24th Regional Symposium on Chemical Engineering (RSCE 2017)

MATEC Web of Conferences Volume 156 (2018)

Semarang, Indonesia
15 – 16 November 2017

Editors:

A.C. Kumoro
Hadiyanto
S.A. Roces
L. Yung
X. Rong
A.W. Lothongkum
M.T. Phong
M.A. Hussain
W.R.W. Daud
P.T.S. Nam

ISBN: 978-1-5108-5976-0

Printed from e-media with permission by:

Curran Associates, Inc.
57 Morehouse Lane
Red Hook, NY 12571

Some format issues inherent in the e-media version may also appear in this print version.

Printed by Curran Associates, Inc. (2018)

For additional information, please contact EDP Sciences – Web of Conferences
at the address below.

EDP Sciences – Web of Conferences
17, Avenue du Hoggar
Parc d'Activité de Courtabœuf
BP 112
F-91944 Les Ulis Cedex A
France

Phone: +33 (0) 1 69 18 75 75
Fax: +33 (0) 1 69 28 84 91

contact-edps@webofconferences.org

Additional copies of this publication are available from:

Curran Associates, Inc.
57 Morehouse Lane
Red Hook, NY 12571 USA
Phone: 845-758-0400
Fax: 845-758-2633
Email: curran@proceedings.com
Web: www.proceedings.com

TABLE OF CONTENTS

The 24th Regional Symposium on
Chemical Engineering
(RSCE 2017)
Theme
"Innovations in Chemical Engineering
Toward The Application in Foods, Energy, and Materials"
DEPARTMENT OF CHEMICAL ENGINEERING, FACULTY OF ENGINEERING
DIPONEGORO UNIVERSITY
November 15th-16th, 2017
Patra Hotel & Convention Semarang, Indonesia
Supported by :
UTM
INDONESIA POWER
PERTAMINA
RSCE 24th

ORGANIZED BY

Statement of Peer review

In submitting conference proceedings to MATEC Web of Conferences, I certify to the Publisher that I adhere to the **Policy on Publishing Integrity** of the **journal** in order to safeguard good scientific practice in publishing.

1. All articles have been subjected to peer review administered by the proceedings editors.
2. Reviews have been conducted by expert referees, who have been requested to provide unbiased and constructive comments aimed, whenever possible, at improving the work.
3. Proceedings editors have taken all reasonable steps to ensure the quality of the materials they publish and their decision to accept or reject a paper for publication has been based only on the merits of the work and the relevance to the journal.

Title, date and place of the conference:

The 24th Regional Symposium on Chemical Engineering (RSCE) 2017
15-16th November 2017
Patra Hotel & Convention, Semarang-Indonesia

Proceedings editor(s):

1. Prof. Dr. Andri Cahyo Kumoro (Indonesia) (Editor in chief)
2. Prof. Dr. Hadiyanto (Indonesia)
3. Prof. Dr. Susan A. Roces (Philippines)
4. Assoc. Prof. Lanry Yung (Singapore)
5. Prof. Dr. Xu Rong (Singapore)
6. Assoc. Prof. Dr. Anchaleepon W Lothongkum (Thailand)
7. Assoc. Prof. Dr. Mai Thanh Phong (Vietnam)
8. Prof. Dr. Ir. Mohammad Azlan Hussain (Malaysia)
9. Prof. Dr. Wan Ramli Wan Daud (Malaysia)
10. Dr. Phan Thanh Son Nam (Vietnam)

Date and editor's signature 08 January 2018

Bioconversion Studies of Methyl Laurate to Dodecanedioic Acid using a Wild-type of *Candida tropicalis*

Rifkah Akmalina, *Ronny* Purwadi, and *Johnner* Sitompul*

Chemical Engineering Department, Institut Teknologi Bandung, Jl. Ganesha 10, Bandung, 40132, Indonesia

Abstract. Production of dodecanedioic acid (DDDA), a platform chemical used as raw material for various commodities and polymers, has been studied through a biological process. This process was conducted by using a wild-type of *Candida tropicalis* which can be obtained easily from natural resources. The aim of this research was to study the characteristics of DDDA production from methyl laurate through batch fermentation process. Growth phase was carried out for 20 h, as the beginning of fermentation, then continued to conversion phase for 5 until 6 days. Utilization of methyl laurate and production of DDDA were analysed using gas chromatography, which proved the ability of *C. tropicalis* in assimilating methyl laurate to convert it become DDDA. The highest value of cells yield ($Y_{x/s}$) and DDDA yield ($Y_{p/s}$) successfully obtained were 0.86 g cells/g methyl laurate and 0.20 g DDDA/g methyl laurate, respectively. This study also showed the possibility of fermentation products accumulation as intermediate, or accumulation of DDDA inside the cells. Thus, this study can be applied as an alternative in addition to the use of mutant microorganism in producing DDDA.

1 Introduction

The long-chain dicarboxylic acids are well known platform chemicals for production of various commodities, such as polymers, perfumes, hot-melting adhesives, and high quality lubricants [1-2]. Dodecanedioic acid (DDDA) is one of dicarboxylic acids produced excessively (>40 million pounds per year) [3]. Currently, crude oil-based DDDAs are still used widely as raw material for polymer production, for example nylon which is produced from crude oil through several steps of chemical reaction [4]. These include limitations in complicated process, generation of unwanted and hazardous byproducts, also dependence on non-renewable petrochemical feedstock [1]. The use of crude oil-based compounds continuously will generate following problem when oil feedstock starts decreasing. A substitution of petrochemical with bio-based synthesis routes is one of alternative [4]. Thus, biotechnology based on renewable resources becomes an interesting process, for example by using fatty acids as raw material.

Several works for dicarboxylic acids production have been studied, with the raw materials or substrates used depend on the long chain of dicarboxylic acid desired. Dodecane [5-6] and lauric acid [7] were widely used as substrates for DDDA production. A direct comparison between utilization of alkanes and fatty acids as substrates has not been studied in detail. Moreover, the use of methyl ester of lauric acid is still rarely conducted. Hence, this study will compare utilization of alkane and fatty acid methyl ester, in particular dodecane and methyl laurate as substrates in the production of DDDA.

In biological process, the success of dicarboxylic acid production really depends on the microorganisms used. Most of dicarboxylic acids could be produced by utilizing genetically engineered (mutant) microorganisms, for example *Candida tropicalis* 7/34 [6], *C. tropicalis* CGMCC 356 [2], *C. cloacae* MR-12 [8], and *C. cloacae* FERM-P736 [7]. A sophisticated and complicated methods or technologies are needed to obtain mutant microorganisms. In the other hand, there is no research found that elucidate clearly the use of wild-type microorganism for producing dicarboxylic acid, especially DDDA. Whereas, a-wild-type yeast can be easily obtained from natural resources, *e.g.* palm kernel oil, palm fiber oil, and ambarella fruit. Thus, this work was performed to study the characteristics of wild-type *C. tropicalis* in producing DDDA on dodecane and methyl laurate substrates in general. In this paper, there is also a brief explanation about mechanism of substrate

* Corresponding author: sitompul@che.itb.ac.id

oxidation for producing dicarboxylic acid. Then it can be related to the result of this work for defining more about DDDA production.

2 Experimental

2.1 Materials

Methyl laurate (>60% purity) was obtained from chemical conversion of palm kernel oil. n-Dodecane (99% purity) was a general purpose reagent and obtained from BDH Chemicals Ltd. Poole England. The strain of *C. tropicalis* wild-type was isolated from ambarella fruit without any treatment of mutation.

2.2 Cells cultivation

One loop of *C. tropicalis* grown on potato dextrose agar (PDA) slant was transferred into 50 ml of preculture medium containing 1% yeast extract, 2% peptone, and 2% glucose. Inoculum was incubated for 24 h at 30°C in a shaking incubator at 150 rpm.

Growth phase was started by transferring 15 ml of inoculum to a 500-ml conical flask containing 150-ml medium which had the following composition per litre: 5 g glucose, 2 g $(NH_4)_2SO_4$, 2 g NH_4NO_3, 1 g K_2HPO_4, 1.5 g KH_2PO_4, 1.5 g $Na_2HPO_4.2H_2O$, 0.2 g $MgSO_4.7H_2O$, 0.2 g KCl, 4 mg $ZnSO_4.7H_2O$, 4 mg $FeSO_4.7H_2O$, and 10 g yeast extract. The culture was aerobically grown at 30°C, 150 rpm in a shaking incubator. Cultivation was carried out until the beginning of stationary phase.

2.3 Dodecanedioic acid production

Methyl laurate was added into the culture at the beginning of conversion phase, followed by the addition of tween 80 (0.01% v/v). Medium pH was raised to 8.0 by adding 1 N NaOH and maintained using phosphate buffer. Fermentation process was continued for 5 until 6 days at 30°C in a shaking incubator at 150 rpm.

Samples (6 ml) were taken manually at regular intervals. Analysis of cells concentration, methyl laurate consumption, and DDDA production was carried out for each sample by these following methods.

2.4 Analysis

Cell growth was determined by measuring of the turbidity. Samples (0.5 ml) were 10x diluted and measured using UV-vis spectrophotometer at 600 nm. The absorbance data obtained were calibrated with cell dry weight data. Fermentation samples (5 ml) were centrifuged at 5000 rpm, 5°C for 10 min. The precipitated cells were dried in oven at 100°C, and weighted.

Glucose concentration in growth phase was measured by phenol-sulphuric method. To 1 ml of diluted free-cell samples, 1 ml of 5% phenol solution and 5 ml of 95-97% H_2SO_4 were added respectively. The absorbance of the solution was measured using UV-vis spectrophotometer after 10 min at 488 nm, and the data were compared with calibration curve of glucose standard solution.

Methyl laurate consumed was determined by gas chromatography (GC). A half ml of free-cell samples was extracted using 1 ml of hexane. The mixture was centrifuged at 5000 rpm, 25°C for 5 min. Aliquot taken from organic phase was injected into a Shimadzu gas chromatograph. The operating conditions were as follow: injection temperature of 280°C, FID-detector temperature of 300°C, and quartz capillary column coated with RTX-5 (crossbond® 5% diphenyl/95% dimethyl polysiloxane). Initial column temperature was set at 70°C for 3 min, followed by 35°C/min temperature increase to 210°C, followed by 3°C/min temperature increase to 240°C with helium carrier gas and column flow of 1 ml/min.

DDDA produced from fermentation process was also determined by GC. To 0.25 ml of free-cell samples added 0.25 ml of 6 N H_2SO_4. DDDA then was extracted into 1 ml of diethyl ether. The mixture was centrifuged at 5000 rpm, 25°C for 5 min. An aliquot of 200 μl from organic phase was transferred to eppendorf tube, followed by the addition of BSTFA (100 μl) for silylation process. Aliquot from the mixture was injected into a Shimadzu gas chromatograph with these following operating conditions: injection temperature of 300°C, column coated with RTX-5 (crossbond® 5% diphenyl/95% dimethyl polysiloxane), temperature of 190°C (hold for 30 min) with helium carrier gas and column flow of 1 ml/min, and FID-detector temperature of 250°C.

3 Results and discussion

3.1 Growth phase of fermentation

When *C. tropicalis* is grown in a medium containing glucose as a carbon source, there was a significant increase in cell concentration. The highest concentration was reached at 20-h-cultivation which produced about 2.8 g dry biomass/l. Further cultivation until 24 h resulted in an equal value, indicating a stationary phase of cell growth (Fig. 1).

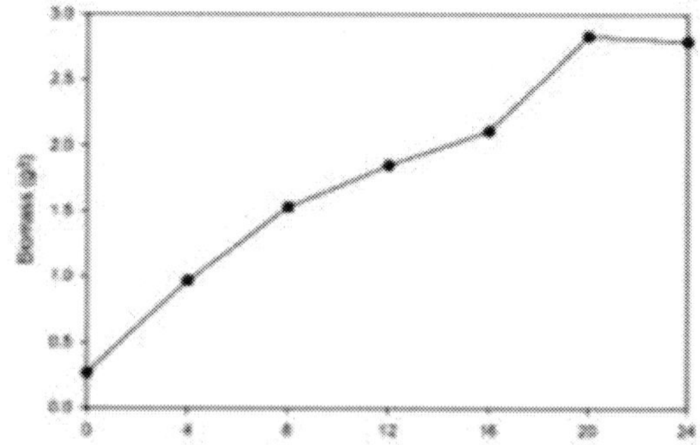

Fig. 1. Cell growth of *C. tropicalis* in medium containing glucose as a carbon source

Table 1. Growth phase in several studies using different medium and carbon source

Yeast	Medium composition	Carbon source	Growth phase	References
C. tropicalis (wild-type)	$(NH_4)_2SO_4$, NH_4NO_3, K_2HPO_4, KH_2PO_4, $Na_2HPO_4.2H_2O$, $MgSO_4.7H_2O$, KCl, $ZnSO_4.7H_2O$, $FeSO_4.7H_2O$, yeast extract	Glucose	20 h	This work
C. tropicalis ATCC 20987	$(NH_4)_2SO_4$, K_2HPO_4, KH_2PO_4, corn steep liquor, $MgSO_4$, NaCl, $CaCl_2$	Glucose	18 h	[3]
C. tropicalis 7/34	$(NH_4)_2SO_4$, NH_4NO_3, K_2HPO_4, KH_2PO_4, $Na_2HPO_4.H_2O$, $MgSO_4.7H_2O$, KCl, $MnCl_2$, $ZnSO_4.7H_2O$, $FeSO_4.7H_2O$, corn steep liquor, polypropylene glycol	Glycerol	24 h	[6]
C. tropicalis CGMCC 356	Yeast extract, corn steep liquor, urea, KH_2PO_4, CH_3COONa, NaCl, $MgSO_4.7H_2O$, polypropylene glicol	Sucrose	18 h	[2]

Growth phase in fermentation process aims to increase cells amount, so that a carbon source which can be easily utilized by microorganism is needed. Glucose is one of carbon source that frequently used for yeast growth [9]. In line with the result given by this study, it showed that glucose was an appropriate carbon source used for increasing cell concentration.

Growth phase can be carried out in slightly different times. It depends on medium composition, type and amount of carbon source, also type of yeast used in fermentation, as shown in Table 1. In general, a mutant and wild-type of *C. tropicalis* have the same capability in assimilating sugar.

3.2 Growth on dodecane and methyl laurate

Candida tropicalis grown in dodecane and methyl laurate resulted in a different tendency. Those substrates could be assimilated and converted into cells (biomass). At the beginning of substrates addition, there was a decrease in cell concentration, indicating an adaptation phase of cell growth (Fig. 2). Microorganisms need to rearrange their enzyme to assimilate dodecane or methyl laurate after assimilating glucose as carbon source. A significant growth was shown after 24-h-fermentation in both substrates, but further fermentation until 72 h showed that *C. tropicalis* could assimilate methyl laurate better than dodecane.

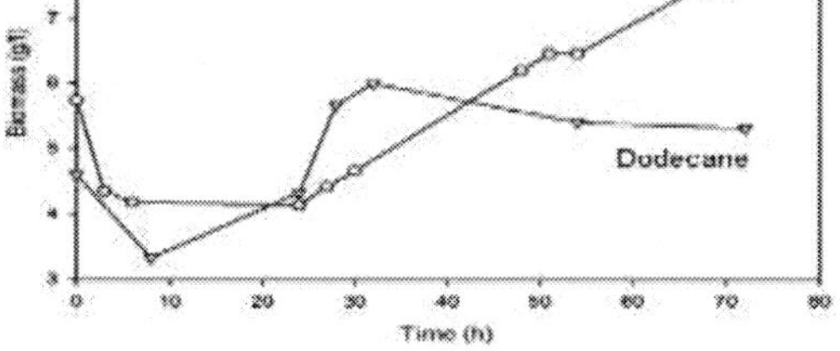

Fig. 2. Growth of *C. tropicalis* in dodecane and methyl laurate during 72-h-fermentation with substrates concentration of 5% v/v respectively

Dodecane is a long-chain alkane that supplies carbon source alternative for growth, while short-chain alkanes can be toxic that inhibit the yeast growth [7]. In addition, short-chain fatty acids can also inhibit the growth of microorganisms by disturbing basal cell functions [10]. The results shown in Fig. 2 are in an agreement with the theories above, in which dodecane and methyl laurate were able to be used as substrates for growth.

However, consumption rate of methyl laurate by *C. tropicalis* was faster than dodecane. It can be seen from Fig. 2 that until 72 h fermentation was carried out, cell concentration kept increasing significantly with methyl laurate as substrate. This study also proved that the use of fatty acid methyl ester (FAME) did not inhibit cell growth. It is presumably because the concentration of methanol contained in FAME is relatively low, so that it supplies more carbon source for yeast growth.

By using dodecane as substrate, the highest cell concentration attained was 6.0 g dry biomass/l at 32-h-fermentation. Further fermentation showed a gradual decrease in cell amount because there was insufficient carbon source to be consumed afterwards.

Theoretically, dicarboxylic acids can be produced from alkanes and fatty acids. Alkanes were oxidized into fatty alcohols, fatty aldehydes, and fatty acids. Further oxidation of fatty acids will produce dicarboxylic acids [11]. This study showed that biooxidation of alkanes was slower than biooxidation of fatty acids, indicating that oxidation step of alkanes into fatty acids was a rate limiting step. Using methyl laurate as substrate can accelerate oxidation process for producing DDDA, which is more advantageous than using dodecane as substrate.

3.3 Production of dodecanedioic acid

Growth of *C. tropicalis* was initially analysed in methyl laurate with concentration of 5 g/l, 10 g/l, and 15 g/l. In general, *C. tropicalis* was able to assimilate methyl laurate in various concentrations and convert it into biomass. At the end of fermentation process, there was a little amount of methyl laurate remained in medium: 0.5, 1.6, and 2.1 g methyl laurate/l from the initial concentration, respectively (Fig. 3). Theoretically, the

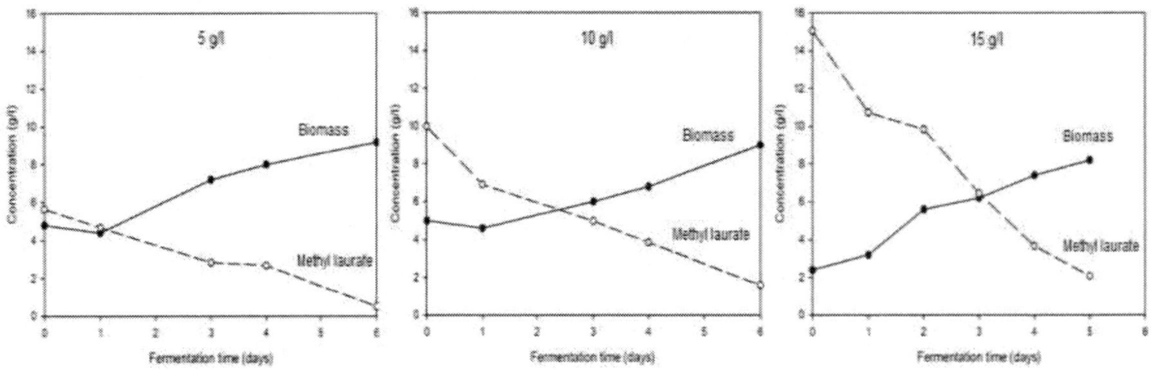

Fig. 3. Assimilation of methyl laurate in various concentrations with the production of biomass

Table 2. Biomass and product yield from various concentrations of methyl laurate

Parameters	Initial concentration of methyl laurate		
	5 g/l	10 g/l	15 g/l
Yx/s (g cells/g methyl laurate)	0.86	0.48	0.45
Yp/s (g DDDA/g methyl laurate)	0.20	0.11	0.08

more methyl laurate consumed, the more biomass produced. In actual, the biomass produced was relatively equal among three variations of concentration (Table 2).

The equal value of biomass yield indicates other products that might also be produced. There was a probability for DDDA produced from fermentation process. Further analysis was conducted to prove the ability of wild-type *C. tropicalis* in producing DDDA. Gas chromatography analysis showed the existence of DDDA in fermentation broth as shown in Fig. 4.

DDDA produced from various concentrations of methyl laurate was relatively equal as fermentation time prolonged (Fig. 4). The highest DDDA concentration that successfully attained was only 1 g/l in approximate. Relating this result with assimilation data shown in Fig. 3 have not yet answered the question related to products obtained beside of biomass. More DDDA should be produced as more methyl laurate was consumed.

When n-dodecane used as a substrate for producing DDDA, 12-hydroxydodecanoic acid, n-dodecanol, and n-dodecanoic acid were excreted and detected in the medium [5] as intermediate products. In this study, fatty acid was used as a substrate. There is a possibility of fermentation products accumulated in the form of hydroxy-fatty acid and fatty acid aldehyde as intermediate products of fatty acid oxidation. Hence, there was insufficient DDDA accumulated in fermentation broth.

Another possibility related to the process occurred in utilization of a wild-type of *C. tropicalis* is the accumulation of DDDA inside the cells (DDDA as intracellular product). This phenomenon might cause in low DDDA concentration in fermentation broth, as shown in Fig. 4. The use of mutant strain successfully

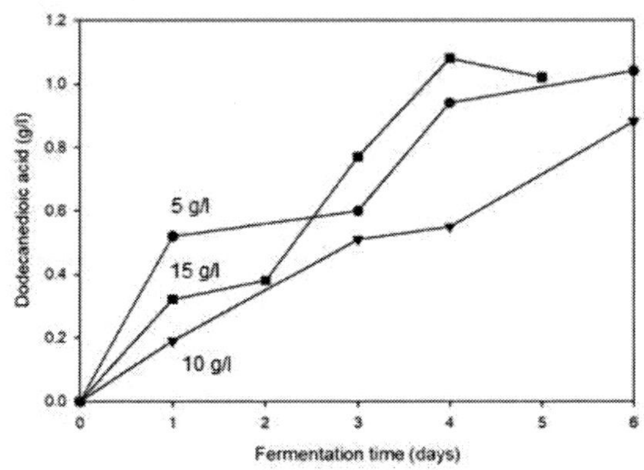

Fig. 4. Dodecanedioic acid produced from several concentrations of methyl laurate

improved the ability to produce and secrete dicarboxylic acid through cell membrane [12]. Extracellular pH is a factor influencing acid secretion. In a higher pH, cells had to secrete more acid to neutralize OH^- in order to maintain the micro-environment balance inside cells [12]. The appropriate pH adjustment is important to make secretion process becomes effective while using a wild-type microorganism.

Comparing biomass and DDDA yield (Table 2), the amount of biomass was about four times higher than DDDA. In other word, methyl laurate was converted to biomass more than DDDA.

Several studies had successfully proved that mutant microorganisms were able to produce dodecanedioic acid (Table 3). Compared with this study, the use of mutant microorganism resulted in higher yield of DDDA in shorter fermentation time. DDDA yield of 0.20 g/g methyl laurate was obtained in 144-h-fermentation from this study. It means that the use of mutant yeast was more advantageous because the fermentation process was more effective than the use of wild-type yeast.

Table 3. Comparison of DDDA yield from several researches using mutant yeasts

Substrate	Microorganism	Fermentation time	Yield (g DDDA/g substrate)	References
Dodecane	*C. tropicalis* 7/34	48 h	0.91	[6]
Dodecane	*C. tropicalis* CGMCC 356	120 h	0.73	[2]
Dodecane	*C. cloacae* MR-12	48 h	0.26	[8]
		72 h	0.41	
		48 h	0.20	
		72 h	0.29	
Dodecane	*C. cloacae* FERM-P736	72 h	0.20	[7]
Lauric acid			0.28	
Dodecane	*C. tropicalis* S-76	36 h	0.23	[5]
		48 h	0.28	

3.4 Mechanism of Fatty Acid Oxidation

The alkane or fatty acid metabolism of *C. tropicalis* is divided into three steps [12]: (1) substrate uptake, (2) metabolism of substrate in *C. tropicalis* cells, and (3) secretion of dicarboxylic acid. Substrate uptake into the cell is the first important step before starting oxidation process. In this study, tween 80 was used as surfactant to emulsify methyl laurate into fermentation medium, thus reducing mass transfer limitation between methyl laurate and aqueous phase of medium.

The different result obtained from using mutant and wild-type yeast in producing DDDA was related to the oxidation mechanism of substrate used. A reaction of β-oxidation might cause in low yield of DDDA when using wild-type *C. tropicalis* in fermentation process.

Fatty acids can be oxidized through reactions of β-oxidation, α-oxidation, and ω-oxidation. Beta-oxidation, which produces acetyl-CoA in peroxisome is the main reaction of fatty acid oxidation. Further oxidation of acetyl-CoA through citric acid cycle generates CO_2, and energy in the form of ATP is produced through oxidative phosphorylation (Fig. 5) [13].

In biological process of dicarboxylic acid production, ω-oxidation is more desired, but it occurs in small amount. Methyl group of fatty acid is hydroxylated into ω-hydroxy-fatty acid. Further oxidation of that compound leads to the production of fatty acid aldehyde and dicarboxylic acid (Fig. 5) [11,13]. Dicarboxylic acids produced from this reaction are only intermediate compounds, and will further be degraded by β-oxidation resulting in shorter-chain dicarboxylic acids.

Naturally, a wild-type yeast will oxidize fatty acids through β-oxidation reaction. This reaction is avoided in dicarboxylic acid production due to CO_2 and energy generation as the main product, so that insufficient amount of dicarboxylic acids is accumulated in fermentation broth.

A genetic engineering through deletion of genes involved in β-oxidation and overexpression of ω-oxidation enzyme can be applied to block β-oxidation and make ω-oxidation as the main pathway for fatty acid oxidation, resulting high yield of dicarboxylic acids [11,14].

Genetically engineered microorganisms might also have improved ability to secrete dicarboxylic acids produced from oxidation process [12], so that dicarboxylic acids could be obtained easily from fermentation broth in sufficient amount. In contrast, a wild-type microorganism needs distinctive condition to make an effective secretion process, for instance by raising extracellular pH [12]. As a result, there are differences in DDDA yield by using mutant and wild-type microorganisms as shown in Table 3.

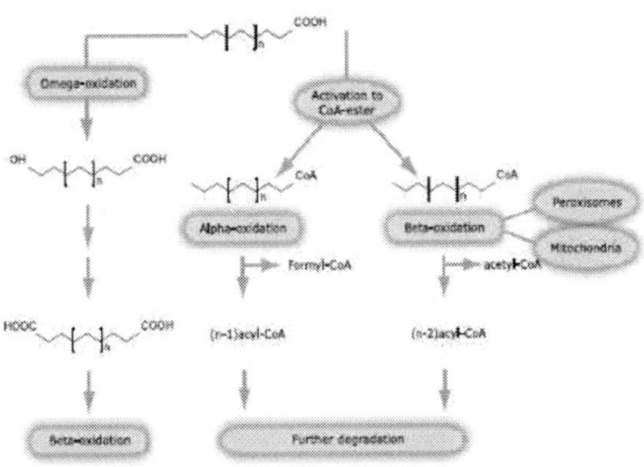

Fig. 5. Mechanism of fatty acid oxidation [15]

4 Conclusions

This study has shown that a wild-type of *Candida tropicalis* is able to produce dodecanedioic acid from methyl laurate. However, the amount of DDDA produced is less than the use of mutant yeast. In addition, dodecane can also be used as a substrate for DDDA production. However, cell growth analysis indirectly showed that oxidation of dodecane was slower than oxidation of methyl laurate.

Various concentrations of methyl laurate used in fermentation process do not affect the biomass and DDDA yield. The possibility of fermentation products accumulation in the form of hydroxy-fatty acid and fatty acid aldehyde may cause the low yield of DDDA. Hence, it is important to do further analysis of compounds contained in fermentation broth using gas chromatography-mass spectrometry (GC-MS). In the other hand, DDDA may also be accumulated inside the cells. Using wild-type microorganisms did bring the disadvantage of poor DDDA secretion from the cells, so that pH adjustment should be performed well to improve the ability of secretion process. Moreover, an experiment for disrupting cells can also be carried out to obtain more DDDA from the cells.

In general, using mutant microorganisms gives more advantages than using a wild-type one. It is related to β-oxidation reaction that may occur, which also produce CO_2 and energy as main metabolism products of wild-type microorganisms. However, this preliminary study could be useful for the following research and might be applied as an alternative in addition to the use of mutant microorganism in producing DDDA.

This work was financially supported by Korea Research Institute of Bioscience and Biotechnology (KRIBB).

References

1. C.S. Prabu, S.K. Lee, J. Agric. Food Chem. **63**, 8199-8208 (2015)
2. S. Li, C. Li, X. Fang, Z. Cao, Enzyme Microb. Technol. **34**, 73-77 (2004)
3. D.P. Mobley, *Biosynthesis of Long-Chain Dicarboxylic Acid Monomers from Renewable Resources* (Final Technical Report, United States, 1999)
4. N. Ladkau, M. Assmann, M. Schrewe, M.K. Julsing, A. Schmid, B. Buhler, Metab. Eng. (2016)
5. Z. Yi, H. Rehm, European J. Appl. Microbiol. Biotechnol. **14**, 254-258 (1982)
6. F.F. Hill, I. Venn, K.L. Lukas, Appl. Microbiol. Biotechnol. **24**, 168-174 (1986)
7. K.D. Green, M.K. Turner, J.M. Woodley, Enzyme Microb. Technol. **27**, 205-211 (2000)
8. R. Uchio, I. Shiio, Agr. Biol. Chem. **36**, 426-433 (1972)
9. M. Saghbini, D. Hoekstra, J. Gautsch, *Methods in Molecular Biology, Two-Hybrid Systems* (2001)
10. Z. Hunkova, Z. Fencl, Biotechnol. Bioeng. **29**, 1623-1641 (1977)
11. S. Huf, S. Krugener, T. Hirth, S. Rupp, S. Zibek, Eur. J. Lipid Sci. Technol. **113**, 548-561 (2011)
12. R. Lin, Z. Cao, T. Zhu, Z. Zhang, Biop. Eng. **22**, 391-396 (2000)
13. D. Voet, J.G. Voet, C.W. Pratt, *Fundamentals of Biochemistry* (John Wiley and Sons, England, 2016)
14. P.C. Williams, *Biological Conversion of Alkanes to Dicarboxylic Acids* (University of Cape Town, 2005)
15. J.A. Wanders, J. Komen, S. Kemp, The FEBS Journal **278**, 182-194 (2011)

Preparation of Simvastatin Hydrogel through Arginine Addition for Drug Delivery System

Niswati Fathmah Rosyida[1], *Teguh* Ariyanto[2*], *Pinandi Sri* Pudyani[1], *Ika Dewi* Ana[3*]

[1]Department of Orthodontic, Faculty of Dentistry, Universitas Gadjah Mada, 55281, Yogyakarta, Indonesia
[2]Department of Chemical Engineering, Faculty of Engineering, Universitas Gadjah Mada, 55281, Yogyakarta, Indonesia
[3]Department of Biomedicine, Faculty of Dentistry, Universitas Gadjah Mada, 55281, Yogyakarta, Indonesia

Abstract. Simvastatin is a lipid lowering agent which has been used recently as drug delivery system for stimulating bone regeneration. Because of low therapeutic efficacy and bioavailability, it is necessary to deliver simvastatin by local administration e.g. by hydrogel system. However, simvastatin has very poor solubility which restricts to prepare hydrogel formulation. The aim of this study is to improve solubility of simvastatin with arginine as co-solvent for developing a controlled released drug delivery system by loading simvastatin into gelatin hydrogel. The solubility study was performed by addition of an excess mass of simvastatin into the specified molar solutions of the arginine. All conical flasks were placed in a mechanical water bath shaker at the temperature of 25, 40, and 50 °C and shaken for a maximum period of 72 hours. The drug concentration was analyzed by UV/Visible spectroscopy at 238 nm. The hydrogel was prepared by a dehydrothermal method. The results showed that simvastatin solubility increases with increasing arginine concentrations and temperature. Characterizations showed a successful preparation of simvastatin-loaded gelatin hydrogel. The arginine simvastatin hydrogel and the gelatin hydrogel (as a blank) exhibited a comparable swelling index (ca. 6.5). Furthermore, microparticles of the material show a narrow particle size distribution in the range between 150-250 μm.

1 1 Introduction

Bone damage due to trauma, osteoporosis, and genetic diseases is a health issue that requires serious treatment. Bone injury can decrease the quality of life, so is needed therapy to accelerate the bone healing process. Without the aid of drugs and bone substitution, healing of bones takes a long time and is less than perfect [1]. Recently, one of the drugs known to have effects on bone healing is simvastatin. Simvastatin is a drug used to lower cholesterol in the blood through inhibition of cholesterol-forming enzymes [2]. Simvastatin could induce bone formation by increasing bone morphogenetic protein expression (BMP-2) and inhibiting bone resorption by reducing RANKL expression [3,4]. However, the bioavailability of simvastatin is only about 5%, since it is metabolized in the liver [2]. Systemic administration of high doses of simvastatin is required to obtain statin effects on bone, but increased doses will also enhance the toxicity of a drug [5]. Therefore, it is necessary to modify the simvastatin formulas given either injections or implants locally to obtain the optimal therapeutic effect on bone.

The hydrogel is a drug carrier polymer that is currently developed due to many advantages. The natural material employed to make the hydrogel is gelatin, which has a biocompatible, and can be degraded [6]. The combination of active substances of simvastatin and gelatin is expected to become the right combination formula for bone healing therapy with local injection. Arginine is one type of amino acid that can be used to improve drug solubility [7]. Furthermore, arginine can accelerate wound healing [8]. Arginine has antibacterial effect and promotes collagen synthesis and cell proliferation [9].

The aim of this study is to improve solubility of simvastatin with arginine as co-solvent for developing a controlled drug delivery system by loading simvastatin into gelatin hydrogel. Material preparation and characterization of drug-loaded hydrogel are also discussed.

2 Material and methods

2.1. Materials

Simvastatin and arginine (pharmaceutical grade, Sigma Aldrich, Singapore) were used as active substance and

* Corresponding author: teguh.ariyanto@ugm.ac.id; ikadewiana@ugm.ac.id

surfactant, respectively. Gelatin type B was obtained from Nitta Osaka, Japan.

2.2 Solubility study of simvastatin

Surfactants in the 0.01-1 mol L^{-1} concentration range were prepared by weighing the right material mass and incorporated into 100 mL of water in Erlenmeyer. The mixture was then stirred to form a homogeneous solution. Excessive amounts of simvastatin were added to each Erlenmeyer containing the specified molar solutions of the arginine. All conical flasks were placed in an incubator shaker under controlled temperature conditions (25 °C, 40 °C, and 50 °C) up to a maximum of 72 h to ensure equilibrium. At the end of the incubation period, the suspensions were filtered and measured its absorbance by UV/Vis spectroscopy at position 238 nm. The typical UV/Vis spectra of simvastatin solution are shown in **Fig. 1**. A linier correlation of absorbance *vs.* simvastatin concentration results.

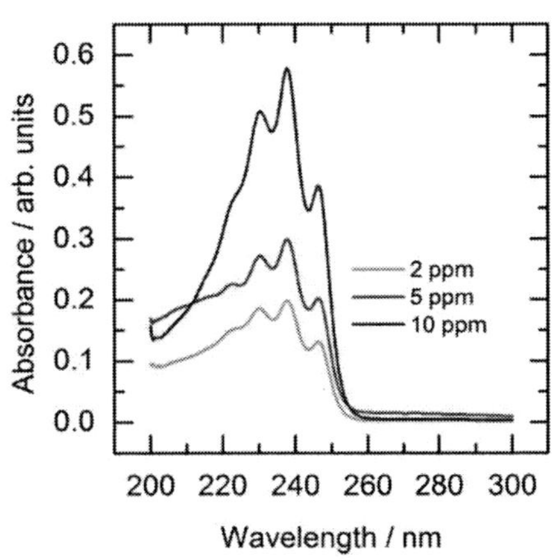

Fig. 1. UV/Vis spectra of simvastatin solution showing a maximum peak at 238 nm.

2.3. Preparation of microparticle of simvastatin-arginine-gelatin hydrogel

Hydrogels are made from gelatin with 7% w/v concentration. Gelatin powder was dissolved in 10 ml of distilled water in a beaker and waited until it was swelling. Gelatin was then stirred on a water bath until homogeneous solution. Simvastatin and arginine were added to gelatin solution and stirred at 40°C. The mixture was cast into polypropylene mold. Specimens were then included in the freeze dryer overnight. Subsequently, the sample was wrapped in parchment paper and aluminum foil and inserted into a vacuum oven until obtaining solid material. The formed hydrogel block matrix was then crushed to obtain microparticles.

2.4 Characterization of simvastatin-arginine-gelatin hydrogel

Image of microparticles was taken by stereomicroscope (Olympus SZX7, Japan). Swelling of the hydrogel index was measured through a hydration process. First, the initial weight of the hydrogel was measured by an electric balance. The hydrogel was then immersed in 100 μL of aquadest. The final weight of the hydrogel was measured at every hour for seven hours. The swelling index was defined as an increased mass after immersion divided by the initial mass of hydrogel.

The Fourier Transform Infrared Spectroscopy (FTIR) test was performed to determine the functional and qualitative groups of hydrogel microparticles and was taken by NICOLET iS10 (Thermoscientific, USA). The sample was mixed with KBr and compacted on the disc. The translucent film was scanned over a range of 4000 to 400 cm^{-1} at ambient temperature.

3 Results and discussion

3.1 Solubility of simvastatin in arginine solution

To investigate the amount of simvastatin that can be dissolved in water solution as solvent in hydrogel preparation, solubility study was carried out by introducing arginine in different concentrations. **Fig. 2** displays solubility of simvastatin as a function of arginine concentration and temperature. The solubility of simvastatin in water is ca. 0.01 mmol L^{-1} at room temperature. This result is in agreement with other studies of simvastatin solubility in pure water [10]. When a higher temperature is used (40 °C), the solubility is a little bit increased. At room temperature, as can be seen, the solubility increases linearly when introducing arginine. At 0.1 M arginine, solubility of simvastatin can be enhanced seven times higher than that of in water. The enhancement of solubility can further be increased at higher solution temperature of 40 °C. However, the study showed that increasing further the temperature to 50 °C does not give a positive impact on solubility of simvastatin. This is due to the arginine is not stable at a high temperature [11].

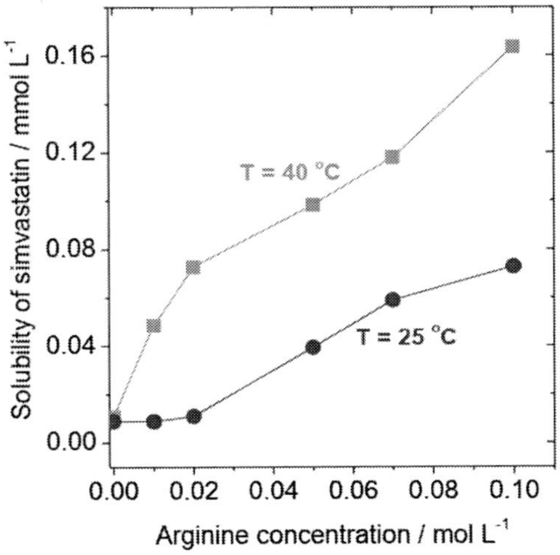

Fig. 2. Solubility of simvastatin as a function of temperature and arginine concentration (pH = 7).

The positive impact of arginine in solubility of simvastatin could give an advantage when dispersing simvastatin as an active substance in hydrogel media.

3.2 Hydrogel microparticle synthesis and characterizations

The active substance of simvastatin was dispersed in hydrogel carrier. After size reduction of hydrogel which contains arginine-simvastatin, particle sizes were observed by using stereomicroscope. **Fig. 3** shows microparticles with a relatively homogenous size ca. 200 µm. This evidences that microparticles are obtained. The hydrogel exhibited the swelling index of more than 4 which indicates the material can adsorb high amount of water.

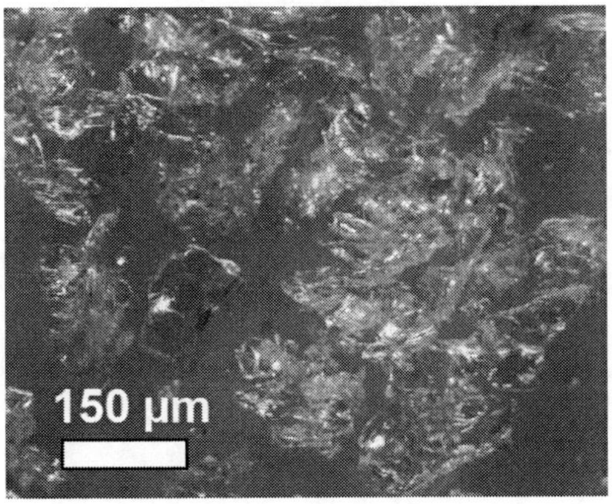

Fig. 3. Hydrogel microparticles observed by stereomicroscope.

Fig. 4 shows the swelling index of gelatin arginine simvastatin and pure gelatin hydrogel. Simvastatin hydrogel showed the higher swelling rather than the pure gelatin hydrogel. Swelling index is influenced by the composition and the degree of cross linking [12]. Swelling index is lower when the degree of cross linking is high. In the present study, there was no significant difference between hydrogel with or without simvastatin.

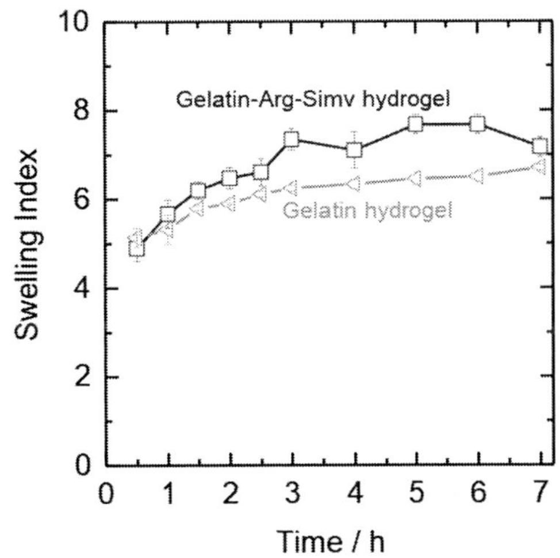

Fig. 4. Swelling index of hydrogel.

To further study the success of preparation of simvastatin-arginine-gelatin hydrogel and functional group involved in the material, FTIR study was carried out. FTIR profiles for gelatin, arginine, simvastatin, and the hydrogel are shown in **Fig. 4**. As individual substances, simvastatin exhibits clearly hydroxyl (-OH) at ca. 3550 cm^{-1} peak, alkyl (-RH) at ca. 3000 cm^{-1} peak and alkene (C=C) at ca. 1950 cm^{-1} peak of functional groups. For arginine and gelatin which are amino acid and protein, it can be seen –NH functional group in the range of 3200-3500 cm^{-1} wavelength. After all of the substances were mixed in the hydrogel system, dominantly of gelatin signal is observed. But a stronger FTIR spectrum is seen. This is likely due to a presence of simvastatin and arginine in hydrogel system.

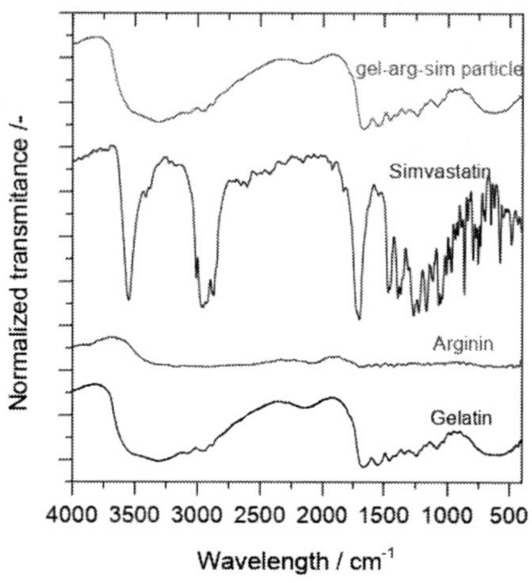

Fig. 4. FTIR signals of hydrogel microparticle and its components.

4 Conclusions

The solubility of simvastatin can be enhanced by addition of arginine in water. The study showed that concentration of arginine and temperature of solution have significant effects on the simvastatin solubility. This positive influence of arginine was further used to formulate simvastatin-loaded hydrogel based on gelatin. Characterizations showed a successful preparation of hydrogel microparticles.

The authors gratefully acknowledge the funding of the Ministry of Research, Technology and Higher Education of the Republic of Indonesia (MENRISTEKDIKTI) through Penelitian Fundamental 2017 (Grant No. 7183/UN1.P.III/DIT-LIT/LT/2017). NFR thanks to Universitas Gadjah Mada for funding of Hibah Peningkatan Kapasitas Dosen Muda 2017.

References

1. Y. Kim, Y. Tabata, Adv. Drug Delivery Rev. **94**, 28-40 (2015)
2. CR. Sirtori, Pharmacol Res. **88**, 3–11 (2014)
3. Z. Wu, C. Liu, G. Zang, H. Sun, Int. J. Oral Maxillofac. Surg, **37**, 170–6 (2008)
4. I. R. Garret, G. R. Mundy, Arthritis Res. **4**, 237-40 (2002)
5. Z. Jia, et al, J. Controlled Release. **200**, 23–34 (2015)
6. J. Zhang, H. Wang, Y. Wang, K. Lai, X. Yang, X. Chen, G. Yang, Scientific Report. **158,** 1–12 (2016)
7. A. Hirano, T. Kameda, T. Arakawa, K. Shiraki, J. Phys. Chem. B. **114**, 13455–62 (2010)
8. S. Sinha, S. C. Goel, Indian J of Orthop, **43**, 328–34 (2009)
9. M. He, A. Potuck, Y. Zhang, CC. Chu, Acta Biomater, **10**, 2482–94 (2014)
10. M. Rúbia, W De. Vargas, FN. Raffin, T. Flávio, A De. Lima, Rev Ciênc Farm Básica Apl, **33**, 497–507 (2012)
11. G. Wu, *Amino Acids: Biochemistry and Nutrition* (CRC Press, 2013)
12. S. Yasasvini, R. Anusa, B. Vedhahari, P. Prabhu, D. Ramyadevi, Mater Sci Eng C, **72**, 160-7 (2017)

Pretreatment of Starch-Free Sugar Palm Trunk (*Arenga pinnata*) to Enhance Saccharification in Bioethanol Production

Kusmiyati[1,*], *Duwi* Maryanto[1], *Ringga* Sonifa[1], *Sabda* Aji Kurniawan[1], and H. Hadiyanto[2,3]

[1]Renewable Energy Research Centre, Department of Chemical Engineering, Faculty of Engineering, Universitas Muhammadiyah Surakarta, Jl. A. Yani Tromol Pos 1, Pabelan, Kartasura 57102, Surakarta, Indonesia
[2]Department of Chemical Engineering, Faculty of Engineering, Diponegoro University, Jl. Prof. Soedarto, SH-Tembalang Semarang 50275, Indonesia
[3]Master Program of Environmental Science, School of Postgraduate Studies, Diponegoro University, Jl. Imam Bardjo, SH-Semarang 50241

Abstract. Starch-Free Sugar Palm Trunk (*Arenga pinnata*) can be utilized to produce bioethanol because of their high lignocellulosic contents. Production of bioethanol from lignocellulosic materials consist of pre-treatment, saccharification and fermentation processes. In this work, conversion of starch-free sugar palm trunk (*Arenga pinnata*) to fermentable sugar and bioethanol was carried out through g pretreatment, saccharification and fermentation processes. The pretreatment was carried out by addition of 1% (v/v) HNO_3 and NH_4OH for 30 min and 60 min, respectively. The saccharification was carried out at enzyme celullase loadings of 10 and 20 FPU/g and substrate loadings of 10 and 20 g for NH_4OH pretreated samples. Fermentation was carried out using two methods i.e. separated hydrolysis and fermentation (SHF) and simultaneous saccharification and fermentation (SSF) techniques. The results showed that pretreatment using NH_4OH was more effective than HNO_3 for 60 minutes. IFurthermore, the results also presented the reduction of the lignin content of 9.44% and the increase of cellulose content to 18.56% for 1% (v/v) NH_4OH 60 min of pretreatment. The increase of enzyme cellulase (20 FPU/g substrate) and substrate loading (20 g) could produce more reducing sugar (17.423 g/L and 19.233 g/L) than that at 10 FPU/g substrate and 10 g substrate (11.423 g/L and 17.423 g/L), respectively. The comparison of SHF and SSF showed that SHF process yielded higher ethanol (8.11 g/L) as compared to SSF (3.95 g/L) and nontreatment process (0.507 g/L) for 72 h..

1 Introduction

World oil reserves continue to decrease, it is estimated that global oil production will decrease from 25 million to 5 million barrels in 2050 [1]. Bioethanol is one of the renewable energy that is used to reduce the dependence on petroleum fuels and environmental impacts [2]. Bioethanol fuel production from starch sources have been widely applied especially in USA and Brazil. The current biofuel production has been moved towards non-food raw materials, such as forestry or agricultural residues. Sugar palm trees (*Arenga pinnata*) or known locally as Aren Palm is one of the potential sources which have been identified for bioethanol production [3]. To increase the use of the potential sugar palm plants, a large scale sugar palm plantations which is as wide as 800,000 to 4,000,000 ha have been developed in Indonesia [4]. The sugar sap from its flower stalks is the most prominent of aren's products. Besides, the sugar palm tree also generates edible fruits, starch from trunk and fibres for building material and household utensils [4], [5]. Flour (starch) is obtained from the extraction of the central part of the stem after the palm tree is no longer producing sap [6]. Starch free sugar palm trunk (SFSPT) is a solid waste derived from sugar palm starch extraction. Dry SFSPT is comprised of lignin (37% w/w), cellulose and hemicellulose (49.5%). Lignocellulose from industrial waste palm flour, SFSPT, (*Arenga pinnata*) can be used to manufacture of bioethanol therefore the environmental damage can be reduced. Bioethanol is produced by fermentation of the fermentable sugars generated by enzymatic hydrolysis of cellulose and hemicellulose which are structural carbohydrates. However, lignin is a complex aromatic polymer that forms a cover surrounding the carbohydrate fractions that are limiting the accessibility of carbohydrates to hydrolytic enzymes. Therefore, delignification is important to improve enzymatic digestibility of lignocellulose [7].

Pretreatment is necessary to degrade lignin in bamboo culms to enable bioethanol production as the bamboo lignin network prevents enzymatic saccharification and fermentation [8]. Tutt et al. [9], reported the effect of pretreatment with a dilute solution

* Corresponding author: kusmiyati@ums.ac.id and psea_ums2014@yahoo.com

of sulfuric acid, hydrochloric acid, nitric acid and potassium hydroxide, followed by enzymatic hydrolysis of pretreated wheat straw and subsequent fermentation to ethanol. The highest glucose concentration of 316.7 g/kg biomass and the lowest (221.3 g/kg) were achieved by pretreatment with nitric acid and hydrochloric acid, respectively. The best ethanol yield of 104.3 g/kg was achieved by samples pretreated with KOH (rinsed). In another study by Jeong and Lee [10] pretreatment of yellow poplar biomass with oxalic acid which showed the structural was changed for pretreated biomassa. The increasing severity of pretreatment resulted an gradual increased of biomass degradation rate which caused the increased crystallinity value and the surface area of the pretreated biomass. Effect of nitric acid on pretreatment for ethanol production of rice straw has been studied by Kim et al. [11], pretreatment conditions were HNO_3 concentration (0.2–1.0%), temperature (120–160◦C) and reaction time (1–20) min. A maximum xylose yield of 86.5% and an enzymatic digestibility of 83.0% were achieved at a condition of 0.65% HNO_3, 158.8°C and 5.86 min. Pretreatment of sorghum fibers with ammonium hydroxide followed by enzyme hydrolysis of pretreated fiber and subsequent ethanol production have been evaluated. The treatments removed 44% lignin and 35% hemicellulose at a ratio of 1:0.14:8 sorghum fibers, ammonia, and water, 160°C for 1 h under 140–160 psi pressure. Ethanol yields were 25 g/100 g dry biomass and 10 g ethanol/100 g dry biomass for treated and untreated biomass, respectively [12].

The objective of this study was to evaluate the effect of pretreatments of starch free sugar palm trunk (SFSPT) using dilute nitric acid and ammonium hydroxide on the hydrolysis and subsequent fermentation using Simultaneous Saccharification and Fermentation (SSF) and Separated Hydrolysis and Fermentation (SHF) methods to produce ethanol. The comparison of fermentation by SHF and SSF were also evaluated.

2 Materials and Methods

2.1 Feedstock Materials

Lignocellulose biomass (*Arenga pinnata*) was taken from solid waste of the palm starch production house located in Klaten, Central Java Indonesia. Sugar palm trunk, usually brought to the plant in pieces about four feet high. Furthermore, palm trunks are milled using a milling machine to get trunks powder. To extract the palm starch from palm trunk powder by pouring the powder with water continuously and subsequent filtered to separate starch and solid waste containing cellulose. The solid waste or biomass fibre of starch-free sugar palm trunk (SFSPT) was dried in the sun to dry (+2-3 days) then it was milled to small particles by a grinding with machine and sieved to obtain a size 40 mesh and subsequent stored in a dry place prior to use. The starch-free sugar palm trunk (SFSPT) was composed of 23.96% cellulose, 26.82% hemicelluloses, and 18.57% lignin based on its dry weight [13].

2.2 Pretreatment

Starch-free sugar palm trunk (SFSPT) substrate added with alkaline solution of 1% (v/v) NH_4OH or acid solution 1% (v/v) HNO_3 at concentration of 50g/500 ml solution was pretreated in an autoclave for 30 min or 60 min at temperature 120°C. The samples were filtered with whatman filter paper to separate solids and liquids, the the pretreated solids was washed repeatedly with distilled water until pH 7 (neutral). The pretreated solid (SFSPT) was dried using an oven at a temperature of 65°C to obtain a constant weight. The pretreated solids (SFSPT) are stored for further use as raw material for hydrolysis.

2.3 Enzymatic Hydrolysis

Non treatment and The pretreated solid SFSPT (NH_4OH 1% (v/v) for 60 min) was hydrolyzed using commercial cellulase enzyme (SQzyme CSP, China). Hydrolysis was conducted in a 500 mL erlenmeyer flask containing 200 mL of 50 mM sodium acetate buffer solution and loaded with 20 g substrate, 0.5 g yeast extract, 1 g peptone. The pH was adjusted to 4.8 using hydrochloric acid. The flasks were autoclaved at 121°C for 20 min. After the flasks was cooled to 30°C, the cellulase enzyme was loading at 10 FPU/g substrate or 20 FPU/g substrate. The hydrolysis was conducted in a shaker incubator for 72 h at 150 rpm and temperature 50°C. Samples were taken for analysis of reducing sugars.

2.4 Inoculum Preparation

2.4.1 Preculture of Aspergillus Niger

Isolates of *Aspergillus niger* was subculture in test tubes slant containing PDA (Potato Dextrose Agar) and incubated at 30°C for 72 h. To enrich amount of cell, so must be preculture. Making pre-cultur in 100 mL consists of 1 g $(NH_4)_2SO_4$; 0.3 g urea; 0.3 g KH_2PO_4; 0.05 g $MgSO_4.7H_2O$; 0.05 g $CaCl_2$, Autoclave 121°C for 20 min, shaker 120 rpm, incubation time 72 h.

2. 4.2 Preculture of Saccharomyces cerevisiae

Isolates of *Saccharomyces cerevisiae* was subculture in 100 mL and divided into 10 reaction tubes slant containing 2 g glucose; 2 g peptone; 1 g yeast extract, and 1.5 g agar. To increase the amount of cells that must be preculture, preculture made in 100 mL containing 0.2 g $(NH_4)_2HPO_4$; 0.64 g urea; 0.2 g KH_2PO_4; 0.1 g $MgSO_4.7H_2O$. Sterilization 121°C for 20 min, shaker 120 rpm, incubation time 72 h.

2.4.3 Main culture of Aspergillus niger

In order to be used *Aspergillus niger* should familiarized with the addition of the substrate. Making 2 main culture in 200 mL and 20 mL plus preculture consists of 4 g substrate which has been in treatment; 2 g yeast; 4 g peptone; 2 g $(NH_4)_2HPO_4$; 0.6 g urea; 0.6 g

KH_2PO_4; 0.1 g $MgSO_4.7H_2O$. Autoclave 121°C for 20 min, main culture incubated for 72 h and 120 rpm shaker.

2.4.4 Main culture of Saccharomyces cerevisiae

Saccharomyces cerevisiae start to familiarized with the addition of the substrate, making main culture in 200 mL (+ 20 mL preculture) consisting of 4 g substrate which has been in treatment; 2 g yeast; 4 g peptone; 0.4 g $(NH_4)_2HPO_4$; 1.2 g urea; 0.4 g KH_2PO_4; 0.2 g $MgSO_4.7H_2O$. Autoclave 121°C for 20 min, main culture was incubation for 72 h and 120 rpm shaker.

2.5 Simultaneous Saccharification and Fermentation (SSF) And Separated Saccharification and Fermentation (SHF)

Solid substrate without pretreatment and previously pretreated with 1% (v/v) NH_4OH autoclave for 60 min at 121°C and hydrolized with 20 FPU and 20 g subtrate for 72 h was fermented with Saccharomyces cerevisiae alone for SHF method. For SSF *Saccharomyces cerevisiae* and *Aspergillus niger* was used. Biomass 50 g in 500 mL solution, mainculture *Saccharomyces cerevisiae* added 1 mL/g and mainculture *Aspergillus niger* 1 mL/ g. Fermentation for 72 h 30°C. Shaker for 72 h, after 72 h is done, then distillation to separate the liquid and the substrate. Ethanol was obtained stored at a temperature of 20°C.

3. Result and Discussion

3.1 Effect of Acids and Alkaline pretreatments on Lignin, Hemicellulose and Cellulose

Amount of Lignocellulosic biomass is very abundant and can be converted to energy using technology [14]. Lignocellulosic biomass contains lignin which envelops cellulose, lignin will inhibit the enzyme accessibility and can decrease the rate of hydrolysis [8]. Biomass palm trunks can not be directly fermented due to very rich content of lignin, so that necessary to delignification treatment with the addition of acids and alkaline. Delignification also affected by pretreatment time and temperature. Long treatment time will efffect to high cellulose [15]. Breakdown of lignin very necessary to remove the cellulose contained.

In Fig 1, can be seen initial lignin content of the raw material, after treatment with HNO_3 and NH_4OH, the lignin content in raw materials has decreased, both with pretreatment using HNO_3 and NH_4OH. The maximum decrease of lignin obtained for pretreatment 1% (v/v) HNO_3 is 8.28% for 60 min treatment and 1% (v/v) NH_4OH is 9.44% for 60 min treatment. while the hemicellulose content varies in each treatment, and the content of cellulase has increased significantly. The maximum increase in the cellulose treatment 1% (v/v) HNO_3 is 17.33% and 1% (v/v) NH_4OH treatment to a maximum is 18.56% for 60 min treatment. This indicates the acidic and alkaline pretreatment can damage lignin [16], so the content of cellulose can be out and will be converted into sugar and then fermented using *Saccharomyces cerevisiae* to produce ethanol.

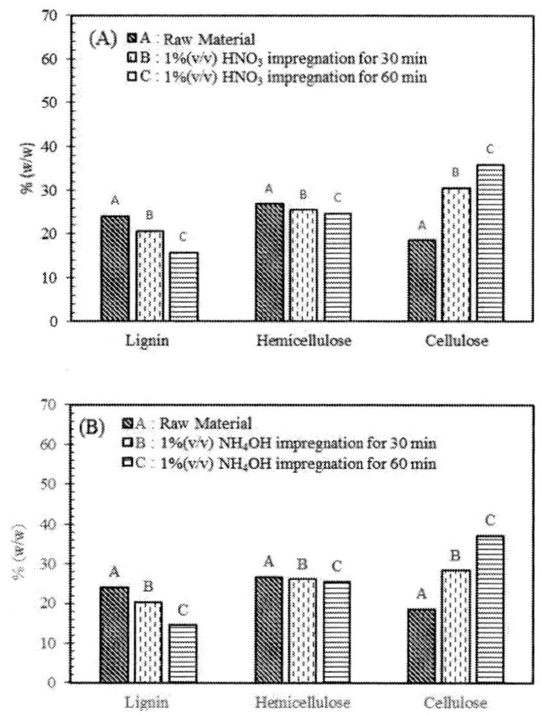

Fig. 1. Lignin after pretreatment using (A) 1% (v/v) HNO_3 for Non-impregnation; Impregnation of 30 min and 60 min. (B) 1% (v/v) NH_4OH to Non-impregnation; Impregnation 30 min and 60 min.

3.2 Relations with reduced sugar content of ethanol after fermentation

When fermented, amount of reduced sugar content greatly affect amount ethanol produced. Levels of bioethanol produced with SFF treatment process using 1% (v/v) NH_4OH produce high levels greater than the levels of ethanol using 1% (v/v) HNO_3 pretreatment SSF. Characteristic of pretreatment alkaline are able to eliminate lignin without having a great effect to other components [17]. The highest ethanol is 3.91 g/L for 60 min 1% (v/v) NH_4OH treatment and 2.73 g/L for 1% (v/v) HNO_3 treatment. Pretreatment of acids and alkaline aim to improve the yield of sugar, for the next can be converted into ethanol using *S. cereviciae*. Pretreatment with alkaline, when compared with the acid or enzymes may increase 8 times and 23 times the yield of fermentable sugars [18]. From the literature, pretreatment on wheat straw with dilute HNO_3 gave an ethanol yield of only 95.0 $g.kg^{-1}$ and highest glucose yield (316.7 $g.kg^{-1}$) [9]. Saha et al. (2013) reported that NH_4OH is the best pretreatment on pteris and could produce 0.642 mg/ml reducing sugar and 0.333 mg/L ethanol with 0.075 gm of yeast at 50 h under the conditions pH 7, 35°C, substrate diameter 45μm-63μm, substrate loading 0.25 gm in 100 ml working volume [19].

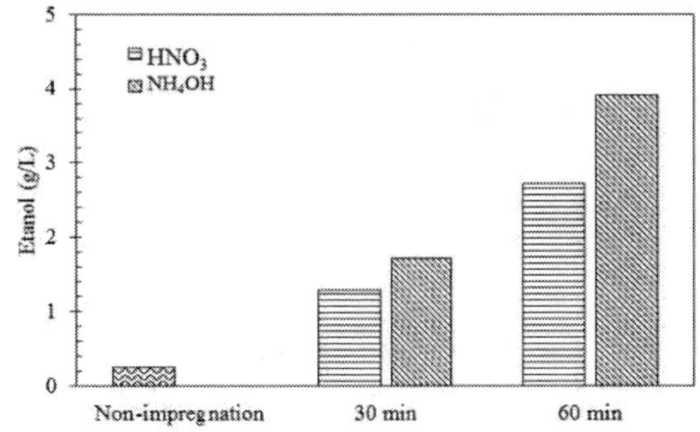

Fig. 2. Bioethanol after fermentation (SSF 72 h) 1% (v/v) HNO_3, 1% (v/v) NH_4OH and non-impregnation; impregnation for 30 min and 60 min with 20 FPU/g substrate enzyme loading and 20 g substrate.

3.3 Correlation between reducing sugar and ethanol in SHF and SSF processes

Enzyme loading for the hydrolysis of biomass palm trunks used 10 and 20 FPU/g celluclast. Raw materials used previously have been treated by NH_4OH 60 menit. As seen in fig 3, the maximum yield achieve of glucose produced by the enzyme loading 20 FPU/g celluclast and 20 g substrate is 19.223 g/L, while the glucose produced with non treatment is 1.718 g/L. Increasing the amount of celluclast and hydrolysis time will be proportional to the increase in the amount of glucose produced. Ayeni et al. [20], reported the highest reducing sugar on shea tree sawdust in 4 day hydrolysis time, 50 FPU/g dry biomass, 45°C, pH 4.8, 40 g/L substrate concentration was 482.40 mg/g dry biomass. In variations of the use of enzymes cellucase SHF process will affect the content of reducing sugar produced, can be seen in graph A, where the variation of cellulase enzymes 20 FPU/g produces (17.423 g/L for 72 h), more reduced sugar content than 10 FPU/g variation (11.423 g/L for 72 h), while the non treatment produced 1.718 g/L for 72 h. To effect substrate amount of raw materials also affect the content of reducing sugar obtained when the hydrolysis process is carried out on raw materials. In the graph (B), shown by the increasing amount of substrate then reducing sugar obtained is also greater. The variation of 20 g substrate could produced 19.233 g/L than 10 g substrate was 17.423 g/L and non treatment was obtained 1.718 g/L for 72 h. In this stage of saccharification enzyme, modified cellulose into cellobiose and further into simple sugars such as glucose, saccharification process using enzymes celluclast, enzyme hydrolysis process takes place at pH 4.8 and the temperature of 45-50°C [21]. From Ganguly et al. [22], and Aveni et al. [20], the addition of biomass loading and hydrolysis time influences increase in reducing sugar, the maximum substrate loading to produces highest reducing sugar is 7% for 27.78 mg/g while the 40 g/L substrate concentration could produce maximum sugar yield for 42.4%.

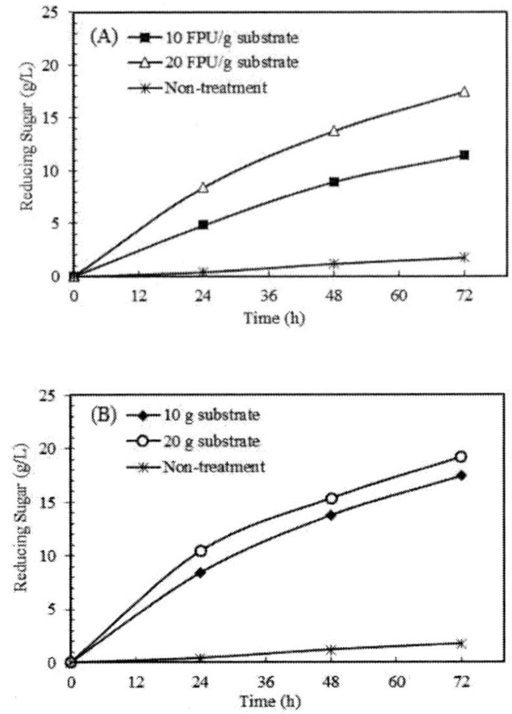

Fig. 3. (A) Graph reduction sugar versus time with 10 g substrate (1% (v/v) NH_4OH treatment for 60 min), the enzyme loading 10 FPU/g substrate and 20 FPU/g substrate, while (B) The relationship of reducing sugar versus time with substrate loading 10 g and 20 g (1% (v/v) NH_4OH treatment for 60 min, 20 FPU/g substrate enzyme loading).

In the use of this type of fermentation process between SHF and SSF, have some differences apart in terms of stages and the time required in the process. SHF process to produce bioethanol is higher, and less energy is required and the production cost can be minimized. SHF process for treatment and nontreatment used 20 FPU/g substrate enzyme loading and 20 g substrate (1% (v/v) NH_4OH impregnation for 60 min).

As It is presented by fig 5, the concentration of bioethanol using the process of SHF was higher as compared to one obtained by using the SSF. Fig 5 also shows that from the graph (A) and (B) it appears that the fermentation time for 72 h, concentrations of bioethanol on the two types of fermentation process has a very significant difference in the SHF obtained biethanol concentration of 8.11 g/L, while the non treatment was obtained 0.57 g/L. The SSF process derived bioethanol concentration of 3.95 g/L, while the non treatment was obtained 0.57 g/L.

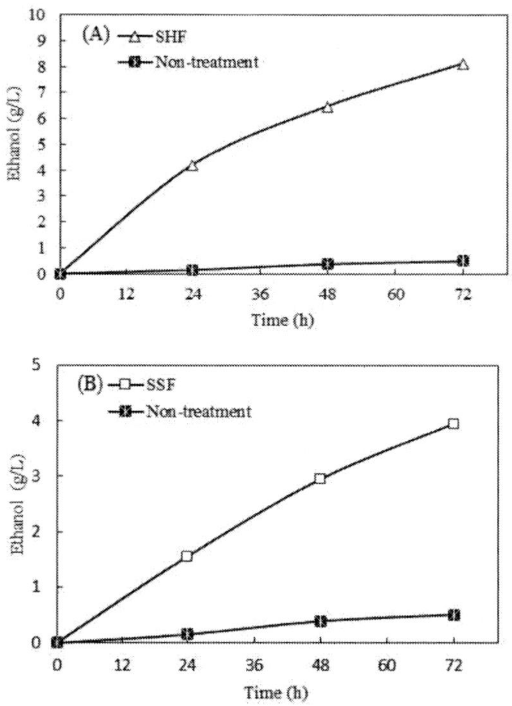

Fig. 4. (A) graph ethanol on fermentation time for the SHF with 1% (v/v) NH_4OH for 60 min, 20 FPU/g substrate cellulase and 20 g substrate biomass loading, (B) graph ethanol on the SSF fermentation process with 1% (v/v) NH_4OH for 60 min, 20 FPU/g substrate cellulase and 20 g substrate biomass loading.

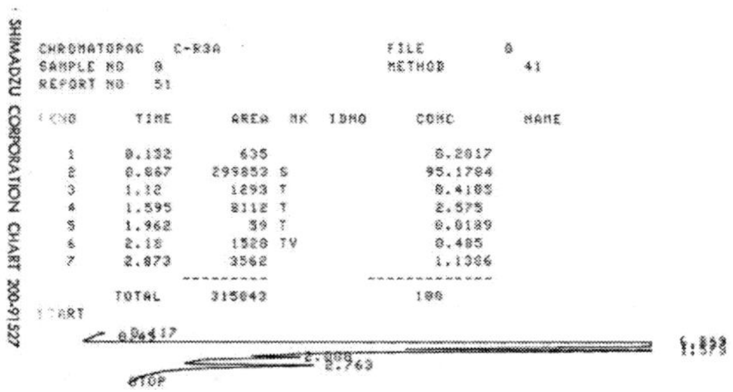

Fig. 5. Graph ethanol GC

GC graph showing measurements of bioethanol for the SHF and SSF.

4. Conclusion

Lignocellulose from industrial waste palm flour, SFSPT, can produce an ethanol with several processes, pretreatment, saccharification and fermentation. Pretreatment SFSPT with NH_4OH was more effective than H_3PO_4 for 60 min impregnation. The more celluclast and substrate loading has a greater reducing sugar (17.3 g/L for 20 FPU/g and 19.2 g/L for 20 g substrate) for 72 hours saccharification. SHF process significantly produce higher ethanol than SSF process for 72 hours (8.11 g/L for SHF and 3.95 g/L for SSF) for 72 h fermentation.

Acknowledgements

The authors would like to thank the Directorate of Research and Community Service, Ministry of Education, Technology and Higher Education Indonesia (Kemenristekdikti) for funding this Research (PB Kompetensi 2017).

References

1. Y. Sun and J. Cheng, *Bioresour. Technol.*, **83**, 1–11 (2002)
2. M. Galbe and G. Zacchi, *Microbiol Biotechnol*, **59**, 618–628, 2002
3. H. Lantemona, A. L. Abadi, A. Rachmansyah, and J. Pontoh, *IOSR J. Environ. Sci. Toxicol. Food Technol.*, **4**, 2, 42–48 (2013)
4. E. Martini, J. M. Roshetko, M. Van Noordwijk, A. Rahmanulloh, E. Mulyoutami, and S. Budidarsono, *Agrofor. Syst.*, **86**, 3, 401–417 (2008)
5. M. R. Ishak, S. M. Sapuan, Z. Leman, M. Z. A. Rahman, U. M. K. Anwar, and J. P. Siregar, *Carbohydr. Polym.*, **91**, 2, 699–710 (2013)
6. M. Lempang, *Info Tek. EBONI*, **9**, 1, 37–54 (2012)
7. J. Xu, Y. Chen, J. J. Cheng, R. R. Sharma-Shivappa, and J. C. Burns, *BioResources*, **6**, 707–720 (2011)
8. H. Suhara, S. Kodama, I. Kamei, N. Maekawa, and S. Meguro, *Int. Biodeterior. Biodegradation*, **75**, 176–180 (2012)
9. M. Tutt, T. Kikas, and J. Olt, *Agron. Res. Biosyst. Eng. Spec. Issue*, 1, 269–276 (2012)
10. S.-Y. Jeong and J.-W. Lee, *Ind. Crop. Prod.*, **79**, 1–6 (2016)
11. I. Kim, B. Lee, J.-Y. Park, S.-A. Choi, and J.-I. Han, *Carbohydr. Polym.*, **99**, 563–567 (2014)
12. D. A. Salvi, G. M. Aita, D. Robert, and V. Bazan, *J. Ind. Microbiol. Biotechnol*, **37**, 27–34 (2010)
13. J. Sahari, S. M. Sapuan, Z. N. Ismarrubie, and M. Z. A. Rahman, *Fibres Text. East. Eur.*, **2**, 91, 21–24 (2012)
14. B. Butkutė, N. Lemežienė, and J. Cesevičienė, *Zamdirbyste-Agriculture*, **100**, 3, 251–260 (2013)
15. X.-D. Hou, T. J. Smith, N. Li, and M.-H. Zong, *Biotechnol. Bioeng.*, vol. xxx, no. xxx, 1–10 (2012)
16. E. C. Bensah and M. Mensah, *Int. J. Chem. Eng.*, **2013**, 1–21 (2013)
17. J. D. McMillan, *Renew. Energy*, **10**, 2/3, 295–302 (1997)
18. L. C. G. Filho, G. A. A. Fischer, N. Sellin, C. Marangoni, and O. Souza, *J. Environ. Sci. Eng.*, **2**, 65–69 (2013)
19. P. Saha, A. C. Baishnab, F. Alam, M. R. Khan, and A. Islam, *Procedia Eng.*, **90**, 504–509 (2014)
20. A. O. Ayeni, J. A. Omoleye, F. K. Hymore, and R. A. Pandey, *Brazilian J. Chem. Eng.*, **33**, 1, 33–45, (2016)

21. E. Hermiati, D. Mangunwidjaja, T. C. Sunarti, and O. Suparno, *J. Litbang Pertan.*, **29**, 4, 121–130 (2010)
22. A. Ganguly, S. Das, A. Bhattacharya, A. Dey, and P. K. Chatterjee, *Indian J. Exp. Biol.*, **51**, 556–566 (2013)

Optimization of the fermentation time and bacteria cell concentration in the starter culture for cyanide acid removal from wild cassava (*Manihot glaziovii*)

Mohamed Hawashi[1], *Tika* Surya Ningsih[1], *Sekar* Bias Tri Cahyani[1], *Kuswandi, Tri* Widjaja[1], and *Setiyo* Gunawan[1,*]

[1]Department of Chemical Engineering, Institut Teknologi Sepuluh Nopember, Surabaya 60111, Indonesia

Abstract. Cassava is one of the most widespread starchy tuberous roots in Indonesia, being one of the typical plants used in the starch market. However, due to the high cyanide content (338.41 ppm), these roots become a poison if they are unsuitably processed. Therefore, a detoxification process is needed to reduce the cyanide level to the safe level for human consumption (10 ppm). This study was focused on *(i)* the investigation of the detoxification potential of fermentation with *Lactobacillus plantarum (L. plantarum)* on the cyanide level of wild cassava tubers (*Manihot glaziovii*) and *(ii)* the optimization of the fermentation time and bacteria cell number in the starter culture. The fermentation was performed for different periods of time (12, 24 and 36 h) and various initial bacteria cell number ($7x10^{10}$, $7x10^{11}$, $1.05x10^{12}$, and $3.5x10^{12}$ *L. plantarum* cells). The results showed a significant decrease of the cyanide level, 97 % of cyanide degradation being noticed after 36 h of fermentation for an initial bacterial cell number of $3.5x10^{12}$ cells. Hence, the strong point of the study consists of a noteworthy reduction of the cyanide content in wild cassava in short periods, whereas the protein content was increased (from 1.5% to 3.5%) in Modified Cassava Flour (MOCAF).

1 Introduction

Due to its high content in starch, one of the key food sources for humans, cassava (*Manihot esculenta*) is a raw material to prepare different food products, along with rice, sago, and corn. Cassava plant has several advantages such as high drought tolerance and grown in poor soil, whereas other staple crops are unable to grow under critical conditions of climate and soil [1]. The estimated production of cassava in Indonesia is more than 23 million tons/year [2]. Also, cassava has high productivity, it is easy to cultivate and can provide a good source of food for a vast majority of the Indonesian population and can increase national food security of Indonesia. Although there are more than 100 cassava cultivars in Indonesia, especially on Java Island, only few of them are commonly used. One cultivar that has not been widely used is wild cassava. This cultivar of cassava has a gnarled root system that develops well and has resistance to drought. The analysis of 45 Indonesian cultivars of cassava roots for cyanogenic potential (CP) showed that CP content usually ranges from 17.8 to 233.2 ppm [3].The wild cassava roots contain high content of hydrocyanic acid (HCN) ranges from 183.2-357.2 mg/kg on dry matter basis (DM) [4]. Its flavor is influenced by the amount of cyanogenic glucosides present in roots, which make it bitter. However, cassava root has a high starch content, which makes it a good source for starch processing with applications in food industry[5]. Generally, starch contains 15-30% amylose and 70-85% amylopectin [6]. Amylose has a strong effect on starch, while amylopectin provides crisp effects in food [7].The utilization of wild cassava root as such for food purposes has few drawbacks, because it contains high toxic cyanogenic glucosides compounds (93% linamarin and 7% lotaustralin) [8]. Cassava roots also contain the enzyme linamarase, which can hydrolyze linamarin to produce glucose and acetone cyanohydrin. Furthermore, it can also decompose to acetone and HCN spontaneously at pH higher than 4 and temperatures greater than 30 °C [9], as shown in Figure 1.

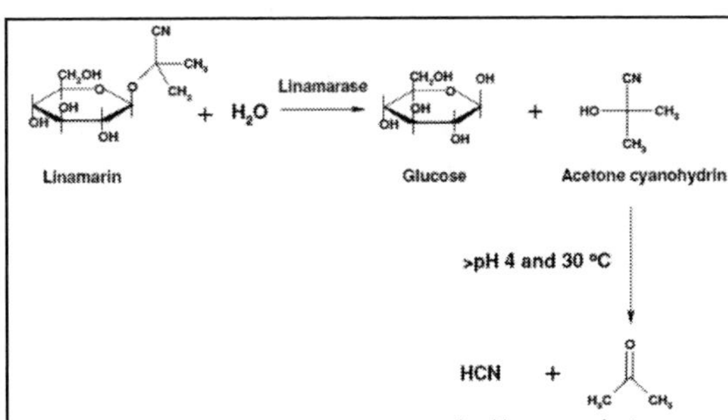

Fig.1 Enzymatic breakdown of the major cyanogenic glucosides (linamarin) in cassava roots [9].

* Corresponding author: gunawan@chem-eng.its.ac.id

 https://doi.org/10.1051/matecconf/201815601004

Hydrocyanic acid is a highly toxic compound for humans. Therefore, the toxic potential of cassava roots should be addressed during cassava root processing before human consumption. The roots have to be detoxified to less than 10 ppm, which is the safe limit proposed by the World Health Organization [10]. The lactic acid fermentation is one of the most used processes for cassava detoxification during the fermentation process the formation of lactic acid decreases the pH below 6. This causes the laminarase enzyme to become non-active, and linamarin does not change into HCN. *L. plantarum* is a lactic acid bacteria commonly used in the fermentation process. These bacteria have antagonistic properties against food-damaging microorganisms, such as Staphylococcus aureus and Salmonella. *L. plantarum* is tolerant to salt, produces acid quickly and has an optimum pH of 5.3-5.6 [11]. Nevertheless, although the fermentation process is widely used, there is a lack of information about optimal parameters, in particular the fermentation time and bacteria concentration, to carry out the process.

Therefore, the objective of this study was to evaluate the effect of the bacteria concentration in the starter culture on the cyanide removal in wild cassava for the production of cassava flour. The fermentation time was also investigated in order to optimize the detoxification process of the roots. The fermentation was performed by a submerged process with the addition of *L. plantarum* bacteria.

2 Materials and methods

2.1 Materials

Wild cassava roots were obtained from a cassava plantation in Blitar, which is located in East Java, Indonesia. The *L. plantarum* bacteria was obtained from the "Chemical Laboratory of Microbiology" of Chemical Engineering Department, Institut Teknologi Sepuluh Nopember (ITS) Surabaya.

2.2 Sample and preparation

Freshly harvested cassava roots were chosen so that to keep the normal level of the proximate content. Cassava roots were peeled out and the skin was washed with running water at 28 °C. After that, the proximate composition of wild cassava was performed. Furthermore, wild cassava was cut into chip form with a thickness of 0.5-1 cm to increase the surface area.

2.3 Production of MOCAF

3.3.1 *Preparation of starter bacteria (L. plantarum)*

The *L .plantarum* inoculum was prepared in an Erlenmeyer flask containing 15 ml of nutrient broth (NB), and 135 mL of distilled water. The mixture was incubated for 16 h to achieve bacterial growth in log phase. The starter volume used for fermentation was adjusted by the variable number of *L. plantarum* bacteria cells. All the tools used for experiments were sterilized previously. The sterilization was carried out in an autoclave at 121 °C for 15 min.

2.3.2 *Fermentation process*

A 150 g of cassava roots were used for the fermentation process. The starter amount of *L. plantarum* cells of $7x10^{10}$, $7x10^{11}$, $1.05x10^{12}$ and $3.5x10^{12}$ were added. The fermentation was performed for 12, 24 and 36 h. The temperature was kept constant at 32 °C during fermentation.

2.3.3 *Milling process*

After fermentation, cassava roots were dried in oven at 45 °C for 4 h, and then milled in a crusher. After milling, cassava flour was sieved to obtain a granulometric fraction lower than 100 mesh. The resulted flour was denoted as (MOCAF)

2.4 Determination of proximate composition

The proximate composition of wild cassava variety was determined for moisture, ash, crude fiber, fat and protein content as described by Association of Official Analytical Chemist (AOAC) [12].

2.5 Determination of cyanide content

The cyanide acid content on fresh and fermented cassava roots was determined as described by [13]. 20 g of sample were transferred into a Kjeldahl digestion flask. Then, 200 mL of distilled water was added to the digestion flask. The flask was stirred to mix the content thoroughly for 2 h, then placed on the heater to recover cyanide acid as distillate. The distillate was collected in a conical flask containing 20 mL of 2.5% NaOH solution. Next, 8 mL of NH_4OH solution and 5 mL of 5% KI solution were added as an indicator. The resulting mixture was titrated against 0.02 N $AgNO_3$ until there was turbidity. A blank was also run through all steps above. The cyanide acid content was calculated from the amount of $AgNO_3$ used for titration.

$$\text{HCN (mg/kg)} = \frac{(V_1 - V_2) \times N \times 27}{(V_3 \times W)} \qquad (1)$$

Where V1, V2, V3, N, and W are the blank titration volume, sample titration volume, distillate volume, $AgNO_3$ normality, and sample weight, respectively.

2.6 Determination of Starch Content

The starch content of the cassava sample was determined as described by [12]. A 2 g of sample was immersed into 50 mL of distilled water and stirred for 1 h on a magnetic stirrer. Next, the solids were separated by filtration and washed with distilled water to produce 250 mL of filtrate, while the residue left on filter paper was washed with 10 mL diethyl ether and then with 150 mL of 10% ethanol. Then, the solid was transferred to an Erlenmeyer flask and cleaned with 200 mL of distilled water. 20 mL of HCl was added and the mixture was heated above its boiling point for 2.5 h. After cooling, the sample was neutralized with 500 ml of 45% NaOH solution and filtered through Whatman paper. Furthermore, the sugar content of the filtrate was analyzed using the Nelson-Somogyi method. From these methods, the glucose content was obtained, and the percentage of starch was determined by multiplying the glucose content by 0.9 (factor number).

3 Results and discussion

3.1 Chemical composition of fresh wild cassava root

According to the proximate composition, the wild cassava used for this study contains 81.57 % starch, 1.25 % protein, 0.39 % lipids, 1.28 % fibers, 0.26 % ash, and 13,74 % moisture. The cyanide acid content of fresh cassava was 338.41 ppm. These results clearly shows that the cassava roots used for this investigation are promising raw materials for starch production but the high content of the cyanide limits its exploitation. Therefore, detoxification process of these roots is a mandatory step before utilization in food industry.

3.2 Microbial growth

The fermentation process was performed by using of *L. plantarum* bacteria. To determine the time necessary to reach the log phase of *L. plantarum*, the growth curve was obtained. The calibration of the growth curve of microorganisms was performed using the counting chamber method, which aims to determine the number of microorganisms used in fermentation. Calculation of cell count was made every 2 h for 24 h. The logarithmic phase of the *L. plantarum* growth curve was obtained from 6 to 16 h (Figure 2). This log phase was used to determine the amount of bacteria by using the number of different starter volumes on each variable of the number of microorganisms.

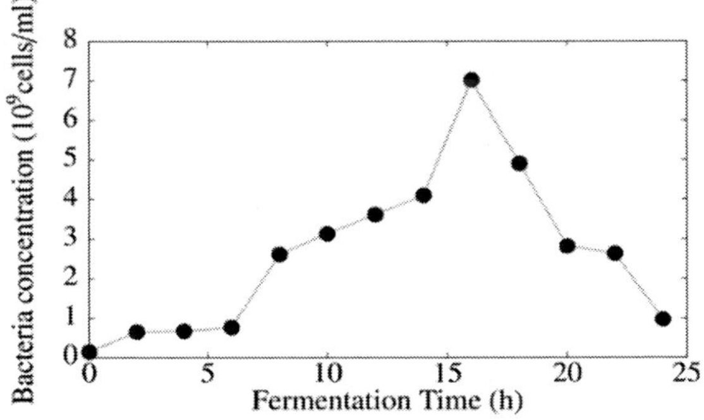

Fig.2. Growth curve of bacteria *L. plantarum*

3.3 Effect of the starter culture concentration on wild cassava fermentation

Two types of fermentation are used for processing of cassava roots, spontaneous and the starter culture, both of them are commonly used in Central Africa and Asia [9]. In this study, fermentation with starter cultures on wild cassava roots was used with the number of microorganisms being increased from $7x10^{10}$, $7x10^{11}$, $1.05x10^{12}$, to $3.5x10^{12}$ *L. plantarum* cells with a fermentation time of 12, 24 and 36 h.

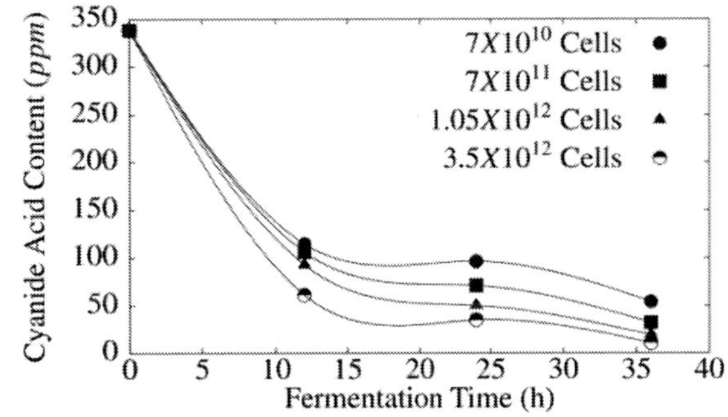

Fig.3. Effect of initial bacteria cell number on cyanide acid (HCN) removal in cassava roots fermentation.

Figure 3 shows that higher concentration of *L. plantarum* cells and longer fermentation periods have a positive effect on the HCN content in cassava. The decrease in HCN levels occurs drastically from ~ 350 to ~ 50 ppm in the first 12 h of fermentation. After 36 h of fermentation and using a concentration of $3.5x10^{12}$ cells, the cyanide content was further decreased up to 10.44 ppm. This remarkable fall could be explained by the enzymatic activity during fermentation, when the pH is lowered at 4, a value at which the enzymatic activity of linamarase, whose optimum pH is 6, is drastically reduced [14]. Due to the fact that linamarin and HCN are very soluble in water, the submerged fermentation process will provide the best conditions for the enzymatic activity aiming to reduce the HCN content in the wild cassava tuber. It is important to mention that a starter culture containing $3.5X10^{12}$ cells decreases the cyanide content up to 94% in the first 24 h, value that is superior to 91% obtained for mixed-starter culture [15]. In the light of this result, the advantage of pure culture such an *L. plantarum* is reproducibility and control to produce MOCAF with low cyanide content.

3.4 Effect of the starter culture concentration on the starch content

Starch is composed of two glucose polymers, i.e., amylose and amylopectin. Amylose is a linear α-(1 → 4) glucose polymer, while the amylopectin consists of short linear α-(1 → 4) linked chains with α-(1 → 6) linked side chains [16]. The mass ratio between these two components is around 1:3. However, from the medical point of view, a starch with higher content in amylose is preferred (the so-called "resistant starch") because it has a good effect on the physiological function of the body, due to its hypoglycemic effect (lowers blood sugar after meals), prebiotic effect, it lowers cholesterol and reduces the risk of colon cancer [6]. Thus, there is a huge interest to prepare such kind of starch. Our study shows that it is feasible to regulate the ratio between amylose and amylopectine in starch by controlling the concentration of the *L. plantarum* cells in the starter culture.

Figure 4 shows the evolution of the starch content over time depending on the concentration of microorganisms during cassava fermentation, when a decrease in the starch content with the increase of fermentation time and cell concentration was observed.

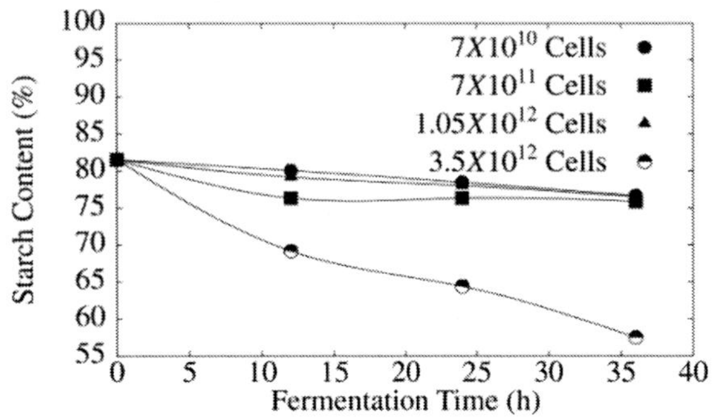

Fig.4. Effect of initial bacteria cell number on starch content.

A dramatic decay was noticed for a concentration of $3.5x10^{12}$ cells. The decrease in starch content is caused by the use of organic matter to meet the energy needs for the growth of microorganisms [6]. It is well known that during the process of fermentation, the starch is hydrolyzed into simpler sugars from oligosaccharides and maltose to glucose [11].

The effect of the cells concentration on the evolution of the amylose and amylopectin content during the fermentation process is depicted in Figures 5 and 6, respectively. It can be noticed that a value of $3.5x10^{12}$ cells was the optimum concentration at which the content of starch components were changed over time in the favour of amylase. These results were in accordance with previous studies [17]. Cassava roots were allowed to ferment enough time to reach an acidic pH value (put the value of pH), as a result of bacterial activity, suitable to start the starch hydrolysis at α- (1.4) bonds, thus increasing amylose groups [6]. According to Syahputri et al, a similar behavior was observed in the amylose content of fermented jali flour, for which the amylose content increased from 26.33% to 29.06% [18]. Enhance of the amylose content is justified by the breaking of the amylopectin branch chain at α- (1.6) glycoside bonds as a result of enzyme activity during fermentation.

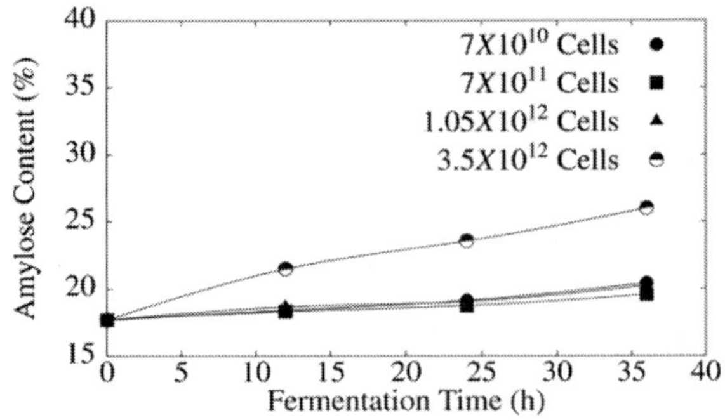

Fig.5. Effect of initial bacteria cell number on amylose content during cassava fermentation.

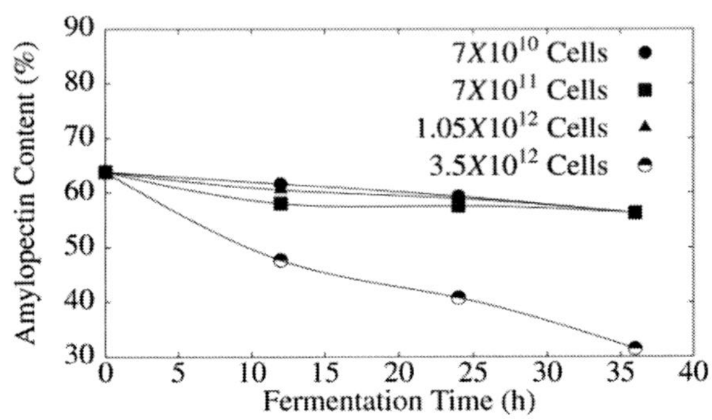

Fig.6. Effect of bacteria concentration on amylopectin content during cassava fermentation.

The highest amount of amylose (26 %) was obtained of fermentation (36 h) for the higher bacteria concentration ($3.5x10^{12}$ cells). As already mention above and according to Aliawati indicated that high amylose content correlates with high resistant starch [19]. Therefore, the product generated by our work has the potential to be used as raw material for resistant starch.

3.5 Effect of the starter culture concentration on the protein content

The current study was mainly focused on the detoxification of the cassava roots. The protein content was also evaluated, since high protein content in MOCAF is essential for nutritional reasons. It was interesting to emphasize that an increase in the protein concentration was noticed for the highest concentration of the cells in the starter culture. This feature is of particular interest knowing that usually, the protein content in MOCAF is improved by adding powdered dried fish debris, oil cake, or soybean flour to the cassava flour [20].The results are illustrated in Figure 7.

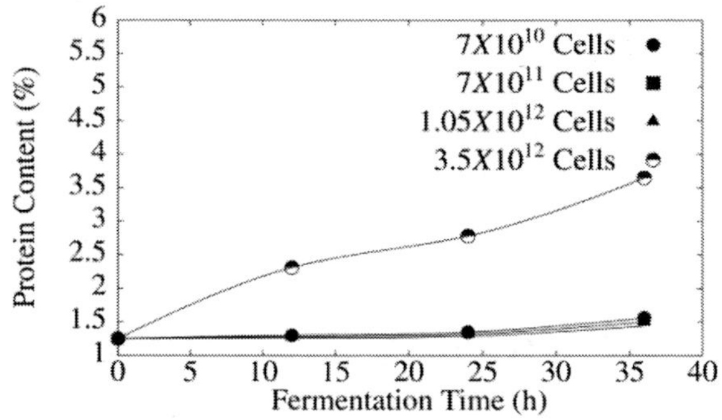

Fig.7. Effect of initial bacteria cell number on protein content during cassava fermentation.

Accordingly, the protein content was below 1.5 % after 36 h of fermentation regardless the concentration of the cells up to a value of 3.5×10^{12} cells. Once the concentration of the cells reached this limit value, a continuous increase of the protein content was noticed right from the beginning of the fermentation process that reached a value of 3.5 % after 36 h of fermentation. This percentage is similar with those reported by other authors in different cassava fermentation 4.07% and 2.19% for 48 h [14, 21]. However, in the present study, a shorter fermentation time was required. It is important to mention that higher protein content in MOCAF is essential for nutrimental reasons. However this is not easy to achieve; in consequence, the supplementation with soybean powders has been suggested by other authors. The present work is focused on the degradation of cyanide in the shorter possible time, but protein content is also an important matter to take into consideration for further studies.

4 Conclusions

This study showed an *L. plantarum* high-performance fermentation process to decrease the cyanide acid content of cassava roots. The toxicity of the roots significantly decreased under optimized conditions concerning the initial bacteria cell number in the starter culture (3.5×10^{12} cells) and fermentation time (36 h). Additionally, an improvement of the protein content in the cassava flour was obtained under optimized fermentation conditions. Therefore, we are confident that the outputs of this work open new ways as concerns the cyanide removal in cassava roots and the enhancement of protein content of MOCAF in a controlled and reproducible manner.

References

1. S. Gunawan, Z. Istighfarah, H. W. Aparamarta, F. Syarifah, I. Dwitasari. *Handbook on Cassava*. Nova Science Publishers, Inc., New York, 271-295 (2017)
2. Food and Agriculture Organization (FAO). Food Outlook: Binnual Report on Global Food Market. Italy (2015)
3. A. Hidayat, N. Zuraida, I. Hanarida. Indones. J. Agric. Sci. **3**, 25-32 (2002)
4. N.M.A. Nassar. Genet. Mol. Biol. **23**, 201-212 (2000)
5. FAO and IFAD international fund for agricultural development food and agriculture organization of the united nations. *The world cassava economy*. Rome, (2000)
6. M.G. Sajilata, S.S. Rekha, R.K. Puspha. Compr. Rev. Food. Sci. Food Saf. **5**, 1-17 (2006)
7. Shu X, Jia L, Gao J, Song Y, Zhao H, Nakamura Y, Wu D. Starch, **59,** 504-509 (2007)
8. Winarno. *Kimia Pangan dan Gizi*. Gramedia Pustaka Utama, Jakarta (1992)
9. J.A. Montagnac, C.R. Davis, S.A. Tanumihardjo. Compr. Rev. Food. Sci. Food Saf. **8,** 17-27 (2009)
10. World Health Organization (WHO). *Aldicarb. Environmental Health Criteria No. 121*. International Programme on Chemical Safety (1991)
11. K.A. Buckle, R.A. Edwards, G.H. Fleet, M. Wooton. *Ilmu Pangan*. Universitas Indonesia Press. Jakarta, (1985)
12. AOAC International. Official methods of analysis of AOAC International. 17th edition. 2nd revision. Association of Analytical Communities, Gaithersburg (2003)
13. SNI standard. Modified cassava flour. SNI 7622. National Standardization Agency of Indonesia, Jakarta (2011)
14. S. Gunawan, T. Widjaja, S. Zullaikah, L. Ernawati, N. Istianah. Int. Food Res. J. **22**,1280-1287 (2015)
15. N.Z Nkoudou, J.J.N Essia, Food Nutr. Sci. **8**,326-333 (2017)
16. M.J.E.C. van der Maarel, B. van der Veen, J.C.M. Uitdehaag, H. Leemhuis, L. Dijkhuizen. J. Biotechnol., **94,** 137-155 (2002)
17. A.L. Lehninger, D.L. Nelson, M.M. Cox. *Principles of Biochemistry*. Worth Publish, New York (1982)
18. D.A. Syahputri, A.K. Wardani. J. Pangan Agroindustri, **3** 984-995 (2015)
19. G. Aliawati. *Teknik analisis kadar amilosa dalam beras*. Buletin Teknik Pertanian. **8,** 82-84 (2003)
20. National Research Council (US) *Applications of Biotechnology to Fermented Foods: Report of an Ad Hoc Panel of the Board on Science and Technology for International Development*. National Academies Press, Washington DC (1992)
21. J. Sulistyo, L.J. Shya, H. Mamat, N.A. Wahab. AIP Conf. Proc. **1744**, 020030-1–020030-9 (2016).

The Effect of Alcohol on Bead Performance of Encapsulated Iron Using Deacetylated Glucomannan

Dyah H. Wardhani [1*] *and Heri* Cahyono[2]

[1]Chemical Engineering Department, Faculty of Engineering, Diponegoro University, Indonesia
[2]MSc Study Program on Chemical Engineering Department, Faculty of Engineering, Diponegoro University, Indonesia

Abstract. The success of encapsulation to protect iron from inhibitor degradation or oxidation depends on many factors including the excipient characteritics. Glucomannan, a neutral heterosaccharide, has a potential for the excipient. To improve the excipient performances, glucomannan is deacetylated to remove the acetyl groups by reacted with Na_2CO_3. This deacylated glucomannan is subject to bead formation after iron loading. The alcohol solution is commonly used in bead forming as dehydration medium during the encapsulation process. The objective of this work was to study the effect of alcohol on the bead performance of encapsulated iron using deacetylated glucomannan. The bead forming was conducted by dropping the excipient into ethanol and isopropyl alcohol (IPA) solution at various concentrations (50, 60, 70, 80 and 90%) and two condition temperatures (27-30° and 7-10°C). The encapsulation samples were subject to yield (YE) and efficiency of encapsulation (EE). The concentration of alcohol showed a positive impact on the yield and efficiency of encapsulation. Ethanol has a better performance compared with that of IPA regarding yield and efficiency of encapsulation. The optimum of yield bead formation (69.67%) and highest EE (66.80%) were obtained at 90% ethanol. Temperature of dehydration did not affect the YE and EE significantly.

1 Introduction

Encapsulation techniques have been developed broadly to accommodate various purposes. Among of the purposes of encapsulation are to protect the sensitive substances from the external environment, to mask the organoleptic properties like color, taste, odor of the substance, to obtain controlled release of the drug substance, for safe handling of the toxic materials, and to get targeted release of the drug [1].

Iron is one of the essential micronutrients that urgently develop into encapsulated form since it is easily oxidized when exposed to air. Its present induces unpleasant organoleptic properties especially odour and metallic taste [2],[3]. Moreover, iron encapsulation leads to increase bioavailability due to excipient protection of the iron from interaction with food components which are inhibitor degradation such as phytates, tannins, and polyphenols [4],[5]. The succeed of encapsulation in protecting the iron from inhibitor degradation or oxidation depends on many factors e.g. the excipient properties and preparation condition [1],[6].

Many compounds have been studied as excipient of iron encapsulation such as modified alginate [3],[7], hydrogenated soybean oil [8], dextrin, cellulose, and sodium hexametaphosphate (SHMP) [9],stearic acid, partially dehydrogenated soybean oil, wax matrix [10].

In recent years, application glucomannan (GM) as an excipient has been explored due to its extraordinary capability in absorbing water and excellent and stable films in both cold and hot water [11],[12]. GM is known as one of the highest viscosity among various gum types [13],[14]. However, the water-absorption of GM which is able to absorb up to 100 g water/g becomes a problem in the excipient application. Hence, this property requires to be controlled. Some methods have been reported to modify the water absorption of GM including thermal treatment [12], high-pressure treatment [15] and chemical modification [16]. Since the presence of acetyl groups in the GM chain inhibits interchain interactions [17], hence deacetylation is considered as one of the most appropriate chemical modification in reducing GM water-solubility. This modification is also expected to minimize iron oxidation due to contact with dissolved O_2 in water [6].

There are various techniques of bead forming in encapsulation process including spray drying, spray chilling or spray cooling, and extrusion coating [18]. An extrusion or dripping method is one of the simple method to obtain the bead [1],[19]. This process requires a dehydrating medium to obtain the desired bead of the encapsulated compound. The bead then is separated from the dehydrating medium [6]. Alcohol is commonly used as dehydrating medium due to its solubility in water

* Corresponding author: dhwardhani@che.undip.ac.id

https://doi.org/10.1051/matecconf/201815601005

[20]. With respect to solvation, the similar in chemical structure to each other depends primarily on their polarity [21]. Moreover, alchohol has high volatility which could help in dehydrating process. Zilberboim et al. reported that temperature of dehydrating medium affects the success in bead forming of Paprika oleoresin encapsulation with gum arabic [22]. Gupta et al. suggested to apply lower temperature in iron encapsulation when using alcohol as dehydrating medium [6].

Ethanol have been reported as a dehydration medium in iron encapsulation [19],[22] but no report isopropyl alcohol (IPA) as dehydrating medium, even though IPA has potential for dehydrating medium. However, to the best of our knowledge, there is lack of information in comparing both dehydrating medium in iron encapsulation using deacetylated GM (DGM). Therefore, the objective of this study was to determine the effect of type of alcohol and temperature of dehydrating medium in iron encapsulation with DGM. The samples were subject to yield and efficiency of encapsulation.

2. Materials and method

2.1. Materials

A. oncophyllus flour was bought from a local farmer at Nganjuk, East Java-Indonesia. GM of A. oncophyllus was extracted following Rahayu method's using IPA [23]. The purified flour contained 76.3% GM.

2.2. Deacetylation of Glucomannan

The deacetylation was conducted based on the method of Chen et al [24]. Ethanol (50%, 100 ml) and GM (5 g) were placed in Erlenmeyer which magnetically stirrer at 250 rpm and 50°C. After 30 min, Na_2CO_3 (0.4M, 25ml) was added to the suspension. The mixture was stirred using a magnetic stirrer (225 rpm) for 24 h. The solid which is separated using filter paper was oven-dried at 40°C for 6 h to obtain DGM.

2.3. Bead Forming

The bead forming was conducted following Wardhani et al. [25]. The dried DGM (2 g) was diluted with 100 ml of distilled water in Erlenmeyer and stirred at 250 rpm to make a suspension. One gram of $FeSO_4.7H_2O$ was diluted into the suspension. After 30-50 min, the solid was precipitated and the liquid was dropped carefully into ethanol (50%, 100 ml) using 10ml-syringe. The bead which formed after 12h was collected and oven-dried at 40°C for 4 hours to obtain dried-iron encapsulate. The yield (YE) of encapsulation was calculated following Eq.1.

$$YE = \frac{A}{B} x100\% \quad [1]$$

where A and B are the weight of final dried-iron bead and dried DGM, respectively.

2.3. Determination of Iron

Iron concentration was determined based on Tamura et al [26] procedure. Dried bead (1 g) was diluted in 100 ml distilled water. Place 0.1 ml sample into a 10 ml measuring flask. The solutions which are 1 ml of the hydroxylamine solution (100g/l), 10 ml of the 1,10-phenanthroline (1g/l) solution, and 8 ml of the sodium acetate solution (1.2 M) were added to the flask as the order. Distilled water was added upto the limit. Read the absorbance of the mixture at 520 nm against the absorbance of the iron standard solution (10-500 ppm). The efficiency of encapsulation was defined as Eq. 2:

$$Efficiency\ of\ encapsulation = \frac{Bound\ iron}{Total\ iron} x\ 100\% \quad [2]$$

3. Result and Discussion

Since iron was sensitive to be oxidized, solvent evaporation was selected as a method to form beads during iron encapsulation using deacetylated glucomannan. Type of alcohol, its concentration and temperature was studied over the yield and encapsulation efficiency.

3.1. Yield of Encapsulation

Alcohol was used in this study since it did not only serve as a precipitate agent for the glucomannan but it also promote dehydration of the forming bead [27]. Figure 1 describes the effect of ethanol and IPA as dehydration medium and its concentration on the yield of encapsulation (YE). In this study, the mixture of DGM-iron sol was syringe-dropped to the alcohol as dehydration medium. As soon as the sol contact with the alcohol, the sol dehydrated and resulted in the beads formation. This bead formation involved dehydration process where water content of the sol reduced. The water of the sol system was relatively in higher concentration toward the water in the alcohol which acted as the desiccant. As a result, dehydration occurred due to water removal of the sol to the alcohol. This principle of phenomenon was similar to the osmotic dehydration in which the water diffused from the the higher osmotic pressure solution to the hypertonic ones through a membrane [28]. In this case the bead of DGM-Iron acted as the membrane while alcohol was the hypertonic solution.

In general, YE of ethanol was higher than that of IPA in the all concentrations (Figure 1). The highest YE (69.67%) obtained using 90% ethanol and the lowest YE (23.97%) obtained from 50% IPA. This phenomenon could be due to different DGM response to the properties of both desiccant solutions including polarity, solubility and volatility.

One of the driving force of this dehydration was solubility of the desiccants. A solute dissolves in a solvent that has most similarity in chemical structure to itself. With respect to solvation, this similar in chemical structure to each other depends primarily on their polarity [21]. This polarity could be expressed with dielectric constan [29]. Dielectric constant of ethanol (25.3) was higher than that of IPA (20.18) [30].This higher solubility of ethanol led to higher driving force for water transfer between the sol and the desiccant. This condition allowed to gain more beads in using ethanol as desiccant. Yeo and Park suggested higher solubility desiccant allowed relatively faster mass-transfer between the excipient and the desiccant. This led to faster precipitation of the polymer [31].

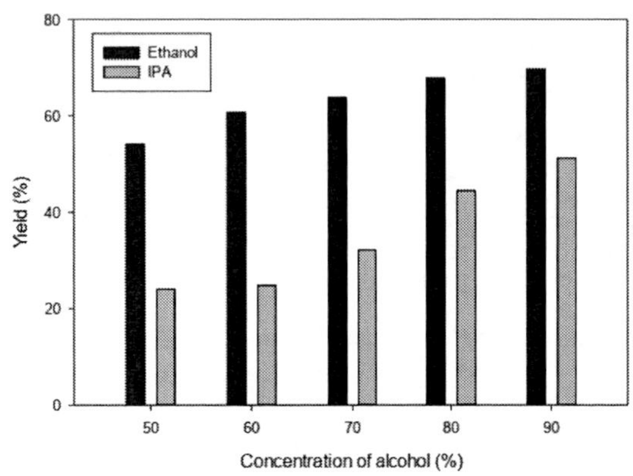

Fig. 1. Effect of type and concentration of alcohol on the yield of encapsulation

IPA has more number of carbon than ethanol, hence it is less volatile than ethanol [32]. As consequent, it could require more time to obtain similar amount of beads than that of ethanol.

Increasing alcohol concentration influenced positively on the yield (Figure 1). The water content of high alcohol concentration was low. This led to create high driving force between the system and more beads were formed in shorter time [19]. Similar result was reported by Zilberboim et al. who found higher alcohol concentration as a dehydrating medium increase yield of encapsulation of oleoresin and aromatic esters of paprika. Lower concentration of alcohol caused particle agglomeration which difficult to be filtered [22].

The effect of temperature on YE was conducted in ethanol solution only. Two temperature were studied *i.e.* 7-10°C and 25-27°C. The results is represented at Figure 2. In general lower temperature showed slightly higher yield than higher temperature. The highest YE (55.04%) was obtained at 90% ethanol of low temperatures. Meanwhile, the lowest YE (20.58%) was achieved at low temperatures using 60% ethanol. Temperature is one of the factors that affect the solubility of the substance. Higher temperature facilitated disruption of the beads and dissolve them [6].

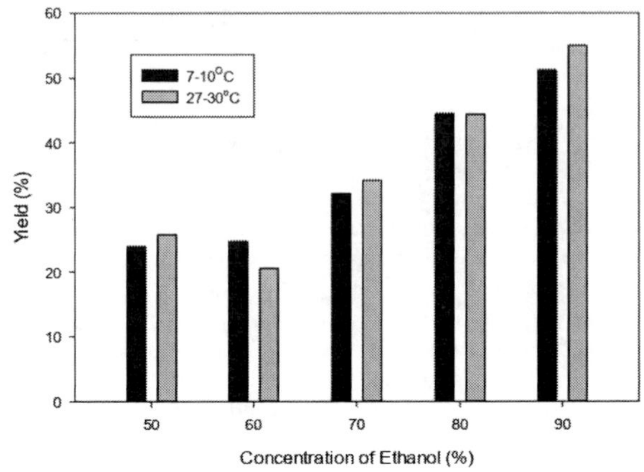

Fig. 2. Effect of temperature of dehydrating medium and concentration ethanol on the yield of encapsulation

3.2. Efficiency of Encapsulation (EE)

Encapsulation efficiency (EE) is one of the success parameters on the encapsulation process. EE determined by comparing the iron content of the beads and the iron added in the encapsulation process. The iron content was measured by spectrophotometric method. The effect of type and concentration of alcohol on the EE is described at Figure 3. EE has a positive correlation with alcohol concentration in both types of alcohol. Ethanol as a hydrating medium showed a higher EE than IPA. The highest EE (51.23%) was obtained at 90% ethanol, while the lowest level of EE (16.41%) was obtained using 50%.

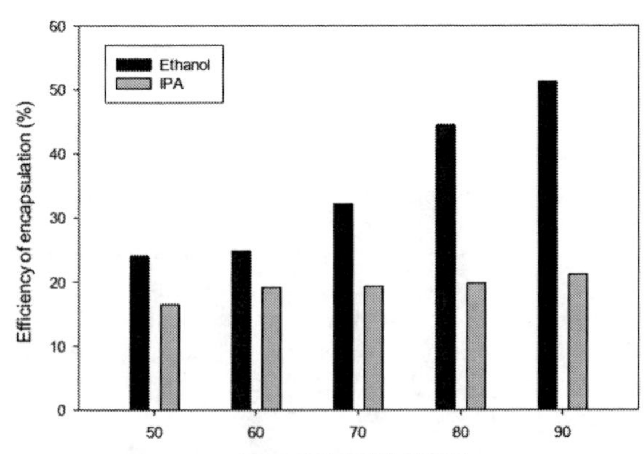

Fig. 3. Effect of type and concentration of alcohol on the efficiency of encapsulation (EE)

In this study, ferrous sulphate heptahydrate ($FeSO_4.7H_2O$) was used as a source of iron. This compound is water soluble (29.51g/100ml, at 25°C) but insoluble in alcohol [33]. Hence, the use of alcohol as dehydration medium in bead forming was an appropriate step.

Ethanol was better in retaining iron in the bead than IPA during the dehydration process. Favourable solubility of iron in the dehydration medium would enhance EE. Low water solubility of desiccant required more time to obtain the beads. However, extended the time allowed to the iron to diffuse to the desiccant [31]. Since IPA has a lower water-solubility than ethanol, hence iron tended to exist on dehydrating medium than bound to DGM excipient.

Alcohol concentration also has an influence on EE on both types of alcohol. Concentration of alcohol showed a positive relation with the EE. High concentration of alcohol led to rapid bead formation due to low water content in the medium. This reduced the possibility of iron loss during beading [19]. This result supported by Gupta et al. which confirmed that the increase in alcohol concentration in bead forming of encapsulated iron could improve the efficiency of iron encapsulation [6].

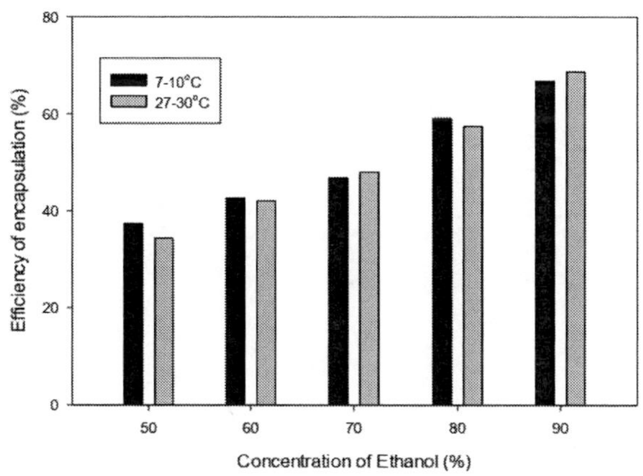

Fig. 4. Effect of temperature of encapsulation and ethanol encapsulation on the efficiency of encapsulation (EE)

Fig. 4 shows increasing ethanol concentration led to increase EE. The highest EE (68.70%) obtained at low temperatures of 90% ethanol. Meanwhile, the lowest EE (34.39%) was found at low temperatures with an ethanol concentration of 50%. The impact of temperature on EE is not significant (Fig.4) which was in line with the yield result. Yeo and Park suggested EE was not affected by solvent evaporation temperature [31].

4. Conclusion

This study confirmed that the ethanol was better as dehydrating medium than IPA in iron encapsulation using deacetylated glucomannan. Higher alcohol concentration was selected in gaining more bead formation. The optimum bead formation (69.67%) and highest EE (51.24%) were obtained at 90% ethanol. Temperature of dehydration did not affect the YE and EE significantly.

References

1. N. V. N. Jyothi, P. M. Prasanna, S. N. Sakarkar, K. S. Prabha, P. S. Ramaiah, G. Y. Srawan., J. Microencapsulation, **27**, 187-197 (2010).
2. C. Liyanage, S. Zlotkin., Food Nutr. Bull., **23**, 133-7. (2002).
3. S. Khosroyar, A. Akbarzade, M. Arjoman, A. A. Safekordi, S. A. Mortazavi., Afr. J. Microbiol. Res. **6**, 2455-2461. (2012)
4. M. B. Zimmermann, R.Wegmueller, C. Zeder, N. Chaouki, R. Biebinger, R. F. Hurrell, E. Windhab., Am. J. Clin. Nutr., **80**, 1283-1290. (2004)
5. K. Raes, D.Knockaert, K. Struijs, J. Van Camp, Trends Food Sci. Technol., **37**, 32-41. (2014).
6. C. Gupta, P. Chawla, S. Arora, S. K. Tomar, A. K. Singh., Food Hydrocoll., **43**, 622-628. (2015).
7. A. N. Bezbaruah, S. S.Shanbhogue, S.Simsek, E. Khan, J. Nanopart. Res., **13**, 6673-6681. (2011)
8. S. Zlotkin, P. Arthur, K. Y. Antwi, G.Yeung,. Am. J Clin Nutr., **74**, 791-795. (2001).
9. K. S. Rao, & B. N. Rao, Food Chem., **17**, 13-23. (1985)
10. R. Wegmüller, M.B. Zimmermann, R.F. Hurrell, J. Food Sci., **68**, 2129-2135. (2003)
11. J. Yang, J. X. Xiao, L. Z. Ding., Eur. Food Res. Technol., **229**, 467-474. (2009).
12. N. Herranz, S. Gallage, M. Mellone, T. Wuestefeld, S. Klotz, C. J. Hanley, A. Georgilis., Nat. Cell Biol, **17**, 1205-1217. (2015)
13. E. I. Yaseen, T. J. Herald, F. M Aramouni, S. Alavi,. Food Res. Int., **38**, 111-119. (2005)
14. Y. Wang, , J. Liu, Q. Li, Y. Wang, C. Wang,. Biotechnol Lett., **37**, 1-8. (2015).
15. H. M. Moreno, , B.Herranz, , A. J. Borderías, , C. A. Tovar., Food Hydrocoll, **60**, 437-444. (2016).
16. D. H. Wardhani, N. Aryanti, K. Haryani, F. Maharani., Carpathian Journal of Food Science & Technology, **9**, 43-56. (2017)
17. X. Du, J. Li, J. Chen, B. Li., Food Res Int, **46**, 270-278. (2012).
18. F. Gibbs, K. Selim, A. Inteaz, N. Catherine, B. Mulligan., Int. J. Food Sci. Nutr., **50**, 213-224. (1999).
19. E. Asghari-Varzaneh, M. Shahedi, H. Shekarchizadeh., Int. J. Biol. Macromol, **103**, 640-647. (2017).
20. Y. Guo, Y.Hao, Y. Zhou, Z. Han, C. Xie, W. Su, H. Hao., J. Chem. Thermodyn., **106**, 276-284. (2017).
21. J. McMurry, *Organic Chemistry,. 6th Ed.*. Brooks Cole, Pacific Grove, CA (2000).
22. R. Zilberboim, I. J. Kopelman, Y. Talmon., J. Food Sci., **51**, 1301-1306. (1986).

23. L. H. Rahayu, D. H. Wardhani, A. Abdullah,. J. Metana, **9**, 45-52. (2013) (in Indonesia)
24. J. Chen, J. Li, B. Li, Carbohydr. Polym., **86**, 865-871. (2011).
25. D. H. Wardhani, A. Abdullah, A. N.Azizah, M. Y. Ananta, AIP Conf. Proc, **1746**, 020039. (2016).
26. H. Tamura, K. Goto, T. Yotsuyanagi, M. Nagayama, Talanta, **21**, 314-318. (1974).
27. F. Lim, R. D.Moss, *Vitamin encapsulation*, U.S. Patent No. 4.389.419. (1983)
28. S. Mizrahi, S. Eichler, O. Ramon, J. Food Eng., **49**, 87-96. (2001).
29. A. D. Smith, ed. *Oxford Dictionary of Biochemistry and Molecular Biology.* 2nd Edition. Oxford: Oxford University Press, (2000)
30. J. A. Dean, *Lange's handbook of chemistry. Material and manufacturing process*, **5**, 687-688. (1990).
31. Y. Yeo, K. Park, Arch Pharm. Res, **27**, 1-12. (2004)
32. L. Berg., Separation of ethanol from isopropanol by azeotropic distillation, U.S. Patent No. 5,415,741 A. (1995)
33. W. F. Linke, A. Seidell, *Solubilities, inorganic and metal-organic compounds: a compilation of solubility data from the periodical literature.* 2nd Vol, 4th ed, Van Nostrand, New York (1965)

Performance Comparison of Commercial Enzymes for The Synthesis of Glucosamine by Chitosan Hydrolysis in The Presence of Surfactant

Nur Rokhati[1], *Heru* Susanto[1], *Titik* Istirokhatun[2], Purwono[2] and *Bambang* Pramudono[1,*]

[1]Department of Chemical Engineering, Faculty of Engineering, Diponegoro University
[2]Department of Environmental Engineering, Faculty of Engineering, Diponegoro University

Abstract. Glucosamine has attracted much attention due to its potential biological, pharmaceutical and nutritional effects. In this research, glucosamine was prepared by enzymatic hydrolysis of low molecule weight chitosan with nonionic surfactant (Tween 80) addition. The change in reducing sugar content was used as an indicator for the hydrolysis reaction rate. The results showed that addition of Tween 80 (1%, w/w) increases the formation of reducing sugars approximately two times higher than without Tween 80 addition. Hydrolysis using a combination of cellulase and β-glucosidase showed higher reaction rate than using a combination of cellobiohydrolase and β-glucosidase or using β-glucosidase only. The combination of cellulase and β-glucosidase showed optimum hydrolysis at reaction time of 24 hours. High-performance liquid chromatography (HPLC) analysis of the glucosamine product showed typical peak distributions the same as those of commercial standard glucosamine hydrochloride.

1 Introduction

Glucosamine, 2-amino-2-deoxy-D-glucose, is a common monosaccharide that is derived from chitosan [1] and the biochemical precursor of all amino sugars in the human body [2]. Glucosamine is receiving much more attention recently because of its has therapeutic potential and is used to prevent or treat a wide variety of diseases. Many biological, pharmaceutical and nutrimental effects of glucosamine have been reported, such as liver protective effect [3], treatment of osteoporosis [1], cancer [4], skin wound healing [5], healing of surgical incisions [5], inflammation [6], bone regeneration [7], and so on.

Glucosamine can be produced by either chemical or enzymatic hydrolysis of chitosan. In the case of commercial production, glucosamine was prepared by complete hydrolysis of chitosan using concentrated HCl. Chemical hydrolysis has some disadvantages, such as low yield, acidic waste by the use of concentrated hydrochloric acid, chemical modifications of glucose ring and so on [1]. Therefore, enhancing of Glucosamine production by enzymatic hydrolysis gets a lot of attention.

Chitosanases are a group of specific complex hydrolytic enzymes that catalyze depolymerization of chitosan. Chitosanase hydrolyzes chitosan to oligoglucosamine chains. Subsequently β-D-glucosaminidase hydrolyzes oligomers to glucosamine monomers. However, Chitosanases are very expensive and unavailable in bulk quantities for commercial application. Xia *et al.*, [8] reported that some non-specific enzymes such as cellulases, lipases, and proteases have the ability to hydrolyze chitosan and those were comparable with the results achieved by chitosanase. However, these enzymes were added at relatively high concentrations.

In this paper, we report the preparation of glucosamine product by hydrolyzing chitosan using cellulase and glucosidase with surfactant Tween 80 addition. The addition of surfactant was expected to be able to decrease the amount of enzyme used. The optimal reaction conditions and the characterization of the glucosamine product were investigated.

2 Materials And Methods

2.1 Materials

Raw chitosan is low molecule weight chitosan was purchased from Sigma-Aldrich. The cellulase enzyme (Cat. 22178), Cellobiohydrolase enzyme (Cat. E6412-100UN), and β-Glucosidase enzyme (Cat. 49290-1G) were purchased from Sigma-Aldrich. Acetic acid glacial (CH_3COOH), sodium carbonate (Na_2CO_3), and

* Corresponding author: pramudono@live.undip.ac.id

surfactant Tween 80 were purchased from Merck Germany. Potassium ferricyanide ($K_3(Fe(CN)_6)$) was purchased from Sigma-Aldrich, Germany.

2.2 Hydrolysis of Chitosan

The method of chitosan hydrolysis followed our previous publication [9]. Briefly, chitosan was dissolved in acetic acid/sodium acetate buffer (0.1 M) at a concentration of 1% (w/v) and was mixed with the surfactant (Tween 80, 1% w/w of chitosan). Chitosan solution (50 mL) was treated using an enzyme with the ratio 1:100 (w/w). The reaction mixture was incubated at 50^0C with constant stirring (100 rpm). After hydrolysis, the enzyme was inactivated at water bath 100 ^{0}C for 10 min. Then the amount of reducing sugar of the filtrate was measured.

2.3 Crystallization of Glucosamine Product

The filtrate from the hydrolysis was filtered using an ultrafiltration membrane and added with 0.02 N HCl solutions. Then it was concentrated by evaporation at low pressure until the final volume was 10% of the initial solution. The concentrate was mixed with an equal volume of ethanol yielding a precipitate. The precipitate was collected by centrifugation, washed with ethanol several times, and dried in a vacuum oven.

2.4 Analysis

Analysis of total reducing sugars was conducted using Imoto method [10]. The absorbance of the mixture solution was measured at 420 nm. Total reducing sugars was calculated based on a standard curve obtained with D-glucosamine HCl.

The glucosamine product was analyzed by using HPLC on Chromolith® Performance NH_2 column (100 x 4,6 mm) at temperature of 30°C. The Mobile Phase is Acetonitrile: Distilled Water (60:40 v/v) with a flow rate was 0.8 ml/min. The detector used was refractive index detector (RID) with injection volume 20 µl at running time 30 min.

FTIR spectra were recorded with KBr powder using the instrument IR Prestige-21 Shimadzu, Japan. Thirty-two scans were performed at a resolution of 4 cm^{-1} and a temperature of 21±1°C over the wavelength range of 500 – 4000 cm^{-1}.

3 Results And Discussion

3.1 Effect of Enzyme Type

Cellulase is an endo-acting hydrolase, which performs random cuts at internal amorphous sites of the polysaccharide chain generating oligosaccharides of various lengths responsible for decreasing degree of polymerization of substrates [11]. Whereas cellobiohydrolase and β-glucosidases are exo-acting hydrolase. Cellobiohydrolase acts in a processive manner from the ends of polysaccharide chains, processing along the polymer chain while liberating disaccharide as the major product. β-Glucosidases attack the substrates from the nonreducing-end, producing monosaccharide [12]. In this study, the effect of enzyme type, i.e., β-glucosidase only, the combination of cellulase - β-glucosidase, and the combination of cellobiohydrolase - β-glucosidase, on chitosan hydrolysis was investigated. The nonionic surfactant (Tween 80) was added in chitosan solution.

The formation of reducing sugar indicates the activity of an enzyme that cleaves glycosidic bond in chitosan polymer [13]. Therefore, the activities of chitosan enzymatic hydrolyzed can also study by measuring reducing sugar content. The reducing sugar formation from hydrolysis chitosan with various enzyme types at reaction time 24 hours showed in Fig. 1. The hydrolysis using a combination of cellulase - β-glucosidase showed higher reaction rate than using a combination of cellobiohydrolase - β-glucosidase or using β-glucosidase only.

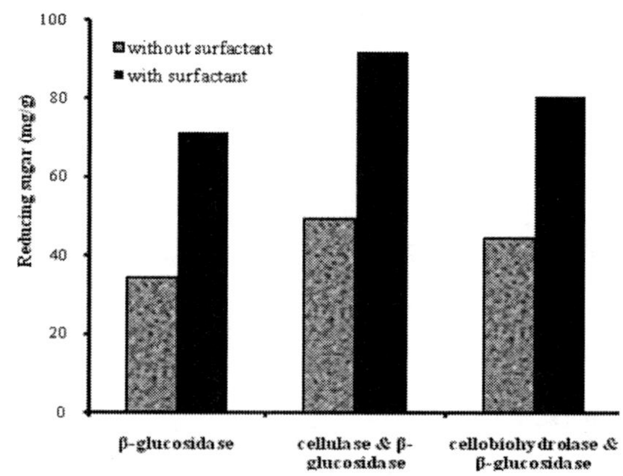

Fig.1. The effect of enzyme type and surfactant addition on the reducing sugar formation. The reaction time and surfactant concentration were 24 h and 1% (w/w), respectively.

β-glucosidases primarily hydrolyze glycosidic bonds of the short chains polysaccharide. Disaccharide has been shown to have a stronger inhibitory effect on the enzymes than monosaccharide, and cellobiohydrolase produces disaccharides as the main product [12]. Mechanism of hydrolysis of chitosan by a combination of cellulase - β-glucosidase can be illustrated as: cellulase hydrolyzes chitosan to chitooligosaccharides chains, and β-glucosidase hydrolyzes oligomers to glucosamine monomers.

Fig. 1 also shows that Tween 80 could increase the formation of reducing sugars approximately two times higher than without Tween 80. Enzymatic hydrolysis was affected by the rates of adsorption and desorption of enzyme to the substrate [14]. The surfactant can improve a desorption of enzyme from a functional group of the substrate and then absorb another substrate functional group. The presence of acetyl groups in chitosan chain caused chitosan has hydrophobic properties [15]. The

https://doi.org/10.1051/matecconf/201815601006

hydrophobicity of the backbone of chitosan chain played an important role in its interactions with surfactants [16].

3.2 Effect of reaction time

Fig. 2 shows the effect of reaction time on the formation of reducing sugar.

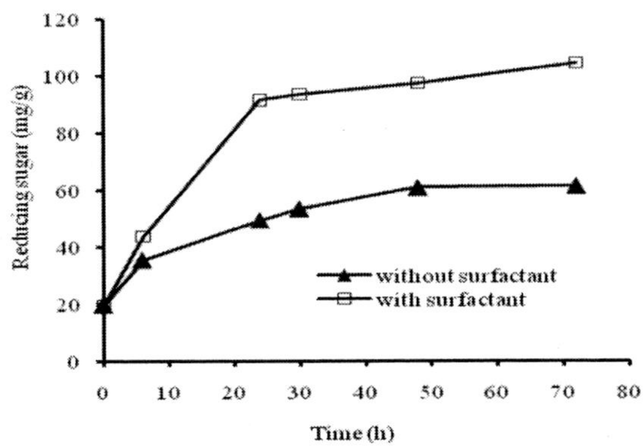

Fig. 2. The Effect of reaction times on the formation of reducing sugar. The enzyme type was a combination of cellulase – β-glucosidase.

During the entire time reaction, hydrolysis of chitosan with Tween 80 addition always possessed higher reducing sugar content than without Tween 80 addition. There was a sharp increase in the formation of reducing sugar within 24 h. After 24 hours, the formation rate of reducing sugar decreased. This phenomenon is probably due to the inhibition of the enzyme activity by the end products. This statement is similar to the previous publications by Li *et al.* [17] and Zhou *et al.* [18].

3.3 Characterization of the glucosamine

The FT-IR spectra of raw and hydrolyzed chitosan (the enzyme type and reaction time were a combination of cellulase – β-glucosidase and 24 h, respectively) are shown in Fig. 3.

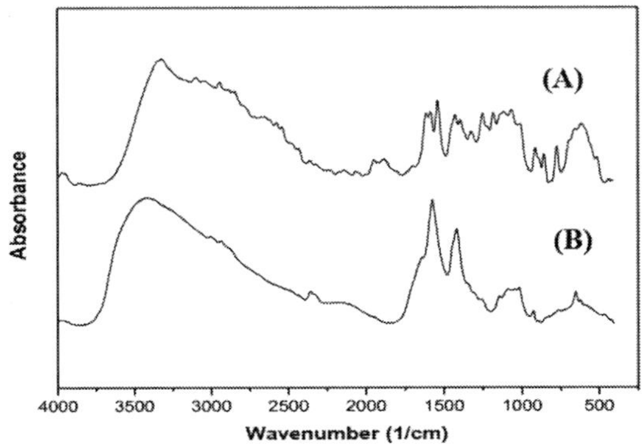

Fig. 3. FTIR spectra of raw chitosan (A) and hydrolyzed chitosan (B)

The spectrum of raw chitosan showed the absorption peaks at 3446.79, 1595.13, and 1423.47 cm^{-1}, which indicates the amino peak characteristic of chitosan. The absorption peak at 1153 cm^{-1} indicates –C-O-C- glycosidic linkage between repeating unit of chitosan. These results are in good agreement with previous publications [19, 20]. The absorption peaks of glycosidic linkage disappear at spectra of hydrolyzed chitosan.

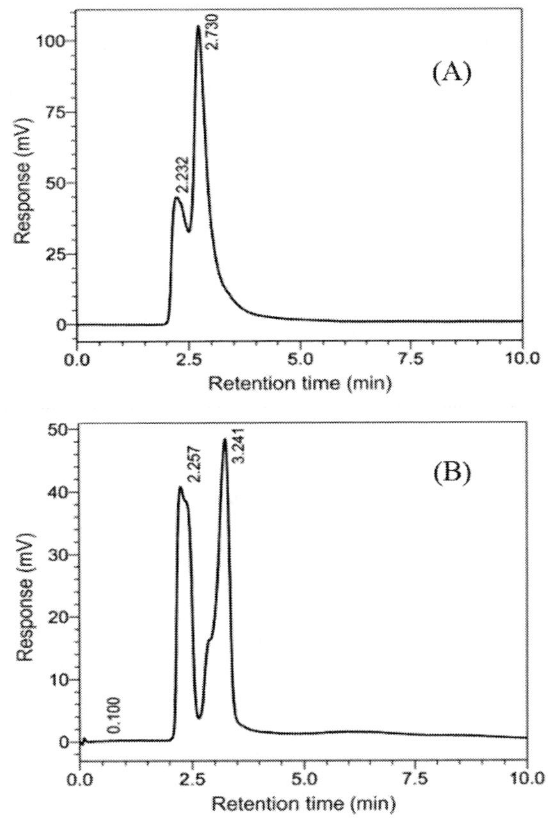

Fig.4. HPLC patterns : (A). Glucosamine hydrochloride and (B). Product of hydrolyzed chitosan

The molecular weight distribution of chitosan has been considered to be one of the most important characteristics affecting the functionality of the polymer [21]. The molecular weight distribution of the hydrolyzed chitosan product was analyzed using HPLC (Fig. 4). The eluted peak of the product of hydrolyzed chitosan similar with glucosamine hydrochloride (the retention time range about 2.4 – 2.7 min). It shows that D-glucosamine can be produced by the enzymatic hydrolysis of chitosan using a combination of cellulase – β-glucosidase.

4 Conclusion

Enzymatic hydrolysis of chitosan using various type enzyme with surfactant addition was investigated. The results of this study showed that addition of non-ionic surfactants Tween 80 could effectively increase enzymatic hydrolysis. Within the range studied, glucosamine could be prepared using a combination of cellulase – β-glucosidase with enzyme:chitosan ratio of 1:100 (w/w) and the addition of surfactant Tween 80 (1%, w/w), at temperature 50 ^{0}C for 24 hours.

References

1. Y. Sun, J. Zhang, S. Wu, S. Wang, Int J Biol Macromol. **61** (2013).
2. R. Dalirfardouei, G. Karimi, K. Jamialahmadi, Life Sci. **152** (2016).
3. S.I. Sal'nikova, S.M. Drogovoz, I.A. Zupanets, Farmakol Toksikol. **53**,4 (1990).
4. H.J. Oh, J.S. Lee, D.K. Song, D.H. Shin, B.C. Jang, S.I. Suh, J.W. Park, M.H. Suh, W.K. Baek, Biochem Bioph Res. Co. **360**, 4 (2007).
5. K.A. Martin Xavier, K.G. Ramachandran, *Standardization of Optimum Conditions for the Production of Glucosamine Hydrochiloride from Chitin,* (Central Institute of Fisheries Technology, 2006)
6. I. Nagaoka, M. Igarashi, J. Hua, Y. Ju, S. Yomogida, K. Sakamoto, Carbohyd. Polym. **84**, 2 (2011)
7. S. Sinha, R.. Kumar, S.R. Dhakate, S. Chand, IPCBEE. **9** (2011).
8. W. Xia, P. Liu, J. Zhang, J. Chen, Food Hydrocolloid, **25**, 2 (2011)
9. N. Rokhati, H.Susanto, M. Waskito, R. Rahmawati, B. Pramudono, Adv Sci Lett. **23**, 6 (2017)
10. Z. Zhang, C. Li, Q. Wang, Z.K. Zhao, Carbohyd. Polym. **78**, 4 (2009).
11. L.R. Lynd, P.J. Weimer, W.H. Van Zyl, I.S. Pretorius, Microbiol. Mol. Biol. R. **66**, 3 (2002).
12. N. Andersen, *Enzymatic hydrolysis of cellulose*. (Bio Centrum, Technical University of Denmark, 2007).
13. S.J. Horn, V.G. Eijsink, Carbohyd. Polym. **56**, 1 (2004).
14. Y. Chen, X. Zhang, S. Zhang, W. Qin, C. Guo, X. Guo, D. Xiao, Turk. J. Biol. **38**, 4 (2014)
15. C. Bangyekan, D. Aht-Ong, K.Srikulkit, Carbohyd. Polym. **63**, 1 (2006)
16. I. Pepić, J. Filipović-Grčić, I. Jalšenjak, Colloid Surface A, **336**,1 (2009).
17. J. Li, Y. Du, H. Liang, Polym Degrad Stabil. **92**, 3 (2007).
18. Y. Zhou, H. Chen, F. Qi, X. Zhao, D. Liu, Bioresource Technol. **182** (2015)
19. H.Y. Zhou, X.G. Chen, M. Kong, C.S. Liu, D.S. Cha, J.F. Kennedy, Carbohyd. Polym. **73**, 2 (2008)
20. I. Prasertsung, S. Damrongsakkul, N.Saito, Polym Degrad Stabil. **98**, 10 (2013)
21. S.B. Lin, Y.C. Lin, H.H. Chen, Food Chem. **116**, 1 (2009)

Polyvinyl Alcohol (PVA) Partially Hydrolyzed Addition in Synthesis of Natural Hydrogel Carboxymethyl Cellulose (CMC) Based from Water Hyacinth

Asep Handaya Saputra[1,*], *Nadia* Huda Apriliana[2]

[1]Department of Chemical Engineering, Faculty of Engineering Universitas Indonesia Depok 16424
[2]Department of Chemical Engineering, Faculty of Engineering Universitas Indonesia Depok 16424

Abstract. There are many benefits from Indonesia's biodiversity. One of them is cellulose of Water hyacinth. Water hyacinth contains 60 % cellulose. Cellulose of water hyacinth can be used as material for polymer product such as hydrogel. Hydrogel is a hydrophilic superabsorbent polymer that can absorb water until 200 times from its weight without dissolved in water. In synthesis of hydrogel, the cellulose should be modified to Carboxymethyl Cellulose (CMC) by alkalization and carboxymethylation process in reaction media ethanol and isopropanol (2:8). The aim of the study is to obtain the hydrogel with high swelling ratio characteristic by adding Polyvinyl Alcohol (PVA) Partially Hydrolyzed with variation of PVA/CMC composition (20:80, 80:20, 50:50, 25:75 and 75:25). The optimum PVA/CMC composition are 20:80 and 25:75 which give results 29% and 25% in their swelling ratio. This low swelling ratio is caused by morphology structure of hydrogel that has a compact structure and small pores. This structures lead to difficulity in water diffusion. In compare with hydrogel from PVA Fully Hydrolyzed addition, hydrogel with PVA Partially Hydrolyzed addition contain vinyl acetate that cause hydrogel less hydrophilic and has a low value on liofil-hidrophilic balance.

1 Introduction

The amount of biodiversity in Indonesia has became a great opportunity to explore its potential. For example, water hyacinth can be utilized to obtain natural polymer in cellulose form. Water hyacinth consists of 60% cellulose, 8% hemicellulose, and 17% lignin[1]. Utilization of water hyacinth may enhanced its value, since water hyacinth is noxious weed due to its congested growth.

Fiber of water hyacinth can be used as the basic material of polymer product in the form of hydrogel. Hydrogel is a superabsorbent polymer with the ability to absorb liquid up to 200 times the weight of itself without getting dissolved in water. This polymer product is environmental friendly for it easy to be dissolved.

Hydrogel can be used in many applications. Recently, extensive research and development has been conducted for hydrogel applications in the scopes of healthcare, biomedicine, agriculture, pharmacy, food packaging, horticulture industry and other scopes. As in healthcare, hydrogel can be used as wound dressing, drug release system, contact lenses, and transdermal drug delivery. In agriculture sector, hydrogel can ensure the availability of water, reduce the frequency of irrigation up to 50% and also applicable to areas that lacks of water supplies. Futhermore, hydrogel can be used as constituent material in diaper and period pads.

Hydrogel is classified into two categories, the natural and the synthetic hydrogel. Hydrogel from water hyacinth cellulose is classified into natural hydrogel, whereas synthetic hydrogel is *Polyhydroxyethyl Methacrylate* (pHEMA) dan *Polyacrylamide*. Synthetic hydrogel is not biodegradable and may increase air pollution, since about 90% of used hydrogel ought to be incinerated or disposed on landfills.[2]

In synthesis of hydrogel, cellulose needs to be modified to become its derivative, Carboxylmethyl Cellulose (CMC), for it has a functional group which causes electrostatic repulsion, that would arise between macromolecules, and water would be taken up.[3]

Moreover, addition of hydrophilic polymer: Polyvinyl Alcohol (PVA) is essential, in order to maximize water absorption. PVA is not only biodegradable and nontoxic, but also has stable both thermal and chemical properties. PVA is classified into 2 groups based on its synthesis process, PVA Partially Hydrolyzed and PVA Fully Hydrolyzed.

PVA fully hydrolyzed has a high degree of crystallinity. Addition of fully hydrolyzed PVA results in

* Corresponding author: nadia.huda@ui.ac.id

reduction of swelling ratio as reported in previous research that was conducted in the same laboratory and using the same instrument.[4]

Polymer with high crytallinity embodies the characteristic of being physically stable. Consequently, it is difficult for water as a swelling medium to enter within the structure. [5] Hence, the synthesis of hydrogel in this study will add PVA Partially Hydrolyzed which has low value in degree of crystallinity. Addition of PVA Partially hydrolyzed is expected to produce hydrogel with higher swelling ratio properties.

The purpose of this study is to obtain optimum composition ratio of PVA and CMC in hydrogel synthesis, based on the properties and characteristics of hydrogel. Addition of PVA Partially Hydrolyzed and its effects on properties and characteristics of produced hydrogel were studied. Differences between addition of PVA Partially Hydrolyzed and PVA Fully Hydrolyzed on synthesis hydrogel were also analysed.

2 Materials and Methods

2.1 Material

This research used water hyacinth from Cipondoh Lake, Tangerang, Indonesia. Several materials that used for isolating cellulose are toluene, ethanol, $NaClO_2$, NaOH, distilled water, acetate acid, and buffer solution. While materials that used for synthesis CMC and hydrogel are Na-MCA, NaOH, ethanol, isopropanol, and PVA Partially Hydrolyzed.

2.2 Methods

2.2.1 Isolation of Water Hyacinth Cellulose

Preparation started with separation of water hyacinth stems (part with highest cellulose content) from roots and leaves. Stem was dried under direct sunlight for 7 days to remove water content. The dried stem was grinded and sieved with Willey Ball Mill until it reached size of 60 mesh. The next step was dewaxing. The water hyacinth powder was wrapped in filter paper, then extracted with Soxhlet Apparatus and mixture of toluene and ethanol (2:1) as solvent.

Prior to lignin content removal, the water hyacinth powder was mixed with 1 %wt solution of $NaClO_2$. The mixture was put into prepared bottles and laid in water bath at 80°C for 3 hours. After lignin removal process, the powder was soaked in NaOH 17.5% solutions for 3 hours to remove hemicellulose. The powder was washed and neutralized with dillistilate water or CH_3COOH and dried in room temperature. This isolation obtained α-cellulose from water hyacinth.

2.2.2 Synthesis of Carboxymethyl Cellulose (CMC)

Synthesis of Carboxymethyl Cellulose (CMC) consists of alkalization and carboxymethylation process. Alkalization process was conducted by mixing 5 grams cellulose powder that obtained from isolation cellulose process with ethanol and isopropanol (2:8) as reaction medium. Then, the solution was put in 3 neck flash with stirrer for 10 minutes and then added with 20 ml of 10% of NaOH. The alkalization process was spent 1 hour at room temperature. The next step was carboxymethylation process which conducted by adding NaMCA as much as 6 grams and stirred it for 3.5 hours at 55 °C.[6] The obtained CMC was neutralized and purified with CH_3COOH and 96% ethanol then dried for one night at room temperature. To determine the degree of substitution, solid CMC was analyzed in BPPK Bandung.

2.2.3 Synthesis of Hydrogel

Synthesis of hydrogel was initiated by mixing CMC and Polyvinyl Alcohol (PVA) Partially Hydrolyed with various compositions (20:80, 80:20, 50:50, 25:75 and 75:25). The solution was mixed for 10 minutes at room temperature. Then the esterification process was conducted by adding 10% citric acid [7] and stirred it for 2 hours until it was homogeneous.

The hydrogel solution was put on petri dish to mould the hydrogel with a thickness of 10mm and a diameter of 90mm. Next, Hydrogel solution was pre-dried for 24 hours at 30°C at vacuum condition, followed by drying process of hydrogel. Hydrogel solution dried in oven at 80°C for 13 hours. The procedure above was repeated with different composition of PVA and CMC.

The hydrogel was analyzed with FTIR, SEM, and swelling ratio to obtain the characteristics of hydrogel. To obtain the swelling ratio, each sample was weighed before and after soaked in distilled water for 1 hour at room temperature. After that, the excess of distilled water on the surface of the hydrogel is removed with filter paper (blotting) and the hydrogel after immersion was weighed.[8] Measurement of swelling ratio can be determined by Eq. (1) :

$$\text{Swelling Ratio} = \frac{W_t - W_o}{W_o} x100\% \quad (1)$$

3 Result and Discussion

3.1 Water Hyacinth α-cellulose

Dewaxing process is a process to remove wax, pectin, and chlorophyll. As a result of dewaxing process, solvent and filter paper show marked colour change. On the other hand, bleaching is lignin removal process that occurs due to bleaching reaction of Chlor Dioxide (ClO_2) in acid media as shown in reaction below.

$$NaClO_2 + CH_3COOH \rightarrow ClO_2 + NaHCO_2 + CH_2 \quad (1)$$

$$ClO_2 + \text{lignin} \rightarrow \text{oxidized lignin} + HClO_3 + HClO_2 \quad (2)$$

Bleaching solution will show a colour change to yellow, owing to the presence of oxidized lignin. This colour change indicatied lignin removal from powder, and caused the colour of the powder change into white. After dehemicellulose process, a pure α-cellulose was obtained and it has white colour powder.

3.2 Carboxymethyl Cellulose (CMC)

Synthesis of Carboxymethyl Cellulose (CMC) started with alkalization process then carboxymethylation process. The process occurs in reaction media of ethanol and isopropanol (2 :8). Polarity of reaction media plays significant role in alkalization process. Reaction media solvent with low polarity will increases reactivity of alkalization reaction, since NaOH will be more reactive to cellulose compared to the reaction medium. Besides the higher concentration of NaOH that exceeds its optimum concentration will cause NaOH more soluble and difficult to substitute -OH group in cellulose as a result cellulose has crystalline structure. The crystalline structure also cause carboxymethylation process occurs hardly. Due to the cellulose still contain the lignin, CMC powder has yellow colour as shown in Figure 1.

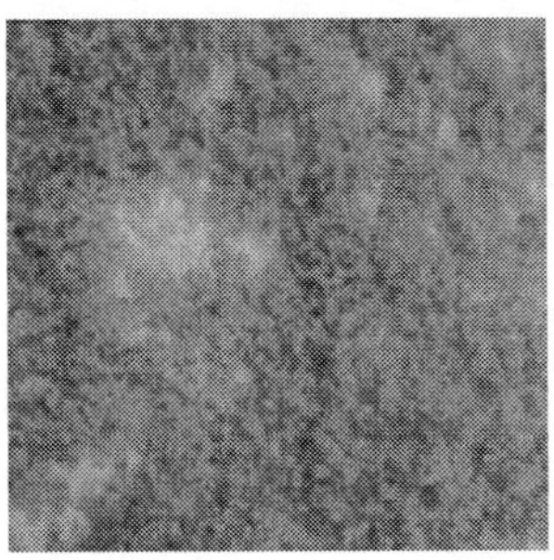

Fig. 1. Carboxymethyl Cellulose (CMC)

The successful synthesis CMC can be indicated by measuring the degree of substitution (DS) that conducted in BPPK Bandung. The result of DS value in this study is 1.64 of 3 (maximum DS value of CMC). This result has relevant value with previous study which uses NaOH 10% and has DS value 1.65 and another study which uses same reaction media has 1.59 of DS value.

3.3 Hydrogel

The hydrogel synthesis in this study utilizes CMC as the main material and PVA Partially Hydrolyzed as the additive. The hydrogel synthesis method employed the chemical crosslinking with citric acid monohydrate and carboxylic group as the cross linker agent.[7] Table 1 shows the variation of the composition between CMC and PVA Partially Hydrolyzed.

Table 1. The Variation of The Composition between CMC and PVA Partially Hydrolyzed

PVA : CMC Composition			Sample
50 : 50	2:2	2 grams : 2 grams	P5C5
25 : 75	1:3	1 grams : 3 grams	P1C3
75 : 25	3:1	3 grams : 1 grams	P3C1
80 : 20	4:1	3.2 grams : 0.8 grams	P4C1
20 : 80	1:4	0.8 grams : 3.2 grams	P1C4

Both polymers are mixed and stirred with 10% citric acid, and results in homogenous hydrogel solution. Then, the solution undergoes two drying processes, the pre-drier process inside vacuum oven for 24 hours and drying process at 80°C for 13 hours. The process of pre-drier aims at cyclic anhydra formation in the citric acid, while the final drying process of hydrogel causes the hydrogel bonds in the hydroxyl group in each of the PVA and CMC to lose, so that both polymers can be connected in the hydrogel structure. This drying process aligns to the reaction mechanism as in Figure 2.

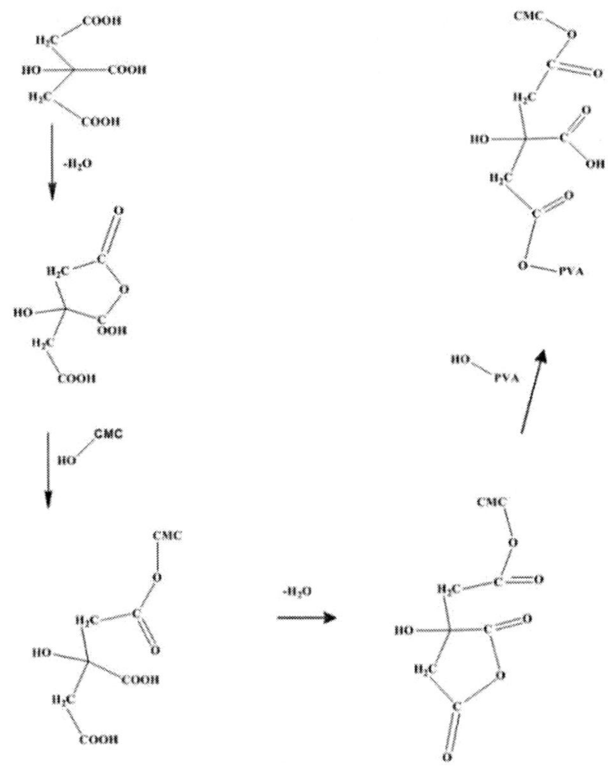

Fig. 2. The Reaction Mechanism of Synthesis Hydrogel

From pictures shown in table 2, it can be concluded that increasing PVA compositions, such as the P3C1-type hydrogel (PVA :CMC 75:25) and P4C1 (PVA:CMC 80:20) will results in more adhesive and soft hydrogels.

This results may be caused by the growth of the number pof acetate group with thermoplastic properties. Thermoplastic property is prone to melting or soften when being heated or cooled as well as is prone to giving rubbery property.[10]

Hence, in synthesis of hydrogel, the optimum composition ratio of PVA and CMC were 20 : 80 (Hydrogel P1C4) and 25 : 75 (Hydrogel P1C3). As-synthesized hydrogels are generated in sheet form, ransparant, and non-adhesive.

Table 2. Hydrogels with Different Composition

P1C4 (PVA:CMC 20 : 80)	P1C3 (PVA:CMC 25 : 75)
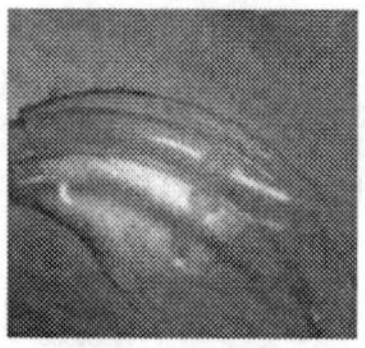	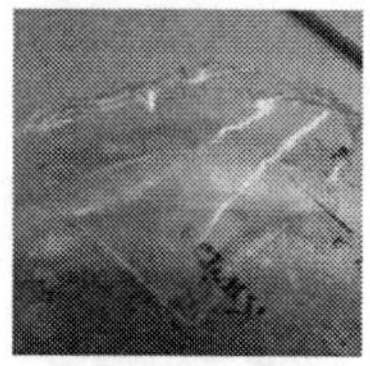
P2C2 (PVA:CMC 50 : 50)	P3C1 (PVA:CMC 75 : 25) P4C1 (PVA:CMC 80 : 20)
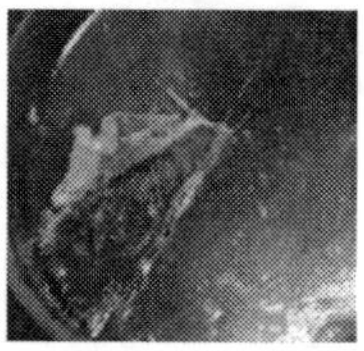	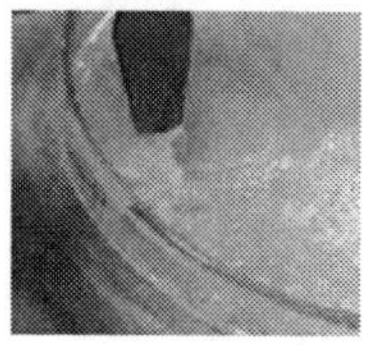

3.4 Hydrogel, PVA Partially Hydrolyzed, and CMC FTIR Analysis

CMC, PVA Partially Hydrolyzed, and Hydrogel P1C3 samples were analyzed by FTIR, in order to determine the functional group content. Figure 3 shows FTIR spectra image of CMC. The functional groups formed out of CMC are hydroxyl group (-OH), ether group (-O-), and carboxyl group (-COO). In CMC, there is a peak in the wave of 1589.17 cm^{-1} which indicates the presence of aromatic group. That occurs because the presence of lignin content in the cellulose which has a polymer structure of aromatic group. The presence of lignin content is also supported by the yellow color produced by CMC.

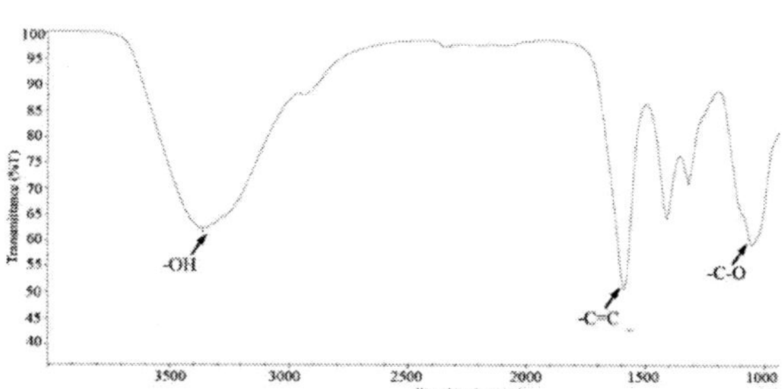

Fig. 3. FTIR spectra of CMC

Figure 4 is the FTIR spectra image on the sample of PVA Partially Hydrolyzed. The functional groups in PVA are hydroxyl group, carboxyl group, and ether group. Carboxyl and ether groups arise due to the fact that the PVA used is the PVA Partially Hydrolyzed, so there are still carboxyl group that has not hydrolyzed into hydroxyl group and still being present in the vinyl acetate.

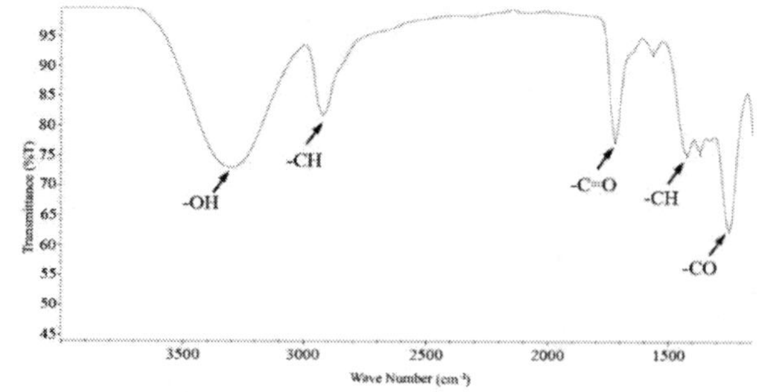

Figure. 4. FTIR spectra of PVA Partially Hydrolyzed

The FTIR spectra of the hydrogel is shown in Figure 5. The hydrogel used is P1C3 with the composition of PVA:CMC 25:75. The hydrogel has hydrophilic group in the form of hydroxyl group, alkane group, and –CO group which show that the formed hydrogel contains ester group of PVA Partially Hydrolyzed. The success of synthesis hydrogel can be seen in the wave number of 1716 cm^{-1} which indicates the presence of carboxyl group.[7] The presence of such group signifies a crosslinking reaction which occurs between PVA Partially Hydrolyzed, citric acid, and CMC in order to form hydrogel tissue.

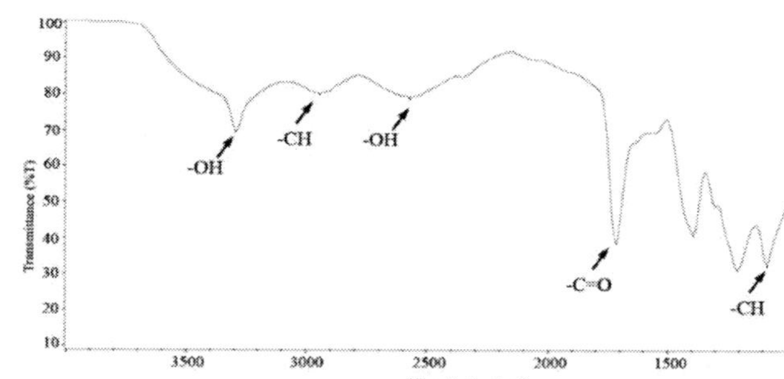

Fig. 5. FTIR Spectra of Hydrogel

The FTIR spectra of CMC, PVA, and hydrogel are presented in Figure 6. From the graph, it can be seen regarding the different transmittance of hydroxyl group. Hydroxyl group (-OH) in CMC has the lowest transmittance value followed by PVA and Hydrogel. This is supported by the fact that formed hydrogel consists of larger composition of CMC compared to the composition of PVA. The transmittance value of hydrogel is the lowest and this shows that the addition of PVA Partially Hydrolyzed does not increase the hydroxyl group content in the hydrogel.

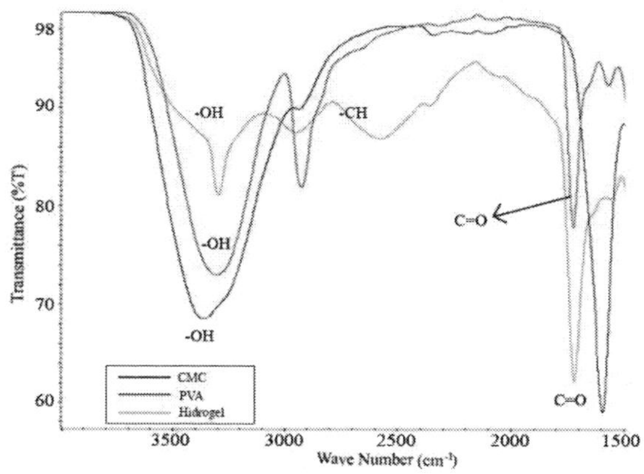

Fig. 6. The FTIR Spectra of CMC, PVA Partially Hydrolyzed, and Hydrogel

3.5 Hydrogel, PVA Partially Hydrolyzed, and CMC Swelling Ratio Analysis

Hydrogel polymers content will have significant role in determining hydrogel swelling ratio. Table 3 displays that the addition of PVA Partially Hydrolyzed decreases the swelling ratio of the hydrogel.

Table 3. Swelling Ratio of Hydrogel

Hydrogel types	PVA:CMC Composition	Swelling Ratio Value (%)
P1C4	20 : 80	29,22
P1C3	25 : 75	25,75

Meanwhile, higher composition of CMC results in higher swelling ratio value. Being associated with the result of FTIR analysis, PVA Partially Hydrolyzed has fewer hydroxyl group than CMC. Hence, this may decrease the hydrophilic characteristic of the hydrogel because PVA Partially Hydrolyzed is not being more hydrophilic than CMC.[11] Additionally, the addition of PVA reduces the double role of swelling in CMC-based hydrogel, as well. The aforementioned swelling is the opening of a hydrogel tissue which allows water to enter. Such swelling occurs due to electrostatic repulsion by a protonated carboxyl group (COO-) of CMC.[3] The highest swelling value is achieved by P1C4 with the composition of PVA:CMC 20:80 and the average selling of 29%, while P1C3 has the composition of PVA:CMC 25:72 and the average selling of 25%.

Low swelling ratio of hydrogel may be induced by the surface morphology of the hydrogel. From Figure 7, it can be seen that P1C4 has larger pores and is less compact in terms of structure. Thus, swelling ratio of this hydrogel is higher than P1C3. Pores have a role in water absorption of hydrogel.[12] Compact surface morphology will act as a barrier to water diffusion.

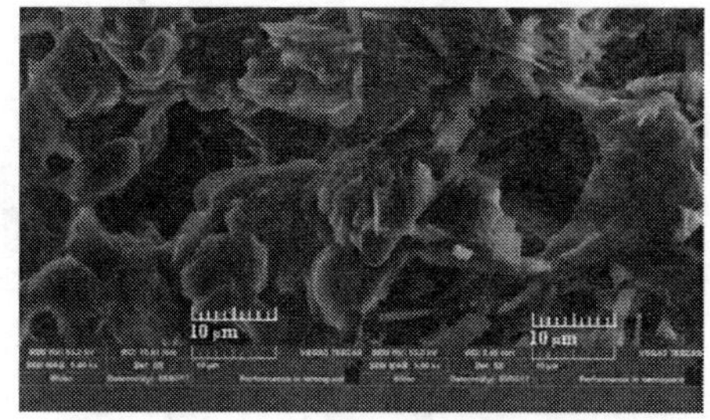

Fig. 7. Microscopic surface of a) Hydrogel P1C3 and b) Hydrogel P1C4

Hence, Hydrogels P1C4 and P1C3 have swelling ratio characteristics 29% and 25%, respectively. Low swelling ratio might be originated from more compact morphology structure and less pores. Consequently, this condition acts as barrier to water diffusion. The characteristics of functional groups in PVA Partially Hydrolyzed hydrogels is characterized by the presence of hydroxyl group (-OH) as water absorbent group and carboxyl group (C=O) as an indication of hydrogel successful crosslinking reaction.

3.6 The Comparison of Hydrogel-PVA Fully Hydrolyzed and Hydrogel PVA-Partially Hydrolyzed

The comparison between hydrogel-PVA Fully Hydrolyzed and hydrogel-PVA Partially Hydrolyzed is based on the FTIR analysis and the morphological structure. Based on the FTIR result between Partially Hydrolyzed PVA and Fully Hydrolyzed PVA, the differences are visible in the transmittance values of C-O and C=O groups. These results imply that the hydrogel derived from PVA Partially Hydrolyzed with the content of vinyl acetate having not been hydrolyzed.

The hydroxyl group in the PVA has hydrophilic properties, while the acetate group is a hydrophobic. Polymers with hydrophobic properties show a low affinity towards water.[14] In addition, the presence of acetate group residue weakens the intra and intermolecular hydrogen bonds from the adjacent hydroxyl group. The hydrogen bond with different molecule is the bond which occurs between hydrogel and water molecule. One of the factors in determining the swelling properties of hydrogel is also the hydrophilic-lipophilic balance (HLB) in the hydrogel. HLB is a function of the amount of functional group, side chain, and ions. Functional group enhancing the hydrophilic properties of gels is polar groups which are ionic and non-ionic at the same time.[15] The ester group (carboxylic group) derived from PVA Partially Hydrolyzed on the hydrogel has low polarity compared to alcohol group (-OH). The comparison of morphological structure is displayed in Figure 8. The hydrogel with the addition of PVA Partially Hydrolyzed has fewer and smaller pores than the hydrogel containing PVA Fully Hydrolyzed. This result confirms the fact that the acetate group affects PVA so that it is prone to having crystal-like structure (solid and tidy) when the PVA is heated.[16]

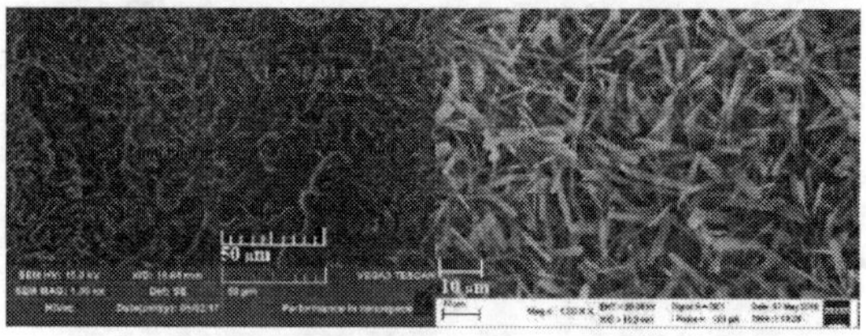

Figure. 8. Microscopic surface of a) Hydrogel- PVA Partially Hydrolyzed and b) Hydrogel-PVA Fully Hydrolyzed

Therefore, Hydrogel with the addition of PVA Partially Hydrolyzed has lower swelling ratio than hydrogel with PVA Fully Hydrolyzed, due to the presence of vinyl acetate.

4 Conclusion

This study showed that hydrogel can be produced from CMC and PVA partially hydrolyzed. The optimum composition ratio of PVA and CMC were 20 : 80 (Hydrogel P1C4) and 25 : 75 (Hydrogel P1C3). On the other hand, the swelling ratio of P1C4 and P1C3 are 29% and 25%, respectively. Low swelling ratio might be originated from more compact morphology structure and less pores. Consequently, this condition acts as barrier to water diffusion. The characteristics of functional groups in PVA Partially Hydrolyzed hydrogels is characterized by the presence of hydroxyl group (-OH) as water absorbent group and carboxyl group (C=O) as an indication of hydrogel successful crosslinking reaction. Hydrogel with the addition of PVA Partially Hydrolyzed has lower swelling ratio characteristic than hydrogel with PVA Fully Hydrolyzed, due to the presence of vinyl acetate.

5 References

1. A.F Abdel-Fattah and M. Abdel-Naby, Carbohydr. Polym., **87,** 2109 (2012)
2. S. Kiatkamjornwong, K. Mongkolsawat and M. Sonsuk, Polym., **43**, 3915 (2002)
3. R. Barbucci, A. Magnani, and M. Consumi, Macromol, **33**, 7478 (2000)
4. E. Ogur, *Materials, Processing and Applications*, **16**, 7 (2005)
5. K. Somwang, T. Enomae, A. Isogai, F. Onabe, Japan TAPPI J, **56**, 866 (2002)
6. A.H Saputra, M. Hapsari, and A.B Pitaloka, Contemp. Eng. Sci., **8**, 1573 (2015)
7. A.H Saputra, M. Hapsari, A.B Pitaloka, P.P.D.K Wulan, JESTECH, **2**, 78 (2015)
8. J. Lu, Y. Li, D. Hu, X. Chen, Y. Liu, L. Wang, Y. Zhao, Biomed Res Int, **2015**, 3 (2015)
9. B.K Barai, R.S Singhai and P.R. Kulkarni, Carbohyd. Polym., **32**, 229 (1997)
10. A.J Kovacs, R.A Stratton, and J.D. Ferby, J. Phys. Chem., **67**, 155 (2016)
11. G. Buhus, M. Popa, C. Peptu, J. Desbrieres, J. Optoelectron. Adv. M., **9**, 3450 (2007)
12. M. Hashem, S. Sharaf, M.M Abd El-Hady, A. Hebeish, Carbohydr. Polym., **95**, 425 (2013)
13. Z. Krol, M. Malik, K. Marycz, A. Jarmoluk, Polym., **8**, 260 (2016)
14. M.L Bruschi, *Strategies to Modify the Drug Release from Pharmaceutial Systems* (Elsevier Ltd.) (2015)
15. N.C Dafader, S. Ganguli, M.A Sattar, M.E Haque, F.Akhtar, MPJ, **4**, 37 (2009)
16. C.M Hassan and N.A Peppas, Polym. Sci,, **153**, 38 (2000)

Effect of Biodiesel Concentration on Corrosion of Carbon Steel by *Serratia marcescens*

Yustina M Pusparizkita[1], *Tjandra* Setiadi[1], and *Ardiyan* Harimawan[1,*]

[1] Institut Teknologi Bandung, Department of Chemical Engineering, 40132 Bandung, Indonesia

Abstract. Biodiesel come into being used as an alternative source of energy as the diminishing of petroleum reserves. This fuel is typically stored in tanks that are commonly made from carbon steel, which is easily corroded by microorganisms. Recent studies have shown that bacteria aside from SRB may also be involved in corrosion. Therefore, this research was aimed to evaluate the effect of biodiesel concentration (15%, 20% and 30% v/v) mixed in diesel oil on the corrosion of carbon steel by *S. marcescens* that dominate biocorrosion on hydrocarbon products. In this study, the corrosion process was investigated by evaluation of biofilm morphology and composition, the rate of corrosion and the corrosion product of carbon steel which was exposed in the mixture of hydrocarbons and the presence of *S. marcescens*. It can be concluded that higher concentration of biodiesel in diesel oil leads to higher growth of bacteria in the biofilm and higher corrosion rate.

1 INTRODUCTION

In response to the petroleum crisis, the existence of biofuels as alternative energy sources are necessary to anticipate the depletion of petroleum fuels. Moreover, Indonesia has natural resources that can be potentially exploited for the production and exported biofuels as biodiesel. In this case, biodiesel can be produced from palm oil, soybeans, and jatropha curcas. Additionally, they are commonly planted and have potential to be developed in Indonesia. Due to the increasing demand for diesel oil in the transportation and industrial sectors, correspondingly, the biodiesel mixed with diesel fuel in Indonesia continues to be investigated. The Indonesian government through Permen ESDM No. 12 2015 expects the use of biodiesel as a mixture of up to 30% of total fuel demand, which can be achieved by 2025.

Petroleum products and biofuels are usually stored in a large tank made of carbon steel. Carbon steel is more widely used by the industry as a construction material as well as material for storage tanks. The use of carbon steel in term of price is due to carbon steel is lower cost than other metals such as stainless steel [1]. However, the carbon steel has a susceptibility to corrosion by microorganisms. Several studies have demonstrated a significant negative impact of biocorrosion occurring in carbon steel. Microbiologically influenced corrosion (MIC) and biofilm may cause pitting because sulphides resulting from SRB metabolism settle on the surface of iron and forming iron sulphide (FeS) at oil field separator and offshore installation [2, 3]. Furthermore, on the crude oil tank, SRB may form a biofilm in the form of a black layer on the bottom of the tank containing crude oil and fuel oil [4].

Petroleum products and biofuels are usually stored in large tanks, then pumped through pipes to be stored and or transported to smaller tanks. In the process of distribution, there is potential for contamination of microorganisms. Factors that contribute to the contamination of hydrocarbon products and the growth of microorganisms are climate, distribution system design, the chemical composition of fuels, methods, and effectiveness of cleaning as well as product quality control [5]. Contamination of microorganisms in products is very difficult to prevent because sterile conditions during transport and fuel storage can not be maintained. Moreover, microorganisms are commonly found in air, water, and soil [6]. The development of these microorganisms contributes to corrosion of the tank.

Corrosion in the tank may occur because microorganisms can degrade hydrocarbons as a source of nutrients. Also, biofuels are a suitable medium for the growth of microorganisms. Products generated through MIC metabolism in the form of extracellular polymeric substance (EPS), organic and inorganic acids will form biofilms that can change the condition of the surrounding environment and trigger corrosion. The corrosion reaction can run faster or slower in the presence of microorganisms as it affects the electron transfer process on the metal surface [1].

Biofilms are formed by the most common microorganism as survival strategies for allowing bacterial species to develop their own microenvironment as part of a defense, food supply and metabolic interactions [1]. Several types of bacteria can live simultaneously in biofilm such as bacteria that produce metabolic products that can support the growth of other

* Corresponding author: ardiyan@che.itb.ac.id

bacteria. The formation of microbial colonies begins with the interaction and attachment of single-cell bacteria on the metal surfaces coated with a thin film layer. Carbon steel is a type of metal that is easy to corrode because of the passive layer of iron oxide is a perfect surface for the attachment of microorganisms [1]. This process is called adhesion or attachment. While electrostatic interactions contribute to the attachment of microorganisms to the surface, the attachment process of the cell is influenced by EPS which causes the strong bonding between microbes to each other and the surface [2]. The bonds between microbes will be strengthened along with the joining of other bacteria (colonization) and accumulated, characterized by increasing the complexity biofilms formed [3]. The next process is early biofilm maturation stage. The ongoing process is the unification of all metabolic processes of colonized cells, including the use of substrates and nutrients in the medium [4] and each bacterium uniting its structure, function, and activity into the biofilm [5]. In this stage, there is a process of cleavage, enlargement, and maturation of polysaccharides which causes the colony to appear enlarged so that the thickness of the biofilm increases [1]. In the final phase, some bacteria will spread and colonize elsewhere [6].

Further, the most common microorganisms involved in the corrosion of the petroleum production system includes the anaerobic SRB [7]. However, the recent studies have shown that other types of bacteria may also be involved in biocorrosion [8-11]. The several species such as *S. marcescens* have the ability to degrade a variety of hydrocarbons [12]. Here, these species after *Bacillus sp.* dominate in the corrosion product of the napha piping transport [13].

With the increasing use of biodiesel mixed with diesel oil, it was necessary to examine the characteristics of carbon steel on biocorrosion including the ratio of the number of bacteria attached to the biofilm, the biofilm morphology and the rate of corrosion in biodiesel and diesel oil blends.

2. Methods

This research consisted of three main stages namely preparation, experimental and corrosion analysis to observed biocorrosion characteristics of carbon steel ST-37 in a mixture of 15%, 20% and 30% v/v biodiesel produced from laboratory and diesel oil from Pertamina containing *S. marcescens* ATCC 27117.

2.1 Preparation

The preparation stage consisted of the culture stage, the test solution preparation, the preparation of metal and the reactor.

2.1.1 Culture stage

The growth medium used for *S. marcescens* was Bushnell Haas (BH). BH medium consisted of $MgSO_4$ 0.2 g/L, $CaCl_2$ 0.02 g/L, KH_2PO_4 1 g/L, K_2HPO_4 1 g/L, NH_4NO_3 1 g/L and $FeCl_3$ 0.05 g/L. To increase the number of *S. marcescens*, these bacteria were cultured in a sterilized BH medium with a working volume of 100 mL until it enters a stationary growth stage. After entering the stationary phase, the acclimatization was done by taking 2 mL inoculated bacteria into 300 mL of BH medium and 1 g of sterilized diesel oil as a sole nutrition in an Erlenmeyer, which was then incubated with the temperature of 27°C and 200 rpm for 30 days in a rotary shaker.

2.1.2 Test solution preparation

The hydrocarbon to be used as the immersion medium was filtered with CA filter membrane with a pore size of 0.45 μm to remove impurities and microorganism that may be contained.

2.1.3 Specimens and reactor preparation

Specimen preparation followed several steps. According to ASTM G 31-72 standard, the specimen of carbon steel was cut into 1 cm x 1 cm. Based on ASTM G 1-81, the specimens were polished using 240-1200 grid abrasive paper and washed using water and ethanol to remove fat and impurities. Once clean, the specimens were dried and stored in a desiccator to keep it dry.

The soaking reactor and the hanging pole for specimens were made of glass, while the reactor cover was made of acrylic.

2.2 Experimental

Types of hydrocarbons which were used as a soaking medium are the mixture of diesel oil with 15%, 20% and 30% biodiesel as variations. Immersion process was done by hanging specimens in pole hangers made of glass and dipped in submersion glass reactor containing 10% v/v culture of *S. marcescens*. Volume of immersion working reactor is 800 ml. All submersion reactors kept sterile from outside influences that can be contaminated by sealing the reactor with wax. The reactor was then stored at the temperature of 25°C and under anaerobic condition for 10 days. At certain time intervals, the reactor was opened to pick up metal specimens and analysed the biofilms.

2.3 Corrosion analysis

Some analysis was done in this study including total plate count (TPC) method to analyse the bacterial colonies on biofilm, biofilm morphology using scanning electron microscopy (SEM) and gravimetric method

 https://doi.org/10.1051/matecconf/201815601008

refers to ASTM G1 to observe the corrosion rate and the influence of *S. marcescens* on corrosion. For each method of the analysis, two specimen replicas were assessed.

3. Results

3.1 The growth of *S. marcescens* in biofilm

By counting the number of bacterial colonies in the biofilm that are formed on the metal surface using TPC analysis, bacterial growth in a biofilm can be observed to determine the ability of *S. marcescens* to live in biofilms. The number of *S. marcesccens* colonies that live was calculated every two days during the immersion period to obtain a growth profile of *S. marcesccens* in the biofilm. The profile of *S. marcescens* colonies on the biofilms can be seen in Figure 1.

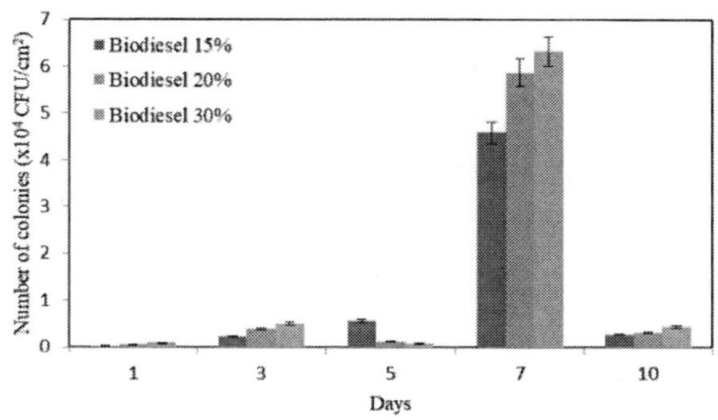

Fig. 1 The growth of *S. marcescens* in biofilm

At 15% biodiesel concentration, the number of bacterial colonies on the biofilm slowly grew from the first day and then increased significantly on the seventh day. On the tenth day, the number of colonies declined drastically. While at 20% and 30% biodiesel concentrations, the number of bacterial colonies on the biofilm increased slightly until the third day. However, the number of bacterial colonies decreased on the fifth day. On the seventh day, the number of colonies increased significantly and dropped dramatically on the final day of immersion. The growth of *S. marcescens* on biofilms is mainly controlled by the concentration of biodiesel in the diesel oil. With the increasing concentrations of biodiesel, the number of bacteria that grew and lived also increased.

3.2 *S. marcescens* biofilm morphology

The observation of biofilm morphology was done by SEM with 5000 times magnification for samples that had been immersed for one, five and ten days. The biofilm morphology formed by the activity of *S. marcescens* for various biodiesel concentrations over ten days can be seen in Figures 2,3, and 4.

Fig. 2 Biofilm morphologies formed in 1 day at 15% (top), 20% (middle) and 30% (bottom)

On the first day, biofilms begin to appear on the metal surface of all variations of biodiesel. The incidence of cracks seen in Figure 2 indicates that bacteria begin to colonize [14]. After five days of immersion, biofilms formed on 20% and 30% biodiesel soaking medium had reached the final stage of maturation and are beginning to enter the final stages. This can be seen due to some biofilms entered lysis processes so the bacteria spread and may colonize to form biofilms elsewhere [6]. On the tenth day, it can be concluded that the bacteria that spread after the final phase form the bonds again with other bacteria on the metal surface. The presence of colonization causes the thickness of biofilm increases [1]. Soaking of metal for ten days also shows the existence of metal damage caused by the activity of *S. marcescens* in the form of pitting.

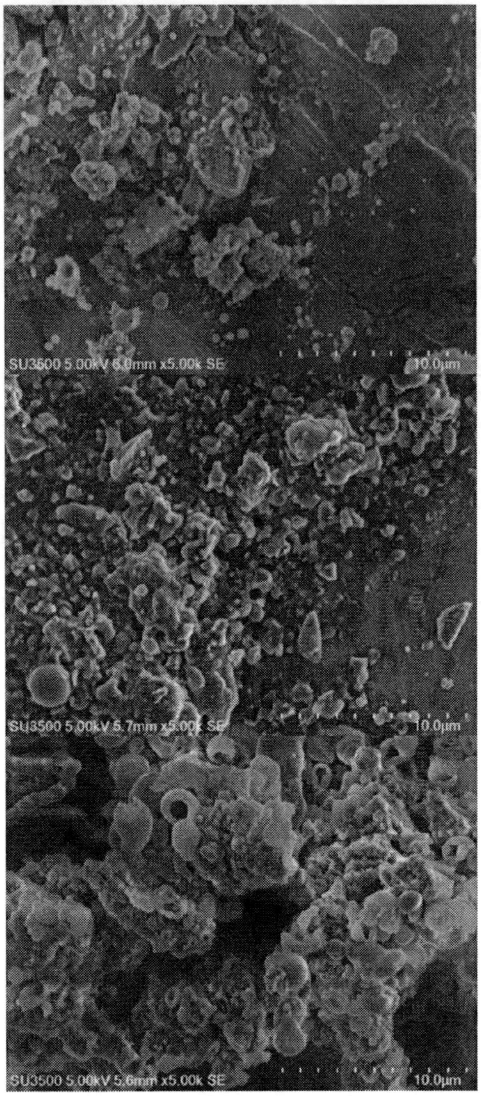

Fig. 3 Biofilm morphologies formed in 5 days at 15% (top), 20% (middle) and 30% (bottom)

3.3 The loss of metal specimens mass

By determining the mass loss occurring in the metal by the gravimetric method, the corrosion rate can be calculated. The result of determining the loss of metal mass at various concentrations of biodiesel can be seen in Table 1.

The average value of carbon steel specimens mass loss in the variation of biodiesel concentration by *S. marcescens* bacteria showed no more than 1.2%. The mean mass loss value of the specimens increases proportional to the biodiesel concentration in the medium. The average mass loss values specimens in the variations in of biodiesel concentration is greater compared to the control loss mass values.

Table 1. Weight loss of control and variation specimens during incubation

Variation	Average	
	Specimen mass loss (%)	Specimen control mass loss (%)
15	0.14±1E-3	0.08±1E-3
20	0,16±2E-3	0.09±1E-3
30	0,38±5E-3	0.13±2E-3

4.1 Growth of Bacteria in Biofilm

From Figure, the concentration of biodiesel in diesel fuel influenced the growth of *S. marcescens* in the biofilm. Increasing the concentrations of biodiesel is directly proportional to the number of bacteria that grow and live. This is because *S. marcescens* can utilize biodiesel as a source of nutrients. Also, the properties of biodiesel provide benefits for microorganisms to grow and multiply. Biodiesel is readily degraded in aerobic and anaerobic conditions compared to diesel fuel [15-18]. Stolz et al. [19] and Sendzikiene et al. [20] reported that the fatty acid methyl esters (FAMEs) components comparing to diesel oil components are more rapidly degraded up to four times due to the high percentage of unsaturated esters. Biodiesel made from palm oil containing 40.2% palmitate (16:0) and 42.4% oleate (18:1) make it easier to degrade and useful for microorganisms metabolic process [21]. Reduction for the number of living *S. marcescens* colonies that occur can be caused by the microorganisms converting the unsaturated fatty acids such as palmitate and oleate into components that can poison the microorganism itself namely saturated fatty acids in anaerobic system [16, 22]. Furthermore, supported by the observation of biofilm morphology the lysis process in the biofilm when bacteria reach the final phase, can decrease the number of living colonies.

4.2 Corrosion Rate

Determination of the corrosion rate of each variation of biodiesel concentration can be obtained from the mass loss value in the Table 1.

The corrosion rate is illustrated in Figure 6. During the first three days, the corrosion rate of *S. marcescens* in 15% biodiesel concentration soaking medium increased slowly to 0.13 mpy. On the fifth day, the corrosion rate declines up to 0.05 mpy. However, there was an increase of corrosion rate up to 0.22 mpy on the tenth day.

A slightly different phenomenon occurs in the variation of 20% biodiesel concentration. The corrosion rate of 0.13 mpy on the first day decreased periodically to 0.06 mpy until the fifth day. After the fifth day, the increase in corrosion rate that occurs reaches 0.27 mpy at the end of immersion process.

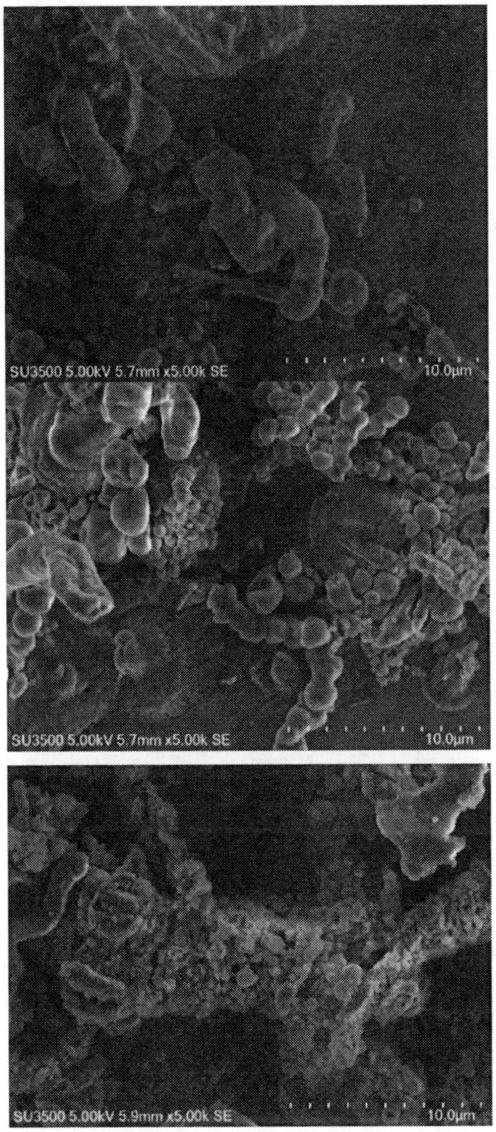

Fig. 4 Biofilm morphologies formed in 10 days at 15% (top), 20% (middle) and 30% (bottom)

In the variation of biodiesel concentration 30% the corrosion rate increase until the third day. Slowly the rate of corrosion decreased on the fifth day, then increased rapidly reached 0.69 mpy at the end of immersion time.

Increasing in corrosion rate occurring in the first three days for all variations in biodiesel concentration is due to the activity of *S. marcescens* and biofilm formation. The SEM analysis results show that since the first day *S. marcescens* has formed a biofilm on the metal surface.

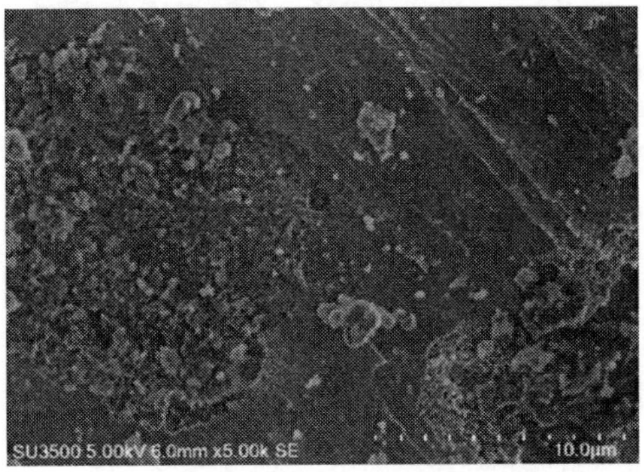

Fig. 5 *S. marcescens* corrosion attack on carbon steel's surface

The biofilms that formed then matured caused the thickening of the biofilm. After maturation on the fifth day some biofilm for variation of biodiesel concentration 20% and 30% experiencing lysis resulting in decreasing the number of *S. marcescens* colonies seen in Figure 1. The biofilm still attached to the metal surface protects the metallic layer from a potentially corrosive environment, resulting in a decrease in corrosion rate [23-25].

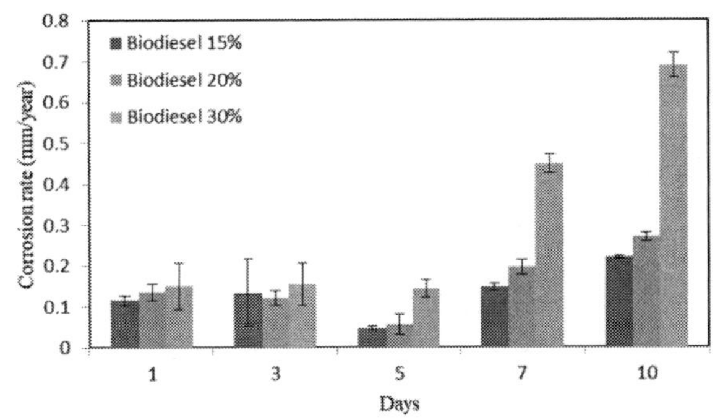

Fig. 6 Carbon steel specimens corrosion rate at various biodiesel concentration

Decreased corrosion rate can also be caused by *S. marcescens* biofilm can minimize the contact of oxygen with the metal surface because the thicker the biofilm formed the oxygen will be difficult to diffuse. Entering the seventh day, the corrosion rate of *S. marcescens* at various concentrations of biodiesel has increased due to the number of colonies that reach the peak point in all variations in the concentration of biodiesel. Many live colonies have the potential to produce many of the metabolites that affect corrosion processes [26-28]. Metabolite products result of biodiesel degradation by microorganisms can decrease the pH value of the environment because of its acidic nature. In addition to the effect on corrosion, acidic environments can be toxic to microorganisms in biofilms resulting in a decrease in the number of colonies until the final day of immersion [16]. Although the number of living microorganisms decreases, the mass loss of metallic specimens and corrosion rate can be affected by acidic environments. This can be seen from the high corrosion rate for various concentrations of biodiesel on the tenth day.

The corrosion rate in the mixture of biodiesel and diesel oil and the presence of *S. marcescens* is given in Table 2. Overall the average corrosion rate of controls as well as by *S. marcescens* has the potential to increase along with the addition of biodiesel concentration in diesel oil. In control, the biodiesel concentration in the diesel influences the final pH of the immersion medium. Increasing in the concentration of biodiesel available in diesel causes a decrease in pH due to the acidic of biodiesel [29, 30].

Table 2. Average corrosion rate of control and *S. marcescens* at various biodiesel concentration

Variation	15%	20%	30%
Average corrosion rate control (mm/year)	0,085±0,057	0,087±0,060	0,130±0,113
Average corrosion rate specimen variation (mm/year)	0,132±0,067	0,155±0,075	0,317±0,223

5. Conclutions

From the research that has been done can be taken some conclusion that is:

1. Bacterial growth in biofilms was influenced by the concentration of biodiesel in diesel oil.
2. In the absence of microorganisms, the addition of biodiesel to diesel oil can increase the corrosion rate.
3. The corrosion rate of mixed diesel and biodiesel medium increased up to 0,317 mm/year due to the activity of *S. marcescens*.

6. Acknowledgments

The authors would like to express their special thanks to PMDSU Scholarship Ministry of Reseach, Technology and Higher Education, which has provided research funding, Dr. Isdiriayani Nurdin and Dr. Hary Devianto who has provided knowledge, especially regarding theories of corrosion, Chemical Engineering Program at the Institute of Technology Bandung which has provided facilities and infrastructure in the implementation of this research. The staff of Microbiology and Technology, Laboratory of Bioprocess for providing the assistance and doing research are also appreciated.

7. References

[1] A. Heyer, F. D'Souza, C. F. L. Morales, G. Ferrari, J. M. C. Mol, and J. H. W. d. Wit, *Ocean Eng.*, **70,** 188-200, (2013).

[2] A. J. Wikieł, I. Datsenko, M. Vera, and W. Sand, Bioelectrochemistry, **97,** 52-60, (2014).

[3] M. W. Dunne jr, Clin.Microbiol. Rev, **15,** 155-166, (2002).

[4] M. T. Martinko, *Brock Biology of Microorganisms*. New York: Pearson Prentice Hall, 2006.

[5] T. S. Rao, in S. Rajagopal, H. A. Jenner, and V. P. Venugopalan, Eds., ed Boston, MA: Springer US, 2012, pp. 95-126.

[6] D. Monroe, PLoS Biology **5,** 307, (2007).

[7] C. A. H. Von Wolzogen Kuehr and L. S. van der Vlugt, (2006).

[8] X. Y. Zhu, J. Lubeck, and J. J. Kilbane, App. Environ. Microbiol., **69,** 5354-5363, (2003).

[9] N. Muthukumar, S. Mohanan, S. Maruthamuthu, P. Subramanian, N. Palaniswamy, and M. Raghavan, Electrochem Commun **5,** 412-426, (2003).

[10] A. Rajasekar, S. Maruthamuthu, N. Muthukumar, S. Mohanan, P. Subramanian, and N. Palaniswamy, Corros. Sci., **47,** 257-271, (2005).

[11] J. Jan-Roblero, A. Posadas, F. Zavala-Dıaz de la Serna, and R. Garcıa, World J Microbiol Biotechnol **24,** 1775-1784, (2008).

[12] P. Wongsa, A. Tanaka M Fau - Ueno, M. Ueno A Fau - Hasanuzzaman, I. Hasanuzzaman M Fau - Yumoto, H. Yumoto I Fau - Okuyama, and H. Okuyama, 20050207 DCOM-20050303.

[13] A. Rajasekar, T. G. Babu, S. T. K. Pandian, S. Maruthamuthu, N. Palaniswamy, and A. Rajendran, J. Ind. Microbiol. Biotechnol., **34,** 589-598, (2007).

[14] T. Lie, J. Periodontal Res., **13,** 391-409, (1978).

[15] X. Zhang, C. Peterson, D. Reece, R. Haws, and G. Moller, Trans. ASAE, **41,** 1423-1430, (1998).

[16] J. S. Lee, R. I. Ray, and B. J. Little, DTIC Document, (2009).

[17] D. F. Aktas, J. S. Lee, B. J. Little, K. E. Duncan, B. M. Perez-Ibarra, and J. M. Suflita, Int. Biodeterior. Biodegrad., **81,** 114-126, (2013).

[18] J. S. Lee, R. I. Ray, B. J. Little, K. E. Duncan, A. L. Oldham, I. A. Davidova, et al., Biofouling, **28,** 465-478, (2012).

[19] J. F. Stolz, P. Follis, G. Floro, R. Donofrio, J. Buzzelli, and W. M. Griffin, (1995).

[20] E. Sendzikienea, V. Makarevicienea, P. Janulisa, and D. Makareviciuteb, Eur. J. Lipid Sci. Technol., **109,** 493-497, (2007).

[21] R. El-Araby, A. Amin, A. K. El Morsi, N. N. El-Ibiari, and G. I. El-Diwani, Egyptian Journal of Petroleum, (2017).

[22] D. Z. Sousa, H. Smidt, M. M. Alves, and A. J. Stams, FEMS Microbiol. Ecol, **68,** 257-272, (2009).

[23] S. G. G. d. Saravia, M. F. L. d. Mele, and H. A. Videla, Corros. Sci., **46,** 302-306, (1990).

[24] Z. Lewandowski, W. Dickinson, and W. Lee, Water Sci Technol, **36,** 295-302, (1997).

[25] J. D. Gu, M. Roman, T. Esselman, and R. Mitchell, Int. Biodeterior. Biodegrad., **41,** 25-33, (1998).
[26] J. W. Sutherland, Trends Microbiol, **9,** 222-227, (2001).
[27] J. S. Lee, R. I. Ray, and B. J. Little, Biofouling, **26,** 623-35, Aug (2010).
[28] T. H. Ching, B. A. Yoza, R. Wang, S. Masutani, S. Donachie, L. Hihara, et al., Int. Biodeterior. Biodegrad., **108,** 122-126, (2016).
[29] M. A. Fazal, A. S. M. A. Haseeb, and H. H. Masjuki, Corros. Sci., **67,** 50-59, (2013).
[30] T. Eryilmaz and M. K. Yesilyurt, Renewable Energy, **95,** 233-247, (2016).

ENZYMATIC HYDROLYSIS OF BITTER CASSAVA AND *GADUNG* STARCHES WITH DIFFERENT COMPOSITIONS AT LOW TEMPERATURE

Hargono Hargono [1*)], Andri Cahyo Kumoro[1], Bakti Jos [1]

[1]Department of Chemical Engineering, Faculty of Engineering, Diponegoro University
Jl. Prof. Soedarto, SH, Tembalang, Semarang 50275, Indonesia, Telp./Fax.: +62247460058 / +622476480675

Abstract. The effect of compositions of bitter cassava (*Manihot glaziovii*) and *gadung* (*Dioscorea hispida* Dennst) starches on reducing sugar during hydrolysis using granular starch hydrolyzing enzyme (GSHE) was studied. All hydrolyses were conducted at concentration of substrate was 200 $g.L^{-1}$, while concentration of enzyme was 1.5 % (w/w), during of hydrolysis time 24 h, at 30°C. Mass compositition of bitter cassava and *gadung* starches were 9:1 to 1:9 The increase *gadung* starch compositions will decrease the reducing sugar. The optimum condition of the process using concentration of substrate 200 $g.L^{-1}$ with compositions of bitter cassava and *gadung* starches was 9:1 at 18 h. It was found that reducing sugar was 50.20 $g.L^{-1}$. The concentration of reducing sugar mainly depend on starch content on bitter cassava, it is much bigger than the *gadung* starch.

1 Introduction

In Indonesia, bitter casava (*Manihot glaziovii*) and bitter yam (*Dioscoreae hispida* Dennst) which locally known as *gadung* are commonly found in secondary forest. They grow under shaded areas or near sreams, not consumed by humans because of the presence of poisonous alkaloids known as cyanogens and dioscorin [1], which can seriously impact the health of the people who consump the tuber. Lambri et al. [2] reported that the total cyanide content of cassava root for sweet white and bitter white were 374 and 442 mg/kg (d.w), respectively. Djazuli and Bradbury [3] reported of 14 samples of cyanogen content of cassava starch in Indonesia. The maximum value of cyanogens content in cassava starch and cassava flour were 12 ppm and 149 ppm, respectively.

The conventional conversion of starch to glucose requires a two-step process, namely, liquification and saccharification. During liquefaction, gelatination of starch is promoted by applying high temperatute (90-100 °C), α-amylase is added in order to convert starch into dextrins, maltose and maltotriose. The resulting mash is then cooled to 60 °C and glucoamylase is added to convert dextrins to fermentable sugars as glucose [4-7]. Granular starch hydrolyzing enzyme (GSHE) is a mixture of α-amylase and glucoamylase which hydrolyzes granular starch directly into fermentable sugars and works at low temperature (30 to 48°C) and pH (4.0 to 4.5) [8]. This process is an energy-intensive, and therefore increases the production cost. It has been estimated that about 30-40% of the total energy is demanded during ethanol production from starch [5]. The use of GSHE presents others advantages, as the GSHE will negate liquefaction and saccharification. Previous researches were commonly conducted via conventional of hydrolysis and hydrolysis without heating a single substrate, for example the hydrolysis of corn, wheat, tapioca, sago, sorghum starches, etc, only few researches were conducted on hydrolysis enzymatic of double substrate. Uthumporn et al. [9] also reported the hydrolysis of maize, sago and sweet potato starches using α-amylase and glucoamylase, at 35°C to 50°C. Li et al. [10] reported the hydrolysis of corn and triticle starches at 30-50°C using α-amylase and glucoamylase. Shariffa [11] investigated hydrolysis of tapioca and sweet potato starches by α-amylase and glucoamylase, at 35-65°C.

The objective of this study is to investigate the effect of bitter cassava and *gadung* starches compositions and on reducing sugar.

2 Materials and method

2.1 Cassava and *gadung* tubers

Ten month old of bitter cassava was called Pandemir (*Manihot glaziovii*) tuber was obtained from Wonogiri district in Indonesia, while *gadung* tuber with 9 month old was obtained in Godean district in Indonesia.

The extraction method applied to the *gadung* is the same as the bitter casava extraction method. The method of extraction of the bitter cassava starch, physicochemical properties of the bitter cassava and *gadung* used in this study were the same as the those used in the previous research conducted by Hargono [12]. Potassium sodium tartrate tetrahydrate and 3,5-Dinitrosalicylic acid, sodium hydroxide (98%, Merck), sodium sulfite

* Corresponding author: hargono@che.undip.ac.id

(98.5%, Merck), sulfuric acid (98.5%, Merck), sodium acetate buffer, glucose (99.5%), were purchased from their authorized distributor. Granular starch hydrolyzing enzyme as Stargen ™ 002, which is a mixture of α-amylase and glucoamylase was produced by Genencor (Palo Alto, USA). The activity and optimal pH range declared by the producer are 570 GAU/g and 4.0-4.5, respectively [8].

2.2 Enzymatic hydrolysis

All hydrolyses were conducted at concentration of starch slurry of 200 $g.L^{-1}$. Mass compositions of bitter cassava and *gadung* starches were 9:1, 4:1, 2.33:1, 1.5:1, 1.25:1, 1:1, 1:9, 1:4, 1:2.33 and 1:1.5. Concentration of Stargen ™ 002 was 1.5 % (w/w) was added into the samples. Samples were then incubated in an incubator shakher at 30±1°C, speed of 110 rpm. After 24 h, hydrolysis was stopped by adjusting the pH 4 (in 50 mM sodium acetate buffer). Samples were periodically withdrawn from the flask at 6 h interval and substantially subjected to reducing sugar analysis. Before the samples were analyzed by Spectrophotometer, they were centrifuged (100Hz, 4°C and 10 min) to obtain the filtrate. All experiments were performed in triplicates.

2.3 Analytical methods

The starch content was determined by AOAC method [13]. The water content in cassava starch was determined by standard drying method in oven at 105 °C to constant mass [14]. The total cyanide analysis by the acid hydrolysis method [15]. During the cassava starch hydrolysis, the content of reducing sugar was measured using dinitrosalicylic acid method [16]. Reagent consisting of aqueous solution of 1% 3,5-dinitrosalicylic acid, 0.05% Na sulfit, 20% Na-K tartrate and 1% NaOH solution was added in the ratio 3:1 to the samples in glass tubes, shaken in incubated in a boiling water bath for 8 min. Reacted samples were cooled in an ice water bath for 5 min, prior to measuring absorbance at 540 nm by using a UV/visible spectrophotometer (UV-160A, SHIMADZU, Kyoto, Japan). Glucose (0 to 60 $g.L^{-1}$) was used as standard, therefore reducing sugar concentrations was reported as $g.L^{-1}$.

3 Result and discussion

3.1 Determination of reducing sugar from bitter cassava and *gadung* starches hydrolysis

The reducing sugars obtained from bitter cassava and *gadung* starches by Stargen ™ 002 with concentration 1.5 % (w/w), and concentration of starch 200 $g.L^{-1}$, at 30°C and pH 4, during of hydrolysing time 0 to 24 hours are given in Figure 1. Over the hydrolysis time from 0 to 24 h, product of the reducing sugar obtained from of bitter cassava starch was higher than *gadung* starch.

Production hydrolysis from 0-18 h showed a maximum reducing sugar concentrations (bitter cassava starch 52.80 $g.L^{-1}$ and *gadung* starch 24.65 $g.L^{-1}$), while production rate showed of bitter cassava starch 2.93 $g.(Lh)^{-1}$ and *gadung* starch 1.37 $g.(Lh)^{-1}$, respectively. Furthermore after 18 hours, the hydrolysis of both of cassava and *gadung* starches decelerated, and then followed by constant rate to 24 h. The differences of the reducing sugar concentrations, it is because the starch content in *gadung* is much lower than the bitter cassava starch, as shown in Table 1 and so probably amylose content in bitter cassava starch is much lower than *gadung* starch. Sharma et al. [17] investigated effect amylose:amylopectin ratios on ethanol production from maize starch using GSHE. They conclused the amylose content in starch increased, the amount of ethanol produced decreased. Wu et al. [18] reported the result showed that the amylose content in starch had a significant effect on the ethanol conversion efficiency. Conversion efficiency increased as the amylose content decreased. So in addition the cyanide content in *gadung starch* is much higher than bitter cassava starch (Table 1), as a consequence the cyanide inhibit the enzyme leading to decrease of enzyme activity.

Table 1. Properties of bitter cassava and *gadung* starches

Parameters	bitter cassava starch	*gadung* starch
Moisture content (%), w/w	10.80	11.12
Starch (%), w/w	82.70	78.85
Cyanide, mg/kg	152.45	186.82

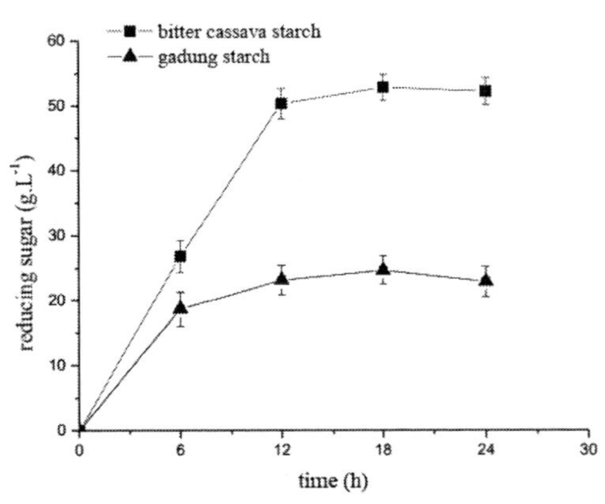

Fig 1. Hydrolysis profile of bitter cassava and *gadung* starches at concentration of substrate 200 g/L^{-1}, concentration of Stargen ™ 002 1.5 % (w/w), pH 4 and 30°C. Each value represents the mean of three independent measurements. Deviation from the mean below 5% for all values displayed.

Hargono et al. [12] investigated of effect of cyanide on reducing sugar from bitter cassava and *gadung* flours, we conclused increasing of cyanide concentration during hydrolysis process, decreased enzyme activity, as a consequence amount of reducing sugar decreased. Shanavas et al. [19] reported the reducing sugar formed from cassava starch by varying level of Stargen, at pH 4.5 and 30°C, it was found as 98.3 g/L could be hydrolyzed by Stargen level 100 mg on 10% (w/v) starch. Also reported by Yussof et al. [20], that the during time 8 to 24 h hydrolysis native tapioca starch was increasing reducing sugar, as indicated by increasing of dextrose equivalent (DE) 18 to 35.7% .

3.2 Effect of different composition bitter cassava and *gadung* starches on reducing sugar

Figure 2 shows reducing sugar release by enzymatic hydrolysis at a substrate concentration 200 $g.L^{-1}$, concentration Stargen ™ 002 was 1.5% (w/w), 30°C and pH 4 at various compositions of bitter cassava to *gadung* starches (9:1 to 1:1) during 24 h. It was found the maximum of concentration of reducing sugars at 18 h were 50.20 $g.L^{-1}$, (9 :1), 46.68 $g.L^{-1}$ (4:1), 43.98 $g.L^{-1}$ (2.33:1), 40.45 $g.L^{-1}$ (1.5:1), and 47.49 $g.L^{-1}$ (1 :1). The production of reducing sugar increased when bitter cassava to *gadung* starches ratio in the starch composition increased.

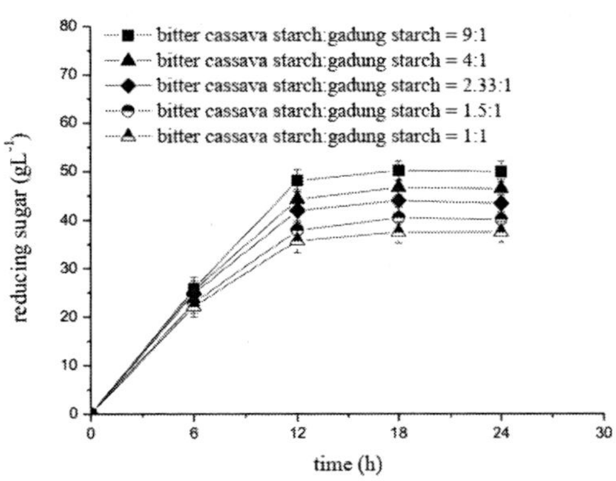

Fig 2. Hydrolysis profile of bitter cassava and *gadung* starches compositions (9 :1 to 1:1) w/w, at concentration of substrate 200 $g.L^{-1}$, concentration of Stargen ™ 002 1.5 % (w/w), pH 4 and 30°C. Each value represents the mean of three independent measurements. Deviation from the mean below 5% for all values displayed.

The result showed that reducing sugar mainly depend on starch content on bitter cassava, it is much bigger than the *gadung* starch. Thereafter the reducing sugar release by enzymatic hydrolysis at a substrate concentration 200 $g.L^{-1}$, concentration Stargen ™ 002 was 1.5% (w/w), 30°C and pH 4 during 24 h at various compositions of bitter cassava to *gadung* starches from 1:9 to 2:3, as shown in Figure 3. The maximum reducing sugar release at 18 h. It was 27.12 $g.L^{-1}$ (1:9), 30.12 $g.L^{-1}$, (1:4), 33.14 $g.L^{-1}$ (1:2.33) and 36.08 $g.L^{-1}$ (1:1.5). Masiero et al. [21] reported a hydrolysis of fresh potato (200 $g.L^{-1}$) with Stargen™ 002 (45 $GAU.g^{-1}$) during 62 h. After the first 20 h, glucose was liberated 30 $g.L^{-1}$. Shanavas et al. [18] investigated a hydrolysis of cassava starch (200 $g.L^{-1}$), by Stargen™ 002 with concentration 0.4 % (w/w) at 24 h and 30°C, reducing sugar was obtained 185 $g.L^{-1}$. The difference in hydrolysis time of our study are much shorter compared to other researchers, because the enzyme concentration is much bigger so that it is more active in degrading starch to reducing sugar.

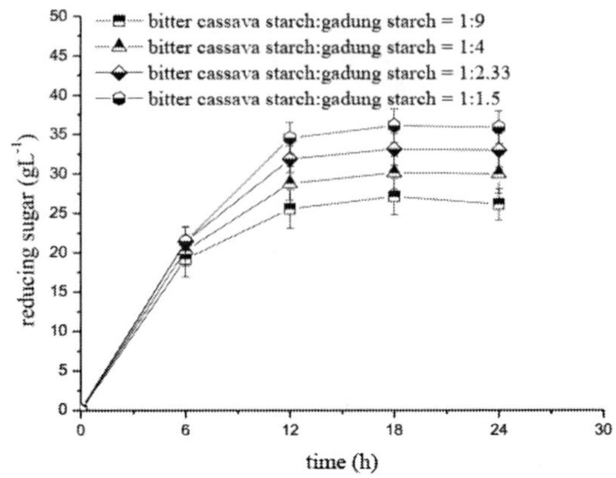

Fig 3. Hydrolysis profile of bitter cassava and *gadung* starches compositions (1:9 to 1:1.5) w/w, at concentration of substrate 200 $g.L^{-1}$, concentration of Stargen ™ 002 1.5 % (w/w), pH 4 and 30°C. Each value represents the mean of three independent measurements. Deviation from the mean below 5% for all values displayed

Conclusion

The experiments conducted an attempt is made to present kinetic expression for the hydrolysis of bitter cassava and *gadung* starches compositions, at concentration of substrate 200 $g.L^{-1}$, concentration of Stargen ™ 002 1.5 % (w/w), pH 4 and 30 °C. The results showed that at various compositions of bitter cassava to *gadung* starches (9:1 to 1:9) during 24 h. It was found the maximum of concentration of reducing sugars at 18 h at ratio 9:1 was 50.20 $g.L^{-1}$, while the minimum of concentration of reducing sugars at 18 h at ratio 1:9 was 27.12 $g.L^{-1}$. The release of reducing sugar increased when bitter cassava to *gadung* starches compositions increased, the concentration of reducing sugar mainly depend on starch content on bitter cassava, it is much bigger than the *gadung* starch.

References

1. M. Nashriyah, Y. Nornasuha, T. Salmah, N. Norhayati, M. Rohaizad, (2010 *Bulletin UniSZA*, **4** (2010)
2. M. Lambri, M.D. Fumi, A.Roda, D.M. Faveri, Afr J Biotechno. **19** (2013)
3. M. Djazuli, J.H. Bradbury, Food Chem. **65** (1999)
4. R.F. Power, *The Alcohol Textbook*, 4th edition, 23-32, (Nottingham University Press, Nottingham, 2003)

5. G.H. Robertson, D.W.S., Wong, C.C. Lee, J.Agric. Food. Chem. **54** (2006)
6. E. Sarikaya, T. Higasa, M. Adachi, B. Mikami, Process Biochem. **35** (2005)
7. S. Govindasamy, C.G. H.A. Oates, Carbohydr. Polym. **18** (1992)
8. Genencor (2009) STARGEN ™ 002: granular starch hydrolyzing enzyme for ethanol production
9. U. Uthumporn, Y.N. Shariffa, A.A. Karim, Appl. Biochem. Biotechnol. **166** (5) (2012)
10. J. Li, T. Vasanthan, D.C. Bressler, Carbohydr. Polym. **87** (2), (2012)
11. Y.N. Sharrifa, A.A. Karim, A. Fazilah, I.S.M. Zaidul Food Hydrocoll. **23**, (2009)
12. H. Hargono, B. Jos, A.C. Kumoro, Bull. Chem. React. Eng. Catal. **12** (2) (2017)
13. AOAC. Official Methods of Analysis of AOAC Intl. 16th ed. Method 991.43. Association of Official Analytical Chemists, Arlington, VA, USA. (1995)
14. B.V. Mc Cleary, V. Solah, T.S. Gibson, J.Sereal Sci. **20** (1994)
15. J.H. Bradbury, S.V. Egan, M.J. Lynch, J. Sci. Food. Agr. **55** (1991)
16. G.L. Miller G.L. Anal.Chem. **31** (1959)
17. V. Sharma, K.D. Rausch, M.E. Tumbleson, V. Singh, Starch/Starke **59** (2007)
18. X. Wu, R. Zhao, D. wang, S.R. Bean, P.A. Seib, M.R. Tuinstra, M. Cambell, A. O Brien, Cereal Chem. **63** (5) (2006)
19. S. Shanavas, G. Padmaja, S.N. Moorthy, M.S. Sajeev, J.T. Sheriff, Biomass and Bioenergy **35** (2) (2010)
20. N.S.Yussof, U.Utra, A.K. Alias, Starch/Starke **65** (2013)
21. S.S. Masiero, A. Peretti, L.F. Trierweiler, J.O. Trierweiler, Biomass and Bioenergy **70** (2014)

Cellulase Enzyme Production From Rice Straw Using Solid State Fermentation and Fungi Aspergillus niger ITBCC L74

Siti Maftukhah[1,*], and *Abdullah* Abdullah[2]

[1]*Master of Chemical Engineering, Diponegoro University, Indonesia, 50275*
[2]*Chemical Engineering Deparment, Diponegoro University, Indonesia, 50275*

Abstract. Rice straw is one of very abundant waste of agricultural and has not utilized maximally. This waste contain cellulose and potential in the manufacture of cellulase enzymes. Research on the production of cellulase enzyme from lignocellulose has been done a lot of enzyme activity is still low. This research using cellulose is 71.95% and conducted with 6 stages. First, the preparation of raw material. Second, the decrease of lignin content with alkali pretreatment. Third, the breeding of fungi *Aspergillus niger* ITBCC L74 . Fourth, incubation in the inoculum. Fifth, the production of cellulase enzyme by solid fermentation method. Finally, the analysis includes protein content, enzyme activity, enzyme characterization and kinetics of enzymatic reactions. The highest enzyme activity of this study is 3.12 U/ml and protein content is 0.34 mg/ml with fermentation time is 4 day and water content is 75%. In enzyme characterization obtained optimum pH and temperature are 4 and 60^{0}C, respectively. And obtained paramatic kinetic are V_{max} and K_m for 40, 50, 60 and 70^{0}C temperature are V_{max} : 6.42; 4.7; 5.82 and 4.46 U/ml and K_m : 1.32; 0.38; 0.32; and 0.12%, respectively.

1 Introduction

One of the most abundant agricultural residues available in the world is rice straw. And usully the word produced rice straw are731 million tons, consists of Africa: 20.9 million tons, Asia: 667.6 million tons, Europe: 3.9 million tons, America: 37.2 million tons and Oceania: 1.7 million tons [1] and overall in the word, rice straw production of 650-975 tons per year [2]. According to data from BPS in 2016 [3], rice production is 75.36 million tons of dried unhulled rice so the potential of rice straw in Indonesia is about 105.80 million tons [20]. Overall, the used of rice straw as fodder, fertilizer of organic and be berned [2,4,5,6]. Production of rice straw is very high and the utilization is not maximal because most of the cellulose content can not be fully utilized [7] makes rice straw very potential to be used as feedstock for biofuel production [2,5].

With the decline of petroleum production, energy sources are needed to meet future needs [8]. From to Pertamina's data, consumption of national fuel during 2000 to 2011 is about 312 to 364 m barrels of oil and the equivalent increases annually. Fuels of fossil is not environmentally friendly because produce a lot of pollution. For that needed a renewable energy source from biomass that environmentally friendly [22]. One of it is the biofuel production from lignocelluloses through hydrolysis and enzymatic fermentation processes, example bioethanol that can use enzymes from fungi that produce glucose, followed by fermentation of glucose into ethanol [9].

Rice straw is waste with cellulose content can be utilized as substrate for production of cellulase enzyme. Cellulase enzyme is an important enzyme because it have a role in the conversion and processing of materials in various fields. Of the total enzyme trade, 20% are cellulase enzymes [10] and increased enzyme demand along with the utilization of cellulase enzymes in the beverage, food, detergent, pulp and paper industries, ethanol, pharmaceuticals and others [9-13]. The use of enzymes on a large scale is still limited by cost factors. Some of the strategies for reducing the cost of cellulose enzymes include: increased enzyme productivity, enzyme production with cheaper substrate, and cellulase production with higher activity [14-16] .

The production of cellulase enzymes from biomass is one of way to decrease the price of cellulase enzymes [16] and increase the value of agricultural waste [5,17]. Production of cellulase enzymes from various agricultural wastes has been widely practiced but the resulting enzyme activity is still low. Cellulase enzymes are synthesized from various microbes so that selection of microbial types also greatly affects enzyme productivity. Abdullah [18] conducted research on water hyacinth, rice straw and corn cob and obtained the highest enzyme activity of rice straw substrate of 2.57 U/ml. Improvement of enzyme productivity in this study started from pretreatment to decrease lignin content, use of fungal strains that have high cellulase enzyme activity, use of appropriate fermentation method and use of optimum fermentation conditions.

* Corresponding author: maftukhah_tuchah@yahoo.com

The purpose of this research is to study the pretreatment effect of rice straw substrate in cellulase enzyme activity, to obtain optimum condition of moisture content and fermentation time to cellulase enzyme activity, characterization the highest activity of crude enzymes by determining the optimum of pH and temperature of the enzymatic reaction and to determine kinetics parameter of enzymatic reaction from the highest activity of crude enzymes.

2 Materials and Methods

2.1 Material

Raw materials in this study used post harvest rice straw obtained from Jepara, Central Java. Fungi *Aspergilus niger* ITBCC L 74 was purchased from Microbiology Laboratory and Bioprocess, Chemical Engineering Department, Bandung Institute of Technology. Aquades obtained from the Reverse Osmosis Unit are available in the Membrane Research Center Laboratory, Diponegoro University. And chemical reagents used with quality analysis (≥9.90% w/w) such as phenol fenol, DNS, BSA, PDA.

2.2 Methods

2.1.1 Preparation of material

Preparation of raw material was done by dried the rice straw in the sun, then cut, crushed and sieved so that obtained by rice straw powder with maximal water content (AOAC, 1990) [19] of 10% and pass 1 mm sieve [17]. Analysis of lignocellulosic content with Cesson Datta Method, protein content with Kjeldahl Method and ash content with thermogravimetri method conducted in food and nutrition laboratories Gajah Mada University..

2.1.2Pretreatment

Pretreatment was done by adding 2.75 % NaOH solution of 1:10 (w/v) ratio of rice straw powder, then the mixture was heated by hot plate stirrer at 100 ° C for 80 minutes. After cold, the mixture was neutralized on the filter paper and vacuum filter using 0.2 N HCl solution up to pH 7. The solid was dried using a oven in 70^{0}C to a constant water content. After cold the substrate was stored in plastic equipped with silica gel [17]. Analysis is done as in rice straw before pretreatment

2.1.3 Cultivated of microbes

Fungi *Aspergillus niger* ITBCC L74 is cultivated on medium of Potato Dextro Agar (PDA). Fungi applied zig-zag on PDA media using ose wire on aseptic condition. Then the fungi were incubated at temperature of 30 ± 2 ° C [8,21].

2.1.4 Inoculation in the inoculum

Inoculum was obtained by adding pure culture of *Aspergillus niger* ITBCC L74 into inoculum of medium solution. Inoculum of medium solution used 50 ml glucose solution 10 g/l, then added 5 ml the solution of nutrition according to Mendels prescription : $(NH_4)_2SO_4$ 1.4 gram; KH_2PO_4 2.0 gram, urea 0.3 gram; $MgSO_4.7H_2O$ 0.3 gram; $ZnSO_4.7H_2O$ 0.0014 gram; $FeSO_4.7H_2O$ 0.005 gram; $MnSO_4$ 0.0016 grams; $CoCl_2$ 0.002 grams; $CaCl_2$ 0.002 grams; Tween-80 2.0 ml and 1.0 gram pepton and aquades added to volume 1 L. Set the pH to 3.5 and sterilization. After cold, added the microbes and the inoculum was incubated at 30^{0}C [8].

2.1.5 Cellulase enzymes production

Production of cellulase enzyme was done by put the sample in to erlen meyer flask, then add the nutrients solution [8,34] and water to the water content of 60%, 65%, 70%, 75% and 80% [8,18,33] And set the pH to pH 3.5 [8].Next done sterilization. After cold, added the incubated inoculum of 1 x 10^6 spores/ml [12]. Subsequently, the mixture was incubated at 30± 2°C with fermentation timeof 3,4,5 and 6 days [8,16,18,33].

2.1.6 Extraction of enzyme

Taking the enzyme started by extracting the fermentation result with a ratio of substrate : aquades was 1: 50 (w/v) using a funnel and filter paper on 250 ml beaker glass. The filtrate was shaken using incubator shaker at 200 rpm at 30°C for 1 hour. Supernatant was obtained after centrifuge at 6000 rpm for 15 min and continued with analysis [16].

2.1.7 Cellulase enzyme assay

According to Abdeshahian et al., [21] the first analysis was a protein content by Lowry method to know the presence of enzymes in sample [23]. And enzyme activity is determined by the initial treatment of taking 1 ml of crude enzyme inserted in erlen meyer. Then adding 1% CMC (w/v) and 0.1 M buffer of sodium citrate solution (pH 4.8). The mixture was incubated at 40°C for 30 minutes. The reducing sugar is measured by DNS method (3,5-dinitrosalicylic acid) [24]. One unit (U) of enzyme activity is the amount of enzyme required to produces 1.0 μmol of glucose per minute under the assay conditions.

The enzyme characterization is performed on the enzyme which has the highest activity by varying the pH and temperature.

To determine the kinetic parameter by altering the substrate concentration (S) in the enzyme activity test of the crude enzyme having the highest enzyme activity performed at 40, 50, 60 and 70^{0}C. Different substrate concentrations resulted in different enzyme activity (V_0). 1/S and 1/V_0 graphs are obtained so that 1/K_m and 1 / Vmax of each temperature are obtained based on the Michaelis-Menten equation.

3 Results and Discussions

3.1 Effect of pretreatment on cellulase enzyme activity

he result of raw material characterization is presented in Table 1.

Table 1. Result of raw material analysis

Parameter	Without pretreatment (%)	With alkali pretreatment (%)
Cellulase	42.26	71.95
Hemicellulose	9.79	13.23
Lignin	11.28	7.08
Protein	5.11	1.14
Ash	4.92	5.12

Pretratment or delignification is a process of altering the chemical structure of lignocellulosic biomass in order to degrade lignin selectivity so as to decompose chemical bonds and other components are kept intact [25]. Cellulose and hemicellulose are carbohydrate structures in rice straw that can be depolymerized by enzymatic hydrolysis to glucose [17]. Pretreatment needs to be done because the presence of lignin in the material can inhibit cellulosic digestion during enzymatic hydrolysis [26] and microbial fermentation [27].

From Table 1. it can be seen that pretraetment causes content of cellulose increased by 70.25%, hemicellulose content increased by 35.14% and lignin content decreased by 59.32%. This is caused because alkaline pretreatment can remove lignin without affecting the other components. Pretreatment using NaOH causes a reduction of linin in the material, increased surface area, increased pore size, and the breaking of bonds between lignin and cellulose and hemicellulose as soon as to decrease the degree of crystallinity [28]. The increase in cellulose during alkaline pretreatment is largely due to the decrease in lignin. Lignin decreases as it dissolves in an alkaline solution [17].

From the Table 1 too. can also be seen that the protein content decreased 3 fold. This is because the protein starts denatured at 45 ° C [29]. Proteins are amino acids that have elements C, H, O, N as nutrients to produce enzymes. Carbon acts as an enzyme cofactor while nitrogen is required in the fermentation process because affect the activity of *Aspergillus niger*. In it growth, carbon and nitrogen are used for the body's fungi cell components [18].

For ash content does not change because NaOH only removes lignin without other components [17].

Thus, when compared with the research of Abdullah [18] the production of cellulase enzymes without pretreatment using the same substrate, fungi and method and using conditions slightly different from this research using pH 4, the highest cellulase enzyme activity increased 21.40%. Or the highest enzyme activity in the study of Abdullah [18] is 2.57 U/ml. While in this research the highest cellulase enzyme activity is 3.12 U/ml at 75% moisture content and 4 day fermentation time. This phenomenon can be seen in Figure 1. It is also revealed by Ilyas [28] that the activity of cellulase enzyme using pretretment increased by 51% compared with without pretreatment on *Vigno mungo* substrate.

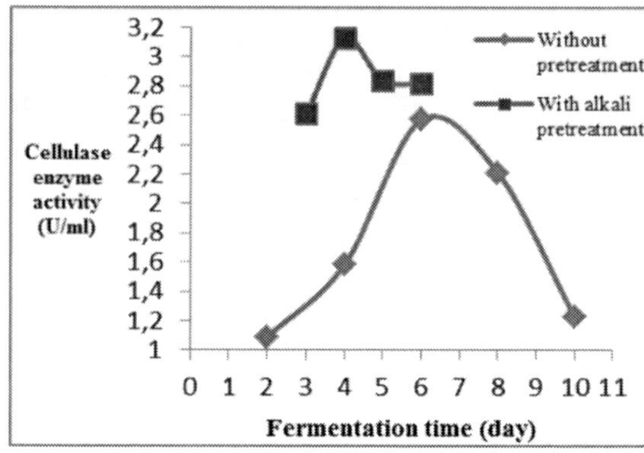

Fig. 1. Cellulase enzyme activity with pretreatment and without pretreatment

3.2 Effect of water content and length of fermentation on cellulase enzyme activity

3.2.1 Effect of water content on cellulase enzyme activity

Growth, biosynthesis and enzyme secretion of microbes affected by water content because it is an important factor in solid state fermentation [30]. The effect of water content to cellulase enzyme activity in this study can be seen in Fig 2.

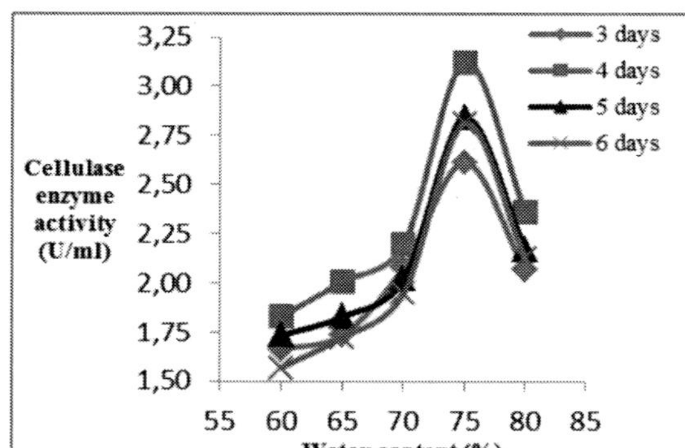

Fig. 2. Effect of water content on cellulase enzyme activity

Enzyme activity will increase with increasing of water content and optimum enzyme activity at 75% water content with 4 days of fermentation time is 3.12 U/ml. However, enzyme activity decreased at 80% water content. This phenomenon is the same for all variations of fermentation time. The water content is low to cause low growth of microbes, as well as low development and accessibility of nutrients. While the water content is high resulted in decreasing porosity of the substrate so reducing the penetration of oxygen to the particles [16].

Abdullah [18] conducted a study on rice straw using *Aspergillus niger* ITBCC L 74 and obtained the highest cellulase enzyme activity at 80% water content. While Ilyas [28] conducted a study of cellulase enzyme production from *Vigno mungo* and fungi *Aspergillus niger*, obtained the highest activity at 70% water content. This is agree with a statement from Ganjdar [31] that *Aspergillus niger* can grow well on water content of 70-80%.

The presence of protein content in a substance indicates the presence of an enzyme in the material. Increased protein content along with the growth of mold (fungi) because the fungus body consists of elements containing nitrogen. And the enzyme produced by the fungus is also a protein. The cell wall of the fungus contains 6.3% protein, while fungal cell membranes with hyphae contains 25-45% protein and 25-30% carbohydrates. In its growth, fungi using carbon and nitrogen for the fungus body cell components. This process occurs because during the fermentation of *Aspergillus niger* ITBCC L74 using nutrients (especially carbohydrates) for growth and increased protein content. [18]. The effect of water content on protein content can be seen in Fig 3.

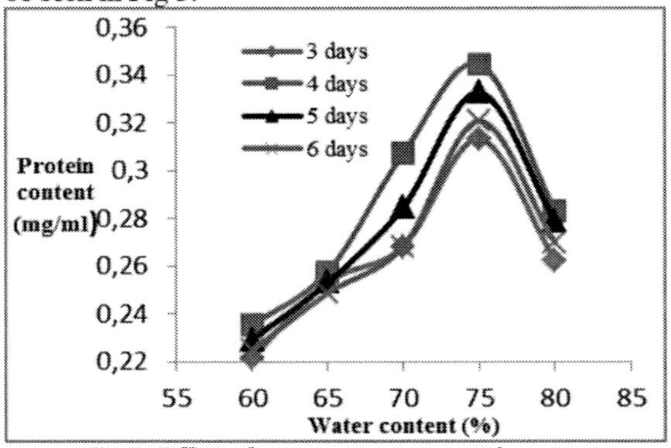

Fig. 3. Effect of water content on protein content

The highest protein content was obtained at 75% water content of 0.34 mg/ml with 4 days of fermentation and the lowest protein content obtained at 60% water content and 3 days fermentation time of 0.22 mg/ml. *Aspergillus niger* in growth media containing complex molecules can secrete extracellular enzymes such as α-amylase, β-amylase, glucoaamylase, protease and cellulase [32]. Based on the results of this study obtained increase in cellulase enzyme activity along with the increase in protein content. This is because most of the proteins in the crude enzyme filtrate consist of cellulase enzymes [9-10].

3.2.2. Effect of fermentation time on the cellulase enzyme activity

The effect of fermentation time on the cellulase enzymes activity can be seen in Fig 4.

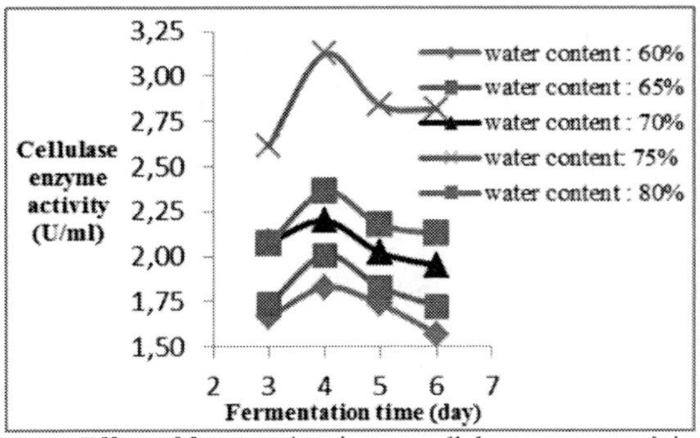

Fig. 4. Effect of fermentation time on cellulase enzyme activity

From the Fig 4. it can be seen that the highest cellulase enzyme activity is obtained at the time of fourth day fermentation for all variations of water content. Enzyme activity increases with fermentation time. However, enzyme activity decreased on the fifth and sixth days. In the beginning fermentation of enzyme activity is still very low. Enzyme activity increases with increasing fermentation time and decreases when available nutrients are depleted [32]. It follows the growth pattern of microorganisms. Spore-forming organisms usually produce enzymes in the post exponential phase. So when the enzyme activity produced is high, then the mold has been in that phase [7].

Ilyas [28] conducted a production study of cellulase enzyme from *Vigno mungo*, using the same method and temperature with this study and the same fungi but different strains, obtained the highest enzyme activity is 0.398 U/ml at 4 day fermentation. And Kang [33] also produced cellulase enzymes with similar methods and substrate types with this study and the same fungi but different strains, and obtained the highest enzyme activity on the fifth to sixth day of fermentation. And Omojasola [34] performed the cellulase enzymes production with pineapple waste substrate and the same conditions of various fungi types, obtained enzyme activity of 1.71 U/ml by *Trichoderma longibrachiatum*, 1.29 U/ml by *Aspergillus niger* and 1.07 U/ml by *Saccharomyces cereviseae* on the third and fifth day of fermentation. Various optimum fermentation periods for cellulase enzyme production due to differences in medium properties, types of microorganisms, nutrient concentrations and physiological conditions of the process used [35].

The effect of fermentation time to protein content is presented in Fig 5.

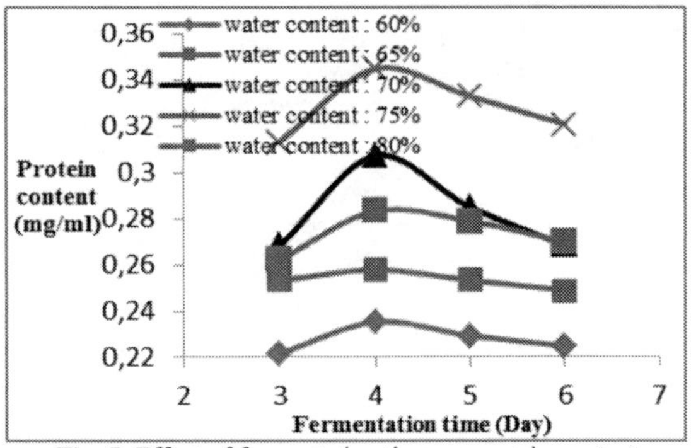

Fig. 5. Effect of fermentation time on protein content

As with enzyme activity, the effect of protein content on fermentation time is obtained by the highest value at 4 days of fermentation time for all variations of water content.

3.3. Effect of pH and temperature on cellulase enzyme activity

3.3.1 Effect of pH on enzyme activity

The active site of the enzyme having the ion group must be in the appropriate form (acid or base) in order to function. The variation in pH of the medium results in ion-formation of the active site and changes in enzyme activity so as to influence the rate of reaction. Changes in pH can also to change the dimensions of the enzyme. Or in other words, the enzyme has an active site with certain groups acting as a catalyst in the formation of the substrate (ES) complex. Changes in pH affect to the ionisation of functional groups, which can cause changes in enzyme conformation and its catalytic properties. For this reason the enzyme is only active in a certain pH range. Maximum reaction rate, K_m and the stability of the enzyme are influenced by the pH of the medium [36].

The effect of pH on enzyme activity with 1% substrate concentration on production of cellulase enzyme having the highest activity is presented in Fig 6.

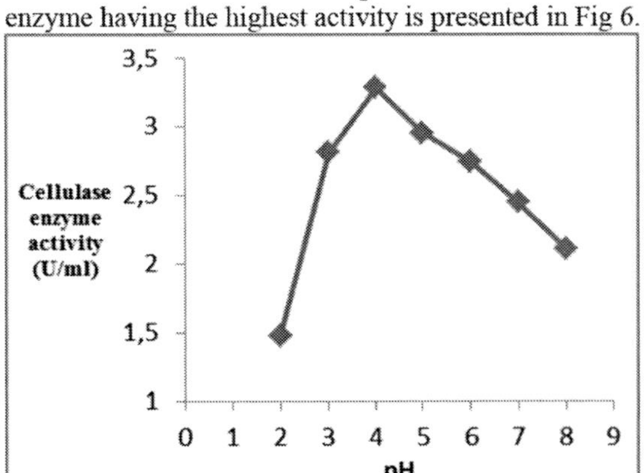

Fig. 6. Effect of pH on cellulase enzyme activity

From the Fig 6. it can be seen that the cellulase enzyme activity at pH 2 is still low and increases at pH 3 and 4. The highest activity is obtained at pH 4 is 3.28 U/ml and its activity decreases up to pH 8, and the decrease is apparent up to 55.45%. According to the statement of Mustafa [13] that optimum cellulase enzyme activity at pH 4 to 6.5. And Liu [37] stated that the optimum cellulase enzyme activity at pH between 3 and 6. Similarly, it is in accordance with a study conducted by Delabona [38] that conducted a study of *Aspergillus fumigates* with wheat bran substrate obtained the highest activity at pH 5. In another study also obtained the activity of the highest cellulase enzyme at pH 4.5 by *Aspergillus niger* [39]. As well as research from *Aspergillus tereus* obtained optimum activity at pH 5.5 [13].

3.3.2 Effect of temperature on enzyme activity

Temperature is one of the environmental factors that influence the development of enzyme activity. The rate of the enzyme-catalyst reaction increases with increasing temperature to some extent. The temperature rise will increase the reaction rate to only a certain range. At first the rate of reaction increases with increasing temperature and and the increase in rate in reaction is caused by the increased kinetic energy of the reacting molecules. Higher energy increases will break the weak hydrogen and hydrophobic bonds that retain the secondary-tertiary structure of the enzyme. In addition, high temperatures may also affect substrate conformation and thus experience resistance to entering the active site of the enzyme. Above a certain temperature, enzyme activity decreases due to enzyme denaturation [36] is changes in the structure of enzyme proteins, changes in ionic bonds and hydrogen bonds, so causing a decrease in reaction velocity catalyzed by the enzyme [40].

The effect between temperature to enzyme activity at pH 4 of the enzyme crude of this study, which has the highest cellulase enzyme activity is presented in Fig 7.

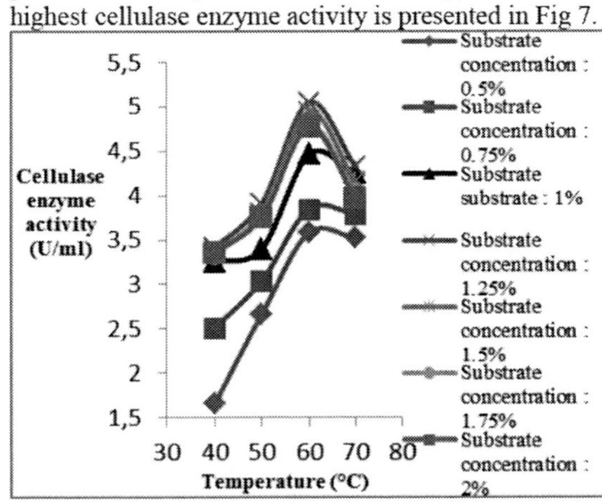

Fig. 7. Effect of temperature on cellulase enzyme activity

From the four variations of substrate concentration the phenomenon that occurs is the same that at first enzyme activity is still low and the increasing temperature, enzyme activity is also increasing. Then the enzyme activity decreases after the temperature of 60^0C. The highest enzyme activity occurred at 60^0C and CMC concentration 1.25% at 5.05 U/ml.

Cellulase enzyme is a thermophilic enzyme with maximum activity between 50 and 65^{0}C. Higher temperatures may increase the formation of products, therefore cellulase enzymes have a potential use in industry [38]. Similarly, other studies such as those conducted by Falkoski [41] is production cellulase enzyme by the plant pathogenic fungus *Chrysoporthe cubensis* obtained the highest enzyme activity at 60°C. While research conducted by Liu [37] from *Aspergillus fumigatus* obtained optimum temperature at 50°C. Research conducted by Mustafa [13] from *Aspergillus Tereus* is optimum at 60^{0}C. As well as research conducted by Saini [42], from *Penicillium oxalicum* obtained the highest enzyme cellulase activity at 50°C.

3.4 Kinetics of Enzymatic Reactions

Determination of maximum the rate of reaction (V_{max}) and constants of Michaelis-Menten (K_m) is an important thing to do to know the enzyme's characteristics [43]. The V_{max} value indicates the saturation level of the enzyme by the substrate, while K_m indicates the catalyst efficiency of the enzyme defined as a substrate concentration at a time when the enzyme's catalytic velocity reaching half of its maximum rate [44]. The K_m value used as an affinity measure E-S also relates to the desosiation balance equilibrium constant of the E-S complex becomes E and S. Adding when the K_m value is small means the E-S complex is steady and the enzyme affinity to the substrate is high, if K_m value of enzyme is large so become low. K_m value varies greatly depending on the type of substrate, the state of the environment and ionic strength[45].

Based on the Michaelis-Menten equation, the effect between substrate concentration and cellulase activity at 40, 50, 60 and 70^{0}C temperature can be presented in Fig 8.

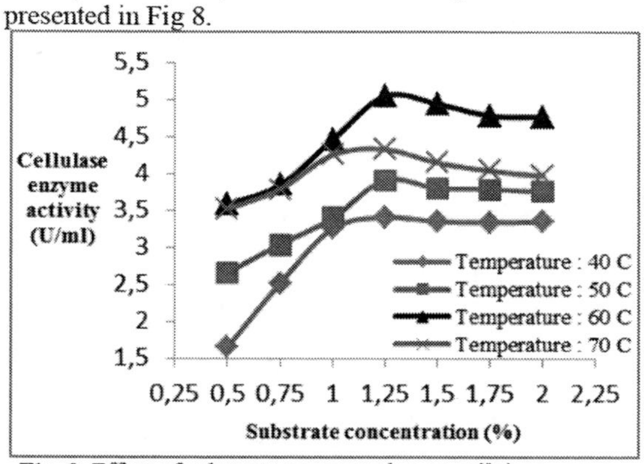

Fig. 8. Effect of substrate concentration on cellulase enzyme activity

Based on the kinetic theory, the higher the substrate concentration, the higher the energy and the frequency of collisions between molecules so that more cellulase enzymes can bind cellulose to form glycosyl-enzyme complexes which will further form glucose products [43]. It can be shown in Fig 8. that as the concentration of substrate increases then the activity of the enzyme is also increased. Nevertheless, the increase in enzyme activity will stop when the enzyme is saturated with the substrate because the molar amount of the substrate has exceeded the molar amount of the enzyme after the maximum limit point (V_{max}) of the enzymatic reaction.

V_{max} and K_m values can be determined by transforming the Michaelis-Menten equation into the Lineweaver-Burk equation and obtain the graph as in Fig 9.

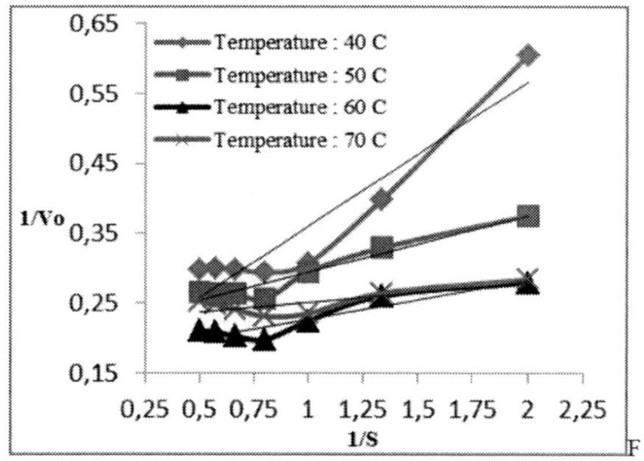

Fig. 9. Correlation between $1/V_0$ with 1/S based on the Lineweaver-Burk equation

Based on the Fig. 9 obtained linear equations, V_{max} and K_m values of each temperature variation presented in Table 2.

Table 2. Parametic of kinetic reaction enzimatic

Temperature (°C)	Linier equestion	R^2	Vmaks (U/ml)	Km (%)
40	y = 0.2053x + 0.1558	0.8949	6.42	1.32
50	y = 0.0813x + 0.2129	0.9364	4.7	0.38
60	y = 0.0553x + 0.1719	0.8757	5.82	0.32
70	y = 0.0271x + 0.2243	0.6076	4.46	0.12

From Table 2. the difference in V_{max} and K_m values at each temperature was due to the reaction of enzyme catalysts with different temperatures causing different reaction rates. An increase in the external temperature will generally increase the rate of chemical reaction of the enzyme, but too high temperature rise will cause enzyme denaturation [36]. Excessive heat energy can increase the enzyme's kinetics to reach the point that can exceed the energy limit required to destroy the covalent bonds that make up the enzyme structure. As a result the polypeptide chain becomes unfolded or denatured and is accompanied by a dramatic decrease in catalytic activity [40].

Mustafa [13] conducted a study of cellulase enzyme production from *Aspergillus terreus* culture filtrate and *Allium cepa* seed sextract. The K_m and V_{max} for cellulase from FCF and OSE were 20 and 8 mg / ml and 10 and 2.78 U/ml, respectively. Saropah [43] conducted a kinetics study of enzymatic reactions of crude extract enzyme cellulase cellulolytic bacteria isolated from bran and obtained values of K_m and V_{max} respectively are 1.694% and 0.0086 ppm/minute. The difference in V_{max} and K_m values is caused because the enzyme extracted from different sources will have different properties, especially its response to environmental conditions, such as: temperature and pH [45].

4. Conclusion

Production of cellulase enzyme from rice straw using fungi *Aspergillus niger* ITBCC L74 and solid state fermentation method successfully done. By doing pretreratment on the substrate used, can increase cellulase enzyme activity is 21.40%. And the variables studied in this research are water content and fermentation length. The highest enzyme activity was obtained at 3.12 U / ml at 75% water content and 4 days of fermentation time.

Enzyme activity is affected by pH and temperature. From the cellulase enzyme produced and the highest activity, the effect of pH and temperature on enzyme activity is studied. And the highest enzyme activity was obtained at pH 4 and temperature 60^0C is 5.05 U/ml.

From the kinetics of the enzymatic reaction at 40.50.60 and 70^0C temperature obtained paramatic kinetic are V_{max}: 6.42; 4.7; 5.82 and 4.46 U/ml and K_m: 1.32; 0.38; 0.32 and 0.12%.

Acknowledgement

We thank you to Bioprocess Laboratory of Chemical Engineering Department in Diponegoro University for all the support and facilities.

References

1. A.Ranjan, V.S. Moholkar. *Fuel,* **112,** 567–571. (2013)
2. G.W. Park, H.N. Chang, K. Jung, C. Seo, Y. Kim, J.H. Choi, H.C. Woo, I. Hwang. *Proc. Biochem.* (2017)
3. Badan Pusat Statistik (2016)
4. D. She, X.N. Nie, F. Xu, Z.C. Geng, H.T. Jia, G.L. Jones. *Cellulose Chem. Technol.*, **46 (3-4),** 207-219. (2012)
5. Ranjan, S. Khanna, V.S. Moholkar. Appl. Energy. 103, 32-38. (2013)
6. T. Kogo, Y. Yoshida, K. Koganei, H. Matsumoto, T. Watanabe, J. Ogihara, T. Kasumi. Biores. Technol., 233, 67–73. (2017)
7. Singh, N. Singh, N.R. Singh. J. Sci. Ind. Res. 69, 232-237. (2010)
8. H. Abdillah, B. Abdullah, S. Priyanto. *Seminar nasional kimia dan pendidikan kimia*, **7**. ISBN : 978-602-73159-0-7.(2015)
9. F.A. Abubakar, O.B. Oloyede. *Inter. J. Sci. Res. Manage.* **1 (5),** 285-291. (2013)
10. B.M. Julia, A.M. Belen, B. Georgina, F. Beatriz. *Biocat. Agric. Biotechnol.*, **6,** 1–8. (2016)
11. P.B. Acharya, D.K. Acharya, H.A. Modi. *African J. Biotechnol.*, **l7 (22),** 4147-4152. (2008)
12. N. Norma, Gamarra, K. Gretty, Villena. *Appl Microbiol Biotechnol*, **87,** 545–551. (2010)
13. F.A. Mostafa, A. Abeer, A. El Aty, R. Eman, B.M. Eid. *Biocat. Agric. Biotechnol.*, **5,** 116–122. (2016)
14. Y. H. P. Zhang, M.E. Himmel, J.R. Mielenz. *Biotechnol. Adv.*, **24,** 452–481. (2006)
15. R.R. Singhania, R.K. Sukumaran, A.K. Pate., C. Larroche, A. Pandey. *Enzyme Microb. Technol.*, **46,** 541–549. (2010)
16. V.H Vu, T.A. Pham, K. Kim. *Mycobiol.*, **39(1),** 20-25. (2011)
17. Singh, N.R. Bishnoi,N.R. *Ind. Crops Prod.*, **44,** 283–289. (2013)
18. Abdullah, S. Maftukhah, F. Faradiba, E. Listyaningrum. *15th International Conference on Quality in Research (QiR 2017)*. (2017)
19. AOAC. Association of Official Analytical Chemist. *Official Methods of Analysis 15 th Edition,* **Vol 2**.
20. M.I. Asy'ari, and M. Sudibandriyo. Thesis of *Universitas Indonesia*, (2014).
21. P. Abdeshahian, N. Samat, A. Hamid, W. Yusoff. *Biotechnol. Bioproc. Eng.*, **16,** 238-244. (2011)
22. V.T. Rosyida, A.W. Indrianingsih , R. Maryana, S.K. Wahono. *Energy Procedia*, **65,** 368 – 371. (2015)
23. O.H. Lowry, N.J. Rosebrough, A.L. Farr, R.J. Randall. *J. Biol. Chem.*, **193**, 265. (1951)
24. G.L. Miller. *Analytical Chem* **31**, 426-428. (1959)
25. R. Singh, A. Shukla, S. Tiwari, M. Srivastava. *Renew. Sustain. Energy Rev.*, **32,** 713-728. (2014)
26. F. Cotana, M. Barbanera, D. Foschinia, E. Lascaro. C. Buratti. *Energy Procedia,* **82,** 389–394. (2015)
27. J.S. Kim, Y.Y. Lee, T.H. Kim. *Bioresour. Technol.*, **199** , 42–48. (2016)
28. U. Ilyas, A. Majeed, K. Hussain, K. Nawaz, A. Shakil, M. Nadeem. *World Appl. Sci. J.*, **8**, 1172-1178. (2011).
29. J.E. Bailey, D.F. Ollis, D. F. Biochemical Engineering Fundamentals (edition 2 th). New York: *McGraw-Hill Book Co.*(1986)
30. S. Mrudula, R. Murugamma. *Brazilian J. Microbiol.*, **42,** 1119-1127. (2011)
31. Gandjar, A. Oetari, W. Sjamsuridzal. Mikologi dasar dan terapan. Jakarta: *Yayasan Obor*.(2006)
32. I.B.W. Gunam, K. Buda, I.M.Y.S. Guna. *J. Biologi XIV*, **1**, 55-61. (2010)
33. S.W. Kang, Y.S. Park, J.S. Lee, S.I. Hong, S.W. Kim. *Bioresour. Technol.*, **91,** 153–156. (2004)
34. P. Omojasola, Falakemi., Jilani, O. Priscilla, S.A. Ibiyemi, S. A. *Nature Sci.*, **6(2),** 64–79. (2008)
35. S. Kumar, H.K. Sharma, B.C. Sarkar. *Food Sci. Biotechnol.*, **20(5),** 1289-1298. (2011)
36. M.L. Shuler, F. Kargi. Bioprosess Engineering Basic *Consept. Prenticehall,Inc.*(1992)
37. D. Liu, R. Zhang, X. Yang, H. Wu , D. Xu, Z. Tang, Q. Shen. *Inter. Biodeterior. Biodegrad.*, **65,** 717-725. (2011)
38. P.S. Delabona, R.D.P.B. Pirota, C.A. Codima, C.G. Tremacoldi, A. Rodrigues, C.S. Farinas. *Ind. Crops Prod,.,* **42** , 236– 242. (2013)
39. C.S. Farinas, M.M. Loyo, A. B. Junior, P. W. Tardioli, V.B. Neto, S. Couri. *New Biotechnol.*, **27,** 810-815. (2010).
40. S. Putri. *UIN Maulana Malik Ibrahim Malang*. Skripsi. (2016)
41. D.L. Falkoski, V.M. Guimarães, M.N. Almeida, A.C. Alfenas, J.L. Colodette, S.T. Rezende. *Bioresour. Technol.*, **130,** 296–305. (2013)
42. R. Saini, J.K. Saini, M. Adsul, A.K. Patel, A. Mathur, D. Tuli, R.R. Singhania. *Bioresour. Technol.*, **vol 188** page 240–246. (2015)
43. D.A. Saropah, A. Jannah, A. Maunatin. *Alchemy*, **2(1)**, 34-45. (2012)
44. Lehninger. (1982). Principles of Biochemistry. Worth Publisher, Inc.
45. G.P.G. Putra. J. Biologi, 8(1), 21 -24. (2009)

Tailoring Properties of Acidic Types of Natural Deep Eutectics Solvents (NADES): Enhanced Solubility of Curcuminoids from *Curcuma zeodaria*

Orchidea Rachmaniah[1,*], Lailatul Jumiati Fazriyah[1], Nurul Hesti Seftiyani[1], and *M.* Rachimoellah[1]

[1]Department of Chemical Engineering, Faculty of Technology Industry, Institut Teknologi Sepuluh Nopember, Surabaya 60111, Indonesia

Abstract. Recently Natural Deep Eutectic Solvents (NADES) show their potential as a promising green solvents at 21th century for extraction of natural products. Rutin, a poorly water soluble flavonoid, was reported better solubilized in NADES than in water as well as a paclitaxel and ginkgolide B, a completely water-insoluble compound. In case of curcuminoids, phenolic compounds from powder of *Curcuma Zeodaria* shown better solubilized in acidic type of NADES such malic acid-sucrose-water (MAS-H_2O = 1:1:11, mole ratio) and citric acid-sucrose-water (CAS-H_2O = 1:2:15, mole ratio) compare to water and ethanol. Indicating that NADES, a water-based solvent, appropriate for extracting curcuminoids. However, the inherent high viscosity of NADES hamper the process extraction. Lowering the viscosity, water content of NADES is varied, adding a certain amount of water up to 60% of water content. All the varied water content of both CAS-H_2O and MAS-H_2O were successfully extracted curcuminoids. Yielded 0.06-0.16 mg curcuminoids/g dry weight. However, 60% of water content of both CAS-H_2O and MAS-H_2O gave more or less similar value of curcuminoids to CAS-H_2O (20% of water) and MAS-H_2O (30% of water). Curcuminoids is more stable in CAS-H_2O compare to MAS-H_2O for 96 h of time duration.

1 Introduction

Turmeric has been found to be rich source of phenolic compounds, namely curcuminoids consisting a natural mixture of three polyphenols: curcumin (C), desmethoxycurcumin (DMC), and bisdemethoxycurcumin (BDMC). In which curcumin as the main constituent. It responsible for the biological activities of curcuma and it yellow to bright orange colour characteristic [1]. Curcumin commonly called diferuloylmethane is a fat-soluble, hydrophobic polyphenol pigment [2]; recognized for their broad spectrum of bioactivity which is safe for use in food and pharmaceuticals industries [1]. It known for its antioxidant, anti-inflammatory, anti-microbial, anti-parasitic, anti-mutagenic, and anti-cancer properties [2, 3] which also used in the treatment of infection with human immunodeficiency virus (HIV) [1]. As an anticancer agent for various cancers such colorectal, pancreatic, breast, prostate, multiple myeloma, lung, head and neck, curcumin has been demonstrated to inhibit cancer cell survival and proliferation [4]. The chemical structure of curcumin takes responsible for its unique physicochemical and biological properties. Moreover, curcumin is highly safe consumed and well tolerated by human up to 12 g/day [3], prevents multi-drug resistance (MDR) and chemo-resistance which are commonly occurred in cancer patients [2].

In spite all of its advantages, until now curcumin not yet been approved as a therapeutic agent. Though those multi-target bioactivities of curcumin makes it a good candidate for a chemical derived anticancer agent from plant. The explanation might be due to low solubility of curcumin in water and its poor bioavailability.

Curcumin shown a low systemic bioavailability after oral consumption, hence only a trace or even negligible amount of curcumin may found in the tissues outside the gastrointestinal tract [2]. Negligible amount of curcumin was found in blood plasma of rats after oral administration of 1 g/kg of curcumin [3]. Practically, curcumin not soluble in water at acidic and neutral pH, ca. pH = 1-7. In contrast with, at aqueous alkaline solution, its solubility is increased. However, at the same time curcumin degradation also takes a place [4, 5].

Several studies have been done to improve both of solubility and bioavailability of curcumin such: (1) adjuvants utilization, e.g. piperine [6], quercetin [7], or genistein [8], which was combined with curcumin in order to block metabolic pathways of curcumin; (2) nanoparticle technology application for triggered drug

* Corresponding author: orchidea@chem-eng.its.ac.id

delivery system containing curcumin. However, the low solubility of curcumin in water yet the major obstacle. In addition (3) a drug delivery system based on liposomes, micelles, and phospholipid complexes were also studied.

Therefore in our preliminary study, a natural deep eutectic solvent (NADES), a recently reported green solvent [9] with wide range of solvent polarity was used. Twelve kind of different NADES were studied; determining the NADES with the highest extractability of curcuminoids from *C. zeodaria* powder. Hence, enhance the solubility of curcumin in water [10]. Though the highest extractability gives by choline chloride-malic-acid-water (CCMA-H_2O) at 1:1:2 mole ratio. It is found that curcumin not stable within 96 h. Meanwhile, curcumin is also well extracted in acidic type of NADES, e.g. citric acid-sucrose-water (CAS-H_2O) and malic acid-sucrose-water (MAS-H_2O). Which is contrary to the fact that curcumin have low solubility at acid solution (pH = 1-7) [5]. In addition, the inherent high viscosity of NADES reduces the mass transfer rate at the extraction process. Therefore, the effect of water content of CAS-H_2O and MAS-H_2O, by means diluting the NADES, to the extractability of curcuminoids from its plant matrix, *C. zeodaria* powder, was currently studied.

2 Materials and methods

Curcuminoids standards was purchased from TCI America. Fine powder of *Curcuma zeodaria* was kindly donated from Sari Herbal (Sukun, Malang, Indonesia). While individual component of NADES: sucrose, citric acid, and malic acid were purchased from Sigma Aldrich (St. Louis, MO, USA). In case of solvents such acetonitrile and acetic acid are in HPLC grade were commercially purchased form Merck.

2.1. NADES preparation

NADES was prepared followed a protocol by [11] with a slight modification; a freeze dry was used instead of a vacuum evaporation. Two individual compounds of NADES were precisely weighted according to its molar ratio; mixed and dissolved in water equipped with water bath (70 °C) and stirring bar. The mixture was agitated till a clear liquid was formed. The resulted liquid called further as a NADES.

In case of water content variation, a determined amount of water, i.e. aquadest, was added to the NADES reaching the specified percentage of water content. CAS-H_2O (1:2:15) and MAS-H_2O (1:1:11) were selected as the lowest content of water, 20% and 30% of water content respectively. Lower value of water content make the extraction process difficult, due to high viscosity of both NADES.

2.2 Curcuminoids extraction with NADES

Extraction was carried out by added 20 mg of the powder of *C. zeodaria* with 5 mL of NADES at amber bottle with a cap, stirring (350 rpm) at 40 °C for 24 h. After the extraction time was ended, 1 mL of each sample was collected and kept in the freeze and dark place for further analysis.

Maceration with acetone using the same plant material for 4 h, at dark place was also conducted. Determining the initial content of curcuminoids in the raw material [12]. A 1:8 (v/w) ratio of *C. zeodaria* powder to acetone was applied, meaning a 37.5 mg of powder with 300 mL of acetone was macerated at room temperature (30 °C) equipped with stirrer.

2.3 Properties tests

Density and viscosity of each NADES was measured in triplicates and the average value was reported. The Density was performed using pycnometer at room temperature (30 °C) with water as a reference; whereas viscosity was determined by Ostwald viscometer.

2.4 Curcuminoids analysis

The extracted curcuminoids was determined by high performance liquid chromatography (HPLC) Agilent Technologies in triplicates with the following column: 3.5 µm Zorbax eclipse XDB C-18 150 mmx4.0 mm ID (Agilent, USA). Chromatograph was equipped with diode array detector (DAD). The detection was performed at 421 nm. All the chromatographic measurements was performed at 35 °C and 1.0 mL/min of flow rate. The elution was carried out with an isocratic solvent system consisted of acetonitrile:acetic acid 1% (50:50, v/v) with the total running time was 15 min.

Prior injected to the HPLC, a ca. 200 mg of sample was accurately weighted, diluted up to 2 mL with methanol; subsequently vortexed, centrifuged and filtered with 0.2 µm nylon membrane syringe filter. Thus 50 µL of filtrate was directly injected to HPLC for quantitatively analysis.

3 Results and Discussion

3.1 Initial content of curcuminoids

Curcuma zeodaria as a raw a material contains 1.96% dry weight of curcuminoids. It was determined by previously described maceration protocol (**subsection 2.2**). In case of selectivity of curcuminoids both using volatile organic solvent and NADES was presented at **Table 1**.

It is clearly shown that acetone is the most suitable solvent for extracting curcuminoids from its plant matrix. Therefore maceration using acetone was applied for determining the initial content of curcuminoids in the *C. zeodaria* powder. It is reasonable, since curcuminoids is relatively non-polar compound. "Like dissolves like", a compound will be easily dissolved in solvent with similar polarity.

In contrast, acetone is volatile and highly flammable solvent with boiling point 56 °C. At high concentration, it is irritating and may depress central nervous system. Moreover, both of NADES and ethanol (EtOH) yielded

https://doi.org/10.1051/matecconf/201815601011

approximately similar amount of curcuminoids. Hence, NADES is a promising solvent for curcuminoids extraction from its plant matrix. In case of lower yield of curcuminoids by EtOH, might be explained that curcuminoids are relatively medium- to non-polar compounds; therefore more suitable with acetone than EtOH. Moreover, it might be degraded considering the length of the maceration time.

Table 1. Curcuminoids yield and selectivity obtained by using different solvent including NADES.

Solvent	Yield curcuminoids (mg/g [α]) (% selectivity [γ])			Total curcuminoids (mg/g±SD% [β])
	BDMC	DMC	C	
Acetone [λ]	2.505 (13)	6.693 (34)	10.402 (53)	19.600±0.35%
Ethanol [π]	0.026 (17)	0.034 (22)	0.092 (61)	0.152±0.43%
CAS-H_2O[κ] (1:2:15)	0.012 (8)	0.042 (27)	0.098 (65)	0.151±0.74%
MAS-H_2O[κ] (1:1:11)	0.005 (3)	0.027 (20)	0.100 (77)	0.131±1.82%

[α] yield curcuminoid expressed as mg curcuminoids/g dry weight of *C. zeodaria*.
[β] deviation standard calculated from triplicate data.
[γ] selectivity was calculated as percentage of total weight of extracted curcuminoids.
[λ] Maceration method was applied following [12].
[π] Maceration extraction was applied till the obtained filtrate was colourless, approximately 15 days.
[κ] Extraction was conducted for 24 h at 40 °C agitated with stirring bar at 350 rpm.

Selectivity of curcuminoids directly dependent to the kind of solvent (**Table 1**). In case of MAS-H_2O (1:1:11), it gave the highest selectivity to curcumin followed with desmethoxycurcumin (DMC) and bisdemethoxycurcumin (BDMC). Whereas CAS-H_2O (1:2:15) gave more or less similar selectivity of curcumin with ethanol. Despite of highest yield of curcuminoids, acetone only shown 53% selectivity on curcumin, hence the lowest one.

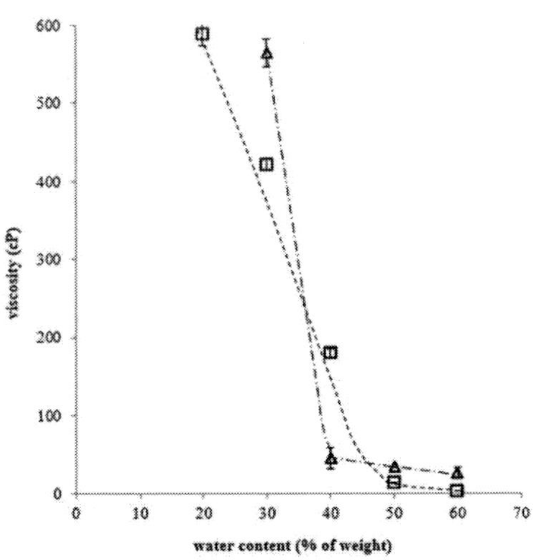

Fig. 1. Relationship between viscosity and percentage of water content of CAS-H_2O (□) and MAS-H_2O (△).

3.2 Physical properties of NADES

Inherent high viscosity of NADES making a limitation especially to the extraction application, hamper mass transfer of target compound form plant matrix to the bulk of solvent, i.e. NADES. Hence, mass transfer resistance is increased.

Viscosity is the major obstacle for the application of NADES. Therefore, a mole ratio of water both at CAS-H_2O and MAS-H_2O are varied. Consequently, physical properties of NADES, e.g. viscosity, density, solubilizing capacity, were altered, leading in some cases a higher solubility of rutin, carthamin, and cinnamic acid [11]. The viscosity and density of NADES were shown at **Fig. 1** and **Fig. 2**, respectively.

Dilution of NADES by adding water was dramatically lowering the viscosity (**Fig. 1**), especially when the water content was >40%. The viscosity of NADES is affected by percentage of water and temperature as well as the density. The density of all tested NADES proved to be higher than that of water (**Fig. 2**). The higher the water percentage, decreasing the density. Lastly, closer to the water density.

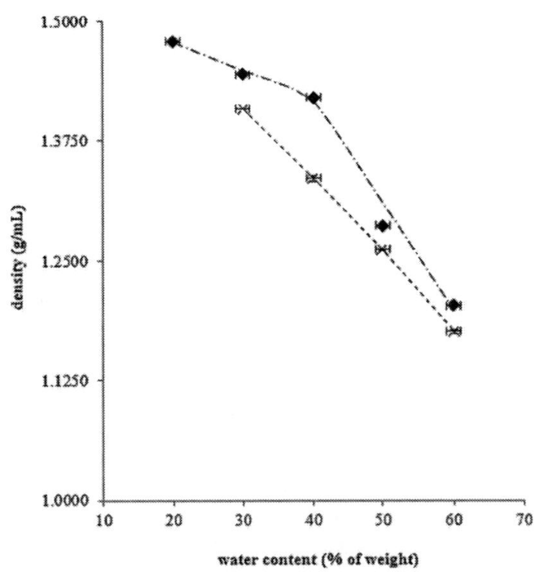

Fig. 2. Relationship between density and percentage of water content of CAS-H_2O (♦) and MAS-H_2O (×).

3.3 Solubilizing ability of NADES

The influence of water content on the solubilizing capacity of both CAS-H_2O (1:2:15, 20% of water content) and MAS-H_2O (1:1:11, 30% of water content) was investigated (**Fig. 3**). Water content of both NADES was variated by adding a specified amount of water, reaching the required percentage of water content.

Dilution of NADES decreasing its solubilizing capacity to curcuminoids. Which is reasonable, considering that curcumin has very low solubility in water at pH = 1-7 [4, 5]. Either CAS-H_2O or MAS-H_2O are an acid solution, pH = 2, measured after 10x diluted with aquadest. However, its native pH cannot be determined due to the viscous property of NADES. This results show that water is an important factor in optimizing NADES.

However, when the percentage of water reached 60%, the obtained yield is comparable to 30% of water. Though, extensive dilution of NADES resulting in the loss of existing hydrogen bonds, consequently the disappearance of superstructure of NADES [11]. In case of CAS-H_2O or MAS-H_2O, the 60% of water content yielded a similar amount of curcuminoids compared to 20-30% of water content. Concluded, both viscosity and density of NADES do not influenced the obtained yield.

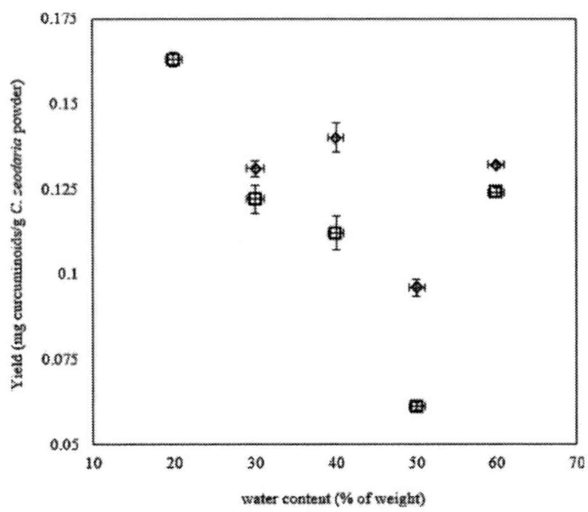

Fig. 3. Extractability of curcuminoids in acidic type of NADES diluted with different percentage of water content (CAS-H_2O (□) and MAS-H_2O (△)).

Exploring whatever yield increase or decrease, the extraction time was prolonged to 96 h (4 days). A prolonged extraction was conducted using CAS-H_2O (20% of water) and MAS-H_2O (40% of water) since both of them gave the highest yield of curcuminoids (**Fig. 3**).

Remarkably, 96 h of extraction time of curcuminoids with CAS-H_2O yielded fourth times higher than 24 h extraction, ca. 0.676 mg curcuminoids/g dry weight. Whereas MAS-H_2O yielded lower value of curcuminoids, ca. 0.083 mg curcuminoids/g dry weight; 50% lower than 24 h of extraction time. Hence, curcuminoids more stable in CAS-H_2O than MAS-H_2O.

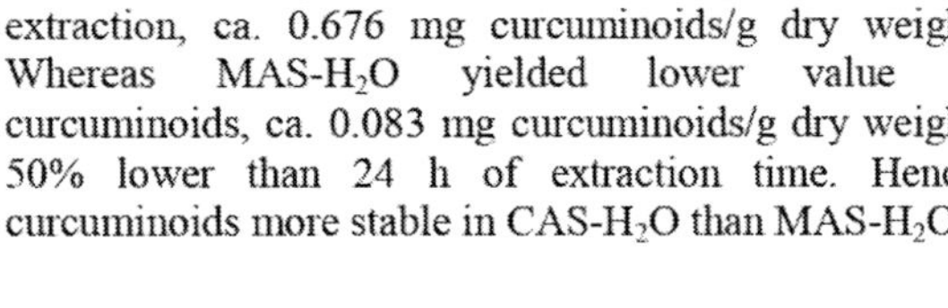

Fig. 4. Chemical structure of citric acid (a) and malic acid (b).

Considering the structure between citric acid and malic acid (**Fig. 4**), the presence of extra hydroxyl or carboxyl groups allowing more hydrogen bond to be formed. Hence increasing the stability of NADES. In case of organic acid with more carboxylic groups, such as citric acid, can form stable liquids than those with less carboxylic groups as in the case of malic acid. The more hydrogen bond are formed, more curcuminoids will be extracted.

Acknowledgement. Sari Herbal (Sukun, Malang, Indonesia) is acknowledged for supplying fine powder of *C. zeodaria*. OR is grateful for a grant No. 695/PKS/ITS/2017 from Lembaga Penelitian dan Pengabdian Masyarakat (LPPM) of Institut Teknologi Sepuluh Nopember, Surabaya, Indonesia, Kementrian Riset, Teknologi dan Pendidikan Tinggi.

4 Conclusions

Water content of NADES significantly influence the physical properties of NADES, i.e. density and viscosity, subsequently also the polarity and extracting capability. In case of acidic type of NADES, i.e. CAS-H_2O and MAS-H_2O, with various water content; all the variated NADES were successfully extracted curcuminoids from *Curcuma zeodaria* powder. Yielded 0.06-0.16 mg curcuminoids/g dry weight. However, 60% of water content of both CAS-H_2O and MAS-H_2O gave more or less similar value of curcuminoids to CAS-H_2O (20% of water) and MAS-H_2O (30% of water). Curcuminoids is more stable in CAS-H_2O compare to MAS-H_2O when it was extracted for prolonged time, ca. 96 h.

References

1. G. K. Jayaprakasha, L. Jagan Mohan Rao, and K. K. Sakariah, J. Agric. Food Chem., **50**, 3668, (2002).
2. G. Bar-Sela, R. Epelbaum, and M. Schaffer, Curr. Med. Chem., **17**, 190, (2010).
3. P. Anand, A. B. Kunnumakkara, R. A. Newman, and B. B. Aggarwal, Molec. Pharm., **4**, 807, (2007).
4. M. Salem, S. Rohani, and E. R. Gillies, RSC Adv., **4**, 10815, (2014).
5. H. H. Tønnesen and J. Karlsen, Z. Lebens. Unters. Forsch., **180**, 402, (1985).
6. G. Shoba, D. Joy, T. Joseph, M. Majeed, R. Rajendran, and P. S. S. R. Srinivas, Planta Med, **64**, 353, (1998).
7. M. Cruz-Correa, D. A. Shoskes, P. Sanchez, R. Zhao, L. M. Hylind, S. D. Wexner, *et al.*, Clinic. Gastro. Hepatol., **4**, 1035,
8. S. P. Verma, E. Salamone, and B. Goldin, Biochem. Biophys. Res. Comm., **233**, 692, (1997).
9. Y. H. Choi, J. van Spronsen, Y. Dai, M. Verberne, F. Hollmann, I. W. C. E. Arends, G. J. Wiktamp, R. Verpoorte, Plant Physiol., **156**, 1701, (2011).
10. S. Zullaikah, O. Rachmaniah, A. T. Utomo, H. Niawati, and Y.-H. Ju, "Green Separation of Bioactive Natural Products using Liquefied Mixture of Solids," in *Green Chemistry*. InTechOpen, 2017, (accepted).
11. Y. Dai, J. van Spronsen, G. J. Witkamp, R. Verpoorte, and Y. H. Choi, Anal. Chim. Acta, **766**, 61, (2013).
12. A. K. Popuri and B. Pagala, Int. J. Innov. Res. Studies, **2**, 289, (2013).

The Effect of Growth Medium Composition on X.*campestris* Metabolism in Producing Xanthan Gum

Nancy Siti Djenar[1] * and *Edi Wahyu Sri* Mulyono[1]

[1] Chemical Engineering Department, Politeknik Negeri Bandung, Bandung 40012, Indonesia

Abstract. In the fermentation industry, the role of medium is very important in increasing the growth rate of microbes. Nutrient Agar (NA) is one of the common medium used for growth and increase regeneration X.*campestris*, while Luria-Bertani medium (LB) is usually used for matters related to X.*campestris* productivity in producing xanthan gum. To increase the productivity of X.*campestris* and xanthan gum rendement, this research used medium variation of Nutrient Broth (NB) added with mineral plus inducer and Luria-Bertani Broth (LBB) with 4% substrate of liquid sugar and 2% maize, respectively. Based on FT-IR analysis, both substrates can produce xanthan gum which qualitatively has been compatible with commercial xanthan gum. The 4% liquid sugar yielded of 34.20% - 57.88%, while 2% maize was 54.93% - 73.61%. The use of NB and LBB produced xanthan gum with viscosity of 28-31 cPs and 15-30 cPs, respectively. These viscosity values indicated that xanthan gum can be used as a thickener in its application in industry. Precipitation using a combination of salt with IPA did not provide significant changes to the xanthan gum gain. This is due to the low number of bonds between the cations of the salt and the anions of the xanthan solution.

1 Introduction

Xanthan gum is one of the natural carbohydrate polymers which are now used as thickening ingredients ranging from the food industry to mining. Xanthan gum belongs to the group of hydrophilic colloidal sap (hydrocolloid) produced by the fermentation process of the *Xanthomonas campestris* bacteria.

In Indonesia, the use of xanthan gum in the food and beverage industry began around the 1980s. Since its inception in the industry until now, xanthan gum is still imported because there is no domestic industry capable of producing the biopolymer [1].

Substrates for the production of gum xanthan are carbohydrate compounds such as sorbitol, starch, potato starch, or hydrolyzate from molasses for carbon sources, while the nitrogen source may use hydrolyzate from casein, soybean waste, or cell hydrolyzate [2]. X.*campestris* fermentation using liquid sugar substrate at certain process conditions can produce xanthan gum between 7.9147 -13.8421g / L, yield 36.52% with highest viscosity 11.1453 cPs [3].

The process of gum xanthan formation is very complicated and not widely known. The mechanism of its formation simply by enzymatic transglycoside reaction. The enzyme responsible for this reaction is cyclodextrin glucanotransferase (1,4-α-D-1,4-gluco-transferase, EC 2.4.1.19) abbreviated to CGTase. CGTase is one of the important industrial enzymes that play a role in the cyclization process, which converts starch and α-1,4 glycosidic to cyclodextrin via intramolecular transglycosylation reaction. In addition, this enzyme can also hydrolyze starch and cyclodextrin becomes simpler [4]. According to D.S Prastiko [5] X. *campestris* produces a high activity CGTase enzyme. Therefore, in this research X.*campestris* cultures were grown in NA medium with an inducer, also grown on Luria-Bertani (LB) medium containing tripton. So it is expected in the process of metabolism X.*campestris* can increase its productivity in producing CGT ase enzyme. Furthermore, this enzyme will perform transglycoside reaction to produce xanthan gum. The objective of this research studied the effect of growth medium composition on *X.campestris* to increase xanthan gum production.

2 Experimental Details

2.1 Microbial Preparation

At this stage regeneration of bacteria X. *campestris* on:

a) Nutrient Agar (NA) medium containing nutrient agar + mineral + inducer and incubated at 28^{0}C.

b) Luria-Bertani (LB) agar medium containing yeast extract, tryptone, NaCl plus agar and incubated at 28^{0}C.

* Corresponding author: nancysitidjenar@yahoo.com

2.2 The production of inculum of X. *campestris*

2.2.1 The production of inoculum of X. campestris in modified Nutrient Broth (NB).

Modified Nutrient Broth medium was prepared by dissolving 0.3 g beef extract, 0.5 g NaCl , 0.5 g of glucose, mineral and 0.1 g inducer in an erlenmeyer that contained 100 ml of distilled water. The medium was sterilized for 20 minutes at 121°C and 1.4 atm. The stock culture X. *campestris* in nutrient agar was inoculated to the modified NB medium and incubated at 28°C for 48 hours and stirred at 150 rpm.

2.2.2 The production of inoculum of X. campestris in Luria-Bertani Broth (LBB).

This Broth medium was prepared by dissolving yeast extract, tryptone, NaCl in an erlenmeyer that contained 100 ml of distilled water. The medium was sterilized for 20 minutes at 121°C and 1.4 atm. The stock culture X. *campestris* in LB agar was inoculated to the LBB medium and incubated at 28°C for 48 hours and stirred at 150 rpm.

2.3 Experimental Work

Aerobic fermentation was done at 250 rpm, 28°C for 72 hours using 4% liquid sugar and 2% maize as substrates with additive substances (Table 1).

Table 1. Variation of medium and substrate on xanthan gum production.

No	Growth Medium		Types of substrate	Types of mineral
	Slant Agar	Inoculum		
1	NA + inducer	NB + inducer	4% liquid sugar	$FeCl_3.6H_2O$, $MgCl_2$, $ZnCl_2$, KH_2PO_4, Na_2SO_4, H_3BO_3, $CaCO_3$
2	LB	LBB	2% maize	$(NH)_2SO_4$, $MgSO_4$, $CaSO_4$, $MnSO_4$, $FeSO_4$, buffer Na-phosphate

Pasteurization. The pasteurization was done at 90°C for 25 minutes. The pasteurized fermentation was separated between its xanthan solution and its cells using centrifugation.

Purification and precipitation. Precipitation of xanthan gum was done by added an organic solvent of isopropyl alcohol (IPA) with a ratio of 1: 1, 2: 1, 3: 1 (v / v) to the supernatant, and with and without salt addition. The obtained precipitate was then dried in an oven at 50°C, weighed, and the yield and rendement were calculated. Analysis of xanthan gum was consisted of viscosity, FTIR and HPLC analysis.

3 Result and Discussion

The first step of this research was the preparation of 4% liquid sugar and 2% maize. Purification of X.*campestris* by regenerating it into two types of growth medium were Nutrien Agar (NA) and Luria-Bertani (LB) using dilution method. To the substrate was added some minerals to support the growth of X. *campestris* during fermentation. The inoculum medium used were Nutrient Broth (NB) and Luria-Bertani Broth (LBB). Fermentation was carried out at 28^{0}C, 250 rpm for 72 hours aerobically within the incubator shaker.

Based on the results of FTIR analysis, the presence of functional groups in the xanthan gum obtained from various fermentation process conditions can be recognized. Figure 1 and 2 show the FTIR spectrum of xanthan gum using modified NB and LBB with 4% liquid sugar and 2% maize, respectively.

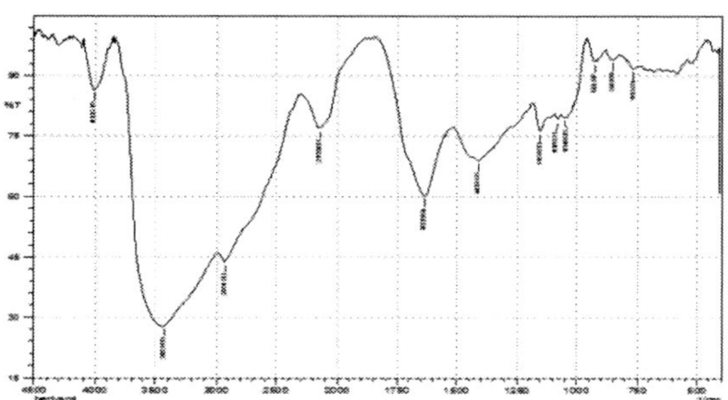

Fig. 1. FTIR Spectrum of Xanthan Gum from X. campestris Fermentation Using a Modified Nutrient Broth Medium with a Substrate of 4% Liquid Sugar.

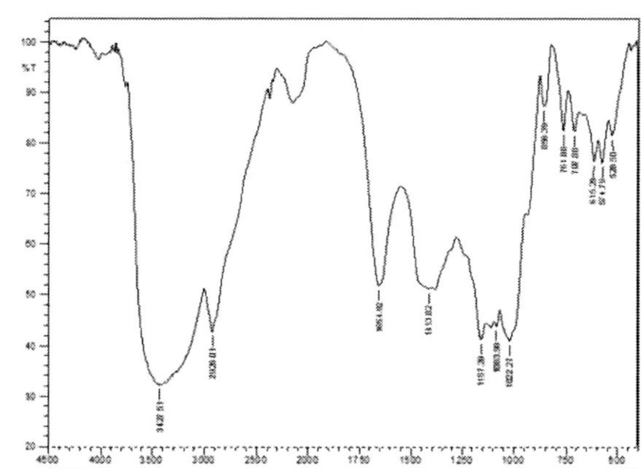

Fig. 2. FTIR Spectrum of Xanthan Gum from X. campestris Fermentation Using a Luria-Bertani Broth Medium with a Substrate of 2% Maize.

Based on FTIR analysis results, it can be seen the existence of functional groups contained in xanthan gum from fermentation 4% liquid sugar and 2% maize. The comparison of functional groups between commercial xanthan gum and xanthan gum of the study using two medium types is shown in Table 2 below.

Table 2 showed the compatibility of functional groups between xanthan gum of the study and commer-

Table 2. The Comparison of functional groups in commercial xanthan gum and in xanthan gum obtained using two types of medium.

Annotation	-OH (3600-3000 cm^{-1})	$-CH_2$ (3000-2800 cm^{-1})	-C=O (1760-1690 cm^{-1})	-C=O from pyruvate cm^{-1}	-COO- cm^{-1}	-C-O-C- cm^{-1}	-C-O-C-O-C acetal cm^{-1}	C_2H from β-glycoside cm^{-1}
Xanthan Gum Commercial	3427,85	2922,59	1733,69	1623,77	1410,67	1277,61	1159,01-1028,84	795,49
Xanthan Gum (NB, using 4% liquid sugar)	3423.65	2931.80		1635.64	1409.96		1049.28-1078.21	848.68
Xanthan Gum using (LBB, 2% maize)	3427,51	2926,01		1654,92	1413,82		1022,27-1157,29	856,39

Table 3. Xanthan gum research result.

Nutrien Broth (NB) + Mineral + Inducer 0,083%					Luria-Bertani Broth (LBB)				
Yield (g/L)	Dry Cells (g/L)	Rendement (%)	Viscosity (cPs)	Solvent	Yield (g/L)	Dry cells (g/L)	Rendement (%)	Viscosity (cPs)	Solvent
23, 35	1,48	57,88	31	IPA (1:2)	22,20	3,16	73,61	15	IPA (1: 3)
23,15	1,48	57,88	31	IPA (1:2) + NaCl 0,1%	22,08	3,16	73,61	15	IPA (1:3). + NaCl 0,1%
13,68	3,36	34,20	± 28	IPA (1:2)	16,48	8,72	54,93	30	IPA (1:3).
22,91	1,67	57,28	± 30	IPA (1:2) + NaCl 0,1%	18,57	7,68	61,90	25	IPA (1:3). + NaCl 0,1%

cial gum xanthan, in particular the presence of pyruvate and acetal groups. This suggests that the fermentation of X.*campestris* using the liquid sugar substrate and maize under certain conditions of process can qualitatively produce xanthan gum.

Overall, the results of this study can be shown in Table 3 as shown above.

In Table 3 it was shown that in X. *campestris* fermentation both of type and medium composition greatly affect cell growth and the amount of product obtained. Based on the acquisition of the product (yield, g / L) from both medium have almost the same value range. In this case either Nutrient Broth (NB) enriched with minerals and inducers or in Luria - Bertani (LB) can increase the growth of X.*campestris* and produced xanthan gum. Other than that from Table 3 it was showed that the dry cell weight and its product meets the average value of xanthan gum industry in general, which contains 10-30 g / L xanthan gum with 1- 10 g / L dry cells [6]. It shows that both of 4% liquid sugar and 2% maize enriched with mineral can be used to support the growth of X.*campestris* and also converted into xanthan gum.

Xanthan gum produced from this study yielded varying rendements. The use of 4% liquid sugar yielded of 34,20%- 57,88%, while 2% Maize was 54,93% - 73,61%. This suggests that most of the 4% liquid sugar and 2% maize may be converted to xanthan gum. While based on the dry cell weight, it shows that the small part of both substrates were used to support the growth of X. *Campestris* [7]. According to Anshor [8], the conversion of glucose substrate to xanthan gum is about 50%. Conversion of glucose into xanthan gum gives a theoretical value of about 70-80%, if fermentation goes well.

Based on its rendement, LBB medium produced a relatively high of 73.61%. From this data it can be mentioned that X.campestris grown in LB medium is better able to convert substrate to its product than grown in NA. The presence of tryptone containing essential amino acids in LB medium will increase the metabolism of X. campestris to produce CGTase enzymes that play a role in the process of starch cyclization and α-1,4 glucan to cyclodextrin [5].

The use of NB and LBB produced xanthan gum with viscosity of 28-31 cPs and 15-30 cPs, respectively. Although the viscosity value was still low but these viscosity values indicated that xanthan gum can be used as a thickener in its application in industry [6]. In general, the viscosity value of xanthan gum depends on temperature, biopolymer concentration and pH. In this study the fermentation temperature between 28-300C with a pH between 6-7. However, in certain conditions, the pH was often reduced to acidic, this will affect the viscosity of gum xanthan obtained.

The most common method for gum xanthan refining is the precipitation method using an alcohol solvent. In this research, the purification of the product was tested on three types of solvent ie IPA, acetone and ethanol with the ratio between the products with each solvent 1: 1, 1: 2, 1: 3. From the experimental results it can be shown that the amount of solvent required depends on the nature of the solvent. The best xanthan precipitation was to use 3 and 2 volumes of IPA added with 1 xanthan volume. The addition of salt to a certain concentration will in addition reduce the water affinity of the polymer but there will also be chemical bonds between the cations of the salt and the anions of the xanthan solution, so that the precipitation will increase. According to Ochoa [6], if xanthan is deposited using a combination of salt with IPA then the required amount of IPA will be lower. However, in this study the addition of salt did not provide significant changes to the gum xanthan gain. In the chemical structure of xanthan gum there is a pyruvate and acetal group which is an anionic polysaccharide producer. According to Kennedy and Bradshaw [9], X. *campestris* can produce xanthan gum with pyruvate and acetal content of 7.1% and 6.5% respectively. From the results of this study the acetal and pyruvate content contained in the xanthan amount is still low, so the bond between the cations of the salt and the anions of the xanthan solution is also low. As a result, the addition of salt to the precipitation process does not provide significant changes.

4 Conclusion

Based on the FT-IR analysis, showed that the compatibility of functional groups between xanthan gum of the study and commercial gum xanthan, in particular the presence of pyruvate and acetal groups. This suggests that the fermentation of X.campestris using modified Nutrient Broth (NB) and Luria-Bertani Broth (LBB) with 4% liquid sugar and 2% maize under certain conditions of process can qualitatively produce xanthan gum.

The use of 4% liquid sugar yielded of 34.20% - 57.88%, while 2% maize was 54.93% - 73.61%. The use of NB and LBB produced xanthan gum with viscosity of 28-31 cPs and 15-30 cPs, respectively. These viscosity values indicated that xanthan gum can be used as a thickener in its application in industry. Precipitation using a combination of salt with IPA did not provide significant changes to the xanthan gum gain. This is due to the low number of bonds between the cations of the salt and the anions of the xanthan solution.

References

1. M.A. Pulungan,. *Kajian Perkembangan Perdagangan Gum Xanthan Sebagai Bahan Pengental Untuk Industri Pangan di Indonesia*. Bachelor Thesis of Institut Pertanian Bogor, (1994)
2. A. Palaniraj and V. Jayaraman, J. Food Eng., **106**, 1–12 (2011)
3. Djenar N.S and Mulyono E.W.S., Advance Science Letters. **23**, 5678-5680 (2017)
4. T. Kometani, Y. Terada, T. Nishimura, H. Takii, S. Okada, J. Biosci. Biotech. Biochem., **58**, 1990-94 (1994)
5. D.S. Prastiko, *Pengujian Aktivitas Transglikolasi Biakan Xanthomonas campestris Pada Berbagai Sumber Karbohidrat,* Bachelor Thesis of Institut Pertanian Bogor, (2011)
6. F. Garcia-Ochoa, VE. Santos , JA. Casas, E. Gómez, Biotechnol Adv., **18**, 549-79 (2000)
7. P. Muhiddin P dan R. Herlina, Bionature, **10** (1): 24-28 (2009)
8. Anshori, *Teknologi Fermentasi Industrial*, Penerbit Arcan, Bogor (1992)
9. JF. Kennedy JF, IJ. Bradshaw IJ, Prog Ind Microbiol, **19**; 319-71 (2004)

Formulation and characterization of nanoemulgel mangosteen extract in virgin coconut oil for topical formulation

Kamarza Mulia*, *Rosalia M.A.* Ramadhan, and *Elsa A.* Krisanti

Chemical Engineering Department, Faculty of Engineering, Universitas Indonesia, Depok 16424, Indonesia

Abstract. A hydrogel of nanoemulsion system containing mangosteen extract of mangosteen rind was developed and characterized for the purpose of topical formulation. Mangostins, the bioactive in the extract, has been known to have biological activities such as antifungal, antibacterial, antioxidant, antiviral and antitumoral. Due to the hydrophobic nature of mangostins, they can not to be directly formulated into a gel. Oil-in-water nanoemulsions were formulated of virgin coconut oil (VCO) as the oil phase and mixed surfactant consisting of Tween 80 and Span 80, using the high-speed homogenization method. Nanoemulgel formed by mixing the nanoemulsion with aqueous solution of xanthan gum, and added phenoxyethanol as a preservative, was a homogeneous milky white gel. The stability test through accelerated centrifugation and freeze-thaw cycle showed that the nanoemulgel would be stable for at least one year. The nanoemulgel penetrated the skin layer up to 12 μg/cm^2 or more than 95% of its total mangostin content, better than the skin penetration result of the nanoemulsion formulation. The present study revealed that VCO-mangostins nanoemulgel formulation is a prospective topical formulation.

1 INTRODUCTION

Cosmetics and skin care products, which composed of numerous chemicals, are recently becoming one of the most-demanded product. Particularly in the skin care category, it has become one of the basic and daily necessities. The products are declared to be safe by the industry as the amount of chemicals added is acceptable. However, regular of these products may result in a chemical accumulation, and hence, increases the chance of unwanted adverse effects. Further development of natural skin care products has to be made in order to reduce the use of harsh chemicals by using the ingredients extracted from plants, such as *Garcinia mangostana* L., commonly known as mangosteen. Mangosteen is a tropical fruit that grows mainly in Southeast Asia, Southwest India, and other tropical area such as Puerto Rico and Florida. The secondary metabolites, known as xanthones contained in fruit and leaves of mangosteen, possess therapeutic activities such as antioxidant, anti-tumoral, anti-inflammatory, anti-allergy, antibacterial, antifungal and antiviral [1]. The major component of xanthones present in mangostana fruit rind is α-mangostin, a compound that has been proven to inhibit certain microbes such as *Bacillius subtilis* and *Staphylococcus aureus*, which justify the topical uses of *G. Mangostana* to treat infection in traditional medicine [2].

To achieve effective topical application of mangostins, these bioactives should diffuse through the skin where the diffusion process is dependent on its aqueous solubility, diffusivity, oil or water partition coefficient, concentration in the formulation vehicle, the surface area of the skin to which it is exposed, and the thickness of the stratum corneum according to Fick's Law. Unfortunately, mangostins contained in the extract, which act as antioxidants, can be easily degraded. Encapsulation of mangosteen extract into nanoemulsion can avoid or delay the degradation, as well as to increase the bioavailability of the of mangostins [3].

Nanoemulsions are emulsions made of oil, surfactant, and water, with droplet size in the range of 20-200nm [4] or from 50-1000 nm [5]. They can present as oil-in-water (O/W) emulsions or and water-in-oil (W/O) emulsions, where the disperse phase of the emulsion is either oil or water, respectively. With its special characteristics, nanoemulsions are ideal vehicles for the purpose of delivery drug or active compounds that lead to enhancement in the therapeutic efficacy by maximizing

* Corresponding author: kmulia@che.ui.ac.id

the drug penetration into the skin, also minimizing the adverse effect and toxic reactions [6]. Nanoemulsions can easily penetrate the pores of the skin carrying the active substance to the target location, resulting in an effective drug delivery system.

Nanoemulsions are considered to be more stable than conventional emulsions because of the small droplet size that would prevent phase separation. The continuous water phase in the (O/W) nanoemulsion can increase hydration of the top layer of the skin (stratum corneum), which leads to a rapid increase in the permeability and then a slow down to a steady-state diffusion [7,8]. It was also reported that the nano-sized droplets resulted in uniform distribution on the skin, larger surface area, modified release and drug carrier properties, better occlusiveness, film formation on the skin, high stability, and gives pleasant aesthetic character and skin feel [9].

In this study, the mangosteen extract-loaded nanoemulsions were prepared using the high speed homogenization method. The virgin coconut oil (VCO) was selected as the oil phase, where the extract was dissolved within. In addition to having a role as oil phase in (O/W) nanoemulsions, VCO has many benefits in skin formulas. It can reduce skin moisture, and also improve the skin's moisture-locking barrier. The fatty acids contained also have antimicrobial effects by disrupting bacterial, fungal, and viral cell membranes [10-12]. Tween 80 and Span 80, as non-ionic surfactants, were used in some mole ratios to obtain the proper value of hydrophilic lipophilic balance (HLB) surfactants that form the most stable nanoemulsion.

The low viscosity of nanoemulsion limits its application in transdermal delivery due to its inconvenient use [14]. It was reported that gels are the best formulation as well as better controlled release, absorption properties as well as bioavailability [14]. Incorporating nanoemulsion into gel makes it a dual-control release system, which possesses both advantages of nanoemulsions and gels. Apart from better penetration ability, it is also thixotropic, greaseless, non-staining, and easily spreadable, unlike any other topical formula which requires excess rubbing [15]. The gelling agent was selected in this study was xanthan gum.

Xanthan gum consists of a repeating unit of penta-saccharide with glucuronic acid residue between two mannose units as the side chain. Higher degree of viscosity can be obtained from xanthan gum solutions even at low concentrations, unlike the other polysaccharide. Its elastic properties make it a great stabilizer, which allows the suspension of solid particles. Xanthan gum is also very resistant to pH-variations, meaning that it is stable in both alkaline and acidic conditions. Xanthan gum has the best thermal stability compared to other water-soluble hydrocolloids, which provides better stability of the formulation, despite being kept in a refrigerator, stored at room temperature, or heated [16].

The objectives of this study was to obtain the physicochemical properties of nanoemulgel that were formed by using VCO as the oil phase in O/W nanoemulsions and xanthan gum as a gelling agent in the water-continuous phase. The α-mangostin content, pH, organoleptic, and viscosity were measured within two-week period. Accelerated stability test of nanoemulgel was performed using centrifugation kecepatan tinggi and freeze-thaw cycle. To study topical application of nanoemulgel, the skin penetration characteristics of nanoemulgel were determined using an in-vitro test using a Franz diffusion cell with abdominal skin of rat as the model skin.

2 EXPERIMENTAL

2.1 Materials

Mangosteen pericarp was obtained from Solo, central Java, Indonesia, and was identified as *G. mangostana* L. by Herbarium Bogoriense, Research Center for Biotechnology, Indonesian Institute of Sciences (LIPI). Ethyl acetate, 96% ethanol, Sodium biphosphate and potassium biphosphate and xanthan gum were bought from Merck, Span 80 and Tween 80 from Sepppic and Virgin Coconut Oil (VCO) were from Kautsarherbal.

2.2 Extraction of mangosteen pericarp and mangostin quantification

The maceration using 96% ethanol for 7 days was done to extract mangostin from the mangosteen fruit pericarp powder, followed by fractionation in a mixture of water - ethyl acetate with a ratio of 1:1. The mangostin-rich ethyl acetate fraction was dried in rotary evaporation to obtain dry paste of mangosteen extract. The quantitative analysis of total mangostins content in the extract was carried out by using a UV spectrophotometer and represented as α-mangostin content. The absorbance of the extract sample solution and standard solution of α-mangostin were measured at a wavelength of 316 nm. A similar procedure was applied for measuring the content of α-mangostin in nanoemulsion and nanoemulgel samples.

2.3 Nanoemulsion and nanoemulgel preparation

The nanoemulsion was prepared using VCO as oil phase, where the mangosteen extract was dissolved in it. The surfactants used were combination of Tween 80 and Span 80 that gave a HLB value of 12. Tween 80 was dissolved in the aqueous phase, while Span 80 was dissolved in the oil phase of the VCO. The oil phase was then added into the aqueous phase under stirring, using the high-speed homogenizer, Ultra Turrax T18, for 15 minutes with speed at 10000 rpm.

Based on procedure reported by Bhura et al. (2015) [17], with slight modification, hydrogel of nanoemulsion or nanoemulgel was obtained by mixing the prepared nanoemulsion with the gel base in 1 to 1 volume ratio, under gentle stirring. The gel base was prepared by dissolving xanthan gum in distilled water. The amount of xanthan gum added into distilled water was varied into 1.0, 1.5 and 2.0 % (w/w). Phenoxyethanol was added lastly to the formed nanoemulgel as a preservative.

Nanoemulgel was characterized visually, and chemically by measuring its pH and mangostin content retained for two-week period. The composition of nanoemulgel with different weight of gel bases were shown in Table 1.

Table 1. Composition of nanoemulgel with HLB 12 surfactant and xanthan gum gelling agent

			Composition (%-w/w) of nanoemulgel #		
			1	2	3
Nanoemulsion	Water phase	Tween 80	11.21	11.21	11.21
Nanoemulsion	Water phase	Distilled Water	27.56	27.54	27.53
Nanoemulsion	Oil phase	VCO	7.12	7.12	7.12
Nanoemulsion	Oil phase	Mangosteen extract	0.0033	0.0033	0.0033
Nanoemulsion	Oil phase	Span 80	4.06	4.06	4.06
Gel base		Xanthan gum	1.03	1.54	2.06
Gel base		Distilled water	48.30	47.80	47.30
Gel base		Phenoxy-ethanol	0.72	0.72	0.72

The stability and physicochemical characteristic of nanoemulgel samples were determined by performing several tests, i.e. measurement of viscosity, pH, accelerated stability, alpha mangostin content, organoleptic and customer evaluation.

2.4 Accelerated stability testing

To determine the predicted stability of nanoemulgel for long period, there were two tests conducted.

2.4.1 Centrifugation

The accelerated stability test was carried out by centrifugation of samples inside a 10-cm centrifuge tube at 3800 RPM for five hours. The stability of samples was observed when there was no phase separation after centrifugation. In his report, Laba (1993) [18] mentioned that centrifugation of sample in this condition was equivalent with the effect of gravity of one year. Thus, all samples that showed no phase separation after centrifugation have shelf life of one year.

2.4.2 Freeze-Thaw cycle

Nanoemulgel prepared was also determined its stability using this test as reported by Jadhav et al., (2014) [19]. The samples were stored at -10 °C for 48 hours, then removed and allowed to thaw up to 25 °C. The parameters measured for stability were the pH and α-mangostin content for each formulation. The whole steps were repeated for three cycles.

2.5 Organoleptic tests

All samples of nanoemulsions and nanoemulgels were observed physically every week for four-week period. The parameters observed were the color, scent and visual phase separation.

2.6 Viscosity and pH measurement

These tests were applied to nanoemulsion samples. The viscosity of nanoemulgel with various content of xanthan gum was measured using Cannon-Fenske capillary viscometer. The stability of nanoemulsion samples was also observed by measuring the pH every 4 days for two-week period. The changes of pH during that period of time might indicate instability of nanoemulsion.

2.7 Mangostins quantity measurement

Quantitative analysis of the mangostin extract was carried out using a UV spectrophotometer. A stock solution was prepared by diluting 2.5 mg mangostin extract in 25 ml of ethanol, followed by further dilution to obtain a variation of the solution concentration. The absorbance of each solution concentration was measured using a UV spectrophotometer at a wavelength of 316 nm. The total mangosteen or xanthone in the extract can be determined as the amount of α-mangostin content measured and calculated based on the standard α-mangostin calibration curve that has been prepared.

2.8 In-vitro skin penetration study

The skin penetration study was performed by Franz diffusion cell with abdominal skin of rat as the model skin. The sample skin was placed between the donor and the receptor compartment, with stratum corneum side facing the donor. The donor medium consisted of sample formulation, while the receptor was phosphate buffer saline (pH 7.4). The receptor medium (0.5 mL) was withdrawn after 0, 10, 30, 60, 90, 120, 180, 240, and 300, 360, 420, 480 minutes and analyzed by UV-spectrophotometer to determine the mangostin content [20].

2.9 Customer testing

This test was conducted only for nanoemulgel samples, which was an additional test to help deciding the best formulation by involving consumer acceptance and

https://doi.org/10.1051/matecconf/201815601013

satisfaction. The formulations were evaluated according to its consistency, color, and odor; with consistency and odor being the more significant parameter.

3 RESULT AND DISCUSSION

The mangosteen extract in the ethyl acetate fraction was obtained and analyzed using UV-spectrophotometer for its α-mangostin content, that represent the amount of total mangostin within the extract. The result showed the total amount of mangostin was 57.2 % (weight) of total dry mangosteen extract. The mangosteen extract of the ethyl acetate fraction was used for preparation of nanoemulsion and nanoemulgel.

3.1 Mangostin extract in ethyl acetate fraction

Maceration of dried mangosteen pericarp was conducted for 7 days with periodic stirring which was aimed to increase the contact between simplicia and the solvent. The thickened solvent was fractionated using ethyl acetate and distilled water (1:1). Phase separation occurred naturally, since ethyl acetate is semi-polar and water is polar solvent. The upper part, which is the ethyl acetate fraction containing mangostins, was then collected and subjected to rotary evaporator to remove the ethyl acetate solvent, leaving the dried paste extract that is rich in mangostins.

3.2 Nanoemulgel characterization

The characteristics of nanoemulgel were observed based on the organoleptic, viscosity, pH determination, mangostin quantitative analysis and stability test.

3.2.1 Organoleptic, viscosity and pH determination

The obtained nanoemulgels were milky-white and homogeneous, with odor comparable to the VCO. The viscosity of the nanoemulgels increases with the increase of xanthan gum concentration within the formulation. Within two-week observation period at room temperature, all of the nanoemulgel formulations with various amounts of xanthan gum remain stable; no phase separation observed, and no changes in consistency or color.

As shown in Fig.1, there were no significant changes of pH within observation time and no significant effect from variation of XG concentration. It was found that the pH measured of nanoemulgel in 12-day observation has minimum pH value of 5.5 and maximum 6.4, which is within the skin pH human skin (4.5 – 6.5) reported by Leonardi, et al., (2002) and Zlotogorski (1987) [21,22]. The relatively constant value of pH within two weeks indicated the stability of nanoemulgel.

The pH measurement was also conducted for all nanoemulgel samples with different amount of xanthan gum (XG) at every four days for two-week observation period.

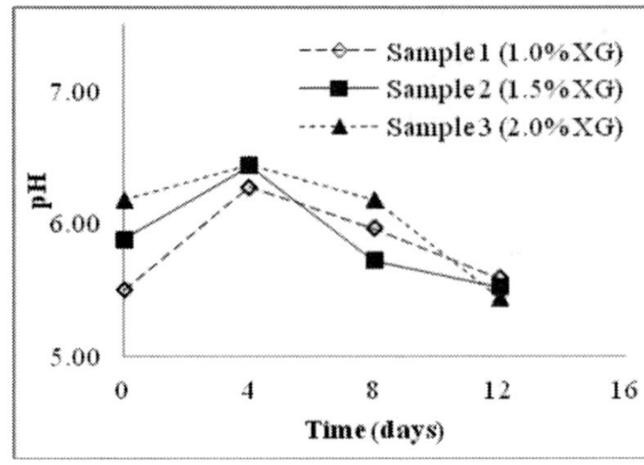

Fig. 1. Plot of pH of nanoemulgels in two-week period with different ratio of Xanthan Gum (XG) in gel-base aqueous solution.

As shown in Fig.1, there were no significant changes of pH within observation time and no significant effect from variation of XG concentration. It was found that the pH measured of nanoemulgel in 12-day observation has minimum pH value of 5.5 and maximum 6.4, which is within the skin pH human skin (4.5 – 6.5) reported by Leonardi, et al., (2002) and Zlotogorski (1987) [21,22]. The relatively constant value of pH within two weeks indicated the stability of nanoemulgel.

3.2.2 Mangostin quantitative analysis

The quantitative analysis of the mangosteen extract in ethyl acetate fraction was carried out using UV spectrophotometer. The absorbance of the extract sample solution and standard solution of α-mangostin were measured at wavelength of 316 nm. The amount of mangosteen in extract of ethyl acetate fraction was measured as 57.9 % (g/g).

The α-mangostin content of each formulation of nanoemulgel was measured every four days for two week period, and no significant change observed during time of observation, i.e around 0.002% (w/w). This experimental data was slightly less than the calculated total mangosteen extract (0.0033%) in the nanoemulgel, as shown in Table 1.

The observation of α-mangostin and pH in two weeks showed that both pH and α-mangostin were stable. This stability was attained due to the chemical structure of xanthan gum as the gelling agent. The glucan backbone is protected by the side chains which lie alongside, making it relatively stable to acids, alkalis and enzymes, factors that may trigger molecules collisions and hence degradation [23]. Regardless of the value of α-mangostin content that were quite small in nanoemulgel, the results showed that the variation of XG content in the gel base showed no effect on the content of mangostin within the nanoemulgel as well as on the pH.

3.2.3 Accelerated stability testing for shelf life

All nanoemulgel samples were centrifuged at 3800 RPM and after five hours there were no phase separation observed. This result indicates the stability of all hydrogel formulations that equals to having a shelf life of one year. The stability of nanoemulgel using these formulations might be due to the presence of xanthan gum that reduce the interfacial tension and tends to form a cohesive interfacial film around the nanoemulsion droplets, hence improving the stability of nanoemulgel [24]. The effect of xanthan gum also can be explained by its ability, as hydrocolloid, to stabilize and slow the thermodynamically breakdown of gel [23].

3.2.4 Freeze-Thaw cycle test for stability detection of nanoemulgel

All of the formulations were also subjected to freeze-thaw testing where the samples are subjected to two extreme temperature changes, which in this case is -10°C and 25°C. There was no sign of phase separation for any formulations, indicating that the nanoemulgel would be stable even under such condition, which could likely be encountered in real life. α-Mangostin content in all formulations that have been subjected to this test, were evaluated using UV-spectrophotometer with similar procedure described previously. The α-mangostin analysis and pH measurement after conduction of Freeze-Thaw test are reported in Table 2.

Table 2. The pH and mangostin content of various nanoemulgel formulations as a function of freeze-thaw cycles.

		Nanoemulgel # (%-w/w XG)		
Parameter	Cycle	1 (1.0%)	2 (1.5%)	3 (2.0%)
pH	0	5.70	5.48	5.28
	1	5.65	5.58	5.71
	2	6.21	6.10	6.20
	3	6.58	6.53	6.52
α-mangostin content (%)	0	0.0019	0.0021	0.0020
	1	0.0018	0.0019	0.0018
	2	0.0018	0.0018	0.0018
	3	0.0017	0.0018	0.0018

Note: XG = xanthan gum, the composition of samples # 1,2 and 3 are described in Table 1. Each cycle is performed at -10 ℃ and 25 ℃.

It is shown in Table 2, the content of α-mangostin in the nanoemulgel in all samples were relatively constant, around 0.0018%, regardless of the differences in xanthan gum amount. The value in three cycles was not very far from the initial concentration of α-mangostin in the emulgel, which was 0.002%. The result indicated that even as low as 1% XG has given effect on stability of nanoemulgel. The stability of nanoemulgel in these formulations might be due to the good thermal stability of xanthan gum, having a rigid helical conformation that may provide better stability despite being kept in a refrigerator [16].

As the nanoemulgel samples frozen, numerous small ice crystals would be generated and the size may increase as the samples cycled through freeze/thaw temperatures. Kuntz (1995) [25] reported that the presence of xanthan gum within the nanoemulgel samples may help in controlling crystal growth by binding the water, keeping it from coalescing into bigger ice crystals. Xanthan gum is also attributed to possible steric blocking of the oil droplet interfaces or inhibiting solute transport at the interface by forming hydrocolloids gel [26].

The pH of each composition was also measured at the end of each cycle and showed an increase as shown in Table 2 and Fig 2. This is due to the reason that two phases would be generated at the end of freezing process; a region with pure ice crystals, and the unfrozen solutes are effectively crowded into a smaller remaining liquid volume. The effective concentration of unfrozen solutes will increase with the number of cycles, that shift the pH of the formulation [27]. Nevertheless, the pH range measured after the cycles was still within the acceptable pH range (4.5-6.5).

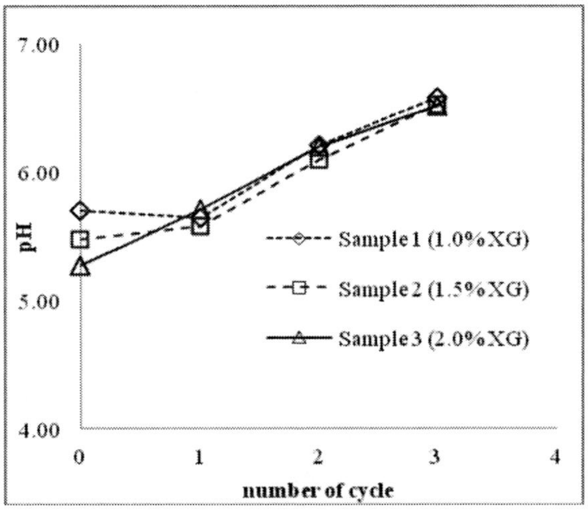

Fig. 2. The pH of nanoemulgels at 3 cycles of freeze-thaw testing with different ratios of xanthan gum in the gel-base aqueous solution

3.3 In-vitro skin penetration study of nanoemulgel

The advantage of delivering active compounds through nanoemulgel was mentioned by Basera, et al. (2015) [28], due to better adhesion to the surface of the skin and the high solubilizing capacity, which leads to larger concentration gradient towards the skin, and hence better skin penetration. Chellapa *et al.* (2015) [29] also reported that upon skin contact nanoemulgel will release the oily droplets from the gel network that then penetrate into the stratum corneum of the skin and directly deliver the drug or active molecules.

The ability of different formulations of the gel to penetrate the skin was studied. At the same time, nanoemulsion was also subjected to the same test as a comparison, as seen in Fig 3. In the first ten minutes the cumulative amount of penetrating mangostin reached about 6 μg/cm^2 or 50% of initial content and it increased slowly to about 12 μg/cm^2 or 95% at about eight hours. The addition of xanthan gum to the nanoemulsion around 1-1.5 % (w) significantly increased the skin penetration of the mangostin extract, however the addition of more xantham gum to 2% (w) lower the penetration. It seems that 1.5 % (w) is the optimum amount of xanthan gum in the nanoemulgel, that showed the highest skin penetration of mangostin, while nanoemulsion without xanthan gum showed the lowest.

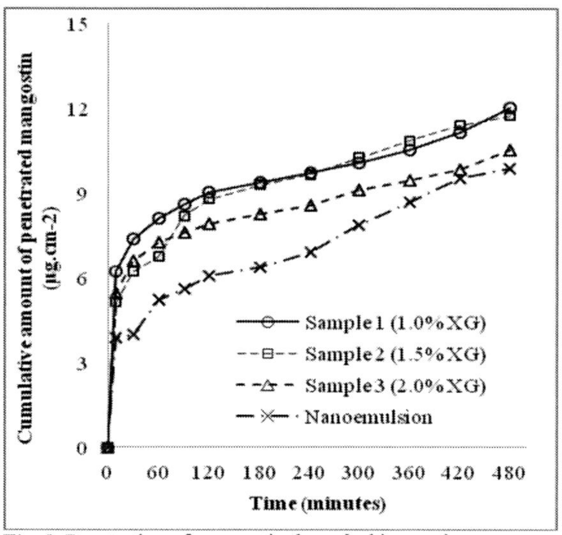

Fig. 3. Penetration of mangostin through skin membrane: Cumulative penetrated mangostin (μg/cm^2) of nanoemulgel and nanoemulsion samples with xantham gum.

Thermodinamic activity of active compound or drug in the applied solution, such as nanoemulgel, affect the penetration of drug through a skin membrane [30], as given in the steady-state Fick's 1st law of diffusion of a drug through skin (J):

$$J = \frac{A_v D}{\gamma_s L} \quad (1)$$

where A_v, D, γ_s and L are the thermodinamic activity of the drug in a formulation, the diffusion coeffient, the activity coefficient of the drug in the skin barrier, and the thickness of the skin barrier, respectively. Salim, et al. (2012) [31] reported that addition of hydrocolloid gum, such as xanthan gum, increased the lipophilicity of the oil droplets, and thus increased the thermodynamic activity of the drug, that lead to the increase of the skin penetration rate. This study showed similar results, i.e. the addition of xanthan gum to the nanoemulgel increased the flux of mangostin that penetrated the skin membrane.

3.4 Customer testing for nanoemulgel

Customer testing was conducted as an additional test to help deciding the best formulation by involving consumer acceptance and satisfaction. The formulations were evaluated according to some parameters, i.e. consistency, color, and odor. Each parameter has different weight according to the level of importance, affecting customer acceptance of the sample product. Consistency and odor are more significant parameter compared to color, so they have a weight of 40% each, while the color has a weight of 20%. This is due to the non-staining and soft color of the nanoemulgel, and hence, the less effect of color on the customer acceptance. Most of the respondents were satisfied with the consistency, color, and odor of emulgel formulation number 1 with 1% of xanthan gum. It is then concluded that the addition 1% of xanthan gum into the formulation will produce the most suitable gel consistency, odor as well as maintain the stability.

4 CONCLUSIONS

A prospective mangostin nanoemulgel for topical formulation using virgin coconut oil as oil phase in making nanoemulsion, and xanthan gum as a gelling agent in aqueous solution, was developed and characterized. Milky-white, homogeneous nanoemulgels were obtained from mixing the gel base (xanthan gum solution), and nanoemulsion with 1:1 weight ratio. The results of accelerated test showed that the stability of all nanoemulgel formulations equals to having a shelf life of one year. The Freeze-Thaw cycle test indicated that as low as 1% (w), xanthan gum has given significant effect on stability of nanoemulgel.

Nanoemulgels were also confirmed to afford better penetration ability than its nanoemulsion (12 μg/cm^2 or 95% cumulative). The skin penetration test showed that addition of xanthan gum to the nanoemulsion as much as 1 % (w) significantly increased the skin penetration of mangostins, however, the 1.5 % (w) is the optimum amount of xanthan gum in the nanoemulgel.

When nanoemulsion was combined with an aqueous gel based to form nanoemulgel, the results showed that the variation of xanthan gum content in the gel base showed no effect on the content of mangostin as well as on the pH of nanoemulgel. Based on customer testing, the nanoemulgel containing 1% of xanthan gum was the most accepted by the respondents.

5 ACKNOWLEDGEMENT

The authors are grateful for financial support from the Universitas Indonesia through the DRPM PITTA Grant 2017.

REFERENCES

1 J. Pedraza-Chaverri, N. Cárdenas-Rodríguez, M. Orozco-Ibarra, and J.M. Pérez-Rojas, Food Chem. Toxicol. **46**, 3227 (2008).

2 S. Al-Massarani, A. El Gamal, N. Al-Musayeib, R. Mothana, O. Basudan, A. Al-Rehaily, M. Farag, M. Assaf, K. El Tahir, and L. Maes, Molecules **18**, 10599 (2013).

3 K.O. Aboalnaja, S. Yaghmoor, T.A. Kumosani, D.J. McClements, Expert Opinion on Drug Delivery. **13**, 1327, 2016.

4 C. Solans, P. Izquierdo, J. Nolla, N. Azemar, And M. Garciacelma, Curr. Opin. Colloid Interface Sci. **10**, 102 (2005).

5 C. Lovelyn and A.A. Attama, J. Biomater. Nanobiotechnol. **2**, 626 (2011).

6 M. Jaiswal, R. Dudhe, and P.K. Sharma, 3 Biotech **5**, 123 (2015).

7 E. Abd, Targeted Skin Delivery of Topically Applied Drugs by Optimized Formula Design, University of Queensland, 2015.

8 A. Vyas, A.K. Sonker, B. Gidwani, The Scientific World Journal, Article ID 276260, 2014.

9 M.N. Yukuyama, D.D.M. Ghisleni, T.J.A. Pinto, and N.A. Bou-Chacra, Int. J. Cosmet. Sci. **38**, 13 (2016).

10 G. Bergsson, J. Arnfinnsson, O. Steingrimsson, and H. Thormar, Antimicrob. Agents Chemother. **45**, 3209 (2001).

11 S. Intahphuak, P. Khonsung, and A. Panthong, Pharm. Biol. **48**, 151 (2010).R. Arora, G. Aggarwal, S.L. Harikumar, and K. Kaur, Adv. Pharm. **2014**, 1 (2014).

12 M. Shilling, L. Matt, E. Rubin, M.P. Visitacion, N.A. Haller, S.F. Grey, and C.J. Woolverton, J. Med. Food **16**, 1079 (2013).

13 R. Arora, G. Aggarwal, S.L. Harikumar, and K. Kaur, Adv. Pharm. **2014**, 1 (2014).

14 L.P. Kaur and T.K. Guleri, Asian J. Biomed. Pharm. Sci. **3**, 1 (2013).

15 R. Wani, M. Patil, P. Dhurjad, C. Chaudhari, and S. Kshirsagar, Int. J. Pharm. Res. Sch. **4**, 397 (2015).

16 C. Qiu, M. Zhao, D. J. McClements, Food Hydrocolloids, **43**, 377 (2015).

17 M.R.G. Bhura, K.A. Bhagat, and S.K. Shah, World J. Pharm. Sci. **3**, 1013 (2015).

18 D. Laba, *Rheological Properties of Cosmetics and Toiletries* (CRC Press, 1993).

19 C. Jadhav, V. Kate, and S.A. Payghan, J. Nanostructure Chem. **5**, 107 (2015).

20 P. Widyanati, M. Jufri, and B. Elya, Int. J. Pharm. Sci. Rev. Res **27**, 1 (2014).

21 G.R. Leonardi, L.R. Gaspar, and P.M.B.G. Maia Campos, An. Bras. Dermatol. **77**, 563 (2002).

22 A. Zlotogorski, Arch. Dermatol. Res. **279**, 398 (1987).

23 M. Chaplin, Xanthan Gum (2016).

24 S.H. Saharudin, Z. Ahmad, and M. Basri, Int. Food Res. J. **23**, 1361 (2016).

25 L.A. Kuntz, Freeze/Thaw-Stability (1995).

26 B.M. Degner, C. Chung, V. Schlegel, R. Hutkins, and D.J. McClements, Compr. Rev. Food Sci. Food Saf. **13**, 98 (2014).

27 J. Brunstein, Med. Lab. Obs. 44 (2015).

28 K. Basera, G. Bhatt, P. Kothiyal, and P. Gupta, World J. Pharm. Pharm. Sci. **4**, 1871 (2015).

29 P. Chellapa, A.T. Mohamed, E.I. Keleb, A.M. Eid, Y.S. Issa, and N.A. Elmarzugi, IOSR J. Pharm. **5**, 43 (2015).

30 H. Ishii, H. Todo, and K. Sugibayashi, Chem. Pharm. Bull. (Tokyo). **58**, 556 (2010).

31 N. Salim, M. Basri, M.B. Rahman, D.K. Abdullah, H. Basri, Int. J. Nanomedicine, **7**, 4739 (2012).

Influence of Soy Protein Isolate on Gelatin-based Edible Film Properties

Iryanti Fatyasari Nata[1,*], *Chairul* Irawan[1], *Lazuardi* Ramadhan[1] and *Muhammad* Rizky Ramadhani[1]

[1]Chemical Engineering Department, Lambung Mangkurat University, 70714 Banjarbaru, Indonesia

Abstract. Gelatin is a biopolymer that can be obtained from the partial hydrolysis of collagen skin tissue of chicken leg. This study aims to utilize gelatin of chicken leg skin in combination with soy protein isolate (SPI) as raw material for edible film production. The experiment was initiated by extracting gelatin and edible film was prepared by mixtures of SPI in ethanol solution and gelatin solution (containing 1% chitosan, w/v) with different weight ratio for SPI : gelatin (1:1, 1:2, 1:3, 1:4, and 1:5) then heated at 70 °C for 45 min and added 10% glycerol. Furthermore printed and dried in oven for 24 h at 60 °C. The yield of preparation of gelatin from chicken leg skin was obtained about 40.3%. Higher concentration of gelatin in solution gives a transparent color and smooth surface of edible film. Base on mechanical properties analysis, the composition of 1:3 has the best result of tensile strength and elongation at break about 4.80 MPa and 113.37%, respectively. The antibacterial activity test showed low antibacterial activity with a 0.3 cm inhibit zone.

1 Introduction

The skin of chicken leg is composed of a dense connective tissue rich in collagen. The utilization of collagen particles by hydrolysis will produce gelatin [1, 2]. Gelatin is widely used as a stabilizer, gelling agent, binder, thickeners, emulsifiers, adhesives and edible food wrappers [3]. Gelatin in Indonesia is imported from Europe and America, the utilization of gelatin application in the field of food about 60%, it mean the production of gelatin will be increased in line with increasing of population. The production of gelatin is commonly from cows by 33% and 27% gelatin derived from pigs [4], so that gelatin is potential alternative material to be developed which is made from chicken leg skin, considering chicken production in Indonesia is abundant. Gelatin is material which can be reversibly changed from the sol form to the gel, swells or expands in cold water and is capable of forming a film [5]. Commonly, gelatin is used as raw material for edible film for food packaging. In addition, the chicken leg skin will be useful where only as a waste of chicken utilization. The advantage of using edible film is to extend the shelf life of the product and not to pollute the environment because the edible film can be eaten.
Previous studies have reported the utilization of edible film from soya protein isolate [6], gelatin [7, 8], and combination of gelatin, protein isolate and materials which have antibacterial activity [9, 10]. However, there has been little discussion about characterization of edible film from combination of soy protein isolate (SPI) and gelatin which produce from chicken leg skin. It is necessary to develop gelatin production made from chicken leg skin for making edible film and expected to overcome the problem of food decay due to the activity of *Staphylococcus aureus* bacteria [11].

This study was focus on preparation of edible films in terms of comparison of SPI and gelatin concentration as a raw material that has the ability to kill bacteria. The aim of this study was characterized the morphological structure of edible film in terms of increasing the concentration of gelatin in the present of SPI through the scanning electron microscope (SEM) test, evaluating the optimum weight composition ratio of SPI:gelatin for edible film production. The characterization and analysis of mechanical properties such as tensile strength, elongation, and antibacterial activity of the edible film were also evaluated.

2 Materials and Method

2.1 Materials

Natrium hydroxide (NaOH), sulfuric acid (H_2SO_4), citric acid ($C_6H_8O_7$), hydrochloric acid (HCl), ethanol (CH_3COOH), and glycerol ($C_3H_8O_3$) were obtain from Sigma-Aldrich and Acros. All other chemicals were analytical grade used without further purification. Chicken leg skin and soy protein isolate were collected

* Corresponding author: ifnata@unlam.ac.id

from traditional local market.

2.2 Preparation of gelatin from chicken leg skin

Chicken legs skin was washed with water flow then drained. The skin (300 g) was immersed in 1.5% acetic acid solution, the skin and solution ratio 1: 8 (w/v) for 3 days. After that, it washed thoroughly with dionized (DI) water until filtrate of solution reach to neutral pH. The residue of sample was extracted in 65% ethanol in a three-neck flask with ratio about 1: 2 (w/v) for 1 h. After extraction was completed, added distilled water in to mixture with ratio 1: 1 (v/v) and keep at 64 °C for 1 h, followed by washing and filtering. The extracted product was dried in an oven at 70 °C and blended and sieved to obtain gelatin powder.

2.3 Edible film preparation

Edible film was prepared by combination of SPI and gelatin in weight ratio is 1: 1, 1: 2, 1: 3, 1: 4 and 1: 5. The gelatin was dissolved in 1% w/v chitosan in acetic acid. Soy protein isolate dissolved with ethanol and each solution was stirred until homogeneous. After that the solution was mixed together and added 10% v/v glycerol with stirring rate 90 rpm at 70 °C for 45 min. The mixture was poured into a Teflon (d = 20 cm). The edible film was formed after sample was dried at 55 °C for 24 h.

2.4 Characterization

Scanning electron microscope (SEM) images of the samples were obtained by scanning electron microscope (JEOL, JSM-6500LV). The tensile strength of edible films was tested by using Lloyd's universal Testing Instrument machine. The elongation is expressed as the maximum force applied to the film.

Staphylococcus aureus grown in LB medium was employed to examine the antibacterial effect of edible film *Staphylococcus aureus* grown overnight at 37 °C were harvested by centrifugation and washed twice with pH 6.5 and 0.9% NaCl solution. *Staphylococcus aureus* solution was uniformly plated on LB nutrient agar plates. A piece of edible film was thick of diameter about 1 cm which placed on the agar plates. The plates were incubated at 37 °C for overnight. The antibacterial activity of the samples was measured based on the width of the clear zone formed around the samples.

3 Results and Discussion

Gelatin was produced by soaking and extraction of chicken leg skin. When it is soaked in 1.5% acetic acid, it is possible to interact between collagen fibrils with acetic acid and result in collagen protein will be swelling. During heating process, the collagen protein will be extracted into gelatin. Yield of gelatin produced from the skin of chicken leg about 40.3%. The chicken leg skin as raw material is shown in Fig.1 (a) and gelatin product can be seen in Fig.1 (b).

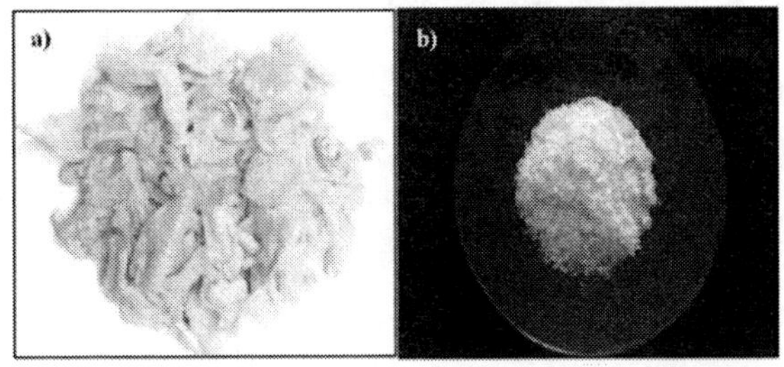

Fig. 1. Chicken leg skin (a) and gelatin product from chicken leg skin (b).

3.1 Edible film production

The edible films resulting from variation weight ratio of SPI and gelatin can be seen in Fig. 2. Visualization test is done by placing an image below edible film. In Fig. 2 there is a clear for edible film which weight ratio 1:1 has brightest/white color; it is because the composition of SPI and gelatin is same and SPI is difficult to dissolve in the solution and not completely dissolve, so that the color of edible film produced was cloudy.

The variation of 1:5 has light transparent color; image below the edible film could be seen clearly. By increasing concentration of gelatin will give transparent color, it is because the high concentration of gelatin facilitates SPI dissolves in the solution.

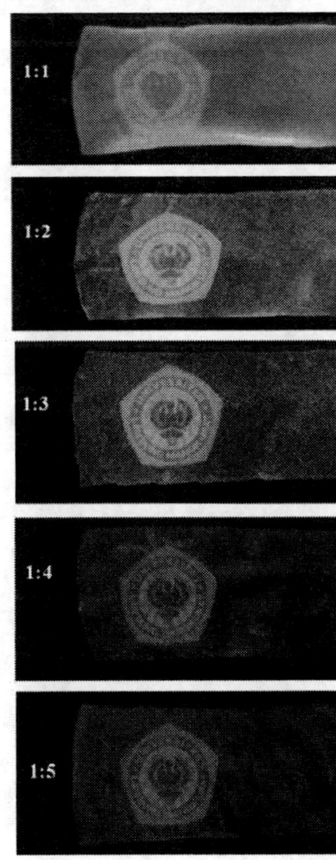

Fig. 2. Edible film product from different weight ratio of SPI : gelatin for 1:1; 1:2; 1:3; 1:4; and 1:5.

https://doi.org/10.1051/matecconf/201815601014

Prior studies that it has noted that the effect of gelatin concentration on edible film product is become transparent. High concentration of gelatin will make the edible film more transparent, homogeneous and not easily broken [12]. In addition, when the mixture of gelatin and SPI under heating condition, the gelatin is form triple helix collagen structure which able to dissolve SPI and the surface product of edible film becomes smoother [13].

3.2 Edible film characterization

The surface of edible film was observed by SEM, it is shown in Fig.3. It is apparent from the weight combination ratio of 1:1 shows the number of aggregates on the surface and its dense of SPI.

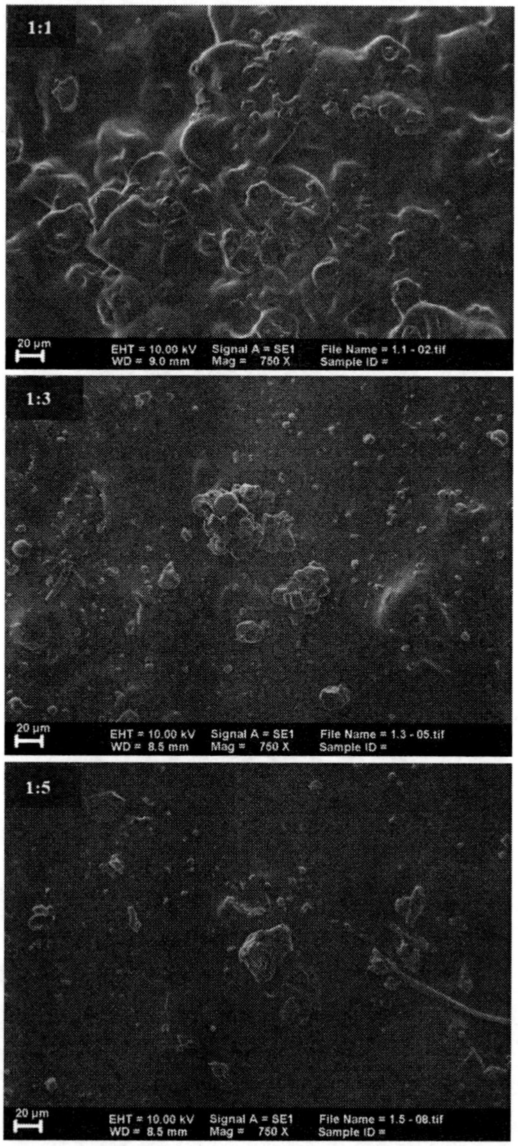

Fig. 3. SEM images of edible film with weight combination ratio of SPI : gelatin for 1:1; 1:3; and 1:5.

Due to limited solvent in the mixture, SIP is not dissolving completely. Effects of this, color of edible film becomes white (Fig. 2). From fig. 3 for composition of 1:3, SEM resulted in the lowest concentration of SPI. It means more SPI was dissolved.

In order to understand how the mechanical properties of edible film, the tensile test was performed to determine the tensile strength of samples and also measure maximum force required to break the sample. This test useful for application of edible film with high tensile strength will be able to protect the product packaged from environment [14]. The result of edible film tensile strength is presented in Fig. 4.

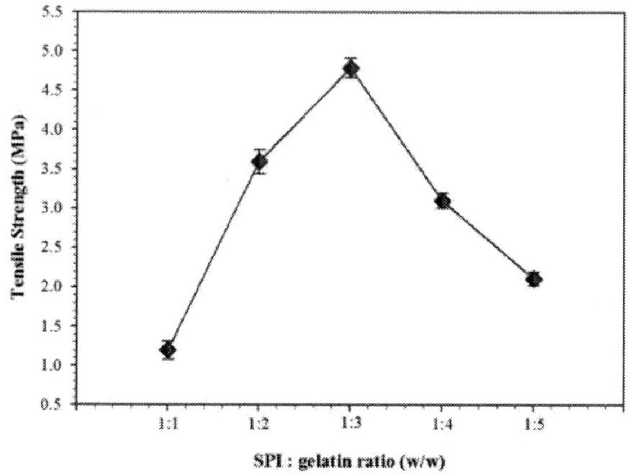

Fig. 4. Tensile strength of edible film product from different weight ratio of SPI : gelatin for 1:1; 1:2; 1:3; 1:4; and 1:5.

The tensile strength test for weight combination ratio 1:1; 1:2; 1:3; 1:4; and 1:5 were 1.20 MPa; 3.60 MPa; 4.80 MPa; 3.10 MPa; and 2.11 MPa, respectively. The most striking result to emerge from the data is that the value of tensile strength test increases with the increasing gelatin concentration, it is due to an increase concentration of collagen in gelatin [13]. However the increasing tensile strength only up to weight ratio of 1:3, for composition 1:4 and 1:5 ratio, the tensile strength were decreased. The decreasing tensile test in this composition because of reduction of intermolecular interactions so that the matrix formed will be less and also excess of gelatin composition have effects to polymerization is not maximum. The tensile strength value of an edible film is determined by the structure of the woven form and the protein content in gelatin containing collagen [15]. In addition, the tensile strength of edible film will decrease which made from soy protein isolate because there is not all SPI particles participate in the matrix film formation [16]. Gelatin has better mechanical properties than that SPI because gelatin is more flexible and elastic [17]. The results of this study show that edible film with weight composition 1:3 has highest tensile strength value about 4.80 MPa. Another research was prepared edible film using SPI and gelatin in weight ratio has tensile strength about 1.50 Mpa [12].

A strong relationship between tensile strength and elongation has been known for edible film. Elongation test for edible film is to determine the resistance of cracking and to know the flexing endurance as well as to show the maximum film length change when getting the tensile force until the film is broken compared to initial length. The elongation test can be seen in Fig. 5.

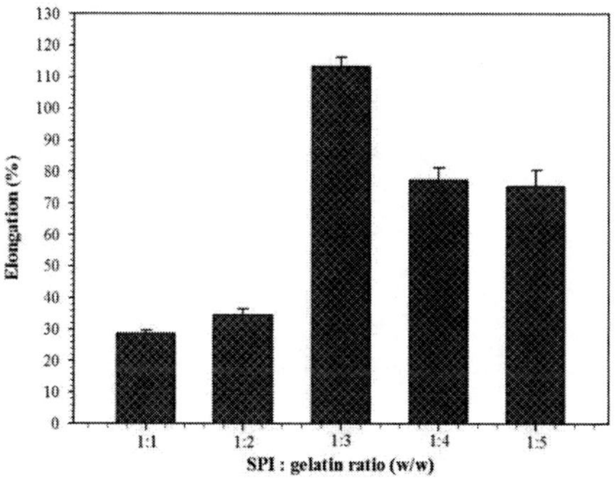

Fig. 5. Elongation of edible film product from different weight ratio of SPI : gelatin for 1:1; 1:2; 1:3; 1:4; and 1:5.

Fig. 5 shows same trend with tensile strength profile (Fig. 4) which elongation of edible film was increased up to 113.37% for weight composition of 1:3. When the gelatin concentration was increased for composition 1:4 and 1:5, the elongation was decreased to 75.52%. The decreasing value of edible film depend on structure of intermolecular the matrix, when excess of gelatin concentration, the polymerization between collagen and SPI are not optimum. The standard elongation value for edible film should more than 50% [3]. This research product as edible film is quite good mechanical properties for elongation about 113.37%. On comparison for mechanical properties of commercial wrapping plastic have value of tensile strength and elongation about 0.86 MPa and 4.8%, respectively.

The edible film was tested for antibacterial activity against the growth of *Staphylococcus aureus* bacteria by the agar diffusion method. Fig. 6 is shown the antibacterial activity for each composition. The antibacterial activity of edible film is shown by a clear zone around the film. If the halo zone does not appear it is assumed that there is no drag zone in the film. The result of the test is presented in Fig. 6, weight ratio of 1:1 was gives antibacterial effects with a clear zone area about 0.3 cm. Edible film composition of 1:2 showed only a little have antibacterial effect. For composition of 1:3; 1:4; and 1:5 have no inhibit zone, this is because SPI and gelatin has no oligopeptide chain, which is suspected to have antimicrobial properties due to the presence of amino groups in the chain. Characteristics of hydrophobic amino acids will make the peptide chain become positively charged so it can enter into the bacterial membrane causing interaction between peptide chains of gelatin with bacteria surface [18].

The existence of chitosan in the edible film have effects to the antimicrobial activity, chitosan will form a polymer membrane on the surface of *Staphylococcus aureus* cells that will inhibit the nutrients into the cell. This is due to the presence of amine groups in chitosan which have a cationic charge which can bind the feeding source of bacteria such as alginate, pectin, protein, and inorganic polyelectrolytes such as polyphosphates. The chitosan antibacterial activity increases with increasing chitosan molecular weight. Organisms will be inhibited when contacted directly with the active side of the chitosan film [19]. In other hand, only chitosan gel or solution form exhibits optimal biological properties, because chitosan in these form the amine group will be easily protonated [11].

In this case, there is different concentration of gelatin containing chitosan which effect to active agent switching that acts as antibacterial. The weigth composition ratio 1:1 has more chitosan contain in the matrix of edible film and shown antimicrobial effect.

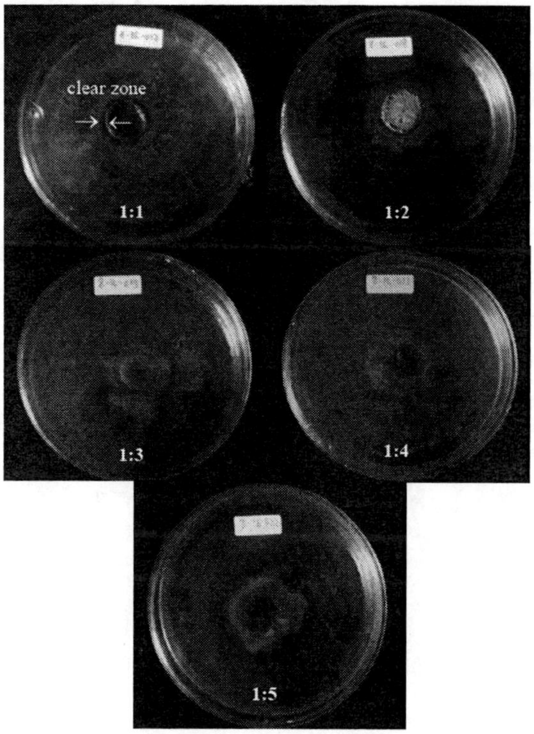

Fig. 6. Antibacterial activity with *Staphylococcus aureus* for edible film product from different weight ratio of SPI : gelatin for 1:1; 1:2; 1:3; 1:4; and 1:5.

4 Conclusions

Edible film can be synthesized by combination weight ratio of soy protein isolate and gelatin from chicken leg skin. The chicken leg skin as a alternative raw material for gelatin production. The optimum value base on mechanical properties of edible film in the weight ratio of SPI : gelatin is 1:3. The value of tensile strength and elongation at break of sample is possible to make it become media for coating or wrapping food.

The authors are grateful for the financial support from Ministry of Research, Technology and Higher Education, Republic of Indonesia, which provided Research Grand (PKM-PE) 2016.

References

1. H. Tavakolipour, World J. of Fish and Marine Sci. **3**, 1 (2011)
2. I. M. S. Negara and I. N. Simpen, J. Chem. **8**, 1 (2014)
3. J. M. Krochta and C. D. Muller-Johnson, CRC Press LCC, Florida (2002)
4. V. S. Wiyono, J. Halal LPPOM-MUI, **36** (2001)
5. R. a. H. G. Schrieber, "Gelatine handbook : Theory and industrial practice," (2007)
6. D. S. Liyanage, M. Tillekeratne and S. Hewage, J. Natn. Sci. Foundation Sri Lanka **29** (2001)
7. M. C. Gómez-Guillén, M. Ihl, V. Bifani, A. Silva and P. Montero, Food Hydrocolloids **21**, 7 (2007)
8. S. Zhang, Y. Wang, J. L. Herring and J. H. Oh, J. of Food Sci. **72**, 9 (2007)
9. A. A. Al-Hassan and M. H. Norziah, Food Hydrocolloids **26**, 1 (2012)
10. N. M. Sarbon, F. Badii and N. K. Howell, Food Hydrocolloids **45** (2015)
11. P. Fernandez-Saiz, J. M. Lagaron, P. Hernandez-Muñoz and M. J. Ocio, Int. J. of Food Microbiol. **124**, 1 (2008)
12. M. Hasdar, Y. Erwanto and S. Triatmojo, , **35**, 3 (2011)
13. H. Chambi and C. Grosso, Food Research Int. **39**, 4 (2006)
14. D. T. H. Suryaningrum, J. Basmal, dan Nurochmawati, Penelitian Perikanan Indonesia **11**, 4 (2005)
15. I. Arvanitoyannis, E. Psomiadou, A. Nakayama, S. Aiba and N. Yamamoto, Food Chem. **60**, 4 (1997)
16. N. Cao, Y. Fu and J. He, Food Hydrocolloids **21, 7** (2007)
17. K. Y. Lee, J. Shim and H. G. Lee, Carbo. Polym, **56**, 2 (2004)
18. D. Mardian, L. Atmaja and M. Najib, Thesis, Sepuluh November Institute of Technology, (2011).
19. J. Gómez-Estaca, López de Lacey, A., LópezCaballero, M. E., Gómez-Guillén, M. C., & Montero, P., Food Microbiol, **27** (2010)

Enzymatic Hydrolysis of Liquid Hot Water Pre-treated Macro-alga (*Ulva lactuca*) for Fermentable Sugar Production

Tri Poespowati[1,*], Ardy Riyanto[1], Hazlan[1], *Ali* Mahmudi[2], and *Rini* Kartika-Dewi[1]

[1]Department of Chemical Engineering, National Institute of Technology, Jl. Bend. Sigura-gura 2, Malang, Indonesia
[2]Department of Electrical Engineering, National Institute of Technology, Jl. Bend. Sigura-gura 2, Malang, Indonesia

Abstract. *Ulva lactuca* is one of green macro-algae that has a significant cellulose content. This study aims to determine the effect of variations in substrate-enzyme ratio and hydrolysis time on the enzymatic hydrolysis process of cellulose extracted from *Ulva lactuca* to produce fermentable sugar or reducing sugar as a raw material for making bioethanol. Firstly, Liquid Hot Water (LHW) pre-treatment process was performed at the temperature of 135°C for 20 minutes; the purpose of this pre-treatment was to reduce the content of hemicellulose and to increase the cellulose content. Secondly, enzymatic hydrolysis process using cellulase enzyme was carried out, in this process citrate buffer was needed in order to stabilize the pH level during hydrolysis process. The process variables were ratio of substrate-enzyme (1:0.05; 1:0.1; 1:1.5; 1:2 and 1:2.5 w/w) and hydrolysis time (24, 48 and 72 hours) under temperature of 45°C and pH level of 5.5. The results shows that the highest reducing sugar yield is 79.7% obtained at a ratio of substrate-enzyme of 1:2.5 (w/w) for 48 hours of hydrolysis time, with the result of reducing sugar concentration is 16.2043 mg/mL.

* Corresponding author: poespowati@lecturer.itn.ac.id

1 Introduction

The world energy demand continues to increase. This increase occurred from the requirement of 90 million barrels per day in 2012 to 100 million barrels in 2020, and raise in volume of 121 million barrels in 2014 [1]. Therefore, in order to fulfil the fuel consumption, alternative renewable energy from biomass sources is necessary; one of them is by converting biomass into bioethanol.

Ethanol is a renewable fuel that can be obtained from fermentation process of mono-sugar. To replace a part of gasoline comes from fossil sources; ethanol fuel has already applied in the US as gasoline fuel additive [2]. This can help to reduce the problem in the decline of fossil fuel reserves and its use as a fuel and chemical raw materials [3]. There are 3 generation types of bioethanol: first generation bioethanol produced from molasses or by-product of refining sugarcane, second generation bioethanol made from lignocellulosic agricultural waste and third generation bioethanol obtained from cellulosic biomass such as algae.

Some researches claim that ethanol made from cellulosic materials is a significant fuel in the future if its technology faces the important improvement [4]. One of the cellulosic natural resources that can be converted into bioethanol is green macro alga *Ulva lactuca*, this alga widely available in nature especially along the coast of the South Island of Java, Indonesia [5]. To obtain mono-sugar for fermentation purposes, polysaccharides in cellulose content of biomass material can be broken down by chemical hydrolysis or enzymatic hydrolysis [6]. Enzymatic hydrolysis process in order to convert cellulose biomass into bioethanol involves three basic steps: first step is pre-treatment process, second step is hydrolysis process, and the third step is fermentation process. By pre-treatment process, lignin content and hemicellulose content in biomass can be removed or decreased, and long chain sugar contained in cellulose can be broken easily into mono sugar. This is due to the increasingly porous of biomass [2].

Liquid Hot Water (LHW) is one of the most pre-treatment procedures to break down hemicellulose from lignocellulosic materials. LHW pre-treatment process can be carried out in temperature range of 120 °C and 200 °C [7] or 120 °C and 260 °C [8], and can be conducted in a simple way by using an autoclave in a ratio of solid sample/water by 1/5 (w/v) [9]. LHW pre-treatment process has several advantages: the process is environmentally friendly because it does not require any chemicals other than water [8, 10], high sugar yields can be obtained; and compared to the pre-treatment process using acid hydrolysis, the LHW pre-treatment process effect provides minimal corrosion in the equipment caused by mild pH levels of the reaction [10]. This pre-treatment is used to enhance the ability of enzymes to degrade cellulose by dissolving hemicellulose, reduce the degradation of sugar products, minimize the formation of HMF (5-hydroxymethyl furfural) and furfural that can be inhibit enzymatic hydrolysis process and yeast fermentation process [11].

Enzymatic hydrolysis of cellulose can be carried out by cellulase enzyme [11]. Reducing sugar and glucose are the products of this enzymatic hydrolysis [2, 11] in relatively high percentage (75%-95%) compared with the usage of acid hydrolysis [12]. Bacteria and fungi can produce cellulase enzymes for hydrolysis of lignocellulosic materials, which trichoderma is the most microbes for cellulase enzyme production [2]. Cellulase is part of hidrolases group containing endo-glucanase, exo-glucanase and beta-glucosidase [11, 13]. Endo-glucanase act on cellulose derivatives by cutting at random and cause breakdown of long chain into the short chain rapidly. Furthermore, exo-glucanase splitting glucose continuously from non-reduced ends of the chain glucan-free and produce cellobiose (disaccharide of glucose). Then beta-glucosidase degrades cellobiose into glucose [14].

The purpose of this research is to investigate the effect of substrate-enzyme ratio and hydrolysis time to enzymatic hydrolysis process of LHW pre-treated green macro-alga (*Ulva lactuca*) to produce high reducing sugar content as a raw material of bioethanol alternative fuel.

2 Experimental Method

2.1. Materials

Green macro-alga *Ulva lactuca* was collected from Kukup beach, Yogyakarta, Indonesia. Chemicals supplied by PT Panadia Malang and CV Makmur Sejati Malang - Indonesia. The chemicals used in this study were fehling A, fehling B, ethanol 98% (Merck), sulphuric acid 98% (Merck), sodium citrate (Merck), citric acid (Merck), dinitrosalisilic acid (Sigma), phenol (Merck), sodium hydroxide (Merck) and sodium sulphite (Merck). All of chemical used was in pure condition, without any further purification process. Cellulase enzyme was supplied by Biotechnology Laboratory, Sidoarjo – Indonesia.

2.1. Sample Preparation

Ulva lactuca was carefully cleaned from sand and other impurities with tap water and then rinsed by distilled water, followed by sun drying then oven drying at 50°C until stable weight was reached [5]. To increase the surface area of mass transfer, dry *Ulva lactuca* was pulverized by using a blender into about 80 meshes in particle size. Content analysis of hot water soluble (HWS), hemicellulose, cellulose, lignin and ash (HHCLA) was conducted on sample prior pre-treatment procedure.

Lignin content and ash content of *Ulva lactuca* result in extremely small concentration and they can be ignored, the most component dominated in this alga were HWS (51%), hemicellulose (27.67%) and cellulose (20.33%). Both of these hemicellulose and cellulose concentration were relevant with the concentration stated by Saini where the concentration of hemicellulose and

* Corresponding author: poespowati@lecturer.itn.ac.id

cellulose of green algae are in the range of (20-50%) and (20-40%) respectively [14].

2.2. Pre-treatment process of Liquid Hot Water (LHW)

During Liquid Hot Water pre-treatment process, 10 gr of dried *Ulva lactuca* was added to 100 ml of distilled water (10% w/v). The mixture was then heated in an autoclave at temperature of 135°C for 20 minutes. After LHW pre-treatment process, Chesson method analysis [15] is needed to be conducted in order to determine the amount of dissolved hemicellulose and other component that might be change such as cellulose, HWS and lignin.

2.3. Enzymatic Hydrolysis

The pre-treatment process explained above was followed by enzymatic hydrolysis process. For this purpose, 100 mL of pre-treatment product was added by 67 mL of buffer citrate (ratio 1:0.67 v/v). Before addition of cellulase enzyme was actuated, pH level of mixture must be maintained at about 5.5 by using sodium citrate or citric acid. Furthermore, cellulase enzyme was added with ratio substrate-enzyme of 1:0.05, 1:0.1, 1:1.5, 1:2 and 1:2.5 w/w; and it was shaken in a water-bath shaker at temperature of 45°C and hydrolysis time of 24, 48 and 72 hours. Hydrolysate was separated from its solid. In turn liquid hidrolysate was analysed on reducing sugar qualitatively using Fehling solution and quantitatively using DNS method [16].

3 Results and Discussion

3.1. Pre-treatment Process

The purpose of pre-treatment process was to increased cellulose content and to reduced hemicellulose content of *Ulfa lactuca*, in turn; cellulase enzymes convert cellulose into reducing sugar easily. Initial analysis of reducing sugar content on *Ulva lactuca* resulted in 3.489 mg/mL.

Cellulose in material structure is covered by hemicellulose and lignin. Hemicellulose in raw material acts as a barrier for the enzymes to degrade cellulose [17-19]. It causes cellulase enzyme hard to reach cellulose and obstruct the way of cellulose become reducing sugar. Lignin also tightly covers cellulose fibres and inhibits enzyme to degrade cellulose [19]. That is why pre-treatment process has to be done in order to dissolve the hemicellulose content. Table 1 provides a summary of the component weight in percentage of *Ulva lactuca* before and after pre-treatment process. And Figure 1 shows a typical history of HHC: Hot water soluble (HWS), hemicellulose, and cellulose content before and after pre-treatment processes.

Table 1. Result of liquid hot water pre-treatment.

Component	Before pre-treatment (% w)	After pre-treatment (% w)
Hot Water Soluble (HWS)	51	47.4
Hemicellulose	27.67	17.8
Cellulose	20.33	33.5
Lignin	1.3	-

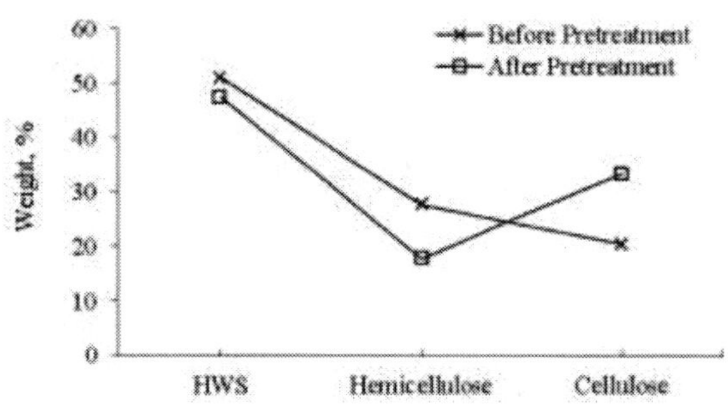

Fig. 1. HHC profile of *Ulva lactuca* before and after pretreatment process.

It can be seen from Table 1 and Figure 1, after the actuation of pre-treatment process, the weight percentages of both HWS and hemicellulose were decreased from 51% to 47.4% and from 27.67% to 17.8% by weight respectively, on the other hand the weight percentage of cellulose was increased from 20.33% to 33.5% by weight. Lignin content on *Ulva lactuca* was quite small (1.3% w), and it can be predicted that lignin was solubilized in total. The less weight of lignin can be ignored, this has in general matching with those reported by Saini [14]. It can be concluded that the pre-treatment temperature of 135 °C able to decrease the weight of HWS and hemicellulose and also able to increase the weight of cellulose in a significant number.

Agreeing with Tutt's experiment, the increasing weight of cellulose after pre-treatment process confirms that the binder of cellulose can be broken down successfully [20]. This is simply because when *Ulva lactuca* was exposed to the high enough of temperature, the cell structure will be broken.

3.2. Hydrolysis Process

Cellulase enzyme activities were examined prior to enzymatic hydrolysis work. The examination of cellulase enzyme activities was conducted twice by using DNS method [16], result in 2.3199 Unit/gram and 2.3128 Unit/gram.

Those two parameters of enzyme activities are much smaller than the parameter that was found by Kumar and

Wyman [21]. The amount of enzyme activity of their experiment was 7.5 Unit/gram, which was done with substrate/enzyme ratio of 1:0.02 (w/w), and producing 90% of sugar yield [21]. In this research, in order to increase the yield of sugar with the low enzyme activity, the ratio of substrate/enzyme should be increased at the range of 1:0.05 and 1:0.25 (w/w). And all results of qualitative analysis of sugar content in solution resulted in brick-red precipitate as a presence of glucose in the solution.

Coding system of hydrolysis work is provided in Table 2.

Table 2. Encoding variables.

Coded level	-2	-1	0	+1	+2
Ratio of substrate/enzyme (w/w)	1:0.05	1:0.1	1:0.15	1:0.2	1:0.25
Hydrolysis time hours)		24	48	72	

Table 3 shows the influences of substrate/enzyme ratio (w/w) and hydrolysis time on reducing sugar concentration and yield of sugar. The volume of applied hydrolysate was 100 ml for every single analysis.

As Table 3 quantified, the highest concentration of reducing sugar (19.6933 mg/mL) was resulted from the ratio of substrate/cellulase enzyme by 1:0.25 (w/w) corresponds to 48 hours of hydrolysis time. On the other hand, the lowest concentration of reducing sugar (4.76 mg/mL) was obtained from the ratio of substrate/cellulase enzyme by 1:0.05 (w/w) corresponds to 72 hours of hydrolysis time. It means that due to obtain the high concentration of reducing sugar; the role of substrate/cellulase enzyme ratio was more significant than hydrolysis time. Table 3 can be visualized as in Figure 2.

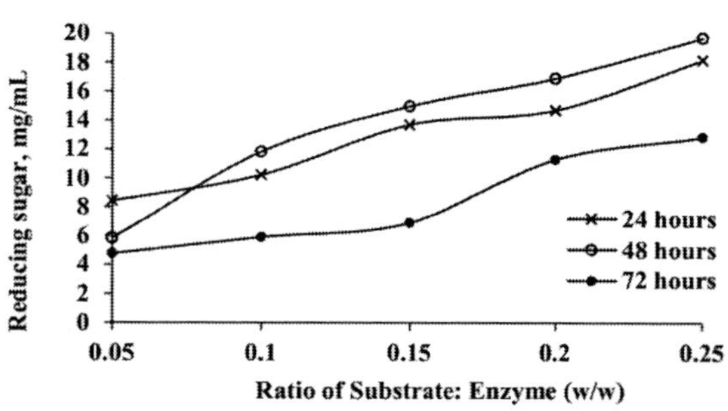

Fig. 2. Reducing sugar concentration versus substrate/enzyme ratio (w/w) corresponds to hydrolysis time

As shown in Figure 2, for all of hydrolysis times, shows that the concentration of reducing sugar increases as the increase of substrate/enzyme ratio. In general, for all ratio of substrate/enzyme, the highest concentration of reducing sugar produced by 48 hours of hydrolysis time, followed by 24 hours oh hydrolysis time, and the less one was obtained by the application of hydrolysis time by 72 hours. Only one point looks unlike that phenomena, which is at ratio of substrate/enzyme of 1:0.05 (w/w) where the reducing sugar concentration increases by the decrease of hydrolysis time. Compared with the two results of reducing sugar concentration by the addition of 1:0.05 (w/w), concentration of reducing sugar resulted by 48 hours of hydrolysis time should be higher than the application of 24 hours of hydrolysis time. This exception result might be caused by the pH level of this point is higher than the others (see Table 3). It means that high pH level affects to the decrease of reducing sugar concentration.

Table 3. Independent variables of substrate/enzyme ratio and hydrolysis time employed of reducing sugar concentration and yield of sugar.

Ratio of substrate/ enzyme (w/w)	Hydrolysis time (hours)	Hydrolisate pH level	Reducing sugar (mg/ml)	Yield (%)
-2	-1	5.4	8.4267	24.3
-1	-1	5.3	10.2267	33.1
0	-1	5.4	13.6933	50.2
+1	-1	5.4	14.6933	55.1
+2	-1	5.4	18.1600	72.2
-2	0	5.9	5.8267	11.5
-1	0	5.3	11.8267	41.0
0	0	5.3	14.9600	56.4
+1	0	5.4	16.8933	65.9
+2	0	5.3	19.6933	79.7
-2	+1	5.3	4.7600	6.3
-1	+1	5.4	5.8933	11.8
0	+1	5.4	6.8933	16.7
+1	+1	5.4	11.2933	38.4
+2	+1	5.4	12.8267	45.9

The experiment result as figured out in Figure 2 can be pictured out by a different graph as shown in Figure 3 that shows the relationship between concentrations of reducing sugar and hydrolysis time with different ratio of substrate/enzyme.

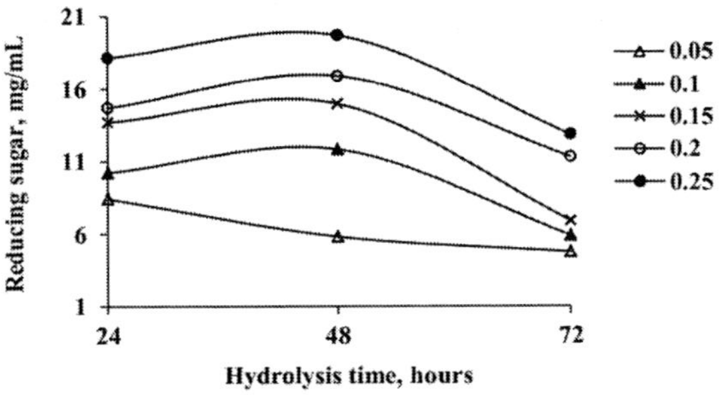

Fig. 3. Reducing sugar versus hydrolysis time corresponds to substrate/enzyme ratio (w/w).

As illustrated in Figure 3, the highest reducing sugar concentration is obtained under 1:0.25 (w/w) ratio of substrate/enzyme corresponds to all hydrolysis times. This concentration is getting smaller in sequences for other 4 ratios of substrate/enzyme (1:0.2 w/w, 1:0.15 w/w, 1:0.1 w/w, and 1:0.05 w/w respectively). In all curves, the lowest number of reducing sugar is at 72 hours of hydrolysis time. This phenomenon is understood as the consequence of enzyme that became inactive at the critical hydrolysis time. In this case, enzyme turned to be self-inhibitor of hydrolysis process. This occurrence was in agreement with Guerard et.al [22]. Except sugar, cellobiose is a by-product of hydrolysis process that can be an inhibitor during the process [23]. As Xu et.al claim that the reducing sugar and the rate of hydrolysis goes down at extending hydrolysis time after optimum hydrolysis time is obtained [24].

The response of substrate/enzyme ratio and hydrolysis time on the yield of reducing sugar is given in Figure 4.

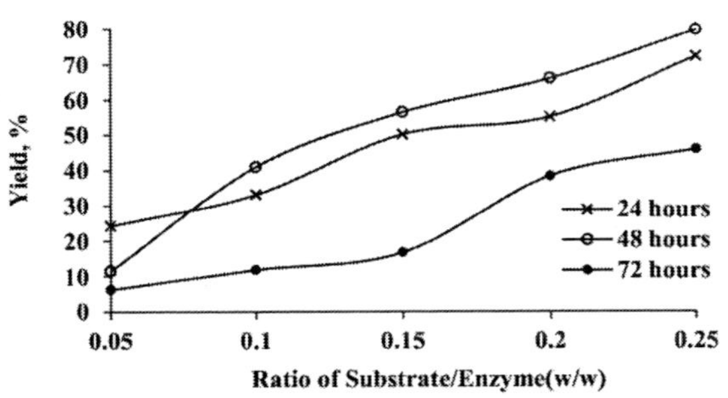

Fig. 4. Reducing sugar yield versus ratio of substrate/enzyme corresponds to hydrolysis times.

The response of substrate/enzyme ratio and hydrolysis time on the reducing sugar yield shown in Figure 4 exposes the high ratio of substrate/enzyme and the modest hydrolysis time tend to enhance yield of reducing sugar (79.7%). The increase of substrate/enzyme ratio from 1:0.05 w/w to 1:0.25 w/w under 48 hours of hydrolysis time improves the yield of reducing sugar production from about 11.5 % to 79.7%. The trends of 3 curves in Figure 4 are quite similar to the trends shown in Figure 2.

This work was supported by Ministry of Research, Technology and Higher Education of the Republic of Indonesia for the scheme of Hibah Kompetensi No. 0299/E3/2016, 27 Januari 2016 with an agreement letter of research from "Institut Teknologi Nasional Malang".
No. ITN.04.189.06/I.LPPM/2016, 25 April 2016.

References

1. U.S. Energy Information Administration, *International Energy Outlook 2016*, (2016)
2. Y. Sun, J. Cheng, Bioresource Technology, **83**, 1-11 (2002)
3. V. Venkatramanan, S. Aravinth, P.C. Santhosh, M. Nithya, B.K. Sathya, Int. J. of ChemTech Res., **6**, 4061-4069 (2014)
4. M. Wang, C. Saricks, D. Santini, 1999, *Centre for Transportation Research, Energy Systems Division,* (Argonne Nat. Lab., 9700 South Cass Avenue, Illinois, 1999)
5. T. Poespowati, A. Mahmudi, R.K. Dewi, Int. J. of ChemTech Res., **8**, 512-518 (2015)
6. J. Payamara, A. Payamara, Int. J. of ChemTech Res., **3**, 1446-1449 (2011)
7. Y. Kim, N.S. Mosier, M.R. Ladisch, Biotechnol. Prog. **25**, 340-348 (2009)
8. C. Wan, Y. Li, J. Biortech, **102**, 9788-9893 (2011)
9. C. Cara, M. Moya, I. Ballesteros, M.J. Negro, A. Gonzalez, E. Ruiz, J. Procbio., **42**, 1003-1009 (2007)
10. B. Bharathiraja, T. Sudharsanaa, A. Bharghavi, G.S. Sowmeya, G. Balaram, Int. J. of ChemTech Res., **6**, 4434-4444 (2014)
11. R. Dhabhai, A. Jain, S.P. Chaurasia, Int. J. of ChemTech Res., **4**, 1497-1502 (2012)
12. Jimmy, T. Poespowati, S. Noertjahjono, Int. J. of ChemTech Res., **8**, 188-193 (2015)
13. A.K. Pandey, G. Edgard, S. Negi, J. Renene, **xxx**, 1-6 (2016)
14. J. K. Saini, L. Tewari., R. Saini, 3 Biotech, **5**, 337-353 (2015)
15. A. Chesson, J. Appl. Bacteriology, **45**, 219-230 (1978)
16. G.L. Miller, Analytical Chemistry, **31**, 426-428 (1959)
17. Y.Y. Lee, P. Iyer, R.W. Torget, Dilute-Acid Hydrolysis of Lignecellulosic Biomass, J. Biotechnology, **65**, 93-115 (1999)
18. M.J. Taherzadeh, K. Karimi, BioResources, **2**, 707-738 (2007)
19. B. Yang, Z. Dai, S.Y. Ding, C.E. Wyman, Biofuels, **2** , 421-450 (2011)

20. M. Tutt, T. Kikas, H. Kahr, M. Pointner, P. Kuttner, J. Olt, Agronomy Research, **12**, 417-424 (2014)
21. R. Kumar, C.E. Wyman, Bioresource Technol., **100**, 4203-4213 (2009)
22. F. Guerard, L. Guimas, A. Binet, J. Molecular Catalysis B: Enzymatic, **19–20**, 489–498 (2002)
23. T.M. Wood, Biochemical Society Transactions, **13**, 407-410 (1985)
24. Z. Xu, Q. Wang, Z.H. Jiang, X. Yang, Y. Ji, Biomass & Bioenergy, **31**, 162-167 (2007)

Optimization of Glucose Production of Cocopeat Using Whole Cell *Trichoderma reesei*

Muhammad Zaki[1*], Suhendrayatna[1], *Misbul* Hadi[1], and *Syukri* Adha[1]

[1]Chemical Engineering Departement, Faculty of Engineering, Syiah Kuala University

Abstract. The high content of cellulose in cocopeat makes this material convertible into glucose. The converting process of cellulose into glucose can be done by hydrolysis. In this research, the coocopeat hydrolyzed enzymatically using cellulose ezyme from *Trichoderma reesei*. The purpose of this study was to obtain optimum conditions of glucose yield and to know the effect of concentration of NaOH, molasses mass, and the effect of hydrolisis time on glucose yield produced. The variabel used was hydrolisis time (0; 124; and 240 hour), NaOH concenteration (1%; 2%; and 3%), and molasses mass (40; 50; and 60 gr/l). The result showed the higest glucose level obtained at 2% NaOH concenteration, molasses mass 60 gram, and hydrolysis time 240 hours, while the predicted resulted of the optimum conditions of glucose level produced using the software Design Expert 6.08 is 776.771 mg/l at NaOH concenteration 1,35%, molasses mass 59.96 mg/l and hydrolisis time 215.62 hours.

1 Introduction

Cocopeat is one type of solid waste from coconut processing industry that commonly found in Indonesia. In 2014, Indonesia has a coconut plantation area about 3.65 million hectares with total production of 3.05 million tons per year [1]. The availability of coconut plantations that make a lot of cocopeat waste abundant in Indonesia, until now the utilization of cocopeat waste is still very limited due to the lack of existing technology. One coconut can produce about 400 grams of coconut husk with 75% fiber and 25% cork composition, so the cocopeat content in one coconut is about 100 grams [2,3].

One of the benefits of waste cocopeat is can be used as raw material for glucose production with involves *Trichoderma reesei* fungus, because its contents of 43.44% cellulose [4]. The converting process of cellulose into sugars can be done by hydrolysis, which can be done either chemically or enzymatically. Chemical hydrolysis process can be carried out by using acid solution with pressure more than 1 atm and temperature above 200°C [4]. Meanwhile, cellulose hydrolysis process using 1% sulfuric acid (H_2SO_4) at a temperature of 215°C with a conversion of 50% [5]. This acidic hydrolysis process is difficult because it's determining difficulty in operating conditions that make this process less effective and requires a lot of energy, also makes the tools to be corrosive, less environmentally friendly, and can produce such furfural disruptive compounds.

Compared to the chemical hydrolysis, enzymatic hydrolysis is more advantageous. The enzymatic hydrolysis process does not degrade the sugar during hydrolysis process, occur at low temperatures, and lower equipment maintenance costs because there is no corrosive as well as environmentally friendly [6]. Enzymatic hydrolysis of cellulose generally uses microbes that can produce cellulose enzymes, Trichoderma *sp* is one of the fungi species that can produce the enzyme [4].

Research that discusses the conversion of cellulose into glucose by using *Trichoderma reesei* fungi has been widely reported. Hydrolysis of coconut husk using *Trichoderma reesei* with coconut fiber concentration of 5 g/ml and the same inoculum concentration can yield 0.8 gr/ml of glucose [4]. In another study using cocopeat as a feedstock yielded the highest glucose concentration at a time of 7 days hydrolysis with glucose levels of 193.25 mg/l [7]. Base on the earlier results research, this research was focused to find the optimum condition of hydrolysis cocopeat process to glucose by using *Trichoderma reseei* fungi.

2 Materials and Methods

2.1 Materials

All chemicals used in the form of grade reagent were obtained commercially from Waco ltd. *Trichoderma reesei* were obtained from the stock culture of School of Life Sciences and Technology, Bandung Institute of Technology. Cocopeat was obtained from coconut farmers around Darussalam, Banda Aceh City. Before use, cocopeat was dried under the sun for 3 days and then cocopeat subsequently shredded and sieved.

* Corresponding author: m.zaki@che.unsyiah.ac.id

2.2 Experimental Procedure

Ten grams of cocopeat was mixed with various NaOH solution (1, 2, and 3%) at a ratio of 1:10 (gr/ml), the sample was then stirred and heated for 2 hours to remove the lignin. Treatment was followed by washing until the pH of wash water become neutral and then drying in the oven. The setting of equipment is shown in Figure 1.

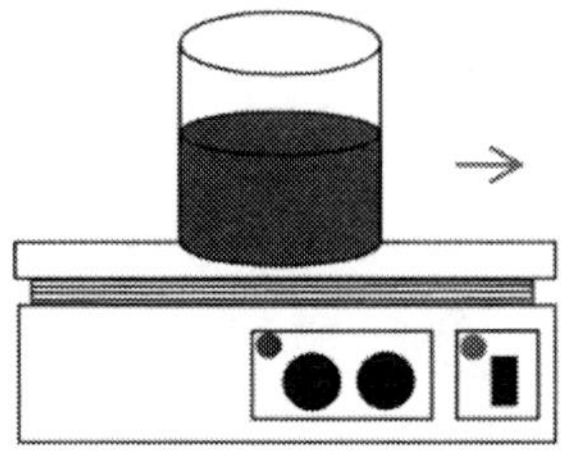

Fig. 1. Setting of equipment

2.3 Glucose Analysis

Glucose content analysis was performed by Di-Nitro salicylates (DNS) method. Two ml DNS reagent mixed with 2 ml of liquid sample and then the mixture was heated to 80-100°C for 5 minutes to develop a red-brown color. The mixture then was added 1 ml of 40% sodium potassium tartrate solution (rochelle salt). After the color of mixture stabilized, the solution was cooled and the absorbance was measured by using UV-VIS spectrophotometer, Shimadzu 1700.

2.3 Data Analysis

Experimental design was carried out using Box-Behnken design with supported by Design Expert 6.0.8 software. The data was processed using ANOVA. Table 1 performs an experimental design using Design Expert 6.0.8 software.

Table 1. Experimental design.

Run	NaOH Conc. (%)	Molasses mass (gr)	Hydrolysis Time (Hours)
1	3	40	120
2	1	50	240
3	3	50	0
4	2	50	120
5	2	50	120
6	2	50	120
7	1	60	120
8	2	60	0
9	1	40	120
10	2	40	240
11	2	50	120
12	2	40	0
13	3	60	120
14	2	60	240
15	1	50	0
16	2	50	120
17	3	50	240

3 Results and Discussion

3.1 Summary of surface response statistics model

The statistical model summary provides suggestions for the selection of appropriate models and compatibility with experimental data. Table 2 shows statistical model for the glucose content analysis resulting from the Box-Behnken design. The suggested model was quadratic and linear.

Table 2. Summary of statistical model.

Cubic	Std. Dev.	R-Squared	Adjusted R-Squared	Predicted R-Squared	Result
Linear	187.55	0.5366	0.4297	0.1208	Suggested
2FI	196.27	0.6097	0.3754	-0.6285	
Quadratic	102.59	0.9253	0,8294	-0,1944	Suggested
Cubic	0	1	1		

Since the quadratic has a value R^2 closer to one (0.9253) and Adj-R^2 is 0.8294, the suggested model is chosen quadratically. If the value of R^2 close to 1 identifies that there is a high correlation between the observed value and the predicted value [8]. The best modeling is focused on Adj-R^2 value and maximum R^2 value [9]. The value of R^2 is the value that describes the ratio of a variation ratio to total variation and the level of conformity of the resulting model. The value of Adj R^2 generated by this research model was 0.8289, which means that the concentration of NaOH and hydrolysis time greatly affect the diversity of resulting response, while the rest was influenced by other factors not used in this study.

The Box-Behnken design used to produce a mathematical model was actually used to predict the value of glucose levels based on the suggested model. Based on the quadratic model, the equations of the mathematical model generated are as follows:

https://doi.org/10.1051/matecconf/201815601016

$$Y = 661{,}49138 + 451{,}33987\ X_1 - 44{,}36996\ X_2 + 5{,}66925X_3 - 80{,}76300\ (X_1^2) + 0{,}45039\ (X_2^2) - 0{,}017623\ (X_3^2) - 1{,}16445\ (X_1.X_2) + 0{,}98519(X_1.X_3) + 0{,}052039(X_2.X_3) \quad (1)$$

Based on the equation model (1) generated above, each variable was the concentration of NaOH (X_1), molasses mass (X_2) and hydrolysis time (X_3). These three variables have a very important influence where the use of these three variables will increase the resulting glucose levels. From the equation (1) obtained results predicted value of glucose levels produced that was tabulated in Table 3.

Table 3. Comparison research and prediction data (Design expert) for glucose levels.

Run	NaOH Concentratio n (%)	Molases mass (gr)	Hydrolysis Time (Hours)	Glucose (mg/l)	
				Research	Prediction (*Desain expert*)
1	3	40	120	506.05	416.39
2	1	50	240	625.26	614.88
3	3	50	0	0	0
4	2	50	120	534.47	534.48
5	2	50	120	534.47	534.48
6	2	50	120	534.47	534.48
7	1	60	120	514.74	604.4
8	2	60	0	0	0
9	1	40	120	435.79	489.42
10	2	40	240	509.21	466.01
11	2	50	120	534.47	534.48
12	2	40	0	0	0
13	3	60	120	538.42	484.8
14	2	60	240	761.84	682.6
15	1	50	0	0	0
16	2	50	120	534.47	534.48
17	3	50	240	149.21	282.12

Comparison of research data with predictive data shows in Table 2. Center point condition was in Run 5 as much as 50 gr of molasses, 2% NaOH concentration, hydrolysis time of 120 hours, and glucose production level of 534.47 mg/l. Table 2 also shows that the highest glucose levels. The plotting data between research results and predicted results of the expert design was shown in Figure 2. This figure shows the relationship of prediction values based on the equations proposed by the response surface design and the actual values generated from the experiments. It can be seen that the points on the chart were spread on a straight line and this condition shows the good accuracy of the model obtained.

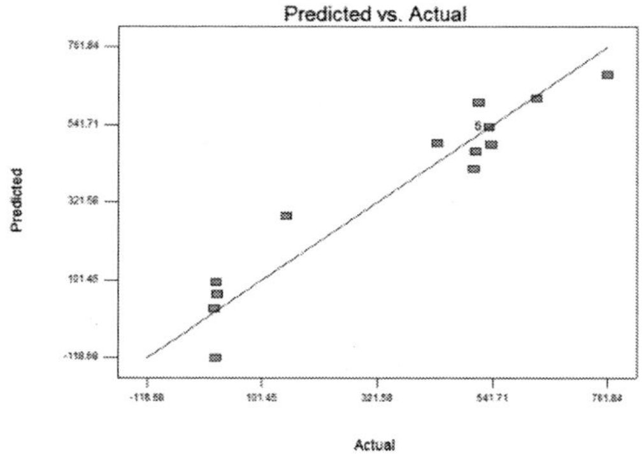

Fig. 2. Relationship between glucose levels of the results of research and prediction.

3.2 Analysis of Variants (ANOVA)

ANOVA is one of the factorials of Box-Behnken design that is useful for generating interaction between process variables and response variables. The components in ANOVA will be used to calculate the F-ratio that serves to determine the effectiveness of a model [10]. Figure 4.2 shows that most of the resulting models from this research were significant. It is proved that the resulting prob. > F value was <0.05. The model acceptable if the P value was less than 5% (0.05) which means the model can represent a significant influence on response [9]. The result of variance analysis showed that both linear factors were significant to the response, whereas for the quadratic, the concentration of NaOH, molasses mass, and hydrolysis time had a significant effect on the resultant response.

https://doi.org/10.1051/matecconf/201815601016

ANOVA for Response Surface Quadratic Model

Analysis of variance table [Partial sum of squares]

Source	Sum of Squares	DF	Mean Square	F Value	Prob > F	
Model	9.132E+005	9	1.015E+005	9.64	0.0034	significant
A	*18553.45*	*1*	*18553.45*	*1.76*	*0.2259*	
B	*16816.77*	*1*	*16816.77*	*1.60*	*0.2467*	
C	*4.942E+005*	*1*	*4.942E+005*	*46.96*	*0.0002*	
A²	*27463.84*	*1*	*27463.84*	*2.61*	*0.1502*	
B²	*8541.10*	*1*	*8541.10*	*0.81*	*0.3976*	
C²	*2.712E+005*	*1*	*2.712E+005*	*25.77*	*0.0014*	
AB	*542.38*	*1*	*542.38*	*0.052*	*0.8269*	
AC	*55906.95*	*1*	*55906.95*	*5.31*	*0.0546*	
BC	*15598.64*	*1*	*15598.64*	*1.48*	*0.2629*	
Residual	73667.55	7	10523.94			
Lack of Fit	*73667.55*	*3*	*24555.85*			
Pure Error	*0.000*	*4*	*0.000*			
Cor Total	9.868E+005	16				

Fig. 3. ANOVA for quadratic model.

3.3 Analysis of Surface Response to Glucose Levels

The model suggested to calculate glucose level in this research was quadratic equation. Model selection was based on several factors, such as the value of R^2. From the model that based on response surface analysis, it used two types of plots, namely contour plots and and 3D plots. This plot was used to facilitate description of the effect of variables on the response. The contour plot was a 2D plot which was the transverse slice of the 3D plot. Contour plots were useful for analyzing the interaction effects among factors on response [11]. The 3D plot shows the effect of two variables on the response to which the variable was made in a fixed state. The model equation suggested by software design expert version 6.0.8 for glucose production was the quadratic model equation. To analyze this model, three-dimensional plots and contour plots can be used. The relationship between the three independent variables to the response can be seen in Figure 4 and Figure 5, respectively.

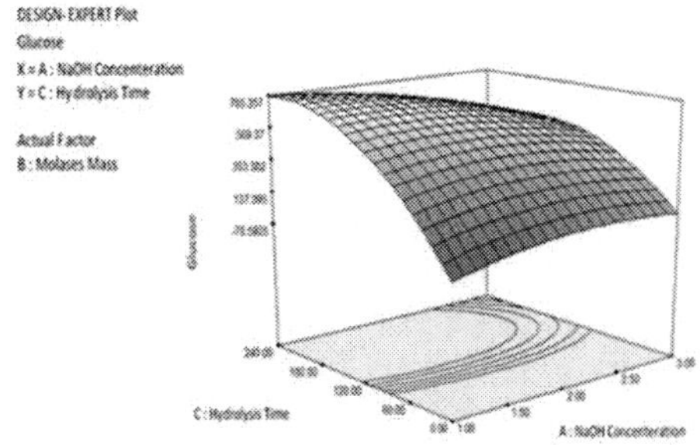

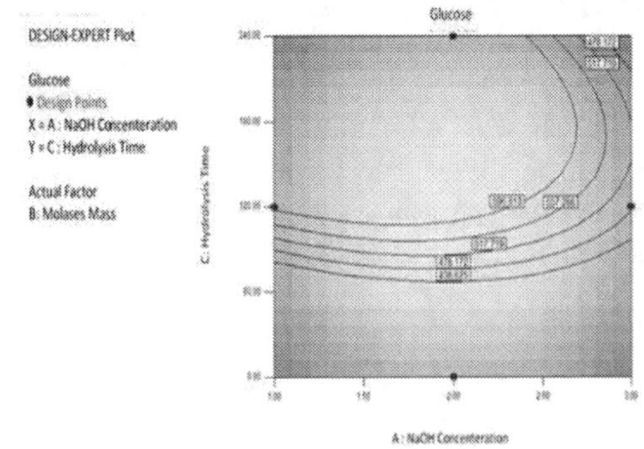

Fig. 4. Effect of NaOH concentration and hydrolysis time on glucose level production (a) 3D Plot (b) Contour Plot.

Figure 4 shows the effect of NaOH concentation on the delignification process on glucose level product. Cocopeat used in this study still contains some impurities such as lignin. The presence of lignin in the cocopeat substrate can decrease the rate of cellulosic decomposition, because lignin can adsorb the cellulase enzyme thus reducing the amount of enzymes adsorbed by cellulose and hemicellulose, therefore before hydrolysis process was occurred, a delignification process was first performed to remove lignin in cocopeat [4]. Delignification is a mass removal process of lignin contained in cocopeat. There are 4 types of delignification method, that is biologically, physically, chemically and physically chemically [12]. The method of delignification used in this research was chemically by addition of Sodium Hydroxide (NaOH) with different concentration (1%, 2%, and 3 %). NaOH was used as a delignifier because NaOH can damage lignin structure. In addition to damage the lignin structure, NaOH can also dissolve lignocellulose, but not able to dissolve cellulose, only some of the cellulose can be dissolved in NaOH due to the degree of polymerization [12]. The difference of NaOH concentration used in this study aims to know the effect of delignification on glucose levels formed.

Based on the results of research and data processing using Design Expert showed that NaOH concentrate greatly affect the resultant glucose concentration, at 1% NaOH concentration (with molasses 40 grams and hydrolysis time 120 hours) yield glucose level became 435.79 mg/l, while at 3% NaOH concentration (with molasses 40 grams and hydrolysis time of 120 h) obtained 506.05 mg/l. However, the highest glucose levels reached 761.84 mg/l that were obtained at 2% NaOH concentration (with molasses 60 g and hydrolysis time of 240 h).

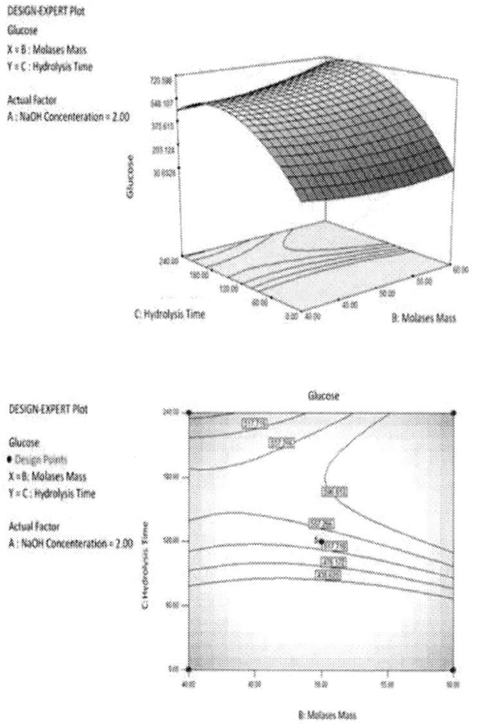

Fig. 5. Effect of molasses concentration and hydrolysis time on the glucose level production (a) 3D Plot (b) Contour Plot

Figure 5 shows the effect of molasses concentration added to the *Trichoderma reesei* culture on the glucose level production. Molasses are a source of nutrients used for the growth of the mycelium of the *Trichoderma reesei* mycelium in which the mycelium may secrete the cellulose enzyme used for hydrolysis process. Better nutrients added to the fungus, more enzymes will be produced by the fungus therefore more enzymes produced, hydrolyzed cellulose will increase into glucose and glucose production will be higher.

Figure 4 and Figure 5 also show the profile of glucose levels produced during the hydrolysis process from 0 to 240 hours. Results of the measurement of glucose levels showed that the treatment that received highest glucose level of 761.84 mg/l occurred at 240 hours. The increase in glucose levels produced shows interaction between cellulose enzyme and cellulose substrate present in cocopeat which will form a complex bond between enzyme and glucose-producing substrate. The interaction between the cellulose enzyme and the substrate at the time of hydrolysis with longer time span causes the reaction to run maximum so the glucose level production becomes higher.

Table 4. Optimum conditions of glucose levels.

Variabel	Value
Molases	59.95
NaOH (%)	1.35
Time (hours)	215.62
Glucose (mg/L)	776.771
Desirability	1

3.4 Optimization Glucose levels

The optimization process was carried out by regulating NaOH concentration, molasses concentration, and hydrolysis time adjusted according to the level range. The response variable as glucose concentration was pointed maximum to obtain the highest yield. Table 4 and Figure 6 show the optimum results of glucose production from cocopeat using *Trichoderma reesei*. The optimum conditions of process will be achieved at molasses mass of 59.95 gr, 1.35% NaOH concentration, and hydrolysis time of 215.62 hours. Figure 6 shows a 3D plot image and contour plot of the optimum conditions suggested to produce the highest glucose levels.

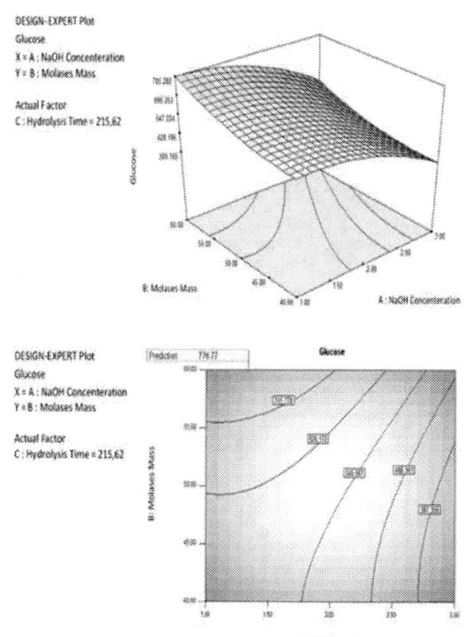

Fig. 6. Model of optimum conditions of glucose levels production (a) 3D Plot (b) Contour Plot for optimum glucose concentration.

4 Conclusion

Based on the results described above, it can be concluded that cocopeat can be hydrolyzed by using

Trichoderma reesei fungus. The higest glucose level obtained at 2% NaOH concenteration, molasses mass 60 gram, and hydrolysis time 240 hours, while the predicted resulted of the optimum conditions of glucose level produced using the software Design Expert 6.08 is 776.771 mg/l at NaOH concenteration 1.35%, molasses mass 59,96 mg/l and hydrolisis time 215.62 hours.

References

1. Directorate General of Plantation, *Statistics of Indonesian Plantation*. Jakarta (2014)
2. I. Isroful, http:// isroful. Wordpress.com/2009/ 10/ 15/ pengolahan-sabut-kelapa-menjadi-papan-partikel-dengan-batang-pisang-sebagai-pelapisnya-pada-interior-bangunan/ (2009)
3. Y. Awang, A. S. Shaharom, R.B. Mohamad and A. Selamat, *American Journal of Agricultural and Biological Sciences* **4**, 1 (2009)
4. S. Sukadarti, S.D Kholisoh, H. Prasetyo, W.P. Santoso, T. Mursini, *Journal of Chemical Technology Development for Indonesia Natural Resources,* D13-1 (2010)
5. Q. Xiang, Applied Biochemistry and Biotechnology, **107**, 1-3 (2003)
6. S. Safaria, N. Idiawati, T.A. Zaharah, J. KK, **2**, 1 (2013).
7. M. Refi, D. Rahmadi, Thesis, Syiah Kuala University, (2016)
8. Q. Zhao, J.F. Kennedy, X. Wang, International Journal of Biological Macromolecules, **49**, 2011.
9. Ulilalbab, T. Estiasih, Varietas, (2012)
10. C. Hanraha, E. Garza, E. Garcia, K. Miller, 2007. Journal of Environmental Informatics **9**, 2 (2007)
11. S.H. Hasan, P. Srivastava, M. Talat, Journal of Hazardous Materials, 168 (2009)
12. A.A. Fajar, Thesis, Bogor Institute of Agricultural (2014)

Bioelectricity Production from Microalgae-Microbial Fuel Cell Technology (MMFC)

Carlito da Costa and *Hadiyanto*[*]

Chemical Engineering Department, Faculty of Engineering, Diponegoro University Jl. Prof. Soedarto, Tembalang, Semarang 50239, tlp/fax : (024)7460058

Abstract. Microbial fuel cell is an ecological innovative technology producing bioelectricity by utilizing microbes activity. Substituent energy is produced by changing the chemical energy to electrical energy through the catalytic reaction of microorganism. The research aims to find out the potency of bioelectricity produced by microalgae microbial fuel cell technology by utilizing the combination of tapioca wastewater and microalgae cultivation. This research is conducted through the ingredients preparation stage – microalgae culture, wastewater characterization, membrane and graphite activation, and the providing of other supporting equipment. The next stage is the MMFC arrangement, while the last one is bioelectricity measurement. The result of optimal bioelectricity production on the comparison of electrode 2 : 2, the power density is 44,33 mW/m^2 on day 6, meanwhile, on that of 1 : 1, 20,18 mW/m^2 power density on day 1 is obtained. It shows that bioelectricity can be produced from the combination of tapioca wastewater and microalgae culture through the microalgae- microbial fuel cell (MMFC) technology.This research is expected to be a reference for the next research particularly the one that observes the utilizing of microalgae as the part of new and renewable energy sources.

1 Introduction

Along with the times, our needs of energy become the priority in mobilizing our daily activities. Energy becomes the important aspect of any activities utilizing the facilities and infrastructures as the support of continuity and success. The economic wheel is driven by energy. Energy that is available and dominant nowadays is fossil fuel energy. The creation of fossil energy takes a really long time and will be exhausted in a certain time, and it is not considered eco-friendly[1].

The use of fossil energy appears some problems, namely the reducing available volume, the difficulty of finding new reserve, the unstable price, and the pollution emerging from the fossil fuel usage[2].

Problems emerged by the use of fossil energy trigger the development of technology which can produce eco-friendly, easy to get, and renewable energy [3].

One energy from aquatic herbs which is potential to be developed, eco-friendly, and competitive is microalgae. The potency of microalgae development is important in the future since microalgae have preeminence as food material, supplement, biodiesel, and bioethanol. It can also be utilized as the electricity producer.

Instead of microalgae cultivation, wastewater produced from tapioca production process, contains starch and hydrosianate acid (HCN), can be elaborated into cyanide, nitrogen, phosphorus, and other organic compounds. If the tapioca wastewater is directly discarded without any treatment, it will endanger the surroundings. [4].

Nowadays, MFC is based on supporting theories, such as utilizing wastewater [5], utilizing waste of alcoholic beverage manufacture [6], and utilizing domestic wastewater [7]. Based on the reports, MMFC technology utilizing microalgae and wastewater is potential to be developed since those two ingredients are easy to find. Both microalgae and wastewater contains nutrition such as carbohydrate, glucose, and fat which are potential to be utilized. Nutrition in the wastewater is a food source for the growth of bacteria while oxygen is produced by microalgae through photosynthesis[8]. The purpose of this research purpose was to investigate the potency of bioelectricity production from microalgae-microbial fuel cell. The evaluated parameters are the COD concentration of tapioca wastewater and the microalgae's growth rate.

1.1 Microalgae

One potential alternative energy source which is renewable and eco-friendly is microalgae [9]. Light and water are used by microalgae as the energy source in

[*] Corresponding author: hadiyanto@live.undip.ac.id

conducting carbon dioxide metabolism to produce biomass as explained on this photosynthesis reaction:

$$CO_2 + H_2O + lights \rightarrow CH_2O + O_2 \quad (1)$$

Oxygen is the side product which is used to form water in the cathode while the transfer ions of electrons and protons of hydrogen atoms occurs. Oxygen is produced by algae through photosynthesis using solar energy while the nutrient and bicarbonate metabolism process is produced by microorganism [10]. The excellence of Microbial Fuel Cell by utilizing microalgae cultivation is because it is efficient and eco-friendly. It naturally changes chemical energy into electricity. It also plays an important role in overcoming air pollution [11].

1.2 Tapioca Wastewater

According to [12], tapioca industry can produce 4 – 12 m^3 of wastewater for each ton of tapioca flour. It contains 4.5 – 5.0 of pH, 10.000 – 15.000 mg/L of COD with 19 – 28 mg/L cyanide compound. The cyanide compound in tapioca wastewater can kill methanogen bacteria and reduce the content of methane. Smith et al, [13] stated the contents of tapioca wastewater are 300 – 7500 mg/L of BOD, 3100 – 20000 mg/L of COD, and 1500 – 8500 mg/L of TSS. Suspended organic compounds in the tapioca wastewater – protein, fat, and carbohydrate which is easily decomposed – can obstruct the respiratory system and inorganic compounds such as cyanide and ammonia (NH_3). It also endangers the marine biota [14].

Organic ingredients consumed by microorganism become new biomass cell and organic substance which produce energy from oxidation reaction and is utilized for metabolism process [15].

2 Microalgae-Microbial Fuel Cell (MMFC) Technology

One way to reduce the pollution caused by tapioca wastewater is Microbial Fuel Cell (MFC). MFC is a mean used to oxidize organic and inorganic ingredients using bacterial activities [16].

Microalgae-Microbial Fuel Cell (MMFC) technology utilizes the result of microalgae photosynthesis as an oxygen in the cathode [17] with the tapioca wastewater as the anode to produce bioelectricity. According to [18], by using dual chamber with the electrolyte of proton exchange membrane (PEM), MFC produces 61,62 W/m^2, 0,57 V and 0,12 A with tapioca wastewater as the anode.

Electrons moving in anode chamber is consumed by the growing microbe cells, while, in cathode chamber, microalgae convert CO_2 into oxygen and biomass via photosynthesis [19]. Oxidized mediator is transformed to the medium by cell. The cycle is repeated. Mediator is reduced by electrons in catholyte [20]. During illumination, biochemical reaction in the anode and cathode is respectively formulated in [21].

$$C_6H_{12}O_6 + 6H_2O \xrightarrow{oxidation} 6CO_2 + 24H^+ + 24e^- \quad (2)$$

$$6CO_2 + 12H^+ + 12e^- \xrightarrow{lights} C_6H_{12}O_6 + 3O_2 \quad (3)$$

3 Materials and Methods

3.1 Materials

The material used as the catholyte is microalgae species spirulina platensis. It is obtained from PT. Neoalgae, Sukoharjo. Meanwhile, tapioca wastewater as the anolyte is obtained from Kedungsari Village RT/RW 01/01, Tayu, Pati Regency, Central Java. The other materials used are nafion 117 membrane with diameter 0.5 inch, while the electrode is a graphite bar [22] with 10 cm length and 7 mm diameter, and 1 meter of copper wire.

3.2 Methods

The research is conducted in the C-BIORE integrated laboratory of Diponegoro University, Semarang. The mechanism of electron transfer from the bacteria to the anode is through the mediator produced by the bacteria [23].

The power density value was calculated based on the influence of the wide of electrode surface and the membrane circle.

The wide of the electrode surface and the membrane circle was calculated by following equation[24]:

$$P = V x I \quad (4)$$

where P is power density (W), I is Current (A), and V is Voltage (V).

Power density can be influenced by the wide of electrode surface and membrane, so the equation is:

$$P = \frac{V \times t}{A_{Total}} \quad (5)$$

3.3 Experiment procedures

Spirulina platensis is diluted (based on cultivation standard), 3 liters are taken and tested – absorbance and wavelength test. The next stage is the tapioca wastewater characterization - the test of sample COD content, the absorbance and the wavelength test of bacterial cell.

The research used four reactors or two pairs of reactor. Each of them contains 3 liters of tapioca wastewater and 3 liters of cultivated microalgae. Each pair is connected by tube with membrane as the media of ion proton transfer and as the separator between anolyte and catholyte. Each reactor respectively filled by graphite electrode 1:1 and 2:2 bars. Both reactors filled by microalgae culture is aerated and 24-hour illuminated for 8 days. The illumination used 15-watt lamps with 20 cm distance. After all tools and materials are set, measure bioelectricity and test the sample every 24 hour. Bioelectricity observation is conducted every 6 hour for 15 minute for each observation, or the measurement result has showed the constant value as the optimum value. The analysis conducted by spectrophotometry is measuring wastewater COD concentration, the density of wastewater bacterial cell and microalgae cell [25].

3.4 MMFC equipments

Tools that are used in MMFC technology are based on modified [26]. MMFC tools design is based on the microbial fuel cell research, namely electrode and membrane usage [27] type of substrate [18], the reactor design [7].

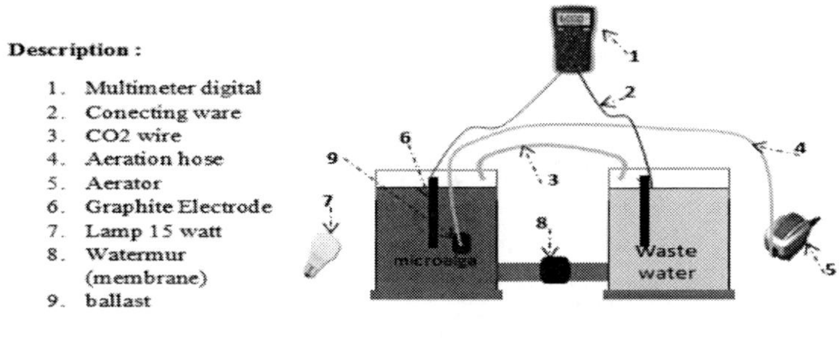

Fig. 1 MMFC series

4 Result and Discussion

Bioelectricity produced from dual chamber on the microalgae-microbial fuel cell (MMFC) technology can be shown by the following:

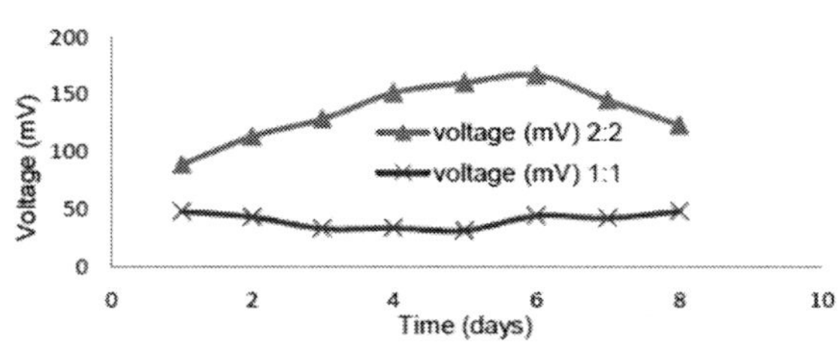

Fig. 2. Bioelectricity Production in Time

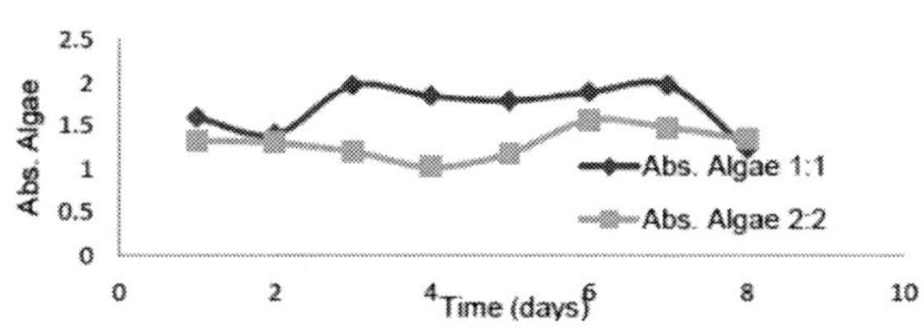

Fig. 3. The Effect of Microalgae Growth in Bioelectricity Production

The results shown in the graph above is the voltage obtained on the difference in the number of electrodes where the use of the number of electrode 2: 2 increased the value of higher voltage that is on day 6 of 166,537 mV with the absorbance value of microalgae cell growth of 1.558 on the same day. While the optimum value achieved on the use of 1: 1 electrode amounted to 48.15 on day 8 with the absorbance value of microalgae cell growth of 1.235.

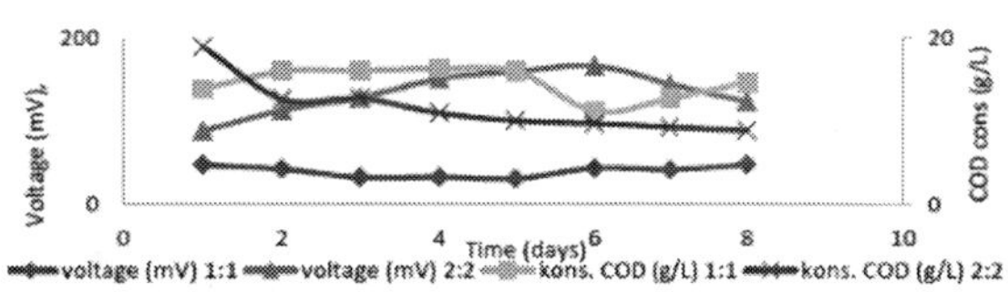

Fig. 4. COD and Voltage Change During MMFC Operation

Data obtained from sample analysis showed that the COD value of tapioca wastewater is significantly decreased (9,7573 g/L) with microalgae absorbance value of 1.558. Sample analysis result of the wastewater optical density showed the deccreasing absorbance value (0,149) on the same day using 1:1 electrodes.

It should be noted that bioelectricity obtained from MMFC technology is small if compared to ([11],[22]), They respectively produce 1926/21.4 mW/m^2 ($8.67/0.10 W/m^3$) and 13.5 mW/m^{-2} power density. It is due to the high carbohydrate content in the tapioca wastewater, so it makes the hydrolysis process slower.

5 Conclusion

It can be inferred that the obtained bioelectricity has not showed the significant result yet. It is due to some factors, namely:

1. The tapioca wastewater is only 2 days old since the production time until it is used in the MMFC technology, so the bacteria growth is obstructed by the high concentration of cyanide The tapioca wastewater was not characterized first. It is shown by its high COD concentration (16,5304 g/L) so it obstructed the growth of bacteria.

After conducting this research, bioelectricity with maximum power density occurred on day 6 (44,33 mW/m^2), with 9,7573 gL^{-1} COD concentration, 2.17 bacteria cell absorbance, and 1.558 microalgae cell absorbance on 2:2 electrodes and 0.5-inch diameter of pipes.

References

1. V. Patil, K.Q.Tran, and H.R.Giselrod, Int. J. Mol. Sci., **9**, 1158-1195 (2008)
2. Ditjenbun. Jakarta : Diretorat Jenderal Perkebunan, Departemen Pertanian, (2006).
3. E. Metcalf., 3rd Edition. Singapore : McGraw Hill Inc., (1991).
4. Mukminin , Wygnyanto and N. Hidayat., J. Teknologi Pertanian Brawijaya, **4**(2): 91-107. (2003)
5. Hyonsoo, I.Moon, I. S. Chang and H.K. Byung., Science Direct. Biores. Technol., **97**, 621-627. (2006)
6. Y. Feng., X. Wang, B.E Logan and H. Lee., Springer- Verlag. Appl Microbiol Biotechnol, **78**,873-880 (2008)
7. H. Liu, and B.E. Logan, Environ. Sci. Technol, pp. **38**(14), 4040-4046. (2004)
8. H. Hadiyanto, and M.M.A. Nur., World Appl. Sci. J., **5**, 959-967. (2014)
9. Y. M. Li, B. W. H. Wang and C. Wu, and Q Lan., Appl. Microbial Biotechnol, pp. **81**, 629-636. (2008)
10. P. Kruzic and J.F. Kreissl., Water Environ. Res., **81**(10):1346-1360. (2009)
11. Y. feng, N. Cui, et al., et al., Energy, pp. **79**, 674-680. (2013)
12. Eva, Oktarina. (2014).
13. Sumestri, Sri and Alaerts. Surabaya Utama : Usaha Nasional, (1984).
14. F.P. Riyanti, Lukitowati and T. Afrilianna., J. Penelitian Sains, **13**(3):34-39. (2010)
15. D.A. Oktavia, et al., Agrointek, **6**(2):65-71. (2012)
16. H.J. Kim, Enzyme and Microbial Technology , pp. **30**, 145-152. (2002)
17. J. Lobato, Applied Energy, **110**, 220-226. (2013)
18. F.N. Cahyani., IPCBEE, **6**, 218-220. Singapore: IACSIT Press. (2011)
19. D.Parlevliet and N.R. Moheimani., Aqua. Biosys., **10**,14. (2014)
20. E.E. Powell and G.A. Hill., Chem. Eng. Res. Des., **87**(9): 1340-1348. (2009)
21. M. Zhou., J. Power Sources, 216-219. (2012a,b),
22. G. Araceli, D. Campo, Journal of Power Sources, **242**, 638-645. (2013)
23. K. Rabaey and W. Verstraete., J. Biotechnol, **23**, 291- 298. (2005)
24. Y.O. Momoh and B. Naeyor., J. Biochem Tech, pp. **2**(4): 216-220. (2010) APHA. 21 st Edition. Washington DC : APHA,AWWA, (2005).
25. B.E. Logan, : A John Wiley & Sons Inc, (2008).
26. S. Cheng, H. Liu and B.E. Logan ., J. Environ Sci. Technol., **40**:364-369. (2006b)

Modification of Cassava Starch Using Lactic Acid Hydrolysis in The Rotary-UV Dryer to Improve Physichocemical Properties

Siswo Sumardiono[1,*], *Bakti* Jos[1], *Denny* Firmansyah[1], *Rahmi* Hidayatunajah[1], and *Isti* Pudjihastuti[2]

[1]Department of Chemical Engineering, Faculty of Engineering, Diponegoro University, Jl.Prof. Soedarto, SH, Tembalang, Semarang, Indonesia 50275
[2]DIII Study Program, School of Vocation, Diponegoro University, Jl.Prof. Soedarto, SH, Tembalang, Semarang, Indonesia 50275

Abstract. Food security should be supported in an effort to utilize local products into import substitution products. Cassava starch has the potential to be developed into semi-finished products in the form of flour or starch which does not contain gluten but can inflate large baking process, potentially as a substitute for wheat flour-the main ingredient for making bread. The characteristic of the starch is influenced by the type of starch composition and structure. Natural starch has physicochemical properties i.e. a long time cooking and pasta formed hard. These constraints allow us to modify cassava starch by a combination of lactic acid hydrolysis and drying with rotary UV system. Modified cassava starch is expected to be used as a substitute for wheat flour. The aim of the research which is a combination of lactic acid hydrolysis and drying using a rotary UV system is to examine the optimum operating conditions in the drying process of starch hydrolysis with parameter the physicochemical and rheological properties of modified cassava starch. The initial process study is to hydrolyze cassava starch using lactic acid. Furthermore, hydrolyzed cassava starch is then dried using UV light in the rotary dryers system. There are a variety of changing variables, i.e. time of irradiation cassava starch-lactic acid hydrolysis products in the rotary UV light and air drying temperature. The research results show that modified starch has a better characteristic than the natural starch. From the analysis, the best point of swelling power, solubility and baking expansion is consequently 15.62 g/g; 24.19 %; 2.21 ml/gr. The FTIR result shows that there is no significant difference of the chemical structure because the starch modification only change the physical characteristics. From the SEM analysis, we can know that the size of the starch's granule changes between the natural starch and the modified starch..

1 Introduction

Starch is a mixture of the two polysaccharides with amylose as straight chain and amylopectin as branched chain [15]. Various kinds of starch have different properties depending on the length of the chain of carbon atoms as well as a comparison between the straight and branched molecules chains [7]. Generally starch contain of 15-30% amylose starch, 70-85% amylopectin and 5-10% others material [26].

Starch that commonly used is a natural starch without treatment modification (native starch) therefore, a natural starch has little additional materials [20]. Natural starch has a limited solubility in water and susceptible starch gel syneresis (separation of water from the gel structure) due to retrogradation starch. Modified starch is a starch that has done physical treatment, chemical or a combination that aims to change one or more of the properties of starch [4]. Starch modification changes the properties of starch which are less profitable so that it can be used more widely. Some properties of starch that can be changed is the gelatinization temperature, gelatinization characteristics during the process, and starch resistance to high temperature.

This experiment focused on the combination of lactic acid hydrolysis process with the rotary drying in addition with UV light on cassava starch, so that the modified product is expected to have properties that almost the same with flour.

Experiment on modification of cassava starch using combination process lactic acid hydrolysis and microwave heating has been conducted with the results that the physicochemical and rheological properties of modified starch changed significantly compared to the native starch [17]. Combination process method of lactic acid hydrolysis and hydrogen peroxide oxidation for cassava starch modification also gives results on improvement in baking expansion of modified starch and showed that the modified starch can substitute wheat flour in optimum condition process [18].

The experiment about combination of chemical and physical modification of cassava starch using lactic acid

* Corresponding author: siswo.sumardiono@che.undip.ac.id

and ethanol under oven and solar drying has also been conducted with results that indicate that the high concentration of lactic acid and ethanol will increase the swelling power of modified starch. It was also determined that starch modified by using esterification and drying using the solar drying method is more effective than using oven drying [19].

The focus of this experiment is to examine the change in physicochemical and rheological properties of modified cassava starch compared to the natural starch by using the combination process of lactic acid hydrolysis and uv-light drying.

2 Materials and Methods

2.1 Materials illustrations

Tapioca starch by the brand of Mount Agung was obtained from Semarang (Indonesia). Other chemicals which were used as additives for process analysis with analitycal grade, were obtained from PT. Merck (Germany).

2.2 Preparation of Lactic Acid Solution

Lactic acid solution was prepared by dissolving lactic acid 1% w / w in 250 ml of water. The solution was left for overnight to make more active lactic bacteria.

2.3 Experimental Procedure

Tapioca starch in accordance with the variable (100 g and 200 g) was dissolved in the lactic acid solution that has been made before. Stir well for 15 minutes at room temperature. After that the solution was filtered using filter paper and cooling it by winding it up until the water content is reduced. The maximum allowable water content contained in the starch before drying is about 40%.

After the hydrolysis process, the drying process is carried out using UV rotary. Before we use, UV rotary tool prepared in advance. Turn the rotary tool, set the temperature (40 ° C, 50 ° C, 60 ° C, 70 ° C) and speed (10 rpm, 20 rpm, 30 rpm, 40 rpm) in accordance with the desired variable. Once ready, the dried tapioca will gradually dry up and in accordance with the expected results.

2.4 Swelling Power

Tapioca starch sample that has been dried taken 0,1gr and dissolved in 10 ml of distilled water. Then the solution is heated using a water bath with a temperature of 60 ° C for 30 minutes. Supernatant was separated using a centrifuge at a speed of 2,500 rpm for 15 minutes. Swelling power calculated by the formula:

$$\text{swelling power}=\frac{\text{weight pasta}}{\text{weight of dry sample}} \tag{1}$$

2.5 Solubility

Tapioca starch sample that has been dried taken 1 g and dissolved in 20 ml of distilled water, then the solution is heated in a water bath at a temperature of 60 ° C for 30 minutes. The supernatant was separated using a centrifuge at 3000 rpm for 20 minutes and then take 10 ml to be dried in the oven and the dry sediment weight recorded.

$$\%\text{sollubility}=\frac{\text{weight of dry sediment}}{\text{volume of supernatant}} \tag{2}$$

2.6 Viscosity

The viscosity measurements carried out simultaneously by means of heating and stirring suspension of starch that is placed on a container on the instrument. The temperature of the sample to be analyzed very decisive change in the viscosity profile of a mixture.

2.7 Baking Expansion

Modified tapioca starch mixed with 12.5 grams of milk, butter, salt, sugar, eggs, water, yeast stirred with a mixer on medium speed for 5 minutes. Suspension (mixed dough) were weighed (50 g) was added to muffin tins and baked in the oven for 25 minutes at a temperature of 135°C. The cold muffin then weighed and measured to determine the specific volume (baking expansion) in ml / gram.

2.8 SEM Analysis

Scanning Electron Microscope (SEM) is a method in which an electron microscopic analysis produced the drawn sample by scanning the sample using a focused beam of electrons. These electrons interact with the electrons in the sample, produced various signals that can be detected as topografy information about the sample surface and also its composition. SEM is able to get the highest resolution smaller than 1 nanometer.

2.9 FTIR Analysis

FTIR is a technique for analyzing organic materials, polymers, coatings, gas, biological samples, inorganic materials, and minerals. FTIR working principle is to emit infrared rays in the sample, after that, it will be shown in graphs consisting of absorbance and also the length of the wave. Cluster analysis contained in the sample was done by interpretation infrared ray absorption spectrum of the results by the sample.

Table 1. The Value of Swelling Power, Solubility and Baking Expansion of Starch as a Result of Lacto Acid Hydrolysis in the Rotary-UV Dryer

No	Variables			Swelling Power (g/g)	Solubility (%)	Baking Expansion (ml/g)
	Playback Speed (rpm)	Temperature (°C)	Capacity (g)			
1	10	40	100	15,62	10,22	1,69
2			200	11,50	15,64	1,59
3		50	100	15,44	11,16	1,67
4			200	12,90	20,22	1,72
5		60	100	12,67	11,16	1,28
6			200	12,61	11,70	1,32
7		70	100	13,70	10,65	1,43
8			200	12,89	10,74	1,06
9	20	40	100	12,19	8,04	1,29
10			200	13,78	13,84	1,62
11		50	100	10,58	10,13	1,61
12			200	12,81	11,72	1,91
13		60	100	11,00	11,67	1,54
14			200	12,02	24,19	1,83
15		70	100	12,73	6,57	2,01
16			200	12,09	19,00	1,97
17	30	40	100	7,50	3,52	2,01
18			200	12,71	19,27	1,81
19		50	100	8,46	6,63	1,82
20			200	13,49	11,36	1,78
21		60	100	7,25	5,39	2,14
22			200	13,09	16,04	1,83
23		70	100	12,01	21,13	1,77
24			200	13,01	18,19	1,56
25	40	40	100	11,50	14,65	2,21
26			200	14,92	9,63	1,71
27		50	100	15,06	12,21	2,00
28			200	11,95	15,01	1,84
29		60	100	10,58	11,32	1,90
30			200	12,79	8,68	1,54
31		70	100	15,73	8,94	1,98
32			200	14,95	13,62	1,68

3 Results and Discussion

3.1. Effect of Playback Speed, Capacity Dryer Feed and Air Temperature on Rotary UV on Swelling Power Value

From Table 1 it can be seen that the average power of modified starch swelling ranged from 7.25 to 15.73 g/g. In general, the swelling power of the modified starch is higher than natural starch which swelling value of 12.51 g/g. The increase in value is due to the swelling power when starch is heated with water media, the starch granules swell and open the resulting presence of heat energy. This heating also resulted in the weakening of the hydrogen bonds that connect between amylose-amylopectin, amylose-amylose and amylopectin-amylopectin [16]. Compactness starch granules will decrease and amorphous regions will be cut off, resulting in the hydration water by the starch granules. Most of amylose molecules will come out of the starch granules and water soluble. According Flenche [6], when the starch molecule is completely hydrated molecule will spread to the media on the outside and the first out is to have a short chain amylose. And the higher of the heating temperature, the more amylose molecules that will come out of the starch granules. In addition, with the increase in temperature of the water molecules have higher kinetic energy so that makes it easier to penetrate the starch granule [9]. Hydrated starch will continue to expand as a result of the entry of water into the starch granules and trapped in the molecular composition of amylose and amylopectin [9].

Shown in Table 1 that there are 12 variables that have a swelling power value is smaller than the value of natural starch swelling power. Impairment swelling power is not due to the leaching of amylose due to heating temperature. Because it can be seen that the temperature of the heater is used instead of above the gelatinization temperature, but the cause that allows the use of a temperature of 40 ° C-70 ° C is the stretch between constituent molecular bonds of starch and starch granules opening. Thus, the drying process occurs retrogradation, where the water that goes into the granules will come out as a result of the drying temperature, so that the bonding between amylose stretched back docked result in swelling power decreased [16].

In addition to the effects of temperature, from table 1 it can be observed on the effect of rotation speed on UV rotary tool used. Swelling power at variable values 4, 12, 20, 28 is equal to 12.9 g / g; 12.81 g / g; 13.49 g / g and 11.95 g / g. Rotational speed proved to be very influential on the value of the modified starch swelling power. This is consistent with research conducted by Bhattarai *et al* [3] which concluded that the rotational speed of rotary no significant effect on heat transfer. This heat transfer effect on the hydration of starch granules. Therefore, the value of swelling power that the difference was not significant in the treatment of different rotational speed.

Subsequent conditions the difference is the capacity of the feed is entered in the UV rotary dryer. Swelling power is strongly influenced by the bonds between the molecules making up the starch. The weight ratio of starch affect the value of swelling power. The highest value of swelling power for 31 variable that is equal to 15.73 g / g. It shows that the smaller the weight ratio of starch that is included as bait, then the reaction will run better [24].

Baking expansion or so-called starch swelling power is incident volume and maximum weight gain experienced by the natural starch in water. In addition, the swelling power is power that characterize the nature and development of a material that is related to the strength of flour [26]. Swelling power is due to the non-covalent bonding between molecules of starch. Swelling power occurs in the amorphous regions of starch granules are influenced by amylopectin. The number of high amylopectin starch resulted in an increase in the value of swelling. In addition, there is a negative correlation between swelling power and amylose content, where the swelling power decreases with increasing amylose content [14]. Amylose can form a complex with lipids and inhibit swelling value. According Moorthy [11] there are several factors that determine the value of the swelling power of starch, amylose-amylopectin ratio is the distribution of molecular weight and chain length and degree of ramifications. When starch is inserted into the cold water, the starch granules absorb water and have swelling. However, a limited amount of water absorbed only about 30% [25].

3.2 Effect of Playback Speed, Capacity Dryer Feed and Air Temperature on Rotary UV on Solubility Value

From Table 1 can be seen that the average solubility of modified starch ranged from 3.52 to 24.19%. In general, the solubility of the modified starch is higher than natural starch which has a solubility value of 10.01%. The increase in solubility of modified starch is caused by the structure of starch which undergo stretching due to the interaction of water and heat. The heat will weaken the hydrogen bonds so that the structure of starch mainly amorphous regions become stretched [16]. Amorphous region is an area that has a lot of branching amylopectin and amylose molecules. Amorphous region is the part that can easily expand and undergo chemical reactions such as hydrolysis by acid or reacted by a functional group [8]. When stretched starch structure, starch will be easier to absorb water and have swelling (swelling power) resulting in increased solubility.

It can be seen from Table 1 that there are 8 variables that have values less than the solubility of natural starch. In variable 9, 15, 17, 19, 21, 26, 30, 31, solubility values of 8.04%; 6.57%; 3.52%; 6.63%; 5.39%; 9.63%; 8.68% and 8.94%. It happened because of retrogadation happened more. At higher temperatures stretching between bonding in amorphous regions are more prevalent. When there retrogadation when drying takes place, it would be a lot closer ties back so that the solubility is lowered [16]. According Olayinka *et al* [13] will increase the drying process chain association between amylose-amylose starch and amylose-amylopectin in amorf area, separate amylose and amylopectin fractions, increasing the cohesiveness of the material in the granules due to the interaction and pressure. In the process of heating the hydrogen bonds that stabilize the double helix structure in crystalline disconnected and replaced by water [12].

In addition to the heating temperature affect the solubility of the modified starch, starch weight ratio also affects the solubility value [24]. The smaller the weight ratio of starch, the reactions were run better and result in starch solubility tends to decrease. In contrast, when the weight of starch bigger then the reaction is unfavourable and solubility tends to increase. This is consistent with the results of research in which of the above table 9 it can be seen that the greatest solubility values contained in 15 variables with the value of the solubility of 24.19%.

According Zulaidah [26] solubility is the ability of a substance to be absorbed in water so as not to form an emulsion. The solubility of starch is one of the most important properties and are useful in a variety of industrial applications both food and non-food. Solubility values is also important to know to determine the suitability for conversion of dextrin in a variety of products. Each starch has a solubility that is different, then the solubility properties is one way to determine the type of starch which one to use. Factors influencing the solubility is the degree of polymerization (DP) and the size of the molecule. Temperature is also an important factor in determining the value of the solubility of starch.

3.3 Effect of Playback Speed, Capacity Dryer Feed and Air Temperature on Rotary UV on Baking Expansion Value

From Table 1 above it can be seen that the average Baking expansion of modified starch ranged from 1.06 to 2.21 ml / g. In general, the solubility of the modified starch is higher than natural starch which has a solubility value of 1.51 ml / g. The increase in the value of this expansion baking indicated by the increasing number of water lost during the baking process. As the starch is gelatinized by heating, the water is absorbed on the starch molecules evaporate and cause high pressure which produces thrust for greater expansion [2]. According Moorthy [11], modified tapioca starch expands the capability greatly influenced by the amylose content. Where amylose can bind water well, so that the higher levels of amylose in the starch obtained dough will be easier to expand. Level developers are also influenced by the ratio of amylose and amylopectin (Martz, 1992). With acid modification, expansion baking results greater than usual tapioca caused a change in the rheological properties and its physicochemical.

Baking process is old technology that is still used and continue to be developed to produce bakery products. The term is not only used in the baking of bread making, but for all products that use flour (flour) as the main ingredients and sources of thermal energy through the process of radiation or convection. Expansion baking measurement process is done by making a dough and then performed the heating process. In the baking process is a physical phenomenon, chemistry, and biology that determines the quality of a product. Heat transfer and mass transfer occurs continuously and interconnected. In the process of baking, heat transfer occurs as a combination of conduction, convection or radiation to the dough (dough). With the heat transfer, mass transfer can then occur based on four stages, namely (1) Evaporation of water, (2) the movement of water into the gas phase, (3) Condensation water and (4) Diffusion of water [22]. In addition to the heat generated by the oven in the baking process, UV radiation on the hydrolysis process also affects the characteristics of modified starch in baking expansion. Reaction of energy UV radiation causes partial depolymerisation of amylose molecules and form a new bond with the hydrogen, which causes water to be more easily absorbed [21]. Water is absorbed at any starch granules will make the starch granules swell and coincident with each other thereby increasing the ability to expand flour.

3.4 Effect of Playback Speed, Capacity Dryer Feed and Air Temperature on UV Rotary for SEM analysis

SEM (Scanning Electron Microscope) is a method in which an electron microscopic analysis produced the first sample by scanning the sample using a focused beam of electrons. SEM consists of an electron absorption which produces an electron beam at an accelerated voltage of 2-30 kV. The electron beam is passed on some electromagnetic lens to produce an image size <10 nm in the samples shown in the form of a photographic film or tube into the screen. These electrons interact with the electrons in the sample, resulting in a variety of signals that can be detected as information on the sample surface morphology and composition. SEM is suitable for use in situations that require observation rough surface with magnification 20 times to 500,000 times.

Observations granules are very important for some applications [23]. According Moorthy [10] the size distribution of starch granules depends on the source and type of starch produced. In general, starch granules are round and oval. Granule size show large variation is about 5-40 lm with round and oval shapes. Starch granule-shaped bowl (cup) and is very compact, but the modification process will break down the starch components into components that irregular shape [11].

Observations granule size is used SEM that shown in 1500 times. The sample used in this analysis is a modified starch and starch that has not been modified. Samples were analysed by using SEM has a variable rotation speed, temperature and UV rotary capacity. It can be seen that the size of the granules and granular form of each sample is greater than the starch that has not been modified. Size and shape of the sample other sample did not have a very significant difference that can be seen in Figure 1 - figure 5.

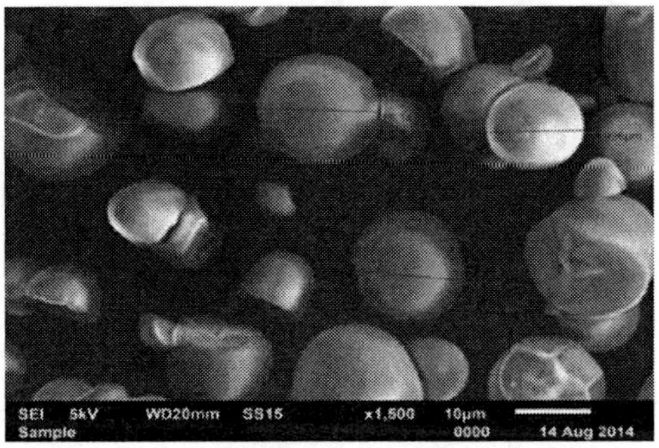

Fig. 1. SEM Analysed (1500x) for starch modified with speed of rotary UV is 40 rpm, temperature 40°C and feed capacity is 100 gr

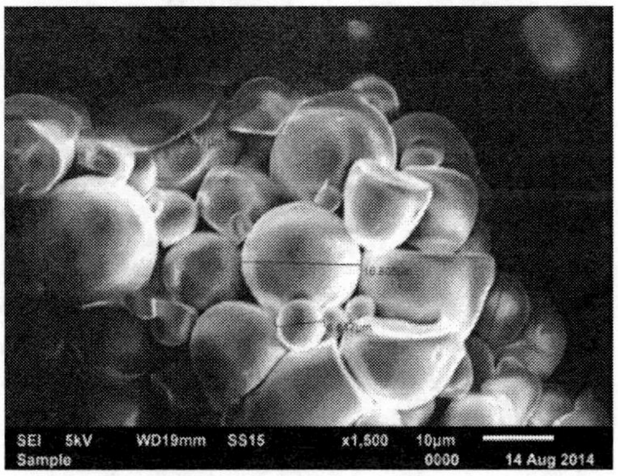

Fig. 2. SEM Analysed (1500x) for starch modified with speed of rotary UV is 30 rpm, temperature 60°C and feed capacity is 100 gr

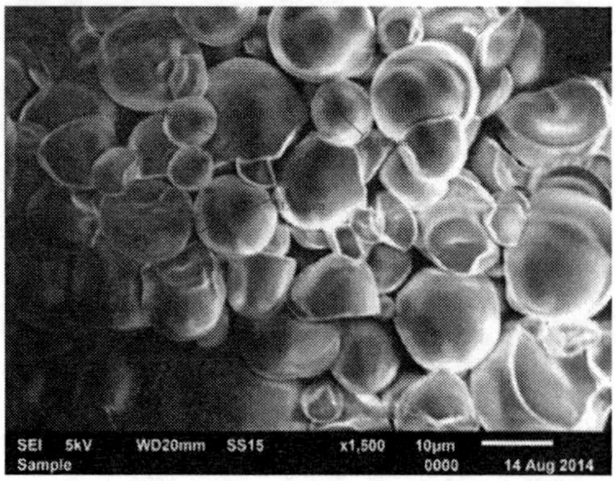

Fig. 3. SEM Analysed (1500x) for starch modified with speed of rotary UV is 20 rpm, temperature 70°C and feed capacity is 100 gr

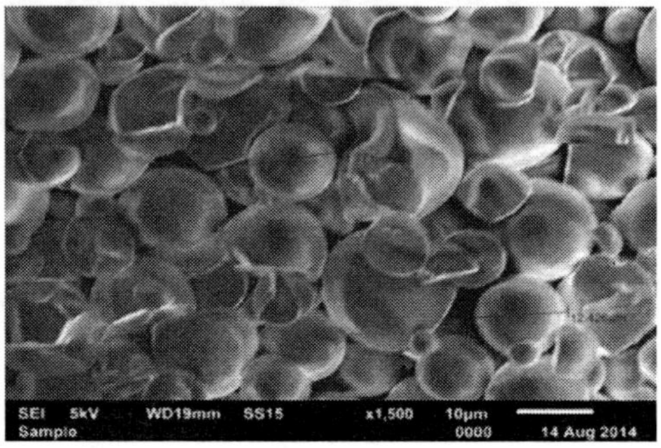

Fig. 4. SEM Analysed (1500x) for starch modified with speed of rotary UV is 40 rpm, temperature 50°C and feed capacity is 100 gr

Fig. 5. SEM Analysed (1500x) for natural starch

3.5 Effect of Modified Tapioca Starch with Lactic Acid Combination and Drying UV Rotary Against sample analysis results using FTIR (Fourier transform infrared spectroscopy)

FTIR analysis is an analysis method using the instrumentation tool that aims to determine the functional groups contained in a sample by using a beam of infrared waves with a specific wavelength. The results analysis is shown by read waveforms with peak absorbance wavelength as indicate a specific functional group. In Figure 6 (a), (b), and (c) the results of analysis of samples using FTIR. In Figure 6 (a) and (b), FTIR analysis using infrared rays with a wavelength between the ranges of 1000-3500 cm-1. In the picture 1 look at specific wavelengths produce specific wavelength absorbance peaks indicating the presence of specific functional groups. From the data results shown that the modified starch and pure flour has the same functional group structure, other than that of Figure 6 (c) looks waveforms and absorbance between pure and modified tapioca nearly equal. The same functional group because the main constituent components of starch are amylose and amylopectin [5] and lactic acid hydrolysis modification process combined with UV radiation causes the breaking of the bond only amylopectin to form amylose [2]. It is this factor that led to the results and interpretation of data from the absorbance images were formed there was no significant change between pure and modified tapioca, in the absence of chemical reactions that form the new product compounds. In addition to UV radiation is generally more cause physical changes such as decreasing levels of water compared to chemical changes as well as a decrease in the degree of crystallization [1].

4 Conclusion

Based on observations and data from the research and experiments, obtained some conclusions below:

1. Increasing drying temperature will results in increase of the solubility and baking expansion meanwhile the swelling power value will decrease.
2. Lowering the dryer feed capacity will decrease the swelling power and solubility value while the baking expansion increase. The smaller the amount of starch inserted, the hydrolysis process will be more effective.
3. Rotational speed is not very influential in changing the physicochemical and rheological properties of the modified starch since it has no significant effect on the mass and heat transfer during drying process.
4. Modified cassava starch has better properties in comparison with the starch prior to modification. Modified starch has lower swelling power value but higher solubility and baking expansion than natural starch. FTIR analysis shows no change in chemical structure of starch while SEM analysis shows change in its physical characteristics..

Acknowledgments

The author would like to thank Kemenristekdikti via PUSNAS research grant 2017 for funding this research.

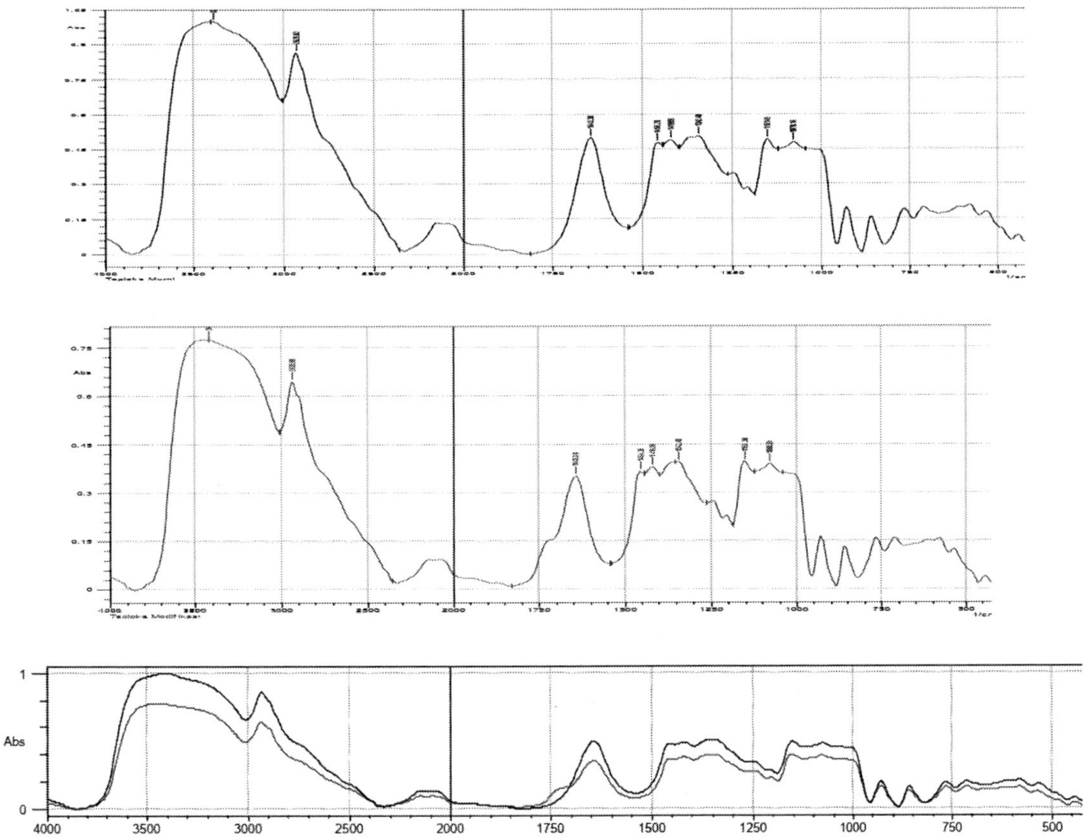

Fig. 6. Results of analysis of samples using FTIR. (a) and (b): FTIR analysis using infrared rays with a wavelength between the ranges of 1000-3500 cm-1; (c): waveforms and absorbance between pure and modified tapioca.

References

1. Bajer, D., Kaczmarek, H., & Bajer, K, Carbohydr. Polym., **98**(1), 477–482. (2012)
2. Bertolini, A. C., Mestres, C., Colonna, P., & Raf, J.. Carbohydr. Polym., **44**, 269–271. (2001).
3. Bhattarai S., Tran, V.H., and Duke, C.C., J. Pharm. Biomed. Anal., **45**, 648-653. (2001)
4. Cui, W., *Food Carbohydrate*, Francise and Taylor, England, (2006).
5. Demiate, I.M., Dupuy, J.P, Cereda, M.P., Wosiaci, G. Carbohydr. Polym. **42**. 149-158. (2000).
6. Fleche, G. *Chemical Modification and degradation of Starch*. In G.M.A.V. Beynum dan J.A Roels (eds.). *Starch Conversion Technology*. Marcel Dekker, Inc., New York. (1985).
7. Koswara, S. Teknologi modifikasi pati. Bachelor Thesis of Diponegoro University. (2006).
8. Liu, H., Yu, L., Xie, F., Chen, L., Carbohydr. Polym., **65**, 357–363. (2006).
9. Meyer, LH. *Food Chemistry*. Textbook Publisher, New York. (2003).
10. Moorthy, S.N. Starch/Starke. **54,** 559-592. (2002).
11. Moorthy, S.N. *Tropical Source of Starch. Di dalam : Ann Charlotte Eliasson (ed). Starch in Food : Structure, Function, and Application*. CRC Press, Baco Raton, Florida. (2004).
12. Muhamed, A. Jamilah B, Abbas KA, Rahman KA and K Roseline, Am. J. Agric. Bio. Sci., **3**(4), 639-646. (2008)
13. Olayinka, O. O., Adebowale, K. O. And Olu-Owolabi, B. I. Food Hydrocolloids, **22**(2), 225-230. (2008).
14. Sasaki T, Masuki J. Cereal Chem. **75**, 525-529. (1998).
15. Shariffa, Y. N., Karim, A. A., Fazilah, A., & Zaidul, I. S. M. Food Hydrocolloids, **23**(2), 434-440. (2009).
16. Siwi, K and Rukmi, W. *Studi Perubahan Sifat Fisik dan Kimia Tepung Ubi Jalar Putih (Ipoema batatas Var Sukuh) sebagai Efek Modifikasi Menggunakan Metode Heat Moisture Treatment*. Bachelor thesis of Diponegoro University, (2010).

17. Sumardiono, S. Pudjihastuti, I. Taufani, M. and Yahya, F. AIP Conf. Proc., **1840**, 060005. (2017).
18. Sumardiono, S. Pudjihastuti, I. Budiyono. Hartanto, H. and Sophiana, I., C. AIP Conf. Proc., **1840**, 060006. (2017).
19. Sumardiono, S. Djaeni, M. Jos, B. Pudjihastuti, I. and Abdallatif, M. Adv. Sci. Lett., **23**, 5792-5795. (2017).
20. Varela, P., & Fiszman, S.M. Food Hydrocolloids, **30**, 477-484. (2013).
21. Vatanasuchart, N, Naivikul, O., Charoenrein, S., & Sriroth, K., Carbohydr. Polym., **61**, 80–87. (2005).
22. Vatanasuchart, Nednapis, & Naivikul, O, Nat Sci, **37**, 334-344. (2003).
23. Wang, JK. 1983. Taro In J.Wang, Taro. Honolulu : Universitas of Hawaii Press. P. 4.
24. Wibowo, A dan Indra, A, .J. Teknol. Kim. Ind., **2**(2), 46-50. (2013).
25. Winarno, F. G. *Kimia Pangan dan Gizi*. Gramedia Pustaka Utama. Jakarta. (2002).
26. Zulaidah, A. *Modifikasi Ubi Kayu Secara Biologi Menggunakan Starter Bimo-CF Menjadi Tepung Termodifikasi Pengganti Gandum*. Bachelor Thesis of Diponegoro University, (2011).

Preliminary study for acetylation of cassava bagasse starch and microfibrillated cellulose of bamboo

Silviana Silviana[1]*, *Siti* Susanti[2], *and Agus* Subagio[3]

[1]Department of Chemical Engineering, Faculty of Engineering, Diponegoro University, Jl. Prof. Sudarto, SH., Kampus Tembalang Semarang, 50275, Indonesia
[2]Department of Food Technology, Faculty of Agriculture, Diponegoro University, Jl. Prof. Sudarto, SH., Kampus Tembalang Semarang, 50275, Indonesia
[3]Department of Physics, Faculty of Science and Mathematics, Diponegoro University, Jl. Prof. Sudarto, SH., Kampus Tembalang Semarang, 50275, Indonesia

Abstract. Bio composite matrixes have been developed from several biomaterials, such as starch. One of potential resources is starch isolated from cassava bagasse still consisting 30-50% of starch. Reinforcement material may be inserted into bio composite to tough and reduce the drawback of the starch-based bio composite or bio plastic. Microfibrillated cellulose of bamboo (MFC) can be used as toughening filler for composite matrix. However, surface modification of material could be employed to alter its properties, such as acetylation of starch-based bio composite and microfibrillated cellulose. The acetylation was executed by using glacial acetic acid (GAA) catalyzed with sodium hydroxide. This paper investigates optimum condition of acetylation for bagasse starch (BS) and bamboo MFC in different weight ratio of GAA to BS or MFC (1:1, 2:1, 3:1, 1:2, 1:3), temperature range of 30°C to 70°C, and pH range of 7 to 11. Data were resulted from degree of susbtitution for each running. The optimum condition of acetylation of BS was obtained at temperature of 50°C (for BS) and 30°C (for MFC), pH of 9, and 2:1 ratio. This acetylation was confirmed by fourier transform infrared spectroscopy and scanning electron microscope.

1 Introduction

Starch, an inexpensive natural resource, has been utilized in several fields, mostly in food area regard with sustainability, economic, and biodegradable. Furthermore, environment issue about replacement commercial plastics is harvesting by raising the demand bioplastic with renewable sources. Starch-based bioplastic have been employed several sources, such as banana [1], potato [2], sago [2-5], rice [6], corn [7], soybean [8], and cassava [9]. Cassava starch has been broadly utilized to be used in bioplastic reinforced with clay nanoparticles [9], thermoplastic chitosan [7], poly lactic acid [10], gelatin [11], bamboo nanofibrils [12], bacterial cellulose [13], and green coconut fiber [14] in order to overcome drawbacks of starch-based bioplastic, such as mostly water-soluble, low resitancy, and low mechanical property. Cassava starch production generates by-product [9], i.e cassava bagasse or pulp consisted a fibrous material containing 30-60% residual starch [15]. Due to low cyanide content with low protein it makes unattractive as ruminant. Furthermore, low ash content denotes broadly usage as biorefinery using microbial cultures [16]. Mostly, this solid is released as wasted resources. By using thermoplastic behaviour of residual starch in cassava bagasse, starch-rich cassava bagasse is promising future feedstock as global bioplastic packaging to replace commercial plastic with high underutilized and inexpensive. In this research it was considered to prepare bioplastic or cassava bagasse starch-based biocomposite reinforced with microfibrillated cellulose of bamboo, bamboo cellulose fiber. Bamboo denotes as potential fiber sources for biocomposites matrix [17-18], as composite filler for polyvinylchloride (PVC) [15], reinforcing nanocomposite starch-polyvinylacetate (PVA) [12], reinforcing polylactic acid (PLA) [20], reinforcing starch-based composite [21]. However, both (cassava bagasse starch and bamboo microfibrillated cellulose) have shortcomings regard with their properties. Therefore, it would be introduced by chemical modification such as acetylation. Acetylation is esterification reaction of acetyl group with hydroxyl group in starch-rich cassava bagasse (BS) and bamboo microfibrillated cellulose (MFC) to form starch acetates and cellulose acetates. This modification is expected to enhance translucency, stability, retrogradation, decrease gelatinasation temperature. Furthermore, dispersing bamboo microfibrillated cellulose in nonpolar or low-polar media can be improved by acetylation of the

* Corresponding author: silviana@che.undip.ac.id

primary hydroxyl groups [22]. This paper reviews investigation for acetylation of cassava bagasse and bamboo microfibrillated cellulose by using glacial acetic acid (GAA) with sodium hydroxide as catalyst. The objectives of this paper are to obtain optimum condition of acetylation such as pH, temperature, and ratio of GAA to BS and GAA to MFC. The assessments of result were confirmed by degree of substitution (DS), fourier transform infrared (FTIR), scanning electron microscope (SEM), and differential scanning calorimetry (DSC).

2 Materials and methods

2.1 Materials

Cassava bagasse was delivered from small entrepreneurship in Pati. Bamboo feedstock was supplied from Sragen, Central Java. Glacial acetic acid was provided from Merck (100%). Sodium hydroxide was supplied from Merck (99%). Hydrochloric acid with 37% was produced from Mallinckrodt Chemicals-USA. Potasium hydroxide pro analysis 99% was provided from Merck. Phenolphthalein also was supplied from Merck. Hydrogen peroxide (H_2O_2) 50% was produced by Evonik Industries, Indonesia and alcohol 96% were produced by PT. Indo Acidatama, Tbk, Indonesia. Toluene 99.5% was provided from Mallinckrodt Chemicals-USA.

2.2 Characterization

Assessment of products were introduced by FTIR Fourier Transform Infrared Spectroscopy. The measurements were obtained using a FTIR model IR Prestige 21, Shimadzu. Spectra were recorded at a spectral range between 400 and 4000 cm−1 at a scan rate of 40 scans and spectral resolution of 4 cm^{-1}. The FTIR spectrum was employed in the absorbance mode. FTIR analyses were performed to study the effect of the acetylation. The matrix of BS and bamboo MFC were observed surface morphology by using scanning electron microscopy (SEM type Inspect S50, FEI). Thermal properties were determined by using differential scanning calorimetry (DSC 60, Shimadzu). Measurements conducted with a refrigerated cooling system (RCS) and a nitrogen DSC cell purge at 30 ml/min was used to achieve temperatures range of 30 °C to 300 °C.

2.3 Methods

Preparation of cassava bagasse was executed by washing and sizing. The bagasse was dissolved in water and filtered a filter cloth. Filtrat was collected and then left a few days. The water at the top of the precipitate was removed, while the precipitate was collected. The sun dried precipitated slurry. After drying, the starch was then sieved to obtain the cassava bagasse starch with fine powder. This powder was homogenized by using high energy milling for one hour. The bamboo stem was cut and chopped prior sizing into mini crusher releasing bamboo powder. Then powdered bamboo stem was sieved to 150 μm size. The bamboo powder was then washed with 90°C hot water three times to reduce the impurities. The isolation process of bamboo MFC was done by extraction using toluene- alcohol (2:1) for 6 hours. The delignification process for the removal of hemicellulose was conducted with a 2% w/v potassium hydroxide (KOH) solution heated at 90°C with magnetic stirrer. Then, delignified bamboo was filtered and dried at room temperature. Bleaching process removing lignin was executed by 4% v/v hydrogen peroxide (H_2O_2) solution heated at 50°C. Then, it was filtered and washed with aquadest up to neutral pH. Furthermore, the bamboo MFC was dried at 105°C for 24 hours. MFC bamboo was homogenized in size by high energy milling. This preparation was conducted with slight modification of previous research [23-24]. Both acetylation was considered to follow Kumoro & Amalia (2015) procedure [25] and also degree of substitution (DS) for BS and bamboo MFC. Starch solution can be prepared by dissolution of BS in aquadest with 4 wt%. The solution was adjusted certain pH by using 1 M NaOH. This solution was then stirred for 30 minutes. Certain mass of GAA was gradually added to the solution, while controlling at constant a pH to certain value of variables. Acetylation reaction would be accomplished at certain time by pouring 1 M HCl having pH of 5.5. The acetylated products were filtrated and washed the product by using aquadest. Washed product was then dried in oven at 50 °C prior to degree substitution analysis [26]. The certain acetylated product was dissolved in ethanol (50 ml of 75 ml ethanol in 100 ml of water) into flask. Afterward, this solution was agitated for 30 min at temperature of 50 °C. Then, the solution was cooled and added 40 ml of 0.5 M KOH. The residual alkali was back titrated with 0.5 M HCl using phenolphthalein as indicator. A blank, each unacetylated substances was used. Calculation of degree substitution uses the following equation.

$$\%Acetyl = \frac{(V_0 - V_n) \cdot MHCl \cdot 0.043 \cdot 100}{M} \quad (1)$$

$$DS = \frac{162 \cdot \%Acetyl}{4300 - (42 \cdot \%\,Acetyl)} \quad (2)$$

Where V_0 is volume of required HCl for blank sample. V_n is volume of required HCl for sample. Normality of HCl is denoted by N. M is dried sample weight.

3 Results and discussion

3.1 Degree of substitution

Observation of optimum condition (pH, temperature, and ratio of GAA to BC or bamboo MFC) based on degree substitution have been conducted. In this research, at beginning the optimum pH was investigated first at temperature of 40 °C, ratio of GAA to BS or bamboo MFC (2:1) and reaction time of 40 minutes. The optimum pH can be attained at 9, while increasing pH can occur decreasing degree of substitution. Acetylated BS obtained DS of 0.22, while acetylated bamboo MFC gained DS of 0.41 at pH of 9. Acetylation percentage denoted the same trending with. It could be expected for increasing pH releasing hydrolysis reaction regard with decreasing acetyl group and increasing residual acetic acid in the bulk liquid [27-28]. Increasing pH may affect starch to undergo partial gelatinization under alkaline resulting high solubility. High solubility may be denoted to weakening of granule structure during modification [29]. This result can be seen in Fig. 1 below.

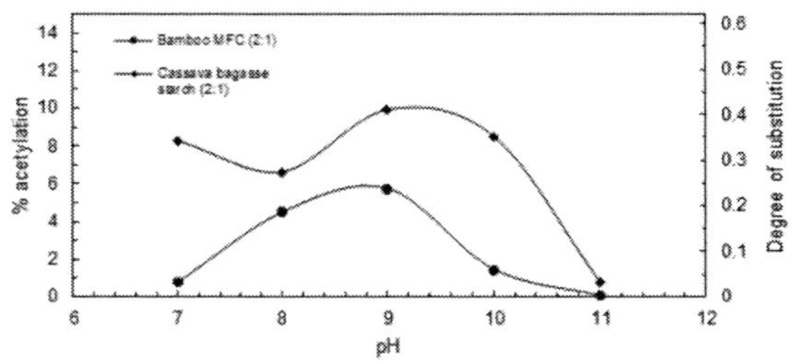

Fig. 1. Effect of increasing pH in acetylation of BS and bamboo MFC with ratio of 2:1 (GAA/BS or bamboo MFC)

By using optimum pH, the acetylation reaction was further conducted to obtain optimum temperature. It can been seen in Fig.2 that optimum temperature can be reached at 60 °C for acetylated bamboo MFC and at 50 °C for acetylated BS. It can be confirmed that due to reversible reaction of acetylation, increasing temperature can increase the solubility of starch (BS). It may be correlated to loss of granular structure and generates of amylase fraction of the starch as the amylase molecules are solubilized and leached from swollen starch (Alam & Hasnain, 2009)

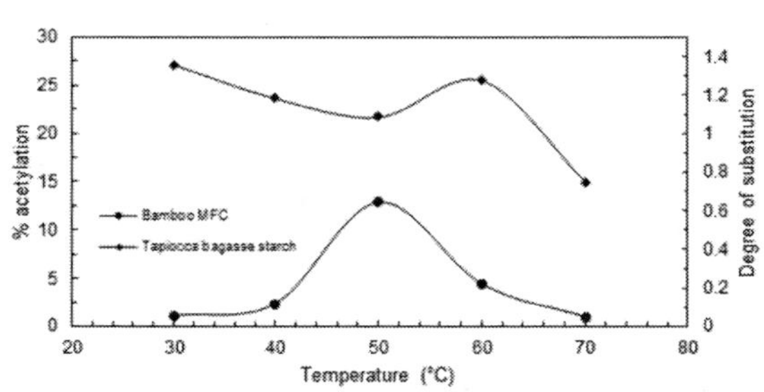

Fig. 2. Effect of increasing temperature in acetylation of BS and bamboo MFC with ratio of 2:1 (GAA/BS or bamboo MFC)

Optimum pH and temperature were used on next observation to obtain optimum ratio of GAA to BS or bamboo MFC. Fig. 3 represented that optimum DS and also % acetylation of GAA over BS or bamboo MFC can occur at 2:1. At beginning, increasing concentration of acetic acid (GAA) released higher DS (0.93 for bamboo MFC and 0.47 for BS), but unfortunately afterward the DS steep down to 0.5 and 0.1 for bamboo MFC and BS, respectively. It can be seen more acid concentration, more residual acid in the bulk phase. In the beginning, the acid can diffuse and withstand adsorption to the BS and MFC. However, for higher amount of BS or MFC did not occur more acetylation regard with less amount of the acid concentration in the bulk phase.

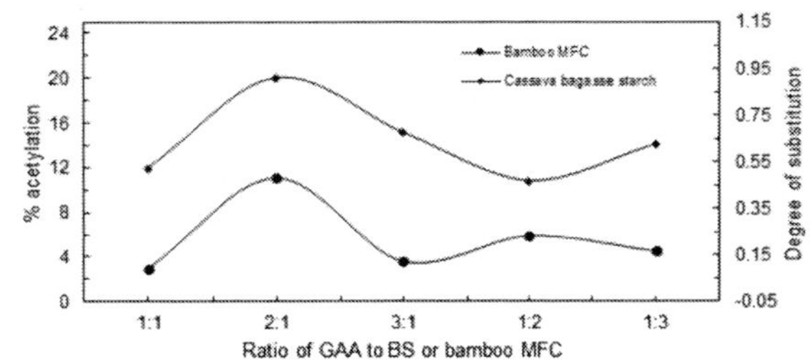

Fig. 3. Effect of increasing reaction time in acetylation of BS and bamboo MFC with pH of 9 and ratio of 2:1 (GAA/BS or bamboo MFC)

3.2 Thermal analysis

Table 1 shows the effect of acetylation on the thermal analysis of the properties of BS or bamboo MFC. The onset (T_o), peak (T_p), and final (T_c) temperature of gelatinization were lower in the presence of acetyl group although the enthalpy (ΔH) was hardly affected by acetylation. This shift to lower temperature for gelatinization indicated that BS or bamboo MFC gelatinized easily with the addition of acetyl group. DSC result of acetylated BS and acetylated bamboo MFC can be seen in Fig.4 and Fig.5, respectively.

Table 1. DSC characteristics of native and acetylated BS or bamboo MFC.

	Endothermal temperature (°C)			ΔH (J/g)
	T_0	T_P	T_C	
Acetylated MFC	65.23	101.27	133.81	173.74
Native MFC	81.15	111.2	142.60	183.44
Acetylated BS	82.15	109.21	155.96	207.15
Native BS	86.3	112.24	159.27	209.34

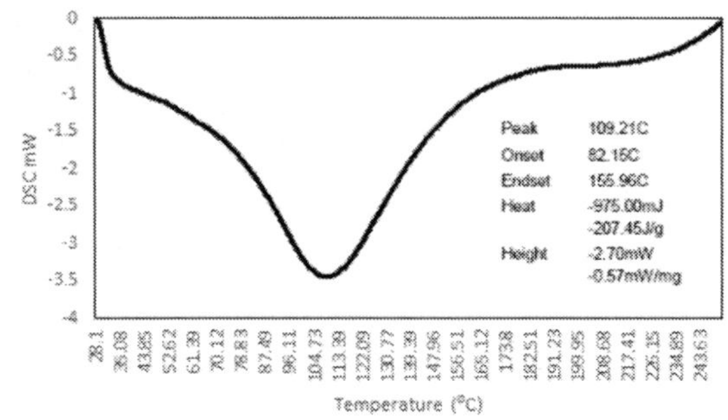

Fig. 4. DSC analysis of acetylated BS

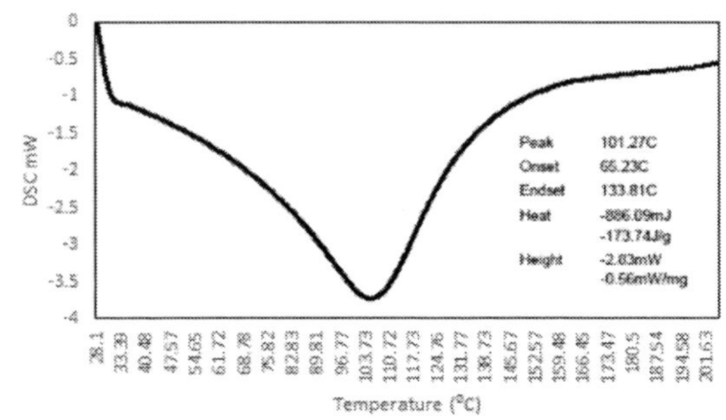

Fig. 5. DSC analysis of acetylated BS

3.3 SEM analysis

Native BS granules have a regular elliptical shape and smooth surface while native MFC have thin flat. After acetylation, both resulted deformation of granules or flat due to the temperature of acetylation as result of the structure reorganization. It is recognized with Fig.6(a) to Fig.6(d). These figures confirm with native bamboo MFC, native BS, acetylated bamboo MFC, and acetylated BS, respectively.

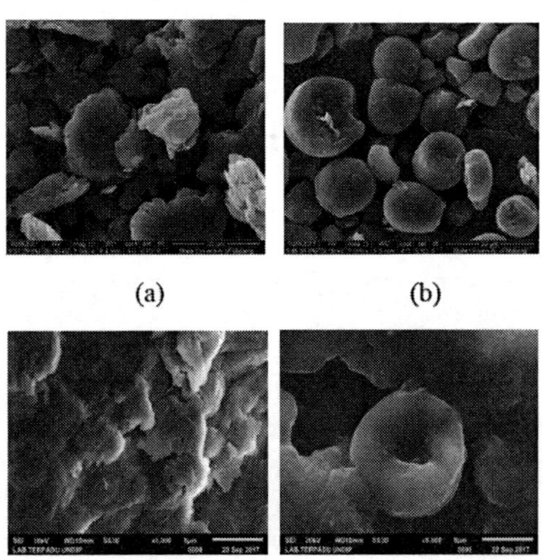

(a) (b)

(c) (d)

Fig. 6. SEM analysis of native bamboo MFC (a), native BS (b), acetylated bamboo MFC (c), and acetylated BS (d) with magnification of 5000x

3.4 FTIR analysis

Acetylated BS/MFC denoted lower absorbance value than that of native BS/MFC expressing decrease of functional group number while in the product did not have significant acetyl group at 1750 cm^{-1} only a slight peak around 1732 cm^{-1} for acetylated MFC and none acetyl group for acetylated BS only decrease in absorption of OH group. These results can be visualized in Fig.7. and Fig.8. for native and acetylated BS, then native and acetylated bamboo MFC, respectively.

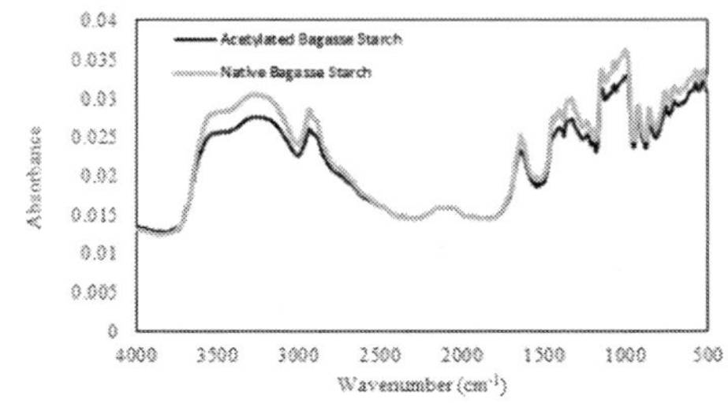

Fig. 7. FTIR analysis of native and acetylated BS

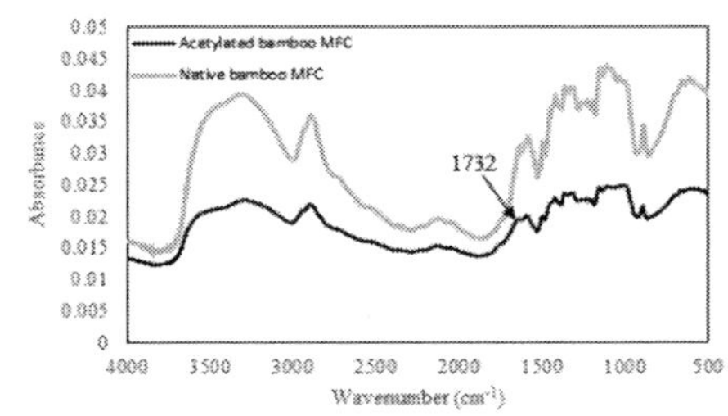

Fig. 8. FTIR analysis of native and acetylated bamboo MFC

3 Conclusion

The optimum condition of acetylation such as pH, temperature, and ratio of GAA to BS and GAA to MFC can be further used to observe the kinetic mechanism of acetylation for BS and bamboo MFC to know rate controlling step of the acetylation, i.e. regime of mass transfer or chemical reaction. The acetylation for BS and bamboo MFC can affect physicochemical, and thermal properties, in terms of the structural, the thermal, and the morphology. It was confirmed by FTIR, DSC, and SEM analysis.

References

1. Pelissari, F. M., Andrade-Mahecha, M. M., Sobral, P. J. D. A., & Menegalli, F. C., Food Hydrocoll. **30**, 2 (2013)
2. Menzel, C. Doctoral Thesis (2014)
3. Al-Hassan, a. a., & Norziah, M. H., Food Hydrocoll. **26**, 1 (2012)

4. Silviana, S. & Anggoro, D. D., Adv. Sci. Lett. **23**, 3 (2017)
5. Silviana, S. & Hadiyanto, H., AIP Conf. Proc. **1885**, 030024 (2017)
6. Dias, A. B., Müller, C. M. O., Larotonda, F. D. S., & Laurindo, J. B., J Cereal Sci. **51**, 2 (2010)
7. Mendes, J. F., Paschoalin, R. T., Carmona, V. B., Sena Neto, A. R., Marques, a. C. P., Marconcini, J. M., Mattoso, L.H.C., Medeiros, E.S., Oliveira, J. E., Carbohydr Polym. **137** (2016)
8. Tajik, S., Maghsoudlou, Y., Khodaiyan, F., Jafari, S. M., Ghasemlou, M., & Aalami, M., Carbohydr Polym. **97**, 2 (2016)
9. Souza, a. C., Benze, R., Ferrão, E. S., Ditchfield, C., Coelho, a. C. V, & Tadini, C. C., LWT-Food Sci Technol. 46, 1 (2012)
10. Teixeira, E. D. M., Curvelo, A. A. S., Corrêa, A. C., Marconcini, J. M., Glenn, G. M., & Mattoso, L. H. C., Ind Crops Prod. **37** (2012)
11. Acosta, S., Jimenez, A., Chafer, M., Gonzalez-Martinez, C., & Chiralt, A., Food Hydrocoll. **49** (2015)
12. Guimarães, M., Botaro, V. R., Novack, K. M., Teixeira, F. G., & Tonoli, G. H. D., Ind Crops Prod. **70** (2015)
13. Martins, I. M. G., Magina, S. P., Oliveira, L., Freire, C. S. R., Silvestre, A. J. D., Neto, C. P., & Gandini, A., Compos. Sci. Technol. **69**, 13 (2009)
14. Lomelí-Ramírez, M. G., Kestur, S. G., Manríquez-González, R., Iwakiri, S., de Muniz, G. B., & Flores-Sahagun, T. S., Carbohydr Polym. **102** (2014)
15. Sánchez, S. A., Lopes, Y., Araújo, R., & Cohim, E., RSER **73** (2017)
16. Pandey, A., Soccol, C. R., Nigam, P., Soccol, V. T., Vandenberghe, L. P. S., & Mohan, R., Bioresour. Technol. **74**, 1 (2000)
17. Abdul Khalil, H. P. S., Alwani, M. S., Islam, M. N., Suhaily, S. S., Dungani, R., H'ng, Y. M., & Jawaid, M., In book: Biofiber Reinforcements in Composite Materials, Chapter 16, Woodhead Publishing USA (2015)
18. Thakur, V. K., & Thakur, M. K., Carbohydr Polym. **109** (2014)
19. Bahari, S. A., & Krause, A., J Clean Prod. **110** (2016)
20. Lu, T., Liu, S., Jiang, M., Xu, X., Wang, Y., Wang, Z., Gou, J., Hui, D., Zhou, Z., Compos Part B Eng. **62** (2014)
21. Liu, D., Zhong, T., Chang, P. R., Li, K., & Wu, Q., Bioresour. Technol. **102** (2010)
22. Bulota, M., Hughes, M., & Paltakari, J., J Appl Polym Sci. 126 (2012)
23. Chen, H., Wang, W., Martin, J. C., Oliphant, A. J., Doerr, P. a., Xu, J. F., Deborn, K.M., Chen, C., Sun, L., ACS Sustain Chem Eng. **1**, 2 (2013)
24. Nguyen, H. D., Thuy Mai, T. T., Nguyen, N. B., Dang, T. D., Phung Le, M. L., Dang, T. T., & Tran, V. M., Adv. Nat. Sci.: Nanosci. Nanotechnol. **4**, 1 (2013)
25. Kumoro, A.D. & Amalia, R., BCREC 10, 1 (2015)
26. Sodhi, N. S., & Singh, N., J Food Eng. **70** (2005)
27. Song, X., He, G., Ruan, H., Chen, Q. Starch/Stärke, **58** (2006)
28. Luo, Z.G., Zhou, Z., Starch/Stärke, **64** (2012)
29. Alam, F., & Hasnain, A., Agric. conspec. sci. **74**, 1 (2009)

Chemical Modifications for Intensity Variation and Spectrum Extension of Brazilein Extract from Sappanwood (*Caesalpinia sappan L.*)

Edia Rahayuningsih[1,*], *Wiratni* Budhijanto[1], *Hana Fitria* Prasasti[2], and *Meyta Tias* Wahyuningrum[2]

[1]Bioresource Engineering Group, Chemical Engineering Department, Universitas Gadjah Mada, Jl. Grafika 2 Yogyakarta, Indonesia
[2]Undergraduate Study Program, Chemical Engineering Department, Universitas Gadjah Mada, Jl. Grafika 2 Yogyakarta, Indonesia

Abstract. Aqueous extract of sappanwood (*Caesalpinia sappan L.*) was treated using acid-base and mordant additions in order to expand the intensity and spectra of the colors. The colorant molecule of sappanwood was brazilein, which gained its color characteristics through its double bonds. The results of this study confirmed three basic mechanisms which formed the fundamental for systematic optimization of natural dye applications. Acid addition induced protonation of the hydroxyl groups in the brazilein structure so that the electrons were distributed in the molecule more evenly. Therefore, the color intensity was reduced. Basic treatment created deprotonation of the hydroxyl groups and hence localized the electron on several spots. This made the color of the extract shifted to deeper tones. The addition of mordant created more significant change in color spectrum through the mechanism of chelation to form coordinated complexes. Transitional metals from group 1 and 2 did not chelate strongly so that the spectrum shift was not very obvious. However, transitional metals from group 3 had strong chelating character and consequently they could strongly bind the dye molecules to increase its intensity and shift the spectrum towards deeper colors.

1 Introduction

The plant of sappanwood or "kayu secang" in Bahasa Indonesia (*Caesalpinia sappan L.*) is considered a common plant in tropical region, including Indonesia. It is known as medicinal plant with various benefits. The claim of its pharmacological effect ranges from anti-inflammatory up to cancer cell inhibitor [1]. The heartwood of this sappanwood (Fig. 1) is known as the source of brazilin (Fig. 2a) and brazilein (Fig. 2b) (the oxidated form of brazilin), which are both the active compounds in the wood to have the pharmaceutical benefits.

Fig. 1. Shaved heartwood of sappanwood

Brazilin is colorless, but brazilein bears natural color widely used in textile dyeing practices as water-soluble natural dye. This natural color had been detected in ancient fabrics to prove its attractiveness as textile dyes. Although it has been known and used for such a long time, its popularity still even increases into the modern era with wider uses as food colorant, pH indicator, and dye-sensitized solar cell [2].

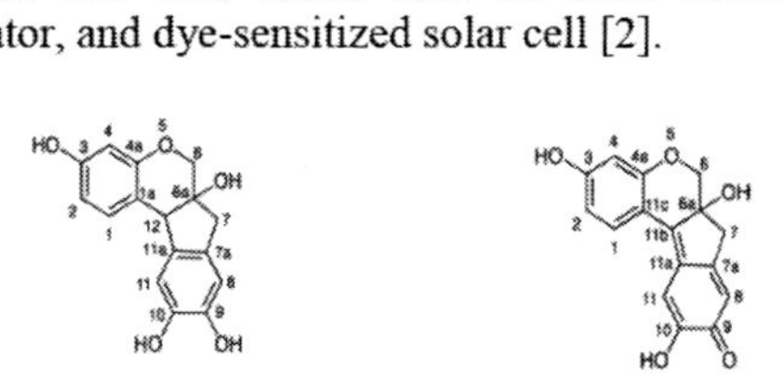

(a) Brazilin (b) Brazilein

Fig. 2. Structure of brazilin and brazilein as the active compounds in sappanwood extract [3]

Brazilein color spectra cover the range of yellow-red depending upon the pH of the solution and the mordanting agents used in the dyeing process [3], [4]. Mordant is the chemicals added in the dyeing process to bind the chromophores into the textile fibers. Besides fixation function, mordants, which are the salts of transition metals, also play important role in changing the depth of the color and creating various shades of the same origin [5]. According to the classic *Reichs-Ausschuß für Lieferbedingungen* (RAL) system for color standardization, the possible colors obtainable from sappanwood extract are presented in Fig. 3. Both the spectrum and the intensity of the colors could be created from the same original extract of sappanwood by changing pH and addition of metal ions of the mordant salts [5], [6].

* Corresponding author: edia_rahayu@ugm.ac.id

Fig. 3. Classic RAL System for color standard

Many studies have been published regarding the application of natural dyes for various uses [3], [6]. Other studies focused on the optimization of the extraction procedure, such as the rigorous methods to separate brazilin and brazilein [1], [2]. Still limited numbers of studies were focused on the color characteristics of the natural dyes. This paper presented a study on the effort to expand the intensity and spectra of the sappanwood extract. The study aimed to define the effect of pH and mordant salts in a systematic way so that the prospective users could decide which procedures to follow for their particular purposes.

2 Research methodologies

2.1 Materials

The raw material of sappanwood (*Caesalpinia sappan L*) was obtained from Beringharjo Market (Yogyakarta) in the form of dried thin shaved wood. Citric acid and acetic acid were used for acid modification while sodium hydroxide was chosen for the basic treatment. All chemicals were obtained from Merck at technical grades. Impurities of the chemicals were not disturbing the accuracy of the color formation because the solutions were quite diluted by water. For other variations of spectra, four chemicals were also tested, i.e. Cu-Acetate ($Cu(CH_3COO)_2$), $MgSO_4$, $FeSO_4$, and $Al_2(SO_4)_3$, which were all analytical grade from Sigma-Aldrich.

2.2 Experimental set up

The experiments were carried out in laboratory scale batch stirred reactor equipped with mantle heater and condenser (Fig. 4).

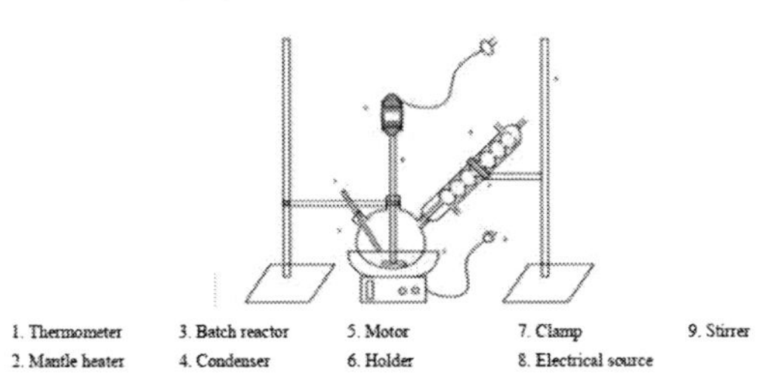

Fig. 4. Experimental set up

The experiments were run in two steps. The first step was brazilein extraction from the shaved sappanwood which was blended first with the solvent using high speed liquid blender. Extraction was run at its optimum conditions [7], which were previously obtained with aquadest as solvent, using the concentration of solid in the solvent of 0.002 g/L, with the mixing speed of 300 rpm, and 70 minutes of extraction at constant temperature of 100°C. The second step was the improvement of the extract, in term of its color spectrum and intensity by chemical modification using additives. Two routes were tested for the chemical modification, which were acid-base reaction and cation reaction.

Chemical modification using acid-base addition was conducted to check the color characteristics under different pH values [8]. For acidic pH, 15 mL of citric acid or acetic acid (both at 0.1 M concentration) was added into 100 mL of original extract, while for basic pH, 15 mL of sodium hydroxide or calcium hydroxide solution (both at 0.1 M concentration) was added into 100 mL of original extract. Other type of additive tested in this study was cation addition. Solutions of Cu-Asetat, $MgSO_4$, $Al_2(SO_4)_3$ and $FeSO_4$ were used as the provider of cations.

After the process with each method, the absorbance of the modified extract was quantitatively measured using UV-Vis spectrophotometer. Spectrophotometer was chosen over chromometer because it offered wider and more flexible measurement range coverage of spectra. The data were recorded as the absorbance measured at the optimum wavelength for each color spectrum, i.e. the wavelength which gave the minimum transmittance for a particular sample. Scanning of optimum wavelength was conducted by measuring the absorbance of each sample at various wavelengths.

3 Results and discussions

3.1 Acid-base modifications

3.1.1 Addition of acids

According to the Classic RAL System, the color of the original extract was categorized as traffic red. Both acid addition, i.e. acetic acid and citric acid, shifted the color category into dahlia yellow. The visual comparison of the colors is presented in Fig. 5. For more objective comparison, Fig. 6 presents the absorbance data of the original and modified extracts.

The absorbance data in Fig. 6 reveals that all extracts exhibits the highest absorbance at the same wavelength, i.e. 440 nm (indicated by vertical line on Fig. 6). This fact translates into more objective interpretation than the visual observation, that acid addition does not change the spectrum of the color, although visually the colors look different. The change caused by acid addition was actually on the intensity of the colors. Fig. 6 shows that the addition of acids significantly lowered the color intensity of the original extract.

https://doi.org/10.1051/matecconf/201815601020

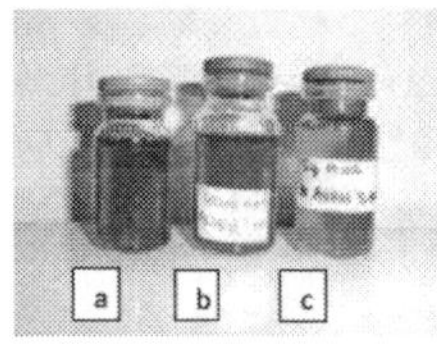

Fig. 5. Visual comparison of original extract (a) with the citric acid added extract (b) and acetic acid added extract (c)

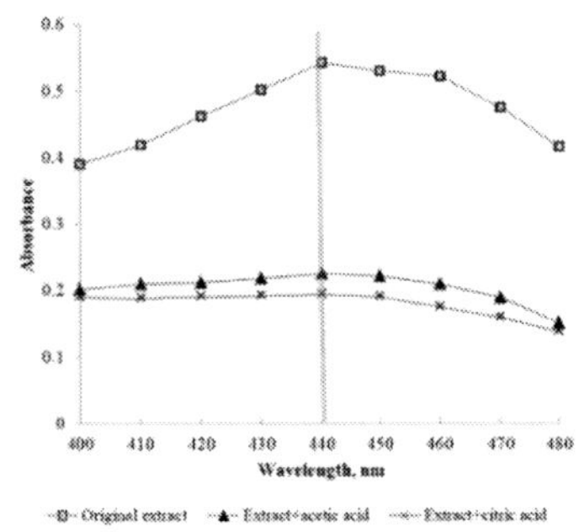

Fig. 6. The change of color intensity by acid addition into original extract of sappanwood based on UV-Vis absorbance data

Both Fig. 5 and Fig. 6 show that different acids (citric acid and acetic acid) did not cause different spectrum or different intensity. This confirms that the change of intensity by acid addition was caused by the H+ protons so that different acyl (RCO-) group arrangement of the acids did not matter. The effect of H^+ proton concentration was then tested by the addition of different amount of citric acids, which gave the variations of color intensity presented in Fig. 7 (for citric acid addition) and Fig. 8 (for acetic acid addition). With the extra H^+ protons available in the solution from acid partial dissociation, the hydroxyl groups of the brazilein became protonated [9]. As a result of that, the initially localized electrons in the π-conjugated orbitals of the double bonds in brazilein molecule became somehow delocalized [2]. Hence, slightly different shade of color was observed.

Slight difference in H^+ proton concentration was created by the addition of different amount of acid solutions, i.e. 0.06 mL, 0.15 mL, and 0.3 mL in 10 mL of original extract. Fig. 7 and Fig. 8 show that color intensity was not very sensitive to H^+ proton concentration as the mechanism of color changes was only due to electron delocalization.

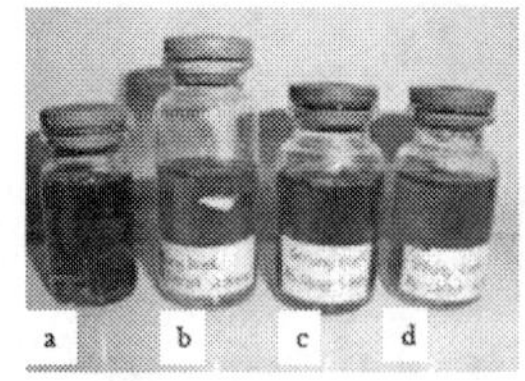

Fig. 7. Visual comparison of original extract (a) with citric acid added extract: 0.06 mL (b), 0.15 mL (c), and 0.30 mL (d) of citric acid into 10 mL of original extracts

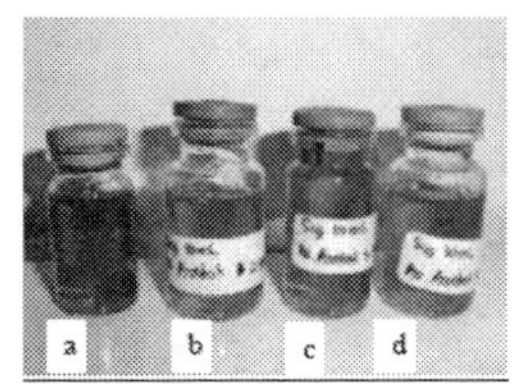

Fig. 8. Visual comparison of original extract (a) with acetic acid added extract: 0.06 mL (b), 0.15 mL (c), and 0.30 mL (d) of acetic acid into 10 mL of original extracts

3.1.2 Base addition

The addition of bases into the original extract of sappanwood gave totally different results from the acid additions. Both bases tested in this experiment, i.e. NaOH dan $Ca(OH)_2$, shifted the originally traffic red of the original extract into ruby red. Fig. 9 presents a visual comparison between the original extract and the base-added extract, which used NaOH as the base. The addition of $Ca(OH)_2$ led to the same color change, however it was not suggested as it turned out that $Ca(OH)_2$ solution induced precipitation of the extract. The absorbance analysis was then only conducted on NaOH added extract and the result is presented in Fig. 10.

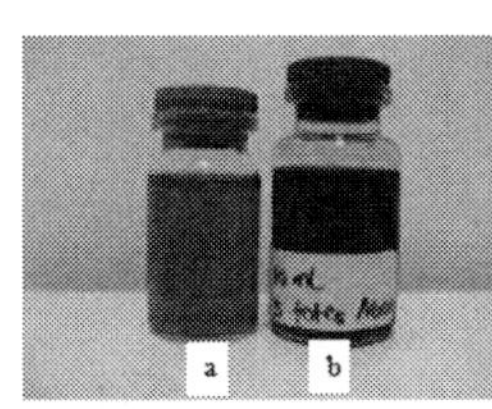

Fig. 9. Visual comparison of original extract (a) with NaOH added extract (b)

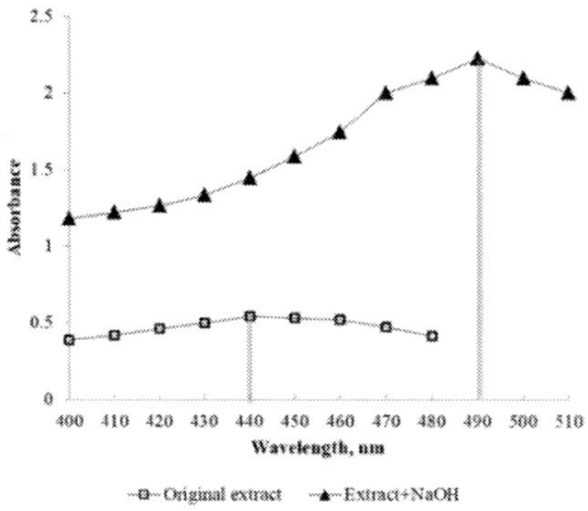

Fig. 10. The change of color intensity and color spectrum by base addition into original extract of sappanwood based on UV-Vis absorbance data

Fig. 10 exhibits that the addition of NaOH solution induced not only the change of color intensity but also shifted the wavelength which gave the maximum absorbance, from 440 nm for the original extract into 490 nm. It means that the color resulted from base addition was of quite different spectrum from the original extract. Such change could only be caused by the change of the the electron distribution caused by the addition of NaOH, which shifted the pH into basic environment.

The hydroxyl ions from the base dissociation created deprotonation of the hydroxyl groups in the brazilein molecules. Consequently, there were spots with high density electrons in some parts of the molecules. This variations in electron localization created different shades of colors. Further tests were conducted to strengthen this speculation, i.e. the addition of different concentrations of NaOH solution (Fig. 11). In Fig. 11, referring to the RAL system of color standardization, the addition of 0.06 mL of NaOH resulted in signal red, 0.15 mL addition of NaOH solution gave ruby red, and 0.3 mL of NaOH addition made pearl ruby red. In contrast to the protonation of hydroxyl groups in the previous section which did not change the color shade, the deprotonation occurred in the brazilein molecules more significantly affected the color spectrum of the extract.

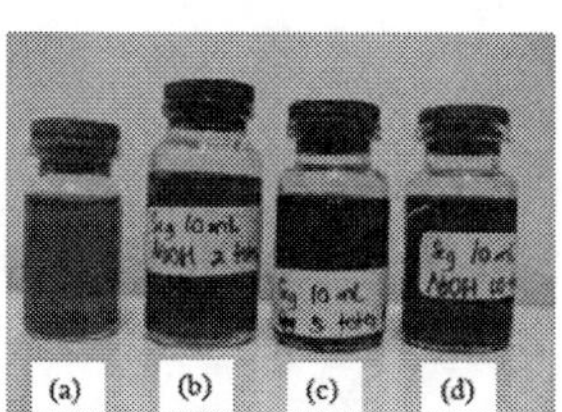

Fig. 11. Visual comparison of original extract (a) with NaOH added extract: 0.06 mL (b), 0.15 mL (c), and 0.30 mL (d) of NaOH solution into 10 mL of original extracts

3.2 Effect of cations from mordant salts

Four transition metal cationic solutions of mordant salts were tested, which were $MgSO_4$, $(CH_3COO)_2Cu$, $FeSO_4$, and $Al_2(SO_4)_3$. As being displayed in Fig. 12, the addition of $MgSO_4$ solution or $Al_2(SO_4)_3$ solution into the original extract did not change the color spectrum as it was still traffic red. Other color, claret violet (which gave optimum absorbance at 510 nm of wavelength), was resulted when $(CH_3COO)_2Cu$ solution was added with dilution. Addition of $FeSO_4$ solution (also with dilution) gave red lilac color (optimally absorbing light at 500 nm of wavelength). The various spectra created by cation addition from the solutions studied in this research were presented in Fig. 13.

The new spectra found by the addition of mordant salts were broader than those obtained in acid-base equilibria explained in the previous section. Transitional metals were known for their ability of chelating, i.e. complex formation with organic compounds. The d and f orbitals of the transition metal were only partially filled so that it was easy to form bonding with any nucleophiles, although the bonding could be weak or strong, depending upon the period of the metal [2]. It can be noticed in Fig. 12 that most intense colors were caused by Fe^{2+} and Cu^{2+} cations, while Al^{3+} and Mg^{2+} did not induce such strong colors (similar to the color with the presence of Na^+ described in the previous section). These results were in good accordance to the theoretical reasons of the chelating strength of different groups of transitional metals [5]. Group 1 cation (Mg^{2+}) and group 2 cation (Al^{3+}) did not have strong tendency to form complexes with the brazilein so that they failed to concentrate the electrons to create deeper colors. On the other hand, group 3 cations (Fe^{2+} and Cu^{2+}) were known for their ability to form coordinated complexes with brazilein. Up to 6 molecules of brazilein could be bond around one cation and hence it induced very deep color.

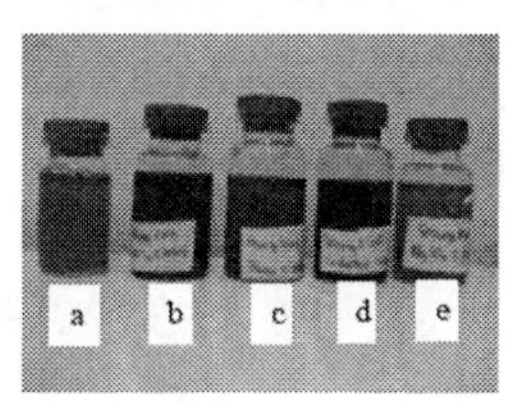

Fig. 12. Visual comparison of original extract (a) and with addition of $FeSO_4$ (b), $Al_2(SO_4)_3$ (c), $(CH_3COO)_2Cu$ (d), and $MgSO_4$ (e)

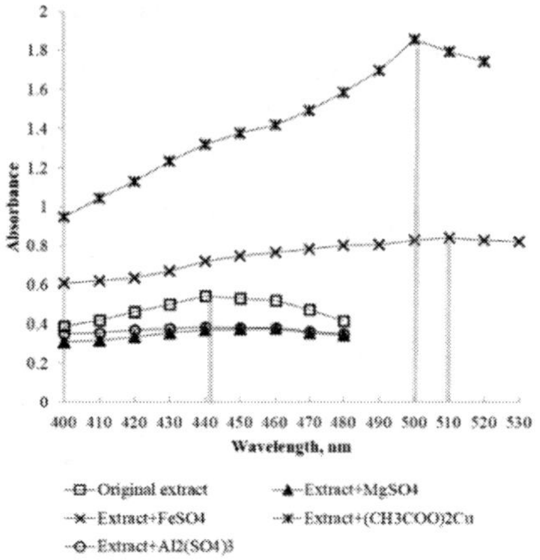

Fig. 13. The change of color intensity and color spectrum by cation addition into original extract of sappanwood based on UV-Vis absorbance data

4 Conclusions

The phenomenological study on the efforts of intensifying the color and expanding the spectra of sappanwood extract raised important understandings as the following:

a. Acid addition changed color intensity but did not affect the spectrum. The intensity variations might be dominantly influenced by the H^+ proton from the dissociated acids, which induced protonation of the hydroxyl groups in the brazilein structure. The protonation created delocalization of electrons so that they were not concentrated on the double bonds and hence the colors of the extract turned lighter. The alkyl group of the organic acids tested in this study had no influence on both color intensity and spectrum.

b. Base addition not only changed the intensity of the color but also expanded the spectra. In this case, the hydroxyl ion from the base caused deprotonation of the hydroxyl groups in the brazilein structure. Consequently, the electrons became localized around the deprotonated hydroxyl groups and induced deeper colors.

c. Different cations created different shades of colors. The sources of cations were usually mordant salts, which were salts from transitional metal. Metals from group 1 and 2 (such as Mg^{2+} and Al^{3+}) might be useful as mordant (to fix the color onto the fiber), however, they did not have the chelating strength to form complexes with the brazilein. They could not concentrate dye molecules to create deeper colors. On the other hand, metals from group 3 (Fe^{2+} and Cu^{2+}) could form strong coordinated complexes with the brazilein. Therefore, Fe^{2+} and Cu^{2+} were not only functioning as the mordant but also could create deeper colors as well.

References

1. N. P. Nirmal, M. S. Rajput, R. G. S. V. Prasad, and M. Ahmad, Asian Pac. J. Trop. Med., **8**(6), 421–430, (2015).
2. G. Richhariya, A. Kumar, P. Tekasakul, and B. Gupta, Renew. Sustain. Energy Rev., **69**, 705–718, (2017).
3. R. Rondão, J. S. Seixas De Melo, J. Pina, M. J. Melo, T. Vitorino, and A. J. Parola, J. Phys. Chem. A, **117**(41), 10650–10660, (2013).
4. A. Manhita, V. Santos, H. Vargas, A. Candeias, T. Ferreira, and C. B. Dias, J. Cult. Herit., **14**(6), 471–479, (2013).
5. A. Moiz, M. Aleem Ahmed, N. Kausar, K. Ahmed, and M. Sohail, J. Saudi Chem. Soc., **14**(1), 69–76, (2010).
6. S. Masood, M. A. Randhawa, M. S. Butt, and M. Asghar, J. Food Process. Preserv., **40**(1), 3–13, (2016).
7. M. T. Wahyuningrum, *Penentuan Kondisi Operasi Optimum Ekstraksi Brazilin dari Kayu Secang*, No Title," Yogyakarta, (2003).
8. H. . Prasasti, *Peningkatan intensitas dan perluasan spektrum warna dari ekstrak kayu secang (Caesalpinia sappan L)*, Yogyakarta, 2013.
9. G. Solomons, C. Fryhle, and S. Snyder, "Org. Chem., 1255, (2014).

Performance of Deacetyled Glucomannan as Iron Encapsulation Excipient

Dyah H. Wardhani[1,*], *Heri* Cahyono[1], *M. Farkhan* H. Dwinanda[1], *Putri R.* Nabila[1], *Nita* Aryanti[1], *Dina* R. Pangestuti[2]

[1]Chemical Engineering Department, Faculty of Engineering, Diponegoro University, Semarang, Indonesia
[2]Public Health Department, Diponegoro University-Semarang, Indonesia

Abstract. Encapsulation protects iron from degradation or oxidation possibilities due to its encapsulation material. Glucomannan (GM) is a neutral polysaccharide consist of D-mannose and D-glucose connected with β-1,4 linkage. Deactylation transforms solubility of glucomannan as well as its gel structure. These properties support for excipient application. The aim of this work was to determine performance of deacetylated glucomannan as iron matrix. Deacetylation was conducted heterogeneously. Deacetylation did not change the backbone of GM. Higher alkali concentration has better ability to encapsulate iron. Extended deacetylation time and alkali concentration affect insignificantly on the performance of encapsulation to protect iron from oxidation. The release of iron from the matrix influences by deacetylation degree.

1 introduction

Iron is an element which is very abundant its availability in nature and one of essentials components of every living creature. In the human body, Iron (Fe) is essential for producing red blood cells, i.e. haemoglobin which contributes in oxygen bonding and then it is transported throughout the body. Iron can be obtained from two types of iron, i.e. heme and non-heme. Under certain conditions, Iron is vulnerable to damage due to inhibitors. Besides fitic acid and tannin, some research revealed the inhibitors of Iron is calcium and peptide from digested protein [1].

The requirement of Fe per day varies for each individual. However, the normal Iron diet can not satisfy the Iron requirement which is caused by low its bioavailability (the ability of the body to absorb Iron), i.e. 14% - 18% [2]. Hence, it is needed an additional Iron intake in the form of supplements in order to satisfy the body requirements of Fe. Iron intake which is required by the body can be obtained by consuming the Iron supplements.

In the recent days, there are many variants of supplements form which are used to supply the Iron, for examples capsules, syrup, and even intravenous Iron therapy. One of the forms that is currently developed is encapsulated Iron. Encapsulation aims for giving the maximum protection from the factors of Iron damage, and also minimalise the less common smells and tastes of Iron [3][4].

The study of Iron encapsulation has been conducted by Bezbaruah et al. with modified alginate as an encapsulation material [5]. They used $CaCl_2$ which contributes as strengthen agent and alginate particles stabiliser to make alginate insoluble in water. Iron encapsulation which employs a gelling agent such as modified alginate reduce organoleptic problems and increase the Iron bioavailability [2]. Bioavailability is the Iron ability to be absorbed by the body. Iron encapsulation can increase the bioavailability due to the coating material protects the Iron from degradation by inhibitors and oxidation.

Glucomannan has high viscosity and the molecular weights ranging from 200 to 2000 kDa. This mannan has ability to form a very good coating, stable in cold and hot water, and the ability to absorb water with high absorption relativity index, i.e. 100 gram water per gram glucomannan [6][7], however, it becomes new problems. The ability of glucomannan to absorb water needs to be controlled or reduced through chemical modification, for example through deacetylation. Glucomannan modification using deacetylation is expected as a proper method to be chosen because it reduces the solubility of glucomannan in water, and also minimalise the Fe^{2+} oxidation into Fe^{3+} as the result of Iron contact with O_2 in the air [8]. This modification method is expected to prevent the degradation by inhibitors. The deacetylation

* Corresponding author: dhwardhani@che.undip.ac.id

alters glucomannan form into stronger, elastic, and stable in extreme pH and high temperature [9]. Deacetylated glucomannan could be more difficult to dissolve in water and form gel structures [10]. By forming a gel structure which is a stable solid phase, encapsulation using KGM is more likely to be performed because it can wrap the active agent, insoluble in water, and has a higher boiling point [11]. The increase of deacetylation degree (DD) will cause the glucomannan insoluble in water [12]. These properties affect on Fe^{2+} oxidation into Fe^{3+} as the result of interaction with water which contains dissolved oxygen [13].

The ability of KOH as deacetylation agent is weaker than NaOH because the ionisation degree of KOH is lower than NaOH in the alcohol medium [14. Deacetylated glucomannan using KOH has higher viscosity compared to that of NaOH [15]. The utilisation of glucomannan of A. oncophyllus itself as an excipient of iron has not much explored. Hence, this work aimed to study the effect of deacetylation conditions of glucomannan of A. oncophyllus on iron encapsulation performance. Effects of concentration of alkali (0.1 and 0.25 M) and deacetylation reaction time (2, 8, 16 and 24 h) on the efficiency of encapsulation of Iron, oxidation protection performance and its release were studied.

2 Materials and Methods

2.1. Materials

A. oncophyllus flour was bought from a local farmer at Nganjuk, East Java-Indonesia. Glucomannan was extracted from A. oncophyllus flour following Rahayu method's using 2-propyl alcohol solution (IPA, 70%) [16]. The glucomannan content of the purified flour was 76.3%. KOH and ethanol (95%) were used for deacetylation reaction and washing the deacetylated glucomannan, respectively.

2.2. Deacetylation of Glucomannan

The deacetylation was conducted based on the method of Chen et al. [17]. Ethanol (50%, 100 ml) and glucomannan (5 g) were placed in beaker glass and stirred 250 rpm for 30 min at 50°C. A particular concentration of KOH was added to the mixture while maintained the stirring (225 rpm) for a time corresponding to the variable. The solid which is separated using filter paper was oven-dried at 40°C to obtain deacetylated glucomannan (DGM).

2.3. Functional Groups and Morphology

The morphology and functional groups of the sample after deacetylation was analysed using FEI Inspect S50 device and IR Prestige Shimadzu. The peaks of sample were assigned by comparison with the literature data.

2.4. Formation of Fe bead

The dried deacetylated glucomannan (2 g) for each variable was placed into a beaker glass containing 100 ml of distilled water. $FeSO_4$ (1 g) was added to the mixture. The mixture was dropped into the ethanol (50%) using a syringe. The bead was collected using filter paper, and oven-dried at 40°C to obtain encapsulated Fe.

2.5. Determination of iron

The sample (0.1 g) is placed in a beaker glass together with HCl 6M (10 ml), then heated for 1 h. The solution is filtered and the filtrate is fed into a 10 ml flask. Added distilled water to the boundary marks and homogenized. To each flask (including the distilled water "blank" and the unknowns), pipet in order 1 ml of the hydroxylamine solution,10 ml of the 1,10-phenanthroline solution, and 8 ml of the sodium acetate solution. Swirled each flask to mix the contents, then dilute each solution to the 100 ml mark and mix thoroughly. The solution was shaken and silenced for 120 min, then measured its absorbance using a UV-Vis spectrophotometer at 520 nm.

2.6. Efficiency of Encapsulation

The determination of the efficiency of encapsulation was calculated as follow:

$$Efficiency\ of\ encapsulation = \frac{Bound\ iron}{Total\ iron} x\ 100\% \quad (1)$$

2.7. Encapsulation Protection

The effect of encapsulation again the oxidation effect was determined at ambient temperature. After 5 days incubation, the samples was analyzed for the iron content. The iron calculations was compared before and after exposure for oxidation determination.

2.8. Release of Iron

The in vitro release of iron from its excipient was conducted at free-enzymatic liquid prepared using 0.1 M HCl to simulate the stomach liquid at pH 1.2.[18] This simulated liquid was incubated at 37°C, 150 rpm. After a certain time, some liquid are took and measured for Fe determination. [19]

3 Results and Discussion

3.1. Morphology and Functional Groups of the Deacetylated Glucomannan

Figure 1 shows the FT-IR spectra comparison of the native glucomannan and the deacetylated sample. In general, the peaks of both samples are very similar but difference in the amount of absorbance. This indicates that glucomannan backbone was not changed during deacetylation. The samples had the broad absorption band at around 3400 cm^{-1} and was attributed to the stretching vibration of O-H groups [20]. This absorption band shifted to a lower wave number with the present of alkali, indicating a gradual increase of intermolecular hydrogen bonding between glucomannan chains. It can be clearly seen that deacetylation decrease the peak at approximate 1730 cm^{-1} which have been assigned to the stretching of C=O of the carbonyl of the acetyl groups, as found in the previous work [21][22].

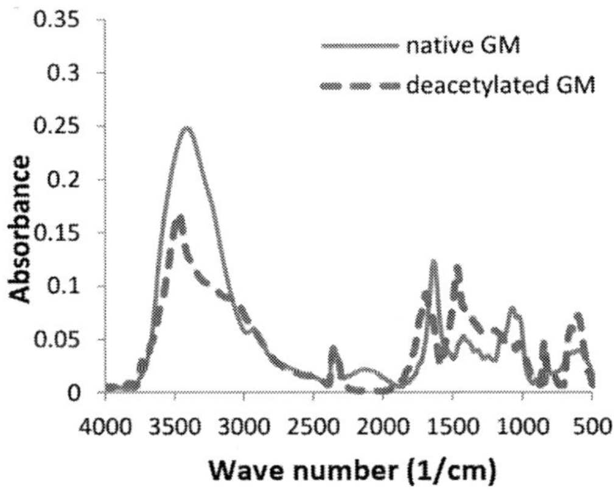

Fig. 1. Comparison of FTIR spectra of native and deacetylated glucomannan.

As a consequence of deacetylation, there was also a significant increase in the intensity of the band in the region of ~1450 and ~880 cm^{-1}, as well as the sift of ~1800 cm^{-1}. This observation suggested that the alkali has been integrated into the sample. A similar phenomenon was reported by Huang et al. in deacetylation using Na_2CO_3 [23].

The morphology of the native and deacetylated is presented in Figure 2. The difference between both samples was insignificantly observed in a 1000 magnification (figure not shown). However, it was detected in 10000 magnificent that the surface of the native sample is rough and covered with the dusk-like which separated each other. While in the same magnificent, the deacetylated sample has a smoother surface in which the dusk-like is connected with some of the melted part.

Fig. 2. Comparison of morphology of native glucomannan (top) and deacetylated glucomannan (bottom) in 1000 magnification.

3.2. Efficiency of iron encapsulation

The efficiency of encapsulation describes ability of the excipient to trap the iron. Effects of KOH concentration and reaction time of deacetylation on the efficiency of encapsulation is presented at Figure 3. Higher concentration of alkali resulted in higher ability to encapsulate the iron. Both concentrations of KOH showed a similar trend in which the highest efficiency obtained after 8 h of deacetylation. However, Figure 3 shows all the efficiencies of encapsulation are less than 50%. This suggests the amount of glucomannan used in this study could be too low to bind the iron added.

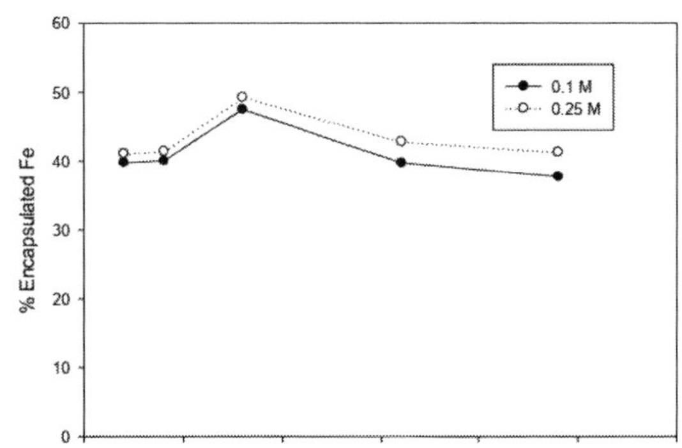

Fig. 3. Effect of time of deacetylation and alkali concentration on Fe encapsulation efficiency.

Deacetylation using alkali is necessary for the gelation of glucomannan. The physical change in gelation due to deacetylation is quite remarkable. The initial glucomannan solution which has not been treated with alkali does not form a gel [24]. One of the changes could be the ability to trap iron in the gels. Wardhani et al. reported deacetylation of glucomannan using higher alkali concentration results in the more deacetylation substitution which led to the higher value of the degree of deacetylation [25]. This influences stronger interchain attractions among the glucomannan molecules as well as the more number of junctions, which results in a denser and more flexible physical network with higher [20]. The junctions could provide sites to trap the iron. Wandrey et al. suggested deacetylated glucomannan has ability to encapsulate active agent by forming a gel which is a stable and soluble matter [11]. Hence, it is expected that higher degree of deacetylation results in better performance in iron encapsulated. However, Figure 3 demonstrates overextended the deacetylation time reduces the performance of the encapsulation. This suggests the forming junctions are not the only condition that affected the iron binding. After reaching certain deacetylation degree, additional acetyl substitution could become a steric hindrance for iron to be trapped in the junctions.

3.3. Ability of Deacetylated Glucomannan to Protect Iron Oxidation

Fe^{2+} of $FeSO_4$ is easily being oxidised into Fe^{3+} during storage, due to contact $FeSO_4$ with oxygen. However, Fe^{3+} has lower bioavailability in the human body [26]. One of the aims of iron encapsulation is to protect direct contact between the iron source with oxygen.

Figure 4 shows the samples of encapsulated Fe2+ using deacetylated glucomannan after incubated for 5 days. The result showed higher alkali concentrations have better ability to protect the oxidation. However, extended deacetylation time did not demonstrate a significant difference in protecting the oxidation. This research has demonstrated that encapsulated iron using deacetylated glucomannan was able to prevent or minimise the oxidation of the iron. All of the encapsulation samples were able to protect >95% of iron oxidation after 5 days. This means deacetylated glucomannan as excipient of iron works very well in protecting iron interaction with oxygen.

3.4. Release of Iron in pH 1.2

It is expected that the encapsulation is able to deliver the iron into the absorption site which is in the small intestine. Hence the excipient is designed to protect the release of iron in the stomach condition by deacetylation in various alkali concentration and time of deacetylation.

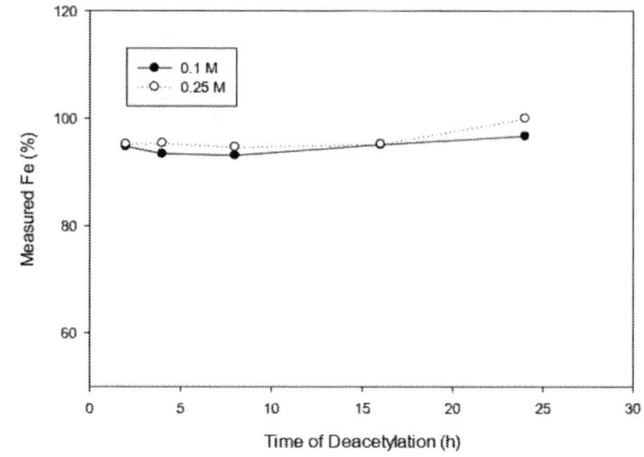

Fig. 4. Effect of time of deacetylation on Fe concentration in encapsulation after 5 days storage.

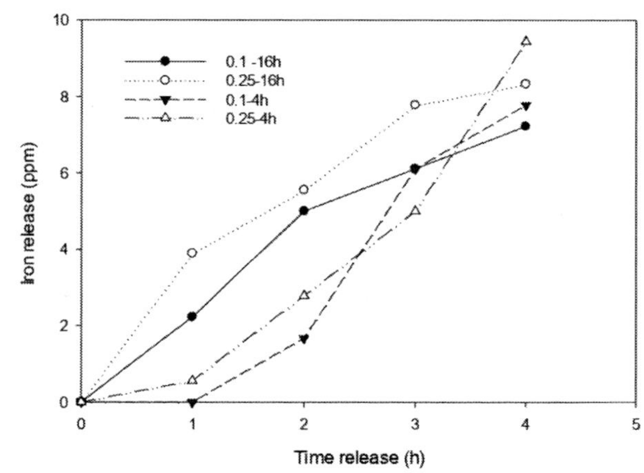

Fig. 5. Profile of iron release at pH 1.2.

Figure 5 shows the release of iron encapsulated in pH 1.2 of HCl solution using a different degree of deacetylated glucomannan as the excipient. Extended deacetylation time resulted in the more iron released. Interestingly, the higher release of iron was observed using excipient with higher degree of deacetylation. This profile supports the result in efficiency encapsulation in the previous discussion.

References

1. P. K. Ashok, K. Upadhyaya, Tannins are astringent. J Pharmacogn Phytochem, **1**. (2012).
2. S. Khosroyar, A. Akbarzade, M. Arjoman, A. A. Safekordi, S. A. Mortazavi, J. Microbiol. Res., **6**, 2455-2461. (2012).
3. C. Liyanage, S. Zlotkin., Food Nutr Bull., **23**, 133-7. (2002).
4. P. M. Schrooyen, R.van der Meer, C. G. De Kruif., Proc Nutr Soc, **60**, 475-479. (2001).

5. A. N. Bezbaruah, S. S. Shanbhogue, S. Simsek, E. Khan., J. Nanopart. Res., **13**, 6673-6681. (2011).
6. J. Yang, J. X. Xiao, L. Z. Ding., Eur Food Res Technol, **229**, 467-474. (2009).
7. Y. Wang, J. Liu, Q. Li, Y. Wang, C. Wang,. Biotechnol Lett., **37**, 1-8. (2015).
8. C. A. Gorski, , & M. M. Scherer., Am. Chem. Soc. 315-343. (2011).
9. V. Dave, M. Sheth, S. P. McCarthy, J. A. Ratto, D. L. Kaplan, Polym. J, **39**, 1139-1148. (1998).
10. K. E. Prasetya, N. Nurgirisia, F. Fadilah, SENATEK Proc. UMP. faculty of Engineering. (2015). (in Indonesia)
11. C. Wandrey, A. Bartkowiak, S. E. Harding., *Materials for encapsulation. In Encapsulation technologies for active food ingredients and food processing*. Springer New York. 31-100. (2010).
12. S. Wang, Y. Zhan, X. Wu, T. Ye, Y. Li, L. Wang, B. Li. Carbohydr Polym., **101**, 499-504. (2014).
13. Z. Pan, K. He, Y. Wang., J. App. Polym. Sci., **108**, 1566-1573. (2008).
14. Z. Pan, J.Meng, Wang, Y. Particuology, **9**, 265-269. (2011).
15. W. Xian, W. Ting, W. Zhiyong, Int. J. Biol. Macromol, **42**, 256-263. (2008).
16. L. H. Rahayu, D. H. Wardhani, A. Abdullah,. J. Metana, **9**, 45-52. (2013) (in Indonesia)
17. J. Chen, J. Li, B. Li, Carbohydr. Polym., **86**, 865-871. (2011).
18. K. Wang, Z. He, Int. J. Pharm, **244**, 117-126. (2002).
19. T.Lee, H. Lim, Y. Lee, J. W.Park, *Use of waste iron metal for removal of Cr (VI) from water.* Chemosphere, **53**, 479-485. (2003).
20. B. Solo-de-Zaldívar, C. A. Tovar, A. J. Food Hydrocolloids., **35**, 59-68. (2014).
21. K. Maekaji. Agric Biol Chem., **38**, 315-321. (1974).
22. H. Zhang, M. Yoshimura, K. Nishinari, M. A. K. Williams, T. J. Foster, I. T. Norton. Biopolymers, **59**, 38-50. (2001).
23. C. K. Huang, P. F. Kerr., Am. Mineral, **45**, 311-324. (1960).
24. Perols, C., Piffaut, B., Scher, J., Ramet, J. P., & Poncelet, D. Enzyme Microb. Technol, **20**, 57-60. (1997).
25. D. H. Wardhani, D.Puspitosari, M. A. Ashidiq, N. Aryanti, A. Prasetyaningrum, AIP Conf. Proc. **1855**. 030020. (2017).
26. C. Valenzuela, V. Hernández, M. S. Morales, A. Neira-Carrillo, F. Pizarro., LWT-Food Sci Technol, **59**, 1283-1289. (2014).

Edible Coating from Green Tea Extract and Chitosan to Preserve Strawberry (*Fragaria vesca L.*)

Dwi Apriyanti[1,3,*], *Nur* Rokhati[1,3], Novia Mawarni[1,3], *Zuroidatul* Khoiriyah[1,3] and *Titik* Istirokhatun[2,3]
[1]Chemical Engineering Department, Engineering Faculty, Diponegoro University
[2]Environmental Engineering Department, Engineering Faculty, Diponegoro University
[3]Membrane Research Centre (Mer-C), Center of Research and Services-Diponegoro University
Jl. Prof Soedharto, Tembalang, Semarang, Indonesia 50275

Abstract. Food quality is an important factor in food sector. One way to increase the quality of food is by applying edible coating to slow down the degradation of foods quality. Combination of chitosan and green tea extract is one of the types of edible coating which many function as food preservative Green tea extract is chosen because of its high antioxidant contents. The goal of this research is to make a film from the composite of chitosan and green tea extract as a food coating and to observe the effect of chitosan concentration and the addition of green tea extract to the phenolic contents and antioxidant activity of the film. Moreover, this research also aims to apply chitosan film and green tea extract to foods, especially strawberries, and observing the effect of chitosan concentration and the addition of green tea extract to the shrinking of the fruit's weight as well as its antimicrobial activities. Strawberries were coated with five different coating formula including one set as blank (uncoated fruit). The result of this research showed that the increasing of chitosan concentration and the volume of green tea extract addition also increase the antioxidant activity on the film. The phenolic contents on the green tea extract addition will also increase its phenolic contents as well. However, the increasing number of chitosan concentration can reduce the phenolic content on the film. Meanwhile, during the application to strawberries, it is found that 3% chitosan concentration which is combined with the addition of 2.5% green tea extract will minimize the shrinking of the fruit's weight. Also 3% chitosan concentration will give a better antimicrobial activity. In the other side the addition of green tea extract will increase antioxidant activity, the addition of green tea extract also causing a decrease in antimicrobial activity.

INTRODUCTION

Edible film is a thin layer of a material that can be consumed and placed between products. Edible films can be either their own sheets or for coatings on food [4,17,21]. This layer serves to protect food products from mechanical damage by reducing the transmission of water vapor, smell, and fat. Edible coatings can be made from materials capable of forming a layer [4,23]. One of the materials that can be used as raw material for edible coating is chitosan

Chitosan can be used as a basic material for edible coating because it is one of the non-toxic and edible materials. The advantages compared to other materials are chitosan has a positive amine group that makes it able to form a good layer and its antimicrobial properties [4,17].

In edible coatings of chitosan can be added with other additives. One of them is by adding green tea extract. Green tea is known to contain many polyphenolic compounds. Polyphenol compounds are known to function as excellent antioxidants. Therefore, green tea extract will give its own value when added in the manufacture of edible coating.

The purpose of this research is (1) Making edible coating of combination of chitosan and green tea extract. (2) to observe the effect of the addition of green tea extract concentration to the content of phenolic compounds, antioxidant activity, weight loss, and antimicrobial activity. (3) observed the effect of adding chitosan concentration to phenolic content, antioxidant activity, heavy shrinkage, and antimicrobial activity.

* Corresponding author: dwi.titik07@gmail.com

MATERIALS AND METHODS

Materials

Fresh strawberry and green tea leaves extract were obtained from local market in Semarang. Glycerol, CH3COOH and DPPH (2,2-dyphenil-1- pikrilhydrazil) were purchased from Merck, (Hohenbrunn,Germany). Commercial chitosan as reference (C6H11NO4)n, was purchased from Biotech Surendo, Cirebon, Indonesia. Distilled water for all experiments was produced from a home-made pure water unit.

Methods

Green tea leaves are heated in aquadest 90^{0}C for 10 minutes at ratio of 1gr to 5ml and filtered using Whatmann No.1 filter paper. Chitosan powders with different concentration (1-3%) were mire mixed with 1% acetic acid solution to dissolve it. Chitosan film-forming solution was prepared according to the procedure of *Siripatrawan, et al* [25]. Each of the prepared chitosan solution (1-3%) was added with glycerol 25% V/V and green tea extract solution according to the variables (0-10%V/V chitosan solution). The solution then stirred until homogeneous.

Fresh strawberries are immersed in the film solution for a while until the film solution coats the entire surface of strawberries evenly. Then the strawberries are removed from the film solution and let stand for a while until it dries. Coating is done on each sample variable.

Analysis

1. Antioxidant activity by DPPH analysis

The antioxidant activity of the chitosan film was evaluated according method from Blois [25]. Each sample solution put into the test tube. Each of these solutions added with 2 ml of 0.1 M DPPH solution. Meanwhile, a control solution was prepared by mixing 1 ml of solvent solution (acetic acid 1%) with 2 ml of 0.1 M DPPH solution. It is also necessary to make a blank solution, using 3 ml of solvent solution (1% acetic acid solution). Each solution that has been made is dissolved left until 30 minutes. After 30 minutes, each solution measured its absorbance using a spectrophotometer at a 517nm wave. The antioxidant activity can be calculated by following equation

$$Aktivitas\ antioksidan\ (\%) = \frac{Abs_{DPPH} - Abs_{sampel}}{Abs_{DPPH}} \times 100\%$$

2. Phenolic content

Total phenolic content in chitosan solution was determined according to the Folin-Ciocalteu method as described bu Singleton,Orthofer, and Lamuela-Raventos [25]. Each 1 ml sample extract was put into the test tube. Then added with 2 mL of10% folin compound and 2.5 ml of 7.5% Na2CO3 solution. The mixed solution is then allowed to stand for 30 minutes. After that, each solution measured its absorbance using a spectrophotometer with 765 nm wavelength.

3. Strawberry weight loss analysis

Strawberries without coating as controls and strawberries that have been coated with the film solution were weighed initially based on method by Kopkhar (1990). All strawberries are again weighed every day for a week. Then compared the value of heavy shrinkage strawberries are dicoating every day with strawberries without coating.

4. Antimicrobial activity *Siripatrawan, et al* [25].

Test microbial activity using total plate count method (TPC). The test stages were performed by weighing each strawberry sample as much as 1 gram which then diluted by dissolving into 9 ml of sterilized aquadest. Each aquadest that has been given strawberries mixed using a vortex. After mixing, the second dilution is done in the same way. After that, a third dilution is done in the same way. Then each diluted sample is taken 1 ml and fed into the petridish. After that the media poured into each petridish that has been filled samples and media made flat and then petridish dittup meeting. In all these processes are done in a sterile state. All petridish that already contains media and samples, are incubated for 24 hours in reverse position to prevent moisture from dripping into the media. After 24 hours, the number of bacteria is calculated.

RESULTS AND DISCUSSION

Phenolic Contents

The Foline-Ciocalteu phenol reagent is used to obtain an estimate of the amount of phenolic compounds contained in the chitosan film. Phenolic compound undergo complex redox reactions with phosphotungstat and phosphomolybdic acids contained in Foline-Ciocalteu reactants. Phenolic compounds have an effect on antioxidant activity [25]. Phenolics consist of one or more aromatic rings containing hydroxyl groups and potentially capable of removing free radicals by forming stable phenoxyl resonance radicals

The results showed that the addition of green tea extract on the base material of chitosan film caused an increase in the content of phenolic compounds. Increased concentrations of green tea extract (GTE) can significantly increase phenolic compound (Figure 1). The highest increase occurred in the addition of 10% GTE of ten times that of chitosan film without the addition of GTE.

Meanwhile, based on the results of the study also known that there is the influence of chitosan concentration on phenolic content. Phenolic content in film with chitosan concentration 3% less compared with phenolic content in film with 2% chitosan concentration. This is due to the increased concentration of chitosan, capable of causing a decrease in phenolic content [2]. Therefore, the phenolic content in the film at 3% chitosan concentration is lower than that of phenolic content in films with chitosan 2 %.

The DPPH test is used to show the antioxidant activity of the film. Chitosan film without GTE has little activity on DPPH (Fig.2). The mechanism of DPPH activity from chitosan is related to the fact that free radicals can react with free amino group residues (NH2) to form stable macromolecular radicals, and the NH 2 group can form ammonium (NH3 +) groups through the absorption of hydrogen ions from the solution [25]. In chitosan films with higher GTE content, the antioxidant activity becomes increasingly increasing. The highest antioxidant activity is shown in movies with a 10% GTE content. The antioxidant activity on chitosan film without GTE addition with chitosan film with GTE addition looks very significant. Differences in antioxidant activity in chitosan films with 10% increase in GTE can reach 10 times greater than the antioxidant activity in chitosan films without GTE addition. Antioxidant activity is largely derived from the content of polyphenolic compounds present in green tea extracts.

Meanwhile, in films with 3% chitosan intended to have greater antioxidant activity compared to fikm with 2% chitosan concentration. From Fig. 2 it is shown that antioxidant activity in film with 3% chitosan concentration is 67,59% bigger than film with 2% chitosan concentration. This is because, antioxidants are not only found in green tea extract. However, in chitosan itself there are antioxidant substances [25]. Therefore, with increasing chitosan concentration in the film can lead to increased antioxidant activity in the film.

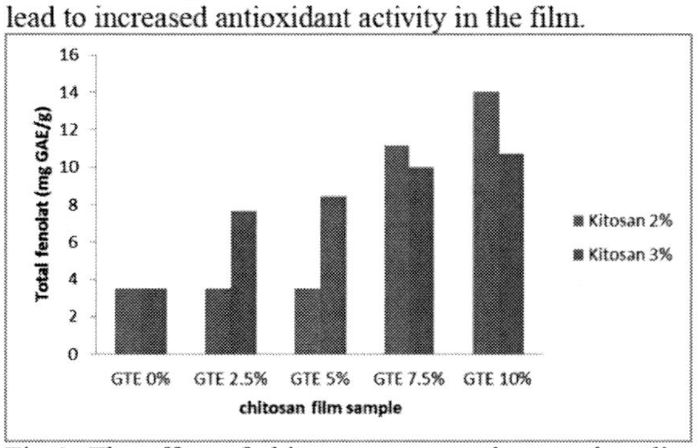

Fig 1. The effect of chitosan concentration on phenolic contain of green tea extract

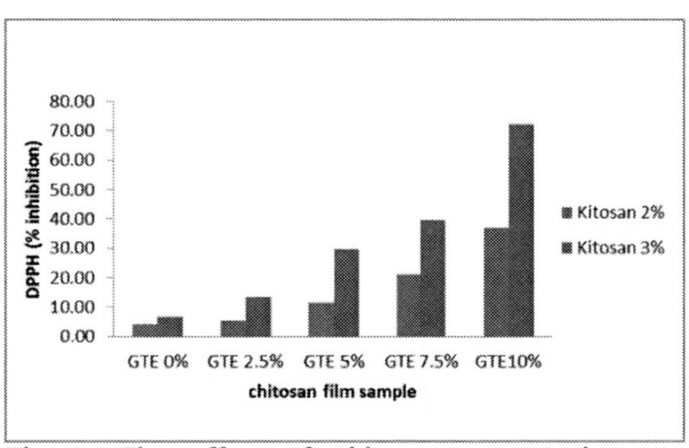

Fig 2. The effect of chitosan concentration on antioxidant activity of green tea extract

Weight Loss

Strowberry fruit is very susceptible to the speed of water loss resulting in shrinkage of fruit and weakening of tissue due to very thin skin of the fruit. Therefore, the weight reduction of fruit during the 4-day storage period was conducted to evaluate the effect of coating. All strawberries weighed heavily during the storage period (Fig. 3). However, after the second day of storage, the weight loss for all coated strawberries was significantly lower than that of strawberries without coating.

Chitosan acts as a physical barrier to loss of moisture, thus reducing dehydration and fruit shrinkage [30]. The addition of green tea extract on the chitosan film can reduce the rate of fruit shrinkage. The lowest shrinkage was obtained in strawberries with 3% incremental chitosan film of 2.5% GTE which lost weight 55% lower than fruit without coating.

In addition, when compared between strawberries with 2% chitosan film coating with 3% chitosan coating film there will be differences. In strawberries with 2% chitosan coating, the weight level of strawberries is greater than the depreciation of strawberries with 3% concentration of chitosan film coating. The amount of weight loss in strawberry fruit with 2% chitosan coating caused by the loss of water in strawberries. Meanwhile, in strawberries with 3% coat kitoasan there is a thicker layer of barrier that protects the fruit so it can suppress the evaporation of water content in strawberries [8]. This can be compared also with strawberry control (without coating) which experienced the greatest weight loss rate. This enormous weight loss is caused by the evaporation of the water content of the strawberries which causes the weight of the strawberries to decrease (shrink). This happens because the control strawberries contain no water barrier to evaporate as in the strawberry fruit that is coated.

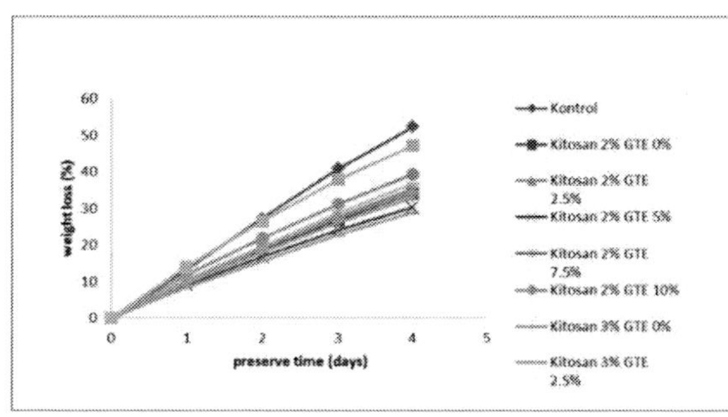

Fig. 3. The effect of chitosan concentration and green tea extract on strawberry weight loss.

Antimicrobial Activity

Chitosan has antimicrobial activity against bacteria, yeast, and fungi. Chitosan is considered a soluble chelating agent and activator because of the positive charge on C-2 of the glucosamine monomer. These characteristics provide good antimicrobial activity. The destruction of protein and intercellular components occurs due to the interaction between the amine groups in the positively charged chitosan molecule and the negatively charged microbial cell membrane [19].

In this study chitosan is used as a coating on strawberries. Analysis of bacterial growth was done by Total Plate Count (TPC) method. Fig. 4. shows the total bacteria of strawberry fruit coated with 2% and 3%

concentration chitosan film with the addition of GTE at various concentrations and has been stored for 6 days. Non-coating sroberries contain bacteria close to the maximum number of colonies that humans can consume 6 x 106 CFU / g. Strawberry fruit with chitosan coating without addition of GTE has the lowest total bacteria (± one tenth of fruit without coating). The addition of GTE causes an increase in total bacteria. The increased concentration of GTE causes the total bacteria to increase. This is caused by the increasing concentration of GTE the amine group content in chitosan becomes decreased so that antimicrobial activity also decreased.

Meanwhile, from Fig. 4. on strawberries with chitosan coating concentrations of 2% and 3% there were antimicrobial antiviral differences. For strawberries with 3% concentration chitosan coatings a higher antimicrobial activity was demonstrated with less total bacteria compared with strawberries with 2% concentration of chitosan coating. This antimicrobial activity difference occurs because antimicrobial activity in chitosan one of them influenced by chitosan concentration [19]. If the chitosan concentration is greater, the antimicrobial activity becomes larger so that fewer microbes will contaminate the strawberry fruit that is coated. Therefore, the number of microbes in strawberries with chitosan coating 3% less than the number of microbes in strawberries with 2% coating chitosan.

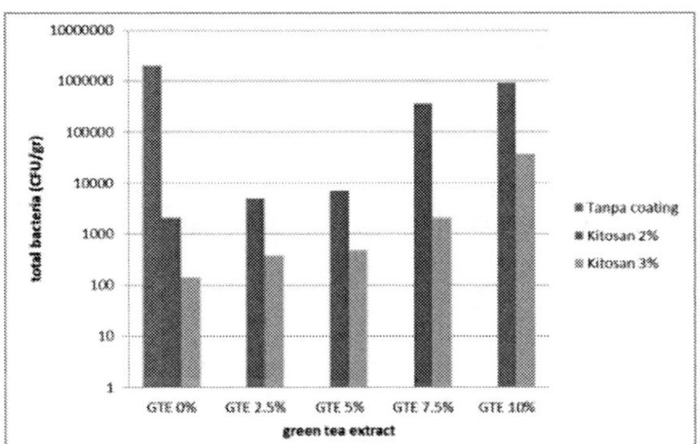

Fig. 4. The effect of blending chitosan concentration and green tea extract on antimicrobial activity.

CONCLUSION

The results showed that the addition of green tea extract can increase total phenolic content, antioxidant activity, and moisture loss barrier. In green tea extract there are antioxidant compounds one of them is a polyphenolic compound. Each addition of green tea extract in chitosan solution will increase total phenolic content and antioxidant activity. From antimicrobial test, it was found that green tea extract can decrease antimicrobial activity from chitosan film because with addition of green tea extract the amine group content in chitosan decreased so that antimicrobial activity also decreased.

Meanwhile, increasing chitosan concentration can cause total phenolic content reduction. However, in addition to decreasing total phenolic content, chitosan concentrations actually increase antioxidant activity, barrier to moisture loss and antimicrobial activity. This is because in chitosan also contain antioxidant substances so that the antioxidant activity increases along with the addition of chitosan concentration. While chitosan serves as a physical barrier to the loss of moisture. From antimicrobial test results obtained at greater chitosan concentrations then increased antimicrobial activity characterized by decreased bacteria that contaminate the test strawberries.

ACKNOWLEDGEMENT

Researcher would like to thank you to the Mer-C Laboratory for its contribution during this research.

REFERENCES

1. Abugoch, L. E., Tapia, C., Villamán, M. C., Yazdani-Pedram, M., & Díaz-Dosque, M. Characterization of Quinoa Protein–Chitosan Blend Edible Films. Food Hydrocolloid, **25**, 879–886.(2011).
2. Ali, A., Maqbool, M., Alderson, P. G., & Zahid, N. Effect of gum arabic as an edible coating on antioxidant capacity of tomato (Solanum lycopersicum L.) fruit during storage. Postharvest Biol Tec, **76**, 119–124. (2013).
3. American Dietetic Association. Antioxidants. American Dietetic Association. (2010).
4. Bourtomm, T. Edible Films and Coatings : Characteristics and Properties. ifrj, **15**. (2008).
5. Carvalho, R. A. De, & Grosso, C. R. F. Properties of Chemically Modified Gelatin Films. Braz J Chem Eng, **23**, 45–53.(2006).
6. Castricini, A., Coneglian, R. C. C., & Deliza, R. Starch Edible Coating of Papaya : Effect on Sensory Characteristics. Ciencia Tecnol Alime, **32**, 84–92. (2012).
7. Chillo, S., Flores, S., Mastromatteo, M., Conte, A., Gerschenson, L., & Nobile, M. A. Del. Influence of Glycerol and Chitosan on Tapioca Starch-Based Edible Film Properties. J Food Eng, **88**, 159–168. (2008).
8. Dávila-aviña, J. E. D. J., Villa-rodríguez, J., Cruz-valenzuela, R., Rodríguez-armenta, M., Espino-díaz, M., Ayala-zavala, J. F.,González-aguilar, G. Effect of Edible Coatings , Storage Time and Maturity Stage on Overall Quality of Tomato Fruits. ajabs, **6(1),** 162–171. (2011).
9. Debeaufort, F., & Voilley, A. Edible Films and Coatings for Food Applications. (M. E. Embuscado & K. C. Huber, Eds.). New York: Springer. (2009).
10. Garcia, L. C., Pereira, L. M., Luca Sarantópoulos, C. I. G., & Hubinger, M. D. Selection of an Edible Starch Coating for Minimally Processed

Strawberry. Food Bioprocess Tech, **3(6)**, 834–842. (2010).

11. Ghasemzadeh, R., Karbassi, A., & Ghoddousi, H. B. Application of Edible Coating for Improvement of Quality and Shelf-life of Raisins. World. (2008).
12. Hassani, F., Garousi, F., & Javanmard, M. Edible Coating Based on Whey Protein Concentrate-Rice Bran Oil to Maintain the Physical and Chemical Properties of the Kiwifruit (Actinidia deliciosa). Trakia Journal of Sciences, **10(1),** 26–34. (2012).
13. Henriques, M., Santos, G., Rodrigues, A., Gomes, D., Pereira, C., & Gil, M. Replacement of Conventional Cheese Coatings by Natural Whey Protein Edible Coatings with Antimicrobial Activity. Journal of Hygienic Engineering and Design, **4**, 34–47. (2013).
14. Hue, S., Boyce, A. N., & Somasundram, C. Antioxidant Activity , Phenolic and Flavonoid Contents in the Leaves of Different Varieties of Sweet Potato (Ipomoea batatas). Aust J Crop Sci, **6(3)**, (2012).
15. Levic, L., Koprivica, G., Misljenovic, N., Filipcev, B., Simurina, O., & Kuljanin, T. Effect of Starch as an Edible Coating Material on the Process of Osmotic Dehydration of Carrot in Saccharose Solution and Sugar Beet Molasses. Acta periodica technologica, **212(39),** 29–36. (2008).
16. Lim, R., Stathopoulos, C. E., & Golding, J. B. Effect of Edible Coatings on Some Quality Characteristics of Sweet Cherries. ifrj, **18(4)**, 1237–1241. (2011).
17. Mehdizadeh, T., Tajik, H., Mehdi, S., Rohani, R., & Oromiehie, A. R. Antibacterial, Antioxidant and Optical Properties of Edible Starch-chitosan Composite Film Containing Thymus kotschyanus Essential Oil. Veterinary Research Forum, **3(3),** 167–173. (2012).
18. Motalebi, A. A., A, H. R., Khanipour, A. A., & Soltani, M. Impacts of Whey Protein Edible Coating on Chemical and Microbial Factors of Gutted Kilka during Frozen Storage. Iran J Fish Sci, **9(2)**, 255–264. (2010).
19. Nadarajah, K. Development and Characterization of Antimicrobial Edible Films from Crawfish Chitosan. Lousiana State Unversity. (2005).
20. Ochoa, E., Rojas-molina, R., Garza, H. de la, & Charles-Rodríguez, A. V. Evaluation of a Candelilla Wax-Based Edible Coating to Prolong the Shelf-Life Quality and Safety of Apples. AJABS, **6(1)**, 92–98. (2011).
21. Pascall, M. A., & Lin, S. The Application of Edible Polymeric Films and Coatings in the Food Industry. J Food Process Technol., 4(2). (2013)
22. Shahidi, F., Arachchi, J. K. V., & Jeon, Y.-J. Food Applications of Chitin and Chitosans. Trends Food Sci Tech, **10**, 37–51. (1999).
23. Shariatifar, M., & EhsanJafarpour. Edible Coating Effects on Storage Life and Quality of Apple. JBASR, **3(6),** 24–27. (2013).
24. Sinija, V. R., & Mishra, H. N. Green tea : Health benefits. Journal of Nutritional & Enviromental Medicine, **17**, 232–242. (2008)
25. Siripatrawan, U., & Harte, B. R. Physical Properties and Antioxidant Activity of an Active Film from Chitosan Incorporated with Green Tea Extract. Elsevier, Food Hydrocolloids, **24(8),** 770–775. (2010).
26. Skurtys, O., Acevedo, C., Pedreschi, F., Enrione, J., Osorio, F., & Aguilera, J. M. Food Hydrocolloid Edible Films and Coatings. Universidad de Santiago de Chile. (2010).
27. Sun, T., Zhou, D., Xie, J., & Mao, F. Preparation of Chitosan Oligomers and their Antioxidant Activity. Shanghai, China.(2006).
28. Vargas, M., Albors, A., Chiralt, A., & González-Martínez, C. Quality of Cold-Stored Strawberries as Affected by Chitosan–Oleic Acid Edible Coatings. Elsevier, Postharvest Biol Tec, **41(2),** 164–171. (2006).
29. Zaveri, N. T. Green Tea and Its Polyphenolic Catechins : Medicinal Uses in Cancer and Noncancer Applications. Els, Life Sciences, **78**, 2073–2080. (2006).
30. Zivanovic, S., Li, J., Davidson, P. M., & Kit, K. Physical , Mechanical , and Antibacterial Properties of Chitosan / PEO Blend Films. Biomacromolecules, **8**, 1505–1510. (2007).
31. Zulkefli, H. N., Mohammad, J., & Abidin, N. Z. Antioxidant Activity of Methanol Extract of Tinospora crispa and Tabernaemontana corymbosa. Sains Malays, **42(6)**, 697–706. (2013).
32. F. De Lillo, F. Cecconi, G. Lacorata, A. Vulpiani, EPL, **84** (2008)

Kinetic of Biomass Growth and Protein Formation on Rice Bran Fermentation Using *Rhizopus oryzae*

Andhika Sukma[1], *Bakti* Jos[2], and *Siswo* Sumardiono[2*]

[1]Master Program in Chemical Engineering, Faculty of Engineering, Diponegoro University, Indonesia
[2]Department of Chemical Engineering, Faculty of Engineering, Diponegoro University, Indonesia

Abstract. Rice bran is the outermost part of the rice grain wrapped in rice husks. Rice bran contains macro and micro nutrients that are bound to fibers that affect psychochemical properties when used in food products. The purpose of this study was to increase the nutritional value of rice bran with solid state fermentation using *Rhizopus oryzae* FNCC 6011 in tray bioreactor at 30°C for 120 h. The results showed an increase in protein, fat and ash content by 58.5%, 124.5%, and 18.6%, respectively, while carbohydrate and fiber content decreased respectively by 25.6% and 51.2%. The deceleration model successfully depicts the profile of biomass growth and improved protein. The kinetic parameters obtained k, A, and Y_{PX} were 0.0536/h, 5.2537, and 0.1821 g/g, respectively.

1 Introduction

Rice bran is a layer of rice grain wrap (layer of aleuron/epidermis) that is released during paddy rice milling process. The total grinding yield of rice bran about 10% [1]. About 90% of the resulting rice bran is used as animal feed at a low price and the remaining 10% is extracted into rice bran oil [2]. Rice bran is a source of protein, fatty acids, minerals, dietary fiber, and vitamin E (tocopherol and tocotrienol) as a source of antioxidants [3]. Rice bran also contains bioactive compounds such as inositol, inositol hexaphospat, ferulic acid, gamma oryzanol, and sterols that are proven to prevent cancer, hyperlipidemia, liver fat, hypercalciuria, kidney stones, and heart disease. With the discovery, rice bran has a huge potential to serve as a functional food [4-7].

Although rice bran is available abundantly, but application in food products is still very limited, this is because the content of macro and micro nutrient of rice bran is still strongly bound with fiber. To overcome these problems, the fermentation technology is required to improve the quality by utilizing microbes, yeasts, and fungi to decompose organic matter in substrates into simpler components. Fermentation technology can increase protein content in agricultural products and agro-industrial wastes through growth and activity of microbial metabolism [8-14]. Fermentation can also improve nutrients, aroma, flavor, and texture through enzymatic activity produced by microbes that can be used as functional food [15,16].

Fermentation can be divided into two methods, submerged fermentation/liquid state fermentation and solid state fermentation. In recent years, many researchers have shown that solid state fermentation has a major impact on productivity, higher yields and improved product characteristics compared to liquid state fermentation [17,18]. Solid state fermentation is defined as fermentation involving solids in the absence of free water. Solid state fermentation benefits include lower energy requirements, less waste water, and environmentally friendly [19,20].

Rhizopus oryzae is a filamentous fungi of the *Zygomycetes* class. Most strains of *Rhizopus oryzae* are isolated as active components in the production of food or alcoholic beverages in Indonesia, China and Japan. *Rhizopus oryzae* has long been used for the production of enzymes (glucoamylase and lipase), organic acid synthesis, and various fermentation applications in foodstuffs [21]. Various types of enzymes produced by various strains of *Rhizopus oryzae* include cellulase which serves to break down cellulose into oligosaccharides, glucose or other compounds. Another enzyme produced is xylanase which catalyzes the hydrolysis of 1-4-beta-D-glycosidic on xylans which is the constituent of hemicellulose, pectinase which serves to break down pectin into simple sugars, and amylase [22-27].

* Corresponding author: siswo.sumardiono@che.undip.ac.id

To develop the potency of rice bran into a larger scale of production, a kinetic model that plays an important role in monitoring and predicting the results of the fermentation process is needed. The unstructured model is simpler because cell biomass is viewed as a single component without the internal dynamics of the cell so that the rate of reaction depends only on the phase conditions in the reactor, while structured models incorporate internal cell factors such as changes in biomass composition, so the model is more complex [28-31]. The kinetic equation is chosen based on the most appropriate regression result. Some kinetic models in solid state fermentation include linear, logistic, exponential, and deceleration models [32-34].

This research aims to determine the characteristics of fermented rice bran and determine the appropriate kinetics model to describe the growth profile of biomass and increase protein content in rice bran by solid state fermentation method for further industrial application.

2 Materials and Methods

2.1 Rice bran

Rice bran was obtained from rice mill in Teter Village, Simo Subdistrict, Boyolali Regency, Central Java, Indonesia. The rice bran is sieved using a 35 mesh sieve (Retsch AS200, Haan, Germany) to separate impurities such as rice husk and rice. The rice bran was packed in polypropylene box and stored at 4°C until used.

2.2 Inoculum preparation

Rhizopus oryzae (FNCC 6011) in the form of ampoule was obtained from Food and Nutrition Culture Collection (FNCC) of Food and Nutrition PAU of Gadjah Mada University, Indonesia. The cultures were transferred in a petri dish that has been filled with potato dextrose agar medium (PDA, Merck KGaA, Darmstadt, Germany) and incubated at 30°C for 7 days [35]. The spores that were formed were taken and dissolved in Tween 80 (0.2%) (Merck KGaA, Darmstadt, Germany). Spore concentration was measured using TPC (total plate count) method. Spore suspension being stored at 4°C.

2.3 Fermentation

Rice bran was fermented in solid state fermentation method using tray bioreactor, with dimensions of 16 cm x 8 cm x 4 cm. The rice bran (100 gr) was placed in tray bioreactor with a thickness of 2 cm substrate layer. The tray bioreactor containing the rice bran was sterilized in an autoclave (Hirayama HVE-50, Tokyo, Japan) at 121°C for 30 minutes. Take a sample of ± 2 g for protein analysis. Then into the rice bran substrate added 45 ml of nutrient solution (KH_2PO_4 2 g/l, $MgSO_4$ 1 g/l, and $(NH_4)SO_4$ 8 g/l in 0.4 N HCl) (Merck KGaA, Darmstadt, Germany) [36].

The *Rhizopus oryzae* spore suspension was added to the substrate with a concentration of 4 x 10^6 spores/g of rice bran [35]. Sterile aquadest was then added until the moisture content of the substrate reached 55% [37]. The rice bran was incubated in an incubator (Memmert IN55, Schwabach, Germany) at 30°C for 120 hours [37,38]. Samples were taken every 24 hours for biomass and protein content analyses

2.4 Proximate composition

The moisture, ash, and fat content were determined by the gravimetric method according to the AOAC (2000) recommendation. The total protein content was determined as total nitrogen by Kjeldahl method [39]. The conversion factor used is 6.25. The crude fibers were determined through gravimetrical measurement of the residue obtained by the acid and basic digestion of the samples according to NDF method.

2.5 Biomass determination

Samples of 1 gr was ground until smooth and then put into a centrifuge tube. Add 5 ml of 15% Na_2SO_4 solution (Merck KGaA, Darmstadt, Germany), then centrifuged at 12000 rpm for 15 minutes (LaboGene 1730R, Alleroed, Denmark). After centrifugation there will be three layers, the top layer is biomass, the middle layer is the Na_2SO_4 solution and the bottom layer is the substrate. The biomass is then separated by a filter paper of Whatman 42 (Merck KGaA, Darmstadt, Germany) that has been known to weigh. The same is done for substrate. Then the filter paper contains biomass and the substrate is dried in an oven (Heraeus T6060, Hanau, Germany) at 40°C until a constant weight is obtained [40]. The results of measurement of dry weight of biomass using the following formula [32],

$$X = \frac{x}{x+S+W} \qquad (1)$$

2.6 Kinetics model

Some models of kinetics of biomass growth in solid state fermentation include linear, exponential, logistic, and deceleration models [32-34].

2.6.1 The linear model

It is the simplest model in which the change in the biomass concentration over time is constant according to the equation

$$\frac{dX}{dt} = k \qquad (2)$$

The result of the integration of the above equation becomes

$$X = X_0 + k\,t \qquad (3)$$

2.6.2 The exponential model

The Exponential model expression of biomass growth is

$$\frac{dX}{dt} = \mu\,X \qquad (4)$$

The result of the integration of the above equation becomes

$$X = X_0\, e^{\mu t} \qquad (5)$$

2.6.3 The logistic model

The logistic model is the most widely used model (75%) to describe the solid state fermentation kinetics, while 25% is the other model [33].

$$\frac{dX}{dt} = \mu X \left(1 - \frac{X}{X_m}\right) \qquad (6)$$

The result of integration of the above equation is obtained

$$X = \frac{X_m}{1 + \left(\frac{X_m}{X_0} - 1\right) e^{-\mu t}} \qquad (7)$$

2.6.4 The deceleration model

The deceleration model of biomass growth as follows,

$$\frac{dX}{dt} = k\, A\, X\, e^{-kt} \qquad (8)$$

The result of the integration of the above equation becomes

$$X = X_0 \exp(A(1 - e^{-kt})) \qquad (9)$$

2.7 Model of protein formation

Changes in protein content that has been formed during the fermentation process is determined using the following equation [32,41],

$$\frac{dP}{dt} = Y_{PX}\frac{dX}{dt} + m_p\, X \qquad (10)$$

3 Results and Discussion

The nutrition content of unfermented rice bran and fermented rice bran are presented in Table 1 below. Carbohydrate, protein, fat, and ash content are calculated on a dry basis.

Table 1. Nutrition content of rice bran

Component	Unfermented Rice bran	Fermented Rice bran
Carbohydrate (%)	68.28 ± 0.15	50.79 ± 0.18
Fiber (%)	26.51 ± 0.11	12.94 ± 0.08
Protein (%)	12.59 ± 0.07	19.95 ± 0.09
Fat (%)	6.21 ± 0.04	13.94 ± 0.06
Ash (%)	12.92 ± 0.04	15.32 ± 0.03

Values are expressed as means ± standard deviation

Based on Table 1 it can be seen that after fermentation for 120 hours there was a decrease of carbohydrate and fiber content respectively 25.6% and 51.2%, this is due to activity of *Rhizopus oryzae* metabolism that grow and developed in rice bran. *Rhizopus oryzae* produces enzymes used to break down polysaccharides such as cellulase, xylanase, pectinase, and amylase to oligosaccharides and glucose, resulting in reduced carbohydrate and fiber content [22,23,25,26,42].

Glucose from polysaccharide splitting as a source of energy in the cell metabolism *Rhizopus oryzae* produces a number of biomass and extracellular compounds such as enzymes that result in increased protein content in rice bran. 10-25% of the fungus constituents are proteins [14,43]. Protein content in fermented rice bran has an increase of 58.5%. Several authors have reported increase in protein in fermented rice bran by 40%, 53.4%, and 49.3% [35,36,38]. The increase in protein content in this study is higher than previous studies because in this study the rice bran used has particles sizes smaller than 0.42 mm, while in previous research using rice bran with particles sizes between 0.35 and 0.60 mm. The smaller particle diameters will produce higher protein content due to the oxygen transfer rate, heat transfer rate, and transfer rate of metabolic gases during the fermentation process are in optimal conditions for biomass growth and excretion of extracellular enzymes [36].

Fat content in fermented rice bran has an increase of 124.5%, this is due to the growing biomass *Rhizopus oryzae* more and more. The constituents of cell wall and plasma membrane of *Rhizopus oryzae* is fat component in the form of phospolipid and lipoprotein so that fat content in fermented rice bran is increased [43]. Increased fat content in fermented rice bran is also reported by 17.3% after 60 hours of fermentation. The results are much lower than this study, this is because the bran used has particles sizes between 0.39 and 1.67 mm so it is still mixed with broken rice and rice husk. This will affect the increase in fat content obtained. The smaller the rice bran particles will produce higher fat content in the fermentation process [44].

While the ash content in fermented rice bran has an increase of 18.6%. This result is consistent with some researchers who reported an increase in ash content by 30%, 47.8%, and 10.5% [35,38,45]. Increased ash content is thought to result from the growth of hypha and mycelium *Rhizopus oryzae* on the rice bran. *Rhizopus oryzae* cell wall contains several minerals such as phosphorus, potassium, sodium and sulfur, resulting in ash content in the fermented rice bran has increased [35,43]. The same phenomenon also occurs in rice bran fermentation using *Rhizopus oligosporus* and *Saccharomyces cerevisiae* where ash content increases by 24.5% after 48 hours of fermentation [46]. Fungal ash content varies from 1% to 29% depending on species of fungus and growth conditions [47].

To develop the potency of rice bran into a larger scale of production, the kinetic model is required. Regression analysis is used to determine the most appropriate kinetic equation and the values of the parameter of this equation that give the best fit to the experimental data. The results of matching experimental data with the model are presented in the Table 2 and Figure 1 below.

Table 2. Biomass growth model

Model	Equation	Sum of square of residuals (SSR)
Linear	$\frac{dX}{dt} = 0.0021$	0.0027
Exponential	$\frac{dX}{dt} = 0.0437\ X$	0.0220
Logistic	$\frac{dX}{dt} = 0.1920\ X\left(1 - \frac{X}{0.1897}\right)$	0.0030
Deceleration	$\frac{dX}{dt} = 0.0536\ x\ 5.2537\ x\ X\ e^{-0.0536t}$	0.0008

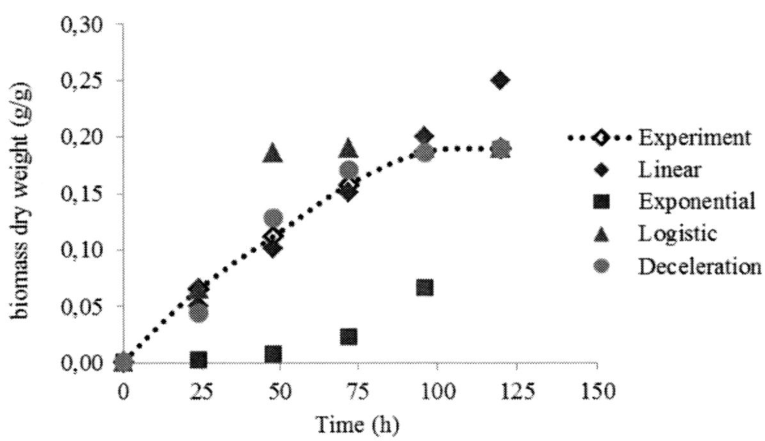

Fig 1. Biomass growth model

From the above kinetic models closest to the experimental data is the deceleration model where each point in the deceleration model coincides with the experimental data, it can also be seen from the smallest sum of square of residuals (SSR) value of 0.0008, so the deceleration model is best suited to describe the biomass growth model of fermented rice bran [32-34]. The kinetic parameters obtained are k (biomass growth rate constant) of 0.0536/h and the pre-exponential factor of 5.2537. The deceleration model for growth kinetics of biomass is also used to describe the growth profile of *Rhizopus oligosporus* using cassava starch substrate in solid state fermentation medium. Model matching results with experimental data resulted in the smallest sum of square of residuals of 0.0626 compared to linear, exponential and logistic models. The kinetic parameter obtained is k (biomass growth rate constant) of 0.044/h. [33,48,49].

The deceleration model on the growth of biomass is used to integrate the product formation equation, so that the following equation is obtained at equation 11. From the calculation result using Ms. excel solver program obtained parameters Y_{PX} = 0.1821 g/g and m_p = 0. The results of experimental protein content curve with the model are depicted in the Figure 2.

$$P = P_0 + \left[Y_{PX}\left(k\,A\left(X_0\ \exp\left(A(1-e^{-kt})\right)\ e^{-kt}\right)\right) + m_p\left(X_0\ \exp\left(A(1-e^{-kt})\right)\right)\right] \qquad (11)$$

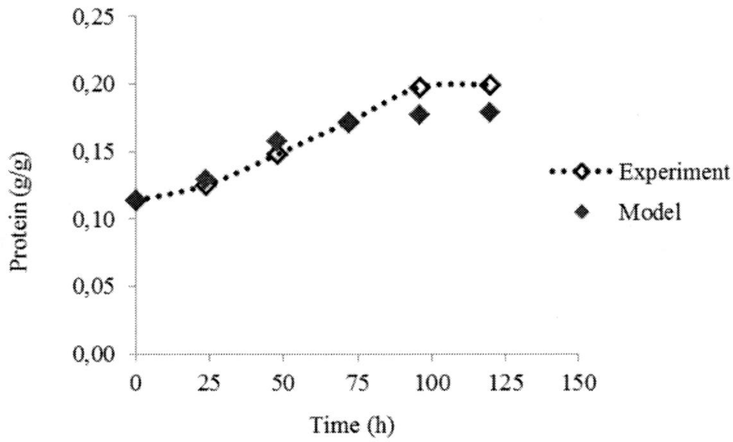

Fig 2. Model increased protein content

The data of experimental results with the model coincide and has a significant proximity between points that the model is acceptable. The kinetic equation of protein formation,

$$\frac{dP}{dt} = 0.1821\ \frac{dX}{dt} \qquad (13)$$

Deceleration model is also used to describe the production of protein from bagasse using *Chaetomium cellulolyticum* through solid state fermentation method. The results of model calculations with experimental data show a significant proximity [50].

4 Conclusions

The protein, fat, and ash content in fermented rice bran have increased by 58.5%, 124.5%, and 18.6%, respectively, while carbohydrate and fiber content decreased respectively by 25.6% and 51.2%. The deceleration model successfully depicts the profile of biomass growth and improved protein well. The kinetic parameters obtained k, A, and Y_{PX} were 0.0536/h, 5.2537, and 0.1821 g/g respectively. This research makes the fermented rice bran has a great potential to be utilized as a functional foodstuff.

Symbols

W	moisture content in the sample (g)
x	dry weight of biomass (g)
S	dry weight of the substrate (g)
X	biomass concentration (g/g)
X_o	initial biomass concentration (g/g)
X_m	maximum biomass concentration (g/g)
t	time (h)
k	growth rate constant of biomass (h^{-1})
μ	specific growth rate of biomass (h^{-1})
A	pre-exponential factor
Y_{PX}	yield product (protein) to biomass (g/g)
m_p	maintenance coefficient of product formation (g/g.h)

Acknowledgments

Authors thank to Department of Chemical Engineering, Diponegoro University for all facilities to do this research.

References

1. W. Hu, J. Henry, T. Shin, I.S. Godber. JAOCS **73**, 1653 (1996)
2. R. Schramm, A. Abadie, N. Hua, Z. Xu, M. Lima, J. Biol. Eng **9,** 1 (2007)
3. Z. Xu, N Hua, J.S. Godber, J. Agric. Food. Chem **49**, 2077 (2001)
4. R.J. Jariwalla, Drugs. Exp. Clin. Res **27**, 17 (2001)
5. S. Khatoon, A.G. Gopalakrishna. JAOCS **81**, 939 (2004)
6. Ardiansyah, H. Shirakawa, T. Koseki, K. Ohinata, K. Hazhizume, M Komai, J. Agric. Food. Chem **54**, 1914 (2006)
7. E.A. Salehi, M. Sardarodiyan, Biochem. Ind. J **10**, 101 (2016)
8. P. Daubresse, S. Ntibashirwa, A. Gheysen, J.A. Meyer, Biotechnol. Bioeng **29**, 962 (1987)
9. C.R. Soccol, I. Iloki, B. Marin, M. Raimbault, J. Food. Sci. Technol **31**, 320 (1994)
10. S.O. Aro, Afr. J. Biotechnol **7**, 4789 (2008)
11. H.S. Hamdy, Roum. Biotechnol. Lett **18**, 7929 (2013)
12. T.C. Do Santos, G.A. Diniz, A.R. De Brito, A.J.V. Pires, M. Fanco, Rev. Caatinga. Mossoró **28**, 248 (2015)
13. C. Sriherwanto, B. Santoso, A. Mahsunah, I. Sujai, S. Toelak, M. Rusmiyati, JBBI **3**, 72 (2016)

14. J.A. Adejuyitan, E.T. Otunola, I.F. Bolarinwa, B.F. Olanipekun, J. Food. Chem. Nanotechnol **3**, 19 (2017)
15. E. Badiale-Furlong, J.L.M. Cacciamani, J. Garda-Buffon, Braz. J. Food. Technol **10**, 233 (2007)
16. C.M. Da Silveira, E. Badiale-Furlong, Braz. Arch. Biol. Technol **52**, 1555 (2009)
17. S. Martins, S.I. Mussatto, G. Martínez-avila, J. Montañez-saenz, C.N. Aguilar, J.A. Teixeira, Biotechnol. Adv **29**, 365 (2011)
18. G.H Hansen, M. Lubeck, J.C. Frisvad, P.S. Lubeck, B. Andersen, Process. Biochem **50**, 1327 (2015)
19. A. Pandey, Biochem. Eng. J **13**, 81 (2003)
20. L. Thomas, C. Larroche, A. Pandey, Biochem. Eng. J **81**, 146 (2013)
21. J.A. Mertens, C.D. Skory, A.S. Ibrahim, Arch. Microbiol **186**, 41 (2006)
22. R.C. Yu, Y.D. Hang, World. J. Microbiol. Biotechnol **6**, 15 (1990)
23. U. Bakir, S. Yavascaoglu, F. Guevenc, A. Ersayin, Enzyme. Microb. Technol **29**, 328 (2001)
24. H.S. Hamdy, IJBT **5**, 284 (2006)
25. P. Chellapandi, H.M. Jani, Braz. J. Microbiol **39**, 122 (2008)
26. M. Karmakar, R.R. Ray, Res. J. Microbiol **6**, 41 (2011)
27. Y.X. Zheng, Y.L. Wang, J. Pan, J.R. Zhang, Y. Dai, K.Y. Chen, Bioresour. Technol **241**, 1138 (2017)
28. J.E. Bailey, D.F. Ollis. *Biochemical Engineering Fundamentals* (McGraw-Hill Book Company, New York, 1987)
29. M.C. Del Barrio, R.B. Demafelis, S.M. Mercado, J.L. Movillon, C.M. Rebancos, Philipp. Agric. Scientist **95**, 129 (2012)
30. J. Almquist, M. Cvijovic, V. Hatzimanikatis, J. Nielsen, M. Jirstrand, Metab. Eng **24**, 38 (2014)
31. J.A.G. Rueda, V.A.L. Gonzales, M.A.O. Supelano, J.A.P. Mendoza, C.G. Luna, L.J.L. Giraldo, Cienc. Tecnol. Futuro **6**, 81 (2015)
32. D.A. Mitchell, N. Krieger, M. Berovic. *Solid-State Fermentation Bioreactor: Fundamental of Design and Operation* (Springer-Verlag, Germany, 2006)
33. G. Viccini, D.A. Mitchell, S.D. Boit, J.C.S Gern, A. da Rosa, R.M. Costa, F.D.H. Dalsenter, F.O. von Meien, N. Krieger, Food. Technol. Biotechnol **39**, 271 (2001)
34. G. Viccini, D.A. Mitchell, N. Krieger, Food Technol. Biotechnol **41**, 191 (2003)
35. M.D. Oliveira, V. Feddern, L. Kupski, E.P. Cipolatti, E. Badiale-Furlong, L.A. de Souza-Soares, Cyta. J. Food **8**, 229 (2010)
36. C.G. Schmidt, E.B. Furlong, Bioresour. Technol **123**, 36 (2012)
37. F.N. Yunus, M. Nadeem, F. Rashid, J. Inst. Brew **121**, 553 (2015)
38. L. Kupski, E. Cipolatti, M. Da Rocha, M.S. Oliveira, L.A. Souza-Soares, E. Badiale-Furlong, Braz. Arch. Biol. Technol **55**, 937 (2012)
39. AOAC. *Official Methods of Analysis 17th* (Washington DC, 2000)
40. A. Augustine, I. Joseph, R.P. Raj, J. Mar. Biol. Ass. India **48**, 139 (2006)
41. D. Mazaheri, S.A. Shojaosadati, Iran. J. Biotechnol **11**, 156 (2013)
42. R. Ravinder, L.V. Rao, P. Ravindra, Food. Technol. Biotechnol **41**, 243 (2003)
43. M.L. Shuler, F. Kargi. *Biopreocess Engineering Basic Concepts* (Prentice-Hall International Inc, New Jersey, 2002)
44. K.C.Massarolo, T.D. De Souza, A.C. Ribeiro, E.B. Furlong, L.A. De Souza Soares, Biocatal. Agric. Biotechnol **8**, 204 (2016)
45. V. Feddern, E.B. Furlong, L. Almeida, S. Soares, Ciênc. Tecnol. Aliment. Campinas **27**, 800 (2007)
46. O.O Odugawa, M.O. Edema, A.O. Ayeni, Bioresour. Technol **99**, 1816 (2008)
47. D.H. Griffin. *Fungal physiology* (John Wiley & Sons, New York, 1994)
48. D.A. Mitchell, D.D. Do, P.F. Greenfield, H.W.A. Doelle, Biotechnol. Bioeng **38**, 353 (1991)
49. L. Ikasari, D.A. Mitchell, D.M. Stuart, Biotechnol. Bioeng **64**, 722 (1999)
50. O. Bravo, A. Ferrer, C. Aieloo, A. Ledesma, M. Davila, Biotechnol. Lett **16**, 865 (1994)

Conversion of Cassava Starch to Produce Glucose and Fructose by Enzymatic Process Using Microwave Heating

Siswo Sumardiono[1,*], *Gita* Budiarti[2], and *Kusmiyati*[3]

[1]Department of Chemical Engineering, Faculty of Engineering, Diponegoro University, 50275, Semarang, Indonesia
[2]Master Program of Department of Chemical Engineering, Diponegoro University, 50275, Semarang, Indonesia
[3]Department of Chemical Engineering, Faculty of Engineering, Universitas Muhammadiyah Surakarta, 57102, Sukoharjo, Indonesia

Abstract. In this study, variation of glycosidase enzyme concentration and saccharification time on enzymatic hydrolysis using microwave have been investigated. Concentration and kinetic parameters rate of glucose and fructose were analyzed. Cassava starch was liquefied and gelatinized by microwave at 80°C. The gelatinized starch was saccharified at 60°C using (0.2;0.4;0.6;0.8;1% (w/v)) glycosidase enzyme for 24, 48 and 72 hours. The glucose which has been saccharified with 1% glycosidase enzyme for 72 hours gave highest conversion 66.23 %. The optimization process by multilevel reaction gave the highest conversion at enzyme concentrations 0.88 %and saccharification time 29 hours that 68.82%. The highest conversion of glucose was isomerized to fructose. The fructose which has been isomerized for 180 minutes gave highest conversion 20.05 %. The kinetics enzymatic reaction was approached and determined by Michaelis - Menten equation, Km and Vmax of reaction for glucose 22.94 g/L; 2.70 g/L hours and for fructose 3.39 g/L; 0.38 g/L. min respectively.

1 Introduction

Sugar is a one of basic needs for the Indonesian people, especially for consumption and food processing. Sugar demand in Indonesia is still dominated by sucrose. Sugar demand in Indonesia is still dominated by sugar (sucrose). Glucose is a monosaccharide with molecular formula $C_6H_{12}O_6$. Glucose is widely used in the food and pharmaceutical industries. Enzymatic hydrolysis produces glucose concentrations higher than the acid hydrolysis [1]. Fructose is the sugar isomer of glucose that is mostly used to produce drinks [2-6].

Raw material for producing glucose is material containing polysaccharide, obtained from starch. Starch is a polymer with the chemical formula anhydrous monosaccharide ($C_6H_{10}O_5$) and with the main constituent amylose and amylopectin [7-10]. Starch for producing glucose and fructose are corn starch, cassava starch, sweet potato starch, and sorghum. The conversion of glucose from cassava starch is greater than corn starch that is equal to 93.56% while the corn starch is only 91.8% [7].

Cassava starch is widely used in the food sector, paper industry, textile industry and chemical industry as a producer of bioethanol [1, 9, 11-13]. Today, the conventional method of acid hydrolysis has been replaced by enzymatic hydrolysis. The enzyme can be derived from bacteria or fungi. The enzyme concentration is one that affects the yield of glucose and fructose. Enzymes contribute 23-40% of the total cost of the production process or in other words the amount of enzyme has a significant impact on the economic viability of the process [11]. This can be overcome by optimizing operating conditions of enzymatic hydrolysis [8-9, 12-13]. Gelatinization heating conventional takes a long time and a lot of energy to reach a temperature of gelatin. Microwave is used to accelerate warming and reduce energy required [7, 14-17]. This study will be focused on the conversion of glucose and fructose from cassava starch by enzymatic process using microwaves. Afterwards, determine the optimal concentration of enzyme saccharification time and kinetics parameter of enzymatic reaction is calculated of each process.

Fructose is a natural sugar that serves as a substitute for cane sugar or sucrose. Advantage fructose is containing triglycerides-resistant, which can reduces cholesterol. Fructose is commonly produced by fruits. Difference fructose and glucose are the optical rotation. Glucose has the optical rotation to the right, while the fructose to the left. Fructose round to the left is what causes the bitter taste of fructose compared to glucose [10].

* Corresponding author: siswo.sumardiono@che.undip.ac.id

2 Materials and method

2.1. Materials

Commercial cassava starch was purchased from a cassava industry in Margoyoso, Pati (Indonesia). Other materials used included α-amylase enzyme, glycosidase enzyme with activities of each 4.000 u/g and 150.000 u/g from Sqzyme, glycoisomerase enzyme from Novozyme and other chemicals agents were analytical grade from Merck (Germany).

2.2 Liquification

In this stage, suspension of starch was prepared by mixing of 200 g starch and aquadest to 500 ml and $CaCl_2.2H_2O$ 40 ppm. Furthermore, the pH of suspension was adjusted to 5.3 or 6 with addition of HCl 5% or 5% NaOH. Then, α-amylase enzyme with concentration of 0.1% (w/v) was added into starch suspension. Starch suspension was heated in a microwave at 80°C for 10 minutes. This stage finish was characterized by a change of the thick white suspension starch solution to brown (caramel) dilute. After the liquefaction process, the enzyme was inactivated with addition of 5% HCl to pH 4.5 [1].

2.3 Saccharification

The results of the previous stages, maltodextrin was divided into five sections. Each of these sections would be used as an ingredient saccharification at 60°C and pH 4.5. Glucosidase enzyme was added into suspension starch with various concentrations (0.2%; 0.4%; 0.6%; 0.8%; 1% (w/v)). The suspension starch was saccharified for 24. 48. 72 hours [11]. The saccharification product was filtered. Filtrate was analyzed dextrose equivalent value using HPLC. The scheme of experiment can be see in Figure. 1.

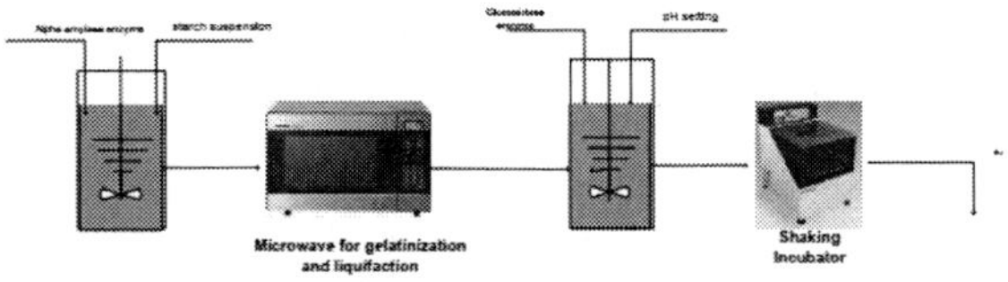

Fig 1. Equipment Series Starch Hydrolysis by Microwave

2.4 Fructose Isomerization

50 mL of the best glucose product was mixed with 16 mg $MgSO_4.7H_2O$. The mixture pH was measured to 7.5. The mixture was added with 50 mg glucoisomerase enzyme. The mixture was incubated in shaking incubator at 60°C for 3 hours. Samples was taken 10 ml every 30 minutes for analysis of enzymatic kinetic reactions [11]. Fructose concentrations were estimated by HPLC.

2.5 Determination of Concentration of Glucose and Fructose by HPLC

Samples was filtered and then was injected into the injector of 20 μL to a passed by port injection which pressurized. Furthermore, by pass tap was opened, so that the sample flowed into the column pressure, from HPLC equipment which using mobile phase aquabidest and detector refraction index [11]. The chromatogram described a retention time of each component in the sample was observed on the printer. Further, interpreted the proportions of the components were compared in the sample. The sample before it was injected into the column, glucose standard solution was injected as the standard comparator. HPLC method was used for quantitative analysis and qualitative at the same time. For qualitative analyed by comparing the sample chromatogram with the chromatogram reference standard based on their retention times. As for the quantitative analyed can be used in equation 1.

$$Cx = \frac{A_X}{A_P} \times C_P \quad (1)$$

Where C_X is the concentration of the sample, C_P is the standard concentration, A_X is the peak area of the sample and the A_P is the peak area of the standard.

2.6 Enzymatic Reaction Kinetics

Kinetics of enzymatic reactions in this study approached by Michaelis-Menten equation. Substrate concentration at time t was obtained from the equation stoichiometry calculations hydrolysis and isomerization reactions. The data obtained by experiments conducted initial substrate [So], the substrate at time t [St] and reaction time (t). The data was estimated by Equation 2 [4].

$$\frac{1}{t}\ln\frac{S_o}{S} = \frac{v_{maks}}{K_m} - \frac{S_o - S}{K_m t} \quad (2)$$

With t is the reaction time (h), Vmax is the maximum reaction rate (mole/L.h) and S is the substrate concentration (mole/L). The equation was linearized to evaluating the value of K_m and V_{max}. $\frac{1}{t}\ln\frac{S_o}{S}$ is the y-axis, $\frac{S_o - S}{K_m t}$ is ax and $\frac{v_{maks}}{K_m}$ is b.

3 Results and Discussion

3.1. Liquifaction

The liquefaction process produces maltodextrin, which has a dextrose equivalent (DE) of less than 20. In this study, liquefaction time 10 minutes using a microwave with a temperature of 80°C. The household microwave does not have temperature indicator, so controlling the temperature was done by power settings. The power is used 100% to get a temperature 80°C.

The principle of microwave heating is the collision solvent and governed by two phenomena that ionic conduction and dipole rotation. Microwave radiation can reduce the solubility and break up the crystalline structure of starch. Heating with microwaves can also affect the concentration of amylase and amylopectin better than conventional heating. The advantage of using microwaves are the liquefaction process becomes faster and more efficient because can reduce heat loss [5].

3.2 Saccharification

Saccharification is a fermentation process which convert maltodextrin into glucose. At this stage it is a glycosidase enzyme used. Mechanism of action of the enzyme α-glycoside is break-1.6 glycoside bond and presented in the Figure 2. Conditions operating specific of glycosidase enzyme are pH 4.5 and a temperature of 60°C. Factors that affect the reaction of glucose are glycosidase enzyme concentration and incubation time [7]. Result from experimental determination of enzyme concentration and time can be seen in Figure 3.

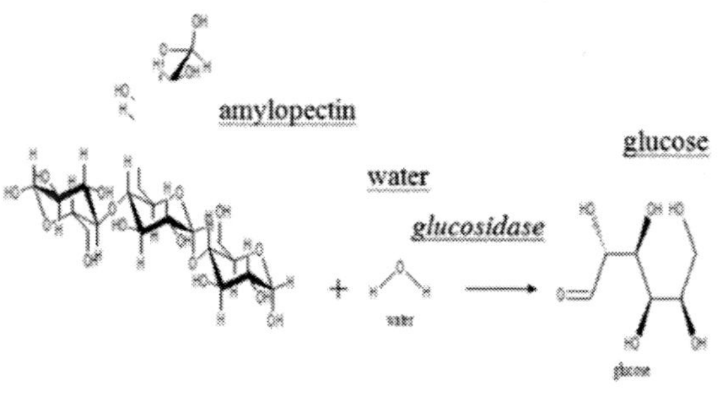

Fig 2. Starch Hydrolysis Reaction

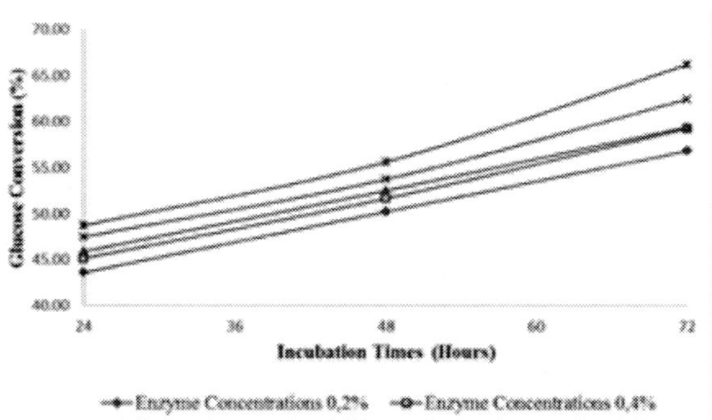

Fig 3. Relation of The Hydrolysis Incubation Time and Conversion Glucose by Various Glucosidase Enzyme Concentration

Figure 3. shows that the highest conversion was obtained 66.23% by the concentration glycosidase enzyme 1% (w/v) for 72 hours. Conversions obtained lower compared to conventional enzymatic hydrolysis [20]. Previous researcher conducted a study of cassava starch hydrolysis with conventional heating, the conversion of glucose were obtained by 95.95% [7]. Conversion of cassava starch into glucose theoretically was at 77 % [6]. The conversion of glucose derived from the hydrolysis is depending on the concentration of cassava starch and the reaction parameters, such as temperature, pressure, substrate concentration, time and catalyst concentration [1].

The reaction in this study was not efficient, because enzyme requirement and reaction time are still high. Enzymes are expensive and difficult biocatalisator to be regenerated, because of soluble solvent. To reduce enzyme requirement reactions carried out multi-stage trial. Multi-stage reaction was obtained with addition a substrate and a little enzyme into the result of glucose from the first reaction [8-9, 13]. Results from a multi-stage reaction experiment are presented in Table 1.

Table 1. Glucose Yield Data on Multi Stage Reaction.

Glucosidase Enzyme Conc. (% w/v)	α-amilase Enzyme Conc. (% w/v)	Starch Conc. (% w/w)	Time (Hours)	Glucose Conc. (% w/w)	Glucose Conversion a (%)
0.80	0.05	24.26	24	11.52	47.48
0.88	0.09	34.26	29	23.57	68.82
0.95	0.12	44.26	36	28.61	64.65
1.01	0.14	54.26	40	32.83	60.50

Table 1. shows that at 24 to 29 hours the reaction increased, but the reaction 36 hours decreased conversion from 68.82% to 64.65%. This is due to the enzyme has reached the maximum conversion so that the addition of substrate. This process is more efficient in terms of cost because it requires little glycosidase enzyme and incubation times are also shorter. Enzymatic hydrolysis using microwaves could be considered for the production of glucose in the future because it can accelerate the process of gelatinization. Glucose conversion obtained from previous research is not reaching the standards in the industrial standard at least 95-96% [21].

3.3 Isomerization of Fructose

Isomerization is reaction of formation fructose from glucose [3, 9, 13, 21]. Isomerization was using enzymes glycoisomerase as biocatalyst. Glycoisomerase was obtained from commercial enzyme artificial Sweetzyme or was isolated from *Lactobacillus brevis*, *Pseudomonas hydrophila*, *Streptomyces phalochromogeneses*. This enzyme requires Mg^{2+} ion as activators in the form of compounds $MgSO_4.7H_2O$. Glycoisomerase enzyme optimal conditions was obtained at pH 7.5 at 60°C. The incubation time should not be too long to avoid arising the presence of dyes [7]. Isomerization fructose reaction conversion data was presented in Figure 4 and Figure 5 respectively.

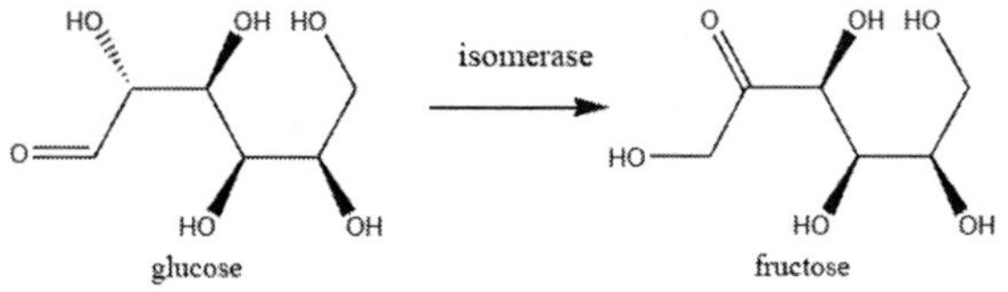

Fig 4. Isomerization Fructose Reaction

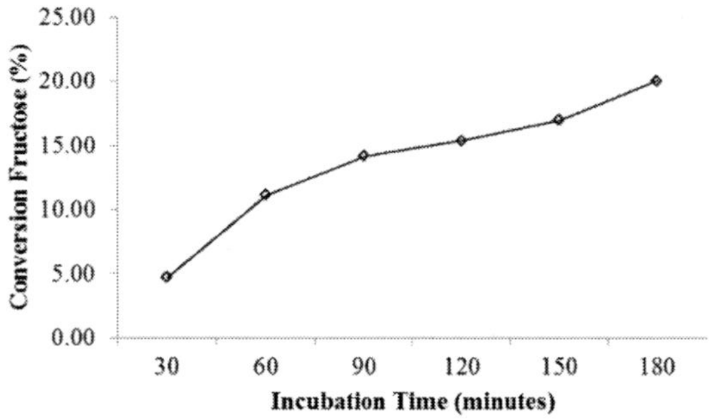

Fig 5. Fructose Conversion about Reaction Time

The highest conversion of fructose 20.05% was reached at 180 minutes. The reaction times affect conversion increase. If the enzyme reaches saturated condition, reaction time and concentration of the substrate does not affect the conversion of fructose. Even the addition of substrate and reaction time can reduce the yield of fructose [20]. Conversion of fructose does not depend on the content of starch and glucose content, but the raw materials used [7].

3.4 Kinetics Enzymatic Reactions

This study obtained kinetic enzymatic hydrolysis graph that shows in Figure 6. From this figure the equation can be calculated Michaelis-Menten constant (K_M) and maximum speed (Vmax). Line equation y = -0.0436x + 0.1177, so the value of K_M is 22.94 g / L and Vmax was 2.70 g/L.h. The result has been tested with a total sum square regression, a significant value of the coefficient of 99.14%. It shows that kinetic coefficient is significant.

Previous researcher conducted a study and calculation of the reaction kinetics of enzymatic hydrolysis of starch to glucose [8]. The calculated result of K_M value was 194.5 g/L and Vmax was 50.8 g/L min. Differences constant value and the maximum speed is obtained due to the starch substrate and reaction temperatures used. To fructose chart is presented in Figure 7.

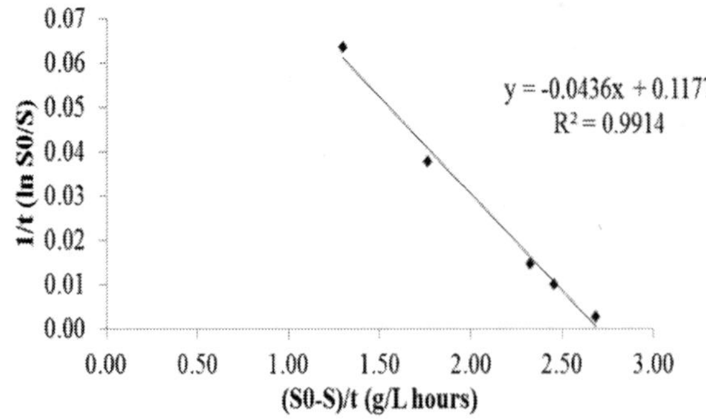

Fig 6. Kinetics Enzymatic Hidrolysis

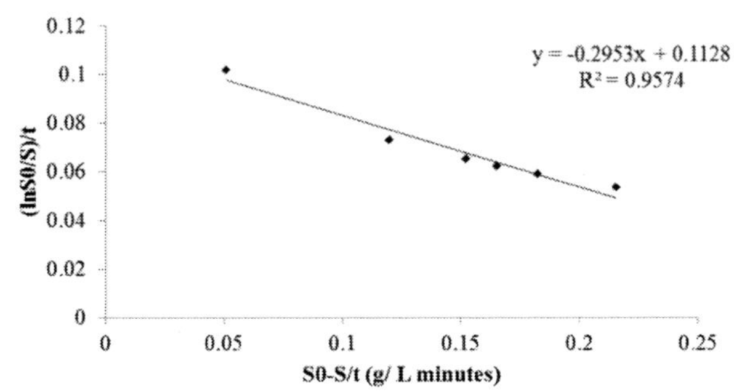

Fig 7. Kinetics Isomerization

The equation of the line in Figure 7. y=-0.2953x+0.1128 obtained K_M and Vmax values of 3.39g/L and 0.38g/Lmin. The result has been tested with a total sum square regression. A significant value of the coefficient of 95.74%. It shows the kinetic coefficient is significant. Previous researcher conducted isomerization kinetics calculations using chicory inulin as substrate, value of K_M amounted to105g/L and Vmax values of 0.67g/L min [2].

4 Conclusions

Based on research that has been done can be conldced as follows : i) glucosidase enzyme concentration affects the conversion of glucose, the greater the concentration of the enzyme. the greater the yield of glucose obtained, ii) incubation time affects the yield of glucose, the longer the incubation time.The higher the yield of glucose produced, iii) glucosidase enzyme concentration was 1% and the optimal incubation time of 72 hours, the yield obtained under these conditions is 66.23%, iv) in order to more optimally use glucosidase enzyme reaction multilevel, optimal glucosidase enzyme concentration of 0,88 % with a reaction time of 29 hours, the resulting conversion of 68.82%, v) the best conversion of fructose 20.05 % was obtained at the incubation time 180 minutes, vi) coefficient of enzymatic reaction kinetics K_M glucose is22.94 g/L and Vmax is 2.70 g/L hours, andvii) coefficient of enzymatic reaction kinetics fructose K_M and Vmax are 3.39 g/L and 0,38 g/L min.

Acknowledgement

The authors thank to Faculty of Engineering, Diponegoro University via Competitive Research Grant for financial support.

References

[1] A.A. Ayoola, O.A. Adeeyo, V.C. Efeovbokhan, Ajileye. O, Journal of Science and Technology, **2**, 6, 353–357 (2012)

[2] S. Curcio, G. Iorio, E. Ricca, V. Calabro, Process Biochemistry. **44,** 466–470 (2009)

[3] A. El-fallal, M.A. Dobara, A. El-sayed, N. Omar, Intech. 459–488 (2012)

[4] H.S. Fogler, *Elements of Chemichal Reaction Engineering Third Edition*. New Delhi. Prentice Hall of India. (2004)

[5] G. Gong, D. Liu, Y. Huang, *Bio. Eng*. **107**, 67-103 (2010)

[6] R. Howeler, *Workshop on Partnership in Modern Science to Develop a Strong Cassava Commercial Sector in Africa and Appropriate Varieties by 2020.Bellagio. Italy. May.*(2006)

[7] R. Johnson, G. Padmaja, International Research Journal of Biology Sciences. **2**,10, 68–75 (2013)

[8] F.J. Lopez, W.K. Wong, Journal of Thermal Biology. **215,** 1, 1-15 (2005)

[9] S. Morales, H. Alvarez, C. Sanchez, Journal of Elsevier. **86**, 25–30, (2008)

[10] I.N. Widiasa, H. Susanto, Journal of Food Engineering. **95**, 3, 423-431 (2009)

[11] S. Chamsart, C. Sawangwon, S. Tungkao, Y. Waiprib, *Proceeding of the 15th Thailand Chemical Engineering and Applied Chemistry. Chonburi*, (2005)

[12] K. Saelim, Y. Dissara, A.H. Kittikum, Songklanakarin Journal of Science and Technology. **30**, 65-71 (2007)

[13] M..U. Ukwuru, S.E. Egbonu, Academia Journal of Food Research. **1,** 1, 001-013, (2012)

[14] A.F. Aderibigbe, A.N. Anozie, L.A. Adejumo, R.U. Owolabi, New Clues in Sciences. 2, 3, 50-58 (2012)

[15] F.H.M. Souza, R.F. Inocentes, R.J. Ward, J.A. Jorge, R.PM. Furriel, Journal of Molecular Catalysis B : Enzymatic. **94**, 119-128, (2013)

[16] V. Singh, A. Tiwari, P. Kumari, S. Tiwari, Journal of Carbohydrate Research. **341**, 2270-2274 (2006)

[17] H.M. Nguyen, S.H. Ha, Y.M. Koo, *Journal of Biotechnology*. **136s**, s356-s401 (2008)

[18] S. Zhu, Y. Wu, Z. Yu, C. Wang, F. Yu, S. Jin, Y. Ding, R. Chi, J. Liao, Y. Zhang, Biosystem Engineering. **93**, 3, 279-283 (2006)

[19] F.J. Izquierdo, E. Penas, M.L. Baeza. R. Gomez, International of Dairy Journal. **18,** 918-922 (2006)

[20] A.M. Omemu, I. Akpan, M.O. Bankole, O.D. Tenida, *African Journal of Biotechnology*. **4**, 1, 19-25 (2005)

[21] J. Pontoh, N.H. Low, Food Research International. **28**, 4, 379-385 (1995)

Improvement the Chemical Properties of Cassava Through Microbial Fermentation Using Rhizopus Oligosporus

Totok Prasetyo [1,*], *Ferry* Ardhianto[2], and *Mayang* Pawitra[2], *Siswo* Sumardiono[2]

[1] Departement of Mechanical Engineering, Semarang State Polytechnic, Postal Code 50275, Semarang, Indonesia
[2] Department of Chemical Engineering, Diponegoro University, 50275, Semarang, Indonesia

Abstract. Indonesia is the world's largest importer of wheat. Because of high wheat imported, Indonesia's researchers has seek an alternative to wheat-based local materials that having equivalent physicochemical properties. Potential alternative replacement of wheat flour is cassava. Cassava can be modified its rheology and physicochemical properties in order to be wheat equivalents. Cyanide contained in the cassava flour cannot be removed immediately by physical treatment. Several efforts have been made, such as soaking with saline solution, the addition of citric acid, and heating, showed a reduction in levels of cyanide in the flour. In this study, reduced levels of cyanide bound by converting it into protein using of Rhizopus oligosporus. While cyanide dissolved with the help of soaking with saline solution and buffer solution. In this case, use a buffer solution such as sodium sulphites which serves to change the form of cyanide dissolved into a form that is less toxic. In addition, a buffer solution pH also controls a range of 6, because in this range of cyanide bound to be easier to dissolve. With the combination treatment is expected to be a reduction in the levels of cyanide and exhibited significantly increased levels of protein in cassava flour. Flour has a protein content of best nutritional protein at 3.68 % dry basis. Cyanide reduction was highest in samples with high levels of *Rhizopus oligosporus* grated 1.5% (w/w) as 72 hours, with a lag of 0.09 (mg/kg) cyanide. Flour with the best nutritional quality is obtained from fermentation with conditions: grated, 12 hours of fermentation as well as using *Rhizopus oligosporus* 3.5% (w/w).

1 Introduction

Wheat (Triticum sp.) is a group of tribes cereals grains are rich in carbohydrates. Wheat flour is made with raw wheat and Indonesia to meet this need by importing 5.4 million tons of wheat and 680 thousand tons of wheat, with a nominal value of imports amounted to 15.5 trillion rupiah. Wheat flour is widely used as raw materials for various food products, such as noodles, cakes, breads, cookies, or other applications. One of the food commodities should be considered to be developed as an alternative use of wheat flour in Indonesia are tubers like cassava. Unlike wheat, cassava grows well in Indonesia. Data show that in Indonesia, cassava is grown around 1.4 million hectares per year, and its roots are used for food, feed and as raw material for starch extraction. Meanwhile, preliminary data obtained from BPS show that Indonesia cassava production in 2010 was 2.080.853 tons [9, 33]. The fact that can be found in the community that has been largely cassava by farmers only sold without being processed, so the price is very cheap. But actually if cassava is processed further to produce modified starch with a higher price [34]. This modification effort can certainly improve cassava farmers and encourage small and medium entrepreneurs as producer's cakes, bread, noodles, and others to develop. Processed cassava flour price is much cheaper than wheat flour from wheat.

Flour made from cassava could potentially be used as a replacement or substitute the use of wheat flour. Excess flour made from cassava flour is having a short chain, so it is easy to digest digestive system, does not cause allergic reactions in the body because it does not contain gluten, as well as cooking oil does not absorb excess oil. The main problem that arises is the amount of cyanogenic glucoside content of compounds in the cassava flour is equal to 62 mg [1] which have a tendency as a poison. In Indonesia, cyanide levels in cassava flour allowed by BPOM is 40 ppm/unit weight. Value levels far above the standard FAO / WHO, 2010 in the amount of 10 ppm / unit weight.

Previous researcher revealed that the bound cyanide contained in the cassava flour cannot be removed immediately by physical treatment. Some efforts have been made, such as soaking with saline solution, the addition of citric acid, and heating, showed a reduction in levels of cyanide in the flour 35 w/w. Other

* Corresponding author: tprasetyo06@yahoo.com

researchers, reveals something else about his research on cassava. His research concluded that by varying the method of drying and drying temperature on fermented cassava cyanide levels lagged obtained a minimum of 30.35 w / w. Then Owuamanam conducted a study with cassava soaking with water at pH 8 for 72 hours with citric acid. This treatment is getting late for 39.28 mg/kg HCN levels. The above research was focused on removing cyanide levels without improving nutrition flour produced.

Cassava can be modified its rheology and physicochemical properties in order to be wheat equivalents. Cassava starch modification, which involves the alteration of the physicochemical characteristics of the native starch to improve better paste clarity, gel stability, increased resistance to retrogradation, increased solubility and improved freeze–thaw stability, which improves their application as stabilizers, fillers, binders, adhesives, and its functional characteristics, can be used to tailor starch to specific food applications [2-4, 8, 13, 18-23, 24, 26, 29, 31-32, 35-36].

The fermentation and sun-drying or UV irradiation processes produce cassava depolymerization [16] which is correlated with expansion ability [6-7, 17]. Sour cassava starch (Polvilho azedo) is a naturalfermented and sun-dried cassava starch that shows expansion properties during baking. However, acidification of cassava starch with lactic, acetic, or butyric acid (10-11, 30], followed by oven drying, failed to produce any great expansion ability after baking. Similarly, amylase depolymerization [10] or natural lactic fermentation [16] did not afford expansion properties to cassava starch. Consequently, it has been suggested that the expansion properties of sour cassava starch result, at least in part, from the combined effect of starch fermentation and sun-drying [12, 15-16, 25, 30].

In laboratory studies, cassava starch acidified with lactic acid and subjected to UV irradiation expanded during baking [5-6, 12, 15, 17]. Thus, expansion could be a result of the photochemical treatment. However, in contrast to UV-irradiated cassava starch, UV-irradiated corn starch showed no expansion during baking [5]. The structural changes involved in the transformation of sour cassava starch dough are not clearly understood. Besides the natural fermentation in tanks which normally used approximately 30 days, soaking in 1% (w/w) lactic acid solution for 15-30 min was also studied [14, 27-28]. Direct irradiated with artificial UV source was applied to the sour cassava starch. The oxidative modification of cassava starch with lactic acid together with either sun dried or UV proved to obtain baking characteristic of an increase specific volume of tested biscuits. However, this behavior did not show in sour cassava starch treated under oven-drying method [30].

In this study, reduced levels of cyanide bound by converting it into protein using of *Rhizopus oligosporus*. While cyanide dissolved with the help of soaking with saline solution and buffer solution. In this case, use a buffer solution such as sodium sulphites which serves to change the form of cyanide dissolved into a form that is less toxic. In addition, a buffer solution pH also controls a range of 6, because in this range of cyanide bound to be easier to dissolve. With the combination treatment is expected to be a reduction in the levels of cyanide and exhibited significantly increased levels of protein in cassava flour.

2 Material and Methods

Cassava has acquired from the estate then peeled the skin and washed under running water. This treatment is in addition to remove dirt still attached, can also eliminate some of cyanide contained in the outer surface of the cassava. Soaking with salt and mixing grated cassava with buffer concentration. Cassava has been washed and soaked with saline solution and a buffer solution (Na_2SO_3). Soaking done in a vessel containing a solution of each salt and buffer to a concentration of 2% w/w. Soaking do in each solution was performed for 1 hour. After soaking cassava is ready to be fermented with *Rhizopus oligosporus*. After a thorough soaking, cassava is then divided into two variables, the form of chips and grated. Then each variable is taken 250 grams for fermented in the fermentor with the help of *Rhizopus oligosporus*. Used *Rhizopus oligosporus* use of three variables, ie 1.5 w/w, 2.5 w/w and 3.5 w/w. Each period of 12 hours, 24 hours, 36 hours, 48 hours, 60, and 72 hours, samples harvested.

Fermentation yields then dried using sunlight. Drying is carried out by using sunlight for 2 to 3 days. Flour and sieving fermented cassava that has been dried is then transformed into flour using a blender or food processor. Results have been obtained and sifted, to obtain flour with size of 60 mesh.

Qualitative analysis of cyanide used in testing, the testing principle of HCN - soluble in water, in the heat and acid HCN will evaporate, and HCN vapor reacts with picric acid to form a red color. Quantitative analysis can be done by spectrophotometryc analysis. The working principle of this method is the cyanide in the sample is converted into cyanogen chloride (CNCl) reacts with Chloramin T at a pH less than 8 hydrolyzed to cyanate. After reacting perfectly, CNCl forming red blue with barbituric acid in pyridine and color happens to read at a wavelength of 578 nanometers.

3 Results and Discussion

Sample form is one of the factors that can influence the process of linamarin breakdown during fermentation. In the experiment, tested forms of chips and grated against the final cyanide levels in cassava flour. Of the three graphs above in the Figure 1, Figure 2 and Figure 3, it can be seen that the cassava flour with grated good form with a concentration of 1.5% by weight *Rhizopus oligosporus* up to 3.5% by weight, resulting in levels of cyanide which is less than the cassava flour in the form of chips. That is because the contact surface on the grated surface area larger than the chips. In the process of solving linamarin with the help of a buffer solution to maintain a neutral pH, the greater the surface area of contact, the easier linamarin to rupture and produce aspartate.

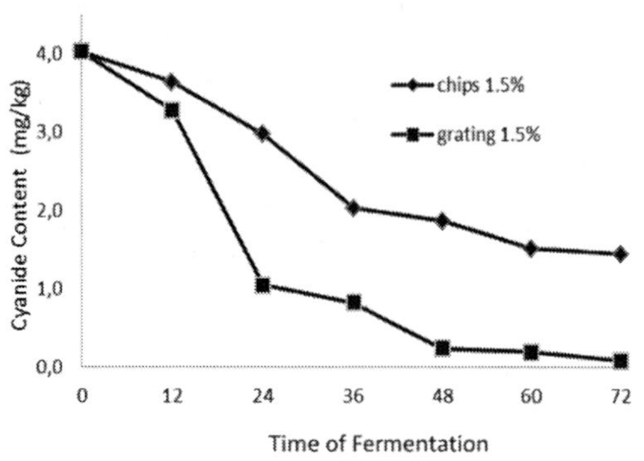

Fig 1. Cyanide content as function of time fermentation with variable chip and grating cassava for *Rhizopus oligosporus* concentration 1.5 % (w/w)

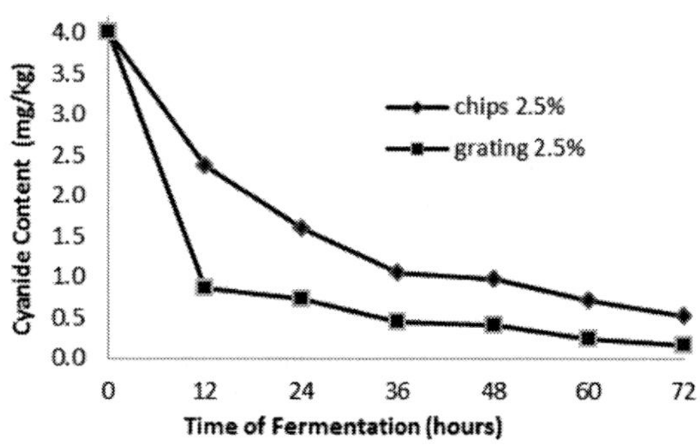

Fig 2. Cyanide content as function of time fermentation with variable chip and grating cassava for *Rhizopus oligosporus* concentration 2.5 % (w/w)

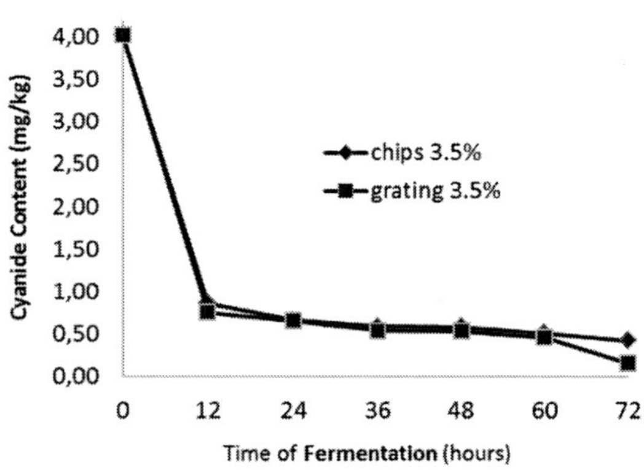

Fig 3. Cyanide content as function of time fermentation with variable chip and grating cassava for *Rhizopus oligosporus* concentration 3.5 % (w/w)

Figure 4 and Figure 5 is the % of *Rhizopus oligosporus* as a factor affecting the final level of cyanide in cassava flour. From the figures can be seen that for both types of cassava flour (in the form of grated and chips) cyanide levels decreased significantly. From graphics chips, it can be seen that the greater percent of *Rhizopus oligosporus*, the lower levels of cyanide were left behind. This phenomenon is also found in cassava flour in a grater. That is because the more weight percent of *Rhizopus oligosporus* is used, the more the enzymes produced in the fermentation. Small levels of cyanide showed that regardless how much cyanide is converted into proteins.

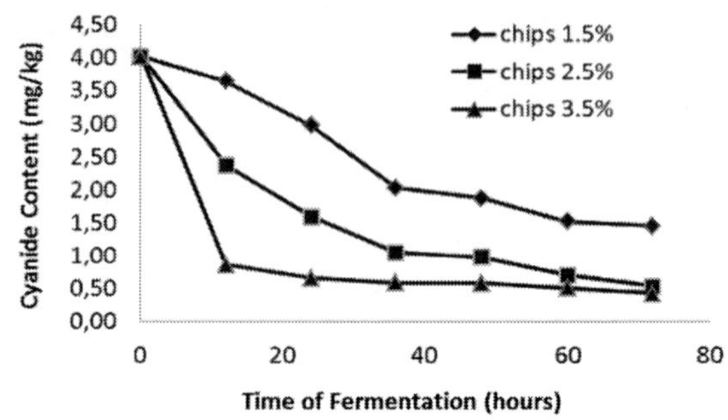

Fig 4. Cyanide content as function of time fermentation of chip cassava with variable *Rhizopus oligosporus* concentration

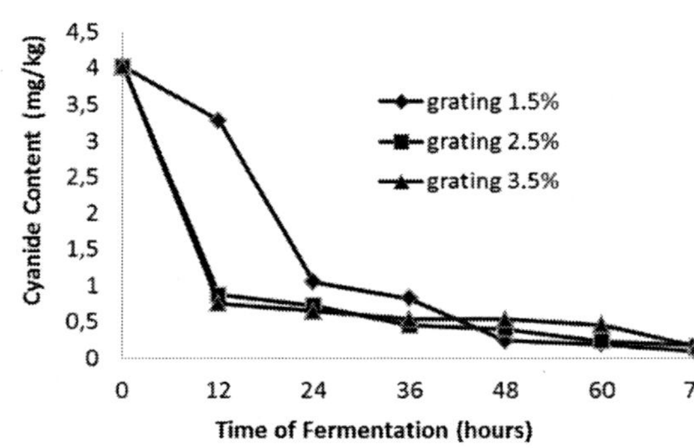

Fig 5. Cyanide content as function of time fermentation of grating cassava with variable *Rhizopus oligosporus* concentration

Mold has a short life cycle, as well as *Rhizopus oligosporus* we use has survival up to 72 hours. If the fermentation is carried out over time, the results will not be too much of a difference. It is caused by a fungus that has entered the phase of stagnation, meaning that the amount of mold that life equal to the number of dead mold. This means that energy use will remain the same after this phase occurs. So the reduction of cyanide after 72 hours will not have much difference. Maximum decrease obtained for the forms of chips is 0.423 (mg/kg), while the form of grated 0.09 (mg/kg).

In the Figure 6, Figure 7, and Figure 8 above, were described relations protein content obtained with different forms of media. The chart shows, the use of media chips produce higher protein than using media grate, either on the use of 1.5% (w / w), 2.5% (w / w) or 3.5% (w / w) *Rhizopus oligosporus*. Grated media has advantages in a more rapid decomposition of cyanide. But in terms of the formation of the protein, the opposite occurs. Media chips are more porous, as there are cavities in the media that facilitates the diffusion of air into and out of the media. This provides an advantage in the metabolism of *Rhizopus oligosporus*, because it is an aerobic organism. Better metabolic enzymes will produce more protein to form cyanide decomposition. Decomposed of cyanide in the media chips is less, but more optimally utilized for conversion into protein, so the protein produced.

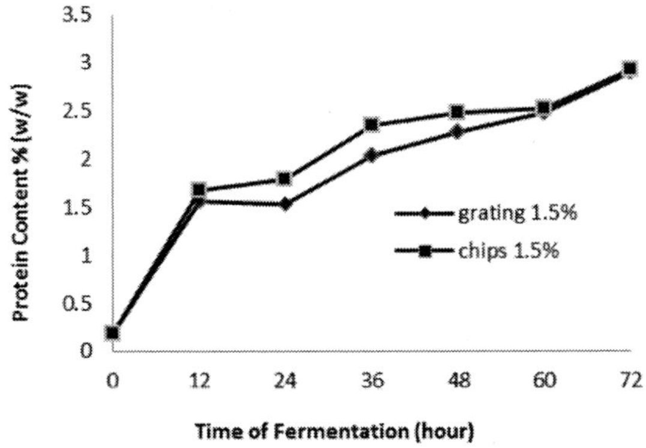

Fig 6. Protein content as function of time fermentation with variable chip and grating cassava for *Rhizopus oligosporus* concentration 1.5 % (w/w)

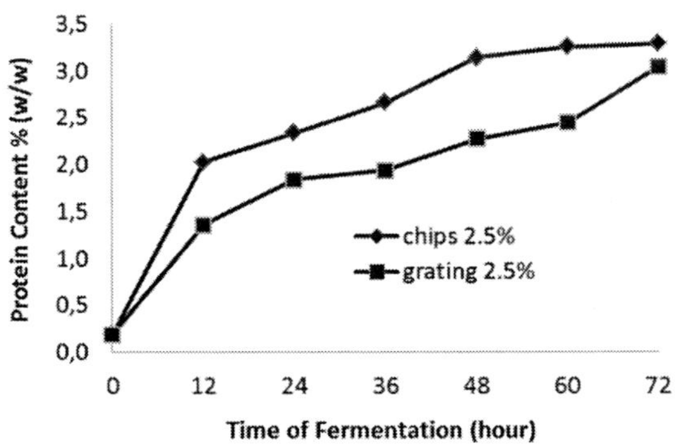

Fig 7. Protein content as function of time fermentation with variable chip and grating cassava for *Rhizopus oligosporus* concentration 2.5 % (w/w)

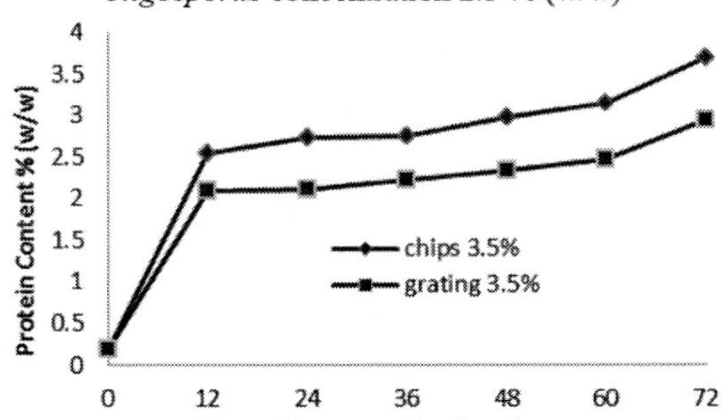

Fig 8. Protein content as function of time fermentation with variable chip and grating cassava for *Rhizopus oligosporus* concentration 3.5 % (w/w)

Rhizopus were inoculated into the substrate affects the proteins produced. It can be seen from the Figure 9 and Figure 10 above, an increase in levels of the protein were obtained. Mold serves as a producer of enzymes that function in converting cyanide that has been released into amino acids [1]. This enzyme breaks down acetone sianohidrin, in order to obtain the final form of the amino acid aspartate and NH_3. Thus, the more molds were added to the substrat, will be the more enzyme is produced, the more protein is formed.

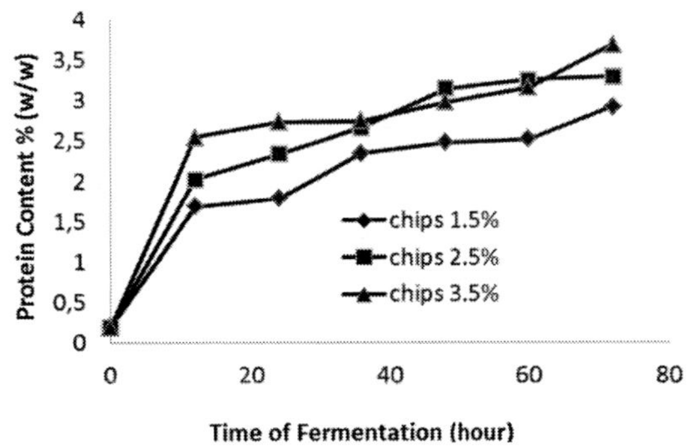

Fig 9. Protein content as function of time fermentation of chip cassava with variable *Rhizopus oligosporus* concentration

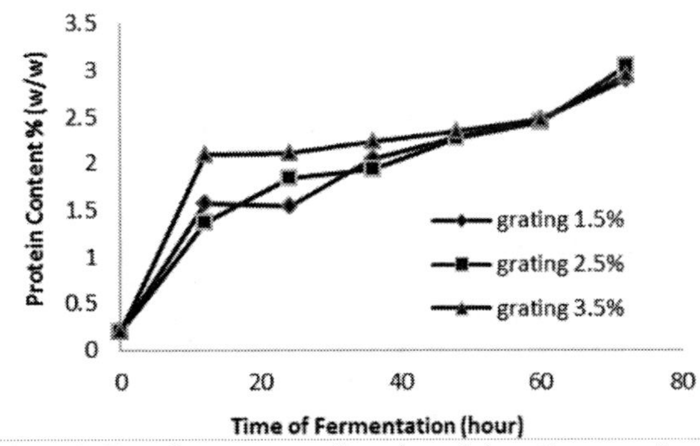

Fig 10. Protein content as function of time fermentation of grating cassava with variable *Rhizopus oligosporus* concentration

Optimum operating conditions can be seen from the data, by comparing the parameters behind the cyanide levels at least as well as the highest protein content in the samples that have been soaked in a solution of salt and buffer before. From the above data in the Table 1 can be concluded that by comparing the nutrition standards in accordance with the WHO reference parameter and flour, the best nutritional quality obtained from fermentation is in the conditions: grated, 72 hours of fermentation, and using 3.5% (w/w) *Rhizopus oligosporus* [17].

Table 1. Comparison cyanide and protein content of various experiments

Cassava Flour	Cyanide Content (mg/kg) dry basis	Protein Content (%) dry basis
Cassava (control)	4.03	0.19
Cassava chip (72 hours fermentation, 3.5% (w/w) R. Oligosporus	0.42	3.69
Cassava grating (72 hours fermentation, 1.5% (w/w) R. Oligosporus	0.09	2.89

4 Conclusions

.Cassava flour with the best protein content has protein content of 3.69 % d.b. The largest reduction of cyanide is in grated samples with Rhizopus oligosporus 1.5% weight level at 72 hours, with 0.09 ppm left. Cassava flour with the optimum result is obtained from fermentation with the conditions: grated with 12 hours of fermentation, and use 3.5 % weight Rhizopus oligosporus.

Acknowledgments

The authors are grateful thank to Diponegoro University for financial support of this research with scheme International Research and Publication Grant.

References

1. M.N. Adindu, F.F. Olayemi, O.U. Nze-Dike, J. Food Composition and Analysis. **16**, 21–24 (2003)
2. A.S. Ahmed, U.E. Igbo, C.C. Igwe, Nigerian Food J. **23,** 85-90 (2005)
3. N.J. Alummoottil, S. Korappati, S.S. Moothdasseri, R. Revamma, N.M. Subbramoney, Starch/ Stärke. **57**, 11, 547-555 (2005)
4. P. Tomasik, C.H. Schilling, Advances in Carbohydrate Chemistry and Biochemistry. **59**, 175-403 (2004)
5. A.C. Bertolini, C. Mestres, P. Colonna, D. Lerner, G.D. Valle, *Comprehensive studies of molecular changes occurring in sour cassava starch. In: Biopolymer Science: Food and Non Food Applications.* INRA: Montpellier, France. **91**, 87-91 (1998).
6. A.C. Bertolini, C. Mestres, P. Colonna, Starch. **52**, 340-344 (2000)
7. A.C. Bertolini, C. Mestres, C. Lourdin, G.D. Valle, P. Colonna, *J. Sci. Food Agric.* (2001)
8. R.N. Tharanathan, Food Science and Nutrition. **45**, 371-384 (2005)
9. Badan Pusat Statistik (BPS). *Tanaman Pangan.* diakses dari http://www.bps.go.id/ . (2011)
10. C. Camargo, P. Colonna, A. Buleon, D. Richard-Molard, J. Sci. Food Agric. **45**, 273-289 (1988)
11. O. S. Cardenas, T.S. Buckle, J. Food Sci. **45**, 1509-1528, (1980)
12. M.P. Cereda, O.L.S. Nunes, A. Westby, *Brazilian fermented cassava starch. I. Production and use. XVI International Carbohydrate Symposium, Paris.* 688 (1992).
13. S.G. Choi, W.L. Kerr, *Effects of chemical modification of wheat starch on molecular mobility as studied by pulsed ^1H NMR, Lebensmittel-Wissenchaft und Technologie.* **51**, 1-8 (2003)
14. I.M. Demiate, N. Dupuy, J.P. Huvenne, M.P. Cereda, G. Wosiacki, Carbohydrate Polymers. **42**, 149-158 (2000)
15. D. Dufour, S. Larsonneur, F. Alarcon, C. Brabet, *Cassava flour and starch; Centro Internacional de Agricultura Tropical: Cali, Colombia.* 133-142 (1996)
16. D. Dufour, *Starch: Structure and Functionality, The Royal Society of Chemistry:Cambridge, U.K.* 42-50 (1997)
17. I. Pudjihastuti, S. Sumardiono, Pengembangan proses inovatif kombinasi reaksi hidrolisis asam dan reaksi photokimia UV untuk produksi pati termodifikasi dari tapioka, Prosiding Seminar Nasional Kejuangan Teknik Kimia UPN, Yogyakarta. (2011)
18. A.M. Hermansson, K. Svegmark, Trends in Food Science and Technology. **7**, 345–353 (1996)
19. J. Jane, J. Macromolecular Science, Part A. **32,** 4, 751-757 (1995)
20. H. R. Kim, A.M. Hermansson, C.E. Eriksson, Starch. **44**, 111-116 (1992).
21. H.R. Kim, P. Muhrbeck, A.C. Eliasson, J. Sci. Food Agric. **61**, 109-116 (1993)
22. C.C. Seow, K. Thevamalar, Starch/Stärke. **45**, 85-88 (1993)
23. H. Liu, L. Ramsden, H. Corke, Carbohydrate Polymers. **40**, 175-182 (1999a)
24. H. Liu, L. Ramsden, H. Corke, Starch/Stärke. **51**, 249-252 (1999b)
25. C. Mestres, O. Boungou, N. Akissoe, N. Zakhia, J. Sci. Food Agric. **80**, 665-672 (2000)
26. J.M.N. Miller, Starch/Stärke. **49**, 4, 127-131 (1997)
27. N. Vatanasuchart, O. Naivikul, S. Charoenrein, K. Sriroth, Kasetsart J. (Nat.Sci.). **37,** 334-344 (2003)
28. N. Vatanasuchart, O. Naivikul, S. Charoenrein, K. Sriroth, Carbohydrate Polymers. **61**, 80-87 (2005)
29. C. Perera, R. Hoover, A.M. Martin, Food Research Int'l. **30**, 235–247 (1997)
30. M. Plata-Oviedo, C. Camargo, J. Sci. Food Agric. **77**, 103-108 (1998)
31. J. Singh, L. Kaur, O.J. McCarthy, Food Hydrocolloids. **21**, 1-22 (2007a)
32. J. Singh, L. Kaur, O.J. McCarthy, Food hydrocolloids. **21**, 1-22 ((2007b).
33. S. Sumardiono, I. Pudjihastuti, B. Jos, M. Taufani, F. Yahya, *Modification Of Cassava Starch Using Combination Process Lactic Acid Hydrolysis And Micro Wave Heating To Increase Coated Peanut Expansion Quality.* AIP Conference Proceedings. **1840**,060005 (2017)

34. S. Sumardiono, I. Pudjihastuti, Budiyono, H. Hartanto, I.C. Sophiana, *Combination Process Method Of Lactic Acid Hydrolysis And Hydrogen Peroxide Oxidation For Cassava Starch Modification*. AIP Conference Proceedings 1840,060006 (2017)
35. S. Sumardiono, R.B. Rakhmawati, Advanced Science Letters. **23**, 6, 5789-5791 (2017)
36. S. Sumardiono, M. Djaeni, B. Jos, I. Pudjihastuti, M. Abdallatif, Advanced Science Letters. **23**, 6, 5792-5795 (2017)

The Effect of Emulsifier and Hydrocolloid on Baking Expansion and Texture of Bread from Modified Cassava

Isti Pudjihastuti[1], *Noer* Handayani[2],and *Siswo* Sumardiono[2,*]

[1]Vocational Program Study of Chemical Engineering Diponegoro University, Jl.Prof. Soedarto, SH, Tembalang, Semarang, Indonesia 50275

[2]Departement of Chemical Engineering, Faculty of Engineering, Diponegoro University, Indonesia, Jl.Prof. Soedarto, SH, Tembalang, Semarang, Indonesia 50275

Abstract. Indonesia has a very abundant cassava that can be used instead of wheat. Bread made from cassava is safe for celiac sufferers, in which cannot tolerate a protein called gluten found in wheat flour. However, bread from cassava has the disadvantage that it cannot inflate perfectly. Our research goal is to study the effect of emulsifier and hydrocolloid concentration as modifying agents on baking expansion and bread texture (hardness). The test level hedonic preference for bread products results from modified tapioca is also necessary to know the level of customer satisfaction. This study were conducted by three main stages, modification of cassava, baking process, and analyses. Modification of cassava starch was applied using combination of lactic acid solution and ultra violet (UV) irradiation. Emulsifier (DATEM) and hydrocolloid (xanthan gum) were used in baking process. The addition of emulsifier and hydrocolloid can improve baking expansion. The addition of 7% emulsifiers on modified cassava can increase the volume of bread, taste, and texture so it can give greater satisfaction to consumers. Hydrocolloid can replace the function of gluten so the bread can inflate perfectly. The optimal composition of modified cassava in bread making is 25% of modified cassava and 75% of wheat flour. The low value of texture (hardness) on bread made from modified cassava indicated a better performance in comparison with native cassava. Baking expansion and texture of the bread is influenced by the modification process. Furthermore, the comprehensive and optimum studies of modification need to be investigated.

1 Introduction

Wheat flour contains about 7-22 % protein and mostly consisted of gluten and gliadin. The existence of high gluten shows that wheat flour has excellent physicochemical properties. This component holds the gas to keep the volume, texture, and viscoelastic form of bread [1,2]. Indonesia's market year wheat import is 8.9 million tonnes. For this reason, the potential of local commodities are urgently required to replace the wheat and apply in the baking fields.

Cassava is one of staple food and is gluten free, thus secure for celiac sufferers [3]. Therefore, it does not have a good texture of bread while use in baking field. Furthermore, modification of cassava starch should be provided in order to obtain the desirable characteristics, i.e. better of paste clarity, gel stability, solubility, and freeze–thaw stability. These properties also could improve their properties as stabilizers, fillers, binders, and adhesives on specific food applications [4-12].

Fermentation, sun-drying, and UV irradiation processes might produce cassava depolymerisation which is correlated with expansion ability [13-19]. Cassava starch (*Polvilhoazedo*) is a natural fermented that shows the expansion properties during baking. However, acidification of cassava starch, using lactic, acetic, or butyric acid and followed by oven drying, was unsuccessful to produce any great expansion ability after baking [20]. Additionally, amylase depolymerisation or natural lactic fermentation also did not provide the better expansion properties [13,19]. In the other hands, some researcher suggested to modify the cassava starch using combination of fermentation and sun drying method for better expansion baking properties [13,20,21].

Acidification process followed by UV irradiation on cassava starch could improve the expansion ability during baking [13-19]. The previous studies used lactic acid solution as acidifying agent [22-24]. The combination processes of UV irradiation and lactic acid solution followed by sun drying process represented the increase of specific volume of tested biscuits [20]. However, these better properties did not show in sour cassava starch treated by oven-drying method [20]. In addition to modification, the use hydrocolloid and emulsifier agents could trap the gas inside the dough thus increase the baking expansion.

* Corresponding author: siswo.sumardiono@che.undip.ac.id

Our research goals are to study the effect of emulsifier and hydrocolloid concentration on baking expansion and bread texture (hardness). The hedonic test for bread made from modified cassava starch is also investigated to know the level of customer satisfaction.

2 Materials and Method

2.1. Materials

The main material in this study, cassava was obtained from local market. Furthermore, DATEM, xanthan gum, wheat, and margarine were purchased from Indrasari, Bratachem, and Multi Kimia Raya, respectively.

2.2 Modification of cassava

Modification of cassava starch was applied using combination of lactic acid solution and ultra violet (UV) irradiation. 500 g of dried cassava starch, 1% lactic acid solution, and 1500 ml aquadest were placed and blended. This mixed solution would be irradiated using UV light for 20 minutes. Filtration process and sun drying would be conducted for final treatment.

2.3 Bread baking process

White bread dough is made with weight composition of 200 g of modified flour (corresponding variable), 0,6 g of bread improver, 3 g of yeast, 20 g of sugar, 4 g of milk, Xanthan Gum appropriate with variable, and 100 mL cold water. The dough is mixed using mixer at medium speed. After that, added 16 g of white margarine, 2 g of salt, and DATEM according to desired variables. The dough is mixed with a mixer again until uniform, then wait for 10-15 minutes. The dough was placed on aluminium pan that has been smeared with white margarine and wait for 1 hour. After that, it is baked in the oven at 180°C for 30 minutes. The composition of DATEM, xanthan gum, modified cassava, and wheat were shown in Table 1.

2.4 Baking expansion analysis

After baking process is finished, bread's volume and mass are measured one hour later. Volume of bread was calculated by deviate the bread volume before and after baking process. Volume bread was measured by the dimension volume of muffin mold. Baking expansion can be determined using formula Equation (1)

$$\text{Baking expansion}\left(\text{cm}^3/\text{g}\right) = \frac{\text{Volume of bread (cm}^3)}{\text{Dough weight (g)}} \quad (1)$$

2.5 Texture analysis

Texture analysis was analyzed using texture profile analyzer (TPA). Hardness will be presented as texture values.

2.6 Preference hedonic analysis

This test is carried out by conducting a quantitative survey toward 10 samples of consumer. The survey includes taste, colour, texture, aroma, and the satisfaction of bread product from modified cassava.

Table 1. Distribution of Experimental Method

Variable	Composition (%)			
	DATEM	Xanthan Gum	Modified Cassava	Wheat
A	3.5%	0.5%	0%	100%
B	3.5%	0.5%	25%	75%
C	3.5%	0.5%	50%	50%
D	3.5%	0.5%	75%	25%
E	3.5%	0.5%	100%	0%
F	3.5%	1%	0%	100%
G	3.5%	1%	25%	75%
H	3.5%	1%	50%	50%
I	3.5%	1%	75%	25%
J	3.5%	1%	100%	0%
K	7%	0.5%	0%	100%
L	7%	0.5%	25%	75%
M	7%	0.5%	50%	50%
N	7%	0.5%	75%	25%
O	7%	0.5%	100%	0%
P	7%	1%	0%	100%
Q	7%	1%	25%	75%
R	7%	1%	50%	50%
S	7%	1%	75%	25%
T	7%	1%	100%	0%

3 Results and Discussions

3.1 The effect of overall parameter to baking expansion

Modification of cassava was conducted using lactic acid solution followed by UV radiation. This modified starch would be used as substitute material of wheat in the baking process. The effect of hydrocolloid and emulsifier on baking expansion depicts on Figure 1.

Generally, the use of 25% substitute modified cassava starch and 7% emulsifier has a better performance on baking expansion properties in comparison with the substitution by modified cassava 50%, 75%, and 100%. However, the application of 100% wheat had the best value of baking expansion (3.50 ml/g). It is in agreement with Gillian Eggleston (1992) that the value of baking expansion is volume 3.98 mL/g [25]. The high percentage of modified cassava composition indicates lower gluten concentration in the

 https://doi.org/10.1051/matecconf/201815601026

dough. Thus, the values of baking expansion were lower than the low percentage. Gluten could provide elastic texture properties which can improve bread's volume.

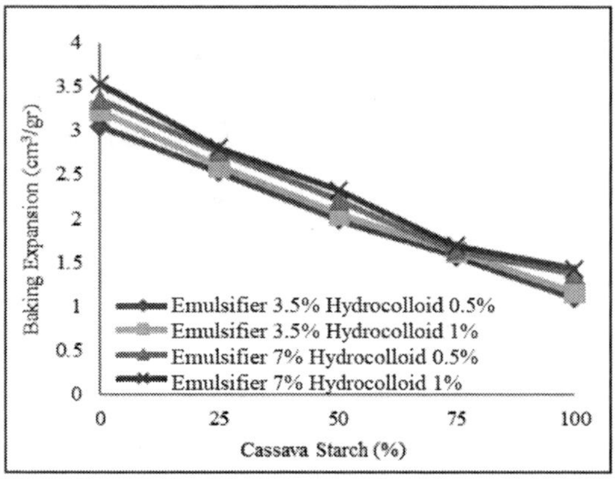

Fig. 1 Correlation between overall parameter to baking expansion

The addition of 7% and 3.5% emulsifier are more efficient for increasing baking expansion. Emulsifier might create a thin layer which can entrap carbon dioxide and strengthen the dough, thus the bread has a strong ability to expand [26]. Hydrocolloid is also has the similar function for improving baking expansion due to its role as a gluten substitute.

3.2 Texture analysis (hardness)

Texture analysis was conducted using texture profile analysis (TPA). Hardness is one of the texture test parameters. It is the force in the form of pressure or stress which is required to change the physical form of the any food material. Table 2 represents the value of hardness from native cassava, modified cassava, and wheat flours. The analyses were taken from the centre and side (wall) of breads.

Table 2 shows that there were a significant difference in rheological properties of native cassava, modified cassava, and wheat. The decreasing value of hardness on bread made from modified cassava indicated a better performance in comparison with native cassava [1,14]. Breads made from modified cassava have a delicate texture. The hardness value of modified cassava on centre and wall of bread were 169.02 and 305.35 gf, respectively. This values approach the hardness level of wheat (Table 2).

Table 2 Hardness values of native cassava, modified cassava, and wheat.

Component	Hardness (gf) (centre)	Hardness (gf) (wall)
Native cassava	213.69	338.03
Modified cassava	169.02	305.35
Wheat	148.64	215.10

3.3 The effect of overall parameter to hedonic test

Hedonic test was conducted on bread made from modified cassava. Ten respondents would participate on this test. The survey includes taste, colour, texture, aroma, and satisfaction of bread parameters (Figure 2-6). The values of each parameter were ranging from 1-10.

Figure 2 shows that the addition of emulsifier and hydrocolloid will improve taste of the bread. The high level of emulsifier and hydrocolloid has the positive respond which is indicate from the higher respondent value. The addition of 7% emulsifier and 1% hydrocolloid has the best value on both modified cassava composition. Figure 2 also indicates that respondent were more appetizing on bread which is made from 100% of wheat or the minimum composition of modified cassava (25%).

Figure 3 depicts that the high level of modified cassava composition has the uninteresting bread's colour. The use of 0% and 100% of modified cassava has the value 10 and 5.79, respectively. Based on this hedonic test, colour of food might be the important parameter which influences consumer appetite.

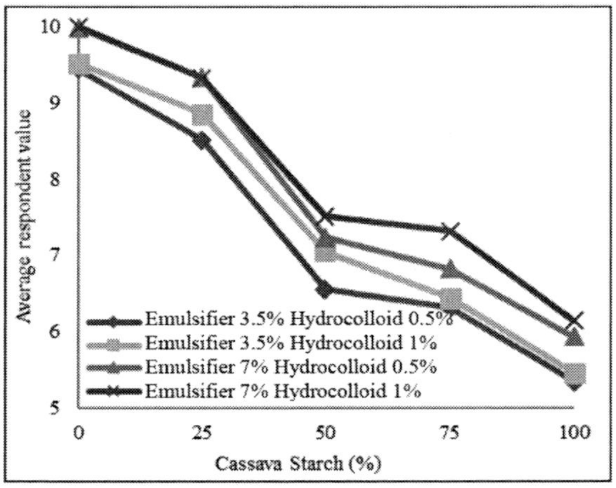

Fig. 2 Correlation between overall parameter to taste

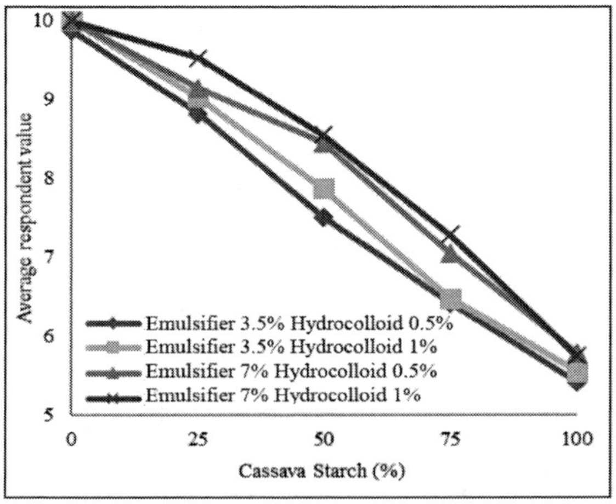

Fig. 3 Correlation between overall parameter to colour

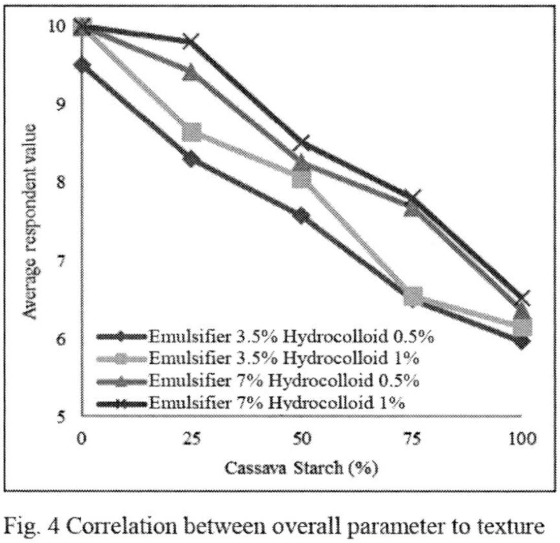

Fig. 4 Correlation between overall parameter to texture

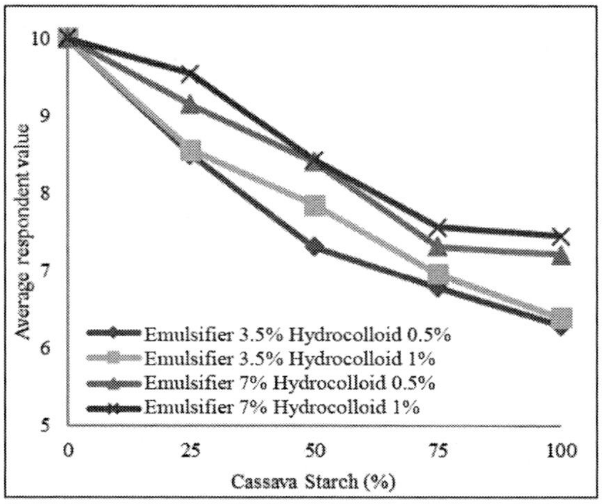

Fig.5 Correlation between overall parameter to aroma

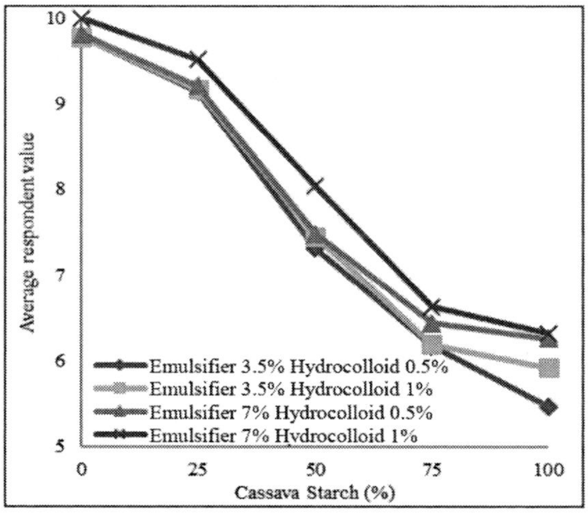

Fig. 6 Correlation between overall parameter to satisfaction

Based on Figure 4, we can see that the bread texture is better and delicate on the addition 7% emulsifier and 1% hydrocolloid. Emulsifiers could improve the texture of bread, while hydrocolloid stabilizes the dough. These results were in agreement with aroma parameter. Aroma of bread is influenced by emulsifier addition and modified cassava composition (Figure5). The hedonic test of 1% hydrocolloid and 7% emulsifiers on 25% modified cassava get average respondent value 9.54. This value is very close with aroma value of bread from 100% wheat with value 10.

Figure 6 shows that modified cassava composition on baking dough determines the bread quality. Bread with 25% modified cassava have similar baking expansion, taste, colour, and aroma parameters with bread from 100% wheat. Based on hedonic test, bread with 25% modified cassava get satisfaction value of consumer above 9 (scale 0-10), while 50%, 75%, 100% of modified cassava get average value under 8. The ideal composition of modified cassava in bread is 25% from the total of flour needed because it can produce bread which very similar with wheat bread. Bread which consist of 50% modified cassava still have a better texture rather than 75% and 100% composition.

Emulsifier and hydrocolloid addition also increase baking expansion, taste, texture, and aroma of bread made from modified cassava. Bread will have higher value of customer satisfaction and getting closer to bread satisfaction from 100% wheat.

4 Conclusions

The addition of 7% emulsifiers on modified cassava can increase the volume of bread, taste, and texture so it can give greater satisfaction to consumers. Hydrocolloid can replace the function of gluten so the bread can inflate perfectly. The optimal composition of modified cassava in bread making is 25% of modified cassava and 75% of wheat flour. Baking expansion and texture of the bread is influenced by the modification process. Furthermore, the comprehensive and optimum studies of modification need to be investigated.

Acknowledgement

Authors thank to Diponegoro University through Fundamental Research Grant Program for funding this research.

References

1. M. Miyazaki., N. Morita., Food Res. Int., **38** (2005)
2. S. G. Gujral., C. M. Rosell., Food Res. Int., **37** (2004)
3. J. E. King., *Mayo Clinic on Digestive Health*, In: Mayo Clinic Proceedings, 2nd ed. Mayo Clinic, USA, p. 126.(2006)

4. S. Sumardiono., I. Pudjihastuti., B. Jos., M. Taufani., F. Yahya., AIP Conf. Proc., 1840,060005 (2017).
5. A. M. Hermansson., K. Svegmark., Trends Food Sci. Technol., **7** (1996)
6. P. Tomasik., S. H. Schilling., Adv. Carbohydr. Chem. Biochem., **59** (2004)
7. A. S. Ahmed., U. E. Igbo., C. C. Igwe., Nigerian Food Journal, **23** (2005)
8. N. J. Alummoottil., S. Korappati., S. S. Moothdasseri., R. Revamma., N. M. Subbramoney, Starch/ Stärke, **57** (2005)
9. J. Singh., L. Kaur., O. J. McCarthy., Food Hydrocoll., **21** (2007).
10. R. N. Tharanathan., Crit. Rev. Food Sci. Nutr., **45** (2005)
11. S. G. Choi., W. L. Kerr., Lebensmittel-Wissenchaft und Technologie, **51** (2003)
12. C. Perera., R. Hoover., A. M. Martin., Food Res. Int., **30** (1997)
13. D. Dufour., S. Larsonneur., F. Alarcon., C. Brabet., Centro Internacional deAgricultura Tropical: Cali, Colombia (1996)
14. A. C. Bertolini., C. Mestres., P. Colonna., Starch, **52** (2000)
15. A. C. Bertolini., C. Mestres., C. Lourdin., G. Della Valle., P. Colonna., J. Sci. Food Agric., **81** (2001)
16. I. Pudjihastuti., S. Sumardiono., Prosiding Seminar Nasional Kejuangan Teknik Kimia UPN, Yogyakarta (2011)
17. S. Sumardiono., I. Pudjihastuti., Budiyono., H. Hartanto., I. C. Sophiana., AIP Conf. Proc., 1840,060006 (2017)
18. S. Sumardiono, M. Djaeni, B. Jos, I. Pudjihastuti, and M. Abdallatif, Adv. Sci. Lett., **23** (2017)
19. S. Sumardiono, R. B. Rakhmawati, Adv. Sci. Lett., **23,** 6 (2017)
20. M. Plata-Oviedo., C. Camargo., J. Sci. Food Agric., **77** (1998)
21. C. Mestres., O. Boungou., N. Akissoe., N. Zakhia., J. Sci. Food Agric., **80** (2000)
22. I. M. Demiate., N. Dupuy., J. P. Huvenne., M. P. Cereda., G. Wosiacki., Carbohydr. Polym., **42** (2000)
23. N. Vatanasuchart., O. Naivikul., S. Charoenrein, K. Sriroth., Carbohydr. Polym., **61** (2005)
24. N. Vatanasuchart., O. Naivikul., S. Charoenrein., K. Sriroth. Kasetsart J. (Nat.Sci.), **37** (2003)
25. G. Eggleston, P. E. Omoaka., D. O. Ihedioha., J. Sci. Food Agric., **59,** 3 (1992)
26. M. Pojić., M. Musse., C. Rondeau., M. Hadnađev., D. Grenier., F. Mariette., T. D. Hadnađev, Food Bioproc. Tech., **9**, 8 (2016)

Effect of pH on Physicochemical Properties of Cassava Starch Modification Using Ozone

Isti Pudjihastuti[1], *Noer* Handayani[2], and *Siswo* Sumardiono[2,*]

[1]Vocational Program Study of Chemical Engineering Diponegoro University, Jl.Prof. Soedarto, SH, Tembalang, Semarang, Indonesia 50275
[2]Departement of Chemical Engineering, Faculty of Engineering, Diponegoro University, Jl.Prof. Soedarto, SH, Tembalang, Semarang, Indonesia 50275

Abstract Nowadays, starch modification is carried out in order to change the native properties into the better ones, such as high stability, brightness, and better texture. The objectives of this study are to investigate the effect of pH on carboxyl content, swelling power, and water solubility of starch. This research was divided into two main stages, i.e. starch modification by ozone oxidation and analysis. The physicochemical properties of modified cassava starch were investigated under various reaction pH of 7-10 and the reaction time between 0-240 minutes. Reaction condition at pH 10 provided the higher value of carboxyl content and water solubility, but the lower of swelling power. This increase in solubility indicates that the modified oxidation starch readily dissolves in water, due to its small size granules and high amylose content. The significant changes of both parameters were achieved in the first 120 minutes of ozone reaction times. The graphic pattern of water solubility was in contrast with swelling power.

1 Introduction

Cassava is a potential material as wheat substitute due to its high level of carbohydrate. However, the baking expansion ability for bread production is low due to gluten free. Nowadays, starch modification is carried out in order to change the native properties into the better ones, such as high stability, brightness, and better texture [1]. Commonly, there are four methods of starch modification to obtain these desired characteristics, i.e. acid, enzyme, oxidation, and crossing bond modifications [1-7]. Among these, chemical and oxidation method is widely used in both food and nonfood industries.

Each modification creates the different characteristics of modified starch. Modification by chemical is provided to increase paste consistency, smoothness, clarity, and the stability in the cold storage [1-3]. Many factors influence the chemical modification, i.e. starch source, reaction condition (concentration, time, pH, presence of catalysts), and type of chemical reagents or substituent. Acetylated starch (with low degree of substitution) is commonly obtained by the esterification using acetic anhydride in the presence of an alkaline catalyst.

Modification by oxidation reaction using hypochlorite was very popular and conducted by some researcher [8-13]. The result indicated that it could reduce viscosity and changed the structure of starch molecule, color and easier in gel formation. In the other hand, it has some drawbacks, such as the formation of residue and chloride during reaction. Oxidation using bromine reduced the enthalpy, viscosity, temperature gelatinization, and changing the structure of starch [14]. Others chemical could also use as oxidizing agent, such as chlorine, perodate, and hydrogen peroxide [15-19]. Some factors also determine the success of this modification, i.e. pH, concentration of oxidizing agent, and temperature reaction. The reaction rate, type and amount of functional groups formed in the starch molecule depend on the pH [20,21].

Recently, the development of ozone application is growing rapidly due to its affordable price and also available in household scale, especially for drinking water processing [14]. Ozone is very popular in water and waste water processing. It used to reduce organic compound and micro pollutant and also inorganic compound [15,16]. Ozone serves to eliminate the odors, taste, color, and particle solubility in the waste water [15]. Ozone is also a strong oxidator which is able to change the structure, physicochemical, and rheological properties of starch. It also produces non residual component in modified starch process. In based on author's knowledge, there is lack of information for the use of ozone (O_3) for starch modification. The objectives of this study are to investigate the effect of pH on

* Corresponding author: siswo.sumardiono@che.undip.ac.id

carboxyl content, swelling power, and water solubility of starch.

2 Materials and Method

2.1 Experimental

The research experiment was conducted in the Food Process Engineering Laboratory, Chemical Engineering Department, Diponegoro University. This research was divided into two main stages, i.e. starch modification by ozone oxidation and analysis. The equipment of this study as depicted in Figure 1.

2.2 Ozone mass transfer in liquid phase

First step on this study is dissolving starch in aquadest 10-40% concentration dry base and then mixed it into a stirred tank reactor. This reactor was immersed in a water bath for control temperature so that keep on desired condition (about 30-50°C). This reactor was equipped with temperature and pressure sensors for identifying the temperature and pressure stability in reactor. Before through to 0.2 micron filter, Air from compressor should be flowed to flow regulator first and filtered before being through to ozoniser as ozone gas generator. Ozone gas will entered pass through under of reactor with sparger distribution. In this research, ozone rates were varied about 0 – 90 mg O_3/g of starch and 7 – 10 of acidity range. Acidity of solution was controlled by adding NaOH or HCl solution. While the oxidizing reaction take place, sample was withdrawn every 5 minutes for measured contents of ozone in solution.

2.3 Determination of carboxyl content

The carboxyl content of ozone-modified cassava starch was determined by the ISO method (ISO-11214, 1996). Starch sample (5 g) was stirred in 0.1 M HCl for 30 min. The slurry was then filtered and washed with distilled water until free of chloride ions. The filtered cake was transferred to 300 ml water and the starch slurry was heated in a boiling water bath with continuous stirring until gelatinized and continue stirring at that temperature for another 15 min. The hot sample was titrated with 0.1 M NaOH using phenolphthalein as an indicator.

2.4 Determination of solubility and swelling power

The water solubility and swelling power analyses were evaluated by the method describe by Kusumayanti (2015) with slight of modification, respectively [25]. These were carried out by weighing 0.1 g of modified cassava starch and 10 ml of distilled water into a reaction tube. The tube was placed it on beaker bath and heated at 60 ° C for 30 minutes. After heating process, the sample was centrifuged (C-5 Test Tube Centrifuge) at 1600 rpm for 15 min. The supernatant was removed from sediment carefully. The swelling power was determined as sediment weight per initial dry weight of sample. 5 ml of supernatant was collected and dried it into a constant weight. The water solubility was expressed as weight percent of the soluble flour of the heated solution.

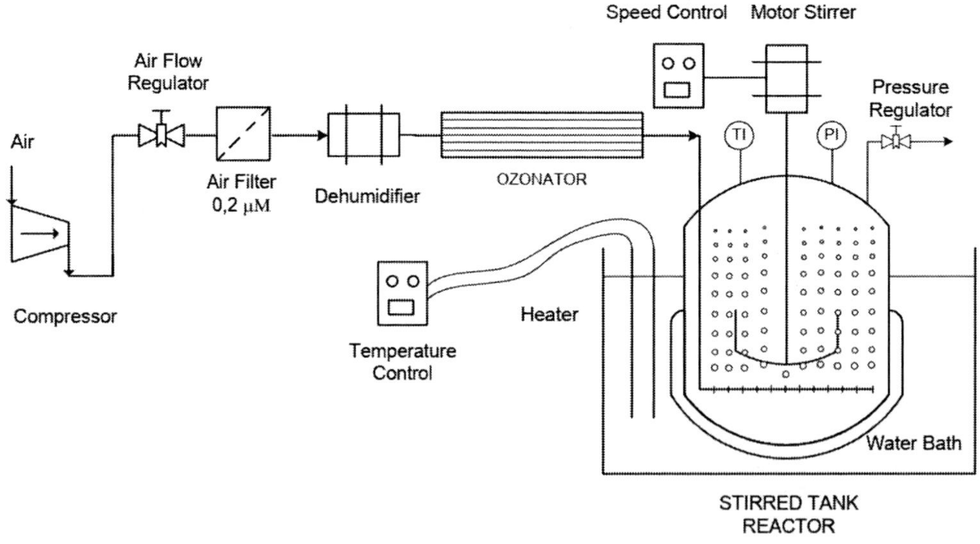

Fig. 3 Experiment apparatus of ozonisation in the stirred tank reactor

3 Results and Discussions

3.1 The effect of pH on carboxyl content (%)

Various pH (7-10) were applied on starch modification using ionization to investigate the carboxyl content (%) of cassava starch. Figure 2 shows the effect of pH on carboxyl group during ozone reaction time (0-240 minutes). The carboxyl content of ozone-modified cassava starch was determined by the ISO method (ISO-11214, 1996) using phenolphthalein as acid-alkaline indicator.

Generally, carboxyl content increased in the first 120 minutes ozone reaction, but it was not significant different in the last 120 minutes. Both oxidized starches had higher carboxyl contents in comparison with unmodified starch. The highest value of carboxyl group (0.092%) was obtained at pH 10 for 120 minutes ozone reaction time. Some factors might influence the complex reaction mechanism of ozone modification, such as type of starch, oxidizing agent, and reaction condition. Modification starch using ozone allows hydroxyl groups replaced by carboxyl and carbonyl groups [26]. The carboxyl and carbonyl content indicates the level of oxidation. According in Figure 2, the higher level of carboxyl content was obtained at alkaline pH.

3.2 The effect of pH on swelling power (%)

The effect of pH on oxidation modification was conducted on cassava starch. Figure 3 depicts the influence of pH (7-10) on swelling power of modified starch. Swelling power signify the water storing capacity of starch, which is generally used as one of parameters in food application, especially in baking process.

Modified starch which was produced by ozone reaction at pH 10 had a highest carboxyl level but lowest of swelling power in comparison with others pH reaction. The significant decreases of swelling power were obtained up to 65% and 81% in the first 30 and 120 minutes of ozone reaction, respectively. This characteristic is influenced by the amylase and amylopectin level. Oxidation reaction could increase the amylose content due to the high of carboxyl group. For this reason, the swelling ability of modified starch decreased, and it might be a disadvantageous for baking process (Figure 3).

3.3 The effect of pH on water solubility (%)

Figure 4 represents the effect of pH on modified starch's water solubility. Commonly, the values of water solubility are in contrast with swelling power. The value of this parameter increased along with ozone reaction time. This result is in agreement with carboxyl content (Figure 2). The highest value of water solubility was achieved at pH 10. This trend was also similar with Kusumayanti (2015) [25]. The graphic pattern of water solubility was in contrast with swelling power. However, there is no straight relation between water solubility and swelling power that could be investigated [27].

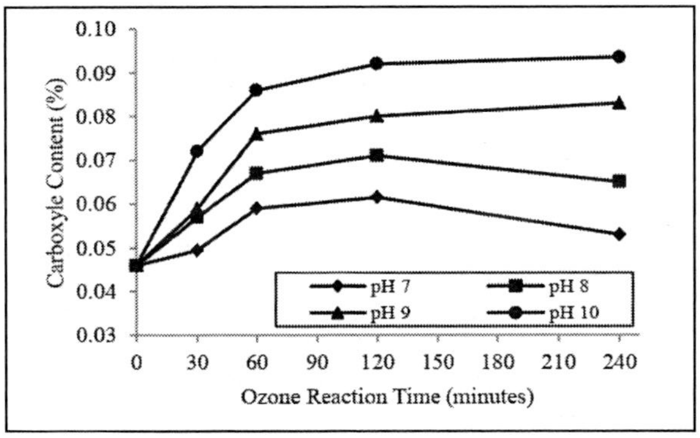

Fig. 2 Carboxyl content of ozone modified cassava starch.

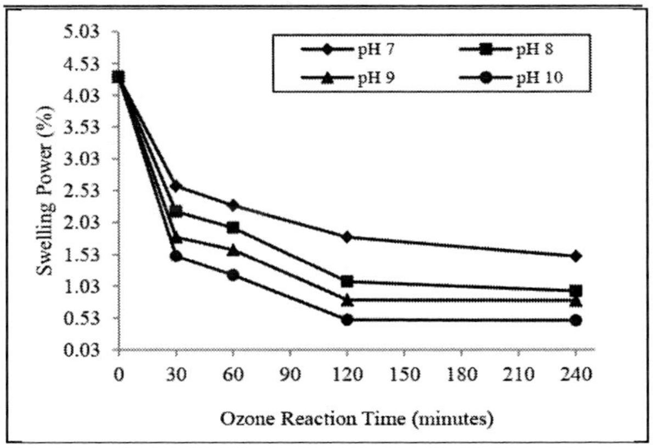

Fig 3. Swelling power of ozone modified cassava starch

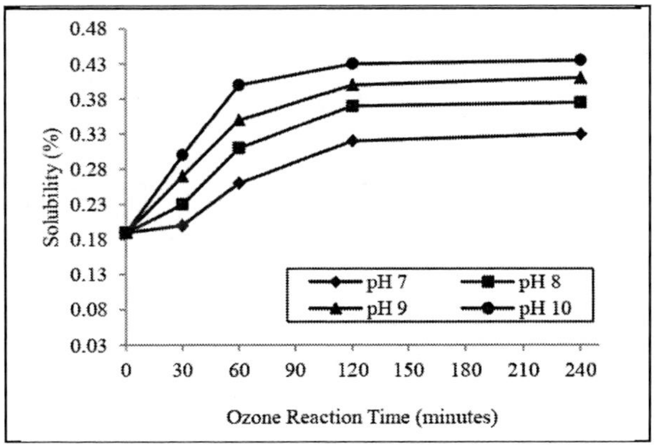

Fig 4. Water solubility of ozone modified cassava starch

4 Conclusions

Modification of cassava starch was conducted using ozone oxidation. Water solubility of oxidized cassava starch increases along with reaction time. Reaction condition at pH 10 provided the higher value of carboxyl content and water solubility, but the lower of swelling power. This increase in solubility indicates that the modified oxidation starch readily dissolves in water, due to its small size granules and high amylose content.

Acknowledgement

Authors thank to Diponegoro University through Fundamental Research Grant Program for funding this research.

References

1. J. J. M. S. Jane., J. Macromol. Sci., Pure Appl. Chem., **32**, 4 (1995)
2. J. M. N. Miller., Starch/Stärke, **49**, 4 (1997)
3. R. N. Tharanathan., Critical Reviews in Food Science and Nutrition, **45**, 371-384. (2005)
4. S. Sumardiono, I. Pudjihastuti, M. Taufani, F. Yahya, AIP Conf. Proc., 1840, 060005 (2017)
5. S. Sumardiono, I. Pudjihastuti, Budiyono, H. Hartanto, I. C. Sophiana, AIP Conf. Proc., 1840, 060006 (2017)
6. S. Sumardiono, M. Djaeni, B. Jos, I. Pudjihastuti, M. Abdallatif, Adv. Sci. Lett., **23** (2017)
7. S. Sumardiono, R. B. Rakhmawati, Adv. Sci. Lett., **23**, 6 (2017)
8. P. Parovuori., A. Hamunen., P. Forssell., K. Autio., K. Poutanen., Starch/Stärke, **47**, 1 (1995)
9. F. Kantouch., S. Tawfik., Starch/Stärke, **50** (1998)
10. D. Kuakpetoon., Y. J. Wang., Starch/Stärke, **53** (2001)
11. E. Bidzinska., K. Dyrek., T. Fortuna, M. Labanowska., S. Pietrzyk., Starch/Stärke, **56** (2004)
12. C. E. C. Murillo., Y. J. Wang., L. A. B. Perez, Starch/Stärke, **60** (2008)
13. K. Sangseethong., S. Lertphanich., K. Sriroth., Starch/Stärke, **61**(2009)
14. P. Muhrbeck., A. C. Eliasson., A. C. Salomonsson., Starch/Stärke, **4**, 11 (1990)
15. A. Hebeish., F. El-Sisy., S. A. Abdel-Hafiz., A. A. Abdel-Rahman., M. H. El-Rafie., Starch/Stärke, **44**, 10 (1992)
16. A. N. Jyothi., K. Sasikiran., M. S. Sajeev., R.Revamma., S. N. Moorthy., Starch/Stärke, **57**, 11 (2005)
17. S. Veelaert., M. Polling., D. de Wit., Starch/Stärke, **47**, 7 (1995)
18. P. Tolvanen, A. Sorokin, P. Mäki-Arvela, S. Leveneur, D. Y. Murzin, T. Salmi, Ind. Eng. Chem. Res., **50**, 2 (2010)
19. K. Sangseethong, N. Termvejsayanon, K. Sriroth,. Carbohydr. Polym., **82**, 2 (2010)
20. O. S. Lawal, K. O. Adebowale, B. M. Ogunsanwo, L. L. Barba, N. S. Ilo, Int. J. Biol. Macromol., **35**, 1-2 (2005)
21. S. Veelaert, D. de Wit, K. F. Gotlieb, R. Verhé,. Carbohydr. Polym., **33**, 2-3 (1997)
22. A. Bohme., Ozone: Sci. Eng., **21** (1999)
23. V. Camel., A. Bermond., Water Res., **32** (1998)
24. U. von Gunten., C. J. Hoign., Environ. Sci. Technol., **28** (1994)
25. H. Kusumayanti, N. A. Handayani, H. Santosa, Procedia Environ. Sci., **23** (2015)
26. A. R. G. Dias, E. da Rosa Zavareze, E. Helbig, F. A. de Moura, C. G. Vargas, C. F. Ciacco, Carbohydr. Polym., **86**, 1 (2011)
27. A. C. Kumoro, D. S. Retnowati, C. S. Budiyati,. Research Journal of Applied Sciences, Engineering and Technology, **4**, 1 (2012)

Effect of Soaking Time in Sodium Metabisulfite Solution on the Physicochemical and Functional Properties of Durian Seed Flour

Andri Kumoro[1*] and *Jefri* Hidayat[2]

[1]Department of Chemical Engineering, Faculty of Engineering, Diponegoro University, Semarang, Indonesia
[2]Master of Chemical Engineering Study Program, Faculty of Engineering, Diponegoro University, Semarang, Indonesia

Abstract. With regard to its high carbohydrate and gum contents, durian seed flour has been used as a substitute to wheat flour in butter cake and cookies manufactures. Unfortunately, processing of fruit seeds into flour may be hampered by discoloration, which may affect the nutrition and sensory quality of the food products. The objective of this study is to investigate the effect of soaking time on the physicochemical and functional properties of durian seed flour. Durian seed chips were soaked in 0.6% w/v sodium metabisulfite solution as anti-browning agent at 30°C for desired periods (40, 60, 80, 100, 120 minutes), followed by drying at 50°C for 17 hours, milling and sieving to obtain flour. The flour was analyzed for its moisture, ash, protein, fat and carbohydrate contents, sulfite residue, yield, gelatinization temperature, and water and oil absorption capacities (WAC and OAC). The results showed that longer soaking time reduced the moisture and ash content of durian seed flour, but increased most of the other studied parameters. The WAC, OAC and fat content were less affected by soaking time. It can be concluded that soaking of durian seed chips in sodium metabisulfite solution can improve the nutrition and functional properties of flour. Based on the residual sulfite content, durian seed flour obtained in this work is safe for consumption.

1. Introduction

Wheat flour has been used for production of noodle and baked products such as breads, cakes, biscuits, and cookies for centuries. This is due to the unique nature and functional properties of wheat flour proteins called gluten [1]. However, wheat cannot be commercially cultivated in the tropical countries such as Indonesia and Malaysia mainly due to climate differences. In order to cover the domestic demand, Indonesia imported wheat flour amounted 97,830 tons or equivalent to US$ 33,963 in 2015 as a reason for the diversification of food products [2, 3]. Unfortunately, individuals with celiac disease may often experience adverse reactions to products containing gluten. Larrosa *et al.* suggested that the elimination of gluten from food products can be done by substitution of wheat flour with specific materials with closely similar physicochemical properties to gluten [4].

Durian (*Durio zibethinus* Murr.) is one of the seasonal fruits in South East Asia and popularly known as *The King of Fruits*. Thailand, Indonesia, Malaysia and the Philippines are the major producers of durian [5]. In 2013, production of durian in Indonesia was approximately 1,818,949 tons and still increased after year [6]. The fruit pulp has a unique taste and strong smell. It is usually eaten fresh or processed into ice cream, pancake and various traditional foods [7]. The edible portion of durian fruit is only about 30-35% of its total weight, whereas the seeds (20–25%) and the shell are usually disposed. It is clear that, the durian seed as by-product of durian fruit production can reach 454,737 tons per year, but it has not been fully utilized. Ripe durian seeds contain 51.1% moisture, 43.6% carbohydrates, dietary fiber, polysaccharide gum (20-25% w/w), and protein (3-5% w/w) [8-9]. The carbohydrate content of whole and dehulled durian seed flour is 73.90% and 76.8%, respectively [10]. The amylose content of durian seed is 22.76 % [11]. Hence, it gives a great potential to substitute the consumption of wheat flour as much as 10.1% with the yield of durian flour by 60%. Unfortunately, the seed also contains cyclopropene fatty acids, namely sterculic acid, dihydrosterculic acid and malvalic acid which may cause shortness of breath as well as cancer in rainbow trout [12] and aortic atherosclerosis in rabbit [13]. Therefore, durian seeds are toxic while consumed raw. Thermal cooking, such as boiling, steaming, burning and roasting were reported to reduce the

* Corresponding author: andreykumoro@che.undip.ac.id

toxicity of the seed and made them safe for consumption [14].

The color intensity is one of the main important factors influencing the appearance and overall acceptability of food products. In order to substitute wheat flour, durian seed flour must have minimum 92.5% lightness [15]. The problem in making durian seed flour is lack of lightness due to the accumulation of brown pigment resulting from non-enzymatic and enzymatic reactions. Sulfite has been used as anti-browning agent because it is a strong inhibitor of polyphenol oxidases and peroxidases, which may inhibit enzymatic and non-enzymatic browning in fruits and vegetables [16]. The objective of this study is to investigate the effect of time on the prevention of browning through soaking durian seed chips in sodium metabisulfite solution by evaluating physicochemical and functional properties of durian seed flour.

2. Material and Methods

2.1 Material

In this study, durian (*Durio zibethinus* Murr.) seed as by-product of durian fruit was collected from durian farmers in Gunungpati, Semarang, Indonesia. Sodium metabisulfite as an anti-browning agent and other chemical reagents for analyses were purchased from an authorized chemicals distributor in Semarang.

2.2 Making of durian seed flour and Browning Investigation

Durian seed flour was produced based on the method developed by Chowdhury *et al.* with some modification [17]. Matured durian seeds were washed thoroughly and peeled to remove outer seed coat and pericarp. Subsequently, they were sliced into small size of 2 cm × 2 cm ×1 cm and were soaked immediately in sodium metabisulfite solution with 0.6% w/v concentration at 30°C for 40, 60, 80, 100, and 120 minutes. The durian seeds were then drained and oven dried at 50°C for 17 hours. To obtain durian seed flour, the dried chips were milled and passed through 80 mesh sieves. Ultimately, the flour was collected and stored in plastic containers at around 8°C till further use. A control durian seed flour sample was prepared from durian seeds without soaking in sodium metabisulfite solution. Flour yield was calculated as the percentage ratio between mass of durian seed flour and initial mass of durian seed.

2.3 Physicochemical analysis of flour

2.3.1 Chemical analysis

Proximate composition of the durian seed flour representative samples were analyzed for crude protein, fat, ash and moisture content according to standard methods of AOAC [18]. The total carbohydrate was calculated by the difference [19]. Sulfite residue was determined by iodometric titration method prescribed in American Public Health Association [20]. All of the analyses were carried out in triplicates and the reported values were the average of each measurement.

2.3.2 Physical analysis

Gelatinization temperature was determined by differential scanning calorimetry method as previously used by Coral *et al.* [21]. The whiteness of durian seed flours were measured by using a Chroma Meter (Minolta Type CR-300, Japan) and considered the parameters L from 0 for black to 100 for white [1].

2.3.3 Functional properties analysis

Water absorption capacity (WAC) was examined according to the methods used by Niba *et al.* [22] and was reported as percentage of volume of water per mass of flour sample. Oil absorption capacity (OAC) was evaluated according to the method developed by Sosulski *et al.* and was expressed as percentage of mass of oil absorbed per mass of flour sample [23].

3. Results and Discussion

3.1 Chemical properties of durian seed flours

Soaking of durian seed chips in sodium metabisulfite solution was mainly expected to prevent the seeds from browning. However, some metabolic reactions can take place during soaking, altering the content of some components in the flour obtained [24]. This effect varied with botanical sources and soaking conditions such as type of soaking solutions, soaking period and temperature [25]. The chemical composition of durian seed flours obtained in this work are presented in Table 1.

Table 1. Chemical properties of durian seed flour

Chemical properties	Soaking time (minutes)					
	0	40	60	80	100	120
Moisture (%)	9.18	8.27	7.62	6.35	4.78	5.17
Ash (%)	3.25	3.22	3.07	2.76	2.52	2.21
Fat (%)	0.64	0.91	0.45	0.73	0.64	0.91
Protein (%)	6.30	6.34	6.84	7.12	7.42	8.24
Carbohydrate (%)	75.90	81.26	81.99	83.04	84.64	83.47
Sulfur residue (ppm)	0	23.12	37.45	50.34	85.71	102.56

The values are expressed as the mean of triplicates.

Moisture content of the durian seed flour decreased significantly as the soaking time increased. This result could be attributed to the effect of sulfites in degrading the seed cells tissue, resulting in larger pore sizes on the cut surface and facilitating easy removal of moisture through internal diffusion and evaporation during drying of samples [26-27]. Low moisture content means low water activity and is an indication of longer shelf life [28]. Similarly, the creation of larger pores induced leaching of minerals from the inner pores to the outside pores of the seed chips and finally to the soaking solution, which caused reduction of ash content in the durian seed flours. Fat content of durian seed flours did not change significantly with the increase of soaking time. However, no regular pattern of the change was observed.

In contrast to the above findings, the protein and carbohydrate content of durian seed flours increased with soaking time. Rom *et al.* [26] reported that protein bodies did not change during soaking in sodium metabisulfite solution suggesting no protein degradation during this process. Therefore, the increase in protein carbohydrate content may likely be the result of reduction of moisture and ash contents. Unfortunately, the protein content of the durian seed flours were only about half of that of commercial wheat flour. This fact suggests that the durian seed flour is not suitable for use in bread making; however, it could be suitable for making cookies or cakes [10].

In general, the chemical properties of durian seed flour obtained in this work are comparable to that reported in the literature [10]. The differences are more likely to be caused by treatment and processing method, type or cultivar, maturity and cultivation method.

As expected, the residual sulfite content increased as the durian seed chips were soaked in sodium metabisulfite solution for longer time. Longer contact time between durian seed chips and sodium metabisulfite solution let the sulfite to penetrate the durian seed chips more extensively and promote the reaction to proceed. However, some of the entrapped sulfite may be left in the durian seed chips after the drying process. The maximum level of sulfite is changeable for different products, such as crustaceous 50 mg/kg, meat products 450 mg/kg, dry biscuit between 30 and 50 mg/kg, vegetables between 50 and 2000 mg/kg and beverages between 20 and 2,000 mg/kg [29]. In particular, the General Standards for Food Additives (GSFA) Codex has set the maximum permissible level (MPL) for sulfites in starch and flour as food ingredients as 200 mg/kg. It is obvious that the residual sulfur in all of the durian seed flours obtained in this work were still in acceptable level. Therefore, the treatment of durian seed chips to produce durian seed flour using sodium metabisulfite solution is safe and the durian seed flours can be used to manufacture bakery products.

Sulfite residue is not the only parameter of food safety. The harmful condition may also come from the compounds originally contained in the food sources. Therefore, extensive investigations on the cyclopropene fatty acids content as the toxic sompound in durian seed flour obtained in this research is still need to be conducted to ensure the safety of its long term consumption.

3.2 Physical properties of durian seed flours

The appearance of durian seed flours obtained from sodium metabisulfite pre-soaked and native durian seed chips are shown in Fig. 1, while the yield and physical properties of durian seed flour are presented in Table 2.

It can be visually observed in Fig. 1 that as the soaking time increased, the durian seed flours obtained are brighter. Bleaching process associated with prevention of the formation of melanoidin compound from the reaction between sulfite and carbonyl group in durian seed flour. Hence, the flours obtained from pre-soaked durian seed chips in sodium metabisulfite solution at longer time exhibited brighter color than that of control sample. The color analysis also revealed that the lightness of durian seed flours increased from 65 to 80 as shown in Table 2. Sodium metabisulfite has played its role as inhibitor of polyophenol oxidases (PPO) and peroxidases and therefore inhibit enzymatic and non-enzymatic browning in durian seed flours [16]. Lambrecht explained that sulfites react with PPO itself, and thus irreversibly inhibiting PPO by modifying the protein structure [30]. Since the lightness of durian seed flours are still below the acceptable value [15], they are still unsuitable for use as substitute to wheat flour.

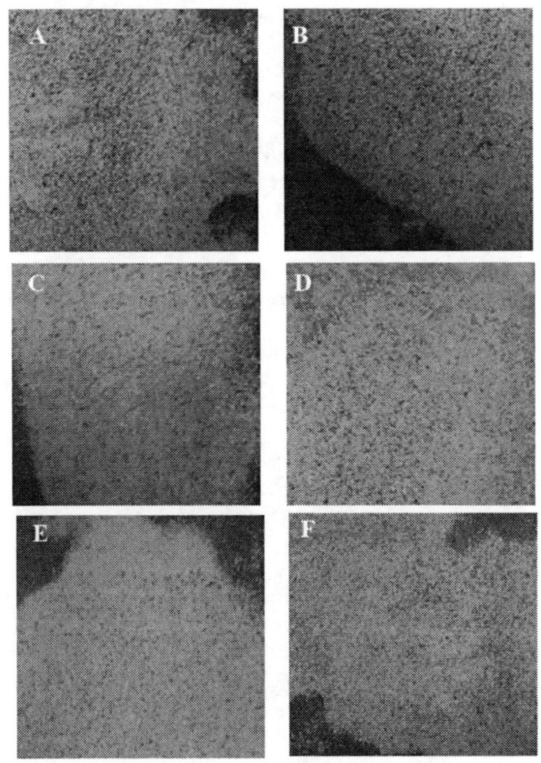

Fig. 1. Visual appearance of durian seed flours

Table 2. Physical properties of durian seed flour

Physical properties	Soaking time (minutes)					
	0	40	60	80	100	120
Y (%)	50.12	50.52	49.25	50.67	51.53	52.97
GT (%)	56.91	58.22	60.54	62.17	65.21	70.89
W (L)	65	70	71	73	75	80
WAC (%)	145.62	147.45	148.17	149.33	150.28	151.77
OAC (%)	24	28	27	27	26	25

Y = yield, GT = gelatinization temperature, W = whiteness, WAC = water absorption capacity, OAC = oil absorption capacity. The values are expressed as the mean of triplicates.

Furthermore, the durian seed flour yields ranged from 49.25% to 52.97%, which is consistent with the literature that fresh durian seeds may contain 51.1%

moisture and the rest being dry matters [8]. It is likely that only very little dry matters were leached out from the durian seed chips during soaking in sodium metabisulfite solution.

The temperature at which gelatinization of starch take place is called as the gelatinization temperature [31]. Generally, the flour which was higher in starch content required lower temperature for gelatinization [32]. Table 2 indicates that the durian seed flours obtained from soaking of durian seed chips in sodium metabisulfite solution at longer time exhibited higher gelatinization temperature. Gelatinization temperature of starch depends on some parameters, which include moisture content, lipid content, heating rate, process applied to starch before gelatinization and amylose/amylopectin ratio of starch [33].

Gelatinization process is not only followed by carbohydrate level due to amylose and amylopectin molecule, but also water absorption which is inside pore gain more water [23]. With lower moisture content and highly porous microstructure the durian seed flour obtained from longer soaking time may adsorb more water. The starch granules become extraordinarily swollen and not necessarily return to their original state. Since carbohydrate and water absorption is getting higher, and then the gelatinization temperature will be higher.

3.3 Functional properties of durian seed flours

Water absorption capacity (WAC) represent the ability of a product to associate with water under conditions where water is limited and assure product cohesiveness [34, 35]. Table 2 shows that WAC of durian seed flours increased as the soaking time increased. A higher WAC value of flour could be attributed to the presence of higher amount of carbohydrates [36]. The other components, such as fiber, protein and amylose content are also the main factors that may influence WAC value [32]. The higher protein content lead to strong hydrogen bond, which subsequently increase the WAC of durian seed flour [37].

The oil absorption capacity (OAC) is the ability of the dry starch to physically bind fat by capillary attraction and it is of great importance that makes the flour suitable in promoting improvement in flavor and mouth feel when used in food preparation [32, 38]. The OAC of durian seed flour increased when the durian seed chips were soaked in sodium metabisulfite solution for 40 minutes to a maximum value, but then leveled off at longer soaking time. Both the protein content and the type contribute to the oil-retaining properties of food materials [39]. Non-polar amino acid side chains of protein can form hydrophobic interaction with hydrocarbon chains of lipid and has implication in functional properties of durian seed flour [40]. Therefore, the lower OAC of flour might be due to lower content hydrophobic proteins which show superior binding of lipids [41]. However, the OAC values obtained in this work did not correspond with the protein content of the respective soaking conditions.

4. Conclusion

The effect of sodium metabisulfite soaking time on physicochemical properties of durian seeds flour has been studied. Soaking of durian seed chips in sodium metabisulfite solution significantly altered the nutrition content, except fat content. This process also changed the whiteness, gelatinization temperature, water absorption capacity and oil absorption capacity of durian seed flour. The residual sulfite levels were varied from 0 to 102.56 ppm. Based on the residual sulfite level, the durian seed flours obtained are safe for consumption, mainly in the manufacture of butter cake and cookies.

5. Acknowledgement

The authors are grateful to dr. Niluh Putu Ratih Restiti for financial support and Chemical Engineering Department, Diponegoro University, Semarang for providing research facilities.

References

1. N. Tharise. E. Julianti, M. Nurminah, Int. Food Res. J. **4**,1641-1649 (2014)
2. Bizteka, *Perkembangan Impor Tepung Terigu* (PT. Citra Cendekia Indonesia, 2016)
3. A. Amin, A. Ahmad, Y. Yin-Yin, N. Yahya, N. Ibrahim, Food Hydrocol. **21**, 273-279 (2006)
4. V. Larrosa, G. Lorenzo, N. Zaritzky, A. Califano, J. Food Sci. **57**, 520-526 (2013)
5. F. Azima, T. Anggraini, D. Syukri, A. Septia, Pak. J. Nutr., **3**, 175-178 (2017)
6. Statistics Indonesia, *Fruits Production by Province (Ton)* (Statistics Indonesia, Jakarta, 2014)
7. S. Ketsa, T. Daengkanit, J. Sci. Hortic., **80**, 181–188 (1999)
8. M. J. Brown, *Durio - A Bibliographic Review* (International Plant Genetic Resources Institute, New Delhi, India, 1997).
9. H. Mirhosseini, B. Tabatabaee Amid, Chem. Cent. J. **7**, 1-7 (2013)
10. A. M. Amin, R. Arshad, Int. J. Postharvest Technol. Innovation, **1**, 367-375 (2009)
11. T. Tongdang, Starch/Starke **6**, 199-207 (2008)
12. R. O. Sinnhuber, J. D. Hendricks, G. B. Putnam, J. H. Wales, N. E. Pawlowski, J. E. Nixon, D. J. Lee, Fed. Proc. (Abs.) **35**, 1652 (1976).
13. T. L. Ferguson, J. H. Wales, R. O. Sinnhuber, D. J. Lee, Food Cosmet. Toxicol. **14**, 15 (1976)
14. S .K. Berry, Lipids **15**, 452-255 (1979)
15. Wheat Associates, *Wheat and Flour Testing Methods: A Guide to Understanding Wheat and Flour Quality* (Wheat Marketing Center Inc., Portland, Oregon, USA, 2007)

16. A. M. Y. Brown, *Understanding Food: Principles and Preparation* (Wadsworth, Thomson Learning Inc., Singapore, 2005)
17. A. Chowdhury, A. Bhattacharyya, P. Chattopadhyay, Indian J. Nat. Prod. Resour. **3,** 347-353 (2012)
18. AOAC, *Official Methods of Analysis* (Washington, D. C., 1990)
19. G. Winarno, *Kimia Pangan dan Gizi* (PT. Gramedia Pustaka Utama, Jakarta, 1995)
20. S. Clesceri, E. Greenberg, D. Eaton, *Standard Methods for the Examination of Water and Wastewater* (APHA American Public Health Association, 1998)
21. D. F. Coral, P. Pineda-Gomez, A. Rosales-Rivera, M. E. Rodriguez-Garcia, J. Physc. Conf. Ser. **167**, 012057 (2009)
22. L. L. Niba, M. M. Bokonga, F. L. Jackson, D. S. Schlimme, B. W. Li, J. Food Sci. **67**, 1701-1705 (2001)
23. F. Sosulski, E. S. Humbert, K. Bui J. D. Jones, J. Food Sci. **41**, 1349–1352 (1976)
24. C. Vidal-Valverde, J. Erias, S. Vulverde, 1992, J. Food Prot. **55**,301–304.
25. A. El-Baltegy, *Effect of home traditional methods on quality aspect of some legumes* (M.Sc. Thesis, Faculty of Agriculture, Minofiya University, Shibin El-Kom, Egypt, 1996).
26. D. L. Rom, J. M. Shull, A. Chandrashekar, A. W. Kirleis, Cereal Chem. **69**, 178-181 (1992)
27. M .S. Rahman, T. P. Labuza, *Water Activity and Food Preservation* (CRC Press, Taylor and Francis Group, New York, 2007)
28. M. B. Go, S. P. Velos, A. V. Minyamin, R. D. Bagsit, R. G. Pableo, Trop. Tech. J. **19**, 1-7 (2015)
29. C. Ruiz-Capillas, F. Jiménez-Colmenero, Food Chem. **112**, 487–493 (2009)
30. H. S. Lambrecht, *Sulfite substitutes for the prevention of enzymatic browning in foods* (American Chemical Society, Washington D.C., 1995)
31. K. M. Sahay, K. K. Singh, Unit *operations of Agricultural Processing* (Vikas Publishing House Pvt. Ltd. New Delhi, 1996)
32. S. Chandra, Samsher, Afr. J. Agric. Res.**8**, 4849-4852 (2013)
33. F. Altay, S. Gunaseakran, J. Agric. Food Chem. **54**, 4235-4245 (2006)
34. U. Singh, J. Food Sci. Technol. **38**,191-199 (2001)
35. P. Housson, G. S. Ayenor, Afr. J. Sci. Technol. **3**, 121-126 (2002)
36. E. I. Adeyeye, P. A. Aye, La Vista Italiana Della Sostanze Grasse **1**, 253-261 (1998)
37. A. R. Olukemi, O. A. Olayiwola, S. S. Abdulsalam, J. Agric. Sci. **1**, 1-5 (2013)
38. M O., Aremu, O. Olaofe, E. T. Akintayo, J. Food Technol. **5**, 109–115 (2007)
39. R. Ravi, N. S. Sushelamma, J. Food Sci. **70**, S539–S547 (2005)
40. M. Eltayeb, A. Ali, A. Abou-Arab, F. Abu-Salem, Afr. J. Food Sci. **5**, 82-90 (2011)
41. J. E. Kinsella, Crit. Rev. Food Sci. Nutr. **5**, 219–225 (1976)

Liquid-liquid Equilibria for Quaternary System of Eugenol (1) + β-Caryophyllene (2) + 1-Propanol (3) + Water (4) at Temperatures 303.15, 313.15, and 323.15 K

Irwan Hidayatulloh[1], *Nurcahyo* Nugroho[1], *Gede* Wibawa[1] , and *Kuswandi* Kuswandi[1*]

[1] Department of Chemical Engineering, Faculty of Industrial Technology, Institut Teknologi Sepuluh Nopember, 60111 Kampus ITS Sukolilo Surabaya, Indonesia

Abstract. In this work, the experimental liquid–liquid equilibrium (LLE) data were measured for quaternary mixtures of eugenol(1) + β–caryophyllene(2) + 1-propanol(3) + water(4) at temperatures 303.15, 313.15, and 323.15 K and at atmospheric pressure. The equilibrium data ware determined by using an equilibrium cell with temperature controlled. Quantitive analysis was performed by using gas chromatography (GC) equipped with thermal conductivity detector and capillary column Rtx-5. The experimental data were then correlated by using the NRTL and the UNIQUAC models. The reliability of these models were tested by compared with the experimental results using a root mean square deviation (RMSD). Based on the RMSD, the NRTL and the UNIQUAC models suited really well and give satisfactory result for the system of eugenol, β–caryophyllene, 1-propanol, and water with the average RMSD of the UNIQUAC and the NRTL models were 0.417% and 0.502%, respectively. Furthermore there was no significant effect of temperatures on the equilibrium composition for the system.

1 Introduction

Clove oil is one kind of essential oils that has many uses, in general, such as aromatherapy, botanical pesticides, additives in the food industry and the tobacco industry. Clove oil is a complex mixture of chemical compounds, the main component of clove oil is eugenol. The content of eugenol in clove oil is approximately 70% - 90% and the largest impurities is β-caryophyllene in the level of less than 10% [1]. Low purity of eugenol which is less than 70% can only be sold at US $ 4.75 / kg, while high purity of eugenol which contain more than 98% can be sold at US $ 7.80 / kg [2].

Extraction is a common process for separating eugenol from their impurities. In the extraction process, the choice of solvent should consider the nature and characteristics of solvent such as polarity, volatility, and toxicity [3]. Alcohol is commonly used as solvent because of the nature of polarity, relatively cheap, and low toxicity. Methanol is the simplest alcohol that can be used as a solvent, but methanol easily corrode to steel and toxic to the human body. Other alcohols such as ethanol is widely used, but ethanol has an azeotropic properties that make phenol and terpen compounds (such as β-caryophyllene) will not completely dissolve in ethanol. Because of that, it's important to find other types of alcohol as a solvent in the eugenol extraction process. One of the best solutions is 1-propanol. Eugenol has high solubility in 1-propanol that is better than methanol and ethanol. Additionally 1-propanol is less toxic to humans and has a relatively high volatility [4].

Liquid-liquid equilibrium data are important in the modeling and design of extraction process for eugenol purification. In previous work, the Liquid-liquid equilibrium data of ternary eugenol-ethanol-water [5] and the quaternary eugenol-β-caryophyllene-ethanol-water [6] at same temperatures 303.15, 313.15 and 323.15 K have been reported. The purpose of this work is to obtain Liquid-liquid equilibrium data for the quaternary system eugenol, β-caryophyllene, 1-propanol, water at temperatures 303.15, 313.15 and 323.15 K and correlating the results of the experiments with NRTL [7] and UNIQUAC [8] models to obtain the binary interaction parameters.

2 Experimental

2.1 Materials

Eugenol and β-caryophyllene were obtained from PT. Indesso Niagatama. with purity of 99.9 mole %, 1-Propanol was supplied by Merck Co. with 99.8 mole % purity. All of those chemicals were used without further purification. Aquabidestilata or double distilled water

* Corresponding author: kuswandi@chem-eng.its.ac.id

used throughout this experimental work was supplied by PT. Ikapharmindo Putramas.

2.2 Apparatus and procedure

The apparatus used in this work consists of an equilibrium cell with 40 cm^3 volume equipped with a heating jacket to keep the temperature remain constant. A magnetic stirrer is used to make sure that the solution was completely mixed. The equilibrium cell is connected to a chiller (water bath). The chiller equipped with a controller and a thermocouple with a rate of accuracy about ± 0.10°C.

Table 1. Experimental data for quaternary system of eugenol + β-caryophyllene + 1-propanol + water at 303.15 K

Organic Phase				Aqueous Phase			
x_1^a	x_2^a	x_3^a	x_4^a	x_1^b	x_2^b	x_3^b	x_4^b
0.0871	0.0850	0.3950	0.4329	0.0005	0.0005	0.0676	0.9314
0.1120	0.0971	0.4089	0.3820	0.0005	0.0005	0.0569	0.9421
0.1733	0.1593	0.3709	0.2965	0.0005	0.0005	0.0609	0.9381
0.1768	0.1815	0.3466	0.2951	0.0005	0.0005	0.0452	0.9538
0.0902	0.0917	0.3951	0.4229	0.0005	0.0005	0.0710	0.9280
0.3587	0.2391	0.2275	0.1748	0.0005	0.0005	0.0476	0.9514
0.1151	0.0165	0.3669	0.5016	0.0005	0.0005	0.0700	0.9290
0.1230	0.0237	0.3792	0.4740	0.0005	0.0005	0.0526	0.9464
0.5808	0.4192	0	0	0	0	0	1

Table 2. Experimental data for quaternary system of eugenol + β-caryophyllene + 1-propanol + water at 313.15 K

Organic Phase				Aqueous Phase			
x_1^a	x_2^a	x_3^a	x_4^a	x_1^b	x_2^b	x_3^b	x_4^b
0.1212	0.0763	0.3945	0.4079	0.0005	0.0005	0.0674	0.9316
0.1508	0.1545	0.3725	0.3222	0.0005	0.0005	0.0688	0.9302
0.1906	0.1729	0.3556	0.2809	0.0005	0.0005	0.0658	0.9332
0.1933	0.1347	0.3636	0.3084	0.0005	0.0005	0.0561	0.9429
0.1204	0.0547	0.3717	0.4533	0.0005	0.0005	0.0600	0.9390
0.2915	0.2765	0.2642	0.1678	0.0005	0.0005	0.0601	0.9389
0.1438	0.0145	0.3526	0.4891	0.0005	0.0005	0.0631	0.9359
0.1493	0.0231	0.3633	0.4643	0.0005	0.0005	0.0640	0.9350
0.5963	0.4037	0	0	0	0	0	1

Table 3. Experimental data for quaternary system of eugenol + β-caryophyllene + 1-propanol + water at 333.15 K

Organic Phase				Aqueous Phase			
x_1^a	x_2^a	x_3^a	x_4^a	x_1^b	x_2^b	x_3^b	x_4^b
0.1060	0.0674	0.4031	0.4234	0.0005	0.0005	0.0587	0.9403
0.1564	0.1259	0.3936	0.3241	0.0005	0.0005	0.0578	0.9412
0.1948	0.1832	0.3519	0.2702	0.0005	0.0005	0.0592	0.9398
0.1897	0.1672	0.3632	0.2800	0.0005	0.0005	0.0514	0.9476
0.1127	0.0848	0.4027	0.3999	0.0005	0.0005	0.0631	0.9359
0.3284	0.2056	0.2730	0.1930	0.0005	0.0005	0.0285	0.9705
0.1130	0.0205	0.3720	0.4946	0.0005	0.0005	0.0687	0.9303
0.1303	0.0173	0.3754	0.4771	0.0005	0.0005	0.0632	0.9358
0.5961	0.4039	0	0	0	0	0	1

To obtain the equilibrium data, the quatenary mixtures of components with known composition, were mixed into the equilibrium cell equipped with a magnetic stirrer and jacket to circulate heating water according to desired temperature. The quaternary mixtures were stirred at a constant temperature (303.15, 313.15, and 323.15 K) at atmospheric pressure for 4 hours, and then the mixture solution was left for 20 hours until the equilibrium was achieved. Once equilibrium was reached, the samples of each phase were taken for analysis by using gas chromatography (GC).

The type of Gas Chromatography used was Shimadzu 2010 with TCD detector using RTX-5 column 30m, 0.32mm ID, 0.25μm. Whereby helium was used as carrier gas with a flow rate of 3 ml / min. GC Analysis was done using temperature program method by setting the oven initial temperature at 343.15 K in 5 minutes hold time, and then temperature was raised until 473.15 K with an increment of $40^{O}C$ / min in 8 minutes hold time, while the temperature of the detector and injector were 603.15 K. The resulting mole fraction from GC analysis was then used to find binary parameters using NRTL and UNIQUAC models.

3 Results and discussion

The experimental liquid-liquid equilibrium data for quaternary system eugenol(1) + β–caryophyllene(2) + 1-propanol(3) + water(4) at temperatures 303.15, 313.15, and 323.15 K are shown in Tables 1-3 and Figs.1-6.

Table 4. Binary interaction parameters and the RMSD values for the NRTL model.

T(K)	Comp. (ij)	NRTL Parameters			RMSD (%)
		α	a_{ij} (J/mol)	a_{ji} (J/mol)	
303.15	1-2	0.2	3817.37	4530.55	0.484
	1-3		-3313.96	1069.65	
	1-4		7045.09	3548.73	
	2-3		-2368.44	2416.38	
	2-4		3928.50	2865.81	
	3-4		-1200.03	862.55	
313.15	1-2	0.2	-1209.21	3021.76	0.468
	1-3		4993.55	4892.45	
	1-4		17687.32	32020.01	
	2-3		349.78	4808.39	
	2-4		10907.27	2643.04	
	3-4		7173.45	1701.00	
323.15	1-2	0.2	4920.82	492.82	0.553
	1-3		-2332.28	1349.87	
	1-4		13904.6	4154.15	
	2-3		10042.0	24188.8	
	2-4		1737.70	1774.12	
	3-4		2143.57	-1821.3	

The experimental data obtained were correlated using NRTL and UNIQUAC models. Those models are used since they can be applied for binary or multi-component systems of liquid liquid equilibrium.
The binary interaction parameter pairs can be determined by minimizing the following objective function (OF) :

$$OF = \sum_{k}^{n} \sum_{j} \sum_{i} \left(x_{ijk}^{exp} - x_{ijk}^{cal}\right)^2 \quad (1)$$

Where x_{ij}^{exp} and x_{ij}^{cal} are the experiment and calculated mol fraction, respectively. k = 1 to n is number of tie-line, subscription i =1-4 is the number of components, and subscription j =I-II denotes the phase (organic and aqueous)[9].

Table 5. Binary interaction parameters and the RMSD values for the UNIQUAC model.

T(K)	Comp. (ij)	UNIQUAC Parameters		RMSD %
		u_{nit} (J/mol)	u_{nit} (J/mol)	
303.15	1-2	364.08	283.15	0.345
	1-3	203.33	123.86	
	1-4	737.41	397.08	
	2-3	147.36	227.94	
	2-4	541.88	296.23	
	3-4	76.64	244.30	
313.15	1-2	537.56	256.55	0.426
	1-3	260.93	178.02	
	1-4	8490.29	411.17	
	2-3	170.68	203.07	
	2-4	480.21	334.02	
	3-4	-71.17	256.97	
323.15	1-2	507.95	240.26	0.480
	1-3	76.67	-20.89	
	1-4	850.43	357.40	
	2-3	145.98	163.85	
	2-4	601.63	336.66	
	3-4	33.10	236.62	

Based on this work the binary parameters for NRTL and UNIQUAC models along with root mean square deviation (RMSD) are given in Tables 4-5. Using those parameters, the experimental data were correlated then compared with the result of calculations by using RMSD in composition of both phase. The RMSD deviation model was defined as :

$$RMSD = 100\% \; x \sqrt{\frac{\sum_{k}^{n} \sum_{j} \sum_{i} \left(x_{ijk}^{exp} - x_{ijk}^{cal}\right)^2}{2\theta N}} \quad (2)$$

Where θ is the number of components and N is number of tie-line [10]

The experimental and calculated data of this system were then plotted in psoudoternary diagram as shown in Fig.2-7, with average RMSD for NRTL and UNIQUAC models are 0,502% and 0,417%, respectively. The result of both models are less than 10%, it was good enough to represent the equilibrium data of this system [11].

Furthermore, from **Fig**. 1-3 and **Fig**. 4-6 it can be observed that the temperature has no significant effect on the equilibrium composition, indicated by the large area under the dome (two phase area) that look almost the same for each temperatures.

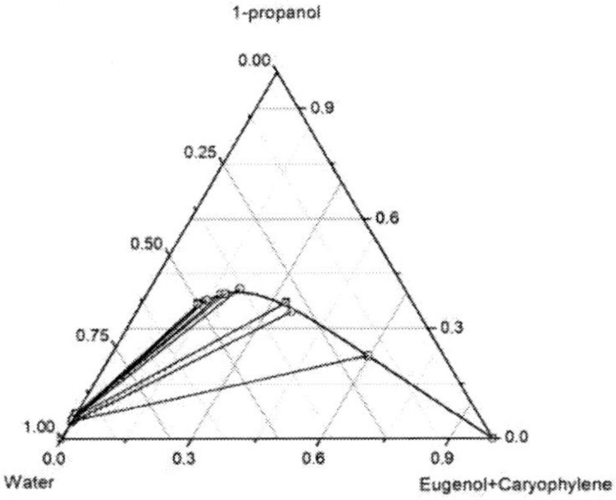

Fig. 1. LLE diagram for eugenol + β-caryophyllene + 1-propanol + water system at 303.15 K. (○) experimental data, (■) calculated data (NRTL), (····) experimental tie-lines, (—) correlated tie-lines using NRTL model.

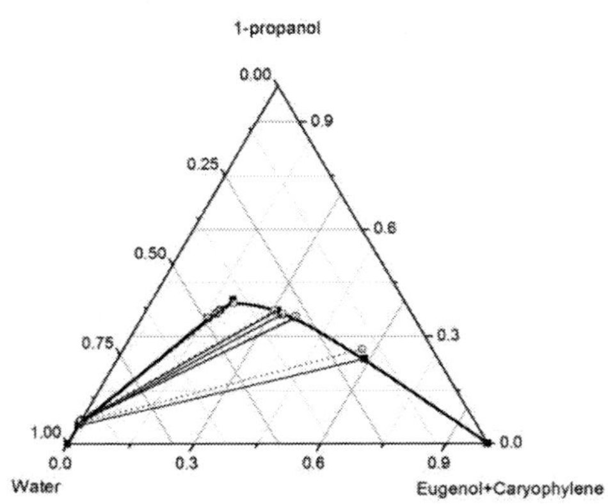

Fig. 2. LLE diagram for eugenol + β-caryophyllene + 1-propanol + water system at 313.15 K. (○) experimental data, (■) calculated data (NRTL), (····) experimental tie-lines, (—) correlated tie-lines using NRTL model.

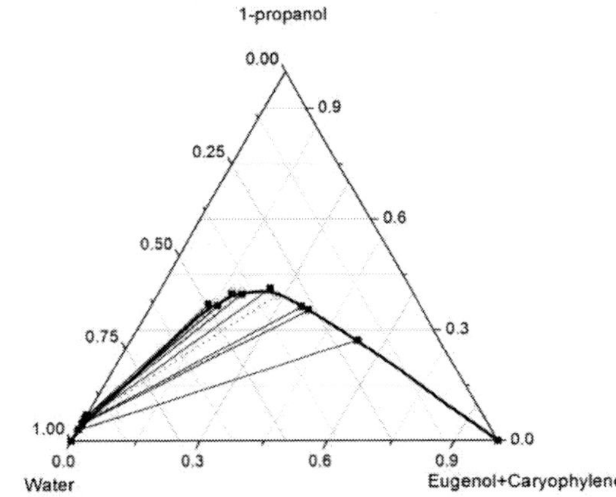

Fig. 3. LLE diagram for eugenol + β-caryophyllene + 1-propanol + water system at 323.15 K. (○) experimental data, (■) calculated data (NRTL), (····) experimental tie-lines, (—) correlated tie-lines using NRTL model.

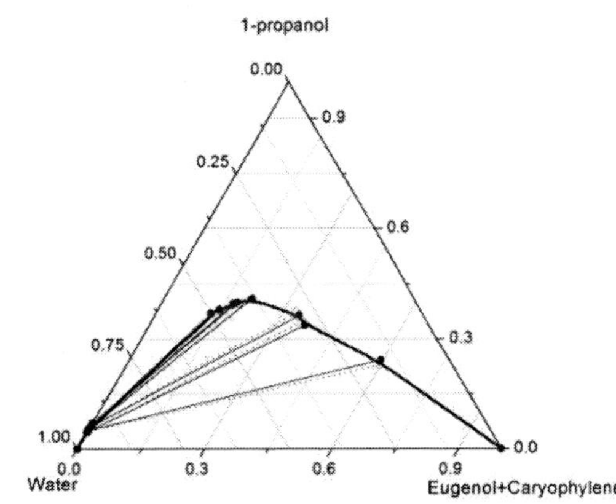

Fig. 4. LLE diagram for eugenol + β-caryophyllene + 1-propanol + water system at 303.15 K. (○) experimental data, (■) calculated data (UNIQUAC), (···) experimental tie-lines, (—) correlated tie-lines using UNIQUAC model.

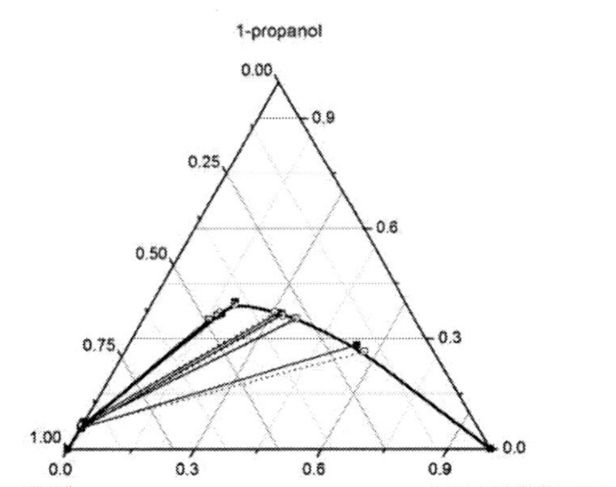

Fig. 5. LLE diagram for eugenol + β-caryophyllene + 1-propanol + water system at 313.15 K. (○) experimental data, (■) calculated data (UNIQUAC), (···) experimental tie-lines, (—) correlated tie-lines using UNIQUAC model.

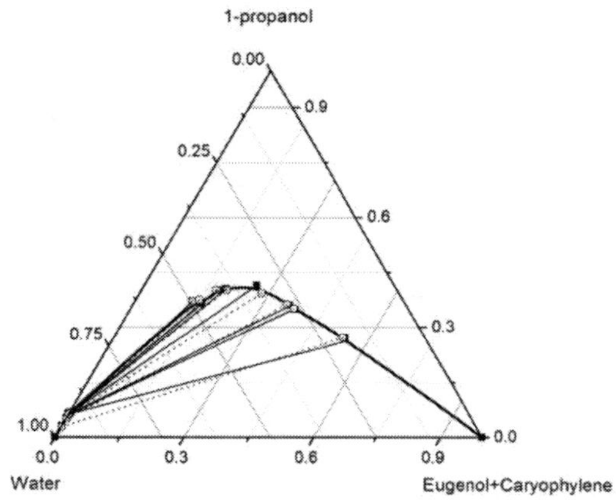

Fig. 6. LLE diagram for eugenol + β-caryophyllene + 1-propanol + water system at 323.15 K. (○) experimental data, (■) calculated data (UNIQUAC), (····) experimental tie-lines, (—) correlated tie-lines using UNIQUAC model.

4 Conclusion

In this study, the liquid-liquid equilibria of quaternary system of the eugenol + β-caryophyllene + 1-propanol + water at temperatures of 303.15, 313.15, and 323.15 K have been measured accurately. Experimental data were well correlated using the NRTL and the UNIQUAC models. The results are satisfactory, with the average RMSD of the NRTL and the UNIQUAC models 0.502% and 0.417%, respectively. In the range of the temperatures studied, it showed that the temperatures are no significant effect to the equilibrium composition.

This research was supported by PT. Indesso Niagatama and the Thermodynamics Laboratory, Chemical Engineering Department, Institut Teknologi Sepuluh Nopember.

References

1. E. Gunther, *Minyak Atsiri Jilid 1*. (1987)
2. J. Towaha. *Indones. Res. Inst. Ind. Baverage Crop*. **11**, 1412-8004 (2012)
3. A. Cháfer, J. de la Torre, R. Muñoz, M. C. Burguet, Fluid Phase Equilib. **238**, 72–76 (2005)
4. Y. -K. Kim, D. -W. Park, J. Ind. Eng. Chem. **14**, 602–607 (2008)
5. K. Kuswandi, E. Naryono, *Int. Rev. Chem. Eng (IReChE)*. **3**, 663–667 (2011)
6. R. Fitriani, K. Kuswandi, *int prosiding seminar nasional pasxasarjana XIII (SNPs)* (2013)
7. H. Renon, J. M. Prausnitz, AIChE J. **14**, 135–144 (1968)
8. D. S. Abrams, J. M. Prausnitz, AIChE J. **21**, 116–128 (1975)
9. I. C. Hwang, S. J. Park, J.S. Choi, Fluid Phase Equilib. **269**, 1–5 (2008)
10. C. Stoicescu, O. Iulian, R. Isopescu, Rev. Roum. Chim. **56**, 3214–3221 (2011)
11. J.M. Smith, H .C. Van Ness, M. M. Abbot, *Chemical Engineering Thermodynamics 6th edition*. (2001)

Effect of Promoter Concentration on CO_2 Separation Using K_2CO_3 With Reactive Absorption Method in Reactor Packed Column

Junety Monde[1], *Tri* Widjaja[1*], *Ali* Altway[1]

[1]Chemical Engineering Department, Industrial Technology Faculty, Sepuluh Nopember Institute of Technology (ITS), Surabaya 60111, Indonesia

Abstract. The presence of carbon dioxide (CO_2) in the gas is not expected because CO_2 can reduce heating value and CO_2 is the major emission contributor into the atmosphere. Various separation technologies can be used to reduce CO_2 content and improve quality of gas. Chemical or reactive absorption is most widely used because it provides higher removal rate. This paper will study the effect of the addition di ethanolamine (DEA) concentration into aqueous 30wt.% potassium carbonate(K_2CO_3) with reactive absorption method in a reactor packed column at temperature from 40^{0}C to 80^{0}C, DEA concentration range of (1% - 3%) and absorbent flow rate (0.5, 0.75 and 1) L. min^1. Contacting the gas and absorbent are countercurrent flow in packed column with 1.5 m high and 50 mm in diameter. The absorption column was randomly packed with a packing material raschig rings 5 mm in diameter. The CO_2 loading in the liquid samples was determined by titration. It is found that the best result of CO_2 loading is 0.065594 mole/mole K_2CO_3 and CO_2 removal 28%. The result show that the loading capacity (mole CO_2/mole K_2CO_3) and CO_2 removal increased with the increase of DEA concentration. **Keyword**: potassium carbonate, promoted and packed column and reactive absorption

1. Introduction

Energy is an important part of human life, because most of human activity need energy. Indonesia is one of the countries that include in category of countries with the largest natural gas reserves in the world. Indonesia also known as a country that produce large natural resource that can be used as raw material to produce biogas. It means that gas is potential to be developed for use as an energy alternative in Indonesia. As an alternative energy, gas quality needs to be improved by reducing the CO_2 component so that methane content increases and the economic value of the gas can increase as well.

The presence of CO_2 in the gas is not expected because CO_2 can reduce heating value and CO_2 is the major contributor emission to the atmosphere. Various separation technologies can be used to reduce CO_2 content and improve quality of gas. Five methods that can be used are adsorption, cryogenic distillation, membrane separation, absorption both physically, and chemically. Since the chemical reaction in the solution affects the liquid phase absorption coefficient as compared with the physical absorption. Potassium carbonate is a solvent commonly used in chemical absorption. The main advantage of potassium carbonate is high solubility of CO_2 in the carbonate or bicarbonate system, ease of regeneration, low solvent cost, low toxicity, and low degradation [1], but potassium carbonate has disadvantages that is low absorption rate. Various technologies can be used to improve the absorption rate of potassium carbonate by adding a number of promoters (activators) or by increasing mass transfer efficiency with equipment such as packing [2]. This study will be focused on adding promoter to potassium carbonate. The reaction of potassium carbonate with CO_2 in general can be written:

$$CO_2 + H_2O + K_2CO_3 \leftrightarrow 2KHCO_3 \quad (1)$$

Amines have been widely used as promoters in CO_2 absorption. Every alkali amine has at least one hydroxyl group and one amine group. The group of hydroxyl used to reduce the vapor pressure and increase the solubility in water while the group of amine supply the necessary alkalinity in the solution to react with the acid gas [3]. One of the amine compound which can be used as a promoter in CO_2 uptake is a secondary amine DEA (di-ethanolamine) with acid gas loading rate of about 0.30-0.35 mole acid gas / mole acid amine [3] and has a

* Corresponding author: papatri2003@yahoo.com

corrosion-resistant advantage over MEA, high reactivity and fast absorption rate.

In the mechanism, the promoter is called homogeneous catalysis that is the promoter first forms intermediates with CO_2 and intermediates then hydrolyzed to produce the final product, bicarbonate.

$$CO2 + promoter = intermediate \quad (2)$$

$$intermediate + OH- = HCO_3^- + promoter \quad (3)$$

In another mechanism, the so-called shuttle mechanism proposed by Danckwerts [4] states that the promoter acts as a carrier to provide additional lines for transporting CO_2 from the liquid gas to the bulk liquid, the carbonate ions then diffuse into the liquid and will react with the OH ion to produce Amine free of CO_2 so that amines can again diffuse into the interface to react with CO_2. DEA reaction with carbon dioxide in general can be written:

$$2(C_2H_5)_2NH + CO_2 \rightarrow (C_2H_5)_2NCOOH_2(C_2H_5)_2 \quad (4)$$

Both these mechanisms can be summarized that the shuttle mechanism can describe the behavior of promotor at ambient temperature and at higher temperature the system is better represented by homogeneous catalysis mechanism [5].

This paper will study the effect of the addition di-ethanolamine concentration into aqueous 30wt.% potassium carbonate using chemical absorption method. The gas and absorbent were contacted with countercurrent flow in column with packing that serves to provide for long time contacted so that enhance CO_2 absorption. The absorption capacity performance of CO_2 into aqueous 30wt% potassium carbonate solute is expressed as CO_2 loading and CO_2 removal.

2.Material and Methods

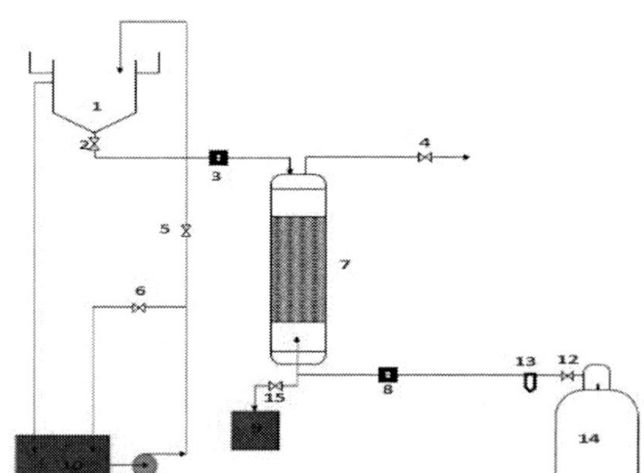

9. Absorbent Tank
10. Valve on absorbent inlet to column
11. Absorbent flowmeter
12. Valve on gas outlet column
13. Valve on discharge of absorbent pump
14. Valve on discharge by-pass of absorbent pump

1. Gas flowmeter
2. Absorbent tank outlet column
3. Overflow tank
4. Absorbent pump
5. Valve on gas inlet to column
6. Manometer on gas inlet to column
7. Gas tank

Fig.1. The experimental apparatus

The experimental apparatus is shown in Fig. 1, the process is simple, the feed gas with the composition of 40% CO_2 and 60% N_2 is flowed with a fixed flow rate at 40 $L.min^{-1}$ from the bottom of the column and absorbent with the DEA composition, aqueous solution of K_2CO_3 and water are flowed from the top of the column with a flow rate of variables of 0.5; 0.75; 1 $L.min^{-1}$ flows through which contains packing that serves to provide for long time contacted so that enhance CO_2 absorption. The principle of absorption is the different solubility of the gas so that only CO_2 gas will be absorbed. The CO_2 content in the liquid samples was determined by acid-base titration using phenolphthalein. The apparatus with column of pyrex glass cylinder of 1.5 m high and 50 mm in diameter. The absorption column was randomly packed with a packing material of raschig rings 5 mm in diameter. The material used is potassium carbonate with purity > 99%, synthetic gas with CO_2: 40% and N_2: 60% composition, in ethanolamine with purity 98% and water. This research will use aqueous 30wt.% Potassium carbonate (K_2CO_3) and DEA Concentration range of (1% - 3%) with operating conditions temperature from 40^0C to 80^0C, absorbent flow rate (0.5, 0.75 and 1) L.min-1, gas flow rate 40 L.min-1 and pressure 1 atm.

3.Results and discussion

The amount of absorbed CO_2 can be expressed in % CO_2 removal and CO_2 loading. Some factors that affect the % CO_2 removal and CO_2 loading are temperature, promoter concentration, absorbent flow rate

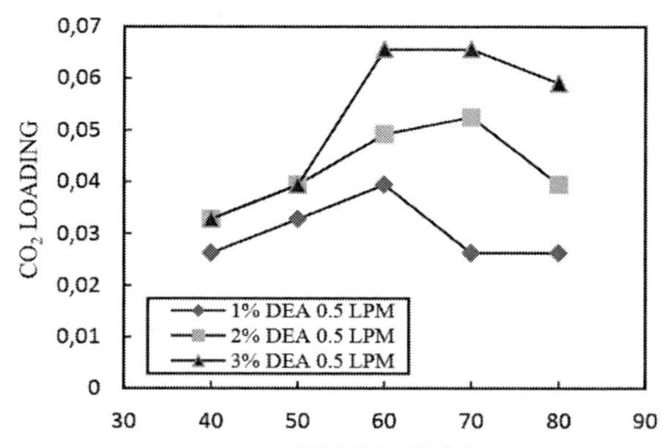

Fig.2. Effect of DEA concentration and Temperature on CO_2 Loading at Flow Rate 0.5 LPM

https://doi.org/10.1051/matecconf/201815602002

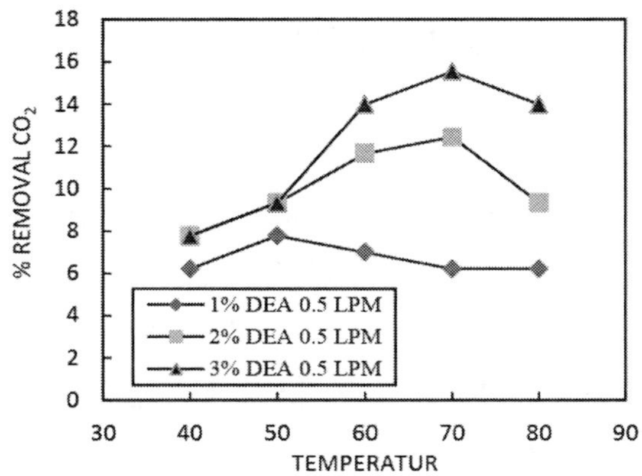

Fig.3. Effect of DEA concentration and Temperature on % Removal CO_2 at Flow Rate 0.5 LPM

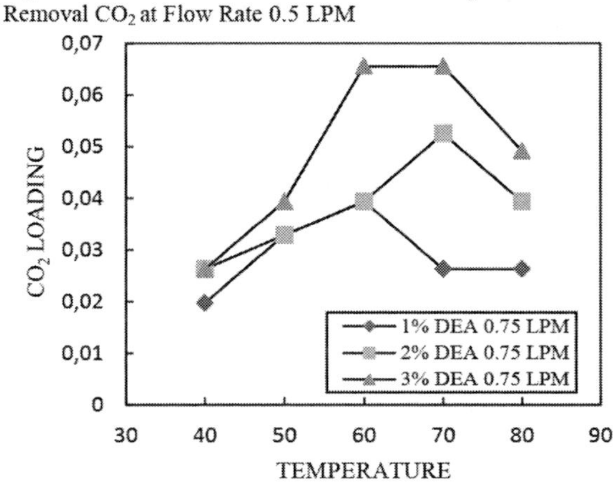

Fig.4. Effect of DEA concentration and Temperature on CO_2 Loading at Flow Rate 0.75 LPM

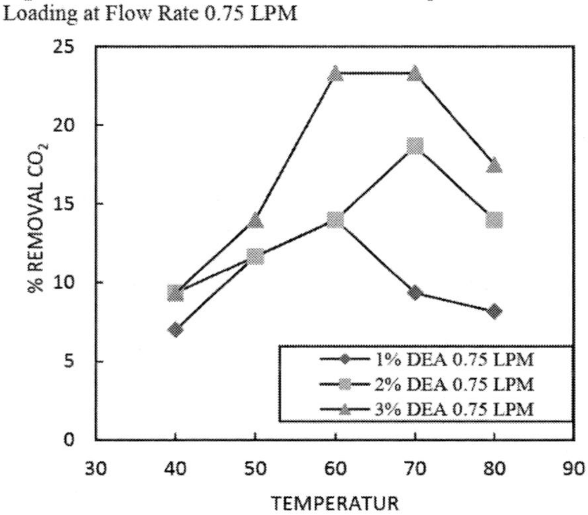

Fig.5. Effect of DEA concentration and Temperature on % Removal CO_2 at Flow Rate 0.75 LPM.

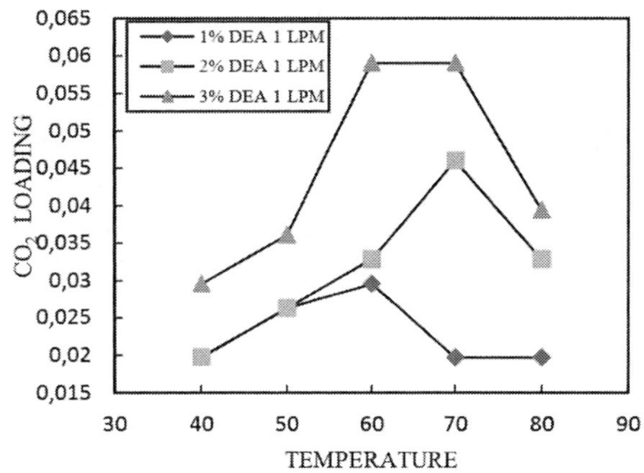

Fig.6. Effect of DEA concentration and Temperature on CO_2 Loading at Flow Rate 1 LPM.

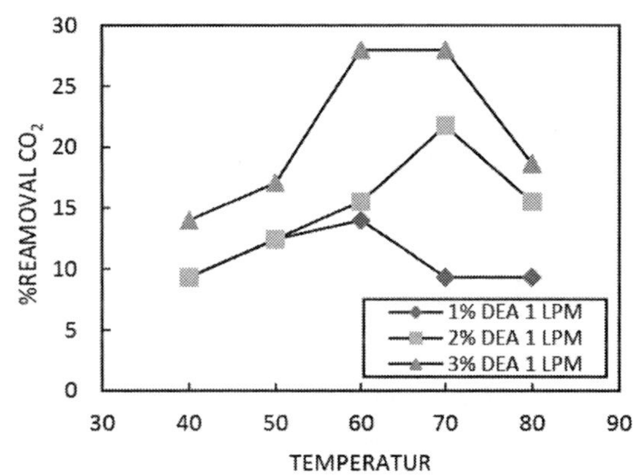

Fig.7. Effect of DEA concentration and Temperature on % Removal CO_2 at Flow Rate 1 LPM.

Table.1. Effect of Concentration, Flow Rate and Temperature on % CO_2 Removal.

T (C)	DEA (WT%)	CO_2 REMOVAL (%)		
		0.5 LPM	0.75 LPM	1 LPM
40		6.218	6.991	9.322
50		7.773	11.652	12.429
60	1	6.995	13.983	13.983
70		6.218	9.322	9.322
80		6.218	8.156	9.322
40		7.768	9.321	9.322
50		9.322	11.652	12.429
60	2	11.652	13.983	15.536
70		12.429	18.644	21.751
80		9.322	13.983	15.536
40		7.768	9.322	13.983
50		9.322	13.983	17.09
60	3	13.983	23.304	27.965
70		15.536	23.304	27.965
80		13.983	17.478	18.644

Table.2. Effect of Concentration, Flow Rate and Temperature on % CO_2 Loading

T (C)	DEA (WT%)	CO_2 LOADING ($L.min^{-1}$) 0.5 LPM	0.75 LPM	1 LPM
40		0.026236	0.019703	0.019678
50		0.032795	0.032839	0.026237
60	1	0.039406	0.039406	0.029517
70		0.026236	0.026271	0.019678
80		0.026236	0.026271	0.019678
40		0.032817	0.026254	0.01969
50		0.03938	0.032817	0.026254
60	2	0.049226	0.03938	0.032817
70		0.052507	0.052507	0.045944
80		0.03938	0.03938	0.032817
40		0.032798	0.026238	0.029517
50		0.039357	0.039356	0.036076
60	3	0.065595	0.0590357	0.059034
70		0.065595	0.065594	0.059034
80		0.059036	0.049195	0.039356

3.1 Effect of Promoter Concentration

The concentration of DEA in K_2CO_3 gives a very important influence in the absorption of CO_2. In Fig.7 to Fig.4,Table.1 and Table 2 it can be seen that when the DEA concentration increases in potassium carbonate, CO_2 Loading and % CO_2 Removal will increase to 0.065594 mole/mole K_2CO_3 and 28% this is because DEA concentration contributes to increased diffusivity and higher mass transfer because the number of DEA molecules is greater in potassium carbonate, this results in more CO_2 absorption and increases CO_2 for contact and reacts with absorbent. The concentration of DEA in a solution of Potassium carbonate (K_2CO_3) causes the larger CO_2 absorbed.

3.2 Effect of Absorbent Temperature

Temperature has a big role in increasing the diffusion and mass transfer so that with increasing temperature will increase CO_2 Loading and % CO_2 Removal as shown in Fig.2 to Fig.7, Table.1 and Table 2. first at temperature 40; 50; 60; 70^{0}C CO_2 loading and % Removal CO_2 up increases to 0.065594 mole/mole K_2CO_3 and 28% at 1 $L.min^{-1}$ and CO_2 Loading moles of CO_2/mole absorbent. In a previous study stated that higher temperatures contribute to higher mass transfer efficiency which can lead to increased CO_2 uptake [6]. In addition to increasing the diffusion and mass transfer of temperature can negatively affect the high temperature of CO_2 solubility will decrease as at temperature of 80^{0}C CO_2 Loading and % CO_2 Removal value decreases to 0.039356 moles of CO_2/mole K_2CO_3 and 15.53% in 1 $L.min^{-1}$. According to previous study by Aroonwilas et al [7] when the temperature is too high can cause a decrease in the efficiency of mass transfer this is caused by the exothermic thermodynamic properties of the process.

3.3 Effect of Absorbent Flow Rate

The absorbent flow rate significantly affects CO_2 reduction in the gas. The relationship between the absorbent flow rate on % CO_2 Removal is shown in fig.9,11,13 and Table.1 when the absorbent flow rate increas % CO_2 Removal will increase to 28% at the flow rate of 1 $L.min^{-1}$. The absorbent flow rate influences the degree of alkalinity of the packing area so that the contact between the gas and the absorbent becomes large and increases the performance of CO_2 absorption. An increase in the flow rate of the solution allows potassium carbonate to absorb more absorbed moles of CO_2. According to Aroonwilas et al [8], increasing flow rate will increase mass transfer in the liquid-side and more absorbent molecules react with CO_2 per unit time. The relationship between the absorbent flow rate on CO_2 loading is shown in fig.8,10,12 and table 2. at flow rate 0.5 to 0.75 $L.min^{-1}$ causing absorption of CO_2 increase so that the CO_2 loading also increase. Increasing of flow rate from 0.75 to 1 $L.min^{-1}$ CO_2 absorbed is not increase significantly so the loading also decreases. The decreased of CO_2 loading indicates that there are free K_2CO_3 or excess K_2CO_3.

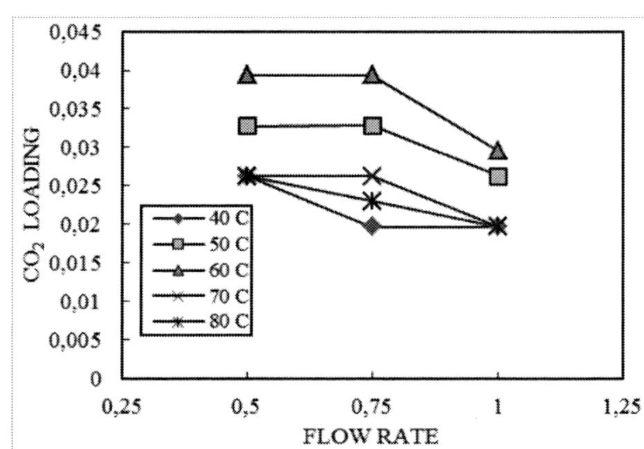

Fig.8. Effect of Flow Rate on CO_2 Loading at concentration 1 Wt% DEA.

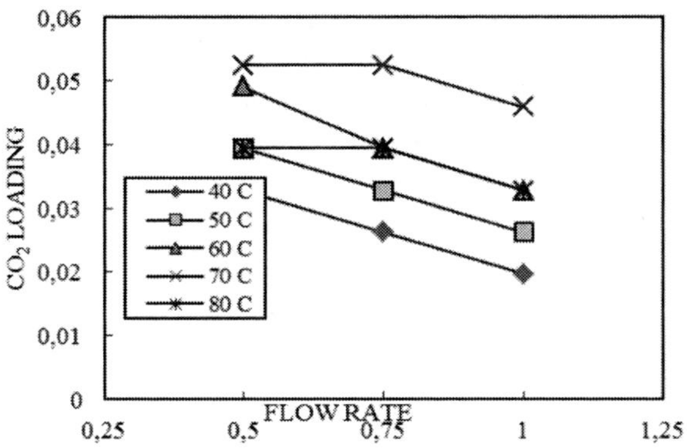

Fig.9. Effect of Flow Rate on % Removal CO_2 at concentration 1 Wt% DEA.

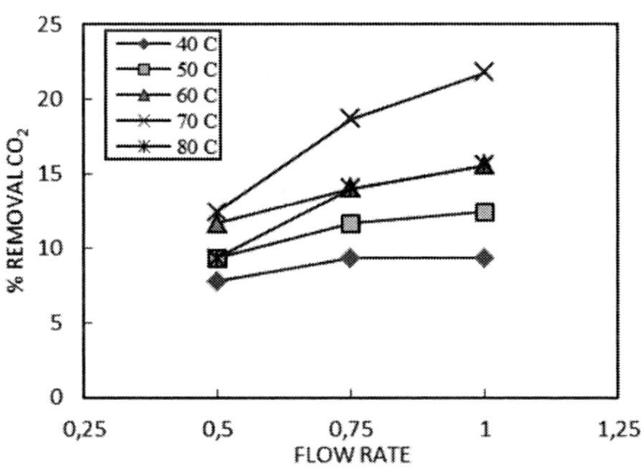

Fig.10. Effect of Flow Rate and Temperature on CO_2 Loading at concentration 2 Wt% DEA

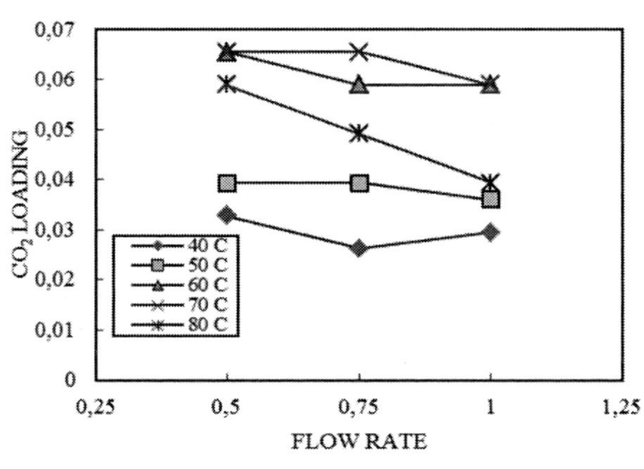

Fig.11. Effect of Flow Rate on % Removal CO_2 at concentration 2 Wt% DEA

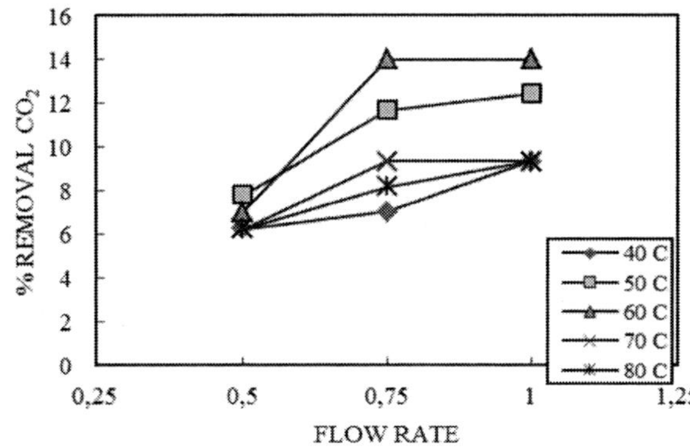

Fig.12. Effect of flow Rate on CO_2 Loading at concentration 3 Wt% DEA.

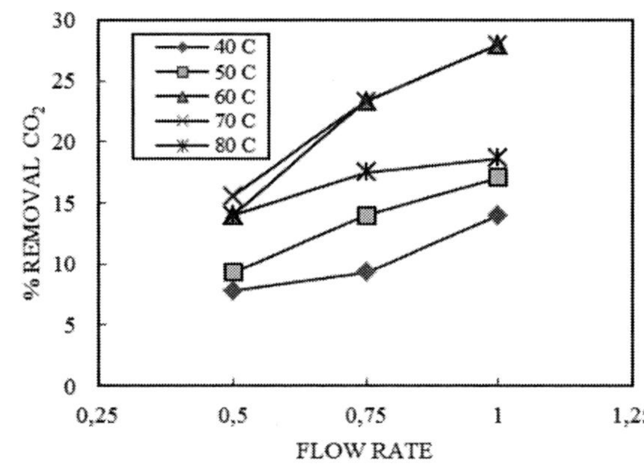

Fig.13. Effect of flow Rate on % Removal CO_2 at concentration 3 Wt% DEA.

4. Conclusions

The absorption of CO_2 by using potassium carbonate with promoter DEA get best result at temperature 70^0C with value of CO_2 Removal and CO_2 loading equal 0.065594 mole/mole K_2CO_3 and 28% at absorbent flow rate 1 $L.min^{-1}$, concentration of promoter 3%, pressure 1 atm, packed column with 1.5 m high and 50 mm in diameter.

5. References

[1] T. N. G. Borhani, V. Akbari, M. K. A. Hamidand Z. A. Manan, *J. Ind. Eng. Chem.*,**22**. 306–316 (2015)

[2] X. Zhao, Y. Wang, Z. Ye, A. G. L. Borthwick, and J. Ni, *Process Biochem.*,**41**,7, 1475–1483, (2006)

[3] A. L. Kohl and R. B. Nielsen, *Gas Purification*, **10** (1997)

[4] A. L. Shrier and P. V. Danckwerts, *Ind. Eng. Chem. Fundam.*, **8**, 3, 415–423 (1969)

[5] M. Hussein, Int. j. eng. sci, **4**, 2, 6-12 (2015)

[6] L. S. Tan, A. M. Shariff, K. K. Lau, and M. A. Bustam, *J. Ind. Eng. Chem,* **18**, 6, 1874–1883 (2012)

[7] A. Aroonwilas, A. Veawab, and P. Tontiwachwuthikul, *Ind. Eng. Chem. Res*, **38**, 5, 2044–2050 (1999)

[8] A. Aroonwilas and A. Veawab, *Ind. Eng. Chem. Res*, **43**, 9,. 2228–2237 (2004)

MATEC Web of Conferences **156**, 02003 (2018) https://doi.org/10.1051/matecconf/201815602003
RSCE 2017

Process Design of Virgin Coconut Oil (VCO) Production Using Low-Pressure Oil Extraction

Patricia Janelle Ferrer[1], *Vanessa Ferl* Quilinguen[1], *Jeremiah* Rosario[1], and *Lola Domnina* Pestaño[1, 2*]

[1]Chemical Engineering Department, Faculty of Engineering, University of Santo Tomas
[2]Research Center for the Natural and Applied Science, University of Santo Tomas
España Boulevard 1015 Manila, THE PHILIPPINES

Abstract. Virgin coconut oil (VCO) has become one of the most prominent high-value coconut product in coconut producing countries because of its versatility. This research attempts to design a fresh-dry process based on the Low-Pressure Oil Extraction Method for the production of VCO to reduce the settling time of the oil after extraction, that usually takes 1-2 weeks. Different parameters, such as drying temperature, centrifuge speed, and centrifugation time were varied and analysed. Three mathematical models were examined to describe the drying behaviour of the grated coconut meat at 65, 70, and 75°C using a tray dryer. A VCO production fresh-dry process based on the Low-Pressure Oil Extraction Method was developed through the employment of a centrifuge. The modified method lessens the settling time while still producing standard quality VCO. As predicted by the Laplace Transform Model, the shortest time for the comminuted coconut meat to reach a moisture content of 11% at which oil from nuts can be extracted using low pressure is at 29.07 minutes using a tray dryer. The best setting of VCO production using the modified method is at a drying temperature of 70˚C and at 2700 RPM and 60 minutes of centrifugation as it produced the clearest oil with a yield of 92.84 % v/v and a recovery of 18.43%. The produced VCO was tested for free fatty acid (FFA), moisture and volatile matter, colour, peroxide value, and iodine value, and the results are 0.03%, 0.11%, 0R/0.3Y, 0, and 5.77, respectively, which all passed the Philippine National Standards for VCO.

1 Introduction

The purest type of coconut oil, virgin coconut oil (VCO), was introduced to the world market at the end of the 20th century. It is considered one of the products of great value derived from the fresh coconut [1]. VCO, the clear, high value oil resulting from the fresh and mature kernel of coconut (*Cocos nucifera L.*), is obtained through mechanical and natural means, with or without the use of heat, without undergoing chemical refining, bleaching or deodorizing, which does not lead to alteration or transformation of the natural characteristics of oil [2]. It is now gaining a worldwide popularity because of its wide range of applications in medicine, food, cosmetics and the like [3].

VCO processing technologies can be generally categorized into fresh-dry process and fresh-wet process. The term fresh-wet is for the VCO process in which the VCO is obtained from the coconut milk by a variety of means after it has been extracted from freshly comminuted coconut kernel. The term fresh-dry on the other hand, is for the VCO process where VCO is obtained directly from the fresh coconut kernel which requires drying of the kernel in comminuted form before the extraction of oil [1].

Under the fresh-dry technologies is the low pressure oil extraction method. This method is common among micro- and village-scale industries and works on the principle that oil from seeds or nuts can be extracted using low pressure at about 460 psi provided that the moisture content of the material is within the range of 10–13% [3].

This process, however, requires at least two weeks of settling time to separate the fine particles of dried kernel from the extracted oil. A centrifuge, which is commonly used for emulsion breaking of coconut milk in fresh-wet VCO processing technologies [4] may be utilized to greatly lessen the settling time while still producing VCO which passes the quality standards set by the Philippine National Standards (PNS) for VCO, as well as the Asian and Pacific Coconut Community (APCC). Other factors,

* Corresponding author: lbpestano@ust.edu.ph

such as drying temperature, centrifugation speed, and centrifugation time may also affect the quality of the VCO produced [5].

The general objective of this study is to lessen the settling time of the low pressure oil extraction method of VCO fresh-dry production through the employment of a centrifuge as replacement to the traditional settling. Different parameters such as drying temperature, centrifugation speed, and centrifugation time were varied and analyzed in order to obtain the optimum setting for the process design.

Specifically, the objectives of the study are:

(1) To design a VCO production fresh-dry process based on the low-pressure oil extraction method that greatly lessens the settling time through the employment of a centrifuge while still producing standard quality VCO; and
(2) to determine the optimum drying temperature and drying time to achieve 11% MC of the comminuted coconut meat, centrifugation speed, and centrifugation time, and the optimum setting that will yield the fastest process and the highest quality of product.

2 Methods

2.1 Raw material collection and preparation

Fresh de-husked coconuts of 10 to 12 months of age were obtained from Oriental Mindoro, Philippines. The coconuts used in this research were of similar sizes, brown in color and did not have haustorium. The nuts were labelled Nut 1 to Nut 10. The fresh mature coconuts were split and comminuted through a manual/motor grater.

2.2 The equipment used

The equipment used in the experimentation were the tray dryer, fabricated manual press and the digital centrifuge DSC-301SD. The tray dryer was readily available in the Unit Operations Laboratory (Lab 4A) of the Chemical Engineering Department in the Roque Ruaño building of University of Santo Tomas, while the digital centrifuge used was from the UST College of Education, located at the Food Technology Laboratory Annex, 2nd floor of the UST Albertus Magnus building. The 4-ton hydraulic jack manual press was fabricated and bought from an external supplier.

2.2.1 Manual press

The hydraulic manual press used in the experiment was pre-fabricated. The manual press' frames and body were made from aluminum while the parts that come in direct contact with the grated coconut meat, such as the tray and the roof, were made of stainless steel. A bottle-type hydraulic jack with a maximum allowable load of 4 tons, manufactured by Hoyoma, Japan Power Tools, was the source of the hydraulic system of the manual press. Figure 1 shows the dimensions and the different views of the fabricated hydraulic jack manual press.

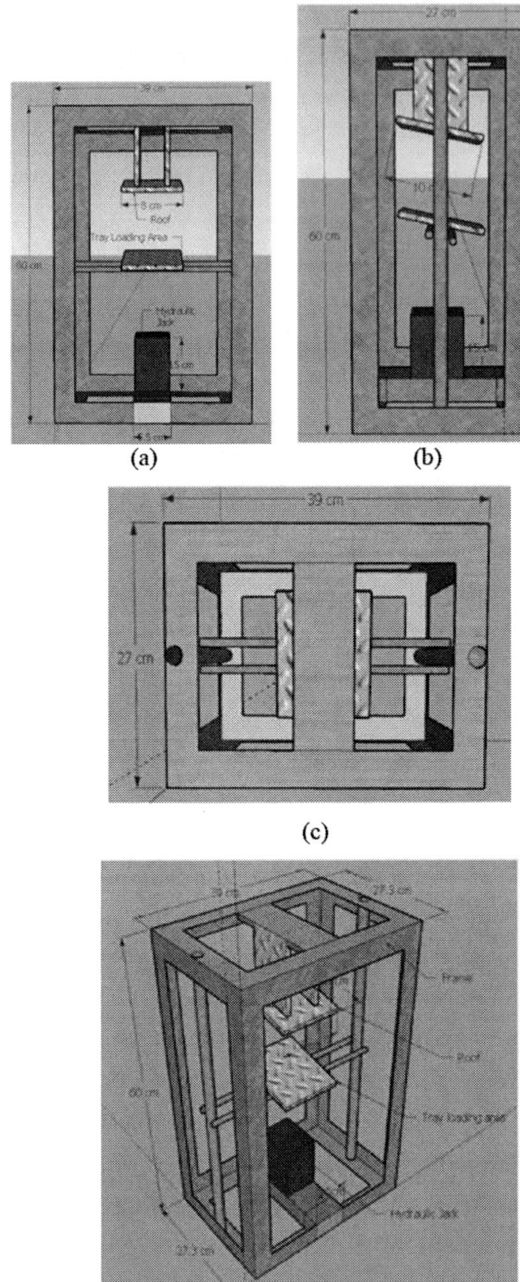

Fig. 1. Fabricated hydraulic manual press: (a) Front view, (b) Right-side view, (c) Top view, (d) Isometric view.

2.2.2 Digital Centrifuge

The digital centrifuge used was Digital Centrifuge Model DSC 301SD developed by Digi system Laboratory Instrument Inc. It has a maximum RPM of 4000 and can contain 4 units of 50 mL corning tubes per run. A specifications sheet of the necessary information is given in Table 1.

Table 1. Specifications sheet of Digital Centrifuge Model DSC-301SD

Speed	300-4000 RPM
Motor	DC Motor
Rotor	SB-071550
Rotor Type	45°Angle
Capacity	16 x (5-7 mL) 12 x (10-15 mL) 4x (50 mL)
Weight	N.W. 19 Kg G.W. 21 Kg
Overall Dimensions	500mm x 390 x 290
Power Supply	AC110V 50/60Hz

2.3 Mass balance

For each nut, the shell, water, and grated coconut meat were weighed using a precision balance with accuracy of 0.01g. Per cent composition of the materials for each nut was determined. Mass balance was performed to account for the losses. The average values for the ten nuts were calculated.

2.4 The Drying Experiment

The grated coconut meat was spread evenly and thinly on a tray with a surface area of 182.88 square centimeters and a thickness of not more than 1 cm. The grated coconut meat was heated at temperatures, 65-75°C at 5°C intervals every 5 min until constant weight is obtained. Before beginning the experiments, the tray dryer system was preheated in order to achieve a desirable steady state condition of temperature.

Weighing of the dried grated coconut meat was done using a precision balance with an accuracy of 0.01 g. Drying experiments for each temperature were conducted in triplicate.

2.4.1 Mathematical Modelling

The parameters that will be used in the modeling are the moisture ratio (MR) and the drying rate (R). The MR during drying was calculated using Equation 1:

$$MR = \frac{M - M_e}{M_i - M_e} \tag{1}$$

where M, Mi and Me are the MC, in g, at any time t, the initial MC and the equilibrium MC, respectively. The R was calculated from the experimental data using Equation 2:

$$R = -\frac{W_S}{A}\frac{d\overline{X}'}{dt} \tag{2}$$

where Ws is the weight in g, of the dry solid, A is the surface drying area in m^2, X' is the bulk MC and t is time. Equation 2 can be rearranged and integrated. The integral form Equation 3 was used to obtain the drying time.

$$\int_0^t dt = -\frac{W_S}{A}\int_{\overline{X}'_1}^{\overline{X}'_2}\frac{d\overline{X}'}{R} \tag{3}$$

The drying data (MR vs. t) at 65, 70 and 75°C were fitted into the three mathematical models: Page Model, Laplace Transform Model and Non-linear Decomposition Model. Drying is a complex process and as a means to simplify the analysis of the drying kinetics of grated coconut meat, empirical expressions are used [6]. Mathematical modeling using thin-layer drying equations are used to evaluate drying time of products based from experimental data. Models are often used to study the variables involved in the process, predict drying kinetics of the product and optimize the operating parameters and circumstances [7]. The values of the correlation coefficient, R^2, for the Page and Non-linear Decomposition Model, the time constant, tau, τ for the Laplace Transform Model and the total error for the three models were computed. The selection of the best model to describe the drying behavior of the grated coconut meat will be based on the least total error and the goodness of fit of a model.

The Page Model was successfully used to describe the drying characteristics of some agricultural products [8, 9, 10]. The Page Model [11, 12] that was used in this study was derived by linearizing the R equation shown in Equation 4.

$$MR = ae^{-kt^n} \tag{4}$$

Eq. (5) is the linearized form of Eq. (4).

$$MR = \ln a - kt^n \tag{5}$$

The Laplace Transform Model and Non-Linear Decomposition Model are used for prediction of the behavior of a system as reduction of moisture proceeds until it reaches equilibrium [13]. The Laplace Transform Model was derived from a material balance of the system in the experiment. The general material balance states that the input minus the output of the system is equal to rate of accumulation as shown in Equation 6.

$$M_i - M_o = \tau \frac{dM}{dt} \quad (6)$$

Equation 6 is then formulated to its Laplace Transform Model [14], integrated and simplified forming Equation 7 used in modeling the drying rate characteristics of the grated coconut meat.

$$M = M_f - M_f e^{-\frac{t}{\tau}} + M_i e^{-\frac{t}{\tau}} \quad (7)$$

The Non-linear Decomposition Model is based on the differential equation for batch decomposition as given in Equation 8.

$$\frac{1}{C^{n-1}} = \frac{1}{C_o^{\ n-1}} + (n-1)kt \quad (8)$$

The determination of the MR is represented by the concentration [15] and is represented below in Equation 9.

$$MR = \sqrt[n-1]{\frac{MR_o^{\ n-1}}{1+(n-1)ktMR_o^{\ n-1}}} \quad (9)$$

2.4.2 Determination of Drying Time to Attain 11% MC

The drying time needed to attain the needed 11% MC was determined through Lagrange interpolation. This method generates nth-order polynomials that pass through (n + 1) points. These polynomials attempt to produce interpolation values of increased accuracy by assuming a curvature in the relationship of the data. Its general equation is represented by Equation 10. The three versions are Equations 11–13.
The Lagrange interpolation:

$$f_n(x) = \sum_{i=0}^{n} L_i(x) f(x_i) \quad (10)$$

First-Order version:

$$f_1(x) = \frac{x-x_1}{x_0-x_1} f(x_0) + \frac{x-x_0}{x_1-x_0} f(x_1) \quad (11)$$

Second-Order version:

$$f_1(x) = \frac{(x-x_1)(x-x_2)}{(x_0-x_1)(x_0-x_2)} f(x_0) + \frac{(x-x_0)(x-x_2)}{(x_1-x_0)(x_1-x_2)} f(x_1) + \frac{(x-x_0)(x-x_1)}{(x_2-x_0)(x_2-x_1)} f(x_2) \quad (12)$$

Third-Order version:

$$f_3(x) = \frac{(x-x_1)(x-x_2)(x-x_3)}{(x_0-x_1)(x_0-x_2)(x_0-x_3)} f(x_0) + \frac{(x-x_0)(x-x_2)(x-x_3)}{(x_1-x_0)(x_1-x_2)(x_1-x_3)} f(x_1) + \frac{(x-x_0)(x-x_1)(x-x_3)}{(x_2-x_0)(x_2-x_1)(x_2-x_3)} f(x_2) + \frac{(x-x_0)(x-x_1)(x-x_2)}{(x_3-x_0)(x_3-x_1)(x_3-x_2)} f(x_3) \quad (13)$$

2.5 Combined low pressure oil extraction and centrifugation method

The dried grated coconut meat was subjected under low pressure oil extraction using a fabricated manual press. A clean cheesecloth was used to hold the dried grated coconut meat together and to filter the undesired solid particles from the extract. The extracted crude coconut oil was collected in corning tubes in preparation for the final method which was the centrifugation method. A digital centrifuge (DSC-301SD) was used to speed up the settling time and to separate the foots or the fine kernel particles entrained in the oil, from the desired liquid.

Specific parameters such as the centrifugation speed and centrifugation time were varied occasionally to determine its effect on the resulting product. For the centrifuge speed - 2000 RPM and 2700 RPM were utilized, due to the limitations of the said equipment used. Centrifugation time was varied for each centrifugation speed using 20, 40, and 60 minutes. The centrifugation method was carried out using fresh samples of extracted crude coconut oil for each time and speed. After centrifugation, decantation was done to obtain the VCO, which was the clear top-most layer of the corning tube after centrifugation. Figure 2 shows the flowchart of the modified and improved low-pressure oil extraction method.

https://doi.org/10.1051/matecconf/201815602003

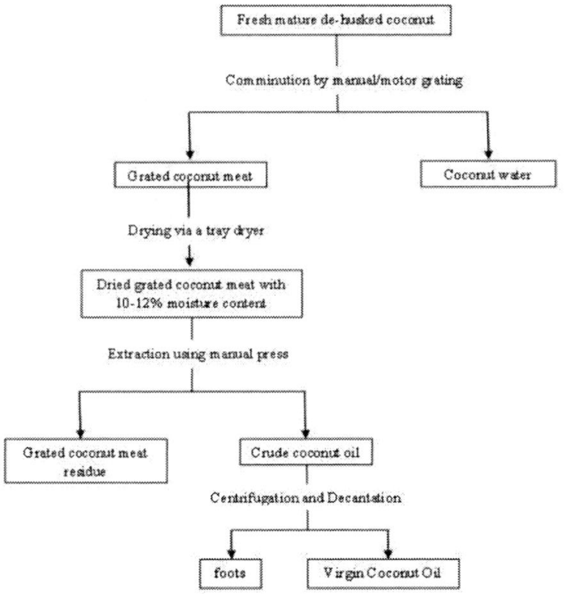

Fig. 2. Flowchart of low pressure oil extraction method for the production of VCO

2.6 Quality testing

The resulting VCO was tested for specific parameters such as average yield per coconut, color, peroxide number, iodine value, moisture and volatile matter, and presence of free fatty acids (FFA). The extracted oil samples were brought to the Philippine Coconut Authority (PCA) for the testing of said parameters. The tests were done at the Research and Development department, Laboratory Services Division of PCA. VCOs from different parameter alterations of numerous trials were tested and compared. The tests were done at the ambient temperature, 25°C.

3 Results and Discussion

3.1 Coconut Mass Balance

The composition and corresponding per cent by weight composition of each nut were calculated. From this data, average values were determined. For an average weight of 1374.01 grams, 520.07 grams coconut water, 312.21 grams shell and 517.27 grams grated coconut meat peel were accounted. The mass balance was done to estimate the oil recovery per coconut and to check the consistency of the weights of the coconuts used in this process. Matured coconut has more meat, resulting to higher residue yield.

An average de-husked coconut used in this process, weighing at least 1,300 grams, contained at least 22.72% shell, 37.85% water, and 37.65% meat by weight and accounting for 1.78% losses. The losses can be attributed to the spillage of coconut water and the fragments of grated coconut meat which were not collected during comminution. Figure 3 shows the average composition of the parts of the 10 coconuts.

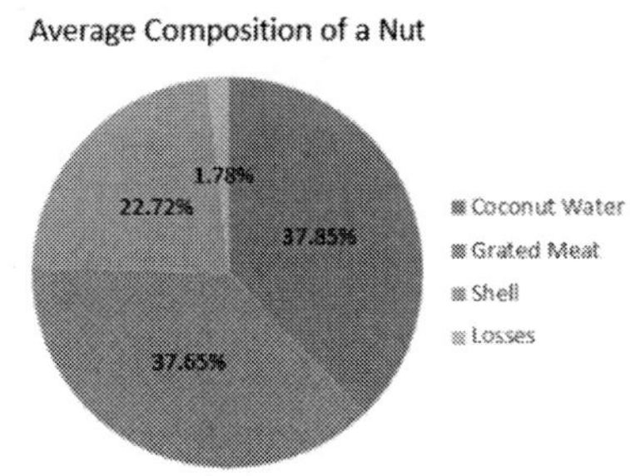

Fig. 3. The average composition (Ave. wt. = 1374 g) of a nut

3.2 Analysis of the Mathematical Models of the Drying of Grated Coconut Meat

The MR data were fitted into the two models. The results, namely, value of the correlation coefficient, R^2, for the Non-Linear Decomposition Model, the time constant, tau, τ for the Laplace Transform Model and the total error for the three models were recorded and are shown in Table 2. The best results based on the least total error were shown by the Laplace Transform Model at the three temperatures studied. The Laplace Transform Model yielded notable accuracy, precisely modeling the data series of grated coconut meat and predicting its future behavior. Using visual inspection, the Laplace Transform Model also gave the best fit as shown in Fig. 4. Hence, the Laplace Transform Model may be assumed to represent the drying behavior of grated coconut meat. Similar findings were reported for the air-drying kinetics of thin slices of coconut meat [14]. Earlier research works reported that the Page Model represent the drying kinetics of coconut meat slices [16, 17]. Figure 4 shows the experimental drying kinetic data as they are fitted with the simulated models at 70°C drying temperature.

Table 2. Summary of Results

Temp.	Page Model			Non-linear Decomposition Method			Laplace Transform Model	
	n	R^2	Total Error	*n*	R^2	Total Error	τ	Total Error
65°C	1.4	0.9787	0.0349	0.7	0.9994	0.00328	897	3.50E-07
70°C	1.5	0.9928	0.0109	0.75	0.9464	0.16354	479	6.50E-07
75°C	1.8	0.9864	0.0274	0.7	0.9951	0.01758	267	4.20E-06

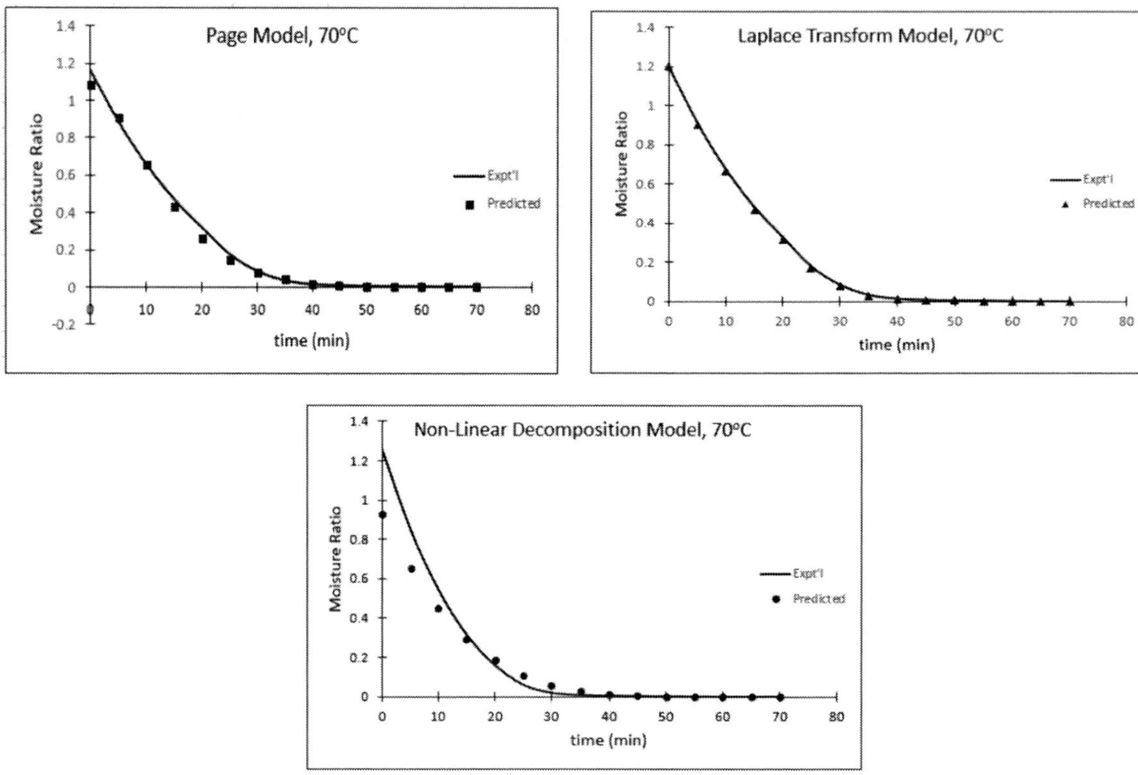

Fig. 4. Comparison of the experimental and predicted values of MR vs. time at 70°C using the three models

Table 3. Calculated drying time at different treatment temperatures as predicted by the Laplace Transform Model

Trial No.	Temperature		
	65°C	70°C	75°C
1	39.26 min	29.70 min	30.51 min
2	54.90 min	34.00 min	27.84 min
3	54.61 min	35.57 min	34.05 min

3.3 Determination of the Drying Time using the Lagrange Method of Interpolation

The drying time for the grated coconut meat to reach the desired MC of 11% at which oil from nuts can be extracted using low pressure [3]. The Laplace Transform Model was used to predict the optimum drying time at which the MC of the grated coconut meat will be 11%. The drying time was computed using Lagrange interpolation. The average drying time to reach 11% MC from the initial range of 54–56% was 49.59, 33.09 and 30.80 min at drying temperatures of 65, 70 and 75°C, respectively. Table 3 shows the calculated heating time as predicted by the Laplace Transform Model. A shorter heating time at 70°C was utilized to attain a 11% MC in the grated coconut meat.

3.3.1 Observation of the grated cocount meat after drying at different temperatures.

The grated coconut meat that was dried at 65°C was too milky that would require a longer centrifugation time, which may cause undesired heating due to the centrifugal force. On the other hand, when 75°C was used as a drying temperature, the grated coconut meat was burnt or scorched. Burnt coconut meat causes yellow oil, which does not pass the standards for VCO.

3.4 Statistical analysis of the computed drying Time using the best fit Mathematical Model

The best condition for drying the grated coconut meat was at a drying time of 29.70 minutes at 70°C, which was computed using the Laplace Transform Model and the Lagrange Method of Interpolation. Table 4 shows the statistical analysis of the drying data gathered at 70°C, using a 95% confidence level.

The standard deviation and the confidence limits were evaluated using Microsoft Excel. Based on the table, for the drying of grated coconut meat at 70°C, the results were 95% confident that the drying time needed to achieve 11% moisture content ranges from 29.65 to 36.53 minutes.

Table 4. Statistical analysis of the best drying temperature at 95% confidence level (α=0.05)

Trial	Drying Time, min	Ave.	Std. Dev.	Conf. Int.	LL	UL
1	29.70					
2	34.00	33.09	3.04	3.44	29.65	36.53
3	35.57					

3.5 Analysis of data from the centrifugation process

Centrifugation time and centrifugation speed were both varied to obtain the optimum condition that will provide the greatest amount of oil yield and greatest initial quality of VCO upon organoleptic evaluation. Two variations of centrifuge speed, 2000 and 2700 RPM, were used. Centrifugation time was varied for each centrifugation speed using 20, 40, and 60 minutes. Figure 5 shows the plot for %VCO yield against centrifugation time for both speeds. The process was done under ambient temperature of 25°C.

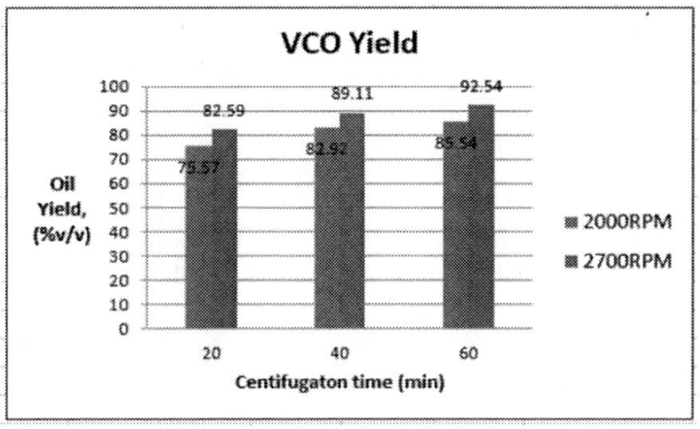

Fig. 5. VCO yield with the effect of centrifugation time for 2000 RPM and 2700 RPM

Based on Figure 5, the highest oil yield occurs at 2700 RPM with 60 minutes of centrifugation time at 92.84 v/v%. Also, upon organoleptic inspection, the clarity of oil increases as centrifugation time and centrifugation speed increases, yielding the clearest oil at 60 minutes of centrifugation for both 2000 and 2700 RPM. This observation is explained by the fact that increasing the centrifuge speed results in increasing the rate of sedimentation and consequently increases the separation of the two phases [4]. Also during centrifugation process, heat is generated by the centrifugal rotation. Density and viscosity, which are temperature dependent properties are altered. When temperature increases, viscosity and density decreases. As a result, the foots and fine particles will settle faster at higher temperatures [18].

3.5.1 Calculation of Oil Recovery

With the density of the virgin coconut oil at room temperature known as 0.903 g/mL, based on literature [5], the oil recovery for the optimum setting was determined using Equation 14.

$$\text{Oil Re cov ery} = \frac{\text{wt. of VCO re cov ered}}{\text{wt. of coconut meat used}} \text{x}100 \quad (14)$$

For each full tray loaded, the average volume of crude coconut oil obtained from 5 trials was 75.20 mL, with 69.82 mL being the VCO obtained, based from the data from the oil yield. The average weight of each full tray from 5 trials was 342.15 g. The oil recovery of the designed process was computed to be 18.43%, while the oil recovery of the traditional low pressure oil extraction is 25 kg / 100 kg meat or 25%.

The amount of VCO that can be obtained from a single coconut can be computed by Equation 15.

$$\text{Amount of VCO per nut} = \left(\frac{\%\text{Re cov ery}}{100}\right)(\text{wt. of meat / nut}) \quad (15)$$

The weight of grated coconut meat was determined by the mass balance in the previous part of the experiment, with a value of 517.27 g. Knowing those values, the weight of oil that can be extracted from a single coconut was computed to be 95.33 g or 105.57 mL.

3.6 Results of Quality Testing

The VCO sample produced at the best condition, 29.70 minutes of drying at 70°C, and 60 minutes of centrifugation at 2700 RPM, was brought to the Laboratory Services Division of the Philippine Coconut Authority and was tested for the FFA, moisture and volatile matter, color, peroxide value, and iodine value. Table 5 shows the result of the laboratory tests, with the Philippine National Standards for Virgin Coconut Oil alongside.

Table 5. Results of quality testing of the sample VCO (70°C drying time, 2700 RPM and 60 minutes of centrifugation)

Parameters	Results	Standards
FFA (Lauric)	0.03%	0.20% max.
Moisture and Volatile Matter	0.11%	0.20% max.
Color: (5 1/4" cell) Red/Yellow	0R/0.3Y	Colorless
Peroxide Value (meq/Kg)	0	3 max.
Iodine Value	5.77	4.11-11.0

Based on the quality testing results, all values of each tested parameter passed the Philippine National Standards for VCO. The color test result suggests that there were 0 red color and 0.3 yellow color detected in the produced VCO sample. The detected yellow value was very small and insignificant; thus, the color does not affect the quality of the VCO produced. Favorably, there was no detection of the red color in the sample, otherwise, the produced VCO sample will already be affirmed as a refined, bleached, and deodorized (RBD) oil, not VCO.

There is no standard set by the Philippine Bureau of Product Standards for the iodine values of virgin coconut oil, however, based on the study of Dayrit and his colleagues on the Standards for Essential Composition and Quality Factors of Commercial Virgin Coconut Oil and its Differentiation from RBD Coconut Oil and Copra Oil in 2007, the Codex Alimentarius standard for coconut oil range for iodine value is 6.3 to 10.6 [19], while the Asian and Pacific Coconut Community (APCC) standards for VCO range from 4.1 to 11.0, making the iodine value of the produced VCO sample in this study pass the both the Codex standard and the APCC standard.

Both the results for the FFA and peroxide value were within the limits set by the Philippine National Standards for VCO. These two parameters are the most critical chemical properties to be met for an oil to be classified as a VCO. VCO is the purest form of coconut oil containing very low amount of FFA and having a low peroxide value.

Naturally, FFAs are recorded to be low for vegetable oils. A high amount of FFA would give the oil an unpleasant taste, which is undesirable for a VCO that has a fresh and raw coconut taste. In addition, an FFA of above 0.5% will cause itchiness in the throat when consumed [20]. This would mean that a change in the oil's composition has taken place and it is corrosive which can be dangerous to the health.

Peroxide value on the other hand is the measurement of the extent of an oil's rancidity. Rancidity denotes the spoilage of a food which results from oil or fat deterioration. VCO has a low content of unsaturated fatty acids making its tendency to be rancid poor. High peroxide value would mean that the oil is unstable and already contaminated [21].

4 Conclusions and Recommendations

4.1 Conclusions

A VCO production fresh-dry process based on the low-pressure oil extraction method was developed through the employment of a centrifuge. As predicted by the Laplace Transform Model, the shortest time for the comminuted coconut meat to reach a moisture content of 11% is at 29.07 minutes using a tray dryer. The best setting of VCO production using the modified method is at a drying temperature of 70°C and at 2700 RPM and 60 minutes of centrifugation. The developed method lessens the settling time while still producing standard quality VCO.

The authors would like to express their deepest gratitude to the University of Santo Tomas Food Technology Department, College of Education, for allowing us to use the equipment in their laboratory.

References

1. D. D. Bawalan, *Process Manual for Virgin Coconut Oil, its Products and By-products for Pacific Island Countries and Territories (2011)*
2. M. Arumugam, M. Raman, & K. Eagappan, Int. J. of Pharmacy & Pharmaceutical Sci. 186-190 (2014)
3. D. D. Bawalan, & K. R. Chapman, *Virgin Coconut Oil Production Manual for Micro- and Village-Processing* (2006)
4. S. Panpanya, & S.P. Pathaveerat, *Study and design manufacture of virgin coconut oil by centrifuge method* (2012)
5. Y. Wong, & H. Hartina, Or. J. of Chem. **30(1)**, 237-245 (2014)
6. Clary C, Farid M, Fasina O, Hui Y, Noomhorm A, & Welti-Chanes J. *Food Drying Science and Technology: Microbiology, Chemistry, Applications.* In: Lypez-Malo A & Rios-Casas L Solar Assisted Drying of Foods, p. 84. (Lancaster: DEStech Publications Inc.(2008)

7. Belessiotis VG & Karathanos VT. Agr. Eng. Resour. 355– 361 (1991)
8. Singh S, Raina GS, Bawa AS, & Saxena DG.. Drying Technol. **24**:1487–1497 (2006)
9. Doymaz I & Ismail O. Food Bioprod. Process; **89**:31–38 (2011)
10. Hassan-Beygi SR, Aghbashlo M, Kinamehr MH, & Massad J. **23**:129–135 (2009).
11. Agrawal YC & Singh RC. *Thin layer drying studies on short grain rough rice* (ASAE Paper No. 77–3531. MI, U.S.A.: St. Joseph (1977)
12. Zhang Q & Litchfield JB. Drying Technol. **9**:383–395 (1991)
13. Charbonnel C, Ghoul M, Guiga W, & Ioannou I. *Food and Bioproducts Processing Frozen mirabelle plum drying: Kinetics, modeling and impact on biochemical properties*. (2010)
14. Pestaño LB.]. In: Third International Conference on Advances in Applied Science and Environmental Engineering (ASEE 2015) (USA: ©Institute of Research Engineers and Doctors). ISBN: 978–1–63248–055–2 (2015)
15. Jose WI & Pestaño LB. J. Adv. Agr. Technol.; **3(1)**:58–62 (2016)
16. Akoy EOM. Int. Food Res. J. **21(5)**: 1911–1917 (2014)
17. Goyal RK, Kingsly ARP, Manikantan MR, & Ilyas SM. J. of Biosystems Eng.; **95(1)**:43– 49 (2006)
18. C. G. Khoo, A. Nour, & A. H. Nour, *Production of Virgin Coconut Oil (VCO) by Centrifuge Method* (2011)
19. F. M. Dayrit et al. Philippine Journal of Science **136 (2)**: 119-129 (2007)
20. A. Albao, *Coco Wonder*. Retrieved from http://www.cocowonder.com.ph/index.php/health-benefitsof-cocont-cacao/how-to-choose-top-quality-vco (1969)
21. O. Ojeh, *Effects of refining on the physical and chemical properties of cashew kernel oil. J Fats Oils Technology* (1981)

Mathematical Modeling of the Drying Kinetics of Thinly-Sliced Saba (Musa Balbasiana) Using Hot-Air Dryer

Lola Domnina B. Pestaño[1,]*, *John Paul T. Bautista*[1], *Reizl JR. H. Leguiab*[1] *and Sean Danielle D. Puri*[1]

[1]Faculty of Engineering, University of Santo Tomas, España Boulevard, 1015 Manila, The Philippines

Abstract. Banana is one of the top produced crops in the Philippines, and among its cultivars is the *Musa balbasiana*, commonly known as *saba*. Due to its high moisture content, *saba* is perishable and one of the methods to increase its shelf-life is drying. The shelf life of *saba* can be extended up to six months when dried to 12% moisture content. The research focuses on the effect of banana maturity and choses a mathematical model which will best fit its drying kinetics. The banana samples, the unripe and ripe *saba*, were bone-dried without pre-treatment using a hot-air tray drier. To produce repeatable data, three trials were done for temperatures 40°C, 50°C and 60°C. Among the three mathematical models used in the study, the treated data for both unripe and ripe *saba* best fit the Laplace Transform Model. Using Lagrange interpolation, the time per trial was computed; unripe *saba* dried at 50°C achieved the ideal moisture content in an average time of 87.5574 minutes for the three trials while that of ripe *saba* dried at 40°C achieved the same moisture content in an average time of 88.8619 minutes for the three trials. With the increase in temperature, the discoloration decreased indicating an enzymatic character ofbrowning.

1 Introduction

Banana, a staple fruit on Filipino tables, is hailed as the Philippines' fourth largest produced crop; the country being able to contribute to 10% of the world supply of banana from the year 1996 up to 2005 [1]. Among the varieties of bananas the Philippines grow, Saba (Musa balbisiana), which is usually eaten fresh as dessert when ripe, leads in market demand and production and can be processed as banana flour, ketchup and the popular banana chips [2].

Having a high moisture content, banana is prone to mold growth [3]. Because of its shelf life, it must be consumed a few days after harvest. Post-harvest losses such as spoilage of banana according to researches account up to 40% around the world [4], not only physical but also in the fruit quality [5] and preservation methods such as dehydration is employed. The most common technique for dehydration of food is air-drying because of its easy set-up [6] and a common way to increase their shelf life [7]. Commercially, banana is dried down to 14-15 % final moisture content on a dry basis (12% on a wet basis) which corresponds to an approximate 70% mass loss. At such a level of moisture content, dried banana has a shelf life of at least 6 months [8].

In drying a perishable fruit like banana, wherein a drying medium such as hot air is passed over the exterior of the material, moisture content is related to time. Mathematical models were constructed to analyze the changes in density and shrinkage of banana during drying by experimentation [9]. Likewise considering shrinkage, Karim and Hawlader, provided a description of the heat and mass transfer properties of banana during drying through the use of a mathematical model [10]. Similar studies use analytical solutions of the diffusion equation by intermittent drying and continuous drying and some of them used the model describing well the drying of bananas and other agricultural products [2].

Mathematical models illustrate the drying kinetics being affected by the temperatures and moisture contents, which in turn affect other superior attributes of banana [2], therefore, analysis of the drying process is of absolute necessity in obtaining quality products. In reference with this, the researchers have come up with the study which involves hot air drying of saba to 12% moisture content which will aid in the its preservation and prolonged shelf-life of up to six months.

2 Methodology

2.1. Preparation of saba

Fresh matured banana (ripe and unripe) grown in a farm located in Laguna, were bought from Trabajo market in Sampaloc. The fruit was peeled and cut into thin slices

* Corresponding author: lbpestano@ust.edu.ph

of 1-2 mm thickness. No pretreatment was done on the samples.

2.2. Drying Procedure

Immediately after slicing, the banana samples were laid out on the area of the perforated tray and weighed using an analytical balance (+ 0.01 g) as shown in Fig. 3.3, and then placed into the drying chamber which was preheated at 60°C. The tray containing the banana sample was weighed every 15 minute time interval throughout the drying period until three constant weights were recorded. Trials were done in triplicates, this procedure was repeated at 40 and 50°C.

2.2 Data Interpolation and Calculation

Three empirical models that best describes the thin-layer drying kinetics of agricultural products were used in the study: Henderson and Pabis' model shows the effect of moisture loss on the system as time increases, Page model, a modification of the Henderson and Pabis' model, best describes the thin-layer drying of whole bananas, and Laplace transform model provides an easier way to solve differential equations derived from the mass balance around the dryer for the prediction of the system's behavior.

Henderson and Pabis model
The drying model of Henderson and Pabis is shown in equation 1.1:

$$MR = ae^{-kt} \quad (1.1)$$

Equation 1.1 is then linearized as shown in equation 1.2:

$$\ln MR = \ln a - kt \quad (1.2)$$

Modified Page model
The Modified Page model equation is shown in equation 2.1:

$$MR = ae^{-kt^{n}} \quad (2.1)$$

This equation was linearized as described by equation 2.2:

$$\ln MR = \ln a - kt^{n} \quad (2.2)$$

Laplace transform model
In deriving the Laplace transform model, a general material balance was formulated stating the rate input minus output in the system equals the rate of accumulation as shown in equation 3.1:

$$M_i - M_o = \tau\frac{dM}{dt} \quad (3.1)$$

To derive the drying rate of banana, Laplace transformation was applied in equation 3.1. The derived equation was represented in equation 3.2:

$$M = M_f - M_f e^{\frac{-t}{\tau}} + M_i e^{\frac{-t}{\tau}} \quad (3.2)$$

2.3 Determination of the moisture ratio

By dividing the difference of the weight of the sample, W, at any time t and the equilibrium weight of saba, Weq, by Weq, the moisture ratio, MR, is determined as shown in equation 4.1:

$$MR = \frac{W - W_{eq}}{W_i} \quad (4.1)$$

2.4 Determination of the drying time

Lagrange method of interpolation was used to calculate for the time needed for dried *saba* to attain 12% moisture as shown in equation 5.1:

$$y = \frac{(x-a)(x-b)(x-c)}{(d-a)(d-b)(d-c)}f(d) + \frac{(x-a)(x-b)(x-d)}{(c-a)(c-b)(c-d)}f(c) + \frac{(x-a)(x-c)(x-d)}{(b-a)(b-c)(b-d)}f(b) + \frac{(x-b)(x-c)(x-d)}{(a-b)(a-c)(a-d)}f(a) \quad (5.1)$$

3 Results and Discussion

3.1 Analysis of mathematical models

A total error of 0.6798, 0.5033 and 0.9306 for 40°C, 50°C and 60°C respectively shown in Table 1 were obtained by comparing the corrected and experimental moisture ratio values of the most appropriate fit from the three trials conducted for each temperature for unripe *saba*. As observed from Figure 1 the Henderson and Pabis model best represented the drying kinetics of unripe *saba* at 50°C. The overlapping curves for the corrected and experimental values for each temperature signified which is the most appropriate fit. The total error recorded from the other trials varies from 0.7984 to 0.8882, 0.5494 to 0.5731 and 0.9888 to 1.2056, correspondingly for each temperature.

Table 1. Henderson and Pabis model

Banana Maturity	*Temperature*	*Trial*	R^2	*Total Error*	*Time (mins) to reach 12% (wet basis)*
Ripe	40°C	1	0.97713	0.7984	78.5705
		2	0.97268	0.8882	73.0914
		3	0.98368	0.6798	83.9978
	50°C	1	0.92699	0.5731	99.8809
		2	0.92306	0.5494	102.5936
		3	0.91941	0.5033	107.0004
	60°C	1	0.98930	0.9306	105.5947
		2	0.97131	1.2056	100.5280
		3	0.98605	0.9888	103.7430
Unripe	40°C	1	0.80191	0.5786	118.6259
		2	0.88216	0.8989	106.3204
		3	0.83769	0.9940	135.1477
	50°C	1	0.98020	0.2637	108.2601
		2	0.97884	0.2518	107.1768
		3	0.96316	0.3554	106.6902
	60°C	1	0.91708	1.8606	119.1495
		2	0.99147	0.1461	128.7729
		3	0.88176	1.8972	116.2403

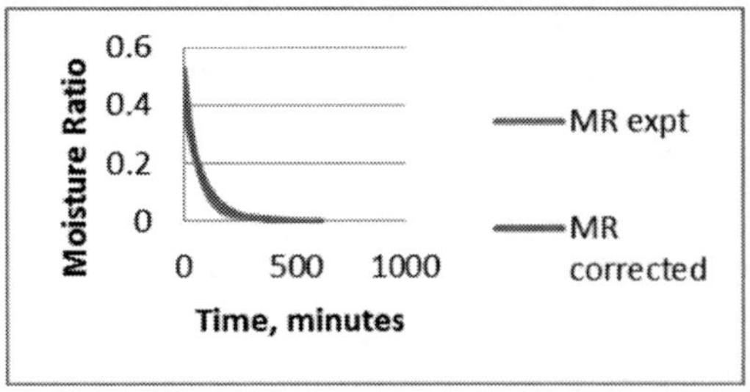

Fig.1. Comparison of Experimental and Corrected Moisture Ratio Values of Unripe *Saba* Using Henderson and Pabis Model (50°C, Trial 3)

Corrected and experimental moisture ratio of unripe *saba* presented a total error of 0.2295 for 40°C, 0.2041 for 50°C and 0.2346 for 60°C for the trial displaying the most appropriate fit as shown in Table 2. Figure 2 shows that Page model best describes the drying kinetics of unripe *saba* at 50°C. The order of reaction and correlation were found to be 0.835 and 0.9245 respectively. The overlapping lines were again observed as to which considerably coincided to choose the most appropriate fit for each temperature. Other trials have total errors varying from 0.2757 to 0.3000, 0.2270 to 0.2945 and 0.2854 to 0.2900 respectively.

Table 2. Modified Page model

Banana Maturity	*Temperature*	*Trial*	*n*	R^2	*Total Error*	*Time (mins) to reach 12% (wet basis)*
Ripe	40°C	1	0.795	0.99551	0.2757	84.5922
		2	0.765	0.99566	0.3000	82.7654
		3	0.83	0.98694	0.2295	87.6941
	50°C	1	0.835	0.92449	0.2041	97.1547
		2	0.855	0.96131	0.2270	100.3673
		3	0.895	0.95365	0.2945	105.3460
	60°C	1	1.21	0.99824	0.2346	111.7294
		2	1.26	0.98601	0.2900	107.5572
		3	1.23	0.99553	0.2854	110.4164
Unripe	40°C	1	0.75	0.78706	0.5963	111.5814
		2	0.78	0.87083	0.4912	102.7607
		3	0.87	0.90892	0.7993	131.3016
	50°C	1	1.06	0.98166	0.0809	109.4937
		2	0.95	0.97885	0.1541	106.0543
		3	0.95	0.96337	0.2040	105.7828
	60°C	1	1.4355	0.96144	0.7665	134.6936
		2	1	0.99147	0.1461	128.7354
		3	1.48	0.93651	0.8395	132.8613

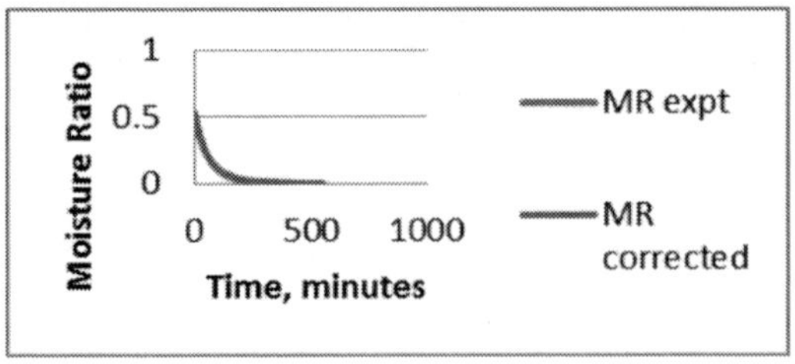

Fig. 2. Comparison of Experimental and Corrected Moisture Ratio Values of Unripe *Saba* Using Modified Page Model (50°C, Trial 1)

Each temperature had three trials and Figure 3 shows the overlapped graphs of corrected and experimental values having the least total error for 60°C as shown in Table 3. The comparison for unripe *saba* at 60°C had the smallest total error equal to 0.1410 and a residence time, τ, of 67.01, while 40°C has a total error of 0.1717 and 50°C, a total error of 0.2138. The other values obtained for the total error ranged from 0.1752 to 0.1876, 0.2280 to 0.2351 and 0.1638 to 0.2214 correspondingly for each temperature.

Table 3. Laplace transform model

Banana Maturity	*Temperature*	*Trial*	*τ*	*Total Error*	*Time (mins) to reach 12% (wet basis)*
Ripe	40°C	1	66	0.1752	92.2472
		2	65	0.1876	90.5289
		3	65	0.1717	90.7879
	50°C	1	60	0.2138	89.0309
		2	60	0.2280	88.7680
		3	60	0.2351	88.7867
	60°C	1	67.01	0.1410	109.6246
		2	67.01	0.2214	108.8106
		3	67.01	0.1638	109.2968
Unripe	40°C	1	60	0.6682	85.7605
		2	62.5	0.4523	88.3711
		3	62	0.6657	88.5406
	50°C	1	67.01	0.1169	106.5511
		2	67.01	0.1168	106.3588
		3	67.01	0.1217	106.5505
	60°C	1	72	0.1302	121.8528
		2	74	0.1316	126.0181
		3	74	0.1316	125.4099

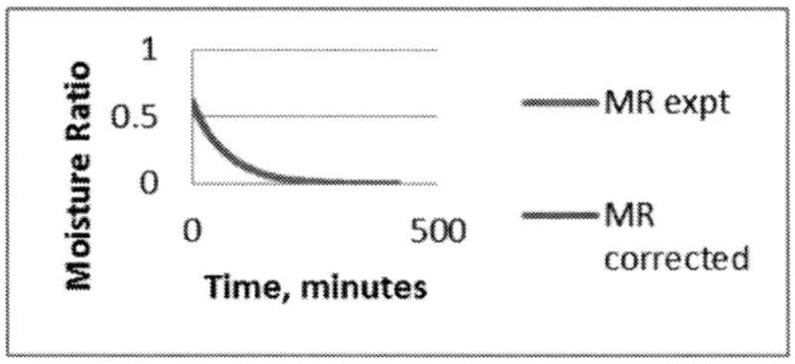

Fig. 3. Comparison of Experimental and Corrected Moisture Ratio Values of Unripe *Saba* Using Laplace Transform Model (60°C, Trial 1)

3.2 Analysis of moisture content and varying temperature

Because of its directly proportional relationship with the drying rate, temperature is one major factor that greatly influence the drying kinetics of *saba*. On the other hand, as the temperature increases, equilibrium moisture decreases, since a greater amount of moisture is removed from the dried banana. This causes a longer drying time to occur. Results from the trials as described by the Laplace Transform Model show that unripe *saba* subjected at 40°C took 90.7879 minutes as compared to 60°C which took 109.6246 minutes and at 50°C, it took 89.0309 minutes.

As for the ripe saba, similar results were observed. At 40°C, it took 88.3711 minutes to reach the desired moisture content of 12% wet basis. 50°C and 60°C took 106.3588 and 121.8528 minutes respectively, showing an increasing trend in relation to the temperature.

Table 4 below shows the comparison of the average total error for each temperature for both ripe and unripe *saba*.

Table 4. Average Total Error

Ripe			***Unripe***		
Henderson and Pabis Model					
40°C	50°C	60°C	40°C	50°C	60°C
0.7888	0.5419	1.0417	0.8238	0.2836	1.3013
Modified Page Model					
40°C	50°C	60°C	40°C	50°C	60°C
0.2684	0.2419	0.2700	0.6289	0.1463	0.5837
Laplace Transform Model					
40°C	50°C	60°C	40°C	50°C	60°C
0.1782	0.22563	0.1754	0.5954	0.1185	0.1311

As evidently shown in Table 4, Laplace Transform model consistently gave the least total error for 40°C, 50°C, and 60°C both for unripe and ripe *saba*. Therefore the Laplace transform, as compared to the Henderson and Pabis Model and Modified Page model is a better fit for the drying kinetics of thinly sliced *saba*.

Table 5 shows the tabulated results of the drying time in minutes to attain the 12% moisture ratio for each temperature for both unripe and ripe *saba* obtained using Laplace Transform model.

Table 5. Drying Time to Attain 12% Moisture Ratio

	Unripe			*Ripe*		
	40°C	50°C	60°C	40°C	50°C	60°C
Trial 1	92.2472	89.0309	109.6245	85.7605	106.5511	121.8528
Trial 2	90.5289	88.7680	108.8106	88.3711	106.5505	126.0191
Trial 3	90.7879	88.7867	109.2968	88.5406	106.3588	125.4099

For a more convenient way to determine which temperature gave the least drying time for both unripe and ripe *saba* the results are tabulated side by side as presented in the table above. The unripe *saba* dried at 50°C gave the least drying time while for the ripe it was at a temperature of 40°C.

3.3 Observation of the Effect of Banana Maturity

After the drying process, visual inspection was done to monitor the effects of maturity on the dried bananas containing 12% moisture. Temperature affects the physical structure of banana, considering even the slight differences observed in color. The thinly sliced fresh samples of the two banana maturities were white-yellowish, and after drying, a yellow-brownish color developed on these samples. With the increase in temperature, the discoloration decreased indicating an enzymatic character of browning as presented by [11].

Ripe *saba* took a longer time to reach 12% moisture content at temperatures 50 and 60°C with an average time of 106.48 and 122.76 mins respectively compared to the average time of unripe *saba* of 91.34 and 109.25 mins respectively. While at temperature of 40 °C ripe *saba* took only an average time of 87.56 mins compared to 91.19 mins of unripe saba. The effect of maturity of *saba* on drying time is that at 40 °C ripe *saba* will be dried faster compared to unripe *saba*, while at temperatures 50 and 60 °C unripe *saba* will be dried faster based on the results obtained.

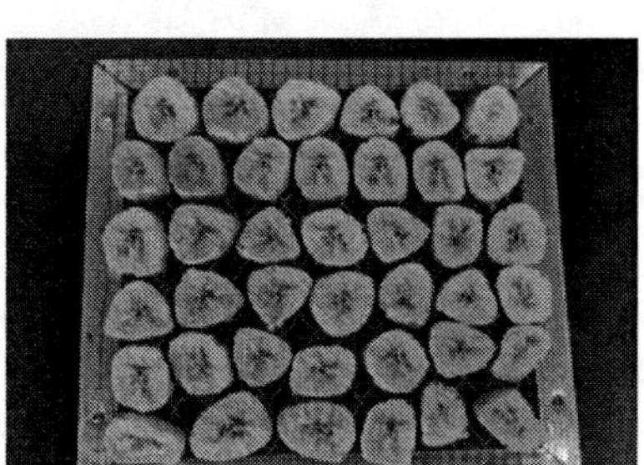

Fig. 4. Dried unripe banana at 50°C

Fig. 5. Dried ripe banana at 50°C

Theoretically, as temperature increases the drying time should decrease but this study shows that as the temperature increases the drying time also increases. This phenomenon could be explained by the large moisture that can be removed from the *saba* samples because the equilibrium moisture decrease as temperature increases. According to Nguyen and Price (2007), the large difference in the initial moisture content could strongly affect the drying rate and this effect was clearly observed at low temperatures. Mass loss was seen to be insignificant for both ripe and unripe bananas as both showed similar results [12]. However, chemical changes such as the increase in sugar of ripe bananas contributed to its mass loss during drying. This would compensate for the more open structure of ripe bananas which favors an increased drying rate.

Presented in Table 6 is the statistical analysis of the maturity with the corresponding temperature that gave the least drying time using 95% confidence level.

Table 6. Statistical Analysis

Unripe (50°C)			Ripe (40°C)		
Trial 1	*Trial 2*	*Trial 2*	*Trial 1*	*Trial 2*	*Trial 3*
85.6705	88.3711	88.5406	89.0309	88.7680	88.7867
Standard Deviation	**Average Drying Time**	**α**	**Standard Deviation**	**Average Drying Time**	**α**
1.5585	87.5574	0.05	0.1467	88.8619	0.05
Confidence Interval	**Lower Limit**	**Upper Limit**	**Confidence Interval**	**Lower Limit**	**Upper Limit**
1.7635	85.7939	89.3209	0.1660	88.6959	89.0279

4 Conclusions and Recommendation

4.1 Conclusions

Among the three models used in this study, the results obtained using the Henderson and Pabis model and Modified Page model weren't good fit to describe the drying kinetics of *saba* because of a large total error. Based on the total error, the Laplace transform model well describes the drying kinetics of thinly sliced saba at all investigated temperatures. Hence, the Laplace transform model can be used to determine mathematical expression for the drying time. The difference in the moisture that can be removed from *saba* has a greater factor as compared to the temperature in determining the drying time.

4.2 Recommendations

This study can be further improved by using the maturity index instead on basing only on the color of the sample for a more accurate data comparison. The use of different models can also be done to enhance this research. It is also recommended to use a more accurate top load balance. This study recommends that a pre-treatment specifically, water bath, for the sample is done to have same initial moisture content.

We would like to express our sincerest gratitude to the Department of Chemical Engineering, Faculty of Engineering, University of Santo Tomas for allowing us to conduct experiments and use their facilities.

References

1. Postharvest Industry: Banana. Philmech.gov.ph. Retrieved 3 February 2016, from http://www.philmech.gov.ph/phindustry/banana.htm (2016)
2. Bezerra, C. V., Amante, E. R., de Oliveira, D. C., Rodrigues, A. M., & da Silva, L. H., Ind. Crops Prod., 241-249 (2013)
3. Saba Banana Production Guide. Retrieved 3 February 2016, from http://cagayandeoro.da.gov.ph/wp-content/uploads/2013/04/SABA-BANANA-PRODUCTION-GUIDE.pdf. (2016)
4. Baini, R., Langrish, T.A.G., J. Food Eng., **79**, no. 1, 330–343 (2007)
5. da Silva, W. P., e Silva, C. M., & Gomes, J. P., J. Food Eng., 417-424 (2013)
6. Hofsetz, K., Lopes, C. C., Hubinger, M. D., Mayor, L., & Sereno, A. M., J. Food Eng., **83**, no. 4, 531-540 (2007)
7. Da Silva, W.P., Silva, C.M.D.P.S., Farias, V.S.O., Gomes, J.P., Drying Tech. **30**, no. 2, 164–174 (2012)
8. Robinson, A. A., *Research Design and development of banana dehydration process* (Food Engineering, Sydney, UNSW, Australia, 1980)
9. Talla, A., Puiggali, J.R., Jomaa, W., Jannot, Y.,. J. Food Eng., **64**, no. 1, 103–109 (2004)
10. Karim, M.A., Hawlader, M.N.A., J. Food Eng., **70**, no. 1, 35–45 (2005)
11. Demirel, D., & Turhan, M., J. Food Eng., **59**, no. 1, 1-11 (2003)
12. Nguyen, M. H., & Price, W. E., J. Food Eng., 200-207 (2007)

Optimization of Supercritical Carbon Dioxide and Co-solvent Ethanol Extraction of Wasted Peanut Skin Using Response Surface Methodology

Nicky Rahmana Putra[1,2], Ahmad *Hazim* Abdul Aziz[1], Lee Nian *Yian*[1], Wan *Diyana* Ramli[1], and Mohd Azizi Che Yunus[1*]

[1] Centre of Lipid Engineering and Apllied Research (CLEAR), Ibnu Sina Institute for Scientific and Industrial Research, Universiti Teknologi Malaysia, 81310, UTM Johor Bahru, Malaysia
[2]Faculty of Chemical and Energy Engineering, Universiti Teknologi Malayia, 81310, Johor Bahru,Malaysia

Abstract. Peanut skin is a waste of peanut industries especially peanut butter industries. Peanut skin contain high antioxidant and high nutritional values. The objective of this study was to optimize the effect of parameters to obtain high extract yield and high antioxidant activity by using supercritical carbon dioxide and co-solvent ethanol. A Box-Behnken design (BBD) was used to optimize the extraction process at the condition temperature (40, 55 and 70 °C), rate of co-solvent ethanol (2.5, 5, and 7.5 % Vethanol/Vtotal), and extraction pressure (10, 20 and 30 MPa). The extraction time was 3 hours. The optimum condition to obtain yield of extraction and antioxidant activity 22.05 MPa, 62.76 °C and 6.03 %($V_{ethanol}/V_{solvent}$) with 15.404 % yield extract and 94.040 % antioxidant activity.

1 Introduction

Peanut is an agricultural product that has been used for many dishes in the asean country such as Indonesia, Malaysia and Philippines. In food industries, peanut skin is always removed from peanut due to taste of product. Peanut skin decreases the quality of product because peanut skin has astringent taste. Astringent taste indicates that peanut skin contains high antioxidant inside of its skin. Previous researcher has already found that peanut skin contains procyianidin, catechin, epicatechin, and anthocyanidin [1,2]. Most of this antioxidant of peanut skin extract reduced risk of cardiovascular diseases and cancers, anti-diabetic, anti-inflammatory [3].

Supercritical carbon dioxide is a green and safe technology to extract oil, bioactive compounds and antioxidant. Furthermore, supercritical carbon dioxide does not have residues after extraction process because carbon dioxide will be naturally separated from the extract. Moreover, supercritical carbon dioxide extraction has many advantages in food process extraction because unexpansive as solvent, and also has small critical temperature and pressure (304 K and 71 Bar). Although, supercritical carbon dioxide has many advantages, the built up the extraction apparatus is expensive but the extract of extraction has high quantity of bioactive compound and antioxidant activity compared with conventional extraction such as soxhlet extraction.

Many researchers has successfully used supercritical carbon dioxide extraction to extract bioactive compounds from herbs and plants effectively. Recently, pipper batle leaves, *Pithecellobium Jiringan* (Jack) prain seeds, palm oil, and rubber seeds oil are successfully extracted by supercritical carbon dioxide extraction [4,5,6,7,8]. Moreover, high selectivity and diffusivity to certain compound and condition in the solute are the significant advantages of using supercritical carbon dioxide. furthermore, the another advantages of supercritical carbon dioxide extraction is the solubility of solvent can be manipulated by pressure, rate of co-solvent and temperature [9].

Ethanol as co-solvent in the extraction process is needed for this extraction process to encourage the polarity of solvent. Therefore, with addition co-solvent ethanol in the supercritical carbon dioxide extraction can extract polar and non-polar compounds inside of the solute. Moreover, Ethanol is safer than other commonly co-solvent such as methanol, propanol and butanol because in the market, ethanol is sold with food grade quality.

To optimize and determine influences of variables and fitted the experimental data on the supercritical carbon dioxide extraction has successfully used by response surface methodology (RSM). Commonly, pressure, temperature and rate of co-solvent are the variables of extraction process. From previous studies shows that extraction of red-fleshed pitaya, piper battle linn leaveas[10], tomato skin[11], Passiflora seed oil,

* Corresponding author: azizi@cheme.utm.my

and palm oil [12] have successfully optimized by response surface methodology.

The aim of this work was to optimize effect parameters to obtain high extract yield and antioxidant activity.

2 Materials and Methods

2.1 Materials

Peanut skin was obtained from G-Tachfood Industries SDN-BHD. For the pre-treatment process, peanut skin was dried in oven at 60°C for 4 hours and blended into powder with average size of particle 425 μm and placed in the freezer until used (-20 °C). Liquid carbon dioxide (99% purity) was purchased from Kras Instrument, Johor Bahru, Malaysia. Food grade of ethanol (99.86%) was obtained from Permula Sdn Bhd, Johor Bahru Malaysia.

2.2 Extraction Process

2.2.1 Supercritical Carbon Dioxide and Co-solvent Ethanol Extraction

A set of supercritical carbon dioxide and co-solvent ethanol extraction apparatus was used for this experiment. The extraction time was 3 hours with every 30 minutes of extraction process, extracts was collected and recorded. The flowrate of carbon dioxide is maintained 3 mL/min by HPLC pump (Lab Alliance, Series II Pump) from the CO_2 tank into the system continuously, and also co-solvent ethanol 98.86% was pumped continuously when the desired temperature was achieved. Temperature of chiller and back pressure regulator (Jasco BP 2080 Plus Automated BPR) was set at 6°C and 50°C. Next, 5 ±0.005 g of peanut skin was placed into extraction vessel.

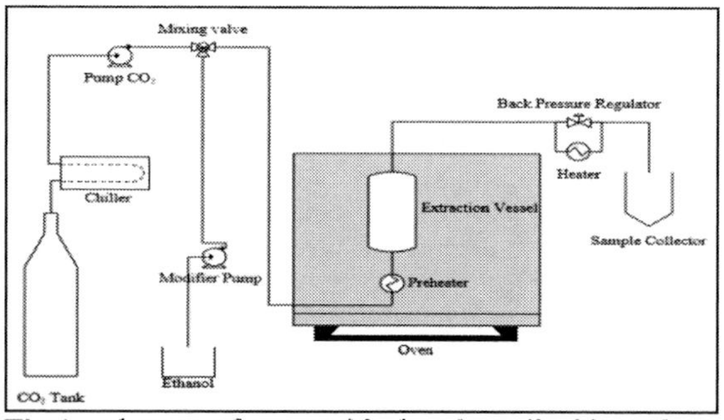

Fig.1. schemes of supercritical carbon dioxide and co-solvent ethanol extraction apparatus

2.2.1 Soxhlet Extraction

Soxhlet extraction is conventional extraction process. Soxhlet extraction gives high extract yield due to exhausting process and small amount of solvents in the extraction process. In this study, a comparison antioxidant activity between supercritical carbon dioxide extraction and soxhlet extraction were revealed. The solvents used was ethanol, n-hexane and water. The particle size was 425 μ and extraction time was 6 hours.

2.3 Determination of Antioxidant Activity

The determination of antioxidant activity by 2,2-diphenyl-1-picryl-hydrazyl-hydrate (DPPH) of peanut skin extract was evaluated as described by previous study[13]. First of all, the peanut skin extract 0.5 μgr/L was mixed with a 0.165 mM ethanoic (DPPH). Furthermore, 30 min incubation at room temperature. Spectrophotometer UV-Vis (Webler) was used to obtain the absorbance values were measured at 516 nm and converted into percentage of antioxidant activity (% AA). The measurement of antioxidant activity is:

$$(\%\ AA) = (A_a - A_b)/A_a \times 100\% \qquad (1)$$

Where %AA is percentage of antioxidant activity, A_a is absorbance of DPPH solution and A_b is absorbance of mixture between DPPH solution and extract solution.

2.4 Experimental Design

Three parameters of this study are pressure (X_1), temperature (X_2) and rate of co-solvent ethanol (X_3) in order to obtain optimum yield extract of peanut skin, Y_1 and percentage of antioxidant activity, Y_2. Box Behnken Design was used for experimental design.

Table 1. Factors and levels for the experimental design

Factors	Low (-1)	Medium (0)	High (+1)
Pressure,(X_1, MPa)	10	20	30
Temperature,(X_2, °C)	40	55	70
Rate of Co-solvent (X_3, %)	2.5	5	7.5

In this study, quadratic model was used to fit the experimental data as follows:

$$Y = B_0 + \sum_{i=1}^{k} B_i X_i + \sum_{i=1}^{k} B_i X_i^2 + \sum_i \sum_j B_{ij} X_i X_j \qquad (2)$$

Y is the predicted response, B_0 is a constant, B_i, B_{ii}, B_{ij} are the interaction of effect, X_i and X_j are coded value of factor. Furthermore, the response surfaces of the variables inside the experimental domain were analyzed by analysis of variance (ANOVA).

3 Result and discussion

3.1 Experimental Result

To optimize the extraction process, the effect of pressure, temperature and rate of co-solvent were the three main variables in order to obtain high extract yield and high antioxidant activity on peanut skin extract. The three variables was investigate by using second order polynomial statistical model. Preliminary studies was needed to determine fixed variables of the optimization. The quadratic effect of the treatments variables, their interactions and coefficients on the response variables were obtained by analysis of variance (ANOVA). Experimental design was made by using box bhenken design shown in table 2.

Table 2. The experimental design for the SC-CO_2

Run	Pressure, X_1 (MPa)	Temperature, X_2 (°C)	Rate of Co-solvent, X_3 (%)	Yield (%)	AA (%)
1	-1	-1	0	7.007	73.857
2	0	+1	-1	8.872	84.892
3	0	+1	+1	14.855	84.143
4	0	0	0	14.176	94.040
5	0	-1	+1	15.267	73.124
6	0	0	0	14.176	94.040
7	0	0	0	14.176	94.040
8	+1	-1	0	13.960	84.802
9	+1	0	-1	9.9087	82.859
10	+1	0	0	14.601	82.780
11	+1	+1	0	15.404	93.077
12	-1	0	-1	7.876	78.333
13	0	0	0	14.176	94.040
14	0	0	0	14.176	94.040
15	-1	+1	0	12.144	90.216
16	-1	0	+1	11.533	93.168
17	0	-1	-1	12.127	79.31

From the table 2 shows that the highest amount of yield extract as experimental data was 15.4048 % with condition 30 MPa, 70 °C and 5% rate of co-solvent. Furthermore, The highest antioxidant activity was obtain at condition 20 MPa, 55 °C and 5% rate of co-solvent ethanol was 94.04%.

3.2 Statistical Data Analysis

The analysis of variance (ANOVA) is presented in Table 3 for extract yield and antioxidant activity. Based on the analysis, second order polynomial equation was chosen and successful to optimize the extract yield and antioxidant activity of peanut. The analysis showed that both models were statistically significant (sig. < 0.05) at 95% confidence. The coefficients of quadratic model are shown in Tables 5 and 6.

Table 3. ANOVA table of supercritical carbon dioxide and co-solvent ethanol extration

Source	Sum of square	df	Mean Square	F	p-value
Extract Yield					
Regression	98.17	9	10.91	4.64	0.0278
Residual	16.47	7	2.35		
Total	114.65	16			
Antioxidant Activity					
Regression	765.95	9	85.11	5.17	0.0207
Residual	115.21	7	16.46		
Total	881.16	16			

Table 4. Coefficient of polynomial quadratic model of extract yield

Extract yield (%)	Coefficient (β)	Standard Error	p-value
B_0	14.18	0.69	
X_1	1.91	0.54	0.0096
X_2	0.36	0.54	0.5234
X_3	2.18	0.54	0.0050
X_1^2	-1.92	0.75	0.0368
X_2^2	-0.12	0.75	0.8737
X_3^2	-1.27	0.75	0.1324
X_1X_2	-0.92	0.77	0.2679
X_1X_3	0.26	0.77	0.7458
X_2X_3	0.71	0.77	0.3850
$R^2 = 0.8563$			

Table 5. Coefficient of polynomial quadratic model of antioxidant activity

Antioxidant Activity (%)	Coefficient (β)	Standard Error	p-value
B_0	94.04	1.81	
X_1	0.99	1.43	0.5111
X_2	5.16	1.43	0.0088
X_3	0.98	1.43	0.5172
X_1^2	-2.32	1.98	0.2796
X_2^2	-6.24	1.98	0.0161
X_3^2	-7.44	1.98	0.0071
X_1X_2	-2.02	2.03	0.3523
X_1X_3	-3.73	2.03	0.1087
X_2X_3	1.36	2.03	0.5246
$R^2 = 0.8692$			

3.3 Effect parameters on the extract yield

Table 3 and 4 shows the response and interaction of extract yield of peanut skin extraction by supercritical carbon dioxide and co-solvent ethanol.

$$Y= 14.18 + 1.91X_1 + 0.36X_2 +2.18\ X_3 - 1.92X_1^2 - 0.12\ X_2^3 - 1.27X_3^3 - 0.92X_1X_2 + 0.26X_1X_2 + 0.71X_2X_3 \quad (3)$$

Due to the coefficient of determination (R^2) 0.853 and p-value of model was 0.0278, the second order polynomial model has been successfully fitted the experimental data of extract yield as shown on the Eq.3. From table 4, for the response extract indicated that pressure and rate of co-solvent ethanol were the most influence variables on the extraction process based on value of coefficient is higher than another variables.

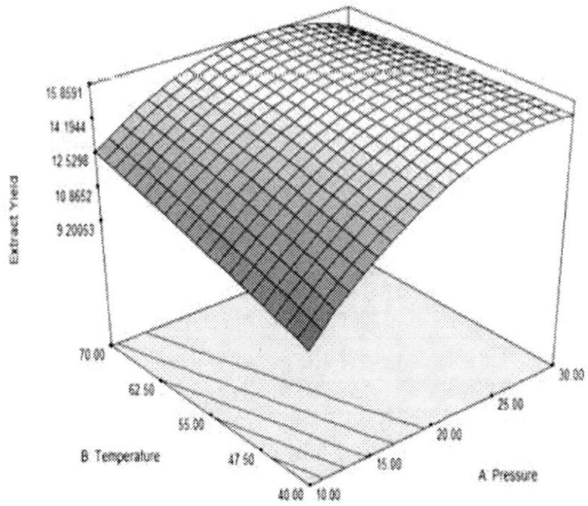

Fig.3. 3D response for the effect of pressure and temperature on the extract yield

Fig. 3 shows that extract yield increase with increasing of pressure (X_1) at constant lowest and highest temperature (X_2). This is because Increasing of solubility and density of CO_2 to extract from the solute will be easily extract went out from the solute[8]. Increasing of pressure on the extraction process increases the density of carbon dioxide may affect of diffusivity of carbon dioxide to carry out of extract from solute[14]. This is similar funding with extraction of castor oil that increasing of pressure will increase the yield extracts[15]. It proofed that in the table 3 p-value of pressure X_1 is lower than p-value ($0.05 > 0.0096$)

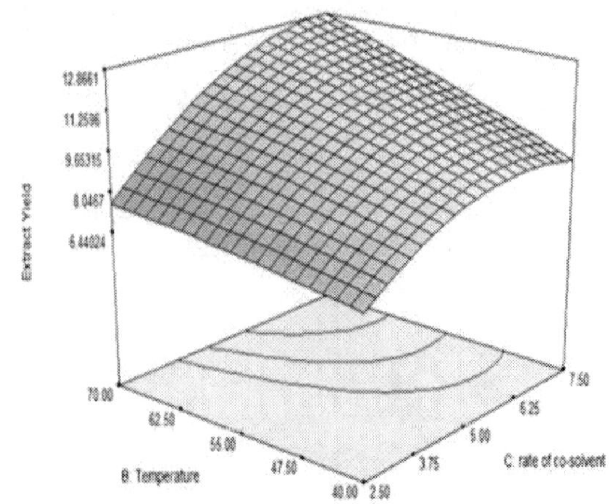

Fig.4. 3D response for the effect of temperature and rate of co-solvent on the extract yield

Increasing of the temperature at lowest pressure condition enhanced the yield extract of peanut skin, but effect of temperature have small effect on the extraction process due to coefficient of quadratic model and p-value temperature is higher than p-value ($0.5234 > 0.05$).

Fig.4 and table 4 shows that the co-solvent ethanol was the main influence variable to obtain the maximum peanut skin extract. Ethanol as co-solvent is a polar solvent that can easily extract the polar compound from the peanut skin due to compositions of peanut skin extract are polar compounds such as catechin, epicatechin, procyanidin and anthocyanidin. The ethanol as co-solvent had a smaller value of significant than pressure ($0.005 < 0.0096$).

The pressure 22.05 MPa, temperature 62.76 °C and 6.03 % rate of co-solvent ethanol are optimum condition to obtain maximum extract yield (15.4043 %). The calculated of coefficient of determination (R^2) of the optimization of extraction with response yield extract was significance because the value is above 0.8. Futhermore, p-value of model on the ANOVA was lower than 0.05. Quadratic model has been successful giving the optimum pressure, temperature and rate of co-solvent ethanol in range of adjustable.

3.4 Effect parameters on the Antioxidant Activity

Table 2 and 5 shows the response of antioxidant activity and interaction of parameters. The coefficient of regression model for extract yield was similar with extract yield model. Second order polynomial was successful to interpret the experimental data as shown in table 3 and can be written on Eq.4

$$Y= 94.04 + 0.99X_1 + 5.16X_2 + 0.98\ X_3 - 2.32\ X_1^2 - 6.24\ X_2^3 - 7.44\ X_3^2 - 2.02X_1X_2\ .\ 3.73X_1X_2 + 1.36X_2X_3 \quad (4)$$

Quadratic model has been successfully described and fitted the experimental data of antioxidant activity due to satisfactory coefficient (R^2) of response was high (0.8692) and p-value model (0.02079). Table 2 and 5 shows temperature were the most influence on the extraction process because it had big values of coefficient compared with others. It is proofed that in the table 5 p-value of pressure X_1 is lower than p-value (0.05 > 0.008).. Quadratic rate-of co-solvent ethanol (X_3^2) was most influence in the extraction process due to p-value and coefficient on the second order polynomial that is the highest value on the equation model.

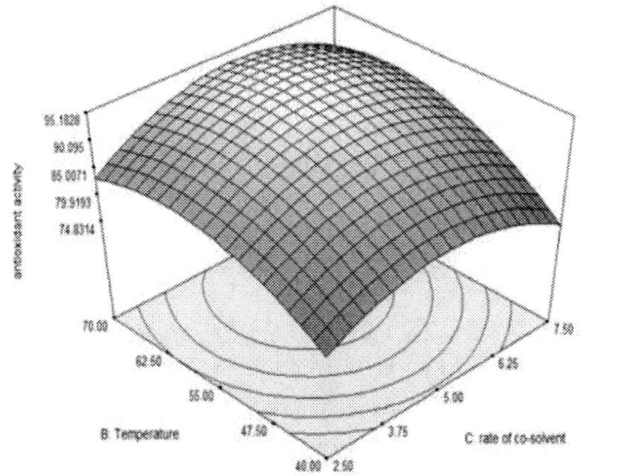

Fig.5. 3D response for the effect of temperature and rate of co-solvent ethanol on the antioxidant activity

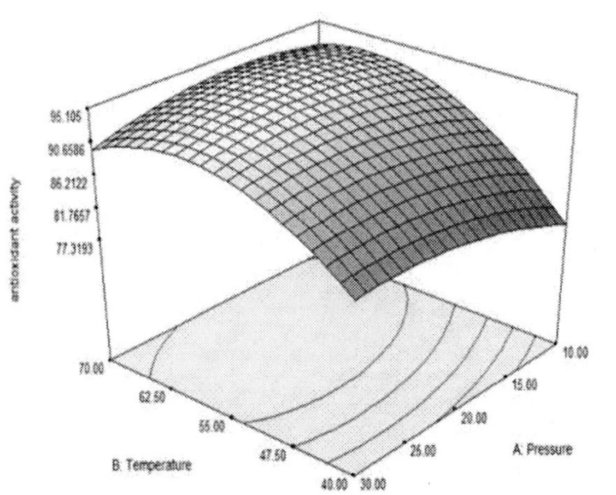

Fig. 6. 3D response for the effect of temperature and pressure on antioxidant activity

Fig 6 shows that increasing rate of co-solvent X_3 at lowest and highest temperature (X_2) reveals the peak optimum condition in the middle of range variables. Increasing of rate of co-solvent will enhance the polarity of solvent from nonpolar that without co-solvent to polar after added ethanol as co-solvent[16,17]. Furthermore, most of bioactive compound inside of peanut skin extract is polar compounds automatically will enhance the antioxidant activity [2].

. From Fig 7, 3D response shows that in the effect of temperature, the model has already success to optimize the condition that optimal temperature was revealed. At the highest temperature condition, temperature will degrade the bioactive compound inside of extract, but in the lowest temperature condition, effect of the solute vapour is not dominant that solvent is difficult to carried out the extract and its bioactive compound from solute. Effect of pressure is not dominant in this extraction process to enhance the antioxidant of extract because increasing of pressure will not degrade the bioactive compound [15,18]. It was proofed that p-value on the coefficient of model was higher than 0.05. Furthermore, effect of vapour solute was revealed on this studies because increasing of temperature will increase the antioxidant activity [19,17,5]. There are more bioactive compound was successfully extracted from solute due to increasing of temperature. This is similar funding with extraction of jack fruit that increasing of temperature will increasing the antioxidant activity and yield of extract[20].

3.5 Comparison antioxidant activity between supercritical carbon dioxide with co-solvent ethanol and soxhlet extraction

In this study, extract of supercritical carbon dioxide with co-solvent ethanol gives the highest antioxidant activity compared with soxhlet extraction with various solvents (ethanol, water and n-hexane).

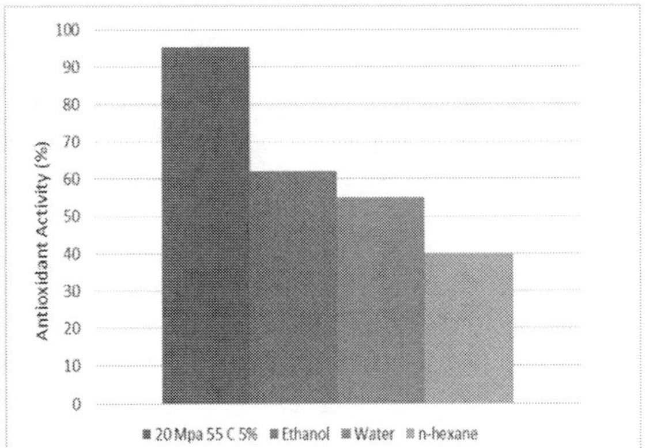

Fig.7. Comparison antioxidant activity between SC-CO_2 and soxhlet extraction with various solvents

Antioxidant activity of soxhlet extraction was lower than supercritical fluid extraction due to temperature of extraction process. Soxhlet extraction uses the high temperature to extract the solute that will degrade the bioactive compound of extract[21]. Antioxidant of extract with n-hexane as solvent is the lowest antioxidant activity indicated that most of compound inside of peanut skin is polar compound because n-hexane is non polar that is difficult to extract polar compound inside of peanut skin such as catechin, procyanidin, anthocyanidin and epicatechin [22],[2],[3].

4 Conclusion

The second order polynomial as quadratic model was chosen to optimize the supercritical carbon dioxide extraction with co-solvent ethanol with response optimum yield extract of peanut skin and antioxidant activity due to p-value of model ($p < 0.05$). The optimum yield peanut skin extract and antioxidant activity was 15.40 % and 94.04 % with conditions at 62.76 °C, 22.05 MPa and 6.03 % rate of co-solvent. Furthermore, the second order polynomial was successful to optimize the effect condition in the range variables adjusted and gave the maximum extract yield and antioxidant obtained. The supercritical carbon dioxide with co-solvent ethanol gave the higher antioxidant activity compared with soxhlet extraction due to temperature of extraction process. High temperature will degrade the bioactive compound and antioxidant activity of extract [23,24].

The authors acknowledge the Centre of Lipid Engineering and Applied Research (CLEAR) and the financial support from Universiti Teknologi Malaysia for professional development research university grant (QJ130000.2509.14H23) and (R.J130000.7909.4F766).

References

1. V. H. Hoang, P. Apoštolová, J. Dostálová, F. Pudil, and J. Pokorný, Czech J. Food Sci., **26**, no. 6, 447–457 (2008)
2. J. Yu, M. Ahmedna, and I. Goktepe, Food Chem., **90**, no. 1–2, 199–206 (2005)
3. V. Nepote, N. R. Grosso, and C. a. Guzman, Grasas y Aceites, **53**, no. 4, 391–395 (2002)
4. N. H. Arsad, M. Azizi, C. Yunus, M. Abbas, and A. Zaini, **12**, no. 4, 741–751 (2016)
5. N. Lee, N. Sharif, A. Ahmad, and A. Khatoon, **17**, no. 10, pp. 46–52 (2013)
6. A. N. Mustapa, Z. A. Manan, C. Y. Mohd Azizi, N. A. Nik Norulaini, and A. K. M. Omar, J. Food Eng., **95**, no. 4, 606–616 (2009)
7. A. N. Mustapa, Z. A. Manan, C. Y. Mohd Azizi, W. B. Setianto, and A. K. Mohd Omar, Food Chem., **125**, no. 1, 262–267 (2011)
8. M. C. Yunus and M. Hasan, J. …, **60**, 21–25, (2013).
9. C. Yi, J. Shi, S. J. Xue, Y. Jiang, and D. Li, Food Chem., vol. **113**, no. 4, 1088–1094 (2009)
10. A. H. A. Aziz, M. A. C. Yunus, N. H. Arsad, N. Y. Lee, Z. Idham, and A. Q. A. Razak, IOP Conf. Ser. Mater. Sci. Eng., **162**, no. 1, 12031 (2016)
11. L. S. Kassama, J. Shi, and G. S. Mittal, Sep. Purif. Technol., **60**, no. 3, 278–284 (2008)
12. M. S. Md Sarip, Y. Yamashita, N. A. Morad, M. A. Che Yunus, and M. K. Abdul Aziz, J. Chem. Eng. Japan, **49**, no. 7, 614–621 (2016)
13. C. S. G. Kitzberger, A. Smânia, R. C. Pedrosa, and S. R. S. Ferreira, J. Food Eng., **80**, no. 2, 631–638 (2007)
14. H. P. Cornelio-Santiago, C. B. Gonçalves, N. A. de Oliveira, and A. L. de Oliveira, J. Supercrit. Fluids, **128**, pp. 386–394 (2017)
15. J. M. Danlami, M. A. A. Zaini, A. Arsad, and M. A. C. Yunus, Ind. Crops Prod., **76**, 34–40 (2015)
16. H. D. Belayneh, R. L. Wehling, A. K. Reddy, E. B. Cahoon, and O. N. Ciftci, JAOCS, J. Am. Oil Chem. Soc., **94**, no. 6, 855–865 (2017)
17. S. C. S. Corzzini, H. D. F. Q. Barros, R. Grimaldi, and F. A. Cabral, J. Food Eng., **194**, 40–45 (2017).
18. T. Sato, F. Fukuda, K. ichi Nihei, and N. Itoh, J. Supercrit. Fluids, **130**, 23–29 (2017)
19. J. M. Danlami, M. A. A. Zaini, A. Arsad, and M. A. C. Yunus, J. Taiwan Inst. Chem. Eng., **53**, 32–39 (2015)
20. M. A. C. Yunus, N. H. Arsad, S. Zhari, Z. Idham, S. H. Setapar, and A. N. Mustaph, J. Teknol. Sciences Eng., **60**, 45–50 (2013)
21. M. C. Yunus and M. Hasan, J. …, **60**, 21–25, (2013)
22. T. S. Ballard, P. Mallikarjunan, K. Zhou, and S. F. O'keefe, J. Agric. Food Chem.,. **57**, no. 8, 3064–3072 (2009)
23. L. P. Cunico, M. C. Acosta, and C. Turner, J. Supercrit. Fluids, (2017)
24. P. Khuwijitjaru, J. Plernjit, B. Suaylam, S. Samuhaseneetoo, R. Pongsawatmanit, and S. Adachi, Can. J. Chem. Eng., **92**, no. 5, 810–815 (2014)

Experimental Study on Pressure Drop and Flow Dispersion in Packed Bed of Natural Zeolite

Petric Marc Ruya, *Herri* Susanto*, and *Mubiar* Purwasasmita

Institut Teknologi Bandung, Department of Chemical Engineering, 40132 Bandung, Indonesia

Abstract. The use of conventional correlation for pressure drop and dispersion coefficient calculation may result in inaccurate values for zeolite packed bed as the correlations are generally developed for regularly shaped and uniformly sized particles. To support the research on the application of modified natural zeolite as tar cracking catalyst, the research on the hydrodynamic behaviour of zeolite packed bed has been conducted. Experiments were carried out using a glass column with diameter of 37.8 mm. Natural zeolite with particle size of about 2.91 to 6.4 mm was applied as packing material in the column, and the bed height was varied at 9, 19 and 29 cm. Air was used as the fluid that flows through the bed and nitrogen was used as a tracer for residence time distribution determination. Air flow rates were in the range of 20 to 100 mL/s which correspond to the laminar-transitional flow regime. The pressure drops through the bed were in the range of 1.7 to 95.6 Pa, depending on the air flow rate and bed height. From these values, the parameters in the Ergun equation were estimated, taking into account the contribution by wall effect when the ratio of column to particle diameter is low. The viscous and inertial term constants in the Ergun equation calculated ranges from 179 to 199 and 1.41 to 1.47 respectively while the particle sphericity ranges from 0.56 to 0.59. The reactor Peclet number were determined to range from 5.2 to 5.5, which indicated significant deviation from a plug flow condition.

1 Introduction

Syngas produced through biomass gasification will generally contains tar which must be removed prior to utilization as tar may cause various problems when it condense in the downstream equipment. Catalytic tar cracking is one of the most promising method for tar removal due to the absence of subsequent waste. Gasification Group of ITB Chemical Engineering Department is currently working on the development of tar cracking catalyst through the modification of domestic natural zeolite. Natural zeolite particles are irregularly shaped and distributed in size. The pressure drop and extent of deviation from plug flow condition on the fluid that flows through the zeolite packed bed will therefore be higher compared to when more regularly shaped and uniformly sized particles are used.

The research on the zeolite particles packed bed hydrodynamic reported in this paper was conducted by initially characterizing the zeolite particles used to determine the particle diameter, roundness, sphericity and void fraction of the resulting packed bed. The measurement of pressure drop and determination of retention time distribution by tracer injection experiment were then conducted by applying the particles in a glass column. The results from the pressure drop measurement were used to estimate the equivalent particle diameter, and both the viscous and inertial term constants in the Ergun equation. The results from the tracer injection experiment were used to calculate for the reactor Peclet number. The reactor Peclet number was subsequently used to predict the effect of dispersion on the reactor performance using the kinetic data of several tar cracking catalyst tabulated by [1]. More details on the particle characterization, pressure drop measurement and tracer injection experiment can be found in the previous work [13].

* Corresponding author: herri@che.itb.ac.id

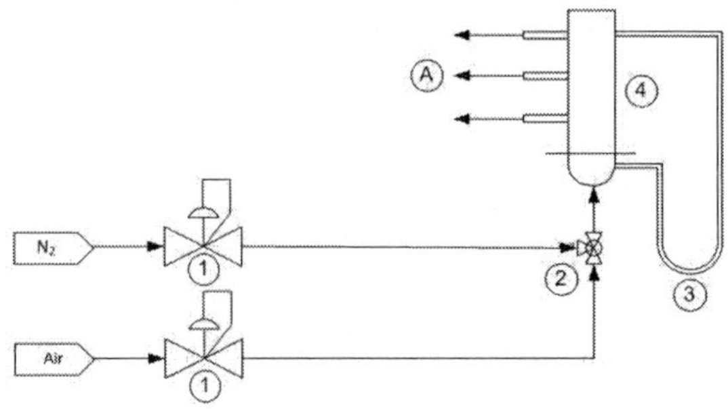

Fig. 1. Experimental set up for pressure drop measurement and tracer injection

Pressure drop measurement results were also used to calculate for the friction factor (f/2), which in turn will be used to calculate for the Le Goff number. Le Goff number is the ratio of the Chilton-Colburn j-factor to the friction factor (f/2). The Le Goff number is defined by [8] to be the amount of mass or heat (energy) transported per unit amount of momentum transferred. The calculation of Le Goff number may enable the comparison of mass or heat (energy) to momentum transport efficiencies at different operating condition and thus assist in determining the most suitable operating condition. According to [15], maximum value of Le Goff number generally exists and the maximum value can be determined when the Le Goff number is plotted as a function of Reynold number. The Chilton-Colburn j-factor used in the calculation was obtained from the work of [2].

2 Materials and Method:

The three types zeolite particles used in the experiment were acquired from domestic sources. The first and second type originated from Sukabumi (West Java) and will be referred to as Zeolite 1A (Z1A) and Zeolite 1B (Z1B) respectively to distinguish the two from Zeolite 2 (Z2) that originates from Klaten (Central Java). Zeolite 1A has been modified for tar cracking purpose prior to the experiment.

The zeolite particles used were initially characterized to determine various parameters, such as equivalent particle diameter, particle roundness, sphericity and void fraction, when the particles were applied in packed column. Particle roundness were measured using the method described by [7].

2.1. Equipment Set Up

The zeolite particles were initially placed inside a cylindrical glass column with inner diameter and height of 3.78 cm and 50 cm respectively. Manometer was integrated with the column for pressure drop measurement as shown in Figure 1. The height of the zeolite packed bed was varied at 9, 19 and 29 cm for pressure drop measurement. Air was blown through the packed bed with the use of air compressor. Tracer injection experiment was performed on the zeolite packed bed by introducing technical grade N_2 into the column. The response from the disturbance was then measured and recorded using Bacharach PCA-3 combustion analyser at every 1 s interval.

2.2 Parameter Estimation

The Ergun equation, a semi empirical model, was initially developed to be used for packed bed with uniformly sized and regularly shaped particles. The values of viscous and inertial terms constants that are generally used are 150 and 1.75 respectively. These values were developed for spherical particles. The increase in pressure drop when non-spherical particles are used as packing material can be taken into account by the inclusion of particle sphericity as shown in Equation 1.

$$\Delta P = 150\frac{(1-\varepsilon)^2}{\phi^2 d_p^2 \varepsilon^3}\mu u_s L + 1.75\frac{(1-\varepsilon)}{\phi d_p \varepsilon^3}\rho u_s^2 L \quad (1)$$

$$d_p = \frac{6}{a_v} \quad (2)$$

$$a_v = \frac{S_p}{V_p} \quad (3)$$

The particle sphericity was obtained through both volumetric measurement and parameter estimation by using the result obtained from pressure drop measurement. The definition of sphericity that is generally used in this context is the ratio of surface area of sphere with the same volume as the particle to the surface area of the particle. However, due to the difficulty in measuring the specific surface area of the zeolite particle, the particle sphericity was instead determined by initially measuring the longest side of several particles and then by measuring the average volume of a particle through water displacement method. The particle sphericity was then calculated as the cube root of the ratio of particle volume to the volume of the circumscribing sphere whereby the average longest side of the particles was taken as the diameter of the circumscribing sphere.

Parameter estimation was not only conducted to calculate for the sphericity but also for both the viscous and inertial term constants of the Ergun equation. The constants were estimated by initially assuming reliable volumetrically measured sphericity and then by allowing the particle sphericity to vary with the constants. The Ergun equation constants are not only affected by the particle shape; the column diameter and the particle size distribution also play significant role. The lower the ratio of column diameter to the particle diameter are, the more significant the wall effect would be. In addition, packed bed that is made using particle with larger particle size distribution will also inflict higher pressure drop on the fluid, despite having the same mean particle diameter [12]. Another approach in the parameter estimation method was also conducted where both the viscous and inertial term constants were taken as a function of equivalent particle diameter. The equivalent particle diameter, in this case, is taken as the product of sphericity with volume-mean diameter. The definition of volume mean diameter was presented by [10] and is shown in Equation 4. Volume-mean diameter can be determined by conducting only volumetric measurement.

$$d_{vs} = \frac{6V_p}{A_{sp}} \quad (4)$$

The equation to calculate for both the viscous and inertial term constants were obtained from the work of [3] and is presented as Equations 5 and 6 respectively.

$$C_1 = K_1 A_w^2 \quad (5)$$

$$C_2 = \frac{A_w}{B_w} \quad (6)$$

$$A_w = 1 + \frac{2}{3\left(D/d_p\right)(1-\varepsilon)} \quad (7)$$

$$B_w = \left[k_1\left(\frac{d_p}{D}\right)^2 + k_2\right]^2 \quad (8)$$

The value of the coefficients used are for the spherical particles, whereby $K_1 = 154$, $k_1 = 1.15$ and $k_2 = 0.87$. The parameter estimation method varies the particle sphericity to calculate for the equivalent particle diameter, viscous and inertial terms constants.

2.3 Fluid Dispersion

The result from the tracer injection experiment was initially used to calculate for the residence time distribution with the results subsequently used to calculate for the variance and mean residence time. The reactor Peclet number was then calculated assuming open-open boundary condition. The reactor Peclet number was then used to calculate for the dispersion coefficient to be used in the axial dispersion model (Eq. 9). Using the calculated reactor Peclet number, tar cracking simulation was performed using the kinetic data tabulated by [1]. The reactor operating temperature of 600 and 700^{0}C was used and toluene was used as the tar model compound in the simulation. The reactor performance simulation was performed on FlexPDE.

$$u\frac{dC_A}{dz} - D_a\frac{d^2C_A}{dz^2} = r_A(\rho_b) \quad (9)$$

2.3 Calculation of Le Goff Number

The calculation of Le Goff number requires both the friction factor (f/2) and Chiton-Colburn j-factor. The results obtained from the pressure drop measurement were used to calculate for the friction factor by using Equations 10 [14] and the j-factor was calculated using the equation presented by [2] which is presented as Equations 11 and 12 respectively.

$$f = \frac{1}{4}\left(\frac{D_p}{L}\right)\left(\frac{P_0 - P_L}{0.5\,\rho v_0^2}\right) \quad (10)$$

$$j = \frac{2.06}{Re^{0.575}\,\varepsilon} \quad (11)$$

$$Re = \frac{\rho u d_p}{\mu} \quad (12)$$

Table 1. Toluene cracking kinetic parameter, [1]

No.	Parameter	Ardy et al. (2015)	Palma (2013)	Agustina (2011)	Swierczynski et al. (2008)
1.	Catalyst type	Ni/a-Al2O3	Ni/Al2O3	Dolomite	Ni/Olivine
2.	Frequency factor (k_0), $m^3/(kg_{cat}.h)$	272	23000	35	3.14×10^{13}
3.	Activation energy (E_a), kJ/mol	11.6	40	13.2	196
4.	Toluene concentration, g/m^3	51	2	3-4	27
5.	Particle size, mm	2	-	1	0.125-0.6
6.	Bulk density, kg/m^3	939	939	2900	1462
7.	Catalyst requirement, $kg/(mol.h^{-1})$	1.167	0.657	4.928	0.082

Table 2. Comparison of parameter estimation results

No.	Method	Parameter	Z1A	Z1B	Z2
1	Method 1[a]	Viscous constant	190	179	199
		Inertial constant	1.45	1.41	1.47
		Sphericity	0.56	0.59	0.58
2	Method 2[b]	Viscous constant	318	890	460
		Inertial constant	4.87	2.19	3.41
		Sphericity	0.693	0.77	0.51
3	Method 3[c]	Viscous constant	150	269	414
		Inertial constant	3.01	1.2	3.23
		Sphericity	0.47	0.42	0.48

[a] Using volume-mean diameter and Equation 5 and 6 for calculating constants
[b] Assuming that the measured sphericity is reliable [13]
[c] Simultaneously estimating the sphericity and the constants [13]

3 Results and Discussions

This section presents the results obtained from parameter estimation, reactor simulation and Le Goff number calculation. Details on the results obtained from particle characterization, pressure drop measurement and tracer injection experiment have been presented and discussed in the previous work [13].

3.1 Parameter Estimation

The results from the parameter estimation using the method that involves the use of volume mean diameter, and Equation 5 and 6 to calculate for the Ergun equation constants is presented in Table 2 along with other methods, presented and discussed in the previous work [13], for comparison.

As can be observed in Table 2, using Method 1 results in Ergun equation constants and sphericity that is relatively closer in value for every type of zeolite particles used. This method takes into account the contribution by wall effect into the Ergun equation constants and varies only sphericity to produce subsequent variation in both the equivalent particle diameter and the Ergun equation constants. The results obtained from this method suggest that the particle sphericity may be relatively similar for each zeolite type and the differences produced when the other methods were applied occurs mainly because of the wall effect and the error in measuring the equivalent particle diameter. The measurement of volume-mean diameter provides an upper limit to the equivalent particle diameter.

3.2 Fluid Dispersion

The reactor Peclet number obtained from the tracer injection experiment has been reported in the previous work [13]. The values of the reactor Peclet number calculated varies at about 5 for the 3 types of zeolite particles, which implies significant deviation from plug flow condition [6]. The results of the simulation is presented as Figure 2 and 3, where PF correspond to the plug flow condition and DF to dispersed flow condition. From the results of the simulation, it can be observed that lower operating temperature would cause lower catalyst activity and thus longer residence time requirement to achieve the same amount of conversion. Longer residence time thus causes the effect of fluid dispersion on the conversion to be more significant. From both Figure 2 and 3, it can be seen that the assumption of plug flow condition may cause significant overestimation of the reactor performance and this can be detrimental, especially when the very low tar content is required for subsequent use of the product gas.

Table 3. Particle Diameter

No.	Parameter	Value (mm)		
		Z1A	Z1B	Z2
1	Volume-mean diameter	5.07	3.77	6.00
2	Equivalent diameter (estimation)[b]	2.82	2.20	3.51
3	Equivalent diameter (measurement)	3.50[a]	2.91	6.40

[a] Averaged according to mass fraction
[b] Volume mean diameter multiplied by sphericity

https://doi.org/10.1051/matecconf/201815602006

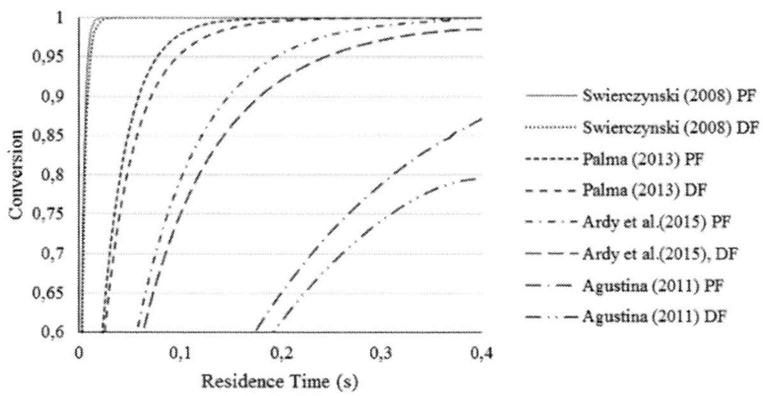

Fig. 2. Tar cracking simulation at operating temperature of 700^{0}C
*PF-Plug Flow, DF- Dispersed Flow

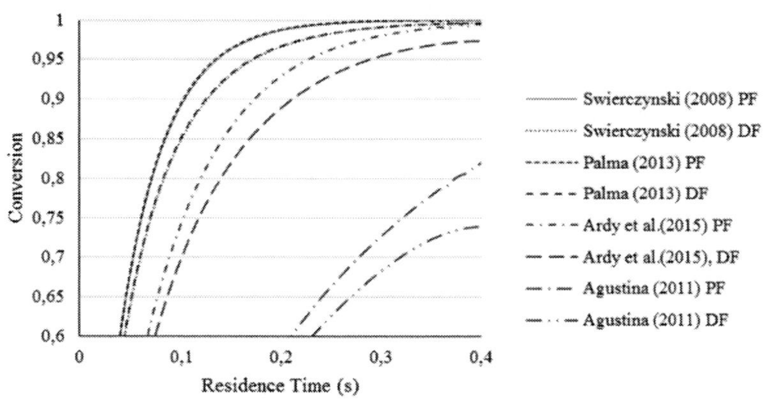

Fig. 3. Tar cracking simulation at operating temperature of 600^{0C}
*PF-Plug Flow, DF- Dispersed Flow

3.3 Le Goff Number

The plots of the Le Goff number for the three types of zeolite packed bed are shown in Figure 4. From Figure 4, the increasing trend of Le Goff number for Z1B and Z2 packed bed can be observed and this indicates that the potential for increased efficiency for mass or energy transferred per unit of momentum transferred.

The Le Goff number for Z1B produces the least scatter from the trend line and this may be attributed to the comparatively uniform particle size. As for Z1A packed bed, the graph has shown maximum point and thus the point of maximum efficiency might have been reached. However, Z1A consist of particles with the widest particle size distribution among the three types of zeolite particles and thus the peak may possibly be caused by the variation in particle size instead. The equation to calculate for the Chilton Colburn j-factor reported by [2] were originally developed for spherical particles and thus the use of this equation for zeolite particles may results in inaccuracies.

The currently reported work deals with hydrodynamic study on zeolite packed bed under low Reynold number operating condition. Further work on higher Reynold number region should be done to determine suitable operating condition and to extend the findings to be applicable at higher Reynold number condition which is prevalent in industrial application.

The reactor Peclet number can also be used in the future work to correct the calculation of mass transfer coefficient, which generally assumes the existence of strictly plug flow condition throughout the cross-sectional area of the packed bed. [9] has applied the axial dispersion coefficient in calculating the mass transfer factor (Chiton Colburn j-factor) and reported that the previously different value obtained between packed bed and dispersed bed were very similar when corrected.

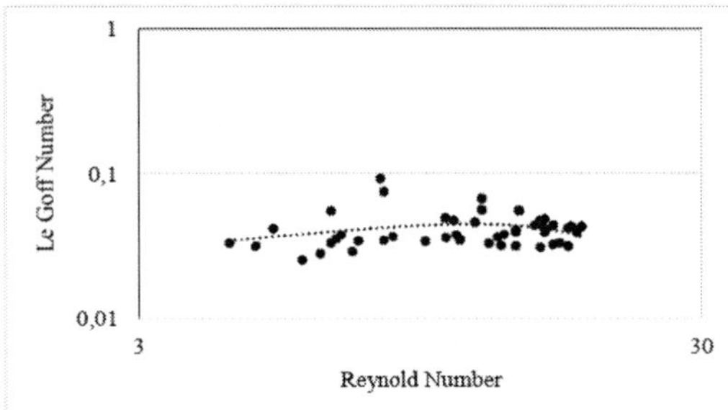

(a) Zeolite 1A

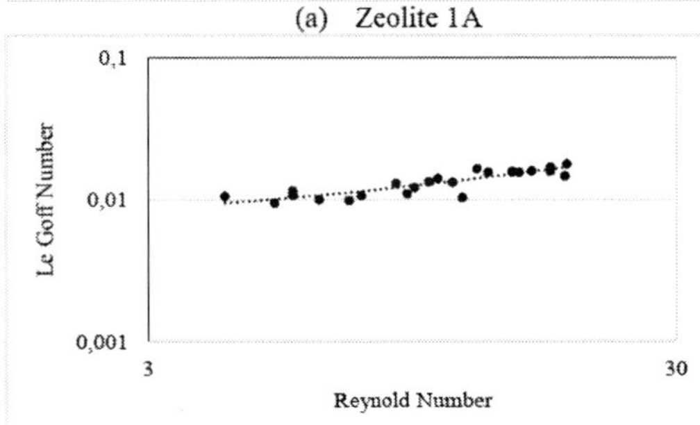

(b) Zeolite 1B

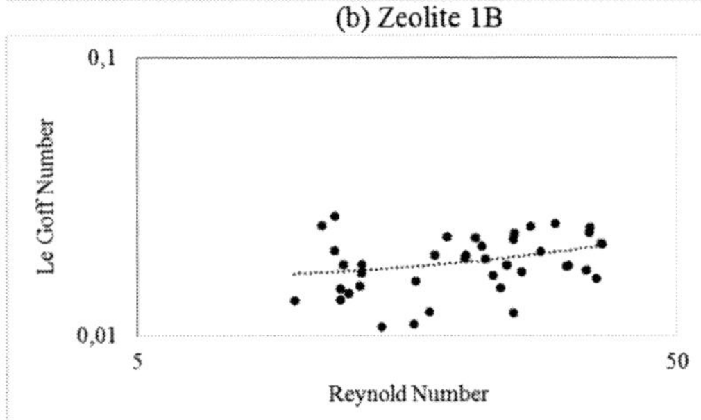

(c) Zeolite 2

Fig. 4. Plot of Le Goff number

4 Conclusions

Parameter estimation to determine particle sphericity and Ergun equation constants was conducted using the data obtained from pressure drop measurement. Different values of particle sphericity and Ergun equation constants were obtained when different assumption and methods were applied. Among the methods mentioned, the method which varies particle sphericity to calculate for the equivalent particle diameter and subsequently the Ergun equation constants produce relatively the most similar value of sphericity and Ergun equation constant. This may indicates the relatively substantial contribution of wall effect and equivalent particle diameter on the parameter estimation. The mentioned method takes into account the contribution by wall effect and further improvement may be obtained if the effect due to particle size distribution is also taken into account.

The reactor Peclet number indicates large amount of deviation from plug flow condition for fluid that flows through the zeolite packed bed. The result of the tar cracking simulation has revealed that the performance of the packed bed reactor may be significantly affected by the fluid dispersion. The effect becomes more significant at lower temperature as longer residence time may be required to achieve a particular amount of conversion.

The calculation of Le Goff number has revealed that the relative efficiency of heat and mass transfer per unit of momentum transferred potentially may still increase at higher Reynold number condition.

More accurate insight might be obtained if the mass or heat transfer experiment were conducted to obtain data applicable for Chilton-Colburn j-factor calculation. The current work on fluid dispersion can also be used in future work to investigate the heat and mass transfer as this will allow the axial mixing factor to be calculated and used to correct the assumption of plug flow condition in calculating the heat or mass transfer coefficient.

This experimental study is a part of the research on the production of bio-synthesis gas and DME synthesis which are funded by Oil Palm Plantation Funds – BPDPKS, Indonesia

Nomenclature

A_w	: Wall effect correction term	
A_{sp}	: Surface area of sphere (equivalent volume as particle)	(m^2)
a_v	: Particle specific surface	(m^{-1})
B_w	: Wall effect correction term	
C	: Nitrogen mole fraction	
C_1	: Viscous term constant	
C_2	: Inertial term constant	
D	: Column diameter	
D_a	: Dispersion coefficient	(m^2/s)
d_p	: Particle effective diameter	(m)
d_{vs}	: Volume-mean diameter	(m)
j	: Chiton-Colburn j-factor	
K_1	: Coefficient of pressure drop correlation	
k_1	: Coefficient of wall effect correction term	
k_2	: Coefficient of wall effect correction term	
ΔP	: Pressure Drop	(Pa)
r_a	: Reaction rate for component a	(mol/m^3.s)
Re	: Reynold number	
S_p	: Particle surface area	(m^2)
t	: Time	(s)
u	: Fluid velocity	(m s^{-1})
u_s	: Superficial velocity	(m s^{-1})
V_p	: Particle volume	(m^3)
z	: Axial length	(m)
ε	: Void fraction	
μ	: Viscosity	(Pa.s)
ρ	: Density	(kg m^{-3})
ρ_b	: Bulk density	(kg m^{-3})
ϕs	: Particle sphericity	

References

1. A. Ardy, *Proceeding of National Chemical Engineering Seminar " Kejuangan"* (Yogyakarta, 2015)
2. A.S. Gupta, G. Thodos, A.I.Ch.E. Journal , **9**, 6, 751-754 (1963)

3. B. Eisfeld, K. Schnitzlein, Chem. Eng. Sci., **56**, 14, 4321-4329 (2001)
4. C. Palma, Appl. Energy, **111**, 129-141 (2013)
5. D. Swierczynski, C. Courson, A. Kiennemann, Chem. Eng. Process., **47**, 3, 508-513 (2008)
6. H.S. Fogler, *Elements of Chemical Reaction Engineering 4th Edition* (Prentice Hall, USA, 2006)
7. H. Wadell, J. Geol., **43**, 250-280 (1935)
8. J.M. Engasser, C. Horvath, Ind. Eng. Chem, Fundamen., **14**, 2, 107-110 (1975)
9. L.J. Petrovic, G. Thodos , Ind. Eng. Chem. Res., **7**, 2, 274-280 (1968)
10. L. Li, W. Ma, Transport Porous Med, **89**, 35-48 (2011)
11. M. Agustina, *Perbandingan uji aktivitas katalis dolomit dan batu kapur lokal dengan katalis nikel komersial untuk dekomposisi toluen sebagai model konversi tar* , Master's Thesis (Institut Teknologi Bandung, Bandung, 2011)
12. M.J. Keyser, M. Coradie, M. Coertzen, J.C. Van Dyk, Fuel, **85**, 1439-1445 (2006)
13. P.M. Ruya, J. Waluyo, Yuono, H. Susanto, Journal of Engineering and Technological Science (in review, pending publication)
14. R.B. Bird, W.E. Stewart, E.N. Lightfoot, *Transport Phenomena 2nd edition* (John Wiley & Sons, Inc., USA, 2002)
15. M. Purwasasmita, *Analogi peristiwa perpindahan massa, panas dan gerak di dalam alat-alat perpindahan dan reaktor heterogen* (Institut Teknologi Bandung, Bandung, 1985) (Unpublished)

Calcium soap from palm fatty acid distillate (PFAD) for ruminant feed: quality of calcium source

Lienda A. Handojo[1,*], *Antonius* Indarto[1], *Dian* Shofinita[1], *Anggina* Meitha[1], *Rakhmawati* Nabila[1], and *Harry* Triharyogi[1]

[1] Department of Chemical Engineering, Faculty of Industrial Technology, Institut Teknologi Bandung, Bandung, Indonesia

Abstract. Calcium soap is potentially used as fat supplements for ruminants since it contains high concentration of fat and calcium that are useful for ruminants. The consumption of calcium soap may increase the yield and the fat content of milk, as well as increase the ruminant's fertility. Calcium soap can be produced from palm fatty acid distillate (PFAD), which is a by-product of crude palm oil (CPO) refining process, and calcium oxide (CaO). In this study, the effect of CaO quality on the acid value of the product has been observed. It was found that the reaction with lower concentration of active calcium of CaO resulted in products with a higher acid value, which indicates a lower reaction conversion. Thus, the produced calcium soap requires further treatment in order to remove the unreacted calcium and free fatty acid. Washing with hexane followed by either vacuum or convection drying has been found to be able to reduce the acid value of the product significantly.

1 Introduction

Calcium soap can be used as supplement for ruminants since it contains high concentrations of fat and calcium that are useful for ruminants. The calcium soap gives benefits for ruminants, particularly regarding the milk productivity and fertility increase. Calcium soap is produced from the reaction between fatty acid and calcium source.

Previous study reported that it was preferable to use the naturally source of fatty acids as raw material to produce calcium soap, such as beef or mutton tallow, palm oil, or lard [1]. Palm fatty acid distillate (PFAD) can be potentially used as raw material for the production of calcium soap since the fat content in PFAD can reach 81.7% [2]. Usage of PFAD with 87.1% free fatty acid content for calcium soap production has been reported [3]. Furthermore, PFAD is a by-product from crude palm oil (CPO) refining process, therefore the utilization of PFAD may add cost benefit for the industry. The product of CPO refining process usually consists of 5% PFAD and 95% refined, bleached, and deodorized (RBD) palm oil [4]. Considering that Indonesia is the largest producer of palm oil with a contribution of approximately 54% to the world total palm oil production in 2016 [5], the study regarding the utilization of PFAD will be interesting.

There are several calcium sources that can be used as raw materials for the calcium soap production. A previous study compared the possibility of using $Ca(OH)_2$ and CaO as calcium sources, and concluded that the use of CaO resulted in a higher reaction conversion [6]. However, the quality of CaO produced from several sources in Indonesia may vary. The production of CaO depends on the effectivity of the $CaCO_3$ combustion, on the condition during transport and storage since CaO is hygroscopic. This study discusses the effect of CaO quality on the acid value of the product, which reflects the reaction conversion. In addition, possible additional treatments, such as washing and drying, to decrease acid value of the product after reaction have also been explored.

2 Methods of research

2.1 Materials

Materials used in this study were CaO (Stone Garden and Padalarang Company) and PFAD (PT Tunas Baru Lampung Tbk).

2.2 Methodology

2.2.1 Actual CaO determination

Calcium oxide used was obtained from the ignition of quick lime as calcium carbonate. The actual amount of CaO should be determined by dissolving it in sucrose solution from ASTM rapid sugar test. The solution of calcium oxide was boiled and cooled at room temperature, then it was added 50% sucrose solution and

* Corresponding author: lienda@che.itb.ac.id

shaken for 30 minutes. Water was added until 250 mL mark in erlenmeyer was reached and filtered through whattman paper no. 1, discharging the first 15-30 mL. Then sulfuric acid titrated the solution. The available lime was calculated as 4 times the volume of 0.357 N sulfuric acid [7].

2.2.2 Saponification reaction

The saponification reaction was carried out using modified reaction method followed by washing and drying process as shown in Fig. 1.

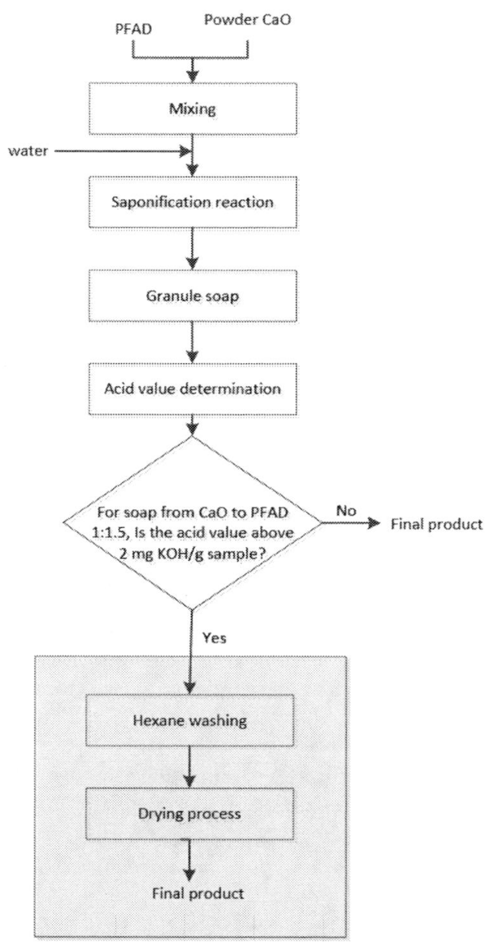

Fig. 1. Block flow diagram of the experiment methodology

The method was chosen since it needs less time for the reaction [8]. In this study, the stoichiometric mole ratio of CaO to PFAD was varied from 1 to 2, the initial temperature of PFAD in the process was 60°C, and percentage of mass of water was about 20% of PFAD. The mixture of molten PFAD and CaO was heated and stirred homogenously. Saponification reaction took place immediately after a certain amount of 60°C hot water was added and stirred for less than 10 minutes, during which the granule soap formed. The soap was relatively dry since the water evaporated as the mixture temperature raised due to the exothermic process of saponification.

2.2.3 Washing and drying process

Since the ruminants requires limited fatty acid content in their feed, the washing process was conducted to remove residual fatty acid present in calcium soap. The washing process was carried out by mixing nonpolar solvent, hexane, and product with ratio 3:1 and stirred it at 60°C for 30 minutes. The calcium soap filtrate obtained was dried in vacuum oven at 40°C, 0.4 atm and another was dried under convection dryer at 50°C for 3-4 hours.

2.2.4 Acid value determination

Acid value was determined by ISO 660: 1990. One gram of calcium soap sample was taken and dissolved in ethanol, the mixture was then heated for 10 minutes. The solution was titrated by 0.01 N potassium hydroxide using phenolphthalein indicator [9].

3 Results and analysis

3.1 The effect of CaO quality on acid value of the products

Two different sources of CaO were used in this experiment. Based on the CaO analysis the result showed that the CaO content was about 40% (CaO 1) and 75% (CaO 2). The CaO production process has a significant influence on the quality of CaO produced. Some factors affecting the quality of CaO are the natural limestone, the calcination process, the method of storage, and the method of transport [10]. The quality of CaO which may vary due to source of production can be determined with titration method using sulfuric acid after CaO is dissolved in a solution of sucrose to calculate the amount of calcium in the form of oxide.

The quality of CaO used as the main raw material is essential and is directly proportional to the quality of the products. Figure 2 shows that the CaO content used had a very significant effect on product quality. The CaO quality in this research may be affected by the storage method that could convert the CaO to $Ca(OH)_2$ due to moisture in the environment. The use of $Ca(OH)_2$ for making calcium soap had lower quality than the product using CaO as calcium source [11]. Another strong factor was likely the presence of calcium carbonate in quicklime used regarding the incomplete combustion process.

A lower acid number was preferred and indicated a higher reaction conversion. In the stoichiometric mole ratio of CaO to PFAD of 1, the CaO 1 (40%) acid value was much higher than that of CaO 2 (75%) which are 31.51 and 10 mg KOH/g sample and are proportional to 14.35% and 4.5% fatty acids as palmitate, respectively.

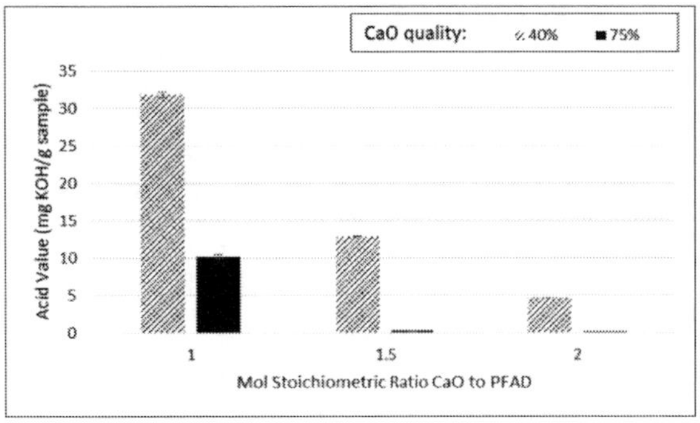

Fig. 2. The acid value of products vs. mol stoichiometric ratio at different CaO quality.

The higher the mole ratio of stoichiometry, the acid value will decrease. In the three variations of the CaO mole ratio, all the acid values of the product of CaO 1 are still above 4 indicating the product was still below the commercial product standard. As for the product of CaO 2, starting from the ratio of CaO to PFAD 1.5, the acid product number has been very low reaching 0.23 mg KOH/g sample. This value was already below to the free fatty acid content claimed by the commercial product [11].

3.2 The effect of washing and drying process on Acid value of the products

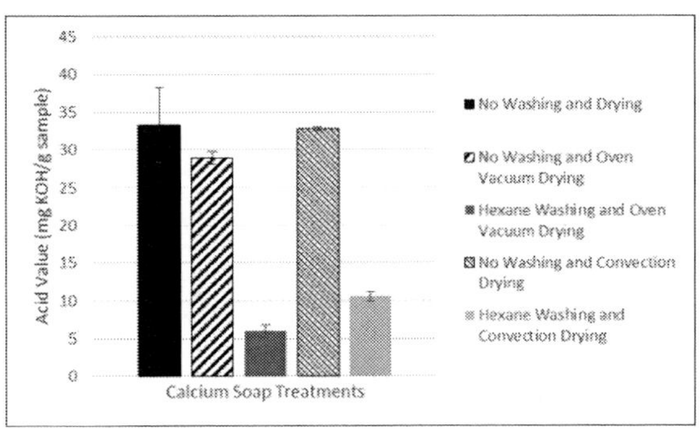

Fig. 3. The acid value of the products after washing and drying process.

The poor quality of CaO leads to a product which requires further processing to reach the desired quality. Figure 3 provides the effect of washing and drying process on the acid value of calcium soap. The soap was obtained from the reaction with a stoichiometric mole ratio of CaO to PFAD 1.5. The hexane wash treatment could reduce the acid value up to 68-82%. The drying process was then necessary to remove hexane in the final product. Hexane was used to extract fatty acids or unsaponified material. Based on Akers and Peters (1984) the organic solvent used to extract the unsaponifiable matter is hydrocarbon with 4-14 C atoms in molecular compounds like hexane [12].

The high decrease in acid value was obtained from hexane washing followed by drying using a vacuum oven, which have an end product acid number of 6 mg KOH/g, compared to hexane leaching and convection drying with an acid number of 10.6 mg KOH / g sample. The convection drying could not remove hexane as good as vacuum oven drying. As a result, hexane left in the product which still carry some unreacted free fatty acid, increases the acid value. The drying process was carried out to remove the hexane from the final product. However, the final acid value was still higher than in the commercial products. Besides, the additional processes required more energy and additional materials, therefore it was relatively uneconomical.

4 Conclusions

CaO utilized as a raw material needs to be analyzed before reacted with fatty acids to produce calcium soap. Higher CaO content yields better calcium soap product which has lower acid value. The washing process followed by drying has an effect on the quality improvement of calcium soap products. The high quality CaO still has a greater effect on the product, compared to the additional treatment applied to the product resulted from low quality CaO.

The authors would like to extend their sincere gratitude to BPDPKS Indonesia for providing the research grant and PT Tunas Baru Lampung Tbk for providing PFAD for this research.

References

1. McAskie, W., 1989. Ruminant Feedstuff, their production and apparatus for use therein. US Patent 485 323 33.
2. A.G.M. Top, Production and Utilization of Palm Fatty Acid Distillate, DOI10.1002/lite.200900070, (2010)
3. L. Handojo, A. Indarto, D. Shofinita, A. Meitha, R. Nabila, H. Triharyogi, Calcium Soap from Palm Fatty Acid Distillate for Ruminant Feed: Reaction Method. JEAS (2017)
4. Zero and Rainforest Foundation Norway, *Palm Fatty Acid Distillate (PFAD) in Biofuels,* (2016)
5. US Department of Agriculture, *Oilseeds: world markets and trade. Foreign Agricultural Service, United State,* (2016)
6. L. Handojo, A. Indarto, D. Shofinita, A. Meitha, R. Nabila, H. Triharyogi, J. Medwell ICABSE (2017)
7. W.E. Stone, F.C. Scheuch, A Method for Determining Calcium Oxide in Quicklime, JACS XVI **11** (1894)
8. L.F. Scott, M. Monrovia, H.D. Strachan, E. Monte, C.M. McCloskey, A. Altadena, C. Calif, Manufacture of Metallic Soap, US Patent 3 803 188 (1974)
9. FSSAI, *Manual of Methods of Analysis of Foods Oils and Fats* (2012)
10. M. Hassibi, *Factors affecting the Quality of Quicklime (CaO) from Mining to Manufacturing Usage.* Chemco System L.P. (2015)

11. L. Handojo, A. Indarto, D. Shofinita, A. Meitha, R. Nabila, H. Triharyogi, J. IJCAET **287** (2017)
12. J.B. Akers, D.C. Peters, , EP 0 015 116 B1 (1984)

Syngas Compositions And Kinetics Of South Kalimantan Lignite Coal Char Gasification With Steam

Yusuf Rumbino[1,2*], *Suryo* Purwono[2], *Muslikhin* Hidayat[2] and *Hary* Sulistyo[2]

[1]Department of Mining Engineering, Faculty of Science and Engineering, Nusa Cendana University , Indonesia
[2]Department of Chemical Engineering, Faculty of Engineering, Gadjah Mada University, Indonesia

Abstract. The aim of this research was to investigate the gasification of a South Kalimantan lignite coal char in the temperature range of 873-1073 °K and steam condition to evaluate the reactions rates and the product gas compositions. Prior to the gasification experiments the raw char was pyrolysed under nitrogen atmosphere and at a temperature of 673 °K. The gasification experiments were conducted in a fixed bed reactor, at atmospheric pressure, isothermal conditions, equipped with cooling system, gas reservoir, and temperature control. Char from coal pyrolysis weighed then gasified at variations of temperature. Gas sampling is done every 15 minutes intervals for 90 minutes. The reactivity study was conducted in the kinetically controlled by the heterogeneous reaction between solid carbon from the char and a gas phase reagent. Two theoretical models were tested to fit the experimental data and the kinetic parameters were determined. It was found that an increase in temperature enhances the reaction rate and also the formation of H_2, CO, CH_4, and CO_2. The results show that higher temperature contributes to more hydrogen production The gasification kinetics was suitably described by the Random Pore Model. Activation energy determination of char gasification reactions by using Arrhenius graph.

1 Introduction

1.1 Indonesian Coal

Coal is considered the cheapest and most widely abundant fossil fuel in the world (Kurt, 2008).Indonesia has the potential coal that potential enough.Total coal from Indonesia reserves is ± 28 Billion tons consisting of probable reserves ± 17 billion and proven reserves ±10 billion tons. In reserves, Indonesia has 3% of world reserves at 28 Billion of 948 Billion tons (Sihite, 2012). The world existing reserves are about 50% is low rank coal (LRC). In terms of world reserves, Indonesia has a 5% low rank coal reserves. For more practical and commercial means, LRC in this paper refers to Indonesia Government version Keppres No.13/2000:

- coal with high moisture content (>35%), and low calorific value (<5100 kcal/kg adb)
- coal rank lignit to sub bituminous C included

Total Indonesian LRCis 49.44% that spread in South Sumatera 38.01%, South Kalimantan 7.68%, East Kalimantan 3.75%. Coal gasifiction in Indonesia stipulated by Keppres No.5/2006. The inititative is to replace the use of natural gas by LRC. (Tirtosoekotjo,2006). Hydrogen is considered as the major energy carrier of the future, so an increase in the demand for hydrogen can be expected. Nowadays, there is increasing interest in lower cost fuels that can be used to produce mixtures of hydrogen and carbon monoxide by means of gasification.

1.2 Gasification Reaction

The most practical technologies for coal conversions and utilizations are carbonization, gasification and combustion (Speight, 2012). Coal gasification is process for converting carbon in coal by using a gasifying agent under restricted amount oxygen into fuel gas which consists hydrogen, carbon monoxide, carbon dioxide and methane. The coal gasification process is divided into two main steps ; pyrolysis and char gasification. Char gasification step is a controlling step for overall conversion process because its reaction rate is much slower than the pyrolysis step. As a result, kinetics of the char gasification step can provide a basic knowledge and crucial for a better understanding and a proper reactor design for the coal gasification proces (Molina, 1998, Bayarsaikhan, 2005).

Usually the following reactions are used in gasification (Highman, 2008) :

a. Boudard reaction :

$$C + CO_2 \leftrightarrow 2CO \quad + 172\ MJ/kmol \qquad (1)$$

b. Water gas reaction

$$C + H_2O \leftrightarrow CO + H_2 \quad + 131\ MJ/kmol \qquad (2)$$

*Corresponding author: yusufrumbino70@gmail.com

c. Methanation reaction

$$C + 2H_2 \leftrightarrow CH_4 \qquad -75\ MJ/kmol \tag{3}$$

d. Water gas shift reaction

$$CO + H_2O \leftrightarrow CO_2 + H_2 \ -41\ MJ/kmol \tag{4}$$

1.3 Chemical reaction kinetic models

Since coal properties differ widely, kinetic data of coal gasification shoud be specific. The design optimization and operation require characterizing the reactivity and kinetics of char gasification. Char-steam kinetics have been widely investigated . Koba (1980) found that rates of steam gasification with lignite and bituminous coal were almost 2-5 times higher than gasification use CO_2.

Considering that the gasification is a single step reaction , the H_2O partial pressure is constant and the reaction rate (*dX/dt*) are expressed by Equation 5 (Jüntgen, 1983; Fermoso *et al.*, 2009):

$$dX/dt = k(T)f(X) \tag{5}$$

where the temperature dependent term is the gasification rate coefficient, (*k*, 1/min), and *f(X)* is a structural term, related to the changes of coal char properties during the reaction. *k(T)* is expressed by means of the Arrhenius relation given by Equation 6,

$$k = A exp(-E_a/RT) \tag{6}$$

where *A* (1/min) is the frequency factor, *Ea* (kJ/mol) is the activation energy, *R* is the constant of gases (8.314 J/mol/K) and *T* (K) is the gasification temperature.

f(X) is provided by the *n*th-order model and in this work, two representative gas-solid models were applied in order to calculate the kinetic parameters.. These are known as single-step gas–solid kinetic models, which have received more attention in recent years.

The first model is the volumetric model (VM) or first order reactionThe Volumetric Model (V) is a simplified gas-char heterogeneous model, since it considers that the reaction is similar to a homogeneous reaction and that the surface area decreases with conversion (Wen, 1968; Ishida and Wen, 1971):

$$\frac{dX}{dt} = k_v(1 - X) \tag{7}$$

The second model is the random pore model (RPM), which was presented for the first time by Bhatia and Perlmutter (1980):

$$\frac{dX}{dt} = (1 - X)\sqrt{1 - \psi\, ln(1 - X)} \tag{8}$$

The Random Pore Model (RP) is the only one that contemplates structural changes in the coal char in the gasification process, by using the parameter Ψ (pore structure parameter) , it is expressed by:

$$\psi = \frac{4\pi L_0(1-\varepsilon_0)}{S_0^2} \tag{9}$$

Where S_o, L_o, ε_0 represent the initial pore surface area, pore length, and porosity

2 Methodology

2.1 Proximate and Ultimate Analysis

Coals samples from South Kalimantan were used in this study. Proximate analysis (ash, volatile matter and fixed carbon) was determined according to the standard ASTM E-1131. It is the simplest means of characterization of coal leading to its evaluation in respect of its quality, rank and type, and in turn broadly indicating its suitability for a particular mode of use. The analysis is done on air dried sample of coal. Ultimate analysis is expressed in terms of percentages of C, H, N, S and O. The elements C, H, N and S are being determined while O is obtained by difference. Ultimate analysis (C, H, N, S and O) and calorific values were determined according to the standards ASTM D-5373 and ASTM D-2015,

Table 1. Proximate analysis of lignite coal

Analytical parameter	Result (%) dry base
moisture (%)	37,743
ash (%)	2,071
volatile matter (%)	24, 656
fixed Carbon	35,530
calori (kal/g)	4303,203

Table 2. Ultimate analysis of char from lignite coal

Analytical parameter	Result (%) dry base
carbon (% adb)	65,70
hidrogen total (% adb)	4.12
nitrogen (%adb)	1,06
sulphur total (%adb)	0.27

2.2 Pyrolisis

The coal sample was grounded, sieved and devolatilised in a fixed bed reactor heated by an electric furnace . Conditions applied during pyrolysis heating rate of 20 °C/min up to the final temperature of 450 °C and residence time of 90 minutes. The resulted chars were cooled down under a flow of nitrogen to room temperature. Chars from low-temperature pyrolysis of lignite coal below 450 °C were more reactive than higher temperature chars (Meng, 2014).

2.3 Surface area measurement and SEM

Porous structure analysis was carried out in liquid N_2 and at 78 K, and surface areas were calculated using the BET method. BET theory introduces the concept of multimolecular layer adsorption of adsorbate on adsorbent. The fundamental assumption of this theory is that the forces active in the condensation of gases also are responsible for the binding energy in multimolecular adsorption. Convention has established that the quantity of gas adsorbed is expressed at S.T.P and the pressure is expressed as a relative pressure (P/P0) which is the actual gas pressure P (mmHg) divided by the saturation

https://doi.org/10.1051/matecconf/201815602008

pressure P0 (mmHg) of the adsorbing gas at the analysis temperature.

Table 3. BET analysis of char from lignite coal

Parameter	Result		Methode
Surface area	5,610	m^2/g	SAA
Pore volume	19,175x10^{-3}	Cc/g	SAA
Pore radius	68,358	Angstrom	SAA

The changes in char morphology was investigated using scanning electron microscopy (SEM). the image shows lignite char coal pyrolysis results at a heating rate of 20 °C / min until it reaches temperature of 450°C and held for 90 minutes. The pore holes appear on the char

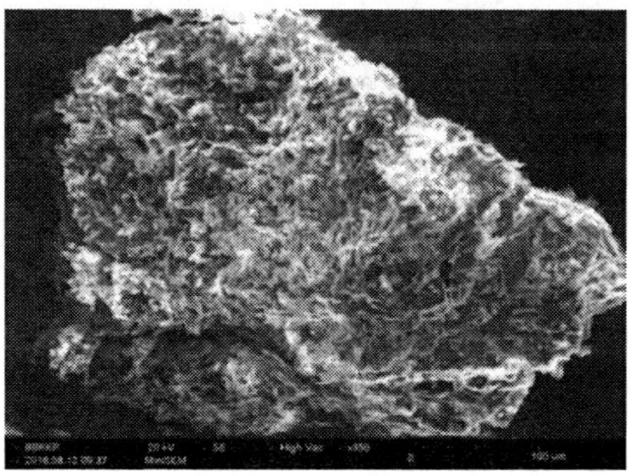

Fig.1. Micro photograph with 350 magnification showing the pores on the char

2.4 Gasification

The charcoal to be gasified is weighed as much as 10 grams each time the gasification is carried out using steam carried out on the fixed bed reactor. a series of gasification tools such as the sketch in Fig. 2.

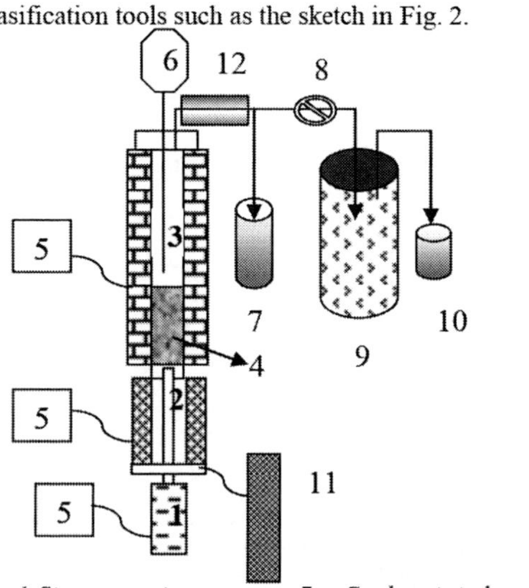

1. Steam generator
2. Super heated steamer
3. Gasification reactor
4. Char coal
5. Power Reguletor
6. Thermocoupel
7. Condensate tank
8. Sampling gas
9. Gas reservoir +water
10. Water reservoir
11. Nitrogen
12. Conndenser

Fig, 2. The schematic of the experimental apparatus

Before gasification begins, it is done flushing using a nitrogen stream. The pressure used in gasification is 1 atmosphere. The steam was discharged in the reactor after the gasification temperature was reached and the gas loading resulting from the reactor was carried out after 15 minutes of 5 ml and fed into the tube which then analyzed the gas composition offline using gas chromatography equipment (Figure 3).

The data generated during the gasification takes place is the amount of gas volume that is collected from the gasification result measured by the addition every 15 minutesA gas analyzer, which can determine the instantaneous contents of CO, CO_2, CH_4, and H_2.

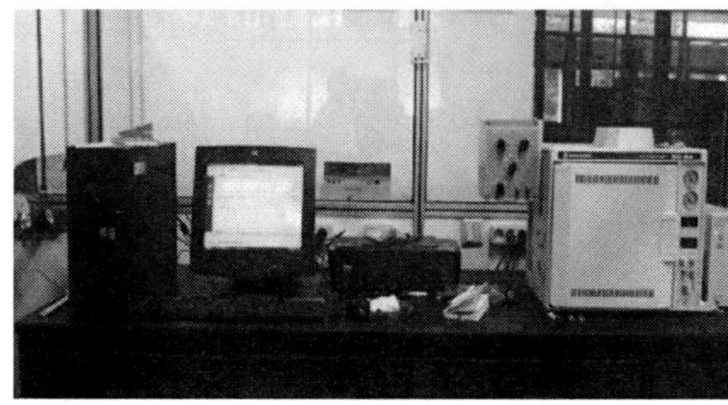

Fig. 3. Gas Chromatography

The result data from the GC tool is the calculation of peak area which is then converted to concentration. The concentration is then used to calculate the number of moles of each gas in each gas volume

3 Result and discussion

3.1. Volume and composition syngas

The gasification reaction is an endothermic reaction, so an increase in temperature will be accompanied by gas formation, it is apparent in an increase in the amount of gas formed each time shown in Figure 4. The resulting gas volume is strongly influenced by the operating temperature of the gasification, which at a temperature of 800 °C produces volume more gas at the same time. At a 90 minute observation resulted in a volume of 15,900 ml of gas at a temperature of 800 °C, 12,150 ml at a temperature of 700 °C and 9,150 ml at a temperature of 600 °C

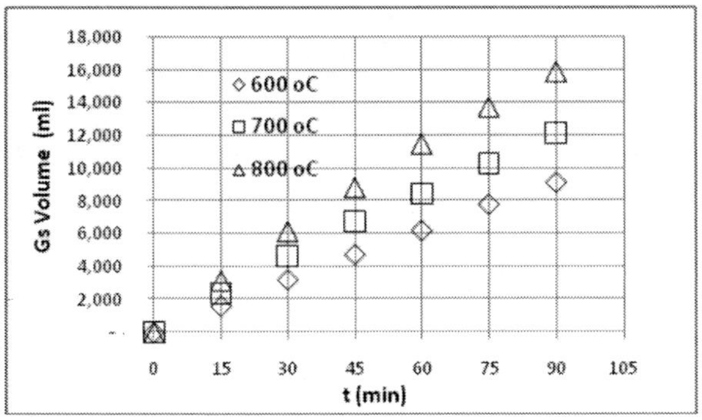

Fig. 4. Gas release rates during the steam gasification of char under 600 °C,700 °C,800 °C,1 atm pressure.

The gas formation is due to the reaction between steam and charcoal, it appears in the reduction of the amount of charcoal weighed at each end of the gasification. The charcoal used in gasification is shown in Table 4.

Table 4. The amount of char used in the gasification

Temperature(°C)	Char (gram)		
	Before	after	used
600	10	7.2936	2.7064
700	10	6.3592	3.6408
800	10	5.3452	4.6548

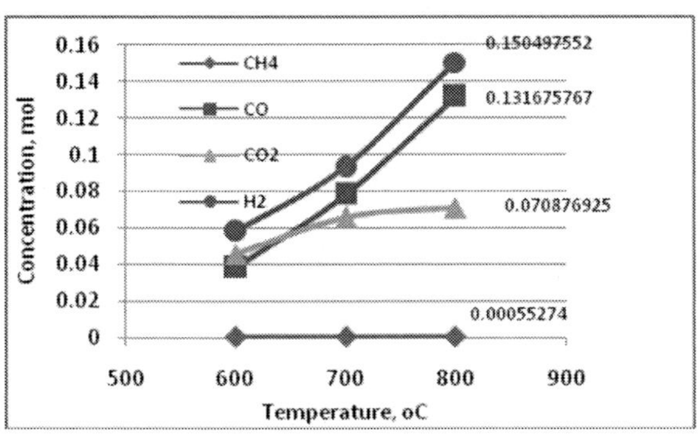

Fig. 5. Gas concentration under different temperature

The Fig. 5 show the tendency of increasing gasification temperature will result in higher H_2 gas that is at temperature of 800°C with concentration reach 15,0497%. At a temperature of 600 °C indicates the amount of CO_2 gas is higher than CO gas but at a temperature of 700 °C and 800 °C CO_2 concentration begins to decrease. The amount of mole concentration of methane gas formed in very small amounts under all conditions of temperature.

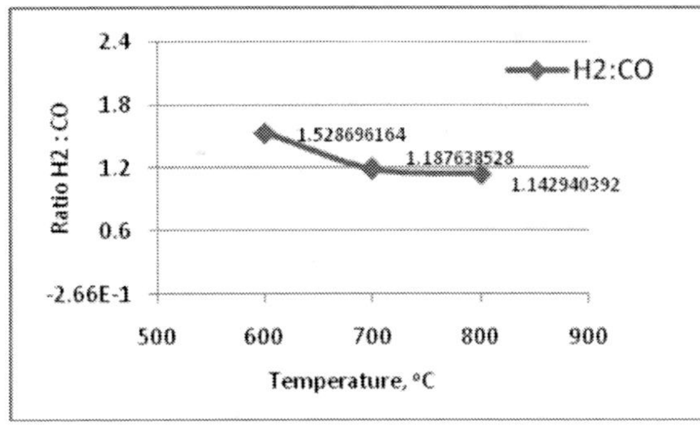

Fig. 6. Ratio H_2/CO under different temperature

At temperature of 600°C producing a ratio of H_2/CO greater than the temperature of 800°C, but the volume of gas produced is lower. Synthesis gas with a stoichiometric ratio (H_2/CO) at 2 or ranging from 1 to 2 is generally used in major synthesis-gas-based chemicals production.

3.2 Char conversion

Carbon conversion was calculated based on moles of CO, CO_2, and CH_4 by using Eq. (10) as follows:

$$X_{char} = \frac{n_{CO}+n_{CO_2}+n_{CH_4}}{n_{C_{char}}} \; x\ 100\% \tag{10}$$

The carbon conversion profiles for the gasification experiments of the coal char are shown in Figure 4. From Fig.7it can be seen that the carbon conversion will increase by time significantly. The final carbon conversion at gasification temperature of 600 °C, 700 °C, and 800°C are 15.42%, 26.47%, and 37.10% respectively. This indicates that temperature is the factors that can increase conversion reaction. In higher temperature, the kinetic energy of molecules will also be greater so that collisions among molecules become more frequent and this causes the reaction occurs faster.

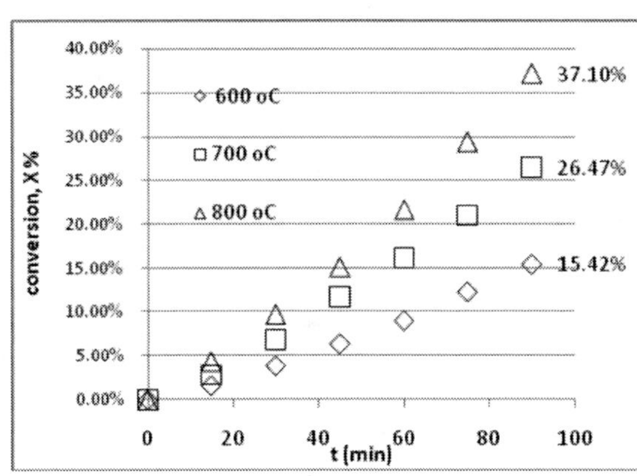

Fig. 7. Carbon conversion for 90 minutes

3.3 Validation of the RPM and Volumetric

The kinetic measurements from this research were tested. The conversion time data for each run were fitted against the Random Pore Model (RPM) and Volumetric Model. The curve fitting method is used in which the value of Sum of Square of Error (SSE) is minimized.

$$SSE = \sum_{i=1}^{m}\left(\frac{X_{data} - X_{calculate}}{X_{data}}\right)^2 \quad (11)$$

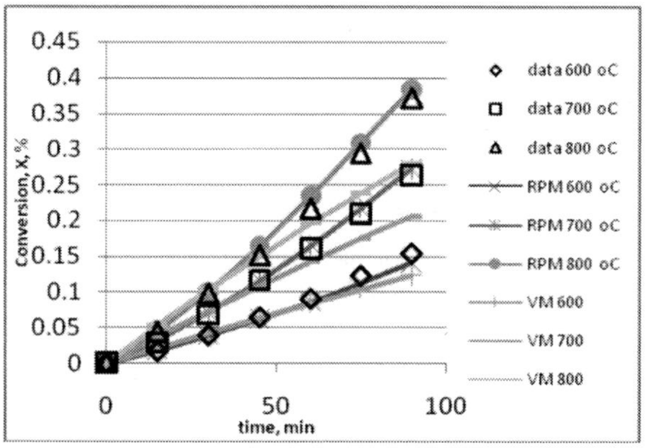

Fig. 8. Comparison betwen measured conversion carbon data with RPM and VM

On the graph shows the value of the data more appropriately approximated by RPM.Calculations using RPM have smaller SSE than using VM. Therefore, the next kinetic parameter calculation uses the RPM approach.

3.4 Calculation of the Kinetic Parameters by Random Pore Model

The Arrhenius plot is a relationship between the apparent rate and its reaction temperature (1/T), and therefore , it can be used to calculate the apparent activation energy (E) as illustrated in Figure 9. From this Arrhenius's plot, the activation energies is observed (Table 5),

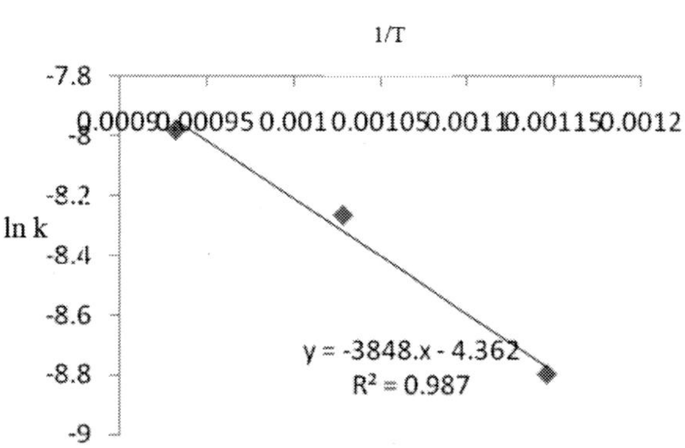

Fig.9. Arrhenius Graph Constant Model RPM

Table 5. Arrhenius constant and parameters on the RPM approach

T	k_{RPM}	SSE	A	E/R	E, kJ/mol
600	0.0151	3.249942%	0.01275	3848	31.9923
700	0.02576	2.77679%			
800	0.03424	2.17269%			

The activation energies is similar with Turkish coal (30,8 kJ/mol) and Indian coal (32 kJ/mol) that calculated by Jayaraman (2013) . Coals from India and Turk have been investigated in the TGA apparatus at a temperature range of 25-1250°C with heating rates of 35 and 45 K/min.

4 Conclusion

The properties of lignite from South Borneo are suitable for applications such as for gasification utility. The char was gasified under steam temperature 600°C , 700°C and 800°C show that the temperature dependence of the reaction .

This research has been financially supported by Ministry ofResearch, Technology, and Higher Education, Republic of Indonesia for funding through Nusa Cendana University (Grant No. 213/UN.15.19/LT/2017)

References

1. Bayarsaikhan, B., Hayashi, J., Fuel, **88**, (2005)
2. Bhatia SK, Perlmutter DD. A random pore model for fluid–solid reactions: I.Isothermal and kinetic control. AIChE J (1980)
3. Fermoso, J., Arias, B., Pevida, C., Plaza, M. G., Rubiera, F., Pis, J.J., J. Thermal Analysis & Calorimetry, **91**, (1998)
4. Highman C., Gasification (2nded.) Elsevier, 2008
5. Jayaraman K., Bonifaci E., High Ash Indian and Turkish Coal Pyrolysis and Gasification Studies in Various Ambiences, Cesme, Izmir, Turkey, September (2013)
6. Jüntgen, H. Fuel, **62**, (1983)
7. Kurt, K.. Coal-to-Gas: The Answer for a Low-Emissions Future? *Nature Reports: Climate Change,* (2008)
8. Molina, Mondrago, F., Fuel, **77**, (1998)
9. Sihite T (Director General Mineral and Coal-Indonesia), Low Rank Coal Utilization in Indonesia, Clean Coal Day in Japan, International Symposium, (2012)
10. Meng F., Energy Fuels, **28**, pp, no. 1, 275–284 (2014)
11. Speight, J. G., *The Chemistry and Technology of Coal* (CRC Press, 2012)
12. Tirtosoekotjo S. (Excecutive Director Indonesian Coal Mining Association-ICMA), Low Rank Coal and Its Contribution to The Energy Development in Indonesia, Presentation at Asia Pacific Lower Rank Coal (2006)

Application of Computational Fluid Dynamics for Modeling of Hydrodynamics and Mass Transfer of Laboratory Scale Crude Palm Oil Degumming Process

Yuswan Muharam*[1], *Aditya* Kurniawan[1], and Andrey S. Wirya[1]

[1]Department of Chemical Engineering, Faculty of Engineering, University of Indonesia, UI Campus,Depok 16424, Indonesia

Abstract. In this research, computational fluid dynamics was applied to model a laboratory reactor for the degumming of crude palm oil (CPO) involving immiscible liquid-liquid mixing and being controlled by mass transfer. The fluid mixing of CPO and phosphoric acid in the reactor was simulated using multiphase mixture model and standard k-ε turbulence model in steady state mode. The simulation predicts the distributions of the drop diameter, the dispersed-phase relative velocity, the drop Reynolds number and the interfacial area density. The mass transfer coefficient from experimental work is correlated using the model as Sherwood number $Sh_d = 0.02576\ Re_d^{0.673} Sc^{0.431}$ with R^2 being 0.91. Finally, the volumetric mass transfer coefficient was calculated and compared to the experiment result.

1 Introduction

Degumming is an important step in crude palm oil (CPO) refining. It reduces or removes gum content from CPO. Gum consists of phospholipid compounds that acts as an emulsifier. The existence of gum disrupts product separation [1].

One of degumming methods, called acid degumming, uses phosphoric acid solution (H_3PO_4) as a reagent [2]. Non-hydratable phospholipids in CPO reacts with the solution and changes into hydratable phospholipids which is easy to remove [3].

The mechanism of CPO degumming involves the phospholipid transfer from the CPO bulk to the H_3PO_4 solution through the interface, followed by the chemical reaction of phospholipids and H_3PO_4. The overall rate of degumming depends on the mass transfer coefficient and the reaction rate coefficient. The volumetric mass transfer coefficient is a function of the total mass transfer coefficient and the interface area which are usually promoted by agitation in a reactor. Therefore, the mixing effectiveness is highly influential to the mass transfer rate.

Several methods have been proposed to correlate mixing dynamics in a reactor to mass transfer rates. Some correlations are merely based on dimensional analysis and thus fully empirical, while others are based on turbulence and relative velocity theories. The latter is built upon the effect of forced convection around particles on mass transfer rate. Other authors derived expressions for relative velocities by solving motion equations of suspended particles in flow fields [4]. This approach uses local parameters (relative velocities between phases) to describe flow dynamics-dependentmass transfers. The approach evolves the differences in flow fields in different geometries of reactors.

Relative velocities can be calculated using semi-analytical methods with some assumptions. Therefore, it is limited by flow geometries of mixing process. Otherwise, it can be determined by solving multiphase flow equations using numerical techniques, which is called computational fluid dynamics (CFD).

CFD uses differential fluid flow governing equations and is solved numerically. It predicts flow parameters of a system, regardless of its geometry. Due to its flexibility, CFD has been widely applied for modeling and designing many kinds of reactors and processes. For liquid-liquid systems, CFD is used by applying multiphase Eulerian models and population balances to predict liquid-liquid mixing and mass transfer in a reactor [5, 6]. A simple mixture model has been used as substitute for more complicated Eulerian model to design a submerged membrane stirred reactor [7].

In multiphase mixture models, continuity equations are solved for each phase. On the other hand, momentum equations are solved for the mixture of continuous and dispersed phases in case of two-phase flows. The model is valid where the local equilibrium between the phases is assumed. When dispersed phase clustering and large slip velocity are possible, the model would be invalid [8]. The two conditions usually arise in mixing with large dispersed phase volume fraction (> 20%).

* Corresponding author: muharam@che.ui.ac.id

In the present study, a mixture model is used to simulate crude palm oil (CPO) degumming process in a laboratory reactor. The simulation result, i.e. the drop diameter distribution, was compared to empirical data to obtain the interfacial area. Mass transfer coefficient was then calculated and correlated to mixing flow parameters. The model was also solved to predict the flow distribution and pattern so that the effect of different internal geometries of the reactor can be quantified.

2 Laboratory reactors

The laboratory data used in the study is obtained from Ristianingsih et al. [9]. The volume of the reactant mixture was 100 mL consisting of CPO as continuous phase and phosphoric acid solution (H_3PO_4 85%) as dispersed phase. The volume fraction of dispersed phase was 1.5%. The reaction took place in a 1000 mL triple-neck round vessel. A magnetic stirrer bar was used to promote agitation. The temperature was kept constant during the reaction. The reaction was run for 120 minutes with initial content of gum in CPO being 1.43%. The parameters deduced from the measurement were the volumetric mass transfer coefficient (k_ca), the reaction rate coefficient (k) and the equilibrium constant (K). The set of experiment data is shown in Table 1, where k_ca becoming the concern.

Table 1. The laboratory experiment data for CPO degumming process.

No.	*rpm*	T(°C)	Re	k_ca (min^{-1})
1.	200	60	148.32	0.00012
2.	300	60	222.48	0.00016
3.	400	60	296.64	0.00022
4.	500	60	370.80	0.00028
5.	350	50	198.41	0.00008
6.	350	60	259.56	0.00012
7.	350	70	339.52	0.00017
8.	350	80	444.05	0.00027
9.	200	50	113.38	0.00006
10.	500	80	634.36	0.00029

The properties of each component used in the simulation is summarized in Table 2. For this study, all properties were not measured. They were calculated using empirical equation available in literatures.

Table 2. Fluid properties in the CPO degumming process.

T (°C)	ρ(kg/m^3)		μ (kg/m s)		D_{AB} (m^2/s)	σ (N/m)
	H_3PO_4	CPO	H_3PO_4	CPO		
50	1,664.0	891.0	0.017	0.023	8.958E-13	0.01738
60	1,656.1	883.1	0.014	0.018	1.114E-12	0.01739
70	1,648.2	875.2	0.011	0.013	1.385E-12	0.01740
80	1,640.3	867.4	0.009	0.010	1.721E-12	0.01742

3 Modeling

3.1. Geometrical modeling

The modeling requires the drawing of the system geometry. The fluid domain of the laboratory reactor was drawn based on its internal configuration. The reactor internal contains only a magnetic stirrer bar submerged in the reactant fluid, as depicted in Figure 1. The fluid is divided into two domains: static and rotating. The rotating domain contains the movable reactor internal.

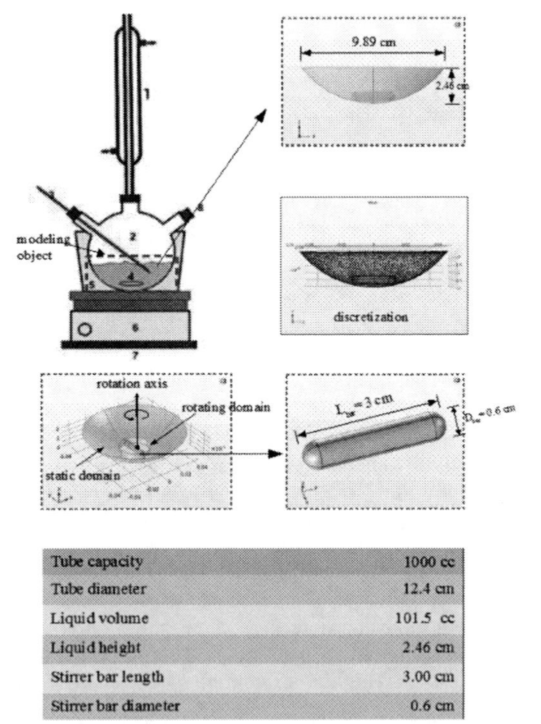

Fig.1. Geometrical modeling of the laboratory reactor.

3.2 Mathematical modeling

The mixing inside the reactor is predicted by solving the differential equations: continuity, momentum balance, mixture model, slip velocity and turbulence k-ε. In order to model the rotating fluid inside the reactor, the multiple reference frames (MRF) equation is applied.

In the mixture model, the continuity and momentum equations are in the form as follow [10]:

$$(\rho_c - \rho_d)\{\nabla \cdot [\phi_d(1-c_d)\mathbf{u}_{slip} - D_{md}\nabla\phi_d]\} + \rho_c \nabla \cdot \mathbf{u} = 0 \quad (1)$$

$$\rho(\mathbf{u}\cdot\nabla)\mathbf{u} = \nabla\cdot\left[-p\mathbf{I} + (\mu+\mu_T)\left(\nabla\mathbf{u} + (\nabla\mathbf{u})^T - \frac{2}{3}(\nabla\cdot\mathbf{u})\mathbf{I}\right) - \frac{2}{3}\rho k\mathbf{I}\right]$$
$$-\nabla\cdot\left[\rho c_d(1-c_d)\left(\mathbf{u}_{slip} - \frac{D_{md}}{1-c_d}\frac{\nabla\phi_d}{\phi_d}\right)\left(\mathbf{u}_{slip} - \frac{D_{md}}{1-c_d}\frac{\nabla\phi_d}{\phi_d}\right)^T\right] \quad (2)$$
$$+\rho\mathbf{g} + \mathbf{F}$$

The velocities of each phase are calculated by using the slip velocity equation. In this case, the Hadamard-Rybczynzki model is used:

$$\frac{3}{4}\frac{C_d}{d_d}\rho_c|\mathbf{u}_{slip}|\mathbf{u}_{slip} = \frac{(\rho_d - \rho_m)}{\rho_m}\left(\mathbf{g} + \frac{\mathbf{F}}{\rho_m} - (\mathbf{u}\cdot\nabla)\mathbf{u}\right) \quad (3)$$

The drag coefficient is calculated as follow:

$$C_d = \frac{24}{Re_t}\left(\frac{1 + \frac{2}{3}\frac{\mu_c}{\mu_d}}{1 + \frac{\mu_c}{\mu_d}}\right) \quad (4)$$

The turbulence model is used to predict the fluid flow instability generated by the stirrer bar rotation. It uses the standard k-ε model. The turbulence is manifested as additional stress in the momentum equation. This stress is calculated by solving two transport equations, i.e. the turbulence kinetic energy and the turbulent dissipation rate:

$$\rho\frac{\partial k}{\partial t} + \rho\mathbf{u}\cdot\nabla k = \nabla\cdot\left[\left(\mu + \frac{\mu_T}{\sigma_k}\right)\nabla k\right] + P_k - \rho\varepsilon \quad (5)$$

$$\rho\frac{\partial \varepsilon}{\partial t} + \rho\mathbf{u}\cdot\nabla\varepsilon = \nabla\cdot\left[\left(\mu + \frac{\mu_T}{\sigma_\varepsilon}\right)\nabla\varepsilon\right] + C_{\varepsilon 1}\frac{\varepsilon}{k}P_k - C_{\varepsilon 2}\rho\frac{\varepsilon^2}{k} \quad (6)$$

The rotating motion inside the reactor is described using MRF equation by introducing additional forces to the momentum equation, i.e. coriolis and centrifugal forces.

In order to evaluate the distributions of the dispersed-phase drop diameter and the interface area, the transport equations for the dispersed-phase volume fraction and the drop number density need to be solved. The transport of the two quantities is determined by the velocity of the dispersed phase obtained from Equation (1) to Equation (6). After several algebraic modifications, the two transport equations become

$$\nabla\cdot\left\{\phi_d\left(\mathbf{u} + (1-c_d)\mathbf{u}_{slip} - D_{md}\frac{\nabla\phi_d}{\phi_d}\right)\right\} = 0 \quad (7)$$

$$\nabla\cdot\left\{n_d\left(\mathbf{u} + (1-c_d)\mathbf{u}_{slip} - D_{md}\frac{\nabla\phi_d}{\phi_d}\right)\right\} = 0 \quad (8)$$

The diameter and the interface area then can be calculated using following equations:

$$d_d = \left(\frac{6}{\pi}\frac{\phi_d}{n_d}\right)^{\frac{1}{3}} \quad (9)$$

$$a = \frac{(4\pi n_d)^{\frac{1}{3}}(3\phi_d)^{\frac{2}{3}}}{1-\phi_d} \quad (10)$$

The two equations take the assumption that the number of the dispersed phase drops is constant throughout the mixing process.

The drop diameter calculated using Equation (9) was compared to that calculated using an empirical equation derived from direct measurement where the data was gathered from published work [11]. The method of developing measurement data-basedempirical equations has been presented by Rueger and Calabrese [12]. The data is better to be classified based on flow regimes, because the dispersion mechanism is different. The effect of dispersed-phase volume fraction has to be also considered.

Based on the plotting of impeller power numbers versus impeller Reynolds numbers, the mixing flow is discovered within laminar regime to intermediate regime. The forces acting on the dispersed phase drops are predominantly in the form of shear stress. Therefore, the empirical equation to estimate the average diameter (d_{32}) of the drops is

$$d_{32}[\mathrm{m}] = 0.17179\frac{C_a(\lambda)\sigma}{\left[\frac{P}{V}\left(\frac{V}{V_{imp}N_p}\right)^{0.42}\right]^{0.55}} \quad (11)$$

C_a is capillary number as the function of viscosity ratio (λ), which can be derived from the plot published by Grace [13].

Equation (11) is applied for a dilute mixture with dispersed-phase volume fraction being $\leq 0.1\%$. To adjust the estimate of average drop diameter due to the effect of volume fraction, the correction factor has been introduced by Rueger and Calabrese [14].

3.3 CFD simulation

The model wassolved using COMSOL Multiphysics. The physics used was Rotating-Machinery Mixture-Model Turbulent Flow. Simulation run in stationary mode (Frozen Rotor). This mode was used to simulate the mixing at steady state. Simulation was conducted for all operating condition mentioned in Table 1.

4 Results and discussion

Several variables solved by simulation are velocity of mixture, pressure, slip velocity, dispersed phase volume fraction, and number density. From this solution, other variables can also be derived, include drop diameter, interfacial area, drop relative velocity, and drop Reynolds number.

4.1 Mixing hydrodynamics

Fig. 2 shows the velocity profile for mixing with the rotating speed of 500 rpm and the temperature of 80 °C. The velocity of the mixture reaches the maximum value near the stirrer bar tip. The high velocity in the front of rotation direction indicates the fluid body acceleration by the impact of the stirrer bar tip towards the wall. The high velocity behinditoccursdue to the void created by rotating motion. The void associates with low pressure zone and thus generates the fluid flow into it.

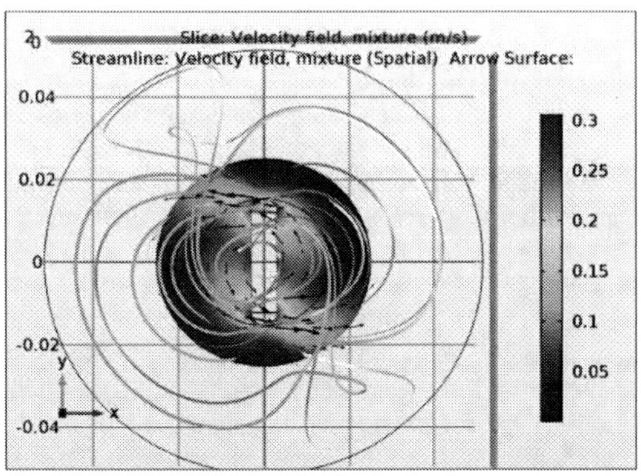

(a)

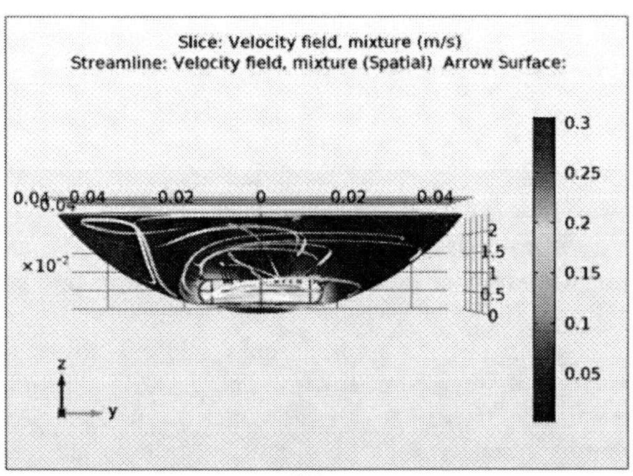

(b)

Fig. 2. Velocity distribution and flow pattern for case 500 rpm, 80 °C (a) horizontal (b) vertical cross section

From the streamline pattern, it seems that solid body vortices are formed following the direction of stirrer bar rotation. The smaller vortex is also generated near the wall. The fluid flow in vertical section, referring to Fig. 2(b), directs toward the reactor wall as the effect ofa centrifugal force and then returns to the center of stirrer bar rotation. A similar flow pattern was also pointed out by previous study, supported by LDV (laser Doppler velocimetry) measurement[15].

The turbulence is not similarly perceived by the drops throughout the reactor. Several authors proposed the usage of drop relative velocityto define local turbulence[4]. This is manifested as drop Reynolds number defined as:

$$\mathrm{Re_d} = \frac{\rho_c |\mathbf{u}_d - \mathbf{u}_c| d_d}{\mu_c} \quad (12)$$

The drop relative velocity andthe drop diameter are available as simulation results. Therefore, the distribution of drop Reynolds number can be calculated. It is showed in Fig. 3giving the picture of local turbulence inside the reactor.

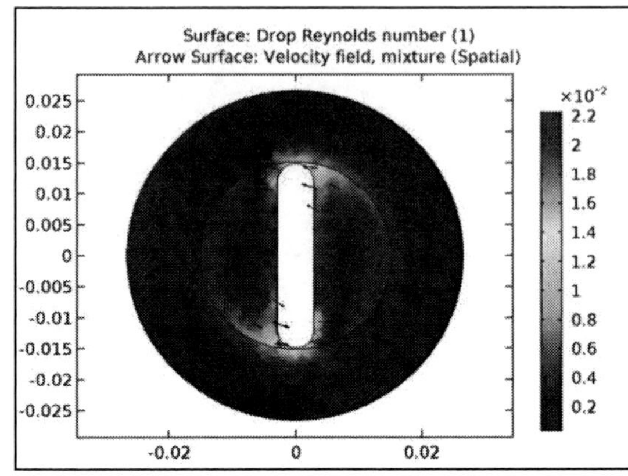

Fig. 3. Distribution of drop Reynolds number for case 500 rpm, 80 °C

The turbulence is more intense around the stirrer bar tip than near the reactor wall. This might be associated with the momentum generated by stirrer bar motion. This momentum is then dissipated away from the rotation zone, also with the turbulence intensity.

4.1 Mass transfer

The purpose of this modeling is to generate a correlation between the local turbulence intensity to the mass transfer coefficient. This correlation, then coupled with interfacial area predicted from simulation, is used to estimate the distribution of the volumetric mass transfer coefficient ($k_c a$). However, due to the availability of the laboratory reactor data as a single value instead of a distribution, the drop Reynolds number value has to be averaged. The averaging method used is the geometrical averaging. This is done by considering the statistical distribution of the variable that approaches the log-normal distribution. Moreover, due to several laboratory experimentsis also performed in different temperature, its effect is also involved in the correlation.

To generate a correlation, each of above factors is represented as dimensionless number, i.e. Sherwood number, drop Reynolds number, and Schmidt number. By applying multivariable regression, an equation to correlate the dimensionless numbers is obtained:

$$\mathrm{Sh_d} = 0.02576\,\mathrm{Re_d^{0.673}}\,\mathrm{Sc^{0.431}} \quad (13)$$

The coefficient of determination (R^2) value of 0.91 shows good correlation between the model and the data asseen in Fig. 4.

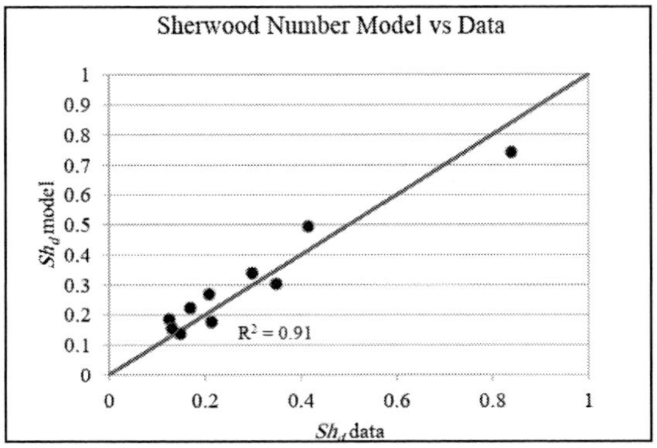

Fig. 4. Comparison between Sherwood number from data and correlation

From Equation (13), due to its higher exponent,the drop Reynolds number has stronger influence to the Sherwood number rather than to Schmidt number. Therefore, the mass transfer coefficienthas more dependence to mixing intensityrather than to temperature.

The second factor affecting the volumetric mass transfer coefficient is interfacial area. This variable is derived from the simulation result by Equation (10). The distribution of the interfacial area density is presented in Fig. 5. The figure describes the distribution at steady state. The dropsare gathered in the front of the stirrer bar tip during the rotation so that the number density and the volume fraction are higher. This leads tohigher interfacial area density. Behind the stirrer bar tip, the effect reverses. This happensbecause the turbulence and therefore the dispersion are more intensive by the existence of trailing vortex[16].

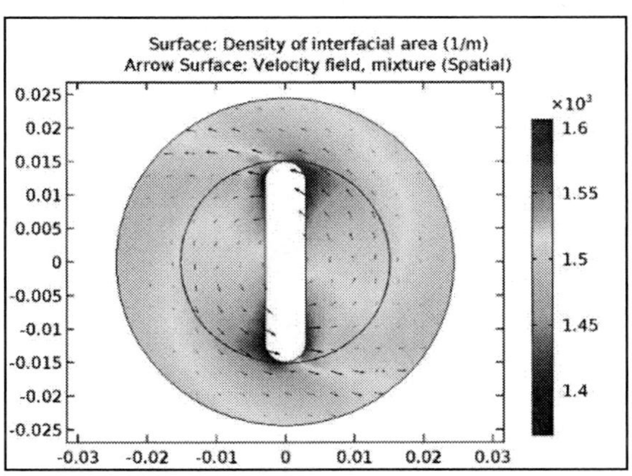

Fig. 5. Distribution of interfacial area density for case 500 rpm, 80 °C

Sherwood number correlation obtained previously is used to calculate the mass transfer coefficient (k_c). By coupling it with the interfacial area density (a), the distribution of volumetric mass transfer coefficient (k_ca) is derived. Fig. 6(a) shows the simulationresults.

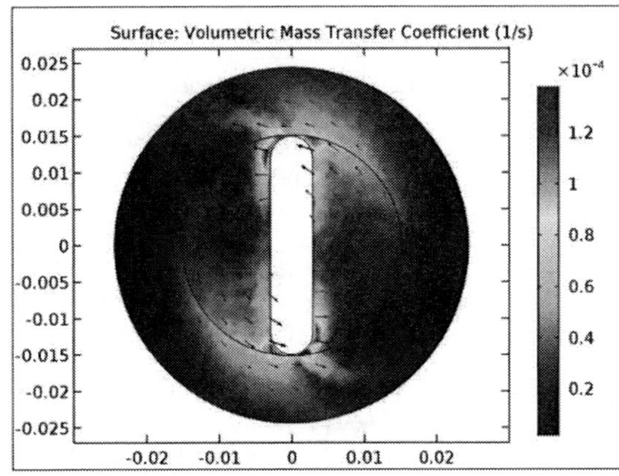

(a)

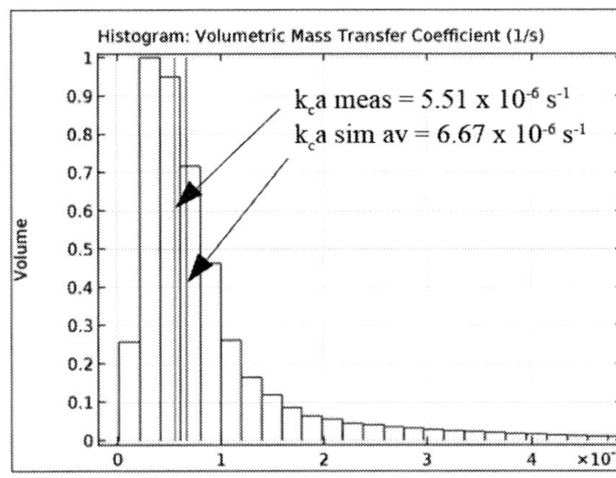

(b)

Fig. 6. (a) Spatial distribution of volumetric mass transfer coefficient (b) statistical distribution and average value of k_ca

The k_ca is higher near the stirrer bar rotation zone, especially near its tip, and is lower approaches the near wall zone. Therefore, the effective zone for the mass transfer to occur is located around the stirrer bar rotation.

Fig. 6 (b) shows the statistical distribution of $k_ca$derived from the simulation. Its average value is compared to the k_ca obtained from the laboratory experiments. The relative error of both values is around 20%. This error is probably related to the error in the Sherwood number correlation. Although the error is rather high, from the modeling point of view, the estimation is quite reasonable.This is because it combines the fundamental equation of fluid flow and the experimental data.Using this modeling approach, value of k_ca for different reactor configurations (e.g. with different impellers or reactor sizes) can be estimated prior to actual experiments. This is useful, for example, if a reactor scale-up is planned.

5 Conclusions

The CFD simulation has been carried out for the laboratory reactor of CPO degumming process. The simulation, combined with the data analysis of the laboratory experiments and the empirical equation to estimate the drop size, is used to generate Sherwood number correlation. The correlation shows the good coefficient of determination (R^2) of 0.91. By applying this correlation and coupled with the value of the interfacial area from the simulation, the distribution of the volumetric mass transfer coefficient ($k_c a$) in the reactor is resulted. The average of this $k_c a$ is comparable to $k_c a$ from laboratory experiments with the relative error being ~20%. This modeling approach is useful in estimating $k_c a$ of other CPO degumming reactor with different sizes or configurations.

We express our gratitude to the University of Indonesia, which funded this research through the scheme of Hibah Publikasi Internasional Terindeks untuk Tugas Akhir Mahasiswa No 863/UN2.R3.1/HKP.05.00/2017.

Symbols

a : interfacial area density(m^{-1})
D_{imp} : impeller diameter(m)
D_{AB} : diffusivity coefficient($m^2 s^{-1}$)
d_{32} : Sauter mean diameter(m)
d_d : drop diameter (m)
ε : turbulence dissipation rate($m^2 s^{-3}$)
$\mathbf{F}$: total force(N)
ϕ_d : dispersed phase volume fraction
$\mathbf{g}$: gravity acceleration($m s^{-2}$)
$\mathbf{I}$: identity matrix
k : turbulence kinetic energy($m^2 s^{-2}$)
k_c : total mass transfer coefficient($m s^{-1}$)
$k_c a$: volumetric mass transfer coefficient(s^{-1})
λ : viscosity ratioof dispersed to continuous phase
μ_c : continuous phase viscosity (kg $m s^{-1}$)
μ_d : dispersed phase viscosity(kg $m s^{-1}$)
N : impeller speed(rpm)
n_d : drop number density(m^{-3})
N_p : power number $N_p = \frac{P}{\rho_c N^3 D^5}$
p : pressure(Pa)
P : impeller output power(W)
ρ_c : continuous phase density(kg m^{-3})
ρ_d : dispersed phase density(kg m^{-3})
ρ_m : mixture density(kg m^{-3})
Re_d : drop Reynolds number
σ : interfacial tension(N m^{-2})
Sc : Schmidt number $Sc = \frac{\mu_c}{\rho_c D_{AB}}$
Sh_d : Sherwood number $Sh_d = \frac{k_c d_d}{D_{AB}}$
$\mathbf{u}$: mixture velocity($m s^{-1}$)
$\mathbf{u}_{slip}$: slipvelocity($m s^{-1}$)
V : vessel fluid volume(m^3)
V_{imp} : impeller swept volume(m^3)

References

1. S. Kanakraj, S. Dixit, Biofuels **7**, no. 5, 537 (2016)
2. R. D. O'Brien, *Fats and oils: formulating and processing for applications*, 3rd ed. (CRC Press,Boca Raton, 2008)
3. E. Deffense, *The AOCS lipid library* (2011) [Online]. Available: http://lipidlibrary.aocs.org/OilsFats/content.cfm?ItemNumber=40322. [Accessed: 01-Jan-2017]
4. V. G. Pangarkar, A. A. Yawalkar, M. M. Sharma, A. A. C. M. Beenackers, Ind. Eng. Chem. Res., **41**, no. 1, 4141 (2002)
5. S. Ibemere, S. M. Kresta, Chem. Eng. Res. Des., **85**(A5), 710 (2005)
6. L. Abu-Faraha, F. Al-Qaessia, A. Schönbuchera, Procedia Comput. Sci., **1**, no. 1, 655 (2010)
7. L. Meng, J.-C. Cheng, H. Jiang, C. Yang, W.-H. Xing, W.-Q. Jin, Chem. Eng. Technol. **36**, no. 11, 1874 (2013)
8. M. Manninen, V. Taivassalo, S. Kallio(1996) [Online]. Available: www.vtt.fi/inf/pdf/publications/1996/p288.pdf
9. Y. Ristianingsih, Sutijan, A. Budiman, Reaktor, **13**, no.4, 242 (2011)
10. COMSOL, *CFD module user's guide* (2015)
11. M. Stamatoudis, L. L. Tavlarides, Chem. Eng. J., **24**, no. 1, 1175 (1985)
12. P. E. Rueger, R. V. Calabrese, Chem. Eng. Res. Des., **91**, no. 11, 2122 (2013)
13. H. P. Grace, Chem. Eng. Commun., **14**, no. 3, 225 (1982)
14. P. E. Rueger, R. V. Calabrese, Chem. Eng. Res. Des, .**91**, no. 11, 2134 (2013)
15. T. Mahmud, J. N. Haque, K. J. Roberts, D. Rhodes, D. Wilkinson, Chem. Eng. Sci., **64**, no. 20, 4197 (2009)
16. S. M. Kresta, R. S. Brodkey, in *Handbook of industrial mixing: science and practice* (John Wiley & Sons, Inc., Hoboken, New Jersey, 2004)

Effect Carrier Agent Formulation in Drying Rate and Antioxidant Activity of Roselle Extract

Febiani Dwi Utari[1*], *Mohammad* Djaeni[1*], *Wahyu Zuli* Pratiwi[1], *Muhammad Alver* Syahputra[1], and *Uma Fadzilia* Arifin[1]

[1]Department of Chemical Engineering, Faculty of Engineering, Diponegoro University Jl. Prof. Soedarto, SH, Tembalang, Semarang, Indonesia 50275

Abstract. Roselle *(Hibiscus sabdarifa L)* contains anthocyanins as the natural colorant and antioxidant. Drying the roselle extract was aims to produce the dry product that easy consumption as antioxidant. The carrier agent was added in roselle extract to improve the drying rate and maintain the nutritional value. This research studied the effect of carrier agent in drying rate and antioxidant activity. The method consists of two step involving roselle extraction using ultrasonic and the drying process. The roselle extraction by ultrasonic use the water as the solvent. The carrier agent (0%,5%,10% of maltodextrin) was added in roselle extract. The mixture was then dried in tray dryer dehumidification using zeolite in drying temperature 50,60, and 70°C. As the response, the moisture content was observed by gravimetry every 15 minutes for 150 minutes. The result showed that Page model was fitted to determine the constant of drying rate. Higher concentration of carrier agent enhanced the moisture evaporation process. Based on the DPPH analysis, the degradation of antioxidant activity in temperature 70°C was 2.14 times higher than in temperature 50°C. As the conclusion, addition of maltodextrin can speed up the drying process and retain the antioxidant activity of.

1 Introduction

The roselle *(Hibiscus sabdariffa Linn.)* calyces has the several function, as the antioxidant [1], antimicrobial activity [2], immunoprotective [3] and natural colorant [4]. The antioxidant compound in roselle stronger than commercial antioxidants such as Butylated Hydroxyanisole (BHA) and α-tocopherol (TOCO) [5]. Anthocyanins and phenols are the main bioactive compounds in Roselle calyces. The concentration of anthocyanin and phenols in roselle depend on the cultivation and extraction process [6].

Study about roselle calyces extraction using water as the solvent resulted 502.33±0.52 mg of anthocyanins [7]. Antioxidant compound including anthocyanin has been understood well as not stable because of the high reactivity and easy to degrade [8,9]. The drying process is the option to produce the stable form of roselle extract and prolong the storage life.

The drying process with addition of carrier agents can reduce the stickiness surface of liquid products [10,11]. The most common carrier agents used in food product is maltodextrins. Maltodextrin is a hydrolyzed starch that gives the advantages such as low cost, good aroma and flavor and low viscosity at high-solids concentrations [12].

Several researcher using maltodextrin in food product drying. Addition 3% maltodextrin in freeze drying improving color stability of roselle anthocyanin powder [13]. Addition 15 % maltodextrin in drying tamarind pulp at drying temperature 70°C had the highest acceptance score [14]. Maltodextrin can retain the color quality of roselle extract drying [11]. In addition, maltodextrin (DE 20) stabilizing the anthocyanin pigments under storage condition [15].

Based on the study, there is a hard information on antioxidant activity degradation of roselle extract dried with carrier agents such as maltodextrin. This work aims to produce the dried roselle extract with the good antioxidant activity. The effect of addition maltodextrin and drying temperature is evaluated. The experimental results are fitted to several thin layer model to describe the drying phenomenon.

2 Materials and Methods

2.1. Roselle extract preparation

The roselle *(Hibiscus sabdariffa Linn.)* calyces variety were red roselle that obtained from a herbal medicine market in Solo, Indonesia. The roselle calyces (dry) was milled until the uniform size (0.25 mm). The roselle extraction based on previously extraction method with several modifications [7]. In this study, the extraction process was carried out using ultrasonic system

* Corresponding author: febianidu@gmail.com, moh.djaeni@live.undip.ac.id

(Branson, 2510 MT). Water (solvent) and roselle calyces mass to water volume ratio were 1:10. The roselle extract was then filtered to separate the filtrate and cake.

2.2 Carrier agent formulation

Carrier agent was added in roselle extract to improve the drying rate. The carrier agent was maltodextrin (food grade) with DE 15. The carrier agent portion was 0%, 5%, 10% of the total mixture. The mixture was stirred using stirrer speed 700 rpm for 5 minutes to make uniform solution.

2.3 Drying process

After the carrier agent formulation, the roselle solution was then dried in tray dryer. The solution was placed in aluminium tray under drying temperature (50^{0}C-70^{0}C) for 150 minutes. The moisture evaporation was determined by gravimetry (weighing the samples every 15 minutes). The drying process took place with superficial velocity of 0.11 m s^{-1}.

2.4 Drying model

Thin layer model was used to determine the constant rate of the drying. The thin layer model correspond to the drying phenomenon (Tabel 1). The good fit model was selected by the highest value of coefficient determination (R^2) and the lowest value of Root Mean Square Deviation (RMSD). The evaluation using POLYMATH 6.1 Educational Free Trial version software.

Table 1. Constant rate of the drying several thin layer model

Model	Model equation	References
1. Newton	$MR = exp(-kt)$ (1)	O'Callaghan et al. [16]
2. Page	$MR = exp(-kt^n)$ (2)	Page [17]
3. Handerson-Pabis	$MR = a\ exp(-kt)$ (3)	Henderson and Pabis [18]
4. Logaritmic	$MR = a\ exp(-kt) + c$ (4)	Yagcioglu et al. [19]

The moisture ratio (dimensionless moisture) could be expressed in equation 5.

$$MR = \frac{M_t - M_e}{M_0 - M_e} \quad (5)$$

where M_t was the moisture content at time t (kg water/kg dry solid), M_0 was the initial moisture content (kg water/kg dry solid), M_e was the equilibrium moisture content (kg water/kg dry solid).

The value of equilibrium moisture content of the solution was determined using equation 6.

$$Me = (XMe)_{roselle} + (XMe)_{maltodextrin} \quad (6)$$

Equilibrium moisture content for roselle and maltodextrin.was previously studied [20,21].

2.5 Antioxidant activity analysis

The capacity of roselle extracts to inhibit DPPH (free radical) was determined according to the method of Brand-Williams et al. with slight modification [22]. The procedure was done by measured the absorbance of the roselle extract and DPPH solution at 517 nm using a spectrophotometer. As the control, the blank (100 μL of ethanol in 1.4 mL of DPPH radical solution) was measured.

Inhibition (%) was used to express the antioxidant activity, as expressed in equation 7.

$$I(\%) = \left[\frac{A_0 - A_s}{A_s}\right. \quad (7)$$

where, A_s was absorbance of the roselle solution and A_o was absorbance of control.

First order reaction was used to describe the anthocyanin and phenolic compound degradation in food product [23]. The anthocyanin and phenolic compound was the source of antioxidant, so the degradation of antioxidant activity following that equation.

$$\ln\left(\frac{I}{I_0}\right) = -k_i t \quad (8)$$

where I and I_0 were antioxidant activity of the sample at time 0 and t. k_i was the constant rate of antioxidant activity degradation(minute^{-1}).

Arrhenius model was used to correlate the dependency of k_i on temperature (T).

$$k_i = k_0 \exp\left(-EA/RT\right) \quad (9)$$

where Ea was the activation energy (J mol^{-1}), and R was the gas constant (8.32 J mol^{-1} K^{-1}) and k_0 was the pre-exponential factor (minute^{-1}).

3 Results and Discussion

3.1 Addition carrier agent

The moisture in roselle extract evaporated during the drying process (Fig. 1.). The addition of maltodextrin could speed up the drying rate. The final moisture content after 150 minutes drying process was 0.62 (gram of moisture per gram of dried product) in addition 10% maltodextrin, while in roselle extract drying without addition of maltodextrin resulted moisture content 2.101 (gram of moisture per gram of dried product). The final moisture content of roselle extract using 10% maltodextrin was 3.9 times lower than without maltodextrin. This research comparable with another research that resulted the constant rate of the drying 1.7 times higher than without maltodextrin [11].

https://doi.org/10.1051/matecconf/201815602010

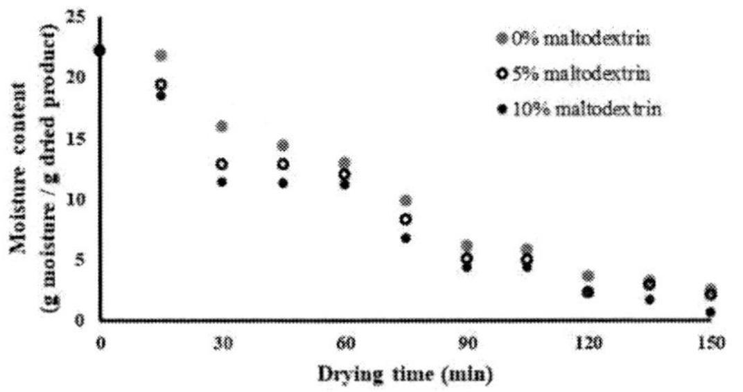

Fig. 1. Effect of addition carrier on moisture content during the drying process

In this research low DE maltodextrin was used as the carrier agent. Addition this carrier agents affected the different final moisture content. The low DE maltodextrin has the less number of ramifications with the hydrophilic groups and not easy to bind with the water molecules from the air [10]. On the other hand, the addition of maltodextrin affected the stickiness surface of roselle extract. The stickiness surface of roselle extract because of low molecular sugar like glucose, sucrose, fructose. Higher concentration of maltodextrin reduces the stickiness surface of the roselle extract [10].

3.2 Drying temperatures

The effect of drying temperatures to the moisture content during the drying process was represented in Fig.2. After 150 minutes drying, the drying process that was conducted in higher drying temperature resulted lower moisture content. Drying roselle extract with the drying temperature 70°C resulted in the final moisture content of 0.163 (gram of moisture per gram of dried product), while the final moisture content in the drying temperature 50°C was 3.8 times higher.

The higher drying temperatures increased constant rate of the drying (Table 2), as also found n the previous studies [24,25]. The constant rate of the drying at drying temperature 70°C was 9.6 times higher than at drying temperature 50°C. As the comparison, another research resulted constant rate of the drying at drying temperature 70°C was 1.6 times higher than at drying temperature 50°C [10]. In higher drying temperature, the relative humidity of air is low, driving force for the drying increase and the sensible heat from the hot air used for water evaporation [25,26,27].

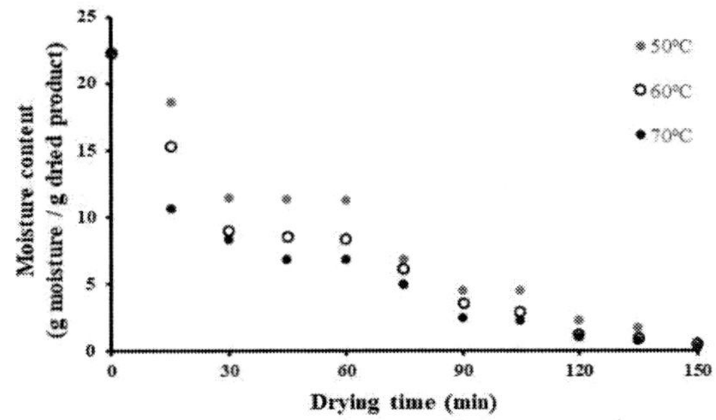

Fig. 2. Effect of drying temperatures on moisture content during the drying process

3.3 Drying rate constant and model

Thin layer model was used to determine the drying phenomenon in roselle extract. Based on the Table 1, all the model resulted the R^2 value range from 0.924 until 0.971 and RMSD value range from 0.014 until 0.021. It was indicated that the thin layer model can well represent the roselle extract drying. The R^2 value in Page model (no 2) was higher than 0.964 with RMSD value was less than 0.017. This model can be used to predict the drying rate of roselle extract drying with addition maltodextrin.

Table 2. Constant rate of the drying several thin layer model

Model	T (°C)	Constant	R^2	RMSD
1	50	k:0.016	0.962	0.018
	60	k:0.021	0.963	0.017
	70	k:0.027	0.924	0.023
2	50	k:0.011, n : 1.083	0.964	0.017
	60	k:0.039, n : 0.852	0.971	0.015
	70	k:0.110, n : 0.646	0.973	0.014
3	50	k:0.017, a : 1.014	0.962	0.018
	60	k:0.020, a : 0.958	0.966	0.016
	70	k:0.024, a : 0.903	0.938	0.021
4	50	k:0.012, a : 1.130 c:-0.145	0.969	0.016
	60	k:0.020, a : 0.960 c:-0.004	0.966	0.016
	70	k:0.030, a : 0.880 c:0.049	0.942	0.020

The drying curves in the form of moisture ratio versus drying time (experiment and model) were depicted in Figure 3. Page model was used to describe the drying phenomenon of food product : bitter melon [28], carrot [29], and seaweed [30]. Page model showed a good fit curves to roselle extract drying with maltodextrin.

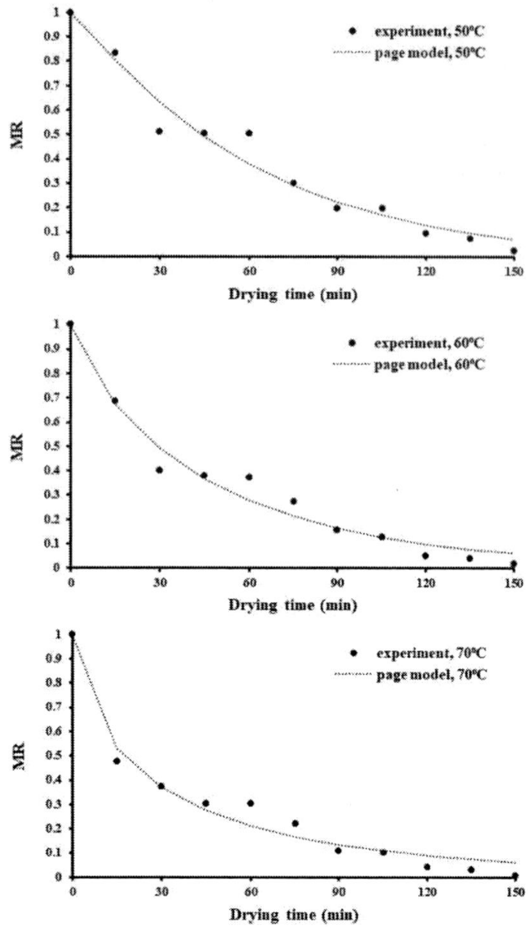

Fig. 3. Comparison of experimental and page model in moisture ratio curves of roselle extract drying at various drying temperature

3.4 Antioxidant Activity Degradation

The antioxidant activity in roselle extract drying with addition 10% maltodextrin was expressed as % inhibition. The degradation of antioxidant activity was faster in higher drying temperature and longer drying time (Figure 4). Anthocyanin as the antioxidant component in roselle extract was thermal sensitive [23,31,32]. The antioxidant activity degradation in roselle extract was caused by the thermal degradation of anthocyanin compound.

The constant rate of antioxidant activity degradation in drying temperature 70°C was 2.14 times higher than in drying temperature 50°C (see Table 3). As the comparison, the constant rate of antioxidant activity degradation in roselle extract drying with maltodextrin was 18 times lower than constant rate of antioxidant activity degradation in beetroot and green pepper at the same temperature (70°C) [33]. Antioxidant activity in roselle extract was reduced 40.1% after thermal process (90°C) for 2 hours [34]. Addition maltodextrin in roselle extract drying could shorten the drying time as well as decreased the antioxidant activity degradation.

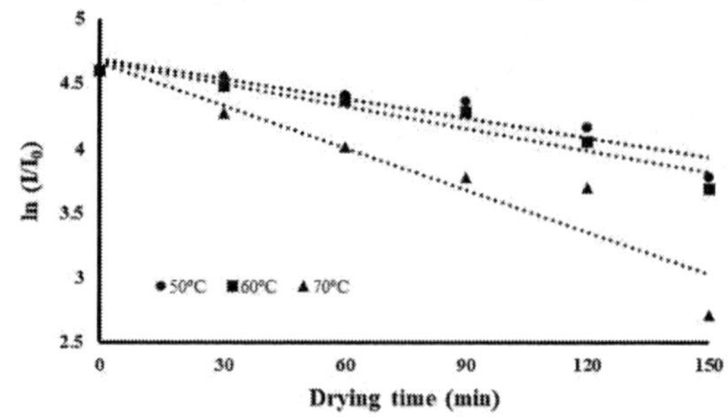

Fig. 4. The antioxidant activity degradation at various drying conditions

Arrhenius model was used to correlate the dependency of k_i on temperature (T). The value of Arrhenius constants were listed in Table 3. Using the value of Ea and K_0, the value of constants rate of the antioxidant activity degradation under various temperature can be predicted.

Table 3. Constant rate of the antioxidant activity degradation

T (°C)	k_i (min^{-1})	R^2	K_0x10^3	Ea (Kj/mol)
50	0.005	0.927	1.932	34.754
60	0.006	0.949		
70	0.011	0.911		

4 Conclusions

The effect of addition carrier agent in constant rate and antioxidant activity were studied. Low DE maltodextrin is not easy to bind with the water molecules from the air and reduce stickiness surface of roselle extract. Page model showed good fit curves with R^2 value in Page model (no 2) was higher than 0.964 with RMSD value was less than 0.017. The values of constant rate for drying at drying temperatures 50°C-70°C ranged from 0.011-0.110 min^{-1}. The degradation of antioxidant activity was faster in higher drying temperature and longer drying time. Addition of maltodextrin could shorten the drying time and decreased the antioxidant activity degradation

This study was supported by Directorate General of Higher Education (DIKTI) under Competency Grant (Hibah Kompetensi) for 2017-2018.

References

1. P. J. Tsai, J. McIntosh, P. Pearce, B. Camden, and B. R. Jordan, Food Res. Int., **35** (4), 351–356 (2002)
2. M. Fullerton, J. Khatiwada, J. U. Johnson, S. Davis, and L. L. Williams, J. Med. Food, **14** (9), 950–956 (2011)
3. T. Okoko and D. Ere, Asian Pac. J. Trop. Med., **5** (1), 33–36 (2012)

4. E. M. Abdou, H. S. Hafez, E. Bakir, and M. S. A. Abdel-mottaleb, Spectrochim. Acta Part A Mol. Biomol. Spectrosc., **115** 202–207 (2013)
5. P.-L. Tee, S. Yusof, and S. Mohamed, Nutr. Food Sci., **32** (1), 17–20 (2002)
6. S. Cid-Ortega and J. A. Guerrero-Beltran, J. Food Res., **3** (5), 83 (2014)
7. P. Chumsri, A. Sirichote, and A. Itharat, J. Sci. Technol., **30** (Suppl.1) 133–139 (2008)
8. M. A. Youssef, A. A. El kady, and K. S. Nagy, Middle East J. Agric. Res., **4** (4), 697–706 (2015)
9. A. N. Al-Baarri, A. T. Suminar, and A. M. Legowo, Potravinarstvo **11** (1), (2017)
10. S. Singh and D. Dixit, Int. J. Appl. Eng. Technol., **4** (1), 1–8 (2014)
11. M. Djaeni, F. D. Utari, and A. C. Kumoro, AIP Conference Proceedings, **1823** 20045–1–20045–6 (2017)
12. M. C. P. de A. Santiago, R. I. Nogueira, D. R. S. F. Paim, A. C. M. S. Gouvêa, R. L. de O. Godoy, F. M. Peixoto, S. Pacheco, and S. P. Freitas, LWT - Food Sci. Technol., **73**, 551–556 (2016)
13. Duangmal, B. Saicheua, and S. Sueeprasan, LWT - Food Science and Technology, **41** (8), 1437–1445 (2008)
14. A. Ekpong, W. Phomkong, and E. Onsaard, Int. Food Res. J., **23** (1). 300–308 (2016)
15. K. A. Selim, K. E. Khalil, M. S. Abdel-Bary, and N. A. Abdel-Azeim, J. Food Sci. Technol., 7–20 (2008)
16. O'Callaghan J R., Menzies D J and Bailey P H J., Agric. Eng. Res **16** (3), 223-244 (1971)
17. G. Page, MS Thesis Submitted to Purdue University West Lafayette, IN, USA (1949)
18. S. M. Henderson and S . Pabis, J. Agric. Eng. Res., **6**, 169-174 (1961)
19. A .Yagcioglu, A. Degirmencioglu and F. Cagatay, Proceedings of the 7th International Congress on Agricultural Mechanization and Energy, 565-569 (1999)
20. Langová, D. Jaisut, R. Thuwapanichayanan, C. Phowong, J. Štencl, D. Jovanovic, J. Fryè, and V. Vlášek, Acta Univ. Agric. Silvic. Mendelianae Brun., **61** (6), 1769–1777 (2013)
21. Y. Roos and M. Karel, Biotechnol. Prog., **7** (20), 49–53 (1991)
22. W. Brand-Williams, M. E. Cuvelier, and C. Berset, **28** (1), 25–30 (1995)
23. Z. Zori, V. Dragovc-Uzelac, S. Pedisic, Z. Kurtanjek, and I. E. Garofulic, Food Technol.Biotechnol, **52** (1), 101–108 (2014)
24. M. Djaeni, S. B. Sasongko, A. Prasetyaningrum, X. Jin, and A. J. van Boxtel, Int. J. Food Eng., **8** (3), 1-21 (2012)
25. M. Djaeni and D. A. Sari, Procedia Environmental Sciences **23**, 2–10 (2015)
26. M. Djaeni, D. Ayuningtyas, N. Asiah, H. Hargono, R. Ratnawati, W. Wiratno, and J. Jumali, Reaktor **14**, (3), 173–178 (2013)
27. M. Djaeni, D. Anggoro, G. W. Santoso, D. Agustina, N. Asiah, and C. L. Hii, Adv. J. Food Sci. Technol., **6** (7), 833–838 (2014)
28. J. Chen, Y. Zhou, S. Fang, Y. Meng, X. Kang, X. Xu, and X. Zuo, Advance Journal of Food Science and Technology **5** (9), 1214–1219 (2013)
29. I. Doymaz, Journal of Food Engineering **61** (3), 359–364 (2004)
30. A. Fudholi, M. Y. Othman, M. H. Ruslan, M. Yahya, A. Zaharim, and K. Sopian, Proceedings of the 4th WSEAS international conference on Energy and development - environment - biomedicine, 129–133 (2011)
31. M. Cisse, F. Vaillant, O. Acosta, D. M. Claudie, and M. Dornier, J. Agric. Food Chem., **57**, 6285–6291 (2009)
32. A. Sinela, N. Rawat, C. Mertz, N. Achir, H. Fulcrand, and M. Dornier, Food Chem., **214**, 234–241 (2017)
33. V. Eyarkai Nambi, R. K. Gupta, S. Kumar, and P. C. Sharma, Journal of Food Science and Technology, **53** (7) 3073–3082 (2016)
34. P.-J. Tsai and H.-P. Huang, Food Research International **37** (4), 313–318 (2004)

Fixed Bed Adsorption of Natural Organic Matter Using Ozonated Carbon Active

Adi Kurniawan[1], *Dian* Listiyani[1], *Jono* Suhartono[1*], and *Suparman* Juhanda[1]

[1]Chemical Engineering Department, Institut Teknologi Nasional
Jl PHH. Mustofa no 23 Bandung 40124, Indonesia

Abstract. Nowadays water, one the most important substance in human life, are being much polluted by not only industrial activities but also caused by human activities and natural polluted material. One of the natural impurities that commonly occur in water is Natural Organic Matter (NOM). NOM are extracted organic compounds from soils, plants, or decomposed animals that infiltrate into the water body when the water contacts with those materials. Thus, the presence of NOM in water body is inevitable. Many techniques have been applied to remove this impurity such as coagulation, flocculation, and filtration. In this research, NOM were controlled using ozonated activated carbon in a fixed bed adsorption system. There were two parameters studied that are activated carbon ozonation time that ranging from 5 - 25 minutes and the fixed bed height of 1 - 7.5 cm. Interestingly, the investigation showed that the optimum condition for the bed height was obtained at 6 cm whilst the optimum carbon active ozonation time was at 5 minute. The experiment was concordant with the conducted BET analysis that showed the highest activated carbon surface area at 5 minute compared to other ozonation time. The ozonated activated carbon was also found to be have slightly better adsorption performance compared to unozonated activated carbon bought from the market with capacity of ozonated activated carbon adsorption was found to be $6.12 \times 10^{-3}\ ^{mg}/_{g}$.

1 Introduction

Clean water has an important role in daily life. However, clean water becomes difficult to find due to varied impurities contained in water. One of the impurities is Natural Organic Matter (NOM), an organic compound that came from decomposition and metabolic reaction of plants and animals. NOM compounds usually consisted of protein, polysaccharide, and aromatic compounds.

NOM is not a single substance but composed of many components. Those components are humin, humic acids, and fulvic acids (Frimmel and Abbt-Braun, 1999; Peuravuori et al, 2002). The difference between those three major NOM compounds may indicate from their solubility through H condition. Humin are insoluble in any pH, fulvic acid are soluble in any pH whilst humic acid are only soluble at pH above 3 (Fulcrum Health Limited, 2004). Humic acid is the largest compound in the NOM constituents, thus humic acid may serve as a parameter content of NOM in the water.

NOM concentration in water must be reduced because it can react with chlorine and chloramines to produce disinfection by-product such as trihalomethanes and haloacetic acid which considered to be carcinogenic and genotoxic (Gora S, 2011).

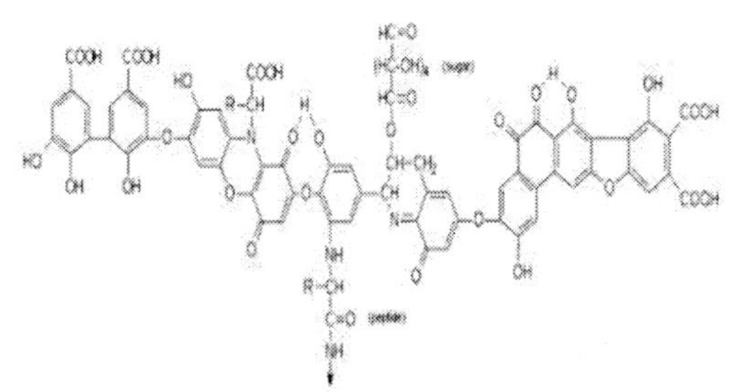

Fig. 1. Structure *humic acid* (Stevenson,1982)

In order to minimise the production of disinfection by-products in water, it must be controlled by removing or reducing the NOM concentration in feed water before disinfection process, thus production of trihalomethanes and haloacetic acids can be reduce. There are many method to control NOM concentration in water such as coagulation - floculation, membrane, adsorption, and ozonation.

Adsorption is a phenomenon when substance adheres to a surface of another substance caused by the attraction of force from each substance. Adsorption process can be conducted either in batch or continuous system. Activated carbon is the most applicable adsorbent uses

*Corresponding author : jono_suhartono@itenas.ac.id

as adsorbent media in the adsorption process because it has a large surface area, easy to produce, cheap, non-toxic, easy and economical to be regenerated.

Activated carbon can be modified by using ozone either to expand their surface area or to modify their surface functionalization. The ozonation of activated carbon may form oxidation bond to the carbon surface such as *carboxylic* (C-OO), *carbonyl* (C=O), and *hidroxyl* (C-OH) through Griegge's mechanism (Ke peng et al, 2011). In this research, NOM were controlled using ozonated activated carbon in a fixed bed adsorption system. To know the efficiency of activated carbon to adsorb the NOM content in water. There were two parameters studied that are activated carbon ozonation time that ranging from 5 - 25 minutes and the fixed bed height of 1 - 7.5 cm. Continuous adsorption process which processed can be observed using breakthrough curves. The output yields can be described by using breakthrough curve between output concentration versus output volume.

2 Material & methods

The materials used in this research were humic acid (Sigma Aldrich, Singapore), NaOH, activated carbon, and HNO_3 were bought from Brataco, Indonesia, Deionisation water. The research stages was conducted was started from preparation of activated carbon, ozonation of activated carbon, preparation of humic acid solution for calibration and feed, adsorption process, and finally the humic acid concentration analysis.

2.1 Preparation of Activated Carbon and Ozonated Activated Carbon

This process is carried out for the preparation of activated carbon before activated carbon used as absorbent bed on adsorption process. Firstly, activated carbon was crushed into a smalled size using a mortar. Then the smaller activated carbon is filtered using a mesh filter to obtained 20/35 mesh particle size. Furthermore, the 20/35 mesh size activated carbon washes using deionisation water to eliminate dust from crushing process. Finally, the activated carbon was modified using ozone for 5, 10, 15, 20, and 25 minutes.

2.2 Preparation of Humic Acid Stock Solution

0.1 grams of NaOH was poured into a 500 mL volumetric flask and dissolved it by 250 mL deionisation water, 0.5122 grams humic acid then was added into the NaOH solution. Stir the humic acid solution for 15 minutes, and add another deionisation water until the marked line. The agitation was extended for 3 hours until homogeneous solution obtained. The next step was filtering the solution using vacuum filter through 0,45 μm membrane filter. This stock solution was stored in a cool and dark place to preserve them for long term utilization.

In order to prepare a 10 mg/L humic acid solution feed, a 20 mL humic acid stock solution was pipetted and diluted using deionisation water to obtain 2L humic acid solution. The solution was then stirred for about 30 minutes to obtain homogeneous solution.

2.3 Adsorption Process

The adsorption process was conducted in a continuous fixed bed adsorption column. The bed column dimension was 2.5 cm diameter to 30 cm height. The solution feed was poured into a feed tank and pumped to a constant head tank. The feed flowrate went into the bed column was controlled using a valve and maintained at 0.5 cm^3/s. Meanwhile, the overflow of constant head tank was returned into the feed tank.

The output of adsorption column was analysed every 2, 5, 10, 20, 35 and 65 minutes to accommodate the breakthrough condition has been yielded. The adsorption process was conducted in a varied height of absorption beds which are 1.5; 3; 4.5; 6; and 7.5 cm. The step was repeated for varied ozonated activated carbon time which ozonated for 5, 10, 15, 20, and 25 minutes.

2.4 Analysis of Humic Acid Concentration and Activated Carbon Pore Size

The humic acid concentration was analyzed using UV-VIS Spectrophotometer (Genesis Thermo Scientific, USA) at 254 nm wavelength (Standard Method). The spectrophotometer was heated for 15 minutes. Then put a blank solution into the spectrophotometer to set up its absorbance value exact to zero before it uses for sample analysis.

The mean pore size of activated carbon was analysed using BET (Brunauer-Emmett-Teller) (Nova 3200e, Quantachrome,UK). Determination of the surface area of the substrate is to be done by physical adsorption of the gas (usually nitrogen). Gas molecules that stick to the surface of the substrate solid (adsorbent) are said to be adsorbed and tend to form a thin layer that covers the entire adsorbent surface.

3. Result

3.1 Standardization Curve

Standardization curve was obtained from measurement of humic acid standard solution which is measured at determined wavelength, i.e 254 nm using variation concentration 2, 5, 10, 15, and 20 mg/L.

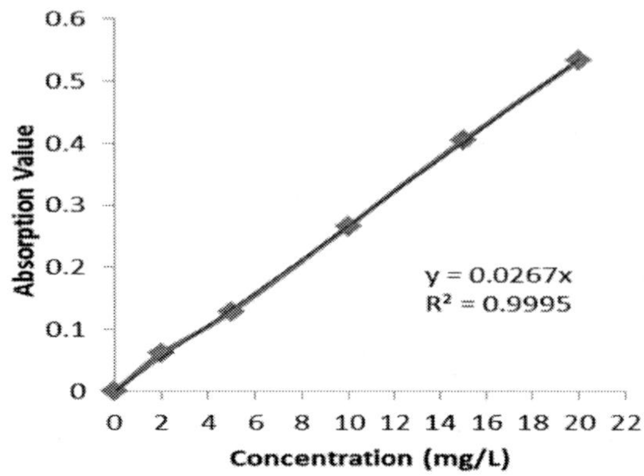

Fig. 2. Standardization curve of humic acid

Figure 2 shows the standardization curve with obtained trend line equation of $y = 0.0267x$ and confidence level of determination, R^2, is 0.9995. This humic acid standardization curve was made to calculate the concentration of humic acid solution used in the feed and the output concentration of humic acid.

3.2 Influence of Ozonated Time

The adsorption process was done using activated carbon which has been modified with different ozonisation time, i.e. 5, 10, 15, 20, and 25 minutes. The activated carbon bed height and NOM concentration feed was set up at 4.5 cm and 10.3 mg/L. The aim was to study the optimum process of modified activated carbon by different ozonisation time, to minimize NOM concentration.

It is shown in fig 3, the influence of ozonisation time against output concentration obtained from 10.3 mg/L of NOM concentration that the optimum modified activated carbon absorption was found at 5 minute of ozonisation time. Figure 3 also shows that longer ozonisation time will decrease activated carbon ability to adsorb NOM. It is indicated by the first 2 minute of adsorption process, NOM removal were 2.63, 1.92, 0.76, 1.61, and 0.60 mg/L for ozonated activated carbon at 5, 10, 15, 20, and 25 minutes, respectively. The concentration getting from plotted absorbance spectro analysis to standarization curve, for each ozonised carbon time of 5, 10, 15, 20, and 25 minutes, respectively.

Figure 3 shows that the longest saturation of carbon bed was obtained by carbon with ozonation time of 5 minutes. However, the saturation time among the varied ozonation time was insignificant and most of the breakthrough point was obtained around 10 minutes. The longest saturation obtained by 5 minutes ozonation time was caused by larger specific surface area of carbon at 5 minutes ozonation time compared to further ozonation time. The specific surface area of 5 minutes ozonation time was 154.71m^2/g whilst further ozonation decrease the carbon surface area to 1.75 after 25 minutes ozonation time. Table 1 shows the specific surface area of ozonated and un-ozonated carbon.

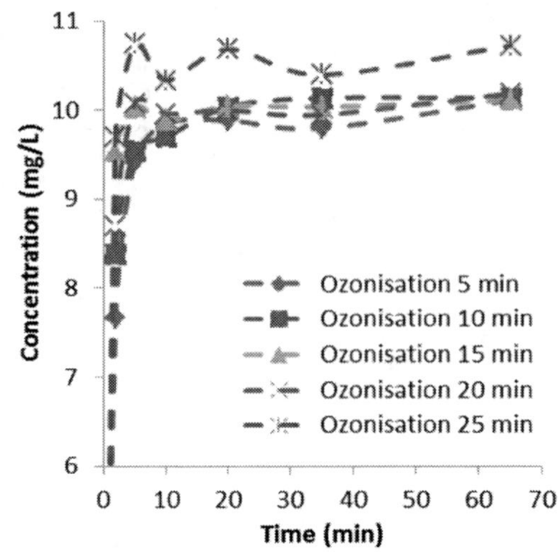

Fig. 3. Continuous adsorption of NOM on humic acid at bed height of 4.5 cm

The surface area at 5 minutes ozonation is larger than 15 and 25 minutes ozonation. This might be caused by the longer ozonisation time apply, the reaction of ozone and the impurity in the carbon pores becomes more intensive. Moreover, oxidation product of carbon impurity pores may cover and reduce activated-carbon surface area.

Table 1. Surface Area of Ozonated and Un-ozonated Carbon

Activated Carbon	Surface Area (m^2/g)
Without Ozonisation	41.723
5 minute	154.713
15 minute	71.079
25 minute	1.746

The reduction of carbon surface area resulted in less ability of activated carbon to accommodate the high NOM solute concentration thus shorter saturation time of the bed as shown in figure 3. The output concentration of 25 minutes ozonation time shows higher value than the initial concentration used. This might be caused by the high amount of exuviating activated-carbon intruding into the output through the adsorption process.

3.3 Bed Height Variation

The adsorption process in this step was using 15 minutes ozonated activated carbon and different height of carbon bed that are 1.5, 3, 4.5, 6, and 7.5 cm.

https://doi.org/10.1051/matecconf/201815602011

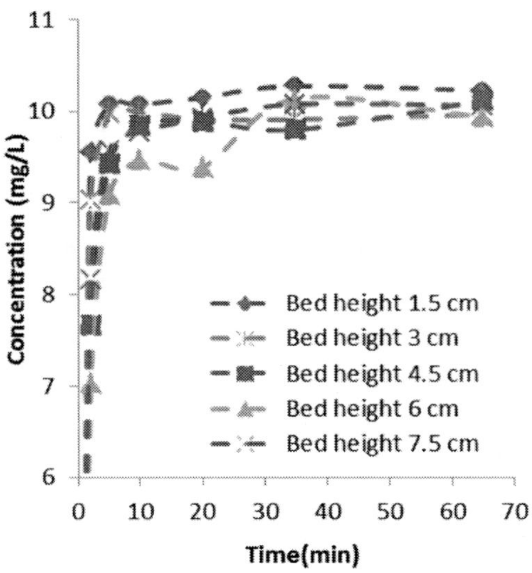

Fig. 4. Continuous adsorption of NOM on humic acid at ozonated time 5 min

The lowest concentration of NOM absorption in the first 2 minutes was obtained at 7.5 cm bed height while the highest concentration was obtained at 1.5 cm bed height as shown in fig 4. The utilization of higher of carbon bed height resulted in higher absorption capability of the carbon to adsorb NOM. However, carbon bed ability to adsorb NOM was declined as process time conducted. This was caused by the increasing of solutes adsorbed in the carbon surface, as shown by the output concentrations which is obtained from the sample at specific intervals on fig 4.

Although the capability of carbon to adsorb NOM increasing as the addition of bed height, the highest absorption was obtained from 6 cm bed height as shown in fig 4. The saturation time of 6 cm bed height was delayed compared to 7.5 cm bed height. This phenomenon might occur due to the influent feed was difficult to pass 7.5 cm bed height of activated-carbon since increasing bed height will also increase the bed retention which generated by higher pressure drop in the bed. Thus, high pressure drop occurs in the bed will result in smaller output rate obtained from the adsorption bed that made NOM retained, adsorbed and accumulated in the activated-carbon longer than shorter activated carbon bed. On contrary, the output concentration of 7,5 cm height is higher than 6 cm bed height since the bed was quickly saturated by the retained NOM.

3.4 Comparison of Activated Carbon from Market and Ozonated Activated Carbon

The performance of optimised ozonated activated carbon process was compared to pristine activated carbon bought from market. The result is presented on the figure 5.

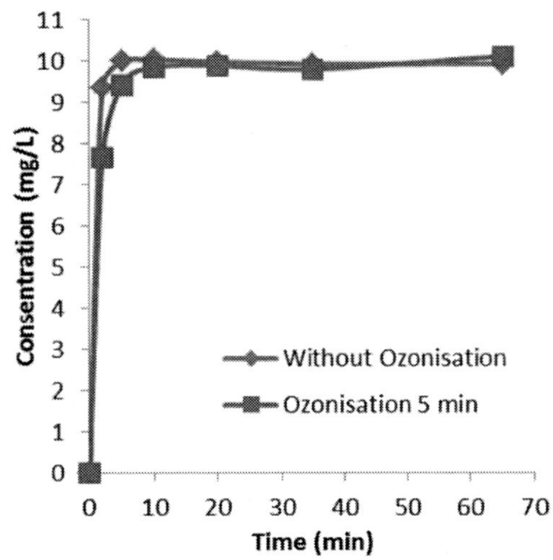

Fig. 5. Breakthrough curve comparison between with and without ozonisation

The data shows that the result obtained from the absorption capability of activated carbon modified by ozone was able to perform higher absorption capability compared to pristine activated carbon bought from market. It can be seen from figure 5 where activated carbon modification and activated carbon from market at the first 2 minute adsorption process. The NOM concentration obtained from output of fixed bed adsorption was 2.559 mg/L for ozone modified activated carbon and 0.861 mg/L for pristine activated carbon where the feed concentration used was 10.3 mg/L. This is caused by the surface area of modified activated carbon were much larger than the pristine activated carbon. Moreover, the introduction of ozone may generate oxidized compounds in the activated carbon surface which makes activated carbon more reactive to adsorb the natural organic matter. According to (Naeimi, 2009) ozonation of carbon will produce oxidation bond in carbon surface area such as *carboxylic* (C-OO), *carbonyl* (C=O), and *hydroxyl* (C-OH). Therefore, the activated carbon which has been modified by the ozone may able to accommodate higher solute absorbed than pristine activated carbon bought from market which was not ozonised.

Figure 5 also shows the comparison of these two activated carbon types used based on saturation time. The breakthrough curve shows that pristine activated carbon had faster saturation time comparing to ozonated activated carbon. The pristine activated carbon was saturated at the first 5 minutes operation whilst ozonated activated has double length of saturation time compare to pristine activated carbon. This may be caused by the activated carbon bought from market has lesser capability to adsorb the solute since their surface area were smaller compared to ozonated activated carbon as shown in table 1. The concentration of NOM in the output of fixed bed adsorption that almost reached to its initial feed value indicates the saturation of activated carbon bed itself.

3.5 Thomas Equations

Thomas equation is the equation that commonly used for approaching a continuous adsorption process in order to find out the mass transfer constant rate (k), and mass transfer capacity of adsorption (qo). A linearization of equation into logarithmic approach

$$\ln\left(\left(\frac{C_0}{C_t}\right)-1\right) \quad (1)$$

to volume of fixed bed adsorption output obtained from the experiment breakthrough curve is demanded to solve this Thomas equation. The linearization is shown in the figure 6.

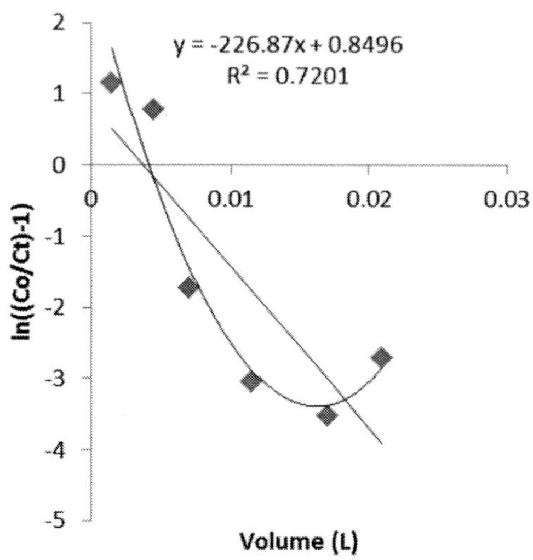

Fig. 6. Linearization curve of Thomas equation at bed height 4.5 cm and ozonated time 5 min

Figure 6 shows the slope and intercept value of the linearization of the breakthrough curve. It is obtained that mass transfer rate (k) value is $\mathbf{1{,}07x10^{-2}}$ L/g.s and the adsorption capacity (qo) is $\mathbf{6{,}12x10^{-3}}$ mg/g. These values indicate that absorbance from activated carbon used in the conducted experiments was small. Absorption capability and small constant rate indicates that mass of adsorbent used were too small compared to the applied NOM feed concentration. The consequence of these, the activated carbon used performed less capability to adsorb the mass of NOM solution. Furthermore, the activated carbon was easily saturated by high concentration of NOM used. In order to obtained better absorption ability of activated carbon, it is required to have a proportional mass of activated carbon used to mass of feed NOM since the concentration of adsorbed NOM is directly proportional to the amount of activated carbon used as the adsorbent.

4 Conclusion

The ozonation process of activated carbon was conducted at 5, 10, 15, 20, and 25 minutes whilst the activated carbon particle size was 20/35 mesh. The optimum condition obtained for this ozonation time is 5 minutes, since longer ozonation time will result in destruction of activated carbon pores and causing exuviating and decreasing of activated carbon pores. Therefore, ozonation after 5 minutes will reduce activated carbon surface area.

The activated carbon fixed bed height was set up at 1.5; 3; 4,5; 6; and 7.5 cm. It is obtained the optimum condition of fixed bed height was at 6 cm. Higher activated carbon bed applied increases activated carbon absorption capacity. Thus, more NOM can be absorbed in the adsorbent. However, whenever bed height was too high, it will cause increasing of retention due to the increasing of pressure drop that may disturb the influent rate and causing activated-carbon becomes quickly saturated.

Natural Organic Matter (NOM) might be processed using adsorption method by activated carbon as the adsorbent materials. However, ozonized activated carbon will perform higher absorption capability than pristine activated carbon. The process of ozonisation affects the surface area of the carbon. On the other hand, if the ozonisation time is taking more than 5 minutes, ozone will damage and decrease the carbon surface area.

Adi and Dian would like to appreciate to Mr Teddy Irnata who has helped the writer to design an instrument that is used in the research, and to Mr Nurcaya who has helped to fund the research.

References

1. Frimmel, F.H. and Abbt-Braun, G. Enviromental International, **25** (2/3), 191-207 (1999)
2. Fulcrum Health Limited. Humifulvate-a natural active ingredients. p. 1–43. Budapest: Humifulvate science, (2004)
3. Gora, S. Study on Characteristics and Removal of Natural Organic Matter in Drinking Water System in Newfoundland and Labor, (2011)
4. Ke peng, et al., Karbon. **49**, 70-76 (2011)
5. Naeimi, H., et al., Applied Surface Science, **256**, 631-635 (2009)
6. Peuravuori, J., et al., Water Research, **36**, 4552–4562 (2002)
7. Stevenson, F.J., *Humus Chemistry. Genesis, Composition, Reactions* (John Wiley and Sons Ltd : N, 1982)

Reuse of Newspaper As An Adsorbent For Cu (II) Removal By Citric Acid Modification

Mardiah, Rif'an Fathoni, Pratiwi Pudyaningtyas, Hamdania Gamu, Rinaldy

Department of Chemical Engineering, Faculty of Engineering, Mulawarman University, Indonesia

Abstract. High Consumption of paper, bring the impact of the waste paper itself. And the utilization of the paper is limited to recycled products and crafts, whereas paper such as newspaper still contains cellulose that can be potential to be used as a heavy metal adsorbent. In this study, newspaper was dissolved in sodium bicarbonate to reduce various impurities and then was reacted with citric acid (CA). The modified adsorbent was characterized by FTIR and was tested for adsorb Cu(II) in artificial solution. After adsorption process, the solution was filtered and analysed using Atomic Absorption Spectrophotometer (AAS). The adsorption experimental data was fitted to Langmuir, Freundlich, Tempkin, and Dubinin-Radushkevich for equilibrium model and was fitted to pseudo first order reaction and pseudo second order reaction for kinetic studies. The result showed that CA-modification newspaper able to remove heavy metals Cu(II) in solution.

1 Introduction

The growth of pulp and paper industry is still increasing. Over 400 million metric ton paper and cardboard are produced worldwide every year and consumtion of paper reached 400% in the last 40 years [1]. From newspaper to paper wrap, paper is still everywhere and mostly ends up in landfills so it will produce accumulated paper waste.

Meanwhile, the utilization of the paper is limited to recycled products and crafts, whereas paper such as newspaper still contains cellulose that can be used as a heavy metal adsorbent. Exposure to heavy metals can cause cancer in humans. For example Copper (Cu), if humans consumed continuously will cause irritation, itching, skin diseases, cancer, liver, kidneys and lungs [2].

Activated carbon as adsorbent has been widely used. However, activated carbon is an expensive material and activated carbon is widely used for adsorption of organic components instead of heavy metals [3].

Old newspapers is one source of cellulose and has the potential to be used as a heavy metal adsorbent. As research conducted by Chakravarty et al [4] used the newspaper that has been modified as adsorbent, capable of reducing the metal content of Zn 9.20 mg / g with initial concentration of Zinc 10.31 mg / lt. Then Chakravarty et al [2] conducted a study with an adsorbent newspaper to absorb Cu (II), obtained a maximum loading capacity of 30 mg / g with initial concentration of Cu (II) 20 mg / l. Dehghani et al [5] using old newspapers that have been modified to absorb chromium of 59.88 mg / g (64%) with a dose of adsorbent 3 g / l and a contact time of 60 minutes. Marshall [6] enhanced the adsorption capacity of soybean hulls for copper uptake by employing base treatment before modification with citric acid (CA). Pitsari et. al [7] showed that Citric Acid- modified newspaper pulp is an effective adsorbent for the removal of Pb^{2+} from wastewater. Lignocellulosic materials exhibit high metal removal ability that is attributed to the sorption properties of cellulose and its affinity towards metal ions. Metals interact with the negatively charged functional groups of cellulose such as carbonyls, carboxyl groups, hydroxyls, phenolics and sulfonic acids [7]. Therefore newspaper needs to be modified. One of treatment for modified is by using citric acid to form esterification [3].

2 Materials and Methods

This research was conducted in two stage: the first was preparation adsorbent from old newspapers. Newspaper was reduced in size, then was dissolved in a solution of 10% of sodium bicarbonate. Then the adsorbent filtered and washed with distilled water until pH 6.5-7. Then it was dried in the oven. After that, it was reacted with 0.5 M citric acid for 90 minutes then filtered and washed with 200 ml of distilled water. The adsorbent was dried in the oven until the weight is constant. The adsorbent was characterized by Fourier

* Corresponding author: mardiah@ft.unmul.ac.id

Transform InfraRed (FTIR) .The second stage, the parameters were tested initial concentration of Cu(II), adsorbent doses, and contacted time.

After the adsorption process was complete, the solution was filtered. Both initial and final concentration of $CuSO_4$. 5 H_2O was analyzed using atomic absorption spectrophotometer.

Percent of Cu (II) removal and adsorption capacity, was calculated using the following equation:

$$\% \text{ removal} = \frac{C_i - C_e}{C_i} \times 100 \qquad [1]$$

Adsorption capacity:

$$q_e = \frac{C_i - C_e}{m} \times V \qquad [2]$$

Information :
Ci: initial concentration of Cu(II) (mg / l)
Ce: the final concentration of Cu(II) (mg / l)
qe: Cu (II) which adsorbed by the adsorbent (mg / g)
V: total volume of solution (liters)
m: dose of adsorbent (g)

Isotherm adsorption Model

Adsorption Model approach which used in this study was the Langmuir, Freundlich, Tempkin, and Dubinin-Radushkevich adsorption. The Langmuir equation model in the form of concentration as follows:

$$q = \frac{QbC}{1 + bC} \text{ ....} (3)$$

Where q is the adsorbate adsorbed mass per unit mass of adsorbent (mg / g), Q is an adsorption capacity (mg / g), C is the concentration of adsorbate in solution and b is the equilibrium constant. The model of the Langmuir adsorption can be expressed in linear form:

$$\frac{C}{q} = \frac{1}{Qb} + \frac{C}{Q} \qquad (4)$$

Where Q and B are constants, and obtained from plotting a graph C / q vs. C. This equation is a straight line with a slope and intercept Q and Q, b.
The well-known Freundlich equation can be written as

$$\text{Log qe} = \log K_f + \frac{1}{n} \text{Log Ce} \qquad (5)$$

where n and k are the Freundlich constant.

The Temkin isotherm has a convenient linear form, which is expressed by the following equation:

$$qe = B \ln A_T + B \ln Ce \qquad (6)$$
$$B = RT/b \qquad (7)$$

where, A_T is Temkin isotherm equilibrium binding constant corresponding to the maximum binding energy (L/g), B is constantly related to the heat of sorption (J/mol), R is the universal gas constant (8.314 J/mol/K), T is absolute temperature at 298 K°, b is Temkin isotherm constant,which indicates the adsorption potential of the adsorbent. Both A_T and B can be determined from a plot qe vs. ln Ce.

Dubinin–Radushkevich (D–R) isotherm :

$$\ln qe = \ln Q_m - k_d\varepsilon^2 \qquad (8)$$

where ε (Polyani potential) is $[RT \ln(1 + 1/Ce)]$, Qm the theoretical saturation capacity(mg g−1), kd a constant related to adsorption energy (mol^2 KJ^{-2}), R is the gas constant (kJ mol^{-1} K^{-1}) and T is the temperature (K). The slope of the plot ln qe versus ε^2 gives the K_d and Q_m values

Adsorption kinetics
The linear form of Lagergren's pseudo-first-order model is generally expressed as follows:

$$\text{Log}\,(q_e - q_t) = \log q_e - \frac{k_1 t}{2.303} \qquad (9)$$

q_t is the amount of Cu (II) adsorbed at time t (min), k_1 is the pseudo first-order rate constant for the kinetic model (1/min).

Ho and McKay in Deghani et.al [5] described pseudo-second-order model as the kinetic process of the adsorption. The equations can be rearranged to obtain linearized as follows :

$$\frac{t}{q_t} = \frac{1}{k_2 q_e^2} + \frac{t}{q_e} \qquad (10)$$

where, k_2 is the rate constant of pseudo-second-order kinetics (g/mg·min).

3 Result and Discussion

3.1 Characterization of adsorbent

The surface functional groups were analysed by Fourier transform infrared spectroscopy (FTIR) in the Treated Newspaper (TN) as describe at Figure 1.

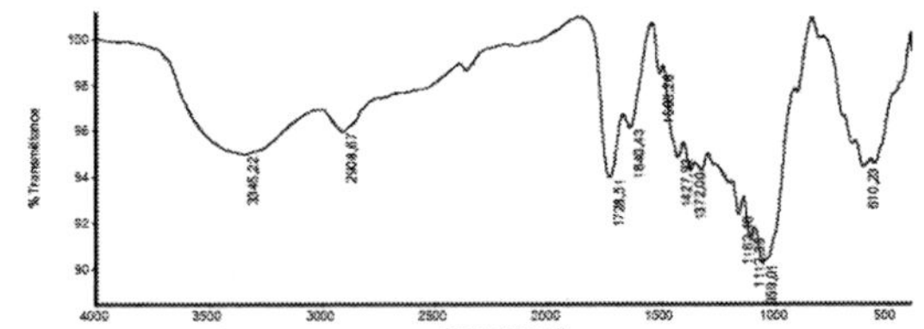

Fig. 1. FTIR spectra of citric-acid modification Newspaper

The FTIR spectra of CA- modified newspaper was in the region 4000- 500 cm^{-1} , showing characteristic cellulose peaks in the range of 1000- 1200 cm^{-1}. The band near 1000- 1100 cm^{-1} corresponds to C- O- C groups from β- (1- 4)- glycosidic bonds in Cellulose. The peaks at 1372 cm^{-1} are assigned to O- H bending and CH_2- wagging vibrations. The band near 3500 cm^{-1} is assigned to hydroxyl vibrations. The wave numbers at 2908 cm^{-1} are characteristic of alkaline groups and the band near 1640 cm^{-1} correspond to bending vibrations of fibres due to the absorbed water. The introduction of additional carboxyl groups to newspaper pulp using CA was further confirmed by characteristic vibration at 1728 cm^{-1} originating from the C=O stretch of ester carbonyl group and carboxylate ions which was the result of CA esterification [7].

3.2 Effect of adsorbent dose

Fig. 2 shows the adsorption of Cu(II) with varying weight of the adsorbent. It indicates that the uptake of Cu(II) increases as the adsorbent dose increases from 0.05, 0.1, 0,25, 0,5 to 1 g which entered into a 50 ml $CuSO_4 . 5H_2O$ with a concentration was 100 mg / l for 60 minutes therefore Cu (II) in solution was decreased.

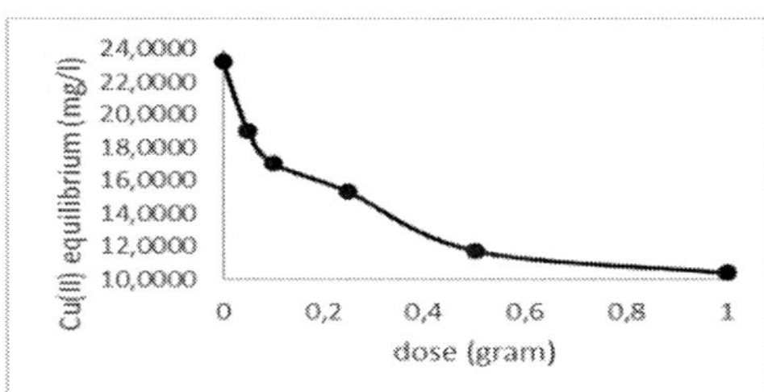

Fig.2. Effect of adsorbent dose

The adsorbent dose of 1 g was able to adsorb 55,30% Cu(II) from 50mL of 23, 27 mgL^{-1} Cu(II) in solution. This increase of Cu(II) removal may be attributed to the active site of the adsorbent surface (Fig.3).

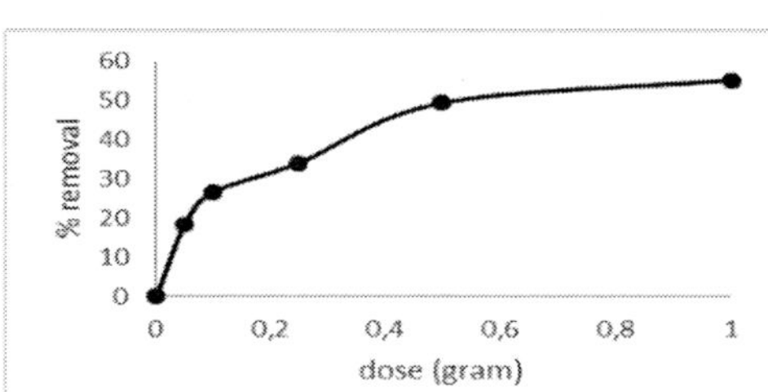

Fig. 3. Removal of Cu(II) as an effect of adsorbent dose

3.3 Effect of initial concentration

The experiments were done with variable initial copper concentration (4.2, 23.27, 35.40, and 42.42 mg/L), with adsorbent dose (1 g/50 mL), contact time (1 h) at room temperature. The percentage of Cu (II) ion uptake on the adsorbent is presented in Fig.4. Fig. 4 shows that by increasing initial Cu (II) concentration, copper removal efficiency is decreased but increased in adsorption capacity (Fig. 5). Thus, results suggest that adsorption capacity and removal efficiency of adsorbent is dependent on the initial concentration of copper [5].

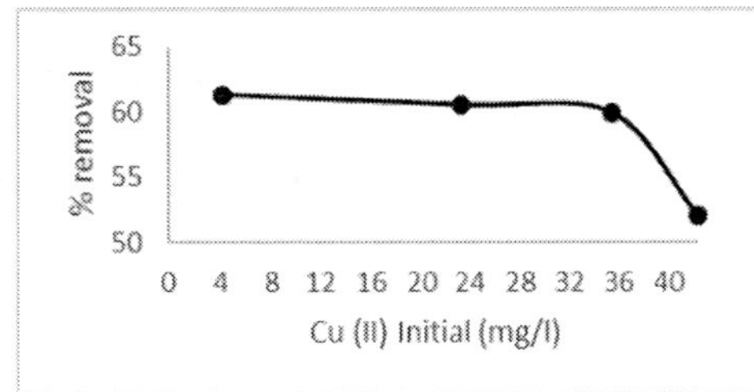

Fig. 4. Removal of Cu(II) as an effect of Cu (II) initial

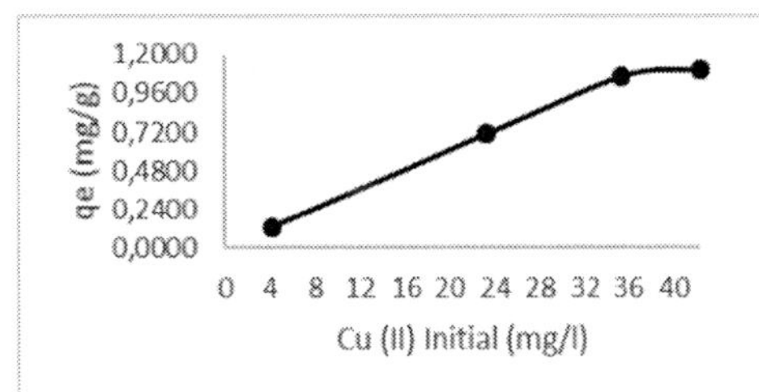

Fig. 5. Adsorption capacity of Cu(II) as an effect of Cu (II) initial

3.4 Effect of contact time

The effect of contact time (15, 30, 60, 90, and 120 minutes) was studied for an initial Cu (II) concentration of 23, 27 mg/L; Adsorbent dose of 1 g/50 mL at room temperature. Fig. 6 shows that Cu (II) in solution was decreased with contact time. The maximum efficiency for removal Cu(II) adsorption was 66,78% at 120 min (Fig. 7).

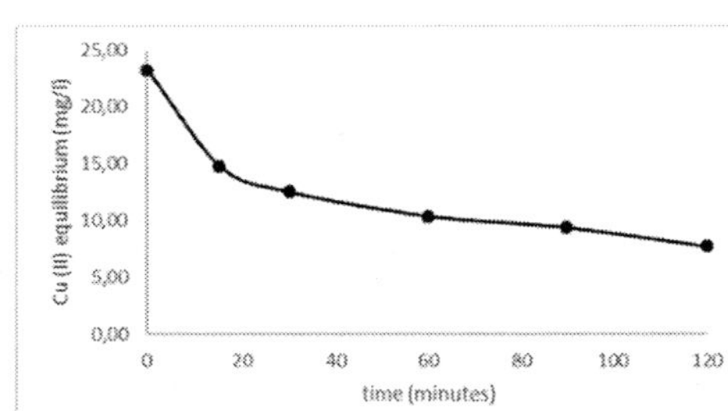

Fig. 6. Effect of contact time

https://doi.org/10.1051/matecconf/201815602012

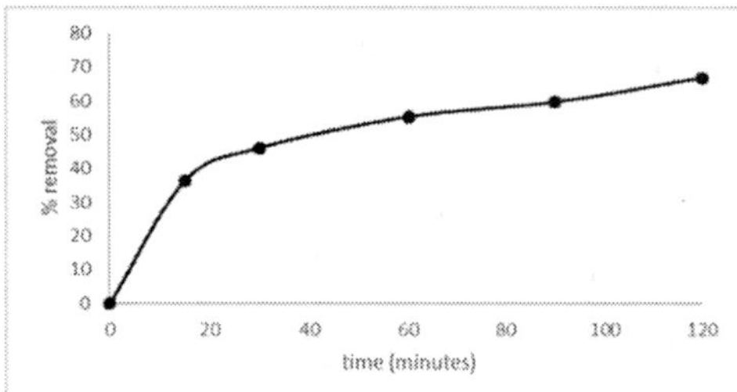

Fig. 7. Removal of Cu(II) as an effect of contact time

3.5 Adsorption Isotherms

The equilibrium adsorption of Cu (II) on the CA-modified newspaper was fitted using adsorption isotherms as discussed before. Fig. 8-11 shows Langmuir, Freundlich, Tempkin, and Dubinin-Radushkevich adsorption.

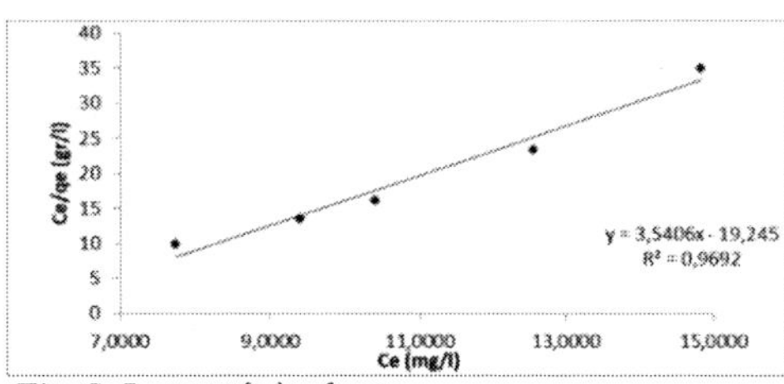

Fig. 8. Langmuir isotherm

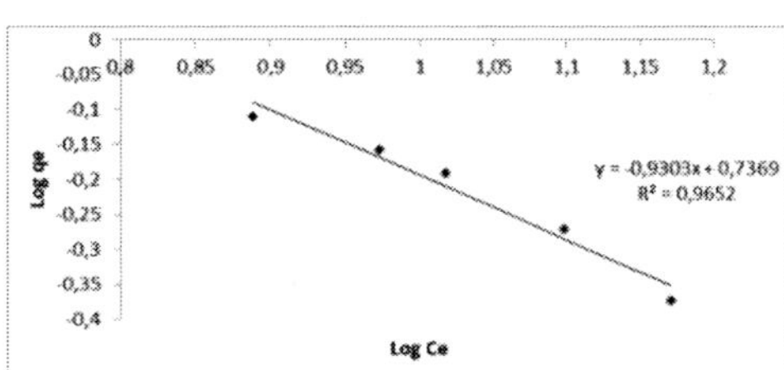

Fig. 9. Freundlich isotherm

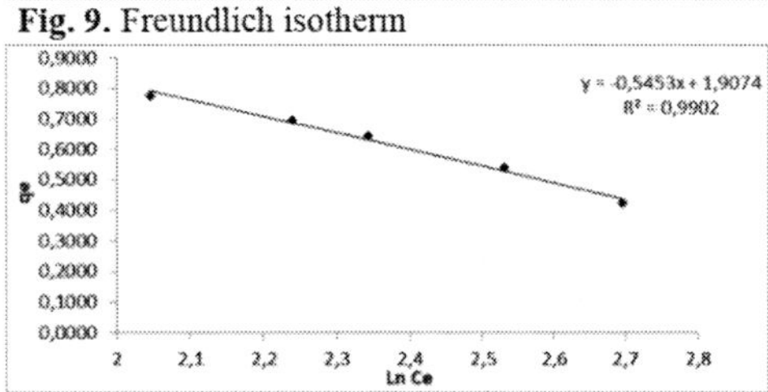

Fig. 10. Tempkin isotherm

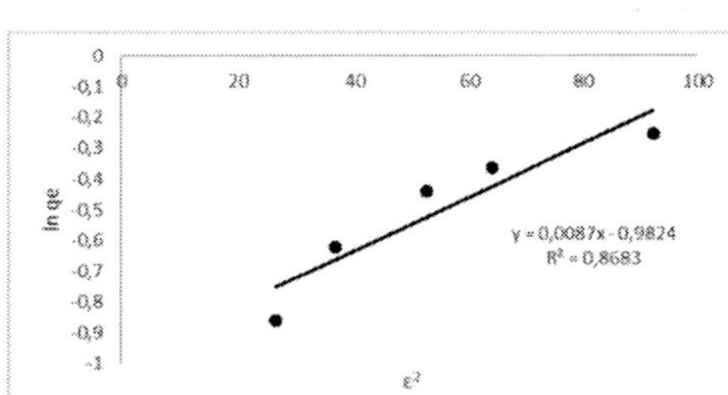

Fig. 11. Dubinin–Radushkevich (D–R) isotherm

Constants values of all four isotherms are presented in Table 1.

Table 1. Isotherm parameters and R^2 values for adsorption of Cu(II) onto CA- modified newspaper

Isotherm	Constants	Value
Langmuir	b (L mg^{-1})	0.1840
	Q_m (mg g^{-1})	0.2824
	R^2	0.9692
Freundlich	n	-1.075
	k_f	5.4563
	R^2	0.9652
Tempkin	A_T (L g^{-1})	0.0298
	B (J mol^{-1})	-0.5453
	R^2	0.9902
Dubinin–Radushkevich	K_d (mol^2/KJ^2)	-0.0087
	Q_m (mg g^{-1})	0.3744
	R^2	0.8683

The results showed that Tempkin gave the best fit for the Cu(II) adsorption by CA-modified newspaper with R^2 = 0.99.

3.5 Adsorption Kinetics

The linear form of Lagergren's pseudo-first-order model shows in Fig.12. and Pseudo- second order shows in Fig.13. at adsorbent dose 1 gram in 50 ml of 23,27 mg/l Cu (II) solution.

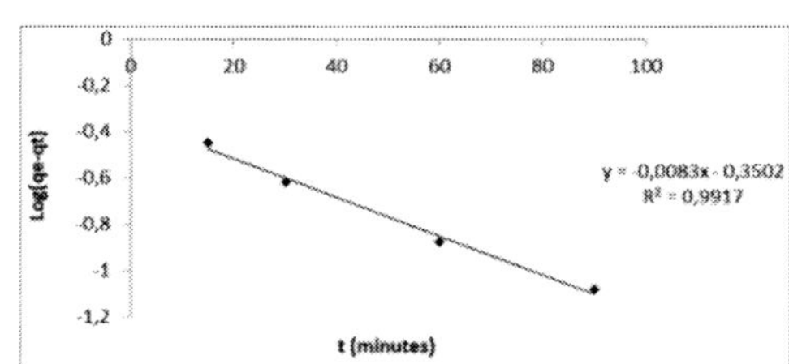

Fig. 12. Pseudo-first-order kinetics plots for adsorption of Cu (II) onto CA-modified newspaper

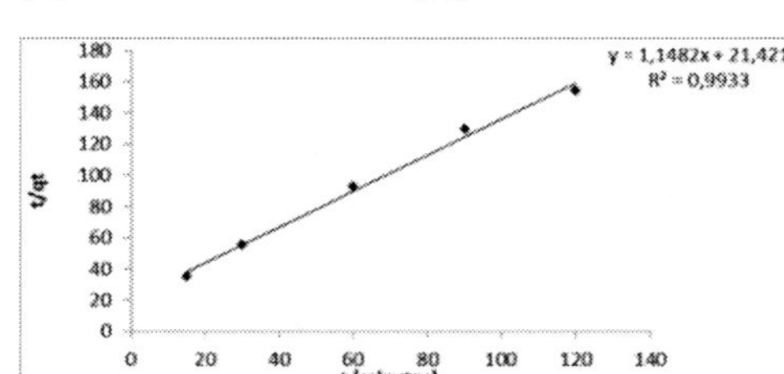

Fig. 13. Pseudo-second-order kinetics plots for adsorption of Cu (II) onto CA-modified newspaper

Kinetic parameters for the removal of Cu (II) by CA-modified newspaper are shows in Table 2.

Table 2. Kinetic parameter for the removal of Cu (II) by CA-modified newspaper

Pseudo 1st order model	k_1	q_e (mg g^{-1})	R^2
	0.8065	0.4465	0.9917
Pseudo 2nd order model	k_2	q_e (mg g^{-1})	R^2
	0.0354	11.482	0.9933

A comparison of Figs. 12 and 13 showed that pseudo-second-order was the best model for the Copper (II) removal onto CA-modified newspaper with a higher correlation coefficient (R^2 = 0.9933) than for pseudo-first-order (R^2 = 0.9917).

4 Conclussion

The initial concentration of the solution, adsorbent dose and contacted time has been tested in this study. Adsorbent from newspapers which modified citric acid can be used to reduce the levels of Cu (II) in the solution.

We acknowledged the LP2M Mulawarman University, Ministry of Research, Technology and Higher Education for funding the research.

References

1. Pratima Bajpai, *Pulp and Paper Industry*, Elsevier (2015)
2. S.Chakravarty , S. Pimple, Hema T. Chaturvedi, S. Singh, K.K. Gupta. Journal of Hazardous Materials, **159**, 396–403 (2008)
3. David William O'Connell, Colin Birkinshaw , O'Dwyer Thomas Francis. Bioresource Technology,**99**, 6709–6724, (2008)
4. S.Chakravarty, S. Bhattacharjee, K.K. Gupta,M. Singh, H.T. Chaturvedi, S. Maity. Bioresour. Technol,**98** , 3136–3141, (2007)
5. Mohammad Hadi Dehghani , Daryoush Sanaei , Imran Ali , Amit Bhatnagar. Journal of Molecular Liquids, **215**, 671–679, (2016)
6. W.E. Marshall, L.H.Wartelle, D.E. Boler, C.A. Toles. Environ. Technol, **21,** 601–607,(2000)
7. S. Pitsari, E. Tsoufakis, M. Loizidou. Chemical Engineering Journal, **223,** 18-30, 2013.

Mathematical modeling of supercritical CO_2 extraction of valuable compounds from *Eucheuma Cottonii* and *Gracilaria Sp*

Dwila Nur Rizkiyah, Nazla, Farah Nadhifah, Siti Machmudah, and Sugeng Winardi*

Department of Chemical Engineering, Sepuluh Nopember Institute of Technology, Surabaya 60111 Indonesia

Abstract. Extraction by using supercritical CO_2 has been extensively developed to extract materials that are easily decomposed at high temperatures. Therefore, in this study extraction of valuable compounds (such as: carotenoids and fatty acids) from macroalgae of Eucheuma cottonii and Gracilaria sp was carried out experimentally with supercritical CO_2 and mathematical model of Broken and Intact Cells (BIC) and Chrastil model. The result with BIC model is compared with the experimental result that has been done. Extraction was conducted at various pressures of 15, 20, and 25 MPa, temperature of 40, 60, and 80 °C, CO_2 flow rate of 6 mL/min, and ethanol concentrations for cosolvent of 5 %, 7.5%, and 10% CO_2 flow rate. The content of extract was analyzed by using a Spectrophotometer UV - Vis and HPLC. The total yield of extract on *Eucheuma cottonii* was about 0.01623-0.03752 gr extract/gr sample at operating conditions 15 MPa pressure, temperature 60 °C, and ethanol flow rate of 10%. The total yield of the extract on *Gracilaria sp* was 0.1982-0.4237 gr extract/gr samples, at operating conditions 15 MPa pressure, 60 °C temperature, and ethanol flow rate 7.5% CO_2 flow rate. The solubility of *Eucheuma cottonii* are greater than *Gracilaria Sp* with constant of Chrastil k, a, and b that are: 2,52; 0,911; and -27,66 at operating conditions. Broken and Intact Cells model could also describe well for extraction yield. The best fitting parameters in BIC model depends on condition at extraction process, such as : 0.45 – 0.46 for *f*, 0.5 - 0.75 for X_c, and 0.15 - 0.4 for *K* value..

1. Introduction

Macroalgae is well known not only as a valuable biofuels source, but also for various other compounds such as carotenoid, caragenant, lipid acid, and phenolic. since the biomass is be able to synthesize lipids, carbohydrates and proteins[1]. Macroalgae is a naturally product that has been applied by some countries. Macroalgae has aplied as suplementary food, medicine, cosmetic, animal feed, and organic fertilizer by chinese and japan since 1670. Algae has some nutrient contents, thats are: proteins, amino acids, minerals, ashes, vitamin A, vitamin C and fats. Some types of algae are include *Eucheuma Cottoni* and *Gracilaria Sp. Eucheuma cottoni* are known as red algae (Rhodophyceae) has founded under the sea. *Eucheuma cottonii* and *Gracilaria Sp* also contains carotenoids and fatty acids[2].

Carotenoids (β-carotene) is serves as a preventative of heart and cancer disease as well as strength the immune of system. This research is about the supercritical fluid extraction of lipids from macroalgae, a new development in recent years whose have purpose to replace conventional solvent extraction. Supercritical fluids are substances that have a critical temperature and pressure. This work is about the supercritical fluid extraction of lipids from macroalgae, a new technique developed in recent years whose purpose is to replace conventional solvent extraction, that has a greater environmental impact mainly due to the use of toxic substances.[3]. The solvent strength of this fluid depends on density. Pressure was increased with increasing of density and solubility was enhanced with enhancing of pressure. Therefore, operating pressure and temperature can be varied to optimize mass-transfer properties like density, diffusivity, viscosity, and, especially solubility. Carbon dioxide is extremely appropriate as a supercritical fluid since its critical conditions are at quite lower pressure : 7.38 Mpa, and temperature : 31.35 °C, [4].

Supercritical CO_2 extraction of lipids from macroalgae exploits the high solubility of fatty acids in supercritical CO_2 and high selectivity for non polar and low polar

* Corresponding author: machmudah@chem-eng.its.ac.id

Nomenclature

a model's constant (vR^2/D_eL)
a_p interfacial area per unit volume of bed (m^{-1})
b model's constant (y_s/x_0)
d_p model's constant (m)
E dimensionless extraction yield (F/Nx_i)
f initial fraction solute in broken cell in particle
F extract (kg)
G model's constant ($k_sa_p/(1-\varepsilon)$
k_f film mass transfer coefficient in fluid(m/s)
k_s film mass transfer coefficient in solid (m/s)
K partition coefficient
$\overline{K}$ partition coefficient in dimensionless (Kx_{10}/y_0)
L bed length (m)
N solid charge in extractor (kg)
r radial coordinate
r_c critical radius of core (m)
R particle radius (m)
S solvent flowrate (kg/s)
t time (s)
v interstitial fluid velocity (m/s)
x concentration of solid phase (kg solute/kg insoluble solid)
x_c transition concentration (kg solute/kg insoluble solid)
x_i solute fraction in untreated solute (kg solute/kg insoluble solid)
x_0 initial solute fraction in particle (kg solute/kg insoluble solid)
x_1 solid phase concentration with broken cells (kg solute/kg insoluble solid)
x_{10} solid phase initial concentration with broken cells (kg solute/kg insoluble solid)
x_2 solid phase concentration with intact cells (kg solute/kg insoluble solid)
x_{20} solid phase initial concentration with intact cells (kg solute/kg insoluble solid)
X dimensionless concentration of solid phase (x/x_0)
X_c dimensionless concentration transition (x_c/x_{10})
X_1 dimensionless concentration broken cells (x_1/x_{10})
X_2 dimensionless concentration intact cells (x_2/x_{10})

Greek letters
ε bed voidage
θ dimensionless time (vt/L)
ρ_f density of fluids (kg/m^3)
ρ_s density of insoluble solid (kg/m^3)
τ dimensionless time (D_et/R_2)
ψ_e dimensionless external mass transfer ($\varepsilon v/k_fa_pL$)
ψ_i dimensionless internal mass transfer ($(1-\varepsilon)v/k_sa_pL$)

compounds. Furthermore, a co-solvent (such as methanol, ethanol, water), can improve the solubility of polar compounds as well [5].

The research are compares both original extraction data for two macroalgae strains (*Eucheuma cottonii* and *Gracilaria Sp*) with a simple approach to modeling the supercritical CO_2 extraction of lipids by a broken and intact cells model and Chrastil model.

2. Materials and method

2.1. Materials

Red algae *Eucheuma cottonii* and *Gracilaria sp* found on Coastal Coast in Pamekasan, Madura. Indonesia. Carbon dioxide (CO_2) liquid with a purity of 99.7% purchased at PT. Samator and ethanol 97% as co-solvent was purchased at UD.Chemical Main Source (SUK).

2.2 Method

Macroalgae *Eucheuma cottonii* and *Gracilaria sp* which have been taken from Madura beach are washed with fresh water which is aim to remove dirt and salt content that still attached to the macroalgae.

The dried macroalgae is then ground to size reduction until the size is approximately 20-35 mesh. It aims to enlarge the area of contact of algae extraction with solvent. At the bottom (inlet) and top (outlet) extractor added glassbead as much as 2.5 gram each. The addition of glassbead aims to prevent channeling. After that, install the extractor in a series of supercritical extraction equipment [6].

Furthermore, the extraction process is modeling for Broken and Intact cell (BIC) model using MATLAB R2013a application with numerical solution by Crank Nicholson method. Experimental data were compared to model analyze.

3. Mathematical Modeling

3.1. Broken and intact cells model

Modeling of the process was established based on mass transfer balances to estimate renewal of extracted compounds: β -karotene and fatty acids. The model was developed about on extraction in the batch system extractor with inside diameter of 21 mm, length of 130 mm and volume of 43 mL. [7].

The model that describes the extraction process is important to optimize the parameters that control and to fix the efficiency. During the extraction process, the oil was move from the macroalgae cells to the solvent quantity, so that the models are focused on the description of mass transfer from a solid matrix to a supercritical fluid. The parameters are the mass transfer coefficients. The model of broken and intact cells was developed by Sovová [8].

Broken and Intact Cells model was chosen because it is illustrates particularly well the macroalgae structure after the pre-treatment. Although it was built at first to describe the supercritical fluid extraction of vegetable oil from plants, it will be shown that it works as well with the extraction of oil from macroalgae[8].

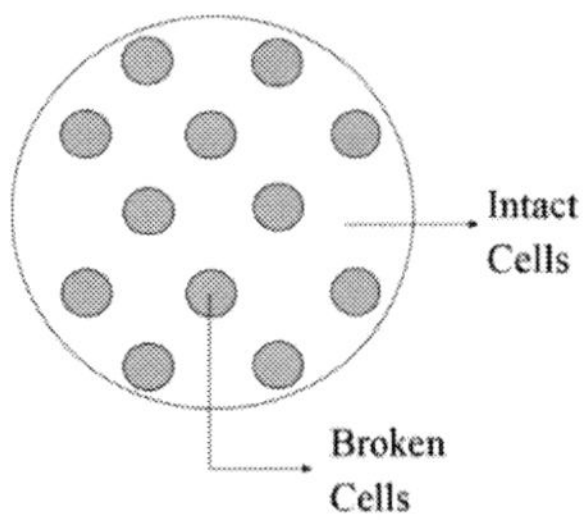

Fig. 1. Illustration of particle for Broken and Intact Cells Models.

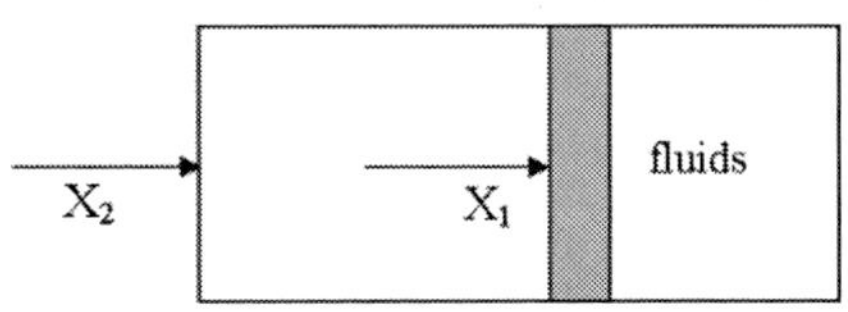

Fig. 2. Simplified representation of solute transfer from the particle to the solvent (fluids).

Therefore, as shown in Figure 1, two regions can be distinguished in the particle: the region of broken cells near the surface with easy accessible solute, and another region in the core as intact cells. The simplified representation of solute transfer from the particle: X_1 as broken cells dan X_2 as intact cells, to the solvent (fluids) can be described in Figure 2 .These condition lead to divide the extraction process in two periods. In the first extraction period, the free solute inside the intact cells (X_2) through the side of broken cells (X_1) goes easily from the particle to the solvent (fluids).

3.1.1 Parameter models

The Sovová model is characterized by equations for the mass balances: the phase equilibrium and the mass transfer that allow to calculate the extraction curve[8]. The equations for the extraction curve are fitted to experimental data to calculate and optimize the parameters X_c, f and K.

The way to solve the mass balance equations are introduce the dimensionless parameters and different equations are formulated for extraction periods, in the cases of plug flow. These equations, published by Sovová in 2005, represent a new general model that extends the group of models based on broken and intact cells [8].

3.1.2 Mass balance equations

The mass balance equations are written for the solute in three phases [8], those are : the fluid phase in equation 1, the solid phase with broken cells equation 2, and the solid phase with intact cells in equation 3, For the mass balance in the fluid phase the accumulation part and the convective part have to be considered :

$$\rho f \varepsilon \left(\frac{\partial y}{\partial t} + \upsilon \frac{\partial y}{\partial z} \right) = js \quad (1)$$

where: ρf is the solvent density, ε is the porosity, y is the solute concentration in the fluid phase, t is the extraction time, υ is the interstitial velocity, z is the axial coordinate, js is the flux from broken cells to solvent.

In the solid phase with broken cells there is no convective term and the flux from intact cells to broken cells is included, that in the previous equation was supposed to be equal to zero :

$$f \rho_s (1-\varepsilon) \frac{\partial x1}{\partial t} = jr - js \quad (2)$$

where: f is the grinding efficiency, ρ_s is the solid density, $x1$ is the solute concentration in broken cells, jr is the flux from intact cells to broken cells, and the others were already defined previously.

Finally, in the solid phase from intact to broken cells only the flux in the solid phase affects the accumulation of solute over time :

$$(1-f) \rho_s (1-\varepsilon) \frac{\partial x2}{\partial t} = -jr \quad (3)$$

where: $x2$ is the solute concentration of intact cells and all the other variables are already defined. To solve the differential equations 1, 2, and 3, the initial concentrations are used as boundary conditions:

$$y|_{t=0} = y_0 = y*; \quad y|_{z=0} = 0;$$

$$x_1|_{t=0} = x_{10}; \quad x_2|_{t=0} = x_{20} = x_i; \left.\frac{dy}{dz}\right|_{z=L} = 0 \quad (4)$$

where: y_0 is the initial solute concentration in the fluid phase while x_1; 0 and x_2; 0 are the initial solute concentrations in broken and intact cells, respectively.

A solute concentration x_c is then defined, that indicates the capacity of the matrix for interaction with the solute. Above this concentration, solubility is equal to phase equilibrium, while below xt it depends on a partition coefficient:

$$y*(x_1) = y_s; \quad for\ x_1 > x_c$$

$$y*(x_1) = Kx_1; for\ x_1 \leq x_c$$

$$where\ Kx_c < y_s \quad (5)$$

where: $y*(x_1)$ is the equilibrium fluid phase concentration, y_s is the thermodynamic solubility of the solute in the solvent and K is the partition coefficient.

The way to solve the mass balance equations, fluxes js and jr have to be defined, being aware that for jf the discontinuity of the phase equilibrium has to be considered. Therefore, the flux from broken cells to the solvent is defined as:

$$js = kfa_p\rho f(y* - y) \quad (6)$$

$$jr = k_s\, a_p\rho_s(x_2 - x_1) \quad (7)$$

The yield of oil extracted, F, collected during the extraction process is calculated as:

$$F = S\int_0^t y\,|_{z=L} dt \quad (8)$$

where: S is the solvent flow rate and L is the length of the extraction bed. Which are some have changed the relationship of equilibrium suggested by Perrut et al [9].

3.1.3 Dimensionless model equations

The model equation are modify into a dimensionless using the dimensionless variable :

$$Z = \frac{z}{L};\ \theta = \frac{vt}{L};\ X_c = \frac{x_c}{x_{10}};\ X_2 = \frac{x_1}{x_{10}};$$

$$Y^* = \frac{y^*}{y_0};\ Y = \frac{y}{y_0};\ X_1 = \frac{x_1}{x_{10}} \quad (9)$$

Dimensionless model equations with initial and boundary conditions according to these notations ,and equation 1 can be written as :

$$\frac{\partial Y}{\partial \theta} + \frac{\partial Y}{\partial Z} = \frac{1}{\Psi_e}(Y^* - Y) \quad (10)$$

The overall dimensionless mass-transport coefficient in equation 2,3 and 4 can be expressed as a function in equation 11, 12, and 13.

$$\frac{\partial X_1}{\partial \theta} = \frac{1}{\Psi_{if}}(X_2 - X_1) - \frac{\Phi}{\Psi_e}(Y^* - Y) \quad (11)$$

$$\frac{\partial X_2}{\partial \theta} = -\frac{1}{\Psi_i(1-f)}(X_2 - X_1) \quad (12)$$

$$E = \frac{F}{N x_i} = \frac{\Phi f}{1+\Phi}\int_0^\theta Y|_{Z=1} d\theta \quad (13)$$

4. Results and discussion

In this experiment was conducted the variation of operating conditions, specifically: temperature, pressure, and ethanol flow rate. The first experiments were carried out at constant pressure and temperature to find the most suitable extraction time and the CO_2 flow rate was kept constant in all the experiments [10]. The extraction time depends mainly on 240 minutes. Afterwards, the role of ethanol was investigated, tempting the different percentages those are: 5; 7.5; and 10%, with CO_2 flow rate of 6 ml/min.. In this research, the results of pressure, temperature, and supercritical CO_2 flow rate of two macro algae on extraction yield were examined, and the experimental and model results were compared [11].

Table 1 and table 2 show the experimental data that were compared with the model of broken and intact cells. Analysis will examine in the following section.

Table 1. *Eucheuma Cottoni*, 15 MPA, 40°

Time (min)	Yield (gr extract/ gr sample)	%recover b-carotene	%recover fatty acid
30	0.1240	0.0562	0.0455
60	0.1428	0.06477	0.05189
90	0.1531	0.06944	0.05635
120	0.1833	0.08311	0.05783
180	0.1937	0.08782	0.06275
240	0.2007	0.09099	0.0656

Table 2. *Gracilaria Sp*, 15 MPA, 40°

Time (min)	Yield (gr extract/ gr sample)	%recover b-carotene	%recover fatty acid
30	0.2123	0.0301	0.0392
60	0.3720	0.0527	0.0756
90	0.3763	0.0533	0.07664
120	0.3875	0.0549	0.07991
180	0.4191	0.0594	0.0871
240	0.4353	0.0617	0.0913

Table 3. Initial concentration in BIC Model.

P (MPa)	T (°C)	y_0 [-]	x_0 [-]
15	40	0.021	0.19
20	60	0.04	0.3
25	80	0.052	0.35

4.1 Effect of Temperature

Broken and Intact Cells (BIC) model were compared into the experimental data. The initial concentration is important for known the values of concentration profile as function in extractor height for BIC model.

The values of concentration profile as function in extractor height for BIC model has shown in Figure 3. The BIC Models has some initial fraction for fitting parameters, those are: f as ground particle, dimensionless transition concentration as X_c, and partition coefficient as *K*.

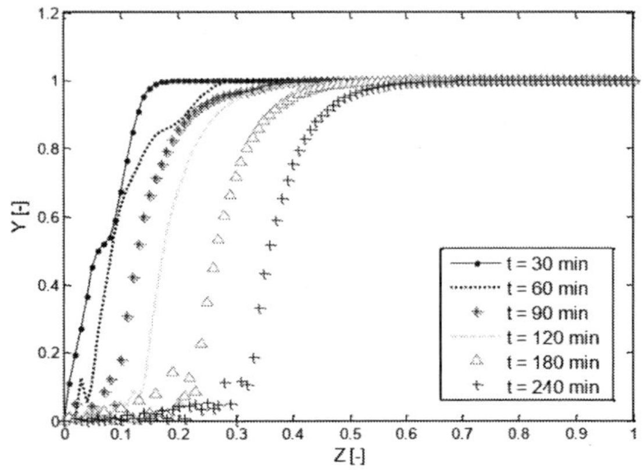

Fig. 3. The values of concentration profile as function for extractor from BIC model.

Figure 3 shows the value of solute concentration (Y) in the fluid phase decrease with increasing of extraction time. In the beginning, solute was certainly dispersed in the fluid phase of extractor. Solute that contained in the particle was placed in the bottom of extractor and was rapidly removed from particle, even particle place at the top of extractor. The solute in the intact cells was simply dispersed into the broken cells and over dispersed solute in the cells of particle. In the BIC model, equilibrium was simply achieved according in the Figure 3.

Table 4. The values of fitting parameters in BIC model.

Condition	*f* [-]	X_c [-]	*K* [-]
P : 15 mpa	0.45	0.5	0.15
P : 20 mpa	0.46	0.6	0.2
P : 25 mpa	0.45	0.7	0.2
T : 40°C	0.46	0.7	0.3
T : 60°C	0.46	0.75	0.25
T : 80°C	0.46	0.75	0.2
Solvent-Flowrate :5%	0.45	0.5	0.4
Solvent-Flowrate :7.5%	0.45	0.5	0.4
Solvent-Flowrate : 10%	0.45	0.5	0.4

The effect of temperature on extraction yield with comparison to BIC Model from macroalgae *Eucheuma cottonii* and *Gracilaria Sp*, at : 15 MPa, and CO_2 flowrate 6 ml/min are shown in Figure 4 and 5. Those figures shows that extraction yield increased with increasing of temperature. In the extraction process, the operating temperature is varied by 40, 60, and 80 °C at supercritical fluid extraction affects into the solvent of density and vapor pressure of the solute to be extracted.

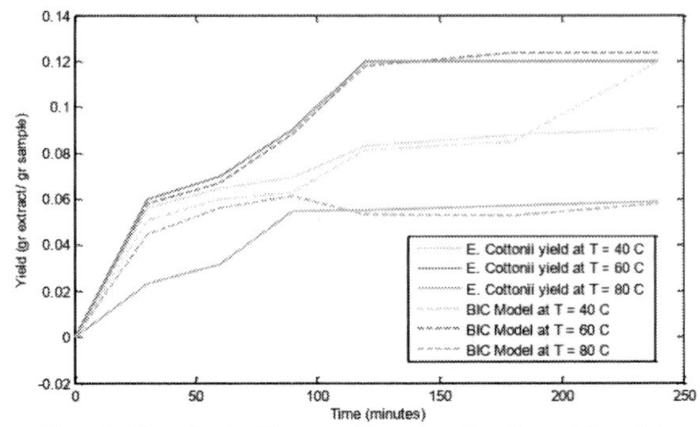

Fig. 4. The effect of temperature on extraction yield with comparison to BIC Model.
(*E. Cottonii*, 15 MPa, CO_2 flowrate 6 ml/min)

As shown in Figure 4, at 30 to 80 minutes the yield of extract at 60 and 80 °C was smaller than the yield of temperature 40 °C, due to the decreased CO_2 density along with the increased in temperature at pressure: 15 MPa.

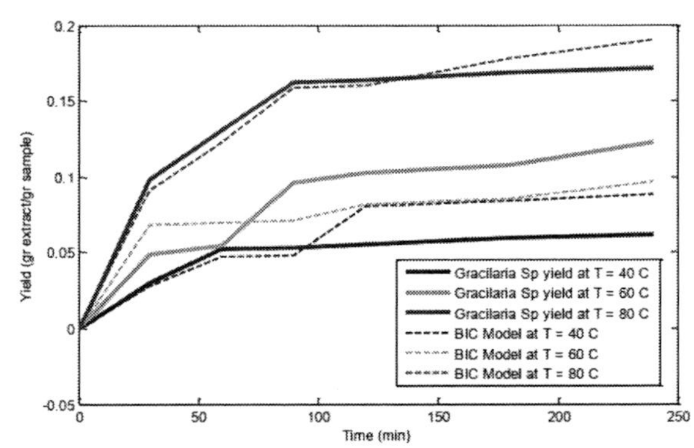

Fig. 5. The effect of temperature on extraction yield with comparison to BIC Model.
(*Gracilaria Sp*, 15 MPa, CO_2 flowrate 6 ml/min).

The different from Figure 5, in time 30 to 90 minutes the yield of extracts at 60 and 80 °C were higer than the temperature of 40 °C due to increased CO_2 density along with the increase in temperature at low pressure: 15 MPa. However, at 240 minutes the yield of extract at 60 and 80 oC was increased slowly so that the accumulation was higher than the extract at 40 °C. Those are causes the vapor pressure of the solute increased and allows the components to diffuse into the solvent. So that the solubility of the solute in solvent was increased, and the total extract produced increases. The total extract at a temperature of 40 °C is more lower than the temperature of 60 and 80 °C because at temperature of 40 °C the influence of solvent density are more dominant[12].

The BIC model could show the experimental data and the comparison of BIC Model had good compatibility with experimental data, but the BIC model could not show the experimental data well at temperature 40 and 60 °C at figure 4 and 5. This might be due to the use of more than one fitting parameters in BIC model. It is important to make the optimum fitting parameter. In addition, the initial concentration of solute in solid (x_0) should also based on maximum yield from experimental data [13].

4.2 Effect of Pressure

The effect of pressure on the extraction process with comparison to BIC Model has been studied at constant temperature of 60 ° C and constant CO_2 flow rate of 6 ml/min with a 5% co-solvent (ethanol) flow rate. To study the effect of pressure in the extraction process, the operating pressure is varied by 15, 20, and 25 MPa. The appropriating extraction yield data for *Eucheuma cottonii* and *Gracilaria Sp* are informed in Figure 6 and 7.

Figure 6 and 7 shows the effect of pressure on the total extract obtained. It can be seen that the increase in pressure is directly proportional to the increase of extract obtained. The most of *Eucheuma cottonii* extracts were obtained at 25 Mpa pressure with weight of 0.00237 gr extract/gr sample.

The addition of extract in Figure 6, at 25 MPa from 30 to 120 minutes has a significant increase but tends to be constant at 120 to 240. This indicates that at the time of extraction for 240 minutes most of the extract has been taken.

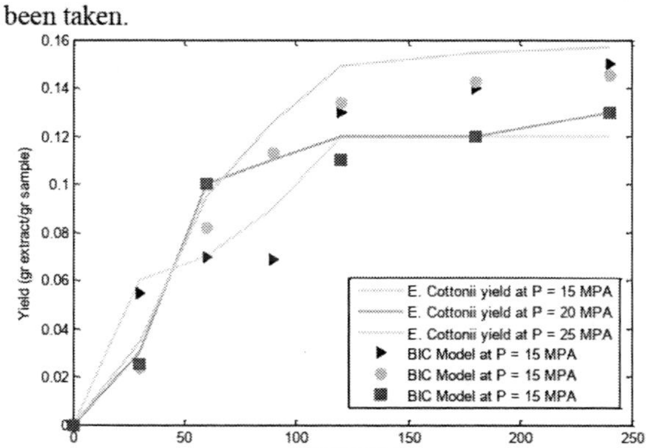

Fig. 6. The effect of pressure on extraction yield with comparison to BIC Model.
(*E. Cottonii*, 60°, CO_2 flowrate 6 ml/min)

Figure 7 has shown the data for *Gracilaria Sp*, that has a middle similar trend to *Eucheuma cottonii*. The total extract that can increase with increasing pressure due to increasing pressure leads to increased of supercritical fluid density, so that the strength of SC-CO_2 as solvent to dissolve the solute also increased.

The contrast yield from *Gracilaria Sp* and *Eucheuma cottonii* caused by the molecular structure of algae *Gracilaria Sp* in the form of flat proved by SEM analysis. So, for the lipid increased difficult to penetrate at high condition shown in the figure 10 :

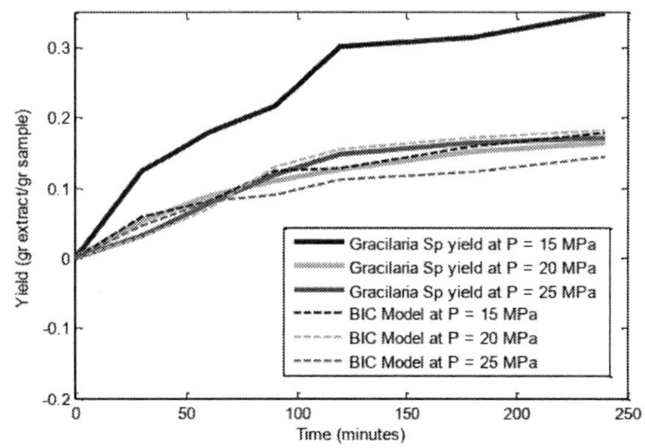

Fig. 7. The effect of pressure on extraction yield with comparison to BIC Model.
(*Gracilaria Sp*, 60°, CO_2 flowrate 6 ml/min)

4.3 Effect of Co-Solvent Flowrate

Figure 8 and 9 shows the effect of co-solvent (ethanol) flowrate on extraction yield with comparison to BIC Model from macroalgae *Eucheuma cottonii* and *Gracilaria sp* were investigated by varying the volume of ethanol that streamed by 5%, 7.5%, and 10%. The addition of ethanol as a co-solvent is needed to increase the solvent power. So that, the extract can be taken maximal.

As shown in figure 8 and 9, extraction yield was increased with increasing of co-solvent flowrate. The higher of co-solvent flowrate, the total yield was increased. The addition of ethanol as a co-solvent flowrate is need to affect the solubility of the extract.

Solvent serves to improve the efficiency of extraction and may affect the properties of supercritical CO_2 by adding ethanol as a co-solvent to help develop biomass pores, by adding supercritical CO_2 interactions with solute matrix cells [14]. BIC Model calculation is explain the satisfactory described the experimental data has shown in figure 8 and 9.

4.4 Solubility and rate extraction

In this study, the extraction rate is increases with increasing of pressure. This is by reason of the CO_2 density increases along with the increasing in operating pressure. Increase in CO_2 density shows an increase in the amount of CO_2 per unit volume that causes more CO_2 molecules to dissolve β-carotene and linoleic acid. The solubility of linoleic acid is better than β-carotene because the amount of linoleic acid in *Eucheuma cottonii* and *Gracilaria sp* is higher than that of β-carotene. Solubility of the experimental results is then correlated with the chrastil equation which results are shown by the table 6 and 7 below:

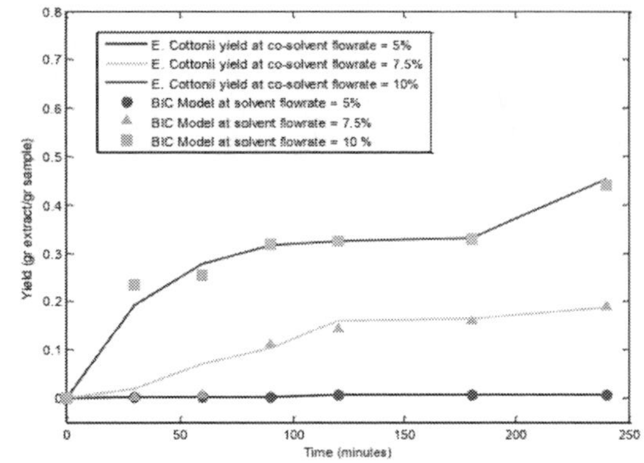

Fig. 8. The effect of co-solvent flowrate on extraction yield with comparison to BIC Model.
(*E. Cottonii*, 60°, 15 MPa)

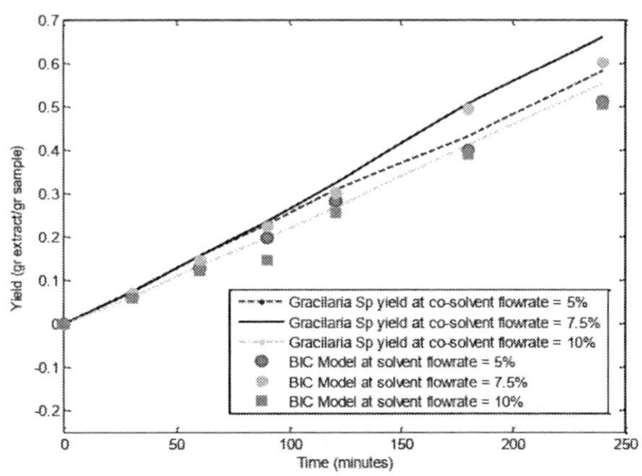

Fig. 9. The effect of co-solvent flowrate on extraction yield with comparison to BIC Model.
(*Gracilaria Sp*, 60°, 15 MPa)

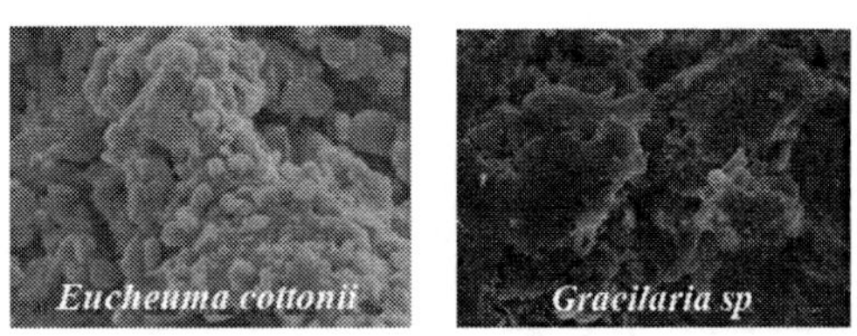

Fig. 10. *Eucheuma cottoni* and *Gracilaria sp* structure of SEM-Analyze.

Table 5. Constant of Chrastil of *Eucheuma cottonii* and *Gracilaria sp* at constant temperature and variation pressure

Macroalgae	Solubility	Constant		
		k	a	b
Eucheuma cottonii	β-carotene	2,52	0,911	-27,66
	linoleic acid	1,82	0,93	-19,84
Gracilaria sp	β-carotene	1,808	-0,066	-22,046
	linoleic acid	-0,758	1,01	-2,44

Table 6. Solubility of *Eucheuma cottonii* at temperature constant

T (K)	P(MPa)	Solubility of Experimental (gr/L)	
		β-karoten	Linoleat
333	15	0,00001	0,00028
	20	1,57E-05	0,000395
	25	1,93E-05	0,000435
T (K)	P(MPa)	Solubility of Chrastil (gr/L)	
		β - karoten	Linoleat
333	15	1,03E-05	0,000282
	20	1,58E-05	0,000384
	25	1,93E-05	0,000443

Table 7. Solubility of *Gracilaria Sp* at temperature constant

T (K)	P (MPa)	Solubility of Experimental (gr/L)	
		B-karoten	Linoleat
333	15	0,000028	0,000665
	20	0,000042	0,000616
	25	0,000044	0,000536
T (K)	P (MPa)	Solubility of Chrastil (gr/L)	
		B-karoten	Linoleat
333	15	2,86E-05	0,000675
	20	3,95E-05	0,000589
	25	4,59E-05	0,00053

5. Conclussion

Oil from macroalgae: *Eucheuma cottonii* and *Gracilaria sp,* were extracted by using SC-CO_2 and ethanol as a co-solvent to consider the effects of temperature, pressure, and ethanol flowrate. Extraction of yield was increased with increasing of pressure and temperature, but straightly increased with addition of ethanol as a solvent. In addition, the experimental data were compared with Broken and Intact Cells (BIC) model in the fluid and solid phase condition. From the comparison of experimental and model calculation, BIC model could describe well for extraction yield. The best fitting parameters in BIC model depends on condition at extraction process. The solubility value of linoleic acid is better than β-carotene because the amount of linoleic acid in *Eucheuma cottonii* and

Gracilaria sp is higher than that of β-carotene correlated with the Chrastil equation.

Acknowledgements,

The authors acknowledge to all member of Fluids and Mechanical Laboratory. Department of Chemical Engineering, Sepuluh Nopember Institute of Technology, Surabaya 60111 Indonesia.

References

1. I. Rawat, R. Ranjith Kumar, T. Mutanda, and F. Bux, Appl. Energy, **103**, 444–467 (2013).

2. P. Mercer and R. E. Armenta, Eur. J. Lipid Sci. Technol., **113**, no. 5, 539–547 (2011).

3. G. Brunner, J. Food Eng., **67**, no. 1–2, 21–33 (2005).

4. L. Fiori, D. Basso, and P. Costa, J. Supercrit. Fluids, **48**, no. 2, 131–138 (2009).

5. H. Sovovà and R. P. Stateva, Rev. Chem. Eng., **27**, no. 3–4, 79–156 (2011).

6. J. C. Germain, J. M. Del Valle, and J. C. De La Fuente, Ind. Eng. Chem. Res., **44**, no. 8, 2879–2886 (2005).

7. S. Machmudah, A. Martin, M. Sasaki, and M. Goto, J. Supercrit. Fluids, **66**, 111–119 (2012).

8. H. Sovová, J. Supercrit. Fluids, **33**, no. 1, 35–52 (2005).

9. M. Perrut, J. Y. Clavier, M. Poletto, E. Reverchon, and M. Modeling, Ind. Eng. Chem. Res., **36**, 430–435 (1997).

10. A. Mouahid, C. Crampon, S. A. A. Toudji, and E. Badens, J. Supercrit. Fluids, **116**, 271–280 (2016).

11. S. Machmudah, K. Kitada, M. Sasaki, M. Goto, J. Munemasa, and M. Yamagata, Ind. Eng. Chem. Res., **50**, 2227–2235 (2011).

12. E. Reverchon, J. Supercrit. Fluids, **10**, 1-37 (1997).

13. A. Tezel, A. Hortaçsu, and O.¨Hortaçsu, J. Supercrit. Fluids, **19**, no. 1, 3–17 (2000).

14. D. M. Triana-Maldonado, S. A. Torijano-Gutiérrez, and C. Giraldo-Estrada, Grasas y Aceites, **68**, no. 1, 1–11 (2017).

Carrageenan Extracted from Eucheuma cottonii Through Variant of Drying Time

Novy Pralisa Putri[1]*, *Ari Susandy Sanjaya*[1], *Neli Kartika Sari*[1], *Reni Puspita Sari*[1], *Yazid Bindar*[2]

[1] Department of Chemical Engineering, Mulawarman University, Jalan Sambaliung 09 Kampus Gunung Kelua, Samarinda 75119, Indonesia E-mail: np.putri@ft.unmul.ac.id
[2] Department of Chemical Engineering, Institute of Technology Bandung, LABTEK X, Jalan Ganesha 10, Bandung 40132, Indonesia

Abstract. Carrageenan is a derivative product of seaweed that used in various fields such as pharmacy, food and cosmetics. One of the most widely used carrageenan from seaweed was *Eucheuma cottonii* sp. The research was conduct to determine the effect of post-harvest seaweed pre-treatment on yields of Carrageenan. The first pre-treatment done by using the fresh seaweed that has been washed clean then directly dried. The second pre-treatment, washed seaweed soaked with clean water for seven days and then dried with a dryer. Drying time for the first and second pre-treatments is varied, then dried seaweed extracted with water to produce carrageenan. The analyses performed were moisture content of seaweed and yields of Carrageenan. The results of the research found that the initial moisture content of seaweed on the second pre-treatment was greater than the first pre-treatment. The variation of drying time on the first pre-treatment did not affect significantly on the yields, but in the second pre-treatment, yield of carageenan increased with accumulation drying time.

1 Introduction

Indonesia is one of the longest coastline countries, and the Indonesian archipelago is separated by oceans. The Indonesian sea is a huge potential for seaweed development. Indonesia is a major producer of Kappaphycus and Euchema seaweed species, these two types are the main contributors of the carrageenan industry [1]. Indonesia has more than five hundred types of seaweed, some of type are used for base material products. Red algae such as types of *Eucheuma, Gelidium, Gracilaria, Hypnea, Sargasum*, and *Turbinaria* which is a type of seaweed that has been widely cultivated in Indonesia [2]. Harvested on seaweed, typically has 90% water content and need to be dried to achieve the quality specifications from the consumers [3]. In Sabah, Malaysia, there are two types of drying method, platforming and hanging [4]. This condition almost the same as Indonesia, drying methods use the platform with wood, bamboo, cement and asphalt. The common conventional technique use in Indonesia for drying seaweed was open drying sun. One of disadvantage of this method, seaweed is not good in quality and also the problem of hygienic. The product gets contaminated from dust, insects, rodents, and other animals, which seriously lowering the food quality [5].

Drying kinetics in seaweeds was also studied under these following pre-treatments: open drying and drying at various temperature from 40-90 °C [6]. From the experiment shows that seaweed drying process that used open sun drying, has took 7 hours to dried, and water content reduce from 94.6% to 8.33% [7]. Moisture content, water activity, colour were affected significantly with

* Corresponding author: novylisa@gmail.com

https://doi.org/10.1051/matecconf/201815602014

temperature of drying [8]. For the appropriate temperature, 50oC was selected as drying temperature.
Carrageenan are sulphated linear polysaccharides of D-galactose and 3,6-anhydro-D-galactose, which have been used in food industry and pharmaceutical industry [9]. Carrageenan from *Eucheuma cottonii* can be taken through four steps, first step was mixed with sodium hydroxide, then extracted which ratio of 1:30, filtered and precipitated [10]. Extraction of carrageenan also can be made with mixed the water and potassium hydroxide [11]. Beside from *Euchema cottonii*, carrageenan can also be obtained from *Euchema spinosum* with ratio solvent 1:20 [12]. However, the aim of the research to determine the effect of pre-treatment and correlation the drying time with characteristics of carrageenan produced

2 Materials and Methods

2.1 Raw material

Eucheuma cottonii were collected from seaweed cultivated in Manggar, Balikpapan. Seaweed was prepared in two pre-treatments. At first treatment, seaweed washed up to clean the sand and salt, then dried. At second treatment, seaweed that has washed up to clean the sand and salt, then immersed with fresh water that replaced every day. Subsequently seven days, seaweed is drained and dried.

Drying

The dried seaweed used tray dryer and heat by humid air. Temperature drying used for was 50 °C. Then observed every variant time of 20, 40, 60, 80, and 100 minutes, especially the of mass losses to determine the water content and the change of colour (physically)

2.2 Carrageenan extraction

After reach the desired time, the dried seaweed was extracted to obtaining the carrageenan. The ratio of dried seaweed-solvent used was 1:15 (g/ml). Carrageenan made by mixed 75 mL solvent with 5 g. then the seaweed soaked in water in 15 minutes. After that filtered then the seaweed extracted. The Volume was kept constant by added hot water. After 3 hours, process of extraction was stopped, then the filtrate from the extraction separated. Filtrate mixed with ethanol 90%. The ratio of the volume of filtrate and ethanol was 3:2 (ml), and so formed hydrocolloid fibres (fibre carrageenan) then wait until 18 hours.

Carrageenan yields (%) were determined by using the equation below

$$Yield = \frac{M_c}{M_{ds}} x100 \quad (1)$$

Where
Mc is the mass of carrageenan extracted (gram)
Mds is the mass of dry seaweed (gram) used for extraction.

3 Result and discussion

3.1 Seaweed properties

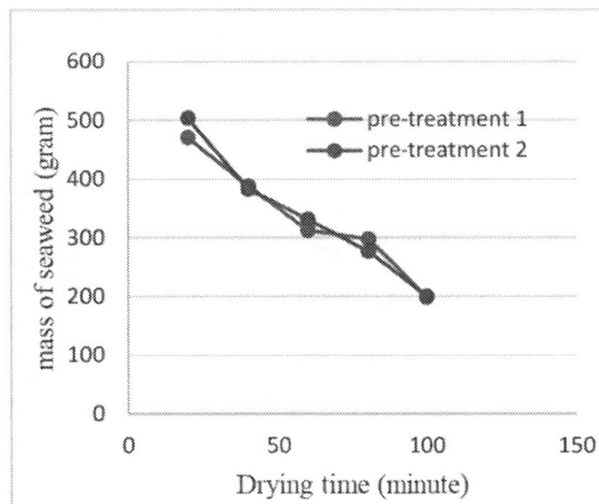

Fig 1. The effect of drying time to mass of seaweed

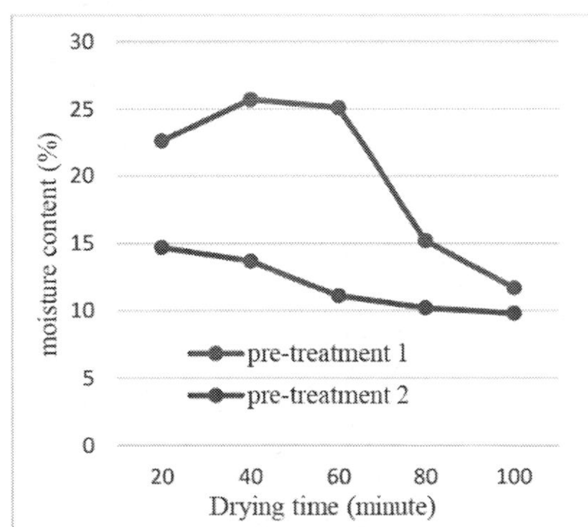

Fig 2. The effect of drying time on moisture content

From Figure 1, it can be found the drying time was inversely proportional to the mass of seaweed, where the higher drying time the lesser mass of seaweed. [7].

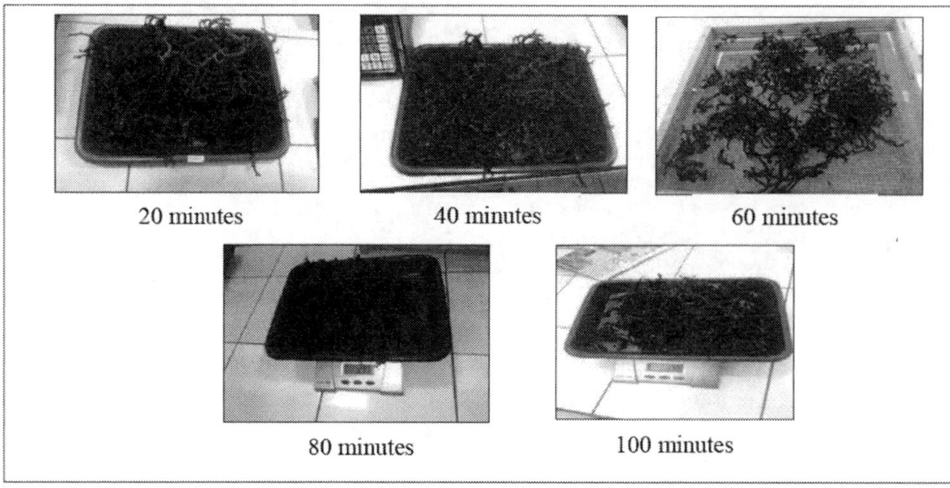

Fig 3. Colour of seaweed after drying (1st pre-treatment)

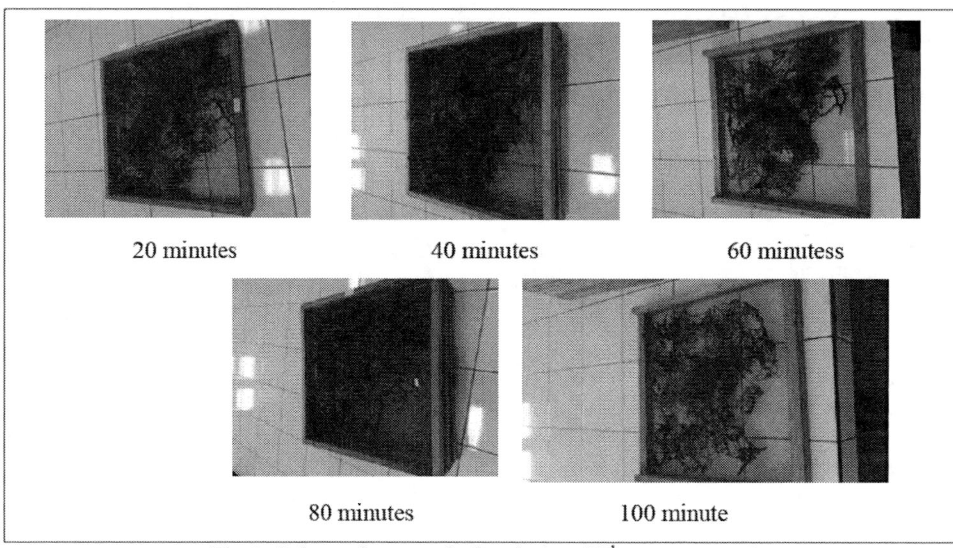

Fig 4. Colour of seaweed after drying (2nd pre-treatment)

The results of Figure 2 also show that mass of seaweed on the second treatment at 20 minutes' look higher than the first pre-treatment. This can be occurred due to the moisture content in the seaweed on the second treatment is higher than first treatment. The initial moisture content of the seaweed for the first treatment was 89% w.b and for second treatment was 90% w.b.

The effect of drying time to moisture content of *Eucheuma cottonii* was shown in Figure 2. The moisture content decrease after 20, 40, 60, 80 and 100 minutes. From the experiment was done above, which have strong correlation with the drying processes of other food products [5, 7, 8,13, 14].

One of the high-quality factor from the consumer's point view was the changing of the colour after dried [14]. From Figure 3 it can be show, on pre-treatment 1, after drying, the colour of seaweed changed to purple brown with strong odours. On Figure 4, for pre-treatment 2, the colour of seaweed turn into light brownish and less pungent odour. The change of the colour during drying is an indication of severe the drying conditions are related with the pigment composition/concentration [15]. The drying rate at below 60 °C can cause less browning as the colour changes. The degradation occurs because need the longer time rather than the action of polyphenol oxidase [16].

3.2 Extraction of Carrageenan from seaweed

From Figure 5, that can be explain, the process of extraction, the results of yield on the first pre-treatment have tendency same with mass of carrageenan for all variant drying time. However, the amount of the carrageenan on first pre-treatment is greater than the second pre-treatment. This can be occurred because in the second pre-treatment most carrageenan was soluble in water when was soaked. It has been reported that carrageenan fractions are water soluble, and being insoluble in oil or fats, and organic solvents [9]

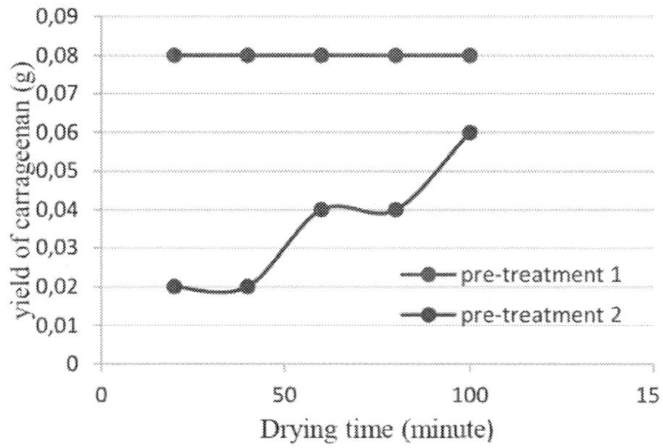

Fig 5. Effect of drying time to mass of carrageenan

4 Conclusion

The research shows that the mass of seaweed was found to decline with drying time. The initial moisture content of seaweed on the second pre-treatment was higher than the first pre-treatment. The variant of drying time on the first pre-treatment did not affect significantly on the yields of carageenan. On the second treatment, yield of carageenan increased with accumulation drying time.

References

1. Harris J. Bixler, H. Porse, A decade of change in the seaweed hydrocolloids industry, J. Appl. Phycol, **23**, 321-335 (2011)
2. E. Nugroho, E. Kusnendar, Seaweed agribusiness, (2015) (in Indonesian)
3. M. Djaeni, D.A. Sari, *Procedia Environmental Sciences*, **23**, 2 – 10 (2015)
4. A. Sade, I. Ali, M.R.M. Ariff, The seaweed industry in Sabah East Malaysia, J. Jati, **11**, 97 – 107 (2006)
5. A. W. Desmukh, M.N. Varma, C.K. Yoo, K.L. Wasewar, Chinese Journal of Engineering, 1 – 7 (2014)
6. G.S. Mattar de Faria, L. Hayashi, A.R. Monteiro, J. Appl Phycol, **26**, 917 – 922 (2014)
7. A. Fudholi, M. Y. Othman, M. H. Ruslan, M. Yahya, A. Zaharim, K. Sopian, Design and testing of solar dryer for drying kinetics of seaweed in Malaysia, *in Recent Research in Geography, Geology, Energy, Environment and Biomedicine*, 119-124 (2011)
8. M.S. Sagrin, G.H. Chong, Industrial Crops and Products, **45**, 430 – 434 (2013)
9. V.L Campo, D.F. Kawano, D. Braz da Silva Jr, I. Carvalho, Carrageenans: Biological properties, chemical modifications and structural analysis-A review, *Carbohydrate Polymers*, **77**, 167 – 180 (2009)
10. Prasetyowati, C. Jasmine A., D. Agustiawan, Journal of Chemical Engineering **15**, 27 – 33 (2008) (In Indonesian)
11. S. Distantina, Fadilah, Rochmadi, M. Fahrurrozi, Wiratni, *Proceedings Seminar of Chemical Engineering and Process*, C-21-1-6 (2010) (In Indonesian)
12. M.I. Hudha, R. Sepdwiyanti, S.D. Sari, J. Periodical Journal Scientific Chemical Engineering **1**, 17 – 20 (2012) (In Indonesian)
13. A. B. Siddique, D. Wright, Asian Journal of Plant Sciences, **2 (13)**, 978 - 982 (2003)
14. S. Kulanthaisami, P. Rajkumar, P. Venkatachalam, P. Subramanian, G.S.V. Raghavan, Y. Gariepy, V. Orsat, Maddras Agriculture Journal, **97(7 – 9)**, 287 – 295 (2010)
15. A. Topuz, H. Feng, M. Kushad, Food Science and Technology, **42**, 1667 – 1673 (2009)
16. D. Argyropoulos, J. Müller, Journal of Applied Research on Medicinal and Aromatic Plants, **1 (1)**, 15-22 (2014)

Mass Transfer Coefficient of Ozone in a Bubble Column

Ratnawati Ratnawati[1,*], *Dyah Arum* Kusumaningtyas[1], *Purbo* Suseno[1], and *Aji* Prasetyaningrum[1]

[1]Department of Chemical Engineering, Diponegoro University, Jl. Prof. Soedarto, SH, Semarang 50275, Indonesia

Abstract. The effect of flow rate of ozone-containing gas and pH on the mass transfer coefficient of ozone through water in a bubble column reactor has been studied. Ozone was generated from air using a corona discharge ozone generator. The flow rate of air was varied from 2 to 5 L min^{-1}, while pH was varied from 4 to 10. The gas containing ozone was bubbled to the reactor containing 1.5 L of 2% KI solution. The temperature was set at 28±1°C. The concentration of ozone was determined using titrimetric method every 5 minutes. The results show that the concentration of ozone increases with time, and it reaches a steady-state concentration after 30 minutes of ozonation. The gas flow rate and pH apparently affect both the concentration and the k_La. The highest k_La of 2.1 ×10^{-2} s^{-1} is obtained at pH 4 with a gas flow rate of 4 L min^{-1}.

1 Introduction

Ozone (O_3) is naturally formed in the stratosphere by chemical reactions involving oxygen with the aid of solar ultraviolet radiation (sunlight) of wavelengths below 240 nm [1]. The reaction can be simplified as Equation (1).

$$3O_2 + h\upsilon\ (\lambda<240\ \text{nm}) \rightarrow 2O_3 \qquad (1)$$

Ozone can be generated using non-thermal plasma corona discharges at atmospheric pressure and room temperature where an extra-high voltage electrical current is passed from an electrode through a neutral fluid, which is usually air [2]. The current is able to ionize the fluid creating plasma around the electrode. Oxygen experiences ionization and the overall reaction that may occur is written as Equation (2) [3].

$$3O_2 + e^- \rightarrow 2O_3 + e^- \qquad (2)$$

Ozone is a strong oxidant with a very high redox potential, i.e. 2.07 V [4]. Ozone is highly reactive that it readily reacts with various organic and inorganic substances [5]. It has been utilized in advanced oxidation process (AOP), without or with H_2O_2 or UV to disinfect microorganism [5], to eliminate pollutants [6, 7], and to control odor problem [8].

To implement AOP in a liquid phase, the ozone must be solubilized in the liquid. The solubility of ozone in water is very low [9]. The solubilization of ozone in water is influenced by mass transfer coefficient (k_La) of ozone in water. The k_La is affected by the ozone concentration in the gas phase, the gas flow rate, temperature, and pH of the liquid phase [10 – 12]. The objective of this work is to study the effect of pH of the water and the flow rate of the gas on the mass transfer coefficient of ozone in water.

2 Materials and Method

2.1. Materials

The materials used in this study included potassium iodide (Merck, Cat. No. 1.05043.0250), sulphuric acid (Merck, Cat. No. 1.12080.1000), sodium thiosulfate (Merck, Cat. No. 1.93248.0521), sodium hydroxide (Merck, Cat. No. 1.06498.050), and hydrochloric acid (Merck, Cat. No. 1.93401.0521). The materials were utilized without any treatment.

2.2 Method

The experiment was conducted by using equipment as depicted in Fig.1, which consisted of a bubble column (a glass cylinder of 10 cm diameter and 50 cm height) equipped with a bubble diffuser, a corona discharge (40 kV) ozone generator (Dipo Technology Indonesia), and a compressor.

Thirty grams of potassium iodide was dissolved in distilled water to form 1500 mL solution with a concentration of 2% (w/w). The pH of the solution was varied (3, 4, 5, 6, 7, 8, 9, and 10) and was adjusted using HCl or NaOH solution. The solution was then placed in the column of which the temperature was kept at 28 ± 1°C. Air was passed the corona discharge ozone generator using the compressor at a certain flow rate (2, 2.5, 3, 3.5, 4, 4.5, and 5 L min^{-1}), then it was bubbled

* Corresponding author: ratnawati.hartono@gmail.com

https://doi.org/10.1051/matecconf/201815602015

into the column. The soluble ozone readily reacted with KI to form I_2 according to Equation (3).

$$O_3 + 2\,KI + H_2O \rightarrow 2\,KOH + O_2 + I_2 \qquad (3)$$

A sample was withdrawn every 5 minutes and was analyzed for concentration of ozone by KI titrimetry. Ozone reacts with KI to form I_2 by a titrimetric method using sodium thiosulfate as the titrant. The concentration of I_2 is equivalent to that of ozone.

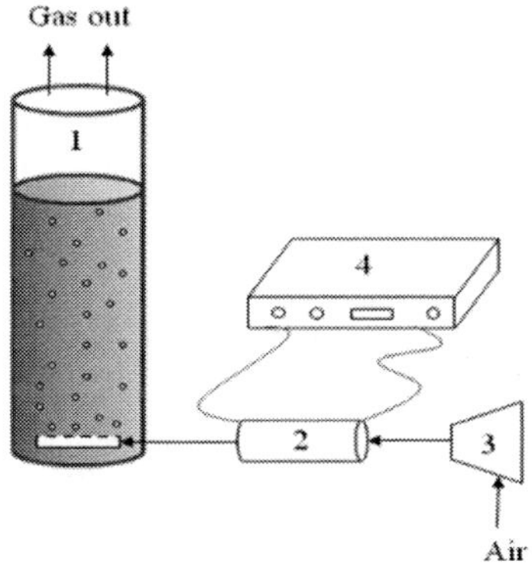

Fig.1. Schematic of equipment for experiment: bubble gas reactor (1), corona discharge ozone generator (2), compressor (3), and power supply (4)

2.3. Mathematical model for mass transfer

According to two-film theory, to be solubilized in liquid water, ozone molecules must diffuse from the gas phase (gas bubble) to the bulk liquid through gas and liquid films [13]. However, as ozone has a very low solubility on water. The absorption of ozone through the liquid film results in a very small concentration difference across the film. The absorption of ozone through the liquid film proceeds very slowly. It is reasonable to assume that resistance to mass transfer through the liquid film is much higher than that through the gas film. The gas-phase resistance of mass transfer is therefore negligible. The mathematical model to describe the mass transfer of ozone from the gas phase to the liquid water can be represented by equation (4) [10],

$$dC/dt = k_La\,(C^* - C) \qquad (4)$$

where k_La is the volumetric mass transfer coefficient of ozone, C and C^* are the concentration ozone at time t and at saturated condition, respectively. Equation (4) is integrated to give Equation (5).

$$\ln(1 - C/C^*) = -\,k_La\,t \qquad (5)$$

3 Results and Discussion

3.1. Influence of time on the concentration of dissolved ozone

The concentration of ozone at various pH and time is depicted in Fig. 2. The results show that the concentration of ozone increases rapidly in the first 10 minutes and slows down afterward. Finally, it reaches a steady-state value after 30 minutes.

A similar result was reported by Egorova *et al.* [14] who passed a mixture of oxygen and ozone through water in a bubble reactor at 20°C. They reported after 10 minutes of bubbling, that concentration of ozone remained constant. The time needed to reach the saturation concentration in this work is much longer than that needed by Egorova *et al.* [14]. It is possibly because this work used air to be fed to the ozone generator, while Egorova *et al.* [14] utilized oxygen. The concentration of ozone in the gas phase in this work must be lower, so it took a longer time to reach the saturation concentration.

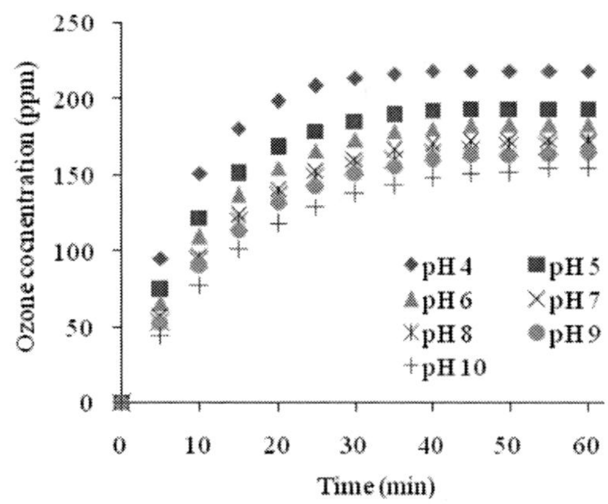

Fig. 2. Effect of time on the concentration of ozone in water at 28±1°C and various pH with a gas flow rate of 4 L min^{-1}

3.2. Influence of gas flow rate on the solubility of ozone

The effect of the flow rate of the ozone-containing gas on the concentration of dissolved ozone is depicted in Fig. 4. All of the data were obtained by bubbling the ozone-containing gas into the reactor at 28±1°C and pH 7 for 30 minutes.

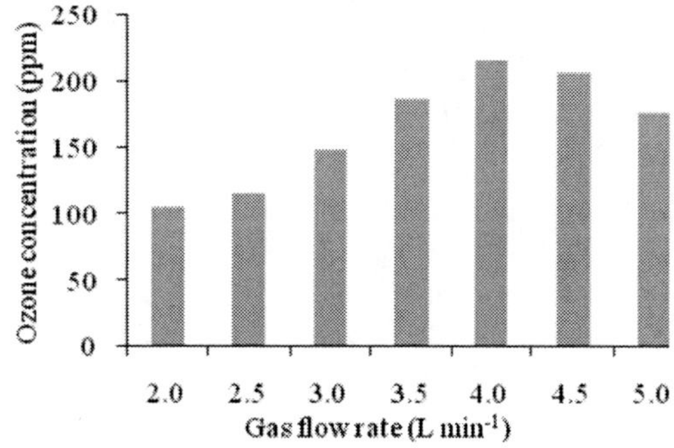

Fig. 4. Effect of gas flow rate on the concentration of dissolved ozone in water at 28±1°C and pH 7 after 30 minutes

Fig. 4 shows that the ozone concentration increases from 105 ppm to 216 ppm as the gas flow rate rises from

2 to 4 L min^{-1}. The ozone concentration then decreases at a flow rate higher than 4 L min^{-1}. As the gas flow rate is higher, the residence time of air in the ozone generator becomes shorter. With less reaction time, less ozone is formed [17]. The same trend was also reported by Boonduang *et al.* [16]. They reported that the optimum ozone concentration was obtained by using a gas flow rate of 2 L min^{-1}.

3.3. Influence of gas flow rate on the k_La

To investigate the effect of gas flow rate on the mass transfer coefficient, a series of experiments at pH 7 with flow rate varied from 2 to 5 L min^{-1} were conducted. The results are depicted in Fig 5. Fig. 5 shows how the k_La of ozone is influenced by the gas flow rate. The k_La increases from 0.4 × 10^{-3} s^{-1} to 1.45 × 10^{-3} s^{-1} as the gas flow rate rises from of 2 L min^{-1} to 4 L min^{-1}, after which the k_La drops with increasing gas flow rate. The maximum mass transfer coefficient is reached with the gas flow rate of 4 L min-1. This flow rate was used to investigate the effect of pH on the mass transfer coefficient.

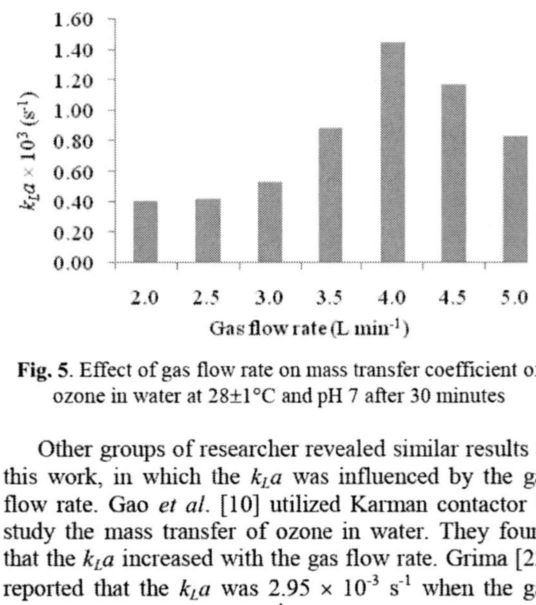

Fig. 5. Effect of gas flow rate on mass transfer coefficient of ozone in water at 28±1°C and pH 7 after 30 minutes

Other groups of researcher revealed similar results to this work, in which the k_La was influenced by the gas flow rate. Gao *et al.* [10] utilized Karman contactor to study the mass transfer of ozone in water. They found that the k_La increased with the gas flow rate. Grima [22] reported that the k_La was 2.95 × 10^{-3} s^{-1} when the gas flow rate was 100 mL min^{-1}, and it increased to 7.47 × 10^{-3} when the experiment was conducted at higher gas flow rate of 600 mL min^{-1}.

When the gas flow rate increases, its superficial velocity (U_G) increases as well. Bouaifi *et al.* [18] explained that in a bubble column, the U_G is related to the k_La in which the k_La increases with increasing U_G. According to Akita and Fumitake [19], the superficial gas velocity of the gas in a bubble column is linearly correlated to the gas holdup (ε_G), which can be related to the mass transfer coefficient through Equation (6).

$$k_La = c_2\, D_L^{0.5}\, \nu_L^{-0.12} (\gamma/\rho_L)^{-0.62} D^{0.17} g^{0.98} \varepsilon_G^{1.1} \quad (6)$$

where c_2 is a constant, D is column diameter, g is gravitational constant, while D_L, ν_L, and ρ_L are liquid phase diffusivity, viscosity, and density, respectively.

However, as it has been previously explained, when the gas flow rate increases, the residence time of the air in the ozone generator is shorter, and less ozone is formed. It eventually decreases the mass transfer coefficient.

3.4. Influence of pH on the k_La

It is shown in Fig. 2 that pH influences the ozone concentration in water. At the same bubbling time, the concentration of ozone increases as the pH decreases. Clever *et al.* [9] reported similar trend, in which at 35°C the solubility of ozone in water increased from 6.8 ppm to 39.6 ppm as the pH decreased from 6.8 to 2.7 when the ozone concentration in the gas was 200 ppm. Roth and Sullivan [20] found that the Henry's constant of ozone in water decreased as the pH dropped. A smaller value of Henry's constant leads to a higher value of solubility. Mizuno and Tsuno [21] found similar results as well.

Egorova *et al.* [13] explained that the solubility of ozone in water decreased with increasing pH. In alkaline solution, the hydroxyl ion (OH^-) catalyzes the decomposition of ozone through the following chain reactions:

$$O_3 + OH^- \rightarrow O_2 + HO_2^- \quad (7)$$

$$O_3 + HO_2^- \rightarrow \bullet O_2^- + \bullet OH \quad (8)$$

$$O_3 + \bullet O_2^- \rightarrow \bullet O_2 + \bullet O_3^- \quad (9)$$

$$H_2O + \bullet O_3^- \rightarrow O_2 + \bullet OH + OH^- \quad (10)$$

Using the data presented in Fig. 2, ln (1 – C/C^*) was plotted versus t for each pH. Only sets of data at pH 4, 7, and 10 are depicted in Fig. 7. Each set of data at respective pH can be approximated by linear lines with excellent correlation coefficients. The k_La can be concluded from the slope of the plot. The results are shown in Table 1, along with those of pH 5, 6, 8, and 9 and respective correlation coefficient. The coefficients of correlation (R^2) of the correlation for all pH are higher than 0.99. The values of k_La obtained by other researchers [12, 22] are presented in Table 1 as well.

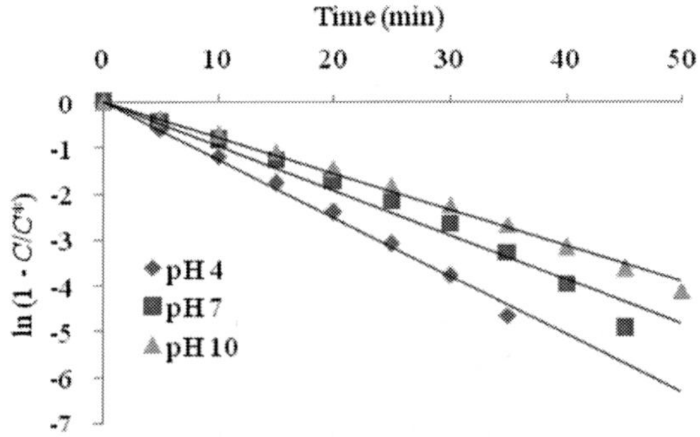

Fig. 7. Plot of ln (1 – C/C^*) versus time

The results presented in Table 1 show that the k_La of ozone drops from 2.10 $\times 10^{-3}$ to 1.22 $\times$ 10^{-3} s^{-1} as pH increases from 4 to 10. Flores-Payán *et al.* [12] revealed a similar trend, in which the mass transfer coefficient of ozone decreased from 6.90 $\times 10^{-3}$ to 4.30 $\times$ 10^{-3} s^{-1} as the pH increased from 1.5 to 11.

Table 1. Mass transfer coefficient at various pH.

pH	k_La (s^{-1})		
	This work	**[12]**	**[23]**
1.5		6.90 × 10^{-3}	
4	2.10 × 10^{-3}	6.30 × 10^{-3}	8.75 × 10^{-3}
5	1.75 × 10^{-3}		
6	1.63 × 10^{-3}		
7	1.45 × 10^{-3}	5.30 × 10^{-3}	7.61 × 10^{-3}
8	1.37 × 10^{-3}		
9	1.35 × 10^{-3}	4.30 × 10^{-3}	6.54 × 10^{-3}
10	1.22 × 10^{-3}		

The decrease of the k_La of ozone in water at alkaline condition is related to the existence of HO• radical. This radical catalyzes the self-decomposition of ozone according to equations (7) – (10) [12].

The values of k_La obtained in this work are lower than those of other groups of researcher [12, 23]. It could be because of different type of reactor, gas diffuser geometry, and gas flow rate. However, the values of k_La are of the same order of magnitude.

Conclusion

The results show that the concentration of ozone increases with time. The saturation concentration is attained after 30 minutes of ozonation. Both the gas flow rate and pH influence the concentration and the k_La. The highest value of k_La of 2.10 × 10^{-3} min^{-1} is obtained at a gas flow rate of 4 L min^{-1} and pH 4.

The authors wish to thank the Department of Chemical Engineering, Diponegoro University for technical support.

References

1. A.K. Dutta, N. Vaval, S. Pal, J. Chem. Theory Comput. **8**, 6 (2012).
2. B. Mennad, Z. Harrache, K. Yanallah, D.A. Aid, A. Belasri, Vacuum **104** (2014).
3. M. Liu, S. Preis, I. Kornev, Y. Hul, C.-H. Wei, J. Environ. Sci. *In press, corrected proof,* (2017).
4. D.R. Lide, W.M. Haynes, *CRC Handbook of Chemistry and Physics 90th Edition* (2010).
5. X. Li, W. Chen, L. Ma, H. Wang, J. Fan, Chemosphere **195** (2018).
6. G. Pak, D.E. Salcedo, H. Lee, J. Oh, S.K. Maeng, K. G. Song, S.W. Hong, H.-C. Kim, K. Chandran, and S. Kim, Environ. Sci. Technol. **14**, 50 (2016).
7. M. Bourgin, E. Borowska, J. Helbing, J. Hollender, H.P. Kaiser, C. Kienle, C.S. McArdell, E. Simon, U. von Gunten, Water Res. **122** (2017).
8. B. Yuan, D. Xu, F. Li, M.-L. Fu, Sep. Purif. Technol. **117** (2013).
9. H.L. Clever, R. Battino, H. Miyamoto, Y. Yampolski, C.L. Young, J. Phys. Chem. Ref. Data **43**, 033102 (2014).
10. M.T. Gao, M. Hirata, H. Takanashi, T. Hano, Sep. Purif. Technol. **42** (2005).
11. A. Al-Abduly, P. Christensen, A. Harvey, K. Zahng, Chem. Eng. Process. **84** (2014).
12. V. Flores-Payán, E.J. Herrera-López, J. Navarro-Laboulais, A. López-López, J. Ind. Eng. Chem. **21** (2015).
13. B.I. Morsi, O. Basha, in M. Solecki (Ed.), *Mass Transfer Advancement in Process Modelling*, (AvE4EvA, 2015).
14. G.V. Egorova, V.A. Voblikova, L.V. Sabitova, I.S. Tkachenko, S.N. Tkachenko, V.V. Lunin, Moscow Univ. Chem. Bulletin **70**, 5 (2015).
15. P.N. Johnson, R.A. Davis, J. Chem. Eng. Data **41** (1996)
16. S. Boonduang, S. Limsuwan, W. Kongsri, P. Limsuwan, Procedia Engineer. **32** (2012).
17. S. Jodpimai, S. Boonduang, P. Limsuwan, J. Electrostat. **74** (2015).
18. M. Bouaifi, G. Hebrard, D. Bastoul, M. Roustan, Chem. Eng. Process. **40** (2001).
19. K. Akita, F. Yoshida, Ind. Eng. Chem. Process Des. Develop. **12**, 1 (1973).
20. J.A. Roth, D.E. Sullivan, Ind. Eng. Chem. Fundam. **20** (1981).
21. T. Mizuno, H. Tsuno, Ozone-Sci. Eng. **32** (2010).
22. N.M.M. Grima, Thesis, School of Engineering Design & Technology, University of Bradford, UK (2009).
23. A. López-López, J.-S. Pic, H. Benbelkacem, H. Debellefontaine, Chem. Eng. Process. **46** (2007).

The Study of EHD Flow:Velocity and electrical power

Sumariyah[1*], *Kusminarto*[2], *Arief* Hermanto[2] *and Pekik* Nuswantoro[2]

[1]*Physics Department, Faculty of Sciences and Mathematics, Diponegoro University, Indonesia*
[2]*Physics Department, Faculty of Mathematics and Natural Sciences, Gadjah Mada University, Indonesia*

Abstract. Electrohydrodynamics (EHD) flow is the yield of the corona discharge. Corona discharge characterized by electrical phenomena that occur together in a gas medium under the influence of an electric field. In corona discharge, the electric field is not homogeneous due to the geometry of the electrode configuration. In this study, a pin-multi concentric ring electrode (P-MRE) and pin-single ring electrode (P-SRE) has been used as electrodes configuration. The objective of this research is to get the velocity of EHD flow and determining the electrical power of the EHD flow. EHD flow generated by a DC high voltage (up to 5 kV). The pin electrode was connected to a positive polarity and electrode ring/multi-concentric rings to negative polarity. Based on our result, we make a comparison of two types of corona configurations. The results were obtained that, EHD flow velocities with the P-MRE was greater than the P-SRE for the same distance and voltage. We found that u-P characteristics of EHD flow with the P-MRE and P-SRE follow relationship as $u \propto P^{1/3}$.

1 Introduction

Electrohydrodynamics (EHD) flow is the yield of corona discharge which is characterized by electrical phenomena. It occurs together in a gas medium under the influence of an electric field that is not homogeneous due to the geometry of the electrode configuration. The electric field will be very high in the area around the active electrode thereby allowing to happen ionization and the formation of high-energy electrons in these area. Around the active electrode the primer ionization process occurs and formed plasma [1]. The plasma surrounds part of the active electrode is usually shaped tapered so that the plasma resembles a corona [2]

The process of the incandescent corona discharge in an electrical field begins with Townsend discharge followed by glow discharge or corona discharge and ends with arc discharge. A Townsend discharge occurs at lower voltages where small number of ionizations occur. While at the time of arc discharge, the raised electric current very quickly followed by the very drop voltage sharply [3].

The prospects of EHD flow in the industry are quite large, such as for drug delivery systems that was developed by Repanas [4] and Sofokleous [5]. The study of EHD flow used as the coolant was carried out by Chen [6] and Hosseinloo [7]. The application of EHD as a drying apparatus was conducted by Bardy [8], Martynenko [9] and Karam [10]. Devices that use corona discharge EHD flow has many advantages such as: free of mechanical vibration and acoustic noise. In addition, the required electric power of EHD corona discharge is quite low [11].

2 Research Procedure

Two configurations of the EHD flow generator were studied, they were a pin-multi concentric ring (P-MRE) and pin-single ring electrode (P-SRE) which was as the control of measurement. Pin electrodes made of stainless steel sewing needle with a diameter of pointed end 0.09 mm. Multi concentric ring electrodes consist of 3 concentric ring electrodes are arranged concentrically interconnected. The third ring of the electrode constituent having diameters in each of 8 mm, 16 mm and 24 mm, respectively. In a single ring electrode, the diameter was 24 mm. All electrode metal rings have the same width and thickness, which were 2 mm. The composition of the generating system of EHD flow in this study includes P-MRE, P-SRE, (high voltage power, set voltage and current measurement tools and a set of measuring tools EHD flow velocity). The air can come from the top and out of the bottom, as shown in Figure.1

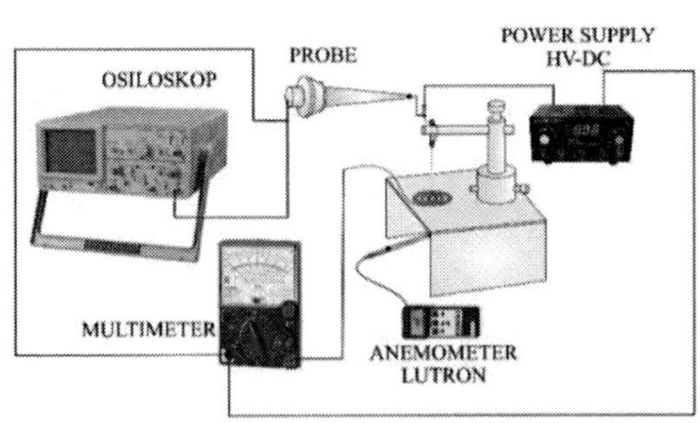

Fig.1. The Composition of a Generating System of the EHD Flow

The EHD flow generated by installing a high voltage DC 10 KV on the electrodes of the corona discharge system with a positive polarity on the pin electrodes and a negative polarity on the ring/ the multi-concentric ring. A high-voltage probe 1000: 1 was used

* Corresponding author: sumariyah@undip.ac.id

to measure the high voltage, which is connected to the CRO. The ion current was measured by using a multimeter. To Measure the velocity of EHD flow, a Hot-Wire Anemometer was placed perpendicular to the pin electrode direction with the distance was 15 mm.

This method is the same as the method of our previous research. They were a study of EHD flow about modeling of electric potential distribution in the zone of EHD flow [12], measurement of the velocity of EHD flow [13] and determination of ion mobility in the zone of EHD flow [14].

3 Results and Discussion

3.1 u-V Characteristic of EHD Flow

Graph of the velocity of EHD flow as a function of voltage (**u**-*V* characteristic) with various distance between the electrodes (*d*=4, 6, 8, 10, 12 and 14 mm) of the electrode pin-multi concentric rings configuration (**UT**) and pin-single ring configuration (**US**) is shown in Figure 2 (a) and (b)

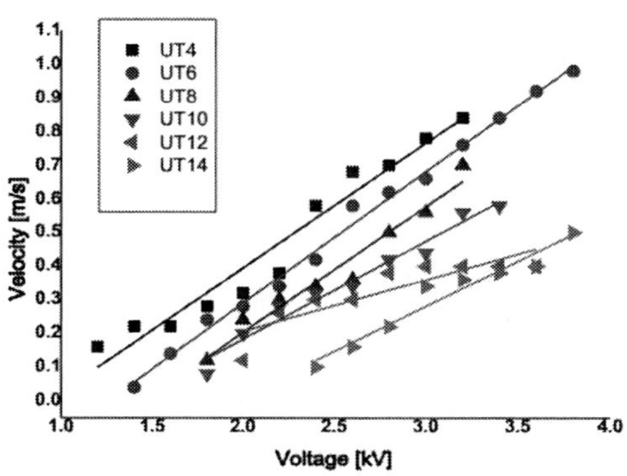

(a)

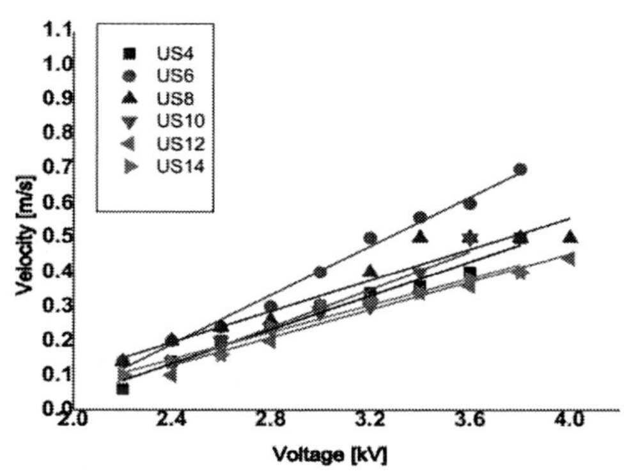

(b)

Fig.2. **u**-*V* Characteristics of (a) pin-multi concentric rings and (b) pin-single ring electrode configuration

In Figure 2.(a) and (b) it appears that in all of the distance between the electrodes (*d*), the greater the applied voltage the greater the velocity of the EHD flow. This is because the relationship between the voltage and the velocity of the EHD flow is a linear function. In accordance with the theoretical formulation by Goodenough et al [15] which shows that **u** ∞ *V*. However, variations in the distance between the electrodes (*d*) at a fixed voltage shows that the greater the distance between the electrodes, the smaller the velocity of the EHD flow. This happens because the electric field strength is a function of the distance between the electrodes. The greater the distance between the electrodes, the electric field strength is getting smaller.

3.2 The Velocity Comparison of the EHD flow

It is very important to know the comparison between the EHD flow velocity for electrode configuration of pin-multi concentric rings (**uT**) with electrode configuration of pin-single ring (**uS**). The EHD flow velocity in pin-multi concentric rings electrodes (**uT**) and in pin-single ring electrode (**uS**) ratio as a function of the distance between the electrodes (*d*) at various the voltages (*V*) is shown in Figure 3. It is shown that the value of ((**uT**)/(**uS**)) at all *d* value is greater than 1 (one).

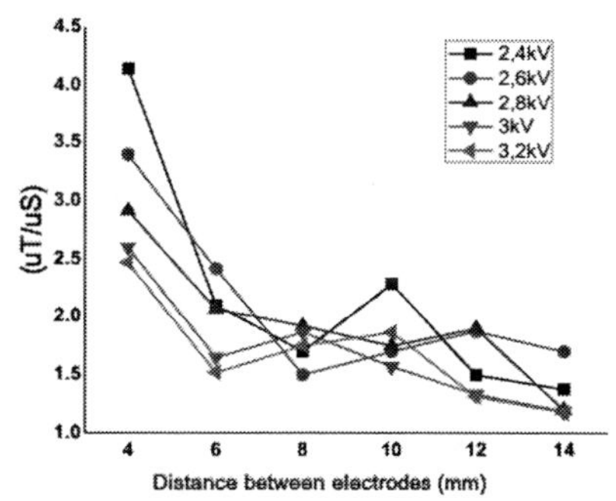

Fig.3. Graph **(uT/uS)** function of the distance between the electrodes (*d*) on voltage (*V*) remain

It can be concluded that at the same electrodes distance *d* the velocity of EHD flow for **uT** is greater than the velocity of EHD flow for **uS** at any voltage. This is due to the electric flux density in the electrode pair of pin-multi concentric rings is greater than that in pin-single ring electrode pair. The large electric flux density will increase the ion density. The ions move towards multi-concentric ring electrodes and collide with neutral air molecules with very high collision frequency. The momentum transfer of positive ions to molecules of gas in the area between the electrodes is very large and in turn this will increase the EHD flow velocity.

3.3 u-P Characteristic of EHD Flow

Figure 4 shows the graph of **u**-*P* characteristics of EHD flow in pin-multi concentric rings electrodes and pin-single ring electrode. The greater the electric power results in the greater velocity of the EHD flow (**u**) produced. This is because the velocity is proporsional to $P^{1/3}$. This is in accordance with the results of studies conducted by Goodenough et al [15] and Komeilia et al [16]. The initial electrical power that produced the initial velocity of EHD flow in pin-multi concentric rings electrodes configuration is smaller than that in the pin-single ring electrode. At the same electrical power, eg at P = 100 mW, the velocity of EHD flow in pin-multi concentric rings electrode configuration will be larger than that in the pin-single ring configuration. This is in accordance with the **u**- V characteristics.

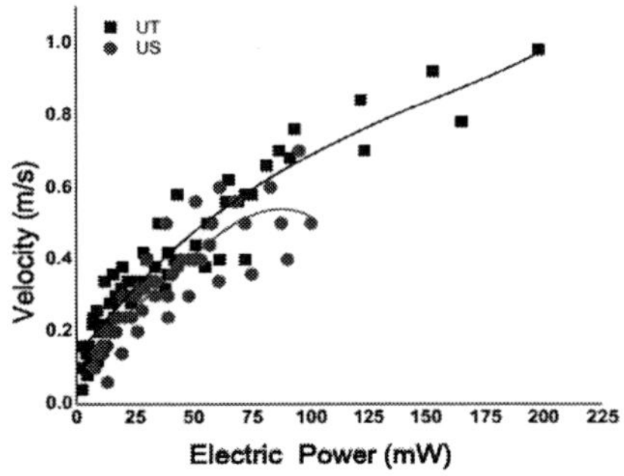

Fig.4. **u**-P Characteristics of EHD flow in pin-multi concentric rings electrodes and pin-single ring electrodes.

Figure 4. represents more obvious that **u**~$P^{1/3}$. The velocity of EHD flow in configuration of pin-multi concentric rings **uT** is always greater than the velocity of EHD flow in pin-single ring configuration **uS**. The value of **uT** can reach 0.9 m/s for a power of 200 mW, before arc discharge which is characterized by extremely high currents and the voltage is reduced to be near zero. In the **uS**, the electric power of 100 mW is still possible to be applied before reaching the arc discharge to produced velocity of EHD flow of 0.7 m/s.

From the aspect of electric power, the EHD flow generator with pin-multi concentric rings configuration can be operated with a higher power than pin-single ring configuration. This is due to the stability (not easily occur arc) of the pin-multi concentric rings.

4 Conclusion

The velocity of EHD flow as a function of the voltage at a certain distance between electrodes showed that the EHD flow velocity is proportional to the voltage and proportional to $P^{1/3}$. At the same distance between the electrodes, the velocity of EHD flow for pin-multi concentric rings is always greater than for pin-single ring at any voltages.

References

1. M. Nur., A. H. Azzulkha, M. Restiwijaya, Z. Muchlisin and Sumariyah, The Study of Electrohydrodynamic and Wind Ions Direction, Produced by Positive Corona Plasma Discharge, *Advances in Physics Theories and Applications* ISSN 2224-719X (Paper) ISSN **Vol. 33**, 2225-0638 (Onlines) (2014)
2. N. St. J. Braithwaite, *Introduction to gas discharges, Plasma Sources Sci. Technol.* **9**, 517-527 (2000)
3. E..D. Fylladitakis, M.P. Theodoridis and A.X. Moronis, *Review on the History, Research, and Applications of of Electrohydrodynamics, IEEE,* 0093-3813 (2014)
4. A. Repanas, S. Andriopoulou, B. Glasmacher, The significance of electrospinning as a method to create fibrous scaffolds for biomedical engineering and drug delivery applications, *Journal of Drug Delivery Science and Technology* **31**, 137-146 (2016)
5. P. Sofokleous, E. Stride, W. Bonfield, M. Edirisinghe, Design, construction and performance of a portable handheld electrohydrodynamic multi-needle spray gun for biomedical applications, *Materials Science and Engineering* **33**, 213–223 (2013)
6. I.Y. Chen, M.Z. Guo, K.S. Yang, C. C. Wang, Enhanced cooling for LED lighting using ionic wind, *International Journal of Heat and Mass Transfer* **57**, 285–291 (2013)
7. A.H. Hosseinloo, S. P. Tan, F. F. Yap, K. C. Toh , Shock and vibration protection of submerged jet impingement cooling systems: *Theory and experiment Applied Thermal Engineering* **73**, 1076-1086 (2014)
8. E. Bardy, M. Hamdi, M. Have, O. Rouaud, Transient exergetic efficiency and moisture loss analysis of forced convection drying with and without electrohydrodynamic enhancement, *Energy* **89**, 519-527 (2015)
9. A. Martynenko, W. Zheng, Electrohydrodynamic drying of apple slices: Energy and quality aspects, *Journal of Food Engineering* **168**, 215–222 (2016)
10. M. C. Karam, J. Petit, D. Zimmer, E.B. Djantou, J. Scher, Effects of drying and grinding in production of fruit and vegetable powders: A review, *Journal of Food Engineering* **188**, 32-49 (2016)
11. J.I. Zhang and and F.C. Lai, Effect of emitting electrode number on the performance of EHD gas pump in a rectangular channel, *Journal of Electrostatics* **69**, 486-493 (2011)
12. Sumariyah, Kusminarto, Arief H., Pekik N., Zaenul M., and Evi S., Modeling of Electric Potential Distribution in EHD Flow Zone Utilizing Pin-Multi Ring Consentric Electrodes, *Procedia Environmental Sciences* **23**, 260 – 265 (2015)
13. Sumariyah, Kusminarto, Arief H., and Pekik N., Velocity Measurement of EHD Flow Produced by Pin-Multi Concentric Ring electrodes Generator, *Applied Mechanics and Materials* **Vol 271**, pp 227-231 (2015)
14. Sumariyah, Kusminarto, Arief H. and Pekik N., Determination of ion mobility in EHD flow zone of plasma generator, *International Conference of Chemical and Material Engineering (ICCME) AIP Conf. Proc.* **1699**, 040019-1–040019-5 (2015)
15. Timothy I.J. Goodenough, Peter W. Goodenough, Sylvia M. Goodenough, The efficiency of corona

wind drying and its application to thefood industry, *Journal of Food Engineering* **80,** 1233–1238 (2007)
16. B. Komeilia, J.S. Changa,G.D.Harvela C.Y. Chingc, D. Brociloa, Flow characteristics of wire-rod type electrohydrodynamic gas pump under negative corona operations, *Journal of Electrostatics* **66**, 342-353 (2008)

MATEC Web of Conferences **156**, 03001 (2018) https://doi.org/10.1051/matecconf/201815603001
RSCE 2017

The Synthesis of Magnesium Soaps as Feed for Biohydrocarbon Production

Meiti Pratiwi[1,*], *Godlief* F. Neonufa[1,2], *Tirto* Prakoso[1], and *Tatang* H. Soerawidjaja[1]

[1]Department of Chemical Engineering, Institut Teknologi Bandung, Bandung 40132, Indonesia
[2]Department of Agriculture Product Technology, Universitas Kristen Artha Wacana, Kupang 85000, Indonesia

Abstract. In previous study, by heating magnesium basic soaps from palm stearine will decarboxylated and produced biohydrocarbon. The frequent method to produced metal soaps from triglyceride in laboratory scale is metathesis. This process is less favored because this method would produced large amounts of salt waste and hard to develop into bigger scale. This study investigated the process and characterization of magnesium soaps from coconut oil and magnesium hydroxide via direct reaction method at 185 °C for 3 and 6 hours. The resulting soaps were washed with water and methanol, then dried. This process yield more than 80%-w metal soaps, acid values lower than 6 mg KOH/g and pH 9.2. Based on Thermogravimetry Analysis (TGA) and SEM results, the initial decomposition temperature of these metal soaps were at 300 °C and have amorphous surface morphology. From decarboxylation test of magnesium basic soaps indicate great potency as feed for biohydrocarbon production.

1 Introduction

Metal soaps, or metal long-chain carboxylates, are alkaline-earth or heavy-metal salts of carboxylic acids with carbon atoms 7-22 [1,2]. Application of metal soaps are diverse such as heat stabilizers for plastics (especially polyvinyl chloride, or PVC), driers in paint or inks, waterproofing agents, fuel additives, cosmetic products, component as lubricating greases, etc [2-6]. Most of these applications are related to the thermal properties of metal soaps [7]. Metal soaps could also use as feed for thermal decomposition to produce bio-hydrocarbon [8,9].

Fatty acids and vegetable oil are the feed for metal soaps manufacture. The metal soaps manufacture processes using fatty acids as saponification feed are precipitation process, or double decomposition; fusion process, direct reaction of fatty acids, or carboxylic acids, with metal oxides, hydroxides, and carbonates; and direct reaction of metals and molten fatty acids [1,4,10]. Fatty acids as saponification feed has greater cost compare to triglycerides derived from vegetable oil, because it is more expensive. Therefore, production metal soaps from vegetable oil is more attractive.

The general method to produce metal soaps derived from vegetable oil is metathesis process [11-13]. This process was started by mixing the vegetable oil and hot ethanol, then stirred the mixture with sodium hydroxide 20% (w/w) solution, become aqueous sodium soaps. This sodium soap is then added the metal aqueous solution (example metal nitrate, acetate, chloride, and sulfate) and yield metal soaps. This process produced sodium nitrate/acetate/chloride/sulfate as waste so it is difficult to develop this process into larger scale. Therefore, this research was carried out by direct reaction of triglyceride and metal hydroxide, as described by [14,16]. The zinc soaps were produced from the reaction mixture of zinc oxide/hydroxide, water, and triglycerides at elevated temperature. The by-product from direct reaction method is glycerine [14], which is valuable.

The metal soaps produced by direct reaction [14] were stoichiometry soaps (formula: $M(RCOO)_2$, if bivalent metals). This experiment aimed to obtained basic soaps (formula: M(RCOO)(OH), if bivalent metal and use hydroxide as saponification feed) because the metal soaps would be the feed for decarboxylation process to produce hydrocarbon (Reaction 1). The decomposition reaction of stoichiometric metal soaps would produce ketones (Reaction 2) [3].

$M(RCOO)(OH) \rightarrow \mathbf{RH} + MCO_3$ (1)
$M(RCOO)_2 + M(R'COO)_2 \rightarrow 2\ \mathbf{RCOR'} + 2\ MCO_3$ (2)

In previous research [12-14], the main product from decarboxylation of magnesium basic soaps from palm stearin and palm kernel oil were green diesel and biokerosene.

* Corresponding author: mei@che.itb.ac.id

One of vegetable oils produced in Indonesia is coconut oil (*Cocos nucifera*). The main fatty acid constituent of coconut oil is lauric acid (C12:0) [11], as palm kernel oil. The preparation and characterization of metal soaps derived from coconut oil was performed by [11] using metathesis process.

The objective of this research was to investigate the characteristics of magnesium basic soaps carried out by direct reaction of coconut oil and magnesium hydroxide. The resulting basic soaps were characterization the acid value, pH, thermal decomposition, morphology, and metal content of magnesium basic soaps. This research also conducted decarboxylation test of magnesium basic soaps to obtain the yield of liquid products.

2 Experimental

Coconut oil (Barco) was used in the experiment with acid value/AV 0,6 mg KOH/g and consisted of caprylic acid (5,92%-mol), capric acid (6,08%), lauric acid (42,39%), myristic acid (20%), palmitic acid (11,41%), stearic acid (3,72%), oleic acid (8,94%), and linoleic acid (1,54%). Magnesium hydroxide was prepared by co-precipitation process of $MgNO_3.6H_2O$ with sodium hydroxide (Merck), as described in [16].

The reaction was carried out at 185 °C for 3 and 6 hours under vigorous stirring, by reacting coconut oil (1 mol), magnesium hydroxide (3 mol), and water (excess). The magnesium soaps were washed with distilled water and methanol, filtered, and dried overnight at 60 °C. The dried soaps was ground into magnesium basic soaps powder. Saponification process was performed in 100 ml stainless steel mini-autoclave. For simplicity, magnesium basic soaps obtained were given code Mg-CO-3 (for 3 h reaction) and Mg-CO-6 (for 6 h reaction).

pH of magnesium soaps was tested by added 2 g soaps into 20 ml distilled water, stirred, then measured the pH with pH-meter digital until the pH was stable. The free fatty acids content of magnesium basic soaps was analyzed using Cunder method [17,18]. Thermogravimetric analysis (Linseis STA PT 1600) of magnesium basic soaps was carried out by heating the sample from 30 until 700 °C at constant heating rate 10 °C/min under nitrogen gas flow at 40 ml/min. A scanning electron microscope (SEM) with energy-dispersive X-ray (EDX) analyses was used to identify the morphology and metal content of soaps. The magnesium basic soaps were coated with gold using the sputtering technique.

The decarboxylation test of magnesium basic soaps was performed in 100 ml glass batch reactor at atmospheric pressure. Decarboxylation reaction was carried out from room temperature up to 370°C and maintained for 5 hours at that temperature. 9-10 g magnesium basic soaps were placed into the reactor. Before heating, the reactor was flushed with nitrogen to remove the remaining oxygen inside. The heat was supplied from heating mantle. The condensed products from decarboxylation reactor, or liquid products, were collected in a beaker and measured. The series of decarboxylation equipment and aparatus used in this experiment is same as [12,13].

3 Results and discussion

3.1. Synthesis and characterization of Magnesium Soaps

The yields magnesium basic soaps carried out by direct reaction of magnesium hydroxide and coconut oil for 3 and 6 hours are 92.23 and 88.23%-w. Both soaps are white and light, similar with magnesium stearate commercial grade. Magnesium basic soaps obtained from this process has similar physical characteristic with magnesium soaps of coconut oil reported by Ek [11]. This is prove that the direct reaction could produce magnesium basic soaps of coconut oil with good yield.

The result acid value and pH of magnesium basic soaps are given in Table 1. These magnesium basic soaps have acid value below 6 mg KOH/g sample, which there were free fatty acids in these basic soaps. The free fatty acids contained in metal soaps indicated the phenomena of direct reaction, triglycerides hydrolysis into free fatty acids then these acids were reacted with Mg^{2+} ion (soluble in water) and formed magnesium soaps. pH magnesium basic soaps about 9.2, higher than magnesium stoichiometric soaps obtained by Ek [11], 8.2. This pH confirmed that direction reaction used in this experiment successfully produce magnesium basic soaps of coconut oil.

Table 1. Acid value and pH of magnesium basic soaps

Sample	AV (mg KOH/g)	pH
Mg-CO-3	5.39	9.20
Mg-CO-6	2.91	9.21

The TGA curve (Figure 1) show that these magnesium basic soaps have similar thermal behavior. These magnesium basic soaps were lose their moisture up to 110 °C, and then stable from 110-350 °C. From 300-500 °C, the basic soaps were decomposed very rapidly and show small weight change during further increased temperature. Thermal decomposition of magnesium basic soaps is written in Reaction 1, which yield magnesium carbonate as residue. Further heating with increased temperature would decompose magnesium carbonate into magnesium oxide and carbon dioxide. The final residue weight of magnesium basic soaps was similar with the weight of magnesium oxide. It is important to studied decomposition behavior so the thermal characteristic and decarboxylation reaction temperature can be predicted.

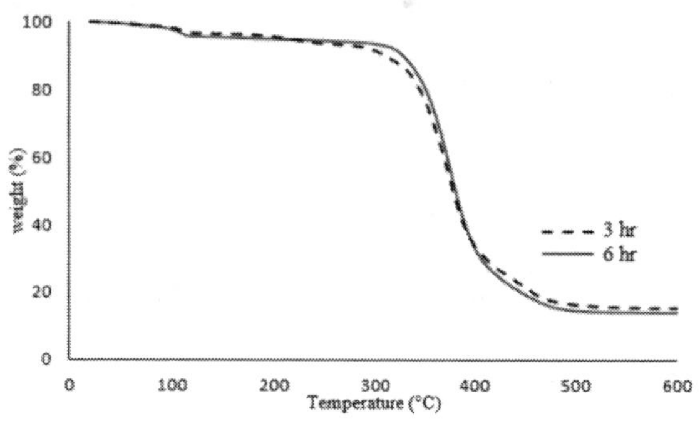

Fig. 1. TGA curve of magnesium basic soaps

The magnesium soaps of coconut oil morphology has not been studied. SEM analysis were conducted to know the morphology of magnesium basic metal soaps. The morphology of magnesium basic soap particles is shown in Figure 2 and 3. The magnesium basic soaps of coconut oil exhibit amorphous morphology, except the particles from Mg-CO-6 were more packed than Mg-CO-3 particles. It might be caused by longer saponification reaction.

Fig. 2. SEM micrographs of Mg-CO-3

Fig. 3. SEM micrographs of Mg-CO-6

EDX is a powerful tool for analyze the elements content, especially for elements with atomic number greater than four [5]. The detector scanned the observed area to identified the elements contained in magnesium basic soaps particles and their spreading. It also detected gold, but neglected because the sample particles were coating with gold. Both soaps contained three common elements : carbon, oxygen, and magnesium. The elements spread almost evenly at magnesium basic soaps surface particles (Figure 4 and 5).

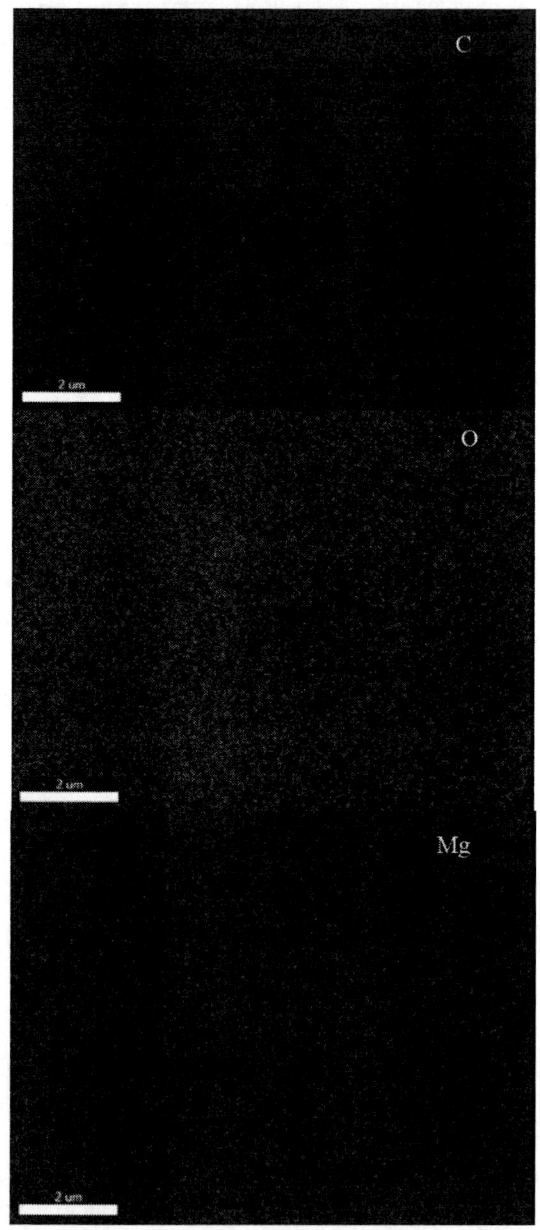

Fig. 4. The element spreading in Mg-CO-3 particles

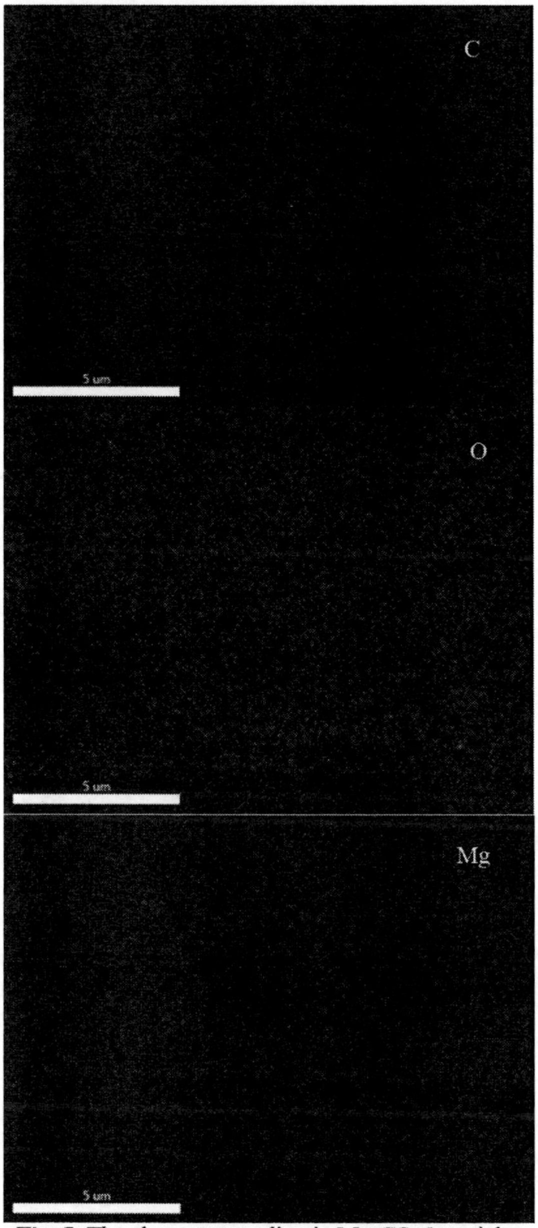

Fig. 5. The element spreading in Mg-CO-6 particles

Table 2 show the EDX measurements of magnesium basic soaps of coconut oil prepared by direct reaction and theoritical values of elements contained, on hydrogen free basis. EDX analysis only give an indication of the element content in the metal soaps since it measured based on the area observed. From the comparison of the experimental results obtained for the soaps prepared using direct reaction method, both magnesium soaps have C and Mg slight higher than theoritical values. A possible explanation for this result could be from free fatty acids, indicated from acid value analysis. The acid value of Mg-CO-3 is slighty higher than Mg-CO-6, so the C content of Mg-CO-3 is higher.

Table 2. The element content in magnesium basic soaps.

Element	%-w		
	Theoritical	Mg-CO-3	Mg-CO-6
C	58.59	64.09	63.6
O	23.40	20.84	22.9
Mg	17.99	15.07	13.5

3.2. Decarboxylation Test of Magnesium Basic Soaps

The products obtained from decarboxylation of magnesium basic soaps were liquid product, water, solid residue, and other (include gas), show in Table 2. Up to 100 ^{0}C, water was collected from beaker. The liquid product color obtained from decarboxylation magnesium basic soaps at 370°C for 5 h is yellow-brown liquid. The yield product was measured by compare the weight of liquid product with theoritical hydrocarbon could obtained. Regardless the liquid product yield still varied (62.75 and 31.51%-w), it indicated that magnesium could be used as feed for bio-hydrocarbon production. When compared with other processes to produce bio-hydrocarbon from vegetable oil, such as via hydrodeoxygenation [19], decarboxylation of metal basic soaps is a potential alternative process to produce bio-hydrocarbon because it performed at atmospheric pressure and not using noble metals as catalyst.

Table 2. Products of Mg-CO decarboxylation

Feed	[%-wt]				Yield liquid product (%)
	Residue	Water	Others	Liquid product	
Mg-CO-3	39.76	0.86	20.6	38.78	62.75
Mg-CO-6	59.49	1.09	19.62	19.8	31.51

4 Conclusion

In this study, the direct reaction of coconut oil and magnesium hydroxide was investigated to obtain magnesium basic soaps. The reaction was carried out at 185 °C in the presence of water, and the processes occurred in this method was hydrolysis of triglycerides hydrolysis into free fatty acids/ffa and saponification reaction between ffa with magnesium ion soluble in water, formed magnesium soaps which insoluble in water. The addition of saponification reaction time did not give significant effect, based on the amount of yield, pH, free fatty acids content, thermal behavior, and surface morphology of magnesium basic metal soaps. Thermal decomposition of magnesium basic soaps was examined in TGA analysis, it concluded that there only occurred one decomposition cycle from 300 °C.

Decarboxylation of magnesium basic soaps at 370 °C for 5 h produced liquid product, predicted as bio-hydrocarbon, as main product. This decarboxylation product need further identification. The decarboxylation process was carried out at atmospheric pressure, moderate temperature, low hydrogen consumption, and not using noble catalyst, such as platina or palladium, indicate the potential of this process as alternative method to produce biohydrocarbon.

We acknowledge the support from P3MI.

References

1. S. B. Elliott, *The Alkaline-Earth and Heavy-Metal Soaps* (Reinhold Publishing Corporation, New York, 1946)
2. Q. Dou, K. M. Ng. Powder Technology 301 (2016)
3. M. S. Akanni, E. K. Okoh, H. D. Burrows, H. A. Ellis, Thermochim. Acta **208** (1992)
4. M. Gönen, D. Balköse, F. İnal, S. Ülkü, Ind. Eng. Chem. Res. **44** (2005)
5. T. O. Egbuchunam, D. Balköse, F. E. Okieimen, Polymer Degradation and Stabillity **92** (2007)
6. D. Balköse, T. O. Egbuchunam, F. E. Okieimen, J. Therm. Anal. Calorim. **101** (2010)
7. R. Sharma, L. C. Heda, A. Joram, Physical Chemistry (2013)
8. C. C. Chang, S. W. Wan, Ind. Eng. Chem **39** (1947)
9. H. Lappi, R. Alen, Journal of Analytical and Applied Pyrolysis **91** (2011)
10. K. S. Markley, *Fatty Acids Their Chemistry, Properties, Production, and Uses* (Interscience Publishers Inc., 1961)
11. O. Ek, J. Appl. Sci. Environ. Manage **18**, 2 (2014)
12. G. F. Neonufa, M. Pratiwi, T. H. Soerawidjaja, T. Prakoso, Proceeding of International Seminar on Chemical Engineering in conjunction with STKSR (2016)
13. G. F. Neonufa, M. Pratiwi, A. N. Istyami, L. Elizabeth, S. S. Dewi, R. Purwadi, Proceeding of IJCAET (2016)
14. J. Blachford, US Patent No 4,316,852 (1982)
15. L. H. Duong, O. Fujita, I. K. Reksowardojo, T. H. Soerawidjaja, G. F. Neonufa, Fuel **185** (2016)
16. M. Pratiwi, P. Ylitervo, A. Pettersson, T. Prakoso, T. H. Soerawidjaja, Proceeding of SOMCHE 2016 (2016)
17. R. E. Laily, dan J. Cunder, US Patent No. 3,476,786. (1969)
18. R. D. Borzelli, dan J. Cunder, US Patent No. 4,307,027 (1981)
19. E. Santilan-Jimenes, M. Crocker, J Chem Technol Biotechnol, **87** (2012)

The Effect of Co-solvent on Esterification of Oleic Acid Using Amberlyst 15 as Solid Acid Catalyst in Biodiesel Production

Iwan Ridwan , *Mukhtar* Ghazali, *Adi* Kusmayadi, *Resza* Diwansyah Putra, *Nina* Marlina, and *Eko* Andrijanto*

Chemical Engineering Department, Politeknik Negeri Bandung, Indonesia

Abstract. The oleic acid solubility in methanol is low due to two phase separation, and this causes a slow reaction time in biodiesel production. Tetrahydrofuran as co-solvent can decrease the interfacial surface tension between methanol and oleic acid. The objective of this study was to investigate the effect of co-solvent, methanol to oleic acid molar ratio, catalyst amount, and temperature of the reaction to the free fatty acid conversion. Oleic acid esterification was conducted by mixing oleic acid, methanol, tetrahydrofuran and Amberlyst 15 as a solid acid catalyst in a batch reactor. The Amberlyst 15 used had an exchange capacity of 2.57 meq/g. Significant free fatty acid conversion increments occur on biodiesel production using co-solvent compared without co-solvent. The highest free fatty acid conversion was obtained over methanol to the oleic acid molar ratio of 25:1, catalyst use of 10%, the co-solvent concentration of 8%, and a reaction temperature of 60°C. The highest FFA conversion was found at 28.6 %, and the steady state was reached after 60 minutes. In addition, the use of Amberlyst 15 oleic acid esterification shows an excellent performance as a solid acid catalyst. Catalytic activity was maintained after 4 times repeated use and reduced slightly in the fifth use.

1. Introduction

Biodiesel is the renewable fuels that not only have great demand, but also pollutants emit less. In biodiesel production, feedstock must be cheap, sustainable and non-edible oil because it can decrease the biodiesel production cost [1-5]. Biodiesel is produced by esterification with acid catalyst and transesterification by the alkaline catalyst. Unfortunately, in transesterification process with an alkaline catalyst, the presence of free fatty acid (FFA) in oil causes saponification reaction between FFA and alkaline catalyst which decrease the biodiesel yield and increase the production cost. Hence, the esterification with acid catalyst must be done before transesterification process to minimize the FFA content in the transesterification reaction. In esterification, the FFA is converted to fatty acid methyl ester (biodiesel). Even though transesterification and esterification are different reactions, it produces the same product namely biodiesel.

Biodiesel production can be influenced by those variables such as co-solvent, amount of catalyst, reaction temperature, reaction time, and the molar ratio of reactant [6-10]. Yunus et al. studied these variables effect to FFA conversion. The FFA conversion using waste cooking oil as a feedstock was 88% under a reaction temperature of 60°C, methanol to oil mass ratio of 18:1, catalyst amount of 3.0 % and stirring speed of 300 rpm [11]. Another study [12] reported that the optimum conditions for the esterification were achieved at 70°C of reaction temperature, 5% of catalyst loading and 6:1 of ethanol to the oleic acid molar ratio by using the zeolite prepared from kaolin and achieved FFA conversion 85% after 60 min [12]. Solid catalysts such as Amberlyst 15 made the highest FFA conversion than the others at catalyst amount (1-2%wt) and temperature 50-60°C in esterification [6].

Another problem in biodiesel production is the oil has very low miscibility in methanol [13]. The low miscibility of methanol in oil causes low overall reaction rate. Co-solvent can decrease the interfacial surface tension and increase the reaction rate. Boocock *et al.* [9] reported that tetrahydrofuran (THF) is the best choice to co-solvent due to its boiling point (67°C) is almost the same with the methanol boiling point (64°C). The loading of THF is for improving the overall reaction rate in both transesterification and esterification [14]. Hence, this study aims to investigate the effects of co-solvent,

* Corresponding author: andrijanto.2002@gmail.com

catalyst amount, reaction temperature and the ratio of methanol to oleic acid for esterification using Amberlyst 15 as a solid acid catalyst in the production of biodiesel.

2 Materials and Methods

2.1. Materials

Oleic acid was purchased from a local supplier (PT. Bratachem), and Methanol (99.8% purity) was purchased from Merck. Amberlyst 15 was used as the solid acid catalyst to improve reaction time and THF as co-solvent to enhance the miscibility of oil and the methanol. Isopropyl alcohol, toluene, potassium hydroxide and phenolphthalein were used to analyze FFA content.

2.2. Methods

2.2.1 Preparation of Catalyst

This preparation aims to activate the pores and remove any impurities that attached to the catalyst. Amberlyst 15 was soaked in methanol and chloride acid (HCl) solution for 1 hour. Afterward, it was heated in the oven at 85°C for 1 hour.

2.2.2 Determination of ion exchange capacity

This step is to measure the acid capacity of the catalyst. Amberlyst 15 was soaked into HCl solution for 1 hour, and after that, the catalyst was dried in the oven at 90°C for 2 h. The dried catalyst was then soaked in NaCl solution for 24 h to add sodium ion to the solution. After that, the 0.01 M NaOH was used to titrate the solution using phenolphthalein as an indicator to determine the titration point. The determination of ion exchange capacity (IEC) is determined by equation [17]:

$$\text{IEC} = \frac{\textit{Volume of NaOH x Concentration of NaOH}}{\textit{Weight of catalyst}} \quad (1)$$

2.2.3 Esterification

The oleic acid esterification in this study was conducted in a batch reactor by mixing oleic acid, methanol, Amberlyst 15 and THF. The reflux condenser was equipped the reactor recovery the remaining methanol. The temperature controller and water bath were equipped to the reactor to control the desired temperature. An agitator was used to mix the solution at 400 rpm. The investigated factors in this study were the co-solvent amount, catalyst loading, the methanol to oil molar ratio, and reaction temperature.
A 50 ml of oleic acid was fed to the reactor for each experiment. Esterification process was carried out in the oleic acid to methanol molar ratio 1:5, 1:10, 1:15, 1:20 and 1:25. The amount of Amberlyst 15 was 2 %w, 4 %w, 6 %w, 8 %w and 10 %w. The temperature varied was 30°C, 40°C, 50°C and 60°C. The amount of co-solvent varied was 2%w, 4%w, 6%w, 8%w and 10%w. Consequently, those variables will be varied to get the optimum condition. Each experiment was conducted for 90 minutes. Sampling was done manually at reaction time every 15 minutes to analyze the FFA concentration.

2.2.4 The Method of Analysis

Determination of FFA was analyzed using American Oil Chemists Society (AOCS) method [14]. The sample was taken from the reactor every 15 minutes. One gram of sample was dissolved in 10 ml of toluene and isopropyl alcohol mixture (1: 1 v/v) as a solvent to ensure the miscibility of sample and titrant. Three drops of phenolphthalein indicator were added to the mixture.
The adding of phenolphthalein was used as an indicator and titrated using KOH (0.25 N) solution until the sample changed the pink color. The percentage of the unreacted FFA was calculated using the equation below [6].

$$FFA\ (\%) = \frac{\textit{Vol. of KOHxNormality KOHx28.2x100\%}}{\textit{Weight of sample}} \quad (2)$$

The FFA conversion was determined by the equation given [6].

$$FFA\ conversion\ (\%) = \frac{(FFA\ i - FFA\ t)}{FFA\ i} \times 100\% \quad (3)$$

where i is initial concentration, t is the final concentration in each reaction time.

2.2.5 Catalyst Reuse

The method of catalyst reuse was adopted from Hykkerud *et al.* [15]. The catalyst used in each experiment was collected. Subsequently, those catalysts were soaked in methanol for 60 minutes and heated in the oven to dryness.

3 Results and Discussion

3.1. Ion Exchange Capacity

The ion exchange capacity demonstrates the ability of the catalyst to increase the conversion of FFA. In a previous study [6], Amberlyst 15 with an ion exchange capacity of 5.2 meq/g yielded FFA conversion of 35-45% [6]. Meanwhile, the catalyst used in this study has an exchange capacity 2.57 meq/g. This result is determined by ion exchange method, and this is slightly lower than expected. This is due to the catalyst was stored for a quite long time. The ion exchange capacity will be very influential in converting FFA into biodiesel [11].

3.2. The Effect of Methanol Molar Ratio

Biodiesel production from oleic acid with molar ratio variation of methanol to oleic acid was done by 5 variations. The differences of methanol to oleic acid molar ratio are 5: 1, 10: 1, 15: 1, 20: 1, and 25: 1. Methanol was chosen because the methanol reactivity is higher than ethanol as a reactant. Fig. 1 shows the changes of the conversion of the FFA every 15 minutes with reaction time for 90 minutes. The esterification reaction of oleic acid is a reversible reaction. Therefore, excessive use of methanol can shift the equilibrium reaction toward the product. It can be seen in Fig. 1 that there was a significant increase in conversion in a process using THF as a co-solvent compared with a reaction without THF. This suggests that the presence of THF helps the oil and methanol mixture into one phase. It facilitates the reaction between oleic acid and methanol and finally increases the overall reaction rate. Fig. 1 explains that the esterification reaction without THF resulted in a maximum FFA conversion of 3.15% whereas with THF the conversion increases to 23.16% at the same reaction time 90 minutes. By comparing with another result [6] for esterification of oleic acid using Amberlyst 15, without using co-solvent, the esterification reaction has reached steady state after 100 minutes while in this research esterification reaction steady state after 75 minutes. This suggests that the THF addition as a co-solvent increase the rate of esterification reaction of oleic acid.

It can be inferred from Fig. 1 that the molar ratio of methanol dramatically influences the conversion of the reaction. The higher mole of methanol causes the higher FFA conversion [12, 16]. At a 5: 1 to 20: 1 methanol molar ratio, the conversion increased sharply and experienced steady state after 75 minutes. In our study with methanol to oil molar ratio of 25:1, produced the highest FFA conversion and steady state is achieved after 60 minutes. The steady state is faster than the previous study, this may due to the role of THF in this reaction by providing single phase system for the reactants although the conversion at this ratio is 23.16% which is almost half of the previously reported. This is due to the ion exchange capacity of Amberlyst 15 used in this study is smaller (2.57 meq/g) compared to the earlier reported by Andrijanto *et al.* which is 4.7 meq/g [17].

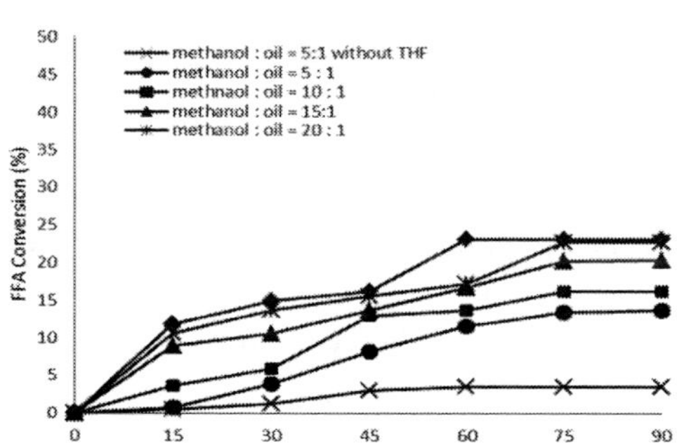

Fig. 1. The Effect of Methanol Molar Ration on FFA conversion

3.3. The Effect of Catalyst Amount

The effects of the amount of Amberlyst 15 used on FFA conversion were investigated. The catalyst amount was varied from 2% to 10%. The temperature of reaction and molar ratio of methanol to oil was kept constant during the experiment. As shown in Fig.2, the FFA conversion changes with time. The conversion increases with the increase of the amount of catalyst.

Fig. 2 indicates the conversion of FFA every 15 minutes at reaction time for 90 minutes. It was clear that the greater the number of catalysts the greater the resulting FFA conversion. The oleic acid esterification is a difficult reaction to occur in the absence of a catalyst, and the conversion is often limited by the nature of the reaction which is reversible. In Fig. 2, it is clear that the oleic acid esterification with 2% and 4% catalyst amount resulted in FFA conversion of about 5.14% to 12.08% respectively. In addition, at catalyst amount of 6%, 8% and 10% there was a significant increase of FFA conversion which is 23.16%, 25.21%, 28.57% respectively.

At the amount catalyst of 2%, the maximum FFA conversion was 5.47%, and by increasing the catalyst amount to 4%, the conversion was almost doubled which was 12.08 %. The steady state condition is reached after 75 min. Further increase the catalyst amount by 6%, 8% and 10% result in faster steady state, and this achieved after 60 min. It means the steady state condition is faster by adding the amount of catalyst. This is due to the more active site presence (hydrogen ion within the Amberlyst 15) in the reaction mixture. The rise in the number of active sites within reaction mixture provided the chance of oleic acid reacted with methanol on the surface of the catalyst and yielded increase reaction rate [6,11,12]. The previous result shows that the increases of the catalyst amount above 10% yield a plateau, which is suggesting that the equilibrium condition is reached.

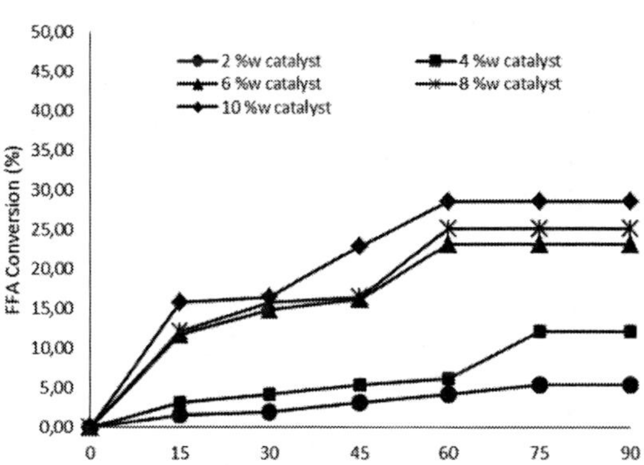

Fig. 2. The Effect of Catalyst Amount on FFA conversion

https://doi.org/10.1051/matecconf/201815603002

3.4. The Effect of Co-solvent Amount

The solubility of methanol in oil is low. The co-solvent is added to the reactor to increase solubility methanol in oil and speed up overall reaction time. In this paper, THF is used as co-solvent for biodiesel production. The ratio of THF to oleic acid was varied at concentrations of 2, 4, 6, 8 and 10%. It is noticeable that by adding THF to the reaction mixture, increase the FFA conversion. The higher the THF concentration, the FFA conversion also increases. This suggests that THF as a co-solvent can increase overall reaction time due to the increase of oleic acid miscibility in methanol which leads to increase overall reaction time [13,16]. The increasing speed of the overall reaction is due to the addition of THF which makes the oil and methanol in one phase as shown in the ternary diagram of Fig. 4.

At THF concentration to the oleic acid of 2% only resulted in maximum FFA conversion of 7.24% while at 4% and 6% THF concentration resulted in FFA conversion of 22.93% and 24.11% respectively. By increasing the concentration of THF to 8% and 10%, the FFA conversion yield 28.57%. This indicates that the THF concentration of 8% is the optimum concentration since the FFA conversion at 10% THF is the same.

At low THF concentration (2%, 4% and 6%) FFA conversion was a steady state at a reaction time of 75 minutes while at THF concentration above 8% FFA conversion had reached steady state at a faster reaction time of 60 minutes. Therefore, the optimum THF concentration was at 8%.

3.5. The Effect of Reaction Temperature

The reaction temperature dramatically influences the conversion of the reaction. The higher the reaction temperature, the collisions that occur between the reactant particles are higher, causing the reaction to be faster and the conversion increases [11, 12]. In this esterification reaction of oleic acid catalyzed by Amberlyst 15, the reaction temperature was varied from 30 °C to 60 °C. The reaction temperature should be below 64 °C because at temperatures above 64 °C the methanol will evaporate, causing contact between methanol and oleic acid to be reduced.

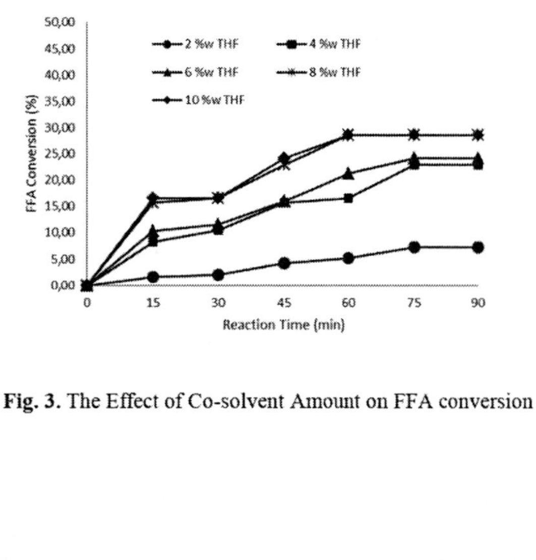

Fig. 3. The Effect of Co-solvent Amount on FFA conversion

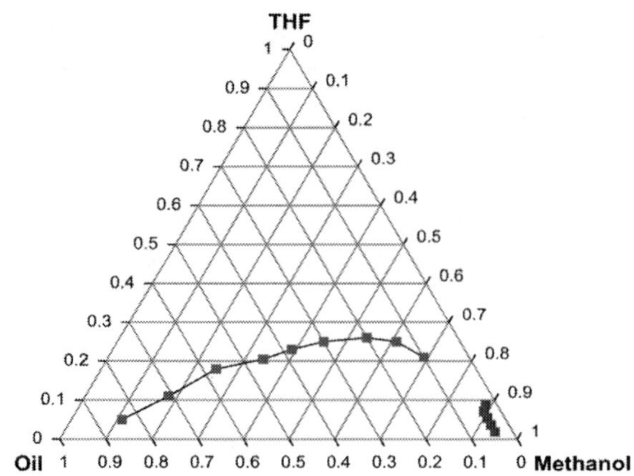

Fig. 4. Terner Diagram of THF, Methanol and Oil. ■ Boocock Research. ■This Research

From Fig. 5, it can be inferred that the reaction temperature greatly affects the conversion of FFA in oleic acid esterification. Fig. 5 illustrates the conversion of FFA with the interval 15 minutes at reaction time for 90 minutes. It is clear that the higher the reaction temperature, the greater the conversion of the FFA. At low temperatures (30 °C), the conversion of FFA is very low at 2.3% under steady state. This indicates that at the low reaction temperature, the oleic acid esterification is difficult. By increasing the reaction temperature to 40 °C and 50 °C, there was an increase of FFA conversion by 9.16% and 17.78% respectively. The highest FFA conversion is achieved at 60 with FFA conversion of 28.57%. In all variations of the reaction temperature, the steady state conditions were reached after 60 minutes.

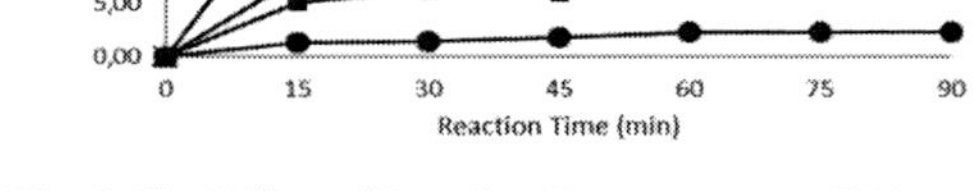

Fig. 5. The Effect of Reaction Temperature on FFA conversion

3.6. The Effect of Catalyst Reuse

One of the advantages of heterogeneous catalysts is an easy separation process and can repeatedly be used. In this study, the heterogeneous catalyst used was Amberlyst 15. The catalyst was repeatedly used in oleic acid esterification to test the Amberlyst 15 performance. In this study, the used catalyst was reused 5 repetitions to see the performance of the catalyst regarding repeated use of FFA conversion [12, 16].

It can be concluded from Fig. 6 that the use of Amberlyst 15 in the esterification reaction of oleic acid shows a good performance. FFA conversion in steady state conditions did not decrease significantly after the fourth usage. After the fifth usage, there was a significant reduction of FFA conversion from 23.07% to 15.42%. This indicates that the Amberlyst 15 must undergo regeneration after being used 4 times. In addition, the decreasing of conversion was possibly due to adsorption of water produced during esterification and blocked the pores of the catalyst and inhibits the catalytic activity. [17].

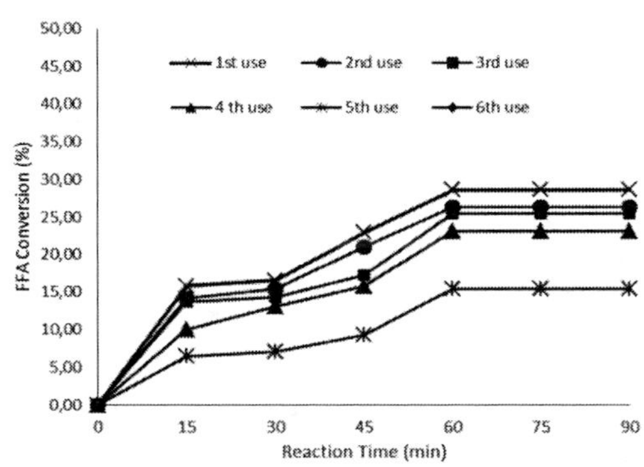

Fig. 6. The Effect of Catalyst Reuse on FFA conversion

4 Conclusions

The conversion of free fatty acid significantly increases by the involvement of THF as a co-solvent. The highest FFA conversion of 28.57% was obtained using molar ratio methanol to the oleic acid of 25:1, the catalyst of 10% w/w, THF concentration of 8%, and a reaction temperature of 60°C. The steady state was achieved after 60 minutes. The catalytic activity was maintained after four repeated use.

We would like to thank to the Ministry of Technology Research and Higher Education of Indonesian Government for financial support to the research.

Reference

1. G. Baskar, R. Aiswarya, Renewable and Sustainable Energy Rev., **57**, 496-504 (2016)
2. B. R. Moser, Dev. Biol. Plant **45**, 229–266 (2009)
3. A. Harvey, J. G. Lee, F. H. Kasim, E. J. Eterigho, J. Sci. Ind. Res., vol. **17**, 822-828 (2012)
4. L.F. Gutierrez, O.J. Sanchez, C.A. Cardona, Bioresour. Technol. **100**, 1227-1237, (2009)
5. P. Nakpong, S. Wootthikanokkhan, Renewable Energy **35**, 1682–1687, (2010)
6. N. Ozbay, N. Oktar, N. A. Tapan, Fuel, **87**, 1789 (2008)
7. D. M. Chesterfield, P. L. Rogers, E. O. Al-Zaini, A. A., Adesina A., Procedia Eng. **49**, 373 – 383 (2012)
8. J. F. Oliveira, I. L. Lucena, R. M. Saboya, M. L. Rodrigues, Renewable Energy **35**, 2581-2584,(2010)
9. D. G. B. Boocock, S. K. Konar, V. Mao, H. Sidi, Biomass Bioenergy **11**, 43–50. (1996).
10. S. Sulaiman, A. R. Abdulaziz, M. K. Aroua, J. Taiwan Inst. Chem. Eng. **44**, 233-238. (2013)
11. N. B. M. Yunus, N. A. B. Roslan, C. S. Yee, S. Z. Abidin, Procedia Eng., **148,** 1274 – 1281 (2016)
12. A. M. Doyle, T. M. Albayati, A. S. Abbas, Z. T. Alismaeel, Renewable Energy, **97**, 19-23 (2016)
13. D. G. B. Boocock, S. K. Konar, V. Mao, H. Sidi, J. Am. Oil Chem. Soc. **73**, 1247–1251, (1996)
14. Y. J. Cho, T. E. Kim, B. Kill, Food and Science Technology, **53**, 517-521(2013)
15. A. Hykkerud, J. Marchetti, Biomass and Bioenergy, **95**, 340-343, (2016)
16. I. F. Nata, M. D. Putra, C. Irawan, C. K. Lee, J. Environ. Chem. Eng. (2017).
17. E. Andrijanto, E. A. Dawson, D. R. Brawn, Appl Catal B **1145,** 115–116 (2012)

Process Stability Identification Through Dynamic Study of Single-bed Ammonia Reactor with Feed-Effluent Heat Exchanger (FEHE)

Tri Partono Adhi[1*], and *Muhammad Iqbal* Prasetyo[1]

[1]Department of Chemical Engineering, Institut Teknologi Bandung, Bandung 40132, Indonesia

Abstract. In ammonia reactor system, a feed-effluent heat exchanger (FEHE) is typically installed to utilize reaction-generated heat to heat the reactor's feed. Utilizing energy from exothermic reaction to the incoming feed stream is often called "autothermic operation". Despite the advantage of FEHE, there is a risk of utilizing FEHE in a reactor system such as instability of process temperature or known as hysteresis. Hysteresis phenomena in chemical process could cause operational problems, for example it could damage the integrity of the equipment's material. This paper aims to evaluate the dynamic behavior of a single-bed ammonia reactor with FEHE, particularly to propose a way to prevent instability within the system. The dynamic simulation of the single-bed ammonia reactor with FEHE was performed with Aspen HYSYS v8.8. The result of the simulation result shows that hysteresis phenomenon in the ammonia reactor system occurs when the feed's temperature is below a certain value. If the feed temperature reaches that value, the temperature of the reactor's outlet oscillates. One of the solution to keep the feed temperature above that critical value is by installing a trim heater within the system. Based on the simulation, trim heater installation within the system is able to prevent hysteresis in the system evaluated.

1 Introduction

Ammonia production process is the one of the most important industrial-scale chemical process. Mostly, ammonia is used commercially in fertilizer and its remainder could be used as raw material for variety product, such as plastics, synthetic fibers and resins, pharmaceuticals, explosives, papers and refrigeration [1]. Despite of its importance, one part of the ammonia synthetic process, catalytic reforming reaction of nitrogen and hydrogen is the most energy consuming process that requires massive amount of heat to ensure the reaction is presented [2]. Therefore, the catalytic reforming reaction is usually supported by feed-effluent exchanger (FEHE) to make energy usage more efficient [2].

Feed-effluent heat exchanger (FEHE) is a kind of heat exchanger that widely used to utilize reaction-generated heat carried by reactor outlet as reactor inlet heating media. The hot reactor effluent is recycled back to feed preheater to provide all or a portion of the energy required to preheat the reactor feed [3]. Using FEHE is the one of the practical ways to optimize energy consumption in chemical plant. This system could be implemented in the process that involve exothermic reaction such as ammonia and methanol synthesis. Releasing energy from exothermic reaction to the incoming reactant stream is often called "autothermic operation" [4].

In steady-state study, designing heat integration in reaction system using FEHE is likely seen as interesting way to reduce energy consumption. However, some dynamic study shown that heat integration that involved FEHE could potentially lead to process instability in certain condition [5,6]. The example of process instability is the oscillation of ammonia reactor outlet temperature that lead to operation failure. The oscillation of ammonia reactor outlet temperature is shown in figure 1. Because of sudden pressure decreasing caused by decrease of fresh feed stream, oscillation of operation temperature in ammonia reactor occurred with the period of 6 minutes and range within 200°C. This oscillation condition could be damaging the catalyst in the reactor and harmful to the process. Therefore, we could investigate the dynamic behavior of ammonia reactor with the aid of simulation application such as Aspen HYSYS.

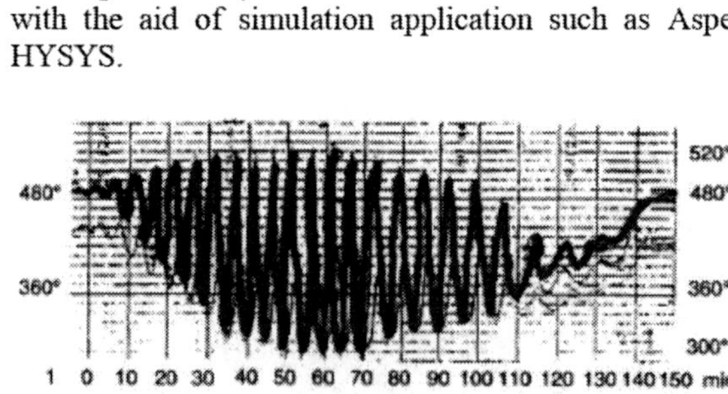

Fig. 1. Example record of industrial-scale ammonia synthesis reactor outlet temperature oscillation [1].

*Corresponding author: tpadhi@che.itb.ac.id

https://doi.org/10.1051/matecconf/201815603003

2 Methodology

Ammonia reactor system is modelled using Aspen HYSYS v8.8 for steady state simulation and then converted to dynamic simulation. Peng-Robinson as fluid package is used for whole simulation. Process flow diagram scheme of the process is shown in figure 2 while simulation of the reactor system scheme is shown in figure 3.

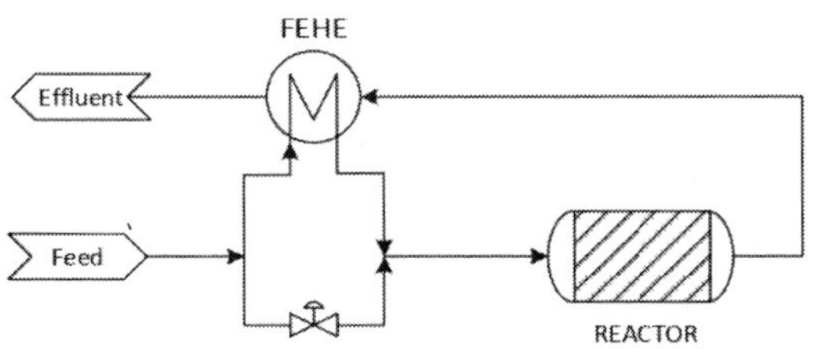

Fig. 2. Process flow diagram of single-bed ammonia reactor system with feed-effluent heat exchanger (FEHE)

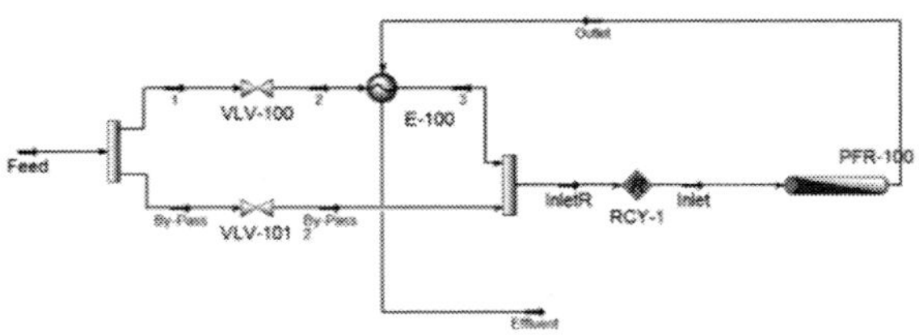

Fig. 3. Simulation scheme of autothermic single-bed ammonia reactor system with Aspen HYSYS v8.8.

Reactor feed is consisted of nitrogen, hydrogen and slightly amount of ammonia which specification is shown in table 1. Feed stream split into two part, one part which consist of 75% from total reactor feed amount is transferred to FEHE to be heated by reactor effluent and the other goes directly to the reactor after mixed with heated stream from FEHE. Specification of heat exchanger used as FEHE is shown in table 2.

Table 1. Ammonia reactor system feed stream specification.

Temperature	250°C
Pressure	163.5 bar
Mass Flow	200 ton/h
Composition	
N_2	75.6 %
H_2	16.3 %
NH_3	8.1 %

Heated feed then entered the ammonia reactor. The reaction undergoes in this reactor is assumed perfectly adiabatic. Reactor effluent that carried heat from exothermic reaction then flows through FEHE. Specification of single-bed ammonia reactor used in this study is based on [6] and presented in table 3.

Table 2. FEHE specification.

TEMA Type	BEM
Tube No.	379
Tube OD	19.05 mm
Tube Thickness	4.19 mm
Tube Pitch	23.08 mm
Tube Pattern	30-Triangular
Baffle Type	Double segmental
Baffle Cut Orientation	Vertical
Shell ID	600 mm
Shell OD	700 mm
Tube length	3500 mm
Baffle spacing	600 mm
Baffle No.	4
Shell in series	1
Shell in parallel	4

Table 3. Ammonia Reactor Specification.

Total Volume	30 m^3
Length	6.11 m
Diameter	2.50 m
Wall Thickness	5.00×10^{-3} m
Void Fraction	0.40

Reaction equilibrium and kinetics parameter, which also referred from [6], are shown in equation (1) and table 4 respectively.

$$N_2 + 3H_2 \rightleftharpoons 2NH_3 \quad (1)$$

Table 4. Reaction Kinetics Parameters.

Parameters	Forward	Reverse
N_2 Order	1.0	0
H_2 Order	1.5	-1.5
NH_3 Order	-1.0	1.0
A	23.62	3.40×10^{13}
E	87090 kJ/kmol	198464 kJ/kmol

A = Arrhenius Constant
E = Activation Energy

3 Result and Discussion

3.1 Instability phenomenon of single-bed ammonia reactor

The stability behavior of the ammonia reactor could be represented by outlet-feed temperature profile. Feed temperature is varied to obtain outlet temperature by using study case feature in Aspen HYSYS in forward step and backward step. Varying reactor inlet temperature is also performed to identify the feed-outlet temperature profile. Feed-outlet temperature profile is shown in figure 4. This relation was generated by varying reactor inlet temperature in such way to get the value of feed temperature and reactor conversion.

https://doi.org/10.1051/matecconf/201815603003

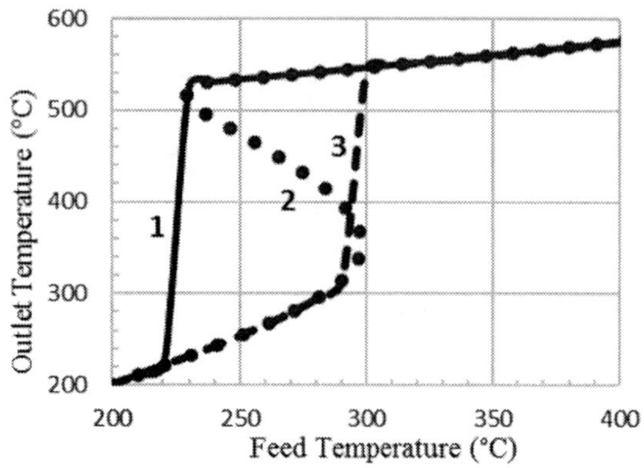

Fig. 4. Ammonia reactor feed-outlet temperature profile.

The 'S' curve, as seen in figure 4, implies that exothermic reactor could operate at several steady states condition under same operating condition depending on initial or feed conditions. There are three main information that could be obtained. First, from feed-outlet temperature profile generated by forward step (line 3), the reaction was activated when the feed temperature is above 290°C. Therefore, when doing the start-up, feed temperature must be set above 290°C to initiate the reaction. This temperature is often referred as "ignition temperature". Second, from feed-outlet temperature profile generated by backward step (line 1), it is shown that the reaction extinguished when feed temperature is below 230°C. The temperature value when the reactor extinguished often referred as "extinction temperature". Third, from feed-outlet temperature profile generated by varying reactor inlet temperature (line 2), it is shown that there is a transition zone between ignition temperature and extinction temperature. When reactor operated in the "upper outlet temperature" of the transition zone, once feed temperature goes below the extinction temperature, the outlet temperature would dramatically fall to "lower outlet temperature". After that moment, increasing the feed temperature back to its original value could not restore outlet temperature directly to "upper outlet temperature" path. However, the outlet temperature would follow "lower outlet temperature" path until feed temperature reach the ignition temperature. This transition zone that has non-reversible behavior could result instability of the process, known as hysteresis.

The stability of autothermic reactor as temperature margin have been introduced as the difference between the temperature of feed and a temperature near the extinction temperature of autothermic reactor [7]. Thus, to make sure that the reactor is safe from extinction, feed temperature should set away from extinction temperature. However, the desired operating temperature should be set to the as low value as possible that inside the region of multiplicity. Reactor operation is held close to the extinction temperature point because of the region is importance as economic optimum in commercial ammonia synthesis reactors [8]. Relation between reactor conversion and feed temperature shown in figure 5. As seen in figure 5, low operation temperature result higher conversion of reaction yet slight change of temperature at this region that may occur during operation could trigger the hysteresis phenomenon in the reactor. Assessment of reactor stability under operation in this region is necessary because slight process disturbances and changes in operating conditions could cause the reactor slide to extinction or to suffer growing oscillations.

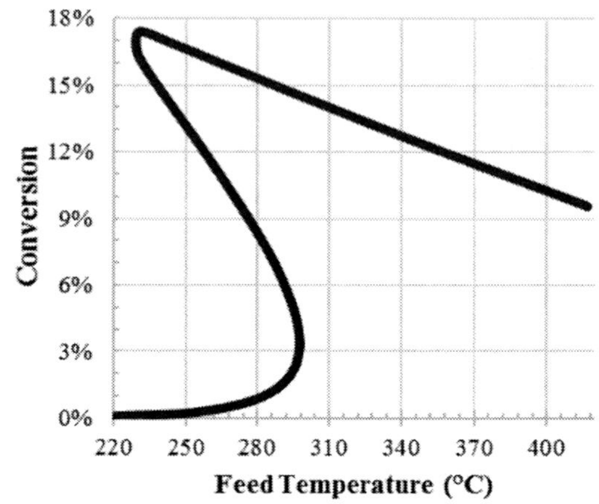

Fig. 5. Ammonia reactor feed-outlet temperature profile.

Furthermore, hysteresis phenomenon could be identified from dynamic simulation. Feed temperature was set at 250°C in steady operation condition and then suddenly dropped down to several temperatures around the extinction temperature which value is 230°C, referred as T_E. The temperature drops performed with a step function after the reactor operated in 1,200 seconds. In this simulation, the final feed temperature would be T_E + 5°C, T_E and T_E - 5°C. Result of dynamic simulations to identify the instability behavior are presented in figure 6.

Reactor still in stable condition when feed temperature dropped from 250°C to T_E + 5°C (235°C) as shown in figure 6(a), although the outlet temperature was slightly oscillating in the beginning of temperature drop. Instability phenomenon could be observed when feed temperature dropped to T_E and T_E - 5°C, as shown in figure 6(b) and 6(c) respectively. After temperature have been dropped, outlet temperature was roughly oscillating in range of 300-570°C before finally extinguished the reaction. As shown in the figure 6(b) and 6(c), the obvious difference of these two instability phenomena is the duration of outlet temperature oscillating behavior. Reactor with final temperature of T_E had experienced oscillating behavior for roughly 9,000 seconds, while the other one was only 5,000 seconds.

Huge difference of oscillating temperature could damage equipment material because of temperature sudden change that could decrease equipment material integrity. Increased temperature within the reactor could change physical property of catalyst's support that affects the performance of the catalyst itself [9,10].

Unstable operation condition also could also cause problem to the next process; the uncertainty of the outlet stream specification from the reactor system may cause problems to the further process.

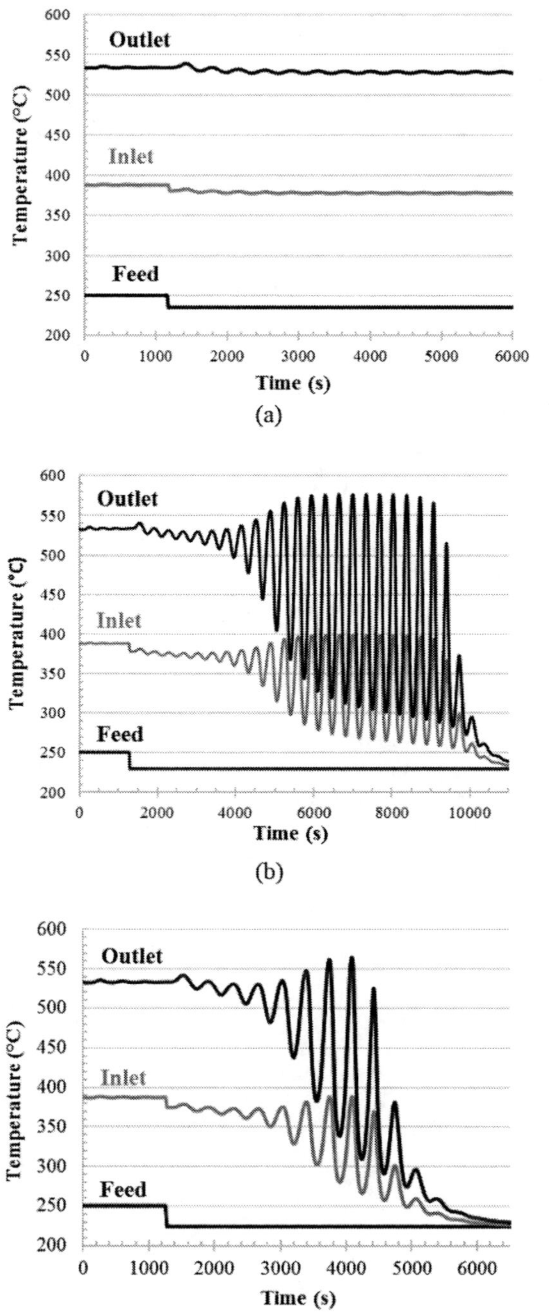

Fig. 6. Reactor temperature dynamic behavior of ammonia reactor system when feed temperature suddenly dropped from 250°C to (a) T_E + 5°C, (b) T_E and (c) T_E - 5°C

3.2 Prevention of hysteresis phenomenon in single-bed ammonia reactor by installing trim heater

Hysteresis phenomenon indicated by the oscillation of outlet temperature value is a dangerous event that could even causing loss or equipment failure to ammonia plant. One of the solution that proposed to prevent this phenomenon is installing trim heater in the reactor system. Trim heater used in ammonia reactor system is a kind of heat exchanger that keep reactor feed temperature above hysteresis temperature during the operation. Trim heater could be installed whether at the feed stream line or reactor inlet stream line. Each of this trim heater scenario illustrated in figure 7.

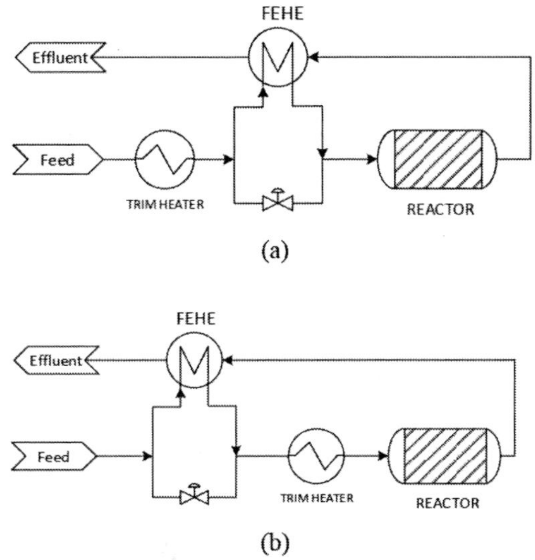

Fig. 7. Trim heater installation scenario (a) at feed stream line, (b) at reactor inlet stream line

The advantage of installing trim heater right before stream entered the reactor is the temperature of reactor inlet directly controlled and hysteresis zone could definitely be prevented, regardless whatever happens with stream in FEHE or its feed temperature. Conversely, installing trim heater before the FEHE only ensure the feed temperature to meet the feed temperature requirement. The inlet temperature still depends on FEHE performance. However, the advantage of installing trim heater before FEHE over before the reactor is that less heat requirement because of lower temperature target. Based on figure 4, to avoid the occurrence of instability behavior, minimum feed temperature required is 230°C, along with minimum inlet temperature requirement which is 400°C.

Usage of trim heater as unit that prevented hysteresis phenomenon is shown in dynamic simulation using Aspen HYSYS v8.8. Two simulation schemes, trim heater before the reactor and trim heater before FEHE, are presented in figure 8. Based on hysteresis zone data on figure 4, temperature control that govern the trim heater are set in 400°C and 240°C, respectively.

https://doi.org/10.1051/matecconf/201815603003

Simulation was performed with initial feed temperature 250°C and gradually decreased to 200°C in 250 minutes with ramp function. Dynamic simulation without the trim heater also performed as a comparison. Result of dynamic simulation of design case, installing trim heater before the reactor and installing trim heater before FEHE are presented in figure 9, 10 and 11, respectively.

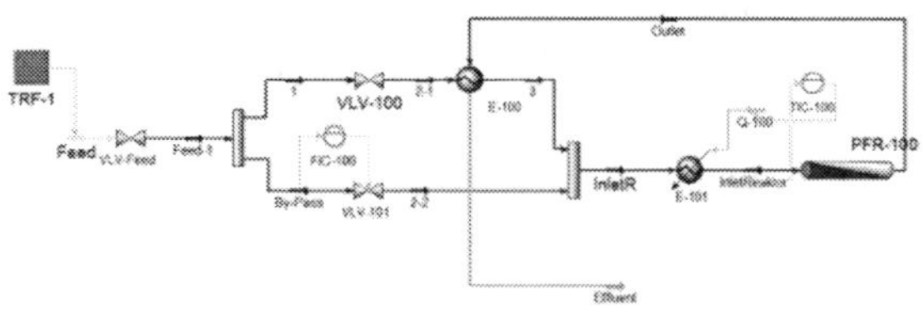

(a)

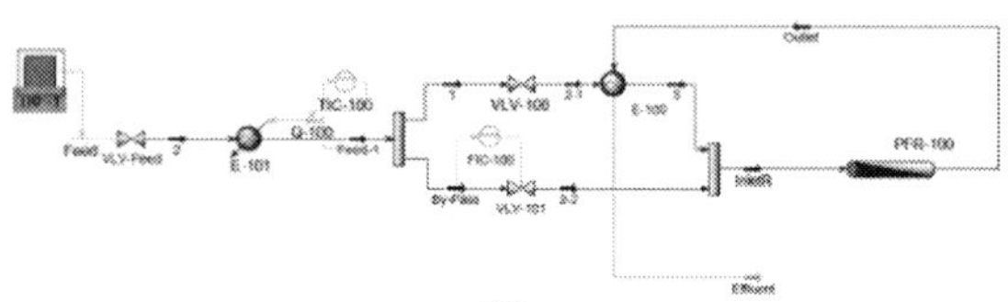

(b)

Fig. 8. Trim heater installation simulation schemes, (a) trim heater before the reactor and (b) trim heater before FEHE.

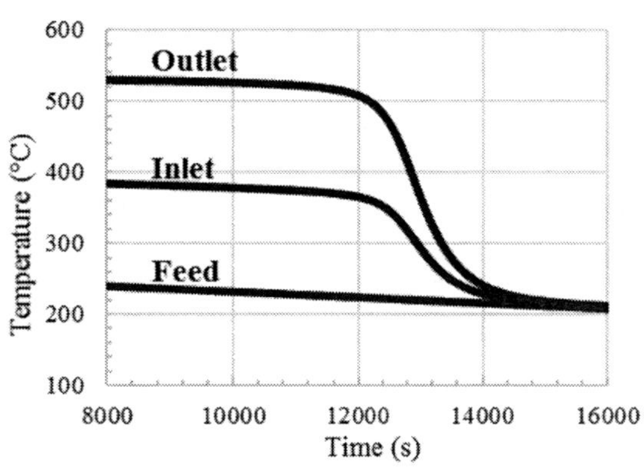

Fig. 9. Design case ammonia reactor temperature profile result.

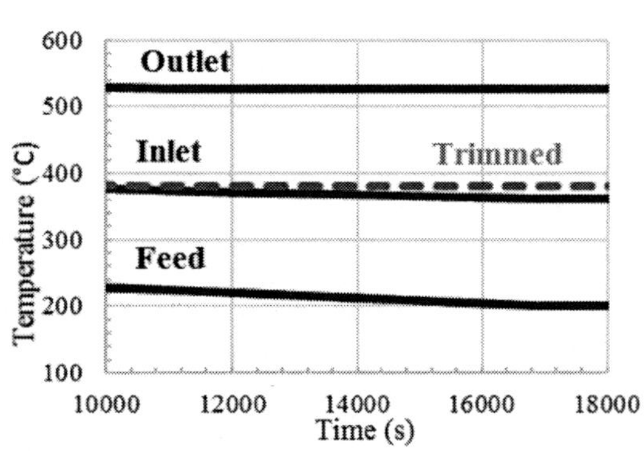

Fig. 10. Trim before the reactor case ammonia reactor temperature profile result.

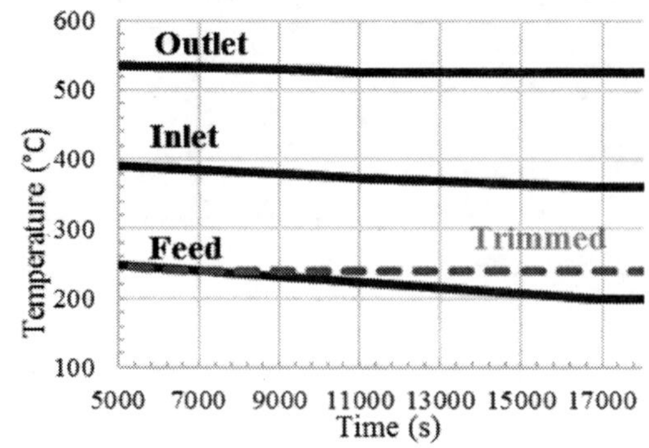

Fig. 11. Trim before FEHE case ammonia reactor temperature profile result.

Based on figure 10 and 11, trim heater is proven to prevent hysteresis phenomenon also avoid the reactor system to become extinguished as if design case in figure 9. Through the regulation of the feed or inlet temperature which value above the hysteresis temperature, the ammonia reactor should remain to work in stable condition without significant temperature fluctuations.

Energy usage of the trim heater from each scheme are analyzed with steady-state condition with feed temperature range of 150-220°C. The result of the energy analysis is presented in figure 12. From that figure, it is shown that case of installing trim heater before the reactor required less energy supply than installing trim heater before FEHE. However, trim heater that installed before the reactor should require more complex utility and tougher material since the stream must be heated until the temperature reach 400°C.

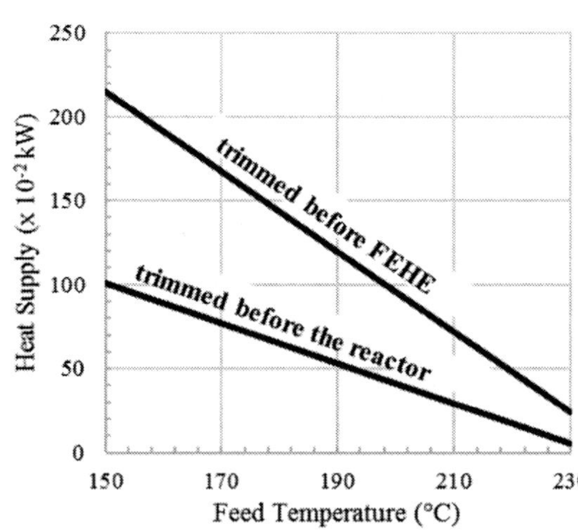

Fig. 12. Energy analysis of trim heater installation in ammonia reactor system.

4 CONCLUSION

Reactor that implemented autothermic configuration, like ammonia reactor, is at risk to become unstable when the reactor is operating. The instability often appears as hysteresis, the oscillation of operating parameter that could lead reactor to extinguish. Hysteresis could occur when reactor inlet temperature reaches certain value that made conversion of the reactor fall down. Hysteresis in single-bed ammonia reactor could be prevented by installing trim heater before stream entered the reactor or in the reactor system feed stream, before entering FEHE.

References

1. M. Appl, *Ammonia: principles and industrial practice,* Wiley-VCH (1999).
2. T. A. N. Nguyen, T. A. Nguyen, T. D. Vu, K. T. Nguyen, T. K. T. Dao, K. P. H. Huynh, Mater. Sci. Eng., **206** (2017)
3. W. L. Luyben, Ind. Eng. Chem. Res., **51**, 8566-8574 (2012).
4. C. G. Hill, *An introduction to chemical engineering kinetics & reactor design,* (John Wiley & Sons, 365, 1997).
5. J. C. Morud, S Skogestad, AIChE J., **44**, 888-895 (1988).
6. G. R. Setyadi, T. P. Adhi, *Simulasi dinamik sistem reaktor eksothermal adiabatik dengan alat penukar panas umpan-keluaran (FEHE),* (Departemen Teknik Kimia ITB, 2005).
7. S. G. Metchis, A. S. Foss, AIChE J., **33**, 8, 1288-1299 (1987).
8. R. F. Baddour, P. L. T. Brian, B. A. Logeais, J. P., Chem. Eng. Sci., **20**, 281 (1965).
9. D. L. Trimm, Stud. Surf. Sci. Catal., **68**, 29-50 (1991).
10. T. Rayment, R. Schogl, J. M. Thomas, G. Ertl, Nature, **315**, 311-313 (1985).

Palm H-FAME Production through Partially Hydrogenation using Nickel/Carbon Catalyst to Increase Oxidation Stability

Elsa Ramayeni[1], *Bambang Heru* Susanto[1,*], and *Dimas Firlyansyah* Pratama[1]

[1] Chemical Engineering Department, Faculty of Engineering, University of Indonesia, Depok 16424, Indonesia

Abstract. One of the methods to improve the oxidation stability of palm biodiesel is through partially hydrogenation. The production using Nickel/Carbon catalyst to speed up the reaction rate. Product is called Palm H-FAME (Hydrogenated FAME). Partial hydrogenation breaks the unsaturated bond on FAME (Fatty Acid Methyl Ester), which is a key component of the determination of oxidative properties. Changes in FAME composition by partial hydrogenation are predicted to change the oxidation stability so it does not cause deposits that can damage the injection system of diesel engine, pump system, and storage tank. Partial hydrogenation is carried out under operating conditions of 120 °C and 6 bar with 100:1, 100:3, 100:5, 100:10 % wt catalyst in the stirred batch autoclave reactor. H-FAME synthesis with 100:5 % wt Ni/C catalyst can decrease the iodine number which is the empirical measure of the number of unsaturated bonds from 91.78 to 82.38 (g-I_2/100 g) with an increase of oxidation stability from 585 to 602 minutes.

1 Introduction

High quality biodiesel will generate many benefit. The advantage of biodiesel is environmentally friendly emissions because it does not contain SO_2 and has a good lubrication effect. Biodiesel is defined as a fuel consisting of mono alkyl esters of long chain fatty acids derived from vegetable oils and animal oils. Biodiesel is produced by a process of transesterification reaction by reacting vegetable oil or animal fat catalytically with short chain aliphatic alcohols that produce FAME (Fatty Acid Methyl Ester) and glycerol as byproducts [1].

Biodiesel is a biofuel that reduces the level of exhaust emissions. In addition, biodiesel mixes perfectly with diesel oil (Solar) and works well on all types of diesel engines. The main biodiesel advantage is that engine modifications are not required to run diesel engines. To ensure the quality of biodiesel, Indonesia has implemented the National Standard SNI 04-7182-2006 which was first ratified in 2006 and revised in 2015 with changes in the value of several parameters.

In fact, biodiesel that has passed through the production process can be degraded into other compounds that result in changes of biodiesel components during the storage process even though the biodiesel meets the established standards. These changes happen because biodiesel is very easily degraded into compounds such as acids, aldehydes, ketones, esters, peroxides and alcohols [2]. Changes in the characteristics of biodiesel caused by the oxidation process include increasing the viscosity value, peroxide number, acid number and decreasing the induction period [3].

Pertamina currently produces Biosolar with 15% biodiesel and 85% of diesel. Furthermore, the government hopes to increase biodiesel percentage by up to 30%. However, this mixture should be further investigated because the mixture results in a low biodiesel oxidation stability effect in which diesel engines are adversely affected by the emergence of deposits and gums that would damage the injection system, pumping systems and storage tanks.

The oxidation process of biodiesel is influenced by long acid chain fatty acids (C_{16} to C_{24}) which contain unsaturated compounds that have lower oxidation stability value. Therefore, to improve the oxidation stability, it is necessary to modify the biodiesel component (FAME). One way that can be done is by a partial hydrogenation reaction in which the reaction will break the unsaturated bond on FAME which is a key component of the determination of oxidative properties. Changes in the composition of FAME will change the oxidation stability, cetane number, and biodiesel viscosity [4].

On the other hand, palm oil is the most reliable as biodiesel feedstock because palm biodiesel has low FFA (Free Fatty Acid) (<5%), therefore in biodiesel synthesis only through transesterification reaction. In addition, the main components of palm oil biodiesel are methyl palmitate (18: 0) and methyl oleate (18: 1), both of these

* Corresponding author: bambanghs@che.ui.ac.id

 https://doi.org/10.1051/matecconf/201815603004

components exhibit very low saturation bonds so that biodiesel is not easily oxidized and has good cetane index. Thus, this study was conducted to prove that the oxidation stability of the product is increased through the partial hydrogenation of the FAME component with the source of the biodiesel feedstock from palm oil.

2 Methodology

2.1 Materials

This study was conducted in two steps namely the synthesis of H-FAME in a hydrogenation reactor and characterization of H-FAME product. The materials needed in this study were: Palm Biodiesel (Wilmar), Ni/C (Pratama), Aquadest (CV. Dwinika), Chloroform (Merck), Wijs (Merck), Sodium Thiosulfate (Merck), Potassium Iodide (Merck), Starch (Merck), Potassium Hydroxide (Merck), Phenolphthalein (Merck).

2.2 Partially Hydrogenated FAME

The reaction was performed in a 250 mL batch stirred reactor integrated with a condenser and furnace. Extensive stirring of the biodiesel and catalyst in the reactor was accomplished using a magnetically driven, stirred with a DC geared motor. Ni/C catalyst was purged using nitrogen for 30 minutes, and then the Ni/C catalyst was activated by hydrogen at 10 bar and 400 °C for 1 hour. Palm FAME was used as feed and Ni/C catalyst with variation ratio of 100:1, 100 :3, 100:5, and 100:10 %wt. Before the hydrogenation, the feed and the catalyst were purged using nitrogen to remove O_2 for 30 minutes. Stirring speed and pressure during the reaction were kept constantly at 200 rpm and 6 bar respectively. The reactions were performed at temperature 120°C for 2.5 hours. In this step, the double bonds that were presented in the FAME were saturated with hydrogen. Figure 1 shows a schematic diagram of the apparatus.

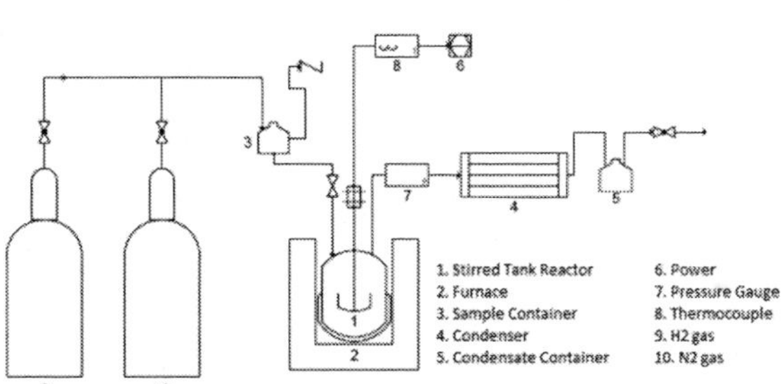

Fig. 1. Scheme of Hydrogenation Reaction

The product is characterized by iodine number test to determine the double bond reduction, acid number test to determine the amount of free fatty acid in the biodiesel, FT IR (Thermo Scientific) to know the effect of reaction to the presence of functional group in FAME, GCMS (Agilent Technologies) to identify components of sample, and oxidation stability test to determine the increased oxidative stability. Furthermore, the product is characterized by the density, viscosity, and cetane index to examine quality of combustion of the product.

3 Result and Discussion

The used catalyst was Nickel/Carbon. In test variations, the catalyst was activated first. A physical properties result of the catalyst that had been used in H-FAME synthesis is attached in Table 3.1.

Table 3.1 Physical Properties of Catalyst Used

Catalyst	Surface Area (m^2/gram)	Pore Size (nm)	Crystal Size (nm)
Ni/C	285.85	2.2	77.78

To facilitate an explanation of the characterization results of the H-FAME samples in the next section, the researcher made a standard naming sample based on the amount of the catalyst used on the reaction, presented in Table 3.2.

Table 3.2 Naming and Sample Operating Conditions

Products	Ni/C (% wt)
H-FAME a	100:1
H-FAME b	100:3
H-FAME c	100:5
H-FAME d	100:10

2.1 Result of Characterization

Result of chemical characterization of H-FAME product can be seen in Table 3.3.

Table 3.3 Chemical Properties of H-FAME

Product	Iodine Number (g-I_2/100g)	Oxidation Stability (minute)	Acidity Number (mg-KOH/g)
B100	91.78	585	0.33
H-FAME a	87.21	-	0.30
H-FAME b	86.19	-	0.30
H-FAME c	82.38	602	0.26
H-FAME d	80.64	-	0.25
SNI 2015	Max. 115	Min. 480	Max. 0.5

Based on Table 3.3, the iodine number decreased with the increasing of the amount of catalyst used. The highest decrease in iodine number was in the catalyst ratio of 100:5 %wt. The decreasing of iodine number showed a reduction in the number of double bonds in the FAME component.

The oxidation stability characteristic was only performed on H-FAME c and Palm FAME (B100). It was based on the decreasing of the highest iodine number. The oxidation stability of B100 were examined to know the initial data before the partial hydrogenation process. Based on the Table 3.3, it was proved that the partial hydrogenation reaction could increase the oxidation stability of biodiesel.

Table 3.4 Identification of Functional Groups on H-FAME Product

Bonds	Functional Groups	Wave Number (cm^{-1})	B100	H-FAME c
$=CH_2$	Alkene	991-723	√	√
C-O	Alcohol	1169-1243	√	√
C=O	Carbonyl	1740	√	√
C-H	Alkane	1360-1459	√	√
	Alkane	2853-2932	√	√

Identification of functional groups using FTIR indicated that the partial hydrogenation reaction did not break the functional group bond but only broke the double bond of the FAME component which was a key component of oxidative properties.

Table 3.5 Component Identification of Sample B100 and H-FAME c

Components	B100	H-FAME c
% Methyl Palmitate (18:0)	33.17	33.29
% Methyl Stearate (18:0)	9.78	9.80
% Methyl Oleate (18:1)	38.63	39.81
% Methyl Linoleate (18:2)	0.28	0.20
% Methyl Linolenate (18:3)	0.23	0.14

According to Table 3.5, the percentage of Methyl Linoleate and Methyl Linolenate components were decreased, while the components of Methyl Oleate, Methyl Stearate, and Methyl Palmitate were increased. These results indicated that the partial hydrogenation reaction had broken the saturated bond on the FAME component into an unsaturated bond. These results also confirmed the result of the decrease in iodine number.

Table 3.6 Physical Properties of H-FAME

Sampel	Density (kg/m^3)	Viscosity (CSt)	Cetane Index
B100	856.96	4.5	52.51
H-FAME a	858.21	-	-
H-FAME b	859.19	-	-
H-FAME c	863.69	4.6	55.59
H-FAME d	864.43	-	-
SNI 2015	850-890	2.3-6.0	Min. 51

Based on the results of physical properties, density of H-FAME product increased, this was because the FAME component bound to hydrogen. While the viscosity of the product increased due to the increasing number of saturated methyl ester components and partial hydrogenation. The breakdown of unsaturated bonds into saturated bonds will increase the oxidation stability of biodiesel but it decreases flow properties, so it is necessary to obtain optimum conditions for both parameters to meet the specification [6]. In addition, the cetane index increased after the partial hydrogenation reaction showed that the more saturated components of the FAME, the more cetane index of the FAME which indicated combustion quality.

Table 3.7 Conversion. Selectivity and Yield of Product

Product	% Conversion	% Selectivity	% Yield
H-FAME c	93.26	10.58	9.87

The calculation of selectivity was based on the amount of addition of the methyl oleate component, since methyl oleate is the desired component by the partial hydrogenation reaction in biodiesel. If the FAME component gets saturated, it will cause a high viscosity of biodiesel thereby imparting the flow properties [6].

4 Conclusion

Palm H-FAME production through partially hydrogenation using Ni/C catalyst under operating conditions of 120 °C and 6 bar with 100:5 % wt catalyst in the stirred batch autoclave reactor decreased the iodine number which was the empirical measure of the number of unsaturated bonds from 91.78 to 82.38 (g-I_2/100 g) with an increase of oxidation stability from 585 to 602 minutes. These conditions with activated Ni/C led to dominant yield of 10.58%, conversion of 93.26%, and selectivity of 9.87% of methyl oleate (18:1).

References

1. A. Demirbas, Prog. Energy Combust. Sci, **33**, 1–18 (2007)
2. J. Xin, H. Imahara, S. Shiro, Fuel. **88 (2)**, 282-286 (2009)
3. Agarwal, A. Kumar, K. Deepak, Fuel Processing Technology, **106**, 447-452 (2013)
4. G. Knothe, Some aspects of biodiesel oxidative stability. Fuel Process. Technol., **88 (7)**, 669-677 (2007)
5. Ministry of Forestry, Internal Report (2015)
6. S. Goto, M. Oguma, N. Chollacoop *EAS-ERIA Biodiesel Fuel Trade Handbook,* 28-38 (2010)

Carbon Capture and Storage System Using Pinch Design Method

Renanto Handogo[1*]

[1] Chemical Engineering Department, Institut Teknologi Sepuluh Nopember, Surabaya 60111, East Java, Indonesia

Abstract. Carbon capture and storage (CCS) have been investigated for a long time. It was intended to reduce carbon dioxide (CO_2) in the atmosphere due to fossil fuel combustion in power generation and industrial processes. CO_2 is captured and stored in various geological formations. The problem here is to match between source and sink such that alternative storage and unutilized storage capacities are minimum. Pinch Design Method as has been proposed by was used in this work. The concept is overwhelming that it can be used other than in the heat exchanger networks, such as in the water system design, mass exchanger networks and many other processes. Initially this concept was applied to carbon capture and storage but with no exact pairing between sources and sinks as proposed in this work using grid diagram as commonly shown in other processes. This work can point out the exact pairing between sources and sinks, and within the given time frame. A four different cases are investigated where the time difference between the starting time of CO_2 generated in the source and the beginning of sink availability. A value of 0, 5, 10 and 15 years are chosen to evaluate the amount of CO_2 that has to be stored and the amount of unutilized storage capacity. The case study has been prepared with 5 sources and 3 sinks. The result shows that the larger time difference the larger alternative storage and unutilized storage capacities. Therefore, having a shorter time difference will be more acceptable in the design CCS system.

1 Introduction

Global warming is an issue that is being discussed in the international world lately. Global warming is characterized by rising global average temperatures on the earth's surface. Studies show that the average global air temperature has increased by about 0.5 - 1.0° F (0.3 - 0.6° C) over the past century. This is the largest rise in surface temperature in the last 1,000 years and scientists predict a larger increase will occur over a century. This heating is mostly caused by increases in greenhouse gases, especially carbon dioxide (CO_2) and methane over the earth's atmosphere caused by the burning of fossil fuels, industrial activities, agriculture and deforestation. One solution that can be applied quickly and on a large scale is to use Carbon Capture and Storage (CCS) technology.

According to the ESDM Ministry, CCS is one of the climate change mitigation technologies that has the potential to reduce large-scale CO_2 emissions resulting from burning fossil fuels. CCS is an effort with a technique approach to capture carbon and store it in a safer form. This technology is a series of activities starting from capturing CO_2 sources such as natural gas processing facilities and power plants, then transporting them to CO_2 storage locations in appropriate geological sinks.

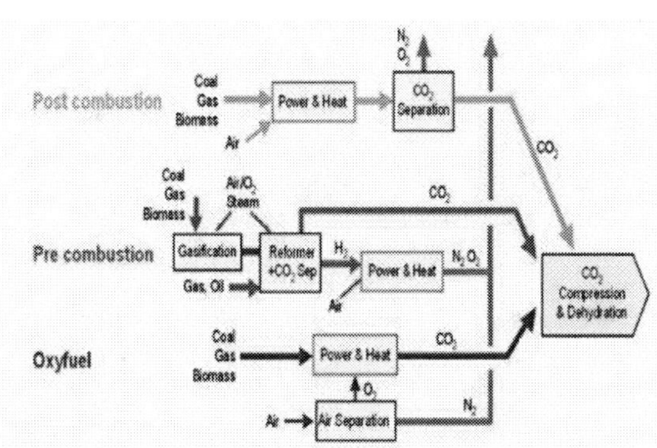

Fig. 1. Ways of catching CO_2 from industries (post combustion, pre combustion, and Combustion) (IPCC, 2005)

In the CCS process, the process of matching between the source and storage of CO_2 are often subjected to various constraints, such as storage capacity availability, the location of the source and storage availability, and the time source and storage availability. These problems

* Corresponding author : renanto@chem-eng.its.ac.id

https://doi.org/10.1051/matecconf/201815603005

can be solved with a design technology that is pinch technology that can provide an minimum alternative and unultilized storage. Pinch technology is built on the laws of thermodynamic heat transfer. This technology is dominated by the design of Heat Exchanger Network (HEN) which provides optimum energy cost savings. The concept is overwhelming that it can be used other than in
the heat HEN, such as in the water system design, Mass Exchanger Networks (MEN) and many other processes. In water design system exactly cooling water system, the performance of a cooling tower is maximized by maximizing the inlet temperature to the cooling tower and minimizing the inlet flowrate. The role of the pinch technology here is for getting minimum cooling water flowrate (Smith, 2005). By MEN, we mean the systematic generation of a cost-effective network of mass exchangers with the goal of transferring certain species from a set of rich streams to a set of lean streams. A mass exchanger can be virtually any countercurrent, direct-contact mass-transfer operation that uses a Mass-Separating Agents (MSA) (eg, absorption, desorption, adsorption, liquid-liquid extraction, leaching, and ion exchange) (El-Halwagi and Manousiouthakis, 1989).

Initially this concept was applied to carbon capture and storage but with no exact pairing between sources and sinks as proposed in this work using grid diagram as commonly shown in other processes (Raymond et.al, 2013). This work can point out the exact pairing between sources and sinks, and within the given time frame. The pinch calculations in this work are performed with composite curve and Carbon Storage Cascade Analysis (CSCA). Just like in HEN, the composite curve CCS will consist of two lines, namely a source line as hot streams and a sink line as cold streams. By shifting the sink line until it gets pinch according to the set time difference it will get the number of alternative storage and unutilized storage.

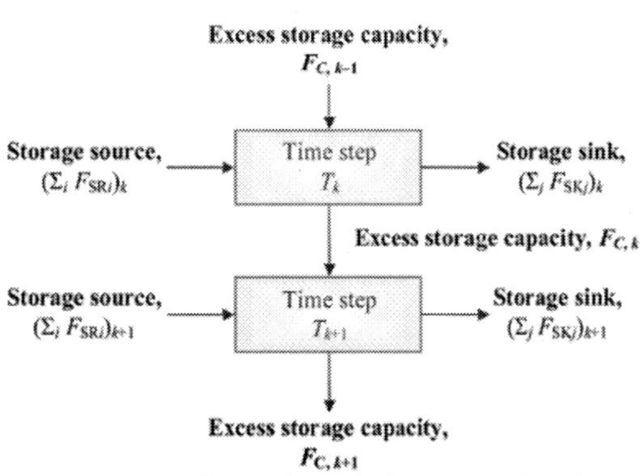

Fig. 2. Concept of Cascade Analysis (Raymond et.al, 2013)

While the cascade analysis as shown in figure 2 is based on the cumulative amount of CO_2 load at a certain time interval from the first year until the last year. The biggest deficit in an interval is as alternative storage and the value is rearranged as surplus value at the beginning of the year so there is no deficit at each interval and will be obtained at a time with a value of 0 called a pinch. The amount of surplus in the last year is as unutilized storage.

2 Study Description

2.1. Data Collections

Data collection is very important with respect to the reality of data in the field, as the success of the research results is influenced by proper data retrieval. Until now, in Indonesia the companies still have not implemented CCS system. Therefore, in order to obtain the data (case) which almost resembles the field situation, it is necessary to study literature as a comparison.

2.2 Create a Grid Diagram on Each Variable

The grid diagram is used to present the results of the Pinch Design Method.

3 Results and Discussion

3.1. Data Collections

From the study of literature, then in this study were given four variables namely dT (time difference) 0 years, 5 years, 10 years, and 15 years. The following data will be used in this research.

Table 1. Sources Data

Source	CO_2 flowrate (Mt/y)	CO_2 load (Mt)	Start time (y)	End Time (y)
1	15	300	0	20
2	5	150	0	30
3	5	150	0	30
4	20	400	20	40
5	10	400	30	70
Total	n/a	1400	n/a	n/a

Table 2. Sinks Data

Sink	Injectivity (Mt/y)	Storage capacity (Mt)	Earliest time available (y)	Characteristic end time (y)
1	20	900	10	55
2	15	750	20	70
3	10	550	30	85
Total	n/a	2200	n/a	n/a

3.2. Optimization Based on Multi Period

At each time interval, each source portion is combined into a composite curve section, by adding the line diagonally. Substitution of the composite part will start from the end of the previous section, which will be converted into a source composite curve. Similarly, the composite sink curve is also done the same as the source composite curve.

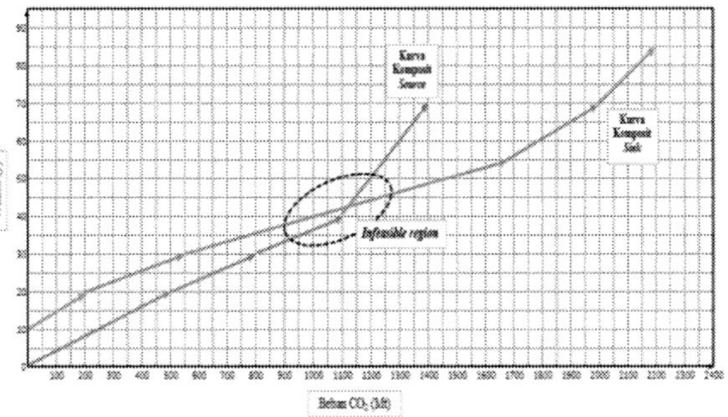

Fig. 3. Merging composite source and sink curves

In the figure 3, it shows that the relative position of the composite curve indicates an overlap between the source and sink composite curves. It is not possible in the real event, so it is called an infeasible region. Part of the composite sink curve is to the left of the source composite curve which means that the sink is only ready once the source is available.

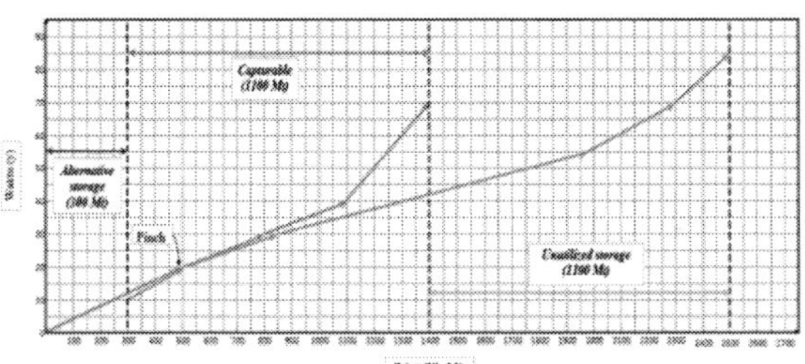

Fig. 4. Merging composite source and sink curves with Pinch 0 year

The deficit may be interpreted as delaying the decision to capture CO_2, until internal sink time (storage) is readily available for use. Areas where the two composite curves overlap each other imply that 1100 Mt of CO_2 can be sent from source to available sinks. The composite curve shows that within 20 years the pinch point with the time difference is 0. In other words, the two composite curve encounters are at the pinch point. Further observation on the pinch diagram is the availability of a storage capacity of 1100 Mt which is not used as a sink from the source. Captured CO_2 of 1100 Mt presented that 50% of the 2200 Mt that can be stored in the 3 sinks. The same is shown by table 3 which is the cascade table for the 0 year difference show the pinch in the 20 year. The cascade table also show the same alternative storage of 300 Mt and unutilized storage of 1100 Mt.

Table 3. Cascade analysis for source and sink with 0 year time difference

t (year)	Source, $S_{i,s}$ (Mt/y)	Sink, $D_{j,s}$ (Mt/y)	Δt	Flowrate CO_2 (Mt/y)	Load CO_2 (Mt)	CO_2 cascade (Mt)	
						Infeasible	Feasible
0	SR1 SR2 SR3					0	300
	15 5 5		10	-25	-250		
10		SK1				-250	50
		20	10	-5	-50		
20	SR4	SK2				-300	0
	20	15	10	5	50		(PINCH)
30	SR5	SK3				-250	50
	10	10	10	15	150		
40						-100	200
			15	35	525		
55						425	725
			15	15	225		
70						650	950
			15	10	150		
85						800	1100

Pinch point on the diagram divides the system into 2 regions. Before pinch point, there are areas where CO_2 requires storage. While after pinch point, there is excess storage. The above scenario is one of the problems that occur in this network, because almost all potential storage locations will experience some delay before the sink receives CO_2 from source. From figure 4 shows below the composite sink curve starts in the 10th year. In other words, alternative storage should be sought in the first 10 years of operation. If no external storage is available, then the decision to capture CO_2 must be suspended in the 10th year, when the internal storage is ready for use.

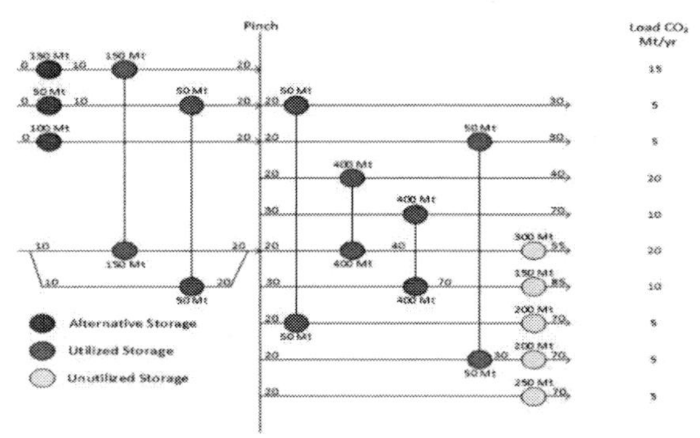

Fig. 5. CCS grid diagrams with 0 year Pinch time difference

Figure 5 shows that sink 1 receives CO_2 from source 1,2, and 4 in the 10 to 40 year timeframe. Meanwhile, CO_2 from partially source 2, and 3 is sent to sink 2 that ends up to year 70. Then sink 3 is receives CO_2 from source 5 in the 30 to 70 timeframe. From the grid diagram obtained the number of alternative and unutilized storage is suitable with the composite curve and cascade analysis calculations so that the pairing done with this grid diagram and resulted the exact pairing. However, in a plant occurring it is not possible to match pinch points with a 0 year time difference. From this research, discuss the optimization network with various time difference variables of 5, 10 and 15 years. This is based on real events in a carbon capture plant that experiences delays in source delivery to sinks or experienced CO_2 escapes from this system.

In the next time difference variable that is pinch point with 5 year time difference. In the figure 6 shows that CO_2 from various sources can be captured only 950 Mt of a total of 1,400 Mt. This leads to an increase in CO_2 deficit captured and requires another alternate storage of 450 Mt. So does the unused storage capacity of 1,250 Mt. In this case there is a pinch point with a 5-year time difference that is at the time of the year 20-25. The captured CO_2 of 950 Mt presents that 43.2% of the 2,200 Mt that can be stored in the 3 sinks. The same is shown by table 4 which is the cascade table for the 5 year difference show the pinch in the 20-25 year. The cascade table also show the same alternative storage of 450 Mt and unutilized storage of 1250 Mt.

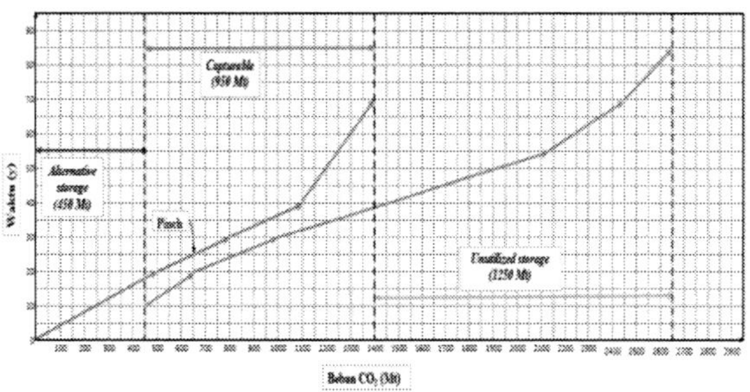

Fig. 6. Merging composite source and sink curves with Pinch 5 year

Table 4. Cascade analysis for source and sink with difference time of 5 years

t (year)	Source, $S_{i,t}$ (Mt/y)	Sink, $D_{j,t}$ (Mt/y)	Δt	Flowrate CO_2 (Mt/y)	Load CO_2 (Mt)	CO_2 cascade (Mt) Infeasible	CO_2 cascade (Mt) Feasible
0	SR1 SR2 SR3					0	450
	15 5 5		15	-25	-375		
15		SK1				-375	75
		20	5	-5	-25		
20	SR4					-400	50
	30		5	-10	-50		
25		SK2				-450	0
		15	5	5	25		(PINCH)
30	SR5					-425	25
	10		5	5	25		
35		SK3				-400	50
		10	5	15	75		
40						-325	125
			20	35	700		
60						375	825
			10	15	150		
70						525	975
			5	25	125		
75						650	1100
			15	10	150		
90						800	1250

Because of the time difference difference, the pairing obtained on the grid diagram is different too as shown in figure 7. Sink 1 receives CO_2 from source 1,3, and 4 in the 10 to 35 year timeframe. Meanwhile, CO_2 from partially source 2, 3, and 5 is sent to sink 2 that ends up to year 65. Then sink 3 available from year 30 to year 85 is an available storage but no injektivity from source. This is because the amount of CO_2 from the source that existed at the time over the pinch has been sufficiently stored in sinks 1 and 2. The grid diagram obtained the number of alternative and unutilized storage which is consistent with the composite curve and cascade analysis calculations, so that the pairing can be done using this grid diagram and resulted in the the exact pairing.

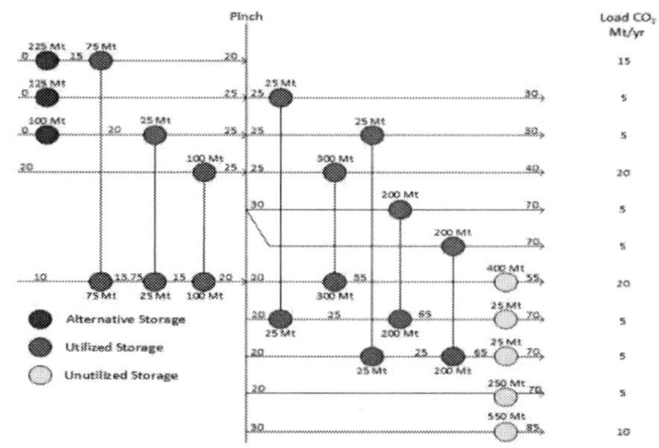

Fig. 7. CCS grid diagram with 5 year Pinch time difference

Table 5. Cascade analysis for source and sink with difference time of 10 years.

t (year)	Source, $S_{i,t}$ (Mt/y)	Sink, $D_{j,t}$ (Mt/y)	Δt	Flowrate CO_2 (Mt/y)	Load CO_2 (Mt)	CO_2 cascade (Mt) Infeasible	CO_2 cascade (Mt) Feasible
0	SR1 SR2 SR3					0	600
	15 5 5		20	-25	-500		
20	SR4	SK1				-500	100
	30	20	10	-10	-100		
30	SR5	SK2				-600	0
	10	15	10	5	50		(PINCH)
40		SK3				-550	50
		10	25	35	875		
65						325	925
			5	15	75		
70						400	1000
			10	25	250		
80						650	1250
			15	10	150		
95						800	1400

In the next time difference variable that is pinch point with 10 year time difference. In this cascade table for the 10 year difference show the pinch in the 20-30 year. The cascade table also show the alternative storage of 600 Mt and unutilized storage of 1400 Mt. The captured CO_2 of 800 Mt presents that 36.4% of the 2200 Mt that can be stored in the 3 sinks. The pairing obtained sink 1 as shown in figure 8 receives CO_2 only from source 4 in the 10 to 30 year timeframe and sink 2 receives CO_2 only from source 5 in the 20 to 60 year timeframe. The grid diagram obtained the number of alternative and unutilized storage is suitable with the cascade analysis calculations so that the pairing done with this grid diagram and resulted the exact pairing.

In the last time difference variable that is pinch point with 15 year time difference. In this cascade table for the 15 year difference show the pinch in the 20-35 year. The cascade table also shows the alternative storage of 750 Mt and unutilized storage of 1550 Mt. The captured CO_2 of 650 Mt presents that 29.5% of the 2,200 Mt that can be stored in the 3 sinks. The pairing obtained sink 1 as shown in figure 9 receives CO_2 only from source 4 in the 10 to 25 year timeframe and sink 2 receives CO_2 only

https://doi.org/10.1051/matecconf/201815603005

from source 5 in the 20 to 55 year timeframe. The grid diagram obtained the number of alternative and unutilized storage is suitable with the cascade analysis calculations so that the pairing done with this grid diagram and resulted the exact pairing.

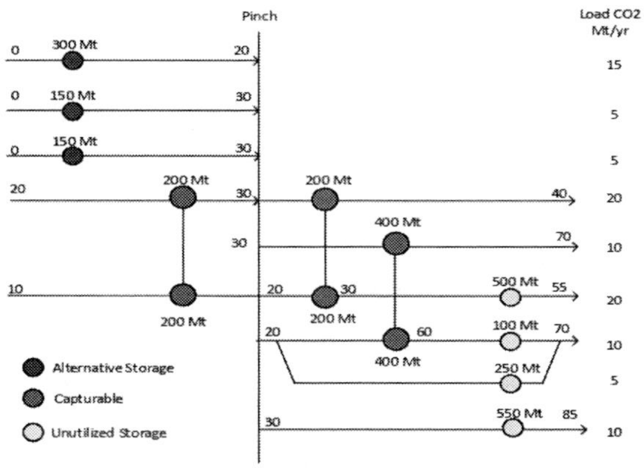

Fig. 8. CCS grid diagram with 10 year Pinch time difference

Table 6. Cascade analysis for source and sink with difference time of 15 years

t (year)	Source, $S_{i,t}$ (Mt/y)	Sink, $D_{j,t}$ (Mt/y)	Δt	Flowrate CO_2 (Mt/y)	Load CO_2 (Mt)	CO_2 cascade (Mt) Infeasible	Feasible
0	SR1 SR2 SR3					0	750
	15 5 5		20	-25	-500		
20	SR4					-500	250
	20		5	-30	-150		
25		SK1				-650	100
		20	5	-10	-50		
30	SR5					-700	50
	10		5	-10	-50		
35		SK2				-750	0
		15	5	5	25		(PINCH)
40						-725	25
			5	25	125		
45		SK3				-600	150
		10	25	35	875		
70						275	1025
			15	25	375		
85						650	1400
			15	10	150		
100						800	1550

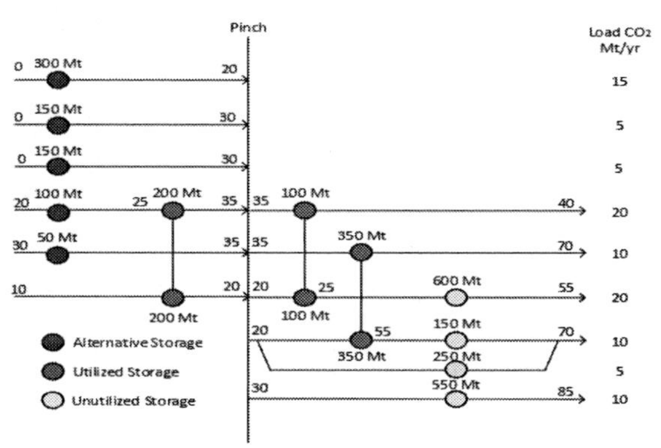

Fig. 9. CCS grid diagram with 15 year Pinch time difference

From some of the above variables, it can be concluded that the best pairing occurs when the time difference variable is 0 years. It can be seen from the percentage of carbon captured with dt 0 that is 50% larger than using other dt as in the table shown in table 7.

Table 7. Results calculation of CSS with pinch diagram and cascade analysis

Δy (year)	Alternative storage (Mt)	Carbon Capturable (Mt)	Unutilized storage (Mt)	% CC
0	300	1100	1100	50
5	450	950	1250	43.2
10	600	800	1400	36.4
15	750	650	1550	29.5

4 Conclusion

The result from each grid diagram obtained shows the exact pairing based on the suitability of alternative and unutilized storage with composite curve and table cascade calculations. The result shows that the larger time difference the larger alternative storage and unutilized storage capacities. Therefore, having a shorter time difference will be more acceptable in the design CCS system.

The author would like to express his gratitude for the financial support from Kemenristekdikti under the contract no 582/PKS/ITS/2017

References

1. IPCC, *IPCC Special Report on Carbon Dioxide Capture and Storage Intergovermental Panel on Climate Change* (Cambridge University Press, New York, 2005)
2. J. A. R. Diamante, R. R. Tan, K. B. Aviso, S. Bandyopadhyay, D. K. S. Ng, D. Y. Foo, *Proceedings of the 6th International Conference on Process Systems Engineering, PSE ASIA*, (2013)
3. M. M. El-Halwagi, V, A.I.Ch.E. J **35**, 1233-1244 (1989)
4. R. E. H. Ooi, D. C. Y. Foo, D. K. S. Ng, R. R. Tan, Che. Eng. Rsch. and Dsgn., **91,** 2721-2731 (2013)
5. R. Smith, *Chemical Process Design and Integration* (John Wiley & Sons, England, 2005)

Study of Acid Hydrolysis on Organic Waste: Understanding The Effect of Delignification and Particle Size

Nadiem Anwar[1,*], *Iman* Mukhaimin[1], *Mining* Harsanti[1], and *Ate* Romli[1]

[1]University of Jendral Achmad Yani, Chemical Engineering Department, Terusan Jendral Sudirman, Cimahi

Abstract. Organic wastes from Swiettenia marcophylla L, Artocarpus heterophyllus L, Mangifera indica L, and Annona muricata L were prepared by grinding into 0.1875, 0.3750, 0.7500 mm of particle size and delignified by 2% NaOH at 80°C for 90 minutes. Acid dilution hydrolysis process with H2SO4 1% was performed at 150°C for 120 minutes in a closed reactor. The effect of particle size and delignification on and reducing sugar concentration were investigated. The result showed (1) leaves that can be used as raw material to produce hydrogen should have 38–49% cellulose and hemicellulose. (2) Reducing sugar concentration increased with particle size reduction and delignification. (3) the best result with the highest reducing sugar concentration was achieved by 0.1875 mm particle size with delignification on Annona muricata L.

1 Introduction

The increasing energy demand, the decreasing production of crude oil, and the fluctuating crude oil prices have an impact on the economies of Indonesia. Similarly, The usage of fossil-based fuel has increased CO_2 accumulation in the atmosphere that increased trigger global warming. Furthermore, there are significant interest and challenge to develop an alternative fuel that is affordable, renewable sources, and eco-friendly such as biomass-based energy.

Indonesia as a tropical country produced large amounts of biomass about 3.9 billion tons of dry plant biomass per year like agricultural waste, plantations, and forestry residues. The usage of agricultural waste, plantations, and forestry residues will minimize environmental impacts because they can be used as feedstocks for biofuel production or increase the value-added chemical. Generally, hydrolysis process can convert dry plant biomass into reducing sugar in some ways, such as chemically [1], fermentatively [2], or enzymatically [3-6]. In previous research, Enzymatic hydrolysis of rice straw with pure or crude enzymes from Aspergillus niger and Trichoderma reesei, showed that hydrolysis in a batch process for seven hours more increased significantly reducing sugar concentration than the usage of commercial enzymes, but required time is too long [7]. This will be an obstacle to produce in large quantities. Anwar [8] has produced the same level reducing sugar concentration from rice straw by acid hydrolysis with a shorter time than enzymatic hydrolysis and it can be converted into hydrogen by the fermentation process.

Other plant biomass that has large quantities especially in University of Jendral Achmad Yani come from falling leaves of trees. The falling leaves have high lignocellulose concentration that can be converted into reducing sugar. Therefore, the utilization of falling leaves as biomass feedstock will be very potential and does not interfere sustainability of food crops

The lignocellulose concentration of plant biomass mainly consists of 40–50% cellulose, 25–30% hemicellulose, 15–20% lignin, and traces of pectin, nitrogen compounds, and inorganic ingredients [9,10]. Cellulose is a polysaccharide with β-1,4-glucoside bonds in a linear chain. The β-1,4-glycoside bond on cellulose fibers can be hydrolyzed to produce glucose monomers by acid or enzymatic hydrolysis but the existence of lignin composition make the conversion lignocelluloses to energy is costly and ineffective up to now [2,11–15]. Besides that, differences of lignocelluloses composition make difference structure of biomass. Thus, it is necessary to design suitable pretreatment to break down the structure of biomass to improve the effectiveness of conversion.[16,17]

Prior to hydrolyzed, to increase reactivity in hydrolysis process and to break down cellulose crystallinity is necessary to remove lignin from the lignocellulose. Because lignin is a polymer formed through the phenylpropane units which have a function to rigidified cellulose microfibril and has physically bond with hemicellulose. Hemicellulose is a group low molecular weight of polysaccharide that can hydrolize into glucose, mannose, galactose, xylose, and arabinose. This step will increase reagent accessibility to polysaccharide in biomass [18-20]. Pretreatment is the

* Corresponding author: nadiemanwar20@yahoo.com

important step to convert lignocelluloses to energy. It is necessary to separate the complex components of lignocellulose which consider the compatibility of feedstocks, enzymes, and organisms. The pretreatment process generally can be classified into physical [21], chemical [22], physical-chemical, biological methods and their combinations.

Krishnan [23] had given attention increasing the accessibility reagent to carbohydrates by chemical pretreatment, ammonia solution, adding ammonia solution had removed lignin and swelled the solid residue. Zhang [24] had pretreated lignocellulosic biomass using the synergistic effect of mechanical crushing and metal salts. Then compare both with different oxidation numbers on polymerization and crystallinity of biomass. Yu [25] had used an alkaline condition on softwood and hardwood to determine the effect of delignification methods on enzymatic hydrolysis of forest biomass. Both delignifications had improved enzymatic hydrolysis, especially for softwood.

J. Yoo [26] had compared the effect of two traditional methods using dilute acid (1% sulfuric acid) and alkali (1% sodium hydroxide) and thermal mechanical extrusion pretreatment on the enzymatic hydrolysis of xylose from soybean. Pretreatment acid, alkali, and hot extrusion had successfully increased the xylose yield on the enzymatic hydrolysis by 69.6%, 128.7%, and 132.2%, respectively. Based on some previous research, alkali pretreatment is the best chemical pretreatment and important to remove lignin from biomass to increase the efficiency of hydrolysis process.

Current studies on pretreatment mainly focus on, identifying, evaluating, and understanding potential approaches on the effect of particle size and delignification, alkaline pretreatment, for subsequent acid hydrolysis with different feedstock biomass from falling leaves.

2 Experimental

1.1 Material

Plant biomass waste, Swietenia marcophylla L, Artocarpus heterophyllus L, Mangifera indica L, and Annona muricata L were collected from University of Jendral Achmad Yani. Natrium hydroxide (NaOH), Sulfuric acid (H_2SO_4), reducing sugar standard, Somogy I, Somogy II, and Nelson chemical reagents were of analytical grade.

1.2 Preparation of feedstock

1.2.1 Reducing particle size

The preparation methods for all the three feedstock was basically with an objective of reducing the particle size to increase the surface area of targeted reactions. Plant biomass waste, falling leaves, were prepared by cutting into ± 1 cm, then dried under sunlight. Dried biomass was ground in an electric mixer to give a fine powder and screened into 0.1875, 0.3750, 0.7500 mm of particle size.

1.2.2 Analysis of Cellulose and Lignin

Analysis of lignin and cellulose followed the method of Chesson [32]. 1 g of dried biomass (a) and 150 mL of aquadest was heated in a water bath at a 90-100°C for 1 hour. The mixture filtered, washed, and dried until the weight was constant (b). The residue was added to 150 mL of H_2SO_4 1 N and heated in the water bath at 90–100°C for 1 hour. The mixture was filtered and washed with 300 mL of aquadest and then the residue was dried (c). The dried residue was soaked with 10 mL H_2SO_4 at room temperature. After that, 150 mL of 1 N H_2SO_4 was added to the mixture and refluxed in the water bath for 1 hour. The solid was washed with 400 mL of aquadest, heated in the oven at 105°C and weighed until the constant weight (d). Finally the solid was heated until becoming ash and weighted (e).The percentage of cellulose and lignin was calculated as follows:

$$\%\ cellulose = \frac{c-d}{a} \times 100\% \quad (1)$$

$$\%\ lignin = \frac{d-c}{a} \times 100\% \quad (2)$$

where: a = sample weight (g), c = the residue weight (g) at the third weighing, d = the residue weight (g) at the fourth weighing, e = the weight of ash (g).

1.3 Alkaline treatment.

A 2500 mL beaker glass reactor was used for the delignification studies. A uniform suspension of the selected sample in distilled water with a by addition of 2% NaOH at 80°C for 90 minutes in a batch-closed mixed flow reactor and the agitation was started using a mechanical stirrer. Before and after process weight of sample was determined to know the effect of chemical pretreatment on delignification,

1.4 Acid Hydrolysis of lignocellulose

Prior to hydrolysis process, organic wastes were divided into two types, with or without delignification. they were hydrolyzed using 1% H_2SO_4 with a ratio of acid-biomass 1: 20. The hydrolysis was performed 150°C, pH 5.5 for 120 minutes in a batch-closed reactor.

The reducing sugar was analyzed by Nelson-Somogy method with Uv-Vis Spectrophotometer. The schematic of Experimental can be seen in **Fig. 1**

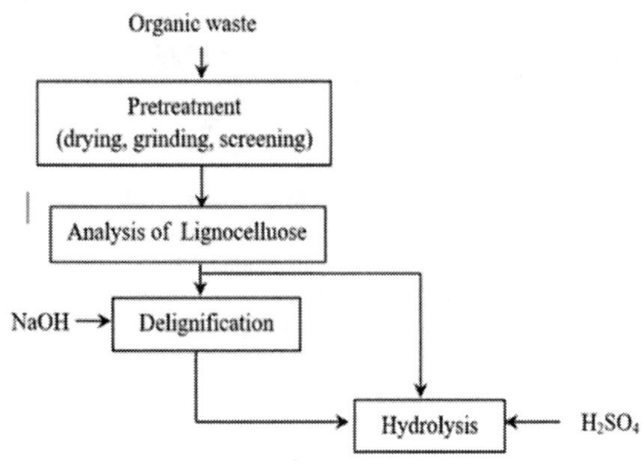

Fig. 1. The schematic of Experimental

3 Result and discussion

The following section analyzes the effect of particle size and delignification on hydrolysis process to produce reducing sugar. This research will compare with or without delignification treatment to produce reducing sugar concentration. The experimental results provide a basis data for the understanding of pretreatment biomass, which resulted in highest concentration of reducing sugar.

3.1 The effect of feedstock type

Different types of leaves as feedstock of had given effect on the process of hydrolysis. The effect was derived from lignocellulose content which is different from each leaf type. The result of lignocellulose analysis in Table 1. showed that leaves contained 14-15% cellulose, 21-35% hemicellulose, 17-29% lignin, with a total potential of the hydrolyzable component to produce reducing sugars such as glucose and xylose 38-49%.

Table 1. the result of lignocellulose analysis

Feedstock	Cellulose	Hemicellulose	Lignin + Ash
Swiettenia marcophylla L	17%	21%	23%
Artocarpus heterophyllus L	25%	21%	29%
Mangifera indica L	14%	35%	22%
Annona muricata L	20%	24%	17%

Based on table 1, different types of leaves as a feedstock have different lignocellulose composition with cellulose content slightly smaller than lignin + ash. It makes the difference in characteristic leaf fiber, microfibril in the cell wall, pore structure on the cellulose surface and strong bond between cellulose, hemicellulose, and lignin in the crystallization zone than others[2]. In addition, these will increase the ability to block the accessibility reagent to cellulose and stability of cell walls.

Artocarpus Heterophyllus L has the highest lignin+ash content. Thus, it will need extra effort to degrade lignin. Because with highest lignin content, it will have the highest stability and ability to protect cellulose. Otherwise, Annona muricata L has the lowest lignin+ash content that makes more easily to degrade lignin from the feedstock. It is proved with some data on **Tabel 2**. that Annona muricata L has the highest weight loss after addition of NaOH.

3.2 The effect of delignification

The result of lignocellulose analysis in Table 2 showed that decreasing particle size causes increasing weight loss from delignification results. A small particle will have larger contact surface area between NaOH and lignocellulose, thus solvent gives more chance to remove lignin. In addition, the weight loss after delignification is greater than the weight of previous lignin concentration. Based on weight loss after the delignification process, *Annona muricata* L is the highest, followed by *Swiettenia marcophylla L*, *Mangifera indica L*, and the last *Artocarpus heterophyllus L*. This showed that addition of alkali on biomass has removed some lignin from lignocellulose.

Tabel 2. The result of alkali pretreatment for delignification

Feedstock	Particle size	Weight (gram)	
		Before delignification	After delignification
Swiettenia marcophylla L	0,7500	50.00	19.15
	0,3750	50.00	16.22
	0,1875	50.00	15.24
Artocarpus heterophyllus L	0,7500	50.00	30.79
	0,3750	50.00	22.26
	0,1875	50.00	20.06
Mangifera indica L	0,7500	50.00	23.49
	0,3750	50.00	23.17
	0,1875	50.00	21.39
Annona muricata L	0,7500	50.00	17.12
	0,3750	50.00	16.30
	0,1875	50.00	13.93

The solubility of lignin in an alkali solution is the key to determine the effectiveness of alkali treatment by saponification reaction between inner molecules that will breakage the aryl ether linkages present in the lignin (contributing to about 50–70% of all the linkages) [33,34]. Consequently, the connecting bond lignocellulose will break and changes the pore structures, microstructure, macrostructure, chemical composition of and crystallinity of lignocellulose[2,35]. Tabel 2. Showed that the highest initial lignin content will give the highest weight feedstock after the delignification process. In other

https://doi.org/10.1051/matecconf/201815603006

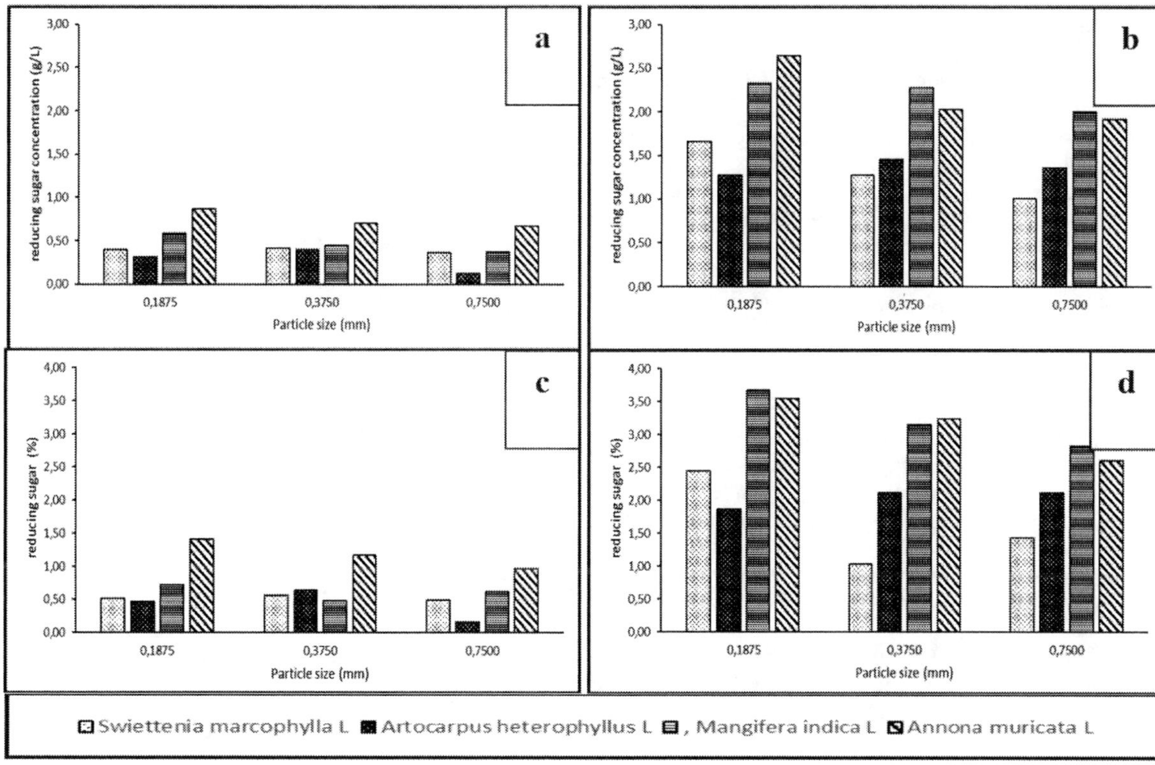

Fig. 2. The result of reducing sugar analysis, a) and c) without delignification, b) and d) with delignification

words, slightly lignin which can soluble in NaOH. Dissolving lignin changes crystalline structure of cellulose and improve the availability to release some hemicellulose and cellulose and slightly soluble in NaOH. This proved with average percent weight loss after delignification, 60.15%, although initial lignin + ash content before delignification only 22.75%. Otherwise, the smaller initial lignin content will give the highest weight loss of cellulose and hemicellulose. It also shows the function of lignin to protect cellulose from the action of reagent chemicals and microorganism.

3.3 Hydrolysis process

The result of this study is consistent, that is proven by increasing production of reducing sugar as a result of acid hydrolysis of plant biomass with increasing lignin removal. Acid hydrolysis was conducted on feed stock with or without delignification using 1% H_2SO_4. Hydrolysis was converted cellulose into glucose and hemicellulose into glucose, mannose, galactose, xylose, and arabinose. Fig. 2a. and 2b showed reducing sugar concentration as g/L and percent weight to investigate the influence of delignification treatment. As a result, plant biomass with delignification almost have higher reducing sugar concentration three times than without delignification especially for Annona muricata L. Hydrolysis with delignification treatment had removed lignin by breakage the lignin bond that one of the major barriers for acid to hydrolyze cellulose and hemicellulose. Removal of lignin with alkaline improves the performance of acid to penetrate cellulose in hydrolysis process. Besides that, Fig 2a-d showed that reducing particle size makes the surface area of leaves more larger and increase chance contact between acid and cellulose. Its makes smaller particle will have higher reducing sugar concentration than the bigger one. The highest reducing sugar concentration is about 0.1875 mm particle leaves size.

There are some factors to understanding the effect of without delignification on hydrolysis process. The first, acid hydrolysis process is limited by lignin that will block the accessibility of acid to penetrate to cellulose. Second, lignin has ability binding with acid. Almost 70% of the total added acid can be bound to lignin after complete hydrolysis. Finally, lignin-carbohydrate limits reaction acid-cellulose on hydrolysis process. This scheme will reduce the concentration of acid to convert cellulose into glucose [25].

The yield of reducing sugar is substantially proportional to its concentration. Fig 2 shows a slightly non-linearity for both with and without delignification leaves. This is due to the difference in the volume of sugar solution that can be separated from the residual solids. In addition, Fig. 2b inform that the highest reducing sugar concentration in this study was 2,642 g / L. It is still too small if the sugar-producing will be used as a fermentation feedstock to produce hydrogen. To produce hydrogen required reducing sugar concentration higher than 1% (10 g/L). For further study higher reducing sugar concentrations can still be obtained by improving operating conditions such as increasing the ratio of solid to Liquid, increasing acid concentration or using other acids such as HCl and others Using higher acid concentrations decreases the pH, thus more chemicals are needed to neutralize the acid. This needs further studied because using large chemicals can increase cations and anions in the solution that will affect the resilience of cell membrane microorganisms in the next fermentation.

4 Conclusion

Pretreatment of lignocellulose and evaluation on its conversion to reducing sugar have been conducted by comparing with and without delignification. The presence of alkali treatment with 2% NaOH changes the structure and removed lignin from lignocellulose in a closed reactor. Thus improve the accessibility of acid by creating pore and breaking the lignin complex. This treatment makes cellulose and hemicellulose can convert more easily than without delignification. Leaves that can be used as feedstock to produce hydrogen should have 38–49% cellulose and hemicellulose. With delignification, 2,642 g/L was produced by acid hydrolysis. Its three times higher than without delignification on Annona muricata L.

Besides that, reducing particle size makes the surface area of leaves more larger and increase possibility contact between acid and leaves particle. It makes smaller particle have higher reducing sugar concentration than the bigger one. The highest reducing sugar concentration is about 0.1875 mm particle leaves size.

Reference

1. Q. Xiang, Y. Y. Lee, P. O. Pettersson, and R. W. Torget, Appl. Biochem. Biotechnol., **105–108**, 505–14, (2003).
2. H. Chen *et al.*, Fuel Process. Technol., **160**, 196–206, (2017).
3. B. O. Aderemi, E. Abu, B. K. Highina, Afr. J. Biotechnol., 7 (11), 1745-1752 (2008)
4. T. De Vrije, G. De Haas, G. B. Tan, E. R. P. Keijsers, and P. A. M. Claassen, Int. J. Hydrogen Energy, **27 (11–12),** 1381–1390, (2002).
5. S. M. Kotay and D. Das, Bioresour. Technol., **98** (6), 1183–1190, (2007).
6. M. D. Dewi, N. Anwar, A. Widjaja, S. Winardi (2011), *Prosiding Seminar Nasional Perkembangan Riset dan Teknologi di Bidang Industri ke-17, Yogyakarta, 16 Mei 2011,* ISBN: 978-979-95620-7-4, Makalah TRTP-31.
7. N. Anwar, A. Widjaja, S., S. Winardi (2010), *Prosiding International Seminar on Fundamental and Application on Chemical Engineering, Dept. of Chemical Engineering - Sepuluh Nopember Institute of Technology, Bali, 3-4 Nopember 2010, Makalah A-005.*
8. N. Anwar and S. Winardi, *Pap. Present. Chem. Eng. RDK 660.284 49 Anw p, 2012,* 2012.
9. T. Mori, Y. Tsuboi, N. Ishida, N. Nishikubo, T. Demura, J. Kikuchi, Sci. Rep. **5**, 1–12 (2015).
10. M. Knauf, M. Moniruzzaman, Int. Sugar J. **106,** 147–150 (2004).
11. V. Menon, M. Rao, Prog. Energy Combust. Sci. **38,** 522–550 (2012).
12. G. Salehi Jouzani, M.J. Taherzadeh, Biofuel Res. J. **2,** 152–195 (2015).
13. S. Wilkinson, D. Greetham, G.A. Tucker, Biofuel Res. J. **3,** 357–365 (2016).
14. K. Zhang, Z. Pei, D. Wang, Bioresour. Technol. **199,** 21–33 (2016).
15. J. K. Kurian, G.R. Nair, A. Hussain, Renew. Sustain. Energy Rev. **25,** 205–219 (2013).
16. R. Chandra, J. Saddler, J. Sci. Technol. For. Prod. Process. **3,** 6–14 (2013).
17. C.E. Wyman, Biotechnol. Prog. **19,** 254–262 (2003).
18. H. W. Blanch, B. A. Simmons, D. Klein-Marcuschamer, Biotechnol. J. **6 (9)**, 1086–1102 (2011).
19. R. El Hage, N. Brosse, P. Sannigrahi, A. Ragauskas, Polymer Degradation and Stability **95 (6)**, 997–1003 (2010).
20. Q. Q. Wang, Z. He, Z. Zhu, Y. H. P. Zhang, Y. Ni, X. L. Luo, J. Y. Zhu, Biotechnol. Bioeng. **109 (2)**, 381–389 (2012b)
21. G. Chatel, K. De Oliveira Vigier, and F. J. Me, Chem. Sus. Chem., **7** (10), 2774–2787, (2014).
22. Z. Liu *et al.*, Bioresour. Technol., **135**, 23-29, (2013).
23. C. Krishnan, L. D. Sousa, M. J. Jin, L. P. Chang, B. E. Dale, V. Balan, Biotechnol. Bioeng. **107** (3), 441–450 (2010).
24. Y. Zhang, Q. Li, J. Su, Y. Lin, Z. Huang, Y. Lu, G. Sun, M. Yang, A. Huang, H. Hu, Bioresour. Technol. **177**, 176–181 (2015).
25. Z. Yu, H. Jameel, H. min Chang, and S. Park, Bioresour. Technol., **102** (19), 9083–9089, (2016)
26. J. Yoo, S. Alavi, P. Vadlani, V. Amanor-Boadu, Bioresour. Technol. **102**, 7583–7590 (2011).
27. F. Cotana, G. Cavalaglio, A. Nicolini, M. Gelosia, V. Coccia, A. Petrozzi, L. Brinchi, Energy Proced. **45,** 52–60 (2014).
28. T. Leskinen, S.S. Kelley, D.S. Argyropoulos, ACS Sustain. Chem. Eng., **3**, 1632–1641 (2015).
29. A.S. Liggenstoffer, N.H. Youssef, M.R. Wilkins, M.S. Elshahed, J. Microbiol. Methods **104**, 43–48 (2014).
30. S. Behera, R. Arora, N. Nandhagopal, S. Kumar, Renew. Sustain. Energy Rev. **36,** 91–106 (2014).
31. Z. Chaula, M. Said, G. John, S. Manyele, C. Mhilu, Smart Grid Renew. Energy **5,** 1–7 (2014).
32. Datta, R., Biotech. and Bioeng. **23** (9), 2167-2170 (1981)
33. Z. Yan, J. Li, S. Chang, T. Cui, Y. Jiang, M. Yu, L. Zhang, G. Zhao, P. Qi, S. Li, Fuel **158,** 152–158 (2015).
34. E. B. Heggset, K. Syverud, K. Øyaas, Biomass Bioenergy **93**, 194–200 (2016).
35. Y.-X. An, M.-H. Zong, H. Wu, N. Li, Bioresour. Technol. **192**, 165–171 (2015).

Solvent-free microwave extraction of essential oil from *Melaleuca leucadendra* L.

Aviarina Widya Ismanto[1], *Heri* Septya Kusuma[1], and *Mahfud* Mahfud[1,*]

[1]Department of Chemical Engineering, Faculty of Industrial Engineering, Institut Teknologi Sepuluh Nopember, Surabaya, Indonesia 60111

Abstract. Cajuput (*Melaleuca leucadendra* L.) oil is one of potential commodity that provides an important role for the country's foreign exchange but the extraction of these essential oil is still using conventional method such as hydrodistillation which takes a long time to produce essential oil with good quality. Therefore it is necessary to optimize the extraction process using a more effective and efficient method. So in this study the extraction is done using solvent-free microwave extraction method that are considered more effective and efficient than conventional methods. The optimum yield in the extraction of cajuput oil using solvent-free microwave extraction method is 1.0674%. The optimum yield is obtained on the feed to distiller (F/D) ratio of 0.12 g/mL with microwave power of 400 W. In the extraction of cajuput oil using solvent-free microwave extraction method is performed first-order and second-order kinetics modelling. Based on kinetics modelling that has been done, it can be said that the second-order kinetic model (R^2 = 0.9901) can be better represent experimental results of extraction of cajuput oil that using solvent-free microwave extraction method when compared with the first-order kinetic model (R^2 = 0.9854).

1 Introduction

Cajuput (*Melaleuca leucadendra* L.) oil is one of potential commodity that provides an important role for the country's foreign exchange. It can be obtained from parts of plants including stems, leaves, or fruits. Cajuput oil can be extracted using several methods like hydrodistillation, water-steam distillation, and steam distillation [1].

Melaleuca leucadendra L. or cajuput is a traditional low-growing tree that the leaves can be produced an essential oil. Cajuput oil is used as an herbal remedy such as antiseptic, analgesic, antireumatic, and anti-inflammatory [2-4]. Furthermore cajuput oil is also used in cosmetic products, perfume and fragrance industries [2, 5-6]. The main product from the cajuput plant is cajuput oil that produced by the distillation of cajuput leaves. The cajuput plant in Java Island has a capacity of 53,760 tons per year for cajuput leaves and total annual production of cajuput oil produced in Java by 300 tons [7]. The consumption for cajuput oil is increasing along with the growing variation of the utilization of cajuput oil. In Indonesia, the annual supply of cajuput oil necessity is 1500 tons [8].

Along with the relatively good prospect of cajuput oil in Indonesia, it should be followed by the development of the cajuput oil distillation technology. The cajuput oil is generally extracted using steam distillation, water-steam distillation and hydrodistillation [1-2, 9]. The disadvantages that used a conventional methods to extract essential oils such as loss of some volatile compounds, high solvent requirements, and long extraction times [10-11]. Microwave energy is an alternative extraction from conventional methods. By using microwave energy, the extraction can be done quickly and reducing solvent consumption. Solvent-free microwave extraction (SFME) is a "green-technology" extraction method that are combining a microwave heat and dry distillation without adding any solvents [12].

In this study the extraction of cajuput oil is examined using solvent-free microwave extraction method. The effect of microwave power on product yield and extraction kinetics model from cajuput oil extraction was also investigated.

2 Materials and Methods

2.1. Materials

Cajuput leaves (*Melaleuca leucadendra* L.) were collected from Mojokerto, East Java. Cajuput leaves were gathered from the same plantation age at approximately 6-12 months of plantation.

* Corresponding author: mahfud@chem-eng.its.ac.id

2.2. Solvent-free microwave extraction

The microwave oven used was Electrolux model EMM-2007X operating at 2450 MHz. The maximum power of the oven was 800 W. Microwave oven was modified by drilling a hole at the top. Sixty grams of fresh cajuput leaves was placed in sample flask in the oven and connected to condensor through the hole. The hole around the neck. During the process, the vapor passed through the condenser outside the microwave cavity where it was condensed. The SFME process was performed for different times and was continued until no more essential oil was obtained. The extraction yield of essential oil was calculated according to Eq. (1).

$$\text{Extraction yield} = \frac{\text{Mass of extraction essential oil}}{\text{Mass of dried material}} \times 100\% \quad (1)$$

2.3. Kinetic Model

2.3.1 First-Order Kinetic Model

Lagergren's kinetic equation for first-order model [13-18] can be written in the following differential form:

$$\frac{dC_t}{dt} = k_1(C_S - C_t) \quad (2)$$

where k_1 (min^{-1}) is the rate of extraction constant for first-order and t (min) is the time of extraction.

Eq. (2) has been obtained can be converted into linear form as follows:

$$\log(C_S - C_t) = \log(C_S) - \frac{k_1}{2{,}303}t \quad (3)$$

The extraction capacity, C_S, and the first-order extraction rate constant, *k*, can be determined experimentally from the slope and intercept by log (C_s – C_t) versus time, t.

2.3.2 Second-Order Kinetic Model

Second-order model in extraction field means that the extraction occurs in two simultaneous processes [16-20]. The rate of dissolution for the essential oil contained in the solid to solution can be described by Eq (4).

$$\frac{dC_t}{dt} = k_2(C_S - C_t)^2 \quad (4)$$

where k_2 is the rate of extraction constant for second-order (L g^{-1} min^{-1}), C_S is the extraction capacity (g L^{-1}) and C_t is the concentration of cajuput oil at any time *t* (min).

A linear form shown in Eq. (5) can be obtained and the extraction rate can be written as Eq. (6).

$$\frac{t}{C_t} = \frac{1}{{C_S}^2 k_2} + \frac{t}{C_S} \quad (5)$$

$$C_t = C_S - \frac{C_S}{1 + C_S k_2 t} \quad (6)$$

The initial extraction rate, *h*, as *Ct*/*t* when *t* approaches 0, can be defined as:

$$h = k\,{C_S}^2 \quad (7)$$

Then ploted between t/C_t with time, t, to get slope and intercept which can be used to determine the initial extraction rate, *h*, the extraction capacity, C_S, and the second-order extraction rate constant, *k*.

3 Results and Discussion

3.1. Cajuput oil yield

The yield of cajuput oil was measured throughout the extraction process. Figure 1 shows the more feed to distiller ratio (F/D) used, the increasing weight of cajuput oil. However the amount of raw material (cajuput leaves) and cajuput oil output are not always positively correlated with the increase of cajuput oil yield. These is because of cajuput oil yield was influence by the factor of the ratio of cajuput oil and the amount of raw material. In Figure 1, the optimum yield in the extraction of cajuput oil using solvent-free microwave extraction method is 1.0674%. The yield was decreased in the feed to distiller ratio of 0.14 g/mL according to the convenience of the raw material (cajuput) used is too much (solid) and almost filled the sample flask. It cause the vapor difficult to penetrate inside the material diffuses out the essential oil of the material. The amount of the material is closely related to the large space between materials. The excessively high and uneven amount of materials cause the formation of vapor lines that can be decreased the yield and quality of essential oils [21].

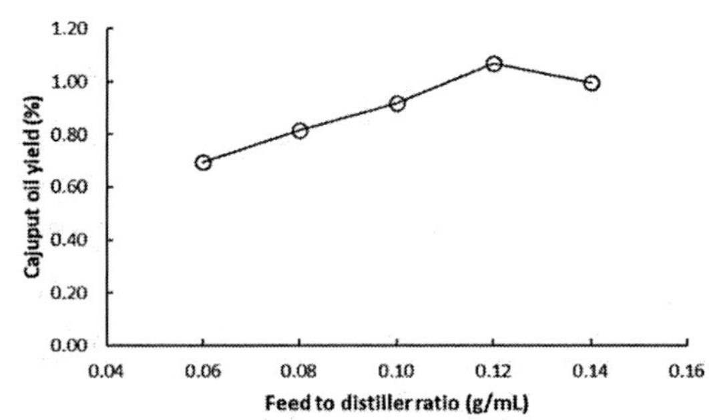

Fig. 1. Yield profile of cajuput oil as function of the feed to distiller (F/D) ratio

Figure 2 shows an increase in product yield when the microwave power is increased. It is known that the microwave power of extraction process will controlled the amount of energy received by plant material, hence to be converted into heat energy. This heat energy helps the process of removal of essential oils from plant material or sample. The power effect on the yield for the solvent-free microwave extraction method can be interpreted as the greater the microwave power used, the greater the energy received by the plant material to be converted into heat which eventually results in lower yields of cajuput oil. This case occurs because of thermal degradation of the material.

In this study, the optimum yield oil extracted by solvent-free microwave extraction was higher when compared with conventional method like hydrodistillation or water-steam distillation. Based on previous research conducted by Pujiarti et al., 2011 [2], during 5 hours of extraction process using water-steam distillation method obtained yield of 0.61 ± 0.11%.

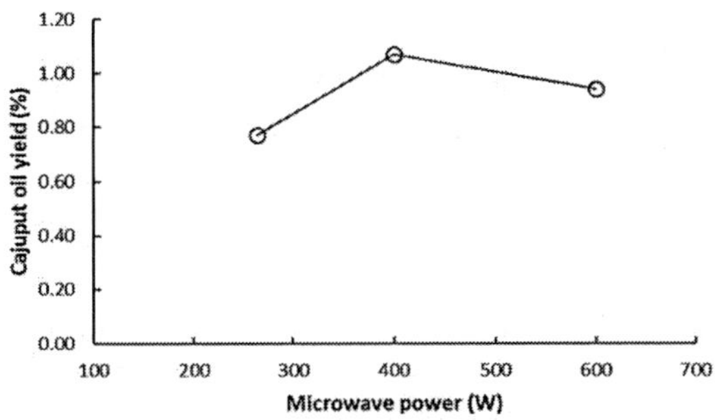

Fig. 2. Yield profile of cajuput oil as function of the microwave power

3.2. Kinetic modelling of cajuput oil extraction with solvent-free microwave extraction method

Figure 3 shows the first-order kinetic model results analyzed using a plot between log (Cs-Ct) versus time, t. In conventional methods of hydrodistilation, heat transfer and mass work in different times. The heat transfer occur from the gland to inside the cell when the temperature increased. Whereas the process of mass transfer occur from inside to outside to the gland when inside the cell was open.

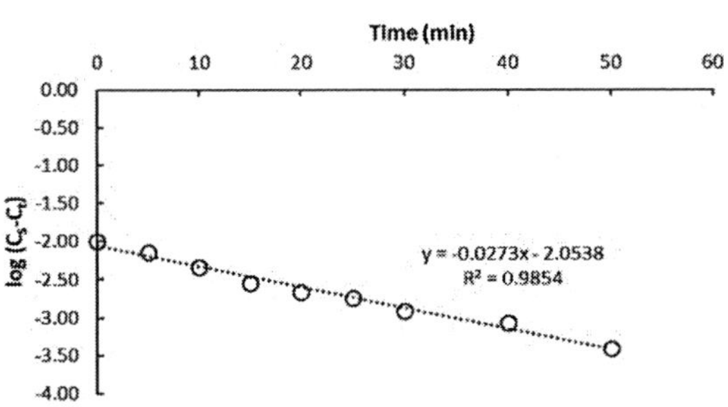

Fig. 3. First-order kinetics model of cajuput oil

Figure 4 shows the second-order kinetic model results analyzed using a plot between t / Ct versus time, t. Based on Figure. 3 and 4, it can be seen that the reaction kinetics corresponding to the SFME method is a second-order kinetics model. This is related with the transfer of heat and mass moving in the same time and direction from the inside to the outside of the gland.

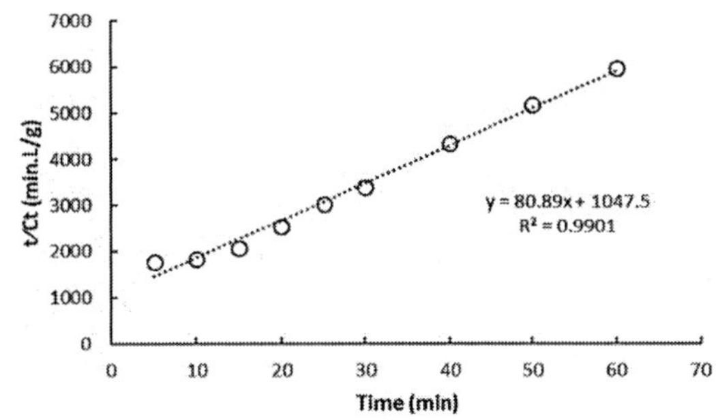

Fig. 4. Second-order kinetics model of cajuput oil

There is the comparison between first-order and second-order kinetics model that shown in table 1. Based on the data showed in Table 1, the coefficient of determination, R^2, second-order kinetics model can be better represent than first-order kinetics model.

Table 1. Comparison between first-order and second-order kinetic model of solvent-free microwave extraction of *Melaleuca leucadendra* L.

Kinetic model			
First-order		Second-order	
k_1 (L $g^{-1}min^{-1}$)	0.0629	k_2 (L $g^{-1}min^{-1}$)	9.3877
Cs (gL^{-1})	0.0101	Cs (gL^{-1})	0.0124
R^2	0.9854	h (g L^{-1} min^{-1})	0.0009
		R^2	0.9901

4 Conclusion

In this study, SFME resulted in a reduced extraction time and a substantial energy saving compared to the conventional method HD technique. The optimum yield in the extraction of cajuput oil using solvent-free microwave extraction method is 1.0674%. Based on kinetics modeling that has been done, it can be said that the second-order kinetic model (R^2 = 0.9901) can be better represent experimental results of extraction of cajuput oil that using solvent-free microwave extraction method when compared with the first-order kinetic model (R^2 = 0.9854).

References

1. A. Widiyanto, M. Siarudin, Jurnal Penelitian Hasil Hutan **31 (4)**, 235-241 (2013).
2. R. Pujiarti, Y. Ohtani, H. Ichiura, J. Wood Sci. **57**, 446-451 (2011).
3. E.S. Angela, W.L. Davis, Altern. Med. Rev. **15 (1)**, 33-47 (2010).
4. U.R. Juergens, U. Dethlefsen, G. Steinkamp, A. Gillissen, R. Repges, H. Vetter, Respir. Med. **97**, 250-256 (2003).
5. J.J. Brophy, E.V. Lassak, Flavour Fragrance J. **3**, 43-46 (1988).
6. M. Yatagai, T. Ohira, K. Nakashima, Biochem. Syst. Ecol. **26**, 713-722 (1998).
7. A. Rimbawanto, M. Susanto, *Prosiding Ekspose Hasil Litbang Bioteknologi dan Pemuliaan Tanaman Hutan*, 83-92 (2004).

8. Muyassaroh, Jurnal Teknik Kimia **10 (2)**, 36-39 (2016).
9. J.A. Pino, E.L. Regalado, J.L. Rodriguez, M.D. Fernandez, Chem. Biodiversity **7**, 2281-2288 (2010).
10. B. Bayramoglu, S. Sahin, G. Sumnu, J. Food Eng. **88**, 535-540 (2008).
11. M. Zill-e-Huma, J.F. Abert-Vian, F.C. Maingonnat, J. Chromatogr. A **1216**, 7700-7707 (2009).
12. A. Farhat, C. Ginies, M. Romdhane, F. Chemat, J. Chromatogr. A **1216**, 5077-5085 (2009).
13. S. Lagergen, K. Sven. Vetenskapsakad. Handl. **24(4)**, 1–39 (1989).
14. Z. Reddad, C. Gerente, Y. Andres, P.L. Cloirec, Environ. Sci. Technol. **36**, 2067-2073 (2002).
15. Y-S. Ho, Scientometrics **59(1)**, 171-177 (2004).
16. H.S. Kusuma, M. Mahfud, RSC Adv. **7**, 1336-1347 (2017).
17. H.S. Kusuma, T.I. Rohadi, E.F. Daniswara, A. Altway, M. Mahfud, Korean Chem. Eng. Res. **55(4)**, 574-577 (2017).
18. H.S. Kusuma, M. Mahfud, Int. Food Res. J. **24(4)**, 1697-1702 (2017).
19. H.S. Kusuma, M. Mahfud, ASEAN J. Chem. Eng. **15(2)**, 62-69 (2015).
20. H.S. Kusuma, P.D. Amelia, C. Admiralia, M. Mahfud, Commun. Sci. Technol. **1**, 15-18 (2016).
21. E. Guenther, *The Essential Oils Volume One: History, Origin in Plants, Production, Analysis*, (1987).

Simulation of the Oxidation and Combustion of Mixed Diesel-Biodiesel Fuel

Yuswan Muharam[1,*], *Danny* Leonardi[1], and *Alisya* P Ramadhania[1]

[1]Department of Chemical Engineering, Faculty of Engineering, University of Indonesia, UI Campus, Depok 16424, Indonesia

Abstract. A comparative simulation-based research has been set up to obtain valid kinetic models of the oxidation and combustion of biodiesel surrogate and diesel surrogate, as well as mixed diesel-biodiesel surrogates which is used to predict their ignition delay times (IDT). The research consists of the development of the detailed kinetic models of the oxidation and combustion of biodiesel surrogate and diesel surrogate, the validation of the two models with the corresponding experimental IDT data, the merging and the validation of the two models for mixed diesel-biodiesel surrogates. The biodiesel surrogate kinetic model was validated with the experimental IDT data of methyl 9-decenoate at 20 atm and three equivalence ratios. The diesel surrogate kinetic model was validated with the experimental IDT data of n-hexadecane at the pressure ranging from 2 atm to 5 atm and the equivalence ratio of 1.0. The diesel-biodiesel surrogate kinetic model was validated with the experimental IDT data of real diesel-biodiesel fuels for four compositions and at 1.18 atm. The validation results of all models show that the models and the experiments are in good agreement.

* Corresponding author: muharam@che.ui.ac.id

1 Introduction

People's needs for fuel keep rising for the last decades as people's living standards have been raised continuously, along with the growth of industrial development. For the last 30 years, one of the sectors which undergoes continuous increase in fuel consumption is the transportation sector with percentages as high as 30% of total energy consumption in the world. In this case, almost all fuel consumed for transportation purposes is derived from fossil sources. With the continuous depletion of crude oil reserves each year, some solutions for alternative energy sources have been suggested and verified, one of it being biodiesel. Biodiesel is a type of fuel equivalent to diesel fuel in terms of usage and properties, but derived from natural sources.

Recently, the development of biodiesel has become the central theme of many researches, due to its renewable properties and less CO_2 emission throughout its life cycle [1]. But in its real application in daily lives, especially in Indonesia, biodiesel is still used in mixtures with diesel fuel due to some of its different physical properties compared to normal diesel fuel. Biodiesel has higher oxygen content, higher viscosity and lower volatility, which contribute to negative effects towards the diesel engine in which the fuel is used [1]. However, biodiesel has higher cetane number compared to normal diesel fuel leading to shorter ignition delay times (IDT). This is considered as an advantage in terms of diesel engine combustion process.

Ignition delay time is defined as the time lapse which is measured between the start of fuel injection and the start of combustion in a diesel engine. Many factors influence IDT, but it can be generally divided into two; physical factor and chemical factor. The physical factors are fuel atomization process, fuel vaporization, and mixing with air in a combustion chamber. The chemical factors include elementary chemical reactions occurring before and during the combustion [2]. This factor depends on the reaction mechanism of the fuel. The difference of molecular structures of fuels differs the reaction mechanisms resulting in different IDTs. One of the methods to approach this problem is simulation of kinetics of reaction mechanisms of these fuels.

Many studies have been carried out to characterize combustion processes involving biodiesel and diesel fuels. Some of these studies were also intended to find out new surrogate compounds or to examine existing surrogates generating similar combustion characteristics to the fuel without simulating the full composition of the fuel itself. In general, surrogate compounds suitable to fit the combustion characteristics of biodiesel are methyl esters with short to medium-length carbon chain. In case of diesel fuel, the surrogate compounds might involve hydrocarbons with long carbon chain (C_{12} or longer).

Herbinet et al. [3] developed a detailed combustion kinetic model for biodiesel surrogates in the form of medium-length methyl esters with a double-bond, such as methyl 9-decenoate and methyl-5-decenoate. Methyl 9-deceonate is chosen because its double bond is at the same position as the one in methyl oleate and at the same location as the first double bond in methyl linoleate and in methyl linolenate. Westbrook et al. [4] developed a combustion kinetic model to simulate normal hydrocarbons with carbon chains being longer than octane, ranging from C_8 to C_{16}. The kinetic model uses reaction routes and classes which is similar to the ones used in the kinetic model of isooctane by Curran et al. [5].

Muharam et al. [6] simulated the IDT of various commercial fuels to find out their octane number. The IDTs of commercial fuels were compared to the ones of the primary reference fuel consisting of n-heptane and isooctane.

The objective of the present work is to have valid kinetic models of detailed reaction mechanisms of the oxidation and combustion of a biodiesel surrogate, a diesel surrogate, and a diesel-biodiesel surrogates to predict their IDTs.

2 Method

The models in the present work are based on the ones developed by Westbrook et al. [4] for diesel surrogate which is n-hexadecane, and by Herbinet et al. [3] for biodiesel surrogate which is methyl 9-decenoate. Both models have been validated by corresponding researchers. We re-validate both models if there are still IDT differences with the experimental results. The point of this is that when the two models are emerged to give a diesel-biodiesel model, the model can reproduce well the experimental real diesel-biodiesel IDT.

The IDT experimental data for the validation are the ones obtained by Wang et al. [7] for methyl-9-decenoate, and by Haylett et al. [8] for n-hexadecane.

The model for a diesel-biodiesel surrogates is the merging of the biodiesel surrogate model and the diesel surrogate model that have been validated. The IDT experimental data for the validation of diesel-biodiesel model is obtained from Hoang et al. [9] using real diesel-biodiesel fuel.

The validation was performed by changing the values of pre-exponential factors (*A*') of several reactions giving high sensitivities to IDT. The determination of high sensitivity reactions is carried out through sensitivity analysis.

3 Model

In chemistry a reaction mechanism is a step by step sequence of elementary reactions by which overall chemical change occurs [10]. If many elementary reactions exist, the net amount of a reactant consumed or a product produced arise from the addition of the contributions of each elementary reaction. In general, species takes part as a reactant in a reaction and as a product in others. The rates of the production and the consumption are a net result of all elementary reactions.

For a mechanism consisting of R elementary reactions of S species, which is given by [11]

$$\sum_{s=1}^{S} \nu_{rs}^{(e)} \mathrm{A}_s \xrightarrow{k_r} \sum_{s=1}^{S} \nu_{rs}^{(p)} \mathrm{A}_s \qquad (1)$$

where r =1, 2, ..., R, $\nu_{rs}^{(e)}$ and $\nu_{rs}^{(p)}$ indicate the stoichiometric coefficients of reactants and products, respectively, the rate of the formation of a species i is obtained from the addition of the whole rate equations where the species i is involved,

$$\left(\frac{\partial c_i}{\partial t}\right)_{\text{chem}} = \sum_{r=1}^{R} k_r \left(\nu_{ri}^{(p)} - \nu_{ri}^{(e)}\right) \prod_{s=1}^{S} c_s^{\nu_{rs}^{(e)}} \qquad (2)$$

where i = 1, 2, ..., S.

The rate laws for a reaction mechanism consisting of R reactions among S species can be written as a system of first order ordinary differential equations,

$$\frac{dc_i}{dt} = F_i\left(c_1, \ldots, c_S, k_1, \ldots k_R\right) \qquad i = 1,2,\ldots,S \qquad (3)$$

with initial conditions being

$$c_i = c_i^0 \text{ at } t = t_0 \qquad (4)$$

For many elementary reactions, a change in the parameters of the system (the rate coefficients) has nearly no effect on the time-dependent solution. Even if the reaction is included explicitly in the mechanism, it does not need a highly accurate rate coefficient. On the other hand, for a few of the elementary reactions, changes in the rate coefficients have really large effects on the outcome of the system. Accordingly, accurate rate coefficients are demanded. These few important reaction steps are rate-determining steps or rate-limiting steps. The dependence of the solution c_i on the parameters k is called sensitivity and sensitivity analysis is a procedure to validate the kinetic model of detailed reaction mechanism.

4 Results and Discussion

3.1 Diesel surrogate

As mentioned in [4] the diesel surrogate mechanism includes both high temperature and low temperature reaction pathways. The validation of the diesel surrogate model was performed at the temperature range of 1100 K – 1400 K, the equivalence ratio of 1, and the pressure of 2 atm – 5 atm.

The IDTs calculated using the diesel surrogate model is initially much lower than the ones from the experiments. Therefore, the values of pre-exponential factors of reactions with positive sensitivity need to be decreased, and/or of reactions with negative sensitivity need to be increased so that the calculated IDTs slow down and approache the experimental IDTs. The changes were made to the reactions with the highest sensitivity to IDT according to their order. In case of the diesel surrogate, we simply changed the reaction with the highest sensitivity to IDT, i.e. $C_3H_5 + HO_2 \leftrightarrow C_3H_5O + OH$, the best fit was obtained, as shown in Figure 1.

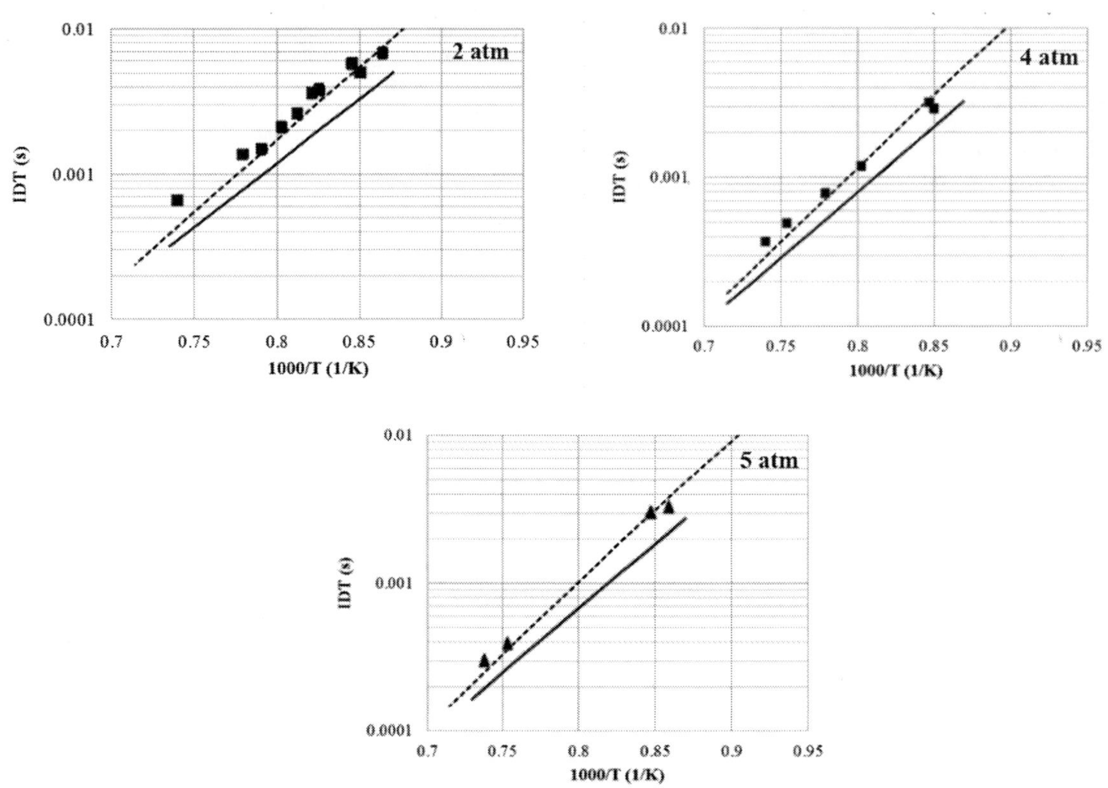

Fig. 1. Comparison of calculated and experimental IDTs of diesel surrogate
(▲ and ■ are experimental data, — is simulation by Haylett et al. [8], and - - - is simulation by present work).

From Figure 1, it can be seen that the calculated IDTs are in good agreement with the experimental ones at 2 atm, 4 atm, and 5 atm. At 5 atm and 1325 K, the calculated IDT is 0.37 ms and the experimental IDT is 0.38 ms. The calculated and experimental results show that IDTs are faster when the initial temperature and pressure increase. The figure exhibits also the calculation results performed by Haylett et al [8].

3.2 Biodiesel surrogate

Validation of the kinetic model of biodiesel surrogate was carried out at three different temperature zones: low, medium and high temperatures, the pressure of 20 atm and the equivalence ratios of 0.5, 1.0, and 1.5. First the simulation results show the deviation of the calculated IDTs against the experimental ones at low-medium temperature (700 K – 900 K). The deviation is negative for all equivalence ratios. Since the deviation values are different among the equivalence ratios, the method of adjusting the kinetic parameters is to find the optimum position between the simulation results and the experimental ones. The optimum position may not precisely match the calculated results with the experimental ones, but at least an approach was carried out such that the simulation results of one of the equivalence ratios may coincide with the experimental results with minimum deviation and good curve shape (the NTC profile at medium temperature remains visible). Therefore, the following optimization steps may be carried out:

1. Low temperature zone: Increasing the values of pre-exponential factors in reactions with positive sensitivity, and/or decreasing the values of pre-exponential factors in reactions with negative sensitivity with the emphasis being on the highest sensitivity reactions in the low temperature zone.
2. Medium temperature zone: Performing the same activities with the above, but with advance notice to the IDT simulation results due to the change in the values of pre-exponential factors in the low temperature zone. When the results are good enough, no changes are needed for sensitive reactions in the medium temperature zone. If the adjustment of pre-exponential factors is necessary, it is done by noting that the reactions with the modified pre-exponential factors in the low temperature zone can not be changed in the medium temperature zone.

https://doi.org/10.1051/matecconf/201815603008

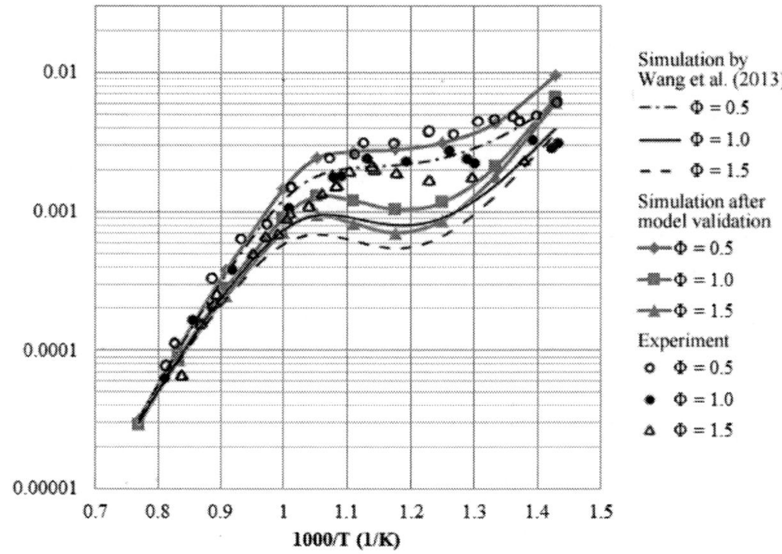

Fig. 2. Comparison between the calculated and experimental IDTs of biodiesel surrogate. (Dots represent experimental data, lines represent simulation results).

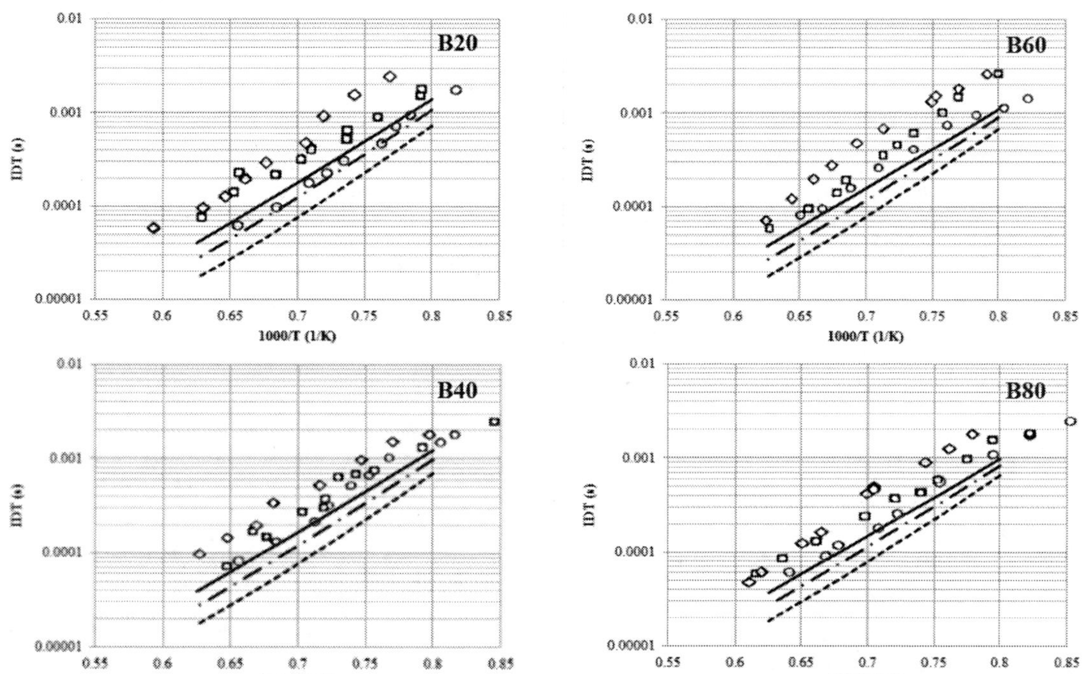

Fig. 3. Comparison between calculated and experimental IDTs of mixed diesel biodiesel fuel before validation (○,□,◊ experimental data for the equivalence ratios of 0.5, 1.0, 1.5;
----, —·, —— simulation results for the equivalence ratios of 0.5, 1.0, 1.5).

The reactions to which the Arrhenius parameters are changed are: H atom abstraction from the fuel by oxygen, peroxy $C_{11}H_{19}O_2$ radical isomerization, O-O scission on peroxy $C_{11}H_{19}O_2$ hydroperoxy radical, and decomposition of peroxy $C_{11}H_{19}O_2$ oxy radical.

Figure 2 shows the calculated IDTs in this study and the experimental ones. From the figure it can be seen that the calculated IDTs for the equivalence ratio of 0.5 have good agreement with the experimental ones, while those for the equivalence ratios of 1.0 and 1.5 are not entirely agreed, especially in the low-medium temperature range

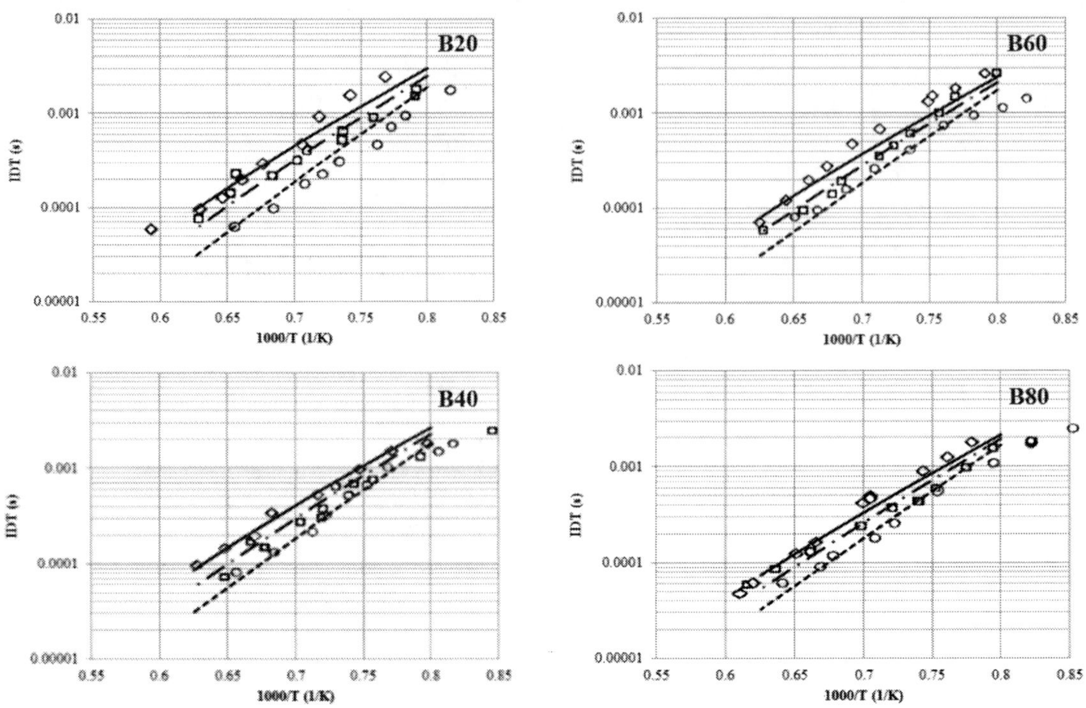

Fig. 4. Comparison between calculated and experimental IDTs of mixed diesel biodiesel fuel after validation (○,□,◊ experimental data for the equivalence ratios of 0.5, 1.0, 1.5; ----, — · , —— simulation results for the equivalence ratios of 0.5, 1.0, 1.5).

(750-950 K). When compared with the simulation results by Wang et al. [7], the IDT results in this study have better agreement. At 900 K, the simulation yields the IDT of 0.37 ms, while the experimental data is 0.38 ms.

The calculated and experimental results show the IDT behavior in three temperature zones. In the low and high temperature zones, IDTs are faster with increasing temperature. In the medium temperate zone, the opposite situation occurs where the IDTs slow with increasing temperature.

3.3 Mixed diesel-biodiesel

The IDT simulation for the mixed diesel-biodiesel surrogates was carried out in the temperature range of 1200 K – 1600 K, the pressure of 1.18 atm, and for the fuel composition of B20 (biodiesel 20%), B40, B60, and B80.

The IDT simulation results are shown in Figure 3. From the figure it is noticed that there is a considerable deviation between the calculated IDTs and the experimental ones. This is because the kinetic model uses diesel and biodiesel surrogates, while the experiment uses real multicomponent diesel-biodiesel fuel.

Since the deviation is negative, it is necessary to adjust the kinetic parameter, i.e. decreasing *A'* values in the reactions with positive sensitivity, and/or increasing *A'* values in the reactions with negative sensitivity. With such a high deviation, it is necessary to change the *A'* values of many reactions as follows:

$CH_3 + HO_2 \leftrightarrow CH_3O + OH$
$C_2H_3 + O_2 \leftrightarrow CH_2CHO + O$
$C_3H_5 + O_2 \leftrightarrow C_2H_3CHO + OH$
$C_3H_5 + O_2 \rightarrow C_2H_2 + CH_2O + OH$
$C_{14}H_{29} + C_2H_5 \leftrightarrow nC_{16}H_{34}$
$C_{11}H_{20}O_2 \leftrightarrow C_8H_{15}O_2 + C_3H_5$

Figure 4 shows the calculated and experimental IDTs for mixed diesel-biodiesel fuel. From the figure it can be seen that in general the calculated IDTs after adjustment of some kinetics parameters have good agreement with the experimental IDTs for each composition. The best match is for B40. At 1350 K and the equivalence ratio 1.0, the simulated IDT is 0.699 ms and the experimental one is 0.69 ms.

4 Conclusions

The kinetic model of detailed reaction mechanism of the oxidation and combustion of diesel surrogate has been validated with the experimental IDT data at the pressure ranging from 2 atm to 5 atm and the equivalence ratio of 1.0. The kinetic model of detailed reaction mechanism of the oxidation and combustion of biodiesel surrogate has been validated with the experimental IDT data at 20 atm and three equivalence ratios. The kinetic model of detailed reaction mechanism of the oxidation and combustion of diesel-biodiesel surrogates has been validated with the experimental IDT data of real biodiesel-diesel fuel for four mixture compositions and at 1.18 atm. In general, the validation results of all models show good aggrement between the models and the experiments.

We express our gratitude to the University of Indonesia, which funded this research through the scheme of Hibah Publikasi Internasional Terindeks untuk Tugas Akhir Mahasiswa No 864/UN2.R3.1/HKP.05.00/2017.

References

1. A.E. Atabani et al., Renew. Sust. Ener. Rev. **16**, 2070 (2012)
2. P.A. Lakshminarayanan, Y.V. Aghav, *Modelling Diesel Combustion* (Springer, 2010)
3. O. Herbinet, W.J. Pitz, C.K. Westbrook, Combust. Flame **157**, 893 (2009)
4. C.K. Westbrook, W.J. Pitz, O. Herbinet, H.J. Curran, E.J. Silke, Combust. Flame **156**, 181 (2009)
5. H.J. Curran, P. Gaffuri, W.J. Pitz, C.K. Westbrook, Combust. Flame **129**, 253 (2002)
6. Y. Muharam, C. Hadiwijaya, Reaktor **14**(2), 109 (2012)
7. W. Wang, S. Gowdagiri, M.A. Oehlschlaeger, Ener. Fuels **27**, 5527 (2013)
8. D.R. Haylett, D.F. Davidson, R.K. Hanson, Combust. Flame **159**, 552 (2011)
9. V.N. Hoang, L.D. Thi, Biosys. Eng. **134**, 1 (2015)
10. J. March, *Advanced Organic Chemistry: Reactions, Mechanisms, and Structure* (Wiley, New York, 1985)
11. M. Nehse, J. Warnatz, C. Chevalier, P. Louessard, H. Melenk, *MOLEC: A Program for the Generation of Chemical Reaction Equations, User's Manual* (Interdisziplinares Zentrum fur Wissenschaftliches Rechnen Universitaet Heidelberg, Heidelberg, 1996)

Biodiesel Production from Wet *Spirulina sp.* by One-Step Extraction-Transesterification

Yano Surya Pradana[1,2,*], *Fariz* Azwar Azmi[1], *Wildan* Masruri[1] and *Muhamad* Hartono[3]

[1]Chemical Engineering Department, Faculty of Engineering, Universitas Gadjah Mada, Jalan Grafika No. 2, Yogyakarta 55281, Indonesia
[2]Center for Energy Studies, Universitas Gadjah Mada, SekipK-1A, Yogyakarta 55281, Indonesia
[3]Chemical Engineering Department, Faculty of Applied Sciences, Delft University of Technology, Mekelweg 2, 2628 CD Delft, Netherlands

Abstract. Microalgae has gained immense interests as the raw material for biofuel production. The lipid content in microalgae can be converted into biodiesel through conventional method which involves separated process of extraction and transesterification. In this study, the production of biodiesel from Spirulina sp. was performed through one-step extraction-transesterification using KOH as base catalyst to simplify the production of biodiesel. The mixture of methanol-hexane was employed as both solvent and reactant in the process. The resulting biodiesel was found to be mainly composed of methyl oleate and methyl palmitate. On the other hand, increasing the reaction temperature and reducing the quantity of methanol in solvent mixture would also increase the yield of biodiesel. The optimum methanol-hexane volumetric ratio and temperature which gave the highest biodiesel yield were 3:7 and 50°C, respectively.

1 Intoduction

The proliferating growth of population followed by the rapid development of technology has resulted in an increase in energy demand from industrial sector, transportation, to household [1]. Until now, the non-renewable fossil fuel has been opted as the main source of energy in many parts of the world with the limitation in availability [2]. The increasing demand which does not correspond with the depleting fossil fuel resources has promoted many attempts to develop renewable energy sources [3]. In the recent years, the search for alternative fuel is growing quite rapidly. This new and renewable energy search is not only intended to reduce dependence on fossil fuels but also to reduce greenhouse gas emissions [4]. One of the alternative renewable energy sources is derived from biomass or commonly referred to as biofuel [5]. Agricultural products such as corn, soybeans, and cassava can be used to produce biofuels such as bioethanol and biodiesel. However, the use of agricultural products in the manufacture of biofuels raises several problems as it can increase global food prices and cause damage to the soil nutrient cycle [6]. Hence, new raw materials for biofuel production are vastly required.

Microalgae is one of the most potent raw materials for biofuel development [7]. These unicellular species contain fat, carbohydrates, proteins and other compounds that can be utilized to produce biodiesel through extraction and transesterification processes [8]. Microalgae have several advantages over other biomasses such as higher photosynthetic efficiency, higher biomass production and rapid growth, high lipid content, and do not require large areas to breed [9]. By those advantages, microalgae are suitable as agent for accelerating the growth rate of fish, which is the similar effect of micro bubble generator application in fish farming [10]. Moreover, microalgae can be used to produce several renewable fuels such as biodiesel, bioethanol, bio-hydrogen, methane, and syn-gas as presented in Fig. 1 [11].

Among the option as seen in Fig. 1, the conversion of microalgae biomass into biodiesel seems attractive. Biodiesel is a mono-alkyl ester of long chain fatty acids from conventional and non-conventional animal oils and fats [12]. Biodiesel has several advantages compared to petro-diesel as it is renewable, non-carcinogenic and non-mutagenic, biodegradable, soluble in petroleum diesel, lubricating ability, high cetane number and absence of aromatic compounds [13]. Biodiesel can also reduce emissions of carbon monoxide (CO), hydrocarbons (HC), and particulates in waste gas compared to diesel petroleum, thus benefiting the environment [14;15]. Biodiesel production from microalgae is usually carried out in two stages. In the first stage, microalgae are extracted by mechanical and chemical means. Subsequently, the microalgae oil undergoes transesterification with alcohol with the aid of an acid or base catalyst to form biodiesel and glycerol as a by-product [16]. The base catalyst is often used to increase reaction rate. Moreover, the conversion is higher than when using an acid catalyst [17].

*Corresponding author: yanopradana@ugm.ac.id

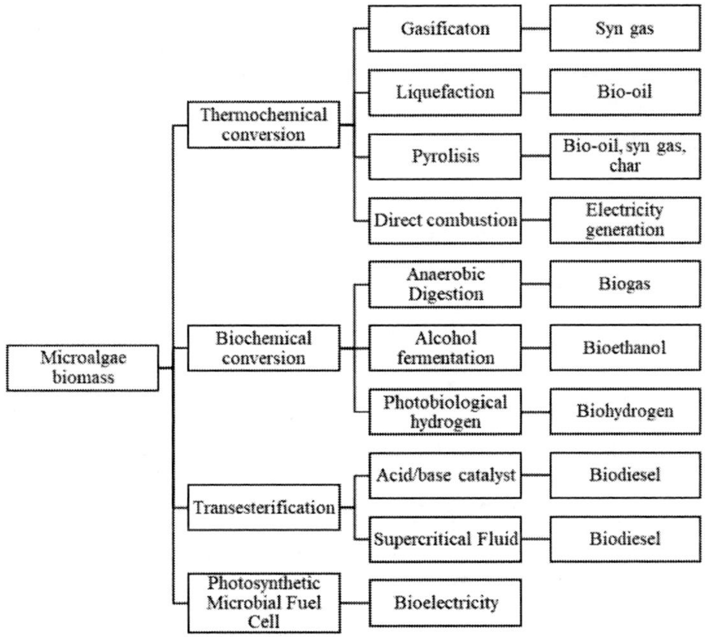

Fig. 1. Framework of microalgae biomass utilization as renewable energy source [10]

In this study, biodiesel is produced from microalgae through a single stage extraction and transesterification process using methanol-hexane mixture. The process employed in this study reduces the number of steps in the preparation of biodiesel in the conventional method, that is by removing the extraction step and that the extracted oil is in direct contact with the alcohol. Hence, alcohol acts both as a solvent and reactant. By using fewer steps, the production of biodiesel using this process will simplify the biodiesel manufacturing process and reduces the manufacturing cost [18].

2 Materials and Methods

2.1. Materials

Spirulina sp. used as raw material was originated from Lamongan, East Java. Methanol (CH_3OH) used as a solvent and reactant in biodiesel synthesis reaction was purchased from CV. General Labora, Yogyakarta. Hexane (C_6H_{14}) used as a solvent and potassium hydroxide (KOH) used as a base catalyst was also purchased from CV. General Labora, Yogyakarta.

2.2 Biodiesel synthesis

Spirulina sp., as much as 150 g, were weighed using digital analytical balance. Methanol and hexane of 500 mL were prepared, with a methanol-hexane volumetric ratio of 1: 1. As much as 0.2 g of KOH catalyst was dissolved with prepared methanol. Methanol, KOH, and hexane which have been prepared were put into a three-necked flask. Microalgae were then also put into a three-neck flask. The cooling water stream and stirrer were switched on and set at 800 rpm. Heater stove was set to the reaction temperature of 45°C. After 2 hours of running, the mixture was cooled to room temperature and then filtered to separate the filtrate from the formed cake. The filtrate, which comprised of 2 layers of glycerol at the bottom and biodiesel at the top layer was evaporated to recover methanol and hexane.

A similar procedure was repeated with reaction temperature of 35 and 50°C and a variation of methanol-hexane volumetric ratio of 3:7 and 7:3 as with a fixed solvent volume of 500 mL.

2.3 Data analysis

Yield of biodiesel was calculated by using Equation (1).

$$\text{Yield (\%)} = \frac{mb}{mm} \times 100\% \qquad (1)$$

Where *mb* denotes mass of biodiesel (gram) and *mm* denotes mass of microalgae (gram). Methyl ester component in biodiesel from *Spirulina sp.* was analysed by using GC-MS to determine the constituent compounds.

3 Results and discussion

The determination of fatty acid present in biodiesel derived from *Spirulina sp.* is a very important parameter to determine since it can define the properties of biodiesel such as molecular weight and cetane number (CN). The result of GC-MS analysis of the biodiesel produced from Spirulina sp. is presented in Fig. 2 and Table 1. It can be seen that biodiesel derived from microalgae mainly composed of methyl oleate with a percentage of 45.10%, followed by methyl palmitate and significantly lower percentage of methyl linoleate and methyl stearate.

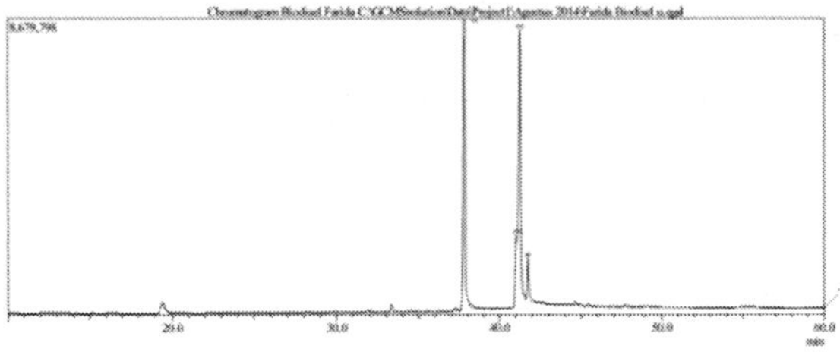

Fig. 2. Result of GC-MS analysis of biodiesel derived from *Spirulina sp.*

Table 1. Fatty acid composition of biodiesel derived from *Spirulina sp.*

Component	Chemical formula	Yield (%)
Methyl Palmitat	$C_{17}H_{34}O_2$	41.03
Methyl Linoleate	$C_{19}H_{34}O_2$	8.34
Methyl Oleate	$C_{19}H_{36}O_2$	45.10
Methyl Stearate	$C_{19}H_{38}O_2$	5.54

Methyl oleate has been used to enrich biodiesel fuels as it can improve biodiesel properties [19;20]. Moreover, Knothe [21] found that the CN of methyl oleate exceeds the minimum standard value in ASTM D6751. The presence of methyl oleate as fatty acid in biodiesel has been shown to reduce regulated hydrocarbon as well as CO emission compared to the other fatty acid methyl ester. However, further studies are required to test the quality and properties of the resulting biodiesel from *Spirulina sp.*

3.1 Effect of methanol-hexane volumetric ratio

The effect of methanol-hexane volumetric ratio on biodiesel yield is presented in Table 2.

Table 2. Effect of methanol-hexane volumetric ratio on biodiesel yield

Methanol-hexane volumeric ratio	Yield (%)
3:7	0.5730
1:1	0.5728
7:3	0.5726

Table 2 shows that the higher amount of methanol added to the methanol-hexane mixture as the solvent decreased the yield of biodiesel, although insignificantly with an order of 10^{-4}. The decline in biodiesel yield is due to the decline of the microalgae oil extracted by the solvent. Non-polar oils are more prone to be extracted by solvents that have the same polarity (like dissolve like) that, in this case, is by hexane. The solvent mixture with higher hexane fraction extracted more oil, which then reacted with methanol. Methanol is not effective as a solvent because it is a polar compound so that it cannot extract microalga oil as efficient as hexane. Hence, the addition of hexane which is a non-polar compound as co-solvent can increase the mass transfer of oil from microalga to solvent mixture [22].

3.2 Effect of reaction temperature

Table 3 presents the result of reaction temperature variations on biodiesel yield. As the process temperature rose, the biodiesel yield would also rise. The highest biodiesel yield of 0.6870% was achieved at 50°C. The increase in yield can be ascribed to higher oil temperatures which allowed solvents to extract more oil. In addition, higher temperatures would also increase the interaction between reactants [18]. However, the temperature was limited to 50°C for the safety of experiment.

Table 3. Effect of reaction temperature on biodiesel yield

Reaction temperature, °C	Yield (%)
35	0.5727
45	0.5728
50	0.6780

4 Conclusions

One-step extraction-transesterification of *Spirulina sp.* using methanol-hexane mixture was able to simplify the production of biodiesel. Methyl oleate and methyl palmitate was found to be the main components of biodiesel derived from *Spirulina sp.* The use of hexane as co-solvent in extraction-transesterification was able to improve the extraction of algal oil hence increasing biodiesel yield. Increasing the quantity of methanol in the solvent mixture reduced the yield of biodiesel. However, increasing the reaction temperature would increase the biodiesel yield. The optimum methanol-hexane volumetric ratio and temperature which gave the highest biodiesel yield were 3:7 and 50°C, respectively.

The authors would like to thank Prof.Arief Budiman for providing suggestion in this study and Chemical Engineering Department, Faculty of Engineering, Universitas Gadjah Mada, Indonesia for financial support through Research Grant No. 451s/H1.17TK/PL/2017.

References

1. Y.S. Pradana, A. Hidayat, A. Prasetya, A. Budiman, Defect Diffus. Forum **382**, 280-285 (2018).
2. Y.S. Pradana, Y. Kusumastuti, F.N. Rahma, N. Effendy, Chem. Eng. Trans. **56**, 1549-1554 (2017).
3. S.S.M. Mostafa, N.Sh. El-Gendy, Arab. J. Chem. **10** (2), S2040-S2050 (2013).
4. M.B. Tasic, L.F.R. Pinto, B.C. Klein, V.B. Veljkovic, R.M. Filho, Renew. Sust. Energ. Rev. **64**, 260-270 (2016).
5. T.M. Mata, A.A. Martins, N.S. Caetano, Renew. Sust. Energ. Rev. **14**, 217-232 (2010).
6. G.A. Nafis, P.Y. Mumpuni, Indarto, A. Budiman,

AIP Conference Proceedings **1699**, Art. No. 030021 (2015).
7. E.A. Ehimen, Z.F. Sun, C.G. Carrington, Fuel **89**, 677-684 (2010).
8. Y.S. Pradana, H. Sudibyo, E.A. Suyono, Indarto, A. Budiman, Energy Procedia **105**, 277-282 (2017).
9. H. Sudibyo, Y.S. Pradana, T.T. Samudra, A. Budiman, Indarto, E.A. Suyono, Energy Procedia **105**, 270-276 (2017).
10. W. Budhijanto, D. Darlianto, Y.S. Pradana, M. Hartono, AIP Conference Proceedings **1840**, Art. No. 110008 (2017).
11. H.T. Chung, L.S. Pau, C. Jo-Shu, C.L. Tau, C.L. John, Biotechnol. Adv. **33**, 1219-1227 (2010).
12. G. Knothe, J. Krahl, J. Van Gerpe, *The Biodiesel Handbook* (American Oil Chemist's Society (AOCS) Press, Urbana, 2005).
13. D.R.Sawitri,Sutijan, A. Budiman, ARPN Journal of Engineering and Applied Sciences, **11** (16), 9951-9957.
14. G. Knothe, Fuel Process. Technol. **86**, 1059-1070 (2005).
15. M. Lapuerta, O. Armas, J. Rodriguez-Fernandez, Prog. Energy Combust. Sci. **34**, 198-223 (2008).
16. R. Zakaria, A.P. Harvey, Fuel Process. Technol. **10**, 53-60 (2012).
17. G.H. Huang, F. Chen, D. Wei, X.W. Zhang, G. Chen, Appl. Energy **87**, 38–46 (2010).
18. P. Nautiyal, K.A. Subramanian, M.G. Dastidar, Fuel **135**, 228-234 (2014).
19. N.A. Bringe, *The Biodiesel Handbook* (AOCS Press: Champaign, IL, 2005).
20. A.J. Kinney, T.E. Clemente, Fuel Process. Technol. **86**, 1137–1147 (2005).
21. G. Knothe, Energ Fuel **22**, 1358-1364 (2008).
22. I.A. Kartika, M. Yani, D. Ariono, Ph. Evon, L. Rigal, Fuel **106**, 111-117 (2013).

The effects of microorganism on coffee pulp pretreatment as a source of biogas production

Sri Rachmania Juliastuti[1,*], *Tri* Widjaja[1], *Ali* Altway[1] , *Vivi* Alvionita Sari[1], *Dessy* Arista[1] and *Toto* Iswanto[1]

[1]Chemical Engineering Department, Institut Teknologi Sepuluh Nopember, Campus ITS Sukolilo Surabaya 60111, Indonesia

Abstract. Coffee pulp waste composition consist of cellulose, hemicellulose, lignin, pectin and caffeine, tannin, and polyphenol as inhibitor substance. The high cellulose compound in coffee pulp can be used for alternative raw materials in the manufacture of biogas. This study aims to define the composition of the mixture of microorganisms of *Pseudomonas putida, Trichoderma harzianum*, and *Aspergillus niger* that are best to use in coffee pulp pretreatment to degrade inhibitor substance. The best result of pretreatment will be applied to biogas production. The first step is to do a pretreatment of the coffee pulp with variable *Pseudomonas putida : Trichoderma harzianum : Aspergillus niger* with a ratio of 1:1:1, 1:2:1, 1:1:2, 1:2:2, 2:1:1, 2:1:2, 2:1:1 (v:v:v), then variables that are most excellent in degrading inhibitor substance are selected. The second step, is doing anaerobic fermentation for 20 days at mesophilic temperature (30-40°C) on a reactor working volume of 4.5 L. In the making of biogas, a varied starter as much as 10% of the total are put into the reactor in the form of a mixture of cow dung : rumen fluid with a ratio of 1:0, 0:1, 1:1, 1:2, 2:1 (w/v). The parameters measured include the decreasing of the inhibitor substance, Chemical Oxygen Demand (COD), biogas (CH_4 and CO_2) and calorific value of combustion (Heating value). This study results a composition of ingredients within the pretreatment process which includes a mixture of microorganisms with a ratio of *Pseudomonas putida : Trichoderma harzianum : Aspergillus niger* is 1:2:1 (v:v:v). For biogas, a mixture of cow dung and rumen fluid can produce higher methane gas is the ratio of cow dung : rumen fluid is 1:2 (w/v) with the concentration of methane (CH_4) formed at 1.825%. At the concentration of methane is 1.825%, the heating value obtained is 76.032 kcal/kg with volume biogas 0.0032 m^3/ kg converted of COD.

1. INTRODUCTION

One of the many industrial waste obtained from processing coffee beans include coffee pulp. Its composition consists of 57.9% cellulose, 21.63% hemicelluloce, 5.21% lignin, 2.28% pectin and inhibitor substance such as 0.86% caffeine, 4.81% tannins, 3.48% polyphenol. Coffee pulp is high in cellulose, which as a raw material, can be generated into an alternative source for the production of biogas.. However, some substances within coffee pulp contains inhibitor, such as caffeine, tannins, and polyphenols which are able obstruct the production of biogas [1].

Degradation of the toxic material contained in coffee pulp such as caffeine, tannins and polyphenols, can be executed by the use of microorganisms [1]. Enzymes in microorganisms is involved within the process of bacterial inhibitor degradation, by breaking down the inhibitor. The process of degradation of inhibitor substance in coffee pulp, helps the microorganism to turn the cellulose in coffee pulp into biogas.

Ethanol and sodium hydroxide are the chemical used in the pretreatment process. Within the pretreatment, lignin is broken down to increase the cellulose in the coffee pulp [2]. The drawback of chemical pretreatment is the expensive ingredients used and the toxicity of the raw material itself.

Whereas in biological pretreatment, non-toxic parts can be produced as the microorganisms grow on the inhibitor substances. In short, biological pretreatment is more environmentally friendly than the chemical pretreatment because harmful materials such as chemicals that will be bad for the environment are not being used.

In degrading caffeine, microorganisms such as *Klebsiella, Serratia*, and *Pseudomonas* found the highest percentage of caffeine degradation in *Klebsiella* with 100% degradation percentage in 10 hours [3]. *Pseudomonas putida* has of 67.2% percentage of caffeine degradation within 48 hours.

* Corresponding author: julizainul@gmail.com

Several microorganisms can degrade Tannins, however fungi species have a specialty in degrading tannins such as Chaetomium, *Fusarium, Rhizoctonia, Clyndrocarpon, and Trichoderma harzianum* [4] and microorganisms that can degrade polyphenols such as *Aspergillus, Fusarium, Coprinus, Penicillium* and the highest degradation rate is *Aspergillus niger* of 8 mg/L/h.

It can be analyzed in previous research that *Pseudomonas putida, Trichoderma harzianum,* and *Aspergillus niger* have special ability to degrade each inhibitor substance. So the three selected microorganisms to degrade the inhibitor substances as a variable in degrading inhibitor substance and then applied in the manufacture of biogas.

2. EXPERIMENT DETAILS

2.1Sampling and coffee pulp treatment

Coffee pulp samples were obtained from PTPN XII plantations, Malang, East Java, Indonesia. The coffee pulp was prepared by separating leaves, rocks, and other materials and leaving only the coffee pulp.

The water content of the coffee pulp is decreased through the process of drying. Then the dried coffee pulp is reduced in size about ± 100 mesh.

2.2 Preparation of medium and microorganism

2.2.1 Medium for Bacteria

4 Grams of nutrient broth, 6.25 grams of agar with 1.5 grams of glucose into 500 mL of boiling water are mixed to create the liquid medium. The autoclave then sterilizes at 121°C, and media was transferred into test tube and cooled until 28°C. Local strain of Pseudomonas putida was moved to agar medium. An incubator was used to inoculate the bacteria and media at 30°C.

2.2.2 Medium for Fungi

19.5 Grams of PDA powder and 6.25 grams of agar into 500 mL of boiling water were mixed as the liquid medium. The autoclave then sterilizes at 121°C, and media was transferred into test tube and cooled until 28°C. Local strain of *Trichoderma harzianum* and *Aspergillus niger* were then moved to agar medium. Fungi and media ware inoculated inside incubator at 30°C.

2.3 Pretreatment Coffee Pulp

Mixing the coffee pulp with water at 1:4 ratio results in substrate. Substrate was prepared by mixing 42.5 grams of coffee pulp and 170 mL of water in Erlenmeyer (Fig 1). Then, a mixture of microorganisms as much as 15% of the amount of substrate around 30 mL was moved from agar medium to substrate. Substrate in the reactor were inoculated inside incubator shaker at 30°C and 100 rpm. An aerobic system is maintained through the substrate and microorganisms in bioreactor. There were 7 bioreactors with different variables (ratio microorganisms *Pseudomonas putida : Trichoderma harzianum : Aspergillus niger* are 1:1:1, 1:2:1, 1:1:2, 1:2:2, 2:1:1, 2:2:1, and 2:1:2 (v:v:v)).

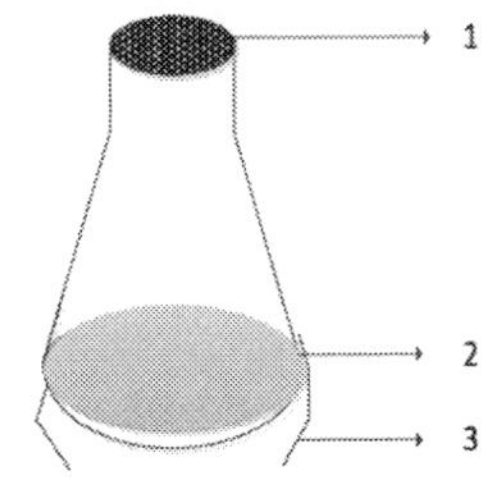

Fig 1. Bioreactor for Pretreatment of Coffee Pulp
1. Cover of Erlenmeyer
2. Substrate and mixture of microorganisms
3. Clamp of shaker

For 7 days, pretreatment of coffee pulp is being executed. Every 2 days, samples are taken from the bioreactors to be analyzed. The analysis include the decrease of inhibitor substance and increase of microorganisms in bioreactors. From the process pretreatment, the best comparison of microorganisms added with the indicated amount of decrease of inhibitor substance the highest is obtained. The best variable will be applied to biogas manufacture.

2.4 Manufacture of Biogas

2.4.1 Preparation of Starter

Cow dung and rumen fluid are used as starter in this experiment. There were 5 biogas reactors with different starter variable. Starter added to the ratio of cow dung: rumen fluid is 1:0, 0:1, 1:1, 1:2, 2:1 (w/v). Based on a study about the manufacturing of biogas by Abdillah Jaka et al in 2013, starter mixed with 10% of the best result of pretreatment coffee pulp and fermented for 5 days. Then, starter is fed into a biogas reactor that containing 90% of total coffee pulp.

2.4.2 Anaerob Fermentation

A separation between substrate and the filtrate is done and the substrate is then dried with an oven for 24 hours and with the temperature of 105°C before mixed with starter. The volume of biogas reactor used is 6 L with a substrate volume of 4.5 L and starter volume of 0.5 L. For 20 days fermentation of biogas is carried out to produce methane (CH_4). The operating state of biogas fermentation is mesophilic temperature (30-40°C) and in anaerobic system (Fig 2).

Every 4 days, samples substrate and gas from fermentation taken from biogas reactors and to analyzed pH, Chemical Oxygen Demand (COD), biogas (CH_4 and CO_2) and caloric value of combustion (Heating Value).

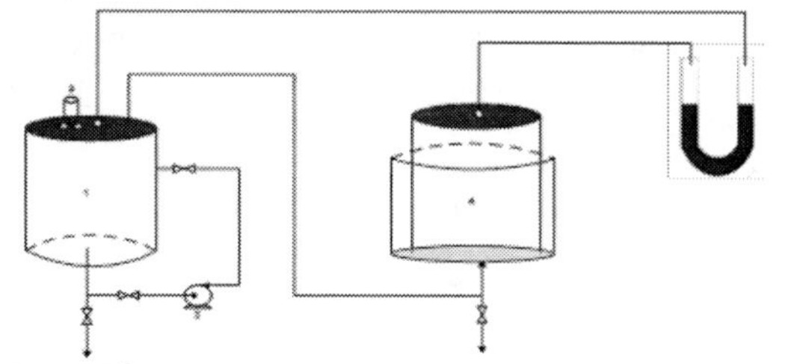

Fig 2. Biogas Reactor

1. Feed Tank
2. Pump
3. Inlet Feed
4. Gas Holder
5. Manometer

2.5 Total Cell Counts

During the degradation of inhibitor substances, total cells counts were determined. The counting method was counting chamber. 1 ml of slurry was diluted by 9 ml aquadest. Dilution was put in haemacytometer. Cell counts were observed and counted under microscope. Cell counts were done everytime for a week. The result was plotted in graphic a form, compared with time and decrase of inhibitor substaces.

2.6 Analysis of Biogas

Biogas sample consist of methane (CH_4) and carbon dioxide (CO_2) from biogas reactor then analyzed at Energy Laboratory, Sepuluh Nopember Institute of Technology, Surabaya, by gas chromatography (GC) method and nitrogen as gas carrier at 200°C.

3. RESULT AND DISCUSSION

3.1 Composition of Coffee Pulp

Table 1. Composition of Coffee Pulp at PTPN XII Malang

Composition Of Coffee Pulp	Percent Composition (%)
Cellulose	57.9
Hemicellulose	21.63
Lignin	5.21
Pectin	2.28
Tannin	4.81
Caffeine	0.86
Polyphenol (Total Phenol)	3.48

That composition is the result of dry coffee pulp analysis before pretreatment. Based on Van Dam of 2003, composition of coffee pulp is different from each other. It is because of the different in source, type, and maturity of raw materials.

3.2 Coffee Pulp Pretreatment

The addition of microorganisms such as *Pseudomonas putida*, *Trichoderma harzianum*, and *Aspergillus niger* spill accelerate the process degradation of inhibitor substance in coffee pulp. The addition of microorganisms was supposed to decrease inhibitor substance such as caffeine, tannins, and polyphenol that was contained in the coffee pulp. The results of the analysis related to the inhibitor substance in seventh bioreactors were shown in Fig 3 to Fig 9.

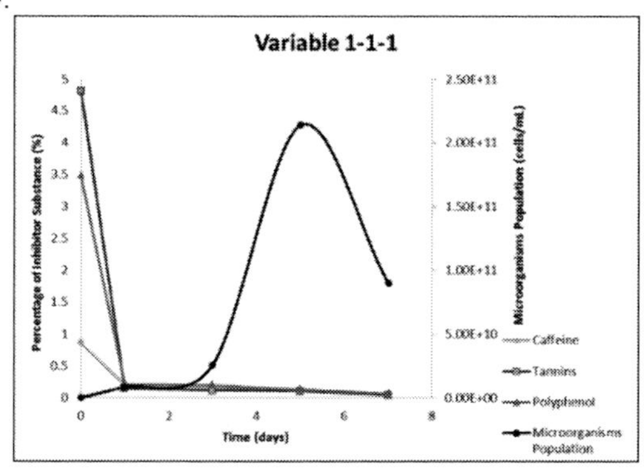

Fig 3. The correlation between the percentage of inhibitor substances, the increase in microorganisms population and time on the ratio of *Pseudomonas putida : Trichoderma harzianum : Aspergillus niger* of 1:1:1

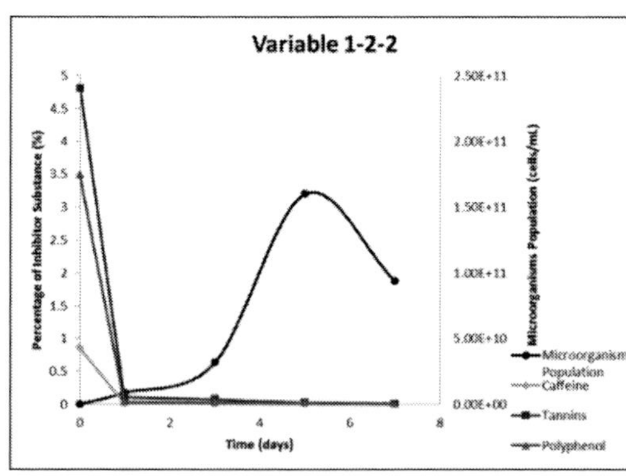

Fig 4. The correlation between the percentage of inhibitor substances, the increase in microorganisms population and time on theratio of *Pseudomonas putida :Trichoderma harzianum : Aspergillus niger* of 1:2:2

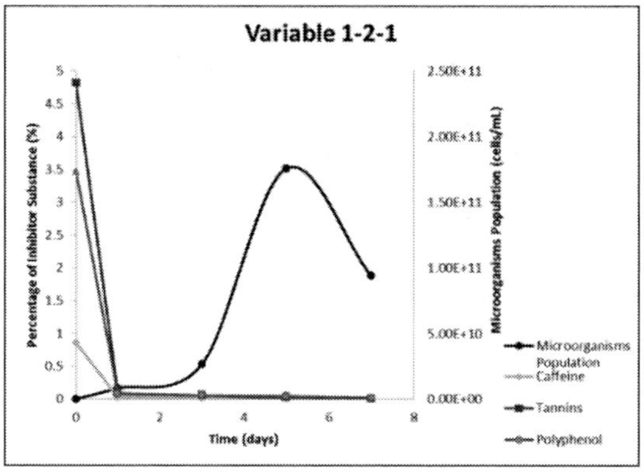

Fig 5. The correlation between the percentage of inhibitor substances, the increase in microorganisms population and time on the ratio of *Pseudomonas putida : Trichoderma harzianum : Aspergillus niger* of 1:2:1

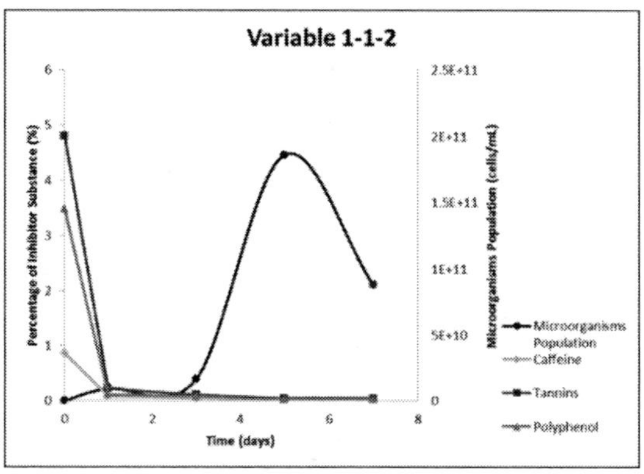

Fig 6. The correlation between the percentage of inhibitor substances, the increase in microorganisms population and time on the ratio of *Pseudomonas putida : Trichoderma harzianum : Aspergillus niger* of 1:1:2

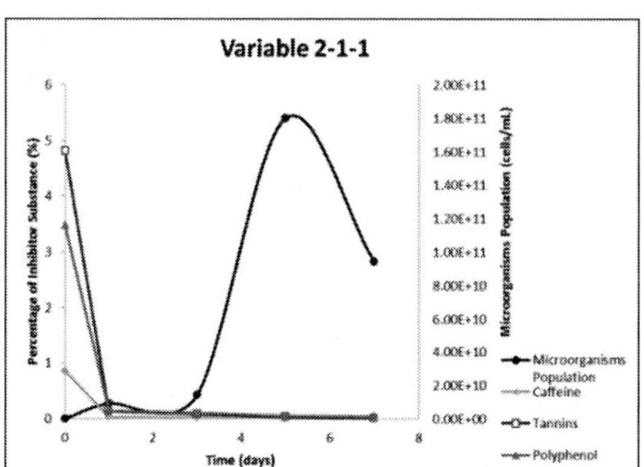

Fig 7. The correlation between the percentage of inhibitor substances, the increase in microorganisms population and time on the ratio of *Pseudomonas putida : Trichoderma harzianum : Aspergillus niger* of 2:1:1

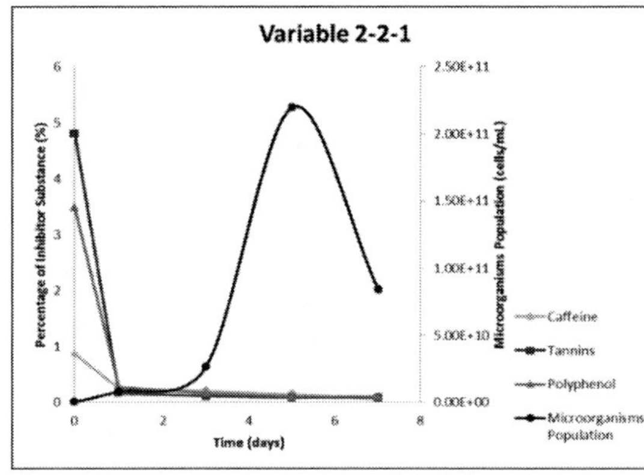

Fig 8. The correlation between the percentage of inhibitor substances, the increase in microorganisms population and time on the ratio of *Pseudomonas putida : Trichoderma harzianum : Aspergillus niger* of 2:2:1

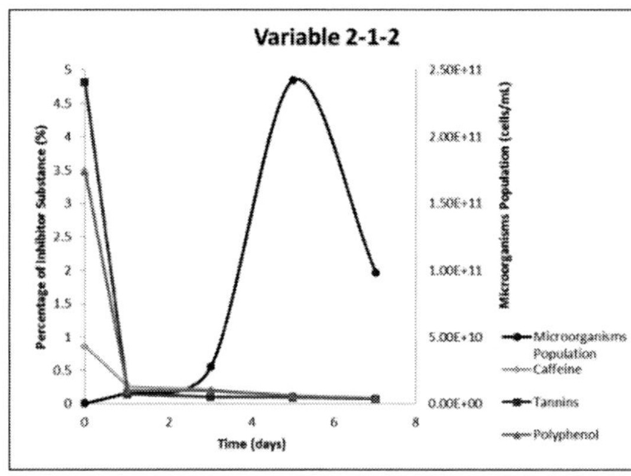

Fig 9. The correlation between the percentage of inhibitor substances, the increase in microorganisms population and time on the ratio of *Pseudomonas putida : Trichoderma harzianum : Aspergillus niger* of 2:1:2

Concentration of inhibitor substance decreased day by day for all bioreactors. It shown that microorganism are growing by being able to adapt on the substrate so that the microorganism would degrade the inhibitor substance. Caffeine compounds are degraded into xanthine and tannins and phenol are degraded into citric acid.

In bioreactor 1, with addition of *Pseudomonas putida, Trichoderma harzianum*, and *Aspergillus niger* with ratio 1:1:1, at the end of process caffeine was decreased from 0.86% become 0.04%, tannins was decreased from 4.81% become 0.05%, and polyphenol was decreased from 3.48% become 0.07%.

In bioreactor 2, with addition of *Pseudomonas putida, Trichoderma harzianum*, and *Aspergillus niger* with ratio 1:2:2, at the end of process caffeine was decreased from 0.86% become 0.005%, tannins was decreased from 4.81% become 0.01%, and polyphenol was decreased from 3.48% become 0.0011%.

In bioreactor 3, with addition of *Pseudomonas putida, Trichoderma harzianum*, and *Aspergillus niger* with ratio 1:2:1, at the end of process caffeine was decreased from 0.86% become 0.001%, tannins was decreased from 4.81% become 0.01%, and polyphenol was decreased from 3.48% become 0.02%.

In bioreactor 4, with addition of *Pseudomonas putida, Trichoderma harzianum*, and *Aspergillus niger* with ratio 1:1:2, at the end of process caffeine was decreased from 0.86% become 0.02%, tannins was decreased from 4.81% become 0.04%, and polyphenol was decreased from 3.48% become 0.03%.

In bioreactor 5, with addition of *Pseudomonas putida, Trichoderma harzianum*, and *Aspergillus niger* with ratio 2:1:1, at the end of process caffeine was decreased from 0.86% become 0.01%, tannins was decreased from 4.81% become 0.02%, and polyphenol was decreased from 3.48% become 0.05%.

In bioreactor 6, with addition of *Pseudomonas putida, Trichoderma harzianum*, and *Aspergillus niger* with ratio 2:2:1, at the end of process caffeine was decreased from 0.86% become 0.06%, tannins was decreased from 4.81% become 0.08%, and polyphenol was decreased from 3.48% become 0.1%.

In bioreactor 7, with addition of *Pseudomonas putida, Trichoderma harzianum*, and *Aspergillus niger* with ratio 2:1:2, at the end of process caffeine was decreased from 0.86% become 0.08%, tannins was decreased from 4.81% become 0.07%, and polyphenol was decreased from 3.48% become 0.08%.

Decrease of caffeine from various variables can be seen in Fig 10. It shown that addition of microorganisms with a ratio 1:2:1 has the highest caffeine decreased of 99.88%.

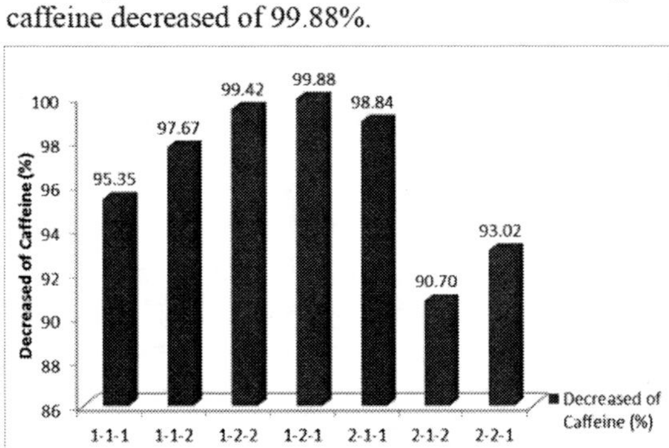

Fig 10. The correlation of Decreased Caffeine (%) on Various Ratio of *Pseudomonas putida : Trichoderma harzianum : Aspergillus niger*

Decrease of tannins from various variables can be seen in Fig 11. It shown that addition of microorganisms with a ratio 1:2:2 and 1:2:1 has the highest tannins decreased of 99.79%. Tej K. Bhat (1998) [5] explained that Trichoderma harzianum has ability to degrade tannins to gallic acid. At that ratio, Trichoderma harzianum has more ratio than Pseudomonas putida and Aspergillus niger, so Trichoderma harzianum is more effective in degrading tannins.

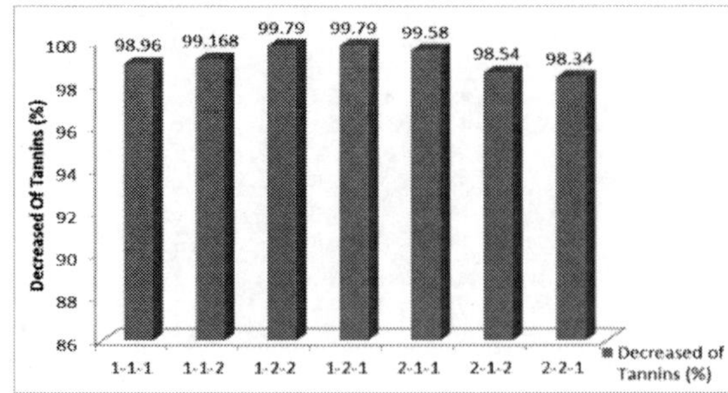

Fig 11. The correlation of Decreased Tannins (%) on Various Ratio of *Pseudomonas putida : Trichoderma harzianum : Aspergillus niger*

While decrease of polyphenol from various variables can be seen in Fig 12. It shown that addition of microorganisms with ratio 1:2:2 has the highest polyphenol decreased of 99.68%.

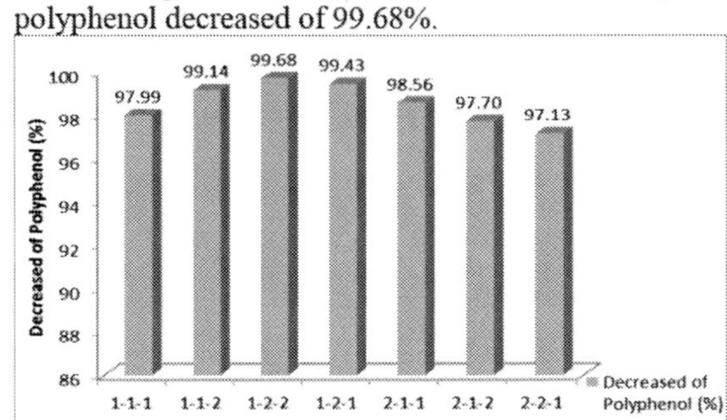

Fig 12. The correlation of Decreased Polyphenol (%) on Various Ratio of *Pseudomonas putida: Trichoderma harzianum : Aspergillus niger*

In addition to being seen in decreasing the amount of inhibitor substances also seen the cellulose produces at the end of the process. It is because cellulose plays an important role in the formation of biogas. Cellulose can be converted into methane gas with help of microorganisms. The amount of cellulose in various variables can be seen in the Fig 13.

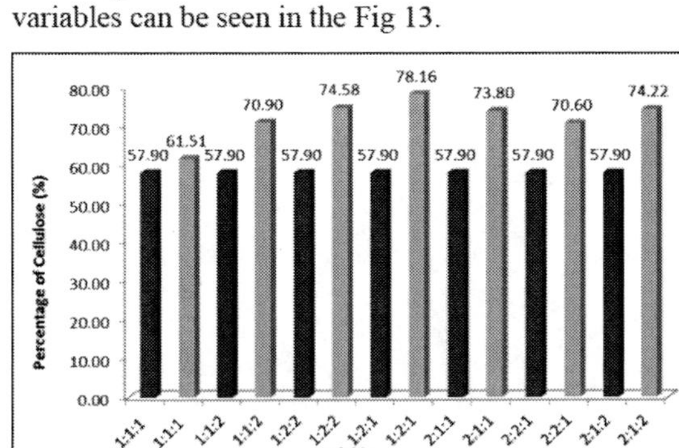

Fig 13. Percentage cellulose on Various Variables on Days 0 and Days 7

Fig 13 showed the increased of cellulose in various variables at the end process. The highest amount of cellulose in the addition of microorganisms with a ratio of 1:2:1 from 57.90% to 78.16%. So the ratio addition of microorganisms Pseudomonas putida : Trichoderma harzianum : Aspergillus niger is 1:2:1 was selected and will be used in biogas manufacture.

3.3 Manufacture of Biogas

In the manufacture of biogas, variable used is the ratio of the addition of cow dung and rumen fluid as starter with the ratio of cow dung : rumen fluid is 1:0, 0:1, 1:1, 1:2, and 2:1.

The results of the analysis related to the increase methane gas concentrations in the five biogas reactors to the fermentation time were shown in Fig 14.

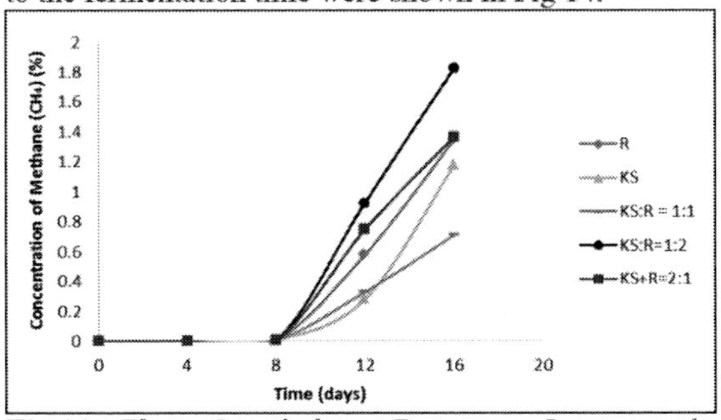

Fig 14. The Correlation Between Increase in Concentration of CH_4 (%) on Time

Fig 14 shows the correlation between methane gas concentration and fermentation time. Starter such as cow dung and rumen fluid use substrate as carbon source to do their metabolism and to reproduce. Fig 14 showed that in biogas reactor, methane gas increased slowly. On days 0 to days 8, methane gas has not formed. It is because microorganisms in cow dung and rumen fluid still adapt to the environment which there are toxic substances such as caffeine, tannins, and polyphenol from coffee pulp [1]. Furthermore, methane gas that has not been formed in the biogas reactor is due to microorganisms still at the hydrolysis and asidogenesis stage [5].

The results of biogas formation for 20 days on variable cow dung : rumen fluid is 1:2 with methane gas concentration of 1.825%. It is because the main microorganisms (Ruminococcus sp) in rumen fluid more and more specifically in degrading cellulose and organic material to methane (CH4) than microorganisms in cow dung [6]

According to previous research (Corro, 2013)[1], biogas that uses coffee pulp as substrate formed methane gas for 1 month fermentation time with methane gas formed very small. Benefit of pretreatment biologically on coffee pulp obtained results 1.825% within 16 days of fermentation. This indicate an improvement that has been made by previous research using coffee pulp without pretreatment to produce biogas.

The accumulated volume of biogas from day 0 to day 20 is shown in Fig 15. Methane gas volume has increased slowly.

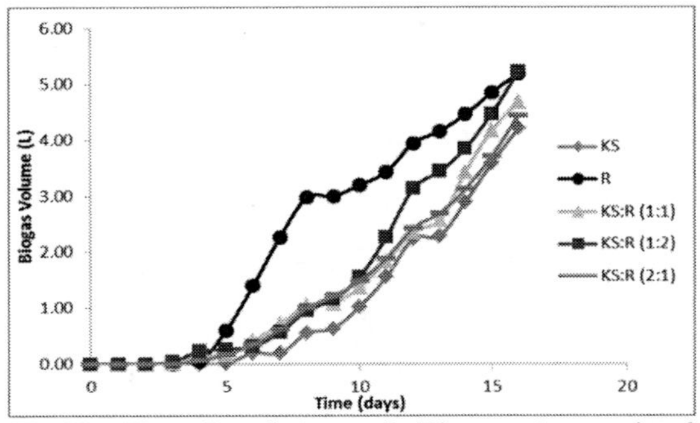

Fig 15. The Correlation of Biogas Accumulated Volume to Time (days) at Various Ratio of Addition Cow Dung : Rumen Fluid

On the days 0 to days 4 can be seen in all variables not yet formed biogas. After day 4, all variables begin to produce biogas. For the result of volume of biogas at the end of process is at ratio cow dung : rumen fluid is 1:0 of 4.61 L/days, at ratio 0:1 of 5.68 L/days, at ratio 1:1 of 5.12 L/days, at ratio 1:2 of 5.71 L/days, and at ratio 2:1 of 4.86 L/days. From the data obtained the highest volume of biogas accumulation on variable cow dung : rumen fluid at ratio 1:2. Thus, it can be calculated methane gas (CH_4) resulting from 0.0032 m^3/kg converted COD.

Tabel 2. Heating value on various variables in biogas reactor

Waktu (Hari)	KS			R			KS : R (1:1)		
	LHV (BTU/ft^3)	Densitas (lb/ft^3)	LHV (kcal/kg)	LHV (BTU/ft^3)	Densitas (lb/ft^3)	LHV (kcal/kg)	LHV (BTU/ft^3)	Densitas (lb/ft^3)	LHV (kcal/kg)
4	0.00	0.00	0.00	0.00	0.00	0.00	0.00	0.00	0.00
8	0.00	0.00	0.00	0.00	0.00	0.00	0.00	0.00	0.00
12	2.293	0.123	10.378	5.069	0.122	23.032	2.715	0.123	12.319
16	10.795	0.122	49.247	12.247	0.122	55.933	6.297	0.122	28.633

Waktu (Hari)	KS : R (1:2)			KS : R (2:1)		
	LHV (BTU/ft^3)	Densitas (lb/ft^3)	LHV (kcal/kg)	LHV (BTU/ft^3)	Densitas (lb/ft^3)	LHV (kcal/kg)
4	0.00	0.00	0.00	0.00	0.00	0.00
8	0.00	0.00	0.00	0.00	0.00	0.00
12	8.253	0.122	37.588	6.608	0.122	30.061
16	16.595	0.121	76.032	8.496	0.122	38.808

In measuring the value of heating value in biogas, composition gas greatly affect the heating value, especially methane gas. If composition of methane gas is high then the heating value obtained will be higher. The highest heating value of biogas in reactor is with variable cow dung : rumen fluid is 1:2 with heating value of 76.032 kcal/kg on end the process.

4 CONCLUSION

From the research that has been conducted, it could be concluded as follows :

1. Biological pretreatment by using microorganisms allows to degrade the inhibitor substance in coffee pulp with the ratio of addition of *Pseudomonas putida : Trichoderma harzianum : Aspergillus niger* of 1:2:1 is best composition for degrading the component. In the variable of 1:2:1, the amount of inhibitor substances decreased by 99.8% for

caffeine, 99.7% for tannins, and 99.4% for polyphenol.

2. Biogas with highest methane concentration was produced by a mixed variable between cow dung and rumen fluid (1:2) of 1.825% with 20 days of fermentation time. The highest heating value was obtained when the highest methane composition was 1.825% in variable cow dung: rumen fluid (1:2) of 76.032 kcal/kg.

We would like to thank the Chemical Engineering Department of Institut Teknologi Sepuluh Nopember Surabaya, especially to the Wastewater Treatment Laboratory crew.

REFERENCES

1. G. Corro, L. Panigua, U. Pal, F. Banuelos, M. Rosas, Energy Conversion and Management. (2013)
2. J. Erwanto, I. T. Hastari, *Studi Perbandingan Produksi Biogas Menggunakan Feses sapi, Cairan rumen dan feses luwak pada co-digestion*. (Jurusan Teknik Kimia. Fakultas Teknologi Industri. Institut Teknologi Sepuluh Nopember Surabaya, 2016)
3. P. Mazzafera, "Degradation of Caffeine By Microorganisms and Potential Use Of Decaffeinated Coffee Husk and Pulp in Animal Feeding". (2002)
4. K. B. Tej, S. Bhupinder, O. P. Sharma. "Microbial Degradation Of Tannin- A Current Prespective", (Regional Station, Indian Veterinary Research Institute Palampur, Biodegradation, 1998)
5. S. Yadvika, T. R. Sreekishnan, S. Kohli, V. Rana, Bioresour. Technol. (2004)
6. A. Takeneka, "The Properties of Rumen Microorganism and Their Contribution to Methane Production", (National Institute of Livestock and Grassland Science, Japan Technol., 2008)

Study on color removal of Sewage Treatment Plant (STP) effluent using granular activated carbon

Anita Nurfida, *I Nyoman* Widiasa*

Chemical Engineering Department, Diponegoro University, Semarang, Indonesia

Abstract Recycling of sewage treatment plant (STP) effluent is one of the attractive solutions to fulfill clean water for hotels and malls in Indonesia. STP effluent has average characteristics as follow: pH 6.8; color (true color) 107 PtCo; A_{254} (UV absorption) 0.36 cm^{-1} and COD 35.9 mg/L. Because of high color and organics contents, STP effluent needs further treatment such as activated carbon adsorption. Batch and continuous treatment of synthetic STP effluent by a commercial granular activated carbon (Filtrasorb 300) were investigated. The results show that adsorption capacity of the Filtrasorb 300 for color body of STP effluent was 184 PCU/g of carbon. Langmuir model is appropriate to describe the isotherm adsorption process in this study, with Q_m: 476.2 PCU/g of carbon. The results of isotherm adsorption model evaluation, SEM photograph and FTIR analysis show that color adsorption occurred was physical adsoption.

1 Introduction

Construction of hotels and malls have been increasing enthusiasticly in major cities in Indonesia such as Jakarta, Bandung, Yogyakarta and Bali. That phenomenon is not complemented with the availability of fresh water supply. Recycling of Sewage Treatment Plant (STP) effluent is one of the solutions in order to fulfill clean water supply for hotels and malls in Indonesia. After being processed, STP effluent are generally used for toilet flushing water [1] and watering the garden, irrigation, fire extinguishing and cooling tower [2]. STP effluent that has characteristics of pH, color (true color), A_{254} (UV absorption), and COD respectively 6.8; 107; 0.36 cm^{-1}; 35.9 mg/L [3], is a potential alternative source of clean water. Since its color and organic content are still high, advanced wastewater treatment is required. Color in STP effluent is caused by humic substances as a result of contact between water and Natural Organic Matter (NOM) [4]. Dissolved ions in water such as iron (Fe) and manganese (Mn) [5] can also contribute the color.

The adsorption process using activated carbon is one of the most frequently method to remove color and organic pollutants in water and wastewater. Mostly, color removal process is applied in the textile wastewater [5–9] since it generates high color waste. Organics removal in the STP effluent using activated carbon adsorption has been widely applied [10–15], but study on its color removal is still slighty done [3]. Color removal process is needed to meet the requirements of water quality set by the Ministry of Health No. 416/MENKES/Per/IX/1990 which the maximum amount is 50 TCU.

The objective of this paper is to determine the contact time and maximum adsorption capacity for color removal in the STP effluent. In the adsorption process, the contact time plays an important role for achieving maximum adsorption capacity, relating to the materials transfer to be absorbed into the pores of activated carbon. In addition, this paper aims to determine the isothermic adsorption models that appropriate with the color removal process in STP effluent.

2 Materials and methods

2.1. Preparation of activated carbon

Granular activated carbon (GAC) was purchased from Calgon (Filtrasorb 300). The GAC was throughly washed using demineralized water in order to clean its impurities. This process was conducted until the turbidity of rinse water below 1. Then GAC was dried at 105°C until the constant weight was achieved. Afterwards, it was stored in a desiccator. Morphological features of carbon were obtained with Scanning Electron Microscope-Electron Dispersive X-Ray (SEM-EDX) Jeol JSM-6510LA, Japan. The specific surface areas was calculated by the Brunauer-Emmett-Teller (BET) method while the pore size distributions (PSD) of the activated carbon particles were obtained using the nitrogen gas adsorption-desorption method with Autosorb-iQ, Quantachrome ASIQC0600-5, USA. The

* Corresponding author: widiasa@undip.ac.id

https://doi.org/10.1051/matecconf/201815603011

functional group of samples were determined by Fourier transform infrared spectroscopy (FTIR) IR Prestige 21, Shimadzu, Japan.

2.2 Preparation of synthetic effluent STP

Synthetic STP effluent was collected from the activated sludge treatment water cultured in wastewater laboratory of Chemical Engineering, Diponegoro University. As analogy of STP effluent, the activated sludge was settled for 30 minutes. The synthetic STP effluent has characteristics of pH, color (true color), and COD respectively 6, 59-115 PCU, and 31-93 mg/L. The effluent water characteristics was vary, depends on the prior biological treatment process (activated sludge).

2.3 Experimental procedure

2.3.1 Batch adsorption tests

Isotherm adsorption experiments were carried out by shaking 100 mL sample solution with activated carbon in 250 mL erlenmeyer flasks. Activated carbon (0,25-2 g) were immediately added to each flasks, and they were shaken on a rotary shaker at a constant temperature and 150 rpm constant velocity for 10-120 minute, which had been previously determined to be the time required to reach adsorption equilibrium [16]. All of the samples were extracted by filtering through 0.45 μm syringe filters to remove carbon particles. Effect of adsorbent dosage was studied with different adsorbent doses (0.25-2 g) and 100 mL of effluent synthetic solution and shaking for equilibrium time. Langmuir and Freundlich equations were employed to study the equilibrium adsorption.

2.3.2 Continuous adsorption tests

The adsorption process was conducted in a single set of apparatus consisting of two storage tanks (feed and product), a single centrifugal pump, and a carbon unit contactor (Fig. 1). Synthetic STP effluent as the feed was passed into the GAC contactors using a centrifugal pump while opening BV 02 and BV-03. The feed flow rate was varied by adjusting BV-01 to determine the contact time of the adsorption process. The product water was obtained by opening BV-05. Afterwards, the color and turbidity of product water were analyzed using Multiparameter Photometer HI 83099.

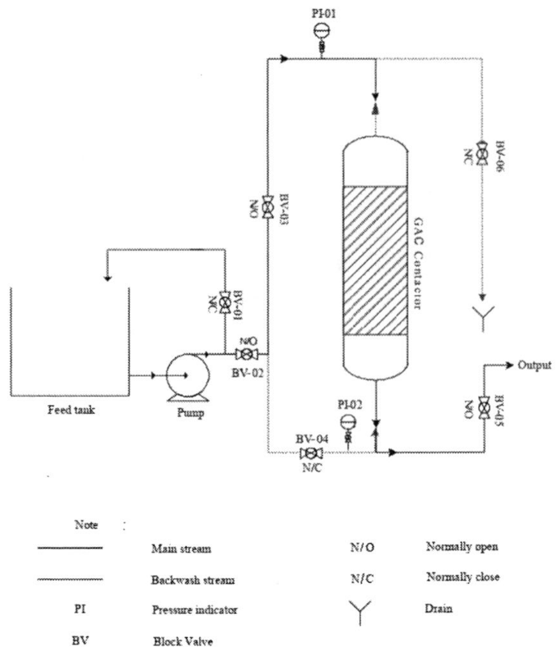

Fig. 1. Scheme of the continuous adsorption test

3 Result and discussion

3.1 Characterization

3.3.1 Scanning electron micrograph (SEM) studies

The aim of activated carbon characterization using SEM was to determine the surface morfology and pore structure of activated carbon. The pore size of activated carbon is divided into 4 types : submicropore (< 0,4 nm), micropore (< 2 nm), mesopore (2-50 nm) and macropore (> 50 nm). Based on the analysis of SEM at a magnification of 5.000x (a) and 20.000x (b), it is obtained that range of the new activated carbon pore size is 89-180 nm, including in the size range mesopore and macropores.

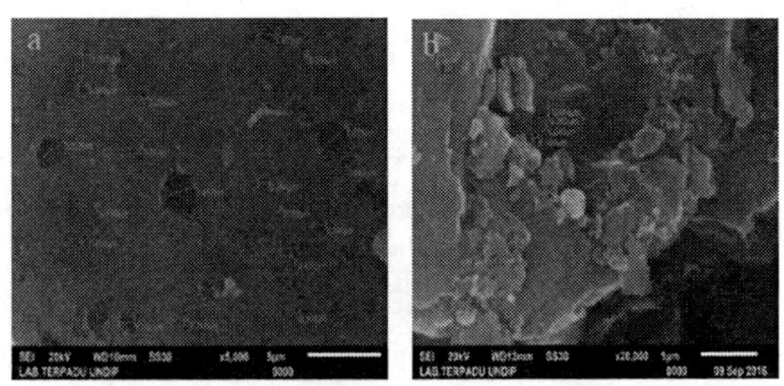

Fig. 2. SEM of new activated carbon

3.3.2 Brunauer-Emmett-Teller (BET) studies

Based on the BET analysis, it is obtained that the surface area of activated carbon is 956 m2/g with a pore radius

1.7 nm and a pore volume of 0.42 cc/g. An average pore diameter of the new activated carbon is 3.4 nm, including the size of the mesoporous. The results of this analysis in accordance with the characteristics of granular activated carbon which has surface area of 700-1300 m^2/g and a pore radius of 16-30 Å [17].

3.3.3 FTIR studies

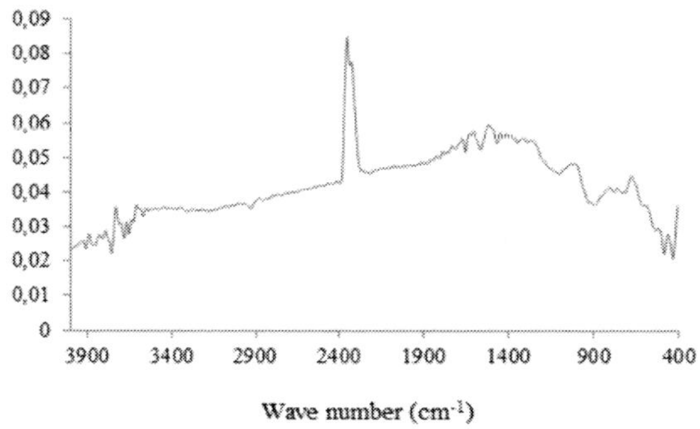

Fig. 3. FTIR new activated carbon

Based on FTIR analysis, the new activated carbon shows sharp peaks around wave number 678-2400 cm-1 (Figure 3). A sharp peak around the wave number of 600-1700 cm-1 (678.89 cm-1) indicates the absorption of the carbonyl functional group, amides and carboxylic. While peak located around the wave number of 950-1170 cm-1 is the strain-O functional group C polysaccharide [18]. The emergence of the peak around the wave number 1200-1280 cm-1 is the group C-O strain and deformation OH COOH (carboxylic) [18,19]. Another peak is observed in the wave number 1517-1590 cm-1 (1519.86 cm-1) relating to a functional group COO and deformation of the N-H bond [18]. Peak at wave number 1600 cm-1 (1608.58 cm-1) is an aromatic functional group C = C [19].

3.2 Batch studies

Batch adsorption process was carried out in a span of 0-120 minutes. During that time span, the adsorption capacity of activated carbon was investigated until its equilibrium and ultimately it was unable to adsorp anymore. Carbon mass is added to the effluent water of 0.25; 0.5; 1; 1.5 and 2 grams, this refers to a previous study using carbon dose of 1-10 grams in 100 ml of sample for color removal application in the form of textile waste water [8].

Results of analysis of the colors on the water output can be seen in figure 4, it appears that the color of the products 58, 34, 18, 12 and 9 PCU with carbon mass 0.25 respectively; 0.5; 1; 1.5 and 2 grams.

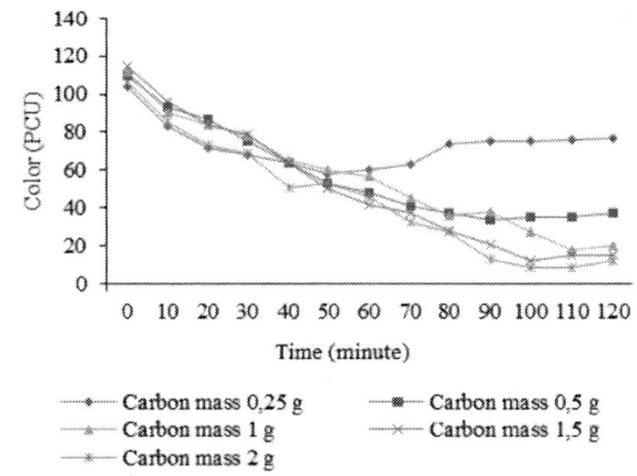

Fig. 4. Water color on various doses of carbon

Water color the resulting product is in conformity with the Ministry of Health less than 50 PCU on carbon mass 0.5-2 grams. At 0.25 gram mass optimum product color only reached 58 PCU, this may be due to the color of the absorbed fraction of excess, while the activated carbon as adsorbent has been exhausted absorptive capacity. A dose of 1 gram of carbon is the most appropriate dose for color removal in this study, it is suspected the fraction of the color components can be absorbed makismal by carbon. Equilibrium occurs at 110-120 minute time span, in which the color components are absorbed not declined against the contact time.

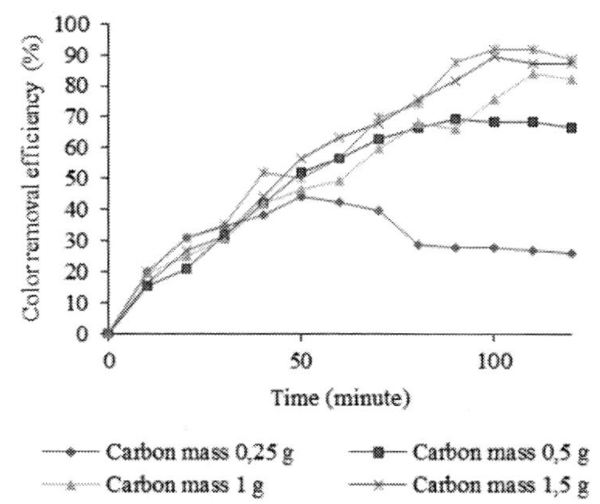

Fig. 5. Color reduction (%) on various doses of carbon

Color removal efficiency of 4.4 consecutive pictures: 44.23%; 69.09%; 83.93%; 86.96% and 91.51% in the carbon mass of 0.25; 0.5; 1; 1.5 and 2 grams. From the description it appears that the most effective color removal efficiency in carbon mass of 2 grams with an efficiency of 91.51%, which feed water color by 112 PCU and after 100 minutes the adsorption process, the color of the water effluent (product) to 9 PCU. This is because the components / fractions of color in the effluent water is absorbed into the pores of the carbon.

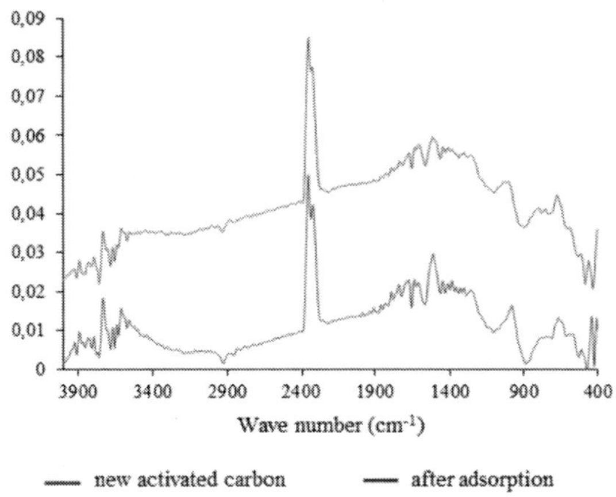

Fig. 6. FTIR new activated carbon and after adsorption

Based on the results of FTIR analysis before and after the activated carbon adsorption (figure 6) can be concluded that the adsorption occurs is physical adsorption, it is not visible from the discovery of new functional groups on the activated carbon after adsorption.

3.3 Equilibrium studies

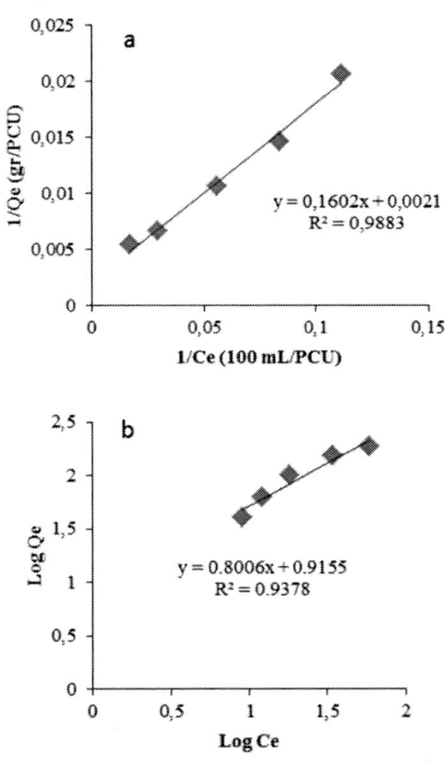

Fig. 7. Langmuir (a) and Freundlich (b) plot for color adsorption

From the equation obtained in the model Langmuir namely 0,1602x + y = 0.0021, we can determine the estimate of the adsorption capacity (Qm) amounted to 476.190 PCU/g of carbon. Langmuir models chosen for this model produced a value greater R^2 is 0.9883, while on the Freundlich models is 0.9378.

3.2 Continuous studies

In the continuous process, using three feed color variants of 59, 77 and 105 PCU, to observe the efficiency and ability of activated carbon in absorbing effluent water color. At the early of the continuous adsorption process, NOM (Natural organic matter) concentration is lower and contributes mainly to the reduction of color caused by micropolutan/organic. Increased flow rate and pressure have a large share in the reduction of color, this is related to the contact time of the adsorption process occurs. Along with the increase in pressure that affects the reduced contact time between the effluent and the adsorbent so as to reduce the efficiency of decreasing color. After a continuous process, pore blocking is more dominant and causes the capacity and kinetics of adsorption to be reduced [20].

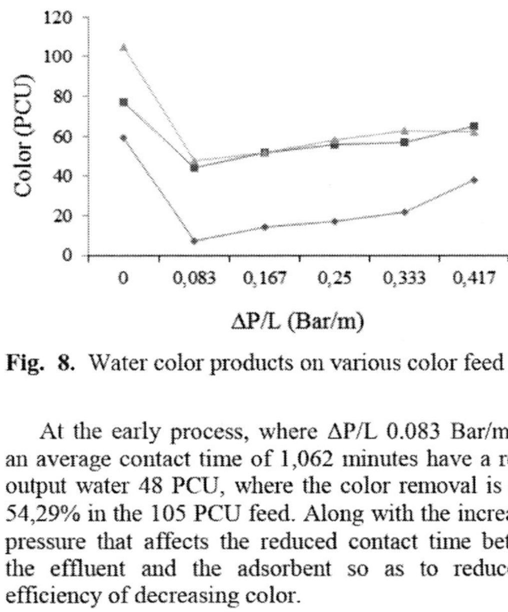

Fig. 8. Water color products on various color feed

At the early process, where ΔP/L 0.083 Bar/m with an average contact time of 1,062 minutes have a results output water 48 PCU, where the color removal is up to 54,29% in the 105 PCU feed. Along with the increase in pressure that affects the reduced contact time between the effluent and the adsorbent so as to reduce the efficiency of decreasing color.

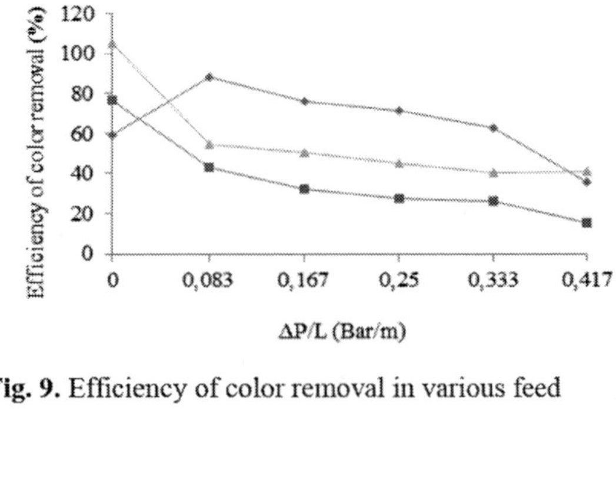

Fig. 9. Efficiency of color removal in various feed

Efficiency of color removal was 88.14%; 42,86% and 54,29%, on feed color variation 59, 77 and 105 PCU can be seen in figure 9. The greatest color removal efficiency in 59 PCU feed is 88,14% with color output reach 7 PCU, this value even according to PERMENKES drinking water quality standard that is maximum 15 PCU). Pretreatment process should be done first to get color feed lower than color for effluent water STP (105 PCU).

4 Conclusion

In the batch process obtained the largest adsorption capacity obtained at 0,5 grams of carbon mass where capacity reached 184 PCU/g of carbon. Langmuir model is most appropriate to describe the adsorption process isothermis, the maximum adsorption capacity of 476,190 PCU/g of carbon. Based on the results of FTIR analysis before and after the activated carbon adsorption can be concluded that the adsorption occurs is physical adsorption, it is not visible from the discovery of new functional groups on the activated carbon after adsorption. With two justifications the adsorption process that occurs in this study is physical adsorption. Increasing the flow rate and pressure has a big share in the reduction of color in a continuous process, it is related to the contact time of the adsorption process occurs. In ΔP/L 0083 Bar/m with an average contact time of 1,062 minutes is the optimum process variables in reducing the color of the synthetic STP effluent water produces 48 PCU.

References

1. M. A. Gual, Moia, and J.G. March. Journal of Desalination, **219**, 81-88 (2008).
2. N. Apostolidis, C. Hertle, and R. Young, Water, ISSN 20173 – 4441 (2011).
3. B. Zhu, and L. Zou, J. Environ. Manage., **90**, 3217-3225 (2009).
4. C. N. Sawyer, L. Perry, and F. P. Gene, *Chemistry for Environmental Engineering and Science*. (**5th** edition. Mc Graw Hill Companies, Inc., 2003).
5. R. Bulan, S. S. Mersi, and Darnianti. Indo. J. Chem. (2010).
6. S. Sirianuntapiboon, O. Sadahiro, and P. Salee, J. Environ. Manage., **85**, 162-170 (2007).
7. S. A. Al-Mamari, and D. G. Rao, Journal International Environmental Application & Science, **8**, 553-540 (2015).
8. A. Yadav, S. Mukherji, and A. Garg, Ind. Eng. Chem. Res.. (2013).
9. K. W. Jung, M. J. Hwang, D. S. Park, and K. H. Ahn, Sep. Purif. Technol., **146**, 154-167 (2015).
10. K. Y. Kim, H. Y. Kim, J. Kim, J. W. Nam, J. M. Kim, S. Son, Desalination, **243**, 132 – 144 (2009).
11. J. Altmann, A. S. Ruhl, F. Zietzchmann, Water Res., 185-193 (2014).
12. J. Altmann, F. Zietzschmann, E-L. Geiling, A. S. Ruhl, A. Sperlich, and M. Jekel, Chemosphere, **125**, 198-204 (2015).
13. J. Altmann, D. Rehfeld, K. Träder, A. Sperlich, and M. Jekel, Water Res., **92**, 131-139 (2016)
14. F. Bonvin, L. Jost, L. Randin, E. Bonvin, and T. Kohn, Water Res., **90**, 90-99 (2016).
15. G. Knopp, C. Prasse, T. A. Ternes, and P. Cornel, Water Res., **100**, 580-592 (2016).
16. Y. Matsui, S. Nakao, A. Sakamoto, T. Taniguchi, L. Pan, T. Matsushita, and N. Shirasaki, Water Res., **85**, 95-102 (2015).
17. Metcalf and Eddy. *Wastewater Engineering Treatment and Reuse*. **4th** edition. New York : Mc Graw-Hill Companies, Inc. (2003).
18. A. W. Zularisam, A. F. Ismail, and R. Salim, Desalination, **194**, 211-231 (2006).
19. K. W. Jung, B. H. Choi, M. J. Hwang, T. U. Jeong, and K. H. Ahn, Bioresour. Technol., **219**, 185-195 (2016).
20. J. Hu, S. Ran, B. Heijman, dan L. Rietveld, J. Env. Manage. **160**, 98-104 (2015).

Advanced Oxidation Processes (AOPs) for Refinery Wastewater Treatment Contains High Phenol Concentration

Alif Nurul Azizah[1] and *I Nyoman* Widiasa[2*]

[1]Chemical Engineering Department, Engineering Faculty,University of Diponegoro, Jln Prof. Soedarto, SH. Semarang, Central Java, Indonesia, 50239, Telp: (+62) 24 7460058
[2] Chemical Engineering Department, Engineering Faculty,University of Diponegoro, Jln Prof. Soedarto, SH. Semarang, Central Java, Indonesia, 50239, Telp: (+62) 24 7460058

Abstract. Petroleum Refinery wastewater is characterized by a high phenol content. Phenol is toxic and resistant to biological processes for treatment of the petroleum refinery wastewater. The combination of an AOP and a biological process can be used for treatment of the refinery wastewater. It is necessary to conduct a study to determine the appropriate condition of AOP to meet the phenol removal level. Two AOP configurations were investigated: H_2O_2 / UV and H_2O_2 / UV / O_3. From each process samples, COD, phenol and pH were measured. The oxidation was carried out until the targeted phenol concentration of treated effluent were obtained. The better result obtained by using process H_2O_2 / UV / O_3 with the H_2O_2 concentration 1000 ppm. After 120 minutes, the final target has been achieved in which phenol concentration of 37.5 mg/L or phenol degradation of 93.75%.

1 Introduction

Currently increasing of fuel consumption of about 5% per year make Indonesia have a very high fuel deficit if there is no additional oil processing capacity in the country. As a state-owned enterprise engaged in the energy sector, Pertamina continues to work hard in sustaining the availability of public fuel and achieving national energy security. Various oil and gas projects were established, one of which is the Residual Fluid Catalytic Cracking project, known as RFCC Project in Cilacap, Central Java. The RFCC refinery will increase premium production to 91,000 barrels per day or up 30,000 barrels per day (Pertamina.com).

The oil refinery consumes a minimum of 75.7 liters to 227.1 liters of clean water to process 1 barrel of crude oil [1]. The amount of waste water produced is about 41% of the amount of water used in the process [2,3]. From several previous laboratory tests it was found that refinery waste water had COD concentrations of 750-1600 mg / L, BOD about 300 mg / L, oil and mud concentrations 500-3000 mg / L, sulphide 13-17 mg / L, ammonia 14 mg / L, phenol around 960 mg / L, temperature around 30^{0}C and pH above 7 [4-6].

The use of AOPs is proven to be effective for wastewater treatment, but it tends to be expensive because intermediate oxidation is formed during the process and makes a long chemical degradation, requiring considerable energy [7]. AOPs technology is one or a combination of several processes such as ozone (O3), hydrogen peroxide (H_2O_2), UV radiation, titaniumoxide (TiO_2-photocatalyst), Fenton process, as well as several other processes that can produce hydroxyl radicals). OH radical is the active species that has the highest relative oxidation potential, which is about 2.8 V, bigger than ozone having an oxidation potential of 2.07 V. This makes the OH very easy to react with other compounds around it [8].

The UV / O_3 / H_2O_2 method as an alternative technology and further development of UV / O_3 and UV / H_2O_2 methods. The AOPs technology was developed to obtain a rapid and nonselective concentration or amount of hydroxyl radicals for processing and removing persistent contaminants, as well as pollutants of EDCs (endocrine disrupting contaminants). Oxidizing agents, H_2O_2 and O_3 are added to the body of the liquid to be treated, and then activated with UV radiation photon energy so that spontaneously generate hydroxyl radical (• OH).

RFCC wastewater has a high phenol content. Phenol is toxic and resistant to biological processes, and required further process in order to be biologically processed in the existing refinery wastewater plant [9]. The use of biological treatment processes will be constrained by high levels of phenol which interfere the nitrification process. Some toxic compounds that can inhibit the nitrification process include: cyanide,

* Corresponding author: widiasa@che.undip.ac.id

thiourea, phenol, aniline, and heavy metals such as silver, mercury, nickel, chrome, copper and zinc.

In order to be biologically processed, wastewater must contain Dissolved Oxygen (DO) about above 2 mg / L, COD 500-3000 mg / L, ammonia 20 mg / L, phenol 50-100 mg / L, temperature around 30^{0}C and pH about 5-10 [10-12]. In dealing with the problem proposed biological system of two stages. There was the addition of aeration tubs that require larger place and longer time [11].

The combination of AOPs processes and biological processes can be used in refinery wastewater treatment [13]. The main drawback of the AOPs process is the high cost due to large energy consumption. Selection of AOPs processes and operating conditions need to take into account the most extreme wastewater characteristics. From the existing wastewater data the process UV / H_2O_2 and UV / H_2O_2 / O_3 process have to be studied. The oxidation of organic pollutants in wastewater by AOP process utilizes the hydroxyl radical (OH) obtained by combining processes such as ozone O_3, UV, and H_2O_2 [14]. The number of OH radicals formed is influenced by the type of reaction and the presence of a strong oxidizer in the form of a mixture of hydrogen peroxide with ultraviolet light, a mixture of ozone with ultraviolet light and a mixture of ozone with hydrogen peroxide.

Phenol is classified in alcohol so can be oxidized to ketones, aldehydes and carboxylic acids [15]. Morkini et al., [16] states that carbon in ortho positions and the phenol compounds are electron donors so in such positions the electrons has high density. Hence the OH radical has a tendency to strike that position forming intermediate compounds of catechol, hydroquinone, and benzokuinon. The end product of the oxidation reaction is carboxylic acid and ketone which can be processed in biological process [17]. The phenol degradation reaction is a first-order reaction, so the oxidizing concentration and reaction time will direct the fast-moving and precise reaction to form carboxylic acids and ketones. The more oxidizing concentrations and reaction time may not necessarily produce the desired product, whereas to obtain low phenol content requires high energy and cost. It is necessary to conduct a study to determine the condition of AOPs process that meets the target.

The experiment will evaluate the processing of AOPs with UV / H2O2 and UV / H_2O_2 / O_3 process and determine the operating conditions of AOPs target effluent with phenol concentration 50-100 mg / L.

2 Experimental

2.1 Materials

This experimental uses synthetic and real wastewater. Synthetic wastewater was made by dissolving phenol PA with deionized water. The real wastewater come from Pertamina RU IV Cilacap. Hydrogen peroxide (technical). UV Lamp Model Viqua Sterilight Silver S12Q-PA UV System 15 gpm Serial 131219515, 110-240v 50/60 Hz. Ozonator model Viqua Sterilight Model S8Q-OZ/2 Serial 13060530, 110-240v 50/60 Hz; Fe_2SO_4; H_2SO_4 and NaOH (*Analytical Grade*) as pH controller.

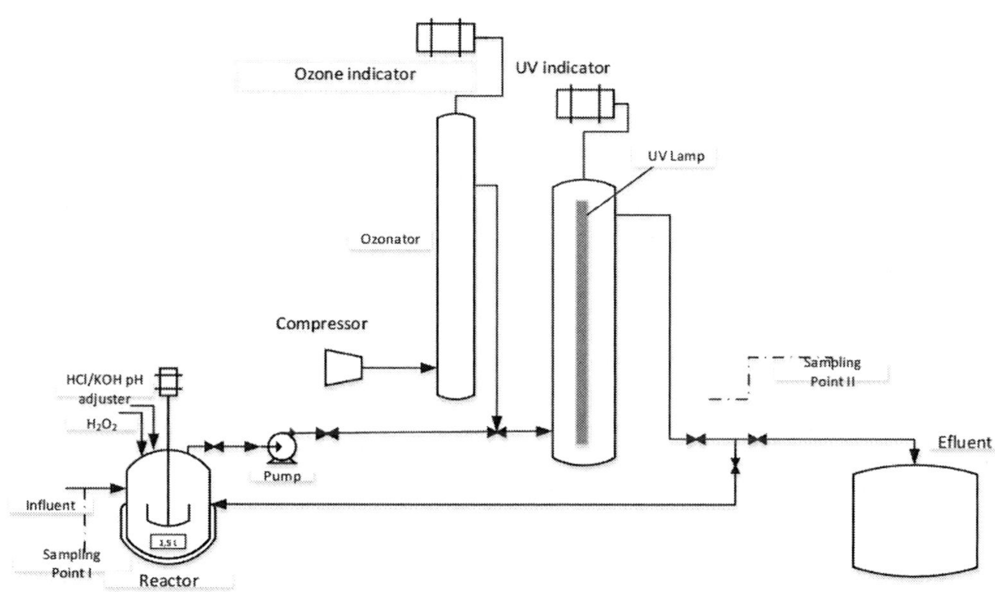

Fig 1. Advanced Oxidation Processess System

2.2 Method

Pertamina refinery waste water is first separated from solid impurities such as sand, stone, and solid waste. Prior to processing, wastewater was carried out by preliminary test, ie COD, phenol, and pH. A total of 20 liters of waste water goes into the reactor for AOPs processing using H_2O_2, ozone, and UV. Wastewater (influent) analyzed the COD, phenol, and pH content first. Then added H_2O_2 on various variables and arranged at pH using HCl and KOH. Influent is pumped into a UV process that is equipped with an ozone generator. (See Fig 1.).

After the process is done the analysis of effluent characteristics. The process is circulated with various time and hydrogen peroxide variables then analyzed effluent characteristics to obtain the required conditions to achieve the biological treatment on existing waste water treatment plant. The experiments were carried out with variations of the combination of H_2O_2 / UV and H_2O_2 / UV / ozone.

For Fenton system, prepare 250 ml wastewater on Schotts Bottle, then adjust pH into 3 by adding H_2SO_4. Add the Fe_2SO_4 catalyst then hydrogen peroxide for each variable. This experiment run with time variable from30 until 180 minutes. The results analyzed as like as the AOPs process.

2.2 Analytical

Influent and Effluent are analyzed Phenol, COD, and pH using Hanna Instruments HI 3864, HI 93754C-25 HR, and HI 8424 in order.

3 Results and discussion

3.1 UV/H_2O_2 process

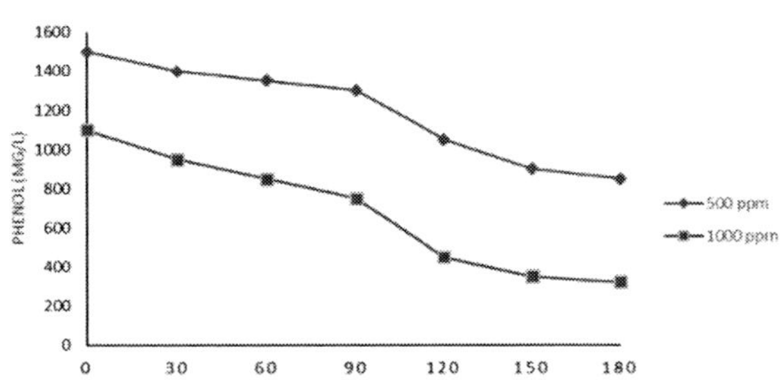

Fig 2. Phenol Degradation for synthetic RFCC wastewater by adding 500 mg/L H_2O_2 and 1000 mg/L H_2O_2

Fig 2. illustrates the experimental values of phenol final concentration in those wastewaters which have been put under advance oxidation process for reaction time 180 minutes. It can be concluded from Fig. that increasing the reaction time and the higher H_2O_2 concentration can lead more phenol degradation and reduction the final concentration of phenol in wastewater. The increase in H_2O_2 concentration increased the phenol degradation and COD removal. Both 500 mg/L and 1000 mg/L H_2O_2 have not reach targetted phenol concentration yet. The initial phenol concentration 1100 mg/L decreases to 320 mg/L and COD removal of 2090 mg/L from initial COD 2186 mg/L.

H_2O_2/UV process was chosen for this purpose because it is known to be simple and effective for the removal of moderate to low amounts of nonbiodegradable organic compounds. Eliminate the ozon process means energy and cost saving, so it will be reliable if H_2O_2/UV process has similar result with H_2O_2/UV/O_3 process. Hence, it can be concluded that the H_2O_2/UV process needs longer time and higher H_2O_2 concentration removed the phenols from the solution to levels below the target.

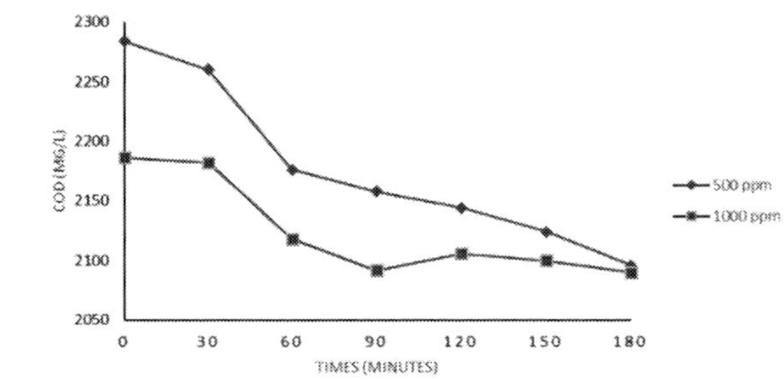

Fig 3. COD removal for synthetic RFCC wastewater by adding 500 mg/L H_2O_2 and 1000 mg/L H_2O_2

Figures 3. We can see that the COD removal show the descending trend againt increasing the H_2O_2 concentration. By adding 500 mg/L H_2O_2 concentration just remove a little amount of COD. While adding more concentration to 1000 mg/L, COD removal reach 2090 mg/L in 180 minutes. From the figures above can be concluded that the COD removal show the descending trend againt increasing the H_2O_2 concentration.

However, when H_2O_2 concentration exceeded the optimum concentration, amount of phenol degradation and COD removal was reduced. Although scavenging effect of hydrogen peroxide towards •OH occurs at higher concentrations [18], this is expected to give asymptotic values for phenol degradation. But reduction in phenol and COD removal observed may be due to the reaction of •OH and H_2O_2 and combination two •OH to form H_2O_2 [19]. In the other hand, the formation of •OOH which are significantly less reactive than hydroxyl radicals also reduces the reaction rate.

3.2 UV/H_2O_2/O_3 process

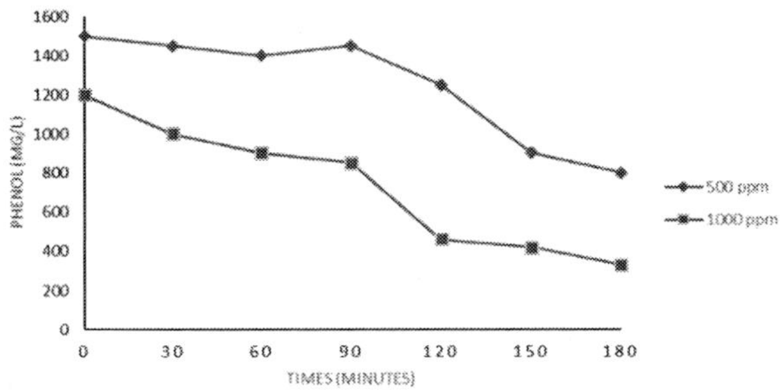

Fig 4. Phenol Degradation for synthetic RFCC wastewater by adding 500 mg/L H_2O_2 and 1000 mg/L H_2O_2

The effect of H_2O_2 concentration on the phenol degradation are despicted in Fig 4. It can be seen from the figures that the phenol concentration show the descending trend againt increasing the H_2O_2 concentration. By adding 500 mg/L H_2O_2 we get phenol degrades from 1500 mg/L to 800 mg/L. Then by adding 1000 mg/L H_2O_2 we get phenol degrades from 1200 mg/L to 330 mg/L In other hand COD removal on synthetic RFCC wastewater a bit lower too. COD decreases from 2360 mg/L to 2130 mg/L with 500 mg/L H_2O_2 and 2360 mg/Lto 2072 mg/L by adding 500 and 1000 mg/L H_2O_2 respectively.

The phenol degradation using the H_2O_2/UV and H_2O_2/UV/O_3 are not having significant result. Phenol degradation using H_2O_2/UV process has efficiency 70,9 % then using H_2O_2/UV/O_3 process has efficiency 72,5 %. Irradiation of H_2O_2 with UV generated •OH radicals with powerful oxidizability, so the time for phenol degradation was higher those for H_2O_2/UV/O_3 The •OH radicals generated by photocatalysis were distributed more widely than the •OH radicals generated from dissolved ozone [20]. Ozone is unstable in solution, and •OH radicals are generated by an autolytic process [21].

$O_3 + OH^- \rightarrow HO_2^- + O_2$ (1)
$O_3 + HO_2^- \rightarrow O_3- + HO_2$ (2)
$O_3^- + H^+ \rightarrow HO_3$ (3)
$HO_3 \rightarrow •OH + O_2$ (4)
$O_3^- + (H_2O) \rightarrow •OH + O_2 + OH^-$ (5)

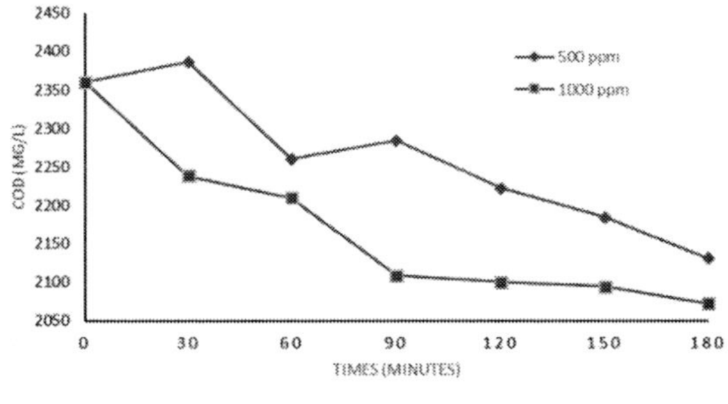

Fig 5. COD removal for synthetic RFCC by adding 500 mg/L H_2O_2 and 1000 mg/L

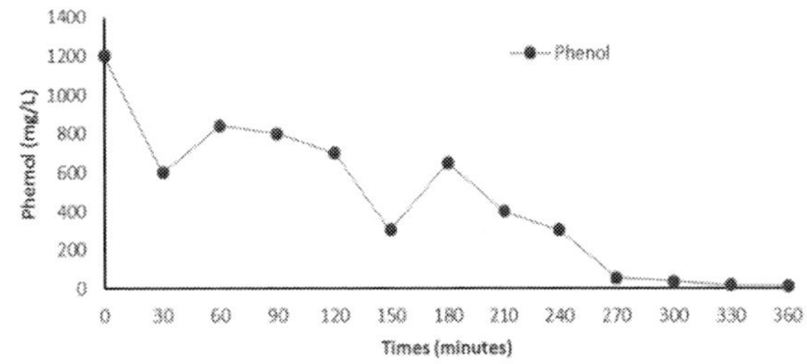

Fig 6. Phenol degradation for real RFCC by adding 500 mg/L H_2O_2

From the figure 6 we can see phenol degradation on real RFCC wastewater mostly symmetrical with the synthetic RFCC wastewater. Although we get a lower degradation there. The reason for this variation could be explained by the following reaction :

$H_2O_2 + hv \rightarrow 2OH•$ (6)
$H_2O_2 + OH• \rightarrow HO_2• + H_2O$ (7)
$O_3 + H_2O_2 + hv \rightarrow 2OH• + 3O•$ (8)

Theorically, if the initial H_2O_2 concentration increases, higher hydroxyl radicals can be produced and phenol can be converted to CO_2 and H_2O more easily.

Fig 6. illustrates the experimental values of phenol concentration in RFCC wastewater which have been put under H_2O_2/UV/O_3 process for different reaction time. It can be conclude from Fig 6. that increasing reaction time will lead more phenol degradation and reducing the final concentration of phenol.

However there is an targeted condition of hydrogen peroxide concentration and reaction time. We see on Fig 6, adding 500 mg/L hydrogen peroxide have not reach target yet. The lowest phenol concentration reach at reaction time 150 minutes with 300 mg/L or degraded 57% from the initial concentration. While, by adding more 500 mg/L hydrogen peroxide on the next reaction reached to 55 mg/L along reaction time 270 minutes. The Ghaly et al. experimental observation confirmed the above statement. It can be clearly seen from Fig 2. that high initial concentration of H_2O_2 consumes the hydroxyl radicals according to desired reaction (7) and reduces the phenol degradation extent (increases the final concentration of phenol as despicted in Fig 6.)

The degradation mechanism of the real RFCC wastewater organic compounds, mostly aromatic-containing structures, is an electrophilic •OH addition to the rings that results in benzene ring conversion to phenol and then to dihydroxybenzene derivatives [22]. Both benzene and phenol produces dihydroxybenzene derivatives, along with quinines and several ring-opened products (namely carboxylic acids groups) [22;23]. This implies that as phenol is degraded, more is generated by the degradation of benzene. The direct reaction of •OH with the phenol molecules also gives rise to phenoxy radical. This radical, as detailed by Hosseini et al. (2007), is in resonance with radical structures in *o*- and *p*- positions [24] and becomes the starting point for the formation of different intermediates. More hydroxylated compounds such as catechol, hydroquinone, benzoquinone or hydroxyl hydroquinone are then produced from subsequent reactions of the intermediates with •OH [25]. So, the degradation rate are less than the synthetic RFCC wastewater.

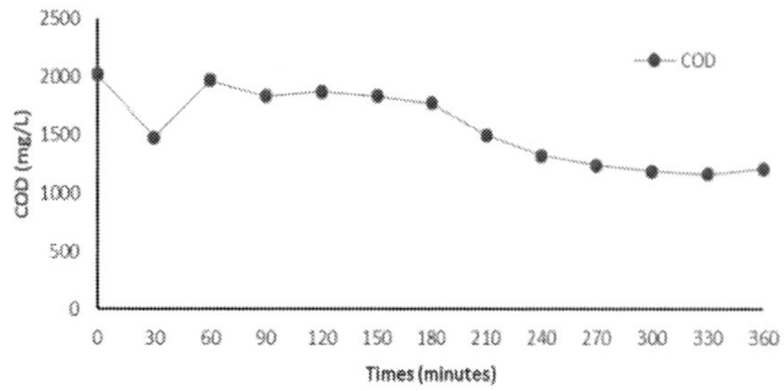

Fig 7. COD removal for real RFCC by adding 500 mg/L H_2O_2

Fig 7. illustrates COD removal as a function of time at different H_2O_2 concentration. COD removal are approximately symmetrical with the phenol degradation. The plots show that increasing reaction time will lead more COD removal and reducing the final concentration COD. By adding 500 mg/L H_2O_2 concentration just remove a little amount of COD typically same with the phenol degradation has not reach the target yet with this condition. While adding more concentration 500 mg/L, COD removal reach 1166 mg/L in 3300 minutes.

From the figures above can be seen from that the COD removal show the descending trend againt increasing the H_2O_2 concentration. Removal effieciencies are reduced at 180 minutes reaction time. It may cause by the alkaline media that dissociated form of hydrogen peroxide (HO_2^-) reacts with hydroxcyl radicals (reaction 2.) more than two orders of magnituted faster than observed hydrogen peroxide, leading loss of carbon oxidation potential which is confirmed by a slight decrease in COD removal under alkaline condition [26].

4 Conclusion

Phenol degradation in RFCC wastewater using AOP was found to be possible. For $UV/H_2O_2/O_3$ process target phenol concentration 55 mg/L easily reached by adding 500 mg/L H_2O_2 along 180 minutes then adding more 500 mg/L H_2O_2 along reaction time 90 minutes. For H_2O_2/UV process both 500 mg/L and 1000 mg/L H_2O_2 have not reach targetted phenol concentration yet. The initial phenol concentration 1100 mg/L decreases to 320 mg/L and COD removal of 2090 mg/L from initial COD 2186 mg/L. It can be concluded that the H_2O_2/UV process needs longer time and higher H_2O_2 concentration removed the phenols from the solution to levels below the target. In other hand COD removal on synthetic RFCC wastewater a bit lower too. This implies that as phenol is degraded, more is generated by the degradation of benzene. The direct reaction of •OH with the phenol molecules also gives rise to phenoxy radical.

The research supported by PT. Pertamina RU IV Cilacap, is greatly acknowledge. The author acknowledge the laboratory equipments provided by Chemical Engineering Departement of Diponegoro University Semarang that have resulted in this article.

References

1. C.H. Guernsey & Company, Water/Energy Sustainability Forum, (2009)
2. T. Doggett,, A. Rascoe, *Global Energy Demand Seen up 44 Percent by 2030,* (2009).
3. C. Marcilly, J. Catalyst. **216**, 47–62 (2003)
4. A. D. al deen A. P. Palaniandy, A. H. Bin Abdul, F. Shaik, J. Env. Chem. Eng., **3** 1117-1124, (2015)
5. A. M. Mansouri, F. Shahrezaei, A. A. L. Zinatizadeh, A. A. Hemati, M. Pirsaheb, K. SharafiJ. Taiwan Inst. Chem. Eng. **45**; 2501-2510 (2014)
6. Z. Wasi, Khan, I. Najeeb, M.Tuiyebayeva, Z.Makhtayeva, Process Saf. Environ. Prot., (2014).
7. I. Muñoz, J. Rieradevall, F. Torrades, J. Peral, X . Doménech, Solar Energy,**79**, 369–75, (2005)
8. C. Gottschalk, JA. Libra, A. Saupe, *Ozonation of water and waste water: a practical guide to understanding ozone and its applications*, John Wiley & Sons, (2009)
9. M. Lapertot, C. Pulgarín, P. Fernández-Ibañez, MI. Maldonado, L. Pérez-Estrada, I Oller, *Water Res*; **40**:1086–94, (2006)
10. G. Bitton., *Wastewater Microbiology*. New York: A John Wiley & Sons, INC., (1994)
11. I. Oller, S. Malato , J.A. Sánchez-Pérez, Sci. Total Environ. **409,** 4141–4166, (2011)
12. Anonim, Data Desain WWTP-Ipal, Pertamina, (2017)
13. V. Sarria, M. Deront, P. Péringer, C. Pulgarin, Appl. Catal. B. Environ., **40**, 231–46, (2003)
14. W. H. Ghaze, J-W. Kang, D.H. Chapin, *The chemistry of water treatment processes involving ozone, hydrogen peroxide and ultraviolet radiation*, (1987)
15. Isyuniarto, W. Usada, Suryadi, A. Purwadi, Degradasi Fenol dalam Limbah Pengolahan Minyak Bumi dengan Ozon, P3TM BATAN Yogyakarta (2003).
16. A. Morkini, D. Oussi, E. Chamarro, S. Esplugas, Photoxidation of Phenol Uv Radiation Aqueous Solution, http://www.photobiology.com/IUPAC98/esplugas/phenol.html. (1998)
17. M. Y. Ghall, G. Hartel, R, Hasender, Photochemical oxidation od p-chlorophenol bu UV/H2O2 and photo-Fenton process. *A comparative study, Waste Management* (New York, N.Y.) **21** 41-7, 11150131, (2001)
18. B. Subramanian, Q. Yang, Q. Yang, A.P. Khodadoust, D.D. Dionysiou, J. Photochem. Photobiol. A: Chem. **192** 114–121, (2007)
19. X.R. Xu, X.Y. Li, X.Z. Li, H.B. Li, Degradation of melatonin by UV, UV/H2O2, Sep. Purif. Technol. **68** 261–266, (2009)
20. Harufumi S., Sadao A., Hideki Y., Journal of Water Process Engineering **7** 54–60, (2015)

21. Bai C., Xiong X., Gong W., Feng D., Xian M., Ge Z., Xu N., Desalination **278** 84–90 (2011).
22. Aranda, E., Marco-Urrea, E., Caminal, G., Arias, M.E., García-Romera, I., Guillén, F., J. Hazard. Mater. **181**, 181–186, (2010)
23. Z. Zhang, Q. Xiang, H. Glatt, K. L. Platt, B. D. Goldstein, G. Witz,, Free Radic. Biol. Med., **18**, 411–419, (1995)
24. S. N. Hosseini, S. M. Borghei, M. Vossoughi, N. Taghavinia, Appl. Catal. B: Environ., **74**, 53–62, (2007)
25. A. M. Peiró, C. Domingo, J. PeralX. Domènech, E. Vigil, M.A. Hernández-Fenollosa, M. Mollar, B. Marí, J.A. Ayllón, Thin Solid Films **483**, 79–83, (2005)
26. K. Kosala, H. Yamada, K. Shisida, S. Echigo, R.A. Minear, H. Tsuno, M. Saburo, Water Res., **35** 3587, (2001)

A decision modelling approach for selection of biological nutrient removal systems for wastewater

Carla Mae Pausta[1], *Ramon Christian* Eusebio[2], *Arnel* Beltran[1], *Aileen* Huelgas-Orbecido[1], and *Michael Angelo* Promentilla[1,*]

[1]Chemical Engineering Department, De La Salle University, Manila City
[2]Department of Engineering Science, University of the Philippines Los Baños, Laguna

Abstract. This paper proposes a decision model built on a hierarchical network for optimal selection of biological nutrient removal systems (BNR) in wastewater treatment plants. BNR is an important component of a sustainable wastewater management wherein resource recovery from wastewater becomes an integral part of the municipal wastewater treatment plants (WTP). However, selection of the most appropriate technology or systems requires a multiple criteria analysis. This study focuses on the following criteria namely 1) Economic aspect; 2) Technical aspect; 3) Environmental Aspect; and 4) Space Requirement. The following alternatives were then evaluated: 1) 3 Stage Pho-redox (A2O); 2) 5 Stage Bardenpho (5BP); 3) University of Cape Town (UCT); 4) Virginia Initiative Plant; 5) Sequencing Batch Reactor (SBR); 6) Membrane Bioreactor (MBR). A fuzzy ANP approach with Monte Carlo simulation was used to derive the overall priorities of these alternatives. This decision modelling approach addresses the uncertainty and complexity involved in the selection of appropriate BNR in Metro Manila's WTP.

1 Introduction

Nitrogen (N) and phosphorus (P) are the major nutrients needed in order to sustain and improve growth of plants both in land and in water. While nutrients in the form of fertilizers create beneficial effects to agricultural land, conversely, excessive nutrient concentration in the water bodies causes adverse effects to the aquatic biodiversity. This leads to eutrophication which is the accelerated growth rate of aquatic plants and algae due to increased availability of nutrients. Algal bloom and excessive growth of aquatic plants limit light penetration and deplete dissolved oxygen which are needed to support aquatic organisms [1]. Common sources of nutrient pollution in the waters are storm water run-off, wastewater effluent and agricultural water run-off. With that, efforts are being made to address this problem by researchers and industries involved. In order to control the nutrient concentration in the wastewater effluent, different technologies were developed including biological nutrient removal (BNR) systems and nutrient recovery systems [2-4]. Several countries have already started the monitoring of nutrient content in water and wastewater discharges especially in their wastewater treatment plants (WTP) [2-3].

The domestic and industrial WTPs in the Philippines are still on the planning stages of implementation of BNR technologies. The Department of Environment and Natural Resources (DENR) released a new Department Administrative Order (DAO) 2016-08 last May 2016 indicating the new water quality guidelines and general effluent standards starting 2016 2016 with the inclusion of ammonia-N (NH3-N), Nitrate-N (NO3-N) and Phosphorus as Phosphates among the monitored parameters [5]. The government gave a grace period of not more than five years to comply with the new DAO. Because of that, WTPs need to upgrade their current system that will integrate BNR technologies in order to comply with the new standards.

The challenge faced by the stakeholders is the selection of the best BNR system to be installed in municipal WTPs located in urban areas. Water and wastewater treatment systems are complex and dynamic since treatment is influenced by interactions of factors such as water quality, regulatory requirements, consumer and environmental concerns, construction challenges, operational constraints, and economic feasibility [6]. The United States Environmental Protection Agency (USEPA) identified the following barriers in the implementation of BNR system: 1) costs; 2) limitations on physical expansion; 3) state resources; 4) increased carbon footprint; 5) advanced operations and control [4]. Industries with WTPs are now faced with a complex decision making problem to evaluate the best BNR alternative considering multiple criteria.

This study will focus on the selection of the optimum BNR system using Fuzzy Analytical Network Process (FANP), a Multi-criteria Decision Analysis (MCDA)

* Corresponding author: michael.promentilla@dlsu.edu.ph

tool, with uncertainty analysis. MCDA is a tool that can provide decision support to complex and multifaceted problems considering the interconnection of various factors [6-8]. Common MCDA methods used in different environmental decision making problems are Multiattribute Utility Theory (MAUT), Analytical Hierarchical Process (AHP) and Analytical Network Process (ANP). AHP utilizes pairwise comparisons of all criteria and alternatives to institute relationships within the hierarchical problem structure [9]. ANP is the generalization of AHP which considers the interactions and dependence between the elements of clusters or networks [10]. A study was made to evaluate the best BNR technology for WTP in urban areas using ANP [11]. However, the ambiguity in the value judgements provided by experts and uncertainties involved were not considered. In this paper, Monte Carlo simulations are done to model the variabilities caused by subjective judgements and coarse quantitative data [12].

2 Methodology

The study proposed a decision modelling approach utilizing the fuzzy concept in Analytical Network Process coupled with Monte Carlo simulations in order to solve the complex decision making problem as summarized in Fig. 1.

Setup the Network Structure.

Obtain the triangular fuzzy numbers through value judgements from experts and data collected from literature.

Calculate the priority vectors using the nonlinear fuzzy preference programming approach

The supermatrix containing the priority vectors will be normalized per cluster through row sum then raise to large powers until it converges to a limit.

Evaluate the uncertainties using Monte Carlo simulations.

Fig. 1. Methodology flow chart

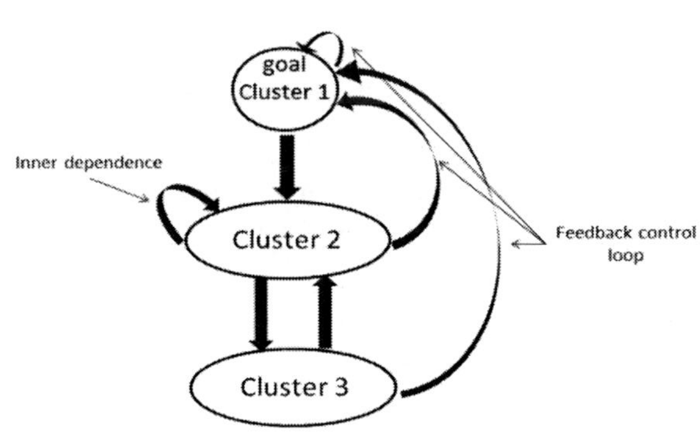

Fig. 2. An example of ANP network structure

The first phase was to setup a network structure to represent the problem as shown in Fig. 2. The network structure has a top to bottom hierarchy followed by feedback from alternatives to criteria, inner dependence of all criteria, and feedback control loop as represented by the direction of the arrows and arc. The feedback control loop indicates that all the elements in the structure are important and influenced by the goal element [13].

The second phase was to gather pairwise comparisons of all interacting elements in the network structure through value judgements from experts and data collected from literature. The relative importance between criterion i and criterion j is represented by a triangular fuzzy number (TFN) which is described by <lower bound, modal value, upper bound> ($<l_{ij},m_{ij},u_{ij}>$) to address the ambiguity of the value judgements (a_{ij}) provided by the experts [12]. The lowest relative judgement provided by an expert was considered as the lower bound (l_{ij}) value while the highest judgement was considered as the upper bound (u_{ij}) value. The modal (m_{ij}) value was obtained by calculating the geometric mean of the relative value judgments. The geometric mean method was used to get the aggregated value of the expert judgments taken from multiple decision makers [14]. The pairwise comparison matrix is presented in eq. (1).

$$\hat{A} = \begin{bmatrix} \langle 1,1,1 \rangle & \hat{a}_{12} & \cdots & \hat{a}_{1n} \\ \hat{a}_{21} & \langle 1,1,1 \rangle & \cdots & \hat{a}_{2n} \\ \vdots & \vdots & \ddots & \vdots \\ \hat{a}_{n1} & \hat{a}_{n1} & \cdots & \langle 1,1,1 \rangle \end{bmatrix} \quad (1)$$

where $\hat{a}_{ji} = \langle l_{ij}, m_{ij}, u_{ij} \rangle$; $\hat{a}_{ji} = \left\langle \frac{1}{u_{ij}}, \frac{1}{m_{ij}}, \frac{1}{l_{ij}} \right\rangle$

The third phase involved the calculation of the priority weights using the nonlinear fuzzy preference programming approach described in the paper of Promentilla et. al. [15]. The approximate ratio-scale weights (w_i) were obtained, wherein the value of λ was maximized. The λ value was used as a measure of consistency where a value of 1.0 means perfect consistency while a value of 0.0 indicates that the fuzzy judgments are only satisfied at their boundaries [16].

In the fourth phase, the normalized priority vectors were arranged in a supermatrix, ***S***, that represents the network structure of the decision making problem to calculate for the overall priority weights. The initial supermatrix or the unweighted matrix is composed of priority weights obtained from the previous step. This was normalized by the row sum. The supermatrix was then raised to large powers until it converged into a limiting value [17]. The resulting answers were normalized per cluster to get the overall priorities.

The last phase was to evaluate the uncertainty in the computation of the overall priority weights using Monte Carlo simulations [18]. This will also address the ambiguity of value judgments provided and probable imprecise quantitative data from different literatures.

3 Case Study

This study evaluates the best biological nutrient removal (BNR) system to be installed in existing wastewater treatment plants (WTPs) located in the urban areas based on the following criteria: (1) economic aspect (EC); (2) technical aspect (TL); (3) environmental aspect (EN); and (4) space requirements (SR). EC considers lower upgrade costs (i.e. construction costs, etc.). The economic aspect data of alternatives to be evaluated are taken from different literatures [19-21]. The cost data are then normalized using the Engineering News Record Construction Cost Index [19]. TL considers the robustness of the processes and the technical capabilities of the assigned personnel. EN is being evaluated by the effluent quality especially but not limited to lower nitrogen and phosphorus concentration in the effluent. The effluent quality must pass the new standards set by the new DAO 2016-08 at the minimum. However, for the comparison of elements with respect to EN, priority must be given to the system with effluent having the least nutrient concentration. SR refers to the physical expansion and upgrade of conventional WTP when the BNR is integrated. In cities, addition of space-consuming technology such as BNR systems must also be considered as an important factor. In this criterion, the priority must be given to the alternative with lesser space requirement.

The following are the six alternatives to be evaluated: 3 stage pho-redox (A2O), 5-stage Bardenpho (5BP), University of Cape Town (UCT), Virginia Initiative Plant (VIP), Sequencing Batch Reactor (SBR) and Membrane Bioreactor (MBR). A2O is a conventional activated sludge system involving anaerobic-anoxic-aerobic zone. In order to enhance denitrification, the nitrate-rich liquor is recycled from the aerobic zone to the start of the anoxic zone. The return activated sludge (RAS) is returned in the anaerobic zone. The 5BP consists of five stages with anaerobic-anoxic-aerobic-anoxic-aerobic zones. The nitrate-rich liquor is recycled to the first anoxic stage from the first aerobic stage while the RAS is recycled to the beginning of the anaerobic stage from the clarifier. UCT also consists of three stages, anaerobic-anoxic-aerobic stage. However, the RAS is returned from the clarifier to the anoxic stage while the nitrate-rich liquor is recycled from the aerobic stage to the anoxic stage [4]. VIP having the same anaerobic-anoxic-aerobic stages recycles its RAS to the anoxic stage where denitrification occurs before entering the anaerobic stage [22]. SBR is a batch operation of anaerobic-anoxic-aerobic stages through adjustment of mixing and aeration [4]. MBR utilizes membranes for the nutrient removal [23].

The complex decision making problem for the optimum selection of BNR system can be summarized in a network structure shown in Fig. 3 which considers the aforementioned set of criteria and six possible alternatives. Since the selection of the BNR system is influenced by a cluster of criteria, it is represented by an arrow from the goal cluster to the criteria cluster, denoted by W_{21} in the supermatrix, ***S***. However, the evaluation of these criteria may also be influenced by other criteria, thus the inner dependence represented by the arc loop must also be considered in the supermatrix as W_{22}. The preferred alternative with respect to each criterion, W_{32}, is also being assessed as represented by the arrow from the criteria cluster to the alternative cluster. On the other hand, the evaluation of each alternative may be influenced by the dominating criteria, thus the feedback is considered as represented by the arrow from the alternative cluster to the criteria cluster in the submatrix W_{23}. Since there are no dependencies between the alternatives, it will be assigned as a null submatrix, W_{33}, in the supermatrix. A null matrix is also assigned to W_{31} because the goal is manifestly not dependent to the each of the alternatives. The feedback control loop is represented by the arcs pointing to the goal cluster from the goal, criteria and alternative clusters. This indicates that all the elements in the structure are influenced by the single controlling element, goal, as indicated by the value of 1.0 in the submatrices, W_{11}, W_{12}, W_{13}, respectively.

$$S = \begin{bmatrix} W_{11} & W_{12} & W_{13} \\ W_{21} & W_{22} & W_{23} \\ W_{31} & W_{32} & W_{33} \end{bmatrix}$$

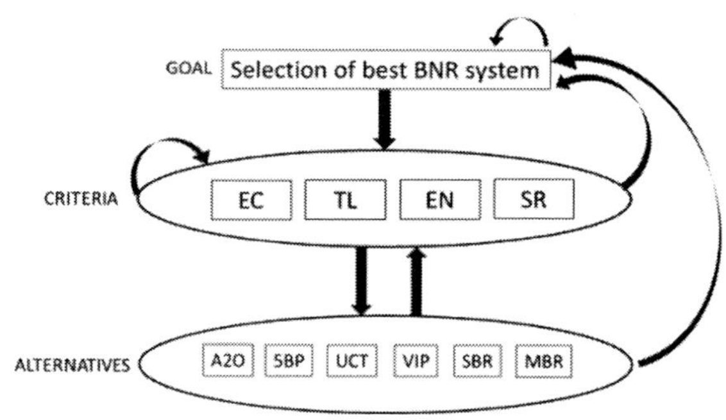

Fig. 3. ANP network structure for the selection of best BNR systems in urban areas

Three experts from the academe participated in the provision of value judgments through pairwise comparisons of the elements in the clusters. The TFN described by $<l_{ij},m_{ij},u_{ij}>$ are composed of the lowest value judgement, geometric mean of all judgements and the highest value judgement respectively. Using the nonlinear fuzzy preference programming approach with the aid of Lingo, the approximate ratio-scale weights (w_i) are obtained while maximizing the value of λ.

The priority vectors from the pairwise comparison matrix for all the interacting elements are populated in an unweighted supermatrix, shown in Table 1. All elements in the submatrices W_{11}, W_{12} and W_{13} have a value of 1.0 that can be attributed to the influence of the goal which is the single controlling element. The submatrix W_{21} shows the resulting priority vectors from the pairwise comparison of the relative importance of each criterion with respect to goal. The submatrices W_{22}, W_{23} and W_{32} are populated by priority vectors considering the inner dependence of evaluation of criteria, feedback of alternatives to criteria, and evaluation of preferred of alternatives considering the criteria, respectively. W_{31} is a null matrix that contains a value of 0.0 for each element that shows the independence of goal from the alternatives. Simultaneously, W_{33} is composed of a null matrix due to the independence of the alternatives amongst each other.

Table 1. Unweighted supermatrix with overall priority weights

	GO	EC	TL	EN	SR	A2O	5BP	UCT	VIP	SBR	MBR	overall priority weights
GO	1	1	1	1	1	1	1	1	1	1	1	**1.00**
EC	0.22	0.56	0.24	0.40	0.58	0.27	0.40	0.26	0.28	0.20	0.31	**0.32**
TL	0.25	0.16	0.58	0.27	0.14	0.28	0.24	0.23	0.27	0.24	0.17	**0.26**
EN	0.33	0.11	0.09	0.21	0.09	0.17	0.16	0.30	0.27	0.29	0.28	**0.23**
SR	0.21	0.17	0.09	0.12	0.19	0.27	0.19	0.21	0.18	0.27	0.24	**0.19**
A2O	0	0.39	0.29	0.05	0.10	0	0	0	0	0	0	**0.24**
5BP	0	0.11	0.23	0.16	0.05	0	0	0	0	0	0	**0.14**
UCT	0	0.05	0.11	0.10	0.04	0	0	0	0	0	0	**0.08**
VIP	0	0.08	0.09	0.13	0.05	0	0	0	0	0	0	**0.09**
SBR	0	0.13	0.21	0.09	0.23	0	0	0	0	0	0	**0.16**
MBR	0	0.24	0.06	0.46	0.53	0	0	0	0	0	0	**0.30**

Presented in Fig.4 is the box-whisker plot of alternatives with their corresponding overall priority weights and variations. The most preferred BNR system to be installed in WTPs located in urban areas is the MBR system followed by A2O. In terms of economic aspect, MBR is now becoming a cheaper technology as it only requires lesser space. Moreover, different MBR configurations are studied in order to optimize and enhance the wastewater effluent quality [24]. The least preferred alternative is UCT which has a very close overall priority weight with VIP. Overall, the most important criterion is the economic aspect followed by technical aspect, environmental aspect and space requirement.

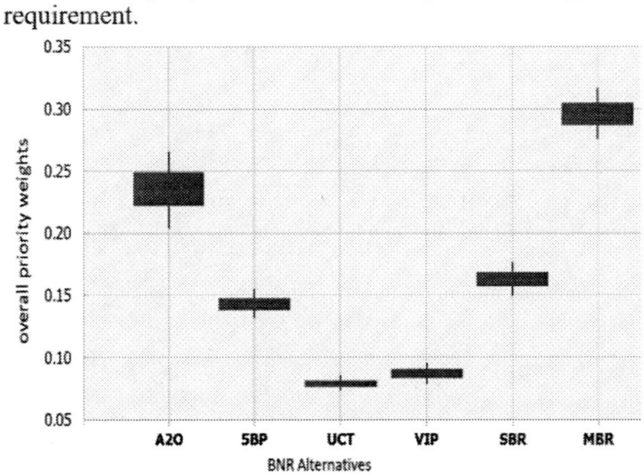

Fig. 4. Box-whisker plot of Monte Carlo simulation results

Furthermore, the uncertainties of the results are evaluated using the Monte Carlo simulation. An iteration of ten thousand times was adopted in order to come up with the distributions. The distributions were converted into a box-whisker plot as shown in Fig. 4. The box plot shows that the decision makers are consistent and certain that VIP and UCT are the least preferred BNR alternatives since the distributions for both alternatives show very low variations. SBR and 5BP, third and fourth in rank respectively, also have closer weights with overlapping whiskers. It can be observed that there is a relatively larger uncertainty and sensitivity in the resulting overall weight of A2O. However, the large variation and whisker length did not affect its rank. Though MBR has the second largest variation among the six alternatives, its rank as the best BNR system for WTPs in urban area is considered stable.

4 Conclusion

The study presents the complex decision making problem of selecting the best biological nutrient removal technology for wastewater treatment plants located in urban areas in a network structure. Fuzzy analytical network process and Monte Carlo simulations were utilized as tools to obtain the overall priority weights of the six alternatives considered. These tools could address the possible subjectivity and uncertainties in the evaluation of all the interacting elements in a systematic approach. The evaluation of the overall priorities shows that the membrane bioreactor is the best option followed by 3-stage pho-redox and sequencing batch reactor. The most important criterion is the economic aspect then technical aspect. Upon utilizing Monte Carlo simulations, the variation for every alternative is shown with respect to the ranking. The results in this study could aid the stakeholders in addressing the decision making problem related to sustainable wastewater management. Future works would extend to considering sub-criteria for each criterion such as dividing economic aspect into upgrade or capital cost and operational cost. This would account a holistic evaluation of alternatives encompassing a more detailed network structure.

The authors would like to acknowledge the Philippine Council for Industry, Energy and Emerging Technology Research and Development (PCIEERD) under the Department of Science and Technology (DOST) for the support given to the Project NexCities. One of the authors, C.M. Pausta, is also supported by the Engineering Research and Development for Technology (ERDT) Program of DOST.

References

1. M. P. Chislock, E. Doster, R. A. Zitomer and A. E. Wilson, Nature Education Knowledge. **4**, no. 4, 10 (2013)
2. R. Kleemann, J. Chenoweth, R. Clift, S. Morse, P. Pearce, and D. Saroj, Resour. Conserv. Recycl., **105**, 347–359 (2015)

3. E. B. Estrada-Arriaga et al., Sci. Total Environ., **571**, 1172–1182 (2016)
4. The Cadmus Group, *Nutrient control design manual: State of technology review report*, 1–104 (2009)
5. Department of Environment and Natural Resources, *Water Quality Guidelines and General Effluent Standards of 2016* (DENR Administrative Order 2016-08, Philippines, 2016)
6. M. A. Hamouda, W. B. Anderson, and P. M. Huck, Water Sci. Technol., **60**, no. 7, 1767–1770 (2009)
7. G. A. Kiker, T. S. Bridges, A. Varghese, T. P. Seager, and I. Linkov, Integr. Environ. Assess. Manag., **1**, (2), 95–108 (2005)
8. J. J. Wang, Y. Y. Jing, C. F. Zhang, and J. H. Zhao, Renew. Sustain. Energy Rev., **13**, no. 9, 2263–2278 (2009)
9. T. L. Saaty, J. Math. Psychol., vol. **15**, no. 3, 234–281 (1977)
10. T.L. Saaty, *Decision Making with Dependence and Feedback: the Analytic Network Process, 2nd ed.* (RWS Publica,ions, Pittsburgh, 2001)
11. C. M. Pausta, A. Huelgas-Orbecido, A. Beltran, R. C. Eusebio, J. J. Ignacio, and M. A. Promentilla, *Selection of Optimum Biological Nutrient Removal (BNR) System for Urban Areas ' Wastewater Treatment Plants using Analytical Network Process (ANP)*, (2017)
12. M. A. B. Promentilla, J. F. D. Tapia, K. B. Aviso, and R. Raymond, **61**, (2017)
13. M. A. B. Promentilla, T. Furuichi, K. Ishii, and N. Tanikawa, **62**, (3), 308–324 (2006)
14. A. H. Orbecido, A. B. Beltran, R. A. J. Malenab, K. I. D. Miñano, M. A. B. Promentilla, Chem. Eng. Trans., **52**, 643-648 (2016)
15. M.A. Promentilla, K. Aviso, R. Tan, Chem. Eng. Trans., **45**, 829-834 (2015)
16. R. Tan, K. Aviso, A. Huelgas, M.A. Promentilla, Process Saf. Environ. Prot., **92**, 467-475, (2014)
17. M. A. B. Promentilla, T. Furuichi, K. Ishii, and N. Tanikawa, Waste Manag., **26** (12), 1410–1421 (2006)
18. M. A. B. Promentilla, J. I. B. Janairo, and D. E. C. Yu, **61**, (2017)
19. United States Environmental Protection Agency, *Biological Nutrient Removal Processes and Costs,* (Office of Water, Washington, 2007)
20. C. W. Randall and E. U. Cokgor, *Performance and economics of BNR Plants in the Chesapeake Bay Watershed , USA*, 21–28 (1995)
21. T. M. Gellner, *Feasible and Cost Effective MBR Applications*, 1–9 (2014)
22. L. R. Mayor, J. V. Camacho, and F. J. F. Morales, Water, Air, and Soil Pollution, **152**, 279–296, (2004)
23. A. F. Silva et al., Appl. Microbiol. Biotechnol., **93**, (5), 2171–2180 (2012)
24. M. Sibag and H. Kim, Bioresour. Technol., **104**,173–180 (2012)

Adsorption of Nickel and Chromium Ions by Amine-Functionalized Silica Aerogel

Sudarat Sertsing[1], *Thanaphat* Chukeaw[1], *Sitthiphong* Pengpanich[2], and *Bawornpong* Pornchuti[1,*]

[1]Chemical Engineering Department, Mahanakorn University of Technology, Bangkok, Thailand
[2]PTT Global Chemical Public Company Limited, Rayong, Thailand

Abstract. In this study, silica aerogel was synthesized by drying at atmospheric pressure and modified further with aminopropyl triethoxysilane (APTES). The amine-functionalized silica aerogel was investigated as an adsorbent for removal of nickel and chromium ions. The effect of contact time, solution pH, and initial solution concentration were studied. The equilibrium was achieved within 60 min. The optimum pH was found to be 4. Adsorption equilibrium data were agreed fairly well with Langmuir isotherm model. Adsorption capacities for nickel and chromium ions were found to be 40.32 mg/g and 46.08 mg/g, respectively.

1 Introduction

Electroplating is one of the most hazardous chemical industries due to its metal-contaminated wastewater. Heavy metals cause environmental problems as pollutants because of their characteristics such as high toxicity, non-biodegradability, and accumulation in food chain [1]. Therefore, electroplating wastewater must be treated before discharge.

Treating electroplating wastewater can be employed by various techniques such as chemical precipitation, coagulation-flocculation, floatation, adsorption, ion-exchange, membrane filtration, and electrochemical processes. Among them, adsorption is one of the most promising techniques when initial heavy metal concentration below 100 ppm [2]. Adsorption process is simple design, easy to use, and flexible. Moreover, it has low operating cost, low fouling problems, and most economic for elimination of heavy metals from wastewater [3].

Many adsorbents have been developed in treatment of effluent comprising heavy metals. Because of high surface area and porous network, activated carbon has been investigated extensively. However, the adsorption capacity is quite low. This may be caused by lacking of surface affinity groups [4].

Aerogels are interesting material. They have extraordinary characteristics such as high porosity, low density, high specific surface area, with tailor-made surface chemistry [1, 4]. Normally, aerogels are prepared by a sol-gel technique followed by suitable drying methods. To conserve the three-dimensional and the highly porous structure, various procedures are employed such as supercritical drying, freeze drying, and drying at atmospheric pressure. In this study, silica aerogel was synthesized by drying at atmospheric pressure which is easy and safe. Furthermore, modification of adsorbents with some chemicals containing N, O, S, P in functional groups increases adsorption capacity of heavy metals [4]. The objective of this work was to study the adsorption of nickel and chromium from aqueous solutions by amine-functionalized silica aerogel.

2 Materials and methods

2.1. Chemicals

Tetraethylothosilicate (Sigma Aldrich), APTES (Sigma Aldrich), Cetyltrimethylammonium bromide (Fluka), isopropanol (Merck), toluene (Merck), hydrochloric acid (Carlo Erba), and ammonia (Labscan) were used without further purification.

2.2 Preparation and modification of the adsorbent

Silica aerogel was synthesized by a procedure adjusted from that of Aravind et al. [5]. Briefly, 4.66 g Tetraethylothosilicate (TEOS), 5.37 g isopropanol, 6.44 g HCL solution were mixed together. Then, gelation was conducted by adjusting pH to 5 using ammonia solution. The resulted hydrogel was kept at 50°C for 1 day. After that, the pore liquid was replaced with isopropanol. The alcogel was then aged further with TEOS solution. After silylation, replace unreacted TEOS with 0.05% CTAB in isopropanol and dry slowly at 70°C.

* Corresponding author: bawornpong@yahoo.com

https://doi.org/10.1051/matecconf/201815603014

Amine-functionalized silica aerogel was prepared according to Parida and Rath [6]. Briefly, 2.00 g of silica aerogel was mixed with 60 mL toluene and then 1.2 mL of APTES was added drop wise. The mixture was refluxed for 8 hours. The resulted solid was washed with ethanol followed by distilled water. Finally, it was dried at room temperature for 12 hours.

2.3 Adsorption experiments

All adsorption experiments were conducted on a shaker using erlenmayer flasks containing 100 mL of single metal solution and 100 mg of amine-functionalized silica aerogel. The effects of several parameters including contact time, pH, and metal concentration were examined. After filtration of the adsorbent, the metal concentration was measured by UV-visible spectrophotometer. The equilibrium adsorption capacity was calculated by Eq. (1).

$$q_e = (C_0 - C_e)V/m \quad (1)$$

where q_e is the equilibrium adsorption capacity (mg/g), C_0 is the initial metal concentration (mg/L), C_e is the equilibrium metal concentration (mg/L), V is the solution volume (L), and m is the adsorbent mass (g).

3 Results and discussions

3.1 Properties of silica aerogel

The synthesis of silica aerogel was comprised of 3 steps: (a) sol-gel process (b) aging process (3) drying at atmospheric pressure. Firstly, reactive monomers were generated from hydrolysis of TEOS. Then, three-dimensional open network structure was constructed via condensation reaction. The gel prepared was strengthened by aging process. Finally, the pore liquid was released by drying. During evaporation of the pore liquid, capillary stress occurred leading to mechanical damage. Therefore, drying process was the critical step. TEOS was applied to change gel surface to be hydrophobic. Moreover, the pore liquid was exchanged to isopropanol to lower surface tension. CTAB was utilized to reduce shrinkage and prevent cracking as well. After drying at ambient pressure, silica aerogel synthesized was shown in Figure 1.

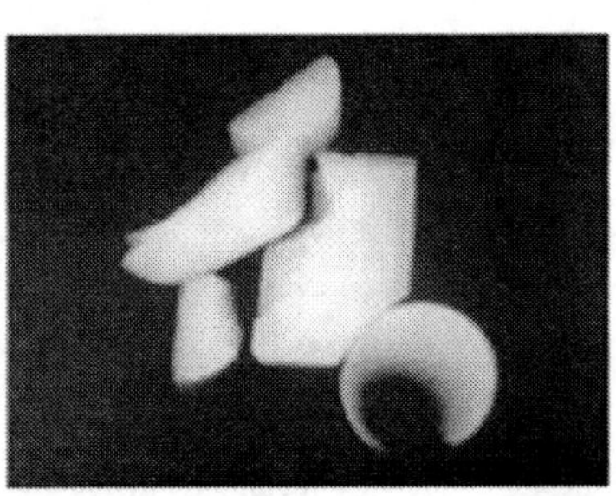

Fig. 1. Silica aerogel.

Figure 2 demonstrates the adsorption-desorption isotherm of N_2 at 77 K for silica aerogel. It was found that silica aerogel produced was mesoporous material because of the existence of hysteresis loop. The hysteresis loop was caused by the capillary condensation in the mesopores. The BET surface area of this synthesized material was 1,401.3 m^2/g while the average pore size was 21.2 nm.

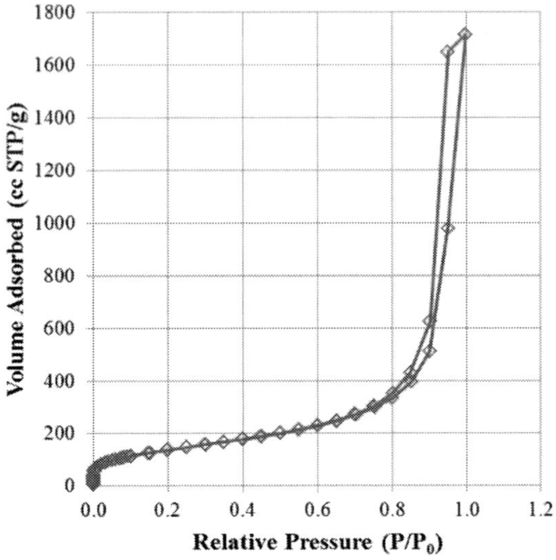

Fig. 2. Adsorption-desorption isotherm of N_2 for silica aerogel.

3.2 Effect of contact time

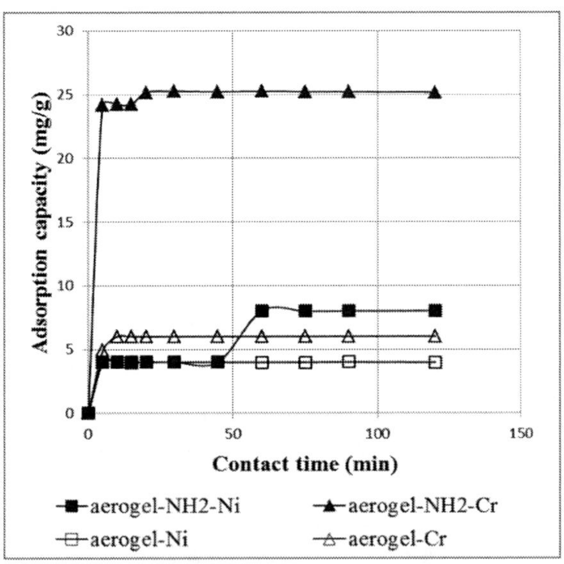

Fig. 3. Contact time effect.

https://doi.org/10.1051/matecconf/201815603014

The effect of contact time on the adsorption of nickel and chromium ions was carried out in the range of 0-120 min as illustrated as Figure 3. The adsorption capacity increased dramatically within the first 5 min. Most of them reached equilibrium within 20 min except for the adsorption of nickel by modified silica aerogel. At the beginning of the adsorption process, there were plenty of adsorption sites resulting in high adsorption rate. The contact time of 60 min was selected for all subsequent experiments. In addition, modification of silica aerogel with APTES improved adsorption capacity of both nickel and chromium ions. The amino groups on the adsorbent surface can form a coordination complex with metal ion [4].

3.3 Effect of pH

One important factor governing adsorption process is the solution pH. The effect of the solution pH on the removal of nickel and chromium ions was investigated at pH 1-4 and the result are presented in Figure 4. It was found that the increase of solution pH resulted in the increase of adsorption capacity. This may be caused by competition between hydrogen ions and metal ions at low pH [1].

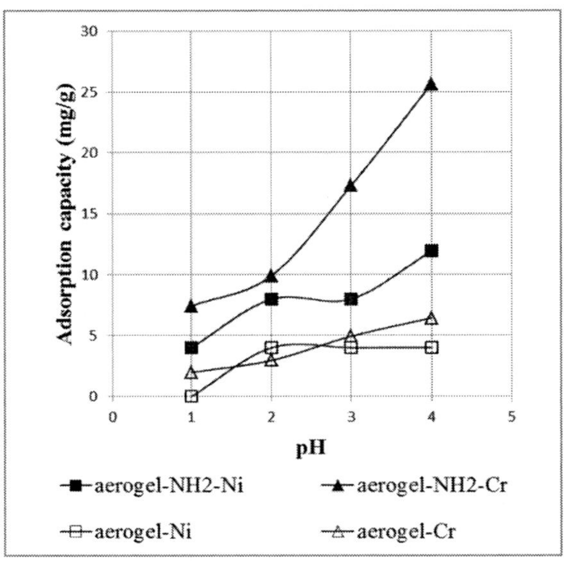

Fig. 4. The effect of pH.

3.4 Adsorption isotherm

As shown in Figure 5, adsorption capacity increased when metal concentration increased due to higher driving force of the metal ions. This figure also indicates that adsorption capacity of chromium ion was higher than that of nickel. This was caused by higher charge to radius ratio (Z/r) of chromium ion [4].

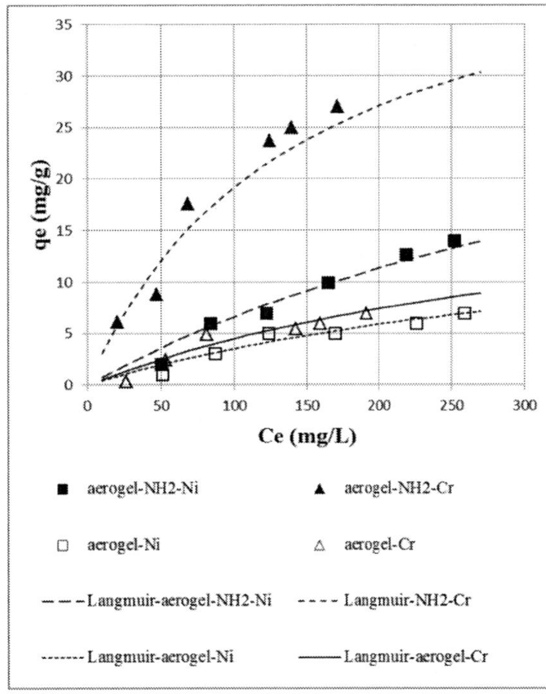

Fig. 5. Metal concentration effect and adsorption isotherm.

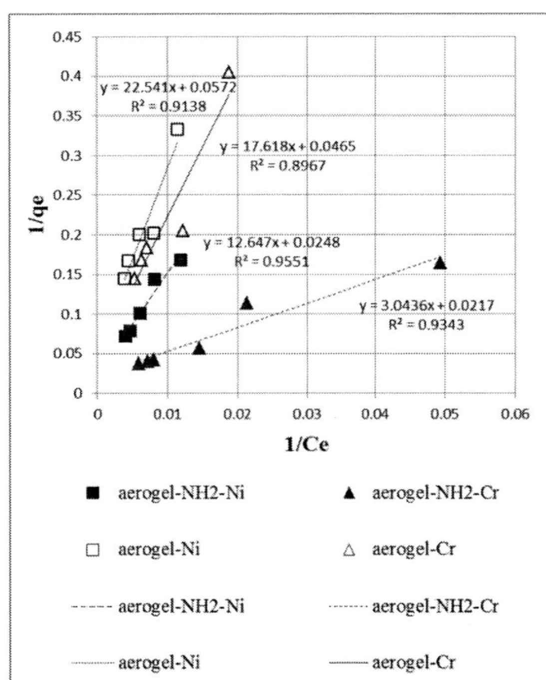

Fig. 6. The linear Langmuir model.

To design any sorption system, adsorption isotherms are the important information. In this study Langmuir isotherm was chosen. It can be shown by equation (2):

$$q_e = Q_m b C_e/(1+bC_e) \quad (2)$$

where q_e is the equilibrium adsorption capacity (mg/g), C_e is the equilibrium metal concentration (mg/L), Q_m is the maximum adsorption capacity (mg/g), b is the equilibrium constant (L/mg). To obtain the model parameters, the linearized Langmuir equation was used as equation (3):

$$1/q_e = (1/Q_m b)(1/C_e) + (1/Q_m) \quad (3)$$

Linear Langmuir plot is displayed in Figure 6 and the obtained isotherm parameters are summarized in Table 1. It can be seen in Figure 5 that the Langmuir isotherm fitted the experimental data fairly well. The maximum adsorption capacities for nickel and chromium ions on amine-functionalized silica aerogel were found to be 40.32 mg/g and 46.08 mg/g, respectively.

Table 1. Isotherm parameters.

Adsorbent/adsorbate	Q_m (mg/g)	b (L/mg)	R^2
Aerogel/Ni	17.48	0.0025	0.914
Aerogel/Cr	21.51	0.0026	0.897
Amine-aerogel/Ni	40.32	0.0020	0.955
Amine-aerogel/Cr	46.08	0.0071	0.934

4 Conclusions

Adsorption of nickel and chromium ions onto amine-functionalized silica aerogel was studied. It was found that adsorption process achieved equilibrium within 60 min. The removal of both metals was optimum at pH 4. The adsorption loading of chromium ion was higher than that of nickel ion. The Langmuir model fitted the experimental data fairly well.

References

1. H. Faghihian, H. Nourmoradi, M. Shokouhi, Pol. J. Chem. Tech., **14**, 50 (2012)
2. H. Faghihian, H. Nourmoradi, M. Shokouhi, Pol. J. Chem. Tech., **14**, 50 (2012)
3. C. F. Carolin, P. S. Kumar, A. Saravanan, G. J. Joshiba, Mu. Naushad, J. Environ. Chem. Eng., **5**, 2782 (2017)
4. H. Maleki, Chemical Engineering Journal, 300, 98 (2016)
5. P. R. Aravind, P. Shajesh, P. Mukundan, K. G. K. Warrier, J. Sol-Gel Sci. Technol., 52, 328 (2009)
6. K. M. Parida, D. Rath, J. Mol. Catal. A: Chem., **310**, 93 (2009)

Solvent Development for Post-Combustion CO_2 Capture: Recent Development and Opportunities

Anggit Raksajati[1,*], *Minh* Ho[2], and *Dianne* Wiley[2]

[1]Department of Chemical Engineering, Institut Teknologi Bandung, Bandung 40132, Indonesia
[2]School of Chemical and Biomolecular Engineering, The University of Sydney, Sydney NSW 2006, Australia

Abstract. Chemical absorption is widely regarded as the most promising technology for post-combustion CO_2 capture from large industrial emission sources with CO_2 separation from natural gas using aqueous amine solvent system having been applied since the 1930s. The use of monoethanolamine (MEA) in CO_2 absorption system possesses several drawbacks, such as high regeneration energy, high solvent loss, and high corrosion tendency. Various solvents have been developed for post-combustion CO_2 capture application including the development of aqueous solvents and phase-change solvents. Some of these alternate solvents have been reported to have better solvent properties, which could improve the CO_2 absorption system performance. This paper reviews key parameters involved in the design improvement of several chemical absorption process systems. In addition, some novel solvent systems are also discussed, for example encapsulated solvents systems. Some of the key solvent parameters that affect the capture performance, such as heat of reaction, absorption rate, solvent working capacity, solvent concentration, and solvent stability, are discussed in this paper, particularly in relation to the economic viability of the capture process. In addition, some guidelines for the future solvent development are discussed.

1 Introduction

A CO_2 reduction scheme that is gaining growing interest is Carbon Capture and Storage (CCS) [1]. Within the CCS process chain, CO_2 capture is the costliest stage and therefore it is important to develop the technologies that can reduce costs. Among all CO_2 capture methods, post-combustion CO_2 capture using chemical absorption has been recognized as the most commercially ready technology. The concept of this technology has been applied, albeit at different feed gas sources, in natural gas industry since the 1930s, where CO_2 is absorbed using aqueous amine solvent system [2]. Many researchers suggested that other CO_2 removal methods, such as membranes and adsorption, are not likely to be competitive because of compression work [2]. The application of physical absorbents in post-combustion CO_2 capture is likely to be more limited than that of chemical absorbents because of the low CO_2 partial pressure in the flue gas [3]. The future development of chemical absorption will be the focus of this paper. This paper aims to review the key parameters involved in the design improvement of several chemical absorption process systems. In particular, the recent updates on solvent development are presented for two solvent classes (aqueous solvents and phase-change solvents). In addition, some novel solvent systems are also discussed, for example encapsulated solvents systems.

2 Post-combustion CO_2 Capture using Chemical Absorption

In CO_2 capture using chemical absorption, weak chemical bonds between CO_2 from emission gases and a solvent solution is formed. Heat is then typically provided in the regeneration column in order to reverse this reaction. A simplified schematic of the post-combustion CO_2 capture process using chemical absorption is shown in Figure 1. Various solvents have been developed for post-combustion CO_2 capture application in order to improve the performance of the absorption system compared to MEA [4]. Typically, these alternate solvents have better solvent properties..

* Corresponding author: anggit@che.itb.ac.id

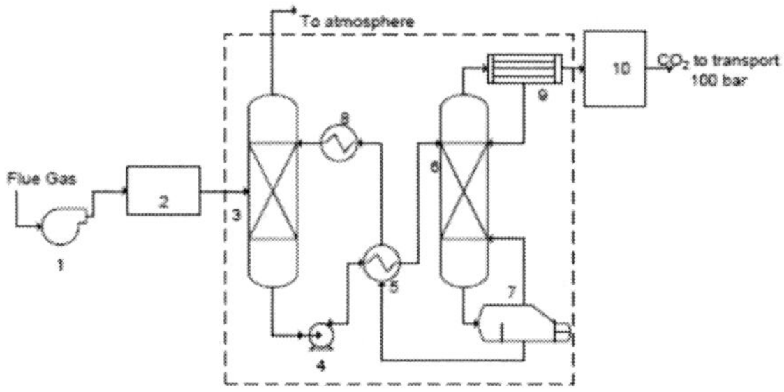

Fig. 1. Simplified process flow diagram for CO_2 capture process using chemical absorption (1) Blower, (2) Pre-treatment, (3) Absorber, (4) Solvent pump, (5) Cross heat exchanger, (6) Stripper, (7) Reboiler, (8) Lean solvent cooler, (9) Condenser, (10) Post-treatment (dehydration and CO_2 transport compressor)

Some of the key solvent parameters that affect performance are:

- Heat of reaction - The use of solvents with a low heat of reaction may decreases the regeneration energy. However, the reduction in regeneration energy for solvents with a low heat of reaction can only be achieved if it is not coupled with high water evaporation in the stripper [5].
- Absorption rate - By formulating solvents with a high absorption rate, the dimensions of the absorber decreases, reducing the capital cost [6].
- Solvent working capacity - The difference between the solvent rich and lean loading is commonly defined as solvent working capacity. Improvement in this property would decrease the solvent circulation rate and potentially reduces the regeneration energy [7-8]. A higher solvent capacity also reduces the dimensions of the absorber, heat exchangers, pumps, and piping, thus lowering the capital cost [8].
- Solvent concentration - High solvent concentrations translate to a low solvent circulation rate, and hence it only requires relatively small dimensions of heat exchangers, absorber, stripper, and pipelines. This also leads to a decrease in the regeneration energy because of the lower amount of water that goes into the stripper [7-8].
- Solvent stability - Solvent degradation accounts for 15 – 25 % of total solvent losses for the MEA system. By formulating a solvent with better solvent stability, the operational cost for solvent make-up cost decreases [9].

Table 1 summarizes the key solvent properties of various solvent groups reported in literature based on their molecular structure. The typical ranges of values for the heat of reaction, absorption rate, and solvent capacity are also given. For the amine-based solvents, the values of absorption capacity and absorption rate are gathered from a study by Puxty et al. [10]. For other solvent groups, the level of information of solvent properties is not as extensive as amine-based solvents. As a consequence, the values presented in Table 1 for amino acid salts, ionic liquids, and deep eutectic solvents may not encompass all solvents within these solvent groups, but are representative of the most prevalent solvents for these groups.

3 Aqueous Solvents

In most industrial applications, only two phases are involved in the CO_2 absorption, the gas phase made up of the gas/gases to be recovered and the solvent liquid phase. In the following section, the key characteristics of different groups of aqueous solvents are discussed.

3.1. Primary Amines

As shown in Table 1, there are various solvents belonging to the primary amines group including MEA, DGA, and EDA. The most widely used is MEA. MEA was first applied and widely developed to capture CO_2 from feed gases with low partial pressure of CO_2 as it possesses fast reaction rate and high working capacity compared to other solvent groups. MEA has several advantages over other amine-based solvents [9, 11-12], such as high reaction rate, high absorption capacity on the mass basis (because of low molecular weight), good thermal stability, and cheap.

However, MEA owns deficiencies such as high energy penalty to regenerate the solvent, corrosive, and high solvent make-up rate [4]. MEA has a high heat of reaction (82 kJ/mole), thus it requires high regeneration duty in the stripper. This is due to the formation of the stable carbamate ion ($MEACOO^-$). The regeneration energy of the generic MEA 30 %-wt. is reported to be between 4.1 to 4.4 MJ/kg CO_2 [8, 13-15]. The high level of solvent loss in an MEA system is due to the vaporization at the absorber. The other major issue regarding the solvent loss is the degradation of solvent, which has several causes (oxidative and thermal reactions). Make-up solvent is required to compensate for the loss that occurs in the process. Some operating strategies that have been implemented to address the issue of solvent loss include limiting the solvent concentration while corrosion effects are dealt with by ensuring proper equipment material selection and utilizing mild operating conditions [8].

3.2 Secondary and Tertiary Amines

CO_2 separation from natural gas using secondary and tertiary amines, such as DEA and MDEA, have been commonly applied commercially for decades. DEA and MDEA have a lower heat of reaction and a higher working capacity compared to MEA and thus require lower regeneration energy. The drawback of these solvents is a much slower absorption rate compared to MEA. Aqueous solvents consisting entirely of secondary and tertiary amines are unlikely to be applied in post-combustion CO_2 capture applications.

Table 1. Solvent properties of various solvent groups

Solvent group	Examples	Working capacity (moles CO_2/mole solvent)	Absorption rate (min^{-1})	Heat of reaction (MJ/kg CO_2)	Solvent class
Primary amines	Monoethanolamine (MEA) Ethylenediamine (EDA)	0.15-0.69	0.003-0.032	70-85	Aqueous
Secondary amines	Diethanolamine (DEA) Diisopropanolamine (DIPA)	0.10-1.0	0.004-0.031	65-75	Aqueous
Tertiary amines	Methyldiethanolamine (MDEA) Triethanolamine (TEA)	0.07-1.1	0.001-0.008	50-65	Aqueous
Hindered amines	Aminomethylpropanol (AMP) 2-piperidineethanol (2-PE)	0.25-1.0	0.003-0.02	50-75	Aqueous
Polyamines	Piperazine (PZ) Hydroxyethylpiperazine (HEP)	0.50-1.8	0.003-0.3	66-78	Aqueous
Alkali carbonates	Potassium carbonate (K_2CO_3) Sodium carbonate (Na_2CO_3)	0.4-1.0	0.001-0.002	30-50	Aqueous and phase-change
Ammonia	Ammonia (NH_3)	0.35	<0.007	27	Aqueous and phase-change
Amino acid salts	Potassium taurate Potassium sarcosine	~0.1-1.0	0.003-0.02	60-80	Aqueous and phase-change
Ionic Liquids	([bmim]-PF_6) ([bmim]-BF_4)	<0.5	<0.001	15-100	Aqueous and phase-change
Deep eutectic solvents	Choline chloride (ChCl)/urea Choline chloride (ChCl)/MEA	<0.2	<0.001	15-100	Aqueous

The absorption rate and kinetics of DEA solution have been reported in a study by Rinker et al. [16] and Cadours et al. [17], while the kinetics for MDEA solution have been investigated in a study by Ko and Li [18]. The other possible drawback of using DEA or MDEA in post-combustion CO_2 capture applications is that they are less suitable for applications where the partial pressure of CO_2 is low (less than 100 kPa) [19]. The rate promoter addition (such as MEA, piperazine, and carbonic anhydrase) or blending with other solvents have been reported to be able to lessen this deficiency [19-22].

3.3 Hindered Amines

Hindered amine compounds are commonly recognized when the nitrogen atom of the amine molecule adds a bulky substitute group. The bulkier substitutes attached to the amine backbone lead to a lower rate of carbamate formation and hence the bicarbonate formation will then be the main reaction mechanism [23-24]. Furthermore, as the reaction mechanism in hindered amines is dominated by bicarbonate formation, the heat of reaction of hindered amines is lower than MEA. A proprietary hindered amine, trademarked as KS-1, has been developed by Mitsubishi Heavy Industries (MHI) in 1990 [25]. The CO_2 capture process using KS-1 is similar to the MEA (Figure 1), except that it incorporates a heat recovery system and absorption inter-cooling. It was reported that the regeneration energy obtained during testing was 2.53 MJ/kg CO_2, which is up to 40 % less than MEA [26]. KS-1 has a larger loading capacity, thus the circulation rate of KS-1 is 40 % lower than MEA.

3.4. Polyamine

The most widely known polyamine solvent for CO_2 absorption is piperazine (PZ), which has been commercially developed as a rate promoter for various solvent types, such as MDEA and potassium carbonate. When PZ is used as a promoter, the concentration used is low, between 0.5 and 2.5 m PZ, because PZ is not highly soluble. The application of CO_2 capture using aqueous concentrated PZ has attracted increasing interest since the early 2000's. It is reported that in order to preserve a liquid solution without precipitation at ambient temperature for 8 m PZ, a CO_2 loading of approximately 0.25 mole CO_2 per mole alkalinity is necessary [27]. Under these conditions, the working capacity of 8 m PZ (0.90 mole CO_2 per kg PZ+H_2O) is twice that of 7 m MEA (0.43 mole CO2 per kg MEA+H_2O), while the rate of CO_2 absorption into 8 m PZ is up to triple the rate of 7 m MEA. The drawback of PZ is that it forms nitrosamines, a class of carcinogenic compounds, at faster rates compared to MEA [28]. This could hinder the use of PZ in CO_2 capture applications because nitrosamines can be released from the absorber units through aerosol formation and accidental spills [28].

3.5 Alkali Carbonate

A hot potassium carbonate process has been used for CO_2 absorption from natural gas and ammonia synthesis since the 1960's. In this process, the CO_2 is captured at high pressure and the concentration of the potassium carbonate is limited to 20-30 %-wt. to prevent solid precipitation. Alkali-carbonate solvents have several benefits compared to other solvents including low heat of reaction and low solvent cost [5, 29-31]. The CO_2 is absorbed through bicarbonate formation, which leads to the significantly lower heat of reaction than that of MEA.

Endo et al. (2011) reported that the regeneration energy of an aqueous potassium carbonate system is 4.9 MJ/kg CO_2, which is 10-20 % higher than that of MEA. Although potassium carbonate has a lower heat of reaction than MEA, the regeneration energy is higher because of the lower working capacity and hence the solvent circulation rate is higher [5, 31]. The other interesting feature of alkali carbonates is the possibility to exclude the FGD and SCR from the pre-treatment facilities [31]. The NO_x and SO_x react with the potassium ion to form potassium nitrate and potassium sulfate precipitates, which can be removed as by-products using reclaiming units (such as ion exchange) prior to entering the absorber column. This potentially reduces the capital cost of the capture plant presumably because the capital cost of the precipitate removal is less than the cost of FGD and SCR.

3.6 Ammonia

Aqueous ammonia has been investigated as a solvent for post-combustion CO_2 capture. The advantages of this system compared to MEA include higher loading capacity, lower heat of reaction, and better resistance to oxidative and thermal degradation [32]. The other potential benefit of this system is the possibility to use carbon steel for equipment because ammonia is significantly less corrosive compared to amine solutions. One of the drawbacks of using aqueous ammonia is the high solvent volatility because of its low vapor pressure. Ammonia slip to the atmosphere must be prevented because of its toxicity. To overcome this, ammonia wash sections are required at the top of the absorber and the stripper, which increases the capital costs by 1-2 % [33].

The CSIRO and Delta Electricity assessed aqueous ammonia technology for a post-combustion CO_2 capture pilot plant at Munmorah Power Station in NSW, Australia [32, 34]. The ammonia solution investigated had concentrations of 0-6 %-wt. The absorber was operated at 15-20 °C and atmospheric pressure, while the stripper was operated at 300-850 kPa. They found that the absorption rate was at least 50 % less than that of MEA [32]. The regeneration energy obtained during the trials was reported to be 3.0-4.2 MJ/kg CO_2 [35].

3.7. Other Solvents

Several other solvents that have been developed for CO_2 capture include ionic liquids (IL) and deep eutectic solvents (DES). IL are defined as salts that is composed by ions with a melting point lower than 100 °C [36]. Ionic liquids have been reported to show excellent properties for CO_2 capture applications due to their tunability, thermal stability, non-flammability, and high CO_2 solubility [36-39]. There are various ionic liquids currently under development for CO2 capture, most of them can be classified into two classes: room temperature ionic liquid (RTIL) and task specific ionic liquid (TSIL). Among these, the imidazolium-based RTIL class is the most widely investigated. However, the use of RTIL is based on physical absorption technology and not widely applicable for low partial pressure flue gas post-combustion CO_2 capture.

A deep eutectic solvent (DES) is a mixture of a substituted quaternary ammonium or metal halide salt with a hydrogen bond donor (HBD), such as an amide or carboxylic acid. This class of ionic liquid is an eutectic mixture that has a quite low melting point [40-41]. DES has been suggested as a low cost alternative to other ionic liquids with interesting characteristics, such as negligible vapor pressure, wide types of liquid range, good stability, and tunability [41]. High purity DES can be easily prepared using biodegradable materials at low cost [42]. However, on a mass basis, the solubility of CO_2 in DES is still significantly lower than that of MEA 30 %-wt., by an order of magnitude, because of the higher molecular weight of DES [43]. The other drawbacks of DES for CO_2 capture applications are their very high viscosity, low absorption rates, and high solvent cost compared to aqueous solvents [44].

4 Phase-Change Solvents

The CO_2 absorption process involves equilibrium reaction, where an increase in the CO_2 absorption capacity (moles of CO_2 per mole of solvent) can be achieved by eliminating one of the reaction products during reaction (Le Chatelier's principle) through precipitation. This could decrease the required regeneration energy up to half that of MEA 30%-wt. [45-47]. However, this removal of reaction product is achieved by precipitation of solids, which can cause plugging and scaling in the gas-liquid contactor. A spray tower has also been suggested to prevent plugging. However, spray tower only has 10-30% surface area compared to that of packed columns. Thus, at the same capacity, a larger spray tower and hence a higher capital cost are required.

The compounds of precipitate in the absorption with phase-change solvents depends on the solubility of the reaction products and the chemical structure of the solvent. Raksajati et al. [48] proposed two classes of phase-change solvents. The S class solvents are those that react with CO_2 to produce bicarbonate precipitate, for example ammonia, alkali carbonates, and amino acids with tertiary or sterically hindered secondary amino group. On the other hand, the L class solvents are those that react with CO_2 and form carbamate ion solution and a zwitterion precipitate. The detailed explanation of reaction mechanism for both phase-change solvent classes is described in Raksajati et al. [48]. For both phase-change solvent classes, the reaction equilibria towards the reaction products when precipitation occurs.

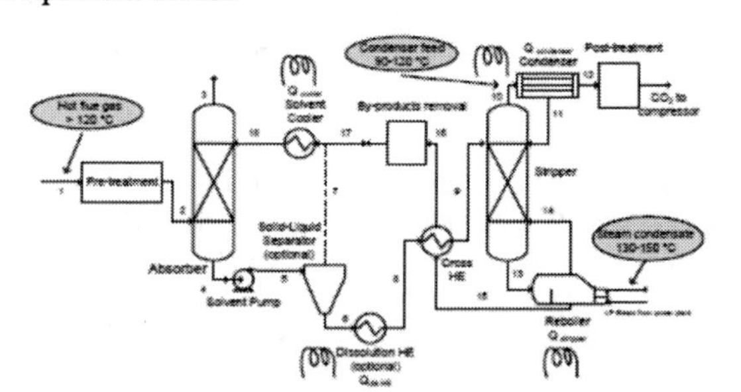

Fig. 2. Simplified process flow diagram for generic phase-change solvent CO_2 capture[48]

Figure 2 shows the simplified process flow diagram of CO_2 absorption using phase-change solvent. The most commonly used solvents for this process are based on three molecular structure groups, namely amino acid salts, alkali carbonates and ammonia.

4.1. Amino Acid Salts

Amino acid salts are one of the phase-change solvents currently under development for CO_2 capture. Some of the attractive characteristics of amino acid salts are high solvent working capacity, relatively fast reaction kinetics, good stability towards degradation, favorable heat of reaction, and low environmental impact [49]. Kumar et al. reported that amino acid salts show negligible volatility because of the ionic nature [50].

The CO_2 solubility in various amino acid salts has been reported in recent studies [50-54]. Portugal et al. (2009) reported that the CO_2 absorption capacities of potassium glycinate and potassium threonate solutions are within the same order of magnitude as that of MEA 30 %-wt. Ma'mun and Kim [53] reported that the CO_2 absorption capacity of potassium sarcosine is 34 % higher than that of MEA. In a recent study by [49], two processes to enhance the absorption and desorption process of a precipitating amino acid solvent system was introduced (DECAB and DECAB Plus). The addition of a solid-liquid separator between the absorber and the stripper may result in a stripper feed stream with a higher solid fraction. Therefore, the pH of the rich solution changes and it consequently enhances the hydrolysis of the carbamate species producing amino acid salts and bicarbonate. This translates to a lower heat of desorption required in the stripper. Sanchez-Fernandez et al. [53] reported that the regeneration energy of DECAB and DECAB Plus are 3.3 and 2.5 MJ/kg CO_2, which are approximately 15 to 40 % lower than MEA 30 %-wt.

4.2 Alkali Salts

Recently, the use of high concentration potassium carbonates for post-combustion CO_2 capture has been investigated. At this concentration range (30-40 %), the solubility limit is exceeded hence bicarbonate precipitates are formed [31] The precipitating alkali carbonate systems have the same advantages as those of aqueous alkali carbonates, such as low heat of reaction, low solvent cost, negligible solvent loss, and a possibility to exclude FGD and SCR [31].

The performance of "UNO Mk 3", a proprietary precipitating potassium carbonate solvent developed by the Cooperative Research Centre for Greenhouse Gas Technologies (CO2CRC), has been reported by [31]. The solution contains potassium carbonate ranging from 35 to 50 %-wt. The CO_2 loading in the solution was reported to be between 0.3 to 0.68 and the regeneration energy of this system was reported to be 3.6 MJ/kg CO_2 [31]. UNO Mk 3 has been evaluated in a pilot plant at The University of Melbourne and at the Hazelwood Power Station [55-56].

4.3 Other Phase-Change Solvents

One example of a carbonate-based solvent is the chilled ammonia process (CAP), which is licensed by Alstom and was installed on a plant at Mountaineer to capture 100,000 tons of CO_2 per year. This project was a co-operation with American Electric Power [57]. Key advantages of CAP include low regeneration energy, good resistance to solvent degradation, high desorption pressure and high solvent working capacity compared to MEA [33, 58-60]. The main drawbacks of this system are the need for refrigeration to support the low-temperature absorption process and the high solvent volatility. An ammonia washing system is typically installed to recover vaporized ammonia at the outlet of the absorber, which contributes to the capital costs of the CAP system. It has also been reported that CAP has a low absorption rate, about 1.5 to 2 times lower than MEA [61]. There are five sections in the CO_2 capture system using CAP: flue gas cooling, CO_2 absorption into solution, NH_3 stripping from the vent gas, CO_2 regeneration, and CO_2 compression [12].

5 Encapsulated Solvent

Encapsulated solvents are one of the new processes currently under development for CO_2 capture [62-65]. In the encapsulated solvent system, the solvent fluid is enclosed in a thin membrane capsule typically 100-600 μm in diameter [65]. The capsule shell can be made from polymers with typical wall thickness of 10-30 μm [65]. The working solvent is isolated within the capsule while the gas is absorbed through the capsule wall. Thus, the membrane capsule needs to be thin enough in order to prevent excessive mass transfer resistance, but also strong enough to maintain its integrity and stability.

While any class of solvents can be used within the encapsulated solvent system, for phase change solvents, encapsulated solvent systems can overcome some of the processing challenges of the traditional process design. If conventional absorption equipment such as a packed column is used, the precipitation of solids in the absorption process raises concerns of plugging [49]. However, if CO_2 absorption is operated in a spray tower to prevent plugging, the capital cost for the system may be uneconomic. The use of encapsulated solvents overcomes this as the precipitates are isolated in the capsule, thus preventing plugging. The other advantage of an encapsulated solvent is that it has significantly higher surface areas compared to conventional packing [63, 65]. Compared to a conventional absorption system, an encapsulated solvent system is likely to possess a drawback in performing heat recovery between the rich and lean sorbent streams. As observed in a solid adsorbent system, exchanging the heat between solid streams is not as straightforward as exchanging heat between liquid streams [66-68].

At the current stage of encapsulated solvent development, the optimal capsule properties such as the capsule diameter, capsule thickness, membrane capsule permeability, and so forth, have not been formulated.

Different process configurations including fixed-bed circulated fluidized bed, and bubbling fluidized bed columns have been proposed as possible equipment for use as the absorber and the regenerator [69]. However, this has not been examined in detail.

6 Discussion

Aqueous Solvents

Raksajati et al. [70] discussed the parameters that have the largest effect on the CO_2 capture cost by using aqueous solvent. The authors reported that solvent stability to SO_x and NO_x, solvent working capacity, solvent concentration, and heat of reaction are the most important variables that affect the CO_2 capture cost significantly. On the other hand, the solvent loss had a relatively small impact on the CO_2 capture cost. Furthermore, in terms of process design improvement, it was found that reduction of capture cost could be achieved by employing novel heat transfer configuration to reduce the temperature difference across stripper, a high solvent regenerator pressure system, advanced column packing with high surface area, and alternate cheaper materials for process vessels.

The costs estimation of CO_2 capture by using aqueous solvent was estimated to be in the range of US$62-US$80 per metric ton of CO_2 avoided and US$44-US$59 per metric ton of CO_2 avoided for capture plant with and without FGD and SCR, respectively [70]. Based on the parameters that are important for aqueous processes, it was reported that capture cost can be reduced significantly for solvents that have good stability to SO_x and NO_x, high working capacity (> 0.35 moles of CO_2/mole of solvent), high concentration (> 50 wt. %), and fast (comparable absorption rate with MEA). This result may suggest that future generations of CO_2 capture processes may be likely to involve a phase change during CO_2 absorption that have higher solvent capacity.

To date, most of the developments for aqueous solvents have focused on amines. However, this presents a difficulty if developments continue to be based on amines, as the degradation of amines due to acid gas impurities will make the possibility of excluding the FGD and SCR very challenging. Continued development of amines solvents would require a novel approach, for example integrating the CO_2 and acid gas impurities removal, such that has been applied in the Cansolv system in Boundary Dam Project [71]. Alternatively, future development of aqueous solvents could focus on developing non-amine solvents [34].

Phase-Change Solvents

Raksajati et al. [48, 72] studied the techno-economic analysis of the carbon capture using phase-change solvents. The authors estimated that the capture costs using phase-change solvents were US$52−108 per tonne CO_2 avoided if the absorber can be operated in a packed column, in comparison to the cost of MEA 30% wt solvent at US$88 per tonne CO_2 avoided. Furthermore, in terms of process design improvement, the use of the additional solid-liquid separator increases the capital costs. However, the separator may reduce the heat duty for the solvents, due to a lower heat of absorption duty and water vaporization duty, or the lower sensible heat duty and water vaporization duty in the stripper[48].

For phase-change solvents, potassium carbonate, ammonia and potassium taurate are at the most advanced stage of development within this solvent class, being at various stages of demonstration field. Potassium carbonate is the main compound used in UNO Technology's UNO MK 3 solvent processes [31], ammonia is the compound used in Alstom's Chilled Ammonia Process (CAP) [58], and potassium taurate is one of the amino acid salt solvents under development by Siemens Energy and TNO [49, 73].

The carbonate phase-change solvents currently tested at demonstration stage do not require FGD and SCR pre-treatment. This is because the SO_x and NO_x are neutralised by alkali hydroxides resulting in sulphate and nitrate salts. For potassium taurate, the behavior of the solvent with SO_x and NO_x has not been reported in published studies. For phase-change solvents, one of the challenges identified is the need to improve absorber designs. This is because current process technologies require the use of spray towers to avoid plugging. During the pilot testing of the Chilled Ammonia Process (CAP) by Alstom power, it was revealed that managing solids in a packed column was unfeasible because of equipment clogging resulting in difficulties in plant operation and control [46]. Therefore, the use of a novel absorber or packing design to allow precipitation during absorption and regeneration is needed to advance the development of phase-change solvents, such as the proposal by CSIRO for their ammonia solvents [34]. Another important process improvement required for this solvent class is the use of internal heat recovery [48].

Encapsulated Solvents

Encapsulated solvents are still in the early development phase and have only been demonstrated at lab scale. Sodium carbonate was identified as an excellent solvent in an encapsulated solvent system following an evaluation of a broad range of solvents by Stolaroff and Bourcier [64] and Vericella et al. [65]. To date, there is no published study that reports the solvent degradation for an encapsulated solvent system, which needs to be investigated in future research. There has also been minimal research of process and equipment designs for encapsulated solvent systems. It is also crucial to improve the current process and equipment designs, especially by developing a novel CFB absorber, such as a multistage counter-current CFB absorber, in order to enhance the equilibrium conditions in the absorber.

Raksajati et al. [74] presented the techno-economic assessment of encapsulated solvent for CO_2 capture using two configurations: (1) multiple fixed bed columns and (2) a circulating fluidized-bed absorber and a bubbling fluidized-bed regenerator. The authors reported that the capture costs using this process were 1.6-2 times more expensive than the conventional MEA solvent system, which is due to the extra membrane resistance in the encapsulated system which increases the regeneration energy required, the higher equipment cost, and the higher capital cost. Further improvement of encapsulated solvent may involve the use of a suitable heat recovery scheme within the process, the use of novel absorber and/or regenerator column designs, the use of solvents encased in very thin capsules, and the use of new solvents with more favourable properties [74].

7 Conclusion

This paper reviews the development and the key parameters involved in the design improvement of several chemical absorption process systems. These processes include chemical absorption using aqueous solvents, phase-change solvents, and encapsulated solvents systems. In addition, some guidelines for the future development has also been presented.

References

1. IEA-GHG, *Energy technology perspectives 2012*, (IEA GHG, 2012)
2. G. T. Rochelle, Science, **325(5948)**, 1652-1654 (2009)
3. GCEP, An assessment of carbon capture technology and research opportunities, (Global Climate Energy Project (GCEP) Stanford University, 2005)
4. M. Wang, , A. Lawal, P. Stephenson, J. Sidders , C. Ramshaw, Chem. Eng. Res. Des., **89(9)**, 1609-1624 (2011)
5. J. Oexmann, A. Kather, Int. J. Greenhouse Gas Control, **4(1)**, 36-43 (2010)
6. K. Smith, U. Ghosh, A. Khan, M. Simioni, K. Endo, X. Zhao, S. Kentish, A. Qader, B. Hooper, G. Stevens, Energy Procedia **1(1)**, 1549-1555 (2009)
7. M. R. M. Abu-Zahra, L. H. J. Schneiders, J. P. M. Niederer, P. H. M. Feron, G. F. Versteeg, Int. J. Greenhouse Gas Control **1(1)**, 37-46 (2007)
8. A. N. M. Peeters, A. P. C. Faaij, W. C. Turkenburg , Int. J. Greenhouse Gas Control **1(4)**, 396-417 (2007)
9. M. T. Ho, *Techno-economic modelling of CO_2 capture systems for Australian industrial sources* (PhD Thesis, University of New South Wales, 2007)
10. G. Puxty, R. Rowland, A. Allport, Q. Yang, M. Bown, R. Burns, M. Maeder, M. Attalla, Env. Sci. Tech., **43(16)**, 6427-6433 (2009)
11. D. G. Chapel, C. L. Mariz, *Recovery of CO_2 from flue gases commercial trends* (Canadian Society of Chemical Engineers Annual Meeting, Saskatoon, Canada, 1999)
12. A. Kothandaraman, *Carbon dioxide capture by chemical absorption: A solvent comparison study* (PhD Thesis, MIT, 2010)
13. J. R. Gibbins, R. I. Crane, D. Lambropoulos, C. Booth, C. A. Roberts, M. Lord. *Maximising the effectiveness of post combustion CO_2 capture systems* (7th International Conference on Greenhouse Gas Technologies, Vancouver, Canada, Elsevier Science Ltd., 2004)
14. A. B. Rao, E. S. Rubin, M. B. Berkenpas. *An integrated modelling framework for carbon management technologies* (Carnegie Mellon University, Center of Energy and Environmental Studies, 2004).
15. P. H. M. Feron, N. A. M. ten Asbroek. *New solvents based on amino-acid salts for CO_2 capture from flue gases* (7th International Conference on Greenhouse Gas Technologies, Vancouver, Canada, Elsevier Science Ltd., 2005)
16. E. B. Rinker, S. S. Ashour, O. C. Sandall, Ind. Eng. Chem. Res. **35(4)**, 1107-1114 (1996)
17. R. Cadours, D. Roquet, G. Perdu, Ind. Eng. Chem. Res., **46(1)**, 233-241 (2007)
18. J. J. Ko, M. H. Li., Chem. Eng. Sci. **55(19)**, 4139-4147 (2000)
19. T. Sema, A. Naami, K. Fu, M. Edali, H. Liu, H. Shi, Z. Liang, R. Idem, P. Tontiwachwuthikul, Chem. Eng. J., **209(0)** (2012)
20. F. Closmann, T. Nguyen, G. T. Rochelle, Energy Procedia **1(1)**, 1351-1357 (2009)
21. N. J. M. C. Penders-van Elk, E. S. Hamborg, P. J. G. Huttenhuis, S. Fradette, J. A. Carley and G. F. Versteeg, Int. J. Greenhouse Gas Control **12(0)**, 259-268 (2013)
22. A. Naami, T. Sema, M. Edali, Z. Liang, R. Idem, P. Tontiwachwuthikul, Int. J. Greenhouse Gas Control **19(0)**, 3-12 (2013)
23. T. Mimura, S. Shimojo, T. Suda, M. Iijima, S. Mitsuoka, Energy Convers. Mgmt. **36**, 397-400 (1995)
24. T. Mimura, H. Simayoshi, H. Suda, M. Iijima, S. Mitsuoka, Energy Convers. Mgmt., **38**, S57-S62 (1997)
25. T. Kamijo, N. Imai, M. Iijima, T. Takashina, H. Tanaka, *Recent technology development of KS-1 CO_2 recovery process* (2004)
26. M. Iijima, T. Nagayasu, T. Kamijyo, S. Nakatani, Mitsubishi Heavy Industries Technical Review, **48(1)**, 26-32 (2011)
27. S. A. Freeman, R. Dugas, D. H. Van Wagener, T. Nguyen, G. T. Rochelle, Int. J. Greenhouse Gas Control **4(2)**, 119-124 (2010)
28. M. J. Goldman, N. A. Fine, G. T. Rochelle, Env. Sci. Tech., **47(7)**, 3528-3534 (2013)
29. M. D. Hilliard, *A predictive thermodynamic model for an aqueous blend of potassium carbonate, piperazine, and monoethanolamine for carbon dioxide capture from flue gas* (PhD Thesis, University of Texas, 2008)
30. J. Oexmann, C. Hensel, A. Kather Int. J. Greenhouse Gas Control **2(4)**, 539-552 (2008)
31. K. Endo, G. Stevens, B. Hooper, S. Kentish, C. Anderson. *A process and plant for removing acid gases* (Patent WO 2011130796 A1, 2011)
32. G. Puxty, R. Rowland, M. Attalla, Chem. Eng. Sci. **65(2)**, 915-922 (2010)

33. P. Versteeg, E. S. Rubin, Int. J. Greenhouse Gas Control, **5(6)**, 1596-1605 (2011)
34. H. Yu, S. Morgan, A. Allport, A. Cottrell, T. Do, J. McGregor, L. Wardhaugh, P. Feron, Chem. Eng. Res. Des. **89(8)**, 1204-1215 (2011)
35. Z. Niu, Y. Guo, Q. Zeng, W. Lin, Fuel Process. Tech. **108(0)**, 154-162 (2013)
36. M. Hasib-ur-Rahman, M. Siaj, F. Larachi, Chem. Eng. Process., **49(4)**, 313-322 (2010)
37. D. Wappel, G. Gronald, R. Kalb, J. Draxler, Int. J. Greenhouse Gas Control **4(3)**, 486-494 (2010)
38. M. Ramdin, T. W. de Loos, T. J. H. Vlugt, Ind. Eng. Chem. Res., **51(24)**, 8149-8177 (2012)
39. Y. Zhang, Y., P. Yu, Y. Luo, Chem. Eng. J., **214(0)**, 355-363 (2013)
40. A. P. Abbott, D. Boothby, G. Capper, D. L. Davies, R. K. Rasheed, J. Am. Chem. Soc., **126(29)**, 9142-9147 (2004)
41. Q. Zhang, K. De Oliveira Vigier, S. Royer, F. Jerome, Chem. Soc. Rev., **41(21)**, 7108-7146 (2012)
42. R. B. Leron, M.H. Li. Thermochimica Acta **551(0)**, 14-19 (2013)
43. E. Ali, M. K. Hadj-Kali, S. Mulyono, I. Alnashef, A. Fakeeha, F. Mjalli, A. Hayyan, Chem. Eng. Res. Des., **92(10)**, 1898-1906 (2014)
44. E. L. Smith, A. P. Abbott, K. S. Ryder, Chem. Rev., **114(21)**, 11060-11082 (2014)
45. J. P. Brouwer, P. H. M. Feron, N. A. M. ten Asbroek, *Amino-acid salts for CO_2 capture from flue gases* (TNO Science & Industry, 2005)
46. D. Sutter, M. Gazzani, M. Mazzotti, Chem. Eng. Sci., **133**, 170-180 (2015)
47. E. Sanchez-Fernandez, E. L. V. Goetheer, Energy Procedia, **4(0)**, 868-875 (2011)
48. A. Raksajati, M. T. Ho, D. E. Wiley, Ind. Eng. Chem. Res., **55(7)**, 1980-1994 (2016)
49. E. Sanchez-Fernandez, E., K. Heffernan, L. V. van der Ham, M. J. G. Linders, E. Eggink, F. N. H. Schrama, D. W. F. Brilman, E. L. V. Goetheer, T. J. H. Vlugt, Ind. Eng. Chem. Res., **52(34)**, 12223-12235 (2013)
50. P. S. Kumar, J. A. Hogendoorn, P. H. M. Feron, G. F. Versteeg, Ind. Eng. Chem. Res., **42(12)**, 2832-2840 (2003)
51. P. S. Kumar, J. A. Hogendoorn, S. J. Timmer, P. H. M. Feron, G. F. Versteeg, Ind. Eng. Chem. Res. **42(12)**, 2841-2852 (2003)
52. A. F. Portugal, J. M. Sousa, F. D. Magalhães, A. Mendes, Chem. Eng. Sci., **64(9)**, 1993-2002 (2009)
53. S. Ma'mun, I. Kim, Energy Procedia **37(0)**, 331-339 (2013)
54. C. C. Wei, G. Puxty, P. Feron, Chem. Eng. Sci. **107(0)**, 218-226 (2014)
55. K. A. Mumford, K. H. Smith, C. J. Anderson, S. Shen, W. Tao, Y. A. Suryaputradinata, A. Qader, B. Hooper, R. A. Innocenzi, S. E. Kentish, G. W. Stevens, Energy & Fuels **26(1)**, 138-146 (2012)
56. K. Smith, G. Xiao, K. Mumford, J. Gouw, I. Indrawan, N. Thanumurthy, D. Quyn, R. Cuthbertson, A. Rayer, N. Nicholas, A. Lee, G. da Silva, S. Kentish, T. Harkin, A. Qader, C. Anderson, B. Hooper, G. Stevens, Energy & Fuels, **28(1)**, 299-306 (2014)
57. F. Kozak, A. Petig, E. Morris, R. Rhudy, D. Thimsen, Energy Procedia, **1(1)**, 1419-1426 (2009)
58. NETL, *Chilled Ammonia-based wet scrubbing for post-combustion CO_2 capture* (DOE USA, 2007)
59. P. M. Mathias, S. Reddy, J. P. O'Connell, Int. J. Greenhouse Gas Control **4(2)**, 174-179 (2010)
60. G. Valenti, D. Bonalumi, E. Macchi, Fuel **101(0)**, 74-83 (2012)
61. Q. Zeng, Y. Guo, Z. Niu, W. Lin, Fuel Process. Tech. **(0)** (2013)
62. R. D. Aines, W. L. Bourcier, C. M. Spadaccini, J. K. Stolaroff, *Polymer-encapsulated carbon capture: Liquids that tolerate precipitation of solids for increased capacity* (WO 2013/063501 A2, 2013)
63. R. D. Aines, C. M. Spaddaccini, E. B. Duoss, J. K. Stolaroff, J. Vericella, J. A. Lewis, G. Farthing, Energy Procedia **37(0)**, 219-224 (2013)
64. J. K. Stolaroff, W. L. Bourcier, Energy Procedia **63(0)**, 2331-2335 (2014)
65. J. J. Vericella, S. E. Baker, J. K. Stolaroff, E. B. Duoss, J. O. Hardin Iv, J. Lewicki, E. Glogowski, W. C. Floyd, C. A. Valdez, W. L. Smith, J. H. Satcher Jr, W. L. Bourcier, C. M. Spadaccini, J. A. Lewis, R. D. Aines, Nat Commun 6 (2015)
66. R. P. Lively, R. R. Chance, W. J. Koros, Ind. Eng. Chem. Res. **49(16)**, 7550-7562 (2010)
67. G. D. Pirngruber, F. Guillou, A. Gomez, M. Clausse, Int. J. Greenhouse Gas Control **14(0)**, 74-83 (2013)
68. H. Liu, B. Liu, L.-C. Lin, G. Chen, Y. Wu, J. Wang, X. Gao, Y. Lv, Y. Pan, X. Zhang, X. Zhang, L. Yang, C. Sun, B. Smith, W. Wang, Nat Commun 5 (2014)
69. H. Krutka, S. Sjostrom, *Evaluation of solid sorbents as a retrofit technology for CO_2 capture from coal-fired power plants* (ADA Environmental Solutions 2011)
70. A. Raksajati, M. T. Ho, D. E. Wiley, Ind. Eng. Chem. Res. **52(47)**, 16887-16901 (2013)
71. D. Shaw (2014), *Cansolv at Boundary Dam - Integrated SO_2 and CO_2 capture for SaskPower. Platts* (8th Annual European CCS, 2013)
72. A. Raksajati, M. T. Ho, D. E. Wiley, Energy Procedia **63**, 2280-2288 (2014)
73. Siemens, *CO_2 capture utilization and sequestration (CCUS) with Siemens's PostCap TM Technology*
74. A. Raksajati, M. T. Ho, D. E. Wiley, Ind. Eng. Chem. Res., **56(6)**, 1604-1620 (2017)

Evaluation of *Ankistrodesmus falcatus* for Bicarbonate-Based Integrated Carbon Capture System (BICCAPS)

Arnel B. Beltran[1,*], *Daniel C.* Gravador[1], *Bea Lutchi O.* Ty[1], *Joceleen Michelle O.* Wu[1]

[1]Chemical Engineering Department, De La Salle University – Manila, Manila, Philippines

Abstract. This study evaluates the performance of alkaliphilic microalgae *Ankistrodesmus falcatus* in the Bicarbonate-based Integrated Carbon Capture and Algae Production System (BICCAPS). The system utilized bicarbonate as carbon source for microalgae production. BICCAPS parameters such as pH, algal biomass productivity and CO_2 utilization (inorganic carbon conversion, C_i) were observed at different sodium bicarbonate ($NaHCO_3$) loading concentration and type of culture media. The highest productivity was observed at 10 g/L of $NaHCO_3$ loading in BRSP medium at 3.5539 mg/L/day. This value is 30% lower compared to the control experiment (continuously aerated bioreactor). The C_i values of the different system ranges from 1.17 x 10^{-4} to 1.51 x 10^{-4} moles/L/day. Both the pH of the BRSP and NPK media at 10 g/L and 30g/L loading of $NaHCO_3$ increased through time. The result shows that *A. falcatus* has a potential in BICCAPS utilization.

1 Introduction

The increasing level of carbon dioxide (CO_2) in the atmosphere had caused severe impact in global temperature. The mean CO_2 concentration in the atmosphere had passed the 400 ppm mark which is beyond the safety limit of 350 ppm [1]. The major culprit for in the increase in CO_2 in the atmosphere is the utilization of fossil fuel for energy generation and trasnportation. Several strategies is been used to reduce or prevent further release of CO_2 such as improvement in the technology, energy conservation, utilization of renewable technology and CO_2 sequestration [2].

Carbon Capture and Sequestration (CCS) system is the most viable techniques to be applied for exisiting industrial plants. In CCS system, the CO_2 gas is captured in an amine-based or carbonate-based adsorbents and stored in a vessel as compressed gas and transported to a place where it could have negligible effect on the environment [3]. Still, these technologies are energy intensive because of the thermal treatment required to regenerate the absorbents rendering it to be expensive [4]. Alternative CCS technology utilized alkaline solutions (e.g. NaOH) and microalgae in photobioreactor [5-6].

The popularity of microalgae as CCS is due to high growth rate and capture carbon capability on top of that it could be used as a potential fuels source. One of the several microalgae that has been evalauted as biofuel feedstock is *Ankistrodemus falcatus* [7-9]. Although, study for it as potential for CCS were limited. One study reported that *A. falcatus* grew in a specialized PHM medium able to havecarbon removal rate of 20% utilizing a 2% CO_2 feed gas stream [10]. *A. falcatus* has reported to contain a lipid of more than 35%, optimal growth length of 0.20 μm and a biomass productivity of 7.9 mg/L-day in a BG-11 medium [8]. Also, it was capable to produce a lipid yield of 67.2%, but it had already undergone a morphology change. With another study, it was known that nitrogen, phosphorus and iron nutrient has a significantly effect in lipid productivity and can obtained biofuel conversion rate as high as 91% [7].

Bicarbonate-based Integrated Carbon Capture and Algae Production System (BICCAPS) is an improvement of CCS system where both chemical and biological systems together is being utlized [4]. The technology captures CO_2 in the form of bicarbonate as alkali solution and utilizes it as carbon source for microalgae. The microalgae should be able to withstand a highly alkaliphilic environment for this system to be viable thus limit the study to some cyanobacteria species.

Liu et al. conducted a study that utilized *Chlorella sp.* and *Scenedesmus obliquus sp.* in a continuous system using bicarbonate-based media and observed that the microalgae was able to grow at a rate of 0.41 g (dry weight) per gram of HCO_3^- [11]. In another study, it was determined that *Euthalothece* ZM001 species has potential for BICCAPS system due to its biomass productivity of 1.21 g/L-day while the *Spirulina* species

* Corresponding author: arnel.beltran@dlsu.edu.ph

has low potential due to its low biomass productivity of 0.15-0.43 g/L-day [4]. One study found out that the used of sodium bicarbonate in culture media is advantageous because of its capability to increase microalgae cells and the amount of lipids in *Tetraselmis suecica* and *Nannochloropsis salina* [12]. However, no studies were reported in utilizing bicarbonate solutions in the growth of *falcatus*.

Thus, this study attempted to evaluate the utilization of bicarbonates for the growth of *A. falcatus*. The solution pH, algal biomass productivity and CO_2 utilization as inorganic carbon conversion (C_i) were determined at different salt loading concentration and type of culture media.

2 Experimental

2.1 Materials

The following reagents and materials were used in the study: sodium bicarbonate ($NaHCO_3$, Himedia, 99.5%) and filter membrane (Pall, 0.45μm) purchased from Yana Chemodities Inc., CO_2 gas (Technical grade) obtained from Linde Philippines. Analytical grade of barium chloride (BaCl), potassium hydrogen phthalate, sodium hydroxide (NaOH) pellets, hydrochloric acid (HCl, Univar, 36%), phenolphthalein indicator and bromocresol green indicator were procured from Crimson Scientific & Laboratory Supply.

2.2 Microalgal strain, culture media and cultivation

Southeast Asian Fisheries Development Center (SEAFDEC) provided the alkaliphilic freshwater microalgae *Ankistrodesmus falcatus* maintained in NPK (14-14-14) commercial fertilizer medium and the BRSP medium (formulation from SEAFDEC) as stock culture. For culture media preparation the following formulation were followed: for BRSP media, a ratio of 1:9 of BRSP concentrated solution to distilled water and for NPK media, a 0.1g $N_{14}P_{14}K_{14}$ fertilizer per liter of distilled water.

Microalgae cultivation is done in a 12h light/12h dark cycle utilizing a light source with intensity of approximately 1 Klux. All runs were done at normal ambient temperature conditions in a 4-L PET bottle photobioreactors at a working volume of 3 L.

A. falcatus is cultivated in an open system as air pumps were used to deliver CO_2 from air. The growth curves of obtained from the experiment were used as standard for biomass production.

2.3 Algal production

Batches of 3.0 L NPK and BRSP media with 10 g/L and 30 g/L loading of $NaHCO_3$ were prepared and inoculated with 1:9 algal ratios to culture media. The microalgae were cultured in the solution for 21 days in a 12h light/12h dark cycle with a light intensity of approximately 1 Klux at normal ambient temperature conditions at a closed system. Optical density and pH readings were measured on a regular basis.

2.4 Analysis

The microalgae growth curves in NPK and BRSP media were determined using UV-vis spectrophotometer (Shimadzu UV-170) at a wavelength of 725 nm (optical density, OD) and dry biomass weight (DBW) for 30 days cultivation. DBW were determined by obtaining the mass of the biomass from the sample by filtering it with 0.45 micron filter and dried at 100°C for 3 hrs.

Double-indicator titration method is used to obtain the bicarbonate and carbonate content of the samples with HCl solution as standard. Volumes of the standardized HCl solution were determined in the titration of 10.0 mL filtered sample (with 0.45 micron filter) until the phenolphthalein endpoint (V_{1HCl}, light pink, carbonate content) and the bromocresol green endpoint (V_{2HCl}, faint yellow, bicarbonate content). The carbonate and bicarbonate concentration were determined using (1) and (2), respectively.

$$C_{CO_3^{2-}} = C_{HCl}V_{1HCl} \times 100 \quad (1)$$

$$C_{HCO_3^-} = C_{HCl}\left[V_{2\text{HCl}} - V_{1\text{HCl}}\right] \times 100 \quad (2)$$

The presence of dissolve CO_2 gas in the samples was obtained using precipitation-titration method. Standardized NaOH solution (10 mL) was added to 10 mL sample which convert the dissolve CO_2 in the solution to CO_3^{2-}. The carbonate is precipitated as barium carbonate ($BaCO_3$)by adding an excess BaCl solution in the prepared sample. Excess hydroxide is titrated with standardized HCl solution to phenolphthalein endpoint (V_{3HCl}). Dissolve CO_2 gas concentration in the solution was obtained using (3).

$$C_{CO_2} = \left[C_{HCl}V_{3\text{HCl}} - C_{HCO_3^-}V_{sample}\right] \times 50 \quad (3)$$

3 Results and Discussion

The growth curve of *A. falcatus* for both BRSP and NPK media for 30 days with air sparging is shown in Fig. 1. The results shows that the stationary phase is evident in NPK media at around day 22 but not yet visible in BRSP media during the cultivation period. Fig. 2 show the the linearized plot of a first order growth rate of *A. falcatus* for both BRSP and NPK media using (4).

$$\ln \text{DBW} = u_{\max} t + \ln \text{DBW}_0 \quad (4)$$

The slope of the line is the maximum specific growth rate (u_{max}) of the algal growth. It was determined that the u_{max} for *A. falcatus* are 0.0515 /day and 0.0349 /day for BRSP media and NPK media, respectively. These results indicates that BRSP is the more favourable growth media for *A. falcatus*. This could be due to the present of essential nutrient in BRSP for algal metabolism such as Biotin, Thiamine, and Vitamin B_{12} [13]. *A. falcatus* biomass productivity (BP) is calculated using (5), where DBW_0 and DBW_f are initial and final DBW of *A. falcatus* and t is the time of cultivation. It was determined that the BP for BRSP media is 5.09 mg/L-day while 3.59 mg/L-day for NPK media.

$$BP = \frac{DBW_f - DBW_0}{t} \quad (5)$$

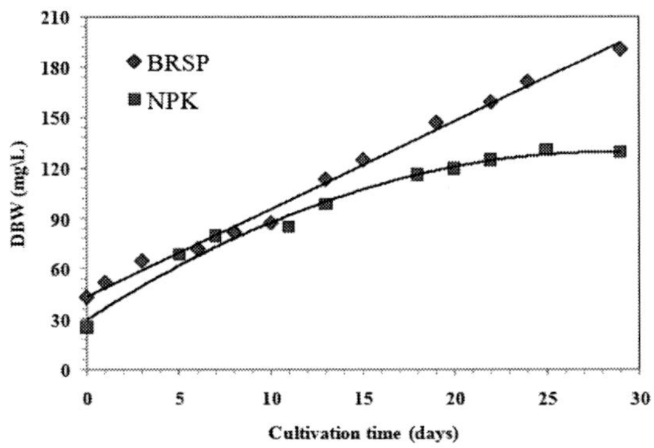

Fig 1. Growth curve of *A. falcatus* in BRSP and NPK culture media.

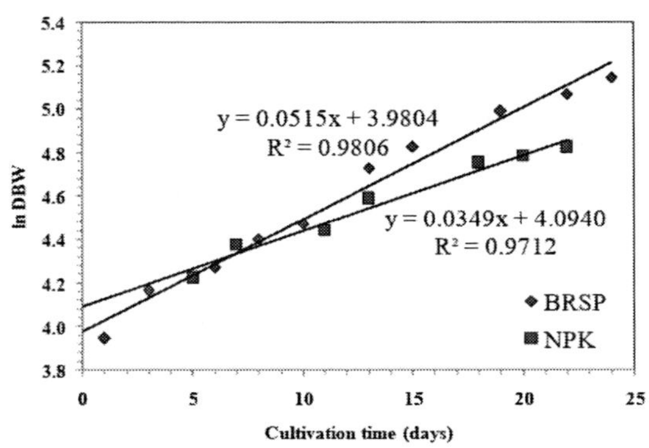

Fig 2. Linearized first order growth rate of *A. falcatus* in BRSP and NPK culture media.

The algal growth and the media physico-chemical properties for bicarbonate-based media at 10 g/L loading were observed for 21 days in a closed system and the results are shown in Fig. 3 to Fig. 7.

From Fig. 3, it could be observe that even though there is no purging of air, the alga was able to grow because of the presence of bicarbonate ions as carbon source. It can be seen that both the growth in BRSP and NPK are still increasing linearly without any trace of stationary phase. Also, the growth at 10 g/L loading of $NaHCO_3$ (Fig. 3a) is much higher than 30 g/L loading (Fig. 3b). This is attributed to higher concentration of Na^+ in the media which is determental to the growth of the algae. Moreover, higher growth is observed for BRSP media with bicarbonate salt in closed system.

Biomass productivities were determined to be 3.554 mg/L-day and 0.954 mg/L-day for BRSP media at 10 g/L and 30 g/L laoding, respectively. While for NPK media, BP values are 1.801 mg/L-day and 0.936 mg/L-day for 10 g/L and 30 g/L laoding, respectively.

Fig. 4 provide the linearized plot of a first order growth rate of *A. falcatus* for bicarbonate-based media in closed system. The u_{max} observed at 10 g/L loading are 0.0604 /day and 0.0428 /day for BRSP media and NPK media, respectively. On the otherhand, u_{max} at 30 g/L loading are 0.0207 /day and 0.0286 /day for BRSP media and NPK media, respectively.

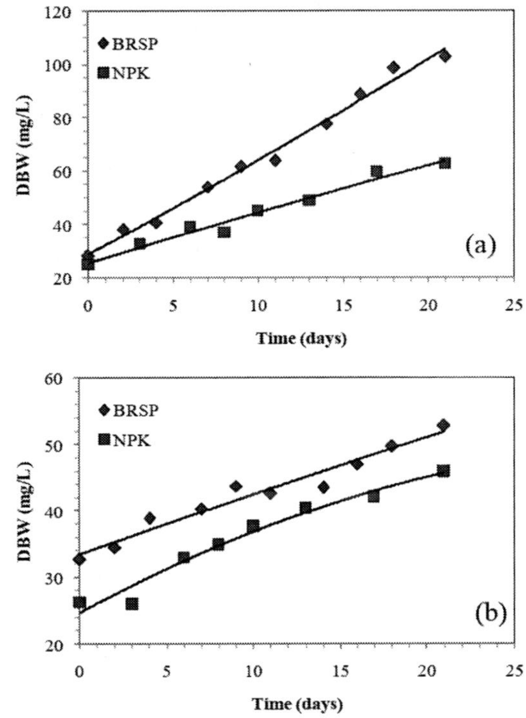

Fig 3. Growth curve of *A. falcatus* in closed BRSP-$NaHCO_3$ solution and closed NPK-$NaHCO_3$ solution media at (a) 10 g/L and (b) 30 g/L loading.

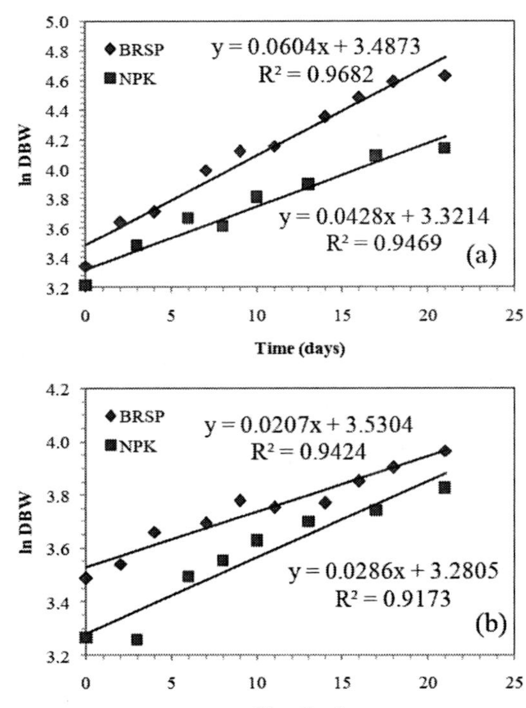

Fig 4. Linearized first order growth rate of *A. falcatus* in closed BRSP-$NaHCO_3$ solution and closed NPK-$NaHCO_3$ solution media at (a) 10 g/L and (b) 30 g/L loading..

At 10 g/L loading, for the bicarbonatete-based BRSP media, the u_{max} was higher by 17% that the airated BRSP media while for the bicarbonatete-based NPK media the u_{max} was higher by 23%. While at 30 g/L loading, the value was reduced by 60% compared to the aerated BRSP media but only 18% reduction is observed for bicarbonatete-based NPK media compared to the aerated one. The higher value for 10 g/L laoding could be due to low agitation observed in the closed system that the aerated system while the lowvalue for 30 g/L loading is caused by Na^+ inhibition.

The pH for both closed systems increases as cultivation time progressed as shown in Fig. 5. The pH increased from 8.54 to 9.07 for the BRSP media while for the NPK, the increased was from 8.64 to 9.06 at 10 g/L loading. At 30 g/L laoding, pH increased from 8.51 to 8.80 in BRSP media while in NPK media the increased was from 8.64 to 8.99.

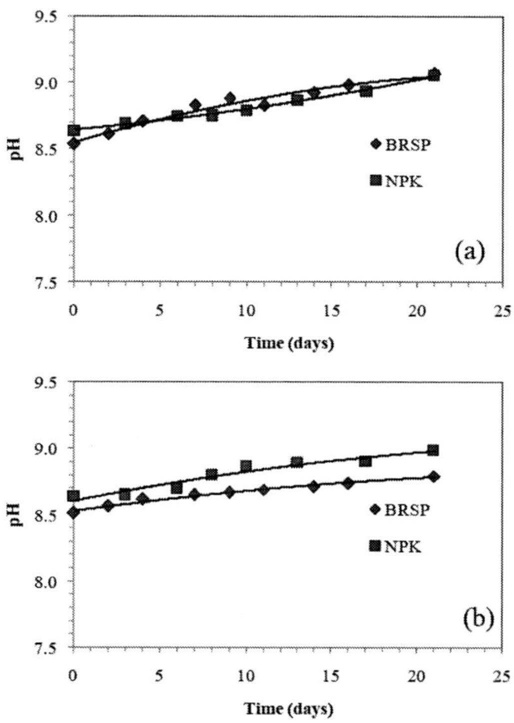

Fig 5. pH profile in closed BRSP-$NaHCO_3$ solution and closed NPK-$NaHCO_3$ solution media during growth of *A. falcatus*.

Fig. 6 and 7 show the HCO_3^- and CO_3^{2-} concentration of the closed system. As the concentration of HCO_3^- in both media goes down it can be observed the CO_3^{2-} concentration have increased. This indicates that the microalgae's growth is due to the uilization of the bicarbonate and convert it to carbonate in the system by the microorganism. Then production of carbonate then thends the pH of the system to goes up.

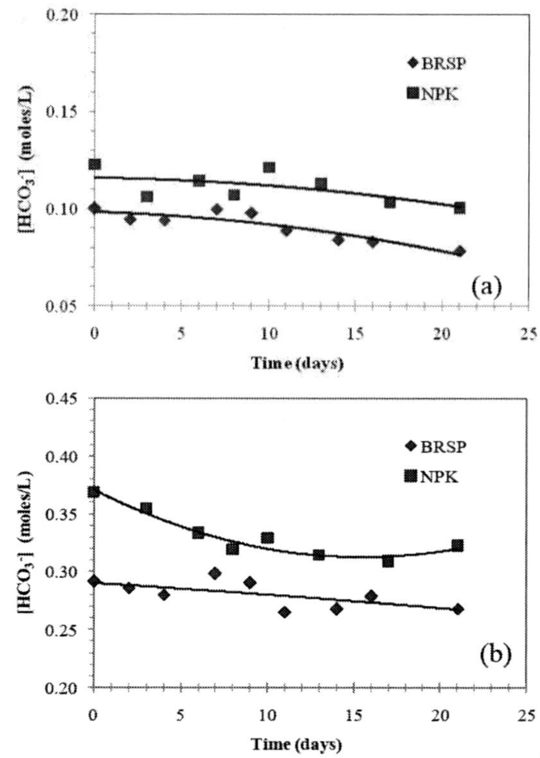

Fig 6. Bicarbonate concentration of closed BRSP-$NaHCO_3$ solution and closed NPK-$NaHCO_3$ solution media during growth of *A. falcatus*.

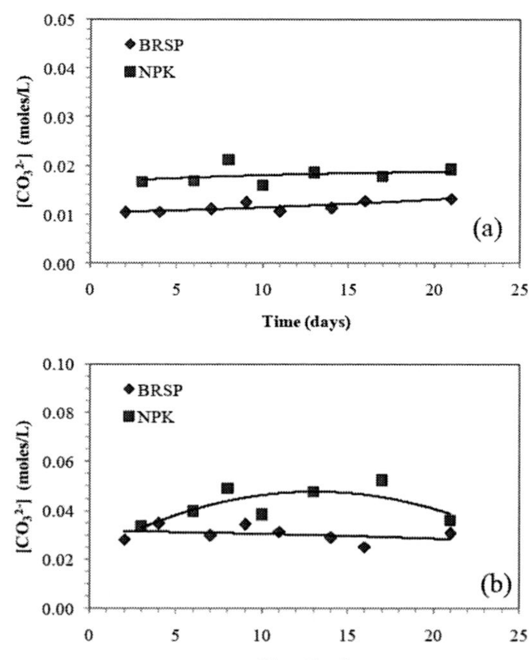

Fig 7. Carbonate concentration of closed BRSP-$NaHCO_3$ solution and closed NPK-$NaHCO_3$ solution media during growth of *A. falcatus*.

Inorganic Carbon (C_i) conversion was defined as the equivalent amount of CO_2 that is converted to biomass per culture volume per time (moles/L-day) and was calculated using (6) where $[CO_3^{2-}]_f$ and $[CO_3^{2-}]_i$ are the initial and final concentration of CO_3^{2-} in the media, $[CO_2]_f$ and $[CO_2]_i$ are the initial and final concentration of CO_2 in the media and t is the time of cultivation. For the bicarbonate-based system, this is the amount of bicarbonate converted to biomass.

$$C_i = \frac{([CO_3^{2-}]_f - [CO_3^{2-}]_i) + ([CO_2]_i - [CO_2]_f)}{t} \quad (6)$$

The calculated C_i for BRSP media at 10 g/L loading was 1.508 x 10^{-4} moles/L-day while 1.260 x 10^{-4} moles/L-day is determined for the NPK media. At 30 g/L laoding, Ci value calculated were 1.432 x 10^{-4} moles/L-day and 1.166 x 10^{-4} moles/L-day for BRSP and NPK media, respectively.

4 Conclusion

The utilization of bicarbonates in solution media for the growth of *Ankistrodesmus falcatus* was investigated. It was observed that growth of A. falcatus is higher in BRSP media than in NPK due to some essential nutrients necessary for algal growth. On the otherhand, bicarbonates laoding at 10 g/L show higher growth than the 30 g/L loading attrbitued to the growth inhibition of Na^+. The pH of the solution media increases as the bicarbonates is utilized and converted to carbonate ions. The result shows that *A. falcatus* can be utilized to in BICCAPS process.

References

1. CO2.Earth. Available at : https://www.co2.earth [Accessed September 7, 2017]
2. B.P. Spigarelli, K. Kawatra, J. CO2 Util. **1**, 69 (2013)
3. J. Gibbins, H. Chalmers, Energy Pol. **36**, 4317 (2008)
4. Z. Chi, Y. Xie, F. Elloy, Y. Zheng, Y. Hu, S. Chen, Bioresour. Technol. **133**, 513 (2013)
5. M. Mahmoudkhani, D. Keith, Int. J. Greenh. Gas Control. **3**, 376 (2009)
6. H. Znad, G. Naderi, H.M. Ang, M.O. Tade, Book Chapter In: Z. Nawaz, eds. *Advances in Chemical Engineering* (2012)
7. P. Singh, A. Guldhe, S. Kumari, I. Rawat, F. Bux, Biochem. Eng. J. **94**, 22 (2015)
8. B. George, I. Pancha, C. Desai, K. Chokshi, C. Paliwal, T. Ghosh, S. Mishra, Bioresource Technol. **171**, 367 (2014)
9. T. Jayanta, K. Chandra, G. Chandra, Int. Res. J. Biol. Sci. **61**, 27 (2012)
10. A. Rinanti, E. Kardena, D. Astuti, K. Dewi, Asian Trans. Sci. & Tech. **3**, 1 (2013)
11. G. Liu, L. Qiao, H. Zhang, , S. Xia, D. Zhao, Enrgy. Proced. **61**, 361 (2014)
12. D. White, A. Pagarette, P. Rooks, S. Ali, J. Appl. Phycol. **25**, 153 (2013)
13. L. Provasoli, A. F. Carlucci, in: W. D. P. Stewart eds. *Algal physiology and biochemistry* (1974)

Biodegradation of Sulphide in Biogas by Biofilm on Salak Fruit Seeds: Accuracy of Quasi-steady-state Approximation

Retno A. S. Lestari[1*], *Wahyudi* B. Sediawan[2], and *Sarto*[2]

[1]Chemical Engineering Department, 17 Agustus 1945 University, Semarang, Indonesia
[2]Chemical Engineering Department, Gadjah Mada University, 2 Yogyakarta 55281, Indonesia

Abstract. This study tried to explore the quantitative description of removal of hydroden sulphide (H_2S) by bio-filtration. H_2S is degraded by bacteria immobilized on the packing materials of Salak fruit seeds inside a column. Two kinetics models are proposed. In both models, the biofilm formed on the packing material is assumed to be thin, so intra-film gradient of H_2S concentration can be neglected. In model 1, material balances of H_2S in the bio-film as well as in the flowing gas are set-up. The growth of the bio-film is modelled by Monod's equation. A set of three simultaneous partial differential equations are obtained. Model 2 is set-up using the same concepts as in model1, but the gas phase is assumed to be quasi-steady-state. This assumption reduces the partial differential equation in model 1 to be an ordinary differential equation which is easier to be solved. The comparisons of the results of model 1 and model 2 can be applied to justify the applications of quasi-steady-state approximation. It turned out that the differences of calculated H_2S concentration results are approximately small, around 6 ppm. Hence, it can be concluded that quasi-steady-state approximation in the gas phase is suggested to be applied.

1 Introduction

Biogas, a potential biomass-based material for alternative energy, is produced by anaerobic decomposition of organic substances such as agricultural waste, animal waste, household waste, waste water, and household waste [1]. Biogas is a side product of biological activity to produce energy by converting organic materials in anaerobic condition. The process is commonly recognized as fermentation. This process actually consists of three major steps, which are hydrolysis, asidogenesis and methanogenesis [2]. The compositions of biogas are typically methane (CH_4) 55-70%, carbon dioxide (CO_2) 30-45% and hydrogen sulphide (H_2S) 0-1.5% [3]. According to Pipatmanomai et al. [4] even though the sulphide compounds contained in biogas are relatively small, they can corrode the materials of equipment utilizing biogas.

Hydrogen sulphide removal from biogas can be achieved by physical and chemical processes such as scrubbing [5], absorption [6], [7], adsorption [8] and reactive absorption. Physical and chemical processes are commonly need high cost and generate other waste.

An alternative method to reduce sulphide compounds in biogas is biochemical process, especially bio-filtration [9]. This process uses a bio-filter containing sulphide degrading bacteria immobilize on the surface of packing material [10].

Bio-filtration is a relatively simple process. Gas containing H_2S was flowed through a pile of packing material which has a surface coated by pollutant-degrading bacteria as a biofilm [11]. In this bio-filtration system, H_2S is absorbed by the biofilm, and then degraded by microorganism to produce elemental sulphur (if oxygen was limited) and $SO4^{-2}$ (if oxygen is in excess), which are non-corrosive [12].

Biofilms are microorganisms that attached and stick together in a polymer matrix [13]. The attachment of microorganisms on the surface of the packing material is very important for the initiation of the development of biofilm [14]. Sulphide-oxidizing bacteria include: bacteria of the genus *Thiobacillus (kemoautotrof/ litotrof)*, green and purple sulphur bacteria (*photoautotrophs*), and types of *heterotrophic* include *Atrhrobacter*, *Bacillus*, *Mikrococcus*, *Mycobacterium and Pseudomonas* [15].

The packing material can be either organic or inorganic compounds. In selecting the packing materials for bio-filter there are several criteria that must be fulfilled include nutrient content (source of C and N), moisture content, pH, porosity, ability to absorb contaminants, lightweight, and cheap [12, 16].

This research used salak fruit seeds (SFS) for bio-filter packing, because SFS easily absorbs water, has a nearly uniform size, containing nutrition and is relatively

* Corresponding author: retnotengaran@gmail.com

rigid. Furthermore, it is largely available and low cost [17]. The utilized sulphide degrading bacteria are isolated from local area, specifically Srandakan, Bantul, Yogyakarta.

Even though a number of studies on sulphide gas removal with bio-filters has been extensively conducted, the use of salak fruit seeds as packing material has never be done. Studies on the removal of H_2S in biogas commonly used a bio-filter with other packing materials and mostly focused on semi quantitative descriptive aspects, namely removal efficiency of H_2S gas (RE). Although there was a number of researcher studied quantitative description, their approaches are still open for improvement.

This study aims to develop a kinetic model that can be applied to design a large scale bio-filter for sulphide elimination from biogas. Special focus in this study is to explore the tolerability of quasi-steady-state approximation for the gas flowing through the column.

This work is a quantitative descriptive study; applying the fundamental concepts of mass transfer and kinetics of biochemical reactions of sulphide compounds in the packed bed. The study included laboratory experiments to generated data for verifications of the kinetic models proposed.

2 Mathematical models

Previous researchers who have done modelling in the process of gas and liquid waste handling using bio-filters include Spigno et al. [18] in the elimination of phenol, Majumder et al. [19] in the process of removal of divalent copper dissolved in water, Jaber et al. [20] in modelling bio-filters for hydrogen sulphide contained in the air, Meena et al. [21] makes the mathematical modelling of the separation of a mixture of hydrophilic (methanol) and hydrophobic (α-Pirene), Santos et al. [22] used a mathematical model for the removal of hydrogen sulphide by oxidation in the bio-filter, Kim and Deshusses [23] calculate the mass transfer coefficient on the packing material used for bio-filter, either in the gas layer or the liquid layer. Furthermore Yang et al. [24] makes modelling of porosity variations of bio-filter equipment used in rotary drum, Ramirez et al. [25] studied the kinetics of microbial growth and biodegradation of methanol and toluene, De Visscher and Li [26] studied the disappearances modelling toluene contained in the air, Agarwal et al. [27] studied the mass balance for phenol degradation events. Modelling in this study was set-up by adopting and combining the basic ideas that emerged in the previous studies.

The mathematical modelling of sulphide removal process with bio-filter columns proposed was based on the most fundamental concepts of mass transfer, phase equilibrium and kinetics of chemical reactions, and tried to avoid the use of special empirical equations. The solution of the mathematical equations formed is also based on common numerical methods, and avoiding special simplification. Two models were proposed with the objective to explore the tolerability of quasi-steady-state approximation for the gas flowing through the column

2.1. Model 1

Assuming that biofilms are relatively thin, model 1 consider that the sulphide concentration in the biofilm at certain axial column position is uniform and the value is equal to the sulphide concentration at the surface of biofilm (Cs = Csi). Furthermore the sulphide concentration in the biogas flowing in the column is considered to be unsteady-state.

Similar to Lestari [28], in the model development, the following simplifications are also applied:

1. The gas phase is plug-flow with no axial dispersions.
2. The biofilm is formed on the surface of the SFS only.
3. The thickness of the biofilm is relatively small compared to size of the SFS, so the geometry of the biofilm can be assumed as a slab.
4. Biodegradation reaction occurs in the biofilm only, and follows Monod's type equation.
5. The biofilm thickness increases as time progresses.
6. The system is isothermal.

Sulphide material balance in the biofilm is as follows:

(The rate of accumulation of sulphide in the biofilm) = (The rate of sulphide mass transfer from the gas to the biofilm) – (The rate of biodegradation in the biofilm)

$$\frac{\partial C_S}{\partial t} = \frac{k_C}{\delta}(C_G - C_{Gi}) - r_A \tag{1}$$

The gas-biofilm phase equilibrium of sulphide follows Henry's type equation:

$$C_{Gi} = H_S * C_S \tag{2}$$

The rate of biodegradation is approximated by Monod's equation:

$$r_A = \frac{\mu_{max} C_s}{C_s + K_S} \tag{3}$$

The thickness increase of the biofilm is caused by the growth of microbes, so it is obtained that:

$$\frac{d\delta}{dt} = \frac{Y_{\frac{X}{S}}}{\rho} . r_A . \delta \tag{4}$$

Since the axial dispersion is to be neglected, material balance of the sulphide in the gas can be set-up as follows:

$$\frac{\partial C_G}{\partial t} = -\frac{G}{\varepsilon S}\frac{\partial C_G}{\partial z} - \frac{k_C a_S}{\varepsilon}(C_G - C_{Gi}) \tag{5}$$

The finite difference approximation (FDA) of equation (1) is as follows:

$$(C_S)_{j+1} = (C_S)_j + \frac{k_C \Delta t}{\delta}\left((C_G)_j - H_S (C_S)_j\right) - (r_A)_j \tag{6}$$

Then, FDA for equation (4) produces:

$$\delta_{j+1} = \delta_j + \frac{Y_{X/S}}{\rho} r_A \delta_j \Delta t \tag{7}$$

Equation (5) is then approximated by:

$$(C_G)_{i,j+1} = (C_G)_{i,j} - \frac{G}{\varepsilon.S}\frac{\Delta t}{\Delta z}\frac{(C_G)_{i+1,j} - (C_G)_{i-1,j}}{2\Delta z} - \frac{k_C a}{\varepsilon}\Delta t\left((C_G)_{i,j} - H_S (C_S)_{i,j}\right) \quad (8)$$

2.2. Model 2

Model 2 is basically similar to model 1, but quasi-steady-state a simplification for the gas phase is applied. Hence equation (1), (2), (3), (4), (6), (7) and (8) are the same, but equation (5) can be simplified since the term $\frac{dC_g}{dt}$ vanishes and the result is:

$$\frac{\partial C_G}{\partial z} = -\frac{k_C aS}{G}(C_G - C_{Gi}) \quad (9)$$

The FDA for equation (9) is as follows:

$$(C_G)_{i+1,j} = \frac{(C_G)_{i,j} + \gamma.H_S.\Delta z.(C_S)_{i+1,j}}{(1+\gamma\Delta z)} \quad (10)$$

where $\gamma = \frac{k_C a}{G}$. (11)

Mas transfer coeficient of sulphide in the gas film is then appriximated by the equation of:

$$k_G = a_0 G^{a_1} \quad (12)$$

in which a_o and a_1 are constants.

3 Research Methodology

3.1 Experimental Data

The experimental data applied to test the accuracies of the mathematical models are from Lestari [29].

3.2. Computation

The FDA for model 1 and 2 are solved using Fortran programs. It turned out that the calculation time for each experimental run using model 1 is approximately 1 minute. Meanwhile the one using model 2 is approximately 3 minute. This comparison shows that significant saving of time is observed.

4. Results and discussions

Simulations applying the two mathematical models proposed were performed using adjustable values of parameters involved until the results are close to the experimental data obtained. The comparison between the calculated results and the experimental data for the models are shown in Table 4.1, Table 4.2, as well as in Fig 4.5 and Fig 4.6. It can be observed that the calculation results using the two models are both close to the experimental data. It means that the two mathematical models proposed are adequate to quantitatively describe H_2S removal from biogas using biofilm on packed bed of SFS.

Table 4.1. Comparison of The results of model 1 and model 2 to experimental data (Debit 500 mL/hr)

Time, hr	Axial position, cm	H_2S concentration, ppm		
		Experimental	Model 1	Model 2
0	0	179.62	176.22	178.97
	20	163.90	157.74	162.26
	40	119.67	130.23	137.19
	60	98.24	103.03	112.78
	80	59.00	85.18	96.57
2	0	163.9	163.90	163.90
	20	138.22	139.23	139.99
	40	96.26	102.95	104.78
	60	39.11	67.62	70.53
	80	23.44	45.39	48.83
4	0	142.48	142.48	142.48
	20	104.40	106.44	107.77
	40	49.00	55.45	58.50
	60	24.15	13.27	15.85
	80	4.06	1.81	1.98

Table 4.2. Comparison of The results of model 1 and model 2 to experimental data (Debit 800 mL/hr)

Time, h	Axial position, cm	H_2S concentration, ppm		
		Experimental	Model 1	Model 2
0	0	87.85	88.02	88.02
	20	82.69	78.24	78.13
	40	75.60	63.81	63.17
	60	53.01	49.70	48.91
	80	41.49	40.60	39.62
2	0	89.99	90	90.00
	20	73.31	76.24	75.81
	40	56.96	56.30	55.29
	60	37.89	37.36	35.90
	80	15.49	25.86	24.19
4	0	89.51	89.51	89.51
	20	60.98	70.02	69.23
	40	45.92	42.64	40.92
	60	22.17	18.85	16.60
	80	5.06	7.79	5.80

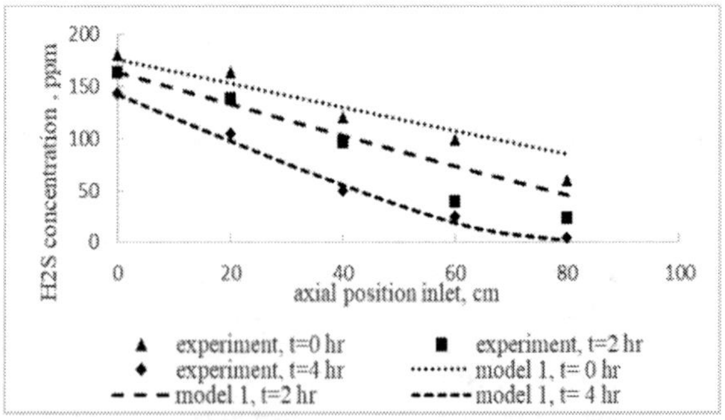

Fig 4.1 H_2S concentration versus axial position at various time at the debit of 500 mL/h for experimental data and model 1

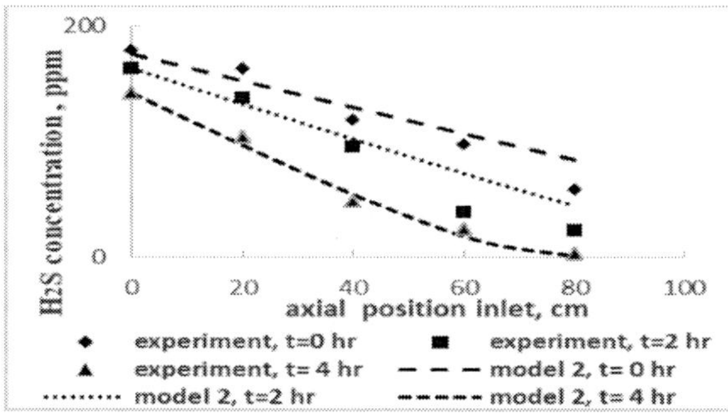

Fig 4.2 H_2S concentration versus axial position at various time at the debit of 500 mL/h for experimental data and model 2

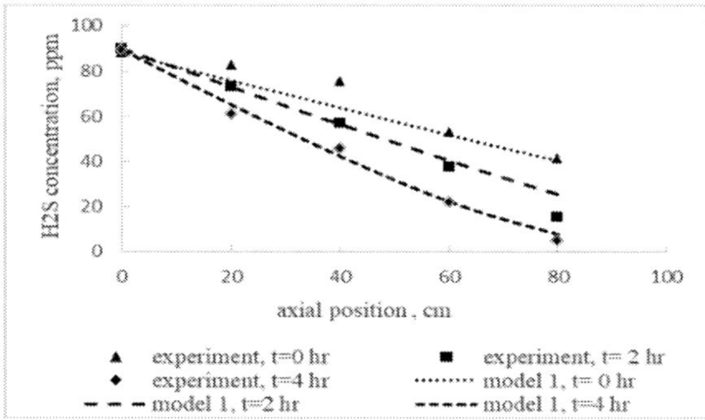

Fig 4.3 H_2S concentration versus axial position at various time at the debit of 800 mL/h for experimental data and model 1

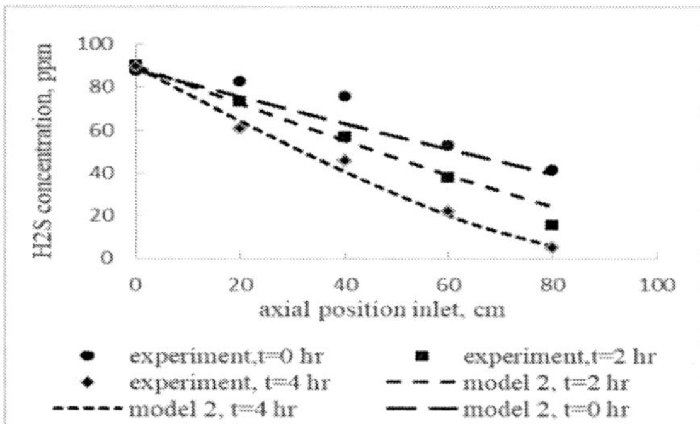

Fig 4.4 H_2S concentration versus axial position at various time at the debit of 800 mL/h for experimental data and model 2

It can also be observed in Table 4.1,Fig 4.1, Fig 4.2 and Table 4.2, Fig 4.3, Fig 4.4 that the results by model 1 are slightly better than the ones of model 2, even though the differences are relatively small. Since the difference between model 1 and model 2 is in the assumption of quasi-steady-state, it can be concluded that simplification of quasi- steady-state in the gas phase gives only minor effects. This phenomenon is logical since the hold-up time of the gas inside the column is relatively very small compare to the time of process. The hold-up time of the process is approximately 0.054 hours, meanwhile time of process approximately 2 hours.

The average absolute errors of model 1 and model 2 are comparable, approximately 6 ppm. All in all, this comparison shows that both models have almost the same accuracies. Model 1 is only slightly better than model 2. Hence quasi-steady-state approximation in the gas phase can be applied with only small deviation.

For model 1 and model 2, the adjustable parameters applied were: the maximum specific growth rate of bacteria (μ_{max}) = 9.3E-7 s^{-1}, K_S = 4.0E-7 g cm^{-3}, H_S = 0.7, $Y_{X/S}$ = 60, and the constants in mass transfer coefficient equation a_0 = 0.038 and a_1 = 0.25.

5 Conclusion

Based on the results of this study it can be concluded that:

1. The accuracies of mathematical models proposed (model 1 and model 2) are turned out to be comparable since the average absolute errors for both models are approximately 6 ppm.
2. Quasi-steady-state approximation in the gas phase gives only minor deviation in biodegradation of sulphide in biogas by biofilm on salak fruit seeds.
3. The computing times for model 1 is approximately 3 times of the one of model 2. Since the accuracies are similar, it is suggested the use of model 2.
4. For model 1 and model 2, the adjustable parameters applied were: the maximum specific growth rate of bacteria (μ_{max}) = 9.3E-7 s^{-1}, K_S = 4.0E-7 g cm^{-3}, H_S = 0.7, $Y_{X/S}$ = 60, and the constants in mass transfer coefficient equation a_0 = 0.038 and a_1 = 0.25.

The authors would like to acknowledge Directorate General of Higher Educations of Indonesia for financial support of this work throug research grant of Hibah Bersaing 2015.

References

1. H.R. Amini, D.R. Reinhart, Waste Manage., **31**, 2020–2026 (2011).
2. T. R. Yadvika, S. Santosh, Sreekrishnan, V. Kohli, Rana, Bioresour. Technol., **95,** 1-10 (2004).
3. M. Syed, G. Soreanu, P. Falletta, M. Béland, Can. Biosyst. Eng. **48** (2006) 2.1–2.14.
4. S. Pipatmanomai, S. Kaewluan, T. Vitidsant, Appl. Energy, **86**, 669-674 (2008).
5. C.C. Lien J.L. Lin, C.H. Ting, Water Scrubbing for, J. Agric. Chem. Environ. **3**, 1-6 (2014).
6. H.B. Boumnijel, H. Amor, N. H. Chekir, C. R. Chim., **30**, 1-8(2016).
7. M. Taheri, A. Mohebbi, H. Hashemipour, A.M. Rashidi, J. Nat. Gas Sci. Eng. **28** 410–417 (2016)
8. M.R.A. Al Mamun, S. Torii, J. Clean Energy Technol., **3,** 6, 428-432, (2015).
8. C. Rattanapan, P. Boonsawang, D. Kantachote, Bioresour. Technol., **100** , 125-130 (2009).
9. Elias, A. Barona, A. Arreguy, J. Rios, I. Aranguiz, J. Penas, Process Biochem., **37**, 813–820 (2002).
10. M.A. Deshusses, G. Hamer, and I.J. Dunn, Environ. Sci. Technol., **29**. 1408–1458 (1995).
11. J.S. Devinny, M.A. Deshusses, and T.S. Webster, Boca Raton. 5–16 (1999).

12. S. Craig, Microbiology: breaking down biofilm. Curr Biol;19:R132–4 (2002).
13. F.P. Chávez, F. Gordillo, C.A. Jerez, Biotechnol. Adv., **24**: 309–20 (2006).
14. P. Edmonds, *Microbiology an environmental Perspective*, Collier Macmillan Publisher. London-New York (1978).
15. M. Hirai, M. Ohtake, and M. Shoda, J. Ferment. Bioeng. (1990)
16. R.A.S. Lestari, W.B. Sediawan and Sarto, *2nd International Conference on Engineering and Technology for Sustainable Development*, Yogyakarta, 13-14 September (2017).
17. G. Spigno, M. Zilli, C. Nicolella, Biochem. Eng. J. 19 (2004).
18. S. Majumder, G. Gangadhar, S. Raghuvanshi, S Gupta, Journal of Water Process Engineering, **6**, 136–143 (2015).
19. M.B. Jaber, A. Couverta, A. Amranea, F. Rouxel, P. Cloireca, L.E. Dumontd, Biochem. Eng. J., **112**, 153–160 (2016).
20. V. Meena, L. Rajendran, S. Kumar, P.G.J. Rani, Egyptian journal of basic and applied sciences, **3**, 94–105 (2016).
21. J.M. Santos, E.S. Lopes, N.C.R. Junior, L. Melo de Sa, N.J. Horand, Water Res., **43**, 3355 – 3364 (2009).
22. S. Kim, and M.A. Deshusses, Chem. Eng. Sci., **63**, 841 – 855 (2008).
23. C. Yang, M.T. Suidan, X. Zhub, B.J. Kimc, G. Zeng, Water Res., **42**,3641–3650 (2008).
24. A.A. Ramirez, S. Bénard, A.G. Fendler, J.P. Jones, M. Heitz, J. Biotechnol. **138**, 88–95 (2008).
25. Visscher, A.D., Li, G.Q., 2008, Toluene removal biofilter modeling: Optimization and case Study, process safety and environment protection 86, 277–282.
26. G.K. Agarwal, A.K. Ghoshal, Bioresour. Technol., **99**, 3765–3773 (2008).
27. R.A.S. Lestari, W.B. Sediawan, and Sarto, *The 2nd International Conference on Engineering and Technology for Sustainable Development* (2017).
28. R.A.S. Lestari, W.B. Sediawan, S. Syamsiah, and Sarto, Eng. J., **4**, 2370–2377 (2016).

Manufacturing Carbon Material by Carbonization of Cellulosic Palm Oil Waste for Supercapacitor Material

Reza Hendriansyah, *Tirto* Prakoso*, *Pramujo* Widiatmoko, *Isdiriayani* Nurdin, and *Hary* Devianto

Institut Teknologi Bandung, Chemical Engineering Department, Jl. Ganesha 10 40132, Indonesia

Abstract. Palm oil waste as biomass resources in Indonesia were not fully utilized. One of the product that can be made from oil palm biomass is activated carbon. Activated carbon characteristic with high porosity and good conductivity, made activated carbon suitable as supercapacitor electrode material. Activated carbon preparation consists of two main steps that are carbonization and activation. In this research carbonization carried out by hydrothermal process while activation conducted by physcal activation. This research focused on manufacturing activated carbon from palm oil waste by hydrothermal carbonization for supercapacitor application. Activated carbon produced from empty fruit bunch have a surface area of 330 - 1181 m^2 / gr, pore volume of 0.19 – 0.69 cm^3/gr, and pore size of 2.1 – 2.3 nm. While activated carbon produced from oil palm shell have surface area of 8 - 451 m^2 / gr, pore volume 0.05 – 1.064 cm^3/gr, and pore size 2.9 – 20.7 nm. The crystallinity of the activated carbon obtained ranged from 46.5 to 51.9%. In this study, the activated carbon is used as a working electrode on an asymmetric hybrid supercapacitor with nickel oxide being used as second electrode. This palm oil-based supercapacitor cell has a capacitance of 1.7554 F / g.

1 Introduction

Indonesia is the largest palm oil producer (Crude Palm Oil) in the world with palm oil production reaches 30 million tons in 2015. Increasing of Indonesia's palm oil production volumes also produced a high amount of palm oil waste especially biomass. Utilization of oil palm biomass for producing new resources is needed to solve the waste management problem. One of the products that can be produced from palm oil waste biomass is activated carbon. Activated carbon is an amorphous carbon material with high porosity and inter-particulate surface area [1]. Activated carbon is usually used as an adsorbent, heterogeneous catalyst support, energy storage applications, etc [2]. Utilization of activated carbon for energy storage applications is used for supercapacitor electrode material due to it has high electrical conductivity, cheap, and easy to produce [2]. Surface area and pore size of activated carbon can be tuned by appropriate method and operating parameters. The pore size of activated carbon which is suitable for capacitor electrode is in the mesopore range (2-50 nm).

Activated carbon preparation consists of carbonization and activation process. The carbonization process is commonly conducted by pyrolysis method by heating the raw material at temperature range 300-650°C with low oxygen condition [3]. Another carbonization process which can be used is a hydrothermal method. Hydrothermal carbonization is a thermochemical conversion process that uses aqueous media to convert biomass into high-value products. The advantage of the hydrothermal process due to it used lower energy than pyrolysis process because the operation temperature is lower than pyrolysis process. Hydrothermal process carried out at a temperature range of 150-350°C [4]. High water content in palm oil biomass waste is suitable for hydrothermal processes because this process does not require the drying process of raw materials [5]. This research was focusing on utilization of palm oil biomass waste to get activated carbon for supercapacitor electrode material using hydrothermal process.

2 Experimental

Hydrothermal carbonization was conducted in a hydrothermal reactor while activation process in a tubular furnace. Empty fruit bunch (EFB) and palm oil shell (POS) were chosen as biomass samples. The samples were gathered from PT. Perkebunan Nusantara VIII in Bogor and palm oil plantation in Riau. EFB and POS were cleaned using water and dried. After that, dried biomass is crushed using disc mill and sieved into coarse granules (35-80 mesh).

* Corresponding author: tirto@che.itb.ac.id

2.1 Hydrothermal

Biomass with the specified ratio of biomass to water (1:10 and 1:3) was fed into hydrothermal reactors containing water mixed with $ZnCl_2$. The ratio of $ZnCl_2$ to the biomass was 2:1 and operating temperature at 275°C. After the reactor reached the desired temperature, it was kept constant for 20-60 minutes. The obtained biochar was then filtered and dried at 105°C for 24 hours.

2.2 Activation

The biochar which recovered from hydrothermal process loaded onto crucible and positioned inside the tubular furnace. The furnace temperature was raised by 10°C/min to reach 800°C under the presence of N_2 at a flow rate of 50 ml/min. After the temperature was reached, N_2 was replaced with CO_2 at a flow rate of 40 ml/min for two hours. Afterwards the furnace was cooled to room temperature in the presence of N_2. Then, activated carbon was stirred for 30 minutes in HCl solution (0,1 M). Furthermore, activated carbon was washed with demineralized water until pH of the wash water reached 7. The activated carbons were washed and dried at 105°C for 24 hours.

2.3 Characterizations

Ultimate and proximate analysis conducted in Laboratorium Pengujian Batubara Tekmira Bandung to determine the chemical and physical characteristic of biomass. Fixed carbon content of biochar determined by gravimetric analysis. FTIR analysis was conducted to get the quantitative and qualitative estimation of oxygen-containing-functional groups. Surface area and pore diameter of activated carbon were analyzed by a gas sorption Quantochrome Instruments Nova 3200e in Laboratorium Instrumentasi Teknik Kimia ITB. The crystalline structure of activated carbon was analyzed by X-ray diffraction (XRD) technique using Bruker D8 Advance with Cu-K radiation (λ = 1.54Å) conducted in Laboratorium Instrumentasi Teknik Kimia ITB. The morphology of raw materials and activated carbon were examined by Scanning Electron Microscopy (SEM) using SEM Hitachi SU3500 conducted in Pusat Penelitian Nanosains dan Nanoteknologi Institut Teknologi Bandung.

2.3 Electrochemical characterizations

The electrochemical analysis was conducted to observed activated carbon performance when assembled as a supercapacitor cells. Cyclic voltammetry (CV), Galvano-static charge/discharge (GCD), and electrochemical impedance spectroscopy (EIS) were examined using Gamry V3000.

The electrochemical examination was performed by assembling supercapacitor cell in an asymmetrical two-electrode configuration. The negative electrode was made from activated carbon which placed in 1 cm^2 of stainless steel mesh. The positive electrode was made from nickel oxide which placed in 1 cm^2 of carbon sheet. Nafion 212 was sandwiched between two electrodes and then infiltrated with an electrolyte solution (6 M KOH).

3 Result and discussion

3.1 Biomass characterizations

Tabel 1. shows the composition of empty fruit bunch and palm oil shell. Biomass composition is depended by where the biomass growth. Based on Table 1, it was the common characteristic of empty fruit bunch and palm oil shell that growth in Indonesia. The ultimate analysis result is observed that both empty fruit bunch and palm oil shell has no significant difference in carbon, hydrogen, nitrogen, oxygen, and sulphur content.

Table 1. The Result of Ultimate and Proximate Analysis of EFB and POS

	EFB	POS
Ultimate Analysis (%)		
C	48.83%	47.38%
H	6.52%	5.94%
N	0.69%	0.47%
O	39.64%	43.54%
S	0.13%	0.05%
Proximate Analysis		
M	5.47%	9.01%
VM	73.17%	68.00%
F C	17.17%	20.37%
A	4.19%	2.62%

Fixed carbon and volatile matter content in empty fruit bunch and palm oil shell is high so that this biomass suitable as a precursor material for activated carbon [6]. The difference of fixed carbon content in empty fruit bunch and palm oil shell will affect the fixed carbon yield that remains in biochar after hydrothermal carbonization process.

3.2 Hydrothermal carbonization

Table 2. shows the result of gravimetric analysis for determining fixed carbon yield of biochar. From Table 2., the holding time and biomass-to-water ratios do not give any significant effect of fixed carbon yield in biochar. Fixed carbon yield of biochar from empty fruit bunch is around 8% while from palm oil shell is around 9%. Fixed carbon yield of biochar from palm oil shell is higher than it from empty fruit bunch. This might occur because palm oil shell has higher fix carbon content as shown in Table 1.

Table 2. *Fixed carbon* yield of bio-char

Raw Materials	Time (min)	Biomass to water ratios	
		1: 10	1 : 3
EFB	20	8.63%	8.70%
	40	9.88%	8.05%
	60	9.30%	8.50%
POS	20	9.15%	9.61%
	40	9.51%	9.46%
	60	9.59%	9.57%

Fig 1. shows FTIR spectrum from empty fruit bunch, palm oil shell and biochar. The intensity of the peak which is around ~3500 cm^{-1} shows stretching vibration of OH-bonds (alcohol, phenol, and carboxyl groups). The intensity of OH-bonds decreased after the hydrothermal process. It is due to dehydration of biomass occurred so that water molecules within the biomass removed [7].

The spectrum of aliphatic (CH) groups present around 2900 cm^{-1}. The intensity weakens after the hydrothermal process which indicates that several aliphatic chains are broke down during carbonization process [7]. The peak around 1600 cm^{-1} attribute to aromatic groups (lignin and aromatic groups) which is formed during carbonization [8]. The intensity around 1000 cm^{-1} shows CO-bonds whose intensity decrease after carbonization because of several CO-bonds are breakdown [8].

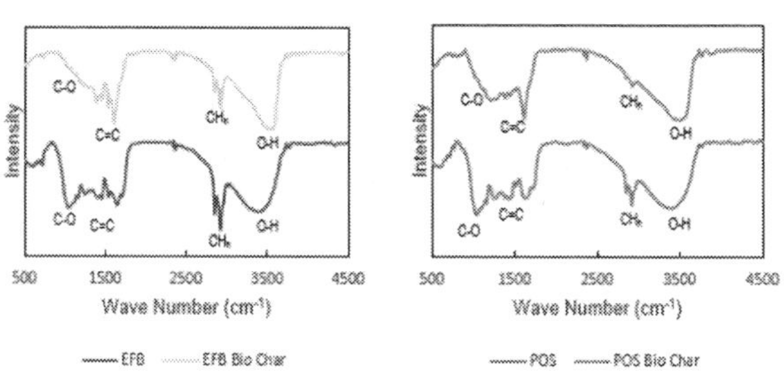

Fig 1. FTIR spectrum of empty fruit bunch, palm oil shell, and biochar

3.3 Activation

The BET surface area of activated carbon which is produced at different process parameters is shown in Table 3. The surface area of activated carbon sample for the biomass-to-water ratio (1:10) increase when holding time increase from 20 to 60 minutes. This may happen because some pores which are formed by the decomposition of organic components in materials increase when holding time of the hydrothermal process is longer. It makes the surface area of activated carbon increased after activation process. The interaction between $ZnCl_2$ and biomass also more intense because of longer holding time during the hydrothermal carbonization. Moreover, $ZnCl_2$ which acts as an activating agent removes the oxygen and hydrogen content in biomass by forming water and could enhance the porosity of the material [9].

The surface area of activated carbon sample increase for residence time 20 minutes whereas biomass-to-water ratio decrease. This may happen because of the concentration of $ZnCl_2$ in the solution increase while biomass-to-water ratio decrease. Thus, the contact between $ZnCl_2$ and biomass is more intense, and resulted the porosity of the activated carbon enhanced.

Table 3. Surface area of activated carbon (m^2/gr)

Raw Materials	Time (min)	Biomass to water ratio	
		(1:10)	(1:3)
EFB	20	330.52	1088.139
	40	383.748	-
	60	644.146	1181.625
POS	20	17.008	451.862
	40	-	89.479
	60	221.407	7.679
Comercial activated carbon			832.967

Average pore size diameter of activated carbon shown in Table.4. Activated carbon from empty fruit bunch and palm oil shell has mesopore size. The activated carbon from palm oil shell has bigger pore size than that from empty fruit bunch and has lower surface area than that from empty fruit bunch. Moreover, the surface area of activated carbon from palm oil shell for the biomass-to-water ratio (1:3) has no tendency increased with elevated residence time. These phenomena might be associated with the decreased solubility of $ZnCl_2$ under subcritical conditions so that diminish its availability for chemical activation [10]. The different raw material which is used for activated carbon manufacturing has also affected the character of activated carbon characteristics [11].

Table 4. Activated carbon pore diameter (nm)

Raw Materials	Time (min)	Biomass to water ratio	
		(1:10)	(1:3)
EFB	20	2.3686	2.1701
	40	2.12972	-
	60	2.232	2.3577
POS	20	12.4039	2.9098
	40	-	4.288
	60	3.128	5.541
Commercial activated carbon			2.29

Average pore size of activated carbon from empty fruit bunch and palm oil shell is on the bottom range of mesopore size. This result affected by $ZnCl_2$-to-raw material ratio. According to Ahmadpour and Do [12],

$ZnCl_2$-to-the raw material ratio which is used in this research drive the micro pore formation and pore widening simultaneously. Even though activated carbon has mesoporous size, it close to micro pore size.

Fig 2. shows the morphology of raw materials and activated carbon after hydrothermal carbonization and activation process. Fig 2a and 2b indicate that both raw materials have a rough surface and closed pores. The morphology of empty fruit bunch and palm oil shell is developed after the heat treatment of hydrothermal carbonization and activation process. More pores are opened on biomass structure creating holes along the biomass surface, so that enhanced the porosity and increased the surface area of activated carbon.

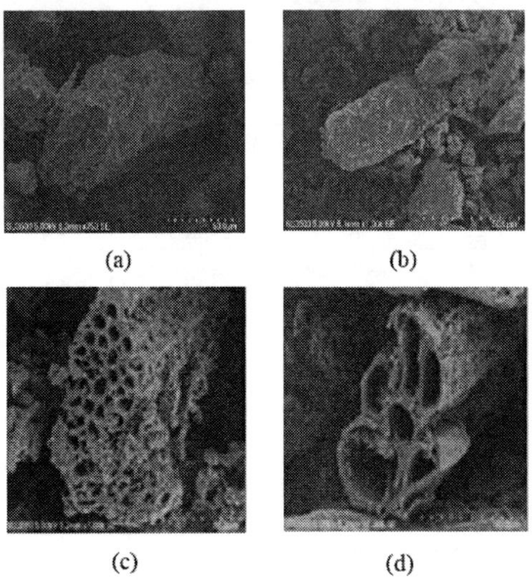

Fig 2. SEM result of: (a) empty fruit bunch; (b) palm oil shell; (c) empty fruit bunch activated carbon; and (d) palm oil shell activated carbon

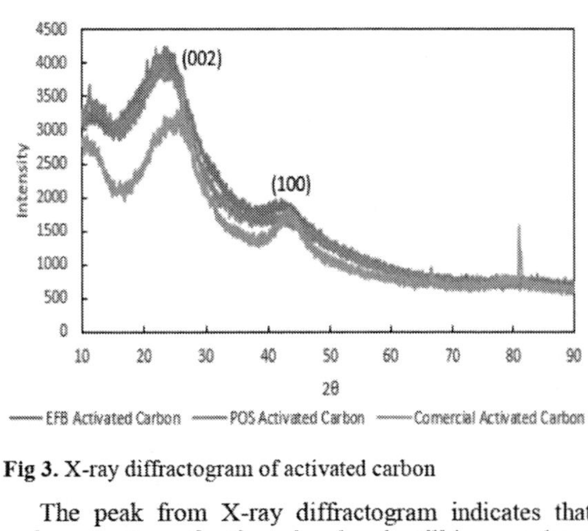

Fig 3. X-ray diffractogram of activated carbon

The peak from X-ray diffractogram indicates that carbon structure of activated carbon is still in amorphous phase. This phase is developed because of the destruction of initial graphite structure during carbonization and activation process. The peak between around 26° and 40° represent disorder carbon structure (002) and (100). This carbon structure is intermediate structure between graphite and amorphous structure which is called turbostratic structure or random layered lattice structure [13]. This structure could be formed due to the presence of local stacking faults, random shifts between adjacent layers, varying interspacing values, unorganized carbons which are not a part of the layer structure, and strain in the layers [14].

3.4 Electrochemical characterizations

Electrochemical characterizations were conducted to analyze the performance of activated carbon while was applied in supercapacitor cell. Activated carbon which produced from empty fruit bunch (biomass-to-water ratio 1:10 and hydrothermal carbonization residence time 40 minutes) is used for this characterization.

Fig 4. shows the voltammogram of supercapacitor cell made by activated carbon produced and commercial activated carbon. The voltammogram has a tendency to form a slope and not in rectangular shape. This shape indicates charging-discharging process through high resistance inside the supercapacitor cell [15]. Moreover from the voltammogram it was suspected that corrosion reaction occurred in electrode due to the dielectric material used was too thin.

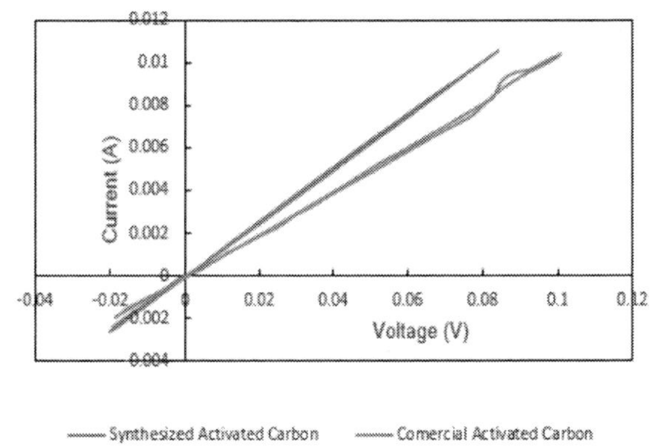

Fig 4. Supercapacitor voltamogram

The specific capacitance of carbon electrode can be calculated from voltammogram according to the following formula [16].

$$C = \sum|I|.\Delta t \ / \ m \, \Delta V \qquad (3.1)$$

where $\sum|I|.\Delta t$ is the area of the current (A) against time (s); m is the mass of active material in the electrode (g); and ΔV (V) is the potential window. Calculation of specific capacitance of carbon electrode shown in Table 5.

According to the capacitance calculation, it shows that the specific capacitance of synthesized activated carbon electrode is greater than commercial activated carbon electrode. The result of specific capacitance shows the contradiction to the surface area of commercial activated carbon. The capacitance is not affected by the surface area of materials. The lower specific capacitance of commercial activated electrode could be caused due to imperfect interphase connection and cell sealing system not optimal [17]. This imperfect cell sealing system could cause the presence of reactive and unstable oxygen functional groups which can interfere the ion transfer process [18]. This result also could be occurred by pore clogging due to some electrolyte ions trapped in the pore and causing the active site of activated carbon for ions storage decreased. In addition, side reactions between carbon, electrolyte and impurities could also decrease cell capacitance [19].

Table 5. Specific capacitance of carbon electrode

Material	Capacitance (F/g)
Synthesized activated carbon	1.7554
Commercial activated carbon	1.5107

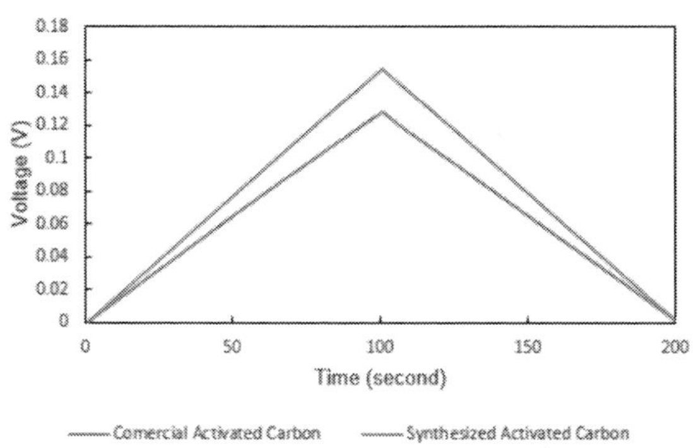

Fig 4. Galvanostatic charge/discharge (GCD) diagram.

Fig 4. shows galvano-static charge/discharge diagram (GCD) of synthesized activated carbon and commercial activated carbon electrode has a triangle and symmetric shape which indicate the supercapacitor cell has good capacitive performance [20]. The GCD of synthesized activated carbon electrode is greater than the that of the commercial activated carbon electrode. This result shows that the capacitance of supercapacitor cells made from synthesized activated carbon is greater than that of commercial activated carbon which is suitable with the results from calculations performed by cyclic voltammetry (CV) analysis.

Electrochemical impedance spectroscopy (EIS) analysis was conducted to estimate the resistance of an electrochemical cell. The result of this analysis is represented in a Nyquist curve which is modelled with equivalent circuits to obtain the impedance value of the supercapacitor cell. The Nyquist curve and equivalent circuit are shown in Fig 5 and Fig 6 respectively.

The Nyquist curve of synthesized activated carbon supercapacitor has one semi circle curve which indicates cell impedance is affected by charge transfer. On the other hands, commercial activated carbon supercapacitor has two semi-circles curves which indicate cell impedance is not only affected by charge transfer but also mass transfer. Moreover, at the end of the semicircle from the Nyquist curve of commercial activated carbon supercapacitor has Warburg line that related to the diffusion layer.

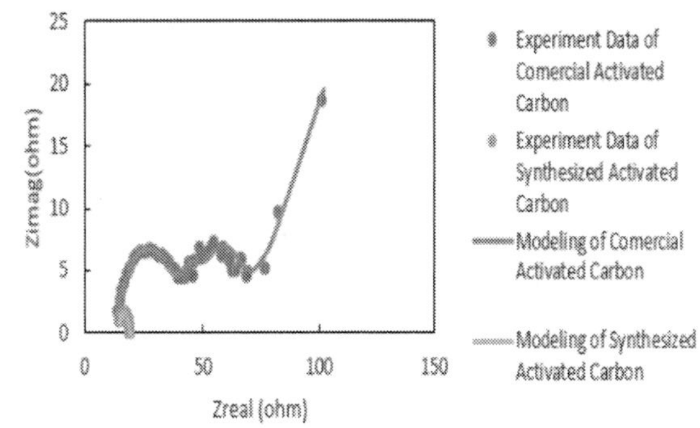

Fig 5. Nyquist curve.

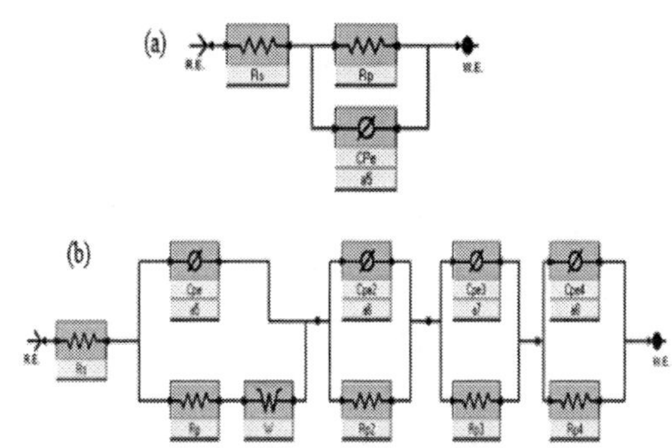

Fig 6. Equivalent circuit: (a) synthesized activated carbon supercapacitor and (b) commercial activated carbon supercapacitor.

Table 6. shows the parameters of the equivalent circuit modelling against the experimental data. Ohmic resistance (Rs) is a parameter which is related to the electrolyte resistance. Both of synthesized activated carbon and commercial activated carbon supercapacitor have a big ohmic resistance number and commercial activated carbon supercapacitor has greater interphase resistance (Rp) than the synthesized activated carbon supercapacitor. Moreover the ohmic and interphase resitance in this research was higher compared to the study conducted by Pramujo et al [17] which ohmic and interphase resistance generated from their supercapacitor cell were 4.9 ohm and 5.4 ohm. This phenomenon might be happen due to the electrolyte dried rapidly because the sealing system was non-optimal. In addition the difficultness of electrolyte diffusion within the entire electrode surface made the both supercapacitor cell from present research have higher ohmic and interphase resistance [17]. Based on the EIS analysis, the interphase resistance of commercial activated carbon supercapacitor causes the capacitance of supercapacitor cell becomes smaller which is suitable with the results from calculations performed by cyclic voltammetry (CV) analysis.

Table 6. Cell impedance (a) synthesized activated carbon supercapacitor (b) commercial activated carbon supercapacitor

(a)

Parameter	Value	Unit
Rs	14.1	ohm
Rp	5.3	ohm
Cpe	0.000043	S*s^a
a7	0.72	

(b)

Parameter	Value	Unit
Rp	4.527	ohm
Rs	12.54	ohm
Cpe	1.848	S*s^a
a5	0.8134	
W	-0.506	S*s^(1/2)
Cpe2	0.000141	S*s^a
a6	0.5693	
Rp2	27.93	ohm
Cpe3	0.02695	S*s^a
a7	0.664	
Rp3	21.74	ohm
Cpe4	0.002883	S*s^a
a8	0.8238	
Rp4	8.52	ohm

4 Conclusions

Activated carbon as an electrode material of supercapacitor can be obtained from palm oil biomass waste through the hydrothermal carbonization and activation process. The capacitance of supercapacitor is 1.7754 F / gr. Decreasing biomass-to-water ratio, as well as increasing residence time during the hydrothermal carbonization process, could increase the surface area of activated carbon.

This work is supported by Badan Pengelola Dana Perkebunan Kelapa Sawit 2016.

References

1. R. C. Bansal, and M. Goyal, *Activated carbon adsorption* (Boca Raton, CRC Press ,2005)
2. E. Frackowiak, Q. Abbas, and F. J. Béguin, Energy Chem. **22(2)**, 226-240 (2013)
3. P. Basu, *Biomass gasification and pyrolysis: practical design and theory* (Academic press ,2010)
4. A. Jain, R. Balasubramanian, & M. P. Srinivasan, Chem. Eng. J **283**, 789-805 (2016)
5. K. Tekin, S. Karagöz, & S. Bektaş, Renew. Sustainable Energy Rev. **40**, 673-687 (2014)
6. A. C. Lua, F. Y. Lau, & J. J. Guo, Anal. Appl. Pyrolysis **76(1)**, 96-102 (2006)
7. A. T. Yuliansyah, T. Hirajima, S. Kumagai, & K. Sasaki, Waste Biomass Valori. **1(4)**, 395-405 (2010)
8. S. Nizamuddin, J. Kumar, N. Subramanian, J. N. Sahu, P. Ganesan, N. M. Mubarak, & S. A. Mazari, Can. J. Chem. Eng. **93(11)**, 1916-1921 (2015)
9. A. Jain, S. Jayaraman, R. Balasubramanian, & M. P. Srinivasan, J. Mater. Chem. **2(2)**, 520-528 (2014)
10. A. Jain, R. Balasubramanian, & M. P. Srinivasan, Microporous Mesoporous Mater. **203**, 178-185 (2015).
11. H. Marsh, and F. Rodriguez, *Activated carbon* (Elsevier Science & Technology Books, 2006).
12. A. Ahmadpour, & D. D. Do, Carbon **35 (12)**, 1723-1732 (1997).
13. B. Manoj, & A. G. Kunjomana, Int. J. Electrochem. Sci **7(4)**, 3127-3134 (2012).
14. B. S. Girgis, Y. M. Temerk, M. M. Gadelrab, & I. D. Abdullah, Carbon letters **8 (2)**, 95-100 (2007).
15. A. Thambidurai, J. K. Lourdusamy, J. V. John, and S. Ganesan, Korean J. Chem. Eng **31 (2)**, 268-275 (2014).
16. A. Elmouwahidi, Z. Zapata-Benabithe, F. Carrasco-Marín, and C. Moreno-Casti, Bioresour. Technol. **111**: 185-190 (2012).
17. P. Widiatmoko, H. Devianto, I. Nurdin, and R. E. Yandra, J. Eng. Technol. Sci. **48 (5)**, 597-613 (2016)
18. W. Gu, N. Peters, and G. Yushin, Carbon **53**: 292-301 (2013).
19. D. Puthusseri, V. Aravindan, S. Madhavi, & S. Ogale, Energy Environ. Sci. *7* **(2)**, 728-735 (2014)
20. A. M. Abioye, Z. A. Noorden, & F. N. Ani, Electrochim. Acta **225**, 493-502 (2017)

Ozonation of Yarn dyed Wastewater in a Continous Stirred Tank Reactor : Kinetic Study and Performance Optimisation

Lieke Riadi[1,2*], *Andrian* Sugiharto[1], and *Hana* Gondokusumo[1]

[1]Departemnet of Chemical Engineering , The University of Surabaya ,Jl.Raya Kalirungkut ,Surabaya 60239, Indonesia
[2]Center for Environmental Studies, The University of Surabaya ,Jl.Raya Kalirungkut ,Surabaya 60239, Indonesia

Abstract. This paper describes the ozonation process in yarn dyed wastewater using continuous stirred tank reactor with the objective to study the kinetic of COD degradation at various volumetric flow rate (30, 50, 70 ml/min) and ozone concentration (2.70 %, 4.25%, 5.86 % mol/mol). The wastewater which was collected from a yarn dying process located in Surabaya area was pretreated by electrocoagulation prior to ozonation process. The electrocoagulation process was carried out to reduce the color intensity and total suspended solid. The pretreated wastewater was then processed using ozonation for 2.5 hours. The result showed that at various concentration of ozone, maximum COD removal was 90.78% which was achieved at 5.86 %mol/mol of Ozon and volumetric flow rate 50 ml/min. It was found that the degradation process followed the pseudo-first order kinetic model. The obtained pseudo-first order rate constants for volumetric rate of 50 mil/min were 0.0307 min^{-1}, 0.0419 min^{-1} and 0.053 min^{-1} for ozon concentration of 2.70 %, 4.25%, 5.86 % mol/mol respectively. The residence time were 41 minutes, 31 minutes and 23 minutes for ozone concentration of 2.70 %, 4.25%, 5.86 % mol/mol respectively. These findings offers an alternative treatment for wastewater containing dyes.

1 Introduction

Yarn dyeing industry generates large amount wastewater with high color intensity, high chemical oxygen demand (COD), suspended solids and some heavy metals. The dyes may be carcinogenic and mutagenic which are harmful for human beings. The dyes have a complex aromatic molecular structure which is difficult to biodegrade [1,2]. Physical methods which are known to be used in the treatment usually used as primary treatment such as precipitation,adsorption, and chemical coagulation. Since the COD content is quite high enough after the primary treatment, it is further treated by other treatment such as chemical degradation or biodegradation. Biological process is slower, requires large area especially in the process of color removal. The electrochemical technique has been used which removed color but not the COD content [3].

Chemical processes are generally more simple in application and need oxidizing agents such as hydrogen peroxide, ozone and hydroxyl radicals. The process is known as Advanced Oxidation Process. Ozone is a strong oxidation agent (E^{o}= 2.07 V) which has been used to oxidize wastewater contained phenols and cyanides [4]. Due to its dipolar nature, ozone can either react as an electrophilic and nucleophilic agent or cause dipolar cycloaddition with most of the compounds with double bonds [5]. As a powerful oxidizing agent, ozone can breakdown the double bonds of dye chromophores and also other functional groups such as complex aromatic rings of dyes [6]. The wastewater used in this study was collected from yarn dyeing industry located in southern part area of Surabaya.

The mechanism of ozonation, follow two main paths: a direct path corresponding to the action of molecular ozone, and an indirect path is a result from the decomposition of ozone to radicals, in which the reaction is initiated by hydroxyl ions (OH^-) [7].The wastewater has a lot of azo dyes which is an aromatic double bonds as can be seen in Figure 1. The wastewater was treated using electrocoagulation in batch system prior ozonation.

The characteristic of wastewater prior and after electrocoagulation is shown in Table 1. The allowable concentration of COD in wastewater to be discharged is 150 ppm as stated in East Java regional government regulation (SK Gub 72/2013). The objectives of this work are to study the effect of ozone concentration, flow rate of wastewater in COD removal efficiency. The pH and conductivity profiles of wastewater during ozonation were also observed. This work also studied kinetic model for COD removal in a Continuous Stirred Tank Reactor (CSTR) mode.

* Corresponding author: lieke@staff.ubaya.ac.id

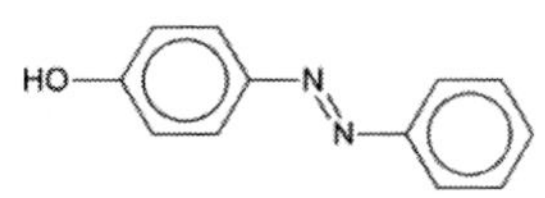

Fig.1. chemical structure of azo dye.

Table 1. Characteristic of wastewater

Parameter	Prior electrocoagulation	After electrocoagulation
Color OD_{498}	1.3117	0.0053
TSS [mg/L]	18,780	2,913
COD [ppm]	745	334
pH	12	8.5
Conductivity [mS/cm]	19.35	13.50

2 Materials and Methods

2.1. Materials

2.1.1 Wastewater

Wastewater used in this study was collected form yarn dyeing industry located in southern part area of Surabaya. Since the fluctuation of wastewater characteristic, we have two types of wastewater with different COD.The wastewater used to study effect of flow rate had a COD content about 1300 ppm prior electrocoagulation and 780 ppm after electrocoagulation. Other experiments for parameter of ozone concentration and pH used wastewater with characteristic as shown in Table 1.

2.1.2. Ozone and oxygen gases

Ozone was generated from oxygen by Ozone generator (Viresco, Singapore). The ozone output was adjusted from 0-50 g/h. The oxygen ouput at 5-15 L/min. The exit port of ozone generator is connected with deep tube and the ozone gas was delivered to the reactor through tube sparger at the bottom of the reactor. The residual ozone was decomposed by trapping excess ozone at the reactor outlet which is connected to 2% potassium iodide (KI) solution trap.

2.2. Combined Electrocoagulation and Continuous Stirred Tank Reactor (CSTR)

An electrocoagulation (EC) reactor with 1.5 L volume was used. The real wastewater was taken form yarn industry located in Surabaya, East Java province, Indonesia. Anode and Cathode Al/Al was used with a distance between two electrodes was 2 cm. The electrodes were connected to DC power supply. The size of Alumunium plate was 8 cm x 8 cm, with a thickness of 2 mm.

The EC process was run for 10 minutes and used a method as described by Riadi *et al.* [8]. The effluent from EC process was then collected and used as influent for ozonation process. A CSTR with 1250 mL volume was developed using a glass vessel connected to collection tank of effluent from electrocoagulation process. Ozone fed to CSTR from ozone generator and diffused through a diffuser in a bottom of reactor.The system used in the experiment can be seen in Figure 2.

2.3. Analysis of wastewater

COD was measured using closed reflux, colorimetric method based on Standard Methods for the Examination of Water and Wastewater [9]. Total Suspended solid was measured using dried method, color measurement was conducted using UV-Vis spectrophotometer at 498 nm wavelength based on Standard Methods for the Examination of Water and Wastewater [9].

The wastewater was also analysed using GC-MS to find out the components of dye in the wastewater. The GC serie is Agilent 7890 and the MS is agilent 5975, the column used was HP5MS, 30 m length and the diameter is 0.2 μm. Ozone concentration trapped in KI solution was measured using sodium thiosulfate tritation procedure. Ozone consumed was calculated as: Ozone produced – Ozone trapped in KI trapper, with an assumption there is no ozone left to the environment.

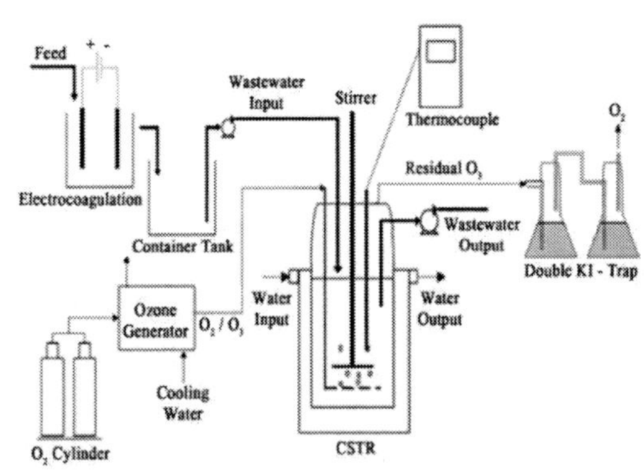

Fig. 2. Scheme of ozonation in CSTR

2.4. Ozonation process

Ozonation process was carried out to evaluate the performance of COD degradation in a CSTR with working volume of 800 mL, and agitation of 200 rpm at 30°C. The parameters used in the experiment were volumetric flowrate of wastewater (30, 50, 70 ml/minutes) with ozone concentration of 4,25 % mol/mol; the effect of ozone concentration (2,70, 4,25, 5,86 %mol/mol) was studied with wastewater flow rate of 50 mL/min. Samples were taken periodically for 2.5

hours. pH and conductivity were measured using pH meter and counductivity-meter from METTLER.

3 Results and Discussions

3.1. Effect of ozone concentration

Experiments were carried out at 2.7 % mol/mol , 4.25 % mol/mol and 5.86 % mol/mol of ozone concentration with the wastewater flow rate (v) of 50 mL/minute. Ozone was assumed dissolved perfectly in wastewater, so the reaction can be considered as homogenous reaction. Since ozone was fed in excess then the reaction (1), is a pseudo first order reaction with z = stochiometri coefficient.

$$O_3 + z\,COD \dashrightarrow \text{degraded pollutant} \quad (1)$$

In a CSTR model [10] , we get :

$$\tau = \frac{V}{v_0} = \frac{Ca_0\, X}{k\, Ca} \quad (2)$$

By rearranging equation (2), we get :

$$k = \frac{Ca_0 - Ca_f}{\tau\, Ca_f} \quad (3)$$

Cao = initial COD concentration, Ca_f = final COD concentration, τ = space time, k= first order kinetic constant.

The pseudo first order model fit with the data at the first 10 minutes of ozonation (Fig.3), hence based on the equation (3), we can calculate the pseudo first order kinetic constant at different ozone concentration as can be seen in Table 2.

Table 2. Pseudo-first order kinetic constant at different [O_3]

Ozone [% mol/mol]	C_{AO} [ppm]	C_{AS} [ppm]	k' [minute^{-1}]
2.70	337.967	225.91	0.031
4.25	345.523	206.65	0.042
5.86	332.930	180.15	0,053

Cao = initial COD concentration, Ca_s = COD concentration at steady state, k'= pseudo first order kinetic constant which is = k x C_{O3}.

The value of k' using 5.86% mol/mol ozone was similar to k' from batch experiment which was 0.0521 min^{-1}. The pseudo first order kinetic constant in batch experiment was also obtained at the first 10 minutes of ozonation and also used 5.86 %mol/mol ozone [11]. Result from the experiment showed that higher concentration of ozone will result in higher COD removal efficiency as can be seen in Fig.3. If we used 150 minutes timeline from the experiments, the percentage removal of COD were 33.17%; 49.57% and 90.78% for 2.7% , 4.25 %, 5.86% mol/mol ozone respectively.

Since the allowable COD concentration to be discharged is 150 ppm (East Java province regulation), then we can recalculate the space time and reactor volume needed using the value of k' as stated in Table 2.

To get 150 ppm of COD concentration in the wastewater at 50 mL/min flowrate (v), we need bigger reactor volume (V) and longer space time (τ) for lower ozone concentration (Table 3).

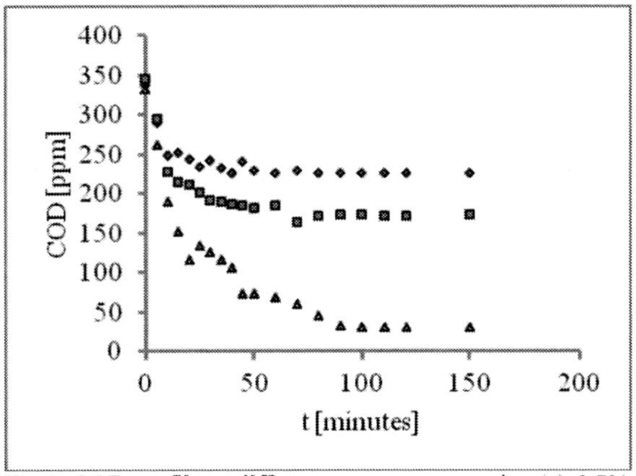

Fig.3. COD profile at different ozone concentration,(♦) 2.7% Ozone , (■) 4.25%, Ozone (▲) 5.86% Ozone.

Table 3. Value of pseudo first order kinetic constant at different ozone concentration

Ozone % mol/mol	C_{AO} (ppm)	C_{AS} (ppm)	k' min^{1}	τ (min)	V (ml)
2.70	337.96	150	0.031	41	2.021
4.25	345.52	150	0.042	31	1.552
5.86	332.93	150	0,053	23	1.151

3.2. Effect of wastewater volumetric flow rate.

The ozone concentration used in the experiment was 4.25% mol/mol, the flow rate variable were 30,50,70 mL/minute. Result from this study showed that higher flow rate of wastewater will result in lower removal of COD as can be seen at Table 4. The experiment result of COD profile was similar to that in simulated model as written in equation (3) as presented in Fig.4. The best removal of COD was at flow rate of 30 mL/min. By working with smaller flow rate at constant volume ,we have bigger space time and we will have higher conversion in a single CSTR.

However, implementation in the field with smaller flow rate may consider the size of reactor which is not favorable from economic point of view. Hence, performance optimization as written in Table 3 gave us an insight that working with high concentration of Ozone is favorable than lower volumetric flowrate. Therefore, a volumetric flow rate of 50 mL/min and 5.86% mol/mol ozone was a suitable parameter for a CSTR with 1150 mL volume to reduce the COD up to an allowable level to be discharged by the regulation (150 ppm). The optimum performance of the reactor to maintain COD level at 150 ppm, was in 1150 L volume of CSTR, 200 rpm of agitation, 50 mL/min volumetric flow rate, 23 minutes of space time and 5.86 % mol/mol of ozone.

Table 4. Removal of COD (%) at different flow rate using 4.25 % mol/mol Ozone

flow rate (v), mL/min	% removal of COD
30	80.09
50	77.6
70	67.84

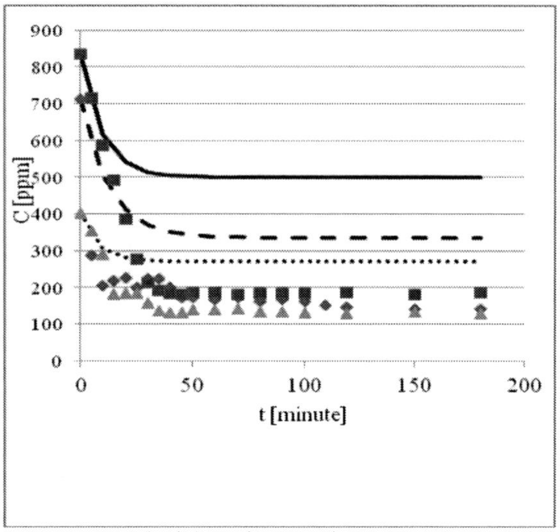

Fig.4. Removal of COD in ozonation process.(♦) 30mL/min experiment,(■)50mL/minexperiment,(▲)70mL/min experiment. (---)30mL/min,model, (–)50mL/min,model, (....)70mL/min,model

3.3 Profile of pH and conductivity during ozonation process

There is no significant change of pH and conductivity during ozonation process, pH and conductivity practically remained constant, it can be seen that pH after ozonation decreased in only 0.4 units (Fig.5).

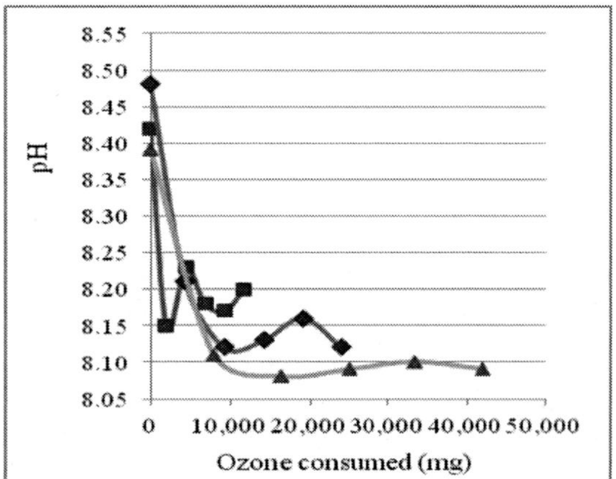

Fig. 5. pH profile during ozonation (■) 2.7% ozone,(♦) 4.25% ozone, (▲) 5.86% ozone

Conductivity after treatment is slightly reduced (Fig.6). It implied that the Total Dissolved Solid (TDS) in the wastewater was stable. Both parameters were determined along the ozonation reaction as a function of mg of ozone/L consumed. This result is similar with the previous work done by Sarasa et.al [12]. Ozonation doesn't have any effect in pH and conductivity, so the treatment doesn't give additional cost to adjust the pH after the treatment. Moreover, ozonation works better in alkaline medium, since the oxidation capacity of ozone may increase due to the formation of highly reactive hydroxyl radicals (HO•) with E° = 2.8eV, which can attack the double bonds of dye chemical effectively.

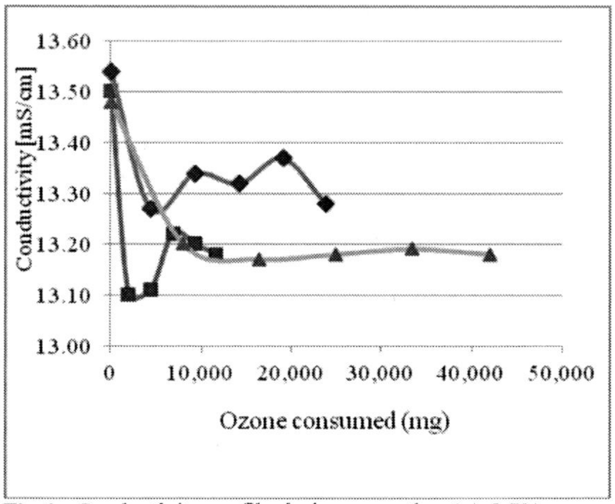

Fig.6. Conductivity profile during ozonation (■) 2.7% ozone, (♦) 4.2% ozone, (▲) 5.86% ozone.

4 Conclusion

Ozone treatment is very effective in removing compounds derived from yarn dyeing manufacture, The optimum performance of ozonation was 23 minutes of space time with 1150 mL reactor volume, 50 mL/min volumetric flow rate and 5.86 % mol/mol of ozone. The pseudo first order kinetic constant was 0.053 min^{-1}.There is no significant change of pH and conductivity during the ozonation process.

We thank to BA yarn dyeing manufacture for financial support for the project and Laboratory of Reaction Engineering staffs for their help in carrying out this study.

References

1. N. Azbar, T. Yonar, K. Kestioglu, Chemosphere **55**,35-43 (2004)
2. R. G. Saratale. G. D. Saratale, J. S Chang, S. P. Govindwar, J. Taiwan Inst. Chem. Eng., **42**,138-157 (2011)
3. G. R. P. Malpass, D. W. Miwa, D. A. Mortari, S. A. S. Macado, A. J. Motheo, Water Res., **41**, 2969-2977 (2007)
4. J. C. Cardoso, G. G. Bessegato, M. V. B. Zanoni, Water Res., **98**, 39-46 (2016)

5. P. R. Gogate, A. B. Pandit, Adv. Environ. Res., **8**, 501-551 (2004)
6. Q. Ruijuan, X. Bingzhe, M. Lingjun, W. Liansheng, W. Zunyao, Water Res., **68**, 316-327 (2015)
7. A. M. Dore, N. Merlet, B. Legube, J. P. Croue , Ozone Sci. Eng., **10**, 153-172(1988)
8. L. Riadi, W. Ferydhiawati, L. D. S. Loeman, Reaktor, **15**, 73-78 (2014).
9. APHA, *Standard Methods for the Examination of Water and Wastewater* (American Public Health Association, 1998)
10. H. S. Fogler, *Essentials of Chemical Reaction Engineering* (Prentice Hall, 2011)
11. L. Riadi, A. Altway, S. M. Vania, A. Widyasayogo, Journal WPT, **12**, 690-697(2017)
12. J. Sarasa, M. P. Roche, M. P. Ormad, E. Gimeno, A. Puig, J. L. Ovelleiro, Water Res., **32**, 2721-2727 (1998)

Low Rank Coal Pre-treatment to Increase Its Reactivity Towards Gasification with Biomass

Jenny Rizkiana*, *Ryzka* Pranata, *Hasna Nisrina* Fauzi, *Winny* Wulandari, *Dwiwahju* Sasongko

Department of Chemical Engineering - Institut Teknologi Bandung, Jalan Ganeca 10 Bandung 40132, Indonesia

Abstract. Coal can be used to produce hydrogen through steam gasification process. Indonesia has abundant coal reserves and thus hydrogen production from coal is very attractive. However, steam gasification of coal usually requires high temperature due to its low volatile content. The use of catalyst, such as alkaline and alkaline earth metal (AAEM) may promote the hydrogen production. AAEM metal can be found in biomass and thus co-gasification of coal and biomass may become the attractive solution as the AAEM may volatilize during gasification and catalyze the coal when it attaches to the coal surface. However, the presence of silicate may decrease catalytic activity of the attached AAEM and thus it needs to be removed by deashing process. This research aims to determine the effect of the solution type, solution concentration, reaction temperature, and reaction time of coal deashing. The results showed that deashing process decreased the ash content of coal to some extent proved by the gravimetric analysis result. The decrease of ash content also affected to the surface morphology of the coal as some pores are formed and thus the surface area of coal increased slightly. The increase of surface area allows more AAEM to be attached to the coal surface so that the coal may become more reactive towards steam gasification.

1. Introduction

Indonesia has abundant coal reserves. Unfortunately, 92% of Indonesian coals are low rank, such as subbituminous and lignite [1]. Low rank coal has high moisture content and low calorific value. Despite the disadvantages, coal is still widely used in Indonesia as main energy sources because of its low price. Generally, coal is used by burning to produce heat as main energy sources, so it causes bad impact to environment. Gasification is considered as an effective way to utilize coal cleanly [2].

Gasification is a thermal conversion process which convert carbon-contain material into synthesis gas. The material may be coal or biomass. Gasification process occurs in several stages, which are drying, pyrolysis, combustion, and gasification [3]. Products of gasification are hydrogen, carbon monoxide, carbon dioxide, and methane [4]. In general, gasification involves the reaction of carbon with gasifying agent at temperature 700°C or more to produce heat or feedstock for chemical industry, liquid fuel or other gaseous materials [5]. In gasification process occurs the following reactions.

$$C + O_2 \rightarrow CO_2 \quad (1)$$

$$C + 0.5\ O_2 \rightarrow CO \quad (2)$$

$$C + CO_2 \rightarrow 2\ CO \quad (3)$$

$$C + H_2O \leftrightarrow CO + H_2 \quad (4)$$

$$CO + H_2O \leftrightarrow CO_2 + H_2 \quad (5)$$

As mentioned above, majority of Indonesian coals are low rank coal. Low rank coal gasification has low efficiency and low carbon conversion [6]. However, gasification of low rank coal still can be accomplished by the addition of catalyst.

Co-gasification of coal with biomass can increase the carbon conversion, especially hydrogen [2]. Catalytic effect on co-gasification process caused by alkali and alkaline earth metal (AAEM) that can be found in biomass ash. In biomass, alkali and alkaline earth metal formed as carbonate. Alkali and alkaline earth metal that formed as carbonate can increase carbon conversion in gasification process through following reaction [7].

$$M_2CO_3 + 2\ C \rightarrow 2\ M + 3\ CO \quad (6)$$

$$2\ M + 2\ H_2O \rightarrow 2\ MOH + H_2 \quad (7)$$

$$2\ MOH + CO \rightarrow M_2CO_3 + H_2 \quad (8)$$

where M is an alkali and alkaline earth metal

However, coal contains silicone in the form of silica, a compound that could deactivate catalyst. Risnes et al. [8] found that silica could deactivate catalytic activity of AAEM by forming silicate, which has no catalytic activity.

Steel et al. [9] proved that silica content in coal could be decreased by leaching method using acid solution. Leaching method is done by mixing coal with acid solution at a certain temperature and time. By reducing

* Corresponding author: jr@che.itb.ac.id

silica content, co-gasification process is expected to be more efficient which resulting higher carbon conversion.

The present study is done to find the suitable condition to deash low rank coal. The conditions include acid solution type, acid concentration, temperature, and residence time. Deashing performance is evaluated by comparing the ash content of raw coal sample with the deashed ones.

2. Material and Methods

2.1. Feedstock preparation

Indonesian low rank coal was used as the feedstock sample. The samples were ground and sieve to particle size of 14-28 mesh (about 0.595-1.19 mm) and dried in the oven at 110°C overnight.

2.2. Coal pre-treatment

The samples were leached with two kinds of strong acid solution, i.e. HF and HCl. The sample mass and solution volume used in all leaching experiments was 10 g and 250 ml, respectively. Experiments were performed at ambient temperature (25°C), medium temperature (47°C) and high temperature which were below the boiling points of the mixtures (70°C). Experiments using HCl were performed in a 500 ml glass beaker, while experiments using HF were performed in 500 ml Teflon beaker. All experiments were mounted on an electric hotplate with magnetic stirring device. There were three variations of reaction residence time of leaching, i.e. 2, 3 and 4 hours. After leaching, the mixtures were immediately filtered by Büchner funnel (vacuum filtration) for experiments using HCl, and it filtered by Polypropylene funnel for experiments using HF. The filtered cakes were washed with distilled water until the wash water became neutral. Filter cake residues were placed in an oven at 110°C overnight to remove water as a final drying step.

2.3. Analysis

The samples were burned in the furnace at 700°C. The ash content of coal was calculated by using the following formula:

$$\text{ash content} = \frac{\text{mass of ash}}{\text{mass of sample}} \times 100\%$$

The ash content of leached samples compared with the ash content of raw sample. The differences of ash content of leached samples and raw sample were used to analyze whether the leaching experiments succeed or not and how was the best condition of leaching experiments should be done.

3. Results and Discussion

Leaching method using acid solution (HCl and HF) could decrease ash content in coal as shown in Fig. 1 and 2. The ash content of raw sample is 8.39%.

Ash content in coal sample could decrease approximately 20-23% weight compared to the ash content of raw sample by leaching method using HCl. Fig. 1(a) shows ash content of raw sample and leached samples at different HCl concentration. However, the increase of HCl concentration in leaching experiments did not affect the ash content significantly. Similar result was also obtained by Steel et al. [9]. They found that dissolution of mineral matter isolated from coal with HCl slightly increased as the increase of HCl concentration. Fig. 1(b) shows ash content of raw sample and leached samples at different temperature. The increase of temperature in leaching experiments with HCl also did not affect the ash content significantly. Steel et al. also did leaching experiments with various temperatures, i.e. 20°C, 70°C, and 100°C. They found that ash content at temperature 20°C and 70°C not much different. However, the dissolution increases slightly when reaction temperature is raised to 100°C. Fig. 1(c) shows ash content of raw sample and leached samples at different residence time. Same with the previous result, the increase of residence time did not affect the ash content significantly.

Ash content in coal sample could be decreased approximately 47-59% weight compared to the ash content of raw sample by leaching method using HF. Fig. 2(a) shows ash content of raw sample and leached samples at different HF concentration. The ash content of sample decreased with the increase of HF concentration. Steel et al. [9] found that the ash content of leached sample decreased as the HF concentration increased. The result of leaching experiments using HF conducted by Steel could decrease the ash content of sample up to 96%. Fig. 2(b) shows ash content of raw sample and leached samples at different temperature. The increase of temperature in leaching experiments with HF also decreased the ash content of sample. Steel et al. also did leaching experiments with various temperatures, i.e. 20°C and 65°C. They found that the increase of temperature could decrease ash content of sample significantly. At 20°C, 96% by weight of sample dissolve using 14 M HF, and at 65°C, 95% by weight of the sample dissolves using only approximately 1.5 M HF. Fig. 2(c) shows ash content of raw sample and leached samples at different residence time. The increase of residence time did not affect the ash content significantly. The ash content of sample at residence time of 3 hours is approximately 4.1% and the ash content did not change at residence time of 4 hours. Accordance with Steel et al. [9], the equilibrium with respect to dissolution of aluminosilicates using HF is established at residence time of 3 hours.

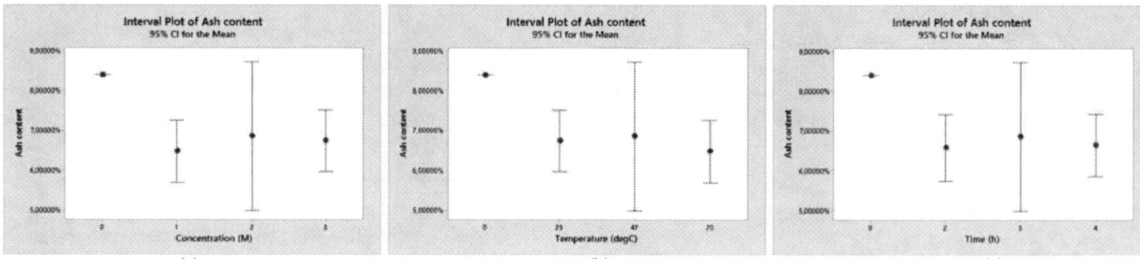

(a) (b) (c)

Fig. 1. Comparison of ash content range of leaching experiments using HCl with various (a) HCl concentration, (b) temperature, (c) residence time.

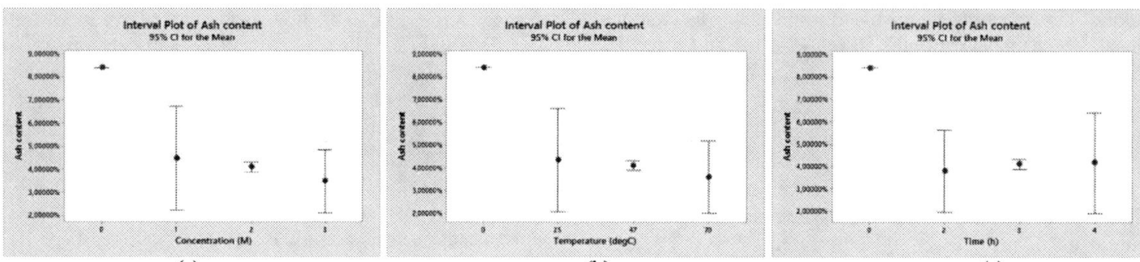

(a) (b) (c)

Fig. 2. Comparison of ash content range of leaching experiments using HF with various (a) HF concentration, (b) temperature, (c) residence time.

Fig. 3 shows comparison of ash content in leaching experiments using HCl and HF 3 M. Fig. 3 shows that leaching experiment using HF more effective to decrease the ash content of sample compared to HCl. The ash content of leached sample using HCl solution is in range 6.0-7.5%, and the average is 6.7%. Meanwhile, the ash content of leached sample using HF solution is in range 2.1-4.9%, and the average is 3.5%. At acid concentration of 3 M, HCl could decrease ash content in the amount of 20%, and HF could decrease ash content in the amount of 59%.

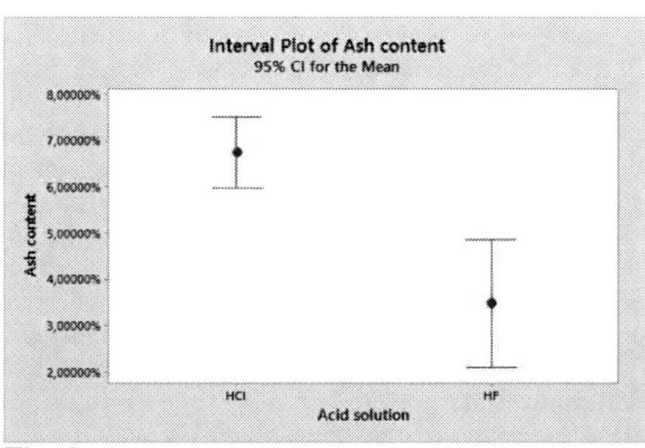

Fig. 3. Comparison of ash content in leaching experiments with acid concentration of 3 M.

Das et al. [10] found that leaching experiments using HCl effective to removes Na, K, Mg, Ca, and Al, while HF removes almost all the ash elements. Steel et al. [9] also found that HCl dissolves essentially all of Na, Ca, Mg, Fe, and some of the Al, K, and Ti. Essentially, no Si is exctracted from the mineral matter by leaching experiments using HCl. Meanwhile, HF is extremely effective at reacting with most of the minerals which occur in coal. In this experiments, the mineral matter that expected to be decreased from coal sample is silica. Si could be removed from sample effectively by HF solution with $SiF_4.2H_2O$ formation in aqueous or gas phase.

4. Conclusion

This paper has shown that HCl and HF could decrease ash content of coal. However, HF is more effective at decreasing the ash content of Indonesian low rank coal. Leaching experiments using HCl with different concentration, temperature, and residence time did not affect the ash content of coal significantly. Meanwhile, leaching experiments using HF could improve the decreasing ash content with the increase of acid concentration and temperature.

Si inhibited the catalytic activity of co-gasification, so it is expected that Si could disappear from the coal in order to increase carbon conversion. However, in this leaching experiments, Si was not the only element that decreased from coal. Hence, the co-gasification of pre-treatmented coal with biomass should be done to prove the catalytic effects so the co-gasification product, i.e. synthesis gas especially hydrogen increased.

References

1. Badan Pengkajian dan Penerapan Teknologi (BPPT), "Outlook Energi Indonesia 2016", Jakarta:Perpustakaan Nasional RI (2016)
2. J. Rizkiana, G. Guan, W.B. Widayatno, X. Hao, X. Li, W. Huang, A. Abudula, Applied Energy, **133**,

* Corresponding author: jr@che.itb.ac.id

282–288 (2014).
3. S. Farzad, M.A. Mandegari, J.F. Gorgens, Biofuel Res. J., **12**, 483-495 (2016).
4. U. Arena, Waste Management **32** (2012), 625-639.
5. A.H. Demirbas, I. Demirbas, Energy Conv. Manage., **48**, 2386-2398 (2007).
6. W. Duan, Q. Yu, J. Liu, T. Wu, F. Yang, Q. Qin, Energy **111**, 859-868 (2016).
7. H. Iwaki, S. Ye, S. Katagiri, K. Kitagawa, Appl. Cat. A: General **270**, 237-243 (2004).
8. H. Risnes, J. Fjellerup, U. Henriksen, A. Moilanen, P. Norby, K. Papadakis, Fuel, **82**, 641-651 (2003).
9. K.M. Steel, J. Besida, T.A. O'Donnel, D.G. Wood, Fuel Process. Technol., **70**, 171-192 (2001).
10. P. Das, A. Ganesh, P. Wangikar, Biomass and Bioenergy, **27**, 445-457 (2004).

Reduction of P_2O_5 and F from Phosphogypsum by CaO addition

Sri Rachmania Juliastuti [1,*], *Nuniek* Hendrianie [1], *Yumna* Dian Pawitra[1], and *Irham* Raditya Putra[1]

[1]Chemical Engineering Department, Institut Teknologi Sepuluh Nopember Surabaya, Surabaya 60111, Indonesia

Abstract. Waste of Phosphoric acid industry, which called phosphogypsum consists of 65% gypsum, 25% water and 10% impurities. Phosphogypsum can only be used up to 15% and the rest is thrown away. The objective purpose of this study is to reduce the content of P_2O_5 and F using best variable CaO and stirring during coagulation process, flocculation and sedimentation. The method of this experiment is conducted by using the ratio of phosphogypsum and variable distilled water 1:1, 1:2 and 1:3, and then put it into a jar test using the variable stirring of 100 and 150 rpm. Then 11, 12 and 13 gram CaO and 50 ml distilled water is added as variables for coagulation – flocculation process. pH is kept at 6 by adding 4% NaOH. Polyalumunium chloride (PAC) 0.1% as much as 1.75 mg/l is added into the mixture and stirred at 30 rpm for 20 minutes. The separation of P_2O_5 and F, is done by the sedimentation process. These experimental results show that variable 1:2 (w/v), 12 gram CaO and 150 rpm stirring able to reduce fluoride and P_2O_5 up to 98.5% and 50% from phospogypsum.

1 Background

Phosphogypsum (PG) is a byproduct of industrial phosphoric acid and gypsum consists of 65-70% gypsum ,25-30% water and 5-10% water impurities, ie phosphoric acid and its salts, hydrofluoric acid and its compounds, R_2O_3 (Al_2O_3 + Fe_2O_3), Quartz, apatite, alkali, organic matter, and others. Some of dirt as P_2O_5 and F can be found in three different forms, namely in the gypsum crystal surface as water-soluble compounds (H_3PO_4, Ca $(H_2PO_4)_2$ • H_2O, H2SiF6), gypsum crystal lattice (solid solution $CaHPO_4$ • $2H_2O$, SrSO4 or Na_2SiF_6) and insoluble compounds, namely apatite and quartz (Al-Hwaiti, 2015).
Only 15% of production worldwide is recycled, while 85% is stored around the factory in coastal areas. PG storage without prior processing can cause serious contamination of soil, water and atmospheric environments (Tayibi, 2008). In further treatment, phosphogipsum can be used for industrial purposes and infrastructure needs, one of which is cement. If used in the manufacture of cement, such impurities can disrupt mechanical properties by altering the gypsum morphology that will harden. Various methods have been suggested to reduce the reactivity of the impurities., including washing simple with water or filter wash solution from phosphogypsum the heating hemihydrate anhydrite followed by washing or neutralization with $Ca(OH)_2$ or the base material, and treating phosphogypsum with a mixture of Sulfuric acid and silica or hot ammonium sulfate solution (Singgh, 1992)

More pure PG without impurities will get more positive results for the environment and surrounding areas. So generally, this study aims to reduce the content of P_2O_5 and F impurities especially in phosphogypsum by performing physical and chemical treatment and determine the effect of such treatment on the set parameters.

2 Methods

In this experiment, three main processes were coagulation, flocculation and sedimentation. In this experiment used a jar test tool to assist the process of blending in the process of coagulation and flocculation. While the process of sedimentation used measuring glass 1 liter to know the best sedimentation process. The process is seen in the picture below

* Corresponding author: juliaz30@gmail.com

 https://doi.org/10.1051/matecconf/201815603021

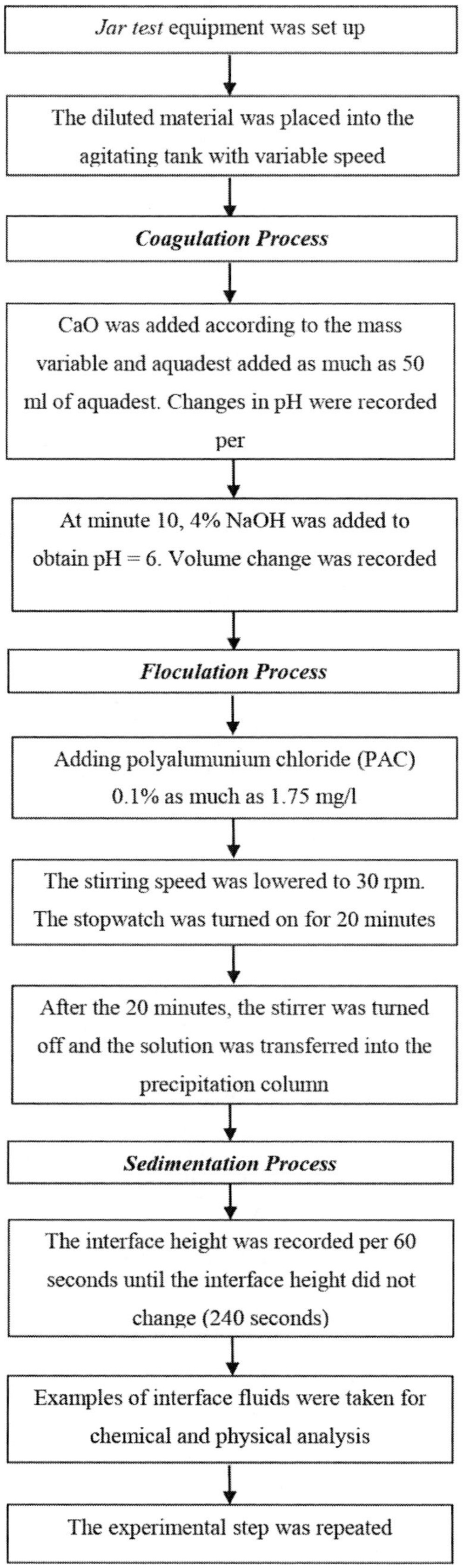

3 Result

3.1 The Effect of Dilution Variables, CaO Mass, and Stirring Rate on Change in pH

In this experiment the pH measurement aims to determined the best pH performance during the coagulation process. The pH changes in the solution occurred every minute, and was measured for 10 minutes in 1: 1 (w / v), 1: 2 (w / v) and 1: 3 (w / v) dilution variables at 100 and 150 rpm, and CaO mass of 11, 12, and 13 gram, could be seen in Fig. 1 and 2. At a speed of 100 rpm, various pH changes occurred in the range of 12.6 to 13.6. This indicated that the phosphogypsum solution was in a high alkaline condition. In the ideal coagulation process the pH of the solution was in the range 7 to 8, which at pH 7 to 8 was the ideal pH for the occurrence of coagulation. However, in this experiment the solution pH was a base in the high pH range, resulting in deviations in the pH value achieved. In this case, it could be seen in the above three samples (Fig. 1) that the best performance of pH occurred at pH 12.6 in sample 2. While at a stirring speed of 150 rpm, the best performance of pH occurred at pH 13 in sample 4 (Fig. 2).

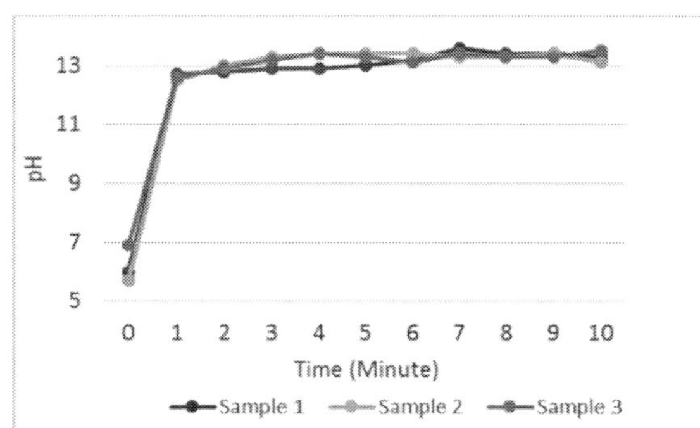

Fig. 1. Change of pH in solution to time at speed of 100 rpm with ratio 1: 1

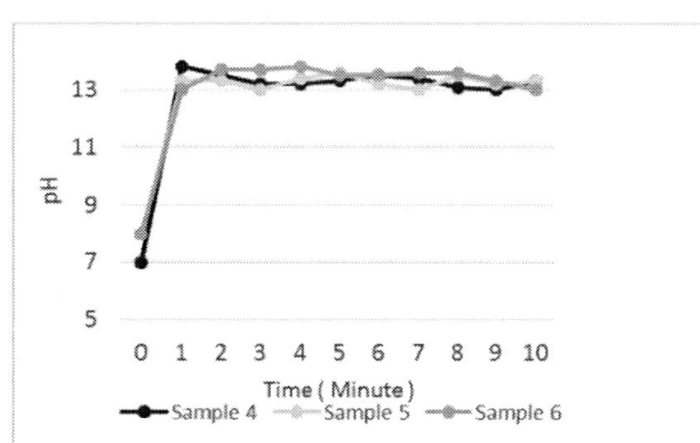

Fig. 2. Change of pH in solution to time at speed of 150 rpm with ratio 1: 1

The best performance of the results was achieved at pH value of 12.5 with stirring speed of 150 rpm. This could be occured because of the influence of several factors that affect the coagulation and flocculation process, such as pH. Coagulation was an important process in this experiment, because coagulant serves to bind impurities (P_2O_5 and F) contained in the solution which will be deposited. The reactions were:

$$2H_3PO_4 + 3CaO \longrightarrow Ca_3(PO_4)_2 + 3\ H2O \quad (1)$$

$$H_2SIF_6 + 3\ CaO \longrightarrow 3CaF_2 + SiO_2 + H_2O \quad (2)$$

$$H_2SO_4 + CaO + H_2O \longrightarrow CaSO_4.2H_2O \quad (3)$$

PH in each variable indicated fairly high base condition, where the conditions achieved exceed the optimum pH at coagulation and flocculation ie pH 7 to 8. This condition was achieved due to the effect of adding CaO as a coagulant in the coagulation process, in which CaO can increase the alkalinity of the solution with an initial pH of 5 to 6. However, the conditions achieved did not meet the optimum conditions, because of the excessive amounts of CaO as coagulants, causing a rise in pH under fairly high base conditions.

Furthermore, on the dilution variables of 1: 2 (w/v) at a rate of 100 rpm and 150 rpm, and the CaO of 11, 12, and 13 gram mass variables were also measured pH on each sample performed every minute for 10 min.

For a speed of 100 rpm in this variable, the pH range occurring, was still in a high alkaline pH condition with a pH value of 12 to 13.6. The best pH performance in this experiment can be achieved, although the ideal pH range occurring in the coagulation process was 7 to 8. The best pH performance at 100 rpm stirring rate, showing the pH value of 12 in sample 7 (Fig. 3), compared with the other sample that have a higher pH value. As for the stirring speed of 150 rpm, the best pH performance occurred in sample 10 with a pH value of 13 (Fig. 4). The pH value in each sample, still under high alcaline conditions, was the same as the 1: 1 (w/v) dilution variables.

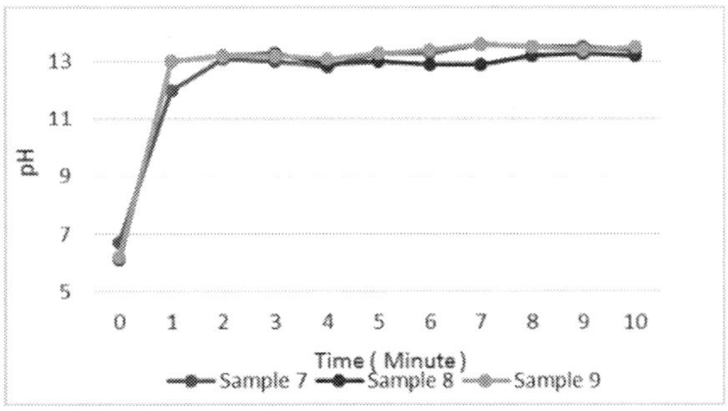

Fig. 3. Change of pH in solution to time at speed of 100 rpm with ratio 1: 2

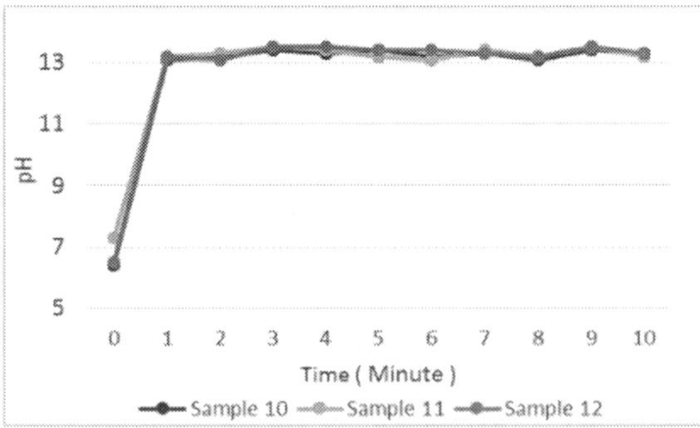

Fig. 4. Change of pH in solution to time at speed of 150 rpm with ratio 1: 2

From the above variables, resulted the best pH performance at pH value 12 that occurred in sample 7. pH became one of the factors affecting coagulation process that occurred in this experiment. Coagulation was a process of destabilizing particles causing the joining of positive and negative ions to form a flock that was easily precipitated, so the coagulation process will also help the sedimentation process. In this case, the pH conditions achieved did not meet the ideal conditions of pH in the coagulation process due to excessive amount of CaO, but at the time of rapid agitation and the addition of CaO, 2 layers of solution indicated that the coagulation process occurred well. Coagulation served to bind the content of P_2O_5 and F in the solution to be easy to precipitate in the sedimentation process. P_2O_5 and F will be bound by the coagulant CaO as Ca_3PO_4 and CaF_2 as salts at the precipitate formed.

In the dilution variables were 1: 3 (w / v) at 100 and 150 rpm, with CaO of 11, 12, and 13 grams, pH was measured every minute at 10 min.

For the speed of 100 rpm resulted the pH value of each sample under high alkaline conditions. In this experiment, there was a pH drift in the coagulation process, where the ideal pH in the coagulation process was 7 to 8. The pH conditions achieved in this experiment were high bases, so in this experiment the best pH performance of this variable was obtained in sample 15 with pH 12.3 (Fig. 5), whereas at a speed of 150 rpm showed the best performance resulted at pH12.1 in sample 17 (Fig. 6). Coagulation was performed on this process using rapid stirring aid, and CaO as coagulant. Aside from being a coagulant, CaO also serves to increase the alkalinity of the starting solution having an initial pH of 5 to 6. Addition of the amount of CaO performed in this experiment was excessive, thus causing an increase in pH value under basic conditions. Although within the pH range of coagulation, the pH achieved was not in ideal conditions, but the coagulation process could occur from the formation of 2 layers during stirring.

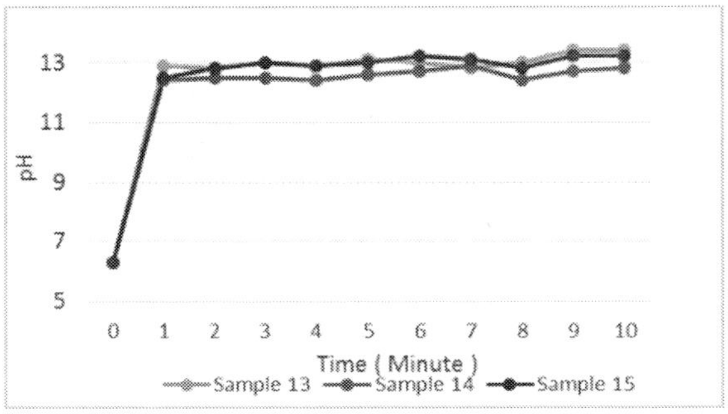

Fig. 5. Change of pH in solution to time at speed of 100 rpm with ratio 1: 3

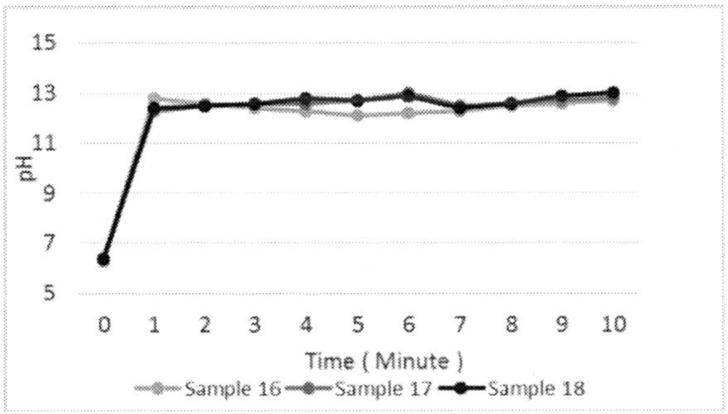

Fig. 6. Change of pH in solution to time at speed of 150 rpm with ratio 1: 3

The suspension formed from this experimental process had properties of electrical properties, which inhibited agglomeration of particles and sedimentation, and the stability of the suspension was caused by the electrostatic forces contained in the suspension. In various industrial waste suspensions, most suspensions had negative ions. So to reduce the electrostatic force of particles it was necessary to neutralize the particles, so that the coagulation process, flocculation, and sedimentation could occur well. With the addition of coagulant, the electro-plastic force became low, and the suspension became unstable due to the presence of positive ions in the coagulant, thus causing the particles binded to form flocks and also affected by rapid stirring aid. In this process also used auxiliary coagulant that was PAC (Poly Aluminium Chloride) which helped to enlarge flock which had happened at previous process, which would collide because of particle agglomeration and felt down because particle weight became heavier. In addition PAC also served to help control the electrostatic force on the suspension. PAC was a heavier molecular weight polyelectrolyte and contained adsorbers which served as flocculant intermediate flock [6].

For pH parameters achieved did not meet the optimum conditions of coagulation and flocculation process, but the coagulation and flocculation process occured well with the formation of 4 layers in the sedimentation process. This was because coagulant reactivity was high CaO. CaO was an alkaline earth metal which was a highly reactive group of IIA elements and can react with water. The reactivity of the alkali metal was closely related to the two valence electrons and the relatively large radius of the atom. As a result, alkali metal easily releases its valence electrons to form a charged cations (+2), and dissolves in water. Therefore CaO could bind impurities (P_2O_5 and F) to easily deposited compounds in the coagulation process, and canceled the negative ion ions present in the suspension, binded each other with the positive ions of the coagulant, due to the high valence, so that the coagulation process proceeds well .

3.2 The Effect of Dilution Variables, CaO Mass, and Stirring Rate on Sedimentation Results

Changes in the sedimentation altitude of the solution per minute, measured for 4 min at 1: 1 (w / v) dilution variables at 100 and 150 rpm, and CaO of 11, 12, and 13 grams, could be seen in (Fig.7) and (Fig. 8)

At a stirring speed of 100 rpm there was altitude change for each sample in the altitude range of 9.1 to 3.8 cm. The height measured was the height of the sludge interface with the clean liquid that occured during the sedimentation process. Sedimentation was the process of separating solid and liquid particles based on the force of gravity. The measured altitude changes showed the sedimentation process that occured, if the rapid precipitation speed, then the altitude changes showed a significant decrease. From the sample that had been measured, the best result of the change in sample 1 which was the first minute to the second minute with the value of 9 cm to 6 cm (Fig.7) with slope 3 cm / minute. While at a speed of 150 rpm, the best altitude change was in sample 4, the first minute to the second minute, with a value of 8 cm to 5.5 cm (Fig. 8) with a slope of 2.5 cm / min.

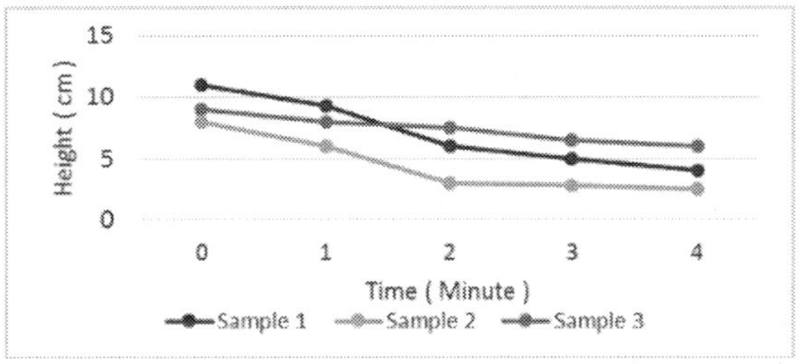

Fig. 7. Change of sedimentation height to time at 100 rpm with ratio 1:1

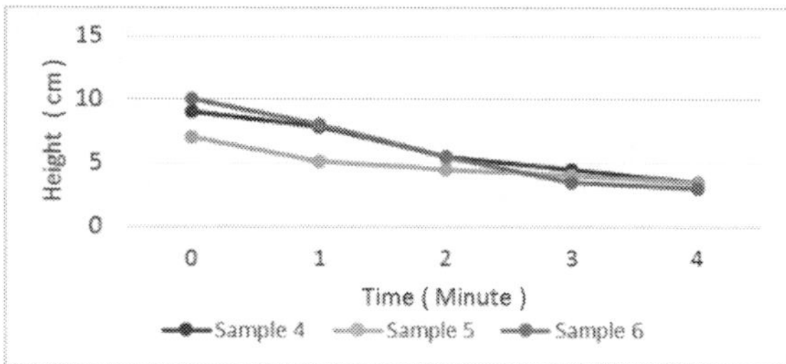

Fig. 8. Change of sedimentation height to time at 150 rpm with ratio 1:1

Sedimentation that occured could not be separated from the influence of coagulation and flocculation processes that have occurred in the previous process. If the coagulation and flocculation process works well, then the sedimentation process would also work well. The coagulation and flocculation process was done with the helped of coagulant ie CaO and PAC. In this process, binding of impurities and the formation of larger flock occurs, so it was easy to precipitate. Later this process would help sedimentation of phosphogypsum solution, so that solid and liquid could be separated. The effectiveness of coagulant usage of CaO and PAC had significant effect. CaO had a high reactivity that helped the opposite ions to bind to form flocks, because of the high cation valency it was easy to release the valence. Although the parameters of pH conditions occured in conditions not ideal, but with reactive CaO properties would help the sedimentation. While PAC was a very effective polyelectrolyte in this process, because the speed of the flock to be fast and produce good sedimentation. Thus the process runs well and forms 4 zones during the sedimentation process.

In the 1: 2 (w/v) dilution variables with CaO mass of 11, 12, and 13 grams, the stirring speed of 150 rpm made changes in altitude measured every minute for 4 min.

For stirring speed of 100 rpm and 150 rpm there was altitude changes during the sedimentation process took place. Sedimentation was done to separate solid and liquid particles based on the force of gravity. In the sedimentation process there was a change in altitude in the solution, the height measured was the height of the sludge interface with clear liquid. This process supported by a good deposition rate. If the speed of settling occured rapidly then the sedimentation process that would work well, and marked by changes in altitude significantly. In this variable, the altitude changes of the best solution in the ninth sample at minute zero to the first minute, with the value of 14.9 cm to 10.1 cm (Fig. 9) with slope value of 4.8 cm/min. As for the speed 150 rpm, the best altitude change of the solution was in sample 10 at minute zero to the first minute, with the value 17 cm to 12.2 cm (Fig. 10) and slope value 4.8 cm/min.

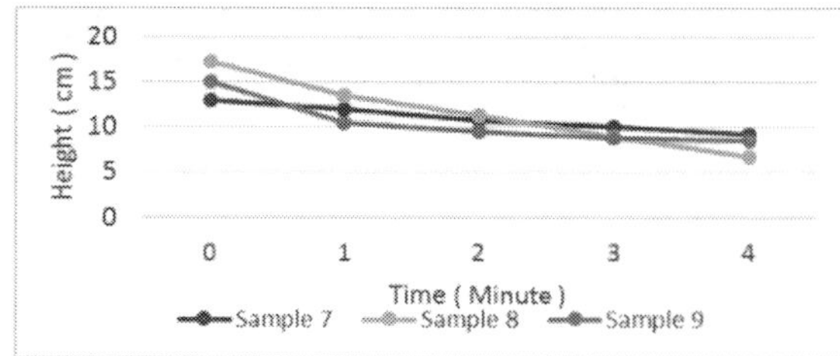

Fig. 9. Change of sedimentation height to time at 100 rpm with ratio 1:2

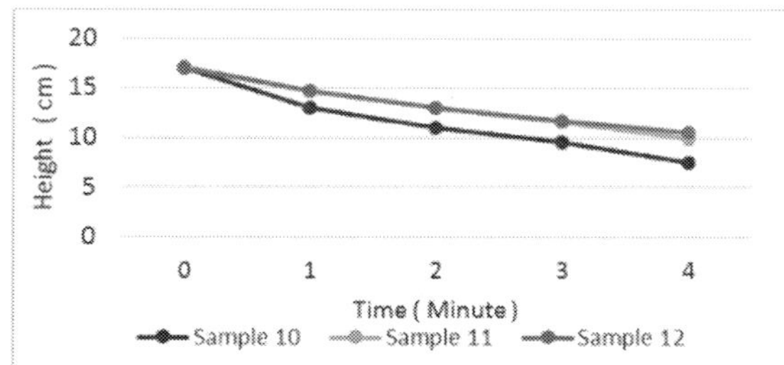

Fig. 10. Change of sedimentation height to time at 150 rpm with ratio 1:2

Sedimentation process was divided into four zones, namely liquid interface zone (layer 1), settling zone (layer 2), transition zone (layer 3), and compression zone (layer 4)

Before the sedimentation process occured, the slurry was at the same concentration. During the beginning of the sedimentation process, sludge settled at the same rate. The settling rate was a function of the initial concentration of the particles in the solution prior to the sedimentation process. As a result of sedimentation, the precipitate deposits of this process reside in the compression zone located at the bottom layer of the settling process occurring at a constant rate. The transition zone would be occured where the settling velocity decreases as a result of increased particle concentration. The solid concentration at the transition zone layer remains constant until the settling interface approaches the transition zone, and a transition zone was formed. Passing through this transition zone, the settling rate would decrease due to increased density and viscosity of the suspension surrounding the particles. If the layer rise of the sediment reached the interface, the compression zone would be formed. Sedimentation effects occured due to several factors, such as the coagulation and flocculation process that had been formed previously which would help the process of sedimentation occurs. In the coagulation process that occurred using coagulant CaO, where the coagulant was very reactive and has the valence of cation (+2) which could easily release the ion to help neutralization, and the ion ion binding into a larger flock, so the role of coagulant was very influential to reach sedimentation conditions Which was good [6].

For 1: 3 (w / v) dilution variables with CaO mass of 11, 12, and 13 grams and stirring speeds of 100 and 150 rpm, measured changes in altitude of the measured solution per minute for 10 min.

In this variable, the sedimentation process that occurs could be seen in the Fig, 11. The measured altitude changes in each sample showed the sedimentation process took place. At such altitude changes, the settling rate affected the altitude changes. If the speed of settling worked quickly, then altitude changes showed significant value and the sedimentation process occured well. Therefore, the best altitude change at 100 rpm was sample 13 from minute zero to the first minute with a value of 22 cm to 18 cm (Fig. 11) with slope value of 4 cm / min, while for 150 rpm the best altitude change occurred in sample 16 at minute zero to the first minute with a value of 20.1 cm to 16 cm (Fig. 12) with a slope value of 4.1 cm / min

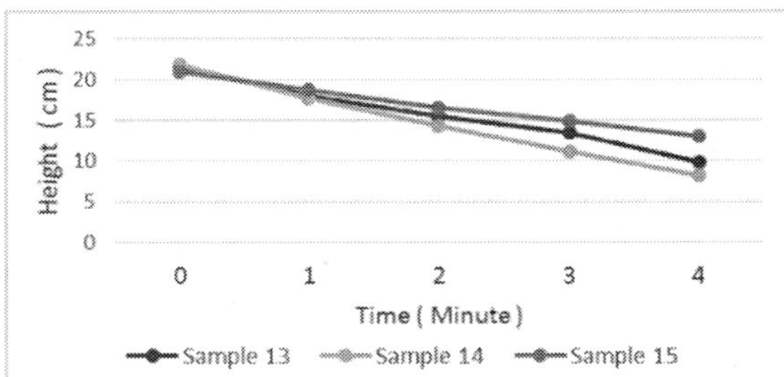

Fig. 11. Change of sedimentation height to time at 100 rpm with ratio 1:3

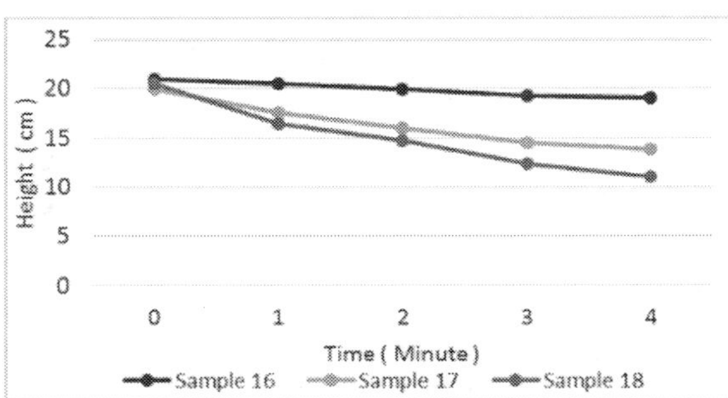

Fig. 12. Change of sedimentation height to time at 150 rpm with ratio 1:3

In the above explanation, it could be seen that there was a relationship that occured in the process of coagulation, flocculation, and sedimentation. The use of coagulant was CaO and PAC could help the process of sedimentation was good. CaO had a very high reactivity and easily released its valence of +2, so it made it easy to bind negative ion ions, and made it a bigger flock. Easily ionic ions were bound, causing the mass density became larger which made the flock felt down due to the influence of gravity force, and finally settles. In addition, because the flock mass became larger, the density or the density of the flock became larger which made it easy to precipitate. With the ion ions merge into flocks, the concentration of the particles becomes larger and would slow down the settling speed. This results in the formation of a larger layer of liquid interfaces, as a result of the rising concentration of the solution and the felt of the flock downward, and the transition zone was formed. Deposition rate in this transition zone slows further due to the large particle concentration, and massive density. This condition would reach a stable state until the interface layer approached the sediment layer. When the layer rise of the sediment reached the interface, this was what drives the compression zone or compaction zone located at the bottom layer.

The forces that affect the sedimentation process were the drag force and impelling force. The force of impelling was the force caused by the gravity of particles or the force of gravity (downward), which caused the big flock to felt down. And buoyant force (bouyant, upward). While the drag force was a force against the impelling force so that the particles were in equilibrium. The direction of this style was upward. The particle mass caused a drag force and was balanced by the impelling force, resulted in a constant particle precipitation speed [6].

From 1 to 18 samples the best results were obtained on the 1: 2 (w / v) dilution variables, CaO of 12 gram mass, and 150 rpm velocity were analyzed to determine the concentrations of P_2O_5 and F at layers 2 and 3, in the measured solution each minute for 4 minutes at the time of sedimentation. The % removal resulted of P_2O_5 and F that had been analyzed at layers 2 and 3 during sedimentation could be seen in the picture with the relation of sedimentation and time height (Fig. 13 and Fig. 14).

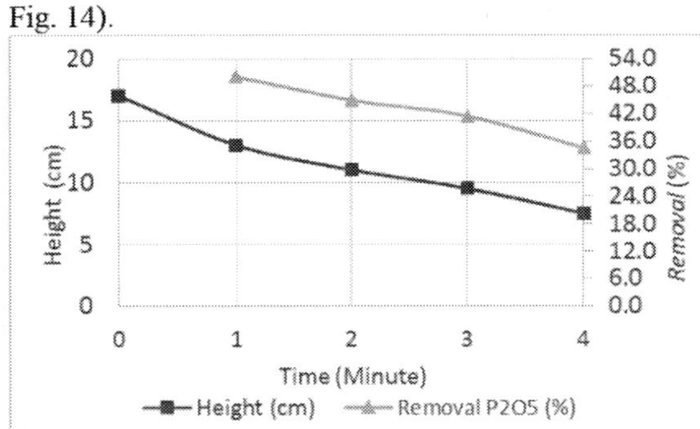

Fig. 13. The relationship between height with% removal of P_2O_5 over time

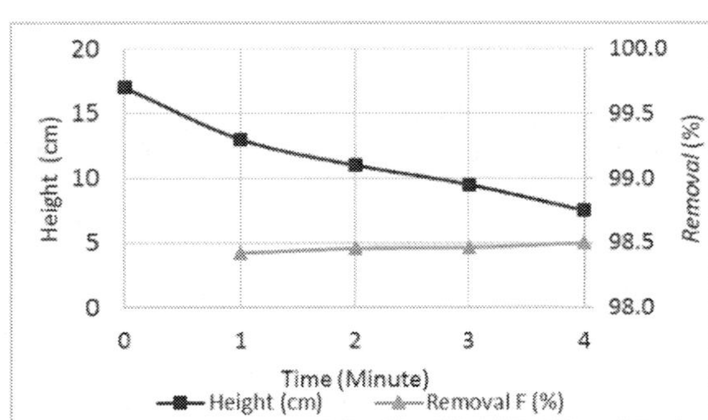

Fig. 14. The relationship between height with% removal of F over time

From the above two figures, different results were obtained on% removal of P_2O_5 and F. The above results showed that the% removal F was greater than the% removal of P_2O_5 (Fig. 15 and Fig .16), there was a decrease of F concentration in layers 2 and 3 , contains a lower concentration of F compared to other layers, whereas for P_2O_5 concentrations, it had a greater concentration than other layers. In Fig. 16 the best% removal occurred in the first minute of 50% with a sedimentation height of 13 cm, and for the best% F removal occurred in the fourth minute of 98.5% with a height of 7.9 cm. In this case there was an anomaly in the removal results for P_2O_5, where there was a decrease in concentration in the first minute and the increase in concentration in the second minute to the fourth minute. This occurs because of the restabilization conditions in the coagulation and flocculation process, caused by the addition of excessive amounts of coagulants, resulting in poor coagulation and flocculation (ASWW, 1998).

This difference in yield could occur due to the effect of the flock size of P_2O_5 and F, which had different sizes. The size of the flock affected the determination of the precipitation characteristics of the flock. In this case the particle diameter becoame an effect on the coagulation process, flocculation and sedimentation. So the particle diameter would determine the location of the particle concentration decrease, ie at the top layer or in the bottom layer of the sedimentation. In addition, other conditions affected the result difference were the effects of interactions between P_2O_5 and F particles which had the same surface area loads produced on electrostatic forces that made them difficult to agglomerate and settle (ASWW, 1998).

In this case the interaction between the particles of P_2O_5 has a larger load surface area than the particle F, so that the particle was difficult to achieve agglomeration and was difficult to precipitate, compared to the more agglomerated F particles and could be precipitated well during the sedimentation process . The result of F concentration decrease was greater than P_2O_5 indicating that the coagulation, flocculation, and sedimentation process works well. Another factor was the size of the particles P_2O_5 and F that affected the occurrence of particle agglomeration. In this case the particle diameter and deposition rate were directly proportional, so the larger the particle diameter the greater the precipitation speed occurs. If the settling speed was faster, the sedimentation process proceeds well, and flocks formed in the coagulation and flocculation process can be well deposited during the sedimentation process. Therefore, F particles more easily precipitate due to larger particle diameter size and better settling rate (ASWW, 1998).

The reactivity of the coagulant also affected the results of % removal of P_2O_5 and F. In this case the coagulant used was CaO. CaO had a cationic valence (+2) and was a highly reactive alkaline earth metal with water, which readily released ions to neutralize particles and helped agglomerate the ion ions to bind. F particles had high reactivity and high electronegativity, so this condition could affect the bonding bonds that occured between the flocks that were formed. As a result of higher F reactivity and electronegativity compared to P_2O_5, ionic ions would be easier to bind and neutralization processes. So in this case the F particles more easily binded to the CaO and form precipitates that were easily precipitated, the flock was bonded and the size becomes larger. F particles had water-soluble properties, where their solubility was high in water, which causes F particles to precipitate easily, compared with P_2O_5 particles, hence F particles would more easily settle than P_2O_5. Thus, the above factor influences the difference in % removal yield between P_2O_5 and F in the best samples, where the% removal occurred on particle F was greater than that of P_2O_5 particles of 98.5%.

Conclusions

1. Phosphogypsum can be used as a cement mixture because the P_2O_5 and F content of the phosphogypsum has decreased, so that the phosphogypsum for the cement mixture can be drawn from layers 2 and 3 at the time of sedimentation.
2. The addition of CaO mass, stirring speed variables, and dilution variables affects the coagulation, flocculation, and sedimentation processes for the pH performance parameters that occur in each experimental sample and CaO reactivity as well as rapid stirring that affected the occurrence of layers during coagulation, flocculation, and sedimentation took place.
3. The percentage decrease of P_2O_5 and F concentration occures at the best condition (layers 2 and 3), with 1: 2 (w / v) dilution variables, CaO of 12 gram mass and stirring speed of 150 rpm. The result is to decrease P_2O_5 concentration by 50%, and decrease F concentration by 98,5%.

References

1. A. Aries, *Makalah Seminar Nasional Soebardjo Brotohardjono*, 1-6 (2007).
2. A. W. Association, *Operational Control of Coagulation and Filtration Process*. America: America Water Works Assocation (1992).
3. N. Z. Association, *Standard For The Supply of Polyaluminium Chloride For Use In Water Treatment*. New Zealand: Water New Zealand (2013).
4. P. Bhawan, *Guideline For Management and Handling of Phosphogypsum Generated From Phosphoric Acid Plants*. Delhi: Ministry of Environment and Forest (2012).
5. P. Borris, Utilization and Disposal of Phosphogypsum. *Proceedings of The International Symposium on Phosphogypsum* (pp. 15-100). Florida: Florida Institute of Phosphate Research (1980).
6. W. Eckenfelder, *Industrial Water Pollution Control Third* McGraw-Hill Companies (2000).

7. I. Nansubuga, Effect of Polyaluminium Chloride Water Treatment. *African Journal of Enviromental Science and Technology*, 146 (2012).
8. Shakhashiri. Lime : Calcium Oxide-CaO. *Chemical of The Week*, 1-2 (2012).
9. M. Singh, *Purifying Phosphogypsum For Cement Manufacture*. India: Central Building Research Institute (1992).
10. H. Tayibi, Enviromental Impact and Management of Phosphogypsum. *Journal of Enviromental Management*, 1-10 (2008).

Temperature, Solid Loading and Time Effects on Recovery of Sugar from OPEFB

Diah Meilany[1,2], *MTAP Kresnowati*[2*], *and Tjandra Setiadi*[2]

[1]Teknik Kimia, Politeknik Negeri Malang, Malang 65141, Indonesia
[2]Teknik Kimia, Institut Teknologi Bandung, Bandung 40132, Indonesia

Abstract. Biorefinery industry used lignocellulosic biomass as the raw material. Oil Palm Empty Fruit Bunch (OPEFB) is one of Indonesian potential lignocellulosic biomass, which consists of hemicellulose with xylan as the main component. Xylitol production via fermentation could use this xylan since it can be converted into xylose. However, the structure of OPEFB is such that hemicellulose is protected in a way that will hinder hydrolysis enzyme to access it. Considering that hemicellulose is more susceptible to heat than cellulose, a hydrothermal process was applied to pre-treat OPEFB before it was hydrolyzed enzymatically. The aim of the research is to map the effect of temperature, solid loading and time of pre-treatment process to obtain and recover as much as possible accessible hemicellulose from OPEFB. The results showed that temperature gave more significant effect than time and solid loading for glucose recovery of OPEFB residues. While xylose recovery varies greatly with temperature, but solid loading and time gave less significant effect.

1. Introduction

OPEFB is one of lignocellulosic biomass that is well known as an industrial waste, especially in palm oil industry. Directorate General of Estate Crops of Indonesian Ministry of Agriculture estimated that 32.0 million tons of palm oil were produced in late 2016[1]. OPEFB as industrial waste reached 34.5 million tons but only about 23% utilized as mulch in palm oil plantation [2]. The rest of it will be pilled around the palm oil plant or plantation. This condition raises many problems in environmental such as methane production.

OPEFB is composed of cellulose 31 – 43%, hemicellulose 23 – 35%, and lignin 11 – 23% [3]–[5]. The hemicellulose content is promising enough to be utilized as one of xylose source for xylitol production. Hemicellulose is a polysaccharide composed mainly of xylan that is built-up of xylopyranose unit, connected by 1,4-β glucosidic linkage [6]. A process to depolymerize hemicellulose into xylose is by hydrolyzing xylan and this research focuses on using the biological way that is hydrolyzing OPEFB enzymatically.

There is a structural barrier in order to gain hemicellulose from OPEFB since hemicellulose is covered by lignin and is attached to cellulose fibril. Lignin cover will hinder enzyme to reach the hemicellulose since it is an amorphous polymer [7]. This makes it important to remove lignin from OPEFB by pre-treatment process.

There have been some efforts to pre-treat OPEFB on variety lignocellulosic biomass. It can be chemical pre-treatment using a dilute acid such as reported applied on OPEFB [8], corn cob [9], *Eucalyptus globulus* chips [10], sunflower stalk and tobacco stalk [11]. Others use dilute base such as reported on OPEFB [12], bamboo [13], and wheat straw [14]. However, both acid and base pre-treatment produce a toxic compound that will disturb xylitol fermentation using wild strain yeast. Instead, a genetically modified strain was developed to omit the detoxification process [15]. Another method is the organosolv method such as used on Oil Palm fronds [16], wheat straw [17] the steam explosion on sugarcane bagasse [18], sugarcane straw [19], wheat straw [20], spruce wood chips [21]. These methods partially hydrolyzes lignin bond with cellulose and hemicellulose but most of the hemicellulose are solubilized [22]. The milder process is using a hydrothermal process such was used on OPEFB [23], prairie cordgrass [24], hardwood [25], seeds of *Euterpe oleracea* [26]. The uses of water as an agent to break the bond between lignin, cellulose, and hemicellulose is having mild effects on OPEFB residue.

Hemicellulose is more susceptible to heat than cellulose; thereby it needs a careful choice of pre-treatment method and condition that will not solubilize hemicellulose. Acid or base will solubilize hemicellulose like a steam explosion. The hydrothermal process is a mild process but temperature chosen must be perfect.

* Corresponding author: kresnowati@che.itb.ac.id

Besides, solid loading and time will affect the success of hemicellulose recovery too. This research study the effect of these three parameters in hemicellulose recovery measured as sugar (xylose) in OPEFB residues.

2. Methods

2.1. Raw material

OPEFB was kindly provided by PTPN VIII, Cigudeg, Bogor. It was cut into 10 – 12 cm, washed with tap water and then dried in 60°C oven for 24 hours. Next, the OPEFB was milled and sieved into -60+80 mesh then stored in a ziplock bag before used.

The enzyme used was xylanase from China, in a powder form with an activity of 22,072.48 U/g. Xylose is purchased from Sigma-Aldrich while glucose is purchased from E-Merck as standard in HPLC analysis. Chemicals for citrate buffer was of p.a. grade, purchased from E-Merck.

2.2. Experimental methods

The experiments were conducted following a Box-Behnken design of experiment as presented in table 1.

Table 1 Box Behnken design of experiment

RUN ORDER	SL	TEMP	TIME
1	5	165	15
2	5	165	60
3	5	165	15
4	5	165	60
5	10	165	37.5
6	10	130	60
7	15	130	37.5
8	10	200	15
9	15	165	60
10	15	130	37.5
11	10	130	60
12	10	200	60
13	5	200	37.5
14	15	165	15
15	10	165	37.5
16	15	200	37.5
17	5	130	37.5
18	10	165	37.5
19	10	200	15
20	10	165	37.5
21	15	165	15
22	5	200	37.5
23	15	165	60
24	10	165	37.5
25	5	130	37.5
26	10	200	60
27	10	165	37.5
28	10	130	15
29	15	200	37.5
30	10	130	15

The pretreatment experiments were conducted in a 100 mL autoclave equipped with paddle stirrer and stirrer cooler as shown in Fig 1.

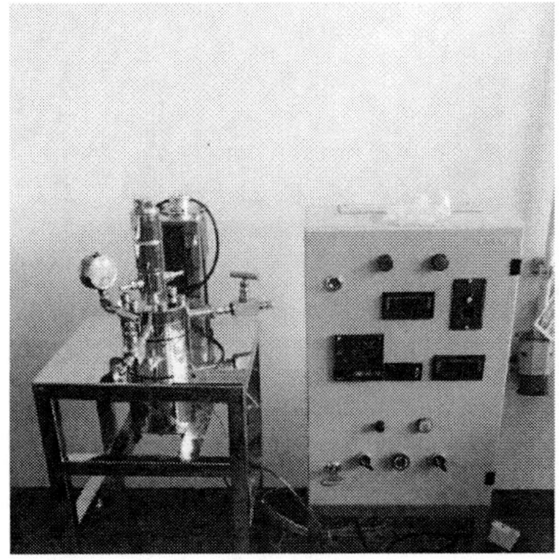

Fig. 1 High-pressure tank

The experiments were aimed to find the effect of temperature, solid loading and time on sugar recovery. The variables are the temperature of 130, 165 and 200°C; solid loading of 5%, 10%, and 15%, and time of 15, 37.5 and 60 minutes. Each pre-treatment experiment used 50 mL of RO water.

After the pre-treatment process, the mixture was screened using fine mesh cloth and the solid residue was left in an oven at 50°C overnight. Before the further process, the supernatant was kept in a freezer to prevent contamination. Hydrolysis for solid residue was performed in a 100 mL Schott bottle at 5% of solid loading in an incubator shaker (Daihan Labtech.Co.Ltd). The citrate buffer solution, 0.05 M pH 5.0, was added up until working volume of 50 mL.

Prior to the addition of enzyme, the mixture of OPEFB and buffer were autoclaved at 121°C for 15 minutes. Incubation was performed for 48 hours at 50°C and 150 rpm. Samples were centrifuged at 6000 rpm for 15 minutes then filtered by 0.22 mm microfilter before sugar concentration analysis of the hydrolyzate.

Sugar concentration in hydrolyzate was analyzed using HPLC (Alliance HPLC System, Waters) using method developed by NREL [27]. Whole experiments were conducted in duplo and the error of each experiments were 15% maximum.

3. Results

The effects of temperature, solid loading and time on sugar recovery are shown on the graphs on fig. 2 to fig. 4 respectively.

3.1. Temperature effect

Temperature effects on sugar recovery are shown in fig. 2.a and 2.b

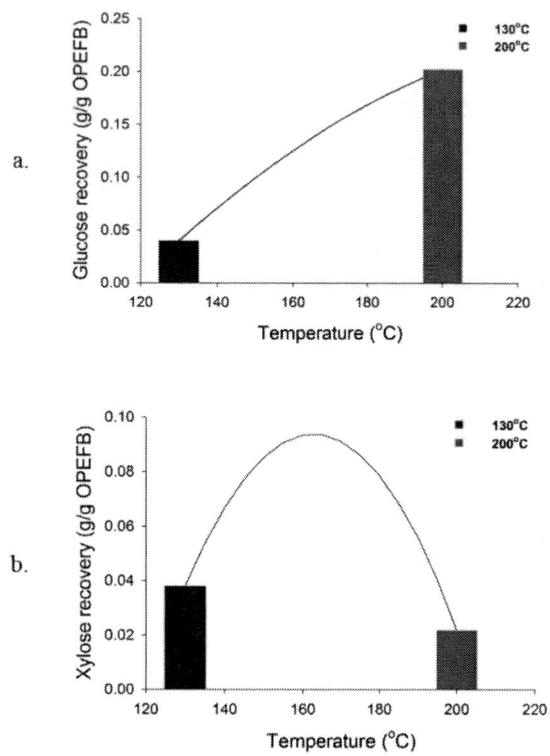

Fig. 2 Effects of temperature on sugar recovery: (a) glucose, (b) xylose. Experiments were conducted at solid loading 10% and 60 minutes.

As can be seen on fig.2.a, Glucose recovery tends to raise as the temperature was increasing, unlike xylose recovery. When the temperature was higher than 130°C, the xylose recovery was lowering significantly. The line curve represents sugar recovery as a function of temperature at 10% solid loading and 60 minutes pre-treatment process. Glucose has not reached the maximum recovery although pre-treatment process had proceeded for 60 minutes, unlike xylose recovery that seems to reach a maximum at 165°C.

This proves that hemicellulose is more susceptible to heat than cellulose. High temperature can solubilize hemicellulose, therefore, xylan cannot be used as xylose source. The high temperature of 200°C had been used in pre-treatment of sugarcane bagasse, sugarcane straw, wood and OPEFB resulted in that hemicellulose is solubilized [18], [19], [28], [29].

3.2.Solid loading (SL) effects

Solid loading effects on sugar recovery are shown in fig.3.a and 3.b.

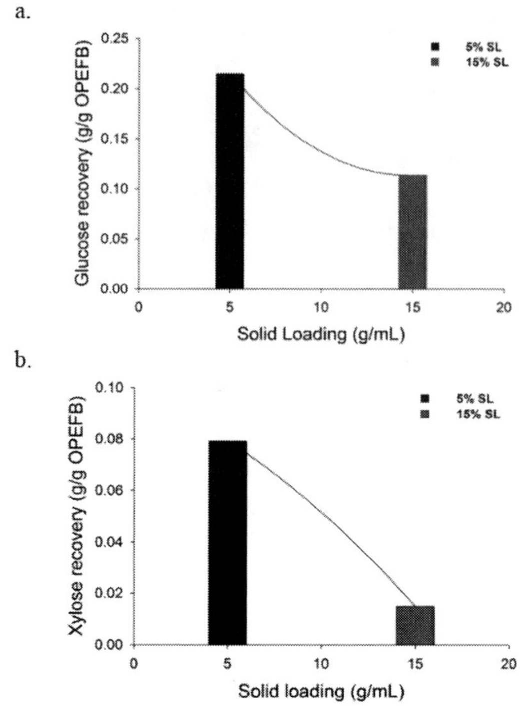

Fig. 3 Effects of pre-treatment solid loading on sugar recovery: (a) glucose, (b) xylose. Experiments were conducted at temperature 165°C and at pretreatment time of 60 minutes

The graphs on fig. 3.a and b showed that the rise of SL will decrease the sugar recovery. This is true since a high solid loading will create a mechanical hindrance.

This experiment was using a 100 mL tank with paddle agitator. The 15% SL in 50 mL mixture formed a compact mixture, makes it difficult to mix thoroughly. This condition leads to lowering glucose recovery more than xylose since there is not enough water molecule to interact with in order to break the lignin bonds with cellulose and hemicellulose. Furthermore, this data was taken at 165°C which is different with fig.2.a. The lower SL (10% SL) at 200°C and 60 minutes gave higher glucose recovery. To overcome this, a specific reactor is needed [30].

3.3. Time effects

Time effects on sugar recovery are shown in fig. 4. Time here means time proceeded while heating the mixture at a certain temperature.

As can be seen in fig.4, a longer pre-treatment time will increase the sugar recovery.

https://doi.org/10.1051/matecconf/201815603022

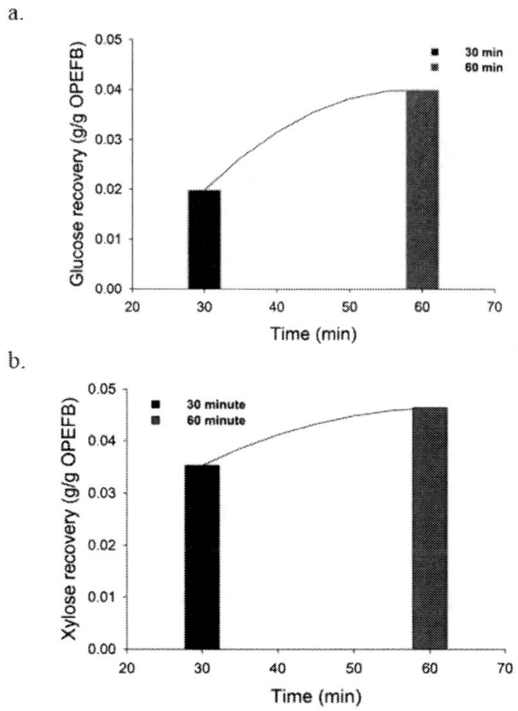

Fig. 4 Effect of pre-treatment time on sugar recovery: (a) glucose, (b) xylose. Experiments were conducted at solid loading 10% and a temperature of 130°C

The graphs in fig. 4.a and 4.b showed that the increase of sugar recovery is affected by time although not significantly. As a comparison, OPEFB which was pre-treated using the hydrothermal process at 127.9°C for 60 and 90 minutes did not give a significant rise of xylose recovery [31].

Another result as shown in fig.2.a. glucose recovery is higher than in fig.4.a. It resulted from pre-treatment process at 200°C;instead 10% SL and 60 minutes time. It seems like temperature gave more significant effect on glucose recovery as to xylose recovery. This result leads to a hypothesis that there are one or more variables that give more significant effect to others, that is to be described in the section below.

3.4. Statistical analysis of effects

Statistical analysis was performed in order to describe the effects of three parameters on sugar recovery. Using main effects plot versus mean value of each sugar recovery, as shown in fig.5.

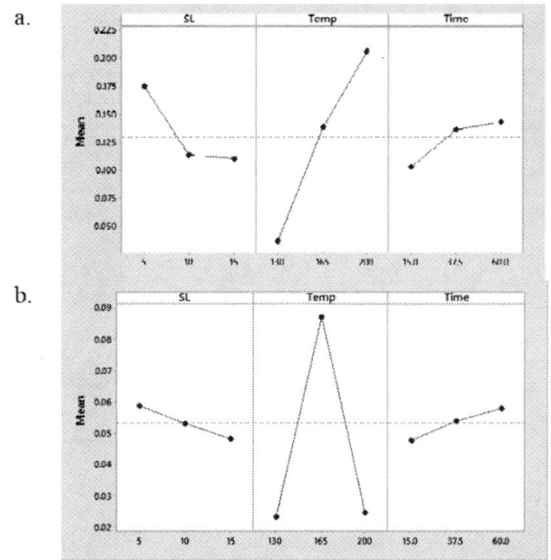

Fig. 5 Main effect plot on glucose (a) and xylose recovery (b)

Fig.5.a and b show that a higher SL will reduce glucose and xylose recovery. The slope of glucose recovery is steeper, this means that SL will affect glucose recovery more significant than to xylose. Temperature shows that a higher temperature will increase glucose recovery. While for xylose recovery shows a great variety of temperature effect. Longer process time shows the same effect as temperature. Since the slope of glucose is steeper means that time gives more significant effect on glucose recovery.

Among these three variables, temperature shows the steepest slope, means that it gives more significant effect on sugar recovery, compared to solid loading and pre-treatment time. There must be one perfect combination of these three variables to achieve a high recovery of sugar.

4. Conclusion

Despite this experiment limitation, the temperature has shown more significant effect than the other two, time and SL. Temperature and time will increase the sugar recovery while SL will decrease the sugar recovery.

References

1. I. M. E. Subiyantoro, & Y. Arianto, Statistik Perkebunan Indonesia 2014 - 2016 Kelapa Sawit (Palm Oil). (2015).
2. R. Omar, A. Idris, R. Yunus, K. Khalid, & M. I. A. Isma, Fuel, **90,** 1536–1544 (2011).
3. S. Octavia, Pengolahan Awal Berbasis amoniak Terhadap Biomassa Lignoselulosik Bahan Mentah Pembuatan Bioetanol. (Institut Teknologi Bandung, 2013).
4. S. H. A. Rahman, J. P. Choudhury, & A. L. Ahmad, Biochem. Eng. J. **30,** 97–103 (2006).
5. M. Shibata, M. Varman, Y. Tono, H. Miyafuji, & S. Saka, J. Jpn. Inst. Energy, **87,** 383–388 (2008).

6. Chen, H. *Biotechnology of lignocellulose: Theory and practice.* (Springer Netherlands, 2014).
7. G. Brodeur, *et al.* Enzyme Research (2011). doi:10.4061/2011/787532
8. S. Duangwang, T. Ruengpeerakul, B. Cheirsilp, R. Yamsaengsung, & C. Sangwichien, Bioresour. Technol., **203,** 252–258 (2016).
9. F.-Q. Wang, *et al.* Bioresour. Technol., **144,** 572–578 (2013).
10. A. Converti, J. M. Domínguez, P. Perego, S. S. da Silva, & M. Zilli, Chem. Eng. Technol., **23,** 1013–1020 (2000).
11. O. Akpinar, O. Levent, S. Sabanci, R. S. Uysal, & B. Sapci, Bio. Resources **6,** 4103–4116 (2011).
12. N. Zulkiple, M. Y. Maskat, & O. Hassan, Mol. Cell. Life Sci. Infect. Dis. Biochem. Struct. Biol. 2015 Conf., **18,** 155–161 (2016).
13. S. X. Chen, Z. D. Wang, K. Fang, G. R. Fan, & P. Wang, Adv. Mater. Res., **1033–1034,** 114–118 (2014).
14. M. C. Coimbra, *et al.* Renew. Energy, **86,** 1060–1068 (2016).
15. E. Hong, *et al.* Biotechnol. Bioprocess Eng., **21,** 612–619 (2016).
16. C. Ofori-Boateng, & K. T. Lee, Fuel, **115,** 170–178 (2014).
17. H. Chen, J. Zhao, T. Hu, X. Zhao, & D. Liu, Appl. Energy, **150,** 224–232 (2015).
18. R. Biswas, H. Uellendahl, & B. K. Ahring, Biomass Bioenergy, **61,** 104–113 (2014).
19. F. M. V. Oliveira, *et al.* Bioresour. Technol. **130,** 168–173 (2013).
20. P. Alvira, M. J. Negro, I. Ballesteros, A. González, & M. Ballesteros, Bioethanol, **2,** 66–75 (2016).
21. T. Pielhop, J. Amgarten, P. R. von Rohr, & M. H. Studer, Biotechnol. Biofuels, **9,** 152 (2016).
22. D. K. Sidiras, & I. S. Salapa, *Biorefinery I: Chemicals and Materials From Thermo-Chemical Biomass Conversion and Related Processes* (ECI Symposium Series, 2015).
23. M. B. Harahap, *Evaluasi Konfigurasi Proses Produksi Xilitol dan Etanol dari Tandan Kosong Kelapa Sawit Secara Terintegrasi.* (Institut Teknologi Bandung, 2016).
24. H. Lei, I. Cybulska, & J. Julson, J. Sustain. Bioenergy Syst., **3,** 250–259 (2013).
25. C. K. Nitsos, T. Choli-Papadopoulou, K. A. Matis, & K. S. Triantafyllidis, ACS Sustain. Chem. Eng., **4,** 4529–4544 (2016).
26. A. R. Oliveira, J., Komesu, A. & Maciel Filho, R. Chem. Eng. Trans., **37,** 787–792 (2014).
27. A. Sluiter, *et al.* Determination of Sugars, Byproducts, and Degradation Products in Liquid Fraction Process Samples in Laboratory Analytical Procedure. (2008).
28. A. P. Nunes, & J. Pourquie, Bioresour. Technol., **57,** 107–110 (1996).
29. J. D. C. Medina, *et al.*, Bioresour. Technol., **199,** 173–180 (2016).
30. J. Zhang, W. Hou, & J. Bao, Bioreactor Engineering Research and Industrial Applications II, 75–90 (Springer, Berlin, Heidelberg, 2015). doi:10.1007/10_2015_307
31. M. B. Harahap & M. T. A. P. Kresnowati, Biomass Convers. Biorefinery, (2017).

Effect of blending ratio to the liquid product on co-pyrolysis of low rank coal and oil palm empty fruit bunch

Siti Zullaikah[1,*], *Ari Suryo* Lenggono[1], *Dennis Farina* Nury[1], and *M.* Rachimoellah[1]

[1]Chemical Engineering Department, Institut Teknologi Sepuluh Nopember, Kampus ITS, Keputih, Sukolilo, Surabaya 60111 Indonesia

Abstract The utilization of Indonesia low rank coal should be maximized, since the source of Indonesia law rank coals were abundant. Pyrolysis of this coal can produce liquid product which can be utilized as fuel and chemical feedstocks. The yield of liquid product is still low due to lower of comparison H/C. Since coal is non-renewable source, an effort of coal saving and to mitigate the production of greenhouse gases, biomass such as oil palm empty fruit bunch (EFB) would added as co-feeding. EFB could act as hydrogen donor in co-pyrolysis to increase liquid product. Co-pyrolysis of Indonesia low rank coal and EFB were studied in a drop tube reactor under the certain temperature (T= 500 °C) and time (t= 1 h) used N_2 as purge gas. The effect of blending ratios of coal/EFB (100/0, 75/25, 50/50, 25/75 and 0/100%, w/w %) on the yield and composition of liquid product were studied systematically. The results showed that the higher blending ratio, the yield of liquid product and gas obtained increased, while the char decreased. The highest yield of liquid product (28,62 %) was obtained used blending ratio of coal/EFB = 25/75, w/w%. Tar composition obtained in this ratio is phenol, polycyclic aromatic hydrocarbons, alkanes, acids, esters.

1 Introduction

Indonesia coal reserves in 2015 reached 126.61 billion tons, and about 80% of coal reserve was low rank coal. Lignite is a type of coal that is classified as low rank coal due to it has high water content and low caloric value[1]. The potential of coal resource in Indonesia is abundant, it's mainly distribute in South Sumatra, Central Java, Papua and Sulawesi[2]. Two conversion processes to convert coal into chemical products are biochemical and thermochemical processes. Biochemical conversion methods are based on the biological activity, whereas thermochemical process includes liquefaction, gasification, pyrolysis and combustion. Coal pyrolysis is a promising process to convert coal into liquid fuels and chemicals[3].

Products on thermochemical conversion are consist of volatile fraction, tar components and char rich of carbon residue. The pyrolysis process consists of a very complex reactions involving the formation of radicals. Pyrolysis is the most widely used to thermochemical processes which is the chemical decomposition of organic material by heating in the absence of oxygen[4].

Since the quantity and quality of pyrolysis products are limited because of the low atom ratio H/C coal[5], so it needs a donor hydrogen from other resources, such as biomass. Biomass can act as donor hydrogen during pyrolysis. Biomass can also generate more liquid than gas product. In addition, biomass contains more volatile matter than coal.

Coal contains monocyclic and polycyclic aromatic structures that represent characteristics of its rank[6]. Coal conversion processes need to be developed further and needs to increased to modern requirements regarding rate of yield, purity and energetic aspects. A study gives an overview on the utilization of lignites for the production of value added chemicals[7]. Other study, which can be found in several of these conversion processes, is the pyrolysis of the higher molecular weight of coal feedstock into lower molecular weight products.

There are several studies focused on coal pyrolysis [7]. The liquids produced during pyrolysis are mixture that contains complex of organic compounds. Generally, the composition in organic compounds depend on the feedstock and pyrolysis conditions, from non-polar to polar and from low to high molecular weight compounds.

* Corresponding author: szulle@chem-eng.its.ac.id, szullehakim@gmail.com

https://doi.org/10.1051/matecconf/201815603023

Hydrogen from biomass can produced more volatiles compounds during co-pyrolysis than that produced from pyrolysis of each sample individually. Hydrogen can prevent recombination and cross-linking reactions during co-pyrolysis process[9]. The use of biomass as an energy source has been known to have environmentally benefits besides the abundance of its biomass. Biomass is considered to be a potential alternative energy source. Especially, biomass from agricultural and forestry industrial wastes that can be used for energy feedstocks.

Biomass is renewable energy resources and more environmentally friendly [8]. An interesting waste from oil palm mill is oil palm empty fruit bunch (EFB). EFB is by-product and one of the top ranked agricultural residues in Indonesia, the availability in Indonesia reaches 7,195,390 tons / year[10].

Several researcher have investigated the effect of blending ratio on pyrolysis products. A study on co-pyrolysis of biomass legume straw (LS) and lignite coal in a drop tube reactor from 500 to 700 °C, the results shown a trend towards linear product distributions at higher temperatures. The results showed that the char yields decreased and consequently the liquid yields increased. The largest degree of synergetic effects on the products yields was observed under 600 °C and the blending ratio of 73 wt.% in the co-pyrolysis of LS/DY. It can be concluded that both the higher blending ratio (around 70 wt.%) and the relatively lower temperature (around 600 °C) are the best conditions during the co-pyrolysis of biomass and coal on free fall reactor[5], [7].

Study of biomass pyrolysis, such as rapeseed were conducted to determine the effect of pyrolysis, such as temperature, heating rate, particle size and gas flow rate on the pyrolysis product yields and their chemical compositions. The results of this experiment showed that the liquid yield product reached a higher of 68% by employing higher heating rate of 300 °C min-1 under temperature of 550 °C and particle size range of 0.6 – 1.25 mm. The proportion liquid products depends very much on the pyrolysis technique used and on the reaction parameters[11].

A research conducted to produce tar from Legume Straw (LS), Sawdust (SD), High-Rank (TF) and Low Rank Coal (DY). Liquid tar was produce more using High-Rank coal than that of Low Rank coal. The best conditions of the blending ratios are 50/50 and 70/30 at temperatures of 600°C and 700°C, respectively. Liquid yields from co-pyrolysis of TF coal are higher than that of DY. Liquid product from the LS/DY co-pyrolysis was analyzed by means of solvent extraction-column chromatography combined with GC-MS. This is because coal levels also greatly spur liquid products from co-pyrolysis[7].

Other study on co-pyrolysis biomass and low rank coal, which produces more tar from low rank coals than pine wood. The best condition of the blending ratio is 70wt% at temperature below 600 °C. Hydrogen donor from biomass can produce more tar, this is because biomass contains higher volatile materials than coal[4], [1].

The effect blending ratio of co-pyrolysis on the product distributions was discussed in detail in this study. Identification of components in tar was done by gas chromatography/mass spectrometry (GC–MS).

2 Experimental

2.1. Raw materials

Low rank coal was obtained from PT. Semen Indonesia Persero (Tbk) and oil palm empty fruit bunch (EFB) was kindly supplied from oil palm plantation in Samarinda, East Kalimantan. The fresh EFB was chopped and dried by sunlight for 3 days before being grinded to reduce its particle size. Samples (coal and EFB) were dried in oven for 1 hour at 105°C and placed in dessicator for 30 minutes. Particle size of samples were distributed in a range of 0.71 – 1.19 mm. The proximate and ultimate analyses of samples are shown in **Table 1.**

2.2. Co-pyrolysis of low rank coal and EFB

Co-pyrolysis was performed in a droptube reactor. It has length of 90 cm, outside diameter of 2 inch, 6 mm of thickness, and it made of *stainless steel* 216 H. Two thermocouples are inserted in the middle of reactor to measure the reactor temperature. Bend heater was installed on the reactor wall to heated up the reactor to the desired temperature. The reactor was heated up to 500°C at rate of 5°C /min under constant N_2 flow rate. A hopper was placed on the top of reactor to fed the sample (100g). Five blending ratios (100/0, 75/25, 50/50, 25/75, and 0/100, % w/w) were employed in this study. Drop tube reactor used in this study was shown in fig. 1.

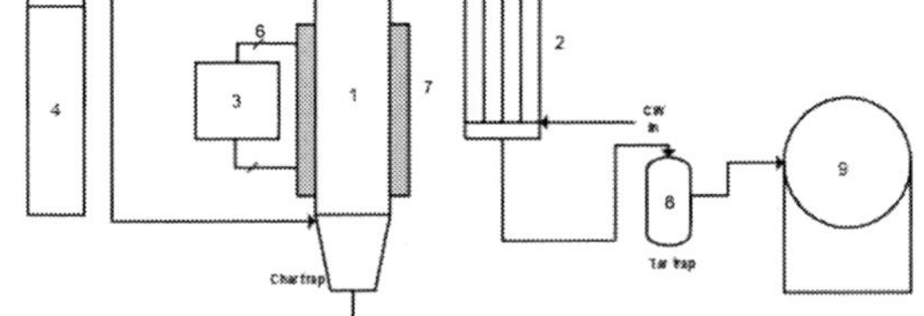

Fig. 1. Droptube reactor :

(1) reactor, (2) condenser, (3) controller, (4) N_2 gas, (5) hopper,(6) thermocouple, (7) heater, (8) tar trap, (9) gas bag

The product gas was cooled through condenser to condensed a condensable gas (liquid product). The collected liquid was separated into tar and water by using a rotary evaporator at 60 °C under vacuum. Tar composition was analyzed qualitatively by using GC-MS. Residual char left in the char trap was measured gravimatrically. The co-pyrolysis was carried out for 1 hour.

The samples were tested for ultimate analysis for

carbon, hydrogen, nitrogen and sulphur contents. The proximate analysis was used to determine the moisture content, volatiles matter, ash content and fixed carbon in both coal and EFB. The HHV (higher heating value) were determined using a bomb calorimeter PARR 1261. TGA (Thermogravimetric analysis) was carried out on the sample of size ~10 mg in order to evaluate the qualitative composition of sample, in particular thermal degradation behavior. TGA was performed with heating rate of 10 °C/min, heated from 25 °C - 900 °C under N_2 gas flow. Schematic diagram is shown in **Fig 1.**

3 Results and discussion

The product yields from the co-pyrolysis of low rank coal and EFB at different blending ratios and yield of tar at temperature 500°C are given in **Fig.3 and 4**, The product yields are calculated by :

$$\text{Yield of char} = \frac{Char\ product\ (g)}{Sample\ (g)} \text{x}100\% \quad (1)$$

$$\text{Yield of liquid} = \frac{Liquid\ product\ (g)}{Sample\ (g)} \text{x}100\% \quad (2)$$

$$\text{Yield of gas} = \frac{Gas\ product\ 9g)}{Sample\ (g)} \text{x}100\% \quad (3)$$

$$\text{Yield of tar} = \frac{Tar\ product\ (g)}{Sample\ 9g)} \text{x}100\% \quad (4)$$

The proximate analysis is analyzed according to ASTM D 5142. Ultimate analysis was performed to determine elemental composition by ASTM D-5373 and ASTM D-4239 to determine of total sulfur.

As shown in Table 1, H/C ratio of EFB is higher than that of coal. It has been reported previously that more volatiles can be produced from coal pyrolysis by adding hydrogen-donorsfrom other organic resources, such as biomass.Higher content of H_2 in EFB may be used as H_2-donor during co-pyrolysis with coal [5].

Table 1 Proximate and Ultimate analysis of samples

Parameter	Sample	
	Coal	EFB
Proximate analysis		
% Moisture content	20.51	7.13
% Fixed carbon	38.50	29.00
% Ash content	5.00	12.00
% Volatile matter	60.00	76.00
Ultimate Analysis		
% Carbon	54.70	42.65
% Hydrogen	5.83	6.45
% Nitrogen	0.75	0.61
% Sulphur	0.24	0.19
% Oxygen	38.48	50.10
H/C	1.28	1.82
HHV (MJ/kg)	26.00	14.50

3.1. Effects of blending ratio on product yields

Fig 2 shown a thermogravimetric (TG) of coal/EFB at any blending ratios. At the blending ratio of coal: EFB= 25/75 and 50/50 are quite similar to 100% EFB. At the blending ratio of coal: EFB 75/25 was similar to that of using coal 100%. From TG profile, it seen that blending ratio of 25- 100% EFB is more active than coal alone.

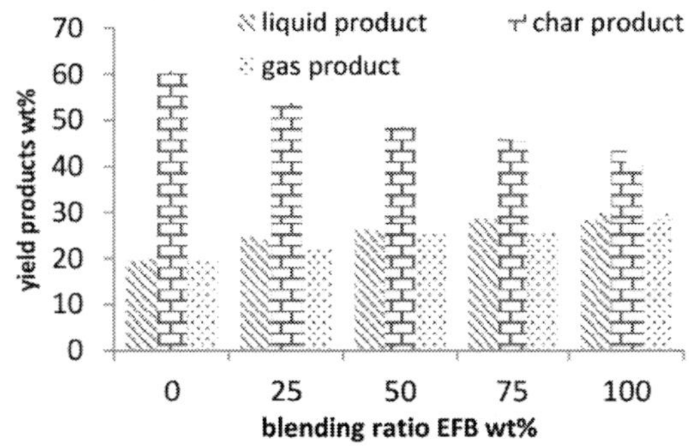

Fig.2. TG of coal/EFB at any blending ratios

Coal decomposition temperature ranges (329.13 – 687 °C),EFB (259.61-362 °C), Coal: EFB 25/75 (259.63-315 °C), Coal: EFB 50/50 (261,18-360 °C) and Coal: EFB 75/25 (246.43-362 °C).

The comparison with the result of TGA conducted [12]on co-pyrolysis subbituminous coal with Malaysia EFB on blending ratioof EFB / Coal 0,20,40,50,60, 80, and 100% for temperature range of 25 -900 °C were obtained two DTG peaks with each T1 = 300-301 °C and T_2 = 425-433 °C. TGA results in this study have lower decomposition temperature range.

The gas, liquid and char product yields co-pyrolysis of coal/EFB was shown in Fig 3. The results of both coal and EFB have the same trend in the acquisition of gas, liquid and char product yields. Increasind of blending ratio, the yield of liquid and gas increased while the yield of char decreased.

The general products trend shows that the amount of gas and liquid products increased, while the char products decreased with blending ratio of coal/EFB. Those results consistent with the previous study on co-pyrolysis of legume straw and lignite in a drop tube reactor from 500 to 700 °C. There was a trend towards linear product distributions at higher temperatures [5]. At the higher blending ratio, hydrogen availability during pyrolysis is much greater. Hydrogen in EFB can be used as hydrogen donor during co- pyrolysis.

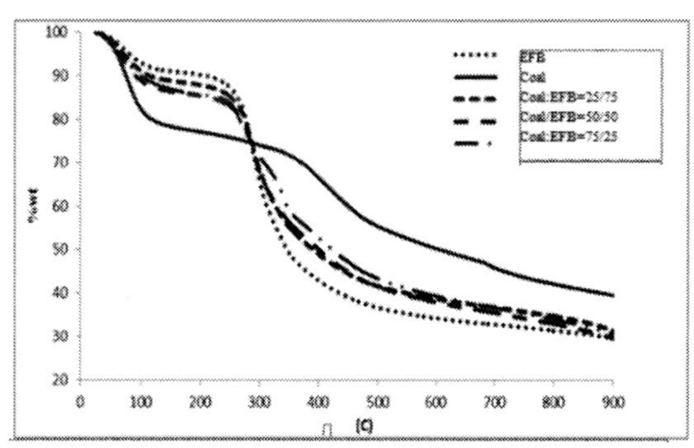

Fig.3. Product yields of coal/EFB at any blending ratio

Hydrogen from biomass may prevent recombination and cross-linking reactions of free radicals that increase char formation [5].

The liquid yield of EFB is greater than that of coal, this can be seen on individual variables, 100/0 and 0/100 respectively. This is because EFB has higher volatile matter than coal which can be seen from proximate analysis in Table 1.

3.2. Physical and chemical properties of liquid product

The liquid product was separated into two phases, tar and water by using a rotary evaporator at 60 °C under vacuum. The yields are given in **Fig 4.** At individual sample, coal/EFB = 100/0 wt% and coal/EFB = 0/100 wt%, the yield of tar were 1.96% and 6.84%, respectively. The yield of tar at blending ratio of 25wt% to 75wt% EFB were increased. The maximum yield of tar (9.05%) was obtained under the condition of blending ratio of 75wt%.

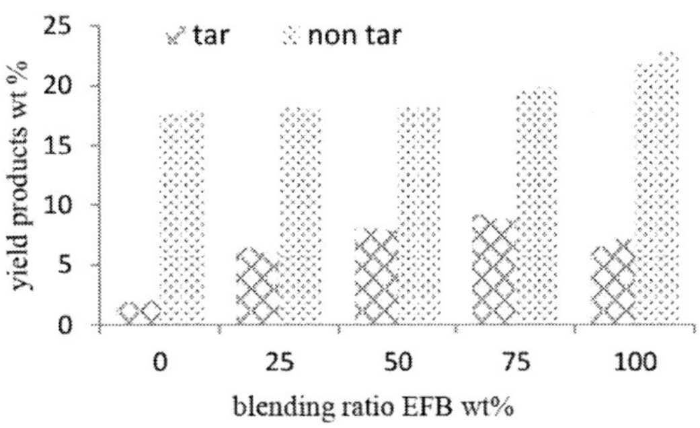

Fig.4. Liquid product yield of coal/EFB at any blending ratio

3.3. Identification of Liquid Compounds by Using GC MS analysis

The liquid product contains an organic phase and a high water content of the aqueous phase. The addition of polar solvents such as methanol or ethanol represents one of the easiest routes [13] and it was estabilished by the authors of this research that the two phases both readily dissolved in methanol giving a homogeneous single phase product with a low viscosity.

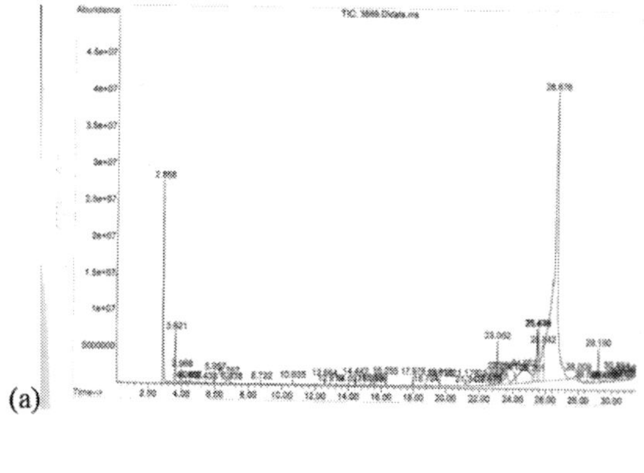

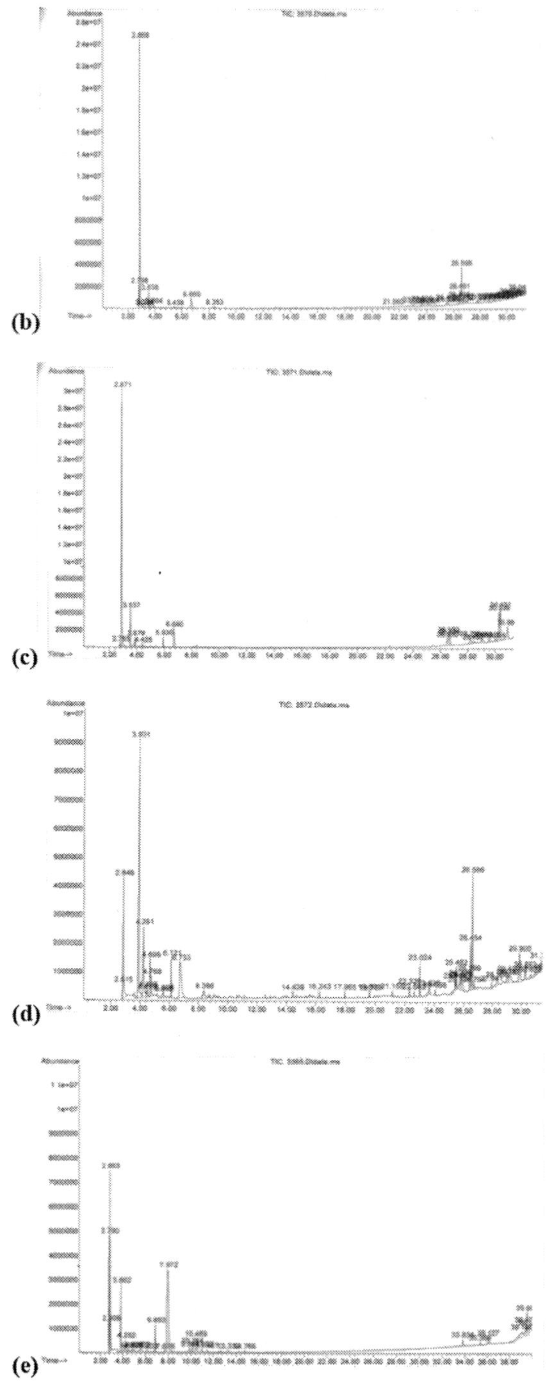

Fig 5. GC MS of liquid products at any blending ratios. (a) EFB 0 wt%; (b) EFB 25 wt%; (c) EFB 50 wt%; (d)EFB 75 wt%; (e) EFB 100 wt%

Tar is consist of many compounds dissolved water. From fig. 5 (a) to (e), we can classified that tar is mainly consist of phenolic compound, acid and ester, alkanes and PAHs (polycyclic aromatic hydrocabons).

The aqueous phase could potentially be co-fired to reduce thermal NOx and enable efficient burning of its dissolved dry matter. As mentioned in the introduction, it might also be possible to obtain phenolic compounds for resin making, as a commercial by-product.

Thermochemical upgrading is another possibility, either via gasification and Fischer-Tropsch, where it may be advantageous to gasify a bio-oil char slurry rather than the biomass. The organic phase may also be used directly in engines, turbines and boilers, if it is first pre-heated to reduce its viscosity, though its thermal instability might limit the temperature it can be brought up to[6], [11].

At blending ratio of EFB 75wt%, tehre are compounds tetradecanoic acid, methyl ester which is the compiler of biodiesel and eicosanol which is an alcohol compound.

In addition, alkanes, alkenes and esters, phenols, benzene, aldehydes, ketones and PAHs were also found at each blending ratio. Percentage of phenols, PAHs decreased, but acid levels increased respectively.

The tar products identified by GC MS analysis, can be utilized in their respective applications. Alkanes compounds can be used as based oil. The compound of acid and ester can be used as feedstuffs for various agnets of food and beverages, textiles, plastic lubricants, paints and its additive, surfactants and pharmaceuticals. Also methyl ester compound which is a biodiesel constituent. PAHs such as naphtalens may be used as an additive.

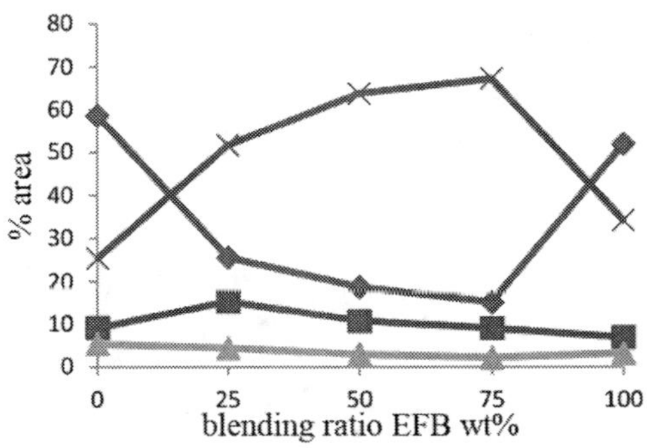

Fig.6 Distributions of liquid product at any blending ratio

Distribution the abundances of the liquid product given in **Fig.6.** Based on GC MS result for EFB 0wt%, the maximum percentage of area are phenols, alkanes and PAHs. This result is accordance which states that liquid coal contains phenolic compounds and polycyclic aromatic hydrocarbons (PAHs) [14].

GC MS analysis results for EFB 100wt%, have found alkanes, acid and ester, phenol and PAHs. Similarly, experiment on pyrolysis of pomegranate seed contains alkanes, phenols, acids, esters and polycyclic aromatic hydrocarbons (PAHs) on liquid products [9].

However, GC MS analysis results for EFB 25wt% up to 75wt% blending ratio, the percentage of phenols, alkanes, polycyclic aromatic hydrocarbons (PAHs) decreased, but the acid and ester increased respectively.

4. Conclusions

Co-pyrolysis of coal with biomass was performed in a droptube reactor working at constant pressure N_2 gas, under operation conditions of temperature of 500°C and 1 hour. The results showed that the higher blending ratio, the yield of liquid product and gas obtained increased, while the char decreased. The highest yield of liquid product (28.62 %) was obtained used blending ratio of coal/EFB = 25/75. Tar composition obtained in this ratio is phenol, polycyclic aromatic hydrocarbons, alkanes, acids, esters. From GC MS analysis, there are found phenol, alkanes, acid and esters, polycyclic aromatic hydrocarbons (PAHs).Methyl ester was also found, a type of fatty acid *ester* that are derived by transesterification of fats with methanol. The molecules are primarily methyl ester, used to produce biodiesel. It can be concluded from this study that co-pyrolysis of low rank coal and EFB, which has a positive effect on liquid product.

The authors would like to express their gratitude to the RISTEKDIKTI on Program Kreativitas Mahasiswa (PKM) for funding this research with grant number: 547/B3.1/KM/2017. The authors would like to thank all members of Biomass Laboratory for their valuable help during this research work. The authors also thank to Chemical Engineering Department Institut Teknologi Sepuluh Nopember (ITS)for the valuable help and support during this research.

References

1. R. M. Soncini, N. C. Means, and N. T. Weiland, Fuel, **112**, 74–82, (2013).
2. D. Supramono, and S. Lusiani, IOP Conf. Ser.: Mater. Sci. Eng., **162** (1), 1-14, (2017).
3. S. Yaman, Chem. In. Form., **35** (5), 651–671, (2004).
4. N. T. Weiland, N. C. Means, and B. D. Morreale, Fuel, **94**, 563–570 (2012).
5. L. Zhang, S. Xu, W. Zhao, and S. Liu, Fuel, **86** (3), 353–359 (2007).
6. O. Onay, E. Bayram, and O. M. Kockar, Energy Fuel, **21** (5) 3049–3056, (2007).
7. W. E. I. Li-gang, Z. Li, and X. U. Shao-ping, J. Fuel Chem. Technol., **39** (10) 728–734, (2011).
8. D. Mohan, C. U. Pittman, and P. H. Steele, Energy Fuel, **20** (3), 848–889 (2006).
9. S. Uçar and S. Karagöz, J. Anal. Appl. Pyrolysis, **84** (2) 151–156, (2009).
10. S. N. Sholikhah, V. S. Bonavia, K. Pertanian, and R. Indonesia,
11. O. Onay and O. M. Kockar, Ren. Energy, **28** (15), 2417–2433, (2003).
12. F. Sulaiman and N. Abdullah, Energy, **36** (5) 2352–2359 (2011).
13. S. Krerkkaiwan, C. Fushimi, A. Tsutsumi, and P. Kuchonthara, Fuel Process. Technol., **115**, 11–18 (2013)
14. N. Abdullah and H. Gerhauser, Fuel, **87**, 2606–2613 (2008).

Effect of Light Intensity, CO_2 Gas Concentration, Culturing Period and Walne Nutrient Concentrations on Biomass and Lipid Productivity of *Chlorella vulgaris* in Sea Water Media

Timotius Candra Kusuma[1], *Anggun* Rindang Pratiwi[1], *Septiandre*[1], *Siti* Zulaikah[1*]

[1] Department of Chemical Engineering, Sepuluh Nopember Institute of Technology,Surabaya

Abstract. The biomass and lipid productivity of Chlorella vulgaris cultured in sea water media were conducted in this study. The effect of light intensity (5000 and 10000 lux), CO_2 gas concentration (0.03%, 1% and 2%), culturing period (7 and 17 days) and walne nutrient concentrations (0%, 0.05%, 0.1% and 0.3%) on biomass and lipid productivity of C. vulgaris cultured in photobioreactor were studied systematically. The biomass and lipid productivity were increased with increasing light intensity and CO_2 gas concentration. Longer culturing period, C. vulgaris produced more biomass and lipid content. However, biomass and lipid productivity at shorter cultured period were higher than longer cultured period. The highest biomass productivity of 139 mg/L/d was obtained under the following condition: light intensity = 10000 lux, CO_2 gas concentration = 2%, culturing period = 7 days, and walne nutrient concentration = 0.3%. The highest lipid productivity of 40.68 mg/L/d was obtained under the following condition: light intensity = 10000 lux, CO_2 gas concentration = 2%, culturing period = 7 days, and walne nutrient concentration = 0.005%. This study shows that a microalga C. vulgaris was a potential candidate as a source of biodiesel production.

1 Introduction

The availability of energy from non-renewable sources was limited. Therefore an innovation is needed to overcome this problem. Microalgae are known as photosynthetic microorganisms that can combine solar, water and CO_2 energy to serve as biomass. Microalgae is a highly efficient biomass capable of taking waste (zero energy) in the form of carbon (CO_2) and converting it into a high density of liquid energy form (petroleum) (Widjaja et al, 2008). According to some estimates, the yield (per hectare) of algae oil can produce 200 times greater than the yield of vegetation or other vegetable oils (Dermibas, 2008). The type of microalga to be used in this study was *Chlorella vulgaris*. Miao and Wu (2006) reported that heterotrophic growth of *Chlorella protothecoides* was able to produce lipid content of 55% and it could also turn into biodiesel (Miao and Wu, 2004, 2006; Miao et al., 2004; Wu et al., 1992, 1994). The effect of CO_2 concentration on lipid productivity can be seen from the previous research which obtained the result that at 50 mL/min CO_2 flowrate, the lipid productivity after 17 days of nutrient deficiency reaches more than 50% which is the highest value when compared to lipid content using a lower CO_2 flowrate (Widjaja et al, 2008).

Therefore, the objective of this research is to study the effect of light intensity, CO_2 gas concentration, culturing period and walne nutrient concentration on biomass and lipid productivity from *Chlorella vulgaris* on sea water medium. The expected benefits of this study are expected to show a comparison of growth rates and lipid production of *Chlorella vulgaris* on optimal amounts of nutrients and light for biofuel production.

2 Materials and methods

2.1. Materials

Chlorella vulgaris microalgae and walne nutrient obtained from Cultivation Centre of Air Payau Situbondo. Technical methanol and aquadest were obtained from local market. Filter paper 2.5μm grade 42 diameter 125mm was obtained from GE Healthcare Life Sciences U. Sea water with salinity of 24.7 ppt (2.47%) was obtained from Centre of decorative fish Gunungsari, Surabaya, Indonesia.

* Corresponding author: szulle@chem-eng.its.ac.id, szullehakim@gmail.com

2.2 Algae cultivation

Chlorella vulgaris (200 mL) obtained from Cultivation Centre of Air Payau Situbondo, Indonesia had cell concentration of ± 5000 cells/ml was added into sterilized seawater (800 mL) with salinity of 24.7 ppt (2.47%). Salinity of seawater was measured use Salinometer EC 10 from ρH on Lab. Cultivation was performed on a lab scale used light intensity of 5000 and 10.000 lux was measure using Lux digital LX1010B from Dekko with a lighting ratio of 12 h : 12 h (dark: light) and added (0%, 0.05%, 0.1% and 0.3% of walne nutrient, where each 1 mL walne nutrient contains 17.26 mg of nitrogen. Aeration was carried out with different flow rate and CO_2 concentration (2 mL/min (0.03%), 62.5 mL/min (1%) and 125 mL/min (2%)).

During the culturing period, each day monitor and maintain the water level of the photobioreactor by adding a sterilized aquadest. The added aquadest volume varies depending on the amount of water lost. UV-VIS spectrophotometer (GENESYSTM 10S, Thermo Scientific USA) was used to measure the absorbance of Chlorella vulgaris culture every 24 hours. Harvesting was done after 7 and 17 days of cultivation

2.4 Extraction and Distillation

The extraction was carried out by soxhlet extractor using methanol as solvent. Algae extraction was done until the paper filters used wrapped microalgae turn back to the original colour. Separation of microalgae oil and methanol was carried out using a rotary vacuum evaporator, Yamato, USA.

2.5 Analysis

Chlorella vulgaris growth analysis was observed daily through absorbance observation using UV-VIS spectrophotometer. The wavelength used was 685 nm. The relationship between absorbance with the number of cells for *Chlorella Sp.*(Hadiyanto, 2010) was:

Number of cells (cell / mL) = $13.4 \times 10^6 \times OD_{685}$

The biomass productivity, lipid content and lipid productivity were calculated as follow:

Biomass productivity (g/L/d) = mass of dry alga (gram)/ (Culture volume (L) * day)

Lipid content (%) = (mass of extracted lipids/dry microalgae mass) x 100%

Lipid productivity (g/L/d) = (Biomass (g/L) x lipid content)/(Number of days of culturing (days)).

Then the oil results was analysed further by using TLC and GC-MS analysis.

3 Results and discussion

3.1. Growth of Microalgae Chlorella vulgaris In Sea Water Media

In this study, the growth of Chlorella vulgaris microalgae was affected by light intensity, CO_2 gas concentration, culturing period and nutrient concentration of walne. Sea water salinity was 24.7 ppt (2.47%). Based on the previous research, it was mentioned that Chlorella vulgaris was tolerant to the environment with salinity content of 0-70 ppt (Chalid et al).

The growth of Chlorella vulgaris was observed in this study by increasing the number of cells. Effect of light intensity, walne nutrient concentration and harvesting time at flow rate of 2mL/min (CO_2 gas concentration= 0.03%) on growth of Chlorella vulgaris microalgae can be seen in Fig 3.1 and 3.2.

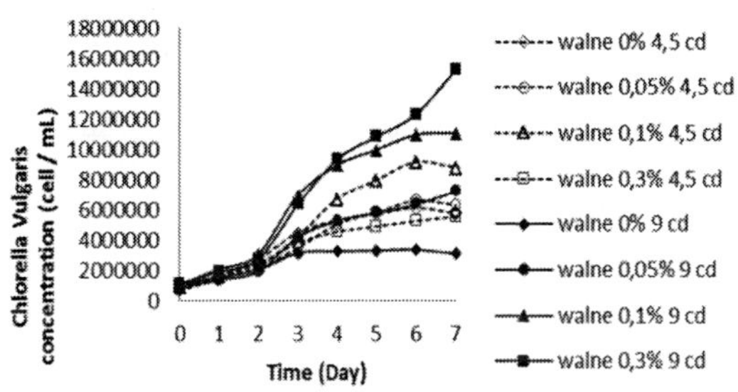

Fig 3.1. *Chlorella vulgaris* concentration in seawater media for 7 days of culturing period with flow rate of 2mL/ min (0.03% CO_2 gas concentration) for various nutrient addition of walne at 5000 lux and 10000 lux.

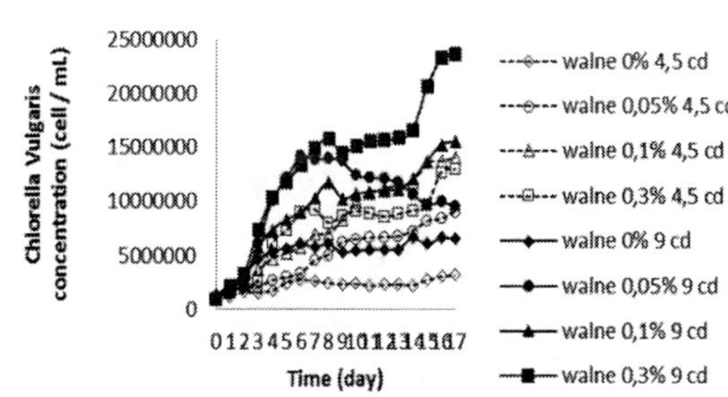

Fig 3.2. *Chlorella vulgaris* concentration in seawater media for 17 days of culturing period with flow rate of 2mL/ min (0.03% CO2 gas concentration) for various nutrient addition of walne at 5000 lux and 10000 lux.

CO_2 used in this study was from the air, it called CO_2 in ambient water. In both Figs, the highest growth was obtained in 0.1% walne nutrient concentration with a light intensity of 10000 lux. From the Fig was known that to increase cell concentration by increasing of walne nutrient concentration. For the trend line formed on growth for 7 days showed a significant increase on the 3rd day to the 7th day. As for the trend line on growth for 17 days showed a tendency to rise until day 8, and then enter the stationary phase during its growth.

Effect of light intensity, walne nutrient concentration and harvesting time at flow rate of 62.5 mL/min (CO_2 gas concentration= 1 %) on growth of Chlorella vulgaris microalgae can be seen in Fig 3.3 and 3.4.

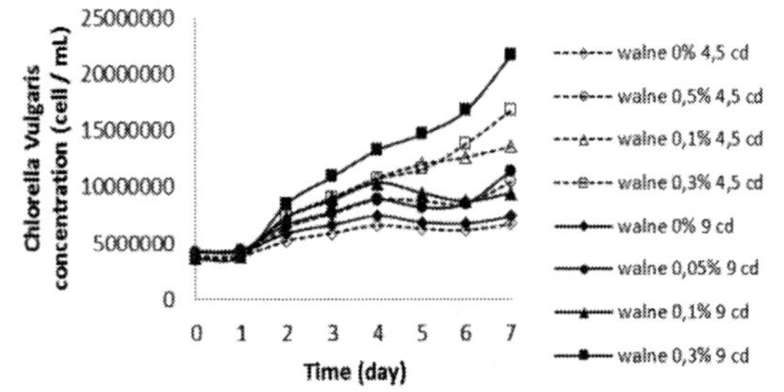

Fig 3.3. *Chlorella vulgaris* concentration in seawater media for 7 days of culturing period with flow rate of 62.5 mL/min (1% CO_2 gas concentration) for various nutrient addition of walne at 5000 lux and 10000 lux.

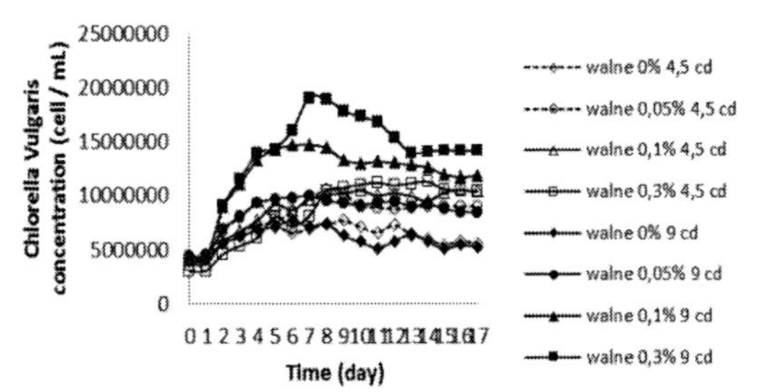

Fig 3.4. *Chlorella vulgaris* concentration in seawater media for 17 days of culturing period with flow rate of 62.5 mL/min (1% CO2 gas concentration) for various nutrient addition of walne at 5000 lux and 10000 lux.

Fig 3.3 and 3.4 shown that the highest cell concentration was obtained by giving walne nutrition concentration of 0.3% and light intensity of 10000 lux. From these can also be observed the tendency of increasing growth along with the increasing number of walne nutrient concentration. In Fig 3.3 it can be seen that the trend line formed on each chart increased on the third day and still continues to increase even though it has been the seventh day (not yet entered the stationary phase). The trend line on each graph is likely to have entered the stationary phase on the day of 5th to day of 17th.

Effect of light intensity, walne nutrient concentration and harvesting time at flow rate of 125 mL/min (CO2 gas concentration= 2 %) on growth of Chlorella vulgaris microalgae can be seen in Fig 3.3 and 3.4..

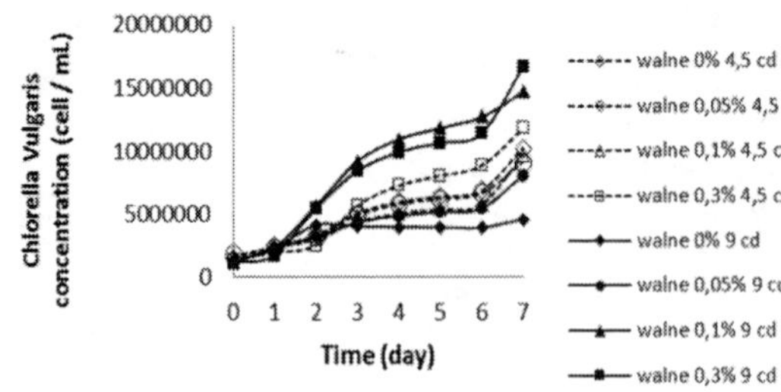

Fig 3.5. *Chlorella vulgaris* concentration in seawater media for 7 days of culturing period with flow rate of 125 mL/min (2% CO_2 gas concentration) for various nutrient addition of walne at 5000 lux and 10000 lux

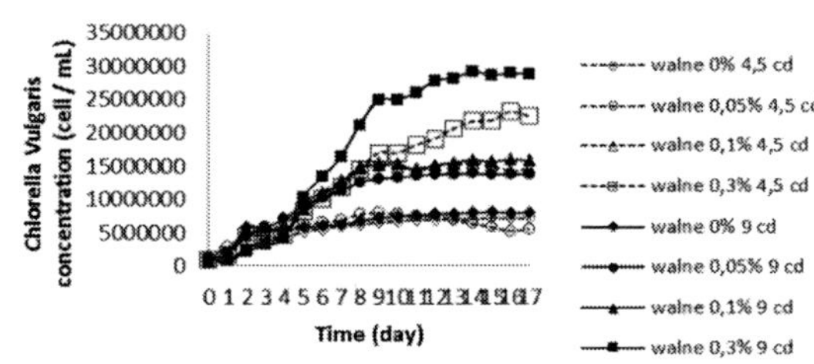

Fig 3.6. *Chlorella vulgaris* concentration in seawater media for 17 days of culturing period with flow rate of 125 mL/min (2% CO_2 gas concentration) for various nutrient addition of walne at 5000 lux and 10000 lux

In both Figs (Figs 3.5 and 3.6) it can be observed that the highest growth was obtained in the nutrient administration of 0.3% and the light intensity of 10000 lux. In Fig 3.5, it can be observed that the growth of microalgae at a light intensity of 10000 lux with the provision of nutrients by 0.1% higher than with the provision of nutrients by 0.3%. While in Fig 3.6, it can be observed also at the light intensity of 5000 lux with walne nutrition by 0.05% experienced a lower growth when compared with walne nutrition by 0%. This can be affected by the aeration in each of these conditions is not good. In addition, temperature and pH also affect each of these conditions.

From Fig 3.1 to 3.6 above can be observed that the growth of the microalgae has increased along with the increasing amount of CO_2 given. In the Figs 3.1 to 3.6 it is also found that the highest growth absorbance was obtained in the 10000 lux, 0.3% walne and 17-day cultures at 125 mL / min CO_2 concentration (2% CO_2).

Microalgae require a light / dark regime for doing photosynthesis, needing light for the photochemical phase to produce Adenosine triphosphate (ATP) and nicotinamide adenine dinucleotide phosphate-oxidase (NADPH). And also need a dark phase for biochemical phase synthesize molecules essential for growth (Al Qasmi et al, 2012). Brautovic learned that Chlorella sp. Which was cultured on 900 lux then changed its exposure to 12,000 lux on the 23rd day of the experiment. The result is that the algae can grow well at higher light intensity stimulate growth until the stable stage is achieved (Brautovic). This is consistent with the results of this experiment shown in Figs 3.1 to 3.6 where at higher light intensities of 10,000 lux (9 cd), the majority of *Chlorella vulgaris* growth is higher than 5,000 lux (4.5 cd). In general, the inhibition of photosynthesis (CO_2 fixation and growth) increases with increasing oxygen concentration and with increasing intensity of light (Salih, 2011). On the other hand, a further increase in light intensity (above 10,000 lux) can make the growth lower. In his research Brautovic also reported that the use of 12,000 lux light from day 0 experiments resulted in a lower growth graph of the use of 6,000 lux intensity.

In his research Brautovic cultured Chlorella sp. With the light intensity of 900, 2000, 4000, 6000, and 12000 lux until stationary phase of growth was achieved, then obtained the result that the shortest period of recorded log phase for algae exposed at 6000 and 12000 lux (on day 18), and longest 2000 lux (On day 26). So from Brautovic's research it can be seen that growth on day 17 will be much higher than day 7 because in that range the linear growth curve is in the log phase. This is in accordance with the experiments shown in Figs 3.1 to 3.6 where the 17-day talk time has a higher growth majority than 7 days.

Basic nutrient growth should be available to maintain proper physiological integration of culture (Salih, 2011). According to BBAP Situbondo, Cultivation of *Chlorella vulgaris* done with the addition of nutrients 1 mL of walne nutrition per 1 liter of culture. In this experiment, however, we tried to use various walne concentrations to find out which variables gave the best growth effect. From the experiments shown in Figs 3.1 to 3.6 it was found that the majority of the best growth trends were obtained when the use of 3 mL walne (0.3%) nutrients.

The effect of CO_2 concentration on growth of *Chlorella vulgaris* can be seen from graph 3.1 - 3.6 above where *Chlorella vulgaris* growth is best found in variable CO_2 with flow rate 125 mL / min (2% CO_2). Chlorella sp microalgae were reported to have a maximum tolerance of CO_2 to 40% (Salih, 2011). On the other hand, in the previous study it was reported that the use of CO2 with a flow rate of 200mL / min (3.33% CO_2) resulted in a lower absorbance graph of CO_2 with a flow rate of 50mL / min (1.67% CO_2) (Widjaja, 2008) . While in another study it is estimated that the rate of CO_2 absorption and ammonia removal of *Chlorella vulgaris* in wastewater is 260 g CO_2 m^{-3} h^{-1} and 0.92 g NH3 m^{-3} h^{-1} (Yun et al., 1997). With the rate of CO_2 absorption it could be one of the reasons that affect the growth of *Chlorella vulgaris* is higher at 125 mL / min.

3.2. Biomass Productivity from *Chlorella vulgaris*

Another quantity observed in this study was the production of biomass from the *Chlorella vulgaris* microalgae. Some of the conditions that influence this research are the light intensity, the amount of nutrients given, the CO_2 that flows and the different harvesting times. The following data show the relationship between the intensity of light, the amount of nutrients, the CO_2 that flows and the time of harvest that is different from the biomass productivity obtained by each of these conditions.

Table 1. Biomass Productivity from *Chlorella Vulgaris*

	CO_2 concentration		
Cultivation Condition	0%	1%	2%
0% 5000 lux 7 days (mg/L/d)	77	91	95
0,05% 5000 lux 7 days (mg/L/d)	80	99	101
0,1% 5000 lux 7 days (mg/L/d)	97	116	119
0,3% 5000 lux 7 days (mg/L/d)	96	119	123
0% 10000 lux 7 days (mg/L/d)	90	101	106
0,05% 10000 lux 7 days (mg/L/d)	107	117	121
0,1% 10000 lux 7 days (mg/L/d)	120	127	134
0,3% 10000 lux 7 days (mg/L/d)	126	134	139
0% 5000 lux 17 days (mg/L/d)	45	52	55
0,05% 5000 lux 17 days (mg/L/d)	56	58	66
0,1% 5000 lux 17 days (mg/L/d)	54	68	74
0,3% 5000 lux 17 days (mg/L/d)	59	78	82
0% 10000 lux 17 days	74	84	89
0,05% 10000 lux 17 days	78	90	95
0,1% 10000 lux 17 days	89	116	120
0,3% 10000 lux 17 days	99	126	133

From the above data can be seen the highest biomass productivity obtained in conditions with a light intensity of 10000 lux, the amount of nutrients given 0.3%, 2% CO_2 and 7 days culturing period.

The effects of light and CO_2 on biomass production are related to photosynthesis, where the reaction is as follows:

$$6\,CO_2 + 6\,H_2O \xrightarrow[\text{chlorophyl}]{\text{light}} C_6H_{12}O_6 + 6\,O_2$$

The more CO_2 and light obtained, the greater the efficiency of photosynthesis so that the biomass production will also increase. The variables exposed to the 10000 lux light intensity resulted in more biomass production than those exposed to a light intensity of 5000 lux. This is in accordance with the literature which states that increasing the intensity of light will also affect the increase in the number of cultivation of existing microalgae. (Al-Qasmi, et al, 2012)

As for the parameters of the amount of nutrients given, biomass productivity is more generated in the provision of nutrients by 0.3%. Biomass productivity is more when compared with the provision of nutrients by 0, 0.05% and 0.1%. As mentioned earlier, Nitrogen is the most needed nutrient for growth as an essential element in the formation of chlorophyll and protein (Kaplan, et al., 1986). Therefore, the more walne is added the more nitrogen and $FeCl_3$ are added, so the growth of the microalgae becomes higher. When growth is higher, then the biomass productivity obtained by the microalgae will also increase.

Another parameter given is the amount of CO_2 that flows. From the data obtained, at 2% CO_2 condition the microalgae have higher production compared with 0% and 1% CO_2. This is consistent with the literature which states that an increase in the amount of CO_2 will cause a significant increase in the growth of the algae (Widjaja, 2008). From the data can be seen also the tendency of significant increase from the provision of CO_2 0% to 1% CO_2. But at 2% CO_2, there was no significant increase in 1% CO_2 to 2% CO_2.

In the literature it is said that giving 50 mL / min CO_2 can increase growth significantly. But on giving 100 and 150 mL / min CO_2 does not provide better growth (Widjaja et al.). In this research, 0% CO_2 is equal to 2 mL / min (ambient air) CO_2 flow while CO_2 of 1% and 2% is equal to CO_2 flow of 62,5 mL / min and 125 mL / min. It can be concluded that in literature that 1% CO_2 (62.5 mL / min) will provide significant increase in the amount of growth and biomass productivity compared to 0% (ambient air) CO_2, but did not provide a significant improvement when compared with 2% CO_2 (125 mL / min).

In this research, culture time is also used as research parameter. It can be seen from the growth chart of each variable (Fig 3.1 - 3.6), in each picture has a tendency to increase on day 6 to day 14. So on the 7th day of the culturing period, the microalgae are still not in the phase of stationer or in the sense still can grow. Therefore, at culturing period, the number of microalgae will be more obtained by 17 days harvesting variable. This applies also to the production of biomass obtained by each parameter. The culturing period of 17 days will result in more biomass production than the 7 days culturing period. However, biomass production is inversely proportional to biomass productivity. This is because biomass productivity has a divisor of the number of cultivation days.

3.3. Lipid Productivity from Chlorella vulgaris

Below is the data of lipid productivity of *Chlorella vulgaris* with several variations of the conditions of light intensity, the amount of nutrients given, the amount of CO_2 that flows and the time of harvesting the microalgae.

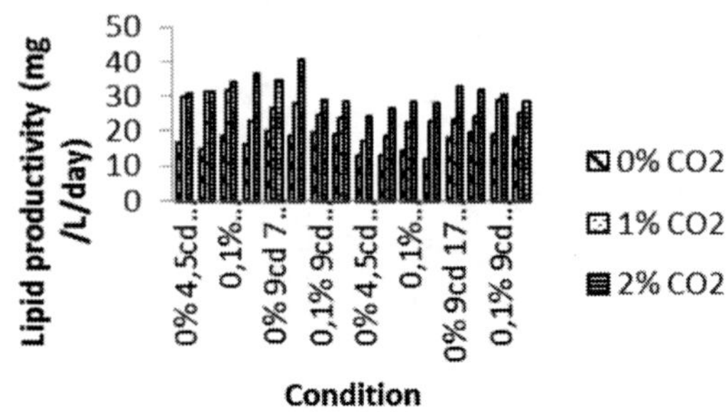

Fig 3.7 Relationship between light intensity, amount of nutrients, CO_2 flowed and culturing period different from lipid productivity from microalgae *Chlorella vulgaris*

From Fig 3.7 can be observed highest lipid productivity obtained at 10000 lux light intensity, walne nutrition 0,05% with 2% CO_2 concentration and 7 day culturing period. In order to enlarge oil productivity microalgae, then the microalgae can be conditioned in a state of stress. Ie by making the microalga nutrient deficient. In principle, nutrients are a source of nitrogen and phosphorus that have a role in influencing the productivity of lipids. This is related to the process of lipid biosynthesis. Lipid biosynthesis in microalgae requires acetil-CoA as the starting point of lipid binding. Acetil CoA carboxylase and several lipid liposuction or pathological enzymes have been used as targets for increased lipid production Schenk et al., (2008). The limitation of nutrients (nitrate and phosphate) in microalgae culture preservation media will increase the activity of ACCase enzyme which is the precursor for lipid formation in microalgae. The lower the concentration of nitrate derived from NaNO3 and the phosphate of NaH2PO4, the total lipid content in microalgae higher (Widianingsih, 2011).

Lipid productivity at culturing period 7 days is higher when compared with lipid production with 17 days culturing period. The longer the microalgae suffer from nutritional deficiencies, the more lipid produced (Widjaja et al, 2008). However, in the production of this lipid has a function of time where if divided per time culture will result in greater value will be available at smaller divisors.

CO_2 concentration also affects the biomass production obtained in this study. The more CO_2 that flows, the higher the production of lipid. This is in accordance with the literature which states that under normal conditions, oil production will increase in line with increased flowrate of CO_2 provided (Widjaja et al., 2008).

In the table below obtained data of biomass production, lipid content and lipid productivity from *Chlorella vulgaris* microalgae. In the table below is given data on the concentration of CO_2 given by 2%. This is because biomass production, lipid content and lipid productivity are highest in this condition.

Table 2. Production Data of Biomass, Lipid Content and Lipid Productivity from *Chlorella vulgaris* at CO_2 concentration of 2%

Condition	Biomass (g/L)	Lipid Content (%)	Lipid Productivity (g/L/day)
0% 5000 lux 7 days	0,67	32,10	30,62
0,05% 5000 lux 7 days	0,71	30,90	31,34
0,1% 5000 lux 7 days	0,83	28,70	34,03
0,3% 5000 lux 7 days	0,86	29,50	36,24
0% 10000 lux 7 days	0,74	32,70	34,57
0,05% 10000 lux 7 days	0,85	33,50	40,68
0,1% 10000 lux 7 days	0,94	21,70	29,14
0,3% 10000 lux 7 days	0,97	20,50	28,41
0% 5000 lux 17 days	0,93	44,20	24,18
0,05% 5000 lux 17 days	1,12	40,10	26,49
0,1% 5000 lux 17 days	1,25	38,40	28,24
0,3% 5000 lux 17 days	1,40	33,90	27,92
0% 10000 lux 17 days	1,52	36,30	32,46
0,05% 10000 lux 17 days	1,61	33,60	31,82
0,1% 10000 lux 17 days	2,04	25,10	30,12
0,3% 10000 lux 17 days	2,26	21,40	28,41

From the above data, it can be known that the highest biomass production is obtained at 10000 lux light intensity condition, walne nutrition given by 0.3% and culturing period of 17 days. The largest lipid content was obtained under conditions of 5000 lux light intensity, without the addition of walne nutrition, and culturing period for 17 days. While the largest lipid productivity obtained at 10000 lux light intensity conditions, the addition of nutrients walne 0.05%, and the cultivation time for 7 days.

From these data it can be concluded that the longer harvesting will be the greater the amount of biomass and lipid content obtained. This is because the longer the harvesting the more the number of microalga contained in the culture so that the production of biomass, lipid content and lipid productivity also more and more. Meanwhile, to increase the amount of lipid production, the microalgae must be conditioned on stress condition. In this study it was done with the restriction of nutrients on the microalgae so that the microalgae suffered from Nitrogen (Nitrogen starvation) deficiency. The limitation of nutrients (nitrate and phosphate) in microalgae culture preservation media will increase the activity of ACCase enzyme which is the precursor for lipid formation in microalgae. The lower the concentration of nitrate derived from NaNO3 and the phosphate of NaH2PO4 then the total lipid content in microalgae is greater (Widianingsih, 2011).

After doing the calculation, then analyzed to one sample in this research that is at the condition of light intensity of 5000 lux, without addition of nutrients (0% walne), without addition of CO_2 (0% CO_2) and 7 days culturing time. In this research, GC-MS (Gas Chromatography Mass Spectrometry) and TLC (Thin Layer Chromatography) were analyzed. Based on the results of GC-MS analysis, it is found that in the microalgae lipid there are characteristics of Hexadecanoid Acid and Octadecanoid Acid. And based on the results of TLC analysis, data obtained that the microalgae oil contains Triglycerides characteristic in it. This may serve as a reference for further research leading to the making of biodiesel from these microalgae oils.

4 Conclusion

The highest biomass productivity was obtained 139 mg/L/d in conditions with 10000 lux light intensity, 0.3% walne concentration, 2% CO_2 concentration and 7 days cultivation days. The highest lipid productivity was obtained at 10000 lux light intensity, walne addition of 0,05%, and 7 days time of 40,68 gr / L / day.

References

1. M. Al-Qasmi. *Proceedings of the World Congress on Engineering 2012 Vol I WCE*, (2012).

2. A. Bajguz, Journal of Plant Physiology, **166** (8), 882 – 886 (2009).
3. E. W. Becker, *Microalgae: Biotechnology and Microbiology*, Baddiley et al. (Eds.). New York: Cambridge University Press, (1994).
4. E. G. Bligh, and W. J. Dyer, Can. J. Biochem. Physiol., **37**, 911, (1959).
5. I. Brautovic, *The influence of light intensity on growth of the marine planktonic alga Chlorella sp. under laboratory conditions*. Institute of Oceanography and fisheries, Dubrovnik, Croatia
6. K. O. Cassidy, *Evaluating Algal Growth that Different Temperature* Theses an Dissertations--Biosystems and Agricultural Engineering. University of Kentucky, (2011).
7. S. Y. Chalid, *Kultivasi Chlorella Sp Pada Media Tumbuh Yang Diperkaya Dengan Pupuk Anorganik dan Soil.* (Program studi Kimia, Fakultas Sains dan Teknologi UIN Syarif Hidayatullah Jakarta).
8. S. Chinnasamy, B. Ramakrishnan, A. Bhatnagar, and K. C. Das, Int. J. Mol. Sci. **10** (2), 518-532. (2009).
9. A. Converti, A. A. Casazza, E. Y. Ortiz, P. Perego, and M. Del Borghi. Chem. Eng. Process. **48** (6), 1146-1151, (2009).
10. V. Endar, Sarjito, J. Hutabarat, B. Prayitno, Journal of Coastal Development, **16** (1), (2012).
11. E. P. Feofilova, Y. E. Sergeeva, and A. A. Ivashechkin. Appl. Biochem. Microbiol. **46** (4), 369 – 378, (2010).
12. Hadiyanto. I. Samidjan, A. C. Kumoro, and Silviana. *Produksi Mikroalga Berbiomasa Tinggi dalam Bioreaktor Open Pond. Pengembangan Teknologi Kimia untuk Pengolahan Sumber Daya Alam Indonesia*. 2010.
13. N. A. Handayani, D. Ariyanti, *Potensi Mikroalga Sebagai Sumber Biomasa dan Pengembangan Produk Turunannya*. **33** (2), ISSN0852 1697.2012 (2012).
14. S. Hiremath, P. Mathad, Journal of algal biomass utilization. Phycospectrum Inc. 2010.
15. http://toddcaldecott.com/herbs/chlorella/
16. http://www.chlorella-vulgaris.eu/
17. http://www.onr.navy.mil/focus/ocean/motion/default .htm
18. A. Isnansetyo, and Kurniastuty. *Teknik Kultur Phytoplankton dan Zooplankton*. Kanisius. Yogjakarta (1995).
19. R. P. John, G. S. Anisha, K. M. Nampoothiri, and A. Pandey, Bioresour. Technol., **102**, 186-193 (1995).
20. J. H. Lee, J. S. Lee, C. S. Shin, S. C. Park, and S. W. Kim. J. Microbiol. Biotechnol. **10** (3), 338-343 (2000).
21. Y. Liang, N. Sarkany, Y. Cui, Springer Science, Business Media B.V. Biotechnol Lett 31:1043–1049 DOI 10.1007/s10529-009-9975-7 (2009).
22. A. W. Mayo, Water Environ. Res. **69** (1), 64-72 (1997).
23. X. Miao, and Q. Y. Wu. Bioresour. Technol., **97**, 841 (2006).
24. J. C. Ogbonna, H. Masui, and H. Tanaka. J. Appl. Phycol. **9** (4), 359-366 (1997).
25. F. M. Salih, J. Environ. Prot., **2**, 648-654 (2011).
26. Soeroso, *Kilang Pengolahan BBM dioptimalkan*, Harian Pagi Jawa Pos, 11 Maret 2005.
27. M. Tsuzuki, E. Ohnuma, N. Sato, T. Takaku, A. Kawaguchi, Plant Physiol., **93**, 851-856 (1990).
28. A. Widjaja, C-C. Chien, Y-H. Ju, J. Taiwan Inst. Chem. Eng., (2008).
29. S. Wirosaputro, *Chlorella Untuk Kesehatan*, Global Teknik Budidaya dan Pengolahan Buku II. Gajah Mada University Press. Yogyakarta (2002).
30. Y. S. Yun, and J. M. Park. Appl. Microbiol. Biotechnol. **55** (6), 765-770 (2001).
31. M. Zhu, P. P. Zhou, and L. J. Yu, Bioresour. Technol., **84,** 93 (2002).

Bioethanol Production via Syngas Fermentation

Irika Anggraini[1], *Made* Tri Ari Penia Kresnowati[1*], *Ronny* Purwadi[1], and *Tjandra* Setiadi[1,2]

[1]Department of Chemical Engineering, Faculty of Industrial Technology Institut Teknologi Bandung, Jl. Ganeesa 10, Bandung 40132, Indonesia
[2]Centre for Environmental Studies (PSLH), Institut Teknologi Bandung, Jl. Sangkuriang 42 A, Bandung 40135, Indonesia

Abstract. Bioconversion of C-1 carbon in syngas through microbial fermentation presents a huge potential to be further explored for ethanol production. Syngas can be obtained from the gasification of lignocellulosic biomass, by which most of carbon content of the biomass was converted into CO and CO_2. These gases could be further utilized by carbon-fixing microorganism such as *Clostridium sp.* to produce ethanol as the end product. In order to obtain an optimum process, a robust and high performance strain is required and thus high ethanol yield as the main product can be expected. In this study, series of batch fermentation was carried out to select high performance strains for ethanol production. Bottle serum fermentations were performed using CO-gas as the sole carbon source to evaluate the potential of some Clostridia species such as *Clostridium ljungdahlii*, *C. ragsdalei*, and *C. carboxidovorans* in producing ethanol at various concentration of yeast extract as the organic nitrogen source, salt concentration, and buffer composition. Strain with the highest ethanol production in the optimum media will be further utilized in the upscale fermentation.

1. Introduction

Syngas bioconversion for ethanol production is one of tools to utilize completely all carbon content of lignocellulosic biomass. Not limited only from gasified lignocellulose, industrial waste gas rich of CO, CO_2, and H_2 or gas waste from municipal solid waste and agriculture and forestry residue can be utilized as the gaseous substrate to produce biofuels and other value added chemicals [1–4]. The conversion process is able to be undertaken by two routes, thermochemical route by Fischer Trop reaction (metal catalyst base process) and biochemical route through fermentation by employing acetogenic bacteria [1,5]. However, compared to other approaches syngas fermentation offers many advantages such as tolerance level of bacteria to several impurities in syngas, bacteria ability in consuming flexible syngas compositions, the applied low temperature and pressure, high specificity of the product, and bacteria ability to use CO_2 in the fermentation process which can be toxic for metal catalyst in thermochemical route [1,5–8].

Some strains of *Clostridium sp.* that have been studied capable of converting syngas into solvent products are *Clostridium ljungdahlii, C. carbodixovorans, C. ragsdalei*, and *C. autoethanogenum* (carboxydotrophic species) [4,9]. These species utilize reductive Acetyl-CoA pathway or Wood-Ljungdahl pathway (WLP) during their growth in the presence of $CO/CO_2/H_2$ and generate acetyl-CoA as the key intermediate product for a diverse metabolites production [10]. These strains are categorized into mesophiles as their optimum growth temperature is between 37°C and 40°C.

C. ljungdahlii and *C. autoethanogenum* are the common model microorganism for basic research and commercial application [4]. From a comparison study of syngas fermentation between *C. ijungdahlii* and *C. autoethanogenum* strains, C. *ljungdahlii* is revealed to be a more superior ethanol producer than *C. autoethanogenum* by ethanol titter produced in low pH condition [11]. Furthermore, some studies also confirm that *C. autoethanogenum* is a low productivity bacteria in ethanol production compared to other carboxidotrophic strains of *Clostridium* sp. despite its genetic modified strain has been developed for ethanol commercial scale production by LanzaTech [1,12–14]. *C. ragsdaleii* and *C. carbodixovorans* are the recent isolated strains which their potential are still in the exploration [15,16]. *C. carbodixovorans* is also a natural butanol producer, beside producing ethanol and hexanol [17]. Meanwhile, capability of *C. ragsdaleii* has been widely studied to produce ethanol under growth on low cost feedstock such as corn steep liquor and cotton seed extract as yeast extract and vitamins substitution in the cultivation and fermentation process [18–20]. Another study also revealed capabilities of *C. autoethanogenum, C. ljungdahli*, and *C. ragsdalei* to produce 2,3-butanediol as another solvent product [21].

There are several important process parameters that may affect the ethanol yield of syngas fermentation [5,8,22]. Low solubility of CO and H_2 which affects to gas to liquid mass transfer is known as the major bottleneck. Other parameters such as gas composition,

* Corresponding author: kresnowati@che.itb.ac.id

pH, temperature, gas pressure, media buffer and nutrient composition may also significantly influence ethanol to acetate ratio. In case of optimum pH, many studies revealed maintaining fermentation medium at pH 4.5 would enhance ethanol production since the condition promotes to solventogenic phase which associated with non-growth or resting cells phase [23]. Metabolically, acetic acid as the product that follows exponential cell growth will be consumed by the cell to produce ethanol as a counteract of internal pH drop caused by the accumulation of undissociated acid [24]. The influence of medium pH has also been studied in ABE fermentation for butanol production which suggested optimum pH for solventogenic *Clostridium* sp. to be in the range of 4.5-5.5 [25]. To give a better pH control in its optimum range, morpholinoethane sulfonic acid (MES) buffer is commonly utilized [20]. However, some studies revealed the absence of MES gave a positive influence to earlier ethanol production driven by a quick drop of medium pH below 5.0 [18,26]. Furthermore, the removal of MES buffer offered a low cost process in upscale fermentation since MES counts for approximately 97% of the standard media cost [20].

The objective of this study is to evaluate the performance of 3 *Clostridium* strains: *C. ljungdahlii*, *C. ragsdalei* (P11), and *C. carbodixovorans* (P7) in converting C1-gaseous compound (CO) into ethanol as the attempt of selection of robust and high performance candidate strain for ethanol production from syngas. As were suggested in many studies that nutrient-limited condition is favorable for ethanol production [18,23], a series of bottle serum fermentations will be performed in media with varied concentration of yeast extract (0.5 g/L and 1 g/L), mineral, and MES buffer (5g/L and 10 g/L). Strain with the highest maximum ethanol production in its favorable condition will be employed in the upscale fermentation. Base composition of fermentation medium is also expected as the output of this preliminary study.

2. Materials and methods

2.1 Microorganism, media, and inoculum Preparation

C. ljungdahlii DSM 13258, *C. carbodixovorans* DSM 15243 was acquired from Deutsche Sammlung von Mikroorganismen und Zellkulturen GmbH culture collection (Braunschweig, Germany) in freeze dried culture while *C. ragsdalei* DSM 15248 in active growing culture. The cultures were activated anaerobically in 10 mL growth medium with 5 g/L fructose as carbon source and further cultivated in 50 mL liquid medium.

Growth media of *C. carbodixovorans* (P7) and *C. ragsdaleii* (P11) were prepared with the same composition of media components based on Rajagolapan [3]. Growth media of P7 and P11 contained (per L) 25-30 mL mineral solution, 10 mL trace metal solution, 10 g 2-mercaptoethanosulfonic acid sodium salt (MESNA), 1 g yeast extract, 0.1 mL resazurin 0.1% w/v solution as redox indicator, 10 mL vitamin solutions and 5 mL cystein-HCl 4% w/v solution as reducing agent (added after medium sterilized).

Growth media of *C. ljungdahlii* was prepared based on DSMZ medium 879 [27]. Medium contained (per L) mineral components that were straightly mixed with other components such as 10 mL trace element, 10 g MES, 1 g yeast extract, and 0.1 mL resazurin 0.1% w/v solution, except 10 mL vitamin solution and 5 mL cysteine 4% w/v added after medium sterilized. Prior to media sterilization at 121°C for 15 min, all media in bottle serums equipped with butyl rubber stoppers and aluminum sealed were purged with ultra-high purity N_2 gas. After sterilization, vitamin and cysteine were added into media then pH of all fresh media were adjusted to 6.0 using 2 M NaOH and 2 M HCl prior to inoculation in anaerobic chamber (N_2 atmospheric). Incubation was at 37°C for 7 days without agitation before transferred to other fresh media for subculturing. After three passages of subculture in fructose media, carbon source was replaced by CO gas. Medium with gaseous substrate was prepared by purging fresh medium with 99% CO-gas at pressure 2 atm (abs) for 1 min prior to inoculation.

2.2 Bottle batch fermentation with variation in MES buffer, yeast extract, and salt concentrations

Batch fermentation using CO gas as the sole carbon source was carried out in 100 mL bottle serum containing 50 mL liquid medium. Fermentation medium of P7 and P11 was prepared based on standard basal medium with 100x salt concentration of growth medium in mineral stock solution and higher concentration of trace elements than in the growth media to promote high ethanol production (Ni, W, and Se) [16]. Standard fermentation medium of *C. ljungdahlii* was prepared in similar concentration of mineral components with growth medium [27]. To study the effect of salt concentration in improving ethanol production, modified medium of P7 and P11 was prepared with dilute salt concentration (referred as modified "10x and 20x NH_4Cl"), while modified medium of *C. ljungdahlii* was in 1.5x and 3x higher salt concentrations (referred as modified media "1.5x NH_4Cl" and "3x NH_4Cl") (Table 1). All modified fermentation media of the three strains were prepared using reduced concentration of yeast extract (0.5 g/L) and buffer MES (5 g/L).

Before inoculated with 10% seed culture grown in the growth media, pH of each fresh medium was adjusted at optimum pH growth 6.0 and purged with CO at 2 atm (abs) for 1 min. Inoculated media was incubated at 37°C in anaerobic chamber with agitation at 190-200 rpm for 24 days. CO-gas in bottle headspace was refilled every 24 h. Two milliliters of fermentation broth were withdrawn at Day 7, Day 11, Day 18 for optical density (OD_{600}) measurement and fermentation product analysis. All experiments were carried out in duplicate.

2.3. Biomass and metabolite product analysis

Cell (biomass) density was determined by optical density measurement using UV-Vis Spectrophotometer at 600 nm (OD_{600}). Ethanol and acetate concentrations as end metabolite products were determined by HPLC (Waters, Milford, MA, USA) with an Aminex HPX 87H column (Bio-Rad). HPLC buffer was 5 mM sulfuric acid in water and the flow rate was 0.6 mL/min. The concentration of metabolites is calculated using external standard of ethanol and acetate calibration curve.

Table 1. Salt composition of modified fermentation medium (g/L)

Component	Standard fermentation media		Modified media P7 and P11		Modified media (*C.ljungdahlii*)	
	P7 and P11	*C. ljungdahlii*	10x NH_4Cl	20x NH_4Cl	1.5x NH_4Cl	3x NH_4Cl
NH_4Cl	3	1	0.3	0.6	1.5	3
NaCl	2.4	0.8	0.24	0.48	1.2	2.4
KCl	0.3	0.1	0.03	0.06	0.15	0.3
KH_2PO_4	0.3	0.1	0.03	0.06	0.15	0.3
$MgSO_4$	0.6	0.2	0.06	0.12	0.3	0.6
$CaCl_2$	0.12	0.02	0.012	0.024	0.03	0.12

3. Results and Discussion

3.1. *C. carbodixovorans* (P7) and *C. ragsdalei* (P11)

3.1.1. Standard growth medium and medium with yeast extract and MES reduction

The effect of media on P7 and P11 cell growth and the production of ethanol and acetate are summarized in Fig 1.

In growth media, standard composition of yeast extract (1 g/L) as organic N source and MES buffer (10 g/L) were applied to promote high cell density of growing culture. By using CO as the sole carbon source, P11 was able to reach final OD_{600} 0.2 after 11 days of incubation and had a stationary phase until Day 18, while P7 final OD_{600} only slightly exceeded 0.1 until Day 18. In media with reduced yeast extract and buffer MES concentration (0.5 g/L and 5 g/L), lower cell densities of P11 during exponential growth (Day 7 - Day 11) were observed in correspond to reduced yeast in medium while P7 could not maintain its cell viability until day 18.

Growth of P11 and P7 in standard growth media was accompanied by ethanol and acetate production (Fig 1). As observed, ethanol and acetate were started to be produced at Day 7 (beginning of exponential cell growth) and continued to increase until Day 18 (data not shown). Final concentration of ethanol at Day 18 reached 0.2 g/L and 0.1 g/L, in P11 and P7, respectively. In P11, significant cell growth in standard medium was also followed by significant ethanol production. In P7, however, acetate accumulation as associated growth product dominated cell metabolite production.

On the other hand, the cell growth and ethanol production were observed to be lower in growth media with yeast and MES reduction. In P11, low ethanol production could only reach 0.12 g/L at Day 18. Interestingly, high concentration of acetate was observed.

The reduction of yeast extract specifically in particular inhibited the growth of P7 as its cell density barely reached final OD_{600} 0.1 throughout the cultivation. Further, no increase in ethanol and acetate production was observed.

Overall, the use of standard media containing rich-nutrient is suitable for ethanol production by P11 in which we observed significant cell growth that was accompanied by high ethanol titter. Media with reduction of yeast and MES is not suitable for ethanol production.

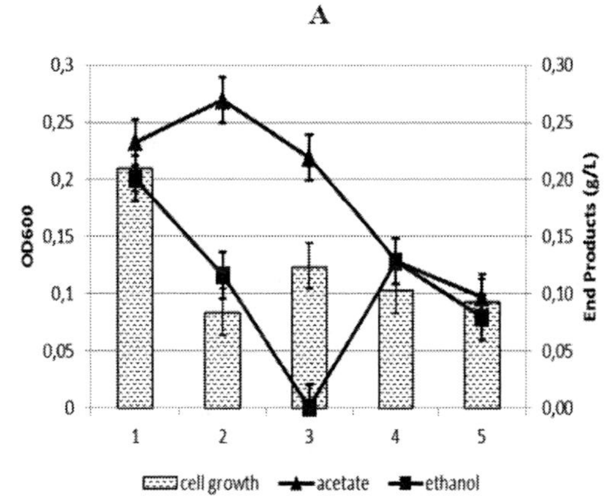

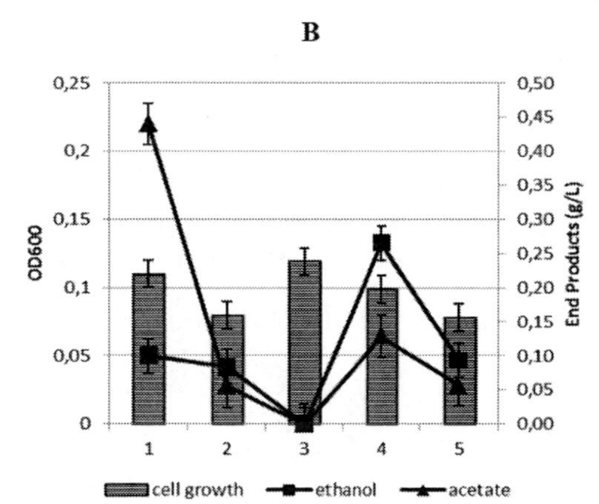

Fig 1. Growth, ethanol, and acetate profile of P11 (A) and P7 (B) after 18 days of incubation in (1) standard growth medium, (2) reduced growth medium, (3)standard fermentation medium, (4) modified $20xNH_4Cl$ medium, (5) modified 10x NH_4Cl medium

3.1.2. Fermentation medium with standard and modified salt concentration

Standard fermentation media of P7 and P11 contained 100x salt concentration higher than in the growth media since yeast extract as organic N source was replaced by NH_4Cl. As were suggested in many studies, nutrient–limited environmental condition induced cell stress condition that initiated resting cell phase which associated with solvent production [23]. Some studies also suggest removal of MES buffer as its absence leads to earlier ethanol production due to a quick drop of medium pH as the result of acetic acid accumulation [26]. In this study, yeast extract and buffer MES were not totally removed but reduced to 0.5 g/L and 5 g/l, respectively.

Compared to the cell growth in rich-nutrient of growth media, a suppressed cell growth of P11 was observed throughout the fermentation. Acetate was detected in comparable concentration as in the growth media, whereas no ethanol was detected. On the other hand, P7 was observed to grow better in this media. However, neither ethanol nor acetate was found in the fermentation broth (Fig 1). It presumed that the presence of excess NH_4Cl as inorganic N source significantly drives cell metabolism to utilize limited CO-gaseous substrate dissolved in media for biomass formation. Thus, it is possible none of significant metabolites produced in P7 and P11 although the fermentation carried out in standard fermentation media.

In other study of P7 batch bottle fermentation which conducted using the same composition of standard media and sole CO-gas supply, maximum biomass concentration (0.13 g/L) had been reached after the first 30 h. Maximum ethanol concentration and acetic acid were able to reach 0.48 g/L and 0.89 g/L, respectively, in 11 days of fermentation [28].

As comparison study, fermentation was also conducted using modified media with more dilute salt concentrations (10x and 20x NH_4Cl of growth medium). Results of the fermentation revealed suppressed cell growth of both P7 and P11 as observed in the reduced growth medium (Fig 1). As observed in both strains, ethanol and acetate was started to be produced at Day 11 when the cell growth started to descend (data not shown). Growth of P7 in medium 20x NH_4Cl was accompanied by high ethanol production which the titer was able to reach the highest value as recorded throughout this study, while the titter in medium 10x NH_4Cl was insignificant (Fig 1).

The result of improved ethanol production in P7 with 20x NH_4Cl medium suggested the bacteria favorable condition for ethanol production. Other P7 study in bottle batch fermentation using formulated minimal media even produced a higher ethanol titter (3.25 g/L) which was also as the effect of gas composition which contained CO and H_2 as a simultaneous electron donor [17]. As observed in this study, unlike P11, acetate production in P7 was significantly suppressed. Based on the overall results of P7 cell growth, the cell density had never been exceeded P11 cell growth, thus it could be a reason why acetate accumulation was not observed. Acetate accumulation in bottle serum fermentation is often linked to high density of biomass that could lead to inhibition of solvent production due to a quick drop of pH or "acid crash" [28]. In bioreactor fermentation, this problem can be avoided by applying pH regulation after medium pH starts to drop [28].

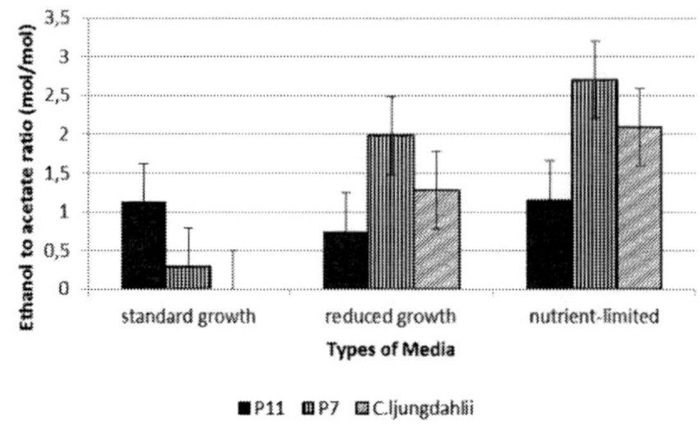

Fig 2. Ethanol to acetate molar ratio of P11, P7, and *C. ljungdahlii* in standar growth media, reduced growth media, and nutrient-limited media after 18 days incubation

In P11 fermentation, ethanol titter reached in both media was not significantly different with dominating acetate over ethanol concentration. However, as the cell density dropped at Day 24, ethanol concentration gradually increased in medium 20x NH_4Cl (0.06 g/L to 0.15 g/L) and followed by increasing ethanol to acetate ratio (data not shown). Overall, stress condition due to the presence of higher salt concentration, unlike P7, did not give positive influence to the enhancement of ethanol production although cell growth was found suppressed. As seen from the result of study in yeast and MES reduction, ethanol production by P11 favors more standard growth medium without excess salt concentration indicated by highest ethanol titer throughout this study although ethanol to acetate molar ratio in average was at the lowest if compared to *C. ljungdahlii* and *C. carbodixovorans* (Fig 2). In other study of bottle serum fermentation by P11, nutrient-limited media also gave insignificant effect to improved ethanol production [18]. Based on references, so far highest ethanol concentration (1.7 g/L) was reached when P11 grown in 20 g/L Corn Steep Liquor (CSL) as low-cost media which also containing 0.32 g/L glucose [19].

By comparing highest ethanol titter and ethanol to acetate molar ratio achieved in each of the two strain's favorable condition, P7 is more superior as ethanol producer than P11 (Fig 2). In addition, P7 fermentation is also able to be performed in minimal media (20x NH_4Cl) with reduction use of yeast and MES buffer which can be the advantage for a low-cost syngas fermentation.

3.2 *C. ljungdahlii*

3.2.1 Standard growth medium and medium with yeast extract and MES reduction

The effect of media on *C ljungdahlii* cell growth and the production of ethanol and acetate are summarized in Fig 3.

In standard growth medium, *C. ljungdahlii* growth reached final OD_{600} of 0.2 after 11 days of incubation and entered stationary growth until Day 18. Cell growth was solely followed by high acetate production (0.63 g/L) as the associated growth product. In media with yeast and MES reduction, a suppressed cell growth was accompanied by ethanol production. At Day 11, ethanol concentration reached 0.18 g/L and continued to increase to 0.21 g/L at Day 18.

3.2.2 Fermentation medium with standard and modified salt concentration

Improved ethanol production was demonstrated in fermentation using 1.5x NH_4Cl of standard salt concentration (Fig 3). In this medium, cells were unable to maintain high cell density as suppressed cell growth was shown as growth in reduced growth medium. Ethanol titer achieved 0.41 g/L with ethanol to acetate molar ratio 2.09 at Day 18. As shown on Fig 2, in spite of lower ethanol to acetate ratio than P7, *C. ljungdahlii* in its favorable condition was able to produce higher ethanol titer. Compared to P7, more acetate was produced by *C. ljungdahlii* yet higher ethanol production was also promoted although the production as the expense of acetate was not observed. Based on references, *C. ljungdahlii* is known to be a superior ethanol production strain even in nutrient-limited media which has been studied to also gave positive effect to the enhancement of ethanol production [29,30].

In medium 3x NH_4Cl, *C. ljungdahlii* showed cell intolerance to high salt concentration, whereas the same salt concentration also used in standard fermentation medium of P7 and P11 (Table 1). It was indicated by a significant drop of cell density after 15 days incubation which was also the lowest final OD_{600} value of the strain throughout this study. The suppressed cell growth was followed by acetate formation yet no ethanol being produced. Too high salt concentrations implicated to unfavorable condition for cells growth and ethanol production. In other study, it was revealed that the presence of 4x salt concentration in medium would implicate to rapid loss of *C. ljungdahlii* cell viability due to bacteria salt stress [31]. Effect of C/N ratio from carbon and nitrogen supply in medium also had influence to *C. ljungdahlii* metabolism in producing metabolite products which can give result to neither ethanol or acetate production as observed in P7 and P11 standard fermentation [23].

3.3 CO as sole carbon source

Low density of cell grown on CO as gaseous substrate was possible as the effect of limited CO supply in bottle serum fermentation. CO-gas mass transfer is also known as the key factor of fermentation performance [32,33]. In general, all the three strains only could reach maximum OD_{600} value 0.2 when being grown on CO, while their cultivation using fructose as carbon source could reach OD_{600} 0.4-0.6 after 7 days incubation (data not shown). Cell growth of the three strains in fructose were also followed with ethanol and acetate production in which acetate accumulation as associated growth product was observed in all strains growing culture (data not shown). However, ethanol titter was insignificant if compared to ethanol titer that achieved in fermentation with favorable condition using CO as sole C-source.

To get high density of cell, cultivation and fermentation can be carried out separately by performing two stage fermentation system using rich medium for cell growth and minimal medium for ethanol production using a continuous suppy of CO-gas [31]. This system has been proved to enhance ethanol production if compared to the performance of single stage fermentation.

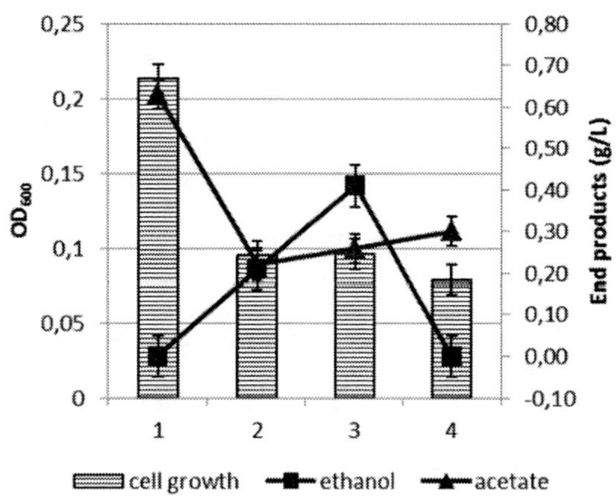

Fig 3. Growth, ethanol, and acetate profile of *C. ljungdahlii* after 18 days of incubation in (1) standard growth medium, (2) reduced growth medium, (3) modified medium 1.5x NH_4Cl, (4) modified medium 3x NH_4Cl

4 Conclusion

From this study, growth and metabolic profile of the three strains [*C. ljungdahlii*, *C. carbodixovorans* (P7), and *C. ragsdalei* (P11)] in standard and nutrient-limited media were able to identify favorable condition of fermentation for each strain. Reduction of yeast extract and buffer MES affected to ethanol production in *C. ljungdahlii* while an appropriate amount of salt concentration used in the modified media of *C. ljungdahlii* and P7 promoted higher ethanol titer. Ethanol production in P11 was favorable to be conducted in standard growth medium.

As shown by the highest maximum ethanol titter achieved in its favorable condition, *C. ljungdahlii* is a superior ethanol producer over the two other strains. The composition of modified medium 1.5x NH_4Cl can be used as base information for upscale fermentation.

This study is part of SATREPS Program (Science and Technology Research Partnership for Sustainable Development), research collaboration between Gunma University, Japan and Bandung Institute of Technology, Indonesia and funded by Japan International Corporation Agency.

Reference

1. M. Kopke, C. Mihalcea, J.C. Bromley, S.D. Simpson, Curr. Opin. Biotechnol. **22,** 320–325 (2011).
2. P. Durre, B.J. Eikmanns, Curr. Opin. Biotechnol. **35,** 63–72 (2015).
3. S. Rajagopalan, R.P. Datar, R.S. Lewis, Biomass and Bioenergy, **23**, 487–493 (2002).
4. B. Molitor, H. Richter, M.E. Martin, R.O. Jensen, A. Juminaga, C. Mihalcea, L.T. Angenent, Bioresour. Technol. **215,** 386–396 (2016).
5. F.R. Bengelsdorf, M. Straub, P. Dürre, Environ. Technol. **34**, 37–41 (2013).
6. D. Xu, D.R. Tree, R.S. Lewis, Biomass and Bioenergy **35**, 2690–2696 (2011).
7. M.R. Wilkins, H.K. Atiyeh, Curr. Opin. Biotechnol. **22**, 326–330 (2011).
8. P.C. Munasinghe, S.K. Khanal, Bioresour. Technol. **10**, 5013–5022 (2010).
9. D. Kennes, N. Abubackar, M. Diaz, C. Veiga, C. Kennes, J Chem Technol Biotechno **91,** 304–317 (2015).
10. S.W. Ragsdale, E. Pierce, Biochim Biophys **1784**, 1873–1898 (2009).
11. M. E. Martin, H. Richter, S. Saha, L.T. Angenent, Biotechnology **113**, 531–539 (2016).
12. J. Abrini, H. Naveau, E. Nyns, Arch. Microbiol. **161**, 345–351 (1994).
13. H. N. Abubackar, M.C. Veiga, C. Kennes, Bioresour. Technol. **186**, 122–127 (2015).
14. Y. Guo, J. Xu, Y. Zhang, H. Xu, Z. Yuan, D. Li, Bioresour. Technol. **101,** 8784–8789 (2010).
15. J. S. Liou, D. L. Balkwill, G. R. Drake, R. S. Tanner, R. S. Tanner, Int. J. Syst. Evol. Microbiol. **55,** 2085–2091 (2005).
16. R. L. Huhnke, R.S. Lewis, R. S. Tanner, Isolation and Characterization of Novel Clostridial Species, US007704723B2, (2010).
17. J. R. Phillips, H. K. Atiyeh, R. S. Tanner, J. R. Torres, J. Saxena, M.R. Wilkins, Huhn, Bioresour. Technol. **190**, 114–121 (2015).
18. J. Gao, H.K. Atiyeh, J.R. Phillips, M.R. Wilkins, R.L. Huhnke, Bioresour. Technol. **147,** 508–515 (2013).
19. P. Maddipati, H.K. Atiyeh, D.D. Bellmer, R.L. Huhnke, Bioresour. Technol. **102**, 6494–6501 (2011).
20. D.K. Kundiyana, R.L. Huhnke, P. Maddipati, H.K. Atiyeh, M.R. Wilkins, Bioresour. Technol. **101,** 9673–9680 (2010).
21. M. Kopke, C. Mihalcea, F. Liew, J.H. Tizard, M.S. Ali, J.J. Conolly, B. Al-sinawi, D. Simpson, Appl. Environ. Microbiol. **77,** 5467–5475 (2011).
22. M.D. Bredwell, R.M. Worden, Biotechnol. Prog. **14**, 31–38 (1998).
23. J.L. Cotter, M.S. Chinn, A.M. Grunden, Bioprocess Biosyst. Eng. **32**, 369–380 (2009).
24. F. Monot, J.-M. Engrasser, H. Petitdemange, Appl. Microbiol. Biotechnol. **19,** 422–466 (1984).
25. I.S. Maddox, K.C. Schuster, J. Mol. Microbiol. Biotechnol. **2,** 95–100 (2000).
26. D.K. Kundiyana, M.R. Wilkins, P. Maddipati, R.L. Huhnke, Bioresour. Technol. **102,** 5794–5799 (2011).
27. G. Najafpour, H. Younesi, Enzyme Microb. Technol. **38,** 223–228 (2006).
28. Á. Fernández-naveira, H.N. Abubackar, M.C. Veiga, Appl. Microbiol. Biotechnol. **100**, 3361–3370 (2016).
29. J.R. Phillips, K.T. Klasson, E.C. Clausen, J.L. Gaddy, Appl. Biochem. Biotechnol. **39**, 0–7 (1993).
30. K.T. Klasson, M.. Ackerson, E.C. Clausen, J.L. Gaddy, Fuel **70,** 605–614 (1991).
31. H. Richter, M.E. Martin, L.T. Angenent, 3987–4000 (2013).
32. P.C. Munasinghe, S.K. Khanal, Biotechnol. Prog. **26**, 1616–1621 (2010).
33. M.D. Bredwell, P. Srivastava, R.M. Worden, 834–844 (1999).

Making Organic Fertilizer Resistant to Bacterial Leaf Blight Disease for Grain Crops

Mohammad Rifki[1], *Luqman* Hanifianto[1],*Sri* Rachmania Juliastuti[1], *Raden Darmawan*[1], and *Nuniek* Hendrianie[1,*]

Departement of [1]Chemical Engineering Institut Teknologi Sepuluh Nopember, Kampus ITS Surabaya 60111, Indonesia

Abstract. Farmers too often use chemical fertilizers to help rice production, but in the long run fertilizer detrimental to land. Xanthomonas Oryzae Pv Oryzae (Xoo) is a microorganism that causes the disease of Bacterial Leaf Blight (BLB) in rice that harms farmers. Pseudomonas flourescens (Pf) is potentially the controller of Xoo and Azotobacter chrococcum (Az) causing rice to be more resistant to disease so as to help reduce the impact of BLB on rice plants in the form of organic fertilizers. The purpose of this research is to utilize corn farm waste into organic fertilizer that is resistant to BLB plant disease (bacterial leaf blight) caused by Xanthomonas Oryzae microorganism. In this research will be made of organic fertilizer from corn farm waste using rotary drum composting method with the help of Bioactivator EM4, Pf and Az bacteria. The results of this study found that the best plant growth in the variable comparison of microorganisms Pseudomonas fluorecense: Azotobacter chroocosum 1: 2 and urea. Plant growth on day 25 is higher (32.8cm) than other variables and the number of grains is 80.

1. Introduction

In the modern era like today the use of fertilizer has become the norm especially inorganic fertilizer to increase nutrient on soil .Many use of fertilizer very affect in agricultural progress in indonesia .Various companies fertilizer urea also increase production to meet agricultural needs in indonesia .But , the use of excessive fertilizer can have a negative impact on the ground and environment . This negative impact it should be stopped or at least reduced , because if they do not dismissed or reduced then as slowly so the structure of the soil be broken . [9]

Besides needs nutrient that must be fulfilled , to increase productivity of rice need to an increase in defense rice of a disease .A disease that there are often in rice is BB (leave Blight) . BB disease caused by bacteria *Xanthomonas oryzae* .The bacteria is can infect of paddy starting from a nursery to harvest . Symptoms of disease BB in plants in seedling , usually characterized by color yellowing on the edge of a leaf that is not easily observed. Symptoms found in phase growth saplings until maturation phase is a symptom of blight until of a line yellowish on the leaves flag .

BB is among the top five malignancy both in the tropics and sub tropics . BB is dangerous not only his potential epidemic , but also his microbe can be transmitted by seeds and can survive in the seed for annuals . Even in india blight microbes can survive until 11 months . In indonesia BB is also one of the most important disease of rice plants , especially on of paddy fields . BB disease can reduce rice productivity to 60 % . [3]

Reduction of various businesses this disease have been conducted , among others by using synthetic chemical ingredients like benzoic acid and nitrites , nor the application of pesticides using basic antibiotic compounds (asman , 1996) .The use of chemicals as fertilizer and pesticides and antibiotics in handling plant disease can cause resistance against bacteria , inflicting residue , and environmental pollution . [10]

Alternatives such as bio control application mikrob controller biodiversity antimikrob produce substance without pollute the environment . Bacteria capable of producing a metabolite compound that has an effect or bakteriostatik bakterisidal to hinder the growth of pathogenic bacteria .A genus of bacteria commonly used and yielding an astringent antimikrob bakteriosin is in the form of *Bacillus sp* [2]

According to [10], *Pseudomonas aeruginosa, Pseudomonas fluorescens* , *Bacillus cereus,* and *Bacillus sp* Have good potential to inhibit pathogen bacterial growth of paddy xanthomonas oryzae . So breeding bacterium Pseudomonas aeruginosa on fertilizer will help improve defense of paddy against blight disease

2. Experimental

2.1 Material

Materials that used in this research are NB Liquid, Agricultural waste corn (Limbah pertanian jagung), bioactivator EM4, *Azotobacter Chroococcum, Xanthomonas oryzae* Pv *Oryzae, Pseudomonas flourescens,* and IR64 Rice.

* Corresponding author: nuniek@chem-eng.its.ac.id

2.2 Procedure

2.2.1 Microbe Preparations.

Inoculum of Pseudomonas flourescens on Nutrient Broth Agar (NBA) media are propagated to the process of generation using scratch method with slant agar . Then moved in liquid media NB with the stab and incubation on the temperature 25 to 30 ^{0}C .The same procedure was uses to treating inoculum of bacteria *Azotobacter chroococcum* and *Xanthomonas oryzae* pv *oryzae*.

2.2.2 Composting Process.

Before start composting process necessary analysis of the N, P,K,C , and the water content of on the agricultural waste corn. Then agricultural waste corn put into rotary drum composter with additional aktivator em4 and bacteria *Azotobacter chrocooccum* and *Pseudomonas flourescens* composed in accordance with a variable that set before. A rotary drum composter drained air about 14 l / minutes using aerator to support the aerobic Process . Performed the measurement of temperature and pH once every day. during composting process the inside drums agitators rotated 3 times a day to compost mature, each rotated as many as 15 times. After compost formed then N, P ,K, C and water contents need to analiyzed.

2.2.3 Test influence fertilizer to intensity attack blight in rice plant.

For rice cultivation , prepared conducive land first and added water to increase moisture land .Land will be prepared size of less more 1 x 2 meters .Rice will be planted is rice has sowing first .Then adding organic fertilizers from the compost above to of paddy newly grown . After that , inject pathogen *Xanthomonas oryzae pv oryzae* with cliping method , as for step :

1. Bacteria pathogen prepared by putting it into a bottle
2. Dipping scissors into a solution that contains bacteria pathogen in bottles 1-2 minutes ~
3. Cut leaves rice with size 1- 2 inches . (IRRI, 2006)

After that, Measuring the intensity of blight desease everyday , and measurement of the height a plant any 2 days.

3. Results and Discussion

Table 1 is the result of analysis of C, N, C/N ratio, P, K of the raw material prior to the composting process. While Table 2 is the result of analysis of C, N, C/N ratio, P, K for variable microorganisms used in Agricultural Waste Corn after the composting process is conducted for 15 days. From table 2 the look changes the levels of C, N, c/n ratio, P, and K on each variable after composting for 15 days.

Tabel 1. The Analysis Result Of C, N, C/n Ratio, P, K On Raw Materials

Compound	Agricultural waste corn	Standart quality of organik fertilizer according to Permentan 2009
C	27,6 %	>12%
N	0,03 %	<6%
C/N ratio	920 %	15-25
P	1,42 %	<6%
K	1,08 %	<6%

Table 2. Analysis Result of C, N, c/n ratio, P, K After 15 days In Composting of agricultural Waste Corn

Ratio Variabel	Compound				
	C (%)	N (%)	C/N Rasio	P (%)	K (%)
Pseudomonas flourescens : *Azotobacter Chroococcum* (1:1)	6,2	0,38	16,32	1,42	1,08
Pseudomonas flourescens : *Azotobacter Chroococcum* (1:2)	7,25	0,41	17,68	1,64	1,38
Ratio Variabel	**Compound**				
	C (%)	**N (%)**	**C/N Rasio**	**P (%)**	**K (%)**
Pseudomonas flourescens : *Azotobacter Chroococcum* (1:3)	6,65	0,36	18,47	1,55	1,57
Pseudomonas flourescens : *Azotobacter Chroococcum* (3:1)	5,38	0,37	14,54	1,39	1,62

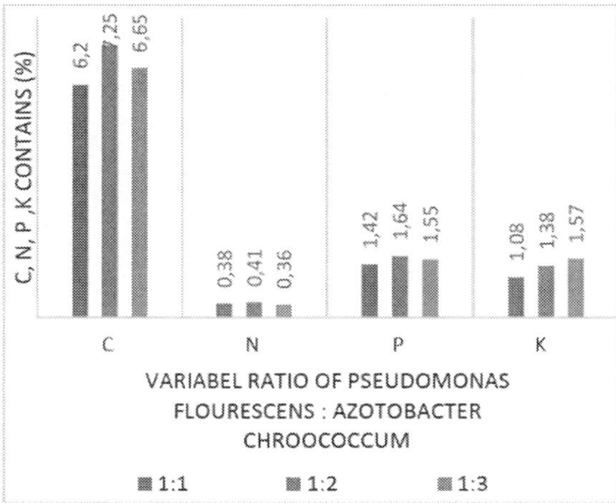

Fig. 1. C, N, P,K contains of ratio Variabel 1:1, 1:2 and 1:3

Can be seen on a fig.1 above that for element nitrogen and phospor, the largest located on the variables of 1: 2. While on the variables of 1: 3 nitrogen and phospor contains is the smallest. This can be caused by decomposition microbes already suffering phase death. The more decomposition microbes used the fast too the process reshuffle done by microbes [7]. So that reshuffle to microbes variable 1: 3 reach maturity before 15 days and a decline N and P contains on the day to 15.

Can be seen in the figur above that element of N, P, and K of the compost is increased on all variables. But the element of Corganik significant decrease, this indicates the elements of C-organic decomposes, it can be seen from the level of C/N Ratio of compost, the lower the value of c/n Ratio showed organic material already decomposes and almost become compost [6].

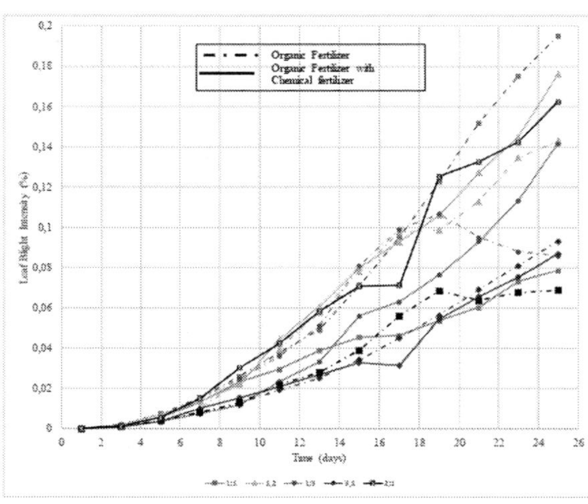

Fig. 2. Level infection of blight desease on each variabel

Fig 2 regarding the intensity of growth of Bacterial leaf blight disease (BB) without using urea addition for all variables. Can be seen fig 1 variable 1:1 is experiencing the highest disease growth compared to other variables. This can be caused by several factors such as the role of microorganisms Pseudomonas fluorecense not working properly, so that its growth is high.

As for the comparison of variable 3:1 the disease experience growth likely to slow, this is due to the role of microorganisms Pseudomonas fluorecense as an inhibitor of Bacterial leaf blight disease (HDB) can work well, because the awarding the more so that the intensity of the disease.

For comparison the variables 2:1 and 1:3 have the intensity of the ride starts the first day to day 18 of the observations, and then descends and nearly constant. This is due to Pseudomonas fluorecense not working on day 1 to 18 to the increase experienced also quite high. For comparison the variable 1:2 keeps rising from the first day until the end of the observation, but on the day of 18 observation experience slightly decreased until the end of the observation, it can be caused on day 18 microorganisms Pseudomonas fluorecense started to work properly.

While the attack blight to rice using additional urea.On a chart it has intesitas attack of illness a bad all the variables, this is because the urea can improve the disease in plants rice, because of an increasing the elements nitrogen on the ground can cause endurance plant against disease decline [8]

4. Conclusion

The best ratio variable of microorganisms *Pseudomonas fluorecense : Azotobacter* **chroococum** is on the ratio 1:2 and urea, evidenced by mild infections blight disease on day 15 that only 0.132% infected.

References

1. A. Asman, *Prosiding Pertemuan Ilmiah Tahunan*; Bogor, 13-14 Mar 1996". Bogor: Research Institute for Spice and Medicinal Crops, pp. 284-290. Bogor (1996)
2. D. Bizani, A. Brandelli, A. Strain, J. Appl. Microbiol., 93: 512-519. (2002)
3. H. R. Hifni, & M. Kardin. . *Risalah Seminar Puslibangtan* (1993)
4. IRRI..*http://www.knowledgebank.irri.org/ricebreedingcourse/Breeding_for_disease_resistance_Blight.htm*. 22.27 (2006)
5. IRRI.2013. *5th ed. Manilla : International Rice Research Institute* (2013)
6. A. Ismayana, N. S. Indrasti & N. Erica, Jurnal Bumi Lestari, **14** (1) (2014)
7. M. Mirwan, F. Rosariawari, *Surabaya: Universitas Pembanguna Veteran* (2010)
8. Suprihanto, A. Guswara & Satoto. *Subang: Balai Besar Penelitian Tanaman Padi* (2008)
9. D. A. Suriadikarta, D. Setyorini. *Bogor:Balai Penelitian Tanah* (2005)

10. Zuraidah. *Banda Aceh : Jurnal Ilmiah Pendidikan Biologi, Biologi Edukasi*, **5** (1) (2013)

Preliminary Evaluation of Potassium Extraction from Bamboo Ash

Tjokorde W. Samadhi[1,*], *Febrizca* Narcia[1], and *Hendro* Amril[1]

[1]Faculty of Industrial Technology, Institut Teknologi Bandung, Jalan Ganesha 10 Bandung 40132, Indonesia

Abstract. Bamboo is a potentially economical fuel crop that has not been utilized at a substantial extent for energy generation in Indonesia. As a thermal conversion waste, bamboo ash is particularly interesting due to its high potassium content. This paper discusses the determination of several key parameters of a simple batchwise extraction process to recover potassium in the form of weak solution from bamboo ash. To produce the ash, black bamboo (*Gigantochloa atroviolaceae*) is charred in a fixed bed combustor. The bamboo char is ground and ashed at 500 °C in an electric furnace. The ash yield is 3.3 %-mass relative to as-received ash, with an ash K_2O content of 12.9 %-mass. The ash is ground until passing 100-mesh standard sieve, and extracted by deionized water on a 2-stage laboratory-scale batchwise extractor battery. Process variables include extractror battery configuration (counter-current and co-current), temperature (nominal setting at 45-80 °C), and contact period of 1-6 hours. The concentration of extracted K_2O increases asymptotically with temperature and contact time. Counter-current extraction yields more than twice the extract K_2O concentration compared to cross-current extraction. The optimum conditions for the counter-current extraction is identified as a temperature of 78 °C and contact time of 4 hours, resulting in a 0.70 %-mass K_2O solution concentration. Spot sampling of commercial liquid fertilizer products in Indonesia indicates an equivalent K_2O content of 0.08-13.6 %-mass, suggesting the potential of the bamboo ash extract as an intermediate for fertilizer product.

1 Background

As the current near- to intermediate-term energy policy of the Indonesian government calls for a 5% contribution of renewable energy sources by the year 2025, it is expected that the production of biomass shall increase substantially in the coming years. With this increase in ash waste volume, it is naturally expected that biomass ash utilization shall also become an increasingly urgent issue for the nation.

Among the widely distributed plants in Indonesia, bamboo is still largely overlooked as a potential energy crop. This potential is driven by its relatively fast growth, and its heating value which is nearly comparable to hardwood. Its poor compostability makes it attractive for thermal conversion processes [1, 2].

As reported by various authors, bamboo stalk ash exhibits a particularly high potassium content relative to other biomass types [2-4]. While the total ash content of bamboo stalk is not particulary high at 1-6 %-mass [1], the high potassium content nevertheless suggests that bamboo ash may serve as an economically viable source of potassium. Works on potassium extraction from biomass ash have been reported in the literature. Wood ash is of course the earliest biomass ash to be exploited for alkali production [5]. Other biomass have also been studied, including wheat straw [6], corn straw [7], banana peels [8], and others. Published works on potassium production from bamboo ash is at best scarce.

This work describes the valorization of bamboo ash through simple, batch-wise, laboratory scale potassium extraction process. As a preliminary work, the objective of this study is to assess the extractability of potassium from bamboo ash with a simple process, and to evaluate the technical feasibility of the extract product for use as a fertilizer intermediate.

2 Experimental

The bamboo species selected for this study is the black bamboo (Gigantochloa atroviolaceae) obtained from local markets in the Bandung area. The bamboo stalk is cut into thin strips roughly 10 cm long, and charred in a simple fixed-bed combustor with natural convection. The bamboo char is then pulverized with mortar and pestle, and ashed in an electric furnace at 500 °C. The obtained ash is then sieved to 100-mesh and stored in a sealed container at ambient condition. The total

* Corresponding author: twsamadhi@che.itb.ac.id

https://doi.org/10.1051/matecconf/201815603027

potassium content of the ash is determined by dissolving all extractables in concentrated nitric acid solution followed by filtration, dilution, and measurement by atomic absorption spectroscopy (AAS).

The extraction is undertaken using 3-neck glass flasks as batch extractors, with deionized water as the solvent. Solid-liquid contact is facilitated by a mechanical stirrer. Two glass beakers are arranged as an extraction battery that can operate in cross-current and counter-current modes alternatively. Contact patterns between solid and liquid flow in these configurations are described in Figures 1 and 2. After the desired contact time is achieved, the stirrer is powered down, the extraction mixture is allowed to settle, and the clear extract is separated by decantation.

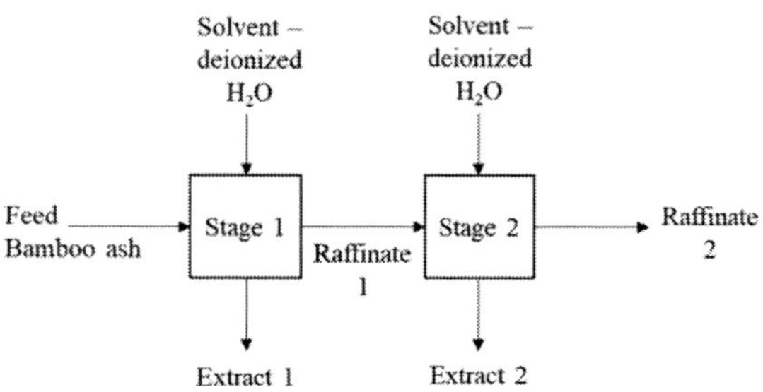

Fig. 1. Materials flow pattern in the cross-current extraction battery configuration.

Based on the total ash feed mass of 33 grams and total extractable K_2O content of 12.9 %-mass on dry ash basis, the total amount of K_2O fed into the extraction battery is 4257 mg. As the numbers in Figure 5 indicate, the total mass of K in the combined extract product streams of the cross-current extraction battery is 898 mg (equivalent to 1082 mg K_2O), which implies that approximately 25.4% of the total K is extracted. The K_2O concentration in the combined extract product is approximately 0.30 %-mass.

In the cross-current extraction, product streams Extract #1 is expected to contain higher K concentration than Extract #2. Both of these streams are mixed into the final extract product. In the counter-current extraction, the process is repeated in three cycles until a steady batchwise process is achieved. At this steady state, the concentrations of the extract product streams are equal. To ensure data consistency however, these product streams are also mixed before analyzed for the K concentration.

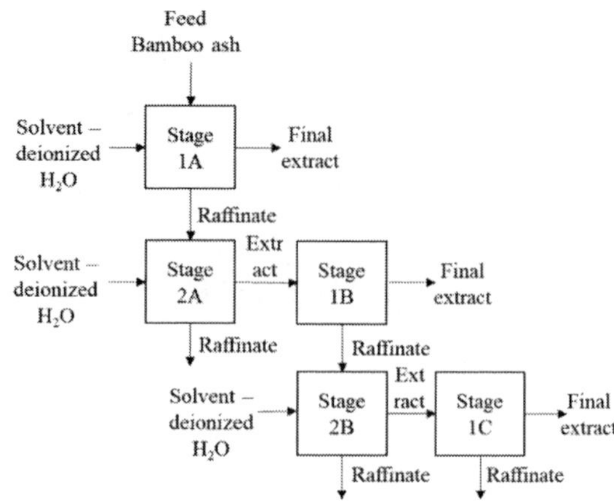

Fig. 2. Materials flow pattern in the counter-current extraction battery configuration.

Experimental variables are varied in an exploratory manner. The extraction temperature is varied in the 45-80°C range, both the cross-current and counter-current flow configuration are tried, and the contact time is varied in the 1-6 hours range. Potassium content in the extract is measured by AAS.

2 Results and discussion

2.2 Temperature and time determination

Mass balance calculations over the entire bamboo combustion and ashing steps indicate an ash content of 3.3 %-mass on dry raw bamboo basis, consistent with the range of values reported in the literature [2]. Total extractable potassium measured by HNO_3 dissolution is 12.9 %-mass (as K_2O) relative to total ash. This value is substantially lower than those reported in the literature, suggesting that the removal of potassium by the acid solution may not be complete, or that the bamboo stalk specimen is of old age, which has been reported to be accompanied by a decrease in its potassium content [9] Since this study is focused more on the relative extractability of the potassium rather than the absolute amount of the extracted material, potential inaccuracy in the total extractable potassium measurement is not considered to be a critical problem.

The first part of the ash extraction experiment addresses the effect of temperature on potassium extractability. Figure 3 presents the K concentration in the extract product of a single-stage batch extraction of the bamboo ash at 45, 60, and 78 °C. While the set point of the latter temperature is 80 °C, the actual measured temperature is 78 °C. Two replicate experiments are run at each temperature; 95% confidence intervals of the average K concentration at each temperature are included.

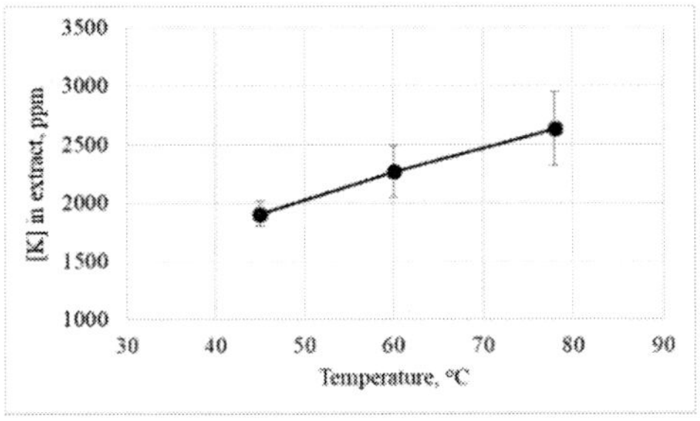

Fig. 3. Effect of temperature on K extractability

Figure 3 clearly indicates the statistically significant increase in K extraction with temperature in the 40-80 °C range. To avoid excessive loss of solvent, subsequent experiments are run at a temperature set point of 80 °C.

In the second extraction experiment, bamboo ash is extracted in a single-stage batch extractor at nominal temperature set point of 80 °C, and contact time varied in the 0.5-6.0 hours range to determine a practical contact time within which the phase equilibrium between extract and raffinate streams has been approached. Two feed to solvent mass ratios are selected, namely 1:5 and 1:250, to assess the presence of mass transfer limitations.

Figure 4 plots the extracted K concentrations against contact time for the two solid/liquid ratios. Incremental gradients between successive time values are computed. The overall trend of extracted K concentration vs. time is similar for both solid/liquid ratios, in which an acceptably low concentration gradient starts to be achieved at contact time between 4 and 5 hours. Therefore, a contact time of 4 hours is selected for subsequent experimental runs.

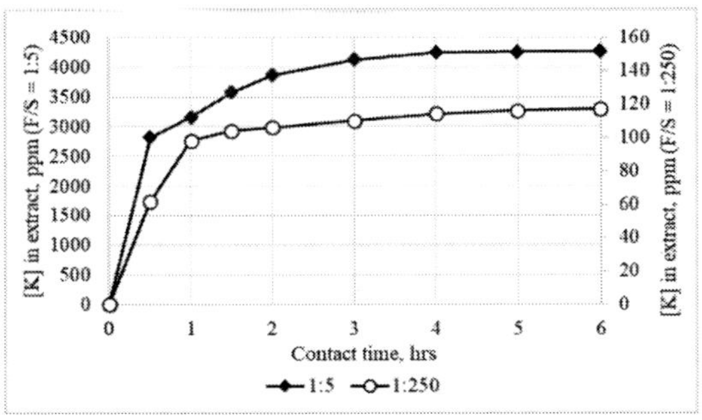

Fig. 4. Extracted K concentrations as functions of contact time at 80 °C, for ash feed to water solvent mass ratios of 1:5 (◆) and 1:250 (O).

2.2 Two-stage extraction

In the final experiment, bamboo ash is extracted in a two-stage batchwise extraction in cross-current and counter-current configurations. Based on the preceding experiments, the extraction temperature is set at 80 °C and contact time at 4 hours. The ash feed mass is 33 grams. The fresh solvent volume is set such that in each extraction stage the fresh feed or raffinate is mixed with 250 g of solvent. The solvent is divided into two parts of 250 grams, each for one extraction stage. Figures 5 and 6 describes the overall mass balance of the two extraction batteries.

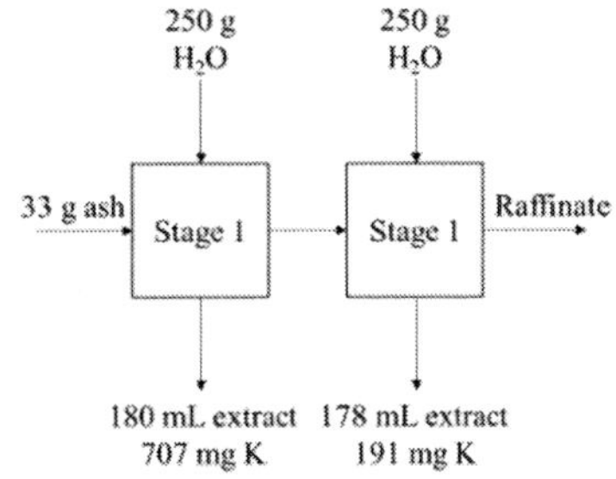

Fig. 5. Overall mass balance of the cross-current 2-stage extraction of bamboo ash.

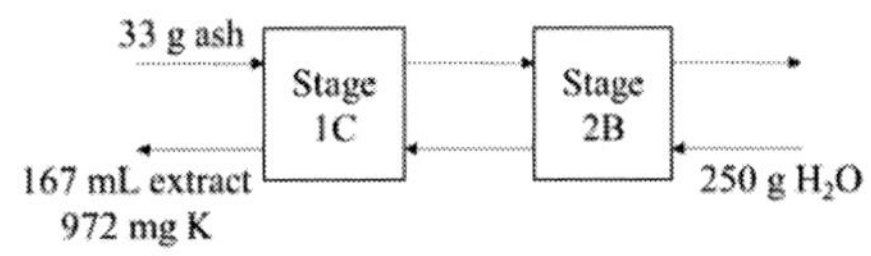

Fig. 6. Overall mass balance of the counter-current 2-stage extraction of bamboo ash

In the counter-current extraction battery, extract from stage 1B (refer to Figure 2 for the designation of the stages) is collected as the final product. This stream contains 972 mg of K (equivalent to 1171 mg of K_2O) dissolved in 167 mL of extract product. Therefore, the extracted K is approximately 27.5% of the total extractable K in the ash. The K_2O concentration in the final extract is approximately 0.70 %-mass, more than twice the concentration produced by cross-current extraction.

To evaluate the characteristics of the bamboo ash extract solution as a potential precursor for fertilizer, spot sampling of commercial organic or NPK fertilizer available in local supermarkets in the Bandung area has been undertaken. Two fertilizer brands were identified, namely Magicgro and GA Premium. The first brand, which is advertised as a liquid organic ready-to-use fertilizer contains 0.08 %-K_2O. The latter brand, which is advertised as a liquid NPK fertilizer, contains 13.6 %-K_2O.

The K concentration of the bamboo ash extract lies in between these two commercial products. Therefore, the bamboo extract is a promising source of potassium for fertilizer. It can be mixed with other macro- and micronutrients to produce ready-to-use liquid organic fertilizer, or concentrated further to render it useful as an NPK fertilizer component.

Both the cross-current and counter-current extraction configurations are able to produce extract with promising K concentrations. Therefore, the choice between these flow configurations is likely to be determined by the suitability of each process to the scale of production. Cross-current extraction is simpler to operate and consumes less ash for small-scale batch production. Counter-current extraction exhibits higher separation efficiency, but is more suitable for larger scale continuous production.

3 Conclusions

A potential route for biomass ash valorization has been identified, which involves the recovery of potassium from bamboo ash through extraction process using water as the solvent. A simple laboratory-scale two-stage batchwise extraction battery is able to recover approximately 25-28% of the total extractable potassium from a black bamboo (Gigantochloa atroviolaceae) ash. Favorable process conditions include an extraction temperature of 78 °C, contact time of 4 hours and solid to liquid mass ratio of 1:5 – 1:7.5 for each extraction stage. The simple batchwise extraction battery process is able to produce extract with K concentration of 0.25-0.58 %. This concentration range compares well to commercial liquid fertilizder products, suggesting that bamboo ash is an attractive source of water-extractable potassium for fertilizer production.

This research is funded by the P3MI grant program for the Chemical Engineering Product Design and Development Expertise Group at the Faculty of Industrial Technology, Institut Teknologi Bandung.

References

1. B. Engler, S. Schoenherr, Z. Zhong, G. Becker, Int. J. Forest Eng. **23**, 2 (2012)
2. J.M.O. Scurlock, D.C. Dayton, B. Hames, Biomass Bioenerg. **19**, 4 (2000)
3. S.V. Vassilev, D. Baxter, L.K. Andersen, C.G. Vassileva, Fuel **89** (2010)
4. T.W. Samadhi, F.G. Gumilar, K.F. Ramadani, Key Eng. Mat. **709** (2016)
5. M.S. Erich, J. Environ. Qual. **20**, 3 (1991)
6. Y. Wang, W. Wang, H. Tan, W. Du, X, Qu, Environ. Prog. Sustain. **34**, 2 (2015)
7. X. Wang, S. Wang, X. Zhao, W. Li, C. Dong, L. Zhao, Adv. Mater. Res. **860-863** (2014)
8. I.O. Olabanji, E.A. Oluyemi, O.S. Ajayi, Afr. J. Biotechnol. **11**, 99 (2012)
9. X.B. Li, T.F. Shupe, G.F. Peter, C.Y. Hse, T.L. Eberhardt, J. Trop. For. Sci. **19**, 1 (2007)

Organic Fertilizer Potential Using *Aspergillus niger*, *Pseudomonas putida* and *Effective Microorganisms* from Coconut Water Waste in Ponorogo, East Java - Indonesia

Raden Darmawan*, *Sri* Rachmania Juliastuti, *Nuniek* Hendrianie, *Setiyo* Gunawan, *Hakun* Wirawasista Aparamarta, *Reynad* D.P. Gultom, *Rillya* Kharisma Prabatiwi

Department of Chemical Engineering, Faculty of Industrial Technology, Institut Teknologi Sepuluh Nopember, 60111 Surabaya, INDONESIA

Abstract. The abundance of coconut water as a waste of household industry activities in Ponorogo can be obtained. There is still organic matter considered to become a liquid fertilizer where it can be expected to substitute for the use of inorganic fertilizers. The research aims to determine the potential of coconut water as a liquid organic fertilizer involving three types of microorganisms, such as *Aspergillus niger, Pseudomonas putida* and *Effective Microorganisms* (EM4). The experiment was conducted using a mixed bacth reactor with 4 L/ minute of aeration rate and initial microorganism`s number of approximately 10^7 cells/ mL for 10 days. The results suggest that the highest reduction of C-organic and the biggest enhancement of N, P, K can be achieved by a combination of *A. niger* and *P. putida* with 29 %; 100 %, 110 %, 552%, respectively.

1 Introduction

Use of inorganic fertilizers excessively can worsen the physical condition of the soil. It can reduce the number of soil's nutrients and minerals. Furthermore, it also can break down the soil structure. In order to improve and maintain the sustainable use of soils, the utilization of organic fertilizer is a way to meet them. The organic fertilizers suppose to heal the texture of soil by resupplying the nutrients and minerals lost.

Ponorogo is an industrial center of micro, small, and medium scale business (UMKM) for several household processed products, including traditional food products (Jenang). In this production, UMKM produces coconut water which is not utilized properly or even thrown away. The utilization of this waste of coconut water needs to be captured as a potential challenge where nutrients present in the wastewater recovered optimally to generate beneficial products to the community. One of an appropriate method is by making organic liquid fertilizer (POC) from wastewater of coconut that can be used as substitutes of synthetic fertilizers. There are economic and ecological advantages found to be higher when the wastewater of coconut fermented become POC. Future perspectives indicate that it will be one of greatest challenges of nutrient variability to utilize the waste become biobased fertilizer products [1]. It is well known that bio-based fertilizers have high variability in nutrient composition [2,3]. The existence of POC can be used to fertilize plants and supply organic materials to improve the physical, chemical, and biological properties of the soil. On the other hand, the use of excessive inorganic fertilizers during this time can worsen the physical condition of the soil, so the need for the development of organic fertilizer.

The research was conducted to find out the used microorganisms appropriately to apply to chilies, tomatoes, and eggplants. *Aspergillus niger, Pseudomonas putida* and bio-activator EM4 were several microorganisms to be variables for generating organic fertilizers from coconut water waste.

2 Materials and Methods

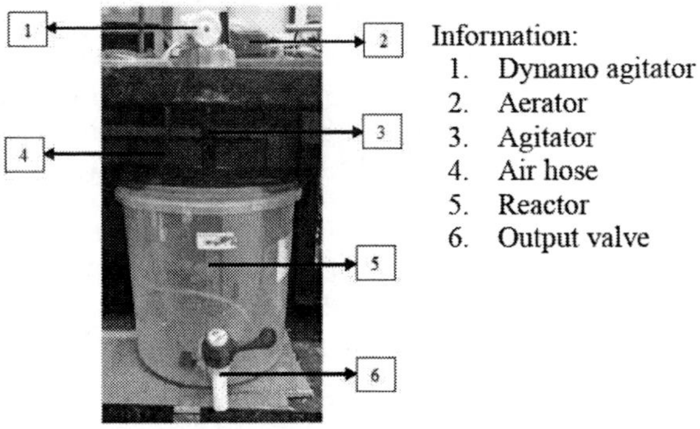

Fig. 1. Mixed Batch Reactor

The organic fertilizers from coconut water waste were produced through fermentation process by implementing

* Corresponding author: rdarmawan@chem-eng.its.ac.id

several microorganisms. They were *Aspergillus niger, Pseudomonas putida* and bio-activator EM4 as a starter with initial number 10^7/ mL. The fermentation was conducted using a mixed batch reactor (Fig.1) with an aeration rate of 4 L/minutes for 10 days. A number of macronutrients (N, P, K, and C-organic) were analyzed before process fermentation started, after the 5th day and the 10th day. The concentration number of macronutrients before the process can be seen in Table 1 as follows.

Table 1. Initial number (%) of N, P, K and C before fermentation periods.

No	Variable	Nutrient (%)			
		N	P	K	C
1	*P. putida*	0.02	0.01	0.018	0.65
2	*A. niger*	0.02	0.01	0.018	0.65
3	EM4	0.02	0.01	0.018	0.65
4	*P. putida : A. niger*	0.02	0.01	0.018	0.65
5	EM4 : *P. putida*	002	0.01	0.018	0.65
6	EM4 : *A.niger*	0.02	0.01	0.018	0.65
7	EM4 : *P. putida : A. niger*	0.04	0.01	0.015	1.19
8	Negative Control	0.04	0.01	0.015	1.19

The produced biofertilizers were applied to manure three type of plants (chilies, tomatoes, and eggplants) for 45 days. The plants were measured and observed for initial height (0 days) and conditions, respectively before spraying the liquid organic fertilizer into the tested plants. The fertilization stage was conducted every two days using 10 mL of each variable then diluted with sterile water up to 20 mL previously. Some physicals development were also investigated such as the hight of plant and the presence and colour of fruit to evaluate the effect of some microorganisms mixtures.

3 Result and Discussion

In this experiment, the macronutrients (N, P, K, and C-organic) were investigated.

Increasing of Nitrogen Element

The fermentation process was carried out for 10 days because all variables implemented had increased at the period especially nitrogen content. As shown in Fig. 2, the mixture of *Pseudomonas putida* and *Aspergillus niger* reported that it had the highest increase of nitrogen compound (100%). Two types of microorganisms believed can improve and have a critical role as the contributors for the nitrogen composition on the plants [4]. *Pseudomonas putida* and *Aspergillus niger* are belonging to diazotroph microorganisms where the microorganisms have an important role to fix nitrogen. The capability of the microorganisms might be the present a nitrogenase enzyme which capable of combining hydrogen and nitrogen [5, 6]. The results also appropriate with [7] that the content of nitrogen can be increased by the fermentation process.

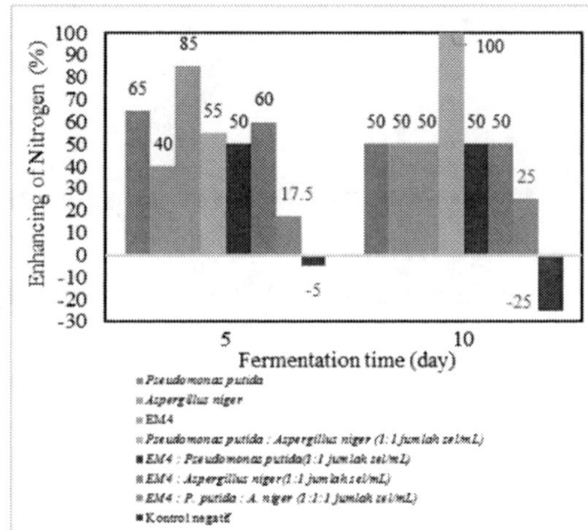

Fig. 2. Percentage increasing of nitrogen element

Increasing of Phosphor Element

The enhancing of phosphor elements were decreased with increasing fermentation time. For 5 days period, the greatest content can be achieved (300 %) then the composition was decreased (110 %) for 10 days fermentation period. The results were shown in Fig. 3 by implementing the combination of *Pseudomonas putida* and *Aspergillus niger*. The microorganisms are not suitable for increasing phosphor element [4].

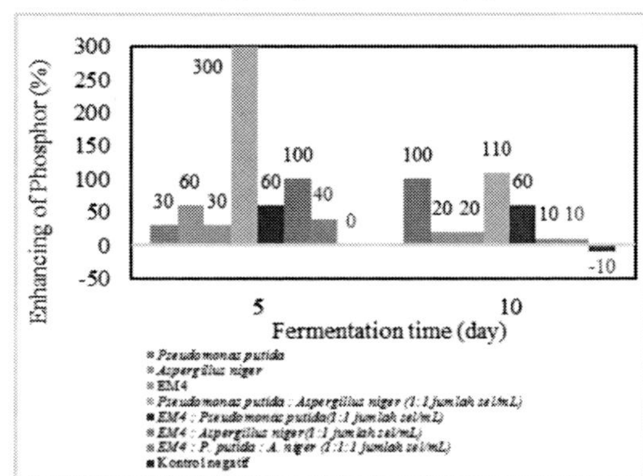

Fig. 3. Percentage increasing of phosphor element

The result is consistent with [4] which present that along with the fermentation process in organic fertilizers, the phosphor contents would also increase although not significantly.

Increasing of Potassium Element

Potassium contents were enhanced along with the fermentation experiment carried out after time 5 days then for several variables were decreased after 10 days period.

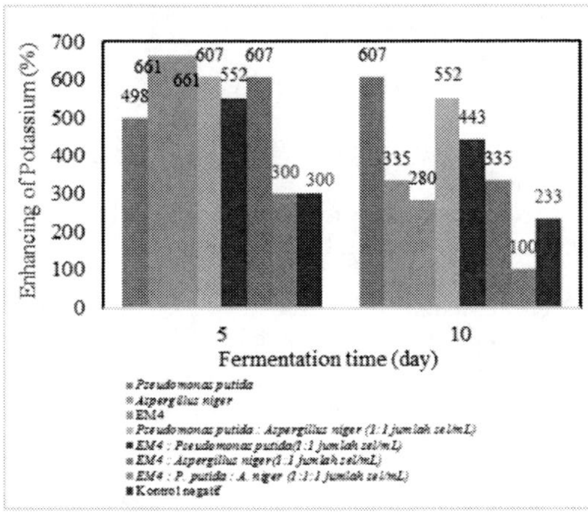

Fig. 4. Percentage increasing of Potassium element

The highest result was obtained by *Pseudomonas putida* (607 %) then followed by the combination of *Pseudomonas putida* and *Aspergillus niger* (552 %). The increasing of K element may be caused by the reaction of hydrolysis and proteolysis that is the loss of K^+ from the origin mineral of organic matter.

Decreasing of Carbon Element

In the experiment, the carbon compounds were breaked down into a simple element with more stable. All variables demonstrated decreased with increasing the fermentation period except the negative control was increased. The condition may be caused by the absence of the supplemented microorganisms that utilize C-organic within coconut water as a carbon source to increase the metabolism of microorganisms.

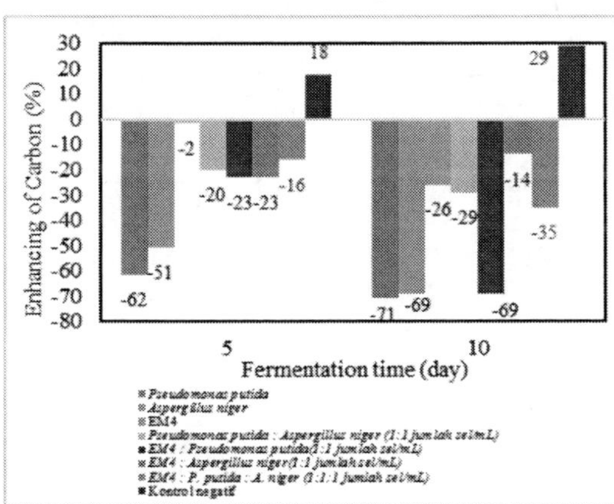

Fig. 5. Percentage increasing of Carbon element

Generally, based on the results of research that the highest levels of macro nutrient are variables using the mixtures of *Pseudomonas putida* and *Aspergillus niger* in which it can increase N, P, and K elements of 100%, 110%, and 552%, respectively.

Application the biofertilizer on the tested plants

Three types of plants were selected to evaluate the produced biofertilizer. The fertilizer was used to manure the plants. The response plants were also observed for the initial time, 15, 30, and 45 days, including high growth (Table 2) and physical development (leaves and fruits) of plants (Fig. 6, 7, 8) using the mixture of *Pseudomonas putida* and *Aspergillus niger*.

Table 2. Difference of height (cm) on the tested plants after manuring for 45 days

No	Variable	$\Delta h = h_{45\ day} - h_{0\ day}$ (cm)		
		Chilli	Tomato	Eggplant
1	EM4	8.67	5	7.33
2	*Pseudomonas putida*	10	5	7.17
3	*Aspergillus niger*	8.33	5.5	7
4	EM4 : *Pseudomonas putida*	8.17	5.5	7.5
5	EM4 : *Aspergillus niger*	9.17	5.5	7
6	*Pseudomonas putida* : *Aspergillus niger*	10.5	6	7.67
7	EM4 : *Pseudomonas putida* : *Aspergillus niger*	8.33	5.33	7
8	Negative Control	6.33	4.5	6
9	Water only	5.67	3.67	5.33

The best of the height difference was obtained using the liquid organic fertilizer with the addition of the mixture of *Pseudomonas putida* and *Aspergillus niger*. From Table 2, it can be summarized that the highest of differences from the tested plants of chilies, tomatoes, and eggplants were 10.5 cm, 6 cm, and 7.67 cm, respectively.

Fig. 6. Testing biofertilizer to manure on chilies for (a) initial; (b) 15 days; (c) 30 days; (d) 45 days

The results indicate that the combination of *Pseudomonas putida* and *Aspergillus niger* is suitable microorganisms for generating the liquid organic

https://doi.org/10.1051/matecconf/201815603028

fertilizer from coconut water waste. On the observation of fruit development, that the tested plants showed the different responses of each variable. The most rapid development was shown by the tested plants using the liquid organic fertilizer with the mixture of *Pseudomonas putida* and *Aspergillus niger* supplemented. The results are confirmed physically in which the development of fruit that has changed color (mature) after 45 days at Fig. 7(d).

Fig. 7. Testing biofertilizer to manure on tomatoes for (a) initial; (b) 15 days; (c) 30 days; (d) 45 days

Fig. 8. Testing biofertilizer to manure on eggplants for (a) initial; (b) 15 days; (c) 30 days; (d) 45 days

The experiment results provide that the microorganisms composition using the mixture of both *Pseudomonas putida* and *Aspergillus niger* can produce the most effective of organic liquid fertilizer from the coconut water waste than using other microorganisms.

4 Conclusion

From the research can be concluded that

1. The bioreactor system described here is a promising alternative method for producing biofertilizer.
2. The coconut waste water can be used as the potential material source for generating liquid organic fertilizer.
3. A combination between *Aspergillus niger* and *Pseudomonas putida* with ratio of 1: 1 is the best supplemented culture condition to make the biofertilizer for manuring to chilies, tomatoes and eggplants.

The authors would like to thank the Institut for Research and Community Services (LPPM) of Institut Teknologi Sepuluh Nopember (ITS) on the Program (Pengabdian Kepada Masyarakat Berbasis Penelitian) for financial supporting with agreement document No: 983/PKS/ITS/2017.

References

1. I. Sigurnjak, C. Vaneeckhaute, E. Michels, B. Ryckaert, G. Ghekiere, F. M. G. Tack, E. Meers, Sci. Total Environ., 599–600 (2017).
2. European Commission, *Study on the behaviour of digestate in agricultural soils*. DG Environment, European Commission (2014).
3. A. Galvez, T. Sinicco, M. L. Cayuela, M. D. Mingorance, F. Fornasier, C. Mondini, Agric. Ecosyst. Environ., **160**, 3–14 (2012).
4. P. Marbun, G. Hardy and P. J. Artha. *Efektifitas Aspergillus niger dan Penicillium sp. dalam Meningkatkan Ketersediaan Fosfat dan Pertumbuhan Tanaman Jagung pada Tanah Andisol.*, Universitas Sumatera Utara (2013).
5. T. T. Yin, U. L. Pin, and A. H. A. Ghazali, Trop. Life Sci. Res., **26**, 1 (2015)
6. S. Issar, and R. K. Gaur, Int. J. Adv. Res., **1**, 8 (2013)
7. M. Day, M. Krzymien, K. Shaw, L. Zaremba, W.R. Wilson, C. Botden, and B. Thomas, Compost Sci. Util. **6**, 2 (1998).

Data Reconciliation and Gross Error Detection for Troubleshooting of Ammonia Reactor

Tri Partono Adhi[1,*] and *Untoro Eko* Saputro[1]

[1]Department of Chemical Engineering, Institut Teknologi Bandung, Jln. Ganeca No. 10, Labtek X, Bandung 40132, Indonesia

Abstract. Data reconciliation (DR) and gross error detection are two common tools used in industry to provide accurate and reliable data, which is useful to analyse plant performance and basis for troubleshooting. DR techniques improve the accuracy of measurements by using redundancies in material and energy balances. This provides reliable information that could help decision making regarding plant operation, which potentially leads to financial benefit. This paper presents the utilization of plant data to perform troubleshooting of ammonia reactor, in particular the profile of catalyst activity. Bad plant data are collected and then analysed using DR to produces reconciled data, which could be used to detect and identify the gross error measurements. The input data for DR and gross error detection were gathered from Aspen HYSYS V8.8 simulations by modelling the single-bed ammonia reactor. The result presents that bad plant data could define actual system condition such as gross error measurements in normal condition or catalyst activity problem. Both conditions are modelled by DR to indicate actual system condition using statistical analysis and to perform troubleshooting. Appropriate troubleshooting could save time and provide financial benefits by avoiding wrong accusation of system problem, specifically in ammonia reactor evaluated in this paper.

1 Introduction

The reliability of plant data could determine the effectiveness of plant performance improvement strategies. If these data are not reliable, decisions might be made based on wrong assumptions and more harm than good may occur [1]. One of the key factors of data reliability improvement is the presence of gross errors in process data. Data reconciliation and gross errors detection could provide efficiency analysis, condition monitoring and operation optimization [2]. These techniques could be beneficial for complex chemical process systems, such as those in an ammonia reactor, where gross errors and data reconciliation are useful to support troubleshooting in this system.

The inaccuracy of measuring instruments could produce gross errors and other problems could appear because of inadequate mathematical models. When a set of data is realistic, then the material and energy is almost balance [3]. Imbalance material and energy might indicate gross errors in data.

Basic issues are whether the estimation of an observed value can be improved by using the other measurements (redundancy), an unobserved value is estimable from the observed ones (determinability) and whether an observed value is a gross error [4]. Process modeling also provide best analysis tools to indicate process condition and decide solution of the problems by comparing redundancy and simulation. Computations and simulation could evaluate the system performance, obtain reconcile data (DR), or to optimize/troubleshoot the system by adjusting some parameters [5].

Good measurement usually provides data redundancy to guarantee data availability. This redundancy could be used to determine the suitable mathematical model and improve the accuracy of plant data by reducing the effect of random errors. Data reconciliation use redundant data and adjust the redundant data according to their estimated standard deviations to obtain the estimation that satisfy system constraints [2].

Data reconciliation adjusts process measurements with random errors by having them satisfy material and energy balance constraints and is a way to improve the quality of the measurements taken from a process via DCS or any other means of data collection. [6] In ammonia plant, plant data is usually provided in logbook and noted by operator, and it is also recorded in DCS system. These plant data could be used to improve the plant performance or to solve plant problems. Monitoring and analyzing bad plant data could trigger wrong troubleshooting solution which eventually could harming plant and environment. This could lead to wrong decisions that results financial loss. Gross errors are the data problems which frequently appear, so it is important to identify and isolate gross errors. Based on gross error data, the decrease plant performance could be identified, so the system could be overhauled to fix the

* Corresponding author: tpadhi@che.itb.ac.id

performance. There may also be a possibility where the system is running well, but there are gross errors in these data because of human/operator errors, inaccurate measurement, or others problems. Bagajewicz explain that some measurements are highly vulnerable to the presence of undetected biases where induced biases are larger than the standard deviation (i.e. precision) many times [7].

There are two main problems in an ammonia reactor system analyzed in this paper, namely the gross error problem and the decreasing of ammonia reactor catalyst activity. This paper presents and demonstrates a brief utilization of plant data to analyze and troubleshooting of ammonia synthesis reactor single bed. Ammonia reactor could be simplified as single bed reactor to ease the analysis or data reconciliation.

2 Methodology

The problem of data reconciliation in plant is to minimize an objective function, e.g. the sum of the squared adjusts pondered by their reciprocal variances, subject to linear or nonlinear restrictions (material balances) [8]. Reconciled process data is used to specify the current status of the plant model and for estimation of the model parameters for plant-model matching [6]. Steady state simulation packages equipped with optimization routines can be used to perform data reconciliation and obtain accurate results. This approach can be used to reconcile linear and nonlinear problems via the minimization of a weighted least squares objective function by varying appropriate flowsheet variables [9]. So, data reconciliation problem simulation is defined as a nonlinear steady-state data by a least sum of square minimization problem as follows:

Error! Reference source not found. (1)

$$f(x_i, y_i) = 0 \quad (2)$$

$\mathbf{Q_{min}}$ is the estimated covariance matrix of measured data; $\mathbf{x^m}$ is the measured values; $\mathbf{x_i}$ is the reconciled values of measured parameters; and $\boldsymbol{\sigma_i}$ represents the estimated error standard deviation. **f(x,y)** represents the steady state system constraint equations; $\mathbf{y_i}$ represents the calculated value for the unmeasured parameters.

Process simulators with optimization capabilities such as Aspen HYSYS can be used to carry out data reconciliation and parameter estimation [9]. Data reconciliation and gross error detection are simulated using Aspen HYSYS V8.8 to achieve efficient calculation including measurement design, gross error detection and elimination, errors propagation, as well as parameters estimation. Data and model are compared to determine an adequate model by formulation as follows:

$$Q_{min} \approx \chi^2_{crit}(v) \quad (3)$$

$\mathbf{Q_{min}}$ represents the random variables with distribution of χ^2, and **v** represents the degree of freedom. If Q_{min} is greater than critical value of distribution χ^2 (σ_c), it indicates that data and model are conflicted which could not be ignored (significant factor). The basis of design of ammonia reactor modeling in Aspen HYSYS uses kinetic data by Temkin equation. Ammonia synthesis is an exothermic reaction with kinetic equation which are defined by following equations:

$$N_{2\,(g)} + 3H_{2\,(g)} \leftrightarrow 2NH_{3\,(g)} \; \Delta H = -92{,}22 \text{ kJ/mol} \quad (4)$$

Error! Reference source not found. (5)

Error! Reference source not found. (6)

Error! Reference source not found.Error! Reference source not found. (7)

Feed composition and conditions are shown in table 1. Ammonia synthesis reactor is presented in table 1. The simulation in Aspen HYSYS V8.8 uses Peng-Robinson Package and the ammonia reactor single bed is modeled as a single adiabatic plug flow reactor.

Table 1. Composition and feed conditions of reactor

Temperature (°C)	437
Pressure (bar)	142.1
Flowrate (kg/h)	203422
Flowrate (kmol/h)	19440
Vapour Fraction	1.00
Composition (%-mol)	
N_2	17.77
H_2	63.77
NH_3	8.81
CH_4	4.82
Ar	4.82

Table 2. Composition and feed condition of reactor

Name	R-100
Volume reactor (m^3)	13.96
Length (m)	6
Diameter (m)	1.721
Void fraction	0.36
Type of reactor	Adiabatic Plug Flow Reactor

Data reconciliation is assumed as single jeopardy. It means that there are no two or more problems and gross errors that are occurred simultaneously. The other assumption is there is no channeling in ammonia reactor system, as the focuses of this paper are gross error measurements and the decreasing of catalyst activity. Based on these data and assumptions, the methodology is divided into 4 main steps, as follows:

1. **Ideal modelling using Aspen HYSYS and operation/plant data collection.** This model is based on design data with standard deviation of measurement devices. Operational data is acquired from operator shift with data redundancy.

2. **Data analysis and gross error identification and isolation.** Data reconciliation uses ideal model and operational/plant data to obtain reconciled data. Error from reconciled data and plant data is compared to standard deviation of measurement devices. If the error is greater than standard deviation, gross error identification and isolation are next step to determine

system conditions. Once a gross error was detected, a new regression was performed and tests were done until the data were free of errors [11].

3. **Troubleshooting with measured data which don't have gross errors.** If there are no gross errors, a new model was generated that is appropriate to the problem in system. Measured data will be checked with new model and produce reconciled data. If error is within the standard deviation range and identical with the new model, thus the system experiences a problem based on the model.

4. **Decision making.** Based on data reconciliation and troubleshooting, solutions are given based on analysis. For example, catalyst regeneration is proposed if there is a catalyst deactivation. The other example of a recommendation to check and calibrate the instrumentation if there are gross errors. The detailed methodology is shown in figure 1.

3 Result and Discussion

Ideal model based on design data from plant is shown in figure 2. Result of streams inlet and outlet is shown in table 3. Specification of measurement devices in reactor system is presented in table 4.

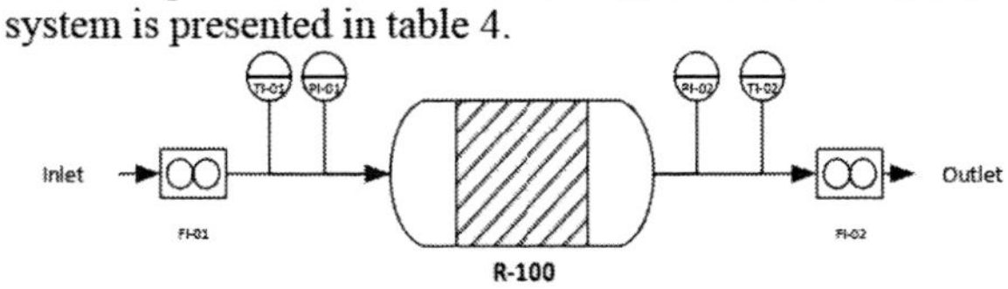

Figure 1. Ideal model of ammonia synthesis reactor

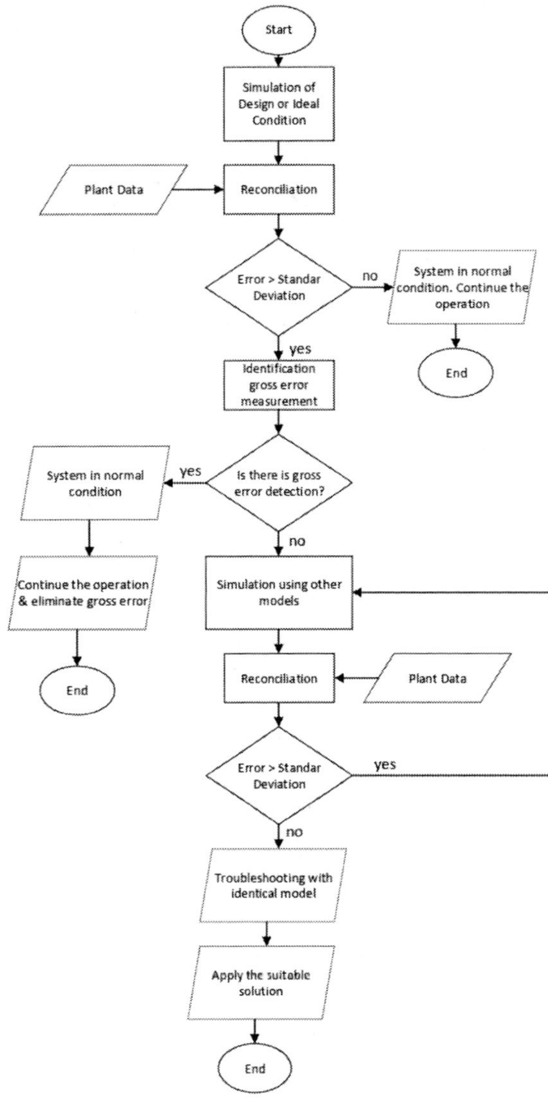

Figure 1. Methodology of simulation data reconciliation for troubleshooting

Table 3. Stream specification of ideal model reactor

Stream	Inlet	Outlet
Temperature (°C)	437	492.3
Pressure (bar)	142.1	141.7
Flow (kg/h)	203422	203422
Flow (kmol/h)	19440	187788
Vapour fraction	1.00	1.00
Composition (%-mol)		
N_2	17.77	16.65
H_2	63.77	60.76
NH_3	8.81	12.61
CH_4	4.82	4.99
Ar	4.82	4.99

Table 4. Measurement devices specification of reactor system

Tag Number	Type	Unit	Min	Max	St. Dev.
FI-01	Online analyser	(kmol/h)	5000	20000	± 100
FI-02	Online analyser	(kmol/h)	5000	20000	± 100
TI-01	Temperature indicator	(°C)	300	500	± 1
TI-02	Temperature indicator	(°C)	300	500	± 1
PI-01	Pressure Indicator	(bar)	125	200	± 1
PI-02	Pressure Indicator	(bar)	125	200	± 1

3.1 Gross Error Detection and Elimination

Hypothetical plant data are acquired as shown in table 5 with 3 redundancies. Data reconciliation result from this plant data is shown in table 6.

Table 5. Plant data from ammonia synthesis reactor

Time	08.00	12.00
FI-01 (H_2) (kmol/h)	12410	12420
FI-02 (H_2) (kmol/h)	17988	17955
TI-01 (°C)	436.1	436.9
TI-02 (°C)	493	492.6
PI-01 (bar)	142.5	142.5
PI-02 (bar)	142.1	141.9

Table 6. Data reconciliation result of plant data

Time	Tag	Plant data	Std. Dev.	Reconciled Data	Design Data	Error/Std. Dev. (%)
08.00	FI-01 (H_2) (kmol/h)	12410	100	12397	12397	12.89
	FI-02 (H_2) (kmol/h)	17988	100	11415	11412	6573.15
	TI-01 (°C)	436.1	1	436.8	437	67.08
	TI-02 (°C)	492.4	1	491.9	492.3	51.09
	PI-01 (bar)	142.5	1	142.1	142.1	37.13
	PI-02 (bar)	142.1	1	141.7	141.7	37.13
12.00	FI-01 (H_2) (kmol/h)	12420	100	12397	12397	22.89
	FI-02 (H_2) (kmol/h)	17955	100	11415	11412	6540.08
	TI-01 (°C)	436.9	1	436.8	437	13.68
	TI-02 (°C)	492.6	1	491.9	492.3	72.25
	PI-01 (bar)	142.5	1	142.1	142.1	38.52
	PI-02 (bar)	141.9	1	141.7	141.7	18.52
Q_{min} (Objective Function)						8599.67
$\chi^2(v) = \sigma_c$						14.10

Table 7. Data reconciliation result of plant data after gross error elimination

Time	Tag	Plant data	Std. Dev.	Reconciled Data	Design Data	Error/Std. Dev. (%)
08.00	FI-01 (H_2) (kmol/h)	12410	100	12397	12397	12.89
	FI-02 (H_2) (kmol/h)	17988	100000	11415	11412	6.57
	TI-01 (°C)	436.1	1	436.8	437	67.08
	TI-02 (°C)	492.4	1	491.9	492.3	51.09
	PI-01 (bar)	142.5	1	142.1	142.1	37.13
	PI-02 (bar)	142.1	1	141.7	141.7	37.13
12.00	FI-01 (H_2) (kmol/h)	12420	100	12397	12397	22.89
	FI-02 (H_2) (kmol/h)	17955	100000	11415	11412	6.54
	TI-01 (°C)	436.9	1	436.8	437	13.68
	TI-02 (°C)	492.6	1	491.9	492.3	72.25
	PI-01 (bar)	142.5	1	142.1	142.1	38.52
	PI-02 (bar)	141.9	1	141.7	141.7	18.52
Q_{min} (Objective Function)						1.78
$\chi^2(v) = \sigma_c$						14.10

Based on table 6, measurement devices FI-02 has high error but others measurement devices still have tolerable error. These data show that the hydrogen content in the product stream is greater than that in the feed stream. The hypothesis is there is gross error measurement in FI-02 and FI-02 measurement is ignored to identify and isolate the gross error. Data reconciliation after gross error elimination of FI-02 measurement is shown in table 7.

Gross error elimination could minimize objective function (Q_{min}) from 8599.67 to 1.78. It indicates that there is gross error measurement in FI-02. So, there is no problem occurred in ammonia reactor based on this analysis because the error from data reconciliation of plant data after gross error elimination is still tolerable.

Action required based on this analysis are instrument FI-02 overhaul and calibration, procedure checking of FI-02 measurement by operator (it could be human error, e.g. value which is measured by FI-02 is not molar flow of H_2 but total molar flow).

3.2 Gross Error Detection for Modelling and Troubleshooting

The decreasing of temperature outlet reactor could be caused by deactivation of catalyst or gross error measurement. Catalyst as one of main component in reactor could determine the conversion and the outlet temperature of reactor. Higher activity catalyst could lead to higher conversion and higher temperature outlet of exothermic adiabatic reactor.

Data reconciliation and gross error detection could analyse the model of system problems. Hypothetical bad plant data were acquired as shown in table 8. It indicates that there were problems in the ammonia reactor because of temperature outlet drop. Temperature measurements have one redundancy because reactor specification and kinetic data is known, so TI-02 would be calculated by model.

Table 8. Plant data of ammonia reactor

Tag	TI-01 (°C)	TI-02 (°C)	FI-01 (kmol/h)	FI-02 (kmol/h)
Design Data	437	492.3	12397	11412
Plant Data 1	436.5	481	12396	11414
Plant Data 2	436.8	480.7	12395	11411

Based on these plant data, the data reconciliation result (using Aspen HYSYS simulation) based on ideal model is shown in table 9.

Table 9. Ideal model data reconciliation

Type	Tag	Plant data	Std. Dev.	Recon Data	Design Data	Error/Std. Dev. (%)
Plant Data 1	TI-01 (°C)	436.5	1	431.4	437	514.2
Plant Data 2	TI-01 (°C)	436.8	1	431.5	437	534.4
Plant Data 1	TI-02 (°C)	481	1	488.9	492.3	789.5
Plant Data 2	TI-02 (°C)	480.7	1	490	492.3	826.0
		Objective Function (Q_{min})				185.55
		$\chi^2(v) = \sigma_c$				3.84

Based on this analysis, the gross error detection should be detected in TI-02 because of gap between plant data and design data, but data reconciliation result show that there were gross error measurements at TI-01 and TI-02. Then, the plant data is actual and there is no gross error measurements after actual checking in field data. Then, data reconciliation result also present that there is a conflict between data and model which could not be ignored because Qmin > σ_c. So, the ideal model is not suitable for this cases.

Further analysis is shown by simulation of other models; the decreasing of catalyst activity model is one suspect of this problems. This model represent the probability of catalyst activity drop. Full activity catalyst (100%) is simulated by default kinetic data, which is shown in table 10.

Table 10. Catalyst activity model of ammonia reactor

Default kinetic data	$k_1 = 36421\exp\left(\frac{-19484}{RT}\right)$ $k_{-1} = 3{,}4585.10^{14}\exp\left(\frac{-38119}{RT}\right)$		
Activity (%)	A_1	A_{-1}	Drop (%)
100	36421	3.4585 x 10^{14}	0
a (model)	a % x 36421	a % x 3.4585 x 10^{14}	100 - a

This catalyst model is modelled using spreadsheet feature in Aspen HYSYS to define the parameter in data reconciliation. Data reconciliation result of decreasing catalyst activity model is shown in table 11.

Table 11. Catalyst activity model data reconciliation

	Tag	Plant data	Std. Dev.	Recon Data	Design Data	Error/Std. Dev. (%)
Plant Data 1	TI-01 (°C)	436.5	1	436.7	437	15.90
Plant Data 2	TI-01 (°C)	436.8	1	436.6	437	15.89
Plant Data 1	TI-02 (°C)	481	1	480.9	492.3	13.18
Plant Data 2	TI-02 (°C)	480.7	1	490.8	492.3	14.76
		Objective Function (Q_{min})				0.09
		$\chi^2(v) = \sigma_c$				3.84

Based on further analysis, gross error detection isn't found at TI-01 and TI-02. Data reconciliation result also present that model is suitable for this cases because of Qmin < σ_c. Case study simulation in Aspen HYSYS analysis is simulated to ensuring the problem of the decreasing catalyst activity. The result of case study between model catalyst activity to outlet temperature reactor is shown in figure 3.

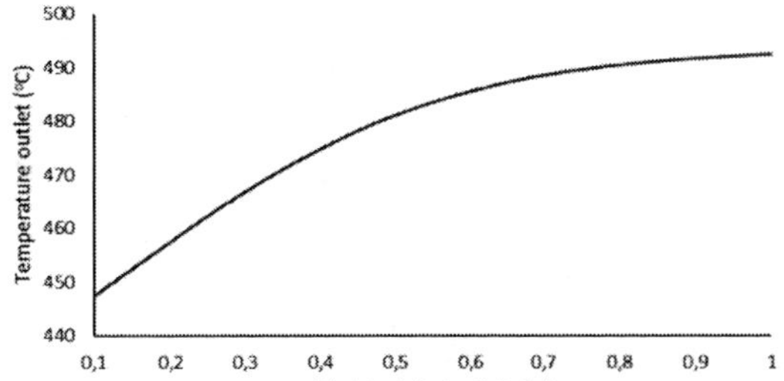

Figure 3. Case study of model catalyst activity to temperature outlet reactor

Based on case study and data reconciliation comparison, the problem of this system is the decreasing of catalyst activity. The data fitting feature shows that the activity of the catalyst is only 50.19% of full activity and it approaches the value of temperature outlet at 50% catalyst activity based on the case study model. These data also match to real condition because catalyst need regeneration in next several months based on plant data collection time. So, the troubleshooting solution of this problem is to regenerate the catalyst in the next turn around (or sooner). This study prove that data reconciliation could provide right troubleshooting based on bad plant data, this study also shows that data reconciliation is particularly true for strategies that rely on first-principles models of the chemical plant, such as model predictive control [12] or real-time optimization [13] and combining with simulation to define faster solution of troubleshooting could save time and financial benefit.

4 Conclusion

In this study, data reconciliation and gross error detection are used as valuable techniques to improve the reliability of plant data, especially for troubleshooting or plant performance improvement purpose. The ammonia synthesis reactor system with redundancy measurements could be analysed by data reconciliation using Aspen HYSYS V8.8 simulations. Bad plant data that could lead to ineffective troubleshooting solutions are fitted with simulation model of data reconciliation. Gross error measurements could be identified and eliminated through data reconciliation and define system condition (in normal or trouble condition). Data reconciliation and gross error detection could provide optimum troubleshooting solutions and may result in operational and financial benefits by mitigating wrong actions due to unreliable data.

References

1. S. Guo, P. Liu, & Z. Li., Applied Energy, Volume **165**, 1037-1051(2016)
2. X. Jiang, P. Liu, & Z. Li. Energy, Volume **75**, 14-23 (2014)
3. V. S. Verneuil, P. Yang, & F.. Chemical Engineering Progress. October 1992 (45-51)
4. R. Maronna & J. Arcas.. Computers and Chemical Engineering **33** (2009) 65-71
5. H. Ijaz, U.M.K. Ati, & V. Mahalec.. Applied Thermal Engineering **52** (2013) 328 - 335
6. D.B. Ozyurt & R.B. Pike.. Computers and Chemical Engineering **28** (2004) 381 – 402
7. M.J. Bagajewicz & D.Q. Nguyen., Computer and Chemical Engineering **32** (2008) 1257-1269
8. J. Placido, A.A. Campos, & D.F. Monteiro. *Data Reconciliation Practice at a Petroleum Refinery Company in Brazil*, 10th International Symposium on Process System Engineering – PSE2009 (2009) 777-782
9. M. Piccolo & P.L. Douglas., Asia-Pasific Journal of Chemcial Engineering Volume **4** Issue 3-4 (1996) 157-182
10. A. Rafiee, & F. Behrouzshad., Journal of Natural Gas Science and Engineering **31** (2016) 538 – 545
11. J.S. Albuquerque, & L.T. Biegler. Process System Engineering, AIChE Journal Vol. **42,** No. 10 (1996) 2841–2856.
12. J.B. Rawlings & B.R. Bakshi. Computer and Chemical Engineering, Volume **30**, Issues 10-12 (2006) 1529 - 1541.
13. J.F. Forbes & T.E. Marlin. Computer and Chemical Engineering, Volume **20**, Issues 6-7 (1996) 717 – 734

Synthesis Study Of Surfactants Sodium Ligno Sulphonate (SLS) From Biomass Waste Using Fourier Transform Infra Red (FTIR)

Slamet Priyanto[1], *Bambang* Pramudono[1,*], *Tutuk Djoko* Kusworo[1], *Suherman*[1], *Hapsoro Aruno* Aji[1], *Edi* Untoro[2], *Puspa* Ratu[2]

[1]Department of Chemical Engineering, Diponegoro University, Jl. Prof. Soedharto, Tembalang, Semarang, Jawa Tengah, Indonesia 50239
[2] Ministry of Energy and Mineral Resources Education and Training Agency (ESDM) Akamigas Cepu Mineral Energy School

Abstract. Lignin from biomass waste (Black Liquor) was isolated by using sulfuric acid 25% and sodium hydroxide solutions 2N. The obtained lignin was reacted with Sodium Bisulfite to Sodium Ligno Sulfonate (SLS). The best result was achieved at 80 ° C, pH 9, ratio of lignin and bisulfite 4: 1, for 2 hours, and 290 rpm stirring rate. The result of lignin formed was sulfonated using Sodium Bisulfite (NaHSO3) to Sodium Ligno Sulfonate (SLS) whose results were tested by the role of groups in peak formation by FTIR and compared to the spectrum of Sodium Ligno Sulfonate made from pure Lignin (commercial) reacted with the commercial Sodium Bisulfite. The result can be seen by the typical functional groups present in the SLS.

1 Introduction

The waste of pulp and paper industry, called black liquor contains 30-45% lignin. Lignin from black liquor (Nurhayati, 1993; Gunawan, 2013), potentially produces Sodium Ligno Sulfonate (SLS) surfactant when synthesized with Sodium Bisulfite (Lubis, 2007; Syahbirin, 2012; Qin, 2015). The development of industry in the world is followed by the increasing need of surfactant. Unfortunately, the need for such surfactants is not matched by increased production.

The purpose of the sulfonation process is to change the lignin hydrophilicity by introducing sulfonate groups as hydrophilic groups. Sulfonation process that occurs is reacting lignin with sodium bisulfite (Denli, 2010). The sulfonate group in lignosulfonate makes lignosulfonate to have an amphipatic structure (surfactant). Sulfonates are known by the general formula R-SO3Na which is a simplification of R-O-SO3-Na sulfate (Kim, 1987). R is a group of C8-C22 aromatic carbon atoms which were hydrophobic groups, while the hydrophilic group comprises carboxylates, sulphonates, phosphates or other acids5. The sodium lignosulfonate surfactant is categorized in an anionic surfactant because it has a sulphonate group and its salt (-NaSO3-) which is an anion (head) and a hydrocarbon group is the tail (Priyanto, 2017, Ma'ruf, 2017, Syahbirin, 2012).

This research studied the influence of temperature, pH, lignin and bisulfite ratio, time and speed of stirring. The purpose of this research is to get the best sulfonation condition of lignin to SLS. Functional group analysis using FTIR to find out the influencing groups in the SLS. The study of each influential variable was applied to find out the combined variables of this synthesis process.

2 Experimental

2.1 Materials

Lignin from Black Liquor was purchased from PT. Indah Kiat Pulp and Paper Factory at Pekanbaru, Riau, West Sumatra, Indonesia. Lignin was isolated from black liquor by acidifying it with 25% sulfuric acid solution. Lignin was synthesized by sodium hydroxide 20% w/w solution with pH 9,5 dried at 60°C for 6 hours and to mesh size of 60 – 80. Sulfuric acid, sodium hydroxide and demineralized water (aquadest) obtained from PT. Indra Sari Semarang, Central Java, Indonesia.

2.1.1 Sulfonation Of Lignin and Synthesis Process of SLS

This phase is the core stage to produce salt lignosulfonates. The reaction occurs between the lignin which has been acidified with sulfite salt (Denli, 2010). There are several types of salt sulphites which can be used in the sulfonation. In US Patent No.4,892,588 sulphite salt used is sodium bisulfite (Rahmawati, 1999; Salminah, 2001).

Two grams of lignin from black liquor was reacted with 2.0 ml 40% bisulfite solution and 60 ml aquadest. Temperature was adjusted from 50 to 95° C, pH from 7 to 10, time from 1 to 4 hours, ratio of lignin and bisulfite from 2 to 5, and the rate of stiring from 250 to 310 rpm. SLS liquid phase was evaporated at a temperature of

* Corresponding author: pramudono2004@yahoo.com

100°C. The concentrated solution formed was then filtered through a buchner funnel using a vacuum pump.

The filtrate obtained is SLS containing lignin and residual bisulfite. The filtrate was mixed with methanol to precipitate an insoluble bisulfite, shaken vigorously, and further filtered with a buchner funnel. The SLS filtrate and residual lignin were evaporated to concentrate the SLS. The obtained concentrated SLS was dried at 60°C to a constant weight, this is the yield to be analysed by the role of the group with FTIR.

2.1.2 Characterization Methods

FT-IR spectrophotometer (SHIMADZU with DRS-8000) was used to analyze the Infrared spectroscopy using a KBr pellets[6]. The KBr pellets consist of 300 mg KBr and 0.1 mg fine powder of SLS sample. Scans were recorded from 400 to 4000 cm^{-1} at a resolution of 16 cm^{-1}.

3 Results and Discussion

3.1 Effect of Process Variables

Figure 1 illustrates the effect of variables temperature, pH, lignin and bisulfite ratio, time and rate of stiring on the yield of SLS. Based on Table 1, it can be seen that the best value for yield at 94% was reached at the temperature of 80°C, pH 9, ratio of lignin and bisulfite 4, time of 2 hours and the rate of stiring of 290 rpm.

Table 1. The best conditions of SLS yield

The best value; Variable: Yield at 94%	
Variables	Best Condition
Temperature, °C	80
pH	9
Ratio, g/g	4
Time, hour	2
Rate of stiring, rpm	290

Table 2 shows the national standardization of Indonesia (SNI) characteristics of SLS from waste Lignin.

Table 2. Characterization of SLS yield

Nu	Parameter	Unit	SLS Ref	SLS from Waste	Method
1	Water content	% w/w	23.96	24.62	SNI 2012-0813-122059
2	Ash Content	% w/w	31.78	32.23	SNI 1247-0442-2009
3	Organic compounds	% w/w	39.97	41.76	SNI 03-2831-1992
4	Volatile Matter	% w/w	4.98	5.14	SNI 13-3999-1995
5	Density (solid)	g/mL	1.09	1.12	SNI 06-2441-1991

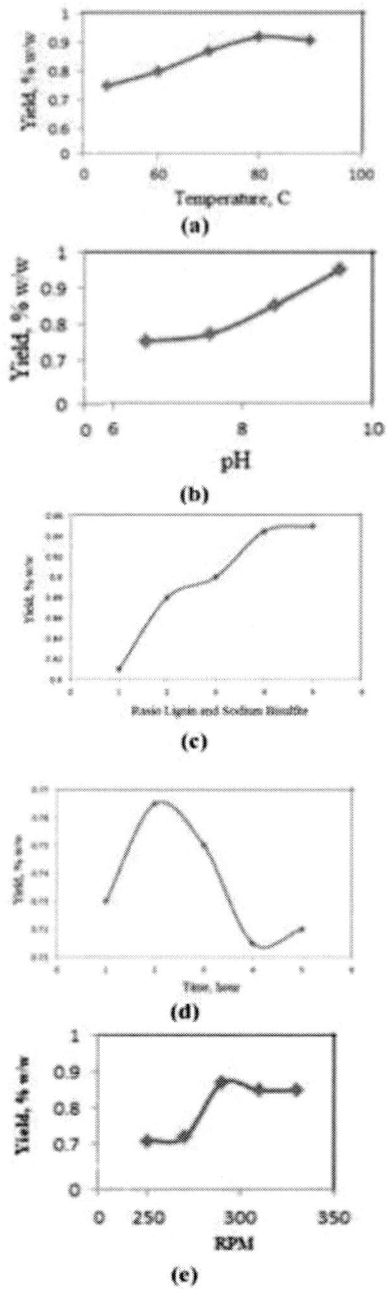

Fig. 1. Effect of temperature (a), pH (b), lignin and bisulfite ratio (c), time (d) and rate of stirring (e) on the yield of SLS

3.2 FT-IR Characteristics of SLS from waste lignin and Pure SLS

Figure 2 shows the spectra of Reference SLS obtained (blue line) and the spectra of SLS from waste lignin (red line). The band between 3425.88-3541.1 cm^{-1} is typical of hydroxyl groups (O-H stretch) in phenolic and aliphatic structures. The peak at 2337.72; 2361 and 2337.72; 2360.87 cm^{-1} indicated a sp^3 C-H MS stretching in methyl ($-CH_3$), methylene ($=CH_2/-CH_2-$) and methoxy ($-OCH_3$) groups[6]. The peak at 1512.19,

1564.34, 1512.19, 1635.14 indicated groups of C-H AS5. In the sulfite (SO_3) region, a weak band is found at 1450.47 cm^{-1} (blue line) and 1450.57 (red line). Furthermore, the S-O, S=O, SO_3, C-H, and OH role of the group in peak formation can be seen in Table-7 (Prakosa, 2017, Ma'ruf, 2017).

Tabel 3. The role of group in SLS

Group	SLS Reference (Blue)		SLS from Waste Lignin (Red)	
	Peak-1	Peak-2	Peak-1	Peak-2
S-O	< 964	-	-	-
S-O	964.41	-	948.9	-
S-O	1033.85	-	1041.56	-
S=O	1111	-	1111	-
SO3	1450.47	-	1450.57	-
C-H AS	1512.19	1564.34	1512.19	1635.14
C-H MS	2337.72	2361	2337.72	2360.87
OH	3425.88	-	3541.1	-
OH	> 3400	-	> 3400	-

The band at 1512.19 cm^{-1} is a typical aromatic skeleton vibration combined with C-H in plane deformations, while 1564.34 and 1635.14 cm^{-1} is of aliphatic C-H stretching in CH_3 (not $-OCH_3$) and phenolic –O-H. The band at 1450.47-1450.57 cm^{-1} shows SO_3. The band at 1111 cm-1 and at 964.41-1033.85 cm^{-1}(blue line) and at 1033.85 Cm^{-1} in SLS reference, at 1041 cm^{-1} in SLS from waste lignin (Prakosa, 2017). FT-IR spectra indicate the spectral features of SLS that are the band at 964.41 and 1450.57 cm^{-1}. The aromatic S-O deformation at 948.9 cm^{-1} and S-O at 1041.56 cm^{-1} appear as aromatic SO_3 at 1450.47 and 1450.57 cm^{-1} or < 964 cm^{-1}.

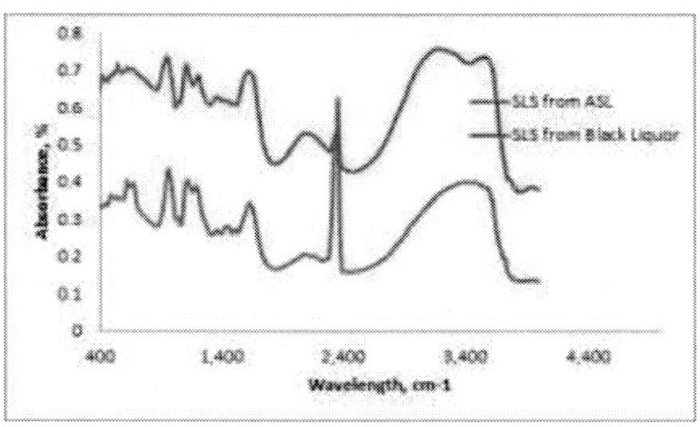

Fig. 2. FT-IR spectra of SLS Reference (blue) and SLS from waste (red)

4 Conclusion

Sodium lignosulfonate (SLS) can be prepared by reacting lignin derived from black liquor biomass waste under best conditions of the temperature of 80°C, pH of 9-9.5, the weight ratio of lignin and bisulfite of 4.00, time of 2 hours, and the rate of stiring of 290 rpm, to obtain the highest yield of SLS 94 %.

This study was supported by Ministry of Research,Technology and High Education Fiscal Year 2017, No:345-22 / UN7.5.1 / PG / 017

References

1. Nurhayati, T., R.A. Pasaribu, Journal of Forest Product Research **11**, 3, 110-116 (1993)
2. Gunawan, D., Scientific Article **1**, 1 (2013)
3. Lubis, A.A., *Isolation of Lignin From Black Liquor Process Cooking Soda Pulp And Pulp Sulphate (Kraft), (Isolasi Lignin Dari Lindi Hitam (Black Liquor) Proses Pemasakan Pulp Soda Dan Pulp Sulfat (Kraft),* Thesis of Faculty of Agriculture, IPB, Bogor (2007)
4. Syahbirin, G., Darwis, A.A., Suryani, A. Syafii, W., 2012, Procedia Chem. **4**, 343 (2012).
5. Qin, Y., Yang, D., Guo, W., and Qiu, X., J. Ind. Eng. Chem. **27**, 192 (2015).
6. Denli, *The Conversion of Lignin to Surfactant (Konversi Lignin Menjadi Surfaktan),* Thesis of Faculty of Agriculture, IPB, Bogor (2010)
7. Kim, H., M.K. Hill, A.L. Fricke, Tappi Journal **12**, 112-115 (1987)
8. Priyanto, S., Suherman, Istadi, Nugroho, A., Aji, H.A., Advanced Science Letters **23**, 6, 5803-5805 (2017)
9. Ma'ruf, A., Pramudono, B., Aryanti, N., Rasayan, J. Chem., 10, 2, 407 (2017)
10. Rahmawati, N., *Lignin Structure of Lumber Leaf Width and Its Influence on Delignification Rate (Struktur Lignin Kayu Daun Lebar dan Pengaruhnya terhadap Laju Delignifikasi),* Thesis of Graduate Program Bogor Agricultural University, Bogor (1999)
11. Salminah, M., *Characteristics of Lignin Result Isolation of Cooking Time Solution Pulp SemiChemical Process at Various Levels of pH (Karakteristik Lignin Hasil Isolasi Larutan Sisa Pemasak Pulp Proses Semi Kimia pada Berbagai Tingkat pH)*, Thesis of Department of Forest Products Technology Faculty of Forestry, Bogor Agricultural University. Bogor (2001)
12. Prakoso, N.I., Purwono, S., Rochmadi, AIP Conf. Proc. **1823**, 1 (2017)

Bioleaching of trace metals from coal ash using local isolate from coal ash ponds

Denvert Pangayao[1,2,*], *Susan* Gallardo[2], *Michael Angelo* Promentilla[2], and *Eric* van Hullebusch[3]

[1]Chemical Engineering Department, Pamantasan ng Lungsod ng Maynila, Manila, Philippines
[2]Chemical Engineering Department, De La Salle University, Manila, Philippines
[3]Department of Environmental Engineering and Water Technology, UNESCO-IHE Institute for Water Education, Delft, The Netherlands

Abstract. Bioleaching of chromium, copper, manganese and zinc from coal ash were investigated using isolates from coal ash ponds particularly *Psuedomonas spp*. Six (6) different coal ash ponds were examined however, after initial screening *Psuedomonas spp.* were only present in three (3) coal ash ponds. Among the three coal ash ponds, results showed that eight (8) putative *Pseudomonas spp.* isolates were present that were identified using the Polymerase Chain Reaction (PCR). Using the eight putative *Pseudomonas spp.* for bioleaching at optimum conditions and 15 days, the pH value ranges from 8.26 to 8.84 which was basic in nature. Moreover, the maximum metal leached were 8.04% Cr, 12.05% Cu, 4.34% Mn and 10.63% Zn.

1 Introduction

Coal is the most widely used and cheap source of fossil resource on earth. The global consumption of coal has increased from 4.92 billion tons in 2002 to 7.43 billion tons in 2012. Coal has been used as a solid fuel to produce heat and electricity via combustion in power plants [1-3]. Upon combustion of coal, 80% is composed of fly ash and 20% is composed of bottom ash. In a global perspective, 40 to 45% of the electric power generated came from thermos power plants amounting to 600 million tons of coal ash are produce annually [2-4]

As of 2016, Philippine's coal consumption is 20.296 million metric tons [5] and more than 47% of the Philippine's power generation came from coal [6]. With the rapid increase in power demand, high quantities of coal ash are being discharge daily. Majority of the coal ash are being disposed in ash ponds and lagoons that can pose environmental and health risks [4, 7-9]. Due to massive amount of coal ash generated, it is considered as secondary material for different applications such as brick production, mine filling, drainage media, replacement for conventional soil and gravel for road side embankment, road construction, structural fill and various cement-based composites including concrete and mortar [10-12].

In the recent years, green technologies are significant in addressing issues on environmental pollution such as biological leaching or bioleaching [10,13]. Bioleaching is an ecological friendly process of using microorganism to extract soluble material from secondary resources via the production of organic or inorganic acids [13-15].

This research used indigenous microorganism from coal ash pond particularly Pseudomonas spp. Pseudomonas spp. is the second most used bioleaching microorganism due to its production of organic acid such as gluconic, oxalic and citric acid [16-17].

The main objective of this research to determine the maximum percent metal leached from coal ash using an indigenous microorganism from coal ash pond under optimum conditions. Moreover, the identified microorganism was Pseudomonas spp., heterotrophic bacterium that produces organic acid. The metals under investigation are chromium, copper, manganese and zinc

2 Materials And Methods

2.1 Coal ash from coal ash pond

A mixture of fly and bottom ash was used in the experiment collected from the ash pond. The coal ash was provided by the private coal fired power plant that uses local and imported coal. Moreover, the power plant uses Circulating Fluidized Bed (CFB) Boiler. The coal ash was dried at 105°C for 24 hours, then sieved passing through particle size 0.075 microns (#200 mesh). The particle size was constant throughout the experiment. The ash is alkali in nature with pH of 10-11.5.

2.2 Determination of Microbial Population and Isolation of the Microorganism from Coal Ash Pond

2.2.1 Coal Ash sample from the Philippines

The ash samples were collected from the different coal power plant in the Philippines namely : PP1 from the partner power plant, PP 2 and PP3 from Central Luzon, PP 4 and PP 5 from Southern Luzon and PP 6 from Visayas. All samples were collected from the oldest part

* Corresponding author: dcpangayao@plm.edu.ph

of the ash pond, one (1) meter below. And stored in a sterile container.

The screening and identification of the microbial population was performed by the Microbiology Department of University of the Philippines – Los Baños (UPLB). Moreover, the

Also, the preservation of the isolated microorganism was done through lyophilization by UPLB National Institute of Molecular Biology and Biotechnology.

2.2.2 Determination of the Microbial Population of the Coal Ash

The initial screening of total heterotrophic bacteria, yeast and molds and detection of *Pseudomonas spp.* was conducted. Initially all coal ash samples were basic in nature with a pH value of 9 to 11. The following procedures were done: (1) heterotrophic bacterial count: plate count agar with 1,000 units/ml nystatin; (2) molds and yeast count: potato dextrose agar with 100 ppm streptomycin and 50 ppm rose bengal, adjusted to pH 9.0 with 5.0 M NaOH and (3) detection of *Pseudomonas spp.* using most probable number (MPN) method on Arginine broth. Heterotrophic bacteria and *Pseudomonas spp.* were counted after 3 days of incubation while yeast and molds were counted after 5 days.

2.2.3 Isolation and Purification of the Microorganism

Based on literature, *Pseudomonas spp.* is the widely used heterotrophic bacteria for bioleaching. This microorganism was isolated in the coal ash samples. The isolated *Pseudomonas spp.* were preserved using lyophilization at UPLB National Institute of Molecular Biology and Biotechnology.

2.3 Polymerase Chain Reaction (PCR) amplification of DNA and DNA Sequencing

The identification of the specific *Pseudomonas* specie was done using TaKaRa PCR thermal Cycler Dice TP600 at the Nakasaki Laboratory of Tokyo Institute of Technology (Japan). The extracted DNAs were amplified using the primers 16SF (5'AGAGTTTGATCCTGGCTCAGGA3') and 16SR (5'GGTTACCTTGTTACG3'). The detected genome sequence was matched using Basic Location Alignment Search Tool (BLAST) database provided by the National Center of Biotechnology and Information (NCBI).

2.4 Bioleaching Experiment

2.4.1 Reviving the Pseudomonas spp

The ampules of the lyophilized *Pseudomonas spp* were opened using a metal file. Then, 1 ml of sterile water was added to the ampule and was swirled and gently tapped for 5 to 10 seconds. The liquid was transferred in a 100 ml fresh nutrient broth media. The *Pseudomonas spp* was incubated at 37°C at 90 rpm for 7 days [18-20].

2.4.2 Growth of Pseudomonas spp

The incubated *Pseudomonas spp* were streaked in a nutrient agar plates. The nutrient agar plates were placed in an incubator at 37°C for 3 days. After which, one colony was place in an inoculating looped then transferred to 100 ml fresh nutrient broth media. *Pseudomonas spp.* was cultured in a Nutrient broth medium containing of 1 g/L yeast, 2 g/L peptone and 5 g/L glucose. 8 grams of nutrient broth medium was dissolved in 1 liter of deionized water. The solution was autoclaved at 121°C for 15 minutes [21]. The resulting solution has a pH of 7.2. Then the inoculum was placed in a water bath shaker (Visio VS-1205SW1) at 90 rpm and 37°C. *Pseudomonas spp* were incubated for 5 days prior to bioleaching [18-20].

2.5 Bioleaching experiment using *Pseudomonas spp*

The bioleaching experiments were conducted in a 250 mL Erlenmeyer flask with the following conditions: 1% pulp density (1 g of ash per 100 ml medium), 90 rpm and 37°C. The solution were autoclaved at 121°C for 15 minutes and cooled. 5 ml of the inoculum was placed in a 100 ml fresh medium with the ash. The bioleaching experiment was for 15 days [22]. The extracted solution was filtered using a 0.45 micrometer cellulous acetate filter.

2.6 Analytical Methods

X-ray Diffraction (XRD) analysis was performed using Bruker D8 Advance diffractometer equipped with copper radiation Sol-X detector (CuKα, λ = 0.15406 nm). The data were recorded with a 1s step time and 0.02° scan step between 2° to 80°. On the other hand, X-ray Fluorescence (XRF) analysis was conducted using Pananalytical Xflourescence spectrometer equipped with 150 eV (Mn Ka) resolution Energy Dispersive Minipal 4 tube (Rh X Ray tube – 30 kV –9W). All samples were dried at room temperature.

Also, metal analysis was done using Shimadzu AA-6300 AAS analysis. Different wavelength were used namely: 267 nm for Cr, 324 nm for Cu, 257 nm for Mn and 213 nm for Zn were used to detect the concentration of the trace metals in the solution, reported as mg/kg or ppm. To preserve the sample, few drops of concentrated nitric acid was added.

3 Results and Discussion

3.1 Mineralogical composition: XRD

Fig. 1 shows the mineral composition of the coal ash namely: quartz-SiO_2, lime-CaO, anhydrite, tricalcium aluminate, periclase, melilite, mullite and rutile. These minerals were also identified by the different researchers [23-29].

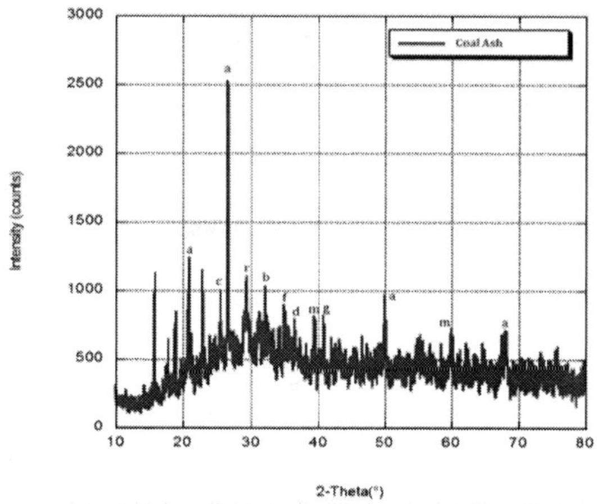

Fig. 1. XRD Analysis for Coal Ash (a: Quartz-SiO2, b: Melilite, c: Anhydrite, d: Tricalcium aluminate, f: Lime-CaO, g: Periclase, m: Mullite, r: Rutile).

3.2 Elemental Composition: XRF

The chemical analyses of the ash samples were performed to determine the composition of the material. Table 1 shows the major and trace elemental composition.

Table 1. XRF analysis of ash samples.

Element	% by weight
Calcium	39.46
Silicon	25.70
Aluminum	10.62
Iron	6.20
Sulfur	5.98
Sodium	3.99
Chlorine	2.11
Potassium	1.83
Magnesium	1.67
Titanium	1.51
Strontium	0.43
Barium	0.22
Zirconium	0.05
Tin	0.035
Vanadium	0.034
Chromium	0.033
Nickel	0.025
Manganese	0.023
Tellurium	0.021
Copper	0.020
Zinc	0.011
Bromine	0.011
Yttrium	0.007

The highest element is calcium which may be originated on the limestone that was being added during the combustion process [24,28,30]. The minerals identified by XRD coincide with the chemical results of the main composition obtained by XRF analysis (Table 2). Moreover, the total metal contents in the ash are 0.96 mg/kg Zn, 4.13 mg/kg Mn, 0.81 mg/kg Cr and 2.22 mg/kg Cu.

3.3 Microbial Population Determination in Coal Ash Samples

Microbial populations were analyzed to identify the possible microorganism that can be used for bioleaching of coal ash. Microbial count was determined by Microbiology Department of UPLB. Based on literature, other than *Acidithiobacillus spp.*, *Pseudomonas spp.* was considered as a possible microorganism for bioleaching [31-35]. Table 2 shows bacterial and molds count and putative *Pseudomonas spp.* present in the coal ash sample.

Table 2. Microbial Count Summary

Power Plant	**Bacteria**	**Molds**	**Putative *Pseudomonas* species**	**Number of *Pseudomonas* isolate**
PP 1	< 2500 cfu/g	-----	< 1 MPN/g	No isolate
PP 2	3.0×10^4 cfu/g	2.0×10^4 cfu/g	< 1 MPN/g	No isolate
PP 3	4.9×10^5 cfu/g	3.3×10^3 cfu/g	233 MPN/g	4 isolate
PP 4	4.9×10^6 cfu/g	1.3×10^5 cfu/g	33 MPN/g	4 isolate
PP 5	2.2×10^3 cfu/g	2.4×10^4 cfu/g	< 1 MPN/g	No isolate
PP 6	2.1×10^5 cfu/g	4.5×10^2 cfu/g	4 MPN/g	1 isolate

cfu: colony forming unit; MPN: Most Probable Number

Table 2 shows that 9 possible Pseudomonas spp. isolate were identified. There are 4 isolates from PP 3 and PP4 and 1 isolate from PP 6. On the other hand, no isolate were detected on PP 1, PP 2 and PP 5.

Table 3. PCR result of *Pseudomonas spp*

Samples	**Most closely related species (Identity: 100%)**
PP 3-1	***P. aeruginosa***, *P. alcaligenes, P. citronellolis, P. delhiensis, P. denitrificans, P. guezennei, P. knackmussii, P. multiresinivorans, P. nitroreducens, P. resinovorans, P. stutzeri*
PP 3-2	***P. putida***, *P. indoloxydans, P. plecoglossicida, P. pseudoalcaligenes*
PP 3-3	***P. aeruginosa***, *P. alcaligenes, P. citronellolis, P. denitrificans, P. knackmussii, P. resinovorans, P. stutzeri*
PP 6- 1	***P. aeruginosa***, *P. alcaligenes, P. citronellolis, P. delhiensis, P. denitrificans, P. guezennei, P. knackmussii, P. multiresinivorans, P. nitroreducens, P. resinovorans, P. stutzeri*
PP 4-1	***P. aeruginosa***, ***P. putida***, *P. alcaligenes, P. brassicacearum, P. denitrificans, P. entomophila, P.monteilii, P. helmanticensis, P.knackmussii, P. lini, P. poae, , P. resinovorans, P. stutzeri, Azotobacter vinelandii*
PP 4-2	***P. aeruginosa***, *P. alcaligenes, P. citronellolis, P. delhiensis, P. denitrificans, P. guezennei, P. knackmussii, P. multiresinivorans, P. nitroreducens, P. resinovorans, P. stutzeri*
PP 4-3	***P. aeruginosa***, *P. alcaligenes, P. citronellolis, P. delhiensis, P. denitrificans, P. guezennei, P. knackmussii, P. multiresinivorans, P. nitroreducens, P. resinovorans , P. stutzeri*
PP 4-4	***P. aeruginosa***, *P. alcaligenes, P. citronellolis, P. delhiensis, P. denitrificans, P. guezennei, P. knackmussii, P. multiresinivorans, P. nitroreducens, P. resinovorans , P. stutzeri*

P: *Pseudomonas*

Pseudomonas spp. were identified using Polymerase Chain Reaction (PCR). Table 3 shows the putative *Pseudomonas spp.* present in the different coal ash.

According to the PCR result, putative *Pseudomonas spp.* isolates are the same for PP 3-1, PP 3-3, PP 4-2 to PP 4-4 and PP 6-1. However, PP 3-2 and PP 4-1 has a different consortium of *Pseudomonas spp. Pseudomonas aeruginosa* is present in 7 samples (PP 3-1, PP 3-3, PP 4-1 to PP 4-4 and PP 6-1) while *Pseudomonas putida* is only present in PP 3-3. In previous researches, these microorganisms were used in bioleaching for extracting metals from secondary resources. These different *Pseudomonas spp.* isolate were used in the bioleaching coal ash. Figure 2 shows the streak plate of the 8 isolate samples.

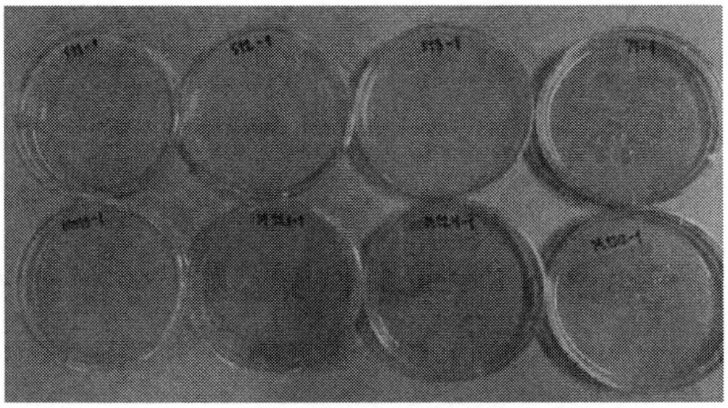

Fig. 2. Different *Pseudomonas spp* isolates (From L to R and T to B: PP 3-1, PP 3-2, PP 3-3, PP 6-1, PP 4-1 to PP 4-4)

3.4 Bioleaching using Pseudomonas spp.

After 15 days of bioleaching experiment, Table 4 shows the percent metal leached for chromium, copper, manganese and zinc for each *Pseudomonas spp.* isolates. Moreover, Figure 3 illustrate the different pH profile for the different *Pseudomonas spp.* isolates.

Table 4. % Metal Leached using *Pseudomonas spp* isolates

Isolate	% Zn	% Cr	% Mn	% Cu
PP 3-1	**10.63**	7.52	**4.34**	10.10
PP 3-2	10.39	**8.04**	2.75	**12.05**
PP 3-3	9.19	8.03	3.90	8.87
PP 6-1	9.61	5.83	3.19	9.47
PP 4-1	9.03	5.99	2.35	9.38
PP 4-2	8.34	7.27	4.14	8.98
PP 4-3	10.41	6.63	3.70	10.24
PP 4-4	9.52	7.90	3.94	11.09

Table 4 shows that the maximum metal leached for chromium is 8.04% and copper is 12.05% from PP 3-2 and 4.34% for manganese and 10.63% for zinc from PP 3-1. Based on the PCR result, PP 3-2 contains *Pseudomonas aeruginosa* and PP 3-1 contains *Pseudomonas putida* in the consortium.

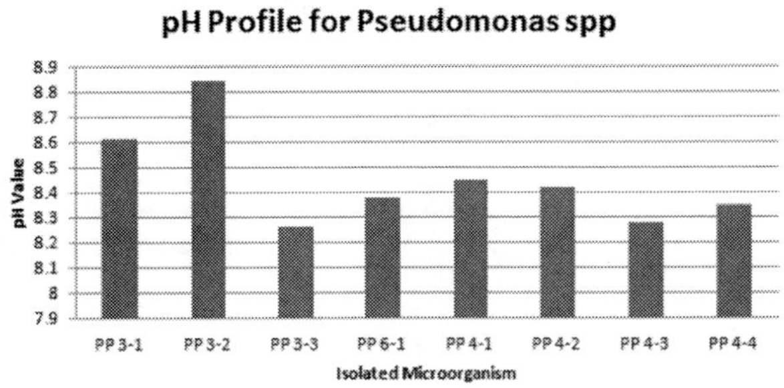

Fig. 3. pH values against the *Pseudomonas* spp isolate

Figure 3 shows the pH values of the *Pseudomonas spp.* isolate after the bioleaching experiment. The pH value of the solution ranges from 8.26 to 8.84 which is basic. However, taking into consideration the pH of the ash which is 11, the decrease in the pH value is an indirect indication of possible formation of organic acid such as oxalic, gluconic or citric acid by the *Pseudomonas spp.* [18-19, 35-36]

Based on literature, bioleaching using *Acidithiobacillus spp.* extracted more metal as compared to *Pseudomonas spp.* The microbial activity of *Pseudomonas spp.* is in the circumneutral pH value of 7.8, because of this the efficiency of metal extraction is lower compared to low pH value. It should be noted that metal extraction is favorable under acidic conditions. Most metals like chromium, copper, manganese and zinc favor metal extraction at lower pH. On the other hand, production of extracellular polymeric substance (ESP) enhances metal solubilization using *Pseudomonas spp.* [31, 37].

4. Conclusion

Bioleaching of coal ash using *Pseudomonas spp.* isolate from coal ash pond was investigated. Based on the result, the total metal content of the coal ash are 0.81 mg/kg Cr, 2.22 mg/kg Cu, 4.13 mg/kg Mn and 0.96 mg/kg Zn. Moreover, the minerals identified were quartz-SiO_2, lime-CaO, tricalcium aluminate and the major elements were calcium, silicon and aluminum. Also, high calcium content results in the alkalinity of the ash sample.

From the initial screening and identification 8 putative *Pseudomonas spp.* were obtained. PCR result shows that 6 isolate has the same *Pseudomonas spp.* consortium namely: PP 3-1, PP 3-3, PP 4-2 to PP 4-4 and PP 6-1. On the other hand, PP 3-2 and PP 4-1 have different consortium. Pseudomonas aeruginosa is present in 7 samples namely (PP 3-1, PP 3-3, PP 6-1 and PP 4-1 to PP 4-4) and Pseudomonas putida in PP 4-2.

After 15 days of bioleaching experiment, the maximum metal extracted were 12.05% Cu and 8.04% Cr from PP 3-2 isolate and 10.63% Zn and 4.34% Mn from PP 3-1 isolate

This research would not be possible without the help of the following institution namely: Engineering Research and Development for Technology (ERDT) – Department of Science and Technology (DOST), Philippines, Campus France, Republic of France, Japan International Cooperation Agency (JICA) Project Office for the ASEAN University Network/Southeast Asia Engineering Education Development Network (AUN/SEED-Net) under the Collaborative Research Program for Common Regional Issues (CRC) with De La Salle University. Furthermore, we would like to thank the partner coal fired power plant, Tokyo Institute of Technology, Tokyo, Japan and Ile de France Region for the ICP-OES analysis.

6. References

1. Y. Deng, Z. Zhang, C. Liu, W. Liu, Y. Cui, X. Du, D. Xu, H. Guo, App. Ener., **200** (2017)
2. J. Xu, J. Dai, H. Xie, C. Lv, C., Ener. Pol. **109** (2017)
3. P. A. Strizhak, M. A. Dmitrienko, Sci. of Tot. Env., **598** (2017)
4. W-S. Ahn, Y-R. Lee, J. T. Soe, S. Zhang, J-W. Ahn, M. B. Park, Chem. Eng. J., **317** (2017)
5. www.doe.gov.ph/overall-statistics-on-coal
6. www.doe.gov.ph/philippine-power-statistcs
7. N. Makul, G. Sua-lam, J. of Clean. Prod., **100** (2017)
8. Y-S. Kim, M. T. Do, H-K Kim, G. Kang, Construc. and Build. Mat., **124** (2016)
9. E. Connors, Utilities Pol., **34** (2016)
10. N. Gupta, V. Gedam, C. Moghe, P. Labhasetwar, Env. Tech. and Innov., **7** (2017)
11. H-K. Kim, Construc. and Build. Mat., **91** (2015)
12. M. L. D. Jayaranjan, E. D. van Hullebusch, A. P. Annachhatre, Rev. in Env. Sci. & Biotech., **13(4)** (2014)
13. M. Diao, E. Taran, S. Mahler, A. V. Nguyen, Adv. in Col. and Interface Sci., **212** (2014)
14. J-C. Lee, B. D. Pandey, Wast. Mang., **32** (2012)
15. F. Bakhtiari, H. Atashi, M. Zivdar, S. Seyedbagheri, M. H. Fazaelipoor, J. of Indus. and Eng. Chem., **17** (2011)
16. S. K. Sen, M. M. Das, P. Bandyopadhyah, R. R. Dash, Eco. Eng., **88** (2016)
17. W. A. G. Hassanien, O. A. N. Desouky, S. S. E. Hussein, Walailak J. of Sci. and Tech., **11** (2013)
18. R. Jujun, Z. Xingjion, Q. Yiming, H. Jian, Wast. Mang., **34** (2014)
19. R. Singh, N. R. Bishnoi, A. Kirrolia, R. Kumar, Biores. Tech., **127** (2013)
20. M. I. Husseiny, M. Abd El-aziz, Y. Badr,M. A. Mahmoud, Spectrochim. Act. Part A, **67** (2007)
21. S. K. Sen, M. M. Das, P. Bandyopadhyah, R. R. Dash, Eco. Eng., **88** (2016)
22. S. M. J. Mirazimi, F. Rashchi, 7th Inter. Chem. Eng. Cong. & Exhib., Kish, Iran, Novmber 21-24, 2011.
23. G. Neupane, R. Donahoe, Fuel, **104** (2013)
24. G. Akar, M. Polat, G. Galecki, U. Ipekoglu, Fuel Process. Tech., **104** (2012)
25. N. Koukouzas, C. Ketikidis, G. Itskos, Fuel Process. Tech., **92** (2011)
26. S. Hareeparsad, L. Tiruta-Barna, C. Brouckaert, C. Buckley, J. of Haz. Mat., **186** (2011)
27. M. Ahmaruzzaman, Prog. in Ener. and Comb. Sci., **36** (2010)
28. G. Skordas, P. Grammelis, M. Prokopidou, E. Kakaras, G. Sakellaropoulos, Fuel, **88** (2009)
29. U. Demir, A. Yamik, S. Kelebek, B. Oteyaka, A. Ucar, O. Sahbaz, Fuel, **87** (2008)
30. B. M. Steenari, S. Schelander, O. Lindqvist, Fuel, **78** (1999)
31. A. Potysz, P. N. L. Lens, J. van de Vossenberg, E. R. Rene, M. Grybos, G. Guibaud, J. Kierczak, E. D. van Hullebusch, Appl. Geochem., **68** (2016)
32. S. M. J. Mirazimi, Z. Abbasalipour, F. Rashchi, J. of Env. Mang., **153** (2015)
33. M. A. Shabani, M. Irannajad, A. R. Azadmehr, M. Meshkini, M. Intern. J. of Min., Metal. and Mat., **20** (2013)
34. V. A. Pham, Y. P. Ting, Adv. Mat. Res., **71-73** (2009).
35. G. Aouad, J-L. Crovisier, V. A. Geoffroy, J-M. Meyer, P. Stille, E. Hutchens, J. Mutterer, Sci. of the Tot. Env., **393** (2008)
36. R. Andreazza, S. Pieniz, L. Wolf, M-K. Lee, F. A. O. Camargo, B. C. Okeke, Sci. of the Tot. Env., **408** (2010)
37. M. Izquierdo, X. Querol, Intern. J. of Coal Geol., **94** (2012)

Synthesis of Green Diesel From Waste Cooking Oil Through Hydrodeoxygenation Technology With NiMo/γ-Al_2O_3 Catalysts

Heri Heriyanto [1]), Sumbogo Murti, SD[2], Septina Is Heriyanti[2], Inayatu Sholehah[1], Ayi Rahmawati[1]*

[1]Chemical Engineering Departement, Sultan Ageng Tirtayasa University, Cilegon, Indonesia, Jl. Jendral Sudirman km. 03 Cilegon 42435
[2]Center of Technology Energy Resources and Chemical Industri, Agency for the Assessment and Application of Technology (BPPT), Gedung Energi 625 Klaster V, Kawasan PUSPIPTEK, Tangerang Selatan 15314

Abstract. Hydrodeoxygenation (HDO) of waste cooking oil (WCO) and trapped grease over sulfide catalysts was examined to produce high quality transportation fuel from low-grade resources. The hydrodeoxygenation of waste cooking oils was carried out in a high pressure of 30 and 60 bar and high temperature of 300 – 400 ^{0}C in a batch reactor autoclave. NiMo/γ-Al_2O_3 catalyst was prepared and for the first time tested in hydroprocessing of waste cooking oil. The content of NiMo/γ-Al_2O_3 in each catalyst was about WCO 5 wt.%. A maximum of 77,97 % green diesel yield was achieved at nearly complete conversion of waste cooking oil using NiMo/γ-Al2O3 at temperature of 400^{0}C, pressure 60 bar and 4 hours of reaction time. The oxygen content was decreased from 14,25 wt.% to 13,35 wt.%, at temperature of 400^{0}C, pressure of 30 bar and 1 hour of reaction time. The Hydrodeoxygenation process was much influenced by temperature, pressure, and time.

1 Introduction

Today energy needs continue to rise but inversely proportional to energy substitutes such as petroleum, natural gas that can not be optimized. The amount of available energy sources currently is sourced from fossil fuels accounted for 94% of the national energy mix mix, comprising 47% of petroleum, 26% coal and 21% natural gas [1]. With an economic growth of 6.3 to 6.8 per year, energy demand is projected to grow around 6% per year until 2014. Fuel oil will continue to be prefered in the final energy demand, because in addition to the fact that fuel-based technologies are more efficient, fuel prices can still continue to compete if compared to other fuels such as coal and natural gas. More over, fuel oil still dominates the use of final energy, especially in the transportation sector, because the use of fuel more practical and efficient and can not be replaced by other fuels.

Considering Indonesia is the largest palm oil producer in the world with oil palm production in 2014 reaching 33 million tons so that the potential of cooking oil produced is quite large [2]. In 2007 used cooking oil (MGB) from various sources as much as 3.9 tons has not been fully utilized. This gives a huge opportunity to make Indonesia as a country that has a strong energy resistance by processing it into greendiesel with hydrodeoxygenation method, in addition to producing high fuel quality compared to biodiesel and petrodiesel, also environmentally friendly [3].

2 Literature Review

2.1 Waste Cooking Oil

Waste cooking oil is waste oil that can be made from various types of cooking oil such as corn oil, vegetable oil, samin oil and so on which when viewed from the chemical composition of the compounds are carcinogenic. In addition to the environmentally destructive wastes the chemical content formed can be harmful to human health. The first sign of cooking oil damage is the formation of acrolein in cooking oil. This acrolein causes itching in the throat when eating fried foods using cooking oil repeatedly. Acreolyn is formed from the hydration of glycerol that forms an unsaturated aldehyde.

2.2 Green Diesel

Green diesel is a straight-chain hydrocarbon fraction similar to that produced by the hydrogenation of triglycerides from vegetable oils [4]. The injection of hydrogen with the support of a bimetallic catalyst for the hydrotreating reaction of the triglyceride compound is capable of breaking the ester or carboxylic group bond of the glycerol group [5,6]. Then followed the breaking of carboxylic groups from the existing fatty acid bonds as well as the saturation reaction of the double chain carbon bonds into straight chain hydrocarbon compounds [7, 8, 9].

2.3 Hydrogenation and Hydrodeoxygenation

Hydrogenation technology or Hydotreatment is a branch of hydroprocessing such as catalytic reactions using hydrogen to eliminate the atoms heterogeneous (heteroatoms) such as sulfur,

* Corresponding author: herfais@yahoo.com

nitrogen, and oxygen, each of which is contained in the compound hydrogen sulfide (H_2S), ammonia (NH_3) and water (H_2O) in a process feedstock [10, 11]. Hydrogenation process usually followed by hidrodeoxygenation reaction (HDO), decarboxylation (DCO), and decarbonylation (DCN) or combination of them [11].

Hydrodeoxygenation (HDO) is a process for eliminating oxygen hydrogenolysis of a material by cutting the carbon-oxygen bond using hydrogen gas [12, 13]. The content of oxygen in organic compounds has a very important role in estimating the characteristics of the fuel to be generated. Oxygen levels in vegetable oils that can reach more than 50% by weight, have an impact on the low calorific value, low thermal stability, corrosive, can not be mixed with fossil fuels and the increasing tendency towards polymerization reaction [14].

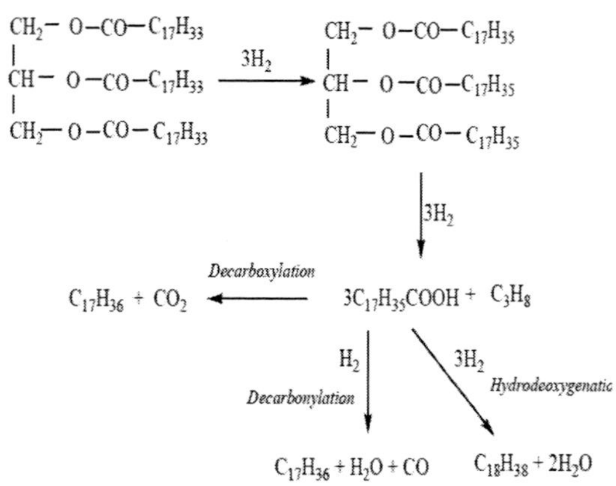

Fig. 1. Reaction pathway of hydrotreatment

3 Methodology

The procedure in this study was including 10 grams of NiMo/γ-Al_2O_3 catalyst passed 200 mesh, 3 grams of sulfur powder and 200 grams of waste cooking oil into the autoclave reactor. Then HDO processes were varied on time, temperature, pressure, and number of reaction stage with rotation speed of 800 rpm. After the HDO process was finished, the product was filtered off and separated between liquid and gas for analysis step. However, in the 2-stage, after the first stage was finished, gas in the reactor was replace by fresh hydrogen, then the reaction was continuoued in the second stage with the same pressure and time variations but with different temperatures. After the reaction process was completed, product gas was analyzed using GC-TCD Shimadzu GC-2014. The liquid product was separated from solid material by filtering. Liquid product was analized using GC-FID. More over the physical and chemical properties of liquid product was also tobe analyzed in accordance to ASTM Standard such as viscosity, flash point and elemental analysis.The operating conditions of each experiment were followed below:

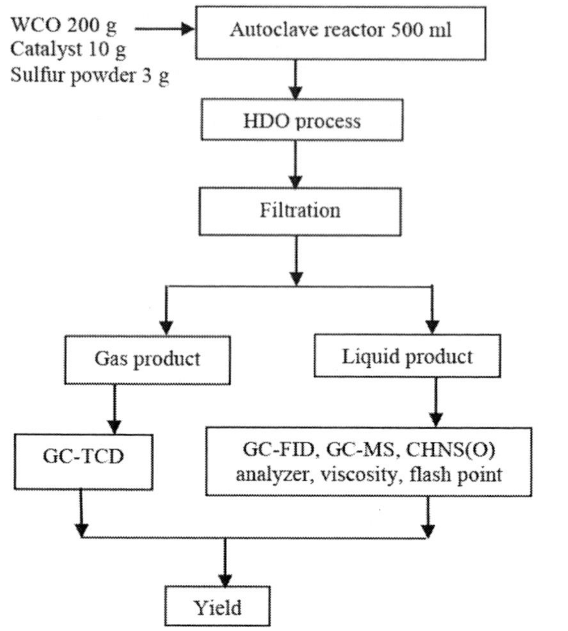

(a)

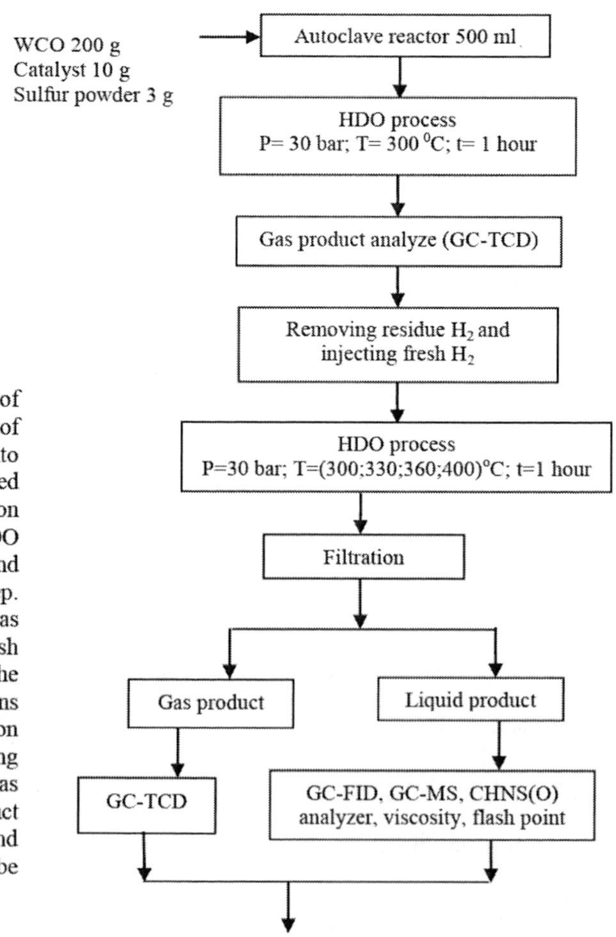

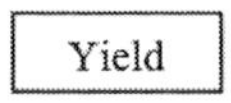

(b)

Fig. 2. Green Diesel Synthesis Scheme (a) Single reaction (b) Two steps reaction

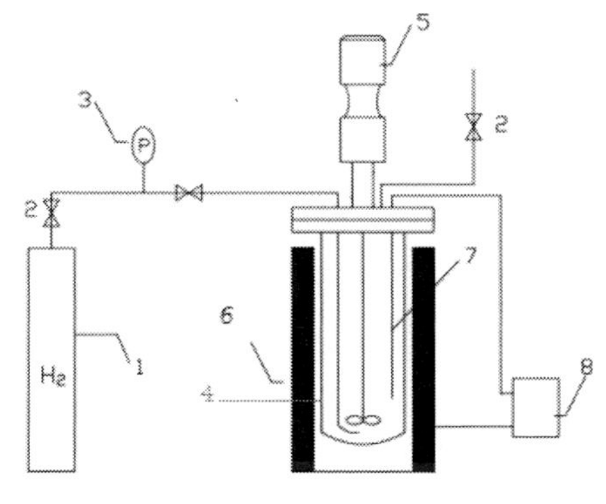

1-steel bottle; 2-valve; 3-pressure gauge; 4-reaktor;5-electrical machine; 6-jacket; 7-thermocouple; 8-controller

Fig. 3. Schematic and Equipments of Experimental

Table 1. Operating Conditions

Run	Operating Condition						
	Stage amont	Temperature (°C)		Initial pressure		Time (hour)	
		T1	T2	P1	P2	t1	t2
1	1	300	-	30	-	1	-
2	1	330	-	30	-	1	-
3	1	360	-	30	-	1	-
4	1	400	-	30	-	1	-
5	1	300	-	60	-	1	-
6	1	330	-	60	-	1	-
7	1	360	-	60	-	1	-
8	1	400	-	60	-	1	-
9	2	300	300	30	30	1	1
10	2	300	330	30	30	1	1
11	2	300	360	30	30	1	1
12	2	300	400	30	30	1	1
13	1	300	-	60	-	4	-
14	1	400	-	60	-	4	-
15	1	300	-	30	-	4	-
16	1	400	-	30	-	4	-

4 Discussion

4.1 Pressure and Temperature Profile

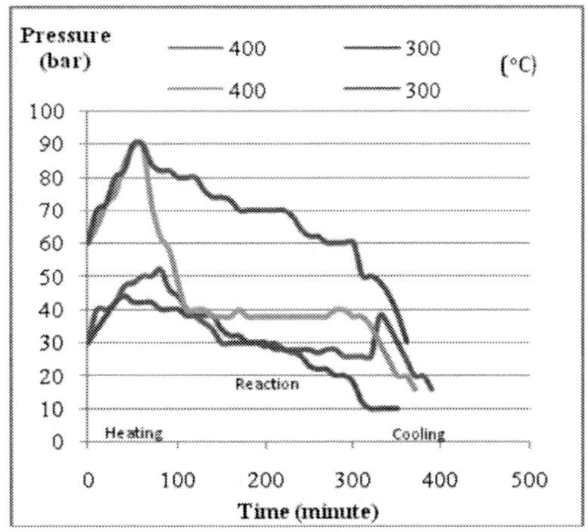

Fig. 4. Pressure profile on the time

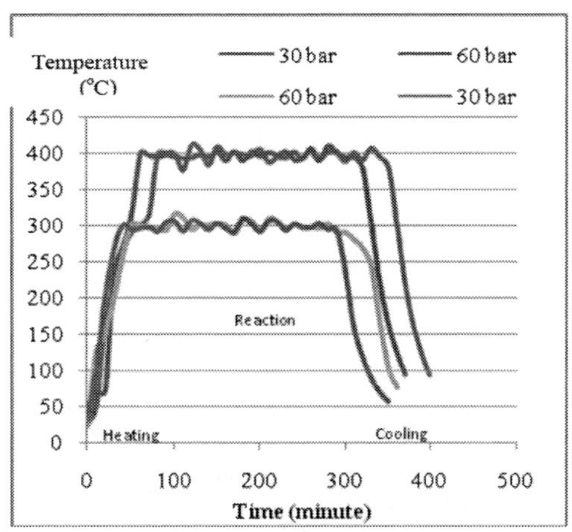

Fig. 5 Temperature Profile on The Time

The pressure and temperature profile of the above reaction was observed at 4 hours reaction time. By controlling the temperature of the coil heater in the reactor it was desirabled that the reaction temperature was immediately achieved and kept constant during the reaction. The observation of the condition of the process pressure was seen fluctuating even pressure drop on the graph above. This pressure drop phenomenon follows the Joule Thompson Effect theory. The hydrogen gas stored in a guarded tube at a pressure of about 200-250 bar to remain in the gas phase was then injected into the reactor at a pressure of 30 or 60 bar. As a result of this pressure drop, the gas became inflate and the rapid

https://doi.org/10.1051/matecconf/201815603032

pressure drop was practically not allowing the occurrence of external heat propagation into the system and vice versa (Q = 0). As a result of this adiabatic development the gas underwent temperature changed both up and down (fluctuations).

4.2 Effect of Temperature on Green Diesel Yield

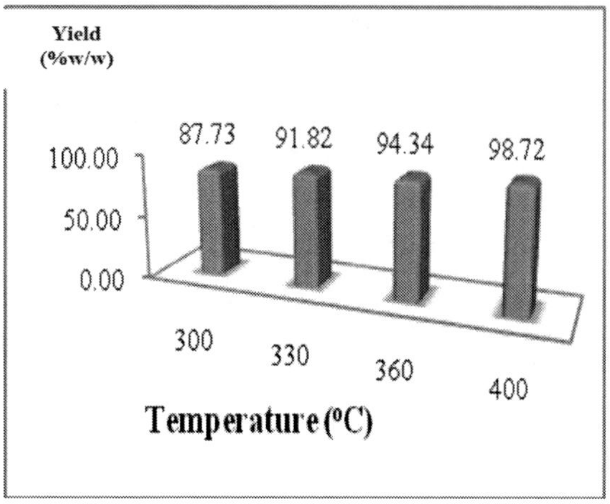

Fig. 6. Effect of Temperature on Green Diesel Yield

At the initial pressure condition of 30 bar and 1 hour reaction time, the highest result is obtained at 400°C operating temperature of 98.72% considering green diesel C_{13}-C_{22}. This is because when the temperature is increased then the cracking process will occur more easily and optimally. It is also caused by the addition of greater heat energy.

4.3 Effect of Initial Pressure Reaction to Green Diesel Yield

At 400 ° C and a reaction time of 1 hour, the green diesel produced at a pressure of 60 bar is lower than the pressure of 30 bar. It is predicted that at higher pressures the incorporation of alkyl in short carbon chains results in long-chain hydrocarbons C23 $^{+}$ which are not green diesel but it was a small part of the wax [15].

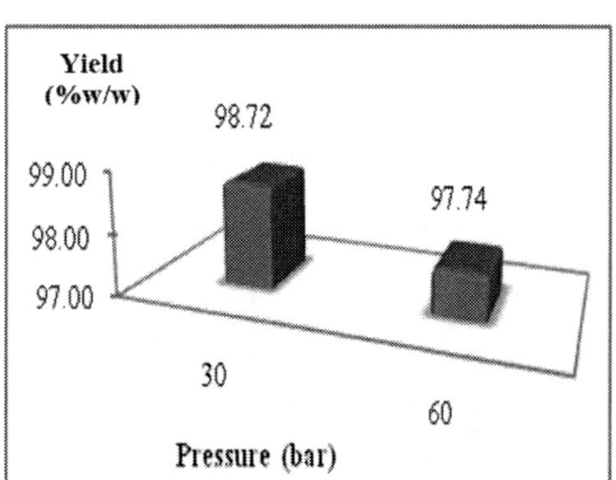

Fig. 7. Effect of Initial Pressure Reaction to Green Diesel Yield

4.4 Effect of Reaction Time to Green Diesel Yield

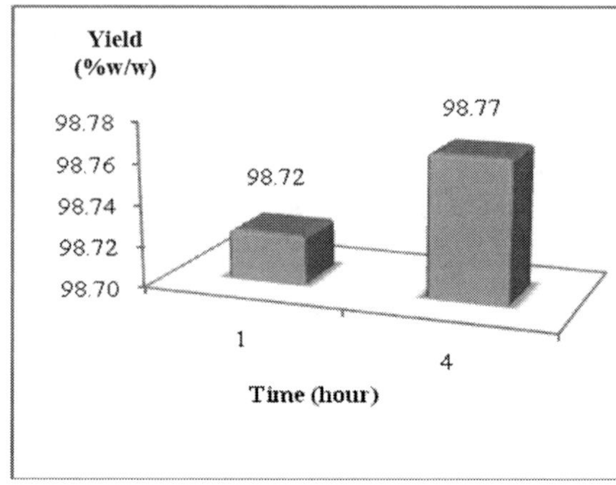

Fig. 8. Effect of Reaction Time on Green Diesel Yield

Results at 4 hours of operating time are higher than the operating time of 1 hour only. The longer the reaction time, the cracking process also occurs more optimally because the contact between reactants and catalysts longer to produce more products that also cause the green diesel increased.

4.5 Effect of Number of Stages of Reaction to Green Diesel Yield

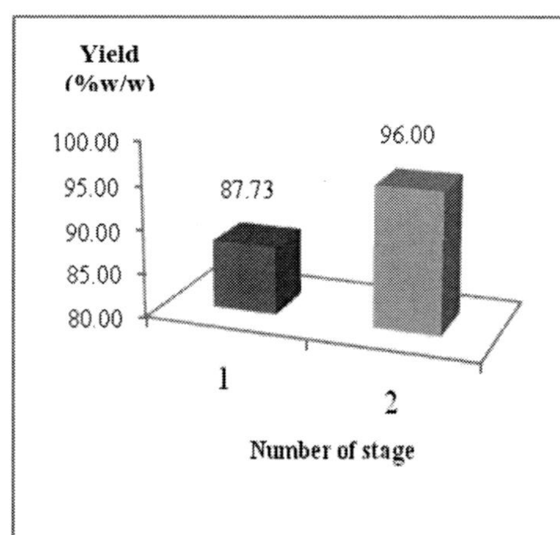

Fig. 9. Effect of Number of Stages of Reaction to Green Diesel Yield

Results at 2 stages are higher than the reaction result of only 1 stage. This is due to the cracking process at 2 stages occurs longer where every 1 stage is given 1 hour so that if 2 stage then happened 2 hour reaction of cracking. In addition, replacement with fresh hydrogen makes the cracking process more optimal.

4.6 Effect of Reaction Temperature on Decreasing Oxygen Content

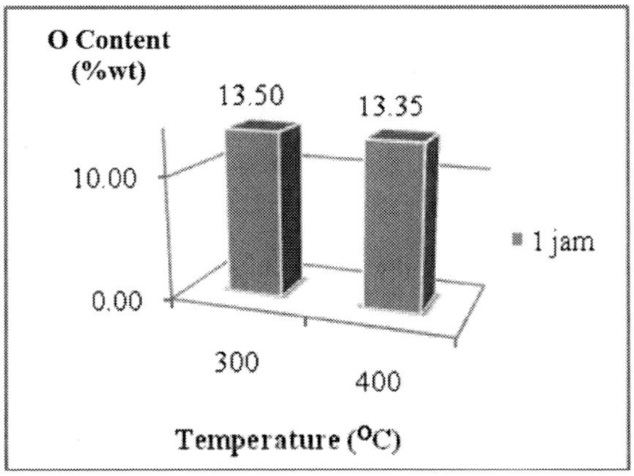

Fig. 10. Effect of Reaction Temperature on Decreasing Oxygen Content

In a reaction that lasted 1 hour, waste cooking oil which originally contained 14.25% oxygen can be reduced to 13.5 and 13.35% oxygen at reaction temperature of 300 and 400oC respectively. Although only less than 1% but with increasing reaction temperature, it is possible to remove oxygen contained in the fatty acid chain in waste cooking oil.

5 Conclusion and Suggestion

5.1 Conclusion

Based on the research that has been done, obtained some conclusions include:

1. The highest yield was obtained under operating conditions with 2 stage reaction, where T_1 = 300 °C, T_2 = 400 °C, $P_1 = P_2$ = 30 bar, $t_1 = t_2$ = 1 hour that was equal to 98,93%
2. Factors affecting green diesel yields were pressure, temperature, number of stages and time.
3. Oxygen levels decreased from 14.25% to 13.35% at T = 400°C, P_1 = 30 bar, t = 1 hour.

5.2 Suggestions

Suggestions that can be submitted for further research include:

1. In order to improve the quality of green diesel product produced, it is necessary to learn more about the hydrodeoxigenation operation condition made from used cooking oil, especially operating pressure and operating temperature need to be increased.
2. Should be further studied related to raw material pretreatment method, product separation, catalyst used and product treatment before analyzed in order to obtain more optimal result.
3. Need further study related to isomerization reaction to get more comprehensive green diesel result.
4. Required product quality analysis such as GCMS, calorific value and FTIR, cetane number measure the success of research that has been done.

Acknowledgments

Financial support from Center of Technology Energy Resources and Chemical Industry - Agency for The Assessment and Applycation of Technology for this research are gratefully acknowledged.

Reference

[1] BPPT, Outlook Energi Indonesia 2015, Jakarta, 2015

[2] Dewajani H., Prosiding Seminar Nasional Energi Alternatif, Politeknik Negeri Malang, (2008)

[3] Dewajani H. dan Budiman A., Prosiding Seminar Nasional Technoin UII, Yogyakarta, (2012)

[4] He Z. Wang X., Catal Sustain Energy (1) 28-52 (2012)

[5] Chew T.L., Bathia S. Bioresour, Technol (03) 009 (2008)

[6] Hilten R., Speir R., Kastner J,. and Das K.C., Bioresour Technol (102) 8288 – 8294 (1992)

[7] Idem R.O., Katikaneni S.P.R., and Bakhshi N.N. ,Energy & Fuels, 10 (6) 1150 – 1162 (1996)

[8] Ketaren S.D.G., Soares V.C.D., Ribeiro E.B., Carvalho D.A., Cardoso E.C.V., Rassi F.C., Mundim K., Rubim J.C., and Suarez P.A.Z., Jounal Anal. Appl.Pyrolysis (71) 987 – 996 (2004)

[9] Maher K.D., and Bressler D.C., Bioresour Technol (98) 2351-2368 (2007)

[10] Ma F., and Hanna M.A., Bioresour Technol (70) 1-15 (1999)

[11] Naveenji Arun, Rajesh V. Sharma Ajay K. Dalai, Renewable and sustainable energy (48) 240-255 (2015)

[12] Bendenzu S. Cid. R., Fierro J.L.G., and Agudo L., Appl.Catal. A: General (197) 47-

60 (2000)

[13] Mohammad M., Kandaramath T., Yaakob Z., Chandra Y., Energy Rev (22) 121-132 (2013)

[14] Perego Carlo, Biomass to Advanced Biofuel, (2015)

[15] Barton D.G., Soled S.L., and Iglesia E., Topics in Catalysis (6) 87 – 99 (1998)

The Relative Humidity Effect Of The Reactants Flows Into The Cell To Increase PEM Fuel Cell Performance

Mulyazmi[1], W. R W Daud[2], *Silvi Octavia*[1], *Maria Ulfah*[1]

[1]Department of Chemical Engineering, University of Bung Hatta Padang, West Sumatera, Indonesia
[2]Fuel Cell Institute, Universiti Kebangsaan Malaysia, 43600 UKM Bangi, Selangor, Darul Ehsan, Malaysia

Abstract. Design of the Proton Exchange Membrane (PEM) fuel cell system is still developed and improved to achieve performance and efficiency optimal. Improvement of PEM fuel cell performance can be achieved by knowing the effect of system parameters based on thermodynamics on voltage and current density. Many parameters affect the performance of PEM fuel cell, one of which is the relative humidity of the reactants that flow in on the anode and cathode sides. The results of this study show that the increase in relative humidity value on the cathode side (RHC) causes a significant increase in current density value when compared to the increase of relative humidity value on the anode side (RHA). The performance of single cells with high values is found in RHC is from 70% to 90%. The maximum current density generated at RHA is 70% and RHC is 90% with PEM operating temperature of 363 K and pressure of 1 atm

1 Introduction

Fuel cells are electro chemical devices that convert chemical energy from fuel (hydrogen) and oxidant (oxygen or air) directly into electrical energy and thermal energy, producing water as a by product [1-3]. One type of fuel cell is the PEM fuel cell, which can be operated at low temperatures to produce electrical energy. Developing the system performance of PEM fuel cells requires a suitable design and appropriate technology. The main purpose is to achieve optimal performance and obtain a low-cost PEM fuel cell system.

Many parameters in the PEM fuel cell system are associated with other parameters in the system. These parameters consist of the operating conditions, the design of the membrane electrolyte assembly (MEA) and the stack PEM fuel cell system [4-5]. In the case of the PEM fuel cells, the system is defined as a group of objects or materials that are integrated and operated together. The performance of the PEM fuel cells is evaluated in the form of the voltage generated, the power density and the efficiency. The optimal performance of the cells is influenced by many internal and external factors, such as the fuel cell design and assembly, the degradation of materials, the operating conditions and the impurities or the contaminants [6]. The design and optimization of a PEMFC system are very complex because they are related to the use of the system, the selection of appropriate components, and the functions of each variable. The purpose of studies on PEM fuel cell systems is to consider the capital and operational costs, efficiency, fuel consumption, flexibility and fuel cell life [7-10]. Other parameters such as the pressure, temperature, composition and utilization of the fuel, and composition and the utilization of oxidant can be varied simultaneously to achieve the desired operating point.

The main challenges facing the commercialization of PEM fuel cells include cost and reliability PEM fuel cell characteristics are highly influenced by the design of an optimal PEMFC system. The PEM fuel cell central system must be supported by providing an oxidant supply (oxygen or air), a fuel supply (hydrogen), heat and water management, process control, equipment and supervision [11-12]. The many failures in PEM fuel cell systems include flooding, the diffusion of water from the cathode side to the anode side and poor of integration heat and mass management during the operation of the system [13-15]. The focus of this research was to develop a fundamental thermodynamic basis for developing a modeling system to optimize the performance of PEM fuel cell systems. The next focus was to determine the relative humidity of the reactants entering the PEM fuel cell stack on the voltage and the current density that was produced by the resulting fuel cell system.

*Corresponding author: mulyazmi@yahoo.com

2 RESEARCH DESIGN

This research design is to determine the performance of a single cell PEM fuel cell based on the process performed by mulyazmi 2013 [16]. Single cell Of PEM fuel cell performance produced using an assumed reactant has passed the external humidifier.The determination of single cell of PEM fuel cell performance was determined by variating relative humidity on the anode and cathode sides, as well as validating results obtained from other studies.The operating parameters used as the basis for the modeling case are listed in Table 1.

Table 1. Basic modeling operation parameters

Parameters	Value
Cell Temperature	353
Thick dry membrane	0.0051 cm for Nafion 112
Catalyst load	0.0002 g cm^{-2}
Thick catalyst	0.0005 cm
Faraday Constant	96,485 C/mol elektron
Stoikhiometric ratio for hydrogen	1.2
Stoichiometric ratio for oxygen	2
Universal gas constant	8.314 J/mol K

3 Mathematics Models

The reaction that occurs in the PEM fuel cell system is:
Anode: $H_2 \rightarrow 2H^+ + 2e^-$
Chatode: $1/2O_2 + 2H^+ + 2e^- \rightarrow H_2O$
Overall: $H_{2(gas)} + 1/2\ O_{2(gas)} \rightarrow H_2O$ + heat + electric

3.1 Water Liquid in the reactant

The water vapor component with hydrogen as a reactant flowing in the PEM fuel cell is (Barbier, F. 2005):

$$\mathcal{N}_{H_2O in H_2} = \frac{S_{H_2} \cdot I}{2F} \cdot \eta_{cell} \cdot \mathcal{Y} \quad (1)$$

The molar humidity ratio of $\mathcal{Y}$ y is as follows::

$$\mathcal{Y} = \frac{\mathcal{N}_v}{\mathcal{N}_a} = \frac{mol\ of\ water\ vapor}{mol\ of\ dry\ gas} \quad (2)$$

The mass and molar relation of humidity is

$$\mathcal{Y}\frac{M_v}{M_a} = x \quad (3)$$

Mole fraction of water vapor can be interpreted as follows[17]:

$$y = \varphi \frac{P_{sat}}{P} \quad (4)$$

φ is relative humidity, P_{sat} is saturated vaporpressure (atm), and P is total pressure (atm).

The saturated vaporpressure p_{H2O} is as follows[18]:

$$P^o_{H_2O}(T) = 6.02724\ x\ 10^{-3} + 4.38484\ x\ 10^{-4}(T - 273.15) + 1.39844\ x\ 10^{-5}(T - 273.25)^2 + 2.71166\ x\ 10^{-7}(T - 273.15)^3 + 2.57731\ x\ 10^{-9}\ (T - 273.15)^4 + 2.82254\ x\ 10^{-11}(T - 273.15)^5$$

(5)

The molar ratio relationship $\mathcal{Y}$ with partial pressure is::

$$\mathcal{Y} = \frac{P_v}{P_a} = \frac{P_v}{P - P_v} \quad (6)$$

P_v is partial pressure of water vapor (atm), P_a is partial pressure gas (atm), P is total pressure.

Based on equation (6), then:

$$\mathcal{Y} = \frac{x.M_a}{M_w} \quad (7)$$

Then the equation becomes

$$\frac{P_v}{P - P_v} = \frac{x.M_a}{M_w} \quad (8)$$

Consequently:

$$\mathcal{Y} = \frac{\varphi.Pv_s}{P - \varphi.Pv_s} \quad (9)$$

So the amount of moles of water vapor in hydrogen is:

$$\mathcal{N}_{H_2O.in\ H_2} = \frac{S_{H_2} \cdot I}{2F} \cdot \eta_{cell} \cdot \frac{\varphi.Pv_s}{P - \varphi.Pv_s} \quad (10)$$

where: P_a is the steam flow pressure in the hydrogen flowing into the anode. While the amount of water vapor in oxygen flowing in PEM fuel cell is:

$$\mathcal{N}_{H_2O,in\ O_2} = S_{O_2}.\mathcal{N}_{O_2}.\eta_{cell}\frac{\varphi.Pv_s}{P_c - \varphi.Pv_s} \quad (11)$$

Where P_c is the pressure of the water vapor flow in oxygen flows into the anode. The amount of water vapor in the air flowing in PEM fuel cell is:

$$\mathcal{N}_{H_2O,in\ air} = S_{O_2}.n_{cell}\frac{\mathcal{N}_{O_2}}{X_{O_2}}\frac{\varphi.Pv_s}{P_c - \varphi.Pv_s} \quad (12)$$

The mass of air flowing out of PEM fuel cell is:

$$\mathcal{N}_{air\ ,out\ steack} = \left[\left(S_{O_2}\frac{1 - XmolO_2}{Xmol\ O_2}\right) + (S_{O_2} - 1)\right]\frac{I}{4F}n_{cell}$$

(13)

3.2 EOD and Back Diffusion

The number of protons that move from the anode to the cathode of the PEM fuel cell ($\mathcal{N}_{drag}$) on the EOD phenomenon can be calculated based on the following equations [19-21]:

$$\mathcal{N}_{drag} = \frac{n_d}{F}I \quad (14)$$

EOD coefficient value (n_d) can be defined based on the following equation:

$$n_d = \frac{2.5\ \lambda pem}{22} \quad (15)$$

Where λpem is the membrane water content that can be calculated based on the following equation [22-24]:

$$\lambda = 0.043 + 17.81a_{H2O} - 39.85\ a^2_{H2O} + 36.a^3_{H2O} \qquad 0 < a_{H2O} \le 1$$

(16)

Membrane water activity a_{H2O} is calculated based on the following equation

$$a_{H2O} = \frac{a_{H2O,a} + a_{H2O,k}}{2} \quad (17)$$

Water diffusion from the cathode side to the anode side PEM fuel cell $\mathcal{N}_{diff,H2O}$ is defined as follows [20, 22, 25]:

$$\mathcal{N}_{diff,H2O} = D_w\frac{(C_{K,v} - C_{A,v})}{x_m} \quad (18)$$

According to [17] that the water diffusion from the anode side to the cathode side of the PEM fuel cell system is as follows:

$$J_w = k_g(y_a - y_c) \quad (19)$$

J_w is humidity flux across the membrane

3.3 PEM Fuel Cell Performance

The value of cell voltage at PEMFC system is [26]

$$V_{operasi} = V_{rev} - V_{irrev} \quad (20)$$

$$V_{irrev} = V_{act} - V_{ohm} - V_{con} \quad (21)$$

3.3.1 .Reversible Cell Voltage (V_{rev})

At standard state the value of PEM fuel cell electric power is [27]:

$E = \frac{-\Delta G_f^0}{nF} = \frac{237.340 \quad Kjmol^{-1}}{2\,mol x\; 96{,}485 \quad Cmol^{-1}} = 1.229V$ At 25°C and pressure 1 atm

The potential of electricity as a function of temperature (E_T) is:

$$E_T = E^0 + \frac{\Delta S}{nF}(T - T_0) \quad (22)$$

The reversible voltage for PEM fuel cell is [27-29]:

$$V_{rev} = E_T + \frac{RT}{nF} ln\left(\frac{P_{H_2}x(P_{O_2}{}^{0.5})}{P_{H_2O}}\right) \quad (23)$$

So the value of PEM fuel cell electric power based on equation (22) is:

$$E_T = 1.229V - 0.85\, x10^{-3}(T - T_0)V/K (24)$$

Assuming that the partial pressure is 1 atm because the water is in the liquid phase, then:

$$V_{rev} = 1.229V - 0.85\, x\, 10^{-3}(T - T_0) + 4.31\, x\, 10^{-5}\, x\, Tx \ln\left(P_{H_2}x(P_{O_2}{}^{0.5})\right) \quad (25)$$

3.3.2 Voltage Activation (V_{act})

In the equilibrium state, the net current (zero down stream) is zero, although the PEMFC response system occurs simultaneously. The consequences are:

$$r = k_f C_{O_2} - k_b C_{H_2} = \frac{i_c}{nF} - \frac{i_a}{nF} \quad (26)$$

The value is expressed as a change in the current density. This condition can be stated as follows:

$$i_c = i_a danr_f = r_b$$

Can be simplified to be the first form:

$$\frac{RT}{F}\left[\frac{d}{dE}\ln(k_b) - \frac{d}{dE}\ln\left(\frac{1}{k_f}\right)\right] = 1 \quad (27)$$

For the reductive symmetry factor expressed by α, then

$$\frac{RT}{F}\frac{d}{dE}\ln\left(\frac{1}{k_f}\right) = \alpha \quad (28)$$

With oxidative symmetry is: $1 - \alpha$

$$\frac{RT}{F}\frac{d}{dE}\ln(k_b) = 1 - \alpha \quad (29)$$

Can be simplified to:

$$\ln(k_b) = \frac{(1-\alpha)FE}{RT} + c \quad (30)$$

k_f value for standard condition is $k_f = k_{f,0}$, $k_b = k_{b,0}$ and $E = E_0$

Can be changed to:

$$k_f = k_{f,0}\exp\left(\frac{-\alpha FE}{RT}\right) \quad (31)$$

And $$k_b = k_{b,0}\exp\left(\frac{(1-\alpha)FE}{RT}\right) \quad (32)$$

Then known as the Butler Volmer equation, then the rate of reaction in the cell is:

$$i = nF\left(k_{f,0}\exp\left(\frac{-\alpha FE}{RT}\right)C_{O_2} - k_{b,0}\exp\left(\frac{(1-\alpha)FE}{RT}\right)C_{H_2}\right) \quad (33)$$

In equilibrium condition it is assumed that $k_{f,0} = k_{b,0} = k_0$

$$i_0 = nFk_0 C_{O_2}{}^{(1-\alpha)} C_{H_2}{}^{\alpha} \quad (34)$$

If this equation is compared to the reference current density $(i_{0,ref})$ is:

$$i_0 = i_{0,ref}\left(\frac{P_R}{P_{R,ref}}{}^{Y}\right) exp\frac{E_c}{RT}\left(1 - \frac{T}{T_{ref}}\right) \quad (35)$$

The i_0 value is detected by the ideal surface reaction area per unit volume of the catalyst layer (a) [30-32]

The equation becomes:

$$i_0 = ai_{0,ref}\left(\frac{P_R}{P_{R,ref}}{}^{Y}\right) exp\frac{E_c}{RT}\left(1 - \frac{T}{T_{ref}}\right) \quad (36)$$

E_c = 76.5 Kj/mol [33]

According to [31] that the value of a a is:

$$a = a_0\frac{m_{pt}}{L}$$

According to [34] the value of a_0 is:

$$a_0 = 10^8(4.4198Y^9 - 27.691Y^8 + 74.206Y^7 - 111.06Y^6 + 101.43Y^5 - 57.841Y^4 + 20.231Y^3 - 4.089Y^2 + 0.39451Y) \quad (37)$$

WithY= % Pt catalysts having value in the range 0 to 100%.

$$a_0 = 10^8(4.4198Y^9 - 27.691Y^8 + 74.206Y^7 - 111.06Y^6 + 101.43Y^5 - 57.841Y^4 + 20.231Y^3 - 4.089Y^2 + 0.39451Y) \quad (38)$$

According to [35] and [31] the reference current density value is:

$$i_{0,ref} = 10^{\left(3.057 - \frac{4001}{T}\right)}$$

The value of the current density based on equation is::

$$i = nF\,k_0\left(C_{O_2}\exp\left(\frac{-\alpha FE}{RT}\right) - C_{H_2}\exp\left(\frac{(1-\alpha)FE}{RT}\right)\right) \quad (39)$$

If the value i i compared i_0 to the equation (35) is:

$$\frac{i}{i_0} = \frac{nFk_0\left(C_{O_2}\exp\left(\frac{-\alpha FE}{RT}\right) - C_{H_2}\exp\left(\frac{(1-\alpha)FE}{RT}\right)\right)}{nF\,k_0\left(C_{O_2}\exp\left(\frac{-\alpha FE_r}{RT}\right)\right)} \quad (40)$$

Nilai $E - E_r = \eta_{act}$, maka nilai $\frac{i}{i_0}$ menjadi:

$$V_{act} = \frac{RT}{2\alpha F}\sin h^{-1}\left(\frac{i}{i_0}\right) \quad (41)$$

The activation voltage value is [36-37]:

$$V_{act} = \frac{RT}{n\alpha F}\ln\left(\frac{i}{i_0}\right) where\, i > i_0 \quad (42)$$

According to [36] the value of i_0 depends on temperature, pressure, type of catalyst used, specific surface area and loading. The value of i_0 is 10^{-2} A to 10^{-8} A. Based on the equation (42), the value of Tafel constant is::

$$b = \frac{RT}{n\alpha F} \quad (43)$$

In the PEMFC system model design, the Tafel constant value is based on research data conducted by [38]is:

$$b = 0.1937\ e^{-2.197(\alpha)} \quad (44)$$

$\alpha$is the exchange coefficient, whose value is [18]:

$$\alpha = \left(0.001552RH_{c,inlet} + 0.000139\right)T \quad (45)$$

$RH_{c,inlet}$ is the relative humidity that flows into the cathode of PEM fuel cell

3.3.3 *Voltage Ohm (V_{ohm})*

Loss caused by ohm resistance occurs in the membrane and catalyst layer and is a function of temperature and humidity [39]. Total voltage loss in PEM fuel cell is a linear function [40]. The total value of the ohm voltage according to [37] is:

$$V_{ohm} = iR_{ohm} \quad (46)$$

While the value of R_{ohm}:

$$R_{ohm} = \frac{t_m}{\sigma_m} \quad (47)$$

The value of the membrane conductivity is [32, 41]:

$$\sigma_m = (0.005139\lambda_m - 0.003260)exp\left[1268\left(\frac{1}{303} - \frac{1}{T}\right)\right] \quad (48)$$

The value of the membrane water content is:

$$\lambda_m = 0.048 + 17.81RH - 39.83RH^2 + 39.85RH^3 \quad (49)$$

a is the average relative humidity in the PEM fuel cell steck..

3.3.4 *Concentric Voltage Kepekatan (V_{conc})*

Loss of concentration capability because the reaction of the electrode is restricted by the reactant mass transfer within the PEM ful cell [40]. The concentration capability value [29] is:

$$V_{conc} = cln\left(\frac{i_L}{i_L - 1}\right) \quad (50)$$

c is the constant of loss by concentration. The value of c is:

$$c = \frac{RT}{nF} \quad (51)$$

$$V_{conc} = \frac{RT}{nF} ln\left(\frac{i_L}{i_L - 1}\right) \quad (52)$$

i_L is the maximum value of the current density generated by PEM fuel cell. The amount of current density limit value by the PEMfuel cell is [42]

$$i_L = nFh_m\left(\frac{c_{O_{2,in}} - c_{O_{2,out}}}{ln\frac{c_{O_{2,in}}}{c_{O_{2,out}}}}\right) \quad (53)$$

$c_{O_{2,indanout}}$ is the concentration of oxygen flowing in and out of PEM fuel cell (mol)

h_m is the convective mass transfer coefficient. The convective mass transfer coefficient value by [43]:

$$h_m = \frac{Sh_f D_{ij}}{H_c} \quad (54)$$

4 ResultS and Discussion

Figure 2 shows the effect that relative humidity has on the single cell performance. The operating temperature was maintained at 353 K. The pressure at the anode and the cathode was 1atm. The hydrogen to oxygen stoichiometric ratio was 1.2 for hydrogen and 2 for oxygen. An increased RHA and an increased RHC value generally improve the single cell performance, as shown by the increases in the voltage and the power density that are produced. The single cell performance can be improved by increasing the relative humidity at the cathode, which increases the voltage and the power density that are produced. This result is reflected by the increased voltage and power density that are produced. Higher single cell performance can be achieved by increasing the RHC value.

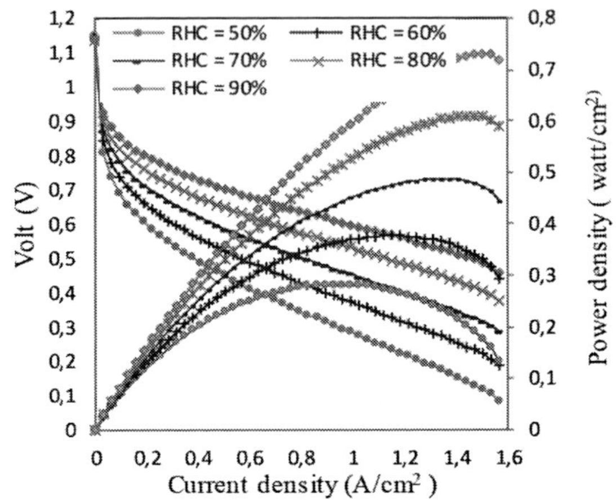

(a)

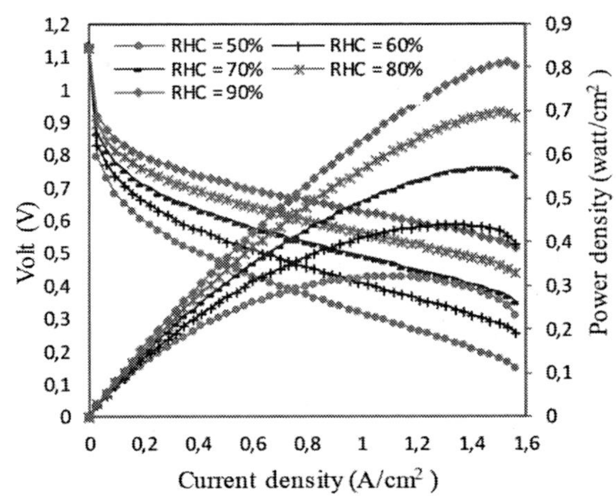

(b)

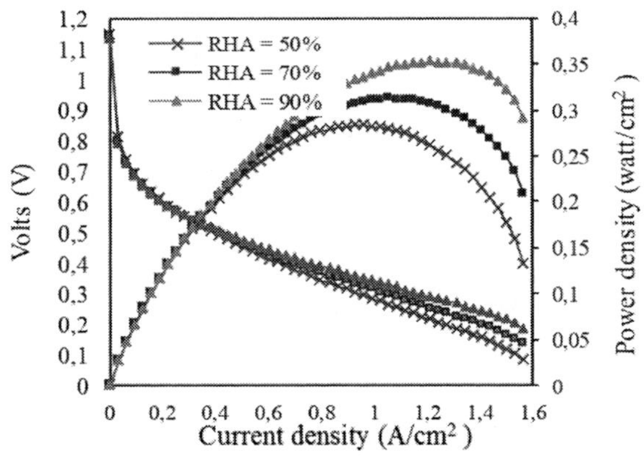

(c)

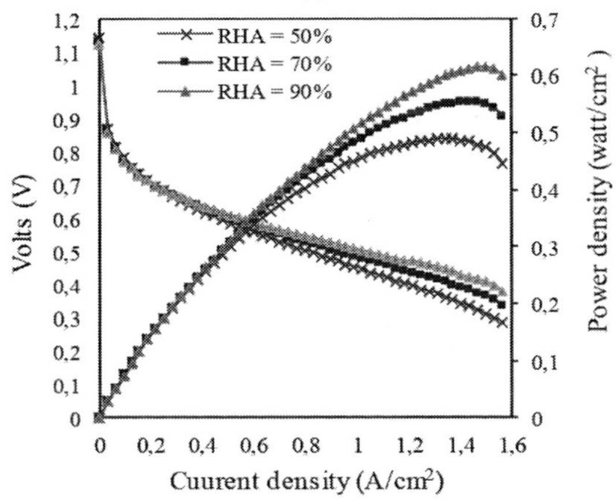

(d)

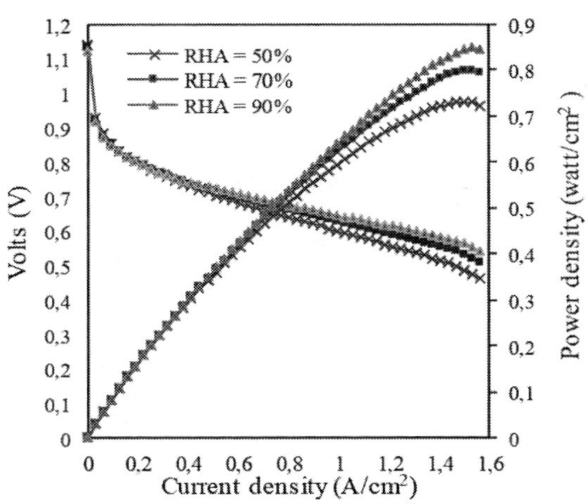

(e)

Fig 2. Effect of relative humidity on single cell performance of PEM fuel cells at temperatures of 353 K, the operating pressure of 1 atm, the stoichiometric ratio is 1.2 for H2 and O2 are 2: (a) RHA is 50% and RHC are 50%, 60%, 70%, 80% and 90%. (b) RHA is 75% and RHA are 50%, 60%, 70%, 80% and 90%. (c) RHA is 50%, 70% and 90%, RHC is 50%. (d) RHA are 50%, 70% and 90%, RHC is 70 %.(e) RHA are 50%, 70% and 90%, RHC is 90%.

Figures 2a and 2c show the effects of varying the relative humidity while all the other conditions are kept constant. The RHA in Figure 2a was maintained at 50% while the RHC was varied from 50% to 90%. In figure 2c, the RHC was maintained at 50% while the RHA was varied from50% to 90%. Figure 2a shows the improvement in the single cell performance with an increase in the RHA. Figure 2c shows that the increase in RHA causes an insignificant improvement in the single-cell performance. Increasing the RHC was more effective than increasing the RHA with regard to improving the single cell performance. The increases in the RHA and the RHC resulted in an increased voltage and an increased power density. Specifically, an increased RHC significantly improved single cell performance. Figures 2c, 2d and 2e show the effect of RHA on the single cell performance; improved performance occurs beyond a voltage of 0.6 V. Figures 2a and 2b show the effect of maintaining the RHA at 50% and 70% while the value of RHC was kept constant from 50% to 90%.

The current density of a single PEM fuel cell at a voltage of 0.6V is shown in table 2.

Table 2. PEM fuel cell performance at 0.6V with different relative humidity at the cathode

Condition	**Current density (A cm^{-2}) at 0.6V with different relative humidity at the cathode**				
	RHC 50%	**RHC 60%**	**RHC 70%**	**RHC 80%**	**RHC 90%**
Condition 1	0.193	0.289	0.449	0.673	0.961
Condition 2	0.193	0.321	0.513	0.801	1.185

Note:
Condition 4: RHA = 50%, PA and PC = 1 atm and T = 353 K
Condition 5: RHA = 75%, PA and PC = 1 atm and T = 353 K

Conditions 1 and 2 in table 2 show that the single PEM fuel cell was able to maintain the current density at the same value when the RHC was 50%, even though the RHA was increased from 50% to 70%. The current density increased when the RHC was increased from 60% to 90%. The same effect was observed when the RHA was increased from 50% to70%, which resulted in an increased current density at a voltage of 0.6V.

Table 3. . PEM fuel cell performance at 0.6V with different relative humidity at the anode

Condition	**Current density (A cm^{-2}) at 0.6V with different relative humidity at the anode**		
	RHA 50%	**RHA 70%**	**RHA 90%**
Condition 3	0.193	0.193	0.193
Condition 4	0.449	0.513	0.545
Condition 5	0.961	1.121	1.281

Note:
Condition 3 : RHC = 50% , PA and PC= 1 atm and T = 353 K
Condition 4 : RHC = 75% , PA and PC= 1 atm and T = 353 K
Condition 5 : RHC = 90% , PA and PC= 1 atm and T = 353 K

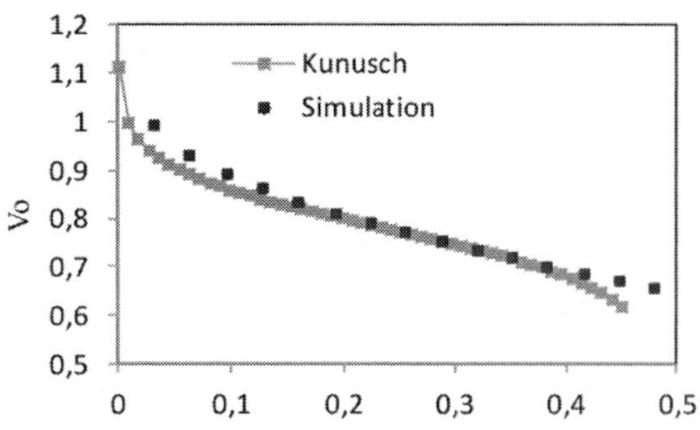

Fig 3. validation of simulation results

Table 3 shows the current density at a voltage of 0.6V when the RHA was varied. When the cell was operated at Condition 3, an increase in the RHA from 50% to 90% did not affect the value of current density at 0.193 A cm^{-2}. However, Conditions 7 and 8 show that an increase in the RHC and the RHA result in an increased current density in the single cell. Tables 3 and 3 show that when the RHC was 50% and the RHA was 50% in Conditions 1, 2 and 3, the current density remained unchanged at 0.193 Acm^{-2}. The increase in the RHA did not significantly increase the current density. However, an increase in the RHC significantly increased the current density. The performance of the single cell was highest at RHC values ranging from 70% to 90%.

The results of the simulation study in figure 3 showed similar trends in the performance of a single cell PEM fuel cell. However, the simulation results in this study did not show a significant decrease in voltage over the current density of 0.45 A cm^2, while the simulation results performed by Khunush showed a decrease in voltage value. The decrease in the value of this voltage is due to insufficient hydrogen and oxygen for the reaction

5 Conclusion

The conclusions that can be drawn from this study are as follows:

1. In the same relative humidity value at one side of the PEM Fuel Cell, the increase in single cell performance is influenced by the increase in RHC value compared to RHA value.
2. The highest single cell performance that occurs at 0.6V is indicated by the RHA value from 70% to 90% and RHC is 90%.

Acknowledgement

The authors would like to thanks the Bung Hatta University (UBH) and RISTEK DIKTI supporting this project through PUPT 2017 research grant. And also thanks to Axel Okta Yuga and Devani Adristi.

References

[1] A. Boudghene Stambouli and E. Traversa, "Fuel cells, an alternative to standard sources of energy," Renewable and Sustainable Energy Reviews, vol. 6, pp. 295-304, 2002.

[2] K. Sopian and W. R. Wan Daud, "Challenges and future developments in proton exchange membrane fuel cells," Renewable Energy, vol. 31, pp. 719-727, 2006.

[3] L.-j. Yu, et al., "Transport mechanisms and performance simulations of a PEM fuel cell with interdigitated flow field," Renewable Energy, vol. 34, pp. 530-543, 2009.

[4] J. H. Lee and T. R. Lalk, "Modeling fuel cell stack systems," Journal of Power Sources, vol. 73, pp. 229-241, 1998.

[5] C.-H. Chen, et al., "Flow distribution in the manifold of PEM fuel cell stack," Journal of Power Sources, vol. 173, pp. 249-263, 2007.

[6] W. Bi, et al., "The effect of humidity and oxygen partial pressure on degradation of Pt/C catalyst in PEM fuel cell," Electrochimica Acta, vol. 54, pp. 1826-1833, 2009.

[7] S. K. Kamarudin, et al., "Design of a fuel processor unit for PEM fuel cell via shortcut design method," Chemical Engineering Journal, vol. 104, pp. 7-17, 2004.

[8] M. G. Santarelli and M. F. Torchio, "Experimental analysis of the effects of the operating variables on the performance of a single PEMFC," Energy Conversion and Management, vol. 48, pp. 40-51, 2007.

[9] N. Rajalakshmi, et al., "Design and development of modular fuel cell stacks for various applications," International Journal of Hydrogen Energy, vol. 33, pp. 449-454, 2008.

[10] C. A. Frangopoulos and L. G. Nakos, "Development of a model for thermoeconomic design and operation optimization of a PEM fuel cell system," Energy, vol. 31, pp. 1501-1519, 2006.

[11] A. D. Le and B. Zhou, "A general model of proton exchange membrane fuel cell," Journal of Power Sources, vol. 182, pp. 197-222, 2008.

[12] J. Wu, et al., "A review of PEM fuel cell durability: Degradation mechanisms and mitigation strategies," Journal of Power Sources, vol. 184, pp. 104-119, 2008.

[13] J. Liu and M. Eikerling, "Model of cathode catalyst layers for polymer electrolyte fuel cells: The role of porous structure and water accumulation," Electrochimica Acta, vol. 53, pp. 4435-4446, 2008.

[14] S. Shimpalee, et al., "Analysis of GDL flooding effects on PEMFC performance," Electrochimica Acta, vol. 52, pp. 6748-6754, 2007.

[15] W.-M. Yan, et al., "Effects of operating conditions on cell performance of PEM fuel cells with conventional or interdigitated flow

field," Journal of Power Sources, vol. 162, pp. 1157-1164, 2006.

[16] Mulyazmi, et al., "Water balance for the design of a PEM fuel cell system," International Journal of Hydrogen Energy, vol. 38, pp. 9409-9420, 2013.

[17] N. S. M. Hassan, et al., "Water management in a single cell proton exchange membrane fuel cells with a serpentine flow field," Journal of Power Sources, vol. 193, pp. 249-257, 2009.

[18] J. Zhang, et al., "PEM fuel cell relative humidity (RH) and its effect on performance at high temperatures," Electrochimica Acta, vol. 53, pp. 5315-5321, 2008.

[19] B. Zhou, et al., "Water and pressure effects on a single PEM fuel cell," Journal of Power Sources, vol. 155, pp. 190-202, 2006.

[20] Y. Zong, et al., "Water and thermal management in a single PEM fuel cell with non-uniform stack temperature," Journal of Power Sources, vol. 161, pp. 143-159, 2006.

[21] X. Yu, et al., "Water and thermal management for Ballard PEM fuel cell stack," Journal of Power Sources, vol. 147, pp. 184-195, 2005.

[22] F. Gao, et al., "Proton exchange membrane fuel cell multi-physical dynamics and stack spatial non-homogeneity analyses," Journal of Power Sources, vol. 195, pp. 7609-7626, 2010.

[23] K. Chu, et al., "A lumped parameter model of the polymer electrolyte fuel cell," Journal of Power Sources, vol. 171, pp. 412-423, 2007.

[24] S. A. J. Erahim Afshari, "Heat and Water Management in a PEM Fuel Cell," Wseas Tctions on Flui Mechanic, vol. 3, p. 6, 1998.

[25] A. J. del Real, et al., "Development and experimental validation of a PEM fuel cell dynamic model," Journal of Power Sources, vol. 173, pp. 310-324, 2007.

[26] S. O. Mert, et al., "Performance investigation of a transportation PEM fuel cell system," International Journal of Hydrogen Energy, vol. 37, pp. 623-633, 2012.

[27] Y.-S. Chen and H. Peng, "A segmented model for studying water transport in a PEMFC," Journal of Power Sources, vol. 185, pp. 1179-1192, 2008.

[28] L. Placca, et al., "Effects of temperature uncertainty on the performance of a degrading PEM fuel cell model," Journal of Power Sources, vol. 194, pp. 313-327, 2009.

[29] H. Shaker, "Analytical modeling of PEM fuel cell i–V curve," *Renewable Energy,* vol. 36, pp. 451-458, 2011.

[30] T. Thampan, et al., "PEM fuel cell as a membrane reactor," Catalysis Today, vol. 67, pp. 15-32, 2001.

[31] D. Song, et al., "Numerical optimization study of the catalyst layer of PEM fuel cell cathode," Journal of Power Sources, vol. 126, pp. 104-111, 2004.

[32] Y. Shan and S.-Y. Choe, "A high dynamic PEM fuel cell model with temperature effects," *Journal of Power Sources,* vol. 145, pp. 30-39, 2005.

[33] B. A. P. Wei Suna, Kunal Karana, "An improved two-dimensional agglomerate cathode model to study the influence of catalyst layer structural parameters " Electrochemica Acta, vol. 50, p. 16, 2005.

[34] D. Song, et al., "A method for optimizing distributions of Nafion and Pt in cathode catalyst layers of PEM fuel cells," Electrochimica Acta, vol. 50, pp. 3347-3358, 2005.

[35] G. Inoue, et al., "Evaluation of the optimal separator shape with reaction and flow analysis of polymer electrolyte fuel cell," Journal of Power Sources, vol. 154, pp. 18-34, 2006.

[36] H. Görgün, et al., "An algorithm for estimation of membrane water content in PEM fuel cells," Journal of Power Sources, vol. 157, pp. 389-394, 2006.

[37] C. Kunusch, et al., "Characterization and experimental results in PEM fuel cell electrical behaviour," International Journal of Hydrogen Energy, vol. 35, pp. 5876-5881, 2010.

[38] R. F. Mann, et al., "Application of Butler–Volmer equations in the modelling of activation polarization for PEM fuel cells," Journal of Power Sources, vol. 161, pp. 775-781, 2006.

[39] K. Scott and M. Mamlouk, "A cell voltage equation for an intermediate temperature proton exchange membrane fuel cell," International Journal of Hydrogen Energy, vol. 34, pp. 9195-9202, 2009.

[40] B. Abdellah, "Modeling and simulation of proton exchange membrane fuel cell systems," Journal of Power Sources, vol. 205, pp. 335-339, 2012.

[41] X. Zhang, et al., "The parametric optimum analysis of a proton exchange membrane (PEM) fuel cell and its load matching," Energy, vol. 35, pp. 5294-5299, 2010.

[42] C. Spiegel, "PEM Fuel Cell Modeling and Simulation Using MATLAB," *Elsevier Inc,* 2008.

[43] R. R. B. Tavakoli, "The effect of fuel cell operational conditions on the water content distribution in the polymer electrolyte membrane," Renewable Energy, vol. XXX, p. 13, 2011.

The Comparison of Different Nutrient Sources in Development of Aerobic Granular Sludge

C. Choerudin[1,2], *Ardiyan* Harimawan[1], and *Tjandra* Setiadi[1,3,*]

[1]Department of Chemical Engineering, Institut Teknologi Bandung (ITB), Jl. Ganesa 10, Bandung 40132, Indonesia
[2]Department of Chemical Engineering, Institut Teknologi Nasional (ITENAS), Jl. P.H.H. Mustofa No.23, Bandung 40124, Indonesia
[3]Center for Environmental Studies (PSLH), Institut Teknologi Bandung (ITB), Jl. Sangkuriang No. 42A Bandung 40135, Indonesia

Abstract. Aerobic granular sludge (AGS) as recent biological wastewater treatment has been developed in a modified sequencing batch reactor (SBR) using real textile wastewater. This experiment focused on comparing the effects of different nutrient sources, i.e. fine chemicals and fertilizer, on the characteristics of the AGS. The development of AGS conducted in a reactor that called as aerobic up-flow fluidized bed (AUFB) reactor. AUFB reactor allows the three-steps SBR operation namely filling-and-discharging, reaction, and settling. The developed AGS was characterized through some parameters i.e. physical characteristics (morphology, size distribution, settling velocity, specific gravity), granulation profile (MLSS and SVI), and removal performances (removal of COD and colour). The results showed that the developed AGS has a slight morphological difference as the effect of each nutrient source. Interestingly, metal ions contained in the nutrient sources affect the granulation profile and removal performances. Higher metal ions in the sources tent towards better profile (higher MLSS and lower SVI) but decreased removal performance (lower colour removal). In conclusions, this experiment suggests that the use of fertilizers as the nutrient source in developing AGS is comparable to the use of fine chemicals.

1 Introduction

Aerobic granular sludge (AGS) become a new standard technology in biological wastewater treatment in the near future [1]. By such a condition, the floccular activated sludge being developed into granular shape. Granular shape, which without carrier material, is very advantageous since it is denser and has very good settling properties. Significant amount of sustainable biomaterial named as alginate-like exopolysaccharides (ALE) can be isolated from the AGS waste [2,3].

AGS technology applied commercially both on municipal and industrial wastewater treatment since first full-scale plant launched in 2005 under the name of Nereda® [1]. The amount of treatment plants (in operation, under construction or in design) on municipal wastewater is much more than on industrial wastewater as informed on its official website [4]. Furthermore, the application on industrial wastewater is limited on slaughterhouse and food processing industry in the Netherlands. However, application on more challenging industrial wastewater is still in research, i.e. textile wastewater.

Textile wastewater can be challenging for AGS technology since it containing synthetic dyes. Textile wastewater usually has high organic carbon and colour compound which is mostly azo dyes type [5]. However, anaerobic-aerobic systems are the most logical concept for dye degradation / colour removal [6]. This concept allows AGS technology being a proper system for the textile wastewater treatment.

Studies on aerobic sludge granulation on textile wastewater are scarce [5,7–11]. All of the study were conducted in laboratory scale and most of them were carried out on the artificial textile wastewater. Besides, most of previous study were conducted in conventional SBR operation which typically 5 steps per cycle namely filling, reaction, settling, discharging, and idling.

The objectives of this study are developing AGS from activated sludge using modified SBR and real textile wastewater. The modified SBR on this research means that the operation of SBR carried out on three steps per cycle: filling-and-discharging, reaction, and settling. The modified SBR designed as the strategy to conduct better granulation process.

As this research using a real textile wastewater, chemicals are used as additional nutrient. It is different to the previous studies that were using artificial wastewater – which are the chemical compositions are well defined. Therefore, this study also aimed to characterizing the AGS properties as the effects of different added nutrient sources.

2 Materials and Methods

This part covered explanation about the reactors, the wastewater characteristics, the nutrient composition, the seed sludge, the experimental setup, and the analytical methods.

* Corresponding author: tjandra@che.itb.ac.id

2.1 Reactors

The design of the reactor was adapted from previous study that is the aerobic up-flow fluidised bed reactor (AUFB) [12]. The reactor consisted of columns and an overhead clarifier. The AUFB was made of acrylic. The columns had diameter 5 cm with 120 cm and 200 cm height. The total working volume was 7.5 L and 9 L. The reactor is illustrated in Fig. 1.

There was also a conventional SBR reactor used as the control. The reactor was a glass column with diameter of 9 cm and height of 100 cm. Feed and air stream were located in the bottom of the reactor. The output stream was located on 70 cm from the bottom. Thus, the working volume of the reactor was 4.5 L.

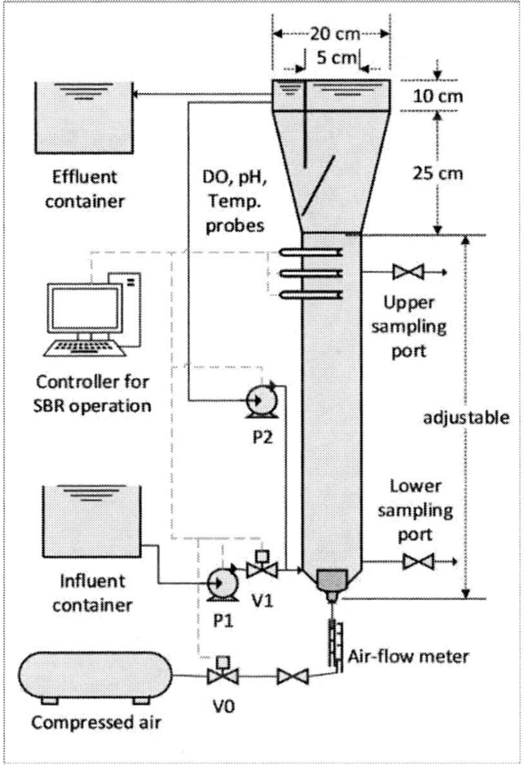

Fig. 1. Schematic representation of the aerobic up-flow fluidized bed (AUFB) reactor.

2.2 Wastewater Characteristics

The wastewater was from local textile industry which is located near Bandung City - Indonesia. The wastewater had COD, BOD_5, and TSS value of 1550 mg/L, 441 mg/L, and 92 mg/L, respectively. The ammonium concentration was <7 mg/L and the pH value was 9.9. Prior to the experiments, the pH value of wastewater set to around neutral by adding HCl and/or NaOH. Then, the wastewater was added with different sources of nutrient that are fine chemicals or fertilizers.

2.3 Added Nutrient Composition

Nutrient from fine chemical sources was referred to previous study [10,13]. The nutrient consisted of CH_3COONa 0.50 g/L, NH_4Cl 0.16 g/L, KH_2PO_4 0.23 g/L, K_2HPO_4 0.58 g/L, $CaCl_2.2H_2O$ 0.07 g/L, $MgSO_4.7H_2O$ 0.09 g/L, $C_{10}H_{14}N_2Na_2O_8$ 0.02 g/L, and trace element solution 1 mL/L. The composition of the trace element solution was H_3BO_3 0.15 g/L, $FeCl_3.4H_2O$ 1.50 g/L, $ZnSO_4.H_2O$ 0.12 g/L, $MnCl_2.4H_2O$ 0.12 g/L, $CuSO_4.5H_2O$ 0.03 g/L, $NaMoO_4.2H_2O$ 0.06 g/L, $CoCl_2.6H_2O$ 0.15 g/L, and KI 0.03 g/L.

Nutrient from fertilizers was taken from NPK 25-7-7 and triple-superphosphate (TSP). The NPK contained 5% N, 7% P_2O_5, 7% K_2O, 1.4% MgO, and 4% CaO. The TSP contained 46% CaH_2PO_4.

2.4 Inoculum

Sludge for inoculation was the activated sludge from a local textile wastewater treatment plant. The sludge acclimatized in anaerobic and aerobic condition in 3 weeks. These two kinds of sludge were mixed up with ratio of 4:1. Anaerobic granular sludge, which was from local pulp and paper wastewater plant, was added 3.5% of the mixture. Then, the inoculum was screened with sieve of 0.425 mm opening and TSS content was measured. Thus, the inoculum for starting up the process had size of < 0.425 mm and MLSS of 8092 mg/L.

2.5 Experimental Setups

Total of three setups had been used based on the type of the reactors – i.e. conventional SBR 4.5 L (CSBR), AUFB 7.5 L (AUFB1) and AUFB 9 L (AUFB2). All reactors were operated in 6 hours cycle with three sequential phases: 15 min filling-and-discharging (anaerobic), 340 min reaction (40 min anaerobic + 130 min aerobic, two times), and 5 min settling. Airflow rate was 0.3-0.8 L/min and volume exchange ratio (VER) was 25% per cycle. At the start-up, inoculum comprised of 50% of CSBR volume, 50% of AUFB1 volume, and 25% of AUFB2 volume. The fine chemicals were mixed with the wastewater and added to CSBR and AUFB1. While, the fertilizers were mixed and added to AUFB2. Comparison of added nutrients is showed in

Table 1.

The change of sludge properties (i.e. MLSS and SVI) along granulation process was defined as the granulation profile. In AUFB1, the granules stability was tested by the change of feeding method that is the separation of wastewater stream from the N and P source stream. The test was as representation of inhomogeneity of system. In AUFB2, the granules stability was tested by increasing the ratio of N and P become 100:10:2. The developed granules were characterized through some physical parameters and also its performance on COD removal and decolourisation.

Table 1. Comparison of added nutrient composition.

Type	Artificial Domestic Sewage		Artificial Textile Wastewater		Real Textile Wastewater	
Ref.	[13]	[14]	[10]	[5,7,9]	AUFB1	AUFB2
COD	100	100	100	100	100	100
N	5.1	10.8	3.4	3.8	5.0	5.0
P	18.1	4.0	12.1	37.3	17.9	1.0
S	1.4	2.6	0.9	0.3	1.4	-
Ca^{2+}	2.2	0.4	1.5	0.8	2.2	0.8
Mg^{2+}	1.0	1.8	0.7	0.2	1.0	0.2
K^{+}	38.1	11.9	25.3	21. 9	37.6	1.1
Na^{+}	0.3	31.1	11.1	29.7	29.1	-

2.6 Analytical Methods

The analytical parameters for the wastewater (i.e. influent and effluent) were COD, BOD, TSS, ammonium, pH, and colour. The colour measurement using a scanning spectrophotometer in range of 400 – 700 nm. For the sludge, the parameters were MLSS, SVI, size distribution, specific gravity, and settling velocity. The morphology of sludge was observed under light microscope with associated camera digital and image processing software. The size distribution of sludge obtained by wet sieve analysis with opening 0.3 mm, 0.425 mm, 0.8 mm, and 2.0 mm. The specific gravity measurement was using pycnometer [15]. The settling velocity was defined as the settling time of single granules in 25 cm of volumetric cylinder. Other parameters that not described were referred to the APHA Standard Methods.

3 Results and Discussion

The results are split into three sub-section. First sub-section is describing the granulation profile in the modified SBR. Second sub-section is describing the developed granules. Third sub-section is covering about the removal performance.

3.1 Granulation profile

The granulation profiles are depicted in Fig. 2. The diagram showed that MLSS in the start-up of CSBR and AUFB1 was 13.4 g/L while in AUFB2 was 8.1 g/L. Most of the sludge had been washed out in the first cycle. All setup has sludge concentration around 2 g/L after 3 days. The MLSS was 1.7 g/L in AUFB1 (11th day) and 2.4 g/L in AUFB2 (18th day). This result is comparable to previous study that is 4.7 g/L in 27 days using conventional SBR on artificial textile wastewater with reducing settling time strategy [5]. But, in this experiment, the MLSS in the CSBR was drop to 0.43 g/L on 11th day. It suggests that the design of AUFB supports the modified SBR operation for granulation process than the conventional SBR.

Initial SVI was 73.4 mL/g in AUFB1 and 121.1 mL/g in AUFB2 based on Fig. 2. After the granulation started, the SVI value was 48.2 mL/g in AUFB1 (11th day) and 71.9 mL/g in AUFB2 (18th day). The SVI were agreed with other result that is 67 ml/g in 27 days [5]. Small SVI value indicated that the granular sludge had been developed.

The stability test in each reactors was affecting the granulation profile, i.e. MLSS decreased and SVI increased. The change of feeding method was applied on the cycle after sampling on day 15th in AUFB1 and on day 18th in AUFB2. The SVI in AUFB1 increases sharply after the test. While, increasing the nutrient ratio in AUFB2 leads the sludge to form biofilm on the column wall. This effect also happened in other experiment using municipal wastewater [14]. The effect in AUFB1 was more dynamic than the effect in AUFB2.

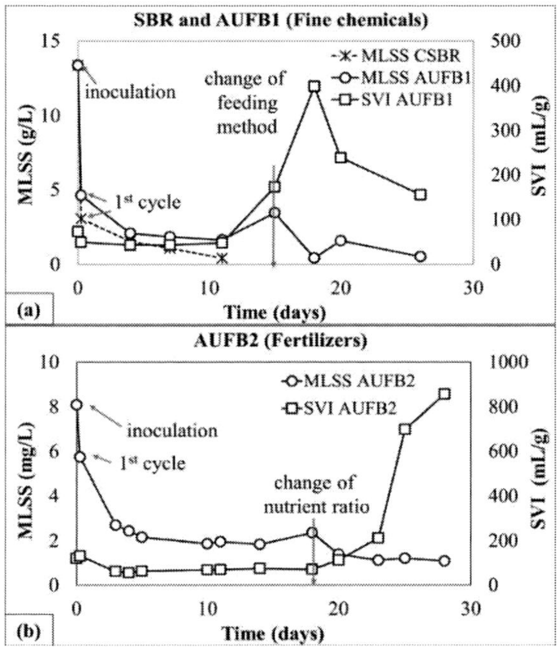

Fig. 2. Aerobic granulation profiles. (□ = MLSS; ○ = SVI)

3.2 Physical Characteristics of Aerobic Granular Sludge

This sub-section talks about physical characteristics of granules including morphology, size distribution, specific gravity, and settling velocity.

3.2.1 Morphology

The morphology of AGS under microscope shown in Fig. 3. It can be seen in Fig. 3a-3b that the inoculum consists of flocs and small granular sludge. After granulation periods, granules were formed and had slightly oval shape. But, granules in AUFB1 was more spherical than in AUFB2 by comparing the granules in

Fig. 3c-3f. As refer to Fig. 3e, parts of anaerobic granular sludge is clumped in the new developed granules as can be seen as black dot in the granules' surface. This finding is agreed with previous study which also using anaerobic granular sludge as the seed [10]. Another thing to note that there were some cavities in the larger granules as seen in Fig. 3d and Fig. 3f.

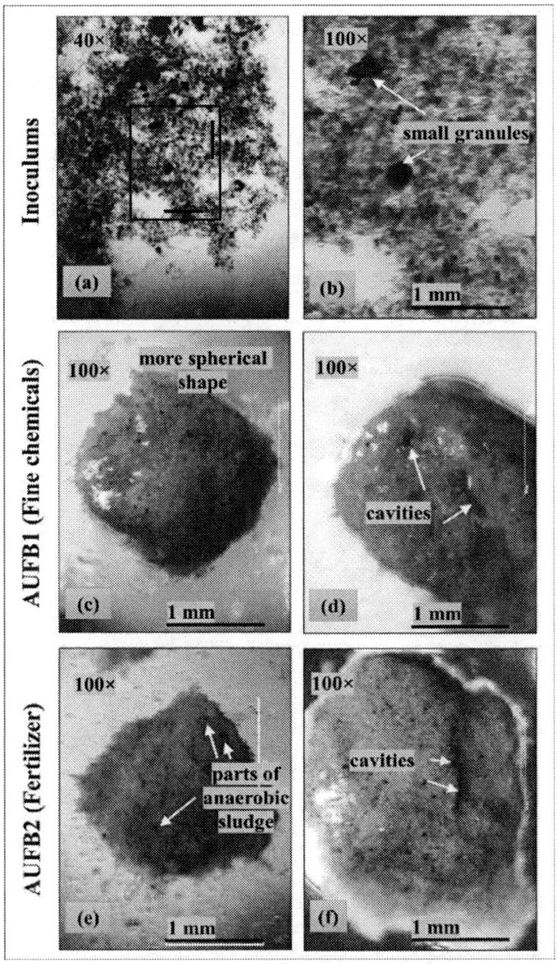

Fig. 3. Appearance of inoculum (a and b) and granular sludge from AUFB1 (b and c) and AUFB2 (e and f) under microscope.

3.2.2 Size distribution

Granular sludge was 4.12% of the total sludge in AUFB1 and 7.78% in AUFB2. It means that both sludge in AUFB1 and AUFB2 had smaller granules portion than flocs. This is agreed with another experiment on the artificial textile wastewater which had 30-45% of granules with size between 0.25-0.65 mm after 77 days [7]. However, full granulation cannot be reached due to short time process. In this experiment, flocs form had larger portion in total sludge because of the flocs had been accumulated in the overhead clarifier instead of washed out.

The distribution of sludge in the reactor is showed in Fig. 4. It shows that granules with size more than inoculum size was 80.6% in AUFB1 and 88.3% in AUFB2. This indicated that small granules in the inoculum were growing and become larger granules. In simply, granulation process was happened. The figure also depicted that granules with range size between 0.4 – 2 mm were stable even after stability test.

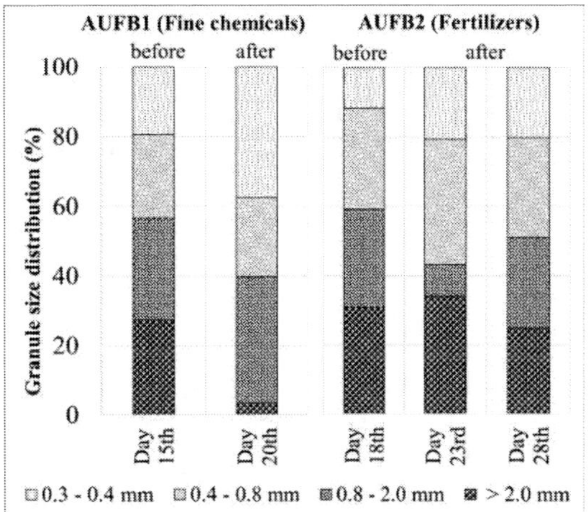

Fig. 4. Size distribution of granules before and after stability.

3.2.3 Specific Gravity

The specific gravity of granules in AUFB1 had range 1.021 – 1.069 (1.039 ± 0.016 in average) while the AUFB2 1.038 – 1.039. Specific gravity of granules in AUFB1 was slightly denser than in AUFB2. This is as the effect of different nutrient composition that is AUFB1 has higher phosphorus and metal ions ratio than AUFB2.

Granule with size range of 0.85 - 2 mm in AUFB1 had average specific gravity of 1.046 ± 0.017. But, granule with size > 2 mm was 1.033 ± 0.013. It means that larger size had lower specific gravity due to cavities which formed in larger granules as described previously.

3.2.4 Settling velocity

Settling characteristics of developed granule is shown in Fig. 5. Granular sludge from AUFB1 had settling velocity minimum of 13.3 m/h and up to 44 m/h. In AUFB2, the minimum settling velocity was 11 m/h and up to 36 m/h. Settling velocity of granule in AUFB1 was slightly better than AUFB2. It also can be seen that granule with larger size is faster to settle not only in AUFB1 but also in AUFB2. It can be concluded that higher ratio phosphorus and metal ions in the nutrient could increases the settling velocity of the granules.

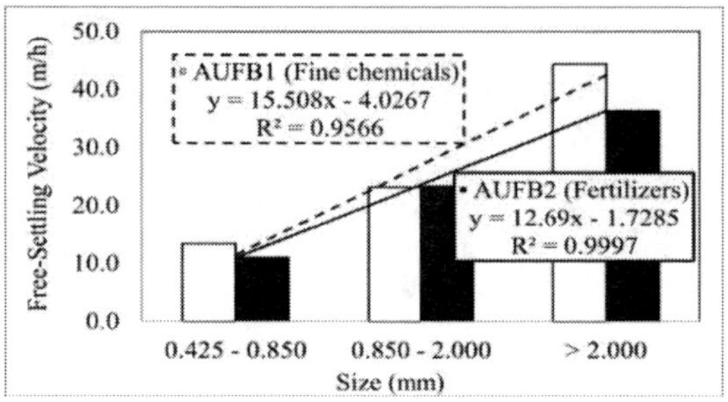

Fig. 5. Free-settling velocity of granules. (empty bar = AUFB1; filled bar = AUFB2; line = linear of AUFB1; dash = linear of AUFB2)

3.3 Removal Performance

The observed performance was the COD removal and the decolourisation. COD removal of AUFB1 and AUFB2 reached up to 84.5% in day 15th and 85.1% in 18th, respectively. It means that there is no significant different between both systems. COD removal performance in this experiment is not that different to other experiments which are 80-85% [5,7,9]. It can be concluded that different sources of added nutrient did not affect the COD removal performance.

Change of feeding method in AUFB1 and nutrient ratio in AUFB2 decreases the COD removal performance. However, after several days, the performance reaches up again to around 80%. This is can be seen in Fig. 6.

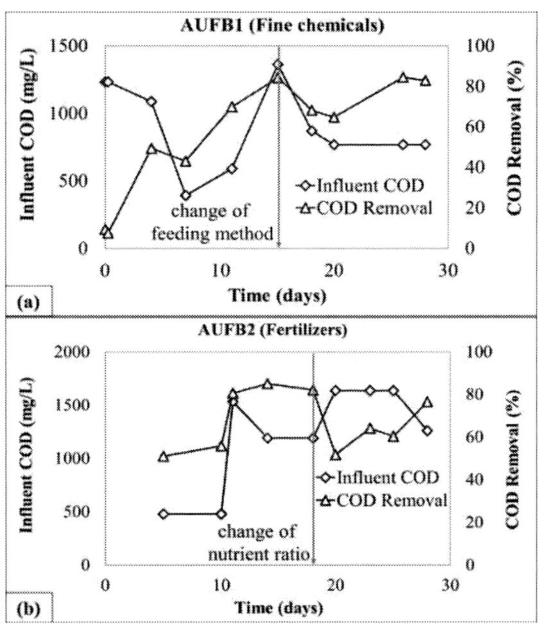

Fig. 6. COD removal performance along granulation.

In decolourisation performance, the AUFB1 has better performance than AUFB2 as the absorbance of effluent in AUFB 2 was below 0.2. In AUFB1, the absorbance of effluent was more than 0.2 while the added nutrients are mixed together in the influent reservoir. But, after the change of feeding methods, the absorbance was below 0.2. While in AUFB2, the effluent had absorbance < 0.2. Interestingly, the absorbance was higher when the nutrient ratio being increased.

The result in Fig. 7 shows that the added nutrient affects the decolourisation performance as described previously. This is happened since metals ion (i.e. bivalent ions) could bind to the azo dyes and make the dye more recalcitrant [16]. This also can be regarded as the metal ions acted as the alternatives terminal acceptor which compete to the azo dyes (the targeted acceptor) [6]. In simply, the AUFB1 had lower decolourisation performance due to the higher metals ion contained in the added nutrients.

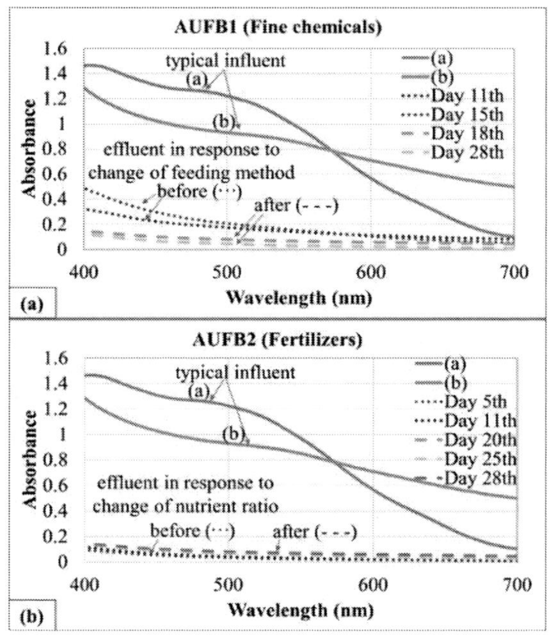

Fig. 7. Decolourisation performance along granulation.

4 Conclusions

This study is to investigate the process of aerobic granular sludge development using modified sequencing batch reactor and the effects of nutrient addition. It can be reported that the three-steps sequencing batch operation can be used for development of the aerobic granular sludge using real textile wastewater.

The main difference of using fine chemical or fertilizer as the added nutrition is the composition of metal ions. This difference affects some physical characteristics of the developed granules. Higher metal ion and phosphorus in the nutrient leads the production of smaller size, higher specific gravity, and better settling velocity. Metal ions in the added nutrition is good for the granulation but less desirable for the decolourisation. However, the use of fertilizer as the added nutrition source for granulation was comparable to the fine chemical source.

The authors want to thank the Ministry of Research Technology and Higher Education of Indonesia (Kementerian Ristek-Dikti) for the scholarships and the program (Program Magister menuju Doktor untuk Sarjana Unggul, PMDSU). The authors would like also to thank the KWEF-AIT Research Grant (KARG) 2016 for a partial funding of this research.

References

1. A. Giesen, L. M. M. de Bruin, R. P. Niermans, and H. F. van der Roest, Water Pract. Technol. **8**, wpt. 2013007 (2013).
2. Y. Lin, M. de Kreuk, M. C. M. van Loosdrecht, and A. Adin, Water Res. **44**, 3355 (2010).
3. Y. M. Lin, K. G. J. Nierop, E. Girbal-Neuhauser, M. Adriaanse, and M. C. M. van Loosdrecht, Sustain. Mater. Technol. **4**, 24 (2015).
4. (n.d.). Accessed on: August 2017. Available at: https://www.royalhaskoningdhv.com/en-gb/nereda/nereda-wastewater-treatment-plants
5. A. M. T. Mata, H. M. Pinheiro, and N. D. Lourenço, Biochem. Eng. J. **104**, 106 (2015).
6. F. P. van der Zee and S. Villaverde, Water Res. **39**, 1425 (2005).
7. R. D. G. Franca, A. Vieira, A. M. T. Mata, G. S. Carvalho, H. M. Pinheiro, and N. D. Lourenço, Water Res. **85**, 327 (2015).
8. T. C. Kee, H. H. Bay, C. K. Lim, K. Muda, and Z. Ibrahim, Desalination Water Treat. **54**, 132 (2015).
9. N. D. Lourenço, R. D. G. Franca, M. A. Moreira, F. N. Gil, C. A. Viegas, and H. M. Pinheiro, Biochem. Eng. J. **104**, 57 (2015).
10. K. Muda, A. Aris, M. R. Salim, Z. Ibrahim, A. Yahya, M. C. M. van Loosdrecht, A. Ahmad, and M. Z. Nawahwi, Water Res. **44**, 4341 (2010).
11. K. Muda, A. Aris, M. R. Salim, Z. Ibrahim, M. C. M. van Loosdrecht, A. Ahmad, and M. Z. Nawahwi, Water Res. **45**, 4711 (2011).
12. N. Kishida, A. Kono, Y. Yamashita, and S. Tsuneda, J. Water Environ. Technol. **8**, 251 (2010).
13. J. J. Beun, A. Hendriks, M. C. M. van Loosdrecht, E. Morgenroth, P. A. Wilderer, and J. J. Heijnen, Water Res. **33**, 2283 (1999).
14. J. J. Beun, M. C. M. van Loosdrecht, and J. J. Heijnen, Water Res. **36**, 702 (2002).
15. M. C. M. van Loosdrecht, P. H. Nielsen, C. M. Lopez-Vazquez, and D. Brdjanovic, *Experimental Methods in Wastewater Treatment*, IWA Publishing, London (2016).
16. Y. Shimizu and T. Takagishi, J. Sericultural Sci. Jpn. **66**, (1997).

High Selectivity of Alkanes Production by Calcium Basic Soap Thermal Decarboxylation

Godlief F. Neonufa[1,2,*], *Meiti* Pratiwi[1], *Tatang H.* Soerawidjaja[1] and *Tirto* Prakoso[1]

[1]ITB, Department of Chemical Engineering, 40132 Bandung, Indonesia
[2]UKAW, Department of Agriculture Product Technology, 85000 Kupang, Indonesia

Abstract. Renewable fuel production from vegetable oil and fat or its fatty acids by direct decarboxylation has been widely reported. An innovative approach to produce drop-in fuel via thermal catalytic decarboxylation of basic soap derived from palm stearin reported in this research. The catalytic effect of the calcium and magnesium metals in the basic soap and its decarboxylation on drop-in fuel yield and product distribution was studied. The catalytic effect was tested in the temperature range up to 370°C and atmospheric pressure for 5 hours in a batch reactor. It has been proved that the calcium basic soap decarboxylation, effectively produce the drop-in fuel in carbon ranges $C_8 - C_{20}$, in which more than 78% selectivity toward alkane. Whereas, only 70% selectivity toward alkane has been resulted from the magnesium basic soap decarboxylation.

1 Introduction

Nowadays, a considerable of attention has been devoted to producing renewable diesel (drop-in fuel) via deoxygenation of vegetable oils and fats or fatty acids. Biodiesel as one of the renewable liquid fuels that have been popular can only be used in a certain amount by blended with petrodiesel as an additive. For instance, a blended fuel comprised of 20% biodiesel and 80% petrodiesel is called B20 [1]. The direct production of liquid biohydrocarbons from vegetable oils and fats via the hydrodeoxigenation method has been reported by many researchers. The disadvantages of the hydrodeoxigenation process of oils and fats are using noble metal catalysts [2] and require fossil fuel-derived hydrogen. So that the process is not 100% renewable.

One promising deoxygenation method for the production of drop-in fuels is basic soap decarboxylation. This method does not require H_2 (when the soap derived from saturated fatty acids) or slightly H_2 (when the soap derived from unsaturated fatty acids), can be operated at atmospheric pressure and only the alkaline earth and transition metals are used to prepare of the basic soap and as a catalyst for decarboxylation of it soap [3].

Decarboxylation of basic soaps (e.g., magnesium soap) into drop-in fuels, illustrated by the equation of reaction 1.

$$\text{(R-COO)M}_g\text{(OH)} \rightarrow \text{RH} + \text{M}_g\text{CO}_3 \quad (1)$$

The factors that determine the success rate of basic soap decarboxylation into liquid biohydrocarbons (drop-in fuels) lies in the level of perfection of basic soap prepared and its hydroxide resistance during decarboxylation [4]. Therefore, the selection of metals for the prepare of basic soaps, highly determined the success rate of the basic soap production and its decarboxylation. Hydroxide and oxide magnesium compound have been reported can be good catalyst for the process of decarboxylation of oils and fats [5-8]. A series of decarboxylation experiments to synthesize drop-in fuels from magnesium basic soaps derived from palm stearin has been studied in our previous study. The drop-in fuel that has been resulted consist of a mixture of biohydrocarbons, which include n-alkane, iso-alkane and 1-alkene fractions, in C_8-C_{20} carbon number. In addition, it also detected a small amount of ketone products.

The presence of the 1-alkene and the ketone component in the fuel product, related to the character of the magnesium basic soaps decomposition. It is decompose more rapidly and at low temperature, i.e., begins at about 300°C, and becomes very rapid between 350° and 400°C [4]. Therefore, the hydroxide of [(R-COO)Mg(OH)] is possible to discharge as water, leaved behind a neutral magnesium soap [(R-COO)Mg(R-COO)], and then the soap lose its basicity before the maximum temperature for decarboxylation is achieved. When the neutral soap decarboxylated, that's result the ketone, and then the decomposition of the ketone would result 1-alkene, as illustrated by the equation of reaction 2.

$$\text{(R-COO)}_2\text{Mg} \rightarrow \text{R-CO-R} + \text{MgCO}_3 \quad (2)$$

Both 1-alkene or ketone, are not expected in all drop-in fuel, because it causes the fuel instability [9]. In addition, the 1-alkenes can reduce cetane number of the green-diesel [10] and become the precursors for oxidation reactions. In principle, the long chain biohydrocarbon fractions, especially n-alkanes, are much

* Corresponding author: godlief17@students.itb.ac.id

more important in green diesel fuel to increase the cetane number than 1-alkene [11]. Therefore, the metal for preparing of the basic soap, must be carefully selected, which have high catalytic and selectivity ability to form n-alkanes during the basic soap decarboxylation.

Sato [12] demonstrated experimentally that calcite, $CaCO_3$, can promote the degradation of fatty acids into n-alkanes. Catalysts with core of calcium metal was reported has the catalytic effects to convert fatty acids into n-alkanes, particularlly in gasoline boiling range [3,13], via the pyrolysis reaction of calcium soap.

Calcium soaps have been reported [4], decompose begin at 350° to 400°C, and becomes rapid at 450° - 500°C. Therefore, hydroxide molecule of the calcium basic soaps may still remain in the soap until their decarboxylation temperature reaches 370°C, and resulting a lot of n-alkanes of long chain during decarboxylation of the basic soap. The purpose of this study is to prove that the decarboxylation of reactants from calcium basic soaps derived from palm stearin, can produce high green diesel content of n-alkanes and free ketones.

2 Experimental

Magnesium acetate tetrahydrate $[Mg(OOC_2H_5)_2.4H_2O]$ and iron acetate dihydrate $[Fe(OOC_2H_5)_3.2H_2O]$ was supplied by Merck (analytical grade), were used to make the soap. The palm stearin oil obtained from PT. Ecogreen, Batam. The magnesium and calcium basic soap were prepared by metathesis process. The process started by mixing oils and hot ethanol (70°C). Then, stirred the mixture with an aqueous solution of sodium hydroxide at 20% (w/w) until it has reached a firm consistency. Magnesium acetate tetrahydrate and iron acetate dihydrate solution in distilled water added slowly to the mixture while continually stirring. The resulting basic soaps were filtered, washed (by hot water) and dried for using in subsequent experiments. Saponification reactions were performed in a glass batch reactors.

Decarboxylation reaction were performed in a glass reactor (100 ml) operating in a semi batch mode and at atmospheric pressure. Decarboxylation proceeded at temperatures up to 370°C for 5 hours. Heat was supplied by a high temperature heating mantle and both liquid (the melted soap) and vapor temperatures were measured with thermocouple. Magnesium and calcium basic soap both from palm stearin oil as a reactant model compound and the reactions were performed without other catalysts.

In a typical semi batch experiment, a number of basic soaps carefully weighed were placed into the reactor. After that, the reactor was flushed with nitrogen in order to remove the remaining oxygen. Then, the reactor was heated to 370°C and maintained for 5 hours at that temperature. The liquid products were collected in a beaker.

The liquid reaction products were analyzed by gas chromatograph (GC) Shimadzu 2010 equipped with a capillary column (rtx-1) with dimensions of 30 m x 0.25 mm x 0.25 μm and flame ionization detector (FID). One microliter of sample was injected into the GC-FID with a split ratio 1:50 and the carrier gas (helium) flow rate was 42.9 ml/min. The injector and detector temperature were 340°C and 340°C, respectively. The following chromatographic temperature program was used for analysis: 40°C (at first) – 300°C (5°C/min) – 340°C (1°C/min, constant 45 min). A number of chemical standards were purchased enabling product identification and calibration.

3 Results and Discussion

The thermal catalytic decarboxylation of two different types of basic soaps, i.e., magnesium and calcium basic soap derived from palm stearin oils, respectively was studied. The reaction was performed at 370°C and atmospheric pressure for 5 hours.

3.1 Product yields from decarboxylation

Table 1 shows various products resulted of Mg and Calcium basic soap decarboxylation.

Table 1. Material balance for the reaction product.

Material	Yield (wt % of feed)	
	Mg base	Ca base
liquid biohydrocarbons	67.21	63.33
water	1.14	1.42
solid residues	28.23	32.93
gas	3.42	2.32

Table 1 shows that the liquid biohydrocarbon resulted of magnesium basic soap is 67.21%-wt, greater than 63.33%-wt of calcium basic soap. The calcium basic soap decarboxylation result the most solid residues product (i.e., 32.93%-wt) than its counterpart magnesium (28.23% -wt).

The gas products by calcium basic soap less than magnesium soap. This implies that magnesium basic soap is faster decomposed to light products than calcium. According to GC-TCD chromatogram, the gases generated from both magnesium and calcium basic soaps contain H_2, CO_2 and CH_4.

3.2 Qualitative analysis of liquid hydrocarbon

Figure 1 shows functional group profiles in liquid biohydrocarbon resulting from decarboxylation of both magnesium and calcium basic soaps, using Fourier transform infrared (FTIR) spectroscopy.

https://doi.org/10.1051/mateccconf/201815603035

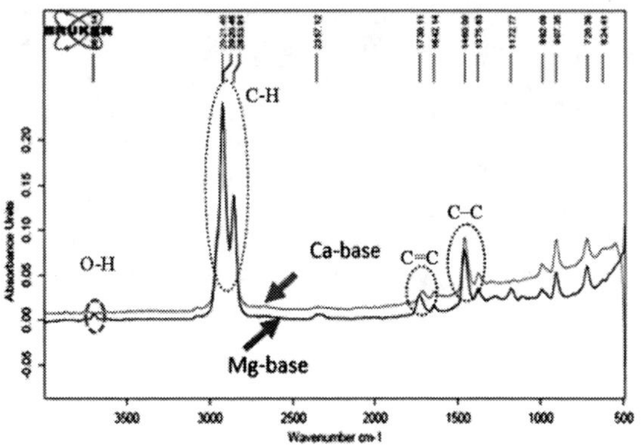

Fig. 1. FTIR of liquid biohydrocarbon from basic soap: red line, Ca base, and blue line, Mg base.

Figure 1 shows that the liquid product was resulted from both the magnesium and calcium basic soap decarboxylation contains the C-C, C = C and C-H functional groups in the range 1375-1460, 642-1730, and 2853-2921 cm^{-1}, respectively. Whereas, the hydroxide (O-H) group at the wave number 3618 cm^{-1}, only detected in the liquid product of the basic soap magnesium. The results mentioned above show that oxygenated compounds still exist in the liquid biohydrocarbon from magnesium basic soap than calcium basic soap.

The GC-MS analysis results, also confirmed that about 2.6% - mole of ketone compounds are present in the liquid biohydrocarbon from magnesium basic soap. Whereas, it is not detected in the liquid biohydocarbon from calcium basic soap.

3.3 Distribution of liquid products profile

The GC-FID chromatogram pattern, related to the distribution of the liquid biohydrocarbon fraction resulting from the decarboxylation of both magnesium and calcium basic soaps derived from palm stearin oil, is shown in Fig. 2. Figure 2 shows that the decarboxylation of magnesium and calcium basic soaps produce liquid biohydrocarbons in the range of C_8-C_{20} carbon number. According to GC-FID chromatogram the n-pentadecane and n-tetradecane was the main product in the decarboxylation of both magnesium and calcium basic soap. However, few of the 1-pentadecene also was formed in the decarboxylation of magnesium basic soap. In case of n-octadecane, significantly were only formed in the decarboxylation of calcium basic soap. It means that calcium can catalyze the polymerization reaction of short-chain into long-chain liquid biohydrocarbon molecules. Figure 2 also shows that the liquid biohydrocarbon product contains alkane (consisting of n-alkanes and iso-alkanes) and alkene compounds. In case of long chain alkanes (e.g., C_{14}, C_{15} and C_{18}) are higher produced in the decarboxylation of calcium basic soap than magnesium basic soap. Whereas, in case of the C_8-C_{13} product, always produced n-alkane and 1-alkenes compounds in the relatively equal peak height in both basic soap.

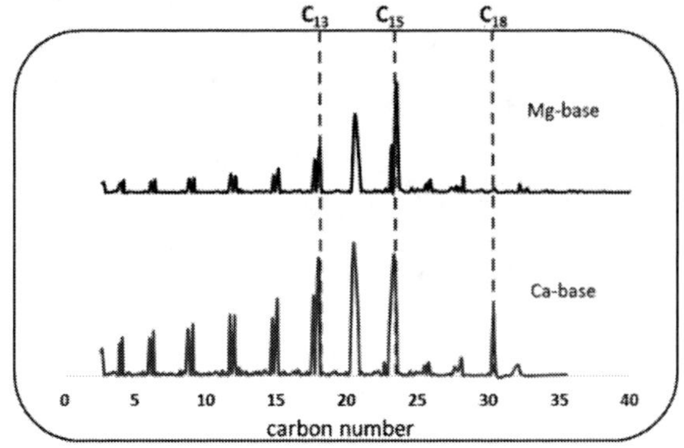

Fig. 2. GC-FID chromatograms patterns for liquid bio-hydrocarbon fractions.

3.4 Quantitative analysis: alkane and olefin

Figure 3 shows the results of the molar fraction analysis of the liquid biohydrocarbons in the range of C_8 to C_{20} carbon number, obtained from the decarboxylation of both magnesium and calcium basic soaps.

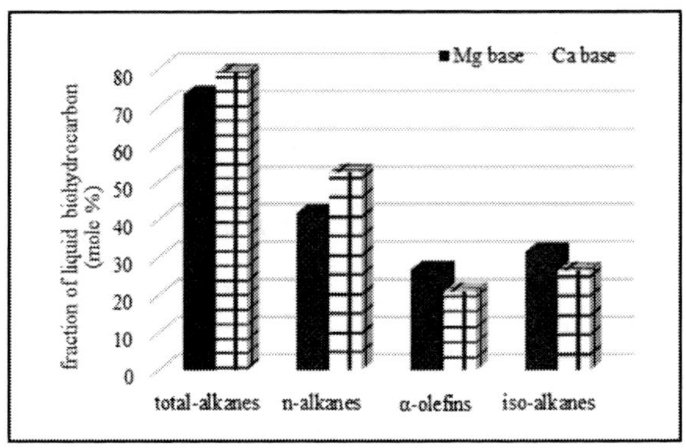

Fig. 3. percentages of C_8-C_{20} total alkanes, n-alkanes, iso-alkanes and α-olefins formed.

Figure 3 shows that, about 78%-mole of liquid biohydrocarbons produced from the decarboxylation of the calcium basic soaps are alkane compounds (i.e., n-paraffin + iso-paraffin), as would be expected in green diesel. The rest (approxymately 20%-mole) are various α-olefin molecules. Whereas, only about 70%-mole of alkanes resulted of magnesium basic soap. Therefore, the resulting α-olefin of magnesium basic soap is high, i.e., about 27%-mole or 7%-mole more than the same product of calcium basic soap.

Figure 4 shows the molar ratio of the n-alkane fraction to 1-alkenes according to C_8-C_{20} carbon number ranges of liquid biohydrocarbon of both magnesium and calcium basic soaps. Figure 4 shows that the molar ratio of n-alkanes to 1-alkene is found to be higher for C_{14}, C_{15}, C_{18} and C_{19} in the liquid product of calcium basic soap. This shows that the decarboxylation of calcium basic soap produces the n-alkane more than 1-alkene fraction for C_{14}, C_{15}, C_{18} and C_{19} molecules. In other words, calcium has high selectivity toward the formation of long chain n-alkane products, as which are expected to green diesel fuel. The more balanced the amount of the n-alkanes and 1-alkenes in the liquid product, its molar ratio close to one, as indicated by the C_8-C_{11} molecules.

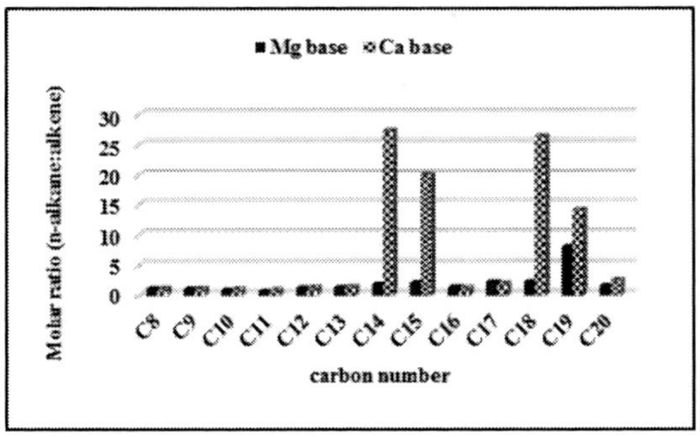

Fig. 4. The molar ratio of n-alkanes to 1-alkenes.

Decarboxylation of magnesium basic soap, yields a low n-alkane to 1-alkenes ratio. The same result occurs almost in all components of the liquid biohydrocarbon fractions, i.e., from the C_8-C_{20} molecule. Therefore, magnesium has low selectivity or is not specific to the formation of n-alkanes. Even magnesium tend to increase of 1-alkene product.

Figure 5 shows the profile of the molar yield of n-alkanes in the range of C_{18}-C_{20} carbon number, resulted from both magnesium and calcium basic soaps.

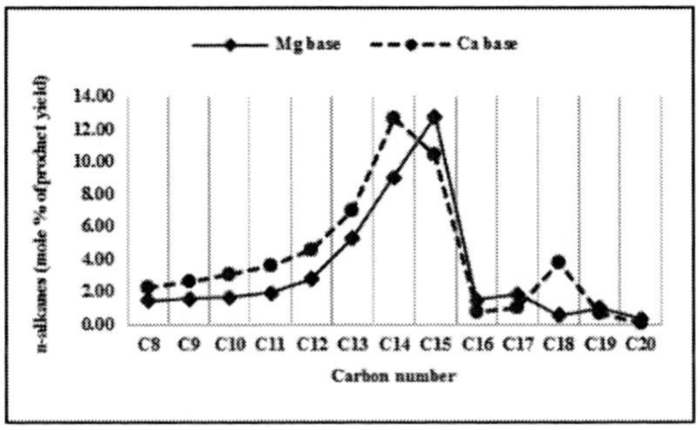

Fig. 5. Molar yield of C_8-C_{20} n-alkanes at 370°C for 5 hours.

Figure 5 shows that about 10%-mole and 13%-mole of 53%-mol total of n-alkane fraction (in the carbon number range C_8-C_{20}) resulting from decarboxylation of a calcium basic soap are n-pentadecane (n-C_{15}) and n-tetradecane (n-C_{14}), respectively. Whereas, only about 13%-mole and 9%-mole of 42%-mol total n-alkanes, generated from magnesium basic soap are n-pentadecane and n-tetradecane, respectively. From the total yield of n-alkane products were resulted from calcium basic soaps, about 29%-mole is a long chain alkane (C_{14}-C_{18}) suitable for green diesel, while only about 26%-mole of magnesium base soap. The n-octadecane product is detected in significant amounts (about 4%-mole), only in the liquid product of decarboxylation of the calcium basic soap. Apparently, the calcium may act as a polymerization catalyst of short chain to long chain biohydrocarbons.

4 Conclusions

Magnesium and calcium basic soaps, can be converted to liquid biohydrocarbon or drop-in fuels, especially green diesel via decarboxylation method. Green diesel generated from calcium basic soaps, contains 78%-mole total alkanes, higher than magnesium basic soap (only about 70%-mole). No ketone compound is detected in green diesel produced by calcium basic soap. The product distribution, in particular the concentration of the saturated C_{14} and C_{15} biohydrocarbons, is significantly higher in the green diesel produced by calcium basic soaps. This proves that, the use of calcium in basic soaps, significantly increases the selectivity to long chain n-alkane products (especially C_{14}, C_{15} and C_{18}), which are expected by green diesel fuel.

References

1. J. J. Yoon, *Advanced Biofuels USA*, 2006
2. I. Kubic˘kova´, M. Sna°re, K. Era¨nen, P. M. Arvela, D. Y. Murzin, Catal. Today **106**, 197-200 (2005)
3. C. Chang, S. Wan, Ind. and Eng. Chem **39**, 1543-1548 (1947)
4. Klare S. Markley, *Fatty Acids,* (InterSciences Publisher INC, New York, 1947)
5. A. Zhang, Q. Ma, K. Wang, X. Liu, P. Shuler, Y. Tang, Appl. Catal. A: General **303,** 103 – 109 (2006)
6. J. G. Na, B. E. Yi, J. N. Kim, K. B. Yi, S. Y. Park, J. H. Park, C. H. Ko, Catal. Today **156**, 44 – 48 (2010)
7. S. A. Khromova, A. A. Smirnov, S. A. Selishcheva, R. G. Kukushkin, V. O. Dundich, L. I. Trusov, V. A. Yakovlev, Biocatal. **5**, 260 – 268 (2013)
8. W. Vermeiren, G. N. Van, Paten WO2011012439A1, 1 – 15 (2011)
9. G. Yasin, M. I. Bhanger, T. M. Ansari, S. M. S. R. Navqi, F. N. Talpur, J of Petro. Techn. and Alternatif Fuels **3**, 29-35 (2012)
10. G. Knothe, A. C. Matheus, T. W. Ryan, Fuel **82**, 971-975 (2003)
11. Ynowitz J., Ratclift M. A., McCormick R. L., Taylor J. D., Murphy M. J., National Renewable Energy Laboratory (2014)
12. M. Sato, J. Soc. Chem. Ind. Japan **30**, 252-260 (1927)
13. H. L. Hsu, J. O. Osburn, C. S. Grove Jr, Ind. and Eng. Chem. **42**, 2141 – 2145 (1950)

The Effect of C/N Ratio and Type of Microbes Sludge to Biogas Production: Combination of Tapioca Industrial Wastewater and Tofu Industrial Wastewater

Agus Hadiyarto[1,*], *Danny* Soetrisnanto[1], *Faishal* Miftahul Huda[1], *Nimas* Amelia Iswara[1]

[1]Chemical Enginering Department of Diponegoro University

Abstract. Biogas will be produced from a single organic substrate and controlled the ratio of carbon nitrogen by added inorganic nitrogen substances. The study has been tried to combine substrates from industrial tapioca industrial wastewater with tofu industrial wastewater. The experiments were carried out in batch anaerobic digesters with a C/N ratio range of 20 - 30 and variation of microbial sludge from cow rumen, anaerobic sludge of tofu wastewater treatment plant (WWTP) and anaerobic sludge of tapioca WWTP. The ratio of C/N in the range of 20-30 didn't significant effect on the volume of biogas production. The microbial from cow rumen gives the best biogas production performance. Biogas yield is 2,199 liter per kg COD.

1. Introduction

Biogas can be produced by the decomposition of one types of substrate organic, but we need chemicals such as Urea, NPK. or nitrate salts for control C/N ratio. However, given the variation of the elemental content present in various substrates or wastes, it is possible to mix two or more types of waste as biogas substrate as well as regulate the composition of its elements. The combination of different types of substrates or wastes to be converted into biogas is known as co-digestion [1].

Co-digestion is an anaerobic processing of mixtures of two or more different organic substrates with the aim of increasing the efficiency of the digestation process [2]. The co-digestion process is usually carried out by mixing the base substrate with an additional substrate in small amounts [3]. Co-digestion has better nutritional balance and digester work so the ability to produce biogas is higher [4]. Many studies have developed co-digestion methods such as biogas production from industrial waste beer and cow dung [5], co-digestion of livestock manure with food and sludge waste in order to increase biogas production [6], co-digestion of Thaihu algae and waste food waste in kitchens [7], anaerobic co-digestion of water hyacinth and sheep dung [8] and many others. Materials used as the main ingredients are generally derived from livestock waste and food industry waste. In this study, mixed tapioca industrial liquid wastes with tofu industry liquid waste. In tapioca industry liquid waste has an organic content rich in carbon. According to J. Fettig (2013) in tapioca industrial liquid waste containing an organic substrate expressed as COD of 20.640 mgO2/l [9], containing carbohydrate 25.37 gr in 100 g tapioca liquid waste, carbohydrates in tapioca industrial liquid waste is 25.37 gr in 100 gr of tapioca liquid waste [10]. The waste also contains 0.46% nitrogen and 39.58% carbon, with a C/N ratio of about 86: 1. In situations where the amount of nitrogen required for bacterial growth is very small compared to carbon it will limit the production of biogas [11]. Nitrogen can be added in inorganic form eg ammonia or in organic form eg urea, manure or food waste [12]. Nitrogen derived from tofu waste water has a nitrogen content of 161.5 mgN / l. [13]. Nitrogen content in this tofu waste can be utilized and combined with tapioca industrial waste to make biogas for C/N ratio is optimal, so it can increase biogas production.

The specific objective of this study was to study the effect of C/N ratio on the combination of liquid industrial liquid tapioca substrate with the tofu industry liquid waste to biogas produced production and to study the effect of active sludge type used to decompose waste on biogas produced production.

2. Methods

For the above purpose an organic substrate prepared from combination of tapioca industrial waste water and tofu industrial waste water filtered through 200 mesh, then stored at 4° C before use. Likewise activated sludge from anaerobic sludge of wastewater treatment plant (WWTP) of tapioca industry at Sidomukti Village, Pati Regency, activated sludge of anaerobic sludge of WWTP of Tofu/Tahu Industry in Semarang City and Cow rumen from slaughterhouse Semarang City. They are all taken and filtered through a 200 mesh before use. The study was conducted by varying the C/N ratio at 20,

* Corresponding author: agushadi55@che.undip.ac.id

25 and 30, system processed batch, in 2 liter biodigester. The substrate to microbial (F/M) ratio was set at 0.5, neutral initial pH. The volume of biogas produced is measured daily with the principle of the volume of water transferred by biogas equal to the volume of biogas produced (water displacement methode). The CODCr content of the substrate combination is also measured daily until the decomposition process is completed for about 60 days. The COD is analyzed by SNI 06-6989.15-2004. Initial COD of tapioca wastewater and tofu wastewater is 11,931 mg/l and 9,256 mg/l. Carbon content by Walkey and Black method is 11,309 mg/l and 8,690 mg/l. Total Nitrogen content by SNI 19-7030-2004 is 150 mg/l and 589 mg/l. The C/N ratio of tapioca and tofu wastewater is 75,1 and 14,75.

3. Results and Discussion

Based on data are presented at Figure 1, can be describe as follow. Before the 20th day, the volume of biogas produced is very small. After that there will be increased continously. The trend pattern of biogas production shown in the figure 1 indicates that the ideal residence time to produce biogas with cow rumen , minimum is on the 20th day. Beside from the figure it can be concluded that the largest accumulative biogas volume was produced at a C/N ratio of 30. However, the C/N ratio 20-30 didn't significant for produced of biogas volume [14]. With statistical analysis (ANOVA), obtained P value on biogas production of 0.821. It means that the ratio of C/N in the range of 20-30 does not give a significant effect on the volume of biogas produced.

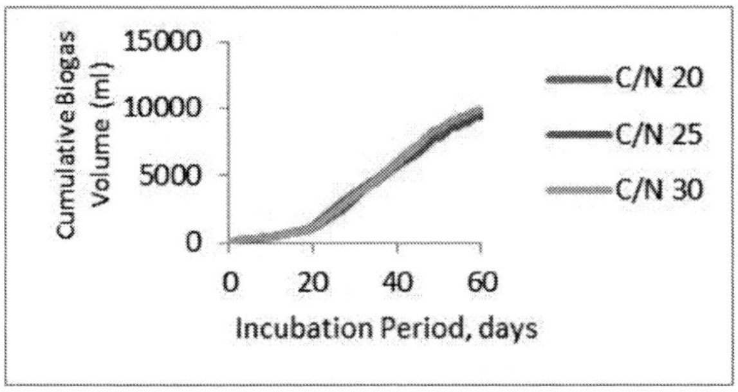

Fig. 1. Biogas production based on incubation period and C/N ratio

The low of C/N ratio (20-25) indicate that nitrogen in the substrate is available in large quantities. The final product of decomposition process is gas and ionic forms, among them is NH3 dan NH4+. Total ammonia nitrogen of 0,17- 14 g/L will be inhibited methanogenic activity to produce methane gas and reduce methane gas until 50% [15, 16]. By the concentration of NH3 in the range of 25-140 mg/l is toxis [17].

Based on biogas production rate, there are four stages , similar to the microbial growth pattern. Table 1 are presented the stages. The first is lag or adaptation stage in the range of 0-7 days incubation periods. Biogas production rate in the range of 43-49 ml/day, The second is growth stage in the range of 7-20 days incubation periods.

Table 1. Average Biogas Production rate, ml per day

Incubation Period, days	C/N 20	C/N 25	C/N 30
0 – 7	43	47	49
8 – 20	73	55	59
20 – 49	371	352	388
49 – 60	137	141	134

Biogas production rate increase slowly in the range of 55-73 ml/day. Biogas production rate increase fastly in third stage or the stationary stage (20-49 days) in the range of 352-388 ml/day. Finally, the fourth stages or death stage, the biogas production rate decrease slowly in the range of 134-141 ml/day.

Table 2. COD removed based on incubation period

Incubation Period, days	COD removed (mg/L)		
	C/N 20	C/N 25	C/N 30
0 – 7	892	537	613
7 - 20	2,685	4,081	4,283
20 - 49	3,848	3,744	4,024
49 - 60	1,204	1,466	1,505

The COD removed based on incubation period can be seen in Table 2. At growth stage of 7-20 days, organic compounds were converted from organic acid to fatty acid (propionic, butiryc and acetic acid), but methane gas hasn't been widely performed. At the stationary stage, the fatty acid were converted to methane gas. Table 3 presented biogas yield as a function of incubation period and C/N.

From the table it can be seen that most biogas production occurs at the stationary stage (incubation period range of 20-49 days) , greater than at other stages. Biogas yield in the range of 2,146-2,200 litre/kgCOD.

Table 3. Biogas Yield, Liter/KgCOD as a function of incubation period and C/N.

Incubation Period, day	C/N 20	C/N 25	C/N 30
0 - 7	0,404	0,735	0,676
7 - 20	0,422	0,212	0,216
20 - 49	2,200	2,146	2,199
49 - 60	1,454	1,270	1,177

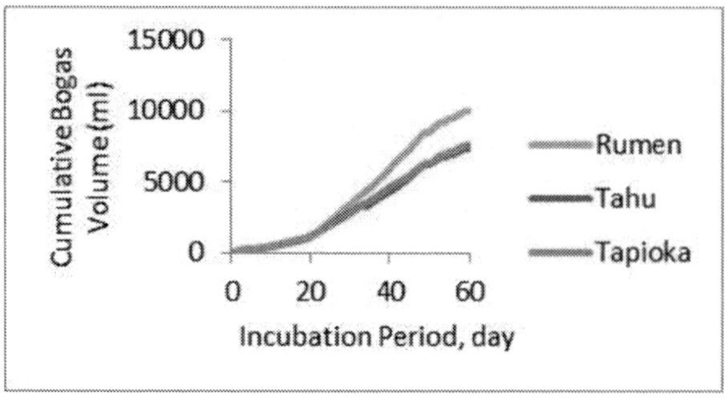

Fig. 2. Biogas production based on incubation period and variation of microbe sludges

The cumulative patern of biogas volume at incubation period value are illustrated in Figure 2 . The rate of biogas production with microbe from cow rumen had a better performance than other microbes. Forty to sixty per cent of microbes in the cow rumen include bacteria and protozoa. They digest non structural carbohydrates of tapioca (amylopectin) respectively.

The biogas production yield based on variation of microbe sludges or activated sludge , C/N ratio 30 , F/M = 0,5 was presented by Table 4.

Table 4. Biogas Yield, Liter/KgCOD as function of incubation period and activated sludge (AS).

Incubation Period, day	Cow Rumen	AS Tofu	AS Tapioca
0 - 7	0,676	0,390	0,319
7 - 20	0,216	0,408	0,474
20 - 49	2,199	1,865	2,051
49 - 60	1,177	2,130	1,688

Table 4 report that the largest biogas yield takes place at a stationary period with rumen microbes (2,199 liter biogas/kgCOD) then with microbe sludge from tapioca (2,051 liter biogas/kg COD) and microbe sludges from tofu (1,865 liter biogas/kgCOD).

4. Conclusions

Variation of C/N ratio range of 20-30 didn't signifi cance difference to biogas production. Microbes derived from cow rumen had better performance than microbes derived from anaerobic sludge WWTP of tapioca industry and activated sludge from WWTP of tofu. In the stationary phase , the biogas yield is 2,199 liters per kg of COD removed.

This research was funded by DIPA Faculty of Engineering Diponegoro University Year 2017 based on Decree of Dean of Faculty of Engineering Number 170/SK/ UN.3.3/V/2017 dated May 15, 2017.

References

1. Rowena T., Romano, Ruihong Zhang, Bioresource Technol. **99**, 3, 631-637 (2008)
2. Alvarez, J.A., Otero, L., Lema, J. M. A, Bioresource Technology **101**, 1153-1158 (2010)
3. Braun R, Wellinger A. Potential of Co-digestion,Task 37-Energy from Biogas and Landfill Gas. IEA Bioenergy (2003)
4. Wu, Wei. *Anaerobic co-digestion of biomass for methane production: recent research achievements* (Iowa State University, 2007)
5. Tewelde S, Eyalarasan K , Radhamani R, Kaliya perumal Karthikeyan. Int. J. Latest Trends Agr. Food Sci **2**, 2, 90-93 (2012)
6. Marañón E, Castrillón L, Quiroga G, Y. Fernández-Nava Y, Gómez L, García M.M., Waste Management **32**, 10, 1821-1825 (2012)
7. Ming –Xing Zhao, Wen-Quan Ruan. Energy Conversion and Management **75**, 21-24 (2013)
8. Fettig J, Pick V, Austermann-Haun U, Blumberg M, Phuoc NV., Water Sci Technol. **68**, 6, 1264-70. (2013)
9. Jagadish H. Patil, MAL AntonyRaj, B.B Shankar, Mahesh Kumar Shetty, M. K., B.P Pradeep Kumar, Energy Procedia **52**, 572–578 (2014)
10. Rahmatul H R., Nurrokhim A., Suwarno N., Nurkhamidah S., Jurnal Teknik Pomits, **2**,1, 1–5 (2013)
11. Sunarso, Siswo Sumardiono. *Pengembangan Teknologi Biogas Dari Limbah Tapioka Dan Limbah Peternakan* (Universitas Diponegoro, 2012)
12. Sterling M.C. Jr., LaceyR.E, Engler C.R, Ricke S.C., Bioresource Technol. **77**, 1, 9-18 (2001)
13. Damayanti A., Hermana J., Masduki A., Jurnal Purifikasi, **5**, 4, 151–156 (2004).
14. Deublein D, Steinhauser A. *Biogas from waste and renewable Resource : An Introduction*. (Wiley-VCH Verlag GmbH & Co. KGaA. Weinheim, 2008)
15. Kayhanian M., Environ.Technol. **20**, 355-365 (1994)
16. Zeshan OP, Karthikeyan, Bioresource Technol. **113** (0), 294-302 (2012)
17. Guerrero L, Omil F, Mondes R, Lema J.M., Bioresource Technol. **61**, 1, 69-78 (1997)

The Digester Modification for Biogas Production from Palm Oil Mill Effluent by Batch System

Martha Aznury[1,*], *Jaksen M.* Amin[1], *Abu* Hasan[1], and *Triadi* Utomo [1]

[1]Department of Chemical Engineering, Politeknik Negeri Sriwijaya, Palembang
Jl. Srijaya Negara Bukit Besar,Palembang 30139, Indonesia

Abstract. The purpose of this research is to determine the optimum production of biogas produced from the tuncated pyramid-shaped modification digester equipment on sedimentation tank and shaped beams on the tank fermentation batch. Pyramid-shaped sedimentation tank aiming for optimal deposition process, While the fermentation tank with beam-shaped used to produce biogas that is optimal. The raw material used is Palm Oil Mill Effluent. In the early stages of process is carried out the sedimentation in the first digester tank at a flow rate of 6 liters/minute and then observing the time stayed for 24 hours. POME flowed into the second digester tank to run into fermentation process by adding active microbial seed to produce biogas. After fermentation is complete, streamed to third digester tank to experience water treatment stage before being released into the environment. COD content test value obtained after the processing of 100 mg/l, while the BOD value is 30.9 mg/l. The percentage of methane gas obtained from starter concentration of 30%, ie 9.82% mol for fermentation time of 10 days, while the fermentation period of 20 days is 15.8 mol%. The production of methane gas obtained for the optimum fermentation period of 30 days ie 33.19% mol. For days 31 through 40, the production of biogas run into highs and caused due to the substrate in the digester began to run out and most of the substrate has not been fermented.

1 Background

The rapid development of the palm oil industry sector has resulted in other impacts. The process of crude palm oil products (CPO) produced solid waste and liquid waste. Liquid waste from industrial of palm oil is palm oil mill effluent (POME). POME is estimated around 63 million ton/year in Indonesia [1].

In the palm oil industry, POME will be processed in a processing pond with an open pond system. Although the pool system is economical, it requires a wider, time-consuming and direct-release of CH_4 into the atmosphere from the breakdown of organic matter occurring in anaerobic ponds. Referring to this condition, modified digester equipment is made.

The modification of the digester was carried out on a change in the shape of previous research that has been done in the past year. According to a final report written by Sari (2015) about comparison of sedimentation in a beam-shaped and a truncated pyramid-shaped [2]. The result of efficiency between beam and truncated-pyramid that were slower compared to the truncated pyramid digester.

According to a previous report, 7.3564% mol CH4 with 30% stater and 4 days for fermentation in beam-shaped digester [3]. The other design was a trucanted.pyramid-shaped that with 30% stater and 4 days fermentation had 6.439% mol CH4. So the design of the beam-shaped digester had been proven that the fermentation process is better than the design with truncated-pyramid digester.

Based on this, the authors conducted a study of the process of wastewater treatment of POME by modifying the digester with trucated pyramid-shaped for sedimentation. The digester for fermentation is a beam-shaped. This design modification is expected to be applicable for industrial palm oil treatment in the potential of POME for capture of biogas. The modified digester design tool is one of the more effective and efficient ways in terms of place, time, and processing costs

2 Methodology

2.1. Raw material

Raw material is Palm Oil Mill Effluent (POME) from PT Minangan Ogan Tbk. Inoculum was get from manure and aclimitation with POME since 1 month before used on biogas production.

2.2 Modification digester

In the design of digester tools consists of several tools that are important components to support the POME for wastewater treatment. The design has trucated pyramid

* Corresponding author: martha_aznury@polsri.ac.id

for sedimentation tank (tank A) and beam-shapeds (tank B and C for fermentation dan collection tanks). Digester also have some supporting components such as centrifugal pump, flowmeter, tedlar bag, swan neck and vacuum pump. The modification of digester shown in Figure 1

2.3 Analytical

Analyzed POME before and after fermentation was used by the American Public Health Association [4]. Gas was analyzed by using using a gas chromatograph (GC-8A, Shimadzu, Kyoto).

.

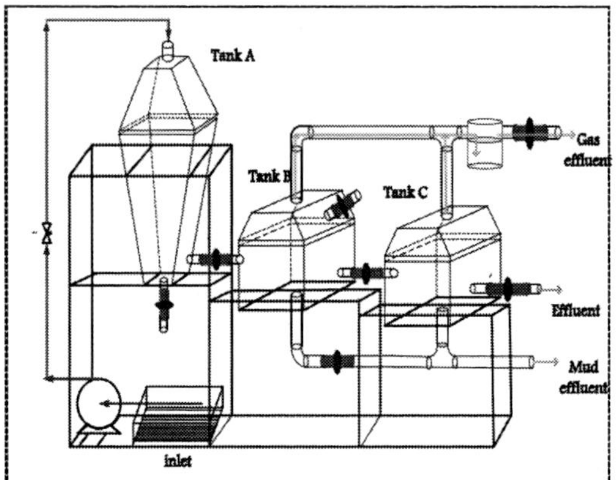

Fig. 1. Modification design of a truncated pyramid-beam Digester

POME is pumped into tank A for sedimentation process for 24 hours for the sedimentation process. Then, from tank A will be directly piped into the second tank B for the fermentation process for 1 month and catch gas in capture bag. Tank C was filled after fermentation process for collection before transfer to river bank or for felitizer process.

3 Results and discuss

3.1. POME after processing

POME using modified digester, the product can be obtained in the form of water and sludge (sludge). Before the water obtained from the processing at the digester dibuangke the water receiving body must first know the chemical oxygen demand (COD) and biological oxygen demand (BOD) because it could not pass the predetermined quality standard to prevent adverse effects on the environment. After analysis, it can be known that COD and BOD water content obtained from the results of this study are as follows in Table 1.

Table 1. Result of POME after Processing

Parameter	After 40 days Processing
COD	100 mg/liter
BOD	30.9 mg/liter
TDS	151 mg/liter
TSS	6200 mg/liter

Chemical oxygen demand (COD)
Biological oxygen demand (BOD)
Total dissolved solids (TDS)
Total suspended solids (TSS

The result of COD is 100 mg/liter where the value of COD is much lower when compared with Fahlevi (2015) with 4 days fermentation had a COD value of 784 mg/liter. From the data value of COD it can be seen that the COD value is determined based on the type of POME itself which means that the COD value of each palm oil liquid waste vary according to the source of the POME and the fermentation time greatly influences the reduced COD content. While for the value BOD after 40 days of fermentation is 30.9 mg/liter which is still much lower than that of previous research has BOD 100 mg/liter [3].

In this research, the results of COD, BOD and TSS content analysis in POME after anaerobic fermentation process for 40 days has met the standard quality standard for which has been established based on the governor regulation of South Sumatra No. 8 of 2012 on the quality standard of liquid waste for the palm oil industry. POME could not directly be discharged into the river or water source dikarana kan still have other contents that have not meet the standard quality standards such as TSS reached 6200 mg/liter, where the content is still too high compared to quality standards by the government of 250 mg/liter.

3.2 Sedimentation Process

The optimum sedimentation process for separating water and sludge from treatment of POME in pellet-shaped digester tool can be determined based on residence time and flow rate by looking at the result of previous research done that is using flow rate 6 L/m which can achieve sedimentation efficiency of 15.6% with the observed deposition period that is at 1, 3, 6, 12, and 24 hours [2].

In this research, there is one idea to optimally treat POME by creating a tool that represents the modeling system in waste treatment ponds in the palm oil industry, which is expected to provide more benefits in treating waste. The principle of processing done by sedimentation and fermentation process, in the hope that the absence of result from waste treatment is thrown away. The relationship between residence time to water volume and sludge at a flow rate of 6 liters/min can be seen in Figure 2.

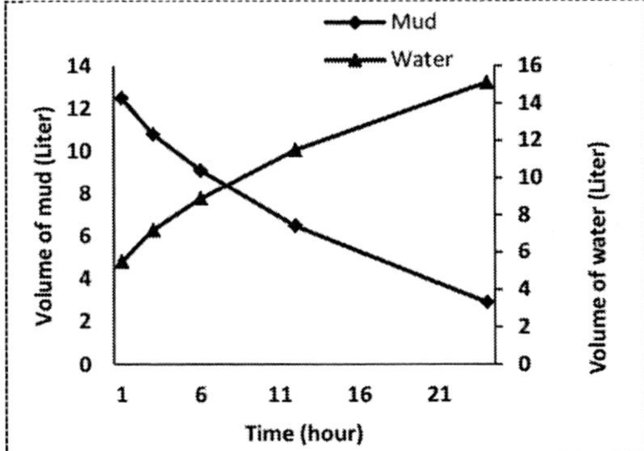

Fig. 2. Separation of Water and Mud Volume at 6 L/m

Figure 2 shown of separation between mud and water. At 1 hour, volume of mud are at 12.5 liters of the total feed volume of 18 liters. Until 24 hours decreased the volume of sludge on the pyramid-shaped digester reaches 2,9 liters. While the volume of water at 1 hour is 5,5 liters and continues to increase until the residence time of 24 hours is 15,1 liters.

This is because the sedimentation process that occurs in the digester takes place well during the residence time, where the longer the residence time the mud will settle completely and solidify on the base of the digester, causing more water to separate from the mud. The residence time is directly proportional to the flow rate, so the faster the flow rate will be the faster the time to stay. The longer the residence time, then the volume of water from the pyramid-shaped digester will be constant because the mud has settled completely [3].

3.3 Biogas production

POME which has been treated with a sedimentation process for 24 hours with a flow rate of 6 liters/minute in tank A, where the treatment is based on Sari research (2015) to obtain the separation of water and sludge with optimum efficiency.

Biogas is clean and renewable energy that can be an alternative to conventional energy sources that can cause problems for the environment and increase the rate of energy depletion in a long time. This biogas formation process requires a special installation called a digester, its function is that anaerobic reshuffle can take place well. The process of organically changing anaerobic material consists of four stages of the process of hydrolysis, acidogenesis, acetogenesis and methanogenesis. Therefore, this research treats POME to reduce the harmful chemical compounds of POME by utilizing the product of the processing of sludge and biogas (methane gas) by using pellet-shaped digester. The graph of biogas measurement results can be seen in Figure 3.

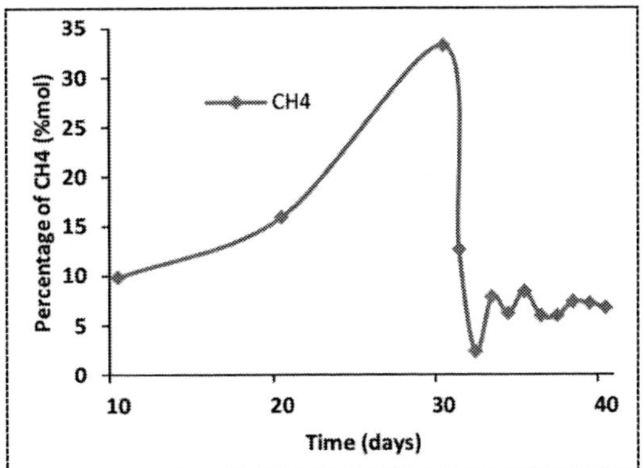

Fig. 3. Time Influence on Biogas Product Quantity

Figure 3 can be observed the effect of fermentation time on biogas produced. At the time of fermentation for 10 days, 20 days and 30 days can be seen that the quantity of biogas produced increased significantly so it can be seen that the bacteria grow and develop very well. The volume of gas increases with increasing fermentation time. The longer the fermentation time is the increased activity of microorganisms to use the substrate so that this will affect the resulting product [5]. At day 30 is the peak of biogas production of 33.1 mol%, it indicates that the substrate is maximally converted completely into biogas. When the number of bacterial populations increases, bacterial activity produces methane gas is also increased so as to produce methane gas with greater composition. In its growth phase, every living creature needs sufficient nutrients and supporting environmental conditions for the growth process, including bacteria [6].

Biogas production decreased in days 31 and 32 to 12.6 mol% and 2.3 mol%, it shows that the fermentation process does not have enough substrate in the digester to be converted into biogas resulting in decreased biogas production. At the time of long fermentation will be less biogas produced due to reduced substrate which is a nutrient for microorganisms [5]. The graph of biogas production on the 33rd day has increased again. This shows that microorganisms do not work maximally in substrate remodeling. Biogas production after the 34th to 40th day has up and down of trend. This indicates that the substrate has not been completely overhauled and there are some that have not been overhauled. The graph shows that the biogas fermentation time can be extended until there is not enough substrate available in the digester so that the graph of biogas production after the 40th day will decrease and eventually no biogas formation occurs.

Biogas production in this research achieved optimal result on 30th day that yield 33.1% mole of CH_4 gas which can be from fermentation process with 30:70% substrate concentration. The production of biogas above is much bigger compared to previous research with the same substrate concentration obtained only 7.3564% mole of CH_4 gas at 4 day fermentation time [7]. This is because in the research Saputri with fermentation within 4 days is not enough to remodel the substrate in the digester optimally. Fermentation time gives a significant difference to gas production, where fermentation time has a direct effect on biogas production [8]. The effect of fermentation on biogas production is due to the content of POME which contains lots of protein and high fat and takes several days to experience the process of fermentation.

3.4 Effect of oxygen on Biogas production

The process of biogas formation is done anaerobically. Bacteria break down organic matter into biogas and organic fertilizer. The process of weathering this organic material is done by microorganisms in the anaerobic fermentation process [9]. Biogas is a flammable gas produced from anaerobic processes at low temperatures and without oxygen [10]. Methane-forming bacteria are bacteria that do not require free oxygen in their metabolism, even the presence of free oxygen can be toxic or affect the bacterial metabolism [11]. The effect of oxygen on biogas production can be seen in Figure 4:

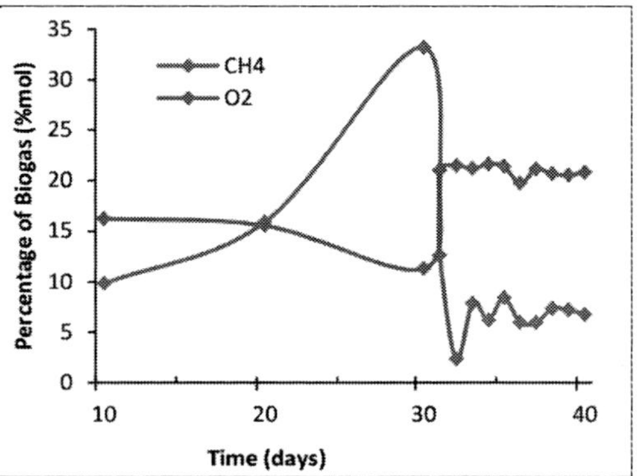

Fig. 4. Effect of Oxygen Content on Methane Gas Production

From Figure 4, the oxygen content on the 30th day is at the lowest position of 12.6 mol%. In this condition, methane gas production at highest condition compared to other day that is 33.1%. While on the 31st day oxygen content rose to 21 mole% which made the methane gas content decreased by 11.3% mole. Thus, it can be analyzed that the oxygen content can have an effect on the determination of CH4 or methane gas. The higher the oxygen content the less likely the formation of CH4 gas.

The amount of oxygen compounds contained in the sample due to environmental factors and the reaction of bacteria in the input composition of raw materials and starter that produce side compounds or byproducts in the form of oxygen gas. High levels of oxygen in biogas greatly affect the formation of methane compounds. This is because bacteria decomposer-forming methane compounds are anaerobic or do not require oxygen. If there is a content in the fermentation process will cause the bacteria to die and will not form a methane compound [12].

3.5 Comparison with previous research on biogas production

Biogas production is obtained from the fermentation process of POME in reactor, where the fermentation system used is batch process. The batch process is fermented by incorporating the media and starter simultaneously into the bioreactor and the product taking is done at the end of the fermentation. Comparison of biogas production results from previous researchers can be seen in Table 2:

Table 2. Comparison of Biogas Production Results with Previous Research

Forms Digester	% vol strater	Ferm.t ime (days)	% mol CH_4	Ref
Beam	10%	4	1.3811	Saputri, 2015
Pyramid	10%	4	1.7972	Fahlevi, 2015
Beam	20%	4	2.0607	Saputri, 2015
Pytamid	20%	4	1.4938	Fahlevi, 2015
Beam	30%	4	7.3564	Saputri, 2015
Pyramid	30%	4	6.439	Fahlevi, 2015
Modification of pyramid and beam	30%	30	33.199	This Research, 2016

From Table 2, it can be seen that previous study produced CH4 optimally at 30% volume of starter in 4 days fermentation time, where Fahlevi yield was 6.439% mole CH4 by using pine digester while CH4 The resulting Saputri is 7.3564% mole CH4 [3,5]. So from the data above, the form digester also affect the process of biogas formation. The beam-shaped digester produces more biogas than the pyramid-shaped diets. But the pellet-shaped digester is more efficient in the sedimentation process than the beam digester.

The biogas production result from previous research is still far below the optimum result of modified digester with % mole CH4 which is 33.199% mole with fermentation time for 30 days [3,5]. Then it can be analyzed that with the process of sedimentation and optimum fermentation process made digester modification tool can produce a more optimal biogas product. In addition, the fermentation time also affects the production of biogas where the longer the fermentation time the more biogas produced. Biogas production is affected by fermentation time because the time of fermentation is directly related to the amount of time required to pass through the methane generation stages of hydrolysis, acetogenesis and methanogenesis.

4 Conclusion

From the results of research on modification of pyramid-shaped digester for sedimentation and block-shaped stage in the fermentation stage in the wastewater treatment of palm oil industry, it can be concluded:

1. The result of analysis in determining COD and BOD of POME after processing in this study contained a small pollutant load compared to the standard quality standard that has been determined based on Governor Regulation of South Sumatera no. 8 of 2012 About the Standard of Liquid Waste for Oil Palm Industry namely:
2. The percentage of biogas production on the 10th, 20th and 30th day was 9.8%, 15.9% and 33.1% mole of CH4, while on the 40th day it was 6.73%. Then the production of biogas with batch system in the long fermentation time will decrease due to the substrate which starts to run out.
3. The optimum biogas production obtained from the digester modifier was produced on day 30 of 33.1% mole CH4

The authors would like to acknowledge the financial suport of **Penelitian Stategis Nasional,** Directorate General of Higher Education provides funding research project grants. NOMOR SPPK: 050/SP211/LT/DRPM /IV/2017. Also, authors acknowledge for PT. Perkebunan Mitra Ogan was a suport of POME.

References

1. [BPS] Badan Pusat Statistik, *Data Luas Lahan Sawit, Produksi, serta Ekspor CPO 2009-2015*,[doawnload 5 april 2016], 2016
2. Sari, Octa Endang, *Perbandingan Design Alat Sedimentasi yang Berbentuk Balok dan Limas pada Pengolahan Air Limbah Industri Minyak Kelapa Sawit,* Thesis of Politeknik Negeri Sriwijaya. Palembang, 2015)
3. Fahlevi, M Reza.. *Pengaruh Laju Alir Terhadap Waktu Tinggal Pada Digester Berbentuk Limas Secara Batch Dalam Mengolah Air Limbah Industri Kelapa Sawit Menjadi Biogas,* Thesis of Politeknik Negeri Sriwijaya (2015)
4. American Public Health Association (APHA), *Standard Methods for the examination of Water and Wastewater*, 19th Ed. (Washington, D.C., 1995)
5. Suryani, Y., Journal of Asian Scientific Research **12**, 789-797 (2013)
6. Budiyanto. A., K.. *Nutrisi Mikroba, Esensi Dasar untuk Kehidupan Mikroba*. froam http://zaifbio. wordpress.com/category/mikrobiolog, [doawnload at 12 June 2016], 2009
7. Saputri, Hardina Apri, *Pengaruh Persentase Volume Starter Dan Waktu Fermentasi Pada POME Terhadap Kuantitas Biogas Menggunakan Digester Balok Sistem Batch,* Thesis of Politeknik Negeri Sriwijaya (2015)
8. Mujdalipah, Siti, S. Dohong, A. Suryani, and A.Fitria,, *Pengaruh Waktu Fermentasi terhadap Produksi Biogas Menggunakan Digester Dua Tahap pada Berbagai Konsentrasi Palm Oil Mill Effluent dan Lumpur Aktif,* Thesis of Universitas Pendidikan Indonesia (2014)
9. Polprasert, C., *Organic Waste Recycling* (John Willey and Sons, Chicester, 1980)
10. Agustine, Restiyana, *Produksi Biogas dari Palm Oil Mill Effluent (POME) dengan Penambahan Kotoran Sapi Potong sebagai Aktivator,* Thesis of Institut Pertanian Bogor (2011)
11. Deublein, D. and Steinhauster, A., *Biogas from Waste and Renewabe Resources* (WILEY-VCH Verlag GmbH & Co. KGaA, Weinheim, Berlin 2008)
12. Noresta, Febrina. Jecika Yavia Nadiaty, dan M. Faizal, Jurnal Teknik Kimia **19**, 1, 2013

Drinking Water Treatment by Inductively Coupled Plasma System to Remove of Microorganisms in River Water

Reni Desmiarti[1,*], *Ariadi* Hazmi[2], *Primas* Emeraldi[2], *Munas* Martynis[1], *Yenni* Trianda[1], and *Ulung* M. Sutopo[1]

[1]Chemical Engineering Department, Universitas Bung Hatta, 25147, Padang, Indonesia
[2]Electrical Engineering Department, Andalas University, 25166, Padang, Indonesia

Abstract. Inductively coupled plasma system was used in drinking water treatment system to kill the microorganisms in water such as total coliforms (TC), fecal coliforms (FC) and other coliforms (OC) from river water. The aim of this study was to investigate the effect of flowrate on removal efficiency (RE), death rate, and death yield and energy consumption of bacteria's. The frequency of the system was set at 4.6 MHz. The results show that the removal efficiencies and death rate of TC, FC and OC decreased with increasing flowrate. Compared to FC, the first-order reactions of TC and OC were lower in the following order: FC > OC > TC. The death yield of TC and OC significantly increased when the removal efficiency increased. The electromagnetic flux varied from 19.44 to 20.55 W/cm^2 and the energy consumption was 0.26, 0.32, and 0.67 with flow rate at 20, 10 and 5 mL/minute, respectively. These results are very necessary to improve drinking water treatment.

1 Introduction

Kuranji River is a raw water source for the water treatment plant in Padang City, West Sumatra, Indonesia. In this river people wash, bathe and defecate. These activities cause concentrations of pathogenic bacteria, such as total coliforms and fecal coliforms are very high in water. Pathogenic bacteria such as total coliforms (TC), fecal coliforms (FC) and other coliforms (OC) have been detected in natural waters [1,2]. Pathogenic bacteria can also enter the surface water and groundwater. The Indonesian Ministry of Health has reported that pathogenic bacteria cause many human deaths in Indonesia through intestinal illnesses such as typhoid and paratyphoid fever.

In drinking water treatment plants, chlorine compounds are used to kill pathogenic bacteria. This process produces tri halo methanes (THMs) as chlorination byproduct from the chlorine reacting with other natural compounds in the water [3]. In the effluent of drinking water treatment plants, THMs concentrations in the range of 80-228 mg/L have been detected [3]. THMs are very carcinogenic, even at low concentrations. Besides, THMs can trigger the production of free radicals in humans, causing cell injury, the leading cause of cancer in the world.

Plasma can generate oxidizing species radicals ($H^\bullet$, $O^\bullet$, and $OH^\bullet$) and molecules (H_2O_2, O_3), electrohydraulic cavitation, shockwaves and ultraviolet light [4]. The removal of pathogenic bacteria has been studied using pulsed high voltage [5], photocatalysis using TiO_2 [6,7,8], anda combination of adsorbents with plasma [8]. The removal efficiency (RE) of microorganisms varied from 50-100%. The RE depends on the amount of bacteria, the flowrate and the current. Compared to using the addition of chlorine, fixed-bed filtration and floating media filter-microfiltration, induction plasma is a new technology for the removal of pathogenic bacteria from water. This technology achieves significantly enhanced disinfection of microorganisms and degradation of organic contaminants due to being environmentally friendly, making no demands on temperature and pressure, reaching higher removal efficiencies, and being unresponsive to pollutants.

In a previous study, the authors have investigated the removal of pathogenic bacteria such as TC, FC and OC using plasma in a batch experiment [9]. The removal of TC and FC was about 93% and 95%, respectively. The current work concentrated on the use of a plasma electric system to study the effect of flowrate on the treatment of pathogenic bacteria. The aim of this study was to investigate the application of inductively coupled plasma to remove pathogenic bacteria from river water. Based on the experimental data, the removal efficiency and disinfection yield were evaluated.

2 Material And Methods

2.1. Source of Water

River water was took from Kuranji river. This river is raw water for water treatment plant in Padang city, West Sumatra, Indonesia. The characteristics of river water is shown in Table 1.

* Corresponding author: renidesmiarti@gmail.com

Table 1. The characteristics of Kuranji river

Parameters	Unit	Values
Electric Conductivity (EC)	μS/cm	150
Total Dissolved Solid (TDS)	mg/L	75
pH	-	8.2
Total coliforms	CFU/100 mL	25.000
Fecal coliforms	CFU/100 mL	5.500
Other coliforms	CFU/100 mL	18.000

2.2. Experiments

The inductively coupled plasma system involved a radio frequency generator and a plasma reactor, as shown in Fig.1. This system was equipped with a current probe (Tektronix P6022 in USA), a high-voltage probe (Tektronix P6015A in USA), a Picoscope 4424 with four channels, and a glass reactor with a volume of 100 ml wrapped in copper wire (∅ 2 mm). To investigate the effect of flowrate on the removal of TC, FC and OC, the flowrate was controlled at 5, 10 and 20 ml/minute using a peristaltic pump. The effluent was checked at 10, 30, 60, 120, 150 minutes. The frequency of the system was set at 4.6 MHz [9]. The oscilloscope, the high-voltage probe and the current probe were used to record the applied voltages and discharge currents.

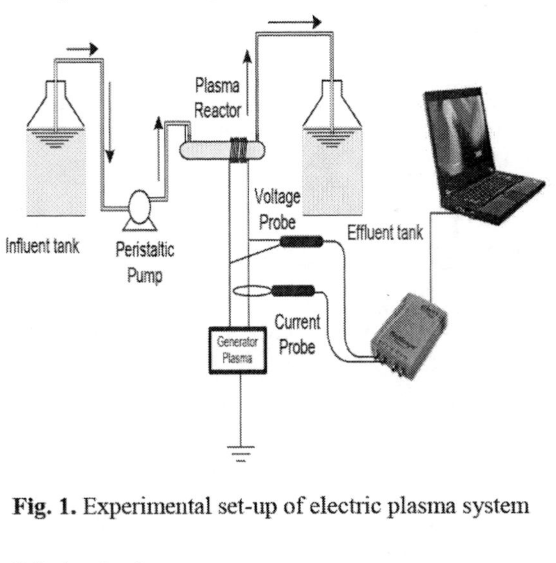

Fig. 1. Experimental set-up of electric plasma system

2.3. Analysis

The number of microorganisms was measured with the plate count method [8,12]. To give 10^1 dilutions, 0.5 mL of treated water was added into a tube containing 4.5 mL of normal saline water. Then, the tube was carefully mixed on a vortex for one minute. Agar plates were prepared by adding 14.05 g of nutrient agar (brilliant green lactose bile 2% (BGLB) broth by Oxoid, UK) to 500 mL of sterilized ultra-pure water.

To analyze the number of pathogenic bacteria, 1 mL of treated water was poured onto a petri dish with a diameter of 5 cm, containing 5 mL of melted nutrient agar and then kept at 37 °C in an incubator for 24 h. The total number of microorganisms was counted by enumerating purple, pink and white colonies formed as fecal coliforms, total coliforms and other coliforms, respectively, in a colony-forming unit (CFU/mL).

3 Results and Discussion

3.1. Death rate of microorganisms

Fig. 2 illustrates the removal efficiency (RE) of TC, FC, and OC when the applied frequency was set at 4.6 MHz and the flowrate at 5, 10 and 20 mL/minute. Fig. 2 shows that the RE of TC, OC and FC in the effluent increased to 44, 54 and 73%, respectively after 30 minutes of treatment time, and after running 60 minutes increased to 74, 89 and 100% with respect to TC, OC and FC, respectively. Compared with previous studies using groundwater the removal efficiency of FC and TC were lower [8,9]. Similar results were also found with the flow rate at 10 and 20 mL/minute (Fig. 2b and 2c). This is due to the high concentration of microorganisms in the river compared to groundwater. In another study, using a combination of floating media filter-microfiltration and coagulation, Chiemchaisri et al. [11] found a RE of FC and TC in the range of 96-98% and 52-98%, respectively. This hybrid system requires high energy and uses $FeCl_3$ and Alum as coagulants.

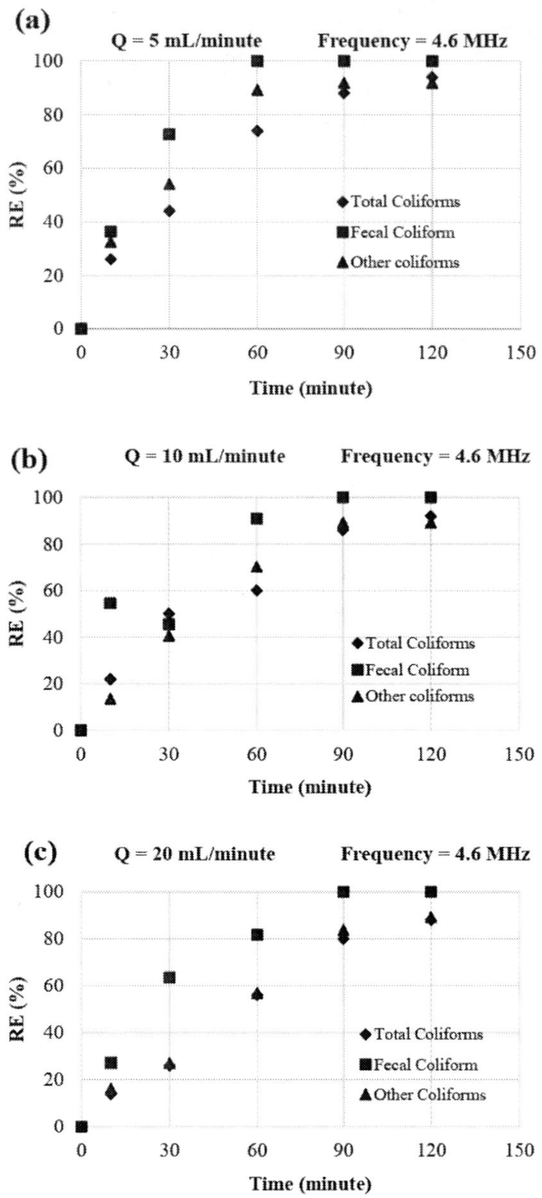

Fig. 2. Profile of pathogenic bacteria during experiment

The death rate kinetic (k) of the pathogenic bacteria was calculated by first-order reaction. k was calculated by plotting ln C/C_0 vs t. As displayed in Fig.3, the correlation coefficients (R^2) were 0.85 ~ 0.97. The k values were calculated from the slopes. The results show that the k values decreased with increasing flowrate, suggesting that the flowrate controls the disinfection of pathogenic bacteria in river water.

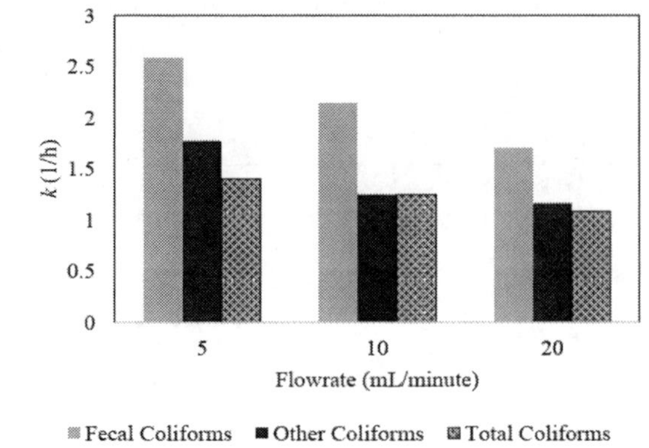

Fig. 3 The death rate of pathogenic bacteria at different flowrates.

3.2. Death yield

The death yield is defined as the number of deaths of pathogenic bacteria per unit of energy consumed in the process to describe the removal efficiency [8,12]. The death yield of bacteria was calculated using the following Eq. (1):

$$Y = \frac{N_0 \; x \; V x \frac{1}{100} \; x \; RE}{P \; x \; t} \tag{1}$$

where Y (CFU/KWh) is death yield (CFU/L), V (L) is volume of water in the reactor, P (KW) is average power, and t (h) is retention time.

The death yields (DY) with varied flowrates are shown in Fig. 4a-4c. Generally, the disinfection yields of FC, TC and OC increased with increasing removal efficiency (RE). The disinfection yield also increased with increasing flowrate. The death yield of TC was higher compared to FC and OC. In the present study a good correlation was obtained for the effect of flowrate on the death yield of TC, FC and OC.

Our previous study found that the death yield of TC was four times higher than that of FC with inductively coupled plasma in a batch experiment to treat ground water [8]. These results showed that the death yield of bacteria using a plasma system depends on energy consumption, flowrate and type of microorganism. In an inductively coupled plasma system, active species can be generated by decomposition of ozone and peroxide and it can also produce shock waves and ultraviolet light [4,5,13,14]. The most important result from the present study is that the flowrate is a key parameter in the removal of all microorganisms from water. Moreover, the results are needed to improve the design drinking water treatment plants in the future.

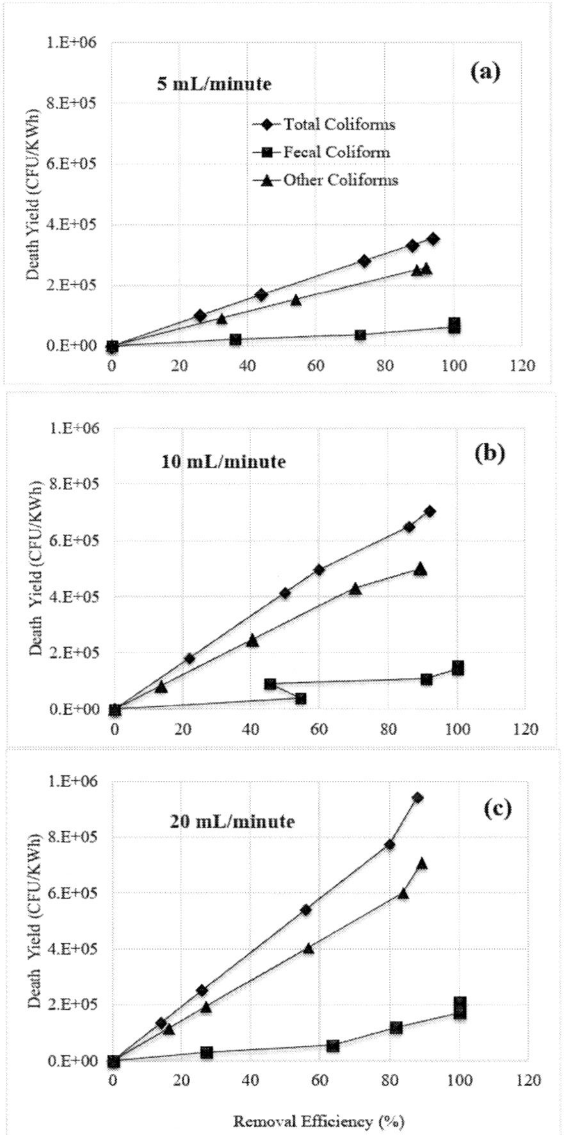

Fig. 4. Relationship between removal efficiency and death yield of pathogenic bacteria.

3.2. Electromagnetic Flux (S) and Energy Consumption (Econs.)

The performance of the inductively coupled plasma system was confirmed using electromagnetic flux (*S*) and calculated using Eq. (2) [6,13].

$$S = 9.94 \cdot 10^{-2} \cdot (I.n)^2 \sqrt{\frac{f}{\sigma}} \quad (2)$$

where S is electromagnetic flux (W/cm^2), I is current (A), n is number of turns per unit length (turn/cm), f is frequency (MHz), and σ is electric conductivity ($ohm^{-1}\ cm^{-1}$). The electromagnetic flux generated in the inductively coupled plasma reactor was quite similar in all runs, from 19.44 ± 1.63 to 20.55 ± 0.23 W/cm^2. The energy consumption was calculated at 0.67, 0.32, 0.26 KWh/L with flowrate at 5, 10 and 20 mL/minute, respectively. These results were achieved with a quite stable current during the experiments.

4 Conclusion

An inductively coupled plasma system was used to remove total coliforms (TC), fecal coliforms (FC) and other coliforms (OC) from river water. The removal efficiency of TC, FC and OC decreased with increasing flowrate. The removal efficiency of TC and OC in all runs was lower than that of FC. The death yields of pathogenic bacteria increased significantly when removal efficiency increased. During the experiments, the electromagnetic flux was quite similar, from 19.44 ± 1.63 to 20.55 ± 0.23 W/cm^2. To remove all pathogenic bacteria in river water, further research is needed to combine plasma system with others processes such as filtration.

This work was supported by Ministry of Research, Technology and Higher Education, Republic of Indonesia (RISTEK DIKTI) in accordance with the Contract of Research Number : 001/KONTRAK-PENELITIAN/K10/ KM/2017 dated April 10th, 2017.

References

1. G. Wilkes, T. Edge, V. Gannon, C. Jokinen. Lyautey, E.; Medeiros, D., Water Research, **43**, 2209-2223 (2009).
2. C. Levatensi, L. Bonadonna. Briancesco R.; Grohmann E.; Toze S.; Tandoi V., **45**, 587-602 (2012).
3. F.B.J. Jackline, S.M. Hiasmyne, C.B. Patricia, J.A. Nelio., International Journal of Food Microbiology, **190**, 9-13 (2014).
4. C. Chiemchaisri, S. Passananon, H.H. Ngo, S. Vigneswaran., Desalination, **234**, 335-343 (2009).
5. A. Hazmi, R. Desmiarti, E.P. Waldi, D. Darwison., *J. Engineering Technological Sciences,* **45,** 1, 1-8 (2013).
6. M. Khraisheh, L. Wu, A.H. Al-Muhtaseb, M.A. Al-Ghouti., Journal of Industrial and Engineering Chemistry, **28**, 369-376 (2015).
7. M. Kharaisheh, J. Kim, L. Campos, H. Al-Muhtaseb, A. Al-Hawari, M. Al Ghouti, G.M. Walker., Journal of Industrial and Engineering Chemistry, **20**, 3, 979-987 (2014).
8. S.N. Hussain, A.P. Trzcinski, H.M.A. Asghar, H. Sattar, N.W. Brown, E.P.L. Roberts., Journal of Industrial and Engineering Chemistry, **44**, 216-225 (2016)
9. R. Desmiarti, A. Hazmi, S. Ellyta, Y. Trianda., Research Journal of Pharmaceutical, Biological and Chemical Sciences, **6,** 1, 889-897 (2015).
10. R. Devi, E. Alemayehu, V. Singh, A. Kumar, E. Mengistie., Bioresource Technology, **99**, 2269-2274, (2008).

11. C. Chiemchaisri, S. Passananon, H.H. Ngo, S. Vigneswaran., Desalination, **234**, 335-343 (2009).
12. B. Van den Akker, T. Trinh, H.M. Coleman, R.M. Stuetz, P. Le-Clech., Bioresource Technology, **55**, 432-437(2014).
13. M. Magureanu, D. Piroi, N.B. Mandache, V. David, A. Medvedovici, V.I. Parvulescu., Water Research, **44**, 3445-3453 (2010).
14. H. Krause, B. Schweiger, J. Schuhmacher, S. Scholl, U. Steinfeld, Chemosphere, **75**, 163-168 (2009).

Characteristics of Nanoparticles in Drinking Water Treatment using Biological Activated Carbon

Reni Desmiarti[1,*], *Yoji* Morishita[2], *Tomonari* Fujisawa[2], *Yasushi* Ishiguro[3], *Toshiro* Yamada[2] and *Fusheng* Li[3]

[1]Chemical Engineering Department, Universitas Bung Hatta, 25147, Padang, Indonesia
[2] Civil Engineering Department, Gifu University, 501-1193, Gifu, Japan
[3] River Basin Research Center, Gifu University, 501-1193, Gifu, Japan

Abstract. Characteristics of nanoparticles in drinking water treatment were performed using five types of biological activated carbon (BAC) columns (BAC1-BAC5) in continuous flow experiments. The BAC was created by covering granular activated carbon (GAC) with attached microorganisms from water samples taken from the Nagara River in Japan. The total running time was about 2000 h. The characteristics of the nanoparticles were investigated based on size distribution and volume distribution measured by Zetasizer Nano. Total dissolved organic carbon (DOC) and ultraviolet absorbance at 260 nm (UV260) were also studied. The important results in this study were that the detached nanoparticles in the effluent were within the size distribution ranges of 0.26~5.62 nm, 0.62~3.62 nm, 0.62~3.12 nm, 0.62~4.19 nm, and 0.62~6.50 for BAC 1, 2, 3, 4 and 5, respectively. The profile of peak size and peak number along the bed depth of the BAC columns was evaluated for better understanding the characteristics of the nanoparticles. This result is very important for improving drinking water treatment using granular activated carbon to remove microorganisms.

1 Introduction

In Japan, water resource development used for domestic and industrial water accounts for 71% of water intake from rivers. The Nagara River, one of three rivers used as water resources in Japan, is used to satisfy the population demand for water in Aichi Prefecture, a region located in the center of Japan [1]. The water is flowed through a treatment plant and then distributed to communities in Aichi Prefecture. One of the technologies used in the drinking water treatment plant is adsorption by activated carbon.

Activated carbon technology is one of the most mature and effective processes in drinking water treatment. Application of activated carbon was demonstrated to remove organic micropollutants such as pharmaceuticals, phenols, Trihalomethane (THMs) and pesticides. The activated carbon was also used to reduce odor and colour in drinking water treatment plant [2,3]. Activated carbons have varied surface characteristics and pore size distributions. These characteristics of activated carbon play an important role in the adsorption of contaminants in water [4,5]. Biological activated carbon (BAC) filtration has been investigated as a pre-treatment for reducing organic fouling, thus allowing a higher sustainable flux for microfiltration [5,6].

Although there have been numerous experimental studies on water adsorption by activated carbons, little is known about the removal of nanoparticles using this method. Exposure to nanoparticles through the ingestion of drinking water may pose a potential direct human health threat or an indirect risk, including DNA damage [7,8]. The characteristics of normal organic contents (NOM) as DOC and UV260 have been studied [9,10,11,12]. The purpose of the present study was to investigate the characteristics of nanoparticles in water taken from the Nagara River using five types of biological activated carbon (BAC) columns (BAC1-BAC5) in continuous flow experiments. The characteristics of the nanoparticles were investigated based on size distribution and volume distribution measured by Zetasizer Nano. Zeta potential was also measured to predict the stability of colloidal nanoparticles in effluent. Furthermore, DOC and ultraviolet absorbance at 260 nm (UV260) were also studied to quantify the content of natural organic matter in water.

2 Materials and methods

2.1. Materials

Water samples were collected from the right side of the Nagara River, located near Nagara Bridge, Gifu Prefecture. The sampled water sample was filtered on site using a sieve of 38 μm pore size to remove relatively large organic matters such as fallen leaves and fallen branches floating in the river water. The sampling was conducted once a week.

* Corresponding author: renidesmiarti@gmail.com

2.2. Preparation of activated carbons

In this study, five activated carbons with different pore distributions were pulverized and used for the experiment. Before using the activated carbon for the experiment, the surface area was cleaned by washing with water and boiled for about 1 hour to remove impurities in the activated carbon pores and then dried at 105 °C to 115 °C (FO810, Yamato Scientific Co., Ltd.). In this study, five types of activated carbon were used; the characteristics of each type are presented in Table 1.

Table 1. Activated carbon pore volume and pore size distribution

Name	Pore volume (cm^3/g)	Pore size distribution (m^2/g)
BAC 1	0.532	1192
BAC 2	0.356	1025
BAC 3	0.483	1088
BAC 4	0.350	1200
BAC 5	0.518	1029

2.3. Continuous flow experiment

The column used for water the flow experiments was composed of a vinyl chloride tube with a diameter of 2.54 cm. The tube was filled with activated carbon to a depth of 15 cm. The flow rate of the sample water was controlled by adjusting the pump so that the residence time in each column was constant (20 min). A schematic figure, similar to that given in Li et al. [13], the granular activated carbon column apparatus is shown in Figure 1.

The effluent was collected twice every week and promptly filtered through a 0.2 μm membrane filter (manufactured by ADVANTEC Corporation, Cellulose Acetate material) for the purpose of analysis.

2.4. Analysis

The effluent from each column was sampled. The sampled water was filtered using a 0.2 μm membrane filter (manufactured by ADVANTEC Corporation, Cellulose Acetate material) before being subjected to analysis. Dissolved organic carbon (DOC) was analysed using a total organic carbon (TOC) analyser (TOC-V, manufactured by Shimadzu). A UV-visible spectrophotometer (U-3210, Hitachi Co., Ltd.) was used for measurement of the UV-absorbance at 260 nm (UV260). The index of UV260 is accomplished of reflecting the present of humic molecules in water and the specific ultraviolet absorption (SUVA), a calculated parameter of UV260 divided by DOC, can thus be used as a humic content indicator.

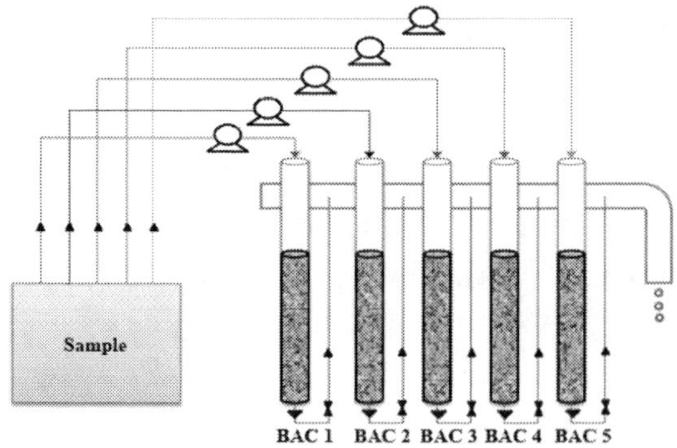

Fig. 1. Schematic figure of granular activated carbon column apparatus.

Three-dimensional fluorescence excitation-emission matrix (RF-EEM) spectroscopy was used to differentiate different types and origin of dissolved organic carbon (DOC) in the effluent. EEM was measured using an RF-5300 PC spectrofluorometer (manufactured by Shimadzu Corporation) using a xenon lamp as light source. For measurement conditions, the slit width was 5 nm for both excitation wavelength and fluorescence wavelength; the measurement range was 220 to 550 nm; the data sampling interval was 5 nm; the sensitivity setting was high; and the response time was 0.02 sec. The fluorescence intensity of the obtained data was calculated by correcting the fluorescence intensity at excitation wavelength 350 nm / fluorescence wavelength 450 nm of the 10 ppb quinine sulfate solution at room temperature to the relative fluorescence intensity as 1 QSU (Quinine Sulfate Unit). Furthermore, nanoparticle size distribution and zeta potential analysis were carried out using a zeta potential meter (Zetasizer Nano, manufactured by Malvern).

3 Results and discussions

3.1. UV absorbance at 260 nm wavelength

The behavior of natural organic matter and nanoparticles in each column over elapsed time from the continuous flow experiment is shown in Figure 2. In each column, the removal rate exceeded 80% at the initial stage of the experiment. However, after over 500 hours, the removal rates of columns BAC 2 and BAC 3, which had relatively large pore volumes, gradually decreased. The performance of columns BAC 1 and BAC 5 remained stable throughout the experiment with an average removal rate of 80%.

This indicates that for all organic components contained in the Nagara river water, UV260 components were excellently adsorbed, while some constituents not detected by UV260 were less favourably adsorbed. This is similar to the result found by Li et al. [13].

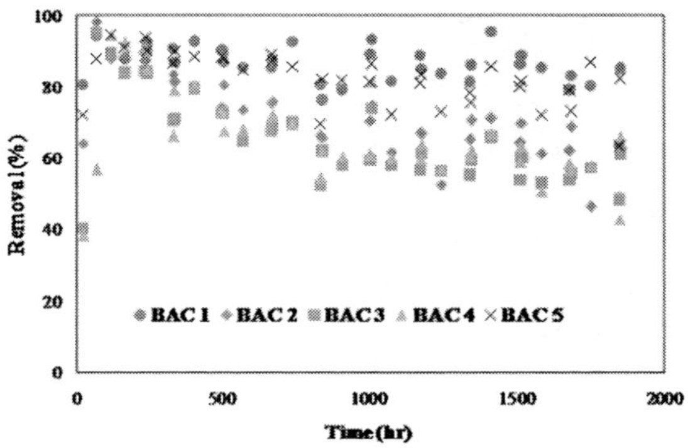

Fig. 2. Removal of organic matter analysis by UV260 absorbance.

3.2. Dissolved organic carbon (DOC)

The dissolved organic matter measured as DOC removal during the experiment is shown in Figure 3.

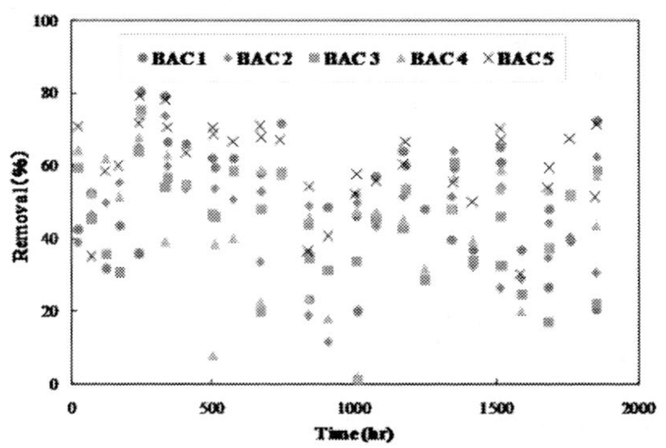

Fig. 3. Removal of DOC removal during continuous flow experiment.

Looking at the changes in DOC removal rate over time for each column, the results are varied and it is difficult to confirm a clear trend, even if the pore sizes and pore size distributions of the granular activated carbons packed in these columns were different, as shown in Table 1. It is thus reasonable to infer that the organic contents of the effluent consisted of those not absorbable by activated carbon. The presence of non-absorbable components was also defined in Nagara river water [13].

3.3. Fluorescence properties

A three-dimensional excitation emission matrix fluorescence spectroscopy (3DEEM) can provide significant information on the chemical nature and source of dissolved organic carbon (DOC) [14, 15]. Figures 4-9 present the 3DEEM characteristics for the DOM fractions for raw water and effluent in each BAC column. Peak 1 (Ex/Em, Ex: 320-350 nm; Em: 435-450 nm) and peak 2 (Ex/Em, Ex: 250-260 nm; Em: 430-460 nm) shown in each figure represent humic-like substances, while peak 3 (Ex/Em, Ex: 260-270 nm; Em: 340-350 nm) represents protein-like substances.

From Figure 5-9 it can be seen that after 2 weeks of experiment only protein-like substances were observed in every BAC column, suggesting the formation of biofilm. The humic-like substance peak did not appear in all columns, indicating that adsorption in the BAC column had occurred. However, this peak was not observed after 4 weeks of continuous flow experiment. Also, there were no differences in the maximum Ex/Em wavelengths of the protein-like fractions, except for column BAC 5. Coble et al [15] has informed that DOM fractions of the blue shift of the Ex wavelength reflect a quite small molecular weight composition (lower intensity), while a red shift reflects a quite larger molecular weight composition (higher intensity). Peak 3 of column BAC 5 displayed the highest fluorescence intensity at 2 weeks, but it could hardly be observed after 4 weeks.

3.4. Size distribution and zeta potential

The nanoparticle size distribution was measured of which the result is shown in Figures 10 and 11. After 2 weeks of experiment, particles of 0.35-10 nm represented the majority of nanoparticle occupied volume.

However, after 4 weeks of experiment, there was a shift in size distribution. The size of nanoparticles found in columns BAC 2 and BAC 3 was increased, as larger sizes of nanoparticles occupying the effluent were detected. This indicates that after 4 weeks, columns BAC 2 and BAC 3 had a limitation in adsorbing the organic matter contained in the river water. For the other columns, the size distribution remained the same. The detached nanoparticles in the effluent were within the size distribution number ranges of 0.26-5.62 nm, 0.62-3.62 nm, 0.62-3.12 nm, 0.62-4.19 nm, and 0.62-6.50 for BAC 1, 2, 3, 4 and 5, respectively.

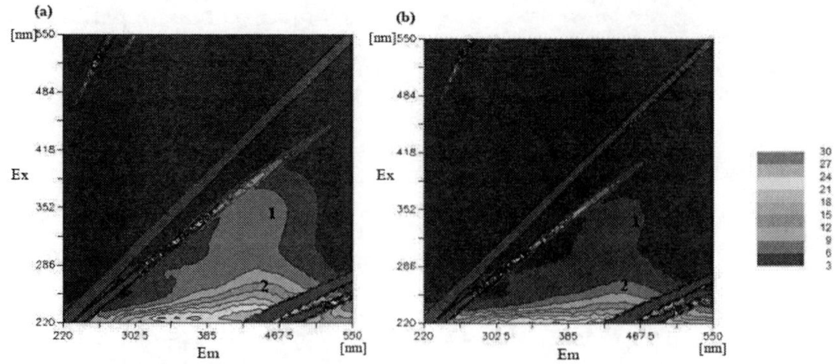

Fig. 4. The 3DEEM plots of raw water (a) at 2 weeks (b) at 4 weeks

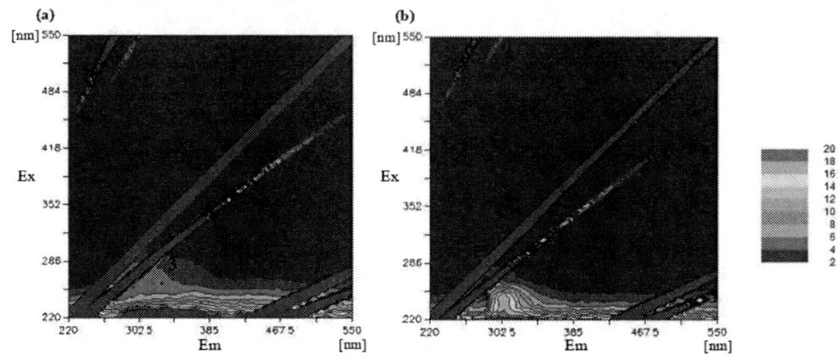

Fig. 5. The 3DEEM plots of BAC 1 (a) at 2 weeks (b) at 4 weeks

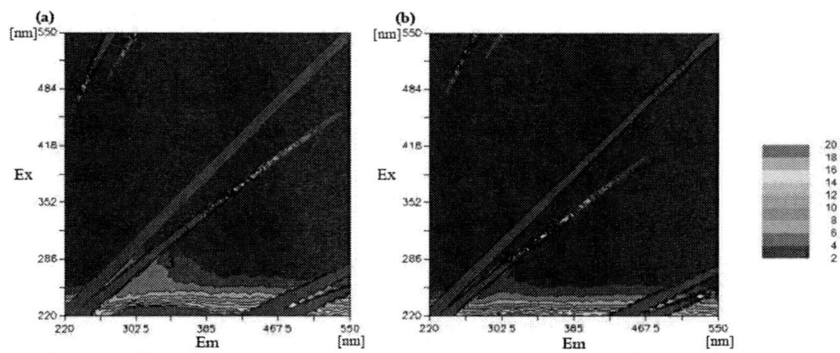

Fig. 6. The 3DEEM plots of BAC 2 (a) at 2 weeks (b) at 4 weeks

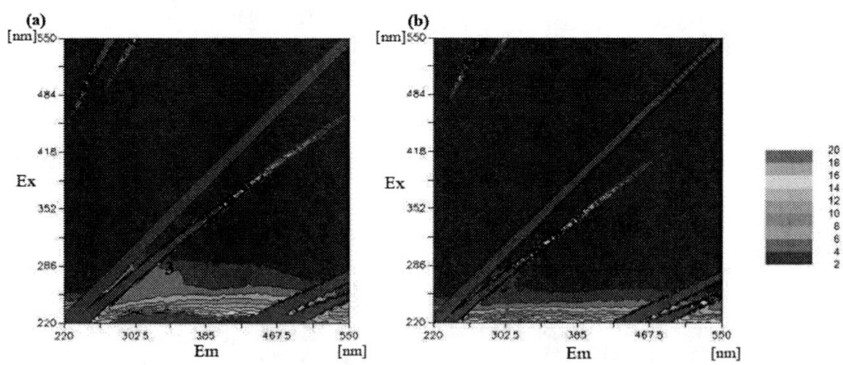

Fig. 7. The 3DEEM plots of BAC 3 (a) at 2 weeks (b) at 4 weeks

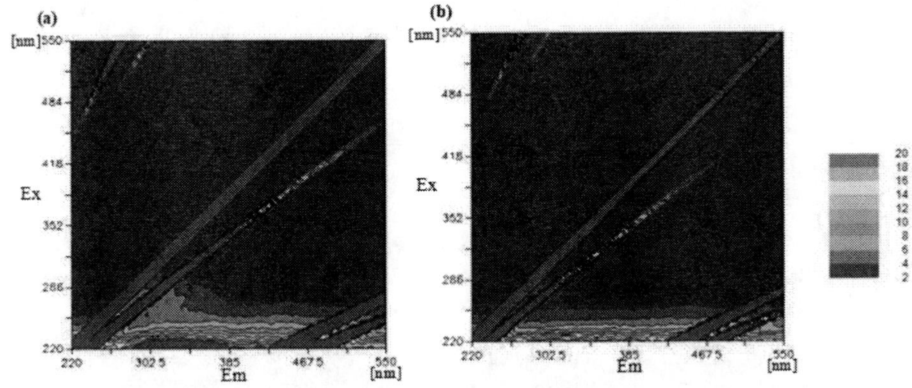

Fig. 8. The 3DEEM plots of BAC 4 (a) at 2 weeks (b) at 4 weeks

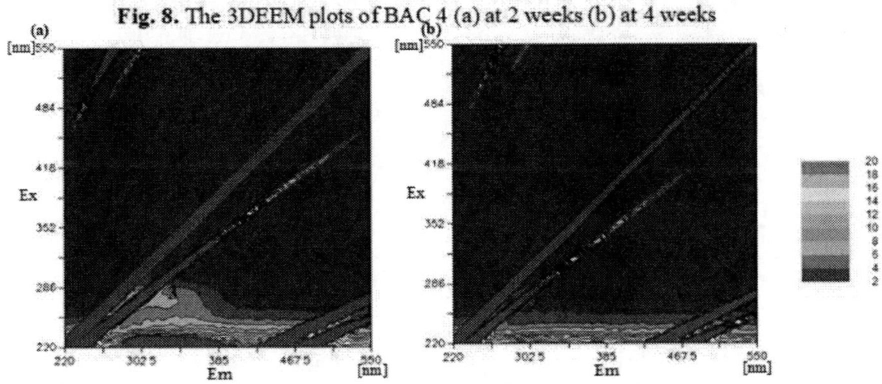

Fig. 9. The 3DEEM plots of BAC 5 (a) at 2 weeks (b) at 4 weeks

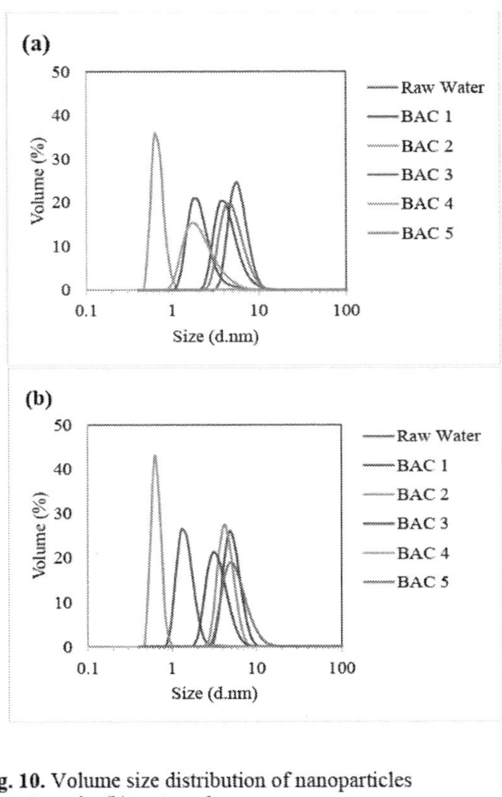

Fig. 10. Volume size distribution of nanoparticles
(a) at 2 weeks (b) at 4 weeks

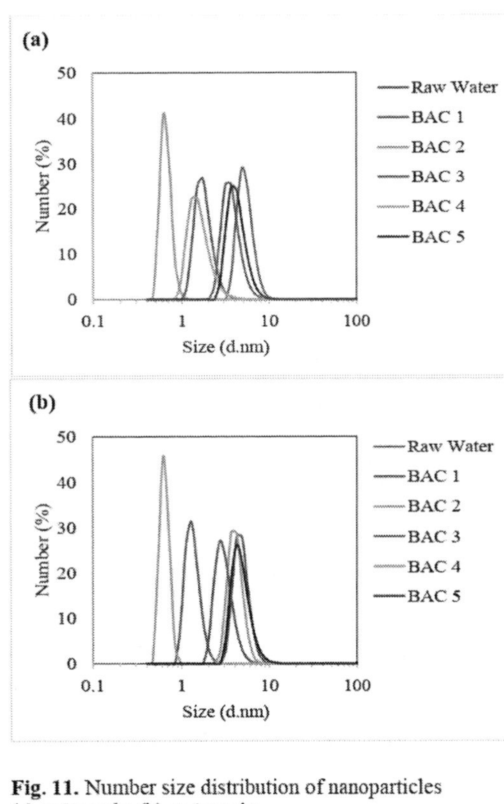

Fig. 11. Number size distribution of nanoparticles
(a) at 2 weeks (b) at 4 weeks

Table 2 summarizes the zeta potential and electrophoretic mobility values obtained for the different continuous flow columns of biological activated carbon. The table also includes the measured conductivity values for these samples. In the results summarized above, the measured mobility values after 2 weeks of experiment were consistent for every BAC column. However, after 4 weeks of investigation, the mobility values became less negative, i.e. the magnitude of mobility decreased. This indicates that the increase of nanoparticles present in the effluent resulted in obscuration of light transmission. Furthermore, the measured zeta potential values were decreased after 4 weeks of experiment. This indicates that the stability of the colloidal nanoparticles was reduced and adsorption by biological activated carbon had occurred.

Table 2. Zeta potential and electrophoretic mobility results for various biological activated carbon (BAC)

Time (weeks)	AC column	Zeta potential (mV)	Electrophoretic mobility ($m^2Vs^{-1} \times 10^{-8}$)	Conductivity ($mS\ cm^{-1}$)
2	BAC 1	-22.9	-1.7980	0.0818
	BAC 2	-21.9	-1.7170	0.0817
	BAC 3	-24.1	-1.8870	0.0851
	BAC 4	-10.3	-0.8048	0.0857
	BAC 5	-11.9	-0.9296	0.0817
4	BAC 1	-4.68	-0.3669	0.0857
	BAC 2	-3.18	-0.2492	0.0844
	BAC 3	-6.05	-0.4746	0.0802
	BAC 4	-2.64	-0.2071	0.0842
	BAC 5	-7.77	-0.6093	0.0850

4 Conclusions

Dissolved organic matter in Nagara River was characterized into two major components: humic-like substances and protein-like substances. After 2 weeks of experiment only protein-like substances were observed in every BAC column. However, the humic-like substance peak did not appear in all columns, indicating that adsorption by BAC had occurred. Organic matter removal rates of 80% were obtained in each BAC column as indicated by UV260 absorbance and DOC measurement. After 4 weeks of continuous flow experiment, the measured zeta potential values were decreased, indicating that the stability of the colloidal nanoparticles was reduced and adsorption by biological activated carbon had occurred. The detached nanoparticles in the effluent were within the size distribution ranges of 0.26-5.62 nm, 0.62-3.62 nm, 0.62-3.12 nm, 0.62-4.19 nm, and 0.62-6.50 for BAC 1, 2, 3, 4 and 5, respectively.

This study was supported the JASSO Follow-up Research Fellowship for the fiscal year 2016 from Gifu University, Japan. This work was also supported by Ministry of Research, Technology and Higher Education, Republic of Indonesia (RISTEK DIKTI) in accordance with the Contract of Research Number: 001/KONTRAK-PENELITIAN/K10/ KM/2017 dated April 10th, 2017

References

1. Kiso River Lower Reaches Works Office, Chubu Regional Construction Bureau, Ministry of Land, Infrastructure and Transport, (2004)
2. E. Worch, *Adsorption technology in water treatment* (Walter de Gruyter GmbH & Co. KG, Berlin/Boston, 2009)
3. R. Desmiarti & F. Li, ITB J. Eng. Sci., **43**, 2, 153-160 (2011).
4. S. Ismadji, S.K Bhatia, Carbon **39**, 1237-1250 (2001)
5. O. Gibert, B. Lefevre, M. Fernandez, X. Bernard, M. Paraira, M. Calderer, X. Martinez, Water Research, **47**, 1101-1110 (2013)
6. J. Niu, I. Kasuga, F. Kurisu, H. Furumai, T. Shigeedaa, Water Research, **47**, 7053-7065 (2013
7. V. Sharma, P Singh, A.K. Pandey, A. Dhawan, Mutat Res **745**, 1-2, 84-91 (2012).
8. Antizar-Ladislao, B., Lopez-Real, J. & Beck, A.J., Chemosphere, **64**, 5, 839–847 (2006).
9. H. Gui, F. Li, Y. Wei, T. Yamada, R. Desmiarti, J Eng. Technol. Sci. **48**, 5, 631-644 (2016)
10. S.A Baghoth, S.K. Sharma, G.A. Almi, Water Research, **45**, 797-809 (2011)
11. Sheng, G.P. & Han, Q.Y., Water Research, **40**, 6, 1233-1239 (2006).
12. W. Chen, P. L. Westerhoff, J. A. Leenheer, K. Booksh, Environ. Sci. Tech. **37**, 24, 5701-5710 (2003)
13. F. Li, A. Yuasa, H. Tanaka, Y. Katamine, Adsorption, **14**, 389-398 (2008).
14. A. Baker, Environ. Sci. Tech. **35**, 5, 948-953 (2001)
15. P.G. Coble, Mar. Chem. **51**, 4, 325-346 (1996)

Utilization of Solar Energy for Air Conditioning System

Juwari Purwo Sutikno[1,*], *Serlya Aldina*[1], *Novita Sari*[1] , and *Renanto Handogo*[1]

[1]Chemical Engineering Departement, Institut Teknologi Sepuluh Nopember, Surabaya, 60111

Abstract. The purposes of this research are to do a system simulation of air conditioning utilizing solar energy with single effect absorption refrigeration method, analyze the coefficient of performance (COP) for each absorbent-refrigerant variable and compare the effectivity of every absorbent-refrigerant variable used. COP is a constant that denotes the effeciency of a refrigeration system, that is ratio of work or useful output to the amount of work or energy input. The higher the number of COP, the more efficient the system is. Absorbent-refrigerant (working fluids) variables used in this research depend on its chemical and thermodynamics properties. Steps in this research are including data collection and tabulation from literature and do a simulation of air conditioning system both commercial air conditioning system (using electrical energy) and solar energy air conditioning system with Aspen Plus software. Next, run the simulation for each working fluid variables used and calculate the COP for each variable. Subsequently, analyze and compare the effectivity of all variables used from COP value and economical point of view with commercial air conditioning system. From the result of the simulation, can be concluded that solar air conditioning can achieve 98,85 % of energy savings than commercial air conditioning. Furthermore, from the calculation of COP, the highest COP value is achieved by solar conditioning system with $LiNO_3$-NH_3 as working fluid where 55% of the composition is the refrigerant and 45% of absorbent.

1 Background

Air Conditioner (AC) is now a sophisticated solution to cope with hot room temperatures, especially in urban areas. Almost all offices, schools or homes in the city can not be separated from the use of air conditioner.
However, adding an energy load to an AC installation will also accelerate global warming and greenhouse effects. Two-thirds of the total energy is used for buildings, where 30-50% of electricity consumption in buildings is dominated by air conditioning systems.
The negative effects of this increase in energy consumption include:

1. The increasing demand of air-conditioning systems will lead to increased consumption of electricity that will be associated with greenhouse gas emissions.
2. In addition, electrical vapor compression air cooling technology uses CFC and HCFC as pollution-causing refrigerants.

Therefore, the existence of alternative technologies capable of creating air conditioning systems without the use of electricity is very important.

Solar air conditioning is a solar energy-based cooling system. Solar energy can be used as an energy source in air conditioning systems in two ways. First, solar energy can be converted into electricity by using photovoltaic cells which are then used to operate vapor compression air conditioning system. Secondly, solar energy can be used to heat the working fluid and regenerate refrigerant in the generator of the vapor sorption (absorption or adsorption) system of air conditioning.

The application of solar energy to air conditioning systems has several advantages, in addition to offering solutions of negative impacts that have been described earlier, this technology has the advantage:

1. The maximum cooling load is equal to the maximum available radiation
2. The working fluids used are harmless, such as water and salt solution

Current technologies for the production of air conditioners using thermal energy are: absorption chiller, liquid-solid desiccant, and solid adsorption.

1.1 Previous Research

Some research on the development of solar-powered air conditioning system has been done. Here is a brief description of previous studies :
Li and Sumathy [1], presented the results of solar-powered air conditioning systems. The system uses a flat plate collector array with a cooling capacity of 4.7 kW. The system is also equipped with two hot water storage tanks. The value of SCR obtained is 0.07. The results of the research indicate that in cloudy weather, the system cannot operate in a commonly used.
Syed et al [2], reported the performance of a single effect LiBr + water absorption refrigeration with a nominal cooling capacity of 35 kW powered by hot water from 49.9 m^2 flat plate collector and 2 m^3 hot water tank during the day. SCR value ranges from 0.11. Maximum

* Corresponding author: joecheits@yahoo.com

cooling capacity of 7.5 kW (21% of nominal) and COP of 0.34-0.42.

Zambrano et al. [3], presented the results of a LiBr + water single effect absorption chiller with a nominal cooling capacity of 35 kW powered by hot water from a 151 m^2 flat plate collector and equipped with an auxiliary gas fired burner and a 2.5 m^3 container tank. Auxiliary gas fired burners are utilized to support the heat of solar radiation when solar radiation heat is not sufficient to drive the system. SCR values obtained by 0.15 and SHF vary from 0.1 to 1 depending on the position of the sun.

1.2 Solar Thermal Collector

One important component of solar air conditioning system is solar thermal collector. According to research conducted by Ariane Rosario [4], the total energy received by solar thermal collector per-day can be formulated as follows:

$$E = 2\,I\,E_0\left[\left(\cos(\varphi)\cos(\delta)\left[\frac{\sin(0.2618\,\tau)}{0.2618}\right]\right) + (\sin(\varphi)\sin(\delta)\tau)\right] \quad (1)$$

Indonesia is one of the tropical countries that have the potential to develop solar thermal energy, including the city of Surabaya. Geographically Surabaya is located at 7.24917 south latitude (806 km from the equator line) and 112.75083 east longitude. If the type of solar thermal collector used is a flat plate collector with 80% efficiency [5] with operating time for 10 hours (07.00 to 17.00) then the average energy produced by solar thermal collector will be 528.4 W / m^2.

2 Methodology

In this study, there are two models of refrigeration system used, there are vapor compression air conditioning for commercial air conditioning and absorption-refrigeration air conditioning for solar air conditioning.

The first step is collecting the required data. The purpose of this step is to achieve a good simulation result approaching to the actual condition. One of the required data is the air temperature of Surabaya. This data will be used to determine the indoor air temperature to be cooled. The data of the air temperature of Surabaya over the year is presented on the table 1.

Table 1. The average of the Surabaya air temperature over the year

Month	Jan	Feb	Mar	Apr	May	Jun	Jul	Aug	Sep	Oct	Nov	Dec
T (°C)	33,9	33,0	34,4	33,6	34,3	33,4	33,1	33,2	33,9	33,4	33,6	32,5

Based on the data, the average of air temperature is 33,5°C.

The next step is simulation process. This step include process modelling and running model by Aspen plus. In this study 4 types of working fluid with 3 different composition as a variable is used. The four working fluids are LiBr-H2O, LiNO3-NH3, H2O-NH3, and NMP-R32 with a ratio of absorbance and refrigerant composition of 45%: 55%, 50%: 50% and 55: 45%.

The last step is analyse the economic and COP for the each model. COP and economic analysis are the important factors that must be taken into account in determining whether solar air conditioning is feasible to be used or not.

3 Vapor Compression Air Conditioning

Commercial air-conditioning systems generally use vapor compression cooling systems. Here is a block diagram of the vapor compression cooling system.

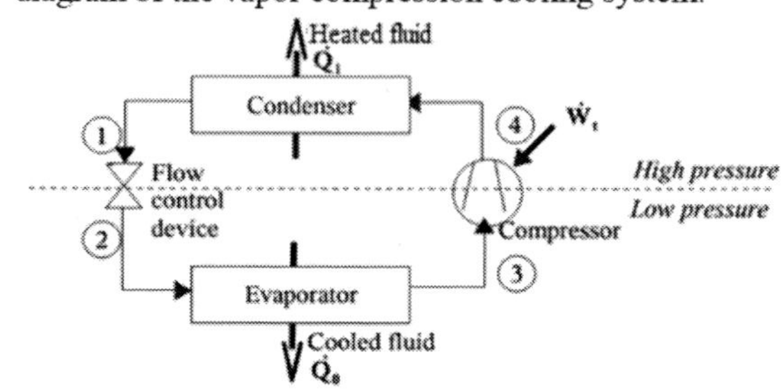

Fig. 1. Block Diagram of Vapor Compression Cooling System

The vapor compression refrigeration system consists of four main processes: compression, condensation, expansion and evaporation [6].

The first stage of the simulation process is the selection of the thermodynamic model to be used. In this study thermodynamic model Peng-Robinson Equation of State selected. This model was chosen because it can be applied over a wide range of conditions, i.e. for temperatures greater than -271°C and pressures of less than 100,000 kPa [7].

In this study, it is selected R-22 refrigerant as working fluid for commercial air-conditioning system. This is because most commercial air conditioners use the R-22 refrigerant as its working fluid [8].

After that, done the modelling using the process blocks that exist in Aspen Plus software. The following is the result of the simulation :

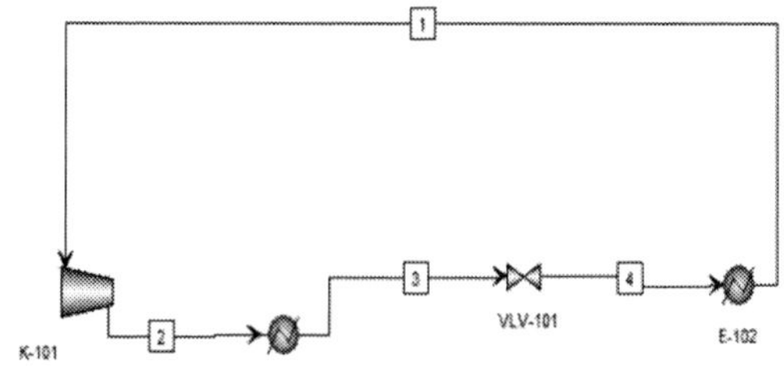

Fig. 2. Vapor Compression Air Conditioning System Block Processing by Using Aspen Plus Software

The first process block is a compressor (K-101). This process aims to increase the pressure of working fluids and circulate them. Flow (1) is a R22 refrigerant with a vapor phase with a mass of 34 kg/h and a charge of 72.327 psig. This pressure is adjusted to the standard R22 vapor compression air conditioner, where Standard

Freon R22 AC commercial pressure is 57.6 psig up to 84.1 psig. For the K-101 (compressor) block the power value is set at 400 watts. This is according to the standard AC ½ PK, where the electric power of the compressor is 320 watts up to 400 watts.

The second process block is a condenser (E-101). In this study the condenser is modelled by using heat exchanger with outdoor air as cooling media. At this stage there is a process of cooling the hot refrigerant vapor from the compressor by air to the condensation temperature, then the vapor is condensed.

The next process block is the expansion process. In this study the expansion process is modelled with a valve (VLV-101). The process of expansion takes place from stream (3) to stream (4). In the process, there is a decrease of the refrigerant pressure from the condensation pressure to the evaporation pressure. When the liquid is expanded through the valve to the evaporator, the refrigerant temperature also drops from the condensate temperature to the evaporative temperature. This process is an adiabatic process, in which the fluid enthalpy does not change throughout the process. The refrigerant in stream (4) is in a mixed of vapor and liquid state.

The final process block is the evaporation process that occurs in the evaporator and takes place at a constant pressure. In this research the evaporation process is modelled by heater. The refrigerant at stream (1) coming out from the evaporator in a saturated vapor state. During this evaporation process occurs, there is absorption of heat from the room that will be cooled. The amount of heat absorbed is set at 5000 Btu/hour.

From the simulation results then calculated the energy requirements for vapor compression air conditioning with a cooling capacity of 5000 Btu/hour. The required energy comes from the compressor's electric power requirement of 0.4 kW.

4 Solar Air Conditioning

Solar air conditioning is a solar energy based air conditioner. Solar thermal energy can be used as an energy source in air conditioning systems in two ways. First, solar energy can be converted into electricity by using photovoltaic cells which are then used to operate the vapor compression air conditioning system. Secondly, solar energy can be used to heat the working fluid and regenerate the refrigerant in the generator of the sorption vapor system (absorption or adsorption) of air conditioning [9].

In this study cooling absorption-refrigerant solar air conditioning system is selected. The absorption-refrigerant air conditioning cooling system consists of five main process units, they are solution pump, generator, condenser, evaporator and absorber [8]. Here is a block diagram of the absorption-refrigerant air conditioning cooling system:

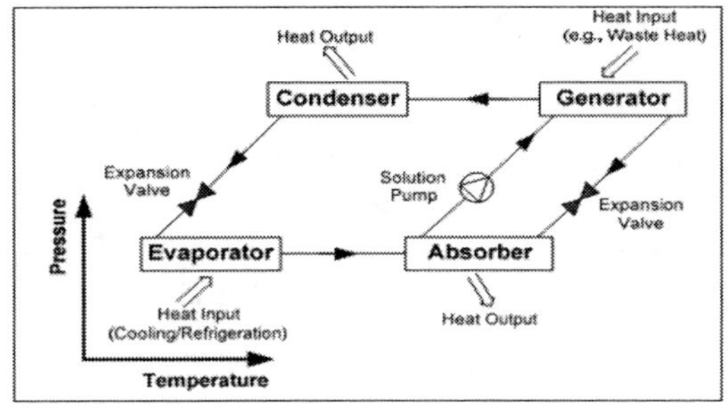

Fig. 3. Block Diagram of Absorption-Refrigerant Air Conditioning System

In this study 4 types of working fluid with 3 different composition as a variable is used. The four working fluids are LiBr-H2O, LiNO3-NH3, H2O-NH3, and NMP-R32 with a ratio of absorbance and refrigerant composition of 45%: 55%, 50%: 50% and 55: 45%.

The first stage of the simulation process is the selection of thermodynamic models to be used. For simulations using working fluids LiBr-H2O and LiNO3-NH3 using ELECNRTL thermodynamic model. This model is used because it is specifically designed for electrolytic solutions such as LiBr-H2O [10]. In contrast with LiBr-H2O and LiNO3-NH3, the ELECNRTL thermodynamic model is not suitable for H2O-NH3 and NMP-R32. Therefore, a different thermodynamic model is chosen. Since there is no specific thermodynamic model for this type of working fluid, a common thermodynamic model is used. The exact thermodynamic model is Peng-Robinson [10].

After that, the modelling using the process blocks that exist in Aspen Plus software is done. The following is the result of the modelling:

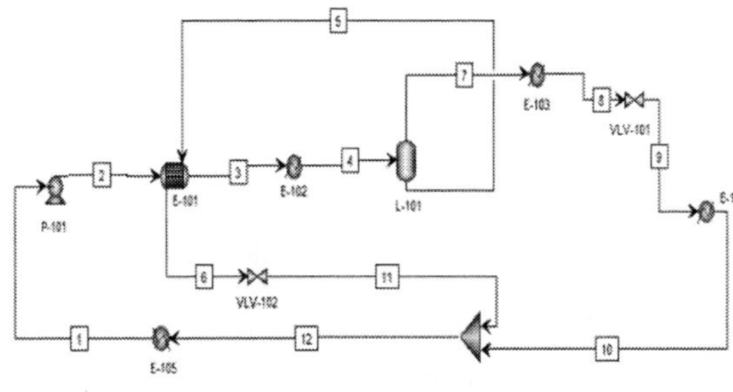

Fig. 4. Absorption-Refrigeration Air Conditioning System Simulation Model Using Aspen Plus V8.8 Software

From the simulation results then calculated the energy requirements for solar air conditioning system with a cooling capacity of 5000 Btu/hour. The energy required comes from the need for electric power of the pump and the energy produced by the solar thermal collector. Here are the details of energy requirements for solar air conditioning systems.

Table 2. Details of Solar Air Conditioning Energy Requirements

Fluid	Ratio	Cooling load (kW)	Energy Required (kW)		
			Pump	Solar Thermal Collector	Total
H_2O/ LiBr	45/55	1,465	1,54E-05	1,810	1,810
	50/50	1,465	1,35E-05	1,800	1,800
	55/45	1,465	1,20E-05	1,794	1,794
NH_3/ H_2O	45/55	1,465	0,005	1,958	1,963
	50/50	1,465	0,002	1,885	1,887
	55/45	1,465	0,001	1,849	1,851
NH_3/ $LiNO_3$	45/55	1,465	0,006	1,790	1,796
	50/50	1,465	0,005	1,781	1,786
	55/45	1,465	0,004	1,774	1,778
R32/ NMP	45/55	1,465	0,026	1,945	1,971
	50/50	1,465	0,020	1,948	1,968
	55/45	1,465	0,016	1,909	1,925

From the simulation shows that the least energy requirement when working fluid used is LiNO3 / NH3 with a ratio of 55% refrigerant and 45% absorbance.

5 Coefficient of Performance Analysis

The performance of the cooling system is generally measured by the value of coefficient of performance (COP). Coefficient of performance (COP) is the ratio between the cooling capacity (per unit of energy required) [11]. The higher the coefficient of performance (COP) the more effective the cooling system is. Coefficient of performance (COP) can be formulated based on the following equation [11]:

$$Coefficient\ of\ Performance\ (\mathrm{COP}) = \frac{Q_C}{W} \quad (2)$$

The absorption refrigerant system is a Carnot cycle, so as a parameter of system effectiveness Carnot COP can be calculated based on the following equation:

$$COP_{carnot} = \frac{T_O}{T_C - T_O}\,\frac{T_H - T_C}{T_H} \quad (3)$$

Where T is in Kelvin

Table 3. COP Calculation Result for Vapor Compression Air Conditioning System

Working fluid	**COP**
R-22	3,663

Table 4. COP Calculation Result for Solar Air Conditioning System

Working fluid	**Ratio**	**COP**	**COP_{carnot}**
H_2O/LiBr	45/55	0,809	1,25
	50/50	0,814	1,25
	55/45	0,817	1,25
NH_3/H_2O	45/55	0,746	3,51
	50/50	0,776	3,90
	55/45	0,792	3,92
$NH_3/LiNO_3$	45/55	0,815	1,27
	50/50	0,820	1,28
	55/45	0,824	1,28
R32/NMP	45/55	0,743	5,21
	50/50	0,744	5,24
	55/45	0,761	5,24

Based on the above result, the COP value for vapor compression air conditioning cooling system with a cooling capacity of 5000 Btu/hour is 3.663, while the value of coefficient of performance (COP) for solar air conditioning with the same cooling capacity ranges from 0.743 to 0.824. Based on these data, it shows that vapor compression refrigeration systems have a higher COP. This is in accordance with the literature which states that the absorption refrigeration system has a range of COP of 0.5 to 1.5, while the vapor compression system has a value of COP more than 3. Although the value of COP absorption refrigeration system lower than vapor compression system, this system is still widely used because this system can be used in low-temperature heat utility (<100°C). In this study, COP for solar air conditioning system cannot be compared with commercial air-conditioning system, because the energy input for solar air conditioning system is obtained free of charge [10]. Therefore, to compare the two systems is done by the calculation of electrical energy savings that can be achieved.

From the results of the COP obtained, for a system with working fluid LiBr / H2O the COP value is 0.809-0.817. While the research done by Al-Ugla, et al [9], for a single absorption refrigeration system with the same working fluid (LiBr / H2O) placed in Dhahran, Saudi Arabia (latitude = 26.16 °N) found a COP value of 0.8567. When compared with these studies, the COP results in this study are relatively lower. This is because Dhahran has a larger average daily total solar radiation energy (5.84 kWh / m2) compared to Surabaya (5,284 kWh/m2).

The solar air conditioning cooling system has a varying coefficient of performance (COP) value, depending on the type of working fluid and its composition. Here are the coefficient of performance (COP) results of these variables in graphical form:

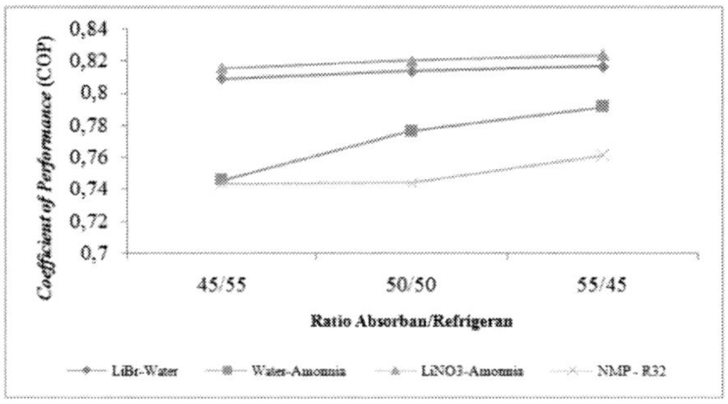

Fig. 5. Graph of COP Relation with Composition

From the graph above shows that LiNO3-H2O has a relatively higher COP value compared to LiBr-H2O. In addition, the use of LiNO3-NH3 as a working fluid can also avoid the occurrence of crystallization processes that can inhibit system performance [12] and the use of NH3 as a refrigerant is more advantageous because it can be evaporated at lower temperatures than water [13].

In addition, from the graph above it can be concluded that for all four working fluid variables, the greater the composition of refrigerant in working fluid, the greater the coefficient of performance (COP) value of the cooling system. This is because to achieve the cooling capacity of 5000 Btu/hour, working fluid with a small refrigerant composition will require more working fluid, so the pump and solar thermal collector load will be greater.

6 Economic Analysis

Economic analysis is one important factor that must be taken into account in determining whether solar air conditioning is feasible to be used or not. In the economic analysis will be analyzed the annual worth of solar air conditioning system and commercial air conditioning.

Previously it was found that based on the simulation results for solar air conditioning system that has been done found that the combination of working fluid LiNO3/NH3 with the composition of each substance 45%: 55% has the highest Coefficient of Performance (COP) so that in this economic analysis, solar air conditioning system analysed is a system with working fluid LiNO3/NH3 with the composition ratio of each substance 45%: 55%.

Next, calculate the capital cost for solar air conditioning by finding the price for each component of the equipment needed to install a solar air conditioning. The solar air conditioning system (absorption - refrigeration) consists of a solar thermal collector, three heat exchanger units, two vessel units and one pump unit. Capacity for each equipment is obtained from simulation result using Aspen plus.

The price of the equipment at any time will change depending on the economic changes. If the price of a device with a certain capacity is known, then the price of equipment with other capacity can be estimated by using Chemical Engineering Plant Cost Index (CEPCI). Similarly, if the price of the tool in a given year is known, then the price of the tool in another year can be estimated [14].

$$C_E = C_B \left(\frac{Q}{Q_B}\right)^{0,6} \quad (4)$$

$$\frac{C_1}{C_2} = \frac{INDEX_1}{INDEX_2} \quad (5)$$

Where C_E = equipment cost with capacity Q

C_B =known base cost for equipment with capacity Q_B

C_1 = equipment cost in year 1

C_2 = equipment cost in year 2

$Index_1$ = cost index in year 1

$Index_2$ = cost index in year 2

Based on calculations that have been done, the average capital cost for solar air conditioning with a cooling capacity of 5000 btu / hour is as follows:

Table 5. Capital Cost for Solar Air Conditioning System

Equipment	Price ($)	Price (Rp.)
Solar Thermal Collector	456	6.081.406
Pump	46	608.095
Heat Exchanger 1	165	2.193.476
Desorber (Storage Tank)	245	3.270.810
Condensor (Fan Coil)	119	1.590.387
Evaporator (Fan Coil)	202	2.698.297
Absorber (absorption chiller)	326	4.347.985
Accecories (pipe, insulation, etc)	117	1.564.888
CAPITAL COST	**$1.677***	**Rp. 22.355.346**

* 1 US $ = Rp 13.300,-

From the table above can be seen that the total investment that must be spent to install one solar air conditioning is Rp. 22.355.346. When compared with the price of vapor compression air cooler with the same cooling capacity sold in the market is about Rp. 2.950.000. It can be seen that the initial investment that must be spent to install one solar air conditioning is much larger.

After calculating the capital cost spent for the installation of a solar air conditioning and commercial air conditioner, calculated also the cost of electricity bills to be issued for each type of air conditioner, either a regular air conditioner or solar air conditioning. This is done to see how big solar air conditioning is able to save electricity energy consumption. Assume air conditioning is used every day for 10 hours per day and electricity cost is calculated for consumption for one year (30 days). Meanwhile, the basic electricity tariff used in the cost calculation is taken from State Electricity Company (PLN) tariff for household R-1/900 VA-RTM group that is Rp. 1.352 / kWh. To calculate the cost of electricity, used the formula:

Electricity Cost = Usage (kWh) x Basic Electricity Tariff

Based on the simulation results, it is found that commercial air-conditioning has a power of 0.4 kilowatts, equivalent to 400 watts. Based on the calculations that have been done, obtained the cost of electricity commercial air-conditioning system for cooling capacity of 1.465 kW is Rp. 5.408, - / per day or equivalent to Rp.1.973.920, - / per year.

Meanwhile, in the same way calculated also the cost of electricity for solar air conditioning, it was found that solar air conditioning has a power of 0.0046 kilowatts or equivalent to 4.6 watts. Based on the calculation that has been done, got the cost of solar air conditioning electricity for cooling capacity of 1,465 kW is Rp. 62, - / per day or equivalent to Rp.22.700, - / per year.

Based on the calculation of the cost of electricity per month to be spent for each type of air conditioner, it is found that by using solar air conditioning can save electricity costs by 98.85% compared to using commercial air conditioning. This shows that solar air conditioning is capable of efficient use of electrical energy that can impact on the cost savings of electricity bills while reducing greenhouse gas (CO2) emissions that can have adverse environmental impacts.

To compare between the two cooling systems, an annual worth (AW) analysis is used as a basis for consideration. The Annual Worth Method (AW) or also called annual equivalent is the method by which cash inflows and cash outflows are distributed in a series of annual money values evenly (equal), any period of time over the life of the investment, at a desired minimum rate of return (MARR) [14].

To perform an annual worth analysis, data on the life time of each system is required. For solar air conditioning system (absorption-refrigeration) can last up to 30 years, while commercial AC can last up to 10 years [6]. Therefore, the life time set for this analysis is 30 years. Here are the data used for the annual worth analysis:

Table 6. Analysis Data for Annual Worth Calculation

Tipe	Capital Cost (Rp)	Operating Cost (Rp)	*Life Time* (year)
Solar Air Conditioning	22.355.346	22.700	30
Commercial Air Conditioning	2.950.000	1.973.920	10

By setting a MARR value of 6.5% (Bank Indonesia deposit interest rate), it is found:

AW of solar air conditioning = -Rp. 1.734.615, -

AW of commercial air conditioning = -Rp. 2.384.278, -

A negative sign at the annual worth value indicates the value of expenditure to be incurred annually. In the analysis of annual worth selected system with the largest annual worth value. Of the two systems annual worth solar air conditioning is higher than the commercial air-conditioning system, which means solar air conditioning system is more economically profitable.

Based on the economic analysis that has been done, it can be seen that solar air conditioning system is more economically advantageous compared with commercial air conditioner.

Conclusion

Based on the study that has been done, it can be obtained the following conclusions:

1. In terms of economic aspects, consumption of electrical energy and the impact on the environment, solar absorption is more effective than commercial cooling systems.
2. Of the four working fluids used for solar air conditioning systems, it is obtained the highest COP value is $LiNO_3$-NH_3 with a ratio of 45% absorbance and 55% refrigerant composition.
3. From the Coefficient of Performance (COP) analysis, $LiNO_3$-NH_3 can be used as a better alternative to LiBr-H_2O working fluid.
4. Of the four working fluids tested, it was found that the higher the composition of refrigerant in working fluid, the higher the value of Coefficient of Performance (COP) for solar air conditioning system.

References

1. Z.F. Li and K. Sumathy, Solar Energy **71**, 285-297 (2001)
2. A. Syed, International Journal of Refrigeration **28**, 859-871 (2005)
3. D. Zambrano, International Journal of Refrigeration **31**, 315-327 (2008)
4. A. Rosario. 2014, Undergraduate Journal of Mathematical Modeling: One + Two. **6**, 1, Article 1
5. F. Struckmann. *Analysis of a Flat-plate Solar Collector* (Project Report MVK 160 Heat and Mass Transport Sweden, 2008)
6. E. Pita. *Air Conditioning Principles and Systems* 4rd ed. (Prentice Hall inc, Sadle River, N.J, 2002)
7. J.G. Ibanez. Prospect **11**, 13-20 (2013)
8. C.C Allgood and C. Lawson. *Performance of R-438A in R-22 Refrigeration and Air Conditioning Systems*. Proceedings of the International Refrigeration and Air Conditioning Conference Purdue, July 12-15 (2010)
9. A.A. Al-Ugla, M.A.I El-Shaarawi and S.A.M. Said. International Journal of Refrigeration **53**, 90-100 (2014)
10. C. Somers, Y. Hwang, R. Radermacher, P. Rodgers and S. Al-Hashimi, Applied Energy **88**, 4197-4205 (2011)
11. R. Smith. *Chemical Process Design and Integration* (John Wiley & Sons, Ltd, New York, 2005)
12. W. Wu, B. Wang, W. Shi, and X. Li, Hindawi Publishing Corporation Advances in Mechanical Engineering Research Article (2013)
13. J.M. Abulateef, K. Sopian, M.A. Alghoul, M.Y Sulaiman, A. Zaharim, and I. Ahmad. *Solar Absorption Refrigeration System Using New Working Fluid Pairs*. Proceedings of the Energy and Environment Conference UK, February 23-25 (2008)
14. M.S. Peters and K.D. Timmerhaus. *Plant Design and Economics for Chemical Engineers*. (Mc Graw Hill, New York, 1991)

Modelling Sulfonation Kinetics On The Sodium Lignosulphonate Synthesis From Black Liquor

Hapsoro Aruno Aji[1], *Bambang Pramudono*[2,*], *Slamet Priyanto*[3]

[1] Master Degree of Department of Chemical Engineering, Faculty of Engineering, Diponegoro University, Jl. Prof. Sudarto, Tembalang-Semarang, 50275, Telp/fax: (024)7460058 / (024)76480675
[2] Department of Chemical Engineering, Faculty of Engineering, Diponegoro University ,Jl. Prof. Sudarto, Tembalang-Semarang, 50275, Telp/fax: (024)7460058 / (024)76480675
[3] Doctor Degree of Department of Chemical Engineering, Faculty of Engineering, Diponegoro University, Jl. Prof. Sudarto, Tembalang-Semarang, 50275, Telp/fax: (024)7460058 / (024)76480675

Abstract. Surfactant is a compound that has the ability to lower the interfacial tension between the face of two phases (liquid). The need of surfactant has increased along with the development of the industrial sector. One example is sodium lignosulfonat (SLS) which is useful to improve oil acquisition in its application for EOR (Enhanced Oil Recovery). SLS is a result of the reaction between lignin, $NaHSO_3$ (bisulfite) and sodium hydroxide as catalyst. The purpose of this research is to determining the reaction mechanism and determining the value of the reaction rate constant (k). The sulphonation kinetics of lignin was carried out in various temperature (50, 60, 70, 80, and 90°C). From this research obtained a positive (linear) response between sulfonation temperature against reaction rate constants. The lowest k values obtained k_{50} = 2,05 x 10^{-4} $minute^{-1}$ and the highest one is k_{90}= 3,35 x 10^{-4}. From graphycally calculation obtained the value of A is 0,015 and the energy activation (Ea) is -11,708 kJ/mol with R^2 = 95,69%. The correlation of the temperature influence to the constant of reaction rate espressed in arhenius equation as follows :

$$k = 0{,}015 \exp^{(-1408{,}3/T)}$$

1. Introduction

Black liquor (BL) is side result of pulp processing with lignin as main component approximately 46% of total solid [1]. In successive year, Black liquor production increase as the growth of paper demand. This solution is stable, complex, smelly and has toxic compound such hidrogen sulfide, methyl mercaptan. It means, dumping to the environment cause pollution. Nowadays, the utilization of BL only for energy source for pulp mill, and use as biofuel[2]. Black liquor has high of density is caused by black liquor also has high of solid content[3].

Lignin is a major wood component beside cellulose and hemicellulose, consists of three dimensional monomer namely p-coumaryl alcohol (H), koniferil alcohol (G) and sinapyl alcohol (S)[4]. Lignin are formed by polymerization of monomers which differ structure depending on plant type. It means that, each plant type have different structure of lignin[5]. Lignin is colorless, amorphorus, renewable[6]. Usually, lignin is used as a cement water reducer, composite materials [7], surfactant and dispersant for cement admixture [8], Flocculants for waste water[9]. Basicly lignin is hydrophobic in nature, so lignin insoluble in water, acid, and hydrocarbon but soluble in alkali solution. It means, lignin from BL can be isolate by acid.

Surfactant are compound that lower the interfacial tension, between two liquids. It compose of polar (hydrophilic) and non polar (lypophilic). The need of surfactant in Indonesia increase a long with the development in the industrial sector. During this time, surfactant commonly used is petroleum-based. This surfactant have good performance, but not eco-friendly, hard to degrade, comes from non-renewable materials[10]. Therefore, it takes an effort to create a new surfactant with the same good or better performance and more sustainable. Development of surfactants is being carried out which one of them is sodium lignosulfonat (SLS) from biomass such as lignin. SLS can be produced from lignin through a sulfonation process. This process aim to substitute hydroxyl groups with sulphonate groups which is more polar (Fig.1). The use of this surfactant such as emulsifying agents, emulsion stabilizer, dye dispersant, floacculants[11], for mud viscosity control during deep oil well drilling, corrosion inhibitor [13].

This research is intended to study the kinetic reaction of sulfonation process in batch reactor. It becomes important to determine the chemical reaction mechanism and determine the value of reaction rate constant (k). The reaction rate is defined as increasing

* Corresponding author: pramudono2004@yahoo.com

the amount of product or decreasing the amount of reactant. Sodium lignosulphonate can be reduced by sulfonation process following reaction :

Lignin + Natrium bisulfate → SLS + water (1)

2. Experimental Method

2.1 Materials

Black liquor obtained from PT. Indah kiat pulp and paper (Riau), lignin isolate, commercial lignin from sigma, sodium bisulfit (~ 40%v/v solution in water) from sigma, sulfuric acid 98% from PT. Indrasari, natrium hydroxide (NaOH) 98% from merck, methanol 99,8%v/v from PT. Indrasari, aquadest, and aquabidest.

2.2 Synthesis of surfactant

2 grams of lignin isolate reacted with 1,77 ml of 40% v/v sodium bisulfit ($NaHSO_3$) in 60 ml aquades. From previous research, obtained that the best of sulfonation condition in ratio (lignin:$NaHSO_3$) is 1 gram of lignin reacted with 0,458 gram $NaHSO_3$, initial pH 8,31 and 79,67°C in temp. Sulfonation temp (T) for reaction kinetic studies is 50, 60, 70, 80, and 90°C. Every 20 min (t), samples are taken to analyze.

2.3 Kinetic Reaction

Experiment performed in batch reactor with fix volume, and temperature condition. Follow as 2nd order reaction.

[in] – [out] + [rate of rx] = [acc]

Batch reactor, so V = constant

0 – 0 – C.V. k = dVC/dt

If conversion of A = Xa, and initial concentration of A and B symbolized as Ca_0, Cb_0 so we get

$$-r_A = k.C_{A0}(1-X_A)(C_{B0}-C_{A0}.X_A) \quad (1)$$

If $C_{b0} = M.C_{a0}$, so $M = C_{b0}/C_{a0}$ and we get

$$-r_A = k.C_A^{\,2}(1-X_A)(M-X_A) \quad (2)$$

$$\ln\frac{M-X_A}{M(1-X_A)} = \ln\frac{C_{A0}.C_B}{C_{B0}.C_A} \quad (3)$$

$$\ln\frac{C_B}{C_A} = (C_{B0}-C_{A0}).k.t + \ln M \quad (4)$$

2.4 Analysis

UV spectrophotometric method is the best suited for determining the concentration of component like lignin, sodium bisulfit, and sodium lignosulfonat (SLS). In this experiment concentration were analyze by UV mini - 1240 spectrophotometer pharco - 300. Each component have different wavelength, for lignin λ = 205 nm [14], for sodium bisulit λ = 288 nm, and for Sodium lignosulfonat λ = 232 nm [15].

3.Result And Discussion

3.1 Reactant concentration vs Temperature

From fig.1 and fig.2, known that at reactant concentration per unit time (t) decrease. It means that, at the longer time more reactants are converted as surfactant. The lowest concentration occurs at 90°C, it is indicate that at the temperature (T) give the best condition in yield of surfactant.

a.

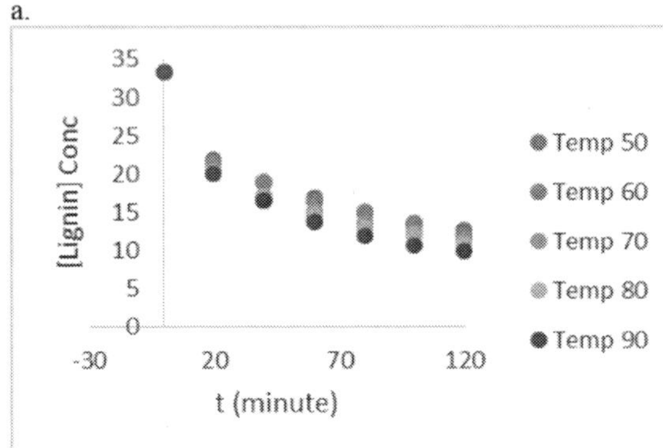

b.

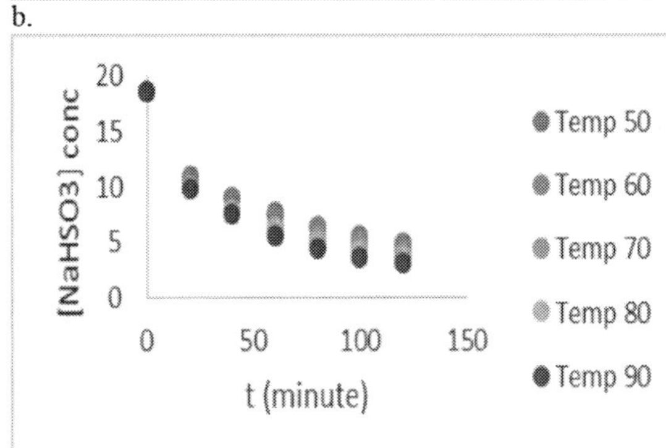

Fig. 1. Reactant concentration (a) Lignin and (b) Sodium bisulfit.

3.2 Reaction rate constant (k) vs Temperature

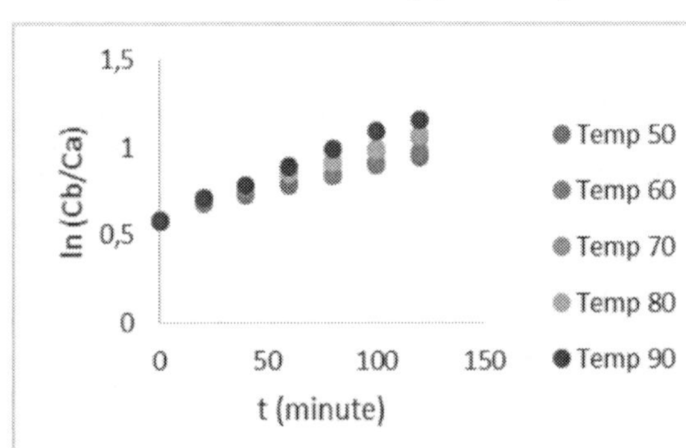

Fig. 2. ln (Cb/Ca) vs time (t)

Table 1. Reaction rate constant vs T

Suhu (°C)	ln $C_{B0} - C_{A0}$ (k)	k
50	0,0030	2,05 x 10^{-4}
60	0,0032	2,19 x 10^{-4}
70	0,0038	2,60 x 10^{-4}
80	0,0040	2,74 x 10^{-4}
90	0,0049	3,35 x 10^{-4}

Based on fig.2 and table 1 known that at the higher temperature of sulfonation process, reaction rate constant also higher. This is due to at the higher temperature (T) , the reactant molecular activation and the collision factor also increase. The reaction rate constant were increase at 1,1 fold every 10oC increment. based on the "rule of thumb" which states that every 10 degree temperature rise, the reaction rate constant (k) will increase twice as much. But in this study, only 1.1 times increased. It's cause the sulfonation reaction rate is affected by the concentration of the lignin active site and the total SO2 ion concentration. This result corresponds to the arrhenius equation, that temperature is proportional to the reaction rate constant (k).

3.3 Determine A and Ea

Ea (activation energy) is the minimum energy required for the reaction occur. Activation energy determined using Arrhenius equation.

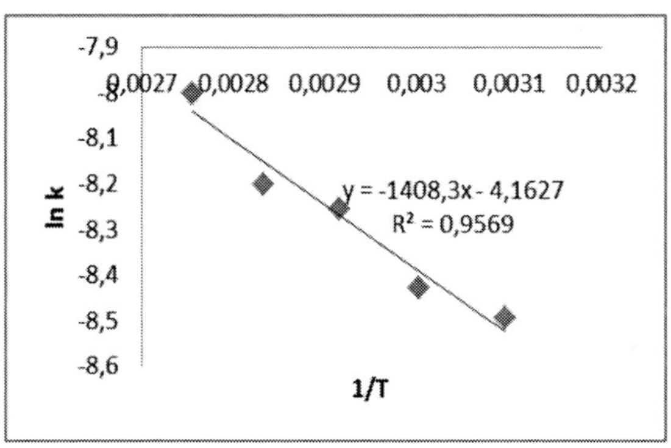

Fig. 3. ln k vs 1/T

Based on fig.3, show that slope representing the value of –Ea/R and the intercept representing ln A. From the calculation, value of –Ea (with R = 8,314 J/K.mol) is 11,708 kJ/mol and the collision factor 0,015. The results showed that the value of Ea in this study was smaller when compared to the ismiyati (2013) in which Ea obtained was 21.27 kJ/mol. It is indicates that the energy required for sulfonation reaction is smaller. For this data experiment, reaction rate constant follow this equation :

$$k = 0{,}0015.e^{-1408{,}3/T}$$

3.2 Validation of reaction kinetics model

Reaction kinetics model used for the calculation of reactor capacity, in order to scale up the capacity of production. The best condition of sulfonation process is 90°C in temperature, with reaction rate constant 3,35x10^{-4}. Validation process is performed by comparing the value of $Ca_{lab\ 90oC}$ and $Ca_{calc\ 90C}$.

$$\ln (Cb/Ca) = (Cb_0\text{-}Ca_0)\ k.t + \ln M \qquad (5)$$

$$Ca = Cb/M.e^{(Ca0\text{-}Cb0).k} \qquad (6)$$

Table 2. Validation of Ca Calculation (at 90°C)

Time (minutes)	Ca lab	Ca calc
0	33,33	33,9603
20	19,95	19,6089
40	16,58	16,6623
60	13,63	13,5826
80	11,86	11,8131
100	10,59	10,5123
120	9,92	10,1249

Based on the data, it can be concluded that reaction kinetics model is acceptable. From the calculation, the average of %error kinetics model is 1,091%.

4. Conclusion

Lignin sulfonation reaction kinetics follow as 2nd order with the highest reaction rate constant occur at 90°C, with k = 3,35x10^{-4} and give 1,1 fold every 10° C increment. The activation energy, collision factor (A) is 11,708 kJ/mol and 0,0015. Validation indicate that laboratory data (Ca_{lab}) and calculation (Ca_{calc}) result in similarity, so the reaction kinetics model is acceptable.

References

1. Sjöström, E. *Wood chemistry fundamentals and application (2nd edition)* (CA:Academic Press, Inc., San Diego, 1993)
2. Bora, M.K., J. Engineering Science and Tehnology: an International Journal **4**, 2 (2014)
3. Alabi, S.B., *Development and implementation of an online kraft black liquor viscosity soft sensor.* (Christchurch, New Zealand, 2010)
4. Fasching, M., Schroder, P., Wollboldt, R.P., Webber, H.K., Sixta, H., Holzforschung **62**, 15-23 (2008)
5. Norgren, M., Edlund, H., Current opinion in colloid and interface science **19**, 5, 409-416 (2014)
6. Radoykova, T., Nenkova, S., Valchev, I., Journal of chemical technology and metallurgy **48**, 5, 524-529 (2013)
7. Kamoun, A., Jelidi, A., Chaabouni, M., Cement and Concrete research **33**, 995-1003 (2003)
8. Kumar, M.N.S., Monhanty, A.K., Erickson, L., Misra, M., J. Bio. Mater. Bioenergy **4**, 1-24 (2009)

9. Ouyang, X., Ke, L., Qiu, X., Guo, Y., Pang, Y., J. Dispersion Sci. Techol, **30**, 1, 1-6 (2009)
10. Wei, Y., Cheng, F., Zheng, H., Carbohydr Polym, **74**, 673-679 (2008)
11. Afif, M.F., *Optimizing composition of Sodium lignosulfonate and Oleic acid to reach IFT 10-3 Mixed surfactant on EOR operation*. (In Indonesia Petroleum Association, 2015)
12. John Wiley & Sons, Inc http://www.scribd.com/doc/30137763/Lignin (Accessed on 10 November 2017)
13. Altwaiq, A., Abdel-Raheem, R., Alshamaileh, E., Al-luaibi, S., Khouri, S., Eurasian J of Anal Chem **10**, 1, (2015)
14. Jahan, M.S., Mun, S.P., J. Of wood chemistry and technology 27 2 (2007)
15. Wesco Technologies. *Typical properties of weschem ammonium lignosulfonat, calcium lignosulfonate, sodium lignosulfonate, zinc lignosulfonate, CA.92674 – 3880* (USA, 1995)

Solid-liquid anaerobic baffled reactor treating food waste

Indro Sumantri[1]*, *Ika* Bagus Priyambada[2], *Hadiyanto* Hadiyanto[1]

1Chemical Engineering Department, Faculty of Engineering, Diponegoro University, Semarang, Indonesia
2 Environmental Engineering Department, Faculty of Engineering, Diponegoro University, Semarang, Indonesia

Abstract. The characteristics of the sauce industrial wastewater is high organic degradable materials, it requires adequate treatment in order to process wastewater is technically and economically feasible. The main treatment to remove high content of organic material is anaerobic process. The presence of a preservative agent in wastewater would lead to disruption of the growth of micro-organisms. Slow the growth of anaerobic bacteria require a solids residence time (solid retention time, SRT) period in anaerobic reactors. This research is to observe the effect of input COD with constant sludge height to the removal of COD. Main equipment process of anaerobic reactor is baffled reactor. Activated sludge used are obtained from the tofu and tempe wastewater treatment unit in Semarang while synthetic wastewater was obtained from dissolution of preservative material product (Del monte sauce). The control parameter was: pH, alkalinity, the ratio of COD: N: P, baffled configuration, wastewater flow rate (20 L/day), addition of micro-nutrient and height of suldge. The variable of reseach was initial COD input. The observed response is: COD removal and pH. The total COD removal for low organic loading (2369.7 mg COD/L) is 80.36 to 98.03 % and for high organic loading (7334 mg COD/L) is 98.15.

1 Introduction

Currently, in Indonesia, there are about 24 companies which produce tomato sauce and/or chili sauce compared to 2003 which only 16 companies. The process production ofthis supplementary food is a simple process and also the availability of abundant raw materials make these products often encountered with various brands and quality which can not be accountable. Even sometimes the sauce product is often forged by using dangerous raw materials [1]. In Indonesia, there is a various range of products which are a good sauce manufacturer both modern process and the local products. Process production in modern industry usually uses materials concentrate [2] while the local products more tend to utilize from its original raw materials because of the abundant raw materials or the use of a low economic value raw materials.

Research conducted by several institutions indicates that the local products of sauce, soy sauce and spicy show that six of the nine brands of chilli sauce researched were obtained sodium benzoate preservative containing more than 1,000 mg/kg [3] as the value of the threshold limit the use of sodium benzoate on the sauce according to regulation of Ministry of Health Number 033 in 2012. While, in a locally sweet soy sauce products, test the content analysis of the preservative sodium benzoate for eight brands of the products examined, six of which contain preservatives exceeding the maximum limit allowed namely 600 mg/kg [3]. Table 1. indicates the levels of sodium benzoate is present in the ketchup and soy sauce products local products.

Table 1. The content of sodium benzoate in the local products of ketchup and soy sauce [4]

No	Brand	Location	Content of sodium benzoat
Soy sauce product			
1	Sumber Agung	Makassar	4.656 mg/kg
2	Cap Cabe Payung	Batam	3.114 mg/kg
3	Tri Sari	Surabaya	2.165 mg/kg
4	Cap AVE	Medan	1.311 mg/kg
5	Sari Nikmat	Jakarta	1.353 mg/kg
6	Cap Captain	Medan	1.231 mg/kg
7	Saus tomat Sumber Jaya	Makassar	1.037 mg/kg
8	Cap Tomato	Surabaya	1.601 mg/kg
Ketchup product			
9	Ayam Panggang	Makassar	802 mg/kg
10	KOKI	Semarang	1.276 mg/kg
11	Ikan Bawang	Surabaya	1.109 mg/kg
12	Burung Bangau	Padang	1.194 mg/kg
13	Panah Sinar Langkat	Medan	2.103 mg/kg

The addition of a food preservative substance is a prevention efforts or hold so that the food does not produce odor, this is done by changing the availability of chemical composition of the foodstuff, the addition of chemicals known as preservatives and anti-oxidants or by certain biochemical processes (fermentation) to

* Corresponding author: indrotekim@yahoo.com

produce preservative compounds. The use of chemical preservatives have been practiced by humans since antiquity. Salt, fogging, the use of acidic vinegar and cooking meats with nitrates are an example of the preservation techniques of those used to be based on the addition of preservatives. Preservation of drinks with alcoholic fermentation and lactic fermented milk with a previous preservation practices are based on the conditions of preservation through biochemical processes [5].

Benzoic acid, C6H5COOH, and its derivatives is a food preservative which is most widely used. Benzoic acid has a low solubility in water, sodium benzoate is a commonly used form. However, only molecules that do not dissosiate have anti-microbial properties. Therefore active benzoic only at low pH, under 4.5. Concentrations commonly used ranges from 500 to 2000 ppm such as: to preserve fruit juices, vegetable juices, vegetable, olive pieces. The use of benzoic acid at low pH is meant to mold and decay fungi but not for bacteria. In fact, the damage because the bacteria are somewhat resistant to benzoate. The use of sodium benzoate in foods on the above mentioned range permitted by many countries [5].

For health, the consumption of excess sodium benzoate can cause stomach cramps, numb fingers taste mouth for people who are tired. This nature accumulative preservative properties could also cause cancer in the long period of time and there were also reports indicating that these preservatives can damage the nervous system [6]. Research on the use of sodium benzoate as a preservative against the effects of genetic damage (genotoxic effects) in a laboratory with a concentration 6.25 – 100 μ g/mL produces that sodium benzoate is compound clastogenic, mutagenic and cytotoxic against lymph cells of man [7].

In the process of food complement production both sauce and soy sauce will be produced wastewater. Wastewater sauce and ketchup are classified in food waste and easy to degrade. Food industry wastewater has characteristics of high levels of COD (16,000 > mg/L) [8], suspended solids, dissolved organic compounds, microorganisms and inorganic salts [9], and the ratio BOD/COD $\geq$ 0.5 [2]. Based on the content of waste organic compounds is high, then the processing treated with biological process. The main biological processes used are anaerobic biological processes because of the high removal efficiency but it still does not meet threshold quality of raw water so enhanced aerobic processes to meet existing quality raw water.

Based on operational considerations between aerobic and anaerobic processes such as organic load, rate of production of sludge, nutrient needs, energy requirements and energy generated [10] as well as the ratio BOD/COD $\geq$ 0.5 [11] then the waste water treatment system with the preservative material selected material is anaerobic. Anaerobic digestion has been known since decades ago, the use of microbes in a condition without the presence of oxygen molecules and there is a large amount of organic compounds. An organic compound is a food source for microbes and convert it into oxidized compounds, new cells, energy, for the life of the process, and some of the results of the compounds in the form of gases such as methane and carbon dioxide.

On the anaerobic process, complex organic waste will be degraded into methane, carbon dioxide and water through three main steps (hydrolisis, acetogenesis and methanogenesis including asidogenesis) in the absence of oxygen. Anaerobic processes have gained rapid progress especially for wastewater treatment based on concept retrieval and utilization of resources as well as the fulfillment of the condition in the control of pollution [12; 13].

The main problem in anaerobic processing is contact between an existing wastewater with sludge in a digester or reactor. The better the contacts then it will be the less time consumes, the less time is required then it will be the smaller the dimension of the digester required to be built. Construction of stirred flow reactor requires a depth of up to 5 m, it will be difficult if its area has a shallow well water sources. Periodical anaerobic baffled reactor (PABR) reactor requires special construction because of the shape of a circle. Therefore the design of anaerobic baffled reactor intended to overcome the lack of contact between the wastewater with sludge and also its construction does not require the high depth.

Advantages of anaerobic baffled reactor (ABR) is : simple design, without mechanical stirring, reduce the occurrence of blockage, and the cost of operation and low maintenance. For biomass used: does not require the presence of granular sludge biomass, requires no special deposition with low growth, solid sludge retention time is high, does not require a fixed media and does not require a specific gas or solid separation. Review of operation: low liquid residence time, allowing operation in intermitten, stable against hydraulic shock loads, long operation without disposal of sludge, high levels of stability against organic shock load.

Application of anaerobic technology for processing wastewater industry depends on the development and utilization of high-rate anaerobic bioreactor. Amongs the high rate of the reactor, anaerobic baffled reactor (ABR,) has been suggested as a promising system for the processing of wastewater industry. Anaerobic baffled reactors already are widely used to process industrial wastewater of palm oil mill [14], swine wastewater [15], leachate of pulp and paper mills [16], azo dye, landfills wastewater, leachate of synthetic wastewater with a content of sulphate and chromium (III) [17], wastewater of distillate liquors [18], wastewater contains nitrogen [19], liquid waste containing sulfate [20], liquid waste of textile dyes [17], wastewater containing p-nitrophenol [21], an alcoholic wastewater [22] and also domestic wastewater [17] and the wastewater of antibiotics [23].

The reactor consists of a number of the UASB reactor connected in series. Wastewater will flow above and below the vertical baffle when the flow enters the reactor. The unique design of the baffle of the ABR was able to reduce the loss of biomass so that it generates the very active biomass content. ABR can quickly recover remarkably from a load of hydraulic and organic shock, requires no special equipment to separate active sludge and gas [24].

Existing configuration ABR allows functions as a two-phase anaerobic treatment system with biomass separation of acidogenic and methanogenic. The reactor has a simple form and specifically does not require equipment of the separation of gas and active sludge and can be applied for almost all the dissolved organic wastewater from low to high organic content. Consideration of the a simple structure and operations, these can be considered as a potential reactor systems for urban waste water processing in tropical and sub-tropical countries for developing countries [25].
Anaerobic baffled reactor, ABR. First developed by McCarty [26] and friends of the researchers, consisting of compartments with divider multilevel, wastewater will flow down contact with active sludge, passes through the baffle and up flow through the blanket of anaerobic sludge. These reactor requires no active to granulated effectively. Very stable to shock load to (changing) because the compartment structure. Other benefits are: does not require biomass with specific properties to deposition, the cost of construction and operations. Based on the design of the reactor then ABR resembles a good mixing with reactor or ABR reactor resemble stopper (plug-flow reactor).

2 Materials and Methods

2.1 ABR design

A lab-scale transparent ABR bioreactor was fabricated using plexi glass sheets and designed has 3-pairs down and upflow with 60 cm of length, 25 cm of wide, 40 cm of height, 10 cm of distance between baffles, 1,5 cm of distance of baffle and bottom reactor and total volume of 60 L. The ABR consisted of three pair equal compartments and each compartment had a pair of baffles of downflow and upflow (hanging and standing baffles) and sample ports.

The temperature difference of inside reactor, due to the changes in the temperature during the day and night wass less than 3oC. Therefore, the reactor did not require cooling or heating system to maintain the temperature. A peristaltic pump was used to feed the ABR with synthetic wastewater.

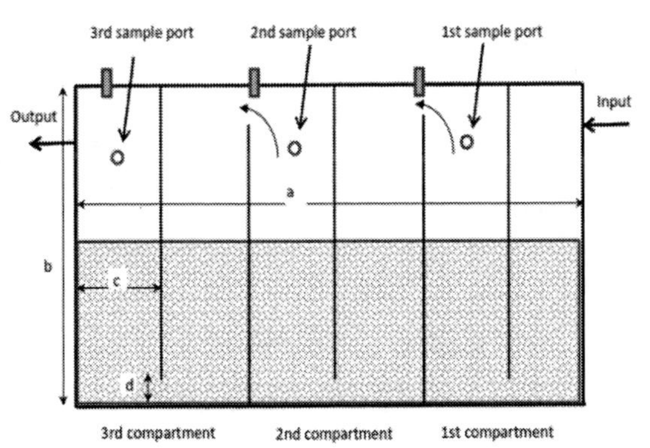

Fig. 1. Anaerobic baffled reactor (ABR) and its baffles with dimensions (cm) a = 60 cm; b = 40 cm; c = 105 cm; d = 1.5 cm; wide: 25 cm **[27]**

2.2 Sludge preparation

Anaerobic sludge obtained from Centralized of Wastewater Treatment plant of small scale tofu industries at Lampertengah, Semarang Municipality. Anaerobic sludge was taken by water sampler in the middle of anaerobic reactor. Sludge was set up into the reactor with certain height 50 % of reactor height. The sludge preparation was conducted by synthetic wastewater of 2 g/L of sucrose, pH (7.5), Ca^{++} ion (40 mg/L), and addition of macro nutrients of N and P with rasio of COD : N : P of 300 : 5 : 1, and micro nutrients with composition of (per g COD) **[28]** : Co (500 μg), Cu (300 μg), Fe (4000 μg), Mo (150 μg), Ni (6000 μg), and Zn (600 μg). This sludge conditioning was run for a month with flow rate of synthetic wastewater was 20 L/day. During this acclimatization, sludge having lighter density would be washed from the reactor and released as suspended solid. While remaining sludge on the reactor will be used for experiments.

2.3 Experimental set up

Sludge after acclimatization period was ready for experiment, first set up of sludge height was 60 % height of reactor,. The flow rate of wastewater was 20 L/day by using peristaltic pump. The synthetic wastewater is set up from the market product with trade mark of Del monte and this synthetic wastewater was made by dissolving a predetermined amount of commercial sauces of Del monte brand in distilled water and followed by analyzing the content of COD to be around 2300 and 7300 mg/L. pH was set up by dilute of sodium hydroxide or nitric acid to the system. This synthetic wastewater was set up for one day and every day had to be prepared equal volume and equal COD to feed to the ABR reactor. The responses of the experiment were the change of pH and COD of each compartment. The pH and COD were analyzed according to the standard method [29].

 https://doi.org/10.1051/matecconf/201815603042

3 Result And Discussion

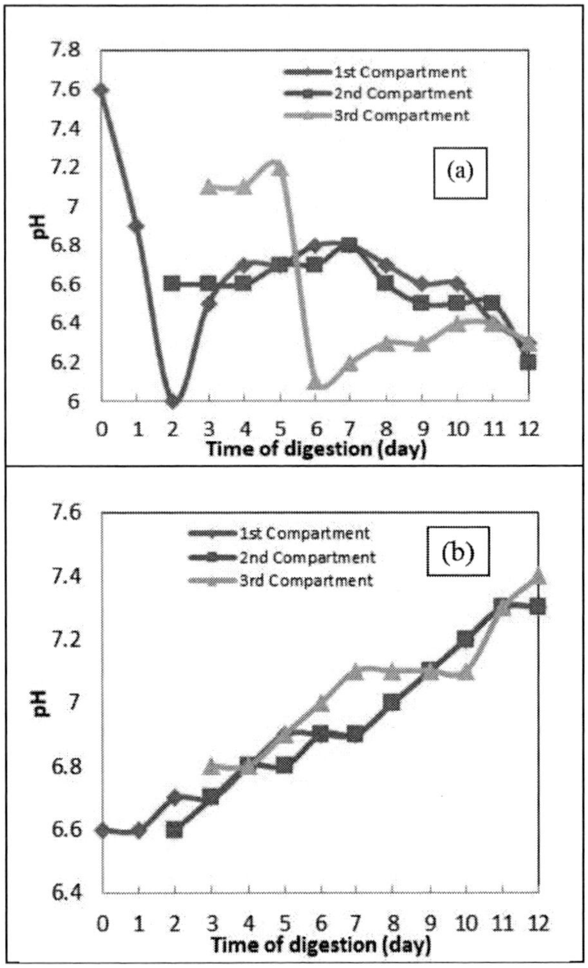

Fig. 2. Result of pH observation in the compartments with 60 % height of activated sludge (a). Initial COD : 2369.7 mg/L and (b). Initial COD : 7334 mg/L

3.1 pH

Observation of pH to the wastewater with initial COD of 2369.7 mg/L COD (Fig. 2a) showed that all of compartment 1, 2 and 3 tend to decrease. The hughest decline at compartment 1 from 7.6 to 6.3, compartment 2 (6.6 to 6.2) and compartment 3 (7.1 to 6.3). On day of 11-12 pH for the third compartment is close to the value of about 6.2-6.3. Normally once it reached a stable pH conditions the the pH will increase [30] due to the formation of alkalinity [31]. Load of the first compartement is the most severe because of the decrease in pH very sharp up to the second day, food wastewater will be easily degraded so that on initial contact with the activated sludge process of hydrolysis occurs and acidogenic with a high VFA result [32; 24]. Acidogenic advanced process occurred in compartment 3. Microorganisms on three early stages of anaerobic process has a broad pH range i.e. 4.0-8.5 [33]. Changes of pH 7.6 to 6.3 still be within range for the occurrence of decomposed process into volatile fatty acids. Start the day of 11, the pH for the third compartment was close the same value, this indicated that the acsidogenesis process is complete, shown in the graph of the COD that in to day of eleven the curve in the form of exponential.

While for wastewater with initial COD of 7334 mg/L (Fig. 2b), the initial pH of the wastewater enters in conditions of acid (6.6), the characteristic of the wastewater that is easily degraded, high content of COD leads to the formation of alkalinity from the substance of protein in foodstuffs is also fastly occurred, the rate of formation of alkalinity is more quickly then causing the rise in pH [34]. Based on the picture above (Fig. 2b), phases of hydrolysis is more dominant than with the stages of acidogenic with increasing pH results. Complementary foods with significant nutrient content will provide degradation is enough to suppress the formation of the VFA until the pH will increase to about 7.3-7.4 on the digestive time 11 and 12 days. The presence of the addition of the calsium ion serves to maintain the alkalinity anaerobic system will neutralize the acids are formed, and also to prevent the formation of regional acid **[34]**.

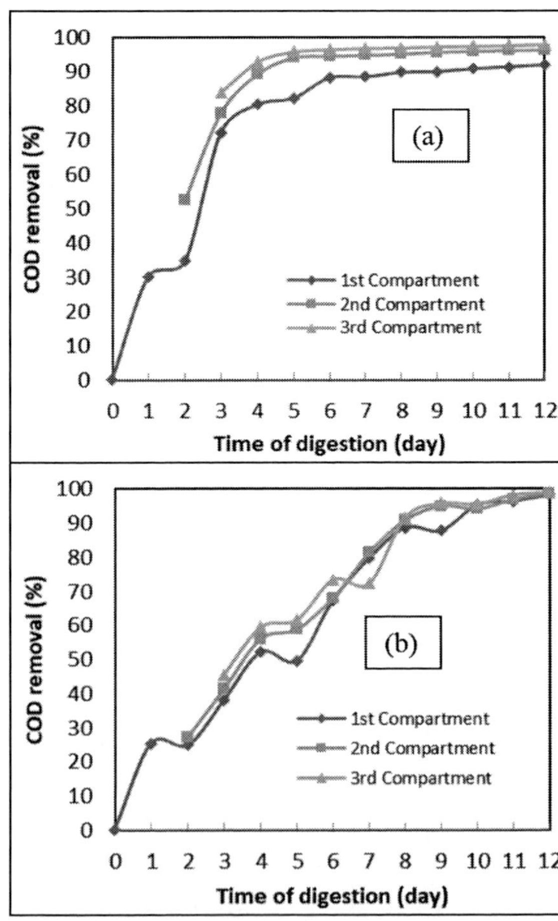

Fig. 3. Result of COD removal in the compartments with 60 % height of activated sludge (a). Initial COD : 2369.7 mg/L and (b). Initial COD : 7334 mg/L

3.2 COD

Observation of COD removal in the ABR system for initial COD of 2369.7 mg/L (Fig. 3a) to the digestion time of 12 days obtained results that total COD removal efficiency of the first, second and third compartments were 91.84%, 96.83% and 98.03%. The COD removal until the fourth day is the highest with COD removal for each compartment was 80.36%, 89.16 % and 92.80 %, while for the successive time addition of only slight its increase of COD removal. Curves for low initial COD expressed as a form of clear exponential curve with a stationary point between the fourth and fifth days. While for wastewater with initial COD of 7334 mg COD/L (Fig. 3b), the performance of the removal COD of each compartment up to digestion time of 12 days was 98.15%, 98.51% and 98.83%. In general the digestion time for 12 days, the total COD removal efficiency in ABR, the higher organic loading the higher COD removal than low content of organic loading of wastewater. The curve of the COD removal to the digestion time is an exponential curve. At a low organic loading, the exponential curve shape is clearly visible at the beginning of the curve, but to a higher organic content, the curve is shifted to the right. For higher loads then the stationary point will shift to the right on the nine days. Based on the stasionary point of the curve, the degradation time is doubled for increasing organic loading of 2369.7 mg/L be 7334 mg/L or 3 times.

The highest COD removal occurred in the first compartment while each successive compartment was only removed small fractions of the remaining COD. As COD decreased in the preceding compartment, a reduction in substrate utilization rate of the microorganisms in subsequent compartments would result to the leading to a lower removal efficiency. This pehenomenon has been well supported by the bacrterial kinetics that lower substrate concentration will cause lower growth rate [35]. Studies that have been done by some researchers such as: If the organic load was increased 2-fold then needed four days to recover to its original condition [30], the efficiency COD removal in ABR reached 97.7% [36], if HRT is increased then the efficiency COD removal is also increased [37], while for molasses with HRT increase then efficiency COD removal to be lower which is about 91.6% [33].

For ABR system, the first compartment, COD was removed through the cytogenesis and gas releases (CO_2 and H_2), this is dominantly by acidogenic bacteria, while significant amount of COD was converted into liquid intermediate prodcuts (e.g. ethanol, butyrate and propionate) and stay in the system. The substrate in second compartment were converted into acetic acid and hydrogen by acetogen and COD was mainly removed through the conversion of intermediate products (e.g. acetic acid) to methane by the methanogenic bacteria in the third and fourth compartments [24].

Food wastewater treatment with anaerobic process in the ABR pointed out that the COD removal for the first compartment plays a very important role while successive compartments of two and beyond just reduce insignificant COD. This is useful if for anaerobic reactor design for food wastewater then it could be selected as single stage ABR similar as UASB reactor.

4 Conclusions

1. Anaerobic process of food wastewater treatment with ABR shows that low levels of COD content resulted the curve the exponential stabilization at 4-5 days digestion whereas for higher load of COD content, the curve the exponential stabilization of shifts to the right.
2. The first compartement is the main reactor to decrease the organic content of wastewater, while for the removal efficiency of the second and third compartments was not significant.
3. The increase in organic loads to 3 times resulted stabilization time 2 times

The authors gratefully acknowledge to the financial support of the Diponegoro University through research scheme of PNBP DIPA Universitas Diponegoro No. SP DIPA-042.01.2.400898/2016

References

1. K. Dede, *Industri Rumahan Saos Sambal Di Bandung Digerebek Polisi*, okezone.com (2015, January 27).
2. PT. Indofood. Laporan UKL-UPL PT. Indofood Fritolay, Semarang (2012).
3. Peraturan Menteri Kesehatan Republik Indonesia Nomor 033 Tahun 2012 Tentang Bahan Tambahan Pangan
4. K. M. Didin, (2011, Juli 24). Bahan Pengawet Saus dan Kecap Lebihi Ambang Batas Halal Guide MUI
5. M. S. Rahman, *Handbook of Food Preservation, Second edition* (CRC Press, Boca Raton, London, New York, 2007).
6. B.S. Lennerz, S.B. Vafai, N.F. Delaney, C.B. Clish, A.A. Deik, K.A. Pierce, Molecular Genetics and Metabolism **114**, 73–79 (2015).
7. N. Zengin, D.Yüzbaşioğlu, F.Ünal, S.Yilmaz, S., Food and Chemical Toxicology **49**, 4, 763–769 (2011)
8. A.S. Cassini, I.C. Tessaro, L.D.F. Marczak, C. Pertile, Journal of Cleaner Production **18**, 3, 260–265 ()(2010)
9. Y. Zeng, C. Yang, J. Zhang, W. Pu. Feasibility investigation of oily wastewater treatment by combination of zinc and PAM in coagulation/flocculation. Journal of Hazardous Materials, 147(3)(2007), 991–996.
10. D. T. Sponza, A. Uluköy, Process Biochemistry **40** 11, 3419–3428 (2005)
11. Metcalf, Eddy, *Wastewater Engineering Treatment and Reuse. 4th Edition* (McGraw-Hill, New York, 2003)
12. Y. Li, S.Y. Park, J. Zhu, J., Renewable and Sustainable Energy Reviews **15**, 1, 821–826 (2011)

13. L. Seghezzo, G. Zeeman, J.B. Liel, H.V.M.. Van, Hamelers, G. Lettinga, Bioresource Technology **65**, 175-190 (1998)
14. T. Setiadi, Husaini, A. Djajadiningrat, Wat. Sci. Tech. **34,** 11, 59-66 (1996)
15. R. Boopathy, Bioresource Technology **64**, 1-6 (1998)
16. R. Grover, S.S. Marwaha, J.F. Kennedy, Process Biochemistry **34**, 6-7, 653–657 (1999)
17. D.J. Batstone, V. Virdis, Current Opinion in Biotechnology **27**, 142–149 (2014)
18. J.C. Akunna, M. Clark, Bioresource Technology **74**, 3, 257–261 (2000)
19. I. Bodík, K. Kratochvíl, E. Gašpariková, M. Hutan, Bioresource Technology **86**, 1, 79–84 (2003)
20. M. Vossoughi, M Vossoughi, M. Shakeri, M. Shakeri, M.I. Alemzadeh, I.Alemzadeh, Analysis **42**, 811–816 (2003)
21. Ö.S.Kuşçu and D.T. Sponza, D. T., Enzyme and Microbial Technology **36**, 7, 888–895 (2005)
22. M.I. Baloch, J.C. Akunna, and P.J. Collier, Bioresource Technology **98**, 9, 1849–1855 (2007)
23. S. Chelliapan, T. Wilby, P.J. Sallis, Water Research **40**, 3 , 507–516 (2006)
24. G.F. Zhu, J.Z. Li, P. Wu, H.Z. Jin, Z. Wang, Bioresource Technology **99**, 17, 8027–8033 (2008)
25. D. Cui, Y-Q. Guo, H-S. Lee, W.-M. Wu, B. Liang, A.-J. Wang,H.Y. Cheng, Bioresource Technology **163**, 254–261 (2014)
26. I.V. Skiadas,H.N. Gavala, G. Lyberatos, Water Research **34**, 15, 3725–3736 (2000)
27. I. Sumantri, P, Purwanto, Budiyono, AIP Conference Proceedings **1699**, 050017 (2015)
28. J. Ma., L. J. Mungoni., W Willy Verstraete., M. Carballa, Bioresource Technology **100**, 3477–3482 (2009)
29. APHA, *Standard Methods for the Examination of Water and Wastewater, 21st ed* (2012)
30. G.D. Ji., T.H. Sun, J.R. Ni, J.J. Tong, Bioresource Technology **100**, 3, 1108-1114 (2009)
31. S. Ozdemir, K. Cirik, D. Akman, E. Sahinkaya, O. Cinar, Bioresource Technology **146**, 135–143 (2013)
32. P. Thanwised., W. Wirojanagud., and A. Reungsang, International Journal of Hydrogen Energy **37**, 15503-15510 (2012)
33. M. Pirsaheb, M. Rostamifar, A.M. Mansouri, A.A.L. Zinatizadeh, K. Sharafi, Journal of the Taiwan Institute of Chemical Engineers **47**, 137–148 (2015)
34. F. Xu., Z. Huang., H. Miao., H. Ren., M. Zhao., and W. Ruan, Journal of Environmental Sciences **25**, 10, 2031-2040 (2013)
35. K. Saritpongteeraka, S. Chaiprapat, Bioresource Technology **99**, 8987-8994 (2008)
36. W. Cao, M. Mehrvar., Chemical Engibeering Research and Design **89**, 1136-1143 (2011)
37. Z. Li,., D. Suzuki,., C. Zhang,., N. Yoshida,., S. Yang, A Katayama, Journal of Bioscience and Bioengineering **116**, 5, 602–609 (2013)

Biogas Production from Palm Oil Fruit Bunch in Anaerobic Biodigester through Liquid State (LS-AD) and Solid State (SS-AD) Method

Bakti Jos[1,*], *Hanif* Farhan [1], *Nadia Dwi* Ayu[1], *Budiyono*[1], and *Siswo* Sumardiono[1]

[1]Department of Chemical Engineering, Faculty of Engineering, Diponegoro University, Jl.Prof. Soedarto, SH, Tembalang, Semarang, Indonesia 50275

Abstract. The crucial problem facing the world today is energy resources. Waste production of palm oil fruit bunch potentially produce as renewable energy resource. Palm oil fruit bunch contains 44% cellulose, 18% lignin and 34% hemicellulose. Organic carbon source is contained in biomass potentially produce biogas. Biogas is one of alternative energy, which is environmentally friendly and has been widely developed. This research is aimed to examine the effect of pretreatment in raw material of waste palm oil fruit bunch for the production of biogas, the effect of time, ratio C/N, and effect of microbial consortium. The variables are total solid (TS) used 10% and 18% with a 40 mesh physical pretreatment, chemical pretreatment with NaOH 8% gr / gr TS, and biology 5% g/vol with microbial consortium. Biogas production process was conducted over 2 months in room temperature, the test response quantitative results in the form of biogas volume every 2 days and also flame test. The result of this research shows that the highest daily production rate of biogas obtained from this study was 5,73 ml/gr TS and the highest biogas production accumulation generated at 58,28 ml/gr TS produced through a 40 mesh sieve of waste oil palm empty fruit bunch, immersion in NaOH, through solid state fermentation and C/N 30. From this research, it can be concluded that the optimum production of biogas formation occurs with the value of C/N 30, physical and biological pretreatment, and solid state method.

1 Introduction

World energy demand is expected to increase rapidly along with population growth and economic development. Dependency on fossil energy particularly fossil fuel, to fulfill high energy consumed of the country, which is 96% (48% crude oil, 18% gas, and 30% coal) of total consumption and to optimize renewable energy utilization [1]. Energy consumption is increased continuously due to population growth [2]. Total fuel consumption for energy production is being increased every year, in the last 5 years, fuel oil consumed increase 18% until 2013 its consumption increase by 197,4 thousand kl/d and total fuel consumption in 2014 is 308 million equal to barrel oil or decrease 0.18% per year.

In order to overcome the increasing of energy demands, renewable energy has remained one of the best alternatives for sustainable energy development [3]. Energy consumption has an effect on economic growth and national defense. One of renewable energy sources which is being pursued for its development nowadays is bioenergy. Bioenergy or biofuel is energy from organic matter (biomass) [4]. Biofuels can be divided into bioethanol, biodiesel and biogas [5]. Biogas is one of alternative energy, which is environmentally friendly and has been widely developed [6]. Biogas can be categorized as bioenergy, because the energy produced from biomass [7].

Biogas production is carried out by microbes in anaerobic condition as known Anaerobic Digestion (AD) process or fermentation from organic matters or biomass [8,9], for instance agricultural waste, organic waste, animal or human manure, industrial waste and domestic waste in anaerobic condition [10-12]. In anaerobic digestion, organic materials will be converted by bacteria into the biogas through four major phases i.e. hydrolysis, acidogenesis, acetogenesis and methanoogenesis [13-15]. In the hydrolysis phase, complex organics (carbohydrate, protein, fat) are converted into simple organics (sugar, amino acid, LVFA) ; in the acidogenesis phase, simple organics are converted into organic acids ; for the acetogenesis phase, organic acids are converted into acetic acid, and the final phase is methanoogenesis that aim for converting acetic acid into CH4 and CO2 [16] Biogas is a renewable energy with a composition consisting of 50-75% methane (CH4), 34-45% carbon dioxide (CO2) and other gases in small amounts such as CO, N_2, H_2, H_2S, and O_2 [17].

The anaerobic digestion (AD) process is generally classified into two categories based on the total solid (TS) content of the mixture in the digester [18]. TS contents of 15% or greater are classified as solid state anaerobic digestion (SS-AD) while TS contents of 15% or lower are classified as liquid anaerobic digestion (LS-

* Corresponding author: baktijos@che.undip.ac.id

AD) [19]. Biogas has 20% lower weight than air and has heat of combustion value between 4800-6200 kcal/m^3. Its value is lower than methane heating value reach 8900 kcal/m^3 [20]. Biogas can be produced through appropriate technology which is relatively simpler [21], so it can be applied either in town or village. According to the Ministry of Energy and Mineral Resources, biogas has applied in both developed and developing countries.

First generation biogas started from 1884, Pasteur attempted to produce biogas from cattle [22]. Biogas is also often produced from other livestock waste such as chicken waste, pig waste, and horse waste. Biogas has developed rapidly until the second generation of biogas [23]. Second generation of biogas can be produced from lignocelluloses material such as agricultural or forestry waste for instance include straw, waste plant corn, palm oil fruit bunch, bagasse, grass, wood and many more [24]. Palm oil fruit bunch is one with huge potential to be developed in Indonesia. Palm oil fruit bunch is waste residue generated from palm oil industries. This country is the world's largest producer and exporter of palm oil and expected to be continuously increased.

According to Directorate General of Plantations, Ministry of Agriculture Indonesia, the average of palm oil area growth rate from 2004 – 2014 reached 7.67 % and palm oil production increases to an average of 11,09 % per year. Palm oil fruit bunch is one of the solid wastes produced in huge volume by palm oil mill, however, it was not fully utilized till then, and therefore it needs to be followed for the extensive utilization of waste palm oil fruit bunch. One of these is to convert waste palm oil fruit bunch into biogas as an alternative energy [25].

Palm oil fruit bunch contains 40-60 % cellulose, 20-30 % hemicellulose, and 15-30 % lignin [26]. Lignocellulose content within palm oil fruit bunch can be converted into biogas, however, lignocellulose biomass has a complex internal structure, thus it needs aggressive pretreatment to yield a substrate easily hydrolyzed by cellulolytic enzymes or enzyme producing microorganisms [27].

Pretreatment of raw material lignocellulose in order to improve the yield of the biogas produced can be done by chemistry, physics, and biology pretreatment [28-30]. Therefore, to achieve increased biogas yield higher, require further study regarding pretreatment method more effective. This research will use a combination pretreatment method that consists of physical, chemical and biological pretreatment. Physical pretreatment is used to disrupt cellulose crystallinity, reduce particle size, and increase the surface area [31], it can be done by milling or grinding the raw material [27]. Afterward, the process followed by adding NaOH for chemical pretreatment, it refers to enhance the biodegradability of cellulose and lignin removal [31].

The final step of this combination pretreatment is biological pretreatment. Biological pretreatment is carried out by microorganisms such as microbial consortium, which mainly degrade lignin, also reduce polymerization degree and crystallinity of cellulose [29]. Thus, pretreatment combination method is expected to increase biogas yield in anaerobic digestion. This research is also aimed to observe the comparison of methane yield in liquid state anaerobic digestion (LS- AD) and solid state anaerobic digestion (SS-AD).

2 Materials and Method

2.1. Materials

Materials used in this study are the palm oil fruit bunch, cow rumen obtained, microbial consortium, NaOH, urea, water, and HCl .

2.2. Preparation Stage

At this stage, the samples of farm waste such as palm oil fruit bunch from Banten, cow rumen is taken from the slaughterhouse in Semarang, microbial consortium from Banyuwangi. Besides, NaOH as lignocellulose degradation agent and HCl are bought in Indrasari chemicals store, Semarang. After that, the palm oil fruit bunch as much as 220 grams (for liquid state variables) and 360 grams (for solid state variables) are reduced according to 40 mesh. Then the raw materials are added with 8% gr/gr TS of NaOH and the pH is set by using HCl for maintaining a close to neutral pH. For biological pretreatment is used 5% v/v microbial consortium to be processed at a further stage.

2.3.Operation Methods

Fig. 1 The Apparatus of biogas production

Substrates that have been homogeneously are mixed with cow rumen as much as 110 grams (for liquid statevariables, 10% total solid) and 220 grams (for solid state variables, 18% total solid) as starter of methanogen microbes. After that, the urea is added as micronutrients as much as 3.92 grams for c/n 25 and 2.65 grams for c/n 30 and water is also added until 2000 ml. the fermentation process is started. The volume of gas hat is formed is measured by water displacement technique every other day for 62 days, as shown in fig 1.

2.4.Results Analysis Stage

The data of biogas volume is analyzed in graphical form as relationship biogas volume versus time.

3 Results and Discussion

3.1. The Effect of Palm Oil Fruit Bunch Size on Physical Pretreatment against Biogas Production

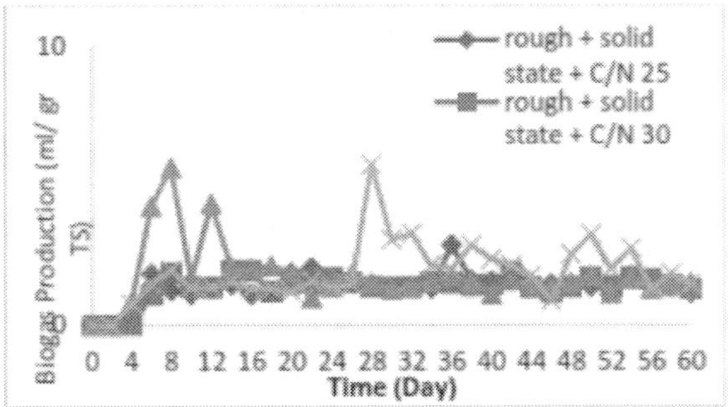

Fig. 2. The rate of the daily biogas production from palm oil fruit bunch in solid state condition with c/n 25 and 30

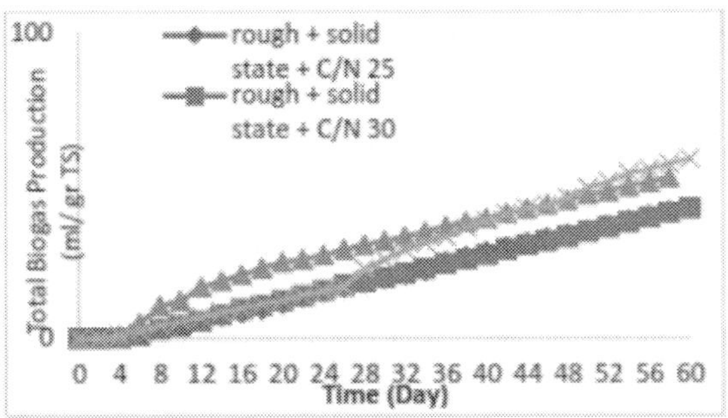

Fig. 3. The total biogas production from palm oil fruit bunch in solid state condition with c/n 25 and 30

Figures 2 and 3 show a comparison of the biogas production of palm oil fruit bunch through size reduction 40 mesh and without being reduced. Total production of biogas in smooth palm oil fruit bunch C/N 25, smooth C/N 30, rough C/N 25 and rough C/N 30 consecutive reaching 53.47 ml/gr TS, 58.28 ml/gr TS, 42.43 ml/gr TS, 42.15 ml/gr TS. The highest total volume of biogas produced in delicate palm oil fruit bunch C/N 30, in the amount 58.83 ml/ gr.

Biogas production on a palm oil fruit bunch through size reduction process has the higher volume than biogas production in palm oil fruit bunch without being reduced. This is due to the particle size reduction that can reduce the crystallinity of cellulose and lignin defense disorders [32]. Thus, it will facilitate the process of hydrolysis resulting in an increased production of bioenergy [33]. In addition, size reduction for palm oil fruit bunch increases the surface area and reduces the degree of polymerization for cellulose [29]. So that, the smaller size of lignocelluloses substrate in this case is palm oil fruit bunch, the higher the production of biogas produced by the crystalline structure due to the increasing number of palm oil fruit bunch degraded to simplify the process of hydrolysis.

The addition of 8% NaOH can help the destruction process of lignin structure, or commonly called delignification. Delignification process can destroy lignin structure and release carbohydrate compound [34]. Delignification process is a method to convert lignocellulose into glucose compound [35]. Delignification is believed as a potential process in preparation method of raw materials [29].

3.2. The Effect of Microbial Consortium Addition against Biogas Production

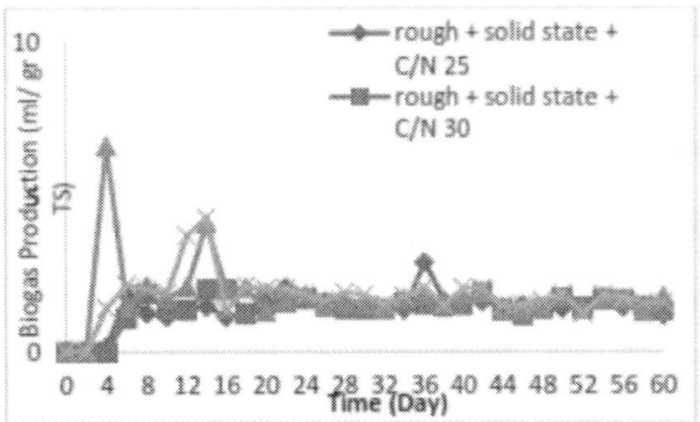

Fig. 4. Daily biogas production rate of palm oil fruit bunch in solid state condition with a C/N 25 and 30

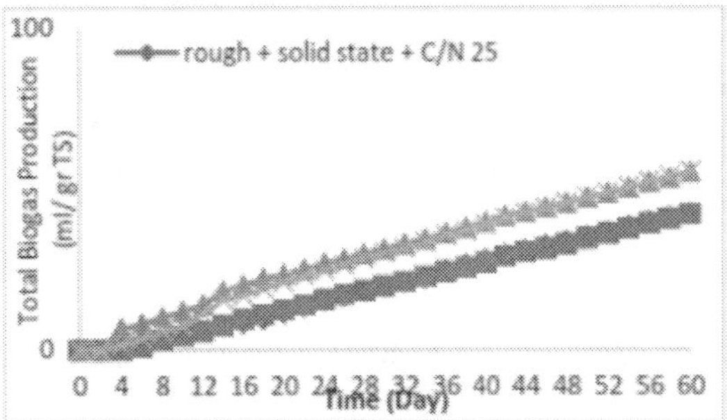

Fig. 5. Total biogas production rate of palm oil fruit bunch in solid state condition with a C/N 25 and 30

From Figure 4 and 5, it can be referred the comparison between biogas production from palm oil fruit bunch in solid state condition with the addition of NaOH at a ratio of C/N of 25 and 30 with the addition of microbial consortium and without microbial consortium. Total production of biogas in the C/N 25 either by the addition of microbial consortium and without microbial consortium respectively for rough palm oil fruit bunch reached 55.11 ml/gr TS and 42.42 ml/gr. While total production of biogas in the C/N 30 either by the addition of microbial consortium and without microbial consortium respectively for rough palm oil fruit bunch reached 55.43 ml/gr TS and 42,15 ml/gr TS. The highest total biogas production on palm oil fruit bunch C/N 30 with the addition of microbial consortium, in the amount 55.43 ml/gr TS.

Biogas production on palm oil fruit bunch with the addition of microbial consortium has the higher volume than without the addition of microbial consortium. The activity of microbial consortium, which consisted by Streptomyces sp., Geobacillus sp., and Trichoderma fungi can enhance the delignification, decrease cellulose polymerization degree, and hydrolysis of hemicellulose. The addition of microbial consortium can change component structure and increase enzymatic hydrolysis, it can also relieve the hydrolysis process of cellulose into glucose [36, 37].

3.3. Influence of C/N and Biodigester Condition against Biogas Production

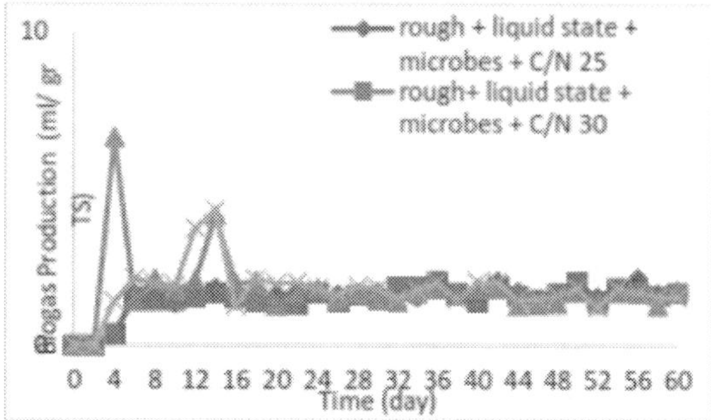

Fig. 6 Daily biogas production rate of palm oil fruit bunch in liquid state condition with a C/N 25 and 30

In the figure 6 and 7 can be seen a comparison of the production of biogas from palm oil fruit bunch in different condition of biodigester (Solid State and Liquid State) with the addition of NaOH and microbes, with differences in C/N, which serves as C/N 25, C/N 30. Total production of biogas in the liquid state C/N 25, liquid state C/N 30, solid state C/N 25 and solid state C/N 30 respectively for palm oil fruit bunch reached 47.28 ml/gr TS, 45.65 ml/gr TS, 55.11 ml/gr TS, 55.43 ml/gr TS. The amount of C/N 25 both liquid state and liquid state have the higher volume than C/N 30, and also for solid state conditions reached the higher volume than liquid state conditions. The highest amount of biogas production on palm oil fruit bunch C/N 30 in solid state biodigester condition, amounting 55.43 ml/gr TS.

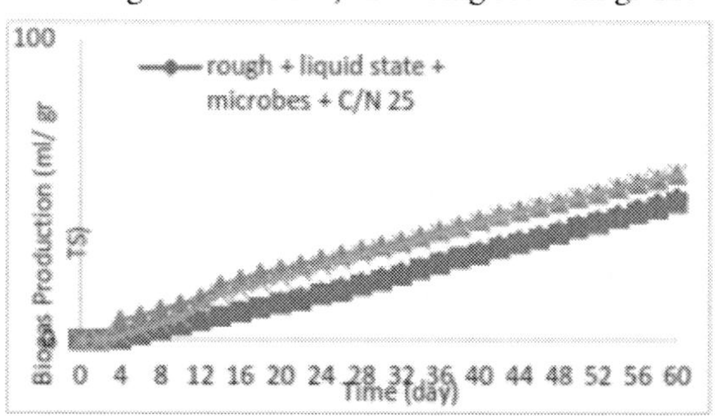

Fig. 7 Total biogas production rate of palm oil fruit bunch in liquid state condition with a C/N 25 and 30

C/N from organic matters greatly influenced microbial activity and biogas production [24,38]. Microbes that play a role in the anaerobic processes require nutrients to grow and develop, in the form of a carbon source and a nitrogen source [39]. Equilibration of carbon and nitrogen in materials used as substrates also requires attention. Therefore, excessive nitrogen concentration can limit bacterial growth, in this case for the material that has very high ammonia contents [40], while the substrate that has slight nitrogen content, the bacteria will not be able to produce the enzyme which is needed to synthesize its compound (substrate) that contains carbon. The optimum production of biogas refers to C/N 30 because the most optimum substrate contains C/N ranges from 20: 1 to 30: 1 [41].

Solid State Fermentation (SSF) holds high productivity level due to the nutrient content per volume that can be much more concentrated. Thus, it results the higher volume of biogas production [42-43].

3.4. The Kinetic of Biogas Production Reaction

By Assuming that the rate of biogas production in batch biodigester is proportional to the specific growth rate of methanogenic microorganisms in biodigester, the rate of biogas production will follow the Gompertz Equation [44]. Originally, Gompertz equation was used to predict bacterial growth rate. Currently, many authors used modified Gompertz equation to predict biogas production rate with assume that biogas production rate had correspondence to methanogenic bacteria growth rate in digesters [45]. This equation is a mathematical model for an observation time series, which is the slowest growth at the beginning and end of the period of observation time and have the general form as follows:

$$y = A\exp\left\{-\exp\left[\frac{Ue}{A}(\lambda - 1) + 1\right]\right\} \tag{1}$$

y = the cumulative production of biogas, L/kg of "t" time.
A = the maximum production of biogas, L/kg
U = the rate of maximum biogas production, (L/kg.day)
λ = the time of lag phase, day
t = the time of cumulative biogas production, day
e = Euler number (e = 2.71828...)

By using non-linear regression method, the data obtained from experiments using biodigester kinetic constants can be obtained as presented in Table 1.

The value of Rm is the lowest value on palm oil fruit bunch in solid state condition and C/N 25 with a value of 1.92 ml/gr TS day. It can be referred that the processes of biogas production in this variable proceed slowly. While the highest Rm value figures on smooth palm oil fruit bunch in liquid state condition and C/N 30 with a value 5.73 ml/gr TS day. Thus, with that size, the addition of microbial consortium, biodigester condition and C/N affect the biogas production process [46].

Table 1. The Kinetics data of biogas production according the calculation of Gompretz equation

Variable	Rm (ml/g TS.day)	P (ml/g TS)	λ (day)
rough + liquid state + C/N 25	4.43	104.40	26
rough + liquid state + C/N 30	5.23	114.08	12
rough + solid state + C/N 25	5.06	123.25	16
rough + solid state + C/N 30	5.09	120.47	14
rough + microbial consortium + liquid state + C/N 25	8.63	145.49	2
rough + microbial consortium + liquid state + C/N 30	6.06	126.51	12
rough + microbial consortium + solid state + C/N 25	4.82	112.70	12
rough + microbial consortium + solid state + C/N 30	4.41	98.46	6
smooth + liquid state + C/N 25	4.82	120.29	10
smooth + liquid state + C/N 30	4.94	131.18	6
smooth + solid state + C/N 25	6.51	154.12	4
smooth + solid state + C/N 30	4.99	120.33	14
smooth + microbial consortium + liquid state + C/N 25	7.71	161.13	4
smooth + microbial consortium + liquid state + C/N 30	4.65	113.63	10
smooth + microbial consortium + solid state + C/N 25	4.29	91.54	20

The value of P from each variable has different value, starting from 38.30 until 58.28 ml/gr TS where the lowest volume of smooth palm oil fruit bunch with the addition of microbial consortium, in solid state condition, and C/N 25, and the highest volume contained on palm oil fruit bunch, in solid state condition and C/N 30. It can be indicated that the presence of a physical pretreatment (milling), biology (the addition of EM-4) and chemical (addition of NaOH) affect the amount of biogas production in each variable. The highest λ value almost consists on each variable, while biogas production started from the second day. The lowest λ value contained in rough palm oil fruit bunch, solid state condition, C/N 30 and on rough palm oil fruit bunch, liquid state condition, C/N 30, where the biogas is formed at day 264. It proves that size does not significantly influence the time of biogas production.

The biogas production kinetics modified Gompertz equation and found that the data predicted by the model are quite close to the experimental data. It also indicates the presence of a physical pretreatment (milling), biology (the addition of microbial consortium), and chemical (addition of NaOH) maximum biogas production (U) and biogas production potential (A) of all improved lignocellulose biomass. Modified Gompertz equation also describes the cumulative gas production as a function of residence time [47]. λ value has an influence on the start time of the first biogas when it is formed or increased [48] Some raw materials originating from livestock manure has a longer adaptation time. In this study, using of cow dung and thus require a longer adaptation period, but the organic matter can decompose perfect views of cumulative production fairly competitive with a variety of other substrates such as in Table 4.2 [49]. By doing so, palm oil fruit bunch combined with cow dung can be used as feedstock for biogas production, but takes time to form a fairly long compared to some raw materials from previous studies.

4 Conclusion

Physic pretreatment (size reduction) and biological pretreatment (the addition of microbial consortium) in palm oil fruit bunch can increase surface area, decrease crystallinity and degree of polymerization for cellulose. Microorganism can be used for lignocellulosepretreatment and increase enzymatic hydrolysis. Biogas production is higher for solid state condition instead of liquid state because of the nutrient content per volume is more concentrated. C/N from organic matters determined microbial activity and biogas production. The optimum production of biogas refers to C/N 30 because the most optimum substrate contains C/N ranges from 20 to 30.

Authors thank to Department of Chemical Engineering, Diponegoro University for all facilities to do this research.

References

1. S. Said. *Indonesian Energy Outlook 2014*. National Energy Board-Indonesian National Energy Council (2014)
2. S. Fletcher, Oil and Gas Journal (2013)
3. P.A. Ukpai. and M.N. Nnabuchi, Advances in Applied Science Research, **3**, 3,1864–1869 (2012)
4. . Basker, S. Basker, R.S. Dhillon. *Biomass Conversion: The Interface of Biotechnology*. Chemistry and Materials Science, Springer (2012)

5. A. Demirbas, Energy Conversion and Management **49**, 2106-2116 (2008)
6. J. Twidell and T. Weir. *Renewable Energy Resources Third Edition*. New York: Routledge (2015)
7. D. Waskito. *Biogas Power Plant Analysis from Cow Dung Manure in Cow Farm*. Thesis of Faculty of Engineering, University of Indonesia (2011)
8. X. Chen, R. Romano, R. Zhang, International Journal of Agricultural and Biological Engineering, **4**, 51－62 (2010)
9. T. Prasetyo, S. Sumardiono, H.A.A. Aji, A.Y. Pratama, Advanced Science Letters **23**, 6, 5810-5814 (2017)
10. K. Zieminski and F. Magdalena, African Journal of Biotechnology **11**, 18, 4127-4139 (2012)
11. K. Doelle and F. Oliveira, Journal Food Process Technol,**6**, 2157-7110 (2015)
12. S. Sumardiono, A.B. Riyanta, H.H.A. Matin, T.D. Kusworo, B. Jos, Budiyono, MATEC Web of Conferences, **101,** 02014 (2017)
13. Batstone, D.J., J. Keller, I. Angelidaki, S.V. Kalyuzhnyi, S.G. Pavlostathis, A. Rozzi, W.T.M. Sanders, H. Siegrist and V.A. Vavilin, Water Sci. Technol. **45**, 557-565 (2002)
14. Esposito, G., L. Frunzo, A. Panico and F. Pirozzi, J. Process Biochem **46**, 557-565 (2011)
15. Budiyono, I. Syaichurrozi, S. Sumardiono, Journal of Applied Sciences, Eng. And Technol. **7**, 13, 2798-2805 (2014)
16. S. Sumardiono, Budiyono, D.T. Mardiani. AIP Conference Proceedings **1699**, 050018 (2016)
17. Budiyono, I. Syaichurrozi, S. Sumardiono, World Applied Sciences Journal **26**, 11, 1464-1472 (2013)
18. Budiyono, I. Syaichurrozi, S. Sumardiono, International Journal of Engineering, Tramsactions B: Applications **27**, 2, 177-184 (2014)
19. D. Brown, J. Shi, Y. Li. *Comparison of solid-state to liquid anaerobic digestion of lignocellulosic feedstocks for biogas production*. US National Library of Medicine National Institutes of Health (2010).
20. I. Mara, Engineering Dynamic, Mechanical Engineering **2** , 1 (2010)
21. B. Rahman. *Biogas, Sumber Energi Alternatif.* Lembaga Ilmu Pengetahuan Indonesia (2005)
22. T. Hoerz, K. Pedro, B. Klingler, C. Kellner, T. Wittur, F. V. Klopotek, A. Krieg, H. Euler. *Biogas Digest,***I**. Information and Advisory Service on Appropriate Technology (ISAT) Germany (2008)
23. F. Streffer, JSM Biotechnol Bioeng **2**, 1, 1023 (2014)
24. A. Teghamar. Chalmers University of Technology (2013)
25. M.M. Kabir, M. Shafiei, H. Zilouei, I.S. Horvarth, K. Karim, Bioresource Technology, **XXX**, 32 (2014)
26. M. Dekker. *Wood and Cellulosic Chemistry*. New York: Madison Avenue (1991)
27. P.F.H. Harmsen, W.J.J. Huijgen, L.M. Bermudez Lopez, R.R.C. Bakker, Food and Biobased Research Wageningen Netherlands (2010)
28. N. Mosier, C. Wyman, B. Dale, R. Elander, Y.Y. Lee, M. Holtzapple, and M. Ladisch, Bioresour. Technol. **96**, 673-686 (2005)
29. M. J. Taherzadeh, K. Karimi, International Journal of Molecular Sci. **9**, 1621-1651 (2008)
30. A.T.W.M. Hendriks and G. Zeeman, Bioresource Technology **100**, 10–18 (2009)
31. P. Kumar, D.M. Barrett, M.J. Delwiche and P. Stroeve, Ind. Eng. Chem. Res. **30**, No. 20, 40-58 (2009)
32. E. Bruni. *Improved Anaerobic Digestion of Energy Crops and Agricultural Residues. PhD Thesis*. Department of Environmental Engineering, Technical University of Denmark (2010)
33. Y. Li, S. Park, J. Zhu, Renew Sustain Energy **15**,821-826 (2011)
34. Y. Zheng, J. Zhao, F. Xu, Y. Li, Progress in Energy and Combustion Science **42**, 35-53 (2014)
35. Budiyono, A. Wicaksono, A. Rahmawan, H.H.A. Matin, L.G. K. Wardani, T.D. Kusworo, S. Sumardiono, MATEC Web of Conference **101**, 02011 (2017)
36. M. Taniguchi, H. Suzuki, D. Watanabe, K. Sakai, K. Hoshino, and T. Tanaka, J. Biosci. Bioeng. **100**, 6, 6737-643 2005)
37. M. Kurakake, N. Ide, T. Komaki, Current Microbiology **54**, 424-428 (2007)
38. K.M. Kangle, S.V. Kore, V.S. Kore, G.S. Kulkarni, Universal Journal of Environmental Research and Technology **2**, 4, 210–219 (2012)
39. M. Yani dan A.A. Darwis. *Biogas Technology Book*. Bogor Institute of Agriculture (1990)
40. Y. Lin, D. Wang, L. Wang, Waste Manag. **28**,800-810 (2010)
41. D. Deublein, A. Steinhauser. *Biogas from Waste and Renewable Resource.* Wiley-VCH Verlag GmbH & Co. KGaA ,Weinheim (2008)
42. D.O. Indriani, L.N. Syamsudin, L. N., F.H. Sriherfyna, A.K. Wardani, Food and Agroindustry Journal **3**, 1405-1411 (2015)
43. S. Merlina, A.D. Lubis, Y. Retnani. *Perubahan Kandungan Nutrient Wheat Bran yang Difermentasi Menggunakan Level Starter Aspergillus niger yang Berbeda.* Thesis of Institut Pertanian Bogor (2012)
44. A. Nopharatana, P.C. Pullammanappalli, W.P. Clarke, Waste Management **27**,595–603 (2007)

45. I. Syaichurrozi, Budiyono, S. Sumardiono, Bioresource Technology **149**, 390-397 (2013)
46. A.N. Ikrimah, A. Pramianshar. *Innovation in Biogas Technology through Anaerobic Digestion from Fish Processing Waste as A Renewable Energy Source.* Thesis of Chemical Engineering Department, Diponegoro University, Semarang (2014)
47. M.D. Ghatak, P. Mahanta, International Journal of Engineering and Advanced Technology (IJEAT) **3**, 2249 – 8958 (2014)
48. Budiyono, I. N. Widiyasa, S. Johari, Sunarso. International Journal of chemical and biological Engineering **3**, 1, 39-44 (2010)
49. Budiyono, I.N. Widiasa, and S. Sumardiono,Waste Technol. **1**, 1, 1-5 (2013)

Livestock Feed Production from Sago Solid Waste by Pretreatment and Anaerobic Fermentation Process

Siswo Sumardiono[1,*], *Aditya* Dwi W N[1], *Farel* Rahman A[1], and *Isti* Pudjihastuti[2]

[1]Department of Chemical Engineering, Faculty of Engineering, Diponegoro University, Jl.Prof. Soedarto, SH, Tembalang, Semarang, Indonesia 50275
[2]DIII Study Program, School of Vocation, Diponegoro University, Jl.Prof. Soedarto, SH, Tembalang, Semarang, Indonesia 50275

Abstract. Food needs in Indonesia is increasing, including beef. Today, Indonesia has problem to do self-sufficiency in beef. The cause of the problem is the quality of local beef is still lower compared with imported beef due to the quality of livestock feed consumed. To increase the quality of livestock is through pretreatment and fermentation. Source of livestock feed that processed is solid sago waste (*Arenga microcarpa*), because in Indonesia that is relatively abundant and not used optimally. Chemical pretreatment process for delignification is by using NaOH solution. The purposes of this research are to study NaOH pretreatment, the addition of *Trichoderma sp*, and fermentation time to improve the quality of sago solid waste as livestock feed through anaerobic fermentation. The variables used are addition or without addition (4%w NaOH solution and Trichoderma sp 1%w) and fermentation time (7, 14 and 21 days), with the response of crude fiber and protein. The result of this research shows that the pretreatment with soaking of NaOH solution, addition of *Trichoderma sp* and 14 days of fermentation was more effective to improve the quality of solid sago waste with decrease of crude fiber from 33.37% to 17.36% and increase of crude protein from 4.00% to 7.96%.

1 Introduction

The Government of Indonesia has made a solution to continue to reduce the famine by making food self-sufficiency policy. Beef is one example of the food. Beef is a kind of food that has multiple roles, both in terms of food security and economy. The increase in beef prices is currently the result of an imbalance between production quotas and the level of public demand for beef [1]. The consequence is that Indonesia must import beef. Based on statistical data on import and export volume of beef during 1996-2015 [2], the import volume of beef in Indonesia tends to increase every year with the highest number obtained in 2014 of 246,000 tons. The data shows that Indonesia is still can't to be self-sufficient in beef. The imported beef in Indonesia comes from the United States has a smooth and soft texture, contains high protein and beneficial to the human immune. This meat also has fat (marbling), as other premium beef meat, such Angus beef. The quality of USA beef is different when compared to the quality of local beef because of differences in the livestock feed [3].

Indonesia's conventional livestock feed products are declining due to weather conditions and fluctuations in price. This reason is causing for seeking alternative sources, environmentally friendly, economically acceptable, and socially acceptable non-conventional feed sources [4]. The livestock feed should also be able to become a mainstay feed in the long term. Potential and useful alternative livestock feed and also available throughout the year are generally from agricultural processing industries, for example, in the production of sago flour [5]. Sago plants (Arenga Microcarpa) grow in damp tropical regions. Currently, the use of sago only focuses on the starch contained in it [6]. The processing of sago flour is usually conducted near water sources such as rivers or creeks edge [7]. The process of sago flour processing in the palm sugar industry in Daleman village, Klaten regency, Central Java produces an average of 200 tons / year of sago flour with 659 tons / year or 2.19 tons / day [8].

Solid sago waste contains lignocellulose rich in cellulose that can be used optimally as a carbon source. The crude protein content of the sago waste is very low, i.e. 2.63 %, while the crude fiber content is 15.90 % [9]. Sanchez [10] states that Cellulose and hemicellulose are also substances of plant tissue composed of different sugars. Cellulose is the main material that encloses the plant skeleton resulting from the process of photosynthesis of plants. This compound is not soluble in ordinary solvents and comprises a plurality of β-D-glucopyranose units connected via β-1-4 bonds to form straight and long chains reinforced by a cross-linked hydrogen bond providing a crystalline structure and arranged in a micro fibril. Hemicellulose is the second largest polysaccharide after cellulose. Hemicellulose is a complex polymer consisting of a mixture of different

* Corresponding author: siswo.sumardiono@che.undip.ac.id

monosaccharide polymers in which the constituent sugars are grouped in the form of hexoses (glucose, mannose, and galactose), pentose (silose, arabinopiranosa, arabinofuranosa), hexuronic acid (glucoronate, methylgucoromate, galactoromat), deoxyhexose (rammosa, fucosa) [11].

Lignin is an anti-nutrient for livestock that interfered with livestock digestion and decrease the nutritional value of animal feed ingredients [12]. Lignin is an undigested component, affecting the digestibility of crude fiber. This compound is an aromatic polymer of phenylpropanoid produced from synthesis of coniferyl, synaphyl, p-coumaryl alcohol. Lignin is highly resistant to any chemical degradation including enzymatic. Bases pretreatment usually use basic solutions such as NaOH, Ca(OH)2 (limestone) or ammonia to remove lignin and parts of hemicellulose, so this process can efficiently increase cellulosic enzyme capability [13]. Generally this pretreatment is conducted under low temperature conditions with long time and high base concentrations [14]. The advantage of using a base as a lignocellulose pretreatment is the presence of lignin destruction wherein lignin inhibits the action of the enzyme during the anaerobic process [15]. Alkali can damage ester bonds between lignin and xylem [16].

Fermentation is one of the ways to improve the quality of the material where in the process there is an overhaul of hard structures physically, chemically and biologically so that the material from complex structures becomes simple, thereby making the digestibility more efficient [17]. According to Oseni and Ekperigin [18], Fermentation of plant waste not only increases the contents of protein, fat or crude fiber, but also increases the content of essential minerals that can be useful to the body's metabolism. The purposes of this research are to study NaOH pretreatment, the addition of Trichoderma sp and fermentation time to improve the quality of solid sago waste as livestock feed through anaerobic fermentation.

2 Materials and Method

2.1 Materials

The material used in this study is solid sago waste from Jepara District (Central Java), Trichoderma sp from Pradipta Paramita Solo, molasses from Ungaran, NaOH, aquadest, and acetic acid.

2.2 Preparation Stage

At this stage, the sago palm waste raw material is collected from sago palm processing industry in Plajan village, Jepara District, Central Java Province. Trichoderma sp was purchased from Pradipta Paramita Solo, molasses from Ungaran, while NaOH and acetic acid from the Indrasari chemical store. After that, preparation is conducted by sieving the solid sago waste and each variable is measured 50 grams.

2.3 Operation Stage

Solid sago waste was soaked with 4% w NaOH solution for 24 hours or heated with 4% w NaOH solution at 60OC for an hour. The pretreatment substrate is reduced to a pH of 5-6 with acetic acid and then fermented for 7, 14 and 21 days with Trichoderma sp and molasses. Fermentation process in this study conducted anaerobically.

2.4 Result Analysis Stage

The analysis was conducted graphically based on the phenomenon for the effect of the independent variables on the contents of crude fiber and crude protein.

3 Results and Discussion

3.1 Effect of pretreatment type on Crude Fiber and Crude Protein Content

Figure 1 and 2 shows the correlation of the pretreatment of NaOH and the crude fiber content of the feed material derived from the solid sago waste. In Figure 1, the solid sago waste variables without the addition of Trichoderma sp on the day 7 samples without pretreatment, heating pretreatment, and soaking pretreatment resulted in crude fiber content of 33.04%; 27.50% and 22.03%, respectively, on the day 14 samples without pretreatment, heating pretreatment, and soaking pretreatment resulted in crude fiber content of 33.01%; 27.49% and 22.63%, respectively, on the day 21 samples without pretreatment, heating pretreatment, and soaking pretreatment resulting in crude fiber content respectively 33.87%; 27.98% and 23.00% respectively. In Figure 2, the solid sago waste variables with the addition of Trichoderma sp on the day 7 samples without pretreatment, heating pretreatment, and soaking pretreatment resulted in crude fiber content of 30.00%; 27.01% and 21.79%, respectively, on the day 14 samples without pretreatment, pretreatment heating and soaking pretreatment resulted in crude fiber content 25.18%; 21.71% and 17.37%, respectively, on the day 21 samples without pretreatment, heating pretreatment, and soaking pretreatment resulted in crude fiber content of 27.62%; 24.53%, and 19.11%, respectively.

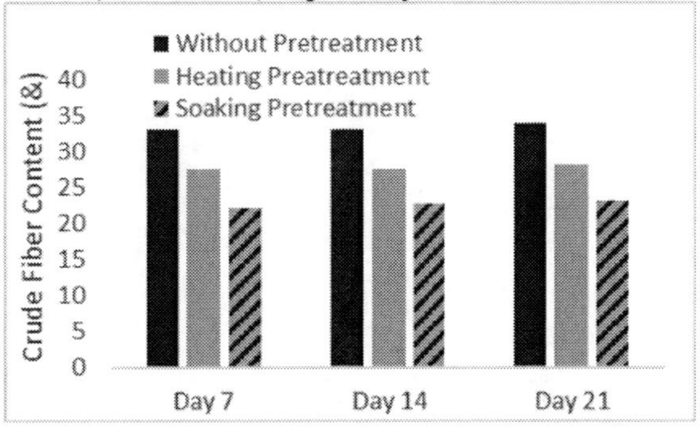

Fig. 1. Correlation of NaOH Pretreatment vs. Crude Fiber Content in Variables without Addition of Trichoderma sp.

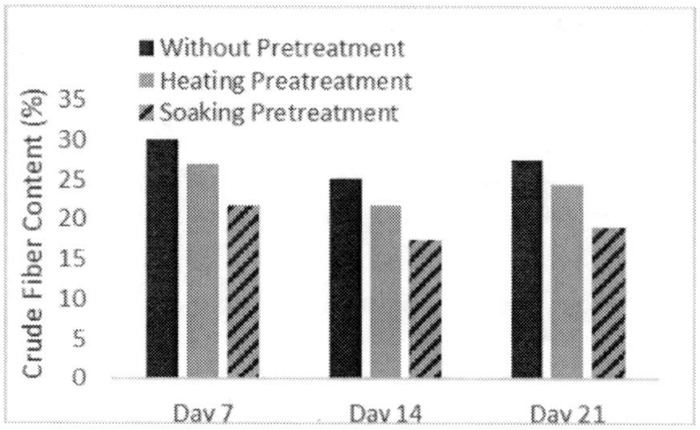

Fig. 2. Correlation of NaOH Pretreatment vs. Crude Fiber Content in Variables with Addition of Trichoderma sp.

From both figures it can be seen that the type of soaking pretreatment is more effective to decrease the crude fiber content in solid sago waste followed by heating pretreatment and without pretreatment. Choosing the right pretreatment method is very important because it determines which hydrolysis and fermentation methods will be used later [19]. The uses of strong base solution in this study aims to make the solid sago waste can undergo a delignification process before going into the fermentation process. Delignification of solid sago waste aims to damage the structure of lignin and a part of hemicellulose and also swelling of cellulose structures [20]. Saponification of ester bonds between lignin and hemicellulose residues makes cellulose more open and easier to interact with enzymes, and can decrease the degree of polymerization and crystallinity of cellulose structures [21]. This causes the crude fiber content that has passed the pretreatment stage to be lower than its original level.

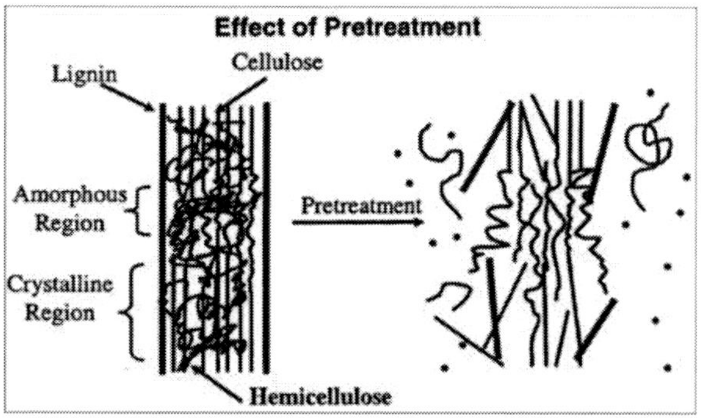

Fig. 3. Schematic process of destruction of lignin structure [22].

Compared with acid pretreatment, the base pretreatment method is more effective as lignin solubilization as presented in Figure 3. Base pretreatment methods in lignocellulose biomass processing are commonly use sodium, potassium, calcium, and ammonium hydroxide. The uses of a base solution cause changes in the structure of lignin by degrading the ester and glycoside side chain. The uses of base solution also cause partial decrystallization of cellulose, partial solubility of hemicellulose and resulting in enlarged cellulose. This process is carried out by soaking the biomass in an alkaline solution at a predetermined temperature and time. The loosened lignin binds to the Na + ion to form soluble sodium phenolate. Dissolved lignin is characterized by dark color in the solution (black liquor) [23,24]. The neutralization stage needs to be conducted before entering the enzymatic hydrolysis stage to remove lignin and inhibitor substances (eg salts, phenolic acids, and aldehydes) [25]. Pretreatment can increase lignocellulose digestibility in feedstuffs inhibited by several factors, such as: lignin content, cellulose crystallinity, polymerization degree, pore volume, acetyl group bound to hemicellulose, surface area and biomass particle size [26]. The factors that affecting the digestibility of feeds are contents of chemical, protein, fat presentations, fiber and minerals [27].

Figure 4 and 5 shows the correlation of the NaOH pretreatment type and the crude protein content of the feed material derived from the solid sago waste. In Figure 4, the solid sago waste variables without the addition of Trichoderma sp on the day 7 samples without pretreatment, heating pretreatment, and soaking pretreatment resulted in crude protein content of 3.96%; 3.12% and 4.05%, respectively, on the day 14 samples without pretreatment, heating pretreatment, and soaking pretreatment yielded crude protein content of 4.04%; 3.18%, and 4.06%, respectively, on the day 21 samples without pretreatment, heating pretreatment, and soaking pretreatment yielded consecutive crude protein content 4.06%; 3.24% and 4.11%, respectively. In Figure 5, the solid sago waste variables with the addition of Trichoderma sp on the day 7 samples without pretreatment, heating pretreatment, and soaking pretreatment resulted in crude protein content of 2.80%; 3.38% and 3.54%, respectively, on the day 14 samples without pretreatment , heating pretreatment and soaking pretreatment resulted in crude protein content of 4.06%; 5.99% and 7.96%, respectively, on the day 21 samples without pretreatment, heating pretreatment, and soaking pretreatment resulted in crude protein content of 4.28 %; 6.14% and 8.07%, respectively.

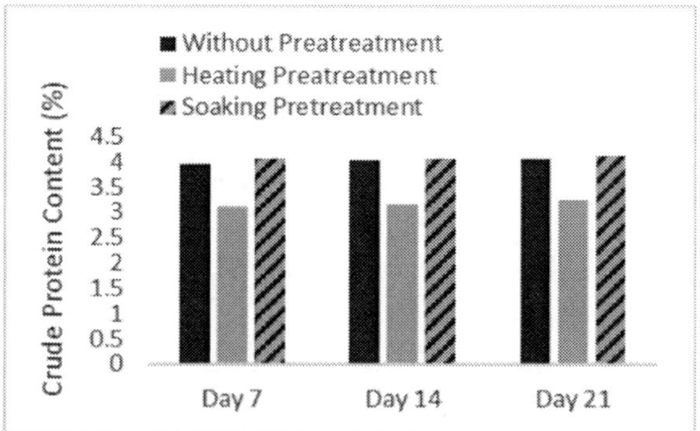

Fig. 4. Correlation of NaOH Pretreatment vs. Crude Protein Content in Variables Without Addition of Trichoderma sp.

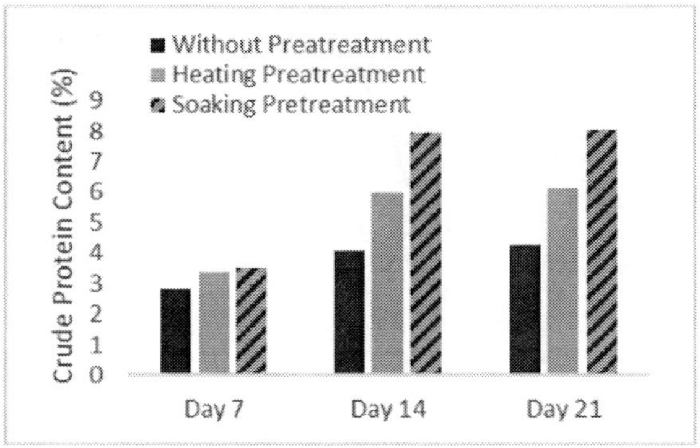

Fig. 5. Correlation of NaOH Pretreatment vs. Crude Protein Content in Variables with Addition of Trichoderma sp.

From both figures it can be seen that soaking pretreatment is best way to help raise the crude protein content in the solid sago waste followed by heating pretreatment and without pretreatment. Delignification of solid sago waste aims to damage the structure of lignin and swelling of cellulose structures [20]. According to Saparianti [28], lignin which is a substance of lignocellulose which is a component of crude fiber will prevent enzyme penetration into substrate. In the variable solid sago waste with the addition of Trichoderma sp, the presence of delignification can help the work of protease enzymes which are one of the enzymes that produced by Trichoderma sp to form protein compounds [29]. Solid sago waste that not added Trichoderma sp and without pretreatment procedure tends not to affect the crude protein content. In the variable with the pretreatment of heating at 60°C without Trichoderma sp there was a decrease in the crude protein content due to the nature of the protein having denaturation at 50 to 80 °C. Each temperature increase of 10 °C, protein denaturation rate increased to 600 fold [30]. The hydrogen bonds and hydrophobic interactions of non-polar ions of proteins can be damaged by heat.

According to Ophar [30] the kinetic energy that increase due to high temperatures can cause protein molecules to move or vibrate faster, thus breaking the bonds of the molecule. In addition, heat energy will lead to disconnection of non-covalent interactions that exist in the natural structure of the protein but do not break its covalent bond in the form of peptide bonds. Increased the content of crude protein in livestock feed consumed by livestock are good for livestock in terms of nutritional fulfillment. The quality of livestock feed depends on the composition of the nutrients contained therein mainly against dry matter, crude protein, crude fat, crude fiber and feed grade digestibility [31].

3.2 Effect of the addition of Trichoderma sp on Crude Fiber and Crude Protein Content

Figure 6, 7, and 8 shows the correlation of the effect of the addition of Trichoderma sp to the crude fiber content of the solid sago waste. Figure 6 shows the correlation of the effect of the addition of Trichoderma sp vs. crude fiber content to the solid sago waste variable without pretreatment. Figure 7 shows the correlation of the effect of the addition of Trichoderma sp vs. crude fiber content on the variable of solid sago waste with heating pretreatment. Figure 8 shows the correlation of the effect of the addition of Trichoderma sp vs. the crude fiber content to the solid sago waste variable with soaking pretreatment. In figure 6, the solid sago waste without pretreatment (without the addition of Trichoderma sp and with the addition of Trichoderma sp) resulted in crude fiber content: on the day 7 samples: 33.04% and 30.00%, on the day 14 samples: 33.01% and 32.52%, on the day 21 samples: 33.87% and 27.62%. In figure 7, the solid sago waste with heating pretreatment (without the addition of Trichoderma sp and with the addition of Trichoderma sp) resulted in crude fiber content: on the day 7 samples: 27.50% and 27.01%, on the day 14 samples: 27.49% and 21.71%, on the day 21 samples: 27.98% and 24.53%. In figure 8, the solid sago waste with soaking pretreatment (without the addition of Trichoderma sp and with the addition of Trichoderma sp) resulted in crude fiber content: on the day 7 samples: 22.03% and 21.79%, on the day 14 samples: 22.62% and 17.36%, on the day 21 samples: 23.00% and 19.11%.

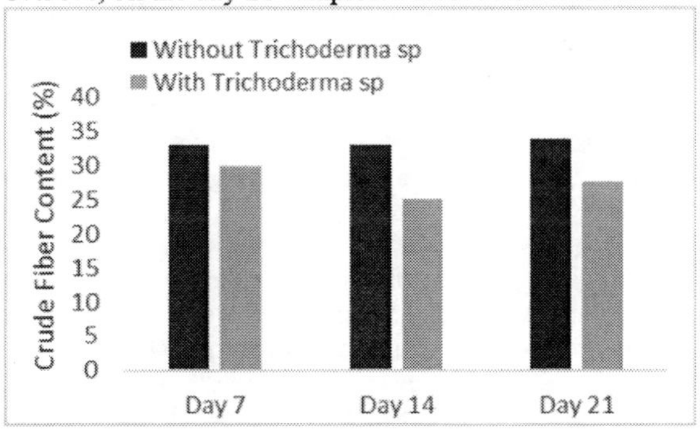

Fig. 6. Correlation of the Addition of Trichoderma sp vs Crude Fiber Content to Variables Without Pretreatment.

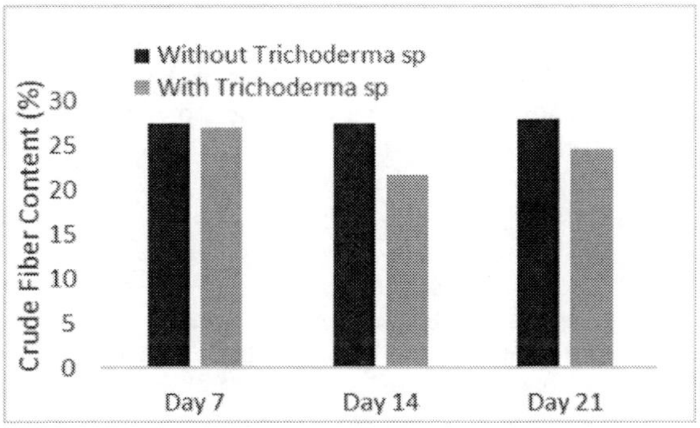

Fig. 7. Correlation of the Addition of Trichoderma sp vs Crude Fiber Content to Variables with Heating Pretreatment.

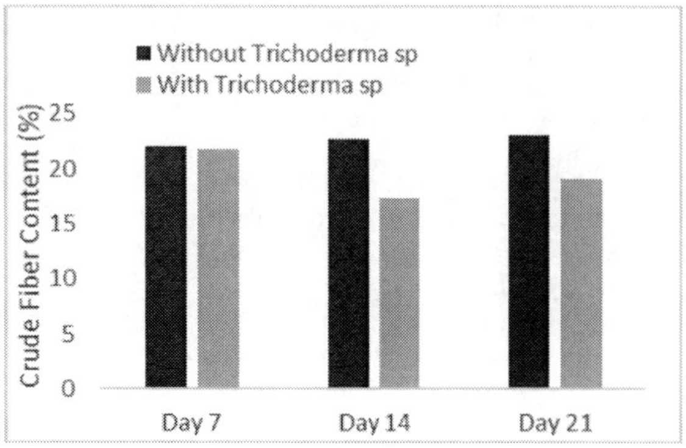

Fig. 8. Correlation of the Addition of Trichoderma sp vs Crude Fiber Content to Variables with Soaking Pretreatment.

From the figure 6, 7, and 8, it can be seen that with the addition of Trichoderma sp is more effective to reduce the crude fiber content in the solid sago waste compared with without the addition of Trichoderma sp. Trichoderma sp is a group of complete cellulase-producing soil fungi as well as other components required for total hydrolysis of crystalline cellulose [32]. Cellulase enzyme contains the largest components of cellobiase and β-1,4-glucan-cellobiohydrolase (C1), and β-1,4-glucan-cellobiohydrolase (Cx) in small amounts. This enzyme is hydrolytic and works either consecutively or simultaneously. Cellobiohydrolase is an enzyme that has an affinity for high level cellulose capable of breaking crystalline cellulose while endoglucanase works on amorphous cellulose. Cellobiohydrolase breaks down cellulose by cutting the hydrogen bonds that cause the glucose chains to be readily hydrolyzed. Further hydrolysis takes place to obtain selobiose and eventually glucose by β-gluconase and β-glucosidase enzymes. This series of cellulose hydrolysis reactions causes the crude fiber content of the solid sago waste to be given Trichoderma sp to be lower [33]. The factors that affecting the digestibility of feeds are contents of chemical, protein, fat presentations, fiber and minerals [27].

Figure 9, 10, and 11 shows the correlation between the effect of the addition of Trichoderma sp and the crude protein content of the solid sago waste. Figure 9 shows the correlation of the effect of the addition of Trichoderma sp vs the crude protein content on the solid sago waste variables without pretreatment. Figure 10 shows the correlation of the effect of the addition of Trichoderma sp vs the crude protein content on the solid sago waste variables with heating pretreatment. Figure 11 shows the correlation of the effect of the addition of Trichoderma sp vs the crude protein content on the solid sago waste variables with soaking pretreatment. In figure 9, the solid sago waste without pretreatment (without the addition of Trichoderma sp and with the addition of Trichoderma sp) resulted in crude fiber content: on the day 7 samples: 3.9578% and 2.8%, on the day 14 samples: 4.0439% and 4.06%, on the day 21 samples: 4.0558% and 4.28%. In figure 10, the solid sago waste with heating pretreatment (without the addition of Trichoderma sp and with the addition of Trichoderma sp) resulted in crude fiber content: on the day 7 samples: 3.12% and 3.38% on the day 14 samples: 3.18% and 5.99%, on the day 21 samples: 3.24% and 6.14%. In figure 11, the solid sago waste with soaking pretreatment (without the addition of Trichoderma sp and with the addition of Trichoderma sp) resulted in crude fiber content: on the day 7 samples: 4.05% and 3.54% on the day 14 samples: 4.06% and 7.96%, on the day 21 samples: 4.11% and 8.07%.

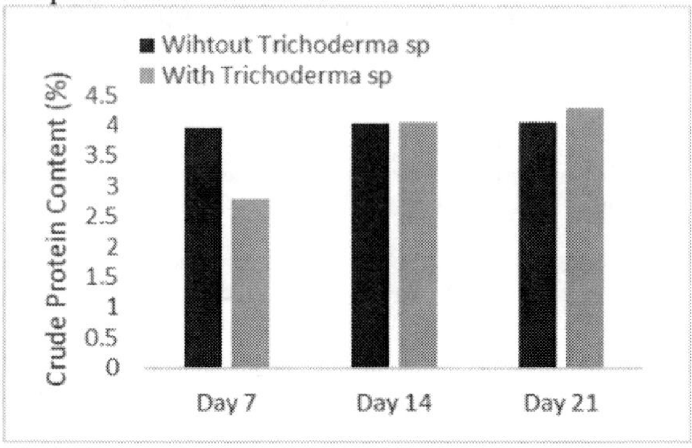

Fig. 9. Correlation of the Addition of Trichoderma sp vs Crude Protein Content to Variables Without Pretreatment.

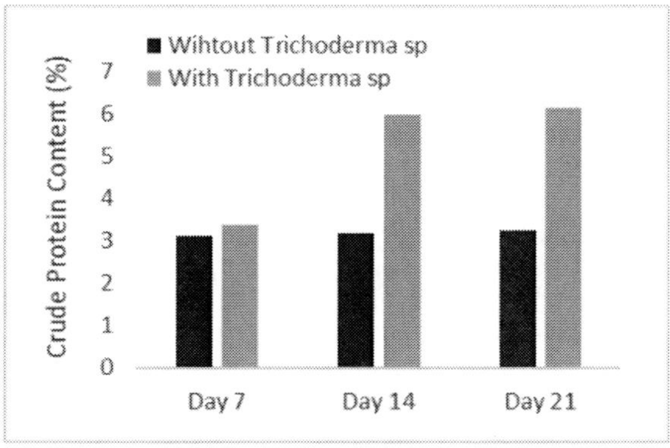

Fig. 10. Correlation of the Addition of Trichoderma sp vs Crude Protein Content to Variables with Heating Pretreatment.

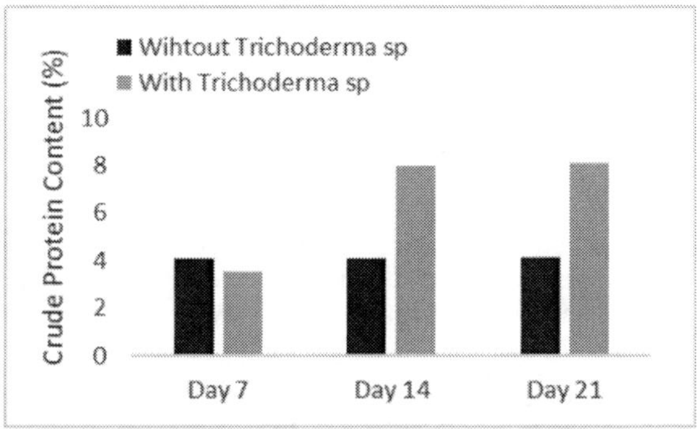

Fig. 11. Correlation of the Addition of Trichoderma sp vs Crude Protein Content to Variables with Soaking Pretreatment.

From figure 9, 10, and 11, it can be seen that with the addition of Trichoderma sp is more effective to raise the crude protein content in solid sago waste than with no addition of Trichoderma sp. The uses of microorganisms can alter the structure of components and increase enzymatic hydrolysis, added microorganisms can help cellulose hydrolysis process into glucose [34, 35]. Trichoderma sp is not only produces cellulose enzymes but also produces protease enzymes that serve to break down proteins [29]. The protein is converted into a polypeptide, and then becomes a simple peptide which ultimately undergoes a further reshuffle into amino acids, which will be exploited by microorganisms to multiply itself. Increasing the number of colonies of microorganisms that are single cell proteins during the fermentation process indirectly increases the crude protein content of the substrate [27, 36]. On the variables without the addition of Trichoderma sp there was no significant change in crude protein content.

3.3 Effect of Fermentation Time on Crude Fiber and Crude Protein Content

Figure 12 and 13 shows the correlation of the fermentation time and the crude fiber content of solid sago waste. Figure 12 shows the correlation of fermentation time vs. the crude fiber content of the solid sago waste without the addition of Trichoderma sp while figure 13 shows the correlation of fermentation time vs. the crude fiber content of the solid sago waste with the addition of Trichoderma sp. From both figures it can be seen that on the day 7, almost all samples except the sample without pretreatment and without the addition of Trichoderma sp had decreased contents of crude fiber from the initial level of 33.37%. Then, on the day 14 all samples with the addition of Trichoderma sp continued to decrease the crude fiber content while all samples without the addition of Trichoderma sp tended not to increase or decrease the crude fiber content. On the day 21, samples with the addition of Trichoderma sp have elevated the crude fiber content while samples without the additionTrichoderma sp remain constant.

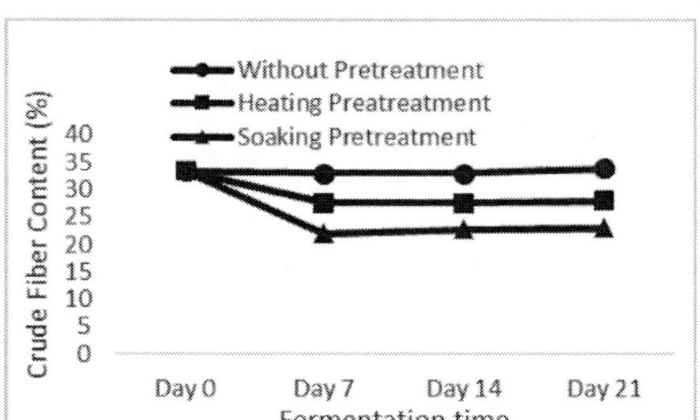

Fig. 12. Correlation of Fermentation Time sp vs Crude Fiber Content to Variables with Addition of Trichoderma sp.

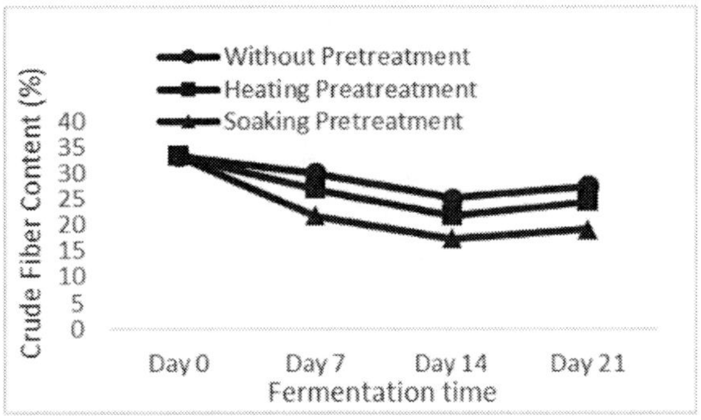

Fig. 13. Correlation of Fermentation Time sp vs Crude Fiber Content to Variables Without Addition of Trichoderma sp.

According to Saparianti [28] the longer the fermentation process takes place, then the glucose content produced from the reshuffle of cellulose is higher. This shows the longer the fermentation, cellulose hydrolytic process becomes more effective in lowering crude fiber content. This statement is reinforced by Detroy et al (1981) in [37] which explain that during the first 15 days of fermentation, cellulase-producing microorganisms can increase 2-3 times of baseline reducing sugar. The longer the fermentation means the total mold produced is also higher. Mold will multiply itself in a certain time interval so that the longer the fermentation then Trichoderma sp opportunity to multiply itself becomes larger so that the number of cells produced higher. The number of cellulose adsorbed by cellulase depends on the surface area and cellulase concentration. The cellulase adsorbtion on the cellulose surface is usually faster than the overall hydrolysis rate [28]. This causes the hydrolytic reaction to occur more quickly so that the crude fiber content in the solid sago waste becomes lower. This phenomenon occurred from day 0 to day 14 of fermentation on variable solid sago waste with addition of Trichoderma sp.

On the day 14 until the day 21 of fermentation there is an increase in the crude fiber content of the solid sago waste with the addition of Trichoderma sp. This is due on after the day 14 Trichoderma sp has entered the death phase. The cell wall of the dead Trichoderma sp is calculated as crude fiber. Meanwhile the solid sago waste variables without the addition of Trichoderma sp, the phenomenon that occurs is a decrease in crude fiber content on the day 7 except in the variable without pretreatment. There was no significant change on the day 14 and 21 compared to the previous day. This may result from the apparent delignification of the sample seen significantly on the day 7 which results in a decrease in crude fiber content.

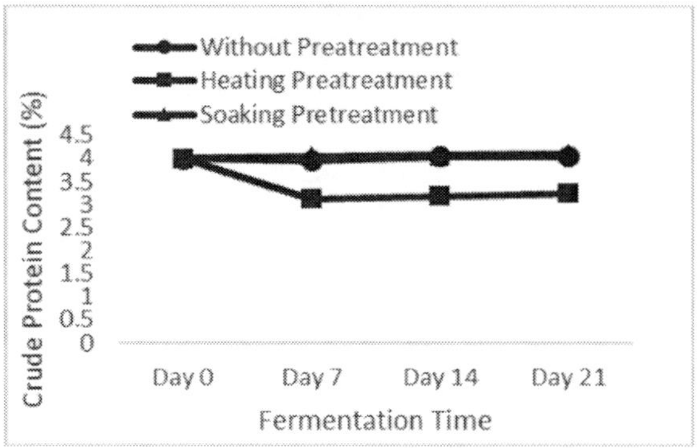

Fig. 14. Correlation of Fermentation Time sp vs Crude Protein Content to Variables with Addition of Trichoderma sp.

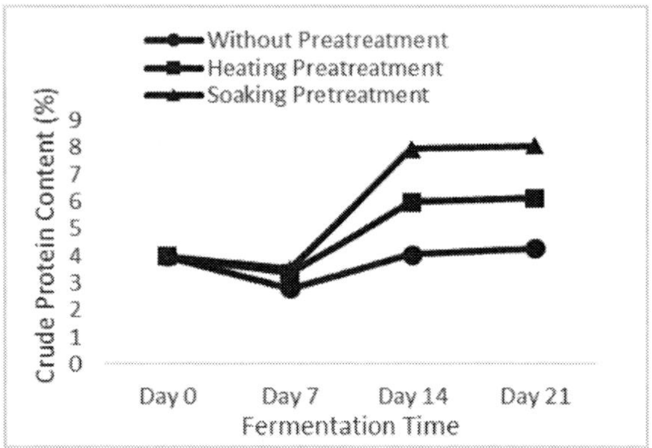

Fig. 15. Correlation of Fermentation Time sp vs Crude Protein Content to Variables Without Addition of Trichoderma sp.

Figure 14 and 15 shows the correlation of the fermentation time and the crude protein content of solid sago waste. Figure 14 shows the correlation of fermentation time vs. the crude protein content of the solid sago waste variables without the addition of Trichoderma sp, while figure. 15 show the correlation of fermentation time vs. the crude protein content of the solid sago waste variables with the addition of Trichoderma sp. From both figures it can be seen that on the day 7, all samples except the sample without pretreatment without addition of Trichoderma sp and the sample through pretreatment by soaking without Trichoderma sp had decreased crude protein content from baseline. Then, on the day 14 all samples with the addition of Trichoderma sp increased significantly in crude protein content while all samples without Trichoderma sp did not increase or decrease in crude fiber content. On the day 21, all samples either with the addition of Trichoderma sp or without the addition of Trichoderma sp did not change the crude protein content.

In the solid sago waste variables with the addition of Trichoderma sp on the day 7, the decrease is due to the protein contained in the solid sago waste consumed by Trichoderma sp for its growth. Then on the day 14, the phenomenon that occurs is the activity of formation of new proteins from Trichoderma sp through protease enzymes. According to Supriyati [17], the best fermentation length is 8-12 days where Trichoderma sp increases the crude protein content of the substrate. The protein is converted into a polypeptide, and then becomes a simple peptide which ultimately undergoes a further reshuffle into amino acids, which will be exploited by microorganisms to multiply itself. Increasing the number of colonies of microorganisms that are single cell proteins during the fermentation process indirectly increases the crude protein content of the substrate [27, 36].

The difference in levels of the crude protein content in sample added with Trichoderma sp was suspected because the crude fiber content of the solid sago waste had been reduced previously through the delignification process. Delignification of sago pulp aims to damage the lignin structure and swell the cellulose structure [20]. According to Saparianti [28], lignin which is a substance of lignocellulose which is a component of crude fiber will prevent enzyme penetration into substrate. The decrease in crude fiber indicates that the enzyme penetration process into the substrate is better and capable of producing higher the crude protein content.

The life phases of microorganisms in the substrate affect the content of crude fiber and crude protein [38]. The phase is generally influenced by the type of microorganisms used and the type of substrate used. The life phase of the organism is divided into 5: adaptation phase, initial growth phase, fast growth phase (logarithmic), stationary phase, and death phase. Adaptation phase is the initial period and is a phase of adjustment (adaptation), so there is no increase in the number of cells and sometimes the number of cells decreases. In the early growth phase there was no population increase. Cells undergo changes in chemical composition and increase in size and the intracellular substance increases. Logarithmic phase is a rapid multiply period. In this period can be observed features of active cells. The generation time on each microorganism can be determined in this rapid phase. At that phase it can be seen that some cells begin to divide, others half divide, and others finish splitting. In the static phase the breeding begins to decrease and some cells die. If the rate of multiply is equal to the rate of death, then the total number of cells remains constant. This may be due to reduced nutrients or the formation of metabolic products that tend to accumulate may be toxic to microorganisms. The death phase is the phase in which the multiply process has stopped. The cells are dead, which will then be followed by a lysis process. If the rate of death exceeds the rate of multiply, then the actual number of cells decreases [38].

4 Conclusions

Livestock feed production using solid sago waste processed through soaking pretreatment and fermentation with Trichoderma sp yields data of crude fiber content decrease from 33,3734% to 17,363% and increase of crude protein content from 4% to 7,955% at 14 days of fermentation time. This indicates the optimum enzyme productivity in the growth phase of Trichoderma sp in 14 days.

Authors thank to Department of Chemical Engineering, Diponegoro University for all facilities to do this research.

References

1. M. Anshar, Tadete, F.H. Elly, L.S. Kalangi, R. Hadju, *District*, **36,** 363-371, (2016)
2. Center for Agricultural Data and Information Systems, *Outlook of Agricultural Commodities Sub-Subsector of Beef Husbandry, Center for Agricultural Data and Information System of the Secretariat General of the Ministry of Agriculture*, (2015)
3. M.F. Asnath, R. Priyanto, E.L. Aditia, M. Baihaqi, M. Ismail, J. Indo. Agricultural. Sci, (2015)
4. R. Salemdeeb, E.K.H.J. zu Ermgassen, M.H. Kim, A. Balmford, A. Al-Tabbaa, J. Cleaner. Production, (2016)
5. Antonius, K. Simanihuruk, J. Sirait, *Use of Sago Waste as a Mixture of Complete Feed for Growth Phase Boerka Goat*, National Seminar on Livestock Technology, (2013)
6. R.S. Singhal, J.F. Kennedy, S.M. Gopalakrishnan, A. Kaczmarek, C.J. Knill, P.F. Akmar, Carbohydrates Polymer **72**, 1-20, Malaysia, (2008)
7. S.K. Thangavelu, A.S. Ahmed, F.N. Ani, J. App. Energy **128**, 277-283, (2014)
8. Sudarsono, S. Huda, M. Yuniwati, *Utilization of Palm Sugar Fiber as a Composite Material - Polyester*. Competitive Grant Research Report, Institute of Science & Technology Akprindo. Yogyakarta, (2013)
9. A. Lay, *Utilization of Sago Baruk as Plant Conservation of Food and Livestock Feed Products, Seminar and Exhibition of Research* Results of Forestry Research Institute of Manado, (2012)
10. C. Sanchez, *Biotechnology Advances* **27**, 185-194, (2009)
11. D. Fengel, G. Wegener, *Wood: Chemistry, Ultrastructure. Reaction.* Walter de Gruyter, Berlin., (1989)
12. E. Musnandar, J. Animal. Sci. **9**, 225-234, (2006)
13. Budiyono, P.S. Sari, Waste Tech **30**, 17-25, (2014)
14. M.J. Taherzadeh, K. Karimi, Int. J. Mol. Sci. **9**, 1621–1651 (2008)
15. Z. Wang, D.R. Keshwani, A.P. Redding, *J. Bioresour. Technol.* **101**, 3583-5, (2010)
16. S. McIntosh, T. Vancov, Bioresour. Technol **101**, 6718-6727, (2010)
17. Supriyati, T. Haryati, I.G.M. Budiarsana, I.K. Sutama, LRC (2010)
18. O.A. Oseni, and M. Ekperigin, Biochemistry **19**, 75-79, (2007)
19. E. Tomas-Pejo, P. Alvira, M. Ballesteros, M.J. Negro, *Pretreatment Technologies for Lignocellulose-to-Bioethanol Conversion. Di dalam Pandey A (ed.),* Biofuels: Alternative Feedstocks and Conversion Processes, 149-176 (2011)
20. W.L. Marsden, P.P. Dan, Gray, Cricital Rev. In Biotechnol **3**, 235-576, (1986)
21. Y. Sun, *Enzymatic Hidrolysis of Rye Straw and Bermudagrass for Ethanol Production,* Ph.D. thesis: NC State University. Raleigh, NC, (2002)
22. P. Kumar, D.M. Barrett, M.J. Delwiche, P. Stroeve, Ind. Eng. Chem. Res. **48**, 3713–3729, (2009)
23. B.E. Dale, M.J. Moreira, Biotechnology and Bioengineering Symposium **12**, 31–43, (1982)
24. S. Safaria, N. Idiawati, T.A. Zaharah, *Effectiveness of Cellulase Enzyme Blends from Aspergillus Niger and Trichoderma ressei in Hydrolyzing Substrates of Coconut Shell.* Thesis of Faculty of MIPA University of Tanjungpura (2013)
25. V. Menon, M. Rao, Progress in Energy and Combustion Science **38**, 522–550 (2012)
26. W.F. Anderson, D.E. Akin, J. Ind. Micro. Biotech, **35,** 355-366 (2008)
27. Anggorodi, *General Livestock Science,* Jakarta: Publisher PT Gramedia (1994)
28. E. Saparianti, T. Dewanti, S.K. Dhoni, Agricultural Tech **5**, 1-10, (2012)
29. Samadi, S. Wajizah, Sabda, *Improving the Quality of Sugar Cane as Animal Feed Through Fermentation by Adding Different Sago Flour Levels*. Banda Aceh, (2015)
30. A. Triyono, *Study of the Effect of Addition of Some Acids on Protein Isolation Processes to Protein Isolate Green Bean (Phaseolus radiatus L.) Flour,* Center for Appropriate Technology Development. Indo, LRC (2010)
31. M. Martawidjaja, I.K. Sutama., I.G.M. Budiarsana, T. Kostaman, *Productivity of PE goat fed with fermented rice,* Indo. LRC. Ciawi. Bogor (2004)
32. C.H. Prayitno, *Supplementation of microminerals in Trichoderma Viride fermented agro-industry wastes reviewed from VFA and N-NH3 concentrations in vitro,* Proceedings of National Seminar on Livestock and Veterinary, Bogor, 761 – 767, (2008)
33. P.R. Chawla, I.B. Bajaj, S.A. Survase, R. S. Singhal, Food. Tech. Biotech. **47**, 107-124 (2009)
34. M. Kurakake, N. Ide, T. Komaki, Curr Microbiology **54**, 424-428 (2007)
35. M. Taniguchi, H. Suzuki, D. Watanabe, K. Sakai, K. Hoshino, T. Tanaka, J. Biosci. Bioeng **100**, 6737-643, (2005)
36. A.S. Agustono, Widodo, W. Paramita, J. Fish and Mar. Scie. **2**, 37-43, (2010)
37. B.J.B. Wood, *Microbiology of Fermented Foods. 2th.* Blackie Academics and Profesional. London, (1998)
38. Wardayanti, *Crude Fiber; Crude Fat and BETN Content of Corn Cocks Inoculated by Fungi Trichoderma sp at different incubation times, Faculty of Animal Husbandry,* Thesis of Hasanuddin University, Makassar, (2015)

Optimized Ultrasound-Assisted Oxidative Desulfurization Process of Simulated Fuels over Activated Carbon-Supported Phosphotungstic Acid

Peniel Jean Gildo[1], *Nathaniel* Dugos[1,*], *Susan* Roces[1] and *Meng-Wei* Wan[2]

[1]Chemical Engineering Department, De La Salle University – Manila, 2401 Taft Avenue, Malate, Manila, Philippines, 0922
[2]Department of Environmental Resources Management, Chia Nan University of Pharmacy and Science, 60 Erh-Jen Rd., Sec.1, Jen-Te, Tainan, Taiwan, 71710

Abstract. Recent technological advancements respond to the call to minimize/eliminate emissions to the atmosphere. However, on the average, fuel oils which is one of the major raw materials, is found to increase in sulfur concentration due to a phenomenon called thermal maturation. As such, a deeper desulfurization process is needed to obtain low/ultra-low sulfur fuel oils. In the present study, the ultrasound assisted oxidative desulfurization (UAOD) processes using the H_2O_2 and HPW-AC oxidizing system applied to simulated fuel (~2800 ppm sulfur in the form of dibenzothiophene, benzothiophene, and thiophene dissolved in toluene), were optimized. After the pre-saturation of the HPW-AC with the simulated fuel, H_2O_2 was added just before the reaction was commenced under ultrasonic irradiation. After the application of both 2^k-factorial design of experiment for screening and Face-Centered Design of Experiment for optimization, it was found that 25.52 wt% of H_2O_2 concentration, 983.9 mg of catalyst dose, 9.52 mL aqueous phase per 10 mL of the organic phase and 76.36 minutes of ultrasonication time would render 94.74% oxidation of the sulfur compounds in the simulated fuel. After the application of the optimized parameters to kerosene and employing a 4-cycle extraction using acetonitrile, 99% of the original sulfur content were removed from the kerosene using the UAOD optimized parameters. The desulfurization process resulted in a low-sulfur kerosene which retained its basic fuel properties such as density, viscosity and calorific value.

1 Introduction

Sulfur oxides (SO_x) and particulate matter (PM) are one of the criteria pollutants set by the United States Environmental Protection Agency that significantly contributes to air pollution. These are particularly emitted by processes utilizing raw materials such as crude oils and metal ores – in which sulfur is prevalent. Legislative efforts have been exerted by various countries and regions to prevent the addition of these criteria pollutants to the atmosphere. Developed countries such as Japan, USA, Canada and the European Union have set a 50 ppm sulfur (low-sulfur oil, Euro IV) limit for its petroleum products while Taiwan has implemented a 10 ppm sulfur limit (ultra-low sulfur oil, Euro V). The Philippines has recently implemented its Euro IV Standard in the mid-2016s.

Although efforts have been done to decrease the emission of sulfur pollutants, scientists and engineers have to continually develop the process they utilize to meet these standards because of thermal maturation – the natural increase in the sulfur concentration of crude oil obtained from sources [1]. Thus, the existing technologies for desulfurization needs to be intensified to be able to handle the increasing sulfur concentration of the crude oil as well as the increasingly stringent legislation against sulfur emissions.

Because of this, the currently applied industrial process of fuel desulfurization, which is hydrodesulfurization (HDS) is operated using extreme conditions – high temperatures and high pressures [2]. Also, the most common sulfur compounds in a fuel, which are dibenzothiophenes (DBT), benzothiophenes (BT), and thiophenes (T) – refractory compounds – were found to be less reactive to HDS [3]. It is for these reasons that HDS needs either an assistance or replacement and intensification.

Oxidative desulfurization (ODS) has gained interest in the recent years because of its potential to answer the concerns in HDS. For one, ODS can be accomplished using ambient conditions and without the use of the expensive hydrogen gas. The use of heteropolyacids (HPAs), particularly phosphotungstic acid, as catalysts in a hydrogen peroxide oxidizing system has proved to be effective (more than 99% efficient) in oxidizing the sulfur compounds [4]. Supporting this HPA can improve its catalytic activity by increasing the effective surface area and making it more economical because less HPW is utilized in the process and the catalyst may be recovered by simpler separation methods.

In this paper, the application of ultrasonication as assistance to ODS process – Ultrasound-Assisted Oxidative Desulfurization or UAOD – was investigated. Ultrasonic irradiation creates fine emulsions that

* Corresponding author: nathaniel.dugos@dlsu.edu.ph

improve the mass transfer between the organic and aqueous phases. In addition, UAOD promotes cavitation that creates local conditions of extreme pressures and temperature that aids the oxidation as well [5,6]. More specifically, the focus of the paper was the optimization of the UAOD process varying parameters such as H_2O_2 concentration, catalyst dose, AP:OP ratio and reaction time under a constant ultrasonication condition. The optimization process using a response surface methodology as well as the use concentrated simulated fuel (~2800 ppm sulfur) has yet to be investigated.

2 Materials and Methods

2.1. Chemicals and Reagents Used

Phosphotungstic Acid ($H_3PW_{12}O_{40}$), Hydrogen peroxide (H_2O_2, 30%w/w), Dibenzothiophene ($C_{12}H_8S$, 98-99% purity), and Benzothiophene (C_8H_6S, >99% purity) were purchased from Merck, Inc. Thiophene (C_4H_4S, ≥99%) was purchased from Sigma Aldrich while Toluene (C_7H_8, AR grade) was purchased from RCI Labscan Ltd and Activated Carbon was purchased from Fluka Analytical.

2.2. Equipment and Instrumentation Used

A Vulcan® Box Furnace with Programmable Controls Model 3-550 was used to dry and calcine the catalysts. Ultrasonic irradiation was provided by an Elmasonic E 120 H Ultrasound Cleaning Unit (37 kHz, 150 W). Immediate separation of organic and aqueous phases of the samples was carried out using Hermele Small Centrifuge Z 206 A at 4000 rpm for 20 minutes. The simulated fuel samples were analyzed using a Perkin Elmer Clarus 500 Gas Chromatograph – Flame Ionization Detector with the oven temperature initially at 100°C, increased at a rate of 5°C/min until 200°C and then at a rate of 10°C/min until 290°C. FID and Injection Temperature both at 300°C and H_2 and Air were set at 45 mL/min and 450 mL/min respectively. The carrier gas was He and N_2 at 1 mL/min.

2.3. Preparation of Catalysts

Activated carbon-supported phosphotungstic acid, thus labelled HPW-AC, was prepared following the proceeding procedures. Based on a previous experiment, the loading of HPW on AC was set to 10%, thus, HPW-AC-10. To synthesize HPW-AC-10, 1 gram of HPW was dissolved in excess deionized water. Ten grams of activated carbon was slowly added to the solution which was then set aside for 24 hours. The mixture was then put in a programmable furnace and dried at 120°C for 12 hours and calcined at 350°C for 4 hours. The catalysts were characterized using FTIR and SEM.

2.4. Desulfurization of Simulated Fuel

The simulated fuel was prepared by dissolving the three most abundant organosulfur compounds (dibenzothiophene, benzothiophene, and thiophene). About 2800 ppm of sulfur with a fixed composition of DBT, BT and T was prepared by dissolving certain amounts of these compounds in toluene. 2800 ppm of sulfur was used to approximate the amount of sulfur in crude diesel. A volume of simulated fuel was initially obtained to be treated by oxidative desulfurization. The catalyst was pre-saturated with the solution for 5 minutes to achieve true catalytic performance [7]. Just before the reaction under sonication is commenced, a certain volume of hydrogen peroxide oxidant with a certain concentration was introduced in the solution. The reaction was commenced under ultrasound and then stopped after a certain period of time. The sample was centrifuged to easily separate the organic phase and the aqueous phase and the residual catalyst in the aqueous phase was separated by syringe filters. The organic phase was then sent to a GC-FID for analysis.

2.5. Design of Experiment and Optimization

To account for the factors that might have interactions in the UAOD process, a response surface methodology (RSM) design of experiment was employed – Face Centered Composite Design. Design Expert v7.0 was used to come up with the sequence of runs and analyzing the data obtained. Table 1 shows the process parameters and its corresponding range of values.

Table 1. Process Parameters and the Levels of Values used in the Design of Experiment

Parameter		Levels		
		-1	0	1
A	H_2O_2 Concentration (%wt)	10	20	30
B	Catalyst Dose (mg)	500	750	1000
C	mL AP per 10 mL OP	1	5.5	10
D	Ultrasonication Time (min)	30	60	90

A 2^k factorial experiment with center points was initially done to investigate the significance of the four parameters and its interaction and if a quadratic curvature is significant ($p < 0.05$). Once confirmed, additional experimental runs were done to complete the FCC design of experiment and optimization can be performed.

2.6. Desulfurization of Actual Fuel and Determination of Fuel Properties

Once the optimized parameters were determined, these were applied to the actual fuel – kerosene. An additional step of actually removing the oxidized sulfur compounds was done after sufficient amount of sample has been processed. The oxidized sulfur compounds was extracted from the sample using a four-cycle extraction using acetonitrile producing the desulfurized fuel. The samples were sent to an external testing laboratory to analyze for

https://doi.org/10.1051/matecconf/201815603045

the sulfur content before and after the desulfurization process. In addition to the sulfur content, fuel properties such as density and viscosity were determined. The higher heating value (HHV) of the desulfurized kerosene was estimated using an empirical formula shown [8]:

$$\mathrm{API} = \frac{141.5}{\mathrm{s.g.}} - 131.5 \quad (1)$$

$$\mathrm{HHV} = 42{,}890 + 93(\mathrm{API} - 10) \quad \mathrm{kJ/kg} \quad (2)$$

3 Results and Discussion

3.1. Characterization of the Catalysts

Figure 1 shows the SEM images of the activated carbon (a), HPW (b), and the HPW-AC-10 catalyst (c). Figure 1b shows the more orderly HPW and the more chaotic Figure 1c indicates the scattering of HPW on the surface of the activated carbon which indicates the efficiency of the impregnation process and one of the reasons why HPW-AC is an efficient catalyst.

Figure 2 shows the FTIR chromatograph of the AC, HPW and HW-AC-10 catalysts. The chromatograms confirmed the presence of the HPW on the surface of AC because of the presence of P–O, W–O, W–O_b–W, W–O_c–W peaks which were present in the HPW chromatograms as well as in previous studies [9].

Fig. 1a. Scanning Electron Microscope Image of the Activated Carbon (AC) at 5,000x Magnification

Fig. 1b. Scanning Electron Microscope Image of the Phosphotungstic Acid (HPW) at 5,000x Magnification

Fig. 1c. Scanning Electron Microscope Image of the Activated Carbon-Supported Phosphotungstic Acid (HPW-AC) at 5,000x Magnification

3.2. 2^k Factorial Design of Experiments

In previous experiments, the proponents have investigated the performance of various HPW-AC loading and it has indicated that the optimal loading was 10%, thus, in this paper and in further experiments, this loading is used. Increasing the loading more than 10% did not increase the percent oxidation and it may be hypothesized that HPW will not attach on the activated carbon due to overcrowding of molecules, though this has yet to be confirmed. Table 2 shows the 2^k factorial experimental runs with their corresponding percent oxidations.

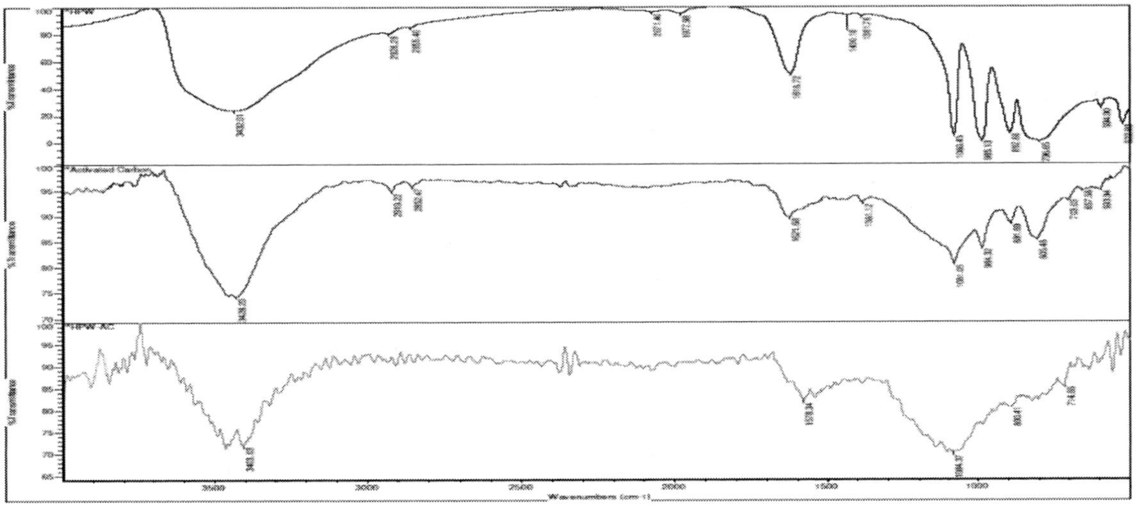

Fig. 2. FTIR Chromatograms of HPW, Activated Carbon and HPW-AC-10

Table 2. 2^k Factorial Runs for UAOD applied to Simulated Fuel with their corresponding Percent Oxidations

Run No.	A	B	C	D	%Oxidation
1	10	500	1	90	70.47
2	30	1000	1	30	78.72
3	10	1000	1	90	67.57
4	30	1000	10	90	96.64
5	30	500	1	30	64.39
6	30	500	1	90	82.37
7	30	500	10	90	86.64
8	20	750	5.5	60	85.47
9	20	750	5.5	60	81.46
10	10	1000	10	30	77.53
11	20	750	5.5	60	84.28
12	30	1000	1	90	88.74
13	10	500	1	30	46.32
14	30	1000	10	30	84.80
15	20	750	5.5	60	82.72
16	10	1000	1	30	59.98
17	10	1000	10	90	86.91
18	10	500	10	30	55.43
19	10	500	10	90	71.75
20	30	500	10	30	68.44

It can be noted that the HPW-AC shows high efficiency in terms of oxidizing the sulfur compounds ranging from a low oxidation of 46.32% and 96.64% as the highest oxidation efficiency. The numbers show that the HPW-AC oxidation system is an efficient oxidation system that can oxidize higher sulfur concentrations of ~2800 ppm S compared to previous literatures which only utilized ~500 ppm S feedstock. This shows the potential of the HPW-AC oxidation system as a stand-alone ODS system rather than an assistance to HDS. It must also be noted that the system could not approach a near 100% oxidation because of the presence of thiophene, the most unreactive of all the sulfur compounds present in the fuel. This oxidation suggests that the less unreactive benzothiophene, was also oxidized by the HPW-AC oxidation system. Thus, it can noted that the organic sulfur compounds were reactive to the HPW-AC oxidation system in the following order: DBT > BT > T, owing to the electron densities of each sulfur compounds.

The data obtained comprises the 16 2^k factorial runs with additional four center points; these were also processed to see the factors and their interactions were significant and to see if the curvature is significant as well. Table 3 shows the significant factors and interactions from the 2^k factorial runs.

Table 3 indicates that all of the individual parameters – H_2O_2 concentration, Catalyst Dose, AP per 10 mL OP and ultrasonication time – were all significant while Interactions AC, BC, BD, and ABC were shown to be significant with p values below 0.05. These factors were retained in the Face Centered Design of Experiment because of their significance. Since the curvature was found to be significant also, a quadratic model is must be used for the analysis of the design space.

Table 3. Significant Parameters for UAOD applied to Simulated Fuels

Source	p-value
Model	< 0.0001
A–H_2O_2 Concentration	< 0.0001
B–Catalyst Dose	< 0.0001
C–AP per 10 mL OP	< 0.0001
D–Ultrasonication Time	< 0.0001
AB	0.9001
AC	0.0087
BC	0.0020
BD	0.0007
ABC	0.0212
Curvature	< 0.0001
Residual	
Lack of Fit	0.4774

* Corresponding author: nathaniel.dugos@dlsu.edu.ph

3.3. Face Centered Design of Experiment

A Face Centered Design of Experiment was employed for further analysis and optimization since this utilizes the runs used in the 2^k factorial experiments. Since there will be no change in the parameters to be considered (because they were deemed significant), this design can be employed by just adding some additional experimental points. Table 4 shows these additional points and their percentage oxidations. Table 5 shows the significant parameters of the RSM FCC Design of Experiment.

The quadratic model utilized in the ANOVA table can now be appropriately explored. The ANOVA shows that all the model and its individual factors were all significant. AC, BC, and BD interactions were also significant. The square of the terms A, C and D were also found to be significant. It should also be noted that the Lack of Fit was found to be not significant at a value of 0.5251. The value of the R^2, which is 0.9761, indicates that the quadratic model was more appropriate for use than the two-factor interaction model which was done in the previous section.

Table 4. Additional Experimental Points for RSM Face Centered Design of Experiment for UAOD applied to Simulated Fuel with their corresponding Percent Oxidations

Run No.	A	B	C	D	%Oxidation
21	20	1000	5.5	60	91.84
22	20	500	5.5	60	75.69
23	20	750	5.5	60	84.28
24	20	750	10	60	85.20
25	10	750	5.5	60	75.41
26	30	750	5.5	60	90.15
27	20	750	5.5	90	89.28
28	20	750	5.5	30	74.14
29	20	750	5.5	60	86.33
30	20	750	1	60	78.96

Table 5. Significant Parameters of the UAOD Applied to Simulate Fuels

Source	p-value
Model	< 0.0001
A–H_2O_2 Concentration	< 0.0001
B–Catalyst Dose	< 0.0001
C–AP per 10 mL OP	< 0.0001
D–Ultrasonication Time	< 0.0001
AC	0.0103
BC	0.0016
BD	0.0004
A^2	0.0371
C^2	0.0120
D^2	0.0064
Residual	
Lack of Fit	0.5251

The insignificant lack of fit supports the adequacy of the model generated since it indicates that the model can predict and approximate the dependent variable (percent oxidation) with great accuracy should one try to repeat the experiments using various combinations of the independent variables. The model that the software generated is shown in Equation 3 with all the terms used in the equation being the actual values of the parameters.

$$\begin{aligned}\%\text{Oxidation} = \\ &-11.43105 + 2.07760A + 0.03380B \\ &+ 2.25346C + 1.00967D - 0.03469AC \\ &+ 0.00179BC - 0.00032BD - 0.02918A^2 \\ &- 0.17852C^2 - 0.00443D^2\end{aligned} \quad (3)$$

3.4. Model Graphs

Figure 3 shows the model graphs of the significant factors and interactions indicated by the ANOVA. Should an individual factor be deemed significant in any of the interactions, single factor variations were not shown.

Fig. 3a shows the interaction plot of the H_2O_2 concentration (A) with the AP per 10 mL OP (C). At high levels of A and C, high oxidation was shown. This increase in oxidation with an increase of A and C can be attributed to the amount of H_2O_2 species which serve two functions: (1) direct oxidation of the sulfur compounds by the increased hydroxyl radicals from the H_2O_2 enhanced further by ultrasonic irradiation and (2) its interaction with the catalyst to form a peroxo-metallate complex which are more reactive to organic sulfur compounds. The increase in AP could entail the same effect (increase in H_2O_2) but to a certain extent only. This might be because of the scattering of the catalyst on the two phases. In the experiments, it was seen that the catalyst had an affinity to both the OP and AP thus increase AP so much would make less contact with the

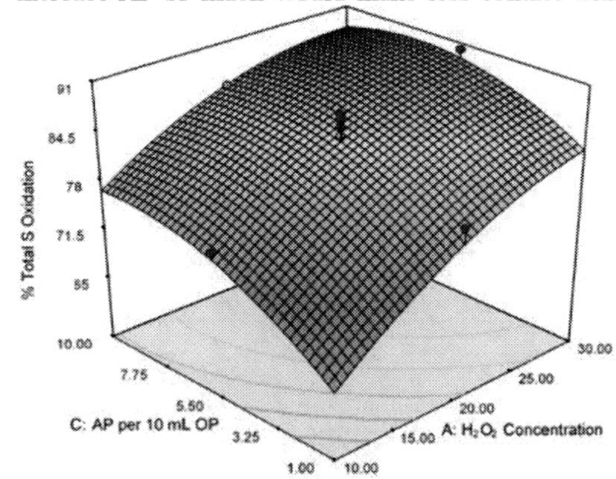

Figure 3a

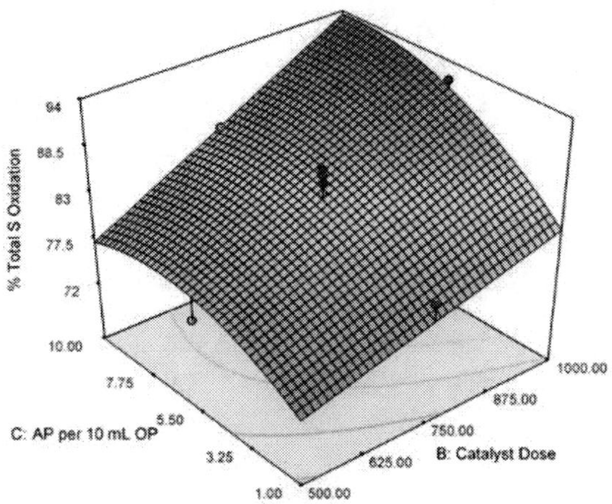

Figure 3b

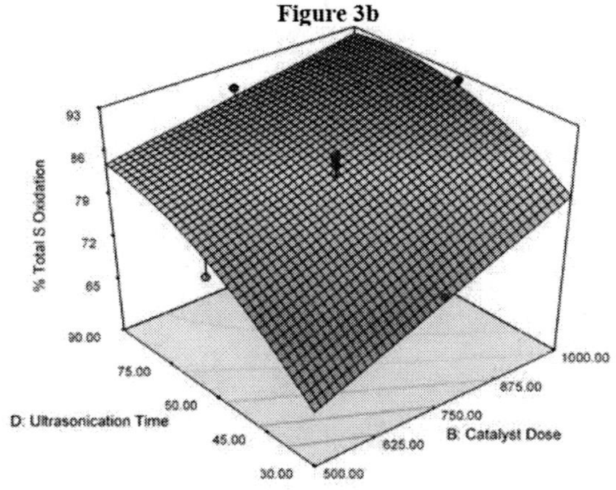

Figure 3c

Fig. 3. Model Graphs of Variables for UAOD Applied to Simulated Fuel: (a) H_2O_2 Concentration and AP per 10 mL OP Interaction Plot; (b) Catalyst Dose and AP per 10 mL OP Interaction Plot; (c) Catalyst Dose and Ultrasonication Time Interaction Plot

OP. As the reaction proceeds with ultrasonication, the catalyst was seen to scatter on both phases. This causes more losses of the catalyst on the AP rather than facilitating the reaction in the OP. Using a sufficient amount of AP to just the right level would provide H_2O_2 for the oxidation as well as sufficient contact with the OP.

Figure 3b shows a steep surface that increases in percent oxidation with the increase of both the catalyst dose and AP per 10 mL OP. As expected, an increase in the percent oxidation was shown with the increase of catalyst dose because of the increase in the active species that would produce the peroxo-metallate complex. At smaller catalyst doses, however, a sufficient amount of AP should be provided at a moderate level only due to the affinity of the activated carbon catalyst to both phases. At this lower level, the system does not have enough catalyst to form the complex and substantially oxidize the sulfur compounds. In addition, the increase in the solid materials in the system increases the nucleation sites for the formation of cavities to occur. These cavities are known to form local conditions of extreme temperatures and pressure thus, increasing the production of chemically active species such as the radicals and complexes [10].

Figure 3c shows the interaction plot of catalyst dose (B) and ultrasonication time (D). Both of these parameters have a positive effect on the oxidation. Ultrasonication and the solid catalyst have a synergistic effect on the oxidation reaction because of the presence of the solid particles. The solid particles provide additional nucleation sites for the cavity formation to occur thus, there is an increase in cavitational intensity and consequently, better sonochemical effects [10]. The intensified cavitation increases the hydroxyl radical production thereby increasing the oxidation of the sulfur compounds. At low levels of catalyst dose, an increase in the ultrasonication time not only forms more radicals but also increases the interfacial mass transfer and provides more contact time for the sulfur compounds and oxidant responsible for its conversion be it the hydroxyl radicals or the peroxo-metallate complex.

3.5. Optimized UAOD Parameters

Based on the model generated, optimized parameters for the ultrasound-assisted oxidative desulfurization applied to simulated fuel were found to be 25.52 wt% H_2O_2 concentration, 983.9 mg of catalyst dose, 9.52 mL per 10 mL OP and 76.36 minutes of ultrasonication time. Table 6 summarizes these optimized values. The table also shows the predicted percent oxidation of the system based on the model at these conditions which is 97.02%. Confirmatory runs were also done to check the validity of the model and an adequate 2.35% error was obtained with the actual oxidation efficiency of 94.74%.

Table 6. Optimized UAOD Parameters applied to Simulated Fuel

Parameter	Value
H_2O_2 Concentration	25.52 wt%
Catalyst Dose	983.9 mg
AP per 10 mL OP	9.52 mL
Ultrasonication Time	76.36 minutes
%Oxidation (Predicted)	97.02%
%Oxidation (Actual)	94.74%
Error	2.35%

3.6. Application to Actual Liquid Fuel

The optimized parameters were then applied to an actual fuel, in this study, it was kerosene. After that, removal of the oxidized sulfur compounds was done a four-cycle extraction procedure using acetonitrile. The sulfur content of the kerosene before the treatment was analyzed as 1370 ppm of S and after the desulfurization process, the sulfur content was reduced to 14 ppm S – a value that is lower than the Euro IV standard of fuels (50 ppm S) – rendering 99.0% sulfur removal.

3.7. Evaluation of the Fuel Properties

The fuel properties (i.e. density, viscosity) were determined using common experimental procedures and the calorific value of the fuel was approximated using Equation 1 and 2. The density of the original kerosene was 0.7801 g/cm^3 while the desulfurized kerosene was found to be 0.7648 g/cm^3. The density of the feed kerosene was well within the typical ranges of density for kerosene – 0.7750 to 0.840 g/cm^3. Though the density of the desulfurized kerosene was well below this range of values, this is of limited significance as fuel quality indication unless correlated to other fuel properties. Although, one may say that the less dense the fuel would make the amount of fuel per unit volume less also.

The density may be correlated to the heating value of the fuel with adequate accuracy. Once the specific gravity of the fuels were determined, the degree API can be calculated immediately followed by the calculation of the heating value. The original heating value was 46.60 MJ/kg while the desulfurized kerosene was found to have 46.94 MJ/kg of heating value. With the decrease in density of the fuel, the estimated heating value increased, though an actual determination of heating value must be done for verification. This is because on the weight basis, the carbon (low-heating value) to hydrogen (high-heating value) ratio increases as specific gravity increases [11].

The viscosity was also determined before and after the desulfurization and was found to be 0.8575 cP and 0.8655 cP. Translating these to kinematic viscosity, in cSt, these values are 1.09902 cSt and 1.1317 cSt, respectively. All these values fall within the typical range of kinematic viscosity of kerosene which is 1.0 and 1.9 cSt [11]. The increase in viscosity shows that after desulfurization, the fuels may be harder to transport using a pump.

From these, it can be said that the desulfurization of the fuel not only reduces the sulfur content of the fuel but also retains the basic properties of the fuel.

4 Conclusions and Future Works

The study shows the effectivity of an HPW-AC oxidation system in oxidizing sulfur compounds to sulfoxides and sulfones high efficiency. Once applied to an actual fuel, it shows that 99.0% of sulfur removal was shown resulting to a low-sulfur fuel (below 50 ppm Euro IV standard). Using a response surface methodology of Face Centered Composite Design of Experiment, the optimal UAOD parameters were found to be 25.52wt% of H_2O_2, 983.9 mg of catalyst dose, 9.52 mL of AP per 10 mL of OP and 76.6 minutes of ultrasonication producing 94.74% oxidation of the 2800 ppm sulfur, a feedstock that contains relatively high concentrations of sulfur compared to typical researches indicating the possibility of applying the ODS in replacing the HDS which were indicated to be less reactive to refractory sulfur compounds or in which process intensification is needed (i.e. higher temperature and pressure). Using kerosene as an actual fuel, the sulfur content was reduced from 1370 ppm S to 14 ppm S, equivalent to 99.0% desulfurization of kerosene.

The ultrasonication device used in the experiment is unable to vary certain ultrasonication parameters such as ultrasonication power and frequency. It may be suggested that further studies investigate on the effect of these parameters to the oxidation efficiency. Previous studies have indicated an HPW–H_2O_2 oxidizing system was shown to follow a first-order kinetics, so in an HPW-AC oxidizing system, kinetics may also be investigated. Finally, a pilot scale design and fabrication of a continuous desulfurization process may be done.

5 References

1. L. Liu, H. Lü, J. Qian, J. Xing, China. Pet. Process. Pe. **12,4,** 1–6 (2010).
2. X. Ma, A. Zhou, C. Song, C. Catal. Today, **123,** 276–284(2007)
3. T.O. Sachdeva, K. K. Pant, Fuel Process. Technol. **91,** 1133–1138 (2010)
4. H. Mei, B. W. Mei, T.F. Yen, Fuel **82,** 405–414 (2003)
5. M.W. Wan, T.F. Yen, Appl. Catal. A-Gen. **319,** 237–245 (2007)
6. Y. Dai, D. Zhao, Y. Qi, Ultrason. Chem. **18,** 264–268 (2011).
7. L. Liu, Y. Zhang, W. Tan, W. Ultrason. Sonochem. **21,** 970–974 (2014).
8. Eugene L. Keating, *Applied Combustion, 2nd ed.* (Taylor and Francis Group, Florida, 2007)
9. M. Prathap Kumar, A. Sivasamy, A. *International Conference on Advances in New materials (ICAN 2014)* **2(1),** 228–233 (2014)
10. A.G. Chakinala, P.R. Gogate, A. E. Burgess, D. H. Bremner, Ultrason. Chem. **14, 5,** 509–514 (2007)
11. Salvatore J. Rand, Significance of Tests for Petroleum Products (ASTM International, 2003)

https://doi.org/10.1051/matecconf/201815603046

Ball-milling effect on Indonesian natural bentonite for manganese removal from acid mine drainage

Widyawanto Prastistho[1,2*], *Winarto* Kurniawan[1], *Hirofumi* Hinode[1]
[1] Department of International Development Engineering, Tokyo Institute of Technology, 2-12-1, O-okayama, Meguro-ku, Tokyo 152-8550, Japan
[2] Department of Environmental Engineering, Universitas Pembangunan Nasional "Veteran", Jl. SWK 104, Sleman, Daerah Istimewa Yogyakarta 55283, Indonesia

Abstract. The influences of mechanical milling on Indonesian Natural Bentonite (INB) characteristics and manganese (Mn) removal from acid mine drainage (AMD) were investigated. The INB characteristics were observed by scanning electron microscope (SEM), X-ray diffraction (XRD), nitrogen adsorption-desorption for specific surface area (SSA) and microporosity measurement, cation exchange capacity (CEC) and particle size distribution (PSD) analyzer. Four minutes milling with frequency 20 Hz on INB caused morphological change which showed more crumbled and destructed particle, lost the (001) peak but still retained the (100) peak that indicated delamination of montmorillonite mineral without breaking the tetrahedral-octahedral- tetrahedral (T-O-T) structure, rose the CEC from 28.49 meq/100g to 35.51 meq/100g, increase in the SSA from 60.63 m2/g to 104.88 m2/g, significant increase in microporosity which described in the t plots and decrease in the mean particle size distribution peak from 49.28 μm to 38.84 μm. The effect of contact time and effect of adsorbent dosage on Mn sorption was studied. Both unmilled and milled samples reached equilibrium at 24 hours and the pH rose from 4 to 7 in first 30 minutes. The Mn removal percentage increased significantly after milling. Using Langmuir isotherm, the maximum adsorbed metals (qmax) also increased from 0.570 to 4.219 mg/g.

1 Introduction

Bentonite, clay material containing montmorillonite minerals including to the smectite group, naturally formed by volcanic ash alteration or hydrothermal alteration, is widely used for environmental purposes such as adsorbent, radioactive barrier, liner disposal and so on due to its absorbability and low permeability [1,2]. Some modifications have been developed to improve bentonite sorption performance such as acid activation, pillaring, and milling [3,4,5]. Milling itself is non-chemical modification technique, usually using spinning jar with stainless still balls inside. The spinning jar causes pounding motion of the balls to the sample. This ball milling process reduces the bentonite particle size, change the morphology and crystal structure, exfoliation, increase at specific surface area (SSA) and cation exchange capacity (CEC), and improve heavy metal sorption performance [6,7]. Continuous milling will gradually affect montmorillonite crystal structural change which finally leads to amorphization. Intensive milling process gradually increased the SSA and CEC and reached the peak which then decreased [8].

This study is utilizing Indonesian natural bentonite (INB), located at Wonosegoro area, Boyolali district, Central Java, Indonesia which has hypothetical resource 58 million tons of natural bentonite [9]. Milling mechanical modification was conducted using small ball miller instead. Simple mechanism moving left right with one stainless steel ball inside at certain frequency that would give straight impact on the sample.

Milling modification of INB is applied for manganese (Mn) removal from acid mine drainage (AMD). One of coal mining company in Jorong area, Tanah Laut district, South Kalimantan, Indonesia showed elevated Mn concentration with range 1.7 mg/L – 27.2 mg/L and pH 3.03-3.04 [10]. Indonesian government set the maximum limit of Mn concentration 0.1 mg/L for drinking water and 4 mg/L for mining waste. The difference is because the usage purposes. Indonesian government divides the water type into several classes considering the usage purpose. For instance, drinking water has different limit parameter with agricultural and fish cultivation purpose [10,11]. Indonesian natural bentonite and milling modification are considered as abundant low-cost material and low-cost treatment for AMD treatment.

2 Material and methods

2.1 Field sampling and sample preparation

Bentonite was taken from Wonosegoro area, Boyolali district, Central Java, Indonesia. Prior to sampling, outer bentonite exposed was dug ±10 cm first in order

* Corresponding author: w.prastistho@gmail.com

to get uncontacted sample with the atmosphere. Sample was crushed using mortar and sieved pass 50 mm in order to get finer and relatively uniform particle size. The sample was then milled using Retsch MM 400. One gram sample was put inside grind jar (size 50 ml) and milled for 4 minutes with vibrational frequency setting was 20 Hz.

2.2 Characterization

Crystal structural changes were observed using x-ray diffraction (XRD) equipment Rigaku Multiflex, operated at 40kV and 20 mA. The XRD analyzer using Jade 5 software. Morphological changes were observed using Scanning electron microscope (SEM) Keyence VE-8800 with 3000 - 5000 times magnifications. Microporosity and SSA were measured using multiple- point nitrogen adsorption-desorption technique, with BELSORP-max. Particle size distribution measurements were made by SALD-2300 laser diffraction particle size analyzer. The CEC was determined using Chapman method by saturating the sample with sodium acetate and replacing sodium with ammonium acetate [12].

2.3 Adsorption study

Effect of contact time was investigated. Two portions 100 mL of Mn solutions with concentration of 27.48 mg/L (diluted from Mn Wako standart solution 1000 ppm) with pH 4 were prepared in 100 mL beaker glasses. 1 gr unmilled and 4 minute-milled bentonite were added to each beaker glass. Magnetic stirrer was used to mix the bentonite and solution. The magnetic stirrer was set to 230 rotations per minute (RPM) at room temperature. The mixtures were sampled at 30 m,1 h, 2 h, 4 h, 6 h, 12 h, 24 h, 48 h and 72 h.

Effect of sorbent dosage was investigated. Various mass of unmilled and 4 minutes milling bentonite were prepared (0.4, 0.6, 0.8, 1 and 1.2 g). The manganese solution was the same condition with the previous solution.

The concentration of all samples were measured using induced coupled plasma atomic emission spectroscopy (ICP-AES) Seiko Instrument, SPS 7800 (II). Adsorbed amount of Mn on INB then calculated using this equation:

$$q = \frac{(Co - Ce)V}{m} \tag{1}$$

q = adsorbed manganese on sample (mg/g)
Co = Initial concentration (mg/L)
Ce = Final Concentration (mg/L)
V = Solution volume (ml)
m = Sample mass (g)

3 Result and discussion

3.1 Characterization

Fig. 1 (SEM photos) shows the morphological change before (a) and after 4 minutes milling (b). The morphology of milled bentonite appears more destructed, crumbled and more blunt angularity. The frequency also affected the change of morphology beside the milling time.

Fig. 1. SEM photo before (a) and after 4 minutes milling (b)

The Jade 5 XRD analyzer software did not exhibit any peak read of (001) montmorillonite after 4 minutes milling (Fig. 2). But the interesting found is it still retained the (100) without any structural change, indicating delamination of montmorillonite mineral without breaking the tetrahedral-octahedral-tetrahedral (T-O-T) structure. It is different from Vdović et al (2010) [8], the decrease in intensity of (001) followed by (100) together which finally both lost peak, indicated transformed to amorphous due to prolonged milling. Different miller and energy would probably give different effect on exfoliation and deformation phases of montmorillonite.

The initial CEC of INB was 28.49 meq/100 g and increased after 4 minutes milling to 35.51 meq/100 g, attributed to disaggregation of INB which leads to more opened exchangeable cation on basal planes of montmorillonite

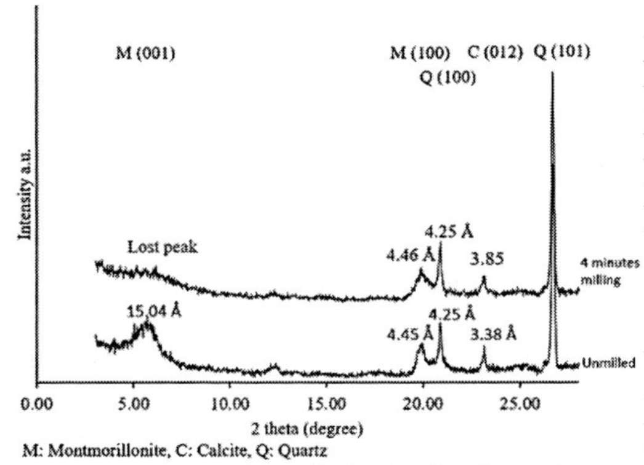

Fig. 2. XRD pattern of unmilled and milled bentonite

The N_2 adsorption-desorption shows hysteresis, indicated the INB is porous material (Fig. 3). This hysteresis shape also confirmed the INB as slit pore shape (type B) [13]. After milling the hysteresis appears narrower. The initial SSA of unmilled INB, using Brunauer, Emmet, and Teller (BET) method, was 60.63 m^2/g, and increased after 4 minutes milling to 104.88 m^2/g.

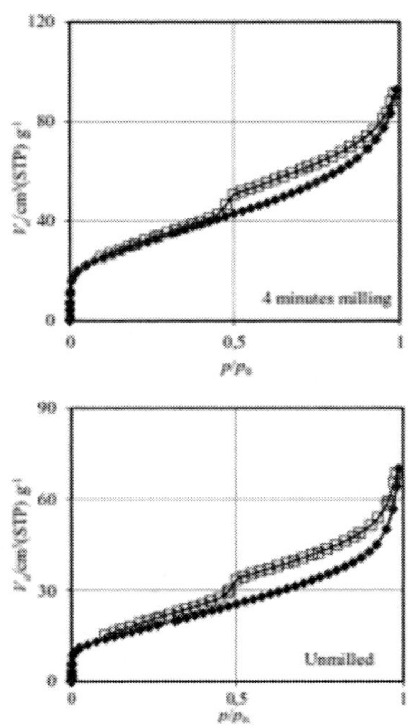

Fig.3. Adsorption-desorption isotherm of unmilled and 4 minutes milling INB

Significant increase in microporosity after milling is described in the t plots (Fig. 4, Table 1). The t-plot method, developed by Lippens and de Boer, is standard isotherm describes relationship between relative pressure and thickness of adsorption layer [14]. The thickness (t) can be obtained from this equation:

$$t = \frac{V_a}{V_m} x 0.354 (nm) \qquad (2)$$

Va = volume of gas adsorbed (mL)

Vm = volume of monolayer gas adsorbed on the sample surface (mL)

The number 0.354 nm is the thickness of monomolecular layer where the nitrogen molecules considered as hexagonal on adsorbent surface, therefore, it is smaller than nitrogen molecule diameter.

The unmilled and milled INB t plots have two slopes that the one sharp and the other steeper, it means the sample is type (II) which indicates it has micropores [14].

The increase of micropore volume and micropore surface area are confirmed the particle disaggregation also opened new pores due to milling. On the other hand, the 2 t did not significantly change. Initially phase of delamination only shifting along the basal planes [15]. Therefore, in this study, is assumed the delamination after 4 minutes milling the basal planes only shifted, and has not caused fully separation between T-O-T planes.

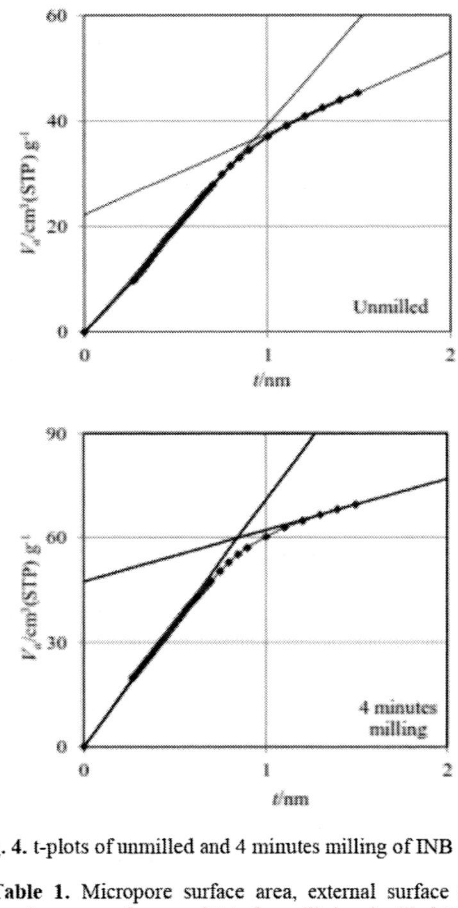

Fig. 4. t-plots of unmilled and 4 minutes milling of INB

Table 1. Micropore surface area, external surface area, micropore volume and 2 t of unmilled and milled INB

Sample condition	Micropore surface area (m^2/g)	External surface area (m^2/g)	Micropore volume (cm^3/g)	2 *t* (nm)
Unmilled	60.73	23.70	0.034	1.85
4minutes milling	109.15	22.63	0.074	1.69

The PSD curve before and after milling shown in Fig. 5. The PSD of unmilled INB, dominated by coarser particle population with mean 49.28 µm. After milling, the coarser population decreased to 38.84 µm. The finer particle population remains unchanged at 9.316 µm.

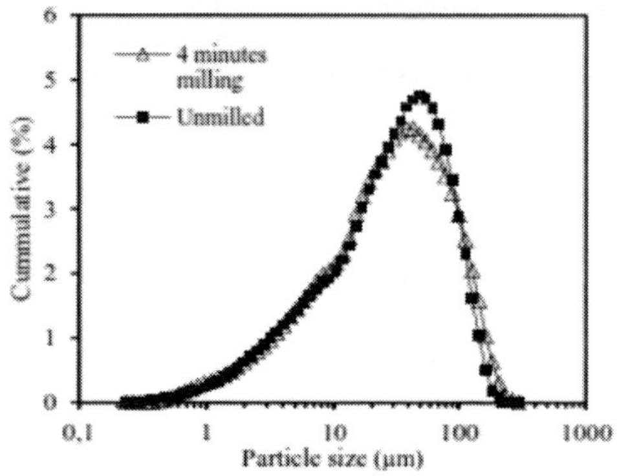

Fig. 5. Particle size distribution of unmilled and 4 minutes milling of INB

3.2 Sorption study

Effect of contact time can be seen in Fig. 6. Amount sorbed respect to time both unmilled and 4 minutes milling of INB. Sorption improvement of Mn shown after 4 minutes milling. Initially, adsorption rate increased rapidly, and the optimal sorption was reached around 24 hours. Further increase was not significant indicated reached equilibrium. The pH has risen became 7 in 30 minutes.

Table 2. Isotherm parameter

Isotherm model	Sample condition	Isotherm parameters		
		q_{max} (mg/g)	K_L (L/mg)	2
Langmuir	Unmilled	0.570	0.029	0.605
	4 minutes milling	4.219	0.457	0.962

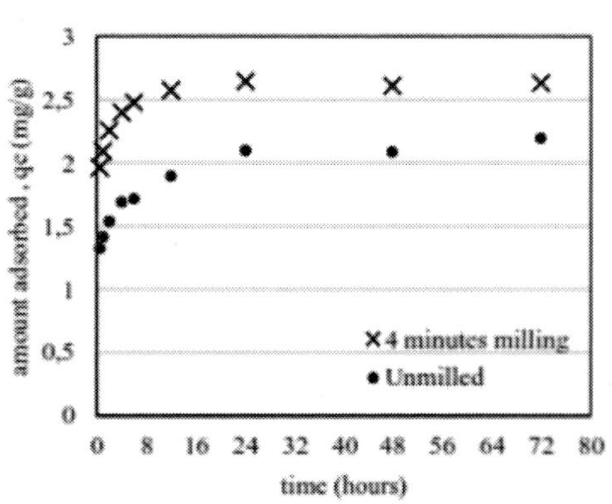

Fig. 6. Effect of contact time

Effect of adsorbent dosage can be seen in Fig. 7. Percentage removal was plot versus adsorbent dosage. Addition of adsorbent dosage slightly increased the percentage removal of Mn. The percentage removal of 4 minutes milling of INB is improved and even almost reached 100 % at 1.2 g dosage.

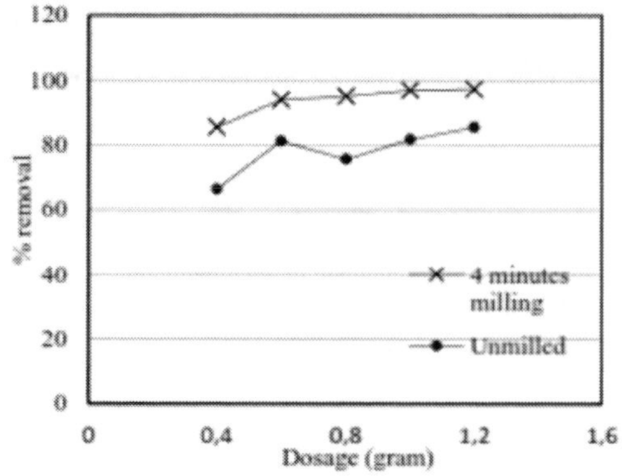

Fig. 7. Effect of adsorbent dosage

Isotherm is relationship between adsorbed amount and final concentration remains in solution. Langmuir isotherms was used to determine the type of sorption between adsorbate and sorbate. Langmuir isotherm is assumed the sorbate attached on surface of adsorbent as monolayer. Langmuir equation can be expressed as:

$$qe = \frac{q_{max} K_L}{1 + K_L c_e} \quad (3)$$

Where :

qe = amount of solute sorbed per gram on sample (mg/gr)

q_{max} = amount of maximum solute sorbed per gram on sample (mg/gr)

K_L = constant related energy of sorption (L/mg)

C_e = the equilibrium concentration of solute (mg/L)

Langmuir isotherm plot is shown in Fig.8. The q_{max} and KL can be calculated from slope = $1/q_{max}$ value and intercept = $1/KL.q_{max}$ value. The calculation result is shown in Table 2. The value of q_{max} is increasing significantly, exhibits milling has brought on better sorption.

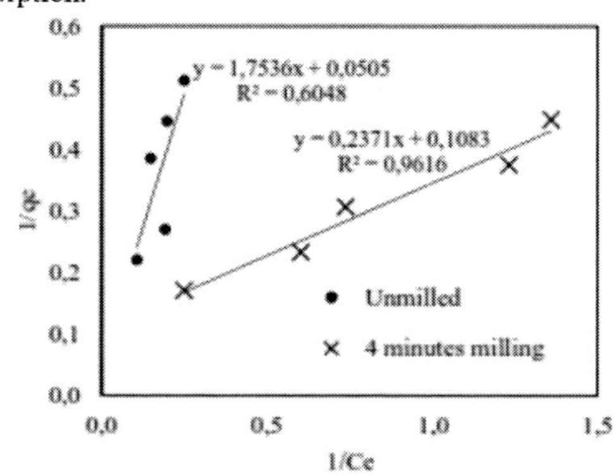

Fig. 8. Langmuir isotherm unmilled and 4 minutes milling of INB

4 Conclusion

Left-right moving milling mechanism with 20 Hz in frequency during 4 minutes deformed the (001) montmorillonite but still remained the (100), indicated delamination without destruct the T-O-T structure. According to the significant increase in microporosity from the t plot, but there was no significant change of t plot, it is assumed the delamination did not fully separated between the T-O-T planes. Further study about equipment and energy effect, and mineral orientation after and before mechanical modification are needed.

The INB adsorption performance for Mn removal from AMD was improved by milling. The removal percentage, maximum adsorbed metal and the pH was successfully increased. However, milling optimization on INB need to be more investigated in order to get more maximum Mn removal and also for the other heavy metals. Milling and INB is promising as low-cost modification and low-cost material remediation.

References

1. R.E. Grim, *Clay Mineralogy*, McGraw-Hill Book Company (1968)
2. T. Al-Ani, O. Sarapää, *Clay and Clay Mineralogy*, Geological Survey of Finland (2008)
3. S. Budsaereechai, K. Kamwialisak, Y. Ngernyen, KKU Res. Jour. **17**, 5, 800 – 810 (2012)
4. M. El miz, S. Salhi, I. Chraibi, A. El Bachiri, M.L. Fauchonnier, A. Tahani, Open Jour. of Phys. Chem. **4**, 98 – 116 (2014)
5. V. Masindi, W. Gitari, Jour. of Env. Chem. Eng. **3**, 2416-2425 (2015)
6. Djukic, U. Jovanovic, T. Tuvic, V. Andric, J.G. Novakovic, N. Ivanovic, L. Matovic, Ceramics Int., **39**, 7173 – 7178 (2013)
7. A.R.Ramadan, A.M.K.Esawi, A.Abdel Gawad, App. Clay Sci. **47,** 196–202 (2010)
8. N. Vdović, I. Jurina, S.D. Škapin, I. Sondi, App. Clay Sci. 48, 575 –580 (2010)
9. Indonesian Ministry of Energy and Mineral Resource, Indonesia Geological Resource: Book 3 : Java Island (2009)
10. Indonesian Government Regulation No. 82, Quality Management and Water Pollution Control (2001)
11. Regulation of Indonesian Ministry of Environment No. 113, Standard quality of waste water for industry and or coal mining activities (2003)
12. H.D. Chapman, *Cation-Exchange Capacity*, in: C.A.Black (Ed.), *Method of Soil Analysis, Part 2: Chemical and Microbiological Properties*, Am.Soc.Agron, Madison, Wisconsin, 891–900 (1965)
13. S. Kondo, T. Ishikawa, I. Abe, *Science of adsorption* (*Kyuutyaku no kagaku*), Maruzen Publisher (2001) (Japanese version only)
14. B. C. Lippens and J. H. de Boer, J. Catalysis 4, 319 (1965)
15. I. Sondi, M. Stubičar, V. Pravdić, Colloids and Surfaces A: Physicochemical and Eng. Aspects **127**, 141–149 (1997)

Dynamic Study of Feed-Effluent Heat Exchanger Addition on Double Bed Configuration Ammonia Reactor System within Varied Quenching Ratio

Tri Partono Adhi[1,*], *Ferdyan Ihza* Akbar[1]

[1]Institut Teknologi Bandung, Chemical Engineering Department, 40135 Bandung, Indonesia

Abstract. Ammonia is one of the most important industrial commodity due to its wide function. Ammonia synthesis reaction is an exotermic reaction. Therefore, Feed-Effluent Heat Exchanger (FEHE) is added to increase thermal efficiency. However, FEHE could lead the process to experience hysteresis phenomenon due to interaction between equipments as one steady state T feed could result several T outlet. Hysteresis phenomenon may result asset losses like explosion, leakage, and loosing material integrity. Double bed reactor configuration allows us to use several operating parameters as variation to overcome hysteresis. In this review, quenching ratio was chosen to be that varied parameter. This study aims to determine how quenching ratio affects hysteresis zone by utilizing Aspen Hysys® V8.8 as simulation tool. Simulation showed that quenching ratio would narrow hysteresis zone yet increased extinction temperature that lower the conversion. Conversion profile showed that 0.2 quenching ratio got the highest conversion for system with bed volume ratio 2:1 while total volume was 30 m^3. However, the feed temperature was fallen at hysteresis zone. Dynamic simulation showed that highest conversion feed temperature (10%ΔTf above extinct temperature) was still able to preserve stability with descending temperature approach. Hysteresis itself started to occur at 1.7%ΔTf above extinct temperature

1 Introduction

Ammonia is one of important industrial commodity due to its wide utilization as chemical product precursor such as urea, nitric acid, and etc. In ammonia production system, Feed-Effluent Heat Exchanger (FEHE) is commonly used that aims to exchange heat between reactor effluent and reactor feed as shown in Fig. 1.

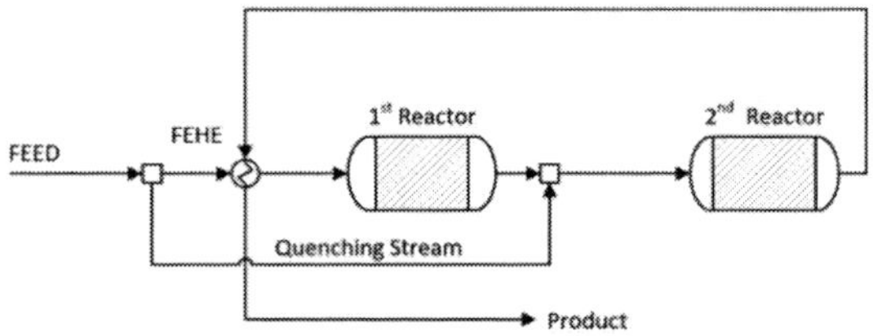

Fig. 1. Ammonia reactor system configuration

Such heat integration system would reduce utility requirement and increase thermal efficiency. However, the integration system could possibly lead the process system to experience significant temperature oscillation as shown in Fig. 2. Oscillations occur as the result of two or more processes that are coupled together with each other [1,7]. In this case, reactors performance is closely related to how the FEHE performs. As the result, reactor possibly undergoes a phenomenon while same operating condition could produce several steady state outputs that lead the process to oscillate [1,6]. This condition could cause several hazards for the equipment material and even operational staff [4,5]. Hazards that occurred could dangerously result disasters such as explosion or poisoneous chemical leakage which causing economic losses.

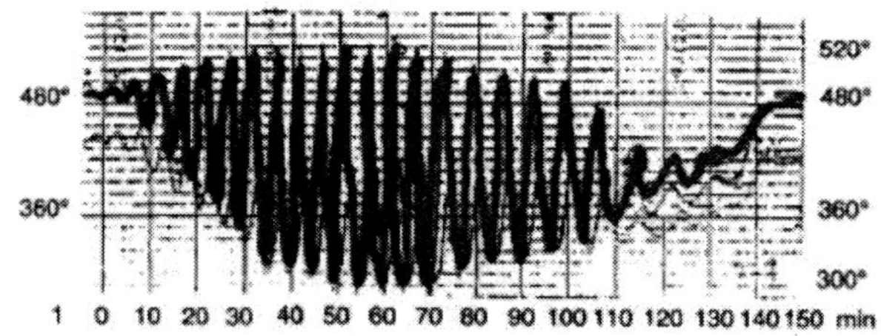

Fig. 2. Hysteresis phenomenon on industrial scale ammonia reactor. [3]

Hysteresis area could be predicted by plotting one of steady state parameter such (T outlet) as result of T feed input as shown in Fig. 3.

* Corresponding author: tpadhi@che.itb.ac.id

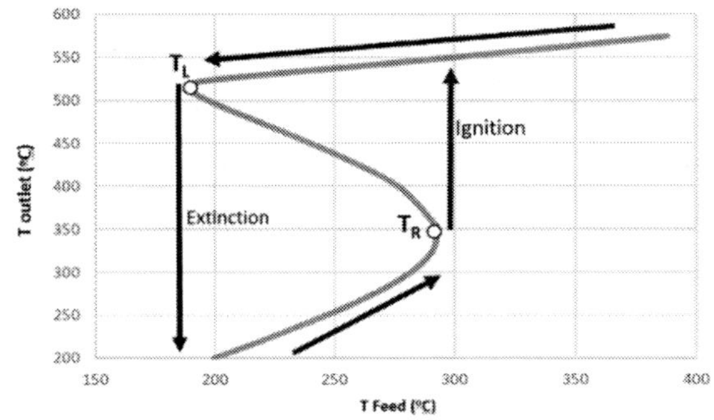

Fig. 3. Curve plot to determine hysteresis zone

The 'S' curve produced showed that within range of T_L and T_R (ΔT_f), each T feed input will result several values of T outlet. Therefore, there will be possibility for the process to experience oscillation while entering this region.

Some efforts to deal with this condition could be done by changing other supporting operation parameter for double bed system such as adding trim heat exchanger, bed volume distribution, or quenching ratio. Therefore, this study would show parametric review and simulation regarding actions to overcome and reduce hysteresis phenomenon by choosing quenching ratio as varied operation parameter. As the quenching ratios were evaluated, conversion profile was also analyzed to map the best temperature at particular quenching ratio to produce optimum reactant conversion. As the matter of limitting the problem, this review focused on observing feed temperature, outlet reactor temperature, quenching ratio, reactant conversion, and starting hysteresis temperature.

2 Methods

Aspen Hysys® V8.8 were used to act as an assistive tools for simulating the dynamic behaviour resulted from altered parameter value. Simulation configuration is shown in Fig. 4.

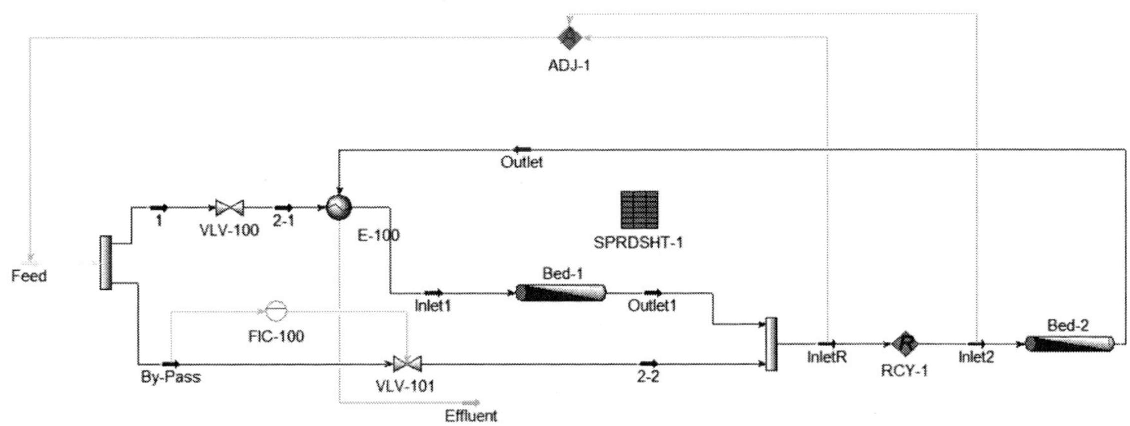

Fig. 4. Simulation Configuration

Fluid package used in the simulation was Peng-Robinson. Ammonia reaction kinetics behaviour for calculation was based on following equation [2] .

$$N_2 + H_2 \leftrightarrow NH_3 \quad (1)$$

$$r = 23.618 \exp\left(-\frac{87{,}090}{RT}\right) \frac{P_{N2} P_{H2}^{1.5}}{P_{NH3}} - 3.39\text{x}10^{13} \exp\left(-\frac{198{,}460}{RT}\right) \frac{P_{NH3}}{P_{H2}^{1.5}} \quad (2)$$

Plug Flow Reactor (PFR) with total volume 30 m^3 was used in the simulation. Quenching process was simulated by using splitter and mixer features of Aspen Hysys® V8.8. Quenching ratio was varied with assumed 1st bed volume at 20 m^3 and 2nd bed at 10 m^3. FEHE simulation was conducted with steady-state-rating mode and sized rigorously by using interactive mode to obtain certain characteristic on heat transfer. FEHE specification sheet is shown in Table 1.

Hysteresis zone identification was conducted by using Adjust feature on Aspen Hysys® V8.8. This process was done by varying the 2nd reactor inlet temperature to adjust the feed temperature on each quenching ratio. This attempt eventually would result plot T Feed vs T Outlet correlation to identify hysteresis zone. Reactant conversion was then identified by using Case Study feature on Aspen Hysys® V8.8. This study was done by varying feed temperature on each quenching ratio or bed volume distribution and observing the mole flow of the ammonia in the outlet stream. At the end, Aspen Hysys® V8.8 Dynamic feature was conducted to observe the process dynamic behaviour at quenching ratio which resulted highest reactant conversion. Eventually, dynamic observation was used to determine starting temperature of oscillation as the percentage of ΔT_f. Dynamic simulation was also used to determine process stability while feed temperature was set at value that resulted highest conversion in chosen quenching ratio.

https://doi.org/10.1051/matecconf/201815603047

Table 1. FEHE Specification Sheet.

Specification	Value	Specification	Value
TEMA	AEL	Shell Diameter (mm)	818.85
Number of Shell per Pass	1	Number of Tube per Shell	872
Number of Shell in Series	1	Tube Pitch (mm)	23.81
Number of Shell in Parallel	2	Tube Layout Angle	Triangl uar (30o)
Tube Passes per Shell	1	Shell Fouling Factor ($^{\circ}C$ m2/kJ)	0.00002 8
Orientation	Horizon tal	Baffle Type	Single
First tube pass flow direction	Counter	Baffle Orientation	Horizon tal
Elevation	0	Baffle Cut (%Height)	20.00
Shell HT coeff. ($kJ/m^2\,^{\circ}C$)	8908	Baffle Spacing (mm)	800.00
Tube HT coeff. ($kJ/m^2\,^{\circ}C$)	12304	Tube OD (mm)	19.050
Overall U ($kJ/m^2\,^{\circ}C$)	2604	Tube ID (mm)	10.668
Overall UA ($kJ/m^2\,^{\circ}C$)	101900	Tube Thickness (mm)	4.191
Shell pressure drop (bar)	0.5855	Tube Length (m)	3.750
Tube pressure drop (bar)	0.1386	Tube Fouling Factor ($^{\circ}C$ m2/kJ)	0.00002 8
Heat transfer area per shell (m^2)	195.7	Themal cond. (W/mK)	45
Tube volume per shell (m^3)	0.2923	Wall Heat Capacity Coeff. (kJ/kg C)	0.473
Shell volume per shell (m^3)	1.043	Wall Density (kg/m3)	7801

3 Results

Unstable process as the result of FEHE utilization in double bed reactor is possible to create instability zone like the one in single bed configuration. Identified instability zone study resulted from varied quenching ratio and bed volume distribution of double bed reactor would be explained in the following section.

3.1 Hysteresis Zone Alteration Due to Varied Quenching Ratio

Fig. 5 shows that generally higher quenching ratio would lead to more narrow hysteresis but higher extinct temperature.

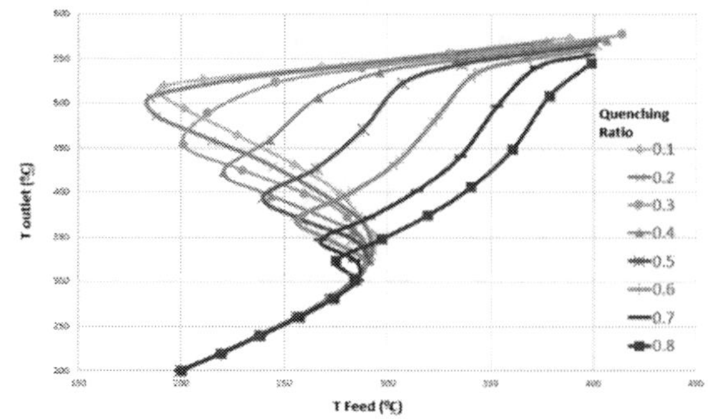

Fig. 5. Hysteresis zone alteration on varied quenching ratio

Higher extinct temperature could possibly lead the reaction to easily extincted and furthermore it would result lower reactant conversion. At bed volume distribution of 2:1, hysteresis zone experienced trend of rise from 0.1 to 0.2 and drop from 0.2 – 0.8. From here, there was an observed reversed trend at 0.2 quenching ratio. This possibly indicates that there was optimum point regarding the process parameter related to feed or outlet temperature. Closest relation to these parameters was conversion profile of the reaction. Reactant conversion was then evaluated from each quenching ratio at steady state simulation as shown in Fig. 6.

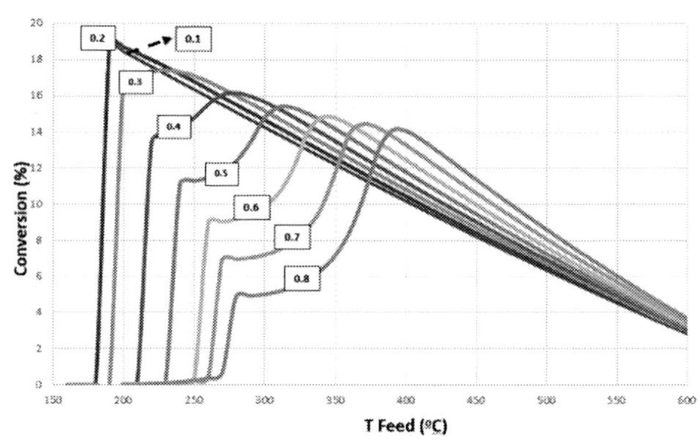

Fig. 6. Conversion profile on varied quenching ratio with 1st bed volume = 20 m^3

Fig. 6 shows that generally lower quenching ratio will increase the conversion and it shows highest overall conversion profile at 0.2 quenching ratio. Optimum point of conversion profile at each quenching ratio was then plotted as shown at Fig. 7.

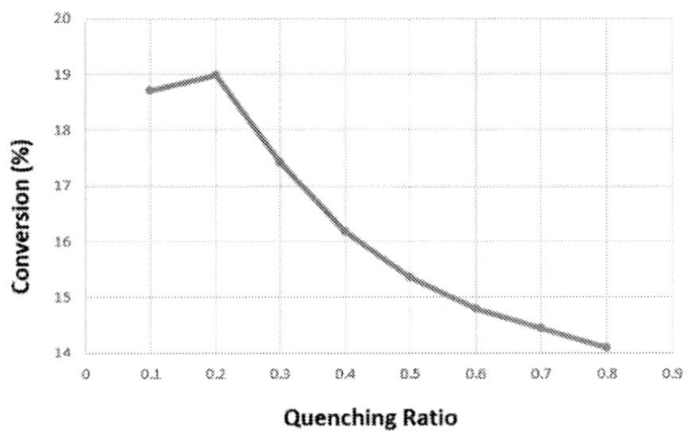

Fig. 7. Highest conversion plot at each quenching ratio

Fig. 7 shows that quenching ratio 0.2 resulted highest conversion which is 18.97% at bed volume distribution 2:1. This is the equilbrium conversion and obtained at T feed = 190°C. However, this temperature was located inside the hysteresis zone and was close to the 0.2 quenching ratio extinct temperature which was 184.69°C. Therefore, dynamic simulation then was conducted to determine the stability behaviour at 190°C as the operation is basically purposed to achieve conversion requirement as high as possible.

3.2 Dynamic Simulation

In this section, Aspen Hysys® V8.8 Dynamic Mode feature was utilized to identify stable/unstable behaviour in real time. From Fig. 5, we know that extinct temperature is 184.69°C at 0.2 quenching ratio. Hence, utilizing Dynamic Mode could possibly show minimum temperature to start hysteresis (as function of ΔT_f = 107.62°C) before the oscillation extinct the reaction. Dynamic Mode was started with ingniting the reaction by set the T feed to 300°C (refer to Fig. 8). T feed was then lowered steppedly to 200°C and then 190°C. Starting from this point, feed temperature was lowered rampedly by 0.1°C to find the hysteresis minimum temperature as accurate as possible. Dynamic simulation showed that oscillation phenomenon started at 186.6°C. This means that at 0.2 quenching ratio and bed volume distribution 2:1, hysterisis mathematically will start approximately at:

$$T\,feed\ hysteresis = T\,extinction + 1.77\%\,(\Delta T_f) \qquad (3)$$

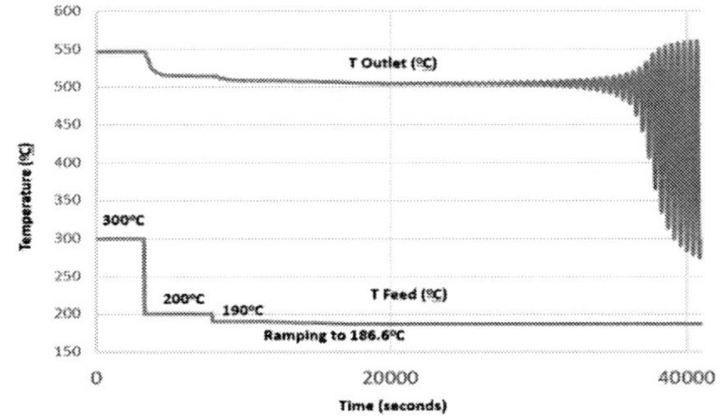

Fig. 8. Minimum hysteresis temperature determination at 0.2 quenching ratio

Highest conversion while applying 0.2 quenching ratio and bed volume distribution 2:1 was obtained at 190°C. Mathematically this point was formulated as:

$$T\,feed\ highest\ conversion = T\,feed + 5\%\,(\Delta T_f) \qquad (4)$$

Therefore, this feed temperature was tested in Dynamic Mode to measure whether the highest conversion could be achieved stably. Result is shown in Fig. 9. Referring to Fig 9, we are able to conclude that highest conversion of the reaction (at bed volume distribution 2:1) could be achieved stably.

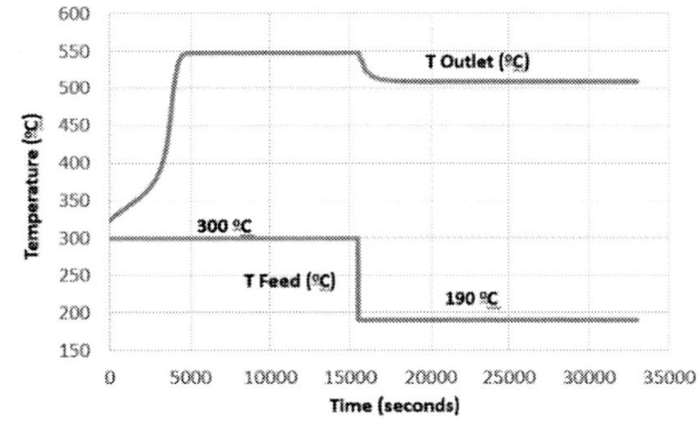

Fig. 9. Stability identification at highest conversion feed temperature (190°C)

4 Conclusion

Quenching ratio alteration could affect how wide the hysteresis area is while plotting T outlet vs T feed. Higher quenching ratio showed narrowing hysteresis yet increase the extinct temperature that could lower the conversion. Highest conversion was obtained at 0.2 quenching ratio (bed volume distribution 2:1) with T feed = 190°C which entering the hysteresis zone. However, this condition was chosen to be analyzed more in the Dynamic Mode to determine whether 190°C would already triggered the oscillation. Dynamic Mode showed that oscillation started at 186.6°C and further extinct the reaction. The highest conversion feed temperature was then tested in Dynamic Mode and resulted stable operation. Therefore, while using bed volume distribution 2:1 (total volume = 30 m^3), ammonia reaction highest conversion of 18.97% (quenching ratio = 0.2) is still able to be obtained at stable operation.

References

1. B. A. Finlayson, ACS Symposium Series, **65**, 1 (1978)
2. G. R. Setyadi & T. P. Adhi, *Dynamic Simulation Adiabatic Extothermal Reactor with addition of FEHE*, Thesis of ITB Chemical Engineering Department (2005)
3. J. C. Morud & S. Skogestad, AIChE.J, **44**, 888-895 (1988)

4. J.R. Rouse & C. Hyde, Materials (Basel) **9**, 1 (2016)
5. P. Laššak, J. Labovský, Ĺ. Jelemenský, Journal of Loss Prevention in the Process Industries, **23**, 2, 280 – 288 (2010).
6. R. K. Herz, *Chemical Reaction Engineering-Part 15-CSTR Thermal Effects*, SS + Dynamic, University of California San Diego, **15,** 5 (2015)

Organic pollutant degradation of tapioca flour industrial waste with photo-fenton reaction

Alin Elita Esther Sarajar[1], *Refa Putri* Ramadhania[1], *and Purwanto* Purwanto[1,2*]

[1]Department of Chemical Engineering, Faculty of Engineering, Universitas Diponegoro, Semarang, Indonesia
[2]School of Postgraduate Studies, Universitas Diponegoro, Semarang, Indonesia

Abstract. Cassava (*Manihot utilissima*) is the main raw material of tapioca flour industry. The tapioca industry waste water still contains high enough BOD, COD, and cyanide, above the quality standard of industrial waste water. Photo-Fenton is one of the most effective alternative methods of degrading organic pollutants. This study aims to determine the effect of Photo-Fenton method on the degradation of organic pollutants such as COD and cyanide. For operating conditions of this study use the addition of concentrations of H_2O_2 and $FeSO_4.7H_2O$ and different UV light intensity. The analysis of COD concentration using SNI method is spectrophotometric. From the research result, it is found that the concentration of COD and cyanide in tapioca waste can be reduced by using Photo-Fenton method. With initial COD content of 6000 mg/L and cyanide level of 0.67 mg/L. And in the operating conditions of the addition of 5 ml/l of H_2O_2, Fe^{2+} 1.25 mg/l, and light intensity of 200 lm, the highest COD and cyanide levels were found to be 900 mg/l for COD and 0.08 mg/l for cyanide. Final levels of COD up to 900 mg/l and cyanide concentrations up to 0.08 mg/l.

1 Introduction

Tapioca industry is an industry that can be managed in many ways. It can be a small scale, medium scale, or even large scale industry. It also has a great potential in penetrating domestic and international market. The greatest domestic demand comes from some parts of Java Island, such as: Bogor, Tasikmalaya, and Indramayu. While the international demand comes from some of ASEAN countries and European countries. As for the industry itself, there are a lot of promising places in Indonesia, they are: Pati, Batang, Temanggung, Wonosobo, and DIY Yogyakarta.

In tapioca production, a lot of water is needed. Each ton of cassava needs 6–9 m3 water. The waste water of this process contains relatively high COD and cyanide. It has 6,000-10,200 mg/l for COD and 0.67 mg/l for cyanide. Based on Ministry of Environment (KLH) decree No. Kep-03/Men-KLH/II/1991, the waste water of the tapioca industry exceeds the standard quality of industrial water waste (which are 400 mg/l for COD and 0.5 mg/l for cyanide). Letting the waste water to a watercourse will reduce the oxygen in the water. As a result, microbes are starting to die and rot, which will pollute the environment with its smell. [1]

There is a poison known as hydro cyan or HCN in tuber flesh and cassava skin [2]. The poison does not stand alone. It is bonded in a glycoside cyanogenic chain. It contains glucose, acetone and HCN. The poison can be reduced, or even eradicated, by producing tapioca flour from the cassava. When the cassava is shredded, cells within the cassava will be broke, therefore the glycoside and HCN will be separated. This kind of method needs a lot of water and the water should be updated frequently. Thus the glycoside and cyanide are disposed as waste.

Fenton's reaction is one of the Advance Oxidation Process (AOPs) and one of the popular waste processing methods[3]. AOPs is a chemical and physical waste processing method which can lower COD and poisonous substance such as phenol and xylene. In addition, AOPs can reduce the levels of organic substance such as pesticide, surfactant, and pharmaceutical waste, which are difficult to be omitted. Whereas Photo-Fenton process is a development of Fenton's Process by adding ultraviolet light to accelerate and increase the formed hydroxyl radical [4].

2 Materials And Methods

2.1. Materials

Tapioca waste water, which is obtained from Tapioca Flour Home Industry in Margoyoso, Pati, Central Java is one of the materials that will be tested. Chemical materials used in this experiments are Fenton reagents such as H_2O_2 analytical grade 30% w/w and $Fe_2SO_4.7H_2O$ analytical grade 99%. Other substances used for pH regulator is H_2SO_4. Self-made COD Digestion Reagent based on Indonesian National Standard is used to analyze COD levels. Another material used in this experiment is 10 watt UV lamp with wavelength of 380-200 nm and its luminous intensity is ±10 lm/watt.

* Corresponding author: p.purwanto@gmail.com

2.2. Methods

All experiments concerning the application of the ultraviolet radiation were performed in a batch reactor. The schematic diagram of the experimental set-up used for these processes is shown in Fig. 1.

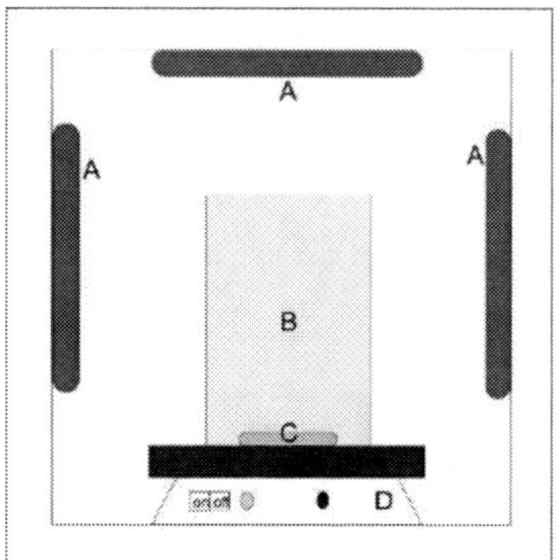

Fig. 1. Series of Experiments Tools. A-UV Lamp, B-Reaction Space (Beaker Glass), C-Magnetic Stirrer, D-Hot Plat Magnetic Stirrer.

Fig. 1 Reactor was cylindrical filled with 0.85 litre of wastewater and was made from quartz glass (used beaker glass) which was available for the transfer of the radiation. Irradiation was achieved by using UV lamp (10 W with wavelength of 200-280 nm and its luminous intensity is ±10 lm/watt) which is placed above or beside the reactor. For Photo-Fenton processes, the reactor was filled with 0.85 liter of wastewater. The desired pH value was adjusted with sulfuric acid before start-up, and then a given weight of iron salt was added. The iron salt was mixed very well with the wastewater before addition of a given volume of hydrogen peroxide. For Photo-Fenton the time at which the UV lamp was turned on was considered the zero time, or the beginning of the experiment that was taking place simultaneously with injection of hydrogen peroxide and adding the dose of iron salt. The amount of reagent $FeSO_4.7H_2O$ and H_2O_2 designed for the experiment was added and homogenized by the magnetic stirrer **[5]**. This experiment investigated the effect of H2O2, Fe2SO4, and light intensity to the concentration of COD and cyanide in tapioca wastewater.

3 Results and Discussions

3.1. The effect of initial H_2O_2 concentration on the degradation of COD concentration

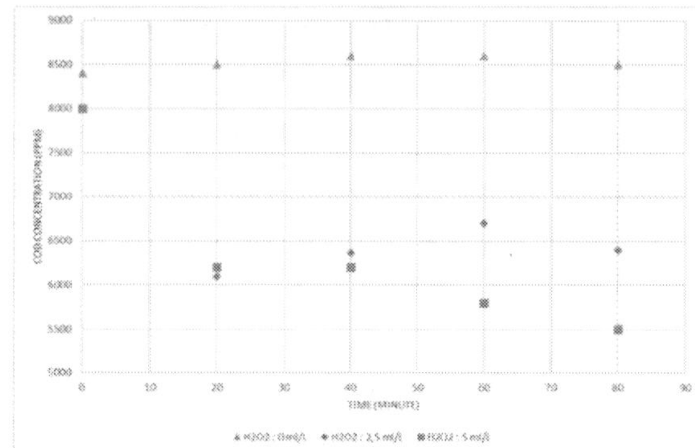

Fig. 2. COD concentration with various addition of hydrogen peroxide (H_2O_2) graphic

Fig.2 shows a decrease of COD over time. From all of the treatments, the highest percentage of RCOD in the concentration of H_2O_2 is 275ppm is 56.25%, which happened in 80-minute treatment. Based on the graphic above, by using the 5 ml/L H_2O_2 the COD level can be reduced, which it appears that the longer the reaction time, the greater the reduction. It means that the Photo-Fenton process can oxidize the organic compound in tapioca waste water.

Chemical oxidation can be defined as a process in which electrons move from one substance to the other [6]. The direction of electron's movement is determined by the oxidation potential. Oxidation potential is an oxidation power which come from oxidation reduction reaction, and it is represented in volt to normalize hydrogen electrode. Besides, hydrogen peroxide (H2O2) is a chemical oxidizing agent with the oxidation potential of 1.77 V which greater than permanganate (1.67 V) and chlorine (1.36 V) [7]. Thus, the oxidation power of hydrogen peroxide can break down organic compounds to simple compounds. The result of the research, the greater the concentration of H2O2 the greater the reduction of COD (%RCOD) is, confirms the theory. However, excessive concentration of H2O2 can make H2O2 reacts with hydroxyl radical which produce perhydroxyl radical (·OOH) that has a lot less oxidation power than hydroxyl radical [8].

3.2. The effect of Fe^{2+} weight ratio on the degradation of COD concentration

The effectivity of Fenton reagents is needed to be observed in terms of Fe2+ concentration. Iron salt such as Fe2+ works as catalyst in Fenton process to convert H2O2 to reactive hydroxyl radical. This phenomenon is shown on Fig. 3.

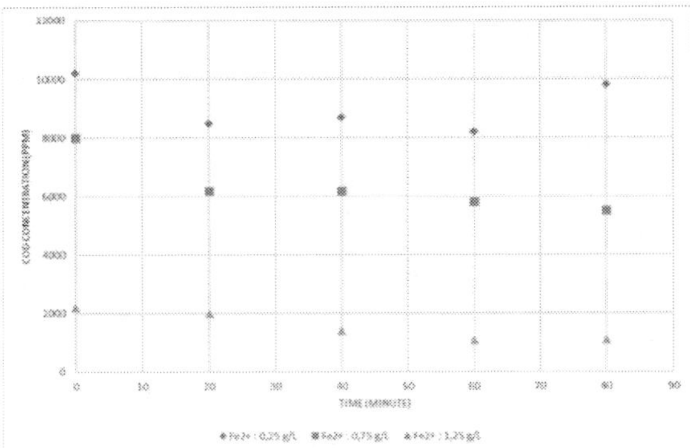

Fig. 3. COD Concentration with various addition of ferrous sulphate catalyst ($Fe_2SO_4.7H_2O$) graphic

Fig.3 shows that the Fe^{2+} concentration of 1.25 g/l can reduce COD concentration by 50%, which is the highest percentage occurred in the experiment. The greater the ferrous sulphate concentration, the more oxidation of pollutants occurred, which is shown by the bigger decrease of liquid waste's COD level. This event happened due to the increasing number of hydroxyl radical formed [9].

Hydroxyl radicals (HO.) react swiftly and nonselective with almost all of electron-rich organic compound which can eradicate organic chemicals effectively. These radicals can mineralize some of the substrates, for instance: it can reduce substrates to CO and H2O [10].

3.3. The effect of luminous intensity on the degradation of COD concentration

Based on Fig. 4 shown above, using 2 UV lamps is proved to reduce the COD concentration with the highest percentage of 70%. There is not much difference between using 3 lamps and using 2 lamps. Hence, in order to reduce the COD concentration of tapioca waste, using 2 UV lamps are enough.

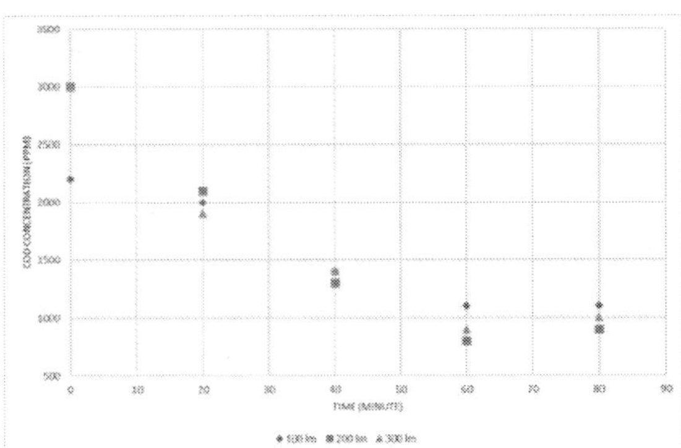

Fig. 4. COD Concentration with various intensity of UV lamp radiation graphic

The function of the combination between UV and oxidation,such as H2O2, or catalyst is to produce hydroxyl radical (·OH) [11]. As shown in the equation (2), a free radical has a very high potential (2.8 V). This radical is the one used to perfectly degrade gas organic pollutants such as CO2 and H2O. The degradation process is done by breaking bonds between atoms in dye molecules so that the molecules can be smaller, colorless and safe. In addition, a point in UV lamp's wavelength (λ = 254 m) will be effective in killing bacteria.

Free radical formation rate in Fenton photo process is 6 times greater than in Fenton process. It means that there is a good catalytic activity and better formation of free radical generation (OH^-) in Fenton photo process.

3.4. The effect of Photo-Fenton process on the degradation of cyanide

The test of cyanide content in tapioca liquid waste water was processed in optimal variable. At first, the cyanide content in the liquid waste is 0.67 mg/l, then after processed by using Photo-Fenton methods, the cyanide content in the liquid waste is 0.08 mg/l. Thereby, the Photo-Fenton methods decreased the cyanide content until 88.06%.

The reason that the reduction seems significant is because the cyanide has reacted with hydrogen peroxide to form ion cyanate.

$$H_2O_2 + CN^- \xrightarrow{Fe^{2+}} OCN^- + H_2O$$

As shown in the reaction above, hydrogen peroxide forms cyanate and water. It is a process that limit the forming of dissolved solids in a solution which being treated. Increasing the addition of hydrogen peroxide will make the greater amount of cyanide which will react to form ion cyanate. Thus, the free cyanide content in the liquid waste will be reduced. This condition occurred when there is a balance between the amount of hydrogen peroxide and the amount of cyanide in the liquid waste [12].

3.5. Reaction Kinetics

Kinetics of COD degradation in tapioca waste water is modelled by first order reaction. The relation between reaction time of Photo-Fenton and $\ln(C_0/C)$ in tapioca liquid waste is shown in Fig. 5, 6, and 7.

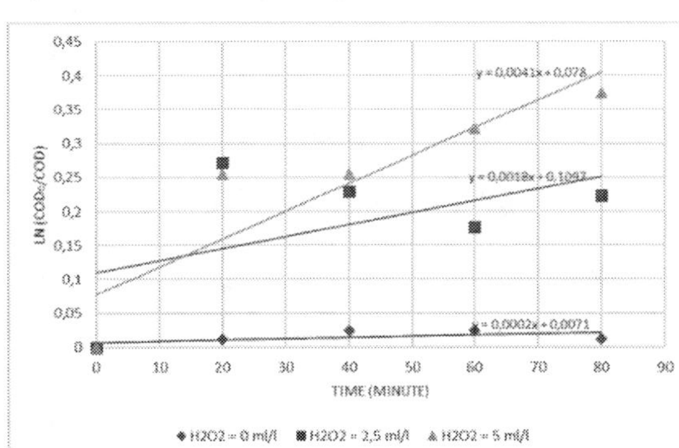

Fig. 5. The relation between reaction rate and ln (C_0/C) in various concentration addition of H_2O_2 graphic

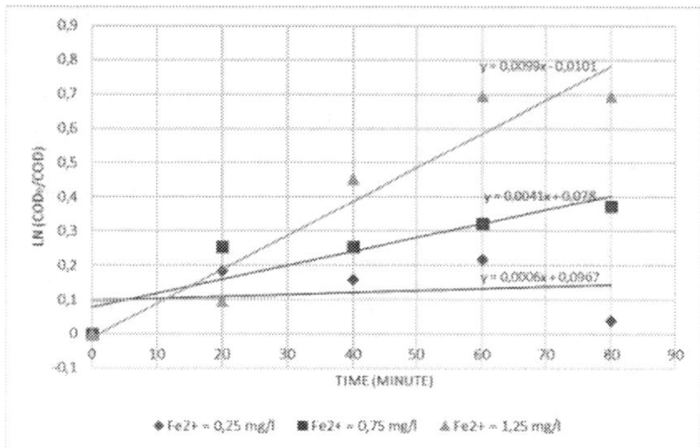

Fig. 6. The relation between reaction rate and ln (C_0/C) in various concentration addition of Fe^{2+} graphic

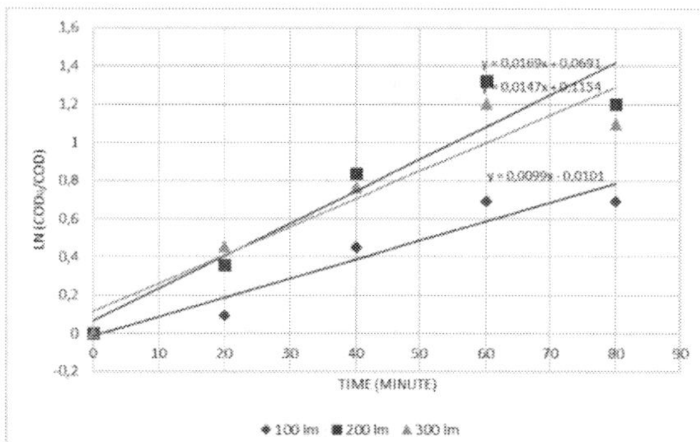

Fig. 7. The relation between reaction rate and ln (C_0/C) in various concentration addition of UV lamp radiation graphic

The rate constant of the Photo-Fenton reaction can be obtained from the graphs in the three figures above through the slope of the linear equation. The constants in each reaction system are shown in the Table below.

Table 1. C value in various operation condition in processing tapioca liquid waste using Photo-Fenton process

Run	Operation Condition			C value (men^{-1})
	C_{H2O2} (ml/l)	C_{FeSO4} (mg/l)	Luminous intensity (lm)	
	0	0.75	100	0.0002
2	2.5	0.75	100	0.0018
3	5	0.75	100	0.0041
4	5	0.25	100	0.0006
5	5	0.75	100	0.0041
6	5	1.25	100	0.0099
7	5	1.25	100	0.0099
8	5	1.25	200	0.0169
9	5	1.25	300	0.0147

As shown in the Table above, the highest constant is 0.0169 men^{-1} which is obtained in the experiment with the addition of 5ml/l H_2O_2, 1.25 mg/l Fe^{2+}, and 200 lm luminous intensity of UV lamp. By adding UV lamp luminous intensity to 300 lm, the constant value is 0.0147 men^{-1}. The lowest constant value occurred on the experiment which has the addition of 0.75 mg/l Fe^{2+} and 100 lm UV lamp luminous intensity.

4 Conclusion

The concentration of COD in tapioca waste water decreased from 6,000 – 10,200 mg/l to 900 mg/l by using Photo-Fenton reaction. Meanwhile the concentration of cyanide diminished by 88.06% from 0.67 mg/l to 0.08 mg/l.

References

1. L.J. Betty Sri, P.R. Winiati, *Waste Management of Food Industry, Food and Nutrition Laboratory,* Thesis of Institut Pertanian Bogor (1993)
2. Lies. Suprapti, *Tapioca Flour: Making and Utilizing* (Kanisius, Yogyakarta, **ISBN 979-21-0854-8**, 2005)
3. L. Zhou, J. Hu, X. Li, Chemical Engineering Research and Design, **90**, 377-382, (2012)
4. M. Tokumura, H. T. Znad, Y. Kawase, Water Research **42,** lB, 4665-4673, (2006)
5. E.E. Ebrahiem, M.N. Al-Maghrabi, A.R. Mobarki, Arabic Journal of Chemistry (2013)
6. M. Rodriquez, Thesis of Master Program, University of Bercelona, Departement of Engineering, (2003)
7. H. B. Hadjltaief, A. Sdiri, W. Ltaief, P.D. Costa, M.E. Gàlvez, M.B. Zina, Comptes Rendus Chimie **xxx** 1-10, (2017)
8. M. Ahmadi, F. Vahabzadeh, B. Bonakdarpour, E. Mofarrah, M. Mehranian, Journal of Hazardous Materials, **B123**,. 187 -195 (2005)
9. G.N. Achilleand L. Yilian, Journal of American Science **6**, 4: 58-66, (2010)
10. J. M. P. Hernández, C. A. M. Huitle, J. L. G. Mar and A. H. Ramírez, Journal of Environmental Engineering Management **19**, 5, 257-265, (2009)
11. S. S. Hutagalung, Jurnal Teknologi Pengolahan Limbah **13**. No. 2, (2010)
12. M.M. Botz, Mining Journal, 28-30 (2001)

Biogas Productivity as Renewable Energy and Performance of Waste Water Treatment in Tofu Small Scale Industry using an AnSBR Reactor

Suparni Setyowati Rahayu[1*], *Totok Prasetyo*[1], *Purwanto Purwanto*[2,3], *Budiyono Budiyono*[2,3]

[1]Mechanical Engineering Department, Politeknik Negeri Semarang, Semarang Indonesia
[2]Doctoral Program of Environmental Science, School of Postgraduate Studies, Universitas Diponegoro, Semarang Indonesia
[3]Department of Chemical Engineering, Faculty of Engineering, Universitas Diponegoro, Semarang Indonesia

Abstract. A research that analyzed the relationship between biogas production as renewable energy and performance of waste water treatment in tofu small industry using AnSBR reactor has been conducted. This research used system operation method in batches in to get a description of how the biogas production system works as it was operated in the batch mode, which is applied in tofu industry. An instrument capable of operation in batch mode using AnSBR reactor was designed as it does not require sludge re-circulation. An AnSBR reactor includes five stages of; filling, reaction, settling, decanting, and stabilization (idling state). The research was run for 24 days using biogas collected in a tank. Biogas was measured for each HRT starting on the second day HRT to 24th day HRT. Biogas characteristics were monitored on day 16th for contents of CH_4, CO_2 and heat value. Results show that CH_4 content in the biogas amounts to 78.26 %, while CO_2 comprises 20.16 %. Meanwhile, the heat value, which depends on acid methane and CO_2 content, is 4,246.55 kkal/kg.

1 Introduction

Application of science and technology for waste water treatment will minimize negative effects of this waste that environmental management can be carried out properly. In order to endorse environmentally friendly sustainable development, the science and technology of waste water treatment needs to be disseminated in the education community, within the general public, and also in business and industry, including the tofu industry.

Tofu waste water can be treated with anaerobic biological process to produce biogas as its inherent characteristic of high concentration biodegradable organic compounds of protein, adequate alkalinity, phosphor, nitrogen and micronutrients suitable for microorganism growth makes anaerobic process the most suitable for its treatment. High content of organic compound is a high energy source in the form of methane that will allow anaerobic bacteria to survive for a long time without the presence of food/substrates. These makes waste water treatment in small scale tofu industry a possibility [1].

An efficient bioreactor should have adequate biomass idling time and facilitate even mixing for contact between cells and their substrates to take place. AnSBR can meet this requirement. Microorganisms in an AnSBR reactor tend to be lifted and re-deposited due to the formation of gas in the process [2].

The most significant advantage of an AnSBR reactor is its ability to separate acid-genesis and methane-genesis longitudinally in its lower part. This facilitates proper growth conditions for different microorganism groups [3].

Other than that, constructing this type of reactor is simple and cheap. There are neither moving parts nor mechanical mixing that bacteria loss in the reactor is small and the outflow is relatively free of biomass solids. This is important as hydrodynamics and mixing level in the reactor affect contact between substrates and microorganisms, and hence influences mass control transfer and reactor performance [4].

It has been mentioned earlier that an AnSBR is simple and easy to construct as there is no need for moving parts or a mixing machine. The design itself helps separate acid-genesis and methane-genesis processes that significant advantages are already inherent. This reactor functions like a two phase system without the need for high cost and does not cause control issues [5].

The performance of an AnSBR reactor in separating COD is around 65-95% and maturation process has to be observed for the first 3 months. Overall, the advantages or an AnSBR reactor compared to any other system are; simple design, no moving parts, nor need for mechanical mixing, inexpensive, low risk of clogging, and low risk of sludge bed expansion, and low operational cost [6].

In terms of the biomass, solid retention time in this reactor is long without the need for media or precipitation room. It takes some time to form mud and resulting mud itself is also only a little. This mud lasts for quite a while that frequent dumping is not necessary, and this reactor does not require microorganisms with certain precipitation abilities [7].

AnSBR is a bioreactor that is both simple and easy to operate with little organic and shock loading hydraulics

* Corresponding author: suparnirahayu@yahoo.co.id

only slightly affect processing efficiency. Considering the benefits of this type of reactor, its use for tofu waste water treatment should be endorsed [8].

The disadvantages of this reactor, as reported by some researchers, include sensitivities to high load anaerobic processing. The other drawbacks, as is the case with the other biological processing, are the need for long Hydraulic Retention Time (HRT), meaning high reactor volume, high concentration of biomass, and the need for biomass wash-out control off the reactor. Therefore, research on anaerobic tofu waste water treatment has to deal with finding the most effective way of holding anaerobic microorganism in the reactor [9].

Tofu waste water treatment in an AnSBR reactor makes use of sewage mud. This mud serves as biological suspension or active microbe mass that degrades dissolved organic materials. This technique takes advantage of microbes that degrade complex organic compounds into stable substances, and hence reduces waste COD down to around 70-95% [10].

Sewage mud can also metabolize and breakdown pollution materials contained in waste water. This mud is an undissolved material; it typically consists of organic fibers rich in celluloses and living microorganisms. Therefore, there is also an interest in figuring out the most effective mud composition for treating tofu waste water, to allow using the mud as active mud [11].

Sewage mud consists of microorganisms in adequate amount to also serve as microorganism germ or oxidation agent in biological treatment of waste. Mud to be used as microorganism germ only comes from sewage of the tofu industry. Hence, development of biomass into active mud cannot be separated from the environment of biomass source. On the other hand, microorganisms coming from polluted waters are usually of high organic content of heterotopic organisms [12].

Adequate supply of food is also needed for microbes to proliferate. This agrees with an earlier research stating that longer germination results in higher microbe load as microbes split and multiply in numbers. Microbes need some time to proliferate, but when required components are available they can multiply fast and hence, does more oxidation on organic compound in tofu waste water. Therefore, the use of sewage mud has great potential to reduce the amount of pollutant in tofu industry [13].

The process in an AnSBR reactor can be categorized as reaction-suspended growth and separation in a single tank. The critical aspect of this process is sludge precipitation rate during precipitation period prior to effluent decanting [14].

2 Method

The waste water used in this research is small scale tofu industry waste. Sample was taken during the working hours from sewage inlet, as these working hours produces maximum quantities of waste water and in terms of quantity, the organic content is highest.

Collected waste water was then added with CaCO3 to raise its pH to an optimum pH level to facilitate anaerobic processes. CaCO3 addition took into account neutralization calculation that will result in pH reaching 7, which is in the rage of optimum pH level of 6.8-7.4.

This stage of research was aimed at understanding the effect of Hydraulic Retention Time (HRT) on the performance of AnSBR bio digester operated in batch mode for biogas production.

Batch mode was chosen in order to get proper description of biogas production system performance, because batch mode is actually operated in tofu making. The research started with designing an instrument capable of operating in batch mode. Some considerations for this design include: (1) AnSBR digester allows solids to be either suspended and floating that it allows effective contact between substrates and microbes; (2) the system comes with an automatic temperature control to ensure control operation at optimum temperature; (3) it is also equipped with a biogas pressure gauge to monitor biogas pressure to prevent damage to the instrument at certain pressure levels; (4) the system also has a microbe inoculum from the sewage. This is one specific factor for biogas production from tofu waste water, the other specific factor being microbe concentration in the digester.

COD and TSS analyses were carried out for each time interval of AnSBR bio digester operation. Analysis for gas composition was conducted at the end of AnSBR bio digester operation period. The effects of sewage mud, COD, TSS were then observed for maximum performance. Optimum conditions for factors of inoculum, HRT and TSS will be optimized and further enhanced for the long terms use of this system.

3 Result and Discussion

When the system was running, TSS concentration was also measured to find out the effect of HRT on effluent quality, based on TSS output concentration during system running.

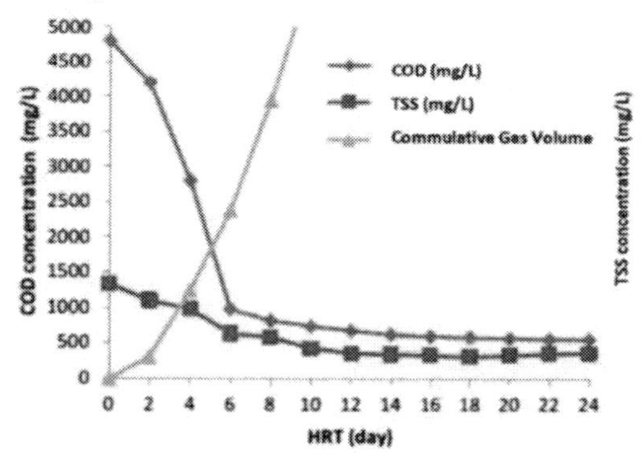

Fig. 1. Decreasing of TSS and COD concentration during system running.

Decreasing TSS concentration measured in each HRT effluent comes along with less COD concentration. This is evident in Figure 1, in which decreasing TSS concentration is in line with lowering COD. Except for the 20th day to the 24th day, TSS concentration is rising.

This increasing TSS is due to the fact that some bacteria were already in dying phase. This dying phase is also marked with COD concentration that is relatively constant. Constant COD concentration on day 20 to day 24 is caused by less substrate available that comes with dead bacteria.

Figure 1 and some earlier analyses show that optimum HRT in TSS separation was obtained on day 16. TSS concentration on the 16th day is still above the standard limit set by Regional Regulation of Central Java Province No. 5 year 2012. Nonetheless, physical conditions have been improved as can be seen with lower TSS concentration in effluent.

System running carried out for 24 days produced biogas stored in a tank. Biogas was measured on each HRT started from HRT 2, onwards to HRT 24. On the 10th day, 5815 mL of biogas was obtained, and it kept on increasing right to HRT 24. This is a significant increase. On day 14, biogas production reached its maximum. And on day 16 onwards, biogas production lowered. Even though day 16 resulted in a biogas volume of 11225 m, biogas production kept on decreasing right to HRT 24. Nonetheless, cumulative volume of biogas production seems to be constant. Biogas production on day 22 to day 24 still lowered, but cumulatively, it has always looked constant since day 16. Therefore, day 16 was an optimum production time.

Constant biogas volume shows that most of the acid substrates have been converted to produce methane. Hence, the phase of optimum HRT has passed. This is in line with Metcalf & Eddy (2003) who stated that 72% of biogas produced comes from substrates, while the remaining figure is from CO2 and H2 substrates.

During system running, characteristic testing for biogas produced up to day 16 include contents of CH4, CO2 and biogas heat value. Examinations show that CH4 content in the biogas amounts to 78.26 %, whereas CO2 content reaches 20.16 %.

Biogas heat value after examination is 4,246.55 kkal/kg. This value depends on the amount of methane acid and CO2 in it. The more the methane content, the higher the heat value is.

4 Conclusion

Some variables that affect the process of biogas production as renewable energy includes 20% sewage mud. The maximum condition for biogas production in an AnSBR bio digester is reached on day 16, with COD concentration of 87.51% and TSS concentration of 75.37%.

Biogas produced on HRT 16 contains 78.26 % CH4 and 20.16 % CO2 and are produced with an AnSBR bio digester operating in batch mode involving states of filling, reaction, precipitation, decanting, and stabilization. Meanwhile, biogas production using conventional bio digester only yields biogas composed of 72.16 % CH4 and 18.12 % CO2.

The researcher wishes to thank DRPM RISTEKDIKTI for the funding of this research, State Polytechnic of Semarang and Diponegoro University for facilitating this cooperative research.

References

1. Alphenaar, P. A., A. Visser, and G. Lettinga Biosource Technology, **43**, 249 (2013)
2. Azimi, A. A. and Zamanzadeh, M, Int. J. Environ. Sci. Tech. **I**, 1, 51-57, Spring (2014)
3. Carucci,G.,F.Carrasco,K.Trifoni,M.Majone, and M.Beccari,. J.Environ. Eng. **131** ,7, 1037-1045 (2015)
4. Demirer,G.N.and S.Chen, Journal of Chemical Technology and Biotechnology, **79**, 1381-1387 (2014)
5. El-Sheikh, Mahmoud A., Hazem I. Saleh, Joeseph R. Flora, Mahmoud R. AbdEl-Ghany. Desalination **276**, 253 -259 (2011)
6. George Nakla, Andre Lugowski, Jounika Patel, Bioresource Technology, **97**, 1, 1-14 (2011)
7. Hang – Qing Yu, Zhen Hu. Hu, Tan – Qiu Hong, Guo – Wei Gu, Bio. Chemistry **38**, 507 – 513 (2008)
8. Pakarinen, 0., A. Lehtomaki, S.Rissanen, J. Rintala, Bioresour. Technol. **99**, 15, 7074-7082 (2013)
9. Pozo, R.D., V. Diez, S. Beltran, Bioresour. Technol. **71**, 2, 143-149 (2015)
10. Rajeshwari, K.V., M. Balakrishnan, A. Kansal, K. Lata, V.V.N. Kishore, Renewable and Sustainable Energy Reviews, **4**, 2, 135-156 (2013)
11. Sarapatka, B, Biomass and Bioenergy, **5**, 5, 387-393 (2013)
12. Van Haandel, A,. Rev. Environ. Sci. Bio Technol, 5-21 (2011)
13. Vavilin,V.A.andI.Angelidaki, Biotechnology and Bioengineering **89** ,1, 113-122 (2015)
14. Yasar, Abdullah dan Tabinda, Amtul Bari,. Polish J. of Environmental Study, **19**, 5 (2010)

Corrosion Inhibitor of Carbon Steel from Onion Peel Extract

Asep Muhammad Samsudin[1] , *Aribella* Samudra Pamungkas[1*], and *Ratih* Estu Nugraheni[1]

[1]Chemical Engineering, Engineering Faculty, Diponegoro University, Indonesia.

Abstract. Carbon steels composed by two main elements, they are iron (Fe) and carbon (C) elements which widely used in industrial because of its resistance and more affordable than stainless steel, but their weakness is they have low corrosion resistance. One way to modify carbon steel is by coating them with antioxidant compounds that can delay, slow down, and prevent lipid oxidation process, which obtained from onion peel extract. Several studies on corrosion inhibitors have been performed. However, the efficiency was not reach the optimum. This study aims to examine the effect of onion peel extract concentration on the efficiency of corrosion inhibitor and characterization of the green corrosion inhibitor from onion peel extract. This research method begins by extracting onion peel to 200 ml solvent which we use aquadest and methanol and mixed with 5 grams of crushed onion peel, then let them be extracted for 60 minutes with room temperature. Once it was filtered and the solution obtained, followed by evaporating process with rotary evaporator to decrease the content of solvent. The product is ready to be used as a green corrosion inhibitor of carbon steel in 1 mol/L HCl. While the analysis used is HPLC qualitative analysis, and electroplatting process. The impedance is measured at a frequency of 100 kHz to 4 mHz with an AC current of 10mV. Inhibitor concentrations are vary between 2 ml and 4 ml of onion peel extract. Electroplatting is done within 30 minutes with 10 minutes each checking time. Furthermore, quantitative analysis was done for the analysis of corrosion rate and weight loss. Based on HPLC analysis, it is known that the extract of onion peel contains 1mg/L of quercetin, which is belong to flavonoid group as green inhibitor. While electroplatting process, aquadest solvent having average efficiency of 99,57% for 2 ml of extract, and 99,60% for 4 ml of extract. Methanol solvent having average efficiency of 99,52% for 2 ml of extract and 99,74% for 4 ml of extract. In line, methanol have more possibilities to gain flavonoid out to act as green inhibitor than aquadest due to its polarity. It is suggested to cover up the extract of onion peel to avoid them against sunlight, to keep the flavonoid not damaged.

1 Introduction

Metal element is one of the basic materials that can not be separated from human life. The metal whom most commonly used is carbon steel with the main content of iron (Fe) and carbon (C). The advantages of carbon steel are more affordable than stainless steel or other steel, but have weaknesses such as not corrosion resistant. The Gross National Product (GNP) survey shows in 2001 corrosion losses could exceed $600 billion. Therefore it is necessary to do special treatment to reduce corrosion, including by coating carbon steel with corrosion inhibitor.

Corrosion inhibitors are chemicals that can decrease the corrosion rate of metals in contact with the environment. The corrosion inhibitor is divided into organic and inorganic inhibitors. Synthetic inhibitors come from hazardous chemicals, they are quite expensive, and are not environmentally friendly. So it is necessary another inhibitor that is safe, easy to obtain, biodegradable, cheap cost, and environmentally friendly [1].

Corrosion inhibitors should have antioxidant compounds that can donate one or more electrons to the prooxidant compound, then convert the oxidant compound into a more stable compound [2]. This compound can be found in nuts, vegetables, fruits, chocolate and tea [3].

onions peel contain active substances such as Allin, Flavonoid, Saponin, Petrin, and Allisin. Flavonols contained in onion peel are 270-1187 mg / kg, indicating that the flavonoid content is greater than garlic skin (89.3-101 mg / kg) and red onion skin (98 mg / kg) [4]. Based on BPS and the Directorate of Horticulture, on 2015 the production of onions in Indonesia reaches about 10,000 tons, where the waste of leather is 10-15% or about 1000-1500 tons of onion peel every year has not been fully utilized.

2 Methodology

2.1 Material

Onion peel, aquadest, hydrogen chloride (HCl), carbon steel, methanol.

2.2 Inhibitor Preparation

200 ml of aquades mixed with 5 grams of onion peel that has been destroyed. Let stand 60 minutes, then separated between the pulp and solvent. Reduce the solvent with rotary evaporator. Perform this step again on methanol solvent.

* Corresponding author: ariesamudra76@yahoo.com

2.3 Material Preparation

Cut carbon steel of the same size (5 cm x 1 cm). Then sanding carbon steel to remove impurities.

2.4 Solution Preparation

Mix 1 mol/L HCl on 100 ml base. Then make an electrolyte solution with an onion peel concentration of 2 ml and 4 ml for each solvent.

2.5 Electroplating Analysis

Calculate weight of carbon steel, clean carbon steel surface, then do electroplating process. In this process, carbon steel is cathode pole, while zinc is anode pole. The electroplating process is carried out at an AC current of 10 mV.

2.6 Efficiency Analysis

Calculate weight of carbon steel before electroplating. Also calculate weight of carbon steel after electroplating for 10 minutes, 20 minutes, and 30 minutes. To calculate the efficiency of inhibitor corrosion can use formula of down below:

$$n = \frac{wo-w}{wo} \text{ x } 100\% \quad (1)$$

Where, n = efficiency (%), Wo = weight of carbon steel before electoplating (gr), and W = weight of carbon steel after electoplating (gr)

2.7 Morphological Analysis

Morphological analysis was performed using HPLC to determine onion peel content.

2.8 Analysis of Corrosion Rate

Calculate different of weight carbon steel and surface area of the electroplating. To calculate corrosion rate use formula:

$$\text{mpy} = \frac{534.W}{D.A.t} \quad (2)$$

Where, W = weight loss (mg), D = carbon steel density (gr/cm^3), A = surface area (in^2), and t = electroplating time (hr)

3 Results And Discussion

3.1 Inhibitors Efficiency

By electroplatting process for around 60 minutes with 1 mol/L HCl of solution, it is shown efficiency :

Table 1. Inhibitors Efficiency

t (time)	without inhibitor	Inhibitors			
		2 ml aquadest	2 ml methanol	4 ml aquadest	4 ml methanol
0	100	100	100	100	100
10	99.74	99.66	99.60	99.66	99.89
20	99.47	99.47	99.36	99.50	99.71
30	99.15	99.16	99.14	99.23	99.37
Average Efficiency					
	99.59	99.57	99.52	99.60	99.74

Table 2. Corrossion Rate (mpy)

Without Inhibitor (mpy)	Inhibitors			
	2ml aquadest	2ml methanol	4ml aquadest	4ml methanol
1,13E-05	1,12E-05	1,17E-05	1,08E-05	8,88E-06

Two table above, it can be seen the value of the efficiency of corrosion rate with and without inhibitors of onion peel extract with various solvents. Efficiency is obtained from the calculation of w / wo of the electroplated metal. The calculation results show that the average efficiency value using aquadest and methanol solvent inhibitors of 4 ml each yields the highest efficiency. This is due to the presence of flavonoid content of the brown onion peel. According to previous research, flavonoids are most commonly found in plant tissues that act as antioxidants [5]. This antioxidative activity of flavonoids stems from the ability to donate its hydrogen atom or through its ability to choke the metal. In addition, the potential as a corrosion inhibitor because it has a pair of free electrons. these free electron pairs will later be adsorbed on the metal surface so as to protect the metal from corrosion attack [6].

3.2 Solvent Used in Extraction

The solvent effect can be shown in fig 1 below. In an inhibitor with a 2 ml variable of methanol solvent and aquadest, the aquadest solvent provides better performance. While the 4 ml variable, methanol give better performance. The components carried on the extraction process are polarized components according to the solvent, so that the type of solvent used may affect the amount of yield produced [7]. Flavonoids are less polar so dissolve in less polar solvents such as methanol and water. Methanol solvent has a small molecular structure capable of penetrating all plant tissues to attract active compounds out [8].

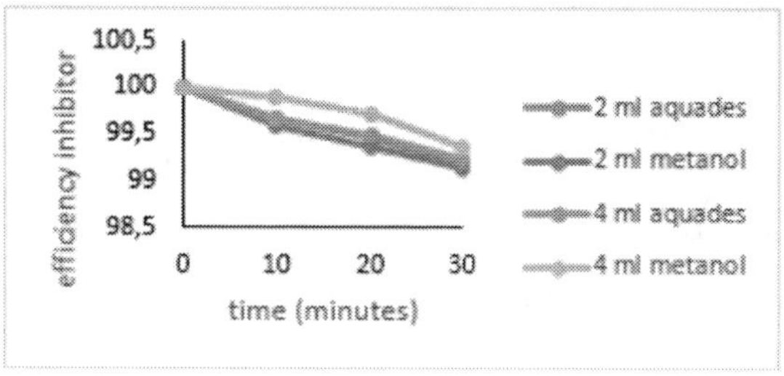

Fig 1. Solvent Extraction Effect

3.3 HPLC Analysis

HPLC Analysis recorded with methanol solvent of brown onion peel extract shows in Figure 2 and Figure 3. HPLC analysis propose to know the concentration of flavonoid in brown onion peel extract. Using the 100 ppm of quercetin standard to get the standard curve as a comparison. The quercetin standard detected at 2,44 s and the brown onion peel extract have quercetin at 2,55 s. Comparing to each area, then get the contration of quercetin in brown onion peel extract is 1 mg/L.

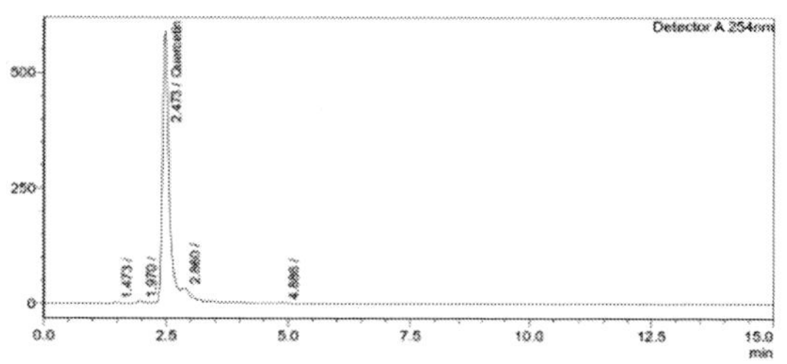

Fig 2. Quercetin Standard

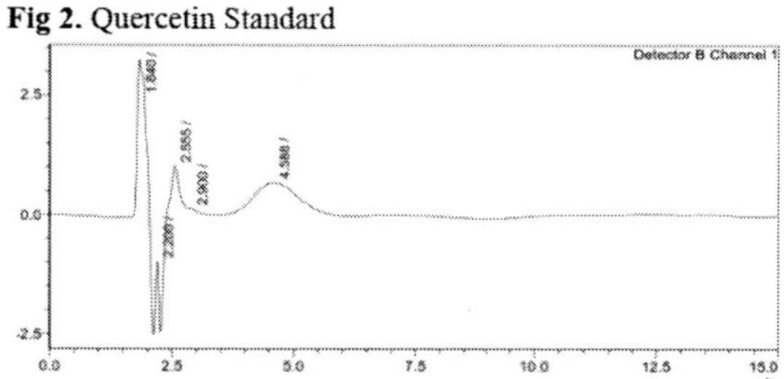

Fig 3. Brown Onion Peel Extract

4 Conclusion

Brown onion peel extract shows good efficiency to be green inhibitors in 1mol/L HCl :

1. In electroplatting process, it is proven that methanol is better than auadest in extracting brown onion due to its polarity.
2. Methanol solvent shows best efficiency with 99,74% in 30 minutes of 1 mol/L HCl solution.
3. HPLC shown that brown onion peel extract contains 1mg/L quercetin which belong to flavonoid group.

Reference

1. Haryono, Gogot. *Ekstrak Bahan Alam sebagai Inhibitor Korosi*. Thesis of UPN "Veteran" Yogyakarta: Yogyakarta (2010)
2. Winarsi, Hery, *Isoflavon Berbagai Sumber, Sifat dan Manfaatnya Pada Penyakit Degeneratif*, (Gadjah Mada University Press, Yogyakarta , 36–43, 2005)
3. Sayuti, Kesuma dan Rina Yenrina. *Antioksidan Alami dan Sintetik.* Thesis of Andalas University Press: Padang (2015)
4. Fossen, Torgils dan Ingunn Molund Vagen . *Onions: A Source of Unique Dietary Flavonoids*. Thesis of University of Bergen (2016)
5. Abdi Redha, *Flavonoid: Structure, Characteristic Antioxidative and the Function in Biologycal System..* Thesis of Pertanian Politeknik Negeri Pontianak (2010)
6. P. B. Raja, M. G. Sethuraman, Materials Letters, **62**, 1, 113–116, 2008.
7. Markham KR. *Distribution of flavonoids in the lower plants and its evolutionary significance.* (Chapman & Hall, London, 427-464, 1988)
8. M. H. M. Y. Yuswan, J. R. Al-Obaidi, A. Rahayu, S. Sahidan, F. Shazrul, D. Fauzi, Advances in Bioscience and Biotechnology, **6**, 4, 320–329 (2015)

Extraction of Phytochemical Compounds from *Eucheuma cottonii* and *Gracilaria sp* using Supercritical CO_2 Followed by Subcritical Water

Dwi Setyorini[1], *Ridlo* Aanisah[1], *Siti* Machmudah[1,*], *Sugeng* Winardi[1], Wahyudiono[2], *Hideki* Kanda[2], *Motonobu* Goto[2]

[1]Department of Chemical Engineering, Institut Teknologi Sepuluh Nopember, Surabaya 60111 INDONESIA
[2]Department of Materials Process Engineering, Nagoya University, Furo-cho, Chikusa-ku, Nagoya 464-8603 JAPAN

Abstract. Extraction of phytochemical compounds (such as β-carotene, linoleic acids, carrageenan, and polyphenols) from algae *Eucheuma cottonii* and *Gracilaria sp* with supercritical CO_2 followed by subcritical water has been investigated. Supercritical CO_2 extraction was carried out at pressure of 25 MPa, temperature of 60°C, CO_2 flowrate of 15 ml/min, and ethanol flowrate of 0.25 ml/min. To determine the content of carotenoids and linoleic acids, the extracted compounds were analyzed using a spectrophotometer UV-Vis. The residue of algae starting material was subsequently extracted by subcritical water at pressures of 3, 5, and 7 MPa, and temperatures of 120, 140, 160, and 180 °C. Carrageenan extracted by subcritical water was analyzed using Fourier Transform Infra Red (FTIR), while the total phenolic compound was analyzed with UV-vis spectrophotometer. Moreover, the antioxidant efficiency of extract was also examined by DPPH assay method. Based on the analytical result, β-carotene and linoleic acid content in *Eucheuma cottonii* were 209.91 and 321.025 µg/g sample, respectively. While β-carotene and linoleic acid content in *Gracilaria sp* were 219.99 and 286.52 µg/g sample, respectively. The optimum condition of subcritical water extraction was at 180°C and 7 MPa. At this condition, the highest TPC content in the extract from *Eucheuma cottonii* and *Gracilaria sp* were 18.51 mg GAE/g sample and 22.47 mg GAE/g sample, respectively; while the highest yield of carrageenan extracted from *Eucheuma cottonii* and *Gracilaria sp* were 61.33 and 65.54 g/100 g dried algae, respectively. At the same condition, the antioxidant efficiency was 0.513 min^{-1} for *Eucheuma cottonii* and 0,277 min^{-1} for *Gracilaria sp*. Based on the results the extraction method effectively separated non-polar and polar compounds, then increased the antioxidant efficiency of extract.

1 Introduction

Seaweed is one of the country's sources of foreign exchange and a source of income for coastal communities. In addition, it can be used as food, drinks and medicines. Alginate and carrageenan are important components in the food and pharmaceutical industries. Most of the seaweed in Indonesia is exported in dry form. To increase the added value of seaweed and reduce the imports of its processed products, the processing of seaweed into carrageenan is needed to be developed [1].

One of the most widely grown algae in Indonesian is *Gracilaria sp*, which is a jelly producer. However, the use of this algae was limited for food and medicinal products, and there was no further development effort on other products that have higher economic value [2]. One of the seaweed contents that is useful in the field of industry is phytochemical compounds. The main component of algae is polysaccharides, which can reach 40-70% dry weight depending on the type of algae and its growing environment. In addition to polysaccharides, algae contain some proteins, fats, minerals, and vitamins.

In order to separate the phytochemical compounds of algae, an extraction method was developed by combination of two extraction methods namely supercritical and subcritical fluids extractions.

Supercritical fluid extraction is a process of extraction using supercritical fluids as solvents. This extraction technology utilizes the solvent strength and physical properties of pure components or mixtures at their critical temperatures and pressures in phase equilibrium [3]. The principle of the supercritical fluid extraction method is the process of separating the component above the critical pressure and critical temperature of a solvent fluid. CO_2 is the most solvent used in supercritical fluid extraction method because it is non-toxic, non-flammable, and easy available. Supercritical CO_2 allows non-polar compounds to be extracted from a natural product [4].

* Corresponding author: machmudah@chem-eng.its.ac.id

In this work, supercritical CO_2 extraction was carried out at constant temperatures and pressures to extract β-carotene and linoleic acid from *Eucheuma cottonii* and *Gracilaria sp* algae. β-carotene and linoleic acid are valuable compounds with high antioxidant activity.

Subcritical water extraction or hydrothermal extraction or pressurized hot water extraction is one such alternative. Water under subcritical conditions is known as a 'natural and green' method for product extraction and has received increased attention as an important alternative to the conventional separation methods. Water under subcritical conditions can be applied to extract polar organic compounds or to decompose lignocellulosic materials to produce valuable compounds such as saccharides and aromatic organic acids. This method has been applied to recover protein, amino acids, and phenolic compounds [5].

In this work, subcritical water extraction was carried out at various temperatures and pressures to extract carrageenan and total phenolic compounds (TPC) from *Eucheuma cottonii* and *Gracilaria sp* algae residue of supercritical CO_2. The effect of pressure and temperature on the yield of carrageenan and TPC was investigated. The antioxidant efficiency of extract was also investigated.

2 Experimental Section

2.1 Materials

Eucheuma cottonii and *Gracilaria sp* algae used for starting material was purchased from Pamekasan, Madura. Prior to the extraction, *Eucheuma cottonii* and *Gracilaria sp* algae were rinsed by using distilled water to remove the impurities and salts. Then, they were ground into fine using a millser. Next, they were dried at 60°C for one day until no standing moisture was visible. The Folin-Ciocalteau's Reagent, 1,1 diphenyl-2-picrylhidrazil (DPPH) and gallic acid ($C_7H_6O_5$) were purchased from Sigma-Aldrich Japan (Tokyo, Japan). Sodium carbonate (Na_2CO_3), methanol (CH_3OH, 99.7%), and ethanol (C_2H_5OH, 99.5%) were supplied by Merck (Germany). Standard carotene (98%) and linoleic acid (98%) were purchased from Wako Pure Chemical Industries Ltd. (Osaka, Japan). They were used without further purification.

2.2 Experimental setup

In this work, the supercritical CO_2 and subcritical water extraction were conducted in a semi-batch process. The main apparatus of both processes consist of a high-pressure pump (200 LC Pump, Perkin Elmer, Germany), heater (Linn High Therm GmbH, model VMK 1600, Germany), reactor (10 mL in volume; Thar Design Inc., USA) and back pressure regulators (BPR; AKICO, Japan). Both sides of the reactor were equipped with removable threaded covers included stainless-steel filters (0.1-1.0 μm). The 1/16 in. stainless-steel tube was used to introduce liquid CO_2 or hot water from the pre-heater to the reactor, which was located in the heater.

2.2.1 Supercritical CO_2 extraction

In Supercritical CO_2 extraction process, initially, 12 g of starting material was loaded into the extractor among glass beads put on the both side of extractor inlet and outlet to prevent channeling. Then the extractor was installed in the heater. CO_2 as a solvent was then pumped at a flow rate of 15 mL/min using an HPLC (High Performance Liquid Chromatography) pump to reach an operating pressure. Besides that, ethanol was then pump at a flow rate 0.25 mL/min using an HPLC pump. The mixture of ethanol and CO_2 was flowed into extractor. The pressure was controlled by adjusting back pressure regulator (BPR) that includes a heater. Furthermore, the CO_2 and ethanol was heated using a preheater and heater to achieve an operating temperature. To ensure the temperature in the extractor according to the desired temperature, temperature in the inlet and outlet of extractor were measured by thermocouples (T_1 and T_2). Extraction solution was collected every 30 min in a collection vial. The CO_2 was flowed into environment after passed through gas flow meter. Extract was then stored in a refrigerator. The schematic diagram of supercritical CO_2 extraction apparatus is shown in Figure 1. The extraction was carried out at temperatures of 60°C and pressures of 25 MPa for 3 hours.

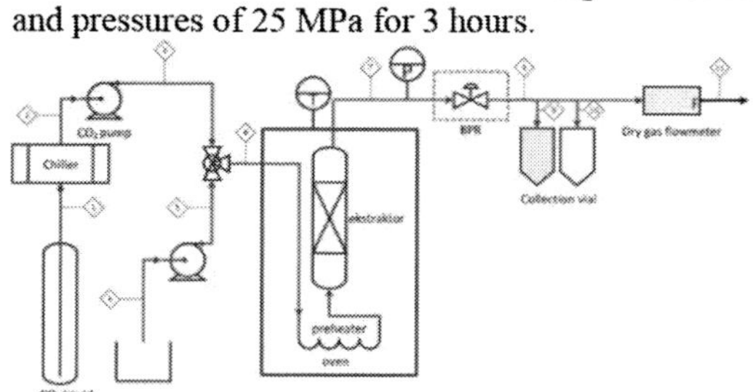

Fig. 1. Apparatus for supercritical CO_2 extraction

2.2.2 Subcritical water extraction

In subcritical water extraction process, initially, 1 g of residue from supercritical CO_2 extraction was loaded into the extractor among glass beads put on the both side of extractor inlet and outlet to prevent channeling. Then the extractor was installed in the heater. Water as a solvent was then pumped at a flow rate of 1 mL/min into the extractor using an HPLC pump to reach an operating pressure. The pressure was controlled by adjusting back pressure regulator (BPR), and was monitored by pressure gauge. Furthermore, the water was heated using a preheater and heater to achieve the operating temperature. To ensure the temperature in the extractor according to the desired temperature, temperature in the inlet and outlet of extractor were measured by thermocouples (T_1 and T_2). The extract solution was cooled by a cooler and then passed the filter and collected in a collection vial every 30 min. Extract solution was then stored in a refrigerator. The schematic diagram of subcritical water extraction apparatus is shown in Figure

2. The extraction was carried out at temperatures of 120 – 180°C and pressures of 1 – 5 MPa for 3 hours.

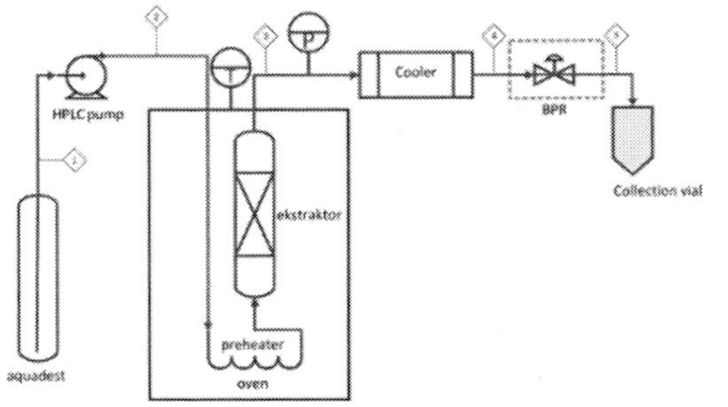

Fig. 2. Apparatus for hydrothermal extraction

2.3 Analytical method

Analysis of phenolic compounds, β-carotene, linoleic acid content, and antioxidant activity in the extracts was conducted using Genesys 10 UV-Vis Scanning Spectrophotometer (Thermo Fisher Scientific, Waltham, MA), allowing spectra between 190 nm to 1,100 nm. Liquid products were analyzed in quartz cuvette with 1 cm path length. The solid products collected at each operating condition were analyzed by a Spectrum One FT-IR Spectrophotometer (Perkin-Elmer, Ltd., England) to determine the structure of the solid products after the subcritical water extraction. The scanning wavenumber ranged was from 4,000 cm^{-1} to 400 cm^{-1}.

2.3.1 β-carotene and linoleic acid Analysis

β-carotene and linoleic acid content in the extract were analyzed using spectrophotometer at a wavelength of 450 nm and 195 nm, respectively. Initially, calibration curve was prepared by measuring a standard solution β-carotene or linoleic acid at concentration of 100 to 500 mg/L. The extract solution was diluted with n-hexane prior to analysis. β-carotene and linoleic acid content in the extract was obtained based on the absorbance of the extract compared with the calibration curve.

2.3.2 Carrageenan analysis

3 ml of the extract was mixed with 90% ethanol in the evaporating disk. The solution then stirred to form a carrageenan fiber (hydrocolloid). And then the mixed solution was leaved for 30 minute at room temperature. After that, it was heated in the heater at 60°C for 24 hours. The results obtained were dry carrageenan sheets. Carrageenan functional group was identified by using FTIR (Fourier Transfer Infra-Red). Equation below is used to calculate the yield of carrageenan:

$$\text{Yield} = \frac{weight\ of\ dried\ carrageenan}{weight\ of\ dried\ seaweed} \text{ x } 100\ \% \quad (1)$$

2.3.3 Total Phenolic Compound (TPC) analysis

TPC content in the extract was determined by Folin Ciocalteu reagent. 1 mL of extract was diluted into 2 mL of aquadest. Then, 1 mL of Folin Ciocalteu reagent was added to the solution followed by the addition of 1 mL Na_2CO_3 solution (7%). And then the mixed solution was leaved for 30 minute at room temperature (in a dark room). Furthermore, the absorbance of the solution was measured with spectrophotometer at the wavelength of 750 nm. TPC content in the extract was determined based on gallic acid standard curve at concentration of 0-200 mg/L. TPC was expressed as mg gallic acid equivalent (GAE)/g sample.

2.4 Antioxidant Efficiency analysis

The efficiency of antioxidant was analyzed using DPPH Assay method. DPPH assay is an easy and accurate method to measure the antioxidant capacity of vegetables, fruits and extracts. DPPH is one of the organic nitrogen (free radical) that is available commercially. The antioxidant efficiency of extract was determined by adding 1000 μL of extract into 2 mL of 25 ppm DPPH in methanol solution. Absorbance of the solution was measured using spectrophotometer at the wavelength of 516 nm every minute until a constant absorbance was obtained. Percentage of the remaining DPPH was calculated with the following equation:

$$\%\text{DPPH}_{rem} = 100\ x\ \frac{[DPPH]_{rem}}{[DPPH]_{t=0}} \quad (2)$$

$[DPPH]_{rem}$ is the absorbance of the extract at a certain time, and $[DPPH]_{t=0}$ is the initial absorbance of DPPH. The efficiency of radical antioxidant was calculated by the following equation:

$$\text{AE} = \frac{1}{EC_{50}\ x\ t_{EC50}} \quad (3)$$

EC_{50} is the concentration of extract that caused 50% decrease in initial DPPH absorbance, and t_{EC50} is time needed to reach steady state at EC_{50} concentration.

3 Result and Discussion

3.1 Extracted compounds from supercritical CO_2 extraction

As the target, the extracted compounds from supercritical CO_2 extraction were β-carotene and linoleic acid. Based on the result of the analysis with UV-VIS spectrophotometer, the content of β-carotene and linoleic acid in the extract from *Eucheuma cottonii* were 209.91 and 321.03 μg/g samples, respectively, while for *Gracilaria sp* the content of β-carotene and linoleic acid in the extract were 219.99 and 286.52 μg/g samples. The extracted carotenoids from *E. cottonii* was higher than the extracted carotenoids reported by Warkoyo and Saati [6]. The reported that the highest carotenoids extracted from 3 types of *E. cottonii* was 1.412 mg/100 g sample extracted from red *E. cottonii* with ethanol. It indicated that supercritical CO_2 extraction could increase the extracted carotenoids from *E. cottonii* as well as from *Gracilaria sp*.

3.2 Functional group of carrageenan

The FTIR analysis is used to determine the presence of molecular functional groups present in a sample, in which the similarity of functional groups present between the standard and the analyzed sample has a group identical to the carrageenan standard groups.

The spectra of the carrageenan product obtained from the extraction using FTIR shows that the carrageenan product has met the standard carrageenan specification, because the functional groups present in the sample spectrum are identical to the standard spectrum of the carrageenan type as shown in Figure 3. In the FTIR spectra of the carrageenan compounds obtained in Figure 4 shows the presence of sulphate ester in the range 1210-1260 cm^{-1}, glycoside bonds in the range 1010-1080 cm^{-1}, 3,6-anhydro-D-galactose in the range 928-933 cm -1, and D-galactose-4-sulfate in the range 840-850 cm^{-1}. Thus it can be concluded that the extracted carrageenan is a kappa-carrageenan.

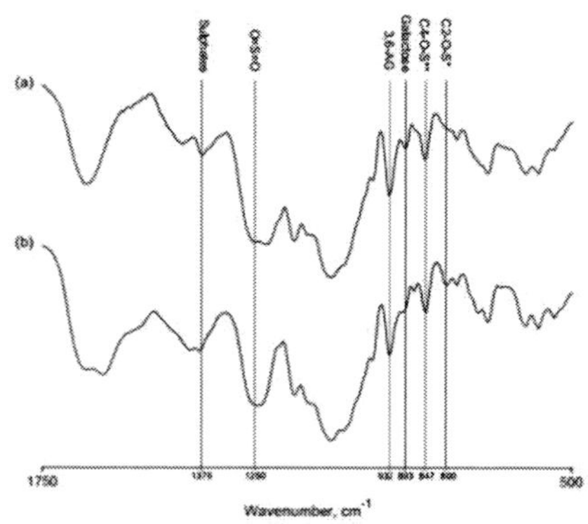

Fig. 3. Standart FTIR spectrum [7]

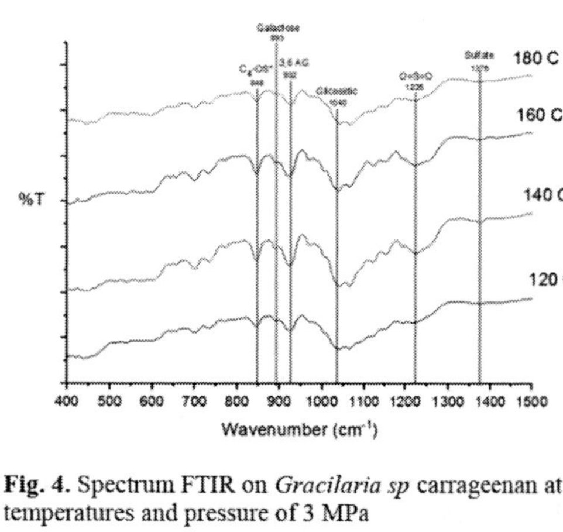

Fig. 4. Spectrum FTIR on *Gracilaria sp* carrageenan at various temperatures and pressure of 3 MPa

3.3 Effect of extraction temperature on the yield of carrageenan

As shown in Figure 5, the carrageenan yields increased as the operating temperature increase. In other words, the solvent can extract optimally when the temperature is increased.

In general, the extracted substance would increase at higher temperatures due to the increasing solubility of the substance at higher temperatures [8]. It can be happened because an increase in operating temperature will result in a decrease in surface tension and the viscosity of the solute, furthermore the extraction efficiency will increase [9]. This is consistent with the experiments which states that the increase of the extraction temperature linear to the yield of carrageenan [10].

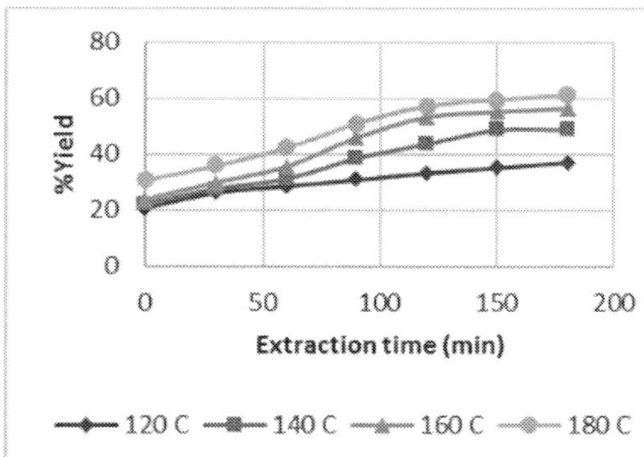

Fig. 5. The Effect of Temperature on *Eucheuma cottonii* Carrageenan at 7 MPa Pressure

3.4 Effect of extraction pressure on the yield of carrageenan

Similar with the temperature, the yield of carrageenan also increased as the increase of the operating pressure. Figure 6 shows the increase of pressure causes the increase of the solubility of the extract in solvent, so the carrageenan will be extracted more. In addition, with the increase in pressure, the number of solvents (subcritical water) per unit volume is greater, and as the result the solubility of the extract in the solvent will increase.

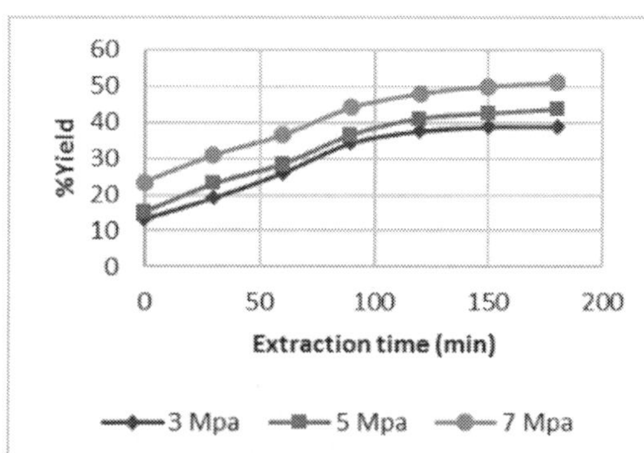

Fig. 6. The Effect of Pressure Against Yield of *Gracilaria sp* Carrageenan at Temperature 140°C

3.5 Effect of extraction temperature on the total phenolic compound (TPC)

Yield of total phenolic compound (TPC) increases as the operating temperature increases as shows in Figure 7.

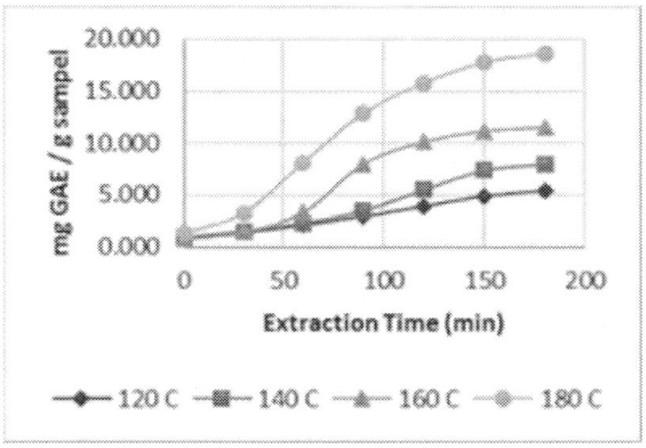

Fig. 7. Effect Analysis of Temperature on Total Phenolic Compound *Eucheuma cottonii* at 5 MPa Pressure

The extracted TPC increased obviously with an increasing extraction temperature. It can be happened because the higher temperature changes affected the solubility of phenolic compounds in water, therefore the solvent (subcritical water) is able to extract seaweed optimally. In addition, As the temperature increases, the permittivity constant of the water falls, so that the phenol becomes more soluble in water.

3.6 Effect of extraction pressure on the Total Phenolic Compound (TPC)

As shown in Figure 8, the effect of operating pressure on the total content of phenol is not very noticeable because the pressure interval taken is too low. In addition, this is also due to the polarity of water that is not influenced by the pressure. For example, at a temperature of 100 ° C, an increase in pressure from 10 MPa to 20 MPa only raises the dielectric constant from 55.9 to 56.2 [11].

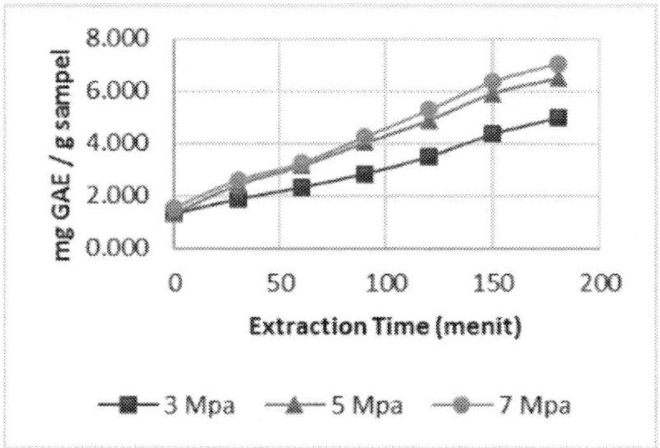

Fig. 8. Effect Analysis of Pressure on Total Phenolic Compound *Gracilaria sp* at Temperature 140°C

3.7 Antioxidant Efficiency

Figure 9 shows %DPPH measurement for antioxidant activity of *E. cottonii*. The decreasing %DPPH in antioxidant activity was caused by the longer incubation time between the extract and DPPH solution as free radical. The seaweed extract acts as an antioxidant that reduce free radical (DPPH) compounds into non-radical compounds. And this can be proved when DPPH concentrations reach 50% of the initial DPPH concentrations.

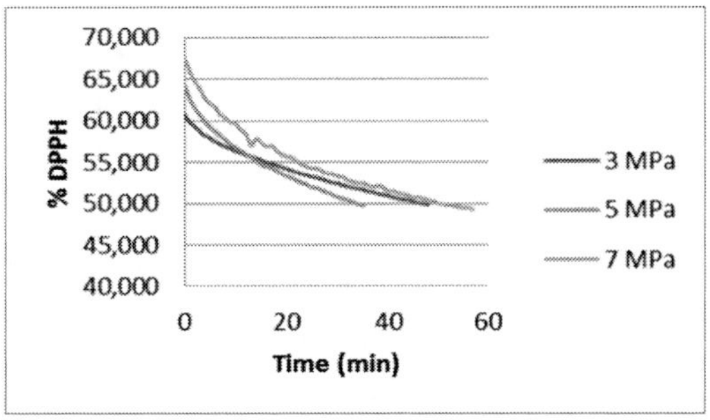

Fig. 9. Decreasing % DPPH Antioxidant Analysis of *Eucheuma cottonii* at 140°C

Eucheuma cottonii contains phytochemicals such as vitamin C, flavonoids, tannins, riboflavin [6]. *Gracilaria sp* are also said to have antioxidant sources such as carotenoids, pigments, flavonoids, terpenes, steroids, tannins, alkaloids, phenols and glycosides in abundant amounts [12].

Table 1 shows effect of subcritical water extraction conditions on the antioxidant efficiency of the extract. The highest antioxidant efficiency (AE) was obtained at the operating temperature of 180 ° C and pressure of 7 MPa. The efficiency of this antioxidant depends on the concentration of antioxidants contained in the extract and the time it takes the extract to reduce the concentration of DPPH per minute. The smaller the concentration of antioxidants needed to react with DPPH, the greater the AE value obtained. The faster the time required by antioxidants to react with DPPH, the greater the AE value obtained. So, the greater the AE value of an extract, the greater the ability of the extract to fight the free radicals.

Tabel 1 Antioxidant Efficiency of Extract at Various Operating Conditions

Seaweed	Temperature (°C)	Pressure (Mpa)	AE (Min^{-1})
Eucheuma cottoni	120	3	0.0016
	140		0.0181
	160		0.0123
	180		0.0636
Gracilaria sp	120		0.004
	140		0.018
	160		0.08
	180		0.0735
Eucheuma cottonii	120	5	0.0018
	140		0.0186
	160		0.049
	180		0.8266
Gracilaria sp	120		0.0056
	140		0.0193
	160		0.2108
	180		0.2984
Eucheuma cottonii	120	7	0.005
	140		0.026
	160		0.1709
	180		4.9636
Gracilaria sp	120		0.0073
	140		0.023
	160		1.2323
	180		1.404

4. Conclusion

Phytochemical compounds from *E. cottonii* and *Gracilaria sp* have been separated and extracted by supercritical CO_2 and subcritical water. The content of β-carotene and linoleic acid in the extract obtained by supercritical CO_2 were 209.91 and 321.025 μg/g sample, respectively, for *E. cottonii,* and 219.99 and 286.52 μg/g sample, respectively, for *Gracilaria sp.* After supercritical CO_2 extraction, the residue was extracted by subcritical water to obtain high concentration of polar compounds (carrageenan and phenolic compounds) with high antioxidant activity. The maximum yield of carrageenan and the content of TPC extracted by subcritical water were 65.54% and 24.74 mg GAE/g sample, respectively. Based on the results the extraction method effectively separated non-polar and polar compounds, then increased the antioxidant efficiency of extract. Furthermore, the extraction method may be applied in pharmaceutical industry.

This research was supported by a grant from Directorate General of Science, Technology and Higher Education, Ministry of Research, Technology and Higher Education of the Republic of Indonesia through a competitive research grant Penelitian Berbasis Kompetensi contract no.: 524/PKS/ITS/2017.

References

1. A. Diharmi, D. Fardiaz, N. Andarwulan, E.S. Heruwati. J. Perikanan dan Kelautan **16**, 117 (2011)
2. A.B Ikrom, Aunurokhim. J. Teknik POMITS **2**, (2013) ISSN: 2337-3539
3. M.V. Palmer, S.S.T. Ting, Food Chem. **52**, 345 (1995).
4. S. Machmudah, K. Kitada, M. Sasaki, M. Goto, J. Munemasa, M. Yamagata, Ind. Eng. Chem. Res. **50**, 2227 (2011).
5. I. Sereewatthanawut, S. Prapintip, K. Watchiraruji, M. Goto, M. Sasaki, A. Shotipruk, Bioresour. Technol. **99**, 555 (2008).
6. Warkoyo, E.A. Saati, Makara J. Technol. **15**, 5 (2011).
7. R. Tuvikene, K. Truus, M. Vaher, T. Kailas, G. Martin, P. Kersen, Proc. Estonian Acad. Sci. Chem. **55**, 40 (2006).
8. M.Z.Ozel, F. Gogus, A.C. Lewis, Food Chem. 82, 381 (2003).
9. H.J. Mockel, G. Welter, H. Melzer, J. Chromatogr. A **388**, 255 (1987).
10. V. Webber, S.M. de Carvalho, P.J. Ogliari, L. Hayashi, P.L. Barreto, Cienc. Tecnol. Aliment. **32**, 812 (2012).
11. Wahyudiono, S. Machmudah, M. Goto, Engineering J. **17**, 1 (2013).
12. K.L. Sreejamole, P.M. Greeshma, Indian J. Nat. Prod. Resourc. **4**, 233 (2013).

Study of Biogas Production from Cassava Industrial Waste by Anaerobic Process

Budiyono[1,*], *Arindra* D. Primaloka[1], *Listi* Ardhannari[1], *Hashfi* H.A. Matin[2], and *Siswo* Sumardiono[1]

[1]Department of Chemical Engineering, Engineering Faculty, Diponegoro University, 50275 Semarang, Indonesia
[2]Master Program of Environmental Science, School of Post Graduate Studies, Diponegoro University, 50275 Semarang, Indonesia

Abstract. Biogas production processes from tapioca wastewater have several problems that cause the biogas production is not optimal, such as pH drop at beginning of the process because the rate of acid formation is too fast and the rate of starch wastewater degradation is too slow. Therefore, to obtain optimal biogas production it is required two-stage reactor. The purposes of this research were to (i) study the influence of one stage fermentation and two stage fermentation on biogas production, (ii) study the effect of buffer Na2CO3 on biogas production, and (iii) study the effect of methanogenic bacteria concentration on biogas production from cassava starch effluent. The first method of our research was hydrolysis process by "Saccharomyces cereviceae" as substrate activator. The second is the arrangement of pH and the last is process of methane production. The results showed that the highest biogas production is achieved at concentration of methanogenic bacteria 20% (v/v) that is equal to 2458 ml. At concentration of 8% (v/v) and 15% (v/v), biogas production was 2105 ml and 2117 ml. The addition of Na_2CO_3 can extend to 16 days with accumulation of 372 ml. While without the addition of buffer, biogas production period was only 9 days with accumulation of 620 ml. In semi continuous process, the analysis carried out every 3 days. Highest biogas production achieved in the variable addition of yeast with the accumulation 9329 ml. Without yeast, accumulation of biogas was 6831 ml. Yeast is use as substrate activator so it can accelerate the hydrolysis process and increased biogas production. The addition of Na_2CO_3 is increase the alkalinity so the pH drop did not occur early in the process.

* Corresponding author: budiyono@live.undip.ac.id

1 Introduction

In the processing of cassava into wet tapioca starch, the tapioca miss rate that comes with wastewater during deposition slurry process slurry is large enough. It's about 1% of total cassava tapioca [1] which directly lost upon environment. Large contain of tapioca in these wastes can be used as raw material for biogas making.

The process of biogas production from tapioca wastes have several problems that led the biogas production not optimal, among others, the pH drop at beginning process because the rate of acid formation is too fast that causing death of microbial methanogens. Besides that, slow degradation rate of tapioca waste caused by the solids content of tapioca waste is still a polysaccharide molecule. Therefore, to obtain optimal biogas production need two stages reactor. The first stage is hydrolysis to produce simpler saccharide compounds (disaccharide and monosaccharide) by Saccharomyces cereviceae. The second stage is formation of methane from the output of first reactor by methanogens bacterial. Meanwhile, pH drop in early process can be prevented by Na_2CO_3 which acts as buffer to maintain the pH. With the addition of Na_2CO_3 and the hydrolysis of polysaccharides is expected to produce a lot of biogas in a short time. Similar with tapioca, many researchers have conducted research on biogas from waste such as agricultural waste is able to produce a lot of gas through an anaerobic process [2-12].

In general, the purpose of the study was to evaluate biogas production with two-stage fermentation of waste water from tapioca industrial with substrate activator. Specifically, the purpose of the research was to study the effect of anaerobic microbial concentration on the rate of biogas formation, study the effect of Na_2CO_3 buffer addition on the rate of formation of biogas, study the effect of fermentation 1 stage and 2 stage in rate of biogas production from starch waste.

2 Materials and Methods

Materials used in the experiment is synthetic wastewater of tapioca manufacture that made by mixing 20 grams of tapioca in 2 litters of water. Yeast (Saccharomyces cereviceae) NKL brand. Water is obtained from Chemical Process Laboratory, Chemical Engineering, Diponegoro University. Urea was obtained from the Waste Treatment Laboratory, Chemical Engineering, Diponegoro University, and Buffer Na_2CO_3 non-PA obtained from chemical shops in Semarang.

The tools used in the experiments is a two-stage anaerobic biodigester which consists of waste and nutrient storage tank, mixing tank, airtight stirred tank, airtight stirred tank and pressure resistant, pH meters, water pump, valve, pipe, pressure gauge, hose with diameter 1 cm, 250 ml measuring cup, water tank, and electric motors. Overall design of equipment used is presented in Figure 1.

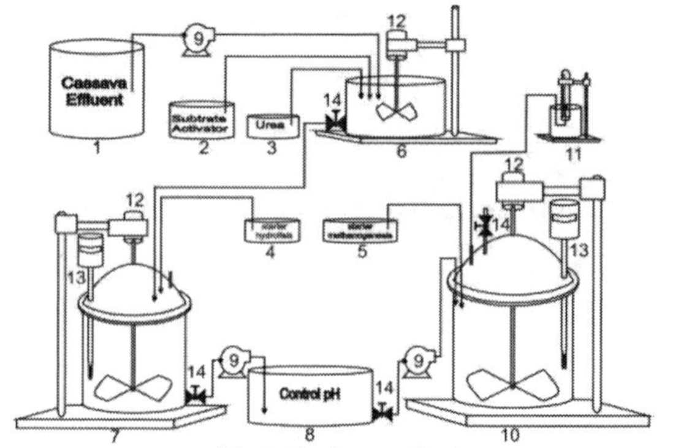

Fig 1. Equipment Design

The fixed variables tested to several respond variables. These variables include: fermentation 1 stage and 2 stage in continuous conditions with Na_2CO_3 buffer, rumen concentration is 10% (v/v), and yeast. In two stage fermentation, when yeast variables of 0.08 w/v, the rumen concentrations is 8%, 15%, and 20% v/v. When yeast variables of 0.15% w/v, the addition of Na2CO3 buffer is batch with variables the presence or absence of and rumen concentration of 15% (v/v) and yeast 0.15% (w/v) in batch conditions.

The process begins by assembling the equipment like in Figure 1. Then mix the tapioca wastes 1% (w/v) with 0.4% urea (w/v). In the second stage of fermentation the mixture is then poured into hydrolysis tank and then add the Saccharomyces cerevisiae 0.08% (w/v), and let stand for 1 day, so polysaccharides will be degraded into disaccharides and monosaccharides. Furthermore disaccharide and monosaccharide produced in hydrolysis tank flowed by pumped to the pH regulator tank. The tank was in the range of pH 7 using Na_2CO_3, and then pumped it into biogas formation tank. In this biogas formation, disaccharide and monosaccharide was mixed with methanogenic microbial starter 10% (v/v) and tapioca wastewater as substrate of 100 ml every 12 days.

Basic experimental of 1 stage and 2 stage fermentation be the same. In 1 stage fermentation, the wastes directly neutralized with buffer and mixed with rumen bacteria with a concentration of 10% (v/v) without yeast, one of its pH tank is not regulated and without addition of substrate, and in each tank was added rumen bacteria with different concentration of 8%, 15%, 20% (v/v) without addition of substrate.

Test results were carried out by analysis amount of methane by using a water displacement technique, temperature and pH measurements was carried out every day. Observations were made continuously until the methane was not formed anymore.

3 Results and Discussions

Biogas is produced depends on composition of feed material. Organic materials that can be used as raw material for biogas production are animal waste, solid waste, sewage sludge, etc. This study used tapioca wastewater with batch and semi-continuous processes. Characteristics of the biogas produced in each process are described below. Experiments were carried out to raise the levels of C/N of 2 liters tapioca waste water tapioca 1% (w/v) in each tank with urea 0.4% (w/v). Experiments with rumen concentration (8%, 15%, 20% v/v), yeast concentration 0.15% (w/v), with Na_2CO_3 buffer batch in 2 stage fermentation system obtained the following results in Figure 2.

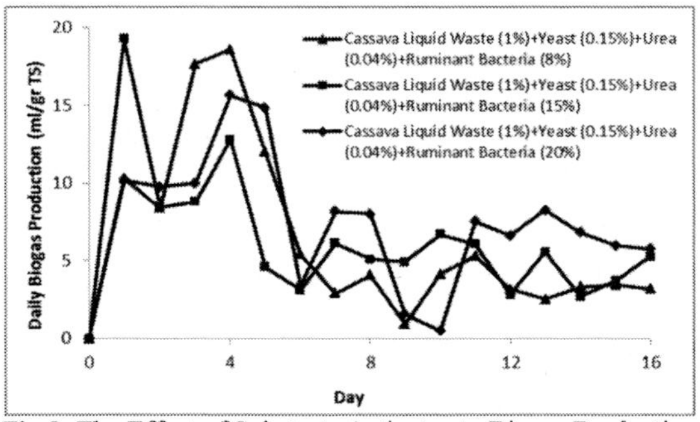

Fig 2. The Effect of Substrate Activator to Biogas Production

Figure 2 describes the relationship of methanogenic bacteria (rumen) concentration in anaerobic fermentation process. It can be seen that the daily biogas production rate is best obtained when the addition of methanogenic bacteria by 8% [1, 13]. This is due to the process of biogas production with tapioca waste material optimum obtained when the number of methanogenic bacteria by 20%.

Bacteria which first worked out in the process of changing such complex carbohydrate polymer is cellulolytic bacteria or other hydrolytic bacteria. Cellulolytic bacteria break or cut the cellulose molecule, the molecule with high weight into cellulobiose (glucose) and free glucose. Glucose is then fermented anaerobically to produce a variety of fermentation products such as acetate, propionate, butyrate, H_2 and CO_2. H_2 results from primary fermentation immediately used by methanogenic bacteria (methanogens) which is the last bacteria used in anaerobic fermentation process. In addition, it is also required for the conversion of acetate to methane in anaerobic fermentation by some methanogenic bacteria.

At this stage of methane gas formation, methanogens bacteria plays the role. Methanogens bacteria will utilize the results of second stage, such as formats, carbon dioxide, and hydrogen as substrates to produce methane, carbon dioxide, trace of gases such as H2S and water. Data shows that the highest accumulation of biogas production obtained when addition of 20% (v/v) methanogenic bacteria is 2458 ml or 122.9 ml/g TS. While the production of biogas for variable concentrations of methanogenic bacteria by 8% (v/v) and 15% (v/v) are 2105 ml or 105.25 ml/g TS and 105.85 ml or 2117 ml/g TS. Biogas production on addition of 20% (v/v) methanogenic bacteria is lower in the beginning of process than the addition of 15% (v/v) methanogenic bacteria. It happened until the fourth day. The characteristics of resulting curve above in accordance with growth curve of bacteria which consisting of four stages, such as lag stage, exponential stage, logarithmic stage, stationary stage, and death stage.

Methanogens bacteria growth in the beginning of process is still going through adjustment to the raw material circumstances in which will be broken down into the biomass, in terms of nutrients, pH, or temperature that corresponds to its place. Furthermore, these bacteria will experience rapid growth process that will produce maximum biogas because the use of good nutrition by methanogens bacteria. The next stage, bacteria begin lack of nutrition in which the number of bacteria that grow as much as the bacteria die so that the biogas produced is relatively constant (fixed). And at the end of the fermentation process the bacteria have started to die, so that biogas production has begun to decline.

There are five generation of methanogenic bacteria that play a role in the formation of methane, namely: Methanobacterium, Methanobacillus, Methanopyrales, Methanococcus, and Methanosarcina. Methanogens bacteria are the most important bacteria in the decomposition anaerobic process. This bacterium is a group of organisms that are very sensitive to oxygen and pH changes in the medium. Bacterial growth rate was slower and more sensitive to changes in the environment when compared with bacterial non-methanogens. Special ability of these bacteria is to remove the excess electrons in an effective way by converting hydrogen into methane which causes the hydrogen concentration in medium remained low. It is important for the presence of cytogenesis bacteria and ultimately led the decomposition process can continue effectively [14].

In the process of anaerobic fermentation for biogas formation used cow rumen as microbes decomposer. There are three groups of bacteria that play a role in anaerobic fermentation, such as:

1. Polymers and monomers bacteria decomposer contained in the waste material and produces mainly acetate and hydrogen, and a number of volatile fatty acids like propionic, butyric and alcohol.
2. Acetogenic obligate bacteria produced hydrogen that covert propionate and butyrate become acetate and hydrogen.
3. Groups of methanogenic bacteria producing methane from acetate or hydrogen.

The final product of cellulose decomposition e by cellulolytic bacteria were succinate, acetate, format or butyrate. Below is a chart of pH and temperature characteristics at various feed composition for variable addition of methanogenic bacteria.

From Figure 3-5, it can be seen that for variable concentrations of methanogenic bacteria (rumen bacteria) showed that in the tank I have temperature range between 27 to 31.20C, tank II between 26 to 32.50C, and tank III between 26 to 33.8^{0}C. From the graph above can be seen temperature fluctuates in each tank for each variable with different ranges. All temperature range is still in the mesophilic temperature range between 28 – 45°C. In addition to mesophilic temperature, anaerobic fermentation process can also occur in the thermophilic temperature, in temperature

https://doi.org/10.1051/matecconf/201815603052

range 45 – 65°C. When the temperature decreased, the activity of microorganisms also decreased so that the production of biogas will be decreased.

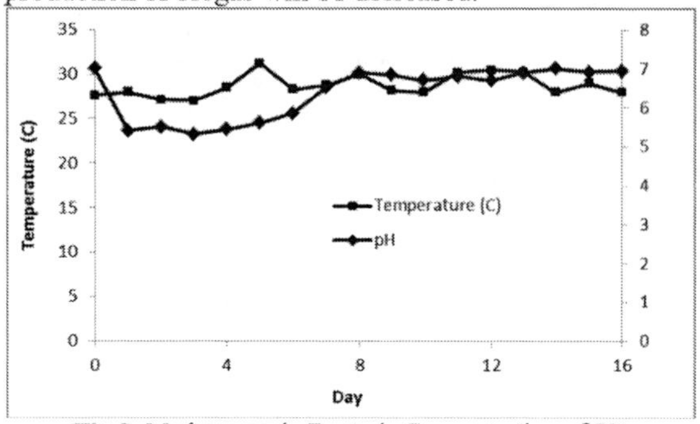

Fig 3. Methanogenic Bacteria Concentration of 8%

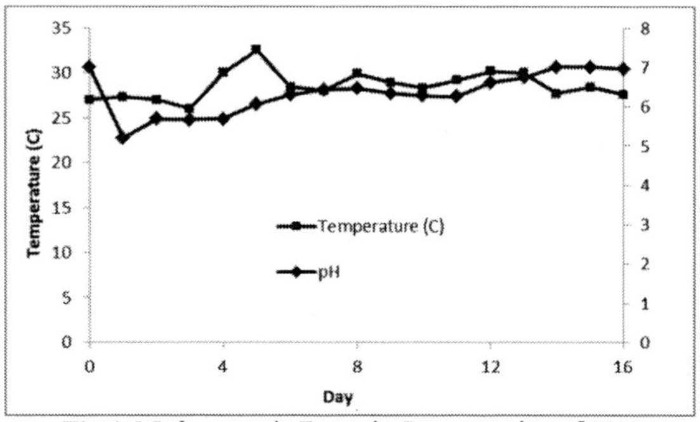

Fig 4. Methanogenic Bacteria Concentration of 15%

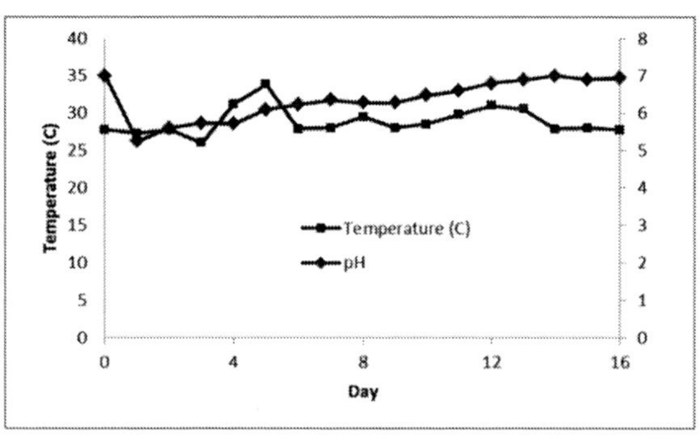

Fig 5. Methanogenic Bacteria Concentration of 20%

In the terms of pH, it can be seen that for variable concentrations of methanogenic bacteria (rumen bacteria) showed that for tank I have pH range between 5.31 to 7, tank II between 5.2 to 7, and tank III between 5.26 to 7. At initial fermentation, acid-forming bacteria will produce acids that can quickly cause pH drop. That acid formation will produce acetic acid, H_2 gas, and some VFA, such as butyric acid and propionate. When the pH value is low, the microorganisms will be in-active state in order to influence the rate of biogas formation, especially methanogenic bacteria. However, it can be handled with the addition of Na_2CO_3 buffer to improve its alkalinity. Methanogenic bacteria were able to grow well in pH range from 6.8 to 7.2 [15]. A good relationship between asidogenesis and methanogens stage is in neutral pH (7) and no dramatically increase in acidity and alkalinity [16].

From Figure 4 above can be seen that the addition of Na_2CO_3 buffer is different each day on each tank. In the tank I, the highest addition buffer is 3.13 grams that achieved on fourth day. Tank II, the highest addition of buffer is 4 grams which occurred on first day. This is because in those days, the pH in the tank II was drop quite dramatically, reaching 4.97 so that the required buffer is much to return it to neutral pH of around 7. While for the tank III, the highest buffer addition is 3.65 grams on first day. In this tank,on first day already showed pH drop reached 5.35.

The addition of Na_2CO_3 buffer every day is different on each tank. This is because the pH drop in each tank also vary every day. Tank with high pH drop will require a lot of buffer, and vice versa. In experiment in variable with or without addition of Na_2CO_3 buffer in 2 stage batch fermentation obtained following results in Figure 6.

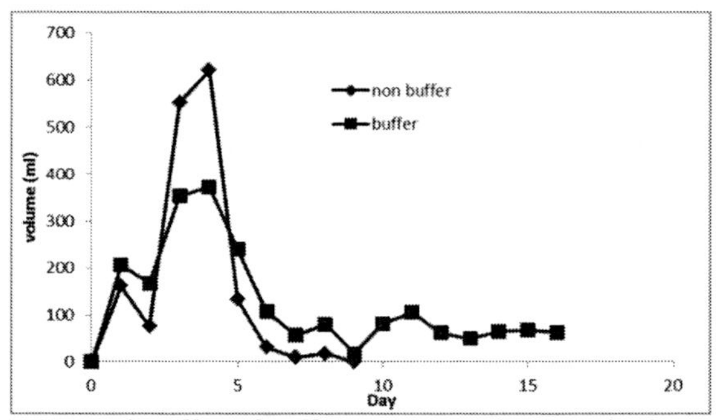

Fig 6. The Effect of Buffer Na2CO3 to Biogas Production

Figure 6 describes the effect of buffers on the rate of biogas formation. It can be explained that the production of biogas by using buffer Na_2CO_3 much less than not using buffer. On day 4, the process of biogas formation by using buffers on feed achieving the highest number of 372 ml with microbial life during 16 days. Rapid formation of acid can cause a number of methanogens bacterial be dead because it is not resistant to acidic conditions. In this case, using Na_2CO_3 aims to maintain pH range so bacteria can survive longer.

While fermentation without buffer, the biogas production is 620 ml. In this pH rumen bacteria have died because the life span of methanogens bacterial was lowest at pH 5, so it can be concluded that the gas methane was not formed, but CO_2. Cessation of biogas production is due to the pH getting down (pH drop) until it reaches 4. These pH conditions cause all the organisms die, so the biogas production ceased.

Figure 7 shows that the highest accumulation of biogas production obtained on the addition of buffer 2117 ml or 105.85 ml/g TS. On the 3rd day of biogas production on fermentation with buffer less rapidly with fermentation without buffer, this is because the gas generated in variable without buffer is not CO_2 but methane. At this time the pH of medium was around 4-5 which optimum pH for yeast growth is. It happened until the 9th day. However, in variable with addition of buffer has higher accumulation. The addition of buffer resulted pH medium is in the range of methanogens bacteria growth, so biogas production can continue until the substrate discharged on 16th day. In variable without buffer the production stopped 9th day. This is due to all microbes die because the pH is too low.

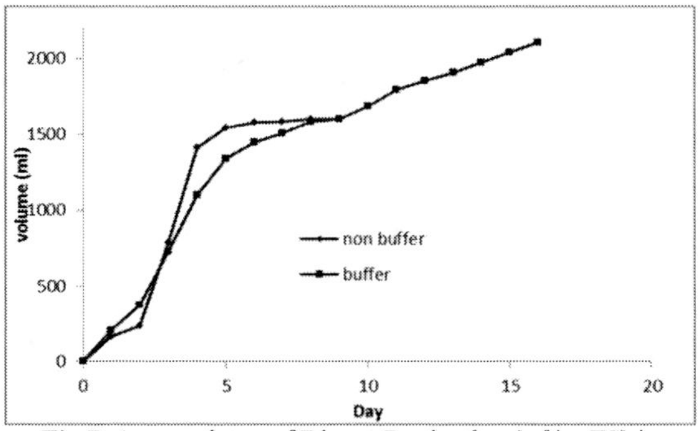

Fig 7. Accumulaton of Biogas Production (ml/gr TS) in Variable with Addition of Buffer and Without Addition of Buffer

In semi-continuous process, which is the feed pumped into tank I consists of 1% (w/v) tapioca waste, urea 0.04% (w/v), and rumen bacteria 10% (v/v). Tank 2 consists of tapioca waste 1% (w/v), urea 0.04% (w/v), yeast 0.08% (w/v), and rumen bacteria 10% (v/v). The addition of 8 grams tapioca waste fresh feed done every 12 days.

Figure 8 shows that tank I consist of tapioca waste 1% (w/v), urea 0.04% (w/v), and rumen bacteria 10% (v/v). Biogas production rate in tank I is still inferior to tank II that consist of tapioca waste 1% (w/v), urea 0.04% (w/v), rumen bacteria 10% (v/v), and yeast 0.08% (w/v). The peak of production in tank II reached on 3rd day that is equal to 625 ml, while tank I was on 6th day that equal to 642 ml. It is because the tank II hydrolyse polysaccharide molecules into simpler molecules is faster with the help of Saccharomyces cerevisiae. So that the substrates used by methanogens bacterial is substrate with a shorter chain so that the process is also more rapid degradation. Therefore, biogas production rate in the tank II is faster than tank I.

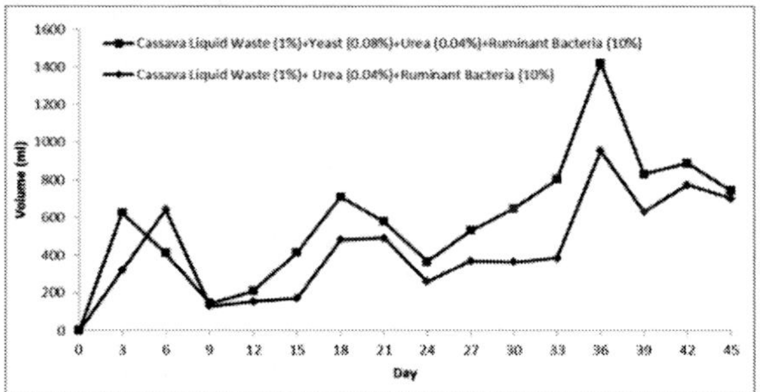

Fig 8. Daily Biogas Production (ml) on Vary Feed Composition

The addition of fresh feed 8 grams tapioca every 12 days can increase the rate of biogas production. It was seen in tank 1 and tank II on day 12 without addition of 8 grams tapioca, the biogas produced only 154 ml in the tank I and 209 ml in tank II. On day 15 with addition of 8 grams tapioca can be seen that production rises equal to 170 ml in tank I and 416 ml tank II but then fell again on day 21. Increased of biogas production is due to the fresh feed able to adding a substrate that will be fermented to produce biogas. Even so, the increase in biogas production occurs only to a certain value, after that it will be decline. That is because the substrate has been exhausted. The decreased biogas production can also caused by environmental factors, especially pH and temperature.

Figure 9 showed the accumulation of biogas production in vary feed composition variable. It can be seen that the accumulation of biogas production in tank 2 is higher than in tank 1. It happens because in the tank 2 contained of yeast (Saccharomyces cereviceae) which acts as substrate activator. Substrate activator can increase the rate of biogas production because it can speed up hydrolysis process by hydrolytic fermentative bacteria that produce simpler compounds for methanogens bacterial substrate. Total production of biogas in tank 1 is 6831 ml or 155.25 ml/g TS, and in tank II is 9329 ml or 212.02 ml/g TS.

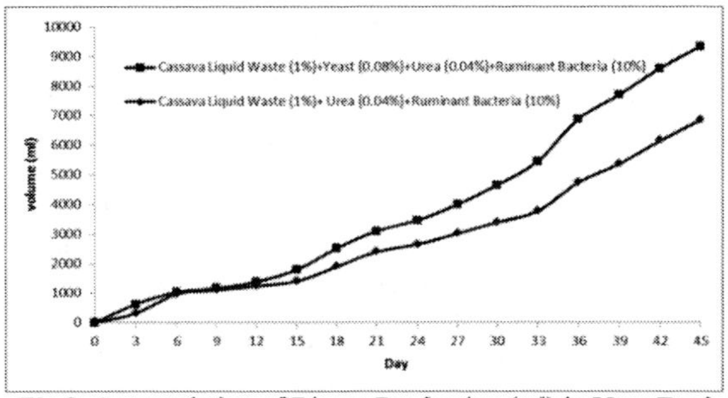

Fig 9. Accumulation of Biogas Production (ml) in Vary Feed Composition

From Figure 10 can be seen that for variable concentrations of methanogenic bacteria (rumen bacteria) showed that tank 1 has pH range between 4.8 to 7, tank 2 between 4.6 to 7. At the time of initial fermentation, acid-forming bacteria will produce acids quickly that caused rapid pH decline. The acid formation will produce acetic acid, H2 gas, and some VFA, such as butyric acid and propionate.

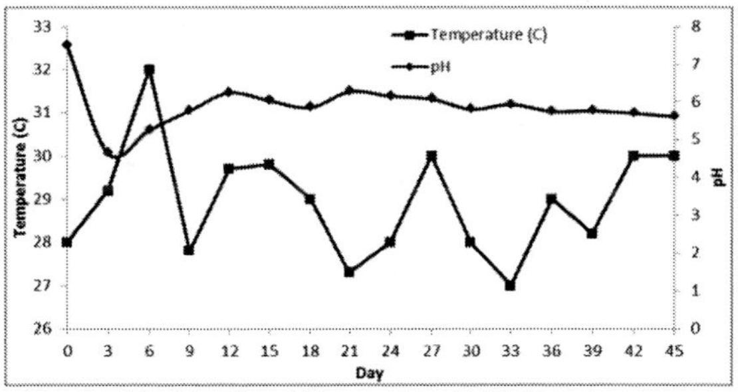

Fig 10. pH and Temperature in Biodigester

Bacteria were first worked out in the process of changing complex carbohydrate polymer is cellolulitic bacteria or other hydrolytic bacteria. Glucose is then fermented anaerobically to produce variety of fermentation products such as acetate, propionate, butyrate, H_2, and CO_2. H_2 resulted from primary fermentation immediately used by methanogenic bacteria which is the last bacteria used in the anaerobic fermentation process. In addition, acetate is also needed for conversion to methane in anaerobic fermentation by some methanogenic bacteria. On the other hand, methanogenic bacteria were able to grow well by addition of buffer solution to increase its alkalinity. The addition of buffer used if the pH drop was above 1. Magnitude of the addition of Na_2CO_3 buffer varies in each tank.

The growth rate of the acid-forming bacteria run faster than the growth of methanogenic bacteria, so the population of methanogenic bacteria is not sufficient to consume the amount of acid production. Excess of acid production in the beginning of the process could lead decrease in pH significantly to below pH neutral (pH = 7). It causes the growth rate of methanogenic bacteria is reduced. However, it can be handled with the addition of Na2CO3 buffer to improve its alkalinity. Methanogenic bacteria were able to grow well on the medium that was initially in range of pH 6.8 to 7.2 with a pH drop is not greater than 1 [16]. From Fig. 10 can be seen that pH drop was greater than 1 at day 21. Good relationship between asidogenesis and methanogens stage is currently in a state of pH drop below 1, because in this state methanogens bacteria do not die in acidity medium levels that drop too fast.

In terms of temperature, from Fig. 10 can be seen that for the variable concentrations of methanogenic bacteria (rumen bacteria) in biodigester has temperature range between 28 – 30°C. The temperature fluctuates in each tank for each variable with different ranges. All of the temperature range is still in the mesophilic temperature range, it is between 28 – 45°C. When the temperature is decreased below 28°C, the microorganism activity decreases which resulting in decreasing biogas production. However, when the temperature rose again, microbial activity will takes place as it should.

4 Conclusions

Biogas production is determined by the concentration of rumen bacteria, the addition of Na_2CO_3 buffer, and the type of fermentation. Greater concentration of rumen will increase accumulation of biogas production because there are more decomposer and rate of acid formation could be offset by formation of methane. The addition of Na_2CO_3 buffer able to providing biogas production period longer and the volume of biogas was greater than without addition of Na_2CO_3 buffer because pH drop in early process can be prevented. More complex fermentation type (2-stage) produce more biogas volume due process hydrolysis is faster.

Acknowledgments

Authors thank to Department of Chemical Engineering, Diponegoro University for all facilities to do this research.

References

1. Soemarno. Rancangan Teknologi Proses Pengolahan Tapioka dan Produk-Produknya (2007)
2. Budiyono, Syaichurrozi, I., & Sumardiono, S. World Applied Sciences Journal, **26(11)**, 1464-1472 (2013)
3. Budiyono, Syaichurrozi, I., & Sumardiono, S. International Journal of Engineering, Transactions B: Applications, **27(2)**, 177-184 (2014)
4. Budiyono, Syaichurrozi, I., & Sumardiono, S. Research Journal of Applied Sciences, Engineering and Technology, **7(13)**, 2798-2805 (2014)
5. Budiyono, Wicaksono, A., Rahmawan, A., Hawali Abdul Matin, H., Gumilang Kencana Wardani, L., Djoko Kusworo, T., & Sumardiono, S. Paper presented at the MATEC Web of Conferences, **101** (2017)
6. Shitophyta, L. M., Budiyono, & Fuadi, A. M. Paper presented at the AIP Conference Proceedings, **1699** (2015)
7. Sumardiono, S., Budi Riyanta, A., Hawali Abdul Matin, H., Djoko Kusworo, T., Jos, B., & Budiyono. Paper presented at the MATEC Web of Conferences, **101** (2017)
8. Sumardiono, S., Budiyono, & Mardiani, D. T. Paper presented at the AIP Conference Proceedings, **1699** (2015)
9. Syafrudin, Dwi Nugraha, W., Hawali Abdul Matin, H., & Budiyono. Paper presented at the MATEC Web of Conferences, **101** (2017)
10. Syafrudin, Nugraha, W. D., Ardinata, I. H., Kencanawardhani, L. G., Matin, H. H. A., & Budiyono. Advanced Science Letters, **23(3)**, 2204-2206 (2017)
11. Syafrudin, Nugraha, W. D., Matin, H. H. A., & Budiyono. Advanced Science Letters, **23(6)**, 5698-5700 (2017)
12. Syafrudin, Nugraha, W. D., Matin, H. H. A., Kencanawardhani, L. G., & Budiyono. Advanced Science Letters, **23(6)**, 5687-5690 (2017)
13. Anunputtikul, and Wantanee. Suranaree University of Technology, Thailand (2004)
14. Adams, K. H. ASEAN Seminar-Workshop on Biogas Technology. Manila (1981)
15. Anunputtikul, W., and Rodtong, S. Abstracts of the 15th Annual Meeting of the Thai Society for Biotechnology and JSPS-NRCT Symposium. Thailand, p.70 (2004)
16. Anunputtikul, W., and Rodtong, S., The Joint International Conference on "Sustainable Energy and Environment (SEE)", 1-3 Desember 2004. Hua Hin, Thailand (2004).

Production of Biogas from Organic Fruit Waste in Anaerobic Digester using Ruminant as The Inoculum

Budiyono[1,*], *Firliani* Manthia[1], *Nadya* Amalin[1], *Hashfi* Hawali Abdul Matin[2], and *Siswo* Sumardiono[1]

[1]Department of Chemical Engineering, Engineering Faculty, Diponegoro University, 50275 Semarang, Indonesia
[2]Master Program of Environmental Science, School of Post Graduate Studies, Diponegoro University, 50275 Semarang, Indonesia

Abstract. Organic waste, fruit waste and vegetable waste are the best substrate to produce biogas. Waste management system for producing biogas can be used as a solution with the waste problem by converting the wastes into biogas. This study is expected to review of the effect of substrate type and substrate composition for the volume of biogas produced. In this study, materials consist of fruit wastes (oranges, apples, papayas, and tomatoes), cow ruminant, urea, cow dung, Na_2CO_3 buffer, NH_4HCO_3 buffer, and distilled water with variations of the substrate materials, F/ W, and the buffer types. The addition of cow manure and Na_2CO_3 buffer with 1:2 of F/W, production of biogas is greater than variable which is used NH_4HCO_3 buffer and without the addition of cow dung. Variables with addition of cow dung with 1:1 of F/W and using Na_2CO_3 buffer, the result is greater than the variable using the same buffer but without the addition of cow dung and variables with 1:1 of F/W with the addition of cow dung and Na_2CO_3 buffer and variables with the same feed and without the addition of cow dung produce more biogases than variable which is the using NH_4HCO_3 buffer, 1:1 of F/W and without the addition of cow dung.

1 Introduction

In recent years, production of oil in Indonesia decreases because of oil derivation spare [1]. Energy consumption on a moderate increase, especially in rapidly developing countries. The overall size of the world energy market nearly doubled between 1971 and 2003, driven by rapid expansion of energy usage in the developing world, where population and energy activity have grown [2]. This case gives effect to the need of oil (BBM) which is an unchangeable source capacity. Therefore, there should be an effort in order to find out the way of solving this problem. In order to minimize the risk of the environment and human health economically feasible solutions are sought for the treatment of solid waste particularly in urban areas. A plan to turn solid organic waste (kitchen waste) into energy through different technology has been possible.

One of the technologies used for recovering the energy from different types of biomass residues is anaerobic fermentation that has as a primary result the production of biogas. Production of biogas provides a versatile carrier of renewable energy, as methane can be used for replacement of fossil fuels in both heat and power generation and as a vehicle fuel [3]. Related with the existing literature, the field of biogas production and applications is developed at a large scale, the existing studies covering different characteristics of this domain. In connection with the process characteristics, [4] presented an extensive study related with the pH effect on anaerobic fermentation on sludge material, [5] conducted a review over the solid state anaerobic digestion from organic waste in accordance with methane production, while [6] presented a new technology for biogas production from solid waste and biomass through means of a double-stage solid–liquid biogas process. Studies were conducted also in the field of biogas production enhancement from solid substrates under different technologies [7].

Biogas technology provides an alternative source of energy mainly from organic waste. Anaerobic digestion [8] is a process of controlled decomposition of biodegradable materials under managed conditions where free oxygen is absent, at temperatures suitable for naturally occurring mesophilic or thermophilic anaerobic and facultative bacteria and archaea species, that convert the inputs to biogas and whole digestive. Most organic wastes produced today originate in municipal, industrial and agricultural sector. Biogas research using agricultural waste (organic waste) has been done and has very good results [9-19]. The advantages of anaerobic digestion include low levels of biological sludge, low nutrient requirements, high efficiency and the production of methane, which can be used as an energy source for on-site heating and electricity [20]. If we observe its composition, fruit waste contents actually has something potential to be produced. The composition and material contents of fruit waste consists of a number of microorganisms that functions as organic-non organic material. One way can be done as the solution of to solve the problem is by changing the fruit waste to be biogas source.

* Corresponding author: budiyono@live.undip.ac.id

According to Food and Agricultural Organization (FAO), the estimated fruit and vegetable wastes percentage for each commodity group in each step of the food supply chain are 15, 9, 25, 10 and 7% in agricultural production, post-harvest handling and storage, processing and packaging, distribution and consumption respectively in South and South-East Asia. Mostly the solid wastes generated largely in traditional markets including fruit and vegetable wastes are disposed in municipal landfill or dumping sites caused environmental problems. Due to their nature and composition, they deteriorate easily and cause foul smell. Considering the high moisture and organic content, these wastes can be treated in biological treatment like anaerobic digestion than other techniques like incineration and composting [21].

Biogas is flammable gas that is produced from biodegradation of organic material by bacteria which are life in a hermetical air condition (anaerobic bacteria) towards digester organic cesspools as well as in the sanitary landfill. Biogas which is produced from organic waste is flammable gas and it is dominated by methane compound (CH_4) and CO_2 compound [22]. The production process happens in two steps which are ready materials preparation and anaerobic biodegradation process by the microorganism to maintain methane gas. Methane gas is gas that contains an atom carbon and four hydrogen atoms (H) which is flammable.

The waste material that is more used as basic biogas material is animals manure such as cattle and chicken, 1 kg cow manure can produce 0.023 – 0.040 m3 biogas [23], and 1 kg chicken manure can produce 0.065 – 0.116 m3 biogas [24]. Besides animal manure, biogas can be produced from compost heap such as the organic waste of fruit and vegetables. This study is expected to provide a review of the effect of substrate type and substrate composition for the volume of biogas produced.

The use of organic waste as basic biogas composition needs some careful methods such as collection, preparation, management, storage way. Methods collection is based on the number and characteristics of basic composition which are varieties. The nature characteristic of basic composition is solid, semi solid, and liquid. The management system is also goes carefully as well and based on the condition where the production is done.

2 Materials and Methods

2.1 Materials

The materials used in this research are organic fruit waste (lemon, apple, papaya, and tomato), ruminant, cow feces, urea, buffer Na_2CO_3, buffer NH_4HCO_3, and distilled water.

2.2 Experimental prosedure

The research is done in some steps such as creating batch digester instruminantt, preliminary experiment, fermentation process, and biogas volume analysis every two days for 30 days. The pre–experiment is done by total solid observation (TS) of basic composition and substrate intermixture.

Biogas production from fruit waste like lemon, papaya, apple, and tomatoes is starting from creating batch digester instrument which will be used. Then, preparing mixed fruit waste which have been rotten and without water. After that, arranging the substrate composition and water based on its variable F/W (1:1;1:2), adding urea as nutrient as the ratio calculation C/N to the substrate, adding animal manure based on the substrate variable, adding 25 ml ruminant on the substrate, set the starting pH that has 7 each by using buffer based on the variable (Na_2CO_3 and NH_4HCO_3), waiting for fermentation process till biogas is formed (one day), then the formed biogas is connected to the liquid displacement than is fill by water using a pipe. Observe its height difference every day that happen in the liquid displacement to find out the formed biogas volume in every 2 days for 30 days.

3 Results and Discussions

The quality and quantity of biogas yield is affected by substrate composition characteristics on each kinds of substrate compound [25]. The substrate used in biogas production will affect its process. Besides substrate, the volatile solid condition, digester temperature, duration, pH system and ratio C/N also need to be considered when producing biogas [26]. This research is conducted to study about the best substrate composition that can be used in producing total solid biogas per gram every day.

In producing biogas, sometimes the pH decrease significantly at the beginning because of it tendency to form acid. Yet, its hydrogen gas form, carbon dioxide, and some VFA such as propionate acid and butyrate. Low pH can cause the microorganism of biogas which is methane genesis bacteria (ruminant bacteria) in inactive condition [27].

Biogas production will be more optimum if the anaerobic fermentation done is really placed without oxygen. In this research, buffer is used to keep its pH so it will be in good and neutral condition in order to anticipate if there is oxygen when pH control is done. Besides, to make the methane genetic can grow and improve better, it needs buffer addition into the digester to improve its alkalinity. Therefore, this research uses kinds of buffer NH_4HCO_3 and Na_2CO_3 which function as buffer to improve its alkalinity as long as the fermentation process is done. From Table 1, it can be seen of notation formula research.

Table 1. Notation Formula Research

Formula	Symbol	Ratio F/W	Biogas Substrate Production
1	♦	1:1	Fruit Substrate + Ruminant + Urea + Cow dung + Na_2CO_3 + water
2	■	1:2	Fruit Substrate + Ruminant + Urea + Cow dung + Na_2CO_3 + water
3	▲	1:1	Fruit Substrate + Ruminant + Urea + Na_2CO_3 + water
4	X	1:2	Fruit Substrate + Ruminant + Urea + Na_2CO_3 + water
5	*	1:1	Fruit Substrate + Ruminant + Urea + NH_4HCO_3 + water
6	●	1:2	Fruit Substrate + Ruminant + Urea + NH_4HCO_3 + water

Figure 1-4 shows biogas accumulative volume per gram total solid (ml/g total solid) from all various orange, apples, papaya, tomato substrate composition. At the beginning fermentation, biogas production is very fast, this happen because of microorganism that work on the early stage is non methanogen microorganism which grows very fast too. This stage is called acid genesis process that is the acid formulation stage process by acid genesis bacteria. Few days after fermentation process done, methanogen bacteria start and be more active in forming methane gas, CO_2, gas residual such as H_2S and water from H_2, CO_2 gas, and acetate acid that is produced on the acid genesis stage [28].

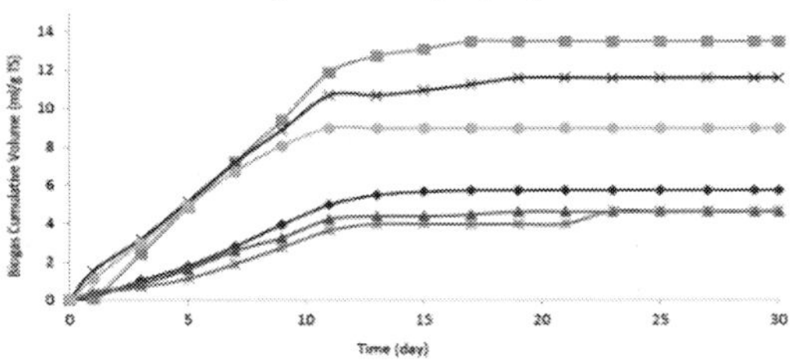

Fig 1. Biogas Accumulative Volume per Gram Total Solid (ml/g total solid) from All Various Orange Substrate Composition

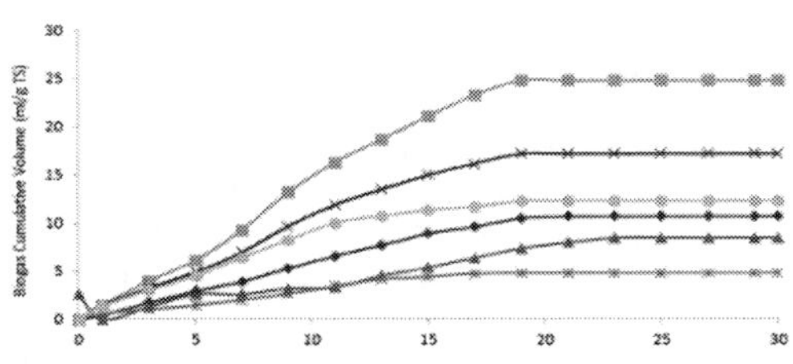

Fig 2. Biogas Accumulative Volume per Gram Total Solid (ml/g total solid) from All Various Apple Substrate Composition

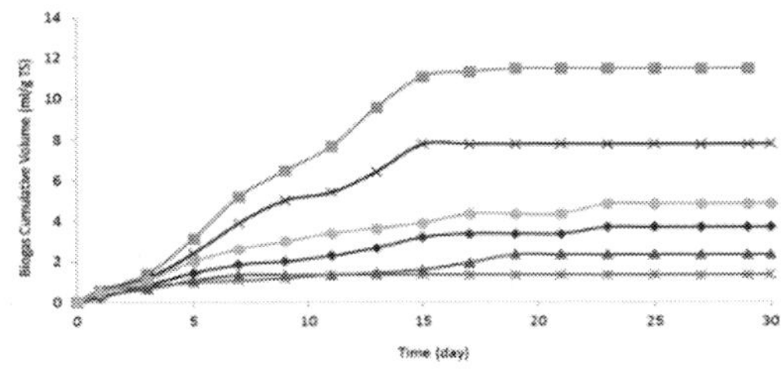

Fig 3. Biogas Accumulative Volume per Gram Total Solid (ml/g total solid) from All Various Papaya Substrate Composition

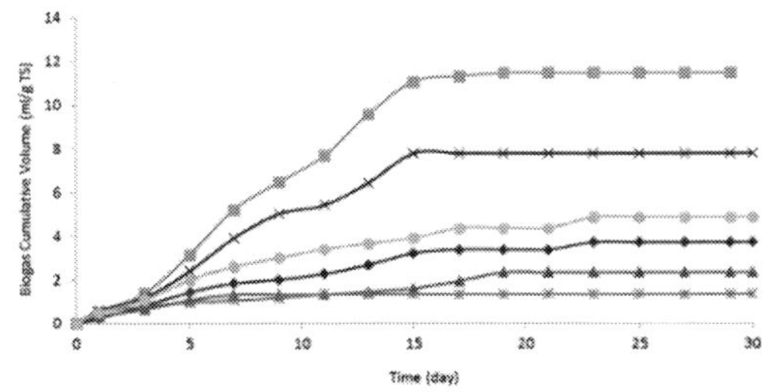

Fig 4. Biogas Accumulative Volume per Gram Total Solid (ml/g total solid) from All Various Tomato Substrate Composition

There are four phases of microorganism growth namely lag phase, exponential phase, stationary phase, and end/death phase. The biogas volume which is formed approximately increase from the 1st day to the 12th day, it happens because of the microorganism is on the exponential phase. In this phase, microorganism is considered in the constant phase (metabolic constant activity). The microorganism concentrates to scatter the organic substance from its substrate. Then the biogas volume decrease till it cannot produce more biogas on at the end of fermentation process. This happen because of the bacteria is on the static phase, where in this phase there is accumulation poisonous product or no nutrient.

This phenomenon is also happen on the use of mixed fruit substrate such as fruit waste mixture as shown on figure 5 which shows daily biogas production per gram total solid on fruit waste mixture substrate between orange – tomato. In the figure 6 shows that daily biogas production per gram total solid on fruit waste mixture substrate between papayas – tomato. Figure 7 shows daily biogas production per gram total solid on fruit waste mixture substrate between papaya – apple.

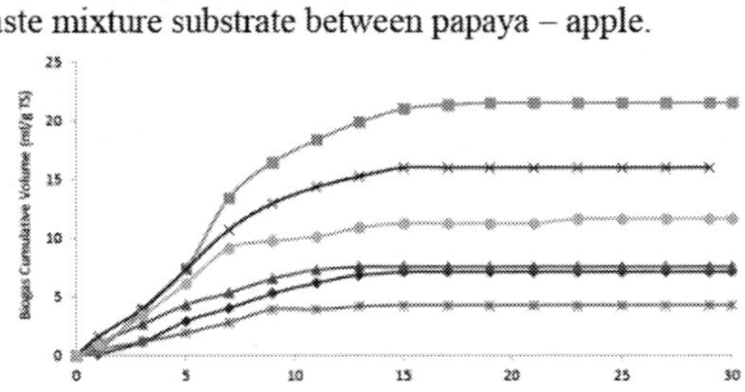

Fig 5. Biogas Accumulative Volume per Gram Total Solid (ml/g total solid) from All Composition Substrates between Tomato and Orange

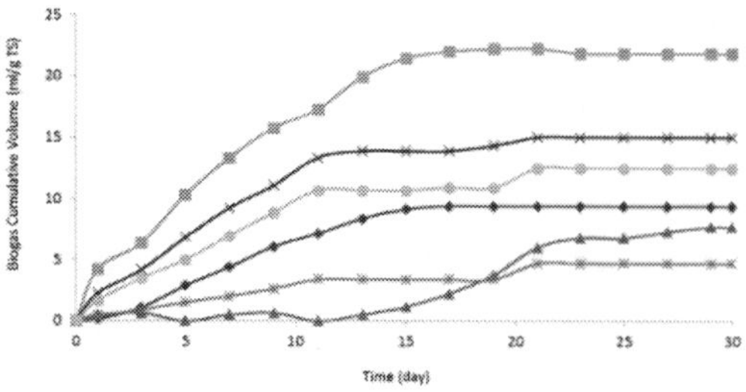

Fig 6. Biogas Accumulative Volume per Gram Total Solid (ml/g total solid) from All Composition Substrates between Papaya and Tomato

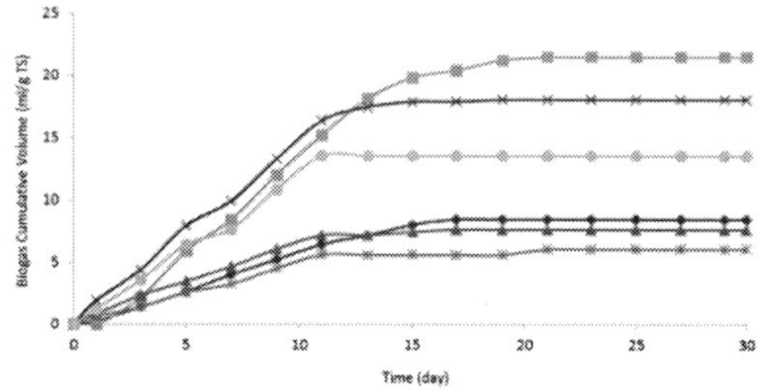

Fig 7. Biogas Accumulative Volume per Gram Total Solid (ml/g total solid) from All Composition Substrates between Papaya and Apple

In producing biogas, there are some processes that directly affect the level and efficiency of biogas including solid volatile, digester temperature, duration, pH system and C/N [26]. As we know that one of the phase in anaerobic organic compound decomposition is acid genesis and asset genesis. In this phase, acid oil which decrease pH in reactor. The whole anaerobic processes happen to pH between 6–8. Even though the methane bacteria former is really sensitive to the pH, but pH reactor is not necessary to be controlled. pH arrangement can be done by keeping the food not too be sour and also controlling its mixture so there is balance reaction between acid genic phase and methane genic phase. When there is no pH balance, then the pH value below 6 of methane bacteria activity will be influenced and if it decreases to 5, 5 there will no bacteria activity. pH concentration in the reactor is most influenced by the number of fatty acid volatile (VFA), ammonia CO_2 and its bicarbonate alkalinity which is produced.

Biological waste process can be differentiated in two types such as aerobic and anaerobic process. The aerobic process happens if only there is oxygen, while the anaerobic process happens the other way around because the oxygen will influence the process. In biogas reactor, there are two kinds of bacteria that play the most important role which are acid genic bacteria and methane genic bacteria. The biogas production will be more optimum if the anaerobic fermentation done is really free from oxygen. That is why this research uses buffer to keep pH so it stays on the neutral condition and to prevent if there is oxygen comes inside the digester when pH controlling done.

In producing biogas, sometimes the pH decrease significantly at the beginning because of its tendency to form acid. Yet, its hydrogen gas form, carbon dioxide, and some VFA such as propionate acid and butyrate. Low pH can cause the microorganism of biogas which is methane genesis bacteria (ruminant bacteria) be in inactive condition [27]. Besides, in order to help methane genic to grow well, it needs buffer addition into the digester to keep up its alkalinity. Therefore, this research uses NH_4HCO_3 and Na_2CO_3 kinds of buffer that function as buffer to increase its alkalinity a long its fermentation process.

From all figure (1 – 7) can be seen that the variable that use feed by formula 2 to produce biogas volume is bigger compared to others. To reach the pH 7 on substrate, bicarbonate ammonium concentration needs to add 0.02 g/ml or 20.000 mg/l. It will be poisonous for the methane genic organism if its ammonium concentration reaches 1500 – 3000 mg/l [29]. It will be poisonous for pH 7.4 if its ammonium concentration reaches 15000 mg/l, and it will be poisonous for all kinds of pH if it ammonium concentration reaches 3000 mg/l [30].

Besides pH, [31], argues that, one of the parameter that influence biogas production is the level of slurry digestion that can be observed from its total solid and its organic composition contents inside the bio – digester. The more Total Solid (TS) it has the easier for pH to decrease and the less TS it has, the les possibility for pH to decrease. This case is as same as the previous research which is done by [32], giving too much substrate into bioreactor causes acidogenic and asetogenic more active and grows faster, and it causes imbalance between these two bacteria. Organic composition that contain less TS is produce easily, makes methane genesis work optimally in producing biogas because of there is less pH reduction (pH controlled). But, containing less TS can reduce the product because its fermentation has substrate limitation that plays role as basic composition.

4 Conclusions

Based on the experiment result, can be conclude that by adding cow dung into oranges waste, apple waste, papaya waste, and tomato waste can increase daily biogas production, also by adding cow dung on fruits substrate can increase daily biogas cumulative production. Using Na_2CO_3 buffer to control F/W 1: 2 ratio can produce daily biogas volume and cumulative compared to F/W 1: 1 substrate ratio composition. For the next experiment, the researcher needs to do test on burning the biogas produced in order to find out the quality of organic fruit waste, besides, needs accurate pH control in which cannot influence the fermentation process.

Authors thank to Department of Chemical Engineering, Diponegoro University for all facilities to do this research.

References

1. Assegaf, F. (Universitas Jenderal Soedirman, Semarang, 2009)
2. Cioablă, A. E.; Ionel, I; Pădurean, I.; Ţenchea, A.; Popescu, F. & Savu, A., Metalurgia Internaţional **XIV**, 3, 40 – 44 (2009)
3. Weiland, P., Appl Microbiol Biotechnol (2009)
4. Wu, H., Yang, D., Zhou, Q., Song, Z., Journal of hazardous materials **172**, 196 – 201 (2009)
5. Li, Y., Park, S.Y., Zhu, J., Renewable and sustainable energy reviews **15**, 821 – 826 (2011)
6. Busch, G., Grobmann, J., Sieber, M., Burkhardt, M., Water Air Soil Pollut: Focus **9**, 89–97 (2008)
7. Karellas, Renewable and sustainable energy reviews **14**, 1273 – 1282 (2010)
8. WRAP. *Specification for whole digestate, separated digestate, separated liquor and separated fibre derived from the anaerobic digestion of source-segregated biodegradable materials*, (Publicly Available Specification PAS 110, U.K, 2010)
9. Budiyono, Syaichurrozi, I., & Sumardiono, S., World Applied Sciences Journal, **26**, 11, 1464-1472 (2013)
10. Budiyono, Syaichurrozi, I., & Sumardiono, S., International Journal of Engineering, Transactions B: Applications, **27,** 2, 177-184 (2014)
11. Budiyono, Syaichurrozi, I., & Sumardiono, S., Research Journal of Applied Sciences, Engineering and Technology **7**, 13, 2798-2805 (2014)
12. Budiyono, Wicaksono, A., Rahmawan, A., Hawali Abdul Matin, H., Gumilang Kencana Wardani, L., Djoko Kusworo, T., & Sumardiono, S., MATEC Web of Conferences **101** (2017)
13. Shitophyta, L. M., Budiyono, & Fuadi, A. M., AIP Conference Proceedings **1699** (2015)
14. Sumardiono, S., Budi Riyanta, A., Hawali Abdul Matin, H., Djoko Kusworo, T., Jos, B., & Budiyono., MATEC Web of Conferences **101** (2017)
15. Sumardiono, S., Budiyono, & Mardiani, D. T., AIP Conference Proceedings **1699** (2015)
16. Syafrudin, Dwi Nugraha, W., Hawali Abdul Matin, H., & Budiyono.,MATEC Web of Conferences **101** (2017)
17. Syafrudin, Nugraha, W. D., Ardinata, I. H., Kencanawardhani, L. G., Matin, H. H. A., & Budiyono., Advanced Science Letters **23**, 3, 2204-2206 (2017)
18. Syafrudin, Nugraha, W. D., Matin, H. H. A., & Budiyono, Advanced Science Letters **23,** 6, 5698-5700 (2017)
19. Syafrudin, Nugraha, W. D., Matin, H. H. A., Kencanawardhani, L. G., & Budiyono, Advanced Science Letters **23** ,6, 5687-5690 (2017)
20. Nallathambi, V., Biomass Bioenergy **33,** 4, 589–596 (2009)
21. Velmurugan B, Alwar Ramanujam R., Int. J. Emerg. Sci. **1**, 478-486 (2011)
22. Hastuti, D., *Aplikasi Teknologi Biogas Guna Menunjang Kesejahteraan Petani Ternak* (Fakultas Pertanian. Universitas Wahid Hasyim, 2009)
23. Budiyono, D.A. Putria, dan R.R. Saputro. *Biogas Production from Cow Manure* (Chemical Engineering Department, Diponegoro University, Semarang, 2012)
24. Setyawan A.H., *Pengembangan Biogas Berbahan Baku Kotoran Ternak* (Magister Studi Pembangunan Institut Teknologi Bandung, Bandung, 2010)
25. Kavacik B., dan Topaloglu B., *Biogas Production From Co-Digestion Of A Mixture Of Cheese Whey And Dairy Manure* (Departement of Environmental Engineering, Turkey, 2010)
26. Putri D.A., Saputro R.R., and Budiyono., Int. Journal of Renewable Energy Development **1**, 2, 61-64 (2012)
27. Arhoun B., Bakkali A., Mail R.E., Rodriguez J.M., dan, Herruzo G. F., Bioresource Technology **127,** 242–247 (2013)
28. Adam, B., *Biogas Markets Amid Fragility and Controversy* (Doctoral School of Organisation and Management Studies PhD Series 17, Denmark, 2010)
29. Kommedal R., Bakke R., Dockery J. and Stoodley P., Wat. Sci. Tech. **43**, 6,129-134 (2011)
30. Nassab A.D.M., T. Amonb., dan H.P. Kaul, *Competition and Yield in Intercrops of Maize and Sunflower for Biogas* (University of Natural Resources and Life Sciences, Department of Applied Plant Sciences and Plant Biotechnology, Austria, 2011)
31. Sjafruddin, R., Jurnal Teknologi Media Perspektif, Riset And Teknologi **11**, 2 (2011)
32. Soeprijanto, Ismail, T., Lastuti, M.D. and Niken, B., Jurnal Purifikasi **11**, 1 (2010)

Utilization of Delignified Sawdust as Raw Material of Biogas Production

Asfarina Zumalla[1*], *Budiyono*[2], *Siswo Sumardiono*[2]

[1] Master Degree of Department of Chemical Engineering, Faculty of Engineering, Diponegoro University, Jl. Prof. Sudarto, Tembalang-Semarang, 50275, Telp/fax: (024)7460058 / (024)76480675
[2] Department of Chemical Engineering, Faculty of Engineering, Diponegoro University ,Jl. Prof. Sudarto, Tembalang-Semarang, 50275, Telp/fax: (024)7460058 / (024)76480675

Abstract. Biogas is one alternative to replace the irreplaceable energy source that has begun to diminish its existence. The raw materials for biogas manufacture are renewable biomass, usually using plantation waste, agriculture, and livestock. Using biogas can also reduce environmental pollution. One of the agricultural waste that has great potential to become the raw material of biogas is teak sawdust. Wood processing industry in Indonesia quite a lot, but wood has a high lignosesluosa content, so it needs the right method to process it. With the delignification of lignin levels on teak sawdust will decrease. Wood sawdust is soaked using NaOH for 1, 2, 3, and 4 days with 4% w / v concentration. The lowest lignin and hemicellulose content was 25.79% and 87.9% in pretreatment for 4 days, while the highest cellulose level was 57, 34%. The accumulated volume of biogas at 1 day pretreatment, ie 709 ml / g TS. Gcms shows the enlarged peak area of methanamine, N-methyl from before pretreatment. The fastest biogas formation (λ) in 4 days pretreatment, 1.60403 days and the largest constant A and U variables at 1 day were 914.5903 ml / g TS and 34.59765 ml / g TS.

1 Introduction

Raw materials with lignocellulosic content are known as potential raw materials to produce environmentally friendly alternative fuels. The timber industry is an industry that produces many lignocellulosic wastes, since wood is a high-lithocellulose raw material [1]. Wood saw waste in Indonesia has not been utilized optimally, causing environmental pollution [2]. Utilizing waste wood industry in Indonesia, is a great potential to become alternative energy raw materials.

Diversification of energy is one way to be able to deal with the scarcity of oil resources and to reduce environmental pollution. Utilization of biogas as an alternative energy is possible to be applied in Indonesia because it has a potential of 49,807.43 MW derived from the activities of forest processing industry, agriculture, plantation, and sewage waste[3]. Biogas is one of the renewable alternative sources of energy that has the potential to replace fossil energy sources[4-6]. Biogas processing using anaerobic digestion process, with methane content (CH_4) is 55-70%, carbon dioxide (CO_2) of 34-45%, and other gas with small quantities are CO, N_2, H_2, H_2S, O_2 [7-9].

The process of Anaerobic Digestion (AD) is the process of transforming organic waste into biogas using microbacteria through conditions without oxygen [10]. There are 4 main processes in the process of Anaerobic Digestion, that are hydrolysis, acidogenesis, acetogenesis, and methanogenesis [11].

Hydrolysis process is the process of changing organic polymer material into monomer. Fat is hydrolyzed into fatty acids or glycerol, the protein is hydrolyzed to asama amino or peptide, and carbohydrates are hydrolyzed into monosaccharides and disaccharides [12]. The process of acidogenesis is the process of forming the result of hydrolysis into organic acid by acid-forming bacteria. The organic acids formed are acetic acid, propionic acid, butyric acid, and valeric acid [13]. The process of Acetogenesis converts organic acids which are oxidized to acetic acid and hydrogen by acetogenic bacteria [14].The process of methanogenesis is the process of converting acetic acid to carbon dioxide and methane by acetropic organisms and by reducing carbon dioxide with hydrogen by hydrogenotropic organisms. The dominant methanogens used in biogas reactors are Methanobacterium, Methanothermobacter, Methanobrevibacter, Methanosarcina and Methanosaeta [12].

Based on its Total Solid (TS) content, the Anaerobic Digestion process is divided into two types: Liquid AD (L-AD) and Solid State AD (SS-AD) [15]. Liquid Anaerobic Digestion (L-AD) is an anaerobic digestion process having a total solid strength (TS) of less than 15%, usually L-AD source from food waste, fertilizer, or animal waste [16]. Solid State Anaerobic Digestion (SS-AD) has a raw material having a total

* Corresponding author: asfarina.zumalla@yahoo.com

solid content (TS) of more than 15% [17]. SS-AD is considered more advantageous than L-AD, because it requires smaller reactor volumes, less energy required for heating, easier handling of raw materials, and less energy loss to be more efficient [18].

The problem for raw materials that have a high TS is to have a high enough lignocellulosic content. Lignocellulose makes the raw material difficult to be processed into biogas, so the need for action to reduce the content of lignocellulose in raw materials to be used [19]. Lignocellulose consists of three main components, namely cellulose, hemerusulose, and lignin. They form a complex chemical bond into the base material of the plant cell wall [20].

Cellulose is the main component of the cell wall. Cellulose is a glucose polymer with β-1,4-glucoside bonds in a straight chain. Build a cellulose base in the form of a selobiose that is a dimer of glucose. The long chain of cellulose is connected together through the hydrogen bonds and van der waals forces. Cellulose molecules are microfibrils of glucose bound to one another forming a very long polymer chain. Cellulose is a polysaccharide consisting of units and has a relatively high molecular mass, composed of 2,000-3,000 glucose. The cellulose molecule formula is $(C_6H_{10}O_5)$ n. Cellulose is a major component of the plant cell wall wall that is glucose polymer composed of β-1,4-glucose units connected with β-1,4 Dllycoside bond. The bonds of β-1,4 glucoside in cellulose fibers can be broken down into glucose monomers by acid or enzymatic hydrolysis. A perfect cellulose hydrolysis will produce a cellulose monomer called glucose. The presence of lignin and hemicellulose around the cellulose is a major obstacle to hydrolyzing cellulose [21-22].

The hemicellulose chain is shorter than the cellulose chain, due to the lower degree of polymerization. In contrast to cellulose, the hemicellulosic polymer is not straight but is a branched polymer and its structure is not crystalline. The hemicellulosic sugar monomer comprises a five-carbon (pentose / C-6), hexuronic and hexose deoxy carbonated monomer. Hemicellulose will undergo oxidation and degradation reactions first than cellulose, due to its shorter and branched chain of molecules. Hemicellulose is hydrophilic (easy to absorb water) resulting in a less orderly structure. Structurally, hemicellulose is similar to cellulose which is a sugar polymer [21-22].

Lignin is one component of the plant. Lignin is a complex molecule composed of phenylphropane units bound in three-dimensional structures, Lignin is very resistant to degradation, either biologically, enzymatically or chemically. Lignin is a three-dimensional polymer comprising a propane phenyl unit through an ete bond (C-O-C) and a carbon bond (C-C) [21-22].

High content of Lignocellulose in biomass causes biomass is difficult to be processed into biogas, so the need for pretreatement. The usual types of pretreatment are pretreatment chemically, physics and biology [23]. Alkali pretreatment is an effective pretreatment method for dissolving lignin and also can neutralize a variety of acidic products from lignocellulosic complex [24-25]. Using NaOH for the delignification process because this solution can damage the lignin structure of the crystalline and amorphous parts, and separate some of the hemicellulose. Hemicellulose has an amorphous structure so that the use of NaOH can remove lignin sekgus extracting hemicellulose. Thus it causes the pretreatment to use alkali more effectively than other prerations [26].

The OH-ions of NaOH will break the bonds from the ligmin base structure, while the Na+ will bind to lignin to form sodium phenolate. This phenolic salt is soluble. Dissolved lignin is characterized by black in a solution called black liquor. After the immersion process, the sample is filtered to remove the dissolved lignin in the solution and then the sample is washed with water to clean the solution still attached to the sample [27].

2 Experimental Method

2.1 Materials

This Biogas using raw material from sawdust, that produced from home industry from Central Java, Indonesia. The rumen obtained from RPH Ungaran. NaOH of 4% w/v used for pretreatment. HCl of 0.2N is used to neutralize ph.

Tabel 1. Sawdust and rumen properties used as biogas feedstock

Parameter	Sawdust	Rumen
TS (%)	88,5	19,88
VS(%)	86	12,46
C(%)	30,9	10
N(%)	0,35	0,56
Rasio C/N	88,3	17,86

2.2 Delignification Process

Sodium hydroxide (NaOH) with 4% w / v concentration is used for the sawdust delignification process. The process is done at room temperature during (1,2,3,4 day) with liquid to solid ratio that is 100 ml liquid per five gram sawdust. After the delignification process, sawdust is washed with aquades and HCl 0,2N, then dried in an oven at a temperature of 100 ° C to a constant weight.

2.3 Anaerobic Digestion Process.

Total Solid (TS) required for making this biogas is 20% . The digester has a volume of 1L. The biogas made has an initial ph 7, because the methanogenic bacteria can work well on ph 6.6-7.6 and it is the best ph for microbial growth. The composition of C / N has a 30/1 ratio, because of the best composition for microbial nutrition in order to develop maximally. Then put in the rumen, water, urea and sample. The digester bottle must be sealed, until no air enters the digester bottle. After that, observed the anaerobic digestion process until 30 days. The product of Anaerobic Digestion analyzed by Gas Chromatography – Mass Spectrometry. The kinetics of gas volume is calculated by Goempertz equation.

3 Result And Discussion

3.1 Results and Discussion Effect of Duration Soaking with NaOH to the content of Lignocellulose

Effect of soaking time on the content of the lignocellulosic NaOH in the sawdust have long immersion variations are 0, 1, 2, 3 and 4 days with NaOH concentration of 4% w / v. The analysis was conducted to determine the content of lignocellulose on sawdust using Chesson analysis [28]. Chesson analysis results can be seen in Figure 1, Figure 2 and Figure 3.

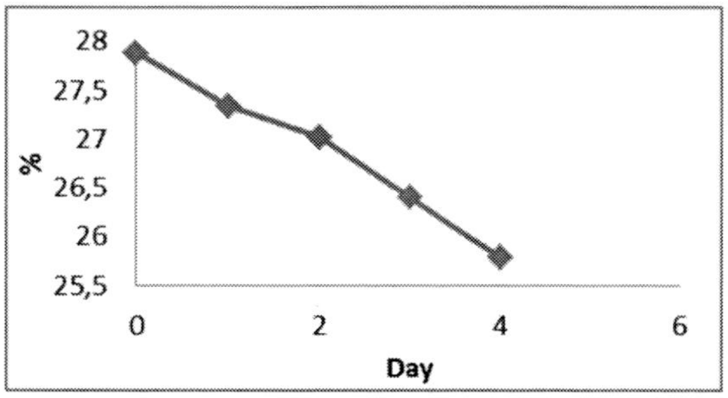

Fig. 1. Lignin on Sawdust For 0, 1, 2, 3, and 4 days pretreatment

Figure 1 can be seen that there is a decrease in the content of lignin every day on sawdust sawdust that has been soaked for 0, 1, 2, 3, and 4 days. The content of lignin at 0 days has 27,88%, the immersion of 1 day have the level 27,34%, 2 day have the content 27,02%, 3 day have the content 26,4%, and 4 day have the content 25,79% From the analysis that has been done the results obtained is 4 days of immersion has the least content, there is 25.79%.

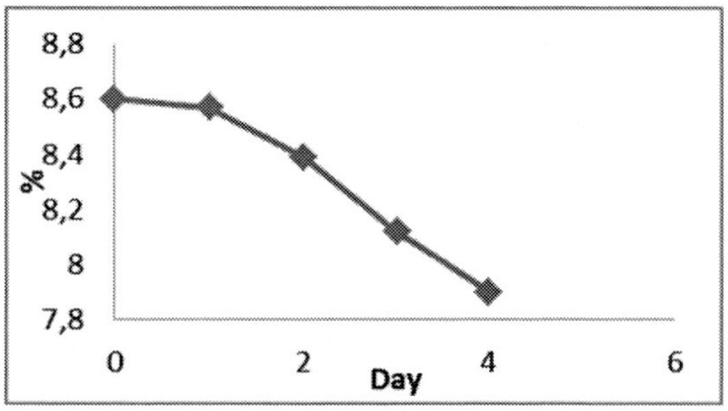

Fig. 2. Hemicellulose on Sawdust for 0,1,2,3 and 4 days pretreatment

Figure 2 shown the hemicellulose content of sawdust that immersed by NaOH 4% w / v at 0, 1, 2, 3, and 4 days. In day 0 immersion, sawdust has 8.6%, on day 1 has 8,57%, day 2 has 8,39%, day 3 has 8,12%, and day 4 has a level of 7.9%. From the graph above, it can be seen that the longer immersion hemicellulose levels decreased. Hemicellulose levels were at least 4 days, that is 7.9%.

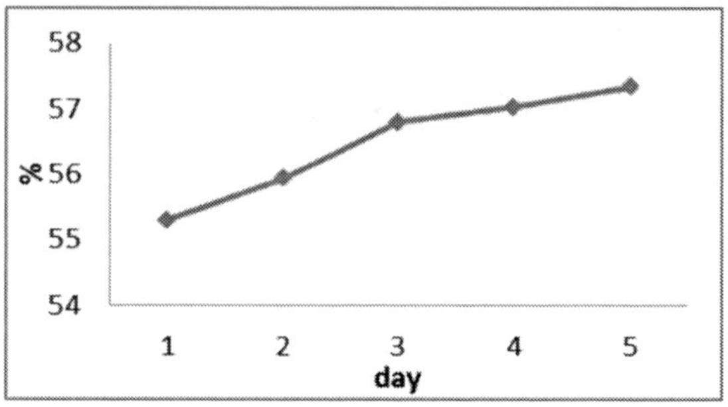

Fig. 3. Cellulose on Sawdust for 0,1,2,3 and 4 days pretreatment

Cellulose content can be seen from Figure 3, an increase from day 0 to day 4. On day 0 concentration of cellulose was 55.3%, and then on day 1 of cellulose content of 55.93%, a day to two levels of cellulose 56, 8%, day 3 level 57.03%, and the highest is cellulose level on day 4, that is 57,34%. This phenomenon occurs due to the occurrence of hemicellulose degradation into xylose, arabinose and glucose [29]. Cellulose can be converted to glucose, and hemicellulose to xylose, arabinose, galactose and glucose [20].

The phenomenon of Figures 1 and 2 decreases the percentage of lignin and hemicellulose content each day in teak sawdust. This decrease is due to the initial treatment of NaOH which can change the chemical composition, chemical structure, and physical characteristics of sawdust during pretreatment, thus increasing the biodegradability of teak sawdust [30]. The content of lignin before immersion using NaOH was 27.88%, then after immersion to 25.79%. Like lignin, hemicellulose also decreased 8.6% before immersion and 7.9% after immersion. This decrease means the conversion of insoluble fiber to soluble component, besides pretreatment of sawdust saw using NaOH causes delignification by breaking the ester bond on lignin and hemicellulose, and increasing porosity of biomass [31].

3.2 Effect of Immersion with NaOH 4% w/v on Biogas Volume

After soaking the teak sawdust with NaOH 4% w / v for 0, 1, 2, 3, and 4 days, the sawdust is used for raw material from biogas manufacture. The biogas made has an initial ph 7, because the methanogenic bacteria can work well on ph 6.6-7.6 because it is the best ph for microbial growth [6]. The C / N composition has a 30/1 ratio because of the best composition for microbial nutrients to develop maximally [32]. The effect of sawdust immersion using NaOH 4% w / v was observed with variations of immersion duration.

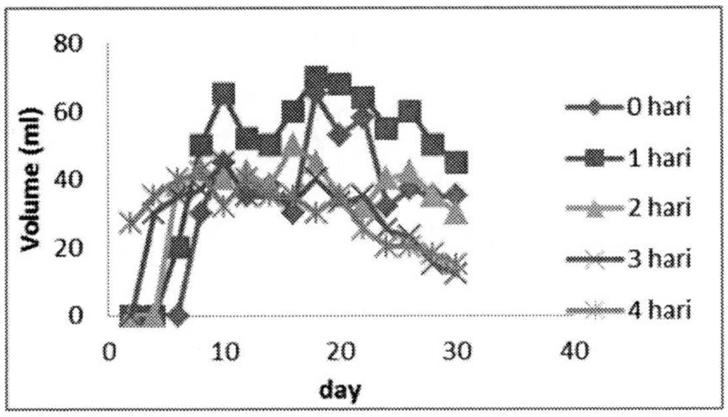

Fig. 4. Biogas Volume Measurement Results

Based on FIG. 4, the first biogas formed is to immerse for 4 days at a volume of 27 ml / g TS. On day 4, biogas begins to form for immersion variables for 3 days with volume 30 ml / g TS. On day 6, biogas begins to form for immersion variables for 2 days and 1 day with consecutive volume of 38 ml / g TS and 20 ml / g TS. On day 8, biogas begins to form for variables without immersion with a volume of 30 ml / g TS.

The phenomenon in Figure 4 is due to the number of lignin in each variable. Chesson analysis conducted before making biogas fermentation to get the result of reduced lignin content the longer done soaking of sawdust teak. The lowest lignin level on pretreatment for 4 days of immersion using 4% w / v NaOH, ie 25.79%. Before pretreatment using NaOH, the lignin content was 27.88 The phenomenon affects the process of the formation of sawdust teak into biogas. Variables without immersion are the longest-producing variables of biogas, because they have high lignin levels, so the fermentation process of biogas formation is quite hampered by the presence of high lignin [33]. Lignin that lies between the cells and inside the cell wall serves as a glue to bond between cells, which can lead to limited rate and extension of hydrolysis, since lignin is protective and complicates digestion of substrate by microbes. Substrate that is difficult to digest by microbes leads to the longer formation of biogas in the fermentation process, so before pretreatment using NaOH there is longer gas formation compared with the dipretreatment sawdust [29]. The most rapid variable produces biogas is 4 days immersion variable, which starts to form after 2 days of fermentation in the digester. The formation of biogas faster than other variables is due to more dissolved and fragmented lignin compared with pretreatment for 1 day, 2 days, and 3 days. Low levels of lignin is evidenced by the chesson analysis that obtained the lowest results on 4 days of immersion using NaOH 4% w / v, ie 25.79%. Low levels of lignin will cause reduced and loosened lignin bonds of cellulose and hemicellulose [34]. Reduced and loosened lignin bonds will accelerate the process of cellulose hydrolysis in sawdust to biogas in fermentation, so the formation of biogas at soaking 4 days faster than immersion for 1 day, 2 days, and 3 days [35].

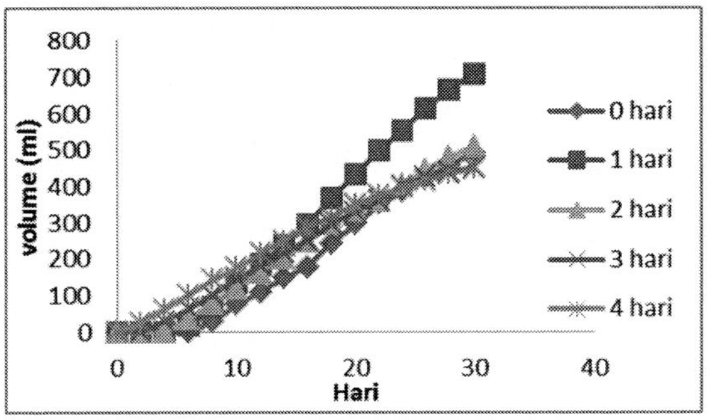

Fig. 5. Accumulation of Biogas Production

Figure 5 shows that the highest cumulative biogas volume is found in 1 day immersion variables with biogas volume reaching 709 ml / g TS. For the 0 days immersion variable, the biogas volume reached 493 ml / g TS. The 2-day immersion variable reaches 510 ml / g TS, 3 days of immersion of 452 ml / g TS and the least is for 4 days of immersion with volume of 449 ml / g TS.

The increase in biogas volume from before pretreatment and after pretreatment for 1 day, due to reduced lignin level before pretreatment. In the chesson analysis obtained lignin levels are reduced from before soaking, ie 27.34%. Reduced lignin levels are caused by the destruction of ester bonds in the LCC (Lignin Carbohydrate Complex) by hydrolysis reaction, so the bonds between units and lignin functional groups, cellulose, and hemicellulose are damaged causing chemical structure changes. Changes in chemical composition, chemical structure, and physical characteristics lead to the easy process of degradation into biogas. Increased cellulose levels can also increase the volume of biogas in pretreatment for 1 day. This phenomenon causes the volume to be formed on immersion for 1 day [36].

Day 2 until day 4 there was a decrease in the accumulation of biogas volume, ie day 2 had accumulated 510 ml / g TS less than 1 day immersion of 709 ml / g TS. The phenomenon is inversely related to the phenomenon of chesson analysis which shows the longer the immersion the less lignin content, should the accumulation of biogas volume increase. The results obtained on the fermentation of sawdust sawdust using 4% w / v NaOH indicated that soaking for 2 days, 3 days and 4 days was getting fewer results than immersion for 1 day. The decrease is due to the more NaOH absorbed in the raw material will cause anaerobic microbial inhibition due to high Na+ concentration and the effect on the lignin decomposer [37-38]. Utilization of NaOH pretreatment may cause inhibition of Na+ ions in methanogenesis [39]. This phenomenon resulted in the accumulation of biogas volume at 2 days, 3 days and 4 days less than 1 day immersion, although in the biogas formation process the fastest occurred in 4 days immersion variable due to low lignin and hemicellulose levels.

3.3 The Influence of Immersion on Biogas Content Characteristics

After immersion and biogas preparation, further analysis is needed to determine the changes of the biogas composition formed after and before immersion.

Table 2. Area and Biogas Component Area Before Immersion

Area	Name
4297054	Methanamine, N-methyl- (CAS)
15257372	Methanamine, N-methyl- (CAS)

Table 2 shows the differences in the area of components found in teak sawdust. The area before immersion NaOH 4% w / v is 4,297,054, whereas after immersion NaOH 4% w / v is 15,257,372. The greater area of the peak area after pretreatment shows that more methanamine N-methyl composition in biogas after pretreatment. The phenomenon is caused by damage to lignin and hemicellulose and the occurrence of de-esterification (saponification) in the bonds of intermolecules ester [31]. It also changes the degree of polymerization of each component, altering the physical characteristics of teak sawdust, ie changes in surface area, porosity, and crystallinity [40], altering chemical composition, and chemical structure [30]. Pretreatment using NaOH causes decreased lignin and hemicellulose content in teak sawdust, thus increasing biodegradability and increased biogas production [36].

3.2 Effect of Immersion with NaOH on Biogas Formation Rate

Biogas production data obtained from the study were matched with the modified Goempertz equation (Equation 1) [41]. In the modified Goempertz equation it assumes the cumulative biogas production in batch digesters is in line with the bacterial growth rate [42-48]. The modified Goempertz equation is as follows:

$$y(t) = A.\ exp\left\{-exp\left[\frac{\mu e}{A}(\lambda - t) + 1\right]\right\} \quad (1)$$

Where P is the production of cumulative biogas (ml), A is the production of potential cumulative biogas (ml / g TS), U is the maximum biogas production rate (ml / g TS days), e is the mathematical constant of 2.71828, λ is the adaptation time bacteria (minimum start time of cumulative biogas production is calculated (days) .The constants A, U, and λ can be determined by using nonlinear regression with the help of POLYMATH software [43-48].

Table 3. Kinetic Constant of Biogas Production

Variabel	A (ml/g TS)	U (ml/gr TS.hari)	λ (hari)	R^2
0 day	621,5533	25,88708	8,52624 8	0,99789 3
1 day	914,5903	34,59765	7,29931 8	0,99598 89
2 day	619,4651	23,66482	5,60889 0	0,99658 8
3 day	528,9307	21,51465	3,84145	0,99733 84
4 day	495,1706	20,56573	1,60403	0,99769 53

The kinetics constants of biogas production from sawdust saw in Table 3 can be seen in saw 2 teak sawdust soaked with NaOH 4% w / v, the highest A and U constant values are immersion for 1 day ie 914,5903 ml and 34 , 59765 ml / gr TS.day, but the lowest λ is 4 days ie 1.60403 days

Pretreatment with NaOH 4% w/v for 4 days is variable that makes the fastest biogas among 4 other variables because the content of lignin and hemicellulose is 25.79% and 7.9%, thus speeding up the hydrolysis process [49-51]. While in the process of hydrolysis is solved by anaerobic bacteria that release a special enzyme that serves to cut the bonds of molecular units, because the molecule is too large [52]. Rapid hydrolysis rate will decrease adaptation phase or minimum start of biogas production (λ) [19], but immersion for 4 days does not get large A and U values. The highest A and U values obtained 1 day immersion variable. This phenomenon is caused by the 4 days of absorbing NaOH more than the 1 day variable. Na + can cause the inhibition of Aerobic Digestion process especially in methanogenesis process [21], so the process of biogas formation in variable 4 days is less than 1 day variable which less NaOH absorption. The formation of biogas on the 1 day variable tends to be slower because the number of lignin and hemicellulose is higher than 4 days, ie 27.34% and 8.57%. High levels of lignin and hemicellulose cause more difficult process of microbial biodegradation more difficult, so the process of hydrolysis runs more slowly.

4 Conclusion

Pretreatment for 1 day produce the highest biogas that is 709 ml / g TS, but it takes a long time to produce it. The longer the pretreatment using NaOH the faster the formation of biogas, this is proven by pretreatment for 4 days to produce the first biogas according to kinetic calculation, which is 1.60403 days because it has low lignin and hemicellulose, that is 25.79% and 7.9 %. The amount of methanamine N-methyl also increases after delignification of NaOH and before immersion.

References

1. Purwanto, D, Riset Industri. **V**, 1, 13-20 (2011)
2. Iphegan, J, O., Evelyn, V, O., IJAIEM. **2**, 6, 91-93 (2013)
3. Haryati, T., Wartazoa. **16**, 3, 160–169 (2006)
4. Budiyono, Widiasa, I, N., Johari, S., Sunarso, International Journal of Basic and Applied Sciences UBAS-IJENS, **10**, 1 (2010)
5. Budiyono, Widiasa, I, N., Johari, S., Sunarso, International Journal of Engineering and Technology. **1**, 3, 109–116 (2009)
6. Putri, D, A, Saputro, R, R., Budiyono, Int. Journal of Renewable Energy Development, **1**, 2, 61-64 (2012)
7. Frigon, J, C., Guiot, S, R., Biofuels Bioprod. **4**, 447-458 (2010)
8. Juanga, J, P., Visvanathandan, C., Tränkler, J., J Waste Manage Res. **25**, 30–38 (2007)
9. Karellas, S, B., Jurnal Renewable and Sustainable Energy Reviews **14**, 1273-1282 (2010)
10. Shitophyta, L, M., Budiyono., Fuadi, A, M., AIP Conference Proceedings. **1699** (2015)
11. Xu, F., Shi, J., Lv, W., Yu, Z., Li, Y., Waste Management **33**, 1, 26–32 (2013)
12. Quintero, M, Castro L, Ortiz C, Guzman C, Escalante, H., Bioresource Technology **108**, 8–13 (2012)
13. Deublein D, Steinhauser, Deggendorf (2008)
14. Zou J, Li Y., Bioresource Technology **218**, 18–26 (2016)
15. Syafrudin, Dwi Nugraha W, Hawali Abdul Matin H, Budiyono, MATEC Web of Conferences **101** doi:10.1051/matecconf/201710102016 (2017)
16. Lin L, Yang L, Xu F, Michel F C, Li Y. Bioresource Technology **169**, 439–446(2014)
17. Shi J, Xu F, Wang Z, Stiverson J A, Yu Z, Li Y., Bioresource Technology **157**, 188–196(2014)
18. Zhu J, Wan C, Li Y., Bioresource Technology. **101**, 19, 7523–7528(2010)
19. Taherzadeh M, Karimi K., Int J Mol Sci. **9**, 9, 1621–51(2008)
20. Zheng Y, Pan Z, Zhang R., Int J Agric and BiolEng. **2,** 3, 51-68(2009)
21. Zheng Y, Zhao J, Xu F, Li Y., Progress in Energy and Combustion Science. **42**, 1, 35–53 (2014)
22. Yang L, Xu F, Ge X, Li Y., Renewable and Sustainable Energy Reviews. **44**, 824–834 (2015)
23. MacLellan J, Chen R, Kraemer R, Zhong Y, Liu Y, Lio W., Bioresource Technology **130**, 418–423 (2013)
24. Hendriks, ATWM., Zeeman, G., Bioresource Technol. **100**, 1, 10–8 (2009)
25. Chen BY, Chen SW, Wang HT., Biomass Bioenergy **39**, 182–91 (2012)
26. Julfana R. JKK. **2**, 1, 52-57(2012)
27. Safaria, S., ISSN: 2303-1077, **2**, 1, 46-51(2013)
28. Aldino, J, G., Fanandy, K., Juliastuti, S, R., Nuniek, H.,Jurnal Institut Teknologi Sepuluh Nopember (2010)
29. Laopaiboon, P., Thani, A., Leelavatcharamas, V., Laopaiboon, L., Bioresource Technology **101**, 3, 1036-1043 (2009)
30. Tian, M., Liu, X., Liu, J., Zhao, Y., T. Chin. Soc. Agri.Eng. **29,** 177-184 (2013)
31. Pang, Y., Liu, Y., Li, X., Wang, K., Yuan, H., Energy Fuel. **22**, 4, 2761-2766 (2008)
32. Budiyono, Syaichurrozi, I., Sumardiono, S., Waste Technology. **1,** 1, 1-5 (2013)
33. Wang, D., Ai, P., Liang, Y.,Tan, Z., Zhang, Y., Science Direct. **132**, 47-55 (2015)
34. Lu, Y, P., Yang, B., Gregg, D., Saddler, J, N., Manfield, S, D., Applied Biochemistry and Biotechnology **98**, 641-54 (2012)
35. Frigon, J, C., Guiot, S, R., Biofuels Bioprod. Bior. **4** , 447-458 (2010)
36. He, Y., Pang, Y., Liu, Y., Li, X., Wang K., Energy and Fuels **22,** 4, 2775-2781 (2008)
37. Bondesson, P, M., Galbe, M., Zacchi, G., Biofuels **6**, 11 (2013)
38. Heiske S, Schultz-Jensen N, Leipold F, Schmidt J E., J. At. Mol. Phys.1-7
39. Sari, F, P., Budiyono, B., Waste Technology **2**, 1, 17–25 (2014)
40. Kim, S, K., Lee, Y, Y., Kim, T, H., Bioresource Technology **199**, 42-48(2016)
41. Lay, M, K., Li, Y, Y., Noike, T., Journal Environment Engineering **124**, 730-36 (1998)
42. Shin, J, D., Han, S, S., Eom, K, C., Sung, S., Park, S, W., Kim, H., Environtment. Eng. Res. **13**, 2, 93-97 (2008)
43. Budiyono, Widiasa, I, N., Johari, S., Sunarso, International Journal of Chemical and Biological Engineering **3**, 1, 39-44(2010)
44. Mudhoo, A., Moorateeah, P, R., Mohee, R., World Academy of Science, Engineering and Technology **69**, 805-10(2012)
45. Patil, J, H., Malourdu, A, R., Mularidhara, P, L., Desai, M, L., Mahadeva, R, G, K., International Journal of Environmental Science and Development **3**, 2 (2012)
46. Adiga, S., Ramya, R., Shankar, B, B., Patil, J, H., Geetha, C, R., International Conference on Biotechnology and Enviroment Management 2nd. **14**, 73-78(2012)
47. Shankar, B, B., Jagadish, H., Patil, P, L., Muralidhara., Ramya, M, C., Ramya, R., Enviromental and Biogical Science **1** (2013)
48. Budiyono., Iqbal, S., Siswo, S., Engineering and Technology **7**, 3, 2798-2805 (2014)
49. Gregg, D., Saddler, J, N, A., Applied Biochemical Biotechnology 711-27 (1996)
50. Grethlein, H, E., Biotechnology. **3**, 155-60 (1985)
51. Palonen, H., Thomson, A, B., Tenkanen, M., Schmidt, A, S., Viikari, L., Applied Biochemical Biotechnology **117**, 1-17(2004)
52. Drapcho, C, M., Nhuan, N, P., Walker, T, H. *Biofuels Engineering Process Technology* (The McGraw-Hill Companies Inc., United States, 2008)

Study of C/N Ratio Effect on Biogas Production of Carica Solid Waste by SS-AD Method And LS-AD

Bakti Jos[1,*], *Fariha* Hundagi[1], *Rizqi* Pindy Wisudawati[1], *Budiyono*[1], and *Siswo* Sumardiono[1]

[1]Departement of Chemical Engineering, Faculty of Engineering, Diponegoro University, 50239, Semarang, Indonesia

Abstract. Biogas is a renewable energy which can be used as an alternative source to replace fossil fuels. Recently, the use of energy has become an important issue because the oil sources and natural gas are depleting. Utilization of carica waste to produce biogas can reduce the consumption of commercial energy sources such as kerosene as well as the use of firewood. Biogas is produced by the process of organic material digestion by certain anaerobic bacterial activity in anaerobic digester. In this study we studied the influence of LS-AD and SS-AD methods, the effect of C / N ratio on biogas yield obtained and kinetics of biogas production reaction. The study was conducted by making a total solid variation of 7%, 9%, 11%, 13%, 19%, 21%, 23% and C/N ratio 25 and 30. The study started with carica waste collection process and examination of the total composition of solids and water content. Thereafter, calculation and determination of variation of C / N ratio by mixing the substrate with inoculum and urea into the reactor. Observe the volume of biogas produced every two-day intervals. The highest biogas production rate of 1.7825 ml/g TS day was obtained from carica solid waste variable by liquid state anaerobic disgestion and C/N 25.

1 Introduction

Energy is a fundamental need for human survival. World oil prices and fuel requirements are increasing while crude oil reserves are depleting, so that sustainable energy supply has become one of the main challenges that humans will face in the future [26]. This moves the current economy towards alternative energy sources [4]. One of the alternatives is the utilization of renewable energy or that can be renewed energy [21]. Renewable energy source is one of environmentally friendly energy.

One of the sustainable biomass utilizations as a renewable energy source is biogas. Flexible renewable energy sources are derived from anaerobic digestion of manure, waste, residue, energy crops, water weeds [26, 38]. Biogas is a carbon-neutral source One of renewable energy, generated from the anaerobic digestion process of biodegradable waste, plant biomass, crop residues and manure. The main compositions of biogas are CH_4 (60-70%) and CO_2 (28-38%) and various gas products such as H_2S and N_2. Biogas production raw materials generally come from agricultural waste such as rice husks [31], rice straw [8,33], wheat straw [15] and livestock waste such as cow dung [24], sugar cane [32], *Eichhornia crassipes* [35], horse manure, pig manure and poultry manure [14]. There are also biogas raw materials derived from industrial waste alcohol [9-12], slaughterhouse wastes [6], fish industry waste [28], cassava starch [5], and tapioca industrial liquid waste [2, 20]. One of the industrial waste that can be processed into biogas is carica waste.

The carica fruit is commonly used for raw materials for making jams, sweets, cold drinks and cocktails [22,29]. From the remain of the production process it also produce another that are considered still can be utilized. One of the byproducts from the processing of candied of carica fruit is carica peels.

Carica fruit peels produced by many industries is still not utilized and become the burden of the industry. In a day, it can be produced as much as 860 kg of carica fruit that will produce carica peels waste as much as 200 kg per day. This means that within a year the peels of carica fruit obtained reaches 73,000 kg [16]. Therefore, these carica peels byproducts need further processing so as to have a positive impact on the industry. To produce biogas with anaerobic digestion, the decomposition of organic waste occurs in four processes simultaneously: hydrolysis, acidogenesis, acetogenesis, and methanogenesis [7, 38-39]. The purpose of this research is to know the effect of total solid and C/N ratio toward biogas yield obtained and to analyze the kinetics of biogas production reaction [34, 39] .

2 Experimental

The materials used in this research are carica waste obtained from carica making industry in Wonosobo area, while rumen cow fluid as inoculum obtained from slaughter house in Pedurungan, Semarang. The tools used are batch digester in the form of 1 liter bottle, rubber cover, hose and clamp, measuring cup, oven, scales, and statif. The independent variables in this study were C / N ratio 25, 30, and total solids 7%, 9%, 11%, 13%, 19%, 21%, 23%. The series of research tools can be seen in Figure 1.

* Corresponding author: baktijos@che.undip.ac.id

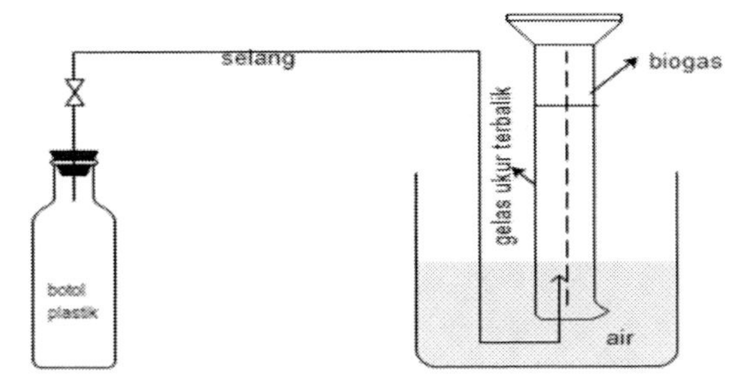

Fig 1. Research Tool Series

The implementation of this research started from the process of making carica waste (biowaste), then examined the total composition of solids and water content [37]. After that, calculation and determination of variation of C/N ratio and also preparation of anaerobic batch reactor. Subsequently, the substrate is mixed with technical inoculum and urea to adjust the C / N ratio. Samples that have been prepared can be put into the reactor, tightly closed to obtain anaerobic conditions, and ready to operate. During the process of processing, the volume of biogas produced each time interval of 2 (two) days is observed. The biogas volume is calculated by passing the gas into a water-filled measuring cup by utilizing the gas characteristic that is by pressing in all directions from the reactor (Boyle's Law), so that when the reactor valve is opened, the gas directly moves into the measuring cup and then being observed the volume of differences. Observation stopped after no more biogas was formed.

To determine the kinetics constant of biogas production was used Gompertz equation [30,36]. the Gompertz equation recently was found to be the most suitable model for describing the growth data of *Lactobacillus plantarum* and *Lactobacillus acidophilus*, which is written as

$$N = \int_0^t r_g \, dt = A \,.\exp\left\{-\exp\left[\frac{\mu_m}{A}(\lambda - t) + 1\right]\right\} \quad (1)$$

The relationship between the bacterial growth rate and the substrate utilization rate can be defined by the following equation:

$$r_g = Y_1 \,.(-r_{su}) \quad (2)$$

On the other hand, the relationship between the substrate utilization rate and the methane production rate is

$$-r_{su} = Y_2 \,. r_m \quad (3)$$

According to (2) and (3) the methane production rate can be defined as follows:

$$r_m = \frac{r_g}{Y_1 . Y_2} \quad (4)$$

Thus, the cumulative methane production M can be defined as

$$M = \int_0^t r_m \, dt = \int_0^t \frac{r_g}{Y_1.Y_2} \, dt = \frac{1}{Y_1.Y_2}\int_0^t r_g \, dt \quad (5)$$

If $\int_0^t r_g \, dt$ is replaced in (1), the resulting expression for the cumulative methane production is

$$M = \frac{A}{Y_1.Y_2} \,.\exp\left\{-\exp\left[\frac{(\mu_m/Y_1.Y_2).e}{(A/Y_1.Y_2)}(\lambda - t) + 1\right]\right\} \quad (6)$$

The term $\frac{A}{Y_1.Y_2}$ can be replaced by the term P, defined as the methane production potential, while $\mu_m/Y_1.Y_2$ can be defined as the maximum methane production rate. Therefore, the resulting final expression is

$$M = P \,.\exp\left\{-\exp\left[\frac{R_m.e}{P}(\lambda - t) + 1\right]\right\} \quad (7)$$

P = Production of biogas cumulative, liter
A = Production of biogas maximum, liter
U =Constant rate of production biogas maximum (liter/day)
λ = *lag phase* time (time minimum to produce biogas), day
t = Time cumulative to produce biogas , day
e = Euler (*e* = 2.71828...)

The equation above illustrates the equation of the biogas production rate at various observation times. Constants U, A, and λ can be determined by non-linear regression method toward data obtained from experimental studies using Polymath 6.0 Program.

3 Results and Discussion

3.1 Effect of Total Solid on C/N ratio 25 by LS-AD method

Daily biogas production volume varies depending on the total solid condition. This can be seen in Figure 2 which shows the daily biogas production rate at C / N ratio 25.

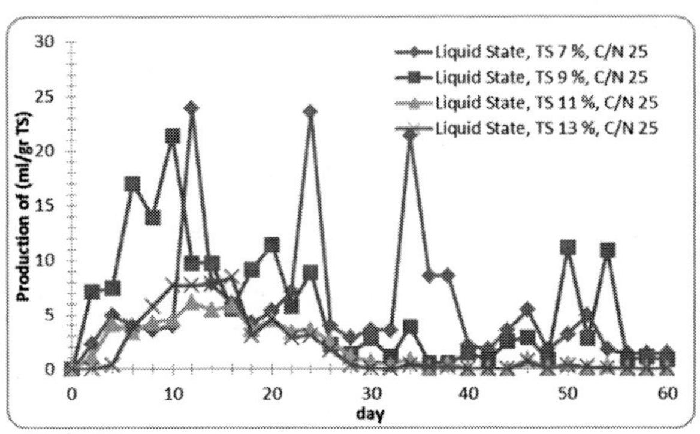

Fig 2. Daily biogas production rate C / N ratio 25

Figure 2 shows the daily biogas yield on carica waste with a C / N ratio of 25. In carica waste with a solid total of 7% fluctuating biogas production tends to increase until day 12. On the 22nd and 34th, days the biogas production went up drastically, but on the next day until the 60th day the fluctuating biogas production tended to decline but still produced biogas. In carica waste with a solid total of 9% biogas production tends to rise until day 10. However on the 12th day the fluctuations tend to decrease until the 60th day still produce biogas. In carica waste with a solid total of 11% biogas production tends to rise until day 12. However, on day 16, fluctuations tend to decrease until day 60 and still producing biogas. In carica waste with a solid total of 13% biogas production tends to rise until the 14th day. However, on day 16, fluctuations tend to decrease until day 60 and still producing biogas.

3.2 Effect of Total Solid on C/N ratio 25 by SS-AD method

Daily biogas production volume varies depending on the total solid condition. This can be seen in Figure 3 which shows the daily biogas production rate at C / N ratio 25.

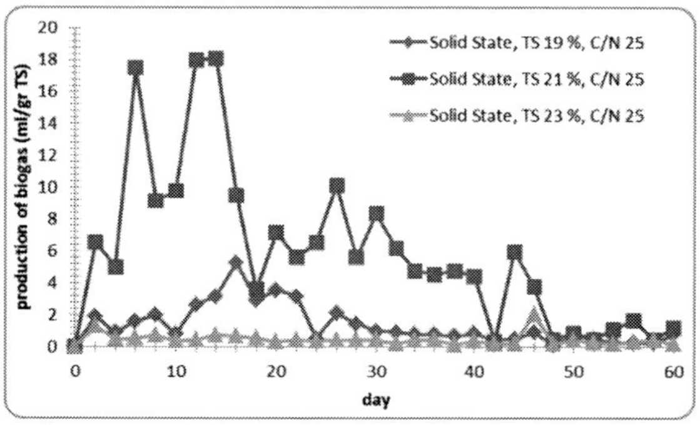

Fig 3. Daily biogas production rate C/N ratio 25

Figure 3 shows the daily biogas yield on carica waste with a C/N ratio of 25. In carica waste with a solid total of 19% biogas production tends to rise until the 16th day. However on the 18th, day the fluctuations tend to decrease until the 60th day and still produces biogas. In carica waste with a solid total of 21% biogas production tends to increase until the 6th day. However, on day 1,6 fluctuations tend to decrease until day 60 and still producing biogas. In carica waste with a solid total of 23% biogas production tends to stagnate or no significant change until day 60. Where the bacteria undergo the last phase, the phase of death (death phase) began to decrease the number of bacteria, until the bacteria die completely until the biogas is not reproduced. Biogas production is most common occured in the total solid 21%. This is evident where on the 12th and 14th day of production biogas reaches 780 ml volume.

3.3 Effect of C/N ratio on Carica Wastes on Biogas Production by SS-AD Method

Daily biogas production volume varies depending on the total solid condition. This can be seen in Figure 4 which shows the daily biogas production rate at C/N ratio 25 and 30.

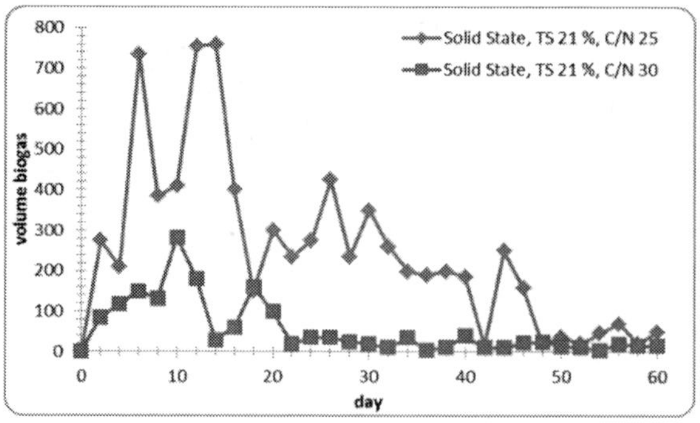

Fig 4. Daily biogas production rate

Figure 4 shows the daily biogas yield in carica solid wastes with C / N ratio 25 and 30. In carica waste with a C/N ratio of 25 fluctuating biogas production tends to increase until the 14th day. On the 16th day the fluctuating biogas production tends to decrease. However, biogas never rises again in harike 46. And then fluctuate down to day 60 still produce biogas. In carica waste the ratio of C/N 30 biogas production increased volatile until day 10. On the 18th day, biogas production fluctuates tends to decline until the 60th day and still produce biogas.

3.4 Kinetics of Biogas Production

By using non-linear regression method, data obtained from experiments can be obtained kinetic constants as presented in Table 4.1 as follows.

From table 1, it can be seen that the lowest Rm (biogas production rate) is a variable with a solid total of 23% C/N 25 ratio with a value of 0.044348 ml / g TS days. Where the highest value of Rm is found in carica waste with a total 9% solid state liquid condition and C/N 25 ratio which value is 1.78 ml / g TS day. Thus the total solid and C/N ratio affects the of biogas production, the time of biogas is formed also influenced by the nutrition factor given to substrates that are used as raw materials because the nutrients are toxic which can increase the pH and can influence the activity of bacteria to remodel the organic compound into biogas [23]. Brown, et al. (2013) also mentioned that significantly lower methane production was generated in SS-AD compared to L-AD [13]. Because a solid total value higher than 15-20% will increase the mass diffusion resistance, it will cause inhibition of hydrolysis and methanogenesis. The lower water content also causes the rapid accumulation of volatile fatty acids (VFAs) that inhibit microbial methanogens [1, 18-19].

Table 1. Kinetics data of biogas production reaction according to calculation of Gompretz equation

Variable	P	Rm	M	M calculation
TS 7 with C/N 25	23.9	1.707	171.2	23.9
TS 9 with C/N 25	21.39	1.782	175.5	21.39
TS 11 with C/N 25	6.13	0.438	61.5	6.13
TS 13 with C/N 25	8.5	0.472	65.7	8.5
TS 19 with C/N 25	5.26	0.292	41.6	5.26
TS 21 with C/N 25	18	1.125	181.45	18.
TS 23 with C/N 25	2.04	0.044	14.4	2.04
TS 7 with C/N 30	2.86	0.11	45.5	2.86
TS 9 with C/N 30	8.6	0.478	75.39	8.6
TS 11 with C/N 30	8.9	0.178	64.59	8.899
TS 13 with C/N 30	6.15	0.181	46.5	6.15
TS 19 with C/N 30	4.21	0.301	40	4.21
TS 21 with C/N 30	6.7	0.67	39.97	6.7
TS 23 with C/N 30	8.04	1.005	48.95	8.04

From table 1 it can be seen that the value of M (cumulative biogas production) between C/N ratio 25 and C/N ratio 30 is greater with C/N ratio 25. Because according to Yani and Darwis (1990), microbes that play an anaerobic need nutrients to grow and develop, in the form of carbon sources and nitrogen sources [3]. The equilibrium of carbon and nitrogen in substrates used as substrates needs attention [27]. Therefore, if too much nitrogen the growth of bacteria will be inhibited, in this case especially the material which ammonian content is very high [25], whereas the substrate is only a few nitrogen, bacteria will not be able to produce the enzymes needed to synthesize the compound (substrate) containing carbon, the most optimum substrate is C:N 25 because the optimum ratio C:N ranges from 20:1 to 30:1 [17].

4 Conclusions

Biogas production rate of carica solid waste is better in LS-AD condition than SS-AD. Optimum biogas production was obtained at C/N ratio 25 compared to C/N ratio 30. Kinetics constant of biogas production from carica waste has the highest production rate (Rm) constant of 1.7825 ml gr TS day with TS 9%.

Authors thank to Department of Chemical Engineering, Diponegoro University for all facilities to do this research.

References

1. A. Abbassi-Guendouz, D. Brockmann, E. Trably, C. Dumas, J.P. Delgenés, J.P. Steyer, R. Escudié, Bioresour. Technol. **111**, 55–61 (2012)
2. L. Appels, J. Lauwers, J. Degrve, L. Helsen, B. Lievens, K. Willems, Renew Sustain Energy Rev. **15,** 9, 4295e301 (2011)
3. M. Yani, A.A. Darwis, *Diktat Teknologi Biogas*. (Pusat Antar Universitas Bioteknologi, Institut Pertanian Bogor, Bogor, 1990)
4. Budiyono, I.N. Widiasa, S. Johari, Sunrso, Int J ChemBiolEng. **3**, 1, 39–44 (2010)
5. Budiyono, T.D. Kusworo, Int'l J. Sci. Eng. **2**, 1, 4-8 (2011)
6. B. Budiyono, I.N. Widiasa, S. Johari, S. Sunarso, Int'l J. Waste Resources (IJWR). **1**, 2, 4-7 (2011)
7. L. Yang, Y. Li, Bioresource Technology. **171**, 233–239 (2014)
8. A. Teghammar, G. Forgacs, I. Sarvari Horvath, M.J. Taherzadeh, Applied Energy. **116**, 125–133 (2014)
9. Budiyono, I. Syaichurrozi, S. Sumardiono, World Applied Sciences Journal. **26**, 11, 1464-1472 (2013a)
10. Budiyono, M.E. Pratiwi, I.N. Sinar, Alchemy Jurnal Penelitian Kimia. **9**, 2, 1-12 (2013b.)
11. Budiyono, I. Syaichurrozi, S. Sumardiono, Int'l J. Eng. **27**, 2, 177-184 (2014a)
12. Budiyono, I. Syaichurrozi, S. Sumardiono, Research Journal of Applied Sci. Eng. Tech. **7**, 13, 2798-2805 (2014b)
13. D. Brown, Y. Li, Bioresource Technology. **127**, 275–280 (2013)
14. M. Carlini, S. Castellucci, M. Moneti, Energy Procedia. **82**, 811-818 (2015)
15. R. Chandra, H. Takeuchi, T. Hasegawa, R. Kumar, J Appl Energy. **94**, 129-40 (2012)
16. Chasanah, Nurul, *Pemanfaatan Kulit Buah Carica Untuk Pektin Di Ud. Yuassa Food Berkah Makmur* (UGM, 2014)
17. D. Deublein, A. Steinhauser, *Biogas from Waste and Renewable Resource* (Wiley-VCH Verlag GmbH dan Co. KGaA ,Weinheim, 2008)
18. A. Schievano, G. D'Imporzato, L. Malagutti, E. Ragalli, G. Ruboni, F. Adani, Bioresour. Technol. **101**, 5728–5732 (2010)
19. F. Di Maria, G. Gigliotti, A. Sordi, C. Micale, C. Zadra, L. Massaccesi, Waste Manage. Res. **31**, 869–873 (2013)
20. H.R. Rahmatul, A. Nurrokhim, N. Soewarno, S. Nurkhamidah, Jurnal Teknik Pomits. **2,** 1 (2013)
21. T. Haryati, J. Wartazoa. **16**, 160 – 169 (2006).
22. H. Idstein, T. Keller, P. Schrejer, J. Agric.Food.Chem. **33**, 4, 663–666 (1985)
23. A.N. Ikrimah, A. Pramianshar, *Inovasi Teknologi Produksi Biogas Secara Anaerobik dari Sludge Limbah Industri Pengolahan Ikan sebagai Energi Terbarukan Masa Depan* (Jurusan Teknik Kimia, Fakultas Teknik, Universitas Diponegoro, Semarang, 2014)

24. D.A. Putri, R.R. Saputro, Budiyono, International Journal of Renewable Energy Development (IJRED). **1**, 2, 61-64 (2012)
25. M. Kumar, J.G. Lin, J. Hazardous Materials. **178**, 1–9 (2010)
26. Patowary, Dipam., W. Helen, C. Michele, D.C. Barurah, *"Biogas Production from Surplus Plant Biomass Feedstock: Some Highlights of Indo-UK R&D Initiative"*. International Conference on Solid Waste Management. (2015)
27. J. Mata-Alvarez, J. Dosta, S. Mace, S. Astals, Crit.Rev.Biotechnol. **31**, 2, 99-111 (2011)
28. M.F. Maulana, Sudarno, I.W. Wardhana, *Pengaruh Pengenceran dan Pengadukan Limbah Industri Ikan Nila terhadap Peningkatan Produksi Biogas with Menggunakan Rumen Sapi sebagai Starter* (Program Studi Teknik Lingkungan Universitas Diponegoro, Semarang 2013)
29. M.A. Moya-León, M. Moya, R. Herrera, Postharvest Biology and Technology. **34**, 211–218 (2004)
30. A. Nopharatana, P.C. Pullammanappallil, W.P. ClarkeWaste Management **27**, 5, 595-603 (2007)
31. O.C. Okeh, C.O. Onwosi, F.J. Odibo, Renewable Energy. 204-208 (2013)
32. S. Sumardiono, A. Budi Riyanta, H. Hawali Abdul Matin, B. Jos, Budiyono, *Increasing Biogas Production From Sugar Cane Baggase By Anaerobic Co-Digestion With Animal Manure.* MATEC Web of Conferences. **101**, 02014 (2017)
33. Budiyono, A. Wicaksono, A. Rahmawan, T. Djoko Kusworo, S. Sumardiono, MATEC Web of Conferences. **101**, 02011 (2017)
34. T. Prasetyo, S. Sumardiono, H.A. Aji, A.Y. Pratama, Advanced Science Letters. **23**, 6, 5810-5814 (2017).
35. S. Sumardiono, Budiyono, D.T. Mardiani, AIP Conference Proceedings. **1699**, 050018 (2015).
36. Budiyono, I. Syaichurrozi, S. Sumardiono, Research J.Appl.Sci.Eng.Tech. **7**, 13, 2798-2805 (2014).
37. Budiyono, I. Syaichurrozi, S. Sumardiono, Int'l. J.Eng. Transactions B: Applications. **27**, 2, 177-184 (2014).
38. Budiyono, I. Syaichurrozi, S. Sumardiono, World Appl.Sci.J. **26**, 11, 1464-1472 (2013).
39. Syaichurrozi, Budiyono, S. Sumardiono, Bioresource Technology **149**, 390-397 (2013).

Production of Biodiesel from Mixed Waste Cooking and Castor Oil

Hadiyanto Hadiyanto[1,2,3*], *Inaya* Yuliandaru [1], and *Rafidha* Hapsari [1]

[1]Chemical Engineering Department, Faculty of Engineering, Universiats Diponegoro
[2]Master program of Environmental Sciences, School of Postgraduate Studies, Universitas Diponegoro
[3]Center of Biomass and Renewable Energy(CBIORE), Universitas Diponegoro

Abstract. Due to increasing population growth, the consumption and needs of energy increase significantly. This leads Indonesia government to search alternative energy to cover the lacks of fossil energy reserves. Biodiesel is one of the prospective alternative energy which are renewable and environmental friendly. A common problem in large-scale biodiesel production is the sustainability of feedstock and the biodiesel stability. Therefore, the purpose of this study was to evaluate the production of biodiesel from two oil sources i.e. waste cooking oil and castor oil. This study examined the effect of mixed oil ratio on yield, biodiesel characteristics and stability. The physical properties included kinematic viscosity, acid number, saponification number, iodine number and cetane number have been evaluated as function of oil ratio. Yield of biodiesel was obtained at 35.07%, 99.2% and 83.69% for jatropha:castor oil ratio of 1: 0, 1: 2 and 2: 1, respectively. Most of these characteristics showed an increase by increasing the oil ratio. The result concluded that at the ratio of 1:1(v/v) was the best characteristic and stability.

1 Introduction

Indonesia is the fourth most populous country in the world after China, India, the United States with a population of 237,641,326 people [1]. The growth of the Indonesian population from the period of 2000-2010 averaged 1.49% per year, this gives a broad impact on energy needs. Energy consumption is still dominated by fuel oil (gasoline, diesel oil, kerosene, aviation fuel and avgas) [2]. Indonesia's oil reserves currently range from 4 billion barrels with an average oil production of 1 million barrels per day, while the demand for fuel oil reaches 1.2 million barrels per day so that the remainder of Indonesia's fuel needs is still imported. Due to this, the government is trying to develop alternative energy in preparation for the fuel oil crisis. One form of effort that has been done by the government is the utilization of biofuel or biodiesel [2].

Biodiesel is a mono-alkyl ester of a long chain of fatty acids having a carbon chain between 6-24 depending on the base material [3]. The presence of oxygen in biodiesel distinguishes it from petroleum diesel whose main component consists only of hydrocarbons. Biodiesel has chemical and physical properties similar to diesel therefore it can be used directly for diesel engines mixed with diesel [4].

Although the use of biodiesel benefits in many ways, which relies solely on single feedstocks, it also raises several problems like competition for food needs and the need for extensive land for the life cycle of biodiesel production [3]. Biodiesel production in Indonesia which still relies on palm oil affects the loss of biodiversity due to deforestation to meet the demand for palm oil.

Multi-feedstock biodiesel can reduce the dependence of biodiesel production on one type of raw material only. Based on research conducted by Elma et al., [5] the production of multi-feedstock and single feedstock biodiesels from coconut oil and cooking oil, the multi-feedstock biodiesel was producing bigger conversion compared with the single feedstock biodiesel.

In the process of making biodiesel, an esterification and / or transesterification stage will be performed. The esterification reaction is a reaction between the carboxylic acid and the alcohol forming the ester. The carboxylic acid derivative forms a carboxylic acid ester. The carboxylic acid ester is a compound containing the -COOR group with R may be either alkyl or aryl. Esterification of acid catalysed and reversible [6]. The esterification reaction converts the free fatty acids contained in the triglycerides into methyl esters. Meanwhile, transesterification is the reaction of an alcohol group exchange of an ester with another ester. The addition of an alkaline catalyst will accelerate the transesterification reaction in the formation of the ester [8]. Alcohols that can be used in the transesterification process of triglycerides are methanol, ethanol, propanol, butanol, and amyl alcohol.

In this research, multi-feedstock biodiesel will be produced, two kinds of oil will be varied composition as base materials. The materials are waste cooking oil and castor oil. The oil mixture will be processed into biodiesel through esterification and transesterification. This research will be focused on the production of multi feedstock biodiesel and biodiesel's stability analysis in order to know the composition of the mixture which able

* Corresponding author: hadiyanto@live.undip.ac.id

to produce biodiesel with a large yield, meet the standards and have the best stability.

2 Materials and Method

2.1. Materials

Waste cooking oil from Warung Burjo around Tembalang and Castor Oil from Chemical Shop Indrasari, Semarang. KOH, NaOH, H2SO4, methanol, distilled water, ethanol 95%, indicator pp, HCl 0,5 N, Na2SO3, Chloroform, Wijs solution, CH3COOH were obtained from Chemical Store in Semarang.

2.2 Raw Materials Pre-Analysis

Preliminary analysis was performed to determine the initial FFA and molecular weight of the oil. If FFA <2.5% esterification was not necessary.

2.3 Esterification

The transesterification process was carried out on a three-neck flask at 60°C with an agitation of 400 rpm. The oil mixture at the top of the funnel separator was introduced into the third neck flask and then KOH mixed with methanol corresponding to the ratio was inserted and reacted for 90 min.

2.4 Purification

Purification was performed 2 stages; first stage is separation by separating funnel. Two layers would be formed and separated, the top layer, a mixture of oil, methanol and methyl esters, was further purified. The bottom layer which is a glycerol was not needed. The second stage was washing by adding 20 mL water 60°C on the top layer in separating funnel, wait for one minute until two layers formed. The washing stage would be done until the washing water is clean.

2.5 Analysis

The biodiesel synthesis results were qualitatively identified using Gas Chromatography Mass Spectrometry (GCMS) and quantitative analyzed by titration method i.e. kinematic viscosity at 40°C, acid number, saponification number, iodine number, and cetane number. This analysis was conducted each week during experiment.

3 Results and Discussion

3.1. Effect of oil ratio to biodiesel yield

From Figure 1 it is shown that the yield of single feedstock of waste cooking oil was very low, about 35.07% as obtained from GCMS analysis. Low yield may be caused by oleic acid content of waste cooking oil. Oleic acid is susceptible to oxidation reactions because it has a low activation energy [9]. The presence of oxidation reactions will reduce the fatty acids that should turn into methyl esters. In contrast, the castor oil contains risinoleic acid which is not easily oxidized. Therefore, a mixture of waste cooking oil and castor oil can produce higher yield biodiesel. The test results showed biodiesel yield with 1:2 oil volume ratios of 99.20% and 2:1 (vol/vol) of 83.69%.

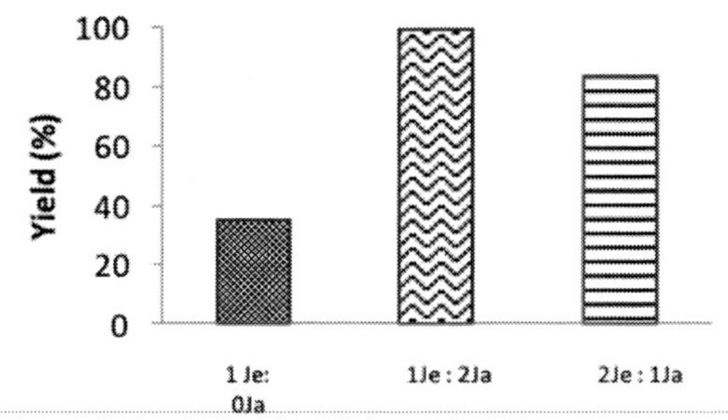

Fig. 1. Yield of biodiesel with variation of ratio of waste cooking oil (Je): castor oil (Ja) (v/v)

3.2. Effect of oil ratio to biodiesel properties

The increase of oil ratio will affect to decrease biodiesel kinematic viscosity at 40°C (Fig 2). Biodiesel produced by castor oil has a high viscosity because its content is dominated by risinoleic acid. The high content of risinoleic acid will provide higher viscosity [10]. Mixing of oil will affect the composition of the fatty acids contained in the oil. Since waste cooking oil is dominated by oleic acid which results in lower viscosity, mixing of waste cooking oil with castor oil can produce biodiesel with lower viscosities. In addition, high viscosity can also be caused by the presence of fatty acids in the product and does not turn into methyl esters[11].

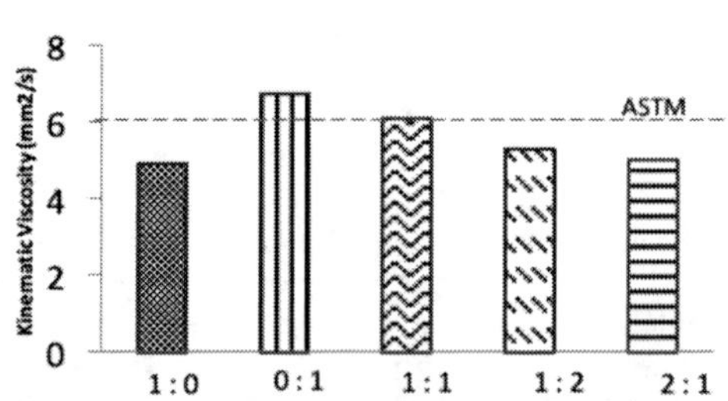

Fig. 2. Kinematic viscosity of biodiesel at various oil ratio (Je : Ja)

Figure 3 shows that biodiesel from waste cooking oil has the largest saponification number. This is due to the dominant oleic acid in waste cooking oil has shorter carbon chains than the dominant fatty acids in castor oil which causes smaller molecular weights. Smaller molecular weights will yield relatively larger saponification number. In waste cooking oil mixed with castor oil, it is found that the saponification number became lower. Castor oil contains the dominant fatty acids in the form of ricinoleic acid which have longer carbon chain. So that the molecular weight is relatively larger which causes the saponification value tends to be lower.

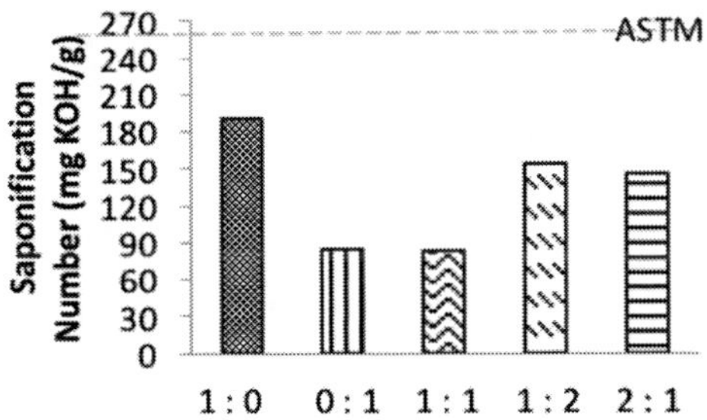

Fig. 3. Saponification number of biodiesel at various oil ratio (Je:Ja)

Based on Fig 4 it is shown that the variables which have higher castor oil ratio resulted higher acid number than the waste cooking oil. High acid numbers can be caused by the oxidation process. Contact with air can also cause oxidation because oxygen can lead aggressive biodiesel reaction and produce degradation compounds such as acids, aldehydes, ketones, dissolved compounds[12]. Acid can be formed because of the water content in biodiesel causing the hydrolysis reaction that converts the ester into alcohol [13].

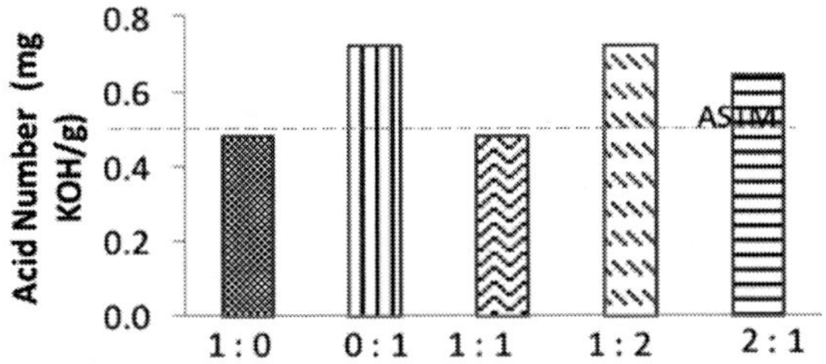

Fig. 4. Acid number of biodiesel at various oil ratio (Je:Ja)

The iodine number is strongly influenced by the double bonds in the dominant fatty acids in the oil. Cooking oil and castor oil composition are predominantly unsaturated fatty acids (double bonded). This leads the mixing of both oils will increases the iodine number (Fig 5). In biodiesel, high iodine is less desirable because biodiesel with high iodine tends to be polymerized so that it is less stable [14].

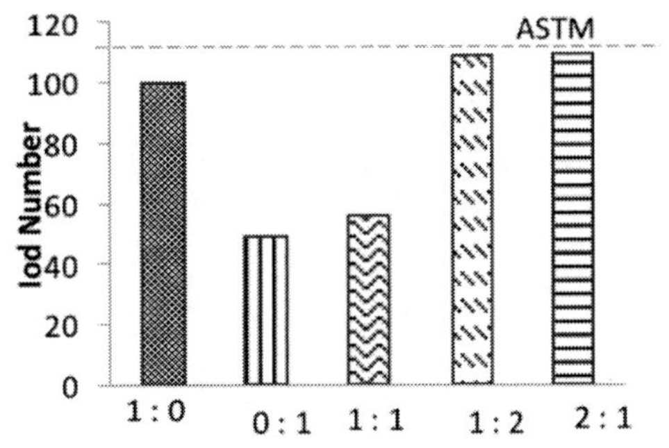

Fig. 5. Iod Number of biodiesel at various oil ratio (Je:Ja)

Cetane numbers are strongly influenced by the chemical structure of fatty acids that composed the biodiesel(Fig 6). The more double bonds and the shorter the carbon chain will decrease the cetane number [15]. Mixing oil produces biodiesel with a higher cetane number. The addition of castor oil that has a fatty acid with fewer double bonds and a longer carbon chain will result in larger cetane numbers.

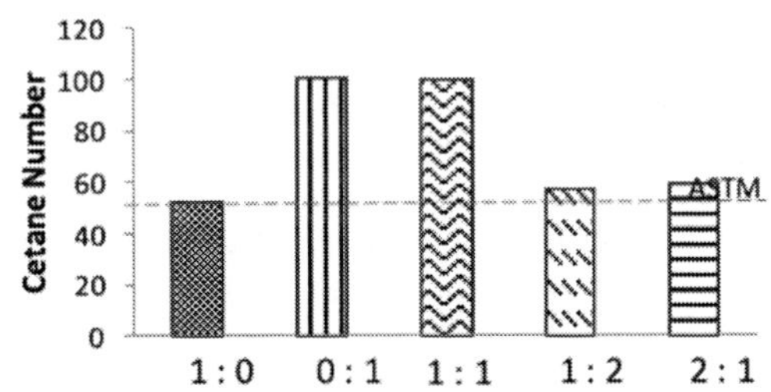

Fig. 6. Cetane number of biodiesel at various oil ratio (Je:Ja)

4 Conclusion

Biodiesel with the highest yield was obtained with the ratio of volume of cooking oil to castor oil at 1: 2 with yield of 99.20%. The characteristics of biodiesel are greatly influenced by the fatty acid content contained in the oil of the manufacture raw material. Mixing of the oils increased or decreased the biodiesel characteristics such as viscosity, density, saponification, and cetane number. Mixing of raw materials did not result in a stable biodiesel than mixture of oil feedstock. The 1Je: 1Ja composition is ratio that gave the best result in terms of biodiesel physical characteristic and the product almost meet the standard.

References

1. BPS (Badan Pusat Statistik). https://bps.go.id/website/pdf_publikasi/Distribusi-Perdagangan-Komoditas-Minyak-Goreng-di-Indonesia-2016.pdf. Accessed on 1 Juli 2016
2. A. Sugiyono, L. Anindhita, M.A. Wahid & Adiarso.(ed), *Outlook Energi Indonesia 2016* (Pusat Teknologi Sumber Daya Energi dan Industri Kimia, Jakarta, 2016)
3. M.I. Jahirul, R.J.Brown, W.Senadeera, I.M. O'Hara, Z.D.Ristovski. Energies. **6**, 3764-3806 (2013)
4. P. Verma and M.P. Sharma. Renewable and Sustainable Energy Reviews. **62**, 1063-1071 (2016).
5. M.Elma, S.A. Suhendra and W.Wahyuddin. Konversi. **5 (1)** (2016)
6. R.J. Fessenden, J.S.Fessenden, and M.W.Logue, Organic Chemistry (2nd edition). Brooks Cole Publishing Company (1982).
7. J. Pullen and S. Khizar. Renewable and Sustainable Energy Reviews **16** 5924-5950 (2012).
8. Istadi, *Teknologi Katalis untuk Konversi Energi : Fundamental dan Aplikasi (1sted.)* (Graha Ilmu, Yogyakarta, 2011)
9. D. Desnelli and Z. Fanani. Jurnal Penelitian Sains **12**, 1(C) (2009)
10. A.Amin, A. Gadallah, A.K. El Morsi, N.N. El-Ibiari, G.I. El-Diwani. 2016. Egyptian Journal of Petroleum. **25 (4),** 509-5014 (2016).
11. T.Kusumaningsih, P.Pranoto and R. Saryoso. Bioteknologi **3(1)**, 20-26 (2006).
12. S.Silviana and L.Buchori. Reaktor, **15(3),** 148-153 (2015)
13. A. Bouaid, M. Martinez , J. Aracil. Fuel, **86,** 2596–2602 (2007)
14. L. Adhani, A. Isalmi, S. Nurbayati and C. O. Okatviana. Jurnal Penelitian dan Pengembangan Ilmu Kimia, **2(1)**, 71-80 (2016).
15. M. Angelovic, Z. Tkac and J. Jablonicky. Animal Science and Biotechnologies, **47 (1)**, 2014

Effect of MWCNT Filler on Properties and Flux of Chitosan/ PEG based Nanocomposites Membranes

Fitri Khoerunnisa[1*], Hendrawan[1], *Dwi Rizki* Primastari[1], and *Riska* Agiawati[1]

[1]Department of Chemistry, Indonesia University of Education, Setiabudi 229 Bandung, Indonesia 40154

Abstract. Biopolymer are expected to be environmentally compatible and to have great potential application as membranes material. The chitosan-poly (ethylene glycol)/PEG based composite membranes was successfully synthesized via inversed phase method. The effect of multiwalled carbon nanotubes (MWCNT) as nanofiller on properties and performances of composite membranes were intensively evaluated. The membrane was prepared by mixing of chitosan and PEG solutions at the same composition ratio while MWCNT amount in the mixture was varied. The synthesized membrane was characterized by means of FTIR spectroscopy, scanning electron microscopy (SEM), contact angle, and tensile strength measurement. The performance of composite membrane on filtration was evaluated in term of flux (permeability) and rejection (rejection) tests. The results showed that the optimum volume ratio of composite membrane solution was found at 30:10:7.5 for chitosan/ PEG/ MWCNT, respectively, as indicated by the largest flux. Insertion of MWCNT nanofiller notably enhanced hydrophilicity, porosity, and mechanical properties of composites membranes that are confirmed by contact angle, SEM images and elongation forces value, respectively. The MWCNT nanofiller remarkably increased both of flux and rejection of composite membranes up to 60 Lm^2h^{-1} and 96%, respectively. The remarkable enhancement of composite membrane performance is attributed to the effective interaction of MWCNT with polymeric matrix.

1 Introduction

The availability of clean water has become a global problem because of the continuously increasing costs of energy and increasing scarcity of water resources [1]. To address the undeniable need of pure water, various water treatment technologies have been proposed and applied at experimental and industrial levels [2]. Filtration techniques by membranes have been deemed more cost and energy effective than other separation methods [1,3]. The ideal membrane system should possess an excellent stability under a wide range of processing conditions, high selectivity and produce a high mass flux with minimal driving force [4-6]. To achieve this goal, both the composition and structure of the membrane are crucial.

Different kinds of membranes from polymers are such potential materials and membrane technology is one such technology at the first line of decontamination and purification techniques [7-10]. During the last few years, continues development of membrane materials for water purification from bio based and renewable polymers, like chitosan have attracted great interests [11] since it is abundantly available as natural sources [12], low cost, compatibility, and has limited impact to the environment during processes and degradation [13].

Application of biopolymer chitosan based membrane has the shortage on water purification due to its low mechanical properties and porosity. Many efforts have been reported on modification of membrane performance. Copolymerization of chitosan with other polymers i.e. poly (ethylene glycol)/ PEG, were frequently reported [14]. On the other hand, use of nanotechnology for membrane fabrication is a new promising method to improve membrane performance through incorporation of nanomaterials [15]. The addition of nanoparticles can alter the membrane morphology, improve permeability and reduce fouling properties of membranes, due to their high specific surface area, chemical stability and feasible functionalization [16,17]. Nanocomposite membranes can be improved by assembling nanoparticles into membranes or blending them with organic or inorganic additives [18]. Some membranes were fabricated using silica, graphite, metal oxide nanoparticle and carbon nanotubes to increase membrane permeability, and fouling resistance [19-21].

Multiwalled carbon nanotubes (MWCNTs), on the other hand, have attracted significant attention as a new type of nanofillers due to their unique structures and properties [22]. The incorporation of MWCNTs into polymeric materials has been examined in many

* Corresponding author: fitri@upi.edu

applications to enhance the mechanical, thermal, and chemical stabilities of polymer matrixes in sever conditions such as high temperature and strongly oxidizing environment,, and also to improve the separation properties of membranes based on the synergism between the organic-inorganic components properties [23]. A theoretical study [24] predicts a rapid mass transport in MWCNTs due to smooth nanotube walls, which may be several orders of magnitude higher than in many other porous fillers, making MWCNTs an ideal candidate filler material for membranes. MWCNT-incorporated membranes have been reported to have extraordinarily high transport rate to liquid water [24,25].

In this study, chitosan/PEG/MWCNT based composite membrane was synthesized by phase inversion method. The effect of MWCNT's concentration in the casting solution during preparation process on the properties and performance (flux) of membrane were intensively studied.

2 Experimental method

2.1 Materials

All chemicals used in the experiments were of reagent grade and all organic solvents were HPLC grade. The MWCNTs (95% of purity) with 20 to 50 nm in outer diameter were obtained from Wako Chemical, Japan. The impurity of MWCNT was removed through acid treatment. Chitosan (molecular weights of 300.000; deacetylation degree of 85%) and PEG/ poly ethylene glycol (mol. wt. of 6000), acetic acid, and sodium hydroxide were purchased from Merck.

2.2 Preparation of chitosan/PEG/MWCNT membrane

Chitosan/PEG/MWCNT filtration membrane was prepared via phase inversion method. The compositions of MWCNT was varied are shown in Table 1. Chitosan (3 *wt%)* and PEG (2 *wt%*) solutions were prepared in acetic acid (0.1 M). MWCNT dispersion (1 *wt%*) was prepared in graphite oxide (GO) colloid. The optimum composition of chitosan/ PEG was followed the preceeding result [14].

Table 1. Cast solutions composition

Sample description	Chitosan (mL)	PEG (mL)	MWCNT (mL)
M0	30	10	0
M1	30	10	2.5
M2	30	10	5.0
M3	30	10	7.5
M4	30	10	10

MWCNT dispersion, chitosan and PEG solutions were homogeneously stirred for 30 min under heating at 50°C. Hereafter, the mixtures were cooling down at room temperature (25°C) for 10 min. The mixtures were further casted on PTFE template and dried at room temperature. The dried membranes was removed from the template and immersed in NaOH solution (1 M) for 3 h, then rinsed with aquadest until obtaining neutral pH, and further dried at room temperature for characterization and performance tests in ultrafiltration processes.

2.3 Membrane characterization

The FTIR spectra of chitosan/PEG/MWCNT nanocomposite membrane with different MWCNT contents were determined between 4000 and 400 cm^{-1} using Shimadzu FTIR spectrometer. Each spectrum was collected by cumulating 40 scans at a resolution of 4 wavenumbers.

A Bruker X-ray diffractometer was used to study the solid-state morphology of the chitosan/ PEG/ MWCNT nanocomposite membranes. The XRD patterns were recorded at a diffraction angle (2θ) of from 5 to 60 to examine the crystal structure.

A scanning electron microscope (SEM-JEOL) directly provides the visual information of the cross-sectional morphology of the membrane.

Membrane hydrophobicity was quantified by measuring the contact angle that was formed between the membrane surface and water. Contact angles were measured using the sessile drop method by using a laboratory-constructed camera apparatus (JVC-TK 1270). One mL water droplets were placed at different positions of the membrane surface for replicates. The average value of at least five measurements was reported.

The overall porosity of all the membranes was determined using a gravimetric method defined in Eq. (1) [26]

$$\varepsilon = \left(\frac{W_{wet} - W_{dry}}{A \times l \times \rho_w} \right) \quad (1)$$

Here W_{wet} is the weight of the membrane immersed in pure water for 3 days, W_{dry} is the weight of the membrane dried in a vacuum oven at temperature 80 °C for 12 h, A is the membrane effective area (m^2), l is the membrane thickness (m) and ρ_w is the water density (0.998 g/cm^3).

2.4 Performance test of membrane

The flux (permeability) and selectivity (rejection) tests were conducted to obtain the information of membrane capability in separation and purification processes with dead-end filtration method. In particular, the flux of membranes was measured by insertion of water into filtration cell (diameter of filtration cell was 4.5 cm, with an effective area of 15.90 cm^2) with a stable stirring speed of 200 rpm. Prior to filtration experiment, each membrane was compacted at 0.20 MPa for 30 min to reach a stable flux, then the pressure was adjusted to 0.10 Mpa. The flux measurement was conducted for 75

 https://doi.org/10.1051/matecconf/201815604001

min. The pure water flux J ($Lm^{-2}h^{-1}$) was measured under steady state condition through measuring the permeate water volume and calculated as the following equation (2) [27]

$$J = \frac{V}{A \times \Delta t} \quad (2)$$

where V (L) is the volume of permeated water, A (m^2) is the efficient membrane area and Δt (h) is the permeation time. The flux of feed solution was recorded as J ($Lm^{-2}h^{-1}$) determined by the water quantity penetrating the membranes.
Meanwhile, the membranes rejection efficiency (selectivity) was measured using the same filtration cell. The stirred filtration cell was refilled rapidly with model foulant feed solution (dyes solution of 300 ppm) and the pressure of 0.1 MPa was applied during measurement the feed and the permeate concentrations was measured via UV spectrophotometer (UV-Vis, Shimadzu) at 644.5 nm and calculated according to equation (3) [28]

$$\%R = \left(1 - \frac{C_p}{C_f}\right) \times 100\% \quad (3)$$

where C_p and C_f are the concentration of foulant in permeate and feed solutions, respectively.

3 Results and Discussion

3.1 Optimum composition of MWCNT filler

The optimum composition of MWCNT filler in membrane matrix was determined from permeability/flux of membrane. Figure 1 shows the flux of nanocomposite membranes at difference MWCNT composition.

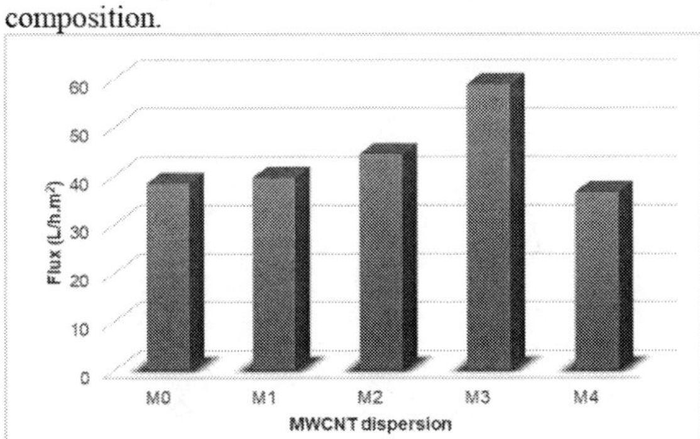

Fig 1. Flux of nanocomposites membrane at different MWCNT composition

The addition of MWCNT proportionally increases permeability of membrane composites where the optimum composition of chitosan/PEG/MWCNT based nanocomposites membrane is achieved at M3 (flux of 59 L/ h.m^2). This may caused by several factors i.e. (1) presence of hydrophilic groups (-OH and –COOH) that enable facilitate the interaction between filler and functional moeities of polymer matrix, (2) enhancement of hydrophilicity of membrane surface thus reduces fouling and consequently increase flux membrane [29], (3) MWCNT effectively inserted into the pore of membrane matrix and created matrix network (nanochannel) through pore connection for water transport [30], and (4) specific interaction of MWCNT with polymer functional groups due to charges modification [31].
However, the addition of MWCNT beyond the optimum composition leads to the reduction in membrane permeability due to aggregation of MWCNT bundles, thus blocking of membrane pore [32].

3.2 Morphological characterization

3.2.1 Fourier-transform infrared spectroscopy (FTIR)

Fourier-transform infrared spectroscopy (FTIR) was used to identify the interaction of functional groups of membrane precursors. Figure 2 demonstrates the FTIR spectra of membranes at optimum composition.

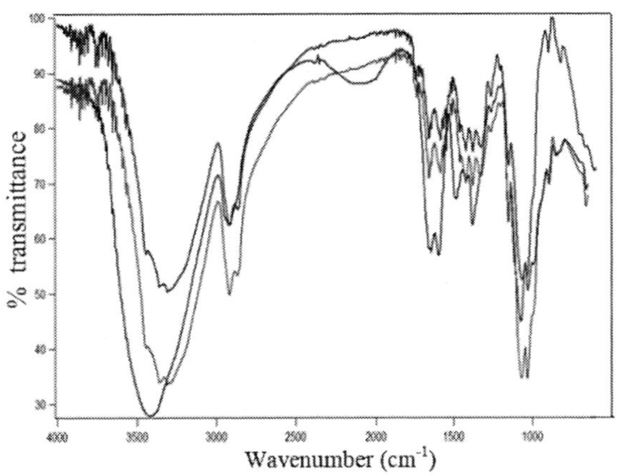

Fig 2. FTIR spectra of chitosan (black), chitosan-PEG (red), and chitosan/PEG/MWCNT nanocomposites membrane at optimum composition (blue).

The FTIR spectra of chitosan/PEG membrane and Chitosan/PEG/MWCNT membrane (Figure 2) shows absorption bands at 3446; 3365; 3299 cm^{-1} that sequencely indicate stretching vibration of hydoxyl O-H, stretching vibration of amine N-H, and N-H/O-H polimerization vibration (Coates, 2000). Absorption bands at 2918 dan 2872 cm^{-1}, respectively, show asymmetric stretching vibration of methylene ($-CH_2$) and symmetric stretching vibration of methyl ($-CH_3$), these finding was also supported by the presence of absorption bands at 1045-1018 cm^{-1} which indicates a contribution of PEG crystal's C-O bending vibration (Wang *et al.*, 2007). Meanwhile, chitosan presence is shown by an abrorption band around 1640-1500 cm^{-1} that refer to amide II (N-H) vibration. There are a little band shifts from wavenumber of 3292 to 3304 cm^{-1} that might be caused by an increase in number of intermolecular hydrogen bonds, as an interaction between (O-H/N-H) of chitosan and functional groups in MWCNT. The decrease in absorption band intensity may be caused by the decrease in energy absorbed by functional groups as a consequence of MWCNT interaction with Chitosan/PEG copolymeric matrix.

3.2.2 X-Ray Diffraction(XRD)

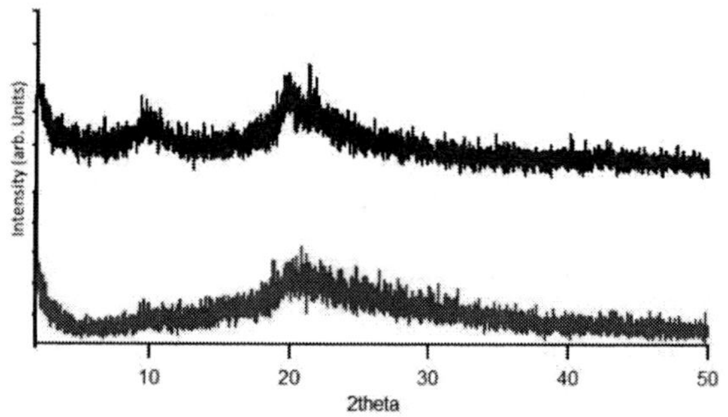

Fig 3. X-ray diffraction patterns of Chitosan/PEG membrane (red) and Chitosan/PEG/MWCNT nanocomposites membrane (black)

The X-ray diffraction pattern of Chitosan/PEG membrane demonstrates the broad peak at 2θ=19.9°-21.8°, indicating the predominant amorf structure of that copolymer. The addition of MWCNT into chitosan/PEG membrane notably increases the peak intensity owing to insertion of MWCNT filler into polymeric matrix. Additional, the new peak at small angle scattering (2θ = 10°) originated from intertube interaction of MWCNT (interlayer distance) bundles (Figure 3).

3.2.3 *Tensile Strength*

MWCNT addition significantly increased Chitosan/PEG membrane mechanical strength (Rahimipour *et al.*, 2012) as indicated by the increasing % elongation at break and maximum force (g) of the membrane after MWCNT addition (Table 3). The increase in mechanical strength of copolymer through addition of MWCNT is due to formation of hydrogen bonds between Chitosan (-NH_2) and MWCNT functional groups (-COOH and –OH).

Sample	Elongation at Break (%)	Maximum force (g)
Chitosan/PEG	52.07	1060,95
Chitosan/PEG/MWCNT	53.40	1103,74

Table 3. Tensile strength of Chitosan/PEG and Chitosan/PEG/MWCNT membranes.

3.3. Surface morphology of membrane nanocomposites

3.3.1 *Cross-sectional SEM images and porosity*

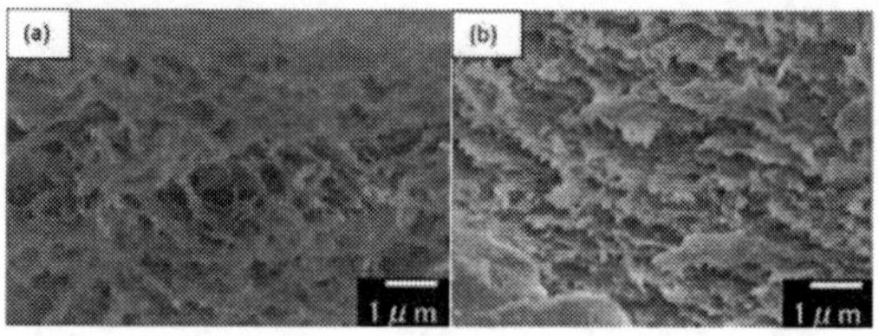

Fig 4. SEM cross-sectional images of (a) Chitosan/PEG and (b) Chitosan/PEG/MWCNT nanocomposite membranes.

Figure 4 shows the cross-sectional images of the membranes with and without the addition of MWCNT. In general, both of membranes have heterogeneous and asymmetrical porous stucture pores. Addition of MWCNT nanofiller induces the expansion of pore size and interlayer distance of copolymeric matrix, due to interconnection of polymer with MWCNT (Tang *et al.*, 2009). The modification of pore size of Chitosan/PEG membrane by MWCNT insertion was verified by measurement of average pore radius (Figure 5). and membrane porosity (Figure 6). The improvement of pore size of copolymer membrane is owing to MWCNT addition into the membrane matrix.

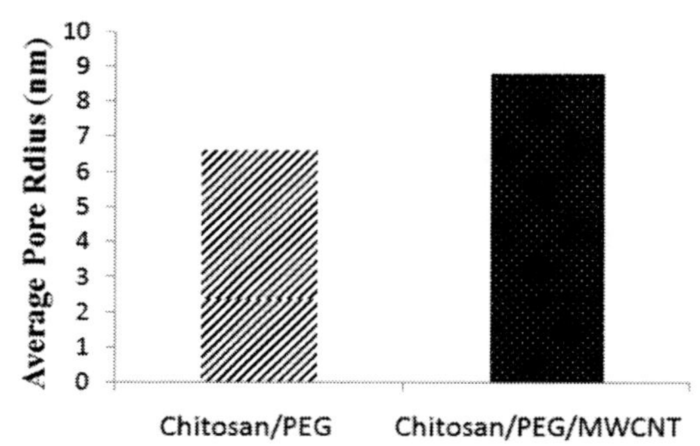

Fig 5. Average pore radius of the membranes (left) Chitosan/PEG and (right) Chitosan/PEG/MWCNT

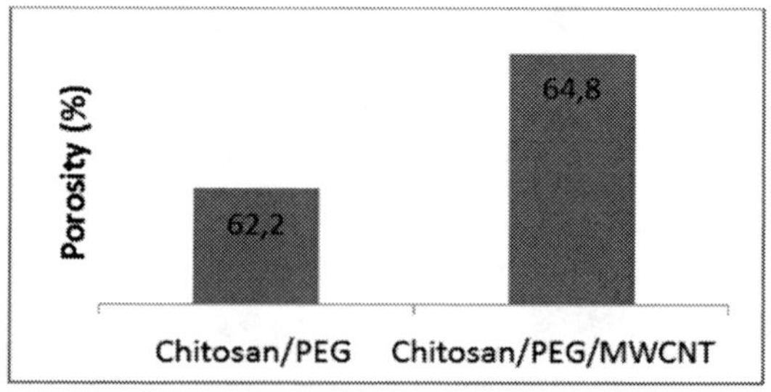

Fig 6. Porosity of Chitosan/PEG (left) and Chitosan/PEG/MWCNT (right) membranes

3.3.2 Contact Angle

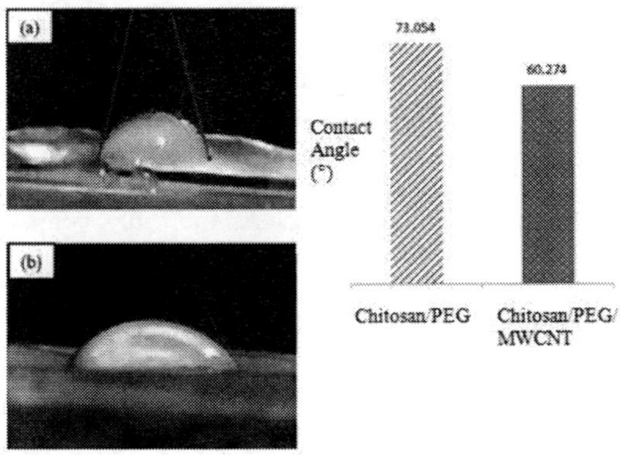

Fig 7. Contact angle of Chitosan/PEG (a) and (b) Chitosan/PEG/MWCNT membranes

Membrane hydrophilicity was tested using sessile drop test to measure the membrane surface contact angle. As shown inFfigure 7 MWCNT addition decreases the membrane contact angle, in other word increases the membrane hydrophilicity. The membrane hydrophilicity was increased as an effect of higher porosity due to MWCNT addition (Figure 6). The functional groups that present in MWCNT (-COOH, -OH) also contribute to increase the membrane hydrophilicity (Vatanpour et al., 2012).

3.4 Permeability and Perselectivity

3.4.1 Pure Water Flux

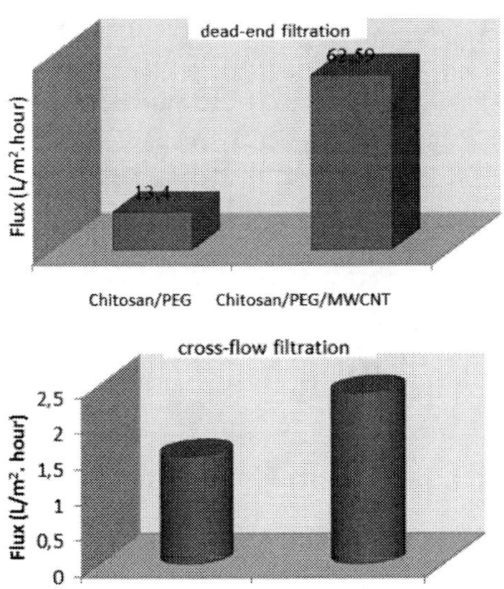

Fig 8. Pure water flux of Chitosan/PEG membrane (left) and Chitosan/PEG/MWCNT nanocomposite membrane (right) at 2 atm and membrane thickness of 0,08mm, using dead-end (top) and cross-flow (bottom) filtation system.

Permeability of the membranes is measured as flux value (L/m^{2}hour). The permeability of Chitosan/PEG/ MWCNT membrane was increased sharply as shown in figure 8. The membrane flux increased because of membrane pore expansion and increase in hydrophilicity due to the addition of MWCNT. Membrane permeability heavily depend on membrane thickness and pressure applied. Figure 9 shows flux value at varied pressure and membrane thickness. Increasing pressure lead to an increase in flux value, while increasing membrane thickness lead to a decrease in flux value (Rahimipour et al., 2012; Zhao et al., 2013; Daraei et al., 2012). The highest flux was obtained using the thinnest membrane (0,07 mm), and at highest pressure (3,6 atm).

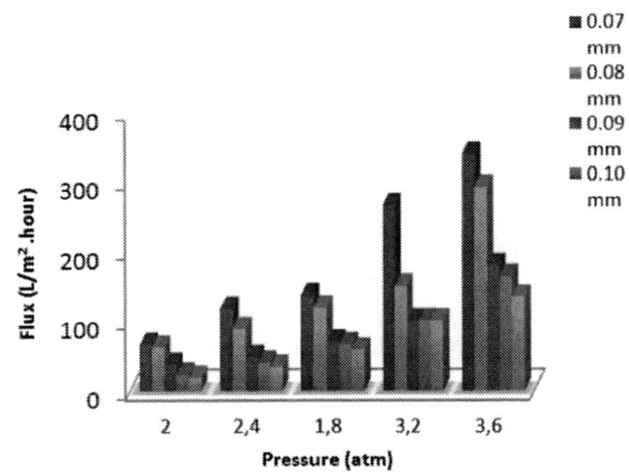

Fig 9. Pure water flux of Chitosan/PEG/MWCNT nanocomposites membrane at varied pressure and membrane thickness using dead-end filtation system.

3.4.2 Membrane selectivity

Selectivity is measured to understand the membrane's ability to reject species of certain size and is written as a % rejection (R).

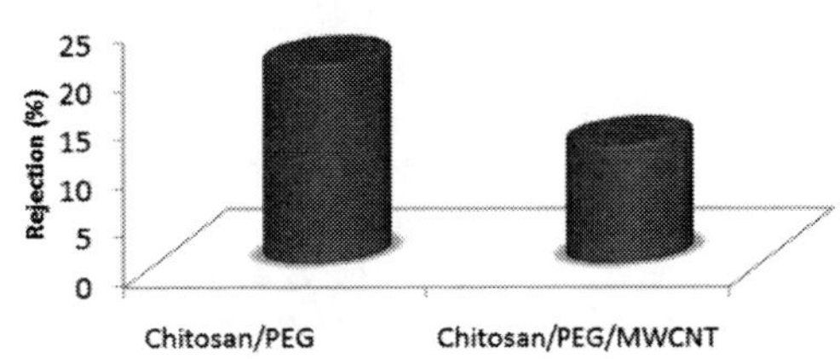

Fig 10. Protein rejection of the Chitosan/PEG membrane and Chitosan/PEG/MWCNT nanocomposites membrane.

Figure 10 shows protein rejection of the membranes. It can be seen that Chitosan/ PEG/ MWCNT nanocomposites membrane. did not reject protein as much as Chitosan/PEG membrane. This slight decrease in protein selectivity is due to the increase in average pore radius in Chitosan/PEG/MWCNT nanocomposites membrane.

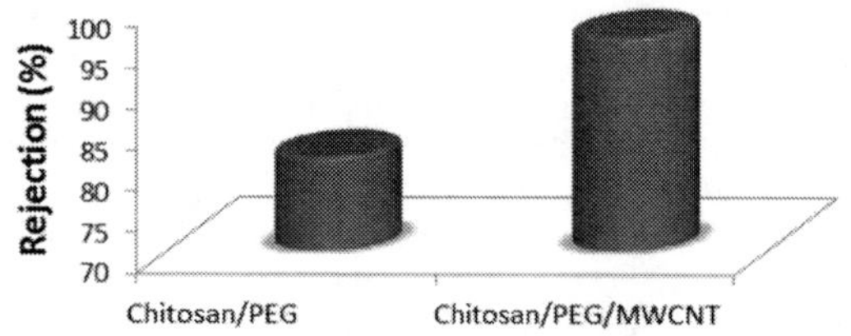

Fig 11. Wantex pigment rejection of the Chitosan/PEG membrane and Chitosan/PEG/MWCNT nanocomposites membrane.

Meanwhile Chitosan/PEG/MWCNT nanocomposites membrane yield higher wantex pigment rejection (Figure 11). This can be explained by the ability of MWCNTs to absorb pigments such as wantex pigment, due to its high surface area and poroeus structure Gupta *et al.* (2013).

4 Conclusion

Based on the results, it can be concluded that Chitosan/PEG/MWCNT nanocomposite membrane had successfully synthesized. The MWCNT successfully incorporated into the membrane matrix shown by FTIR, XRD, and SEM characterization. MWCNT addition to Chitosan/PEG membrane sharply increases membrane premeability and selectivity against pigments as well as increasing its mechanical properties. The nanocomposites membrane can be applied at high pressure to be classified as an ultrafiltration membrane.

References

1. M.A. Shannon, P.W. Bohn, M. Elimelech, J.G. Georgiadis, B. J. Mari˜nas and A. M. Mayes, Nature, 452, 301 (2008)
2. V.K. Gupta, I. Ali, T.A. Saleh, A. Nayak, S. Agarwal, RSC Advances **2** (16), 6380–6388 (2012).
3. M. Elimelech and W. A. Phillip, Science, 333, 712 (2011).
4. D. Li and H. Wang, J. Mater. Chem., **20**, 4551 (2010).
5. M.M. Pendergast, E.M. Hoek, Energy Environ. Sci. **4** (6), 1946–1971, (2011).
6. R. Khalil, E. Sheha, T. Hanafy, & O., Al-Hartomy, *Materials Express*, **4**(6), 483–490 (2014).
7. S. He, Y. Lin, Z. Wei, L. Zhang, J. Lin, & S. Nazarenko, *Polymers for Advanced Technologies*, **26**(4), 300–307(2015).
8. C.S. Ferreira, S.G. Caridade, J.F., Mano, & N.M. Alves, *Polymers for Advanced Technologies*, **25**(12), 1492–1500 (2014).
9. N. Akther, A. Sodiq, S. Giwa, H. Daer, A. Arafat, & S.W. Hasan, *Chemical Engineering Journal*, 281, 502–522 (2015).
10. J. Li, J., Gao, G. Sui, L. Jia, C. Zuo, & Q. Deng, *Materials Express*, **4**(6), 491–498 (2014).
11. V.K. Thakur, A.S. Singha, & B.N. Misra, *Journal of Applied Polymer Science*, ***122***(1), 532–544. (2011).
12. S.I., Voicu, A.I.V. MuhuletAntoniac, & M.S. Corobea. *Key Engineering Materials*, ***638***, 27–30. (2015).
13. F. Khoerunnisa, Hendrawan, Y. Sonjaya, A. Putri, Synthesis and Characterization of composites filtration membranes based chitosan-poly(ethylene glycol), AIP conference proceedings 1699, 040009 (2015)
14. Z. Fan, Z. Wang, M. Duan, J. Wang, S.Wang, J. Membrane Sci. **310** (1-2), 402–408 (2008).
15. T. Wu et al,. RSC Adv., **5**(11), pp.7880–7889. (2014).
16. E. Celik, L. Liu & H. Choi, Water Research, **45**(16), 5287–5294. (2011).
17. X. Qu, P.J.J. Alvarez, Q. Li, Water Res. **47**, 3931–3946, (2013).
18. Liang S. et al., J. Membr. Sci. **394**, pp. 184–192. (2012).
19. A.H. Cui et al. J. Membr. Sci. **360**, pp. 259–264. (2010)
20. A .Bottino G.A. Capannelli Desalination **146**, pp. 35–40. (2002).
21. S. Kar, R.C. Bindal, P.K. Tewari, Nano Today **7**, 385-389 (2012)
22. D.Q. Vu, W.J. Koros, J. Membrane Science 211, 311–334 (2003).
23. A.I. Skoulidas, D.M. Ackerman, J.K. Johnson, D.S. Sholl, Rapid transport of gases in carbon nanotubes, Phys. Rev. Lett. **89**, 18590 (2002)
24. J.K. Holt, H.G. Park, Y. Wang, M. Stadermann, A.B. Artyukhin, C.P. Grigoropoulos, A. Noy, O. Bakajin, Science 312, 1034–1037 (2006)
25. M. Majumder, N. Chopra, R. Andrews, B.J. Hinds, Nature 438, 44 (2005).

Synthesis of Rice Straw as Biosorbent for Colour Removal in Batik Industrial Waste which is Represented by Methylene Blue

Nakesamas Putut Prilininta[1,*], and *Oktrianto*[1]
[1]Department of Chemical and Green Process Engineering, Surya University, Tangerang 15810, Indonesia

Abstract. Use of rice straw as a biosorbent for water treatment in batik industry is not fully developed. Liquid waste for the colour's adsorption using rice straw biosorbent represented by methylene blue solution. Rice straw biosorbent has been activated by HNO_3 1 M and NaOH 0,75 M. Then characterized using Scanning Electron Microscopy with Energy Dispersive Spectroscopy (SEM EDS) dan Fourier Transform Infrared (FTIR) Spectroscopy. The variation of dosage and drying temperature applied to know the optimum dosage for utilization of biosorbent and optimum drying temperature for biosorbent synthesis. The equilibrium were fitted to Langmuir isotherm and Pseudo Order 2 kinetics. The optimum adsorption obtained in 1,67 gr/L biosorbent with 100°C drying temperature in 4 ppm methylene blue waste simulation solution with 95.27% adsorption capacities and 4 ppm batik industrial waste solution with 30.59% adsorption capacities.

1 Introduction

Batik is one of the world cultural heritage derived from Indonesia. Batik becomes an important part in industrial growth in Indonesia, especially in batik producing area. Productivity of batik does not only affect the increase of regional income, but also impact on other problems, especially environmental problems. One of the environmental problem is caused by batik production process that produced liquid waste which comes from residual of batik colouring process. The liquid waste by some industries directly discharged into the river without treatment [1].

Rivers have the ability to clean themselves (self purification). But if the waste is discharged continuously and exceeds the self purification capability of the river, it will result in the occurrence of river pollution [2]. Colour is the easiest parameter to determine whether the river is clean or not. The clean river has a clear flow of water. If the river flow is not clear, it can be indicated that the river is not suitable for daily needs. Rivers in batik industrial areas have been polluted dyes derived from residual of batik colouring process [3]. Based on these parameters can be indicated that the river is not feasible for the continuity of the river ecosystem and the needs of water to surrounding communities.

There are several alternatives to handle the waste problem. The adsorption method becomes one of the ways to absorb the dye on batik industrial waste. This study also focused on using natural material adsorbent (biosorbent) derived from agricultural waste as an alternative to conventional adsorbents. Agricultural wastes are selected as the basic ingredients for the synthesis of biosorbents because of their abundant availability. In addition, its use can be optimal because it does not compete with anything [4]. There are various agricultural wastes, such as coconut shells, rice husks, rice stalks, corncobs, palm oil bunches and others. Agricultural waste is a material that has the potential to be processed into biosorbent [5]. Agricultural waste used in this research is rice straw. Rice straw is chosen because its availability is very abundant and rarely reused. After the harvest season, rice straws are directly burned without being processed first [6]. Rice straw contains C, O, H and N content of 87% of its total weight and Si (4.6%), K (2.5%), P (0.5%), Ca (0.3%) And Mg (0.3%) (Wang, 2007, p.315). These elements content are beneficial to adsorption process, so that rice straw is potentially processed into biosorbent [7].

This study made biosorbent from rice straw that activated using HNO_3 1 M and NaOH 0.75 M. HNO_3 and NaOH were selected become media activation based on Rocha's research, et al [8]. Both were able to make pore of the rice straw have a greater absorbence to adsorbate. Then the biosorbent will be applied to absorb the dissolved dyes in the liquid waste of batik industrial. This research uses methylene blue to represent liquid waste of batik industrial. Methylene blue is a cationic dye commonly used by textile industry [9].

The objective of this work were to convert rice straw to biosorben with high sorption capacity, and to examine potential application of this new biosorben in batik's colour wastewater treatment.

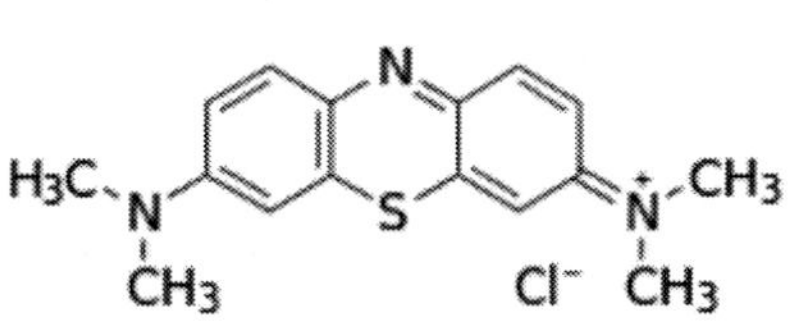

Fig. 1. Chemical structure of Methylene Blue.

* Corresponding author: nakesamasprilininta@gmail.com

2 Materials and Methods

2.1 Materials

Rice straw and batik industrial waste were collected from Pekalongan City, Central Java, Indonesia. Methylene blue (Merck), NaOH (Merck) and HNO_3 (65%, Pudak Scientific). All solution were prepared by using distilled water.

2.2 Preparation of modified biosorbent

Rice straw washed by water, dried in an oven at 60°C for 8 hours. And then crushed to obtain a small size. This material called as SMP. 10 g of SMP was activated by 50 mL of HNO_3 1 M and agitated for 1 hour at room temperature. Next, the material washed by water to removed the excess acid. After the pH was neutral, 10 g of SMP activated by 100 mL of NaOH 0,75 M and agitated for 1 hour at room temperature. The SMP washed again by water until the pH was neutral and dried in an oven with the variation of drying temperature for 4 hours. Rice straw that has passed through this stage is called BMP. BMP crushed and sieved with the mesh 75 (193.5 μm).

Methylene blue is used as a waste simulation solution. Methylene blue solution was prepared by dissolving 4 mg of methylene blue into 1 L of distilled water. Analysis of dissolved methylene blue concentration using UV-VIS spectrophotometer with maximum wavelength of 668 nm.

2.3 Experimental methods and measurements

2.3.1 Drying temperature

Drying temperature variation is carried out in the second drying process. The temperature variation used are 60°C, 100°C and 140°C. The second drying was done by using the oven for 4 hours.

2.3.2 Biosorbent dosage

Variation of dosage were performed by adding biosorbents of 0.1, 0.5, and 1 g to 300 ml of methylene blue solution. The initial concentration of methylene blue solution used was 4 mg/L. The adsorption studies was performed using a batch system in a 600 mL beaker glass that was stirred using a magnetic stirrer at pH 6 and room temperature.

2.3.3 Equilibrium studies

The equilibrium studies is carried out by contacting the rice straw biosorbent with a methylene blue solution. The length of contact time is 2 hours with sampling every 5 minutes. The adsorption study was performed on a 600 mL beaker glass using a magnetic stirrer. Initial and equilibrium concentration were measured using a UV-VIS spectrophotometer with a wavelength of 668 nm. The value of the adsorption equilibrium is calculated by the following formula.

$$q_e = \frac{(Co - Ce)\,v}{W}\ x\ 100 \qquad (1)$$

Where C_o and C_e are initial concentration and equilibrium concentration, v is the volume and m is the dose of the biosorbent. Percent removal is calculated using the following formula.

$$\text{Removal percentage} = \frac{Co - Ce}{Co}\ x\ 100 \qquad (2)$$

2.3.4 Kinetics studies

The kinetics studies is carried out by contacting 300 mL of methylene blue solution with different biosorbent doses. The studies was performed using a 600 mL beaker glass and magnetic stirrer at room temperature and pH 6.

2.3.5 Characterization of biosorbent

There are two characterization performed in this research. The pore morphology on the surface of rice straw was observed using a Scanning Electron Microscopy with Energy Dispersive Spectroscopy (SEM EDS FEI type Quanta 650, Oxford). Then, functional groups on the surface of rice straw was analysed using a Fourier Transform Infra Red (FTIR Tensor 27, Brucker). FTIR spectra were recorded over the wavenumber range of 600-4000 cm^{-1}.

3 Results and Discussion

3.1 Scanning Electron Microscopy with Energy Dispersive Spectroscopy (SEM EDS)

SEM characterization results were performed to analyze the pores which formed on the surface of the biosorbent shown in Fig. 2. Based on SEM test results, the pores formed on the surface of the biosorbent are not uniform. Pores are various in size, ranging from 14.08 to 20.57 μm. Pores on the surface of the biosorbent cause dye adsorption. The adsorbate particles will be trapped into the pores so that the adsorbate content in the waste simulation solution will decrease [10].

The element content on the biosorbent surface is detected by using EDS and presented in Table 1. The results obtained surface of the biosorbent are dominated by elements C, O, Si with a value of 58.07% atoms, 37.44% atoms and 3.35% atoms, respectively. These elements are involved in the adsorption process [11].

https://doi.org/10.1051/matecconf/201815605001

3.2 Fourier Transform Infra Red (FTIR) Spectroscopy

The result of fungsional group identification of biosorbent is shown in Fig. 3. Based on Fig. 3, peaks appeared at 1058 cm^{-1} and 1350 cm^{-1}, which are represented C-O stretching group and C-C group. The absorbance band at 1650 cm^{-1} was assigned to C=O stretching group. Then, the absorbance band at 2810 cm^{-1} and 3440 cm^{-1} was attributed to C-OH stretching group and O-H group, respectively. The results of the FTIR studies are summarized in Table 2.

3.3 Influence of drying temperature

The ability of rice straw biosorbent in dye absorbing is shown in Fig. 4. Biosorbent with 100°C and 140°C drying temperature has sorption capacities 95.27% and 94.39%. Both types of biosorbents also reached a state of equilibrium at 30 minutes. While the biosorbent with a 60°C drying temperature, has a lower adsorption capacities of 88.47%. The equilibrium state for biosorbent with 60°C drying temperature was reached at 95 minutes. If the drying temperature is higher, then the pores that formed on the surface of the biosorbent are more numerous as the amount of H_2O and other volatile substances that can be evaporated is also more.

This relates to the number of active sites formed on the surface of the biosorbent [12]. The optimum drying temperature is specific for each type of biosorbent. Biosorbent with 100°C drying temperature has a higher %sorption capacities when compared by biosorbent with 140°C drying temperature. This is because at a drying temperature of 140°C some of the biosorbent pores begin to damage, this is proven by the sorption capacities for biosorbent with a drying temperature of 140°C lower than the sorption capacities for biosorbent with a drying temperature of 100°C but not far adrift.

3.3 Influence of Biosorbent Dosage

The correlation between ability of rice straw biosorbent in absorbing methylene blue based on adsorbent dose is shown by figure 5. In this analysis, a 100°C drying temperature biosorbent is used. As the adsorption studies proceeded, the adsorption capacities of rice straw biosorbent to the methylene blue is increased. The graph also explains that the more doses of biosorbent added, the more adsorbat absorbed. This is related to the number of active sites available [13]. Adsorption capacitiess for biosorbent at doses of 0.33 g/L, 1.67g/L and 3.33g/L were 92.78%, 95.27% and 95.42%, respectively. Based on the value of sorption capacities achieved, difference of sorption capacities value is not significant in the increase of biosorbent dose. This is because the availability of active sites at small doses is sufficient.

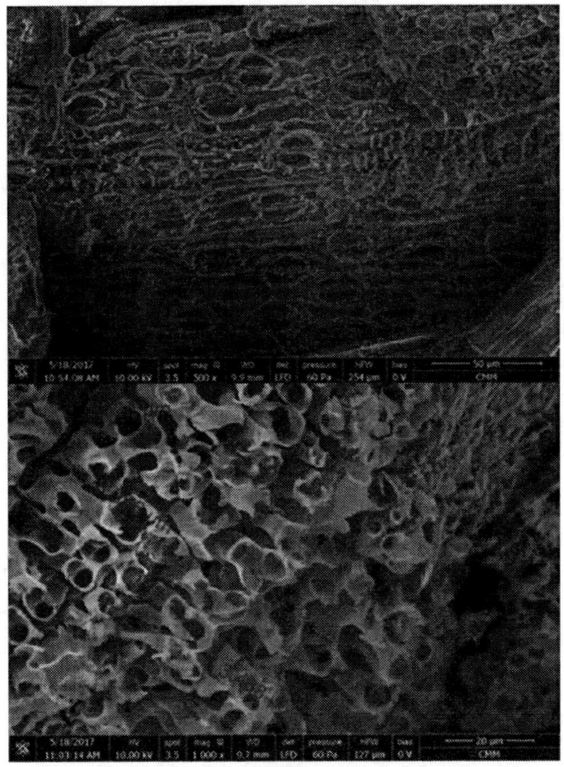

Fig. 2. SEM study of rice straw biosorbent in 100°C drying temperature with 500x magnification (above) and 1000x magnification (below).

Tabel 1. EDS study of rice straw biosorbent of 100°C drying temperature on spectrum 1.

Element	Apparent Concentration	Wt%	Atomic %	Standard label
C	12.01	47.53	56.51	C Vit
O	21.64	44.15	39.41	SiO2
Si	4.34	7.33	3.73	SiO2
Ca	0.56	0.99	0.35	Wollastonite
Total		100	100	

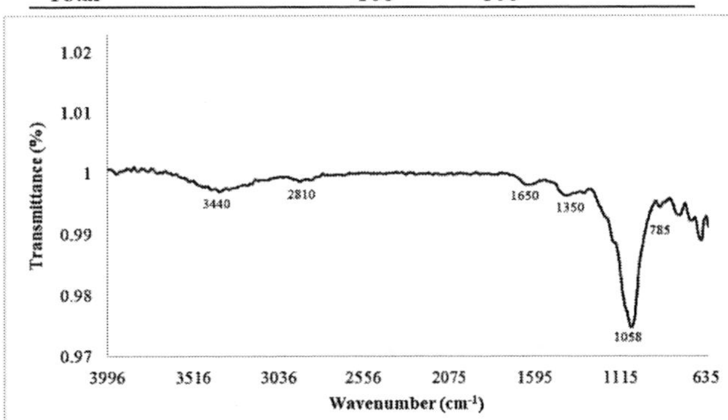

Fig. 3. FTIR study of rice straw biosorbent 100°C drying temperature.

Tabel 2. The result of functional group analysis on rice straw biosorbent in 100°C drying temperature.

IR peak	Frequency (cm^{1})	Functional groups	Type of bond	Intensity
1	785	Si-O		
2	1058	C-O	stretching	Strong
3	1350	C-C		Weak
4	1650	C=O	stretching	Strong
5	2810	C-OH	stretching	Weak
6	3440	OH		

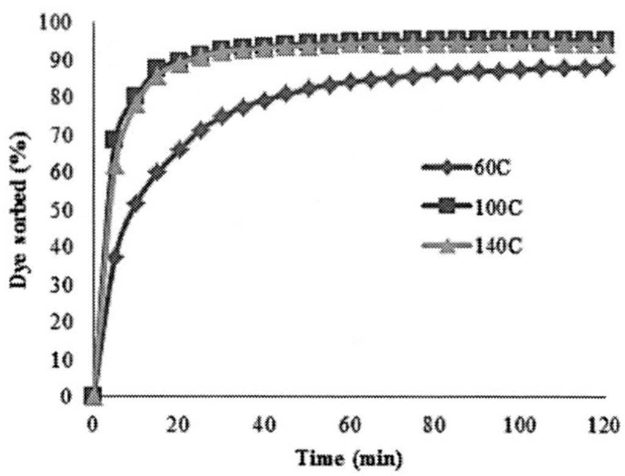

Fig. 4. Adsorption capacities of each rice straw biosorbent with a dose of 1.67 g/L to 300 mL methylene blue.

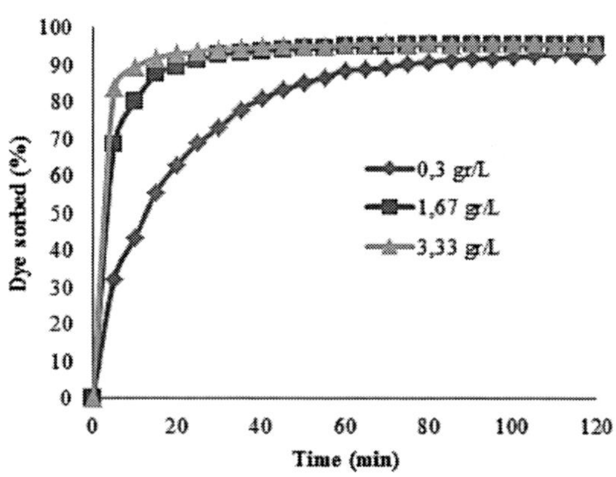

Fig. 5. Adsorption capacities of each rice straw biosorbent dose of 100°C drying temperature to 300 mL of methylene blue.

3.5 Adsorption isotherm studies

There are two isotherms equations used, Langmuir isotherm and Freundlich isotherm. Determination of appropriate isotherms based on the value of R^2 which obtained.

Isotherm Langmuir explains the assumption that optimal adsorption occurs when the solute molecules formed a monomolecular layer on the biosorbent surface, constant adsorption energy, and no mass transfer of adsorbate molecules on the biosorbent surface. Isotherm Langmuir is expressed by the following equation.

$$\frac{Ce}{qe} = \frac{1}{qm\,KL} + \frac{1}{qm}\,Ce \quad (3)$$

q_m and K_L are Langmuir constants which represent the maximum adsorption capacity for solid phases and energy constants associated with the adsorption heat. While q_e and C_e describe the equilibrium condition.

Freundlich isoterm expresses an empirical relationship that explains the adsorption of solutes to the surface of the adsorbent and takes the assumption that the adsorbate formed a multimolecular layer on the biosorbent surface. Freundlich isotherm illustrates the relationship between the amount of adsorbat adsorbed per unit of adsorbent mass and the concentration of adsorbate at equilibrium.

$$\log q_e = \log K_f + \frac{1}{n} \log C_e \quad (4)$$

K_f and n are Freundlich constants, K_f and n are indicators of adsorption capacity and adsorption intensity. While q_e and C_e describe the equilibrium condition.

The result of the analysis of R^2 value for each biosorbent is shown in Tabel 3. Based on the calculation, biosorbent is more suitable with Langmuir isotherm equation. Langmuir isotherm curve can be seen in figure 6. The results of the Langmuir isotherm analysis show that the active surface of biosorbent has homogeneous properties and adsorbate is absorbed on the biosorbent surface forming a monomolecular layer.

3.6 Adsorption kinetics studies

The colour adsorption kinetics that occur between the adsorbate and the rice straw biosorbent progress quickly. At first the adsorption process that occurs is rapid adsorption. Then the adsorption process slows down over time and then enters a state of equilibrium.

Kinetic studies were analyzed using the kinetic equation of Pseudo Order 1 and the kinetic equation of Pseudo Order 2. Determination of corresponding kinetic based on the value of R^2 which achieved. Kinetic equation of Pseudo Order 1 is shown in the following equation.

$$\log(q_e - q_t) = \log q_e - \frac{k1}{2.303}\, t \quad (5)$$

k_1 is a constant on the kinetic equation of Pseudo Order 1. q_t and q_e are the amount of adsorbate at t and at equilibrium.

The kinetic equation of Pseudo Order 2 is shown in the following equation.

$$\frac{t}{qt} = \frac{1}{k2qe^{\wedge}2} + \frac{1}{qe}\, t \quad (6)$$

k_2 is a constant on the kinetic equation Pseudo 2. q_t and q_e is the amount of adsorbate at time t and at equilibrium.

Based on the value of R^2 in table 4, the adsorption phenomenon that follows the kinetic equation of Pseudo order 2, which means that adsorption is chemical adsorption. Pseudo order 2 kinetics curve can be seen in figure 7. The results of this study are similar to Kazembeigi's research [14].

Tabel 3. The value of R^2 in the isotherm analysis of each biosorbent.

Biosorbent		R^2	
T (C)	Dose (gr)	Freunlich	Langmuir
60	0.1	0.9308	0.876
	0.5	0.8818	0.9361
	1	0.9041	0.9748
100	0.1	0.8274	0.9123
	0.5	0.8961	0.9912
	1	0.9504	0.9984
140	0.1	0.8402	0.8729
	0.5	0.8895	0.9867
	1	0.9507	0.9971

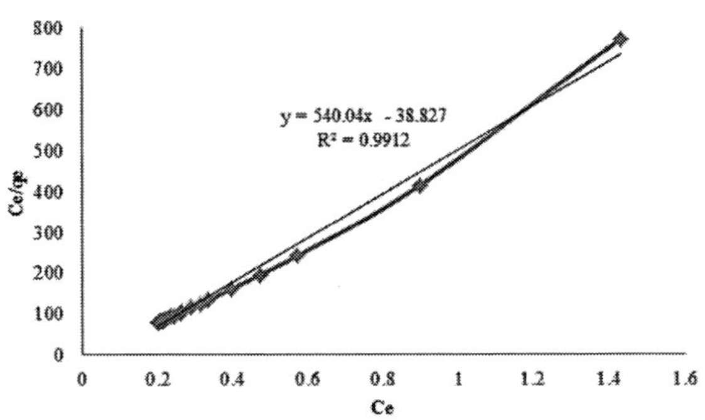

Fig. 6. Langmuir's isotherm curve for adsorption study with rice straw biosorbent 100°C drying temperature dose 0.5 g in 300 mLmethylene blue.

3.7 Aplication of the biosorbent studied for water waste treatment

The adsorption studies for batik industrial waste is carried out under the same conditions as using methylene blue, by using rice straw biosorbent temperature of 100°C and a dose of 0.5 g in 300 mL waste solution. Based on these studies can be seen the ability of biosorben rice straw of batik waste adsorption. The test results showed the dye sorbed increased. Sorption capacities achieved by 30.59%.

Before the adsorption studies, pretreatment is performed for batik liquid waste. Pretreatment performed include sedimentation, filtration and centrifugation. The adsorption capacities of these studies is not as good as when using methylene blue solution. This is because the substances dissolved in batik liquid waste waste are not only dyes, but many impurities such as azo dye, salts, waxes, and other impurities [15]. Salt contained in the waste solution is useful as a color amplifier, so that the color attached to the fabric does not quickly fade.

Tabel 4. The value of R^2 in the isotherm analysis of each biosorbent.

Biosorbent		R	
T (C)	Dose (gr)	Pseudo order 1	Pseudo Order 2
60	0.1	0.9136	0.9996
	0.5	0.9878	0.9983
	1	0.9601	0.9641
100	0.1	0.992	1
	0.5	0.9807	0.9999
	1	0.9743	0.9959
140	0.1	0.9888	1
	0.5	0.9671	0.9998
	1	0.8502	0.9913

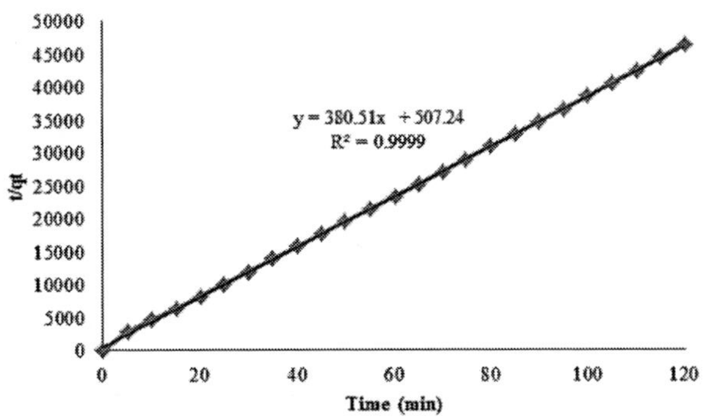

Fig. 7. Pseudo order 2 kinetic curve for adsorption study with rice straw biosorbent of 100°C drying temperature dose 0.5 g in 300 mLmethylene blue.

4 Conclusions

This study confirmed that rice straw was excellent biosorbent for colour removal in liquid waste of batik industries which is represented by methylene blue. The optimum adsorption was in a 100°C drying temperature biosorbent at dose of 0.5 g in 300 mL of a 4 ppm methylene blue solution. The achieved of sorption capacities was 95.27% and reach equilibrium state after 30 minutes of adsorption. Sorption capacities that achieved for the adsorption studies using batik liquid waste was 30.59%. The isotherm and kinetic equation of rice straw biosorbent adsorption are following the Langmuir isotherms and the kinetic Pseudo order 2.

The authors thank BPPT (Badan Pengkajian dan Penerapan Teknologi) and BATAN (Badan Tenaga Nuklir Nasional) for analysis support.

References

1. T. Noorimotlagh, R.D.C. Soltani, A.R. Khataee, S. Shahriyar, H. Nourmoradi, J. Chin. Inst. Chem. Eng, (2014).
2. N. Colin, A. Maceda-Veiga, N. Flor-Arnau, J. Mora, P. Fortuno, C. Vieira, N. Prat, J. Cambra, A. de Sostoa, Ecotoxicol. Environ. Saf. **132**, 295-303, (2016).
3. R.M. Jain, K.H.Mody, J. Keshri, B. Jha, Mar. Pollut. Bull. (2014).
4. Y. Wu, Y. Fan, M, Zhang, Z. Ming, S. Yang, A. A, P. Fang, Biochem. Eng. J. **105**, 27-35, (2016).
5. S. Palsan, W.N.A.W. Jusoh, H. Subramanian, MATEC Web Conf. **87**, (2017).
6. W.C.Li, F.Y. Law, Y.H.M. Chan, Environ. Sci. Pollut. Res. (2015).
7. W. Cao, Z. Dang, X.Q. Zhou, X.Y. Yi, P.X. Wu, N.W. Zhu, G.N. Lu, Carbohydr. Polym. **85**, 571-577, (2011).
8. C.G. Rocha, D.A.M. Zaia, R.V.S. Alfaya, A.A.S. Alfaya, J. Hazard. Mater. **166**, 383-388, (2009).
9. S.A. Umoren, U.J. Etim, A.U. Israel, J. Mater. Environ. Sci. **4**(1), 75-86, 2013.
10. A.A. El-Bindary, A.Z. El-Sonbati, A.A. Al-Sarawy, K.S. Mohamed, M.A. Farid, Spectrochim. Acta, Part A, **136**, 1842-1849, (2015).
11. S.L.Wang, Y.M. Tzou, Y.H. Lu, G. Sheng, J. Hazard. Mater. **147**, 313-318, (2007).
12. F.B. Scheufele, C. Ribeiro, A.N. Modenes, F.R.E. Quinones, R. Bergamasco, N.C. Pereira, Fiber Polym, **16**, 1646-1656, (2015)
13. N. Nasuha, B.H. Hameed, A.T. Mohd Din, J. Hazard. Mater. **175**, 126-132, (2010).
14. F. Kazembeigi, H.R.S. Arezoomand, H. Faraji, S. Mazloomi, F. Mohammadi, F.M.A. Moghadam, R. Khoshneyat, A. Nikonahad, H. Nourmoradi. Veliger. **53**, 1-7, (2014).
15. M. Imran, D.E. Crowley, A. Khalid, S. Hussain, M.W. Mumtaz, M. Arshad, Rev. Environ. Sci. Biotechnol. **14**, 73-92, (2015).

Kaolin as a Source of Silica and Alumina For Synthesis of Zeolite Y and Amorphous Silica Alumina

Endang **Sri Rahayu**[1,*], *Gatot* **Subiyanto**[1], *Arief* **Imanuddin**[1], *Wiranto*[1], *Sabrina* **Nadina**[1], *Rista* **Ristiani**[1], *Suhermina*[1], **and** *Endang* **Yuniarti**[1]

[1]*Chemical Engineering Department, Bandung State Polytechnic, Indonesia*

Abstract. Kaolin is the clay mineral which containing silica (SiO_2) and alumina (Al_2O_3) in a high percentage, that can be used as a nutrient in the synthesis of zeolites and amorphous silica alumina (ASA). The objective of this research is to convert the Belitung kaolin into silica and alumina as nutrients for the synthesis of zeolites and amorphous silica alumina, which are required in the preparation of the catalysts. Silica and alumina contained in the kaolin were separated by leaching the active kaolin called as metakaolin, using HCL solution, giving a solid phase rich silica and a liquid phase rich alumina. The solid phase rich silica was synthesized to zeolite Y by adding seed of the Y Lynde type, through the hydrothermal process with an alkaline condition. While, the liquid phase rich alumina was converted into an amorphous silica alumina through a co precipitation method. Characterization of zeolite and ASA were done using XRD, surface area and pore analyzer and SEM. The higher of alumina in liquid phase as a result of the rising molar of HCL in the leaching process was observed, but it didn't work for its rising time. Products of ASA and zeolite Y were obtained by using liquid phase rich alumina and solid phase rich silica, respectively, which resulted through leaching metakaolin in 2.5 M HCl at temperature of 100° C for 2 hours.

1 Introduction

Kaolin is the clay mineral which containing silica (SiO_2) and alumina (Al_2O_3) which are inert and inactive, which can be activated through the process of calcination. When the calcination of the kaolin is taken place, the regular crystal structure of kaolinite with inactive Si-O and Al-O bonds is converted to metakaolin with an irregular amorphous alumino silicate structure, with active Si-O and Al-O bonds.

Zeolite and Amorphous Silica Alumina (ASA) are built by atoms of Al, Si and O, therefore these materials can be synthesized with raw materials containing silica and alumina compounds. Accordingly, kaolin is a cheap natural material that containing silica and alumina in a high percentage. However, kaolin is inert and inactive, so it requires activation to form the active kaolin, that can be done through the process of calcination. Metakaolin is a relatively inexpensive active kaolin, in which the silica and alumina content can be separated through leaching by acid, that will produce silica-rich solids and alumina-rich liquids. The Silica-rich solids with a certain silica/alumina molar ratio are required in the synthesis of zeolites [1]. Both of the process conditions of calcinating kaolin and leaching metakaolin will affect the results of the separation of silica and alumina. When metakaolin is leached, the Al sheet would be dissolved, that causes the different environment of the Si atoms to form microporous silica [2].

There are few papers reporting on leaching metakaolin for using silica-rich solid as a nutrient in the synthesis of zeolites. Process of leaching metakaolin using H_2SO_4 5.2N at temperature of 90°C within 2 hours, resulting in silica-rich solids having a surface area of 400 m^2 /g and a pore size of 0.7 nm, as a source of silica for the synthesis of zeolite Y [3]. Ajayi (2010) reported that leaching metakaolin using H_2SO_4 > 50%-w in the starting material will result in rasio of Si/Al > 5 w/w [4]. Mamani (2014) has successfully separated silica and alumina in metakaolin using HCl 3M at temperature of 115°C during 2.5 hours, which produces a silica-rich solid with molar ratio of silica/alumina between 20-40 mol/mol, that is suitable for nutrient of the synthesize ZSM5 [5]. Leaching of metakaolin by using H_2SO_4 has also been done by Adeoye (2017), which produces silica-rich solids with a molar ratio between 3-8 mol/ mol [6] .

The main purpose of this paper is to improve efficiency of the synthesis process of zeolite Y from kaolin that has been done previously [7-8], by substituting the supply of porous silica through the process of leaching metakaolin rather than calcinating of kaolin at high exothermic temperature. Furthermore, alumina-rich liquids were be used as nutrients for synthesis of ASA.

2. Experimental

2.1 Materials and methods

The material for the synthesis of Zeolite Y comprises of the silica-rich solids resulting in leaching metakaolin, Sodium hydroxide, demineral water and crystal seed of Lynde Type Y wherein contain Merck sodium silicate, Aldrich sodium aluminate, technical sodium hydroxide and demineral water. The material for ASA synthesis comprises of mother liquor resulted of leaching metakaolin, and ammonia or sodium hydroxide.

2.1.1 Metakaolin

* Corresponding author: endang.srirahayu@polban.ac.id

 https://doi.org/10.1051/matecconf/201815605002

Metakaolin is an active kaolin of Belitung kaolin which is required to provide Si-O and Al-O actives, which are the source of silica and alumina for the synthesis of zeolite Y and ASA. Metakoline was prepared by calcinating of kaolin in Nabertherm GmbH furnace at 700°C for 2 hours, where the kaolin dehydroxylation process takes place to form metakaolin completely, through endothermic reaction.

This temperature is determined by kaolin thermal characterization, using a DSC / TGA.

2.1.2 Silica and Alumina

Preparation of the active silica and alumina was done by leaching process using HCl between 2-3 M at temperature 100°C, with the weight ratio of HCl/ metakaolin solution was 10: 1 w / w, for 2-4 hours. The leaching product of metakaolin is filtered to obtain silica-rich solids, and alumina-rich solutions.

The silica-rich solids are rinsed to neutral pH, dried in an oven at temperature of 110°C for 8 hours, and characterized by AAS. The alumina-rich solution was analyzed by gravimetric method. The condition of the leaching process is adjusted to obtain a silica-rich solids having silica / alumina molar ratio in between 7-10 mol / mol.

2.2. Synthesis Zeolite NaY

Stages of the process in the synthesis of zeolite Y include: preparation and aging of crystal seed Lynde Type Y, preparation and aging starting materials, and hydrothermal synthesis.

2.2.1 Seed

Crystalline seed was prepared by adopting Lynde Type Y of Ginter et al (1992), by stirring the starting material consisting of: 2.72 grams of Na_2SiO_3 solution, 2.09 grams of NaAlO2, 4.07 grams of NaOH and 19.95 grams of H_2O at moderate speeds for 1 hour, and continued by aging at room temperature for 24 hours [9]

2.2.2 Starting Material

The starting material for the synthesis of Zeolite Y from kaolin was prepared by adopting of Rahayu et.al. (2014) with modify by using micropore silica resulted in leaching metakaolin.

The molar ratio of silica / alumina in starting materials is adjusted for having to about 10 mol / mol, whereas the amount of silica is a combination of silica from silica-rich solids and from seed, as well as for the amount of alumina.

The content of silica-rich solids in a starting material solution is set in between 5-10% -w. Fuhermore, the starting material was aged at room temperature for 11 hours and pH 13 to produce gel form, before being processed hydrothermally.

2.2.3 Synthesis of NaY Zeolite

The Gel of starting material was synthesized hydrothermally in a stirred reactor made of teflon, which is equipped with reflux, at temperature between 91-93°C by using oil bath, to form zeolite Y. Sampling of synthesis product was started at 12nd hours and filtered and rinsed to pH 8.

The synthesis product is dried in an oven with a temperature of 110C for 8 hours, prior to characterize of mineralalogi by XRD Rigaku Smartlab tool, morphology with SEM JEOL tool, and pore structure with NovaWin Quantachrome.

2.3 Synthesis of ASA

The synthesis of ASA was carried out by using a co-precipitation method with stages of a neutralization of alumina acid solution with an ammonia or sodium hydroxide solution, which it fed into reactor with flow in parallel for 1 hour at a temperature of ± 65 ° C to form the sol, followed by stabilization for 30 minutes before injection of sodium solution silicates, then done aging for 1-5 hours with pH 8-9 to form a gel that grows to form an amorphous phase.

The product was filtered and rinsed to a neutral pH, dried in an oven with a temperature of 110C for 8 hours, and analyzed by a NovaWin Quantachrome for pore structure , mineralogy analysis with XRD Rigaku Smartlab tool, and morphology analysis with SEM JEOL tool

3. Results and Discussion

3.1 Thermal Analysis

Figure 1 show the result of thermal analysis of the kaolin Belitung. The DSC curve forms a downward peak at temperatures between 480°C – 600°C, where indicating the area of an endothermic reaction in which the kaolin is converted to metakoline.

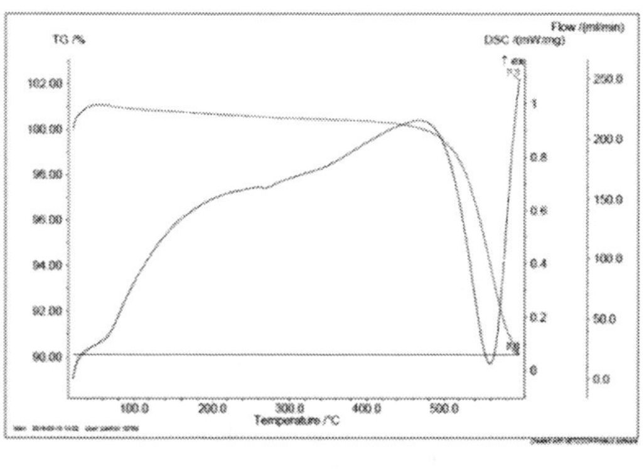

FIG 1 DSC/TGA Curves of Kaolin Belitung

Kaolin is converted to metakaolin through dehydroxylation process mechanism, which accompanied of release of water as in the equation of the reaction 3.1, and as shown in the TGA curve, Furthermore, a temperature of 700°C is set as calcining temperature of the kaolin Belitung to maximize releasing of bonded water in kaolin.

$$\underset{\text{Kaolin}}{Al_2Si_2O_5(OH)_4} \longrightarrow \underset{\text{Metakaolin}}{Al_2Si_2O_7} + 2H_2O \quad (1)$$

3.2 Dealumination Metakaolin By Leaching

Table 1 shows that is important to determine the process conditions of leaching acurately, to obtain silica-rich solids having an appropriate SiO_2/Al_2O_3 molar ratio for using as a raw material for the synthesis of zeolite NaY.

Table 1. Influence of leaching condition to SiO_2/Al_2O_3 ratio in silica-rich solid

Concentration of HCl (M)	Time of Leaching (Hours)	Ratio of SiO_2/Al_2O_3 (mol/mol)
1	2	3,0
	3	2,9
	4	3,1
2	2	5,8
	3	6,0
	4	6,2
2.5	2	10,5
	4	9,9
3	2	23,6
	3	24,5
	4	24,7

When leaching of metakaolin was carried out using HCl 2.5 M at temperature of 100°C during 2 hours, a good results were obtained. That resulting in silica-rich solids having a suitable SiO_2/Al_2O_3 molar ratio for NaY formation, ie between 7-10 mol / mol [1].

The concentration of HCl solution greatly influenced the molar ratio of silica / alumina in solid product, while the processing time did not have much effect. The alumina concentration in the solution of leaching product in between 25-28 g / L was obtained by using HCl 2.5 M at temperature of 100C, which is used as an alumina source in ASA synthesis.

3.3 Synthesis of NaY Zeolite from Kaolin

3.3.1 Diffractogram of Results of Synthesis NaY Zeolite From Kaolin

Synthesis of NaY Zeolite were performed in a teflon stirred reactor equipped with a reflux, using a heating oil bath to achieve a reaction temperature in between 91-93°C, pH ±13, and starting material with a varied of NaOH, SiO_2/Al_2O_3 and H_2O , where these materials were provided as a combined of rich-silica solids with seed. Table 2 and FIG. 2 - 5 show the results of NaY zeolite synthesis, using variations of the starting material formulations.

Table 2. Effect of Synthesis Process Conditions on Zeolite NaY Product

Synthesis Process		Zeolit NaY	
Starting Material (mol/mol)	Time of Syntesis (hour)	Ratio of SiO_2/Al_2O_3 (mol.mol)	Other Zeolite
$1{,}5Na_2O{:}Al_2O_3{:}10{,}4SiO_2{:}774\,H_2O$	15	3,61	v
$2{,}6Na_2O{:}Al_2O_3{:}10{,}4SiO_2{:}694H_2O$	12	3,43	v
	15	3,46	v
	18	3,70	v
	21	3,58	v
$1{,}5Na_2O{:}Al_2O_3{:}9{,}9SiO_2{:}375H_2O$	15	3,08	v
$2{,}5Na_2O{:}Al_2O_3{:}9{,}9SiO_2{:}350H_2O$	15	3,02	-

Table 2 shows that the synthesis of zeolite Y from kaolin requires a high mineralizer solution with a molar ratio of Na_2O/H_2O is not more than 140 mol/mol, and the starting material concentration has a molar ratio of 2.5 Na_2O: Al_2O_3:9,9 SiO_2:350 H_2O suitable to produce zeolite NaY. This starting material has the molar ratio of SiO_2 / Al_2O_3 as recommended by Murat 1992 of about 10 mol / mol, with high SiO_2 and Al_2O_3 concentrations that will increase reaction velocity, also increase the greater Si-O and Al-O collision to build skeleton towards NaY zeolite formation. Synthesis time of 15 hours can prevent the occurrence of further reaction to form zeolite P.

FIG. 2 is a difractogram of the NaY synthesis produced by starting material with the formulation of Starting Material 1.5: Na_2O Al_2O_3: 10.41 SiO_2: 774 H_2O, containing silica with SiO_2 / Al_2O_3 ratio of about 10 mol / mol, from leaching process in HCl 2.5 M solution during 2 hours called silica-rich solids, and seeds.

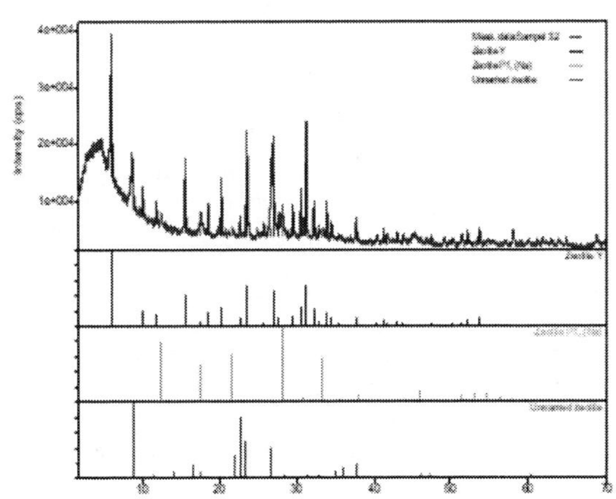

FIG. 2. Product of Zeolit NaY,15 Hours, $1.5Na_2O{:}Al_2O_3{:}10.41SiO_2{:}774H_2O$

How much silica-rich solids can be dissolved by the mineralizer under the conditions of the synthesis process, can be expected to be identical to how high purity zeolite NaY is in the crystals formed in the synthesis product.

The Diffractogram in FIG. 2 shows that the synthesis using a starting material formulation of $1.5Na_2O{:}Al_2O_3{:}10.41SiO_2{:}774H_2O$ produces a side reaction which forms zeolite P and another zeolite, which are probably. The results of parallel reaction. Increasing the quantity of Na_2O and decreasing H_2O, as well as the extension of reaction time will be studied its effect on the ability of silica-rich solids mineralization in the reaction system by increasing Na_2O in the starting material formulation $2.6Na_2O{:}$ $Al_2O_3{:}10.41SiO_2{:}$ $694H_2O$, and the extension of reaction time. This formulation produces synthesis products with difractogranes as shown in FIG. 3.

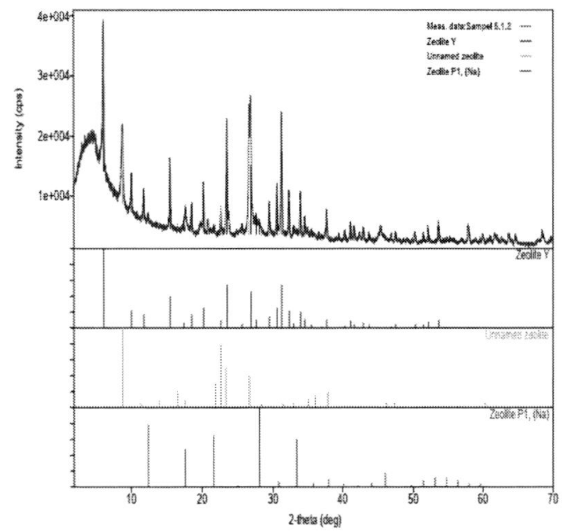

FIG. 3. Product of Zeolit NaY,15 Hours, $2.6Na_2O{:}Al_2O_3{:}10.41SiO_2{:}694H_2O$

The result of synthesis Zeolit NaY using the starting material formulation of $2.6Na_2O{:}Al_2O_3{:}10.41SiO_2{:}694H_2O$ is presented in FIG. 3, and showing a slightly higher the relative intensity in NaY zeolite. The addition of molar Na_2O to the starting material, which does not increase the molar ratio of Na_2O/ H_2O significantly, so that it does not significantly affect the quality of the synthesis product significantly was observed. Accordingly, the concept of the collision frequency required to increase the reaction rate was involved in the design of this study, by increasing the SiO_2 and Al_2O_3 concentrations, and the starting material formulation being $1.5Na_2O{:}Al_2O_3{:}9.9SiO_2{:}375H_2O$ with higher of SiO_2 and Al_2O_3 concentrations, but lower of Na_2O, and $2.6Na_2O{:}Al_2O_3{:}9.9SiO_2{:}350H_2O$ with higher of SiO_2 and Al_2O_3 concentrations,also higher of Na_2O, and giving diffractogram on FIG. 4 and FIG. 5 respectively.

Diffractogram of these synthesis result shown in FIG. 5 has a much higher NaY purity than in FIG. 4. This result proves that high concentrations in a reaction greatly affect the reaction rate, but there is a diffractogram of the synthesis shown in FIG. 5 having a much higher NaY purity than in Figure 4. These results prove that high concentrations in a reaction system are highly advantageous in terms of quantity and product quality, if the mineralizer is still capable of dissolving high solids.

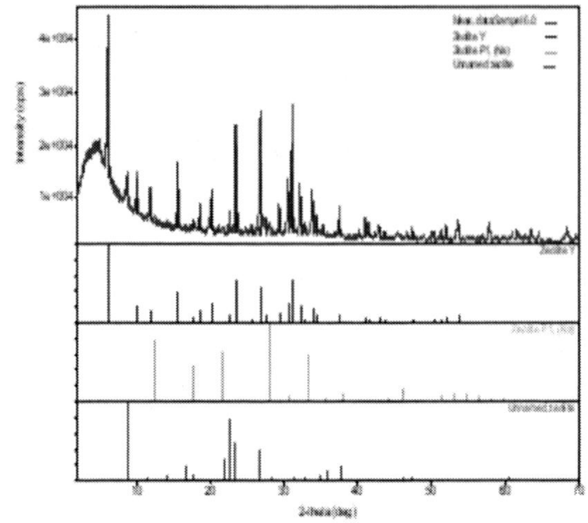

FIG. 4. Product of Zeolit NaY,15 Hours, $1.5Na_2O{:}Al_2O_3{:}9.9SiO_2{:}375H_2O$

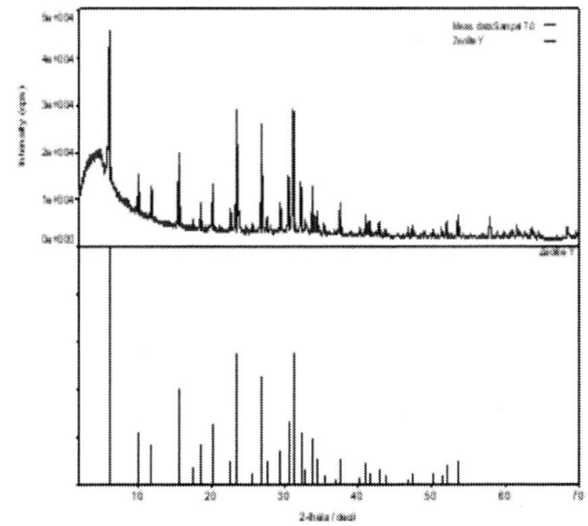

FIG. 5. Product of Zeolit NaY,15 Hours, $2.6Na_2O{:}Al_2O_3{:}9.9SiO_2{:}350H_2O$

Diffractogram of the synthesis result shown in FIG. 5 has a much higher NaY purity than in FIG. 4. This result proves that high concentrations in a reaction greatly affect the reaction rate, but there is a diffractogram of the synthesis shown in FIG. 5 having a much higher NaY purity than in FIG. 4. These results prove that high concentrations in a reaction system are highly advantageous in terms of quantity and product quality, if the mineralizer is still capable of dissolving high solids.

This study was conducted by using the ratio of seed Lynde Type Y/silica rich solids = 0.5 - 1.0 w/w. For comparison, the ratio of seed / active kaolin in Rahayu

(2014) = 4.4 w / w, while the ratio in Brown is 2*.4 b / w and pH = 14, and yields 20% -b NaY* [7]; [10].

3.3.2 Micrograph of Results of Synthesis NaY Zeolite From Kaolin

The zeolite NaY with highest purity in the product was analyzed using SEM, JEOL JSM-6510A Analytical Scanning Electron Microscope, shown in FIG. 6 (a) and FIG. 6 (b) with successive magnification of 5000 and 10,000 times.

(a)

(b)

FIG. 6. Micrograph of Zeolit NaY,15 Hours, $2.6Na_2O:Al_2O_3:9.9SiO_2:350H_2O$
(a) Mikrograph of NaY zeolite with magnification 5000 times
(b) Mikrograf of NaY zeolite with magnification 10,000 times

The micrograph shown in FIG 6 shows the mixture of the zeolite crystalline phase Y and the amorphous phase in the synthesis product. The white / gray crystal plates are zeolite Y crystals, while dark-colored clumps are amorphous.

In comparison to similar processes, the synthesis of zeolite Y from kaolin that Rahayu has done (2014) with a reaction time of 15 hours and the same process conditions, but using a mixture of 3 types of active kaolin: K700C, K1013C, K1050C, the composition is arranged to form a starting material formulation $6.4Na_2O:Al_2O_3:10.4SiO_2:932H_2O$, the Zeolite NaY with molar ratio of SiO_2/Al_2O_3 of 4.8 mol/mol and the purity of zeolite Y 86% in crystals was obtained. While the crystalinity of product was 46% [7]

3.4 Synthesis of ASA from Kaolin

The synthesis of ASA using alumina from alumina-rich solution of leaching metakaolin was done by co-precipitation method, with process conditions at temperature of 60-65°C and pH in between 7-8 at neutralizing stage to form sol and temperature of 50-55°C and pH 8-9 at aging stage process to form gel. A 10% ammonia and 2 M NaOH are solution used for neutralizing acid in neutralization stage. The ASA synthesis results were analyzed their pore structure with the NovaWin Quantachrome tool, shown in Table 3, difractogram with XRD Rigaku Smartlab, shown in FIG.7, and the morphology with SEM, JEOL JSM-6510A, shown in FIG 8.

Table 3. Effect of Process Conditions on Pore Structures

Synthesis Process			Pore Structure of ASA		
Neutralizer	Rasio Si/Al, (w/w)	Aging Time (Hours)	Average Pore Diameter (Å)	Specific Surface Area (m^2/g)	Pore Volume (mL/g)
NH_4OH	0,11	1	57,7	437	1,26
		3	59,3	428	1.27
		5	58,5	432	1.26
	0,22	1	62,5	439	1,37
	0,33	1	62,4	430	1.34
NaOH	0,11	1	52,6	372	0,98

The results of the ASA synthesis shown in Table 3, appear to have a good pore structure, especially on high surface area and total pore volume, which will be used as a hydrocracking catalyst or matrix. But, average pore diameter slightly lower than desired (70 Å).

Surprisingly, the addition of silica or an increase in Si/Al ratio does not have a significant effect on the pore structure. As it is known that silica is required to provide acidic properties to ASA, via a silanol form attached to Al such as: [(H) O-Si-O-Al]. However, the addition of Si to the formation of ASA may destroy the pore structure, thus affecting mainly the pore specific surface area. In addition, the replacement of the Al atomic site by the Si atom can excessively remove the Bronsted acid-forming group (Si-OH-Al), which will decrease the Bronsted acid concentration (Chizallet, 2000).

Rahayu (2014) has also done the synthesis of ASA from kaolin by various procedures, but still has not achieved ASA with a high specific surface area and sufficient pore diameter. Furthermore, ASA was synthesized by using synthetic materials of aluminum sulphate $[Al_2(SO_4)_3.xH_2O]$ and Sodium Aluminate ($NaAlO_2$). The best ASA with a specific surface area of

383.5 m^2/g, average pore diameter = 80.7 Å, and total pore volume = 0.78 ml / g was obtained.

The success of study of the ASA synthesis is also supported by ASA diffraction result, that similar to the commercial ASA as shown in Figure 7, and the ASA micrograph in Figure 8.

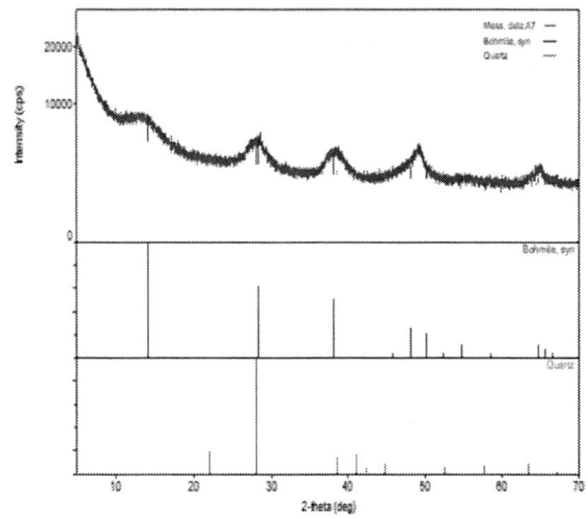

FIG.7. Diffractogram of ASA

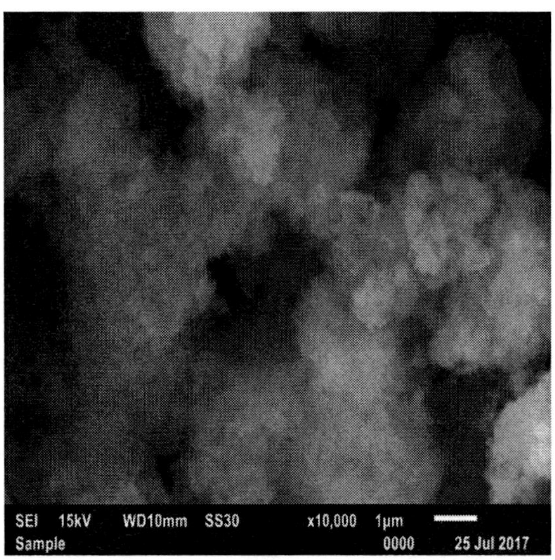

FIG. 8. Micrograph of ASA with magnification 10,000 times

4. Conclusion

Research on the synthesis of NaY and ASA zeolites from kaolin, using silica and alumina from the results of leaching metakaolin has been successfully performed, with high purity NaY zeolite, and ASA having high surface area, which is suitable for use as hydrocracking catalyst component. The silica and alumina were obtained by leaching in a solution HCl 2.5 M, at temperature of 100°C for 2 hour.. Overall, the process undertaken is more efficient and cheaper compared to what has been done before (Rahayu, 2014)

5. Reference

1. M. Murat, A. Amokrane, J. P. Bastide, L. Montanaro, Synthesis f Zeolits From Themally Activated Kaolinite. Some Observation On Nucleation and Growth, *Clay Minerals*, **27**, 119-130 (1992).
2. C. Belver, A.B.M. Miguel, A.V. Miguel, Chemical Activation of a Kaolinite under Acid and Alkaline Conditions. Spain (2002).
3. H. Hosono, Y. Mishima, H. Takezoe, K.J.D. MacKenzie, Nanomaterials: Research Towards Applications, Elsevier Ltd. The Boulevard, Langford Lane, Kidlington, Oxford OX5 IGB UK, (2006)
4. A.O.Ajayi, A.Y. Atta, B.O. Aderemi, and S.S. Adefila, Novel Method of Metakaolin Dealumination -Preliminary Investigation. *Journal of Applied Sciences Research*, 6(10): 1539-1546 (2010).
5. W.A., Mamani, G. Gustavo, H. Jonas and M. Johan and leached diatomaceous earth as raw materials for the synthesis of ZSM-5. *Springer Plus*, 3:292. (2014)
6. J.B. Adeoye, J.A. Omoleye. Development of Zeolit Y From Arobieye Mined Kaolin, 3rd International Conference on African Development Issues (CU-ICADI), 2016
7. E.S. Rahayu, Subagjo, T.W. Samadhi, M.G. Laniwati, Development of Hydrocracking Catalyst Support from Kaolin of Indonesian Origin, *Advanced Materials Research*, 896, 532-536, (2014)
8. Subagjo, E.S. Rahayu, T.W. Samadhi, M.L. Gunawan, Synthesis of NaY Zeolite Using Mixed Calcined Kaolin, [2016]
9. D.M. . Ginter, A. T. Bell, C.J. Radke,: Synthesis Of Microporous Materials, *Molecular Sieves*, Van Nostrand Reinhold, New York, **1**, p 6, (1992)
10. S.M. Brown, dan G.M.Woltermann, "Zeolitized composite bodies and manufacture thereof" *US Patent4235753* (1980).
11. C. Chizallet, and P. Raybaud, Chem. Phys. Chem. **11**, 105. (2010)

6. Acknowledgement

Authors would like thanks to Directorate General of Research Reinforcement & Development for funding this research through the National Competition Grants Program, 2016.

Characterization and acid resistance test of one-part geopolymer from fly ash and water treatment sludge

Aileen Orbecido[1]*, *Vuong* Ho[1], *Hirofumi* Hinode[2], *Winarto* Kurniawan[2], *Long* Nguyen[1]*, and *Michael Angelo* Promentilla[1]

[1]Chemical Engineering Department, Gokongwei College of Engineering, De La Salle University, Manila, Philippines
[2]Department of Transdisciplinary Science and Engineering, Tokyo Institute of Technology, Tokyo, Japan
[3]Chemical Engineering Department, Ho Chi Minh University of Technology, Ho Chi Minh City, Vietnam.

Abstract. Development of geopolymers from wastes or by-products introduces a sustainable approach to replace ordinary Portland cement (OPC)-based concrete with an eco-material of lower green-house gases emissions. However, safety concerns related to the conventional two-part geopolymer has limited large-scale applications of the product. In this context, a novel one-part geopolymer from coal fly ash and water treatment sludge has been presented. The transformation of raw materials to geopolymer was observed by FTIR, SEM and XRD analyses. Acid resistance test has proved that the new binder had great durability against sulphuric acid attack. After 28 days immersion in 5% H_2SO_4 solution, weight of all samples was hardly changed. Compressive strength, on the other hand, has not decreased but significantly increased as curing time increased. The properties were also compared to those of control samples cured in water. It was demonstrated that strong acid immersion did not create any noticeable effect on the weight and strength of one-part geopolymer system developed from coal fly ash and water treatment sludge.

1 Introduction

Since the first attempt of Davidovits to create geopolymer in 1972, numerous efforts have been dedicated to develop an advanced material for wide range of applications [1]. In construction field, geopolymer studies aim in gradually replacing Ordinary Portland Cement (OPC) and can be attributed to its superior strength and low greenhouse gas emission [2,3]. Conventionally, aqueous alkali solution is used to dissolve aluminate and silicate species from starting materials so as to form a three-dimensional network [1]. To avoid the usage of corrosive and viscous alkali solution, recent researches has explored the feasibility of one-part geopolymer using solid activators [4].

In this study, coal fly ash (FA) and water treatment sludge (WTS) were used as raw materials for the production of one-part geopolymer. WTS is the residue from water treatment process that consist significant fraction of precipitated alum. The FA and WTS mixtures were activated by sodium hydroxide and sodium aluminate. The durability of the formed- geopolymer against acid attack is investigated in this study.

2 Materials and Method

2.1 Materials

The coal FA was obtained from a power plant in Luzon, Philippines. It was used without any pre-treatment. The wet WTS was collected from a water treatment plant in Manila, Philippines. The wet sludge was dried at 105 °C then ground using LA abrasion machine for 10,000 cycles. Anhydrous sodium hydroxide (99% NaOH, Mega Source Scientific Supplies, Philippines) purchased from a local supplier in the Philippines and anhydrous sodium aluminate (50-56% Al_2O_3; 40-45% Na_2O) from Sigma-Aldrich were used as activators. Concentrated sulphuric acid (96.7%) was purchased locally and diluted to 5% concentration for the acid resistance test.

2.2 Geopolymer synthesis

Anhydrous sodium hydroxide (99% NaOH) was used as solid activator form fly ash based one-part geopolymer. Two sources of aluminates were examined including ground WTS and sodium aluminate (50-56 % Al_2O_3; 40-

*Corresponding author: aileen.orbecido@dlsu.edu.ph

Table 1. Compositions of two geopolymer mixtures.

Component (%)	GP1	GP2
FA	72.7	93
WTS	22.3	0
Sodium aluminate	0	2
NaOH	5	5

45 % Na_2O, Sigma-Aldrich). Proportion of each mixture GP1 and GP2 is presented in Table 1. Water is added to each mixture and stirred by mechanical mixer for 10 minutes to obtain homogeneous pastes.

The paste was poured into 50x50x50 mm moulds. Steel moulds were greased beforehand in order to be demoulded easier. The samples were hardened for 7 days at ambient temperature and pressure. They were then removed from cubic moulds and kept at room temperature for another 14 days.

2.3 Compressive strength test

Unconfined compressive strength test was conducted in accordance with ASTM C109 standard [6].

Dimensions of specimens were firstly measured. Then, 50-mm cubes were placed on the holder put in specimen chamber of the Compression and Flexural Testing Machine in STRC, De La Salle University. Both surfaces exposed to the force were polished to avoid high deviation during the testing process. The loading speed was fixed at 0.35 kPa/s. A computer was connected with the testing machine to record the result. Average of three replicates of each run was calculated.

2.4 Sulphuric acid resistance test

Following the curing period, samples GP1 and G2 were subjected to acid resistance test conforming to modified ASTM C267 standard [7].

Deterioration of specimens before and after acidic attack indicated acid resistance. It was determined by measuring the changes of compressive strength and weight. Three (3) days prior to acid immersion, samples were all submerged in water [8]. After that, weight and compressive strength of each sample were taken as initial values.

The samples were then immersed in 5% sulphuric acid solution [9, 10]. The solution to specimen volume ratio was kept constant at four throughout the test [8]. Weight and compressive strength were recorded after 7 and 28 days of immersion.

2.5 Microstructure characterization

Chemical composition of precursors was recorded by Energy-dispersive X-ray fluorescence (XRF) using S2 Puma machine in the Center of Viet Nam - German Technology. Morphology of structures was observed by a JEOL 5310 Tungsten Scanning Electron Microscope (SEM). The ground samples were put on the flatting metal circle pellet noted with number (from 1 to 4). The pellet was coated with a thin gold layer. Structural bonds were tested by Nicolet 6700 Fourier Transform Infrared Spectroscopy (FTIR) Spectrometer. Samples were ground and mixed with KBr powder. Then, the mixture was pressed into circle tablet in the circle steel mould. The outputs were recorded and saved in a computer connect with the system.

3 Results and discussion

3.1 Characterization of raw materials

Table 2 shows the elemental composition of FA and WTS in form of oxides as determined by XRF analysis. Both sludge and fly ash compose of high proportions of Si and Al., The percentages of SiO_2 and Al_2O_3 in raw materials are similar with the range of these materials from various sources of WTS (Waijarean et al., 2014; Suksiripattanapong et al., 2015; Huang and Wang 2013) and FA (Blissett and Rowson, 2012). Specifically, WTS

Table 2. Elemental compositions of FA and WTS using XRF analysis.

Element as oxide form	Composition (wt.%)	
	WTS	FA
SiO_2	42.0	33.9
Fe_2O_3	19.3	26.5
Al_2O_3	30.0	13.5
CaO	2.46	13.6
TiO_2	1.38	1.01
K_2O	0.784	1.19
MgO	1.45	7.92
SO_3	0.425	0.864
Na_2O	0.196	0.357
Other	2.005	1.159

is a rich source of Al (30% Al_2O_3), thus it is a prospective precursor for zeolithic formation. Also, CaO content in FA is relatively high so that it can create a strong alkaline media for geopolymerization.

The microscale images of raw materials in Figue 1 show that FA consists of typical microspheres from 1 to 20 μm [11], meanwhile, sludge does not contain any sphere but flaky phases.

Micrographs at magnification x200 imply that FA generally are small size. Whereas WTS contains larger grains probably due to hydroxyl groups of $Al(OH)_3$ presented in non-calcined sludge [12].

FTIR spectra of starting materials are presented in Figure 2. The bands at around 1000-1100 cm^{-1} are assigned to T-O-Si asymmetric stretching vibration that identifies the presence of polysialates of aluminum network. The existence of bands at around 1630 cm^{-1} and 3420 cm^{-1} in both WTS and FA spectra are due to H-O-H bending vibration of aluminosilicates. Especially, WTS has an additional peak at wavenumber 914 cm^{-1} indicating symmetric Al-O-Si stretching vibration of kaolin. A peak at 537 cm^{-1} implies the Si-O-Al bending vibration; however, this peak might be overlapped with absorption band of hematite mineral [13].

3.2 Effect of water treatment sludge on fly ash based one-part geopolymer matrices

The compressive strength values of both combinations sastify the strength requirement of nonloadbearing masonry unit which is 4.14 MPa according to ASTM C129 [14].

Fig. 1. SEM images of WTS and FA at magnifications of x7500 and x200

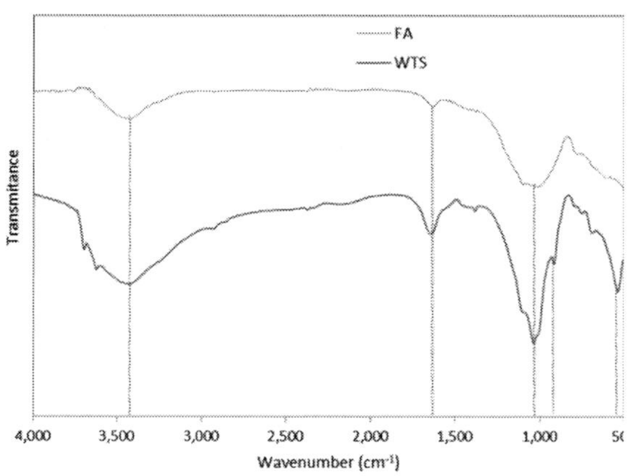

Fig. 2. SEM images of geopolymers GP1 and GP2

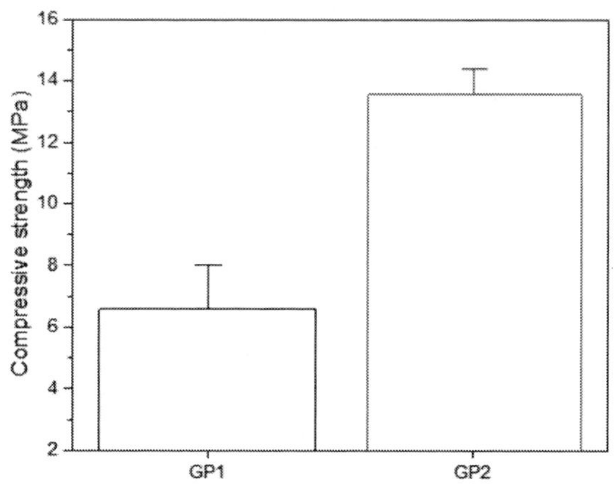

Fig. 3. Compressive strength values of GP1 and GP2, the bars represent standard deviations

Fig. 4. SEM images of geopolymers GP1 and GP2

One-part geopolymer sample using sodium aluminate (GP2) has higher strength than the sample using WTS (GP1). It could be implied that WTS contains Al in inactive phases. Therefore, addition of WTS into the mixture could inhibit the reaction rate. Previous study explained that non-calcined sludge composed of colloidal $Al(OH)_3$ which is the inert octahedral coordinated aluminium, particularly for geopolymer formation [12].

Geopolymer transformation was indicated by SEM micrographs in Figure 4. The microspheres typical for FA almost disappeared in the images of two specimens. Instead, SEM images present fine flakes and particles of ground product. This is associated with the attack of sodium hydroxide that weakened or destroyed the spheres.

3.3 Durability against acid attack

As presented in Table 3, one-part geopolymer binders are very durable with strong acidic environment. The strength values significantly increase after immersed in 5% sulphuric acid solution. After 28 days, sample GP2 gained additional 9 MPa that is close to the increased strength of control samples. However, strength of GP1 slightly decreased. It indicates that utilization of sludge in one-part geopolymer inhibit the acid resistance capacity of product. Mass weights were nearly kept constant through the period of immersion. For both specimens, mass losses are smaller than 2%.

Table 3. Compressive strength and weight of samples before and after immersed in 5% H_2SO_4 solution for 7 and 28 d.

		Acid immersed runs			Control runs
Duration (days)		0	7	28	28
Strength (MPa)	GP1	6.9	7.8	5.7	8.2
	GP2	16.0	20.1	25.0	27.6
Weight (wt.%)	GP1	100	98.6	98.3	100
	GP2	100	98.6	99.5	100

4 Conclusions

Characterization of raw materials using XRF analysis indicated that FA and WTS have suitable elemental and mineral compositions for geopolymer synthesis. SEM micrograph presented that FA consists of spherical forms while WTS contains flaky glassy phases. Although, WTS has high over Al content (30 % Al_2O_3), utilization of this Al-source material in one-part geopolymer significantly reduced compressive strength.

One-part geopolymers have shown very high resistance against sulphuric acid attack. Compressive strength and weight of 28-day acid immersed samples were not significantly different with those of water immersed control runs.

Acknowledgement

The authors acknowledge the financial support of Japan International Cooperation Agency (JICA) through the AUN/SEED-Net scholarship.

References

1. J. Davidovits, *Geopolymer 2002 Conference* (2002)
2. C.D. Atis, E.B. Görür, O. Karahan, C. Bilim, S. Ilkentapar, E. Luga, *Construction and Building Materials,* **96,** 673–678 (2015)
3. G. Habert, C. Billard, P. Rossi, C. Chen, N. Roussel, *Cement and Concrete Research,* **40,** 820–826 (2010)
4. F. De Lillo, F. Cecconi, G. Lacorata, A. Vulpiani, EPL, **84** (2008)
5. H. Choo, S. Lim, W. Lee, C. Lee, *Construction and Building Materials,* **125,** 21–28 (2016)
6. American Society for Testing Materials. Standard Test Method for Compressive Strength of Hydraulic

Cement Mortars (Using 2-in. or [50-mm] Cube Specimens) (ASTM C109/109M – 16a). *ASTM International*, (2016)

7. American Society for Testing Materials. Standard Test Methods for Chemical Resistance of Mortars, Grouts, and Monolithic Surfacings and Polymer Concretes (ASTM C267-01). *ASTM International*, (2012)
8. S. Thokchom, P. Ghosh, S. Ghosh, *ARPN Journal of Engineering and Applied Sciences*, **4**, 1 (2009)
9. V. Kannan, K. Ganesan, *Construction and Building Materials*, **51**, 225–234 (2014)
10. T. Bakharev, *Cement and Concrete Research, 35,* 658–670, (2005)
11. A.S. Vargas, D.C.C.D Molin, A.C.F. Vilela, F.J. Silva, B. Pavão, B., H. Veit, *Cement & Concrete Composites,* **33**, 653–660 (2011)
12. M.A. Tantawy, *Materials Research Bulletin, **61**,* 415–421 (2014)
13. J. Davidovits, *Geopolymer chemistry and application* (4th ed.). Institut Géopolymère – France (2011)
14. American Society for Testing Materials. Standard Specification for Nonloadbearing Concrete Masonry Units (ASTM C129-14a). *ASTM International*, (2015)

Natural Dye Extraction From Teak Leves *(Tectona Grandis)* Using *Ultrasound Assisted Extraction* Method for Dyeing on Cotton Fabric

Lailatul Qadariyah[*], *Mahfud* Mahfud, *Endah* Sulistiawati, *and Prima* Swastika
Department of Chemical Engineering, Institut Teknologi Sepuluh Nopember, Sukolilo 60111 Surabaya Indonesia

Abstract. Synthetic dye waste of textile industry has contaminated the water and aquatic living organism. One of the solution is used teak leaves as the natural dye. Common conventional extraction process is soxhletation which need more time and more solvent. One of extraction method that has been developed is ultrasound assisted extraction (UAE). The objectives of this research were to study the effect of solvent, extraction time, pH, and extraction yield. The conclusion of this research was optimum yield of teak leaves extract about 42,79% at pH 3 and extraction time of 20 minutes with aquadest as the solvent using UAE method. If natural dye of taek leaves was used on cotton, the color will change depend on fixator that was added. In GC-MS analysis, some compound that were detected are decanoic acid methyl ester and methyl tetradecanoate. Both of the compound have chromophore cluster.

1. Introduce

In April 2016, Rancaekek people in Bandung complained about textile industrial waste which contaminated their agriculture lands, the lost was 11,3 trillion Rupiah. Textile industrial waste which was thrown to river contaminated water and aquatic living organism. The contamination of textile industrial waste was shown by physical condition of river surface, the colors changing, turbidity of the water, the smell of the water and agriculture lands that was broken near the river.

One of the solutions to reduce the impact of industrial textile waste so it is not dangerous anymore for the environment or living organism is to use the natural dye for textile. Teak leaves *(Tectona grandis)* can be used as natural dye. Teak leaves contain anthocyanin which is source of brown color [1].

Conventional extraction process that commonly used in research before to extract teak leaves are soxhlet extraction, reflux extraction method, and maseration which is need more time and more solvent. According to those reasons inovation of extraction technology is needed that can save time of extraction process and solvent and also environmental friendly. One of extraction method which be improve recently is Ultrasound-assisted Extraction (UAE). Ultrasound-assisted extraction (UAE] offers an inexpensive, environmentally friendly, less time consuming and efficient alternative to conventional extraction techniques. The enhancement in extraction obtained by using ultrasound is mainly attributed to the effect of acoustic cavitation produced in the solvent by the passage of an ultrasound wave [2,3]. Ultrasound also offers a mechanical effect allowing greater penetration of solvent into the sample matrix, increasing the contact surface area between the solid and liquid phase, and as a result, the solute more rapidly diffuses from the solid phase into the solvent [4,5].

This study was to know about parameters that effect yield of teak leaves extract, those parameters are solvent, extraction time, and pH. After that is to determine the optimum yield, and qualitative analysis of pigment in yield, the chromophore cluster using GC-MS, and will be applied on fabric.

[*] Corresponding author: lailatul_2008@yahoo.com

2. Methods

2.1. Materials

Dry teak leaves with moisture content approximately 10% were purchased from Surabaya, ethanol 60%, HCl, NaOH and aquadest.

2.2. Procedure

First, teak leaves were grinded inside the blender and then sieved until have size 35 mesh – 60 mesh. After that 1 gram of teak leaves and 200 ml of solvent were put into three neck round flask. Then, three neck round flask was put inside the ultrasound cleaning bath. Extraction process was done at temperature 60°C in atmospheric pressure, 200 ml of solvent, 40 Hz frequency, ratio teak leaves with solvent was 0.005 g/ml. The extraction process has done in various variable such as time (5, 10, 15, 20, 30, 40, 50, 60 and 70 minutes), solvent (ethanol 60% and aquadest), and pH (3, 4, 5, 7, 9 and 10). Adding HCl and NaOH to make condition become acid or base. Here is the ultrasound assisted extraction equipment:

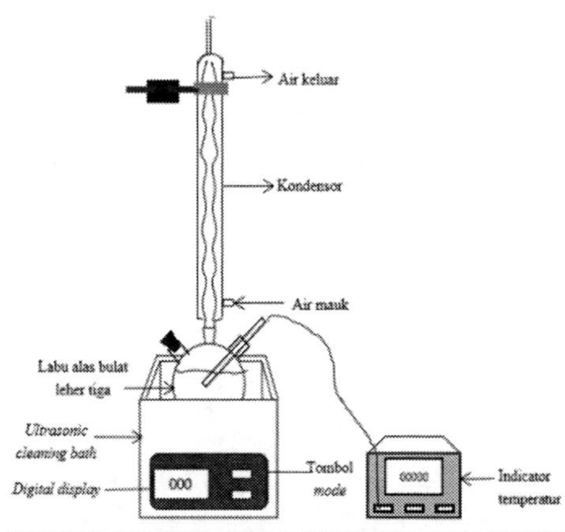

Fig 1. ultrasound assisted extraction equipment

Extraction process will be stopped when the extraction time was completed. After the extraction process then the filtrate will be filtered using vacuum pump to separate it from impurities. Natural dye that mixed with the solvent were separated using hot plate, the solvent will be vaporized and the natural dye powder will stay on the bottom of beaker glass. The natural dye was analyzed by GC-MC and will be applied on cotton fabric.

3. Results and Discussion

3.1 Effect of Solvent

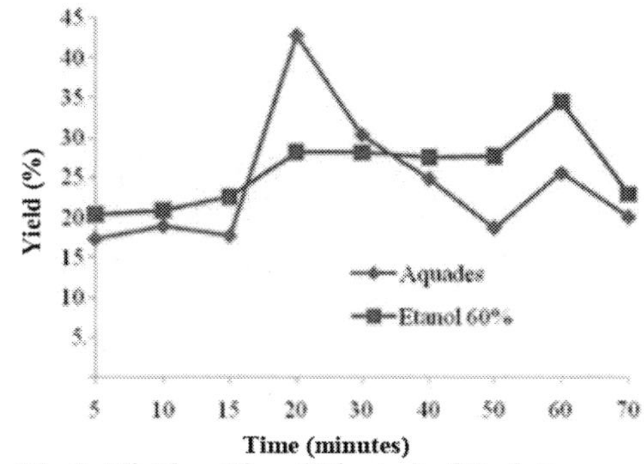

Fig 2. Yield vs Time (Minutes) of Teak Leaves extraction in pH 3

Fig 2 shows the effect of reaction time on teak leaves extract. Extraction using aquadest produce teak leaves extract greater than using ethanol as the solvent. Anthocyanin is polar molecule which has aromatic ring. Anthocyanin polarity influence the extraction process [6]. Commonly polar organic solvent suitable for anthocyanin extraction. The optimum yield of teak leaves extraction with aquadest as solvent in pH 3 was 42,79% at 20 minutes. Meanwhile the optimum yield of teak leaves extraction with ethanol 60% as solvent in pH 3 was occur at 60 minutes. Based on **fig 2**, teak leaves extract was decreasing after reach the optimum condition. It's occur because of vaporization process using hot plate make yield of natural dye also vaporize with the solvent.

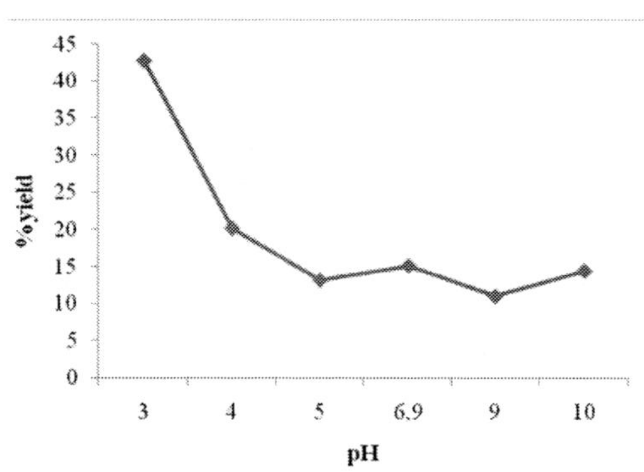

Fig 3. Effect of pH to yield of extract teak leaves

Fig 3. shows that optimum yield of teak leaves in acid extraction condition. The yield of teak leaves extract decreased with increasing of pH. Possibly due to the decreasing of pH makes hydrogen ion density increase and obstruct releasing hydrogen ion [7].

3.3 Effect of Extraction Time

(a) **(b)**

Fig 3. Effect of extraction time to yield of extract teak leaves in minutes (a) 20 (b) 30

Fig 3. shows that color of teak leaves extract in extraction time of 30 minutes was darker than extraction time of 20 minutes, extraction time have effect for the color of the teak leaves extract, from that can be concluded that the more time needed for extraction process, the more material extracted [8].

3.4 GC-MS Analysis

Based on GC-MS analysis, the main compound of teak leaves extract were decanoic acid methyl ester (41,16%), methyl tetradecanoate (22,84%), and hexadecanoic acid, methyl ester (15,17%). Some of that compound consist of chromopore cluster, it is substance which bring colors. **Fig 4.** and **fig 5.** Shows the compound structure of decanoic acid, methyl ester and methyl tetradecanoate. Chromophore cluster can be found in their compound structure which is ketone cluster (C=O) and carboxylate cluster (COOH).

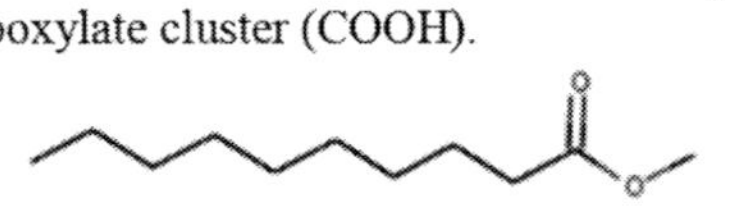

Fig 4. Structure of Decanoic acid, methyl ester

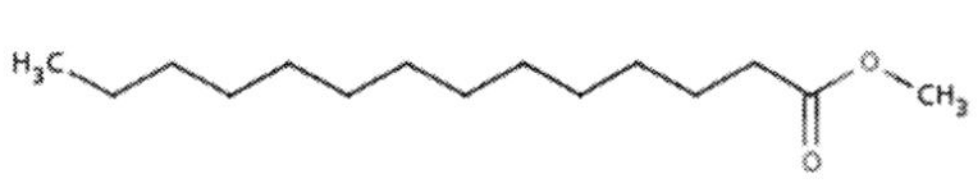

Fig 5. Structure of Methyl tetradecanoate

Table 1. Compounds of Teak Leaves Extract based on area of Teak Leaves Extract

No	Retention Time	% Area	Name
1.	3,076	4,81	*Decanoic acid, methyl ester*
2.	4,894	41,16	*Decanoic acid, methyl ester*
3.	7,020	22,84	*Methyl tetradecanoate*
4.	9,094	15,17	*Hexadecanoic acid, methyl ester*
5.	10,632	0,26	*Heptadecene-(8)carbonic acid-(1)*
6.	10,746	0,56	*Pyridine-3-carboxamide, oxime, N-2-trifluoromethylphe nyl*
7.	10.798	8,53	*9-Octadecanoic acid, methyl ester,(E)-*
8.	11,015	4,02	*Heptadecanoic acid, 16-methyl-, methyl ester*
9.	12,284	1,74	*Thiosulfuric acid*
10.	13.124	0,92	*8-Acetyl-3, 3-epoxymethano-6, 6, 7-trimethylbicyclo*

4. Application on Cotton Fabric

Application of teak leaves extract on cotton fabric was done by adding some fixator such as $CaCO_3$ and $Al_2(SO_4)_3$. Some test were done to know the tenacity of the color such as dipped in TRO (Turkey Red Oil) about 24 hours, TRO has function as dispersing agent which can disperse color to the cotton fabric entirely. Besides, TRO can omit chemical substances or impurities which stick on the cotton fabric so the colors can stick on cotton fabric sturdily. After that the cotton fabric was washed by water then dye into liquid of teak leaves extract and dry in the sun.

Fig 6. Application Teak Leaves Extract on cotton fabric

Fig 6. shows the result of teak leaves extract which was applied on cotton fabric. The precipitation of natural dye will be increase after cotton fabric was dipped in fixator, because the metal grained inside cotton fabric can precipitate natural dye in a big amount compared without fixator [9].

5. Conclusion

Natural dye from teak leaves can be extracted using ultrasound assisted extraction at 60°C in atmospheric pressure, 40 Hz frequency, extraction time were 5, 10, 15, 20, 30, 40, 50, 60 and 70 minutes, solvent ethanol 60% and aquadest, and pH 3, 4, 5, 7, 9 and 10. Optimum yield of teak leaves extract is 42,79% at pH 3 and extraction time 20 minutes with aquadest as the solvent. If natural dye of taek leaves are used on cotton, the color will change depend on fixator that be added.

6. References

1. Yuliana, Sri Kumalaningsih. Pembuatan Pewarna Bubuk Alami dari Daun Jati (Tectonagrandis Linn f.) (Kajian Jenis dan Konsentrasi Filler). Jurusan Teknologi Industri Pertanian. Fakultas Teknologi Pertanian. Universitas Brawijaya. Malang (2012).
2. Y. Ma, X. Ye, Y. Hao, G. Xu, D. Liu, Ultrason.Sonochem. **15**. 227–232. (2008)
3. D.T. Velickovic, D.M. Milenovic, M.S. Ristic, V.B. Veljkovic, Biochem. Eng. J. **42**. 97–104. (2008)
4. M. Toma, M. Vinatoru, L. Paniwnyk, T.J. Mason, Ultrason. Sonochem. **8.** 137–142 (2001)
5. T.J. Mason, Ultrason. Sonochem. 10 175–179 (2003)
6. J. B. Harborne. and R. J. Grayer,. The Flavonoids. p. 1-20. Chapman and Hall. New York. (1998)
7. Fathinatullabibah. Stabilitas Antosianin Ekstrak Daun Jati (Tectona grandis) terhadap Perlakuan pH dan Suhu. Surakarta: Universitas Sebelas Maret. (2014).
8. Putri A.R.W.dan F.C. Nisa. Extraction Anthocyanin from the Sorted Red Rose (Rosa damascene Mill) with Microwave Assisted Extraction. Jurnal Pangan dan Agroindustri. **3**. 701-712. (2015).
9. Rosyida, Ainur dan Didik Achandi. 2014. Pemanfaatan Daun Jati Muda Untuk Pewarnaan Kain Kapas Pada Suhu Kamar. Sukoharjo: Akademi Teknologi Warga Surakarta.

Non-invasive Detection of Human Body Liquor Intake Based on Optical Biosensor

Christian Jhoart Abaoag[1], *Ralph* Alforque[1], *John Benedict* Ordoño[1], and *Edna* Quinto[1,2,*]

[1]Department of Chemical Engineering, University of Santo Tomas, 1008 Manila, Philippines
[2]Research Center for the Natural and Applied Sciences, University of Santo Tomas, 1008 Manila, Philippines

Abstract. Alcohol-related incidents are increasing despite the implementation of RA 10586, the Anti-Drunk or Drugged Driving Act of 2013. Conventionally, blood alcohol content (BAC) is tested by gas chromatography or breathalyzers. This work aims to design and fabricate a paper-strip based sensor for the detection of alcohol using saliva as biomedium. The sensor will act as an alternative alcohol detection platform, which will provide low cost analysis of BAC. Different tests were undergone using p-nitrophenol, PNP, as recognition element, which include stability, repeatability, and sensitivity. In order to establish the change of PNP in the presence of alcohol, a UV-Vis spectrophotometer was used. For stability and repeatability, an RSD of 1.24% and 4.61% were obtained, respectively. Furthermore, the sensitivity to alcohol concentration was found to have an R^2 value equal to 0.987. For the paper-strip application, 10 mm x 60 mm optical sensing membrane (OSM) with immobilized PNP was prepared. It is analysed through the use of digital image captured by a smartphone camera. The RGB values are then measured using the ImageJ software application. The stability, repeatability, and sensitivity of biosensor are 1.85% RSD, 2.23% RSD, and 0.9731 coefficient of linearity, respectively. This method is a promising alternative to breathalyzer.

Introduction

Being charged with DUI or DWI (Driving Under the Influence or Driving While Intoxicated) is a very serious offense. If you are found guilty, it will significantly impact your ability to drive (with suspended or revoked license), increase your insurance rates, and may even include a fine and jail time. A guilty court verdict can further affect your employment options, financial future, and ability to travel internationally. [1]

Field Sobriety test and chemical blood test are two of the most commonly performed by policemen in order to detect if a person is under the influence of alcohol.

According to Badugu, *et. al.*, blood sampling is impractical especially in today's generation, since other bodily fluids, such as sweat and saliva, give far better results compared to the blood tests. Using the latter fluid, different components of the blood can be determined: uric acid, lactic acid, alcohol concentration, glucose level, etc. [2]

According to Philippine National Police-Highway Patrol Group (PNP-HPG), road accidents are one of the leading causes of death in the Philippines. In 2012, there was an average of 227 road accidents per day according to MMDA. [3] Vehicular accidents associated with drunk driving is steadily increasing in number according to HPG. In 2012, 375 car accidents were recorded nationwide involving drunk drivers; 390 in 2013, and 498 in 2014. [4]

Breathalyzer units were distributed and have been in disposal of officer in NCR, Metro Davao, and Metro Cebu since March 12, 2015. [5] The device uses a platinum fuel cell that reacts specifically to presence of alcohol, and will not be affected by external or environmental factors. It is also already calibrated according to the Philippines' legal blood alcohol content (BAC) limit of 0.05%. For non-pofessional drivers, their BAC should not exceed 0.05%. Professional drivers and drivers of public utility vehicles (PUVs), however, must have no trace of alcohol at all- anything higher than 0% will be subject to penalty. [6]

To date the Philippine government will need 3,000 breathalyzers if one is to one ratio per precinct is the target as there are 3,000 precincts throughout the country. [7]

To solve this problem, the researchers intend to design a low-cost, portable, user-friendly version of breath analyser by finding a suitable reagent that is capable of detecting quantitatively both the presence and concentration of ethanol present inside the body through human saliva. The presence of ethanol at varying concentration in saliva will reflect different intensities of visible color change on the biosensor.

* Corresponding author: ecquinto@gmail.com

The study would show that the device/method possess repeatability, stability of the signal, sensitivity of the reagent to changes in the alcohol content of the saliva sample. It should show the limit of detection of the reagent and the response time. A real sample would be tested using the biosensor.

The primary goal for this prototype is to develop an alternative to breathalyzers that are currently commercialized in the market by fabricating a low-cost and portable or compact and easy to carry paper-strip based biosensor with higher efficiency and sensitivity. The development of this biosensor will be advantageous to (1) the country's economy, (2) other professionals, and (3) the general public.

The proposed method uses non-invasive sampling, which may be applicable to determine BAC instead of using the conventional method of gas chromatography or breathalyzer using blood and breath as sample, respectively. However, the study will cover only the determination of alcohol in saliva based on the standard reactions of ethanol. It will not include any other organic substances, such as glucose, glycoproteins, amylase, etc., and inorganic substances, i.e. fluoride, calcium ions, and inorganic phosphates, present in saliva nor will it do interference studies.

The presence of alcohol in human body may exist up to very small quantities when compared to other constituents of human body fluids, which include saliva. For this reason, there is a need for an indicator that has the capacity to detect alcohol even in small quantities. Hence, it will serve as the sensing reagent for alcohol.

Preparation of indicator and alcohol standard solutions

The indicator was prepared according to its solubility to water. 1.6 g of PNP (RDH, Spain) was dissolved in 100 mL of distilled water.

Alcohol solutions of standard concentrations were prepared from ethyl alcohol (J.T. Baker, USA). The stock solution is a 0.14% (v/v) alcohol solution. Other concentrations of alcohol, 0.02% (v/v) to 0.12% (v/v), were derived from the stock solution by serial dilution. These alcohol concentrations are such that the limit alcohol percentage, 0.05% (v/v), is between the suitable ranges. It is to clearly see the variations of the readings in the upper and lower boundaries of the target concentration.

The reaction of the p-nitrophenol with ethanol produces phenolate. The color change is brought upon by the pH adjustment carried out by such reaction. The p-nitrophenol is colorless from pH 5 and has the most intense yellow color at pH 7. [8]

Spectrophotometric analysis of p-nitrophenol

Instrumentation system

The instrumentation system for the analysis of p-nitrophenol is composed of a personal computer connected to a UV-Vis spectrophotometer, Perkin Elmer LAMBDA™ 25.

The sample was placed inside a cuvette and was then introduced into the spectrophotometer. The UV-Vis spectrophotometer was interfaced with a personal computer for processing. Data acquisition and off-line analysis was performed using Microsoft Excel (Microsoft Corporation, USA). Spectra readings were taken in the visible region. All spectroscopic experiments were performed at room temperature.

Determination of the working wavelength

Working wavelength was chosen by reading the absorbance of the blank PNP and comparing it with the spectra of a PNP solution with a few drops of 0.14% ethanol solution.

The absorbance readings obtained in the two conditions were plotted against the wavelength range considered. The two resulting spectra were compared. The wavelength that gave the largest difference between the two absorbance readings was the working wavelength used in the succeeding sensor tests.

Figures of merit of the sensor

Stability determines the invariability of the sensor against other atmospheric elements. Ideally, the sensor should not react with other atmospheric elements or compounds other than the target analyte. Signal stability was determined by reading and plotting the absorbance with time. Signal stability was based on the Relative standard deviation of the Absorbance readings done in 2 hours.

The repeatability test determines precision of the results, this is done by alternately reading the absorbance of the blank solution and that of the blank with 0.08% ethanol.

The sensitivity test determines the response of the PNP to varying alcoholic concentrations. Increasing concentrations of alcohol solution was used within the range of 0.04% (v/v) to 0.12% (v/v) ethanol.

Testing of the PNP immobilized on paper-strip by photo-monitoring

Immobilization of the reagent

A paper-strip (Hard Copy, substance: 20/70 gsm, ultra-white color) of size 10 mm by 60 mm was used as matrix where the reagent was immobilized to create an optical sensing membrane for alcohol detection. Immobilization was carried out by dipping in order to equally distribute the reagent on the surface of the paper strip and allowed to dry.

Instrumentation system

The immobilized p-nitrophenol on paper strip, IPP needs to be photographed under direct lamp light (3W) to ensure consistent lighting. A smartphone camera (Samsung A5 2016, 13 MP, f/ 1.9, 28 mm) was used to take pictures of the IPP and these images were subjected to treatment using ImageJ, a computer application with extensive photography capabilities, for their red, green, and blue (RGB) measurements. [9]

A section of the photo was selected for treatment. The section was carefully chosen so that there would be no air gaps or misdistribution of the liquid along the surface of the section chosen. The goal is to achieve a uniform distribution of the liquid sample on the chosen section.

The uniformly selected section of each picture gave particular data points corresponding to the intensity of the color. All photographic experiments were performed at room temperature and at a height 13 cm from the paper to the lens of the camera.

Figures of merit of the IPP

In the determination of the stability, the IPP was read and results based on RGB measurement. Stability was determined by determining the RSD of the RGB readings. A graph of RGB versus time is plotted.
The response time is the time at which the IPP can sense the presence of ethanol right after the alcohol is introduced to the matrix.

The response time of the biosensor was obtained from the plot of RGB values with and without alcohol against time. The RGB measurements of the paper-strip based biosensor were measured using photographs taken for 20 minutes every 10 seconds.

To determine the degree of repeatability of the sensor, the RGB measurements of the IPP using photographs taken for 1 minutes every 10 seconds. These were subjected to ImageJ data software. The IPP was exposed to equal concentrations of alcohol every trial. Repeatability is presented as RSD value.

To measure the sensitivity of the IPP, it was exposed to different alcohol concentrations. The data obtained are plotted to give a chart of RGB values versus concentration of alcohol.

Testing IPP using real sample

The IPP was tested on an individual who has taken a single shot of alcoholic beverage. The procedure that was employed was the same as getting the response time of the sensor. The sample saliva was dropped on the IPP

Each trial succeeded in seconds after taking an alcohol shot.

Results and discussion

Spectrophotometric analysis of p-nitrophenol

Series of tests was done using the UV-VIS spectrophotometer. The tests include the determination of the working wavelength, stability, response time, repeatability, and sensitivity.

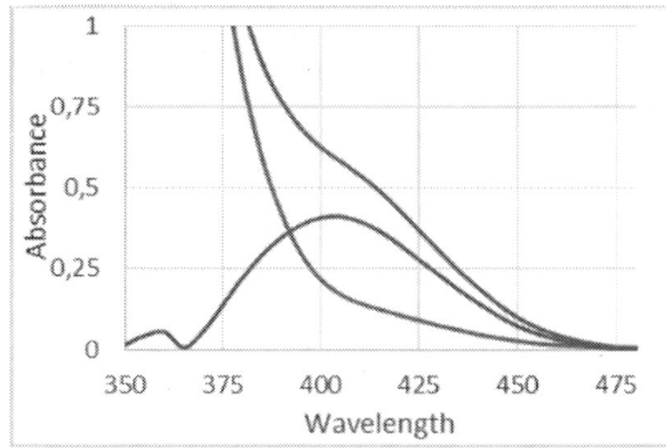

Figure. 1. Spectra of p-nitrophenol in blank medium and 0.14% ethanol solution

The spectrum in Figure 1 showed the response of p-nitrophenol to two different environments: blank (water) and 0.14% ethanol solution. The third curve is a plot of the difference of the absorbance readings between 0.14% ethanol concentration and the blank. The result showed that the peak of the curve of the difference occurred at 405 nm. For analytical reading purposes, all readings were done at 405 nm. This is the working wavelength for the succeeding tests.

The stability of p-nitrophenol was a measure of the absorbance for an hour. Analysis of the data resulted to an RSD of 1.24% This behavior indicated fair stability of the p-nitrophenol at 405 nm working wavelength.

Figure 2 showed the response time of p-nitrophenol to 0.08% ethanol solution. Ethanol solution was introduced after taking the absorbance of the blank. There was a significant increase in the absorbance 0.29 seconds after the introduction of ethanol solution.

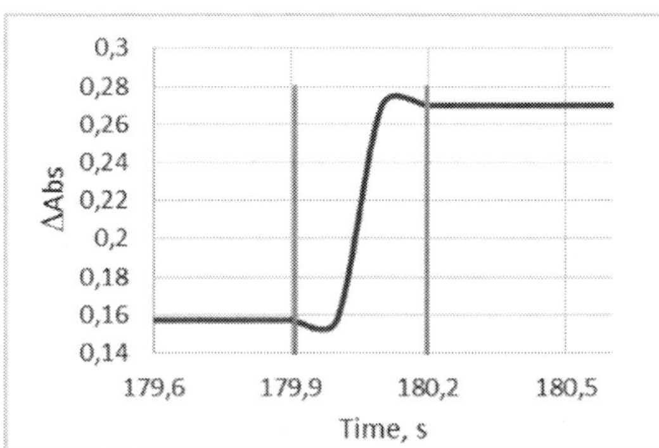

Figure 2. Response time of p-nitrophenol to 0.08% ethanol solution

Using a 0.08% solution of ethanol in different runs, repeatability test yielded results shown in Figure 3 with an RSD of 4.61%.

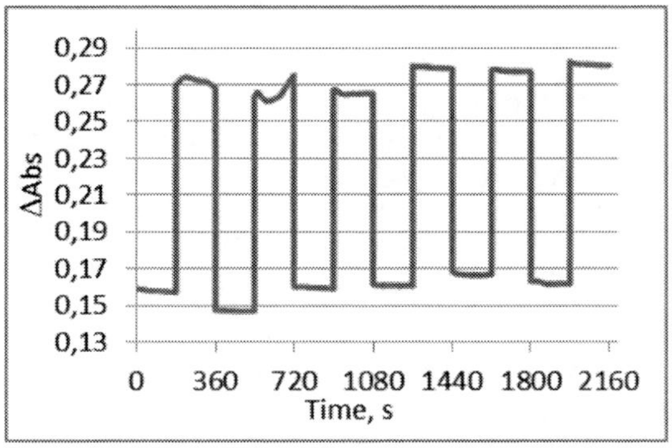

Figure 3. Change in absorbance vs. time using 0.08% ethanol solution (n=6)

The response of the sensor with varying ethanol concentrations in the range of 0.04 - 0.1% (v/v) is presented in Figure 4. The response showed a direct relationship between change in absorbance and ethanol concentration.

Figure. 4. Sensitivity of the p-nitrophenol to varying ethanol concentration

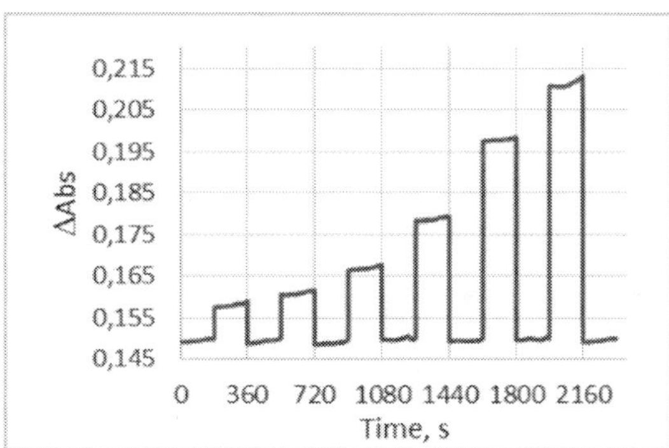

Figure 5 showed the relationship between the change in absorbance readings to the different concentration of alcohol. The sensitivity is 0.56 change in absorbance per (v/v) % ethanol with a coefficient of linearity of $R^2 = 0.987$. The limit of detection of the sensor is 0.02% ethanol solution.

Figure 5. Calibration curve for the sensitivity of p-nitrophenol indicator (n=3)

Testing of paper-strip biosensor by photo monitoring

The applicability of the sensor was tested as an immobilized p-nitrophenol using a paper-strip. The same tests were done on the paper-strip which include stability, repeatability, and sensitivity. Using a computer software, ImageJ, the histogram of the solution in the paper was directly observed and analyzed. The RGB (Red, Green, and Blue) color system was utilized and changes in the colors were determined.

Stability is often the most cited parameter in biosensor publications according to Gibson. [10] Furthermore, it relates to the operating lifetime and reusability of the device. The operational stability of a biosensor is defined as the retention of the activity of an indicator when it is immobilized to a medium. The latter investigated if p-nitrophenol, the alcohol indicator, will react to paper, environment, and any other external factors other than the analyte that it should detect. In this experiment, the stability is shown by Figure 6.

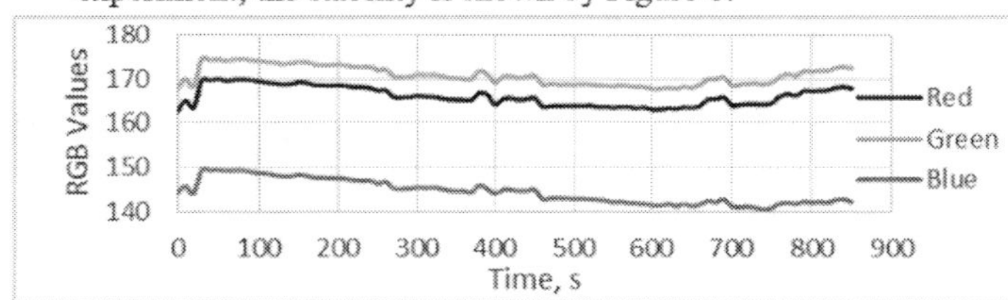

Figure 6. RGB values reading vs. time for stability test

The computed over-all RSD for the signal stability of p-nitrophenol to paper was computed to be 1.31%, 1.23%, and 1.85% for red, green, and blue respectively. With these RSD values, IPP is stable to paper and can detect the alcohol with a minimal reaction in the environment.

Response time is shown in figure 7. For a 95% change, t_{95} of the response time is about 270 seconds. In this test, the "blue" color was chosen as the significant indicator.

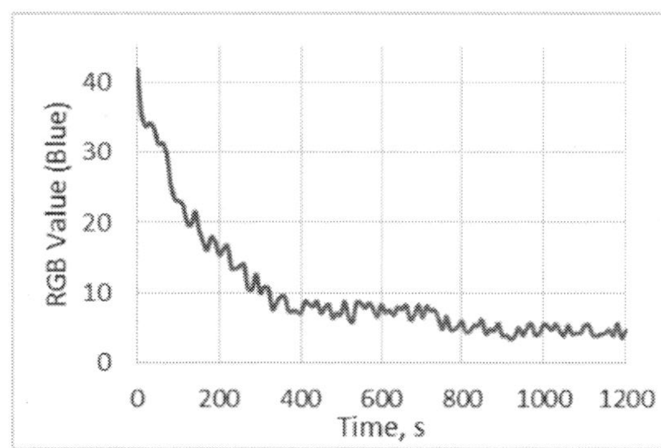

Figure 7. RGB value reading for color blue vs. time

The repeatability of IPP sensor using the same method in the UV-VIS, employed the 0.09% ethanol. Thiswas prepared based on the 0.05% minimum allowable ethanol concentration in the human body. The RSD value of the repeatability test is 2.23%. The low RSD value indicates that the biosensor will produce the same result for the same alcohol concentration.

The sensitivity of paper-strip based biosensor is defined by the calibration curve shown in Figure 8. In this test, concentrations of ethanol in the range of 0.04 to 0.1% (v/v) were used. Figure 8 showed the reading for color blue. Alcohol concentration of unknown samples can be detected using the calibration curve $y = -191.63x + 118.76$, The actual color response is from yellow to lighter or pale yellow when viewed on camera.

As seen in Figure 8, the IPP sensor has a lot to improve. Light scattering and light tightness needs to be improved

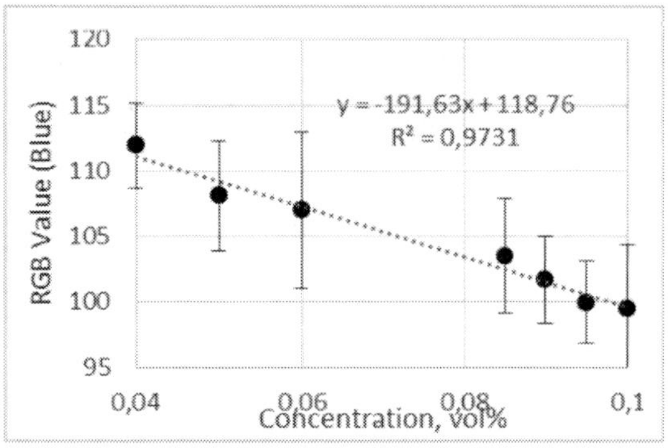

Figure 8. RGB value reading for color blue vs. volume per volume ethanol concentration (n = 3)

Limit of detection or (LOD) for this study is obtained by dividing thrice of the standard deviation of the stability data to the slope of the calibration curve. The computed LOD of the biosensor using the color "blue" is 0.03% per RGB value.

Testing the IPP sensor using real sample

For the analysis of real sample, six readings were also taken into consideration. Fresh saliva was extracted from the primary source, and immediately tested. Mean of the data from the photo analysis were computed and compared with the calibration curve in order to determine the alcohol concentration. The obtained RGB value for color blue of the real sample is 113.4435. Using the calibration curve formula, the alcohol concentration of the saliva sample was found to be around 0.028% (v/v) ethanol. Extrapolation was done as the resulting RGB value is not within the range of the curve.

Based on this limit, the sensitivity of the p-nitrophenol biosensor offers a promising detection capability for the quantification of alcohol in human fluid.

Conclusion and recommendation

In the standardization of the p-nitrophenol as a sensing agent of alcohol, UV-Vis spectrophotometer was used as the main equipment. The absorbance of the solution was utilized in establishing the figures of merit of p-nitrophenol.

The study showed that the reagent is stable when read at 405 nm with an RSD value of 1.24% and a response time of 0.29 seconds. The repeatability test showed a 4.61% RSD. Furthermore, the reagent is sensitive with a linear coefficient of 0.987 and a sensitivity of 0.56 % ethanol/change in absorbance. The limit of detection is 0.02%.

The IPP sensor photographs of the immobilized PNP on paper-strip corresponded to a significant change in RGB values specifically for color "blue". Color changes of the paper-strip from light to intense yellow were observed. The photographs showed a correlation of the response of the paper-strip sensor with alcohol concentration.

The biosensor was analyzed based on its response to alcohol presence. It was found to give a stable signal with an RSD of 1.85% in the blue region. The response time t_{95} is 270 s. The repeatability test showed an RSD value of 2.23%. The biosensor is also sensitive to different alcohol concentrations. It showed that higher concentrations of alcohol intensify the color of the indicator. Alcohol concentration and RGB values were found to have an inverse relationship of proportionality. The slope of the linear behavior has a sensitivity of -191.63 RGB value per (v/v) % alcohol. The limit of detection is 0.03% per RGB value of color blue.

The paper-strip based alcohol biosensor using PNP as an indicator may serve as a low-cost alternative to breathalyzers.

The paper-strip biosensor was tested as an alcohol detector. Further researches on other matrices may be used. Researches on the system response under various conditions such as optimum pH, maybe done to determine the effects on immobilized indicator. Further development may include interference studies since the sample may vary in terms of pH, biological components such as glucose, uric acid, etc.

The paper-strip biosensor may also be improved by using a more controlled environment such as a black box to eliminate other light sources.

Further researches may also include comparison of actual data to those gathered by the breathalyzer and further compare its stability, repeatability and sensitivity to such device to produce a benchmark analysis.

References

1. Official Gazette of the Republic of the Philippines. (2013, May 27). Republic Act No. 10586. Retrieved December 4, 2016 from the Website of the Official Gazette of the Republic of the Philippines, ttp://www.gov.ph/2013/05/27/republic-act-no-10586/
2. G. G. Guilbault, G. Palleschi, G. Lubrano, Biosensors and Bioelectronics, *Non-invasive biosensors in clinical analysis*, **10.** 379-392 (1995)
3. Moneymax. (2014, August 10). Top causes of road accidents in the Philippines. Retrieved September 15, 2015, from Moneymax Weblog, http://moneymax.ph/blog/top-causes-road-accidents-philippines/
4. C.S. Felipe (2015, April 6). Drunk driving cases rising. Retrieved March 28, 2017, from The Philippine Star Website, http://www.philstar.com/metro/2015/04/06/1440677/drunk-driving-cases-rising

5. R. Ramirez, M. Frialde (2015, March 7). LTO to start breath analyzer tests next week. Retrieved September 15, 2015, from Philstar Weblog, http://www.philstar.com/metro/2015/03/07/1430838/lto-start-breath-analyzer-tests-next-week
6. K. Francisco (2015, March 11). Traffic enforcers train in using breathalyzers. Retrieved September 15, 2015, from Rappler Weblog, http://www.rappler.com/nation/86367-lto-breath-analyzer-anti-drunk-driving
7. P. E. Tadeo (2013, July 25). How many breathalyzers will the government need? Retrieved January 25, 2017, from TopGear Philippines Website,http://www.topgear.com.ph/news/motoring-news/how-many-breathalyzers-will-the-government-need
8. Merck Millipore. (2017). 4-Nitrophenol. Retrieved March 28, 2017, from Merck Millipore Website, http://www.merckmillipore.com/INTL/en/product/4-Nitrophenol,MDA_CHEM-106798
9. ImageJ. (2016, September 20). ImageJ Ecosystem. Retrieved November 20, 2016, from ImageJ Website, http://imagej.net/Welcome
10. T. D. Gibson, Biosensors, *Biosensors: The stability problems,* **27** 630-638 (1999)

Chemical and Spectral Characterization of The Ozonation Products of κ-Carrageenan

Aji Prasetyaningrum[1]*, *Bakti* Jos[1], *Yudhy* Dharmawan[2], *Ratih* V. Octaviani[3], *Ratnawati* Ratnawati[1]

[1] Department of Chemical Engineering, Faculty of Engineering, Diponegoro University
[2]Departement of Public Health, Faculty of Public Health, Diponegoro University
[3] Medical Faculty, Diponegoro University

Abstract. Kappa (κ-) carrageenan oligomers are known to have several biological activities. Recent progress in the development of modified κ-carrageenan has resulted low molecular of κ-carrageenan. Ozone is a powerful oxidant and considered for depolymerization of κ-carrageenan. However, few studies have investigated the changes in κ-carrageenan properties associated with ozone treatment. This study would investigate on the changes in chemical structure after ozonation process. The experiments were carried out in a glass reactor equipped with an ozone bubble diffuser. Ozone with concentration of 80 ± 2 was bubbled into the solution. The ozone treatment was conducted at different times, i.e., 0 (control), 5, 10, 15, and 20 minutes. The experiments were conducted at pH 7 and constant stirring speed (200 rpm). Ozone-treated κ-carrageenan was dried at 60 °C for 24 h in a forced air oven. The chemical and spectral analyses of κ-carrageenan after ozonation process were carried out using UV-Vis and FT-IR spectroscopy. These changes are seen in the UV spectra as a high intensity of absorbance peak at 290 nm. It is shows that ozonation of κ-carrageenan leads to some chemical changes such as the formation of carbonyl, carboxyl or double bonds.The FT-IR spectra reveals that the chemical structure of degraded κ-carrageenan, in term of sulfate content, is only slightly affected by the ozone treatment.

Keywords: κ-carrageenan, ozonation, spectral characterization

1 Introduction

κ-Carrageenan is a considerable gelling sulfated polysaccharide extracted from *Kappaphycus alvarezii* (Rhodophyceae). It comprises of repeating disaccharide units of alternating (1/3)-α-D-galactose-4 sulphate, (1/4)-β-3 and 6-anhydro-D-galactose residues [1-4]. The size or degree of polymerization of κ-carrageenan plays a critical role in determining their application. Large molecular size of κ-carrageenan have limited applications as it is insoluble in water and it forms viscous solution. Converting the High Molecular Weight Fraction of carrageenan (HMWF) into Low Molecular Weight of carrageenan (LMWF) will construct the bioavailability and enlarge the potential uses of carrageenan in pharmaceutical and biomedical [5-21].

Currently, LMWF of carrageenan has been produced by several different techniques, such as thermal depolymerization [20], acid hydrolisis [6, 16, 21 - 25], enzymatic hydrolysis [12, 13, 15, 26, 27], microwave-assisted depolymerization [28], sonication [29 - 31], irradiation [17, 32], and oxidation using H_2O_2 [21, 33]. To produce a series of the carrageenan oligomers, acidic hydrolysis has been considered as a common and rapid method. However, the use of chemicals causes an increasing of the enviromental pollution level. The enzymatic method, either using specific or non-specific enzymes, is a relatively not recommended. The enzymatic method need *kappa carragenase* enzyme that relatively high-priced and complex process [27]. Due to the high oxidation potential, ozone can be an alternative method to attaining degradation of organic and anorganic compounds [34 - 53].

Polysaccharide depolymerization with ozonation process is widely developed because the ozonation

* Corresponding author: ajiprasetyaningrum@gmail.com

method provides a convenient, inexpensive, and easily controllable for producing low molecular of polysaccharide [27]. Ozone has been proven to be able to degrade macromolecules. Ozone has been used to depolymerize for many polysaccharides such as starch [36], chitosan [37], gelatin [43] and cellulose [44].

During ozonation process, may occur the change physicochemical and rheological process. . It is has been known that depolymerization of polysaccharides by ozonation process leads to some chemical changes such as the formation of carbonyl, carboxyl or double bonds [6, 10, 15]. In the degradation of chitosan, oxidation of functional groups by hydrogen peroxide give the results of ring-opening reaction, the formation of carboxyl groups and deamination [27]. There was limited reports of carragenan depolymerization treatment by ozonation The chemical changes which occur in polysaccharides are similar to those with simple carbohydrates. For this reason, our study was to analyze the effect of ozonation for chemical changes of κ-carrageenan.

2 Material and Methods

2.1. Materials

Refined κ-carrageenan was acquired from CV. Karaindo, Center of Java, Indonesia. κ-carrageenan was dilluted in hot distilled water (70°C). Carragenan solution was filtered for separate with the impruities. Purified κ-carrageenan was obtain by precipitation with isopropyl alcohol and dried at 60^{0}C for 24 hours.

2.2 Methods

Water distillation was used to dissolve the purified carrageenan. Purified κ-carrageenan was completely dissolved in distilled water to form 1 % (w/v) solution. The ozonation were carried out at pH 7 and 29 ±1 ^{0}C °C. The sample inserted glass reactor that equipped with an ozone bubble diffuser. The sample was then ozonated with different time. Ozone gas with concentration of 80±2 ppm was produced by an ozone generator (*dielectric barrier discharge*). Ozone is generated by passing air between two electrodes with 30 kV high potential. Ozone was bubbled into the solution with a constant flow rate of 3 L min^{-1}. This experiments were stirred with constant speed (200 rpm). The pH of κ-carrageenan solution was adjusted with HCL and NaOH. The measuring of pH during ozonation process using pH meter (Hanna Instruments HI 98107). The ozone treatment was carried out at different times, i.e., 0 (control), 5, 10, 15, and 20 minutes. Ozone-treated κ-carrageenan was collected, rinsed with distilled water, and filtered with Buchner funnel. The temperature used to dry the residu is at 60^{0}C for 24 h in a forced air oven for UV-VIS and FT-IR analysis.

2.3 UV-Vis Analysis

The spectra analysis using Shimadzu spectrophotometer. The spesification was UV-265 FW and wavelength range (200 - 600 nm). The perform UV-Visible spectroscopy of carrageenan solutions at ambient temperature and at 0.025% (w/v) concentration.

2.4 FT-IR Spectroscopy

The measurement of FT-IR spectra of native and ozone–oxidized of κ-carrageenan samples was using an FT-IR Prestige-21, Shimadzu, in the wavenumber range of 4000–400 cm^{-1}. Pellets were prepared by mixing the sample with KBr. The ratio used is 1: 100 (Sample: KBr).

3 Results and Discussion

The chemical changes that occur in sulfated galactan polysaccharides are the same as those simple carbohydrates. Reducing groups (carbonyl) and acid groups appear and carbon-carbon scission occurs. The chemical changes of κ-carrageenan before and after ozonation process were analyzed with UV-Vis and FT-IR spectra.

3.1. UV-Vis Spectra

Figures 1 show that the peak intensity increases with increasing of ozonation time. The absorbance peak of κ-carrageenan at around 290 nm after ozonation process (5, 10, 15 and 20) minutes ozonation were (0.637; 0.642; 0.774 and 0.777), respectively. For comparison, the absorbance peak of κ-carrageenan before ozonation was 0.587. The intensity of absorbance was increase 32.36 % during 20 minutes of ozonation.

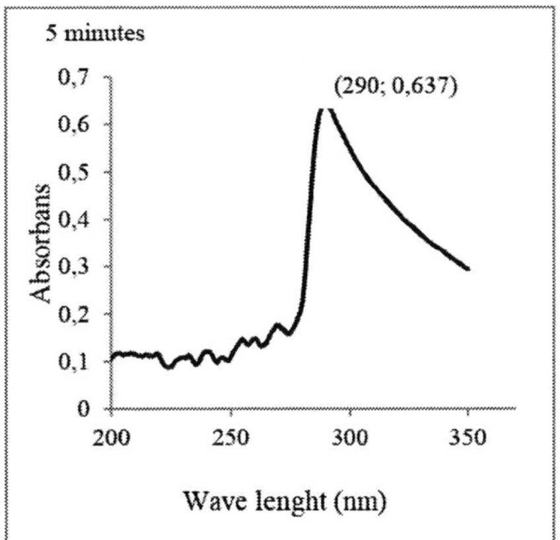

Fig. 1. UV-Vis spectra of 5 minutes ozonated κ-carrageenan

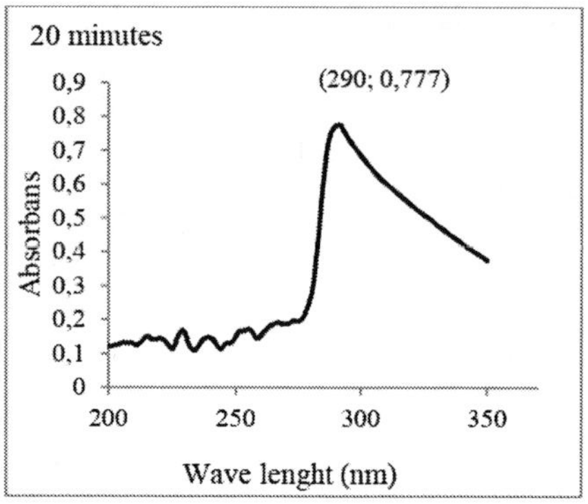

Fig. 2. UV-Vis spectra of 20 minutes ozonated κ-carrageenan

3.2. FT-IR Spectra

The sulfate groups contained in κ-carrageenan before and after ozonation process was identified by FT-IR spectra. Sulfate group in κ-carrageenan recognized in the area of 1230-1270 cm^{-1}. The band at 840-850 cm^{-1} shows C4 position (within the galactose ring). The band at 928-933 cm-1 is connected with a coupling of the C–O stretching vibrations of 3,6-anhydro bridges [45, 46]. This study shows that absorption peaks at 1257 and 845 cm-1 contribute to S=O of sulfate esters and C–O–S of axial secondary sulfate on C–4 of galactose, respectively.

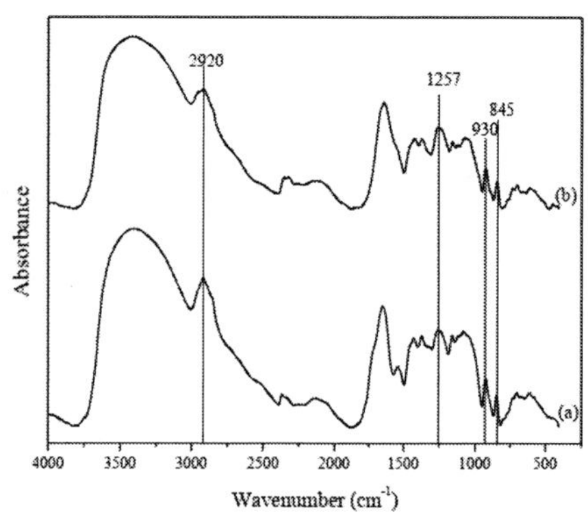

Fig. 3. FT-IR spectra of (a) native and (b) ozonated κ-carrageenan

Meanwhile, the band at 930 cm-1 is characteristic of C–O of 3,6-anhydro-D-galactose. FT-IR spectra are indicative of the structural integrity of κ-carrageenan in the process of depolymerization.The biological activity of LMWF of carrageenan relies upon the presence of sulfate groups. This study indicates that even after the process of ozonation, there is sulfate group existence in the κ-carrageenan.

Conclucions

Chemical and spectral characterization of κ- carrageenan during ozonation process has been investigated. The ozonation process gives no significant effect on the reduction of sulfated content of κ-carragenan. The UV-Vis spectra shows that the formation of double bonds increase after 20 minutes of ozonation. The FT-IR spectra shows that there are no enormous changes in the functional groups of κ-carrageenan after ozone treatment. This finding suggests that the sulfate groups are only slightly removed during ozonation process.

Acknowledgement

This research was funded by the Directorate General of Higher Education, Ministry of National Education The Republic of Indonesia for financial support of through *Hibah Penelitian Terapan Unggulan Perguruan Tinggi* (PTUPT) 2018.

References

1. J. Necas, L. Bartosikova, Vet. Med., **58,** 4 (2013)
2. G. Jiao, G. Yu, J. Zhang, H.S. Ewart, Mar. Drugs, **9**, 2 (2011)
3. V.L. Campo, D.F. Kawano, D.B.D Silva, I. Carvalho, Carbohydr. Polym., **77,** 2 (2009)
4. M. Sen, E.N. Erboz, Food Res. Int., **43** (2010)
5. H. J. Bixler, H. Porse, J. Appl. Phycol., **23** (2010)
6. V.D. Prajapati, P. M. Maheriya, G.K. Jani, H.K. Solanki, Carbohydr. Polym, **105** (2014)
7. W. Wang, P. Zhang, G.L.Yua, C.X. Li, C. Hao, X. Qi, L.J. Zhang, H.S. Guan, Food Chem., **133** (2012)
8. I. Wijesekara, R. Pangestuti, S.K. Kim, Carbohydr. Polym., **84** (2011)
9. Gomez-Ordonez, E., Jimenez-Escrig, A., Rupérez, P, Bioact. Carbohydr. Diet. Fibre **3** (2014)
10. de Araújo, C.A., Noseda, M.D., Cipriani, T.R., Goncalves,A.G., Duarte, M.E.R., Ducatti, D.R.B, Carbohydr. Polym., **91** (2013)
11. Silva, F.R.F., Dore, C.M.P.G., Marques, C.T., Nascimento, M.S., Benevides, N.M.B., Rocha, H. A.O., Chavante, S.F., Leite, E.L, Carbohydr. Polym, 79 (2010)
12. Yamada, T., Ogamo, A., Saito, T., Uchiyama, H., Nakagawa, Y, Carbohydr. Polym, **41** (2000)
13. Yao, Z., Wu, H., Zhang, S., Du, Y, Carbohydr. Polym, **101** (2014)
14. Haijin, M., Xiaolu, J., Huashi, G, J. Appl. Phycol., **15** (2003)
15. Raman, R., Doble, M, J. Funct. Foods, **15** (2015)
16. Kalitnik, A.A., Barabanova, A.O.B., Nagorkaya, V.B., Reunov, A. V., Glazunov, V. P., Solov'eva, T. F., Yermak, I. M. J. Appl. Phycol, **25** (2013)
17. Chiu, Y.H., Chan, Y.L., Tsai, L.W., Li, T.L., Wu, C.J.: *Antiviral Res.*, **95** (2012)

18. de Souza, L.A.R., Dore, C.M.P.G., Castro, A.J.G., de Azevedo, T.C.G., de Oliveira, M.T.B., Moura, M.F.V., Benevides, N.M B., Leite, E.L. Biomedicine & Preventive Nutrition, **2** (2012)
19. Pomin, V.H.: Glycoconjugate Journal, **27** (2010)
20. Qi, H., Zhang, Q., Zhao, T., Chen, R., Zhang, H., Niu, X., Li, Z, Int. J. Biol. Macromol., **37** (2005)
21. Lai, V. M.F., Lii, C.Y., Hung, W.L., Lu, T.J. :, *Food Chem.*, **68** (2000)
22. Sun, Y., Yang, B., Wu, Y., Liu, Y., Gu, X., Zhang, H., Wang, C., Cao, H., Huang, L., Wang, Z.: , *Food Chem.*, **178** (2015)
23. Tang, F., Chen, F., Li, F, J. Appl. Polym. Sci.,**127** (2013)
24. Yuan, H., Song, J.: *J. Appl. Phycol*, **17** (2005)
25. Yu, G., Guan, H., Ioanoviciu, A. S., Sikkander, S. A., Thanawiroon, C., Tobacman, J. K.,Linhardt, R. J, *Carbohydr. Res.*, **337** (2002)
26. Wu, S.J.:. *Carbohydr. Polym*, **89** (2012)
27. Duan, F., Yu, Y., Liu, Z., Tian, L., Mou, H.: *Algal Res.*, **15** (2016)
28. Zhou, G., Yao, W., Wang, C.: *Carbohydr. Polym*, **64** (2006)
29. Ratnawati, R., Prasetyaningrum, A., Wardhani, D.H.:, *Bull. Chem. React. Eng. Catal.*, **11** (2016)
30. Taghizadeh, M. T., Abdollahi, R.:, *Int. J. Biochem. Biophys.*, **3** (2015)
31. Yamada, T., Ogamo, A., Saito, T., Uchiyama, H., Nakagawa, Y.: *Carbohydr. Polym*, **4** (2000)
32. Abad, L.V., Kudo, H., Saiki, S., Nagasawa, N., Tamada, M., Fub, H., Muroya, Y., Lin, M., Katsumura, Y., Relleve, L. S., Aranilla, C. T., DeLaRosa, A. M.: *Nucl. Instrum. Methods Phys. Res., Sect. B*, **268** (2010)
33. Zúñiga, E., Matsuhiro, B., and Mejías, E.: *Carbohydr. Polym*, **66** (2006)
34. Loures, C.C.A., Alcântara, M.A.K., Filho, H.J.I., Teixeira, A.C.S.C., Silva, F.T., Paiva, T.C.B., Samanamud, G.R.L: *Int. Rev. Chem. Eng.*, **5** (2013)
35. Seydim, Z. B., Greene, A. K.:, *LWT-Food Science and Technology*, **37 (2004)**
36. Sandhu, H.P.S., Manthey, F., Simsek, S.: *Carbohydr. Polym*, 87 **(2012)**
37. Seo, S., King, J. M., Prinyawiwatkul, W.:, *J. Food Sci.*, **72** (2007)
38. Klein, B., Vanier, N. L., Moomand, K., Pinto, V. Z., Colussi, R., da Rosa Zavareze, E., and Dias, A. R. G.:, *Food Chem.*, **155** (2014)
39. Dodgson, K.S., Price, R.G.:, *Biochemical Journal*, **84** (1962)
40. No, H. K., Kim, S. D., Kim, D. S., Kim, S. K., Meyers, S. P, *Journal of Chitin and Chitosan Science*, **4** (1999)
41. Chan, H,T., Leh, C, P, Bhat, R, Senan C, Williams P, A., Karim, A, A.: *Food Chem.*, **126** (2011)
42. Tiwari, B.K., Muthukumarappan, K., O'Donnell, C.P., Chenchaiah, M., Cullen, P.J *Food Res. Int.*, **41** (2008)
43. Cataldo, F.:. *Int. J. Biol. Macromol.*, **41** (2007)
44. Wang, Y., Hollingsworth, R. I., and Kasper, D. L.:, *Carbohydr Res.*, **319** (1999)
45. Prado-Fernández, J., Rodrıguez-Vázquez, J.A., Tojob, E., Andrade, J. M.: *Anal. Chim. Acta***480** (2003)
46. Rochas, C., Lahaye, M., Yaphe, W.:, *Bot. Mar.*, **XXIX** (2015)
47. Karlsson, A., Singh, S.K.:, *Carbohydr. Polym*, **38** (1999)

Comparison Drying Behavior of Seaweed in Solar, Sun and Oven Tray Dryers

Suherman Suherman*, *Moh.* Djaeni, *Andri* C. Kumoro, *Rizky* A. Prabowo, *Sri* Rahayu and *Sufrotun* Khasanah
Departement of Chemical Engineering, Faculty of Engineering, Diponegoro University, Jl. Prof. Soedarto, SH Tembalang, Semarang, Indonesia 50239

Abstract. The main objective of this study is to analyze and compare the drying behavior of seawage experimentally in solar dryer, sun dryer and oven dryer. Seaweeds with moisture content of 70% have been cleansed of impurities attached and washed with fresh water. The solar dryer tool is made of glass and iron shaped dryer and box, where the solar collector is made of black painted iron plate. The results showed that the air temperature in solar dyrer was higher than the ambient. The air humidity relative in the solar dryer is lower than the ambient. Drying seaweed by solar dryer method can dry the material up to 12.2% water content within 16 hours, and faster than sun dryer method. Energy utilization per kg of material for solar dryer, sun dryer and oven dryer is 1493 kJ / kg, 1338 kJ / kg, and 1620 kJ / kg, respectively. The quality of the product by the solar dryer method produces the best dried seaweed products in terms of color, aroma and contaminants in the product

1 Introduction

Drying is the transfer of heat and water vapor simultaneously, which requires heat energy to evaporate the water content removed from the surface of the material, which is dried by the media which is usually a hot dryer. One of the materials can be dried seaweed because it has a high water content. Seaweed is classified as low-level plant, does not have roots, stems and true leaves, but only resemble rods called thallus, growing in the wild to attach themselves to rocks, mud, sand, rocks and other hard objects. Indonesia is one country with a great potential sea grass, where Indonesia has approximately 555 species of 8642 species of seaweed found in the world (Bosse, 1899). Although it has great potential but the seaweed development program potential is still very low because of the production is still using the conventional drying process is drying with the help of sunlight. This drying is done 2-3 days depending on weather conditions. Drying method It also causes the product to be contaminated with dirt from the outside air. Drying method is one of the potential development program seaweed one using solar drying method. Solar drying is a method or way of drying a food using solar radiation energy, this method is the most inexpensive and easily as using direct heat from the sun. Solar drying is assisted simple tools and a drying chamber as a place. Solar drying is shaped like a closed room with transparent walls. With the closing of this tool, it can avoid the contaminants from the surrounding environment. With the closing of this tool, it can avoid the contaminants from the surrounding environment. At the bottom will be placed a heat collector rectangular painted black. Drying seaweed using Solar drying shows that the mathematical model of the drying curve is a variable measured during the drying process and shows the effectiveness of the drying method (Fudholi et al., 2013).Solar drying method is expected to improve the quality of dried seaweed.

The research objective of drying seaweed with methods of solar drying is to determine the profile of the temperature and humidity, know the profile of the curve of drying the seaweed with a method of solar dryers, electric dryer and sun dryer, as well as analyzes of economic deals with the analysis of the energy efficiency of products produced from three various methods namely solar dryers sun electric dryer and dryer.

2 Materials and Methods

2.1 Materials

The method used in this seaweed drying are solar drying method. Preliminary research needs to be conducted to determine the initial water content in the ingredients seaweed. Initial preparation of washing the seaweed. Then the fresh seaweed was cut with dimensions of 3cm x 0.5 cm x 0.2 cm. Once cut to the desired dimensions, materials incorporated into solar dryers dryers.

* Corresponding author: suherman.mz@che.undip.ac.id

2.2 Drying Equipment

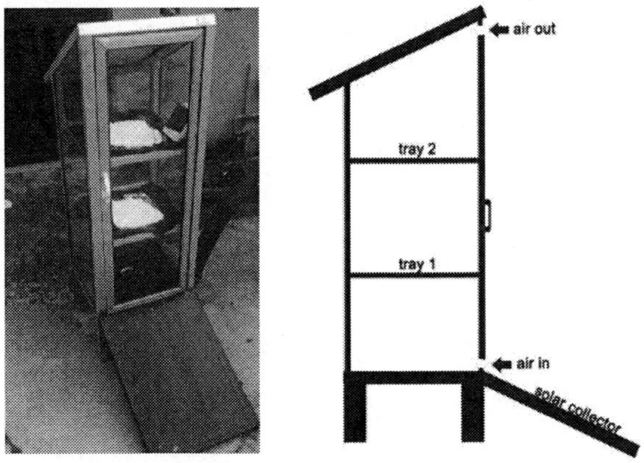

Fig. 1. Solar Tray Dryer

The prototype of Solar dryer has a dimension 40x40x150 cm. The materials used in the manufacture of the prototype of Solar dryer has some special characteristics. First, choose materials affordable to expectations in its application were able to save cost from the price. Second, lightweight materials to facilitate the operation of the device itself, for example if you want to be moved or taken to another place. Third, the selected material has the properties of heat collectors. With these properties then the heat will accumulate inside the tool so that it can speed up the drying process. This tool is also made of a material that is not easily broken, broken or porous so as to reduce the risk of damage and loss. Aluminum is used as the framework is a good conductor of heat, light and affordable.

Drier is made in the form of shelves in order to have a drying capacity of the lot. The roof is sloped so that when it rains, the water can drip down easily so it will not wet the dried material. In addition, the slope of the roof is water vapor attached to the roof produced during drying is expected to easily flow out through the roof and walls so that the dried material will be protected from water vapor that condenses droplets. On the roof there is a hole as a water vapor expenditure, the purpose is that water vapor formed during the drying process may soon be out of the appliance. *Solar dryer* is shaped like a closed room with transparent walls. With the closing of this tool, it can avoid any contamination from the surrounding environment. At the bottom of the tool will be placed a heat collector rectangular painted black.

3 Results and Discussion

3.1 Profile Temperature, Humidity and Solar Radiation Average on Drying Solar Dryer

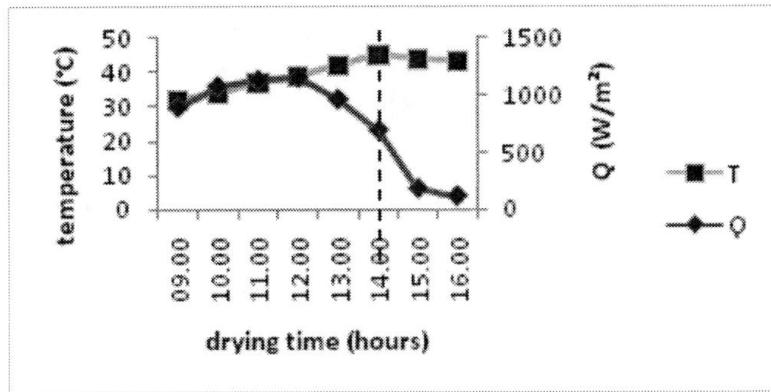

Fig. 2. The average solar temperature and radiation curve curve with the drying time of the solar drying method

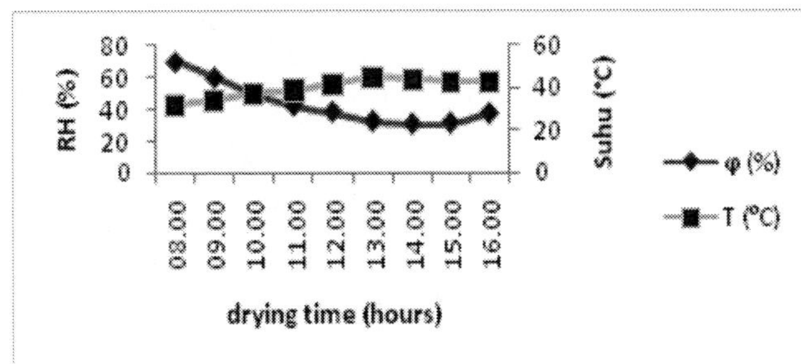

Fig. 3. Average temperature and humidity relation curve with drying method drying time

Based on Figure 3.1 shows that in times of high solar radiation, temperatures in the dryer will increase and vice versa when the solar radiation began to decrease the temperature in the dryer will be reduced, the solar dryers.Their thermal collectors will store heat so that the current decline in solar radiation were not significant air temperature in the drier solar dryers remain high, the potential at the time of solar radiation maximum and the air temperature dryers are high is used to lower the water content of seaweed on the methods of solar drying.

In addition to the temperature and solar radiation, solar drying method is also affected by the humidity in the dryer, by drawing 3.2 it can be seen that when the solar radiation is low, the drying air temperature is low so the drier the air humidity is high. Air humidity will decrease with increasing solar radiation and drying air temperature, so that it can be concluded that the value of the air humidity is inversely proportional to the solar radiation and air temperature dryer (Yuwana, 2003).

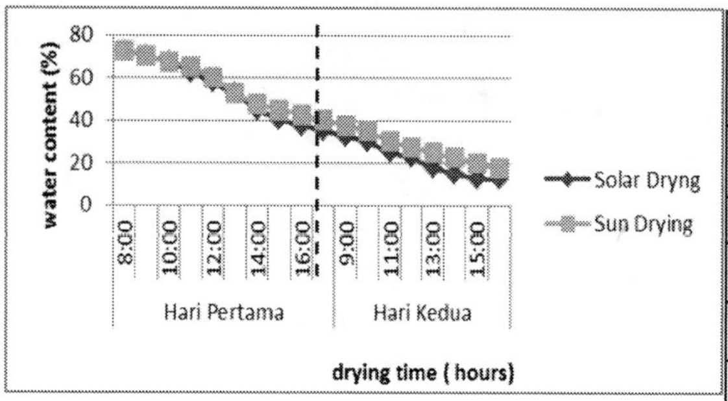

Fig. 4. Curve relation with the water content of the drying time of solar dryers and sun dryer metods

Drying with a solar dryer capable of lowering the water content of seaweed quickly than the sun dryer, because of differences in temperature dryer. Temperature in the collector is higher compared to the wild open. The temperature difference is due to drier solar dryers were able to raise outside air temperature average 130C, in addition to the effects of angina open can increase the humidity in the environment dryers while in solar dryers shape of the instrument closed on several sides so as angina not significant in raising the humidity in dryer.

3.2 Comparison of Drying Tray Dryer and Oven

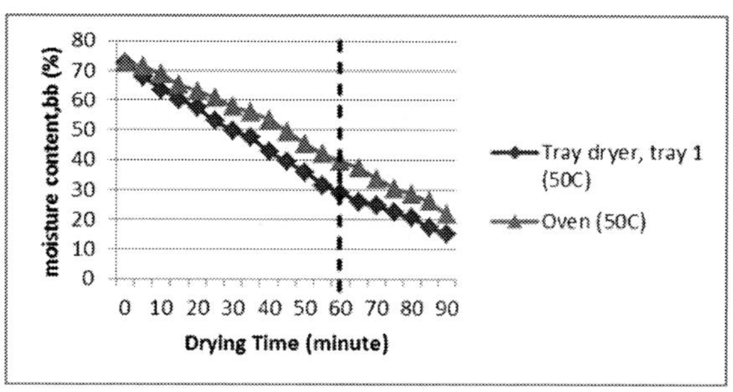

Fig. 5. Curve moisture content relation with electric dryer when drying method

Based on Figure 3.4 shows that a decline in the water content of seaweed is not significant spike at any given time.This drying using a tray dryer.Drying is able to reduce the water content of seaweed quickly than using the oven, is due to the direct contact with the seaweed, so the water will evaporate on the surface of the material.

3.3 Analysis of Energy Efficiency Drying Seaweed

Method / Analysis	Solar dryer/ 400gram	Sun Dryer / 200 gram	Electric dryer (KJ)	
			Tray dryer / 200 gram	Oven / 100 gram
Water content	12,6	17,6	14,31	21,82
Energy	597,8 kj	267,6 kj	297,98 kj	162,4
Efficiency of drying	9,5 %	9,1 %	4,3%	3,7%
Efficiency of solar collector	27,7%	-	-	-
Efficiency of drying*	27%	-	-	-
Efficiency of solar collector*	35%	-	-	-

Fig. 6. Dryer Efficiency and Drying Time of Solar Drying Method

Based on Table 3.1 it can be seen that the solar dryer to produce dry seaweed with a water content of 12.6% smaller than the sun method dryer and electric dryer. Drying efficiency is directly proportional to the amount of heat that is utilized where the efficiency increases if the heat given are largely utilized to dry some more materials to achieve the desired moisture content. whereas in drier tray dryer with the amount of energy used is greater than the method of sun dryer but the value of the efficiency of drying is lower, it is because of the difference in temperature in the dryer, where the temperature in the dryer tray dryer is higher, should the heat generated can be utilized to drain more ingredients that will increase the drying efficiency.
The efficiency of the solar collector solar dryers method at 27.7% efficiency solar collector where the results showed a smaller value of research fudholi et al, 2013 at 27% and 35% this is because in this study, the design of solar dryers used tool is simple yet equipped with a heat supplier.

4 Conclusion

Air humidity will decrease with increasing solar radiation and air temperature dryers, drying seaweed solar dryers method can dry ingredients until the water content of 12.2% faster than the methods of sun dryer, drying energy analysis method for solar dryers that 597.8 kJ / 0, 4kg material, sun dryer 267,6kj / 0,2kg of material, the efficiency of solar dryers drying method, namely, sun dryer, respectively 9.5%, 9.1%, where the value of solar dryers dryer efficiency is 27.7%

References

1. B. M. A. Amer, M. A. Hossain, , & K. Gottschalk. *Energy conversion and management*, **51**, 813-820. (2010).
2. W.L. Badger, and J.T. Banchero, Introduction to chemical Engineering. McGraw Hill, Kogakusha,Ltd, Tokyo. (1995).
3. Bono, A., Farm, Y. Y., Yasir, S. M., Arifin, B., & Jasni, M. N. Production of fresh seaweed powder using spray drying technique. (2011).

4. A.A. El-Sebaii, ,and S.M. Shalaby., Renewable and Energy Reviews. **16**. 37-43. (2012).
5. A. H. M. A. D. Fudholi, M. H.Ruslan, , M. Y. Othman, A. Zaharim, , & K. Sopian, In Proceedings of the 6th WSEAS Int. Conf. on Renewable Energy Sources (RES'12) 36-41. (2012).
6. S.Gupta, , S.C ox, , & N. Abu-Ghannam,. *LWT-Food Science and Technology*, **44,** 1266-1272. (2011).
7. S. Hirao., Overseas Technical Cooperation Agency Goverment of Japan, **148** . (1971).
8. Juraev, E.T., and J.S, Akhatov. Study of Mixed-Mode Forced Convection Solar Drier. Applied Solar Energy, **45**,. 61-64. (2009).
9. Mc. Ketta. John.. Encyclopedia Chemical Process and Design. Marchell Dekker Inc., New York (1983)
10. N. S. Rathore Renewable and Energy Reviews, **16**, 37-43. (2010).
11. A.Nonclercq, Food Sec, **1**, 197-205. (2009).
12. A.Reyes,. Energy Conversion and management, **83**, 241-248. (2014).
13. Yuwanadan Silvia Evanila. 2003. Application of YSD-UNIB12 Model Solar Dryer for Red Pepper, Mustrad and Cassava Leaf Drying. JurusanTeknologiPertanian.FakultasPertanian,,Universitas Bengkulu.

Performance Analysis of Solar Tray Dryer for Cassava Starch

Suherman Suherman*, *Mohammad* Djaeni, *Dyah* H. Wardhani, *Mukhtar* Dzaki R, and *Muhammad* N. Bagas F

Department of Chemical Engineering, Faculty of Engineering, Diponegoro University, Jl. Prof. Soedarto, SH Tembalang, Semarang, Indonesia 50239

Abstract. The main objective of this study is to analyze the performance of solar tray dryyer for cassava starch. The solar tray dryer is made of glass and iron shaped dryer and box, where the solar collector is made of black painted iron plate. The initial moisture content of cassava starch is 50% wet bases. The experimental result show the moisture content of starch decrease rapidly in 8 hours of drying the first day until 14% moisture content. Further drying able to dry starch to 8% water content. The water temperature in solar dryer can reach 60°C was higher than the ambient of 32°C. Furthermore, the air humidity relative in the solar dryer can drop dramatically to 30% is lower than the ambient 80%. Maximum drying rate can reach up to 0.2 g/h. Thermal efficiency of the dryer can reach 40%. The performance of this solar dryer is much better than the sun dryer which veing piloted in this study

1 Introduction

Drying cassava flour using solar drying have been developed in several countries such as Thailand, while in Indonesia itself is still using conventional drying is dried by relying on sunlight , Solar drying is a method or way of drying a food using solar radiation energy, this method is the most inexpensive and easily as using direct heat from the sun. Solar drying is assisted simple tools and a chamber as a place of the drying. Solar drying is shaped like a closed room with transparent walls. With the closing of this tool, it can avoid the contaminants from the surrounding environment. At the bottom will be placed a rectangular heat collector painted black.

There is a lot of research on drying cassava flour that has been done before with a wide range of prototype devices. Mixed flow dryer (Olawale et al., 2015); pneumatic conveying dryer (Aichayawanich S. et al., 2011); rotary dryer (FT Ademiluyi et al., 2010); tray dryer (Ndubisi et al., 2014); spray dryer (Monica and Anibal, 2014); convection solar dryers (Tarin et al., 2005); indirect solar dryers (Azimi et al., 2012). This study investigated the drying characteristics of cassava flour with a variety of different treatments by the drying temperature and drying efficiency. However, the earlier study relied more prototype tool that tends to be more expensive. There's drying solutions is by sun drying or natural drying whose treatment is only necessary to drying under the sun, but is inefficient and requires a long drying time.

In research conducted by Anhalt (2003) entitled "The use of Renewable Energy in the Production of Goods: Seaweed" indicates that the natural drying process usually takes a long time because the temperature and energy dependent on sunlight. In addition, the effects of weather, season, and the change of day and night to make this process more and more limited. If the weather conditions do not support the increasingly high water content so that it can be mold and mildew growth media. Moreover, if the drying is done in the open. This will trigger the presence of contaminants that degrade the quality of tapioca starch products, such as dust, dirt or foreign objects are not desired. Based on the problems it is necessary to have a technology development efforts that are efficient, effective and efficient in drying starch. These efforts are expected to reduce the water content in accordance default, the drying process is faster, does not depend only on the energy of the sun, the weather and climate of the local area.

2 Materials and Methods

2.1 Materials

Materials (samples) were used and tested is starch (wet basis), was taken from the SME starch in Pati, Central Java. For drier *(solar dryers)*, the material used in the form of metallic aluminum conductors, transparent Acrylic, zinc, iron, rubber and bolts. The equipment used during the testing period, among others timepiece, digital weight scale, a thermometer, and an oven

* Corresponding author: suherman.mz@che.undip.ac.id

2.2 Drying Equipment

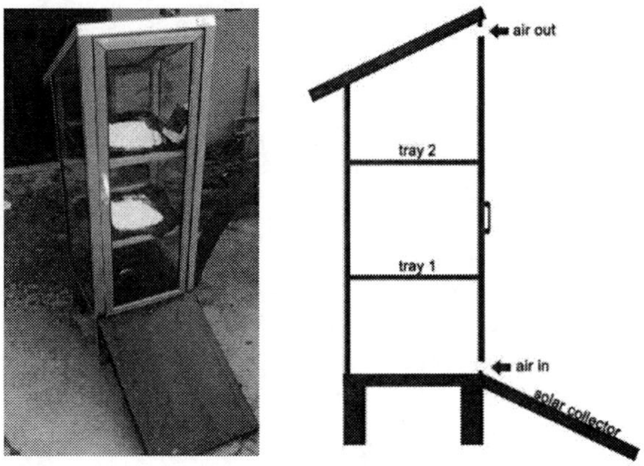

Fig. 1. Solar Tray Dryer

The prototype of Solar dryer has a dimension 40x40x150 cm. The materials used in the manufacture of the prototype of Solar dryer has some special characteristics. First, choose materials affordable to expectations in its application were able to save cost from the price. Second, lightweight materials to facilitate the operation of the device itself, for example if you want to be moved or taken to another place. Third, the selected material has the properties of heat collectors. With these properties then the heat will accumulate inside the tool so that it can speed up the drying process. This tool is also made of a material that is not easily broken, broken or porous so as to reduce the risk of damage and loss. Aluminum is used as the framework is a good conductor of heat, light and affordable.

Drier is made in the form of shelves in order to have a drying capacity of the lot. The roof is sloped so that when it rains, the water can drip down easily so it will not wet the dried material. In addition, the slope of the roof is water vapor attached to the roof produced during drying is expected to easily flow out through the roof and walls so that the dried material will be protected from water vapor that condenses droplets. On the roof there is a hole as a water vapor expenditure, the purpose is that water vapor formed during the drying process may soon be out of the appliance. *Solar dryer* is shaped like a closed room with transparent walls. With the closing of this tool, it can avoid any contamination from the surrounding environment. At the bottom of the tool will be placed a heat collector rectangular painted black.

2.3 Research Procedure

Research tapioca starch drying is done according to the procedure of the first is the early preparation stage. Initial preparation phase aims to wet starch by preparing samples of starch (wet basis) and weigh the wet base sample (100 grams). Then the drying step is done by setting up a solar dryer tray dryer and prepare 100 grams of starch. Put the tapioca flour (wet basis) on a tray in a tray dryer solar appliance. Please note the change of tapioca flour weight every hour for 8 hours within 2 days. Take measurements of temperature, moisture content, drying rate, dryer efficiency, relative humidity.

2.4 Experimental Data Analysis

2.4.1 Temperature measurement

Each 100 grams sample is placed on each tray on the solar dryer. Measuring the temperature every 1 hour on each tray for 8 hours, measure the temperature at the inlet, outlet, ambient temperature and the temperature of the drying chamber operating solar dryers. Then be concluded associated with operating time. We get the graph of temperature relation with time

2.4.2 Analysis of moisture content

To determine the moisture content, we have to get the data of weight per hour at each tray by weighing the sample on each tray every 1 hour then obtained graph relation of moisture content with time, for example:

$$M_c\ (wet\ basis) = [\ (M_i - M_d) \div M_i\ ]\ x\ 100\% \quad (1)$$

(Mi) is the mass of the sample after drying for 1 hour, (Md) is the mass of dry basis.

2.4.3 Analysis of drying rate

To obtain the drying rate, it requires data sample of weight on each tray. By using equation (Dhanushkodi Saravanan, 2004). Then the drying rate obtained graph with time.

$$R_d = (M_i - M_d) \div t \quad (2)$$

Dimana : Rd = drying rate
Mi = initial mass
Md = final mass
t = drying time (s)

2.4.4 Analysis of relative humidity

To determine the relative humidity, using a hygrometer or use the measurement method Tw and Td. Tw is the wet bulb temperature, which is measured by measuring the temperature of the wet with wet cotton stick on the tip of the thermometer. While Td is dry bulb temperature, which is measured by measuring the temperature of the usual dry. By connecting Td and Tw on psychometric chart will be obtained per cent RH.

2.4.5 Analysis of dryer efficiency

To measure dryer efficiency, using the equation :

$$\eta_c = [\ mC_p\ (T_o - T_i)\ ] \div I\ x\ A\ ] \quad (3)$$

Dimana : To = outlet temperature
Ti = inlet temperature

M = air flowrate
I = solar intensity, from the reference
A = area of heat collector

The temperature is measured every hour and graphed the relation of dryer efficiency with time

3 Results and Discussion

3.1 Temperature profile and humidity of Solar Dryer

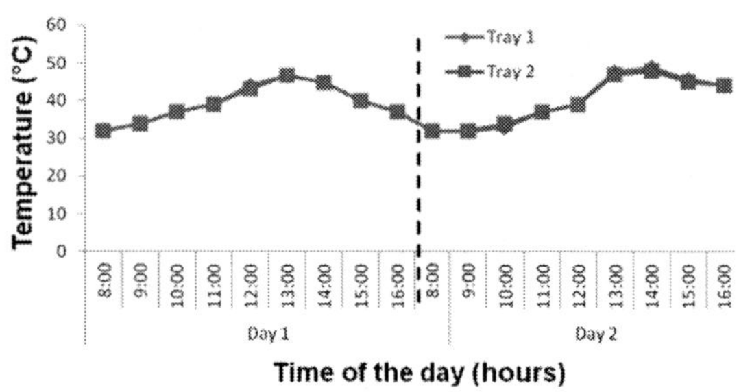

Fig. 2. Temperature and Drying Time of Solar Drying

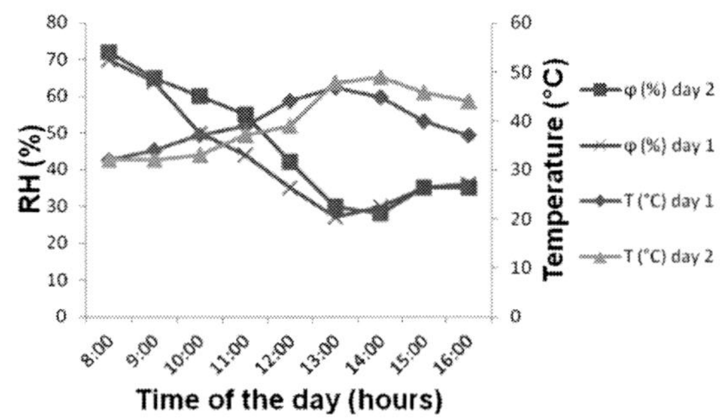

Fig. 3. Temperature and Humidity Average with Drying Time in Solar Drying Method

Based on the figure (2) shows the fluctuation of dryer temperature. The maximum dryer temperature on the first day was 44.8°C at 14:00 pm and dryer temperature minimum on the first day was 32°C at 08.00 pm. The maximum dryer temperature of the second day was 49°C on a tray 1 and 48°C at the tray 2 when at 14.00 pm. The minimum temperature dryer on the second day was 32°C at 08.00 pm. This is influenced by the weather and time of drying. In fine weather the sun is absorbed by optimal because no clouds blocked. In addition it will be effective and optimal drying if done during the day, particularly at 11:00 am until 14:00 pm. At the time of the emitted light intensity is strong enough. In the tray 1 and tray 2 having a bit of difference in temperature at which the tray 1 is located at the bottom and the tray 2 is at the top. This causes the heat from the heat collector leads directly to the tray 1 first and then passed to tray 2 so that the temperature of the tray 1 higher. This tool borrows the concept of green house that aims to trap heat. Heat from the sun is absorbed by the tool but is not directly reflected off the tool, but is reflected back by the wall of the new tool further repeatedly reflected out. As a result, Solar Dryer trapping heat from the sun, causing the temperature inside the appliance increases.

Based on the figure (3) can be known to the fluctuation in the value of air moisture with the length of drying time. This is directly affected by changes in temperature also fluctuates. When the comparison is made, it turns out the air inside the instrument has a relative humidity lower than outside. When moisture begins to evaporate from the material decline in relative humidity so it will reduce the water content (Rathore, 2010). This affects the drying speed levels are higher. In addition to the temperature, weather and light intensity, drying with solar drying method is also affected by the humidity in the dryer. In Figure 4.2 it can be seen that at the time of the drying air temperature is low, moist air at high dryer appliance. Air humidity will decrease with increasing temperature of the drying air, so that it can be concluded that the value of the air humidity is inversely proportional to the drying air temperature. (Chramsa, 2013)

3.2 Moisture Content with Drying Time of Solar Dryer

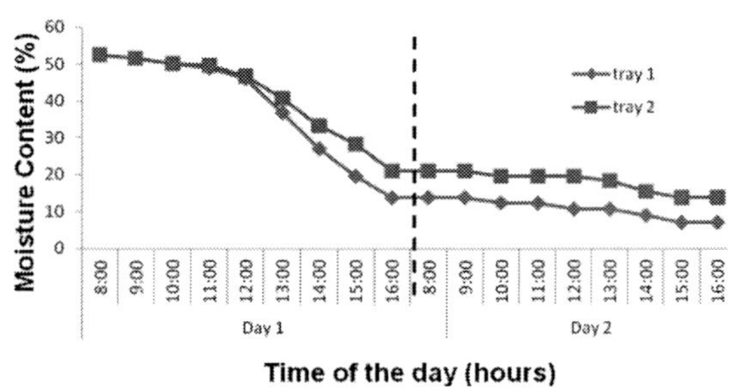

Fig. 4. Moisture Content with Drying Time of Solar Drying Method

Based on the figure (4) shows that there is a significant drop in water levels on the tray 1 from 49.03% to 27.07% occurred at 11.00 pm until 14.00 pm on the first day of testing and from 12.22% to 8.84% on the second day of testing that occurred at 11.00 pm to 14.00 pm. On drying starch tray 1 by using solar dryers second day of testing at 15:00 pm moisture content of materials amounted to 7.05%. In tray 2 a decline in water levels significantly from 49.57% to 33.23% occurred at 11.00 pm until 14.00 pm on the first day of testing and from 19.66% to 15.35% on the second day of testing, which occurred at 11.00 am to 14.00 pm. On drying starch tray 2 by the method of solar dryers second day of testing at 15:00 pm moisture content of materials amounted to 13.81%. This indicates that the products meet the standard of a maximum moisture content in accordance with SNI for tapioca starch product that is less than 15% (Taihua, 2017).

3.3 Drying Rate with Drying Time of Solar Dryer

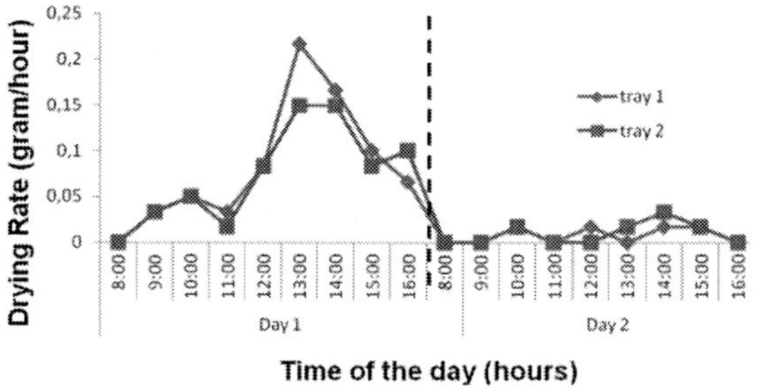

Fig. 5. Drying Rate and Drying Time of Solar Drying Method

Based on the figure (5) shows the comparison of drying rate with respect to time. Fluctuations can be seen in the rate of drying by the method of the solar dryers. In tray 1 drying rate increased significantly when the 11:00 am until 13:00 pm which is 0.03 g / h to 0.21 grams / hour on the first day and remained constant at 11.00 pm until 14.00 pm which is 0.016 grams / hour on a second. In tray 2 actually increased drying time of 11:00 am until 14:00 pm that is 0,016 grams / hour up to 0.15 grams / hour on the first day and on the second day from 13:00 am until 14:00 pm that is 0,016 grams / hour up to 0,033 grams / hour , So the longer the drying time will make increasing the drying rate is directly proportional to the weight loss that occurred (Dhanushkodi, 2014)

3.4 Dryer Efficiency with Drying Time of Solar Dryer

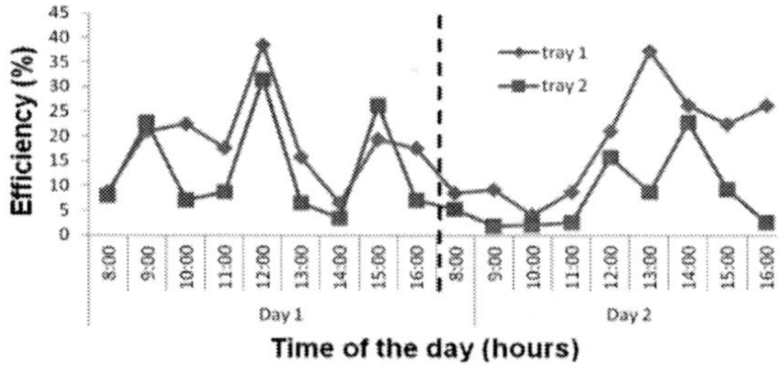

Fig. 6. Dryer Efficiency and Drying Time of Solar Drying Method

Based on the figure (6) can be seen on the tray 1 maximum efficiency is 38.5% on the first day and 37.25% on the second day of drying. In the second tray is a maximum efficiency of 31.5% on the first day and 22.7% on the second day of drying. The maximum drying efficiency occurred at 12.00 pm on the first day and 13:00 pm until 14:00 pm on the second day where the intensity of the sun is high and the optimal temperature dryer at that hour. The drying efficiency solar dryers follow the same pattern with the intensity of the sun. So the optimum drying efficiency occurred at 12.00 pm due to the high intensity of the sun affected. (Dhanushkodi, 2014)

4 Conclusion

Results of the drying of starch with methods of solar dryer can be concluded that the air humidity will decrease with increasing solar radiation and air temperature dryer, it can be concluded that the value of the air humidity is inversely proportional to the solar radiation and air temperature dryer. Drying tapioca flour with solar dryers method can dry ingredients until the water content of 7.05% in a faster time. Analysis of temperature and relative humidity where the maximum achievable temperature is 49°C. The humidity will decrease with increasing temperature of the drying air, so that it can be concluded that the value of the air humidity is inversely proportional to the drying air temperature. The maximum drying rate in this study was 0.21 grams / hour, where the value of solar dryers dryer efficiency is 38.5%. Tapioca starch drying method produces solar dryers to dry tapioca starch product more desirable in terms of color, aroman and contaminants in products rather than drying tapioca with sun conventional methods. Therefore, solar dryers can be an alternative means of dryers in food substance drying.

References

1. J. D. Anhalt,. World Climate and Energy Event, Rio de Janeiro, Brazil, 44-53. (2003).
2. A. Azimi, T. Tavakoli, H.K. Beheshti, A. Rahimi, Iran. J. Energy Environ. **3.** 347-353. (2012).
3. S. Dhanuskodi, V. Wilson, & S. Kumarasamy, Facta Universitatis, Series: Mechanical Engineering, **12**, 277-288. (2014).
4. M.Ordoñez, & A. Herrera., Powder technol. **253**, 89-97. (2014).
5. N. A.Aviara, L. N. Onuoha, O. E. Falola, & J. C. Igbeka, Energy and exergy analyses of native cassava starch drying in a tray dryer. Energy, **73**, 809-817. (2014).
6. N. S. Rathore, N. L. Panwar. Renewable Sustainable Energy Rev., **16**, 37-43.(2010).
7. O. U. Dairo, A. A.Aderinlewo, O. J., Adeosun, I. A. Ola, & T. Salaudeen., Acta Technologica Agriculturae, **18,** 102-107. (2015).
8. S. Aichayawanich, M. Nopharatana. A. Nopharatana, & W. Songkasiri,. Carbohydr. polym., **84**, 292-298. (2011).
9. Standar Nasional Indonesia 3451:. Tapioka. www.bsn.go.id (2011)
10. Taihua Mu, Hongnan Sun, Miao Zhang dan Cheng Wang. *Sweet Potato Processing Technology*. Beijing : Academic Press Chinese Academy of Agricultural Sciences. (2017).
11. E.Tarigan, P.Tekasakul, Proceedings of the Australia and New Zealand Solar Energy Society Annual Conference. 1–9. 2005,
12. W. Chramsa-Ard, S. Jindaruksa, C. Sirisumpunwong & S. Sonsaree, Energy Procedia, **34**.189-197. (2013).

Investigation on mixture design of one-part geopolymer from fly ash and water treatment sludge

Vuong Ho[1], *Aileen* Orbecido[1] and *Michael A.* Promentilla[1*]

[1]Chemical Engineering Department, Gokongwei College of Engineering, De La Salle University, Manila, Philippines

Abstract. This study presents a one-part geopolymer system from coal fly ash and water treatment sludge. Geopolymer is typically produced from two parts namely the aluminosilicate solids, which is typically sourced out from industrial by-product, and an alkali activator solution which reacts with aluminosilicate solids to form an inorganic polymeric network. For a one-part geopolymer system, the solid binder with activators will just be added with water to address the drawback of corrosive and viscous alkali activator solution. Formulation of the proportion of geopolymer precursors with the two solid alkali activators namely sodium hydroxide and sodium aluminate was conducted using statistical mixture design. Effects of each components as well as interactions between them were evaluated by step-wise regression analysis. It was found that high alkali content decreased the compressive strength of binder. Meanwhile, the incorporation of sludge in this system helps reduce the unit weight of samples. Multiple response surface analysis that maximized compressive strength and minimized unit weight resulted in the optimal combination of 18.9% sludge, 76.1% fly ash and 5.0% NaOH.

1 Introduction

The shortage of raw materials for Ordinary Portland Cement (OPC) production and the stress of global warming urged many cement companies to find sustainable alternatives. In recent years, many efforts have been dedicated to replace OPC with coal fly ash (FA) in cementitious concrete. FA is a by-product from coal-fired thermal power plant. It contains large proportions of reactive silicate and aluminate that can be used as a geopolymer precursor which reacts with alkali activator solutions to form inorganic polymeric network [1]. The FA-based geopolymers have found to have superior strength for construction applications [2,3].

Water treatment sludge (WTS) is residue from water treatment plant. It is classified as non-hazardous waste and usually disposed to landfills. However, WTS has potential applicability on the manufacture of geotechnical materials [4,5]. For example, WTS was used as lightweight aggregate in FA-based geopolymer [6].

In this study, FA and WTS mixtures were activated by solid sodium hydroxide and sodium aluminate. The usage of solid activators creating a one-part geopolymer system could address the drawbacks of using corrosive and viscous alkali activator solution in a two-part geopolymer system [7]. Note that geopolymer is typically produced from two parts namely the aluminosilicate solids, which is typically sourced out from industrial by-product, and an alkali activator solution which reacts with aluminosilicate solids to form the hardened binder.

2 Materials and Method

2.1 Materials

Coal fly ash was collected from a certain power plant in Luzon, Philippines and used without any pre-treatment. The wet WTS was collected from a water treatment plant in Manila, Philippines. Wet sludge was dried at 105°C overnight then ground using an abrasion machine for 10,000 cycles. Anhydrous sodium hydroxide (99% NaOH) purchased from a local supplier in the Philippines and anhydrous sodium aluminate (50-56% Al_2O_3; 40-45% Na_2O) from Sigma-Aldrich were used as activators.

2.2 Geopolymer synthesis

Firstly, FA was mixed with WTS in mechanical mixer. Solid sodium hydroxide (NaOH) and sodium aluminate ($NaAlO_2$) were added to raw materials and continuously blended for a minute. After that, water was poured into the solid mixture. The mixing was done for 10 to 30 minutes using a mechanical mixer. The paste was poured into 50x50x50 mm molds. Steel molds were greased beforehand for easier demolding. To prevent the

*Corresponding author: michael.promentilla@dlsu.edu.ph

Table 1. Compositions of two geopolymer mixtures.

Component	Lower bound, %	Upper bound, %
FA	65	95
WTS	0	30
$NaAlO_2$	0	10
NaOH	5	15

superficial carbonation of alkali activated samples, all of cubes were kept in plastic bags. Those were cured at ambient condition for 21 days before tested for compressive strength and unit weight.

2.3 Design of Experiment – Mixture design and Multiple response surface analysis

Mixture design is an optimization design of experiment in which all factors are the components of a mixture. Proportion of each component was varied within ranges as shown in Table 1. Sum of percentage of $NaAlO_2$ and NaOH shall not exceed 15% based from the preliminary test. In the region of interest, 30 different combinations were chosen with an aim to fit the full cubic model:

$$E(y)=\sum_{i=1}^{p}\beta_i x_i+\sum\sum_{i<j}^{p}\beta_{ij}x_i x_j+\sum\sum_{i<j}^{p}\delta_{ij}x_i x_j\left(x_i-x_j\right)+\sum\sum_{i<j<k}\sum\beta_{ijk}x_i x_j x_k \tag{1}$$

To identify the point with the highest strength and lowest weight, the multiple response surface analysis using desirability function was applied [8].

Desirability function d_i is a transformation of the responses variable to a 0-to-1 scale. Response of 0 represents a completely undesirable response and 1 represents the most desirable response. If all responses are evaluated equally, the overall desirability function is calculated as follow:

$$D=\sqrt[m]{d_1 d_2 \ldots d_m} \tag{2}$$

2.4 Compressive strength test

Unconfined compressive strength test was conducted in accordance with ASTM C109 standard [9]. Dimensions of specimens were firstly measured. Then, 50-mm cubes were tested using the Compression and Flexural Testing Machine at STRC, De La Salle University. Both surfaces exposed to the force were polished to avoid high deviation during the testing process. The loading speed was fixed at 0.35 kPa/s. A computer was connected with the testing machine to record the result. Average of three replicates of each run was calculated.

3 Results and discussion

3.1 Engineering properties results

Figure 1 shows the trend of compressive strength and unit weight in correlation with variety of fly ash proportions. Generally, higher FA resulted in stronger

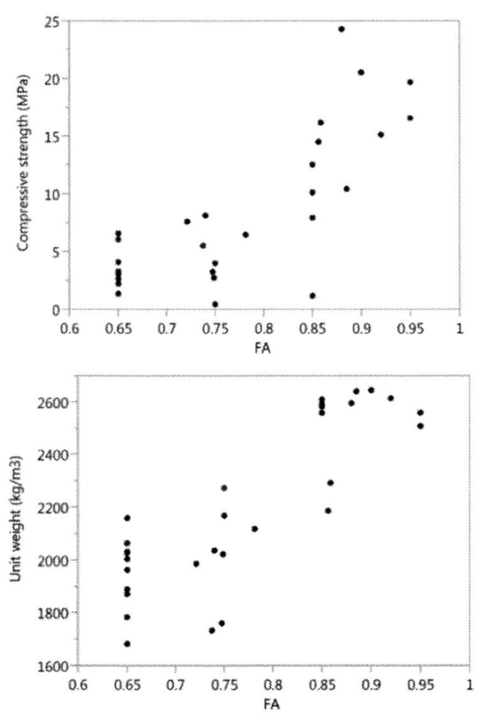

Fig. 1. Correlation between FA content and compressive strength and unit weight

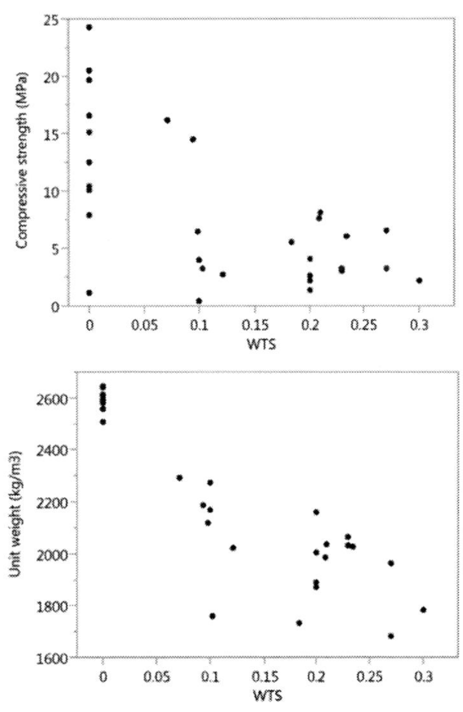

Fig. 2. Correlation between WTS content and compressive strength and unit weight

one-part geopolymers with larger bulk density. Note that FA is a by-product of coal burning process, so it had been already thermally activated at high temperature. Therefore, higher dosage of FA increased proportion of reactive phases which increase the geopolymerization reaction rate and strengthen the geopolymer matrix. However, density of binder from FA is increased with concentration of precursor. On the other hand, the effect of WTS to the mixture is decreasing the bulk density of the geopolymer product. As noticed in Figure 2, the increase of WTS proportion leads to the reduction of unit weight of geopolymer. However, higher percentage of WTS reduced compressive strength of product. Without WTS, the binder could reach compressive strength above 20 MPa. Meanwhile, addition of 30 % of sludge resulted in low-strength sample (< 5 MPa). It appears that WTS is relatively inactive and acts as lightweight aggregate in fly ash based gepolymer as observed in previous study [6].

3.2 Multiple response analysis

Experimental data in mixture design were fitted into regression models as follow:

$$\begin{aligned} \text{Compressive strength} &= 2.8854*\left(\frac{WTS}{0.3}\right)+19.3089*\left(\frac{FA-0.65}{0.3}\right)-54.3640*\left(\frac{NaAlO_2}{0.3}\right) \\ &-2.1453*\left(\frac{NaOH-0.05}{0.3}\right)+90.3054*\left(\frac{WTS}{0.3}\right)*\left(\frac{NaAlO_2}{0.3}\right) \\ &+111.5347*\left(\frac{FA-0.65}{0.3}\right)*\left(\frac{NaAlO_2}{0.3}\right)*\left(\frac{FA-0.65}{0.3}-\frac{NaAlO_2}{0.3}\right) \end{aligned} \quad (3)$$

$$\begin{aligned} \text{Unit weight} &= 1823.769*\left(\frac{WTS}{0.3}\right)+2633.647*\left(\frac{FA-0.65}{0.3}\right)+2603.340*\left(\frac{NaAlO_2}{0.3}\right) \\ &+2160.139*\left(\frac{NaOH-0.05}{0.3}\right)-1056.761*\left(\frac{WTS}{0.3}\right)*\left(\frac{FA-0.65}{0.3}\right) \\ &-41703.112*\left(\frac{NaAlO_2}{0.3}\right)*\left(\frac{NaOH-0.05}{0.3}\right)*\left(\frac{NaAlO_2}{0.3}-\frac{NaOH-0.05}{0.3}\right) \end{aligned} \quad (4)$$

Based on these predictive formulas, the prediction profiler was presented in Figure 3. Multiple response analysis resulted in the optimal composition of geopolymer which is 76% FA, 19% WTS, 5% NaOH. Optimal composition has compressive strength value of 8.98 MPa that satisfies the requirement for non-load bearing units (4.14 MPa) according to ASTM C129 standard [10]. Bulk density of the product is within the range for moderate-weight brick.

4 Conclusions

This study demonstrates that coal fly ash (CFA) and water treatment sludge (WTS) could be utilized in a one-part geopolymer binder for construction applications. Mixture design of experiment shows that FA is a good base material for geopolymerization as increased percentage of FA led to increase in compressive strength. Addition of WTS into the mixtures reduced unit weight of the binders but also reduce the compressive strength of the product. Multiple response analysis in which compressive strength was maximized and unit weight was minimized has given the optimal composition of FA = 76%, WTS = 19%, NaOH = 5% and sodium aluminate = 0. The compressive strength and unit weight of optimal combination are 8.98 MPa and 1877.7 kg/m3 respectively.

Acknowledgement

The authors acknowledge the financial support of Japan International Cooperation Agency (JICA) through the AUN/SEED-Net scholarship.

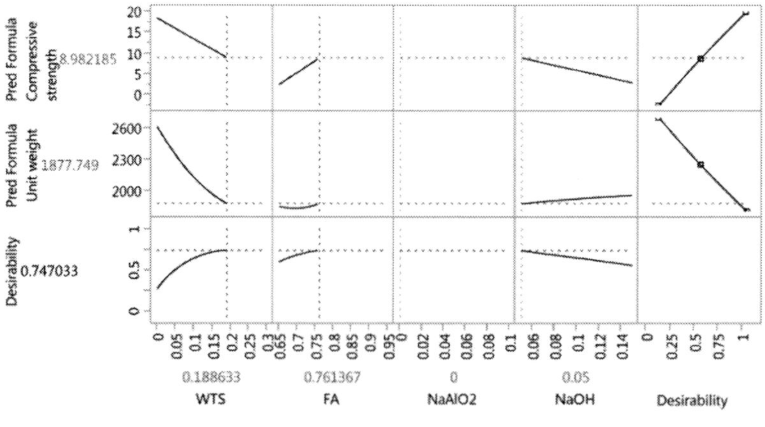

Fig. 3. Prediction profiler

References

1. J. Davidovits, *Geopolymer 2002 Conference* (2002)
2. C.D. Atis, E.B. Görür, O. Karahan, C. Bilim, S. Ilkentapar, E. Luga, *Construction and Building Materials,* **96,** 673–678 (2015)
3. G. Habert, C. Billard, P. Rossi, C. Chen, N. Roussel, *Cement and Concrete Research,* **40,** 820–826 (2010)
4. S. Horpibulsuk, C. Suksiripattanapong, W. Samingthong, R. Rachan, A. Arulrajah, J. Mater. Civ. Eng, (2015)
5. E. Nimwinya, W. Arjharn, S. Horpibulsuk, T. Phoongernkham, A. Poowancum, *Journal of Cleaner Production*, **119**, 128-134 (2016)
6. C. Suksiripattanapong, S. Horpibulsuk, S. Boongrasan, A. Udomchai, A. Chinkulkijniwat, A. Arulrajah, *Construction and Building Materials,* **94**, 807–816 (2015)
7. B. Nematollahi, J. Sanjayan, F.U.A Shaikh, *Ceramics International*, (2015)
8. D.J. Obermiller, (n.d.). Multiple Response Optimization using JMP. The Dow Chemical Company, Midland, MI. Retrieved from http://www2.sas.com/inforvis/paper178
9. American Society for Testing Materials. Standard Test Method for Compressive Strength of Hydraulic Cement Mortars (Using 2-in. or [50-mm] Cube Specimens) (ASTM C109/109M – 16a). *ASTM International*, (2016)
10. American Society for Testing Materials. Standard Specification for Nonloadbearing Concrete Masonry Units (ASTM C129-14a). *ASTM International*, (2015)

Synthesis and Characterization of Non-Halogenic Fire Retardant Composite with Epoxy Resin and Additive Combination $Al(OH)_3/Mg(OH)_2$

Asep Handaya Saputra[1*] and *Farhan* Fathurrahman[1]

[1] Chemical Engineering Department, Faculty of Engineering, University of Indonesia, Depok, 16424, Indonesia

Abstract. Epoxy has many advantages over other resin. However, in certain usage, additive is needed to raise the thermal resistance of the composites, while lowering its negative effect on health and environment. One of the most common additives used for thermal resistance is halogen. Halogen gives negative effect on health and environment because of the release of toxic gas following its combustion. An alternative for halogen substitution is using $Al(OH)_3/Mg(OH)_2$. Therefore, in this research, the synthesis and characterization of non-halogenic fire retardant composite with epoxy resin and various concentration of additive $Al(OH)_3/Mg(OH)_2$ was conducted. The characterization of this research is the fire retardancy, morphology dispersion, and mechanical properties of the synthesized composite, such as tensile strength and hardness. The result of this research is that epoxy resin 50% with 50% of additive $Al(OH)_3$ gives the best flame retardancy behavior and mechanical properties. This composite gives flammability rating V-0 with T_{max} 364,3 °C, MLR 12,51 %/menit, total mass loss 57,26%, tensile strength 11,7 MPa, and hardness 79.

1 Introduction

Composite is a material that is widely applied in various fields. Epoxy resins, one of the components commonly used in the manufacture of composites has limitation in terms of fire resistance.

Epoxy is widely used in the manufacture of composites because of its advantages, such as its wide range of properties, relatively small shrinkage among other thermosets, resistance towards chemical, oil and water, and has excellent mechanical and electrical properties. Due to these advantages, epoxy resins are widely applied in various fields. However, due to its flaws that are easy to burn, it is necessary to improve the fire retardant capability of the material better, so it will be safer to use.

Composite materials that have such properties are called fire retardant composite. Currently, fire retardant composite is available and is used for applications of materials that are susceptible to temperature rise or even burning. Halogen additive can improve the composite's fire retardancy but nowadays, the use of halogen additive has been minimized due to the release of halogen radical and halide acid which are harmful to the human health during burning[1]. Alternative materials which are more environmental friendly and can be used as replacement for halogen additive are magnesium hydroxide ($Mg(OH)_2$) and aluminium trihydroxide ($Al(OH)_3$) [2].

Both materials, aluminium trihydroxide and magnesium hydroxide, are known to be able to improve the fire retardancy of the composite material through decomposition mechanism that produce water vapour, absorb the heat, and form protective layer in the material surface.[3] $Al(OH)_3$ will begin to decompose at 200°C, whereas $Mg(OH)_2$ will decompose at a higher temperature than $Al(OH)_3$, at a temperature of 300°C [4].

The effect of using $Mg(OH)_2$ in the manufacture of fire retardant composite has been done by Tang et al. in 2013 using unsaturated polyester resin. Increasing the concentration of $Mg(OH)_2$ to some extent provides a better fire retardancy effect on the composite material[5]. In addition, the combination of $Mg(OH)_2/Al(OH)_3$ and

*Corresponding author: asephandaya@gmail.com

microencapsulated red phosphor (MRP) has been researched by Liang et al. in 2014 using polypropylene resins. The results obtained show that the use of a combination of three is able to increase the capability of fire retardancy of the material [6]. Then, the combination of $Al(OH)_3/Mg(OH)_2$ and the carbon black filler was investigated by Arfiana in 2016 using unsaturated polyester resins. The results obtained show that the addition of carbon black filler can improve the hardness and thermal stability [2].

Eventhough epoxy resin is one of the most common resin and applied in various fileds, most research that has been done rather use unsaturated polyester resins, while there is currently little research on fire retaradancy on composites that utilize epoxy resins. Therefore, through this research fire retardancy of epoxy resin can be studied and further improved. In addition, epoxy resins also have a wide range of properties and good mechanical and electrical properties. Moreover, epoxy resins have an advantage over unsaturated polyester resins, wherein epoxy resins have a lower shrinkage during curing and have faster curing time compared with unsaturated polyester resins.

Therefore, research on epoxy resin is required to improve its fire retardancy properties. In this study, this was conducted by using additives combination of $Al(OH)_3$ and $Mg(OH)_2$ on epoxy resins. The choice of these two additives is due to their properties that can produce water vapor when decomposed with fire and does not produce toxic compounds when caught on fire as occurs in halogen compounds.

2 Methodology

The composition of composites this study was varied as shown in Table 1.

Table 1. Varied Composition of Resin and Additive

Sample Code	% wt		
	Epoxy	**$Al(OH)_3$**	**$Mg(OH)_2$**
A0	50	0	0
A1	50	50	0
A2	50	35	15
A3	50	25	25
A4	50	15	35
A5	50	0	50

Composites with code A0 are neat resins, namely composites with epoxy content in the absence of additives. This neat resin is used as a standard in comparing the synthesized composite characteristics with the addition of $AlOH)_3/Mg(OH)_2$ as additive. Meanwhile, the composite with additive $Al(OH)_3/Mg(OH)_2$ comprises several variations of additive concentration to see and determine the epoxy resin composite with the best additive composition in terms of fire resistance, as well as to assess and determine which additive is more dominant in terms of improving fire retardancy in composite. Variations of the resin and additive compositions used can be seen in Table 1, where the column for epoxy consists of epoxy resin and hardener with 2:1 ratio.

Materials used in this study were epoxy resin and epoxy hardener (provided by PT Justus), aluminium hydroxide, and magnesium hydroxide. In the manufacture of these composites, stainless steel plate mould is used with a size that has been adjusted for testing flammability with UL94 standard and tensile strength with ASTM D638 type I standard. Composite with various additive concentrations which has been synthesized is then tested for fire resistance, thermal stability, mechanical strength, and its morphology is also observed.

The properties of fire retardancy were tested by flammability test and TGA-DSC, while the tensile strength and hardness were performed to see the mechanical properties of the composites, and SEM-EDX is used to observe the dispersion of additives within the composite.

Flammability test was performed based on the UL-94V standard procedure which resulted in four levels of flammability rating; V-0, V-1, V-2 and NR (non-rated) for the specimen that was completely burned. From this test, parameters obtained are time to ignition (t_i), time taken to retard the fire (t_r), and burning time (t_b).

For each concentration variation, testing required 5 samples. Each sample was burned for 10 seconds with a maximum repetition of 3 times if the sample had not burned or died after burning before the third repetition. From the test, the value of time to ignition (t_i), burning time (t_b) and time taken to retard the fire (t_r) are recorded. Burning material droplets are also observed to see if they are able to ignite the cotton located at the bottom of the test apparatus.

Flammability test UL-94V produces data that is used to see fire retardancy of the composite. Those data are time to ignition, time taken to retard the fire, burning time, and flammability rating. Time to ignition (t_i) is seen from the time required by the sample to be burned. Burning time value (t_b) is the sum of afterflame time ($t_1 + t_2$) and afterglow time (t_3). Afterflame time is the duration of the burnt sample for the first and second burning experiments, while the afterglow time is the length of time the sample burns on a third fuel trial. While time taken to retard the fire (t_r) is calculated when the sample begins to burn until the fire can be extinguished by itself. If the sample is burned completely then the comparable data for each resin/additive composite code is the value of t_i and t_b. Whereas if the fire can be extinguished after being burned then t_r data can be obtained.

Thermal stability test was performed by using Thermal Gravimetry Analysis (TGA). The result of this test was a graph that showed the temperature when the mass of material started to decrease (T_{onset}), the mass loss rate (%/min), and the maximum temperature (T_{max}) when the mass loss rate reached its highest point.

Mechanical properties observed in this study are tensil strength and hardness. The tensile strength was characterized based on ASTM-D638 type I. The hardness was based on ASTM D2240-15.

3 Results & Discussion

3.1 Flammability Analysis

Based on the parameters calculated in Table 2, which shown the tabulation data of resin/additive composite sample test results, code A1 has the greatest t_i value with the lowest t_r value and the best flammability rating (V-0) based on the UL-94V test scale. Thus, the addition of 50% $Al(OH)_3$ additive to epoxy resin is capable of producing a fire retardant composite with best fire retardancy capability compared to composites with other concentration variation of $Al(OH)_3/Mg(OH)_2$ additive.

Table 2. UL-94V Flammability Test Results

Sample Code	t_i (second)	t_r (second)	t_b (second)	Rating
A0	10	-	129.6	NR
A1	30	0	205	V-0
A2	22	25	153.6	V-0
A3	20	-	209.6	NR
A4	20	-	210.4	NR
A5	20	-	197.4	NR

From the flammability testing conducted, it is known that the sample neat resin with the code A0 burned with a value of t_i of 10 seconds. The neat resin data is the standard data that will be used as a benchmark for other composite code flammability tests. The time to ignition data for the resin/additive composite in Table 2 is made into graph form shown in Figure 1.

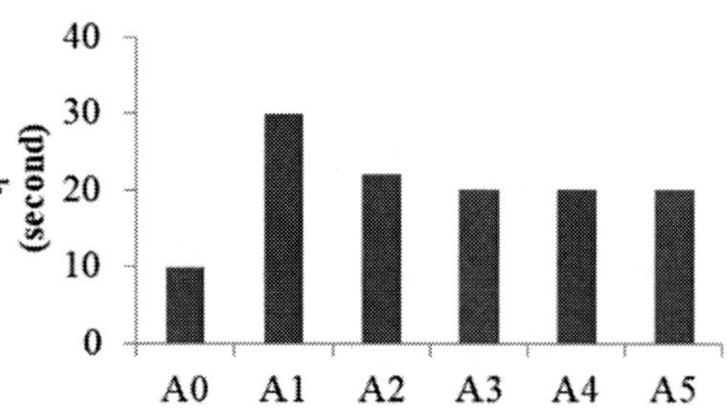

Figure 1. Comparison of Time to Ignition

Based on the time to ignition graph, figure 1, it can be seen that t_i of neat resin increases with the addition of additives, both $Al(OH)_3$ and $Mg(OH)_2$. In addition, this increase is most significant in code A1 with an additive content of $Al(OH)_3$ 50%. In code A1 there is an increase in the value of ti as much as 3-fold from t_i of neat resin. While in other run codes, t_i decreases with the decreasing of $Al(OH)_3$ and increasing of $Mg(OH)_2$. However, the A5 code with $Mg(OH)_2$ 50% and doesn't contain any $Al(OH)_3$ still has a higher t_i value than neat resin.

This suggests that the two additives, $Al(OH)_3/Mg(OH)_2$, are capable of increasing time to ignition of epoxy composite neat resin. However, the additive $Al(OH)_3$ is capable of significantly increasing the value of t_i rather than $Mg(OH)_2$ additive. The $Mg(OH)_2$ additive has a coagulation effect when added to the epoxy resin. The higher the concentration of $Mg(OH)_2$, the thicker the texture of the resulting mixture. This condition causes stirring more difficult and causes the distribution of $Mg(OH)_2$ in the mixture to be less evenly distributed, whereas the additive distribution may affect the flammability property of the material. In contrast to $Mg(OH)_2$, the addition of $Al(OH)_3$ results in a mixed texture that is sufficiently fluid, thus allowing the $Al(OH)_3$ additive to spread more evenly in the mixture.

According to Hapuarachchi et al. (2009), the endothermic decomposition reaction due to the increase in temperature of the hydroxide group fire additive is capable of releasing moisture and absorbing heat in the material, which can slow down the process of material decomposition [7]. In addition, the phenomenon also causes the material temperature to be under the ignition temperature for a certain time when the additive decomposition process occurs so that time to ignition material increases. Thus, the addition of $Al(OH)_3/Mg(OH)_2$ additive is able to give a positive effect on composite fire resistance marked by the increase of time to ignition value.

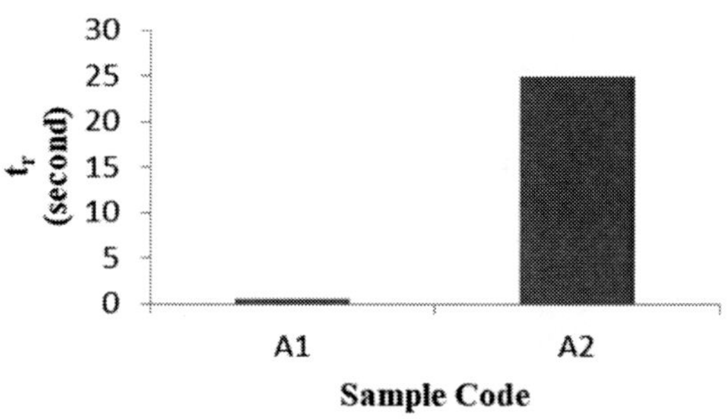

Figure 2. Comparison of Time Taken to Retard the Fire

Time taken to retard the fire, t_r, is the time it takes material to be able to extinguish the flame. The resin/additive composites in this study which demonstrate the ability of fire retardancy are composites with codes A1 and A2. In both of these codes, the results obtained are not uniform, where from the five test samples, two samples are able to extinguish the flame and some burn. This may be due to an uneven distribution of additives due to thickening, especially in code A2 when added with an $Mg(OH)_2$ additive. The t_r value for the resin/additive composite can be seen in Figure 2.

The lower the t_r value indicates better fire retardancy capability because the flame can be extinguished faster. The data in Table 2 shows the t_r values for the composite codes A1 and A2 are 0 and 25 seconds respectively. The additive composition which gives the smallest t_r value is obtained in a combination of $Al(OH)_3$ 50% additive (code A1), where for this composition, the composite sample is not burned by fire. As for neat resin (code A0), composite resin/additive with composition of $Al(OH)_3$ 25%/$Mg(OH)_2$ 25% (code A3), $Al(OH)_3$ 15%/$Mg(OH)_2$ 35% (code A4), and $Mg(OH)_2$ 50% (code A5), the entire sample burns.

From the graph it can be seen that the use of $Al(OH)_3$ 50% still gives better results when compared with the use of $Mg(OH)_2$ with the same concentration. This occurs because of the thickening effect that occurs when the mixture is added with an $Mg(OH)_2$ additive that reduces the homogeneity of the additive particle distribution in the mixture so that the fire retardancy is reduced.

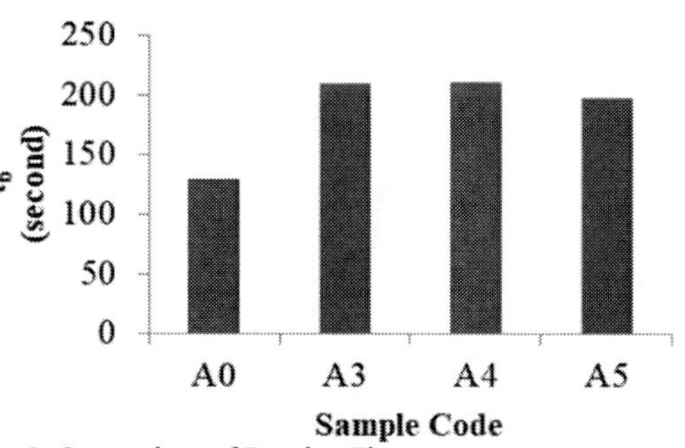

Figure 3. Comparison of Burning Time

For burned composite samples, codes A0, A3, A4, and A5, the parameter which is compared is burning time or t_b values. The value of t_b indicates how long the sample burns after the source of the flame is removed. Thus, although the composite samples with the above codes have not demonstrated the capability of fire retardancy, the increase in t_b value can be an indication that there is an increase in the resistance of the material. The burning time graph for resin/additive composites can be seen in Figure 3.

From the burning time graph, it can be seen that for the four varied composite compositions burned, burning time increases with the increase of additive concentration of $Al(OH)_3$ and the decrease of additive concentration $Mg(OH)_2$. This shows that the influence of $Al(OH)_3$ additive on the material's fire resistance is more dominant than $Mg(OH)_2$ additive. The effect is caused because the additive $Mg(OH)_2$ tends to produce a thick mixture and difficult to stir. Therefore, when the $Mg(OH)_2$ additive concentration is lower, the mixture becomes more homogeneous so that the fire retardancy of the material is better which is hinted by increased burning time for burned composite codes in the flammability test.

3.2 Thermal Stability Analysis

The thermal stability and decomposition pattern of the fire retardant composite material is known through thermal gravimetry analysis (TGA) with inert gas atmosphere conditions and heating rate of 10°C/min to a temperature of 800°C. The inert gas atmosphere conditions were chosen to determine the effect of heat degradation on the decomposition of the material, because if using air atmosphere then the influence of heat degradation is difficult to observe due to the influence of thermal oxidation. The results of thermal stability testing are TGA, DTG, and DSC graph that can explain the phenomenon of decomposition that occurs in the material due to heat degradation.

Table 3. Thermal Stability (TGA-DTG-DSC) Test Results

Sample Code	T_{onset} (°C)	T_{max} (°C)	MLR (%/min)	Total Mass Loss (%)
A0	301.5	362.4	20.47	73.3
A1	260	364.3	12.51	57.26
A2	256.9	340.4	14.86	53.84
A3	231.4	362.5	21.89	59.44
A4	282.6	362	25.49	64.37
A5	365.2	352.8	14.68	61.39

From the TGA test temperature of onset, total mass loss, mass loss rate value, and maximum temperature is known. Further analysis between neat resin and other resin/additive composite compositions is easier to be done by comparing the value of decomposition temperature (T_{onset}), maximum temperature (T_{max}), mass loss rate (MLR), and total mass loss. The temperature of the onset is the temperature at which there is a phase change in the test material. From several peaks that appear on each DTG chart, the maximum temperature (T_{max}) taken is the temperature at which the mass loss rate (MLR) reaches the highest value. Total Mass Loss is the total mass lost at the end of the test. The details of the tabulation of the thermal stability test results for neat resins and resin/additive composites can be seen in Table 3.

From the result of TGA-DSC test for neat resin (A0), it can be seen that the decomposition process started to occur at temperature around 301.5 °C (temperature onset) with maximum temperature at 362.4 °C where mass loss rate (MLR) reached its highest value, 20.47% /min. In addition, the total mass loss for neat resin reached 73.3%. This value indicates that most of the resin is lost due to temperature rise, where the sample mass reaches 26.06% at the end of the test at a temperature of about 800°C.

While the resin/additive composite composition with the best flammability rating (A1), generally exhibits better thermal stability properties than the neat composite resin (A0). This is indicated by the increase of the maximum temperature value (T_{max}), decrease of the MLR value, and also decrease of the total mass loss value. Thus, the best resin/additive composite with $Al(OH)_3$ 50% increased the thermal stability by increasing Tmax to 364.3 °C, decreasing MLR to 12.51% /min and decreasing total mass loss to 57.26%.

For composite resin/additive codes A1 through A5 decomposition occurs gradually with different intermediates. The A1 to A5 resin/additive composite DTG graphs also show more than one peak due to the addition of $Al(OH)_3$ and or $Mg(OH)_2$ additives. The presence of more than one peak confirms the TGA graph that decomposition occurs in several stages. As for the DSC graph for the resin/additive composite there is a difference, where for codes A1 and A5 there is only one peak, whereas in the A2, A3, and A4 codes there are two peaks. The difference in the number of peaks is due to the presence of different additive content. In code A1 and A5 there is only one additive, $Al(OH)_3$ in A1 and $Mg(OH)_2$ in A5. While in code A2, A3, and A4 there is a combination of additive $Al(OH)_3/Mg(OH)_2$ which decomposes at different temperatures causing the existence of two different peaks.

The TGA test results for resin/additive composites in Table 3 shows that the onset temperature has a fairly varied trend. Compared to the T_{onset} of neat resin, the T_{onset} of resin/additive composite values for the A1, A2, A3, and A4 codes decreased. The largest decrease in T_{onset} occurred in A3 composite, which is 23%. While the resin/additive composite code A5 increased the value of T_{onset} by 21%. The same thing also occurs in the value of T_{max} where the trend is quite varied. When compared with neat resin (code A0), there is a decrease in T_{max} value in A2, A4, and A5 codes with the largest decrease in code A2 which is 6%. On the other hand, in composite A1 and A5, there is no significant increase of T_{max} value.

From the DTG graph and the TGA-DSC test results data in Table 3 above it can be seen that when compared with neat resin (A0), mass loss rate (MLR) for varied additive concentration A1, A2, and A5 has lower value. When compared with MLR of neat resin, MLR of the best flammability rating resin/additive composite (A1) experienced the greatest decrease, from 20.47% /min to 12.51% /min. This MLR value of A1 is the lowest value compared to other codes of resin/additive composite. In addition, when compared between the use of single additive $Al(OH)_3$ with $Mg(OH)_2$, MLR for composites with a single additive $Al(OH)_3$ has smaller value compared to composites with a single additive $Mg(OH)_2$. This suggests that the additive $Al(OH)_3$ has a more dominant influence on MLR of composites. Addition of $Al(OH)_3$ and $Mg(OH)_2$ additives is also capable of decreasing the total mass loss of composite by about 8.93-19.46% by weight when there is an increase in temperature. The addition of $Al(OH)_3$ and $Mg(OH)_2$ additives tend to increase the thermal stability of the composite material especially when viewed from the decrease in total mass loss value.

3.3 Mechanical Strength Analysis

Tensile strength and hardness tests were performed on neat resin composite (A0) and resin/additive combination composite with $Al(OH)_3$ 50% which have the best flammability rating (A1). The graph of tensile strength test results can be seen in Figure 4.

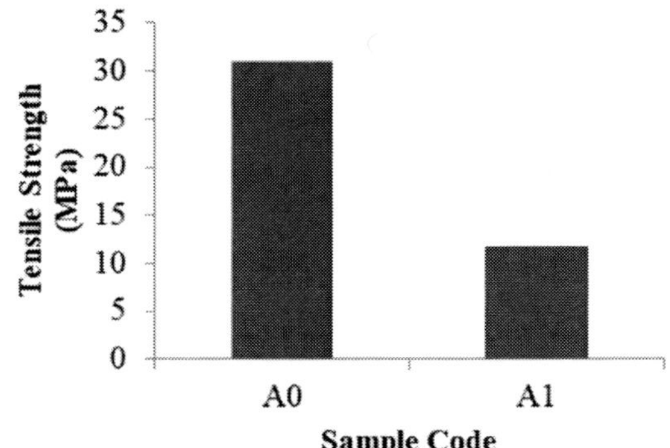

Figure 4. Tensile Strength Comparison

From the graph above it can be seen that the addition of the additive lowers the resulting material tensile strength. Addition of $Al(OH)_3$ 50% additive caused a decrease in tensile strength from 31 MPa (neat resin) to 11.7 MPa, or a decreased by 62.4%.

According to Wang, et al. (2015), the polarity difference between the additive and the resin may cause the resulting composite compatibility to be low and the additive particles can not be dispersed evenly in the matrix, especially when the additive concentration is high [8]. Meanwhile, according to Devendra et al. (2013), the increase of additive concentration in composites can reduce the strength of interfacial bonds between additives and matrix [9]. These things tend to cause a decrease in the mechanical properties of the synthesized material. In addition, according to Reddy et al. (2015), the addition of additives up to 60% can lead to an increase in viscosity that triggers the formation of agglomerates which can then become a central point of stress and cracking [10].

As for the graph of hardness test results can be seen in Figure 5. From the graph of hardness, the effect of additive to the resulting composite hardness can be seen. The combined use of $Al(OH)_3$ additives by 50% can increase hardness by 21.22%, from 65.5 to 79. According to Flores et al. (2000), the increase in composite hardness with additive additions may be caused by the additive particles that is evenly distributed resulting in a structure that can harden the material [11].

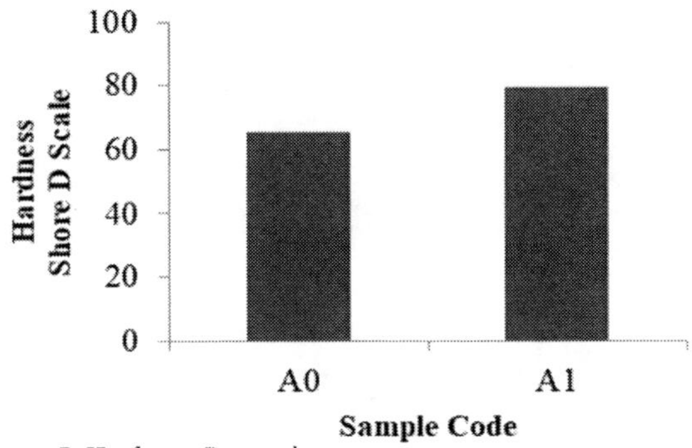

Figure 5. Hardness Comparison

3.4 Morphology Analysis

The dispersion of additive particles as well as the interfacial interactions of particles with the matrix are two important factors affecting the resulting composite material [12]. To see the dispersion of the additive particles, as well as the possible formation of aggregates and/or agglomerates in the composite, the tests performed using SEM-EDX tools include morphological tests and percentage analysis of elements.

The morphological and elemental percentage tests were performed for neat resin (A0) as well as resin/additive composite with the best flammability rating (A1). In addition, the test was also performed for composites with an additive of $Mg(OH)_2$ 50% to see the homogeneity of the composite. The test is performed on the composite surface. The dispersion pattern of additive particles of $Al(OH)_3/Mg(OH)_2$ in epoxy and morphological matrices for neat resin is shown in Figure 6.

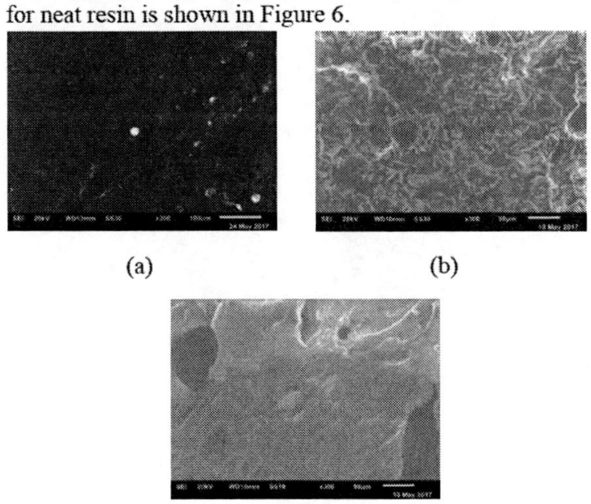

Figure 6. Composite Surface (a) Neat Resin, (b) A1 ($Al(OH)_3$ 50%), (c) A5 ($Mg(OH)_2$ 50%)

Based on the image, it can be seen that on the surface of neat resin composite almost no air bubble exists. Bubble should be avoided in the manufacture of composites as they may negatively affect material characteristics. From general morphological testing it can be seen that adding additives tends to trigger agglomerates where additives accumulate in several places. In the resin/additive composite with concentration of additive $Al(OH)_3$ 50% (A1) formed agglomerates of varying sizes but relatively smaller compared to agglomerates in resin/additive composite with concentration of additive $Mg(OH)_2$ 50% (A5). This suggests that the additive $Al(OH)_3$ is more easily dispersed in epoxy resins than with the $Mg(OH)_2$ additive.

The morphological results show that the addition of $Mg(OH)_2$ to the epoxy resin composite mixture tends to trigger the formation of larger size agglomerates which can lead to the formation of a material structure less resolved by the resin thus affecting the bond between the matrix (resin) and the additive particles. The more agglomerate formed with the addition of $Mg(OH)_2$ will affect the composite fire resistance characteristics produced, where if the additive is not dispersed properly then the fire resistance properties obtained will also decrease. This is also supported by flammability test results that tend to decrease with the increasing concentration of $Mg(OH)_2$. Therefore, it can be said that the morphological results show that $Mg(OH)_2$ additive is difficult to dispersed well on the epoxy resin, so it tends to form agglomerates which can decrease the fire and mechanical resistance properties of the material.

4 Conclusions

This research showed that the best fire retardancy for epoxy resin with additive $Al(OH)_3/Mg(OH)_2$ was obtained from the composite with concentration of $Al(OH)_3$ 50%. The addition of $Al(OH)_3$ 50% to epoxy resin composite could also improve the thermal stability of composite by reducing MLR (mass loss rate) to 12.51% /min and total mass loss to 57.26%. On the other hand, the tensile strength of composite decreased to 11.7 MPa but the hardness improves to 79 based in hardness shore D scale. Overall, the addition of both $Al(OH)_3$ and $Mg(OH)_2$ to epoxy resin can improve the fire retardancy of composite eventhough $Al(OH)_3$ have more dominant effect in increasing the fire retardancy of composite and both can be used as an alternative to synthesise fire retardant composite.

5 Acknowledgements

Great gratitude to PT Justus for providing materials needed in this research and PITTA Research Fund of Universitas Indonesia for the research financial support.

6 References

1. C. Gerard, G. Fontaine, S. Bourbigot, *Journal of Material 3*. 4476-4499. (2010)
2. A. H. Saputra, Arfiana. *Chemical Engineering Transactions*. **56**. 1759-1764. (2017)
3. M. Rakotomalala, S. Wagner, and M. Doring. *Recent Developments in Halogen Free Flame Retardant for Epoxy Resins for Electrical and Electronic Applications*. (2010)
4. E. D. Weil, S. V. Leivchik, and P. Moy. *Flame and Smoke Retardant in Vinyl Chloride Polymers – Commercial Usage and Current Development*. (2006)
5. H. Tang, X. B. Zhou, and X. L. Liu. *Effect of Magnesium Hydroxide on Flame Retardant Properties of Unsaturated Polyester Resin*. (2013)
6. J. Z. Liang, J. Q. Feng, C. P. Tsui, C. Y. Tang, D. F. Liu, S. D. Zhang, and W. F. Huang. *Composite: Part B*, **71**. 74 – 81. (2014)
7. T. D. Hapuarachchi, T. Peijs, *eXPRESS Polymer Letters 3*, 743-751. (2009)
8. B. B. Wang, H. B. Sheng, Y. Q. Shi, W. Z. Hu, N. N. Hong, W. R. Zeng, H. Ge, X. J. Yu, L. Song, and Y. Hu. *Polymer Degradation and Stability*. **113**. 96 – 109. (2015)
9. K. Devendra, and T. Rangaswamy. *Journal of Minerals and Materials Characterization and Engineering*. **1**. 353 – 357. (2013)
10. B. H. N. Reddy, K. R. V. Mahesh, V. V. Lakshmi, and M. A. Kumar. *Advanced in Polymer Science and Technology: An International Journal*. **5**. 11 – 17. (2015)
11. A. Flores, M. E. Cagiao, A. Ezquerra, and F. J. B. Calleja. *Applied Polymer Science*. **79**. 90 – 95. (2000)
12. Xin Wen, Y. J. Wang, Jiang Gong, Jie Liu, Nana Tian, Y. H. Wang, W. J. Zhi, and Jian Qiu. *Polymer Degradation and Stability*. **97**. 793 – 801. (2012)

 https://doi.org/10.1051/matecconf/201815605011

Biosorption of Lead (II)–containing Sasirangan Textile Wastewater using Nanocomposites of *Eleocharis dulcis* Fibers with Iron (III) Nanoparticles as Adsorbent

Chairul Irawan[1*], *Iryanti* F. Nata[1], *Meilana* D.Putra[1], *Yuli* Ristianingsih[1]

[1] Chemical Engineering Study Program, Engineering Faculty, University of Lambung Mangkurat, Banjarbaru, South Kalimantan 70714, Indonesia.

Abstract. This research focuses on the study of biocomposite nanoparticles of *Eleocharis dulcis* (ED) as potentials biosorbent to reduce the concentration of lead (II) ion containing Sasirangan textile industry wastewater. *Eleocharis dulcis*, locally named as Purun Tikus, has been developing becomes the biocomposites nanomaterial and valuables material in this research. Batch experiments were carried out to considering the kinetic of biosorption of lead onto the adsorbent, evaluating the effects of lead ion equilibrium concentration, equilibrium pH, and temperature on the adsorption of lead (II). Kinetic data of lead (II) biosorption onto EDB and EDB-MH revealed that equilibrium time was reached within 2 h, and the isotherm data showed that the Langmuir maximum adsorption capacity of the EDB-M and EDB-MH at pH_e of 6±0.2, room temperature were 150.43 mg/g and 180.92 mg/g, respectively. The thermodynamic of lead (II) biosorption onto the adsorbent implied the biosorption was spontaneous and endothermic indicating by increased in temperature would increased in adsorption capacity.

1 Introduction

Heavy metals are present harmful, carcinogenic and non-biodegradable characteristics when they appeared in the aquatic environment. Recently, from the point of view the heavy metals categories, lead has gained an increasing attention due to its numerous occurrence in wastewaters and its high toxicity impacts to human health [1, 2]. Lead (Pb) is considered detected in outflow approaching from smelters industry and refineries, battery manufacturing, glass manufacturing, alloy steel industry, and textile industry even from dyestuff and printing process [3, 4]. Then, lead is considered detected in Sasirangan textile wastewater. In Sasirangan textile wastewaters, Pb-ion concentrations approach of 2–5 mg/L. This concentration is quite high in relative to provisional guidelines value, and Pb-ion concentration of the wastewaters must be decreased to below of 0.01mg/L before its discharging into water ways body.

A number of technologies already used prior to diminish Pb-ion concentration in textile wastewater including chemical precipitation, coagulation–flocculation, electrochemical methods, ion exchange, membrane separation, reduction, adsorption, etc [5]. In recent years, biosorption method, a part category of adsorption, has gain concern for treatment lead–containing wastewater due to its high efficiency, easy operation, low cost, abundant amount in nature, mitigating heavy metal removal in low metal concentration (< 100 mg/L), reduced quantity of chemical and sludge handling.

Nowadays, biosorption using agricultural biomass-derived magnetic as adsorbent offer various attractive features, such as highly adsorption capacity, porous, inexpensive, non-toxic, biodegradable and biocompatible [6]. Sugarcane bagasse [7], oak wood and oak bark [8], pine bark [9], bagasse [10], and pine wood [11] are a samples of agricultural biomass-derived magnetic for lead adsorption. The agricultural biomass-derived magnetic as adsorbent would be easy separated after adsorption process from solution by using magnet kits. This research proposed *Eleocharis dulcis* (ED), locally named as Purun Tikus, becomes the biocomposites nanomaterial with magnetic modifiers and valuables material as an adsorbent for treating lead-containing Sasirangan textile wastewater. The contact time effect, equilibrium pH, equilibrium Pb-ion concentration, and temperature, related to kinetic and isotherm adsoprtion models, and thermodynamic parameters, such as entalphy, free energy change, and entropy were considered in batch model experiment for a more appropriate understanding of the mechanism of

* Corresponding author: cirawan@unlam.ac.id

biosorption process of lead (II)-containing Sasirangan textile wastewater.

2 Experimental Details

2.1 Materials

Lead containing Sasirangan textile wastewater samples of the typical high density of dyes were collected from the factory discharge point without any treatment in Banjarmasin, South Borneo, Indonesia. The samples were kept stored at below 4°C. $Pb(NO_3)_2.4H_2O$ as an artificial lead solution. HCl and NaOH were utilized to adjust the pH value of the solution, used in the present work were obtained from MERCK. All others chemicals employed in this work were of pure grade analytical without any further treatment. EDB-M and EDB-MH as adsorbent were prepared by author according to [12].

2.2 Batch model adsorption experiments of Lead containing Sasirangan textile wastewater

Batch model adsorption experiments were accomplished by placing 50 mL solution of certain amount of lead containing Sasirangan textile wastewater into 100 mL glass bottle and arranging its pH value using 1M NaOH or 1M HCl. Afterwards, weighed amount of the biosorbent, EDB-M and EDB-MH was taking into glass bottle, respectively. The mixtures was then placed in a water bath shaker (Firstek Scientific) at 150 rpm at room temperature for a desired contact time. The adsoprtion was carried out while shaking at room temperature for 6 hours. Afterward, besides separation of the solution from adsorbent assisted by using magnet kits. At the last of the experiment, the mixed solution are centrifuged and filtered through 0.2 μm PVDF membrane (Advantec). Finally, the lead concentration in the initial and filtrate solution were measured by using inductively coupled plasma optical emission spectrophotometer (HORIBA Jobin Yvon 2000-2 ICP-OES). Adsorbed Pb-ion was determined from the difference between the initial and equilibrium adsorbate concentrations. The batch adsorption experiments were done triplicate, where as the average value was taken.

The equilibrium adsorption capacity, *qe* (mg/g) was determined by the following equation:

$$q_e = (C_o - C_e)V/M \quad (1)$$

Where C_o (mg/L) was the initial lead concentration, C_e (mg/L) was the effluent or equilibrium lead concentration, V (L) was the volume of the lead solution and M (g) was the mass of the biosorbent used.

The biosorbent, EDB-M and EDB-MH, were used to investigate the factor affecting lead adsorption such as pH, initial lead concentration, and temperature. The adsorbent was separately used for adsorption isotherm studies and finding out the correlation between the adsorbent content that has a major role in the lead adsorption mechanism.

In addition, to regenerated the adsorbent by suitable desorption proses, the EDB-M and EDB-MH loaded metal ions were desorbed by shaking in 0.1 N HCl for 24 hours for repeated uses. After washing properly with deionized water , the regenerated adsorbent was charged for the next cycle of lead adsorption process. The cycles of such adsorbent were repeated for three times in triplicate analysis.

3 Results and Discussion

3.1 Biosorption of Lead containing Sasirangan textile wastewater by using ED as biosorbent

The ED biocomposite with iron (III) nanoparticles (EDB-M) with diameter size around 30–50 nm is a promising precursor to be used in the biosorption process for the reduction of lead from Sasirangan industrial textile wastewater. Clearly as shown in **Fig. 1**, the adsorbent easily separate from the solution after adsorption process by using magnet kits.

Fig. 1. The samples of lead adsorption onto EDB-M a. before separate adsorbent from the solution, b. separate adsorbent assisted by using magnet kits

3.2 Kinetic studies of biosorption of Lead-containing Sasirangan textile wastewater

The contact time effect on the Pb-ion adsorption onto EDB-M was conducted at room temperature, pH_e of 6.2±0.2, shaking rate of 150 rpm, initial Pb-ion concentration of ca. 100 mg/L, and dose of biosorbent of 2.5 g/L.

Fig. 2 shows that during the first 15 minutes, Pb-ion uptake capacity significantly increased from 0 mg/g to 21.90 mg/g, which was apparently due to the diffusion taking place into the vacant pores and or adsorption onto the surface of the adsorbent. Initially, the Pb-ion concentration gradient was quite high diffusing into and through the vacant porous of the adsorbent, then the lead uptake capacity are increased slightly to lead uptake capacity of 24.56 mg/g and being constant until reached equilibrium condition within 2 hours. The equilibrium adsorption experiments were conducted within 2 h of reaction time.

The kinetic models of pseudo-first-order and pseudo-second-order are adapted to fit the adsorption reaction model of lead adsorption onto the EDB-M. The Lagergren pseudo-first-order expression is given as [**13**]:

$$\frac{dq_t}{dt} = k_1 (q_e - q_t) \quad (2)$$

where q_e and q_t are the amount of Pb-ion adsorbed onto biosorbent per mass unit of the biosorbent (mg/g) at equilibrium and contact time t, respectively, and k_1 is the rate constant of biosorption (1/min). Integrating of equation (2) for boundary conditions $t = 0$ to $t = t$ and $q_t = 0$ to $q_t = q_t$ resulted in the following:

$$q_t = q_e(1 - \exp(-k_1 t)) \quad (3)$$

The model of pseudo-second order is considered as [**14**]

$$\frac{dq_t}{dt} = k_2 (q_e - q_t)^2 \quad (4)$$

where q_e is the maximum adsorption capacity (mg/g); k_2 the rate constant of the pseudo-second order equation (g /mg.min); q_t are the amount of lead adsorbed onto biosorbent per unit mass of the biosorbent (mg/g). Integrating the equation for boundary conditions $t = 0$ to $t = t$ and $q_t = 0$ to $q_t = q_t$ gives:

$$q_t = q_e \left(\frac{q_e k_2 t}{1 + q_e k_2 t} \right) \quad (5)$$

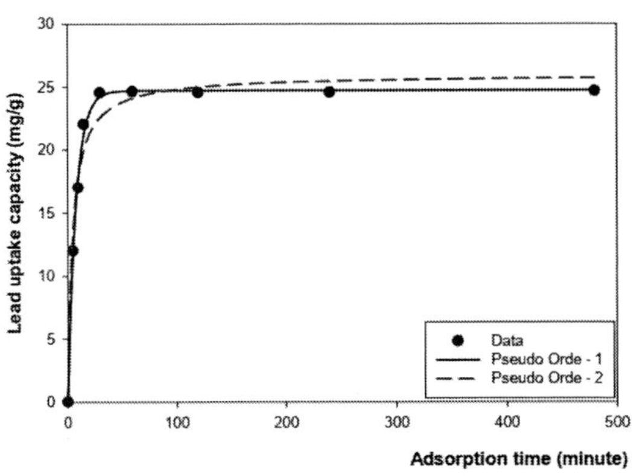

Fig. 2 Kinetic plot of lead biosorption onto the EDB-M at room temperature, pH_e of 6.2±0.2, shaking rate of 150 rpm, initial Pb-ion concentration of ca. 100 mg/L, and dose of biosorbent of 2.5 g/L.

The comparison figure of the different kinetic model for lead adsorption onto the EDB-M was shown in Fig. 2. The resulting of the parameter and the correlation coefficients achieved by using non-linear regression was calculated and listed in Table 1. The values of the coefficient related to the Lagergren kinetic model for pseudo-first order was found to be 0.997, which was more representative fitted than the kinetic model for pseudo-second order. Based on the composition of EDB-M, it was likely several reaction type or mechanisms are involved in the fixation of lead onto EDB-M. However, uses of the kinetic models tend to suggest that the rates of this reaction could be adjacent to pseudo-first-order kinetic model. A better correlation of kinetic reaction rate data in this model justified the physical biosorption mechanism occured.

Table 1. Kinetic parameters of lead biosorption onto the EDB-M at room temperature

Kinetic Model	Parameter	Value
	q_e (mg/g)	24.722
Pseudo-first order	k_1 (1/min)	0.1293
	R^2	0.9969
	q_e (mg/g)	25.948
Pseudo-second order	k_2 (g/(mg.min)	0.0085
	R^2	0.9756

3.3 Equilibrium pH studies of biosorption of Lead-containing Sasirangan textile wastewater

Plotted in Fig. 3 shown that the lead adsorption envelops for the adsorbent was bell-shaped like typical oxyanions sorption curves. The lower value of lead adsorption capacity when the pH was low (pH around 4) could be due to the excess of hydrogen ions surrounding the binding sites making adsorption process was unfavorable. Lead and H^+ ions compete with the active adsorption site and account for the lower adsorption uptake capacity.

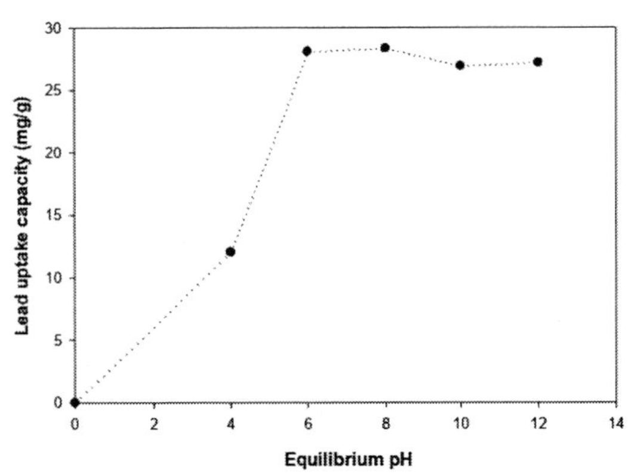

Fig. 3 The equilibrium pH effect on lead biosorption onto the EDB-M at equilibrium adsorption time of 2 h, room temperature, shaking rate of 150 rpm, initial Pb-ion concentration of ca. 100 mg/L, and dose of biosorbent of 2.5 g/L.

Thus, lead biosorption capacity will increase with increasing pH around 6–8 up to 27.18 mg/g for EDB-M as biosorbent. The increase in lead uptake capacity as equilibrium pH increases can be interpreted by a decrease in competition between protons (H^+) and positively charged metal ions at the surface sites. At higher pH (pH > 8) due to metal solid hydroxide precipitation of the lead, several hydroxil low-soluble such as $Pb(OH)_2$ maybe form, then the lead uptake capacity seems remaining constant [15].

3.4 Effect of temperatures studies of biosorption of Lead-containing Sasirangan textile wastewater

The effect of temperature on lead adsorption was investigated in order to figure out the behavior of lead biosorption process, and also used to predict the thermodynamic parameters. The temperature effect was studied by varying the temperature from ambient temperature up to 60°C at equilibrium adsorption time of 2 h, pH_e value of 6.2 ± 0.2, shaking rate of 150 rpm, and dose of biosorbent of 2.5 g/L.

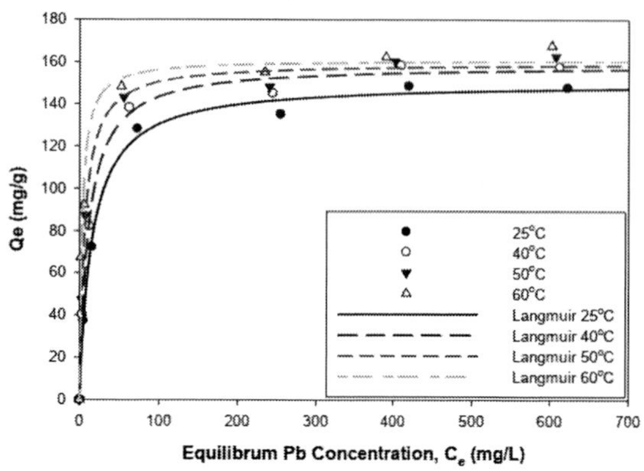

Fig. 4 Langmuir isotherm model plot at various temperatures for lead biosorption onto the EDB-M at equilibrium adsorption time of 2 h, pH_e value of 6.2 ± 0.2, shaking rate of 150 rpm, and dose of biosorbent of 2.5 g/L.

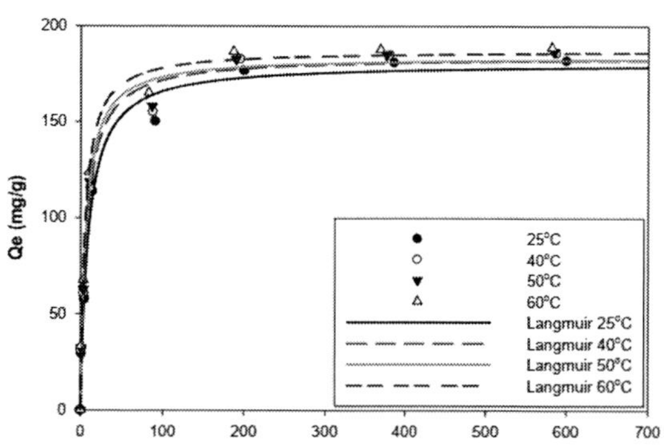

Fig. 5 Langmuir isotherm model plot at various temperatures for lead biosorption onto the EDB-MH at equilibrium adsorption time of 2 h, pH_e value of 6.2 ± 0.2, shaking rate of 150 rpm, and dose of biosorbent of 2.5 g/L.

Fig. 4 shows that lead uptake capacity was slightly increased with increasing temperature. Lead adsorption onto the surface of the EDB-M seems was favored at higher temperature. The Langmuir isotherm model was illustrated to predict the thermodynamic parameters. The linearity of the data was derived from regression analysis and it was found that the correlation coefficient are average better than 0.996 at temperature 25, 40, 50, and 60°C. The tendency was also same shown in Fig. 5.

3.5 Adsorption isotherm models studies

The adsorption isotherm model studies were conducted by varying initial lead concentration of c.a. 50 to 1000 mg/L at room temperature, equilibrium adsorption time of 2 h, pH_e value of 6.2 ± 0.2, shaking rate of 150 rpm, and dose of biosorbent of 2.5 g/L.

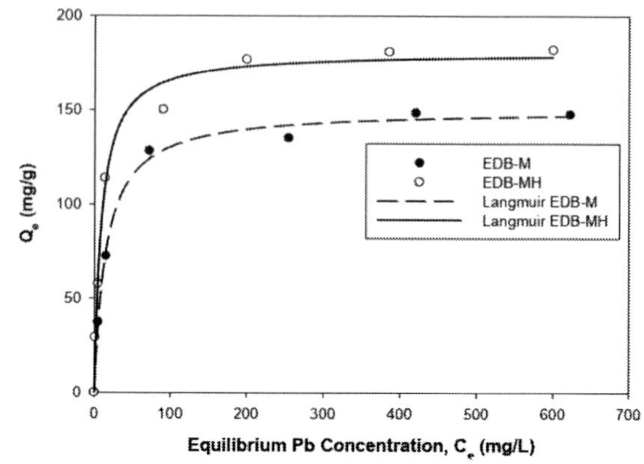

Fig. 6 Langmuir isotherm plot for lead adsorption onto EDB-M and EDB-MH at room temperature, equilibrium adsorption time of 2 h, pH_e value of 6.2 ± 0.2, shaking rate of 150 rpm, and dose of biosorbent of 2.5 g/L.

Fig. 6 shown that lead adsorption density increased up to 147.72 mg/g using EDB-M, then up to 179.51 mg/g using EDB-MH as adsorbent. The Langmuir model described the adsorption data satisfactorily (R^2 value average = 0.996), and the batch maximum adsorption capacities were found as 150,43 and 180,92 mg/g using EDB-M and EDB-MH, respectively at pH value of 6.2 ± 0.2. The maximum adsorption densities compared with previously reported adsorbents shown in Table 2.

3.6 Thermodynamic biosorption studies

Entalphy, Gibbs free energy and entropy changes are considerations into the thermodynamic biosorption studies, Its should be taken into account to determine which process will take place spontaneously.

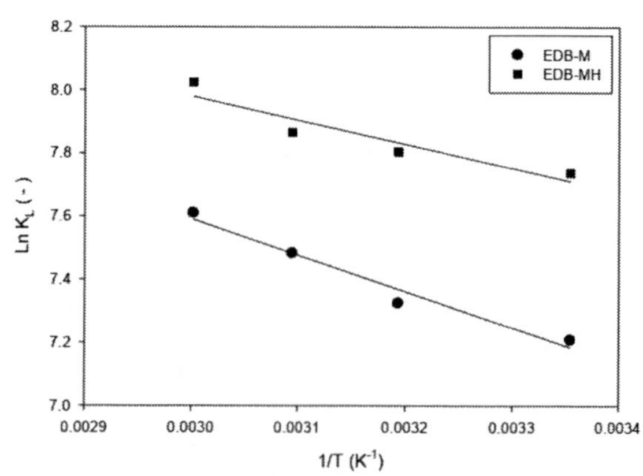

Fig. 7 Van't Hoff plots at different temperature, equilibrium adsorption time of 2 h, pH_e value of 6.2 ± 0.2, shaking rate of 150 rpm, and dose of biosorbent of 2.5 g/L for EDB-M and EDB-MH as biosorbents, respectively.

The values of thermodynamic parameters are subtantial indicator for practically of process applied, where the thermodynamic parameters should be calculated at various temperatures [2]. For this reason, the amounts of lead adsorbed capacity at 25, 40, 50, and 60°C were calculated to obtain the thermodynamic properties such as enthalpy change (ΔH°), Gibbs free energy change (ΔG°), and entropy change (ΔS°). The molar free energy change of the adsorption process is related to the equilibrium constant (K_L) and calculated from the following equation is shown below :

$$\Delta G^0 = -RT \ln K_L \quad (6)$$

where R is the gas constant in the ideal gas equation (8.314 J/mol.K), T is the absolute temperature, and the free energy change is calculated at 25◦C. K_L is Langmuir constant (L/mol)

Enthalpy change value, ΔH°, and entropy change value, ΔS°, of biosorption of lead can be predicted by using the Van't Hoff equation is shown below:

$$\mathrm{Ln}\, K_L = \frac{\Delta S^o}{R} - \frac{\Delta H^o}{RT} \quad (7)$$

A plot of ln K_L against 1/T renders a straight line, as shown in Fig. 6. The slope of the plot is equal to −ΔH°/R and its intercept value is equal to ΔS°/R. It was well known that these parameters can evaluate the orientation and feasibility of the physicochemical adsorptive reaction [2].

The thermodynamic parameters were calculated and tabulated in Table 2. The general explanation accounts for the thermodynamic study of lead adsorption onto EDB-M and EDB-MH were probably generally accompanied by release of energy, that is, most adsorption processes are endothermic in nature [**2**]. This adsorption process was favor under high temperature condition. The negative values of ΔH° (9.56 and 6.34 kJ/mol) confirmed the endothermic nature of biosorption of lead. The ΔH° value is quite high (9.56 and 6.34 kJ/mol), suggesting it was physical adsorption and chemical reaction mechanism, since the change of enthalpy value in physical adsorption process was usually of 5–40 kJ/mol [**16**]. Biosorption of lead was a spontaneous process, therefore its the change of free energy value is negative (ΔG < 0). The positive values of ΔS° (91.79.15 and 85.38 J/mol K) show the increased randomness at the solid/liquid interface during lead biosorption onto the biosorbent, which was accorded with fact that the mobility of adsorbates on adsorbent surface become more restricted as compared with that of those in the solution, and also reflect the affinity of the adsorbent material for lead removal.

Table 2. Langmuir adsorption isotherm model constants for the lead biosorption onto the EDB-M at various temperature

Isotherm Model	25°C	40°C	50°C	60°C
Langmuir				
q_{max} (mg/g)	150.43	160.79	165.72	170.75
K_L (L/mg)	0.0652	0.0928	0.1598	0.3228
R_L^2	0.996	0.997	0.996	0.988

Table 3. The maximum lead adsorption capacity and pH optimum of some bioadsorbent as reported in the literature.

Adsorbent	Max.adsorption capacity (mg/g)	pH Optimum	References
Rapeseed biomass	21.29 (20°C)	around 5	[2]
Phytolacca americana L.	10.83 (PAL) 12.66 (HPAL)	6	[17]
Flax fibres	10.741	6	[18]
Peanut shell	38.91	5.5±0.2	[15]
Cyclosorus interruptus	46.25	5.5	[19]
Magnetic Co-Fe NP	44.58 (25°C)	-	[20]
EDB-M	150.43 (25°C)	6.2±0.2	Present study
EDB-MH	180.92 (25°C)	6.2±0.2	

4. Conclusions

The study of biocomposite nanoparticles of *Eleocharis dulcis* (ED) as promising biosorbent for the treatment of lead (II) ion containing Sasirangan textile industry wastewater was investigated. Kinetic data of lead (II) biosorption onto EDB and EDB-MH revealed that equilibrium time was reached within 2 h, and the isotherm data showed that the Langmuir maximum adsorption capacity of the EDB-M and EDB-MH at pH_e of 6±0.2, room temperature were 150.43 mg/g and 180.92 mg/g, respectively. The thermodynamic of lead (II) biosorption onto the adsorbent implied the biosorption was spontaneous and endothermic indicating by increased in temperature would increased in adsorption capacity. Even the ΔH° value is quite high (9.56 and 6.34 kJ/mol), suggesting it was physical biosorption since the value of enthalpy change in physical adsorption process was usually of 5–40 kJ/mol, but chemical reaction mechanism probably included that indicating by favorable in high temperature.

ACKNOWLEDGEMENT

This research was supported by a grant of Directorate of Research and Community Service, The Ministry of Research, Technology and Higher Education of Indonesia, Project No: 070/SP2H/LT/DRPM/IV/2017, the fund for Applied Research University Grant.

REFERENCES:

1. M. D. Meitei and M. N. V., J. of Environ Chem. Eng. **1,** 3 (2013)
2. I. Morosanu, C. Teodosiu, C. Paduraru, D. Ibanescu and L. Tofan, New Biotechnol (2016).
3. V. V. Basava Rao and S. Ram Mohan Rao, Chem. Eng. J. **116**, 1 (2006)
4. L. Largitte and P. Lodewyckx, Microporous and Mesoporous Materials **202** (2015)
5. F. Fu and Q. Wang, J. of Environ Manag **92**, 3 (2011)
6. N. M. Noor, R. Othman, N. M. Mubarak and E. C. Abdullah, J. of the Taiwan Instit. of Chem. Engine **78** (2017).
7. O. Karnitz, L. V. A. Gurgel, J. C. P. de Melo, V. R. Botaro, T. M. S. Melo, R. P. de Freitas Gil and L. F. Gil, Biores Technol **98**, 6 (2007)
8. D. Mohan, H. Kumar, A. Sarswat, M. Alexandre-Franco and C. U. Pittman, Chem Engine J. **236** (2014)
9. D. H. K. Reddy and S.-M. Lee, Adv. in Colloid and Interface Sci. **201** (2013)
10. S. A. Saad, K. M. Isa and R. Bahari, , Desalination **264**, 1 (2010)
11. H. Tahir, M. Sultan, N. Akhtar, U. Hameed and T. Abid, J. of Saudi Chem. Society **20** (2016)
12. C. Irawan, I. F. Nata, M. D. Putra, M. Elma and K. F. Hanisa, Int. J. on Adv. Scie, Engine and Infor. Technol **7**, 4 (2017)
13. H. Yuh-Shan, Scientometrics **59**, 1 (2004)
14. Y. S. Ho and G. McKay, Water Resear **34**, 3 (2000)
15. K. Zhou, Z. Yang, Y. Liu and X. Kong, J. of Environ. Chem. Engine **3**, 3 (2015)
16. D. D. Do, *Adsorption analysis: Equilibria and kinetics*, Imperial College Press: London, 1998.
17. G. Wang, S. Zhang, P. Yao, Y. Chen, X. Xu, T. Li and G. Gong, Arabian J of Chem. (2015).
18. B. Abbar, A. Alem, S. Marcotte, A. Pantet, N.-D. Ahfir, L. Bizet and D. Duriatti, Process Safety and Environ. Protection **109** (2017)
19. Ş. Taşar, F. Kaya and A. Özer, J. of Environ Chem Engine **2**, 2 (2014)
20. S. P. Mun, Z. Cai and J. Zhang, , Materials Let, **96** (2013)

Hydrothermal Carbonization of Seaweed For Advanced Biochar Production

Tirto Prakoso[1]*, *Rima* Nurastuti[1], *Reza* Hendriansyah[1], *Jenny* Rizkiana[1], *Gede* Suantika[2], *Guoqing* Guan[3]

[1] Departement of Chemical Engineering Institut Teknologi Bandung 40132 Indonesia
[2] School of Life Sciences and Technology Institut Teknologi Bandung 40132 Indonesia
[2] North Japan Research Institute for Sustainable Energy, Hirosaki University, Aomori 030-0812 Japan

Abstract. Seaweed such as *Eucheuma Cottonii* is a potential source of biomateriallts high moisture content makes it suitable for hydrothermal conversion process since it doesn't need to utilize dry feedstock. The aim of this study is to convert the biomass of red seaweed *Eucheuma Cottonii* into alternative fuels and high value biomaterials using hydrothermal process. The hydrothermal process seaweed *Eucheuma Cottonii* produce two types of products, liquid product and char (solid). This research focus on the char product. The char from hydrothermal process was then activated using the tubular furnace. The yield for activated char is 7.5 % and results of SEM analysis of activated char showed the formation of allotropes carbon include carbon micro spheres, carbon micro fibres and graphene. These structures have encountered application in a wide range of technological fields, such as adsorption, catalysis, hydrogen storage or electronics.

1 Introduction

United Nation Convention on the Law of the Sea (UNCLOS) in 1982 reported that Indonesia marine teritority ocupied 5.8 million km^2 and it contains 27.2% of all species flora and fauna from all over the world. Seaweed or known as macroalgae is one of the most abundant biological resources in Indonesian, which is about 8.6% from the total biota in the sea [1]. The habitat of seaweed in Indonesia reaches 1.2 million hectares and it is considered as the largest seaweed habitat in the world [1].

Abundant aviablity of seaweed in Indonesia should be optimized by turning this resources into other high value product. Some of product that can be obtained from seaweed are biomaterial and biofuel. Seaweed conversion into biomaterial and biofuel by hydrothermal process has drawn much attention. Hydrothermal process take place in water at high temperature ranging at 250-347°C and pressure ranging at 4-22 MPA. The most advantages of hydrothermal process is it can use feedstock with high moisture content without the need of pre-dying [2]. Therefore, hydrothermal process is suitable to be used for seaweed processing since it has high moisture content.

Hydrothermal process produced three main products such as bio-gas, bio-oil, and bio-char. These products can be obtained by tuning the temperature of hydrothermal process. Bio-char as a main product from hydrothermal process obtained when temperature process tuned at 150-275°C. Meanwhile for bio-oil purposes hydrothermal temperature tuned above 275°C. Furthermore in this research is focused on development of seaweed into biomaterial and biofuel products via hydrothermal process.

2 Experimental

In this present research, hydrothermal carbonization was performed in hydrothermal reactor equipped with heater. Eucheuma Cottonii was chosen as biomass sample and then dried biomass were choped into 1 cm long.

Biomass precursor was mixed with water in mass ratios of 1:20, 1:10 and 1:5. Hydrothermal reactor was then heated to 250-275°C and held for 30-90 minutes. After cooling process, bio-char and bio-oil were separated by filtration and bio-char dried at 105°C for 24 hours.

The bio-char recovered from hydrothermal process was then treated by CO_2 activation method to activate the char. Bio-char was loaded onto crucible and positioned inside the tubular furnace. Temperature of furnace ramped to 800°C at a rate 10°C/min in the presence of N_2 at flow rate of 50 ml/min. After the final temperature was reached, N_2 then replaced by CO_2 at a flow rate of 40 ml/min and held for 2 hours.

The solid product was characterized by several methods. Ultimate and proximate analysis were

* Corresponding author: tirto@che.itb.ac.id

 https://doi.org/10.1051/matecconf/201815605012

conducted in Laboratorium Pengujian Batubara tekMIRA Bandung to determine the chemical and physical characteristics of biomass. FT-IR analysis were carried out in quantitative and qualitataive estimation of oxygen-containing functional groups. Surface area of bio-char and activated carbon were obtained by using a gas sorption Quantochrome Instruments Nova 3200e conducted in Laboratorium Instrumentasi Teknik Kimia ITB. The morphology of activated carbon were examined by Scanning Electron Microscopy (SEM) using SEM Hitachi SU3500 conducted in Pusat Penelitian Nanosains dan Nanoteknologi Institut Teknologi Bandung. As for the liquid products, it was evaporated in a vacum evaporator to separate the water from the bio-oil prior characterization using GC-MS.

3 Result and discussion

3.1 Seaweed characterizations

Table 1 shows the proximate and ultimate analysis of *Eucheuma cottonii*. The elemental composition of the seaweed obtained from ultimate analysis. It is observed that the seaweed had high oxygen and carbon content. The seaweed had a biger proportion of volatile matter than fixed carbon, moisture and ash. Moreover, the seaweed had a high ash content comparing to other biomass due to the alkali content from this seaweed was high [3]. Since it has high ash content, the thermochemical process such as direct combustion, pyrolysis, and gasification was not suitable for this seaweed. However this characteristics was more acceptable when hydrothermal process being used [4].

Table 1. Proximate and Ultimate analysis of Eucheuma Cottonii.

Proximate (%)	
M	11.2
VM	47.4
FC	27.82
Ash	13.58
Ultimate (%)	
C	21.11
H	4.26
N	0.35
O	57.73
S	2.97

3.2 Solid product (bio-char)

Table 2 shows the bio-char obtained after hydrothermal process. In General io-char obtained from hyrothermal process decreased when holding time of hydrothermal process was set longer. The longer holding time increases reaction severity so that bio-char obtained after hydrothermal process decreased [5]. Furthermore, the decrease of bio-char was attributed to devolatilization of volatile matter within the biomass precursor [6].
However, some samples showed the opposite. Longer residence time resulted in a higher bio-char produced. it was expected due to secondary char formation of polyaromatic char from non-dissolved lignin content. It can be seen that bio-char produced after hydrothermal process decreased while hydrothermal process temperature increased.

The decreased of bio-char produced was expected due to primary decomposition of biomass and secondary decomposition of char residue at higher temperature [7]. Some of samples observed produced more bio-char while hydrothermal process temperature increased. This phenomena was expected due to secondary char formation caused by extensive condensation/intermolecular dehydration and formation of polymers [8]. This phenomena was also found when the biomass to water ratio decreased, causing the bio-char produced increased. The increased of bio-char produced was expected due to polimerization becomes dominant so that the dissolved monomer of decomposed products formed solid phase products [8].

Table 2. Bio-char produced after hydrothermal process (g).

Biomass:water ratio	1 : 20		
Holding (min)	30	60	90
250 °C	8.6	5.72	6.7
275 °C	10.78	5.89	4.64
Biomass:water ratio	1 : 10		
Holding (min)	30	60	90
250 °C	13.35	13.49	13.09
275 °C	12.89	12.99	9.69
Biomass:water ratio	1 : 5		
Holding (min)	30	60	90
250 °C	24.55	24.645	25.145
275 °C	23.935	19.04	16

Figure 1. shows the comparison of FT-IR spectrun from raw seaweed, bio-char and bio-oil which obtained from hydrothermal process. Peak around ~3500 cm^{-1} represents O-H bonds from alcohol, fenol and carboxyl functional groups. The intensity of O-H bond in bio-char decreased after hydrothermal process compared to raw seaweed due to dehydration of the raw seaweed occured so that water molecules within the biomass decreased [9]. Meanwhile, the intensity of O-H bond in bio-oil sample greater than the others due to it has high amount of water content. The peak around ~2900 cm^{-1} attributed to aliphatic groups (CH_n). The intensity of this peak in bio-char decreased compared to raw seaweed after hydrothermal process, indicating some of aliphatic long

chain broke down [9]. Peak around ~1600 cm^{-1} represnted the C=C bonds was attributed to aromatic ring of lignin and aromatic groups formed during hydrothermal proces [10]. The decreased of intensity around ~1600 cm^{-1} indicated the removal of inorganic carbonate from raw materials [10]. The peak around ~1000 cm^{-1} represented the C-O bond, the intensity of this bond decreased after hydrothermal process due to breaking C-O linkages during hydrothermal process [10].

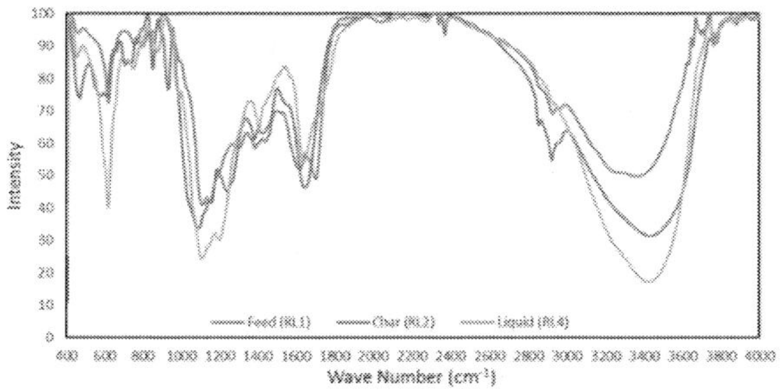

Fig. 1. FT-IR spectrum of raw seaweed and bio-char.

These bio-char could be further applied for various applications such as solid fuel, soil enrichment, catalysis, water purification, energy storage, and CO_2 sequesterian. As solid fuel and soil enrichment these bio-char could applied directly without further modification. Commonly for catalysis, water purification, energy storage and CO_2 sequesterian applications these bio-char further treated by thermal, chemical, or pyhsical activation to improved the characteristic of the carbonaceous materials, by turning these bio-char into activated carbon. The desired characteristic of activated carbon furthermore could be tuned by changing the parameter process of activation process [11].

3.3 Liquid product (bio-oil)

Bio-oil from hydrothermal process at 275°C and holding time 30 minutes was chosen for GC-MS analysis. Liquid product or bio-oil obtained after separation process was evaporated to reduce the water content of the bio-oil. The residue from evaporation were water and concentrated bio-oil. The concentrated bio-oil afterwards diluted on aceton for bio-oil aceton ratio was 1 to 100. This diluted bio-oil then injected to GC-MS to analyze the composition of bio-oil. The result from GC-MS analysis shown in Table 3.

It can be seen that most of the compounds indentified from bio-oil were alkanes, aromatic, alcohol and organic acid. Toluene was the compound of highest content, contributing 63,01% of the total area followed by benzene which contributing 14,53% of the total area. Only one organic acid, levulinilic acid and one alcohol, diacetone alcohol were detected. During the hydrothermal process lignocelullosic biomass first hydrolyzed and leads to form polyscharides, afterwards this polyscharides hydrolyzed into glucose and some might go further to form fructose. The decomposition of theses glucose and fructose leads to the formation of organic acids such as levulinic acid [8,12]. The aromatics compounds detected in bio-oil such as benzene and toluene was expected formed via intermolecular dehydration of aromatic compounds derived from lignin [8]. The present of lipid in algae contains saturated and unsaturated fatty acids, which is derived from triglyceride hydrolysis. It was expected that the present of small amount of alkanes in bio-oil were formed from decarboxylation reaction of fatty acid presented in algae [13,14].

Table 3. Chemical detected in bio-oil from GC-MS analysis.

Number	Retention Time	Area %	Name
1	2.18	14.53	Benzene
2	2.307	1.31	Cyclopentane
3	2.341	1.39	Heptane
4	2.541	6.28	Methylcyclohexane
5	2.821	3.18	2,5-Dimethylhexane
6	2.889	63.01	Toluene
7	2.981	1.82	cis-1,3 Dimethylcyclohexane
8	3.17	2.45	Cyclohexane
9	3.623	4.71	Diacetone alcohol
10	6.39	1.31	Levulinic acid

3.4 Post modification of bio-char

Bio-char from hydrothermal process at 250°C and holding time 30 minutes was chosen for activation process. Solid product or bio-char obtained after separation process was loaded into tubular furnace and heated at 800°C for two hours in CO_2 atsmosphere. Figure 2. Shows the comparison of SEM images from raw seaweed and activated carbon. SEM image of Fig 2. (a) shows the morphology of raw seaweed before sequentally treatment. It can be seen that morphology of raw seaweed has irregular shape like block and closed surface. Fig 2. (b) and (c) shows the differences of raw seaweed and activated carbon. After actvation process it observed that activated carbon has some morphological shapes. Fig 2. (b) and (c) shows that activated carbon composed of agglomerates of spherical particles like microspheres and long structure like fiber. This structure was probably generated as a consequence of the sequentially treatment that given to the raw seaweed and also affected from characteristics of raw seaweed.

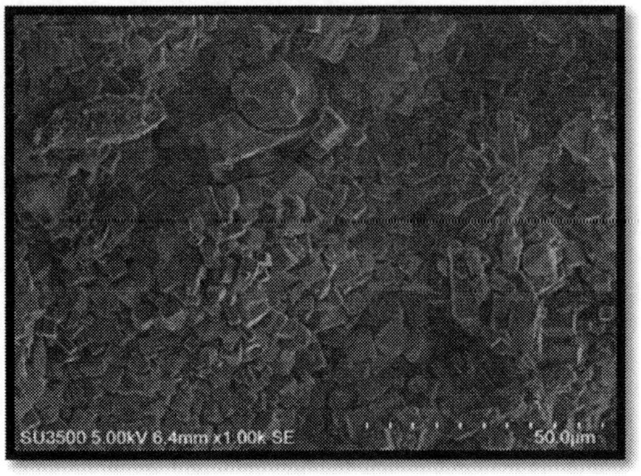

(a)

(b)

(c)

Fig. 1. SEM images of (a) raw seaweed (b) activated carbon (c) activated carbon.

Table 4 shows the comparison of surface area of activated carbon obtained from this research and commercial activated carbon. It can be seen that comercial activated carbon has greater surface area than synthesized activated carbon and also the gaps between these samples was very significant. Adding activating agent during hydrothermal process could enhance the porosity of the material. Sevilla et al. [16] added KOH as activating during hydrothermal process to produce high-surface area carbons up to 2700 m^2/gr from furfural, glucose, starch, cellulose and eucalyptus sawdust. Meanwhile Wang et al. [17] demonstrated the production of high-surface area carbons that can be reached up to 2610 m^2/gr using H_3PO_4 as activating agent during the hydrothermal process. Based on those results, the improving the porosity of the activated carbon could be done by adding activating agent during the hydrothermal process.

Table 4. Surface area of synthesized activated carbon and comercial activated carbon.

Sample	Surface Area (m^2/gr)
Synthesized Activated Carbon	12.35
Comercial Activated Carbon	832.967

From the result obtained, carbon microspheres and carbon nano fibers could be applied as materials for supercapacitors electrode. Carbon microshperes offered several advantages such as have regular morphology, adjustable porosity, and particle size causing the resistance of ion diffusion decrease and improved the electrochemical performance while applied as materials for supercapacitor electrode [18] (Zhu et al., 2015). Meanwhile carbon nano fiber gives several advantages such as tunable surface functionalities, dimensional stability, and transport property [19] (Jang et al., 2005).

Conclusion

Utilization of seaweed by hydrothermal process has a great potential to be an alternative process for production of wide variety products. Furthermore, the product obtained from hydrothermal process such as bio-char and bio-oil could be applied for various applications. However the further investigation is still needed for product and process development so that the characteristic of product can be specialized to specific applications. From the result achieved of this research, the bio-char obtained should be more developed by adding chemical on hydrothermal process to enhanced the porosity of activated carbon and the activated carbon produced could be applied as material for supercapacitor electrode.

Acknowledgment

This work is supported by Program Insentif Riset Sistem Inovasi Nasional **(Insinas)** 2016.

References

1. Suparmi, Sahri, Achmad., Sultan Agung Vol XIV **96**, 118 (2009)
2. K. Tekin, S.Karagöz, & S. Bektaş, , Renew. Sustainable Energy Rev., **40**, 673-687 (2014)
3. Smith, M. Aidan, Ross, B. Andrew,Algal Res, **16**, 1-11 (2016)
4. Q. Xu, Q.Qian, A. Quek, N. Ai, G. Zeng, , J. Wang, ACS Sustainable Chem. Eng **1**, 1092-1101 (2013)
5. A. Funke, & F. Ziegler, Biofpr **4**, 160-177 (2010)
6. He, C., Giannis, A., & Wang, J. Y., Appl. Energy **111**, 257-266 (2013)
7. N. M. Nor, L. C. Lau, K. T. Lee, & A. R. Mohamed, , JECE **1**, 658-666 (2013).
8. A. Jain, R. Balasubramanian, & M. P. Srinivasan, Chem. Eng. J. **283**, 789-805 (2016)
9. A. T. Yuliansyah, T.Hirajima, Waste and Biomass. Valori. **1**, 395-405 (2010)
10. S. Nizamuddin, J. Kumar, , Subramanian, N., Sahu, J. N., Ganesan, P., N. M. Mubarak, & S. A. Mazari,, Can. J. Chem. Eng. **93**, 1916-1921 (2015)
11. M. M. Titirici, R. J. White, C. Falco, & M. Sevilla, Energ. Environ. Sci. **5**, 6796-6822 (2012)
12. M. T. Reza, J. Andert, B.Wirth, D. Busch, J. Pielert, , J. G. Lynam, & Mumme, J., Appl. Bioenerg. , **1** (2014)
13. Y. Guo, T. Yeh, W. Song, D. Xu, & S. Wang, , Renew. Sustainable Energy Rev. **48**, 776-790 (2015)
14. L.Hadhoum, M. Balistrou, G.Burnens, K.Loubar, & M. Tazerout, , Bioresour. Technol. **218**, 9-17 (2016)
15. P. Matanjun, S. Mohamed, N.M. Mustapha,, Muhammad, K., J of Appl Phycol **21**, 75-80 (2009)
16. M. Sevilla, A. B.Fuertes, & R. Mokaya, Energ. Environ. Sci. **4**, 1400-1410 (2011)
17. L.Wang, Y Guo, B. Zou, C.Rong, X.Ma, Y.Qu, & Z. Wang, Bioresour. Technol, **102**, 1947-1950 (2011)
18. D. Zhu, Y.Wang, L.Gan, Liu, M., K Cheng, Y. Zhao, & D. Sun, Electrochim. Acta, **158**, 166-174 (2015)
19. J. Jang, J. Bae, M. Choi, & S. H. Yoon, Carbon, **43**, 2730-2736 (2005)

Mass Balances and Thermodynamics Study of Thermal Triglyceride Hydrolysis

Astri Nur Istyami[*], *Tatang Hernas* Soerawidjaja, and *Tirto* Prakoso

Department of Chemical Engineering, Institut Teknologi Bandung, Indonesia

Abstract. Triglyceride hydrolysis is a process to convert triglyceride into fatty acids and glycerol, which are important precursors in oleochemical industry. Commercial technology of triglyceride hydrolysis established at the present time is thermal hydrolysis (a.k.a. Colgate-Emery process), which operates in robust condition (250°C, 50 bar). Although this technology has been the most preferable process to produce fatty acid for a century, but information published about its process parameters is limited. In this study, an analysis of mass balances and thermodynamic aspects of thermal hydrolysis of triglyceride was performed based on literature review. Composition of input and output process stream shows some indication of polymerization and/or hydrogenation reaction of linoleic and linolenic acid, and also geometric isomerization of oleic acid. Thermodynamic feasibility of reaction was compared between high temperature and room temperature. In the near future, more energy-efficient and less-side reaction technology to produce fatty acids seems to compete with this conventional process.

1 Introduction

Oleochemical products are important commodities in plant-oil producing countries, such as Indonesia, Malaysia, Canada, China, etc. Plant-oil processing mainly begins with conversion of triglyceride into fatty acids and fatty alcohols. At the present time, the common process to convert triglyceride into fatty acids in industries is thermal hydrolysis. This technology has been used for almost a century, but the details of hydrolysis process parameters are rarely found in scientific publications. This paper describes several aspects of thermal triglyceride hydrolysis that might be useful for development of conventional or novel production technology of fatty acids.

2 Triglyceride Hydrolysis Principle and Development

Reaction of triglyceride hydrolysis is described as

$$Triglyceride + 3\,H_2O \rightleftharpoons 3\,Fatty\,Acids + Glycerol \quad (1)$$

Mechanism of this reaction is similar to other ester hydrolysis, where hydroxide ion from H_2O bonds with carbon atom from electrophile carboxylic group, conforming tetrahedral intermediate. The glycerol group then separate with the ester, results in free fatty acids.

Development of triglyceride hydrolysis technology was described in reference [1]. It was stated that in 1813, fat was discovered as a substance consist of glycerol and one, two, or three fatty acids. It was also known that reaction with water will separate fats into its components. In 1853, Tilghman claimed that the fat-water reaction into fatty acids and glycerol can be performed with assistance of heat and pressure. During that time, robust technology was limited by material construction and process technology. In 1898, Twitchell offered a simpler method of triglyceride hydrolysis involving organosulphate acid as catalyst [2]. This method was widely used for a long period because it can be performed in atmospheric pressure and temperature of 100°C, but the waste processing remained a problem. As an alternative, high temperature, non-catalytic process of triglyceride hydrolysis was initiated in 1939 [3], developed in 1949 [4] and utilized by Colgate-Palmolive-Peet and Emery Industries (thus known as Colgate-Emery process). It requires higher temperature and pressure (250°C, 50 bar) than Twitchell process, but it was more economical due to high yields and minimum waste processing. The thermal hydrolysis remains the commercial process used for fatty acid production until the present time [1]. Fig 1 shows process flow diagram for thermal triglyceride hydrolysis.

[*] Corresponding author: anistyami@che.itb.ac.id

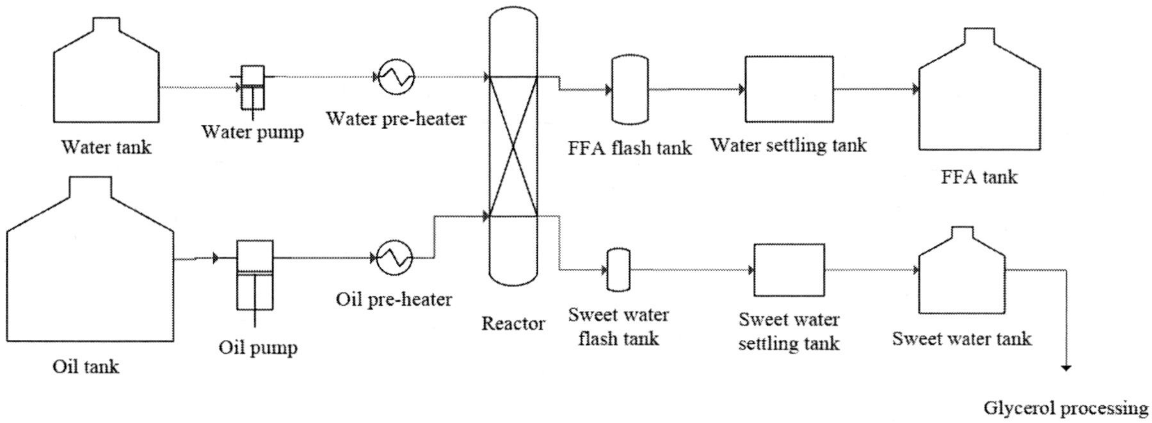

Fig. 1. Process flow diagram of thermal triglyceride hydrolysis

3 Mass Balances Aspect of Thermal Triglyceride Hydrolysis

Thermal triglyceride hydrolysis takes place in column where oil and water are contacted in countercurrent mode. Reactor feed and product scheme is shown in Fig 2. The main literature reviewed in this section [5] reported thermal hydrolysis with oil/water ratio 1:2, 1:3, and 1:4, with fatty acid yield respectively 92%, 93%, and 89%. It was not clearly stated if the value includes soluble fatty acids in sweet water and if the process involves fatty acid separation from sweet water. In fact, fatty acid was fairly soluble in water, especially in high temperature (above 200°C). In this study, mass balances was calculated based on several assumption:

- Triglyceride mass flow (M), fatty acid mass flow (F), and fatty acid product composition was taken from reference [5]
- Oil/water ratio is 1:4
- Oil feed composition was determined based on typical composition of each oil type [6]
- Amount of soluble fatty acids in sweet water was determined based on stearic acid solubility in water (250°C)
- Acid value of oil feed is 1 mg KOH/gr oil
- Oil feed contains no water
- Water feed contains no glycerine

Mass balances is shown in Table 1. Calculation of soluble fatty acid products in sweet water results in higher yield than reported by reference [1]. Table 2 shows comparison of product yield from reference [1] and this study, in terms of product yield compared to oil feed (based on triglyceride mass) and theoretical product yield (based on theoretical fatty acid product mass). Fatty acid composition in oil feed and products is shown in Table 3.

Hydrolysis of canola, camelina, and algal oil shows several tendencies. First, linoleic and linolenic acid content decreases along with hydrolysis reaction. This is similar to linseed oil (with more than 50% linolenic acid content), which was reported to have a lower iodine value after thermal hydrolysis [1]. It was also reported that iodine value of fish oil (which typically contain high polyunsaturated fatty acid content) decreases 20-40 points after thermal hydrolysis. Other reference reported that declining of iodine value of menhaden oil from 168 to 142 gr I_2/100 gr oil was simultaneous with distillable product mass, which concluded with the occurrence of double- and triple-bond polymerization [7]. It is a disadvantage in industries where products are desired to have high content of double and triple bond fatty acids, such as linoleic and linolenic acid, which is valuable in food industries. Several negative effects of thermal triglyceride hydrolysis have been reported before [8], such as thermal degradation, isomerization, dehydration, polymerization, and colour darkening.

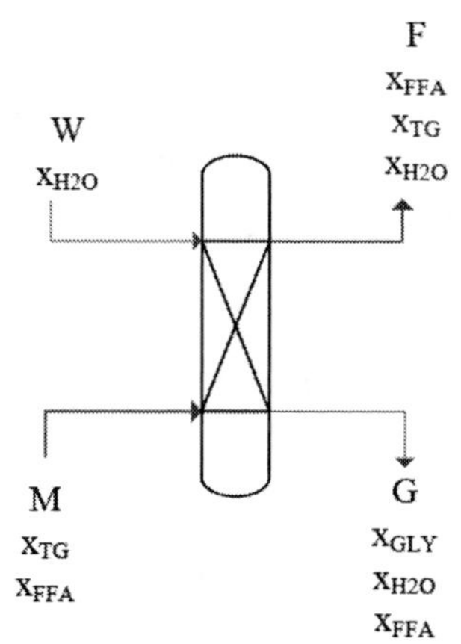

Fig 2. Mass balances of thermal triglyceride hydrolysis

Table 1. Mass balances of thermal triglyceride hydrolysis

	Flow			
	M	W	F	G
Canola oil				
Mass (gr)	2459.4	9837.6	2385.0	9912.0
Mass fraction				
x_{TG}	99.49%	0.00%	1.07%	0.00%
x_{H2O}	0.00%	100.00%	2.49%	97.18%
x_{FFA}	0.51%	0.00%	96.44%	0.28%
x_{GLY}	0.00%	0.00%	0.00%	2.50%
Total	100%	100%	100%	100%
Camelina oil				
Mass (gr)	2260.7	9042.8	2192.7	9110.8
Mass fraction				
x_{TG}	99.49%	0.00%	1.92%	0.00%
x_{H2O}	0.00%	100.00%	0.77%	97.60%
x_{FFA}	0.51%	0.00%	97.31%	0.28%
x_{GLY}	0.00%	0.00%	0.00%	2.50%
Total	100%	100%	100%	100%
Algal oil				
Mass (gr)	904.2	3616.8	878.1	3642.9
Mass fraction				
x_{TG}	99.49%	0.00%	2.34%	0.00%
x_{H2O}	0.00%	100.00%	4.39%	96.81%
x_{FFA}	0.51%	0.00%	93.27%	0.28%
x_{GLY}	0.00%	0.00%	0.00%	2.42%
Total	100%	100%	100%	100%

Table 2. Comparison of yield calculated in reference [1] and this study

	Oil		
	Canola	**Camelina**	**Algal**
Product yield to oil feed			
Reference [1]	92%	93%	89%
This study	94%	94%	91%
Theoretical product yield			
Reference [1]	96%	96%	90%
This study	98%	99%	93%

Table 3. Fatty acid composition in hydrolysis feed and product with oil/water ratio 1:4

	Mass (gram)					
	Canola oil		Camelina oil		Algal oil	
	Before	After	Before	After	Before	After
C 14:0	1.5	0.8		1.2		0.2
C 16:0	92.2	71.6	119.8	105.2	60.6	31.7
C 16:1^c	5.2	38.9		63.6		26.2
C 17:0	1.0	2.3		0.0		0.9
C 18:0	46.0	35.8	67.8	48.4	22.6	13.8
C 18:1^c	1534.9	1571.7	422.8	534.1	429.5	453.6
C 18:1^t		34.3		33.6		14.5
C 18:2^c	494.8	253.8	361.7	329.0	325.5	142.2
C 18:3^c	205.9	83.9	861.3	502.3	66.0	31.2
C 20:0	15.7	144.8	31.6	166.4		86.5
C 20:1^c	37.9	19.1	262.2	259.2		5.5
C 20:2^c	2.7	0.0		0.0		0.0
C 22:0	8.6	7.6		17.8		0.7
C 22:1^c	0.0	0.0	56.5	34.2		0.0
C 20:5^c		3.0		10.1		0.7
C 24:0	6.6	3.0		1.8		0.7
C 24:1^c	6.4	1.5		0.9		0.5
Others			76.9			
Total	2459.4	2272.2	2260.7	2107.8	904.2	808.7

^{c} = cis, ^{t} = trans

The second tendency is the forming of trans oleic acid, which was possibly produced by isomerization of cis oleic acid. This is also an undesirable product in food industries. Trans fatty acids are rarely found in nature, but the forming activation energy is low (125 kJ/mol), makes it easy to be found in elevated temperature of oils, significantly around 230°C, and exponentially above 240°C [9, 10].

Based on relative isomerization rate of linolenic, linoleic, and oleic acid (100:10:6), geometric isomerization of linolenic and linoleic acid occur faster than oleic acid. Nevertheless, trans linolenic and linolenic acid were not found in mass balances product result in this study. It was concluded that polymerization or hydrogenation dominates the reaction, causing absence of isomerization products. Moreover, oleic acid content is much higher than the other two, makes the isomerization product of oleic acid shown significantly.

Beside polymerization, hydrogenation and isomerization, there is also tendency of increasing content of heptadecanoic acid, which is only found as trace in ruminant fats. It might be relevant to declining content of stearic acid, which decarboxylates in high temperature. This phenomenon is relatively insignificant compared to the others, and might occur due to different reference between reactant and product composition.

4 Thermodynamics Aspect of Thermal Triglyceride Hydrolysis

In this study, reaction enthalpy and Gibbs free energy for thermal hydrolysis process is evaluated in condition 1 (25°C, 1 bar) and condition 2 (250°C, 50 bar). Gibbs free energy and reaction enthalpy relation is described as

$$\Delta G = \Delta H - T\Delta S \quad (2)$$

Actual enthalpy and entropy is described as

$$H = H^{ig} + H^{d} \quad (3)$$

$$S = S^{ig} + S^{d} \quad (4)$$

Where H^{ig} is ideal gas enthalpy, S^{ig} is ideal gas entropy, H^{d} is departure or residual enthalpy and S^{d} is departure entropy. Reaction enthalpy in condition 1 is defined as

$$\Delta H_1 = H_{product} - H_{reactant} \quad (5)$$

$$\Delta H_1 = H^{ig}{}_{product,1} + H^{d}{}_{product,1} - H^{ig}{}_{reactant,1} - H^{d}{}_{reactant,1}$$

$$\Delta H_1 = \Delta H^{ig}{}_1 + \Delta H^{d}{}_1 \quad (6)$$

Similar to equation (3), reaction enthalpy for condition 2 is stated as

$$\Delta H_2 = \Delta H^{ig}{}_2 + \Delta H^{d}{}_2 \quad (7)$$

$$\Delta H_2 = \Delta H^{ig}{}_1 + {}_{T1}\int^{T2}\Delta Cp.dT + \Delta H^{d}{}_2 \quad (8)$$

Reaction enthalpy of ideal gas in condition 1 can be estimated by bond dissociation energy (Ei) of each product and reactant.

$$\Delta H^{ig}{}_1 = \sum E_{i,reactant} - \sum E_{i,product} \quad (9)$$

H^{d} and S^{d} was determined with Peng-Robinson's method [11]. Accentric factor, critical temperature, critical pressure, and heat capacity of involved molecules are based on reference [12].

Standard entropy for various triglyceride is unavailable in references, thus triglyceride is simplified as triolein. Calculated reaction enthalpy, entropy, and Gibbs free energy for each condition is shown in Table 4.

Table 4. Enthalpy, entropy, and Gibbs free energy of thermal triglyceride hydrolysis reaction

	25°C, 1 bar	250°C, 50 bar
ΔH	129,99 kJ/(mol.K)	-320,61 kJ/(mol.K)
ΔS	-0,11 kJ/(mol.K)	-0,52 kJ/(mol.K)
ΔG	161,84 kJ/(mol.K)	-47,42 kJ/(mol.K)

In room temperature, it was confirmed that thermal hydrolysis reaction is thermodynamically unfeasible. Feasibility is gained at 250°C, where the Gibbs free energy is slightly more negative than the minimum value of common reaction in industries (-10 kkal/mol.K). The necessity of high temperature (also high pressure to keep the water in liquid phase) is confirmed. This causes requirement of high amount of energy and expensive materials, unless catalyst is used.

5 Triglyceride Hydrolysis Technology in the Future

The shortcomings in conventional triglyceride hydrolysis has encourage engineers to develop alternative technology. Hydrolysis in moderate temperature and reaction time with acidic ionic liquids as catalyst was reported to yield more than 95% [13], and subcritical hydrolysis was reported to result in 97% conversion in 15-20 minutes [14].

Another potential technology in the near future is utilization of lipase. The enzyme enables triglyceride hydrolysis in room temperature, preventing side reactions such as polymerization, thermal degradation. Microbial and animal lipase activity has been reported for a long time and enhanced in several technique [15, 16]. Plant lipase, though encountering few challenges, is also potential source might be utilized in the future [17,18]. Though the technology is under development to produce high yield, it might compete the conventional triglyceride hydrolysis in the future.

References

1. H.L. Barnebey and A.C. Brown, J. Am. Oil Chem. Soc., **25**(3), 95-99 (1948)
2. E. Twitchell, *US601603* (1898)
3. V. Mills, *US2156863* (1939)
4. H. Martin, *US2458170* (1949)
5. W.C. Wang, T.L. Turner, L.F. Stikeleather, W.L. Roberts, Chem. Eng. Process., **57-58**, 51-58 (2012)
6. R.C. Zambiazi, R. Przybylski, M.W. Zambiazi, C.B. Mendonça, B. CEPPA, Curitiba, **25**(1), 111-120 (2007)
7. H.D. Allen, Chem. Eng. Prog., **43**, 8 (1947)
8. D.J. Anneken, S. Both, R. Christoph, G. Fieg, U. Steinberner, A. Westfechtel, *Ullmann's Encyclopedia of Industrial Chemistry*, **14**, 73-116 (2012)
9. M. Kellens, *Current Developments in Oil Refining Technology* (Technical Report De Smet-Belgium, Antwerp, Belgium), 35-48 (1997)
10. R.L. Wolff, J. Am. Oil Chem. Soc., **70**(4), 425-430 (1993)
11. D.Y. Peng and D.B. Robinson, Ind. Eng. Chem. Fundam., **15**(1), 59-64 (1976)
12. W.C. Wang, R.H. Natelson, L.F. Stikeleather, W.L. Roberts, Comput. Chem. Eng., **58**, 144-155, (2013)
13. H. Luo, K. Xue, W. Fan, C. Li, G. Nan, Z. Li, Ind. Eng. Chem. Res., **53**, 11653-11658 (2014)
14. R.L. Holliday, J.W. King, G.R. List, Ind. Eng. Chem. Res., **36**, 932-935 (1997)
15. P. Adlercreutz, Che. Soc. Rev., **42**, 6406-6436 (2013)

16. F.X. Malcata, H.R. Reyes, H.S. Garcia, C.G. Hill, Jr., C.H. Amundson, J. Am. Oil Chem. Soc., **67**(12), 890-910 (1990)
17. M. Barros, L.F. Fleuri, G.A. Macedo, Braz. J. Chem. Eng., **27**(1) 15-29 (2010)
18. S. Seth, D. Chakravorty, V.K. Dubey, S. Patra, Protein Expr. Purif., **95**, 13-21 (2014)

Synthesis of Ceria Zirconia Oxides using Solvothermal Treatment

Siti Machmudah[1,*], Widiyastuti[1], *Sugeng* Winardi[1], Wahyudiono[2], *Hideki* Kanda[2], and *Motonobu* Goto[2]

[1]Department of Chemical Engineering, Institut Teknologi Sepuluh Nopember, Surabaya 60111, Indonesia
[2]Department of Materials Process Engineering, Nagoya University, Furo-cho, Chikusa-ku, Nagoya 464-8603, Japan

Abstract. Ceria oxide (CeO_2) is widely used as catalyst with high oxygen storage capacity at low temperature. The addition of zirconia oxide (ZrO_2) to CeO_2 can enhance oxygen storage capacity as well as thermal stability. In this work, ceria zirconia oxides has been synthesized via a low temperature solvothermal treatment in order to produce ceria zirconia oxides composite with high oxygen storage capacity as electrolyte of solid oxide fuel cells (SOFC). Under solvothermal conditions, solvent may control the direction of crystal growth, morphology, particle size and size distribution, because of the controllability of thermodynamics and transport properties by pressure and temperature. Water, mixed of water and ethanol (70/30 vol/vol), and mixed of water and ethylene glycol (70/30 vol/vol) were used as solvent, while $Ce(NO_3)_3$ and $ZrO(NO_3)_2$ with 0.06 M concentration were used as precursor. The experiments were conducted at temperature of 150 °C and pressure for 2 h in a Teflon-lined autoclave of 100 mL volume. The synthesized products were dried at 60 °C for 6 and 12 h and then calcined at 900 °C for 6 h. The particle products were characterized using SEM, XRD, TG/DTA, and Potentiostat. The results showed that the morphology of particles formed were affected by the solvent. Solid plate shaped particles were produced in water, and tend to be pore with the addition of ethylene glycol. The addition of ethanol decreased the size of particles with sphere shaped. The XRD pattern indicated that ceria-zirconia oxides particles are uniformly distributed in the structure to form a homogeneous solid solution. Based on the electrochemical analysis, ceria zirconia oxides produced via solvothermal synthesis had high conductivity ion of 0.5594 S/cm, which is higher than minimum conductivity ion requirement of 0.01 S/cm for SOFC electrolyte. It indicated that ceria zirconia oxides produced via solvothermal synthesis is suitable for SOFC electrolyte.

1 Introduction

The central part of a Solid Oxide Fuel Cell (SOFC) is electrolyte, which is sandwiched between cathode and anode. The oxygen ions (O^{2-}) reduced on cathode or proton (H^+) oxidized on anode sides is transported through the electrolyte materials and reacts with hydrogen or oxygen, respectively, to form water on either side of the cell. The electrons (e^-) are generated due to the reaction pass in the opposite direction through the outer circuit [1]. There are two types of materials: oxide ion and proton conducting electrolytes. Oxide ion conducting electrolytes cover a wide range of materials which basically include fluorite-, perovskite-, brownmillerite-structured materials. Fluorite-structured electrolyte is adopted by ceria at room temperature and zirconia at high temperature. Basically zirconia and ceria based materials are widely investigated for its application as electrolyte in solid oxide fuel cell (SOFC) [2].

To increase performance of SOFC, electrolyte should be able to store and transport oxygen in large quantities to be reacted with hydrogen. In this case, oxygen storage/release capacity (OSC) is an important feature for electrolyte fabrication. Ceria has high OSC based on the reversible redox reaction ($CeO_2 \leftrightarrow CeO_{2-x} + x/2\ O_2$; $x = 0 - 0.5$) [3]. However, the OSC performance and durability of pure CeO_2 are still inadequate for its use at high temperature. Furthermore, it is necessary to mix it with ceramic material that has stability at high temperature. Zirconia oxide (ZrO_2) is a polymorphic oxide that is stable at high temperature [4] and exists in three different crystal structures: (i) cubic from 2367 °C up to its melting point of 2677°C, (ii) tetragonal from 1167°C up to 2367°C and (iii) monoclinic below 1167 °C [5].

Nagai et al. [3] has prepared CeO_2-ZrO_2 mixed oxides by precipitation method. This method is simple, but the morphology and size of particles are difficult to be controlled. Synthesis of 8 mol% yttria-stabilized zirconia doped with 10 mol% ceria has also been

* Corresponding author: machmudah@chem-eng.its.ac.id

conducted by Bekale et al. [6] via wet chemical process. This process is expensive, takes long time and needs various chemicals to be used. Solvothermal method is one of the simple methods for synthesis of ceria zirconia oxides particles. Similar to the hydrothermal method, in solvothermal method particles are formed in an organic solvent or water at temperature above boiling point of solvent and pressure above vapour pressure of solvent. Solvothermal process has many advantages in addition to being a green process: (1) it is energy efficient because of low temperature synthesis, (2) it is an environmentally friendly process because of closed system conditions, (3) unused components can be recycled, (4) precipitants are not always needed, (5) high-purity products can be synthesized, (6) metastable and new phases can be accessed, and (7) crystal size, morphology, composition, and polymorphism of the synthesized phases are easily controlled [7, 8].

This work is focused on the synthesis of ceria zirconia oxides using solvothermal method at various solvent compositions. The solvent used was water, ethanol, and ethylene glycol with various compositions. The morphology, oxygen storage capacity, and electrical property of particles were also investigated.

2 Experimental Method

2.1. Synthesis of ceria zirconia oxides particles

Ceria zirconia oxides particles were prepared by a solvothermal process in a batch reactor. All chemicals were of analytical grade and used without further purification. $Ce(NO_3)_3.6H_2O$ (Sigma Aldrich, Germany) and $ZrO(NO_3)_2$ (Wako, Japan) were used as precursor. Water, mixed of water and ethanol (70/30 and 50/50 vol/vol), and mixed of water and ethylene glycol (70/30 vol/vol) were used as solvent. Firstly, solution of 0.06 M cerium nitrate and zirconium oxynitrate were prepared separately in the solvent, and mixed with ratio of 1:1 (vol/vol). Then, the certain volume of solution was moved into a Teflon-lined autoclave of 100 mL volume. The autoclave reactor was placed into an electric furnace (Linn High Therm GmbH, model VMK 1600, Germany) and quickly heated to 150°C. The temperature in the reactor was measured by a thermocouple (K-type) inserted in the reactor cover. The time required to heat the reactor from room temperature to 150°C was around 10 min and thereafter the reactor temperature was constant. The difference between the furnace temperature and the reactor temperature was approximately 7°C. The pressure was autogenously generated when the reactor was sealed and heated. After 2 h (including the heating time of about 10 min), the reactor was removed from the electric furnace and quickly quenched in a water bath at room temperature. After cooling, the reactor was opened and the resultant precipitate was filtered and washed repeatedly with deionized water. The obtained particles were dried at 60°C for 6 and 12 h and then calcined at 900°C for 6 h.

2.2 Characterization of particles

The morphologies of the calcined particles were observed using SEM (JEOL JSM-6390LV). It is well known that SEM has the capability to detect particle diameters and morphologies, but the resolution is lost at extreme magnifications. In addition, SEM requires the samples to be electrically conductive, therefore, for most of the particles produced, a gold or platinum coating must be applied that may alter the diameter readings at higher magnifications. XRD patterns were determined to confirm that the crystal structure of ceria-zirconia oxides was formed via this synthesis method. A Rigaku RINT 2100/PC XRD machine (40 kV and 200 mA) equipped with a h–h wide-angle goniometer and scintillation detector was used for all XRD measurements, using Cu Ka radiation (k = 1.5406 A°). The composition of ceria zirconia oxides particles was analyzed using Energy Dispersive X-ray spectroscopy (EDX). The oxygen storage capacity (OSC) was determined by thermo-gravimetric analysis [3]. Initially, ceria zirconia oxides samples loaded on the platinum plate were fed with N_2 and heated from room temperature to 600°C at temperature rate 50°C/min to release free oxygen in the sample, then the temperature was reduced. The samples were oxydized with O_2 and and heated from room temperature to 600°C at temperature rate 50°C/min to determine oxygen storage and stability of samples at high temperature. The weight loss and gain were measure using a thermo-gravimetric analysis (TGA-50, Shimadzu, Japan). The value of the weight change between the releasing and oxydizing corresponds to total OSC. To determine electrical performance of particles product, the impedance spectra was measured by Potentiostat VersaStat 4.

3 Result and Discussion

Morphology of ceria zirconia oxides particles produced at various solvents at drying times is shown in Figure 1. Ceria zirconia oxides was produced using water, mixed water/ethanol (70/30 and 50/50 (v/v)), and mixed water/ethylene glycol (70/30 (v/v)) as solvent at 150°C for 2 h with drying times of 6 and 12 h. In general, particles were composed of small sphere particles agglomerated into plate like shape. Particles produced using water as solvent were agglomerated into solid plate like shape with smooth surface and big size. By addition ethanol into the solvent, the agglomerated particles size became smaller and tend to be spherical shape. Moreover, by addition ethylene glycol into the solvent, the spherical particles formed agglomerated particles with large amount of pore. These results indicated that morphology particles were highly influenced by the type of solvent. It can be explained that every solvent has a specific dielectric constant that influences reaction occurred during the solvothermal synthesis. Moreover, the addition of chemicals into the water in high temperature particles synthesis could control the size and shape of particles formed [9].

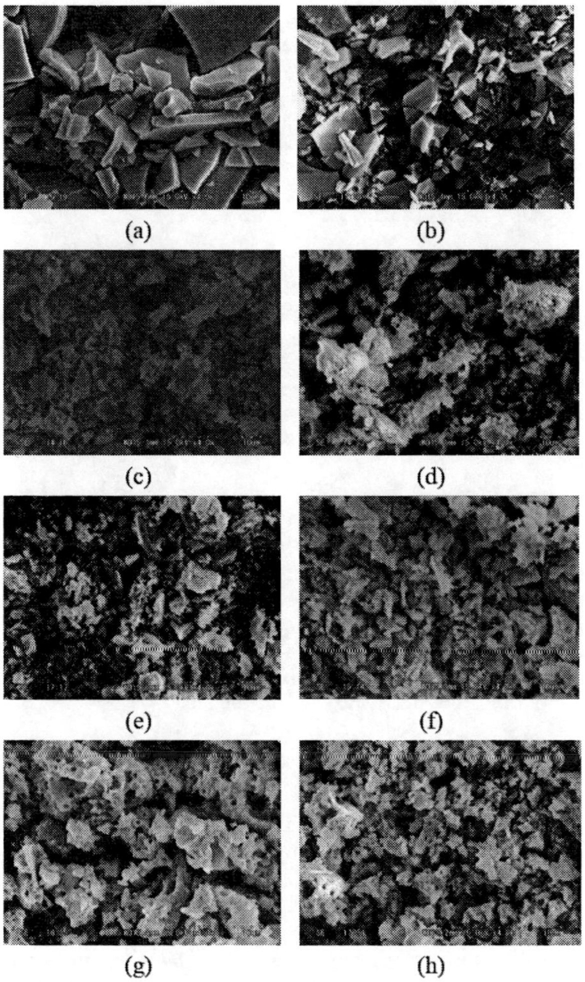

Fig. 1. SEM images of ceria zirconia oxides particles produced at various solvents and drying times. (a) water, 6 h drying time; (b) water, 12 h drying time; (c) 70/30 (v/v) water/ethanol, 6 h drying time; (d) 70/30 (v/v) water/ethanol, 12 h drying time; (e) 50/50 (v/v) water/ethanol, 6 h drying time; (f) 50/50 (v/v) water/ethanol, 12 h drying time; (g) 70/30 (v/v) water/ethylene glycol, 6 h drying time; (h) 70/30 (v/v) water/ethylene glycol, 12 h drying time.

The drying time also influenced the morphology of particles. For water and mixed water/ethylene glycol (70/30 (v/v)) as solvent, the longer the drying time, the smaller particles were obtained. In contrast, by addition ethanol into the water, the bigger particles were formed for longer drying time. In water and mixed water/ethylene glycol, the drying process reduced the agglomeration effect on the particles due to slower evaporation. However, for mixed water/ethanol, the property of solvent became more volatile than water itself, and it resulted in fast evaporation of the particles, and agglomeration was still occurred. Vehring et al. [10] explained that longer drying proses resulted in slower evaporation and bigger particle size. Longer drying time, it proposed the agglomeration of particles [11].

Scanning result of EDX analysis for ceria zirconia oxides particles produced with mixed water/ethylene glycol (70/30 (v/v)) and 12 h drying time is depicted in Figure 2. It was confirmed that produced particles were composed of cerium, zirconium, and oxygen atom. This result also confirmed that composite of ceria zirconia oxides particles were produced. The result was in agreement with the previous work [11] that ceria-zirconia oxide composite could be produced in continuous hydrothermal treatment.

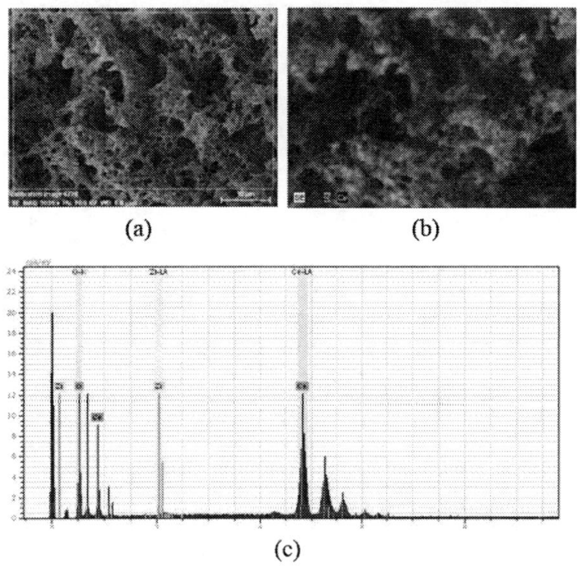

Fig. 2. EDX analysis of ceria zirconia oxides particles produced with 70/30 (v/v) water/ethylene glycol solvent and 12 h drying time.

XRD is a powerful method for the study of nanomaterials, especially to characterize crystallographic structure for bulk, nano and thin film materials. Figure 3 shows XRD pattern of ceria zirconia oxides particles produced with various solvents at 6 h drying time. All particles produced with all types of solvents showed in crystal form. The addition of organic solvent in the water solvent induced the crystal particles formation. The diffraction lines at 2θ = 28.25°, 34.20°, 49.30°, 57.52° and 59.07° correspond to the composite ceria-zirconia oxides. These figures also indicated that the diffraction patterns of ceria-zirconia oxides particles are uniformly distributed in the structure to form a homogeneous solid solution. Under hydrothermal and solvothermal conditions, an increase in temperature will increase both the nucleation rate and the linear growth rate, hence, the crystallinity of the samples normally increases in time. With the same reaction time, at higher temperature leads to the crystallization of denser products [12]. At the same temperature, the addition of chemicals including organic solvent in the hydrothermal synthesis increased the crystallinity of the particles. Therefore, as shown in Figure 3, the higher volume of ethanol resulted in higher peak intensities while pure water resulted in lower peak intensities.

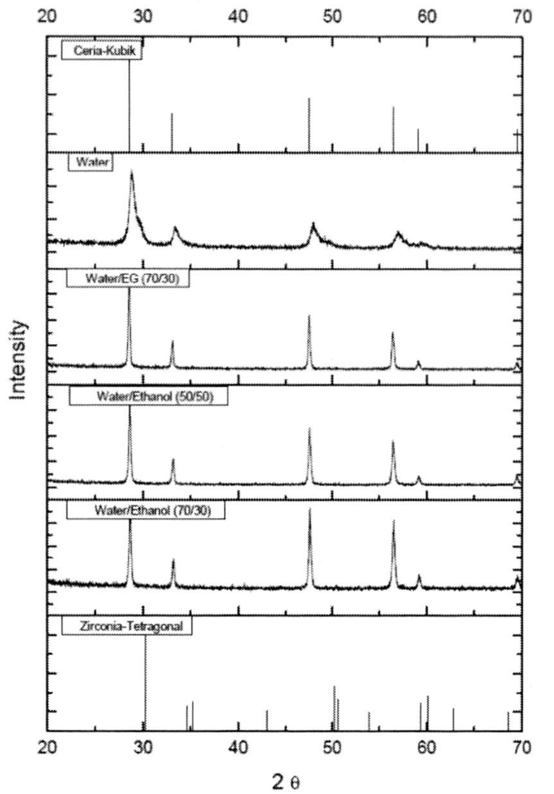

Fig. 3. XRD patterns of ceria zirconia oxides particles produced by various solvents at 6 h drying time.

Table 1. Total oxygen storage capacity (OSC) of ceria zirconia particles produced by various solvents at 6 h drying time.

Solvent	Temperature Maximum (°C)	OSC (%)
Water	410	1.82
Water/Ethanol (70/30)	580	2.09
Water/Ethanol (50/50)	585	51.76
Water/E. Glycol (70/30)	480	0.61

Table 1 shows total oxygen storage capacity (OSC) of ceria zirconia oxides particles produced with various solvents and 6 h drying time. OSC represents the ability of particles to store oxygen to be reacted with hydrogen in SOFC. The addition of ethanol into the solvent increased the stability and OSC of the particles. With the addition of 50% volume of ethanol into the solvent resulted stability of particles up to 585°C with the highest OSC 51.76%. It can be confirmed with the SEM image of particles obtained with water/ethanol (50/50 (v/v)) that smaller particles were obtained at this condition. The smaller particles caused the larger surface area and larger area to contact with the oxygen atom. In the previous work [13], the supercritical synthesis could lead to ceria–zirconia mixed oxides with higher thermal stability and better OSC. At supercritical synthesis, the operating temperature was much higher than the hydrothermal synthesis, and the process may result in smaller particle size.

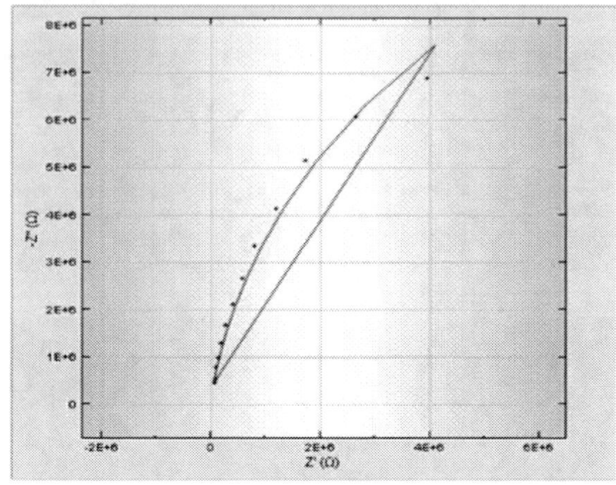

Fig. 4. Nyquist plot of ceria zirconia oxides particles produced by water/ethanol (70/30) at 6 h drying time.

Figure 4 shows Nyquist plot of ceria zirconia oxides particles produced by water/ethanol (70/30) at 6 h drying time. The Nyquist plot is a plot of Z' (real value of impedance) versus –Z" (imaginer value of impedance) composed of two curve parts, power part and linear part. The power curve part performs the characteristic of material sample to be measured and the interface resistant of electrode material, while the linear curve part performs the resistant of material sample. Based on the curve, the obtained resistant |Z| of material sample (ceria zirconia oxides particles) is 0.2669 Ω. Furthermore, the conductivity of material sample (σ) was calculated using Equation (1):

$$\sigma = 1 / R \times A \quad (1)$$

where σ is conductivity (S/cm), l is sample thickness (cm), R or |Z| is sample resistant (Ω), and A is surface area of sample (cm^2).

In this work, the impedance measurement was conducted with 0.3 cm sample thickness and 2.0096 cm2 surface area of sample. Based on the calculation, conductivity of ceria zirconia oxides particles was 0.5594 S/cm. This result is higher than the conductivity of zirconia based oxide material reported by Lv et al. [14]. They reported that zirconia doped with yittria (Y_2O_3), scandia and ytterbia (Yb_2O_3) achieved maximum electrical conductivity of 0.18 S/cm, and zirconia doped with yittria (Y_2O_3), dysprosia (Dy_2O_3) and ytterbia (Yb_2O_3) had maximum electrical conductivity of 0.16 S/cm. Moreover, ceria zirconia oxides particles produced with water/ethanol (70/30 (v/v)) as solvent meets the requirement of SOFC electrolyte. The minimum ion conductivity requirement of SOFC electrolyte is 0.01 S/cm.

4 Conclusion

Ceria zirconia oxides particles has been prepared using solvothermal synthesis at various solvents. The synthesis was carried out at temperature of 150°C for 2 h in a Teflon-lined autoclave of 100 mL volume. The synthesis solvents influenced the morphology and size of ceria zirconia oxides particles formed. Ceria zirconia oxides produced via solvothermal synthesis had high conductivity ion of 0.5594 S/cm, which is higher than minimum conductivity ion requirement of 0.01 S/cm for SOFC electrolyte. It indicated that ceria zirconia oxides produced via solvothermal synthesis is suitable for SOFC electrolyte.

Acknowledgement

This research was supported by a grant from Directorate General of Science, Technology and Higher Education, Ministry of Research, Technology and Higher Education of the Republic of Indonesia through a desentralisasi research grant Penelitian Unggulan Perguruan Tinggi contract no. 616/PKS/ITS/2017.

References

1. A.B. Stambouli, E. Traversa, Renew. Sustain. Energy Rev. **6**, 433 (2002).
2. T. Hibino, H. Tsunekawa, S. Tanimoto, N. Sano, J. Electrochem. Soc. **147**, 1338 (1990).
3. Y. Nagai, T. Nonaka, A. Suda, M. Sugiura, R&D Rev. Toyota CRDL **37**, 1-20 (2002).
4. M. Fernandez-Garcia, A. Martinez-Arias, J.C. Hanson, J.A. Rodriguez, Chem. Rev. **104**, 4063 (2004)
5. S. Machmudah, O.P. Prastuti, Widiyastuti, S. Winardi, Wahyudiono, H. Kanda, M. Goto, Res. Chem. Intermed. 42, 5367 (2016).
6. V.M. Bekale, C. Legros, C. Haut, G. Sattonnay, A.M. Huntz, Solid State Ion. **177**, 3339 (2006).
7. T.R.N. Kutty, J.A.K. Tareen, B. Basavalingu, B. Puttaraju, Mater. Lett. **1**, 67 (1982).
8. S. Komarneni, D. Li, B. Newalkar, H. Katsuki, A.S. Bhalla, Langmuir **18**, 5959 (2002).
9. K. Kanie, Y. Seino, M. Matsubara, M. Nakaya, A. Muramatsu, New J. Chem. **38**, 3548 (2014).
10. R. Vehring, W.R. Foss, D. Lechuga-Ballesteros, J. Aeros. Sci. **38**, 728 (2007).
11. J.-R. Kim, K.-Y. Lee, M.-J. Suh, S.-K. Ihm, Cat. Today **185**, 25 (2012).
12. S. Machmudah, M.A. Hadian, L. Samodro, S. Winardi, Wahyudiono, H. Kanda, M. Goto, Modern Appl. Sci. **9**, 134 (2015).
13. J.-R. Kim, W.-J. Myeong, S.-K. Ihm, Appl. Cat. B: Environ. **71**, 57 (2007).
14. Z.G. Lv, P. Yao, R.S. Guo, F.Y. Dai, Mat. Sci. Eng. A **458**, 355 (2007).

The Current Density on Electrosynthesis of Hydroxyapatite with Bipolar Membrane

Adrian Nur[1,*], *Arif* Jumari[1], *Anatta W* Budiman[1], *Nazriati* Nazriati[2], *Fauziatul* Fajaroh[2], *Jeihan H* Fariza[1], and *Tovaliery L* Anisa[1]

[1]Chemical Engineering Department, Faculty of Engineering, Universitas Sebelas Maret, 57126 Surakarta, Indonesia
[2]Chemistry Department, Faculty of Mathematics and Science, Universitas Negeri Malang, 65111 Malang, Indonesia

Abstract. Synthesis of hydroxyapatite by electrochemical method was has been successfully done. The novelty of this research is used of the bipolar membrane to separate electrolysis chamber. The bipolar membrane is used to keep the cations still around the cathode and react to form hydroxyapatite. The aim of this paper was to compare the current density on electrosynthesis of hydroxyapatite with and without bipolar membrane and the effect of current density on electrosynthesis. The electrosynthesis was performed at 2 hours at 400 to 600 mA/cm^2 at room temperature. The bigger the current density, the more pure HA formed. The electrosynthesis of hydroxyapatite with bipolar membrane more effective than without bipolar membrane. The formation of HA is very effective in the cathode chamber. HA can be formed at low current density. In electrosynthesis with the bipolar membrane, particles of HA are nanosheet flower-like. The nanosheet flower-like HA growth at 1.6 A to 2 A.

1 Introduction

Hydroxyapatite (hydroxylapatite, HA) is the most potential material to be used as a substitute for human bone and teeth [1]. Actually, the formula of HA is $Ca_5(PO_4)_3(OH)$ but usually written as $Ca_{10}(PO_4)_6(OH)_2$. HA has been used for the biomedical application [2], scaffolding for new bone formation [3], and as matrices for drug release control [4]. HA is also used for the nonbiomedical application such protein adsorption [5], gas sensors [6], hydrogen production [7], biodiesel production [8] and catalysts [9].

There are several methods for HA synthesis, e.g. mechanochemical method [10], emulsion method [11], the sol-gel process [12], chemical precipitation method [13], and hydrothermal technique [14, 15]. Mechanochemical treatment is suitable for large-scale hydroxyapatite production but the disadvantage of this method is the possibility of being contaminated. Emulsion method is a simple process method with the good product that is high crystallization without high temperature and crystalline size with the narrow distribution. The sol-gel method for HA production can process in low temperatur and results high product purity. Chemical precipitation method is simple, low cost, and the low-temperature but difficult to control particle size. The particles have the wide particle size distribution. The hydrothermal methode need the high temperature for the high degree of crystallinity. The alternative method for synthesis of HA is the electrochemical method. The electrochemical methods were originally used for HA coatings on metals. The electrochemical method is known to assemble fine particle. Synthesis of nanosized hydroxyapatite by electrochemical method explained in reference [16].

The synthesis of hydroxyapatite by electrochemical method has been developed using pulsed direct current [17]. In this method, the pulsed direct current is periodically interrupted. When the current is interrupted, HA particles around cathode move away to bulk solution. This condition is to avoid high concentration of HA particle and agglomeration process. This method requires the longer process.

In another our work [18, 19], we conclude that water reduction at cathode to produce hydroxide ions is very important process in HA formation. The problem in the electrochemical method is the concentration of OH^- ions produced by water reduction difficult increase due they move away from the cathode and react with H^+ ions produced by water oxidation in the anode. In this work, we used the bipolar membrane to separate chamber system in the electrochemical cell.

The novelty of this work is used of the bipolar membrane to separate electrolysis chamber at hydroxyapatite synthesis by the electrochemical method. The bipolar membrane is used to keep the cations still

* Corresponding author: adriannur@staff.uns.ac.id

around the electrode and react to form hydroxyapatite. The effect of current density on electrosynthesis of HA with the bipolar membrane was studied.

2 Experimental

A bipolar membrane was installed in th electrochemical cell (Fig. 1.). The bipolar membrane is Fumasep FBM from FUMATECH BWT GmbH. The bipolar dimension is 2×5 cm and 0.2 – 0.25 mm in thickness.

The solution used for the HA synthesis by electrochemical method was consisted of $Na_2H_2EDTA.2H_2O$ (Merck, reagent grade), KH_2PO_4 (Merck, reagent grade) and $CaCl_2$ (Merck, reagent grade). The concentration of $Ca^{2+}/EDTA/PO_4^{3-}$ as homogeneous solution was 0.25/0.25/0.15 M. All chemicals were used without further treatment. The electrolysis was done in a electrochemical cell. There are two carbon sheet with dimensions of (2×1) cm as the anode and the cathode. The acrylic material was used to the electrochemical cell. The anode and the cathode were within 3 cm and immersed in the solution at a depth of 2 cm. The anode and the cathode were connected to a DC power supply (Zhaoxin PS-3005D). The bipolar membrane was installed between the anode and cathode. The solution was electrolysed at 2 hours at 400 to 600 mA/cm^2 at room temperature and under ultrasonic cleaner to void agglomeration.

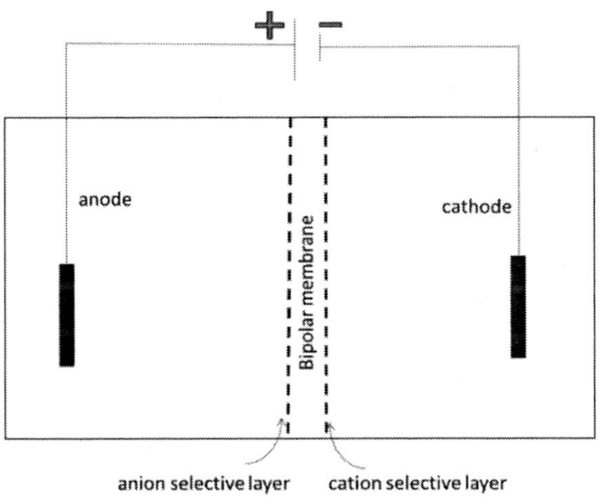

Fig. 1. The electrochemical cell for HA synthesis with bipolar membrane

The suspension produced was filtered. The particle was washed with demineralized water and dried at 110°C for 2 days. The X-ray diffraction (XRD) pattern of the particles was detected using an X-ray diffractometer (Shimadzu 6000). The morphologies of the particles analysis were observed using scanning electron microscopy (Inspect S40, FEI). Fourier transform infrared spectroscopy (FTIR) (Shimadzu 8400s) was performed to evaluate the functional groups of specimens. The FTIR spectra were obtained over the region 500-4,000 cm^{-1}.

3 Result and Discussion

The XRD patterns of particle resulted in electrosynthesis of HA without and with bipolar membrane at various current density were compared. Fig. 2 show the XRD patterns of particles obtained at various the current density at 2 hrs without the bipolar membrane. Fig. 3 show XRD patterns at the same condition with the bipolar membrane at cathode chamber and Fig. 4 at the anode chamber.

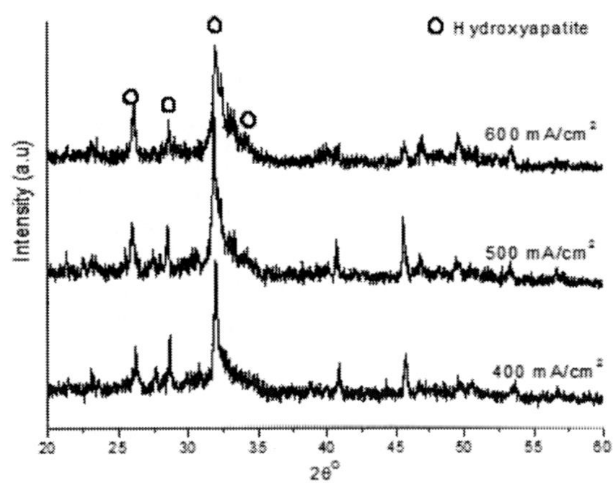

Fig. 2. The effect of current density at XRD patterns for particles prepared without bipolar membrane

At electrosynthesis of HA without bipolar membrane, the particle products are HA. There are four characteristic peaks at 26° (002), 29° (210), 32° (300), and 34° (202). These peaks match with the standard pattern of HA (JSPDS 09-0432). The 34° (202) peak is not clear at 400 mA/cm^2 and is clear at 500 and 600 mA/cm^2. The bigger the current density, the higher purity of HA product. The crystalline properties of the HA particles increase with the increase of current density. The final pH of solution shows the phase of calcium phosphate. If final pH of solution < 8, the phase product is brushite. If final pH of solution > 8, the phase product is HA [18]. In this work, we find that the final pH of solution for 400, 500, and 600 mA/cm^2 are 8.1, 9.1, and 9.5.

At the cathode of electrosynthesis of HA with the bipolar membrane (Fig. 3), the particle products are HA. There are four characteristic peaks at 26° (002), 29° (210), 32° (300), and 34° (202). They match with the standard pattern of JSPDS 09-0432. All of product is pure HA. The final pH solution for 400, 500, and 600 mA/cm^2 are 12.5, 13.4, and 13.7.

https://doi.org/10.1051/mateconf/201815605015

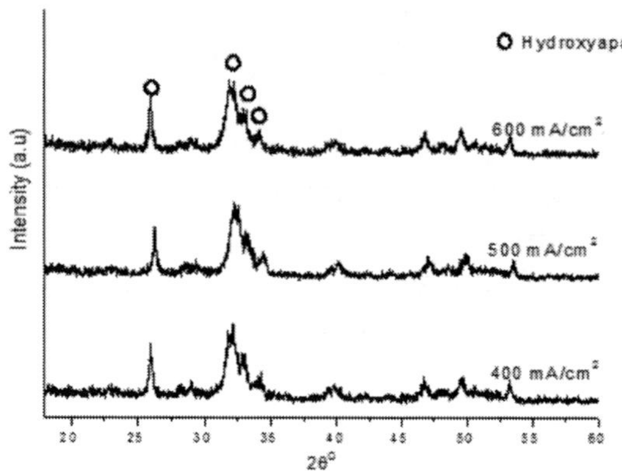

Fig. 3. The effect of the current density at XRD patterns for particles prepared with the bipolar membrane at the cathode

At the anode of electrosynthesis of HA with the bipolar membrane (Fig. 4), the particle products are brushite (at 400 and 500 mA/cm^2) and hydroxyapatite (at 600 mA/cm^2). At current density 400 and 500 mA/cm^2, the patterns have 7 characteristic peaks at 21°, 25°, 27°, 29°, 32°, 37°, and 42° which the standard of brushite. The final pH solution at anode for 400, 500, and 600 mA/cm^2 are 4.5, 5.6, and 7.8.

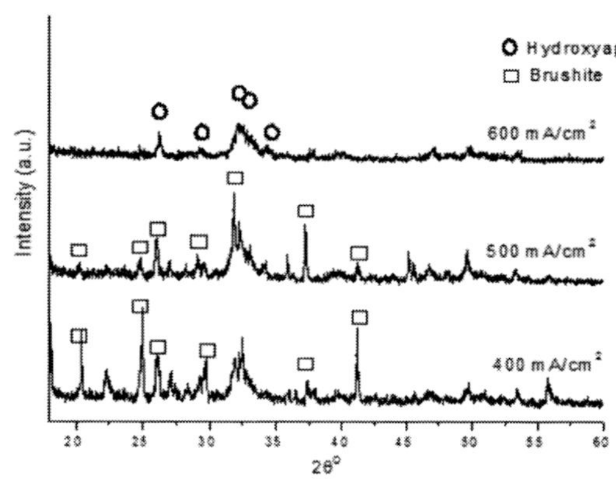

Fig. 4. The effect of the current density at XRD patterns for particles prepared with the bipolar membrane at the anode.

The reactions occurred in the cell without bipolar membrane can be explained in Fig. 5. The reduction of water in cathode produce OH$^-$ ions. The hydroxyde ions is very important in the formation of HA. The hydroxyde ions change the equilibrium of phosphate and EDTA-Ca. In this case, hydroxyde ions produced by reduction of water move away from the cathode and react with H$^+$ ions produced by oxidation of water in the anode. We will find HA formation if we can produce OH$^-$ ions more than react with H$^+$.

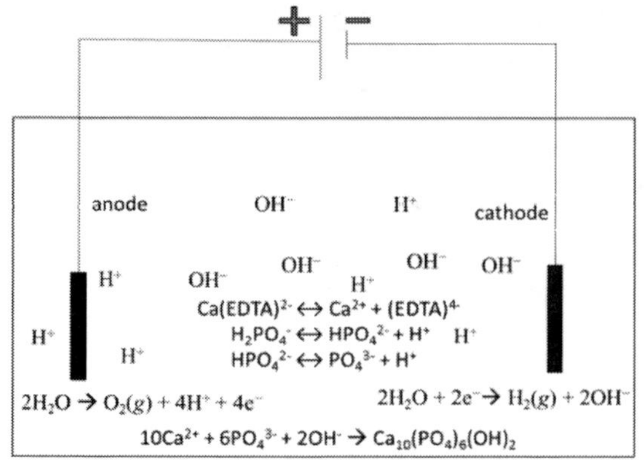

Fig. 5. The reactions occurred in the cell electrochemical for HA synthesis without the bipolar membrane

Fig 6. shows the movement of ions when the bipolar membrane installed in the cell electrochemical.

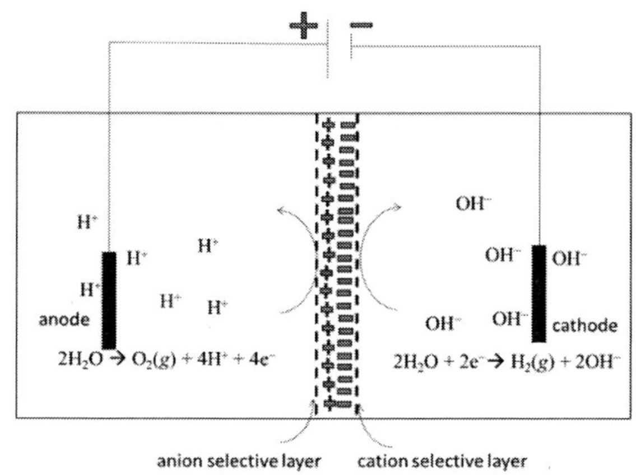

Fig. 6. The movement of ions in cell electrochemical for HA synthesis with bipolar membrane

In the cathode chamber, OH$^-$ ions do not move away from the cathode because the bipolar membrane prevents OH$^-$ ions to the anode chamber. The pH solution rapidly increases and the formation of HA increase. The formation of HA is very effective in this chamber. HA can be formed at low current density and low time electrolysis. In the anode chamber, H$^+$ ions do not move away from the anode chamber to the cathode chamber. In this chamber, pH solution increases very slowly. The pH solution still can increase due to the equilibrium condition from ions in solution. HA can be formed if the pH solution more than 8.

The FTIR analysis used to identify the HA product. The Fig. 7 shows the FTIR spectra of the particle with and without the bipolar membrane. According of the FTIR spectra, there is no significant difference of HA product with and without the bipolar membrane. The spectra were analyzed in the 4,000 to 500 cm^{-1} wavenumber.

The FTIR spectra of particle prapred without and with the bipolar membrane show that there are the phosphate and hydroxide groups. The characteristic bands for phosphate was seen at 493 cm^{-1}, 559 cm^{-1}, 976 cm^{-1}, and 1,032 cm^{-1}. The broadband of low intensity in the range 3,000–3,500 cm^{-1} can be attributed to traces of water.

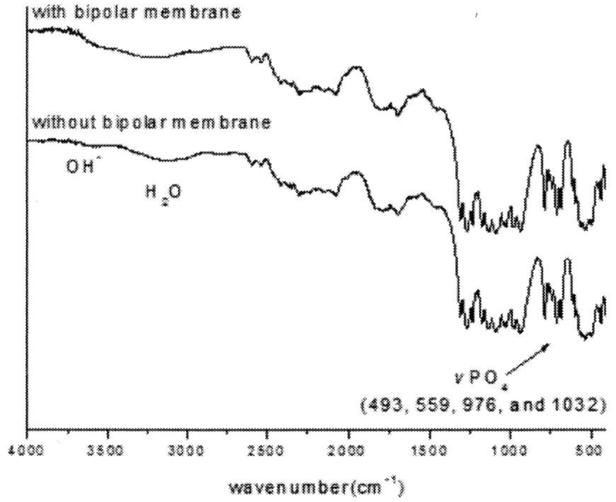

Fig. 7. The comparation of FTIR spectra of the HA particles without and with the bipolar membrane

Fig. 8. The morphologu of HA particles by SEM without the bipolar membrane. (a) 1.6 A and (b) 2 A

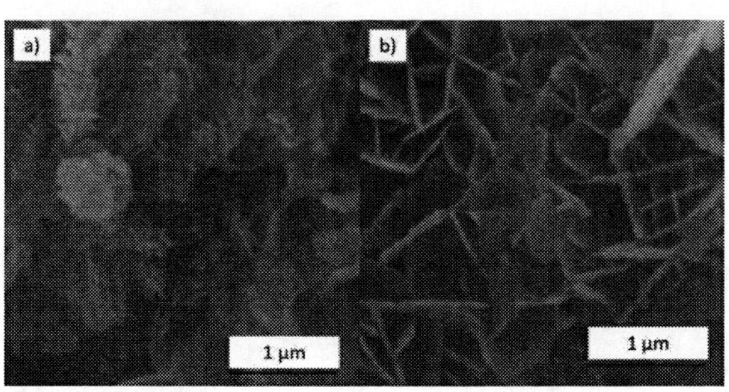

Fig. 9. The morphology of HA particles by SEM with the bipolar membrane at cathode chamber. (a) 1.6 A and (b) 2 A

The SEM images of HA product with and without the bipolar membrane at high magnification is provided in Fig. 8 and Fig. 9. The particles of HA without the bipolar membrane at 1.6 A (Fig. 8 (a)) are the spherical-like shape. The growth of particles at 2 A (Fig. 8 (b)) forms plate sheet-like. In electrosynthesis with the bipolar membrane, particles of HA are nanosheet flower-like. The nanosheet flower-like HA growth at 1.6 A to 2 A (Fig. 9).

4 CONCLUSION

The synthesis of HA was performed by electrochemical method at 2 hours at 400 to 600 mA/cm^2 at room temperature. We compare the current density on electrosynthesis of hydroxyapatite with and without the bipolar membrane. The increase of current density results higher purity of HA. The electrosynthesis of hydroxyapatite with bipolar membrane more effective than without bipolar membrane. The bipolar membrane is used to keep the cations (OH^- ions) still around the electrode and react to form hydroxyapatite. The formation of HA is very effective in the cathode chamber. HA can be formed at low current density. In electrosynthesis with the bipolar membrane, particles of HA are nanosheet flower-like. The nanosheet flower-like HA growth at 1.6 A to 2 A

Acknowledgment

The authors would like to thanks to KEMENRISTEKDIKTI and Sebelas Maret University for Hibah PU UNS 2017.

References

1. S.V. Dorozhkin, Acta Biomater. **6**, 715 (2010)
2. H. Zhou and J. Lee, Acta Biomater. 7, 2769 (2011)
3. G. Tripathi and B. Basu, Ceram. Int. **38**, 341 (2012)
4. M. Kamitakahara, R. Imai, and K. Ioku, Mater. Sci. Eng., C **33**, 2446 (2013)
5. T. Nagasaki, F. Nagata, M. Sakurai, and K. Kato, Journal of Asian Ceramic Societies **5**, 88 (2017)
6. L. Huixia, L. Yong, L. Lanlan, T. Yanni, Z. Qing, and L. Kun, Mater. Sci. Eng., C **59**, 438 (2016)
7. K. Yaemsunthorn and C. Randorn, Int. J. Hydrogen Energy **42**, 5056 (2017)
8. S.H.Y.S. Abdullah, N.H.M. Hanapi, A. Azid, R. Umar, H. Juahir, H. Khatoon, and A. Endut, Renew. Sustainable Energy Rev. **70**, 1040 (2017)
9. M. Zahouily, Y. Abrouki, B. Bahlaouan, A. Rayadh, and S. Sebti, Catal Commun **4**, 521 (2003)
10. B. Nasiri-Tabrizi, P. Honarmandi, R. Ebrahimi-Kahrizsangi, and P. Honarmandi, Mater. Lett. **63**, 543 (2009)
11. A. Banerjee, A. Bandyopadhyay, and S. Bose, Mater. Sci. Eng., C **27**, 729 (2007)
12. M.H. Fathi and A. Hanifi, Mater. Lett. **61** 3978 (2007)

13. M.A. Martins, C. Santos, M.M. Almeida, and M.E.V. Costa, J. Colloid Interface Sci. **318** 210 (2008)
14. M. Tomozawa, and S. Hiromoto, Acta Mater. **59** 355 (2011)
15. R. Xin, F, Ren, and Y. Leng, Mater. Des. **31** 1691 (2010)
16. M.S. Djosic, V.B. Miskovic-Stankovic, S. Milonjic, Z.M. Kacarevic-Popovic, N. Bibic, and J. Stojanovic, Materi. Chem. Phys. **111**, 137 (2008)
17. A. Nur, A. Rahmawati, N.I. Ilmi, S. Affandi, and A. Widjaja, *AIP Conference Proceedings* **86**, 1586 (2014)
18. A. Nur, H. Setyawan, A. Widjaja, and I. W. Lenggoro, BCREC, 9(3) 168 (2014)
19. A. Nur and H. Setyawan, J. Chem. Eng. Jpn., 49(2) 144 (2016)

Thermal, Morphological and Physic-Mechanical Properties of Natural Rubber - $CaCO_3$ Composites Using Jatropha Oil as Softener.

Nasruddin[1*], and *Tri Susanto*[1]

[1]Palembang Institute for Industrial Research and Standardization, Ministry of Industry, Indonesia

Abstract The urgency of green technology in rubber compounding has become a critical issue recently. In this research, the effect of using renewable resources in rubber compounding has been studied. Commercial Calcium Carbonate, Silica and Jatropha Oil were used in natural rubber composite. The research was designed by varying the types of commercial filler namely $CaCO_3$ (47-51) phr, silica (47-51) phr and Jatropha Oil (4-6) phr in natural rubber composites (SIR-20). The formulas were intentionally designed for rubber tips vulcanizates. The samples were characterized by the determination of physic-mechanical, thermal (TGA) and morphological (SEM) properties. From the measured results, there is no significant effect on the tensile strength, specific gravity, and hardness on the loading of commercial $CaCO_3$ and Silica in natural rubber composites using Jatropha Oil. However, a slight difference in elongation at break and abrasion resistance could be detected. Compared to the commercial rubber tips, the rubber tips produced in this research have higher tensile strength, elongation at break and abrasion resistance. Due to the usage of commercial $CaCO_3$ and Silica, the SEM micrographs show rough surface because of the agglomeration. The thermogram shows clearly the compositional analysis of the rubber tips vulcanizates consist of Jatropha Oil and natural rubber, $CaCO_3$, ash and other filler residues such as Silica.

1 Introduction

Fillers play an important role in rubber composites to enhance physic-mechanical properties and reduce the material costs since the second highest portion of the manufacturing of rubber products [1, 2]. Fillers are grouped into two types reinforcing and non-reinforcing fillers. Fang, Song [3], [4] mentioned that the reinforcing such as Carbon Black (CB), SiO_2 has been used to improve the modulus and strength of rubber products, whereas, non-reinforcing have not significant effect on the rubber properties.

CB consumption of world rubber industry has been increasing recently. Meanwhile, the increasing of rubber industries development is in the same way with the increasing price of CB per year [5]. On the other hand, CB is obtained from derivative oil industries [6], while the petroleum reserves in the world are getting less [5]. Therefore, it is a need to find CB substituents that are renewable, low cost, abundant and environmentally friendly. Considering its availability from natural resources, Calcium Carbonates ($CaCO_3$) is an prospective filler to be developed for rubber and plastics industry.

$CaCO_3$ is a low-cost commercially available and is able to be non-reinforcing filler that only upsizing the volume of the rubber products [7, 8]. Furthermore, $CaCO_3$ is also able to impart significant improvements on polymer properties, such as abrasion resistance, stiffness and tensile strength as well as thermal stability [9] and Fang, Song [3], [8, 9] reported that to improve the mechanical properties of the reinforced natural rubber-$CaCO_3$ composites. According to Avella, Errico [10], the benefits of blending polymer with the $CaCO_3$ reduce cost without loosing the tensile properties significantly. [9], Bartczak, Argon [11] stated that the selection of appropriate size distribution and surface treatments with hydrophobic agents such as stearic acid and silane.

Fang, Song [3] observed that rubber composite filled chain shape Nano Calcium Carbonates (NCC) has positive benefits on Mullin effects, Payne effects, and its dynamic characteristics. While Lazzeri, Thio [12], [13] showed that the surface modified Calcium Carbonates also exhibited better processing capability than that Carbon Black in SBR vulcanizates. In addition, the improvement on mechanical properties, however, is always limited because Calcium Carbonates with high surface energy tends to agglomerate [8, 9]. Qu, Chen [14] found that when the amount of co-precipitated Nano Calcium Carbonates and bulk NCC is identical, the mechanical properties of the former can achieve high tensile strength (13,38 MPa) which was superior over the later.

In rubber industry, Calcium Carbonates is also commonly used as the filler for SBR composites [13, 14], Acrylonitrile-butadiene rubber [15], and Chloroprene composites [16, 17]. Fang, Song [3], [9, 14] stated that addition NCC can produce outstanding stiffness, toughness, and dimensional stability rubber compound. Hence, the deep understanding of the relationship between particle size and mechanical properties are important to improve the end-user properties of rubber composite.

Focusing on the industrial application (end-use rubber composite), this research applied commercial precipitated $CaCO_3$, which is finer and

* Corresponding author: nas.bppi@gmail.com

high purity (400 mesh) along with the commercial Silica for natural rubber composite.

Bearing in mind the green technology approach, the importance of searching new alternative is not only for CB as filler but also mineral plasticizers [18-21]. With regard to some weakness of using mineral plasticizers such as high cost, low availability, corrosive and toxic chemicals [21, 22], in this research would apply commercially natural plasticizer namely Jatropha Oil (Ricini Oil). Gogoi, Boruah [23] has conducted research on the use of jatropha seed oil and other vegetable oils as coupling agents as well as plasticizers in the range of 1-3 phr. The other research showed that it had a significant effect on mechanical properties such as modulus, tear strength, and hardness [22-24]. Based on the previous researchers, Jatropha seed potentially could be used as plasticizers to substitute the mineral oil.

Figure 1. Rubber tips products

Thus, the current study focuses on the incorporation of commercial fillers such as Silica and Calcium Carbonate into natural rubber composites using natural based plasticizers. The tensile properties, physic-mechanical properties, thermal and morphological properties were investigated. The usage of the NR composites was intentionally designed as manufacturing of rubber tips vulcanizates (end-up products) as shown in Figure 1. The product of this research was commonly used for furniture accessories, canes including quad canes and tripods. In order to observe the product specification compliance with this research, the commercial rubber tip was also examined. Consequently, the research would contribute new information about the essential properties rubber tips vulcanizates that using commercial $CaCO_3$, Silica, Jatropha Oil in natural rubber composites. Thus, this technology could be applied for rubber tips manufacturing and rubber industries.

2 Experimental

2.1. Materials

Natural Rubber/ NR (Standard Indonesian Rubber/SIR 20) was used as polymer. The other compounding ingredients including Zinc Oxide (NC 105 Global Chemical Co, Ltd), Stearic Acid (SA 1806, Oleochemicals Industry, PT Sumi Asih), Carbon Black (CB N330, Vulcan, CABOT), Calcium Carbonate (400 mesh, SK Product), Precipitated Silica powder (Type SD Grade 185, Chemisil, PT Supersilica Indo Semesta), TiO_2, Sulphur. While, Jatropha Oil (Oleum Ricini), PEG-4000, TMTD, Paraffinic Wax, N-(1,3-dimethylbutyl)- N'-phenyl-p-phenylenediamine 6 PPD were obtained from Brataco Chemical.

2.2 Methods

2.2.1. Rubber Compounding

The rubber compounding was carried out using a laboratory two roll mill XK-160, Shanghai Rubber- Machine Worker, China. NR was masticated in two roll mills for 2-3 minutes. The next process is adding the compounding ingredients as listed in Table 1, so that the 5 vulcanized sheets were obtained. This process was conducted based on ASTM D3182 Standard Practice for Rubber (Materials, Equipment, and Procedures) for Mixing Standard Compounds and Preparing Standard Vulcanized Sheets. The vulcanized sheets were pressed for 12 minutes at 125 °C. The formulas given in Table 1 were developed for rubber tips products (end –up). Therefore, in this research, the rubber tips in market (F6) would be tested to compare its quality. In order to obtain the accurate data, the rubber compounding and rubber tips pressing were conducted as 3 replications for each formulas.

Table 1. Formulation of Rubber Compounding for Rubber Tips

Material	Formula				
	F_1	F_2	F_3	F_4	F_5
NR (SIR 20)			100		
TiO_2			3.65		
Carbon Black			0.85		
ZnO			4.65		
Stearic Acid			2.44		
Silica	47	48	49	50	51
Calcium Carbonate	51	50	49	48	47
Jatropha Oil	4.5	5.00	5.50	6.00	6.50
PEG			3.85		
TMTD			1.58		
Paraffin Wax			0.82		
6.PPD			1.25		
Sulfur			2.75		

Note : F6 is the commercial rubber tips

2.2.3. Physical-Mechanic Properties Testing

The five formulas for solid tyres were tested for parameters as follows hardness Shore A (ASTM D 2240-05 (ra 2010); tensile strength (ASTM D 412); modulus 300% (ASTM D 412); specific gravity (ASTM D 297-93 (ra 2002); abrasion resistance (ASTM D 5963-04 (ra 2010). Each replication was tested for those parameters, the data presented in this results are the average of 3 replications.

2.2.4. Scanning Electron Microscope and Thermal Gravimetry Analysis

Thermal Gravimetric Analysis was conducted using LINSEIS STA PT 1600 based on Standard Test Method for Rubber – Composition Analysis by Thermogravimetry – ASTM D 6370-99 (2003). Heat from 30-800 °C, 10 °C/min. While, the morphological analysis of NR-Calcium Carbonates composites was examined using SEM JEOL JSM 6510.

3 Results and Discussion

3.1 Tensile Properties

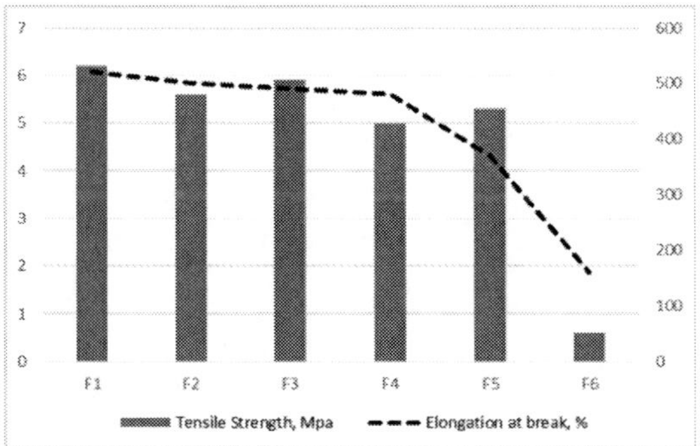

Figure 2. Tensile Properties of NR-$CaCO_3$ Composites Using Jatropha Oil

The relationship between tensile strength and loading of precipitated $CaCO_3$ and Silica in NR vulcanizate is presented in Figure 2. It could be seen that the tensile strength of F1 (47 phr Silica and 51 phr $CaCO_3$ loading) has the highest number about 6,2 MPa. The graph illustrates the fillers loading (F1, F2, F3, F4, and F5) may not influence the tensile strength of NR vulcanizate significantly. The tensile properties is influenced by the interaction between the fillers and rubber within the vulcanizate along with the cross linking density [4, 25]. Even though the rheological properties were not examined in this study, the using $CaCO_3$ as non-reinforcement filler in a high proportion would affect on cross-linking interaction between fillers and rubber matrix negatively [7, 13, 14]. The using of 400 mesh $CaCO_3$ and commercial Silica as fillers which were very large particle size probably could not produce more interfacial bonding and causes the molecular mobility to increase resulting in the stiffening of the matrix. Furthermore, the precipitated $CaCO_3$ in very large number possesses a low surface area and initiate to form agglomerate in the rubber filler interaction [14, 25]. Therefore, when it incorporates into natural rubber composites cause severe reduction of the elongation at break [7, 10, 11]. The decreasing of elongation at break is shown in Figure 2. It indicates that the more loading $CaCO_3$ (non-reinforcement filler) causes the worse interaction between filler and natural rubber matrix in the composites.

However, compared to the tensile strength of F6 (commercial rubber tips), the NR vulcanisates in this formula has higher tensile properties (tensile strength and elongation at break). The nature of NR has very high tensile properties among other polymers [26].

3.2. Physical and Mechanical Properties

Table 2. Physical and Mechanical Properties of NR-$CaCO_3$ Composites Using Jatropha Oil

Formula	Specific Gravity, g/cm^3	Hardness, Shore A	Abrasion Resistance, DIN, mm^3
F1	1.392	55	233.8
F2	1.338	50	226.8
F3	1.540	56	324.8
F4	1.395	55	204.8
F5	1.391	56	195.4
F6*	1.598	73	778.8

**) commercial rubber tips*

Table 2 lists the physic-mechanical properties of rubber tips vulcanized namely specific gravity, hardness and abrasion resistance. The changing hardness with different loadings of $CaCO_3$ and Silica in NR vulcanizate is shown in Table 2. It was observed that the fillers used in this composites have no significant influence on hardness of the vulcanizate. As listed in Table 1 that the quantity of reinforcement filler (Silica) and Calcium Carbonates as non-reinforcement filler is not significantly different. This very slight variation could not give significant changing on hardness and specific gravity as well. Aforementioned in tensile properties, the particle size, shape and surface area of fillers has a significant role on the hardness of vulcanizate [11, 14, 18, 25]. Therefore, using very fine reinforcement filler in natural rubber composites such as Carbon Black (CB) is suggested to increasing the rigidness and stiffness of the vulcanizates [12, 25]. The higher hardness due to the very good dispersion between filler and rubber in polymer matrix that causes increasing the cross linking density of rubber composites [7]. In this research, the fillers $CaCO_3$ and silica do not have the other advantages of CB like the functional groups of CB could form sulfuric cross-linking and also Van Der Walls interaction in the rubber matrix.

Compared to the commercial rubber tips, the 5 formulas have smaller specific gravity and lower hardness. The hardness of rubber tips commercial is about 73 Shore A, while the 5 formulas (F1, F2, F3, F4, and F5) only approximately 55 Shore A. The quantity of reinforcement filler should be added in a certain amount to improve its physical properties.

The physic-mechanical properties namely abrasion resistance was examined and evaluated. Table 2 illustrates the value of abrasion resistance each formula and the commercial rubber tips. The test method (ASTM D 5963-04) is subject to abrasive/ frictional wear in actual service. This test was conducted to evaluate the product specification compliance. By comparing test results of the designed formula (listed in Table 1) in this research to the commercial rubber tips in the market, some recommendations could be reported. The interpretation of the testing is expressed as volume loss in cubic millimeter, a smaller number indicates better abrasion resistance. As listed in Table 2, the abrasion resistance of commercial rubber tips 778,8 mm^3. In contrast for other formulas (F1, F2, F3, F4 and F5) have considerably small volume loss (200-300 mm^3), which indicates has a better abrasion resistance.

3.3 Morphological Properties

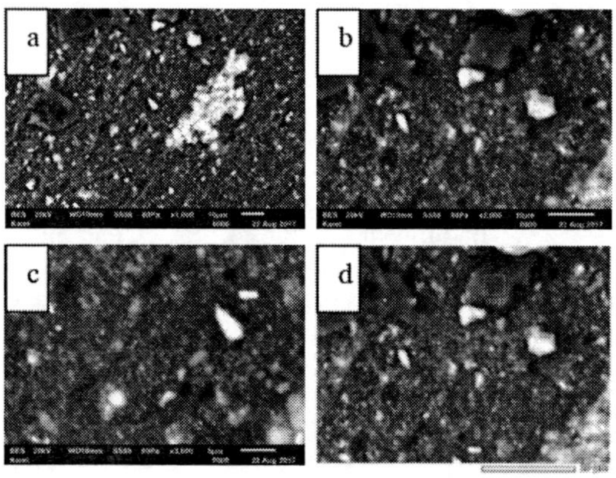

Figure 3. SEM Micrographs for Vulcanized of Natural Rubber- $CaCO_3$ Composites (100:47) using Jatropha Oil a) 1000x, b) 2000x, c) 3000x magnification. d) analysis on 20 um

The SEM micrographs for Vulcanized Rubber of NR-$CaCO_3$ 100:47 phr is shown in Figure 3. The magnification of the composites figure were 1000X, 2000X, 3000X a, b, c respectively while the micrographs were magnified for quantitative analysis.

It could be seen clearly that the micrographs show an irregular surface which is not so smooth and very rough. It indicates that the dispersion between $CaCO_3$ in the NR rubber matrix using Jatropha Oils is not homogenous so that the surfaces are not smooth (irregular surface). Since the particle size of $CaCO_3$ (400 mesh) is not Nano-sized, the agglomeration could not be hindered, and the flocculation appear on the surface, so that the sample has thicker tear lines and rough surface. Fang, Song [3] stated that using Nano-calcium carbonates (NCC) with spherical and cube shape has a positive effect to the Mullins effects and the filler networking in the rubber matrix. Therefore the micrographs show clear and smooth surfaces. Consequently, using the smaller NCC has the outstanding mechanical and tensile properties that resulted from good dispersion in the NR composites as observed in SEM micrographs [3, 8, 9, 18]. Conversely, the large size of $CaCO_3$ using in this research, the non-homogenous dispersion of $CaCO_3$ and Jatropha Oil in the rubber matrix contributed to the flocculation which had negative effects on the mechanical properties as shown in table 2.

The micrographs indicate that the rubber blends undergo ductile failure. So that, the failure of the tensile test (as shown in table 1) could be explained by the phenomena. Compared to the composite using Carbon Black which has finer particle size rather than $CaCO_3$, the surface has the smoother surface and better tensile properties [8, 18, 27]

3.4 Thermal Properties

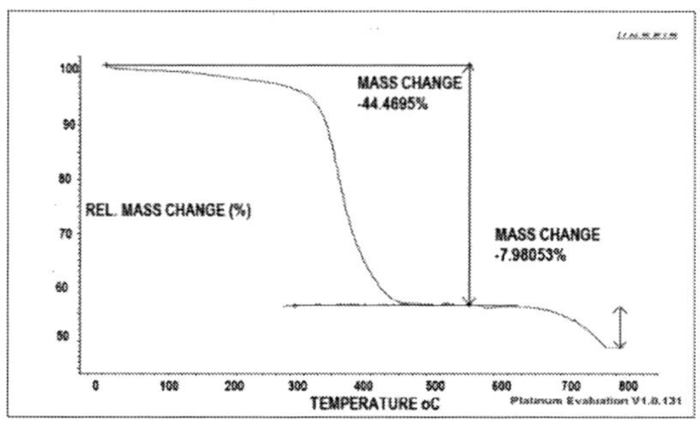

Figure 4. Thermogram for Vulcanized of Natural Rubber-$CaCO_3$ Composites (100:47) using Jatropha Oil.

The thermogram as shown in Figure 4 shows the curve of relative mass loss of the polymer by the increasing of temperature. The polymer as shown in table 1 consists of the major quantity of NR, filler SiO_2, and $CaCO_3$ 100, 51, and 47 phr respectively and very small quantity of Jatropha oil 6.5 phr. TGA was examined to observe the weight changing to changes in temperature (thermal stability), components in the polymer, organic and inorganics existed in the polymer, and residues percentage [28]. In order to obtain accurate weight loss, the derivative curve was used with the high-resolution dynamic method. Firstly, Nitrogen gas was purged with the flow 75 ml/min for sample, heating until 50 °C and allow to equilibrate for 2 minutes. Secondly, hear from 50 °C to 560 °C at 10 °C/min. Then, cool at about 300 °C and allow to equilibrate for 2 minutes. The purge gas changed to air at 75 ml/min, and heat from 300 °C to 800 °C at 10 °C/min. This changes due to the composites contain $CaCO_3$ as filler which decomposed at 680 °C - 800 °C, thus oxygen was applied in order to see the decomposition clearly [29, 30]. In the presence of Oxygen, the $CaCO_3$ was decomposed into CaO and CO_2 [29]. As shown in Figure 4, at the 300 °C, the decreasing weight is about 5%, probably, this is the loss of Jatropha oil and other oils ingredients in the sample. At the 550 °C, the natural rubber and all organic materials were burn, which is about 45 % of composites weight. While, the 7.9% mass change in the range 600 °C – 800°C is the decomposition of $CaCO_3$. The residual which about 48% weight loss at the temperature above 800°C indicated the ash and filler residue. As listed in Table 1, the number of SiO_2 and TiO_2 is 51 and 3.65 phr respectively, so that relative mass change at 800°C is considerably high. Similarly, the thermogram typical and its compositional analysis (Figure 4) which using $CaCO_3$ as filler was also obtained for SBR vulcanizate [31], for NR and SBR blending [29]. Furthermore, to obtain the more accurate decomposition and compositional analysis, the ending temperature should be about 1300 °C.

Conclusion

In conclusion, there is no significant improvement in the tensile strength, specific gravity, and hardness on the loading of commercial $CaCO_3$, Silica in natural rubber composite using commercial Jatropha Oil. However, a very slight difference in elongation at break and abrasion resistance could be detected due to the increasing loading of $CaCO_3$. The rough surface appears in SEM micrograph indicates the non-homogenous dispersion between commercial CaCO3, silica in natural rubber composites. Consequently, the vulcanisates have a slightly lower physic mechanical properties rather than commercially rubber tips. Based on the thermogram analysis, about 44% weight of the jatropha oil and other organics components completely loss at 500 °C, while 8% weight of the $CaCO_3$ was decomposed into CaO and CO_2 at 680 °C – 800 °C. In addition, the residual (48% weight) is expected an organic materials, ash and filler residues (Silica and TiO_2).

Acknowledgement

This research is under the project of Palembang Institute for Industrial Research and Standardization, Ministry of Industry, Indonesia. The authors would thank to the chief of for the financial and technical support, crumb rubber factories for providing the SIR 20, and PT Shima Prima Utama for the rubber tips manufacturing. We are very grateful to the research team for the hard work in conducting the research.

References

1. Leblanc, J.L., Prog. Pol. Sc. **27**(4), 627 (2002)
2. Nan, C.-W., Y. Shen, and J. Ma, Annu. Rev. Mater. Res. **40**, 131 (2010)
3. Fang, Q., et al., Compos Eng. 60, 561 (2014)
4. Fröhlich, J., W. Niedermeier, and H.-D. Luginsland, Compos Part A **36(4)**, 449 (2005)
5. Fang, Y., M. Zhan, and Y. Wang, Mater. Desig. **22(2)**, 123 (2005)
6. Wigmore, F.-P.R., *Carbon black production process*. 1972, Google Patents.
7. Ismail, H., R. Nordin, and A.M. Noor, Polym-Plast. Tech. Eng. **41(5)**, 847 (2002)
8. Leong, Y., et al., J. Appl. Polym. Sci. **91(5)**, 3315 (2004)
9. Yaling, L., et al., Synth. Rub. Ind. **6**,154 (2004)
10. Avella, M., M.E. Errico, and E. Martuscelli, Nano Lett. **1(4)**, 213 (2001)
11. Bartczak, Z., et al., Polymer. **40(9)**, 2347 (1999)
12. Lazzeri, A., Y. Thio, and R. Cohen, J. Appl. Polym. Sci. **91(2),** 925 (2004)
13. Mishra, S., N. Shimpi, and U. Patil, J. Polym. Res. **14(6)**, 449 (2007)
14. Qu, G.-L., X.-M. Chen, and X.-S. Ma, *Mater. Mec. Eng **35(1)**, 79 (2011)*
15. Zhang, A., et al., Polym Compos. **31(9)**, 1593 (2010)
16. Das, A., et al., Eur. Polym. J. **44(11)**, 3456 (2008)
17. Younan, A., S. Abd- El- Messieh, and A. Gasser, J. Appl. Polym. Sci. **70**(10), 2061 (1998)
18. Arroyo, M., et al., *Compos. Sco. Tech.* **67**(7), 1330 (2007)
19. Kanking, S., et al., *Mater. Des.* **41,** 74 (2012)
20. Sae- Oui, P., C. Rakdee, and P. Thanmathorn, J. Appl. Polym. Sci. **83**(11), 2485 (2002)
21. Vieira, M.G.A., et al., Eur. Polym. J. **47(3)**, 254 (2011)
22. Petrović, Z.S., et al., Rubber Chem. Technol **86**(2), 233 (2013)
23. Gogoi, P., et al.. ACS Sust Chem Eng **3**(2), 261 (2015)
24. Kukreja, T., et al., J. Appl. Polym. Sci. **87**(10), 1574 (2003)
25. Fu, S.-Y., et al., Compos Part B: Eng, **39**(6), 933 (2008).
26. Ku, H., et al., Compos Part B: Eng, **42**(4), 856 (2011)
27. Teh, P., et al., J. Appl. Polym. Sci. **94**(6), 2438 (2004)
28. Jin, F.-L. and S.-J. Park, Mater. Sci, Eng. **478**(1), 406 (2008)
29. Fernández-Berridi, M.J., et al.,. Thermochim. Acta **444**(1), 65 (2006)
30. Li, Z.-s., et al., Energy Fuels **19**(4), 1447 (2005)
31. Zhang, Q., et al., Appl. Clay Sci **65**, 134 (2012)

Synthesis of a one-part geopolymer system for soil stabilizer using fly ash and volcanic ash

April Anne S. Tigue[1], *Jonathan R.* Dungca[2], *Hirofumi* Hinode[3], *Winarto* Kurniawan[3], and *Michael Angelo B.* Promentilla[1*]

[1]Chemical Engineering Department, De La Salle University, Manila 1004, Philippines
[2]Civil Engineering Department, De La Salle University, Manila 1004, Philippines
[3]Department of Transdisciplinary Science & Engineering, Tokyo Institute of Technology, Tokyo 152-8550, Japan

Abstract. A novel approach one-part geopolymer was employed to investigate the feasibility of enhancing the strength of in-situ soil for possible structural fill application in the construction industry. Geopolymer precursors such as fly ash and volcanic ash were utilized in this study for soil stabilization. The traditional geopolymer synthesis uses soluble alkali activators unlike in the case of ordinary Portland cement where only water is added to start the hydration process. This kind of synthesis is an impediment to geopolymer soil stabilizer commercial viability. Hence, solid alkali activators such as sodium silicate (SS), sodium hydroxide (SH), and sodium aluminate (SA) were explored. The influence of amount of fly ash (15% and 25%), addition of volcanic ash (0% and 12.5%), and ratio of alkali activator SS:SH:SA (50:50:0, 33:33:33, 50:20:30) were investigated. Samples cured for 28 days were tested for unconfined compressive strength (UCS). To evaluate the durability, sample yielding highest UCS was subjected to sulfuric acid resistance test for 28 days. Analytical techniques such as X-ray fluorescence (XRF), X-ray diffraction (XRD), and scanning electron microscope/energy-dispersive X-ray spectroscopy (SEM/EDX) were performed to examine the elemental composition, mineralogical properties, and microstructure of the precursors and the geopolymer stabilized soil.

1 Introduction

Coal fly-ash, a by-product of coal-fired power plant, has known to have a natural pozzolanic property which can be utilized as substitute for ordinary Portland cement. In Philippines, Department of Energy (2015) has reported a coal consumption of 8.76 million metric tons in 2000 to 22.0 million metric tons in 2015 [1]. The tremendous increase in consumption pattern can be accounted to the use of coal as precursor for power generation, a major contributor to the country's total generation. In 2015, based on coal consumption statistics, 79.77% of coal is consumed by the power energy sector [2]. With the remarkable coal dependency for power generation, significant surge in the production of coal fly ash as waste by-product is expected. A significant amount of 1.14 MMT of coal fly ash generation has been reported by five major coal-fired powerplants in the Philippines [3]. Currently, a huge proportion of this waste go directly to landfill disposal. However, landfill disposal may pose threat as coal fly ash has been known to leach toxic heavy metals under certain conditions [4-6]. Similarly, volcanic ash or lahar which comprises volcanic material fragments such as pulverized rocks and minerals has also a natural pozzolanic property. Massive amount of volcanic material deposits was noted on the slope of Mt. Pinatubo amounting to approximately 5 to 6 cubic kilometres during its 1991 eruption [7]. These deposits being washed off downstream during heavy rain fall or typhoon is a major area of concern in the provinces of Central Luzon [8-9]. In fact, it has been reported that lahar flooding affects the river channels around the area. Due to a huge amount of by-product production and potential threat of lahar flooding, a need for potential utilization of these materials were studied.

Fly-ash and volcanic ash, a material rich in alumina silica, are promising precursors which can be used for soil stabilization through the process called geopolymerization. Geopolymerization involves dissolution of raw feedstock rich in aluminosilica in the presence an alkali activator to increase the reactivity. The aluminosilicate ions formed in the dissolution process will undergo polycondensation forming a gel matrix and will turn into a hardened molecule network. Factors such as sodium hydroxide concentration, alkali activator ratio, ash content, curing temperature and curing time are found to have influence in geopolymerization. The mechanism of geopolymerization can be described into three stages namely destruction-coagulation, coagulation-condensation, and condensation-crystallization [10].

* Corresponding author: michael.promentilla@dlsu.edu.ph

However, the traditional geopolymer synthesis which uses alkali activator solution is an impediment to geopolymer stabilized soil commercial viability. Hence, the researcher explored the one-part geopolymer synthesis. This kind of synthesis, instead of utilizing alkali activator solution, employs solid alkali activator. The synthesis involves dry mixing of solid alkali activators and geopolymer precursors, and addition of water. With the aid of 'just add water' concept like in the case of ordinary Portland cement to start the hydration process, its commercial viability will greatly increase [11-12].

In this study, the potential utilization of waste materials such as fly ash and volcanic ash, and use of solid alkali activator for soil stabilization were explored. The influence of amount of fly ash, addition of volcanic ash, and ratio of alkali activator on the unconfined compressive strength (UCS) of geopolymer stabilized soil were investigated using factorial design. To evaluate the durability, sample yielding highest UCS was subjected to sulfuric acid resistance test. X-ray diffraction (XRD), and scanning electron microscope/energy-dispersive X-ray spectroscopy (SEM/EDX) were performed to further analyse the structural and chemical changes after acid resistance test.

2 Materials and Method

2.1 Sources of raw materials

Fly ash samples were obtained from a coal fired power plant located at the central region of Luzon, Philippines and were used as received. The soil samples were obtained from a construction site located at the southern part of Manila, Philippines. Volcanic ash samples were collected from central region of Luzon, Philippines. The soil and volcanic ash samples were immediately sieved using Mesh No. 4 and stored in a container. Alkali activator used in the study are analytical grade of sodium silicate with composition 44% SiO_2, sodium hydroxide micropearls with 99% purity, sodium aluminate with 50-56% Al_2O_3/ 40-45% Na_2O, and sulfuric acid with 95% purity were purchased from a local company.

2.2 Characterization of raw materials

Chemical analyses of fly ash, volcanic ash, and soil were conducted using techniques such as X-ray fluorescence dispersive (XRF), and X-ray diffractometer (XRD). Elemental analysis was performed by irradiating the sample with x-ray beams generated from a voltage of 50 kV and current of 35 A using X-ray fluorescence (XRF) instrument (Shimadzu Model EDX-720). The crystalline phase of raw materials was determined using Multiflex Rigaku Automated Powder X-ray Diffractometer ($\lambda_{Cu\ K\alpha}$ = 1.54 Å, Voltage=40kV, Current=30.0 mA) with a measuring angle of 5°-60°. Particle size analysis was performed in accordance with ASTM D-422 using sieves to determine the particle size distribution of soil and volcanic ash [13]. In compliance with the standard method giving uniform spacing, a set of sieves 3-in, 1½-in, ¾-in, 3/8-in, No. 4, No. 8, No. 16, No. 30, No. 50, No. 100, and No. 200 were used. For particle size analysis of fly ash, dynamic light scattering was considered with single exposure and a total exposure time of 10 seconds using sodium lauryl sulfate as solvent. Atterberg limit, which includes plastic limit and liquid limit of soil, was performed in accordance with ASTM D-4318 [14]. The soil type was determined based on unified soil classification using the data obtained from Atterberg limit and particle size distribution.

2.3 Research design

The influence of amount of fly ash, addition of volcanic ash, and ratio of alkali activator SS:SH:SA were explored in this study. For this reason, factorial design was chosen. The response variable considered in this experiment was unconfined compressive strength (UCS) of geopolymer stabilized soil. The controllable factors that were kept constant during the experiment were alkali activator/ash content (30% w/w), amount of water (8% of the soil), curing temperature (ambient temperature), and curing time (28 days) [15-17].

2.3.1 Factor and levels

The research design consisted of 2 sets of 6 factorial combinations to determine the influence of amount of fly ash (% w/w), and alkali activator ratio (SS:SH:SA) and to determine the effect of adding volcanic ash (%/w/w) on unconfined compressive strength (UCS) were investigated. Other parameters used in this experiment are shown in Table 1-2 [18-23].

Table 1. Design factors range and levels to determine the influence of amount of fly ash, and alkali activator ratio.

Design Factor	(-)	(0)	(+)	Type of Factor
Amount of FA	15%	-	25%	Numerical
Alkali activator ratio	50:20:30	33:33:33	50:50:50	Categorical

Table 2. Design factors range and levels to determine effect of adding volcanic ash.

Design Factor	(-)	(0)	(+)	Type of Factor
Amount of VA	0%	-	12.5%	Numerical
Alkali activator ratio	50:20:30	33:33:33	50:50:50	Categorical

2.3.2 Statistical treatment of data

The statistical technique used was ANOVA (analysis of variance). This can determine whether there is statistically significant difference in the obtained data of various experimental runs. p-value was chosen to determine the statistically significant difference. When p-value<0.05, amount of fly ash, addition of volcanic ash, and ratio of alkali activator are significant. Parameters such as coefficient of determination, adjusted coefficient of determination, predicted coefficient of determination, coefficient of variation, predicted residual sum of squares, and regression data were used to evaluate how well the model fits the data with the aid of Design Expert Software for better analysis.

2.4 Soil stabilization

Dry mixing of materials (mixture of fly ash, volcanic ash, and alkali activator) were performed according to the runs generated by the design considering the parameters listed in Table 1-2. Amount of water equivalent to 8% of the soil was added on the dry formula followed by mixing for at least 5 minutes. The geopolymer paste binder, a mixture of dry formula and water, was completely mixed with soil. After stabilization, geopolymer stabilized soil was compacted in a cylindrical mold 50 mm diameter by 100 mm height for 3 layers of about 25 tamps per layer. The molded specimen was sealed and cured for 28 days at ambient conditions. After curing, the unconfined compressive strength was determined using Digimax Plus automatic Uniframe compression tester.

2.5 Sulfuric acid resistance test

Specimen, prepared based on the parameters yielding highest UCS, was compacted in a cylindrical mold 50 mm diameter by 100 mm height and allowed to cure for 28 days. After curing, sample was immersed in 5% (w/w) sulfuric acid solution [17, 24]. The compressive strength was measured after 28 days of immersion. Soaked sample was further analysed using XRD and SEM/EDX. The sample for XRD analysis were pulverized in powdered form and were sent to Tokyo Institute of Technology. The samples for SEM/EDX analysis were obtained from the broken samples resulted from the destructive compressive test and were then sent to DOST-Admatel. Samples, coated with gold, were analysed using Field-emission scanning electron microscope (voltage=2.0 kV and beam current=43 pA) equipped with Energy-dispersive X-ray spectroscopy with a voltage of 15.0 kV and beam current of 0.69 nA.

Table 3. Chemical composition of soil, fly ash and volcanic ash.

Chemical Compound	Soil (% wt)	Fly Ash (%/wt)	Volcanic Ash (%/wt)
SiO_2	38.87	24.65	56.75
Fe_2O_3	34.06	44.63	13.00
Al_2O_3	12.53	8.55	11.34
CaO	6.82	16.04	10.93
TiO_2	3.09	1.32	1.35
K_2O	1.87	1.18	5.01
SO_3	-	1.63	-

3 Results and discussion

3.1 Raw material characterization

3.1.1 Soil

Soil used in this work was classified as brown silty sand (SM) in accordance with unified soil classification system. The results of liquid limit, plastic limit and plasticity index showed that the fines are non-plastic and had a specific gravity of 2.54. Soil particles has a cloudy and an irregular honeycomb like structure as presented in Figure 1. The particle size distribution is shown in Figure 2. This illustrates that the median diameter (D_{50}) is 0.70 mm. The elemental analysis of soil in oxides form, conducted by ITDI-DOST is shown in Table 3. It is primarily composed of SiO_2, Fe_2O_3, Al_2O_3, and CaO. The mineralogical phase also showed presence of Si and Al as shown in Figure 3a wherein it contains quartz (SiO_2), cristobalite (SiO_2), montmorillonite ($(Na,Ca)_{0.3}(Al,Mg)_2Si_4O_{10}(OH)_2 \cdot nH_2O$), and magnetite ($Fe_3O_4$).

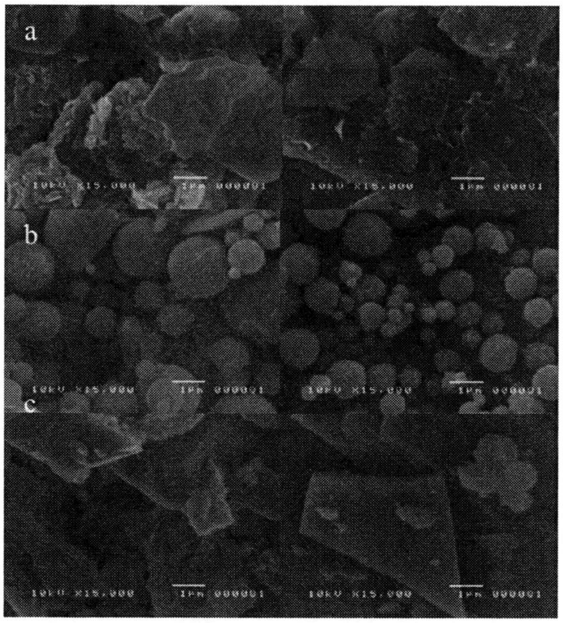

Fig. 1. SEM images of a) soil, b) fly ash, and c) volcanic ash.

3.1.2 Fly ash

Fly ash is classified as ASTM class F fly ash or low-calcium fly ash considering that the total SiO_2, Al_2O_3, and Fe2O3 is more than 70%. It is primarily composed of SiO_2, Fe_2O_3, CaO, and Al_2O_3 with median particle size of 1.92 microns as shown in Figure 2. Morphological characteristics of fly ash showed non-uniform spherical particle as illustrated in Figure 1. XRD pattern of fly ash revealed the presence of quartz (SiO_2), mullite ($2Al_2O_3SiO_2$), hematite (Fe_2O_3), and magnetite (Fe_3O_4). as shown in Figure 3b.

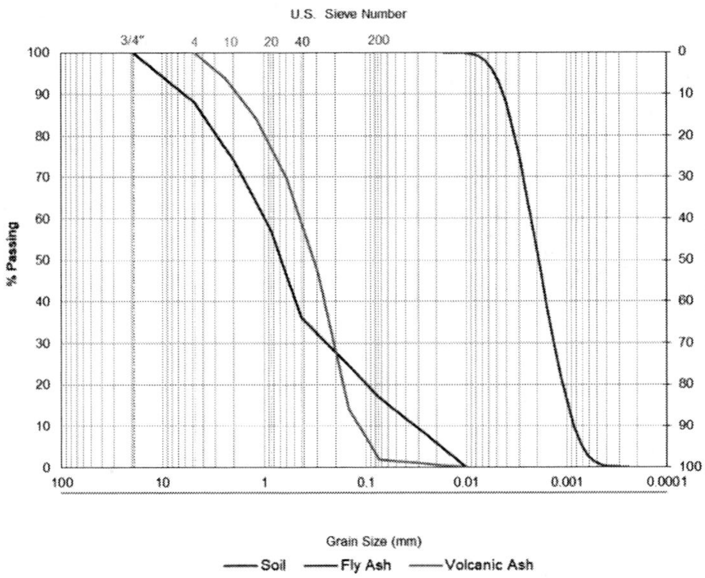

Fig. 2. Particle size distribution of soil, fly ash and volcanic ash.

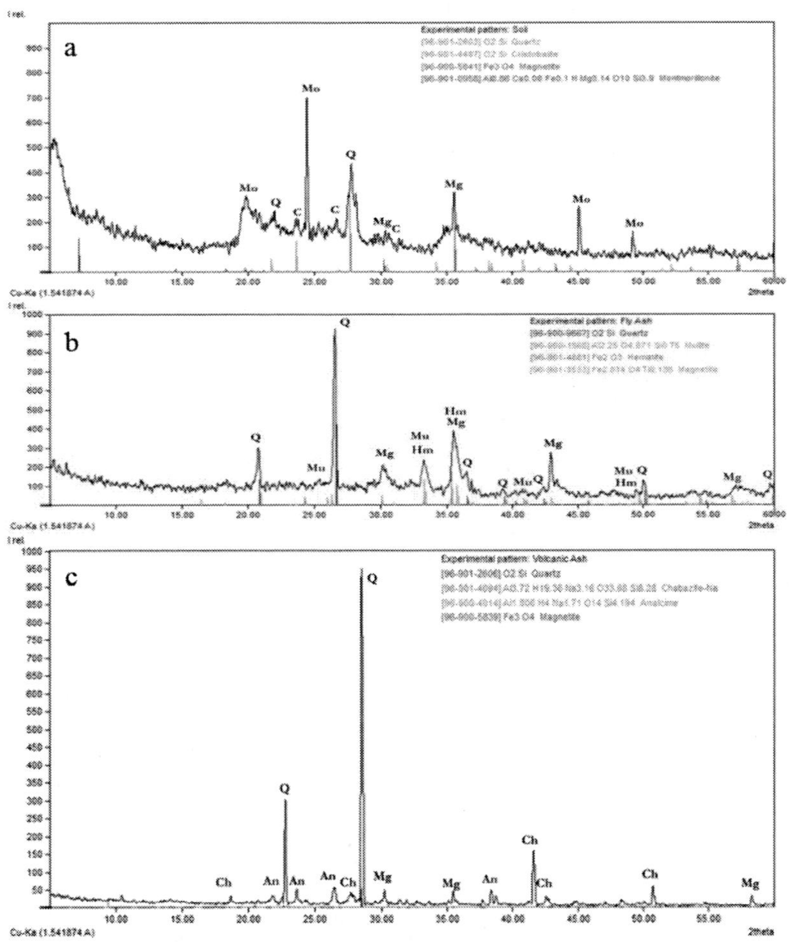

Fig. 3. XRD pattern of a) soil (Mo- Montmorillonite; Q- Quartz; C- Cristobalite; Mg- Magnetite); b) Fly Ash (Q- Quartz; Mu- Mullite; Hm-Hematite; Mg- Magnetite); and c) Volcanic Ash (Q- Quartz; A- Analcime; Ch- Chabazite; Mg- Magnetite).

3.1.3 Volcanic ash

Volcanic ash primarily comprises SiO_2, CaO, Al_2O_3, and Fe_2O_3. The particle size distribution of volcanic ash is shown is Figure 2. It is observed that 50% of the particles has a diameter of 0.33 mm. Morphology of volcanic ash particles consists irregular flaky with angular edges as illustrated in Figure 1. The result of XRD also showed presence of crystalline phases such as quartz (SiO_2), analcime ($NaAlSi_2O_6 \cdot H_2O$), chabazite ($(Ca,Na_2,K_2,Mg)Al_2Si_4O_{12} \cdot 6H_2O$), and magnetite ($Fe_3O_4$) (Figure 3c).

3.2 Unconfined compressive strength

Table 4-5 summarize the measured unconfined compressive strength of stabilized soil geopolymer at different combination of raw materials and alkali activator. The unconfined compressive strength of the stabilized soil geopolymer was observed to range from a minimum 45 kPa to a maximum 971 kPa. The stabilized soil with 25% fly ash/75% soil and alkali activator with 50% SH/50% SS was measured to yield the highest value of unconfined compressive strength. As per requirement of National Structural Code of the Philippines, a compressive strength of at least 75 kPa can be used for possible application in structural fill [25].

Table 4. Measured response (unconfined compressive strength) at varying amount of fly ash and alkali activator.

Amount of Fly Ash (%)	Alkali Activator Ratio (SS:SH:SA)	UCS (kPa)
15	50:50:0	455
25	50:50:0	971
15	50:20:30	45
25	50:20:30	348
15	33:33:33	149
25	33:33:33	623

Table 5. Measured response (unconfined compressive strength) at varying amount of volcanic ash and alkali activator.

Addition of Volcanic Ash (%)	Alkali Activator Ratio (SS:SH:SA)	UCS (kPa)
0	50:50:0	971
12.5	50:50:0	745
0	50:20:30	348
12.5	50:20:30	242
0	33:33:33	623
12.5	33:33:33	421

To statistically measure the influence of amount of fly ash, addition of volcanic ash, and alkali activator ratio using varying combination of parameters, experimental data shown in Table 4-5 were analysed using ANOVA (analysis of variance). Results from the statistical analysis showed that the amount of fly ash, and ratio of alkali activator have significant effect on unconfined compressive strength as shown in Table 6-7 (p-value of 0.0221 and 0.0445). Additionally, presence of volcanic ash in the geopolymer matrix is found to significantly affect the strength of geopolymer stabilized soil having a p-value of 0.0399. The measured response fitted significantly, with a p-value of 0.0336 and 0.0162, in both models. Furthermore, the resulting adjusted coefficient of determination of the models were calculated to be 94.36% (amount of fly ash) and 97.29% (addition of volcanic ash), respectively.

Table 6. Measured p-value at varying amount of fly ash and alkali activator.

Source	p value (Prob>F)	
Model	0.0336	significant
Amount of fly ash (FA)	0.0221	significant
Alkali Activator Ratio	0.0445	significant

Table 7. Measured p-value at varying amount of volcanic ash and alkali activator.

Source	p value (Prob>F)	
Model	0.0162	significant
Amount of volcanic ash (VA)	0.0399	significant
Alkali Activator Ratio	0.0124	significant

The final equation for the unconfined compressive strength were found to be:

Alkali Activator ratio (50SS:50SH:0SA)

$$UCS = -149 + (43.10*FA) \quad (1)$$

$$UCS = 947 - (14.24*VA) \quad (2)$$

Alkali Activator ratio (50SS:20SH:30SA)

$$UCS = -665 + (43.10*FA) \quad (3)$$

$$UCS = 384 - (14.24*VA) \quad (4)$$

Alkali Activator ratio (33SS:33SH:33SA)

$$UCS = -476 + (43.10*FA) \quad (5)$$

$$UCS = 611 - (14.24*VA) \quad (6)$$

The coefficients show the magnitude effect of factors on unconfined compressive strength. Equations 1, 3, and 5 show that high levels of amount of fly ash will result in an increase in compressive strength while equations 2, 4, and 6 show that high levels of amount of volcanic ash will result in the reduction of compressive strength. An increase in the amount of fly ash relative to the amount of volcanic ash could result to an increase in compressive strength. This may be attributed to the reactivity of the precursor in which particle size is a major factor [17, 26]. Finer particles of fly ash may be an advantage as compared to the particle size of volcanic ash. Thus, increasing the amount of fly ash, more reactive source of aluminosilica in the precursor, could lead to a higher value of compressive strength.

* Corresponding author: michael.promentilla@dlsu.edu.ph

https://doi.org/10.1051/matecconf/201815605017

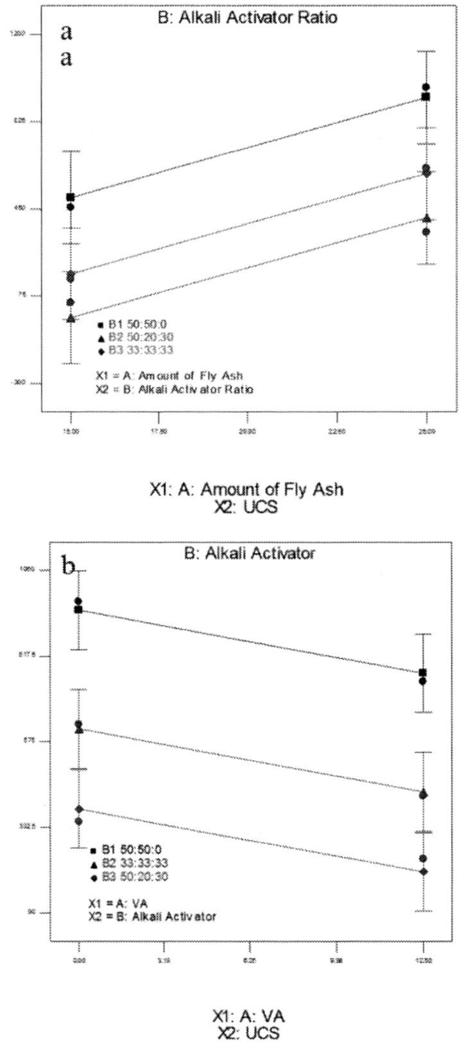

Fig. 4. Contour plot of a) fly ash and alkali activator ; b) volcannic ash and alkali activator.

The increase in compressive strength of geopolymer stabilized soil have been remarkably noticed in alkali activator containing 50%SS:50%SH as shown in Figure 4. The model suggests that as the amount of sodium hydroxide increases, the strength of the stabilized soil tends to increase. Low amount of sodium hydroxide in the alkali activator, <20%, at varying amount of fly ash and volcanic ash could lead to a low strength stabilized soil. The production of low strength stabilized soil could be attributed to a lower dissolution rate, less leaching of alumina and silica ions from the precursor, thus lower value of compressive strength [16].

On the other hand, controlling the amount of sodium aluminate could produce a high strength stabilized soil considering higher amount of fly ash is present in the precursor and as long as more sodium hydroxide is present in the alkali activator. A greater proportion of sodium aluminate relative to the amount of sodium hydroxide in the alkali activator could lead to a low strength stabilized soil geopolymer. This may be due to the increase in amount of alumina source and decrease in amount of silica source since at low amount of sodium hydroxide, less leaching of silica is anticipated [23]. Additionally, geopolymer produced with more silica tend to have higher values of compressive strength [27]. Thus, for this case, low amount of silica relative to the alumina could produce a low strength stabilized soil geopolymer.

3.3 Sulfuric acid resistance

The strength of geopolymer stabilized soil was observed to increase when exposed in acidic medium. The unsoaked specimen was measured with strength of 971 kPa while the compressive strength of soaked specimen exposed in 5% (w/w) sulfuric acid solution for 28 days was measured with strength of 2620 kPa. The increase in strength may be attributed by the influence of curing time considering that the gel matrix seems to be not affected by acid attack. A longer soaking time could mean an increase also with the curing time prior unconfined compressive test. The unreacted aluminosilica and alkali activator might continue to undergo geopolymerization resulting to an increase of strength over time [13]. Thus, the increase in strength of soaked specimen showed that geopolymer stabilized soil is resistant to acid attack.

3.4 Mineralogical properties

Raw materials such as soil and fly ash were observed to have sharp diffraction peak of quartz as shown in Figure 5. Significant mineralogical change has been demonstrated between soil, fly ash, unsoaked specimen, and soaked specimen. The notable decrease in peaks of quartz as depicted in Figure 5 shows the dissolution of crystalline phase results to an increase in reactivity. On the other hand, formation of new crystals such as anorthite (calcium tecto-dialuminodisilicate) and dolomite (calcium magnesium carbonate) for unsoaked specimen was observed. Also, presence of an iron rich crystals, magnetite, was also expected given that the soil and fly ash contain a significant amount of iron as revealed by the XRF result. After immersion in sulfuric acid solution, soaked specimen showed a change in mineralogical property where sharp peaks of gypsum were observed at 11.60° and 20.70° 2θ, respectively. Presence of gypsum may be related to the reaction between the free calcium in the geopolymer matrix and the sulfuric acid solution [17]. The peaks of gypsum clearly demonstrate that unreacted calcium compounds can be easily attacked by an acid. Additionally, disappearance of anorthite peaks and decrease in peak intensity of quartz in unsoaked specimen may be attributed to the dissolution of crystalline phases. This results to an increase in amorphous state thereby increasing the reactivity of these compounds for further geopolymerization [28].

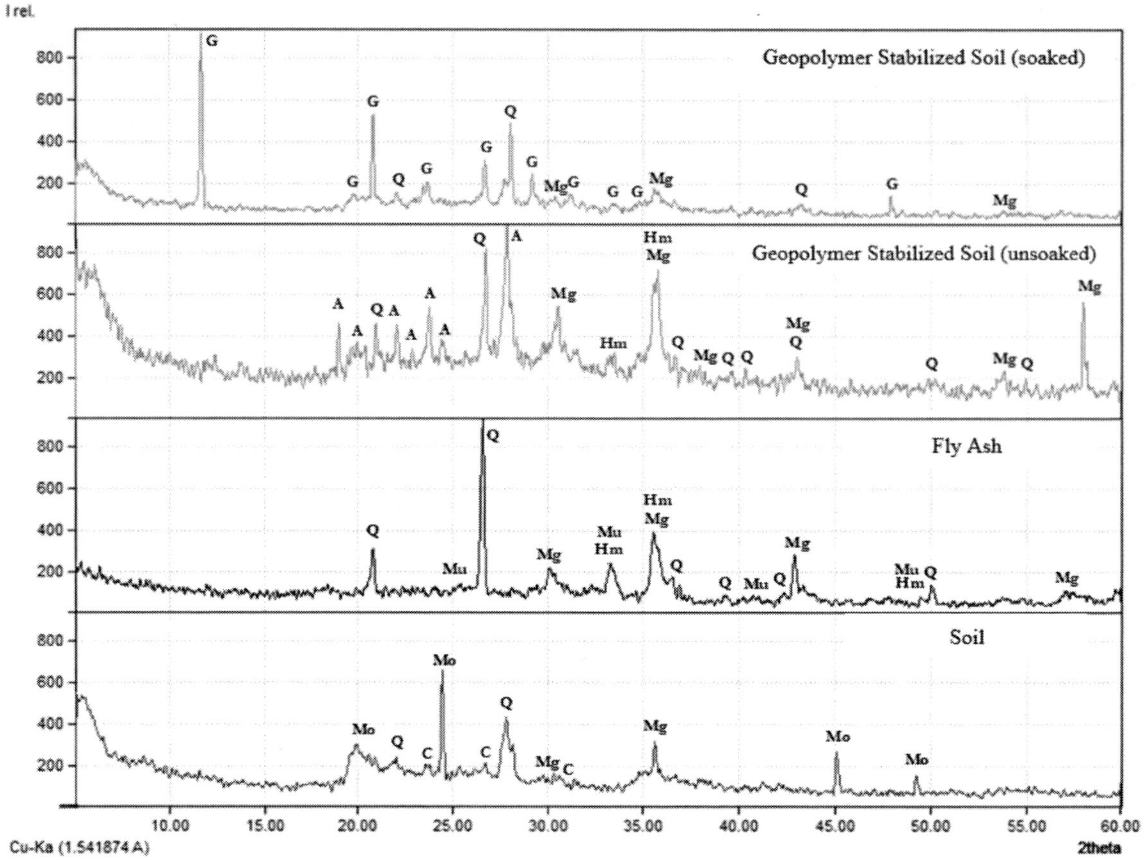

Fig. 5. XRD comparative analysis of raw materials and stabilized soil geopolymer (Q- Quartz; Hm-Hematite; Mu-Mullite; Mg- Magnetite; Mo- Montmorillonite; C- Cristobalite; G- Gypsum; A-Anorthite).

3.5 Microstructure properties

Morphological characteristics of soaked and unsoaked stabilized soil geopolymer was carried out using SEM/EDX analysis as shown in Figure 6. Geopolymeric gel microstructure were also investigated for both samples by comparing its composition based on EDX spectra. The SEM images of samples unsoaked specimen was clearly observed to contain a number of spherical particles similar to morphology of fly ash. This may correspond to unreacted reacted fly ash in the sample. The specimen as shown in Figure 6 was also observed to be partially polymerize due to the partial covering of fly ash particles in the matrix. The particles were also noted to have large voids and pores. Additionally, presence of fibrous crystals and flower like structures are clearly seen. Small spongy particles were also visible covering the spherical particles. This structure resembles the morphology of magnetite. This is consistent with the XRD pattern of unsoaked specimen which showed diffraction peaks of magnetite in Figure 5.

On the other hand, soaked specimen evidently contains lesser amount of unreacted fly ash resulting to a homogenous phase structure as illustrated in Figure 7. The sample also appears to be well compacted and cemented leading to smaller voids and pores on the morphology. A needle like structure from a broken sphere was also observed. This may indicate partial dissolution of fly ash, thus exposing the smaller particles. Moreover, a hexagonal shape structures were also evident in the sample as shown in Figure 8. As confirmed by the EDX analysis, these structures are rich in calcium and sulfur. Formation of crystal rich in calcium and sulfur is an indication of gypsum in the matrix. The formation of gypsum is a result of reaction between free Ca^{2+} ions in the specimen and the free SO_4^{-2} ion in the acidic medium. The free Ca^{2+} ions in the specimen was noted to be directly attacked by the acidic medium and the matrix rich in Na gel was found to be resistant with no indication of sulfur in the EDX analysis. The evidence of gypsum formation was also observed in the result of XRD pattern of soaked specimen in Figure 5.

Fig. 6. SEM images of unsoaked pecimen.

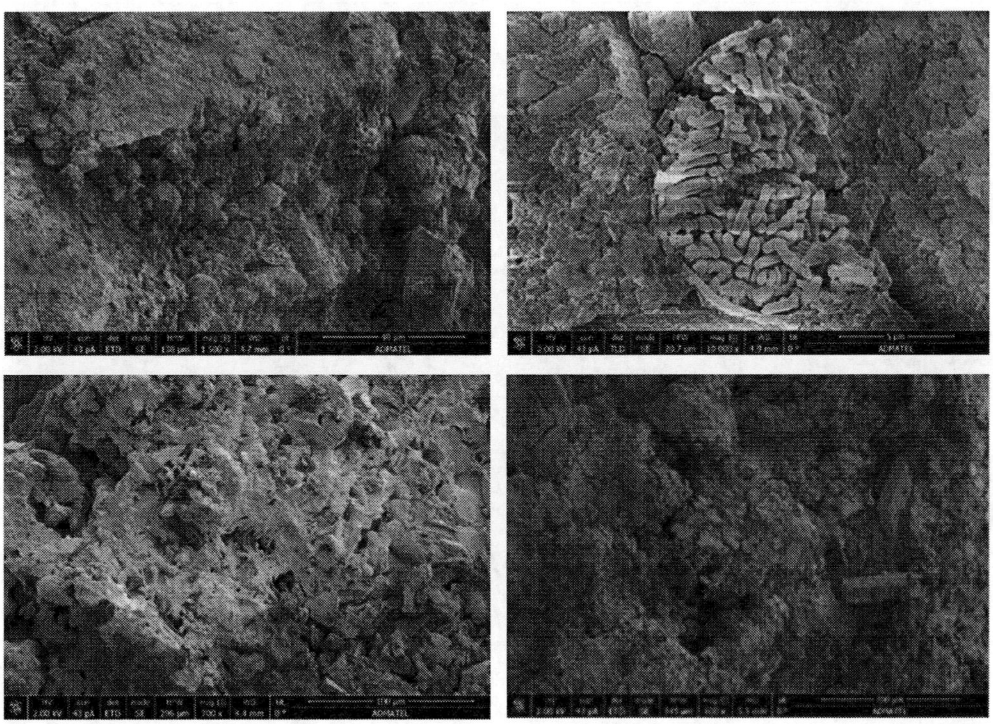

Fig. 7. SEM images of soaked specimen.

* Corresponding author: michael.promentilla@dlsu.edu.ph

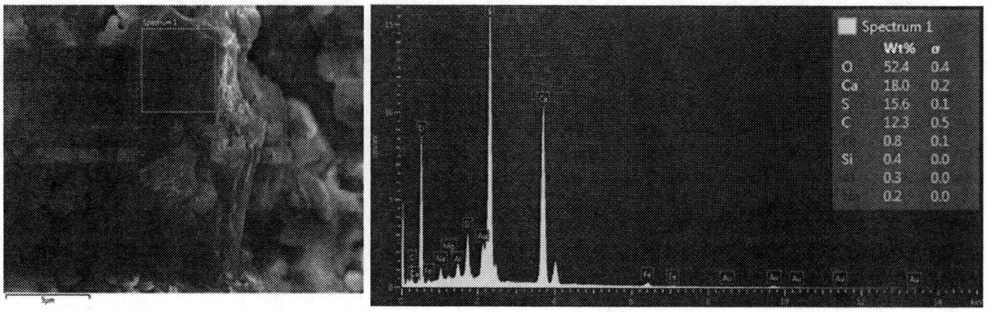

Fig. 8. SEM/EDX analysis of gypsum in soaked specimen.

The gel formation EDX analysis of stabilized soil geopolymer for both unsoaked and soaked specimen are shown in Table 8 and Figure 9. The elemental analysis of both specimen predominantly consists of Na, Ca, Fe, Si, Al, O, and C. The molar ratios are calculated to be Na/Al=2.05, Si/Al= 2.64, Na/Si=0.78, Ca/Si=0.11, and Fe/Si=0.57 for unsoaked specimen. Similarly, the composition of the gel of soaked specimen in terms of molar ratios of Na/Al, Si/Al, Na/Si, Ca/Si, and Fe/Si are 1.11, 2.83, 0.39, 0.06, and 0.17, respectively. These composition of gel resembles a typical poly(ferro-sialate-disiloxo) type of binder with a formula relatively close to [Ca,Na,K,Mg]–[–Fe–O–]$_x$–[Si–O–Al–O–]$_{1-x}$–[–Si–O–] having values of x less than or equal to 0.5, and a value of y between 0 and 25 [29, 30]. This is an evidence that the iron component of the raw materials participated in geopolymerization. This is consistent with the result of XRD analysis wherein peak intensities of iron rich crystals was decreased.

Table 8. EDX Analysis of gel formation.

Specimen	Composition (wt%)							Molar Ratio				
	Na	Ca	Si	Al	Fe	O	C	Na/Al	Si/Al	Na/Si	Ca/Si	Fe/Si
G1	6.3	1.5	9.9	3.6	11.3	38.8	26.8	2.05	2.64	0.78	0.11	0.57
G2	5.5	1.4	17.1	5.8	5.8	48.8	13.6	1.11	2.83	0.39	0.06	0.17

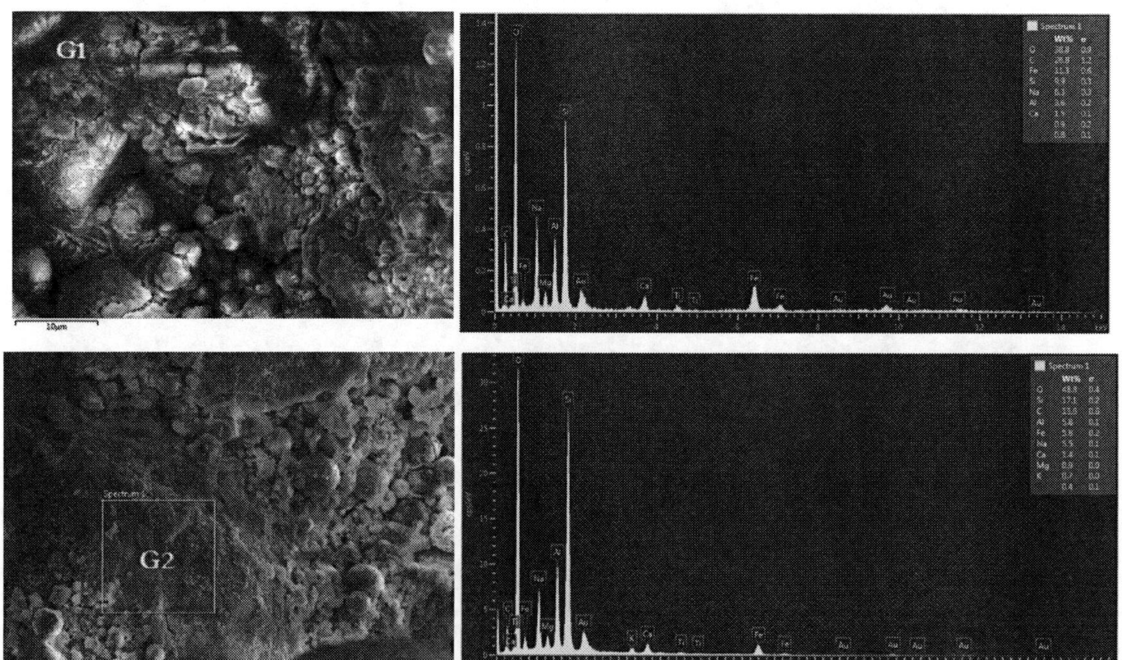

Fig. 9. EDX analysis of gel formation of unsoaked specimen (G1) and soaked specimen (G2).

4 Conclusion

This study has explored the use of geopolymer precursors such as fly and volcanic ash with solid alkali activators such as sodium silicate, sodium hydroxide, and sodium aluminate. The strength of geopolymer stabilized soil was observed to be affected by the amount of fly ash, addition of volcanic ash, and the ratio of alkali activator. Increasing the amount of fly ash resulted in an increase in compressive strength while addition of volcanic ash in geopolymer matrix resulted in the reduction of compressive strength. The significant difference in the morphological images between unsoaked specimen and soaked specimen acts as support on the increase in mechanical strength of stabilized soil geopolymer after immersion in acid solution. The latter which appears to be well compacted and cemented gives a higher value of unconfined compressive strength while the former evidently contains large voids and pores and a number of spherical particles corresponding to unreacted fly ash gives a lower value of unconfined compressive strength. Moreover, the gel matrix of soaked specimen has shown to be acid resistant as confirmed by the EDX analysis with no indication of sulfur. A hexagonal shape structures were also evident in the sample indicating gypsum formation as confirmed by EDX and XRD pattern. Also, the composition of gel matrix formed resembles a typical poly(ferro-sialate-disiloxo) type of binder. The decrease in peaks of XRD pattern was found to be a manifestation of geopolymerization.

The authors would like to acknowledge the financial support from Engineering Research and Development for Technology (ERDT), and URCO Research Project. We are also thankful for the laboratory facilities and support of International Development Engineering of Tokyo Institute of Technology (TIT), Mechanical Engineering Laboratory, Chemical Engineering Laboraroty, and Geotechnical Laboratory of De La Salle University (DLSU), Advanced Device and Materials Testing Laboratory (ADMATEL) of Department of Science and Technology. Dr. Promentilla also acknowledges the JSPS Bridge Fellowship Program (BR171401) for the financial support during the conduct of the study TIT, Japan

References

1. Department of Energy. Coal Statistics. Available online: http://www.doe.gov.ph (accessed on 19 September 2016).
2. Department of Energy. Power Situation. Available online: http://www.doe.gov.ph (accessed on 19 September 2016).
3. S. Gallardo, J. Dungca, R. Gallardo, M. Kalaw, Sustainability Issues Due to Coal Ash from Coal Fired Power Plants Phase1: Impact Assessment of Coal Ash Dumping in a Typical Power Generating Facility; Interdisciplinary Research Final Report, (2012).
4. Ibrahim, Water Sci. **29**, 109 (2015).
5. G. Akar, M. Polat, G. Galecki, U. Ipekoglu, Fuel Process. Technol. **104**, 50 (2012).
6. N. Koukouzas, C. Ketikidis, G. Itskos, Fuel Process. Technol. **92**, 441 (2011).
7. T. Orejas, Phivolcs: Pinatubo lahar threats not over, Inquirer.Net, 1(2015) Available online: http://newsinfo.inquirer.net (accessed on 19 September 2016).
8. I. O. Flora, Mudflow threat still present in Pampanga, Sunstar Pampanga, 1(2017) Available online : http://www.sunstar.com.ph (accessed on 30 June 2017).
9. Phivolcs, Phivolcs volcano monitoring, Philippine Institute of Volcanology and Seismology, 1(2009) Available online : http://www.phivolcs.dost.gov.ph (accessed on 30 June 2017).
10. A. Fernández-Jiménez, A. Palomo, in *Geopolymers*, edited by J. L. Provis and J. S. J. van Deventer (Woodhead Publishing, 89, 2009).
11. P. Duxson, in *Geopolymers*, edited by J. L. Provis and J. S. J. van Deventer (Woodhead Publishing, 37, 2009).
12. B. Singh, G. Ishwarya, M. Gupta, S. K. Bhattacharyya, Constr. Build. Mater. **85**, 78 (2015).
13. *Standard Test Method for Particle-Size Analysis of Soils* (ASTM International, West Conshohocken, PA, 2007).
14. *Standard Test Methods for Liquid Limit, Plastic Limit, and Plasticity Index of Soils* (ASTM International, West Conshohocken, PA, 2010).
15. P. Pavithra, M. S. Reddy, P. Dinakar, B. H. Rao, B. K. Satpathy, A. N. Mohanty, J. Clean. Prod. **133**, 117 (2016).
16. C. Phetchuay, S. Horpibulsuk, A. Arulrajah, C. Suksiripattanapong, A. Udomchai, Appl. Clay Sci. **127–128**, 134 (2016).
17. J. N. Y. Djobo, A. Elimbi, H. K. Tchakouté, S. Kumar, Constr. Build. Mater. **124**, 606 (2016).
18. A. Fauzi, U. J. Fauzi, W. M. Nazmi, Procedia Eng. **54**, 675 (2013).
19. R. A. Mozumder, A. I. Laskar, Comput. Geotech. **69**, 291 (2015).
20. M. Zhang, M. Zhao, G. Zhang, P. Nowak, A. Coen, M. Tao, Appl. Clay Sci. **108**, 199 (2015).
21. M. Z. N. Khan, F. U. A. Shaikh, Y. Hao, H. Hao, Constr. Build. Mater. **125**, 809 (2016).
22. A. A. Aliabdo, A. E. M. A. Elmoaty, H. A. Salem, Constr. Build. Mater. **121**, 694 (2016).
23. A. Hajimohammadi, J. L. Provis, J. S. J. van Deventer, Ind. Eng. Chem. Res. **47**, 9396 (2008).
24. M. Jin, Z. Zheng, Y. Sun, L. Chen, Z. Jin, J. Non. Cryst. Solids **450**, 116 (2016).
25. Association of Structural Engineers of the Philippines, National structural code of the Philippines 2015: Volume I building, towers, and other vertical structures. (Association of Structural Engineers of the Philippines, Inc., 1, 2015).

* Corresponding author: michael.promentilla@dlsu.edu.ph

https://doi.org/10.1051/matecconf/201815605017

26. M. Zhang, M. Zhao, G. Zhang, D. Mann, K. Lumsden, M. Tao, Constr. Build. Mater. **124**, 373 (2016).
27. F. Pacheco-Torgal, J. Castro-Gomes, S. Jalali, Constr. Build. Mater. **22**, 1315 (2008).
28. L. Chen, Z. Wang, Y. Wang, J. Feng, Materials (Basel). **9**, (2016).
29. J. Davidovits, Geopolymer Inst. Libr. 1 (2013).
30. S. Kumar, J. N. Yankwa Djobo, A. Kumar, S. Kumar, J. Build. Eng. **8**, 172 (2016).

Development of Abaca Fiber-reinforced Foamed Fly Ash Geopolymer

Ngo, Janne Pauline S.[1*], *Promentilla,* Michael Angelo B.[1,2]

[1]Chemical Engineering Department, De La Salle University, Manila 0922, Philippines
[2] National Research Council of the Philippines, Taguig, Metro Manila 1631, Philippines

Abstract. The growing environmental and economic concerns have led to the need for more sustainable construction materials. The development of foamed geopolymer combines the benefit of reduced environmental footprint and attractive properties of geopolymer technology with foam concrete's advantages of being lightweight, insulating and energy-saving. In this study, alkali-treated abaca fiber-reinforced geopolymer composites foamed with H_2O_2 were developed using fly ash as the geopolymer precursor. The effects of abaca fiber loading, foaming agent dosage, and curing temperature on mechanical strength were evaluated using Box-Behken design of experiment with three points replicated. Volumetric weight of samples ranged from 1966 kg/m^3 to 2249 kg/m^3. Measured compressive strength and flexural ranged from 19.56 MPa to 36.84 MPa, and 2.41 MPa to 6.25 MPa, respectively. Results suggest enhancement of compressive strength by abaca reinforcement and elevated temperature curing. Results, however, indicate a strong interaction between curing temperature and foaming agent dosage, which observably caused the composite's compressive strength to decline when simultaneously set at high levels. Foaming agent dosage was the only factor detected to significantly affect flexural strength.

1 Introduction

Sustainability has been an important drive for research for the past years. Increasing consumptions of resources and materials, as well as the immoderately massive generation of waste, has pulled the research and development of modern materials towards greener alternatives. An important factor to consider is the construction industry to which over 40% of the global energy consumption and excessively high quantity of carbon dioxide emission is due [1]. This sector has nonetheless seen growth in demand continually and inevitably. Hence, the development of sustainable construction materials to possibly substitute these commodities and reduce environmental footprint should be sought after. Considering this, one of the most promising and fast-growing research in the past two decades is on geopolymer. Geopolymer, an inorganic binder formed by alkaline activation of solid alumina- and silica-containing materials which are often sourced from industrial waste, has emerged as a potential concrete binder – an alternative to ordinary Portland cement (OPC). Geopolymer production is reported to curb CO_2 emission of traditional cement clinker by 60 to 80% [2]. Moreover, it finds increasing applications in other areas due to its unique properties such as thermal and fire resistance.

Previous studies have reported foamed concrete's brittleness to be unsuitable for bending and squash loads [3]. Fiber reinforcement is commonly employed to enhance mechanical strength by allowing load transfer and crack bridging. Commonly reinforced fibers in cementitious materials include steel and polymeric fibers [4]. The use of natural fiber has recently been explored, particularly for its inherent mechanical properties, low density, low cost, abundance and biodegradability.

1.1 Abaca Fiber

Abaca fiber which is native and abundant in the Philippines is widely used for its commendable mechanical properties and utility for polymer reinforcement. An annual report in 2015 reveals that 70,400 metric tons of the fiber was produced [5].The large volume of harvested high-grade abaca in turn generates a considerable amount of waste or low-grade fibers [6]. As much as three quarters of the abaca is considered agricultural waste and left in plantations to wither after the high-grade fiber is harvested [7]. About 328.5 million kilograms of abaca waste was reported to be generated annually in 2014. The industry, however, is foreseen to grow even more in the succeeding years. In 2015, it was reported that the outbound shipments of abaca rose by 39.8% from the previous year [8]. In 2016,

* Corresponding authors: janne_ngo@dlsu.edu.ph; michael.promentilla@dlsu.edu.ph

farmers have increased their production of abaca by 5.7% (from 58,665.8 MT to 62,008.6 MT) from the previous year to meet the demand for the crop in the Philippine's export market [9]. It was reported in 2015 that the about 87.7% of the profit gained from abaca came from finished goods such as fabrics, cordage and fiber crafts, while raw abaca fibers accounted for the remaining exports. However, revenue from raw abaca fiber has been tripling year after year [8]. Raw abaca fiber's application in various industries has indeed broadened. As of present, the most common innovative use of abaca-reinforced composites has been in under-floor protection of automotive [10]. With the high volume of agricultural waste involved in this growing industry, it would therefore be advantageous for the environment if these wastes or scrap fibers can be treated and utilized in advanced materials. With that said, this study aims to explore the use of waste abaca fiber as reinforcement in geopolymer with foaming. Particularly, this study aims to develop an abaca-reinforced fly ash geopolymer foamed with hydrogen peroxide (H_2O_2) and to assess its compressive and flexural strength under various fiber loading, H_2O_2 dosage and curing temperature.

1.2 Fiber Pretreatment

Often, reinforcement is employed in order to improve the material's mechanical properties. Fibers are used as load-bearing materials imbedded in another's matrix, thus providing rigidity, strength and flexibility in many cases [4]. The resulting junction is called a composite. Concrete is often reinforced to address premature cracking that develop with shrinkage during the hardening of the paste. This shrinkage, called plastic shrinkage, occurs due to loss of moisture [11]. Fiber reinforcement has been effective preventing crack propagation by bridging cracks and by allowing load transfer. However in some cases, fiber loading can be counterproductive and lead to compressive strength reduction due to factors like improper distribution and weak interfacial interaction. When the fiber acts more as an interference in the matrix than an aid or reinforcement, mechanical properties may become less desirable [11-12].

Stress sharing and transfer between the fiber and the matrix depend on fibre-fibre and fibre-matrix adhesion and interactions. Effective reinforcement relies on compatibility of components. Likewise, incompatibility may only result to decline in mechanical strength and promote deterioration. This has been reported by prior studies to be a disadvantage of natural fibers. Natural fibers typically consist of lignin, pectin, cellulose, hemicellulose, moisture and other compounds such as fats and waxes. It is basically a composite in itself of polymers: cellulose, hemicellulose and lignin [13]. Due to its hydrophilic nature and thus, tendency for water absorption, natural fibers may exhibit weak interfacial bonding with hydrophobic matrix [12, 14-17]. This is often observed in polymer composites. Furthermore, 'swelling' may occur when a hydrophilic fiber within a hydrophobic resin absorbs moisture, making it susceptible to weakening and even deterioration [18]. Consequently, softened fibers will result to reduced strength.

The most economical and well-studied treatment to overcome this disadvantage is alkali treatment [16]. Soaking fibers in alkali solution, commonly sodium hydroxide (NaOH), strips them of impurities which contribute to water absorption and incompatibility. Parts of the amorphous components lignin, hemicellulose and pectin are removed, resulting to smaller fibers with increased effective surface area. The remaining fiber after these components come off will reveal a rough surface, thereby increasing the fiber's resistance to being pulled out of the matrix [16].

2 Methodology

In this study, alkali-pretreatment was performed on the waste abaca as follows: fibers were initially cut to approximately 1 cm strands. A solution of 6 wt% NaOH was prepared from 12M NaOH available. The waste abaca fibers were then submerged in the alkali solution following the proportion 70-80 mg fiber per 100 mL solution for 2 hours. The fibers were subsequently washed with water until they are sufficiently neutralized to about pH 7. Then, the fibers were air-dried for 24 hours. This procedure was reported to yield the best mechanical strength improvement of abaca reinforced in epoxy [17]. **Fig. 1** shows the untreated and treated abaca fibers.

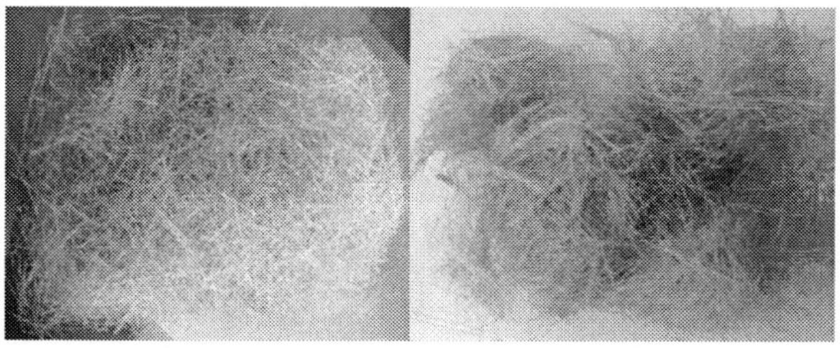

(a) (b)
Fig. 1. (a) Untreated and (b) treated abaca fibers

Preparation of geopolymer composites consists of two parts: (1) preparation of alkali activator solution and (2) preparation of the composites. For the alkali activator solution, waterglass to sodium hydroxide solution (WGS/NaOH) was fixed at mass ratio 1.66. The activator solution was mixed thoroughly with a magnetic stirrer. After 15 minutes, 50 wt% hydrogen peroxide was added at 0.16, 0.24 and 0.32 weight % of the fly ash used to the activator solution and mixed for 5 more minutes.

To prepare the specimen with fiber reinforcement, the cut, treated and dried abaca fibers were dry-mixed with sieved fly ash (2mm) to ensure uniform distribution. Fiber loading was designated to 0%, 0.25% and 0.5 wt% of the weight of fly ash used. Meanwhile, the prepared alkali activator was incorporated with the fly ash and abaca following mass ratio NaOH/FA = 0.2. The slurry was manually mixed first to ensure the desired consistency. Then, the mixture was thoroughly mixed in a mechanical mixer for about 15 minutes until fiber strands were well-distributed and that homogeneity was achieved. The geopolymer paste was then casted on moulds, allowed to dry for 24 hours and removed from the moulds. Samples were then placed in air-tight plastics and cured at various temperatures, 30°C, 52.5°C and 75°C, for 2 hours, followed by 24-hour hardening at room temperature.

Scanning electron microscopy (SEM) images of treated and untreated abaca fibers were taken and analysed. Thermal gravimetric analyses were also performed on the fibers. Subsequently, the prepared geopolymer composites were characterized by SEM image analysis, volumetric weight and water absorption. Finally, compressive strength and flexural strengths of the composites were measured.

3 Results and Discussion

3.1. Scanning electron microscopy (SEM) analysis of abaca

Scanning electron microscopic image analyses were carried out using Field-emission scanning electron microscope (FESEM Dual Beam Helios Nanolab 600i) equipped with electron-dispersive X-ray spectroscopy. Images were taken at accelerating voltage = 2.0 kV and beam current = 43 pA. SEM images of the fiber samples at 500x and 5000x magnification are presented in **Fig. 2. Fig. 2a** and **Fig. 2c** shows the untreated fiber whose surface evidently has less uniformity. "Impurities" which could likely be the weak and dislodged hemicellulose are visible. It is observed in **Fig. 2b** and **Fig. 2d** that these impurities were stripped off, revealing a more uniform but rougher surface. After treatment, corrugation on the fiber surface was observed. Corrugation and roughening of the surface indicate a better fiber–matrix interface by virtue of mechanical interlocking. This surface modification indicates higher frictional bond between the two components, which has previously shown to be effective in resisting shear and bending stress

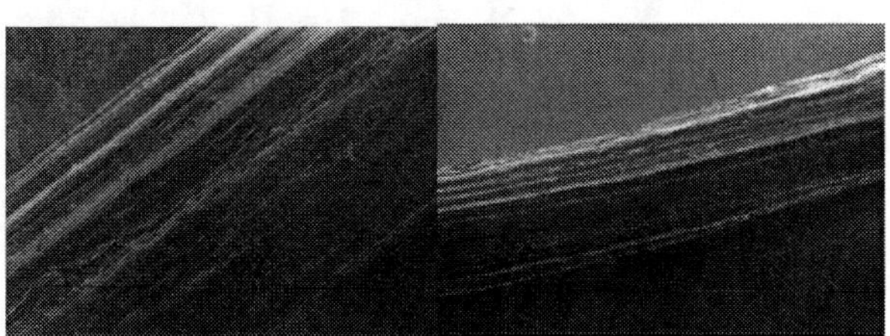

a. untreated, 500x magnification b. treated, 500x magnification

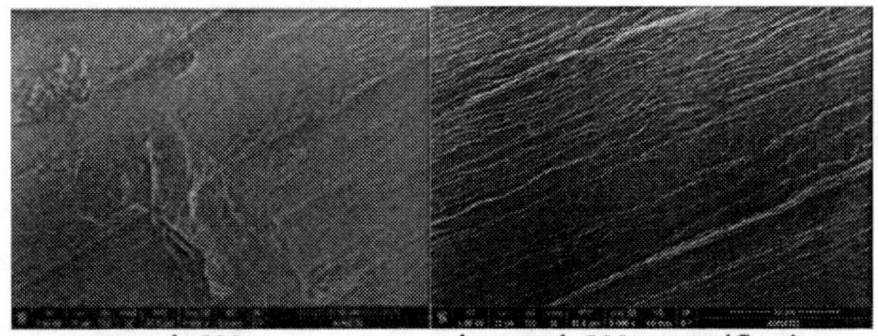

c. untreated, 500x magnification d. treated, 500x magnification

Fig. 2. SEM images of waste abaca before and after treatment

3.2. Thermogravimetric analysis (TGA) of abaca

Percent mass loss during the ramp increase in temperature was analyzed using TGA. This analysis allows evaluation of the fiber's composition. According to previous studies, cellulosic fibers' thermal decomposition commonly comprises of loss of moisture, degradation of hemicellulose, cellulose and lignin. **Table 1** summarizes this with each component's degradation temperature [19].

Table 1. Decomposition Temperatures of Natural Fibers' Components.

Components	Decomposition Temp, °C	T_{max}, °C
Moisture	30 – 100	80
Hemicellulose	160 – 350	245 - 298
Cellulose	240 – 365	335
Lignin	300 – 500	337

Fig. 3 shows the thermograms of both the untreated (a) and treated (b) abaca samples. Meanwhile, **Table 2** summarizes the observed thermal degradation of the untreated and treated abaca samples. T_{max} is defined as the temperature at which the fastest rate of degradation was observed, and it was taken using differential thermogravimetry (DTG). The first stage of decomposition primarily removed free and absorbed moisture from the abaca fibers. For the untreated fiber, decomposition observed between 200°C and 370°C resulted to a mass loss equivalent to 78.2%. T_{max} was observed at 345°C wherein the maximum rate of decomposition was 1.1% per °C. This is mainly accounted to the decomposition of hemicellulose. At the same decomposition stage, the treated fiber experienced a lower % mass loss (71.2%), which indicates its lower hemicellulose content. Beyond this is the onset of the decomposition of cellulose and lignin. This was allowed to proceed up to 900°C, where 87.1% and 79.0% were the measured mass losses for the untreated and treated fiber samples, respectively. This signifies that the treated samples contained lower combined cellulose and lignin content. The untreated fiber's residual mass was 12.9% while that of the treated sample was 21%. These results

signify that the treatment was able to remove the weak and dislodged components off of the abaca fiber.

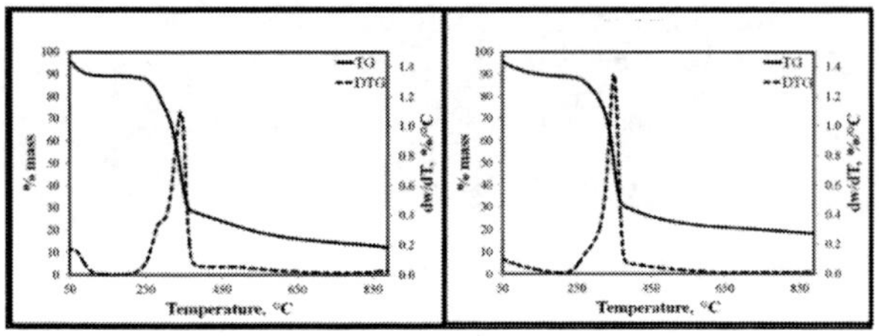

Fig. 3. Thermograms of (a) untreated and (b) treated abaca

Table 2. Observed Decomposition Stages of Abaca Samples

fiber	Decomposition 1			Decomposition 2			% mass loss up 900°C
	T_{max} °C	% mass loss	Rate, % /°C	T_{max} °C	% mass loss	Rate, %/°C	
Raw	62	8.9	0.3	345	78.2	1.1	87.1
Treated	53	7.8	0.1	350	71.2	1.3	79.0

3.3. Density and water absorption of geopolymer composites

The resulting volumetric weight of each treatment was taken as the average of three cubic samples from each run. Each sample's dimensions were measured using caliper and weighed subsequently. Pristine geopolymer samples (0% fiber loading, 0% H_2O_2 dosage, cured at 52.5°C) were measured to have a mean 2149.1 kg/m^3 volumetric weight. The resulting volumetric weights of the samples ranged from 1966.38 kg/m^3 to 2249.52kg/m^3. The mean is 2099.687 kg/m^3. The average value for the foamed samples is relatively high compared to that of a typical foam concrete whose volumetric weight are at most 1000kg/m^3 [20]. Since the abaca fiber's density is significantly lower than that of fly ash and other raw materials, it is reasonable for the composite's volumetric weight to decrease with abaca loading. Likewise, increasing the foaming agent dosage reduced the volumetric weight due to formation of void spaces within the composite. Increasing the curing temperature also resulted to lower volumetric weight which could be due to higher moisture loss during curing.

Water absorption was measured following ASTM D570. The mass of a cubic sample from each run was measured. The samples were then submerged in distilled water for 24 hours, wiped dry at the surface before being weighed again. Volume changes were observably negligible throughout the experimental procedure. The mass percent moisture absorbed is taken as the water absorption, computed using **Eq. 1**.

$$WA = \frac{W_{sat} - W_{dry}}{W_{dry}} x\, 100 \quad (1)$$

where WA = *water absorption* ; W_{sat} = *mass at saturated state* ; W_{dry} = *mass at dry state*

The pristine geopolymer sample's measured water absorption value was 4.499%. The values ranged from 3.3821% to 13.9730% with a mean value of 6.6228%. The values measured for the experimental samples are found to be in the same range as reports on hardened OPC composites, which can range between 2-15%. A study of coir fiber-reinforced epoxy reported that higher fiber content resulted to increased water absorption due to the present hydroxyl groups [21]. In this study, however, the effect of the fiber loading was statistically insignificant to the composite's water absorption. This is favorable since one of the main concerns of natural fiber reinforcement is the tendency for the organic fiber to draw water in due to its hydrophilic nature. The lack of significance of fiber loading effect to water absorption in this study suggests that the alkali treatment has sufficiently reduced the hydrophilic components in the fiber.

Measured volumetric weight and water absorption are summarized in **Table 3**.

Table 3. Volumetric weight and water absorption of composites

Fiber Loading (%)	H_2O_2 Dosage (%)	Curing Temp. (°C)	Volumetric Weight (kg/m^3)	Water Absorption (%)
0	0.16	52.5	2249.52	4.26
0	0.32	52.5	2063.17	7.18
0	0.24	30	2178.76	13.97
0	0.24	75	2142.64	5.68
0.25	0.16	30	2155.76	8.95
0.25	0.16	75	2066.80	5.53
0.25	0.32	30	1986.12	5.25
0.25	0.32	75	2016.05	4.94
0.25	0.24	52.5	2190.26	3.26
0.5	0.16	52.5	2179.78	6.86
0.5	0.32	52.5	1966.38	7.09
0.5	0.24	30	2129.43	6.33
0.5	0.24	75	1976.08	6.78

3.4. Scanning electron microscopy (SEM) analysis of geopolymer composites

SEM images of selected samples at 2000x magnification are shown in **Fig. 4**.

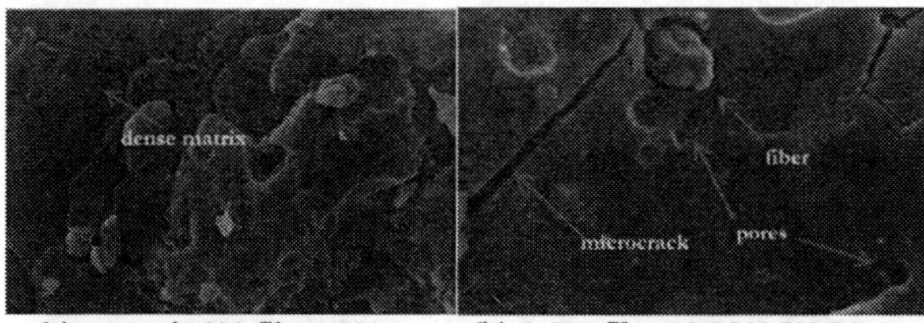

(a) control: 0% fiber, 0% H2O2, cured at 52.5°C (b) 0.5% fiber, 0.32% H2O2, cured at 52.5°C

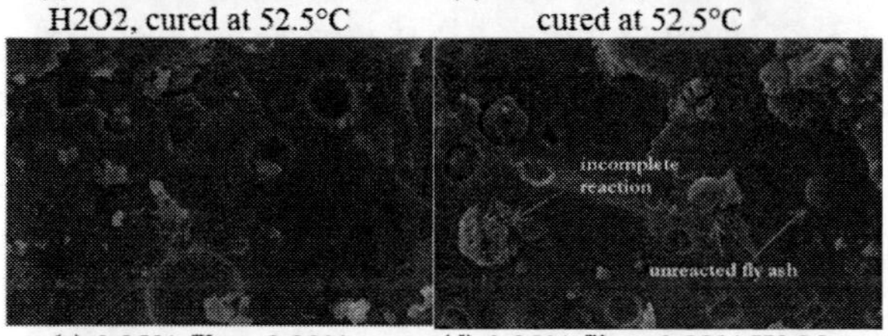

(c) 0.25% fiber, 0.32% H2O2, cured at 75°C (d) 0.25% fiber, 0.32% H2O2, cured at 30°C

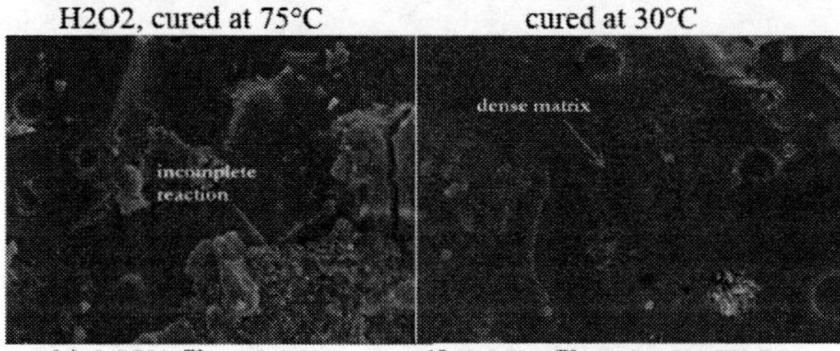

(e) 0.25% fiber, 0.16% H2O2, cured at 30°C (f) 0.25% fiber, 0.16% H2O2, cured at 75°C

Fig. 4. SEM Images of geopolymer composites

SEM analysis can provide clues and qualitative validation of premises about the water absorption, and provide later on, the mechanical properties and thermal properties of the composite samples. Unreacted fly ash and incomplete reactions were observed largely in samples that were cured at low temperature as shown in **Fig. 4d** and **Fig. 4e**. Samples cured at higher temperatures formed more dense matrices, exhibiting less unreacted fly ash and incomplete reactions. Pores were visibly bigger and more evenly distributed in **Fig. 4c** and **Fig. 4f** which were cured at 75°C. Microcracks were also observed in most of the samples but were more evident in **Fig. 4b** and **Fig. 4c.** It can be observed in Figure 5.4b that the fiber-matrix gap seemed to have been the site for microcrack formation. In **Fig. 4d**, in which the sample was cured at low temperature, microcracks were not present along the abaca fibers. This verifies the premise that microcracks tend to form at high fiber loading and high curing temperature. Curing at low temperature, however, can result to incomplete reaction [22], as evidenced by **Fig. 4d** which shows more unreacted fly ash and less dense matrix. This can translate to lower strength, as well. Microcracks present in **Fig. 4c** could be due to the high H_2O_2 dosage and high curing temperature.

3.5. Mechanical properties of geopolymer composite

UNIFRAME stand-alone universal compression/flexural tester was used to measure the compressive and flexural strength of the samples. Compressive strength test was performed following ASTM C109. Meanwhile, the three-point flexural test (ASTM C348) was performed to calculate for the flexural strength of the composites.

3.5.1 Compressive strength of geopolymer composites

Compressive strength is a measure of the structure's resistance or ability to withstand compression stress. This parameter, which will be considered as a response in the experiment, is a key mechanical property for construction materials. Standard values of materials acceptable for use as structural 'residential concrete' are at least 17 MPa. Moreover, 28 MPa is the minimum compressive strength for materials finding application as commercial structures [23]. 50mm cubic samples were prepared, cured for 28 days and tested for compressive strength. In addition to the 13 treatments, a pristine geopolymer sample and 3 replicated treatments were prepared for the test. Samples were subjected to gradual load with rate 0.4 MPa/s. The maximum loads were recorded at failure. Three replicates were tested for each treatment. The mean ultimate load of the three replicates was taken as the compressive strength for each treatment. **Fig. 5** shows the set up for the compressive strength test.

a. compression test set-up b. sample after testing

Fig. 5. Compressive strength test set-up

The average of the compressive strengths of three specimens was taken as the measured value for each run. Pristine geopolymer samples (0% fiber loading, 0% H_2O_2 dosage, cured at 52.5°C) were measured to have a mean compressive strength of 24.04 MPa. The compressive strength values ranged from 19.56 MPa to 36.85 MPa. The mean value is 29.26 MPa. Compressive strength requirements are at least 17 MPa and 28 MPa for residential and commercial concrete, respectively [24]. Compressive strength was analyzed as response to the three factors and a 2nd degree model was generated. The measured R^2 is 0.8275. **Table 4** and summarizes the compressive strength data. **Table 5** shows the ANOVA for compressive strength.

Table 4. Compressive strength of the abaca-reinforced geopolymer composites

Fiber Loading (%)	H_2O_2 Dosage (%)	Curing Temp. (°C)	Compressive strength (MPa)
0	0.16	52.5	27.58
0	0.32	52.5	23.70
0	0.24	30	21.54
0	0.24	75	34.50
0.25	0.16	30	24.96
0.25	0.16	75	35.54
0.25	0.32	30	28.46
0.25	0.32	75	19.56
0.25	0.24	52.5	32.25
0.5	0.16	52.5	33.27
0.5	0.32	52.5	30.97
0.5	0.24	30	28.11
0.5	0.24	75	34.27
*0	0.24	75	2.23
*0.25	0.32	30	1.82
*0.5	0.16	52.5	2.96

*replicated treatments

Table 5. Compressive strength summary of fit and Analysis of Variance

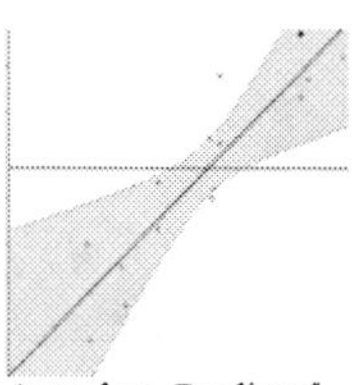

Actual vs. Predicted

Source	DF	Sum of Squares	Mean Square	F Ratio
Model	7	342.15	48.8788	6.8917
Error	8	56.74	7.0925	Prob > F
C. Total	15	398.89		0.0071

Table 6 summarizes the parameter estimates of the model for compressive strength. The abaca reinforcement was found to have improved the compressive strength of the composites. This signifies that the fibers have sufficiently been treated and have formed an effective interaction with the matrix. Foaming agent dosage expectedly decreases compressive strength since more voids are expected to form. Higher curing temperature also improved the compressive strength. This agrees with previous reports [19]. It is said that heat curing facilitates in the reaction and formation of dense geopolymer matrix. The interaction between H_2O_2 dosage and curing temperature, however, resulted to weaker composites. As evidenced by the SEM images and qualitative results of preliminary preparation of samples, it is likely due to the microcrack formation when H_2O_2 dosage and curing temperature are simultaneously set at high levels.

Table 6. Compressive strength model parameter estimates

Term	Estimate	Std Error	t Ratio	Prob>|t|
Intercept	32.220605	1.723139	18.7	**<.0001**
Fiber(0,0.5)	2.5907439	0.899289	2.88	**0.0182**
H2O2(0.16, 0.32)	-2.663281	0.901261	-2.96	**0.0161**
CT(30,75)	2.7500156	0.901261	3.05	**0.0138**
H2O2*CT	-4.752168	1.277011	-3.72	**0.0048**

Fiber - % Fiber Loading; H2O2 – % H_2O_2 Dosage; CT – Curing Temperature °C

3.5.2 Flexural strength of geopolymer composites

Flexural strength, which is also referred to as bend strength, is the mechanical property which measures the stress in the material right before it yields or deforms in a bending test. It is the highest stressed loaded to the material at point of failure. Foam concrete's flexural stress was typically measured to be in the range 15% to 35% of its own compressive strength [25]. Very few studies have considered this parameter. A specimen with dimensions 40x40x160-mm was prepared from each treatment for this test. The test was performed after 28 days of curing. Each sample was subjected to rate of load 0.3 MPa/s; flexural strength was calculated following ASTM C348. **Fig. 6** shows the set up for flexural test.

Fig. 6. Flexural strength test set-up

A pristine geopolymer sample (0% fiber loading, 0% H_2O_2 dosage, cured at 52.5°C) was measured to have 5.81 MPa flexural strength. As shown in **Table 7**, the flexural strength values ranged from 2.415 MPa to 6.254 MPa, with a mean value of 4.323 MPa. These values are in the same range as previous reports on geopolymer's flexural strength [26-27]. **Table** 7 summarizes the measured flexural strength for each treatment.

Table 7. Flexural strength of the abaca-reinforced geopolymer composites

Fiber Loading (%)	H_2O_2 Dosage (%)	Curing Temp. (°C)	Flexural Strength (MPa)
0	0.16	52.5	6.25
0	0.32	52.5	4.23
0	0.24	30	5.39
0	0.24	75	2.41
0.25	0.16	30	5.18
0.25	0.16	75	5.98
0.25	0.32	30	3.70
0.25	0.32	75	3.87
0.25	0.24	52.5	4.95
0.5	0.16	52.5	5.23
0.5	0.32	52.5	3.84
0.5	0.24	30	4.01
0.5	0.24	75	2.82
*0	0.24	75	2.51
*0.25	0.32	30	3.64
*0.5	0.16	52.5	5.13

*replicated treatments

A second-degree model was fit with the data, and R^2 = 0.5362. The model itself was not very significant as shown in **Table 8**. The table summarizes the model's fit and ANOVA for flexural strength.

Table 8. Flexural strength summary of fit and Analysis of Variance

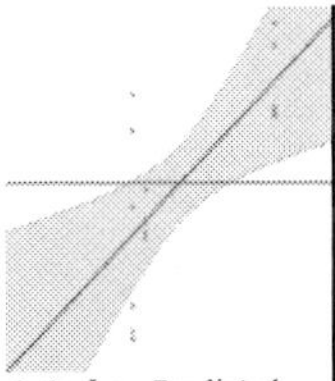

Source	DF	Sum of Squares	Mean Square	F Ratio
Model	2	11.17650	5.58825	7.5135
Error	13	9.668871	0.74376	Prob > F
C. Total	15	20.84537		0.0068

Among the factors, foaming agent dosage was found to be a significant factor, which expectedly caused flexural strength to decline due to formation of void spaces in the matrix. This relationship was observed from Table 9, which shows the parameter estimates of the model.

Table 9. Compressive strength model parameter estimates

Term	Estimate	Std Error	t Ratio	Prob>|t|
Intercept	3.6819481	0.352079	10.46	<.0001
H2O2 (0.16,0.32)	-0.850624	0.27272	-3.12	**0.0081**
H2O2*H2O2	1.0251364	0.445349	2.3	**0.0385**

H2O2 – % H_2O_2 Dosage

The negative coefficient associated with the first-degree factor H2O2 (foaming agent dosage) signifies increasing H_2O_2 dosage results to lower flexural strength. However, the model generated has found significance in the second-degree factors of the H_2O_2 dosage. Intriguingly, the second-degree parameter was associated with a positive coefficient. This signifies that a curve exists and that the flexural strength can possibly increase at a threshold H_2O_2 dosage. The voids and pores formed from added foaming at an optimal or desired dose, can possibly serve as a "shock absorber" during impact. This could explain how foaming can improve flexural strength as suggested by the positive coefficient associated with this parameter.

In this study, the effect of natural fiber reinforcement to flexural strength remained inconclusive. Although statistical analysis presents no indication of any significant effect of fiber reinforcement to flexural strength, it was qualitatively observed that samples subjected to bending load were somehow held together by the fibers at the point of failure, as shown in **Fig. 7**.

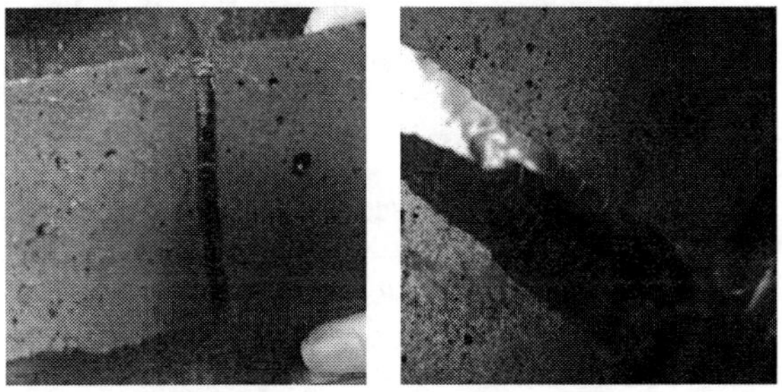

Fig. 7. Bridging fibers at composite's point of failure

4 Conclusions

Treated fibers appeared rougher and more uniform in SEM images; this, in theory, facilitated in interlocking of fiber with the matrix and could therefore enhance the composite's performance. The alkali treatment has sufficiently removed hydrophilic components and impurities as evidenced by TGA results and the lack of significance of fiber loading's effect on the composite's water absorption.

Compressive strength of the samples ranged from 19.56 MPa to 36.85 MPa. The mean value is 29.26 MPa – comparable to the standards for residential and commercial structural concrete. Abaca reinforcement and heat curing both improved the compressive strength. Foaming resulted to weaker composites. The interaction of H_2O_2 dosage and curing temperature significantly affected compressive strength. Compressive strength declined when these factors are set at high levels at the same time due to microcrack formation as observed in the samples' SEM images.

Flexural strength values ranged from 2.415 MPa to 6.254, with a mean value of 4.323 MPa. Addition of foaming agent was observed to weaken the composite's flexural strength, as expected. The correlation of the factors to flexural strength remains inconclusive due to the model's lack of fit. Fiber aggregation during the hardening period could have contributed to the experimental errors It was, however, qualitatively observed that the abaca fibers bridged across the point of failure during the performance of the flexural test.

References

1. L. Perez-Lombard, J. Ortiz, C. Pout. Energy Build. **40**, 394-398 (2008)
2. Z. Zhang, J.L. Provis, A. Reid, H. Wang. Cem. Concr. Compos. **62**, 97-105 (2015)
3. M.W. Ibrahim, N. Jamaludin, J. Irwan, P. Ramadhansyah, A.S. Hani. AMR. **911**, 489-493 (2014)
4. S. Taj, M. Munawan, S. Khan. *Proceedings of the Pakistan Academy of Sciences*. **44**(2), 129-144 (2007).
5. Philippine Statistics Authority. Selected Statistics on Agriculture 27, 13-18 (2016) Retrieved May 6, 2017, from https://psa.gov.ph/sites/default/files/Selected%20Statistics%20on%20Agriculture%202016.pdf
6. R. Dungani, M.S. Karina, A. Sulaeman, D. Hermawan, A. Hadiyane. Asian J. Plant Sci. **15**(1), 42-55 (2016).
7. R. Araral. Turning Abaca Wastes Into Export Grade Handicraft (2004) Retrieved August 13, 2017, from http://sntpost.stii.dost.gov.ph/frames/OctToDec04/pg41_exportGradehandicraft.htm
8. C. Feliciano. Abaca shipments up nearly 40% by value in 11 months to November. (2015). Retrieved on August 13, 2017 from http://www.bworldonline.com/content.php?section=Economy&title=abaca-shipments-up-nearly-40%-by-value-in-11-months-to-november&id=106608
9. J. Arcalas. PHL abaca production grew 5.7% in 2016. (2017). Retrieved August 13, 2017 from http://www.businessmirror.com.ph/phl-abaca-production-grew-5-7-in-2016/
10. E. Agung, S. Sapuan, M. Hamdan, H.D. Zaman, U. Mustofa. Int. J. Phys. Sci. **6**(8), 2100-2106 (2011)
11. G. Islam, S. Gupta. Int. J. Sust. Built Environ. **5**(2), 345-354 (2016)
12. P. Ramadevi, D. Sampathkumar, B. Bennehalli, P. Badyankal, S. Venkateshappa. Int. J. Chem. **35**(2), 1699-1706 (2014)
13. O. Faruk, A. Bledzki, H. Fink, M. Sain. Prog. Polym. Sci. **37**(11), 1552-1596 (2012)
14. P. Ramadevi, S. Venkateshappa, R. Sampathkumar, D. Bennehalli. Bioresour. **7**(3), 3515-3524 (2012)
15. A. Jabbar, J. Militký, J. Wiener, M.U. Javaid, S. Rwawiire. Indian J. Fibre Text. Res. **41**, 249-254 (2016)
16. O. Onuaguluchi, N. Banthia. Cem. Concr. Compos. **68**, 96-108 (2016).
17. M. Cai, H. Takagi, A.N. Nakagaito, Y. Li, G.I. Waterhouse. Composites Part A. **90**, 589-597 (2016)
18. M.M. Kabir, H. Aravinthan, H. Wang, T. Cardona, K.T. Lau. *Energy, Environment and Sustainability. EddBE2011 Proceedings*. 2011, 94-99 (2011)
19. J. Zhang, L. Feng, D. Wang, R. Zhang, G. Liu, G. Cheng. Bioresour. Technol. 153, 379-382 (2014)
20. V. Ducman, L. Korat. Mater. Charact. **113**, 207-213 (2016)
21. G. Das, S. Biswas. Mater. Sci. Eng. **115**(2016)
22. M. Abdullah, K. Hussin, M. Bnhussain, K. Ismail, Z. Yahya, R. Razak. Int. J. Mol. Sci. **13**(2012)
23. M. Jones, A. Mccarthy. Fuel. **84**(11), 1398-1409 (2005)
24. National Ready Mixed Concrete Association [NRMCA]. Concrete in Practice (2003). Retrieved June 29, 2017, from http://welschreadymix.com/wp-content/uploads/2016/04/CIP35-Testing_Compresive_Strength_of_Concrete.pdf
25. Y. Amran, N. Farzadnia, A.A. Ali. Constr. Build. Mater. **101**, 990-1005 (2015)
26. K. Korniejenko, E. Frączek, E. Pytlak, M. Adamski. *Procedia Engineering*. **151**, 388-393 (2016)
27. A. Bhutta, P.H. Borges, C. Zanotti, M. Farooq, N. Banthia. Cem. Concr. Compos. **80**, 31-40 (2017)

Preparation of Graphene Oxide Sand Composites as Super Adsorbent for Water Purification Application

Eko Andrijanto [1,*], *Gatot* Subiyanto[2], *Nina* Marlina[3], *Hanifa* Citra[4], *Cahya* Lintang[5]

[1-5] Chemical Engineering Department, Politeknik Negeri Bandung, Indonesia

Abstract. This paper describes a method to synthesize a graphene oxide sand composites (GSC) as filter media (absorbent) for water purification. Graphene oxides is synthesized from graphite using modification of Hummer's method. The graphene oxide sand composites is prepared through solution method at 100 °C. The graphene oxide is analyzed using XRD, FTIR to confirm its formation. The FTIR spectrum and XRD diffraction pattern confirmed that the graphene oxide synthesized by this method is able to convert graphite into graphene oxide. Performance tests were conducted using a column to purify contaminated water which was mimicked using dyes such as rhodamine B, methylene blue and methyl orange.The initial concentration for all dyes were set for 5, 10, 25, 50 and 100 ppm. The color removal for methylene blue was 100% at all concentrations. However, for the rhodamine B and methyl orange, the color removal achieved 100% for the first three concentration 5, 10 and 25 ppm. The higher concentration of 50 and 100 ppm, the removal were slightly reduced. For the 50 ppm, the color removal of rhodamine B was 98% and for methyl orange 87% respectively. At 100 ppm, the color removal for rhodamine B drops to 92% and for the methyl orange was only 77% respectively. The GSC was very effective to remove methylene blue dyes at any concentration followed by rhodamine B and methyl orange. This GSC composite material is potential to be applied for water purification.

1 Introduction

Numerous use of nanomaterial in the decontamination of air, water and soil has recently acquired many attentions. In the early work, the application of nanomaterial focused on the removal of pesticide [1]. However, recent progress on water purification and water treatment also utilized the use of nanomaterial including carbon nanotubes [2] and graphene oxide [3]. Water purification technologies have also used activated carbon for removing contaminant including color and heavy metals due to high surface area and large iodine number and this material which is derived from plant sources nowadays become the most used adsorbent for water purification and treatment [3-4]. This material is also considered the most efficient and affective due to low cost and practicality especially for developing countries. Other method for water purification which is more advanced such as membrane filtration and ion exchange had been used for sometimes. However, the high cost of investment and operation limits the large scale application the use of the method especially in some countries.

The new member of carbon namely graphene and graphene oxide, the one atom thick sheets of carbon derived from graphite or other carbon materials, become fascinating in recent years due to its remarkable chemical

* Corresponding author: andrijanto.2002@gmail.com

and physical properties [5]. Many carbon materials have been used for water purification such as charcoal, activated carbon [6] and carbon nanotubes [7]. Those carbons mentioned have been used in the water treatment for long time especially charcoal and activated carbon and just recently carbon nanotubes was introduced for this application. Another new carbon material which has interesting properties are graphene and graphene oxide.

Graphene is a truly two dimensional material honeycomb structure of carbon with single carbon thickness and this material has just recently reported and awarded Geim and Novozelov a Noble Prize in Physics in 2010 [5]. Production of graphene is carried out by oxidizing graphite into graphene oxide and then reduce graphene oxide into graphene. This process is initially well known as Hummer Methods [8]. Since the introduction of graphene by Geim in 2010, research to find wide range application of graphene has been enormous including in water purification. Graphene oxide (GO) also attracts many attentions since this intermediate material of graphene has attractive properties for water treatment. GO is soluble in nonpolar and polar solvents including water due to hydrophilicity of this material.

If we compare activated carbon and charcoal, graphene oxide are produced at relatively low temperature, it is likely to be the most cost effective and efficient used for adsorbent. This material is fascinating in term of properties and relatively inexpensive to produce for large scale application.

This study, we show the synthesis of graphene oxide from graphite using modified Hummer Methods [8], and introducing how to prepare composites between graphene oxides with silica based material such as river or beach sands. This composites is later being used as adsorbent for removing any color or other contaminants such as pesticide, heavy metal or even viruses in contaminated water. In this report, we demonstrate a simple method to convert a regular beach sand into graphene oxide-sand composites (GSC).

This study also reported the use of this composites to remove any color contaminants in contaminated water. The color in the contaminated water is mimicked using methylene blue, rhodamine B and methyl orange dyes solutions. The performance of the composites was evaluated based on how much the color is being reduced during adsorption process, and the analysis was done using spectrophotometer method.

* Corresponding author: andrijanto.2002@gmail.com

2 Materials and Methods

2.1. Materials

Graphite oxide was prepared according to the method developed by Macano et al. [9]. This method is modification from Hummers Method which is safer. Graphite powder was supplied from Sigma Aldrich. Concentrated HNO_3, H_3PO_4 , H_2O_2 and $KMnO_4$ were purchased from Sigma Aldrich. and The dyes were purchased from Sigma Aldrich. The detail of this modified procedure is presented below.

2.1. Methods

2.1.1. Graphene oxide preparation

The improved method, a 9:1 mixture of concentrated H_2SO_4/H_3PO_4 (360 mL:40 mL) was added to mixture of graphite powder (3.0 g, 1 wt equiv) and $KMnO_4$ (18.0 g, 6 wt equiv), producing a slight exothermic to 35°C. The reaction is the heated to 50°C and stirred for 12 h. The reaction is then cooled to rt and poured onto ice bath (approximately 400 mL) with 30 % H_2O_2 (3 mL). The end of the addition should be noted the yellow precipitate which is known as graphene oxide particle. The mixture was left overnight to obtain maximum precipitate, the clear water is then decanted and the yellow slurry was then centrifuged for 20 min with the speed of 3000 rpm. The clear solution is again decanted and the precipitate was washed three times with DI water and followed by ethanol washed another three times. The solid obtained on the centrifuged glass was then taken out and dried in the oven at 40 °C overnight.

2.2.2. Preparation of GSC (Graphene Oxide Sand Composites)

GSC was prepared according to the procedure developed by Gao et al. [10]. This procedure is basically combined graphene oxide with beach sand in old fashioned way. The detail of this procedure is fully explained as follow.

Beach sand was purchased from local fish company. It was firstly washed with 10 % HCl before use and continued washing by adding DI water till neutral. A 30 gram of clean sand was put in a 500 beaker glass contains 50 mL 3 % of GO/DI water dispersion, and the mixture was heated up to 105 °C to evaporate the DI water while concentrate the GO/DI mixture. The mixture was heated for further 3-4 h to mix the GO and sand till all the sand particle coated with GO to form GSC. The final product was a black sand adsorbent called GSC. The GSC is then further heated in an oven for further two

https://doi.org/10.1051/matecconf/201815605019

hours at 150 °C. The beach sand and the finish GSC product is illustrated at Figure 1 below.

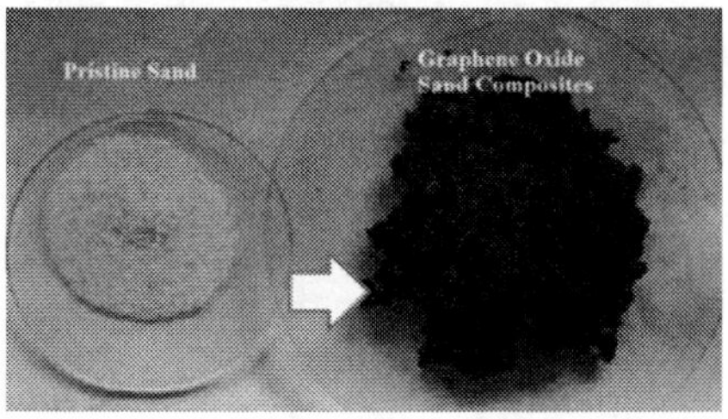

Fig. 1. Conversion graphite into GSC adsorbent

2.2.3. Characterization and Analysis

The GO particle was characterized by XRD and FTIR to ensure the formation of grapene oxide and to differ from the precursor (graphite). Color analysis was performed by spectrophotometer method (Spectronic 20).

2.2.4. Adsorption experiment

Adsorption experiment was conducted in glass tube (drop pipette). The water contaminant was simulated and mimicked using methylene blue, rhodamine B and methyl orange dyes. The flow rate of sample was control to 1 mL per minute. The filtered sample was then analyzed using UV-Vis spectrophotometer (Spectronic 20). The remaining color concentration can be calculated based on the original dyes initial concentration. The initial dyes concentration was varied from 5, 10, 25, 50 and 100 ppm. The wavelength used during the experiments for methylene blue, rhodamine B, methyl orange were set to 660, 555 and 460 nm respectively.

3 Results and Discussion

3.1 Graphene oxide characterization

Graphene oxide synthesis was performed by oxidizing graphite precursor using potassium permanganate as oxidizing agent. The product was graphene oxide which then characterized by FTIR and XRD techniques.

Figure 2 illustrates the FTIR spectra of graphene oxide (GO) and its precursor graphite. The GO spectrum shows the peak at 1288 cm^{-1} and confirming the present of C-O-C bending. The peak at 1066 cm^{-1} is attributed as C-O stretching and C-OH bending is also confirmed at at 1587 cm^{-1}. The peak at 1724 cm^{-1} confirmed the carbonyl group present in GO as C=O stretching and finally a broa

peak at 3448 cm^{-1} is attributed for O-H stretching vibration of the C-OH groups in GO structure.

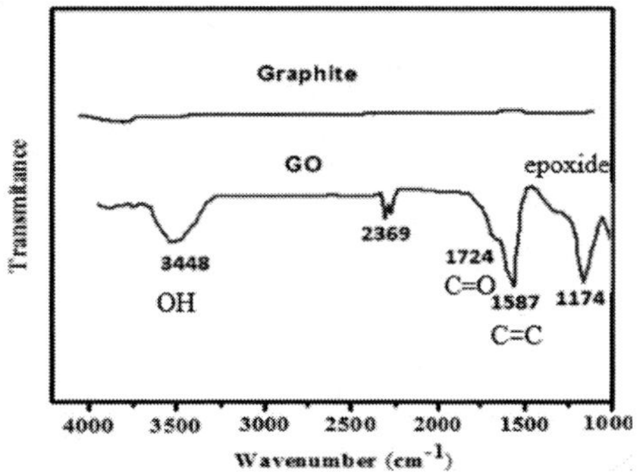

Fig. 2. FTIR spectra of graphite and graphene oxide

Where graphite spectrum shown in Fig.2 do not shows any significant peak observed. This due to graphite is as two dimensional structure consisting of parallel layers consisted of hexagonal rings of carbon atoms hybridized in sp^2 form. Graphite shows almost no peaks detected in FTIR spectrum since in pristine graphite have no functional groups present. Sometimes a few peaks detected due to the present of water adsorbed on the surface. This reason explained that the peaks of graphite in Fig.2 was absence.

3.2 XRD spectra

The existence of graphite and graphene oxide (GO) was characterized using X-ray diffraction method and the result is illustrated in Figure 3. The samples were scanned from 10 ° to 80 ° of angle of 2 theta. Figure 3 shows the peaks of diffraction pattern of the graphite and graphene oxide. Those two diffraction pattern show different in the peak of 2 theta. Pristine graphite peak can be observed at 28° and graphene oxide peak shifted to 15°. This indicates that the interlayer spacing increases after graphite was oxidized. This two patterns show the formation of graphene oxide.

* Corresponding author: andrijanto.2002@gmail.com

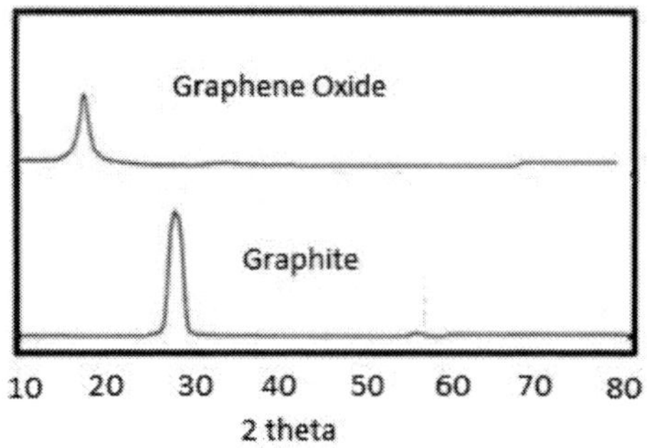

Fig. 3. XRD diffraction pattern of graphite and graphene oxide

3.3 Adsorption experiments

The performance of the adsorbent (GSC) was tested in glass column (drop pipette) in which the sample was run through the column with rate of 1 mL/min. Figure 4 shows the experiment conduted in glass column with the diameter of 0.5 cm and the column was packed with GSC adsorbent with adsorbent height of 5 cm. The color removal of samples can be observed visually or by measuring absorbance changes using spectrophotometer visible at wavelength depending on the dyes. Three different dyes were used in this experiments (methylene blue, rhodamine B and methyl orange). The absorbance of methylene blue, rhodamine B and methyl orange were measured at wavelength of 660, 555 and 460 nm respectively. The initial concentration of each dye were varied from 5, 10, 25, 50 and 100 ppm.

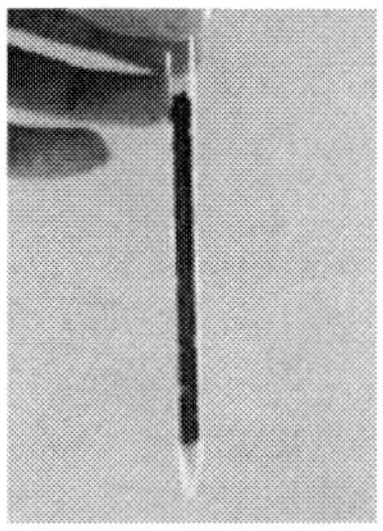

Fig. 4. Adsorption column (drops pipette).

Twenty mililiter of solution after passing through the adsorbent was collected for analysis as illustrated at Figure 5.

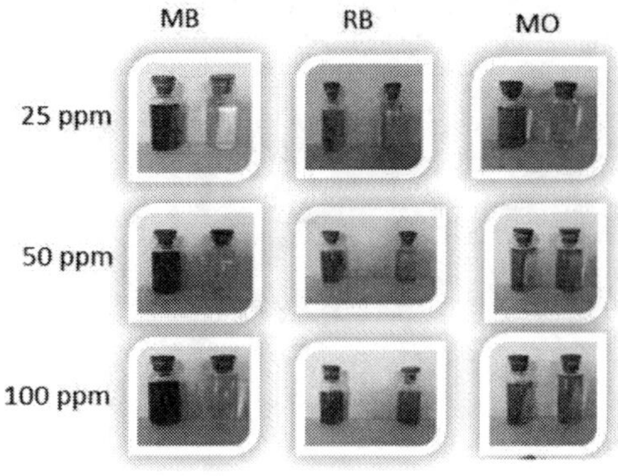

Fig. 5. Dyes adsorption at different concentration.

The percentage of color removed by the adsorbent was measured for the three dyes studied using spectrophotometer method. And the results are illustrated in Figure 6.

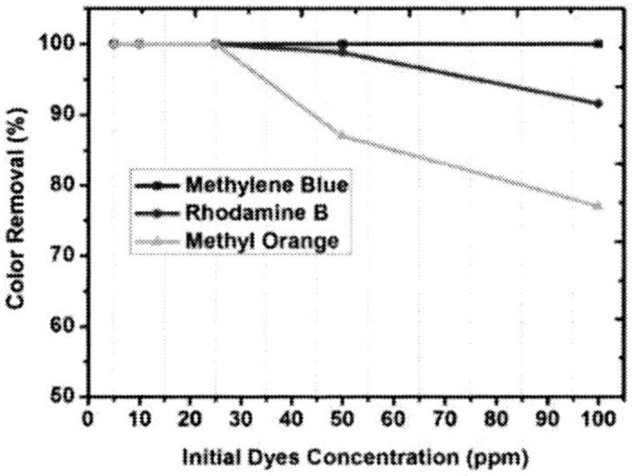

Fig. 6. Color removal for different dye concentration.

Figure 6 shows that the adsorbent is very effective in removing methylene blue dye and followed by rhodamine B and methyl orange. Methylene blue shows 100 % color removal at any different concentration, this can be explained that this adsorbent compatible and effective to remove methylene blue dye. However, rhodamine B show slighly less effective compared to methylene blue. At low initial dye concentration (5, 10 and 25 ppm), removal of rhodamine B is 100 %, but by increasing the concentration to 50 and 100 ppm, the removal decreased to 98 % and 92 respectively. This value is still considered very high and the adsorbent is assumed to be effective in removing rhodamine dye.

Figure 6 shows the ability of synthesized adsorbent in removing methyl orange dye. The same value was obtained as shown by rhodamine B. At low initial dye concentration 5, 10 and 25 ppm, the dye

* Corresponding author: andrijanto.2002@gmail.com

removal by the adsorbent is 100 %. Increasing the initial concentration to 50 and 100 ppm, the color removal decreases to 87 and 77 %. This number is still acceptable.

Performance test of adsorbent was conducted at flexiglass plastic filter, with column diameter of 1.5 cm and column length of 20 cm. Total adsorbent used was 12 g and positioned in the middle of the filter. The solution was run through the column by gravity and the flow rate was set at 2.5 ml/min. Two hundreds milliliters of solution were withdrawn with interval of ten milliliter. This test is to investigate the capability of adsorbent to remove a dye solution. The result of this test is illustrated in Figure 7.

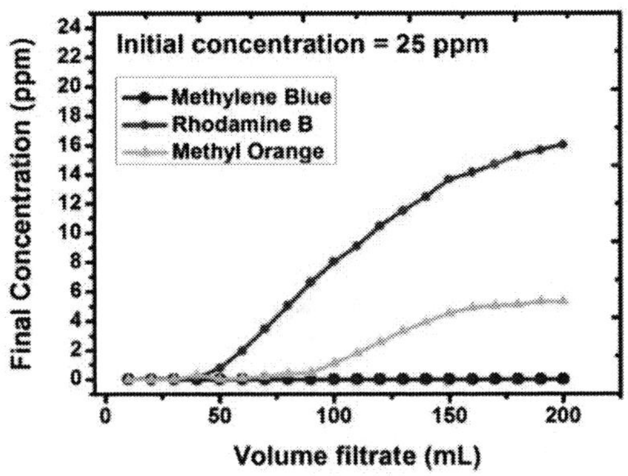

Fig. 7. Performance test of GO-sand adsorbent.

The performance test shows that the ability of the adsorbent towards methylene blue dye is better compared to rhodamine dan and methyl orange. Methylene blue dye was adsorbed almost one hundred percent after 200 mL filtrate collection and for methyl orange, the increasing of the final concentration occurred after 70 mL volume of filtrate collected. However, the increasing of filtrate concentration is limited untill 5 ppm and and this means 80% color removal is retained after 200 mL filtrate collection. The rhodamine B is the only suffering from the decreasing of the efficiency and this can be seen from the increase of the final color concentration very sharply after 40 mL filtrate collection. The color removal decreased to only 25 % after 200 mL filtrate collection. The conclusion that the color removal efficiency of the synthesized adsorbent is increased from methylene blue, methyl orange and the least is rhodamine B.

* Corresponding author: andrijanto.2002@gmail.com

4 Conclusions

Graphene oxide is successfully synthesized and it has been shown its structure by FTIR and XRD. The graphene oxide composite sand showed the ability to remove different dyes molecule from colored water. The removal of methylene blue by the adsorbent close to 100% and slightly reduced for rhodamine B and methyl orange. However, the performance test indicated that the methylene blue is the most easily adsorbed into the synthesized adsorbent and followed by methyl orange and rhodamine B. The ability of synthesized adsorbent in removing dyes from waste water simulated the potential application of this material to be used for water purification.

References

1. S. Sen Gupta, T.S. Sreeprasad, S.M. Maliyekkal, S.K. Das, T. Pradeep, Graphene from Sugar and its Application in Water Purification, ACS Applied Materials & Interfaces, **4**,4156-4163 (2012).
2. Wang, Ng, Wang, Li, Hao, Chem. Eng. J. **197**, 34–40 (2012).
3. G.K. Ramesha, A.V. Kumara, H.B. Muralidhara, S. Sampath, Graphene and grapheme oxide as effective adsorbents toward anionic and cationic dyes, J. Colloid Interface Sci.**361**, 270-277 (2011).
4. M. Goyal, M. Bhagat, R. J. Dhawan, Hazard. Mater., **171**,1009-1015 (2009).
5. A. K Geim,. K. S. Novoselov, Nat. Mater., **6**, 183-191 (2007).
6. A. J. Demirbas, Hazard. Mater., 167,1-9. (2009).
7. S.Yang, J. Li,; D. Shao, J. Hu, X. Wang, J. Hazard. Mater.,166, 109-116 (2009).
8. Hummers, S. William, Offeman, E. Richard. Journal of the American Chemical Society. 80 . 1339 (1958).
9. D. C. Marcano , D. V. Kosynkin , J. M. Berlin , A. Sinitskii , Z. Sun , A. Slesarev ,
10. L. B. Alemany , W. Lu , J. M. Tour , ACS Nano, 4 , 4806 (2010).
11. Gao, W., Majumder, M., Alemany, L.B., Narayanan, T.N., Ibarra, M.A., Pradhan, B.K., Ajayan, P.M.,. Engineered graphite oxide materials for application in water purification. ACS Applied Materials & Interfaces, 3, 1821-1826 (2011)

Characterization Nanofillers from Agriculture Waste for Polymer Nanocomposites Reinforcement

Bastian Arifin[1], *Sri* Aprilia[1*], *Pocut* Nurul Alam[1], *Farid* Mulana[1], *Amri* Amin[2], *Dilla* Mars Anaska[1], and *Dwita* Eka Putri[1]

[1]Chemical Engineering Department Engineering Faculty, University of Syiah Kuala, Banda Aceh, 23111, Indonesia
[2]Mechanical Engineering Department, Engineering Faculty, University of Abulyatama, Lampoh Keude, Aceh Besar, Indonesia

Abstract. The current development of the packaging industry is increasing as well as the dependence of non-renewable oil-based materials encouraging researchers to look for alternative polymeric strengthening materials from biomass. Especially used from agricultural waste because it is cheap and widely available in nature and it can be renewed. In this study, agriculture waste used were rice husk and rice husks ash that prepared as organic nanofillers for the development of polymer nanocomposites. XRF analysis showed that rice husk ash has the highest silica (SiO_2) content of 89. 835%, while rice husk has SiO_2 contents of 82.540%. From XRD analysis on 2 theta there is a crystalline silica region at 22° and this analysis shows the sample is amorphous. FTIR analysis showed Si-H at peak 2339 cm^{-1} in rice husk and 2129 cm^{-1} for rice husk ash.

1 Introduction

The development of bionanocomposite materials has attracted much attention from scientists, researchers and industries, especially to obtain environmentally friendly materials and the use of materials for tomorrow's sustainability [1, 2]. The addition of this nanofiller aims to improve good performance on bionanocomposite materials in various application areas such as: automotiv, construction, electronics and packaging as well, because they have low density, good stiffness, good thermal and mechanical properties [3, 4]. The use of nanofillers as reinforcement in composite polymers has many advantages such as low cost, renewable natural resources, special properties, low density, easy preperation, require low energy for process, biodegradable, widespread and relatively non-abrasive use compared to strengthening with traditional materials [5, 6].

Research on the utilization of rice straw and rice husk in Indonesia is still very limited. These waste is used only as animal feed and energy or fuel or as adsorption on heavy metals [7]. While the utilization of polymer technology is still very rarely reported. Silica (SiO_2) is a basic raw material that can be used to strengthen matrix polymers and is usually used in polymeric industries for electronics, ceramics. Many silica applications can be used in technology applications because silica starch particles have properties as tixotropic agents, thermal insulators, and composite fillers [8]. Silica-based manufacturing products have high strength at high temperatures. Because of this nature, the wastes of rice can be added to the matrix polymer as a bionanofiler and also as an alternative solution to environmental problems, thereby increasing the added value in agricultural waste.

Research has been done using rice husk and rice husk ash. Characterization of both samples has been done to determine the silica content by using X-ray fluorescence (XRF) spectra, fourier transform infrared (FTIR) and X-ray diffraction (XRD) analysis.

2 Materials and methods

Rice husk is light yellow colour fiber. The fiber was ground up from Aceh Province, Indonesia. Rice husk ash was got from the burning of rice husk. The samples were collected from rice plant Aceh Besar, Indonesia.

Rice husk and rice husk ash were washed and dried in an oven overnight. The dried samples were chopped by blender and smoothed using ball mill for 12 h and then screnning with a 325 mesh.

2.1. XRF analysis

X-ray fluorescence (XRF) spectrometer was used for measured the purity of silica by a PANalytical AXIOS. The silica content of surface area was measured on a TriStar II 3020. The samples for surface area measurement were dried at 250 °C over 8 h prior to the measurement.

2.2. FTIR analysis

Fourier transform infrared (FTIR) spectrocopy analysis has used to employed of characterization of the silica

* Corresponding author: sriaprilia@unsyiah.ac.id

content in absorption spectra to get information the organic structure. The analysis was used by using a FTIR spectroscopy (Shimadzu FTIR Spectrometer). The sample was analyzed into the Nicolet Avatar for FTIR Spectrometer. The analyisi of spectra were recorded in the range of wave length about 4000-400cm^{-1}.

2.3. XRD analysis

Analysis of X-Ray diffraction were purposed to determine the crystallinity index and crystal size of samples. XRD analyses were on an X-Ray diffractometer with measurement condition are 40 kV and current 30 mA. The scattering angle range (2θ) was from 10° to 80° with the scan rate of 10°/min.

2.4. Particle size analyzer

Malven Instruments was used for analysis of distribution of particle size on rize husk and rize husk ash. The instrument of Malven with the long Benc Mastersizer S and fitted with the Qspec Dry Powder Feeder. This analysis, the sample weighted in (two) 2 grams. The number of distribution at 10%, 50% and 90% cut-off point was done with well as the number of mean diameter.

3 Result and disscussion

3.1. Chemical compounds

From XRF analysis used for chemical compounds of the rice husk and rice husk ask. The elemental components of rice husk and rice husk ash come in the form of chemical compounds. Chemical compounds of silica in the sample rice husk and rice husk ask shown in Table 1. Silica content in the form of bulk SiO_2. In addition to silica as the main component, many other components exist in the sample.

From Table 1 it can be seen that there are Na_2O, MgO, Al_2O_3, P_2O_5, SO_3, K_2O, CaO, MnO, Fe_2O_3 and SrO contents. The SiO_2 as a main chemical composition. The silica content (SiO_2) in rice husk ash is higher than the rice husks, 89.835% and 82.540%, respectively. The other compounds for the rice husk ask low, it is show that after burning of rice husk, all impurities remove. Liu et al, 2013 [9] has mentioned that removing organic matter occured only heat at moderate temperatures to avoid the fusion of the SiO_2 product. In biological system, the six metals common come such as Na, Mg, Al, K, Ca, and Fe, and also Si, were detected simultaneously with the mass spectrometer.

3.2. Functional group

FTIR analysis for rice husk and rice husk ash are show in Fig. 1 and 2. From Fig. 1 and 2 shows that many chemical compounds found with FTIR analysis. The stretching area is seen at 3266 cm^{-1} and 2895 cm^{-1} peaks for rice husk and 3207 cm^{-1} and 2837 cm^{-1} for rice husk ash. At the peak of about 3200 cm^{-1} there is an O-H stretching hydrogen bond. The hydrophilic area was absorbed at 3100-3700 cm^{-1}, this area being a group of -OH in aliphatic compounds or aromatic alcohols. The peak around 2929 cm^{-1} is due to stretching C-H from compounds of aliphatic saturated. The two peaks of this stretch correspond to the aliphatic parts in cellulose and hemicellulose. In the triple bond region, the peaks are at 2351 cm^{-1} and 2142 cm^{-1}. At peak 2351 cm $^{-1}$ is expressed as P-H stretching and P-O-H stretching. At the peaks around 2129 cm^{-1} shown as Si-H stretching. It can be deduced that the wave length in this region represent multivalent inorganic hydrides or oxides. The C=O strectch at the double bonding region at 1723 cm^{-1} in both spectra is assigned as acetyl and uronic ester.

Table 1. Chemical compositions

Chemical compounds	Rice husk (%)	Rice husk ash (%)
Na_2O	0.254	0.268
MgO	1.055	1.734
Al_2O_3	0.174	0.446
SiO_2	82,540	89.835
P_2O_5	0.501	0.661
SO_3	0.314	0.213
K_2O	1.115	3.490
CaO	1.509	1.498
MnO	1.490	0.512
Fe_2O_3	11.019	1.330
SrO	0.028	0.012

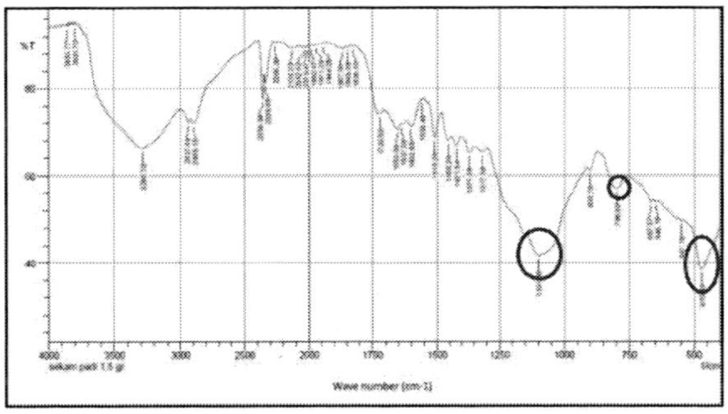

Fig. 1. FTIR of the rice husk

Amide is shown at the peak 1643 cm^{-1} for rice husk and 1649 cm^{-1} for rice husk ash. The amide peak I presents 80% of C=O stretching at the amide group, connected with N-H bending and C-N stretching. In the secondary structure of protein, the vibration depends on the nature of the hydrogen bonds involved in the C=O and N-H groups. In the area of fingerprint, there are many sharp peaks coul be shown from both of spectra. At the range 1450-1370 cm^{-1} of the symmetrical C-H and asymmetric deformation. The area sircle red at 1101, 795 and 470 cm^{-1} for rice husk and 1103, 802 and 466 cm^{-1} for rice husk ash are due to the symmetric, asymmetric, and bending modes of SiO_2, respectively.

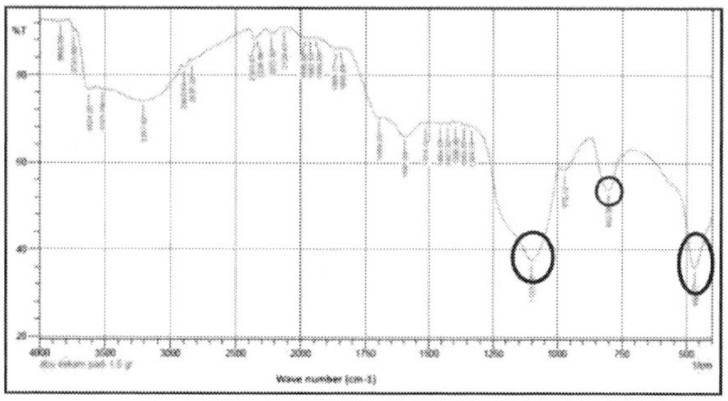

Fig. 2. FTIR of the rice husk ash

3.3. Crystallinity region

The XRD patterns of silica sample from rice husk and rice husk ash shown in Fig. 3. The two patterns are close to each others, where both are seen crystalline silica peak at 2 theta at 22°. The peak of the crystalline silica is consistent with the results, it was shown by Le et al., 2013 [10]. Chen et al., 2013 [11] in his research stated that the same two sample patterns show that both samples are amorpous.

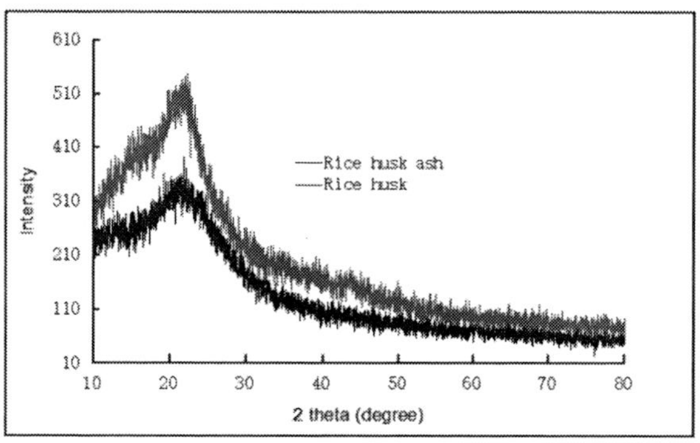

Fig. 3. XRD patern of rice husk and rice husk ash

Depend on Saceda and Leon, 2011 [12], that the received rice husk ash have defined Bragg's peaks at 2θ value of 21.9° and several distinct peaks at 20.9, 26.6 and 36.0°.

3.4. Particle size

The particles size distribution of rice husk and rice husk ash particles has been marked of the base and the level of aggregation and agglomeration form. The used of Malven Mastersizer S Analyzer to predicted particle size and distribution analysis. The results obtained are shown in Fig. 4 and Fig. 5. The range of distribution carbonized particle from rice husk and rice husk ash after sieve pass 325 mesh.. The size of particulate silica at pass trought 325 mesh show in size of distribution range is below 45.06 μm. Particle size of rice husk and rice husk ash in the size of distribution particle shown that in the range because it has come form aglomeration of particles.

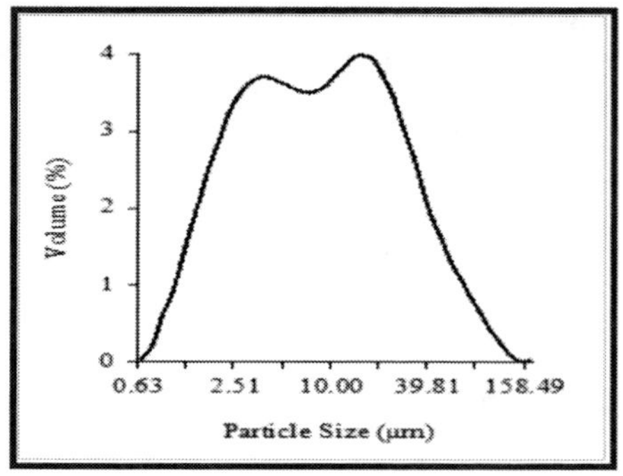

Fig. 4. Particles size of rize husk

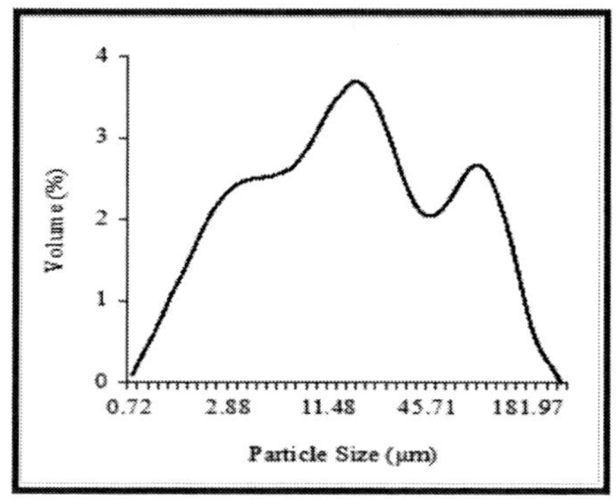

Fig. 5. Particle size rice husk ash

4. Conclusion

Rice husk and rice husk ash have successfully performed characteristics. Bionano filler from rice husk and rice husk ash has been done by passing the particles on the screening of 325 mesh about < 45 μm. The largest silica content is obtained on rice husk ash. Si-H is present at the peak of 2339 cm-1 in rice husk and 2129 cm-1 for rice husk ash. From the XRD analysis for both samples of crystalline silica there are at 2 theta 22 degrees and both samples are in the amophouse.

Acknowledgment: The authors would like to thank the Institute for Research and Community Service Unsyiah and Directorate of Research and Community Service and Directorate General of Research and Development, Ministry of Research, Technology and Higher Education who have funded this research in accordance with the Research Contract No. 105/ SP2H / LT / DPRM / IV /2017, April 03, 2017.

References

1. H.P.S. Abdul Khalil, Rus Mahayuni A.R., Irsyad-ul-haq Bhat, Rudi, D., Almulali, M.Z. and Abdullah, C.K., Bioresources, **7**, 5771-5780 (2012).
2. H.P.S. Abdul Khalil, N.A.Sri Aprilia, Bhat, A.H., Jawaid, M., Paridah, M.T., Dungani, R, Renewable and Sustainable Energy Rev. **22**, 667–685 (2013)
3. N.A. Sri Aprilia, H.P.S. Abdul Khalil, A.H. Bhat, R. Dungani, Hossain, Md.S., Bioresources, **9 (3),**

4888-4898 (2014).
4. K. Majeed, M. Jawaid, and A. Hassan, Abu Bakar, A., Abdul Khalil, H.P.S., Salema, A.A., Inuwa, I., Mater. Des., **46**, 391–410 (2013).
5. K.S. Kim,H. Choi, S.-W., Lee,B.-H., Kim,S., Kim, H.-J., Cho C. W. and Cho, D., J. Therm. Anal. Calorim., **89 (3)**, 821–82 (2007)
6. T. Mekonnen, Mussone, P., Alemaskin, K., Sopkow, L., Wolodko, J., Choi, P. and Bressler, D., J. Mater. Chem. A, **1,** 13186–13196 (20013).
7. N.S. Awwad, Gad, H.M.H, Ahmad, M.I., Aly, H.F., Biointerfaces **81**, 593–599 (2010)
8. V.B. Carmona, Oliveira, R.M., Silva, W.T.L., Mattoso, L.H.C., Marconcini, J.M., Ind. Crops Prod. **43**, 291– 296 (2013).
9. X. Liu, X.dong Chen, L. Yang, H. Chen, Y. Tian, Z. Wang 2016, Res. Chem. Intermed. **42 (2),** 893–913 (2016)
10. V. H. Le, C. N. H. Thuc and H.H. Thuc, Nanoscale Research Letters, 1-10 (2013)
11. H. Chen, W. Wang, J. C. Martin, A. J. Oliphant, P. A Doerr, Jeffery F. Xu, Katelyn M. DeBorn, Caixing Chen, and Luyi Sun, ACS Sustainable Chem. Eng. **1**, 254–259 (2013).
12. J-J. F. Saceda and R.L. De Leon, Quim. Nova, **34**,1394-1397 (2011)

Silicon Conversion From Bamboo Leaf Silica By Magnesiothermic Reduction for Development of Li-ion Baterry Anode

Silviana Silviana[1]*, *Wardhana J.* Bayu[2]
[1] *Department of Chemical Engineering, Faculty of Engineering, Diponegoro University ,Jl. Prof. Sudarto, Tembalang-Semarang, 50275, Telp/fax: (024)7460058/(024)76480675*
[2] *Master Degree of Department of Chemical Engineering, Faculty of Engineering, Diponegoro University, Jl. Prof. Sudarto, Tembalang-Semarang, 50275, Telp/fax: (024)7460058/(024)76480675*

Abstract. Silicon (Si) is a promising alternative material for the anode Lithium ion Battery (LIB). Si has a large theoretical capacity about 3579 mA hg -1, ten times greater than the commercial graphite anode (372 mA hg -1). Bamboo is a source of organic silica (bio-silica). Most part biogenetic content of SiO_2 is obtained in bamboo leaves. This paper aims to investigate the synthesis nano Si from bamboo leaves through magnesiothermic reduction after silica extraction using sol–gel method and to observe nano Si of bamboo leaf as mixed material for lithium ion baterry. Silica and silicon content was determined using XRF. Silica product has 96,3 wt. % yield of extraction from bamboo leaf, while silicon yield was obtained 61.2 wt. %. The XRD pattern revealed that silica and silicon product were amourphous. The extracted silica and silicon from bambo leaf has spherical shape and agglomerated form. As anoda material for LIB, silicon product achieved 0,002 mAh capacity for 22 cycle.

1 Introduction

Silicon, the second most common element in the earth's crust have attracted so many attention in the energy storage application because of their high potential capacity. Silicon has been considering most promosing candidate to replace graphite anode. They have large theoretical capacity about 3579 mAh g^{-1}, almost 10 times higher theoretical specific capacity than graphite anode(372 $mAhg^{-1}$) [1]. Silicon usually has obtained from mineral in rock and sand, however, silicon also can be obtained in organic material such as risk husk, bamboo leaf, wheat husk and another agricultural waste.

Bamboo is a source of organic silica (bio-silica). The bamboo absorbs silicic acid from the water and then convert it into SiO_2 hydrate in the root branches and leaves [2].Most part biogenetic content of SiO_2 is obtained in bamboo leaves. Bamboo leaves have a silica content of 17 -23% by weight, higher than the silica content in rice husk (9.3-13.5% by weight)[3]. Bamboo leaf ash has a large SiO_2 content of 75.90-82.86 wt% [4] [5].

The sol gel technique mainly used for extracting silica from agricultural material like rice husk [6], sugar cane bagasse [7-8], wheat husk [9], corn cob [10], and wheat staw [11]. The sol-gel process is widely used to produce pure silica particles due to its ability to control the particle size, size distribution and morphology through systematic monitoring of reaction parameters [12].

Silica can be reduced to silicon by using several method: carbothermic reduction, aluminothermic reduction, calciothermic reduction, and magnesiothermic reduction [13]. Conventional carbothermic reduction was usually used by industrial apllication to reducing silica into silicon. Oxygen is removed by reaction sequences of heterogeneous-homogeneous at a temperature of about 3000 °C [14]. In the present work, magnesium as a reducing agent, was introduced to reduce the reaction temperature until 650°C [15]. Silica can be reduced by magnesium (magnesiothermic reduction) to produce silicon through following reaction:

$$2Mg\,(g) + SiO_2\,(s) \rightarrow 2MgO(s) + Si\,(s) \qquad (1)$$

or

$$2Mg\,(s) + SiO_2\,(s) \rightarrow 2MgO(s) + Si\,(s) \qquad (2)$$

If Mg excess,

$$2Mg\,(s) + Si\,(s) \rightarrow Mg_2Si\,(s) \qquad (3)$$

The formation of Mg_2Si is affected by the excess Mg and reduction temperature. Decreasing magnesium amount resulted in decreasing Mg_2Si and with increasing temperatures at fixed ratio of Mg to SiO_2, the formation of Mg_2Si increased [16].

* Corresponding author: silviana@che.undip.ac.id

https://doi.org/10.1051/matecconf/201815605021

2 Experimental Method

2.1. Silica Extraction

Bamboo leaves was collected from Sragen, Central Java To remove alkali impurities, the bamboo leaf were crushed and leached by 1M HCl for 2 hours, after washing in water and drying the bambo leaves were calcined at 650°C for 2 hours in muffle furnace to remove organic impurities.

Silica was extracted from Bamboo Leaf Ash, (BLA) using sol-gel Method to produce silica xerogel [7]. In this method, 60 ml NaOH 2M were added to 10g BLA and boiled for 1 h with constant stirring to extract the silica and produce sodium silicate solution. The solution were filtered trough whatman No.41. The filtrate solution was sodium silicate, cooled to room temperature and the pH solution was reduced with 1N HCl to 7.0 under constant stirring to produce silica gels. When the solution gel was formed ,it was aged for 18 h. After aging, the soft gel gently broken by adding 100 ml destilled water and centifuged at 3000 rpm to make a slurry. The slurry was filtered and washed. The supernatant was discarded and the gel was dried at 80 at 12 h to produce white powder silica (xero-gel).

2.2 Silica Reduction

Synthesis of silicon from the as-produced silica was conducted by a magnesiothermic reduction process. Silica (xero gel) mixed with magnesium powder with the SiO_2:Mg ratio of 1:0,4. The powder was heated in a furnace under Ar gas flow and held at 650°C for 2 hours. After the reaction the powders were immersed in 1M HCl solution for 6 hour to remove MgO, MgSi and unreacted Mg [15].

3 Result And Discussion

3.1. Silica Extraction Composition

The silicas's purity can be determined using XRF method. It shown in table 1, sol–gel method can produced silica with purity of 96.3% wt.%. The remaining elements include metallic impurities such K, Ca, Fe, Ni, Cu, In, Yb. This high yield of silica indicated that during sol-gel extraction, only anmorphous silica was synthesized. Silica dissolved from BLA in sodium hydroxide and produce sodium silicate. The reaction as follows:

$$x\ SiO_2 + 2\ NaOH \rightarrow Na_2O.\times SiO_2 + H_2O \qquad (4)$$

The sodium silicate solution reacts with hydrochloric acid to produce silica gel. The reaction is:

$$Na_2O.\times SiO_2 + HCl \rightarrow \times SiO_2\,(gel) + NaCl + H_2O \qquad (5)$$

The presence of ion Na+ have important rule in this reaction. Interaction between Na^+ and negatively silicate would speed up the reaction. Ion Na + promoted the gel formation by screening the surface charge of particles that favors their agregation. [7].

Table 1. Silica product from sol gel method.

Compound	Si	K	Ca	Fe	Ni	Cu	In	Yb
Concentrati on Unit	96.3	0.2	1.8	0.42	0.52	0.16	0.4	0.1

The SEM image of silica xerogel BLA was shown in Fig.1. The extracted silica was spherical an agglomerated form. The particle has various size in nano scale.

Fig.1. The scanning electron micrograph (SEM) image of silica xerogels from BLA

An XRD pattern of prepared silca gel from bamboo leaf is shown in Fig.2. The diffused peak at 2θ=22,6 indicated that silica gel were solid amorphous [3][6]. It was confirmed by the absence of any ordered crystalline structure. This XRD pattern also proved that sol-gel process completely eliminates most of crystalline structured impurities in BLA, ie K_2O, Fe_2O_3, CaO [17].

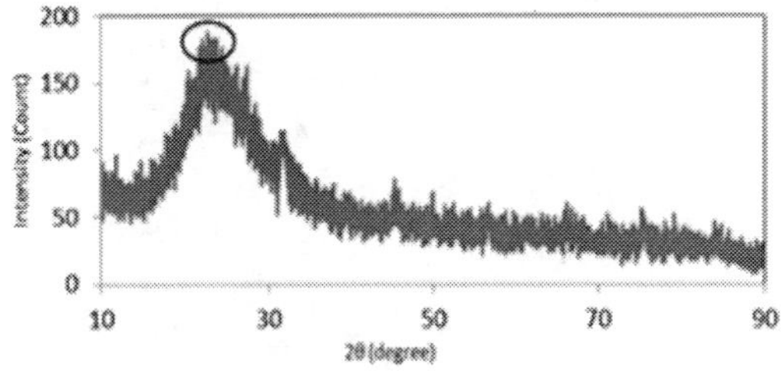

Fig. 2. XRD diagram of Silica BLA

The amorphous, nano scaled silica have many advantages, it can be easily dissolved , have high porosity and high surface area that provide higher activities reaction [18].

3.2. Silicon Reduction

The component of silicon from magnesiothermic reduction were characterized using XRF. The main component was silicon (61.2 %.wt) and Magnesium (31 %.wt). It was also confirmed from XRD Diffractorams (**Fig 3**). There is an another strong peak beside of silica that indicate the presence of magnesium.

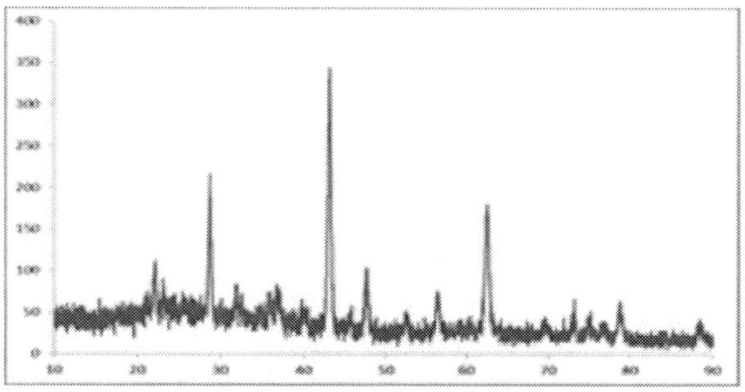

Fig.3. X-ray diffraction pattern of silicon produced from magnesio-thermic reduction.

It is suspected that in this experiment, there was magnesium excess in reaction, and as a result, following Equation 3, the silicon product was reduced [19]. Leaching process method taking big role to obtain high purity silicon. It is need to optimize the condition and variable of leaching process due the low purity silicon product.

From the morphology characterization by using SEM, it was shown in Fig.4., that nano silicon has spherical particle shape and agglomerated. Hence there is no significant difference between silica and silicon product in a term of morphology.

Fig.4. The scanning electron micrograph (SEM) image of silicon product from magnesiothermic reduction

3.3. Silicon as anoda material for LIB

Lithum ion batteries performance. with silicon BLA as anoda material was tested by the galvanostatic cycling method. As the result (Fig.6.) the maximum capacity of electroda was 0.002 mAh at 22 cycles.

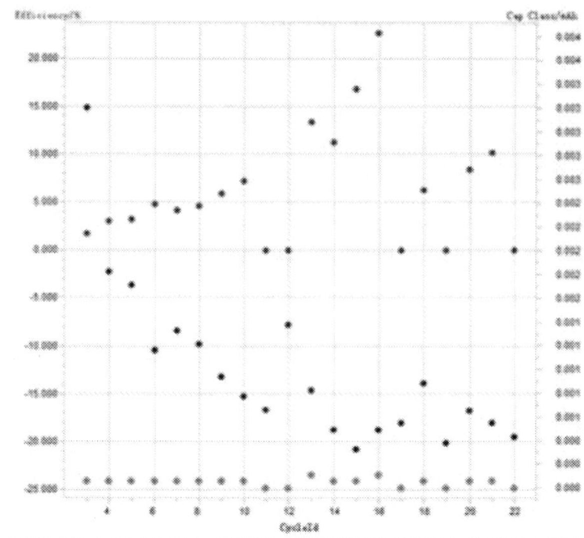

Fig.6. Galvanostatic cycles

The present of MgO has significant effect for the low electroda capacity. MgO is insulator material because of its wide band gap (~7.8 e), that may decrease the elctroda capacity [20].

Conclusion

Nano silica has been suscesfully extracted from bamboo leaves by sol gel technique and then reducted into silicon using magnesiothermic reduction. Silica from bamboo leaves has 93.6 % %.wt purity while silicon was 61.2 % .wt. From XRD and SEM analysis reveals that silica and silicon BLA were amorphous with spherical shape and agglomerated form. Silica performance as LIB anoda material was tested using eight channel baterry. The electroda's capacity was 0.002 mAh for 22 cycle.

References

1. Tian, H., Xin, F., Wang, X., He, W. and Han, W., J. Mater. **1,** 3 (2015)
2. Currie, H. and Perry, C., Ann. Bot, **100**, 7 (2007)
3. Kow, K. W., Yusoff, R., Aziz, A. R. A. and Abdullah, E. C., J. Non. Cryst. Solids, **386** (2014)
4. Dwivedi, V. N., Singh, N. P., Das, S. S. and Singh, N. B., Int. J. Phys. Sci., **1**, 3 (2006)
5. Mohapatra, S., Sakthivel, R., Roy, G. S., Varma, S. and Singh, S. K. Mater. Manuf. Process. **26** (2011)
6. Kamath, S. R. and Proctor, A., Cereal Chem. **75** (1998)
7. Affandi, S., Setyawan, H., Winardi, S., Purwanto, A., and Balgis, R., Adv. Powder Technol. **20**, 5 (2009)
8. Mupa, M., Hungwe, C. B., Witzleben, S., Mahamadi, C., and Muchanyereyi, N., Afr. J. Pure Appl. Chem. **9**, 2 (2015)

9. Javed, S. H., Shah, F. H., and Mansha, M., J. Fac. Eng. Technol. **18**, 0092 (2011)
10. Okoronkwo, E. A., Imoisili, P. E., and Olusunle, S. O. O., Chem. Mater. Res. **3**, 4 (2013)
11. Patel, K. G., Misra, N. M., and Shettigar, R. R., IJCEA **7**, 5 (2016)
12. Rahman, I. A. and Padavettan, V., J. Nanomater. **2012** (2012)
13. Sadique, S. E., Thesis (2010)
14. Aminullah, Rohaeti, E., and Irzaman, Procedia Environ. Sci. **24** (2015)
15. Liu, N., Huo, K., McDowell, M. T., Zhao, J., and Cui, Y., Sci. Rep. **3** (2013)
16. Kalem, T., Thesis (2004)
17. Rani, K. M., Palanisamy, P. N., and Sivakumar, P., Int. J. Adv. Technol. Eng. Sci. **2**, 10 (2014)
18. Awizar, D. A., Othman, N. K., Jalar, A., Daud, A. R., Rahman, I. A., and Al-Hardan, N. H., Int. J. Technol. **7** (2015)
19. Andriayani, Raja, S. L., Herlince, S., and Sofyan, N., Int. J. Technol. 7 (2015)
20. Habibah, Z., Arshad, A. N., Ismail, L. N., Bakar, R. A., and Rusop, M., Procedia Eng. **56** (2013)

Experimental Study of Paddy Grain Drying in Continuous Recirculation System Pneumatic Conveyor

Totok Prasetyo [1,*], *Listiyana* Riska[2], *Rahim* Arlanta[2], and *Siswo* Sumardiono[2]

[1] Departement of Mechanical Engineering, Semarang State Polytechnic, Postal Code 50239, Semarang, Indonesia
[2] Department of Chemical Engineering, Diponegoro University, 50239, Semarang, Indonesia

Abstract. Nowadays, paddy drying using conventional method is irrelevant to used. The increasing need of grain each year is one of the main factor that pushes scientist to invent new methods of drying paddy grain. The old method is unusefull since it takes up space, needed the sun as the drying media which causes a long drying time. One of the alternative of drying paddy grain is by using mechanic dryer. By using this dryer type, the process of drying takes faster comparing to the old method and the distribution of grain is supplied continuously. The dryer which uses pneumatic conveyor can be one of the solution due to its less energy consumption, a better drying result, and high drying capacity. The variable used to identify drying performance is temperature and loading weight. From the experiment, it is resulted that the best temperature for drying paddy grain is 60^{0}C with the weight loading of 150 gram.

1 Introduction

Indonesia is the largest rice consuming country in ASEAN with the numbers 132 kg per capita per year. In 2006, demand for rice in Indonesia increased by 1.6 % per year [7]. The amount of demand for rice in Indonesia requires the development of quality and quantity of rice product. High humidity causes paddy easy to deteriorated, especially in tropical areas [8]. In the rainy season paddy drying can not run optimally, causing decay and decrease the quantity of quality rice.

During this process the grain dried sorted performed in a conventional manner, by direct heating under sunlight on the floor. Conventional drying spend 3-8 days and greatly depends on the intensity of solar radiation [1]. Continuous recirculation drying technology conveyor pneumatic type is a mechanical dryers as an alternative to the previous dryers. This type advantage is recirculation during drying of grain moving at constant speed. Thus the grain to be dried and the dried grain can be maintained at a consistent drying rate of the drying capacity. Recirculation systems that run continuously can help prevent over-drying, drying time is short, high percentage of head rice, grouts bit, and can homogenize paddy drying results [3]. In addition, the tool is not dependent on the season, the capacity of drying has a uniform moisture content, grain condition is not damaged due to sudden changes in temperature (thermal shock), as well as small operating costs and small area required [1].

This research studied the effect of drying air temperature and drying load (load) on the rate of grain drying system with a pneumatic conveyor dryer recirculation type. By reviewing these two variables, expected to be known the optimum conditions that can be used in a predetermined range of values. In addition drying rice profile obtained by a pneumatic conveyor dryer recirculation type.

2 Materials and methods

The raw material used is grain obtained from paddy fields in the area Meteseh, Semarang, central Java, Indonesia. The series of recirculating dryers pneumatic conveyor type equipped by temperature sensors and humidity gauges. Scheme of dryers illustrating in figure 1. The experimental procedure as follows : Wet grain crops incorporated into the winnowing machine, cleaned, and weighed as much as 20 kg of each variable. Turning on the heater blower and continuous recirculation dryer pneumatic conveyor type then heated to the desired temperature variable is obtained. Then enter the appropriate grain into the hopper in desired mass variable. Running circulation of grain in the dryer for the desired time variable. During the process, temperature and flow of the material is kept constant and measured the temperature of the drying air out, air humidity, and temperature of rice grains every 5 minutes. Once the process is completed the tool is switched off and the grain removed. Measured the weight of the grain. The level of air moistrure is checked by Kett moisture tester. Repeating the steps above for the other variables.

* Corresponding author: tprasetyo06@yahoo.com

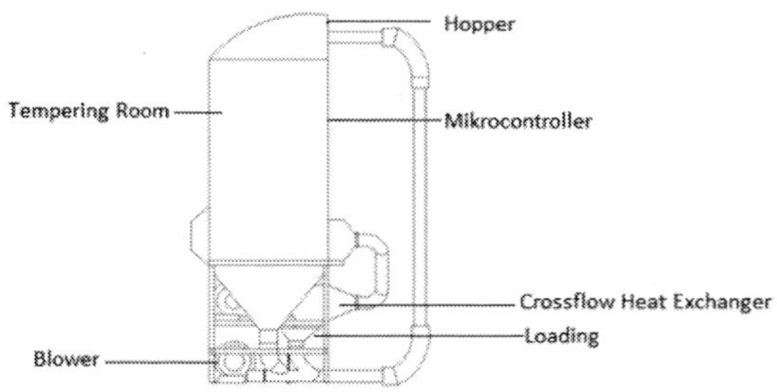

Fig 1. Scheme of paddy drier resirculation type

3 Result and discussion

3.1. Effect of inlet air temperature to the moisture content of grain

From figure 2, it can be seen that the water content of the material (x) will decrease with the decreasing drying rate and constant rate drying over time. Longer drying time will entrained amount of water from material more and more. According to Mujumdar 2004, the mechanism of water evaporation in the material occurs through heat and mass transfer process simultaneously, among others [5] :

1. Convection heat transfer from the air to the grain surface to evaporate the water in the grain surface.
2. Heat conduction from the surface of the grains into the grain.
3. Mass transfer of water from the granules to the surface diffusion or capillary.
4. Mass transfer of water from the surface of the grain to the drying air.

Decrease rate of water content of material is affected by mass and heat transfer process of moisture grain surface. Water displacement in grain surface occurs due to temperature difference on the surface of grains with air conditioning, causing driving force to move the water content in surface materials to the drying air in the form of moisture. The rate of decline relatively quickly and occurred for a short time. It is because during this period, only temperature difference with the material which affects and aren't influenced by thick of medium drying materials and water diffusivity in the material. At the same drying air temperature, it appears that the increase of drying loads number caused moisture content release of the material become smaller. It's because for the same amount of heat, the amount of water vapor that is released from the material will be greater. With the increase of quantity of water vapor, the amount of water vapor released per unit of time will be smaller.

This phenomenon also occurs in the drying material with the same load on the temperature variation of 40, 50, 60 ^{0}C. Seen in the figure 2 that as the temperature rises, the amount of water vapor that is released from the material per unit time was greater. As the temperature increases, the amount of heat used in the drying process the greater. The greater amount of heat will result in water diffusivity of material towards dryers media increased. This is consistent with the proposed drying mechanism of Mujumdar (2004), that the water diffusivity in the solid material is a function of temperature and moisture content [5-6].

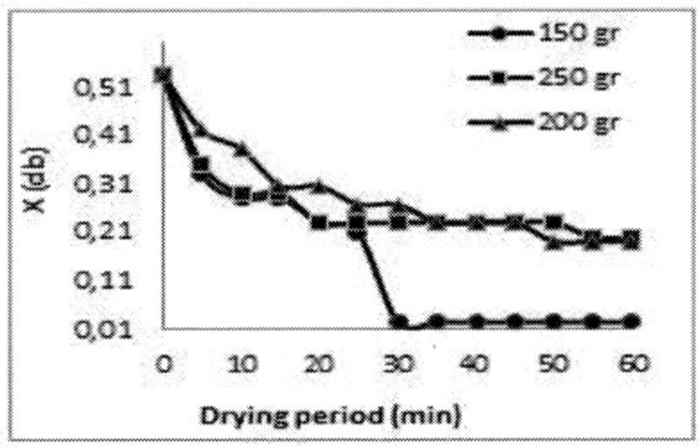

(a)

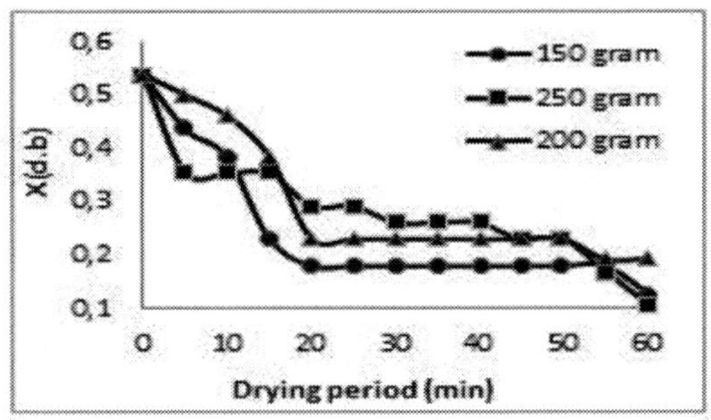

(b)

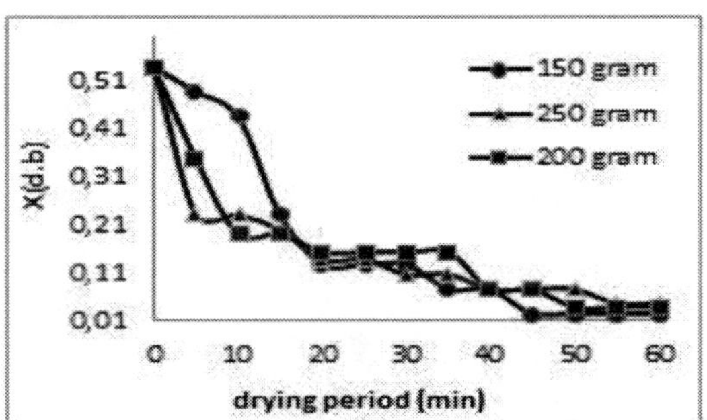

(c)

Fig 2. Grain drying grain moisture content (x) at various drying load with drying air temperature (a) 40^{0}C, (b) 50^{0}C, (c) 60^{0}C

3.2 Dynamics of drying out the air temperature

In figure 3, it is seen that drying air temperature will increase over time. When examined further, at the same temperature with different drying load variation, more mass of material drying caused exit air drying temperature smaller. It can be seen in temperature difference every minute for air conditioning in each drying load variations. For comparison, at 60^{0}C appears that the value of exit air drying temperature at 5 minutes early with a load of 150 grams is 59.23^{0}C while 250 gram load is 58.09^{0}C, decrease of 1.14 ^{0}C.

More material drying load has more moisture content to be evaporated, so the heat required will also greater and exit drying air temperature will be smaller.

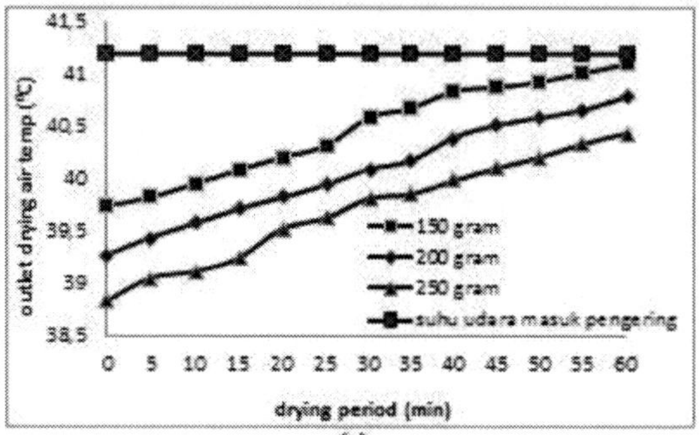

(a)

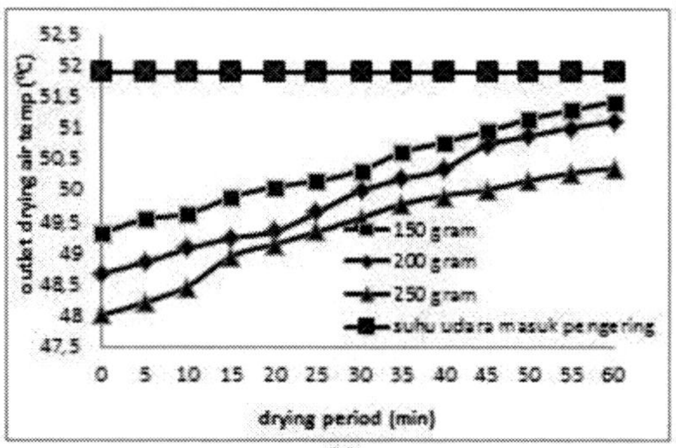

(b)

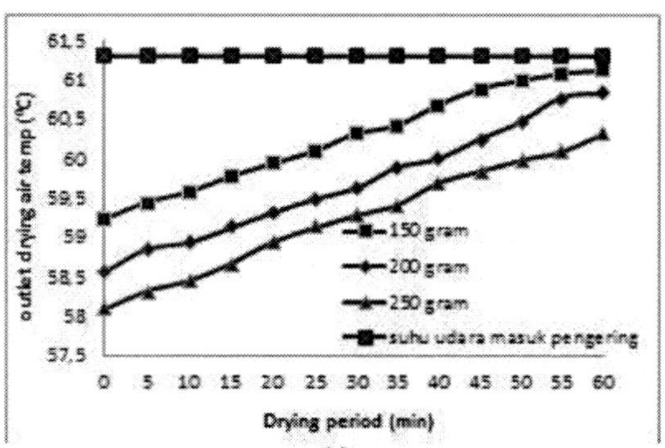

(c)

Fig 3. The results of grain drying for dynamic of outlet drying air temperature on various drying loads with drying air temperature (a) 40^{0}C, (b) 50^{0}C, (c) 60^{0}C

In the drying temperature variations with the same drying load seen that the higher air temperature entering dryer, the exit air temperature will rise to near dryer temperature when empty yet filled materials. It due to longer drying time will reduced water content in materials so that less heat is used. When heat is needed less, the exit dryer temperature will rise closer to the air temperature in the dryer when empty. In the drier, temperature distribution in the material will uniform so can accelerate the rate of heat transfer from the air to the material. Moreover, it also can be seen that at the beginning of drying process, occurs temperature gradient between temperature of the empty dryer and temperature when filled materials. Gradient occurs because the heat provided by drying air is used to evaporate water in the material and heat transfer mechanism occurs by convection, and conduction causing the exit air drying temperature be lower than currently empty.

3.3 The dynamics of granular temperature drying results

Based on figure 4 can be seen that temperature of particles on the grain in temperature and load variations will increase along increasing drying time and air temperature. Particle temperature will rise faster from initial temperature in the beginning of drying. It is because at beginning of drying water that vaporized on the surface particle decreases more and more until the end of drying. The reduced of water vaporized caused by water content on the surface have been unable to form a continuous film layer, so that heat conduction used by materials to provide driving force in order to water inside material diffuses out. At the same temperature with drying load variations, it is seen that the temperature will decrease with increasing particle drying load. It happens because the same amount of heat used to dry larger mass of material so the heat transfer process will result in decrease particle temperature.

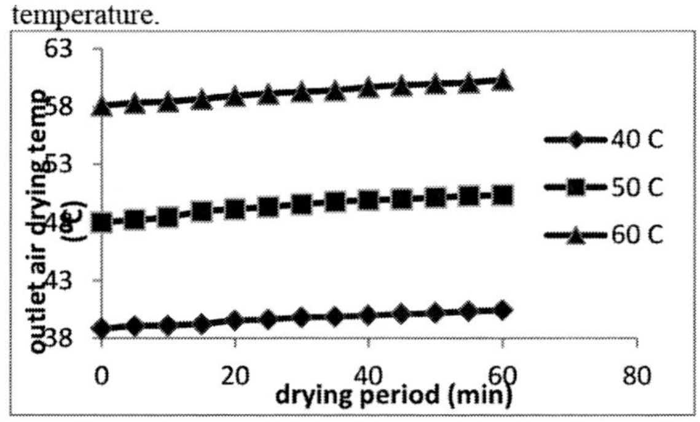

Fig 4. The results for grain drying for dynamic of outlet drying air temperature on drying loads 250 grams with various inlet air temperatures

In the first 5 minutes of drying process with same temperature and load variations drying temperature gradients obtained are quite high, but when 50 minutes the temperature gradient begins to diminish. For example drying temperature 60^{0}C, in the first 5 minutes occurred temperature gradient 10.33^{0}C and at 50 minutes become 15.77^{0}C. This phenomenon can be explained by basic mechanism of drying. At the beginning of drying process where the surface moisture content begin vaporized, the particle temperature will rise rapidly due to mass and heat transfer mechanisms that will form a water vapor continuous film on the surface. With the formation of water film, the heat will be used to release water on the surface and the measured temperature is it's film temperature. Along with increasing drying period, the particles temperature will be nearly constant because the main process is water vapor diffusion in heading out material. This process utilizes heat conduction of the material itself.

Therefore, the particle temperature gradient progressively smaller and tends towards constant .

Another point to be observed is that in the same drying load with dryer air temperature variations, it is seen that drying air temperature increases will further increase particles temperature. An increase of 27% in 5 minutes early looks at 40°C towards 60°C. The temperature rise in accordance with the basic drying law that drying is a function temperature and increasing drying air temperature used will cause heat transferred to material greater and rise the material temperature itself.

3.4 Effect of inlet air temperature of the air humidity

In figure 5 shows that over drying process the % RH will decrease. Decreased % RH is influenced by increase in air temperature [4]. At the beginning of drying process the air humidity is high because large amount of entrained water from the particles. Air humidity will decrease until the end of drying process because entrained water from particle fewer. Moreover, greater air temperature will decrease water vapor content of air due to water evaporation capacity will be greater while water contained in the air constant anyway.

At the same temperature with drying load variations, it appears that the increasing drying load will increase relative humidity drying air. Its look from %RH gradient at each load. For example in the first 5 minutes, the value of %RH at 40°C is 35.45 %RH and at 50 °C the value is decreased by 53.8 % become 16.37 %RH, then at 60°C the value of %RH become 8.49 %. It is because more drying load will release more water vapor from material to air conditioning. Thus, the air conditioning will acquire additional water vapor from water vapor material that has been released and will add value of drying air relative humidity.

In figure 6 shows that at temperature variations with same drying load the higher air temperature dryer used will decrease the relative humidity. In addition, in line with increase in temperature, relative humidity of the air will be smaller. This is consistent with the relationship of air - water systems in the psychometrics diagram. With rising temperatures, the ability of the air to evaporate water in the material will be even greater. Furthermore, the amount of vaporized water vapor in the air will change into vapor so the moisture content in the air will be less when compared to increase ability to evaporate moisture and % RH is getting smaller. When compared with air at room temperature, with the same amount of water, air with higher temperature will be able to get more water vapor [2].

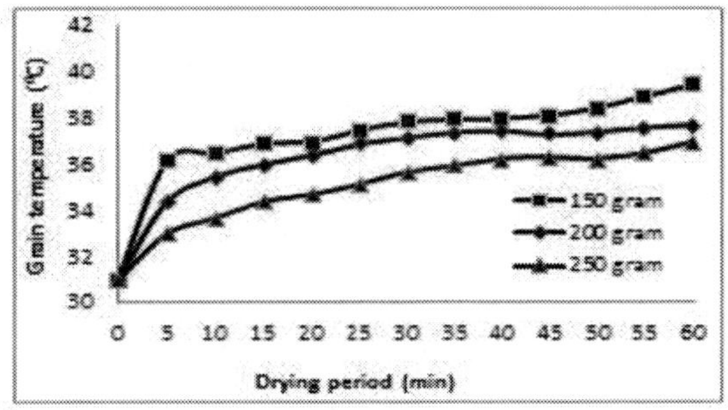

(a)

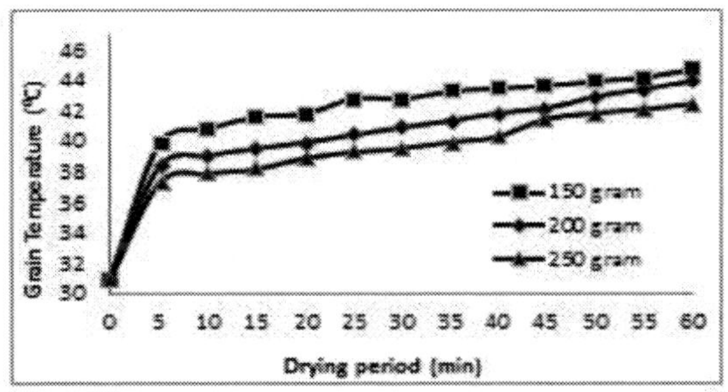

(b)

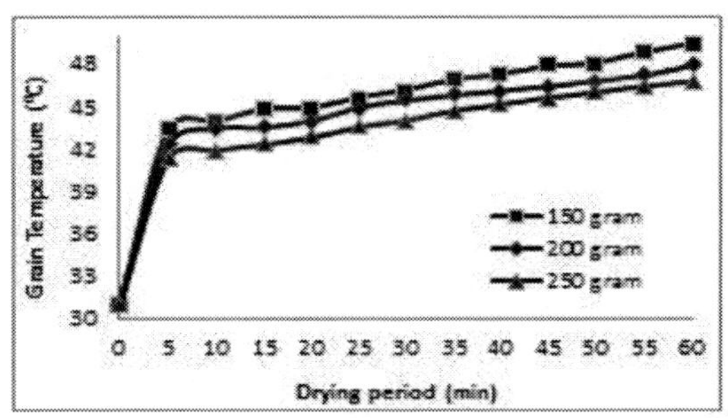

(c)

Fig 5. Results for grain drying for dynamic of grain temperature at various drying loads with air temperature (a) 40°C, (b) 50°C, (c) 60°C

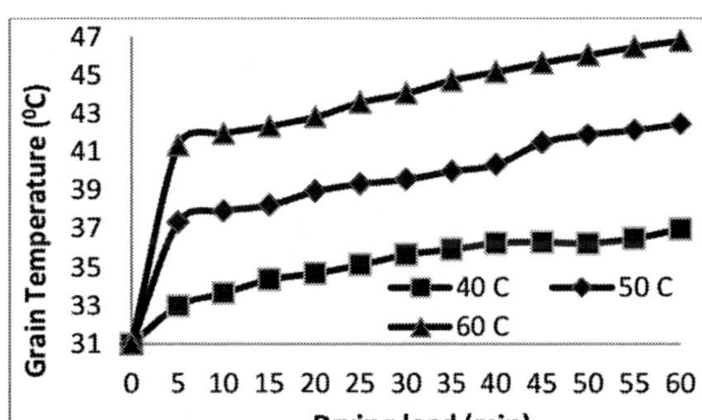

Fig 6. The result of grain drying for dynamics of grain drying temperature on the drying loads 250 grams with different inlet air temperatures

[17].

4 Conclusions

Total drying time to result moisture content 14%, is faster then sun drying that took 2-3 days and it depends on weather. Drying air temperature about 60 ^{0}C obtained 72 -74.3% head rice rendement, while direct drying resulted head rice rendement of 68 – 72 %, in bright weather.

Acknowledgments

Authors thank to Department of Chemical Engineering, Diponegoro University for all facilities and Ministry of Agriculture via KKP3N for funding this research.

References

1. S. Athajariyakul, T. Leephakreeda, J. Food Eng. **75**, 104-114 (2006)
2. P. Bunyawanuchakul, G. J. Walker, J.E. Sargison, P.E. Doe, Biosystem Engineering. **96**, 3, 335-344 (2007)
3. Kamaruddin, CREATA-IPB. (2007)
4. M.S. Khanali, A. Rafiee, S.H. Jafari, Hashemabadi, A. Banisharif, J.Agricultural Tech. **8,** 3, 795-810 (2012)
5. Mujumdar, S. Arun, *Guide to industrial drying principles. Equipment And New Developments*. IWSID: Mumbai, India. (2004)
6. S. Sumardiono, I. Pudjihastuti, A. Riyanto, and M.S. Siswadi, Physicochemical Properties of Analog Rice from Composite Flour: Cassava, Green Bean, and Hanjeli, World Applied Science Journal. 2014, Volume 32(6): pp 1140-1146 (2014).
7. M. Sidik, *Prospect of rice production and food security in east asia. Di dalam Peningkatan Daya Saing Beras Nasional Melalui Perbaikan Kualitas*. Lokakarya Nasional. Jakarta, 13-14 September (2006)
8. S. Soponronarit, P. Luangmalawat, S. Prachayawarakon, A. Nathakaranakule, J. Food Egineering, *LWT*. **41**, 716-723 (2007)

Performance Evaluation of Pneumatic Dryer for Aren (*Arenga piñata*) Flour

Suherman Suherman[1]*, and *Nur* Hidayati[2]

[1] Chemical Engineering Department, 50275 Diponegoro University, Indonesia
[2] Master of Chemical Engineering, 50275 Diponegoro University, Indonesia

Abstract. Aren (*Arenga pinata*) tree is one type of palm plant that produces fruit, sap and starch or flour in the trunk. Industries aren flour usually in small and medium industries is dried by conventional method with dried under the sun. The main constraint is the low drying speed. Therefore, in this study aren flour with a water content of 40% was dried using a pneumatic dryer to moisture content below 15% (wet base). The aim of this research is to analyze the influence of drying air temperature, drying air flow rate and feeding speed to drying speed and energy analysis. Energy analysis was performed to determine the performance of pneumatic dryers. The energy analysis itself is done in the form of energy utilization and energy efficiency. The results show that the increase of dryer temperature from 60 to 100 will increase the utilization of energy from 0.61 to 1.39 J / s, energy efficiency is 33.4-45.1%. the results of a proximate analysis, an ash content of 0.39 grain fiber 0.18 white degree 92.93% which already fulfill SNI standard of Aren flour.

1 Introduction

Aren tree is one type of palm plant. All parts of the tree can be used and have economic value. starting from the physical parts (roots, trunk, leaves, fibers etc.) as well as the results of its production (nira, flour and fruit). Aren including Aracaceae tribe (aracea nut). The tree is not spiny, unbranched, the height can reach 25 meters and the diameter of the tree can reach 65 cm [1].

This plant is almost similar to the coconut tree. Coconut tree trunked clean with old leaves that easily loose, aren tree trunk is very dirty because the stems are clad in fibers, so that the old leaves bark is difficult to be taken or out of the stem. Therefore, aren tree trunks are often overgrown by many types of ferns [1].

Empulur stem aren powder 48.9%. However, each aren tree produces varied flour. In Indonesia from every aren tree trunk can be obtained flour between 60-70 kg. aren flour is widely used for foodstuffs such as cakes, cendol, meatballs, noodles, vermicelli and hun kwe [2]. The standard quality of aren flour in Indonesia is contained in the Indonesian National Standard SNI 01 - 3724 - 1995. In the production process of aren flour, one of the important stages is the drying the product [3].

The pneumatic dryer is one of the mechanical drying machines that utilizes high-speed air and high temperature. The use of high-speed air and high temperature will cause the product to dry in a short time. The pneumatic dryer includes continuous dryer with a granule solid feed [4,5]. The solids are driven to duct drying by the hot air produced by the heater with the air from the blower and then the product passed to the cyclone. this type of dryer can be used for granular materials and non-abrasive solids[6].

Pneumatics can also produce a particle size reduction (friction) because the feeder is a screw conveyor. Pneumatic dryers are also particularly suitable for heat sensitive, explosive, degradable or flammable materials processing, otherwise the average residence time of solids is relatively short (usually a few seconds) [7].

Drying is a process of evaporation of moisture content in materials involving heat and mass transfer between the surface of the product and the environment [8,9]. Drying is widely used in various thermal applications such as food, perennial products and other industrial products [10] The purpose of drying is to degrade the moisture content in the solid can be removed by evaporation. In addition to extending shelf life as well to simplify storage simplify as well as in material transport [11,12].

Research on batch drying system has been done [13], [14] while continuous drier still little information. Energy analysis can be interpreted as a calculation of the flow of energy in a production process, usually for the process to be economical. Energy analysis is based on the first law of thermodynamics[15]. where the first thermodynamic law assumes that energy cannot be created and destroyed so that all incoming energy will be converted into other forms of energy [16].

Some researchers have conducted energy analysis on drying, especially on food products and agricultural products, among others, potatoes drying [11], olive drying [13], drying of palm cocoa [14], carrots drying [8] Mulberry drying [16], beans and nuts drying [17], wheat drying [18], and drying of seaweed [19].

* Corresponding author: suherman.mz@che.undip.ac.id

Information on energy analysis on food drying has been widely described, but the use of energy analysis in continuous system drying is rare.

Energy analysis is a calculation on systems that consume energy that aims to know the balance of energy use, energy conversion equipment efficiency, specific energy consumption [20]. The aims of this study are the experimental research on the drying of aren flour in a pneumatic dryer, to know the performance of the aren flour drying tool by calculating the energy analysis. Energy efficiency calculations can be used to evaluate the performance of a drying machine.

2 Materials and Methods

2.1. Materials

Material used in this experiment were Aren flour, which taken from Klaten, Central Java. The equipment used in this experiment are Pneumatic dryer, moisture keet meter, humidity meter, oven, digital scales, porcelain cup, and anemometer. Drying time is 60 minutes for each variable with moisture content measured every 5 minutes. Two hundreds grams of Aren flour were used on each variables. The variabels were drying temperature (60,70,80,90 and 100°C), feed flow rates (10, 20 and 30 gram/minutes) and air flow rate (2,8; 3,7; 4,1; and 5,1 m/s)

2.2 Drying Equipment

Pneumatic consists of temperature controller, blower, heater, feeder, screw conveyor, duct dryer, cyclone and output product. The Pneumatic dryer must be calibrated first before the drying process. The calibration is done by turning on the dryer and measure the temperature and the air flow rate for 1 minute until constant value

Fig. 1. Pneumatic dryer.

2.3 Research Procedure

The solid feed flow rate of 10 g/minute is calculated manually and continuously by introducing a certain amount of flour mass by weight and increments gradually over a predetermined time. The drying process was carried out by varying the dryer air temperature is 60, 70, 80 90 and 100°C. Raw materials enter through the feeder and then forwarded to the duct dryer Aren flour. Before the experiment began, air temperature and air flow rate set first. After turning on the blower, then set the desired temperature. Wait a while until the air dryer from the blower reaches the desired temperature. Feed aren flour is inserted after the condition of the dryer is constant. The feed from the feeder will be pushed upward through a cylindrical drying tube with a total length of 170 cm and an inner diameter of 12 cm. During drying, the feeder continuously feeds the wet aren flour into the drying duct. Out of the drying duct, the dry aren flour will be passed to the cyclone to separate the solid from hot air. Furthermore, dried aren flour was collected for product analysis

2.4 Calculation of Dryer Efficiency

The energy analysis method follows the formula derived by [21]. EU (Energy Utilization) can be known by applying the first thermodynamics law

$$EU = Ma\ (hai - hao) \qquad (1)$$

Where hai is the air dryer enthalpy at the inlet temperature (J/kg) and hao is the air dryer enthalpy at the exit temperature (J /kg) and the EU is Energy Utilization (J / s). The mass flow rate can be calculated using the equation

$$Ma = \rho a\ Va \qquad (2)$$

Where ρa is the density of dry air in kg/m^3, Va is the volumetric air dryer rate in m^3/s. The drier air enthalpy at the entry and exit temperature of hi and ho can be calculated by the equation equation

$$h = Cpa\ Tda + W\ hsat \qquad (3)$$

Where Cpa is the specific heat of dry air in J / (kg°C), T is the drying air temperature in °C, W is the ratio of the drying air humidity (kg H_2O/kg da) and is the drying air and hsat is the saturated vapor enthalpy in J/kg The specific heat of the dryer air can be calculated by the equation

$$Cpa = 1.0029 + 5.4\ x\ 10\text{-}5\ Tda \qquad (4)$$

The ratio of energy utility can be calculated using equation

$$EUR = (Ma\ (hai\text{-}hao)) / (Ma\ (hai\text{-}ha\alpha)) \qquad (5)$$

Where EUR is the ratio of energy utilization and haα is the air dryer enthalpy of environmental temperature (J/kg). The energy efficiency (%) can be evaluated and the ratio of energy used to the given energy

https://doi.org/10.1051/matecconf/201815605023

$$\eta E = [(Ei - Eo) / Ei] * 100 \qquad (6)$$

ηE is the energy efficiency, Ei is the incoming energy and Eo is the out energy, where

$$Ei - Eo = Ma\ (hai\text{-}hao) \qquad (7)$$

$$Ei = Ma{*}hai \qquad (8)$$

3 Results and Discussion

3.1 Effect of temperature on the drying curve

The drying air temperature is defined as the mean air temperature used to dry the amount of material measured in the drying chamber [15]. During the drying process, the temperature very important in the evaporation process of water, whether on the surface of the material or inside the material. The dryer air temperature should be adjusted so that the dried material can be as dry as possible so that it meets the SNI standard for acceptance in the market. The drying curve shows changes in the water content in the aren flour throughout the drying chamber under different drying conditions. In the experiment, the operating temperature of the dryer air was varied at 60, 70, 80, 90, and 100°C. Within every 5 minutes it measures the moisture content along the dryer duct

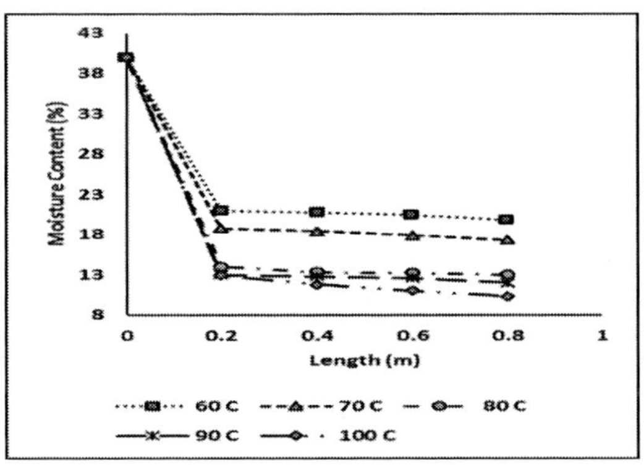

Fig. 2. Effect of dryer air temperature on the moisture content profile of aren flour each the dryer duct (10 g / min).

The average initial moisture content of wet aren before being put into a pneumatic dryer is 40%. The influence of drying air temperature along the drying chamber to the Aren drying curve is shown in Fig 2. The results show that there is a decrease in the drying curve along the drying duct (0.2 - 0.8 m /length) at each drying air temperature. The water content of aren flour at each temperature decreased a significant curve at 0.2 m duct until constant drying occurred on duct 0.8 m. The decrease occurred at 19.8% at 60°C, while at 70, 80, 90 and 100°C, the water content decreased to 17.3%, 13%, 12% and 10.3%. The drying curve shows a decrease in the water content of aren flour along with the dryer duct until the small decrease is almost constant, the solid feed flow rate is 10 g / min and the inlet airflow rate of 0.25 is due to the larger drying air temperature, water in the product. This result is in accordance with previous study which in drying tapioca by means of a continuously fluidized bed fluidization dryer at a drying distance divided into 3 zones of decreasing moisture content throughout the zone and at the drying effected in cassava starch[20,21].

3.2 Effect of air flow rate

Air flow rate is an important factor to be considered in drying by using pneumatic dryers. In this study, variations were made to the air velocity of the dryer as a heat carrier medium. Air velocity variation is carried out by arranging the air inlet opening a blower, is by using a cover plate with 4 openings variation is openings 0.25 (2.8m /s) 0.5 (3.7m/s), 0.75 (4.1m/s) and 1 (5.1m/s)

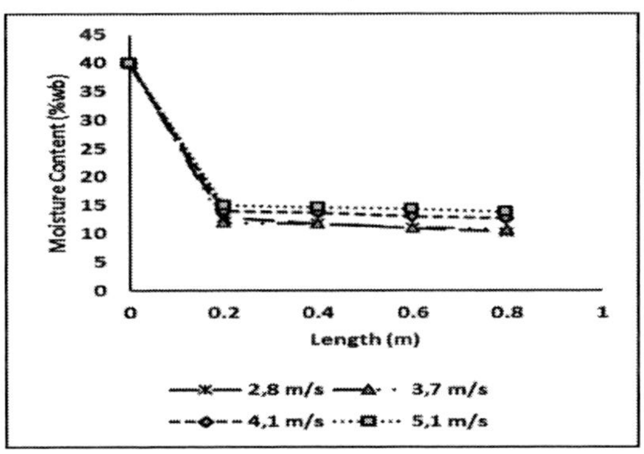

Fig. 3. Effect of air drying inlet openings on the moisture content profile of aren flour along the dryer duct (10 g / min)

The determination of the air velocity to be used to dry a material must be observed. The larger the air inlet opening, the air velocity exhaled by the blower will be higher. To achieve an efficient drying process, the dryer airflow rate used should be greater than the minimum speed required to remove the material. The air velocity exhaled by the blower must be greater than the free fall velocity of the particles to be dried (wet material). The results showed that with the higher air velocity, the water content in solids is also higher. The solid water content at 2.8 m / s is 10.3%, at a speed of 3.7 m / s is 10.8%, at a speed of 4.1 is 13% and at a speed of 5.1 m / s is 14.4%. The speed of air flow during the drying process by using pneumatic (flash) dryer at a very low speed, the material particles cannot be lifted by the air flow, so the drying process cannot run perfectly. While the flow of air velocity is too high will cause the heat contact between dry air with the material will become too short, consequently the drying process becomes ineffective, because the water is vaporized only slightly, and the final water content of the product is usually still higher, this is in accordance with previous research on drying cassava with pneumatic dryer that the higher the rate of air flow rate then the moisture content is getting bigger, this is because the contact between the hot air and the solid becomes shorter so that drying is not optimal [22].

3.3 Effect of feed flow rate

Aren drying research is implementing a continuous system. Aren Flour with water content ± 40% is feed continuously into the dryer. The flow rate of sugar palm is varied by 10 gr/min, 20 gr/min and 30 gr/min

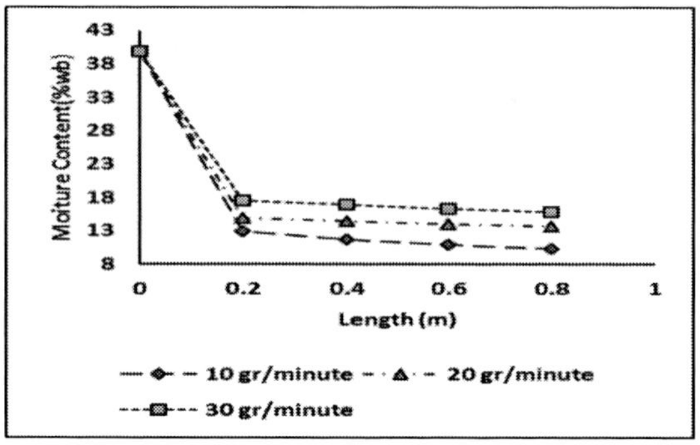

Fig. 4. Effect of solid feed flow rate on the water content profile of the material along the dryer duct (100 ° C).

The drying curve indicates that the Aren feed incorporated every minute in the pneumatic drying apparatus may affect the change in moisture content along the drying duct in the dry palm sugar product. Fig 4 shows that more aren flour is fed to the dryer, causing a decrease in water content longer to reach a constant. This is due to the faster saturation of dryer air. More moisture content material makes the dryer air faster saturated with the same air dryer capacity. This saturation will make the air dryer longer to remove the water content in aren flour. This trend is similar to the results in study monitoring and control of fluid-bed drying of tea where feed flow rate control is necessary to regulate the final moisture content of the product [23]

3.4 Volumetric water evaporation rate

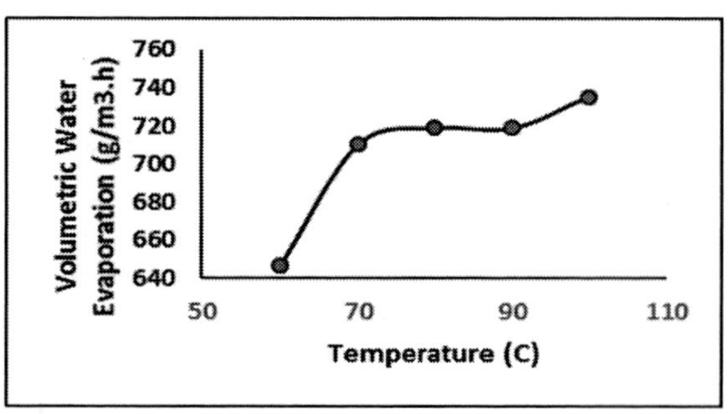

Fig. 5. Effect of temperature on volumetric water evaporation rate.

The rate of volumetric water evaporation increase with increasing air temperatures because higher air temperatures lead to greater differences between air temperature and starch surface temperature, which is the driving force for heat transfer[24]. Changes in air temperatures lead to changes in clearer moisture levels, increased feed flow rates at constant temperature causing a volumetric increase in the rate of water evaporation. Despite using a lower drying temperature, the moisture obtained by using a pneumatic dryer system is considerable

3.5 Energy Utilization

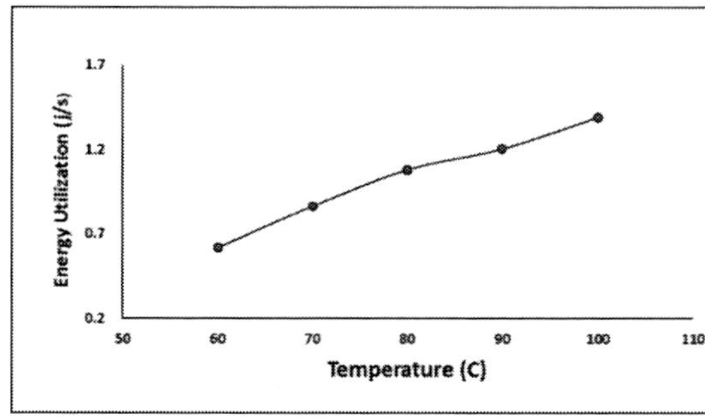

Fig. 6. Effect of dryer air temperature on energy utilization of Aren flour

Energy analysis uses the data obtained from the experiment. The images show the effect of temperature on energy utilization where energy use increase with increasing drying air temperature in pneumatic dryer. The higher the temperature causes a lot of moisture content to be evaporated. This is due to the high temperatures causing heat transfer and also the higher the period to increase the use of energy in reducing the product water content, obtained the average of energy usage at 60-100 C temperature variation increased from 0.61 to 1.39 J /s. The energy used in this study shows a relationship that is directly proportional to the drying air temperature. The results of this study are similar to the results of the tapioca drying research using a tray dryer [21,25] and drying carrots using a continuous band dryer [8]

3.6 Energy efficiency

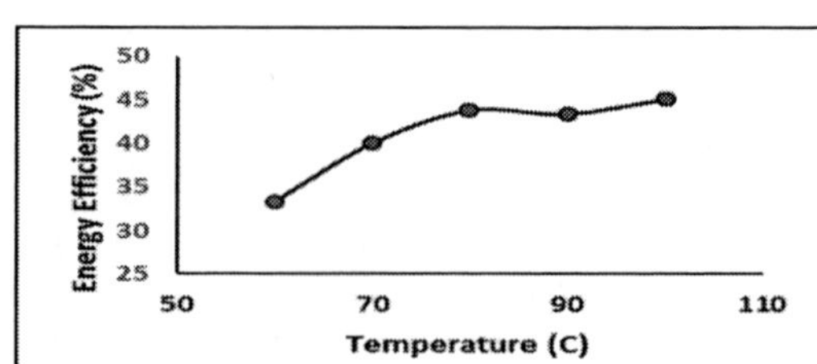

Fig. 7. Effect of dryer air temperature on energy efficiency.

The effect of increased drying air temperature on energy efficiency is shown in Fig 7. At a temperature increase of 60-100°C the energy efficiency to dry the Aren increases from 33.4% to 45.1%. At energy efficiency, the increased energy use of each temperature causes the efficiency of energy use to dry the Aren increase. The effectiveness of energy use is influenced by the energy used from the available energy. The research [21] and [18] also showed a similar effect between increased drying air temperature and energy efficiency. The same trend is shown by all types of dryers (convective dryers, microwave dryers, vacuum dryers, solar dryers, and dryer joints) that energy efficiency increases with increasing drying air temperature

3.7 Product Quality Aren

The quality of Aren flour is determined by the standard requirements for the resulting

Table 1. Results of Quality Analysis of aren flour.

Test Criteria	Unit	Requirement	Test Result
Color	-	White *	white.
Water content (b/b)	%	Maks 13*	±10.3
Ash (b/b)	%	Maks 0.5 *	0.39
Coarse fiber	%	Maks 0.1*	0.18
Starch	%	Min 60 *	60.78
White Degree	-	Min 91 *	92,93
Protein	b/b		0.49
Fat	b/b		0.18

*) Indonesian National Standard 01-3724-1995

Table 1 presents the results of the feasibility test of the quality of aren flour. The quality criteria of aren flour seen from the results of the physical analysis (color), proximate analysis (ash content, starch content, protein content, fat content), physicochemical analysis (white degree). The results show that all test criteria have met the quality standards of aren flour as a food ingredient, for crude fiber the results showed still higher than required because of the fibers are still included in the flour because the content of aren flour is very high in which aren flour obtained from tree trunks. Aren flour is made by extracting the tree trunk. Where the flour is taken if the tree is no longer produces nira (aren sugar). The extracted aren flour will be brownish. Based on the degree of white, the white Aren flour is also better quality. The test results indicate that the white degree value of the Aren product has fulfilled the requirement of 92. One of the processing of Aren flour which can cause the difference of ash content value is the starch extraction stage. The minerals contained in the stems can be wasted along with the pulp from the extraction process of aren flour, so that the measured ash content becomes lower, in addition to the amount of fiber contained due to aren tree trunks. The amount of fiber and impurities affect the quality of sugar palm. The more fiber and impurities it contains, the lower the quality. The test results showed that the level of more palm fiber by 0.18

CONCLUSION

The result of Aren flour proximate analysis used in drying research using pneumatic dryer showed that aren tress has starch content 60.78%, whiteness 92.93%, fiber content 0.18. The energy analysis of the drying process shows that (i) the energy usage increases with increasing temperature. The energy efficiency increases from 33.4-45.1%, (ii) the higher the temperature followed lower the moisture content. The lowest moisture content is obtained at a temperature of 100°C with water content of 10.3% (iii) the speed of the air flow rate is very important because if it is too low, can be lifted by the airflow, if the flow of airspeed is too high will cause the hot contact between dry air and the material will become too short (iv) the higher the temperature the increased energy utilization

References

1. M. Lempang, Info Tek. EBONI, **9**, 37–54, (2012)
2. Purnawan, J. Tekno. Technoscientia, **4**, 28–36, (2011)
3. S. Suherman, M. Djaeni, D. H. Wardhani, and A. C. Kumoro, CSAFS Conf. Proceedings. 430–438, (2017)
4. I. Borde and A. Levy, "*Pneumatic and Flash Drying*," (2006)
5. L. R. Dragiša Tolmač, Zvonimir Blagojević , Slavica Prvulović, Jasna Tolmač, J., Mech. Eng., **8**, 89–96, (2010)
6. S. Suherman, A. C. Kumoro, T. D. Kusworo, and N. Hidayati, Adv. Sci. Lett., **23**,. 5650–5652, (2017)
7. J. Baeyens, D. Van Gauwbergen, and I. Vinckier, J. Powder Technol., **83**, 139–148, (1995)
8. M. Aghbashlo, M. H. Kianmehr, and A. Arabhosseini, J. Food Eng., **91**, 99–108, (2009)

9. S. M. El-behery, W. A. El-askary, M. H. Hamed, and K. A. Ibrahim, J. Comput. Fluids, **68**, 159–167, (2012)

10. S. Suherman and R. Trisnaningtyas, AIP Conf. Proc., **1699**, 2015 (2015)

11. M. Aghbashlo, H. Mohammad, and H. Samimi-akhijahani, J. Energy Convers. Manag. **49**, 2865–2871, (2008)

12. A. M. E. Tsotsas., *Modern Drying Technologi*. (2012)

13. N. Colak and A. Hepbasli, " J. Food Eng., **80**, 1188–1193, (2007)

14. A. V. Otoniel Corzo, Nelson Bracho and A. P. ´squez, J. Food Eng., **86**, 151–161, (2008)

15. S. Suherman, A. C. Kumoro, and T. D. Kusworo, Int. Conf. Chem. Mater. Eng, 050001-1 -050001-6, (2015)

16. A. Akbulut and A. Durmus, J. Energy, **35**, 1754–1763, (2010)

17. İ. Karagüzel, E. Tekİn, and A. Topuz, ,J. Sci. Res. Essays, **7**, 3961–3973,(2012)

18. M. R. Assari, H. B. Tabrizi, and E. Najafpour, Int. J. Therm. Sci., vol. **64**, 213–219, (2013)

19. A. Fudholi, K. Sopian, M. Y. Othman, and M. H. Ruslan, Energy Build., **68**, 121–129, (2014)

20. S. Suherman and Rona., J. Reaktor, **16**, 24–31, (2016)

21. N. A. Aviara, L. N. Onuoha, O. E. Falola, and J. C. Igbeka, J. Energy. 73, 809–817, (2014)

22. J. N. W. K, N. Bintoro, and J. T., Pros. Semin. Nas. Perteta, 97–104, (2012)

23. S. J. Temple, S. T. Tambala, and A. J. B. Van Boxtel, J. Control Eng. Prctice. **8**,165–173, (2000)

24. C. Nimmol and S. Devahastin, J. Appl. Therm. Eng., **30**, 2204–2212, (2010)

25. N. A. Aviara, J. C. Igbeka, and L. M. Nwokocha, J. Int Agrophys, **24**, 219–225, (2010)

26. M. Djaeni, N. Asiah, S. Suherman, A. Sutanto, and A. Nurhasanah, J. Renew. Energy Dev., **4**, 20–24, (2015)

Effect of Impeller Position and Baffle Width on Uniformity of Fe Concentration for Fortification of Baby Instant Porridge

Herry Santosa , *Noer* Handayani*, *Kristinah* Haryani, *Bontor* Sitorus, and *Muhija* Mawarni

Departement of Chemical Engineering, Faculty of Engineering, Diponegoro University, Prof. Soedarto, SH, Tembalang, Semarang, 50275, Indonesia

Abstract. Iron deficiency (ID) is considered to be the most micro- nutritional problem, and the main cause of anaemia, called Iron Deficiency Anaemia (IDA). Nowadays, iron fortification is the most practical and priority solution for IDA alleviation. For toddler, iron fortification could be performed on complementary food of breast milk, such as instant porridge. Iron fortification was conducted using mixing and agitation method to prepare instant porridge from purple sweet potato flour. The aims of this work were to investigate the effect of impeller position and baffle width on the uniformity of iron concentration. Preparations of purple sweet potato flour and iron fortification were conducted using drying and mixing agitation method, respectively. Two independent variables, i.e. impeller position (1, 1.8, 2.6, 3.2 cm from the bottom of stirred tank) and baffle width (1.1, 1.4, 1.8, 2.3 cm), were used to investigate the uniformity of iron. Iron concentration was determined by atomic absorption spectrophotometer (AAS). The results depict that impeller position 1.8 cm from the bottom of the stirred tank and baffle width 1.1 cm provide a relatively good operating condition to obtain the high level of iron uniformity.

1 Introduction

Vitamin A, iron, zinc and iodine have important roles to maintain body metabolism, improved health, and well-being of humans. Furthermore, iron deficiency (ID) is considered to be the most micro- nutritional problem, and the main cause of anaemia, called Iron Deficiency Anaemia (IDA) [1,2]. Children under 5 years old and women of child bearing age are the most susceptible subjects of this micronutrient malnutrition. IDA brings adverse consequences, such as lower cognitive capability and reduced immune functions of young generation in Indonesia, particularly [3,4]. In accordance with these serious impacts, IDA prevention has to be priority by the Government and Health Organization.

Nowadays, iron fortification is the most practical and priority solution for IDA alleviation [5,6]. For toddler, iron fortification could be performed on complementary food of breast milk, such as instant porridge. Purple sweet potato flour is one of potential ingredients for instant porridge due to its interesting colour, high level of beta carotene and prolonged storage time. Beta carotene is very useful for children as the main dietary source of vitamin A, essential for growth, and immune function [7].

Some researchers did iron fortification for many food products using various methods [8-10]. Among those, mixing and agitation is commonly procedure and widely used for food fortification is. It is also regarded as the simplest and cheapest method in food industrial [11]. Previous works were also conducted to investigate micronutrient fortification using mixing and agitation [12-18]. However, it has some difficulties to ensure the iron uniformity. Several variables should be controlled and regulated carefully to confirm it, so that the effective and efficient method would be achieved. Therefore, further studies to increase the mixing and agitation performance are needed.

In this study, iron fortification was conducted using mixing and agitation method to prepare instant porridge. The main substance of instant porridge was purple sweet potato flour. The aim of this work was to investigate the effect of impeller position and baffle width on the uniformity of iron concentration.

2 Materials and Method

2.1. Materials

Purple sweet potatoes were obtained from local market in Indonesia. $FeSO_4$. $7H_2O$ (p.a), as iron source, was purchased from Merck. Demineralized water was produced by Processes Laboratory at Chemical Engineering Department, Diponegoro University.

2.2 Preparation of purple sweet potato flour

Purple sweet potato flours were conducted by drying method. The sorted purple potatoes were sliced (± 1 mm)

* Corresponding author: nora@undip.ac.id

to produce thin chips, and soaked in a sodium bisulphite solution to prevent browning. The chips were dried using oven with a temperature of 40°C for 10 hours. The dried chips were then fed into the grinder to form smaller shape. Sweet potato flours were then obtained by sieving it with a size of 60 mesh.

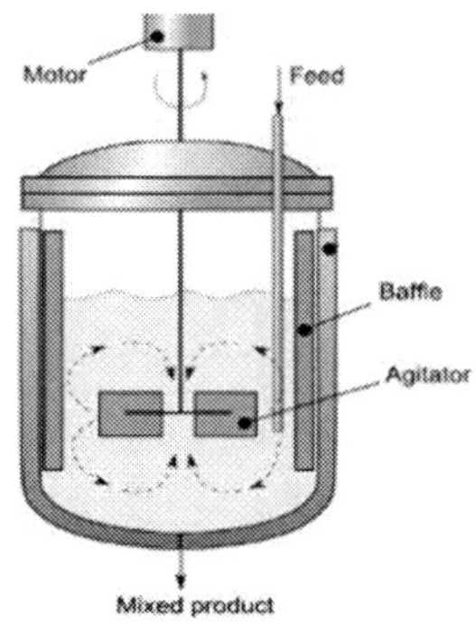

Fig. 1 Mixing tank

2.3 Iron fortification and preparation instant porridge

Fortification was made by using mixing agitation procedures. Iron source $FeSO_4 \cdot 7H_2O$ was added into flours and dissolved in 100 ml of free mineral water with ratio 1:2 water in a mixing tank with particular conditions for 15 minutes until the mixture becomes homogeneous and form slurry (Figure 1). The slurry was heated 70–80°C and stirred at 300 rpm. This process was carried out before gelation occurs. The impeller position (1.0, 1.8, 2.6, 3.2 cm from the bottom of stirred tank) and ratio of baffle width to tank diameter (1.1, 1.4, 1.8, 2.3 cm) were investigated as independent variables. The height of slurry and the diameter of the tank were 9 cm and 10 cm, respectively. The mixtures were dried using oven with temperature of 60°C for 5 hours to forms flakes. Instant porridges were obtained after size reduction process, further sieved with 60 mesh sieve.

2.4 Iron determination

Fe concentration was analysis using atomic absorption spectrophotometer (AAS). The results will be presented in the total Fe concentration.

3 Results and Discussions

3.1 The effect of impeller position on uniformity of iron concentration

Variation of impeller position was carried out due to its affect on the flow pattern of slurry [19]. The pattern and motion of slurry phase influence the acceleration of diffusion mechanism to produce uniformity of concentration [20]. The effect of impeller position on iron concentration was investigated in the stirred tank. Impeller was placed at 1, 1.8, 2.6, 3.2 cm from the bottom of tank. Table 1 shows the iron concentration in different sampling locations (side, top and bottom of the vessel).

Table 2. The uniformity of iron concentration with the different impeller positions

Position impeller from the bottom (cm)	**Sampling position**	**Fe concentration (mg/l)**
1	Side	0.072
	Top	0.073
	Bottom	0.104
1.8	Side	0.110
	Top	0.120
	Bottom	0.103
2.6	Side	0.018
	Top	0.080
	Bottom	0.029
3.2	Side	0.038
	Top	0.016
	Bottom	0.080

Impeller which is placed 1.8 cm from the bottom of stirred tank provide a better uniformity in comparison with others positions. Three sampling locations of the tank (side, top, and bottom) showed a relatively uniform of iron concentration (0.100, 0.120, 0.103 mg/l). This result is in a good agreement with the theory proposed by Dickey (1984) [21]. Impeller should be placed at 1/6 of total fluid height. The height of total slurry in the stirred tank is 9 cm, so that impeller could be put at 1.5 cm from the bottom of the tank. Thus the impeller position 1.8 cm (from the base of the tank) had the best performance. Impeller position which is placed too far from base will cause iron congregate on the bottom of the stirred tank (see Table 1 on impeller position at 3.2 cm from the base tank).

3.2 The effect of baffle width on iron concentration

Baffle, in the mixing systems, has a function as vortex breaker which could make an axial pattern [20]. Those, the effect of baffle width (1.1, 1.4, 1.8, 2.3 cm) on iron uniformity needed to be investigated. In this study, we use number of baffle 4, since it showed the best performance in Santosa. et al (2017) [18].

Table 2 presents the iron concentration in different sampling locations (side, top and bottom of the tank). Baffle width 1.1 cm had a better iron homogeneity in compare with others. The samples were taken in three different locations of the tank (side, top, and bottom) depicted a relatively uniform of iron concentration (137.5, 132.5, 131.0 mg/l). The proper baffle width for mixing is 10 to 21% of diameter of the tank. This study used stirred tank with the diameter of 10 cm, so that the appropriate diameter for iron uniformity is 1.1 cm.

Generally, the larger width of the baffle creates the turbulence flows. However, the slurry movement could be hampered while using widely baffles and the imperfect iron distribution will be obtained (see on Table 3).

Table 3. The uniformity of iron concentration with the different baffle width

Baffle width (cm)	Sampling position	Fe concentration (mg/L)
1.1 (B1)	Side	137.5
	Top	132.5
	Bottom	131.0
1.4 (B2)	Side	106.0
	Top	126.5
	Bottom	129.0
1.8 (B3)	Side	139.5
	Top	198.0
	Bottom	121.0
2.3 (B4)	Side	153.0
	Top	157.5
	Bottom	125.5

4 Conclusions

Preparation of baby instant porridge using mixing and agitation method was conducted in this study. The results depict that impeller position 1.8 cm from the bottom of the stirred tank and baffle width 1.1 cm have the optimal condition to obtain the high level of iron uniformity.

Acknowledgement

Authors would say thank you for financial support from the Fundamental Research from Ministry of Research, Technology, and Higher Education.

References

1. C. R. Titaley., M. J. Dibley., Asia Pas J Clin Nutr, **24**, 1 (2015)
2. B. Tripathi., K. Platel., Food Chem. **126** (2011)
3. S. Akhtar., F. M. Anjum., M. A. Anjum., Food Res. Int., **44** (2011)
4. C. Gupta., P. Chawla., P. Arora., S. K. Tomar., A. K. Singh., Food Hydrocoll., **43** (2015)
5. N. F. Krebs., M. Domellof., E. Ziegler., J. Pediatr., **167**, 4 (2015)
6. J. Berger., F. T. Wieringa., A. Lacroux., M. A. Dijkhuizen., Nutr. Rev., **69** (2016)
7. E. G. Donhowe., F. Kong., Food Bioproc. Tech., **7**, 2 (2014)
8. Y. Wei., M. J. I. Shohag., F. Ying., X. Yang., C. Wu, Y. Wang., Food Chem., **138** (2013)
9. W. Hei., M. J. I. Shohag., Y. Wei, Y. Feng., X. Yang., Food Chem., **141** (2013)
10. N. A. Handayani,, M. Luthfansyah, E. Krisanti, S. Kartohardjono, K. Mulia., AIP Conf. Proc., **1904**, 0200531-0200538 (2017)
11. E. Betoret., N. Betoret., D. Vidal., P. Fito, Trends Food Sci. Technol., **22** (2011)
12. A. Kyritsi., C. Tzia., V. T. Karathanos.. LWT - J. Food Sci. Technol,, **44** (2011)
13. M. L. Samaniego-Vaesken., E. Alonso-Aperte., G. Varela-Moreiras., J. Food Compost. Anal., **23** (2010)
14. F. Jimenez-Colmenero., F. J. Sanchez-Muniz., B. Olmedilla-Alonso., Food Chem., **123** (2010)
15. J. L. Chavez-Servın., A. I. Castellote., M. C. Lopez-Sabater., Food Chem., **106** (2008)
16. G. Puccio, C. Cajozzo, F. Meli., F. Rochat., D. Grathwohl., P. Steenhout., J. Nutr., **23** (2007)
17. P. Alliet., P. Scholtensc., M. Raes., K. Hensen., H. Jongen., J. –L. Rummens., J. Nutr., **23** (2007)
18. H. Santosa, K. Haryani, N. A. Handayani, H. Cahyono, Muhija, Bontor., Adv. Sci. Lett.,, **23** (2017)
19. A. Kaćunić., M. Akrap., N. Kuzmanić., Chem. Eng. Res. Des., **91**, 2 (2013)
20. K. J. Myers., M. F. Reeder., J. B. Fasano., Chem. Eng. Prog., **91**, 12 (2002)
21. D. S. Dickey., In Handbook of Chemical Engineering Calculations, (N.P. Chopey and T.G. Hicks Eds.), McGraw-Hill, New York, (1984)

Production of Biofuel via Catalytic Hydrocracking of Kapuk (Ceiba pentandra) Seed Oil with NiMo/HZSM-5 Catalyst

I Gede Andy Andika Parahita[1], *Yustia Wulandari* Mirzayanti[1], *Ignatius* Gunardi[1], *Achmad* Roesyadi[1], and *Danawati Hari* Prajitno[1*]

[1]Department of Chemical Engineering, Industrial Technology Faculty, Sepuluh Nopember Institute of Technology, Surabaya, 60111, Indonesia

Abstract. Biofuel is one of alternative energy that is being developed today to solve the problem of limited fossil fuel as an energy source. The goal of this study is to produce biofuel from kapuk (*Ceiba pentandra*) seed oil (KSO) through catalytic hydrocracking process using NiMo/HZSM-5 catalyst. NiMo/HZSM-5 catalyst was obtained by impregnation of nickel and molybdenum as metallic precursors on HZSM-5 catalyst as support using incipient wetness impregnation method. It was found that the surface area of the catalyst was 222.1350 m^2/g, the pore diameter was 3.0148 nm and the pore volume was 0.1674 cm^3/g. The diffraction peaks of nickel oxide phase and the metallic phase of nickel were observed at 2θ of 62.5102° and 51.7283°. Molybdenum oxide phases were observed at 2θ of 53.5674° and 60.4682°. The catalytic hydrocracking process was performed using slurry pressure batch reactor at the temperature of 350°C for 2 h. The obtained liquid product was analyzed using GC-MS in order to determine the organic content. It has been found that the highest compounds were the palmitic acid with 23.14 area%. Besides, the hydrocarbon composition consisted of 33.93 area% (i.e. 4.34 area% cycloparaffins, 16.02 area% n-paraffins, 12.26 area% olefins, and 1.30 area% of aromatics) and 58.73 area% of carboxylic acid. Thus, it can be concluded that NiMo/HZSM-5 catalyst can convert KSO into biofuel through catalytic hydrocracking process at the temperature of 350°C for 2 h.

1 Introduction

Biofuel production as renewable energy is promising alternatives to substitute fossil fuel from the depletion of world oil reserve [1]. Biofuel is referred to enriched-energy chemicals derived from such plants, microalgae, and bacteria that used to fulfill the global energy demand. However, approximately 95% of the biofuels produced today are derived from edible vegetable oils. It caused an ongoing debate regarding the use of agricultural lands for fuel purposes. This resulting the exploration of biofuel production from inedible vegetable oil such as kapuk (*Ceiba pentandra*) seed oil (KSO).

Kapuk (*Ceiba pentandra*) is a tropical tree belongs to Malvales order and Malvaceae family. It was originally coming from Southeast Asia and grown naturally through tropical region [2]. KSO has the capability as a sustainable biodiesel feedstock because of simple cultivation and short harvest time (4-5 months) [3]. KSO contains a unique fatty acid (malvalic acid) which are more reactive than the double bond carbon (polyunsaturated) in reaction by atmospheric oxygen. The hydrocarbon chain has the ability to reduce oxidation stability of bio-oil [4].

Biofuel consists of hydrocarbon developed by many technologies such as hydrodeoxygenation, thermal pyrolysis, and transesterification [5]. Two main routes to converting vegetable oil into hydrocarbon were catalytic cracking using zeolites and hydrodeoxygenation using metals or metal-sulfides-based catalyst [6]. Many study had applied cracking vegetable oil using HZSM-5 zeolite to produce hydrocarbons because of its large surface area, crystallinity, and high acidity [7]. HZSM-5 zeolite has acidic site favoring hydrogen transfer that able to produce high aromatic content and some polyaromatic [8] In another study, the renewable diesel produced through hydroprocessing using NiMo/titania catalyst has a high cetane number when it was compared to conventional diesel and hence has better combustion

*Corresponding author: dana@chem-eng.its.ac.id

characteristics [9]. But it still contains high oxygen content.

Nowadays, the production of biofuel with inedible vegetable oil raw material has been developed through catalytic hydrocracking process to reduce the oxygen content in the long chain of trigliserides. Hydrocracking may be preferred conversion route of vegetable oil into biofuel using hydrogen to selectivity remove the oxygen atoms [10]. The selective catalyst in catalytic hydrocracking process was very important to obtain the appropriate conversion. Ni and Mo transitions metals used as active compounds and promoters to enhance the active site on the H-ZSM-5 catalyst used. Ni catalyst can promote the dehydrogenation, hydrogenation and hydrocracking reaction. Ni also can be used in more aggressive role of aromatic hydrocracking process because of its good hydrogenating properties [11]. Chen et al reported that when Ni/HZSM-5 used in catalytic hydroprocessing of FAMEs, the n-paraffin (C_{17}-C_{18}) and isoparaffin (C_5-C_{16}) compounds were increased as long as the increase of temperature. [12] In another case Mo catalyst can promote the hydrodeoxygenation reaction [13]. The impregnation of Mo into HZSM-5 can promote the cyclization reaction and aromatization reaction in cracking of vegetable oil process [8]. In another study, the NiMo catalyst selectivity for renewable diesel source (C_{15}–C_{18} range hydrocarbon) using temperature at 320°C for 1 h reached 91%. The yield of paraffin C_{15} –C_{18} range hydrocarbon increased due to the cracking of heavier molecules of triglycerides into diesel range hydrocarbon as the increasing of temperature from 320 to 400°C but it slightly decreased at 420°C due to the cracking of C_{18} and other triglyceride molecules into smaller molecules [9]. Thus, it is interesting to produce biofuel (C_{15} – C_{18} range hydrocarbons) from kapuk (*Ceiba pentandra*) seed oil through catalytic hydrocracking process at average temperature reaction of 350°C.

2 Material and methods

2.1 Materials

Kapuk seed oil (KSO) was provided from Pandaan, East Java, Indonesia as reported in previous study [14]. The commercial ammonium-ZSM-5 zeolite powder was purchased from Zeolyst International (SiO_2/Al_2O_3 Mole Ratio = 80). Nickel (II) nitrate hexahydrate and ammonium heptamolybdate tetrahydrate with 98% purity was purchased from Merck.

2.2 Catalyst Preparation

As the previous study, the HZSM-5 catalyst was prepared by calcined ammonium-ZSM-5 at 550°C for 5 h under at air atmosphere [14]. NiMo/HZSM-5 was prepared by incipient wetness impregnation method using 1.1247 M of nickel nitrate and 1.3761 M of ammonium heptamolybdatetetrahydrate. It was dissolved in the volume of the solution according to the total pore volume of the HZSM-5 support (0.2712 cm^3/gram) [13]. The solution was pouring into HZSM-5 and stirred for 3 h to obtained homogenous metal impregnated catalyst. Then, the catalyst dried under the temperature at 120°C for 12 h and calcined at 400°C with air flow for 5 h. At last, the catalyst reduced with hydrogen flow at 450°C for 2 h to obtain activated site catalyst.

2.3 Catalyst Characterization

The pore volume, specific surface area, and average pore diameter were obtained by an automatic specific surface area measurement (Quantachrome Instruments version 10.01) with the N_2 adsorption-desorption isotherm at 77K. The total pore volume, specific surface area, and average pore diameter were calculated using the Brunauer-Emmet-Teller (BET) method. The phase analysis of the catalyst was obtained using an X-ray diffractometer (Philips X'PERT MPD) at 30 mA and 40 kV with Kα radiation. Morphology catalyst, actual metal content, and dispersion were performed by scanning electron microscopy-energy dispersive X-ray (SEM-EDX) at 20 kV.

2.4 Catalytic hydrocracking

Catalytic hydrocracking of KSO was performed in a slurry pressurized batch reactor system equipped with a mechanical stirrer (Parr USA 4563) as reported by the previous study [14]. About 200 mL of KSO and the catalyst approximately 0.56 wt% to KSO were loaded into the reactor. Furthermore, the reactor was purged with nitrogen to remove dissolved air in the oil. The operating condition of the experiment carried out at a temperature 350°C for 2 h with reactor pressure range of 10-15 bar after hydrogen flow for 1 h. The obtained liquid product was characterized using gas chromatography-mass spectrophotometry (GC-MS).

3 Results and discussion

3.1 Catalyst characterization

Fig. 1. showed the XRD patterns of NiMo/HZSM-5 catalyst. As reported by the previous study [14], the main peaks of HZSM-5 zeolite observed at 2θ of 7.87°, 8.74°, 23.04°, 23.24°, 23.64°, 212.26° and 24.31° indicate that HZSM-5 have a framework type MFI. Besides, after the impregnation process of nickel and molybdenum into the HZSM-5, the structure of NiMo/HZSM-5 catalyst still similar to HZSM-5. It indicates that metals impregnation did not change the main structure of HZSM-5 [15]. The diffraction peaks intensity of NiMo/HZSM-5 catalyst showed that the nickel metallic phase was formed at 2θ of 51.7283°, nickel oxide phase was formed at 2θ of 62.5102°, whereas for molybdenum oxide phase for peaks intensity were formed at 2θ of 53.5674° and 60.4682°.

https://doi.org/10.1051/matecconf/201815606001

The actual metal content of NiMo/HZSM-5 was observed by EDAX analysis as showed in Fig.2. It showed that the Ni and Mo loading were approximately 0.26%wt and 0.93%wt.

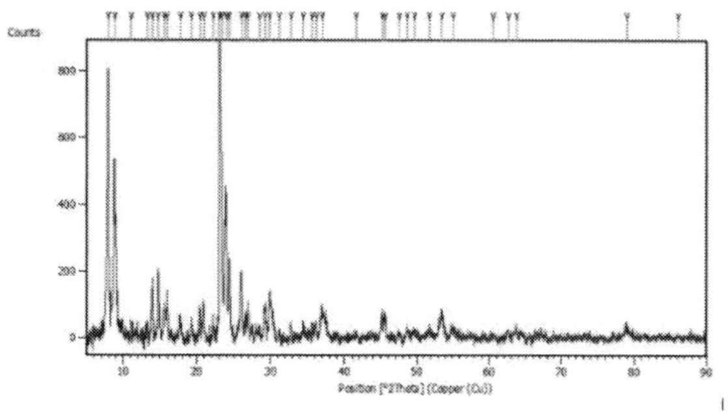

Fig. 1. XRD pattern of NiMo/HZSM-5 catalyst

The SEM image of NiMo/HZSM-5 in Fig. 2. showed that Ni and Mo metals are homogeneously dispersed over the microporous HZSM-5 support. Table 1 shows the textural properties of HZSM-5 and NiMo/HZSM-5 catalyst from BET analysis.

Table 1. The catalyst properties

Catalyst	BET surface area, (m^2/g)	Average pore diameter, (nm)	Total pore volume, (cm^3/gram)
HZSM-5	375.121	2.4900	0.2712
NiMo/HZSM-5	222.135	3.0148	0.1674

The total surface area of NiMo/HZSM-5 is 255.653 m^2/g, which is lower than the total surface area of HZSM-5 (375.121 m^2/g). It seems that the metal content blocking some micropores, clogging the external mesopores and decreasing the total area of the parent zeolitic support [16]

3.2 Product analysis

The obtained liquid product compounds were shown in Fig. 3. The most abundant compound in the liquid product produced by KSO catalytic hydrocracking process using NiMo/HZSM-5 catalyst at 350°C was the palmitic acid with 23.14 area%. The formation of carboxylic acid in the liquid product shows that triglyceride contained in KSO is cracked into the form of free fatty acids. Free fatty acids that form trough hydrogenation reaction in hydrocracking process that cracked the triglycerides bond [14]. The hydrocarbon compounds observed of 33.93 area% (i.e. 4.34 area% cycloparaffins, 16.02 area% n-paraffins, 12.26 area% olefins, and 1.30 area% of aromatics). The aromatic content that found in the liquid product such as pentylbenzene of 1.22 area% was affected by the presence of polyunsaturated fatty acid in the raw KSO. Furthermore, the presence of olefin of 12.26 area% indicates that the hydrogenation reaction occurring in the hydrocracking process was followed by cyclization process.

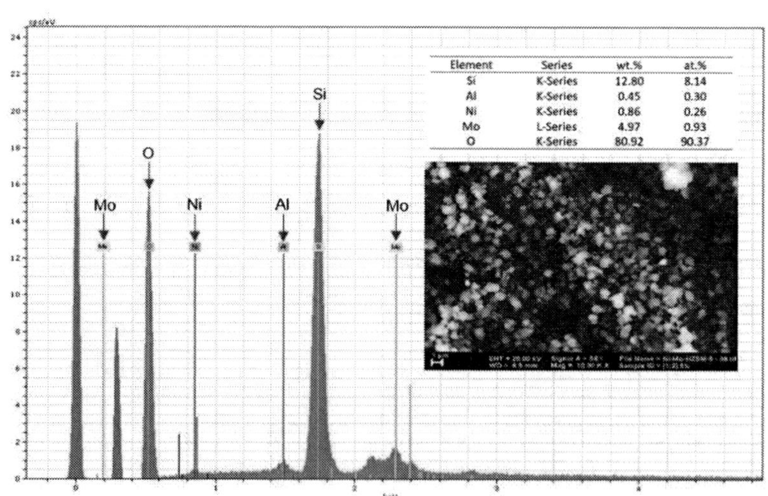

Fig. 2. SEM image and EDX pattern of NiMo/HZSM-5 catalyst

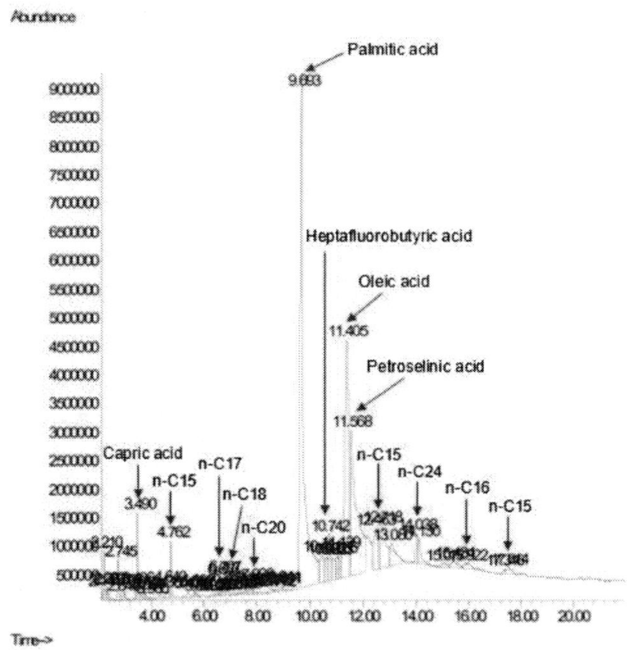

Fig. 3. GC-MS spectra of hydrocarbon liquid product using NiMo/HZSM-5 catalyst at 350°C

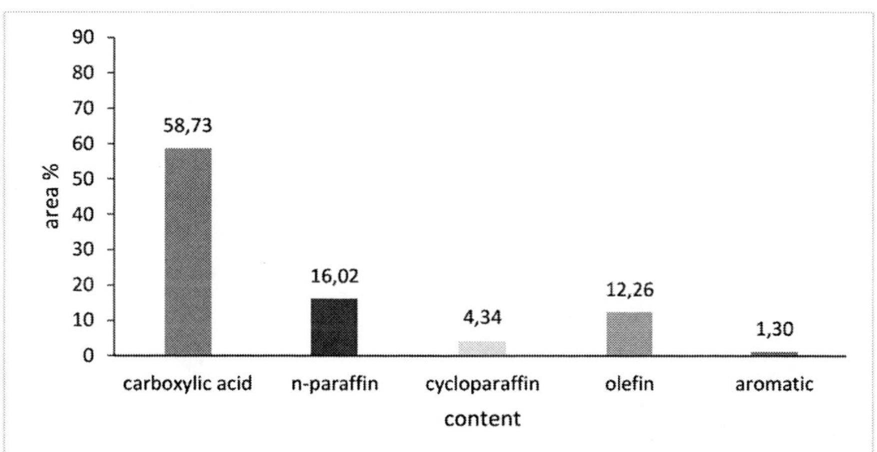

Fig. 4. The compounds of obtained hydrocarbon liquid product produced using NiMo/HZSM-5 catalyst at 350°C

Fig. 4 showed carboxylic acid is quite abundant that was equal to 58.73 area% consisting mostly of palmitic acid proves that triglyceride in raw KSO has begun cracked through the catalytic hydrocracking process into saturated fatty acids. This indicates that the unsaturated fatty acids present in the raw material have begun to crack into saturated fatty acids. The contribution of Ni metal in the cracking process transforms the unsaturated carboxylic acid compounds into saturated carboxylic acid through hydrogenation reaction. The presence of Ni can reduce the barrier of C = O bonds [17]. After the saturated fatty acid is formed, the subsequent reaction is the decarbonylation reaction, but at this stage the decarbonylation reaction has not occurred maximally because the cracking activity has required higher reaction temperature because temperture increase will increasing the conversion [18,19]. Furthermore, hydrocarbon compounds such as olefin (12.26area%), n-paraffin (16.02 area%), cycloparaffin (4.34 area%) and aromatics (1.30 area%) are formed indicating that the successful breaking of double bonds in the long chain fatty acids, although still in small amounts. The olefin compound formed is obtained by the dehydrogenation reaction of n-paraffin [20]. Table 2. showed that the most abundant biofuel fraction was gasoil-range alkanes (C_{15}-C_{18}) with 9.96 area%. Furthermore, another fraction was kerosene-range alkanes (C_{10}-C_{14}) with 5.09 area%. Temperature above 350°C needed to crack more free fatty acid into biofuel fraction.

Table 2. Various hydrocarbon contents of liquid product over NiMo/HZSM-5 catalyst

Various hydrocarbon contents of liquid product (area%)				
nC_5-nC_9	nC_{10}-nC_{14}[a]	nC_{15}-nC_{18}[b]	$nC_{>18}$	Cycloparaffin, olefin, and aromatic
0	5.09	9.96	0.00	16.90

[a]*kerosene range alkanes*
[b]*gasoil range alkanes*

4. Conclusion

Catalytic hydrocracking of kapuk (*Ceiba pentandra*) seed oil (KSO) with NiMo/HZSM-5 catalyst can be applied to biofuel production. The NiMo/HZSM-5 catalyst surface area was 222.1350 m^2/g, the pore diameter was 3.0148 nm and the pore volume was 0.1674 cm^3/g. the biofuel contained hydrocarbon composition consisted of 16.64 area% (i.e. 4.34 area% of cycloparaffins, 16.02 area% of n-paraffins, 12.26 area% of olefins, and 1.30 area% of aromatics) and 58.73 area% of carboxylic acids. The higher temperature needed to convert more carboxylic acids into hydrocarbons fraction.

References

1. S. Dharma, H. C. Ong, H. H. Masjuki, A. H. Sebayang, A. S. Silitonga, Energ. Convrs. Manag. **128** 66-81 (2016)
2. A. S. Silitonga, H. C. Ong, T. M. I. Mahlia, H. H. Masjuki, W. T. Chong, Fuel. **108** 855-858 (2013)
3. M. Yuniwati, Tek. Tech. Sci. 202-212 (2012)
4. X.H. Yu, R. Rawat, J. Shanklin, BMC. Plant. Bio. **11** 1471-1481 (2011)
5. Z-X. Xu, P. Liu, G-S Xu, Q. Liu, Z-X. He, Q. Wang, Energ. **133** 666-675 (2017)
6. J. Horacek, D. Kubicka, Fuel. **198** 49-57 (2017)
7. D. P. Botas, A. Garcia, J. de Vicente, R. Ramos, Catal. Today. **195** 59-70 (2012)
8. R. Frety, J. G. A. Pacheco, M. R. Santos, J. F. Padilha, A. F. Azevedo, S. T. Brandao, L. A. M. Pontes, Anal. App. Pyro. **109** 56-64 (2014)
9. R. Kumar, S. A. Farooqui, M. Anand, R. Kumar, R. Joshi, A. Khan, A. K. Sinha, Catal. Com. **98** 102-106 (2017)
10. J. K. Satyarthi, D. Srinivas, Energ. Fuel. **25** 3318-3322 (2011)
11. L. Chen, H. Li. J. Fu, C. Miao, Lv. Pengmei, Z. Yuan, Catal. Today. **259** 266-276 (2016)
12. H. Morelo, C. Galarrga, F. Feng, E. Hernandez, V. Birss, P. Pereira, Catal. Lett. **132** 402-209 (2009)
13. M. F. Ali, B. M. El Ali, J. G. Speight, *Handbook of Industrial Chemistry*, Mc Graw-Hill (2005)
14. Y. W. Mirzayanti, A. Roesyadi, D. H. Prajitno, 7th *Annual Basic Science International Conference* (2017)
15. S. Wang, Q. Yin, J. Guo, B. Ru, L. Zhul, Fuel. **108** 597-603 (2013)
16. G. Vitale, H. Morelo, E. Hernandez, S. Aquino, V. Birss, P. Pereira-Almao, App. Catal. **452** 75-87 (2013)
17. C. Dupont, R. Lemeur, A. Daudin, P. Raybaud, Catal. **279** 276-286 (2011)
18. H. Zhang, H. Lin, Y. Zheng, App. Catal. B. **160-161** 415-422 (2014)
19. J. Liu, K. Fan, W. Tian, C. Liu, L. Rong, Hydro. Energ. **37** 17731-17737 (2012)
20. S. K. Kim, S. Brand, H. Lee, Y. Kim, J. Kim, Chem. Eng. **288** 114-123 (2013)

The Effect of Metal Loading and Alkaline Treatment on The Zeolites Toward Catalytic Pyrolysis of ɣ-Valerolactone

Carolus Borromeus Rasrendra*, *Muhammad* Luthfi, *Vika* Asriani, *Jenny* Rizkiana, *Muhammad* A Kariem, *Dwiwahju* Sasongko

Department of Chemical Engineering, Faculty of Industrial Technology, Institute of Technology Bandung, 40132 Bandung, Indonesia

Abstract. ɣ-Valerolactone (GVL) is a glucose derivatives and its utilization as organic solvent has been developed to extract lignin from biomass. Both GVL and lignin can be used for producing aromatic chemicals (Bezene, Toluene, Xylene – BTX) via catalytic pyrolysis. We present a study focused on the catalyst modification of the zeolites to produce BTX from GVL. The catalysts were modified using wet impregnation method and alkaline treatment using NaOH solution. The addition of metal oxides were examined. Catalyst performance test were carried out in a fixed bed reactor. The feed was flowed at flowrate of 0.2 ml/min in 50 ml/min N_2. The operating temperature was maintained at 500°C with catalyst-to-GVL ratio of 1:1.5. The results showed that adition of Fe-metal improves zeolites activity compared with parent zeolites. Alkaline treatment had negative impact on HZSM-5 activity because of the change of silica-alumina-ratio (SAR) and average pore diameter.

1 Introduction

Indonesia become the world's largest producer of crude palm oil with recorded a substantial growth over the years, from about 20 million tonnes in 2008 to about 30 million tonnes in 2015 [1]. At conservative estimate, for every tonnes of palm oil another four-thirds tonnes of empty fruit bunches (EFB) are produced. Abundant amount available of EFB provides opportunity and challenge as useful resources [2][3]. Renewable approaches for making aromatics from EFB will reduce the Indonesia dependence on fossil petroleum and mitigate the negative impacts of climate change. One promising approach to utilize EFB is depolymerizing EFB fraction into aromatic compounds such as benzene, toluene, and xylene (BTX) using zeolite as the catalyst.

Benzene, toluene, and xylene (BTX) are considered as important and valuable aromatic chemicals in petroleum industry [4]. These aromatics serve as the most important aromatic platform molecules for the development of high-end products such as plastics, solvents, additives, and other products [5].

There have been a number of studies focusing on modifying the HZSM-5 catalyst to produce aromatic hydrocarbon in higher yield and/or enhance the selectivity to particular target compounds [6-10]. The presence of transition metals on the modified catalyst such as FeHZSM-5 and MgHZSM-5 may impact catalytic behaviour and could improve the aromatics yields [11]. Recently, ZSM-5 with hierarchical structure, which contains mesopores and micropores, was developed to minimize the diffusion limitation of reactant and to prolong the lifetime of zeolite catalyst in some extent due to the enlarged coke capacity [12] by dealumination and desilixation [13].

As part of the study to produce BTX from EFB, we here present our study to convert a biomass derivative solvent from EFB, γ-valerolactone (GVL) to produce BTX. In the present study we have investigated the effect of transition metals (Mg and Fe) over HZSM-5 and HUSY on the aromatics (BTX) yields. Moreover, we evaluate the effect of NaOH concentration in alkaline treatment of zeolite on the BTX yields.

2 Experimental

2.1. Materials

Gamma-Valerolactone (GVL) solvent was purchased from Diaozhen Industrial Park-Jinan City-Shandong-China and was used without further purification. ZSM-5 and USY zeolites were purchased from Pingxiang Naike Chemical Industry Co. Ltd, China.

2.2 Catalyst Preparation

The metal-zeolite catalysts with Fe and Mg content of 2 wt% were prepared by the conventional impregnation method. The stoichiometric quantities of

* Corresponding author: cbr@che.itb.ac.id

ion nitrate [(Fe(NO_3)$_3$.9H_2O) and (Mg(NO_3)$_2$.6H_2O)] were dissolved in distilled water. Then the zeolites (HZSM-5 and HUSY) was then added at 30°C and stirred to obtain a suspension. After being dried at 120°C for 6h, the powders were calcined in air at 550°C for 3h.

Hierarchical zeolites obtained in the presence of strong inorganic bases, NaOH. The NaOH concentration were varied from 0.3 M to 0.8 M. ZSM-5 was mixed with NaOH solution. The ZSM-5 suspension is then dried overnight. HZSM-5 were obtained by the ion-exchange with ammonium sulphate solution at 85°C for 1 h, prior to calcination.

Pyrolysis of GVL was carried out in a tubular reactor (1.5 cm id) made of quartz glass. As shown in Fig.1, 1.0 g zeolites was loaded in the heating zone with quartz wool before the reaction. GVL was vaporized at 200°C before entering the catalyst bed. During the pyrolysis, GVL was feed into the heating zone by syringe pump (12 ml/h) and purge with nitrogen (50 ml/min). The liquid product is condensed and collected in condenser trap.

Liquid samples were analyzed by GC-FID (Shimadzu GC-2010) equipped with a RTX-5 capillary column. Split injection is performed at a split ratio of 50 using helium (99.999%) as a carrier gas.

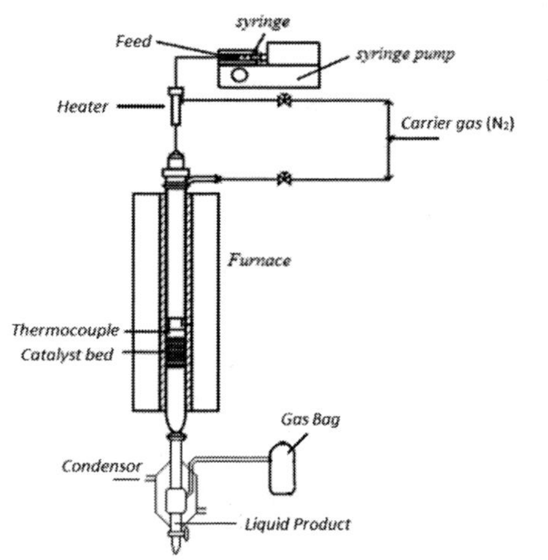

Fig. 1. Schematic diagram of the experimental apparatus for the catalytic pyrolysis

3 Results and Discussion

3.1 Catalyst Characterization

The HZSM-5 modification with 0.3 M and 0.5 M NaOH were examined using XRD. The XRD result, presented in Fig. 2, shows that all the investigated samples preserved their surface structure of zeolite (MFI). However, the degree of crystallinity in the mesopore samples was qualitatively influence by the alkaline treatment. The qualitative prediction based on the relative intensities of the reflection at 23° 2θ, taking the crystallinity of the parent sample (HZSM-5) to be 100%. The results indicate that the intensity of the diffraction peak of HZSM-5 slightly increases over NaOH concentration 0.3M and decreases over NaOH concentration 0.5M. It is possibly attributed to the local Si extraction from the zeolite lattice without significant destruction of the framework, thus preserving long range ordering and microporous structure in the material [14].

The observed difference in the pore structure changes among the samples with different silica to alumina ratios is evident from the textural data presented in Table 1. NaOH on alkaline treatment caused the Si atom apart form the tetrahedral framework so that affect the silica to alumina ratio and also make the new pore, so called mesopore. The greater the NaOH concentration makes the silica to alumina ratio decreases (from 36 to 21) and the average pore size larger than the parent HZSM-5 (15.71 Å to 26.50 Å).

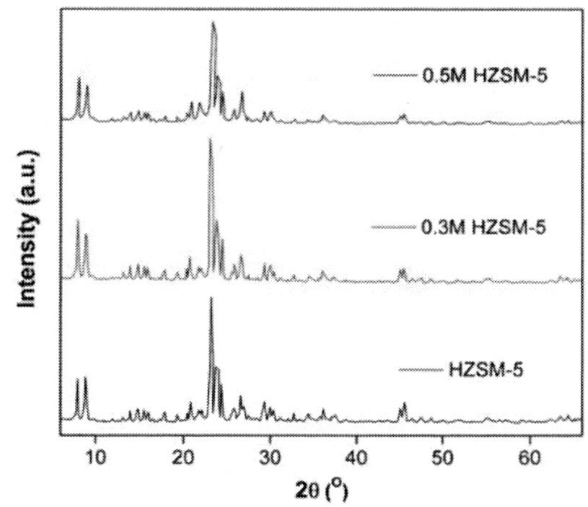

Fig. 2. XRD patterns of NaOH desilication-modified HZSM-5 after calcination

Table 1. Textural properties of the catalyst.

NaOH Concentration	Surface area (m^2/g)[a]	Total pore volume (cm^3/g)[a]	Average pore size (Å)[a]	SiO_2/Al_2O_3[b]
parent	315	0.17	15.71	36.13
0.3M	341.04	0,18	16,87	36
0.5M	314.06	0.27	26.50	21.71

[a] BET method
[b] XRF method

Fig. 3 shows NH_3-TPD profiles of modified-HZSM-5 using NaOH alkaline treatment. Basically, all samples show two peaks centered around 137°C and 461°C, corresponding to NH_3 eluted from the weak and strong acid sites, respectively. Notably, HZSM-5 modified with 0.3M and 0.5M NaOH shows broader peaks of weak acid sites and narrower peaks of strong acid sties compared to HZSM-5. The acid sites distribution are summarized in Table 2.

The amount of weak acid sites increased with higher NaOH concentration, from 0.032 cc/g to 0.041 cc/g. On the contrary, the amount of strong acid sites decreased with higher NaOH concentration, from 0.0074 cc/g to 0.0023 cc/g.

3.2 Catalyst Performances

The result of catalyst activity test which has been treated with NaOH alkaline treatment is presented in Fig. 4 and Fig. 5. Such treatment leads to a decrease in the yield of BTX product when the NaOH concentration was increased. The BTX yield obtain over HZSM-5 catalyst was 2.88%-C, 0.3M HZSM-5 0.87%-C, 0.5M HZSM-5 1.02%-C, and 0.8M HZSM-5 0.02%-C. Meanwhile, unlike the case with selectivity, increasing NaOH concentration provide improving selectivity of BTX product. The decrease in the yield and the increased in selectivity of BTX is due to the liquid product yield.

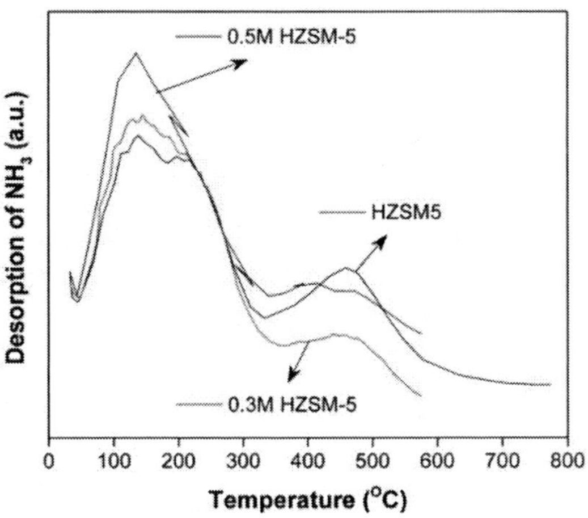

Fig. 3. NH_3-TPD profiles of NaOH desilication-modified HZSM-5

Table 2. Acid sites distributions in NaOH desilication method.

NaOH Concentration	Acidity level		Total acid sites (cc/g)
	Low temp. sites (cc/g)	High temp. sites (cc/g)	
parent	0.032	0.0074	0.0394
0.3M	0.038	0.0016	0.047
0.5M	0.041	0.0023	0.026

The greater the NaOH concentration then the liquid product yield decreases, as shown in Table 3. This indicates that the more a product gas is generated due to the forming of Bronsted Acid sites. The Bronsted acid site is very active so that it tends to lead to cracking reactions [15]. Moreover, the greater the NaOH concentration causing the average pore diameter of the catalyst larger, leading the pores are less selective towards the formation of the desired BTX product.

Fig. 5 shows product distribution. The largest product is occupied by toluene, followed by m-xylene, benzene, and o-xylene. In addition, the varying of NaOH concentration did not cause changing of product distribution.

Table 3. The effect of NaOH concentration on liquid product.

Katalis	HZSM-5	0.3M HZSM-5	0.5M HZSM-5	0.8M HZSM-5
Liquid product (g)	0.512	0.335	0.287	0.129

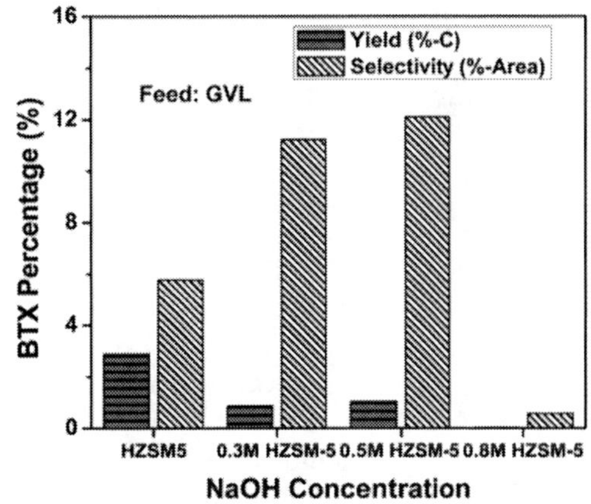

Fig. 4. The effect of NaOH concentration on BTX product yield and selectivity (500°C; 1 bar; catalyst : feed = 1:1.5; feed rate = 12 ml/h; N_2 flow = 50 ml/h)

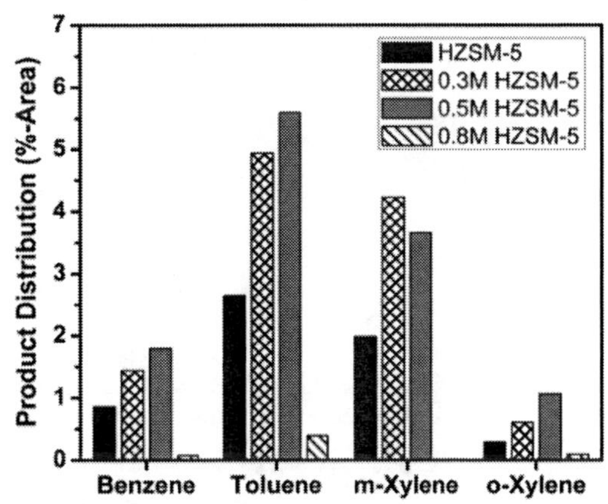

Fig. 5. The effect of NaOH concentration on BTX product distribution (500°C ; 1 bar ; catalyst : feed = 1:1.5 ; feed rate = 12 ml/h ; N_2 flow = 50 ml/h)

Zeolite HZSM-5 and HUSY has shown to be very active in catalytic biomass pyrolysis and also to enhance the selectivity towards aromatic in the liquid product [16]. Modification of zeolite by metal is considered as one way to improve catalyst activity and help to reduce the coking. The use of iron and magnesium to modify the zeolites would be an attractive option because iron is inexpensive. The product yield and selectivity in the catalytic pyrolysis for the various catalyst of our study are summarized in Fig 6. and Fig. 7, while the product distribution of the pyrolysis liquids is presented in Fig 8.

Fig 6 shows that the deposition of the metal on the zeolite can improve the zeolite activity in producing BTX products. In HZSM-5, iron is the best metal because it can improve the yield of BTX product from

2.88%-C to 4.56%-C otherwise in HUSY, magnesium is the best metal because it can improve the yield of BTX product from 0.21%-C to 0.67%-C.

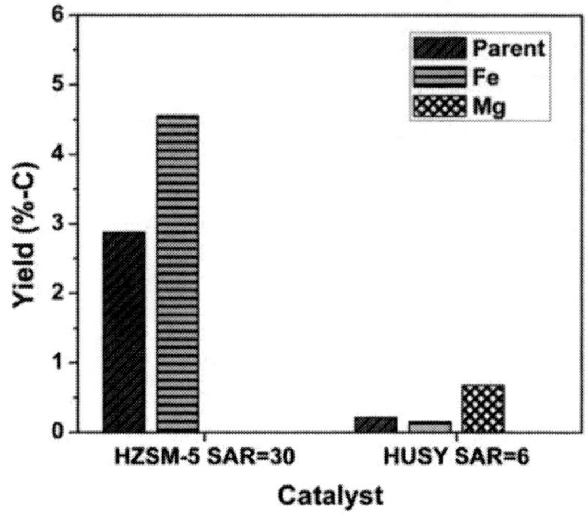

Fig. 6. The effect of metal on BTX yield (500°C; 1 bar; catalyst : feed = 1:1.5; feed rate = 12 ml/h; N_2 flow = 50 ml/h)

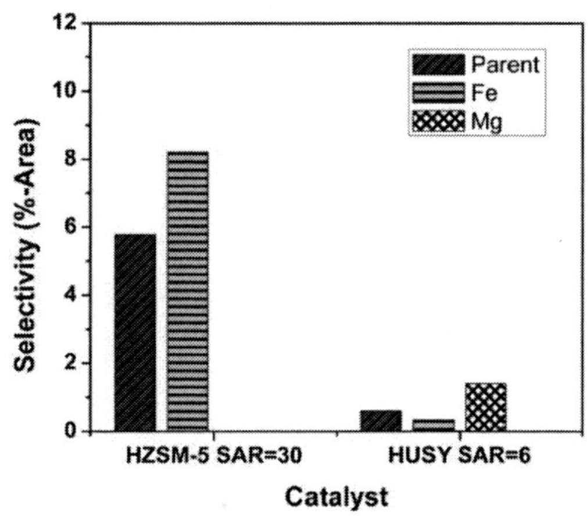

Fig. 7. The effect of metal on BTX selectivity (500°C; 1 bar; catalyst : feed = 1:1.5; feed rate = 12 ml/h; N_2 flow = 50 ml/h)

Iron is known to have ability in facilitating oxygen transfer redox reactions since it has more than one stable state, as such Fe-zeolites (FeHZSM-5 and FeHUSY) catalyst can be used either for selective oxidations or reductions [17]. Iron could find a role in facilitating removal of oxygen in reactions of biomass pyrolysis vapors. Rizkiana *et al* reported that the coke amount formed on Mg loaded zeolite (Mg/USY) is much lower while the hydrocarbon content in the pyrolytic oil is significantly higher [18]. They suggested that the deoxygenation process is mainly occurred by decarboxylation and dehydration reactions over Mg loaded zeolite. One of the possible explanation for the high coke resistance ability of Mg/USY is as follows: Mg species in the form of MgO can easily absorb CO_2 to form carbonate and/or bicarbonate (CO_3^{2-}/HCO_3^-) due to its slight basicity [19] and the formed carbonate/bicarbonate species might react with hydrogen atom generated from the decomposition of pyrolytic vapor and create formate/carboxylate intermediate on the surface of the zeolite [20].

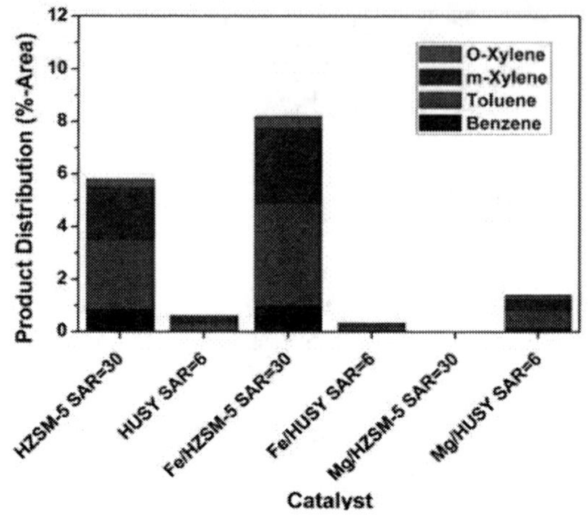

Fig. 7. The effect of metal on BTX product distribution (500°C; 1 bar ; catalyst : feed = 1:1.5 ; feed rate = 12 ml/h ; N_2 flow = 50 ml/h)

Fig. 7 shows product distribution. The largest product is occupied by toluene, followed by m-xylene, benzene, and o-xylene. In addition, the varying of metal did not cause changing of product distribution.

4 Conclusions

The catalytic performance of alkaline treatment and impregnation on the treated HZSM-5 has been studied.

The result of catalyst activity test in HZSM-5 treated with NaOH alkaline treatment shows that the greater the NaOH concentration in the alkaline treatment, the BTX product decreases. The greater the NaOH concentration causing the average pore diameter of the catalyst larger, leading the pores are less selective towards the formation of the desired BTX product. Modified-zeolite HZSM-5 and HUSY has been shown to be very active in catalytic biomass pyrolysis and also to enhance the selectivity towards aromatic in the liquid product. In HZSM-5, iron is the best metal because while in HUSY, magnesium is the best metal. Iron could find a role in facilitating removal of oxygen in reactions of biomass pyrolysis vapors.

Acknowledgement

This research was supported by “Grant Riset Sawit K16-1 BPDP Sawit”.

References

1. K. Azduwin, M. J. M. Ridzuan, A. R. Mohamed, S. M. Hafis, Applied Mechanics and Materials, **695**, 228-231 (2015)

2. K. C. Sembiring, N. Rinaldi, S. P. Simanungkalit, Energy Procedia, **65**, 162-169 (2015)
3. N. Abdulla, F. Sulaiman, H. Gerhauser, Journal of Physical Science, **22 (1)**, 1-24 (2011)
4. L. Shun, X. Zhang, L. Chen, B. Zhao, S. Yang, X. Xie, Journal of Analytical and Applied Pyrolysis (2016)
5. J. F. Zhu, J. C. Wang, Q. X. Li, Chinese Journal of Chemical Physics, **26**, 477-483 (2013)
6. S. Kelkar, C. M. Saffron, K. Andreassi, Z. Li, A. Murkute, D. J. Miller, T. J. Pinnavaia, R. M. Kriegel, Applied Catalysis B : Environmental, **174**, 85-95 (2015
7. A. Galadima, O. Muraza, Energy Conversion and Management, **105**, 338-354 (2015)
8. T. R. Carlson, T. O. Vispute, G. W. Huber, ChemSucChem, **1**, 397-400 (2008)
9. D. J. Mihalcik, C. A. Mullen, A. A. Boateng, Journal of Analytical and Applied Pyrolysis, **92**, 224-232 (2011)
10. Y. Wei, H. Lei, L. Zhu, X. Zhang, Y. Liu, G. Yadavalli, X. Zhu, M. Qian, Di Yan, Fuel, **178**, 77-84 (2016)
11. M. H. Nilsen, E. Antonakou, A. Bouzga, A. Lappas, K. Mathisen, M. Stocker, Microporous and Mesoporous Materials, **105**, 189-203 (2007)
12. C. Yang, M. Qiu, S. Hu, X. Chen, G. Zeng, Z. Liu, Y. Sun, Microporous and Mesoporous Materials (2016)
13. Ke, Zhang, M. L. Ostraat, Catalysis Today (2015)
14. V. Rac, V. Rakic, Z. Miladinovic, D. Stosic, A. Auroux, Thermochimica Acta, **567**, 73-68 (2013)
15. E. Koohsaryan, M. Anbia, Chinese Journal of Catalysis, **37**, 447-467 (2016)
16. E. F. Iliopoulou, S. D. Stefanidis, K. G. Kalogiannis, A. Delimitis, A. A. Lappas, K. S. Triantafyllidis, Applied Catalysis B: Environmental, **127**, 281-290 (2012)
17. C. A. Mullen, A. A. Boateng, ACS Sustainable Chem. Eng (2015)
18. J. Rizkiana, G. Guan, W. B. Widayatno, J. Yang, X. Hao, K. Matsuoka, A. Abudula, RSC Adv, **6**, 2096-2105 (2016)
19. P. Ferreira-Aparicio, I. Rodr'iguez-Ramos, J. A. Anderson, A. Guerrero-Ruiz, Appl. Catal. A, **202**, 183-196 (2000)
20. I. H. Son, S. J. Lee, I. Y. Song, W. S. Jeon, I. Juang, D. J. Yun, D.-W. Jeong, J.-O. Shim, W.-J. Jang, H.-S. Roh, Fuel, **136**, 194-200 (2014)

Diethyl Ether Production as a Substitute for Gasoline

Riza Alviany[1], *Arifuddin* Wahyudi[1], *Ignatius* Gunardi[1], *Achmad* Roesyadi[1*], *Firman* Kurniawansyah[1], *Danawati* Hari Prajitno[2]

[1]Chemical Engineering Department, Industrial Technology Faculty, Sepuluh Nopember Institute of Technology (ITS), Surabaya 60111, Indonesia
[2]Chemical Engineering Industry Departement, Vocation Faculty, Sepuluh Nopember Institute of Technology (ITS), Surabaya 60111, Indonesia

Abstract. Diethyl ether is one of alternative fuel that could be used as a significant component of a blend or as a complete replacement for transportation fuel. The aim of this research is to produce diethyl ether through dehydration reaction of ethanol with fixed bed reactor using nanocrystalline γ-Al_2O_3 catalyst. Nanocrystalline γ-Al_2O_3 catalyst was synthesized by precipitation method using $Al(NO_3)_3.9H_2O$ as precursors and NH_4OH as the precipitating agent. Dehydration reaction was performed at temperature range of 125 to 225°C. The result shows that synthesized γ-Al_2O_3 catalyst gave higher ethanol conversion and diethyl ether yield than that of commercial Al_2O_3 catalyst. The use of synthesized γ-Al_2O_3 catalyst could reach ethanol conversion as high as 94.71% and diethyl ether yield as high as 11,29%.

1 Introduction

Diethyl ether (DEE) is an important chemical for the laboratory and industry uses. It has been widely used as a solvent or extractant, reaction medium, fragrance, explosive materials, pharmaceutical chemical processes, fuels and fuel additives [1]–[6]. As a fuel, DEE has been used as starting fluid in countries with cold climates because of its high volatility and low flash point [6]. In addition, cetane and octane number, reasonable energy, high oxygen content, and high miscibility in diesel fuel and bioethanol become DEE available advantage as blending material to improve the ignition [3]. DEE has a potential as an alternative transportation fuel to replace an excessive use of fossil fuels as major energy sources.

The volumetric heating value and octane number are very important parameters for an alternative fuel of gasoline. The latent heat of ethanol is about 3.7 times higher than that of gasoline, which causes lower temperatures during combustion in the engine. A decrease in engine temperature also causes a formation of less amount of NOx. Ethanol with an octane number of 113 and low Reid vapor pressure is considered as an excellent alternative to gasoline but it also causes some problems in cold-start of the engine. Fuel regulations mandate that motor vehicle fuels should have sufficient vapor pressure to cold-start, as low as -30 °C [5]. Combustion of alcohols and ethers produce much less unburned hydrocarbons and CO and CO_2 than conventional gasoline [3], [5], [6]. Consequently, ethanol-DEE mixtures are considered as excellent gasoline alternatives which diethyl ether can be produced by dehydration of ethanol over acidic catalysts.

Diethyl ether is generally prepared by the dehydration process of ethanol (Barbet process) using sulfuric acid catalyst (homogeneous catalyst). The downside of this process is the difficulty of catalyst separation and catalyst corrosiveness. Corrosive catalysts make investments in equipment quite expensive. Thus, homogeneous catalysts are replaced by heterogeneous catalysts. heterogeneous catalyst that has long been known for ethanol dehydration is Al_2O_3. One of the phase types of Al_2O_3 is γ- Al_2O_3. γ- Al_2O_3 which is widely used as catalysts and catalyst supports. γ-Al2O3 has been known for having a large surface area with acidic and alkaline active sides. The active sides have resulted from the release of water molecules [7]. Primary alcohol dehydration reaction catalysis by using γ-Al_2O_3 catalyst can produce ether compounds and alkenes. While the reaction of the secondary alcohol with an acid catalyst can produce alkenes and base catalyst produces ketones through dehydrogenation reaction [8]. To maximize the effectiveness of the catalyst, a study was carried out by varying technology of catalyst manufacturing.

Because of an excessive use of fossils fuel, it is important to produce desirable, higher value chemicals and fuels from clean, cheap, abundantly available feed to

* Corresponding author: aroesyadi@yahoo.com

contribute to the solution of the two major problems of the next century, namely global warming and depletion of oil reserve.

This research was carried out the synthesis of nanocrystalline γ-alumina from $Al(NO_3)_3.9H_2O$ by precipitation method. Then, to test the activity of prepared γ-Al_2O_3, it was applied in 96% ethanol dehydration reaction into diethyl ether and compared its activity with γ-Al_2O_3 commercial catalyst.

2 Experimental

Some of the steps being taken in this study were γ-Al_2O_3 catalyst synthesis by precipitation method, γ-Al_2O_3 catalyst characterization, and DEE production.

2.1. Materials

The commercial Al_2O_3 used in this study was purchased from Merck Company. Aluminium nitrate nonahydrate (Merck), ammonium hydroxide (Sigma-Aldrich Co. LLC), distilled water (Sumber Ilmiah Persada), hydrochloric acid (37% from Sigma-Aldrich Co. LLC), molecular sieve, nitrogen gas (PT. Aneka Gas Industri) were also employed.

2.2 Catalyst Preparation

Synthesis of γ-Al_2O_3 was performed by precipitation method. Firstly, aluminum nitrate nonahydrate with different concentration (0.2, 0.3, 0.4, 0.5, and 1 M) was dissolved in a solution of HCl (0.5 M) and distilled water. Subsequently, the precipitation was carried out by adding ammonium nitrate to the solution by stirring at several temperatures (60, 70, 80, 90, 100°C) by careful pH adjustment to 9. The precipitate was stirred for different time (3, 4, 5, and 6 hours) and was maintained at the same temperature just like in precipitation step. Then, precipitate filtered and washed by distilled water and ethanol and then dried at 75°C for 24 hours. The solid precipitate was calcined at 600°C under air flow for 6 hours to obtain the synthesized γ-Al_2O_3 catalyst.

2.3 Catalyst Characterization

The synthesized γ-Al_2O_3 catalyst were characterized using X-ray diffraction (XRD) at Material and Metallurgy Engineering Departement, Sepuluh Nopember Institute of Technology (ITS) to determine the crystallinity and the crystal size of the catalyst. Size of the crystal was calculated by using Scherrer formula based on the data of XRD analysis:

$$D = \kappa\, \lambda / \beta cos\theta \qquad (1)$$

Where D is the diameter of the crystal, κ is the shape factor with a value of 0.9-1.4 (generally used ~ 0.9), λ is the wavelength of X-ray (0.15406 nm), θ is the diffraction angle (radians), and β is the FWHM (Full Width Half Maximum) of the diffraction peak at half maximum intensity (radians).

The surface area, pore size distribution and pore volume were determined using a nitrogen adsorption-desorption method at Chemical Engineering Departement, Sepuluh Nopember Institute of Technology (ITS).

2.4 DEE production

DEE production process was performed by adsorption process in adsorption column and followed by dehydration of ethanol in a reactor. The dehydration of ethanol was carried out in a fixed-bed microreactor with an inside diameter of 0.5 in. and length of 30 cm. In the experiment, 10 gram of molecular sieve was loaded into the adsorption column and 3 gram of catalyst was loaded into the reactor. The liquid ethanol was vaporized in a flowing of nitrogen at a flow rate 200 ml/min. the reaction was carried out at several temperatures (125, 150, 175, 200, 225°C). The products condensed by a condenser were analyzed by HP 6890 gas chromatograph with FID using crosslinked methyl siloxane column. Helium (22.98 psi) was used as carrier gas in GC using the temperature detector at 250°C. The arrangement of equipment used for DEE production is shown in Fig. 1 below.

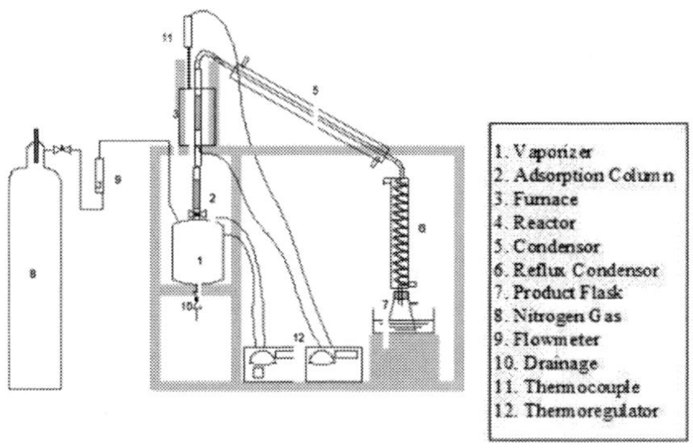

Fig. 1. DEE production equipment

3 Result and discussion

3.1 Nano-crystallite Al_2O_3 preparation

The synthesis/preparation of γ-Al_2O_3 was performed using precipitation method in which aluminum nitrate nonahydrate ($Al(NO_3)_3.9H_2O$) was used as an aluminum precursor, hydrochloric acid (HCl) and distilled water as a solvent, and ammonium hydroxide (NH_4OH) as a precipitating agent. In this research, a variation of the type of solvent used, the concentration of precursor solution, stirring time and temperature of the solution during the precipitation process, has been studied.

For the synthesis process of γ-Al_2O_3 includes several stages, these are [9]:

First stage

The formation of aluminum hydroxide occurs when NH_4OH added into the aluminum precursor solution. Equations 2 and 3 are used for reactions using HCl solvents, whereas reaction 4 is used for reactions using distilled water solvent. This stage is illustrated by the reaction as follows:

$$Al(NO_3)_3 + 3HCl \rightarrow AlCl_3 + 3HNO_3 \quad (2)$$

$$AlCl_3 + 3NH_4OH \rightarrow Al(OH)_3\downarrow + 3NH_4Cl \quad (3)$$

$$Al(NO_3)_3 + 3NH_4OH \rightarrow Al(OH)_3\downarrow + 3NH_4NO_3 \quad (4)$$

Second stage

The solution containing precipitate of aluminum hydroxide in the first stage transformed to boehmite crystal precipitate after stirring at a constant temperature (aging with stirring). The reaction is as follows:

$$Al(OH)_3 \rightarrow AlOOH\downarrow + H_2O \quad (5)$$

Third stage

The precipitate of boehmite crystals from the second stage then washed with water and ethanol sequentially to remove impurities in the form of anions such as nitrate and chloride ions [10]. After that, the boehmite crystal precipitate filtered, dried and then calcined at 600°C for 6 hours. The dried precipitate of boehmite crystal is calcined to form γ-Al_2O_3. The following reactions occur:

$$2AlOOH \rightarrow Al_2O_3 + H_2O \quad (6)$$

3.1.1 Effect of solvent in precipitation process

Two different types of solvents are used to determine the effect of solvent on the diameter of the resulting Al_2O_3 crystals. The operation was performed at temperature 80°C with 5 hours stirring time.

The results in Fig. 2 show that in all precursor concentration variations, using 0.5 M HCl solvent produces the smaller size of Al_2O_3 crystal than using distilled water solvent. This occurred due to the reaction between HCl and NH_4OH will produce NH_4Cl. The presence of NH_4Cl in the solution would increase the viscosity of the solution, decreasing the molecular mobility and dielectric matrix constant of mother solution. It also increased the distance between particles of aluminum hydroxide thus preventing aggregation of aluminum hydroxide particles during the process of nucleation and growth of particles [10].

Based on Fig. 2, it can be seen that the crystal size of the resulting γ-Al_2O_3 is proportional to the amount of precursor used. The lower the concentration of precursor used, the resulting crystal size is also smaller. A lower concentration of precursor make the distance between the particles of aluminum hydroxide higher where can reduce the aggregation of particles of aluminum hydroxide on the stage of nucleation and growth of particles. The size of the smallest crystal produced at 0.2 M concentration of precursor and 0.5 M HCl solvent. The crystal diameter is 8.62248 nm, while the size of the largest crystals produced at 1 M concentration of precursor with distilled water solvent which has diameter 23.6881 nm.

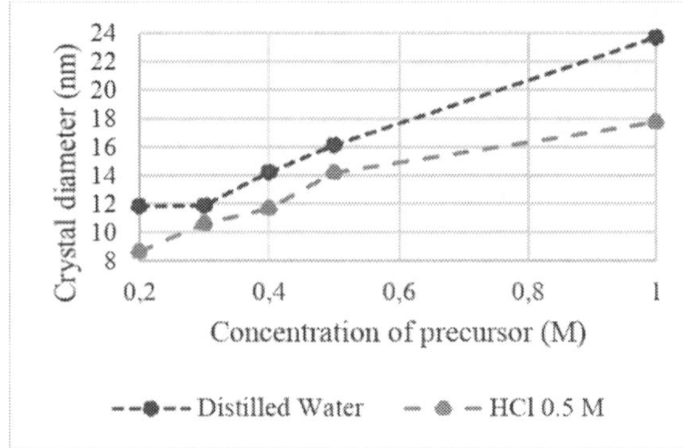

Fig. 2. Crystal diameter of γ-Al_2O_3 at different solvent type and precursor concentration

3.1.2 Effect of stirring time in precipitation process

The operation was performed at the varied stirring time (3, 4, 5, 6 hours) while the concentration of precursor and precipitation temperature was kept constant at 1 M and 80°C respectively. The solvent used was 0.5 M HCl.

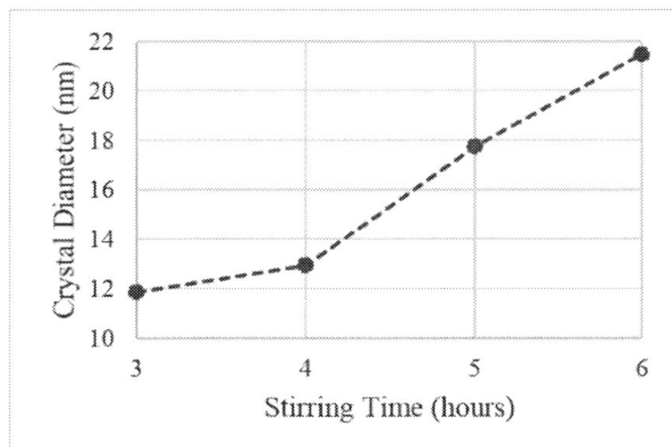

Fig. 3. Crystal diameter of γ-Al_2O_3 at different stirring time

In Fig. 3 it has been shown that the stirring time on precipitation processes influence a crystal diameter of γ-Al_2O_3 produced significantly. The longer the stirring time, the greater the diameter of the resulting crystals. This happens because of the longer the stirring time the mobility of particles in the solution lasts longer Longer duration of stirring time cause particle mobility and collision intensity between the particles in solution increased, so aggregation that occurs will be greater and cause a growth of the crystal particle size of γ-Al_2O_3 greater. The smallest particle size is produced by stirring for 3 hours which is 11.85287 nm.

3.1.3 Effect of temperature in precipitation process

The operation was performed at several operating temperatures (60, 70, 80, 90, 100°C) while the concentration of precursor and stirring time was kept constant at 1 M and 3 hours respectively. The solvent used was 0.5 M HCl.

https://doi.org/10.1051/matecconf/201815606003

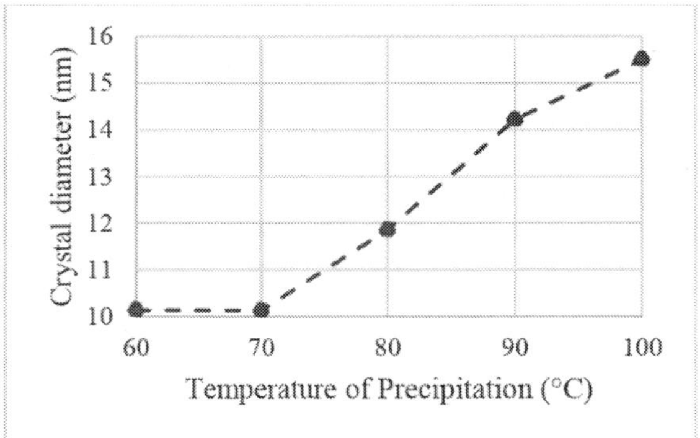

Fig. 4. Crystal diameter of γ-Al_2O_3 at different precipitation temperature

Fig. 4 shows precipitation temperature is a parameter that influences the size of crystals γ-Al_2O_3. Generally, the size of γ-Al_2O_3 crystals produced increases with the rising temperature of the precipitation process. An increasing temperature at precipitation process decrease the viscosity of the solution and increase the particle movement so at higher temperature the aggregation of the particle bigger. The optimum condition was 70°C which produce the smallest crystal size, i.e. 10.12907 nm.

3.2 Catalyst characterization for catalytic test

For the catalytic test, the catalyst which used was the catalyst that prepared using 1 M precursor concentration and 0.5 M HCl as a solvent stirred for 3 hours at 70°C (PR-C) while commercial γ-Al_2O_3 catalyst as a comparison. Those variables are the optimum variable which produce high yield of γ-Al_2O_3 and γ-Al_2O_3 crystal size is still in a nano size.

3.2.1 Phase analysis

The XRD patterns of the synthesized and commercial γ-Al_2O_3 are shown in Fig. 5. In that figure, both of catalyst has three peaks at 2θ = 38°, 2θ = 45.7° and 2θ = 66.7°, this pattern is suitable with the data (JCPDS) Joint Committee on Powder Diffraction Standards in Fig. 6.

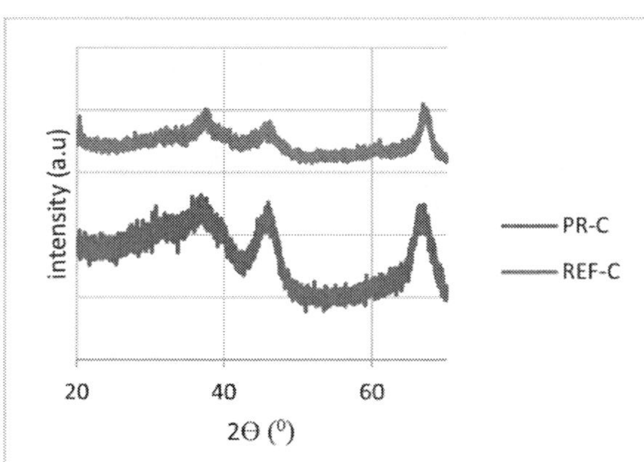

Fig. 5. XRD patterns of synthesized (PR-C) and commercial γ-Al_2O_3 (REF-C)

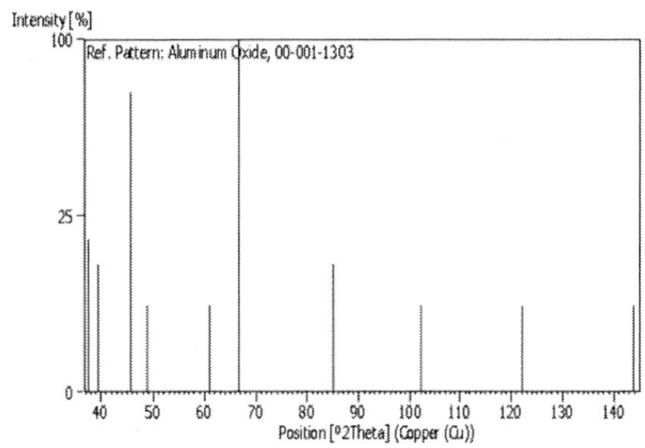

Fig. 6. XRD patterns of γ-Al_2O_3 from JCPDS

The crystallite sizes were calculated using Scherrer formula. Results of calculation are listed in Table 1.

Table 1. Crystallite sizes of the catalyst

	D (nm)
PR-C	10.1183
REF-C	80.3913

3.2.2 N_2 adsorption-desorption (BET) analysis

Physical properties of catalyst (specific surface area, total pore volume, and average pore diameter) are presented in Table 2.

Table 2. Physical properties of the catalyst

Catalyst	Specific Surface Area (m^2/g)	Pore Diameter (nm)	Pore Volume (cc/g)
PR-C	241.437	3.839	0.241
REF-C	133.420	6.185	0.253

The result shows that synthesized γ-Al_2O_3 (PR-C) have a higher specific surface area and smaller pore diameter than commercial catalyst γ-Al_2O_3 (REF-C) which may decrease higher hydrocarbon and byproducts leading to increasing of main product selectivity and ethanol conversion. In addition, the high surface area would affect to the increased catalytic activity of ethanol dehydration [11].

Result at Table 1 and Table 2 show a correlation with an assumption that solids in the form of particles, so the surface area for a solid with a certain mass is greater if the particle size is smaller. By defining the specific surface area as the ratio of the total area of the solid surface to the mass, the specific surface area is greater if the particle size is smaller.

3.3 DEE process production

Diethyl ether production process in this research combining two steps, they are adsorption process and the previous process. For this purpose, it will be compared the activity of γ-Al_2O_3 catalyst which prepared by precipitation method (PR-C) with γ-Al_2O_3 commercial catalysts (REF-C).

Adsorption process aimed to reduce the water content in the ethanol feed. To determine the achievement of the adsorption process, an analysis of the ethanol feed before entering the adsorption column and the output product of adsorption column was done. Ethanol was detected at a retention time of 3.27 minutes and water was detected at a retention time of 3.82 minutes. Based on the results of the GC analysis, there is increasing of ethanol concentration from 95.3% (before entering the adsorption column) to 97.77% (after passing through the adsorption column). An increase in ethanol concentration is happened because of some water in the ethanol solution are adsorbed by molecular sieve. Basically, a molecular sieve adsorbent using in any practice can adsorb sufficient water contained in ethanol [12], [13].

In the ethanol conversion process, the reaction which may occur is a dehydration reaction of ethanol (formation
of diethyl ether compounds, and the formation of ethylene compounds) [4]. Fig. 7 and 8 shows catalytic reactions. The catalytic test aimed to determine the performance of γ-Al_2O_3 catalyst that was prepared in chromatographic analysis of the liquid product reaction at a temperature of 175°C using PR-C catalyst (Fig. 7) and REF-C catalyst (Fig. 8). From chromatogram, diethyl ether and residual ethanol compounds were detected, while the water was not detected in the GC chromatogram. Ethylene compounds might be formed in the gas phase products, but in this research, chromatographic analysis was not performed for gas product. Diethyl ether was detected at a retention time of 6.052 minutes, ethanol was detected at a retention time of 5.3 minutes, and methanol was detected at a retention time of 4.5 minutes. Methanol formation process might exist due to the reaction between methane with water (Eq-7).

$CH_4 + H_2O \leftrightarrows CH_3OH + H_2$ $\Delta G_{398.15}$ = 117,437.85 J (7)

Another process might happen as the effect of catalytic activity of alumina ie cracking, isomerization and dehydration process [14]. The methanol formation might also occur through the reaction:

$CO + 2H_2 \leftrightarrows CH_3OH$ $\Delta G_{398.15}$ = -2,347.41 J (8)

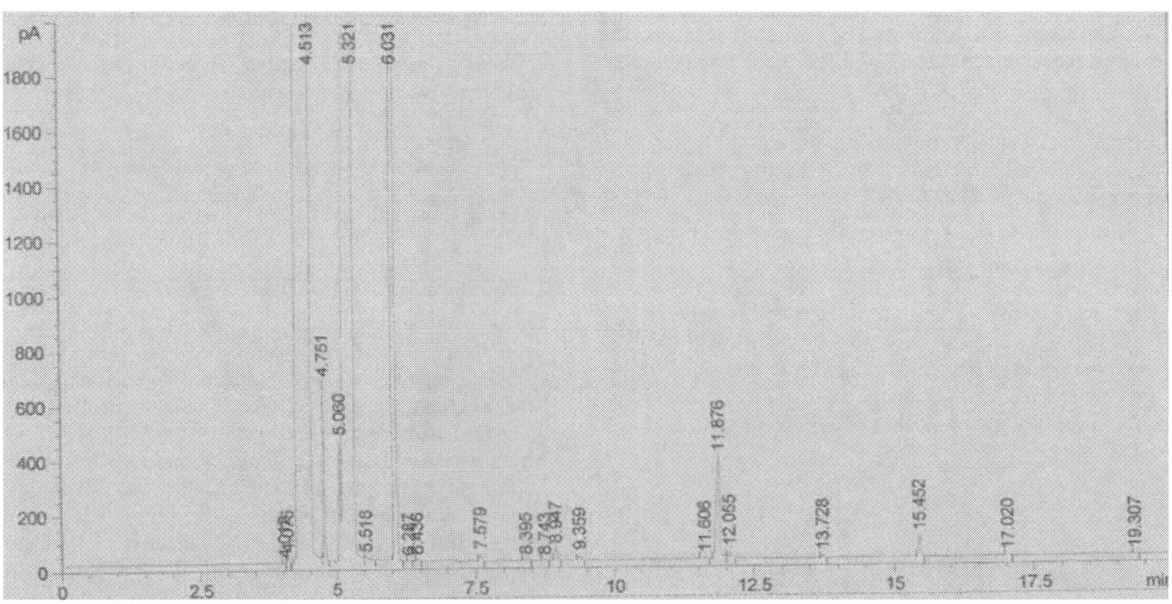

Fig.7. Chromatogram of reaction product using PR-C catalyst at 175°C of reaction temperature

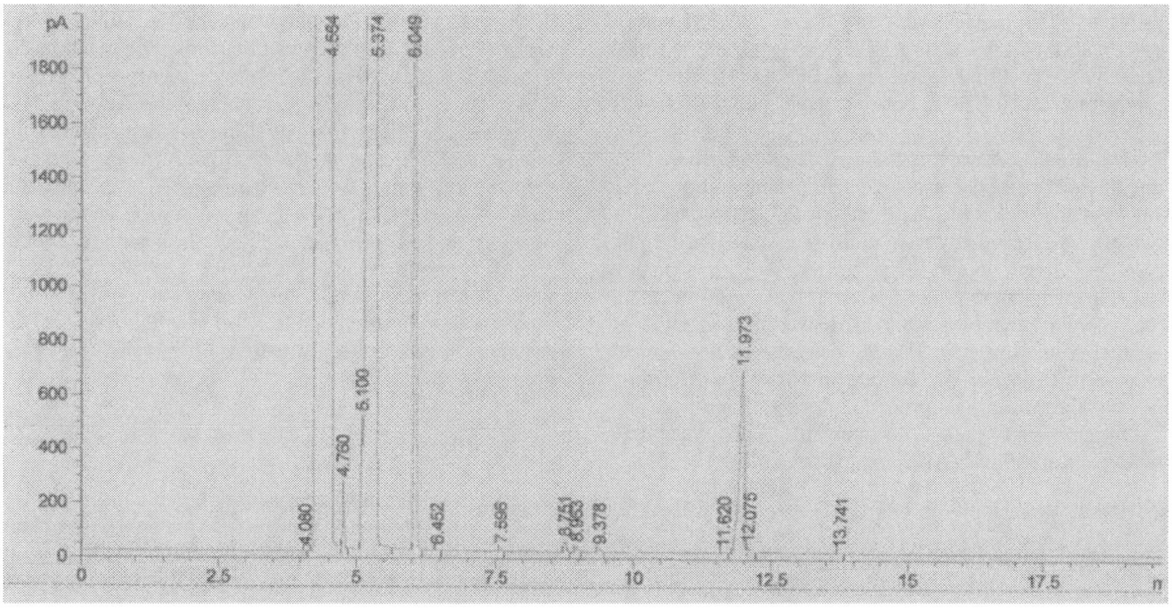

Fig.8. Chromatogram of reaction product using REF-C catalyst at 175°C of reaction temperature

Based on thermodynamics concept, the reaction can be predicted using thermodynamic parameters. A chemical reaction can happen if the value of the Gibbs energy change of a reaction at a specific operating temperature is negative (ΔGreaction<0) [15]. Judging from the value of ΔGibbs energy, the possible reaction for the formation of methanol compounds in the present study is reaction (8). It can be seen at the operating conditions (reaction temperature) in this research, the value of the Gibbs energy change in reaction (8) is negative ($\Delta G_{398.15}<0$), while the reaction (7) has positive Gibbs energy change ($\Delta G_{398.15}>0$).

To know the effectiveness of catalyst, calculation of diethyl ether yield and ethanol conversion are necessary as a reaction parameter.

Fig. 9 shows the result of the conversion of ethanol dehydration reaction at various operating temperatures. From the fig. 9, either with the use of γ-Al_2O_3 catalyst which was prepared by precipitation method (PR-C) or with γ-Al_2O_3 commercial catalyst (REF-C) has increasing conversion of ethanol as increasing of the reaction temperature. Increased temperatures can increase the conversion of ethanol according to the Arrhenius equation. In this equation, the increase of temperature will increase the reaction rate constants that will increase the rate of reaction [5].

The uses of PR-C catalysts generally produced ethanol conversion greater than using REF-C catalyst. This happened because it has been known that the PR-C catalyst has higher surface area than the REF-C catalyst. Higher surface area will increase the conversion of ethanol because it can increase the opportunity for the occurrence of reaction (adsorption and conversion process) [14].

In addition to the ethanol conversion, in this research also calculated the diethyl ether yield which produced by the reaction using catalyst PR-C and REF-C. The diethyl ether yield is shown in the following figure.

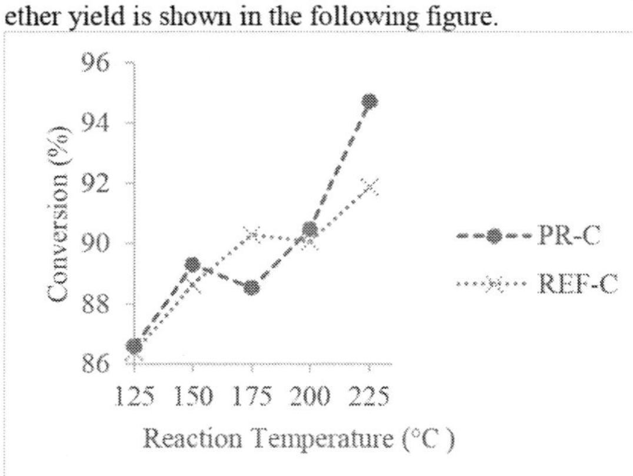

Fig.9. Ethanol coversion vs reaction temperature of ethanol dehydration process using PR-C dan REF-C catalyst

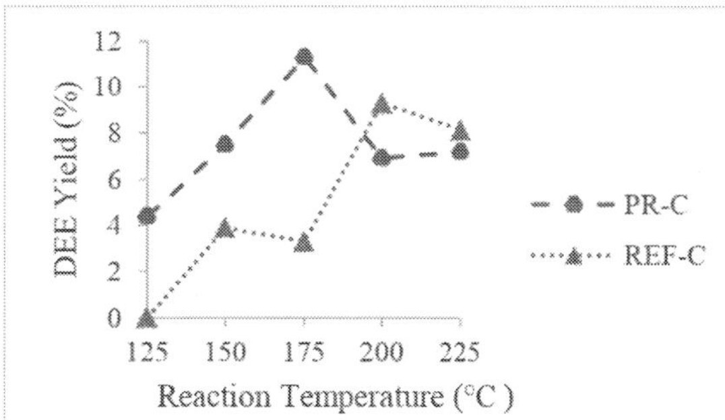

Fig.10. DEE yield vs reaction temperature of ethanol dehydration process using PR-C and REF-C catalyst

From Fig. 10, we can see that using PR-C catalyst has a greater yield of diethyl ether than using REF-C catalyst. That happens because PR-C catalyst has a higher surface area than REF-C catalyst. The greatest DEE yield is 11.29% was produced by using PR-C catalyst while the REF-C catalyst produced DEE yield as 9.28%.

Fig. 10 also shows that the increasing of DEE yield is not linear to the increase of reaction temperature. This suggests that the reaction occurs in the catalytic test is not a single reaction but a multi-reaction where some reactions occur other than diethyl ether formation reaction of ethanol. Other reactions that might occur in the catalytic reaction process is dehydration reaction of ethanol to diethyl ether (Eq-9); dehydration reaction of ethanol to ethylene (Eq-10); reaction of steam reforming (steam reforming) to produce carbon monoxide and hydrogen (Eq-11); and decomposition of ethanol to methane, carbon monoxide and hydrogen (Eq-12) [11].

$$2C_2H_5OH \leftrightarrows C_2H_5OC_2H_5 + H_2O \ \Delta G_{398.15} = -57{,}822.3 J \quad (9)$$

$$C_2H_5OH \leftrightarrows C_2H_4 + H_2O \ \Delta G_{398.15} = -4{,}363.3 \ J \quad (10)$$

$$C_2H_5OH + H_2O \leftrightarrows 2CO + 4H_2 \ \Delta G_{398.15} = 77{,}331.5 \ J \quad (11)$$

$$C_2H_5OH \leftrightarrows CH_4 + 2CO + 4H_2 \ \Delta G_{398.15} = -42{,}577.7 \ J \quad (12)$$

From some reactions above, theoretically (based on the value $\Delta G_{reaction}$) reactions which occured in this catalytic reaction of ethanol is ethanol dehydration reaction (Eq-9 and 10) and the decomposition of ethanol (Eq-12), while the steam reforming reaction (Eq-11) at the operating temperature of this research did not occur which can be seen from the value of the Gibbs energy changing has positive value.

Fig. 10 shows that the yield of diethyl ether decreased at high temperatures. This could happen because the catalytic test process at this research which operated at high temperature can trigger reactions to produce a larger gas product and decreasing the rate of the reaction towards the formation of diethyl ether. This can be explained by analyzing changes in the Gibbs energy of the reactions (Eq-9, Eq-10, and Eq-12) at each reaction temperature. Value of Gibbs energy change for each reaction is shown by the following picture.

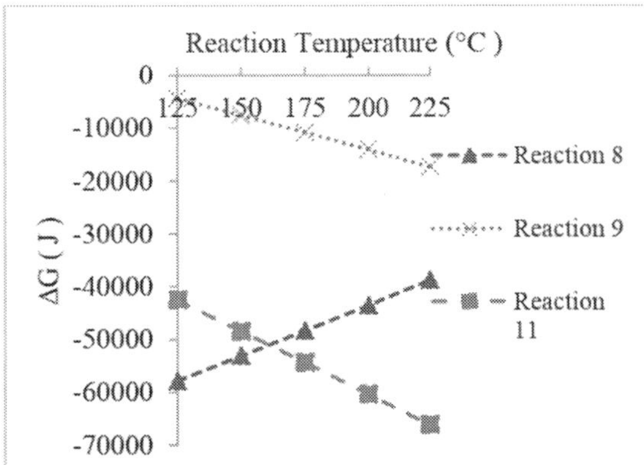

Fig.11. Value of $\Delta G_{reaction}$ vs reaction temperature

Fig. 11 shows that the value of the Gibbs energy change ($\Delta G_{reaction}$) in diethyl ether formation reaction (Eq 9) are more positive (go up) with the increase of reaction temperature. It indicates that the reaction towards the formation of diethyl ether decreases with increasing temperature, due to more minimum (more negative) value of the Gibbs energy change ($\Delta G_{reaction}$) make the reaction easier take place [15]. Instead of gas-forming reaction (Eq-10 and Eq-12) has the value of Gibbs energy change ($\Delta G_{reaction}$) are more negative (go down) with increasing reaction temperature. This shows that with increasing temperature the reaction rate of the reaction towards the formation of gas will be even greater.

4 Conclusion

Based on this research, it may be concluded that: γ-Al_2O_3 nanocrystallite can be produced by using precipitation method with $Al(NO_3)_3.9H_2O$ as an aluminum precursor, NH_4OH as precipitating agent and 0.5 M HCl and distilled water as solvent. Concentration of aluminum precursors, type of solvent, time of stirring, temperature of precipitation process influences the size of γ-Al_2O_3 crystal produced. Generally, the catalyst which prepared by precipitation method (PR-C) have a greater ethanol conversion and DEE yield than γ-Al2O3 commercial catalysts because of its high surface area. The biggest conversion reaction is resulted by using PR-C catalyst at 225°C of reaction temperature as 94.8216%. The biggest DEE yield is resulted by using PR-C catalyst 175°C of reaction temperature as 11.04%.

The authors would like to thank Ministry of Research Technology and Direktorat Riset dan Pengabdian Masyarakat – Direktorat Jenderal Penguatan Riset dan Pengembangan – Kementerian Riset, Teknologi, dan Pendidikan Tinggi Republik Indonesia for financial support under grant PTUPT Scheme for the year of 2017 granted through Institute for Research and Community Services Sepuluh Nopember Institute of Technology.

References

1. M. Sakuth *et al.*, *Ullmann's Encyclopedia of Industrial Chemistry* (Wiley-VCH Verlag GmbH & Co. KGaA,Weinheim, 2010)
2. T. K. Phung and G. Busca, Chem. Eng. J. **272**, 92–101 (2015)
3. A. Ibrahim, Appl. Therm. Eng. **107**, 853–862 (2016)
4. T. Kito-Borsa, D. a. Pacas, S. Selim, and S. W. Cowley, Ind. Eng. Chem. Res. **37**, 97 (1998)
5. T. Doğu and D. Varişli, Turkish Journal of Chemistry **31**, 5 (2007)
6. B. Bailey, J. Eberhardt, S. Goguen, and J. Erwin, SAE transactions, 972-978 (1997)
7. W. Wibowo, S. Sunardi, and I. Yulia, Bull. Chem. React. Eng. Catal. **2**, 21 (2007)
8. B. C. Shi and B. H. Davis, Journal of Catalysis**157**, 2 (1995)

9. O. Rahmanpour, A. Shariati, M. Reza, and K. Nikou, Int. J. Chem. Eng. Appl. **3**, 2 (2012)
10. J. Kong, B. Chao, T. Wang, and Y. Yan, Powder Technol. **229**, 7–16 (2012)
11. Y. Chen *et al.*, J. Ind. Eng. Chem. **16**, 5 (2010)
12. V. J. Chopade, Y. P. Khandetod, and a G. Mohod, Int. J. Emerg. Technol. Adv. Eng. **5**, 11, (2015)
13. M. Simo, Sep. Purif. Technol. Biorefineries, 503–512 (2013)
14. H. Scott Fogler, *Elements of chemical reaction engineering* (Prentice Hall, New Jersey,1987)
15. J. M. Smith, H. C. Van Ness, and M. M. Abbott, *Introduction to Chemical Engineering Thermodynamic* (McGraw-Hill Education, New York, 2005)

Production of Lactic Acid from Empty Fruit Bunch of Palm Oil Using Catalyst of Barium Hydroxide

Apsari Puspita Aini, *Hyung Woo* Lee, *Johnner Parningotan* Sitompul, and *Carolus Borromeus* Rasrendra*

Department of Chemical Engineering, Faculty of Industrial Technology, Institute of Technology Bandung, 40132 Bandung, Indonesia

Abstract. Lactic Acid as a platform chemical has broad application in various industries, especially in the production of Poly Lactic Acid (PLA) for biodegradable plastic. Empty fruit bunch (EFB), abundant by product from palm oil mill industry, is one of potential feedstock to be used in the production of lactic acid from lignocellulose biomass. EFB contains high cellulose and hemicellulose about 37– 59.7% w/w and 16–28% w/w, respectively. The aim of this paper is to study the effects of the operating conditions, such as temperature, reaction time, biomass loading, and catalyst concentration on the yield of lactic acid using barium hydroxide as alkaline catalyst. EFB pretreatment with steam explosion was applied to remove lignin content. The results showed that pretreatment reduced the lignin content from 22.66% to 9.69% w/w. Meanwhile, hemicellulose and cellulose increased from 14.40% to 16.40% w/w and 29.37% to 63.57% w/w, respectively. The highest yield of lactic acid was 21.57% C-mol, achieved by using 0.25 M $Ba(OH)_2$ as the catalyst, with 5% w/v biomass loading, temperature 240°C, during 4 h reaction times. The yield was approximately equal to yield of lactic acid (~ 20%) compared with Pb^{2+} as the catalyst for EFB conversion although the later catalyst produced fewer by products during conversion.

1 Introduction

Palm oil is one of the plantation crops, the source of non-oil foreign exchange for Indonesia. Other than the production of palm oil, margarine, and the cosmetics industry, the palm oil industry also produces liquid and solid waste. About 20 percent of oil palm empty fruit bunch (EFB) is produced from the processing of fresh fruit bunch [1]. The oil palm empty fruit bunch is one of lignocellulosic raw material. Lignocellulose is a major part of the plant cell wall, which consists of cellulose, hemicellulose, and lignin, and comprise a small amount of pectin, protein, extractive, and ash. Cellulose, hemicellulose and lignin content in EFB are about 37-59.7%, 16-28%, and 10-32% [2]–[5]. Limited utilization of EFB for fertilizer on oil palm plantation areas has forced many researchers to use EFB as raw material for various applications, one of which is lactic acid, the raw material for the production of Poly Lactic Acid (PLA).

Lactic acid is one of the chemical platforms that can be applied in chemical industry, food, the pharmaceutical industries, and cosmetics. Lactic acid has a strong market growth. It is included in the top 30 basic chemicals with significant market growth potential in the PLA and solvent industries. Global lactic acid is estimated to grow from 714.2 kilotons in 2013 to 1960.1 kilotons by 2020 [6]. Utilization of lactic acid as a primary material for PLA synthesis to produce biodegradable plastics is expected to be a replacement for plastic produced from petroleum. Due to the increase of awareness among consumers and plastic producers about the environment, PLA's market growth as biodegradable plastic will increase. It also means that the demand for lactic acid, a PLA raw material will increase as well.

Various substrates such as triose, hexose, cellulose or lignocellulose can be used as feedstocks in the production of lactic acid [7]–[12]. The production of lactic acid can be carried out by chemical synthesis, fermentation, and catalytic conversion pathways. The production of lactic acid with chemical synthesis pathways has some limitations because the raw materials depend on the by-products of other processes, Lacto nitrile, and the inability to produce L-lactic acid stereoisomers [8], [13]. The production of lactic acid with fermentation process was more advantageous than with chemical synthesis. This is because it can produce the desired stereoisomer, L-(+)-lactic acid or D-(-)-lactic acid [8]. Although more than 90% of lactic acid production is produced from fermentation, this method has some disadvantages such as reactor control, temperature, pH, and microorganism control. Otherwise, it takes a longer reaction time to produce lactic acid. The production of lactic acid by catalytic conversion process is expected to improve the disadvantages in chemical synthesis and fermentation process.

* Corresponding author: cbr@che.itb.ac.id

Nowadays, the high percentage of lactic acid yield can be achieved with triose as raw material, such as dihydroxyacetone, and glyceraldehyde [7], [9], [14], [15]. It is because triose is a short chain of sugar, which leads to a shorter reaction pathway of lactic acid production when it is compared with other raw materials. The use of other raw materials such as cellulose and lignocellulose has not been able to produce a very high yield of lactic acid than those obtained with triose.

Research conducted by Wang *et al.*, 2013, Sitompul *et al.* 2014, and Chin *et al.*, 2016, suggested that Pb^{2+} is selective catalyst for the reaction of lactic acid production. The yield of lactic acid from the cellulose-ball mill was 68% using Pb^{2+} as a catalyst with the addition of N_2 2MPa [11]. Meanwhile, the raw material of EFB the yield of lactic acid was 8.33% m/m and 6.21% m/m using $AlCl_3$ [12].

Esposito and Antonietti (2013) conducted research on the production of lactic acid with glucose as raw material. The yield of lactic acid with the temperature of 220°C and 12 h reaction time, using 0.05 M $Ca(OH)_2$, $Sr(OH)_2$, and $Ba(OH)_2$ were 49%, 50%, and 53%, respectively. Fewer product yields such as formic acid (4.4-5.9%) and acetic acid (5-7.6%) in their study showed that $Ca(OH)_2$, $Sr(OH)_2$, and $Ba(OH)_2$ is a highly selective catalyst in the formation of lactic acid. In addition to producing a good yield, this catalyst is also safer and less environmentally benign compared to other catalysts. Furthermore, Chin *et al.* (2016) suggested that the yield of lactic acid achieved for EFB as raw materials without pre-treatment and with the pretreatment using ball mill are 24.8% mol and 46.5% mol, respectively. In addition to shrinking size, ball mill can also reduce the crystallinity of EFB. This was resulted in the process of hydrolysis of EFB into a sugar monomer that is easier than EFB without ball mill treatment. So that, the higher yield of lactic acid can be achieved. Due to the high level of Pb toxicity, alternative catalysts may be required to replace Pb.

This study was expected to improve the conversion process of EFB into lactic acid, to overcome the obstacles, such as biomass structure recalcitrance that arise in previous studies. The objectives of this study were to study the effect of alkaline catalysts utilization such as NaOH, $Ca(OH)_2$ and $Ba(OH)_2$, conversion of pre-treated EFB to lactic acid, and to evaluate the effect of several operating parameters on lactic acid obtained. In order to reduce existing lignin content, EFB needs to be treated as early as possible using steam explosion [16]–[19].

2 Materials and Methods

2.1 Materials

The Oil Palm Empty Fruit Bunch (EFB) was collected from Riau Province. The microcrystalline cellulose was purchased from Kuraray Co., Ltd. The chemicals used in this experiment were D-(+)-glucose (99%, Merck), D-(+)-xylose (99%, Merck), lactic acid (90%, Merck), Formic acid (98%, Merck), Acetic Acid (99%, Merck), and Levulinic Acid (Merck).

The EFB was rinsed with water to remove other contaminants, then dried using sunlight. Size reduction of EFB carried out using scissor and disc mill. It was sieved into sizes 60-80 mesh and dried by oven at a temperature of 100 ° C until achieved a constant weight.

To know the content of cellulose, hemicellulose, and lignin, the EFB was characterized by using NREL (National Renewable Energy Laboratory) method. The determination of lignin content includes acid insoluble lignin (AIL) and acid soluble lignin (ASL). The content of cellulose and hemicellulose were known by analyzing the filtrate of hydrolysis using HPLC (High Performance Liquid Chromatography). The analysis began with the preparation of several sugar standard calibration solutions, sugar standard calibration verification solutions, and sample solutions. Standard calibration solutions were prepared from D-(+)-glucose and D-(+)-xylose.

2.2 The EFB pretreatment

Steam explosion method was used to reduce the lignin content in EFB. 50 grams of EFB 60 mesh and NaOH 0.5 M were input into the reactor, with 500 ml reaction volumes. The operating conditions of pretreatment was 180°C , ± 12 bar and 1 h of reaction time. After pretreatment, the EFB fiber was washed in order to neutralised the pH, and was dried in 100°C until it achieved a constant weight. To determine the cellulose, hemicellulose, and lignin content after pre-treatment, the EFB was characterized using NREL (National Renewable Energy Laboratory) method.

2.3 Conversion of EFB to lactic acid

Pre-treated EFB and cellulose microcrystalline were used as the raw material in this experiment. The conversion of EFB to lactic acid was carried out by hydrothermal process using high-pressure stainless steel autoclave with a stirrer to homogenize the raw material and the catalyst during the reaction.

This experiment consist of the variation of alkaline catalyst ($Ba(OH)_2$, $Ca(OH)_2$, NaOH), temperature (160-240°C), reaction time (1-6 h), catalyst concentration (0.05-0.25 M) and biomass loading (2.5-7.5% w/v). The reaction volume of each run was 80 ml. The highest yield was achieved from variation of the catalyst then be used to examine the effect of some parameter above.

2.4 Analytical method

Lactic acid was the product of hydrothermal process and other byproducts, such as formic acid, acetic acid, and levulinic acid were analysed by HPLC (High Performance Liquid Chromatography), equipped with an RI detector and Phenomenex Rezex ROA-Organic Acid H^+ column (300 x 7.8 mm) and 0.005 M H_2SO_4 as the mobile phase. HPLC was operated at 60°C, with flow rate 0.6 ml/min. To determine the concentration of the product, a standard curve was necessary. The concentration of the product was calculated on the basis of C-mol product compared to C-mol of the raw material.

3 Result and Discussion

3.1 Characterization and pretreatment of EFB

The pre-treatment of EFB was aimed to remove the lignin content. Various pre-treatment on lignocellulose biomass as follows: physical, chemical, biological, and physicochemical treatment, have been done by previous research. Physical treatment involves mechanical comminution, pyrolysis, and radiation with high energy. Chemical treatment is treatment using acid, alkaline, oxidative delignification, ozonolysis, and organosolv, ionic liquids pretreatment. In biological treatment, microorganisms (several white-rot fungi) have been examined on lignocellulosic biomass and showed the high yield of delignification. Physico-chemical pretreatment involves steam explosion, ammonia fiber explosion, CO_2 explosion, liquid hot water, wet oxidation, etc [16], [17], [19], [20].

Considering the advantages and disadvantages of each process, steam explosion pretreatment with alkaline was the most appropriate process to apply in this experimental. The results of an experiment conducted by Jeon *et al.* (2014) showed that enzymatic digestibility of EFB using Changhae Ethanol Multiexplosion (CHEMEX), 83.6%, was higher than the other pretreatment processes. The high yield of enzymatic digestibility is required for bioethanol fermentation process on a further step. The amorphous structure of EFB after CHEMEX treatment causes the enzyme to easily access the cellulose and hemicellulose chains, convert them into glucose and was used in bioethanol fermentation. In addition to being easily accessible by enzymes, amorphous structure of EFB will also greatly affect chemical conversion using a catalyst.

Table 1. The chemical composition of Oil Palm Empty Fruit Bunch

Composition	Before pretreatment (%w/w)	After pretreatment (%w/w)
Cellulose	29.37	63.57
Hemicellulose	14.40	17.30
Lignin	22.66	9.69

The characterization of EFB by NREL method showed that EFB pretreatment reduced the lignin content from 22.66% to 9.69% w/w. Meanwhile, hemicellulose and cellulose increased from 14.40% to 16.40% w/w and 29.37% to 63.57% w/w, respectively. The reduced lignin content causes the catalyst more easily accesses cellulose and hemicellulose chains to be converted to lactic acid.

3.2 Catalytic Conversion of EFB into Lactic Acid

The yield of lactic acid as main product from EFB catalytic conversion was determined by using HPLC analysis with a retention time of 30 minutes. The organic acid that was showed in HPLC chromatogram consists of lactic acid, formic acid, acetic acid, levulinic acid, and other byproducts. The 5-hydroxymethyl furfural (5-HMF), furfural, or other intermediate products have not been identified. It was seen from the brown color of the liquid product of the reaction [21].

The alkaline catalysts used in the production of lactic acid were NaOH, $Ba(OH)_2$, and $Ca(OH)_2$. It was found that $Ba(OH)_2$ is the best catalyst to produced lactic acid. The highest yield of lactic acid using $Ba(OH)_2$ at 200°C was 7.40% C-mol. The highest yield of lactic acid using $Ba(OH)_2$ in this experiment was in accordance with previous research conducted by Esposito and Antonietti (2013) and Li (2014).

$Ba(OH)_2$ was able to produce a high yield of lactic acid due to the size of its ionic radius, which can catalyze a 1.2-hydride shift. The size of ionic radius for metal ions are rBa^{2+}: 0.136 nm; rCa^{2+}: 0.100 nm; and rNa: 0.102 nm. The larger ionic radius, the better catalytic effect occurs. It also occurs in the production of lactic acid from glycerine by using $Mg(OH)_2$ and $Ca(OH)_2$ catalysts, where the yield of lactic acid obtained is 18.3 %-mol using $Ca(OH)_2$ and 15.7 %-mol using $Mg(OH)_2$ [22].

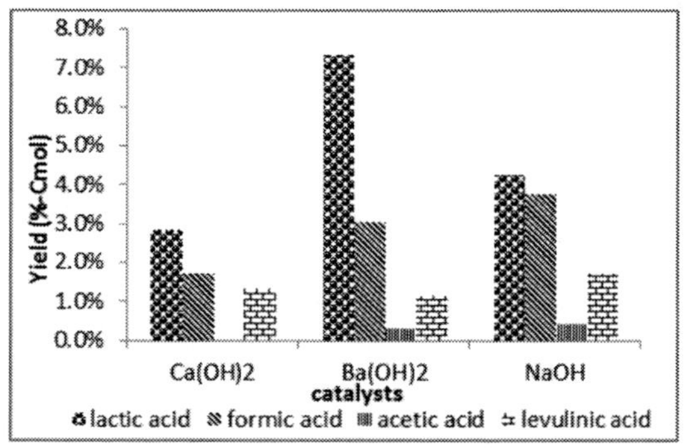

Fig. 1. The yield of lactic acid with alkaline catalyst at temperature of 200°C, 5%w/v biomass, and catalyst concentration 0.1 M

The temperature variations of 160°C, 200°C, 240°C were carried out to examine the effect of temperature on catalytic conversion of EFB into lactic acid. The increased yield of lactic acid occurs because the increase of temperature. It was due to the increased activity of Ba^{2+} and the production rate of lactic acid also increases. The effect of temperature on the yield of lactic acid was also found in the previous research [23] with the raw material of glucose using $Ba(OH)_2$, and research carried out by Wang *et al.*, 2013 with the raw material of glucose and fructose using Pb^{2+} 2mM. The concentration of lactic acid increased at the temperature of 170-230°C.

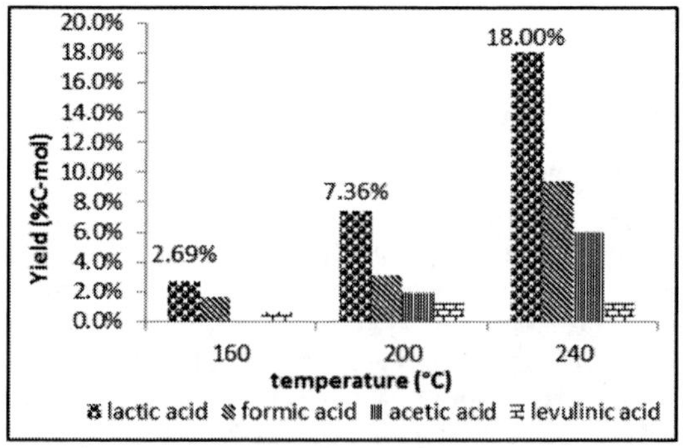

Fig. 2. The yield of lactic acid with variation of temperature, using $Ba(OH)_2$ 0.1 M, 5%w/v biomass

The yield of levulinic acid and formic acid as a byproduct were lower than acetic acid. Formic acid and levulinic acid were formed from 5-HMF. 5-Hydroxy methyl furfural (HMF) can be formed through the dehydration process of $C_6H_{12}O_6$ by eliminating three molecules of H_2O. Formation of levulinic acid and formic acid occurs through HMF rehydration and not directly from sugar. Meanwhile the formation of other by-products of acetic acid occurs through the process of oxidation of levulinic acid [9], [24]–[26]. Beside pretreatment EFB, microcrystalline cellulose, another raw material was also used in this experiment. Figure 3 represents the yield of lactic acid using microcrystalline

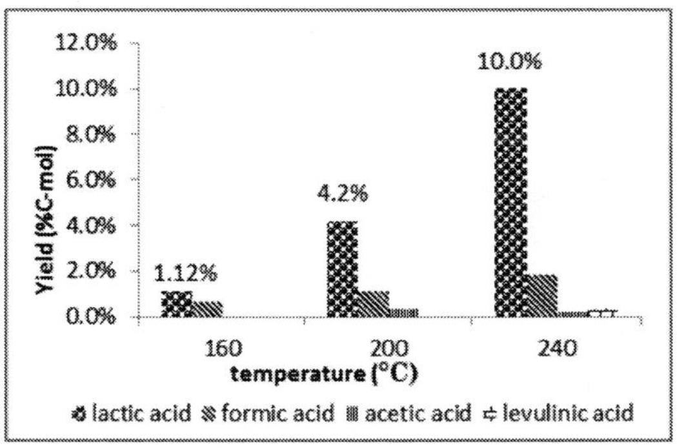

Fig. 3. The yield of lactic acid with variation of temperature, using $Ba(OH)_2$ 0.1 M, 5%w/v microcrystalline cellulose

as the raw material. The highest yield of lactic acid 10% C-mol, was achieved at a temperature of 240°C. Yield of lactic acid from microcrystalline cellulose was lower than yield of lactic acid from EFB, although EFB has more complex structure than cellulose. The lower lactic acid yield was due to its crystalline structure on cellulose that caused cellulose hardly to be hydrolyzed to form sugar component compared with amorphous structure on steam explosion EFB [21]. The comparison between crystalline and amorphous structure in the production of lactic acid was also shown in research conducted by Chin *et al.* 2016. The yield of lactic acid using EFB untreated and ball mill EFB were 24.8% mol and 42.5% mol, respectively. Wang *et al.* 2013 showed that longer reaction time was needed to produced lactic acid in similar yield (~60%) when used microcrystalline cellulose (15 h) rather than ball mill cellulose (4 h).

The variation of reaction time was carried out to found the optimal reaction time on the production of lactic acid. This experiment was conducted at temperature of 240°C using 0.1 M $Ba(OH)_2$ and 5%w/v biomass.

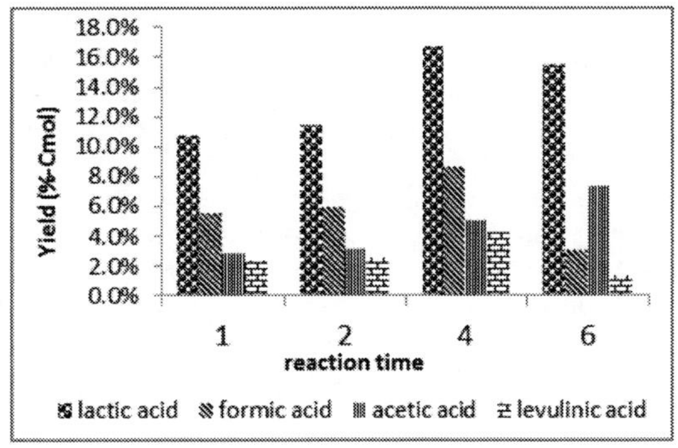

Fig. 4. The yield of lactic acid with variation of reaction time, using $Ba(OH)_2$ 0.1 M, 5%w/v biomass, temperature of 240°C

The highest yield of lactic acid (16.77% C-mol) was achieved at 4 h reaction time. The addition of reaction time to 6 h did not show an increase in yield of lactic acid. It was indicated that degradation of lactic acid occurred into other products. The oxidation process was predicted to occur during the reaction, which is also indicated by the increase of a concentration of acetic acid, and formic acid inasmuch as the increase of reaction time [27]. The highest acetic acid yield (7.40% C-mol) was obtained at 6 h reaction time. This phenomenon was in accordance with research conducted by Chin *et al.* (2016) in the formation of lactic acid from EFB using Pb^{2+}.

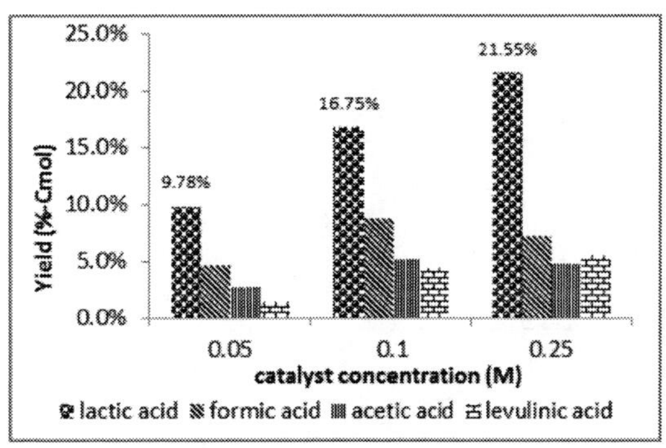

Fig. 5. The yield of lactic acid with variation of $Ba(OH)_2$ catalyst concentration , 5%w/v biomass, temperature of 240°C

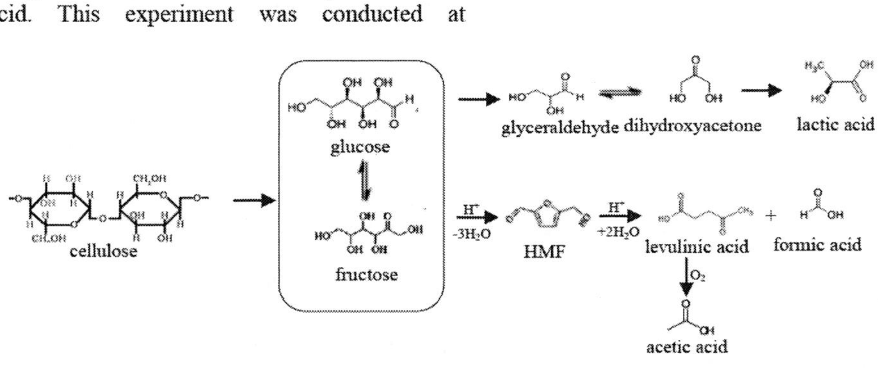

Fig. 6. Possible reaction pathway from the conversion of cellulose to lactic acid and by products

The variations of catalyst concentration on lactic acid production were shown in Figure 5. The highest lactic acid yield in this experiment was 21.57% C-mol, with catalyst concentration 0.25 M. The research concerning the conversion of glucose to lactic acid conducted by Esposito and Antonietti (2013) using $Ba(OH)_2$ also showed the same phenomena. The yield of lactic acid using 0.1 M $Ba(OH)_2$ was higher (57% mol) rather than 0.05 M catalyst (43% mol). The same phenomena were obtained by Shen *et al.* (2009) on the production of lactic acid from glycerine using various alkaline catalysts with concentration 0.025-3M. The increased of the yield of lactic acid was inasmuch as the concentration of the alkaline catalyst. The highest lactic acid achieved in that research was 90% mol using KOH as a catalyst, with concentration of 1.25 M. However, when the catalyst concentration was higher than 1.25 M, there was no significant effect on the yield of lactic acid obtained [22].

Increasing in the amount of biomass has caused decreasing in the yield of lactic acid, because the effectiveness of stirrer and catalyst performance during the reaction were also decreased. It can cause the formation of humin through dehydration of carbohydrate intermediates [9], [11], [28], [29]. The highest yield of lactic acid (19.32% C-mol) was obtained with a biomass concentration of 2.5% m/v.

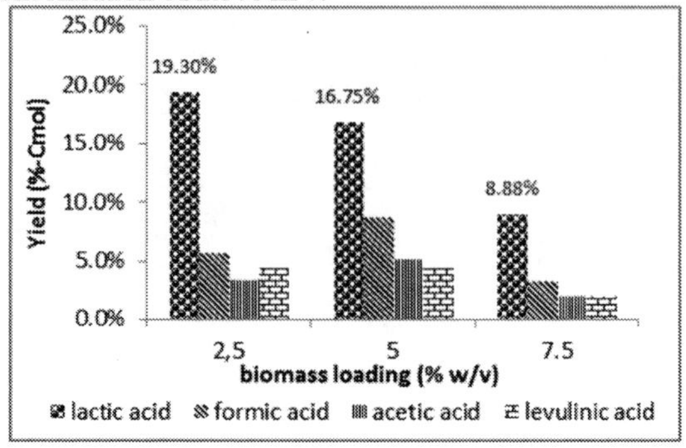

Fig. 7. The yield of lactic acid with variation of reaction time, using $Ba(OH)_2$ 0.1 M, 5%w/v biomass, temperature of 240°C

Table 3. The production of lactic acid using various biomasses and catalysts

Raw materials		Catalyst	Catalyst Concentration	Temperature (°C)	Yield of Lactic Acid (% C-mol)	Yield of byproduct (% C-mol)				
Biomass	(% m/v)					formic acid	acetic acid	levulinic acid	HMF	Furfural
a* Steam Explosion EFB	5	$Ba(OH)_2$	0.1 M	240	21.57	7.16	14.7	6.6	NA	
a* Steam Explosion EFB	5	$PbCl_2$	0.1 M	240	23.67	1.5	4.94	1.033	NA	
b*EFB	2.5	$AlCl_3$	0.1 M	230	13.38	NA				
c*Ball mill EFB	0.5	$PbCl_2$	2 mM	210	21.25	-	v	-	-	-
d*Cellulose	5	$PbCl_2$	0.05 M	220	23.10	-	-	-	NA	

a* This experiment. Reaction time 4 h.
b* Sitompul et. al, (2014). Reaction time 3h, EFB without pretreatment.
c* Chin et. al, (2016). Reaction time 2 h, EFB with ballmill treatment.
d* Alisyahbana and Asrizal (2014). Catalyst concentration 0.05 M, reaction time 4h.

The results of this experiment were also in accordance with previous research conducted by Li (2014) on the production of lactic acid using alkaline catalysts with raw materials of glucose, fructose, mannose, and xylose. The research showed that the increasing of fructose concentration from 1g/L to 8g/L, the yield of lactic acid decreased from 70% C-mol to 57% C-mol.

Further, table 3 showed the yield of lactic acid using alkaline and other lewis acid catalysts from the conversion of EFB. Lactic acid produced from EFB using $PbCl_2$ of 21.25%C-mol is the highest yield compared to other catalysts with 2 h reaction time [21]. Differences in reaction temperature, percentage of the amount of raw material and the treated or untreated EFB can affect the yield of lactic acid. However, considering the toxicity of Pb^{2+}, $Ba(OH)_2$ as the alkaline catalyst can be one of the best options. Further research with the larger scope of operating conditions needs to be carried out so it can be compared accurately.

4 Conclusion

The EFB pretreatment with steam explosion could reduce 57.24% w/w lignin content, increase 53.79% w/w cellulose, and 16.95% w/w hemicellulose. The highest yield of lactic acid 21.57% C-mol was achieved using 0.25 M $Ba(OH)_2$, 5% m/v biomass, and temperature of 240°C, during 4 h reaction time. Production of lactic acid with $Ba(OH)_2$ and Pb^{2+} as catalyst gave approximately equal yield of lactic acid (~ 20%). The use of Pb^{2+} catalysts for the production of lactic acid is more advantageous because it produces fewer by products. $Ba(OH)_2$ could be a good choice for the development of versatile products from EFB.

5 Acknoledgement

This research is funded by Indonesia Estate Crop Fund for Palm Oil through the Ministry of Finance and Research Incentive Program : The National Innovation System through the Ministry of Research and Technology. Dr. H.W. Lee was supported by the National Research Foundation of Korea (NRF Korea) through the TPC program for this work.

References

1. A. Wijono, *Simp. Nas.* **III**, (2014)
2. N. Abdullah, F. Sulaiman, J. Phys. Sci **2**, 2 (2013)
3. A.S. Baharuddin, A. Sulaiman, D.H. Kim, M.N. Mokhtar, M.A. Hassan, M. Wakisaka, Y. Shirai, H. Nishida, Biomass Bioenergy **55**, (2013)
4. Y. Sudiyani, D. Styarini, E. Triwahyuni, Sudiyarmanto, K.C. Sembiring, Y. Aristiawan, H. Abimanyu, M.H. Han, Energy Procedia **32**, (2013)
5. S.A.M. Saman, Y. Uemura, S. Yusup, Procedia Chem **9**, (2014)
6. G. V. Research, *Market research and consulting,* (2014)
7. M. Bicker, S. Endres, L. Ott, H. Vogel, J. Mol. Catal. A: Chem **239**, 1–2 (2005)
8. A. Onda, T. Ochi, K. Kajiyoshi, K. Yanagisawa, Appl. Catal., A **343**, 1–2 (2008)
9. C.B. Rasrendra, B.A. Fachri, I.G.B.N. Makertihartha, S. Adisasmito, H.J. Heeres, ChemSusChem **4**, 6 (2011)
10. Z. Tang, W. Deng, Y. Wang, E. Zhu, X. Wan, Q. Zhang, Y. Wang, ChemSusChem **7**, 6 (2014)
11. Y. Wang, W. Deng, B. Wang, Q. Zhang, X. Wan, Z. Tang, Y. Wang, C. Zhu, Z. Cao, G. Wang, H. Wan, Nat. Commun **4**, (2013)
12. J.P. Sitompul, R.F. Simangunsong, A.A. Asrizal, H. Alisyahbana, H.W. Lee, C.B. Rasrendra, Procedia Chem **9**, (2014)
13. R. P. John, A. G.S., K. M. Nampoothiri, A. Pandey, Biotechnol. Adv **27**, 2 (2009)
14. E. Taarning, S. Saravanamurugan, M.S. Holm, J. Xiong, R.M. West, C.H. Christensen, ChemSusChem **2**, 7 (2009)
15. F.F. Wang, C.L. Liu, W.S. Dong, Green Chem **15**, 8 (2013)
16. P. Alvira, E.T. Pejo, M. Ballesteros, M.J. Negro, Bioresour. Technol **101**, 13 (2010)
17. Y. Sun, J. Cheng. Bioresour. Technol **83,** 1 (2002)
18. P. Harmsen, W. Huijgen, L.M.B. Lopez, R.R.C. Bakker, ECN (2010)
19. P. Kumar, D.M. Barrett, M.J. Delwiche, P. Stroeve, Ind. Eng. Chem. Res **48**, 3713-3729 (2009)
20. P. Bajpai, Green Chem Sustain **34,** (2016)
21. S. X. Chin, S. M. Tasirin, H. Chan, H. Chia, BioResources **11,** 1 (2016)
22. Z. Shen, F. Jin, Y. Zhang, B. Wu, A. Kishita, K. Tohji, H. Kishida, J. Ind. Eng. Chem. Res **2,** (2009)
23. D. Esposito, M. Antonietti, ChemSusChem **6,** 6 (2013)
24. F. Jin, J. Yun, G. Li, A. Kishita, H. Enomoto, Green Chem. **10,** (2008)
25. R. Van Putten, J. C. Van Der Waal, E. De Jong, C.B. Rasrendra, H.J. Heeres, J.G. De Vries, Chem. Rev **113**, (2013)
26. Y. Wang, G. Yao, F. Jin, Green Chem. Sustain. Technol (2014)
27. L. Yang, J. Su, S. Carl, J.G. Lynam, X. Yang, H. Lin, Appl. Catal. B **162,** (2015)
28. H. Alisyahbana, A.A. Asrizal, Institut Teknologi Bandung (2014)
29. M. Li, Auburn University (2014)

Biodiesel Production from Dry Microalga Biomass by Microwave-Assisted In-Situ Transesterification

Lailatul Qadariyah*, *Mahfud*, *Renova* Panjaitan, *Fatkhulloh Mujaddid*, and *Ummu* Kalsum

Department of Chemical Engineering, Institut Teknologi Sepuluh Nopember, Sukolilo 60111 Surabaya Indonesia

Abstract. Microalga is one of the potential feedstocks in the manufacture of biodiesel because it contains high oil content. In this study, *Chlorella sp.* was selected because its high oil content about 28-32% of oil (based on its dry weight) and its presence is abundant among other green algae. In situ transesterification was carried out in round neck flask under microwave irradiation. Microwave irradiation can facilitate the in situ transesterification by extracted the lipid of microalga and simultaneous convert to FAME. The purposes of this study are to investigate the effect of acid catalyst concentration, microwave power, reaction time and the addition of co-solvent (n-hexane) on the yield of biodiesel, to get optimum operating conditions and to know the fatty acid compounds of biodiesel from *Chlorella sp.* The results of oil extraction and biodiesel were analyzed by GC-MS analysis. Based on the experiment, the yield of microalga oil was 11.37%. The optimum yield of biodiesel by in-situ transesterification was 75.68%. It was obtained at the microwave power of 450 watts, the reaction time of 60 minutes, an acid catalyst concentration of 0,2M of H_2SO_4, and the co-solvent addition of 10 ml.

1 Introduce

Nowadays, a lot of research was done on produce alternative fuel called biodiesel. Biodiesel is made from vegetable oil and animal fat. However, the lack of produce biodiesel from vegetable oil is it competes with food and has a long harvesting time [1]. Whereas, biodiesel from animal fat has less quality [2]. One of the natural resources that potential be the raw material of biodiesel is microalga. Microalga has fast growth and contains a lot of lipids. Harvesting of microalga can be done every day [3]. Microalga is the fastest organism in photosynthesis and contains high lipid content. It can produce yield up to 200 times more than other plants [4].Triglycerides which is contained in microalga can easily be extracted and transesterified to get biodiesel [5]. This research used microalga *Chlorella sp.* as the main material. *Chlorella sp.* contains 28-32 % of % lipid based on its dry weight [3].

Generally, biodiesel usually produced using ex situ transesterification, that is extraction lipid of microalga continued by transesterification process. But, this process is uneconomical and take a long reaction time. This process also needs higher of temperature and has a slow rate of reaction. Besides it, this process needs the higher amount of solvent and impossible to do a scale-up [6]. To resolve it, now there has been developed a non-conventional method called in situ transesterification. In situ transesterification is a simple method to produce biodiesel. This process cuts the extraction and purification step of microalga oil [7]. Microwave irradiation is one of the new technology in extract the lipid of microalga. It has been reported that microwave is a simple and effective technology for extraction the lipid of microalga. Heating by microwave can help to speed up the extraction and simultaneously turn it into FAEE or FAME [6]. Microwave irradiation has been applied in the transesterification of *Chlorella protothecoides* oil by Azcan *et al.*[8]. Transesterification by conventional heating was done in the molar ratio of oil to methanol of 1:8, catalyst concentration about 1%, the reaction time of 210 minutes and temperature reaction of 65°C. 96.94% of % biodiesel conversion was obtained by this condition while the transesterification by microwave in the molar ratio of oil to methanol of 1:8, catalyst concentration about 1%, the reaction time of 5 minutes was 96,54%. According to this research, use microwave can decrease the reaction time and save the energy. Another experiment has been explained by Dianursanti *et al.*[9]. They confirmed that the addition of co-solvent is also considered as a way to increase the yield on in situ transesterification because co-solvent can increase the solubility of alcohol and speed up the in situ transesterification.

This paper concerns with the development of in situ transesterification of *Chlorella sp.* using microwave

* Corresponding author: lailatul_2008@yahoo.com

irradiation. The purposes of this experiments were to investigate the effect of acid catalyst concentration, microwave power, reaction time and the addition of co-solvent (n-hexane) on the yield of biodiesel, to get optimum operating conditions and to know FAME compounds of biodiesel from *Chlorella sp*. This work was expected to produce a high yield of biodiesel and expected that methyl ester compounds of biodiesel are suitable for making a good biodiesel.

2. Methods

2.1. Materials

The main materials that use in this research were powder of microalga *Chlorella sp.*, which was collected from BBAP Jepara, Methanol 96%, n-hexane 96 % and the acid catalyst was H_2SO_4 98% of % concentration.

2.2. Equipment

In situ transesterification was performed in EMM2308X Electrolux microwave unit. The experimental equipment of in situ transesterification by microwave is illustrated in Fig. 1.

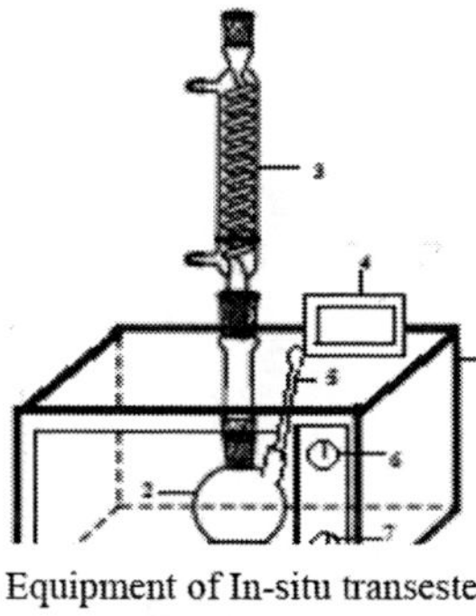

Fig. 1. Equipment of In-situ transesterification

2.2. Experimental Procedure

2.2.1. Extraction step of Microalga Oil by Microwave

The experiment was carried out in round neck flask. The loaded amount of microalga powder was 10 gram for microalga oil extraction and the mixed solvents of hexane and methanol (3:2 v/v) were 50 ml. Microalga powder and solvent were mixed in round neck flask. The extraction process was done at the microwave power of 600 watts, the reaction time of 60 minutes. After the extraction process by microwave irradiation was completed, the product was cooled down and separated by filter vacuum, the filtrate and residue were obtained. The residue was dissolved in the mixture of hexane and methanol to get the remaining oil. The filtrate was put into a separatory funnel and washed using 40°C of aquadest. After that, the filtrate was distilled to separate the solvent and microalga oil. Microalga oil was analyzed by gas chromatography-mass spectroscopy (GC-MS).

2.2.2. Step of Biodiesel Production from Microalga Chlorella sp. by in situ Microwave-Assisted Transesterification

The experimental equipment of in situ transesterification by microwave is illustrated in Fig. 1. In situ transesterification was took place in round neck flask. The loaded amount of microalga powder was 10 gram and methanol was 150 ml for all experiments. Microalga powder and methanol was mixed in round neck flask. In situ transesterification was done at the microwave power of 300, 450, 600 watt, reaction time of 30, 60, 90, 120 minutes, the acid catalyst concentration of 0.2; 0.35; 0.5 M of H_2SO_4. In situ transesterification with co-solvent (n-hexane) was performed in the addition of co-solvent (n-hexane) was 10 ml. This process was done at the microwave power of 450 watts, acid catalyst concentration of 0.2 M of H_2SO_4, and reaction time of 30, 60, 90 and 120 minutes. After the in situ transesterification was completed, the product was cooled down and separated by filter vacuum so that, the filtrate and residue were obtained. The residue was dissolved in the mixture of hexane and methanol to get the remaining FAME. The filtrate was put into a separatory funnel then n-hexane was added into the funnel (1:1 v/v), so it formed two layers. The top layer was washed using 40°C of aquadest then it was distilled to separate the n-hexane from crude biodiesel (product). Crude biodiesel was analyzed by gas chromatography-mass spectroscopy (GC-MS).

3 Results and Discussion

Experiments were carried out in round neck flask under the microwave irradiation. In-situ transesterification was done with and without the addition of co-solvent (n-hexane) using H_2SO_4 as the acid catalyst. The use of H_2SO_4 in this process will give several advantages, those are not sensitive to the content of FFA and water in oil, esterification and transesterification can take place simultaneously, and saponification can be avoided. Biodiesel yield was defined as the mass of crude biodiesel divided by the mass of microalga oil.

3.1 GC-MS Characterization of Microalga Oil

Based on the experiment, the yield of microalga oil extraction was 11.37% (based on its dry weight). GC-MS analysis shows that fatty acid compounds of microalga oil consist of saturated fatty acid group such as palmitic acid (C16:0), stearic acid (C18:0); monounsaturated fatty acid group such as oleic acid (C18:1); and polyunsaturated fatty acid grup such as linoleic acid (C18:2) with total of area reach 22.53% of % area of the sample. Table 1. shows the fatty acid compounds of microalga oil based on the GC-MS analysis.

 https://doi.org/10.1051/mateconf/201815606005

Table 1. Fatty acid compounds of microalga oil based on 22.53% of % area of the sample

Fatty Acid	Formula	Area (%)
Palmitic Acid	$C_{16}H_{32}O_2$	43.44
Linoleic Acid	$C_{18}H_{30}O_2$	6.35
Oleic Acid	$C_{18}H_{34}O_2$	44.33
Stearic Acid	$C_{18}H_{36}O_2$	5.86
Total of SFA	49.30	
Total of MUFA	44.33	
Total of PUFA	6.35	

This analysis fits with the research by Harrington (1986) [10]. He has been reported that all plants oil which is used for biodiesel production must contain C16 and C18 of fatty acids. Based on this result, *Chlorella sp.* is a potential feedstock for biodiesel production.

3.2 Effect of Reaction Time and Microwave Power

Fig. 2 shows the effect of reaction time and microwave power on yield of crude biodiesel at various acid catalyst concentration. At the various concentration of 0.2; 0.35; 0.5M, in situ transesterification was done effectively at the microwave power of 450 watts. Microwave power affects in the increase and the stability of temperature. The most instability of temperature occurred at the microwave power of 300 watts. According to the experiments, the microwave power of 600 watts resulted in the highest temperature reaction compared to microwave power of 300 and 450 watts. Increasing of temperature will help to the breakdown of microalga cell wall, but excessive using of power and heating result in the overheating for reaction and little oil will burn [11]. Besides it, at higher temperature methanol will evaporate and can disturb the transesterification [12]. Addition, three of graphics above have almost the same tren. At 300 watt, the yield of crude biodiesel increased at the reaction time range of 30-90 minutes. At 450 watt, the yield of crude biodiesel increased at the reaction time range of 30-60 minutes. At 600 watt, the yield of crude biodiesel increased at the reaction time range of 30-60 minutes and at 0,35 and 0,5 M of catalyst concentration while the yield of crude biodiesel increased at the reaction time range of 30-90 minutes and catalyst concentration of 0,2M. The highest yield of crude biodiesel was 65%. It was obtained at the microwave power of 450 watts and the reaction time of 60 minutes. This experiments also show that effective microwave power for in-situ transesterification was 450, 600 and 300 watt, respectively.

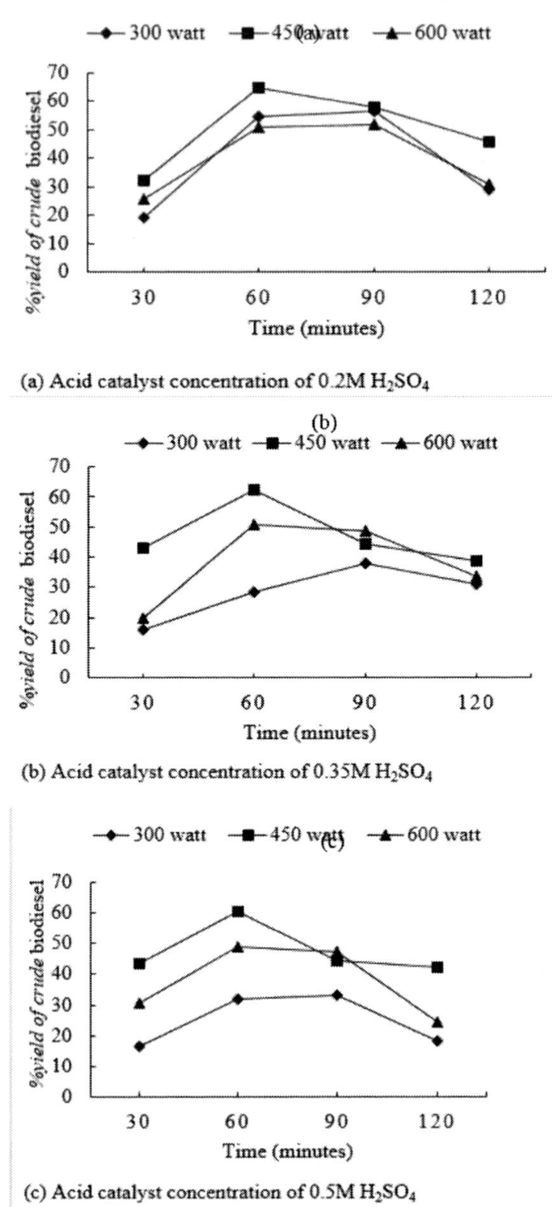

(a) Acid catalyst concentration of 0.2M H_2SO_4

(b) Acid catalyst concentration of 0.35M H_2SO_4

(c) Acid catalyst concentration of 0.5M H_2SO_4

Fig. 2. The effect of reaction time and microwave power on yields at catalyst concentration (a) 0.2 M H_2SO_4; (b) 0.35 H_2SO_4; (c) 0.5M H_2SO_4.

3.3 Effect of catalyst concentration and reaction time

In addition, Fig. 3 shows the effect of the reaction time of 30, 60,90 and 120 minutes on the yield of crude biodiesel at the acid catalyst concentration of 0.2; 0.35;

0.5M of H_2SO_4 and effective microwave power of 450 watts. Acid catalyst concentration and reaction time give varied influence on in situ transesterification. At the reaction time of 30 minutes, the yield of crude biodiesel increased with the higher catalyst concentration. At the reaction time of 60, 90, and 120 minutes, and catalyst concentration of 0.35M and 0.5M, the yield of crude biodiesel was lower compared to catalyst concentration of 0.2M. This occurred because of the negative effect of acid catalysts with high concentrations that can cause side reaction which is called an alcohol dehydration[13].

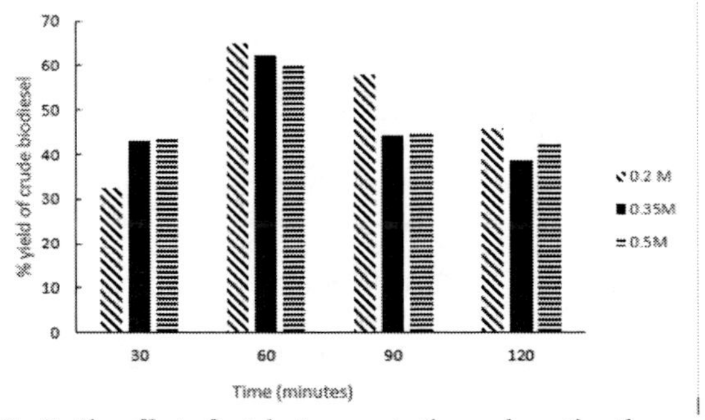

Fig. 3. The effect of catalyst concentration and reaction time on *yield* at the microwave power of 450 watts.

3.4 Effect of addition of *co-solvent*

Fig. 4, shows a comparison of yield with and without the addition of co-solvent in in-situ transesterification. The yield of crude biodiesel increased with the addition of co-solvent (n-hexane). It occurs because co-solvent can increase the solubility of oil in methanol [14]. Co-solvent also can reduce the *interfacial surface tension* of oil and methanol leading to form one phase of the liquid mixture. This condition increases the contact between reactants [15]. The highest yield of crude biodiesel was 75.68%. It was obtained at the microwave power of 450 watts and the reaction time of 60 minutes, the catalyst concentration of 0.2M and the addition of co-solvent was 10 ml.

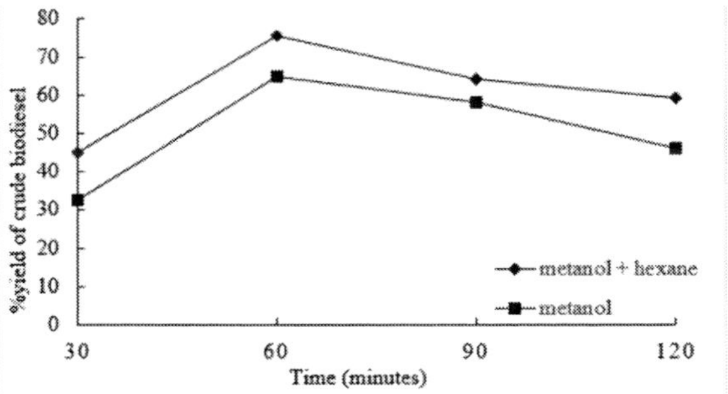

Fig. 4. The comparison of in situ transesterification with and without of co-solvent. The operation conditions at the microwave power of 450 watts and catalyst concentration of 0.2 M H_2SO_4.

3.5 FAME compounds of *Chlorella sp.*

Based on GC-MS analysis, the main compounds of methyl ester of biodiesel were palmitic acid (C16:0), palmitoleic acid (C16:1), myristic acid (C14:0), eicosapentaenoic acid (C20:5) and oleic acid (C18:1) and the total area reach 88,94% of % FAME normalization. Total of % FAME normalization on biodiesel sample was 35.09%. FAME compounds of biodiesel are presented in Table 2.

Table 2. Methyl ester compounds of *Chlorella sp.* based on 35,09% of % FAME area of biodiesel sample

Methyl Ester (FAME)	Formula	Area (%)
lauric acid methyl ester	$C_{13}H_{26}O_2$	1.08
E-11-tetradecanoic acid methyl ester	$C_{15}H_{28}O_2$	0.49
myristic acid methyl ester	$C_{15}H_{30}O_2$	11.10
pentadecanoic acid methyl ester	$C_{16}H_{32}O_2$	0.65
palmitoleic acid methyl ester	$C_{17}H_{32}O_2$	25.50
palmitic acid methyl ester	$C_{17}H_{34}O_2$	35.04
Z-7- hexadecanoic acid methyl ester	$C_{17}H_{32}O_2$	1.18
margaric acid methyl ester	$C_{18}H_{36}O_2$	0.49
8, 11-octadecanoic acid methyl ester	$C_{19}H_{34}O_2$	1.53
oleic acid methyl ester	$C_{19}H_{36}O_2$	6.33
11-octadecanoic methyl ester	$C_{19}H_{36}O_2$	0.60
stearic acid methyl ester	$C_{19}H_{36}O_2$	2.11
10-nonadecanoic acid methyl ester	$C_{20}H_{38}O_2$	0.49

arachidonic acid methyl ester	$C_{21}H_{42}O_2$	2.43
Eicosapentaenoic acid methyl ester	$C_{21}H_{32}O_2$	10.97
Total of SFA		52.91
Total of MUFA		34.59
Total of PUFA		12.50

4 Conclusions

In situ transesterification was studied by microwave irradiation at microwave power of 300, 450, 600 watt, reaction time of 30, 60, 90, 120 minutes, acid catalyst concentration of 0.2; 0.35; 0.5 M of H_2SO_4 while in situ transesterification with co-solvent at microwave power of 450 watt, acid catalyst concentration of 0.2 M of H_2SO_4, and reaction time of 30, 60, 90 and 120 minutes. The highest yield of crude biodiesel was 75.68%. It was obtained at the microwave power of 450 watts, the reaction time of 60 minutes, the catalyst concentration of 0.2M and the addition of co-solvent was 10 ml. The yield of biodiesel was influenced by microwave power, reaction time, acid catalyst concentration and co-solvent. A longer of reaction time and a greater of power provided a higher of biodiesel yield. However, when the condition reaches in optimum condition, the longer of reaction time and the greater of power, provide the lower of biodiesel yield. The addition of co-solvent (n-hexane) in in-situ transesterification gives the higher of biodiesel yield. The main compounds of methyl ester were palmitic acid methyl ester and palmitoleic acid methyl, respectively.

5 References

1. J. Basmal. Squalen **3(1)** (2008)
2. V.Feddern. *Embrapa Swine and Poultry* (Brazil, 2015)
3. Christi,Y. Advances **11**, 294–306 (2007)
4. M. Latief. ResearchGate. (to be published)
5. K.K. Sharma, H. Schuhman, P.M. Schenk. Energies **5 (5),** 1532-1553 (2012)
6. J. Liu, Z. Sun, H. Gerken. *OMICS Group eBooks* (USA, 2016)
7. S. Yucel, P. Tergiozglu. 2013. *Recent Res. Devel. Lipids* **9**, 101-124 (2013)
8. N. Azcan, O. Yilmaz. *Proceedings of the World Congress on Engineering and Computer Science* Vol II (San Francisco, USA, 2012)
9. Dianursanti, P. Religia, A. Wijanarko. (2015). Procedia E.S **23**, 412 – 420 (2015)
10. K.J Harrington. 1986. Biomass **9**, 1-17 (1986)
11. P. Li, X. Miao, R. Li, and J. Zhong. Hindawi **2011**, 1-8 (2011)
12. Y. Tang, Y. Zhang, J. Rosenberg, M. Betenbaugh, F. Wang. Appl. Sci **6**, 343 (2016)
13. P.M. Ejikeme, I.D. Anyaogu, C.L. Ejikeme, N.P. Nwafor, C.A. Egbuonu, K. Ukogu, J.A. Ibemesi. J. Chem **7(4)**, 1120-1132 (2010)
14. V.T. Wyatt, M.J. Haas. J. Am. Oil Chem. Soc **86**, 1009-1016 (2009)
15. D.S. Khang, F.R. Luis, F.M. Cynthia, R.T. Raymond. Chem. Eng. R & D **92**, 1512-1518 (2014)

Optimization of Preparation of Zeolite Y Dealuminate Catalysts for Glycerol Conversion to Glycerol Mono Laurate

Didi Dwi Anggoro*, *Luqman* Buchori, *Istadi* Istadi, Fadhil R. P., and Antonio G.

Department of Chemical Engineering, University of Diponegoro, Semarang 50275, Indonesia

Abstract. This Research has been done to analyze Zeolite Y CBV 712 as a catalyst. Zeolite Y CBV 712 was dealuminated by treating it with H2SO4 solution, with 3 – 8 M concentration at 40 – 60 oC. The dealumination treatment runs for 4 hours. Then, dealuminated zeolite was calcinated with a furnace for 3 hours with temperature range between 500 – 600 oC. Dealuminated zeolit Y then tested to synthesize Glycerol Mono Laurate (GML). By Analysis of Variance (ANOVA) analyze, indicated that the higher Glycerol Monolaurate yield, X2 factor is the most significance factor, indicating with high F-value (29.60768) and low P-value (0.001600). Objective function from this tests, is used to determine the optimum value between dealumination temperature (X1), acid concentration used (X2) and calcinating temperature (X3). The relation between each operating variables and the responses is shown in mathematical equation below :

$$Yield\,(\%) = -304{,}099 + 14{,}083X_1 - 0{,}079X_1^2 + 32{,}836X_2 - 2{,}219X_2^2 - 0{,}309X_3 + 0{,}001X_3^2 + 0{,}83X_1X_2 - 0{,}013X_1X_3 - 0{,}020X_2X_3$$

The optimum condition obtained is 43.25oC for dealumination temperature, 5.45 M for acid concentration and 614oC for calcination temperature. According to the optimum condition obtained, the glycerol monolaurate yield obtained is 59.470%.

1. Introduction

One of the glycerol derivative compound is Glycerol Monolaurate which used in food additives, surfactant, medicine, cosmetics and others [1] As a non-ionic surfactant which consists of hydrophilic and hydrophobic group, Glycerol Monolaurate can be used as a nutrition supplement [2]. In the making of Glycerol Monolaurate, catalysts is used to accelerate the reaction and increasing the yield of Glycerol Monolaurate. One of the catalysts that had been used is Zeolite Y [3]. Dealumination is used to increase the acidity of the zeolites [4] Characterization is used to determine the characteristics of the Zeolit Y that had been dealumination [5]. Zeolite Y is a kind of crystalline aluminosilicate with a microscale cavity which has a pore size of about 0.74 nm framework [6,7]. The zeolite have the largest application in catalysis and classified as faujasites [8]. Faujasites is a kind of the zeolite mineral group which is a silicate mineral [9].

The previous results [10] showed that the increasing crystallinity percentage represents the higher acid conversion. The higher crystallinity defined decreasing the conversion to GML and GDL from 45.17 to 34.98 for Glycerol Monolaurate (GML) and 31.68 to 20.32 for Glycerol Dilaurate (GDL). Conversion to Glycerol Trilaurate (GTL) had different results. The more crystallinity caused the increasing conversion to GTL from 9.07 to 20.32. The results indicated that the increasing of crystallinity of zeolite caused the decreasing of the conversion to Glycerol Monolaurate and Glycerol

*Corresponding author : anggorophd@gmail.com

Dilaurate, except for Glycerol Trilaurate which equal to the percentage of the crystallinity.

In this study, using three stages of chemical treatment processes / dealumination, washing, drying / calcination process. The objective of this study are to determine the optimum value between dealumination temperature (X_1), acid concentration used (X_2) and calcinating temperature (X_3) by using Analysis of Variance (ANOVA) method and to modified Zeolit Y catalysts for produced the Glycerol Monolaurate which higher GML yield.

2. Experimental Details

Zeolite Y CBV 712 (supplied by Zeolyst International in NH_4 form; mole ratio SiO_2 / Al_2O_3 = 12; surface area = 730 m^2/g) was dealuminated by treating it with H_2SO_4 solution. The dealumination treatment was carried out in a three neck flask with stirrer for mixing the sulphuric acid with Zeolite Y, and dealuminated for 4 hours. Then, zeolite that have been dealuminated, calcinated with a furnace for 3 hours. Dealuminated Zeolite Y then, tested to synthesize glycerol monolaurate using a glycerol and lauric acid, as described at previous paper [10].

3. Result and Discussion

The experimental design matrix including the experimental results which is designed using central composite design of response surface methodology are presented in Table 1. The result consists of 16 sets of coded conditions expressed in the natural values. The design consists of eight factorial points, six axial points, three central points.The sequence of experiment was randomized in order to minimize the effects of uncontrolled factors.

Analysis of Variance (ANOVA) used to analyze the Glycerol Monolaurate yield as shown in Table 1. The significance between each factor in Table 1 was tested using F-value and P-value. High F-value and P-value lower than 0.05 indicating that the variables, significantly influencing the response studied. F-value showing the ratio between Mean Square of Factor (MSF) and Mean Square of Error (MSE). The effect of operating variables is shown in Table 1.

Table 1. Analysis of Variance

Variant	Koefisien	F-Value	DF	p-value
X_0	-304,099			
X_1	14,083	9,51471	1	0,021536
X_1^2	-0,079	8,91520	1	0,24452
X_2	32,836	3,686619	1	0,103274
X_2^2	-2,219	29,60768	1	0,001600
X_3	-0,309	0,74971	1	0,419839
X_3^2	0,001	0,95074	1	0,367187
X_1X_2	0,083	0,94272	1	0,369078
X_1X_3	-0.013	9,30101	1	0,022519
X_2X_3	-0,020	1,42125	1	0,278206

As shown in ANOVA table above, for higher Glycerol Monolaurate yield, X_2 factor is the most significance factor, indicating with high F-value (29.60768) and low P-value (0.001600). Objective function from this tests, is used to determine the optimum value between dealumination temperature (X_1), acid concentration used (X_2) and calcinating temperature (X_3). The relation between each operating variables and the responses is shown in equation below:

$$Yield\ (\%) = -304{,}099 + 14{,}083X_1 - 0{,}079X_1^2 + 32{,}836X_2 - 2{,}219X_2^2 - 0{,}309X_3 + 0{,}001X_3^2 + 0{,}83X_1X_2 - 0{,}013X_1X_3 - 0{,}020X_2X_3$$

Order of this equation is chosen by comparing the coefficient of determination, which the highest coefficient of determination is the most recommended one. Due to this experiment, the highest coefficient of determination (R^2) for the mathematical equation shown above is 91.587.

With high value of coefficient of determination (91.587) indication a match between predicted value and experimental data shown in Figure 1. The linier line in Figure 1 called a regression line. Regression line shows the best prediction between dependent variables (Y) against independent variables (X). However, nature is rarely (if ever) perfectly predictable, and usually there is substantial variation of the observed points around the fitted regression line (Figure 1). The deviation of a particular point from the regression line (its predicted value) is called the residual value.

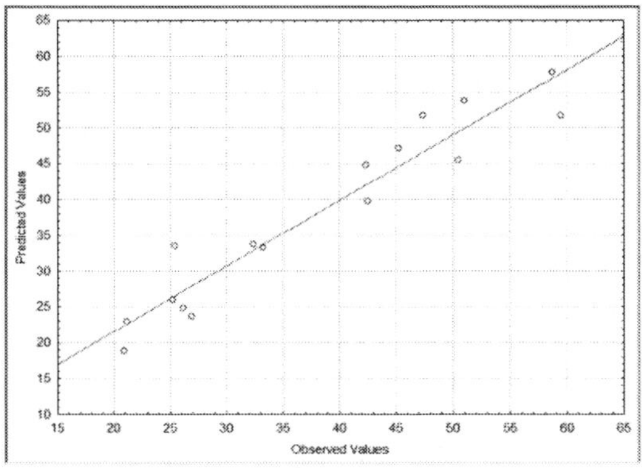

Fig 1. Relation of predicted and observed value of Glycerol Monolaurate yield

Contour plot are the graphical representations of the fitted model that can be used to study the effect of process variables on glycerol monolaurate yield. Figure 2 showed increase in acid concentration used resulting in increasing yield of glycerol monolaurate, increasing acid concentration caused an increasing of glycerol monolaurate yield in the beginning and then achieve an optimal state and then down again. As we can see in Figure 2, the optimal concentration is between 5-6 M, and dealumination temperature between 46-54°C. Further explanation is given following this paragraph.

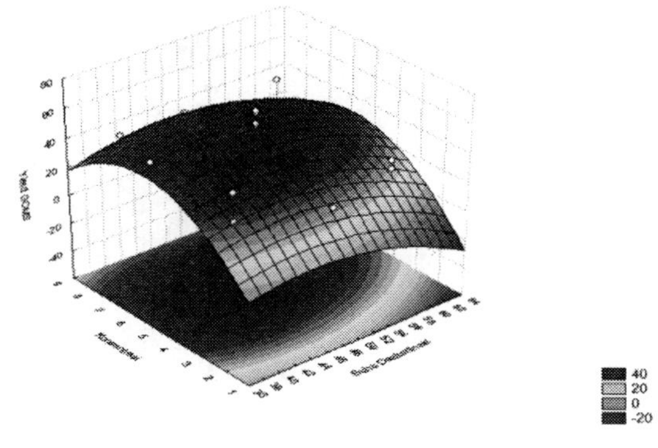

Fig. 2. Contour Plot of Glycerol Monolaurate Yield

Dealumination process is increasing of Si/Al ratio in Zeolite Y, and increasing the catalysts ability to synthesize derivate of glycerol becoming glycerol monolaurate (GML), to reach the optimum yield possible. The dealumination temperature, acid concentration and calcination temperature are needed to be control by limiting the operating condition variables.

Optimizing the most significance variables in this research is achieved by using desirability profiling in statistica software. Figure 3 determined that the optimum condition are dealumination temperature is limited within 37-62°C, acid concentration within 2.2 to 8.7 M and calcination temperature within 485-614°C.

According Figure 3, the optimum condition obtained is 43.25°C for dealumination temperature, 5.45 M for acid concentration and 614°C for calcination temperature, which the glycerol monolaurate yield obtained is 59.470% is obtained.

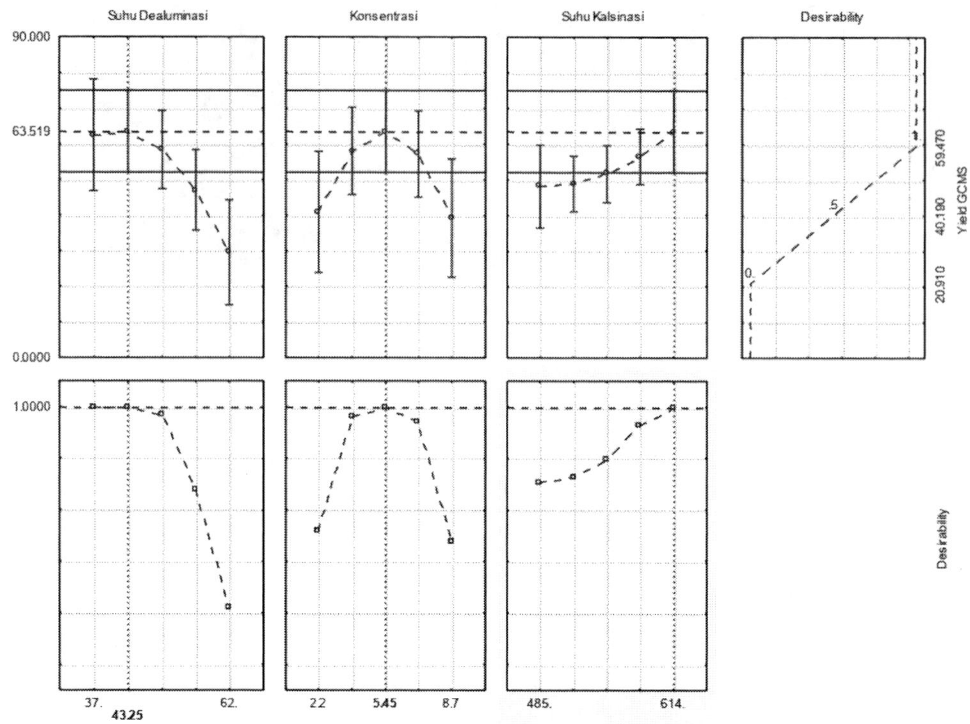

Fig 3. Optimizing Using Responses Desirability Profiling in Statistica 6.0

4. Conclusion

The optimum condition obtained is 43.25°C for dealumination temperature, 5.45 M for acid concentration and 614°C for calcination temperature. According to the optimum condition obtained, the glycerol monolaurate yield obtained is 59.470%.The relation between each operating variables and the responses is shown in mathematical equation below :

$$Yield\ (\%) = -304{,}099 + 14{,}083X_1 - 0{,}079X_1^2 + 32{,}836X_2 - 2{,}219X_2^2 - 0{,}309X_3 + 0{,}001X_3^2 + 0{,}83X_1X_2 - 0{,}013X_1X_3 - 0{,}020X_2X_3$$

References

1. T. Y. Wibowo, A. Z.Abdullah, & R. Zakaria, Appl. Clay Sci. **50**. 280-281. (2010).
2. R. Nakamura, K.Komura, & Y. Sugi,. Catal. Commun. **9.** 511-515. (2008).
3. Y. Wang, R. Otomo, T. Tatsumi, & T. Yokoi,. Microporous Mesoporous Mater. **220**. 275-281. (2015).
4. W. Xu, L. Y. Li, & J. R. Grace,. Chemosphere. **111**. 427-433. (2014).
5. M. d. S. Machado, et al. Appl. Catal. **203**. 321-328. (2008).
6. B. C. Gates, Catalytic Chemistry. New York: John Wiley & Sons, Inc. (1997).
7. Zhao, L. et al. J. Membr. Sci. **498**. 1-13. (2015).
8. C. H. Tempelman, et al. Fuel Process. Technol. **139**. 248-258. (2015).
9. W. Ratanathavorn, C. Samart, & P. Reubroychareon,. Mater. Lett. **159**. 135-137. (2015).
10. D. A. Didi, B. S. Wahyu, W. Triyogo, B. Luqman, R. P. Fadhil and G. Antonio, Adv. Sci. Lett. **23**, 5602–5604 (2017)

A Review of Bio-lubricant Production from Vegetable Oils Using Esterification Transesterification Process

Arianti N. Annisa[1,*] and *Widayat* Widayat[1,2]

[1]Department of Chemical Engineering, Faculty of Engineering, Diponegoro University, Semarang
[2]Laboratory of Material Technology, Centre of Research and Service Diponegoro University, Semarang

Abstract. Since long time ago, petroleum oil has been used as a lubricant in motor vehicles. But, the uses of lubricants petroleum can pollute the environment and disrupt marine ecosystems and terrestrial. On the other hand, the use of lubricants in the world in 2008 reached 46 million kiloliters/year and increased by 2% every year. The future prospect the uses of lubricant petroleum on the vehicle's engine is predicted to have a bad prospect. Thus began research to identify suitable materials to replace petroleum based lubricants. It can produce from vegetable oils and animals oils by chemical modification. But, bio-lubricant properties are easily damaged so we need more research to improve the shelf life of the bio-lubricant. The research must be done to improve the characteristics by adding a bio-lubricant additive.

1 Introduction

In 2025, Indonesia's energy mix is expected to be realized by the use of oil less than 20% and uses of renewable energy more than 5% [1]. It aims to improve the management of natural resources and the environment to support the quality of life and to improve the maintenance and use of biodiversity. On the other hand, the automotive industry and the machine began to focus on the improvement of environment-friendly technologies and energy efficient. Technology developed toward the use of fuel with low pollution combustion and exhaust as well as the efficiency of the vehicle is expected to reduce environmental problems [2]. To achieve an environmentally friendly vehicle conditions requiring lubrication in the engine parts so as to reduce friction between each other. By reduced friction, it can reduce energy loss in the engine.

For a long time oil has been used as a lubricant in motor vehicles. Nonetheless, the uses of lubricants petroleum can pollute the environment and disrupt marine ecosystems and terrestrial. On the other hand, the use of lubricants in the world in 2008 reached 46 million kiloliters/year and increased by 2% every year. In Indonesia, the use of lubricants in 2012 amounted to 2,988,265 barrels and increased 2% - 8% annually [3]. However this is not in line with the availability of petroleum which has declined by 3% annually [3]. Based on Tung and McMillan [4] research, the future prospects the uses of lubricant petroleum on the vehicle's engine is predicted to have a bad prospect. Thus began research to identify suitable materials to replace petroleum based lubricants.

Their depletion of crude oil reserves and demands for protecting the environment by reducing pollution generated from the lubricating oil led to an interest to develop and use alternative lubricants. Bio-lubricant felt can be used as an alternative to replace lubricants from petroleum as it has suitable properties and can be renewed. Compared with a lubricant, bio-lubricant has a high degree of lubricity, high viscosity index, high flash point and low volatility [5-10].

2 Raw Material of Bio-lubricant

Bio-lubricant interpreted as a lubricant obtained from natural raw materials both vegetable and animal oils, renewable and non-toxic to humans and other living things, as well as environmentally friendly. Vegetable oil used for the production bio-lubricant can be obtained from plant seeds, such as vegetable oil that can be consumed or which cannot be consume. Some of the vegetable oil can be used as bio-lubricant i.e. castor oil [11], karanja, neem, rice bran, rapeseed, castor, linseed, mahua [12], palm oil [13], sunflower oil, coconut, soybean, olive and canola [14].

The main content of vegetable oils that are important in making bio-lubricant is triglycerides. Triglycerides are glycerol molecules that have three long chain polar fatty acids attached at the hydroxyl groups via ester linkages [15]. But vegetable oils derived from plants have unstable nature. It is influenced by the free acid content in vegetable oil. Therefore, it should be known fatty acid

* Corresponding author: ariantinuura@gmail.com

content in vegetable oil. The fatty acid composition of vegetable oils were listed in Table 1.

Table 1. Fatty acids composition of vegetable oils [15-16]

Vegetable oils	Palmitic (16:0)	Stearic (18:0)	Oleic (18:1)	Linoleic (18:2)	Linolenic (18:3)
Sunflower oil	6.1	5.3	21.4	66.4	-
High oleic sunflower oil	3.5	4.4	80.3	10.4	-
Safflower oil	6.4	2.5	17.9	73.2	-
High oleic safflower oil	4.6	2.2	77.5	13.2	-
High linoleic safflower oil	6.7	2.6	14.6	75.2	-
Soybean oil	6.0	5.2	20.2	63.7	5.0
High oleic soybean oil	6.2	3.0	83.6	3.7	1.7
Corn oil	10.6	2.0	26.7	59.8	0.9
Cottonseed oil	18.0	2.0	41.0	38.0	1.0
Rapeseed oil	3.5	0.9	19.4	22.3	8.2
Canola oil	2.5	1	64.4	22.2	8.2
Peanut oil	10	3	50	30	-
Linseed oil	5	3	22	17	52
Olive oil	14	2	64	16	2
Coconut oil	9	2	7	1	-
Palm oil	42	5	41	10	-

By knowing the composition of fatty acids in vegetable oil can affect the characteristics to be generated bio-lubricant. Vegetable oils that have more double chains will have a better pour point and low oxidation stability, whereas the vegetable oils those have less double chains. Lubricants have utility for reducing friction between the surfaces in mutual contact and are used in various applications such as agriculture, forestry, transportation, mining, automobile, fishing, chainsaw oils, transmission oils, serving as engine oils and hydraulic oils [17]. Based on that, it must be known to the comparison between the properties of commercial lubricants and vegetable oils to produce bio-lubricant. Properties of commercial lubricant and vegetable oils listed in Table 2.

Table 2. Characteristics of commercial lubricant and vegetable oils [17]

Lubricant requirement	Viscosity 40°C (cSt)	Viscosity 100°C (cSt)	Viscosity index	Pour point (°C)	Flash point (°C)	Oxidative stability (min)	Coefficient of friction	Wear scar (mm)
ISO VG32	>28.8	>4.1	>90	-6	204	-	-	-
ISO VG46	>41.4	>4.1	>90	-6	220	-	-	-
ISO VG68	>61.4	>4.1	>198	-6	226	-	-	-
ISO VG100	>90.0	>4.1	>216	-6	246	1670.26	-	-
Paraffin VG95	95	10	102	-	-	-	-	-
Paraffin VG460	461	31	97	-	-	-	-	-
R150	150.04	-	-	-	195	931.16	-	-
SAE20W40	105	13.9	132	-21	200	-	0.117	0.549
AG100	216	19.6	103	-18	244	-	-	-
75W-90	120	15.9	140	-48	205	-	-	-
75W-140	175	24.7	174	-54	228	-	-	-
80W-140	310	31.2	139	-36	210	-	-	-
Vegetable oil								
Soybean	28.86	7.55	246	-9	325	-	-	-
Sunflower	40.05	8.65	206	-12	252	-	-	-
Passion fruit	31.78	-	-	-	228	7.5	-	-
Moringa	44.88	-	-	-	204	28.27	-	-
Castor	220.6	19.72	220	-27	250	-	-	-
Rapeseed	45.60	10.07	180	-12	240	-	-	-
Jatropha	35.4	7.9	205	-6	186	5	-	-
Coconut	24.8	5.5	169	21	325	-	0.101	0.601
Rice bran	40.6	8.7	169	-13	318	-	0.073	0.585
Palm	52.4	10.2	186	-5	-	-	-	-
Lesquerella	119.8	14.7	125	-21	-	-	0.045	0.857
Pennycress	40.0	9.3	226	-21	-	-	0.054	0.769

3 Production Process of Bio-lubricant

Bio-lubricant is a lubricant made from vegetable oil that can reduce environmental pollution [18]. Bio-lubricant produced by modifying the chemical structure of the fatty acid. The method can be performed is lists in Table 3.

Table 3. Chemical Modifications on Fatty Acids [19]

Modification	**Catalyst**
Modification of Carboxyl Groups	
- Esterification/transesterification	Acids, Bases
- Estolides	$HClO_4$
Modification of fatty acid chains	
- Selective hydrogenation	Copper catalysts
- Dimerisation/oligomerisation	Aluminosilicates
- Diels-Alder cycloaddition	-
- Telomerisation	H_2O vapor
- Co-oligomerisation	$RhCl_3.3H_2O$
- Hydroformylation (oxo-synthesis)	$Rh(Ph_3P)_3$
- Friedel-Crafts alkylation	$Et_3Al_2Cl_3$
- Aminoalkylation	$[Rh(COD)Cl]_2$
- Friedel-Crafts acylation	$EtAlCl_2$
- Ene-reaction	$EtAlCl_2$, SnO_4
- Radical addition	$Mn(OAc)_3$, Cu
- Acyloxylation	Nafion/SiO_2
- Cyclopropanation	CH_2I_2, Zn(Cu), Et_2Zn
- Metathesis	Grubbs catalyst
- Epoxidation	Peracids, chemo-enzymatic
- Ozonation	-
- Oxidative cleavage	Pb_3O_4

4 Esterification/Transesterification

Esterification process is a method to produce ester (RCOOR') by reacting a compound of acid and alcohol by eliminating the water formed in the reaction product. The molecular structure that reacts affect reaction rates, in particular molecular structure that acts as a free radical. The acid catalysts are often used is a strong mineral acid, such as hydrochloric acid and sulfuric acid. Also, it can use other materials such as silica gel, and cation resin exchanger. Biopelumas formation mechanism esterification reaction can be seen in Figure 1.

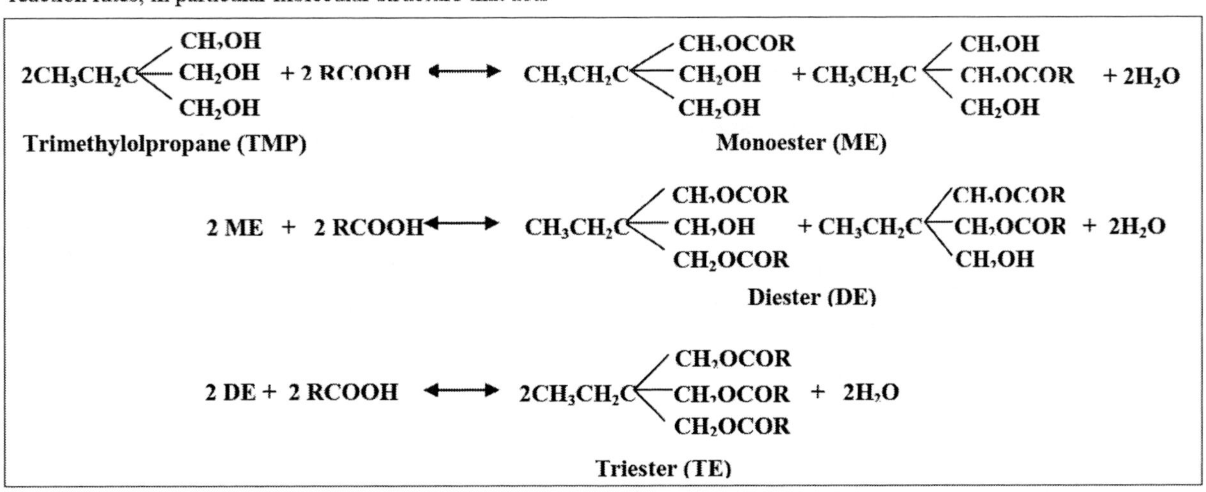

Fig. 1. Esterification of fatty acids [20]

While the process of converting vegetable oil (triglyceride ester) into bio-lubricant (triesters) known as transesterification process, the reaction between triglycerides as a component of vegetable oil with alcohol octavalen, using a base catalyst. The mechanism of the production biopelumas transesterification reaction is shown in Figure 2.

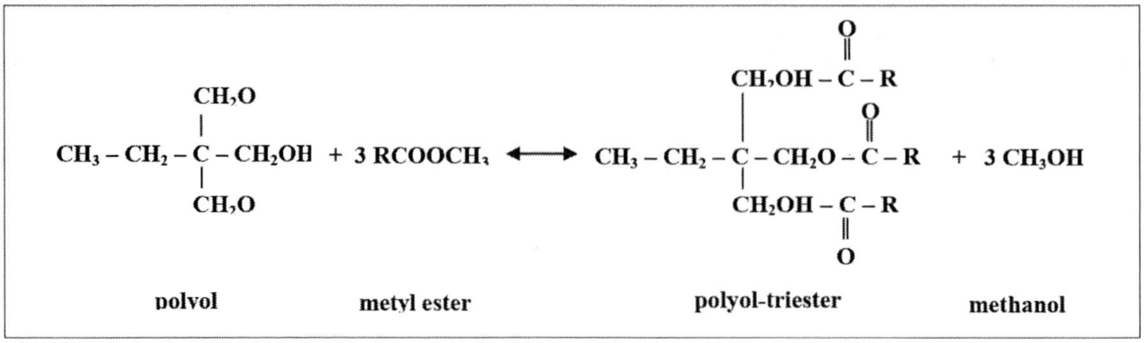

Fig. 2. Transesterification reaction Bio-lubricant [21]

The transesterification process is influenced by the catalyst, contact time, temperature, flow rate, water content in alcohol, the amount of alcohol excesses and free fatty acids in raw materials. The water content in alcohol will result in the formation of soap by consuming a catalyst, and this will reduce the efficiency of the catalyst. To get the maximum conversion, the use of anhydrous alcohol is preferred. Excess alcohol consumption will shift the reaction equilibrium towards the right and the greater the production of methyl ester. Another factor affecting the transesterification process is a free fatty acid content in the raw material. If the free fatty acid content exceeds 0.5 %, the saponification reaction will occur in the form of an emulsion. Soap can be physically separated from the former vegetable oils. While the esterification reaction mechanism on biopelumas briefly transesterification contained in Figure 3 below.

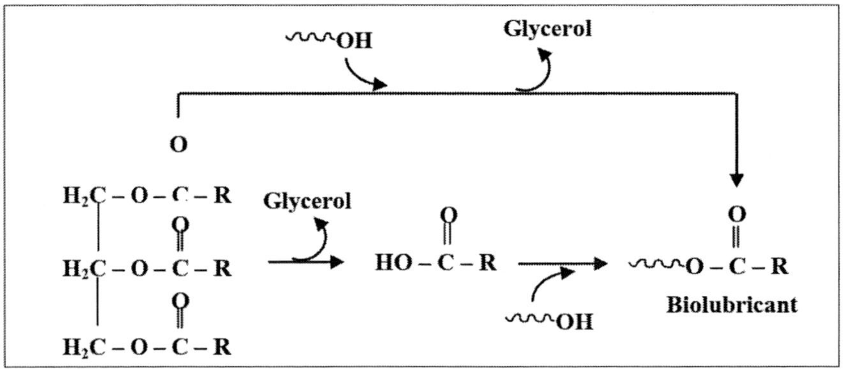

Fig. 3. Esterification Process Scheme (A) and transesterification (B) Bio-lubricant Production [22]

4.1. Transesterification using sodium methoxide catalyst

Aziz et al., [23] using raw materials palm oil methyl ester (POME) to produce a bio-lubricant. In this study used sodium methoxide catalyst at a concentration of 0.5 wt%, 0.75 wt%, 1.0 wt%, 1.25 wt% and 1.5 wt%. Experiments carried out by reacting palm oil methyl ester with pentaerythritol (PEE) and stirring with a speed of 700 rpm. Furthermore, the catalyst is added to the mixture and reacted until the designated time. Then the results obtained were separated from the soap and catalyst.

The existence of bio-lubricant products showed by the composition pentaerythrityol tetraoleate (PETO) and trimethylolpropane ester (TMPE) which is the result of a reaction between PEE and POME [24]. The optimal reaction conditions were obtained, namely at a temperature of 158°C, the catalyst concentration of 1.19% and a molar ratio of 4.5:1 with a reaction time of 60 minutes. From these optimal conditions, the results obtained by the yield of 37.56%, the viscosity at 40°C and 100°C. Zulkifli et al [25] testing the lubrication of biolubricant trimethylolpropane (TMP) and pentaerythritol ester (PE) obtained from POME (palm oil methyl ester) and of the friction value is obtained then this biopelumas fit for use in vehicle engines. Similarly, the results of research by using glycerol [26] and using trimethylolpropane ester [27]. However, to applied bio-lubricant commercially need to increase oxidative stability, thermal and hydrolytic.

4.2 Transesterification using oscillatory flow reactor and sodium methoxide catalyst

Koh et al., [28] undertake the production of bio-lubricant of palm methyl ester (PME) and trimethylolpropane (TMP) using sodium methoxide catalyst ($NaOCH_3$). Then do the transesterification reaction in the oscillatory flow reactor at 110°C to 150°C. An optimal reaction condition was obtained at a temperature of 140°C to be reacted for 25 minutes. Bio-lubricant obtained 94.6% by the concentration diesters 14.7% and 79.9% trimesters. However, the lower yield obtained from the research Yunus et al. [29] 2003) with a concentration of 4.6% diesters and trimesters 93.2%.

Although the results of a study by Koh et al., [28] obtained a high yield, kinematic viscosity and viscosity index which is not accordance with ISO. Bio-lubricant produced has a kinematic viscosity 47.1 cSt at 40°C, 9.0 cSt at 100°C and a viscosity index of 176. Meanwhile, based on ISO VG100 [30] lubricant should have a kinematic viscosity >90 cSt at 40°C, >4.1 cSt at 100°C and a viscosity index >216.

Reeves et al. [31] conduct research on the effect of fatty acid composition of the viscosity index contained in the avocado seed oil, canola oil, corn oil, olive oil, peanut oil, safflower oil, sesame oil and soy oil. The result showed that the fatty acid composition affects the viscosity grades, with the highest viscosity in peanut oil for 70.24 cP. Based on Zubaidah et al. [32] research used palm oil derivative compounds (trimetylolpropane) as bio-lubricant raw material. Production trimetylolpropane performed using a vacuum reactor with transesterification reaction. From this study showed that the highest conversions were achieved by 66% at a temperature operating conditions 120°C for 2 hours. Syaima et al. [33] produce bio-lubricant from palm oil mill effluent (POME). Production of bio-lubricant performed using enzymatic hydrolysis and esterification. From the research, optimum reaction conditions reached at 40°C, pH 7, stirring speed of 650 rpm, the composition of the enzyme 20 U/ml and 50% (v/v) POME.

The research results bio-lubricant using processes esterification/transesterification is shown in Table 4.

Table 4. Characteristics of Bio-lubricant using esterification/transesterification process [17]

Reactants	Product	Catalyst	Reaction conditions	Viscosity 40°C (cSt)	Viscosity 100°C (cSt)	Viscosity index	Pour point (°C)	Oxidative/ Thermal stability	Yield (%)
Jatropha and TMP	TMP triesters	Sodium methoxide	150°C, 10mbar, 3 h	43.90	8.71	180	-6	-	>80
Jatropha ME and TMP	TMP triesters	Sodium methoxide	150°C, 55 min	42.57	9.37	183	-6	325°C Degradation temp	-
High oleic palm ME and TMP	TMP triesters	Sodium methoxide	120-150°C, 0,3 mbar, 45 min	45.5 – 50.7	9.2 - 10	183 - 200	(-37) to (-9)	-	-
Palm ME and TMP	TMP triesters	Sodium methoxide	140°C, 25 mbar, 25 min, oscillatory flow reactor at 1,5 Hz with 20 mm amplitude	47.1	9.0	176	-2	355°C Degradation temp	94.6
Palm ME and TMP	TMP triesters	Calcium methoxide	180°C, 50 mbar, 8 h	-	-	-	-	-	92,38
Canola biodiesel ME and TMP	TMP triester	Sodium methoxide	110°C,1 mbar, 5 h	40.5	7.8	204	-66	Induction time: 0.74 h	90.9
Castor biodiesel and TMP	TMP triester	Dibutyltin dilaurate	170°C, 0.01 bar	287.2	26.13	119	-27	RPVOT: 43 min (Butylated hydroxytoluene added)	89.7
Castor biodiesel and TMP	TMP triester	Amberlyst 15 ionic exchange resin	120°C, 0.01 bar	20.94	4.467	127	-	-	-
Castor biodiesel and TMP	TMP triester	Sodium methoxide	120°C, 0.01 bar	11.28	3.100	141	-	RPVOT: 150 min (Butylated hydroxytoluene added)	-
WCO ME and TMP	TMP triester	KOH	128°C, 200 Pa, 1.5 h	38.60	8.44	204	-8	FP: 240°C	85.7

Soybean oil and various alcohols	n-alcohol-esters	Sulfated zirconia catalysts	140°C, 4 h	10.3 – 432.7	3.0 – 34.4	45 - 195	-	-	>80
Sunflower oil and octanol	FA-n-octyl esters	Fe-Zn double-metal cyanide (DMC) complexes	170°C, 8 h	7.93	2.74	226	-3	23 min (RBOT)	98
Pentaerythritol and oleic acid	Pentaerythritol tetraoletae ester	Ion exchange resin, Indion-130	110°C, 6 h toluene solvent	63.08	12.00	190	-24	-	-
Valeric acid TMP	Valeric acid TMP ester	Silica-sulphuric acid	70°C molar ratio of 3:1, toluene	9.5	2.5	80	-75	-	-
Rubber ME and NPG/TMP/PE	NPG/TMP/PE triesters	p-Toluensulphonic acid	135-140°C, until theoretical reaction complete	23,1-62,6	5.9-12.6	206-222	(-15) to (-3)	10-15 min (RBOT) FP: 266 - 308°C	94.5-96.5
Rapeseed ME and NPG/TMP/PE	NPG/TMP/PE triesters	C Antarctica lipase	150, 200, 50 h	7.8 – 38.2	2.7-8.4	205-224	(-31.3) to (-18)	Δv: 90.1-147.1 ΔAc: 2.9-7.7	98
Thumba ME, xylene and NPG/TMP/PE	NPG/TMP/PE triesters	p-Toluensulphonic acid	135-140°C, until complete	20.65-60.26	5.45-11.89	209-220	(-12) to (-3)	10-15 min (RBOT) FP: 270-318°C	89-95

5 Conclusion

Increasing the amount of environmental pollution can be solved by replacing petroleum lubricant with bio-lubricant those have made from vegetable oils. This is because the use of petroleum lubricant gives big influence on environmental pollution because of their lubricant spills which can damage the ecosystem. However, the production of bio-lubricant has not done commercially. Besides bio-lubricant properties are easily damaged so we need more research to improve the shelf life of the bio-lubricant. In addition, research must be done to improve the characteristics by adding a bio-lubricant additive.

References

1. Precidential Decree Republic of Indonesia No.5. *National Energy Policy*, (2006).
2. Y. Gerbig, S.I.U. Ahmed, F.A. Gerbig, H. Haefke. Journal of Synthesis Lubricant, **21**, 177-191 (2004).
3. BPS. *Produksi Bahan Bakar Minyak*, (2014).
4. S.C. Tung, M.L. McMillan. Tribology International Journal, **37**, 517-536 (2004).
5. S. Asadauskas, J.M. Perez, J.I. Duda. Lubricant Engineering, **52**, 877-82 (1996).
6. S.Z. Erhan, S. Asadauskas. Journal of Industrial Crops and Products, **11**, 277-282 (2000).
7. N.H. Jayadas, K.P. Nair, G. Ajithkumar, G. Tribology International Journal, **40**, 350-354 (2007).
8. S.M. Alves, B.S. Barros, M.F. Trajano, K.S.B Ribeiro, E. Moura. Tribology International Journal, **65**, 28-36 (2013).
9. P. Lathi, B. Mattiasson. Journal of Applied Catalyst B: Environment, **69**, 207-212 (2007).
10. P. Nagendramma, S. Kaul. Journal of Renewable and Sustainable Energy Review, **16**, 764-774 (2012).
11. M. Shahabuddin, H.H. Masjuki, M.A. Kalam, M.M.K. Bhuiya, H. Mehat. Journal of Industrial Crops and Products, **47**, 323-330 (2013).
12. K. Saroj, S.R.K. Padhi. Journal of Chemical Pharmacy, **3**, 39-49 (2011).
13. H.C. Ong, T.M.I. Mahlia, H.H. Masjuki, R.S. Norhasyima. Journal of Renewable and Sustainable Energy Review, **15**, 3501-3515 (2011).
14. R.V. Sharma, K.D. Ajay. Journal of Applied Catalyst B: Environment, **142**, 604-614 (2013).
15. D. Kania, Y. Robiah, O. Rozita, A.R. Suraya, M.J. Badrul. Journal of Petroleum Science and Engineering, **135**, 177-184 (2015).
16. Z.E. Sevim, A. Atanu, K.S. Brajendra. Journal of Synthetic Lubrication, **17**, 22 (2006).
17. J. McNutt, S.H. Quan. Journal of Industrial and Engineering Chemistry, **36**, 1-12 (2016).
18. Department of Ecology.. *Biolubricants. Environmentally Preferable Purchasing Fact Sheet,* Publication No. 13-07-011, (2011).

19. J.C.J. Bart, G. Emmanuele, C. Stefano. *Biolubricants Science and Technology*, (2013).
20. M. Hajar, V. Farzaneh. Journal of Industrial Crops and Products, **59**, 252-259 (2014).
21. M.F.M.G. Resul, T.I.M. Ghazi, A. Idris. Journal of Industrial Crops and Products, **38**, 87-92 (2012).
22. J. Oh, Y. Sungeun, K. Chanyeon, C. Inchang, H.K. Jae. Journal of Applied Catalysis A: General, **455**, 164-171 (2013).
23. N.A.M. Aziz, Y. Robiah, R. Umer, M.S. Azhari. Industrial Crops and Products, **62**, 305-312 (2014).
24. R. Yunus, A. Fakhru'l-Razi, T.L. Ooi, R. Omar, A. Idris. Industrial Engineering Chemical, **44**, 8178-8183 (2005).
25. N.W.M. Zulkifli, S.S.N. Azman, M.A. Kalam, H.H. Masjuki, R. Yunus, M. Gulzar. Tribology International, **93**, 555-562 (2016).
26. G.S. Dodos, D. Karonis, F. Zannikos, E. Lois. Journal of Industrial Crops and Products, **75**, 43-50 (2015).
27. H.A. Hamid, R. Yunus, U. Rashid, T.S.Y. Choong, A.H. Al-Muhtaseb. Chemical Engineering Journal, **200-202**, 532-540 (2012).
28. M.Y. Koh., I.M.G. Tinia, I. Azni. Journal of Industrial Crops and Products, **52**, 567-574 (2014).
29. M.S. Silva, E.L. Foletto, S.M. Alves, T.N.C. Dantas, A.A.D. Neto. Journal of Industrial Crops and Products, **69**, 362-370 (2015).
30. L.R. Rudnick, S.Z. Erhan. *Natural oils as lubricants*, (2006).
31. C.J. Reeves, L.M. Pradeep, J. Tien-Chien, R.L. Michael. Tribology International Journal, **90**, 123-134 (2015).
32. S.S. Zubaidah, A.C. Luqman, A. Fakhru'l-Razi. Journal of Applied Energy, 7 (15): 2002-2005 (2007).
33. M.T.S. Syaima, K.H. Ong, M.N. Ishenny, M.I.M. Zamratul, S.A. Brahim. Journal of Renewable and Sustainable Energy Reviews, 44: 669-675 (2015).

Performance of Dolomite Calcination in a Bench-Scale Rotary Kiln

Winny Wulandari*[,1], Subagjo[1], *Anggoro Tri* Mursito[2], *Fadly Junico* Juanjaya[1], and *Muhammad Faqih* Alwi[1]

[1] Department of Chemical Engineering, Faculty of Industrial Technology, Institut Teknologi Bandung, Jalan Ganesa No.10 Bandung 40132, Indonesia
[2] Research Centre for Geotechnology, Indonesia Institute of Sciences (LIPI), Jl. Cisitu Bandung 40135, Indonesia

Abstract. Dolomite calcination is one of process steps to prepare calcined dolomite for raw materials in magnesium production. Calcination of dolomite involves heating the raw material at sufficient temperature in order to release the carbon dioxide from its carbonate minerals. This process is commonly conducted in a rotary kiln. There have been a number of calcination studies in a laboratory scale, but the study of dolomite calcination in a larger scale is very scarce. This research is aimed to study the performance of dolomite calcination in a bench-scale rotary kiln with 500 gram of feed. The effect of various parameters, including temperatures, feed rate, rotating frequency, and particle size were examined. The temperature of rotary kiln was varied between 700 and 1000 °C, while the particle size of dolomite was varied between 0.149 – 0.297 mm and up to 10 – 15 mm. The temperature distribution inside the rotary kiln was also measured. It is obtained that a conversion of 92% was attained at operation temperature of 1000 °C, which is at a higher temperature compared to the laboratory scale, where a conversion of 100% was obtained at 900 °C. This imply that the effect of heat transfer also plays important role in the calcination of dolomite especially at a larger scale.

1 Introduction

Dolomite, $CaCO_3.MgCO_3$, is a mineral found in several places in Indonesia with an estimated reserve of 2,396,707 thousand tons, based on the data from the Ministry of Energy and Mineral Resources in 2010. Dolomites which are found in Gresik are relatively pure, which contains 40-45% of $MgCO_3$ and 55% of $CaCO_3$, with some impurities include SiO_2 (0,2%), Fe_2O_3 (0,65%), and Al_2O_3 (0,03%).

Dolomite calcination is an important step in the processing of dolomites into other materials of higher value such as fluxes and additives in steel processing smelters, kieserite ($MgSO_4.H_2O$) fertilizer, refractory bricks, sulfur dioxide capture medium in Flue Gas Desulfurization (FGD), and as the raw materials for magnesium production.

According to Samtani *et al* [1], dolomite begins to decompose into its oxide in one step in an inert atmosphere, as in the following reaction:

$$CaMg(CO_3)_{2(s)} \rightarrow CaO.MgO_{(s)} + 2CO_{2(g)} \quad (1)$$

The mechanism is different when dolomite decomposes in atmosphere containing CO_2, as it decomposes with two step reactions as in the following:

$$CaMg(CO_3)_{2(s)} \rightarrow CaCO_{3(s)} + MgO_{(s)} + CO_{2(g)} \quad (2)$$

$$CaCO_{3(s)} \rightarrow CaO_{(s)} + CO_{2(g)} \quad (3)$$

The study on dolomite decomposition in a TGA using nitrogen as the inert atmosphere by the authors [2] showed that decomposition of dolomite is a one step process that began at 700°C and finished at 850 °C with activation energy of 149,5 kJ/mol. This study also showed that the maximum conversion of dolomite calcination in a laboratory scale tubular furnace is 100% at 900 °C. Meanwhile, study on dolomite decomposition in tubular furnace using atmosphere consisting of air and CO_2, as an atmospheric replication in rotary kiln, shows that decomposition of dolomite occurs in two steps, starting at 650°C until 820°C with 162 kJ/mol of activation energy [3]. In addition, the maximum conversion of 80 – 90% in this atmosphere was achieved at 900 °C.

Even though the laboratory scale study shows that the conversion of dolomite decomposition reaching 99% at 900 °C, the rotary kiln in commercial scale operates at 1100 °C to 1200 °C, which means there is unnecessary energy losses. Often the rotary kiln is equipped with preheater to preheat the feed prior to the rotary kiln using the flue gas from combusted gas in the rotary kiln, and cooler to reduce the temperature of calcined materials. Only few literatures that shows the knowledge of rotary kiln process in the scale that larger than laboratory scale. The discrepancy in literature between bench-scale data and full-scale data is the motivation behind this research.

This research aims to understand the effect of temperature, atmospheric condition during reaction,

* Corresponding author: winny@che.itb.ac.id

particle size of dolomites, and rotating frequency of rotary kiln on dolomite conversion and distribution of residence time and temperature in rotary kiln.

2 Methodology

The experiment on calcination was carried out in four steps: sample preparation, determination of initial mass and composition of the sample, calcination, and determination of final mass and composition of the sample. It was carried out in rotary kiln at the specified temperature. Figure 1 shows the photograph of the bench scale rotary kiln. Several variables were varied in order to get rotary kiln performance data on dolomite calcination. The remaining sample at the end of calcination was weighed and analyzed to determine the conversion of dolomite. A final analysis was also performed to determine the final composition of the sample. Residence Time Distribution (RTD) of dolomite was determined by retaining samples exiting the rotary kiln over a certain interval. Actual temperature distribution inside the rotary kiln was determined by real time measurement using a thermocouple.

Fig. 1. Bench-scale rotary kiln

2.1 Procedures

2.1.1 Sample Preparation

Four categories of dolomite particle size were used: 0.149 – 0.297 mm; 1.19 mm – 2.83 mm; 2.83 – 6.73 mm; and 10 – 15 mm. Dolomites were crushed using jaw crusher and ball mill, then separated using mechanical sieve to obtain the desired particle size. As much as 500 g of sample was used for the next step.

2.1.2 Initial Composition Determination

Composition analysis was carried out qualitatively using X-Ray Diffraction (XRD). The output given by XRD were in the form of crystal structure data and constituents of dolomite. Dolomite sample were also analyzed using TG (thermogravimetry) to determine the mass reduction when heated to 950°C.

2.1.3 Measurement of Residence Time Distribution

The prepared sample with mass of 500 grams was fed into the rotary kiln. The sample exiting the kiln every 2 minutes were collected and weighed. The average residence time of the sample can be inferred from the residence time distribution.

2.1.4 Calcination

The kiln temperature is set to be constant at the specified temperature for each variation: 700°C, 800°C, 900°C, and 1000°C. The sample is then fed into the rotary kiln. The experiment is carried out in atmosphere consisting of ambient air and carbon dioxide (CO_2). CO_2 is fed into the kiln using a hose. The flow rate of CO_2 is set at 7 L/minute. The flow is measured using a flowmeter. Rotary kiln rotation frequency is also varied at 18.25; 29.75; 40.87; and 52.10 rpm. Calcined sample exiting the kiln was collected in a container and then weighed to determine the conversion of dolomite calcination.

The conversion of dolomite calcination is calculated by the amount of carbon dioxide that is released from the feed, which is assumed from the weight of initial feed subtracted by the weight of final sample.

2.1.5 Actual Temperature Distribution in Rotary kiln

The actual temperature distribution inside the rotary kiln is measured directly using thermocouple. Temperature measurement was carried out axially and radically. Axial temperature measurement is carried out every 10 cm until 60 cm from the edge of the rotary kiln. Radial temperature measurement was carried out in two points, in the center of the rotary section and 2 cm from the middle.

3 Results and Discussions

3.1 Analysis of the Feed

Figure 2 shows the X-ray diffraction analysis of dolomite. Dolomite contains mineral dolomite ($CaCO_3.MgCO_3$), calcite ($CaCO_3$), with impurities such as hematite (Fe_2O_3) and feldspar (silica minerals).

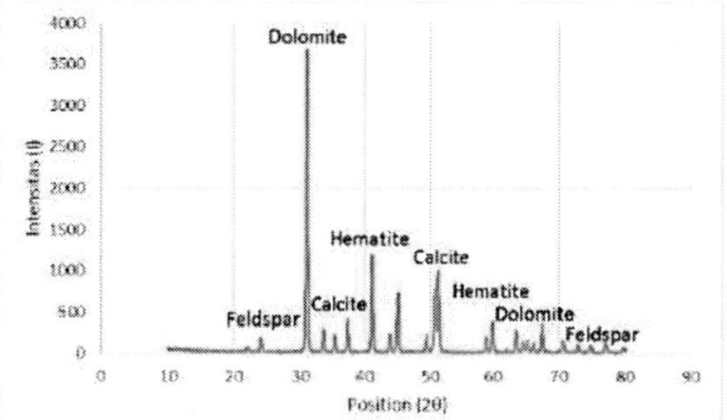

Fig. 2. XRD Analysis of Dolomite

The thermogravimetric analysis has been carried in the previous study by the authors [2] and shown in Figure 3.

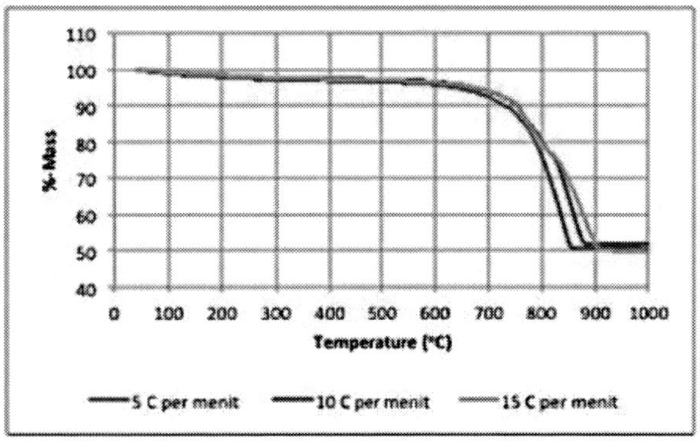

Fig. 3. TGA of Dolomite [2]

Dolomite in inert atmosphere started to decompose at temperature of 600 °C and finished the decomposition at temperature lower than 900 °C. The total mass loss of dolomite is approximately 50 %, which take accounts for the mass of carbon dioxide that is released from this carbonate minerals. For this reason, the variation of temperature in the bench experiments is from 700 to 900 °C.

3.1 Effect of Temperature

The effect of temperature on the conversion of dolomite calcination at various particle sizes is shown in Figure 4. The rotation frequency used in the experiment is 29.75 rpm with the inclination of 3°. The calcination experiments were conducted using various particle sizes of feed material.

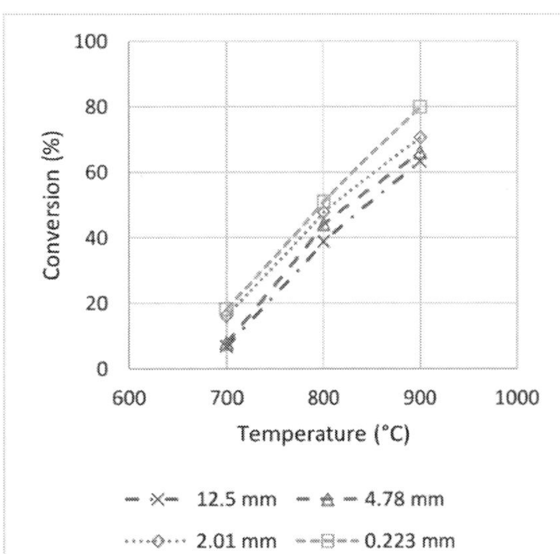

Fig. 4. Effect of temperature on calcination

Based on Figure 4, the conversion of dolomite calcination is ranging from 6 to 19% at 700 °C. The recovery increases to 62-80% at 900 °C. Compared to the laboratory scale, the recovery at 700 °C is much lower, only 20%. This may be caused by the materials is still in the heating zone, as the feed material enter the rotary kiln in a room temperature. Thus, at the rotary kiln temperature of 700 °C, the part of rotary kiln only heating the feed and when calcination begin to occur, the feed material has started to enter the outlet of the rotary kiln.

At a higher temperature, for example at 900 °C, the preheating of material feed occur faster, thus the calcination recovery is higher. The higher the reaction temperature of calcination, the higher the conversion of the reaction. The primary cause is more reactants has gained enough activation energy from higher operating temperature to allow it to undergo calcination. The reaction rate also increases with increasing temperature thus making the resulting calcination conversion higher. Temperature has the most significant effect compared to other process variables.

The effect of temperature on conversion of the reaction is studied further at 1000°C using sample with particle size of 1.19 – 2.83 mm. Figure 2 shows the change in conversion when temperature is set up to 1000°C.

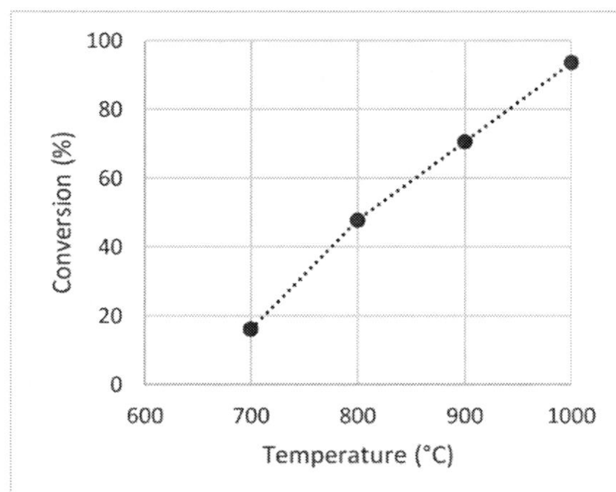

Fig. 5. Calcination conversion at 1000°C

Figure 5 shows that higher temperature up to 1000°C results in higher conversion. Compared to laboratory scale, the conversion of dolomite calcination at bench scale is obtained at higher temperature. For example, using inert gas, dolomite calcination conversion is 100% at 900 and by using CO_2 containing atmosphere, the conversion is 80-90% at 900 °C, while the calcination conversion at bench scale at 900 °C is 72 %. This discrepancy shows clearly that at bench scale, with larger capacity, will result in less recovery. The possibility of the cause may be from the heat transfer of the heat source to the center of particles, mass transfer, etc. This should be studied further.

https://doi.org/10.1051/matecconf/201815606008

3.2 Effect of Particle Size

The effect of dolomite particle size on reaction conversion is shown in Figure 6. The experiments were performed at rotation frequency of 29.75 rpm, 3° in inclination, and atmosphere consisting of air at 700°C, 800°C, and 900°C.

It can be seen that as the particle size decreases, the resulting conversion increases. This is in accordance with the theory that the diminution of particle size in the same mass causes the overall surface area of the particle to be larger. As the surface area of the particle increases, the incoming heat rate increases, thus with the same residence time, the greater the incoming heat. As the heat increases, more particles reach the activation energy need for calcination and are converted to calcium and magnesium oxide.

Particle size affects conversion less significantly compared to temperature because the heat transfer is due to heat transfer into dolomite particles is well under way. Rotary kiln rotation helps distribute the incoming heat to all the particles inside the kiln.

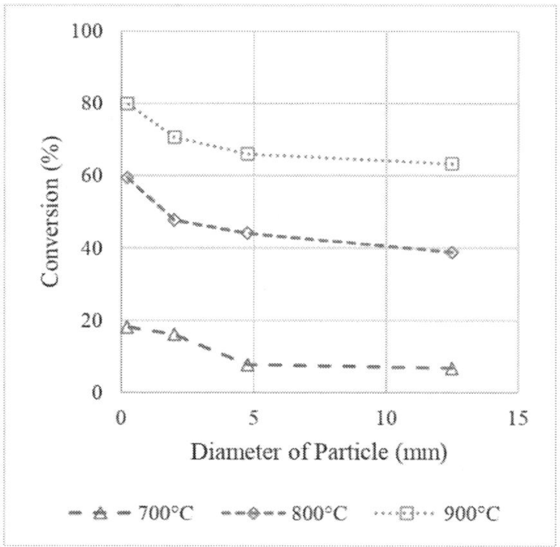

Fig. 6. Effect of particle size on calcination

3.2 Effect of Residence Time Distribution (RTD)

Dolomite sample RTD inside the kiln with a total mass of 500 grams is shown in Figure 7.

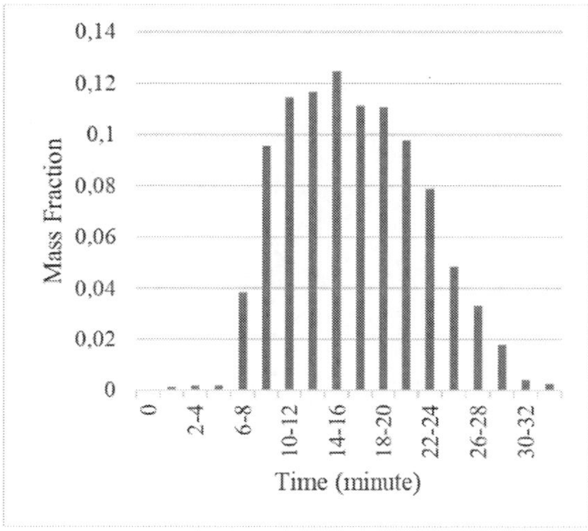

Fig. 7. Dolomite RTD inside the kiln

It can be inferred from Figure 4 that dolomite particles generally have a residence time 8 – 30 minutes. The average residence time obtained from the RTD is 17.5 minutes. The RTD pattern resembles normal distribution (Gaussian distribution). Average mass rate of dolomite is 28.4 grams/min based on the residence time.

3.3 Temperature Distribution profile

Temperature distribution measurements were carried out at three different temperature settings: 700°C, 800°C, and 900°C. Figure 8 and 9 shows the temperature distribution measurements at the center of rotary kiln (r = 0 cm) and at a radial distance of 2 cm from the center of the rotary kiln (r = 2 cm).

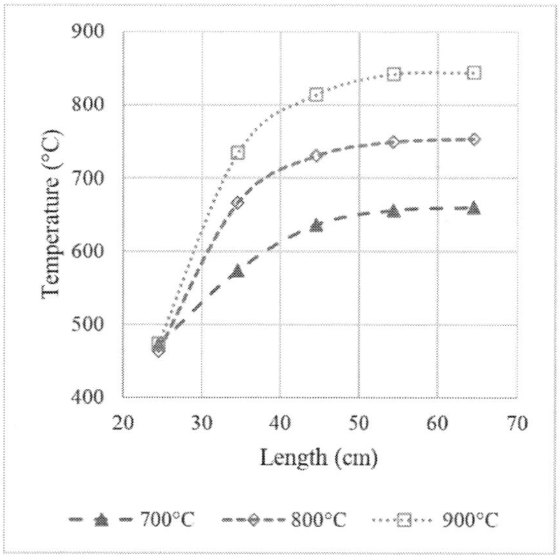

Fig. 8. Temperature distribution at the center of the kiln

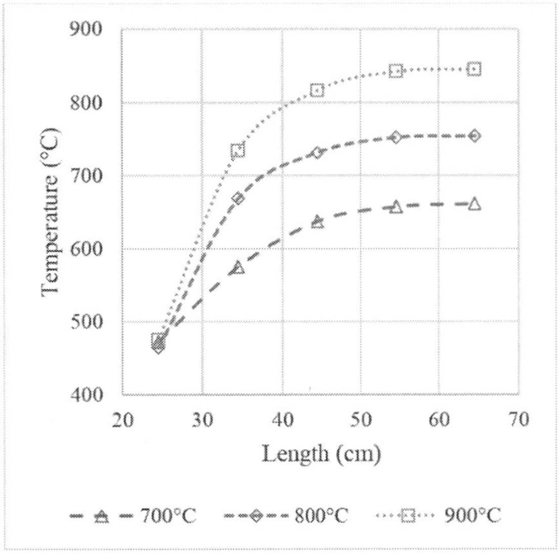

Fig. 9. Temperature distribution 2 cm from the center of the kiln

At the same radial length, the temperature increases over the axial length. At greater axial length, the thermocouple is placed further from the entry point which is at ambient temperature, thus making the temperature higher. At 50 cm from entry point, the temperature is starting to become constant, thus further measurement is unnecessary.

The radial temperature distribution is relatively unchanged compared to the axial temperature distribution. This is due to the small kiln diameter compared to the length of the kiln. The largest radial temperature difference is only 2°C.

According to the actual temperature distribution from Figure 5 and Figure 6, there is a temperature difference from the set point (roughly 50°C). This explains the phenomenon when the temperature set point is increased to 1000°C, the calcination conversion still increases up to 90% which is supposed to be achieved at roughly 950°C and should not go any higher according to the TG analysis.

3 Summary

The study of dolomite calcination at bench scale has been carried out. The effect of various parameters, including temperatures, feed rate, rotating frequency, and particle size were examined. The temperature of rotary kiln was varied between 700 and 1000 °C, while the particle size of dolomite was varied between 0.149 – 0.297 mm and up to 10 – 15 mm. The temperature distribution inside the rotary kiln was also measured. The experiments at bench scale enable us to obtain data that is differ from the data at the laboratory scale, and may act as a bridge to understand the phenomena of dolomite calcination at the commercial scale. Of many variables that has been examined, the conversion of dolomite calcination at bench scale reaches 72% at 900 °C and 93% at 1000 C. The lower conversion of dolomite calcination is postulated to be caused by larger capacity, with affect the heat transfer from the heat source to dolomite particles. Further research suggests to examined this phenomena into heat transfer modelling.

The authors would like to thank the Ministry of Research, Technology, and Higher Education of the Republic of Indonesia to fund this research through Advanced Higher Education Applied Research Program (Program Desentralisasi Penelitian Terapan Unggulan Perguruan Tinggi) 2017.

References

1. Samtani, M.; Dollimore, D.; Alexander, K.S., 2002. Thermochim. Act. **393**, (2002)
2. Subagjo, Wulandari, W., Fajrin, A., Aditama, P.M., AIP Conf. Proc. **1805,** (2017)
3. Wulandari, W., Subagjo, Rio, A, Istiadi, P., Seminar Nasional Teknik Kimia Indonesia (2015)

Methyl Butanoate Adsorption on MoS_2 Surface: A Density Functional Theory Investigation

Wahyu Aji Eko Prabowo[1, 2], *Supriadi* Rustad[1], *T.* Sutojo[1], Nugraha[2], Subagjo[3], *Hermawan Kresno* Dipojono[2, 4]*

[1]Dian Nuswantoro University, Department of Computer Science, 50131, Semarang, Indonesia
[2]Bandung Institute of Technology, Department of Engineering Physics, 40132, Bandung, Indonesia
[3]Bandung Institute of Technology, Department of Chemical Engineering, 40132, Bandung, Indonesia
[4]Research Center for Nanoscience and Nanotechnology, Bandung Institute of Technology, 40132, Bandung, Indonesia

Abstract. Methyl butanoate is one of the compound which is obtained from triglyceride molecule. It has hydrocarbon components and hence may produce hydrocarbon through hydrodeoxygenation (HDO) or decarbonylation (DCO) processes. The first step to uncover the underlying mechanism of HDO or DCO is to find the active site of methyl butanoate adsorption over the catalyst. This study attempts to investigate the active site of methyl butanoate adsorption on MoS_2 surface. Stable bonding configuration for methyl butanoate adsorption on MoS_2 is investigated by using density functional theory (DFT). This investigation consists of geometry optimisation and adsorption energy calculations. The stable configuration of methyl butanoate adsorption on MoS_2 surface is found to be on top of Mo atom in Mo-edge surface.

1 Introduction

Hydrotreating is a catalytic hydrogenation process which is aimed to remove the content of sulfur, nitrogen, oxygen and metals [1]. The hydrotreating catalyst consists of at least 2 metals (Mo or W), and (Co or Ni) as promoters.

Molybdenum disulfide (MoS_2) catalysts are well known as catalyst for refinery processes, especially for hydrodesulfurization (HDS) reactions. In HDS catalytic reactions, Mo-based catalyst reduces sulfur-containing compound from the feed stock. In advanced application, MoS_2 with Co / Ni as a promoters is used for production processes of hydrocarbon from feed stock with ultralow sulfur content for transportation fuels [2-7].

HDS process has a negative effect, especially air pollution and acid rain. It also cause the depletion of fossil fuel reserves. The solution for this problem is to find the alternative fuels. One of the alternative is palm oil based biofuel which is produced from organic material.

In order to improve the catalysts, a detailed understanding of the active site of MoS_2 surface and its interactions with typical molecules is a necessity. Experimental and theoretical studies of the catalysts has been done by many scientist [8-18]. Nevertheless, to the best of our knowledge, the catalytic interactions of MoS_2 with methyl butanoate have never been reported before.

Methyl butanoate ($C_5O_2H_{10}$) is an adsorbate derived from triglyceride (palm oil compound). It can be transformed into hydrocarbon with different cetane number, by hydrodeoxygenation (HDO) and decarbonylation (DCO) reactions. The main problem in hydrotreating process is catalytic activity. The first step to unveil further evidence is to investigate the interaction between methyl butanoate and MoS_2 surface. This study will be a first effort to do so. The density functional theory (DFT) [19], [20] based on ab initio computational method will be used for this purpose.

2 Computational details

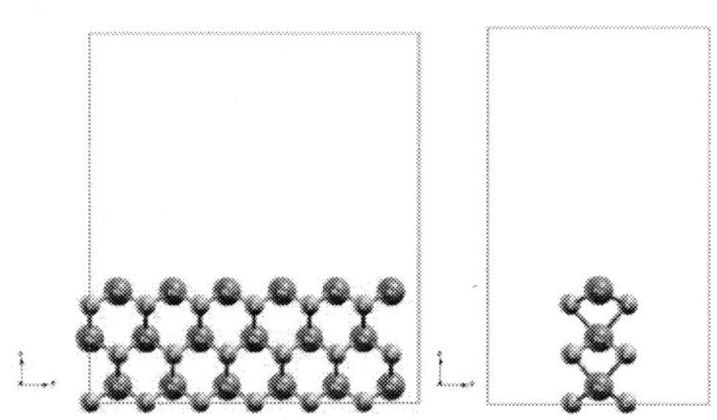

Fig. 1. Mo-edge MoS_2 surface. Mo atom (grey), S atom (gold)

* Corresponding author: dipojono@tf.itb.ac.id

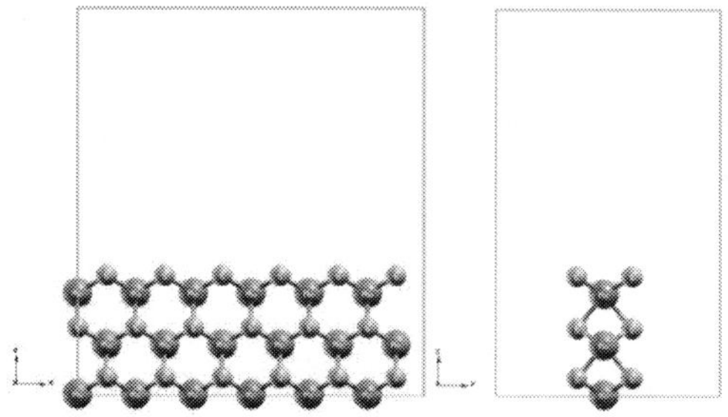

Fig. 2. S-edge MoS_2 surface. Mo atom (grey), S atom (gold)

The calculations are implemented in the open Source Package for Research in Electronic Structure, Simulation and Optimization (Quantum ESPRESSO) [21]. The ultrasoft pseudopotential method is employed to describe the interaction between ion cores and electrons. The electron exchange correlation is treated by a generalized-gradient approximation (GGA) based on Perdew, Burke, and Ernzerhof (PBE) functional [22]. The planewave basis set with a cut-off energy of 500 eV is used for all calculations. Supercell for this study is (18.96 x 12.29 x 21.28) $Å^3$. The Monkhorst-Pack method [23] is used to sample k-point by using 3 x 1 x 1 grid.

Figures 1 and 2 show the structure of MoS_2 surface. We use Mo-edge and S-edge (0001) model from Lauritsen et al [18]. In our model, the surface consists of 54 atoms; namely 18 Mo atoms and 36 S atoms. There are 6 layers in the surface. The difference between M-edge and S-edge is in the top of layer. In Mo-edge, the top of layer consist of Mo atoms, while in S-edge, the top of layer consist of S atoms. To prevent the interaction between molecules and periodicity of the system, we use vacuum space of ~16 Å.

Based on chemical formula, methyl butanoate consists of 17 atoms; namely 5 carbon atoms, 2 oxygen atoms, and 10 hydrogen atoms. Figure 3 shows the methyl butanoate model.

Fig. 3. Methyl butanoate model. C atom (yellow), O atom (red), H atom (light blue)

Interaction between MoS_2 surface and methyl butanoate are displayed in Figures 4 and 5. We choose this adsorption configuration because of the possibility of adsorption of methyl butanoate.

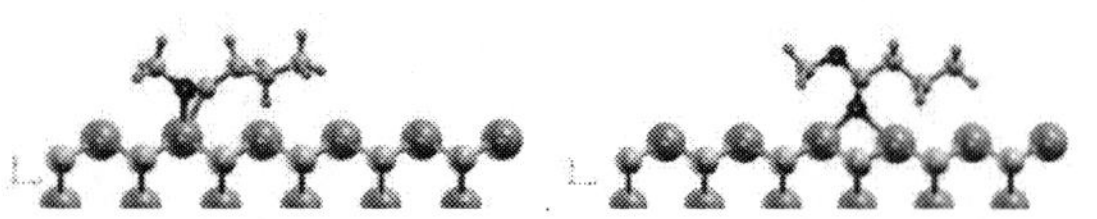

Fig. 4. Methyl butanoate adsorption on Mo-edge MoS_2 surface. Left: Methyl butanoate on the Top of Mo atom. Right: Methyl butanoate on the bridge of two Mo atom.

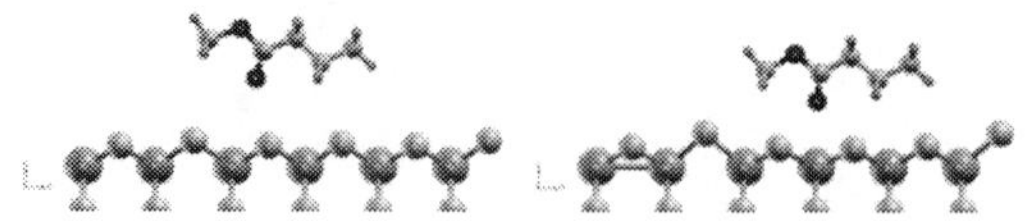

Fig. 5. Methyl butanoate adsorption on S-edge MoS_2 surface. Left: Methyl butanoate on the Top of S atom. Right: Methyl butanoate on the hollow of four S atom.

3 Results and discussion

3.1. Geometry optimisation

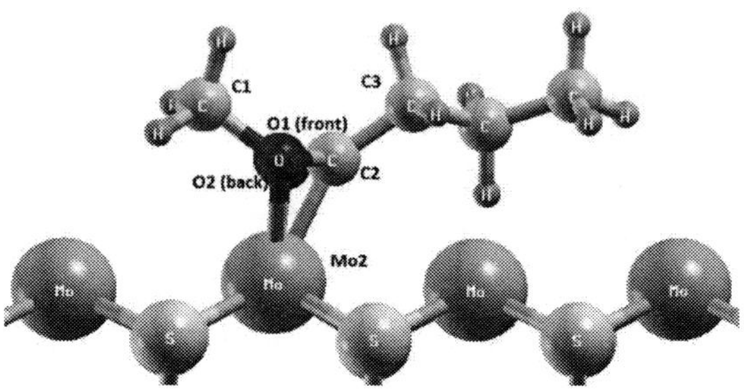

Fig. 6. Final position of methyl butanoate on Mo-edge surface (Top Mo site).

Figure 6 shows the final position of top Mo site. After the interaction with MoS_2 surface, the methyl butanoate molecule structure is slightly changed. In this figure, Mo_2 stands for the second Mo atom on the top of layer. For methyl butanoate, C_1, C_2, C_3 stands for the first, the second and the third C atoms, while O_1 and O_2 stand for the front and the back O atoms respectively.

Table 1. Geometry optimisation of Top Mo site (Mo-edge).

Bonding	Initial distance (Å)	Final distance (Å)	Δr (Å)
C_1-O_1	1.4552	1.4515	-0.0037
C_2-O_2	1.3061	1.2901	-0.0160
C_2-C_3	1.5183	1.5077	-0.0106
O_1-Mo_2	2.4122	2.2460	-0.1662
O_2-Mo_2	2.1942	2.2119	0.0177

Note: (+) bond's weakening

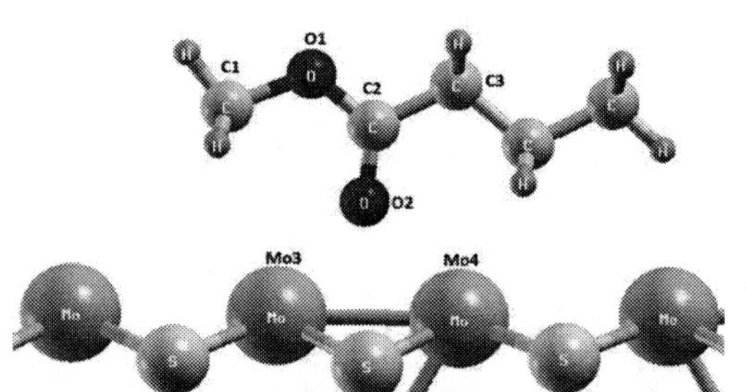

Fig. 7. Final position of methyl butanoate on Mo-edge surface (Bridge Mo site).

Figure 7 shows the final position of bridge Mo site. Similarly after interaction with MoS_2 surface, the methyl butanoate molecule structure is slightly changed. In this figure, methyl butanoate adsorp in the bridge of two Mo atoms, with label Mo_3 and Mo_4. The position of those two Mo atoms in the top of layer are in the third and fourth place from the left.

Table 2. Geometry optimisation of Bridge Mo site (Mo-edge).

Bonding	Initial distance (Å)	Final distance (Å)	Δr (Å)
C_1-O_1	1.4314	1.4564	0.0250
C_2-O_2	1.2233	1.2537	0.0304
C_2-C_3	1.5117	1.5021	-0.0096
O_2-Mo_3	2.0224	2.3451	0.3227
O_2-Mo_4	2.1731	2.4510	0.2779

Note: (+) bond's weakening

In top Mo site, only O_2-Mo_2 is elongated, indicating bond's weakening while other bonds are shortened after the interaction with the surface as shown in Table 1. Meanwhile, in bridge Mo site, only C_2-C_3 is shortened while other bonds are elongated as observed in Table 2. The O_2-Mo_3 bond length is stretched from previous value of 2.0224 Å to 2.3451 Å, indicating the presence of an active site on the surface and the occurrence of charge transfer or electron flow. Further discussion in the next section will clarify this observation.

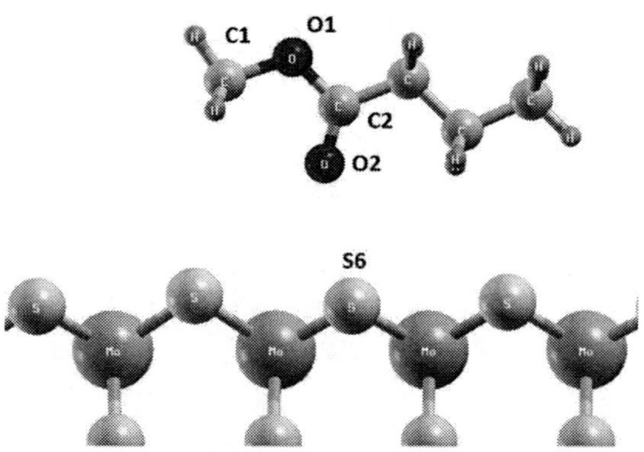

Fig. 8. Final position of methyl butanoate on S-edge surface (Top S site).

Table 3. Geometry optimisation of Top S site (S-edge).

Bonding	Initial distance (Å)	Final distance (Å)	Δr (Å)
C_1-O_1	1.4314	1.4462	0.0148
C_2-O_1	1.3662	1.3626	-0.0036
C_2-O_2	1.2233	1.2156	-0.0077
C_1-C_2	2.3661	1.5094	-0.0023
O_2-S_6	1.7002	3.3458	1.6456

Note: (+) bond's weakening

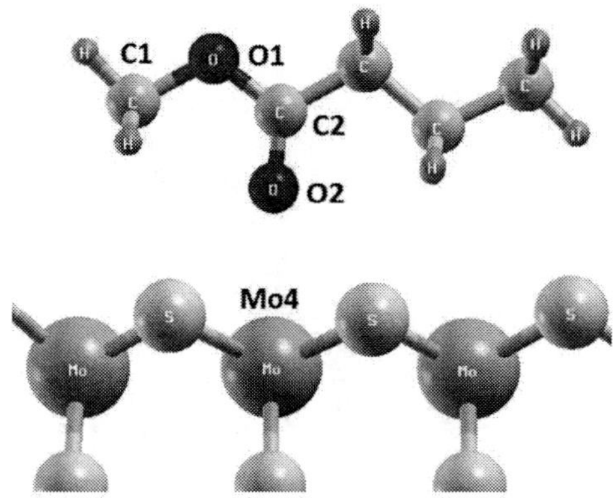

Fig. 9. Final position of methyl butanoate on S-edge surface (Hollow S site).

Table 4. Geometry optimisation of Hollow S site (S-edge).

Bonding	Initial distance (Å)	Final distance (Å)	Δr (Å)
C_1-O_1	1.4314	1.4498	0.0184
C_2-O_1	1.3662	1.3528	-0.0134
C_2-O_2	1.2233	1.2187	-0.0046
C_1-C_2	2.3661	2.3518	-0.0143
O_2-Mo_4	2.0941	2.8486	0.7545

Note: (+) bond's weakening

Figure 8 and 9 shows the final position of methyl butanoate adsorption on S-edge surface. After the interaction with S-edge surface, methyl butanoate molecule away from the surface. There are no bond between the methyl butanoate and surface. With oxygen atom as a reference, we can measure atomic distance between O atom in methyl butanoate with sulphur atom in surface. For Top S site, the distance between O atom and S atom is around 1.64 Å, while in hollow site, the distance between O atom and S atom is around 0.75 Å.

3.2. Adsorption Energy

Adsorption energies are computed by optimized gas-phase methyl butanoate, optimized MoS_2 surface and optimized MoS_2/methyl butanoate according to:

$$E_{ads} = E_{MoS2/methyl\ butanoate} - (E_{MoS2} + E_{methyl\ butanoate}) \quad (1)$$

E_{ads} = adsorption energy, $E_{MoS2/methyl\ butanoate}$ = total energy of methyl butanoate on MoS_2 surface, E_{MoS2} = total energy of MoS_2 surface and $E_{methyl\ butanoate}$ = total energy of isolated methyl butanoate.

Table 5. Adsorption energies of methyl butanoate on MoS_2.

Surface	Site	Adsorption energy (eV)
Mo-edge	Top Mo	-2.79
	Bridge Mo	-1.29
S-edge	Top S	-0.31
	Hollow S	-0.47

From our calculations for Mo-edge surface, methyl butanoate prefer to adsorp in Top Mo. The adsorption energy on top Mo site is 2.79 eV, while in bridge Mo site is 1.29 eV. For S-edge surface, methyl butanoate prefer to adsorp in hollow site, with the adsorption energy in this site is -0.47 eV.

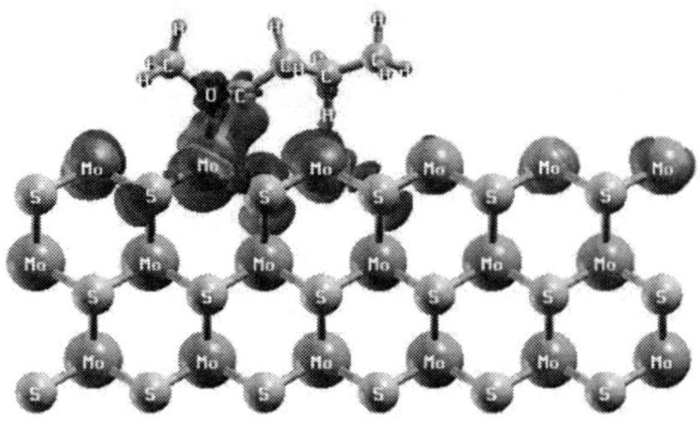

Fig. 10. Charge density difference at top Mo site. Red: charge accumulation. Blue: charge depletion. Isosurface = 0.01

Table 6. Charge density in specific atom at top Mo site

Atom	Initial Charge (e)	Final Charge (e)	Charge Difference (e)
O_1	4.6658	4.6961	0.0303
C_2	2.5539	2.6950	0.1411
Mo_2	6.7726	6.8713	0.0987
Mo_3	6.7725	6.8864	0.1139

Note: (-) transfer electron, (+) receive electron.

In Figure 10, the charge density concentrated near the bonding of Mo_2-O_2 and Mo_3-C_2. There are electron charge accumulation at that site, represent by red contour. There is an accumulation of charge between O_1-Mo_2 atoms. This indicates the presence of bonding between methyl butanoate molecules with Mo atom from MoS_2 surface. Methyl butanoate transfer charge around 0.04e to the surface. This is facilitated by the interaction of d orbitals from the surface with p orbitals of the methyl butanoate molecule.

Table 7. Charge density in specific atom at bridge Mo site

Atom	Initial Charge (e)	Final Charge (e)	Charge Difference (e)
O_2	4.7139	4.7046	-0.0093
Mo_3	6.7725	6.8580	0.0855
Mo_4	6.7725	6.8480	0.0755

Note: (-) transfer electron, (+) receive electron.

Figure 11 shows the electron charge density of bridge Mo site. In this case, we can't see the red contour near surface and methyl butanoate. This indicates that the bonding between methyl butanoate and MoS_2 surface at this site are weaker than that at top Mo site. These results are consistent with Table 7.

Figures 12 and 13 show the electron charge density of top S site and hollow S site. Both of them show that there are no bonding between methyl butanoate and surface. There is no significant difference for the charge density between before and after adsorption.

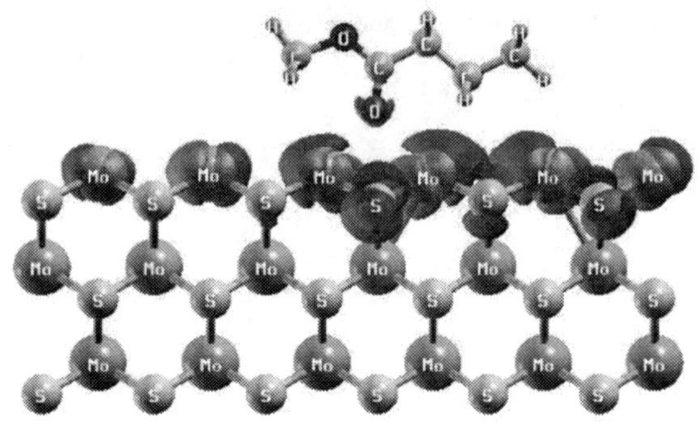

Fig. 11. Charge density difference at bridge Mo site. Red: charge accumulation. Blue: charge depletion. Isosurface = 0.01

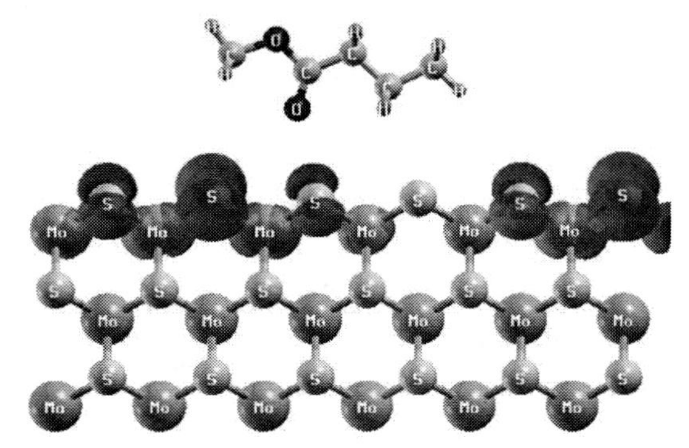

Fig. 12. Charge density difference at top S site. Red: charge accumulation. Blue: charge depletion. Isosurface = 0.01

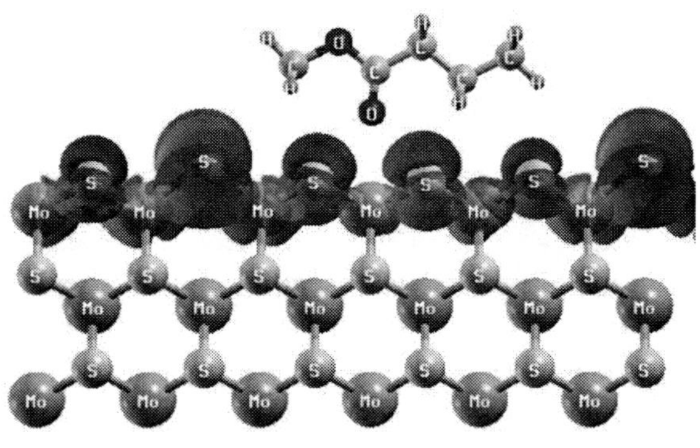

Fig. 13. Charge density difference at hollow S site. Red: charge accumulation. Blue: charge depletion. Isosurface = 0.01

4 Conclusion

Methyl butanoate prefers to adsorp on top Mo site, with the adsorption energy is around -2.79 eV. In this site, methyl butanoate forms a bond with the surface, and yields some shortened chemical bonds. Charge density plots show that there are some charge accumulations around Mo_2 of the surface, and at the vicinity of O_1 and C_2 of the methyl butanoate. At bridge site, charge density plots show that there is charge depletion between atom Mo_3 and atom O_2. Although it has charge depletion between molecule and surface, but there are no bond's formed between the molecule and the surface. The adsorption energy in this site is around -1.29 eV. This value does not only correspond to charge density, but also corresponds to Coulombic interaction between atom O_2 and atom Mo_3. For S-edge surface, the presence of S atoms on the top of layer make the molecule away from the surface because of the Coulombic interaction between S atom and O_2 atom.

Acknowledgement

The authors would like to express their gratitude to the full financial support from Directorate General of Higher Education of Republic of Indonesia through PUPT Research Grant Project. Wahyu Aji Eko Prabowo would like also to thanks and acknowledge the computer facility from Research Center for Nanoscience and Nanotechnology, Bandung Institute of Technology.

References

1. R. Prins, G. Ertl, H. Knzinger, *Handbook of Heterogeneous Catalysis*, Vol. 4, Wiley-VHC Verlagsgesellschaft, (1997).
2. I. Mochida, K. Sakanishi, X. Ma, S. Nagao, T. Isoda, Catal. Today **29** (1996) 185.
3. D. D. Whitehurst, T. Isoda, I. Mochida, Adv. Catal. **42** (1998) 345.
4. J. W. Gosselink, Cattech **4** (1998) 127.
5. H. Topsøe, K. G. Knudsen, L. S. Byskov, J. K. Nørskov, B. S. Clausen, Stud. Surf. Sci. Cata l. **121** (1999) 13.
6. T. Kabe, A. Ishihara, W. Qian, *Hydrodesulfurization and hydrogenation – Chemistry and Engineering*, Wiley-VCH, Kodansha, (1999).
7. E. Derouane, Cattech **5** (2001) 214.
8. R. Prins, V.H.J. de Beer, G.A. Somorjai, Catal. Rev.-Sci. Eng. **31** (1989) 1.
9. B.S. Clausen, B. Lengeler, R. Candia, J. Als-Nielsen, H. Topsøe, Bull. Soc. Chim. Belg. **90** (1981) 1249.
10. T. G. Parham, R.P. Merril, J. Catal. **85** (1984) 295.
11. S. Tauster, T. Pecoraro, R. Chianelli, J. Catal. **63** (1980) 515.
12. H. Topsøe, R. Candia, N.-Y. Topsøe, B.S. Clausen, Bull. Soc. Chim. Belg. **93** (1984) 783.
13. L.S. Byskov, J.K. Nørskov, B.S. Clausen, H. Topsøe, Catal. Lett. **64** (2000) 95.
14. P. Raybaud, J. Hafner, G. Kreese, S. Kasztelan, H. Toulhoat, J. Catal. **189** (2000) 129.
15. L.S. Byskov, M. Bollinger, J.K. Nørskov, B.S Clausen, H. Topsøe, J. Mol. Catal. A **163** (2000) 117.
16. H. Schweiger, P. Raybaud, G. Kreese, H. Toulhoat, J. Catal. **207** (2002) 76.
17. H. Orita, K. Uchida, N. Itoh, Applied Catalysis A : General **258** (2004) 115-120.
18. J. V. Lauritsen, M.V. Bollinger, E Lægsgaard, K.W. Jacobsen, J.K. Nørskov, B.S. Clausen, H. Topsøe, and F. Besenbacher, Journal of Catalysis **221** (2004) 510-522.
19. P. Hohenberg and W. Kohn, Phys. Rev. **136**, B 864, (1964).
20. W. Kohn and L.J. Sham, Phys. Rev. **140**, A1133, (1965).
21. P. Giannozzi, S. Baroni, N. Bonini, M. Calandra, R. Car, C. Cavazzoni, D. Ceresoli, G. L. Chiarotti, M. Cococcioni, I. Dabo, A. Dal Corso, S. Fabris, G. Fratesi, S. de Gironcoli, R. Gebauer, U. Gerstmann, C. Gougoussis, A. Kokalj, M. Lazzeri, L. Martin-Samos, N. Marzari, F. Mauri, R. Mazzarello, S. Paolini, A. Pasquarello, L. Paulatto, C. Sbraccia, S. Scandolo, G. Sclauzero, A. P. Seitsonen, A. Smogunov, P. Umari, R. M. Wentzcovitch, *QUANTUM ESPRESSO: a modular and open-source software project for quantum simulations of materials* J. Phys.: Condens. Matter, **21**, (2009) 395502.
22. J.P Perdew, K. Burke, M. Ernzerhof, Phys. Rev. Lett, **77**, 3865, (1996).
23. Monkhorst, H. J., Pack, J. D. Phys. Rev. B **13** (1976) 5188.

Synthesis of Mercapto Ethyl Ester of Palm Fatty Acid Distillate

I Dewa Gede Arsa Putrawan[1,*], *Adli* Azharuddin[1], *Kartini Ratna* Arum[1], *Dendy* Adityawarman[2], and *Dicka Ar* Rahim[3]

[1]Research Group on Chemical Product Design and Development, Institut Teknologi Bandung, Jl. Ganesha 10, Bandung, Indonesia
[2]Research Group on Chemical Process Design and Development, Institut Teknologi Bandung, Jl. Ganesha 10, Bandung, Indonesia
[3]Research Group on Energy and Processing System, Institut Teknologi Bandung, Jl. Ganesha 10, Bandung, Indonesia

Abstract. This research is aimed to study the synthesis of mercapto ethyl ester of fatty acid (MEFA) using palm fatty acid distillate (PFAD). The synthesis was carried out in a jacketed stirred vessel. Water as a byproduct was removed by using vacuum evaporation. p-Toluene sulfonic acid was used as catalyst. Temperature, mercaptoethanol excess, and catalyst amount were varied during the experiments. The variations ranges were 60-80°C, 5-25 %-mole, and 0.8-3.2 %-mole, respectively. The accomplishment of reaction was evaluated by measuring the mercaptan content and yield of MEFA. The mercaptan content and yield of MEFA were in the ranges of 6.4-8.0% and 64-86%, respectively. It was found that mercaptan content and yield increased with temperature. The optimum temperature was found to be 70°C, considering the mercaptan content, yield, and appearance of the product obtained. A full 2^2 factorial design at a reaction temperature of 70°C showed that mercaptoethanol excess and catalyst amount had significant effects on the mercaptan content and yield of MEFA.

1 Introduction

Mercapto ethyl ester of fatty acid (MEFA) is used as raw material to produce organotin thermal stabilizers for poly vinyl chloride (PVC). Thermal stabilizer is added to PVC resin to prevent degradation due to heat treatment ([1]-[4]). Without thermal stabilizer, PVC degrades when heated even at 70°C. The degradation process is also known as hydro-chlorination since it produces hydrogen chlorides. Thermal degradation changes PVC properties and color, from yellow to black, depending on the number of hydrogen chlorides liberated.

MEFA is produced from fatty acid and mercaptoethanol via the following reaction:

$$RCOOH + HSCH_2CH_2OH \rightleftharpoons RCOOCH_2CH_2SH + H_2O$$

The above is a liquid phase, equilibrium reaction. The esterification products are called reverse-esters as they are derived from the reaction between mercapto-alcohols and fatty acids instead of between alcohols and mercapto-fatty acids [5]. Although it has long been mentioned by Litser and Hecker [6] and Kegule [7], studies on synthesis of MEFA are very limited. Wang et al. [8] investigated the synthesis of MEFA from commercial oleic acid and found that vacuum evaporation is the best method to remove water. They also found that the optimum reaction conditions include mercaptoethanol molar excess of 20%, temperature range of 70~75°C, and catalyst amount of 0.5% of total weight. Knežvić et al. [9] studied the reaction between mercaptoethanol and methacrylic acid and found that the esterification products consist of O-esters as main products and S-esters as side products.

For industrial applications, the common source of fatty acids for MEFA is tall oil fatty acid (TOFA) and the reverse-esters produced are called mercapto ethyl tallate (MET). TOFA is a byproduct of the wood pulp industry and resulted when pine wood chips are digested, under pressure, with an alkaline solution of sodium sulfate or an acidic solution of sodium sulfite ([10], [11]). The alkaline digestion of pine woods produces soap which changes to fatty acid on acidification. The obtained fatty acids are refined by distillation to remove rosins. The refined fatty acids are called TOFA with mainly consists of oleic and linoleic acids.

Recently, TOFA is difficult to obtain due to environmental problems. An alternative, therefore, needs to find. Putrawan et al. [12] have synthesized MEFA from rice bran oil fatty acid and found that fatty acid derived from rice bran oil could be converted into MEAL with satisfied results. Although rice bran oil is possible to be available in the form of free fatty acids, its acid value is not quite high to be used as raw material for MEFA. It needs hydrolysis as an additional preparation step before converted into MEFA.

Palm fatty acid distillate (PFAD) is a byproduct of palm oil refining industry. Fig. 1 shows the schematic of palm oil refining process. PFAD is produced during distillation to remove free fatty acids and odors after

* Corresponding author: idewa@che.itb.ac.id

degumming and bleaching processes. PFAD mainly consists of free fatty acids. Like TOFA, PFAD is available in the form of free fatty acid and is produced as a byproduct. Its fatty acid composition, however, is different from that of TOFA. TOFA is dominated by unsaturated fatty acid, up to 90% ([10], [11], [13]) but PFAD has comparable contents in unsaturated and saturated fatty acids. PFAD is currently used by soap and fuel industries ([14], [15]).

This paper is aimed to study the utilization of PFAD as a substitute for TOFA in the production of MEFA as raw materials for PVC thermal stabilizers. The reaction was carried out in a batch stirred vessel under vacuum. The varied variables included temperature, mercaptoethanol molar excess and catalyst amount. The reaction performances were evaluated by measuring the mercaptan contents and yields of the product, which is named mercapto ethyl palmate (MEP) in this paper.

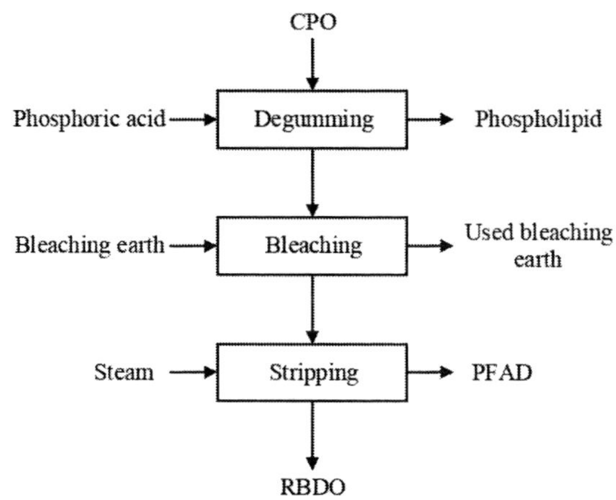

Fig. 1. Palm oil refining process.

2 Experiments

2.1 Materials

PFAD was obtained from a local palm oil refinery. Mercaptoethanol was purchased from Merck (>99% purity) and para-toluene sulfonic acid (pTSA) as catalyst from Fluka (>98% purity) and used as accepted.

2.2 Procedures

A batch jacketed glass beaker with a capacity of 1 L was used as a reactor. The top of the reactor was connected to a vacuum condenser. The temperature of the reaction mixture was controlled by circulating hot water with controlled temperature. The apparatus was arranged as shown in Fig. 2. PFAD and mercaptoethanol were mixed in the reactor and stirred rigorously by using a magnetic stirrer. After the reaction temperature achieved the desired value, catalyst was then added into the reactor. The reactor was vacuumed, started at 200 mmHg and decreased 50 mmHg at every 30 minutes. At the end of the reaction, the reactor was heated and hold at 90°C for 30 minutes. The product obtained was filtered, weighted for yield determination and sampled for measuring mercaptan content and acid value.

A full 2^k factorial design was applied to study the effects of mercaptoethanol excess and catalyst amount on the yield and mercaptan content of the product. The levels of each factor are shown in Table 1.

Table 1. Factor levels.

#	Factor	–	+
1	Mercaptoethanol excess (%-mole)	5	25
2	Catalyst amount (%-mole)	0,8	3,2

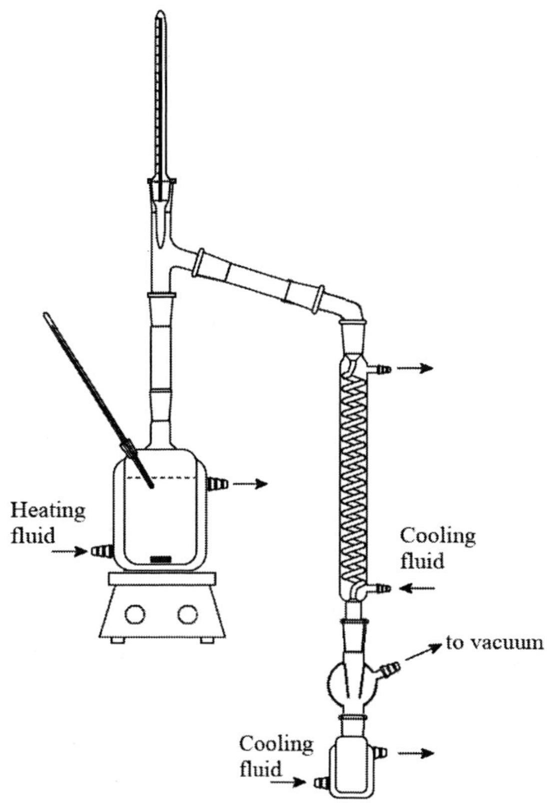

Fig. 2. Schematic of apparatus.

2.3 Analysis

Acid number was determined by direct titration (with KOH solution) of 1 g sample dissolved in 10 ml of neutral ethanol in presence of phenolphthalein as indicator. The end points for the esterification products were not clear by observing color. Because of that, the pH of solution during titration was also monitored using a YSI pH meter pH100 model. The pH curves obtained was utilized to determine titration end points. Saponification value was determined using KOH solution under reflux followed by titration using HCl.

https://doi.org/10.1051/matecconf/201815606010

Iodine value was determined using Wijs method. Titer value was determined according to AOAC Official Method 942.18. The mercaptan contents were determined by iodine titration [16]. IR Spectra was measured by a Shimadzu IR-Prestige-21 Spectrophotometer.

3 Results

3.1 PFAD characters

The PFAD sample used had acid, saponification, and iodine values of 184 mg KOH/g, 207 mg KOH/g, 54 mg I_2/g, respectively. As expected, the acid value was high showing the high content of free fatty acid. The saponification value is higher than the acid value, indicating that PFAD still contains neutral oil. The contents of main fatty acids, are given in Table 2 which also gives typical ranges for TOFA. The titer of PFAD used here was found to be 47°C. It is solid at room temperature. PFAD contains about the same portions of saturated and unsaturated fatty acids. The saturated fatty acids of PFAD are mainly palmitic and stearic acids having melting points of 63°C and 70°C, respectively ([11]). On the other hand, TOFA is approximately 90% unsaturated fatty acid, mainly oleic and linoleic acids having melting points of 16°C and -5°C, respectively ([11]). These are the reasons for PFAD to be solid and for TOFA to be liquid at room temperature.

Table 2. Composition of fatty acids.

Fatty acid	$C_{16.0}$	$C_{18.0}$	$C_{18.1}$	$C_{18.2}$
PFAD (this work)	33	6	31	15
TOFA ([1])	2,5		30-46	36-45

3.2 MEFA characters

Fig. 3 shows the IR spectra of mercaptoethanol and PFAD. Mercaptoethanol showed stretching OH and SH frequencies at 3357 and 2557 cm^{-1}, respectively. PFAD, as carboxylic acids, showed a characteristic peak at 1709 cm^{-1} representing C=O bond. The stretching OH frequency was showed as a broad band in the range of 2500~300 cm^{-1}. A peak at 2924 cm^{-1} representing a double bond C=CH, indicating the existence of unsaturated fatty acids. Fig. 4 shows the IR spectra of MEP. The IR spectra obtained revealed the characteristic peaks for MEFA structure, specifically at 2994 cm^{-1}, 2923 & 2853 cm^{-1}, 2570 cm^{-1}, 1740 cm^{-1} and 1235 cm^{-1}, for the functional groups of C=CH, CH_2, SH, C=O, and CO, respectively. Compared to the spectra of mercaptoethanol, the formation of SH functional group was found at 2570 cm^{-1}. The OH peak still appeared but much shorter. It is due to the formation of S-ester in which the mercaptan site of mercaptoethanol reacted with the acid site of PFAD. Fig. 4 also shows the IR spectra of commercial MET. The IR spectra of MEP and MET are found to be identical. The peak for a double bond can be seen more clearly in the spectra of MET than that of MEP. It was caused by the high content of unsaturated fatty acids in TOFA. Compared to PFAD, the peak for C=CH in MEP became weaker. This occurred as a part of double bound experience thiol-ene reaction with mercaptan ([17]). This side reaction is also possible to occur.

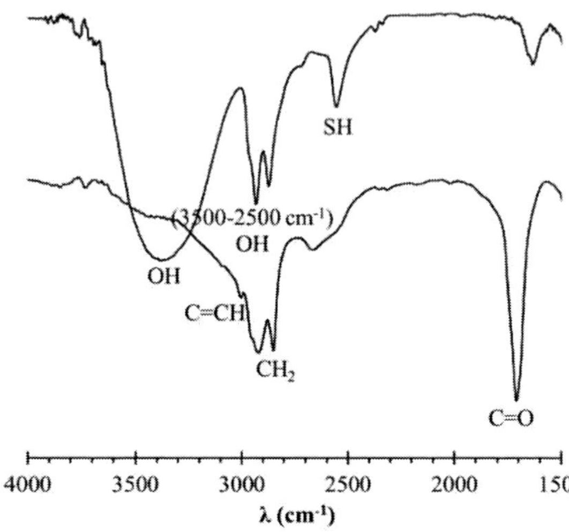

Fig. 3. IR spectra of ME and PFAD.

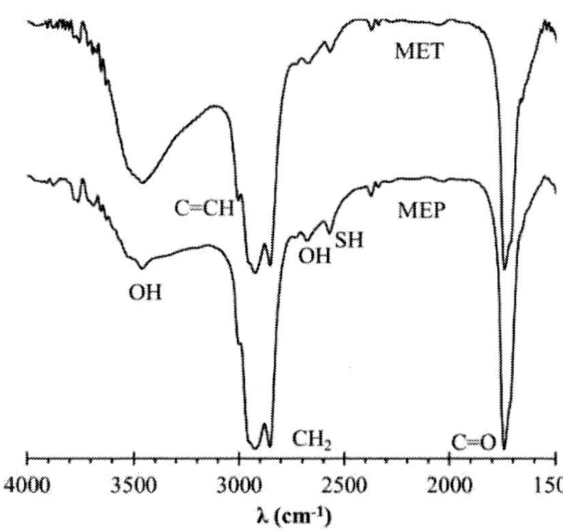

Fig. 4. IR spectra of ME and PFAD.

The product consisted of solid and liquid phases. Compared to PFAD, a decrease in titer was found. The addition of mercaptan functional group decreased melting point so that the product obtained was not totally solid at room temperature. In contrast, most of the product was liquid. It is commonly known that esterification lowered melting point. Palmitic acid, for example, has a melting point of 63°C but its methyl ester has a melting point of 31°C ([11]). Such decrease in melting point also found in the formation of reverse-ester. It occurs because the hydrogen bond of mercaptan is weak. The hydrogen bond of carboxylic acid, however, is strong.

3.3 Effects of temperature

Fig. 5, Fig. 6, and Fig. 7 show the effects of temperature on acid value, mercaptan content, and yield of MEFA. The measurement of acid value in MEFA to represent the residual fatty acid is not accurate since mercaptan also reacts with strong base. However, it could be used as a rough indicator to show significant decreases in fatty acid content have occurred, compared to PFAD.

Acid value decreased with temperature but mercaptan content and yield increased with temperature. Increasing temperature shifted the reaction to the product side as the higher temperature gives the molecules more kinetic energy. This means that the molecules move faster and be more likely to collide and to react successfully. In addition, increasing the temperature of reaction increased the volatility of water as a byproduct. When water evaporate faster, the reaction will shift to the product side. In the same reaction time, more product will be obtained at higher temperature.

The theoretical mercaptan content is 9.7% which corresponds to a yield of 100%. The mercaptan content found here was in the range of 7% to 8%, lower than the theoretical mercaptan content. About the same maximum mercaptan content was also found by Wang et al [8] using oleic acid. It was due the formation of S ester through the following reaction ([9], [18], [19]).

$$RCOOH + HSCH_2CH_2OH \rightarrow RCOSCH_2CH_2OH + H_2O$$

Mercaptoethanol contains two active sites, i.e., mercaptan and hydroxyl. However, hydroxyl is more active than mercaptan. MEFA (O-ester) is formed when the hydroxyl site of mercaptoethanol reacts with the carboxyl site of fatty acid. S-ester is formed when the mercaptan site of mercaptoethanol reacts with the carboxyl site of fatty acid. With a strong acid as catalyst, such as pTSA, the formation of S-ester could not avoided because strong acids have limited selectivity. The formation of this side product has limited the mercaptan content and yield of MEP.

Organotin thermal stabilizers stabilize PVC by binding the escaping hydrochloric acid [1]. They substitute unstable chlorine atoms in the polymer chain and prevent the formation of polyene. The sulfur site of MEFA has important role in stabilizing PVC. When neutralizing the hydrochloric acid, it will be attached onto polyene sequences and thereby disrupting the conjugated double bond system and improve the color of the PVC object. The higher the mercaptan content of MEFA the better for MEFA to use as raw materials for organotin stabilizers. As can be seen, the highest mercaptan content was obtained at 80°C. However, MEP obtained at this temperature solidified after aging. The solid found in MEPs which were synthesized at 60°C and 70°C could be easily melted by heating them. These solids were the saturated fractions of MEP. A part of solid found in MEP obtained at 80°C could not be melted and left as precipitates. Precipitates were also found by Wang et al [8], even using oleic acid, an unsaturated acid, at temperatures 80°C or above. It might be caused by the formation of polymers. The thiol–ene reaction between thiol and unsaturated bonds of oleic acid is a polymerization reaction [20]. Unsaturated fatty acids can also undergo self-condensation reactions to form high molecular weight dibasic and polybasic acids [21]. Thus, 70°C was considered as the optimum temperature for PFAD which is in the range of optimum temperature found by Wang et al. [8].

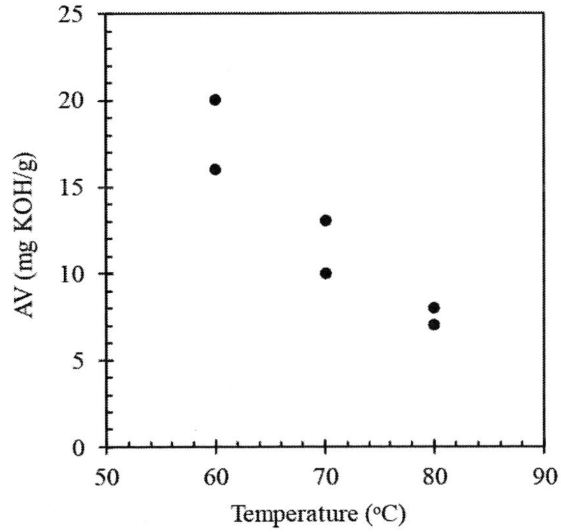

Fig. 5. Effect of temperature on acid value.

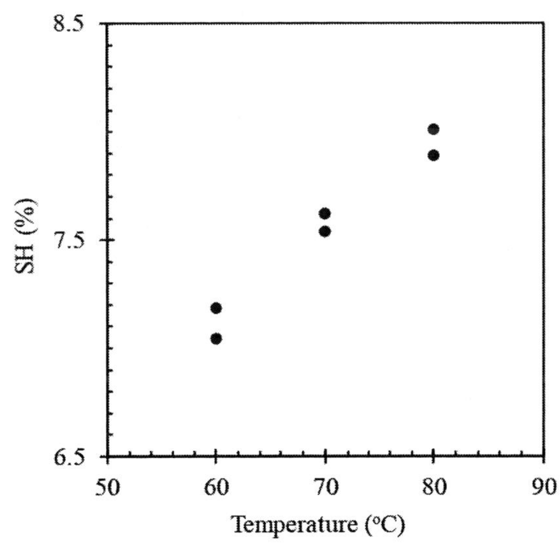

Fig. 6. Effect of temperature on mercaptan content.

3.4 Effects of mercaptoethanol and catalyst

A full 2^2 factorial design with center points was used to study the effects of mercaptoethanol excess and catalyst amount on the mercaptan content and yield of MEFA. The effects of catalyst amount and mercaptoethanol excess on these two responses are presented in Fig. 8 and Fig. 9, respectively. Both mercaptoethanol excess and catalyst amount had positive effects on mercaptan and yield. Paying attention to the slopes of response lines, Fig. 8 at a glance shows that mercaptoethanol excess and catalyst amount had no interaction effect. On the other hand, Fig. 9 shows that an interaction occurred between mercaptoethanol excess and catalyst amount.

Table 3 and Table 4 present the analysis of factorial designs for mercaptan content and yield as responses, respectively. In general, a model for relating responses and factors can be written as ([22]):

$$y = \beta_o + \beta_1 \cdot x_1 + \beta_2 \cdot x_2 + \beta_{12} \cdot x_1 \cdot x_2 \quad (1)$$

where y is response, x is factor, and β is parameter.

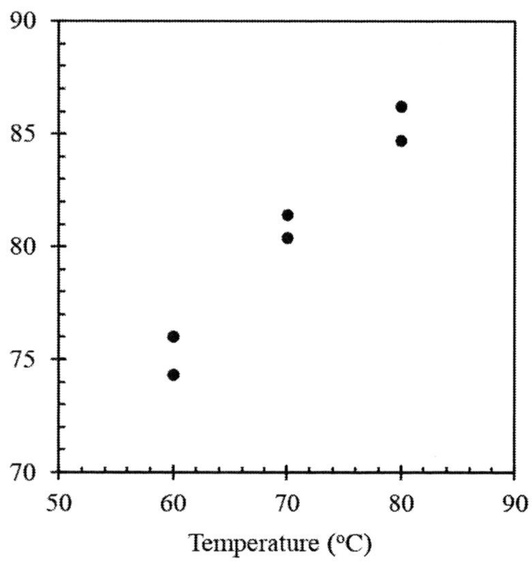

Fig. 7. Effect of temperature on yield.

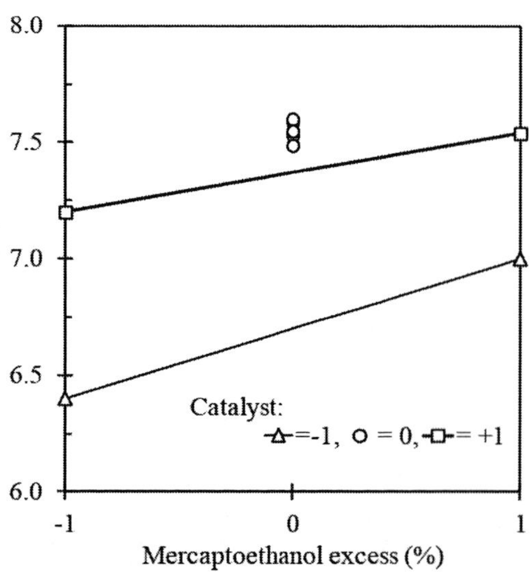

Fig. 8. Effects of mercaptoethanol excess and catalyst amount on mercaptan content.

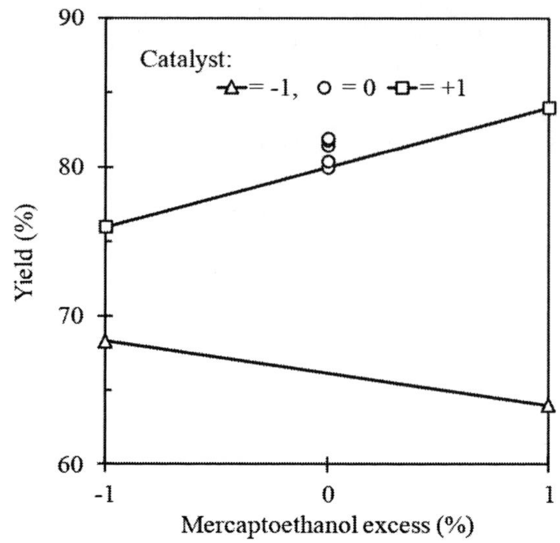

Fig. 9. Effects of mercaptoethanol excess and catalyst amount on yield.

The coefficients of the model in equation (1) are confined by the following confidence intervals:

$$CI_{(\beta i)} = \beta_i \pm t_{n-1}\, s/\sqrt{n} \quad (2)$$

where t_{n-1} is t distribution, s is standard deviation, and n is degree of freedom. From the center point data, standard deviations for mercaptan and yield were found to be 0.04 and 0.89, with four degrees of freedom. For a confidence level of 95%, the confidence intervals for mercaptan content and yield are given bellow.

Mercaptan content:

$$CI_{(\beta i)} = \beta_i \pm 0.06 \quad (3)$$

Yield:

$$CI_{(\beta i)} = \beta_i \pm 1.10 \quad (4)$$

Table 3. Analysis of factorial design for mercaptan content.

Run	M (%)	C (%)	M*C	SH (%)
1	-1	-1	1	6.40
2	1	-1	-1	7.00
3	-1	1	-1	7.20
4	1	1	1	7.54
5	0	0	0	7.53
6	0	0	0	7.49
7	0	0	0	7.59
8	0	0	0	7.60
9	0	0	0	7.55
Σ (-)	13.60	13.40	14.20	
Σ (+)	14.54	14.74	13.94	
Effect	0.47	0.67	-0.12	

Table 4. Analysis of factorial design for yield.

Run	M (%)	C (%)	M*C	Y (%)
1	-1	-1	1	68.3
2	1	-1	-1	64.0
3	-1	1	-1	76.0
4	1	1	1	84.0
5	0	0	0	80.0
6	0	0	0	80.4
7	0	0	0	81.5
8	0	0	0	81.8
9	0	0	0	82.0
Σ (-)	144.30	132.30	140.00	
Σ (+)	148.00	160.00	152.30	
Effect	1.85	13.85	6.15	

It was found that both mercaptoethanol excess and catalyst amount as single factors significantly affected mercaptan content and yield. The interaction factor between mercaptoethanol excess and catalyst amount had no significant effects on mercaptan content but had significant effects on yield. The models for mercaptan content (SH) and yield (Y), therefore, are given as:

$$SH = 7.03 + 0.29M + 0.26C \quad (5)$$

$$Y = 72.40 + 1.86M + 6.70C + 3.05MC \quad (6)$$

where M and C stand for mercaptoethanol excess and catalyst amount, respectively, in coded factors.

Higher mercaptoethanol excess and higher catalyst amount increased mercaptan content and yield. This indicated that increasing mercaptoethanol or catalyst amount shifts the reaction to the MEFA side. Increasing mercaptoethanol means increasing reactant. For equilibrium reaction, this will shift the reaction toward the product side. As hydroxyl is more active than mercaptan, this also means increasing the possibility for the formation of O-ester, i.e., MEP in this case. An increase in MEP was observed as an increase in mercaptan content and yield. The nucleophilic acyl substitution of carboxylic acids with alcohols proceeds via two steps [18]: the attack of the alcohol's electron-rich oxygen atom on the electron-deficient carbonyl carbon, forming a tetrahedral intermediate with a delocalized charged transition state, and the loss of the carboxylic acid's -OH leaving group, resulting in the esterified carbonyl group and water. Increasing catalyst amount increased the rate of reaction steps. As mentioned earlier, reaction pressure was decreased gradually. As each pressure was hold in the same time interval, higher reaction rate resulted in higher yields. Mercaptan content also increased with catalyst amount because pTSA as catalyst favored the formation of O-ester, as mentioned by Knežvić et al [9]. F-test showed that the effects of mercaptoethanol excess and catalyst amount exhibited a curvature. A central composite design is therefore necessary to elaborate the models.

4 Conclusions

Mercapto ethyl ester (reverse-ester) of palm fatty acid distillate has been synthesized in a jacketed stirred vessel under vacuum using pTSA as catalyst. It was found that mercaptan content and yield increased with temperature. A full 2^2 factorial design showed that catalyst amount and mercaptoethanol excess increased mercaptan content and yield significantly. The mercaptan content and yield of the reverse-ester obtained were in the ranges of 6.4-8.0% and 64-86%, respectively.

This research was supported by the Indonesian Estate Crop Fund for Palm Oil (BPDPKS) under GRS 2016.

References

1. M. Schiller, PVC Additives (Hanser Publisher, Munich, 2015).
2. E. Arkiş, D. Barköse, Polymer Degrad. and Stability, **88**, 46-51 (2005).
3. W. H. Starnes, Procedia Chem., **4**, 1-10 (2012).
4. S. Hope, I. Sewell, *South African Vinyls Association Conference*, Johannesburg, South Africa (2014).
5. G. F. Beekman, L. R. Price, Eur Pat 0850987A1 (1999).
6. W. E. Leistner, A. C. Hecker, US Pat 2870182A (1959).
7. T. G. Kugele, US Paten 4 120 845 (1959).
8. A. Wang, J. Li, J. Li, X. Zhong, S. Long, Chinese Plastic Additives, **91** (1), 34-37 (2012).
9. M. Knežvić, L. Katsikas, I. G. Popović, Hem. Ind., **59**(11-12), 321-323 (2005).
10. F. D. Gunstone, *Chemistry of oils and fats* (CRC Press, Oxford, 2004)
11. F. Shahidi, *Bailey's Industrial Oil and Fat Products, Edible Oil and Fat Products: Chemistry, Properties, and Health Effects* (John Wiley & Sons, New Jersey, 2005).
12. I. D. G. A. Putrawan, D. K. Bestari, C. A. Wicaksana, *Chemical Engineering Seminar* (Parahyangan University, Bandung, 2016).
13. R. L. Logan, JAOCS, **56** (11), 777A-779A (1979).
14. N. H. Embong, G. P. Maniam, M. H. A. Rahim, K. T. Lee, D. Huisingh, J Clean. Prod., **116**, 244-248 (2016).
15. A. G. M. Top, Lipid Technol., **22** (1), 11-13 (2010).
16. G. H. Jeffery, J. Bassett, J. Mendham, R. C. Denney, *Vogel's Textbook of Quantitative Chemical Analysis,* 5th edn (Longman Scientific & Technical, England, 1989).
17. S. K. Sharma, A. Mudhoo, *A Handbook of Applied Biopolymer Technology: Synthesis, Degradation and Applications* (Royal Society of Chemistry, Rajasthan, India, 2011)
18. D. L. Compton, M. A. Jackson, JAOCS, **88**, 1799-1805 (2011).

19. P. E. Sonnet, G. G. Moore, Lipids **24**(8), pp. 743-745 (1989).
20. J. Samuelsson, M. Jonsson, T. Brinck, M. Johansson, J. Poly. Sci., **42**(24), 6346-6352 (2004).
21. D. J. Anneken, S. Both, R. Christoph, G. Fieg, U. Steinberner, A. Westfechtel, *Ullmann's Encyclopedia of Industrial Chemistry: Fatty Acid* (Wiley-VCH Verlag GmbH & Co, Weinheim, 2006).
22. M. C. Montgomery, *Design and Analysis of Experiments* (John Willey & Sons, New York, 2001).

Microwave-assisted Extraction of Natural Dyes from *Coleus atropurpureus* Leaves: The Effect of Solvent

Selfina Gala, *Sumarno* Sumarno, and *Mahfud* Mahfud*

Department of Chemical Engineering, Institut Teknologi Sepuluh Nopember, Surabaya, Indonesia
*Corresponding authors: mahfud@chem-eng.its.ac.id (M. Mahfud); selfinagala90@gmail.com (S. Gala)

Abstract. A microwave-assisted extraction system was used to extract natural dyes from *Coleus atropurpureus* leaves. The effects of water solvent of the temperature at 100°C and 80°C as well as ethanol solvent 96% were investigated. This is done at the various material to solvent ratio (0.001, 0.0025, 0.005 g/mL) and microwave power (264, 400, 600 watt). The optimal conditions of the extraction were obtained on a water solvent of the temperature at 100°C with material to solvent ratio 0.001 g/mL, irradiation power 400 watt, extraction time 10 min, material size 35-60 mesh, the optimum yield of 44.48%. Identification of groups of pigments contained in the *Coleus atropurpureus* leaves is known categories tannins and flavonoids. Fourier Transform Infrared Spectroscopy was used to identify the major chemical groups in the extracted dyes.

Keywords: ***Coleus atropurpureus*, extraction, natural dye, microwave, solvent**

1 Introduction

The dye is an organic compound of natural or synthetic sources which are used to give color to the object. Natural dye (pigment) is a dye that is naturally present in plants, animals and minerals. Until now, the main source of natural dyes derived from plants [1-2]. The dye used in the textiles, food, beverages, pharmaceuticals, cosmetics, ink printers, home appliances and others. The use of dyestuffs is inevitable in the interests of the industry because color has a meaning and a very important role where the color is the appeal of a product also as a quality attribute, especially on textile products.

Currently, the needs of textile products increased thereby increasing the use of synthetic dyes in the textile industry. Synthetic dye is widely used because it is more easily obtained and practical use and the resulting color is more diverse. In addition the use of a small amount is sufficient to provide the desired color. But the use of synthetic dyes pose a new problem that environmental pollution. Disposal of synthetic dye waste without first processing will pollute the environment, especially the surrounding aquatic environment.

Natural dyes become widespread with the increasing awareness of the negative impact of synthetic dyes on the environment and the wearer. Advantages of natural dyes are non-toxic, non-carcinogenic, renewable resources, environmentally friendly, reducing pollution and no allergic reaction on the skin. Some natural sources produce really beautiful dyes for textile applications and economical price compared to synthetic dyes [3-5]. In addition, the plant-based natural dyestuffs also function as antibacterial, antioxidant and anticarcinogenic due to the presence of a pigmental group containing phenolic components [6-10].

One of the plants that can be explored for natural dyes is *Coleus atropurpureus.* This plant is widely grown in the tropics, especially in Asia including Indonesia and has a variety of colors on the leaves (red, yellow, pink, green and maroon). The name of these plants vary widely and have different names with different subspecies such as *C. blumei, C. atropurpureus, C. scutellarioides, C. bicolor, C. versschaffelti and C. hybridus.* However, in some publications use the name *C. blumei* [11]. The pigment content present in this coleus leaf is tannin [11-12], flavonoid and tannin [13].

At first all natural coloring materials obtained by traditional extraction method is by boiling raw materials and water as a solvent. For laboratory scale is usually done by conventional methods such as reflux and soxletation. This method requires a long time, the relatively large amount of solvent to extract low yield [5], [14-18]. It is encouraging technology innovation extraction process performed by using microwaves as an

* Corresponding author: mahfud@chem-eng.its.ac.id; selfinagala90@gmail.com

energy source with optimal results in a relatively short time. This method also includes "green techniques" for natural dye extraction process.

In this study, the solvents used are water and ethanol that are environmentally safe, cheap and have relatively large dieletric constants. Therefore, this study aims to determine the effect of solvent on the results of microwave-assisted extraction.

2 Materials and Metods

2.1 Materials and chemicals

The material used in this research, *Coleus atropurpureus* leaves was obtained from South Sulawesi, Indonesia. The powdered *Coleus atropurpureus* leaf were passed through >35 mesh filter. Analytical grade of ethanol (96% v/v) and distilled water as a solvent.
Hydrochloric acid, ferric chloride, sodium chloride and Mg powders is used for phytochemical test. All chemicals used were obtained from domestic distributors in Indonesia.

2.2 Microwave-assisted extraction (MAE)

The tool used a domestic microwave oven Electrolux model EMM2007X with 2450 MHz Magnetron frequency, maximum delivered power of 800 W, 220 V voltage. Oven dimension with length 46.1 cm, width 28.0 cm and height 37.3 cm. The extraction process is carried out in a three round neck flask extractor equipped with a condenser placed at the top of the oven. The temperature inside the extractor flask was measured with K 1/16 inc thermocoppel type fitted to a microwave oven.

Raw materials was milled, sifted to a certain size (35-60 mesh). Powder and solvent was fed into extractor. Water flowed on the cooling system (condensor). After the extraction process in the oven is complete, the extract then filtered with filter paper (whatman no.1, 90 mm diameter) using a vacuum filter.The obtained extract solution was concentrated and dried at a temperature of 60-80°C to a constant weight.

2.3 Determination of extraction yield of *Coleus atropurpureus* leaves (CAL)

The yield of Sample produced from MAE, extractions were calculated by:

$$y = \frac{S}{W} \times 100 \quad (1)$$

where y is extraction yield (%, w/w), S is mass of extracted yield (g), and W is mass of *Coleus atropurpureus* leaves (g).

3 Results and discussion

One important factor that affects the extraction results with a microwave is the choice of solvent. Selection of suitable solvents can make the extraction process work more efficiently. In general, the capacity of the solvent to absorb the microwave energy will be high when the solvent used has a high dielectric constant values [19]. The value of the dielectric constant indicates the ability of the solvent to be polarized by an external electric field and can be regarded as a relative measure of microwave energy density [20]. The dielectric constant also plays an important role in determining the interaction between the electric field to the matrix, so the higher the dielectric constant value that is owned by the solvent, the solvent will be getting better at absorbing microwave energy.

3.1 Effect of solvent on extraction

In this study used ethanol and distilled water as solvent. Water solvent is an environmentally friendly solvent because it is not toxic, non-mutagenic, and is not carcinogenic [21]. While the ethanol solvent selection is based on its polarity and has a high dielectric constant value of about 24.3 [22], which is quite good in microwave absorption. Ethanol is the safest alcohol used because it has at least toxic properties that can be used for the consumer product industry [23]. Also important is that because at atmospheric conditions the boiling point of ethanol is 78°C and does not degrade the pigment (natural dye).

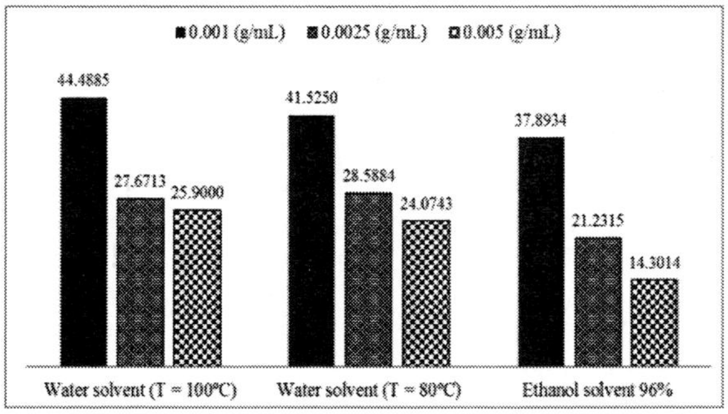

Fig 1. Effect of solvent type on the yield of CAL extraction on various material to solvent ratio (microwave power of 400 watt, time of 10 min, material size of >35 mesh)

Figure 1 shows the best condition for the use of water solvent with temperature in extractor controlled 100°C, material to solvent ratio at 0.001 g/mL with an optimum yield of 44.4885%. This can happen because the water has polarity properties and high dielectric constant is approximately 80.4 [22], so it is very good in microwave absorption. In addition, it is also possible because the leaves contain lot of bioactive including pigment dyes that are more soluble in water than ethanol.

Figure 1 also shows that the ratio of the material to the smaller solvent in the extraction system can cause the solvent to be more in contact with the material. This leads to more dye that can be extracted. If the use of the volume of solvent that more will cause cell swelling or materials so that the pressure pushes the cell wall, stretch and facilitate the breakdown of cell walls and the release of the dye into the solvent [24-25].

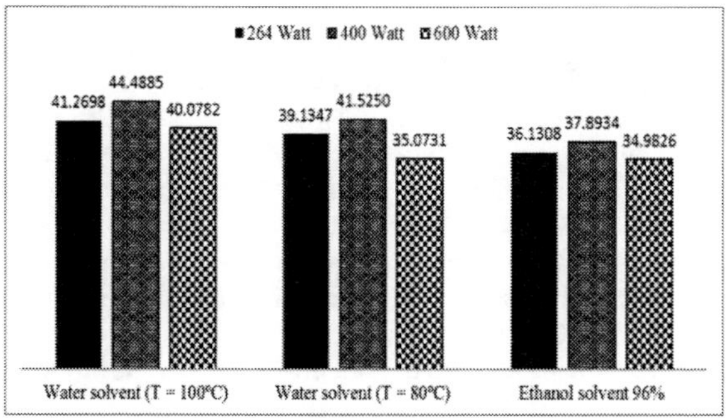

Fig 2. Effect of solvent type on the yield of CAL extraction on various microwave power (material to solvent ratio of 0.001 g/mL, time of 10 min, material size of >35 mesh)

Figure 2 shows that the power of 600 watt was obtained a smaller yield than to power 264 W and 400 W. This is due to the greater the power of microwave used then the system temperature of the extraction process will be more rapid increase and reached the boiling point of the solvent used. Achievement of the boiling point in faster time results in faster solvent evaporation resulting in reduced solvent volume to penetrate into the material.

3.2 Qualitative analysis extracted dyes

This research is conducted qualitative test pigment component which is found in extracts by phytochemical test (pigment identification) and Fourier Transform Infrared Spectrometry (the identification functional groups of extracted dye).

3.2.1 Phytochemical test

Phytochemical test performed on extracts produced by the method of cold maceration (extraction without the provision of heat) and the MAE (The temperature in the extractor around 100°C). The result of phytochemical test identified that flavonoids, tannins, and quinones are found in *Coleus atropurpureus* leaves. This can be shown in table 1.

Table 1. Pigment identification of *Coleus atropurpureus* leaves

Types of material	Types of pigment	Reagents	Observation	Test results	
				Maceration (2 days)	MAE (10 min)
Coleus atropurpureus leaves	Flavonoid	Mg powder + HCl	Yellow	+	++
	Tannin	NaCl + $FeCl_3$	Blackish green	+	+
	Quinone	NaOH	Yellowish green	-	-

Note: - = Not detected
\+ = Strong
++ = Stronger

3.2.2 Characterization of the extracted dye with FTIR Spectrometer (Fourier Transform Infra Red)

Qualitative analysis of functional groups on the raw material and extract the dye after extraction with MAE is done by interpreting the absorption peaks of the infrared spectrum. In Figure 3. shows the FTIR spectra of raw materials and extract, where the results of these spectra are used to observe changes in the structure or functional group of substances after extraction with microwave.

The results show a ribbon stretching extract (vibration strain) -OH, but the -OH group on raw materials (Figure 3a) indicate the presence of free -OH groups (provide absorption with greater frequency with sharp intensity). While in Figure 3b shows the formation of group -OH bound (the hydrogen bonds) are on the uptake with lower frequency and width and a shift in the absorption band.

(a)

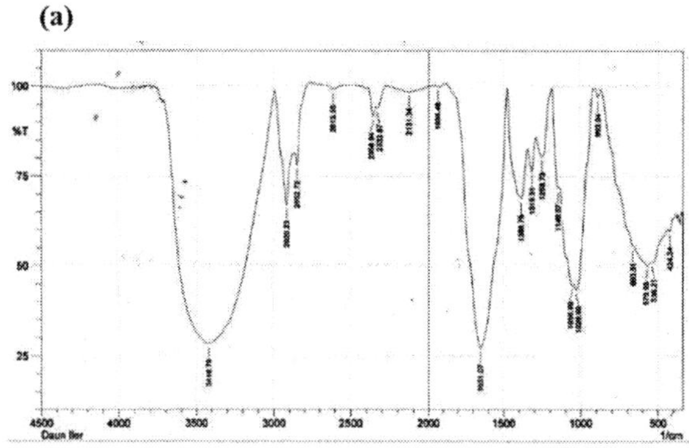

(b)

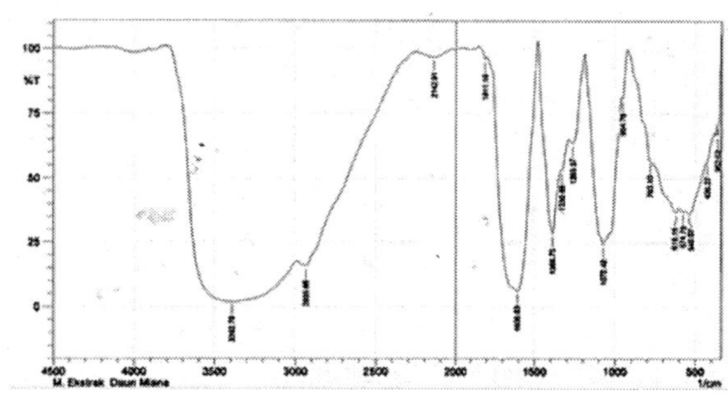

Fig 3. FTIR spectra of *Coleus atropurpureus* leaves: (a) before the extraction, (b) After the extraction with MAE (microwave power of 400 watt, extraction time for 10 min)

The FTIR spectra analysis of the extracted dye after the extraction with MAE show distinct peaks at 3392, 2935, 2142, 1811, 1606, 1330 –1388, 1263, 1072, 954 cm^{-1} respectively (Fig 3). The broad and strong band at 3392 cm^{-1} can be attributed to bonded –OH groups. The peak at 2935 cm^{-1} is indicative of C–H stretching vibration, –C $\equiv$ N stretching of 2142 cm^{-1}, C=O stretching of 1811 cm^{-1}, C=C stretching of 1608 cm^{-1}, C–H bending of 1330-1388 cm^{-1}, O–H *bending* of 1263 cm^{-1}, C–O *stretching* of 1072 cm^{-1}, C–C *stretching* of 954 cm^{-1}.

In FTIR spectrometry analysis, it is identified that there is a functional groups of the chromophore such as carbonyl groups (C=O) and ethylene groups (C=C). Functional groups of auxochrome, hydroxyl groups (-OH) **[27].**

Conclusion

Extraction of natural dyes from *Coleus atropurpureus* leaves have been successfully performed using MAE. The optimum condition is obtained on the use of water solvent with temperature control of 100°C, material to solvent ratio of 0.001 g/mL, microwave power of 400 watt with extraction yield of 44.49%. The identification of pigments with phytochemical test shows the presence of pigment tannin and flavonoid on the extract dyes. In FTIR spectrometry test, it is identified that there is a functional group of the chromophore such as carbonyl groups (C=O) and ethylene group (C=C). Group of auxochrome, hydroxyl (-OH).

Acknowledgement

This work was financially supported by Ministry of Technology Research and Higher Education of Indonesia (1523/K9/KT.03/2017).

References

1. H. Goodarzian, E. Ekrami, World Appl Sci J, **8**, 11 (2010)
2. L.G. Angelini, I. Pistelli, P. Belloni, A. Bertoli, S. Panconesi, Rubia Tinctorum a source of natural dyes: Agronomic evaluation, Ind. Crops Prod., **6** (1997)
3. S. Ali, T. Hussain, R. Nawaz, *J. Cleaner Prod.*, **17**, 1 (2009).
4. P. Leitner, C. Fitz-Binder, A. Mahmud-Ali, T. Bechtold, *Dyes Pigm.*, **93**, 1-3 (2012)
5. K. Sinha, P.D. Saha, S. Datta, Ind. Crops Prod., **37**, 1, (2012)
6. R.M. Selvam, G. Athinarayanan, A.U.R. Nanthini, A.J.A. Ranjit Singh, K. Kalirajan, P.M. Selvakumar, Ind. Crops Prod., **70** (2015).
7. M.P. Sathianarayanan, N. Bhat, S.S. Kokate, V.E. Walunj, *Antibacterial finish for cotton fabric from herbal product* (India Fiber Text Res, 2010).
8. F. Anwar, M. Ali, A.I. Hussain, M. Shahid, Flavour Fragrance J., **24**, 4 (2009)
9. A.P. Kulkarni and S.M. Aradhya, Food Chem. **93**, 2 (2005)
10. M. Riaz, N. Rasool, I.H. Bukhari, K. Rizwan, F. Javed, A.A. Altaf, H.M.A. Qayyum, Oxid. Commun. **36**, 1 (2013)
11. A. Nagpal, B. Singh, S. Sharma, G. Rani, V.G. Singh, *Medicinal and Aromatic Plant Science and Biotechnology*, (Global Science Books, 2008)
12. D. Rizal and W.D.R. Putri, Journal of Food and Agro-Industry, **2**, 4 (2014)
13. Y. Ridwan and Y.Q. Ayunita, Journal of Protein, **14**, 1 (2007)
14. M.F.B. Vazquez, I.R. Comini, R.E. Martini, M.S.C. Nunez, S. Bottini, J.L.Cabrera, Ultrason. Sonochem. **21**,2 (2014)
15. S. Umale and P.A. Mahanwar, Int. J. Basic Appl. Sci. **12**, 1 (2012)
16. W. Xiao, L. Han, B. Shi, *Sep. Purif. Technol.* **62** (2008)
17. K. Sinha, S. Chowdhury, P.D. Saha, S. Datta, Ind. Crops Prod., 41(2013)
18. I.V. Karabegovic, S.S. Stojicevic, D.T. Velickovic, N.C. Nokolic, M.L. Lazic, World Academy of Science, Engineering and Technology, International Journal of Biological, Biomolecular, Agricultural, Food and Biotechnological Engineering. **6**, 11 (2012)
19 G. Spigno, and D.M. De Faveri, J. Food Eng., **93** (2009)
20 G.G. Raju, *Dielectrics in Electric Fields* (Dekker, New York, 2003)
21 Material safety data sheet, *Water MSDS*,(2005).
22 A.C. Metaxas, *Foundations of Electroheat: A unified Approach* (Wiley, New York, 1996)
23. Material safety data sheet, *Ethyl Alcohol*, (2001)
24 I. Calinescu, C. Ciuculescu, M. Popescu, S. Bajenaru, G. Epure, RICCCE, **12**, (2001)
25 J.P. Maran, V. Sivakumar, K. Thirugnanasambandham, R. Sridhar, *Carbohydr. Polym.*, **97**, 2, (2013)
26 U. Supratman, *Structure elucidation of organic compounds: spectroscopic methods for structure determination of organic compounds*, (Publisher Widya Padjadjaran, Bandung, 2010)

Effect of Catalyst Pellet-Diameter and Basicity on Transesterification of Soybean Oil into Biodiesel using $K_2O/CaO-ZnO$ Catalyst over Hybrid Catalytic-Plasma Reactor

I. Istadi, *Luqman* Buchori*, *Brigitta B.T.* Putri, and *Henrikus I.A.* Hantara

Department of Chemical Engineering, Faculty of Engineering, Diponegoro University, Semarang, Indonesia

Abstract. This research is aimed to study the effect of catalyst pellet-diameter and catalyst basicity on the transesterification process of soybean oil into biodiesel over a hybrid catalytic-plasma reactor. Various catalyst diameters (3, 5, and 7 mm) were tested in this reaction system. Catalyst basicity was also examined by comparing fresh and used catalyst as well as with and without K_2O promoter. All catalysts testing were performed in a hybrid plasma-catalytic reactor (dielectric barrier discharge – DBD type). From the results, the synergistic effects roles of the catalyst and the plasma in the transesterification process are important, in which the energetic electrons within plasma assist the reaction on the catalyst surface by an exciting bonded electron. The catalyst basicity was influenced by the composition of CaO on the catalyst as well as roles of the alkaline K_2O promoter. Catalyst basicity is important in producing biodiesel with high performance. Yield of fatty acid alkyl ester (FAAE) or biodiesel is slightly influenced by the catalyst diameter within the range of diameter studied.

Keywords: biodiesel; catalyst pellet-diameter; DBD plasma reactor; $K_2O/CaO-ZnO$ catalyst

1 Introduction

Currently, energy consumption is increasing as economic and population growth globally. Fossil energy as a non-renewable energy source is the world's major energy source. The serious problem faced by developing countries is the limited fossil fuel, while the need for fossil fuels is increasing [1], resulting in an energy crisis. To overcome the energy crisis and to support the government programs to achieve national energy security, the alternatives to fossil fuel sources from renewable fuel sources are needed. One of the energy alternatives is biodiesel which is consisting of fatty acid alkyl ester (FAAE), biodegradable, non-toxic, and has low emissions [2].

Conventionally, biodiesel was produced by transesterification of triglyceride from vegetable oils or animal fats with alcohols using a homogeneous or a heterogeneous catalyst. The heterogeneous catalyst can be used several times with easier end product separation, can be recycled, minimize costs of raw material and production, more environmentally friendly, and can be applied to batch or continuous processes without the need of purification [3-4]. However, the homogeneous catalysts were more widely used due to high conversion and fast reaction. The heterogeneous catalysts can simplify the production process and refining processes, reduce the amount of waste water, reduce the size of process equipment, reduce environmental problems and process costs [4].

Among the alkaline earth metal oxides, CaO is the most heterogeneous base catalyst used for transesterification due to capable of producing 98% FAME yield in the first reaction cycle [6], but it needs longer reaction time than the homogeneous one. ZnO is more commonly used in the transesterification process, because it exhibits a good yield of biodiesel [7,8]. When the ZnO catalyst was impregnated with alkaline earth metals, it showed good catalyst activity for the soybean oil transesterification process [9]. To increase basicity of the alkaline earth-based heterogeneous catalysts (CaO, BaO, MgO, SrO), a catalyst promotor is required. The alkali metal (Li, Na, or K) can be used as a promotor of the catalyst to increase its basicity by impregnation method [4,10]. The mixed of ZnO and CaO metal oxides prepared with co-precipitation followed by impregnation by alkali oxide (K_2O) may increase the surface area and catalyst basicity [4].

The transesterification process by continuous process is expected due to the following advantages, i.e.: easier reaction control, relatively small space requirements, can be used for large capacity, produce high yield and easy scaling for large-scale production [11]. Some additional technologies were applied in order to enhance the catalytic reactor performance, i.e. microwave irradiation, ultrasonic, electro-catalysis method [12], and a hybrid plasma-catalytic method [4]. The hybrid plasma-catalytic

* Corresponding author: luqman.buchori@che.undip.ac.id

process is promising because the high voltage plasma provides high energetic electrons to assist the catalytic process by exciting the bonded electrons of the reactant molecules in the catalyst surface [4,12]. In addition, the presence of a catalyst can increase the contact surface area between the reactant molecules and the high-energetic electrons so that the breakdown of chemical bonds becomes more efficient. Therefore, particle size of the catalyst within discharge zone is important to be studied. This research is aimed to study the effect of catalyst pellet-diameter and basicity of catalyst on the transesterification process of soybean oil into biodiesel over a hybrid catalytic-plasma reactor.

2 Materials and Method

2.1 Materials

This experimental works used Happy™ soybean oil and methanol (Merck, 99.9%) as the raw materials of biodiesel production. $K_2O/CaO\text{-}ZnO$ catalyst was used which was prepared from precursors of calcium nitrate tetrahydrate ($Ca(NO_3)_2.4H_2O$) Merck, 99%), zinc nitrate tetrahydrate ($Zn(NO_3)_2.4H_2O$) (Merck, 98,5%), potassium nitrate KNO_3 (Merck, 99%), sodium carbonate (Na_2CO_3) (Merck, 99%), and sodium hydroxide (Merck, 99%).

2.2 Preparation of $K_2O/CaO\text{-}ZnO$ catalyst

Preparation of 5%wt $K_2O/CaO\text{-}ZnO$ catalyst was initiated by the synthesis of CaO-ZnO catalyst using co-precipitation method followed by impregnation of K_2O. The CaO-ZnO catalyst (CaO:ZnO of 3:1 mole ratio) was prepared by dissolving calcium nitrate ($Ca(NO_3)_2$) 1 M and zinc nitrate ($Zn(NO_3)_2$) 1 M in distilled water. Further, the solution was dropped wisely by 2 M Na_2CO_3 solution (10 mL/min) and was stirred until a white-gel was formed. The resulted gel was added by 1 M NaOH to adjust the pH to be 10. The mixed gel was stirred for 24 hours at a temperature of 60 °C. The resulting solid was washed several times by aquadest until alkaline free. The filtered solid was dried in a Memmert oven at 110 °C for overnight. The catalyst was then calcined in a Ney Vulcan box furnace at temperature of 800 °C for 3 hours.

The resulted CaO-ZnO catalyst was impregnated by solution of 1 M KNO_3 while stirred for 1 hour. The wet resulted solid was then dried in the Memmert oven at temperature of 110 °C for overnight. Subsequently, the dry solid was calcined in the box furnace at temperature of 300 °C for 5 hours to produce a $K_2O/CaO\text{-}ZnO$ catalyst. Thereafter, the catalyst was pelletized to obtained diameter of 3, 5, and 7 mm.

2.3 Catalyst basicity testing

Catalyst basicity was analyzed by titration method according to Tanabe and Yamaguchi method [13]. In this method, the solid catalyst was suspended in benzene and was titrated with benzoic acid. One-half gram of the catalyst was put into an Erlenmeyer flask. Twenty mL benzene and 1 mL indicator solution (128 mg of bromthymol blue (BTB) in 100 mL of benzene) was poured into the Erlenmeyer flask. The catalyst suspension was indicated by change of yellow dye to green-blue color. After that, the 0.1 N benzoic acid was added dropwise from the micro burette into the suspension. The green color of the solid particle suspension gradually disappeared. The endpoint of titration was determined when all the green color disappeared. The basicity was calculated from the titer of 0.1 N benzoic acid required as shown in Eqn. (2), where V denotes the volume of benzoic acid solution (mL), N denotes normality of benzoic acid solution (mgeq.mL^{-1}), and W represents a weight of catalyst sample (g) [4].

$$Basicity\ (mmol.g^{-1}) = \frac{(VxN)_{benzoic\ acid}}{W_{catalyst}} x100\% \qquad (2)$$

2.4 Catalyst testing for biodiesel production

Experimental rig for 5wt%$K_2O/CaO\text{-}ZnO$ catalyst testing in the transesterification process of soybean oil over a hybrid plasma-catalytic reactor (Dielectric Barrier Discharge – DBD type) was depicted in Fig. 1.

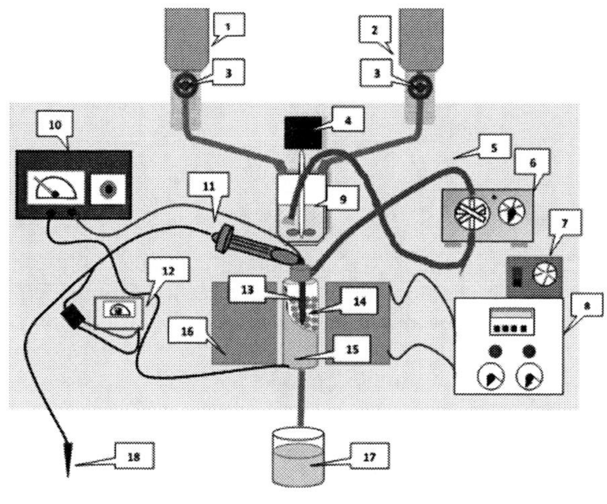

Fig. 1. Experimental rig for transesterification process using a hybrid fixed bed plasma-catalytic reactor: (1) methanol tank, (2) oil tank, (3) valve, (4) agitator, (5) board, (6) dosing pump, (7) regulator stirrer, (8) thermostat & controller, (9) mixing tank, (10) high voltage power supply, (11) high voltage probe, (12) multimeter, (13) electrode rod, (14) catalyst pellet, (15) glass tube, (16) heater jacket, (17) container tank, and (18) ground.

The catalytic-plasma reactor consisted of a high voltage (HV) electrode, a ground electrode, glass barrier, and a packed of catalyst bed placed within HV discharge zone. The HV electrode was made of the copper rod with a 0.5 cm in diameter and 30 cm in length placed in the center of reactor tube. The ground electrode was made of stainless steel plate rounded around the HV copper rod with a 3 cm in tube diameter. The glass barrier placed between both electrodes was acted as a dielectric-like element which made of a glass tube with a thickness of 0.1 cm and a diameter of 2.8 cm. The gap between electrodes, a distance between the high voltage electrode and the ground electrode, was 1.27 cm. A packed of catalyst pellet (certain diameters) was placed within a discharge zone.

The high voltage applied on the catalytic-plasma reactor was supplied by a high voltage power supply (up to 18 kV and 330 W DC-type), and equipped with a high voltage probe (1000X, maximum 40 kV), and an electric split tube furnace. The temperature of the reactor was controlled and measured by a temperature controller (5 °C) connected to the electric tube furnace. The transesterification process was operated at 7 kV voltage, 65 °C reaction temperature, and 1,186 min^{-1} WHSV.

Analysis of the reaction products (mainly of FAAE components) was done by using Gas Chromatography-Mass Spectrometry (QP2010S SHIMADZU, DB-1 column). The samples were analyzed by programming the GC as follows: heated up to 50 °C oven temperature (hold for 5 min) and ramped 10 °C.min^{-1} to 260 °C and held at later temperature for 33 minutes. The yield (%) of FAAE were determined using Eqns. (1).

$$Yield\ (\%) = \frac{\rho_{biodiesel} \times V_{biodiesel} \times \%FAAE\ GC\ Area}{m_{soybean\ oil}} \times 100\% \quad (1)$$

3 Results and Discussion

3.1 Effect of catalyst pellet-diameter on performance of plasma-catalytic process

Production of biodiesel using the hybrid plasma-catalytic process is influenced by synergistic effect between plasma role of DBD reactor and catalytic process with 5wt%K_2O/CaO-ZnO as a solid catalyst. The synergistic effect involves roles of the catalyst and the plasma simultaneously in the transesterification process in which the energetic electrons within plasma assist the reaction on the catalyst surface by exciting bonded electrons.

In the packed bed reactor system within discharge zone, surface area of contact between the energetic electrons and the catalyst surface is important. The intimate contact could be covered when the surface area of catalyst is increased including void fraction of the packed bed. The high void fraction includes void inside catalyst pore and void within packed bed. From Fig. 2, the yield of FAAE is slightly influenced by the catalyst diameter within the range of diameter studied. This means changes of catalyst diameter only slightly affected on reaction performance.

Optimum diameter catalyst was found to be 5 mm that gives a maximum yield of FAAE relatively. In the catalytic-plasma process, active species that consist of free radicals and metastable compounds were generated within the plasma discharge zone and catalyst pores. When the catalyst diameter is smaller than the optimum diameter, the transesterification reaction only occurs predominantly on the surface of catalyst without adsorption-desorption process within internal pores of the catalyst. Also, the smaller the catalyst diameter, the smaller void fraction within the catalyst bed, so that the FAAE yield decreased. The larger the catalyst diameter, the larger void fraction in the catalyst, so that discharge void fraction would be increased.

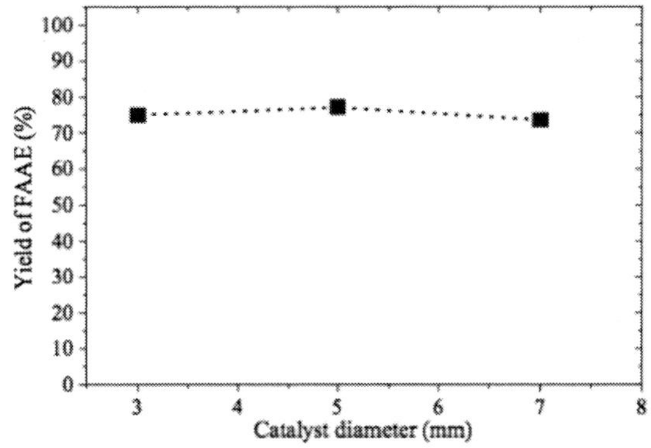

Fig. 2. Effect of catalyst diameter on FAAE yield

3.2 Effect of catalyst basicity on performance of plasma-catalytic process

Basicity of the catalysts (fresh and used) was presented in Table 1. The table indicated an increased basicity due to promoting the CaO-ZnO with K_2O. The high basicity is due to higher basic strength of calcium oxide, and promoted by an impregnation of K_2O. The basicity of the catalyst could also be increased with good dispersion of CaO on the ZnO surface.

The zinc oxide generates strong basic sites that providing a strong interaction between CaO and ZnO. It leads to the electron transfer from metal zinc oxide support interface to CaO as reported by Taufiq-Yap *et al.* [15]. Improving the catalyst basicity could also be done by impregnating with alkaline components as a promoter. In this research, the CaO-ZnO mixed metal oxide was impregnated by potassium nitrate (KNO_3) solution to improve the catalyst basicity as compensation of the basicity decrease due to the presence of ZnO component and the leached CaO content by methanol. From Table 1, the increased basicity of catalyst leads to higher FAAE yield.

Table 1. Effect of basicity of catalyst on FAAE Yield

Catalyst	Basicity (mmol/g)	FAAE yield (%)
CaO-ZnO	0.662	75.58
5%K_2O/CaO-ZnO (fresh)	1.090	77.19
5%K_2O/CaO-ZnO (used)	1.000	-

3.3 Characterization of FAAE as biodiesel product

Characterization of fatty acid alkyl esters (FAAE) as biodiesel product of the hybrid plasma-catalytic reactor is presented in Table 2. From the table, it can be suggested that the reactor product fulfill the basic specification of biodiesel according to the SNI standards.

The measurement of biodiesel characteristics used ASTM method as stated in Table 2.

Table 2. Characterization of biodiesel product (1: ASTM D 1298; 2: ASTM D 445; 3: ASTM D 664)

Parameter	Biodiesel / FAAE at Various Catalyst Diameter (mm)			SNI Standard	Soybean Raw Material
	3	5	7		
Density (g/mL) [1]	0.88	0.90	0.90	0.85-0.89	0.92
Kinematic Viscosity (40 °C) (mm^2/s) [2]	3.84	4.01	4.15	2.3-6.0	8.46
Acid Number (mg KOH/g) [3]	0.34	0.34	0.22	< 0.8	0.56
FFA concentration (% oleat)	-	-	-	-	0.28 %

4. Conclusion

Production of biodiesel over a hybrid plasma-catalytic process is strongly influenced by a synergistic effect between plasma role of DBD reactor and catalytic process with 5wt%K_2O/CaO-ZnO as a solid catalyst. The synergistic effects involve roles of the surface of catalyst and the plasma in the transesterification process in which the energetic electrons within plasma assist the reaction on the catalyst surface by the exciting bonded electrons. The yield of FAAE as biodiesel product is slightly influenced by the catalyst diameter. In the packed bed reactor system within discharge zone, the surface area of contact between the energetic electrons and the catalyst surface is important. The intimate contact between reactant molecules and the energetic electrons from plasma could be covered when the surface area of the catalyst is increased including the void fraction of the packed bed. The higher catalyst basicity is due to higher basic strength of calcium oxide, and promoted by impregnation of alkaline K_2O. Increased basicity of catalyst leads to higher FAAE or biodiesel product yield.

Acknowledgment

The authors thank to the Directorate General of Research and Development, Ministry of Research, Technology, and Higher Education, the Republic of Indonesia for the financial support received under the research project of Hibah Kompetensi Year 2015-2017.

References

1. A. Budi, Abdullah, Gunawan, *Research Report*, Fakultas MIPA, Universitas Lambung Mangkurat, (2009).
2. B. Haryanto, *Research Report*, Universitas Sumatera Utara, (2002).
3. L. Buchori, I. Istadi, P. Purwanto, *Bull. Chem. React. Eng. Catal.*, **11**, 406–430 (2016).
4. L. Buchori, I. Istadi, P. Purwanto, *St. Cerc. St. CICBIA*, **18** (3), 303-318, (2017).
5. J. Zhang, S. Chen, R. Yang, Y. Yan, *Fuel*, **89**(10), 2939-2944, (2010).
6. O. Stamenkovic, V. Veljkovic, Z. Todorovic, M. Lazic, D. Skala, *Fuel*, **88**, 1554-1562, (2009).
7. M. Serio, R. Tesser, L. Pengmei, E. Santacesaria, *Energy Fuels*, **22**(1), 207-217, (2008).
8. S. Semwal, A. Arora, R. Badoni, D. Tuli, *Biores. Technol.*, **102**(3), 2151-2161, (2011).
9. Z. Yang, W. Xie, *Fuel Process. Technol.*, **88**(6), 631-638, (2007).
10. L. Meher, D. Sagar, S. Naik, *Renew. Sust. Energ. Rev.*, **10**(3), 248-268, (2006).
11. A. Buasri, N. Chaiyut, V. Loryuenyong, C. Rodklum, T. Chaikwen, N. Komphan, *Appl. Sci.*, **2**, 641-653, (2012).
12. I. Istadi, A.D. Yudhistira, D.D. Anggoro, L. Buchori, *Bull. Chem. React. Eng. Catal.*, **9**(2), 111–120, (2014)
13. K. Tanabe, T. Yamaguchi, *Journal of the Research Insitute for Catalysis*, **11**(3), 179-184, (1964).
14. G. Kombe, A. Temu, H. Rajabu, H. Mrema, J. Kansendo, K. Lee, *Proceeding of Second International Conference on Advance in Engineering and Technology*, 176-182, (2007).
15. Y.H. Taufiq-Yap, H.V. Lee, M.Z. Hussein, R. Yunus, *Biomass Bioenergy*, **35**(2), 827- 834, (2011).

Synthesis and Characterization of Co/Ni/CoNi-ZSM-5 Catalyst for Hydrogen Production

Widayat Widayat[1,2*], *Arianti* Nuur Annisa[1], *Hantoro* Satriadi[1] and *Syaiful* Syaiful[3]

[1]Department of Chemical Engineering, Faculty of Engineering, Diponegoro University, Semarang
[2]Laboratory of Material Technology, Centre of Research and Service Diponegoro University, Semarang
[3]Department of Mechanical Engineering, Faculty of Engineering, Diponegoro University, Semarang

Abstract. Nickel is commonly used as a catalyst in hydrogen production. However, the use of nickel catalysts in the steam reforming process has the disadvantage of coke formation and high cost. Therefore, in this research, Ni/ZSM-5 catalyst synthesis was used to reduce production cost and an addition of cobalt (Co) metal to avoid coke formation. The method consists of a synthesis of ZSM-5 catalyst using hydrothermal process. Furthermore, the crystalline product was impregnated with the metal cobalt, nickel and combination of cobalt-nickel as much as 2% by weight metal/weight of the catalyst. Then the XRD and EDX characterization of Co/ZSM-5, Ni/ZSM-5, and CoNi/ZSM-5 was done followed by catalytic testing in the production of hydrogen from glycerol using steam reforming process. From XRD characterization results showed that Co/ZSM-5 catalyst has a crystallinity of 78.69%, Ni/ZSM-5 catalyst has 70.04% crystallinity and CoNi/ZSM-5 catalyst has 76.99% crystallinity. Catalytic testing on hydrogen production showed that CoNi/ZSM-5 catalyst produced the highest hydrogen concentration of 1,756.33 ppm while Ni/ZSM-5 catalyst produces 1,240 ppm and Co/ZSM-5 catalyst produces 491 ppm.

1 Introduction

The increased world economy was accompanied by increased energy consumption, especially fuel. During this time the fuel used is sourced from petroleum which cannot be renewed. By the increase of energy consumption, there is depletion of petroleum reserves. Therefore, the fuel was produced using renewable raw materials derived from plant or animal oil. On the other hand, the use of renewable energy produces lower pollution and environmentally friendly. Biodiesel is one of the most mass-produced renewable energy. Sales of biodiesel have increased rapidly in 2014 from 14.2 billion tons in 2008 to 29.1 billion tons [1]. In the production of biodiesel produced glycerol as by product. The production of 10 kg biodiesel can produce 1 kg of glycerol or 10% by weight [2], so by increasing production of biodiesel the amount of glycerol product also increases.

Glycerol is a versatile compound which can be used as a raw material for producing other derivative compounds. Industries that use glycerol as raw materials are food and beverages, personal care, oral care and tobacco [3]. At this time began to develop the use of glycerol to produce new energy. Hydrogen is one of the most widely used compounds in industry and as a fuel with 95% of hydrogen production using fossil raw materials. By the increasing amount of glycerol, it used as raw materials in hydrogen production. Steam reforming is one of the most widely used processes in hydrogen production [4]. The production of hydrogen by steam reforming process produces the highest number of moles of hydrogen [5,6].

In hydrogen production, the commonly used catalyst is alumina, but alumina catalyst (Al_2O_3) has a thermal deterioration that is unstable at high temperatures [7]. The process of hydrogen production that is widely applied in the industry is steam reforming using high temperature. Therefore, the use of zeolite catalysts is preferred because it has micropore structure, high area, high thermal stability and high CO_2 affinity. In addition, the ZSM-5 catalyst is stable at high temperatures (1000°C) and at low pH (pH = 3), so it is appropriately used in the steam reforming process [8]. However, a zeolite is a non-metallic mining product, so its availability is limited. In addition, the composition of natural zeolite varies so that its properties and characteristics not optimal if used as a catalyst [9]. By knowing this it is necessary to make the synthesis of zeolite.

Zeolite type is quite diverse, ZSM-5 is one type of zeolite which is often used as a catalyst [8]. The ZSM-5 catalyst synthesis (Zeolite Socony Mobil-5) was first performed by Argauer and Landolt patented by Mobil Oil Corporation in 1972 [10]. The ZSM-5 catalyst can be used to produce various products through a catalytic

* Corresponding author: widayat@live.undip.ac.id

cracking process. The type of product which produced by ZSM-5 catalyst was influenced by the physical and catalytic properties of the constituent material. Some parameters that may affect the nature of the ZSM-5 catalyst include the silica and alumina sources, the template and the crystallization conditions [11], so that in the preparation of ZSM-5 catalysts can be modified according to the desired product type.

Based on Pastor-Pérez and Sepúlveda-Escribano [4] research, the use of $NiSn30CeO_2/C$ catalyst on GSR process is capable of producing the best hydrogen yield of 47%. The presence of Ni metal is capable of breaking the C-C bond on glycerol thereby increasing the yield of hydrogen produced. Zamzuri et al. [12] and Yurdakul et al. [13] performed hydrogen production using a combination of Al_2O_3, ZrO_2, La_2O_3, olivine, SiO_2 and MgO catalysts. In their research used catalyst support in the form of metal Ni and Pd. The highest hydrogen selectivity was 86% using Ni/Al_2O_3 catalyst at a temperature condition of 692°C [12]. In the Yurdakul et al. [13] research, the highest yield was 74% using $Ni\text{-}Pd/Al_2O_3\text{-}ZrO_2$ catalyst at operating temperature 800°C.

In this research, the synthesis of ZSM-5 catalyst was done using hydrothermal process. Furthermore, ZSM-5 synthesis was impregnated using metals to produce hydrogen. Hydrogen production will be done through steam reforming process. The metal support to be used is metal cobalt (Co) and nickel (Ni). This is because the metal is easy to obtain, cheap and activated at steam reforming operating temperature at 450°C-800°C [14,15].

2 Materials and methods

2.1. Materials

The materials used were $KAl(SO_4)_2.12H_2O$ (for analysis, Merck), H_2SO_4 (98%, Merck), Na_2SiO_3 (extra pure, Merck), NaOH (for analysis, Merck), $Ni(NO_3)_2.6H_2O$ (for analysis, Merck) and $Co(NO_3)_2.6H_2O$ (for analysis, Merck). Glycerol 98% was purchased from Indrasari chemical store. Nitrogen gas (N_2) with 99.9% purity was purchased from PT. Samator Gas. Distilled water was provided by Membrane Research Centre Laboratory, CORES-DU.

2.2 Catalyst synthesis

The catalyst synthesis was performed based on the Latourette et al. [16] and Widayat and Annisa method [17]. The procedure of making catalyst is done by making solution A by mixing $KAl(SO_4)_2.12H_2O$, H_2SO_4 and H_2O. Then the A solution was mixed with sodium silicate to homogeneous. Then 40% NaOH solution was added and homogenized. Molar ratio of mixed solution are 0.056 $KAl(SO_4)_2.12H_2O$: 0.57 H_2SO_4 : 11.5 H_2O : 3.5 Na_2SiO_3 : 1.4 NaOH.

The catalyst feedstock mixture was crystallized using a hydrothermal system with 8 hours and 24 hours crystallization time and crystallization temperature 90°C, 120°C and 150°C. After the crystallization process is complete, the crystal product was cooled to room temperature for 24 hours. Next, the crystal product was washed using distilled water and followed by filtration. The crystal product was dried using an oven at 110°C for 12 hours. Then the crystal was calcined at 600°C for 5 hours.

2.3 Catalyst impregnation

The impregnation of the catalyst is carried out by using nickel metal (Ni), cobalt (Co) and nickel and cobalt metal alloys. The impregnation was carried out by adding 2% weight metal/weight catalyst. The impregnation process was carried out at 60°C and stirred for 6 hours, then stays for 48 hours. Furthermore, the filtration process was done followed by catalyst drying at 110°C for 6 hours. Then the catalyst was calcined at 600°C for 5 hours with nitrogen gas.

2.4 Catalyst characterization

The ZSM-5 crystal product was characterized by composition analysis and crystallinity analysis. The crystallinity analysis was performed using X-Ray Diffractometer, XRD-7000S type, Shimadzu brand with Cu-targeted tube X-ray tube, 30 kV, 30 mA and Kα radiation. The XRD analysis results data was processed by using software PCXRD and PDF4 +. The catalyst composition was analysed using Energy Dispersive X-Ray Fluorescence Spectrometers, EDX-7000 type, Shimadzu brand. The process of catalyst characterization was done in the Diponegoro University Integrated Laboratory.

2.5 Hydrogen production

Hydrogen production was carried out by a steam reforming process using a fixed bed reactor. The procedure is to load the catalyst by filling the centre reactor with 2 grams of catalyst. Then a glass wool was added as a catalyst holder. Loading feed is done by filling vaporizer vessel with a feed consisting of glycerol and water at a ratio of 1: 9 of 1000 ml. Evaporation of the feed was carried out at a temperature of 180°C. Then steam feed that has been formed is driven by the flow of nitrogen gas to the reactor with an operating temperature of 500°C. The hydrogen production process was carried out for 10 minutes. The gas product was streamed online to the hydrogen sensor and gas analyser. In addition, a portion of the gas was accommodated for analysing the composition using Gas Chromatography Mass Spectrometry (GCMS). The steam reforming instrument was showed in Figure 1.

https://doi.org/10.1051/matecconf/201815606013

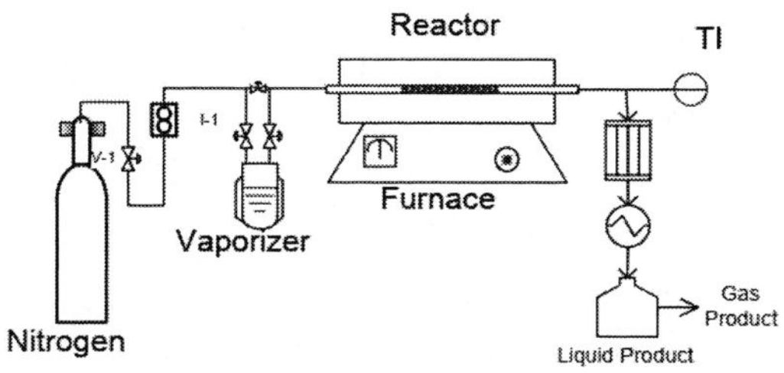

Fig. 1. S*team reforming* instrument (dimension of reactor: inside diameter = 1,5 cm; outside diameter = 2,0 cm; thickness of catalyst = 2 cm)

The product from the steam reforming process of glycerol comprises liquid and gas products. The liquid product was analysed using High Performance Liquid Chromatography (HPLC) to determine the composition of glycerol in the product. The gas product was analysed using an online hydrogen sensor for 10 minutes to determine the amount of hydrogen gas produced. Then in the early and final minutes of the gas product were analysed online by using a gas analyser to know the composition of by products in the form of CO, CO_2 and hydrocarbon gas. Meanwhile, to know the type of hydrocarbon gas produced was analysed by using Gas Chromatography Mass Spectrometry (GCMS) on the gas product that has been accommodated.

3 Result and discussion

3.1 Catalyst synthesis and characterization

3.1.1 XRD Analysis

The crystallization process of ZSM-5 catalyst synthesis was carried out for 8 hours at 200°C. The catalyst product obtained was characterized using X-Ray Diffractometer (XRD) to determine the structure and composition of the catalyst. Diffractogram of XRD analysis results was showed in Figure 2.

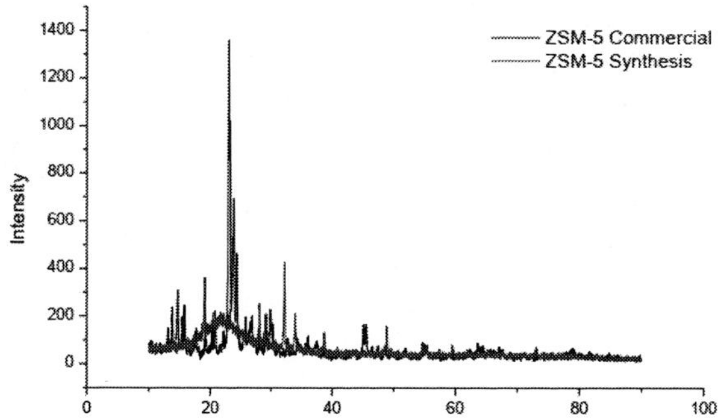

Fig. 2. ZSM-5 diffractograms

From the XRD difractogram shows there is a difference between ZSM-5 synthesis and commercial ZSM-5. ZSM-5 synthesis has the highest peak at 2θ 19°; 28° and 32°, but the commercial ZSM-5 has the highest peak at 2θ 13°; 23° and 45°. However, ZSM-5 synthesis also has a peak at 13°; 23° and 45° with low intensity. Peak at 2 theta 23° can be increased by adding the concentration of NaOH solution [18]. The crystallinity of ZSM-5 synthesis is 44.09% and the commercial ZSM-5 has a crystallinity of 76.06%. This is because ZSM-5 synthesis has an amorphous crystal structure so the degree of crystallinity is low [19].

The ZSM-5 catalyst was impregnated using cobalt and nickel metal. The impregnated catalyst were analysed using XRD and EDX. XRD Diffractogram was showed in Figure 3.

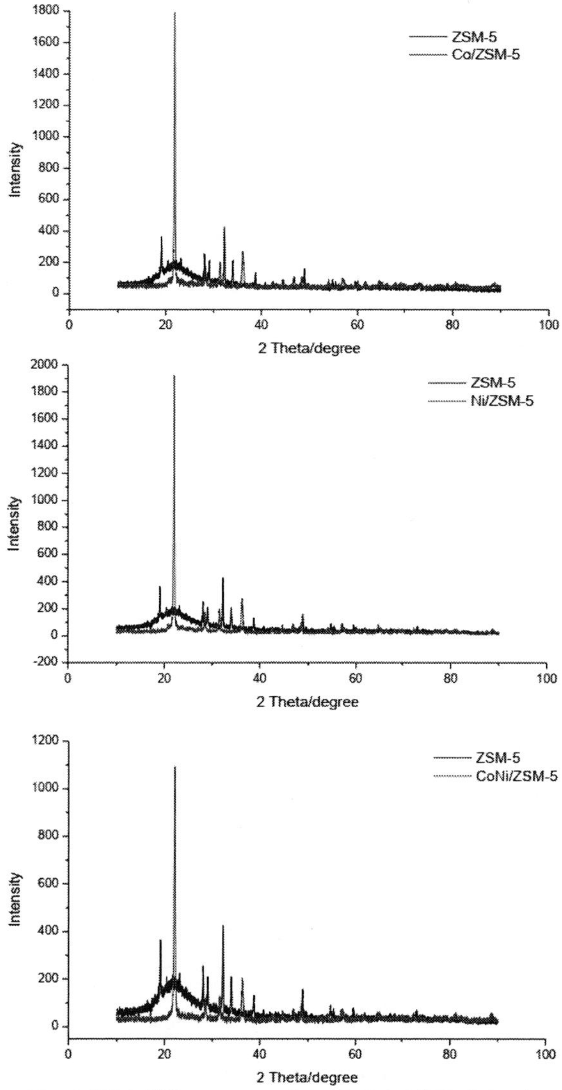

Fig. 3. XRD diffractograms of Co/ZSM-5, Ni/ZSM-5 and CoNi/ZSM-5

From Figure 3 shows that the impregnated catalyst has a higher intensity than ZSM-5 synthesis. Co/ZSM-5, Ni/ZSM-5 and CoNi/ZSM-5 were recalcined at 600°C. Crystallinity of Co/ZSM-5; Ni/ZSM-5 and CoNi/ZSM-5 were 78.69%; 70.03% and 76.98%. Calcination can reduce catalyst impurities and increase crystal formation, so the crystallinity of the impregnated catalyst becomes increases.

3.1.2 EDX Analysis

The catalysts were characterized using EDX to know the metals composition. Figure 4 and Table 1 were showed the EDX analysis result.

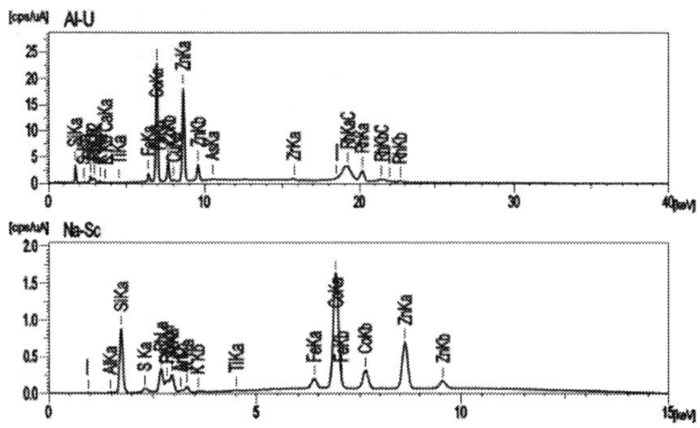

a. Co/ZSM-5

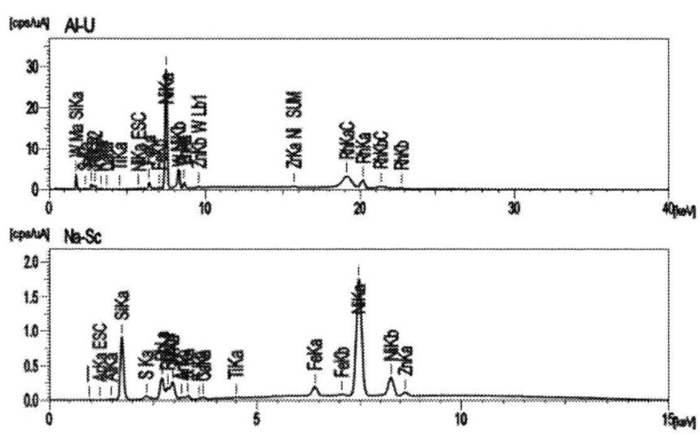

b. Ni/ZSM-5

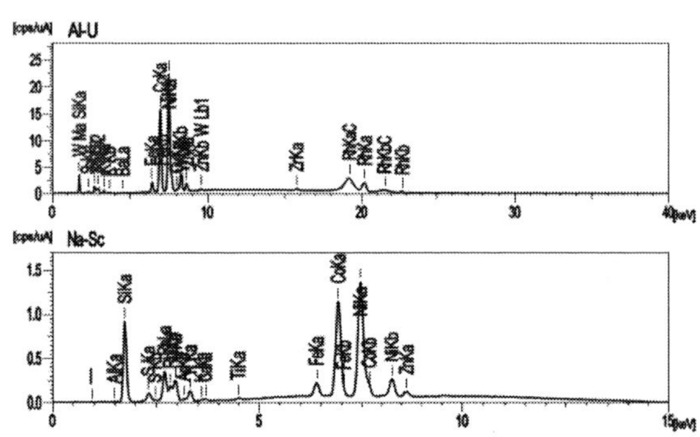

c. CoNi/ZSM-5

Fig. 4. EDX analysis result

Table 1. Metals compound in impregnated catalyst

Catalyst	Metals Compound	Result
Co/ZSM-5	Co_2O_3	1.472%
Ni/ZSM-5	NiO	1.335%
CoNi/ZSM-5	Co_2O_3	1.041%
	NiO	0.935%

From EDX analysis result shows that the metals did not totally attach in the catalyst surface. It caused by the low interaction between metals and support due to imprecise calcining temperatures [20]. The cobalt oxide and nickel oxide compounds formed are caused by an imperfect reduction process at the time of calcination.

3.2 Hydrogen production

The catalysts were used to produce hydrogen gas using glycerol feedstock. The gas product is analysed using a hydrogen sensor to know the hydrogen yield and gas sensor to known composition of other gas products. From the analysis showed that the catalyst can produce hydrogen, carbon monoxide, carbon dioxide and hydrocarbon gas product. Figure 5, 6 and 7 below are shows the product composition.

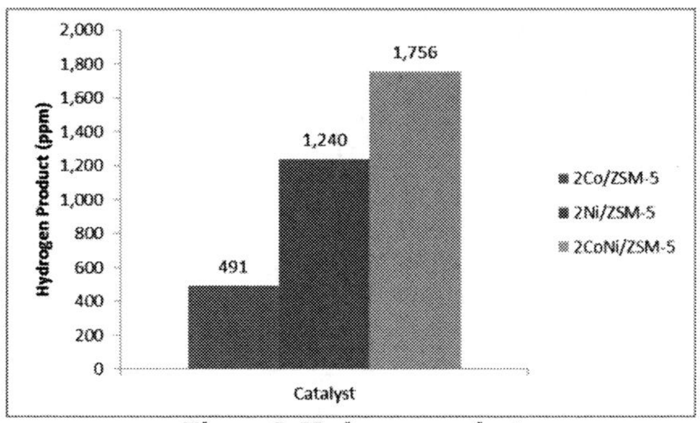

Figure 5. Hydrogen product

The highest hydrogen product obtained was 1756 ppm using CoNi/ZSM-5 catalyst. The Co/ZSM-5 catalyst produced 491 ppm hydrogen and Ni/ZSM-5 catalyst produced 1240 ppm. The impregnation of cobalt and nickel metal on the ZSM-5 catalyst can increase the activity of the catalyst. The hydrogen product produced from the CoNi/ZSM-5 catalyst has higher purity because no carbon formation occurs so no carbon monoxide and carbon dioxide by-products were formed [20]. Therefore, by using the CoNi/ZSM-5 catalyst obtained the highest yield of hydrogen. From this research can be concluded the best metal activity is CoNi > Ni > Co. The Co/ZSM-5 catalyst produces the lowest hydrogen yield because there was a carbon formation on the active site of the catalyst and catalyst sintering [1, 2]. By the coke formation on the surface of the catalyst, it can trigger the formation of carbon dioxide and carbon monoxide so the hydrogen yield becomes low.

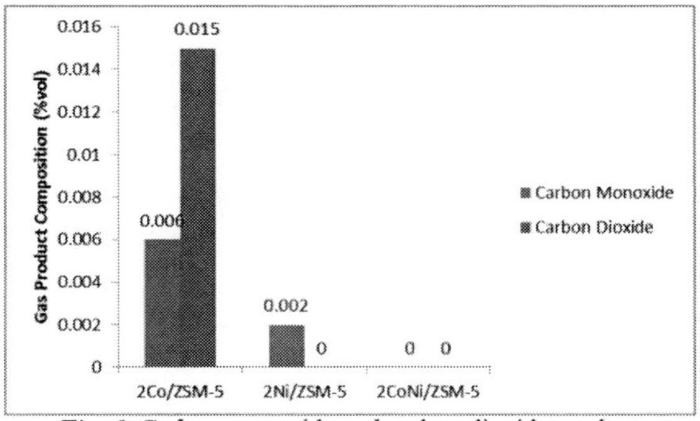

Fig. 6. Carbon monoxide and carbon dioxide product

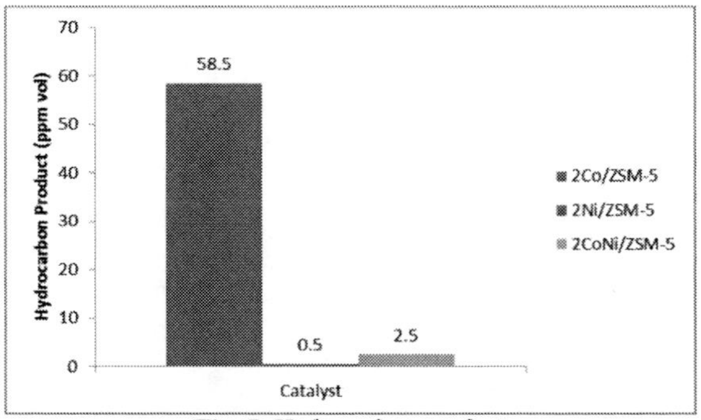

Fig. 7. Hydrocarbon product

From the results, it can be seen that the type of product formed is varies. This suggests that the type of metal may affect the reaction that occurs in the steam reforming process. Glycerol steam reforming by used nickel metal produces H_2, CO and hydrocarbon, on the other hand cobalt metal produces H_2, CO, CO_2 and hydrocarbon. By combined nickel and cobalt metals only produces H_2 and hydrocarbon. The used of cobalt metal produced the highest CO_2 and CO. Using a cobalt metal will produce CO_2 and CO at the beginning of the reaction (for 2 hours), after that there will decrease in CO_2 and CO production as H_2 production increases [21]. The use of nickel and cobalt metal combinations did not produce CO_2 and CO. This is because by doing a combination of metal nickel and cobalt can reduce carbon formation [20].

4 Conclusion

Crystallinity of ZSM-5 synthesis can improve by recalcination after catalyst impregnation because calcination can reduce the impurity of the catalyst and improve the crystal formation. The XRD characterization results showed that the Co/ZSM-5 catalyst had a crystallinity of 78.69%, the Ni/ZSM-5 catalyst had 70.04% crystallinity and the CoNi/ZSM-5 catalyst had 76.99% crystallinity. But, the metals did not totally attach in the catalyst surface. It caused by the low interaction between metals and support due to imprecise calcining temperatures. Catalytic testing on hydrogen production showed that the CoNi/ZSM-5 catalyst produced the highest hydrogen concentration of 1,756.33 ppm while the Ni/Zeolite catalyst produce 1,240 ppm and the Co/ZSM-5 catalyst produce 491 ppm.

References

1. M. Menor, S. Sayas, A. Chica, Fuel **193**, 351-358 (2017)
2. N. D. Charisiou, G. Siakavelas, K. N. Papageridis, A. Baklavaridis, L. Tzounis, K. Polychronopoulou, M. A. Goula, International Journal of Hydrogen Energy **42**, 13039-13060 (2017)
3. S. Adhikari, S. D. Fernando, A. Haryanto, Energy Conversion and Management **50**, 2600-2604 (2009)
4. L. Pastor-Pérez, A. Sepúlveda-Escribano, Applied Catalysis A: General **529**, 118-126 (2016)
5. S. Adhikari, S. D. Fernando, A. Haryanto, Renewable Energy **33**, 1097-1100 (2008)
6. D. H. Chun, J. C. Park, S. Y. Hong, J. T. Lim, C. S. Kim, H-T Lee, J-I Yang, S. J. Hung, H. Jung, Journal of Catalysis **317**, 135-143 (2014)
7. Y. Vafaeian, M. Haghighi, S. Aghamohammadi, Energy Conversion and Management **76**, 1093-1103 (2013)
8. van der Gaag, Technische Universiteit Delft (1987)
9. J. Weitkamp, Solid State Ionics **131**, 175-188 (2000)
10. S. Abbasian, M. Taghizadeh, International Journal Nanoscience Nanotechnology **10** **(3)**, 171-180 (2014)
11. J. Li, S. Liu, H. Zhang, E. Lu, P. Ren, J. Ren, Chinese Journal of Catalysis **37**, 308-315 (2016).
12. N. H. Zamzuri, R. Mat, N. A. S. Amin, A. Talebian-Kiakalaieh, International Journal of Hydrogen Energy **xxx**, 1-12 (2016)
13. M. Yurdakul, N. Ayas, K. Bizkarra, M. El Doukkali, J. F. Cambra, International Journal of Hydrogen Energy **41**, 8084-8091 (2016)
14. F. Rahmani, M. Haghighi, Y. Vafaeian, P. Estifaee, Journal of Power Sources **272**, 816-827 (2014)
15. J. Majewska, B. Michalkiewicz, International Journal of Hydrogen Energy **41**, 8668-8678 (2016)
16. B. Latourette, Le Raincy, C. Magnier, US Patent **4,891,199**, (1990)
17. W. Widayat, A. N. Annisa, AIP Conference Proceedings **1904**, 020061 (2017)
18. W. B. Widayatno, IOP Conf. Series: Journal of Physics: Conf. Series **817**, (2017)
19. W. Widayat, A. N. Annisa, IOP Conf. Series : Materials Science and Engineering **214**, 012032 (2017)
20. W. Nabgan, T. A. Tuan Abdullah, R. Mat, B. Nabgan, S. Triwahyono, A. Ripin, Applied Catalysis A: General **527**, 161-170 (2016)
21. E. A. Sanchez, R. A. Comelli, International Journal of Hydrogen Energy **39**, 8650-8655 (2014)

Cleaning Schedule Operations in Heat Exchanger Networks

Hairul Huda[1], *Renanto* Handogo[1,*], and *Totok Ruki* Biyanto[2]

[1]Chemical Engineering Department, Institut Teknologi Sepuluh Nopember, Surabaya, Indonesia
[2]Engineering Physics Department, Institut Teknologi Sepuluh Nopember, Surabaya, Indonesia

Abstract. Heat exchanger networks have been known to be the essential parts in the chemical industries. Unfortunately, since the performance of heat exchanger can be decreasing in transferring the heat from hot stream into cold stream due to fouling, then cleaning the heat exchanger is needed to restore its initial performance periodically. A process of heating crude oil in a refinery plant was used as a case study. As many as eleven heat exchangers were used to heat crude oil before it was heated by a furnace to the temperature required to the crude unit distillation column. The purpose of this study is to determine the cleaning schedule of heat exchanger on the heat exchanger networks due to the decrease of the overall heat transfer coefficient by various percentage of the design value. A close study on the process of heat exchanger cleaning schedule in heat exchanger networks using the method of decreasing overall heat transfer coefficient as target. The result showed that the higher the fouling value the more often the heat exchanger is cleaned because the overall heat transfer coefficient decreases quickly.

1 Introduction

In a refinery plant one can generally find a lot of heat exchangers which are used in series-parallel to heat cold crude oil to a required temperature before it flows to a column furnace to a higher temperature in order that a somewhat perfect separation can be obtained. There can be many configurations from a set of heat exchangers to construct. If we only use hot and cold utilities to exchange the heat between steam/cooling water and process fluid, then we can imagine the huge amount of energy needed to operate the process. A heat integration concept is aimed at reducing energy consumption from hot and cold utilities by exchanging heat between hot streams and cold streams. By using pinch design method, then maximum energy recovery gained from the process will reduce the need of hot and cold utilities, and will also reduce energy cost as well. Figure 1 shows the process in a refinery plant where a crude oil is being heated through a series of heat exchangers [1]. There is neither hot utilities nor cold utilities used in the process. There are hot streams available as they are side stream products from pipe still of crude distillation unit. We have one cold stream and seven hot streams which are being heat-matched. Eleven heat exchangers process to process are needed without any heaters and coolers. A delta T minimum of 10 °C is given as a constraint due to the physical heat exchangers used. From Figure 1 we have a heat exchanger networks (HEN) where there are four independent heating processes (HE-01, HE-07, HE-09 and HE-10) and seven integrated heating processes (HE-02, HE-03, HE-04, HE-05, HE-06, HE-08 and HE-11). In order to get more understanding how the streams are matched between hot streams and cold streams, a grid diagram is proposed as shown in Figure 2. However, using complete heat integration one can have zero hot and cold utilities with eleven heat exchangers. Since the energy cost outweighs the annual capital cost, the configuration for HEN as given in Figure 2 will be the base case. A fouling problem arises when a heat exchanger operates continuously over time.

2 Process Descriptions

In this process, crude oil from the storage is pumped to HE-01 as cold stream. Its inlet temperature is 32–51°C and its outlet temperature is 72–93°C. The inlet temperature of hot stream is 128–150°C and leaving HE-01 at 61–110°C. For HE-02, the cold stream enters at 72–93°C and leaves at 83–105°C while the hot stream enters at 146–211°C and leaves at 82–115°C. The hot stream of HE-02 is the outlet of HE-06 hot stream. For HE-03, the cold stream enters at 83–105°C and leaves at 99–126°C while the hot stream enters at 137–170°C and leaves at 93–150°C. The hot stream of HE-03 is the outlet of HE-05 hot stream. The cold stream of HE-04 enters at 99–126°C and leaves at 108–139°C while the hot stream of HE-04 enters at 159–240°C and leaves at 110–200°C. The hot stream of HE-04 is the outlet of HE-

* Corresponding author: renanto@chem-eng.its.ac.id

08 hot stream. After leaving HE-04, crude oil is sent to desalter to remove salt from the crude oil.

Crude oil from desalter is sent to HE-05 as cold stream. Its inlet temperature is 103–135°C and its outlet temperature is 127–151°C. The inlet temperature of hot stream is 154–187°C and leaving HE-05 at 137–170°C. For HE-06, the cold stream enters at 127–151°C and leaves at 131–157°C while the hot stream enters at 162-225°C and leaves at 146–211°C. For HE-07, the cold stream enters at 131–157°C and leaves at 135–163°C while the hot stream enters at 254–281°C and leaves at 145–200°C. The cold stream of HE-08 enters at 135–163°C and leaves at 149–188°C while the hot stream of HE-08 enters at 269–296°C and leaves at 189–217°C. The hot stream of HE-08 is the outlet of HE-11 hot stream. For HE-09, the cold stream enters at 149–188°C and leaves at 164–213°C while the hot stream enters at 269–296°C and leaves at 189–217°C. For HE-10, the cold stream enters at 164–213°C and leaves at 175–241°C while the hot stream enters at 314–338°C and leaves at 209–271°C. The cold stream of HE-11 enters at 175–241°C and leaves at 186–253°C while the hot stream of HE-11 enters at 314–337°C and leaves at 228–313°C. After leaving HE-11, crude oil is sent to furnace. The outlet temperature of furnace is 350°C. In this process, there are four independent heating processes (HE-01, HE-07, HE-09 and HE-10) and seven integrated heating processes (HE-02, HE-03, HE-04, HE-05, HE-06, HE-08 and HE-11).

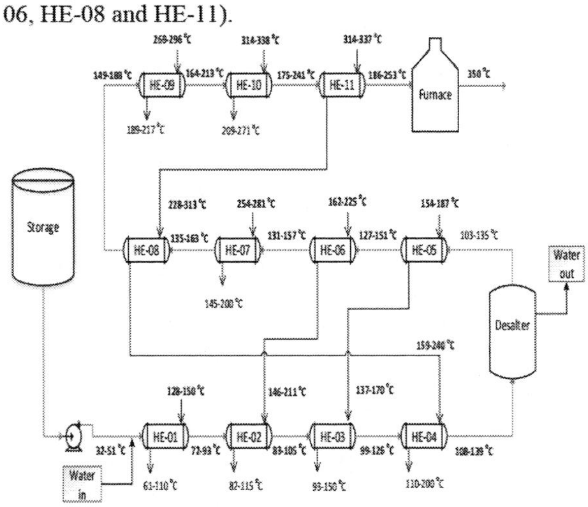

Fig. 1. The schematic diagram of the HEN in crude refinery.

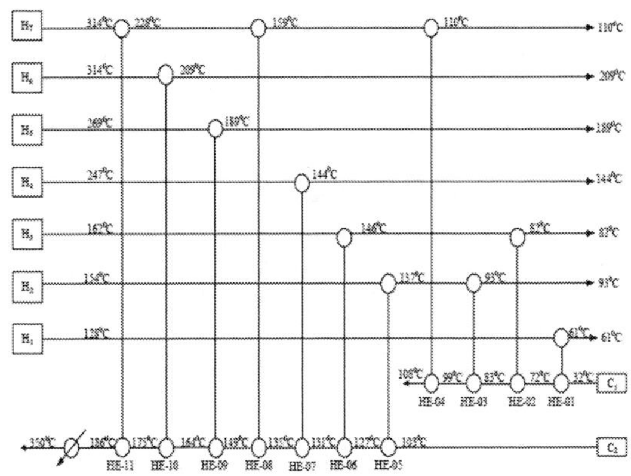

Fig. 2. Grid Diagram of Heat Integration with ΔTmin 10°C.

3 Heat Exchanger Networks

In any process flow sheet, there are several streams that need to be heated and there are some that need to be cooled before entering the next step. For that reason, cooling water is needed to lower the hot process streams while steam is needed for heating the cold process streams. If we only use hot and cold utilities to exchange the heat between cooling water/steam and process fluid, then we can imagine the huge amount of energy needed to operate the process.

A heat exchanger networks (HEN) is an arrangement of heat exchangers; in which cold and hot process streams and hot and cold utility streams interchange energy. The purpose of the HEN is to recover energy from the hot process streams to heat up cold process streams using the least amount of hot and cold utility streams, while achieving specified outlet target temperatures of the process streams [2].

4 Heat Exchanger

Heat exchangers are devices that provide the transfer of thermal energy between two or more fluids at different temperatures [3]. Traditionally, the heat exchanger performance analysis and simulation are performed using steady-state energy balance across the heat exchanger. The energy balance on the hot and cold fluid together with the heat transfer equation constitute the model of heat exchangers. Under the assumptions that there is no heat loss to the surrounding and the heat loss of hot streams shall be equal to heat gained by cold streams, therefore

$$Q_c = Q_h = m_c\, c_{p,c}\, (T_{c,o} - T_{c,i}) = m_h\, c_{p,h}\, (T_{h,i} - T_{h,o}) \quad (1)$$

Where:

m_c: mass flow rate of the cold fluid (crude oil) (kg/hr)
m_h: mass flow rate of the hot fluid (kg/hr)
$c_{p,c}$: specific heat of the cold fluid (W hr/kg °C
$c_{p,h}$: specific heat of the hot fluid (W hr/kg °C)
$T_{c,o}$: outlet temperature of the cold fluid (°C)
$T_{h,o}$: outlet temperature of the hot fluid (°C)
$T_{h,i}$: inlet temperature of the hot fluid (°C)

For the performance of another heat exchangers use equation below:

$$Q = U\,A\,\Delta T_{LMTD} \tag{2}$$

where:
Q: heat duty (W)
U: overall heat transfer coefficient ($W/m^2\,°C$)
A: heat transfer surface area (m^2)
ΔT_{LMTD}: log mean temperature difference (^oC)
ΔT Log Mean Temperature Difference (ΔT_{LMTD}) is defined as

$$\Delta T\ \Delta T_{LMTD} = \frac{(T_{h,o} - T_{c,i}) - (T_{h,i} - T_{c,o})}{ln\left(\frac{T_{h,o} - T_{c,i}}{T_{h,i} - T_{c,o}}\right)} \tag{3}$$

5 Fouling resistance and Overall heat transfer coefficient

Fouling is a major unresolved problem of significant interest in the field of heat transfer [4]. Fouling is the accumulation of undesired material at the heat transfer surfaces. It is transient process that begins with a clean heat transfer surface and continues until the point where the surface become fouled to the point where the heat exchanger can no longer be used effectively. At this point, the heat exchanger must be taken out of service and cleaned [5].

The correlation between fouling resistance and overall heat transfer coefficient can be described as

$$\frac{1}{U_{f(t)}} = \frac{1}{U_c} + R_{f(t)} \tag{4}$$

$$U_{f(t)} = \frac{U_c}{1 + U_c \cdot R_{f(t)}} \tag{5}$$

Where:
R_f: fouling resistance ($m^2\,°C/W$)
U_f: overall heat transfer coefficient dirt ($W/m^2\,°C$)
U_c: overall heat transfer coefficient clean ($W/m^2\,°C$)
t: time (day)
Fouling resistance, R_f, as follows:

$$R_{f\,(t)} = R_{f\infty}\left(1 - exp^{-\beta t}\right) \tag{6}$$

Where:
$R_{f\infty}$: asymptotic fouling resistance ($m^2\,°C/W$)
β: constant fouling formation with time (1/day)

The characteristic decrease in overall heat transfer coefficient dirt (U_f) follows the exponential function as given below:

$$U_{f(t)} = A\,e^{-Bt} \tag{7}$$

The characteristic and relation between U_f and t should be investigated. The value of U_f varies by t, where t is production period. Variable A is determined at initial condition and variable B is determined at final condition [6].

So we can determine the characteristic of overall heat transfer coefficient (U) for one period. The characteristic decrease in heat transfer coefficient (U) follows the exponential function as given below:

$$U_{f(t)} = \frac{U_c}{1 + U_c \cdot R_{f(t)}} + \left(U_c - \frac{U_c}{1 + U_c \cdot R_{f(t)}}\right)e^{-\beta t} \tag{8}$$

The characteristic and relation between U_f and t should be investigated. The value of U_f varies by t, where t is production period. Variable U_c, β are known from data. For the process cleaning scheduling was done on decrease by 20% U_c from each HE. In order to get more understanding how the streams are matched between hot streams and cold streams, a grid diagram is proposed with ΔT_{min} is equal to 10°C.

Table 1. Data.for Uc, R_f dan β

HE	U_c	$R_{f\infty}$	β (1/day)
HE-01	678	$9{,}57\times10^{-04}$	$1{,}99\times10^{-02}$
HE-02	483	$7{,}14\times10^{-03}$	$2{,}26\times10^{-02}$
HE-03	583	$8{,}03\times10^{-03}$	$2{,}96\times10^{-02}$
HE-04	359	$6{,}23\times10^{-03}$	$1{,}74\times10^{-02}$
HE-05	487	$7{,}74\times10^{-03}$	$2{,}14\times10^{-02}$
HE-06	649	$3{,}72\times10^{-03}$	$1{,}94\times10^{-02}$
HE-07	148	$2{,}61\times10^{-03}$	$1{,}67\times10^{-02}$
HE-08	357	$3{,}69\times10^{-03}$	$1{,}48\times10^{-02}$
HE-09	540	$3{,}42\times10^{-03}$	$1{,}51\times10^{-02}$
HE-10	640	$6{,}04\times10^{-03}$	$1{,}22\times10^{-02}$
HE-11	465	$4{,}48\times10^{-03}$	$3{,}46\times10^{-02}$

6 Result and Discussion

This section discusses the fouling resistance model, the total of cleaning and the overall heat transfer coefficient model each heat exchanger.

Table 2. Present of result.

HE	R_f at t = 360	Total of Cleaning	U_f at 80%	U_f at t = 360
HE-01	$9{,}56\times10^{-4}$	6	544,32	411,510
HE-02	$7{,}14\times10^{-3}$	15	390,89	108,674
HE-03	$8{,}03\times10^{-3}$	20	471,41	102,675
HE-04	$6{,}22\times10^{-3}$	10	289,46	111,521
HE-05	$7{,}74\times10^{-3}$	14	390,76	102,328
HE-06	$3{,}71\times10^{-3}$	13	523,32	190,784
HE-07	$2{,}61\times10^{-3}$	3	118,56	106,908
HE-08	$3{,}67\times10^{-3}$	7	285,84	155,475
HE-09	$3{,}40\times10^{-3}$	9	434,75	191,761
HE-10	$5{,}97\times10^{-3}$	10	512,49	139,191
HE-11	$4{,}48\times10^{-3}$	16	365,57	150,930

R_f fouling resistance at each heat exchanger is different. In Figure 3 it can be observed that there is an increase in fouling resistance or an increase in the level of impurities on the heat exchanger. Where for the highest end value of the highest or dirtiest fouling resistance at HE-03 is $8{,}03\times10^{-3}$ (m^2 °C/W) and the lowest fouling resistance value at HE-01 $9{,}56\times10^{-4}$ (m^2 °C/W).

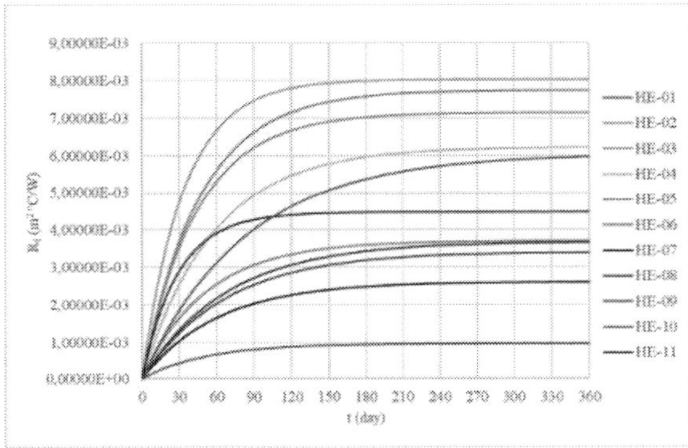

Fig. 3. Fouling resistance all heat exchanger.

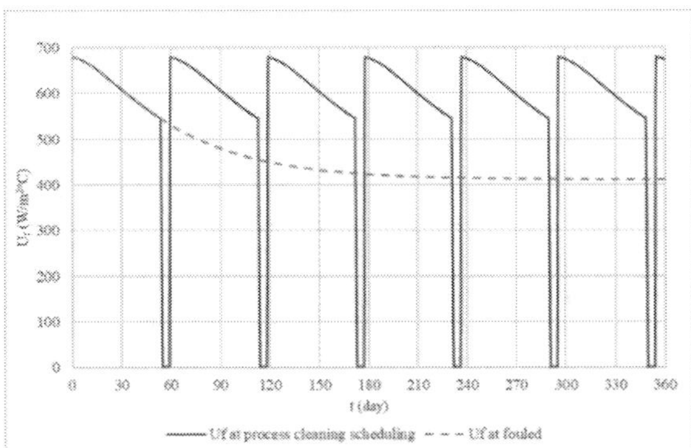

Fig. 4. Cleaning scheduling at HE-01.

In the HE-01 with initial value U_c 678 (W/m^2 °C), U_f at 360 days fouled condition value was 411,51 (W/m^2 °C) and U_f at 80% value was 542,4 (W/m^2 °C). The total of process cleaning 6 times in 360 days, U_f at day 54 value was 544,3245 (W/m^2 °C) and first process cleaning at day 55 in the HE-01.

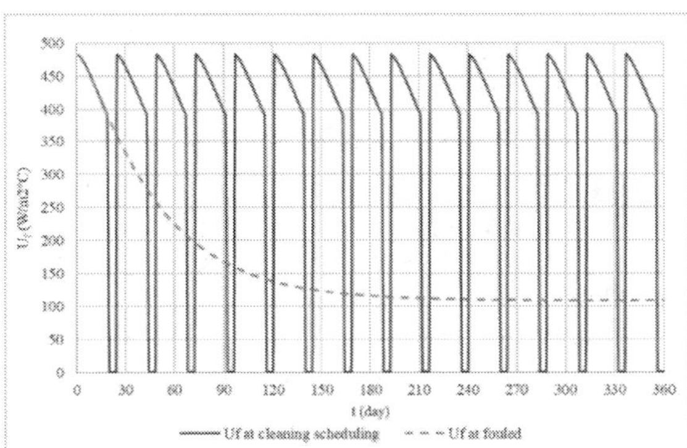

Fig. 5. Cleaning scheduling at HE-02.

In the HE-02 with initial value U_c 483 (W/m^2 °C), U_f at 360 for fouled condition 108,674 (W/m^2 °C) and U_f at 80% value was 386,4 (W/m^2 °C). The total of process cleaning 15 times in 360 days, U_f at day 19 value was 390,8917 (W/m^2 °C) and first process cleaning at day 20 in the HE-02.

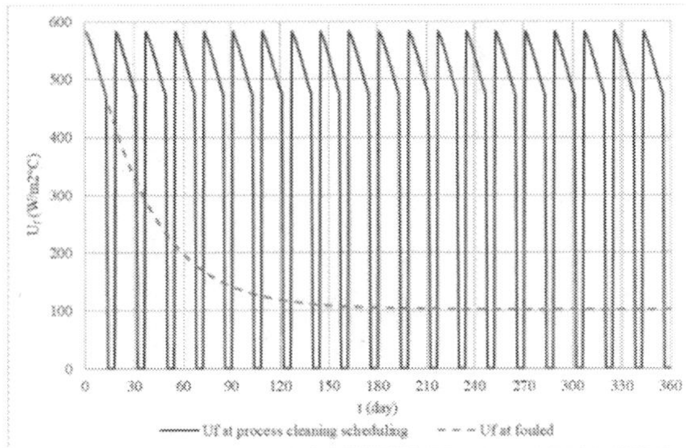

Fig. 6. Cleaning scheduling at HE-03.

In the HE-03 with initial value U_c 583 (W/m^2 °C), U_f at 360 for fouled condition 102,675 (W/m^2 °C) and U_f and at 80% value was 466,4 (W/m^2 °C). The total of process cleaning 20 times in 360 days, U_f at day 13 value was 471,4086 (W/m^2 °C) and first process cleaning at day 14.

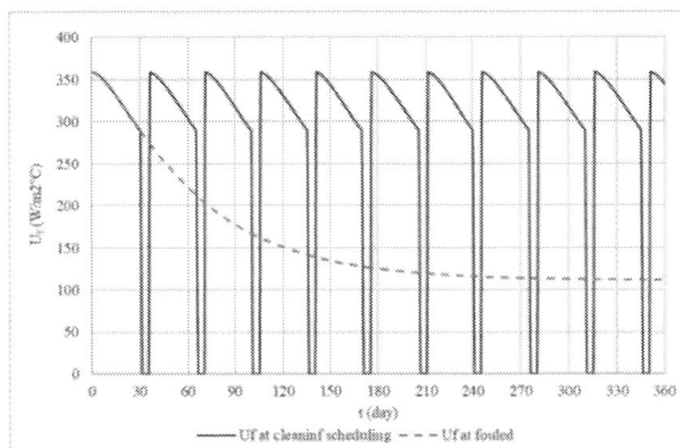

Fig. 7. Cleaning scheduling at HE-04.

In the HE-04 with initial value U_c 359 (W/m^2 °C), U_f at 360 for fouled condition 111,521 (W/m^2 °C) and U_f and at 80% value was 287,2 (W/m^2 °C). The total of process cleaning 10 times in 360 days, U_f at day 30 value was 289,461(W/m^2 °C) and first process cleaning at day 31 the page

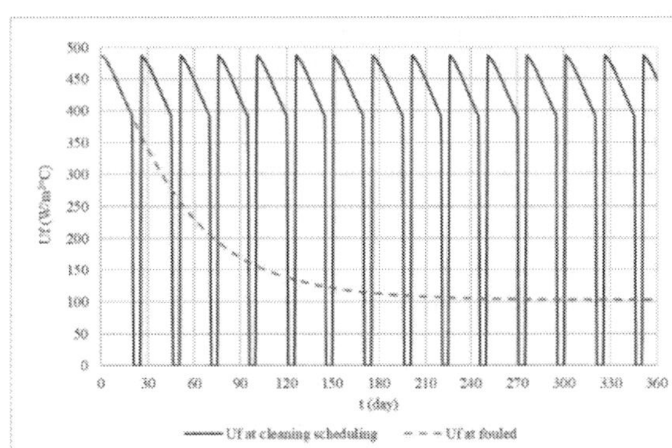

Fig. 8. Cleaning scheduling at HE-05.

In the HE-05 with initial value U_c 487 (W/m^2 °C), U_f at 360 for fouled condition 102,328 (W/m^2 °C) and U_f and at 80% value was 389,6 (W/m^2 °C). The total of process cleaning 14 times in 360 days, U_f at day 20 value was 390,760 (W/m^2 °C) and first process cleaning at day 21.

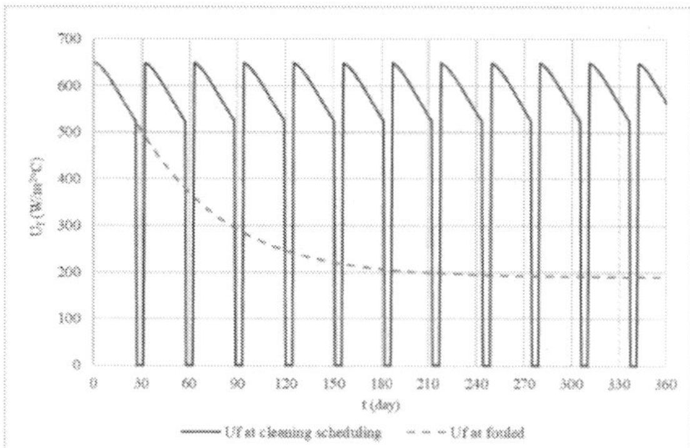

Fig. 9. Cleaning scheduling at HE-06.

In the HE-06 with initial value U_c 649 (W/m^2 °C), U_f at 360 for fouled condition 190,784 (W/m$_2$ °C) and U_f and at 80% value was 519,2 (W/m$_2$ °C). The total of process cleaning 11 times in 360 days, U_f at day 26 value was 523,321 (W/m2 °C) and year first process cleaning at day 27.

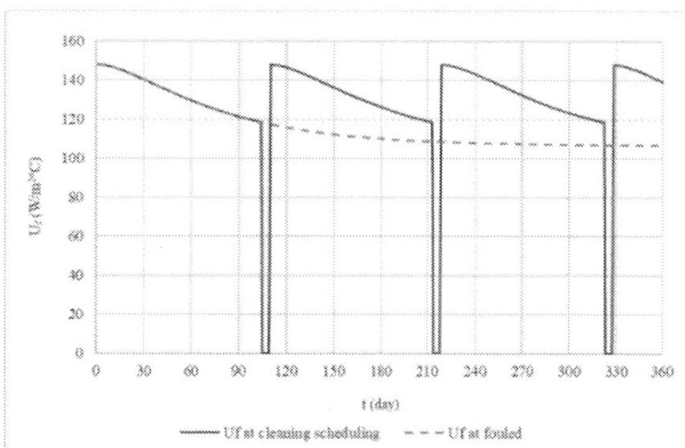

Fig. 10. Cleaning scheduling at HE-07.

In the HE-07 with initial value U_c 148 (W/m^2 °C), U_f at 360 for fouled condition 106,908 (W/m^2 °C) and U_f and at 80% value was 118,4 (W/m^2 °C). The total of process cleaning 3 times in 360 days, U_f at day 104 value was 118,560 (W/m^2 °C) and first process cleaning at day 105.

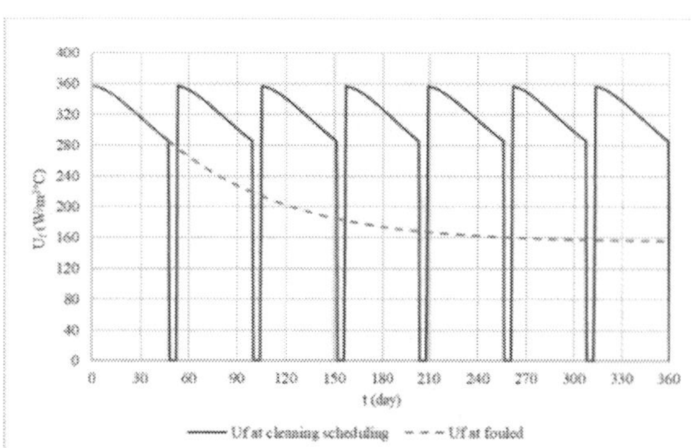

Fig. 11. Cleaning scheduling at HE-08.

In the HE-08 with initial value U_c 357 (W/m^2 °C), U_f at 360 for fouled condition 155,475 (W/m^2 °C) and U_f and at 80% value was 285,6 (W/m^2 °C). The total of process cleaning 7 times in 360 days, U_f at day 47 value was 285,838 (W/m^2 °C) and first process cleaning at day 48.

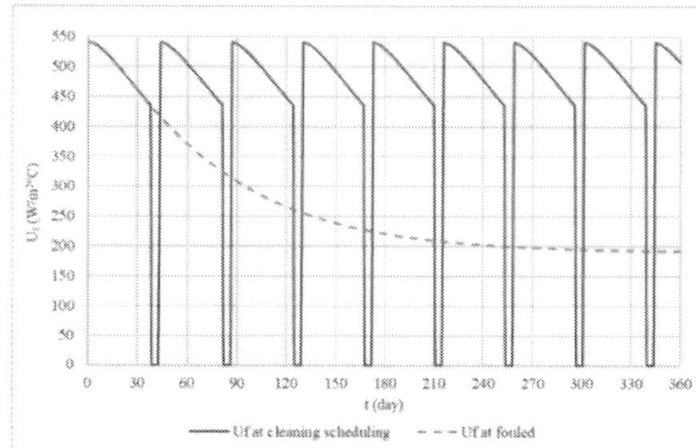

Fig. 12. Cleaning scheduling at HE-09.

In the HE-09 with initial value U_c 540 (W/m^2 °C), U_f at 360 for fouled condition 191,761 (W/m$_2$ °C) and Uf and at 80% value was 432 (W/m$_2$ °C). The total of process cleaning 8 times in 360 days, U_f at day 38 value was 434,750 (W/m^2 °C) and first process cleaning at day 39.

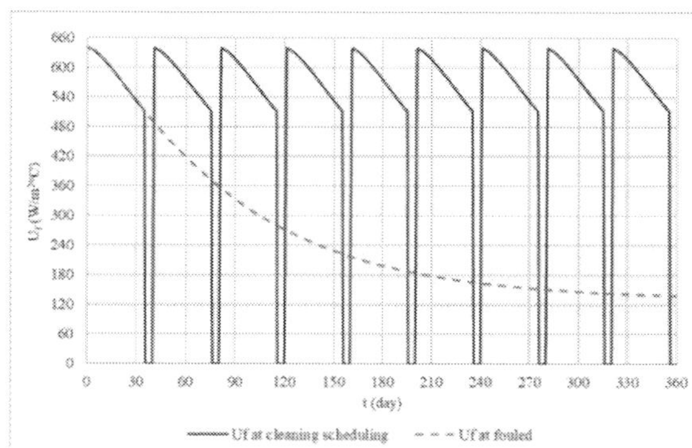

Fig. 13. Cleaning scheduling at HE-10.

In the HE-10 with initial value U_c 640 (W/m^2 °C), U_f at 360 for fouled condition 139,191 (W/m^2 °C) and U_f and at 80% value was 512 (W/m^2 °C). The total of process cleaning 9 times in 360 days, U_f at day 35 value was 512,489 (W/m^2 °C) and year first process cleaning at day 36.

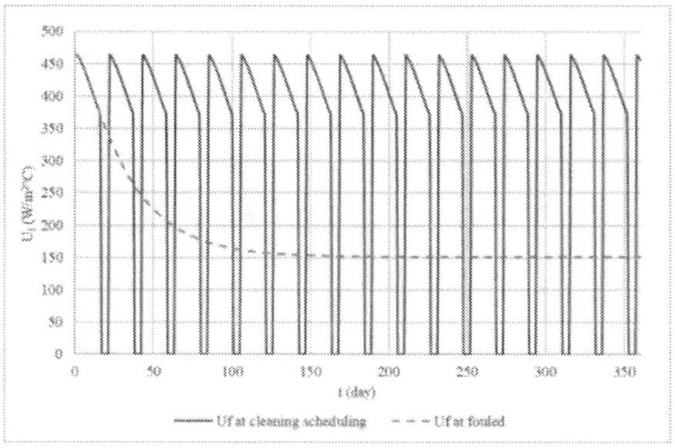

Fig. 14. Cleaning scheduling at HE-11.

In the HE-11 with initial value U_c 465 (W/m^2 °C), U_f at 360 for fouled condition 150,93 (W/m^2 °C) and U_f and at 80% value was 372 (W/m^2 °C). The total of process cleaning 17 times in 360 days, U_f at day 16 value was 372,154 (W/m^2 °C) and first process cleaning at day 17.

7 Conclusions

Scheduling for cleaning is required by looking at the decrease of U over time especially when the value of B is high enough that U will decrease very fast. Smaller values of B is then required when heat exchangers are kept in operation. In other words, infrequent cleaning operation can be implemented. Process cleaning scheduling by decrease Uc at value 80% can be implemented and give result process cleaning frequently.

We thank DIKTI for the financial support through (Penelitian Terapan Unggulan Perguruan Tinggi) PTUPT Project No. 582/PKS/ITS/2017.

References

1. T. R. Biyanto, H.Y. Fibrianto, M. Ramasamy, J. App. Therm. Eng. **108**, 1436-1450 (2016).
2. R. Tellez and B. Young, *Heat Exchanger Network Optimization and Controllability Using Design Reliability Theory*, (Proceedings of the Canadian Design Engineering Network Conference, McGill University, Montreal, Quebec. 2004).
3. S. Kakaç, H. Liu, *Heat Exchangers: Selection, Rating, and Thermal Design* (CRC Press, Boca Raton, FL, 1998).
4. Z. Jin., X. Chen, Q. Dong. and M. Liu, J. Het. Trans. As. Resch., **41**, 33–42 (2012).
5. R. Smith, *Chemical Process Design and Integration* (John Wiley & Sons Ltd, The Atrium, Southern Gate, Chichester, West Sussex PO19 8SQ, England, 2005).
6. D. Licindo, R. Handogo, and J.P. Sutikno, J. Chem. Eng. Trans., **45**, 835–840 (2015).

Fuzzy-GMC Control of Gas-Phase Propylene Copolymerization in Fluidized Bed Reactor

Nazratul Fareha Salahuddin[1], *Ahmad* Shamiri[2], *Mohd Azlan* Hussain[1,*] and *Navid* Mostoufi[3]

[1]Chemical Engineering Department, Faculty of Engineering, University Malaya, 50603 Kuala Lumpur, Malaysia.
[2]Chemical & Petroleum Engineering Department, Faculty of Engineering, Technology and Built Environment, UCSI University, 56000 Kuala Lumpur, Malaysia.
[3]Multiphase Systems Research Lab. School of Chemical Engineering, College of Engineering, University of Tehran, P.O. Box 11155/4563, Tehran, Iran.

Abstract. A modified two-phase model for gas phase propylene and ethylene copolymerization was chosen to represent the process in a fluidized bed reactor. This model considered the entrainment of solid particles in the reactor, as a modification to the original two-phase model assumptions. The non-linearity of this process makes it difficult to control just by using conventional controller such as PID. A hybrid control strategy (a simple designed fuzzy logic controller (FLC) integrated with generic model control (GMC)) is designed to control the temperature of the reactor. This advanced control system was compared with the GMC and conventional PID controller. The simulation results showed that the hybrid controller (Fuzzy-GMC) performed better than both GMC and PID in terms of both servo and regulatory control.

1 Introduction

Polymerization is a process of linking monomer units by chemical reaction to form long chains called polymer. The process is very difficult to measure let alone to control due to its complex reaction kinetics and physical mechanism [1]. There are many phases involve in the reaction hence lead to various heat and mass transfer mechanism in every phase and also the unusual flow characteristics that need to be considered in order to obtain a good model. All these complexity results to high non-linearity of the process dynamics and this are where the process control challenges come in. The difficulty also found in the on-line monitoring of the product (polymer) quality because of the main quality variables such as composition and molecular weight distribution cannot be measured on-line. If it is at all possible, there are still other problems associated with the measurement such as sampling problems, large dead times and/or high noise levels [2].

This present work will focus on the copolymerization of propylene and ethylene in fluidized bed reactor (FBR) with two sites Ziegler-Natta catalyst and triethyl aluminium as co-catalyst. The utilization of FBR in producing polypropylene (PP) is widely preferable in the polymerization industry. The strong interaction between the reactor variables gave additional complexity to the fundamental control problem. Advanced control strategies are needed to cope these problems of which conventional controller are incapable to handle [3]. Although the studies on the modeling and control of polymerization process in fluidized bed reactor are available in the literature [2, 4-6], but minimal work has been done on the polymerization of propylene, specifically copolymerization of propylene and ethylene, for both modeling and control.

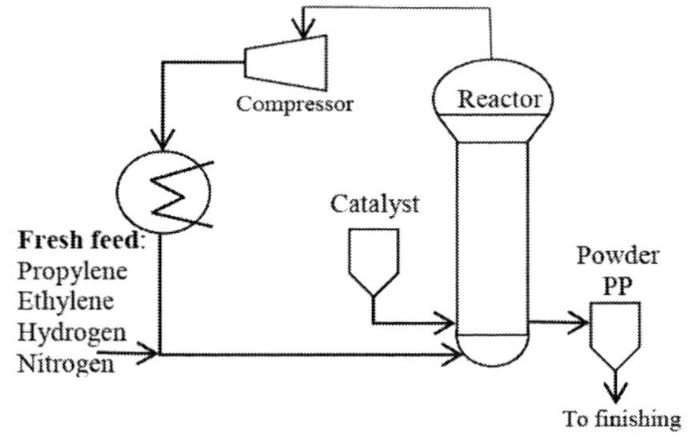

Fig 1. Simplified schematic diagram of fluidized bed propylene and ethylene copolymerization reactor

* Corresponding author: mohd_azlan@um.edu.my

Figure 1**Error! Reference source not found.** shows the simplified schematic of fluidized bed propylene and ethylene copolymerization reactor. The feed gas stream (propylene, ethylene, hydrogen and nitrogen) will help to fluidize the growing polymer particles inside the reactor through the distributor along with the injected catalyst. The solid free gas exits from the top of the reactor, compressed and chilled before combined with fresh feed and fed back to the gas distributor. This recirculation of gas will help to remove the heat of polymerization process. The polymer product is withdrawn continuously from the reactor. The heat removal is crucial since the process is highly exothermic. Hence, it is important to stabilize the reactor temperature above the dew point in order to achieve not only optimal production rate but also to prevent agglomeration of polymer and reactor thermal runaway.

A good model will still need a good controller in order to provide a better environment for the system, especially when controlling the reactor temperature to avoid any explosion in extreme cases. Since polymerization process is very well known for its non-linearity, the searching for the most suitable controller are still going on from time to time. There are many factors influenced the dynamic and steady-state of the polymerization reactor including superficial gas velocity, monomer concentration, monomer feed rate, feed gas temperature and catalyst feed rate among others.

For a polyethylene fluidized bed reactor, Dadebo et al. (1997) [7] used a non-linear error trajectory control (ETC) with SISO system to control the temperature of the reactor. The controller exhibited greater performance and was able to maintain the stability over a wide range of operating conditions compared to optimally tuned PID controller. E. Ali et al. (2003) [8] presented two control schemes in the paper to study the multivariable control of polyethylene in fluidized bed reactor where both non-linear and linear MPC are used and compared. It was found that non-linear MPC was better than the latter in the sense providing a good performance even in the presence of moderate modeling error. Ghasem (2006) [9] had done a study on fuzzy-logic control for temperature regulatory in polyethylene fluidized bed reactor. Both Takagi-Sugeno (TS) and Mamdani FLC, along with PI controller and hybrid of TS-PI and Mamdani-PI are compared. The results showed that hybrid controllers performed better than advanced controller (FLC) and even better than conventional PI controller. In order to control a liquid pool polypropylene hollow-shaft reactor, M. A.-H. Ali et al. (2007) [2] suggested a modified GMC. The modified controller gave better performance for set point and disturbances changes when compared to GMC and PI-Smith predictor controller. Ahmmed S. Ibrehem et al. (2008) [4] studied the performance of neural-network based predictive controller for temperature control of ethylene copolymerization fluidized bed reactor. By using modified three-phase model, the neural-network based MPC displayed greater performance in handling servo and regulatory control than PID. To control molecular weight distribution (MWD), E. M. Ali and Ali (2010) [10] designed non-linear MPC to manipulate hydrogen flow and catalysts feed rate. The non-linear predictive control managed to give a good performance even when there is modeling errors. Adaptive Predictive Model-based Control (APMBC) is proposed by Ho et al. (2012) [3] to control both emulsion temperature and production rate of a polypropylene fluidized bed reactor. The controller used two input two output (TITO) system, and has demonstrated a better ability to track set-point changes when compared to IMC-tuned and Z-N tuned PID controllers. Shamiri et al. (2013) [11] utilized multiple input multiple output (MIMO) centralized model predictive control (MPC) to control polyethylene fluidized-bed reactor. Centralized MPC has proven to be able to control the reactor temperature and production rate successfully and effectively handled disturbance rejection and set point changes compared to PI controller.

As according to Ghasem (2006) [9], a hybrid controller (FLC-PI) was superior than advance controller (FLC) in controlling the polymerization reactors. Fuzzy control exhibited superior performance in transient state while PI has the advantage in controlling the steady state. The combination of two controllers was able to tackle the control problem effectively plus showing an excellent controller performance. Taking that into consideration, to control the reactor temperature, a Fuzzy-GMC hybrid controller is proposed in the current study to cope with the complexity of the non-linear process. A comparative study has been done with the conventional PID controller as well as GMC controller to compare their performance in both servo and regulatory control. An open loop analysis is also done to verify the model.

2 Mathematical Model

In this study, a modified model developed by Shamiri et al. (2015) [12] is chosen to represent the copolymerization process. It is a modified form of the two-phase model developed by Shamiri et al. (2011) [13]. Solid entrainment at the top of the reactor has been taken into account in this model. Generally, only small particles are regarded to be entrained and removed from the system while coarse particles will just stay within the system. But, there exists a situation whereby the velocities are much higher than the terminal velocity that could cause the coarse particles to be entrained from the reactor and the small particles stay. This phenomenon which is known as elutriation is also considered along with solid entrainment in the modified two-phase model.

The assumptions for this model are summarized as below:

a) The emulsion phase does not maintain at the minimum fluidization temperature, and it is assumed to be well-mixed.

 https://doi.org/10.1051/matecconf/201815607002

b) Polymerization reaction occur in both emulsion and bubble phase.
c) Constant bubble size is considered.
d) Heat and mass transfer resistances between solid and gas in both phases are negligible.
e) Uniform particle size throughout the bed.
f) Solid entrainment is considered at the top of the reactor

Based on the assumptions, the material balances for the system are as follows:

For emulsion phase:

$$[M_i]_{e,in}U_eA_e - [M_i]_eU_eA_e - R_v[M_i]_e + ([M_i]_b - [M_i]_e)V_e\left(\frac{\delta}{1-\delta}\right) - (1-\varepsilon_e)R_{i,e} - \frac{K_eV_e\varepsilon_eA_e[M_i]_e}{W_e} = \frac{d}{dt}(V_e\varepsilon_e[M_i]_e) \quad (1)$$

For bubble phase:

$$[M_i]_{b,in}U_bA_b - [M_i]_bU_bA_b - R_v\varepsilon_b[M_i]_b - K_{be}([M_i]_b - [M_i]_e)V_b - (1-\varepsilon_b)\frac{A_b}{V_{PFR}}\int R_{i,b}\,dz - \frac{K_bV_b\varepsilon_bA_b[M_i]_b}{W_b} = \frac{d}{dt}(V_b\varepsilon_b[M_i]_b) \quad (2)$$

The energy balances are as below:

For emulsion phase:

$$U_eA_e(T_{e,in} - T_{ref})\sum_{i=1}^{m}[M_i]_eC_{p,i} - U_eA_e(T_e - T_{ref})\sum_{i=1}^{m}[M_i]_eC_{p,i} - R_v(T_e - T_{ref})\left(\sum_{i=1}^{m}\varepsilon_e[M_i]_eC_{p,i} + (1-\varepsilon_b)\rho_{pol}C_{p,pol}\right) + (1-\varepsilon_b)R_{p,e}\Delta H_R - H_{be}V_e\left(\frac{\delta}{1-\delta}\right)(T_e - T_b) - V_e\varepsilon_e(T_e - T_{ref})\sum_{i=1}^{m}C_{p,i}\frac{d}{dt}([M_i]_e) - \frac{K_eA_e}{W_e(T_e-T_{ref})}\left(\sum_{i=1}^{m}\varepsilon_e[M_i]_eC_{p,i} + (1-\varepsilon_b)\rho_{pol}C_{p,pol}\right) = V_e\left(\sum_{i=1}^{m}\varepsilon_e[M_i]_eC_{p,i} + (1-\varepsilon_b)\rho_{pol}C_{p,pol}\right)\frac{d}{dt}(T_e - T_{ref}) \quad (3)$$

For bubble phase:

$$U_bA_b(T_{b,in} - T_{ref})\sum_{i=1}^{m}[M_i]_bC_{p,i} - U_bA_b(T_b - T_{ref})\sum_{i=1}^{m}[M_i]_bC_{p,i} - R_v(T_b - T_{ref})\left(\sum_{i=1}^{m}\varepsilon_b[M_i]_bC_{p,i} + (1-\varepsilon_b)\rho_{pol}C_{p,pol}\right) + (1-\varepsilon_b)\frac{A_b\Delta H_R}{V_{PFR}}\int R_{p,b}\,dz - H_{be}V_b(T_e - T_b) - V_b\varepsilon_b(T_b - T_{ref})\sum_{i=1}^{m}C_{p,i}\frac{d}{dt}([M_i]_b) - \frac{K_bA_b}{W_b(T_b-T_{ref})}\left(\sum_{i=1}^{m}\varepsilon_b[M_i]_bC_{p,i} + (1-\varepsilon_b)\rho_{pol}C_{p,pol}\right) = V_b\left(\sum_{i=1}^{m}\varepsilon_b[M_i]_bC_{p,i} + (1-\varepsilon_b)\rho_{pol}C_{p,pol}\right)\frac{d}{dt}(T_b - T_{ref}) \quad (4)$$

Where

$$K_e = 23.7\rho_gU_o\frac{A}{W_e}exp\left(\frac{-5.4U_td_p}{U_o}\right) \quad (5)$$

$$K_b = 23.7\rho_gU_o\frac{A}{W_b}exp\left(\frac{-5.4U_td_p}{U_o}\right) \quad (6)$$

$$W_e = AH(1-\varepsilon_e)\rho_{pol} \quad (7)$$

$$W_b = AH(1-\varepsilon_b)\rho_{pol} \quad (8)$$

$$U_t = U_t^*[\mu\rho_g{}^{-2}(\rho_{pol} - \rho_g)g]^{1/3} \quad (9)$$

$$U_t^* = [18(d_p^*)^{-2} + (2.335 - 1.744\emptyset_s)(d_p^*)^{-0.5}]^{-1}, \quad (10)$$

$$\text{for } 0.5 < \emptyset_s \leq 1$$

$$d_p^* = d_p[\mu^{-2}\rho_g(\rho_{pol} - \rho_g)g]^{1/3} \quad (11)$$

Equations (1) to (4) are solved by using MATLAB with the following initial conditions:

$$[M_i]_{b,t=0} = [M_i]_{in} \quad (12)$$

$$T_b(t=0) = T_{in} \quad (13)$$

$$[M_i]_{e,t=0} = [M_i]_{in} \quad (14)$$

$$T_e(t=0) = T_{in} \quad (15)$$

All correlations and equations related to this modified two-phase model are listed in Table 1 below.

Table 1. Correlations and equations used in the mathematical model

Parameters	Formula	Reference
Minimum fluidization velocity	$Re_{mf} = [(29.5)^2 + 0.375Ar]^{1/2} - 29.5$	[14]
Bubble velocity	$U_b = U_o - U_{mf} - U_{br}$	
Bubble rise velocity	$U_{br} = 0.711(gd_b)^{1/2}$	[15]
Emulsion velocity	$U_e = \frac{U_{mf}}{\varepsilon_{mf}(1-\delta)}$	
Bubble diameter	For Geldart B: $d_b = d_{b0}[1 + 27(U_0 - U_e)]^{\frac{1}{3}}(1 + 6.84H)d_{b0} = 0.0085$	[16]
Mass transfer coefficient	$K_{be} = \left(\frac{1}{K_{bc}} + \frac{1}{K_{ce}}\right)^{-1}$ $K_{bc} = 4.5\left(\frac{U_e}{d_b}\right) + 5.85\left(\frac{D_g^{1/2}g^{1/4}}{d_b^{5/4}}\right)$ $K_{ce} = 6.77\left(\frac{D_g\varepsilon_eU_{br}}{d_b}\right)$	[15]

Heat transfer coefficient	$H_{be} = \left(\frac{1}{H_{bc}} + \frac{1}{H_{ce}}\right)^{-1}$ $H_{bc} = 4.5\left(\frac{U_e \rho_g C_{pg}}{d_b}\right) +$ $5.85\left(\frac{(U_e \rho_g C_{pg})^{1/2} g^{1/4}}{d_b^{5/4}}\right)$ $H_{ce} = 6.77(\rho_g C_{pg} k_g)^{1/2}\left(\frac{\varepsilon_e U_{br}}{{d_b}^3}\right)^{1/2}$	
Bubble phase fraction	$\delta = \frac{U_0 - U_{mf}}{U_b}$	
Emulsion phase porosity	$\varepsilon_e = \varepsilon_{mf}$	
Volume of polymer phase in the emulsion phase	$V_p = AH(1-\varepsilon_e)(1-\delta)$	[15]

3 Control Strategy

As mentioned in the previous studies on the temperature control of the polymerization process, in order to tackle this non-linearity issue, a hybrid controller is suggested to be better than other types of controller when it comes to coping with the complex system. In this work, a fuzzy-GMC hybrid controller is designed to control the reactor temperature by manipulating the cooling water valve rate, F_c (%).

3.1 Generic Model Control

GMC is very highly dependent on the model and is only limited for the systems with linear in control.
GMC control algorithms are given by:

$$K_1\varepsilon(t) + \int_0^{t_f} K_2\varepsilon(t)dt = \frac{\partial H(X)}{\partial x} \times F(X,U,t) \tag{16}$$

Where H(x) is the controlled variable.
For any non-linear system with linear control,

$$F(X,U,t) = F(X) + G(X)U \tag{17}$$

Therefore,

$$K_1\varepsilon(t) + \int_0^{t_f} K_2\varepsilon(t)dt = \frac{\partial H(X)}{\partial x} \times [F(X) + G(X)U] \tag{18}$$

In this case, $\frac{\partial H(X)}{\partial x} = 1$.
Hence, the GMC formulation is given by,

$$U = \frac{K_1\varepsilon(t) + \int_0^{t_f} K_2\varepsilon(t)dt - F(X)}{G(X)} \tag{19}$$

In implementing the above strategy for the propylene copolymerization process, $F(X)$ and $G(X)$ are mass and energy balances of the emulsion phase in Equation (3), U is the coolant flow (F_c), which is the manipulated variable to control the reactor temperature, and $\varepsilon(t)$ is the difference between the emulsion temperature and the desired set point. Hence, for the current process, GMC formulation becomes as follows:

$$F_c = \frac{\propto [K_1\varepsilon(t) + \int_0^{t_f} K_2\varepsilon(t)dt] - F(X)}{G(X)} \tag{20}$$

where

$$\propto = V_e\left(\sum_{i=1}^{m} \varepsilon_e[M_i]_e C_{p,i} + (1-\varepsilon_b)\rho_{pol}C_{p,pol}\right) \tag{21}$$

$$F(X) = \begin{array}{c} (1-\varepsilon_b)R_{p,e}\Delta H_R + H_{be}V_e\left(\frac{\delta}{1-\delta}\right)(T_e - T_b) \\ +U_e A_e (T_e - T_{ref})\sum_{i=1}^{m}[M_i]_e C_{p,i} \\ +R_v(T_e - T_{ref})\left(\sum_{i=1}^{m}\varepsilon_e[M_i]_e C_{p,i} + (1-\varepsilon_b)\rho_{pol}C_{p,pol}\right) \\ +V_e\varepsilon_e(T_e - T_{ref})\sum_{i=1}^{m} C_{p,i}\frac{d}{dt}([M_i]_e) \\ +\frac{K_e A_e}{W_e(T_e - T_{ref})}\left(\sum_{i=1}^{m}\varepsilon_e[M_i]_e C_{p,i} + (1-\varepsilon_b)\rho_{pol}C_{p,pol}\right) \\ +(333.15 - T_{ref})(U_e A_e \sum_{i=1}^{m}[M_i]_e C_{p,i}) \end{array} \tag{22}$$

$$G(X) = \frac{Ke^{-\tau_D s}}{\tau s + 1} U_e A_e \sum_{i=1}^{m}[M_i]_e C_{p,i} \tag{23}$$

GMC performance specifications, K_1 and K_2 is obtained from the generalized GMC profile specification graph by Lee and Sullivan (1988) [17]. These values are selected based on the desired response suitable for the process. For the copolymerization process, $K_1 = 9 \times 10^{-3}$ and $K_2 = 2.25 \times 10^{-6}$.

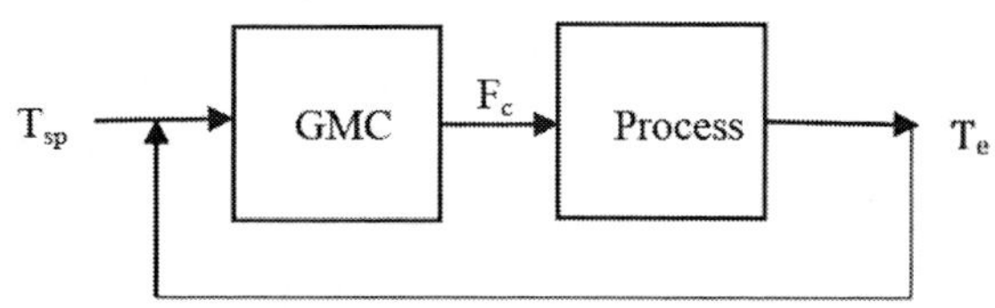

FIGURE 2. GMC control scheme

3.2 Fuzzy-GMC Hybrid Control

The purpose for the hybrid control scheme is to modify the outcome from GMC, in order to reduce the overshoot problem. A fuzzy-logic controller is used to do the modification then coupled with GMC to control the reactor.

FLC is strongly based on its fuzzy sets, linguistic variables and the approximate reasoning. A set of fuzzy rules are set based on the input and output variables. Its input variables are error, *e* (the difference between set-point and reactor temperature) and $\Delta T_e/F_c$ (quotient of the changes of the reactor temperature and the associated coolant valve rate). The output variable is a factor, f where this value is used to compensate the outcome from GMC ($\Delta F_c = f \times F_c$).

This is a simple designed fuzzy logic controller by using Mamdani method. The membership functions are in triangular form.

TABLE 2. Linguistic terms for process variables

Process variables		Linguistic terms
Input	e	Positive (P)
		Negative (N)
		Zero (Z)
	$\Delta T_e/F_c$	Small (Sm)
		Medium (Med)
		Large (L)
Output	f	Negative large
		Negative medium
		Negative small
		Positive large
		Positive medium
		Positive small
		Zero

The fuzzy rules are:

1. IF (error is negative) AND (ΔT_e /F_c is small) THEN (f is negative large)
2. IF (error is negative) AND (ΔT_e /F_c is medium) THEN (f is negative medium)
3. IF (error is negative) AND (ΔT_e /F_c is large) THEN (f is negative small)
4. IF (error is positive) AND (ΔT_e /F_c is small) THEN (f is positive large)
5. IF (error is positive) AND (ΔT_e /F_c is medium) THEN (f is positive medium)
6. IF (error is positive) AND (ΔT_e /F_c is large) THEN (f is positive small)
7. IF (error is zero) THEN (f is zero)

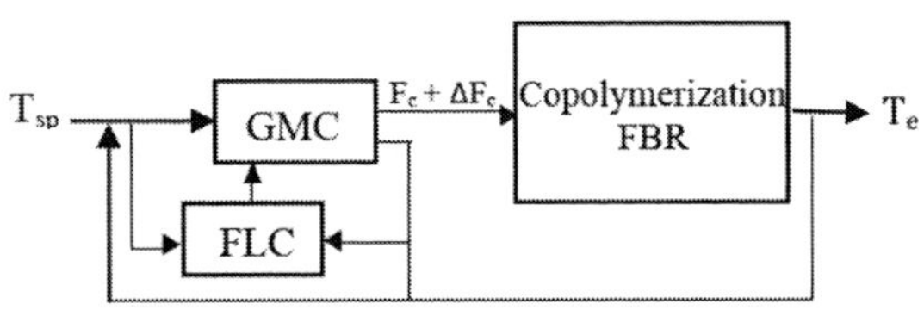

Fig 3. Fuzzy-GMC hybrid control scheme

4 RESULTS AND DISCUSSION

4.1 Open-Loop Analysis

To verify the constructed mathematical model, the process is simulated in MATLAB. Several operating conditions (superficial velocity, propylene and ethylene feed concentration, coolant valve rate) were subjected to step changes and Figure 4 to 7 shows the effect of step change in each parameter to the emulsion temperature.

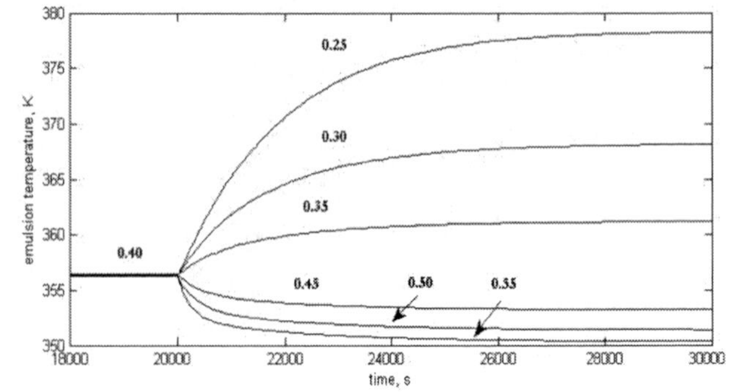

Fig 4. Effect of step change in superficial velocity, U0 on the emulsion temperature.

Based on Figure 4, the step change in superficial velocity shows a significant impact on the emulsion temperature. The nominal value of superficial velocity is taken as 0.4 m/s. In order to demonstrate the non-linearity, step changes are taken 0.05 m/s larger or smaller than the nominal value. Emulsion temperature increases with decreasing the superficial velocity. As shown in the figure, even with a very small change in the superficial velocity, the impact on the emulsion temperature was large and this indicates that it is crucial to monitor the superficial velocity when we are controlling the temperature of the reactor. Since the figure above shows a non-symmetry response, this shows that the relationship is non-linear.

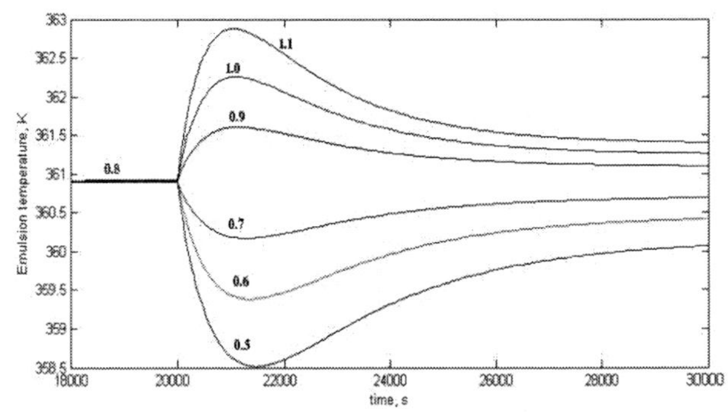

Fig 5. Effect of step change in the propylene concentration on the emulsion temperature.

Effect of monomer concentration on the emulsion temperature is depicted in Figure 5. It can be seen in this figure that changes in the propylene concentration has a significant impact on the reactor temperature when the concentration is changed by 0.1 mol/L from the nominal value. The non-symmetrical pattern in this figure indicates that the reactor temperature changes non-linearly with change of propylene concentration. Hence, it can be concluded that the propylene concentration is among the important parameters when controlling the reactor temperature.

https://doi.org/10.1051/matecconf/201815607002

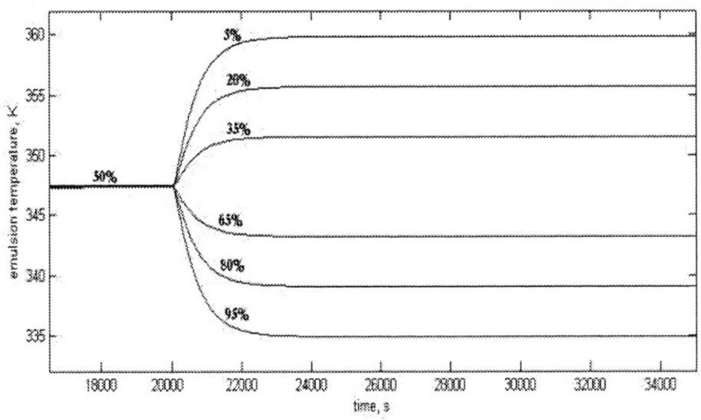

Fig 6. Effect of step change in coolant flow (%) on emulsion temperature

Figure 6 illustrate the effect of step change in coolant valve rate (%) on the emulsion temperature. Change in valve rate should have linear relationship with emulsion temperature in order to directly change the temperature when needed. As shown in figure, the symmetrical pattern shows the linear relationship.

Based on these analysis, it can be said that the model is reliable, and a dependable representative of the copolymerization process.

4.2 Control System Analysis

Conventional controller, PID is also designed to compare its performance. By using Ziegler-Nichols method, the PID setting are as follows:
$K_c = 24.24$
$\tau_I = 129.5$
$\tau_D = 32.37$

4.2.1 Set-Point Tracking

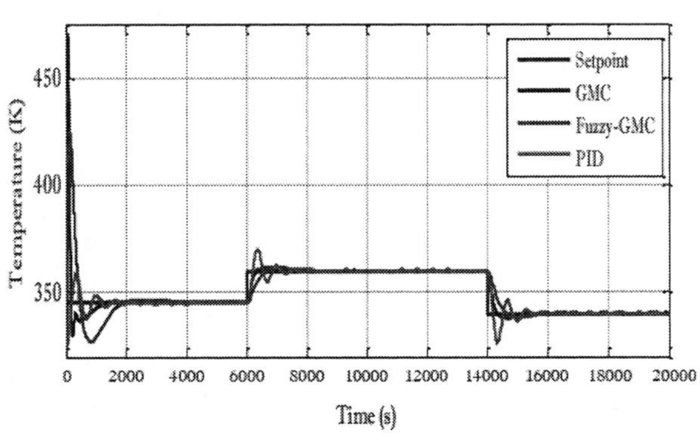

Fig 7. Performance comparison of GMC, Fuzzy-GMC Hybrid and PID controller in set-point tracking.

Figure 7 shows the performances of the controllers tracking the set-point changes. At the first part, GMC shows a very high overshoot and a very good set-point tracking. However, fuzzy-GMC controller manages to track the change in set-point in less than 1000 secs and also reduces the overshoot. PID on the other hand, gives low overshoot but exhibit poor performance in tracking the set-point changes compared to the other two controllers. Hence, hybrid controller exhibit better performance than GMC and PID controllers (low overshoot and very fast response).

4.2.2 Disturbance Rejection

Two disturbances are introduced to the system to analyze the disturbance rejection capability for each controller. Changes in superficial velocity, U_0 from its nominal value (0.35 m/s) are imposed to the system. At 4000 secs, U_0 is increased to 0.55 m/s and then decreased to 0.3 m/s at 8000 secs. Changes in ethylene concentration is also applied to the process. At 6000 secs, the concentration is increased from its nominal value of 0.05 mol/L to 0.15 mol/L and then decreased to 0.03 mol/L at 14000 secs.

Based on Figure 8, PID controller is unable to cope with the disturbance since the response did not stop oscillating. Although the overshoot is low, the oscillating response did not exhibit the capability of the PID to reject the disturbance well. Both GMC and fuzzy-GMC hybrid controller exhibit good performance in rejecting the disturbances. The slight difference is observed that is fuzzy-GMC hybrid controller's response is much stable after the disturbance is introduced compared to GMC controller.

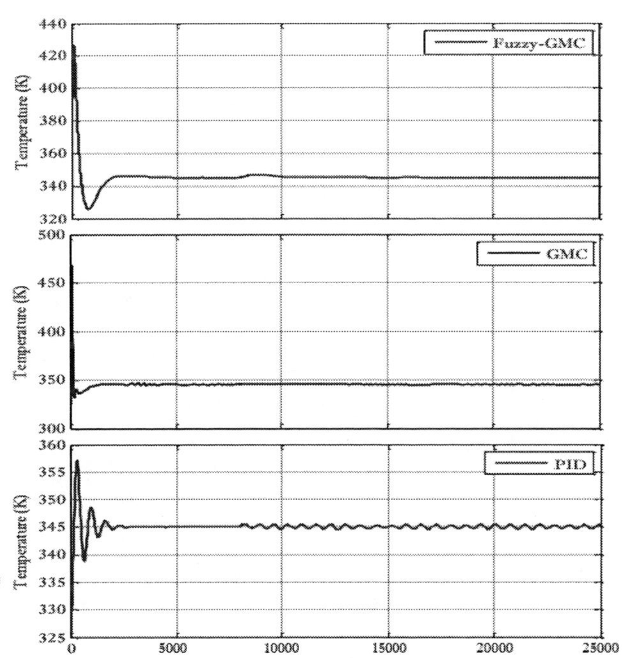

Fig 8. Controllers' performance in disturbance rejection.

5 Conclusion

A modified two-phase model that considered solid entrainment in the fluidized bed reactor is utilized in the study to represent the copolymerization process. The model is then verified by using open-loop analysis. The analysis showed that the model is reliable and a dependable representation of the process.

A model based controller, GMC is designed to control the temperature of the reactor. GMC gave a very high overshoot in its response. A hybrid controller is then designed to reduce the overshoot. Results show that fuzzy-GMC hybrid controller is able to reduce the overshoot problem and provide the best response in set-point tracking as well as disturbance rejection compared to GMC and PID controller. Future study will be needed to further tune the hybrid controller to improve the initial overshoot problem.

The authors acknowledge the funding from FRGS (FP064-2015A) University of Malaya.

References

1. Richards, J. R., & Congalidis, J. P., Comput. Chem. Eng., ***30***(10–12), 1447-1463 (2006).
2. Ali, M. A.-H., Betlem, B., Weickert, G., & Roffel, B., Chem. Eng. Process, ***46***(6), 554-564 (2007).
3. Ho, Y. K., Shamiri, A., Mjalli, F. S., & Hussain, M. A., J. Process Contr., ***22***(6), 947-958 (2012).
4. Ibrehem, A. S., Hussain, M. A., & Ghasem, N. M. Chinese J. Chem. Eng., ***16***(1), 84-89 (2008).
5. Salau, N. P. G., Neumann, G. A., Trierweiler, J. O., & Secchi, A. R., Ind. Eng. Chem. Res., ***47***(16), 6058-6069 (2008).
6. Vahidi, O., Shahrokhi, M., & Mirzaei, A., Iran. Chem. Chem. Eng., ***27***(3), 87-101 (2008).
7. Dadebo, S. A., Bell, M. L., McLellan, P. J., & McAuley, K. B., J. Process Contr., *7*(2), 83-95 (1997).
8. Ali, E., Al-Humaizi, K., & Ajbar, A., Ind. Eng. Chem. Res., ***42***(11), 2349-2364 (2003).
9. Ghasem, N. M., Chem. Eng. Res. Des., ***84***(2), 97-106 (2006).
10. Ali, E. M., & Ali, M. A.-H., ISA T, ***49***(1), 177-187 (2010).
11. Shamiri, A., Hussain, M. a., Mjalli, F. s., Mostoufi, N., & Hajimolana, S., Chinese J. Chem. Eng., ***21***(9), 1015-1029 (2013).
12. Shamiri, A., Wong, S. W., Zanil, M. F., Hussain, M. A., & Mostoufi, N., Chem. Eng. J., ***264***, 706-719 (2015).
13. Shamiri, A., Azlan Hussain, M., Sabri Mjalli, F., Mostoufi, N., & Saleh Shafeeyan, M., Chem. Eng. Sci., ***66***(6), 1189-1199 (2011).
14. Lucas, A., Arnaldos, J., Casal, J., & Puigjaner, L., Ind. Eng. Chem. Process Des. Dev., ***25***(2), 426-429 (1986).
15. Kunii, D., & Levenspiel, O., *Fluidization Engineering*: Butterworth-Heinemann (1991).
16. Hilligardt, K., & Werther, J., Chem-Ing-Tech, ***57***(7), 622-623 (1985).
17. Lee, P. L., & Sullivan, G. R. (1988). Generic model control (GMC). *Computers & Chemical Engineering, 12*(6), 573-580.
18. Alizadeh, M., Mostoufi, N., Pourmahdian, S., & Sotudeh-Gharebagh, R., Chem. Eng. J, ***97***(1), 27-35 (2004).
19. Chatzidoukas, C., Perkins, J. D., Pistikopoulos, E. N., & Kiparissides, C., Chem. Eng. Sci., ***58***(16), 3643-3658 (2003).
20. Choi, K. Y., & Ray, W. H., Chem. Eng. Sci., ***40***(12), 2261-2279 (1985).
21. Cui, H., Mostoufi, N., & Chaouki, J., Chem. Eng. J., ***79***(2), 133-143 (2000).
22. Cui, H., Mostoufi, N., & Chaouki, J., Powder Technol, ***120***(1–2), 12-20 (2001).
23. Hatzantonis, H., Yiannoulakis, H., Yiagopoulos, A., & Kiparissides, C., Chem. Eng. Sci., ***55***(16), 3237-3259 (2000).
24. Ibrehem, A. S., Hussain, M. A., & Ghasem, N. M., Aust. J. Basic & Appl., ***3***(4), 3954-3974 (2009).
25. Ibrehem, A. S., Hussain, M. A., & Ghasem, N. M., Chem. Eng. J., ***149***(1–3), 353-362 (2009).
26. Ibrehem, A. S., Hussain, M. A., & Ghasem, N. M., Iran. Chem. Chem. Eng., ***31***(4), 91-117 (2012).
27. McAuley, K. B., & Macgregor, J. F., Aiche J., ***39***(5), 855-866 (1993).
28. McAuley, K. B., Talbot, J. P., & Harris, T. J., Chem. Eng. Sci., ***49***(13), 2035-2045 (1994).
29. Shamiri, A., Hussain, M. A., Mjalli, F. S., & Mostoufi, N., Chem. Eng. J., ***161***(1–2), 240-249 (2010).
30. Shamiri, A., Hussain, M. A., Mjalli, F. S., & Mostoufi, N., Comput. Chem. Eng., ***36***(0), 35-47 (2012).

Improving the Student Competency of the "Modeling and Computation Process" Course by Using Open Source Application.

Setia Budi Sasongko[1,*], and *Luqman* Buchori[2].

[1, 2] Computational Process Laboratory, Department of Chemical Engineering, Faculty of Engineering, Diponegoro University, Indonesia

Abstract. One of the student outcome based on ABET is the ability to use the techniques, skills, and modern engineering tools (in this case is engineering software) necessary for engineering practice. The problem is how the process of student centered learning so that the learning outcome of the "Modeling and computation process" subject can be achieved especially analysis process using modern engineering software. The first problem is commercial software applications have limitations due to high prices and limited amounts and usage time. The second problem is pasive student caused of one-way learning. The first problem is solve by utilisation of the open source software, that is "scilab" and "COCO". The blended learning method is used to minimize the second problem, that is combination of face-to-face learning in the class, distance learning which use the "edmodo", and accompaniment learning by senior student in the computation laboratory. The results of the active learning are increasing the ability of the student to apply equipment design and process system, improving soft skill outcome such as team work, communication, time management, thinking criticaly and analysing effectively.

1 Background

The change of education from lecturer-centered learning toward student-centered learning is not easy. One way to improve active student is to use software engineering applications in the learning process. This is in line with one of ABET student outcomes, that is the ability to use the techniques, skills, and modern engineering tools (in this case is engineering software) necessary for engineering practice.

The first problem to meet previous performance indicator is commercial software applications have limitations due to high prices and limited amounts and usage time. The second problem is generally learning process based on the lecture activity or one-way learning.

2 Learning Methodology

In the Information Age, the chemical engineering education should shift to the use of industrial software so that students will be easy to apply theory of basic chemical engineering to the design of chemical industry. The limitation of the commercial software (first problem) can be overcome by using open source software that is scilab and COCO. The Scilab (Matlab-like) is free open-source software for the numerical computation and scientific applications[1]. COCO (CAPE-OPEN to CAPE-OPEN) is a free of charge CAPE-OPEN (Computer-Aided Process Engineering) compliant steady state simulation environment [2]. The COCO is developed and maintained by AmsterCHEM which is collection of software component consist of four main component there are COFE the flow-sheeting engine, TEA (Thermodynamic for Engineering Application), COUSCOUS (collection of unit operations) and CORN (reaction numerical package).

The blended learning is applied in this subject to overcome teaching centered learning (second problem). Blended learning (figure 1) means the combination of three learning methods that are first face-to-face learning in the class, second distance learning which use the "edmodo", and accompaniment learning by senior student in the computation laboratory. The Edmodo is teaching tool modeled which designed for teacher/student interaction with emphasis on communication, assignment sharing [3].

The "Modeling and computation process" subject in the Chemical Engineering Department, Diponegoro University is taken as an active learning sample. The

* Corresponding author: sbudisas@live.undip.ac.id

learning outcomes of this subject is the student will have an ability to apply engineering fundamental using computer programming or simulation.

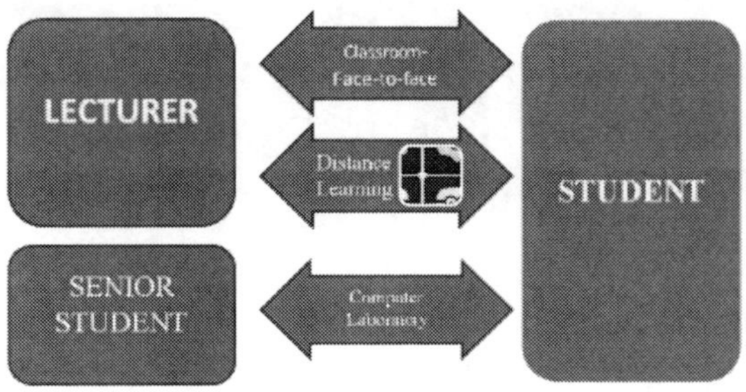

Fig. 1. Blended learning method

In the classroom (face-to-face learning), the learning process implemented tutorial applied program directly, because most students bring their laptops. Requirement specification of the "Scilab" and "COCO" are not high, therefore notebook can be used for the learning process in the class. Lecturer gives the simple problem, than student solve it under lecturer supervision. In this case, learning process consist of small group discussion, problem based learning. Application of Information Technology (IT) in education will improve three C's: Computing, Communication and Collaboration [4]. Therefore, the outcomes of this process are team work, communication, and critical thinking. In general, the relationship between open source software with the learning process can be seen in Figure 2.

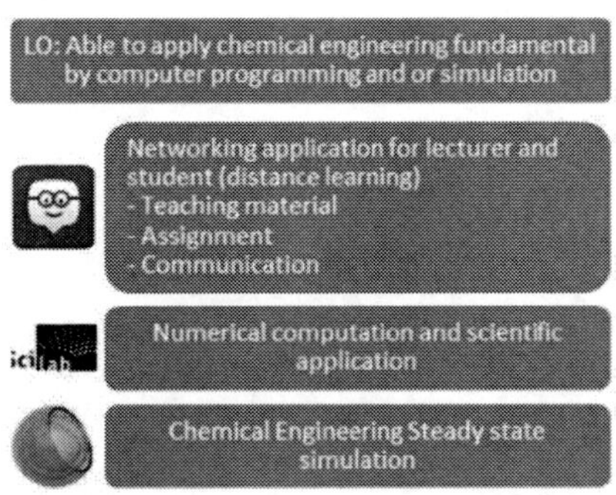

Fig. 2. Relationship between learning outcome (LO) with open source software in the "Modeling and computation process" subject.

Both software (COCO and scilab) in the learning process have different approach. COCO is used to study the overall process system (black-box system). Scilab is used to study each unit (section) in detail. In this case, students in a group get the assignment to design the reactor using scilab. Every group has a different problems, as shown in table 1. Each alphabet has a meaning as shown in table 2.

Table 1. Several type of assignment

Group	Assignment	Group	Assignment
1	A C E G I	7	A C E H I
2	B C E G I	8	B C E H I
3	A D E G I	9	A D E H I
4	B D E G I	10	B D E H I
5	A C F G I	11	A C F H I
6	B C F G I	12	B C F H I

Table 2. Classification code for the group assignment

No.	Classification, based on	Classification code	
1	Type of reactor	**A**	**B**
		CSTR $\frac{X}{V}=\frac{-r_A}{F_{Ao}}$	PFR $\frac{dX}{dV}=\frac{-r_A}{F_{Ao}}$
2	Influence of heat to the environment	**C**	**D**
		Adiabatic $Q=0$	Non adiabatic $Q\neq 0$
3	Heat of reaction	**E**	**F**
		Exothermic $\Delta H_R = -$	Endodermis $\Delta H_R = +$
4	Equilibrium	**G**	**H**
		Irreversible $A \rightarrow B$	Reversible $A \Leftrightarrow B$
5	Molecular of reactant	**I**	**J**
		Mono molecular $A \rightarrow D$	Bimolecular $A + B \rightarrow D$

After completing the assignment, students can work in teams, communicate, share knowledge, and be more creative.

The development of a process system (imitating Coco) by merging excel as a data base with Scilab GUI (Graphical User Interface) is also done by the students as shown in Figure 3.

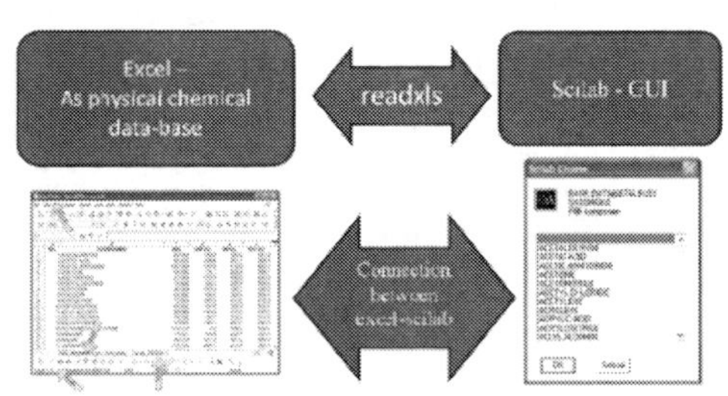

Fig. 3. Relationship between excel with scilab (GUI) to replicate property package in COCO simulator (Hysys-like).

Coco is used to develop the design of chemical plants as shown in Figure 4.

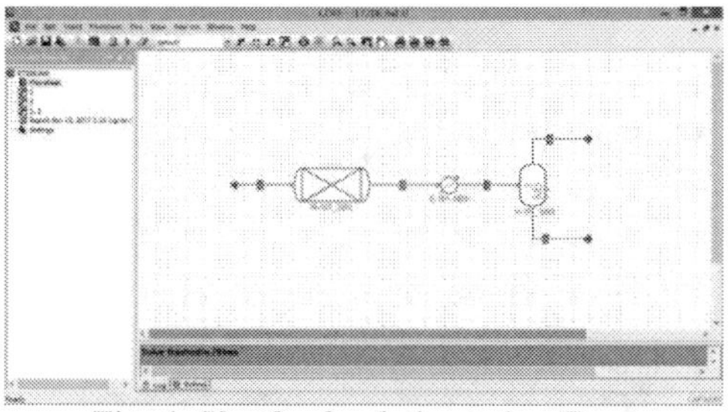

Fig. 4. Simple simulation using Coco

3 Conclusion

The development of learning based on the information technology will improve the student learning outcomes, both knowledge, skills and affective. Open source software can overcome the limitations of expensive commercial software without reducing the function of learning and software development.

References

1. Campbell S.L. Chancelier J, *Modeling and Simulation in Scilab/Scicos*, Available at: http://www.cse.iitb.ac.in/~nitinwagmode/seminar/scicos_paper/book.pdf.
2. www.cocosimulator.org
3. www.edmodo.com
4. Edgar T., *Chemical Engineering Education and the Three C's: Computing, Communication, and Collaboration. AIChE Annual Meeting, LA,* Available at: http://cache.org/site/news_stand/fall00/fall2000_chemengedu.pdf [Accessed December 14, 2014]

Rotating Membrane Emulsification for Producing Single and Multiple Emulsions

Nita Aryanti[1,*], and *Richard* A. Williams[1]

[1]Department of Chemical Engineering, Kampus Undip Tembalang, Semarang, 50278
[2]Heriot-Watt University, Edinburgh EH14 4AS

Abstract. Membrane emulsification is a technique utilising a novel concept of generating droplet 'drop by drop' to produce emulsions. The technique has several distinctive advantages over the conventional emulsification techniques.This paper concerns on the development of membrane emulsification (Rotating Membrane Reactor, RMR) which utilizes rotating tubular membrane to initiate droplet detachments. The RMR uses a rotating stainless steel tubular membrane with laser drilled pores (100 μm pore diameter) and a syringe pump to drive the dispersed phase through the membrane at a given flow rate. O/W formulations were prepared with low viscosity of paraffin wax, two types of emulsifiers, different membrane rotation rate and dispersed phase flow rate. The emulsion droplets exhibited a coefficient of variation of 9% and 81μm droplet size. In this research, the pore size/droplet size ratio could achieve 0.8. This value was below than other membrane emulsification processes. The effects of principal system operating parameters on both the average droplet diameter and droplet uniformity were discussed. In addition, a multiple (W/O/W) emulsion formulation was investigated as well.

1 Introduction

Existing methods for manufacturing emulsions involve droplet break-up using shear or impact stresses. These have some problems relating to emulsion production such as inefficiency of energy used, lack of droplet size and size distribution control and poor equipment reproducibility [1]. In addition, the behaviour of any pair of immiscible phases normally cannot be predicted. The high of energy utilization leads to high cost of manufacture. Furthermore, the quality of the product can vary from one manufacturing vessel design to another.

In order to solve some of these problems, membrane emulsification has been developed as an alternative method for producing emulsions. This method is claimed to be superior in many respects to those existing methods. By using membrane emulsification, size and size distribution of droplets can be carefully controlled through the selection of porous membrane, efflux rate of discontinuous phase, cross flow velocity of continuous phase, etc [1]. It means that the membrane emulsification process can produce a narrow droplets size distribution [2,3], potential to produce monodispersity of emulsions [4] and require lower surfactant [3]. In addition, this equipment has lower apparent shear stress compared to the conventional emulsification process [2,4]. Membrane emulsification also generally require lower energy input (104 – 106 J/m^3) compared with the conventional emulsification (106-109 J/m^3) [2,3,4].

Several recent development reports in the membrane emulsification technology are available in the literature [2, 3, 5]. Due to the advantages of membrane emulsification, this method has been successfully applied for preparation of complex particulate products such as double/multiple emulsions [6]. In more particular, beetroot juice as well as procyanidin-rich extract encapsulated in water-in-oil-in-water (W/O/W) emulsions were prepared using premix emulsification system [7,8]. In addition, food-grade W1/O/W2 emulsions containing resveratrol or vitamin B_{12} and garlic extracts W1/O/W2 emulsions were prepared by membrane emulsification and stirred cell membrane emulsification, respectively [9,10]

This paper investigates development of membrane emulsification system. Droplet detachment is initiated by rotating tubular membranes. In this system, the shear force was developed by the rotating membrane rather than the flowing continuous phase. By applying this system, pump circulation of continuous phase in the system can be eliminated. This has a potential advantage of eliminating the need for pump circulation of the

* Corresponding author: nita.aryanti@che.undip.ac.id

continuous phase in the system, which can be particularly beneficial when fabrications of coarser emulsion droplets or fragile structured multiphase emulsion and capsule products are concerned, as those delicate structures can be easily destroyed during the circulation. The effects of the membrane rotational speeds, and concentrations of emulsifiers and stabilisers on the droplet size and monodispersity in the production of single phase O/W emulsions is discussed. Some preliminary experimental results in using the rotating membrane reactor (RMR) for the fabrication of multiphase W/O/W emulsions is also be presented.

2 Experimental

2.1. Experimental Setup

Fig. 1 and Fig. 2 illustrate a photographic image and schematic diagram of the experimental apparatus respectively.

Fig. 1. Photographic image of rotating membrane emulsification device.

The RMR consisted of a tubular membrane made from stainless steel having diameter of 10 mm and effective length of 85 mm. The membrane was mounted on a digital overhead stirrer (IKA Eurostar), which enables the membrane to rotate in a stationary vessel. The pores of membrane (fabricated via laser drilling) were arranged in a cubic array. The membrane having 100 µm mean pore size and 500 µm pore-pore spacing, which correspond to membrane porosity of 3%. In the current work, a standard 100 ml measuring cylinder (diameter 25 mm) was used as the stationary vessel.

2.2 Single Emulsion

In the single phase emulsion investigation, low viscosity paraffin wax (28mPa.S, Fluka) was used as the disperse phase. As emulsifier, either Tween 20 (polyoxyethylene sorbitan monolaureate) (Fisher Chemical, UK) or sodium dodecyl sulphate (SDS) (Fluka, UK) was used as the emulsifier. Carbomer (Carbopol ETD 2050) (Surfachem Ltd., Leeds, UK) was used as the stabiliser. The dispersed phase was introduced inside the membrane tube by mean of a constant flow syringe pump with adjustable flow rate between 7-211 ml/hr.

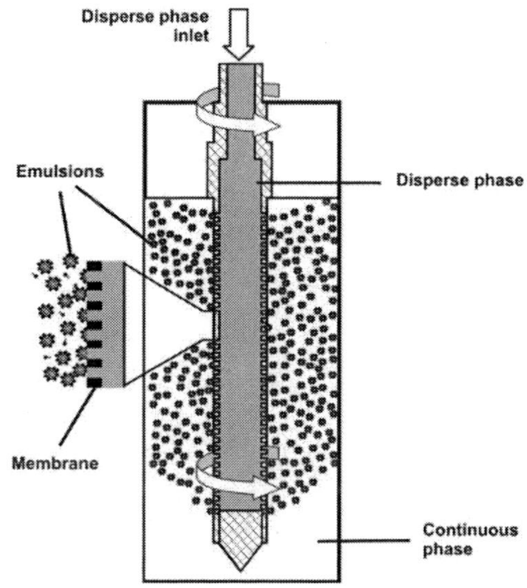

Fig. 2. Schematic illustration of rotating membrane emulsification device, insert showing a microscopic view of the stainless steel membrane.

2.3 Multiple W/O/W Emulsions

In the multiphase emulsion investigation, a rapeseed oil was used as the oil phase. Deionised water was employed as both the inner and outer water phase. The W/O/W emulsions were prepared by a two-step emulsification method. In the first step, a single phase W/O emulsion was prepared through homogenisation (Polytron PT2100) at 19,000 rpm for 3 minutes, which produced approximately 1-10 µm water droplets in the rapeseed oil. A lipophilic surfactant, Polyglycerol polyricinoleate (PGPR) (Danisco, Denmark), was added to the oil phase as the emulsifier. The concentration of deionised water in the rapeseed oil was kept constant at 10% (w/w). The W/O emulsion was then used as the disperse phase for the second step emulsification by RMR, in which the W/O emulsion was carefully injected into the membrane tube using a syringe pump (Razel A99FMZ, from Fisher Scientific), and dispersed into the outer deionised water phase containing 2% Tween 80 (polyoxyethylene sorbitan monooleate) (Across Organic) as the emulsifier.

https://doi.org/10.1051/matecconf/201815608001

2.4 Droplet Characterization

Droplets were examined directly using a digital microscope (Nikon model SMZ800) with 10-63 times magnification. For multiple emulsions, a further analyses was examined using confocal microscope (Nicon Eclipse TE2000-U). The images were then intrepreted by image processing software (Image Pro Plus) for number based distribution. The monodispersity of the droplets was evaluated in terms of the coefficient of variation (CV) which is defined as:

$$CV = \frac{S_d}{D_{avg}} \times 100\% \qquad (1)$$

where Sd and D_{avg} represent the standard deviation of droplet diameter and the average droplet diameter respectively. It was suggested that droplets with a CV no higher than 25% should be considered as monodispersed.

3 Result and Discussion

3.1. Single Emulsion

Fig. 3 presents the effect of membrane rotation on droplet average diameter at carbomer concentration of 0.1% at different type of emulsifier. It can be seen that the average droplet diameter initially decreases sharply with the increase of membrane rotation speed no matter the type of emulsifiers used. This phenomenon is likely du to the shear force at the membrane surface, which is the main droplet detaching force, increases with the increase of the membrane rotational speed [11]. In the meantime, higher membrane rotational speeds also result in shorter droplet growth time. The continuous increase of the membrane rotation speed may force the droplet growth time rapidly approaching its limit, causing the droplet average diameter changes to decrease [12].

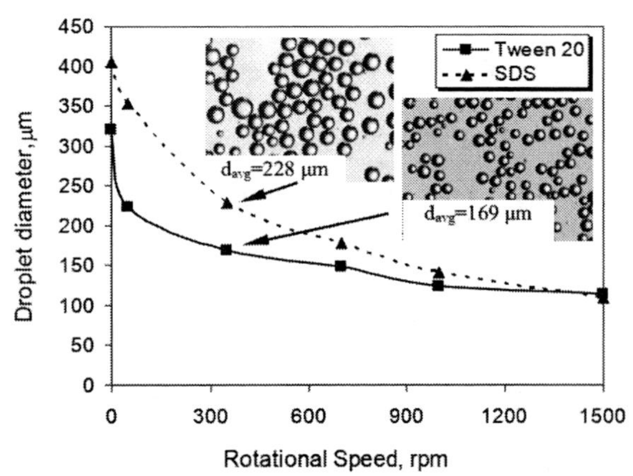

Fig. 3. Effect of membrane rotational speed on droplet size using different type of emulsifier at 0.1 % (w/w) carbomer concentration, insert showing typical micrograph images at the specific condition.

The type of emulsifier also influences the average droplet diameter. At 0.1% (w/w) carbomer concentration, Tween 20 produced smaller droplet diameter than SDS. This result is reversed with Schroder et al.'s investigation which is showed that SDS produced smaller droplet than SDS because of the ability of SDS to lower the interfacial tension very quickly [13] This could be attributed to the fact that both SDS and carbomer are anionic surfactants. The migration of SDS molecules from the bulk solution to the droplet surface, therefore, may be prevented by the electrostatic repulsion due to the high level of presence of the carbomer molecules at the interface. Tween 20, on the other hand, is a non-ionic surfactant, therefore, will not be affected.

The effect of dispersed phase flow rate on droplet diameter is shown in Fig. 4, with the CV is in the range of 9% to 25%.

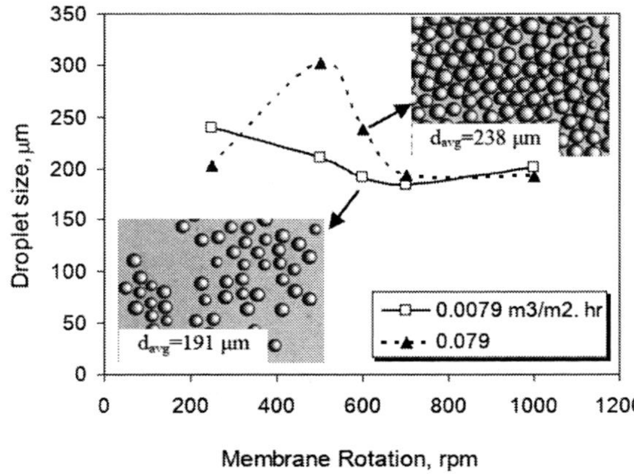

Fig. 4. Effect of membrane rotational speed on droplet size at 0.1 % (w/w) carbomer concentration with different dispersed phase fluxes, insert showing typical micrograph images at specific condition.

In general, there is a tendency for droplet diameter to decrease with the increase of membrane rotation. However, at different dispersed phase flow rates, the results are slightly different. The increase of dispersed phase flow rate correlates with the increase of inertial force which is play a role in additional detaching force and leading to faster droplet detachment. In the case of higher membrane rotation, the drag force will also increase as the membrane rotation raise. Further detaching force certainly will make the droplet detach more rapidly and droplet detachment period is shorter than in the case of low membrane rotation. This is due to the larger inflow velocity of the dispersed phase in the pore neck which is creating a Bernoulli under-pressure drop that arises as a result of the necking process [13]. The Bernoulli under-pressure drop affects the transmembrane pressure (TMP) of membrane emulsification and resulted in smaller droplets.

On the other hand, at higher rotation rate, larger droplet diameter was generated with the increase of dispersed phase flow rate. This was due to the rate of emulsifier to stabilise the interface was not fast enough and hence larger droplets were produced. In addition, with the increase of dispersed phase flowrate, a viscous pressure contribution is created. This pressure maintain the pore neck into open longer and the droplets detachment period is finally retarded [14]. As a consequence, the larger droplets are produced. However, when the membrane rotation is higher, the effect of necking process can be diminished by the shear stress of rotation and smaller droplet diameter can be obtained.

3.2 Multiple W/O/W Emulsions

Only the preliminary experimental results are available at this stage. Fig. 5 illustrates the effect of the membrane rotational speed on the multiple emulsion droplet size at two levels of inner phase emulsifier concentrations respectively.

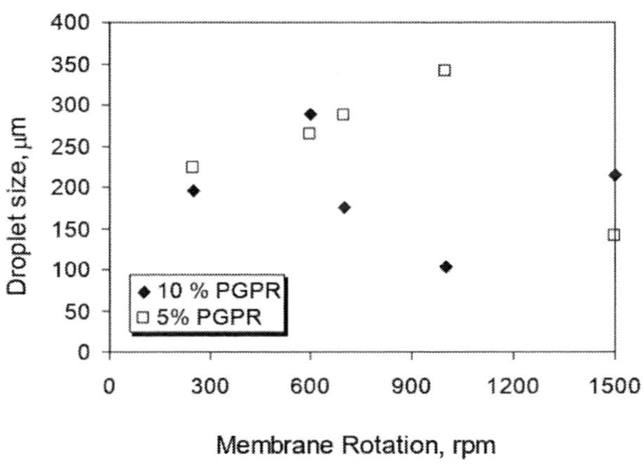

Fig. 5. Effect of inner phase of emulsifier concentration on droplet size.

Fig.5 confirms that application of higher membrane rotation tend to generate smaller droplets as found by previous reports [6]. Preparation of double emulsion by membrane emulsification should have advantage, producing higher encapsulation efficiency compared to higher shear device [15].

The experiments were conducted with the concentration of the outer phase emulsifier (Tween 80) being kept constant at 2%. The coefficient of variation (CV) for 10% PGPR concentration is in the range of 18% to 28%, while the CV for 5% PGPR concentration is much higher, which is in the range of 19% to 48%. The result obviously indicates that the concentration of the inner phase emulsifier has a rather strong influence on both the multiple droplet size and monodispersity. PGPR is soluble in oil and usually applied for stabilizing W/O emulsions. However, the concentration of PGPR did not have effect on the droplet size of inner emulsions [16]. Addition of PGPR reduced instability of the processes due to a lower collision efficiency. Experimental works with RMR on W/O/W emulsion production (10% (w/w) PGPR as inner emulsifier and 2% (w/w) Tween 80 as outer emulsifier) provide multiple emulsions in the range of 102.2 μm to 290.3 μm. Some images of multiple emulsion and droplet size distribution are showed in Fig. 6 and Fig. 7, respectively.

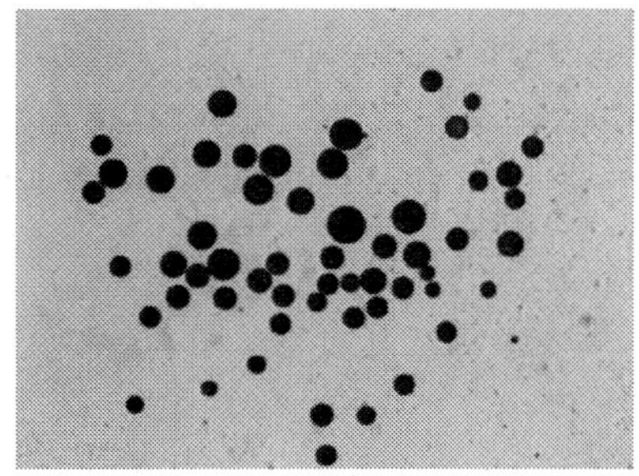

(a)
Davg = 102.2 μm, CV = 20.8%
(1000 rpm, 0.026 m^3/m^2.hr)

(a)
Davg = 290.3 μm, CV = 24.3 %
(700 rpm, 0.053 m^3/m^2. hr)

Fig. 6. Images of W/O/W emulsion droplets using RMR with 10% (w/w) of PGPR as inner emulsifier.

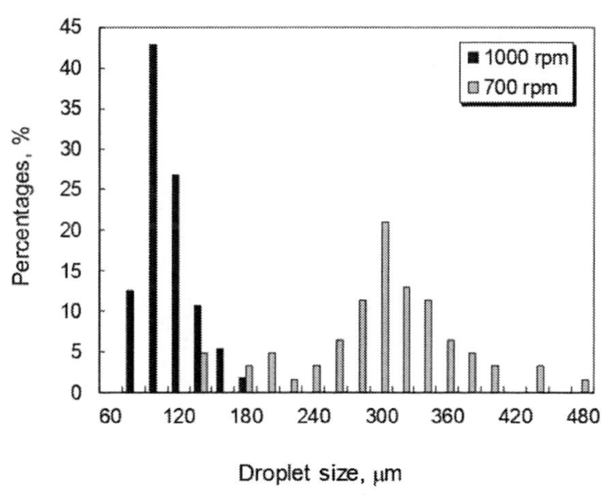

Fig. 7. Droplet size distribution of images at Fig. 6

In a multiple emulsification system, such influences would have to be minimised. Stability of double emulsions was influenced by colaescence inner droplets with the outer droplets. Effect of inner droplet coalesce was reported by Gaitzsch et al. [17] and Schuh et al. [18,19]. The interaction between the inner and outer phase surfactants will induce a redistribution of the surfactant molecules on the two interfaces, which may lead to rupture of the oil layer in the W/O/W droplets, and subsequently result in the loss of the internal W/O droplets [20]. Fig. 8 shows a photographic image of the droplets produced. The multiple phase structure of the droplets is clearly discernable.

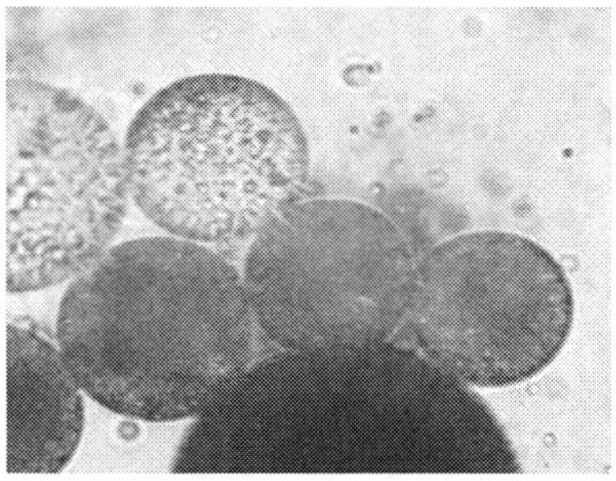

Fig. 8. Micrographic images of the multiphase WO/W droplets produced at membrane rotation rate of 700 rpm and dispersed phase flow rate of 0.053 m^3/m^2.hr taken using: Confocal microscope (Nicon Eclipse TE2000-U) at 20× magnification.

4 Conclusion

Single (O/W) emulsions was successfully produced through a rotating membrane emulsification system. In this system, Tween 20 and SDS were used as emulsifier and carbomer was employed as stabilizer. The average droplets sizes at various stabilizer yielded average droplet size in the range of 80 to 210 μm with CV between 10 to 20%. O/W emulsion droplets with average diameter of 130 μm to 300 μm with coefficient of variation (CV) from 9% to 25% were produced at various dispersed phase flow rate. The droplet size and CV reduced when the dispersed phase flow rate increased. The single phase emulsification experiments indicate that RMR is capable of producing emulsion droplets, which are close to or even smaller than the membrane pore size. This preliminary work has demonstrated the potential application of RMR in manufacturing multiphase W/O/W emulsions. However, the stability of the emulsion droplets need further improvement through a judiciary selection of surfactants, and careful control of the process parameters.

Reference

1. S. J. Peng, R.A. Williams, Trans IchemE. **76**, 894 (1998)
2. Y. Charcosset, I. Limayem H. Fessi, J. Chem. Technol. Biotechnol. **79**, 209 (2004)
3. S.M. Joscelyne, G. Trägådh, , J. Membr. Sci. **169**, 107 (2000)
4. A.J. Gijsbertsen-Abrahamse,. A. van der Padt, R.M. Boom, J. Membr. Sci. **230**, 149 (2004)
5. E. Piacentini, E. Drioli, L.Giorno. J. Membr. Scci. **468**, 410 (2014)
6. S.van der Graaf, C.G.P.H.Schroën, R.M.Boom, J. Membr. Sci. **251**, 7 (2005)
7. V. Eisinaite, D. Juraite, K. Schroën, D. Leskauskaite, Food Chem. **206**, 59 (2016)
8. R. Berendsen, C. Güell, M. Ferrando, Food Hydrocoll. **43**, 636 (2015)
9. M. Matos, G. Gutiérrez, O. Iglesias, J. Coca, C. Pazos, J. Food Eng. **166**, 212 (2015)
10. J. D. Ilić, B. G. Nikolovski, L. B. Petrović, P. S. Kojić, J. S. Petrović, J. Food Eng. **205**, 1 (2017)
11. R.D. Hancocks, F. Spyropoulos, I.T. Norton, J. Membr. Sci. **497**, 29 (2016)
12. Q. Yuan, Q., R. A. Williams, N. Aryanti, Adv. Powder Technol. **21(6)**, 559 (2010)
13. V. Schröder, O. Behrend, H. Schubert, J. Colloids Interface Sci. **202** , 334 (1998)
14. S. Holzapfel, E. Rondeau, P. Muehlich, E. Windhab, Chem. Eng. Tech. **36(10)**, 1785 (2013)
15. A.K. Pawlik, I.T. Norton, J. Membr. Sci. **415-416**, 459 (2012)
16. E. Panagopoulou, V. Evageliou, N. Kopsahelis, D. Ladakis, A. Koutinas, I. Mandala, Coll. Surf.: A. Psycochem. Eng. Aspects **522**, 445 (2017)
17. F. Gaitzsch, A. Gäbler, M. Kraume, Chem. Eng. Sci. **66(20)**, 4663 (2011)
18. A. Schuch, L. G. Leal, H. P.Schuchmann, Colloids Surf., A: Physicochem. Eng. Asp. **461**, 336 (2014)
19. A. Schuch, L. G. Leal, H. P.Schuchmann, Coll. Surf. A: Physicochem. Eng. Asp. **461**, 344 (2014)
20. A. T. Florence and D. Whitehill, Int. J. Pharm. **11**, 277 (1982).

Enhancement Performance of Hybrid Membrane Zeolite/PES for Produced Water Treatment With Membrane Modification Using Combination of Ulta Violet Irradiation, Composition of Zeolite and Thermal Annealing

Tutuk Djoko Kusworo*, *Annizah* Rahmatya Gerhana and *Noor* Hanifah Angga Putra

Chemical Engineering Department, Engineering Faculty, Diponegoro University, Jl. Prof. Sudharto, Tembalang, Semarang, 50239, Indonesia Tel./Fax: +62-24-7460058

Abstract. Produced water is a wastewater from oil production that must be treated well. Membrane is one alternative of water treatments technology based on filtration method. However, in the use of membrane, there's no exact optimal variable that influences membrane performance.This underlying research to assess factors that influences membrane performance to obtain optimal condition. Therefore, the objectives of this study are determining the effect of variable process in membrane fabrication and several modification techniques on membrane performance. The membranes were fabricated via dry-wet phase inversion method. The process variables of this experiment are varying the Zeolite concentration by low level 1% weight and 3% weight, UV irradiation time low level 2 minutes and high level 6 minutes, thermal annealing temperature low level 160°C and high level 180°C. The experiment runs were designed using central composite design. From the research that has been perfromed, PES/Zeolit membrane has a higher permeability after being irradiated by UV light and denser pore after heating and the longer of annealing time.

1 Introduction

Produced water is waste water from crude oil production process. Produced water come out with oil that going up to surface and containing much of dangerous chemicals and contaminants. Before produced water is thrown away, it should be given some treatments so that the produced water is appropriate with valid standard [1]. Produced water can be used for pulling out the oil. Beside that, in some wells, produced water used to pressure maintenance during crude oil production process [2]. The benefit of using this method is to keeps the environment clean. It can reduce the danger contaminant that can affect human, plants, and the others. There are so many pre-treatment methods that can be used like air flotation, hydrocyclone, coalescer bed, and filtration. From all of the methods above, filtration technology is the one of other alternatives that can be developed because it has many advantages [3].

Membrane is one of many water treatment technologies. Where membrane can separate chemical components specifically, can separate in low temperature, work continuously, energy saving, the process is not destructive for the separated chemical components, and didn't give negative effect for environment. Membrane is a selective obstacle between two phases. Membrane has various thickness, there is tin membrane, thick membrane, homogene membrane, and heterogene membrane. Membrane separates material based on molecule size and shape, sustain the component from feed that is having larger size than membrane pores, and release the component that is having smaller size.

In other side the uses of membrane for water treatment still have some disadvantages such as rejection decreasing along with flux increasing, and fouling. Therefore, hybrid membrane with polyether- sulfone material/Zeolit will be developed in this research. Membrane will be made with phase inversion for producing ultrathin membrane, while the combination of UV irradiation and thermal annealing are expected to make the membrane more improving against fouling and enhancing membrane performance such as flux value

* Corresponding author: tdkusworo@che.undip.ac.id

and percent rejection of membrane. The performance of the membranes include membrane permeability and selectivitity. The performance of the membrane influenced by the structure of its morphology, where the structure of a membrane that is formed is a result of the influence of several factors, including the composition of the membrane forming material and the addition of additives, the mistreatment at the grafting and treatment at the time of post treatment.

2 Materialls and Methods

The materials used in this research include: PES from Sigma Aldrich Chemie GmbH Steinheim Germany used as a membrane mix matrix material forming , Zeolite frome Indrasari Chemical Store Semarang, and distilled water UPT Integrated Laboratory of the University of Diponegoro. Original produced water samples from PT. Pertamina E & P Cirebon. This study were divided into three stages. The Variable using Run from Statistica 6. Software.

Table 1. Formulation Variable Factor

Run	A (%wt)	B (mnt)	C (°C)
1	1.00	2.00	160.00
2	1.00	2.00	180.00
3	1.00	6.00	160.00
4	1.00	6.00	180.00
5	3.00	2.00	160.00
6	3.00	2.00	180.00
7	3.00	6.00	160.00
8	3.00	6.00	180.00
9	2.00	4.00	170.00
10	0.24	4.00	170.00
11	3.76	4.00	170.00
12	2.00	0.47	170.00
13	2.00	7.53	170.00
14	2.00	4.00	152.36
15	2.00	4.00	187.64
16	2.00	4.00	170.00

The first step is the of production of cellulose acetate membranes, followed by optimization using RSM for membrane applications in the processing of produced water, and phase characterization. At this stage, fabrication of cellulose acetate membrane starts with making the solution a print composed by variable in Table.1 as the solvent. Casting membranes was using phase inversion method. This method was done by casting the membrane on the glass plate using a casting knife. Furthermore, the membrane was immersed into the coagulation bath with distilled water as non-solvent for 1 hour, followed by immersion in a different coagulation bath at ambient temperature (30 ± 2°C) for 24 hours. After that the membrane application will be examined for produced water treatment. Rejection and permeate water flux measurement are conducted using dead-end separation system. The membrane effective area in the module was determined to be 12.57 cm^2. Before performing the permeability test, membrane compaction is done first using distiled water for 30 min so the membrane polymer chains arrange themselves. After the compaction process, distiled water was replaced with produced water and maintained at 30 ± 2°C. Produced water flux determination is done by measuring the volume of produced water for every 10 min. Membrane rejection was performed by determining the concentration of Cl^-, S^-, TDS, COD , Turbidity before and after passing through the membrane barrier. Determination of TDS was performed using a TDS meter, Turbidity using Turbidimeter, the analysis of Cl^-, S^-, and COD ion were used titrimetric method. Figure 1 illustrated the simple process diagram of a dead-end cell apparatus for produced water separation process using prepared PES membrane. The permeate water flux was calculated by equation 1 [4]:

$$J = \frac{V}{P \ . \ A \ . \ t} \qquad (1)$$

Where, J = Flux (L.m^{-2}bar^{-1}h^{-1}), V = permeate volume (Liter), P = pressure (bar), t = time (hour) A = membrane effective area (m^2). Determination of the coefficient of rejection is done by analyzing the concentration of pollutants in upstream and downstream of the membrane as in the equation 2.

$$R = \left(1 - \frac{C_p}{C_f}\right) . \ 100\% \qquad (2)$$

Where, R is percent of rejection, C_p is permeate concentration and C_f is feed concentration [5].

3 Result and Disscussion

3.1 Independent Variable Influenced towards Membrane Performance

3.1.1 Zeolite Concentration Influenced towards Membrane Flux

Performance membrane PES can be seen from the flux membrane. The measurement of flux membrane performed with filtration tool for measuring the rate of alir permeat membrane per unit area per unit time. From figure 4.1 by increase time, the value of flux membrane tends to go down. It was because the longer the time of surgery, pressure through the process of filtration has been increased. Pressure has indicates fouling occurring to membrane. Fouling will increase, to cover pores membrane, who makes work membrane be the more severe and resulting in a reduction number of permeat. Because flux permeat is directly proportional to the volume of permeat at a particular time, so the decreasing volume of permeat, resulting in flux membrane declining.This is in the overall run where flux decreased along with increase in time.

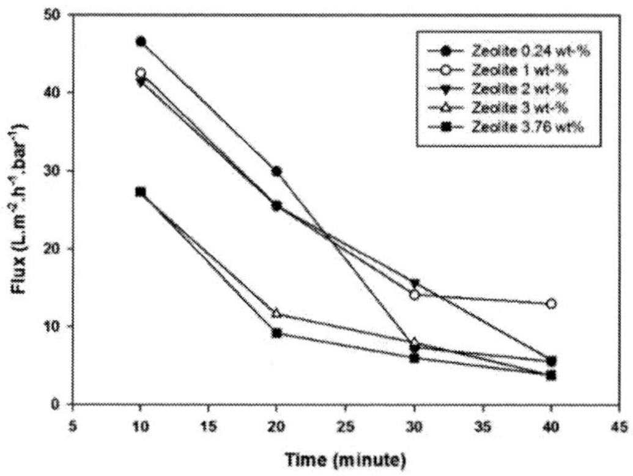

Fig. 1 The Influenced of Zeolite Concentration Toward Flux Membrane

According to Figure 1 , through the increasing of zeolite concentration, the flux on membrane getting down , it was because the more the concentration of a zeolite who adds into a solution of dope will improve the ability of the permeability and transport of a substance through a zeolite. Membrane swelling will barrier by charging on material zeolite matrix membrane PES. A stable zeolite structure is not susceptible to membrane swelling in zeolite area. PES chain polymer who was in the region close zeolite , controlled by interactions among advance .Hence , zeolite particles serves as cross-linker for chains of a polymer around it .This is what causes flow of particles through a matrix membrane being obstructed that result in flux value that down if the number of zeolite the more.

3.1.2 The Influenced of UV Ray Period towards Flux

UV irradiation on the polymer has been studied could affect on structural changes and chemical structural reaction. The type of structural changes strongly influenced by irradiation time. The longer UV irradiation time on the membrane surface, the more complex both structural and chemical changes of the membrane surface. Membrane structure plays an important role in membrane separation performance. Therefore, the determination of optimal irradiation time is required in employing UV light to modify membrane structure. Figure 2 shows the effect of UV irradiation time on permeate fluxes in function of time.

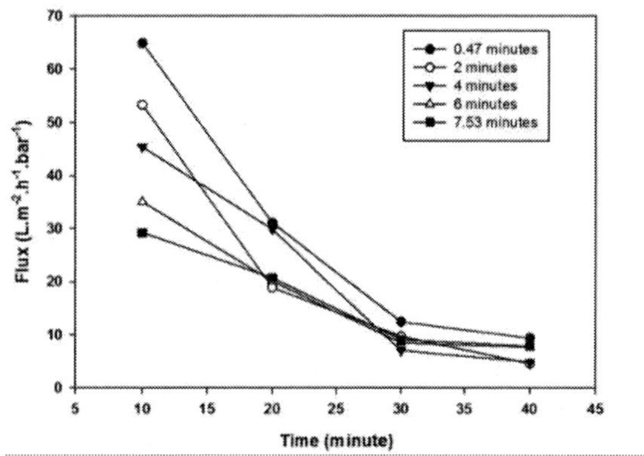

Fig. 2 Flux Membrane in different Period of UV Ray

Based on Figure 2, flux membrane that illuminated UV is decline. This because the delay between membrane after casting and before put in a tub coagulation (delayed mixing) that causes the formation of membrane non pores [6]. When solvent evaporated ,polymers solution that formed liquid move to fill pores for produce tightly pore without evaporation [7].

3.1.3 The Influenced of Thermal Annealing towards Flux Membrane

Polymeric membrane structure can be altered by giving a thermal treatment. the effect of annealing temperature on the permeate flux of PES/zeolite membrane is shown in Figure 3.

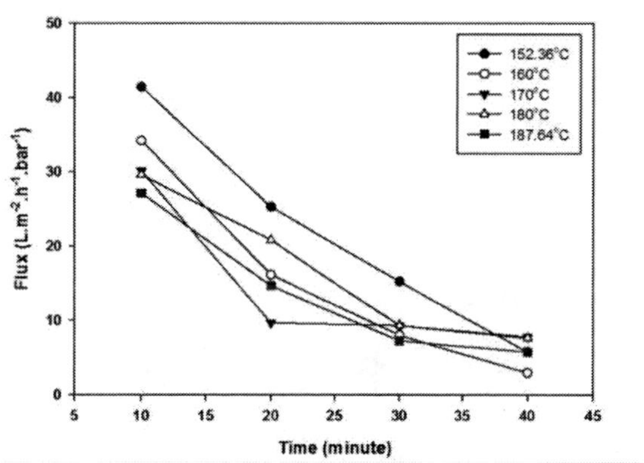

Fig. 3 Flux Membrane in different Thermal Annealing

From Figure 3 can be seen that the value of flux membrane PES by annealing lower than values flux membrane pes without annealing .Treatment annealing to membrane will cause adjustment structure chain polymers in the membrane surfaces and membranous be more stable. Pore structure in the membrane surfaces adjust equilibrium thermodynamics changed due to heat given [8]. Adjustment structure polymer chains cause the morphology change of membrane [6,9,10]. Flux membrane that heated at temperature 187.64°C lower than to the flux the membrane that is heated at temperature 152.36 °C. It was because at a temperature 187.64°C close to transition glass polymer temperature

PES (Tg = 220 °C) (Wibowo et al.,2010) . Warming temperatures glass approaching transition to break molecules PES ties between polymer become more tightly for membrane structure that have sparse skin layer thicker. These results obtained is consistent with the previous studies [11].

Treatment warming or annealling would cause adjustment structure chain polymers in the membrane surfaces. Structure pores in the membrane surfaces adjust equilibrium thermodynamics changed due to heat given. Adjustment structure chains are causing morphological change's membrane [12]. As a result , pores formed into more stable and meetings. Pores more tightly membrane down flux obtained. The higher temperatures annealing apply to membrane , so the membrane will be more tightly so flux obtained becoming progressively lower [13].

3.1.4 The Influenced of Zeolite Concentration towards Rejection of Ion Cl^-, S^{2-}, COD, TDS

Figure 4 shows that greater concentration of Zeolite , rejection is generated. This is due to increase in the concentration of Zeolite raise viscosity of casting film and lose its diffusivity among the components in the system during the process of compacting the solution eventually cast and further hamper the process of precipitation and makes the surface of the membrane has pores smaller [14]. Thus, increasing the composition of the Zeolite will lead to a rise in the value of rejection.

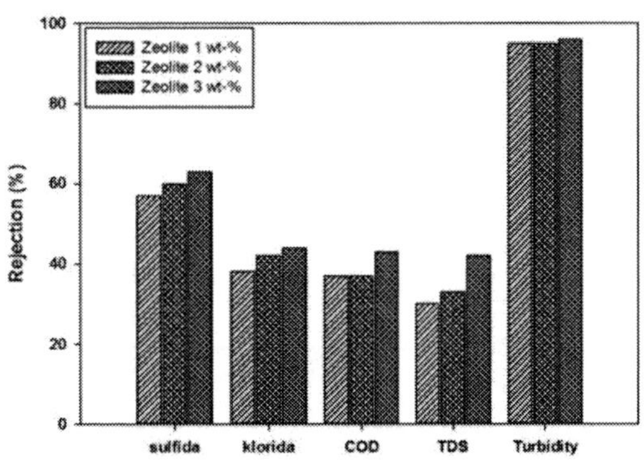

Fig. 4 Membrane rejection in different Zeolite loading and solutes

We obtained that the value of turbidity rejection is most of the influence of various zeolite concentrations. From the view of particle size, Turbidity have a particle size of 2 μm, the particles cause turbidity average molecular size > 5 μm depending on the value of the turbidity of a liquid, ion Cl^- ion has diameter of 194 pm and ion S^2-particle diameter of 145 pm. Rejection value is the value of many substances that can't passing through the pores of membrane. The rejection value of turbidity is around 95 % , then membranes are functional as good to hold macromolecules .While on a rejection level of ions Cl^- having a level rejection at 40 % and for rejection of ions S^{2-} as much as 65 % .So we can conclude that a membrane polyether sulfon with hybrid nanozeolite having the ability most either in merejeksi macromolecules in the form of ions can cause cloudiness

3.1.5 Influenced of UV Ray Periods toward Rejection of Ion Cl^-, S^{2-}, COD, TDS

From Figure 5 in general, show an increase of rejection value in each time filtration. Increasing number was also indicated by each rejection indicators where the longer time irradiating so performance rejection membrane is increased. Long irradiating 6 minutes produce higher rejection than the irradiating for 2 minutes and 4 minutes , the average value of 57 % to Sulphide , 43 % to Chloride , 42 % to the COD, 33 % for TDS and 98 % for Turbiditas. An increasing 0f UV irradiating will result in part sparse layer to a membrane the back so that performance of the inner membrane of rejection is better [15]. This is further strengthened by research formerly practiced by Mansourpanah and Habili [16] where UV irradiating to prove rejection increasing in membrane to salt NaCl and Na_2SO_4 significantly.

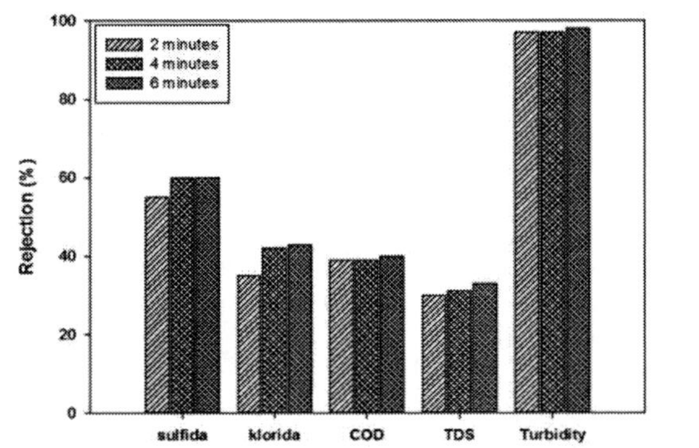

Fig. 5 Membrane rejection in different UV irradiation and solutes

In terms of rejection indicators, hybrid membrane PES-Zeolit known has all the different levels in rejected every components. It is affected by the size of every molecule who filtered out. Almost all turbidity is rejected which the figures 98 % is highest than another rejection indicators. Besides it, Sulphide rejection is higher than the chloride on every long irradiating membrane. [17] claim that irradiating UV to membrane fabrication process will cause distribution membrane pores more uniform in every sections so minimize the arrival of void potential to pass molecules with a bigger size. Hence turbidity will be reduced significantly due to turbidity own the parameter cloudiness is caused by the massive particles dispersed in in the water .This also applies to sulphide , chloride and other rejection indicators .

3.1.6 The Influenced of Thernal Annealing towards Rejection of Ion Cl^{-}, S^{2-}, COD, TDS and Turbidity

The effect of thermal annealing on the membrane structure are increasing the density of the membrane and alterate the polymer become more crystalline. The effect of annealing process on the solutes rejection performance is depicted in Figure 6.

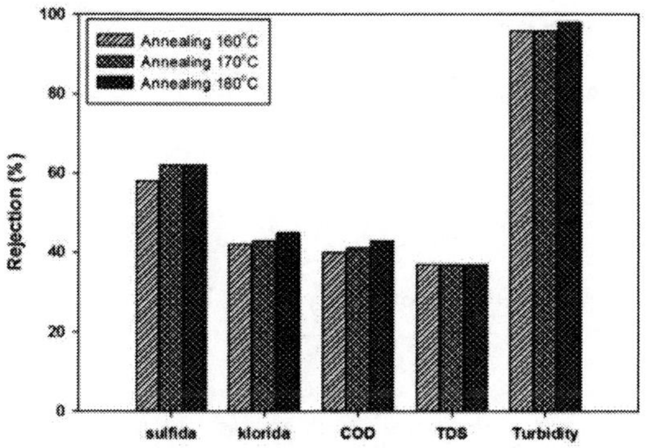

Fig. 6 Membrane rejection in different annealing temperature and solutes

Figure 6 shows the influence of temperature on thermal annealing of membrane rejection. It appears that rising temperatures will also increase the ability of membrane rejection in various components. Analogous to the previous discussion that the existence of thermal treatment on membrane fabrication will affect the structure of its morphology. Myeong and Bhattacharyya [12] reported that at the time of the thermal treatment, given membrane molecules in polymer chain will gain energy and move , so it can fix its structure. Thus the membrane polymer chains will be arranged more regularly resulting in more compact membrane layer. This is confirmed also by Bosko et al. [18] in his research using membrane Pd-Ag/PSS where annealing temperature increase would cause a uniform deposition in each segment lining membrane.

4. Conclusions

The hybrid membrane PES/Zeolite was successfully fabricated in this work. The modification methos seems significantly enhanced membrane performance. The increasing of zeolite concentration, the flux on membrane getting down , it was because the more the concentration of a zeolite who adds into a solution of dope will improve the ability of the permeability and transport of a substance through a zeolite. Membrane swelling will barrier by charging on material zeolite matrix membrane PES. The flux membrane that UV irradiated is decline and permeability is arised. This is because the delay between membrane after casting and before put in a tub coagulation (delayed mixing) that causes the formation of membrane non pores. Treatment annealing to membrane will cause adjustment structure chain polymers in the membrane surfaces and membranous be more stable. Pore structure in the membrane surfaces adjust equilibrium thermodynamics changed due to heat given.

References

1. F.R. Ahmadun, A. Pendashteh, L.C. Abdullah, D.R.A Biak, S.S Madaeni, Z.Z. Abidin. J. Haz. Mat., **170**, 530–551. (2009).
2. A.F. Ismail, Hassan, A.R.; Ng, B.C. J. Sci. Techn. Membr. Sci. & Tech., **24**. 879-889. (2002).
3. Kaeselev, Bozena; Kingshott, Peter; Jonsson, Gunnar, Desalination, **146**, 265-271. (2002).
4. B.E. Kirk, and D.F. Othmer, The Interscience Encyclopedia Inc, New York. (1996).
5. F. Macedonio, A. Ali, T. Poerio, E. El-Sayed, E. Drioli, M. Abdel-Jawad, Sep. Purif. Tech., **126**, 69-81. (2004).
6. M. Mulder, Kluwer Academic Publishers, Netherland. 1996.
7. H. Zhang, Zhong, W. Xing, Desalination, **309**, pp. 84-90. (2013).
8. I.C. Kim, H.G. Yun, K.H. Lee, J. Membr. Sci., **199**, 75-84. (2002)
9. S. Mulyati, Reintek J. Ilmu Pengetahuan Tek. Terapan **3**(2), 2. (2008).
10. A.P. Siswanto, and A.D. Aufa, J. Penelitian Membran Teknik Kimia Undip. 2011.
11. T.D. Kusworo, Budiyono, J. Supriyadi, D.C> Hakika. Intern. J. Waste Res. **4**(1), 1-9. (2014).
12. Myeong, Jin-han, and Bhattacharyya, D., Desalination, **101**, 195-200. (1994).
13. T.D. Kusworo, A.F. Ismail, Budiyono, World Appl. Sci., 28(1), pp. 9-19. (2013).
14. A. Motta, C. Borges, K. Esquerre, A. Kiperstok, J. Membr. Sci., **469**, 371-378. (2014).
15. P.A. Nur, D.K. Sari, H. Susanto, J. Tek. Kim. Ind., **2**, 189-197. (2013).
16. Y. Mansourpanah, and E.M. Habili,. J. membr. sci., **430**, 158-166. (2013)
17. M. Sangermano, A. Chiolerio, G.P. Veronese, L. Ortolani, R. Rizzoli, F. Mancarella, and V. Morandi. Macromol. Rap. Comm., **35**(3), 355-359. (2014).
18. M.L. Bosko, J.B. Miller, E.A. Lombardo, A.J. Gellman, and L.M. Cornaglia. J. Membr. Sci., **369**(1), 267-276. (2011).

Hydrophylicity Enhancement of Modified Cellulose Acetate Membrane to Improve the Membrane Performance in Produced Water Treatment

Tutuk Djoko Kusworo*, *Danny* Soetrisnanto, *Cynthia* Santoso, *Tyas Dwi* Payanti, *and Dani Puji* Utomo

Chemical Engineering Department, Faculty of Engineering, University of Diponegoro, Jl. Prof. Sudharto, Tembalang, Semarang, 50239, Indonesia

Abstract. Produced water is a wastewater generated from petroleum industry with high concentration of pollutants such as Total Dissolved Solid, Organic content, and Oil and grease. Membrane technology has been currently applied for produced water treatment due to its efficiency, compact, mild and clean process. The main problem of produced water using membrane is fouling on the membrane surface which causes on low permeate productivity. This paper is majority focused on the improvement of anti-fouling performance through several modifications to increase CA membrane hydrophilicity. The membrane was prepared by formulating the dope solution consists of 18 wt-% CA polymer, acetone, and PEG additive (3 wt-%, 5 wt-%, and 7 wt-%). The membranes are casted using NIPS method and being irradiated under UV light exposure. The SEM images show that parepared membrane has asymmetric structure consist of dense layer, intermediete layer, and finger-like support layer. The filtration test shows that PEG addition increase the membrane hydrophilicity and the permeate flux increases. UV light exposure on the membrane improves the membrane stability and hydrophilicity. The imrpovement of membrane anti-fouling performance is essential to achieve the higher productivity without lowering its pollutants rejection.

1 Introduction

Produced water is a by-product of the petroleum industry, where during production every barrel of obtained oil produces 3 barrels of produced water and will increase with the age of the oil well [1]. Produced water contains organic and inorganic materials such as Total Dissolved Solid (TDS) of 4380-49.987 mg/L, Magnesium 396.1-678 mg/L, Sulfate 6.2-621.3 mg/L, Calcium 987-2669 mg/L [2]. In the condition of untreated produced water, it can affect the environment and human health [3].

Today, membrane technology is widely used for manufactured water treatment. Membrane technology is a beneficial technology because of its high separation efficiency and the addition of a small chemical [4]. In addition, the design of membrane equipment is also simple with low initial investment costs, low operational energy, and easy maintenance. The nanofiltration membrane is a type of membrane filtrations which effectively reduces the chemical content such as inorganic salts and small organic microsolutes present in produced water, and has a higher flux yield than the reverse osmosis membrane [5]. The nanofiltration membrane is a membrane having a pore size <2 nm.

Many studies have been performed to treat the produced water using membranes. Xi et al. [6] examined the surface-modified Poly (ether sulfone) membrane with electrophoresis - UV grafting, in which the amount of foulant can be reduced. Homayoonfal et al. [7] also modified the surface of the nanofiltration membrane by using UV-grafting Acylic acid on the surface of the Polysulfone membrane with the PEG additive where increasing surface irradiation time of the membrane will cause fouling decrease. However, those modification methods still produced fouled membranes. Fouling can cause a decrease in flux and shorten the life of the membrane used. Fouling on the membrane is the adsorption or accumulation of certain components in the produced water that occurs on the surface of the membrane (external fouling) or within the pores of the membrane (internal fouling).

According to Mondal and Wickramasinghe [8] and Alzahrani et al. [9], to overcome fouling, with increased hydrophilic properties in nanofiltration membranes can increase water permeability and increase rejection of both organic and inorganic components. In this study,

* Corresponding author: tdkusworo@che.undip.ac.id

surface modification was done by mixing poly ethylene glycol additives (PEG). PEG plays a role in forming pores, increasing the number of pores on the nanofiltration membrane thus increasing membrane permeability, suppressing the amount of macrovoids and providing hydrophilic properties of the membrane. UV light exposure is also required to improve the performance and characteristic of cellulose acetate non-fouling nanofiltration membranes in treated water treatment.

2 Materials and Methods

2.1. Materials

The materials used in this research include: produced water from Pertamina Oil and Gas Company, Cirebon, cellulose acetate (CA) from MKR Chemicals, PEG 1000 and PEG 4000 from Sigma-Aldrich Chemie GmbH Steinheim Germany, 99.75% Acetone from Mallinckrodt Chemicals, and distilled from UPT Diponegoro University Integrated Laboratory.

2.2 Fabrication of Cellulose Acetate (CA) membranes

Membrane preparation begins with the preparation of a polymer solution (casting) consisting of CA, PEG and Acetone polymers. The membrane was made with a fixed CA composition of 18 wt%. The PEG types used were PEG 1000 and 4000, while the composition of variations of each type of PEG were 3 wt%, 5 wt%, and 7 wt%. The CA was inserted with Acetone into a closed erlenmeyer flask and stirred for 8 hours with a magnetic sirer until all the CA was dissolved, then PEG was added. Furthermore, the solution is allowed to stand for 1 day to remove air bubbles and ready to be printed by phase inversion technique.

The casting process was carried out by phase inversion technique, the dope solution was poured onto the glass plate that the edges are duct tape. The casting knife is then moved downward to form a thin layer on the glass plate, then irradiated with UV light with variations of time for 30, 60, and 90 seconds, then inserted into a coagulation bath containing aquades for 1 hour followed by immersion in another coagulation bath for 24 hours. Then the membrane is dried in a 40-50 °C oven for 24 hours.

2.3 Membrane characterization

2.3.1 Flux and Rejection of fabricated membrane

The flux value is determined a cross-flow filtration cell as shown in Figure 1. In the filtration cell is placed filter paper and membrane to be measured its permeability value. The distilled water at a volume of approximately 150 mL is fed into the filtration cell and tightly closed, then supplied with air pressure of 4 bar. The compacting process is carried out first in order to replace the polymer for 20-30 minutes. After the compaction process, the distilled water in the filtration cell is replaced by the water produced, the measurement of the produced water flux is carried out by measuring the volume of water produced which can be accommodated for a certain time interval at 15 minute intervals. The flux value is calculated by the ratio of permeate volume per unit time as in (Eq.1). The determination of the rejection coefficient is determined by determining the concentration before and after passing through the membrane as in (Eq.2).

$$J = \frac{V}{P.A.t} \qquad \text{(Eq.1)}$$

$$R = \left(1 - \frac{C_i}{C_0}\right) \times 100\% \qquad \text{(Eq 2)}$$

J is permeate flux ($L.h^{-1}m^{-2}bar^{-1}$), P is a upstream pressure (bar), A is an effective membrane area (m^2), t is an operating time (second). R is a rejection efficiency, C_i is the concentration of TDS in permeate, C_0 is the concentration of TDS in feed.

Figure 1. Membrane filtration cell

2.3.2 Membrane structure morphology by Scanning Electron Microscopy

The determination of membrane morphological structure is done by SEM (Scanning Electron Microscopy) with the following procedure. First the membrane is dried first, then the membrane is immersed in liquid nitrogen for a few seconds until it solidifies. Before the shooting, the membrane is lifted and broken both ends with tweezers. This piece of membrane is coated with pure gold (coating) that serves as a conductor. Furthermore, transverse sections and membrane surfaces are photographed with certain magnification.

2.3.3 FTIR analysis (Fourier Transform Infra Red)

Membrane characterization using FTIR was used to determine functional groups on the membrane. The test was performed to confirm the presence of cellulose acetate and PEG, as well as the influence of the type of PEG used and the irradiation of UV light on the membrane.

3 Results and Discussions

3.1 The effect of PEG Molecular Weight on membrane performance

The performance of cellulose acetate membranes can be known from the membrane flux profile. Flux is the number of permeate volumes obtained at separation operations using membranes per unit area of membrane and time unit. Measurement of membrane flux is done by means of dead end filtration. Measurements and flux profiles of cellulose acetate membranes for treated water treatment with variations of the PEG additive type are shown in Figure 2.

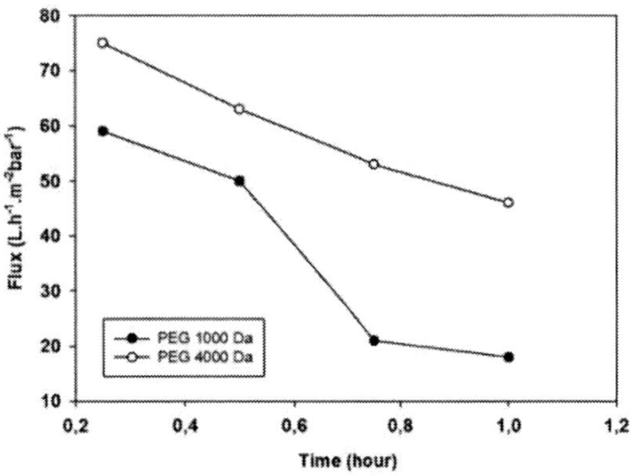

Figure 2. Fluxes profiles of fabricated membrane with different PEG molecular weights.

In the data presented in Figure 4.1. it can be seen that the flux will decrease over the time of filtration due to the fluctuation of the solute contained in the water produced on the surface of the membrane. The membrane with the PEG 4000 additive has a higher flux than the PEG 1000. This phenomenon corresponds to previous studies investigating the effects of 3 different PEG types (400 Da, 6000 Da and 20,000 Da) on membrane transport and morphology [10]. PEG is an additive substance in the membrane that acts as a pore forming on the membrane and has hydrophilic properties. The molecular weight of PEG has an important role in the process of membrane formation rate. Increasing the molecular weight of PEG causes more PEG molecules trapped in the membrane due to reduced PEG mobility after immersion in the coagulation bath. At the same time, by increasing the molecular weight of PEG with the same concentration, the molecular quantity of PEG is decreased which will lead to the formation of larger pores, but fewer.

Membrane performance can also be determined by the value of membrane rejection. TDS rejection eficiency was determined by measurement of TDS concentration in both of upstream and downstream membrane filtration process. The effect of the PEG type on the TDS rejection efficiency is shown in Figure 3.

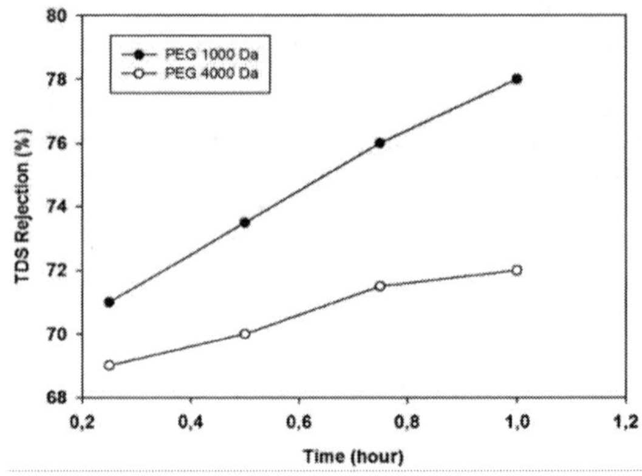

Figure 3. Rejection efficiencies of fabricated membranes with different PEG's MW

CA membrane with 1000 Da PEG doping has a higher TDS rejection efficiency than 4000 Da PEG doping. This could be due to the membrane with lower PEG additive has a lower membrane pore size. The PEG contributes as a porogent in membrane fabrication, the lower MW is well known gives smaller pore size.

3.2 The effect of PEG concentration on CA membrane performance

The flux profile of cellulose acetate membrane for treated water treatment with variation of PEG additive concentration is shown in Figure 4.

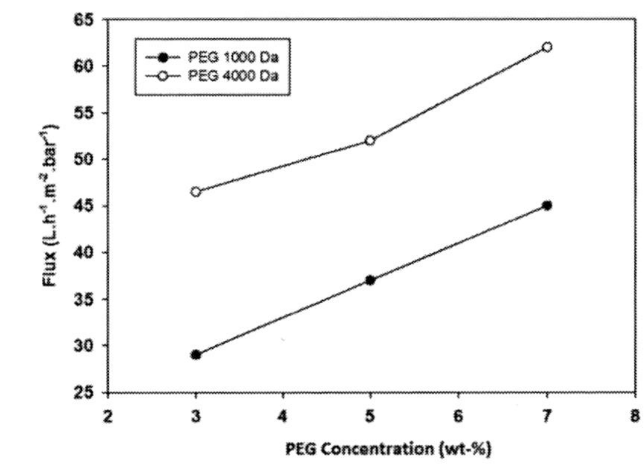

Figure 4. The average flux of membranes with various PEG concentration

The presence of PEG in the membrane can cause two effects: (1) The addition of PEG causes a decrease in solvent in the dope solution resulting in a comparison of polymer concentration with lower solvent concentration and an increase in the viscosity of the dope solution. Therefore, the dope solution becomes less stable, resulting in instantaneous demixing when the dope solution mold is immersed in the coagulation bath. Instantaneous demixing means that the membrane will soon form when the dope solution mold is inserted into a non-solvent coagulation bath. (2) PEG has a natural hydrophilic property. This hydrophilic property helps the exchange of water as non-solvent with acetone in the membrane during the phase inversion process and affects the speed of membrane formation and membrane morphology [11].

PEG acts as a pore-forming that is completely mixed in dope solution and will dissolve with water during the coagulation process. When PEG dissolves in water, the pores begin to form into finger-shaped structures or macrovoid holes. Increased PEG concentrations will lead to an increase in the number of pores on the membrane, so the resulting flux will increase as well.

The values of flux and rejection are directly related to the amount of pore and pore mass of the membrane. From the data presented in Figure 4 shows that the greater the concentration of PEG added to the dope solution the greater the membrane flux obtained. One of the most important things in membrane applications is the ability of membranes to reject certain solutes. In this study, rejection test was conducted on TDS concentration. The data of membrane test results in rejecting the solute are shown in Figure 5.

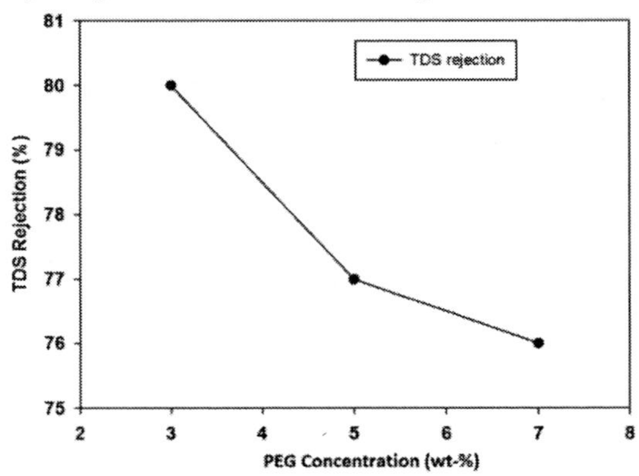

Figure 5. TDS rejection profile of CA membranes with various PEG concentration

In Figure 5, it is seen that the more PEG concentrations are added the rejection will decrease. This is because the more PEG concentrations added to the dope solution will increase the amount of pore [12]. PEG is an additive substance that is hydrophilic and easily soluble with water. The PEG mixed in the dope solution will dissolve in water at the membrane coagulation stage so that it will cause macrovoid in the membrane structure. The greater the percentage of PEG in dope solution, will lead to more pores. If the pores on the membrane are more numerous, the solute present in the produced water will have a chance of escaping during filtration causing a decrease in the membrane rejection value.

3.3 The effects of UV irradiation on CA membrane performance

UV irradiation on the membrane is a pre-treatment prior to the coagulation process. The effect of UV treatment on the membrane can be determined by comparing the membrane performance with UV treatment and without UV treatment on the same type and PEG concentration as shown in Figure 6

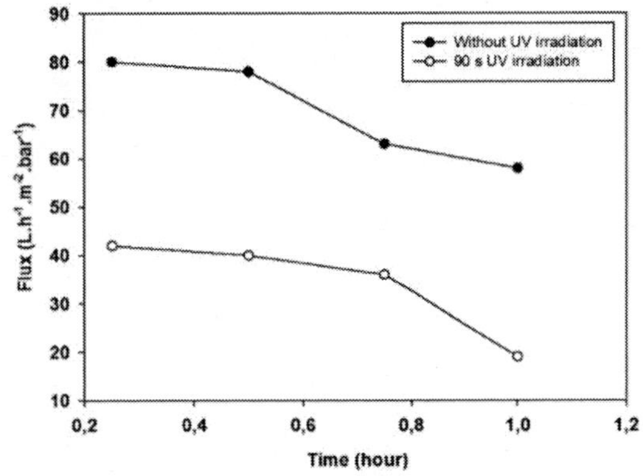

Figure 6. Flux profiles of UV irradiated membranes

Figure 6 shows the performance of cellulose acetate membrane flux with PEG 1000 of 7% without UV irradiation and with UV irradiation for 90 seconds. Membrane with UV irradiation treatment has a lower flux result. The resulting lower flux results from changes in the membrane structure caused by UV light exposed on the membrane surface.

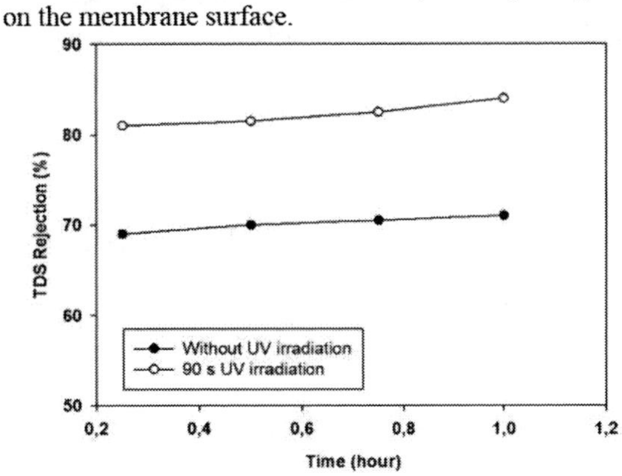

Figure 7. TDS rejection efficiency of UV irradiated CA membrane

Membrane without UV irradiation has a larger flux, but when viewed with regard to the rejection value of the membrane, the UV membrane has a better rejection value. Rejection profiles of Total dissolved solid (TDS) is shown in Figure 7. UV irradiation on the membrane serves for chain scission and crosslinking [13]. Furthermore, UV rays will cause polymerization of the membrane. Polymerization occurs due to the formation of free radicals in the polymer due to UV irradiation, which in turn decreases pore size and increases the hydrophilic properties of the membranes.

3.4 Membrane Characteristics

3.4.1 Membrane structure morphology

The membrane struture prophology was observed using Scanning electron microscopy (SEM). Figure 8 shows the images of surface and cross-sectional membrane structures.

(a)

(b)

Figure 8. SEM images of fabricated membrane (a) surface section, (b) cross-section.

Figure 8.a shows a flat smooth membrane surface of CA membrane. At this magnification (5.000x), no voids were observed. The white spot in the membrane surface could be due to the agglomeration of CA polymer. This phenomenon occured because of un-spontaneous demixing in whole flat sheet membrane during coagulation process in non-solvent soaking. The Figure 8.b shows a finger-like porous structure of cross-sectional membrane. This structure contributes as a support and maintain the permeability of membrane. The membrane consists of three layer: top layer is a dense layer which plays an important role in selectivity performance, the intermediate layer is a porous layer which acts as support and maintaining the membrane flux.

3.4.2 FTIR analysis

Characterization of cellulose acetate membranes using FTIR analysis was performed to determine the functional groups present in the membrane. Figure 8 shows the FTIR characterization of a 7% PEG 1000% cellulose acetate membrane with variation without UV irradiation and with UV irradiation for 90 seconds.

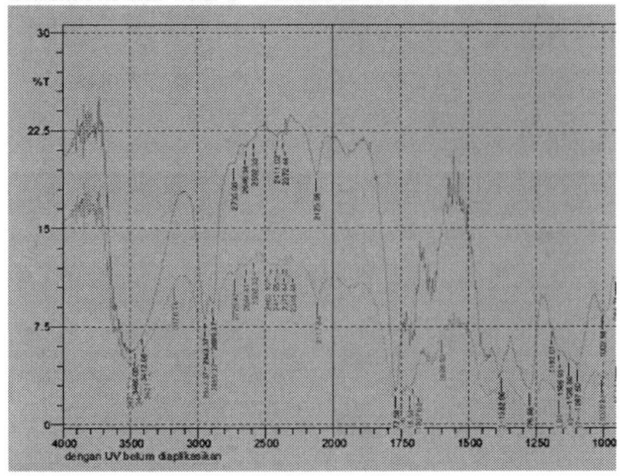

Figure 9. FTIR spectra of Unirradiated and UV irradiated CA membranes.

FTIR spectra showed that cellulose acetate membrane has C-O-C-CH_3-COOH and C=O groups. Cellular cellulose acetate membranes with UV irradiation have higher transmittance values in all clusters compared with UV-free membranes. The C-O-C group on the UV less membrane lies at a wavelength of 1095.57 cm^{-1} and on the UV irradiated membrane lies at a wavelength of 1097.50 cm^{-1}. While the C=O group on the membrane without UV lies at wavelength 1751.36 and on the membrane with UV irradiation located at wavelength 1772.58 cm^{-1}. The occurrence of these wavelength shifts indicates that UV irradiation treatment effect on membrane morphology structure.

In membrane treatment with UV irradiation, there are chain scission, crosslinking and then polymerization process on the membrane. This is illustrated in Figure 9, which is the FTIR result of two membranes without and with UV treatment for 90 seconds. Appears peak at the wave number 1050-1300 cm^{-1} which may indicate the presence of a C-O group of Alcohol / ether / carboxylic acid / ester usually present in the wave number. Appears on a wave number 2500-2700 cm^{-1} which may indicate the presence of an O-H group of carboxylic acids with the hydrogen bonds normally present in the wave number. Appears on top of the wave number 3200-3600 cm^{-1} which may indicate the presence of an O-H group of hydrogen / phenol bonding alcohols which usually appear on the wave number.

There is a wide difference of peaks on both membranes where FTIR results from cellulose acetate membranes treated with UV light irradiation have larger area area than non-UV irradiated membrane. This wide area difference is due to the denser membrane structure. The denser structures are formed because the interactions between molecules of the membrane become larger due to UV irradiation resulting in the absorbance of the membrane against Infrared light from FTIR equipment that increases significantly.

4 Conclusion

Cellulose-based nanofiltration membranes have been successfully manufactured. The cellulose acetate nanofiltration membrane is an asymmetric membrane consisting of three layers of dense layer, intermediate layer and support layer. The filtration test shows that PEG addition increase the membrane hydrophilicity and the permeate flux increases. UV light exposure on the membrane improves the membrane stability and hydrophilicity. The imrpovement of membrane anti-fouling performance is essential to achieve the higher productivity without lowering its pollutants rejection.

The authors would thank to Waste treatment laboratory of Chemical Engineering Department UNDIP for supporting facilities.

References

1. F.R. Ahmadun, A. Pendashteh, L.C. Abdullah, D.R.A. Biak, S.S. Madaeni, Z.Z. Abidin, J. Hazard. Mater. **170**, 530-551 (2009).
2. S. Alzahrani, A.W. Mohammad, N. Hilal, P. Abdullah, O. dan Jaafar, *Sep. Purif. Technol.* **118** 324–341 (2013).
3. H.I. Safitri, F. Ryanitha, and N. Aryanti, J. Teknol. Kimia Industri, 2 (4), 205-211 (2013).
4. M.J. Han, and S.T. Nam, J. Membr. Sci. 202, 55-61 (2002)
5. A.W. Mohammad, Y.H. Teow, W.L. Ang, Y.T. Chung, D.L.O. Radcliffe, N. Hilal, Desalination. **356**, 226-254 (2014).
6. W. Xi, W., Rong, L. Zhangsheng, A.G. Fane, J. Membr. Sci. **273**, 47-57 (2006).
7. M. Homayoonfal, A. Akbari, and M.R. Mehrnia, Desalination, 263, 217-255 (2010).
8. S. Mondal, and S.R. Wickramasinghe, J. Membr. Sci. **322**, 162-170 (2008).
9. S. Alzahrani, A.W. Mohammad, N. Hilal, P. Abdullah, and O. Jaafar, J. Environ. Chem. Eng., **1**, 1341-1349 (2013).
10. B. Chakrabarty, A.K.A Ghoshall, and M.K. Purkait., J. Membr. Sci. **309**, 209-221 (2008).
11. Y. Ma, F. Shi, J. Ma, M. Wu, J. Zhang, and C. Gao, Desalination. **272**, 51-58 (2011).
12. T.D. Kusworo, Budiyono, J. Supriyadi and D.C. Hakika. Inter. J. Waste Resource, **4**(1), 1-9 (2014).
13. N. Addina, K.S. Dini, and S. Heru, J. Teknol. Kimia Industsri, **2**(4), 189-197 (2013).

Desalination of Sea Water Using Polymer Inclusion Membran (PIM) With Aliquat 336-TBP (Tributhyl Phosphate) as Carrier Compound

Muhammad Cholid Djunaidi[1,*], *Henry* Fauzi[1], and *Rum* Hastuti[1]

[1]Departement Chemistry Faculty Sains and Mathematica Diponegoro University, Indonesia

Abstract. Desalination of Sea Water using Polymer Inclusion Membrane has been done. PIM is known has high stability membrane to overcome the instability of liquid membrane. PIM was placed among two phase : source phases was sea water and feed phases was aquadest. Efficiency of desalination is known by determining salinity concentration and ion Na^+ in feed phases and stripping phases using refractometer and AAS, while membrane characterization is done using FTIR, SEM and UV-Vis spectroscopy. The PIM membrane that is produced has thin, transparent, clear and flexible properties. The result showed that PIM transport for 24 hours give the highest of salinity transport, there are 92.68% is transported from feed phases and 84.87% in stripping phases. Membrane characterization result by FTIR and UV spectroscopy showed that PIM membrane is stable enough. .

1 Introduction

Sea water as a water resource can be utilized to meet clean water needs. Rapid population, industrial growth and expansion led to an increase in water demand. Not to mention the existence of water pollution by industries and households that can reduce the supply of clean water available. With the management of sea water into fresh water and ready-to-drink water, it can increase the supply of clean water needs which has an impact on the reduction of poverty rate and is an important aspect in future industrial development. Good quality water can be met, namely, by processing dirty water into clean water or removing salt from seawater (desalination) [1].

Desalination techniques have long been known and carried out, i.e. distillation (thermal processes) and membrane processes such as reverse osmosis, ion exchange and electrodialysis. The use of liquid membrane offers a new alternative to seawater desalination process. Liquid membrane has attracted the attention of many researchers because of its high selectivity in the presence of carrier substances contained in the membrane. Djunaidi et al (2017) [2] recovered chromium (VI) metal ions from electroplating waste using Supported Liquid Membrane (SLM) with results of more than 60% chromium transported to the receiving phase using a single carrier compound of Aliquat 336. Applications of liquid membranes for desalination have been widely undertaken, one of which was conducted by Maryati (2008) [3]. The results showed that decreasing salinity level for Aliquat 336 single carrier compound was 96.696% and for Aliquat 336-TBP mixed carrier compounds was 100%. However, their lack of stabilization dissosiates the liquid membrane from applications / commercializations in the process of separating and purifying the industry. This instability is caused by the loss of the solvent and the carrier into the solution caused by the formation of the emulsion droplets or caused by the pressure difference present in the membrane [4]. Liquid membranes commonly used are supported liquid membrane, emulsion liquid membrane (ELM) and bulk liquid membrane (BLM). Nevertheless, the they still have weaknesses: BLM has interface surface area and low mass transfer rate, while the breakage of emulsion is a problem in ELM. SLM has major weakness in low stability, this weakness is also owned by ELM and BLM [5].

In order to overcome these problems, Sugira et al. (1987) [6] managed to obtain a stable membrane, by entangling the liquid membrane into the cellulose acetate membrane added with plasticizer, which became a new generation of Supported Liquid Membrane. This type of facilitated transport membrane is called Polymer Inclusion Membrane (PIM). PIM shows the advantages than other liquid membranes especially SLM. This type of membrane is very high stability, lost carrier during process can be ignored, more efficient carrier used, higher flux and more versatile [4,7].

Djunaidi, et al (2017) [8] used PIM with Aliquat-TBP carrier compound (4: 1) for the recovery of Cr (VI)

* Corresponding author: choliddjunaidi@live.undip.ac.id

ions from electroplating waste and the result was good. The Cr (VI) ions transferred from the feed phase was 99.24% and the transferred in the receiving phase was 85.88%.

Looking at the advantages of PIM as mentioned previously. in this research. desalination was conducted by using PIM. In this study. PIM used consisted of Aliquat 336-TBP carrier compound. tetrahydrofuran (THF) as solvent. dibutylether (DBE) as plasticizer and PVC as base polymer. The mixture of Aliquat and TBP carrier compounds provided synergistic effect in extracting $ZrCl_6^{-2}$ [9] and $HCrO_4^-$ [10]. PIM performance was influenced by membrane surface area (membrane thickness). feed phase concentration. and n times of the use of membrane. Therefore. in this research variations of those parameters were considered to determine the effect on salinity transport process occuring in both feed and receiving phases. The solution used in the feed phase was sea water while the receiving phase was aquadest.

2 Experimental Method

2.1 Instruments

A set of PIM tool cells. glassware laboratory. glass print (petridish with diameter 9.5. 11.5 and 13.5 cm). analysis scale. Refractomter H-UR-201. AAS (Atomic Absorption Spectroscopy). FTIR and UV- Vis spectroscopy. Scanning Electron Microscope (SEM).

2.2 Material.

Sea Water was taken from Logending Beach Cilacap, aqua deminiralisated (Bratachem), and chemicals were bought from Sigma Aldrich i.e polyvinyl chloride (PVC), Aliquat 336, TBP, dibutyleter (DBE) and THF.

2.3 Experimental Procedures

2.3.1 Membrane Fabrication

A set of PIM tool cells. glassware laboratory. glass print (petridish with diameter 9.5. 11.5 and 13.5 cm). analysis scale. Refractomter H-UR-201. AAS (Atomic Absorption Spectroscopy). FTIR and UV- Vis spectroscopy. Scanning Electron Microscope (SEM).

2.3.2 Sea Water Desalination using PIM Feeding Phase (FP) Solution Preparation

The feed solution used was seawater with dilution of 100. 50 and 10 times.

2.3.3 Preparation of the Receiving Phase (FP) Solution

The receptor solution preparation used in the process was aquadest

3 Results and Discussion

This research was conducted for desalination of seawater using Polymer Inclusion Membrane (PIM).

3.1 Fabrication of Polymer Inclusion Membrane (PIM)

Polymer Inclusion Membrane (PIM) was produced by mixing PIM components consisting of PVC (polyvinyl chloride) base polymer. Aliquat 336:TBP (4: 1) as carrier compounds and DBE (dibutylether) as plasticizers dissolved in THF solvent to produce PIM membrane thin. clear. non-fragile and effectively capable of desalination.

3.2 Membrane Characterisation before Use

3.2.1 FTIR spectra analysis

The determination of active groups of resulting PIM membrane was done using FTIR spectrophotometer. The result of membrane analysis before stirring is shown in the figure 1 below:

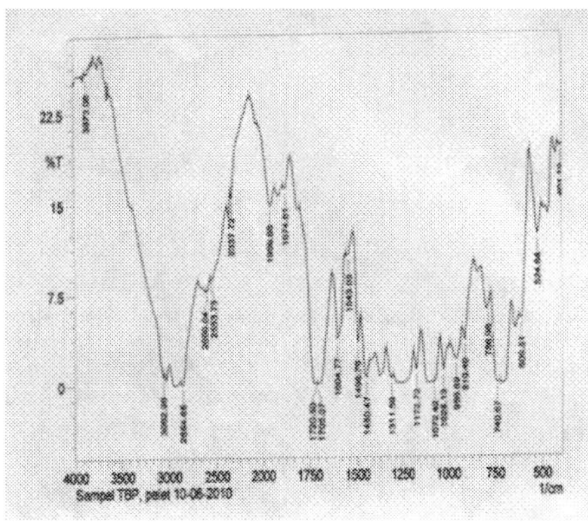

Fig. 1. FTIR Spectrum of the PIM membrane prior to stirring

The FTIR spectrum of membrane prior to stirring showed a sharp absorption band in the 3062.96 cm^{-1} region indicating the presence of an unsaturated CH (=CH-) group. 2854.65 cm^{-1} showing the presence of an alkyl group ($-CH_3$) reinforced by the band at 1311.59 cm^{-1} and 1450.47 cm^{-1}. In the meantime. the absorption at 1720.5 cm^{-1} region shows the carbonyl group (C=O). the absorption band at 1026.13 cm^{-1} region indicates the presence of ether (CO). absorption at 910.4 cm^{-1} assigns to the presence of an alkyl CH_2 group. absorption band at 740.67 cm^{-1} produced a series of methylene. while vibration mode at 609.51 cm^{-1} indicated the presence of C-Cl group [11]. This results confirm that there are compounds of PIM membrane constituent components in the membrane.

3.2.2 Scanning Electron Microscopy (SEM) Analysis

The surface morphology and cross section thickness of PIM membrane with Aliquat 336-TBP carrier compounds can be identified by SEM analysis (Scanning Electron Microscopy). Results of SEM analysis of Aliquat 336-TBP membrane are presented in the following figure (2):

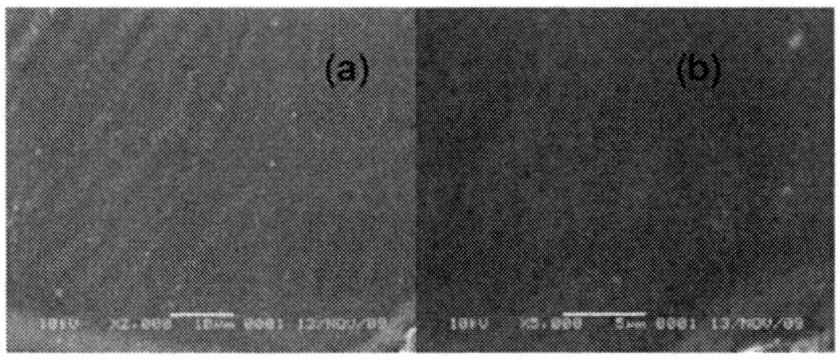

Fig 2. Morphological images of the PIM membrane,
(a) the surface of membrane with the 2000x magnification,
(b) surface of membrane with the 5000x magnification

Analysis of membrane using Scanning Electron Microscopy (SEM) (Fig. 2) shows that surface morphology of PIM membrane with 5000x magnification is not porous. Non-porous PIM membrane can be used to salinity transport occured by diffusion through an intermediate of Aliquat 336-TBP carrier compound (4: 1) [12].

3.3 PIM Membrane Transport Mechanism

The salinity transport process took placed through several stages. Initially the saline diffused into the membrane-feed phase interface. then binding to the carrier compound to form ion pair. In this case the carrier compound used was Aliquat 336-TBP. The reactions occuring in sea water desalination can be written as follows [12]:

$$nR_4N^+Cl^-_{mem} + OH^- \leftrightarrow (R_4N^+.OH^-)_n + Cl_{mem}$$
$$(R_4N^+.OH^-)_n + Cl^-_{aq} \leftrightarrow (R_4N^+.Cl^-_{aq})_n + OH^-$$
$$(R_4N^+.Cl^-_{aq})_n + nTBP \leftrightarrow n(R_4N^+.Cl^-_{aq}).nTBP$$
$$n(R_4N^+.Cl^-_{aq}.TBP) + OH^- \leftrightarrow n(R_4N^+.OH^-) + TBP + Cl^-_{aq}$$

The Aliquat-Cl-ion pair diffused to the membrane-receptor phase interface. In the membrane-receptor phase interface. the Aliquat-Cl^- ion pair will be solvated by the TBP carrier compound so that Cl^- would be released and was exchanged with OH^- in the receiving phase. This process was continuous and reversible.

3.4 Effect of Feed Phase Concentration

In the feed phase. concentration variation of 10x. 50x and 100x samples dilution was used in membrane B and stirring had been conducted for 24 hours. Results obtained as follows (table 1):

Table 1. Effect of variation of sample concentration on % desalination

Variation concentration (dilution)	% desalination in FP	% Desalination in RP
10x	50.24	36.58
50x	72.13	57.37
100x	74.19	58.06

Table 1 shows that salinity transport at 100x dilution using Aliquat-TBP membrane shows a large percentage of desalination. whereas the increasing concentration of seawater samples at 50x and 10x dilutions indicates the decrease of desalination percentage in the receiving phase. This is due to the greater the concentration of feed phase causes the lower the salinity transported from the feed phase to the receiving phase. The ability of membrane performance decreased. the greater the concentration of the sample. the stronger the intermolecular force causing the difficulty of salinity transport through the membrane to the receiving phase.

3.5 Effect of Membrane Thickness

The thickness of membrane was varied by varying the diameter of petridish membrane. namely. 9.5 cm. 11.5 cm and 13.5 cm. Results obtained as follows (table 2).

Table 2. Effect of membrane thickness variation on % desalination

Variation of Membrane Thickness (μm)	% Desalination in FP	% Desalination in RP
25 (9.5 cm)	50.24	36.58
20 (11.5 cm)	83.34	75.60
10 (13.5 cm)	92.68	84.87

Table 2 shows that the thinner the membrane (the larger the diameter of the membrane) the more percentage of desalination in the feed phase and the better the transport to the receiving phase. Conversely. the thicker the membrane the less salinity transported to the receiving phase because the diffusion rate was slower. This can be explained by Fick law $J = - D\ dc/dx$ stating that the diffusion rate is inversely proportional to the thickness of the interphaselayer/flux distance [13]. From this equation it can be seen that the thicker the membrane the longer the flux distance traveled by the Aliquat-TBP ion pair with salinity to get to the receiving phase which causes the low diffusion rate so that the percentage of desalination to the receiving phase was reduced. Other researchers [8] experienced the same thing in transporting Cr (VI).

3.6 Influence n Membrane Usage

In Aliquat 336-TBP membrane for n times use with 3x membrane use. the following data were obtained:

Table 3. Effect of n times the use of membrane on % desalination

Variation of n times of Membrane use	% Transport in FP	% Transpor in RP
1	87.87	84.94
2	83.40	84.81
3	78.94	55.01

As seen in Table 3 it can be concluded that there is decrease in membrane transport capacity (as seen from the decreasing percentage of desalination in the feed phase) due to the use of membrane n times as well as the decreasing percentage of desalination in the receiving phase. This indicates decrease in the ability of the membrane to be used more than 1 times the use. This is likely due to the closed surface of the membrane after the stirring of white flakes prevented the transport of salinity to the receiving phase.

3.6 Use of Blank Membranes

The making of a blank membrane without a carrier compound is performed as a comparison of the PIM membrane with the Aliquat 336-TBP carrier compound. Desalination using a blank membrane with stirring for 24 hours shows a desalination percentage of 0 (no metal transport) either from the feed phase or the receiving phase. this is due to the absence of the carrier compound in the blank membrane. From these results it can be concluded that salinity transportability occurs due to the role of carrier compounds in the membrane.

3.7 Sodium Ion Transport

In addition to transporting Cl^- (salinity). the mixed Aliquat-TBP compound carrier also transported the content of the main constituent cations of salinity. especially sodium (Na^+) ions. This was likely to occur because of the role of the TBP carrier compound as well as the neutralizing reaction resulting from the transfer of Cl^- ions into the receiving phase [12]. The amount of sodium ion transported from the feed phase can be seen in the table below.

Table 4 shows that Na^+ ions contained in seawater were also transferred from the feed phase to the receiving phase in the transport process of sea water salinity. Based on this result. it can be seen also that the use of membrane with thickness of 10 µm in the feed phase. sodium ion transported was 96.89% whereas in the receiving phase. the sodium ion transported was 38.57%.

Table 4. Sodium content transported at feed phase and receiving phase after stirring for seawater using AAS analysis

Thickness of Membrane (µm)		10 (13.5 cm)
Concentration of Na^+ before desalination (ppm)	Feed Phase	109,400
	Receiving Phase	0
Concentration of Na^+ after desalination (ppm)	Feed Phase	3,400
	Receiving Phase	42,200
Na^+ level Transported from feed phase (%)		96.89
Na^+ level Transported to receiving phase (%)	Feed Phase	38.57
	Receiving Phase	

3.7 Membrane Characterization after Stirring

3.7.1 FTIR spectra analysis

FTIR analysis on the PIM membrane after the stirring process is used to determine the presence of functional groups possible missing from the membrane.

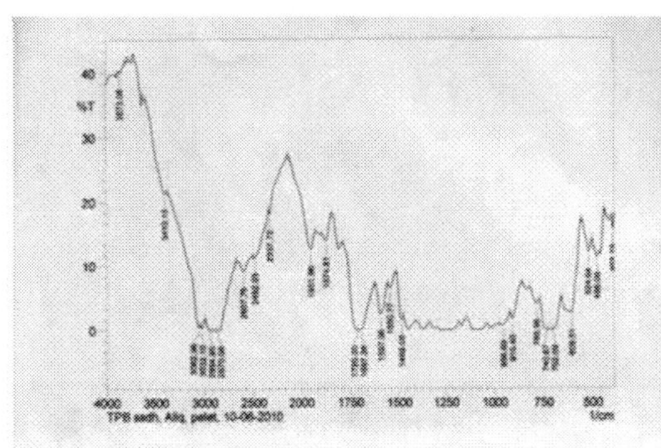

Fig 3. FTIR spectrum of PIM membrane after stirring

FTIR spectrum of PIM membrane after stirring (Fig 3) shows the same spectrum of PIM membrane before strirring in which compounds of constituent components of PIM membrane were present. The peak shift that occured may be the effect of ion pair formation happening between salinity and the Aliquat-TBP carrier compound taking place on the membrane.

3.7.2 Stability Analysis of PIM Membrane

In order to determine the PIM membrane stability. UV-Vis analysis was performed on the membranes. feed and receiving phases before and after stirring in seawater desalination process. Membrane stability is indicated by the absence of organic compounds both in the feed phase and the receiving phase shown in the UV-Vis spectrum (Fig 4-5). Analysis of membrane using UV-Vis spectrophotometer resulted in the following datum:

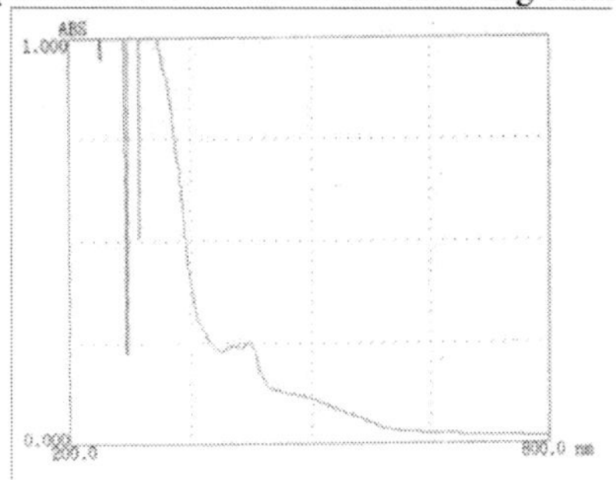

Fig 4. UV-Vis spectrum of feed phase before stirring

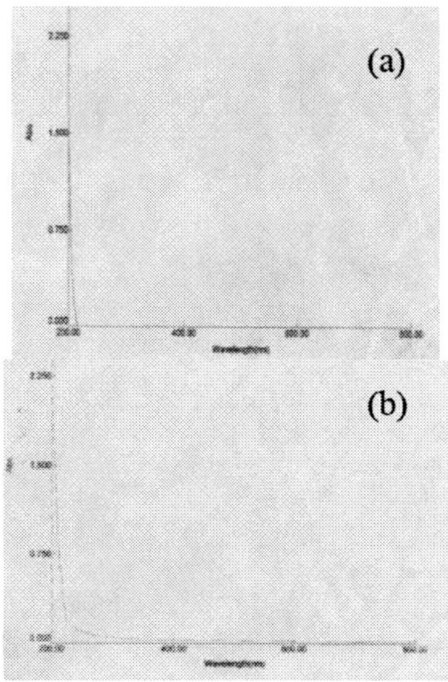

Fig 5. UV-Vis spectrum of feed phase
(a) before stirring
(b) after stirring

The UV-Vis spectrum of the feed phase before and after the stirring (Fig 5) did not show any peak assigned to organic compounds which are constituents of membrane. While the UV-Vis spectrum of the receiving phase after stirring is as follows (Fig 6):

Fig 6. UV-Vis spectrum of the receiving phase after stirring

The above UV-Vis spectrum (Fig 6) shows no new peak of the organic compounds of membrane constutuents. This result proved that there was no organic compound of PIM membrane constituent that dissolved in the feed and the receiving phases. so it can be concluded that the PIM membrane used was quite stable.

CONCLUSION

Based on the above discussion it can be concluded that PIM membrane was produced with the preoperties: thin. clear. and flexible so that PIM method can be used for desalination of sea water. The highest salinity transport process from the feed phase was 92.68% and the transport at the receiving phase was 84.87%. The results of membrane characterisation using FTIR and UV-Vis spectroscopy confirmed that the PIM membrane was quite stable.

REFERENCES

1. W. A. Wardana.. *Impact of pollution of Environmental.* (Andi. Yogyakarta 2004)
2. M. C. Djunaidi. Khabibi. R. Nurfitriana. R.. Alchemy Jurnal Penelitian Kimia. **13**. 119-132 (2017)
3. Maryati.. *Supported Liquid membrane (SLM) With Anionic Carrier for Desalination Sea Water.* Thesis. Diponegoro University Semarang (2008)
4. S. Kusumocahyo. K. Sumarua. T. Iwatsubo. T. Shinbo. T. Kanamori. H. Matsuyamab and M. Teramoto. J. Membr. Sci. **350** 143-152 (2006)
5. L. Nghiem. M. Patrick M.. D. Ian . M. Jilska. Perera. W. Robert . Cattrall. and S. Kolev.. J. Membr. Sci. **281** (2006) 7–41 (2006)
6. M. Sugira. M. Kikkawa. and S. Urita. J. Membr. Sci. **42**. 47-55 (1989)
7. J. Gardner. P. Quinn. O. Jedediah. Walker. D. Bryce. . A. Bibhutosh. G. Roger and J. Lamb. J. Membr. Sci. **121 (9)** 103-107 (2005)
8. M. Cholid Djunaidi. R. Ariadi Lusiana. M. damayanti Rahayu. *The 2nd International Conference on Materials Engineering and Nanotechnology IOP Publishing IOP Conf. Series* (2017)
9. B. M. Misra. *Supported Liquid Membranes in Metal Separation.* C.S. Symposium Series 642. Chemical Separation with Liquid Membrane. American Chemical Society; Washington. pp 361-375 (1996)
10. M. Rahayu. D. Separation of Cr(VI) using Supported Liquid membrane. Thesis (2009)
11. H. Sastrohamidjojo. *Spektroskopi.* Liberty. Yogyakarta (2001).
12. S. M. Khopkar. *Basic Concept of Analytical Chemistry.* Penerbit UI. Jakarta (1990)
13. M. Mulder. *Basic Principles of Membrane Technology.* Kluwer Academic Publisher. London. 10-12 (1996)

Oilfield Produced Water Reuse and Reinjection with Membrane

Utjok W.R. Siagian[1], *Setyo* Widodo[2], Khoiruddin[3], *Anita K.* Wardani[3], and *I Gede* Wenten[3,*]

[1]E Department of Petroleum Engineering, Institut Teknologi Bandung, Jl. Ganesha 10, Bandung 40132, Indonesia
[2]PPPTMGB LEMIGAS, Balitbang KESDM, Jl. Ciledug Raya, Kav. 109, Cipulir, Kebayoran Lama, Jakarta 12230, Indonesia
[3]Department of Chemical Engineering, Institut Teknologi Bandung, Jl. Ganesha 10, Bandung 40132, Indonesia

Abstract. Produced water has become a global environmental issue due to its huge volume and toxicity that may pose detrimental effects on receiving environment. Several approaches have been proposed to provide a strategy for produced water handling such as reinjection, reuse, or discharge. With various advantages, membrane technology has been increasingly used in produced water treatment replacing the conventional technologies. However, fouling is a major drawback of membrane processes in this application which needs to be controlled. This paper gives an overview and comparison of different produced water management. Special attention is given to produced water treatment for reuse purpose. Furthermore, the use of membrane processes in produced water reuse including performance, challenges, and future outlook are discussed.

1 Introduction

Hydrocarbon exploration and exploitation activities pose a serious environmental problem due to the production of produced water stream, which is also the largest portion of the waste streams [1,2]. Produced water can be defined as trapped water under a geological structure that comes along to the surface as oil and gas are being drilled out [3]. Components of produced water vary from one location to another depending on geological formation, geological condition and types of hydrocarbon produced [4]. Most of the constituents include dispersed oil, organic and inorganic contaminants, treatment chemical (from oil production process such as emulsion breaker, corrosion and scaling inhibitor, coagulant, etc.), produced solids, bacteria, metal and naturally occurring radioactive material (NROM) [1,3]. As the major constituent, oily substances and grease [3] with alkanes, alkenes, alkynes aromatic and complex hydrocarbon compounds have become the primary concern in produced water treatment [5]. On the other hand, a high salt concentration of produced water also becomes a problem when the onshore operation takes place [3].

Produced water is usually managed by several ways such as production avoidance, produced water injection, reuse and consumption of treated produced water for beneficial usages [6]. According to the reported studies on produced water treatment, discharge is the most common practice of produced water treatment **(Fig. 1)**. The hazardous constituents in produce water should be removed before being discharged directly [2]. Studies on searching a better method for produced water treatment is increasing due to a more stringent regulation of produced water discharge. Meanwhile, other efforts are conducted to find a cost-effective method for produced water reuse since it is considered as an unconventional water source that may be used for beneficial usages such as industrial water, agriculture, potable water, and reinjection into a nearby formation well for enhancing oil recovery. In the reuse water purpose, it is important to significantly reduce the contaminants by choosing a proper treatment to meet the requirement of water quality. In addition, there are two general objectives in produced water treatment: removal of undesired constituents (excessive hardness, dissolved solids, microorganisms, light hydrocarbon, etc.) and effluent adjustment when it is going to be re-used such as for irrigation [6].

Four methods of produce water treatment have been proposed, i.e. physical, chemical, biological and membrane-based treatment [1]. With the ability of producing a high quality product along with relatively lower cost, lower energy and chemical consumption, smaller footprint, more intensive process, easy to scale-up, and flexible to operate [7–13], membrane is considered as an attractive technology in various processes including in water and wastewater treatment [14–19]. In produced water treatment, the number of publications on membrane applications in the last two decades is increasing which indicates the gaining attention on this technology **(Fig. 1)**. This increasing attention is due to the ability of membrane in producing a high quality effluent that meet the required standard either for discharge and reuse purposes.

* Corresponding author: igw@che.itb.ac.id

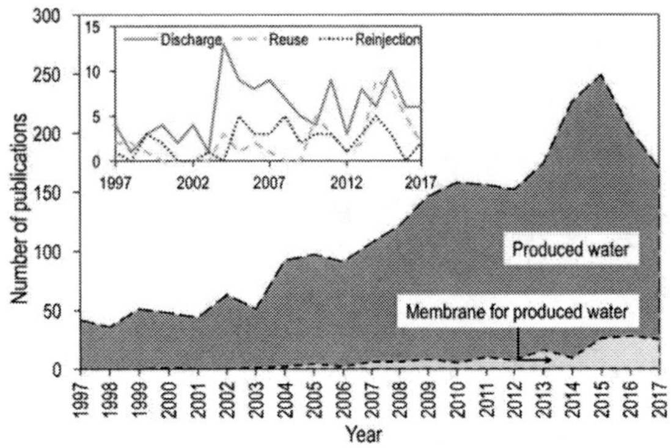

Fig. 1. Publications of produced water in the last 20 years indexed by Scopus (query: TITLE ("terms"))

Several papers reviewing the performance and applications of membrane technology in produced water treatment can be found [20–24]. In this paper, membrane technology for produced water treatment is reviewed focusing on reuse purpose. Current progress and challenges of membrane technology in this application are discussed.

2 Produced water treatment purposes

2.1 Produced water characteristics

Typical characteristics of produced water treatment reported in some studies are tabulated in Table 1. Produced water contains a mixture of inorganic and organic substances, solids, a small fraction of dissolved hydrocarbons, and gasses together with chemicals used for enhanced oil recovery [25,26]. These contaminants can be classified as follows: high total dissolved solids, suspended solid, dispersed oil, dissolved and volatile organic compounds, heavy metals, radionuclides, dissolved gases, and bacteria [27]. Both produced water volume and characteristics highly depends on formation well or reservoir including its geological characteristics, lifetime, and the hydrocarbon produced [28,29]. The volume of produced water increases along with increasing reservoir lifetime. The characteristics and the large volume are challenging for produced water handling. Inappropriate treatment may result in negative effect when it is discharged to a receiving environment. On the other hand, the large volume of produced water may be considered as a potential unconventional water source that can be used for various beneficial purposes [30]. For this purpose, a large fraction of components should be removed to meet the specifications of desired water quality. Standard quality of produced water discharge, reuse, and reinjection have been summarized by Li and Lee [31] as follow:

- Offshore disposal: suspended solid <10 mg/L; oil <5 mg/L;
- Reinjection: Suspended solid <1 mg/L; particle size of solid < 1 μm; Oil content <5 mg/L;
- Irrigation: Oil and grease <35 mg/L; TDS = 500-2000 mg/l;
- Cooling tower: TDS <2700 mg/L;
- Chemical process: TDS <1000 mg/L;

2.2 Produced water discharge

Generally, produced waters are discharged or reused. Produced water is usually discharged into the sea (for offshore), local river (for onshore), and estuarine or coastal water (for terminal or oil refinery) [26]. Due to a more stringent regulation of produced water discharge, the effective treatment technologies are required. Before disposal, produced water undergoes several steps of treatment including (i) removal of suspended solids and dispersed hydrocarbon and (ii) removal of oil and total carbon content, so the oil contents below 10 mg/L [32]. The other standards are stated by OSPAR (Convention for the Protection of the Marine Environment of the North-East Atlantic) and EPA (USA Environment Protection Agency). The standards are as follow [33]: (1) in OSPAR: Dispersed aliphatic oil = 30 mg/L; oil and grease = n/a and (2) in EPA: Dispersed aliphatic oil = n/a; oil and grease = 42 mg/L. This standard is focused on offshore discharge practice. It should be noted that dispersed hydrocarbon is not the only one component which becomes the environmental concern [34]. For onshore, many more parameters should be taken into account [35]. The dissolved solids, COD, BOD, heavy metals and other toxic contaminants should also be removed to avoid pollution in receiving water bodies [35].

2.3 Produced water reuse and reinjection

Since the source of fresh water starts to diminish, the large volume of produced water has been considered as a potential alternative which could be reclaimed for industrial purposes and directly reduces freshwater withdrawals [36]. This will also reduce the discharge of waste effluent to environment thus minimizing the environmental risk. Produced water reuse provides more beneficial economic and environmental impacts rather than discharge. Most of the produced water is reused for water injection. The purpose of reinjection of produced water into formation wells is to enhance oil recovery [37]. Reinjection is also aimed for disposal purpose. The attractive goals of both saving fresh water usage along with reducing waste disposal or near-zero discharge made reinjection as an example of smart water source management which turns waste into resource [38]. It is also viewed as a more cost-effective alternative rather than discharge and conventional waterflooding [39,40]. According to Robinson [41], the advantages of produced water reinjections include increasing hydrocarbon production to around 40% and removing an extensive surface water treatment system. In addition, using produced water as injection also offers some advantages, such as available in oil production field, compatible, near zero bacteria, and solve disposal problem [41].

Table 1. Typical characteristics of produced water reported in several studies.

TDS (mg/L)	Conductivity (μS/cm)	COD (mg/L)	BOD (mg/L)	TOC (mg/L	Oil and grease (mg/L)	TSS (mg/L)	pH	Ref.
-	47,600	588	-	-	-	-	7.1	[42]
80,470	-	1622	695	386	220	167	6.8	[43]
-	20,000-80,000	1000-9000	-	-	200-5000 [a]	-	6.0 – 8.0	[32]
-	-	380 – 490	-	-	50-200	30-160	8.0-9.0	[44]
6000	-	-	-	120	50	-	7.0	[45]
5500-7000	-	-	-	-	0-60 [b]	-	6.0-8.0	[46]
98,800 [d]	100,000	156 [c]	-	113	-	60	6.6	[47]
3100-5800	-	450-1280	-	-	12.5-84.4 [e]	7.9-89.2	7.2-8.1	[48]
6000	-	-	-	120	20	-	7.0	[49]
-	20,000-80,000	-	-	200-2000	32-180 [a]	-	7.0-7.8	[28]
28,980	46,000	1440	-	403	14	-	-	[36]
96,472	-	-	-	268.2 [f]	120.4 [g]	20.2	8.7	[50]
12,900	-	3000	-	-	28	-	6.77	[51]
10,000-17,500	-	-	-	48-164	3-220 [e]	6-110	8.3	[52]

[a] Dispersed oil; [b] Free oil; [c] 3*DOC (equation from dowac.custhelp.com); [d] Salt concentration; [e] Oil; [f] Dissolved organic including hydrocarbon and soluble oil; [g] Free oil droplets; - Data not available;

Nevertheless, the treated produced water should be compatible with the reservoir. For instance, the presence of oil and suspended solid smay cause reservoir interface plugging [39]. Therefore, the component which causes plugging should be removed. Specification of injection water may vary depending on formation characteristics. For low-permeability formations, the injection water should have <1 mg/L of total hydrocarbons and <10 mg/L of suspended solids, and <1 μm of particle size [31].

Moreover, for designing an injection water, some the following parameters should be taken into account: (i) size distribution of oil droplet, (ii) oil content, (iii) size distribution of suspended solids, (iv) total suspended components, (v) dissolved gas, (vi) bacteria, and (v) scaling potential [34].

The large volume of produced water is also viewed as potential source of water for various usages such as irrigation, potable water, and industrial process water. However, the complex mixture of pollutants presents the challenges for achieving treatment steps that both cost effective and high efficiency [27,53]. For example, the high-quality standard and stringent regulation for drinking water applications lead to the complex treatment steps which result in a high treatment cost [54]. Therefore, numerous efforts were made to develop more economical and advanced treatment technologies to meet various end-use requirements.

3 Membrane in produced water reuse and reinjection

The use of membrane technology in produced water treatment is increasing since the conventional technologies are no longer capable to meet the more stringent regulation in produced water treatment. Membrane process may be combined with conventional processes. Membrane-based integrated processes are also reported by several studies to obtain a high quality reclaimed water.

* Corresponding author: igw@che.itb.ac.id

Generally, produced water treatment includes several steps namely pre-treatment, main treatment, polishing treatment, and tertiary treatment [27]. Coarse particles, oil, and gas are removed in pre-treatment step. The main step is usually used to remove small hydrocarbon droplets and small particles resulting in dispersed hydrocarbon content to below 40 mg/L. Further, produced water may undergo polishing and additional treatment depending on the final effluent quality needed.

Microfiltration (MF) and ultrafiltration (UF) membranes are used in several studies to reduce the oil content to an acceptable level. Performance of MF and UF membrane in produced water treatments are shown in Table 2 and Fig. 2. As indicated by Fig. 2, MF and UF can achieve >99% oil removal from produced water. Furthermore, UF membrane with pore size of 0.05 μm can remove 99% of oil with 39% of TOC [28].

Integrated MF/UF/NF or RO process is used when a high quality reclaimed water is required. Typical examples of the integrated process are summarized in Table 3. Ebrahimi et al [28] reported a study on produced water treatment using integrated MF, UF, and nanofiltration (NF). This integrated process successfully removed oil up to 99% with acceptable TOC removal. This study also suggested MF as a first efficient step for oil removal. Khedr [50] investigated the performance of integrated chlorination, coagulation, and NF membrane for produced water treatment. Coagulation can produce effluent with a low concentration of suspended solids, organic, and oil. Meanwhile, the introduction of NF as polishing step help to remove metal cations and to further remove suspended solids, bacteria, and organics. This polishing step is expected to inhibit scale and biofilm formation during injection.

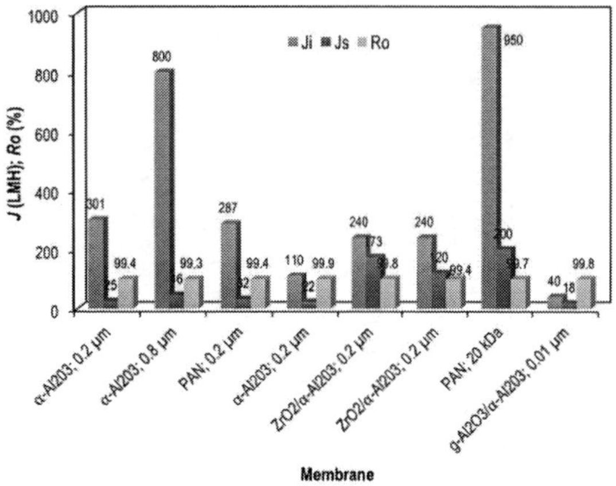

Fig. 2. Initial and stabilized flux (*Ji*, *Js*) and oil removal (*Ro*) of several MF and UF membrane during produced water and oily waste water treatment (data from refs. [55–58])

Integrated UF/NF/RO system was studied by Hayatbakhsh et al [59] to treat produced water from steam-assisted gravity drainage (SAGD). SAGD is generally used to ehance oil recovery for bitumen extraction from oil sands. The treatment purpose is to reclaim produced water for a high quality produced water wherein salt, silica, and dissolved organic matter are the concern. They found the UF membrane was able to remove up to 30% of salt and silica and 50% of dissolved organic matter. Tight NF membrane could remove >86% of salts, silica, and dissolved organic matter.

Besides pressure driven membrane processes (MF, UF, NF, and RO), membrane bioreactor (MBR), a hybrid membrane/bioreactor process, is also a potential membrane process for produced water treatment. Submerged MBR equipped with PVDF membrane was used by Sun et al [60] to produce water injection from produced water. The effluent of MBR contained <1 mg/L of crude oil, <1 mg/L of NH_4^+-N, and 1 NTU of turbidity. This effluent was able to achieve the grade A1 criteria of Recommendation Standard and Analysis Method for the Clastic Rock Reservoir Injection Water Quality (SY/T 5329-94, China).

Table 2. Performance of MF and UF membrane in produced water treatment

Membrane; Pore size or MWCO.	Operating conditions	Feed water	Results	Ref.
Ceramic membrane (93.32% SiO_2)	1 bar	Synthetic oil/water; C_{oil} = 200 mg/L.	Oil in permeate = 2 mg/L (99,8 % rejection)	[65]
Mixed cellulose esters; 0.1 μm	0.5 bar; CFV= 0.43 m/s;	Produced water	Removal: COD = 35%; TOC = 25%; O&G = 92%; Phenol = 35%.	[43]
Al_2O_3; 0.2 μm	1 bar; 60°C;	Produced water	Oil removal = 93%;	[66]
ZrO_2 (active layer) TiO_2/Al_2O_3	1.5 bar; 45°C; CFV= 3 m/s; Pilot scale;	Produced water	Oil in product <5 mg/L; SS = not detected; Perm. = 620-1200 LMH/bar	[31]
α-alumina; 0.8 μm.	0.7 bar; 40°C; 0.91;	Synthetic produced water; C_{oil}= 250 mg/L;	Oil removal = 99.3%;	[55]
α-alumina; 0.2 μm.	0.7 bar; 40°C; 0.24;	Synthetic produced water; C_{oil}= 1000 mg/L;	Oil removal = 99.4%;	[55]
PAN; 0.1 μm.	1.4 bar; 40°C; 0.24;	Synthetic produced water; C_{oil}= 250 mg/L;	Oil removal = 99.8%;	[55]
Cysteic acid functionalized α-alumina; 0.22 μm.	0.55 bar; Single-pass closed loop batch system.	Produced water; TOC = 43,550 mg/L;	TOC removal = 98.5%;	[67]
α-alumina; 1 μm.	1-1.5 bar; CFV= 3-5 m/s.	Synthetic oil/water mixture; C_{oil} = 5000 mg/L;	Oil removal = 94.3%; (<10 mg/L in permeate).	[56]
α-alumina; 0.2 μm.	1-1.5 bar; CFV= 3-5 m/s.	Synthetic oil/water mixture; C_{oil} = 5000 mg/L;	Oil removal = 99.9% (<10 mg/L in permeate).	[56]
ZrO_2/α-alumina; 0.2 μm.	1-1.5 bar; CFV= 3-5 m/s.	Synthetic oil/water mixture; C_{oil} = 5000 mg/L;	Oil removal = 99.8% (<10 mg/L in permeate).	[56]
ZrO_2/α-alumina; 0.2 μm.	1.1 bar; CFV= 2.56 m/s; 25°C; Flocculation (3530S, 40 mg/L)	Oily wastewater produced from refinery process; C_{oil} = 6000 mg/L; COD = 3 g/L; Solid = 5%-wt.	Oil removal = 99.8% (8.762 mg/L in permeate); COD in permeate = 108;	[57]
γ-alumina/α-alumina; 0.01 μm.	1-1.5 bar; CFV= 3-5 m/s.	Synthetic oil/water mixture; C_{oil} = 5000 mg/L;	Oil removal = 99.8% (<10 mg/L in permeate).	[56]
TiO_2; 0.05 μm	1 bar; 60°C;	Produced water	Oil removal >99%; TOC removal = 13.6%;	[68]
Al_2O_3; 40 nm	0.5 bar, CFV = 2 m/s; Total recycle;	Produced water	Oil removal >99.5%;	[32]
PAN; 20 kDa	TMP: 3 bar; T: 40oC; CFV: 1 m/s.	Oily wastewater; O&G = 78 mg/L; TSS = 60 mg/L; TOC = 81 mg/L.	Final effluent: O&G = 0.2 mg/L; TSS = trace; TOC = 19 mg/L.	[58]

Table 3. Produced water treatment schemes and purposes

No.	System configuration	Purpose	Remarks	Ref.
1	A primary sedimentation + oil/water separator + DAF system + 1 μm cartridge filter + 0.2 μm ceramic MF + Activated carbon + RO;	Discharge	Feed: Devecatagi oil well produced water; Product: COD =108 mg/L; TDS = 1726 mg/L; pH = 7;	[42]
2	MF + biological treatment;	Discharge	Feed: Campos basin oilfield wastewater; Final effluent: COD: 230 mg/L, TOC: 55 mg/L, respectively.	[43]
3	Induced gas floatation + walnut shell filter + heat exchanger + DOX stripper + chemical addition + flocculation + heat exchanger + multimedia filter + cartridge filter + RO (two-pass) + constructed wetlands; 79,493 m^3/d.	Surface discharge to local ground water	Feed: Produced water from San Ardo water reclamation facility; Final effluent: TDS = 180 mg/L; Na = 50 mg/L; Cl = 11 mg/L; Sulfate: 125 mg/L; Nitrit = non-detectable; Boron = 0.10 mg/L;	[46]
4	UF + RO + MED; Pilot plant.	Reuse for irrigation	Feed: coal seam gas (CSG) produced water. Clean water recovery= 95%; A final brine containing mostly sodium bicarbonate up to of 48 g/L. RO recovery = 76%; MED recovery= 80%. RO permeate and MED distillate were of high quality and could be blended with UF filtrated CSG produced water for irrigation.	[69]
5	MF + Activated carbon + MD;	Reuse	TDS removal >99.6%; TOC removal >90%; Recovery = 70%	[70]
6	Aeration tank + air floatation + sand filter + UF; Pilot scale.	Discharge or reinjection	Feed: Produced water from Daqing oil field. Oil in product <0.5 mg/L; SS contents <1.0 mg/L; The effluent can meet discharging or injection water quality standard.	[44]
7	Warm Softening (pH 9.5; 71°C) + Cooling (40°C) + Equalization Storage + Pressure Filtration + Cartridge Filtration + Reverse Osmosis (pH 9.5; 40°C) + Disinfection + Sludge Handling Reject + Disposal; Pilot scale.	Reuse for irrigation	Feed: Produced water from Placerita Canyon Oil Field, Los Angeles County, CA. Warm softening removed ~95% of hardness. RO removed >95% TDS. About 90% boron removal was achieved at a pH ≥10.5. An 80% ammonia removal was obtained at a pH ≤8.7.	[45]
8	Oil/water separator + warm softener + sand filters + ion exchange softener + cartridge filter (1.0 μm) + RO; Pilot scale;	Reuse for irrigation	Remineralization (calcium, magnesium) and addition some nutrients are needed for irrigation purpose.	[71]
9	Cartridge filter (10 μm + 5 μm + 1 μm) + Cation exchanger + cartridge filter (0.45 μm) + RO + constructed wetlands; Pilot scale.	Reuse for irrigation	Conductivity reduction= 95%; TDS removal = 94%;	[72]
10	Induced gas flotation + Nut-shell filter + cooking + induced gas flotation + warm lime process + media filter + softener + cartridge filter + two-pass RO + pH adjustment + sodium absorption ratio adjustment + constructed wetlands + recharge basins; 79,493 m^3/d;	Reuse for irrigation	Final effluent: TDS = 422 mg/L; Alkalinity = 92 mg/L; pH = 7.0; Sodium absorption ratio = 17.5; TOC = 15.6 mg/L;	[73]
11	Three-phase separator + gravity separator + dissolved gas flotation + metal removal unit + sand filters + VSEP membrane system + ammonia absorption column + Cartridge filter (5 μm) + RO; Feed flow rate 90 m^3/day (design).	Reuse for Irrigation	Final effluent: TDS = 160 mg/L (TDS in feed = 35,803 mg/L); Recovery = 45%;	[74]
12	MF + Activated carbon + MD; Estimated for a 100 m^3/h designed capacity.	Reuse for irrigation	TDS removal >99.6%; TOC removal >90%;	[70]
13	Warm Softening (pH 9.5; 71°C) + Cooling (40°C) + Equalization Storage + Pressure Filtration + Cartridge Filtration + Reverse Osmosis (pH 10.8; 40°C) + pH Adjustment + Ion-Exchange for Ammonia (pH 7.7; T ambient) + Disinfection + Sludge Handling + Reject Disposal; Pilot scale.	Reuse for drinking water	Feed: Produced water from Placerita Canyon Oil Field, Los Angeles County, CA. Warm softening removed ~95% of hardness. RO removed >95% TDS. About 90% boron removal was achieved at a pH ≥10.5. An 80% ammonia removal was obtained at a pH ≤8.7.	[45]
14	Warm Softening + Cooling + MBR + RO (two-pass); Conceptual design of 79,493 m^3/d;	Ruse for potable water	RO product: TDS = 1.22 mg/L; pH = 8.0; Boron = 0.12 mg/L; O&G = 0.0;	[75]
15	Microfiltration (Zirconia oxide membrane)	Reinjection	TSS and oil in product <1 mg/L;	[47]

* Corresponding author: igw@che.itb.ac.id

Another membrane process which is potential for produced water reuse is membrane distillation (MD). Performance of MD in produced water treatment is summarized in Table 4. MD can recover pure water while rejecting the other components. Moreover, the performance of MD is not limited by the osmotic pressure of the solution. Therefore, it is possible to treat a feed water containing a high dissolved solid with a high recovery factor. MD can also be combined with RO membrane [61] to improve the overall water recovery during produced water treatment, especially for reuse purpose. In addition, since MD is a thermally-driven process, it can be directly used to treat a hot produced e.g. from SAGD without cooling [62]. MD can be integrated with a waste heat source to drive the separation process even to reach a concentration of brine solution of 30% [63]. Tavakkoli et al [64] reported an economic analysis on MD for produced water treatment and showed that by using a waste heat source, the cost of produced water treatment can be significantly reduced from \$5.70/m^3_{feed} to \$0.74/m^3_{feed}.

Among MD configuration, direct contact MD (DCMD), vacuum MD (VMD), sweeping gas MD (SGMD), and air gap MD (AGMD), DCMD is preferred since it does not require external condenser and it requires the simplest equipment and operation [70,76].

Even though membrane shows an excellent performance in treating produced water for reuse purpose, fouling is still the major drawback [28]. This fouling results in a considerable reduction of membrane flux (Fig. 2). Fouling phenomena are inevitable in membrane operation. Strategies for fouling mitigation has been widely studied and can be found in the literature [77–79]. In produced water treatment, several studies used hydrophilic membranes. For example, Xu et al used a hydrophilized PVDF UF membrane for produced water reinjection purpose [82,83]. The membrane exhibited an average 75 LMH. The flux recovery was about 95%. The membrane effectively removed oil, turbidity, and TSS to 1 mg/L, 1 NTU, and <2.5 mg/L (below detection level), respectively, which was able to meet the highest reinjection criteria in China. In addition, the membrane showed an antifouling property. Alzahrani et al [84] investigated the performance of commercial hydrophilic NF and RO membranes for produced water treatment reuse (potable water). The membranes have contact angles of ~23° and ~37°. It was reported that more hydrophilic and negatively charged membrane shows a less susceptible to fouling [59]. Xu et al [85] investigated UF membrane performance in produced water treatment for reinjection purpose.

Polyethersulfone UF membrane was used in a dead-end filtration mode. The membrane showed a high oil removal (>95%) with an oil content of lower than 5 mg/L. They also successfully recovered up to 95% of membrane flux by combining cleaning solution method of citric acid solution, sodium hydroxide solution, and SDBS solution (membrane is soaked in each solution containing 0.5 wt. % chemical).

In MD, besides fouling, wetting still becomes the major drawbacks. The treated produced water may still contain impurities which could result in more severe fouling and wetting when fed to MD system. It was reported by Kim et al [86] that the presence oil and grease as well as scaling induced by multivalent ions can enhance the wetting phenomenon. Results of their study suggest a combination of MD and crystallization to reduce scalant content and effectively prevent wetting. Scaling can also be removed by a simple intermittent cleaning operation using diluted acid solution (e.g. 0.32 wt.% for 10 min every 24 h) which was successfully performed by Zhong et al [87] to maintain permeate flux in 3 cycles operation. Another approach to mitigate membrane wetting is by using a membrane which has an anti-wetting property [88]. For instance, a graphene/PVDF membrane studied by Woo et al [61] showed a stable normalized flux (85% final flux after 10 days continuous operation). For a produced water containing dissolved organic matter and surfactant, optimized pretreatment and surface membrane properties are required to minimize flux reduction due to fouling [89].

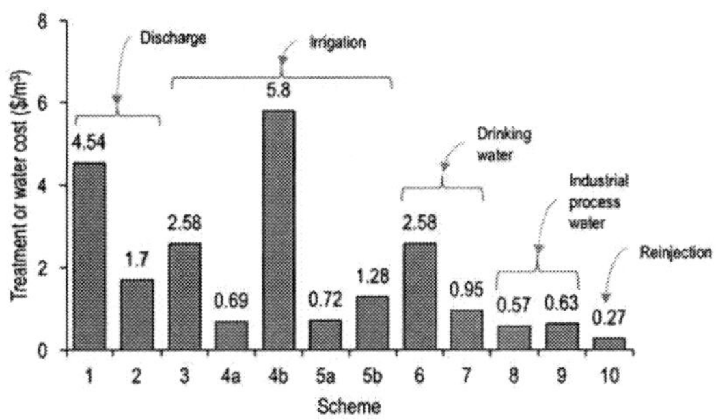

1	A primary sedimentation + oil/water separator + DAF system + 1 μm cartridge filter + 0.2 μm ceramic MF + Activated carbon + RO.
2	DAF with coagulation, flocculation polymer + granular activated carbon fluidized bed biological reactor + ED.
3	Warm Softening (pH 9.5; 71°C) + Cooling (40°C) + Equalization Storage + Pressure Filtration + Cartridge Filtration + Reverse Osmosis (pH 9.5; 40°C) + Disinfection + Sludge Handling Reject + Disposal.
4a	Three-phase separator + gravity separator + dissolved gas flotation + metal removal unit + sand filters + VSEP membrane system + ammonia absorption column + Cartridge filter (5 μm) + RO; TDS in feed = 35,803 mg/L;
4b	Three-phase separator + gravity separator + dissolved gas flotation + metal removal unit + sand filters + VSEP membrane system + ammonia absorption column + Cartridge filter (5 μm) + RO; TDS in feed = 106,670 mg/L;
5a	MF + Activated carbon + MD; Estimated for a 100 m^3/h designed capacity. T_{feed} = 50°C.
5b	MF + Activated carbon + MD; Estimated for a 100 m^3/h designed capacity. T_{feed} = 20°C.
6	Warm Softening (pH 9.5; 71°C) + Cooling (40°C) + Equalization Storage + Pressure Filtration + Cartridge Filtration + Reverse Osmosis (pH 10.8; 40°C) + pH Adjustment + Ion-Exchange for Ammonia (pH 7.7; T ambient) + Disinfection + Sludge Handling + Reject Disposal.
7	Warm Softening + Cooling + MBR + RO (two-pass);
8	Warm Softening (pH 7.7; 71°C) + Equalization Storage + pH Adjustment + Sludge Handling.
9	Warm Softening (pH 8.6; 71°C) + Equalization Storage + pH Adjustment + Sludge Handling.
10	Microfiltration (Zirconia oxide membrane)

Fig. 3. Total cost of schemes used in produced water treatment (data from refs. [42,45,47,70,74,75,95])

* Corresponding author: igw@che.itb.ac.id

Table 4. MD for produced water treatment

Process	Membrane	Operational conditions	Results	Ref.
MD	PTFE (0.03 μm)	DCMD; Tf = 128 °C; Tc = 20 °C; 2-3 atm; Feed = Synthetic PW (NaCl = 3000; Phenol = 45; Cresol = 45; Naphtenic acid = 10; all in mg/L);	Flux = 195 LMH	[62]
MD	PTFE (0.2 μm)	AGMP; Tf = 40-80 °C; Tc = 5-25 °C; Feed = Produced water from ARAMCO (TDS = 187,440 mg/L);	Flux = 7-18 LMH; Salt rejection >99%; Organic rejection >98%;	[90]
MD	PP	DCMD; Tf = 50-85 °C; Tc = 20 °C; Feed = Produced water from Chevron (TDS = 7622 (S1); 12,040 (S2); 35,000 (S3));	Flux = 15-20 LMH (T_f = 80°C); Recovery = 80%; Salt rejection >99%;	[91]
MD	PP (0.2 μm)	DCMD; Tf = 50-70 °C; Tc = 25 °C; Feed = PW treated by MF and AC (TDS = 247,900; TC = 40.7; all in mg/L);	Flux = 6-14 LMH; R_{TDS} = 99.7%; R_{TC} = 92.9%;	[70]
MD	PVDF (0.29 μm)	DCMD; Tf = 50-70 °C; Tc = 25 °C; Feed = PW treated by MF and AC (TDS = 247,900; TC = 40.7; all in mg/L);	Flux = 4-11 LMH; R_{TDS} = 99.8%; R_{TC} = 95.0%;	[70]
MD	PVDF (0.43 μm)	DCMD; Tf = 50-70 °C; Tc = 25 °C; Feed = PW treated by MF and AC (TDS = 247,900; TC = 40.7; all in mg/L);	Flux = 3-7 LMH; R_{TDS} = 99.8%; R_{TC} = 90.6%;	[70]
MD	MWCNT/PP (0.846 μm)	VMD; Tf = 55 °C; Tc = 15 °C; Feed = Oilfield produced wastewater (TDS = 230,000 mg/L);	Flux = 19.6 LMH; R_{TDS} = 99.9%;	[92]
MD	SiNP/PVDF (0.45 μm)	DCMD; Tf = 60 °C; Tc = 20 °C; Feed = prefiltered shale produced wastewater (TDS = 101,000; O&G = 2; Hardness = 5910 all in mg/L);	Flux = 13.6 LMH;	[93]
MD	PVDF (0.22 μm)	DCMD; Tf = 60-80 °C; Tc = 20 °C; Feed = synthetic $NaCl/NaHCO_3$ solution (1 mM NaCl/1 mM $NaHCO_3$ and 500 mM NaCl/25 mM $NaHCO_3$)	Flux = 18.8-41.9 LMH; R = >99.5% and >99.8%;	[94]

Moreover, membrane technology also faces other challenges including a high permeability and product, easy to clean, and chemical and thermal stability [32]. Chemical stability is needed since produced water contains hydrocarbon which may result in membrane damage, especially for a polymeric membrane. Ceramic membrane is used in produced water treatment due to its high chemical stability. This is beneficial in either filtration and cleaning process since a polymeric membrane is generally sensitive to hydrocarbon in produced water as well as chemical used in cleaning. The higher chemical stability will lead to a longer membrane lifetime. Ceramic membrane may also undergo a harsh chemical cleaning resulting in a better membrane regeneration. Therefore, the ceramic membrane can be potential material in produced water treatment.

Weschenfelder et al [25] used zirconia multichannel UF membrane for removal of suspended solids and oil. The membrane can remove suspended solids completely with an oil content of <5 mg/L. This shows the potential applications of ceramic UF membrane in produced water treatment for reinjection. Li and Lee reported that ceramic MF membrane can remove oil and suspended solids from produced water effectively [31]. Ebrahimi et al used ceramic hollow fiber membrane (Al_2O_3; d_{in} = 2 mm, d_{out} = 4 mm) with pore size diameter of 40 nm for produced water treatment [32]. By using this membrane, >99.5% of oil and up to 94% of TOC can be removed from produced water dewatering tank samples. The performance of the membrane is better than membrane used in their previous study [66] which only reached 49% TOC removal with comparable oil removal. MFI-type zeolite membrane can be used to remove dissolved solid replacing the polymeric RO membrane [96]. MFI zeolite membrane gives more than 99% ion rejection with water flux around 10 $l.m^{-2}.h^{-1}$ [31]. This type of membrane is also able to remove dissolved organic content [97]. However, the cost of the inorganic membrane is higher than the polymeric membrane. Therefore, producing a low-cost inorganic membrane is needed for commercial scale plant.

Fig. 3 shows the total cost of different schemes and purposes used in produced water treatment including discharge and reuse (for industrial purposes, drinking water, irrigation, and reinjection). It is obvious that the scheme and the final purpose of the treatment determine the treatment cost. The treatment cost increases along with the increasing quality of the end product. This is due to the more complex system used. In addition, additional nutrient may be required, for example, for irrigation purpose. Moreover, produced water characteristic also determine the overall treatment cost. This effect is showed by scheme 4 a and b and 5 a and b (Fig. 3). A higher TDS content results in an increased treatment cost involving RO. Meanwhile, a higher produced water temperature fed to a system involving MD may reduce the treatment cost. Among the treatment purposes, reinjection may be an interesting option. Since the produced water characteristic is almost similar to the formation water, a simpler treatment system may be used to achieve an acceptable water injection quality rather than the other reuse purposes. Therefore, the treatment cost is lower. Moreover, reinjection is also beneficial for both economic and environmental aspects.

4 Future challenges

Produced water has become a global environmental issue due to its huge volume and toxicity that may pose detrimental effects on receiving environment. Several approaches have been proposed to provide a strategy for produced water handling such as reinjection, reuse, or discharge. Among the reuse purposes, reinjection is the most attractive option. This is due to the attractive goal of both saving fresh water usage along with near-zero discharge. It is also viewed as a more cost-effective alternative rather than discharge and other reuse purposes. With various advantages, membrane technology has been increasingly used in produced water treatment replacing the conventional technologies. However, fouling is a major drawback of membrane processes in this application which needs to be controlled. Various strategies have been conducted to solve this problem such as choosing an appropriate cleaning procedure and using a membrane with an anti-fouling property. For MD, an anti-wetting membrane is also needed to maintain a stable flux for a long-term operation. Since produced water contains hydrocarbon, a chemically stable membrane such as inorganic membrane was recommended by several studies to avoid membrane degradation. This stability is also important to obtain a better membrane regeneration by a harsh chemical cleaning. However, a lower cost inorganic membrane may be needed.

References

1. A. Fakhru'l-Razi, A. Pendashteh, L.C. Abdullah, D.R.A. Biak, S.S. Madaeni, Z.Z. Abidin, *J. Hazard. Mater.*, **170**, 530 (2009).
2. B. Gargouri, O.D. Gargouri, B. Gargouri, S.K. Trabelsi, R. Abdelhedi, M. Bouaziz, *Chemosphere*, **117**, 309 (2014).
3. J.A. Veil, M.G. Puder, D. Elcock, R.J. Redweik Jr, *A white paper describing produced water from production of crude oil, natural gas, and coal bed methane.*, Argonne National Lab., IL (US), (2004).
4. C.E. Clark, J.A. Veil, *Produced water volumes and management practices in the United States.*, Argonne National Laboratory (ANL), (2009).
5. G.T. Tellez, N. Nirmalakhandan, J.L. Gardea-Torresdey, *Adv. Environ. Res.*, **6**, 455 (2002).
6. J.D. Arthur, B.G. Langhus, C. Patel, *All Consult. LLC, Tulsa, OK*, (2005).
7. N.F. Himma, S. Anisah, N. Prasetya, I.G. Wenten, *J. Polym. Eng.*, **36**, 329 (2016).
8. Khoiruddin, I.N. Widiasa, I.G. Wenten, *J. Food Eng.*, **133**, 40 (2014).
9. M. Purwasasmita, D. Kurnia, F.C. Mandias, Khoiruddin, I.G. Wenten, *Food Bioprod. Process.*, **94**, 180 (2015).
10. I.G. Wenten, P.T. Dharmawijaya, P.T.P. Aryanti, R.R. Mukti, Khoiruddin, *RSC Adv.*, **7**, 29520 (2017).
11. I.G. Wenten, Khoiruddin, *J. Eng. Sci. Technol.*, **11**, 916 (2016).
12. I.G. Wenten, I.N. Widiasa, *Desalination*, **149**, 279 (2002).
13. I.G. Wenten, H. Julian, N.T. Panjaitan, *Desalination*, **306**, 29 (2012).
14. D. Ariono, M. Purwasasmita, I.G. Wenten, *J. Eng. Technol. Sci.*, **48**, 367 (2016).
15. P.T.P. Aryanti, M. Sianipar, M. Zunita, I.G. Wenten, *Membr. Water Treat.*, **8**, 463 (2017).
16. Khoiruddin, D. Ariono, Subagjo, I.G. Wenten, *J. Appl. Polym. Sci.*, (2017).
17. I.G. Wenten, Khoiruddin, *Desalination*, **391**, 112 (2016).
18. I.G. Wenten, Khoiruddin, F. Arfianto, Zudiharto, *Desalination*, **314**, 109 (2013).
19. A.K. Wardani, A.N. Hakim, Khoiruddin, I.G. Wenten, *Water Sci. Technol.*, **75**, 2891 (2017).
20. S. Munirasu, M.A. Haija, F. Banat, *Process Saf. Environ. Prot.*, **100**, 183 (2016).
21. S. Alzahrani, A.W. Mohammad, *J. Water Process Eng.*, **4**, 107 (2014).
22. J.M. Dickhout, J. Moreno, P.M. Biesheuvel, L. Boels, R.G.H. Lammertink, W.M. de Vos, *J. Colloid Interface Sci.*, **487**, 523 (2017).
23. A.R.P. da Motta, C.P. Borges, A. Kiperstok, K.P. Esquerre, P.M. Araujo, L.P.N. Branco, *Eng. Sanit. E Ambient.*, **18**, 15 (2013).
24. S. Mondal, *Rev. Chem. Eng.*, **32**, 611 (2016).
25. S.E. Weschenfelder, A.M.T. Louvisse, C.P. Borges, E. Meabe, J. Izquierdo, J.C. Campos, *J. Pet. Sci. Eng.*, **131**, 51 (2015).
26. D. Robinson, *Filtr. Sep.*, **47**, 14 (2010).
27. Society of Petroleum Engineers, Challenges in Reusing Produced Water, http://www.spe.org/industry/challenges-in-reusing-produced-water.php .
28. M. Ebrahimi, K.S. Ashaghi, L. Engel, D. Willershausen, P. Mund, P. Bolduan, P. Czermak, *Desalination*, **245**, 533 (2009).
29. E.T. Igunnu, G.Z. Chen, *Int. J. Low-Carbon Technol.*, **9**, 157 (2012).
30. J.E. Horner, J.W. Castle, J.H. Rodgers, *Ecotoxicol. Environ. Saf.*, **74**, 989 (2011).
31. L. Li, R. Lee, *Sep. Sci. Technol.*, **44**, 3455 (2009).
32. M. Ebrahimi, S. Kerker, S. Daume, M. Geile, F. Ehlen, I. Unger, S. Schütz, P. Czermak, *Desalin. Water Treat.*, **55**, 3554 (2015).
33. D. Robinson, *Filtr. + Sep.*, **50**, 20 (2013).
34. R.C. Evans, Developments in Environmental Protection Related to Produced Water Treatments and Disposal (Produced Water Re-Injection), in: SPE Heal. Saf. Environ. Oil Gas Explor. Prod. Conf., Society of Petroleum Engineers, Jakarta (1994).
35. D. Robinson, *Filtr. + Sep.*, **50**, 40 (2013).
36. A. Hussain, J. Minier-Matar, A. Janson, S. Gharfeh, S. Adham, Advanced technologies for produced water treatment and reuse, in: Soc. Pet. Eng. - Int. Pet. Technol. Conf. 2014, IPTC 2014 Unlocking Energy Through Innov. Technol. Capab., ConocoPhillips, United States (2014), pp. 1621–1631.
37. E.J. Sullivan Graham, A.C. Jakle, F.D. Martin,

Water Int., **40**, 809 (2015).
38. R.R. Nair, E. Protasova, T. Bilstad, S. Strand, Reuse of produced water by membranes for enhanced oil recovery, in: Proc. - SPE Annu. Tech. Conf. Exhib., Department of Mathematics and Natural Science, United States (2016).
39. J. Ochi, D. Dexheimer, P. V Corpel, *SPE Prod. Oper.*, **29**, 192 (2014).
40. R.W. Paige, L.R. Murray, Re-injection of produced water - Field experience and current understanding, in: Rock Mech. Pet. Eng. 29-31 August, Delft, Netherlands, Society of Petroleum Engineers, (1994), pp. 731–738.
41. D. Robinson, *Filtr. Sep.*, **50**, 36 (2013).
42. M. Çakmakce, N. Kayaalp, I. Koyuncu, *Desalination*, **222**, 176 (2008).
43. J.C. Campos, R.M.H. Borges, A.M. Oliveira Filho, R. Nobrega, G.L. Sant'Anna, *Water Res.*, **36**, 95 (2002).
44. X. Qiao, Z. Zhang, J. Yu, X. Ye, *Desalination*, **225**, 113 (2008).
45. R. Funston, R. Ganesh, L.Y.C. Leong, *Gr. Water Prot. Counc. (GWPC), Color. Spring, US*, 16 (2002).
46. C.H. Webb, L. Nagghappan, G. Smart, J. Hoblitzell, R. Franks, Desalination of Oilfield-Produced Water at the San Ardo Water Reclamation Facility, CA, in: SPE West. Reg. Meet. 2009, Society of Petroleum Engineers, San Jose, California (2009), pp. 565–585.
47. S.E. Weschenfelder, C.P. Borges, J.C. Campos, *J. Memb. Sci.*, **495**, 242 (2015).
48. X.-H. Zhen, S.-L. Yu, B.-H. Wang, H.-F. Zheng, *J. Environ. Sci. (China)*, **18**, 1077 (2006).
49. G.F. Doran, F.H. Carini, D.A. Fruth, J.A. Drago, L.Y.C. Leong, Evaluation of technologies to treat oil field produced water to drinking water or reuse quality, in: Proc. - SPE Annu. West. Reg. Meet., ARCO Western Energy (1997), pp. 811–822.
50. M. Gamal Khedr, *Desalin. Water Treat.*, **55**, 3460 (2015).
51. Q. Helmy, E. Kardena, *J. Pet. Environ. Biotechnol.*, **6**, 1 (2015).
52. J. Kus, R. Card, *J. Can. Pet. Technol.*, **23**, 66 (1984).
53. J. Cao, S. Monroe, H. Sun, L. Li, Cost-efficient and improved treatment technology for produced water reuse, in: Proc. - SPE Annu. Tech. Conf. Exhib., Baker Hughes Incorporated, United States (2015), pp. 974–982.
54. F. Hum, P. Tsang, A. Kantzas, T. Harding, Is it possible to treat produced water for recycle and beneficial reuse?, in: SPE/PS-CIM/CHOA Int. Therm. Oper. Heavy Oil Symp. Proc., TIPM Laboratory, Canada (2005), pp. A135–A148.
55. J. Mueller, Y. Cen, R.H. Davis, *J. Memb. Sci.*, **129**, 221 (1997).
56. C. Yang, G. Zhang, N. Xu, J. Shi, *J. Memb. Sci.*, **142**, 235 (1998).
57. J. Zhong, X. Sun, C. Wang, *Sep. Purif. Technol.*, **32**, 93 (2003).
58. A. Salahi, M. Abbasi, T. Mohammadi, *Desalination*, **251**, 153 (2010).
59. M. Hayatbakhsh, M. Sadrzadeh, D. Pernitsky, S. Bhattacharjee, J. Hajinasiri, *Desalin. Water Treat.*, **57**, 14869 (2016).
60. Y. Sun, W.-X. Shi, J. Xu, M. Yu, L.-Z. Jia, *Adv. Mater. Res.*, **113–116**, 1541 (2010).
61. Y.C. Woo, Y. Kim, W.-G. Shim, L.D. Tijing, M. Yao, L.D. Nghiem, J.-S. Choi, S.-H. Kim, H.K. Shon, *J. Memb. Sci.*, **513**, 74 (2016).
62. D. Singh, K.K. Sirkar, *J. Memb. Sci.*, **389**, 380 (2012).
63. O.R. Lokare, S. Tavakkoli, G. Rodriguez, V. Khanna, R.D. Vidic, *Desalination*, **413**, 144 (2017).
64. S. Tavakkoli, O.R. Lokare, R.D. Vidic, V. Khanna, *Desalination*, **416**, 24 (2017).
65. O.A. Al-Harbi, M. Mujtaba Khan, C. Özgür, *J. Aust. Ceram. Soc.*, **53**, 883 (2017).
66. M. Ebrahimi, D. Willershausen, K.S. Ashaghi, L. Engel, L. Placido, P. Mund, P. Bolduan, P. Czermak, *Desalination*, **250**, 991 (2010).
67. S.J. Maguire-Boyle, J.E. Huseman, T.J. Ainscough, D.L. Oatley-Radcliffe, A.A. Alabdulkarem, S.F. Al-Mojil, A.R. Barron, *Sci. Rep.*, **7**, (2017).
68. M. Ebrahimi, Z. Kovacs, M. Schneider, P. Mund, P. Bolduan, P. Czermak, *Desalin. Water Treat.*, **42**, 17 (2012).
69. L.D. Nghiem, C. Elters, A. Simon, T. Tatsuya, W. Price, *Sep. Purif. Technol.*, **146**, 94 (2015).
70. F. Macedonio, A. Ali, T. Poerio, E. El-Sayed, E. Drioli, M. Abdel-Jawad, *Sep. Purif. Technol.*, **126**, 69 (2014).
71. M. Melo, H. Schluter, J. Ferreira, R. Magda, A. Júnior, O. de Aquino, *Desalination*, **250**, 1016 (2010).
72. C. Murray-Gulde, J.E. Heatley, T. Karanfil, J.H.R. Jr., J.E. Myers, *Water Res.*, **37**, 705 (2003).
73. J.E. Myers, Chevron San Ardo Facility Unit (SAFU) beneficial produced water reuse for irrigation, in: Soc. Pet. Eng. - SPE Int. Conf. Heal. Saf. Environ. 2014 Journey Contin., PE, Chevron, United States (2014), pp. 687–697.
74. V. Piemonte, M. Prisciandaro, L. Mascis, L. Di Paola, D. Barba, *Desalin. Water Treat.*, **55**, 565 (2015).
75. P.B. Tsang, C.J. Martin, Economic Evaluation of Treating Oilfield Produced Water for Potable Use, in: SPE Int. Therm. Oper. Heavy Oil Symp. West. Reg. Meet., Society of Petroleum Engineers, (2004), p. SPE 86984.
76. E. Drioli, A. Criscuoli, E. Curcio, *Membrane contactors: fundamentals, applications and potentialities*, Elsevier, (2011).
77. H. Lin, W. Peng, M. Zhang, J. Chen, H. Hong, Y. Zhang, *Desalination*, **314**, 169 (2013).
78. N. Hilal, O.O. Ogunbiyi, N.J. Miles, R. Nigmatullin, *Sep. Sci. Technol.*, **40**, 1957 (2005).
79. M.F.A. Goosen, S.S. Sablani, H. Al-Hinai, S. Al-Obeidani, R. Al-Belushi, D. Jackson, *Sep. Sci. Technol.*, **39**, 2261 (2005).
80. T.D. Kusworo, N. Aryanti, M.M.H. Firdaus, H. Sukmawati, Surface modification of cellulose

acetate membrane using thermal annealing to enhance produced water treatment, in: AIP Conf. Proc., (2015).

81. X. Yang, R. Wang, A.G. Fane, C.Y. Tang, I.G. Wenten, *Desalin. Water Treat.*, **51**, 3604 (2013).
82. J. Xu, W.-X. Shi, S.-L. Yu, W.-M. Qu, *Adv. Mater. Res.*, **113–116**, 1266 (2010).
83. J. Xu, C. Ma, B. Cao, J. Bao, Y. Sun, W. Shi, S. Yu, *Process Saf. Environ. Prot.*, **104**, 564 (2016).
84. S. Alzahrani, A.W. Mohammad, P. Abdullah, O. Jaafar, *J. Environ. Chem. Eng.*, **1**, 1341 (2013).
85. J. Xu, C. Ma, Y. Sun, W. Shi, S. Yu, L. Wang, *Desalin. Water Treat.*, **61**, 42 (2017).
86. J. Kim, H. Kwon, S. Lee, S. Lee, S. Hong, *Desalination*, **403**, 172 (2017).
87. W. Zhong, H. Li, Y. Ye, V. Chen, *Desalination*, **393**, 52 (2016).
88. N.G.P. Chew, S. Zhao, C.H. Loh, N. Permogorov, R. Wang, *J. Memb. Sci.*, **528**, 126 (2017).
89. M. Malmali, P. Fyfe, D. Lincicome, K. Sardari, S.R. Wickramasinghe, *Sep. Sci. Technol.*, **52**, 266 (2017).
90. A. Alkhudhiri, N. Darwish, N. Hilal, *Desalination*, **309**, 46 (2013).
91. D. Singh, P. Prakash, K.K. Sirkar, *Ind. Eng. Chem. Res.*, **52**, 13439 (2013).
92. K. Okiel, A.H.M. El-Aassar, T. Temraz, S. El-Etriby, H.A. Shawky, *Desalin. Water Treat.*, **57**, 10995 (2016).
93. C. Boo, J. Lee, M. Elimelech, *Environ. Sci. Technol.*, **50**, 12275 (2016).
94. D.-W. Cho, H. Song, K. Yoon, S. Kim, J. Han, J. Cho, *Water (Switzerland)*, **8**, (2016).
95. J.A. Miller, A.W. Lawrence, R.F. Hickey, T.D. Hayes, (n.d.).
96. N. Liu, J. Lu, A. Li, R. Lee, Factors determining the reverse osmosis performance of zeolite membranes on produced water purification, in: Proc. - SPE Int. Symp. Oilf. Chem., (2007), pp. 399–405.
97. N. Liu, L. Li, B. McPherson, R. Lee, *J. Memb. Sci.*, **325**, 357 (2008).

Synthesis and Characterization of Nano Hybrid Membrane PES-TiO_2 for Biogas Purification: Combination Effect of Ultra Violet and Cross-Linking

Tutuk Djoko Kusworo[1,*], *Budiyono*, *Qudratun*[2], *Dani Puji* Utomo[1], *Iqbal Ryan* Ramadhan[1], and *Indriyanti*[1]

[1]Department of Chemical Engineering, Diponegoro University, Semarang, Indonesia
[2]Nano Center Indonesia, South Tangeran Indonesia

Abstract: One of the alternative energy offered and can be used as a substitute for fossil energy is renewable energy. Examples of renewable energy are very abundant and can be developed in Indonesia is biogas. Biogas mostly contain CH_4 and CO_2. Membrane separation is one of the most suitable process to make biogas more pure. So, it must be advanced to improve the membrane performance. The objectives of the current research are to investigate the effect of combined UV irradiation and cross linking on nano hybrid membrane PES-TiO_2 on their performance to separate CO_2 that contain in biogas. The nano hybrid membrane PES-TiO_2 was fabricated by preparing the dope solution containing PES, NMP as solvent, and nano TiO_2. The membrane was casted by using NIPS method, with expose under UV lights for 1, 2, 3 min and followed by immersion in acetone-ethanol mixture for 24 hours. UV irradiation and cross linking treatments increased the selectivity and permeability of nano hybrid membrane PES-TiO_2 on CO_2/CH_4 gas separation. The effect either UV irradiation or cross linking addition, make the void larger than before, so the selectivity and permeability can increase. With the addition of cross linking ethanol-acetone solution can reduce agglomeration in the membrane. The combination of UV irradiation and cross linking treatments can increase significantly in nano hybrid membrane PES-TiO_2 performance for CO_2/CH_4 gas separation.

1 Introduction

One of the current global problems is the imbalance between consumption and available energy sources. The main factors causing the problem are the dependence on fossil energy and the lack of public attention to the current energy crisis. According to the National Energy Board of the Republic of Indonesia (2014), dependence on fossil energy, especially petroleum in the fulfilment of domestic consumption is still high at 96% (petroleum 48%, gas 18% and coal 30%) of total consumption and efforts to maximize utilization renewable energy cannot run as planned. On the other hand, Indonesia tends to face a decline in fossil energy reserves that continue to occur and cannot be matched by the discovery of new reserves [1]. This condition causes Indonesia to import energy for domestic needs.

The scarcity of fossil energy requires another energy source that can be used as an alternative to help meet global energy consumption. One of the alternative energies offered and can be used as a substitute for fossil energy is renewable energy. An example of a very abundant renewable energy that can be developed in Indonesia is biomass in the form of biogas.

Biogas is gas produced from the process of fermentation of organic materials by anaerobic bacteria that are resistant to airtight conditions. Biogas can be produced from organic waste which is a combustible gas dominated by methane and CO_2 gas. All types of organic materials containing carbohydrate, protein, and fat compounds can be processed to produce biogas [2]. In principle, making biogas is very simple, that is by entering the substrate in the form of animal or human waste into the digestive unit (digester) and then closed tightly and some time will form a gas that can be used as a source of energy [3]. The composition of biogas according to Seadi et al., consists of methane (50-75%), carbon dioxide (25-45%), and small amounts of water vapour, oxygen, nitrogen, ammonia, hydrogen, and hydrogen sulphide [4]. The purity of methane from the biogas product becomes important because it can affect the value of the resulting heat. According to Mustafa et al., the use of biogas also still has limitations due to the impurities of biogas containing large amounts of CO_2 (20-45 Vol%) and H_2S (0-1 Vol%) other than CH_4 main component (55-80 Vol%) [5]. The presence of CO_2 in CH_4 Very undesirable, this is because the higher CO_2 levels in CH_4 further decrease the CH_4 heat value

* Corresponding author: tdkusworo@che.undip.ac.id

indicated by the yellowish red colour of the resulting flames [6]. In addition CO_2 is also corrosive and acidic with its presence with water compounds present in storage tanks or biogas distribution [7]. Reduced CO_2 levels will significantly increase biogas products.

Many technologies have been developed for the purification of biogas from CO_2, such as CO_2 absorption, purification by using membranes, and CO_2 fixation by biological and chemical methods. From the several solutions offered, membrane technology for gas separation was chosen because of low energy requirements, low maintenance costs, and conditions that are not extreme.

There are three aspects to note in the membrane application for the separation of CO_2, the first aspect of membrane performance that includes the selectivity and permeability of the gas, the second aspect is the membrane resistance to the applied operating conditions such as temperature and pressure critical to have plastic properties, and the third aspect is availability material and cost to be used. The most widely used membrane material for CO_2 separation in the industry is the polymer. Polymers are chosen because they are relatively inexpensive and the manufacturing process is relatively easy compared to other materials. The most common types of membrane polymers used in gas separation processes are polyimide, cellulose acetate, and polyether sulfone (PES). PES was chosen because it has a higher resilience than polyimide and has a fairly high plastization pressure, which is 34 bar [8]. The PES membrane matrix, mostly uses a mixture of zeolites, carbon, silica, carbon nanotubes, or metal oxides as inorganic materials. One of the metal oxides used as inorganic materials in membrane manufacturing is TiO_2. TiO_2 was chosen because of its abundant presence in Indonesia, always stable when working at room temperature, and non-toxic [9].

The development of the fabrication and application of MMMs (Mix Matrix Membranes) containing inorganic particles in the polymer matrix for gas separation is still considered quite low compared to the polymer membrane, providing an opportunity for future membrane development. Success in improving membrane performance with mixed matrix concepts is limited to its use for gas separation, and morphological studies of prevailing membranes (e.g. asymmetric membranes, thin films of composite membranes).

Limited information in inorganic materials on the manufacture of polymer membranes makes the selection of materials in the membrane-making process into a problem that needs to be developed. Therefore, there is a need for better approaches and standards to identify the properties of inorganic materials, especially in terms of their ability as a separator [10]. So it is necessary to develop the ability of the membrane itself because basically membrane technology has a prospect that ensures for the future.

Therefore, in this research, the effort that can be done is by combining UV irradiation with cross linking using polyvinyl pirrolidone (PVP) and also mixture of ethanol and acetone and treatment by using thermal annealing on nano hybrid membrane PES - nano TiO_2 using material polyether sulfone (PES) to determine membrane characterization by determination of flux and selectivity, Scanning Electron Microscopy (SEM), and Fourier Transform Electron (FTIR) to obtain the best membrane performance in CO_2 purification in biogas.

2 Experimental

2.1. Materials Selection

Polyethersulfone (PES) powder was purchased from Solvay Advance Polymer USA.; N-methyl-pyrrolidinone (NMP) from Merck was used as the solvent due to its low toxicity. TiO_2 nanoparticle were prepared in Nano Center Indonesia with the average nanoparticles size is 50 nm.

2.2 Fabrication of asymmetric flat sheet mixed matrix membrane

The polymer solution was consisted of 20 wt-% PES, 80 wt-% NMP and 0.5 wt-% of nano TiO_2 in total solid. The homogeneous polyethersulfone was prepared according to the following procedure; 0.1 gram of TiO_2 nanoparticles and 20 gram PES were added to 80 mL NMP. This solution was further agitated by stirring at high speed for at least 4 hours. The membrane was casted by using NIPS method. The flat sheet membrane was prepared according to the dry/wet phase inversion technique. The solution was poured onto a clear, flat and smooth glass plate that was placed on the trolley. Stainless steel support casting knife was used to spread the solution to a uniform thickness. The casted membrane was exposed under UV lights for 1, 2, 3 min. The glass plate with the membrane film was then immersed in the water bath. During this process, solvent exchange occurred and solidified the membrane film to a complete membrane structure. To ensure all of the solvent in the membrane structure is removed, membranes were immersed in an aqueous bath for 24 hours, followed by cross linking added in ethanol-acetone (1:3) for 24 hours and air-dried for 60 minutes at room temperature.

2.3 Gas Permeation Testing

The permeation test involved the use of gas permeation cell in which the membrane was placed on a sintered metal plate and pressurized at the feed side. Gas permeation rates were measured by a constant pressure system using a soap bubble flow meter. Fig. 1 illustrates the gas permeation cell set up. The cross-membrane pressure difference was maintained 1 bar. Pressure normalized gas permeation flux or permeance for gas I, (P/l)I, in (GPU), can be calculated as follows:

$$(P/l)i = Qi/\Delta P*A \quad (1)$$

Where Qi is the volumetric flow rate of gas i, ΔP pressure difference across the membrane (cmHg), A is a membrane affective surface area (cm^2) and l is membrane skin thickness (cm). The ideal separation factor α i/j can be calculated by using the following equation:

$$\alpha\ i/j = (P/l)i : (P/l)j \quad (2)$$

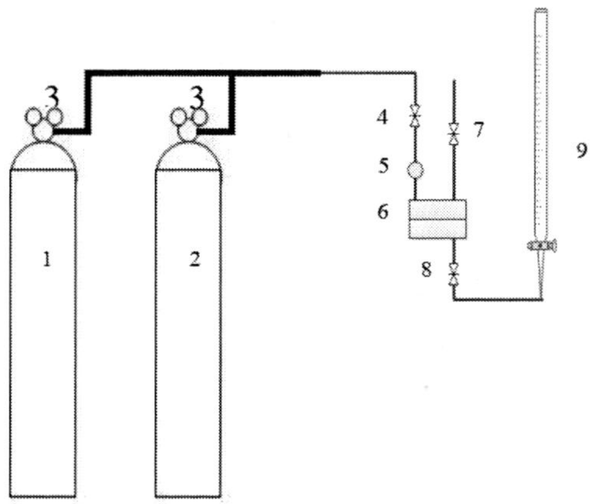

Fig 1. Instrumentation for Biogas Permeation

1. CO_2 tube
2. CH_4 tube
3. Regulator
4. Needle Valve
5. Pressure Gauge
6. Membrane module
7. Retentate
8. Permeate
9. Soap bubble flowmeter

3 Results and Discussion

3.1 Effect of UV irradiation on Nano Hybrid Membrane PES-TiO_2 for Biogas Purification

To observe the effect of the combination of the nano TiO_2 addition and UV modification, the membrane was fabricated with 0.5 wt-% nano TiO_2 loading and was exposed using UV light for 1, 2, and 3 minutes. Then the membrane is used to treat gas separation to evaluate the permeability and selectivity.

The addition of UV light to the PES + TiO_2 membrane increases the permeability of CO_2 and selectivity. CO_2 permeability and selectivity respectively are 104.10 GPU and 4.94 rise to 531.26 GPU and 7.53. This is because the UV-exposed PES + TiO_2 membrane undergoes a chain scission and crosslinking process. This phenomenon can cause membrane pores to form larger. Thus, raising the value of CO_2 permeability and selectivity [11].

The selectivity of the non-irradiated PES + TiO_2 membrane and UV light irradiance was obtained at values 4.94. Whereas in previous studies, PES polymer membranes without UV light irradiation had a selectivity of 8.33 [12]. This phenomenon is caused by the effects of UV light exposed to the polymer membrane. The UV lightinduced the polymer chains to restructure and grafting resulting on the denser polymer. The longer exposure of UV light on the membrane surface, it leads to the polymer chain degradation of polyethersulfone polymer [13-15]. As the consequences, the membrane pores will be larger, and the UV treatment increases the permeability. But this shows that the selectivity still obtained is far from standard.

According to Adib et al. the phenomenon occurs because there is a selective hole in the membrane layer which induces for the Knudsen diffusion mechanism resulting in low selectivity and high permeability [15]. This can be seen in Table 1. which shows that the permeability of CH_4 that escaped very large. While the desired mechanism in the separation of CO_2 and CH_4 is the mechanism of Molecular Sieving or Solution-Diffusion, where only or more CO_2 passes through the membrane whereas CH_4 is not expected to pass.

Table 1. CO_2 permeability and CO_2/CH_4 selectivity for different mixed matrix membranes before and after adding UV irradiation

Type	CO_2 Permeance (GPU)	CO_2/CH_4 Selectivity
PES + TiO_2	104.10	4.94
PES + TiO_2 + UV 1 min	531.26	7.53

In Table 2. We can see the performance of PES membrane for the separation of CO_2 from CH_4. The permeability values of CO_2 and selectivity tend to increase with increasing duration of irradiation. This is because the UV-exposed PES membrane undergoes a chain scission and crosslinking process. This phenomenon can cause membrane pores to form larger. Thus, raising the value of CO_2 permeability and selectivity [11].

However, the selectivity obtained in the experiments was very low when compared with previous experiments on PES membranes without irradiated UV light, ie, 8.33 [12]. While the selectivity shown in Table 2, selectivity does not reach a value greater than 3.75. Adib et al. assessed that the low selectivity of the research results was due to a selective hole in the membrane causing the gas flow mechanism occurring in the membrane to be Knudsen diffusion [15]. With the occurrence of Knudsen diffusion will increase CH_4 permeability and will reduce the selectivity. This can be seen in Table 1. which shows the permeability of CH_4 passes. While the desired mechanism in the separation of CO_2 and CH_4 is the mechanism of Molecular Sieving or Solution-Diffusion, where only or more CO_2 passes through the membrane whereas CH_4 is not expected to pass.

Table 2. CO_2 permeability and CO_2/CH_4 selectivity for different mixed matrix membranes before by adding UV irradiation

Type	CO_2 Permeance (GPU)	CO_2/CH_4 Selectivity
PES + TiO_2 + UV 1 min	531.26	7.53
PES + TiO_2 + UV 2 min	309.33	8.43
PES + TiO_2 + UV 3 min	1194.20	13.18

3.2 Effect of Cross Linking on Nano Hybrid Membrane PES-TiO_2 for Biogas Purification

Table 3. CO_2 permeability and CO_2/CH_4 selectivity for different mixed matrix membranes before by adding cross linker ethanol-aseton

Type	CO_2 Permeance (GPU)	CO_2/CH_4 Selectivity
PES + TiO_2 + UV 3 min	1194.20	13.17
PES + TiO_2 + UV 3 min + cross linking	1644.82	15.64

Based on Table 3, membrane with the addition of ethanol and acetone as cross linker or co-solvent and also internal coagulant together with water will lead to a denser and smoother membrane on its surface [16]. Addition of acetone in the casting solution with a ratio 3:1 with ethanol, results in the formation of a dense upper coating and a decline in the size and number of finger-like pores on the sub layer.

The immersion of membrane material in the organic solvent has been reported to be able to give effects to membrane transport properties [13]. This provides lower permeability and higher selectivity [17]. And also with the presence of acetone as the co-solvent can enhances the bridge formation between the compounds [18]. Based on the presence of ethanol and acetone as cross linker with ratio 1:3 can improve the membrane performance, so that the membrane surface will be denser, smaller pore, and the selectivity will increase.

3 Conclusion

Membrane nano hybrid PES with 0.5 wt-% nano TiO_2 as additive materials with 3 minutes of ultraviolet irradiation for biogas separation can increased the permeability of CO_2 and the selectivity of CO_2/CH_4. This phenomenon is caused by the effects of UV light exposed to the polymer membrane. Variations in ultraviolet irradiance from 1,2,3 minutes, while UV irradiation for 3 minutes has the highest permeability of CO_2 and selectivity of CO_2/CH_4. Besides that, combination between ultraviolet irradiance and immersion of membrane nano hybrid PES-nano TiO_2 into non-solvent solution ethanol-acetone (1:3) can increased the permeability of CO_2 and selectivity CO_2/CH_4 because the presence of acetone as the co-solvent enhances the bridge formation between the compounds.

References

1. Dewan Energi Nasional Republik Indonesia, *Outlook Energi Indonesia 2014* (Dewan Energi Nasional Republik Indonesia, 2014)
2. G. Dong, H. Li, and V. Chen. J. Membr. Sci. **369** 206-220 (2011)
3. Naqiibatin and Triwikantoro. Jurnal Mekanika. **3** 13-25 (2014)
4. T.A. Seadi, D. Rutz, H. Prassl, M. Kottner, T. Finsterwalder, S. Volk, and R. Janssen. *Biogas Handbook* (University of Southern Denmark, Esbjerg 2008)
5. A. Mustafa, T.D. Kusworo, A. Busairi, A.F. Ismail, and Budiyono. Internat. J. of Waste Resour. **2** 22-24 (2012)
6. D. Deublein and A. Steinhauser. *Biogas form Waste and Renewable Resources.* (Wiley – VCH Verlag GmbH & Co. KGaA, Weinheim 2001)
7. S. Li, J.L. Falconer, and R.D. Noble. J. Membr. Sci. **241** 121-135 (2004)
8. R. Guan, H. Dai, C. Li, J. Liu, J. and Xu. J. Membr. Sci. **277** 148-156 (2006)
9. G.A. Karim and A.S. Hanafi. J. Energy Resour. Technol **114** 152-157 (2008)
10. T.S. Chung, L.Y. Jiang, Y. Li, S. Kulprathipanja. Prog. Polym. Sci. **32** 483-507 (2007)
11. H. Susanto, A.P. Nur, and D.K. Sari. Jurnal Teknologi Kimia dan Industri **2** 189-197 (2013)
12. M.S. Suleman, K.K. Lau, and Y.F. Yeong, Chem. Eng. Technol. **39** 1-14 (2016)
13. T.D. Kusworo, A.F. Ismail, N. Aryanti, Qudrotun, D.P. Utomo. Jurnal Teknologi. **79** 129-140 (2017)
14. T.D. Kusworo, D.P. Utomo, N. Aryanti, Qudrotun, J. Environ. Chem. Eng. 5 (2017) 3290–3301.
15. T.D. Kusworo, N. Aryanti, Widayat, Qudrotun, D.P. Utomo. Adv. Sci. Lett. **23** 5744-5747 (2017).
16. H. Adib, S. Hassanajili, D. Mowla, and F. Emaeilzadeh. J. Supercrit. Fluids. **97** 6-15 (2015)
17. Q.F. Alsalhy. Eng. & Technology **25** (2007)
18. S.S. Madaeni, A. Rahimpour, and J. Barzin. 9th Iran. Natl. Chem. Eng. Congr., 8th 4309-4314 (2004)
19. I. Ahmed, A. Idris, M.S. Khan, and S. Chowdhury. Appl. Mech. Mater. **625** 545-548 (2014)

SO_2 Removal from the flue gas by hollow fibre membrane contactor

Danu Ariono[1], *Utjok W. R.* Siagian[2], *Anita K.* Wardani[1], *Ahmad N.* Hakim[1], and *I Gede* Wenten[1,*]

[1]Department of Chemical Engineering, Institut Teknologi Bandung, Jl. Ganesha 10, Bandung 40132, Indonesia
[2]Department of Petroleum Engineering, Institut Teknologi Bandung, Jl. Ganesha 10, Bandung 40132, Indonesia

Abstract. The combustion of fossils fuels is the main source of the environmental problems, such as acid rain, photochemical smog, and ozone depletion. To solve these problems, the appropriate technology to remove acid gasses such as SO2 is urgently required. In this work, SO2 removal was conducted using polypropylene and polysulfone-based hollow fiber membrane contactors. Water and sodium sulfite were used as sorbents. The results showed that the use of sodium sulfite as a sorbent provided 2 times higher SO2 permeation than water. Moreover, permeate flux of polypropylene membrane was higher than polysulfone membrane since the hydrophobic membrane is more suitable as membrane contactor for flue gas cleaning. In addition, a comparison between membrane contactor and conventional scrubber showed that membrane contactor can reduce the contact volume up to 27 times smaller than those used in the conventional scrubber.

1. Introduction

Most of the world energy production is made up from the combustion of fossil fuels such as oil, coal, and gas. The combustion of these fuels is the main source for environmental problems of major concern such as acid rain, photochemical smog, ozone depletion, and greenhouse effect [1, 2]. To cope with these problems, the appropriate technology to remove acid gasses such as SO_2 is urgently required.

The available SO_2 capture technology including physical and chemical absorption, cryogenic distillation solid adsorption, and membrane technologies. Among these, chemical absorption with conventional scrubbing tower is the most used technique to capture SO_2 from flue gas streams. However, this technique is characterized by huge space requirements and related high capital costs. The conventional wet scrubbing techniques also suffer from a number of deficiencies such as loading, flooding, and entrainment limitation [3].

The membrane processes is a promising alternative that rapidly gaining recognition as efficient, energy saving and compact separation technologies [4-6]. Microporous membranes can be used as gas-liquid contractors to remove gas components from gas phase [2, 7, 8]. The major advantage of the membrane contactor lies in the drastic reduction of the contactor volume and the related reductions in the investment and operating costs. Furthermore, due to the principles of contacting, many disadvantages present in conventional contacting equipment such as flooding, loading, and entrainment limitations can be overcome [9]. Hence, the operation of the membrane contactor can easily be adjusted to meet changes in plant load. In addition, due to their compactness and modular nature, membrane contactors are well suited for retrofit applications.

However, the majority of membrane contactor application is limited by membrane wetting phenomena which deteriorates its performance. Membrane wetting is caused by penetration of liquid absorbent into the pore structure. The presence of liquid absorbent inside of the pores add an additional mass transfer resistance causing a dramatic performance drop. Therefore, a material with the higher hydrophobicity is more recommended to assure long-term operation stability [10].

Membrane wetting problem can be overcome by increasing membrane hydrophobicity. The hydrophobic nature of membrane material will prevent penetration of liquid absorbent into the pore structure.

Polypropylene (PP), polyvinylidene fluoride (PVDF), polyethylene (PE), and polytetrafluoroethylene (PTFE) are the hydrophobic material which commonly used in porous membrane contactor fabrication, where polypropylene is considered as the most low-cost material with moderate hydrophobicity [11, 12].

The comprehensive review of the development of superhydrophobic membrane has available in the literature [13]. The coating method is a low-cost and simple approach to improve membrane hydrophobicity. The modification of low cost hollow fiber polypropylene membrane to improve its hydrophobicity by dip coating method was investigated by Himma et.al. [14]. The superhydrophobic membrane with a contact angle of 151.3° can be obtained by using methyl ethyl ketone as nonsolvent.

* Corresponding author: igw@che.itb.ac.id

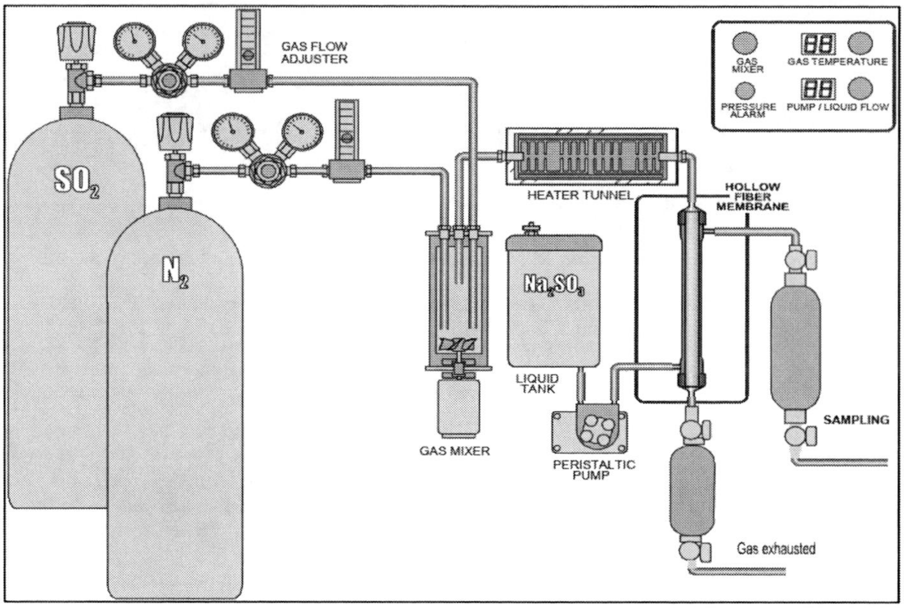

Figure 1 Schematic illustration of hollow fibre membrane contactor

In this work, SO_2 removal was conducted using polypropylene and polysulfone-based hollow fibre membrane contactors. Meanwhile, water and sodium sulfite were chosen as sorbents. Membrane characteristic, structure, and stability were investigated. Finally, the performance of membrane contactors was compared to the conventional scrubber.

2. Methodology

In this work, the comprehensive study of membrane characterization used for membrane contactor has been carried out experimentally and theoretically. Commercial polypropylene (MemTec) and polysulfone (self-preparation) were used in this work. Membrane was packed into units called modules. All experiments were conducted using hollow fibre modules due to the high packing densities obtained in such modules. Schematic illustration of hollow fibre membrane contactor can be seen in Figure 1.

The membrane parameters controlling the transport properties of the membrane such as wettability, porosity, wall thickness, pore size, pore area, and pore tortuosity has been assessed. Information about pore structure, pore area, etc. was obtained from scanning electron microscopy, measurement of the bubble point pressure, and measurement of mercury porosimetry. Chemical stability was studied by immersing membranes into any level pH of sodium sulfite solution for a month. Meanwhile, to measure its thermal stability, membrane contactor was operated at high temperatures.

3. Result and Discussion

3.1 Sorbent and Membrane Selection

In this work, both water and an aqueous solution of sodium sulfite have been used as sorbents. Comparison of the two sorbents is shown in Figure 2. As can be seen in this figure, SO_2 permeation is relatively same at the low gas flow rate. However, in high gas flow rate, an aqueous solution of sodium sulfite provides 2 times higher permeation of SO_2 than water. These results show that aqueous solution of sodium sulfite is more suitable than water. Figure 2 indicates that chemical reaction is controlling the overall process. In addition, an aqueous solution of sodium sulfite has a high absorption capacity for SO_2 so that mass transfer resistance in the liquid phase is negligible by proper selection of operating conditions.

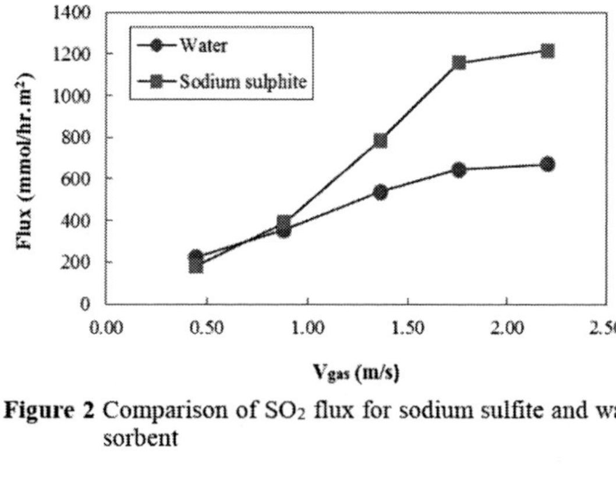

Figure 2 Comparison of SO_2 flux for sodium sulfite and water sorbent

Furthermore, the effect of membrane type is shown in Figure 3. This figure shows that the polypropylene membrane gives higher SO_2 flux than the polysulfone. It proves that the polypropylene membrane mass transfer coefficient is greater than the polysulfone since hydrophobic membrane is more suitable as membrane contactor for gas cleaning [14-17].

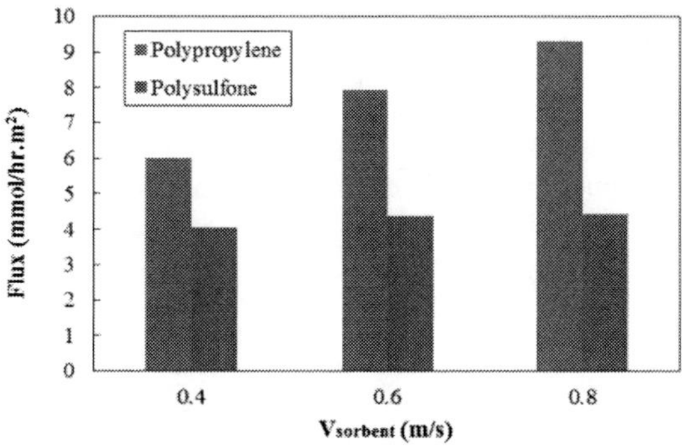

Figure 3 Comparison of SO_2 flux for polypropylene and polysulfone membrane (sorbent flow rate 0,001 m/s and sorbent concentration 0.26 M)

3.2 Membrane Characteristic and Structure

The key concern for selection of membranes used in membrane contactors is transport properties and membrane stability with respect to wettability, chemical, thermal, and mechanical properties. Parameters affected on transport properties are porosity, wall thickness, pore size, and tortuosity. The membrane wettability is influenced by the type of membrane used for membrane contactor. A hydrophobic microporous membrane will generally prevent liquid from penetrating into the pores, and hence the pores will be gases filled. In case of hydrophilic microporous membranes, the liquid will penetrate the pores. A commercial polymer such as polytetrafluoroethylene (PTFE), polypropylene (PP), polyvinylidene fluoride (PVDF) and polysulfone) provide hydrophobic characteristics.

As regards to the thermal stability, PP membranes can be continuously operated in the range 50-90°C (depending on the grade of PP), while temperature ranges for PVDF and PTFE are 100-155°C and 130-260°C, respectively [18]. Typical flue gas temperatures after the dust removal system at coal fire boiler facilities are 110-150°C. For wet scrubbing systems the temperature is normally reduced by conditioning prior to scrubbing, thus the thermal stability of the membrane may not be a critical issue. However, a PVDF and PTFE do have a significance advantage in that an accidental exposure to higher temperatures will not be disastrous.

It is suggested to use a high porosity membrane as membrane contactor because it gives a large contact area between gas and sorbent. Two membranes, i.e. polypropylene and polysulfone have been used in this study. Porosities of the membranes used in this study are so high that they are good enough to be used as a membrane contactor. The experimental results of their characteristics are shown in Table 1. Based on their characteristics, both of them are microporous membranes with the pores sizes is in the range of 0.01-1 μm, the porosity between 0.3-0.85 % and membrane thickness between 25-500 μm.

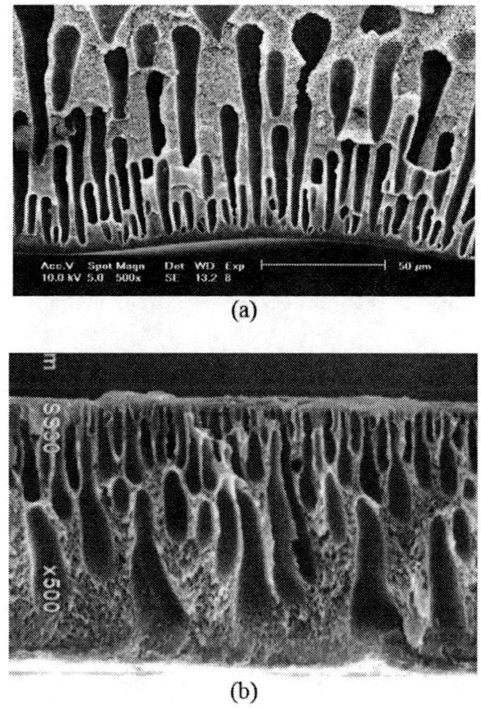

Figure 4 SEM image of (a) polypropylene and (b) polysulfone membrane

Table 1 Membrane characteristics

	Polypropylene	Polysulfone
Pore diameter (μm)	0.2	0.2
Porosity (%)	70	82
Outside diameter (μm)	541	672
Membrane thickness (μm)	150	153

Furthermore, chemical stability of membranes was tested by pre-soaking polypropylene membranes in sodium sulphite solution in various pH for a month. Mercury porosimetry analysis results are listed in Table 2.

Table 2 Stability assessment of PP membrane

pH	Diameter (μm)
Initial	0.2
1	0.1
3	0.2
10	0.2

The results show that pore diameter remains stable before and after the treatment. At extreme acid condition, the membrane seems to shrink, as the pore size becomes smaller. However, the effect is not significant. This proves that extreme pH environment will not significantly influence membrane characteristic.

3.3 Comparison of Membrane Contactor Performance with Conventional Scrubber

Generally, the performance of membrane contactor is not enhanced by the increase of mass transfer coefficient but rather by increasing contact surface area. Absorption in many cases relies on providing a high interfacial area between the gas and liquid phase to achieve high overall rate of gas mass transport. Many commercial devices based on packed columns, plate columns, spray columns, venturi scrubbers, etc. are available. In many cases equipment design can be restricted by limitations in the relative flows of the fluid streams. The use of membrane, in the form of low cost hollow fibres gives a contact surface area between gas and liquid phases, which is independent of gas and liquid flows. The small dimensions of the hollow fibre membranes give high specific surface areas which are significantly larger than most conventional absorbers. Commercially available hollow fibre modules used for filtration duties can be used as absorbers with gas flow through the fibres.

In order to explain advantages of membrane contactor with conventional scrubbers, performance of both contactors is quantitatively compared. One of conventional scrubber i.e. packed column need 3.105 m on diameter and height 19.9 m for separation rate 0.5 kg/s. Membrane contactor with specifications pore size 0.2 µm, porosity 70%, fibre diameter 0.5 µm, module length 30 cm, module diameter 1.5 cm and number of fibre 100 is compared with the same separation rate. The comparison of both contactors is expressed using analytical dimension as shown in Table 3. Table 3 shows that flux of conventional scrubber is higher than hollow fibre contactor. However, membrane contactor volume is smaller than conventional scrubber for the same absorption capacity (27 times lower contactor volume).

Table 3 Analytical dimensions for membrane contactor and conventional scrubber

	Conventional Scrubber	Hollow Fiber Contactor
Capacity (kg/s)	0.5	0.5
Flux (mol/hr.m^2)	250	10
Contactor volume (m^3)	151	5.6

4. Conclusion

Theoretical and experimental investigations to explore the potential application of membrane contactor for flue gas cleaning were carried out. In the case of SO_2 absorption, aqueous solutions of sodium sulphite as listed in the commercial Wellman-Lord and Dual Alkali processes was selected for this study. The performance of the sorbent was compared with water. The result showed high absorption capacity for SO_2 by using aqueous solution of sodium sulphite as a sorbent. Moreover, hydrophobic membranes were more suitable as membrane contactor for flue gas cleaning. These membrane materials allow gas filled pores mechanism to govern the process.

Examination of membrane characteristic before and after treatment showed that pore size remained stable in wide pH range. Furthermore, the improved bubble point method could determine the pore size distribution and permeability of membrane. The nominal pore size obtained using this method was in agreement with the membrane data from the manufacturer. The pore size distribution obtained for each membrane was relatively narrow and shows good distribution.

Furthermore, comparison between membrane contactor and conventional scrubber showed that hollow fibres membrane contactor had higher contact surface area between gas and liquid phases. The small dimensions of hollow fibre membranes gave high specific surface areas which significantly larger than most conventional scrubbers. Analytical dimension were used to compare advantages of hollow fibre membrane contactor with conventional scrubbers quantitatively. The results showed that scrubber conventional had higher flux than membrane contactor, but membrane contactor could reduce the contact volume up to 27 times smaller than those used in conventional scrubber.

References

1. J. Yang, X. Yu, J. Yan, S.-T. Tu and E. Dahlquist, Appl. Energy **112**, 755-764 (2013).
2. Z. Wang, M. Fang, H. Yu, Q. Ma and Z. Luo, Energy Fuels **27**,11, 6887-6898 (2013).
3. J. Albo, P. Luis and A. Irabien, Desal. Water Treat. **27**,1-3, 54-59 (2011).
4. P. T. P. Aryanti, S. R. Joscarita, A. K. Wardani, S. Subagjo, D. Ariono and I. G. Wenten, J. Eng. Technol. Sci. **48**,2 (2016).
5. A. K. Wardani, A. N. Hakim, Khoiruddin and I. G. Wenten, Water Science and Technology **75**,12, 2891-2899 (2017).
6. D. Ariono, P. Aryanti, S. Subagjo and I. Wenten, presented at the AIP Conference Proceedings, 2017 (unpublished).
7. R. Klaassen and A. E. Jansen, Environ. Prog. **20**,1, 37-43 (2001).
8. S. Karoor and K. K. Sirkar, Ind. Eng. Chem. Res. **32**,4, 674-684 (1993).
9. M. Stanojević, B. Lazarević and D. Radić, FME Transactions **31**,2, 91-98 (2003).
10. A. F. Ismail and A. Mansourizadeh, Journal of Membrane Science **365**,1, 319-328 (2010).
11. M. Rezaei, A. F. Ismail, S. A. Hashemifard and T. Matsuura, Chemical Engineering Research and Design **92**,11, 2449-2460 (2014).
12. F. Geyer, C. Schönecker, H.-J. Butt and D. Vollmer, Advanced Materials **29**,5, n/a-n/a (2017).
13. H. Nurul Faiqotul and I. G. Wenten, Journal of Physics: Conference Series **877**,1, 012010 (2017).

14. N. F. Himma, A. K. Wardani and I. G. Wenten, Polym. Plast. Technol. Eng. **56**,2, 184-194 (2017).
15. N. F. Himma, S. Anisah, N. Prasetya and I. G. Wenten, J. Polym. Eng. **36**,4, 329-362 (2016).
16. Y. Lv, X. Yu, S.-T. Tu, J. Yan and E. Dahlquist, Appl. Energy **97**, 283-288 (2012).
17. N. F. Himma and I. G. Wenten, presented at the AIP Conference Proceedings, 2017 (unpublished).
18. M. Mulder, (Kluwer Academic Publishers, 1996).

Hydrogenation of Maltose in Catalytic Membrane Reactor for Maltitol Production

I.G.B.N. Makertihartha[1], Khoiruddin[1], *Ahmad N.* Hakim[1], *P.T.P.* Aryanti[2], and *I.G.* Wenten[1,*]

[1]Chemical Engineering Department, Institut Teknologi Bandung, Jalan Ganesha 10, Bandung 40132, Indonesia
[2]Chemical Engineering Department, Universitas Jenderal Achmad Yani, Jl. Ters. Jend. Sudirman, Cimahi 40285, Indonesia

Abstract. Maltitol is one of the low-calorie sweeteners which has a major role in food industries. Due to its characteristics of comparable sweetness level to sucrose, maltitol can be a suitable sugar replacement. In this work, catalytic membrane reactor (CMR) was examined in maltitol production through hydrogenation of maltose. Commercial ceramic membrane impregnated with Kalcat 8030 Nickel was used as the CMR. The reaction was conducted at a batch mode operation, 95 to 110°C of temperature, and 5 to 8 bar of pressure. In the range of working conditions used in this study, up to 47% conversion was achieved. The reaction conversion was significantly affected by temperature and pressure. Results of this preliminary study indicated that CMR can be used for hydrogenation of maltose with good performance under a relatively low operating pressure.

1 Introduction

Low calorie sweetener usage shows a significant increase recently. This is due to an improved paradigm in order to achieve a healthy lifestyle. Maltitol is one of these calorie sweetener that has a major role in food industry. Since maltitol has a comparable characteristic and sweetness level to sucrose, it makes maltitol suitable as sugar replacement. Moreover, maltilol can be used as an intermediate product for pharmaceuticals industry [1].

Generally, maltose hydrogenation is carried out in a batch slurry reactor. Studies on hydrogenation reactions of maltose in conventional batch slurry reactors can be found in literature [1–8]. The drawback of this conventional reactor is a poor contact between reactant and catalyst. Therefore, a high operating pressure, e.g. 20-30 bar, is usually used. Consequently, it usually requires a relatively high investment cost due to system complexity. This necessitates the improvement of hydrogenation reactor to overcome the gas-liquid interface constraint as well as to simplify the hydrogenation reaction design.

Nowadays, membrane technology is used in almost industrial processes and starts to replace the competing conventional technologies due to its interesting features [9–19]. Membrane offers the advantages of relatively lower foot print, lower cost, easy to scale-up, and more energy intensive [20–23]. Membrane also provides a possibility to combines a several steps or processes into a hybrid process thus an intensified process can be obtained [24–29]. CMR is a hybrid membrane process which facilitates catalytic reaction and contact between reactants. CMR was used by several studies to replace the conventional hydrogenation reactor. Veldsink [30] used CMR for hydrogenation of sunflower seed oil. The hydrogenation reaction was not affected by interfacial transport resistance and intraparticle diffusion limitation. Bottino et al [31] reported a study of hydrogenation-isomeration of methylenecyclohexane in a CMR and compared with the conventional reactor. They also found that the hydrogenation reaction in a catalytic membrane reactor was controlled by kinetic regime and thus the reaction conversion could be increase while avoiding the high pressure operation. In this work, performance of CMR in hydrogenation of maltose is investigated. A ceramic membrane is used as CMR and commercial nickel catalyst is impregnated in the membrane pores. The effect of temperature and pressure on reaction conversion is analyzed and discussed.

2 Materials and method

Maltose used in this study has a 60° Brix concentration. Kalcat 8030 Nickel was used as catalyst. Technical grade citric acid and demineralized water were used to prepare catalyst suspension. High purity hydrogen (99.9%) was used in hydrogenation reaction. Tubular ceramic membrane was used as membrane reactor (Al_2O_3, 100 kDa MWCO, 19 channels, 4 mm inside diameter, and 0.3 m length).

Experimental set up for catalyst impregnation and hydrogenation reaction are shown in Fig. 1 a and b. Catalyst impregnation was conducted by circulating catalyst suspension which contained 25 g catalyst in 500 mL citric acid solution (20%). After circulation, the membrane was drying in oven at 80°C for 6 h and then calcinated at 200°C for 3 h. Hydrogenation was

* Corresponding author: igw@che.itb.ac.id

conducted for 6 h at various operating temperatures and pressures. Between each reaction, membrane module was flush with demineralized water. Subsequently, the membrane was flushed with nitrogen. Every 45 min. reduced sugar content was analyzed by Eynon-Lane titration method in Ref. [32].

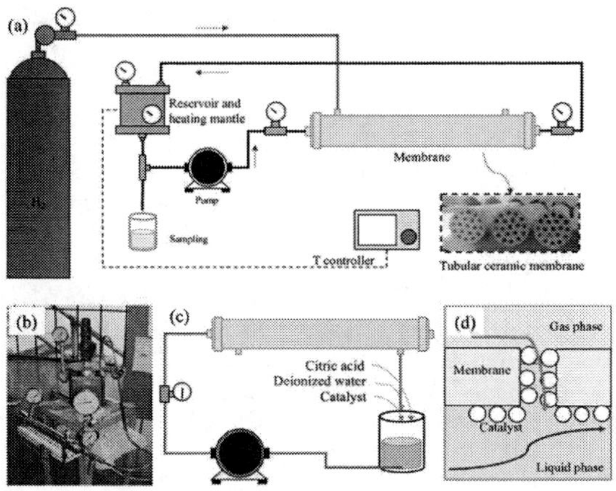

Fig. 1. Experimental set up. (a) Schematic of CMR, (b) photograph of experimental set up, (c) catalyst impregnation, and (b) hydrogenation reaction in CMR.

3 Results and discussion

The effect of temperature and pressure on hydrogenation reaction conversion and profile are shown in Fig. 2 (a) to (b). It can be observed that both operating parameters have the significant effect on hydrogenation reaction conversion. Temperature is one of the parameters that influence the kinetic rate. As the temperature increases, the reaction rate is increased. Operating pressure has a similar effect on reaction conversion. The reaction conversion increases with the increase of pressure. A drastic increase of reaction conversion is observed from 5 to 6 bar. It may be due to the minimum reaction condition was not achieved at 5 bar. In addition, it can be observed that reaction conducted at 6 bar has a conversion reaction which closes to that at 8 bar. It may be implied that contact between the multiple phases of reactants in the catalytic membrane reactor is good enough and no more depended on the dispersion of gas phase in the reactants mixture.

Reaction scheme, C/Co, reaction conversion at various reaction time at 6 bar and 110°C can be observed in Fig. 3. According to the reaction scheme, the nth-order reaction rate equation can be writen as I = $dC/dt = -kC^n$. By assuming that the reaction rate follows the first order mechanism (n = 1), the integration of the reaction rate will result in a linear correlation between ln (C/Co) and t with the slope of the linear plot is the negative value of rate constant, $-k$. Linear regression of the experimental $ln(C/Co)$ vs t plot gives high R-square value of 95.13% and the rate constant of the hydrogenation reaction, k, can be estimated to be 3.3 x 10^{-5} s^{-1}. The high R-square value proves that the assumption of first order mechanism is satisfied.

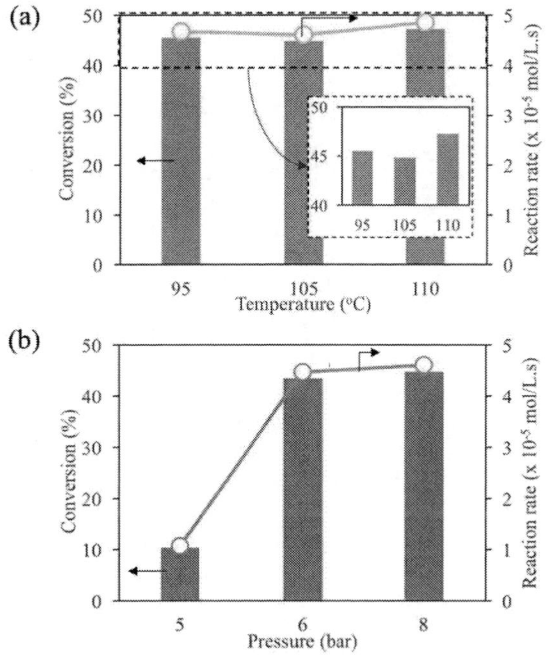

Fig. 2. Reaction conversion and reaction rate at various temperature (a; at 8 bar) and pressure (b; at 110°C).

(a) $$C_{12}H_{22}O_{11} \xrightarrow[\text{Catalyst}]{H_2} C_{12}H_{24}O_{11}$$

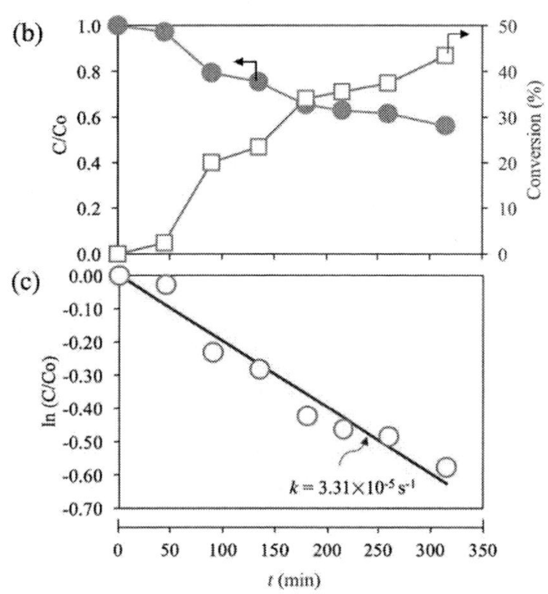

Fig. 3. (a) Reaction scheme of maltose hydrogenation, (b) C/Co vs t (at 8 bar and 110°C), and (c) ln C/Co vs t (at 6 bar and 110°C).

In a conventional reactor, a high hydrogen pressure is usually used to overcome the mass-transfer or diffusion limitation in a three-phase mixture. In this study, at a maximum pressure of 8 bar, reaction conversion of 47%

can be obtained. Even at lower pressure, 6 bar, the conversion is almost similar. This operating pressure is far below the operating pressure that usually used in conventional reactors (see Table 1). The reaction rate is also higher compared to the conventional reactors. According to Bottino et al [31], hydrogenation in catalytic membrane reactor offers a better control on operating condition rather than the conventional slurry bath reactor since the reaction is only a kinetic-controlled regime. Meanwhile, for in conventional reactor, the diffusion-controlled regime was also observed. Benefiting from its better diffusion and contact between hydrogen gas, liquid maltose, and solid catalyst, catalytic membrane reactor requires a relatively low hydrogen pressure to achieve a high conversion reaction. In addition, the reaction rate can be increased further by increasing the reaction temperature. This is expected to reduce the complexity of the reactor unit as well as the investment cost.

Table 1. Conversion of maltose hydrogenation from several reported studies

Operating conditions	Reactor; Catalyst	Conv. (%)	Ref.
50 mL; 40%-wt. maltose; 30 bar; 120°C; 4h.	Slurry batch; Co-B	33.7	[1]
50 mL; 40%-wt. maltose; 20 bar; 90°C; 2h.	Slurry batch; Ni-P	17.2	[7]
50 mL; 40%-wt. maltose; 30 bar; 100°C; 50 min.	Slurry batch; Raney Ni	5.2	[8]
50 mL; 40%-wt. maltose; 30 bar; 100°C, 3 h.	Slurry batch; Ru-B-C	73	[2]
50 mL; 40%-wt. maltose; 30 bar, 100°C; 0.5 h.	Slurry batch; Raney Ni	2	[6]
600 mL; 60%-wt maltose; 6 bar; 110°C; 6 h.	CMR; Kalcat 8030 Ni	47.2	This work

In catalytic membrane reactor, reaction occurs only when the solution contacts with catalyst inside the membrane phase. Meanwhile, during the circulation (in the reservoir and in piping), reaction does not occur. This is one of the disadvantages of CMR compared to slurry reactor batch. Therefore, it is suggested to use a cascade membrane system in order to minimize the inactive reaction time.

4 Conclusions

In this work, catalytic membrane reactor (CMR) was examined in maltitol production through hydrogenation of maltose. Commercial ceramic membrane impregnated with Kalcat 8030 Nickel was used as the CMR. The reaction was conducted at a batch mode operation, 95 to 110°C of temperature, and 5 to 8 bar of pressure. In the range of working conditions used in this study, up to 47% conversion was achieved. The reaction conversion was significantly affected by temperature and pressure. Results of this preliminary study indicated that CMR can be used for hydrogenation of maltose with a good performance under a relatively low operating pressure.

References

1. H. Li, P. Yang, D. Chu, H. Li, *Appl. Catal. A Gen.*, **325**, 34 (2007).
2. H. Li, Y. Wang, Q. Zhao, H. Li, *Res. Chem. Intermed.*, **35**, 779 (2009).
3. D. Chu, H. Li, H. Li, *Shiyou Huagong/Petrochemical Technol.*, **35**, 1125 (2006).
4. E.M. Sulman, M.E. Grigorev, V.Y. Doluda, J. Wärnå, V.G. Matveeva, T. Salmi, D.Y. Murzin, *Chem. Eng. J.*, (2015).
5. Y. Ren, W. Li, Y. Han, M. Zhang, K. Tao, *Shiyou Huagong/Petrochemical Technol.*, **37**, 29 (2008).
6. Y. Wang, L. Xu, L. Xu, H. Li, H. Li, *Cuihua Xuebao/Chinese J. Catal.*, **34**, 1027 (2013).
7. H. Li, D. Chu, J. Liu, M. Qiao, W. Dai, H. Li, *Adv. Synth. Catal.*, **350**, 829 (2008).
8. Q. Meng, H. Li, H. Li, *J. Phys. Chem. C*, **112**, 11448 (2008).
9. K. Khoiruddin, A.N. Hakim, I.G. Wenten, *Membr. Water Treat.*, **5**, 87 (2014).
10. N.F. Himma, S. Anisah, N. Prasetya, I.G. Wenten, *J. Polym. Eng.*, **36**, 329 (2016).
11. M. Purwasasmita, D. Kurnia, F.C. Mandias, Khoiruddin, I.G. Wenten, *Food Bioprod. Process.*, **94**, 180 (2015).
12. D. Ariono, M. Purwasasmita, I.G. Wenten, *J. Eng. Technol. Sci.*, **48**, 367 (2016).
13. D. Ariono, Khoiruddin, Subagjo, I.G. Wenten, *Mater. Res. Express*, **4**, 24006 (2017).
14. P.T.P. Aryanti, R. Yustiana, R.E.D. Purnama, I.G. Wenten, *Membr. Water Treat.*, **6**, 379 (2015).
15. I.N. Widiasa, P.D. Sutrisna, I.G. Wenten, *Sep. Purif. Technol.*, **39**, 89 (2004).
16. Khoiruddin, I.N. Widiasa, I.G. Wenten, *J. Food Eng.*, **133**, 40 (2014).
17. Khoiruddin, D. Ariono, Subagjo, I.G. Wenten, *J. Appl. Polym. Sci.*, (2017).
18. N.F. Himma, A.K. Wardani, I.G. Wenten, *Mater. Res. Express*, **4**, 54001 (2017).
19. P.T.P. Aryanti, M. Sianipar, M. Zunita, I.G. Wenten, *Membr. Water Treat.*, **8**, 463 (2017).
20. I.G. Wenten, P.T. Dharmawijaya, P.T.P. Aryanti, R.R. Mukti, Khoiruddin, *RSC Adv.*, **7**, 29520 (2017).
21. I.G. Wenten, Khoiruddin, *Desalination*, **391**, 112 (2016).
22. I.G. Wenten, Khoiruddin, F. Arfianto, Zudiharto, *Desalination*, **314**, 109 (2013).
23. A.K. Wardani, A.N. Hakim, Khoiruddin, I.G. Wenten, *Water Sci. Technol.*, **75**, 2891 (2017).
24. M. Purwasasmita, E.B.P. Nabu, Khoiruddin, I.G. Wenten, *J. Eng. Technol. Sci.*, **47**, 426 (2015).
25. I.G. Wenten, H. Julian, N.T. Panjaitan, *Desalination*, **306**, 29 (2012).
26. I.G. Wenten, I.N. Widiasa, *Desalination*, **149**, 279

(2002).
27. I.G.B.N. Makertihartha, P.T. Dharmawijaya, M. Zunita, I.G. Wenten, *Adv. Sci. Lett.*, **23**, 5726 (2017).
28. A.N. Hakim, Khoiruddin, I.G. Wenten, *Adv. Sci. Lett.*, **23**, 5640 (2017).
29. I.G. Wenten, Khoiruddin, N.F. Himma, *Adv. Sci. Lett.*, **23**, 5768 (2017).
30. J.W. Veldsink, *J. Am. Oil Chem. Soc.*, **78**, 443 (2001).
31. A. Bottino, G. Capannelli, A. Comite, A. Del Borghi, R. Di Felice, *Sep. Purif. Technol.*, **34**, 239 (2004).
32. C.A. Browne, F.W. Zerban., *Physical and chemical methods of sugar analysis : a practical and descriptive treatise for use in research, technical, and control laboratories*, 3rd ed., J. Wiley & sons, inc., New York (1941).

Do ZnO and Al_2O_3 Nanoparticles Improve the Anti-Bacterial Properties of Cellulose Acetate-Chitosan Membrane?

Titik Istirokhatun[a,b,*], *Ulva* Yuni[a], *Pertiwi* Andarani[a], *Heru* Susanto[b,c]

[a]Department of Environmental Engineering, Diponegoro University, Indonesia
[b]Membrane Research Center(Mer-C), Center of Research and Services-Diponegoro University
[c]Department of Chemical Engineering, Faculty of Engineering, Diponegoro University

Abstract. Biofouling, due to the existence of bacteria in water, becomes one of the barriers in cellulose acetate (CA) membrane applications. Although chitosan has been incorporated into CA membranes, its antimicrobial activity has not been improved significantly yet. In this study, cellulose acetate-chitosan membranes were prepared by modification using ZnO and Al_2O_3 nanoparticles during phase separation method. The membranes were then characterized in term of water permeability, and surface morphology. The anti-bacteria property was examined by using gram-negative bacteria. Modification of cellulose acetate-chitosan membranes with ZnO and Al_2O_3 nanoparticles have not shown optimal results where no clear zones around the membrane are visible.

1. Introduction

Membrane technology has been proposed as promising technology in the last few centuries to remove various contaminants from water and waste water [1-3]. Cellulose acetate (CA) is considered as an attractive raw material for the membrane preparation due to its raw material availability in large quantities that commonly found in nature. CA membrane has been widely developed due to its superior characteristics such as nontoxic, inexpensive, biodegradable (therefore, they are susceptible to fouling) [4, 5].

Fouling membrane is a critical problem that inhibits membrane performance in water treatment processes and until now it has not yet handled. The main consequence of biofouling is the decrease of flux during filtration process [6,7]. Membrane fouling can be distinguished by organic fouling, inorganic fouling, and biological fouling (i.e., biofilm formation) [8]. Between these types of fouling, membrane biofouling, the accumulation of microorganisms and extracellular polymers on the membrane surface, is the major problem in water treatment. Membrane biofouling is formed when adhesion occurs from one or more bacteria on the membrane surface. Then it followed by growth and multiplication of sessile cells, which then form biofilms on the membrane surface [9,10]. Unlike physical and chemical fouling, biofouling is more difficult to be removed and it causes decrease in permeability, resulting in irreversible damage to membrane [11].

Chitosan (CS), also known as the most important derivative of chitin, which is its existence very abundant in Indonesia. Chitosan has been already used in membrane production due to its interesting characteristics such as hydrophilicity [12], non-toxicity [13], and antimicrobial activity [14].

Previous research has added chitosan to improve anti-biofouling performance on cellulose acetate membranes. Unfortunately, it resulted low antimicrobial activity because the membrane loses its positive charge [15]. Another researcher stated that chitosan increased the membrane resistance against microbial fouling, but its microbial activity could only be seen via membrane surface visualization, and could not be seen by inhibition method. Furthermore, addition of chitosan also reduced the hydrophilicity of CA membrane [16]. Therefore, alternatives are needed for preventing biofouling, increasing hydrophilicity, and in the same time improving antibacterial activity of CA-CS membrane. Membrane modification using additives such as inorganic nanoparticles, is one of the alternative ways to control biofouling by improving basic polymer properties such as hydrophilicity and fouling resistance (perm-selectivity) [17,18].

In this study, Al_2O_3 was used to enhance the mechanical resistance and hydrophilicity of the membrane [19]. Al_2O_3 nanoparticle also improves membrane anti-adhesive capability thus reducing the biofilm layer and in the same time increasing membrane hydrophilicity [20].

Zink oxide is another material which is completely hydrophilic and possesses antibacterial nature. Moreover, ZnO also have ultravioresistant property which might potential benefit the antifouling performance of membrane. However, little attention is paid to the application of ZnO on membrane fabrication [21]. Therefore, the ZnO was also used in this study to improve

* Corresponding author: titik.istirokhatun@live.undip.ac.id

hydrophilicity and antifouling properties of membranes [22-25].

In this research, modifications of cellulose acetate-chitosan membranes were prepared by adding Al_2O_3 (0-4%), and ZnO (0-4%) to improve the hydrophilicity and anti-biofouling properties of membranes.

2. Materials and Methods

2.1 Materials

Commercial cellulose acetate (($C_6H_7O_2(OCOCH_3)$)x was purchased from from Sigma Aldrich, Germany. Nanoparticles Al_2O_3 and ZnO were obtained from Merck, Germany. Chitosan ($C_6H_{11}NO_4$)n was purchased from Biotech Surendo, Indonesia. Acetic acid (CH_3COOH) was obtained from Merck, Germany. Distilled water produced from home-made pure water unit was used for all experiment. Escherichia coli bacteria cultures were obtained from Microbiology, Diponegoro University.

2.2 Methods

2.2.1 Membrane preparation

Preparation of membrane was conducted by wet phase inversion. Membrane casting solution was made from 12% cellulose acetate as membrane base material, a mixture of acetic acid and distilled water (3:1) as solvent, 0.1% chitosan, nanoparticles Al_2O_3 and ZnO with concentration of 0-4% as additives.

First of all, chitosan was introduced into the solvent and stirred using a magnetic stirrer until all chitosan dissolved. Then the nanoparticles were added to the solution with concentration variations of 0-4% until dissolved. The final step was addition cellulose acetate then stirring until all the ingredients were homogenously mixed. After completion, casting solution was left without stirring until no bubble was observed.

Membrane molding process was carried out by following the method which has been conducted by Istirokhatun et al. [26] as follows, the solution was firstly cast on a glass substrate using a casting knife, then left in room temperature. Thereafter, the cast membrane was solidified in a coagulation bath containing water for 30 min to remove remained acidity. The resulting membrane was washed and rinsed by soaking in water for at least 24 h before drying.

2.2.2. Membrane Characterizations

The water permeability of the membranes was evaluated through a series of filtration using dead end stirred cell filtration system (Amicon model 8010), which was pressurized by nitrogen gas. Prior to each filtration membrane were pre-compacted by filtration of pure water at high pressure for at least 0.5-1 h to avoid effect of compaction. Thereafter, the pressure was lowered to the desired pressure for water permeability measurements (1 bar). The next step was measuring the permeate weight to know the volume. This permeate volume was then used as basis for calculating water flux. The equations used to calculate the pure water flux are:

$$J = \frac{Q}{A.t} \quad (1)$$

Where: Q : The volume of fluid passing through the membrane (permeate) (L)
t : Time (hour)
J : Flux (L/m^2.hour)
A : Membrane Area (m^2)

2.2.3. Antibacterial activity test

The antibacterial activity of the membrane was examined by an inhibition method, wherein gram-negative bacteria Escherichia Coli was used. This method has already been described by Istirokhatun et al and Ma et al [16,27]. To determine antibacterial activity, all membranes were sterilized in autoclave at 121°C, 2 bars for 15 min. Sterilized membranes were then placed on the surface nutrient agar (NA) containing Escherichia Coli solution and then incubated for 24 hours at 37 °C. The diameter of the clear zone formed around the membrane was noted as an indicator of antibacterial activity.

3. Result and Discussion

3.1 Membrane Characterization

3.1.1 Effect of nanoparticles concentration on water permeability

Membrane permeability can be determined by measuring flux. The results of the pure water flux test from Cellulose Acetate-Chitosan (CA-CS) membrane modified with ZnO and $Al_2O_3$are shown in Figure 3.1.

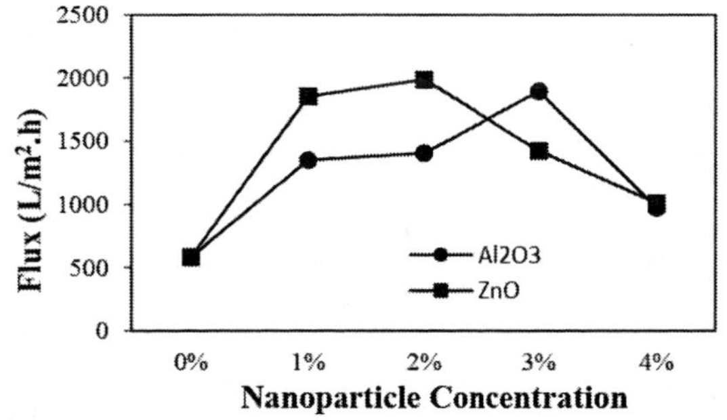

Figure 3.1 Water permeability of CA-CS membranes as function of nanoparticles concentration

It is obviously seen that compared with 0% concentration of nanoparticles, addition of both ZnO and Al_2O_3has significant effect on promoting membrane permeability. These phenomena occurred might be due to complete hydrophilicity of ZnO and Al_2O_3. CA-CS membrane with addition of ZnO reached the highest flux (1,984.71 L/m^2.hour) at 2% of ZnO concentration, while addition of Al_2O_3 achieved the highest flux (1,891.98 L/m^2.hour) at 3% of Al_2O_3 concentration. It is also important to note that overdoses of both nanoparticles could decrease flux [21], as can be observed in the decrease of flux when percent concentration of ZnO and Al_2O_3 were 3% and 4% respectively. These results are consistent with studies conducted by Yan et al. [20] which conclude the increase of Al_2O_3 concentration along with the increase of permeate flux to certain concentration level. According to Ivars et al. [28], membrane permeability increased along with the increasing of nanoparticles in the casting solution. It caused by the affinity of nanoparticles, therefore, the porosity and pore size of modified CA-CS membranes were higher than unmodified membranes. However, the results of this study differ from research conducted by (He and Hong, [29] which states that the addition of nanoparticle the PVDF membrane reduces the flux of pure water and pore size of the membrane.

The presence of nanoparticles in the membrane matrix enhances strength and stiffness, thermal stability, permeability, hydrophilicity, flux recovery and antifouling properties of membranes [30-32]. The presence of nanoparticles improves the membrane hydrophilicity due to high surface energy thus increasing the absorption of water molecules on the membrane surface [33]. In addition, the hydrophilicity of nanoparticles also affects membrane flux. Increased hydrophilicity of the membrane surface and membrane pore could decrease the membrane fouling as it strongly interacts with water molecules and resulting greater flux [34,35].

However, higher concentration of nanoparticles could negatively affect the membrane permeability due to agglomeration of the inorganic nanoparticles on the membrane surface during membrane preparation. These clumping nanoparticles as consequence of Van der Waals forces between nanoparticles caused by high surface energy to achieve stable conditions. It may clog some pores and possibly desrease the membrane hydrophilicity, thus causing decrease in the water flux [34,36].

3.1.2. Scanning Electron Microscopy (SEM)

Figure 3.2 show SEM results from the surface of CA-CS and CA-CS/Al_2O_3 membranes. It was observed qualitatively that addition of Al_2O_3 influences surface morphology of each membrane. On the surfaces of CA-CS/Al_2O_3 membranes, both 1% and 3% concentrations appear agglomerations of nanoparticle covering the membrane pores. The increased concentration of Al_2O_3 resulted thicker agglomeration. This SEM result supports preceding explanation on the results of water permeability that the presence of nanoparticles causing clumps that covered the membrane pores.

Yan et al. [20] stated in his research that the addition of Al_2O_3 nanoparticles in PVDF membranes does not affect thesurface structure, cross section and pore size of membranes. Maximous et al. [37] also stated that there was no significant pore size difference with the addition of Al_2O_3 nanoparticles in the PES membrane. However, based on the results obtained in this study the pore size of membrane modification of Al_2O_3 nanoparticles was greater than the membranes without modification of nanoparticles. This result is in accordance with the research conducted by Liu et al. [19] which states that higher concentration of Al_2O_3 nanoparticles resulting larger pore size of membrane.

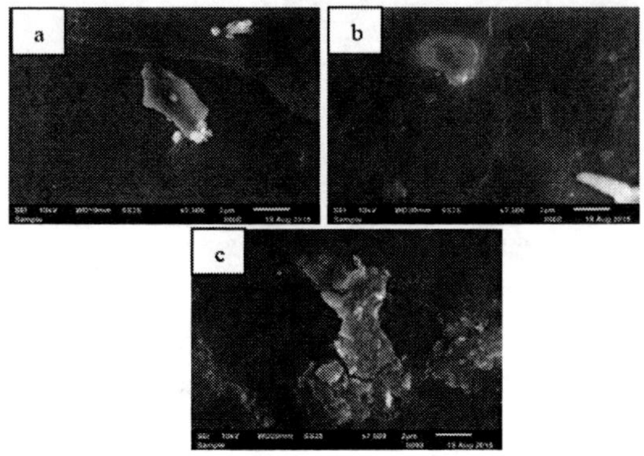

Figure 3.2 SEM of membrane surface (a) CA-CS (b) CA-CS/Al_2O_3 1% (c) CA-CS/Al_2O_3 3% with magnification 7.500x

The surface morphology of the CA-CS/ZnO membrane is shown in Figure 3.3. As well as the addition of Al_2O_3, the addition of ZnO also resulted lumps of nanoparticles on the membrane surfaces. It is also observed that CA-CS/ZnO membranes have smaller pores than CA-CS/Al_2O_3 membranes. This could be caused by the presence of oxide metal nanoparticles in the polymer to increase the viscosity of the solution thereby the rate of solvent diffusion into the non-solvent during the phase inversion process decreased, which eventually led to the formation of tightly pore membranes [38].

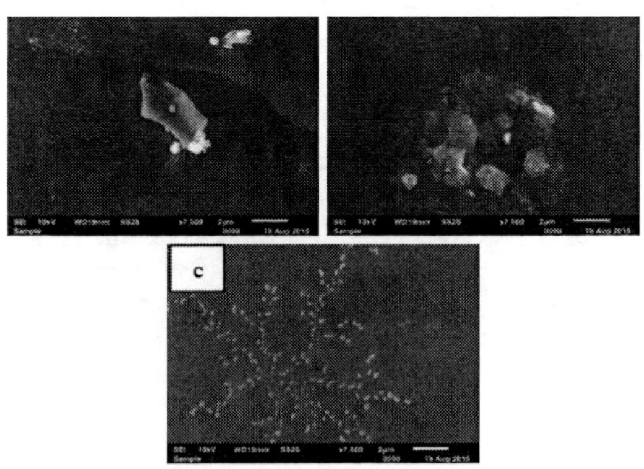

Figure 3.3 SEM of membrane surface (a) CA-CS (b) CA-CS/ZnO 1% (c) CA-CS/ZnO 2% with magnification 7.500x

3.2 Antibacterial Activity Test

The anti-bacterial test was examined to find out the anti-biofouling properties of CA-CS membranes after the addition of Al_2O_3 and ZnO nanoparticles. The antibacterial test was performed by measuring the diameter of the clear zone formed near membrane after *E. coli* incubation. The results are presented in Figure 3.4.

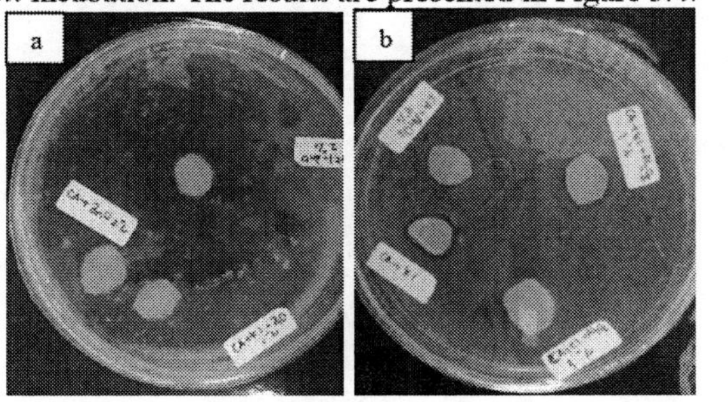

Figure 3.4 Anti-bacterial test on (a) CA-CS/ZnO membrane and (b) CA-CS/Al_2O_3 membrane

In the CA-CS membrane (Figure 3.4.b), the width of the clear zone formed around the membrane was 3 mm. The width of the clear zone formed by CA-CS-nanoparticle membrane was 2 mm and 1 mm for Al_2O_3 1% and Al_2O_3 2% respectively. Meanwhile, the clear zone widths formed by CA-CS membrane with ZnO (Figure 3.4.a) were 1 mm and 0.5 mm at concentrations of 1% ZnO and 2% ZnO respectively. These results suggest that the effect of addition ZnO on membrane matrix resulted in smaller clear zone compared to CA-CS membrane which modified by Al_2O_3, while the modified CA-CS membrane with the addition of nanoparticles has a smaller clear zone than the CA-CS membrane.

Antibacterial activity can be seen from the diameter of the clear zone. The clear zone is an effective way for membrane to demonstrate antibacterial activity [39]. The greater the clear zone formed around on the membrane the better antibacterial properties. From this study, it can be concluded that the addition of nanoparticles further reduced the clear zone around membrane. In other words, the addition of nanoparticles during membrane preparation has not been successful in improving antibacterial properties. These results differ from those stated by Al-naamani et al. [24] and Malini et al. [25] that modified the chitosan membrane with the addition of ZnO and found that the addition of ZnO increased the antibacterial capacity of the membrane significantly. In another study [40], Shenghan and Ruichang, reported that the membrane of cellulose/nanoparticles produces better anti-bacterial properties than the cellulose/chitosan membrane. However, in these results, it is seen that higher nanoparticle concentrations resulted smaller inhibition zones. This might be due to the accumulation of nanoparticles added to the membrane which further decreased their antibacterial activity.

4. Conclusion

Modification of CA-CS membrane by adding ZnO and Al_2O_3 increased flux. The maximum flux of the modified membrane was reached by 2% ZnO (1,984.71L.m^{-2}.h^{-1}) and 3% Al_2O_3 (1,891.98 L.m^{-2}.h^{-1}). In addition, the presence of nanoparticles also enhanced hydrophilicity of CA-CS membranes. The SEM results showed that addition nanoparticle during phase separation process has not occurred perfectly, evidenced by the presence of the agglomerations of nanoparticles on membrane surfaces. The higher the concentration of nanoparticles, the more obvious the appearance of the lumps. An antibacterial test using inhibition zone method was performed. Nevertheless, it produced unideal result, where no clear zone around the membrane were visible.

The authors would like to thanks to Engineering Faculty, Diponegoro University for the financial support of this study.

References

[1] G. Kang, Y. Cao. J. Membr. Sci. **463**, 145-165 (2014)
[2] J. Hong, Y. He. Desalination **332**, 67-75 (2014)
[3] J. Kim, B.V. Bruggen. Environ. Pollut. **158**, 2335-2349 (2010)
[4] D. Bhongsuwan and T. Bhongsuwan, 2008. J. Nat. Sci. **42**, 311 - 317 (2008)
[5] J. Puls, S.A. Wilson, D. Holter.J. Polym. Environ. **19**, 152-165 (2011)
[6] J. Hong, Y. He. Desalination **302**, 71-79 (2012)
[7] L.H. Kim, M.S. Shin, S. Kim, C. Kim, K. Chae, In. S. Kim. J. Membr. Sci. **477**, 86-92 (2015)
[8] L. Dong, H. Yang, S. Liu, X. Wang, Y.F Xie. Desalination **365**, 70-78 (2015)
[9] C.X Liu, D.R Zhang, Y. He, X.S. Zhao, R. Bai. J. Membr. Sci. **346** .121-130 (2010)
[10] N. Hilal, V. Kochkodan, L. Al-Khatib, T. Levadna. Desalination. **167**, 293-300 (2004)
[11] H. Rajabi, N. Ghaemi, S.S. Madaeni, P. Darael, B. Astinchap, S. Zinadini, S.S. Razavizadeh. Appl. Surf. Sci. **349**, 66-77 (2015)
[12] T.C. Yang, C.C. Chou, C.F. Li. Int. J. Food Microbiol. **97**, 237-245 (2005)
[13] G. Ma, D. Yang, Y. Zhou, M. Xiao, J.F. Kennedy, J. Nie. Polym. **74**, 121-126 (2008)
[14] H. Mellegard, S.P. Strand, B.E. Christensen, P.E. Granum, S.P. Hardy. Int J. Food Microbiol **148**, 48-54 (2011)
[15] C. Qin, H. Li, Q. Xiao, Y. Liu, J. Zhu, Y. Du. Carbohyd. Polym. **63**, 367-374 (2006)
[16] T. Istirokhatun, N. Rokhati, D. Nurlaeli, N.N. Arifianingsih, Sudarno, Syafrudin, H. Susanto. Int. J. Environ. Sci. Technol. **10**, 56-67 (2017)
[17] F. Kara, E.A Aksoy, Z. Yuksekdag, N. Hasirci, S. Aksoy. Carbohyd. Polym. **112**, 39-47 (2014)
[18] J.G. Ivars, M.I. Iborra-Clar, M.I. Alcania-Miranda, B.V. Bruggen. J. Membr. Sci. **492**, 709-722 (2015)
[19] F. Liu, M.R.M. Abed, K. Li. J. Membr. Sci. **366**, 97-103 (2011)
[20] L.Yan, Y.S. Li, C.B. Xiang, S. Xianda. J. Membr. Sci. **276**, 162-167 (2006)
[21] S. Liang, K. Xiao, Y. Mo, X. Huang. J. Membr. Sci. **394-395**, 184-192 (2012)
[22] D. Sharma, J. Rajput, B.S. Kaith, M. Kaur, S. Sharma. Thin Solid Films. **519**, 1224-1229 (2010)
[23] G.S. Dhillon, S.Kaur, S.K. Brar. Int. Nano. Lett. **4**, 107 (2014)
[24] L. Al-naamani, S. Dobretsov, J. Dutta, J.G. Burgess. Chemosphere. **168**, 408-417 (2017)

[25] M. Malini, M. Thirumavalavan, W.Y. Yang, J.F. Lee, G. Annadurai. Int. J. Biol. Macromol. **80**, 121-129 (2015)
[26] T. Istirokhatun, R. Rachmawaty, M. Merriyani, N. Rokhati, H. Susanto. IOP Conf. Ser.: Earth Environ. Sci. **55** (2017)
[27] Y. Ma, T. Zhou, C. Zhao. Carbohydr. Res. **343** (2008)
[28] J.G. Ivars, M.I.A. Miranda, M.I.I. Clar, J.A.M. Roca, L.P. Alcaniz. Sep. Purif. Technol. **128**, 45-57 (2014)
[29] Y. He, and J. Hong. Adv. Mat. Res. **311-313**; 1818-1821 (2011)
[30] M.K. Shinha, and M.K Purkait. J. Membr. Sci. **437**, 7-16 (2013)
[31] A.Razmjou. J. Mansouri, V. Chen. J. Membr. Sci. **378**, 73-84 (2011)
[32] T.A. Saleh, and V.K. Gupta. Sep. Purif. Technol. **89**, 245-251 (2012)
[33] S. Balta, A. Sotto, P. Luis, L. Benea, B.V.D. Bruggen, J. Kim. J. Membr. Sci. **389**, 155-161 (2012)
[34] Y.S. Li, L. Yan, C.B. Xiang, L.J. Hong. Desalination **196**, 76-83 (2006)
[35] L. Ahmad, A.A. Abdulkarim, B.S. Ooi. Chem Eng Jour **223**, 246-267 (2013)
[36] J.M. Arsuaga, A.Sotto, G.del Rosario, A. Martinez, S. Molina, S.B. Teli, J. de Abajo. J. Membr. Sci. **428**, 131-141 (2013)
[37] N. Maximous, G. Nakhla, W. Wan, K. Wong. J. Membr. Sci. **341**, 67-75 (2009)
[38] Y. Wang, Q. Zhang, C.L. Zhang, P. Li. Food Chem **132**, 419-427(2012)
[39] Z. Cao, Z. Shen, X. Luo, H. Zhang, Y. Liu, N. Cai, Y. Xue, F. Yu. Carbohyd Polym. DOI. 10.1016/j.carbpol.2017.03.012 (2017)
[40] G. Shenghan and G. Ruichang. Chem. Res. Chin. Univ. DOI:10.1007/s40242-017-6423-4

Innovative Design of Solar-Powered Desalination (SPD) System using Vacuum-Multi Effect Membrane Distillation (V-MEMD) Process

Achmad Chafidz[1,*], *Faisal* RM[1], *Esa* D. Kerme[2], *Irfan* Wazeer[3], *Saeed* M. AlZahrani[3]

[1]Department of Chemical Engineering, Universitas Islam Indonesia, Yogyakarta 55584, Indonesia
[2]Mechanical Engineering Department, King Saud University, Riyadh 11421, Saudi Arabia
[3]Chemical Engineering Department, King Saud University, Riyadh 11421, Saudi Arabia

Abstract. This research focused on the development of an innovative design of solar-powered desalination (SPD) system which was expected to solve the water and energy problem simultaneously. We have developed a portable and hybrid solar-powered desalination (SPD) system for producing potable water from saline water. It is a self-contained and integrated system which combines solar-thermal collector and solar-photovoltaic for its operation, and thus the system can operate to produce water by only using solar energy. Therefore, the system is highly suitable to be implemented in remote arid and coastal areas without infrastructures or connection to the grid (water and power), but blessed with abundant solar irradiation, like in Saudi Arabia. A Memsys Vacuum Multi-Effect Membrane Distillation (V-MEMD) unit was used as the core of the SPD system. A heat pump was also integrated into the SPD system for energy recovery and to improve the performance of the system. The system could be considered as sustainable and "green" desalination technology, which will be very useful for the Kingdom of Saudi Arabia. To study the performance of the system, small-scale tests have been carried out at the Engineering College - King Saud University, Saudi Arabia. Based on the experimental results, the system has run successfully by only utilizing solar energy.

1 Introduction

Fresh water is one of the most important human needs besides food and energy. Without three of them (i.e. water, food, energy) human cannot live normally. Additionally, three of them also interdependent with each other. Energy crisis will lead to water crisis. This is because to produce water requires a lot of energy, for example: for seawater desalination, water treatment, water distribution (by pump), etc. In reverse, to produce energy also require a large amount of water, for example: cooling, power plant, oil production, etc. On the other hand, water crisis will inevitably lead to food crisis. The food crisis cannot be denied will also lead to a decline in the standard of human life, which will affect the performance of their work, e.g. in the energy industrial sector. The relationship/interdependence between these three is what is called by "*The Nexus of Water, Energy, and Food*".

It is already known that most of the Middle East and North Africa (MENA) countries, such as Kingdom of Saudi Arabia (KSA) has limited amount of natural fresh water resources. In contrary, Indonesia, an archipelago country has abundant natural fresh water resources, such as: lake, river, underground water, rain, etc. The most promising method and maybe the only option for a country such as Saudi Arabia to be able to meet the fresh water needs for its citizens is by seawater desalination. It is because seawater resources is very abundant, about 96% of earth's total water distribution is seawater. To meet the water demand, Saudi Arabia has constructed numbers of large scale desalination plants, which accounted for about 24% of world total capacity. The number of the desalination plant will definitely increase with increasing its population. According to figures from Electricity & Cogeneration Regulatory Authority (ECRA) of Saudi Arabia, most of desalination plants in Saudi Arabia are using conventional technologies, such as Multi-Stage Flash process (MSF), Reverse Osmosis (RO), and Multi-Effect Distillation (MED). These conventional processes required significant amount of energy and most of them are driven by fossil fuel. As one of the top oil producers in the world, maybe Saudi Arabia does not have a problem with the water supply through desalination plant driven by fossil fuel. However, in the future, when the oil resources are depleted then the energy crisis will inevitably fall upon this kingdom, which will then lead to water crisis. One the most promising alternative to solve the water and energy problem simultaneously is by utilizing renewable energy sources to power seawater desalination process. One of the most potential renewable energy in Saudi

* Corresponding author: achmad.chafidz@uii.ac.id

Arabia is solar energy. Therefore, utilizing solar energy to drive seawater desalination process may become the best alternative for Saudi Arabia to meet the domestic water demand as well as reducing the oil consumption.

Membrane distillation (MD) is one among desalination technologies that is compatible to be coupled with the solar energy. The MD process can be described as combined process between membrane and thermal distillation proceses. It can also be called a "trans-membrane evaporation" process [1-3]. The basic working principle of MD process is very simple. Heat is applied in the MD process to generate vapor from the saline water. The MD uses microporous hydrophobic membrane for its separation process. Due to its hydrophobic nature of the membrane, only vapor can pass through the membrane, and thus separating the (pure) water from the saline water. There are several advantages of MD process, such as: the process is very simple, it can be used for any range of saline water, less fouling tendency, the heat used can be low grade heat, it can produce water with high quality, it can be run with small temperature difference [1, 2, 4, 5]. Based on the mechanism, the MD process can be classified into several different configurations. Figure 1 shows the schematic illustration of three different configurations of MD process. In Vacuum membrane distillation (VMD), a vacuum is applied to the permeate side and the vapor condensation occurs outside the membrane. This vacuum will help the vaporization process to be started at low temperature. In Direct contact membrane distillation (DCMD), inside the permeate side is a condensing liquid, which in direct contact with the membrane. Whereas, in AGMD, the condensing (cold) surface is separated from the membrane by an air gap [3].

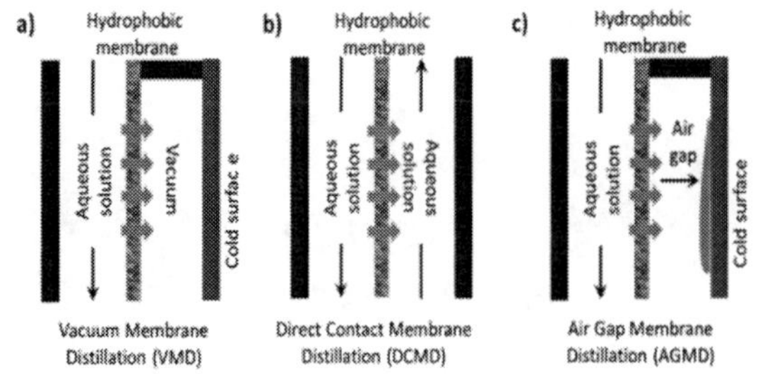

Fig. 1. Illustration of three different configurations of membrane distillation process: a) VMD; b) DCMD; and AGMD. Adopted from the literature [1].

2 The SPD system configuration

The solar-powered desalination (SPD) system is a self-contained and integrated system which combines solar-thermal collector and solar-photovoltaic for its operation. Thus, the system can operate to produce water by only using solar energy. For producing the water, a vacuum-multi effect membrane distillation (V-MEMD) unit (Memsys, Germany) was used. Figure 2 (Adopted from the literature [1]) shows the schematic diagram of the SPD system. The SPD system consists of three main sub-systems, which are: 1) Solar-photovoltaic (PV) system, to harness solar energy and converts into electricity for powering the equipment; 2) Solar-thermal collector system, to harness solar energy and converts into thermal energy to supply the heat required by the Memsys unit via a heat exchanger; and 3) Memsys V-MEMD system for producing the water by utilizing the low grade heat generated by the thermal collector. As seen in Fig. 2, the SPD system was also equipped with a

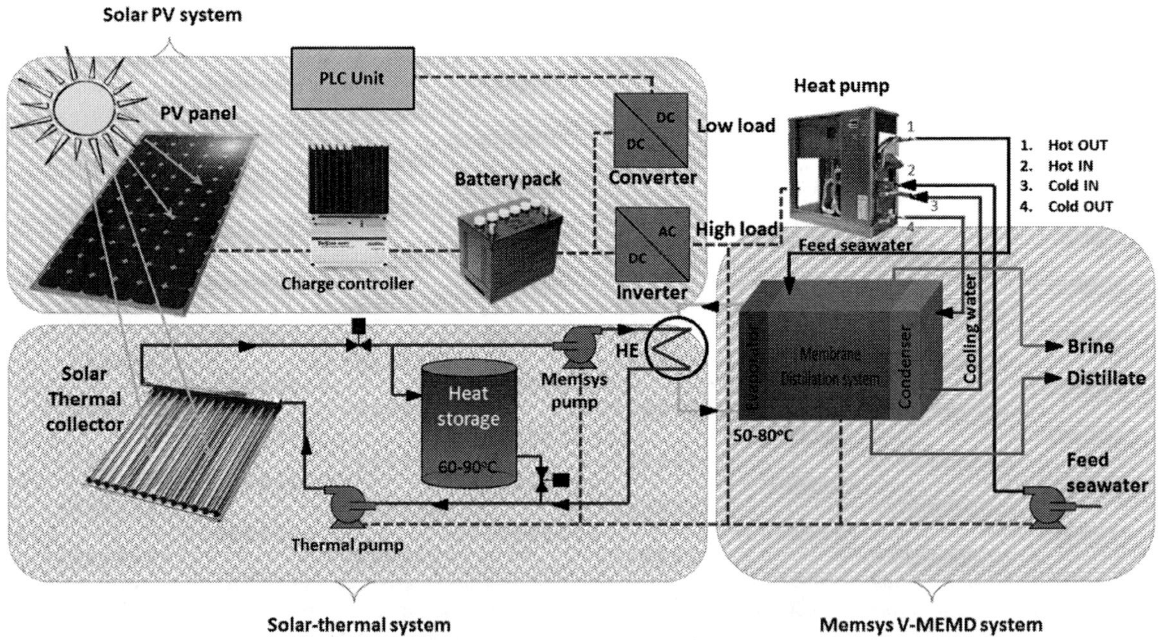

Fig. 2. Schematic diagram of the SPD system.

heat pump, which was used for energy recovery by cools the cooling water (in the *condenser* loop) as well as preheats the feed water (e.g. seawater, brackish water) during operation of the SPD system. The more details of the SPD system could be found in our previous work [1, 6].

Additionally, the system was also designed to be able to operate in arid remote areas. This advantage is coming from the design of the system which utilizes a typical modified container (6.0 m x 2.4 m x 2.6 m) to house all the equipments, thus it is portable and can be deployed at anytime. The seawater tank is also housed in one side of the container. This seawater is used for feed as well as for cooling for the Memsys V-MEMD unit. The seawater was prefiltered using an ultrafiltration cartridge when filling the seawater tank. Figure 3 shows the photograph of the SPD system. Currently the SPD system was placed on the rooftop of College Engineering building, at King Saud University, Riyadh, Kingdom of Saudi Arabia.

Fig. 3. Photographs of the SPD system

The Memsys Vacuum Multi-Effect Membrane Distillation (V-MEMD) unit is the core of the SPD system. The main purpose of this unit/sub-system is to produce a clean water by utilizing the low grade heat generated by the solar-thermal collector. The Memsys unit is desalination process that combines both the thermal process (i.e. multi-effect distillation) and membrane separation process. In general, the working principle of the Memsys unit is based on the Vacuum Multi-Effect Membrane Distillation (V-MEMD) process. According to the literature [7], the Memsys V-MEMD process has advantages due to some reasons, such as: internal heat recovery, high distillate flux, minimum heat loss, low specific heat transfer surface, low mechanical stress due to less pressure difference on membrane from feed stream and negative pressure, modularity. The last point is important to compete with the many advantages of Reverse Osmosis (RO). In which, the Memsys V-MEMD unit, due to its modularity, it has wide range of applications, from small to large scale [7]. Figure 4a shows the photograph of the complete Memsys unit. The unit consists of three major parts: a *steam raiser*, multiple *effect (evaporation-condensation) stages*, and a *condenser*. Please see also Fig. 5b, which shows the configuration of the Memsys unit.

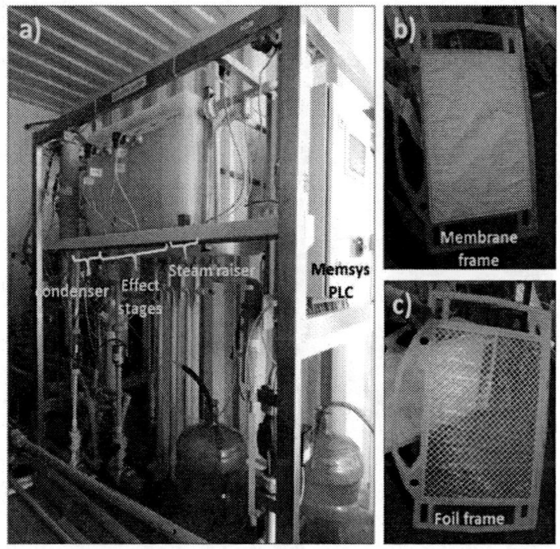

Fig. 4. Photographs of the Memsys unit: **a)** The complete modul of the Memsys unit; **b)** The Memsys membrane frame; and **c)** The Memsys foil frame.

The purpose of the *steam raiser* is to create steam (from fresh water) which will acts as driving thermal energy source for the operation of Memsys process. To produce steam, the *steam rasier* utilizes heat from the thermal tank that is transferred through a plate heat exchanger. Due to the vacuum pressure (< 200 mbar), the evaporation process could happen with low temperature. In general, the Memsys process can run with temperature in the range of 50-80 °C. Whereas, the *evaporation-condensation stages*, or can be called the *effect stages* is where the distillate is produced. The thermal energy from the *steam raiser* evaporates the seawater and the water vapor generated get separated by the microporous hydrophobic membrane, then condensed by the polypropylene (PP) foil, and finally collected at the bottom of *effect stages*. There is a multiple recycle of thermal energy, and thus it can reduces the amount of required energy and increases thermal efficiency of the system. In this work, the Memsys unit consists of four *effect stages*. Each stage consists of four membrane frames and PP foil frames (see Fig. 4b and 4c), with the total effective membrane area of 5.12 m^2. Apparently, this membrane area is smaller than other SPD systems [1]. Figure 5a shows the schematic illustration of how the feed and vapor flows inside one *effect stage* of the Memsys V-MEMD unit. The last part is a condenser, which its purpose is to condense the water vapor being produced at the last *effect stages*. A fresh water with temperature of 20-40 °C is used as the coolant. Additionally, the temperature difference between the *steam raiser* and the *condenser* is the driving force for the Memsys process. Therefore, the greater the temperature difference, the greater the amount of distillate could be produced by the Memsys unit. The more details of the Memsys unit can be found in our previous work [1].

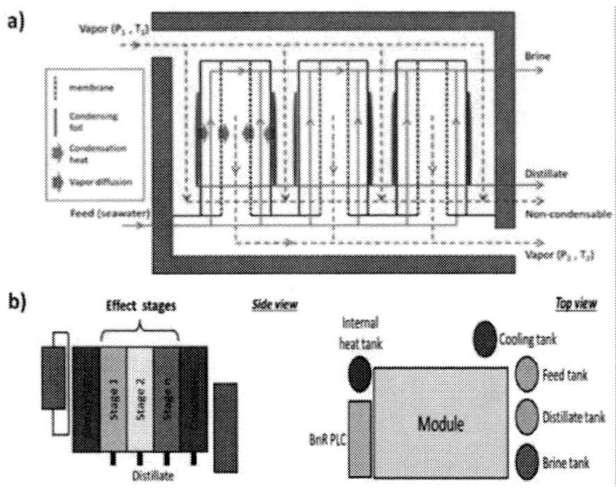

Fig. 5. a) Schematic illustration of how the Memsys V-MEMD process works (inside one *effect stage*) – Adopted from the literature [7], and **b)** The schematic configuration of the Memsys unit (side view and top view).

3 One-day test of the SPD system

The SPD system was place on the roof-top of college engineering building, King Saud University, Riyadh, Saudi Arabia. It is worth to note that the system was a prototype of solar powered desalination system with a small capacity of water production. Small-scale tests of the system were conducted to observe the performance of the SPD system. Figure 6 shows a one-day operation of the SPD system. The test was conducted on 21 Feb 2013 (spring season). The system was run from 9:00 AM until 4:00 PM (about 7 hours). The system was run in manual mode. The solar irradiation level was monitored all the time using a Rotating Shadowband Radiometer (Irradiance Inc., USA). As seen in Fig. 6, the range of global irradiation level on that day was from 308.5 to 870 W/m^2. The peak level of 870 W/m^2 was occurred at 12:22 AM. The Memsys unit started to produce the water at 9:26 AM. The first half an hour was used to stabilize the SPD system, especially the thermal energy required by the Memsys unit. The amount of time needed to stabilize the system depends on the thermal tank temperature. The higher the initial temperature of the thermal tank, the less amount of time needed for the Memsys unit to start producing the water. In the current test, the initial temperature of thermal tank was already at 72 °C (see Fig. 6). This temperature was high enough to provide a driving force for the Memsys unit to produce water. During the test, the thermal tank temperature did not change siginificantly. The temperature was only in the range of 72 – 76.3 °C.

The feed used in this test was a brackish water with conductivity of 2490 μS/cm. This conductivity was offline measured using a Multi-Parameter Analyzer (DZS-708), Cheetah Scientific Ltd, Malaysia. The feed flowrate was kept constant at 69 L/h during the test. Based on the previous tests performance of the system (see our previous work [6]), this feed flow rate was found to be the optimum rate to produce maximum distillate. Additionally, as seen in Fig. 6, the distillate production rate was in the range of 11.75 – 20.29 L/h, with average distillate production rate of 15.91 L/h. As expected the maximum distillate production rate was occurred at noon, around 12:00 AM, where the global irradiation level also reached its peak. Whereas, the total distillate output gained on that day was approximately 102.9 L of fresh water with approximate conductivity of 4.7 μS/cm. This water is clean water and safely to drink. This total volume of distillate output can be considered lower than the other systems (see in the literature). This is due to small capacity of the Memsys unit, which only has a total membrane area of 5.12 m^2. Nevertheless, this is not a problem, since the Memsys unit capacity can be easily increased due to modularity of the Memsys module.

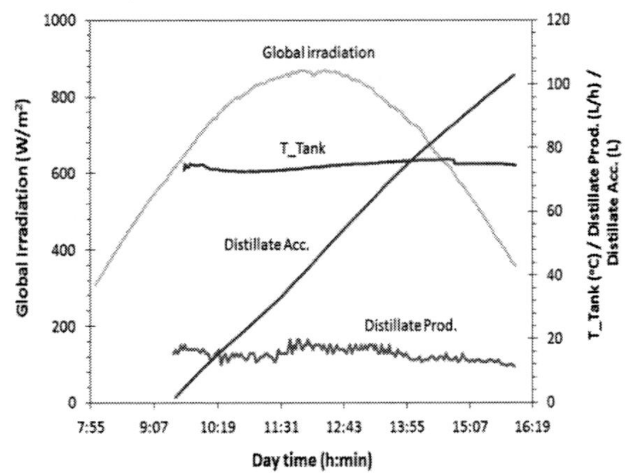

Fig. 6. A one-day performance test of the SPD system.

Additionally, there is one parameter that can be used to compare the performance of the SPD system with the others, proportionally. The parameter is called disitillate flux (J) with the unit of L/m^2.h, which can be defined by Eq. (1) as follows:

$$J = \frac{V}{S.t} \qquad (1)$$

where V is the volume of the distillate produced (L), S is the total effective membrane area (m^2), and t is the running time of the Memsys unit. The total effective membrane area of the Memsys module was approximately 5.12 m^2. Figure 7 shows the distillate flux profile of the Memsys unit on the same day of operation (i.e. 21 Feb 2013). As seen in the figure, the profile of distillate flux was following the trend of global irradiation level, which reached its peak at the noon time. The distillate flux was found in the range of 2.29 – 3.96 L/m^2.h, with the approximate average value of 3.12 L/m^2.h. This distillate flux rate was comparable or even higher than some of other SPD systems [1].

https://doi.org/10.1051/matecconf/201815608010

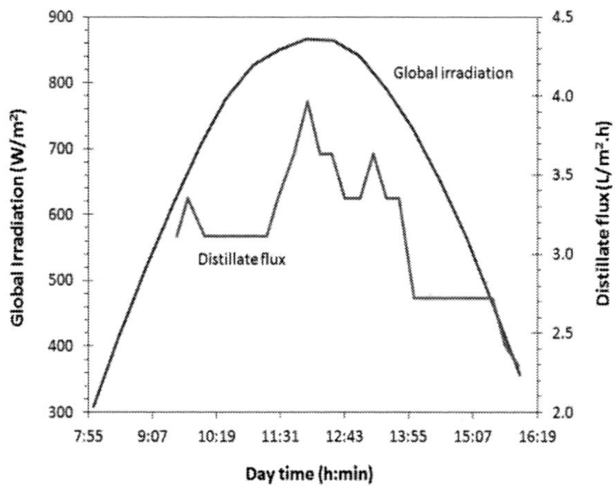

Fig. 7. Hourly distillate flux of the SPD system during one day performance test.

4 Conclusion

An innovative solar-powered desalination (SPD) system which is a stand-alone, integrated, and portable has been designed and successfully developed. It is a hybrid system which combines solar-thermal collector and solar-photovoltaic for its operation. In this work, a one-day operation test of the SPD system has been conducted on 21 Feb 2013 (spring season in Saudi Arabia). The test results showed that the SPD system has run successfully by only utilizing solar energy. The system was run from 9:00 AM until 4:00 PM (about 7 hours). The range of global irradiation level on that day was from 308.5 to 870 W/m^2 with the peak level occurred at 12:22 AM. The thermal tank temperature was in the range of 72 – 76.3 °C. The feed (saline water) was a brackish water with conductivity of 2490 μS/cm. The feed flowrate was kept constant at 69 L/h. The distillate production rate was in the range of 11.75 – 20.29 L/h, with average distillate production rate of 15.91 L/h. The total distillate output gained on that day was approximately 102.9 L with approximate conductivity of 4.7 μS/cm. The distillate flux was found in the range of 2.29 – 3.96 L/m^2.h, with the approximate average value of 3.12 L/m^2.h. This distillate flux rate was comparable or even higher than some of other SPD systems.

This solar-powered desalination (SPD) system has been developed by a joint collaboration project between King Saud University (KSU), Kingdom of Saudi Arabia and Nanyang Technological University (NTU), Singapore. The authors would like to thank all the contributors for this project: Choo F. H., Tan F.L., Darryl, M. Prabu, Naeem.

References

[1] A. Chafidz, E. D. Kerme, I. Wazeer, Y. Khalid, A. Ajbar, S.M. Al-Zahrani, Journal of Cleaner Production, **133** (2016) 631-647.

[2] M. Safavi, T. Mohammadi, Chem. Eng. J. (Lausanne), **149** (2009) 191-195.

[3] J.-P. Mericq, S. Laborie, C. Cabassud, Chem. Eng. J. (Lausanne), **166** (2011) 596-606.

[4] H. Chang, C.-L. Chang, C.-Y. Hung, T.-W. Cheng, C.-D. Ho, International Journal of Environmental Research and Public Health, **11** (2014) 12064-12087.

[5] A. Hausmann, P. Sanciolo, T. Vasiljevic, M. Weeks, M. Duke, Chem. Eng. J. (Lausanne), **211–212** (2012) 378-387.

[6] A. Chafidz, S. Al-Zahrani, M.N. Al-Otaibi, C.F. Hoong, T.F. Lai, M. Prabu, Desalination, **345** (2014) 36-49.

[7] W. Heinzl, S. Büttner, G. Lange, Desalination and Water Treatment, **42** (2012) 177-180.

Optimum parameters for treating coolant wastewater using PVDF-membrane

Erna Yuliwati[1,*], *Ahmad Fauzi* Ismail[2], *Amrifan Saladin* Mohruni [3]and *Agung* Mataram[3]

[1]Department of Chemical Engineering, Faculty of Engineering, Universitas Sriwijaya, Indralaya, Indonesia
[2]Advanced Membrane Technology Research Center, Universiti Teknologi Malaysia, 81310 UTM, Skudai Johor, Malaysia.
[3]Department of Mechanical Engineering, Faculty of Engineering, Universitas Sriwijaya, Indralaya, Indonesia

Abstract. Chemical oxygen demand (COD removal on ultrafiltration membrane were studied from treating of coolant wastewater. The membranes were fabricated via the dry-jet wet spinning method. The experiment was conducted using synthetic coolant wastewater as effluent, and an experimental set-up consist of membrane reservoir was used throughout the investigation. Deposition and accumulation of suspended solids on the membrane surface during filtration were prohibited with continuous aeration. Membrane performance was measured as well as flux and COD removal efficiency. The RSM was performed to investigate the optimised process conditions for treating coolant wastewater. Results using RSM-approximation have demonstrated the improvement in COD removal of 99.63%. This improvement is achieved by optimised process conditions of mixed liquor suspended solids (MLSS) concentration, air bubble flow rate (ABFR), hydraulic retention time (HRT) and pH at 6.00 g/L, 3.05 ml/min, 280.50 min, and 6.50 respectively.

1 Introduction

Wastewater streams typically contain many regulated inorganic and organic contaminants that can restrict its use or disposal thereof. Standards promulgated by the state agency that regulates the maximum content of contaminants in wastewater streams disposed into publicly owned treatment works or discharged into waste injection wells have become increasingly more strict. For instance, in Indonesia, the effluent discharged from industrial sectors should comply with the national primary regulator of the discharged standard from Standard Environmental Regulation, 2015 [1]. For attaining of these limits, the palm oil wastewater treatment needs additional tanks and equipment, but in many cases, the refineries are located in populous areas with little space for expansion. Technologies, which can treat large quantities of wastewater with relatively small requirements are essential. The developed membrane technology can completely retain biomass and operate with high suspended solids concentration.

The removal of organic wastes from wastewater becomes essential increasingly, and submerged ultrafiltration finds its application in this area [2]. The membrane is promising to be used in water and wastewater treatment due to its high packing density and ease of module manufacture and operation. In this device, hollow fibre membranes are directly immersed in the feed reservoir with the withdrawal of permeate through the fibres by the application of vacuum on the outlet of the fibre lumen [3-5]. According to the reports, machining oil wastewater was characterised by the presence of several substances, namely, oil and grease, chemical oxygen demand (COD), total organic carbon (TOC), sulphide, and total suspended solids (TSS) [6]. The process efficiency in the membrane filtration is generally affected by many factors such as aeration flow rate; mixed liquor suspended solids (MLSS) concentration, pH, and hydraulic retention time (HRT). As well, separation efficiency, flux, and COD are considered as results of this study.

Table 1. The composition of synthetic coolant wastewater with standard deviation (S.D) and standard for wastewater and water discharge [1].

Constituent, unit	Concentration (S.D)	Standard
pH	6.7	5.5-9
COD, mg/L	555.0 (0.25)	200
TOC, mg/L	29.1 (1.02)	
Suspended Solids, mg/L	20	100

* Corresponding author: erna.yuliwati@unsri.ac.id

2 Experimental

The properties of PVDF membranes used in this work have been described in the previous study in detail [5]. As a semi-crystalline polymer, PVDF generally exhibits more complicated phase separation behaviour than amorphous polymer. The addition of SiO_2 affected as a modifier for PVDF membranes since they are chemically and thermally stable with larger surface area, highly miscible, fine suspendability in aqueous solution and relatively environmentally inert. Researchers have been carried out on their addition to PVDF matrix have been found to a promising modifier to improve the permeability and selectivity of PVDF membrane [7-9]. Synthetic coolant wastewater was prepared to be used as feed solution in ultrafiltration experiments as shown in Fig 1.

2.1. Experimental setup and procedure

The lab-scale experimental set-up shown in Figure 1 was used in this work. The submerged membrane separation system consisted of a feed reservoir of the 14 L volume, hollow fibre bundles, a peristaltic pump, a permeate flow meter, and a permeate collector.

Fig. 1. Ultrafiltration membrane unit

The filtration experiments were conducted at room temperature and under vacuum on the permeate side (0.5 bar abs). The membrane was created using a peristaltic pump (Master flex model 7553-79, Cole Palmer) with the permeate being withdrawn from the open end of fibres and a constant transmembrane pressure (TMP) of 0.5 bar was maintained to let water permeate from outside to the inside of the hollow fibre. The continuous aeration produced a turbulent flow, which could decrease the cake layer thickness and the average particle size [7].

2.2 Analytical methods

Membrane performance of Flux was tested by membrane bundle that having 10.48 dm^2 of membrane surface area. The pure water flux was measured at 100 kPa after the flux was steady, then calculated as

$$F = V/At \qquad (1)$$

where F is the pure water flux (L/m^2h), V the permeate volume (l), A the membrane surface area (cm^2) and t is the time (h).

Average pore size, r_m was determined by the filtration velocity method, in which pure water flux of the wet membrane was measured by applying pressure (0.1 MP) for a limited period (20 h). It represents the average pore size along the membrane thickness (ℓ) that was the difference between the external radius and an internal radius of the hollow fibre membrane. According to Guerout-Elford-Ferry equation, rm could be calculated by Yu et al., 2006 [8].

$$r_m = \sqrt{\frac{(2.9-1.75\varepsilon)\times 8\eta\ell Q}{\varepsilon \times A \times \Delta P}} \qquad (2)$$

where η is water viscosity (8.9 x 10^{-4} Pa), ℓ is the membrane thickness (m), ΔP is the operating pressure (0.1MPa), ε is the porosity of the membrane (%), Q is the volume of permeate water per unit time ($m^3 s^{-1}$), A is the effective area of membrane (m^2).

COD concentrations were measured using a spectrophotometer (DR 5000, HACH) in accordance with the standard procedures of method 8006 (Photometric method) and method HR TNT 10031 (Salicylate method), respectively. During the operation with high organic loading rates, the concentrations were evaluated daily, and sampling was carried out three times a week again. The COD removal efficiencies were calculated by Eqs. (1)

$$COD\ removal\ (\%) = \frac{COD_0 - COD}{COD_0} \times 100 \qquad (3)$$

where COD_o and COD are the initial COD concentration of the feed synthetic palm oil wastewater and the COD concentration of permeate produced.

A field emission scanning electronic microscope (FESEM, Hitachi Model S-520 Japan) was employed to observe a cross-section of the prepared PVDF membranes [9]. Prior to analysis, the membranes were cryogenically fractured in liquid nitrogen and then spattered with platinum. The FESEM micrographs were taken at certain magnifications.

2.3. Response surface methodology (RSM)

Response surface methodology (RSM) is a collection of mathematical and statistical techniques, commonly used for improving and optimising processes. It can be used to evaluate the relative significance of several affecting factors in the presence of complex interactions. When a combination of several independent variables and their interactions affects desired responses, RSM is an effective tool for optimising the process [10]. RSM uses an experimental design such as the central composite design (CCD) to fit a model by least squares technique. This methodology optimises COD removal in submerged hollow fibre membrane process. The Design Expert software is used for the statistical design of experiments and data analysis and performed in duplicate [11]. The four most important operating variables (factors), air bubble flow rate (x_1), hydraulic retention time (x_2), mixed liquor suspended solids concentration (x_3), and pH (x_4), are optimised. The study ranges (levels) are chosen as coded values for x_1, x_2, x_3, x_4 are set at 5 levels -2, -1 (minimum), 0 (central), +1 (maximum), and +2. Four dependent parameters are analysed as responses. Namely, COD removals, where achieved permeate flux was 145.7 L/m^2 h, in order to obtain optimum process condition.

In RSM, a model with the form of Eq. (3) is fitted to experimental data and, by optimisation, the coefficients in the model are calculated. To identify the right model that can fit the data, it can be started with the simplest model forms like first- and second-degree Scheffe's polynomial [12]. After testing these models for adequacy of fit, they were augmented to simplex centroid and special quartic models by adding the appropriate terms. In this study, the quartic model used for predicting the optimal point is as follows

$$y(x) = \sum_{i=1}^{N} e_i x_i^4 + + \sum_{ij(i<j)}^{N} e_{ij} x_i^3 x_j + \sum_{ij(i<j)}^{N} e_{ij} x_i^2 x_j^2 + \sum_{ijk(i<j<k)}^{N} e_{ijk} x_i^2 x_j x_k + \sum_{ijkl(i<j<k<l)}^{N} e_{ijkl} x_i x_j x_k x_l \quad (3)$$

where y is the response variable, e_i, e_{ij}, e_{ijk}, and e_{ijkl} are the polynomial coefficients of the model, x_i, x_j, x_k and x_l are the coded levels of the independent variables [10]. All these coefficients variables are analysed by multiple regression analysis, and the response contour plot is generated using the software Design-Expert. The validity of the selected model used for optimising the process parameters has to be tested using analysis of variance (ANOVA) that is determined by performing the Fisher's statistical test.

3 Results and discussion

3.1. Membrane structure and morphology

Properties of the membrane can be shown in Table 2. Based on analysis of structural and characterisation of PVDF membrane.

Table 2. Membrane properties

Parameter	Membrane
Membrane configuration	Hollow fibre
Membrane material	PVDF/SiO_2
Outer diameter (mm)	1.2
Inner diameter (mm)	0.6
Pore size (nm)	35.2
Membrane area (dm^2)	10.48

The membranes were obtained based on the size of outer and inner diameter, average pore size and membrane surface area. It showed that SiO_2 had modified the nanoporous membrane pore size. Improvement of membrane morphology is observed by adding of SiO_2. The result shows that the permeate flux of the membrane was better (flux value of 92.4%) as compared to the neat PVDF membrane (flux value of 84.3%). This improvement was caused by the change of membrane morphology of PVDF membrane. It is observed for adding SiO_2 particles, may affect the interaction between SiO_2 particles and PVDF chains. Generally, the addition of nanofillers into PVDF matrix is due to the agglomeration phenomena. This phenomenon is owing to the natural characteristic of the nanofillers (such as small size and high surface energy) and the poor compatibility with hydrophobic OVDF bulk. At present, the dispersion of nanofillers for the preparation of PVDF-inorganic composite membrane, usually achieved by strong mechanical stirring. They have a unique ability to bond polymers with dissimilar materials such as silica. The bond thus formed has good initial strength as demonstrated by the failure of the composite by polymer rupture and the bond exhibits excellent retention of strength even after severe environmental ageing.

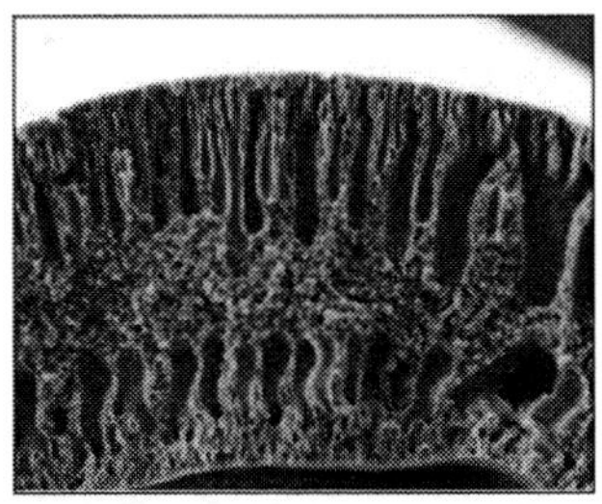

Fig. 2. FESEM images of the cross-section (mag 500x) of PVDF/SiO_2 membrane

Fig.2 illustrated that SiO_2 affects the mass transfer during the phase inversion process. The cross-sectional images of hollow fibres consist of finger-like macrovoids extending from both inner and outer walls of the hollow fibre, and an intermediate sponge-like layer. This phenomenon can be explained by the kinetic effect on the rate of solvent-nonsolvent exchange in the phase inversion process. At lower SiO2 concentration (1%), an increase in the amount of SiO_2 tends to draw more water into the polymer dope, resulting in an increase in the length of finger-like macrovoids and decrease in the thickness of the intermediate sponge-like layer.

3.2. Statistical analysis

The Design Expert software was used to design the experiments and together with their experimental and predicted values of COD removal. The second-order polynomial regression model containing 4 linear, 4 quadratic, and 6 interaction terms plus 1 block term was employed by using RSM. This model was found to be significant with $R^2 = 0.99$. Joglekar and May [10] also suggested that for a good fit of a model, R^2 should be at least 0.80. The R^2 for these response variables were higher than 0.80, indicating that the regression models explained the reaction well. However, its lack of fit was significant ($P=0.0229$), suggesting that this model not accurately represent the data in the experimental region. Therefore, each factor had five levels, up to quartic terms could be included in the model [11]. An empirical relationship between the response and the variables expressed by the following fitting the fourth equation degree. The experimental results were evaluated and approximating function of flux and COD removal was obtained in the final equation in terms of a coded factor are:

$$Y_2 = -208.48050 + 0.46\,X_1 + 0.79\,X_2 + 8.51\,X_3 + 0.28\,X_4 + 0.034\,X_1X_2 - 0.095\,X_1X_3 - 0.19\,X_1X_4 - 0.77\,X_2X_3 - 0.56\,X_2X_4 - 0.57\,X_3X_4 + 0.067\,X_1^2 - 0.46\,X_2^2 - 3.40\,X_3^2 - 2.96\,X_4^2 + 0.11\,X_1X_2X_3\ 0.15\,X_1X_2X_4 + 0.11\,X_1X_3X_4 - 0.18\,X_2X_3X_4 - 0.074\,X_1^2X_2 + 3.82\,X_1^2X_4 - 0.29\,X_1X_2X_3X_4 - 5.81\,X_1^2X_2^2 \quad (4)$$

Empirical model equations are mathematical correlation models that can be employed to predict and optimise the COD removal within the range of variable factors of this experiment.

3.3 Effect of pH-MLSS on COD

Figure 3 shows that the pH exerts a remarkable influence on the biological removal of COD. The increase of pH values with increasing MLSS value resulted in the COD removal in the range 88-91%. The average COD removal efficiencies had slightly different under pH conditions. pH had slightly affected the fatty acid composition in coolant wastewater. It does not lead to a sufficient supply of dissolved oxygen in acid or base conditions [12-13]. The oxygen is needed by aerobic microorganisms as electron acceptor.

COD removal increased slightly with increasing MLSS concentration. It was indicated that the lower MLSS concentration (3.00 g/L) the microbial metabolism and COD removal were quite low. Addition to MLSS concentration to 8 g/L, the sludge increased and COD removal also increased caused by proper organic matter removal was achieved in the process regarding the characteristic of wastewater.

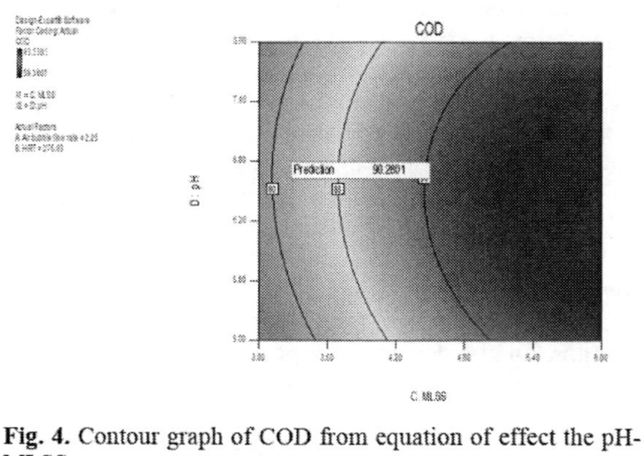

Fig. 4. Contour graph of COD from equation of effect the pH-MLSS

Place the figure as close as possible to the point where it is first referenced in the text. If there is a large number of figures and tables, it might be necessary to place some before their text citation. If a figure or table is too large to fit into one column, it can be centred across both columns at the top or the bottom of the page.

4 Conclusions

Full factorial design and central composite design of response surface methodology can be used to determine the significant variables and optimum condition for submerged ultrafiltration of coolant wastewater concerning the COD removal. Experimental results showed that a submerged ultrafiltration process using modified PVDF membranes has an excellent potential for refinery produced wastewater treatment. The quartic equation developed in this study shows the presence of a high correlation between observed and predicted values. Interestingly, 3-dimension response surface plots can be an excellent driven approach for visualising the parameter interaction. The optimum factor conditions were satisfied at ABFR of 2.25 ml/min, HRT of 276.93 min, the MLSS concentration of 4.50 g/L, and pH of 6.50 to resulted in COD removal of 90.28 %.

The approach of the statistical method has been testified to be a powerful tool for studying membrane ultrafiltration process. Further pilot-scale studies are required, and detailed study is needed to explore the remediation of coolant wastewater filtration.

This work was financially supported by the Ministry of research and technology and higher education of Republic Indonesia.

References

1. A.W. Zularisam, A.F. Ismail, R. Salim, M. Sakinah, T. Matsuura, Desalination **212,** 1–3 (2007).
2. A. Mansourizadeh, A.F. Ismail, Chem. Eng. J. **165,** 3 (2010).
3. A. Bottino, G. Capannelli, A. Comite, R. Mangano, Desalination 245 1-3 (2009).
4. M.L. Hami, M.A. Al-Hasyimi, M.M. Al-Doori, Desalination **216,** 1-3 (2007).
5. M. Kobya, E. Demirbas, M. Bayramoglu, M. Sensoy, Water Air Soil Pollut, **215,** 1-4 (2011).
6. I. Khouni, B. Marrot, R.B. Amar, Chem. Eng. J. **156,** 1 (2010).
7. K.J. Howe, M.M. Clark, Environ. Sci. Technol. **36,** 16 (2012).
8. E. Yuliwati, A.F. Ismail, T. Matsuura, M.A. Kassim, M.S. Abdullah, **283,** Special Issue (2011).
9. D.M. Wardrop, R.H. Meyers, J. Statist. Plann. Inference **25, 1** (1990).
10. Design-Expert® Software Version 6 User's Guide, 2001.
11. Design-Expert® Software Version 8.0.5.2, 2011.
12. W.J. Lau, A.F. Ismail, Membr. Water. Treat. **1,** 1 (2010).
13. E. Yuliwati, A.F. Ismail, T. Matsuura, M.A. Kassim, Desalination, **283,** Dec. (2011).

CO_2 desorption from activated DEA using membrane contactor with vacuum regeneration technology

Nidia Intan Listiyana, *Yeni* Rahmawati*, *Siti* Nurkhamidah, *Hafan* Rofiq Syahnur, and *Yusuf* Zaelana

Chemical Engineering Department, Faculty of Industrial Technology, Institut Teknologi Sepuluh Nopember Surabaya, Kampus ITS Keputih, Sukolilo, Surabaya

Abstract Carbondioxide (CO_2) content in natural gas must be removed because it inhibits liquefication process of natural gas. CO_2 gas separation technology using membrane contactor has been developed, however solvent regeneration using membrane contactors are still rare because it requires a larger energy. The regeneration process by using membrane vacuum technology was put forward to reduce the regeneration energy consumption. In this work, arginine, piperazine (PZ), and potassium carbonate ($K2CO_3$) as activators were added into diethanolamine (DEA) solution to form aqueous solutions of activated DEA. The experiment of CO_2 desorption from activated DEA was carried out in hollow fibre membrane contactor (HFMC). The solvent with rich CO_2 at 30-70°C was flowed in the lumen of the hydrophobic polypropylene HFMC, and the shell side was maintained at a reduced pressure by a vacuum pump at 20 kPa. The effect of solvent temperature and activators were investigated to get CO_2 desorption flux and regeneration efficiency. Experimental result shows that increasing of solvent temperature could enhance CO_2 desorption flux and regeneration efficiency. Instead of that, the activated DEA also give better result compared with non-activated DEA. Among three activators, K_2CO_3 give the best result for desorption flux and regeneration efficiency.

1 Introduction

Natural gas is the third largest natural resource in the world after coal and petroleum. Natural gas processing industry is an important sector in supplying energy or fuel. Compare to traditional coal, combustion of natural gas only produce half of CO_2 and 33% compared to oil. The high potential of natural gas resources increasing efforts to develop more effective and efficient processing in terms of operations, economics, and environment.

In natural gas processing processes where the main component is a hydrocarbon must be separated from impurity components such as CO_2, H_2S, N_2, and water vapor as they may affect the quality and processing of natural gas further. CO_2 as an impurity in natural gas can cause corrosion of piping and equipment, otherwise CO_2 is able to decrease the heating value of natural gas [15]. In the process of Liquefied Natural Gas (LNG) processing, CO_2 content must be removed to prevent icing during natural gas liquefaction process, because the process runs at very low temperatures of -161 ° C while the freezing point of CO_2 around -78.4°C [14].

Contactor membrane technology is an improvement of CO_2 removal technology combining gas absorption process with solvent and membrane technology in order to set the pattern of flow rate for avoiding flooding, loading, foaming, and entrainment that usually occurs in conventional absorption technology. The advantages of membrane contactors compared to conventional absorption technology are high surface area per unit contactor volume, independent control of gas and liquid flow rates without any operational problems, small size device, known gas–liquid interfacial area, modular and being easy to scale up or down. [9].

Feron et al. [3] conducted a study in CO_2 gas separation process using a conventional hybrid membrane contactors for absorption and regeneration column for the desorption process with amine solvents. They compared the performance of the membrane absorber and packed column absorber. The results showed that performance CO_2 absorption in membrane absorber was same compared to conventional absorber column, but the size of absorber can be reduced ten-times using a membrane absorber. Hollow fiber gas-liquid membrane contactor as CO_2 absorber has been comprehensively studied by several researchers. Those experimental and theoritical result have been reported [2, 8, 12, 16].

Research and study about membrane technology for CO_2 removal is increasing. However, it is important to

* Corresponding author: yeni_18des@yahoo.com

notice that there are only a few research on CO_2 desorption through membrane contactors. The solvent regeneration operation unit requires greater energy, it was estimated that the energy requirement for CO_2 desorption reaches 15-30% of coal burning power plant production [1]. Therefore, it is necessary to do a lot of research on CO_2 desorption through membrane contactors with low operating costs and capital costs.

CO_2 desorption using membrane contactor can be done by vacuum or sweep gas method. Kosaraju et al. [9] conducted a study using polypropylene membrane contactor by combining CO_2 absorption and desorption process. They performed a process of CO_2 absorption and desorption on two continuously separated membrane contactor modules for 55 days using novel solvent and helium gas as stripping gas. The result showed that the performance of the membrane contactor was stable and the PP membrane material was not wetted by the solvent.

Mansourizadeh and Ismail [13] also conducted a similar study using a polivinildenefluorid membrane material (PVDF) with two separate modules for absorption and desorption using water as solvent and N_2 as stripping gas. The results showed that the temperature of the solvent also affects the CO_2 desorption efficiency, which higher by increasing the temperature of the solvent. While increasing the flow rate of solvent could decrease desorption flux.

CO_2 desorption for solvent regeneration using membrane contactor vacuum was performed by Fang et al. [1]. MEA with a lot of CO_2 content was flowing to the membrane lumen and membrane shell was pressed by vacuum pump. Low temperature steam flowed in shell side of membrane module which acts as sweep gas improve desorption performance in order to avoid liquid loss. The results showed that CO_2 desorption using membrane vacuum technology represented good performance and reduce energy consumption because it was operated at low temperatures.

Proper choice of solvent plays an important role in determining the performance of gas liquid membrane contactor for CO_2 regeneration. Various solvent such as pure water, aqueous solution of NaOH, KOH, alkanolamines and ammonia have been studied experimentally. Many studies of CO_2 absorption-desorption with membrane contactors still use a single solvent without any mix of activators, such as Khaisri et al. [7], Lv, et al. [10], and Fang, et al. [1] who use MEA as a solvent. Then Mansourizadeh and Ismail [13], Karoor and Sirkar [6], used water, and several other studies using NaOH, MDEA (methildiethanolamine), or DEA [16]. Some amines have some advantages, such as primary and secondary group amines whose reaction rate are faster than the tertiary amine group, but the tertiary amine group has a higher absorption capacity than the primary and secondary amines [11].

Recently, the blended alkanolamines or addition of activator have been presented by several researches in order to improve the performance of CO_2 separation. Yeon, et al. [18] conducted CO_2 separation using membrane contactor for absorption and stripper column for solvent regeneration by mixing PZ and triethanolamine (TEA) as solvent. The results showed that TEA with low absorption capacity could be increased by addition of PZ. TEA is a tertiary amine with high volatility can prevent hollow fiber membrane from wetting phenomenon. Then Hedayat, et al. [4] conducted a study for the absorption of CO_2 and H_2S gases from natural gas using membrane contactors by comparing MDEA solvents mixed with MEA and MDEA with DEA. The results showed that the addition of MEA into the MDEA improved the efficiency of CO_2 separation, while the addition of DEA did not give significant results. Rahim, et al. [15] used four different amino acid salt (AAS) solvents and compared with aqueous solutions of MEA and NaOH. The results showed that AAS improved the performance of CO_2 absorption-desorption when compared with MEA.

In this work, arginine, PZ, and K_2CO_3 as activators were added into diethanolamine (DEA) solution to form aqueous solutions of activated DEA. The activated mechanism were presented to explain the activation phenomenon. The experiment of CO_2 desorption was carried out in polypropylene hollow fiber membrane contactor (HFMC). The use of such amine solvent mixture is expected to increase CO_2 regeneration efficiency and reduce the cost of solvent usage so that it is more economical.

2 Theory

2.1. Characteristic of Activators

Three activators, arginine, PZ, and K_2CO_3 were chosen to form the three DEA-activated aqueous solutions : DEA-Arginine-H_2O, DEA-PZ-H_2O, and DEA-K_2CO_3-H_2O. Arginine is a primary amino acid that can form alkaline amino acids salt. This amino acid salt is predicted as one of the proper substances in CO_2 absorption because it has functional groups that are identical with alkanolamines. While PZ has a special molecular structure, the cyclic structure of a symmetrical diamino. PZ is an efficient and highly successful activator used for high CO_2 separation capacity in industrial processes. Then K_2CO_3 is alkaline salt which has a high CO_2 loading capacity and low heat of reaction. Therefore, it's to be a good activator for CO_2 absorption.

2.2 Reaction Mechanism

The mass transfer of CO_2 gas to the liquid phase is influenced by the diffusivity of the CO_2 into the solvent. If the pore is hydrophobic and membrane is not wetted, then it is assumed that the pore membrane is only fully filled by gas. Diffusion begins in the bulk gas phase by the diffusion of the boundary layer and the chemical reactions on the fluid-membrane surface. The reaction between a CO_2 and DEA is as follows :

$$CO_2 + R_1R_2NH \leftrightarrow COO^- + R_1R_2NH^+ \qquad (1)$$

Where R_1R_2NH is the primary or the secondary amine. The formation of carbamate proceeds as follows :

- Formation of zwitterion (intermediate reaction)

$$CO_2 + R_1R_2NH \leftrightarrow R_1R_2NH^+COO^- \quad (2)$$

- Removal of the proton by a base B

$$R_1R_2NH^+COO^- + B \leftrightarrow R_1R_2NCOO^- + BH^+ \quad (3)$$

Where the base B, could be the amine itself, water molecules, or hydroxyl ions that reacted with DEA solution. The corresponding reactions are as follows :

$$R_1R_2NH^+COO^- + R_1R_2NH \rightarrow R_1R_2NCOO^- + R_1R_2NH^+ \quad (4)$$

$$R_1R_2NH^+COO^- + H_2O \rightarrow R_1R_2NCOO^- + H_3O^+ \quad (5)$$

$$R_1R_2NH^+COO^- + OH^- \rightarrow R_1R_2NCOO^- + H_2O \quad (6)$$

If there is an activator, the reaction that occured not only CO_2-DEA but also CO_2-activator take place in the system. Here is the reaction between DEA-Arginine-H_2O with CO_2 :

$$H_2NCHR'COOH + CO_2 \leftrightarrow H_2NCHR'COOHCOO^- \quad (7)$$

$$H_2NCHR'COOHCOO^- + B \leftrightarrow HNCHR'COOHCOO^- + BH^+ \quad (8)$$

$$HNCHR'COOHCOO^- + H_2O \leftrightarrow HCO_3^- + H_2NCHR'COOH \quad (9)$$

For the system DEA-PZ-H_2O, will give different reactions which will directly form carbamic ions without going through intermediate reactions, as follows :

$$HN(C_4H_8)NH + CO_2 \leftrightarrow HNC_4H_8NCOO^- + H^+ \quad (10)$$

Then the carbamate will convert CO_2 into bicarbonate ion and PZ will regenerate as Eq. (11)

$$HN(C_4H_8)NCOO^- + H_2O \leftrightarrow HCO_3^- + HN(C_4H_8)NH \quad (11)$$

The overall reaction for the absorption of CO_2 using DEA-K_2CO_3-H_2O is described by reaction below :

$$CO_2 + K_2CO_3 + H_2O \rightarrow 2KHCO_3 \quad (12)$$

Since potassium carbonate and bicarbonate are strong electrolytes, it can be assumed that the metal is present only in the form of K^+ ions and Eq. (12) can be represented as :

$$CO_2 + CO_2^{-3} + H2O \rightarrow 2HCO^{-3} \quad (13)$$

Reaction (13) proceeds according to the following sequence of elementary steps :

$$CO_2 + H_2O \rightarrow HCO^{-3} + H^+ \quad (14)$$

$$CO_2 + OH^- \rightarrow HCO^{-3} \quad (15)$$

$$H_2O \rightarrow H^+ + OH^- \quad (16)$$

Reactions (14) and (15) are both followed by subsequent instantaneous reactions as follows :

$$H^+ + CO_2^{-3} \rightarrow HCO^{-3} \quad (17)$$

$$H_2O + CO_2^{-3} \rightarrow HCO^{-3} + OH^- \quad (18)$$

The reaction sequence (14), (16) and (17) are known as the acidic mechanism. The contribution of the acidic mechanism to the overall rate is negligible unless the pH of the liquid solution is very low. Almost all cases of industrial absorption is held at high pH (generally, pH > 8) and as such the acidic mechanism can be neglected. Therefore, reaction (15) is the rate controlling step for absorption of CO_2 into potassium carbonate solution as reactions (16) and (18) are instantaneous reactions.

2.3. Data Analysis

In this study, CO_2 desorption flux (J_{CO_2}) and CO_2 loading (α) were used to analyze the CO_2 desorption performance. They can be expressed by the following equations :

$$J_{CO_2} = \frac{(C_{in} - C_{out}) \times Q_l}{A} \quad (19)$$

$$\alpha = \frac{mole(CO_2)}{mole(solvent)} = \left[\frac{\frac{(V_{gas} - V_{HCl})(P)(273K)}{(101325Pa)(T)(22{,}4L/mole)}}{C_1V_1} \right] \quad (20)$$

C_{in} and C_{out} are the liquid phase inlet and outlet CO_2 concentration. Q_l is volumetric flow rate of solvent. A represent the liquid-gas contact area in the membrane module.

CO_2 concentration in the aqueous solution which written in Eq.20 was determined by the apparatus adapted from Chittick CO_2 analyzer apparatus [15]. From Eq.20 , V_{gas} is volume of displaced solution in the manometer tube (mL), V_{HCl} is volume of titrant (mL), P is pressure (Pa), T is room temperature (K), C_1 is solution concentration (mol/L), and V_1 is sample volume (mL).

Regeneration efficiency was calculated to evaluate the regeneration performance of rich solvent by using membrane vacuum regeneration technology in this study. Regeneration efficiency is defined by the ratio of amount of CO_2-rich solution, which can be calculated by the following equation :

$$\eta = \frac{C_{in} - C_{out}}{C_{in}} x100\% \quad (21)$$

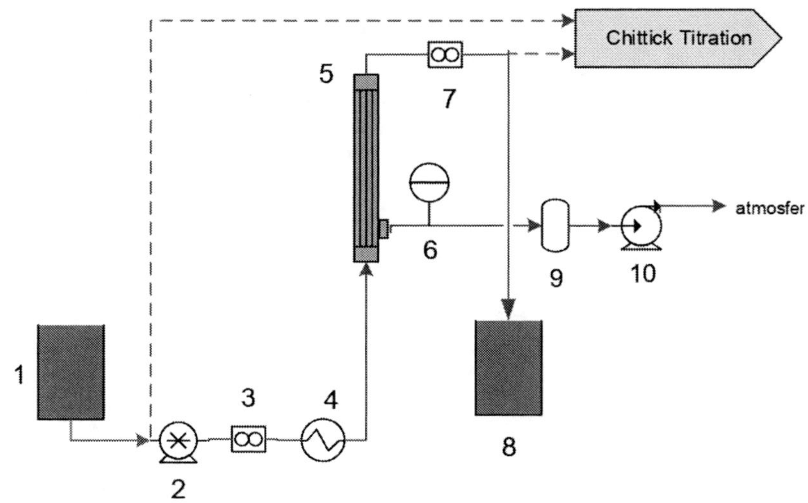

Fig. 1. Experimental setup for CO_2 desorption. 1. Rich solvent tank; 2. Liquid pump; 3. Rich solvent flow meter; 4. Heater; 5. Membrane contactor; 6. Pressure gauge; 7. Lean solvent flow meter; 8. Lean solvent tank; 9. Buffering tube; 10. Vacuum pump

3 Experimental

3.1. Material

In this experiment, the material of membrane is a hydrophobic polypropylene membrane from GDP Filter Indonesia membrane industry with its characteristics shown in Table 1.

Table 1. Specification of polypropylene hollow fiber membrane

Parameter	Specification
Inside Diameter (mm)	0,35
Outside Diameter (mm)	0,5
Pore Diameter (μm)	0,2
Fiber Length (mm)	100
Number of fibers	3302
Membrane porosity (%)	65
Membrane turtuosity	2,3669
Membrane area (m^2)	0,0413

Solvent used is diethanolamine (DEA) were obtained from PT. Pertamina, Balikpapan, Indonesia. CO_2 gas cylinder with 40% volume and 60% volume N_2 balance purchased from PT. Aneka Gas, Sidoarjo, Indonesia. PZ was purchased from Shenzen Lodi Chemical Co.Ttd. While Arginine and K_2CO_3 were purchased from Sigma Aldrich Chemical Co. Ltd. In this study, concentration of solvent used was 30% volume with 1% vol. activator and 69% vol. H_2O.

3.2. Experimental Procedure

The experimental set-up for CO_2 desorption was shown in Fig.1. The solvent containing CO_2 gas from the absorber was prepared with CO_2 loading 0,22 mole/mole and was added into the tank (1), then flows into coil heater water bath (3) by diaphragm pump (2) before entering the desorber (4). This aim of heating was to reduce the solubility of CO_2 in the solvent so CO_2 can easily released from the solvent. The temperature range in this study was 30-70°C. After pre-heating, CO_2-rich solution continuosly pumped into tube side of the hollow fiber membrane contactor with flow rate 100 mL/min. Liquid flow rate was controlled by liquid flow meter. During the regeneration operation, shell side of the module membrane contactor was kept at 20 kPa absolute by vacuum pump (8) that controlled with pressure gauge (5). Buffering tube (7) used to capture water vapor from output gas desorber before entering vacuum pump. The dissolved CO_2 in the solvent in the tube section diffused outward through the membrane pores which was driven by the vacuum pressure. Finally, CO_2 gas enriched and could be collected from the vacuum pump. The solvent exit desorber is called lean solvent (6) then analyzed with chittic titration.

4 Result and Discussion

4.1. Effect of Temperature on Desorption Flux

The effect of rich absorbent temperature on CO_2 desorption flux was shown in Fig.2.

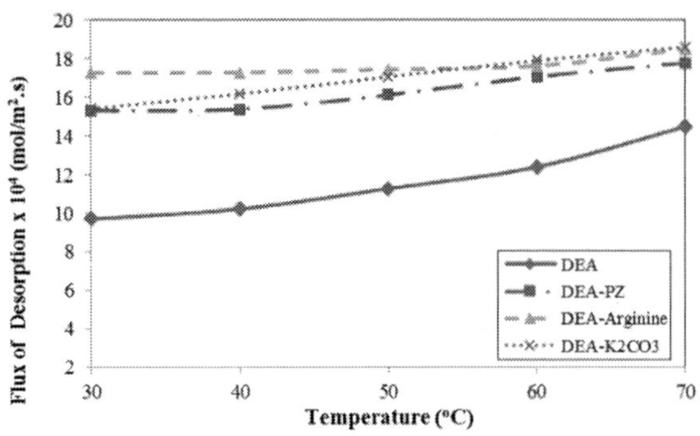

Fig.2 Effect of liquid temperature on CO_2 desorption flux (Q_l = 100 mL/min ; P = 20 kPa ; C_{DEA} = 30% vol. ; $C_{activator}$ = 1% vol. ; α – rich solvent = 0,22 mol/mol)

It is shown that desorption flux increased when temperature of rich absorbent was increased. This phenomenon due to the solubility of CO_2 decreases very rapidly with the increasing of temperature. The increasing of absorbent's temperature increased the CO_2 equilibrium partial pressure; thus also increased the driving force of CO_2 desorption. Moreover, the reversible reaction equilibrium constant and CO_2 diffusivity in DEA solution increased with the solution temperature, which leads to the improvement of desorption performance as well. The effect of temperature on these factor was studied by several researchers through experiments and mathematical modeling also which reported the similar result [2, 5, 13, 15, 17,]. However, the membrane stability must be considered against high temperature operation due to susceptibility to chemical and thermal effect.

4.2. Effect of Temperature on CO_2 Loading of Lean Absorbent

The effect of the rich absorbent temperature on CO_2 loading was shown by Fig.3.

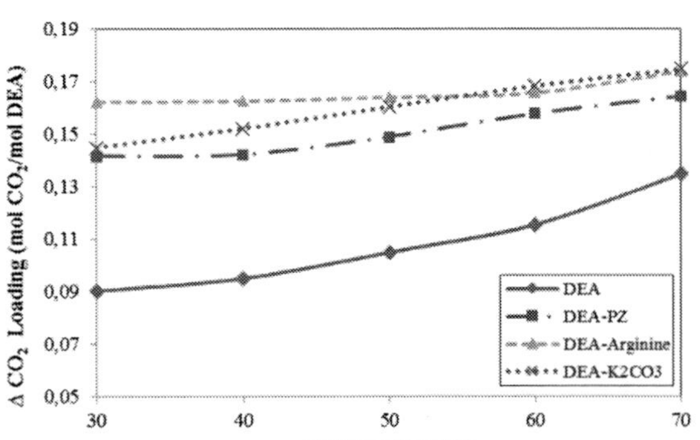

Fig.3 Effect of liquid temperature on CO_2 loading (Q_l = 100 mL/min ; P = 20 kPa ; C_{DEA} = 30% vol. ; $C_{activator}$ = 1% vol. ; α – rich solvent = 0,22 mol/mol)

It is shown for each type of absorbent with the same CO_2 loading of rich amine value of 0.22 mol CO_2/mol DEA, CO_2 loading of lean absorbent decrease with increasing absorbent temperature. In other words, the higher the temperature of the solvent, the more the amount of CO_2 released from the solvent and cause the amount of CO_2 contained in the lean amine solvent decreases. The parameter α is the content of CO_2 dissolved in the solution from both physical and chemical absorption. It is expressed as moles of CO_2 per mole of amine.

4.3. Effect of Activator on Desorption Efficiency

Value of desorption efficiency expresses the performance of CO_2 regeneration process. It is shown in Fig.4.

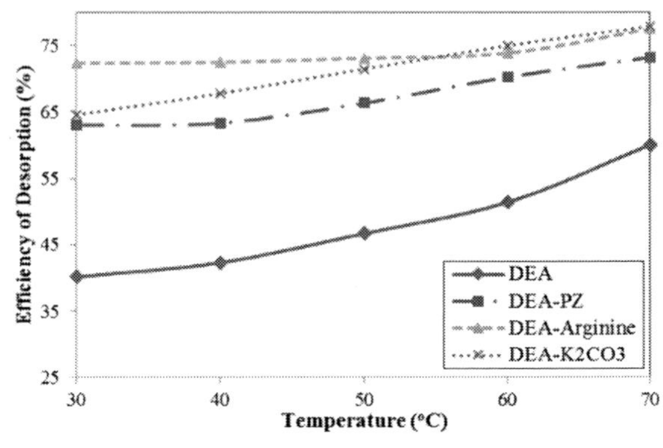

Fig.4 Effect of liquid temperature on efficiency of desorption (Q_l = 100 mL/min ; P = 20 kPa ; C_{DEA} = 30% vol. ; $C_{activator}$ = 1% vol. ; α – rich solvent = 0,22 mol/mol)

At constant liquid flow rate of 100 mL/min and pressure of 20 kPa, efficiency increased with the increase of liquid temperature for any given absorbent. It was caused by the chemical reaction kinetics and CO_2 diffusion coefficient. At higher temperatures, the dissolved CO_2 concentration in the solvent moved to the low-concentration side and increased CO_2 mass transfer driving force. Many researchers had conducted research about the effects of temperature on desorption through experiments and mathematical modeling which reported similiar results [2, 13, 15, 17].

The experimental results showed that the DEA-K_2CO_3 solvent was the best desorption efficiency. The K_2CO_3 solution has an ionic structure, and bicarbonate and carbonate ions were formed by reacting with the CO_2 in the solution. Therefore, bicarbonate and carbamate ions in reaction was unstable which caused the flux of desorption increase and the partial CO_2 equilibrium pressure higher.

5. Conclusion

Desorption of CO_2 from activated DEA using membrane contactor with vacuum regeneration technology was achieved in polypropylene hollow fiber membrane contactor using membrane vacuum regeneration technology. Arginine, PZ, and K_2CO_3 as activators were added into DEA solution to form aqueous solutions of activated DEA. The three activator was compared at any temperature operation and 20 kPa pressure vacuum in order to determine the effect of temperature and chemical structures of absorbent on vacuum regeneration performance.

The experimental results showed that the increasing of solvent temperature could enhance the CO_2 desorption flux and reduce CO_2 loading from rich solvent. Among three solvent in this study, the best solvent for desorption of CO_2 is DEA-K_2CO_3 with a ratio of separation efficiency 1.3 times higher than that of unactivated solvent. Flux of desorption with DEA unactivated at 70°C was 14.49 x 10^{-4} mol/m^2.min, while DEA-K_2CO_3 was 18.62 x 10^{-4} mol/m^2.min. Therefore, the best performance for CO_2 desorption in this study is using K_2CO_3 as activator with efficiency of 77.83%.

The financial support of this work from LPDP Thesis Scolarship (No : PRJ-1651 /LPDP.3/2017) is gratefully acknowledged.

References

1. M. Fang, S. Yan, Z. Luo, M. Ni, and K. Cen, K. Energy Proced., **1** (1), 815–822. (2009)
2. M. Fang, Z. Wang, S. Yan, Q. Cen, and Z. Luo, Int. J. Greenhouse Gas Control, **9**, 507–521. (2012)
3. P. H. M. Feron, and A. E. Jansen, Energy Conversion dan Management, **36** (6–9) : 411–414. (1995)
4. M. Hedayat, M. Soltanieh, and S. A. Mousavi, J. Membr. Sci., **377**: 191–97. (2011)
5. G. L. Jian, F. Z. You, D. C. Min, and J. W. Lian, J. Membr. Sci. **289 :** 138–149. (2007)
6. S. Karoor, and K. K. Sirkar, Ind. Eng. Chem. Res. **32** : 674–84. (1993)
7. S. Khaisri, D. M. David, T. Paitoon, and R. Jiraratananon. Chem. Eng. Res. Des.. (2011)
8. K. Young-Seok and Y. Seung-Man. Separation and Purification Technology **21** (1–2): 101–9. (2000)
9. P. Kosaraju, S. Kovvali, Korikov, and K. K. Sirkar, Ind. Eng. Chem. Res., **44** (5), 1250–1258. (2005)
10. L. Yuexia, Y. Xinhai., S. T. Tu, J. Yan, and E. Dahlquist, Appl. Energy **97**: 283–88. (2012)
11. B. P. Mandal, A. K. Biswas, and S. S. Bdanyopadhyay, Chem. Eng. Sci., **58** (18), 4137–4144. (2003)
12. A. Mansourizadeh, and A. F. Ismail, J. Hazard. Mater. **171,** 1–3 : 38–53. (2009)
13. A. Mansourizadeh, and A. F. Ismail, *Desalination*, **273**, 386–390. (2011)
14. R. H. Perry, and D. W. Green, *Perry's Chemical Engineering Handbook* (2008)
15. N. A. Rahim, N. Ghasem, and M. Al-marzouqi, J. Nat. Gas Sci. Eng. **26**, 108–117 (2015)
16. H. A. Rangwala, J. Membr. Sci. **112** (2): 229–40. (1996)
17. Y. Shuiping, F. Mengxiang, W. Zhen, and L. Zhongyang, Appl. Energy **98** : 357–367. (2012)
18. S. Yeon, B. Sea, and Y. Park, Sep. Sci. Technol. **39** (14), 3281–3300. (2004)
19. W. Zhen, F. Mengxiang, P. Yili, Y. Shuiping, and L. Zhongyang. Chem. Eng. Sci. **98** : 238–249. (2013)

Development of Nano-hybrid Cellulose Acetate/TiO_2 Membrane for Eugenol Purification from Crude Clove Leaf Oil

Tutuk Djoko Kusworo*, *Danny* Soetrisnanto and *Dani Puji* Utomo

Chemical Engineering Department, Faculty of Engineering, University of Diponegoro, Jl. Prof. Sudharto, Tembalang, Semarang, 50239, Indonesia

Abstract. Chemical separation and purification are the important part of the chemical industry which consumes up to 70% energy cost. The separation technology such as distillation and absorption are well known in essential oil purification. The purification of clove leaf oil needs an attention because the current technology still consumes high energy and produces chemical wastes. The employment of membrane separation for clove leaf purification is a novel concept that needs many improvements. The main problem of polymeric membrane utilization is eugenol ability to dissolve the polymer membrane. Cellulose acetate is one of membrane polymers that is insoluble in eugenol. This paper reveals the performance of nanohybrid CA/TiO_2 membrane for eugenol purification. The stability of produced membrane as an organic solvent nanofiltration (OSN) is evaluated in this study. The SEM image result shows that fabricated membrane has an asymmetric structure of membrane sub-layer. The different nano-particles loading shows the variation of permeate fluxes, the increase of nano-particles in polymer blend tends to increase the permeability. Thus, this study provides an overview of the potential CA/TiO_2 for OSN development by incorporating inorganic nano-particles in membrane polymers for eugenol purification that can be integrated in upstream separation process.

1 Introduction

Chemical separation and purification are the important part of the chemical industry which specify the quality of the product. Generally, the separation process consumes up to 70% of energy costs for a chemical plant [1]. The separation technology such as distillation and absorption which requires a solvent removal by heating process are highly applied for essentials oils purification for example crude clove leaf purification [2]. The crude clove leaf oil is produced from leaves and twigs of the clove tree through distilation process. Crude clove leaf oil has lower quality than clove bud oil due to its eugenol content is bellow the quality standard of clove oil. The eugenol concentration in crude clove leaf oil ranged from 60 – 70% and needs to be purified to increase the economic value [3]. However, the recent technologies for crude clove leaf oil purification inflict the other problem such as high energy consumption, chemical waste generation, and needs a certain condition e.g vacuum, high temperature, etc [4]. Membrane separation technology can be proposed as an alternative separation technology because it is currently known as an inexpensive technology, low energy consumption, can be conducted in a mild operating condition, and easily to be scaled up [5].

Membrane separation is well used in the separation of gasses, water, proteins, and viruses. Several membranes have been developed for organic chamicals separation that is also known as an Organic Solvent Nanofiltration (OSN) membrane [6]. The field of OSN membranes refers to the separations of chemicals with molecular weights from 100 to 2,000 g.mol^{-1} [7]. The main concerns in this study are developing the membrane material which is suitable for crude leaf oil purification and determining the best operating condition. The quality of clove leaf oil is defined by the content of eugenol and the main impurity of the crude clove leaf oil is Caryophyllene [3]. These chemicals are challenging to be separated using a membrane due to very few literature which reports about this separation concept.

Eugenol has molecular weight of 164 g.mol^{-1} while the caryophyllene has a higher molecular weight of 204 g.mol^{-1}. Figure 1 shows the molecular structure of eugenol and caryophyllene, where there is a difference in their chemical potentials. Eugenol has hydroxyl and ether group which refer a hydrophilic properties, while caryophyllene has only methyl group. These different properties indicates that eugenol and caryophyllene are potential to be separated using membrane. Nasution et al. studied the purification of the clove oil using cellulose-

* Corresponding author: tdkusworo@che.undip.ac.id

chitosan membrane in a dead-end membrane filtration cell [8]. They found that the highest flux and eugenol quality of the permeate are 0.239 $L.m^{-2}.s^{-1}$ and 58.21% respectively. However, the concentration of eugenol in the permeate product is far bellow the desired standard and also there is not enough scienctific approach in the separation process. In the WO Patent that was invented by Boam et al. they disclosured in general relates to a process for reducing impurities present in an essential oil using selective membrane [9]. They reported that the membrane should gave MWCO at least 400 – 1500 $g.mol^{-1}$ and the essential oil should be dissolved in organic solvent before being filtrated using selective membrane. In their invention, there are no information about the specification of the selective membrane and opertaing condition for this purpose.

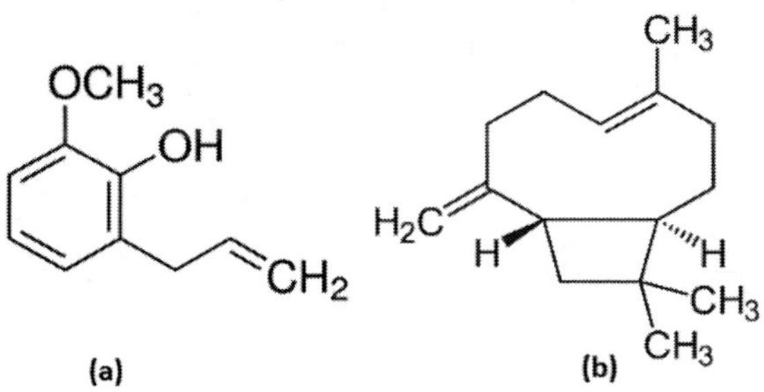

Fig. 1. Molecular structure of (a) eugenol, (b) caryophyllene

In this study, cellulose acetate (CA) polymer is used as membrane material due to its insolubility in eugenol. Other polymeric membrane such as polyethersulfone (PES), polyamide (PA), and polysulfone (PSf), can be used in this application because they are highly soluble in eugenol. The nano-TiO_2 is incorporated as an inorganic nanomaterial filler in the CA membrane fabrication to enhance its mechanical strength. The performance of CA/nano-TiO_2 membranes are discussed in this study and the final product quality of clove oil permeate are also reported.

2 Materials and Methods

2.1. Materials

Cellulose acetate (CA) polymer was purchased from Alpha Chemika, Mumbai, India. The polymer average molecular weight was 30 kDa and the acetic content was 53-56%. Acetone as polymer solvent was purchased from Merck. Nano-TiO2 was supplied by Nano Center Indonesia with average particles diameter were 128 nm. Original crude clove leaf oil was purchased from clove farmers in Semarang, Indonesia. The characteristics of crude clove leaf oil sample was shown in Table 1.

2.2 Fabrication of Nanohybrid CA/nano-TiO_2 Membrane

Flatsheet cellulose acetate (CA) membranes were fabricated by preparing the dope solution containing polymer CA (17 wt-%), nano-TiO_2 (1 and 2 wt-%), and

Table.1 Characteristics of crude clove leaf oil

No.	Parameter	Value
1.	Color and odor	Yellow brownish and typical clove
2.	Density 20°C ($g.mL^{-1}$)	1.014
4.	Solubility in 70% Alcohol	Completely soluble
5.	Eugenol content (% v/v)	69.23 ± 0.71

acetone as solvent. An appropriate amount of nano-TiO_2 was poured into the acetone and then agitated for 1 h. Subsquently, the nano-TiO_2 mixture was added with powdered CA polymer gradually to avoid polymer aggregation The polymer mixture was mixed for 24 h. The CA dope solution was allowed to stand for another 24 h with gentle mixing to remove air bubbles and to prevent the precipitaion of nano-TiO_2. The membranes were casted using dry-wet phase inversion method. The CA dope solution was poured on the flat glass plate and then was casted using mechanic casting knife. The casted film was immediately soaked in distilled water for demixing process. The membranes were handled in storing tank containing distilled water at 30 ± 2°C for 24 h. Wet membranes were dried in the drying chamber at 40-50°C for 24 h [10].

2.3 Membrane characterization

The surface and cross-sectional structure images were analyzed using Scanning Electron Microscope (SEM) JEOL JSM-6510-LA. The membrane sample was immersed in liquid nitrogen and then fractured. The sample was sputtered with a gold layer before analyzing and then scanned at certain magnification.

The molecular weight cut-off (MWCO) of the membrane was determined to estimate the size of pore radius. The determination was performed in a dead-end filtration cell by applying the solution containing PEG with different molecular weight (400; 600; 1,000; 4,000; 6,000; and 10,000 Da), the permeate sample was analyzed using visual handheld refractometer (Atago). The MWCO was determined at 90% of solute rejection. The value of MWCO was used to estimate the pore size radius using Eq. 1 [11].

$$r_m(cm) = 16.73 \times 10^{-10} (MWCO)^{0.557} \quad (1)$$

Porosity of the membrane was performed by weighing the dried membrane and wetted membrane. The different weight of the membrane was assumed as void fraction. The porosity of membrane was calculated using Eq. 2.

$$\varepsilon = \frac{wt_0 - wt_i}{\rho_{water} A l} \quad (2)$$

Where, ε is the membrane porosity, wt0 and wti are the membrane weights before and after drying, respectively. A is the membrane surface area, l is the

https://doi.org/10.1051/matecconf/201815608013

membrane thickness and ρwater is water density. While water contact angle of membrane was determined to observe the hydrophilicity and the surface tension of membrane. Distilled water was used as a probe liquid and then dropped on the membrane surface. The water contact angle was immadiately average measured thress times at two points of positions.

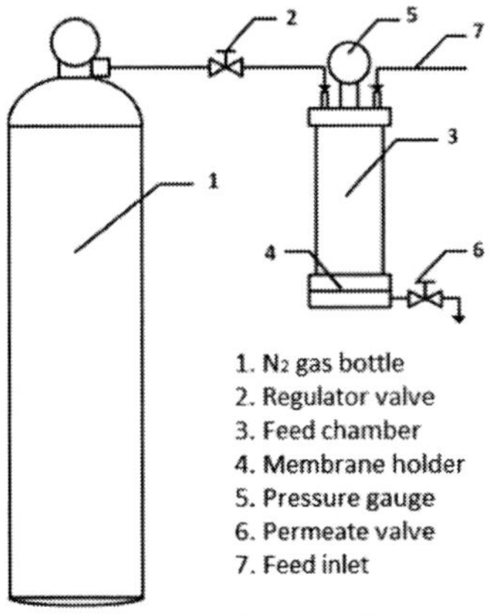

Fig. 2. Dead-end membrane filtration cell

2.4 Performance test of fabricated membrane

The performance test of fabricated membranes were carried out in a dead-end membrane filtration cell (Figure 2). The membrane with effective filtration area pf 12.56 cm^2 was conditioned in hexane solvent for 3 h and then mounted on membrane holder. Clove leaf oil sample was poured in filtration chamber with capacity of ± 100 mL. N_2 gas was streamed into filtration chamber as a gas suppressor and the pressure was maintained at 2 bar-gauge. The permeate flux was determined using Eq. 3. The permeate and consentrate of clove leaf oil were analyzed according to the quality parameter i.e density, and total eugenol content. The analysis methods refer to Indonesian National Standard (SNI 06-2387-2006) [12].

$$J = \frac{V}{t.A.P} \tag{3}$$

Where J is permeate flux (L.h^{-1}m^{-2}bar^{-1}), V is permeate solume at certain operation time (L), t is operation time (h), A is effective area of membrane (m^2), and P is pressure given to membrane (bar).

3 Results and Discussions

3.1 Membrane characterization results

Figure 3 shows the SEM image of CA membrane and CA/nano-TiO2 membrane both of surface and cross-section. The surface image of CA membrane is shown in Figure 3.a, the surface is flat and smooth, there is not observed any void at 20,000x magnification. It indicates that the membrane pores are in nanoscale. Figure 3.b displays the surface image of CA/TiO2 membrane, the polymer matrix is well attached on the nanoparticles surface. The cross-sectional SEM images (Figure 3.c and 3.d) show assymetric structure of membrane consists of dense layer and support layer of finger-like pore structure. Dense layer contributes on the membrane selectivity and the finger-like pore structure plays on the permeation performance.

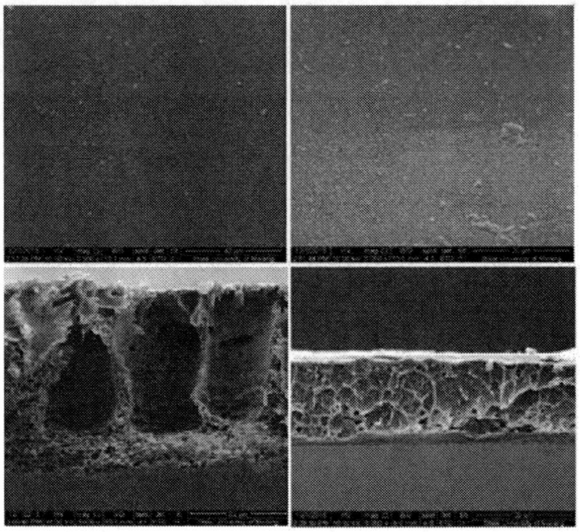

Fig. 3. SEM image of fabricated membrane

Table. 2 Characteristics of fabricated membrane

Membranes	MWCO (Da)	pore radius (nm)
CA	4839 ± 147	1.89 ± 0.27
CA/TiO_2 1%	4387 ± 298	1.79 ± 0.40
CA/TiO_2 2%	4253 ± 185	1.76 ± 0.31

Membranes	Porosity	Contact angle (θ°)
CA	0.51 ± 0.06	67.43 ± 1.2
CA/TiO_2 1%	0.43 ± 0.08	59.77 ± 1.5
CA/TiO_2 2%	0.39 ± 0.11	53.21 ± 1.1

Table 2. shows the characteristics of fabricated membrane which play important role in membrane performance. MWCO indicates the molecular weight of solute which is 90% rejected by membrane, it relates with pore size diameter. The increase in nano-TiO_2 loading concentration will decrease the MWCO value and pore diameter. This is due to the nano-particles fill the nano-voids in the membrane structure. Porosity of the membrane also decrease with the increase of nano-TiO_2 loading. The nano-fillers devide thelarge void into smaller voids without creating any dead-end void. The contact angle represents the hydrophilicity and surface tension of membrane. The lower contact angle of membrane indicates more hydrophilic membrane. The addition nano-TiO_2 in the membrane matrix increase significantly the hydrophilicity of the membrane. This properties is very usefull to enhance the caryophyllene rejection.

3.2 Clove oil permeation of prepared membrane

The permeation tests of clove oil through fabricated membrane were performed in dead-end filtration cell. Figure 4. shows the flux profiles of clove oil for each fabricated membrane. The flux declines with the filtration time due to membrane compaction and fouling. The initial flux of clove oil in CA membrane is 23.4 $L.m^{-2}h^{-1}bar^{-1}$ and then sharply declines until 13 $L.m^{-2}h^{-1}bar^{-1}$ after 25 minutes filtration. CA/nano-TiO_2 membranes show higher permeate fluxes with the initial fluxes are 30 and 34 $L.m^{-2}h^{-1}bar^{-1}$ for nano-TiO_2 concentration 1 wt-% and 2 wt-% respectively. The incorporated membranes are able to achieve higher fluxes because the nano-fillers are avoiding dead-end pores formation and maintain the structure of membrane from compaction as the result of pressure-driven filtration process. This indicates that CA/nano-TiO_2 membrane is a favorable structure for this process. This work revealed that 2% incorporated nano-TiO_2 in CA membrane enhances the permeate flux up to 52% compared with conventional CA membrane.

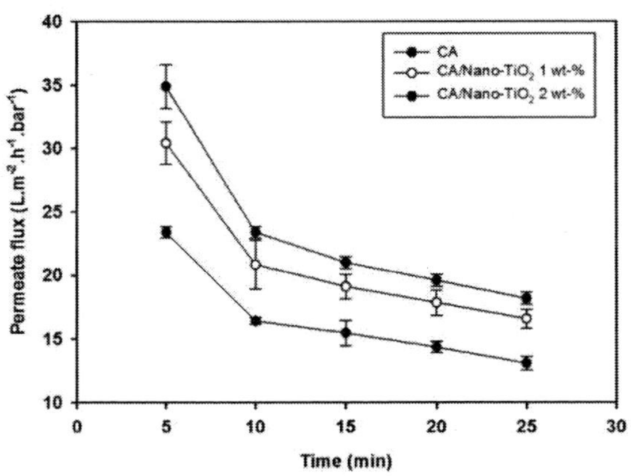

Fig. 4. Flux profiles of clove leaf oil permeate for each membrane

3.3 Separation performance of membrane and permeate product quality

Selectivity of the membrane is representing the separation performance of membrane. In this work the separation efficiency between eugenol and caryophyllene is studied. To observe the separation performance, the eugenol content and clove oil density of feed and permeate are analyzed. The density of pure eugenol and caryophyllene are 1.060 $g.mL^{-1}$ and 0.905 $g.mL^{-1}$, respectively. In other words the density of clove oil will increase as the increase of eugenol content in clove oil. Figure 5 shows the eugenol content and density of clove oil as feed and permeate of each membrane.

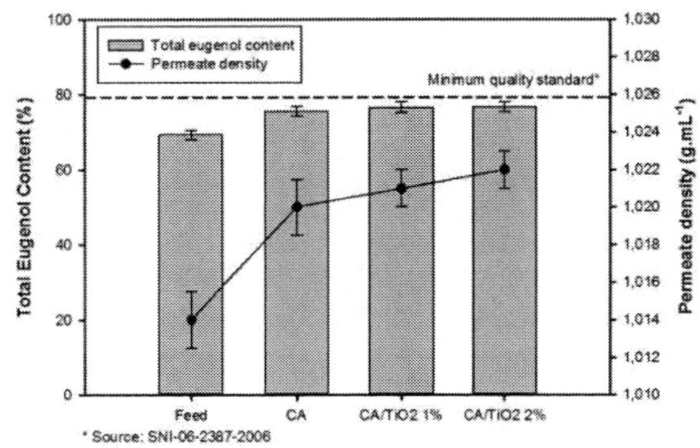

Fig. 5. Eugenol content and density of feed and permeates clove leaf oils

Figure 5 shows that eugenol content of feed is 70% and the eugenol content of clove oil permeate of CA, CA/nano-TiO_2 1%, and CA/nano-TiO_2 2% membranes are 75.5%, 76.4%, and 76.6% respectively. The filtration using conventional CA membrane is able to increase the eugenol content up tp 7.8%, while the nanohybrid membranes of CA/nano-TIO_2 are able to increase the eugenol content up to 9.1 – 9.4%. Nanohybrid membrane exhibits higher separation efficiency than conventional CA membrane. This could be due to the membrane hydrophilicity is enhanced by incorporating the nano-TiO_2 in the membrane matrix. Caryophyllene molecule will be rejected more when a very hydrophilic membrane is employed. The density of clove oil permeates also increase as the increase of nano-TiO_2 loading in membrane matrix. The density profile corresponds with eugenol content in clove oil since the density of clove oil depends on the major component contained in clove oil. Even though the eugenol content in clove oil permeate is bellow the accepted standard, however these results give important information that nanohybrid CA/nano-TiO_2 is suitable to be applied in crude clove leaf oil purification.

4 Conclusion

Nanohybrid CA/nano-TiO_2 flat sheet membranes are successfully fabricated using dry-wet phase inversion method. The MWCO values of fabricated membrane ranging from 4200 – 4800 Da which are equivalent with 1.7 – 1.9 nm of pore radius. The incorporation of nano-TiO_2 in CA membrans are proven to improve the hydrophilicity of membrane, where the 2 wt-% loading in membrane polymer is able to decrease the contact angle from 67.43° to 53.21°. The conventional 17 wt-% CA membrane has been observed capable to increase the eugenol content up to 7.8%, while 2 wt-% nano-TiO_2 loading in CA membrane gives 9.4% eugenol content enhancement. The nanohybrid CA/TiO_2 2 wt-% exhibits the flux enhancement up to 52%. These results inform the potential of CA membrane for crude clove leaf oil purification and nano-TiO_2 incorporated CA membrane is more favorable for this work.

The authors would thank to Waste treatment laboratory of Chemical Engineering Department UNDIP for supporting facilities. Nano Center Indonesia as research partner which supplied the nanomaterials. And this research was supported by a grand from Directorate of Research and Community Service, Directorate General of Research and Development Reinforcement, Ministry of Research, Technology and Higher Education, Subject to Contract Research, No: 007/SP2H/LT/DRPM/IV/ 2017.

References

1. C. Gilmer and N. B. Bowden, ACS Appl. Mater. Interfaces. **8**, 1-30 (2016)
2. J. L. N. R. D. Falconer and D. P. Sperry, Marcel Dekker, (1993).
3. Widayat, M. Hilman, B. Bagus, and A. Rahmawan, Simposium Nasional RAPI XIII, (2014)
4. T. Marwati, M. Syahbanna, E. Noor, and E. Mulyono, J. Pascapanen, **2**, 45-52 (2005).
5. T. D. Kusworo, N. Aryanti, R. A. Anggita, T. A. D. Setyorini, and D. P. Utomo, J. Environ. Sci. Technol. **10,** 35-43 (2017)
6. K. Werth, P. Kaupenjohan, and M. Skiborowski, Sep. Pur. Technol. **182,** 185-196 (2017).
7. H. Siddique, E. Rundquist, Y. Bhole, L.G. Peeva, and A. G. Livingston, J. Membr. Sci. **452**, 354-366 (2014).
8. I. K. Nasution, B. Susilo, and W. A. Nugroho, J. Keteknikan Pertanian Tropis dan Biosistem, **2**, 9-14 (2014).
9. A. Boam, A. Meniconi, and X. Wu, WO Patent No. 2013167307 A1, (2013).
10. T. D. Kusworo, H. Hadiyanto, D. Deariska, and L. Nugraha. Chem. Ind. Chem. Eng. Q. **23**, (2017).
11. C. J. Davey, Z. X. Low, R. H. Wirawan, and D. A. Patterson, J. Membr. Sci. **526**, 221-228 (2017).
12. Indonesia Nastional Standard (SNI), No. 06-2387-2006, (2006)

Physical and mechanical properties of membrane polyvinilidene flouride with the addition of silver nitrate

Agung *Mataram*[1], S. *Rizal*[b*], Estu *Pujiono*[1]

[1]Mechanical Engineering Department, Engineering Faculty, Universitas Sriwijaya, Sumatera Selatan, Indonesia
[2]Politeknik Negeri Sriwijaya, Mechanical Engineering Department, Sumatera Selatan, Indonesia

Abstract. The membrane is a layer formed from fine fibers are used as a filter. In this research, measured the tensile strength and pore size of the membrane that is made in three variations of a specimen with a mixed composition of different Polyvinilidene Fluoride 17.5%, 20%, 22.5%., membrane made using Polyvinilidene Fluoride polimer granules and *N,N-Dimethylformamide* as solvent and silver nitrate. Membrane molded into standardized test specimens and tested with a device Adhesion Tearring Strength Tester. The results show an increase in the value of the tensile test of 404.8356 KPa to the composition of 17.5%, 507.3598 KPa for composition of 20% and 603.7218 KPa to a composition of 22.5%. For testing the microstructure using a Scanning Electron Microscopy with the resulting increase in pore diameter, for the range of 558.4 nm-781.8 nm are void to the composition of 17.5%, range 670.1 nm - 781.8 nm without voids for the composition of 20% and 1.117 µm-1.228µm without voids on a composition of 22.5

1 Introduction

Microbiological contamination of some water sources that exist at the moment is very worrying to the public, according to some researchers, there are many different species of bacteria which are available in pure water and wastewater [1,2].
This application requires a membrane that can improve water quality. The membrane is a layer formed from the composition or the tangle of fine fibers used as filter or barrier for a variety of applications in the industrial world. polymer has been the material of interest to be analyzed and developed its potential as a membrane filter, application examples that are of concern at this time is a membrane filter water [3].
The main objective to be achieved in this discussionare to develop a new method in the production of membran,andTo determine the mechanical properties of the membran of Polyvinilidene Fluoride (PVDF) against tensile testing.
The membrane is a layer formed from the composition or the tangle of fine fibers used as filter or barrier for a variety of applications in the industrial world.
Membrane Polymer produced using Polyvinilidene Fluoride as fiber polymer and DMF as a solvent, the process of printing the membrane by the method of a flat sheet, the membrane of this type is typically used in ultrafiltration process, due to higher porosity and pore structure are interconnected, resulting in permeability is higher for water filtration [4]. be used to determine the characteristics of the membrane there are a number of ways, but a commonused permeability and permeselectivity..

2 Methodology

2.1 Membran Preparation

Dominated by the carbon fiber Polyvinilidene Fluoride fibers due to a combination of the polymer, has good mechanical properties, especially tensile strength [3]. The advantage of the PVDF fiber, high degree of molecular orientation and a higher melting point (PVDF) fiber tends to decompose below the melting point, Tm of 317-330° C), and the results were better than carbon fiber [5], Polyvinilidene Fluoride used Writer (Figure 1)

Figure 1 Polyvinilidene Fluoride (PVDF)

Polyvinilidene Fluoride can to be form thermally stable molecular structure of highly oriented while getting a low temperature heat treatment, which is not significantly impaired during treatment carbonization at a higher temperature, this means the fibers produced have excellent mechanical properties [6,7].

Silver is a chemical element in the periodic table that has the emblem of Ag and atomic number 47. arms is derived from the Latin Argentum. A transition metal is soft, white, shiny, silver has the highest electrical conductivity and heat throughout the metal and there dimineral and in free form. This metal is used in coins, jewelry, equipment tables, and photography. Silver including precious metals such as gold [8].

Figure 2 Silver Nitrate

Chemical Properties Silver Nitrate (figure 2):
Very stable in pure air and water so it does not react with air and water at normal temperature. Less reactive than copper, except ozone, sulfur and hydrogen sulfide, which is rapidly discolor the silver surface. Not soluble in dilute acid solution, but can be dissolved in nitric acid and sulfuric acid.

N,N-Dimethylformamide (DMF) supplied by MERCK (figure 3) and is used as a solvent without further purification, thisis a strong solventfor PES polymer, because they have desirable properties such as low volatility, non-flammable, and relatively low toxicity.

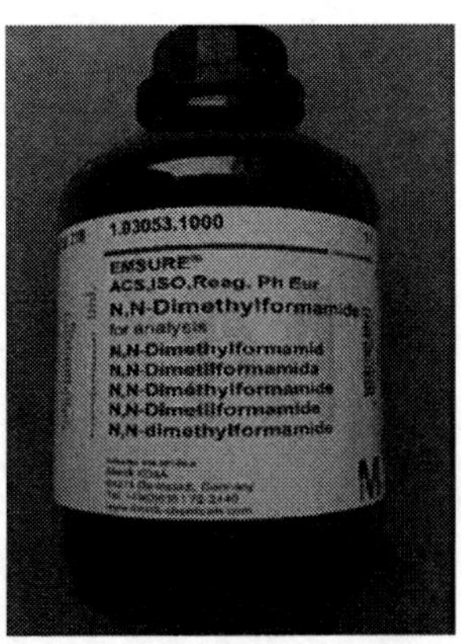

Figure 3 N,N-Dimethylformamide

2.2. Preparation of Polymer Membranes

Captions should be typed in 9-point Times. They should be centred above the tables and flush left beneath the figures.

2.2 Positioning

Membranes are made in this study using two materials with different variations of the mixture on each specimen, materials used is PVDF and DMF. This mixing of the two materials using a Magnetic Stirrer, this tool will mix the two ingredients with the rotation system so that the mixture becomes homogeneous [9].

Methods flat sheet is one way in the process of making the membrane author uses glass plates and duct tape as a template membrane, membrane PVDF prepared by phase inversion method immersion in a tub of coagulation of water, the result of printing membranes with these methods form a flat sheet [9]

Tensile test is a method used to test the strength of a material by providing load coaxial style. This experiment used to measure the mechanical strength of a material is the force of static given slowly, the results of tensile testing is essential to engineering and product design for yield strength data material tensile testing done to complete the design information is basic mechanical strength of a material and as a data support for material specifications.

Microstructure Test tool SEM (Scanning Electron Microscope), which uses an electron microscope. SEM testing is done in stages on each membrane composition. SEM testing is used to view and analyze surface and texture, shape and size of objects to analyze the sample and analyze the composition of the surface of objects both quantitatively and qualitatively. At this time SEM Test, used to see the size of the pores, voids or holes in the membrane and the bond that exists every pore in the membrane.

Research conducted procedure comprising the steps - the steps below:

Preparation of the membrane material, Polyvinilidene Fluoride and *N, N- Dimethylformamide* and prepare materials stirrer (magnetic stirrer).

Considering mixture of PVDF and DMF prior to the mixing process takes place.

Stir the two ingredients for 6 hours under on each specimen. Print the membrane solution that has been homogenized using glass mold that has been provided.

Prepare a tensile testing machine tearing adhesion strength test (figure 4). After that perform appropriate testing procedures. Prepare test equipment SEM, once ready to do appropriate testing procedure.

https://doi.org/10.1051/matecconf/201815608014

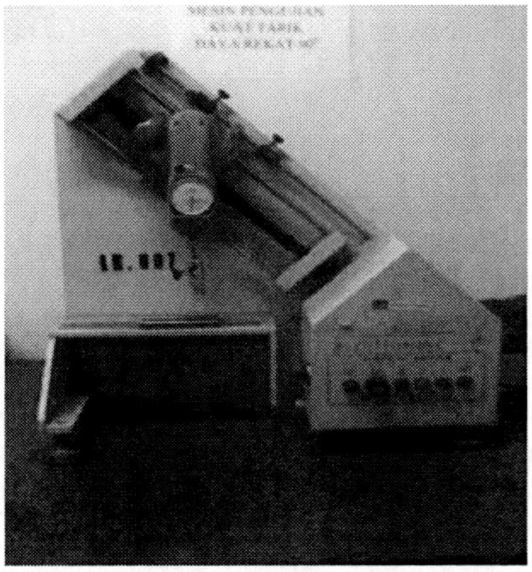

Figure 4 Adhesion Tearring Strenght Test.

3 RESULTS AND DISCUSSION

The test results of the research shows the mechanical properties of the material PVDF membrane material, each specimen using a membrane tensile testing machine tearing Adhesion strength test (figure 6). This test is generally aimed at determining the mechanical properties of a material, specifically on this study aims to determine the tensile resistance of the membrane, and to add the test results do also test Scanning Electron Microscopy (SEM). Results SEM micrograph of the surface of PVDF fiber is shown in Figure 5, when the PES fiber content is at 17.5%, 20% and 22.5% PVDF nanoparticle dispersions relatively simple homogeneous, but with the added amount (%.PES) in a solution with a composition of 22.5%, occur agglomeration excessive resulting solution is more difficult to be homogeneous [10,11].

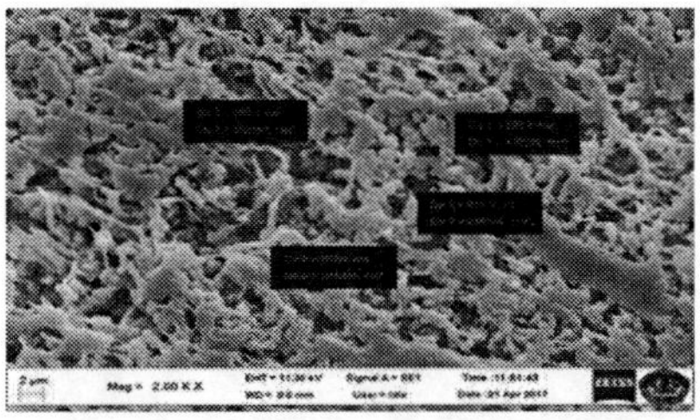

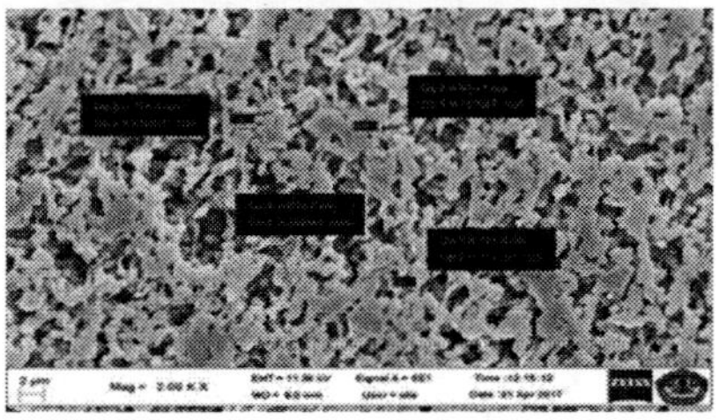

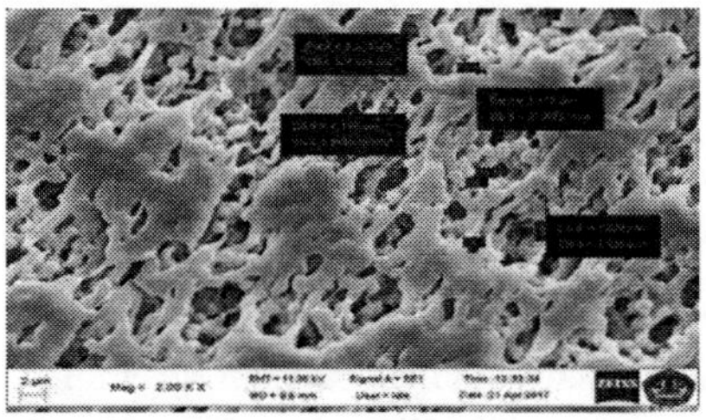

Figure 5 SEM comparison PVDF fibers with different variations 20% by weight, 22.5% by weight 25% by weight.

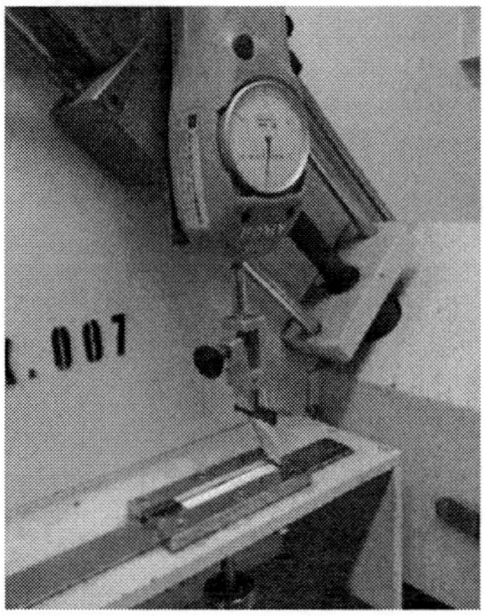

Figure 6 Process Polyvinilidene Fluoride membrane tensile testing

Tensile test specimens having a composition respectively 17.5%, 20%, 22.5% PVDF. The tensile strength of the specimens showed increases in the value of 404.8256 kPa to 603.7218 kPa when the specimen with a composition of PVDF fiber increased from 20% to 22.5%, the value of the tensile strength of the fiber membranes with composition (weight %) 25% PVDF also showed an increase up 603.7218 Kpa. PVDF on each specimen resulting fiber braid structure of the membrane increasingly meeting [12].

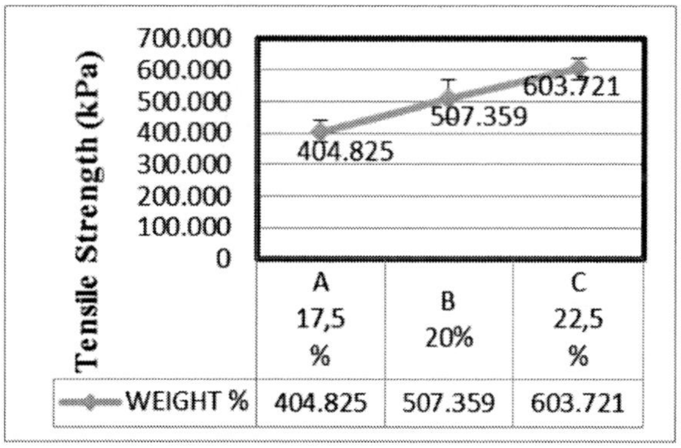

Figure 8 Graph tensile strength (A) of the PVDF composite fiber with changes in The composition (weight % . PVDF) on each specimen.

4 Conclusion

The length of time and the rotation speed of the mixing process of membrane material influence on the results of the homogeneity of the membrane solution.
The temperature at the time of stirring is also important to remove residual solvent and trapped air that can cause bubbles on the membrane
Results of testing the best there is in the composition of 20% PVDF because it can withstand loads of up to 312.2881 kPa

References

[1] S. Agustina, S. Pudji, T. Widianto, A. Trisni. *Penggunaan Teknologi Membran Pada Pengolahan Air Limbah Industri Kelapa Sawit*, Workshop Teknologi Industri Kimia dan Kemasan, Jakarta (2008)
[2] V. Bonnélye, L. Gueya. And J. Del-Castillo, Desalination. **222**, 59–65 (2008)
[3] A. E. Kalyuzhny, *Handbook Of Elispot: Methods and Protocol*. Humana Press. Totowa : New Jersey, p 37-49 (2005)
[4] M. Mulder, *Basic Principles of Membrane Technology*. (1st ed.) London: The Netherlands, Kluwer Academic Publisher. p 12-17, p 138-140 (1991)
[5] M. Nasir, *Sintesi dan Karakteristik Nanokomposit Konduktif Nanofiber*, Pusat Penelitian Kimia Lembaga Ilmu Pengetahuan Indonesia, Bandung, (2013)
[6] M. Schwatz, *Encyclopedia Of Smart Materials Volume 1 and Volume 2*. John Wiley & Sons. Inc : New York, p 807-824 (2002)
[7] S. Redjeki, *Proses Desalinasi Dengan Membran*. Direktorat Jendral Pendidikan Tinggi Departemen Pendidikan Nasional (2011)
[8] H. V. Tiwary, P. D. Mangalgiri, A. R. Upadhya, A. Selvaraja, *Smart Material, Structures And Systems*. Allied Publisher Limited. New Delhi : India (1999)
[9] Y. Wu, *Lithium-ion Batteries: Fundamentals and Applications*, CRC Press, Boca Raton (2015)
[10] S. Zemansky, *Fisika untuk Universitas I: Mekanika, Panas, Bunyi*. Bandung: Binacipta (1994)

Preparation and Characterization of Porous Ceramic Membranes for Micro-Filtration from Clay/CuZn Using Extrusion Methods

Muh Amin[1*] and Muhammad Subri[1]

[1]Mechanical Engineering Department, Universitas Muhammadiyah Semarang, Indonesia

Abstract. In this study, fabrication and characterization of ceramic membranes preparation was carried out. Porous ceramic membranes were fabricated by extrusion process from different percentage composition of CuZn on (80 wt% Clay, 10 wt% TiO_2, 5 wt% Carbon and 5 wt% PVA). The fabricated membranes were sintered at 900°C for 1 hour in an electrical box furnace with heating rate 1oC/min and holding time for 1 hour. Apparent density and porosity were determined by standar methods for ceramic materials. Phase composition of the ceramic support was established by X-Ray Diffraction analysis. SEM studies of the membranes added at different CuZn were carried out.

1 Introduction

Ceramic membranes were used in waste water technology and succesfull and possible applications cover all industries where media fillered, likely: chemical industry, metal industry (surface engineering), textiles (pupl and paper industry), food and beverages, recycling and environment [1]. Membrane application as the filtering of bacteria [2] and water treatment [3].

Commercialized porous ceramic membranes from alumina, zirconia, titania, and silica are not suitable for large scale application because these membranes are very expensive. In recent yeras, the preparation and potensial application of porous ceramic membrane on the basis of naturally occurring minerals such as zeolite, apatite, clay kaolinite, dolomite, have increasingly attracted intention for its low cost. The development of naturally mineral-based ceramic membranes can lead to a critical new technological revolution that existed through out the world [4].

Ceramic membranes are granular membranes and its are made from inorganic materials such as Al_2O_3, TiO_2, ZrO_2, SiO_2, etc. or combination of these materials. Production of micro porous ceramic membrane filters also can be made by compressing and sintering of pyrex powder.

In contrast with polymeric membranes, ceramic membranes have excellent thermal and chemical stability wich makes them usable in high temperature membrane operations and aggresive environments. Also its were very stable mechanicals for many pharmaceutical industries or in water and waste water processing [5].

Fabrication of nanometer membranes from local kaolin and limestone from Algerian were processed by extrusion [6]. In these studies the dried tubular samples were sintered at various temperatures ranging between 800°C and 1100°C following this program:

$$25°C \xrightarrow{2°C/min} 250°C\ (2\ h) \xrightarrow{2°C/min} 800 - 1100°C\ (3\ h)$$

The preparation of the natural zeolite membrane from Kralevo's deposits can be produced into multilayer ceramic membranes for microfiltration at low sintering temperature of 800°C. The obtained by dip-coating membranes are characterized with good adhesion between the support and layers, high porosity of the support membrane and very uniform pore distribution [7].

Ceramic membranes made from alumina have been successfully performed [8]. Its were processed by using the extrusion method. Starting alumina dough was consisted of 71 wt.% alumina powder mixed with 29 wt.% of processing additives (hydroxypropylmethyl cellulose, strach dispersant, plasicizer (PEG, polyethylene glycol), lubricant and water). The alumina dough was extruded and dried. Then, the green extruded were sintered in the temperature range of 1300°C - 1500°C for 2 hours. The increasing of sintering temperature, there was a slight change in particle morphology leading to particle coarsening and densification (less porosity).

Researcher use various porous ceramic membrane to support fabrication methods. Most membrane supports (commercial and home made) were fabricated by powder pressing, past processing, and collodial processing methods. A detailed classification is shown in Figure 1 [9]. Slip casting and dry pressing methods are widely used in lab scales, while the method in industry is extrusion [10].

* Corresponding author: amin@unimus.ac.id

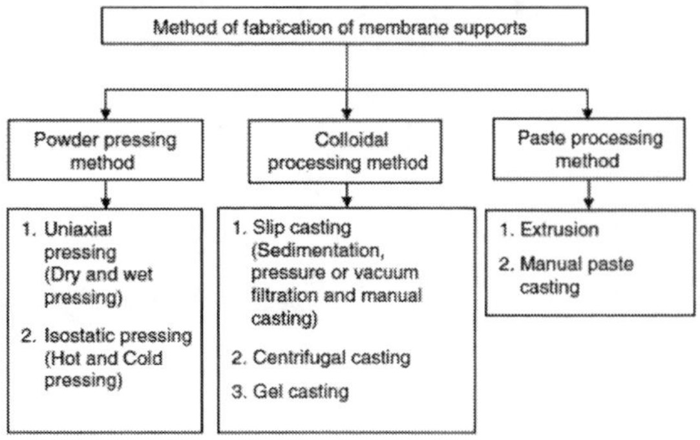

Fig. 1. Classification of the membrane support production methodes based on the precurcor aggregate state

The aim of this work is toward the development of low cost ceramic membrane. In order to obtain a low price in the process of making ceramics, one solution is to use local materials that have a high abundance. The use of the extrusion method is also easy and cost-effective.

2 Method

2.2 Raw materials

The raw materials were used for in this study are Clay, TiO_2, Black Carbon (BC), Polyvinyl Alchohol (PVA) and CuZn. The raw materials are obtained from local Central Java, Indonesia, except PVA from China. The CuZn was obtained from industrial waste from CuZn craftsmen in Juwana, Central Java, Indonesia.

2.2 Preparation of raw materials for obtaining of ceramic membranes

The raw materials were used for this study were has been tested to obtain related information. All the studed materials were tested chemical composition in order to obtain chemical content information. All materials were prepared in the powder shape that passes filtered with 80 mesh, except PVA with 100 mesh.

2.3 Membrane preparation

Manufacturing of green body was done by preparing of 80 wt.% Clay, 10 wt.% TiO_2, 5 wt.% BC and 5 wt.% PVA becoming dough. Then, the additive of CuZn was added to the dough with various of 0 wt.%, 10 wt.%, 20 wt.%, 30 wt.% and 40 wt.%. It was processed to become the samples as shown at the following Table 1. The fabrication of the green body was printed by extrusion method of pressure 25 MPa. The details of the processing step of the membranes are displayed in Figure 2.

Prior to the extrusion process, all materials were mixed at the rotational speed of 64 rpm for 30 min. After that, its were continued for blending process with water addition of 20 wt.% and 5 wt% oil. The addition of oil has a purpose to facilitate the extrusion process and to avoid cracking on the surface of the created membrane material.

Table 1. Percentage composition of the raw materials

Sample	Clay (wt.%)	TiO_2 (wt.%)	BC (wt.%)	PVA (wt.%)	CuZn (wt.%)
A	80	10	5	5	0
B	80	10	5	5	10
C	80	10	5	5	20
D	80	10	5	5	30
E	80	10	5	5	40

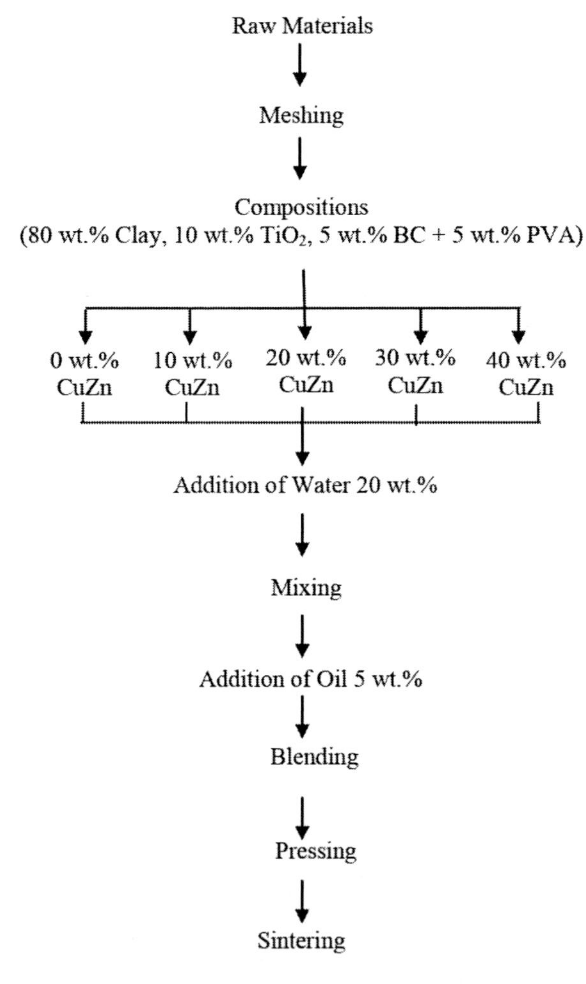

Fig. 2 Squential path for membranes preparation

Sintering process was done after the green body ceramic placed in the room temperature for 24 hours in order to obtain stability. The roles of sintering processed was done according to Figure 3. The process of sintering green body was done on the two stages. The first stage, sintering has done at temperatur 400°C by heating rate 5°C/min and holding time for 0.5 hours. After that, the second sintering was carried out at a temperature of 900°C with heating rate 5°C/min and holding time for 1 hour. The decreasing on tempetarure was done by turn off the button on furnace.

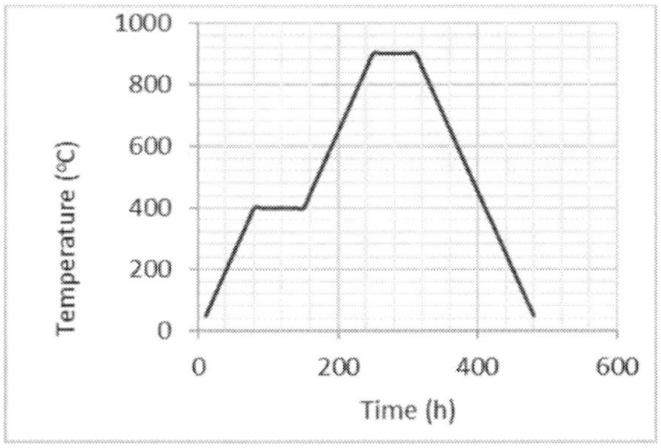

Fig. 3. Sintering temperature profile

2.4 Characterization methods

The result of Clay/CuZn membranes was tested for chemical composition, density, porosity, X-Ray Diffraction (XRD) and Scanning Electron Microscopy (SEM). The chemical composition of the row material was analyzed by ZAF Methode Standarless Quantitive Analysis.

The density test was calculated according to the Archimedes method by weighing the samples in the air and in the water. Weighing results in air and in water are used to determine density and porosity by using the following formula [11]:

$$\text{Bulk Density}\left(\frac{g}{ml}\right) = \frac{\text{Mass in air (g)}}{\text{Volume of water displaced (ml)}} \quad .....(1)$$

$$\text{Apparent Porosity (\%)} = \frac{\text{Soaked mass (g)} - \textit{Mass in air (g)}}{\text{Soaked mass (g)} - \text{Suspended mass (g)}} \quad ....(2)$$

3 Result and discussion

The density test of pore ceramic membranes were calculated by the Archimedes method according to equation (1) wich was prepared with samples of 5 mm x 5 mm x 50 mm. Based on figure 4 it can be seen that the increasing content of CuZn can be increase value of density. The highest density was achieved by ceramic membrane at more CuZn content because CuZn has higher density than all raw materials used ie 8.7 gr/cm^3.

In addition of CuZn 10 wt.% showed a significant increase in density ie 4.59 g/cm^3, after which it will increase linearly as CuZn content increases. Material pore ceramic membrane without CuZn content sintered at 900°C with 1 hour holding time has a density of 1.96 gr/cm^3. So, in other words there is an increase of more than 100% by adding CuZn into the ceramic pore membrane material.

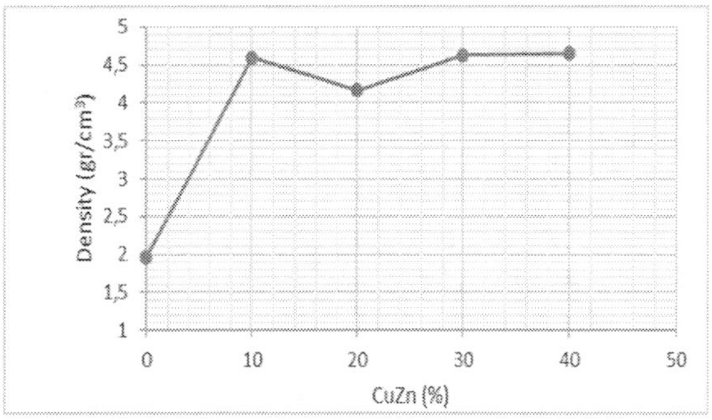

Fig. 4. Density of Clay/CuZn membrane at sintering 900°C

Based on the porosity calculation by the equation (3) that increasing CuZn content on porous ceramic material can be increase the amount of material porosity. As in Figure 5, there is a significant increase of CuZn without content up to 10 wt.% CuZn content of 44.58% ie from 44.58% to 66.67% or an increase of 49.52%. After that, with increasing CuZn content, porous ceramic materials decreased porosity to 55.13 on CuZn content of 40 wt.%. so the highest porosity with the addition of CuZn content is 66.67% on the addition of CuZn 10 wt.%.

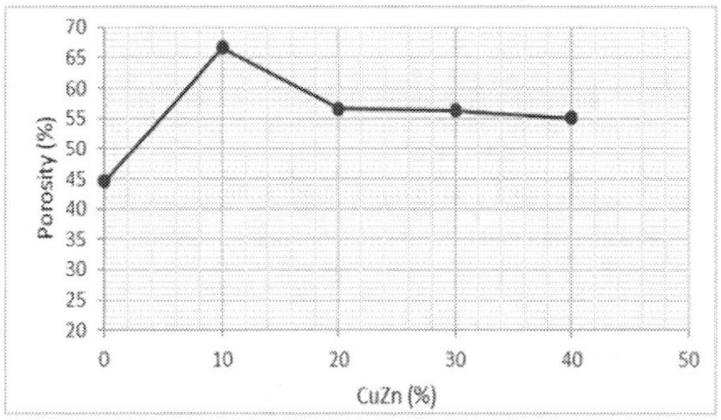

Fig. 5. Porosity of Clay/CuZn membrane at sintering 900°C

XRD Test results were performed with the following parameters: instrument = 6510 (LA), Acc (Figure 6). Beratage = 20.0 kV, probe current = 1.0 nA, PHA mode = T3 and Real Time = 51.34 seconds, Live time = 50.0 seconds, Dead time = 2%, Counting rate = 300 cps and energy range = 0 - 20 keV.

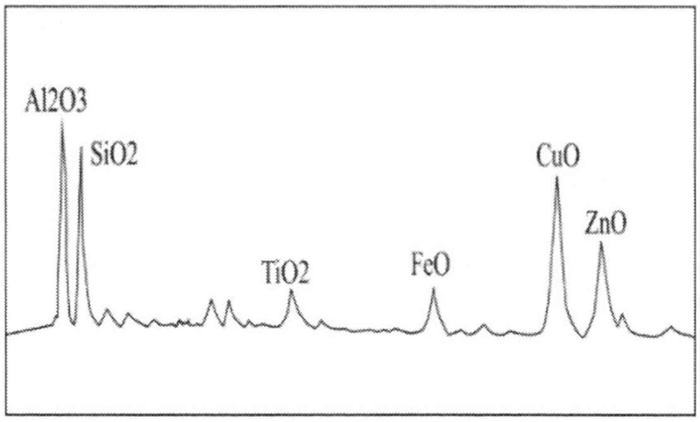

Fig. 6. X-ray diffraction pattern for membrane 10 wt.%

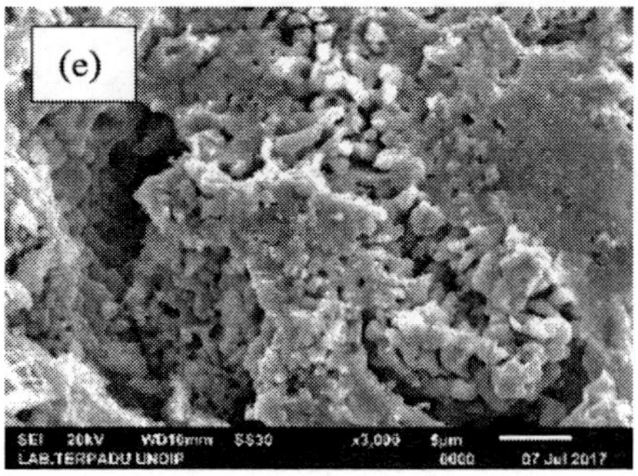

Fig. 7 . SEM micrographs of (a) CuZn 0 wt.%, (b) CuZn 10 wt.%, (c) CuZn 20 wt.%, (d) CuZn 30 wt.% and (e) CuZn 40 wt.%.

Based on ZAF Method Standardless Analysis for chemical composition test obtained that CuO = 32,05%, ZnO = 23,58%, C = 10,34%, Al_2O_3 = 10,27%, SiO_2 = 10,08% FeO = 4.49%, SnO_2 = 3.33%, TiO_2 = 3.07% and ZrO_2 = 1.36%.

According to the observation by SEM as shown at Figure 7. The porous ceramic membranes at 0 wt.% CuZn shows a small porosity if compared with other samples. In all the addition of CuZn on porous ceramic membranes will change the appearance of ceramic surface becomes more porous. This data was supported with the porosity data in Figure 5. Its can be seen in the CuZn content of 10 wt.% indicated the largest amount of porosity and gradually. While the adding more CuZn and slighlty causing the low porosity.

Our special gratitude to the DRPM KemenristekDikti and Mechanical Engineering at Universitas Muhammdiyah Semarang on the facilities provided to us by Research at PENELITIAN PRODUK TERAPAN (PPT) in 2017.

References

1. Manohar, IJMER. **2,** 4, 1492 – 1506 (2012)
2. D. S. Biron, J. Bortoluz, M. Zeni, C. P. Bergmann and V. Santos, Mater. Res., **19,** 3 (2016)
3. F. T. Owoeye, A. P. Azoda and S. B. Udo. J. Mech. Eng. Sci. **7**, 1115 – 1126 (2014)
4. P. Hristov, A. Yoleva, S. Djambazov, I. Chukovska and D. Dimitrov. J. Univ. Chem. Tech. and Metal., **47**, 4 pp. 476–480 (2012)
5. A. A. Hosseini, A. Sadigzadeh, S. Mohammadi and R. Soltani, Iranian Journal of Materials Science & Engineering, **8**, 3 (2011)
6. A. Harabi, A. Guechi and S. Condom. Proced. Eng. **33** 220-224 (2012)
7. P. Hristov, A. Yoleva, S. Djambazov, I. Chukovska and D. Dimitrov, Journal of the University of Chemical Technology and Metallurgy, **47**, 4 pp. 476–480 (2012)
8. C. Mangkolkachit, S. Wanakitti and P. Aungkavattana, Journal of Metal, Materials and Minerals, **20**, 3 pp. 123–125 (2010)

9. P. Monash, Pugazhenthi and P. Saravanan, 2013, Rev. Chem. Eng., **29** (5) pp.357–383 (2013)
10. P. Fan, K. F. Zhen, Z. Y. Zan, Z. Chao, Z. Jian and J. Z. Yun, Chem. Eng. Trans., **55**, 277-282 (2016).
11. F. T. Owoeye, A. P. Azoda, and S. B. Udo, J. Mech. Eng. Sci. **7**, pp.1115–1126 (2014)

Enhancing Properties and Performance of Cellulose Acetate/Polyethylene Glycol (CA/PEG) Membrane with the addition of Titanium Dioxide (TiO_2) by Using Surface Coating Method

Siti Nurkhamidah*, *Yeni* Rahmawati, *Ignatius* Gunardi, *Pitsyah* Alifiyanti, *Krisna* Dimas Priambodo, *Ryanda* Luthfi Zaim, and *Wahyuni* Eka Muqni

Department of Chemical Engineering, Faculty of Industrial Technology, Institut Teknologi Sepuluh Nopember, Kampus ITS Sukolilo, Surabaya 60111, Indonesia

Abstract. In this study, cellulose acetate/polyethylene glycol (CA/PEG) membrane with composition 80/20 was prepared by phase inversion method. Titanium dioxide with different number has been added by using surface coating. Hydrophilicity, morphology, flux permeate and salt rejection of membranes has been studied. The hydrophilicity is determined by Fourier-Transformed Infra-Red (FTIR) spectra and contact angle analysis. Surface and fractured morphology are identified by using Scanning Electron Microscopy (SEM). The experiment results show that hydrophilicity of CA/PEG membrane increases with the addition and the increasing of TiO2 contents. However, with further increasing of TiO2, hydrophilicity of CPT membrane decreases. The optimum membrane is CA/PEG/TiO2 80/20/1,25 g/L solvent (CPT 3) with flux permeate of 111,82 L.m-2h-1 and salt rejection of 48,30%.

1 INTRODUCTION

Water is essential for the survival of all forms of life on earth. On an average, a human being consumes about 2 L of water every day [1]. Water accounts for about 70% of the weight of a human body [2]. Today, access to clean water is becoming a difficult task in many regions of the world. According to the World Health Organization, 1.2 billion people lack access to sufficient amounts of clean fresh water and 2.6 billion lack adequate sanitation [3]. Poor sanitation combined with unhealthy water quality accounts for the largest single cause for disease and death in the world. About 75% part of the earth is covered with water and out of which 97% is saltier and only 3% is available for drinking, agriculture, domestic, and industrial consumption [4]. Fresh water is about 2.8% of the total water in the world, whereas among fresh water, only 0.6% is available for use while the rest is locked up in oceans as salt water, polar ice caps, glaciers, and underground reservoirs [5].

The need for fresh clean water is growing rapidly due to the world population growth that imposes larger demands of water supply for domestic use, agriculture, and industry. Another reason is deterioration of fresh water supplies: aquifers, the largest fresh water resource, are being contaminated constantly by industrial and agricultural activities, as well as by intrusions of seawater or saline water due to overuse. Rivers and lakes (surface water resources) are also in threat. Hence, there is a strong need to increase fresh water availability either by recycling waste water or by production of fresh water from seawater [6]. The need to increase fresh water supply and more extensive water treatment drove the advancement of new water technologies and the maturity of existing ones, in all fields of water: desalination and ion removal by reverse osmosis (RO), disinfection techniques by catalysts and by biological treatment, decontamination, new filtration techniques, and monitoring of water quality. Desalination is a general term for methods to remove salt from salty water to produce fresh water.

The common material used for membrane desalination is cellulose. Cellulose acetate/polyethylene glycol (CA/PEG) membrane has been observed by Ahmad et al. and the result shows that CA/PEG with composition 80/20 has the best performance in salt rejection and flux permeate [7]. The pore number and flux permeate of the cellulose membrane can be increased by the addition of PEG and it is hydrophilic material [8]. However, if flux permeates increase, salt rejection will be decrease. Adding materials to increase salt rejection should solve this problem.

* Corresponding author: nurkhamidah@chem-eng.its.ac.id

In the previous study, nanomaterials such as silica [9], silicon dioxide (SiO_2) [10], zinc oxide (ZnO) [11], titanium dioxide (TiO_2) [12], alumina (Al_2O_3) [13], and carbon nanotube (CNT) [14] have been added to increase the membrane's performance. For the addition of silica into CA/PEG membranes, the results show that hydrofilicity increases when the particle size of silica decrease [9]. The addition of TiO_2 into polyamide membrane has been study by Lee et al. resulting high and stable salt rejection [12]. In this study, CA/PEG membrane is modified with the addition of TiO_2 nanoparticles to increase its performance.

2 Materials and Methods

2.1. Materials

Cellulose acetate (CA, Mw 3000 Da, acetyl content 39%), polyethylene glycol-200 (PEG, Mw 200 Da) are purchased from Sigma Aldrich. Titanium dioxide (TiO_2) with particle size of ≤ 25 nm, surface area of 45 – 55 m2/g, and purity of 99.7% was purchased from Sigma Aldrich. Aceton was used as solvent and water as non-solvent.

2.2 Membrane Preparation

2 gram of CA and 0.5 gram of PEG were dissolved in 17 mL of acetone. This solution was heated at 130°C up to homogen. Then, solution was cast at room temperature and dip into water for 15 min. After immersion process in water, membrane was removed at room temperature from the mold resulting CA/PEG membrane. TiO_2 nanoparticles were self-assembled on the surface of prepared pore-filled membranes by dipping the membranes in the TiO_2 colloidal suspension of 0.05 wt.% TiO_2 distilled water. Before dip-coating of membrane, the TiO_2 suspension was sonicated about 15 min to avoid the agglomeration of nanoparticles in water. UV lights radiated the coated membranes with 160 W lamp for 15 min. TiO_2 colloidal suspension was varied by 0, 0.5, 0.75, 1, and 1.25 g/(L solvent) and named as CPA, CPT1, CPT2, CPT3, and CPT4, respectively.

2.3 Membrane Characterizations

CA/PEG/TiO_2 membrane is characterized for their hydrophilicity, surface and fractured morphology. The hydrophilicity is determined by Fourier-Transformed Infra-Red (FTIR) spectra and contact angle analysis. Surface and fractured morphology areidentified by using Scanning Electron Microscopy (SEM).

2.4 Membrane Performance

The permeate flux (J) represents the amount of pure water collected per unit time and per unit area at variable pressures. It was calculated by: $J = Q/t \times A$, where J is the permeate flux (mL/h.m2), Q is the amount of permeate(mL), t is the time and A is the area (m2). Percentage of salt rejection is the efficiency of membrane and its ability to remove contaminates and calculate by using this equation: $R=(1-C_p/C_f) \times 100\%$, where R is the percentage of salt rejection, C_p (in ppm) is the salt concentration of the permeate, C_f (in ppm) is the salt concentration in the feed water.

3 Results and Discussion

3.1 Fourier Transformed Infra Red (FTIR) Characterization

FTIR result shows the increasing of –OH group with the addition of TiO_2 and the increasing of it contents as shown by the increasing of area in the wavenumber of 3300 cm-1 as shown in the **Figure 1**. The increasing of –OH group up to CPT 3 and decrease at CPT 4. This result shows that hydroxyl group increases up to the addition of TiO_2 is 1 g/L solvent. The increasing of hydroxyl group in TiO_2 is caused by photo catalysis process.

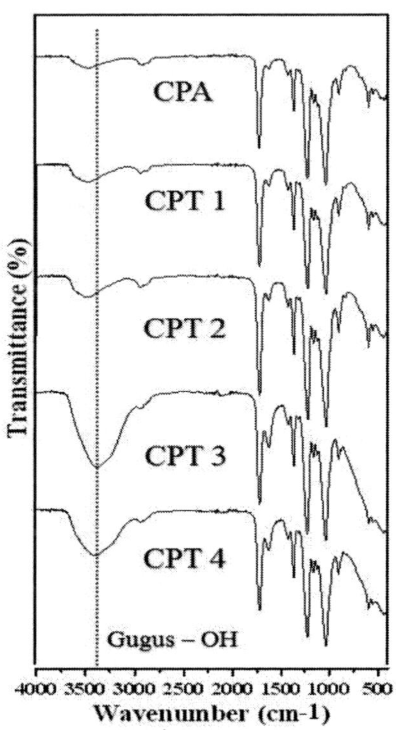

Figure 1 FTIR graphs for CA/PEG membrane with the addition of TiO_2

In the formation process of CA/PEG membrane with the addition of TiO_2, there is a step; that is: dipping the membrane into TiO_2 solution. After that, the membrane was contacted with UV light for 15 minutes. Photo catalytic reaction occurs when the electrons are resulted due to the interaction between membrane and UV light. Those electrons interact with photon and resulting free photoelectrons and photoholes [16].

TiO_2 is semiconductor material and UV light caused the hole and electrons. Anion of super oxide radical is formed from the interaction between photo-generated electrons with oxygen from the environment. However, when photo-generated holes react with water, -OH radical is formed [15].

3.2 Contact Angle Analysis

Another method to identify hydrophilicity of the membrane is by measuring its contact angle. **Figure 2** shows contact angle of membrane CA/PEG with the addition of TiO_2. With the addition of TiO_2, contact angle of membrane CPT decreases up to CPT 3 and increases at the CPT4. The decreasing of contact angle shows the hydrophilicity of membrane increases. Contact angle measurement and FTIR results are the same. Hydrophilicity of CA/PEG membrane increases with the increasing of TiO2 contents and the maximum is at CPT3, with the addition of 1 g TiO2/L solvent. With further increasing of TiO_2 content, hydrophilicity of CA/PEG membrane decreases.

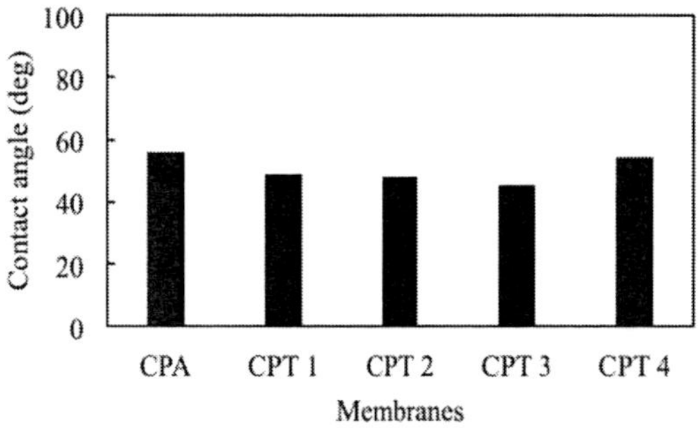

Figure 2 Contact angle of CPT membrane

3.3 Scanning Electron Microscopy (SEM) Characterization

Morphology of the membrane on the top surface of cross section was observed by using SEM. **Figure 3** shows SEM images show the Top surface and cross section of membrane (a) CPA, (b) CPT 1, (c) CPT 2, (d) CPT 3, (e) CPT 4. The pore size of CPT membrane is higher than that of CPC membrane. However, from the pore size both of them are microfiltration membrane because the pore size is in the range 100 nm – 10.000 nm [17].

3.4 Salt rejection and Flux Permeate

Performance of the membrane for desalination process was observed from salt rejection and flux permeate of the membranes. **Figure 4** shows the increasing of salt rejection with the increasing of TiO_2 contents and achieved maximum at CPT3 and decreases with further increasing of TiO_2 content (CPT4). Salt rejection of CPT3 is 48,302% which is higher than that of CPA, 34,053%. Compared to the previous result, salt rejection of CA/PEG membrane with the addition of TiO_2 is higher than that with the addition of silica. Salt rejection of CA/PEG with the addition of silica is 36.88% [9].

Figure 5 shows the increasing of flux permeate of CPT membrane. The result shows that flux permeate increases with the increasing of TiO_2 content and decreases with further decreasing of TiO_2 content. This phenomenon is in agreement with the pore size of CPT membrane as can be seen from SEM images. CPT3 has the maximum flux permeate (111,819 L.m-2h-1) because it has the biggest pore size.

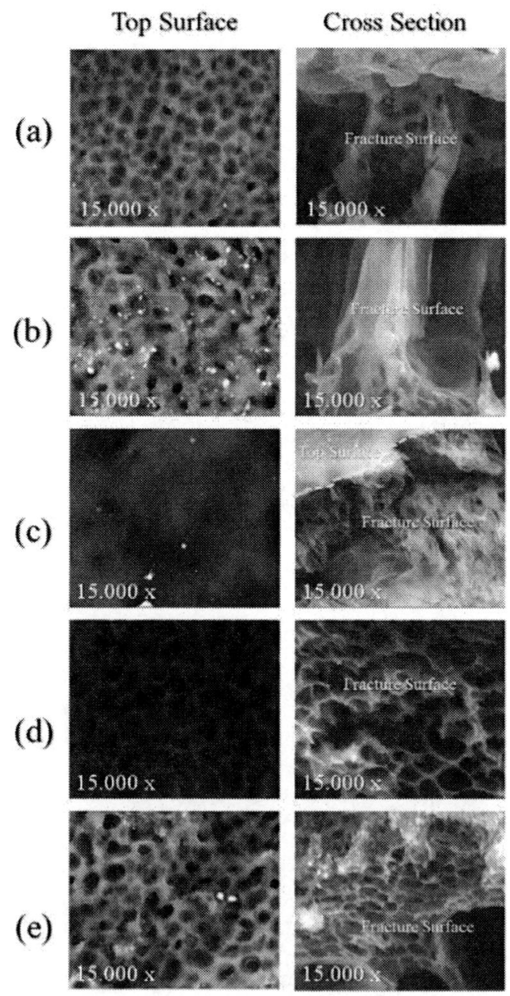

Figure 3 SEM images show the Top surface and cross section of membrane (a) CPA, (b) CPT 1, (c) CPT 2, (d) CPT 3, (e) CPT 4

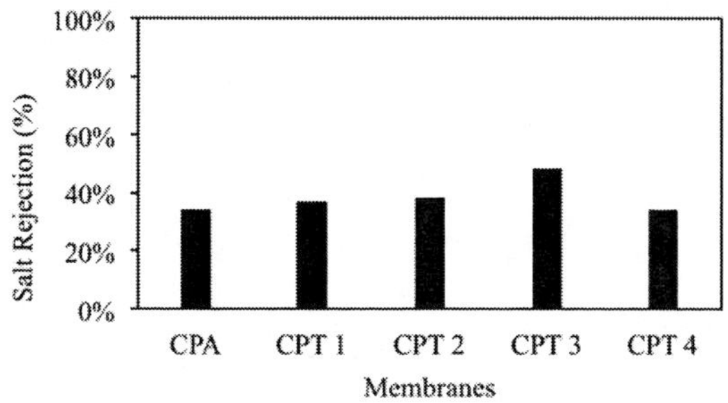

Figure 4 Salt rejection of CPT membranes

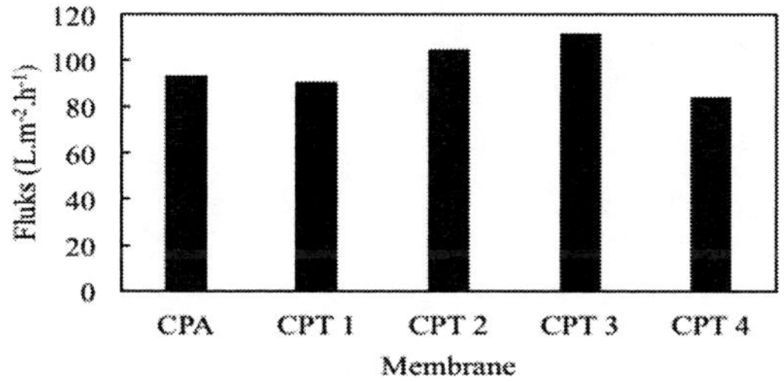

Figure 5 Flux permeate of CPT membrane

4 Conclusions

Hydrophilicity of CA/PEG membrane increases with the addition and the increasing of TiO_2 contents as can be seen from FTIR results and contact angle measurements. However, with further increasing of TiO_2, hydrophilicity of CPT membrane decreases. The optimum membrane is membrane with the composition of CA/PEG/TiO_2 80/20/1,25 $^{g}/_{L\ solvent}$ (CPT 3) with flux permeate of 111,82 $L.m^{-2}h^{-1}$ and salt rejection of 48,30%.

This work has been financially supported by ministry of Research, Technology and Higher Education of the Republic of Indonesia with contract's number 01775/IT2.11/PN.085/2016 and local funding from Institut Teknologi Sepuluh Nopember with contract's number 841/PKS/ITS/2017 to which the authors express their sincerest gratitude.

References

1. M.A. House, D.H. Newsome, Water Sci. Technol. **21**,1137–1148 (1989)
2. S.B. Jonnalagadda, G. Mhere, Water Res. **35**, 2371–2376 (2001)
3. S.M. Liou, S.L. Lo, C.Y. Hu, Water Res. **37**, 1406–1416 (2003)
4. A.J. Melloul, M. Collin, J. Environ. Manage. **54**, 131–142 (1998)
5. J.W. Nagel, Water Sci. Technol. **43**, 285–292 (2001)
6. S.S.S. Loeb, Sea water demineralization by means of an osmotic membrane **38**, 117–132 (1963)
7. A. Ahmad, S. Waheed, S. M. Khan, S. e-Gul, M. Sha, M. Farooq, K. Sanaullah, T. Jamil. Desalination **355**, 1-10 (2015)
8. G. Arthanareeswaran, D. Mohan, M. Raajenthiren. J. Membr. Sci. **350**, 130-138 (2010)
9. S. Nurkhamidah, Y. Rahmawati, F. Taufany, I. M. P. A. Merta, D. D. D. Putra, E. M. Woo. AIP Conf. Proceed., **1699** (2015)
10. H. Wu, J. Mansouri, V. Chena. J. Membr. Sci. **433**, 135-151 (2013)
11. S. Balta, A. Sotto, P. Luis, L. Benea, B. Van der Bruggen, J. Kim. J. Membr. Sci. **389**, 155-161 (2010)
12. H. S. Lee, , S. J. Im, J. H. Kim, H. J. Kim, J. P. Kim, B. R. Min., Desalination **219**, 48-56 (2008)
13. X. Wang, X. Li, K. Shih. J. Membr. Sci. **368**, 134-143 (2011)
14. C. P. Leo, N. H. Ahmad Kamil, M. U. M. Junaidi, S. N. M. Kamal, A. L. Ahmad., Sep. Purif. Technol. **103**, 84-91 (2013)
15. L. Dong, H. Yang, S. Liu, X. Wang, Y. F. Xie. Desalination **365**, 70-78 (2015)
16. Rahimpour, A.; Madaeni, S. S.; Taheri, A. H.; Mansourpanah, Y. J. Membr. Sci. **313**, 158–169 (2008)
17. E. Bet-moushoul, Y. Mansourpanah, K. H. Farhadi, M. Tabatabei. Chem. Eng. J. **283**, 29–46 (2016)
18. R.W. Baker. *Membrane Technology and Application 2nd edition*. California:Mc Graw-Hill (2004)

AUTHOR INDEX

AUTHOR INDEX

AUTHOR INDEX

AUTHOR INDEX

AUTHOR INDEX

EDP Sciences – Web of Conferences
17, Avenue du Hoggar
Parc d'Activité de Courtabouf
BP 112
F-91944 Les Ulis Cedex A
France

ISBN 978-1-5108-5976-0

9 781510 859760

CURRAN ASSOCIATES INC.
proceedings.com